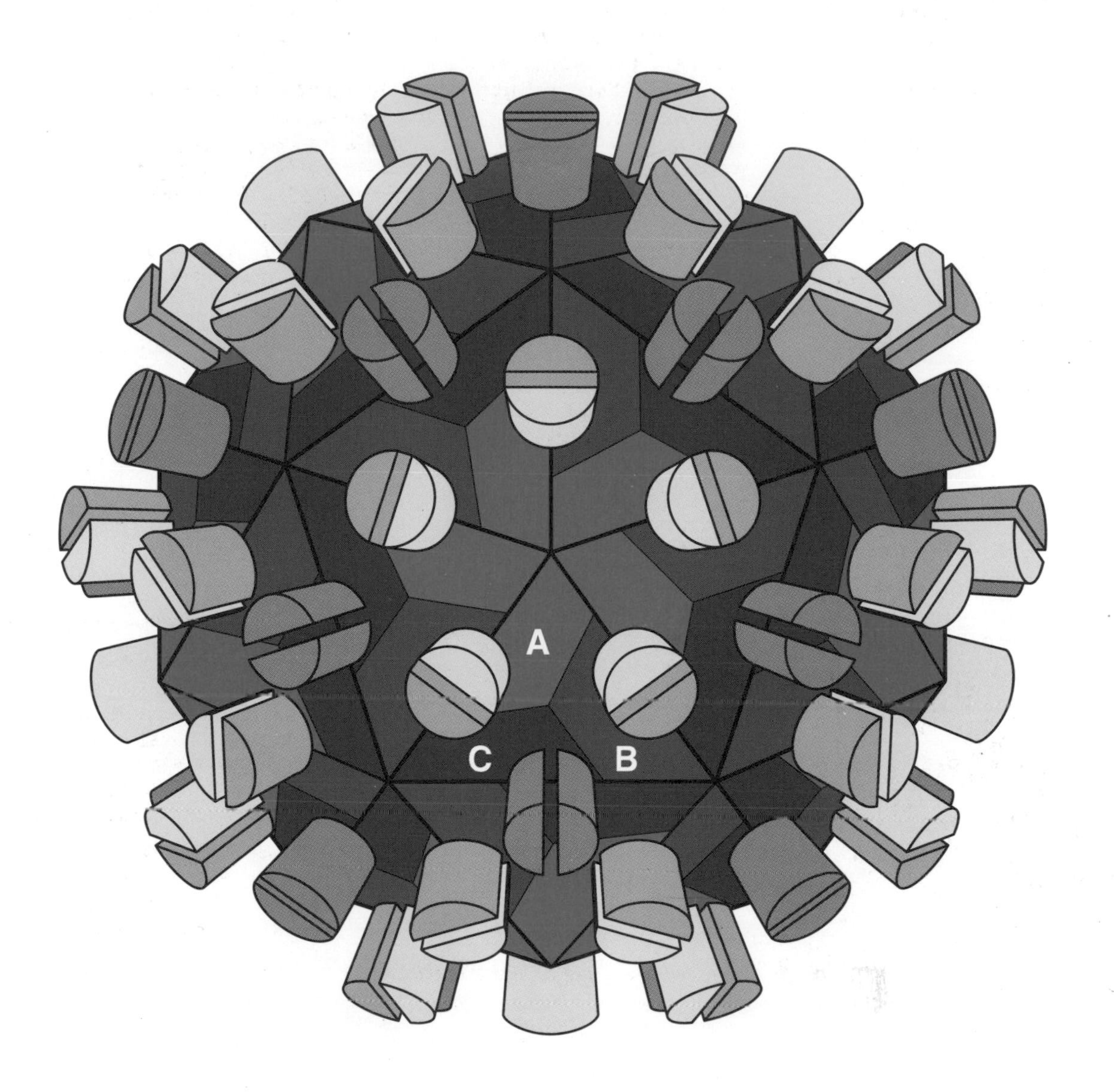

Figure 2.50 Cartoon representation of the quaternary structure of the tomato bushy stunt virus (TBSV) capsid. Subunits labeled A, B, and C have an identical sequence but are present in three different conformations.

Third Edition

TEXTBOOK OF BIOCHEMISTRY

With Clinical Correlations

Edited by

Thomas M. Devlin, Ph.D.

Professor and Chairman
Department of Biological Chemistry
Hahnemann University School of Medicine
Philadelphia, Pennsylvania

WILEY-LISS

A JOHN WILEY & SONS, INC., PUBLICATION

New York • Chichester • Brisbane • Toronto • Singapore

Address All Inquiries to the Publisher
Wiley-Liss, Inc., 605 Third Avenue, New York, NY 10158-0012

Printed in the United States of America.

Recognizing the importance of preserving what has been written, it is a policy of John Wiley & Sons, Inc. to have books of enduring value published in the United States printed on acid-free paper, and we exert our best efforts to that end.

Cover Illustration: (Front cover) An artist's conception of a hormone receptor in a mammalian cell membrane. The signal initiated by the binding of the hormone (messenger) is transmitted across the membrane by a change in the shape of the receptor; this activates an enzyme-catalyzed synthesis of a "second messenger" inside the cell. (Back cover) A channel facilitates the movement across the membrane of ions such as K^+ or Ca^{2+} which can serve as activators of cellular process.

Illustration Concept: Cecile Duray-Bito
Design: Patrick O'Connor

Library of Congress Cataloging-in-Publication Data

Textbook of biochemistry: with clinical correlations/edited by
Thomas M. Devlin. — 3rd ed.
p. cm.
Includes bibliographical references and index.
ISBN 0-471-51348-2 ISBN 0-471-58818-0 (pbk.)
1. Biochemistry. 2. Clinical biochemistry. I. Devlin, Thomas M.
[DNLM: 1. Biochemistry. QU 4 T355]
QP514.2.T4 1992
612'.015—dc20
DNLM/DLC 91-47576
for Library of Congress CIP

To Marjorie

and to the three individuals who most influenced my career,

Britton Chance

Albert L. Lehninger

George E. Boxer

Contributors

Stelios Aktipis, Ph.D.
Professor
Department of Molecular and Cellular Biochemistry
Stritch School of Medicine
Loyola University of Chicago
2160 S. First Avenue
Maywood, IL 60153

Carol N. Angstadt, Ph.D.
Assistant Professor
Department of Biological Chemistry, MS 411
Hahnemann University School of Medicine
Broad and Vine
Philadelphia, PA 19102-1192

William Awad, Jr., M.D., Ph.D.
Professor
Departments of Medicine and Biochemistry
University of Miami School of Medicine
P.O. Box 016960
Miami, FL 33101

James Baggott, Ph.D.
Associate Professor
Department of Biological Chemistry, MS 411
Hahnemann University School of Medicine
Broad and Vine
Philadelphia, PA 19102-1192

Stephen G. Chaney, Ph.D.
Associate Professor
Department of Biochemistry and Biophysics and Department of Nutrition
The University of North Carolina at Chapel Hill School of Medicine CB# 7260
Faculty Laboratory Office Building
Chapel Hill, NC 27599-7260

Joseph G. Cory, Ph.D.
Professor and Chairman
Department of Biochemistry
East Carolina University School of Medicine
Brody Medical Sciences Building
Greenville, NC 27858-4354

David W. Crabb, M.D.
Professor
Departments of Medicine, and Biochemistry and Molecular Biology
Indiana University School of Medicine
Medical Research and Library Building 1B424
975 West Walnut Street
Indianapolis, IN 46202-5121

Thomas M. Devlin, Ph.D., Editor
Professor and Chairman
Department of Biological Chemistry, MS 411
Hahnemann University School of Medicine
Broad and Vine
Philadelphia, PA 19102-1192

John E. Donelson, Ph.D.
Professor
Howard Hughes Medical Institute and Department of Biochemistry
University of Iowa College of Medicine
300 Eckstein Medical Research Building
Iowa City, IA 52242

Robert H. Glew, Ph.D.
Professor and Chairman
Department of Biochemistry
The University of New Mexico
School of Medicine
Basic Medical Science Building
Albuquerque, NM 87131

Dohn G. Glitz, Ph.D.
Professor
Department of Biological Chemistry
UCLA School of Medicine
10833 Le Conte Avenue
Los Angeles, CA 90024-1737

Robert A. Harris, Ph.D.
Professor and Chairman
Department of Biochemistry and Molecular Biology
Indiana University School of Medicine
635 Barnhill Avenue
Indianapolis, IN 46202-5122

Ulrich Hopfer, M.D., Ph.D.
Professor
Department of Physiology and Biophysics
School of Medicine
Case Western Reserve University
2119 Abington Road
Cleveland, OH 44106

Michael N. Liebman, Ph.D.
Adjunct Associate Professor
Department of Molecular and Cellular Biochemistry
Stritch School of Medicine
Loyola University of Chicago
2160 South First Avenue
Maywood, IL 60153

Gerald Litwack, Ph.D.
Chairman, Department of Pharmacology and
Deputy Director, Jefferson Cancer Institute
Jefferson Medical College
Thomas Jefferson University
1020 and Locust Street
Philadelphia, PA 19107

Bettie Sue Siler Masters, Ph.D.
The Robert A. Welch Foundation Professor in Chemistry
Department of Biochemistry
University of Texas Health Science Center at San Antonio
7703 Floyd Curl Drive
San Antonio, TX 78284-7760

J. Denis McGarry, Ph.D.
Professor
Departments of Internal Medicine and Biochemistry
University of Texas
Southwestern Medical Center at Dallas
5323 Harry Hines Boulevard
Dallas, TX 75235-8858

Alan H. Mehler, Ph.D.
Professor
Department of Biochemistry
Howard University College of Medicine
520 W Street, N.W.
Washington, D.C. 20059

Richard T. Okita, Ph.D.
Associate Professor
College of Pharmacy
105 Wegner Hall
Washington State University
Pullman, WA 99164-6510

Merle S. Olson, Ph.D.
Professor and Chairman
Department of Biochemistry
University of Texas Health Science Center at San Antonio
7703 Floyd Curl Drive
San Antonio, TX 78284-7760

Francis J. Schmidt, Ph.D.
Associate Professor
Department of Biochemistry
University of Missouri-Columbia
M121 Health Sciences Building
Columbia, MO 65212

Richard M. Schultz, Ph.D.
Professor and Chairman
Department of Molecular and Cellular Biochemistry
Stritch School of Medicine
Loyola University of Chicago
2160 South First Avenue
Maywood, IL 60153

Nancy B. Schwartz, Ph.D.
Professor
Departments of Pediatrics, and Biochemistry and Molecular Biology
University of Chicago
5825 S. Maryland Avenue, Box 413
Chicago, IL 60637

Thomas E. Smith, Ph.D.
Professor and Chairman
Department of Biochemistry and Molecular Biology
College of Medicine
Howard University
520 W Street, N.W.
Washington, DC 20059

Gerald Soslau, Ph.D.
Professor
Department of Biological Chemistry, MS 411
Hahnemann University School of Medicine
Broad and Vine
Philadelphia, PA 19102-1192

*__Marilyn S. Wells, M.D.__
Adjunct Professor
Veterans Administration Medical Center (128)
1201 N.W. 16th Street
Miami, FL 33125

J. Lyndal York, Ph.D.
Professor
Department of Biochemistry and Molecular Biology
College of Medicine
University of Arkansas for Medical Science
4301 W. Markham Street
Little Rock, AR 72205-7199

* Deceased July 29, 1991

Preface

The **purposes** of the third edition of the *Textbook of Biochemistry With Clinical Correlations* remain the same as for the first two editions: to present a clear discussion of the biochemistry of mammalian cells; to relate the biochemical events at the cellular level to the physiological processes occurring in the whole animals; and to cite examples of deviant biochemical processes in human disease.

The explosive advances in knowledge in the last five years, particularly due to the techniques of molecular biology, required a critical review and evaluation of the entire **content** of the textbook. The topics were selected to cover the essential areas of general biochemistry and physiological chemistry, and to meet the needs of upper-level undergraduate and graduate courses, especially courses presented in medical schools. Since the application of biochemistry is so important to human medicine, the text has an overriding emphasis on the biochemistry of mammalian cells. Significant additions of new material, clarifications, and some deletions were made throughout, and several chapters were totally rewritten by new contributors. Topics such as proteins, molecular biology, and hormones were totally reorganized, and new chapters on genetic engineering, molecular cell biology, and metabolism of xenobiotics were added. As with the previous editions, each contributor included up-to-date information that has been substantiated, but avoided observations so new that time has not allowed adequate evaluation.

The textbook is written such that any sequence considered most appropriate by an instructor can be presented. It is not formally divided into major sections, but related topics are grouped together. After an introductory chapter on cell structure, chapters 2–5 concern the **Major Structural Components of Cells,** that is proteins and their many functions, and cell membranes and their major roles. **Metabolism** is discussed in the next nine chapters, starting with the conservation of energy, then the synthesis and degradation of the major cellular components, and concluding with a chapter on the integration of these pathways in humans. The next section of five chapters covers **Information Transfer and Its Control,** describing the structure and synthesis of the major cellular macromolecules, that is DNA, RNA, and protein. A separate chapter on biotechnology is included because information from this area has had such a significant impact on the development of our current state of biochemical knowledge. The section concludes with a chapter on the regulation of gene expression, in which the mechanisms in both procaryotes and eucaryotes are presented. The fourth major section represents **Signal Transduction and Amplification** and includes two chapters on hormones that emphasize their biochemical functions as messengers. A new chapter titled Molecular Cell Biology describes four major mammalian signal transducing systems. The textbook concludes with six chapters on topics that comprise **Physiological Chemistry,** including cytochrome P450 enzymes and xenobiotic metabolism, iron and heme metabolism, gas transport and pH regulation, digestion and absorption, and human nutrition from a biochemical perspective.

A major change from previous editions is the addition of many original **illustrations.** All figures were reviewed and new drawings were prepared to illustrate the narrative discussion. In many cases the adage "A picture is worth a thousand words" is appropriate and the reader is encouraged to study the illustrations because they are meant to illuminate often confusing aspects of a topic.

The relevancy of the individual topics to human life processes are given in selected **Clinical Correlations** that describe the aberrant biochemistry of disease states rather than specific case reports. Each correlation has a reference to a current review that allows the reader to explore the topic in more detail. In some cases the same clinical correlation is presented in different chapters, each from a different perspective. All pertinent biochemical information is presented in the main text, and an understanding of the material does not require a reading of the correlations. In a few cases, clinical discussions are part of the principal text because of the close relationship of some topics to clinical conditions.

Each chapter concludes with a set of **questions and answers;** the multiple-choice format was retained as being valuable to students for self-assessment of their knowledge and as preparation for taking national examinations. All questions were reviewed and many new ones added. The questions cover a range of topics in each chapter, and each has an annotated answer, with references to the page in the textbook covering the content of the question. The appendix, **Review of Organic Chemistry,** was designed as a reference for the nomenclature of organic groups and compounds and some chemical reactions; it is not intended as a comprehensive review. The reader might find it valuable to become familiar with the overall content and then use the appendix as a reference for specific topics when reading related sections in the main text.

Our experience with the second edition reinforced our commitment to a multicontributor textbook as the best approach to have the most accurate and current presentation of biochemistry. Each author is involved actively in teaching biochemistry in a medical or graduate school and has an active research interest in the field in which he or she has written. Thus, each has the perspective of the classroom instructor, with the experience to select the topics and determine the emphasis required for students in a course of biochemistry. Every contributor, however, brings to the book an individual writing style, leading to some differences in presentation. This should not be an impediment to the reader's ability to understand the material, as students are accustomed to learning from a variety of resources. As editor, I have worked to smooth out differences and to remove unwanted redundancies. Repetition of selected topics was retained because some reiteration should be helpful to the reader.

The individual contributors were requested to prepare their chapters for a **teaching textbook.** The book is not intended as a compendium of biochemical facts or a review of the current literature, but each chapter contains sufficient detail on the subject to make it useful as a resource. Each contributor was requested not to refer to specific researchers; our apologies to those many biochemists who rightfully should be acknowledged for their outstanding research contributions to the field of biochemistry. Each chapter contains a bibliography that can be used as an entry point to the research literature.

In any project one person must accept the responsibility for the final product. The decisions concerning the selection of topics and format, reviewing the drafts, and responsibility for the final checking of the book were entirely mine. I welcome comments, criticisms, and suggestions from the students, faculty, and professionals who use this textbook. It is our hope that this work will be of value to those embarking on the exciting experience of learning biochemistry for the first time and for those who are returning to a topic in which the information is expanding so rapidly.

THOMAS M. DEVLIN

Acknowledgments

Without the encouragement and participation of many people, this project would never have been accomplished. My personal and very deep appreciation goes to each of the contributors for accepting the challenge of preparing the chapters, for sharing ideas and making recommendations to improve the book, for accepting so readily suggestions to modify their contributions, and for cooperating throughout the period of preparation. To each I extend my sincerest thanks for a job well done.

The contributors received the support of associates and students in the preparation of their chapters, and, for fear of omitting someone, it was decided not to acknowledge individuals by name. To everyone who gave time unselfishly and shared in the objective and critical evaluation of the text, we extend a sincere thank you. In addition, every contributor has been influenced by former teachers and colleagues, various reference resources, and, of course, the research literature of biochemistry; we are deeply indebted to these many sources of inspiration.

A very special thanks is extended to two friends and colleagues who have been of immeasurable value to me during the preparation of the third edition. My gratitude goes to Dr. James Baggott, who patiently allowed me to use him as a sounding board for ideas and who unselfishly shared with me his suggestions and criticisms of the text, and to Dr. Carol Angstadt, who reviewed many of the chapters and who gave me valuable suggestions for improvements. To each I extend my deepest gratitude.

I extend my sincerest appreciation and thanks to the members of the staff of the Wiley-Liss Division of John Wiley & Sons who participated in the preparation of this edition. Special recognition and thanks go to Dr. Brian Crawford, Publisher, who as a visionary suggested many excellent new additions and who gave me his unqualified support. I am indebted to Joseph Gill, Production Editor, who so meticulously oversaw the production, kept the flow of activities reasonable, patiently sought answers, made many valuable suggestions, and was always available to answer my questions. I also extend my gratitude to Eileen Cudlipp, Director of Production & Manufacturing, Production Editor Joseph Vella, and the entire production staff of Wiley-Liss, who often put aside their regular tasks to facilitate the production of this textbook. My appreciation is extended to Jeanette Stiefel, copyeditor, and Lillian Rodberg, indexer, both of whom did an excellent job and who willingly shared with me their suggestions for improvement. A significant improvement in this edition is the addition of many original illustrations. My most heartfelt thanks go to Michael O'Connor, head of the Illustrations Department at Wiley-Liss, and his staff, and to artists Cecile Duray-Bito and Elizabeth Morales, who transformed the rough drawings of the contributors into meaningful illustrations, and to computer artist Kitty Ryan of Precision Graphics who rendered the final artwork. I extend my thanks to Larry Graup, who revised the illustrations from the second edition, and to Patrick O'Connor, whose design implemented the concept for the cover. No book is successful without the activities of a Marketing Department; special thanks are due to Reed Elfenbein, David Stier, Merry Aronoff and their colleagues at Wiley-Liss for their efforts. I extend my deepest appreciation to everyone.

I am very indebted to my own staff, Colleen Healy and Christa Hess, who assumed many other responsibilities during the preparation of the book, who faithfully and efficiently completed the multitude of little chores, and who were patient with me when my other responsibilities were not completed on time.

Finally, a very special thanks to my supportive and considerate wife, Marjorie, who had the foresight to encourage me to undertake this project, who supported me during the days of intensive work, and who created an environment in which I could devote the many hours required for the preparation of this textbook. To all my sincerest thank you.

THOMAS M. DEVLIN

Contents in Brief

1 EUCARYOTIC CELL STRUCTURE 1

2 PROTEINS I: COMPOSITION AND STRUCTURE 25

3 PROTEINS II: STRUCTURE–FUNCTION RELATIONSHIP OF PROTEIN FAMILIES 91

4 ENZYMES: CLASSIFICATION, KINETICS, AND CONTROL 135

5 BIOLOGICAL MEMBRANES: STRUCTURE AND MEMBRANE TRANSPORT 195

6 BIOENERGETICS AND OXIDATIVE METABOLISM 237

7 CARBOHYDRATE METABOLISM I: MAJOR METABOLIC PATHWAYS AND THEIR CONTROL 291

8 CARBOHYDRATE METABOLISM II: SPECIAL PATHWAYS 359

9 LIPID METABOLISM I: UTILIZATION AND STORAGE OF ENERGY IN LIPID FORM 387

10 LIPID METABOLISM II: PATHWAYS OF METABOLISM OF SPECIAL LIPIDS 423

11 AMINO ACID METABOLISM I: GENERAL PATHWAYS 475

12 AMINO ACID METABOLISM II: METABOLISM OF THE INDIVIDUAL AMINO ACIDS 491

13 PURINE AND PYRIMIDINE NUCLEOTIDE METABOLISM 529

14 METABOLIC INTERRELATIONSHIPS 575

15 DNA: THE REPLICATIVE PROCESS AND REPAIR 607

16 RNA: STRUCTURE, TRANSCRIPTION, AND POSTTRANSCRIPTIONAL MODIFICATION 681

17 PROTEIN SYNTHESIS: TRANSLATION AND POSTTRANSLATIONAL MODIFICATIONS 723

18 RECOMBINANT DNA AND BIOTECHNOLOGY 767

19 REGULATION OF GENE EXPRESSION 805

20 BIOCHEMISTRY OF HORMONES I: PEPTIDE HORMONES 847

21 BIOCHEMISTRY OF HORMONES II: STEROID HORMONES 901

22 MOLECULAR CELL BIOLOGY 927

23 BIOTRANSFORMATIONS: THE CYTOCHROMES P450 981

24 IRON AND HEME METABOLISM 1001

25 GAS TRANSPORT AND pH REGULATION 1025

26 DIGESTION AND ABSORPTION OF BASIC NUTRITIONAL CONSTITUENTS 1059

27 PRINCIPLES OF NUTRITION I: MACRONUTRIENTS 1093

28 PRINCIPLES OF NUTRITION II: MICRONUTRIENTS 1115

Appendix: Review of Organic Chemistry 1149

Index 1159

Contents

1 EUCARYOTIC CELL STRUCTURE 1
Thomas M. Devlin
1.1 Cells and Cellular Compartments 1
1.2 Cellular Environment—Water and Solutes 4
1.3 Organization and Composition of Eucaryotic Cells 13
1.4 Functional Role of Subcellular Organelles and Membranes 16

CLINICAL CORRELATIONS
1.1 Blood Bicarbonate Concentration in a Metabolic Acidosis 13
1.2 Lysosomal Enzymes and Gout 19
1.3 Zellweger Syndrome and the Absence of Peroxisomes 21

2 PROTEINS I: COMPOSITION AND STRUCTURE 25
Richard M. Schultz and Michael N. Liebman
2.1 Functional Roles of Proteins in Humans 26
2.2 Amino Acid Composition of Proteins 27
2.3 Charge and Chemical Properties of Amino Acids and Proteins 34
2.4 Primary Structure of Proteins 44
2.5 Higher Levels of Protein Organization 49
2.6 Other Types of Proteins 61
2.7 Folding of Proteins From Randomized to Unique Structures: Protein Stability 72
2.8 Dynamic Aspects of Protein Structure 76
2.9 Methods for the Study of Higher Levels of Protein Structure and Organization 78

CLINICAL CORRELATIONS
2.1 Use of Amino Acid Analysis in Diagnosis of Disease 42
2.2 Differences in the Primary Structure of Insulins Used in the Treatment of Diabetes Mellitus 49
2.3 A Nonconservative Mutation Occurs in Sickle Cell Anemia 50
2.4 Symptoms of Diseases of Abnormal Collagen Synthesis 61
2.5 Hyperlipidemias 67
2.6 Hypolipoproteinemias 68
2.7 Functions of Glycoproteins 70
2.8 Glycosylated Hemoglobin, HbA_{1c} 71

3 PROTEINS II: STRUCTURE–FUNCTION RELATIONSHIP OF PROTEIN FAMILIES 91
Richard M. Schultz and Michael N. Liebman
3.1 Overview 92
3.2 Antibody Molecules and the Immunoglobulin Superfamily 93
3.3 Proteins With a Common Catalytic Mechanism: The Serine Proteases 102
3.4 DNA Binding Proteins 116
3.5 Hemoglobin and Myoglobin 119

CLINICAL CORRELATIONS
3.1 The Complement Proteins 96
3.2 Functions of the Different Antibody Classes 96
3.3 Immunization 97
3.4 Fibrin Formation in Myocardial Infarct and the Action of Recombinant Tissue Plasminogen Activator (rt-PA) 104
3.5 Involvement of Serine Proteases in Tumor Cell Metastasis 105

4 ENZYMES: CLASSIFICATION, KINETICS, AND CONTROL 135
J. Lyndal York
4.1 General Concepts 136
4.2 Classification of Enzymes 137
4.3 Kinetics 141
4.4 Coenzymes: Structure and Function 149
4.5 Inhibition of Enzymes 157
4.6 Allosteric Control of Enzyme Activity 162

4.7 Enzyme Specificity: The Active Site 167
4.8 Mechanism of Catalysis 171
4.9 Clinical Applications of Enzymes 179
4.10 Regulation of Enzyme Activity 186
4.11 Amplification of Regulatory Signals 190

CLINICAL CORRELATIONS
4.1 A Case of Gout Demonstrates Two Phases in the Mechanism of Enzyme Action 146
4.2 The Physiological Effect of Changes in Enzyme K_m Value 147
4.3 Mutation of a Coenzyme Binding Site Results in Clinical Disease 149
4.4 A Case of Gout Demonstrates the Difference Between an Allosteric Site and the Substrate Binding Site 163
4.5 Thermal Lability of G6PD Results in a Hemolytic Anemia 179
4.6 Alcohol Dehydrogenase Isoenzymes With Different pH Optima 180
4.7 Identification and Treatment of an Enzyme Deficiency 181
4.8 Ambiguity in the Assay of Mutated Enzymes 183

5 BIOLOGICAL MEMBRANES: STRUCTURE AND MEMBRANE TRANSPORT 195
Thomas M. Devlin
5.1 Overview 196
5.2 Chemical Composition of Membranes 197
5.3 Micelles and Liposomes 206
5.4 Structure of Biological Membranes 208
5.5 Movement of Molecules Through Membranes 215
5.6 Channels and Pores 220
5.7 Passive Mediated Transport Systems 223
5.8 Active Mediated Transport Systems 225
5.9 Ionophores 231

CLINICAL CORRELATIONS
5.1 Liposomes as Carriers of Drugs and Enzymes 208
5.2 Abnormalities of Cell Membrane Fluidity in Disease States 214
5.3 Diseases Due to Loss of Membrane Transport Systems 232

6 BIOENERGETICS AND OXIDATIVE METABOLISM 237
Merle S. Olson
6.1 Energy-Producing and Energy-Utilizing Systems 238
6.2 Thermodynamic Relationships and Energy-Rich Components 241
6.3 Sources and Fates of Acetyl Coenzyme A 247
6.4 The Tricarboxylic Acid Cycle 253
6.5 Structure and Compartmentation of the Mitochondrial Membranes 261
6.6 Electron Transfer 270
6.7 Oxidative Phosphorylation 283

CLINICAL CORRELATIONS
6.1 Pyruvate Dehydrogenase Deficiency 252
6.2 Mitochondrial Myopathies 268
6.3 Cyanide Poisoning 282
6.4 Hypoxic Injury 284

7 CARBOHYDRATE METABOLISM I: MAJOR METABOLIC PATHWAYS AND THEIR CONTROL 291
Robert A. Harris
7.1 Overview 292
7.2 Glycolysis 293
7.3 The Glycolytic Pathway 297
7.4 Regulation of the Glycolytic Pathway 309
7.5 Gluconeogenesis 324
7.6 Glycogenolysis and Glycogenesis 337

CLINICAL CORRELATIONS
7.1 Alcohol and Barbiturates 307
7.2 Arsenic Poisoning 309
7.3 Fructose Intolerance 311
7.4 Diabetes Mellitus 312
7.5 Lactic Acidosis 316
7.6 Pickled Pigs and Malignant Hyperthermia 317
7.7 Angina Pectoris and Myocardial Infarction 318
7.8 Pyruvate Kinase Deficiency and Hemolytic Anemia 324
7.9 Hypoglycemia and Premature Infants 325
7.10 Hypoglycemia and Alcohol Intoxication 336
7.11 Glycogen Storage Diseases 340

8 CARBOHYDRATE METABOLISM II: SPECIAL PATHWAYS 359
Nancy B. Schwartz
8.1 Overview 360
8.2 Pentose Phosphate Pathway 361
8.3 Sugar Interconversions and Nucleotide Sugar Formation 365
8.4 Biosynthesis of Complex Carbohydrates 372
8.5 Glycoproteins 373
8.6 Proteoglycans 378

CLINICAL CORRELATIONS
8.1 Glucose-6-Phosphate Dehydrogenase: Genetic Deficiency or Presence of Genetic Variants in Erythrocytes 362
8.2 Essential Fructosuria and Fructose Intolerance: Deficiency of Fructokinase and Fructose 1-Phosphate Aldolase 366

8.3 Galactosemia: Inability to Transform Galactose Into Glucose 368
8.4 Pentosuria: Deficiency of Xylitol Dehydrogenase 368
8.5 Glucuronic Acid: Physiological Significance of Glucuronide Formation 369
8.6 Blood Group Substances 373
8.7 Aspartylglycosylaminuria: Absence of 4-L-Aspartylglycosamine Amidohydrolase 374
8.8 Heparin as an Anticoagulant 375
8.9 Mucopolysaccharidoses 383

9 LIPID METABOLISM I: UTILIZATION AND STORAGE OF ENERGY IN LIPID FORM 387
J. Denis McGarry
9.1 Overview 388
9.2 The Chemical Nature of Fatty Acids and Acylglycerols 389
9.3 Sources of Fatty Acids 392
9.4 Storage of Fatty Acids as Triacylglycerols 401
9.5 Methods of Interorgan Transport of Fatty Acids and Their Primary Products 404
9.6 Utilization of Fatty Acids for Energy Production 407
9.7 Characteristics, Metabolism, and Functional Role of Polyunsaturated Fatty Acids 417

CLINICAL CORRELATIONS
9.1 Obesity 404
9.2 Genetic Abnormalities in Lipid-Energy Transport 405
9.3 Genetic Deficiencies in Carnitine or Carnitine Palmitoyl Transferase 409
9.4 Genetic Deficiencies in the Acyl CoA Dehydrogenases 410
9.5 Refsum's Disease 414
9.6 Diabetic Ketoacidosis 417

10 LIPID METABOLISM II: PATHWAYS OF METABOLISM OF SPECIAL LIPIDS 423
Robert H. Glew
10.1 Overview 424
10.2 Phospholipids 424
10.3 Cholesterol 438
10.4 Sphingolipids 449
10.5 Prostaglandins and Thromboxanes 461
10.6 Lipoxygenase and the Oxy-Eicosatetraenoic Acids 466

CLINICAL CORRELATIONS
10.1 Respiratory Distress Syndrome 428
10.2 Treatment of Hypercholesterolemia 445
10.3 Atherosclerosis 446
10.4 Diagnosis of Gaucher's Disease in an Adult 461

11 AMINO ACID METABOLISM I: GENERAL PATHWAYS 475
Alan H. Mehler
11.1 Overview 475
11.2 General Reactions of Amino Acids 477
11.3 Reactions of Ammonia 480
11.4 The Urea Cycle 481
11.5 Regulation of the Urea Cycle 485
11.6 Alternative Reactions for Elimination of Excess Nitrogen 488

CLINICAL CORRELATIONS
11.1 Hyperammonemia and Hepatic Coma 483
11.2 Deficiencies of Urea Cycle Enzymes 487

12 AMINO ACID METABOLISM II: METABOLISM OF THE INDIVIDUAL AMINO ACIDS 491
Alan H. Mehler
12.1 Overview 492
12.2 Glucogenic and Ketogenic Amino Acids 492
12.3 Metabolism of Individual Amino Acids 493
12.4 Folic Acid and One-Carbon Metabolism 516
12.5 Nitrogenous Derivatives of Amino Acids 518
12.6 Glutathione 522

CLINICAL CORRELATIONS
12.1 Diseases of Metabolism of Branched-Chain Amino Acids 492
12.2 Diseases of Propionate and Methylmalonate Metabolism 496
12.3 Nonketotic Hyperglycinemia 498
12.4 Diseases of Sulfur Amino Acids 503
12.5 Phenylketonuria 506
12.6 Disorders of Tyrosine Metabolism 508
12.7 Parkinson's Disease 509
12.8 Histidinemia 512
12.9 Diseases Involving Lysine and Ornithine 515
12.10 Folic Acid Deficiency 517
12.11 Diseases of Folate Metabolism 518
12.12 Clinical Problems Related to Glutathione Deficiency 523

13 PURINE AND PYRIMIDINE NUCLEOTIDE METABOLISM 529
Joseph G. Cory
13.1 Overview 530
13.2 Metabolic Functions of Nucleotides 530
13.3 Chemistry of Nucleotides 532
13.4 Metabolism of Purine Nucleotides 536
13.5 Metabolism of Pyrimidine Nucleotides 547

13.6 Deoxyribonucleotide Formation 551
13.7 Nucleoside and Nucleotide Kinases 555
13.8 Nucleotide Metabolizing Enzymes as a Function of the Cell Cycle and Rate of Cell Division 557
13.9 Nucleotide Coenzyme Synthesis 559
13.10 Synthesis and Utilization of 5-Phosphoribosyl 1-Pyrophosphate 563
13.11 Compounds That Interfere With Cellular Purine and Pyrimidine Nucleotide Metabolism: Use as Chemotherapeutic Agents 566
13.12 Purine and Pyrimidine Analogs as Antiviral Agents 569
13.13 Biochemical Basis for the Development of Drug Resistance 570

CLINICAL CORRELATIONS
13.1 Lesch-Nyhan Syndrome 543
13.2 Immunodeficiency Diseases Associated With Defects in Purine Nucleotide Metabolism 544
13.3 Gout 546
13.4 Orotic Aciduria 548

14 METABOLIC INTERRELATIONSHIPS 575
Robert A. Harris and David W. Crabb
14.1 Overview 576
14.2 Starve–Feed Cycle 577
14.3 Mechanisms Involved in Switching the Metabolism of the Liver Between the Well-Fed State and the Starved State 586
14.4 Metabolic Interrelationships of Tissues in Various Nutritional and Hormonal States 593

CLINICAL CORRELATIONS
14.1 Obesity 576
14.2 Protein Malnutrition 577
14.3 Starvation 577
14.4 Reye's Syndrome 582
14.5 Hyperglycemic, Hyperosmolar Coma 584
14.6 Hyperglycemia and Protein Glycosylation 585
14.7 Insulin-Dependent Diabetes Mellitus 597
14.8 Complications of Diabetes and the Polyol Pathway 598
14.9 Noninsulin-Dependent Diabetes Mellitus 600
14.10 Cancer Cachexia 602

15 DNA: THE REPLICATIVE PROCESS AND REPAIR 607
Stelios Aktipis
15.1 Biological Properties of DNA 608
15.2 Structure of DNA 609
15.3 Types of DNA Structure 627
15.4 DNA Structure and Function 641
15.5 Formation of the Phosphodiester Bond In Vivo 649
15.6 Mutation and Repair of DNA 653
15.7 DNA Replication 662
15.8 DNA Recombination 675
15.9 Sequencing of Nucleotides in DNA 677

CLINICAL CORRELATIONS
15.1 Diagnostic Use of Probes in Medicine 626
15.2 Mutations and the Etiology of Cancer 658
15.3 Diseases of DNA Repair and Susceptibility to Cancer 662
15.4 The Nucleotide Sequence of the Human Genome 677

16 RNA: STRUCTURE, TRANSCRIPTION, AND POSTTRANSCRIPTIONAL MODIFICATION 681
Francis J. Schmidt
16.1 Overview 681
16.2 Structure of RNA 683
16.3 Types of RNA 686
16.4 Mechanisms of Transcription 695
16.5 Posttranscriptional Processing 708
16.6 Nucleases and RNA Turnover 717

CLINICAL CORRELATIONS
16.1 Staphylococcal Resistance to Erythromycin 692
16.2 Antibiotics and Toxins Targeting RNA Polymerase 699
16.3 Involvement of Transcriptional Factors in Carcinogenesis 707
16.4 Thalassemia Due to Defects in Messenger RNA Synthesis 714
16.5 Autoimmunity in Connective Tissue Disease 717

17 PROTEIN SYNTHESIS: TRANSLATION AND POSTTRANSLATIONAL MODIFICATIONS 723
Dohn Glitz
17.1 Overview 724
17.2 Components of the Translational Apparatus 724
17.3 Protein Biosynthesis 736
17.4 Protein Maturation: Modification, Secretion, and Targetting to Organelles 746
17.5 Organelle Biogenesis and Targetting 753
17.6 Further Posttranslational Protein Modifications 755
17.7 Protein Degradation and Turnover 761

CLINICAL CORRELATIONS
17.1 Missense Mutation: Hemoglobin 732
17.2 Disorders of Terminator Codons 732

17.3 Thalassemia 733
17.4 I-Cell Disease 753
17.5 Familial Hyperproinsulinemia 755
17.6 Defects in Collagen Synthesis 760

18 RECOMBINANT DNA AND BIOTECHNOLOGY 767
Gerald Soslau
18.1 Overview 768
18.2 Restriction Endonucleases and Restriction Maps 769
18.3 DNA Sequencing 770
18.4 Recombinant DNA and Cloning 773
18.5 Selection of Specific Cloned DNA in Libraries 778
18.6 Techniques for Detection and Identification of Nucleic Acids 781
18.7 Complementary DNA and Complementary DNA Libraries 784
18.8 Bacteriophage, Cosmid, and Yeast Cloning Vectors 785
18.9 Techniques to Further Analyze Long Stretches of DNA 788
18.10 Expression Vectors and Fusion Proteins 788
18.11 Expression Vectors in Eucaryotic Cells 790
18.12 Site-Directed Mutagenesis 793
18.13 Polymerase Chain Reaction 797
18.14 Applications of Recombinant DNA Technologies 798
18.15 Concluding Remarks 802

CLINICAL CORRELATIONS
18.1 Restriction Mapping and Evolution 771
18.2 Restriction Fragment Length Polymorphisms Determine the Clonal Origin of Tumors 783
18.3 Site Directed Mutagenesis of HSV I gD 795
18.4 Polymerase Chain Reaction and Screening for Human Immunodeficiency Virus 799
18.5 Transgenic Animal Models 800

19 REGULATION OF GENE EXPRESSION 805
John E. Donelson
19.1 Overview 806
19.2 The Unit of Transcription in Bacteria: The Operon 807
19.3 The Lactose Operon of *E. coli* 808
19.4 The Tryptophan Operon of *E. coli* 815
19.5 Other Bacterial Operons 821
19.6 Bacterial Transposons 824
19.7 Inversion of Genes in *Salmonella* 827
19.8 Organization of Genes in Mammalian DNAs 828
19.9 Repetitive DNA Sequences in Eucaryotes 829
19.10 Genes for Globin Proteins 833
19.11 Genes for Human Growth Hormone-Like Proteins 837
19.12 Genes in the Mitochondrion 839
19.13 Bacterial Expression of Foreign Genes 840
19.14 Introduction of the Rat Growth Hormone Gene into Mice 842

CLINICAL CORRELATIONS
19.1 Transmissible Multiple Drug Resistances 825
19.2 Prenatal Diagnosis of Sickle-Cell Anemia 836
19.3 Prenatal Diagnosis of Thalassemia 837
19.4 Leber's Hereditary Optic Neuropathy (LHON) 840

20 BIOCHEMISTRY OF HORMONES I: PEPTIDE HORMONES 847
Gerald Litwack
20.1 Overview 848
20.2 Hormones and the Hormonal Cascade System 849
20.3 Major Polypeptide Hormones and Their Actions 854
20.4 Genes and Formation of Polypeptide Hormones 858
20.5 Synthesis of Amino Acid Derived Hormones 862
20.6 Inactivation and Degradation of Hormones 868
20.7 Cell Regulation and Hormone Secretion 868
20.8 Cyclic Hormonal Cascade Systems 876
20.9 Hormone–Receptor Interactions 881
20.10 Structure of Receptors: β-Adrenergic Receptor 884
20.11 Internalization of Receptors 886
20.12 Intracellular Action: Protein Kinases 889
20.13 Oncogenes and Receptor Functions 897

CLINICAL CORRELATIONS
20.1 Testing the Activity of the Anterior Pituitary 852
20.2 Hypopituitarism 854
20.3 Lithium Treatment of Manic Depressive Illness: The Phosphatidylinositol Cycle 876

21 BIOCHEMISTRY OF HORMONES II: STEROID HORMONES 901
Gerald Litwack
21.1 Overview 901
21.2 Structures of Steroid Hormones 902

21.3 Biosynthesis of Steroid Hormones 905
21.4 Metabolic Inactivation of Steroid Hormones 907
21.5 Cell–Cell Communication and the Control of Synthesis and Release of Steroid Hormones 910
21.6 Transport of Steroid Hormones in Blood 915
21.7 Steroid Hormone Receptors 916
21.8 Receptor Activation and Up- and Down-Regulation 920
21.9 Specific Examples of Steroid Hormone Action at the Cellular Level 921

CLINICAL CORRELATIONS
21.1 Oral Contraception 912
21.2 Programmed Cell Death in the Ovarian Cycle 922

22 MOLECULAR CELL BIOLOGY 927
Thomas E. Smith
22.1 Overview 927
22.2 Stimuli Recognition and Transmembrane Signaling 928
22.3 Nervous Tissue: Metabolism and Function 929
22.4 Thc Eyc: Metabolism and Vision 939
22.5 Muscle Contraction 954
22.6 Mechanism of Blood Coagulation 966

CLINICAL CORRELATIONS
22.1 Myasthenia Gravis: A Neuromuscular Disorder 935
22.2 Chromosomal Location of Genes for Vision 953
22.3 Classic Hemophilia 970

23 BIOTRANSFORMATIONS: THE CYTOCHROMES P450 981
Richard T. Okita and Bettie Sue Siler Masters
23.1 Overview 981
23.2 Cytochrome P450: Nomenclature and Overall Reaction 982
23.3 Cytochrome P450: Multiple Forms 984
23.4 Inhibitors of Cytochrome P450 986
23.5 Cytochrome P450 Electron Transport Systems 987
23.6 Physiological Functions of Cytochromes P450 990

CLINICAL CORRELATIONS
23.1 Consequences of Induction of Drug-Metabolizing Enzymes 986
23.2 Deficiency of Cytochrome P450–21-Hydroxylase 994
23.3 Steroid Hormone Production During Pregnancy 996

24 IRON AND HEME METABOLISM 1001
Marilyn S. Wells and William M. Awad, Jr.
24.1 Iron Metabolism: Overview 1001
24.2 Iron-Containing Proteins 1002
24.3 Intestinal Absorption of Iron 1005
24.4 Molecular Regulation of Iron Utilization 1006
24.5 Iron Distribution and Kinetics 1007
24.6 Heme Biosynthesis 1009
24.7 Heme Catabolism 1017

CLINICAL CORRELATIONS
24.1 Iron-Deficiency Anemia 1008
24.2 Hemochromatosis and Iron-Fortified Diet 1009
24.3 Acute Intermittent Porphyria 1012
24.4 Neonatal Isoimmune Hemolysis 1019

25 GAS TRANSPORT AND pH REGULATION 1025
James Baggott
25.1 Introduction to Gas Transport 1025
25.2 Need for a Carrier of Oxygen in the Blood 1026
25.3 Hemoglobin and Allosterism: Effect of 2,3-Diphosphoglycerate 1029
25.4 Other Hemoglobins 1030
25.5 Physical Factors that Affect Oxygen Binding 1032
25.6 Carbon Dioxide Transport 1033
25.7 Interrelationships Among Hemoglobin, Oxygen, Carbon Dioxide, Hydrogen Ion, and 2,3-Diphosphoglycerate 1038
25.8 Introduction to pH Regulation 1039
25.9 Buffer Systems of Plasma, Interstitial Fluid, and Cells 1039
25.10 The Carbon Dioxide–Bicarbonate Buffer System 1041
25.11 Acid–Base Balance and Its Maintenance 1045
25.12 Compensatory Mechanisms 1049
25.13 Alternative Measures of Acid–Base Imbalance 1052
25.14 The Significance of Na^+ and Cl^- in Acid–Base Imbalance 1054

CLINICAL CORRELATIONS
25.1 Cyanosis 1029
25.2 Chemically Modified Hemoglobins: Methemoglobin and Sulfhemoglobin 1031
25.3 Hemoglobins With Abnormal Oxygen Affinity 1032

25.4 The Case of the Variable Constant 1042
25.5 The Role of Bone in Acid–Base Homeostasis 1046
25.6 Acute Respiratory Alkalosis 1051
25.7 Chronic Respiratory Acidosis 1052
25.8 Salicylate Poisoning 1053
25.9 Evaluation of Clinical Acid–Base Data 1055
25.10 Metabolic Alkalosis 1056

26 DIGESTION AND ABSORPTION OF BASIC NUTRITIONAL CONSTITUENTS **1059**
Ulrich Hopfer
26.1 Overview 1060
26.2 Digestion: General Considerations 1062
26.3 Epithelial Transport 1067
26.4 Digestion and Absorption of Proteins 1073
26.5 Digestion and Absorption of Carbohydrates 1077
26.6 Digestion and Absorption of Lipids 1081
26.7 Bile Acid Metabolism 1088

CLINICAL CORRELATIONS
26.1 Electrolyte Replacement Therapy in Cholera 1073
26.2 Neutral Amino Aciduria (Hartnup Disease) 1077
26.3 Disaccharidase Deficiency 1079
26.4 Cholesterol Stones 1086
26.5 A-β-lipoproteinemia 1087

27 PRINCIPLES OF NUTRITION I: MACRONUTRIENTS **1093**
Stephen G. Chaney
27.1 Overview 1093
27.2 Energy Metabolism 1094
27.3 Protein Metabolism 1096
27.4 Protein–Energy Malnutrition 1100
27.5 Excess Protein–Energy Intake 1101
27.6 Carbohydrates 1103
27.7 Fats 1104
27.8 Fiber 1105
27.9 Composition of Macronutrients in the Diet 1106

CLINICAL CORRELATIONS
27.1 Vegetarian Diets and Protein–Energy Requirements 1098
27.2 Low-Protein Diets and Renal Disease 1099
27.3 Providing Adequate Protein and Calories for the Hospitalized Patient 1100
27.4 Fad Diets: High Protein–High Fat 1102
27.5 Carbohydrate Loading and Athletic Endurance 1103
27.6 High-Carbohydrate–High-Fiber Diets and Diabetes 1104
27.7 Polyunsaturated Fatty Acids and Risk Factors for Heart Disease 1106
27.8 Metabolic Adaptation: The Relationship Between Carbohydrate Intake and Serum Triacylglycerols 1108

28 PRINCIPLES OF NUTRITION II: MICRONUTRIENTS **1115**
Stephen G. Chaney
28.1 Assessment of Malnutrition 1116
28.2 Recommended Dietary Allowances 1117
28.3 Fat-Soluble Vitamins 1118
28.4 Water-Soluble Vitamins 1126
28.5 Energy-Releasing Water-Soluble Vitamins 1127
28.6 Hematopoietic Water-Soluble Vitamins 1132
28.7 Other Water Soluble Vitamins 1136
28.8 Macrominerals 1137
28.9 Trace Minerals 1139
28.10 The American Diet: Fact and Fallacy 1143
28.11 Assessment of Nutritional Status in Clinical Practice 1144

CLINICAL CORRELATIONS
28.1 Nutritional Considerations for Cystic Fibrosis 1120
28.2 Renal Osteodystrophy 1122
28.3 Nutritional Considerations in the Newborn 1126
28.4 Anticonvulsant Drugs and Vitamin Requirements 1126
28.5 Nutritional Considerations in the Alcoholic 1128
28.6 Vitamin B_6 Requirements for Users of Oral Contraceptives 1131
28.7 Diet and Osteoporosis 1138
28.8 Calculation of Available Iron 1140
28.9 Iron-Deficiency Anemia 1141
28.10 Nutritional Considerations for Vegetarians 1145
28.11 Nutritional Needs of the Elderly 1145

Appendix: Review of Organic Chemistry **1149**
Carol N. Angstadt

Index **1159**

CHAPTER QUESTIONS AND ANSWERS

The questions at the end of each chapter are provided to help you test your knowledge and increase your understanding of biochemistry. Since they are intended to help you strengthen your knowledge, their construction does not always conform to principles for assessing your retention of individual facts. Specifically, you will sometimes be expected to draw on your knowledge of several areas to answer a single question, and some questions may take longer to analyze than the average time allowed on certain national examinations. Occasionally, you may disagree with the answer. If this occurs, we hope that after you read the commentary that accompanies the answer to the question, you will see the point and your insight into the biochemical problem will be increased.

The *q*uestion *t*ypes conform to those commonly used in objective examinations by medical school departments of biochemistry. For each question, the *q*uestion *t*ype is given in parentheses after its number; for example, 5.(QT2) indicates that question 5 is of type 2. Keys for the question types are given below:

QT1: Choose the one *best* answer

QT2: Answer the question according to the following key:
- A. If 1, 2, and 3 are correct
- B. If 1 and 3 are correct
- C. If 2 and 4 are correct
- D. If only 4 is correct
- E. If all four are correct

QT3: Answer the question according to the following key:
- A. If A is greater than B
- B. If B is greater than A
- C. If A and B are equal or nearly equal

QT4: Answer the question according to the following key:
- A. If the item is associated with A only
- B. If the item is associated with B only
- C. If the item is associated with both A and B
- D. If the item is associated with neither A nor B

QT5: Match the numbered statement or phrase with one of the lettered options given above.

TEXTBOOK OF

BIOCHEMISTRY

With Clinical Correlations

1

Eucaryotic Cell Structure

THOMAS M. DEVLIN

1.1 CELLS AND CELLULAR COMPARTMENTS 1
1.2 CELLULAR ENVIRONMENT—WATER AND SOLUTES 4
Hydrogen Bonds Form Between Water Molecules 4
Water Has Unique Solvent Properties 6
Some Molecules Dissociate With the Formation of Cations and Anions 6
Weak Electrolytes Partially Dissociate 7
Water Is a Weak Electrolyte 7
Many Biologically Important Molecules Are Acids or Bases 8
The Henderson–Hasselbalch Equation Defines the Relationship Between pH, and the Concentration of a Conjugate Acid and Base 10
Buffering Is Important to Control pH 11
1.3 ORGANIZATION AND COMPOSITION OF EUCARYOTIC CELLS 13
The Chemical Composition of Cells 14
1.4 FUNCTIONAL ROLE OF SUBCELLULAR ORGANELLES AND MEMBRANES 16
The Plasma Membrane Is the Limiting Boundary of a Cell 16
The Nucleus Is the Site of DNA and RNA Synthesis 17
The Endoplasmic Reticulum Has a Role in Many Synthetic Pathways 18
The Golgi Apparatus Is Involved in Sequestering of Proteins 18
Mitochondria Supply Most of the Cell's Needs for ATP 18
Lysosomes Are Required for Intracellular Digestion 19
Peroxisomes Contain Oxidative Enzymes Which Involve Hydrogen Peroxide 21
The Cytoskeleton Organizes the Intracellular Contents 21
The Cytosol Contains the Soluble Cellular Components 22
Conclusion 22
BIBLIOGRAPHY 22
QUESTIONS AND ANNOTATED ANSWERS 23
CLINICAL CORRELATIONS
1.1 Blood Bicarbonate Concentration in a Metabolic Acidosis 13
1.2 Lysosomal Enzymes and Gout 19
1.3 Zellweger Syndrome and the Absence of Peroxisomes 21

1.1 CELLS AND CELLULAR COMPARTMENTS

By a process not entirely understood and in a time span that is difficult to comprehend, elements such as carbon, hydrogen, oxygen, nitrogen, and phosphorus combined, dispersed, and recombined to form a variety of molecules until a combination was achieved that was capable of replicating itself. With continued evolution and the formation of ever more complex molecules, the environment around some of these self-replicating molecules was enclosed by a membrane. This development gave these primordial molecules a significant advantage in that they could control to some extent their own environment. A form of life had evolved and a unit of space, a cell, had been established. With the passing of time a diversity of cells evolved, and their chemistry and structure became more complex. Eventually, they were capable of extracting nutrients from the environment, chemically converting these nutrients to either sources of energy or to complex molecules, of controlling chemical processes they catalyzed, and of replicating themselves into other cells. The challenge of biochemical research is to unravel the chemical mechanisms behind the organized and controlled manner in which cells carry out their function.

One of the most important structures of all cells is the limiting outer membrane, termed the **plasma membrane,** which delineates the intracellular and extracellular environments. The plasma membrane separates the variable and potentially hostile environment outside the cell from the relatively constant milieu within the cell and is the communication link between the cell and its surroundings. It is the plasma membrane that delineates the space occupied by a cell.

Based on both microscopic and biochemical differences, living cells are divided into two major classes, the procaryotes and the eucaryotes. **Procaryotic cells,** which include bacteria, blue-green algae, and rickettsiae, lack extensive intracellular anatomy (Figure 1.1). Intracellular structures of cells are due to the presence of macromolecules or membrane systems, which can be visualized in a microscope under appropriate conditions The deoxyribonucleic acid (DNA) of procaryotes is often segregated into discrete masses, nucleoid region, but it is not surrounded by a membrane or envelope. The plasma membrane is often observed to be invaginated, but there are no definable subcellular organized bodies in procaryotes.

In contrast to the procaryotes, **eucaryotic cells,** which include yeasts, fungi, plant, and animal cells, have a well-defined membrane surrounding a central nucleus and a variety of intracellular structures and organelles (Figure 1.1*b*). The intracellular membrane systems establish a number of distinct subcellular compartments, as described in Section 1.4, permitting a unique degree of subcellular specialization. By compartmentalization of the cell, different chemical reactions requiring different environments can occur simultaneously. As an example, reactions requiring different ionic conditions or reactants can occur in different compartments of the same cell because the membranes surrounding a cellular compartment can control the environment in the compartment by regulating the movement of substances in and out.

Membranes are lipid in nature, and many compounds that are soluble in a hydrophobic lipid environment will concentrate in cellular membranes. Many biochemical reactions occur in specific membranes of the cell; thus, the extensive membrane systems of the eucaryotic cell create an additional environment that the cell can use for its diverse functions.

Besides the structural variations, as presented in Figures 1*a* and *b*, there are significant differences in the chemical composition and biochemical activities between procaryotic and eucaryotic cells. Some of the major differences between the cell types are the lack in procaryotic cells of a class of proteins, termed histones, which in eucaryotic cells complex with DNA; major structural differences in the ribonucleic acid–protein complexes involved in the biosynthesis of proteins; differences in transport mechanisms across the plasma membrane; and a host of differences in enzyme content.

Even though there are structural and biochemical differences between the cell types, the many similarities are equally striking. The emphasis throughout this textbook is on the chemistry of eucaryotic cells, particularly mammalian cells, but much of our knowledge of the biochemistry of

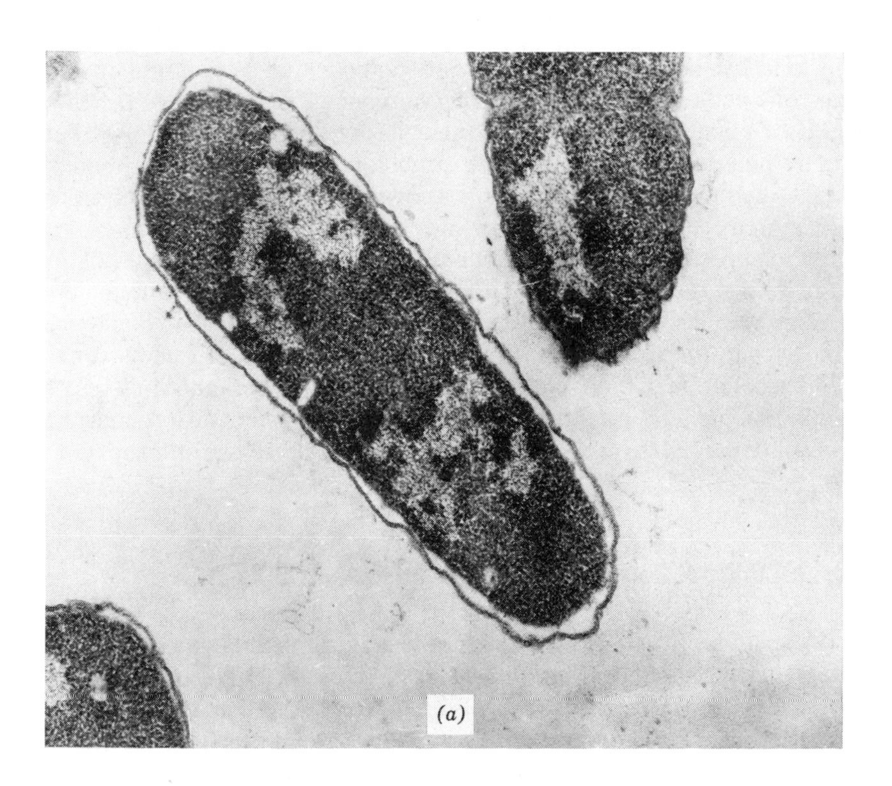

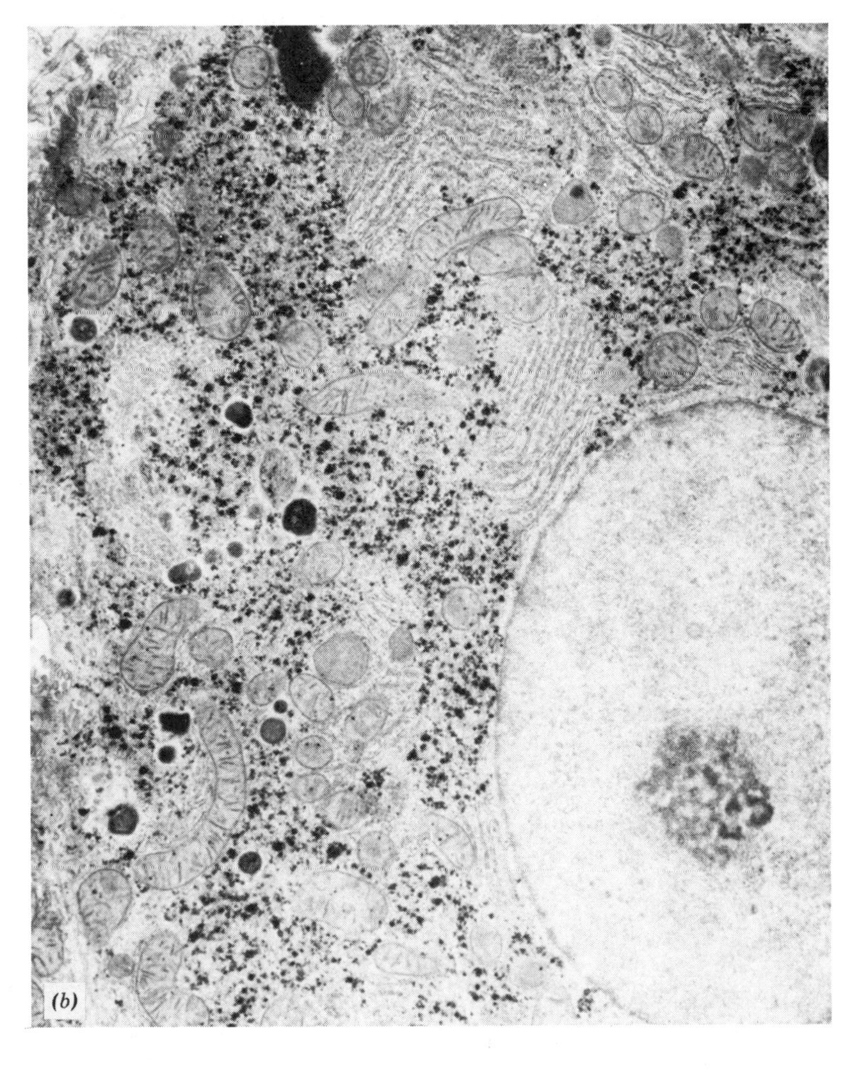

FIGURE 1.1
Cellular organization of procaryotic and eucaryotic cells.
(*a*) An electron micrograph of an *Escherichia* (*E. coli*), a representative procaryote; approximate magnification 30,000×. There is little apparent intracellular organization and no cytoplasmic organelles. The chromatin is condensed into a nuclear zone but is not surrounded by a membrane. Procaryotic cells are smaller than eucaryotic cells. (*b*) An electron micrograph of a thin section of a rat liver cell (hepatocyte), a representative eucaryotic cell; approximate magnification 7500×. Note the distinct nuclear membrane, the different membrane bound organelles or vesicles, and the extensive membrane systems. The various membranes create a variety of intracellular compartments.

Photograph (*a*) generously supplied by Dr. M. E. Bayer, Fox Chase Cancer Institute, Philadelphia, PA; photograph (*b*) reprinted with permission of Dr. K. R. Porter, from Porter, K. R. and Bonneville, M. A. in *Fine Structure of Cells and Tissues*, Philadelphia: Lea & Febiger, 1972.

living cells has come from studies of procaryotic and from nonmammalian eucaryotic cells. The basic chemical components and many of the fundamental chemical reactions of all living cells are very similar. The availability of certain cell populations, for example, bacteria in contrast to human liver, has led to an accumulation of knowledge about some cells; in fact, in some areas of biochemistry our knowledge is derived exclusively from studies of procaryotes. The universality of many biochemical phenomena, however, permits many extrapolations from bacteria to humans.

Before we dissect and reassemble the complexities of mammalian cells and tissues in the following chapters, it is appropriate to review some of the chemical and physical characteristics of the environment in which the various biochemical phenomena occur. It is important to recognize the constraints placed by the environment on the activities carried out by cells.

1.2 CELLULAR ENVIRONMENT—WATER AND SOLUTES

There is a general uniformity in nature and all biological cells contain essentially the same building blocks and types of macromolecules. The general classes of substances in cells are presented in Table 1.1. There are differences in the concentration of specific components in different cells; in eucaryotic cells, there are even differences between intracellular compartments. It is also considered that there are microenvironments created by the cellular macromolecules in which the composition is distinct from the surrounding areas. All biological cells depend on the environment for nutrients, which are requisites for replacement of components, for growth, and for their energy requirements. The composition of the external environment can vary significantly and cells have a variety of mechanisms to cope with these variations. In addition, the different intracellular compartments also have different biochemical and chemical compositions. The one common characteristic of the different environments is the presence of water. **Water,** the solvent in which the substances required for the cell's existence are dissolved or suspended, has an important role in the well-being of all cells. The unique physicochemical properties of water make life possible.

Hydrogen Bonds Form Between Water Molecules

Two hydrogen atoms share their electrons with an unshared pair of electrons of an oxygen atom to form the water molecule. This deceptively simple molecule, however, has a number of unusual properties. The oxygen nucleus has a stronger attraction for the shared electrons than the

TABLE 1.1 Chemical Components of Biological Cells

Component	Range of molecular weights
H_2O	18
Inorganic ions Na^+, K^+, Cl^-, SO_4^{2-}, HCO_3^-, Ca^{2+}, Mg^{2+}, etc.	23–100
Small organic molecules Carbohydrates, amino acids, lipids, nucleotides, peptides	100–1200
Macromolecules Proteins, polysaccharides, nucleic acids	50,000–1,000,000,000

hydrogen, and the positively charged hydrogen nuclei are left with an unequal share of electrons, creating a partial positive charge on each hydrogen and a partial negative charge on the oxygen. The bond angle between the hydrogens and the oxygen is 104.5°, making the molecule electrically asymmetric and producing an **electric dipole** (Figure 1.2). Water molecules interact because the positively charged hydrogen atoms on one molecule are attracted to the negatively charged oxygen atom on another, with the formation of a weak bond between the two water molecules, as in Figure 1.3*a*. This bond, indicated by a dashed line, is termed a **hydrogen bond.** Five molecules of water form a tetrahedral structure (Figure 1.3*b*), since each oxygen can share its electrons with four hydrogen atoms and each hydrogen with another oxygen. A tetrahedral lattice structure is formed in ice and it is the hydrogen bonding between molecules that gives ice its crystalline structure. Some of these bonds are broken as ice is transformed to liquid water. Each hydrogen bond is relatively weak compared to a covalent bond, but the large number of hydrogen bonds between molecules in liquid water is the reason for the stability of water. Even at 100°C liquid water contains a significant number of hydrogen bonds, which accounts for its high heat of vaporization; in the transformation from a liquid to a vapor state the hydrogen bonds are disrupted.

Water molecules hydrogen bond to different chemical structures. Hydrogen bonding also occurs between other molecules and even within a molecule wherever an electronegative oxygen or nitrogen comes in close proximity to a hydrogen covalently bonded to another electronegative group. Some representative hydrogen bonds are presented in Figure 1.4. Intramolecular hydrogen bonding occurs extensively in large macromolecules such as proteins and nucleic acids and is the basis for their structural stability.

Breaking and forming hydrogen bonds occurs more rapidly than covalent bonds because of their low energy; it is estimated that hydrogen bonds in water have a half-life of less than 10^{-10} s. Liquid water actually has a definite structure due to the hydrogen bonding between molecules,

FIGURE 1.2
Structure of a water molecule.
The H—O—H bond angle is 104.5°. Both hydrogen atoms carry a partial positive and the oxygen a partial negative charge, creating a dipole.

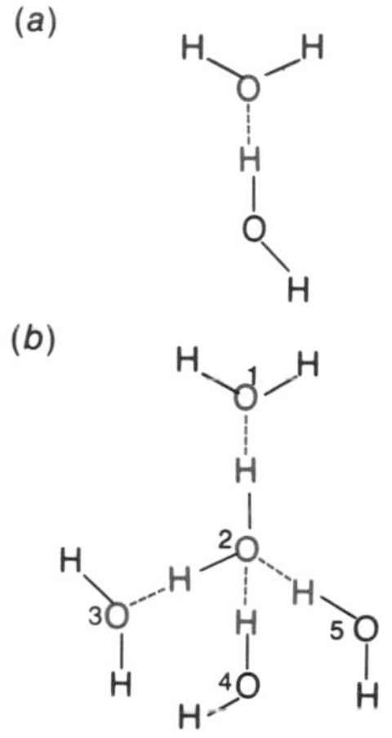

FIGURE 1.3
(*a*) Hydrogen bonding, as indicated by the dashed lines, between two water molecules. (*b*) Tetrahedral hydrogen bonding of five water molecules. Water molecules 1, 2, and 3 are in the plane of the page, 4 is below, and 5 is above.

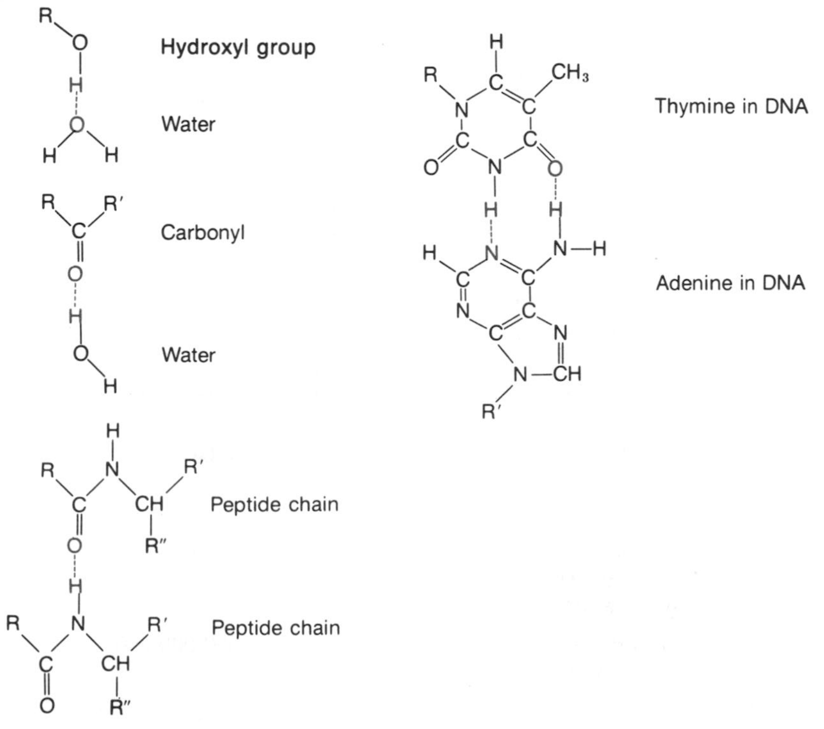

FIGURE 1.4
Representative hydrogen bonds of importance in biological systems.

but the structure is in a dynamic state as the hydrogen bonds break and reform. A similar dynamic interaction also occurs with substances present in the liquid, which are capable of hydrogen bonding. Many models have been proposed for the structure of liquid water, but none adequately explains all of its properties.

Water Has Unique Solvent Properties

The polar nature of the water molecule and the ability to form hydrogen bonds are the basis for its unique solvent properties. Polar molecules are readily dispersed in water. Salts in which the crystal lattice is held together by the attraction of the positive and negative groups dissolve in water because the electrostatic forces in the crystal can be overcome by the attraction of the charges to the dipole of water. Sodium chloride (NaCl) is an example where the electrostatic attraction of the Na^+ and Cl^- is overcome by interaction of Na^+ with the negative charge on the oxygen, and the Cl^- with the positive charge on the protons. A shell of water surrounds the individual ions. The number of weak charge–charge interactions between water and the Na^+ and Cl^- ions is sufficient to separate the two charged ions.

Many organic molecules containing nonionic but weakly polar groups are also soluble in water because of the attraction of the groups to water molecules. Sugars and alcohols are readily soluble in water for this reason. **Amphipathic molecules,** that is, compounds containing both polar and nonpolar groups, will disperse in water if the attraction of the polar group for water can overcome possible hydrophobic interactions of the nonpolar portions of the molecules. Very hydrophobic molecules, such as compounds containing long hydrocarbon chains, however, will not readily disperse in water but interact with one another to exclude the polar water molecules.

Some Molecules Dissociate With the Formation of Cations and Anions

Substances that dissociate in water into a cation (positively charged ion) and an anion (negatively charged ion) are classified as **electrolytes.** The presence of these charged ions, partially prevented from interacting with one another because of the attraction of water molecules to the individual ions, facilitates the conductance of an electrical current through an aqueous solution. Sugars or alcohols, which readily dissolve in water but do not carry a charge or dissociate into species with a charge, are classified as **nonelectrolytes.**

Salts of the alkali metals (e.g., Li, Na and K), when dissolved in water at low concentrations, dissociate totally; at high concentrations, however, there is an increased potential for interaction of the anion and cation. For biological systems in which the concentrations of salts are low, it is customary to consider that such compounds are totally dissociated. The salts of the organic acids, for example, sodium lactate, also dissociate totally and are classified as electrolytes. The anion that is formed reacts to a limited extent with a proton to form the undissociated acid (Figure 1.5). These salts are also electrolytes. It is important to remember that when such salts are dissolved in water the individual ions are present in solution rather than the undissociated salt. If a solution has been prepared with several different salts (e.g., NaCl, K_2SO_4, and Na acetate) the original molecules do not exist as such in the solution, only the ions (e.g., Na^+, K^+, SO_4^{2-}, and acetate$^-$).

Many acids, however, when dissolved in water do not totally dissociate but rather establish an equilibrium between the undissociated compound

(1) CH_3—CHOH—COONa ⟶
Na lactate
Na^+ + CH_3—CHOH—COO^-
Lactate ion

(2) CH_3—CHOH—COO^- + H^+ ⇌
Lactate ion
CH_3—CHOH—COOH
Lactic acid

FIGURE 1.5
Reactions occurring when sodium lactate is dissolved in water.

and two or more ions. An example is lactic acid, an important metabolic intermediate, which partially dissociates into a lactate anion and a H^+ as follows:

$$CH_3—CHOH—COOH \rightleftharpoons CH_3—CHOH—COO^- + H^+$$

Because of their partial dissociation, however, such compounds have a lower capacity to carry an electrical charge on a molar basis when compared to a compound that dissociates totally; they are termed **weak electrolytes.**

Weak Electrolytes Partially Dissociate

In the partial dissociation of a weak electrolyte, represented by HA, the concentration of the various species where A^- represents the dissociated anion can be determined from the equilibrium equation

$$K'_{eq} = \frac{[H^+][A^-]}{[HA]}$$

The brackets indicate the concentration of each component in moles per liter (mol L^{-1}). The activities of each species rather than concentration should be employed in the equilibrium equation, but most of the compounds of interest in biological systems are present in low concentrations, and the value of the activity approaches that for the concentration; the **equilibrium constant** is indicated as K'_{eq} to indicate that it is an apparent equilibrium constant. The term K'_{eq} is a function of the temperature of the system, increasing with increasing temperatures. The degree of dissociation of an electrolyte will depend on the affinity of the anion for a proton; if the weak dipole forces of water interacting with the anion and cation are stronger than the electrostatic forces between the H^+ and anion, there will be a greater degree of dissociation. From the dissociation equation above it is apparent that if the degree of dissociation of a substance is small, K'_{eq} will be a small number, but if the degree of dissociation is large, the number will be large. Obviously, for compounds that dissociate totally, a K'_{eq} cannot be determined because at equilibrium there is no remaining undissociated solute.

Water Is a Weak Electrolyte

Water also dissociates as follows:

$$HOH \rightleftharpoons H^+ + OH^-$$

The proton that dissociates will interact with the oxygen of another water molecule forming the **hydronium ion,** H_3O^+. For convenience, the proton will not be presented as H_3O^+, even though this is the chemical species actually present. At 25°C the value of K'_{eq} is very small and is in the range of about 1.8×10^{-16}:

$$K'_{eq} = 1.8 \times 10^{-16} = \frac{[H^+][OH^-]}{[H_2O]} \quad \textbf{(1.1)}$$

With such a small K'_{eq} there is an insignificant dissociation of water, and the concentration of water, which is 55.5 M, will be essentially unchanged. Equation 1.1 can be rewritten as follows:

$$K'_{eq} \cdot [H_2O] = [H^+][OH^-] \quad \textbf{(1.2)}$$

TABLE 1.2 Relationships between $[H^+]$ and pH and $[OH^-]$ and pOH

$[H^+]$ (M)	pH	$[OH^-]$ (M)	pOH
1.0	0	1×10^{-14}	14
0.1 (1×10^{-1})	1	1×10^{-13}	13
1×10^{-2}	2	1×10^{-12}	12
1×10^{-3}	3	1×10^{-11}	11
1×10^{-4}	4	1×10^{-10}	10
1×10^{-5}	5	1×10^{-9}	9
1×10^{-6}	6	1×10^{-8}	8
1×10^{-7}	7	1×10^{-7}	7
1×10^{-8}	8	1×10^{-6}	6
1×10^{-9}	9	1×10^{-5}	5
1×10^{-10}	10	1×10^{-4}	4
1×10^{-11}	11	1×10^{-3}	3
1×10^{-12}	12	1×10^{-2}	2
1×10^{-13}	13	0.1 (1×10^{-1})	1
1×10^{-14}	14	1.0	0

The value of $K'_{eq} \times [55.5]$ equals the product of H^+ and OH^- concentrations and is termed the ion product of water. The value at 25°C is 1×10^{-14}. In pure water the concentration of H^+ equals OH^-, and substituting $[H^+]$ for $[OH^-]$ in the equation above, the $[H^+]$ is 1×10^{-7} M. Similarly, the $[OH^-]$ is also 1×10^{-7} M. The equilibrium among H_2O, H^+, and OH^- exists in dilute solutions containing dissolved substances; if the dissolved material alters either the H^+ or OH^- concentration, such as on addition of an acid or base, a concomitant change in the other ion must occur.

Using the equation for the ion product, the $[H^+]$ or $[OH^-]$ can be calculated if the concentration of one of the ions is known.

The importance of the hydrogen ion in biological systems will become apparent in subsequent chapters. For convenience $[H^+]$ is usually expressed in terms of **pH,** calculated as follows:

$$\text{pH} = \log \frac{1}{[H^+]} \qquad \textbf{(1.3)}$$

In pure water the concentration of hydrogen ion and hydroxyl ion are both 1×10^{-7} M, and the pH = 7.0. The OH^- ion concentration can be expressed in a similar fashion as the pOH. For the equation describing the dissociation of water, $1 \times 10^{-14} = [H^+][OH^-]$, and taking the negative logarithm of both sides, the equation becomes 14 = pH + pOH. Table 1.2 presents the relationship between pH and H^+ concentration.

The pH of different biological fluids is presented in Table 1.3. In blood plasma, the hydrogen ion concentration is 0.000,000,04 M or a pH of 7.4, whereas the concentrations of other cations are between 0.001 and 0.10 M. An increase in hydrogen ion to 0.000,000,1 M (pH 7.0) leads to serious medical consequences and is life threatening; a detailed discussion of the mechanisms by which the body maintains intra- and extra-cellular pH is presented in Chapter 25.

TABLE 1.3 pH of Some Biological Fluids

Fluid	pH
Blood plasma	7.4
Interstitial fluid	7.4
Intracellular fluid	
Cytosol (liver)	6.9
Lysosomal matrix	5.5–6.5
Gastric juice	1.5–3.0
Pancreatic juice	7.8–8.0
Human milk	7.4
Saliva	6.4–7.0
Urine	5.0–8.0

Many Biologically Important Molecules Are Acids or Bases

The definitions of an acid and a base proposed by Lowry and Brønsted are the most convenient in considering biological systems. **An acid is a proton donor** and **a base is a proton acceptor.** Hydrochloric acid (HCl) and sulfuric acid (H_2SO_4) are strong acids because they dissociate totally and the OH^- ion is a base because it will accept a proton shifting the equilibrium,

$$OH^- + H^+ \rightleftharpoons H_2O$$

When a strong acid and OH^- are combined, the H^+ from the acid and OH^- interact, participating in an equilibrium with H_2O; since the ion product for water is so small, a neutralization of the H^+ and OH^- occurs.

In dilute solutions strong acids dissociate totally; the anions formed, as an example Cl^- from HCl, are not classed as bases because they do not associate with protons in solution. When an organic acid, such as lactic acid, is dissolved in water, however, it dissociates only partially, establishing an equilibrium between the acid, an anion, and a proton as follows:

$$\text{Lactic acid} \rightleftharpoons \text{lactate}^- + H^+$$

The acid is considered to be a weak acid and the anion a base because it can accept a proton and reform the acid. *The combination of a weak acid and the base that is formed on dissociation is referred to as a conjugate pair;* examples are presented in Table 1.4. Ammonium ion (NH_4^+) is an

TABLE 1.4 Some Conjugate Acid–Base Pairs of Importance in Biological Systems

Proton donor (acid)		Proton acceptor (base)
CH_3—CHOH—COOH (lactic acid)	$\rightleftharpoons$	$H^+ + CH_3$—CHOH—COO^- (lactate)
CH_3—CO—COOH (pyruvic acid)	$\rightleftharpoons$	$H^+ + CH_3$—CO—COO^- (pyruvate)
HOOC—CH_2—CH_2—COOH (succinic acid)	$\rightleftharpoons$	$2H^+ + {}^-OOC$—CH_2—CH_2—COO^- (succinate)
$^+H_3NCH_2$—COOH (glycine)	$\rightleftharpoons$	$H^+ + {}^+H_3N$—CH_2—COO^- (glycinate)
H_3PO_4	$\rightleftharpoons$	$H^+ + H_2PO_4^-$
$H_2PO_4^-$	$\rightleftharpoons$	$H^+ + HPO_4^{2-}$
HPO_4^{2-}	$\rightleftharpoons$	$H^+ + PO_4^{3-}$
Glucose 6-PO_3H^-	$\rightleftharpoons$	H^+ + glucose 6-PO_3^{2-}
H_2CO_3	$\rightleftharpoons$	$H^+ + HCO_3^-$
NH_4^+	$\rightleftharpoons$	$H^+ + NH_3$
H_2O	$\rightleftharpoons$	$H^+ + OH^-$

acid because it dissociates to yield a H^+ and ammonia (NH_3), an uncharged species, which is the conjugate base. Phosphoric acid (H_3PO_4) is an acid and PO_4^{3-} is a base, but $H_2PO_4^-$ and HPO_4^{2-} can be classified as either base or acid, depending on whether the phosphate group is accepting or donating a proton.

The tendency of a conjugate acid to dissociate can be evaluated from the K'_{eq}; as indicated above, the smaller the value of the K'_{eq}, the less the tendency to give up a proton and the weaker the acid, the larger a K'_{eq}, the greater the tendency to dissociate a proton, and the stronger the acid. Water is a very weak acid with a K'_{eq} of 1×10^{-14} at 25°C.

A convenient method of stating the K'_{eq} is in the form of pK', which is defined as

$$pK' = \log \frac{1}{K'_{eq}} \tag{1.4}$$

Note the similarity of this definition with the definition of pH; as with pH and $[H^+]$, the relationship between pK' and K'_{eq} is an inverse one, and the smaller the K'_{eq}, the larger the pK'. K'_{eq}, and pK' values for representative conjugate acids of importance in biological systems are presented in Table 1.5.

A special case of a weak acid important in medicine is carbonic acid (H_2CO_3). Carbon dioxide when dissolved in water is involved in the following equilibrium reactions:

$$CO_2 + H_2O \overset{K'_2}{\rightleftharpoons} H_2CO_3 \overset{K'_1}{\rightleftharpoons} H^+ + HCO_3^-$$

Carbonic acid is a relatively strong acid with a pK'_1 of 3.77. The equilibrium equation for this reaction is

$$K'_1 = \frac{[H^+][HCO_3^-]}{[H_2CO_3]} \tag{1.5}$$

Carbonic acid is, however, in equilibrium with dissolved CO_2 and the equilibrium equation for this reaction is

$$K'_2 = \frac{[H_2CO_3]}{[CO_2][H_2O]} \tag{1.6}$$

TABLE 1.5 Apparent Dissociation Constant and pK' of Some Compounds of Importance in Biochemistry

Compound		K'_{eq}M	pK'
Acetic acid	(CH_3—COOH)	1.74×10^{-5}	4.76
Alanine	(CH_3—CH(NH_3^+)—COOH)	4.57×10^{-3}	2.34 (COOH)
		2.04×10^{-10}	9.69 (NH_3^+)
Citric acid	(HOOC—CH_2—COH(COOH)—CH_2—COOH)	8.12×10^{-4}	3.09
		1.77×10^{-5}	4.75
		3.89×10^{-6}	5.41
Glutamic acid	(HOOC—CH_2—CH_2—CH(NH_3^+)—COOH)	6.45×10^{-3}	2.19 (COOH)
		5.62×10^{-5}	4.25 (COOH)
		2.14×10^{-10}	9.67 (NH_3^+)
Glycine	(CH_2(NH_3^+)—COOH)	4.57×10^{-3}	2.34 (COOH)
		2.51×10^{-10}	9.60 (NH_3^+)
Lactic acid	(CH_3—CHOH—COOH)	1.38×10^{-4}	3.86
Pyruvic acid	(CH_3—CO—COOH)	3.16×10^{-3}	2.50
Succinic acid	(HOOC—CH_2—CH_2—COOH)	6.46×10^{-5}	4.19
		3.31×10^{-6}	5.48
Glucose 6-PO_3H^-		7.76×10^{-7}	6.11
H_3PO_4		1×10^{-2}	2.0
$H_2PO_4^-$		2.0×10^{-7}	6.7
HPO_4^{2-}		3.4×10^{-13}	12.5
H_2CO_3		1.70×10^{-4}	3.77
NH_4^+		5.62×10^{-10}	9.25
H_2O		1×10^{-14}	14.0

Solving Eq. 1.6 for H_2CO_3 and substituting for the H_2CO_3 in Eq. 1.5, the two equilibrium reactions are combined into one equation:

$$K'_1 = \frac{[H^+][HCO_3^-]}{K'_2[CO_2][H_2O]} \quad \textbf{(1.7)}$$

Rearranging to combine constants, including the concentration of H_2O, simplifies the equation and yields a new combined constant, K'_3, as follows:

$$K'_1K'_2[H_2O] = K'_3 = \frac{[H^+][HCO_3^-]}{[CO_2]} \quad \textbf{(1.8)}$$

It is common practice to refer to the dissolved CO_2 as the conjugate acid; it is the acid anhydride of H_2CO_3. The term K'_3 has a value of 7.95×10^{-7} and p$K'_3 = 6.1$. If the aqueous system is in contact with an air phase, the dissolved CO_2 will also be in equilibrium with CO_2 in the air phase; a change in any of the components will cause a change in each of the components, that is CO_2 (air), CO_2 (dissolved), H_2CO_3, H^+, and HCO_3^-.

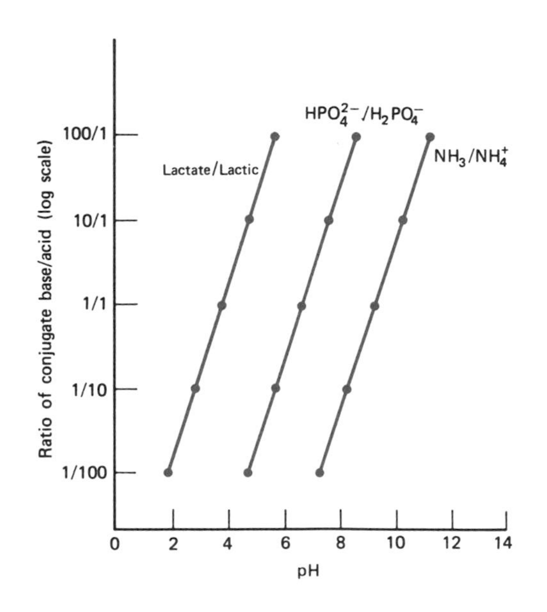

FIGURE 1.6
Ratio of conjugate base/acid as a function of the pH.
When the ratio of base/acid is 1, the pH equals the pK' of the weak acid.

The Henderson–Hasselbalch Equation Defines the Relationship Between pH, and the Concentration of a Conjugate Acid and Base

Changing the concentration of any one component in the equilibrium reaction necessitates a concomitant change in every component. An increase in [H^+] will decrease the concentration of conjugate base with an equivalent increase in the conjugate acid. This relationship is conveniently expressed by rearranging the equilibrium equation and solving for

H^+, as shown for the following dissociation:

$$\text{Conjugate acid} \rightleftharpoons \text{conjugate base} + H^+$$

$$K'_{eq} = \frac{[H^+][\text{conjugate base}]}{[\text{conjugate acid}]} \tag{1.9}$$

Rearranging Eq. 1.9 by dividing through by $[H^+]$ and K'_{eq} leads to

$$\frac{1}{[H^+]} = \frac{1}{K'_{eq}} \cdot \frac{[\text{conjugate base}]}{[\text{conjugate acid}]} \tag{1.10}$$

Taking the logarithm of both sides gives

$$\text{Log}\,\frac{1}{[H^+]} = \log\frac{1}{K'_{eq}} + \log\left(\frac{[\text{conjugate base}]}{[\text{conjugate acid}]}\right) \tag{1.11}$$

Since $\text{pH} = \log 1/[H^+]$ and $pK' = \log 1/K'_{eq}$, Eq. 1.11 becomes

$$\mathbf{pH = pK' + \log\left(\frac{[\text{conjugate base}]}{[\text{conjugate acid}]}\right)} \tag{1.12}$$

Equation 1.12, developed by **Henderson and Hasselbalch,** is a convenient way of viewing the relationship between the pH of a solution and the relative amounts of base and acid present. Analysis of Eq. 1.12 demonstrates that when the ratio is 1/1, the pH equals the pK' of the acid because $\log 1 = 0$, and, thus, pH = pK'. If the pH is one unit less than the pK', the [base]/[acid] ratio = 1:10, and if the pH is one unit above the pK', the [base]/[acid] ratio = 10:1. Figure 1.6 is a plot of the ratios of conjugate base to conjugate acid versus the pH of several weak acids; note the ratios are presented on a logarithmic scale.

Buffering Is Important to Control pH

When NaOH is added to a solution of a weak acid, the ratio of [conjugate base]/[conjugate acid] will change because the OH^- will be neutralized by the existing H^+ to form H_2O, thus decreasing the $[H^+]$. This will cause a further dissociation of the weak acid to comply with requirements of its equilibrium reaction. The amount of weak acid dissociated will be nearly equal (usually considered to be equal) to the OH^- added. Thus the decrease in the amount of conjugate acid is equal to the amount of conjugate base that is formed. Titration curves of two weak acids are presented in Figure 1.7. When 0.5 equiv OH^- are added, 50% of the weak acid is dissociated and the [acid]/[base] ratio is 1.0, and the pH at this point is equal to the pK' of the acid. The shapes of the individual titration curves are similar but displaced due to the differences in pK values. As OH^- ion is added, initially there is a rather steep rise in the pH, but between 0.1 and 0.9 equiv OH^-, the pH change is only ~2. Thus a large amount of OH^- is added with a relatively small change in pH. This is called **buffering,** and defined as *the ability of a solution to resist a change in pH when acid or base is added.*

The best buffering range for a conjugate pair is in the pH range near the pK' of the weak acid. Starting from a pH one unit below to a pH one unit above the pK', ~82% of a weak acid in solution will dissociate, and therefore an amount of base equivalent to about 82% of the original acid can be neutralized with a change in pH of 2. The maximum buffering

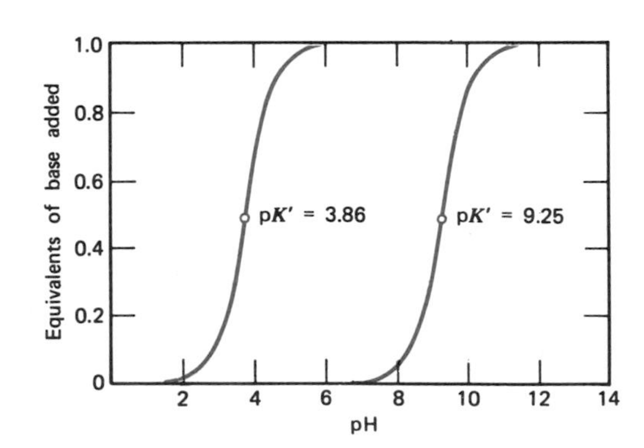

FIGURE 1.7
Acid–base titration curves for lactic acid (pK' 3.86) and NH_4^+ (pK' 9.25).
At the value of pH equal to the respective pK' values, there will be an equal amount of the acid and base for each conjugate pair.

range for a conjugate pair is considered to be between 1 pH unit above and below the pK'. A weak acid such as lactic acid with pK' = 3.86 is an effective buffer in the range of pH 3–5 but has little buffering ability at a pH of 7.0. The $HPO_4^{2-}/H_2PO_4^-$ pair with pK' = 6.7, however, is an effective buffer at this pH. Thus at the pH of the cell's cytosol (~ 7.0), the lactate–lactic acid pair is not an effective buffer but the phosphate system is.

The buffering capacity also depends on the concentrations of the acid and base pair. The higher the concentration of conjugate base, the more added H^+ with which it can react, and the more conjugate acid the more added OH^- can be neutralized by the dissociation of the acid. A case in point is blood plasma at pH 7.4. The pK' for $HPO_4^{2-}/H_2PO_4^-$ of 6.7 would suggest that this conjugate pair would be an effective buffer; the concentration of the phosphate pair, however, is low compared to the HCO_3^-/CO_2 system with a pK' of 6.1, which is present at a 20-fold higher

FIGURE 1.8
Typical problems of pH and buffering.

1. Calculate the ratio of $HPO_4^{2-}/H_2PO_4^-$ (pK = 6.7) at pH 5.7, 6.7, and 8.7.

   Solution:

   $$\text{pH} = \text{p}K + \log [HPO_4^{2-}]/[H_2PO_4^-]$$
   $$5.7 = 6.7 + \log \text{ of ratio; rearranging}$$
   $$5.7 - 6.7 = -1 = \log \text{ of ratio}$$

   The antilog of −1 = 0.1 or 1/10. Thus, $HPO_4^{2-}/H_2PO_4^-$ = 1/10 at pH 5.7. Using the same procedure, the ratio at pH 6.7 = 1/1 and at pH 8.7 = 100/1.

2. If the pH of blood is 7.1 and the HCO_3^- concentration is 8 mM, what is the concentration of CO_2 in blood (pK' for HCO_3^-/CO_2 = 6.1)?

   Solution:

   $$\text{pH} = \text{p}K + \log [HCO_3^-]/[CO_2]$$
   $$7.1 = 6.1 + \log 8 \text{ mM}/[CO_2]\text{; rearranging}$$
   $$7.1 - 6.1 = 1 = \log 8 \text{ mM}/[CO_2].$$

   The antilog of 1 = 10. Thus, 10 = 8 mM/[CO_2], or [CO_2] = 8 mM/10 = 0.8 mM.

3. At a normal blood pH of 7.4, the sum of [HCO_3^-] + [CO_2] = 25.2 mM. What is the concentration of HCO_3^- and CO_2 (pK' for HCO_3^-/CO_2 = 6.1)?

   Solution:

   $$\text{pH} = \text{p}K + \log [HCO_3^-]/[CO_2]$$
   $$7.4 = 6.1 + \log [HCO_3^-]/[CO_2]\text{; rearranging}$$
   $$7.4 - 6.1 = 1.3 = \log [HCO_3^-]/[CO_2].$$

   The antilog of 1.3 is 20. Thus, [HCO_3^-]/[CO_2] = 20. Given, [HCO_3^-] + [CO_2] = 25.2, solve these two equations for [CO_2] by rearranging the first equation:

   $$[HCO_3^-] = 20\ [CO_2]$$

   Substituting in the second equation,

   $$20\ [CO_2] + [CO_2] = 25.2$$

   or

   $$CO_2 = 1.2 \text{ mM}$$

   Then substituting for CO_2, 1.2 + [HCO_3^-] = 25.2, and solving, [HCO_3^-] = 24 mM.

concentration and accounts for most of the buffering capacity. In considering the buffering capacity both the pK' and the concentration of the conjugate pair must be taken into account. Most organic acids are relatively unimportant as buffers in cellular fluids because their pK values are more than several pH units lower than the pH of the cell, and their concentrations are too low in comparison to such buffers as the various phosphate pairs and the HCO_3^-/CO_2 system.

The importance of pH and buffers in biochemistry and clinical medicine will become apparent, particularly when reading Chapters 2, 4, and 25. Figure 1.8 presents some typical problems using the Henderson–Hasselbalch equation, and Clin. Corr. 1.1 is a representative problem encountered in clinical practice.

1.3 ORGANIZATION AND COMPOSITION OF EUCARYOTIC CELLS

As described above, the eucaryotic cell is organized into a series of compartments, each delineated by a **membrane** (Figure 1.9). In many cases these membranes define specific cellular organelles such as the nucleus, mitochondria, lysosomes, and peroxisomes. Other cellular membranes are part of a tubulelike network throughout the cell enclosing an interconnecting space or cisternae, as is the case for the endoplasmic reticulum or Golgi complex. As described in Section 1.4, every cellular compartment has clearly delineated functions and activities.

The *semipermeable nature* of cellular membranes prevents the ready diffusion of many molecules from one side to the other. Very specific transport mechanisms in membranes for the translocation of both charged and uncharged molecules allow the membrane to modulate the concentrations of the transported substances in various cellular compartments. Macromolecules, such as enzymes and nucleic acids, do not cross biological membranes unless the membrane contains a specific mechanism for the movement or the membrane is damaged. Because of the controls exerted by cellular membranes, the fluid matrix of each of the various cellular compartments has a distinctive composition of inorganic ions, organic molecules, and macromolecules. Partitioning of activities and components in the membrane compartments and organelles has a number of advantages for the economy of the cell, including the sequestering of substrates and cofactors where they are required, and adjustments of pH and ionic composition for maximum activity of a biological process.

The activities and composition of many of the cellular structures and organelles have been defined by a variety of techniques, including histochemical, immunological, and fluorescent staining methods with intact tissues and cells. Newer approaches permit the continuous observation in real time of cellular events of intact viable cells; such is the case of studies involving the changes of ionic calcium concentration in the cytosol by the use of fluorescent calcium indicators. In addition, individual organelles, membranes, and components of the cytosol can be isolated and analyzed by various techniques following disruption of the plasma membrane. Methods have been developed to alter the plasma membrane to permit the release of subcellular components, including osmotic shock of cell suspensions, homogenization of tissues, where shearing forces cleave the plasma membrane, and chemical disruption with the use of detergents. In an appropriate isolation medium, the cellular membrane systems can be separated from one another by centrifugation because of differences in size and density of the organelles. A general outline for the separation of cell fractions is presented in Figure 1.10. Chromatographic procedures

CLIN. CORR. 1.1 BLOOD BICARBONATE CONCENTRATION IN A METABOLIC ACIDOSIS

Blood buffers in the normal adult control the blood pH at a value of about 7.40; if the pH should drop much below 7.35, the condition is referred to as an acidosis. A blood pH of near 7.0 could lead to serious consequences and possibly death. Thus in acidosis, particularly that caused by a metabolic change, it is important to monitor the acid–base parameters of the person's blood. The values of interest to the clinician include the pH, HCO_3^- and CO_2 concentrations. The normal values for these are pH = 7.40, $[HCO_3^-]$ = 24.0 mM, and $[CO_2]$ = 1.20 mM.

The blood values of a patient with a metabolic acidosis were pH = 7.03, and $[CO_2]$ = 1.10 mM. What is the patient's blood HCO_3^- concentration and how much of the normal HCO_3^- concentration has been used in buffering the acid causing the condition?

1. The Henderson–Hasselbalch equation is

$$pH = pK' + \log([HCO_3^-]/[CO_2])$$

The pK' for $[HCO_3^-]/[CO_2]$ is 6.10. Substitute given values in the equation.

2. $7.03 = 6.10 + \log([HCO_3^-]/1.10\text{ mM})$

or

$$7.03 - 6.10 = 0.93$$

$$= \log([HCO_3^-]/1.10\text{ mM})$$

The antilog of 0.93 = 8.5; thus,

$$8.5 = [HCO_3^-]/1.10\text{ mM}$$

or

$$[HCO_3^-] = 9.4\text{ mM}$$

3. Since the normal value of $[HCO_3^-]$ is 24 mM, there has been a decrease of 14.6 mmol of HCO_3^- per liter of blood in this patient.

It should be noted that if much more HCO_3^- is lost, a point would be reached when this important buffer would be unavailable to buffer any more acid in the blood and the pH would drop rapidly. In Chapter 25 there is a detailed discussion of the causes and compensations that occur in such conditions.

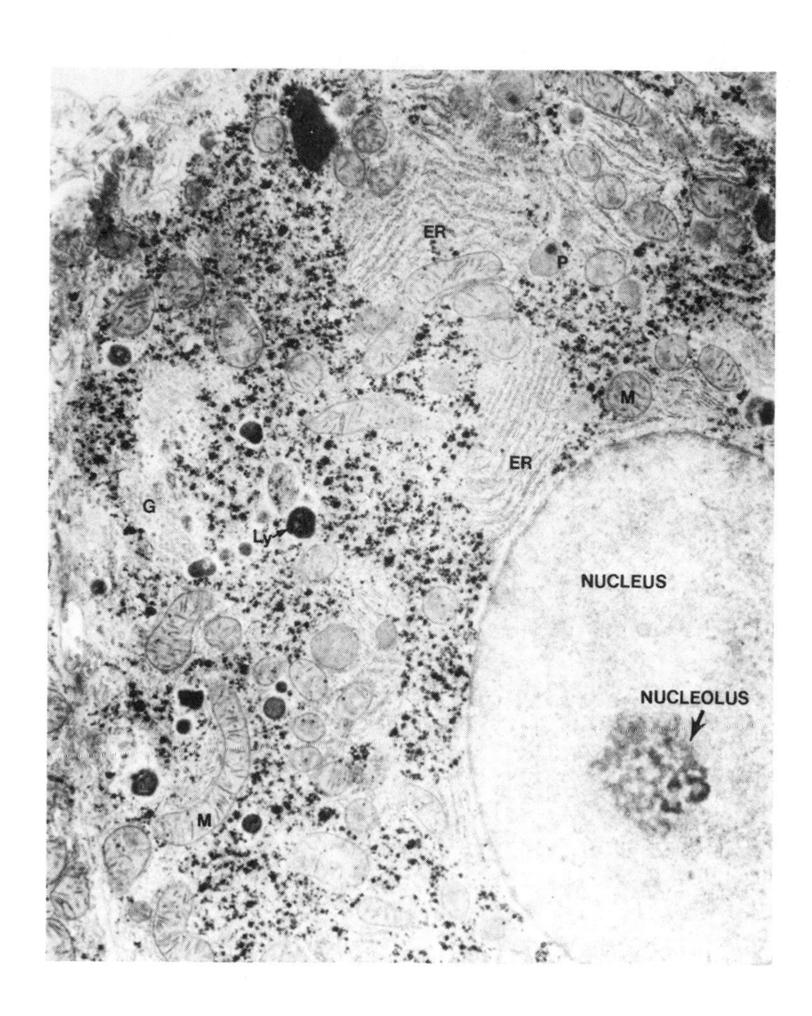

FIGURE 1.9
An electron micrograph of a rat liver cell labeled to indicate the major structural components of eucaryotic cells.
Note the number and variety of subcellular organelles and the network of interconnecting membranes enclosing channels, that is, cisternae. All eucaryotic cells are not as complex in their appearance, but most contain the major structures shown in the figure. ER, endoplasmic reticulum; G, golgi zone; Ly, lysosomes; P, peroxisomes; M, mitochondria.
Photograph reprinted with permission of Dr. K. R. Porter (see Figure 1.1).

have also been employed for the isolation of individual cellular fractions and components. These techniques have permitted the isolation of relatively pure cellular fractions from most mammalian tissues. In addition, components of cellular organelles, such as nuclei and mitochondria, can also be isolated following disruption of the organelle membrane. By these isolation techniques it is possible to study individual organelles.

In many instances the isolated structures and cellular fractions appear to retain the chemical and biochemical characteristics of the structure in situ. But biological membrane systems are very sensitive structures, subject to damage even under very mild conditions, and alterations can occur during isolation, which can lead to a change in the composition of the structure. The slightest damage to a membrane can alter significantly its permeability properties allowing substances that might normally be excluded to traverse the membrane barrier. In addition, many proteins are only loosely associated with the lipid membrane and easily dissociate when the membrane is damaged.

Not unexpectedly, there are differences in the structure, composition, and activities of cells from different tissues due to the diverse functions of the tissues. The major biochemical activities of the cellular organelles and membrane systems, however, are fairly constant from tissue to tissue. Thus, results of a study of biochemical pathways in liver are often applicable to other tissues. The differences between cell types are usually in the specialized activities that are distinctive to a particular tissue.

The Chemical Composition of Cells

Each cellular compartment contains an aqueous fluid or matrix in which are dissolved various ions, small molecular weight organic molecules,

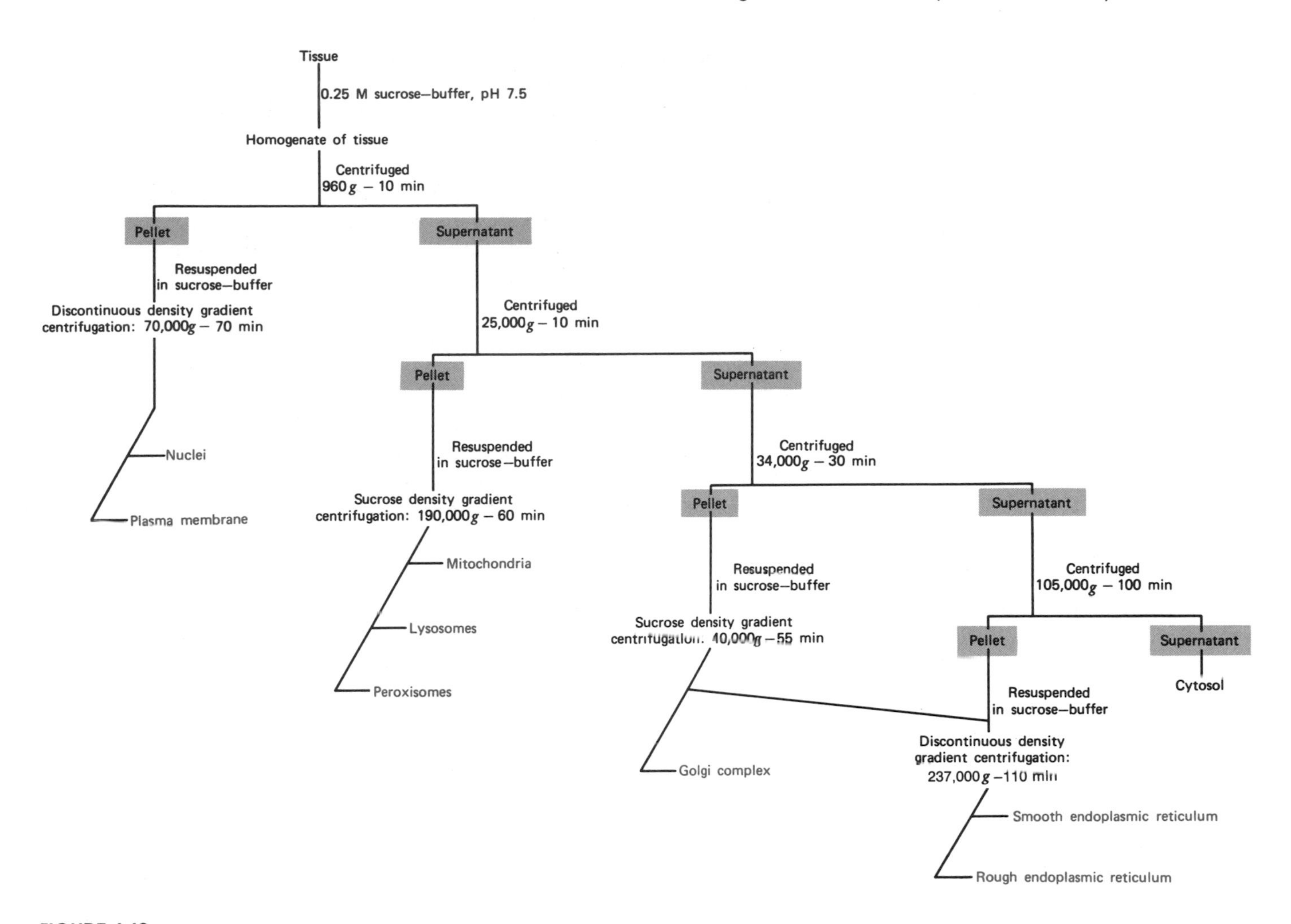

FIGURE 1.10
Diagram of a general procedure for the isolation of subcellular components from rat liver by differential and density gradient centrifugation.
This is an oversimplified scheme; to simplify the diagram, the composition of the various buffers and of the density gradients are not included. The gravitational fields and times are given only to indicate the range of each required for the isolation. There are steps in the procedure, particularly the washing of pellets, that have also been omitted.
Details of the procedure are found in the article by S. Fleischer and M. Kervina, Subcellular fractionation of rat liver. *Methods Enzymol.* 31:6, 1974.

different proteins and nucleic acids. The localization in the cell of specific macromolecules, such as enzymes, has been defined but the exact ionic composition of the matrix of cell organelles has still to be determined; however, it is certain that each has a distinctly different ionic composition and pH. The overall ionic composition of intracellular fluid compared to blood plasma is presented in Figure 1.11. The major extracellular cation is Na^+, with a concentration of ~140 meq L^{-1}; there is little Na^+ in intracellular fluids. The K^+ ion is the major intracellular cation. The Mg^{2+} ion is present both in extra- and intracellular compartments at concentrations much lower than Na^+ and K^+. The major extracellular anions are Cl^- and HCO_3^- with lower amounts of phosphate and sulfate. Most proteins have a negative charge at pH 7.4 (Chapter 2), and thus are anions at the pH of tissue fluids. Inorganic phosphate, organic phosphate compounds, and proteins are the major intracellular anions. In addition, there are other ions, such as Fe^{2+} and Mn^{2+}, which are present in concentrations well below the milliequivalent per liter level. The total anion concentration

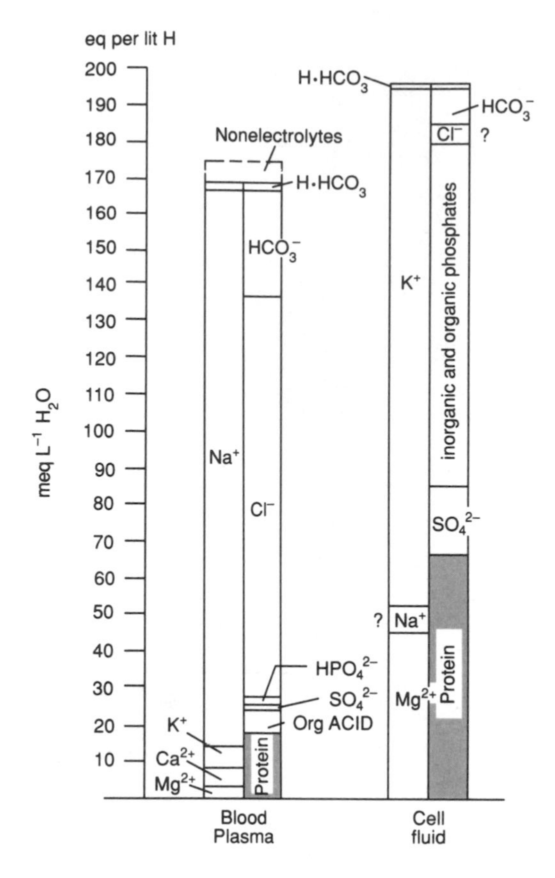

FIGURE 1.11
Diagram showing major chemical constituents of blood plasma and cell fluid.
Height of left half of each column indicates total concentration of cations; that of right half, concentrations of anions. Both are expressed in milliequivalents per liter (meq L^{-1}) of fluid. Note that chloride and sodium values in cell fluid are questioned. It is probable that, at least in muscle, the cytosol contains some sodium but no chloride.
Modified from Gamble. From Magnus I. Gregersen, in *Medical Physiology*, 11th ed., Philip Bard (Ed.), St. Louis, MO: Mosby, 1961, p. 307.

equals the total cation concentration in the different fluids, in that there cannot be a significant deviation from electroneutrality.

The concentration of most small molecular weight organic molecules, such as sugars, organic acids, amino acids, and phosphorylated intermediates, all of which serve as substrates for enzymatic reactions in the various cellular compartments, is considered to be in the range of 0.01–1.0 mM, but in some cases they have significantly lower concentrations, depending on the cell and individual organelle. Coenzymes, organic molecules required for the activity of some enzymes, are in the same range of concentration. In contrast to inorganic ions, substrates are present in relatively low concentrations when considering their overall cellular concentrations but can be increased by localization in a specific organelle or cellular microenvironment.

It is not very meaningful to determine the molar concentration of individual proteins in cells. In many cases they are localized with specific structures or in combination with other proteins to create a functional unit. It is in a restricted compartment of the cell that the individual proteins carry out their role, whether structural, catalytic, or regulatory.

1.4 FUNCTIONAL ROLE OF SUBCELLULAR ORGANELLES AND MEMBRANES

The subcellular localization of various metabolic pathways will be described throughout this textbook. In some cases an entire pathway is located in a single cellular compartment, but many metabolic sequences are divided between two locations, with the intermediates in the pathway moving or being translocated from one cell compartment to another. In general, the organelles have very specific functions. Specific enzymatic activities of individual organelles are used as an identifying characteristic during isolation.

The following describes briefly some of the major roles of the eucaryotic cell structures; the descriptions are not inclusive but rather are meant to indicate the level of complexity and organization of the cell. Table 1.6 summarizes the variety of functions of eucaryotic cells and the division of labor within the cell. Further details of some cellular compartments are presented in later chapters during discussions of their function in metabolism.

The Plasma Membrane Is the Limiting Boundary of a Cell

The limiting membrane of every cell has a unique role in the maintenance of a cell's integrity. One surface is in contact with the variable external environment and the other with the relatively constant environment of the cell's cytoplasm. As will be discussed in Chapter 5, the two sides of the plasma membrane, as well as all other cellular membranes, have different chemical compositions and functions. A major role of the plasma membrane is to permit the entrance of some substances but exclude many others. It contains the mechanism for phagocytosis and pinocytosis and the unique chemical structures for cell recognition. The plasma membrane with the cytoskeletal elements is involved in the shape of the cell and in cellular movements. Through this membrane cells communicate; the membrane contains multiple specific receptor sites for chemical signals, such as hormones (Chapter 20), released by other cells. The inner surface of the plasma membrane is also the site for attachment of a variety of enzymes involved in different metabolic pathways. Plasma membranes from a variety of cells have been isolated and studied extensively; details of their structure and biochemistry are presented in Chapter 5.

TABLE 1.6 Summary of Eucaryotic Cell Compartments and Their Major Functions

Compartment	Major functions
Plasma membrane	Transport of ions and molecules Recognition Receptors for small and large molecules Cell morphology and movement
Nucleus	DNA synthesis and repair RNA synthesis
Nucleolus	RNA processing and ribosome synthesis
Endoplasmic reticulum	Membrane synthesis Synthesis of proteins and lipids for cell organelles and for export Lipid synthesis Detoxication reactions
Golgi apparatus	Modification and sorting of proteins for incorporation into organelles and for export Export of proteins
Mitochondria	Energy conservation Cellular respiration Oxidation of carbohydrates and lipids Urea and heme synthesis
Lysosomes	Cellular digestion: hydrolysis of proteins, carbohydrates, lipids, and nucleic acids
Peroxisomes	Oxidative reactions involving O_2 Utilization of H_2O_2
Microtubules and microfilaments	Cell cytoskeleton Cell morphology Cell motility Intracellular movements
Cytosol	Metabolism of carbohydrates, lipids, amino acids, and nucleotides Protein synthesis

The Nucleus Is the Site of DNA and RNA Synthesis

The early microscopists divided the cell into a **nucleus,** the largest membrane-bound compartment, and the cytoplasm, the remainder of the cell. The nucleus is surrounded by two membranes, termed the perinuclear envelope, with the outer membrane being continuous with membranes of the endoplasmic reticulum. The nucleus contains a subcompartment, clearly seen in electron micrographs, termed the nucleolus. The vast majority of cellular deoxyribonucleic acid (DNA) is located in the nucleus in the form of a DNA–protein complex termed chromatin. Chromatin is organized into chromosomes. Deoxyribonucleic acid is the repository of the cell's genetic information and the importance of the nucleus in cell division and for controlling the phenotypic expression of genetic information is well established. The biochemical reactions involved in the replication of DNA during mitosis and the repair of DNA following damage (Chapter 17), the transcription of the information stored in DNA into a form that can be translated into proteins of the cell (Chapter 18), are contained in the nucleus. Transcription of DNA involves the synthesis of ribonucleic acid (RNA), which is processed following synthesis into a variety of forms. Part of this processing occurs in the nucleolus, which is very rich in RNA. The nucleus may synthesize some of the proteins required for nuclear function, but this activity is small in comparison to the very active protein synthetic activity of the cytosol and endoplasmic reticulum.

The Endoplasmic Reticulum Has a Role in Many Synthetic Pathways

The cytoplasm of most eucaryotic cells contains a network of interconnecting membranes enclosing channels, that is, cisternae, that thread from the perinuclear envelope of the nucleus to the plasma membrane. This extensive subcellular structure, termed the **endoplasmic reticulum,** consists of membrane structures with a rough appearance in some areas and smooth in other places. The rough appearance is due to the presence of ribonucleoprotein particles, that is, ribosomes, attached on the cytoplasmic side of the membrane. Smooth endoplasmic reticulum does not contain ribosomal particles. During cell fractionation the endoplasmic reticulum network is disrupted, with the membrane resealing into small vesicles referred to as **microsomes,** which can be isolated by differential centrifugation. Microsomes per se do not occur in cells.

The major function of the ribosomes on the rough endoplasmic reticulum is the biosynthesis of proteins for export to the outside of the cell and enzymes to be incorporated into cellular organelles such as the lysosomes. The endoplasmic reticulum is also involved in membrane lipid synthesis and contains an important class of enzymes termed cytochromes P450, which catalyze hydroxylation reactions in the biosynthesis of steroid hormones and of other compounds including exogenous toxic substances. These reactions are described in detail in Chapter 22. The endoplasmic reticulum also has a role with the Golgi apparatus in the formation of other cellular organelles such as lysosomes and peroxisomes.

The Golgi Apparatus Is Involved in Sequestering of Proteins

The Golgi apparatus is a network of flattened smooth membranes and vesicles responsible for the secretion to the external environment of a variety of proteins synthesized on the endoplasmic reticulum. Golgi membranes catalyze the transfer to proteins of carbohydrate and lipid precursors to form glycoproteins and lipoproteins. Glycosylation of proteins is required for transport of proteins across the plasma membrane. In addition, the complex is a major site of new membrane formation. Membrane vesicles are formed from the Golgi apparatus in which various proteins and enzymes are encapsulated, which can be secreted from the cell after an appropriate signal. The digestive enzymes synthesized by the pancreas are stored in intracellular vesicles formed by the Golgi apparatus and released by the cell when needed in the digestive process. The role in membrane synthesis also includes the formation of intracellular organelles such as lysosomes and peroxisomes.

Mitochondria Supply Most of the Cell's Needs for ATP

Mitochondria have been studied extensively because of their role in cellular energy metabolism. The studies have been facilitated by the ease with which mitochondria can be isolated in a relatively intact state from tissues. In electron micrographs **mitochondria** appear as spheres, rods, or filamentous bodies; they are usually about 0.5–1 μm in diameter and up to 7 μm in length. The internal matrix is surrounded by two membranes, distinctively different in appearance and biochemical function. The inner membrane convolutes into the matrix of the mitochondrion to form cristae and contains numerous small spheres attached by stalks on the inner surface. The outer and inner membranes contain distinctly different sets of enzymes, which are used for identification of mitochondria during isolation. The components of the respiratory chain and the mechanism for ATP synthesis are part of the inner membrane, described in detail in Chapter 6. In addition to membrane-bound enzymes, the space between the two membranes and the internal matrix also contain a variety of enzymes. Major metabolic pathways involved in the oxidation of carbohydrates, lipids, and amino acids, and special biosynthetic pathways involving urea and heme synthesis are located in the mitochondrial matrix space. The outer membrane is relatively permeable, but the inner membrane is highly selective and contains a number of transmembrane transport systems.

Mitochondria also contain a specific DNA, containing genetic information for some of the mitochondrial proteins, and the biochemical equipment for limited protein synthesis. The presence of this biosynthetic capacity indicates the unique role that mitochondria have in their own destiny.

Lysosomes Are Required for Intracellular Digestion

Hydrolysis of a variety of substances in the cell occurs inside the structures designated as **lysosomes.** These cellular organelles have a single limiting membrane, capable of maintaining a lower pH in the lysosomal matrix than in the cytosol. Encapsulated in lysosomes is a group of enzymes (hydrolases), which catalyze the hydrolytic cleavage of carbon–oxygen, carbon–nitrogen, carbon–sulfur, and oxygen–phosphorus bonds in proteins, lipids, carbohydrates, and nucleic acids. As in the process of digestion in the lumen of the gastrointestinal system, the enzymes of the lysosome are able to split molecules into simple low molecular weight compounds, which can be utilized by the metabolic pathways of the cell. The enzymes of the lysosome have a common characteristic in that each is most active when the pH of the medium is acidic, that is, pH 5. The relationship between pH and enzyme activity is discussed in Chapter 4. The pH of the cytosol is close to neutral, pH 7.0, and the lysosomal enzymes have little activity at this pH. Thus for the lysosomal enzymes to carry out the digestion of various substances the intralysosomal pH must be significantly lower than the cytosolic pH. A partial list of the enzymes present in lysosomes is presented in Table 1.7.

The enzyme content of lysosomes of different tissues varies and depends apparently on specific needs of individual tissues to digest different substances. Intact lysosomes isolated from other cellular components do not catalyze the hydrolysis of substrates until the membrane is disrupted, demonstrating that the lysosomal membrane is a barrier to the ready access of cellular components to the interior of the lysosome. The lysosomal membrane can be disrupted by various treatments, leading to a release of the lysosomal enzymes, which can then react with their individual substrates. The activities of the lysosomal enzymes are termed "latent." The membrane of a lysosome is impermeable to both small and large molecules and mounting evidence suggests the presence of specific protein mediators in the membrane for translocation of substances. Disruption of the membrane in situ can lead to cellular digestion; a variety of pathological conditions have been attributed to release of lysosomal enzymes, including arthritis, allergic responses, several muscle diseases, and drug-induced tissue destruction (Clin. Corr. 1.2).

CLIN. CORR. 1.2
LYSOSOMAL ENZYMES AND GOUT

The catabolism of purines, nitrogen-containing heterocyclic compounds found in nucleic acids, leads to the formation of uric acid, which is excreted normally in the urine (see Chapter 13 for details). Gout is an abnormality in which there is an overproduction of purines leading to excessive uric acid production. A consequence of the disease is an increase in uric acid in blood and deposition of urate crystals in the joints. Uric acid is not very soluble. Some of the clinical symptoms of gout can be attributed to the damage done by the urate crystals. The crystals are phagocytosed by cells in the joint and accumulate in digestive vacuoles that contain the lysosomal enzymes. The crystals cause physical damage to the vacuoles, causing a release of the lysosomal enzymes into the cytosol of the cell. Even though the pH optima of the lysosomal enzymes are lower than the pH of the cytosol, they have some hydrolytic activity at the higher pH, which causes digestion of cellular components, release of substances from the cell, and autolysis. The consequences are clinical manifestations in the joint, including inflammation, pain, swelling, and increased temperature.

Weissmann, G. Crystals, lysosomes and gout. *Adv. Intern. Med.* 19:239, 1974.

TABLE 1.7 Representative Lysosomal Enzymes and Their Substrates

Type of substrate and enzyme	Specific substrate
POLYSACCHARIDE-HYDROLYZING ENZYMES	
α-Glucosidase	Glycogen
α-Fucosidase	Membrane fucose
β-Galactosidase	Galactosides
α-Mannosidase	Mannosides
β-Glucuronidase	Glucuronides
Hyaluronidase	Hyaluronic acid and chondroitin sulfates
Arylsulfatase	Organic sulfates
Lysozyme	Bacterial cell walls
PROTEIN-HYDROLYZING ENZYMES	
Cathepsins	Proteins
Collagenase	Collagen
Elastase	Elastin
Peptidases	Peptides
NUCLEIC ACID-HYDROLYZING ENZYMES	
Ribonuclease	RNA
Deoxyribonuclease	DNA
LIPID-HYDROLYZING ENZYMES	
Esterase	Fatty acid esters
Phospholipase	Phospholipids
PHOSPHATASES	
Phosphatase	Phosphomonoesters
Phosphodiesterase	Phosphodiesters

Lysosomes are involved in the normal digestion of both intra- and extracellular substances that must be removed by the cell. By the process of endocytosis, external material is taken into the cell and encapsulated in a membrane-bound vesicle (Figure 1.12). Formed foreign substances such as microorganisms are engulfed by the cell membrane of some cells by the process of phagocytosis, and extracellular fluid containing suspended material is taken up by pinocytosis. In both processes the vesicle containing the external material interacts with a lysosome to form a cystolic organelle containing both the material to be digested and the enzymes capable of carrying out the digestion. These vacuoles are identified microscopically by their size and often by the presence of partially formed structures in the process of being digested. Lysosomes in which the enzymes are not as yet involved in the digestive process are termed primary lysosomes, whereas secondary lysosomes are organelles in which digestion of material is under way. The latter are also referred to as digestive vacuoles and will vary in size and appearance.

Cellular constituents are synthesized and degraded continuously and lysosomes have the responsibility of digesting the cellular debris. The dynamic synthesis and degradation of cellular substances includes proteins and nucleic acids, as well as structures such as mitochondria and the endoplasmic reticulum. During the normal self-digestion process, that is autolysis, cellular substances are encapsulated within a membrane vesicle that fuses with a lysosome to complete the degradation. The overall process is termed autophagy and is also represented in Figure 1.12.

The products of the normal lysosomal digestive process are able to diffuse across the lysosomal membrane and be reutilized by the cell. Indigestible material, however, accumulates in vesicles referred to as residual bodies; the contents of these vesicles are removed from the cell by exocytosis, where the membrane of a vesicle interacts with the plasma membrane. In some cases, however, the residual bodies, which contain a high concentration of lipid, persist for long periods of time. The lipid is oxidized and a pigmented substance, which is chemically heterogeneous and contains polyunsaturated fatty acids and protein, accumulates in the cell. This material, termed **lipofuscin,** has been called the "age pigment" or the "wear and tear pigment" because it accumulates in cells of older

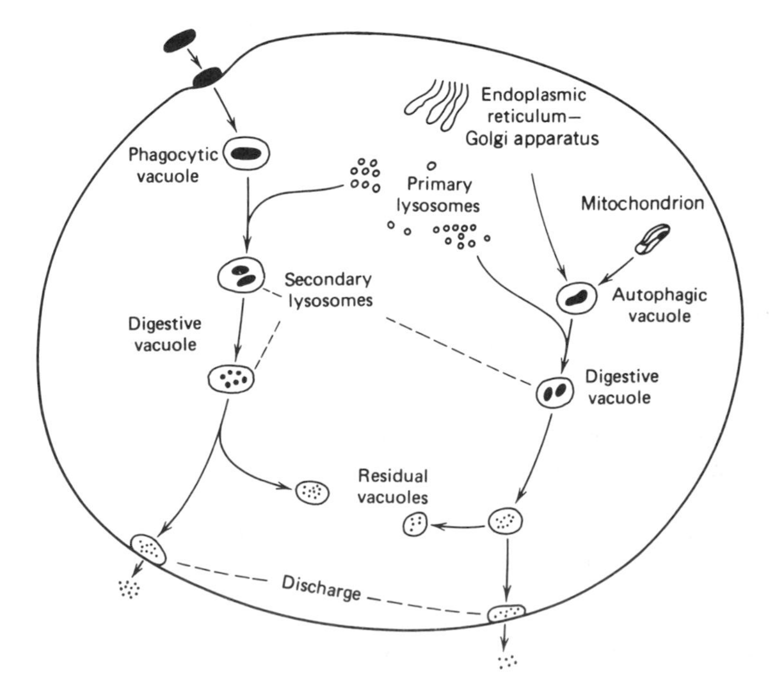

FIGURE 1.12
Diagrammatic representation of the role of lysosomes in the intracellular digestion of substances internalized by phagocytosis (heterophagy) and of cellular components (autophagy).
In both processes the substances to be digested are enclosed in a membrane vesicle, which is followed by interaction with either a primary or secondary lysosome.

individuals. It occurs in all cells but particularly in neurons and muscle cells and has been implicated in the aging process.

Lysosomal enzymes in some cells also participate in the secretory process by hydrolyzing specific bonds in precursor protein molecules, leading to the formation of active proteins that are secreted from the cell. Under controlled conditions lysosomal enzymes are secreted from the cell for the digestion of extracellular material; an extracellular role for some lysosomal enzymes has been demonstrated in connective tissue and prostate gland and in the process of embryogenesis in which they have a role in programmed cell death.

The absence of specific lysosomal enzymes has been demonstrated in a number of genetic diseases; in these cases there is an accumulation in the cell of specific cellular components that cannot be digested. Lysosomes of the affected cell become enlarged with undigested material, which interferes with normal cellular processes. A discussion of lysosomal storage diseases is presented in Chapter 10.

(1) $2H_2O_2 \longrightarrow 2H_2O + O_2$

(2) $RH_2 + H_2O_2 \longrightarrow R + 2H_2O$

FIGURE 1.13
Reactions catalyzed by catalase.

Peroxisomes Contain Oxidative Enzymes Which Involve Hydrogen Peroxide

Most eucaryotic cells of mammalian origin, as well as of protozoa and plants, have a defined cellular organelle, which contains several enzymes that either produce or utilize hydrogen peroxide (H_2O_2). Frequently referred to as microbodies, the designation **peroxisome** is now more widely accepted. The organelles are small (0.3–1.5 μm in diameter), spherical or oval in shape, with a granular matrix and in some cases a crystalline inclusion termed a nucleoid. Peroxisomes contain enzymes that oxidize D-amino acids, uric acid, and various 2-hydroxy acids using molecular O_2 with the formation of H_2O_2. The enzyme catalase, also present in the peroxisome, catalyzes the conversion of H_2O_2 to water and oxygen; it can also catalyze the oxidation by H_2O_2 of various compounds (Figure 1.13). By having both the peroxide-producing and utilizing enzymes in one cellular compartment, the cell protects itself from the toxicity of H_2O_2.

Peroxisomes also contain enzymes involved in lipid metabolism, particularly the oxidation of very long-chain fatty acids and the synthesis of glycerolipids and glycerol ether lipids (plasmalogens) (see Chapter 10). Clin. Corr. 1.3 describes Zellweger syndrome in which there is an absence of peroxisomes.

The peroxisomes of different tissues contain different complements of enzymes, and the peroxisome content of cells can vary depending on different cellular conditions.

CLIN. CORR. 1.3
ZELLWEGER SYNDROME AND THE ABSENCE OF PEROXISOMES

Zellweger syndrome is a rare, autosomal recessive disease characterized by abnormalities of the liver, kidney, brain, and skeletal system. It usually results in death by age 6 months. A number of seemingly unrelated biochemical abnormalities have been described including decreased levels of glycerol–ether lipids (plasmalogens), increased levels of very long-chain fatty acids (C-24 and C-26), and increased levels of cholestanoic acid derivatives (precursors of bile acids). These abnormalities are due to the absence of peroxisomes in the afflicted children. Peroxisomes are responsible for the synthesis of glycerol ethers, for shortening very long-chain fatty acids so that the mitochondria can completely oxidize them, and for oxidation of the side chain of cholesterol needed for bile acid synthesis. This disease can be diagnosed prenatally by assaying amniotic fluid cells for peroxisomal enzymes or analyzing the fatty acids in the fluid.

Datta N. S., Wilson G. N., and Hajra A. K. Deficiency of enzymes catalyzing the biosynthesis of glycerol–ether lipids in Zellweger syndrome. *N. Engl. J. Med.* 311:1080, 1984; and Moser A. E., Singh, I., Brown, F. R., Solish, G. I., Kelley, R. I., Benke, P. J., and Moser, H. W. The cerebrohepatorenal (Zellweger) syndrome. Increased levels and impaired degradation of very long chain fatty acids and their use for prenatal diagnosis. *N. Engl. J. Med.* 310:1141, 1984.

The Cytoskeleton Organizes the Intracellular Contents

Eucaryotic cells contain both **microtubules** and **actin filaments** (microfilaments), which are part of the cytoskeletal network. The cytoskeleton has a role in maintenance of cellular morphology, intracellular transport, cell motility, mitosis, and meiosis. The microtubules consist of a polymer of the protein **tubulin,** which can be rapidly assembled and disassembled depending on the needs of the cell. A very important cellular filament occurs in striated muscle and is responsible for muscular contraction (see Chapter 22). Three mechanochemical proteins, **myosin, dynesin,** and **kinesin,** which convert chemical energy into mechanical energy for movement of cellular components, have been identified. These molecular motors are associated with the cytoskeleton; the actual mechanism for the energy conversion, however, has not been defined completely. It is believed that dynesin is involved in ciliary and flagella movement, whereas kinesin is the driving force for the movement of vesicles and organelles along the microtubule.

The Cytosol Contains the Soluble Cellular Components

The least complex in structure, but not in chemistry, is the remaining matrix or **cytosol** of the cell. It is here that many of the multiplicity of chemical reactions of metabolism occur and where substrates and cofactors of various enzymes interact. Even though there is no apparent structure to the cytoplasm, the high protein content precludes the matrix from being a truly homogeneous mixture of soluble components. Many reactions may be localized in selected areas of the cell, where the conditions of substrate availability are more favorable for the reaction. The actual physicochemical state of the cytosol is poorly understood. A major role of the cytosol is to support the synthesis of proteins catalyzed by the rough endoplasmic reticulum by supplying cofactors and enzymes. In addition, the cytosol contains free ribosomes, often in a polysome form, for synthesis of intracellular proteins.

Studies with isolated cytosol suggest that many reactions are catalyzed by soluble enzymes, but in the intact cell some of these enzymes may be loosely attached to one of the many membrane structures and are released upon cell disruption.

Conclusion

The eucaryotic cell is a complex structure whose purpose is to replicate itself when necessary, maintain an intracellular environment to permit a myriad of complex reactions to occur as efficiently as possible, and to protect itself from the hazards of its surrounding environment. The cells of multicellular organisms also participate in maintaining the well being of the whole organism by exerting influences on each other to maintain all tissue and cellular activities in balance. Thus, as we dissect the separate chemical components and activities of cells, it is important to keep in mind the concurrent and surrounding activities, constraints, and influences. Only by bringing together all the separate parts, that is, reassembling the puzzle, will we appreciate the wonder of a living cell.

BIBLIOGRAPHY

Water and Electrolytes

Dick, D. A. T. *Cell water*. Washington, DC: Butterworths, 1966.

Eisenberg, D. and Kauzmann, W. *The structures and properties of water*. Fairlawn, NJ: Oxford University Press, 1969.

Morris, J. G. *A biologist's physical chemistry*. Reading, MA: Addison-Wesley, 1968.

Stillinger, F. H. Water revisited. *Science* 209:451, 1980.

Cell Structure

Alberts, B., Bray, D., Lewis, J., Raff, M., Roberts, K., and Watson, J. D. *Molecular biology of the cell*. New York: Garland, 1989.

Alterman, P. L. and Katz, D. D. (Eds.). *Cell biology*. Bethesda, MD: Federation of American Societies of Experimental Biology, 1975.

DeDuve, C. Blueprint for a cell. Carolina Biol. Supply: Burlington, 1991.

Dingle, J. T., Dean, R. T., and Sly, W. S. (Eds.). *Lysosomes in biology and pathology*. Amsterdam: Elsevier (a serial publication covering all aspects of lysosomes).

Fawcett, D. W. *The cell*. Philadelphia: W. B. Saunders, 1981.

Hers, H. G. and Van Hoof, F. (Eds.). *Lysosomes and storage diseases*. New York: Academic, 1973.

Holtzman, E. and Novikoff, A. B. *Cells and organelles*, 3rd ed. New York: Holt, Rinehart & Winston, 1984.

Loewy, A. and Siekevitz, P. *Cell structure and function*, 2nd ed. New York: Holt, Rinehart & Winston, 1970.

Porter, K. R. and Bonneville, M. A. *Fine structure of cells and tissues*. Philadelphia: Lea & Febiger, 1972.

Vale, R. D. Intracellular transport using microtubule—based motors. *Annu. Rev. Cell. Biol.* 3:347, 1987.

QUESTIONS

J. BAGGOTT AND C. N. ANGSTADT

*Q*uestion *T*ypes are described on page xxiii.

A. Eucaryotic cell C. Both
B. Procaryotic cell D. Neither

1. (QT4) DNA is distributed uniformly throughout the cell.
2. (QT4) Contains histones.
3. (QT4) Lacks organized subcellular structures.

4. (QT2) Factors responsible for the polarity of the water molecule include:
   1. differences in electron affinity between hydrogen and oxygen.
   2. the tetrahedral structure of liquid water.
   3. the magnitude of the H—O—H bond angle.
   4. the ability of water to hydrogen bond to various chemical structures.
5. (QT2) Hydrogen bonds form only between electronegative atoms such as oxygen or nitrogen and a hydrogen atom bonded *to:*
   1. sulfur.
   2. carbon.
   3. hydrogen.
   4. an electronegative atom.

6. (QT1) Which of the following is *least* likely to be soluble in water?
   A. Nonpolar compound
   B. Weakly polar compound
   C. Strongly polar compound
   D. Weak electrolyte
   E. Strong electrolyte
7. (QT1) Which of the following is both a Brønsted acid and a Brønsted base in water?
   A. $H_2PO_4^-$
   B. H_2CO_3
   C. NH_3
   D. NH_4^+
   E. Cl^-

Refer to the following information for Questions 8 and 9.

A. Pyruvic acid $pK' = 2.50$
B. Acetoacetic acid $pK' = 3.6$
C. Lactic acid $pK' = 3.86$
D. β-Hydroxybutyric acid $pK' = 4.7$
E. Propionic acid $pK' = 4.86$

8. (QT5) Which weak acid will be 91% neutralized at pH 4.86?
9. (QT5) Assuming that the sum of [weak acid] + [conjugate base] is identical for buffer systems based on the acids listed above, which has the greatest buffer capacity at pH 4.86?

10. (QT1) All of the following subcellular organelles can be isolated essentially intact *except:*
    A. nuclei.
    B. mitochondria.
    C. lysosomes.
    D. peroxisomes.
    E. endoplasmic reticulum.

11. (QT2) Biological membranes may:
    1. prevent free diffusion of ionic solutes.
    2. release proteins when damaged.
    3. contain specific systems for the transport of uncharged molecules.
    4. be sites for biochemical reactions.

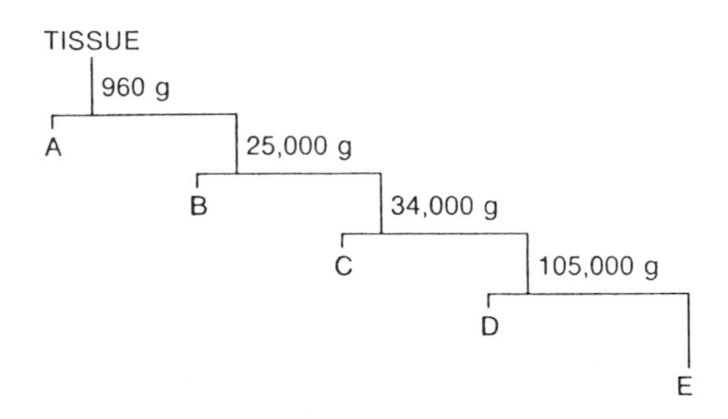

12. (QT1) In separating subcellular components of rat liver by differential centrifugation as shown schematically above:
    A. fraction A contains the Golgi apparatus.
    B. fraction B contains the mitochondria.
    C. fraction C contains the endoplasmic reticulum.
    D. fraction D contains the plasma membrane.
    E. fraction E contains the peroxisomes.
13. (QT1) Analysis of the composition of the major fluid compartments of the body shows that:
    A. the major blood plasma cation is K^+.
    B. the major cell fluid cation is Na^+.
    C. one of the major intracellular anions is Cl^-.
    D. one of the major intracellular anions is phosphate.
    E. plasma and the cell fluid are all very similar in ionic composition.

A. Mitochondria C. Both
B. Nucleus D. Neither

14. (QT4) Contains DNA.
15. (QT4) Enclosed by a single membrane.
16. (QT4) Connected to the plasma membrane by a network of membranous channels.

17. (QT2) Lysosomes may:
    1. combine with phagocytic vacuoles to become digestive vacuoles.
    2. combine with autophagic vacuoles to become digestive vacuoles.
    3. secrete their enzymes from the cell for digestion of extracellular material.
    4. combine with peroxisomes in order to add oxidative capabilities to their catalytic armamentarium.

ANSWERS

1. D The DNA of eucaryotes is confined to the nucleus and mitochondria; the DNA of procaryotes is segregated into discrete masses, the nuclear zone (p. 2).
2. A Histones are unique to eucaryotes; they form complexes with the DNA (p. 2).
3. B Eucaryotic cells typically contain mitochondria, a nucleus, and other organized structures; procaryotic cells do not (p. 2).
4. B 1 and 3 true. Water is a polar molecule because the bonding electrons are attracted more strongly to oxygen than to hydrogen. The bond angle gives rise to asymmetry of the charge distribution; if water were linear, it would not be a dipole (p. 5). 2 and 4 are consequences of water's structure, not factors responsible for it.
5. D Only 4 true. Only hydrogen atoms *bonded to* one of the electronegative elements (O, N, F) can form hydrogen bonds (p. 5). A hydrogen participating in hydrogen bonding must have an electronegative element on both sides of it.
6. A In general, compounds that interact with the water dipoles are more soluble than those that do not. Thus ionized compounds and polar compounds tend to be soluble. Nonpolar compounds prefer to interact with one another rather than with polar solvents such as water (p. 6).
7. A $H_2PO_4^-$ can donate a proton to become HPO_4^{2-}. It can also accept a proton to become H_3PO_4. B and D are Brønsted acids; C is a Brønsted base. The Cl^- ion in water is neither (p. 8).
8. C If weak acid is 91% neutralized, 91 parts are present as conjugate base and 9 parts remain as the weak acid. Thus the conjugate base/acid ratio is 10 : 1. Substituting into the Henderson–Hasselbalch equation, $4.86 = pK + \log (10/1)$, and solving for pH gives the answer (p. 11).
9. E The buffer capacity of any system is maximal at pH = pK (p. 12). Buffer concentration also affects buffer capacity, but in this case concentrations are equal.
10. E Gentle disruption of cells will not destroy A–D. The tubelike endoplasmic reticulum, however, is disrupted and forms small vesicles. These vesicles, not the original structure from which they were derived, may be isolated (p. 13).
11. E All statements are true. 1 and 3 (p. 13); 2 (p. 14); 4 (p. 2).
12. B Fraction A: nuclei and plasma membrane. Fraction B: mitochondria, lysosomes, peroxisomes. Fraction C: Golgi apparatus. Fraction D: microsomes. Fraction E: cytosol (p. 15, Figure 1.10).
13. D A, B, and E: plasma and cell fluid are strikingly different. The Na^+ ion is the major cation of plasma. C and D: phosphate and protein are the major intracellular anions; most chloride is extracellular (p. 16, Figure 1.11).
14. C Most of the DNA is in the nucleus, but the mitochondria also contain DNA (pp. 17, 19).
15. D Both the mitochondria and the nucleus are surrounded by double membranes (pp. 17–18).
16. B This describes only the nucleus (p. 18).
17. A 1, 2, and 3 describe the activities of lysosomes (pp. 20–21). Peroxisomes contain peroxide-producing enzymes and catalase, which destroys peroxide. This arrangement is thought to protect the rest of the cell from the damaging effects of H_2O_2 (p. 21).

2

Proteins I: Composition and Structure

RICHARD M. SCHULTZ
MICHAEL N. LIEBMAN

2.1 FUNCTIONAL ROLES OF PROTEINS IN HUMANS 26
2.2 AMINO ACID COMPOSITION OF PROTEINS 27
Proteins Are Polymers of α-Amino Acids 27
The Common Amino Acids Have a General Structure 27
The Side Chains Define the Structures of the Different Amino Acids 27
The Amino Acid Side Chains Can Have Polar and Apolar Properties 30
Amino Acids Have an Asymmetric Center 31
Amino Acids Are Polymerized Into Peptides and Proteins 31
Cystine Is a Derived Amino Acid 33
2.3 CHARGE AND CHEMICAL PROPERTIES OF AMINO ACIDS AND PROTEINS 34
The Ionizable Groups of Amino Acids and Proteins Are Important for Their Biological Function 34
The Ionic Form of an Amino Acid or Protein Can Be Determined at a Given pH 36
Titration of a Monoamino Monocarboxylic Acid: Determination of the Isoelectric pH 36
Titration of a Monoamino Dicarboxylic Acid 37
The Charge Properties of Amino Acids and Proteins Is Determined by the pH 37
Amino Acids and Proteins Can be Separated Based on pI Values 38
Amino Acids Undergo a Variety of Chemical Reactions 41
Techniques are Available for Determining the Amino Acid Composition of Proteins and Physiological Solutions 42
2.4 PRIMARY STRUCTURE OF PROTEINS 44
Techniques Used to Determine the Amino Acid Sequence of a Protein 44
Knowledge of Primary Structure of Insulin Aids in Understanding Its Synthesis and Action 46
2.5 HIGHER LEVELS OF PROTEIN ORGANIZATION 49
Proteins Have a Secondary Structure 50
Proteins Fold Into a Three Dimensional Structure Called the Tertiary Structure 55
A Quaternary Structure Occurs When Several Polypeptide Chains Form a Specific Noncovalent Association 60
2.6 OTHER TYPES OF PROTEINS 61
Fibrous Proteins Include Collagen, α-Keratin, and Tropomyosin 61
Lipoproteins Are a Complex of Lipids With Protein 67
Glycoproteins Contain Covalently Bound Carbohydrate 70
2.7 FOLDING OF PROTEINS FROM RANDOMIZED TO UNIQUE STRUCTURES: PROTEIN STABILITY 72
Noncovalent Forces Lead to Protein Folding and Contribute to a Protein's Stability 72
Denaturation of Proteins Leads to Loss of Native Structure 75
2.8 DYNAMIC ASPECTS OF PROTEIN STRUCTURE 76
2.9 METHODS FOR THE STUDY OF HIGHER LEVELS OF PROTEIN STRUCTURE AND ORGANIZATION 78
X-Ray Diffraction Techniques Are Used to Determine the Three-Dimensional Structure of Proteins 78
Various Spectroscopic Methods Are Employed in Evaluating Protein Structure and Function 81
Proteins Can Be Separated and Characterized Based on Molecular Weight or Size 85
Chromatographic Techniques Can Separate Proteins on the Basis of Charge and Reactive Groups 87
General Approach to Protein Purification 87

BIBLIOGRAPHY 88
QUESTIONS AND ANNOTATED ANSWERS 89
CLINICAL CORRELATIONS
2.1 Use of Amino Acid Analysis in Diagnosis of Disease 42
2.2 Differences in the Primary Structure of Insulins Used in the Treatment of Diabetes Mellitus 49
2.3 A Nonconservative Mutation Occurs in Sickle-Cell Anemia 50
2.4 Symptoms of Diseases of Abnormal Collagen Synthesis 61
2.5 Hyperlipidemias 67
2.6 Hypolipoproteinemias 68
2.7 Functions of Glycoproteins 70
2.8 Glycosylated Hemoglobin, HbA_{1c} 71

2.1 FUNCTIONAL ROLES OF PROTEINS IN HUMANS

Proteins perform a surprising variety of essential functions in mammalian organisms. These functions may be grouped into two classes: dynamic and structural. Dynamic functions of proteins include transport, metabolic control, contraction, and catalysis of chemical transformations. In their structural functions, proteins provide the matrix for bone and connective tissue, giving structure and form to the human organism.

One of the important groups of dynamic proteins is the enzymes. Enzymes act to catalyze chemical reactions, converting a substrate to a product at the enzyme active site. Almost all of the thousands of different chemical reactions that occur in living organisms and involve covalent bond formation or cleavage require a specific enzyme catalyst for the reaction to occur at a rate compatible with life. Thus the characteristics and functions of any cell are based on the particular chemistry of the cell, which in turn is generated by the specific enzyme makeup of the cell. Genetic traits are expressed through the synthesis of enzymes, which catalyze the chemical reactions that establish the trait to be expressed.

Another dynamic function for proteins is in transport. Particular examples discussed in greater detail in this text are hemoglobin and myoglobin, which transport oxygen in blood and in muscle, respectively. Transferrin transports iron in blood. Other important proteins act to transport hormones in blood from their site of synthesis to their site of action. Many drugs and toxic compounds are transported bound to proteins.

Proteins can also function in a protective role. The immunoglobulins and interferon are proteins that act against bacterial or viral infection. Fibrin is a protein that is formed where required to stop the loss of blood on injury to the vascular system.

Many hormones are proteins. Protein hormones include insulin, thyrotropin, somatotropin (growth hormone), luteinizing hormone, and follicle stimulating hormone. There are many diverse protein-type hormones that have a low molecular weight (<5000), referred to as **peptides.** In general, the term **protein** is used for molecules composed of over 50 component amino acids and the term peptide is used for molecules of less than 50 amino acids. Important peptide hormones include adrenocorticotropin, antidiuretic hormone, glucagon, and calcitonin.

Some proteins have roles in contractile mechanisms. Of particular importance are the proteins myosin and actin, which function in muscle contraction.

Other proteins are active in the control and regulation of gene transcription and translation. These include the histone proteins closely associated with DNA, the repressor and enhancer proteins that control gene expression, and the proteins that form a part of the ribosomes.

Whereas the above proteins are "dynamic" in their function, other proteins have structural, "brick-and-mortar" roles. This group of proteins includes collagen and elastin, which form the matrix for bone and ligaments and provide structural strength and elasticity to the organs and the vascular system. α-Keratin has an essential structural role in epidermal tissue.

It is obvious that an understanding of both the normal functioning and the pathology of the mammalian organism requires a clear understanding of the structure and properties of the proteins.

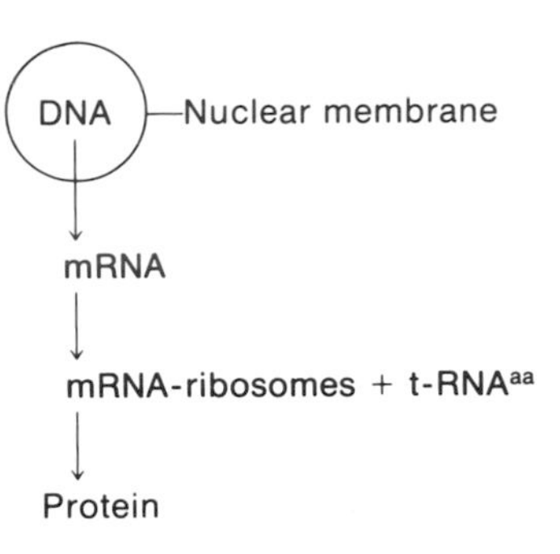

FIGURE 2.1
Genetic information is transmitted from a DNA sequence through RNA into the amino acid sequence of a protein.
DNA, deoxyribonucleic acid; mRNA, messenger ribonucleic acid; tRNA^{aa}, aminoacyl transfer ribonucleic acid (see Chapters 15–19).

2.2 AMINO ACID COMPOSITION OF PROTEINS

Proteins Are Polymers of α-Amino Acids

It is notable that all the different types of proteins are initially synthesized as polymers of only 20 amino acids, known as the common amino acids. The *common amino acids are defined as those amino acids for which at least one specific codon exists in the DNA genetic code.* There are 20 amino acids for which DNA codons are known. The process of the reading of the DNA code, resulting in the polymerization of amino acids of a specific sequence into a protein based on the DNA code, is the basis of a later chapter (Chapter 15). The relationship between the gene organization and the protein for which it codes will be discussed for three examples in Chapter 3. In this chapter we will discuss only the protein product of this genetically controlled synthetic process (Figure 2.1).

In addition to the common amino acids, derived amino acids are found in proteins. *Derived amino acids in proteins are formed from one of the common amino acids,* usually by an enzyme-facilitated reaction, *after the common amino acid has been incorporated into a protein structure.* An example of a derived amino acid is cystine (page 33). Other derived amino acids are desmosine and isodesmosine found in the protein elastin (see Section 2.6), hydroxyproline and hydroxylysine found in collagen (see Section 2.6), and γ-carboxyglutamate found in prothrombin.

The Common Amino Acids Have a General Structure

The common amino acids have the general structure depicted in Figure 2.2. They contain in common a central **alpha (α)-carbon atom** to which a carboxylic acid group, an amino group, and a hydrogen atom are covalently bonded. In addition, the α-carbon atom binds a side chain group, designated R, that in this case uniquely defines each of the 20 common amino acids.

In the structure depicted in Figure 2.2 the ionized form for a common amino acid that is present in solution at pH 7 is shown. The α-amine is protonated and in its ammonium ion form; the carboxylic acid group is in its unprotonated or carboxylate form.

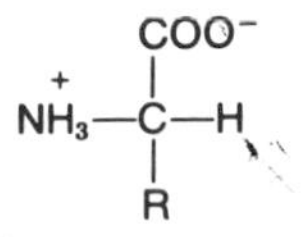

$$\overset{+}{NH_3}-\overset{\displaystyle COO^-}{\underset{\displaystyle R}{C}}-H$$

FIGURE 2.2
General structure of the common amino acids.

The Side Chains Define the Structures of the Different Amino Acids

The structures for the common amino acids are shown in Figure 2.3. In the category of alkyl amino acids are glycine, alanine, valine, leucine, and isoleucine. **Glycine** has the simplest structure, with R = H. **Alanine** contains a methyl (CH_3—) R group and **valine** an isopropyl R group (Figure 2.4). The leucine and isoleucine R groups are butyl alkyl chains that are structural isomers of each other. In **leucine** the branching methyl group in the isobutyl side chain occurs on the *gamma* (γ)-carbon of the amino acid. In **isoleucine** the butyl side chain is branched at the β carbon.

In the category of aromatic amino acids are phenylalanine, tyrosine, and tryptophan. In **phenylalanine** the R group contains a benzene ring,

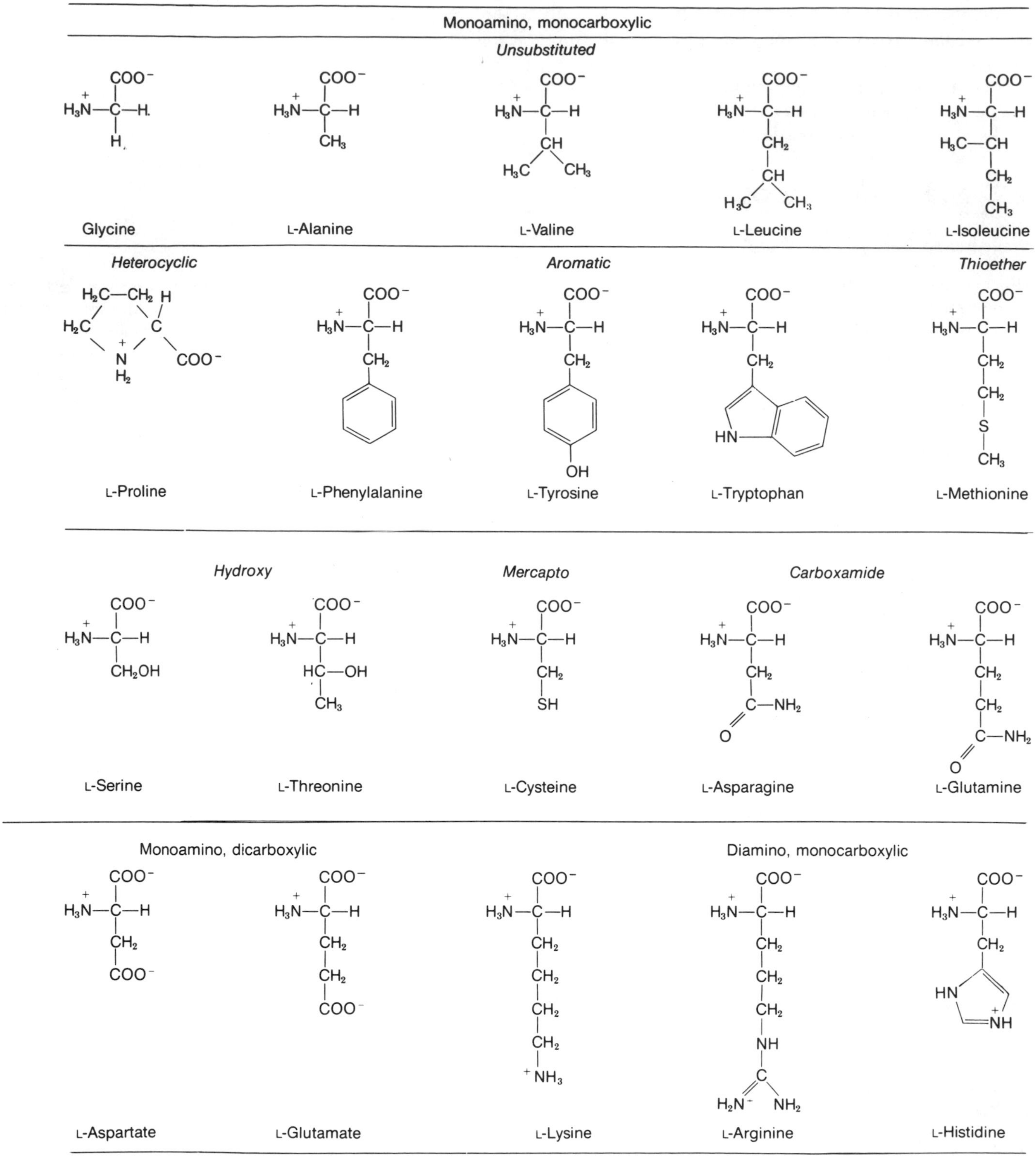

Monoamino, monocarboxylic

Unsubstituted

Glycine | L-Alanine | L-Valine | L-Leucine | L-Isoleucine

Heterocyclic

L-Proline

Aromatic

L-Phenylalanine | L-Tyrosine | L-Tryptophan

Thioether

L-Methionine

Hydroxy

L-Serine | L-Threonine

Mercapto

L-Cysteine

Carboxamide

L-Asparagine | L-Glutamine

Monoamino, dicarboxylic

L-Aspartate | L-Glutamate

Diamino, monocarboxylic

L-Lysine | L-Arginine | L-Histidine

FIGURE 2.3
Structures of the common amino acids. Charge forms shown are those present at pH 7.0.

tyrosine contains a phenol group, and the **tryptophan** R group contains a heterocyclic structure known as an indole. In the three aromatic amino acids the aromatic moiety is attached to the α carbon through a methylene (—CH_2—) carbon (Figure 2.3).

The two sulfur-containing common amino acids are cysteine and methionine. In **cysteine** the R group is thiolmethyl ($HSCH_2$—). In **methionine** the side chain is a methyl ethyl thiol ether ($CH_3SCH_2CH_2$—).

There are two hydroxy (alcohol)-containing common amino acids, serine and threonine. In **serine** the side chain is a hydroxymethyl moiety ($HOCH_2$—). In **threonine** an ethanol structure is connected to the α carbon of the amino acid through the 1 position of the ethanol, resulting in a secondary alcohol structure for R (CH_3CHOH—).

In **proline** the side chain group, R, is unique in that it incorporates the α-amino group in the side chain. Thus proline is more accurately classified as an α-imino acid, since its α-amine is a secondary amine, rather than a primary amine. In subsequent sections we discuss the ramifications of the incorporation of the α-amino nitrogen into a five-membered ring that constrains the rotational freedom around the N—C_α bond in proline to a specific rotational angle. This has important structural consequences for the protein structures in which proline participates.

The categories of amino acids discussed so far contain uncharged side chain R groups at physiological pH. The next category of amino acids, the dicarboxylic monoamino acids, contain a negatively charged carboxylate R group at pH 7. In **aspartate** the side chain carboxylic acid group is separated by a single methylene carbon (—CH_2—) from the α carbon (Figure 2.5). This differs for **glutamate** (Figure 2.5), in which the γ-carboxylic acid group is separated by two methylene (—CH_2—CH_2—) carbon atoms from the α carbon of the generalized structure (Figure 2.2).

Dibasic monocarboxylic acid structures are present in lysine, arginine, and histidine (Figure 2.3). In these structures, the R group contains a nitrogen or nitrogen atoms that may be protonated to form a positively charged side chain. In **lysine** the side chain is simply *N*-butyl amine. In **arginine** the side chain group contains a guanidino group (Figure 2.6) separated from the α carbon by three methylene carbon atoms. Both the **guanidinium group** of arginine and the amino group of lysine are predominantly protonated at physiological pH (pH ~7) and are in their charged forms. In **histidine** the side chain R group contains a five-membered heterocyclic structure known as an imidazole group (Figure 2.6). The pK_a' of the imidazole group of histidine is approximately 6.0 in water, and physiological solutions will contain relatively high concentrations of both the basic (imidazole) and acidic (imidazolium) forms of the histidine side chain at physiological pH (see Section 2.3).

The last two common amino acids are glutamine and asparagine. These two amino acids contain an amide moiety in their side chain R group. **Glutamine** and **asparagine** may be considered structural analogs of glutamic acid and aspartic acid, respectively, with their side chain carboxylic acid groups amidated. However, DNA codons exist for glutamine and asparagine separate from those for glutamic acid and aspartic acid. The amide side chains of glutamine and asparagine cannot be protonated and are uncharged in the range of physiological pH.

In order to easily communicate the sequence of amino acids in a protein, three-letter and one-letter abbreviations for the common amino acids have been established (Table 2.1). These abbreviations will be used throughout the book. The three-letter abbreviations of aspartic acid (Asp) and glutamic acid (Glu) should not be confused with those for asparagine (Asn) and glutamine (Gln). In experimentally determining the sequence of amino acids in a protein, there may be a difficulty in differentiating be-

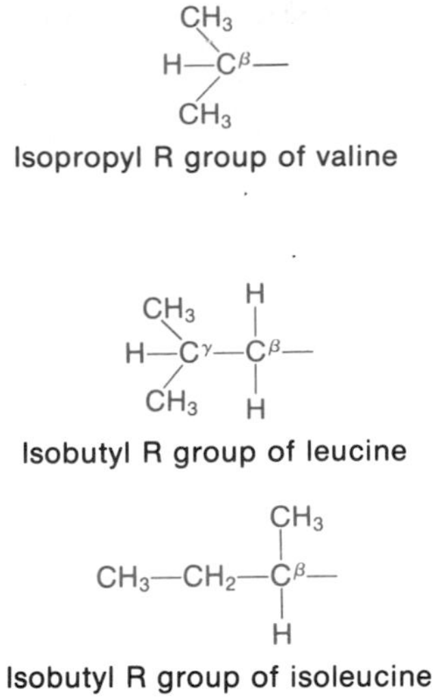

Isopropyl R group of valine

Isobutyl R group of leucine

Isobutyl R group of isoleucine

FIGURE 2.4
Alkyl side chain groups of valine, leucine, and isoleucine.

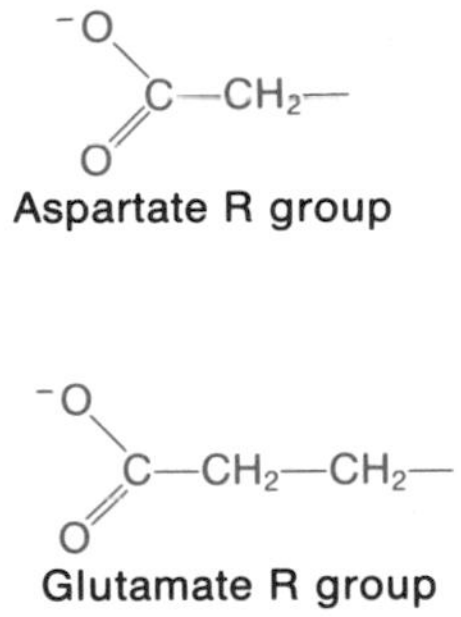

Aspartate R group

Glutamate R group

FIGURE 2.5
Side chain groups of aspartate and glutamate.

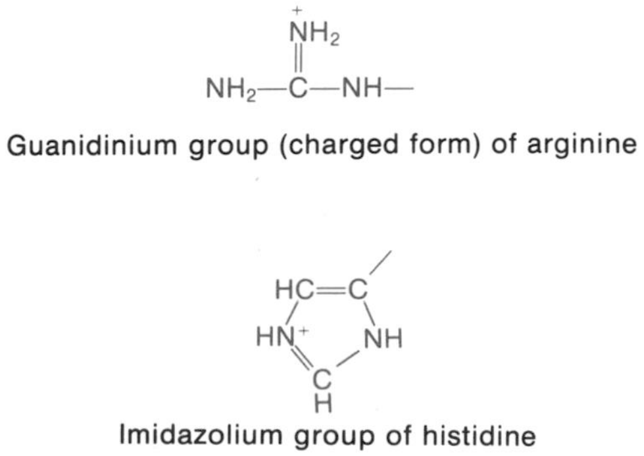

Guanidinium group (charged form) of arginine

Imidazolium group of histidine

FIGURE 2.6
Guanidinium and imidazolium groups of arginine and histidine.

TABLE 2.1 Abbreviations for the Amino Acids

Amino acid	Abbreviation	
	Three letter	One letter
Alanine	Ala	A
Arginine	Arg	R
Asparagine	Asn	N
Aspartic	Asp	D
Asparagine or Aspartic	Asx	B
Cysteine	Cys	C
Glycine	Gly	G
Glutamine	Gln	Q
Glutamic	Glu	E
Glutamine or Glutamic	Glx	Z
Histidine	His	H
Isoleucine	Ile	I
Leucine	Leu	L
Lysine	Lys	K
Methionine	Met	M
Phenylalanine	Phe	F
Proline	Pro	P
Serine	Ser	S
Threonine	Thr	T
Tryptophan	Trp	W
Tyrosine	Tyr	Y
Valine	Val	V

tween Asn and Asp, or between Gln and Glu because the side chain amide groups in Asn and Gln are hydrolyzed to the free carboxylic acids, Asp and Glu, by the chemical procedures utilized in the determination (see Section 2.4). In these cases the ambiguity is reported using the symbols of Asx for Asp or Asn, and Glx for Glu or Gln.

The Amino Acid Side Chains Can Have Polar and Apolar Properties

The relative **hydrophobicity** of an amino acid side chain is considered to play an important role in defining a specific amino acid's contribution to a protein's structure and function. Table 2.2 contains values of relative hydrophobicity of the 20 common amino acids based on the partition between water and a nonpolar solvent. The scale is based on glycine being zero. The amino acid side chains that preferentially dissolve in the nonpolar solvent have a positive hydrophobicity value with the more positive value the greater its preference for the nonpolar solvent (leucine, isoleucine, phenylalanine, tryptophan, tyrosine, and valine are the most positive). In the last column of Table 2.2 are values of the fraction of times an amino acid is found buried within the interior of a protein structure. The data is based on the observed three-dimensional structures of 12 proteins as determined by X-ray diffraction analysis (see Section 2.9). The volume

TABLE 2.2 Some Physical and Observed Properties of the Amino Acids

Amino acid	Relative hydrophobicity[a] (kcal mol^{-1})	Fraction of time found buried in protein structure[b]
Alanine	0.5	0.38
Arginine	(−11.2)	0.01
Asparagine	−0.2	0.12
Aspartic acid	(−7.4)	0.15
Cysteine	(−2.8)	0.40[c]
Glycine	0	0.36
Glutamine	−0.3	0.07
Glutamic acid	(−9.9)	0.18
Histidine	0.5	0.17
Isoleucine	(2.5)	0.60
Leucine	1.8	0.45
Lysine	(−4.2)	0.03
Methionine	1.3	0.40
Phenylalanine	2.5	0.50
Proline	(−3.3)	0.18
Serine	−0.3	0.22
Threonine	0.4	0.23
Tryptophan	3.4	0.27
Tyrosine	2.3	0.15
Valine	1.5	0.54

[a] Hydrophobicities measured by the distribution of the amino acid between a nonpolar solvent, either ethanol or dioxane, and water. Negative values indicate a preference for water and positive values a preference for nonpolar solvent. Values in parenthesis were calculated as described by G. Von Heijne and C. Blomberg, *Eur. J. Biochem.* 97:175, 1979 and the others by Y. Nozaki and C. Tanford, *J. Biol. Chem.* 246:2211, 1971.

[b] Amino acid residue must be at least 95% buried within the interior of the protein structure in order to be counted as buried. Average for 12 proteins, from C. Chothia, *J. Mol. Biol.* 105:1, 1976.

[c] Value is for Cys in thiol form. Value for disulfide form is 0.50.

of a "buried" amino acid must be at least 95% within the interior of the protein structure. Those residues preferring nonpolar solvents (i.e., positive hydrophobicity) are also those amino acids found buried in folded protein structures. This correlation is not exact, however, due to the amphoteric nature of many of the hydrophobic amino acids that place the more polar portions of their structure near the surface. In addition, contrary to expectation, a significant number of nonpolar residues will not be buried and are exposed to the water solvent as a result of the three-dimensional folding of the protein. However, these hydrophobic side chain groups on the surface appear generally dispersed and isolated among polar and ionized surface residues. The clustering of two or more nonpolar side chain groups may occur in small regions of the protein surface and are usually associated with a function of the protein such as providing a hydrophobic binding site for substrate or ligand molecules to the protein. Amino acids of intermediate polarity include glycine, tyrosine with its polar OH group, threonine, serine, and proline. These amino acid side chains are found both in the interior and on the solvent protein interface in significant proportions. In contrast, the polar amino acids glutamine and asparagine and the amino acids containing charged R groups at pH 7 (lysine, arginine, histidine, glutamate, and aspartate) are predominantly found on the surface in globular proteins where the charge is stabilized by the water solvent. As a result of the complex nature of protein folding, specific interactions can cause the stabilization of charged residues in the nonpolar interior of a protein. The rare positioning of a charged side chain into the interior of a globular protein is usually correlated with an essential structural or functional role for the "buried" charged side chain group within the nonpolar interior of the protein.

Amino Acids Have an Asymmetric Center

The common amino acids with the general structure in Figure 2.2 have four substituents (R, H, COO^-, NH_3^+) covalently bonded to the α-carbon atom in the α-amino acid structure. A carbon atom with four different substituents arranged in a **tetrahedral configuration** is asymmetric and exists in two enantiomeric forms. Thus each of the amino acids (*except glycine,* in which R = H and thus two of the four substituents on the α-carbon atom are hydrogen) exhibits optical isomerism.

The **absolute configuration** for an amino acid is depicted in Figure 2.7 using the Fischer projection to show the direction in space of the tetrahedrally arranged α-carbon substituents. The α-COO^- group is directed up and back behind the plane of the page, and the side chain group, R, is directed down and also back behind the plane of the page. The α-H and α-NH_3^+ groups are directed toward the reader. An amino acid held in this way projects its α-NH_3^+ group either to the left or right of the α-carbon atom. By convention, if the α-NH_3^+ is projected to the left, the amino acid has an L absolute configuration. Its optical enantiomer, with α-NH_3^+ projected toward the right, has a designated D absolute configuration. In mammalian proteins only amino acids of L configuration are found.

As the amino acids in proteins are asymmetric, the proteins into which the amino acids are polymerized also exhibit asymmetric properties.

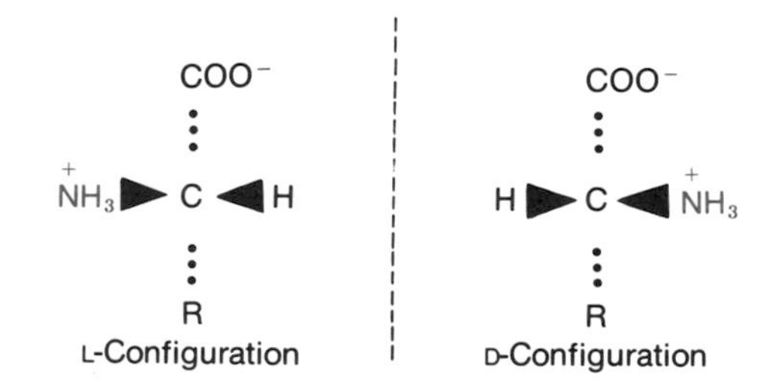

FIGURE 2.7
Absolute configuration of an amino acid.

Amino Acids Are Polymerized Into Peptides and Proteins

The polymerization of the 20 common amino acids into polypeptide chains within cells is catalyzed by enzymes, requires RNA, and occurs on the ribosomes (Chapter 17). Chemically, the polymerization of amino acids into protein is a dehydration reaction. The chemical rationale of the

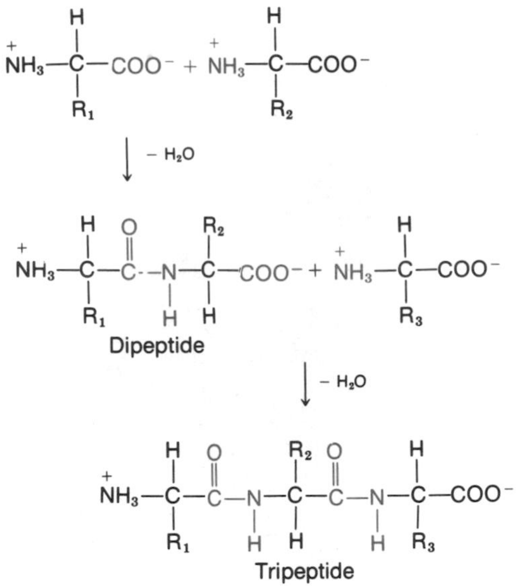

FIGURE 2.8
Peptide bond formation.

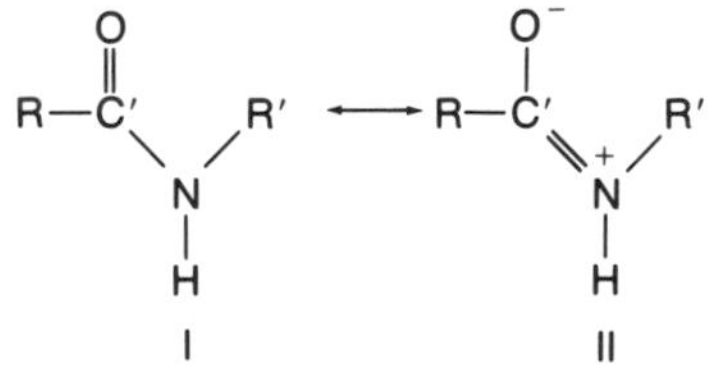

FIGURE 2.9
Electronic isomer structures of the peptide bond.

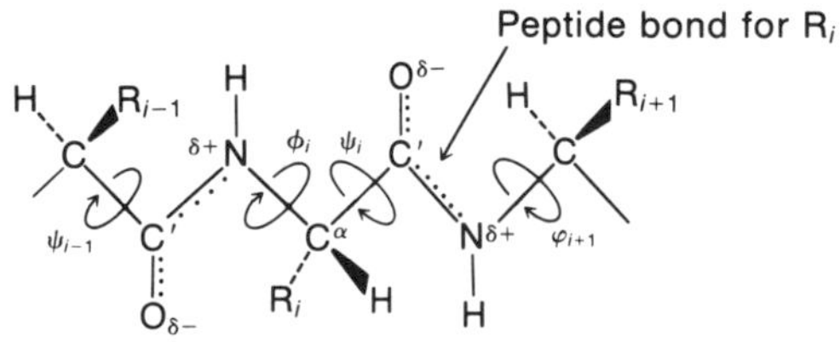

FIGURE 2.10
Each amino acid residue of a polypeptide chain contributes two single bonds and one peptide bond to the polypeptide chain. The single bonds are the bonds between the C_α and the C′, and the C_α and N. See p. 50 for definition of ϕ, and ψ.

reaction is shown in Figure 2.8. The figure shows that the α-carboxyl group of an amino acid with side chain R_1 may be covalently joined to the α-NH_2 group of the amino acid with side chain R_2 by the elimination of a molecule of water to form a type of amide bond known as the **peptide bond.** The dipeptide (two amino acid residues joined by a single peptide bond) can then form a second peptide bond through its terminal carboxylic acid group to the α-amine of a third amino acid (R_3), generating a tripeptide (Figure 2.8). Repetition of this stepwise dehydration process will generate a **polypeptide** or protein of specific amino acid sequence (R_1—R_2—R_3—$R_4 \cdots R_n$). The specific amino acid sequence of a natural polypeptide is determined from the genetic information (Chapter 17). The amino acid sequence of the **polypeptide** chains in a protein is known as the **primary structure** of the protein. It is the primary structure (amino acid sequence) that gives a protein its physical properties and causes a polypeptide chain to fold into a unique structure giving it a characteristic function and role.

The chemical nature of the peptide bond can be represented using two **resonance isomers** (Figure 2.9). In one of the isomer structures (Structure I), a double bond is located between the carbonyl carbon and carbonyl oxygen (C′=O), and the carbonyl carbon to nitrogen (C′—N) is depicted as a single bond. In the alternate isomer (Structure II), the carbonyl carbon to nitrogen bond is shown as a double bond ($C'{=}\overset{+}{N}$) and the carbonyl carbon to oxygen bond as a single bond. In Structure II there is a formal negative charge on the oxygen and a positive charge on the nitrogen. Actual peptide bonds are a resonance hybrid of these two electron isomer structures. The carbonyl carbon to nitrogen bond (C′—N) has equal characteristics of a double and single bond (50% double bond character). The hybrid bond structure of the peptide bond is supported by both spectroscopic measurements and X-ray diffraction studies, which show that the C′—N bond length (1.33 Å) is approximately half-way between that found for formal C—N single bonds (~1.45 Å) and C=N double bonds (1.25 Å). A consequence of this partial double bond character is that *free rotation does not readily occur around the C′—N bond at physiological temperatures*. As a result the atoms attached to C′ and N that form the double bond, $\begin{matrix} O^- & & C— \\ & C'{=}\overset{+}{N} & \\ —C & & H \end{matrix}$, all lie in the peptide bond plane. Thus a polypeptide chain can be viewed as a polymer of peptide bond planes interconnected at the α-carbon atom by single bond linkages. Each **amino acid residue** gives three covalent bonds (two single bonds and a peptide bond, Figure 2.10) to the polypeptide chain. The term *residue* refers to the atoms contributed by an amino acid to a polypeptide chain, including the atoms of the side chain group.

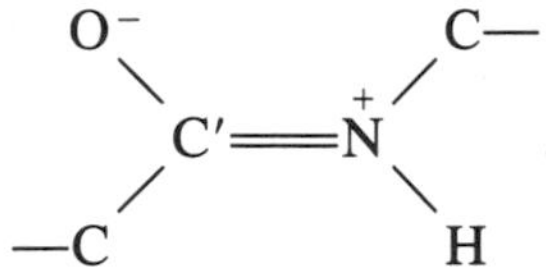

The peptide bond in Figure 2.11*a* shows a **trans conformation** between the oxygen (O) and the hydrogen (H) atoms of the peptide bond. This is the most stable conformation for the peptide bond as the two side chain groups (R and R′) on the amino acids joined by the peptide bond are also trans. Thus they have a lower probability for unfavorable steric interactions than in the alternative cis conformation. In the **cis conformation** (Figure 2.11*b*) the two side chain groups are close together on the same side of the C′—N partial double bond, where repulsive steric forces are generated between the two side chain (R) groups. Because of the steric repulsive forces between side chain groups in the cis peptide bond, trans peptide bonds are the predominant conformation found in proteins except

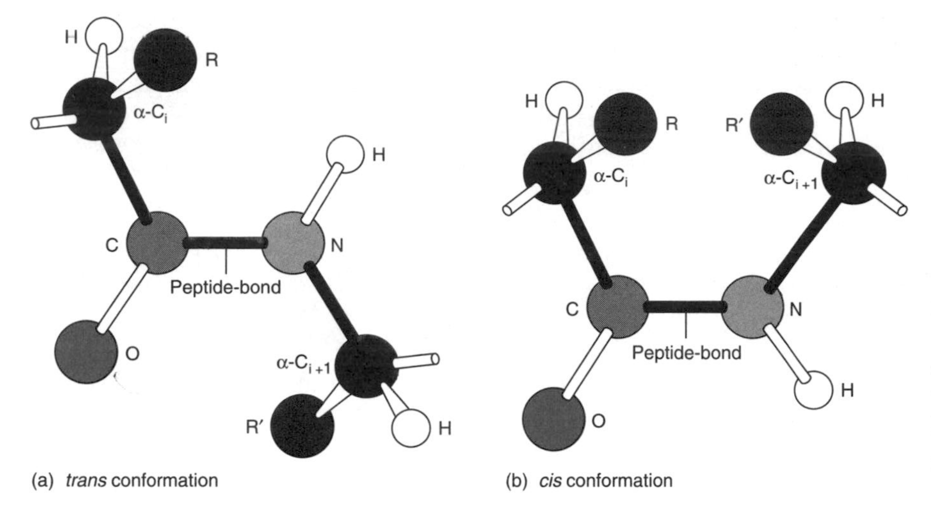

FIGURE 2.11
(*a*) Trans peptide bond, and (*b*) the rare cis peptide bond. The C′ and N have a partial double bond character.
Modified from G. E. Schulz and R. H. Schirmer, *Principles of Protein Structure,* N.Y., Springer-Verlag, 1979, p. 18.

where there are proline residues. In proline the side chain is linked to its α-amino group, and cis- and trans-type peptide bonds with the proline α-imino have near equal energies. Both cis and trans peptide bonds are found in proteins for peptide bonds made with the α-imino group of proline. The conformation of peptide bond actually found for proline in a protein will depend on the forces generated by the folded three-dimensional structure of a protein molecule on the proline peptide bonds within its structure.

One of the largest natural polypeptide chains in humans is that of apolipoprotein B-100 found in the plasma low density lipoprotein (LDL) particles. This protein contains 4536 amino acid residues in a single polypeptide chain. Chain length alone, however, does not determine the function of a polypeptide. Many small peptides with less than 10 amino acids perform important biochemical and physiological functions in humans (Table 2.3).

Primary structures are written in a standard convention and numbered from their NH_2-terminal end toward their COOH-terminal end. Accordingly, for thyrotropin-releasing factor the glutamic acid residue written on the left of the sequence is the NH_2-terminal amino acid of the tripeptide and is designated amino acid residue 1 in the sequence. The proline is the COOH-terminal amino acid in the structure and is designated the third amino acid residue in the sequence. The defined direction of the polypeptide chain is from Glu → Pro (NH_2-terminal amino acid to COOH-terminal amino acid).

Cystine Is a Derived Amino Acid

A derived amino acid found in many protein structures is cystine. **Cystine** is formed by the oxidation of two cysteine thiol side chain residues, which are joined to form a disulfide covalent bond (Figure 2.12). The resulting disulfide amino acid is the derived amino acid cystine. Within proteins covalent disulfide links of cystine formed from cysteines, separated from one another in the primary structure, have an important role in stabilizing the folded conformation of proteins. Cystines are formed after the free SH-containing cysteines are incorporated into the protein's primary structure and after the protein has folded.

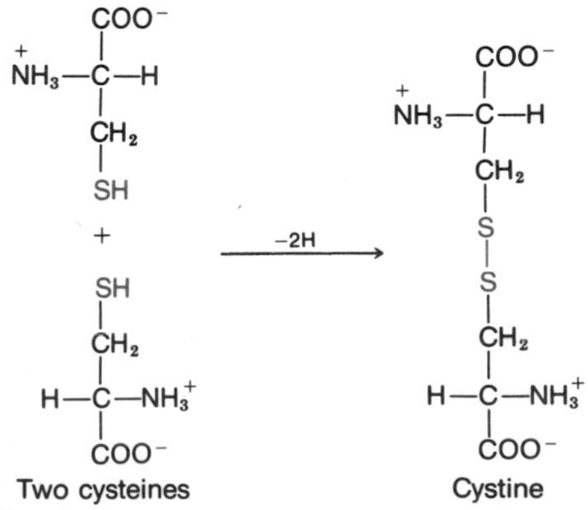

COO^-
$\overset{+}{N}H_3$—C—H
CH_2
SH
\+
SH
CH_2
H—C—NH_3^+
COO^-
Two cysteines

—2H →

COO^-
$\overset{+}{N}H_3$—C—H
CH_2
S
S
CH_2
H—C—NH_3^+
COO^-
Cystine

FIGURE 2.12
Cystine bond formation.

TABLE 2.3 Some Examples of Biologically Active Peptides

Amino acid sequence	Name	Function
1 3 pyroGlu-His-Pro(NH_2)[a]	Thyrotropin releasing factor	Secreted by hypothalamus and causes pituitary gland to release thyrotropic hormone
1 9 H-Cys-Tyr-Phe-Gln-Asn-Cys-Pro-Arg-Gly(NH_2)[b] S ——— S (Cys-1 to Cys-6)	Vasopressin (antidiuretic hormone)	Secreted by pituitary gland and causes kidney to retain water from urine
1 5 H-Tyr-Gly-Gly-Phe-Met-OH	Methionine enkephalin	Opiate-like peptide found in brain that inhibits sense of pain
1 10 pyroGlu-Gly-Pro-Trp-Leu-Glu-Glu-Glu-Glu-Glu- 11 17 Ala-Tyr-Gly-Trp-Met-Asp-Phe(NH_2)[a,c] SO_3 (on Tyr)	Little gastrin (human)	Hormone secreted by mucosal cells in stomach and causes parietal cells of stomach to secrete acid
1 10 H-His-Ser-Gln-Gly-Thr-Phe-Thr-Ser-Asp-Tyr- 11 20 Ser-Lys-Tyr-Leu-Asp-Ser-Arg-Arg-Ala-Gln- 21 29 Asp-Phe-Val-Gln-Trp-Leu-Met-Asn-Thr-OH	Glucagon (bovine)	Pancreatic hormone involved in regulating glucose metabolism
1 8 H-Asp-Arg-Val-Tyr-Ile-His-Pro-Phe-OH	Angiotensin II (horse)	Pressor or hypertensive peptide; also stimulates release of aldosterone from adrenal gland
1 9 H-Arg-Pro-Pro-Gly-Phe-Ser-Pro-Phe-Arg-OH	Plasma bradykinin (bovine)	Vasodilator peptide
1 10 H-Arg-Pro-Lys-Pro-Gln-Phe-Phe-Gly-Leu-Met(NH_2)	Substance P	Neurotransmitter

[a] The NH_2 terminal Glu is in the pyro form in which its γ-COOH is covalently joined to its α-NH_2 via amide linkage; the COOH terminal amino acid is amidated and thus also not free.
[b] Cysteine-1 and cysteine-6 are joined to form a cyclohexa structure within the nonapeptide.
[c] The Tyr 12 is sulfonated on its phenolic side chain OH.

2.3 CHARGE AND CHEMICAL PROPERTIES OF AMINO ACIDS AND PROTEINS

The Ionizable Groups of Amino Acids and Proteins Are Important for Their Biological Function

An understanding of proteins requires a knowledge of the **ionizable side chain** groups of the common amino acids. These ionizable groups common to proteins and amino acids are shown in Table 2.4.

The acid forms of the ionizable groups are on the left of the equilibrium sign and their conjugate bases on the right side. In forming their conjugate base, the acid form dissociates a proton. In reverse, the base form will associate with a proton to form the respective acid. The proton dissociation from a particular acid is characterized by an acid dissociation constant (K_a') with its corresponding pK_a' value ($pK_a' = \log_{10}(1/K_a')$). As shown in Table 2.4, each acid group has a range of pK_a' values, depending on the environment in which an acid group is placed. For example, when a

TABLE 2.4 Characteristic pK_a' Values for the Common Acid Groups in Proteins

Where acid group is found	Acid form		Base form	Approximate pK_a range for group
NH_2 terminal residue in peptides, lysine	$R—NH_3^+$ Amino	⇌	$R—NH_2 + H^+$ Amine	7.6–10.6
COOH terminal residue in peptides, glutamate, aspartate	R—COOH Carboxylic acid	⇌	$R—COO^- + H^+$ Carboxylate	3.0–5.5
Arginine	$R—NH—\overset{+}{C}(\cdots NH_2)—NH_2$ Guanidinium	⇌	$R—NH—C(=NH)—NH_2 + H^+$ Guanidino	11.5–12.5
Cysteine	R—SH Thiol	⇌	$R—S^- + H^+$ Thiolate	8.0–9.0
Histidine	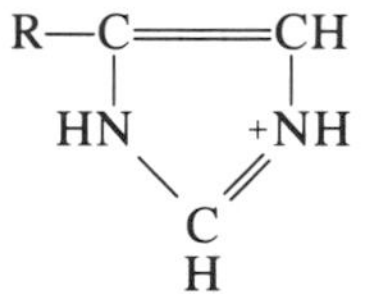 R—C=CH, HN, +NH, C, H (ring) Imidazolium	⇌	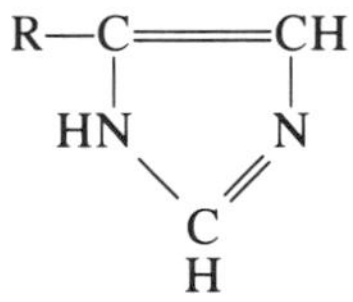 R—C=CH, HN, N, C, H (ring) $+ H^+$ Imidazole	6.0–7.0
Tyrosine	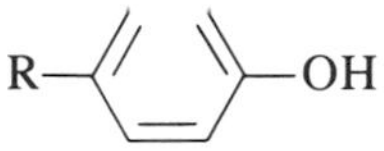 R—(benzene ring)—OH Phenol	⇌	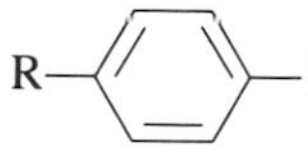 R—(benzene ring)—$O^- + H^+$ Phenolate	9.5–10.5

positive charged amino group ($—NH_3^+$) is placed near a negatively charged group within a protein structure, the negative charge will stabilize the positively charged amino group making it more difficult to dissociate its proton. It will be necessary to go to a higher pH to dissociate the proton from the positively charged amino group in the ion pair with the negatively charged group, and the pK_a' will have a higher value than normally found for an amino group. Factors other than charge that can affect the pK_a' of an acid group include the polarity of the environment, stabilization or destabilization effects transmitted by the atoms of the molecule that the acid group is a part, the absence or presence of water, and the corresponding absence or presence of hydrogen bonding. The terminal α-COOH or α-NH_3^+ typically have lower pK_a' values than those of the side chains. In the protein ribonuclease, the amino terminal α-NH_3^+ has a pK_a' of 7.8, while the *epsilon* (ε)-amino group of the lysine side chains have pK_a' values that average 10.2. Similarly, the carboxyl terminal α-COOH acid will have a pK_a' value of lower value (3.8) than that found for the carboxylic acid groups in the side chains of aspartic acid and glutamic acid (p$K_a' \sim 4.6$).

The amino acids whose R groups contain nitrogen atoms (Lys and Arg) are known as the **basic amino acids,** since their side chains have high pK_a' values and function as good bases. They are usually in their acid forms and positively charged at physiological pH. The amino acids whose side chains contain a carboxylic acid group have relatively low pK_a' values and are called the **acidic amino acids.** They are predominantly in their base forms and are negatively charged at physiological pH. Proteins in which the ratio (Σ Lys + Σ Arg)/(Σ Glu + Σ Asp) is greater than 1 are referred to as basic proteins. Proteins in which the above ratio is less than 1 are referred to as acidic proteins.

$$\text{pH} = \text{p}K_a + \log \frac{[\text{conjugate base}]}{[\text{conjugate acid}]}$$

or

$$\text{pH} - \text{p}K_a = \log \frac{[\text{conjugate base}]}{[\text{conjugate acid}]}$$

FIGURE 2.13
Henderson–Hasselbalch equation. (For a more detailed discussion of this equation, refer to Chapter 1.)

TABLE 2.5 Relationship Between the Difference of pH and Acid pK_a' and the Ratio of the Concentrations of Base to Its Conjugate Acid

$pH-pK_a'$ (Difference between pH and pK_a')	Ratio of concentration of base to conjugate acid
0	1
1	10
2	100
3	1000
−1	0.1
−2	0.01
−3	0.001

The Ionic Form of an Amino Acid or Protein Can Be Determined at a Given pH

From a knowledge of the pK_a' for each of the ionizable acid groups in an amino acid or protein and the Henderson–Hasselbalch equation (Figure 2.13), the ionic form of the molecule can be calculated at a given pH. This is an important calculation since a change in the ionization of a protein with pH will give a molecule different functional properties at different pH values.

For example, an enzyme may require a catalytically essential histidine imidazole in its basic form for catalytic activity. If the pK_a' of the catalytically essential histidine in the enzyme is 6.0, at pH 6.0 one-half of the enzyme molecules will be in the active basic (imidazole) form and one-half in the inactive acid (imidazolium) form. Accordingly, the enzyme will exhibit 50% of its potential activity. At pH 7.0, the pH is one unit above the imidazolium pK_a' and the ratio of [imidazole]/[imidazolium] is 10 : 1 (Table 2.5). Based on this ratio, the enzyme will exhibit $10/(10 + 1) \times 100 = 91\%$ of its maximum potential activity.

Titration of a Monoamino Monocarboxylic Acid: Determination of the Isoelectric pH

$$\overset{+}{N}H_3-CH(COOH)-CH_2-CH(CH_3)-CH_3 \underset{+H^+}{\overset{+OH^-}{\rightleftharpoons}} \overset{+}{N}H_3-CH(COO^-)-CH_2-CH(CH_3)-CH_3 \underset{+H^+}{\overset{+OH^-}{\rightleftharpoons}} NH_2-CH(COO^-)-CH_2-CH(CH_3)-CH_3$$

	I	II	III
Charge	+1	0	−1
pH	pH < 2.4	2.4 < pH < 9.6	9.6 < pH

FIGURE 2.14
Ionic forms of leucine.

An understanding of a protein's acid and base forms and their relation to charge is more clearly obtained after following the titration of the ionizable groups for the simple case of an amino acid. As presented in Figure 2.14, leucine contains an α-COOH with $pK_a' = 2.4$ and an α-NH_3^+ group with $pK_a' = 9.6$. At pH 1.0 the predominant ionic form (form I) of leucine will have a formal charge of +1 and leucine ions will migrate toward the cathode in an electrical field. The addition of base in an amount equal to one-half of the moles of leucine present in the solution will half-titrate the α-COOH group of the leucine, that is, $[COO^-]/[COOH] = 1$. The pH of the solution after the addition of the 0.5 equiv of base is equal to the pK_a' of the α-COOH of the leucine (Figure 2.15).

Addition of 1 equiv of base will completely titrate the α-COOH. In the predominant form (II), the negatively charged α-COO^- and positively charged α-NH_3^+ cancel each other and the net charge on this ionic form is zero. Form II is the zwitterion form of leucine. The **zwitterion form** is that ionic form in which the positive charge from positively charged ionized groups is exactly equal to the negative charge from negatively ionized groups of the molecule. Accordingly, the net charge on a zwitterion molecule is zero, and a zwitterion molecule will not migrate toward either the cathode or anode in an electric field.

The further addition of a 0.5 equiv of base to the zwitterion form of leucine will half-titrate the α-NH_3^+ group. At this point in the titration, the ratio of $[\alpha\text{-}NH_2]/[\alpha\text{-}NH_3^+] = 1$, and $pH = pK_a'$ for NH_3^+ (Figure 2.15).

The addition of a further 0.5 equiv of base (total of 2 full equiv; Figure 2.15) will completely titrate the α-NH_3^+ group. The solution pH is greater than 11, and the predominant species has a formal negative charge of −1 (form III).

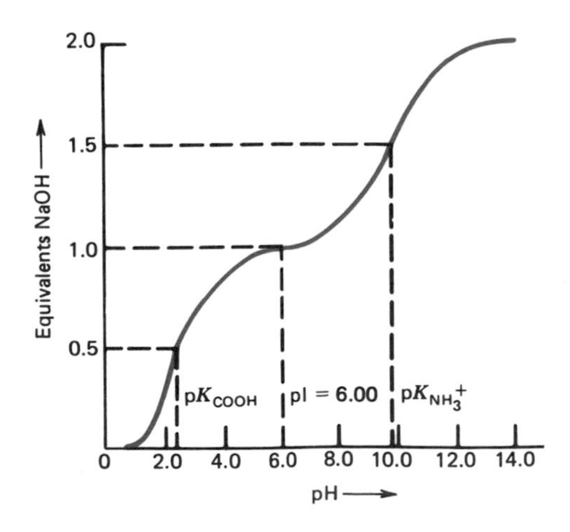

FIGURE 2.15
Titration curve for leucine.

It is useful to calculate the exact pH at which an amino acid is electrically neutral and in its zwitterion form. This pH is known as the **isoelectric pH** for the molecule, and the symbol is **pI.** The pI value is a constant of a particular compound at specific conditions of ionic strength and temperature. For simple molecules, such as amino acids, the pI is the average of the two pK_a' values that form the boundaries of the zwitterion form. Leucine has only two ionizable groups, and the pI is calculated as follows:

$$pI = \frac{pK_a'COOH + pK_a'NH_3^+}{2} = \frac{2.4 + 9.6}{2} = 6.0$$

COOH		COO^-		COO^-		COO^-
$\overset{+}{N}H_3$—C—H	$\underset{+H^+}{\overset{+OH^-}{\rightleftharpoons}}$	$\overset{+}{N}H_3$—C—H	$\underset{+H^+}{\overset{+OH^-}{\rightleftharpoons}}$	$\overset{+}{N}H_3$—C—H	$\underset{+H^+}{\overset{+OH^-}{\rightleftharpoons}}$	NH_2—C—H
CH_2		CH_2		CH_2		CH_2
CH_2		CH_2		CH_2		CH_2
COOH		COOH		COO^-		COO^-
Charge +1		0		−1		−2
pH < 2.2		2.2 < pH < 4.2		4.3 < pH < 9.7		9.7 < pH

FIGURE 2.16
Ionic forms of glutamic acid.

At pH > 6.0 leucine will assume a partial negative charge that formally rises at high pH to a full negative charge of −1 (form III) (Figure 2.14). At pH < pI, leucine will have a partial positive charge until at very low pH it will have a formal charge of +1 (form I) (Figure 2.14). The partial charge at any pH can be calculated from the Henderson–Hasselbalch equation or from extrapolation from the titration curve of Figure 2.15.

Titration of a Monoamino Dicarboxylic Acid

A more complicated example of the relationship between molecular charge and pH is the example of glutamic acid. Its ionized forms and titration curve are shown in Figures 2.16 and 2.17. In glutamic acid the α-COOH $pK'_a = 2.2$, the γ-COOH $pK'_a = 4.3$, and the α-NH_3^+ $pK'_a = 9.7$. The zwitterion form is generated after 1.0 equiv of base is added to the low pH form, and the isoelectric pH (pI) is calculated from the average of the two pK'_a values that form the boundaries of the zwitterion form:

$$\text{pI} = \frac{2.2 + 4.3}{2} = 3.25$$

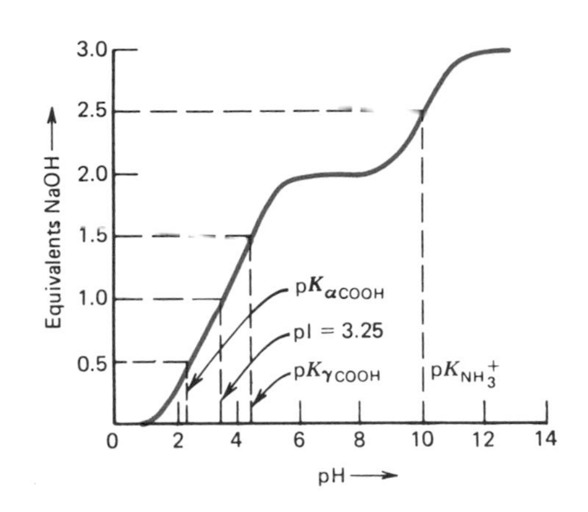

FIGURE 2.17
Titration curve for glutamic acid.

Accordingly, at values above pH 3.25 the molecule will assume a net negative charge until at high pH the molecule will have a net charge of −2. At pH < 3.25 glutamic acid will be positively charged, and at extremely low pH it will have a net positive charge of +1.

The Charge Properties of Amino Acids and Proteins Is Determined by the pH

An analysis of the charge forms present in the other common amino acids shows that the relationship found between pH and the respective pI constants for leucine and glutamate is generally true. That is, *at a solution pH less than the pI of the amino acid, the amino acid is positively charged. At a pH greater than the pI, the amino acid is negatively charged.* The degree of positive or negative charge is a function of the difference between the pH and the pI value of the amino acid and is calculable for an amino acid by the Henderson–Hasselbalch relationship.

Proteins contain multiple ionizable side chain groups and the pI value characteristic of a protein will depend on the relative concentrations of the different acidic and basic R groups. As a protein contains many ionizable residues, calculation of its isoelectric pH from pK'_a values would be difficult. Accordingly, the pI values for proteins are almost always experimentally measured by determining the pH value in which the protein does not move in an electrical field. The pI values found for some representative proteins are given in Table 2.6.

As with the amino acids, *at a pH greater than the pI, the protein will have a net negative charge. At a pH less than the pI, the protein will have*

TABLE 2.6 pI Values for Some Representative Proteins

Protein	pI
Pepsin	ca. 1
Human serum albumin	4.9
α_1-Lipoprotein	5.5
Fibrinogen	5.8
Hemoglobin A	7.1
Ribonuclease	7.8
Cytochrome *c*	10.0
Thymohistone	10.6

pH > pI, then protein charge negative
pH < pI, then protein charge positive

FIGURE 2.18
Relationship between solution pH, protein pI, and protein charge.

a positive net charge (Figure 2.18). *The magnitude of the net charge of a protein will increase as a function of the difference between pH and pI.* For example, human plasma albumin contains 585 amino acid residues of which there are 61 glutamates, 36 aspartates, 57 lysines, 24 arginines, and 16 histidines. The titration curve for this complex molecule is shown in Figure 2.19. The albumin pI = 4.9, at which pH the net charge is zero. At pH 7.5 the imidazolium groups of histidine have been partially titrated and albumin has a formal negative charge of −10. At pH 8.6 additional groups have been titrated to their basic form, and the formal net charge is approximately −20. At pH 11 the net charge is approximately −60. On the acid side of the pI value, at pH 3, the approximate net charge on the albumin molecule in solution is +60.

Amino Acids and Proteins Can Be Separated Based on pI Values

The techniques of electrophoresis, isoelectric focusing, and ion-exchange chromatography are some of the more important techniques for the study of biological molecules based on charge.

In **electrophoresis,** an ampholyte (protein, peptide, or amino acid) in a solution buffered at a particular pH is placed in an electric field. Depending on the relationship of the buffer pH to the pI of the molecule, the molecule will either move toward the cathode (−) or the anode (+), or remain stationary (pH = pI).

An example of a classical apparatus for protein electrophoresis is shown in Figure 2.20. The apparatus consists of a U tube in which is placed a protein solution, followed by a buffer solution carefully layered over the protein solution. The migration of the protein is observed with an optical device that measures changes in the refractive index of the solution as the protein migrates toward the anode (Figure 2.20). This apparatus historically led to the separation and operational classification of the proteins in human plasma. For the **plasma protein** separation, the solution is buffered at pH 8.6, which is at a pH substantially above the pI of the major plasma proteins. The proteins are negatively charged and move toward the positive pole. The peaks in order of their rate of migration,

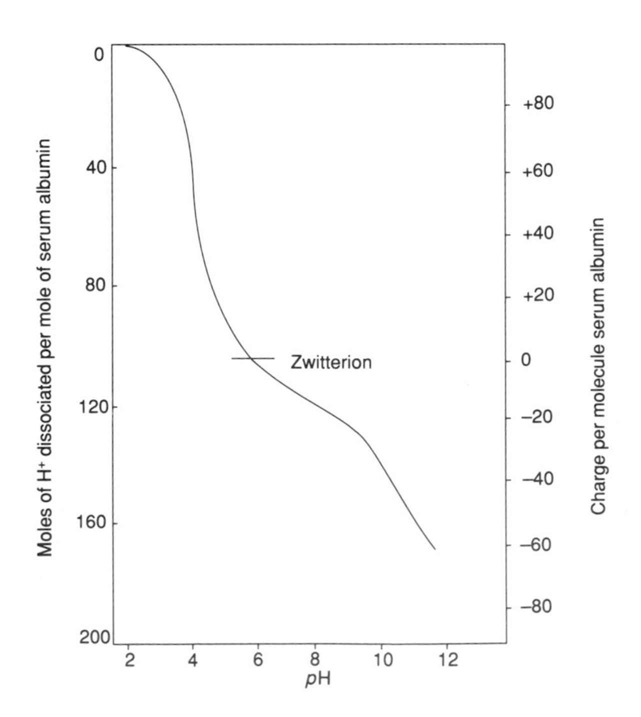

FIGURE 2.19
Titration curve of human serum albumin at 25°C and an ionic strength of 0.150.
Redrawn from C. Tanford, *J. Am. Chem. Soc.* 72:441, 1950.

which is related to the order of their pI values, are those of albumin, α_1-, α_2-, and β-globulins, fibrinogen, and γ_1- and γ_2-globulins. Some of these peaks represent tens to hundreds of individually different plasma proteins that have a similar migration rate to the anode at pH 8.6. Their rate of migration under these experimental conditions is widely used for purposes of their identification and classification (Figure 2.21).

More sophisticated procedures for electrophoresis use polymer gels, starch, or paper as support. The inert supports are saturated with buffer solution, a sample of the proteins to be examined is placed on the support, an electric field is applied across the buffered support, and the proteins migrate in the support toward a charged pole.

A commonly used high-resolution polymeric support is a polyacryl-

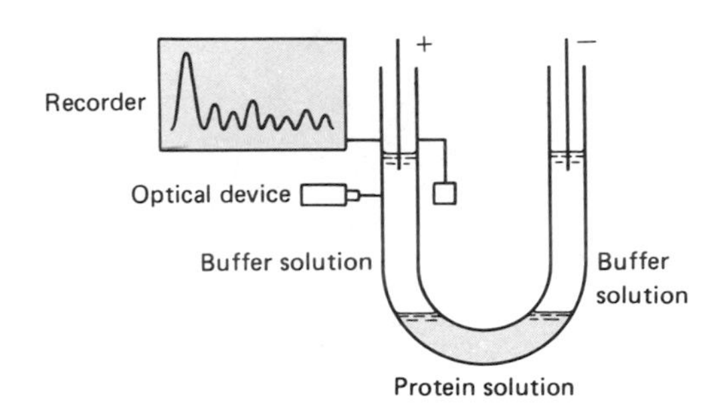

FIGURE 2.20
Classical electrophoresis apparatus.

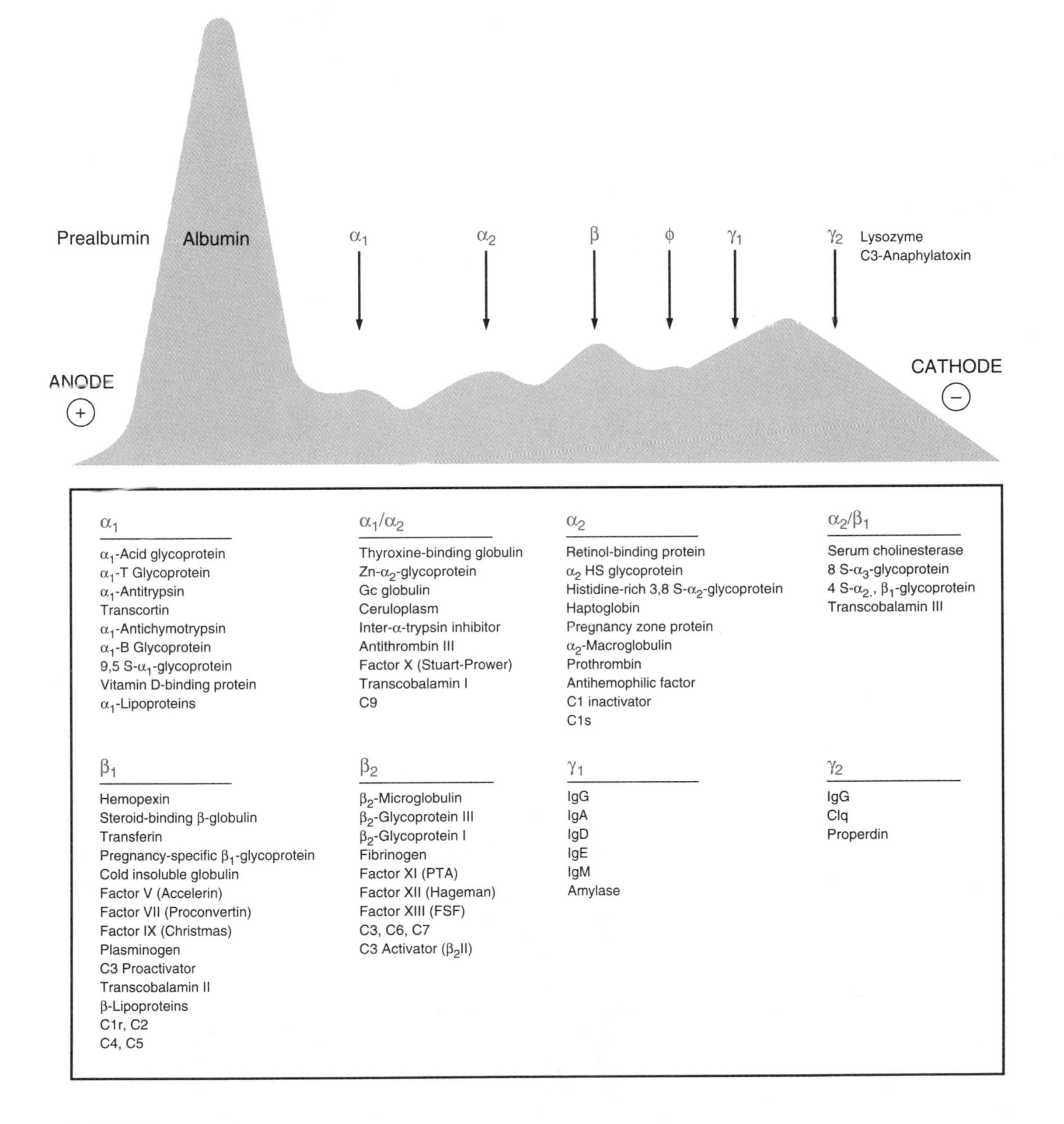

FIGURE 2.21
Classical (Tiselius) electrophoresis pattern for plasma proteins at pH 8.6.
The different major proteins are designated underneath the peaks. The direction of migration is from right to left with the anode (+) at left.
Reprinted with permission from K. Heide, H. Haupt, and H. G. Schwick, in *The Plasma Proteins*, 2nd ed., Vol. III, F. W. Putnam (ed.), New York: Academic Press, 1977, p. 545.

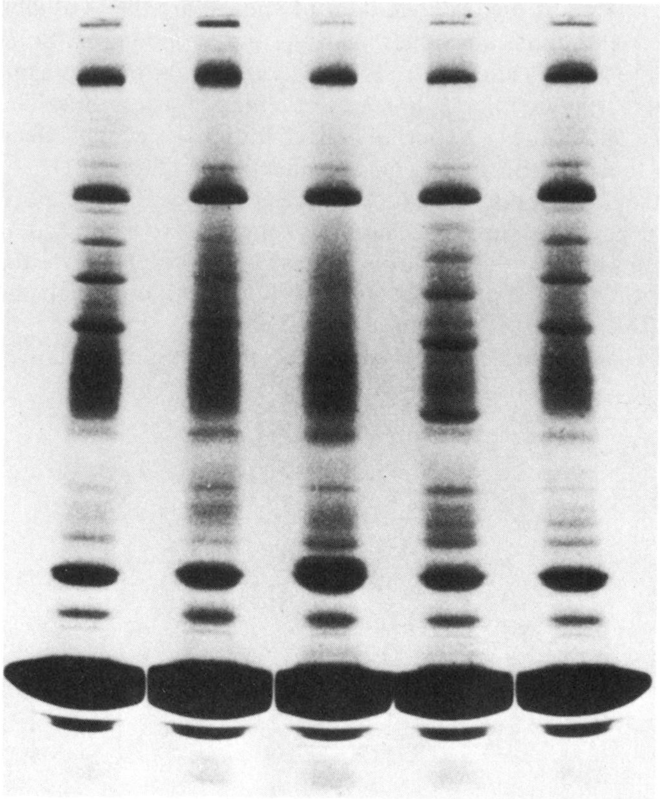

FIGURE 2.22
Serum protein separation by polyacrylamide gel electrophoresis.
Albumin peak is heavy brand near the bottom and slower moving proteins (γ-globulins) are near the top.
Reprinted with permission from P. C. Allen, E. A. Hill, and A. M. Stokes. The plasma proteins: analytical and preparative techniques. Blackwell, Oxford, 1977.

amide cross-linked gel. With the polyacrylamide gel technique, the seven peaks observed in the U-tube electrophoresis may be resolved into a greater number of distinct bands (Figure 2.22). A common *criterion for purity* of a protein is the observation of a single sharp band for a protein in a polyacrylamide gel electrophoresis experiment (Figure 2.23).

A type of electrophoresis with extremely high resolution is the technique of **isoelectric focusing** in which mixtures of polyamino-polycarboxylic acid ampholytes with a defined range of pI values are used to establish a pH gradient across an applied electric field. A charged protein will migrate through the pH gradient in the electric field until it reaches a pH region in the gradient equal to its pI value. At this point the protein

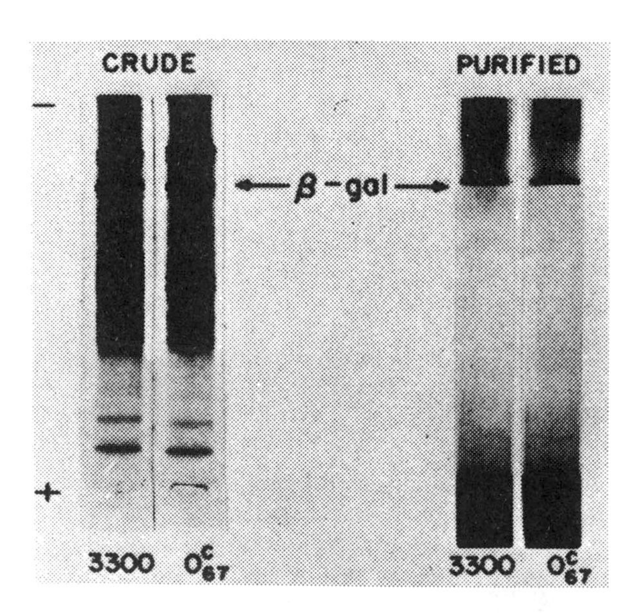

FIGURE 2.23
Purification of the protein β-galactosidase from crude extracts of two bacteria strains, designated 3300 and O^c_{67}.
Reprinted with permission from E. Steers, G. R. Craven, and C. B. Anfinsen, *Proc. Natl. Acad. Sci. U.S.A.* 54:1174, 1965.

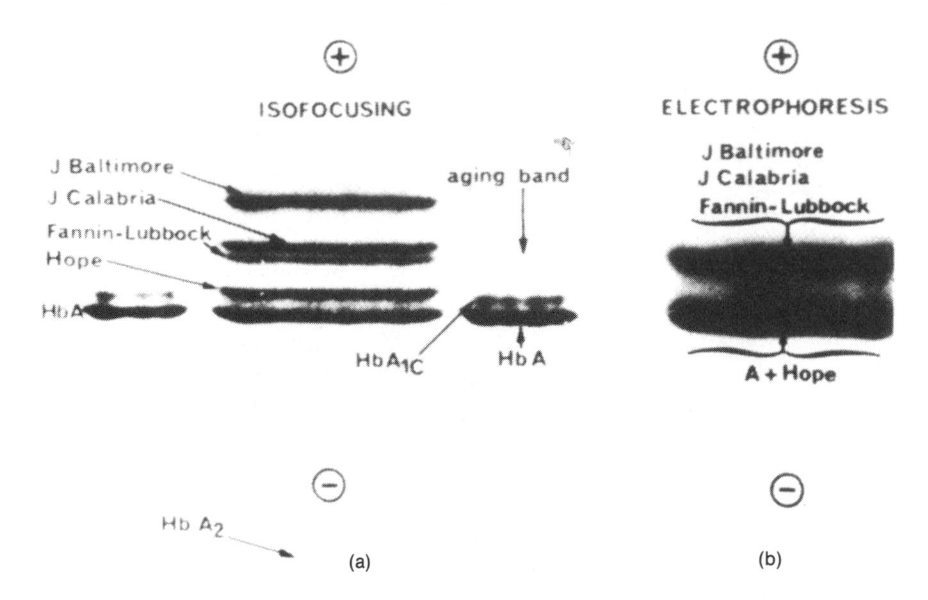

FIGURE 2.24
Migration patterns for a mixture of four variant hemoglobins (*a*) in isoelectric focusing in a pH gradient range of 6–9. Hb–A is normal adult hemoglobin. The variants all have replaced one of the glycines in the HbA sequence by an aspartic acid and have close pI values. (*b*) In conventional electrophoresis on cellulose acetate only two bands are observed.
Reprinted with permission from P. Basset, F. Branconnier, and J. Rosa, *J. Chrom.* 227:267, 1982.

becomes stationary in the electric field and may be visualized or eluted from the column in preparative quantities (Figure 2.24). Proteins that differ by as little as 0.0025 in their pI values can be separated on the appropriate pH gradient.

Separation of proteins by **ion-exchange resins** in a chromatography column is a third important technique for the separation and characterization of proteins by charge. Ion-exchange resins are prepared of insoluble materials (agarose, polyacrylamide, cellulose, and glass) that contain negatively charged ligands (e.g., $—CH_2COO^-$, $—C_3H_6SO_3^-$) or positively charged ligands (e.g., diethylamino) (Figure 2.25) covalently attached to the insoluble resin. Negatively charged resins bind cations strongly and are known as *cation-exchange resins*. Similarly, positively charged resins bind anions strongly and are referred to as *anion-exchange resins*. The degree of retardation of a protein or amino acid by a resin will depend on the magnitude of the charge on the protein at the particular pH of the experiment. Molecules of the same charge as the resin are eluted first in a single band, followed by proteins with an opposite charge to that of the resin, in an order based on the protein's charge density (Figure 2.26). In situations where it is difficult to remove a molecule from the resin because of the strength of the attractive interaction between the bound molecule and resin, systematic changes in pH or in ionic strength may be used to weaken the interaction.

For example, an increasing pH gradient in the eluent buffer through a cation-exchange resin with cationic proteins bound will reduce the difference between the solution pH and the respective molecular pI values. This decrease between pH and pI reduces the magnitude of the net charge on the proteins and thus decreases the strength of interaction between the proteins and the resin.

An increased gradient of ionic strength in the eluting buffer will also decrease the strength of charge interactions.

$R—CH_2—COO^-$

Negatively charged ligand: carboxymethyl

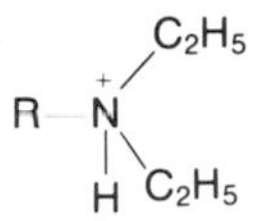

$R—\overset{+}{N}H(C_2H_5)_2$

Positively charged ligand: diethylamino

FIGURE 2.25
Two examples of charged ligands used in ion-exchange chromatography.

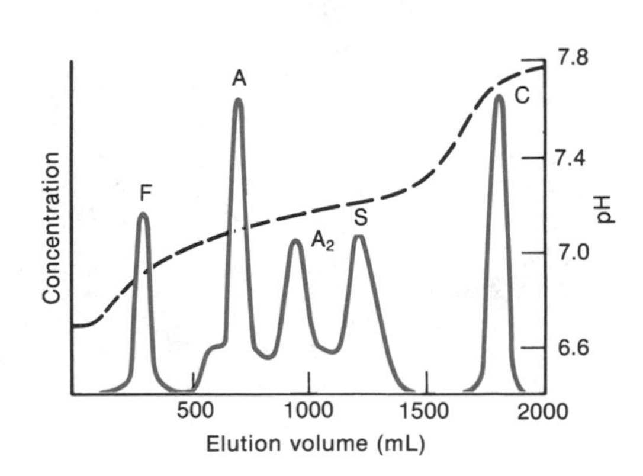

FIGURE 2.26
Example of ion-exchange chromatography.
Elution diagram of an artificial mixture of hemoglobins F, A, A_2, S, and C on Carboxymethyl-Sephadex C-50.
From A. M. Dozy and T. H. J. Huisman, *J. Chromatog.*, 40:62, 1969.

Amino Acids Undergo a Variety of Chemical Reactions

Amino acids in proteins undergo a variety of chemical reactions with chemical reagents that may be used to investigate the function of specific side chain groups. Some of the common chemical reactions are presented in Table 2.7. Chemical reagents have also been synthesized to bind to specific sites in a protein structure, like the substrate binding site in an

TABLE 2.7 Some Chemical Reactions of the Amino Acids

Reactive group	Reagent or reaction	Product
Amine (—NH_2) groups	Ninhydrin	Blue colored product that absorbs at 540 nm[a]
	Fluorescamine	Product that fluoresces
Carboxylic acid groups	Alcohols	Ester products
	Amines	Amide products
	Carbodiimide	Activates for reaction with nucleophiles
—NH_2 of Lys	2,3,6-Trinitrobenzene sulfonate	Product that absorbs at 367 nm
	Anhydrides	Acetylates amines
	Aldehydes	Forms Shiff base adducts
Guanidino group of Arg	Sakaguchi reaction	Pink-red product that can be used to assay Arg
Phenol of Tyr	I_2	Iodination of positions ortho to hydroxyl group on aromatic ring
	Acetic anhydride	Acetylation of —OH
S atom of Met side chain	CH_3I	Methyl sulfonium product
	$[O^-]$ or H_2O_2	Methionine sulfoxide or methionine sulfone
—SH of Cys	Iodoacetate	Carboxymethyl thiol ether product
	N-Ethylmaleimide	Addition product with S
	Organic mercurials	Mercurial adducts
	Performic acid	Cysteic acid (—SO_3H) product
	Dithionitrobenzoic acid	Yellow product that can be used to quantitate —SH groups
Imidazole of His and phenol of Tyr	Pauly's reagent	Yellow to reddish product

[a] Proline imino group reacts with ninhydrin to form product that absorbs light at 440 nm (yellow color).

enzyme. Modifying reagents that are **"active site directed"** may contain structural features of the enzyme's natural substrate. Its reaction with a side chain group of an amino acid in the enzyme active site serves to identify an amino acid as being in the substrate binding site and helps identify its role in the catalytic mechanism of the enzyme.

Techniques Are Available for Determining the Amino Acid Composition of Proteins and Physiological Solutions

Determination of the amino acid composition of a protein is an essential component in the study of its structure and physiological properties. In addition, analysis of the amino acid composition of physiological fluids (i.e., blood and urine) can be utilized in the diagnosis of disease (see Clin. Corr. 2.1). A protein is hydrolyzed to its constituent amino acids by heating the protein at 110°C in 6 N HCl for 18–36 h, in a sealed tube under vacuum; the vacuum prevents degradation of oxidative-sensitive amino acid side chains by oxygen in the air. Because tryptophan is destroyed in this method, alternative procedures are used for its analysis. Asparagine and glutamine side chain amides are hydrolyzed to form free carboxyl side chains and free ammonia; thus asparagine and glutamine will be counted within the glutamic acid and aspartic acid content in the analysis. A common procedure for amino acid identification uses cation-exchange chromatography to separate the amino acids, which are then reacted with

CLIN. CORR. 2.1 USE OF AMINO ACID ANALYSIS IN DIAGNOSIS OF DISEASE

There are a number of clinical disorders in which a high concentration of amino acids are found in plasma or urine. An abnormally high concentration of an amino acid in urine is called an *aminoaciduria*. *Phenylketonuria* is a metabolic defect in which patients are lacking sufficient amounts of the enzyme phenylalanine hydroxylase, which catalyzes the transformation of phenylalanine to tyrosine. As a result, large concentrations of phenylalanine, phenylpyruvate, and phenyllactate accumulate in the plasma and urine. Phenylketonuria occurs clinically in the first few weeks after birth, and if the infant is not placed on a special diet, severe mental retardation will occur (see Clin. Corr. 12.5). *Cystinuria* is a genetically transmitted defect in the membrane transport system for cystine and the basic amino acids (lysine, arginine, and the derived amino acid ornithine) in epithelial cells. Large amounts of these amino acids are excreted in urine. Other symptoms of this disease may arise from the formation of renal stones composed of cystine precipitated within the kidney (see Clin. Corr. 5.3). *Hartnup disease* is a genetically transmitted defect in epithelial cell transport of neutral-type amino acids (monoamino monocarboxylic acids), and high concentrations of these amino acids are found in the urine. The physical symptoms of the disease are primarily caused by a deficiency of tryptophan. These symptoms may include a pellagra-like rash (nicotinamide is partly derived from tryptophan precursors) and cerebellar ataxia (irregular and jerky muscular movements) due to the toxic effects of indole derived from the bacterial degradation of unabsorbed tryptophan present in large amounts in the gut (see Clin. Corr. 26.2). *Fanconi's syndrome* is a generalized aminoaciduria associated with hypophosphatemia and a high excretion of glucose. Abnormal reabsorption of amino acids, phosphate, and glucose by the tubular cells is the underlying defect.

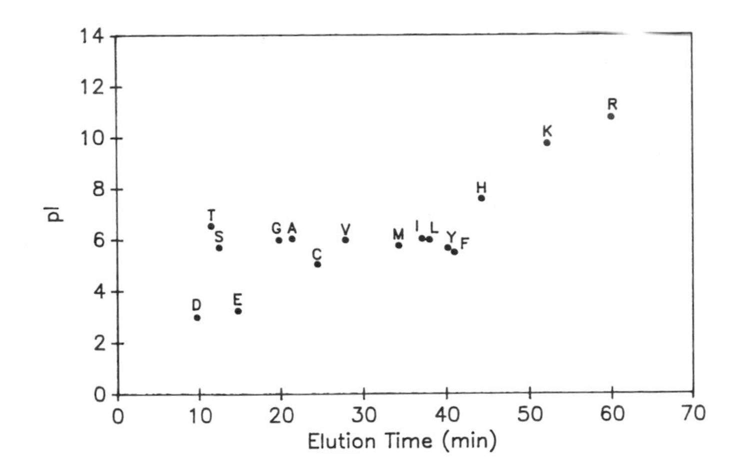

FIGURE 2.27
Plot of amino acid pI versus elution time from a cation exchange resin.
Amino acid analysis utilized a 0.32 × 15-cm bed of Durrum DC-5A resin. Eluent linear flow rate was 2.4 cm min^{-1}; column temperature was 45°C from 0 to 22 min and 65°C after 22 min. Four buffers (0.2 M buffer component) at pH 3.25 (25 min), pH 4.25 (8 min), pH 5.25 (4 min), and pH 10 (28 min) was applied sequentially for the time duration indicated.
Source of experiment: J. R. Benson, *Methods Enzymol.* 47:19, 1977.

ninhydrin, fluorescamine, dansyl chloride (Table 2.7 and Section 2.4), or similar chromophoric or fluorophoric reagents to quantitate the separated amino acids.

Cation-exchange chromatography is carried out at a pH below the pI of the common amino acids. Accordingly, the amino acids with higher pI values are retarded more strongly by the cation-exchange resin. There is also some retention of amino acids due to hydrophobic interactions with the resin, but the charge interaction is the primary cause for the separation of the amino acids by the resin. Figure 2.27 plots the pI versus elution time for the common amino acids obtained from a particular laboratory in chromatography over a sulfonic acid cation-exchange resin in a gradient of buffer pH values.

Amino acids can also be separated on a chromatographic column that solely uses the hydrophobic differences between the amino acid side chain groups (Figure 2.28). The chromatography column uses closely packed insoluble resin beads coated with hydrophobic alkyl groups through which the amino acids are eluted in a mixed organic–aqueous

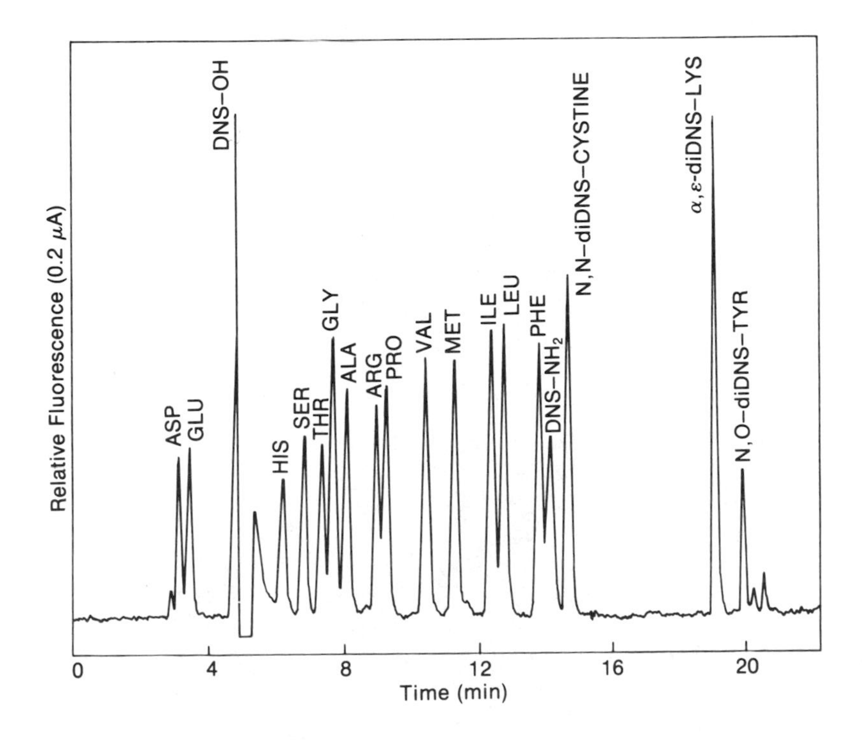

FIGURE 2.28
Separation of amino acids as dansyl (DNS) derivatives by reverse-phase HPLC.
DNS is δ-(dimethylamino)-1-naphthalene sulfonate.
Reprinted with permission from M. W. Hunkapiller, J. E. Strickler, and K. J. Wilson, *Science* 226:304, 1984. Copyright © 1984 by the Amer. Assoc. Adv. Sci.

solvent under high pressure. The chromatography procedure is known as reverse-phase **high performance liquid chromatography** (HPLC) (see Section 2.9). The amino acids are derivatized with chromogenic or fluorogenic reagents either before or after their placement on the column. With some types of derivatization, amino acids can be identified at concentrations as low as 0.5×10^{-12} mol ($\frac{1}{2}$ pmol).

2.4 PRIMARY STRUCTURE OF PROTEINS

Techniques Used to Determine the Amino Acid Sequence of a Protein

Knowledge of the primary structure (amino acid sequence) of a protein is required for an understanding of a protein's structure, its mechanism of action on a molecular level and its interrelationship with other proteins in evolution. With the ability to clone the genes for many proteins the sequencing of the gene is a much faster method to obtain the amino acid sequence of a protein. The sequencing of the actual protein, however, is required for the study of protein modifications that occur after the synthesis of the protein (see Chapter 17).

The primary structure of a protein is the most commonly used data base to predict the similarity in structure and function between two proteins. Such comparisons typically require aligning the two sequences to maximize the number of identical residues while minimizing the number of sequence insertions or deletions that must be used to achieve this alignment. Two sequences are frequently termed **homologous** when their sequences are highly correlatable. In its correct usage **homology** only refers to proteins that have evolved from the same gene (e.g., hemoglobins and immunoglobulins). **Analogy** is used to describe the sequences from two proteins that are similar but for which no evolutionary relationship has been proven.

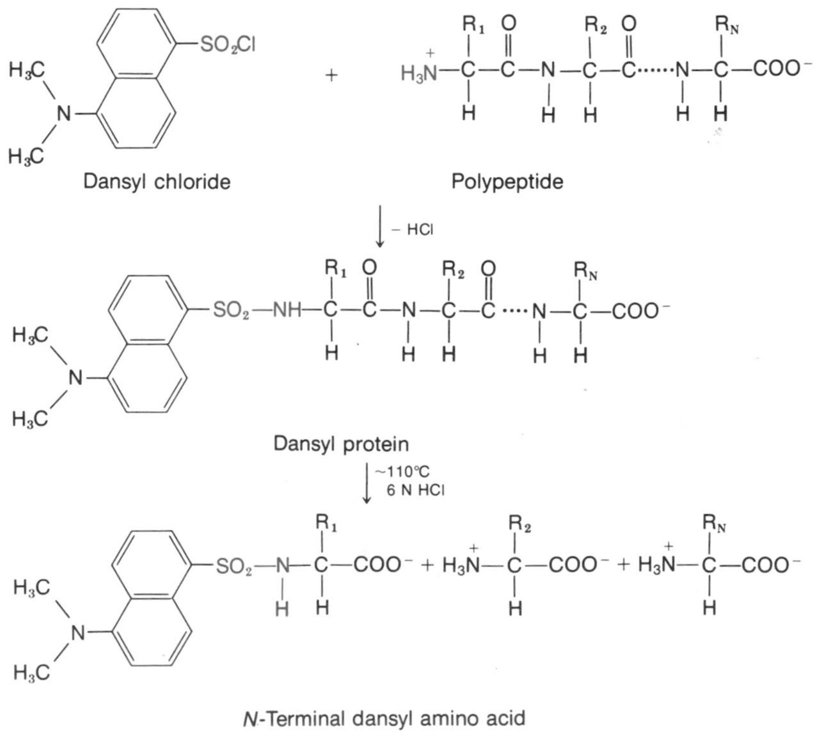

Dansyl chloride

Polypeptide

− HCl

Dansyl protein

~110°C
6 N HCl

N-Terminal dansyl amino acid

FIGURE 2.29
Reaction of a polypeptide with dansyl chloride.

The determination of the primary structure of a protein requires a purified protein. A variety of purification techniques are available including those using charge (Section 2.3), size, or molecular weight (Section 2.9). Many proteins contain several polypeptide chains and it is necessary to determine the number of chains in the protein. The protein can be denatured (see Section 2.7) and then treated with a reagent, such as dansyl chloride (Figure 2.29), that forms a covalent bond with the NH_2-terminal α-amino group of each polypeptide chain. The tagged protein can then be hydrolyzed to its constituent amino acids followed by analysis of the amino acid hydrolysate by chromatographic procedures (Figure 2.30). After determining the number of polypeptide chains, the individual chains are purified by the same techniques used in purification of the whole protein. Since the disulfide bonds of cysteine may covalently join the chains, these bonds have to be broken (Figure 2.31).

Polypeptide chains are most commonly sequenced by the Edman reaction (Figure 2.32). In the **Edman reaction,** the polypeptide chain to be sequenced is reacted with phenylisothiocyanate, which, like the dansyl chloride, forms a covalent bond to the NH_2-terminal amino acid of the chain. However, in this derivative, acidic conditions catalyze an intramolecular cyclization that results in the cleavage of the NH_2-terminal amino acid from the polypeptide chain as a phenylthiohydantoin derivative. This NH_2-terminal amino acid derivative may be separated chromatographically and identified against standards. The polypeptide chain minus the NH_2-terminal amino acid is then isolated, and the Edman reaction is repeated to identify the next NH_2-terminal amino acid of the chain. This series of reactions can theoretically be repeated until the sequence of the entire polypeptide chain is determined.

The repetition of Edman reactions under favorable conditions can be carried out for 30 or 40 amino acids into the polypeptide chain from the NH_2-terminal end. At this point in the analysis, impurities generated from incomplete reactions in the reaction series make further Edman cycles unfeasible. Since most polypeptide chains in proteins contain more than 30 or 40 amino acids, they have to be hydrolyzed into smaller fragments and sequenced in sections.

Both enzymatic and chemical methods are used to break polypeptide chains into smaller polypeptide fragments. For example, the enzymes

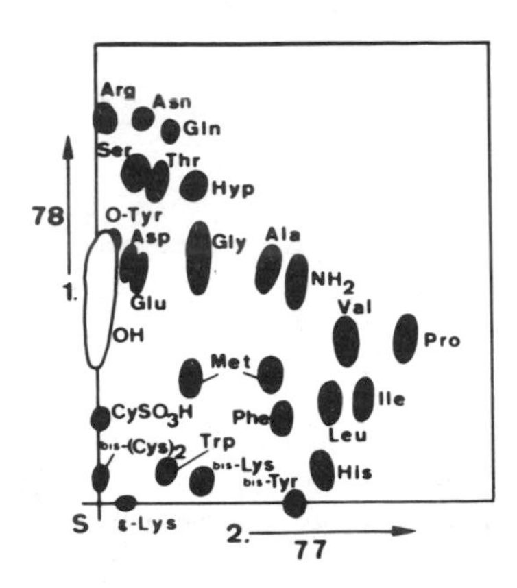

FIGURE 2.30
Identification of dansyl amino acids by thin-layer chromatography (TLC).
Solvent 1: water–90% formic acid (100 : 1.5 v/v); solvent 2: benzene–glacial acetic acid (9 : 1 v/v) on polyamide paper (Cheng-Chin Trading Co.).
According to Dr. R. Woods and K. T. Wang, *Biochem. Biophys. Acta,* 133:369, 1967, from A. Niederwieser in *Methods Enzymol.* 25:60, 1972.

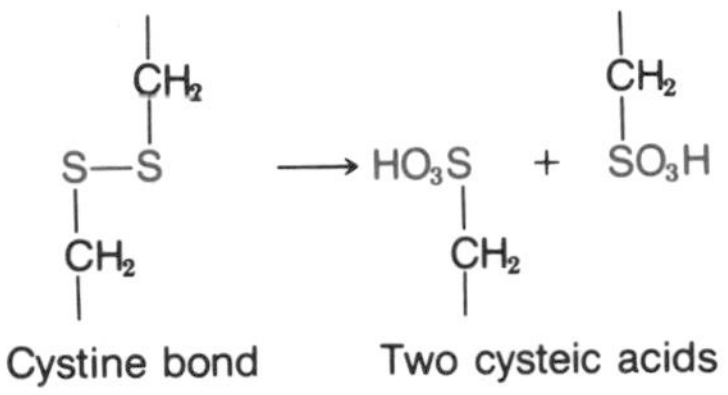

Cystine bond → Two cysteic acids

FIGURE 2.31
Breaking of disulfide bonds by oxidation to cysteic acids.

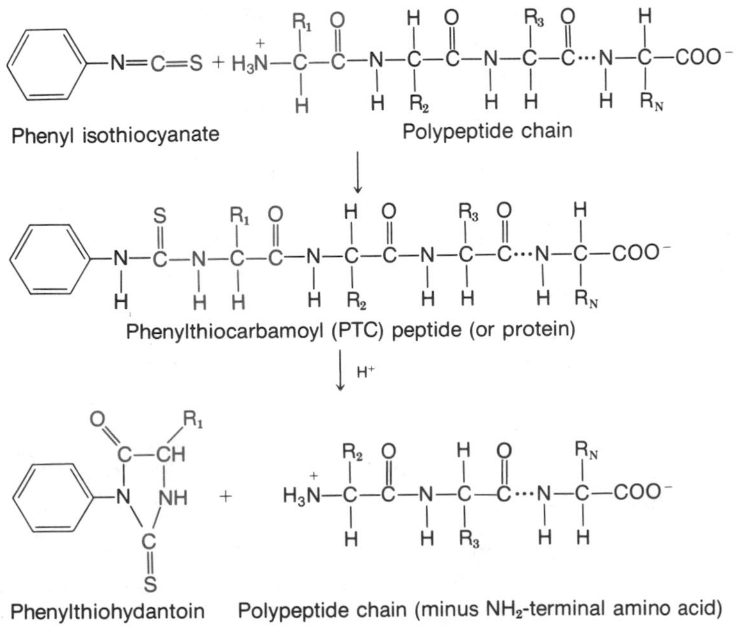

Phenyl isothiocyanate + Polypeptide chain
↓
Phenylthiocarbamoyl (PTC) peptide (or protein)
↓ H^+
Phenylthiohydantoin + Polypeptide chain (minus NH_2-terminal amino acid)

FIGURE 2.32
Edman reaction.

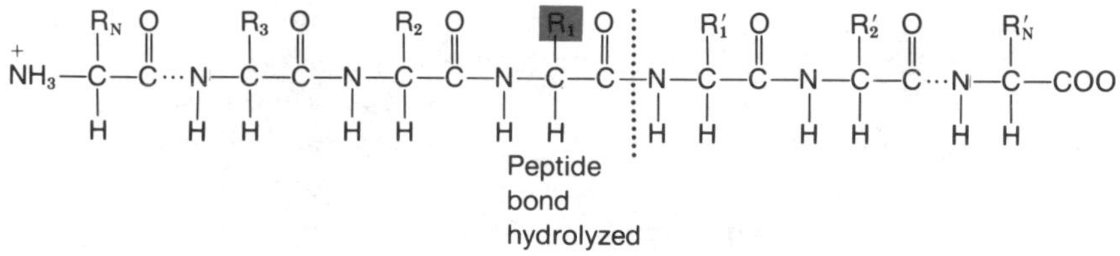

Peptide bond hydrolyzed

R_1	Reagent
Phe, Tyr, or Trp	Chymotrypsin
Arg, Lys	Trypsin
Met	Cyanogen bromide
Trp	*o*-Iodosobenzoic acid
Glu	Staphylococcus aureus endoprotease V8

FIGURE 2.33
Specificity of some polypeptide cleaving reagents.

trypsin and chymotrypsin are *proteolytic enzymes* that are commonly used for partial hydrolysis of polypeptide chains in sequencing. The enzyme trypsin preferentially catalyzes the hydrolysis of the peptide bond on the α-COOH side of the basic amino acid residues of lysine and arginine within polypeptide chains. Chymotrypsin hydrolyzes peptide bonds on the α-COOH side of amino acid residues with large apolar side chains. Other proteolytic enzymes are available that cleave polypeptide chains on the COOH terminal side of glutamic acids and aspartic acids (Figure 2.33). The chemical reagent **cyanogen bromide** specifically cleaves peptide bonds on the carboxyl side of methionine residues within polypeptide chains (Figure 2.33). Thus, to establish the amino acid sequence of a large polypeptide chain, the chain is subjected to partial hydrolysis by one of the specific cleaving reagents, the polypeptide segments are separated, and the amino acid sequence of each of the small segments is determined by the Edman reaction.

The molecular weights of purified peptides generated by partial hydrolysis can be determined by **fast atom bombardment mass spectrometry** (FABMS). The technique is also very useful in showing modified amino acids in the sequence and the structure of any attached carbohydrate.

To order the sequenced peptide segments correctly into the complete sequence of the original polypeptide, a sample of the original polypeptide must be subjected to a *second partial hydrolysis* by a specific hydrolytic reagent different from that used initially. The sequence of this second group of polypeptide segments gives overlapping regions for the first group of polypeptide segments. This leads to the ordering of the initially sequenced peptide segments (Figure 2.34).

Proteolytic enzyme digests of a protein are often used as an analytical tool for protein identification. Figure 2.35 shows the trypsin digest of hemoglobin A and hemoglobin S chromatographed in one dimension and electrophoresed in the other dimension. Close examination of the patterns shows that they are identical except for peptide 4, which contains the genetically determined amino acid substitution that causes sickle cell anemia. The chromatography pattern of such an enzymic digest is known as the protein's "fingerprint." In place of two-dimensional chromatography for separation of peptide fragments, reverse-phase HPLC chromatography (Section 2.9) has been employed.

Knowledge of Primary Structure of Insulin Aids in Understanding Its Synthesis and Action

The study of the primary structure of insulin illustrates the value of knowledge of the primary structure of a protein. Insulin is initially produced in pancreatic islet cells as a single chain precursor known as **proinsulin.** The proinsulin is a substrate for a proteolytic enzyme that catalyzes

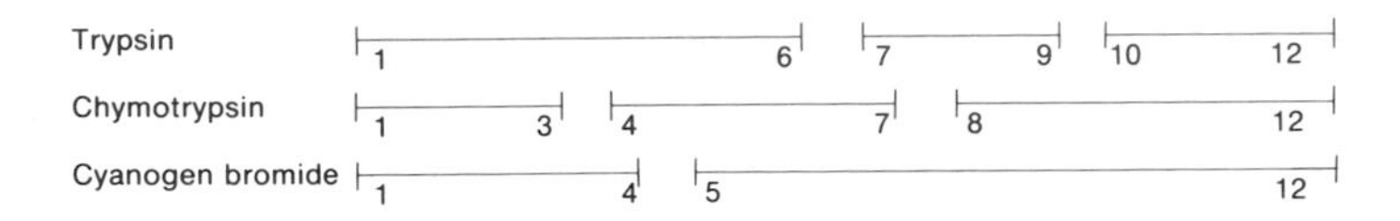

FIGURE 2.34
The ordering of peptide fragments from overlapping sequences produced by specific proteolysis of a peptide.

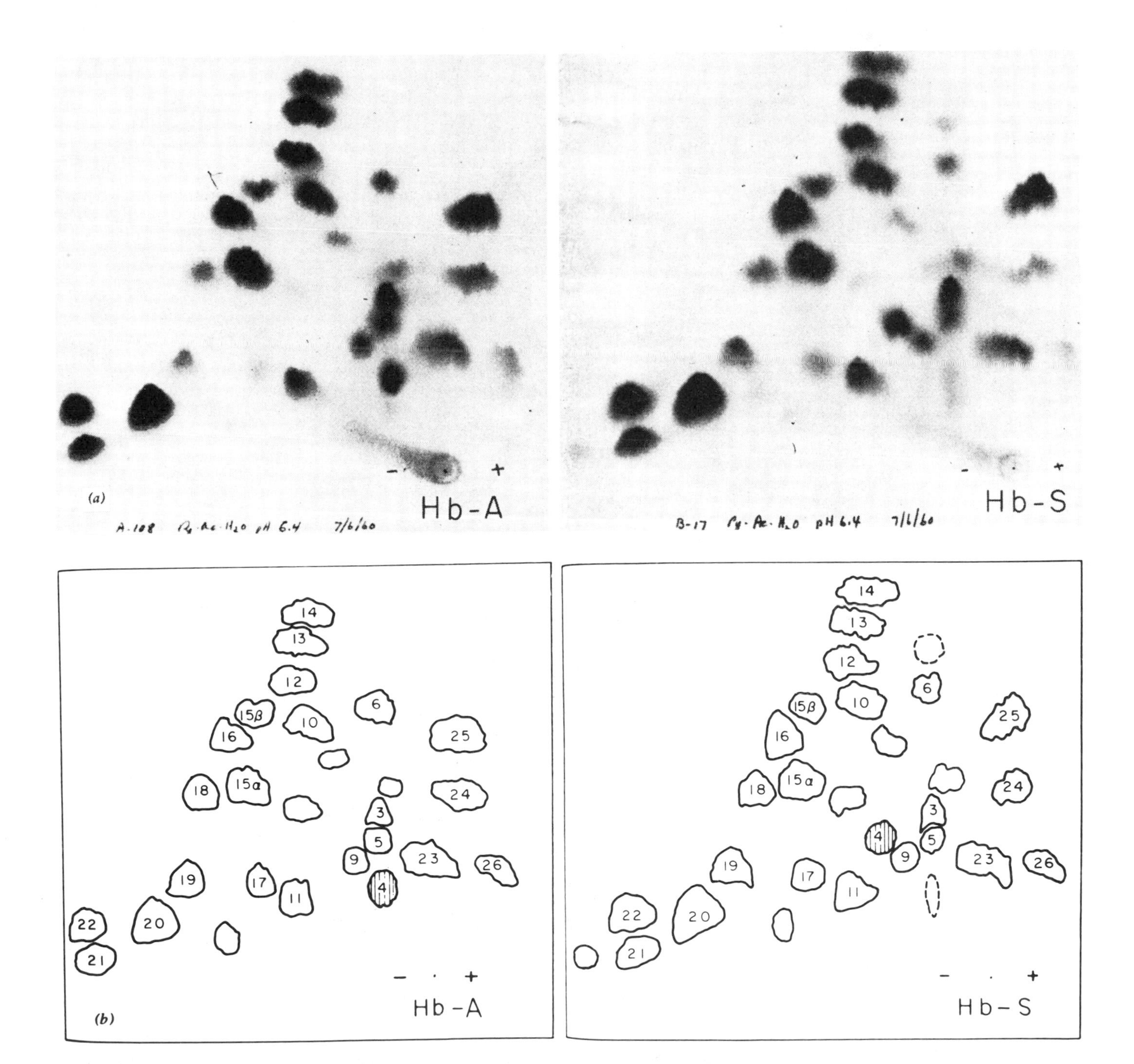

FIGURE 2.35
Trypsin digest of HbA (normal) and HbS (sickle-cell hemoglobin).
(*a*) "Fingerprint" of digest visualized after two-dimensional chromatography. (*b*) Tracing of spots for peptides observed above. Dashed line tracing indicates spots that become visible only on heating. Spot 4 contains mutated amino acid that give HbS its sickle-cell properties; otherwise all other peptides are identical.
Reprinted with permission from C. Baglioni, *Biochim. Biophys. Acta* 48:392, 1961.

the hydrolysis of two intrachain peptide bonds, resulting in the cleavage of a 35 amino acid segment (the **C-peptide**) from the polypeptide chain (Figure 2.36). The remainder is active insulin, which consists of two polypeptide chains (A and B) covalently joined by disulfide bonds of cystine residues formed from cysteines within the A and B chains (Figure 2.36). In insulin the cystine bonds are *inter*chain bonds, whereas these same bonds were *intra*chain cystine bonds in the precursor proinsulin molecule. The C-peptide is further processed in the pancreatic islet cells by an enzyme that hydrolyzes a dipeptide from the COOH terminal and a second dipeptide from the NH_2 terminal. The modified C-peptide is secreted into the blood with the active insulin.

The essential or nonessential function of particular amino acids in the structure of active insulin has been studied by comparing the primary structures of insulin from different animal species. The proinsulin molecules contain from 78 amino acids (dog) to 86 amino acids (human, horse, and rat). The sequences when aligned show an identity in most amino acid positions except for residues 8, 9, and 10 of the A chain and residue 30 of

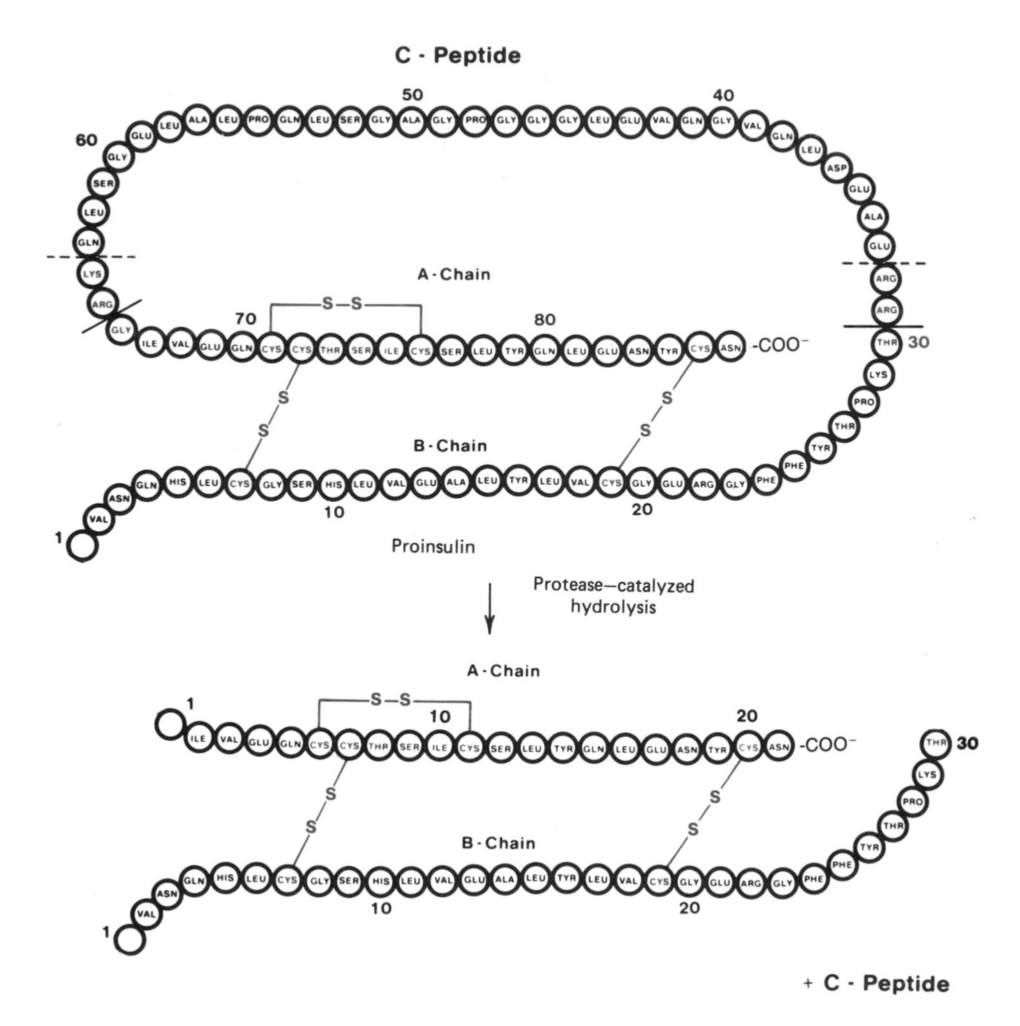

FIGURE 2.36
Amino acid sequence of human proinsulin and insulin.
The B-peptide extends from Phe at position 1 to Thr at position 30, the C-peptide from Arg at position 31 to Arg at position 65, and the A-peptide from Gly at position 66 to Asn at position 86. Cystine bonds from positions 7 to 72, 19 to 85, and 71 to 76 are found in proinsulin.
Redrawn from G. I. Bell, W. F. Swain, R. Pictet, B. Cordell, H. M. Goodman, and W. J. Rutter, *Nature (London)* 282:525, 1979.

the B chain, which appear to be widely variable (Table 2.8). Variations at these positions do not appear to affect the biological properties of the insulin molecule (see Clin. Corr. 2.2) significantly. The other amino acid residues of the primary structure do not vary as much among the different insulins and it has been assumed that these positions have a more critical role in the function of insulin.

An amino acid substituted by another amino acid of similar polarity (i.e., Val for Ile in position 10 of insulin) is termed a **conservative substitution.** This description of substitution is commonly used in the comparison of amino acid sequences of the same protein among different animal species. For example, the mitochondrial protein cytochrome *c* has been sequenced in over 67 plant and animal species, from yeast and fungus to humans. The number of **invariant residues** in the sequence is only 28 out of 104 amino acids. Thus, 76 positions are variable, mostly in a conservative manner. Each of these sequences produce a protein with a similar structure and function. A **nonconservative** change occurs on the substitution of an amino acid by another of dramatically different polarity. This usually leads to severe changes in the properties of the resultant protein (see Clin. Corr. 2.3).

TABLE 2.8 Variation in Positions A8, A9, A10, and B30 of Insulin

Species	A8	A9	A10	B30
Human	Thr	Ser	Ile	Thr
Cow	Ala	Ser	Val	Ala
Pig	Thr	Ser	Ile	Ala
Sheep	Ala	Gly	Val	Ala
Horse	Thr	Gly	Ile	Ala
Dog	Thr	Ser	Ile	Ala
Chicken[a]	His	Asn	Thr	Ala
Duck[a]	Glu	Asn	Pro	Thr

[a] Positions 1 and 2 of B chain are both Ala in chicken and duck; whereas in the other species in the table, position 1 is Phe and position 2 is Val in B chain.

2.5 HIGHER LEVELS OF PROTEIN ORGANIZATION

The *primary structure* of a protein, discussed in preceding sections, refers to the covalent structure of a protein. It includes the amino acid sequence of the protein and the location of disulfide (cystine) bonds. Higher levels of protein organization refer to noncovalently generated conformational properties of the primary structure. These higher levels of protein conformation and organization are customarily defined as the secondary, tertiary, and quaternary structures of a protein.

Secondary structure refers to the local conformations of the polypeptide chain in the protein. The polypeptide chain in this context refers to the covalently interconnected atoms of the peptide bonds and α-carbon linkages that string the amino acid residues of the protein together. Side chain (R) groups are not included at the level of secondary structure. For example, secondary structures of polypeptide chains may form noncovalently generated conformations that are helical (i.e., α helix).

Tertiary structure refers to the total three-dimensional structure of the polypeptide units of the protein. It includes the conformational relationships in space of the side chain groups to the polypeptide chain and the geometric relationship of distant regions of the polypeptide chain to each other.

Quaternary structure refers to the structure and interactions of the noncovalent association of discrete polypeptide subunits into a multisubunit protein. Not all proteins have a quaternary structure.

Proteins generally assume a unique secondary, tertiary, and quaternary conformation for their particular amino acid sequence, known as the **native conformation.** The folding of the primary structure into the native conformation appears to occur in most cases spontaneously, under the influence of noncovalent forces. This unique conformation is that of the lowest Gibbs free energy kinetically accessible to the polypeptide chain(s) for the particular conditions of ionic strength, pH, and temperature of the solvent in which the folding process occurs. Cystine bonds are made after the folding of the polypeptide chain occurs, and act to stabilize the native conformation covalently.

The higher levels of protein organization are individually discussed in the following sections.

CLIN. CORR. 2.2 DIFFERENCES IN THE PRIMARY STRUCTURE OF INSULINS USED IN THE TREATMENT OF DIABETES MELLITUS

Both pig (porcine) and cow (bovine) insulins are commonly used in the treatment of human diabetics. Because of the differences in amino acid sequence from the human insulin, some diabetic individuals will have an initial allergic response to the injected insulin as their immunological system recognizes the insulin as foreign, or develop an insulin resistance due to a high antiinsulin antibody titer at a later stage in treatment. However, the number of diabetics who have a deleterious immunological response to pig and cow insulins is small; the great majority of human diabetics can utilize the nonhuman insulins without immunological complication. The compatibility of the cow and pig insulins in humans is due to the small number of changes and the conservative nature of the changes between the amino acid sequences of the insulins. These changes in primary structure do not significantly perturb the three-dimensional structure of the insulins from that of the human insulin. Pig insulin is usually more acceptable than cow insulin in insulin-reactive individuals because it is more similar in sequence to human insulin (see Table 2.8). Human insulin is now available for clinical use. It can be made using genetically engineered bacteria or by modifying pig insulin.

Brogdon, R. N. and Heel, R. C. Human insulin: a review of its biological activity, pharmacokinetics, and therapeutic use. *Drugs* 34:350, 1987.

CLIN. CORR. 2.3 A NONCONSERVATIVE MUTATION OCCURS IN SICKLE-CELL ANEMIA

Hemoglobin S (HbS) is a variant form of the normal adult hemoglobin in which a nonconservative substitution occurs in the sixth position of the β-polypeptide chain of the normal hemoglobin (HbA_1) sequence. Whereas in HbA_1 this position is taken by a glutamic acid residue, in HbS the position is occupied by a valine. Consequently, individuals with HbS have replaced a polar side chain group on the molecule's outside surface with a nonpolar hydrophobic side chain group (a nonconservative mutation). Through hydrophobic interactions with this nonpolar valine residue, HbS in its deoxy conformation will polymerize with other molecules of deoxy-HbS, leading to a precipitation of the hemoglobin within the red blood cell. The precipitation of the hemoglobin gives the red blood cell a sickle shape, an instability that results in a high rate of hemolysis and a lack of elasticity during circulation through the small capillaries, which become clogged by the abnormal red blood cells.

Only individuals homozygous for HbS exhibit the disease. Individuals heterozygous for HbS have approximately 50% HbA_1 and 50% HbS in their red blood cells and do not exhibit symptoms of the sickle-cell anemia disease unless under extreme hypoxia.

It is of interest that individuals heterozygous for HbS have a resistance to the malaria parasite, which spends a part of its life cycle in the red blood cell. This is a factor selecting for the HbS gene in malarial regions of the world and the reason for the high frequency of this homozygous lethal gene in the human genetic pool. It is known that approximately 10% of American blacks are heterozygous for HbS, and 0.4% of American blacks are homozygous for HbS and exhibit sickle-cell anemia.

Sickle hemoglobin is detected by gel electrophoresis. Because sickle hemoglobin lacks a glutamate, it is less acidic than HbA. Hemoglobin HbS, therefore, does not migrate as rapidly towards the anode as does HbA. It is also possible to diagnose sickle-cell anemia by recombinant DNA techniques.

Embury, S. H. The clinical pathophysiology of sickle-cell disease. *Annu. Rev. Med.* 37:361, 1986.

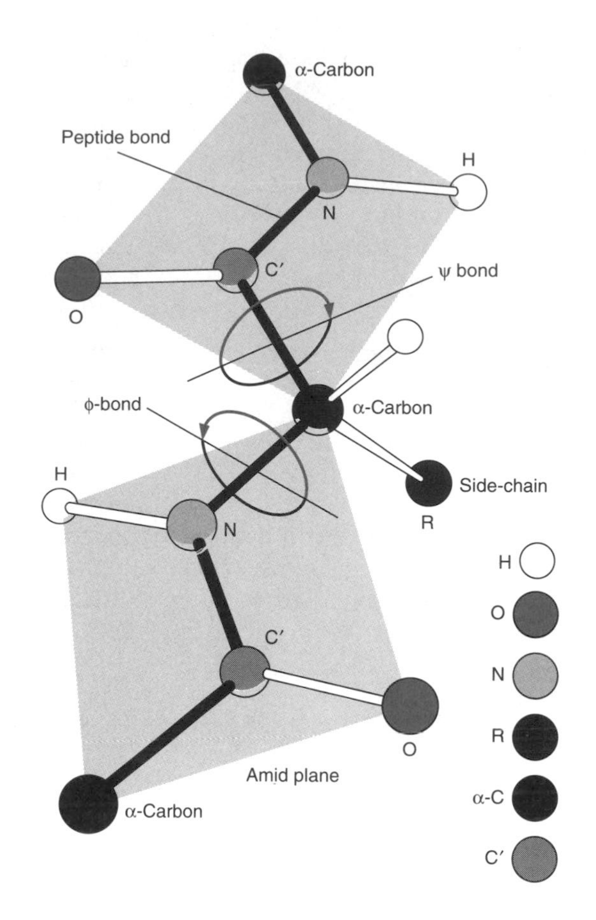

FIGURE 2.37
Polypeptide chain showing φ, ψ, and peptide bond for residue R_i within chain. δ^+ and δ^- indicate partial charge due to delocalization of bonding electrons between carbonyl oxygen and peptide bond nitrogen (see text).
Redrawn with permission from R. E. Dickerson and J. Geis, *The Structure and Action of Proteins*. Menlo Park, Ca.: W. A. Benjamin, 1969, p. 25.

Proteins Have a Secondary Structure

The conformation of a polypeptide chain can be described by the rotational angles assumed about the covalent bonds that interconnect the amino acids. These are the bonds in one amino acid between (1) the nitrogen and α carbon, and (2) the α carbon and the carbonyl carbon, and (3) the peptide bond between the carbonyl carbon and the nitrogen of the adjacent amino acid. The first of these is designated the **phi (φ) bond** and the second the **psi (ψ) bond** for an amino acid residue in a polypeptide chain (Figure 2.37). The third bond contributed by each of the amino acids to the polypeptide chain is the peptide bond. As previously discussed, due to the partial double bond character of the C′⋯N bonds there exists a significant barrier to free rotation about this bond. The atoms of the bond lie in a common plane (see Figures 2.9 and 2.37) and the peptide bond tends to be in a trans conformation except for those involving the α-imino nitrogen of proline, which may be either cis or trans.

Regular secondary structure conformations in segments of a polypeptide chain occur when all the φ bond angles in that polypeptide segment are equal to each other, and all the ψ bond angles are equal. The rotational angles for φ and ψ bonds for common regular secondary structures are given in Table 2.9.

TABLE 2.9 Helix Parameters of Regular Secondary Structures

Structure	Approximate bond angles (°) ϕ	ψ	Residues per turn, n	Helix pitch,[a] p, (Å)
Right-handed α helix [3.6$_{13}$-helix)	−57	−47	3.6	5.4
3_{10}-helix	+49	−26	3.0	6.0
Parallel β strand	−119	+113	2.0	6.4
Antiparallel β strand	−139	+135	2.0	6.8
Polyproline type II[b]	−78	+149	3.0	9.4

[a] Distance between repeating turns on a line drawn parallel to helix axis.
[b] Helix-type found for polypeptide chains of collagen.

The ***α* helix** and ***β* structure** conformations for polypeptide chains are generally the most thermodynamically stable of the regular secondary structures. However, particular amino acid sequences of a primary structure in a protein may support regular conformations of the polypeptide chain other than α helical or β structure. Thus, whereas α helical and β structures are found most commonly, the actual conformation is dependent on the particular physical properties generated by the sequence present in the polypeptide chain and the solution conditions in which the protein is dissolved. In addition, in most proteins there are significant regions of unordered structure in which the ϕ and ψ angles are not equal.

Proline can interrupt α helical conformations of the polypeptide chain since the pyrrolidine side chain group of proline sterically interacts with the side chain group from the amino acid preceding it in the polypeptide sequence when the preceding amino acid has a ψ angle of −47° as required for the α-helical structure. This repulsive steric interaction can prevent formation of α-helical structures in regions of a polypeptide chain in which proline is found.

The helical structures of polypeptide chains are geometrically defined by the number of amino acid residues per 360° turn of the helix (n) and the distance between α-carbon atoms of adjacent amino acids measured parallel to the axis of the helix (d). The **helix pitch** (p), defined by the equation below, measures the distance between repeating turns of the helix on a line drawn parallel to the helix axis.

$$p = d \times n$$

An alternative method for representation of the three-dimensional structure of a polypeptide or protein is the **distance matrix.** The representation utilizes the ordering of the primary structure to generate the x and y axis. The number of amino acids, N, is used to generate a square matrix of $N \times N$ elements. Each element, ij [i and j vary from 1 to N], is the distance in space between each of the α-carbon atoms ($C_{\alpha i}$ and $C_{\alpha j}$) of the amino acid residues in the sequence (see Figure 2.38). The diagonal elements (where $i = j$) are all zero, as these elements represent the distance between the same amino acid in the sequence. The matrix is also symmetrical ($ij = ji$). This distance matrix representation has unique properties specific to the three-dimensional structure of a specific molecule, and its contents are invariant to rotating or translating the polypeptide in space. In an inverse process, the matrix of distances can be directly used to generate the three-dimensional structure of the original polypeptide. The distance matrix form of representation is introduced here and its utility will be developed in later sections.

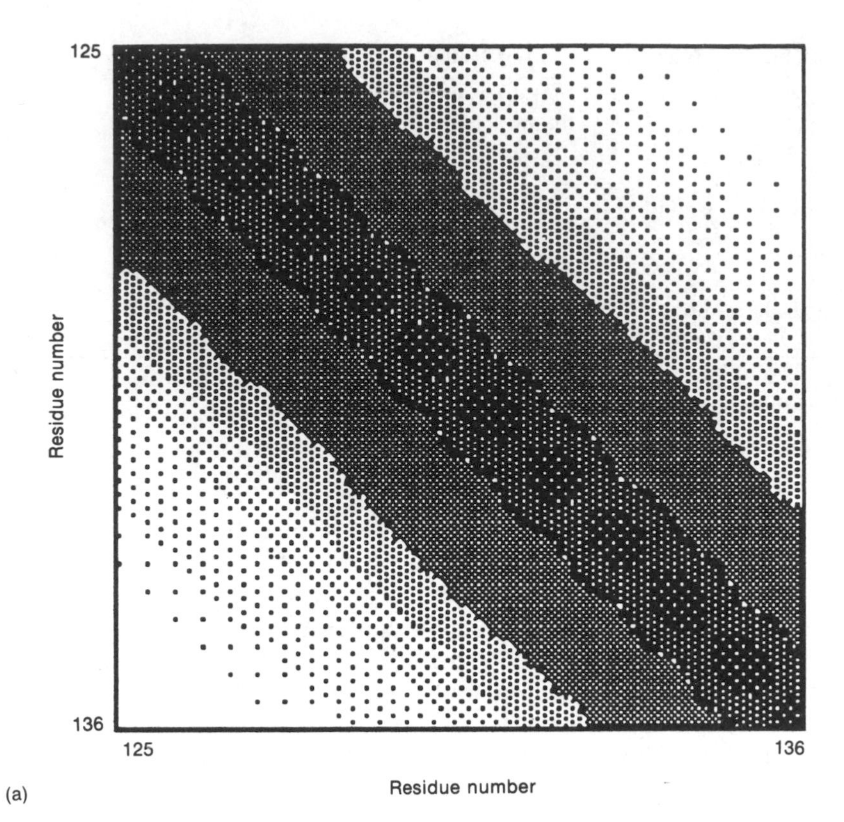

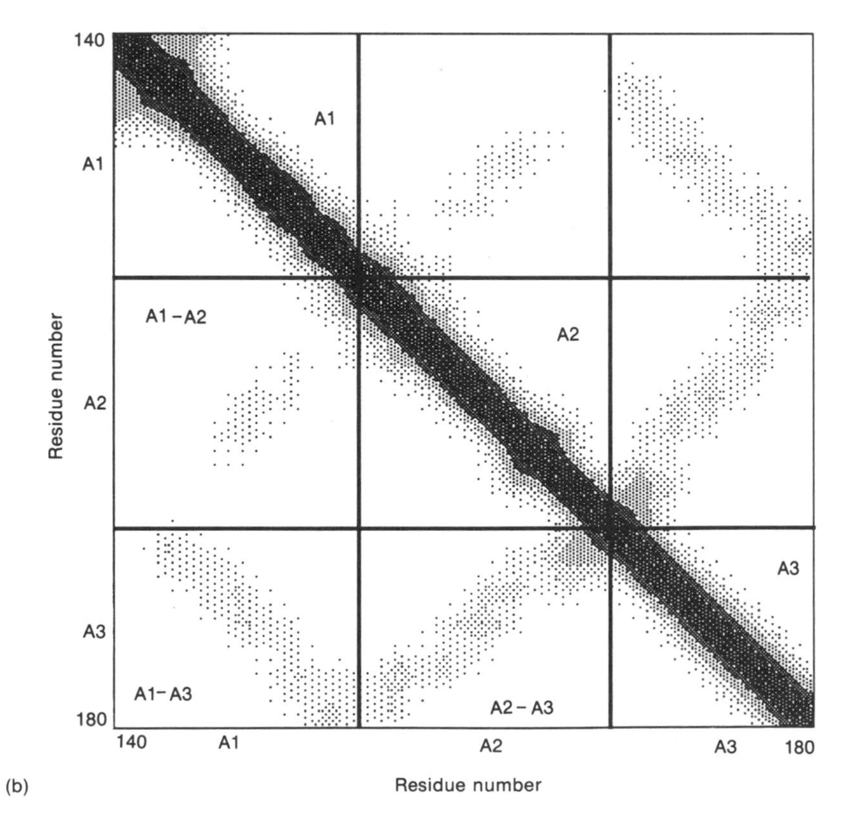

FIGURE 2.38
Interresidue distance matrix shaded to represent four equally spaced distance levels for C_α distances of between 0 and 15 Å.
The darker the shading the closer the distance between the two residues. The examples are (*a*) the H helix of human deoxy-Hb (residues 125–136) showing α-helical structure (reference should also be made to the distance matrix for the full hemoglobin α chain in Figure 3.33) and (*b*) Residues 140–180 of the heavy chain variable domain of antibody Mopc603 revealing three separate β-strand segments, A1, A2, and A3. Beta conformation is shown by broadening along the diagonal for each of the chain segments with the antiparallel β-structural interactions perpendicular to the diagonal and parallel β-structure interaction offset from but parallel to the diagonal. Therefore, polypeptide segments A1 and A2 are antiparallel β structure, A2 and A3 are also antiparallel, and A1 and A3 are parallel β structure. Reference should also be made to the full distance matrix for the Mopc603 V_H and CH_1 domains in Figure 3.10.

α-Helical Structure

A polypeptide chain in an **α-helical conformation** is shown in Figure 2.39. Characteristic of the α-helical conformation are 3.6 amino acid residues per 360° turn ($n = 3.6$). The plane of the peptide bonds in the α helix are approximately parallel to the axis of the helix. In this geometry each peptide forms a hydrogen bond to the peptide bond of the *fourth* amino acid above and the peptide bond of the *fourth* amino acid below in the

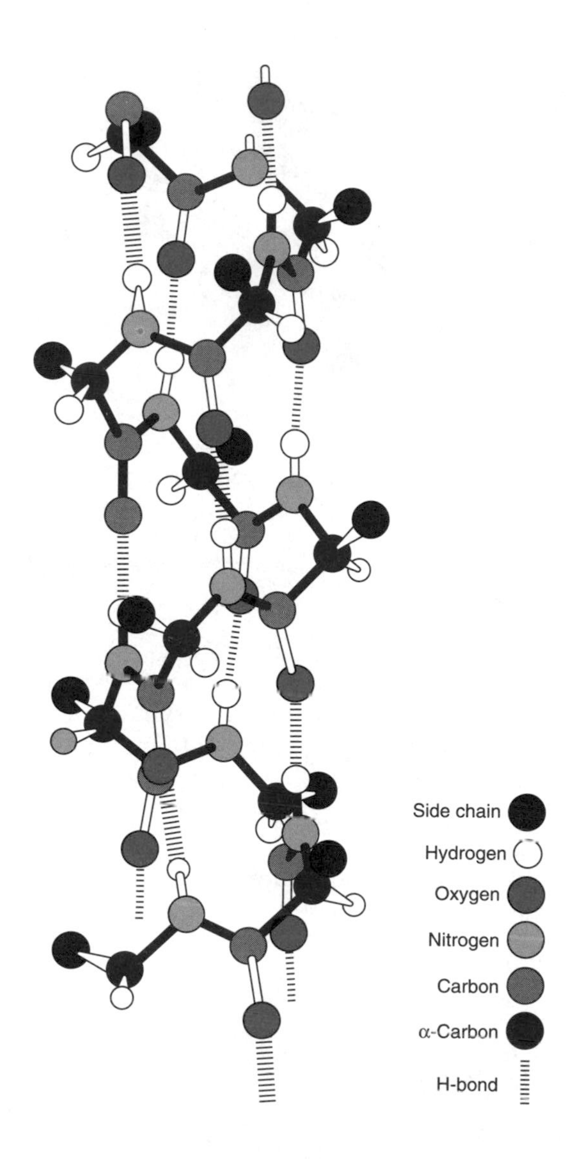

FIGURE 2.39
An α helix.
Redrawn with permission, based on figure from L. Pauling, *The Nature of the Chemical Bond,* 3rd ed., Ithaca, NY: Cornell University Press, 1960.

primary structure. This hydrogen-bonding pattern places 13 interconnected atoms in the intrachain hydrogen-bond loop characteristic of the α helix. Based on these characteristics of 3.6 residues per turn and an intrachain hydrogen-bond loop containing 13 atoms, the α helix is designated a 3.6_{13}-type helix. Other helix parameters such as the pitch (p) of the helix are given in Table 2.9.

In the hydrogen bonds between the peptide groups of an α-helical structure, the distance between the hydrogen-donor atom and the hydrogen-acceptor atom is 2.9 Å. Also, the donor atom, acceptor atom, and hydrogen atom are approximately colinear, that is, they lie on a straight line. This is close to an optimum geometry and distance for maximum hydrogen-bond strength (see Section 2.7).

The side chain groups in the α-helix conformation are perpendicularly projected from the axis of the helix on the outside of the spiral structure generated by the polypeptide chain. Due to the characteristic 3.6 residues per turn of the α helix, the first and every third and fourth R group in the amino acid sequence of the helix comes close to the other. If these R groups have the same charge sign or are branched at their β carbon (valine and isoleucine), their interactions will destabilize the helix structure.

The distance matrix generated for a 10 residue length of α helix is shown in Figure 2.38 and the repetitive nature of the helix is contrasted with that of a β strand.

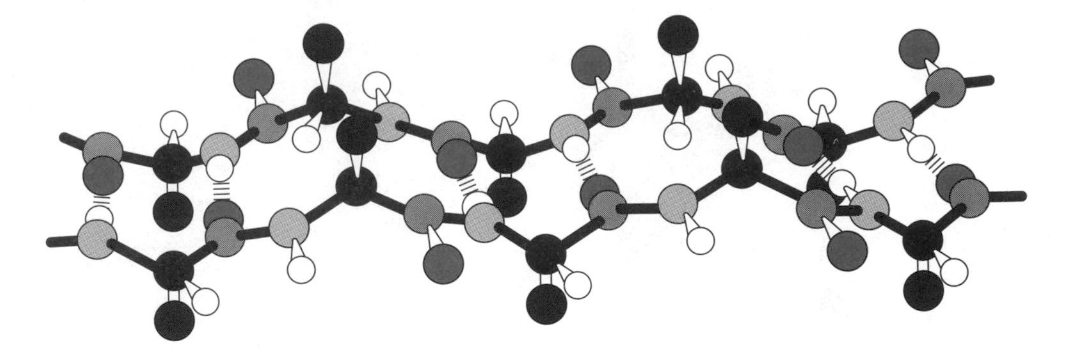

FIGURE 2.40
Two polypeptide chains in a β structure conformation.
Additional polypeptide chains may be added to generate more extended structure.
Redrawn with permission from A. Fersht, *Enzyme Structure and Mechanism,* San Francisco: Freeman, 1977, p. 10.

The α-helix can form its spiral in either a left-handed sense or right-handed sense, giving the helical structure asymmetric properties and a correlated property of optical activity in solution. In the structure shown, the more stable right-handed α helix is depicted.

β-Structure

A polypeptide chain in a β strand conformation is shown in Figure 2.40. In β structure, segments of a polypeptide chain are in an extended helix with

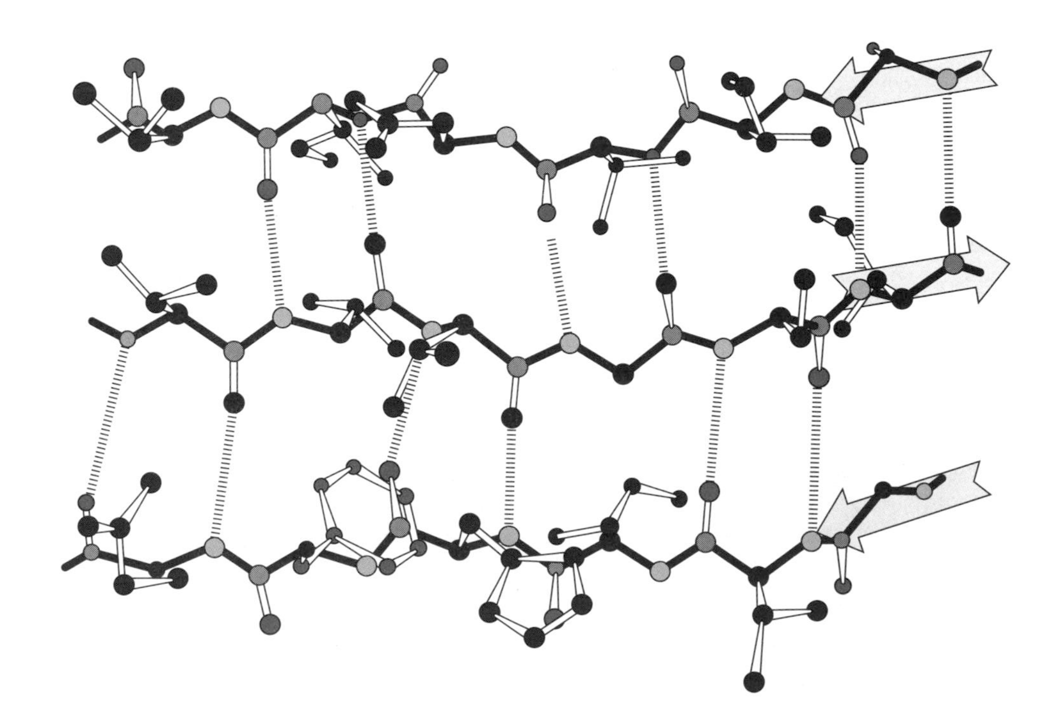

FIGURE 2.41
An example of antiparallel β structure (residues 93–98, 28–33, and 16–21 of Cu, Zn superoxide dismutase).
Dashed line shows hydrogen bonds between carbonyl oxygen atoms and peptides nitrogen atoms; arrows show direction of polypeptide chains from N-terminal to C-terminal. In the pattern characteristic of antiparallel β structure, pairs of closely spaced interchain hydrogen bonds alternate with widely spaced pairs.
Redrawn with permission from J. S. Richardson, *Adv. Protein Chem.* 34:168, 1981.

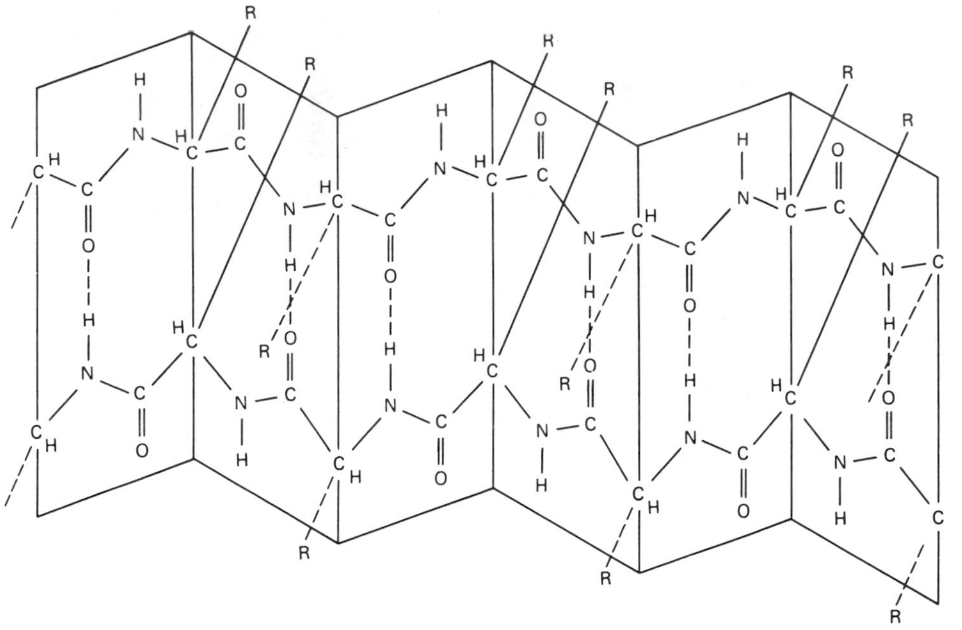

FIGURE 2.42
β-Pleated sheet structure between two polypeptide chains.
Additional polypeptide chains may be added above and below to generate more extended structure.

$n = 2$. The strand of one polypeptide chain or segment is hydrogen bonded to *other* strands in β-structure conformations. The polypeptide segments in a β structure are aligned either in a parallel or antiparallel direction to its neighboring chains (Figure 2.41). Large numbers of polypeptide chains interhydrogen bonded in a β-type structure give a pleated sheet appearance (see Figure 2.42). In this structure the side chain groups are projected above and below the planes generated by the hydrogen-bonded polypeptide chains.

Super-secondary Structures

Certain specific orderings of secondary structure occur in more than one protein and may have a functional role. These orderings are frequently referred to as **structural motifs** and include β-α-β, helix-turn-helix (see p. 116), the leucine zipper (see p. 118), the calcium binding EF hand (see p. 229), and the zinc finger (p. 117). Even longer orderings may take on the structure of a domain (see below) and include the β barrel and the immunoglobulin fold. The significantly longer sequence lengths of secondary structure that fold in patterns observed in more than one protein, such as a motif or folding domains, are referred to as **super-secondary structures.**

Proteins Fold Into a Three Dimensional Structure Called the Tertiary Structure

The tertiary structure of a protein refers to the total three-dimensional structure of the protein, including the geometric relationship between distant segments of the primary structure and the relationship of the side chain group with respect to each other in three-dimensional space. As an example of a protein's tertiary structure, the structure for trypsin is shown in Figure 2.43. In (*a*) of this figure the protein's side chain groups are not shown in order to make the general polypeptide conformation clear. Accordingly, the ribbon diagrammatically shows the relationship in

(a)

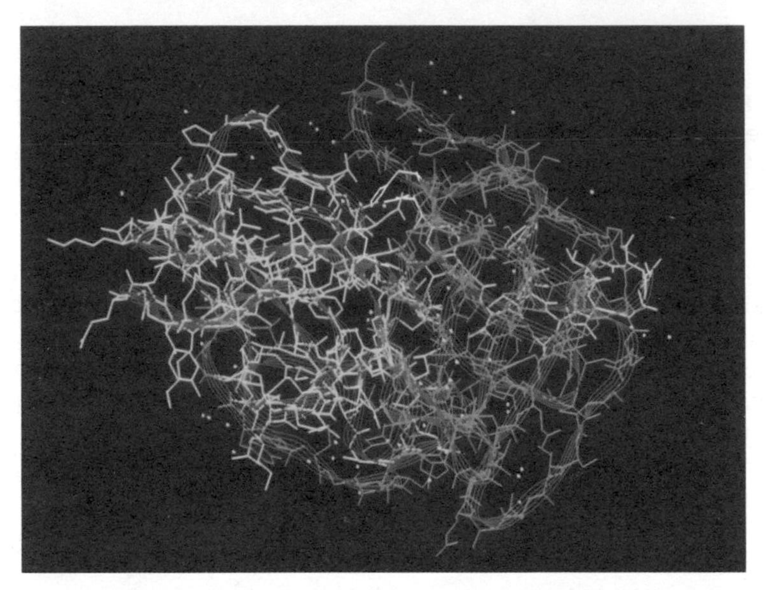

(b)

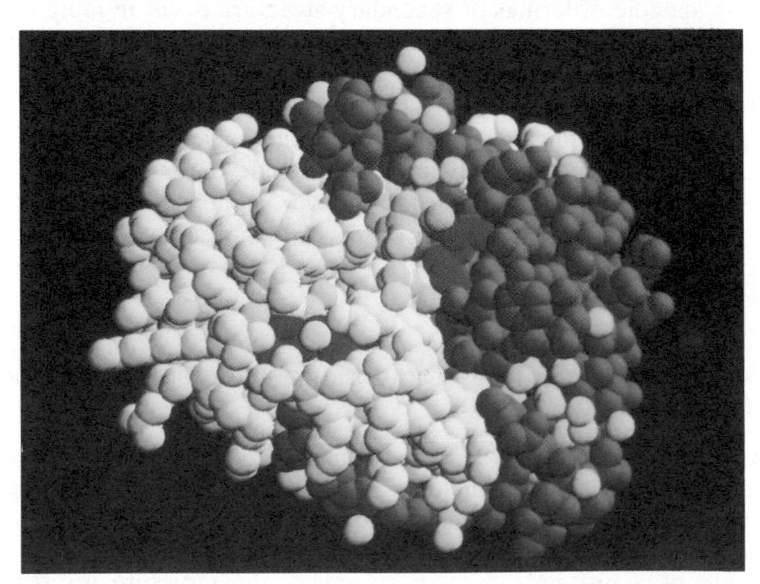

(c)

FIGURE 2.43
Tertiary structure of trypsin.
(*a*) Ribbon structure outlines the conformation of the polypeptide chain. (*b*) Structure shows side chain groups including active site residues (in yellow) with outline of polypeptide chain (ribbon) superimposed. (*c*) Space-filling structure in which each atom is depicted as the size of its van der Waals' radius. Hydrogen atoms are not shown. The different domains are shown in dark blue and white. The active site residues are in yellow and intrachain disulfide bonds of cystine in red. The light blue spheres represent water molecules associated with the protein. This structure shows the density of packing within the interior of the protein.

space of the polypeptide segments with respect to each other. The side chain residues in the catalytic site are shown (*b*), which are the serine side chain hydroxymethyl (residue 177 in the sequence), the histidine side chain imidazole (residue 40), and the aspartate side chain carboxylate (residue 85). Even though these catalytic residues are widely separated in the primary structure, the folded tertiary structure brings them together to form the catalytic site. The polypeptide chains are mostly in β-structure conformation giving an extensively hydrogen-bonded network to the protein.

The tertiary structure of trypsin conforms to the general rules of folded proteins discussed previously (Section 2.2). The hydrophobic side chains are generally in the interior of the structure, away from the water interface. Ionized amino acid side chains are found on the outside of a protein structure, where they are stabilized by water of solvation.

Within the protein structure (not shown) are buried water molecules, noncovalently associated, in specific arrangements. In general, a few noncovalently bound water molecules are buried in the interior and a large number form a solvation layer around the protein.

A long polypeptide strand of a protein will often fold into multiple compact semi-independent folded regions or **domains,** each domain having a characteristic spherical geometry with a hydrophobic core and polar outside very much like the tertiary structure of a whole globular protein. In fact, on separation of a domain unit from its polypeptide chain by hydrolysis of the polypeptide chain at a point just outside the domain's structure, the isolated domain's primary structure will potentially have an ability to fold by itself into its native domain conformation in isolation. The domains of a multidomain protein are often interconnected by a segment of the polypeptide chain lacking regular secondary structure. Alternatively, the dense spherical folded regions can be separated by a cleft or less dense region in the tertiary structure of the protein (Figure 2.44).

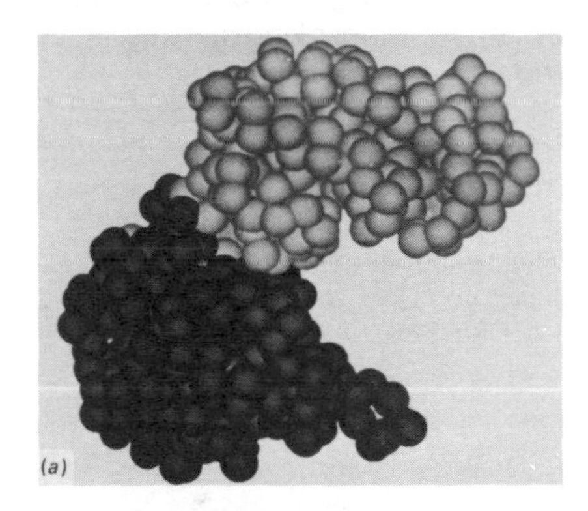

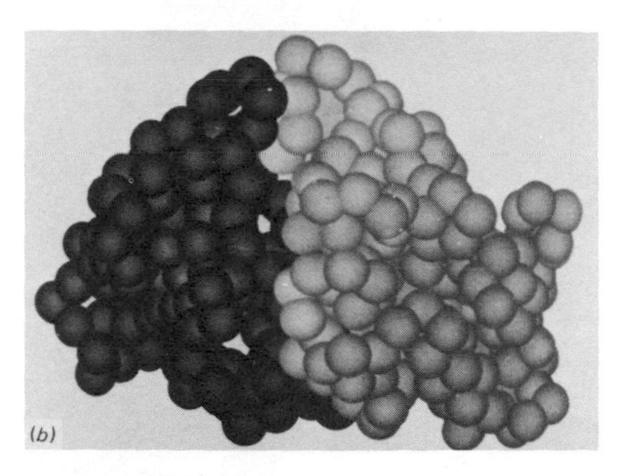

FIGURE 2.44
Globular domains within proteins.
(*a*) Phosphoglycerate kinase has two domains with a relatively narrow neck in between. (*b*) Elastase has two tightly associated domains separated by a narrow cleft. Each sphere in the space-filling drawing represents the α-carbon position for an amino acid within the protein structure.
Reprinted with permission from J. S. Richardson, *Adv. Protein Chem.* 34:168, 1981.

There are two folded domains in the trypsin molecule with a cleft between the domains that includes the substrate binding–catalytic site of the protein. This characteristic that the active site of a **multidomain enzyme** occurs within an interdomain interface appears to be a generalized attribute of these enzymes. Besides achieving an optimal size for folding and energetic stability at ~100 amino acid length, domain organization within a protein is thought to be utilized functionally.

The different domains within a protein can exhibit a relative motion with respect to each other, and these motions can have a functional purpose. The enzyme hexokinase (Figure 2.45), which catalyzes the phosphorylation of a glucose molecule by adenosine triphosphate (ATP), has the glucose substrate-binding site at a cleft between two domains. When the glucose molecule binds in the active site cleft, the surrounding domains close over the substrate to trap it for phosphorylation (Figure 2.45). In enzymes with more than one substrate or allosteric effector sites (Chapter 4), the different binding sites are often located in different domains. In multifunctional proteins, the different domains can each perform a different task.

A protein adopts a unique range of conformations for a particular sequence. Even though each native structure is unique, a comparison of tertiary structures of different proteins shows similar patterns in the arrangement of the secondary structure in the tertiary structures of domains. Thus proteins unrelated by function, exact sequence, or evolution can generate similar patterns of arrangement of their secondary structure.

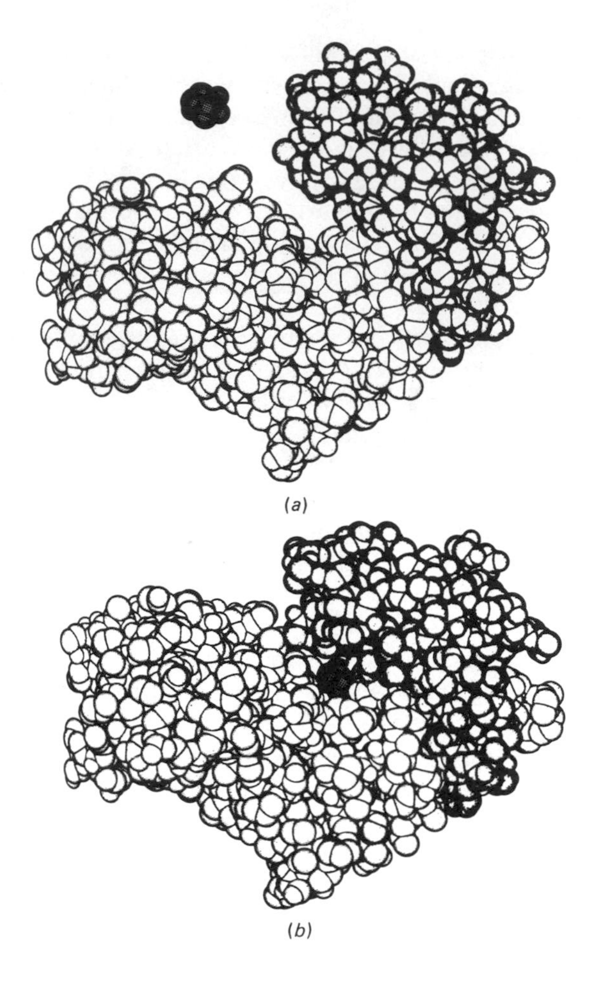

FIGURE 2.45
Drawings of (*a*) unliganded form of hexokinase and free glucose and (*b*) the conformation of hexokinase with glucose bound.
In this space-filling drawing each circle represents the van der Waals radii of an atom in the structure. Glucose is cross-hatched, and each domain is differently shaded.
Reprinted with permission from W. S. Bennett and R. Huber, *CRC Rev. Biochem.* 15:291, 1984. Copyright © CRC Press, Inc., Boca Raton, FL.

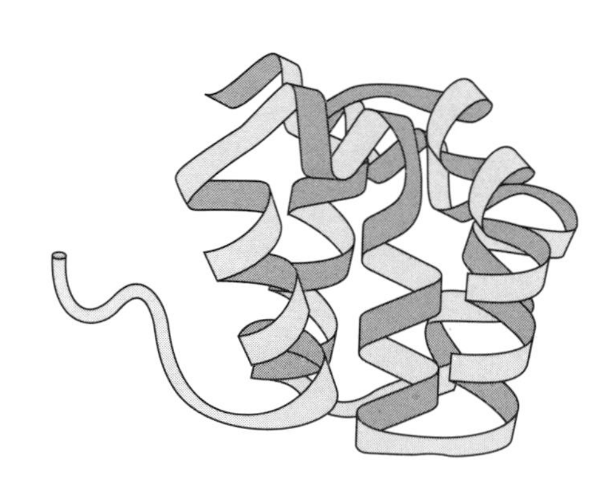

FIGURE 2.46
An example of an all α-folded domain.
In this drawing and the drawings that follow (Figures 2.47–2.49), only the outline of the polypeptide chain is shown. β-Structure strands are shown by arrows with the direction of the arrow showing the N→C terminal direction of the chain; lightning bolts represent disulfide bonds, and circles represent metal ion cofactors (when present).
Redrawn with permission from J. S. Richardson, *Adv. Protein Chem.* 34:168, 1981.

A classification system has emerged for secondary structure patterns that places most proteins or domains within proteins into four categories. An *all-α-structure* for domain 2 of one form of the enzyme lysozyme is shown in Figure 2.46. Other examples of an all-α-structure are found in myoglobin and the subunits of hemoglobin, whose structures are extensively discussed in Chapter 3. In this folding pattern, sections of α helices are joined by smaller segments of polypeptide chains that allow the helices to fold back upon themselves to form a spherical or globular mass.

An example of an **α,β structure** is shown by triose phosphate isomerase (Figure 2.47) in which the β-structure strands (designated by arrows) are wound into a supersecondary structure called a **β barrel.** Each β-structure strand in the interior of the β barrel is interconnected by α-helical regions of the polypeptide chain on the outside of the molecule. A similar supersecondary structure arrangement is found in pyruvate kinase (Figure 2.47). A different type of α,β structure is seen in domain 1 of lactate dehydrogenase and domain 2 of phosphoglycerate kinase (Figure 2.48) in which the interior polypeptide sections participate in a classical twisted β structure. The β-structure segments are again each joined by α-helical regions positioned on the outside of the molecule to give a characteristic α,β folding pattern. It is interesting to note that this structural class contains many of the enzymes that participate in glycolysis.

An *all-β-structure* pattern is observed in Cu, Zn superoxide dismutase, in which the antiparallel β structure strands form a tertiary pattern called a Greek key β barrel (Figure 2.49). A similar pattern is seen in the folding of each of the domains of the immunoglobulins, discussed in Chapter 3. Concanavalin A (Figure 2.49) also shows an all-β-structure in which the

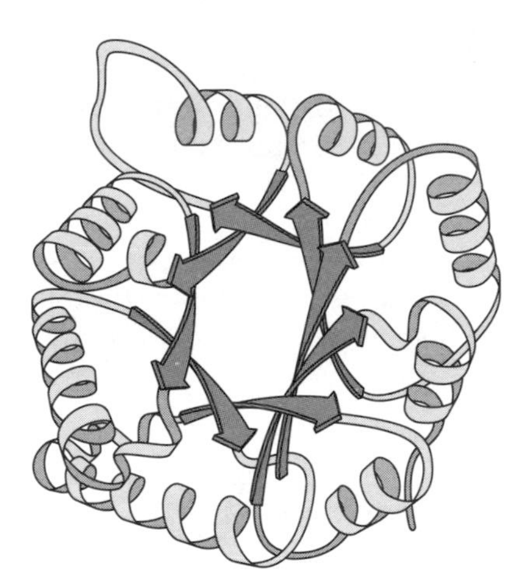

Triose Phosphate Isomerase

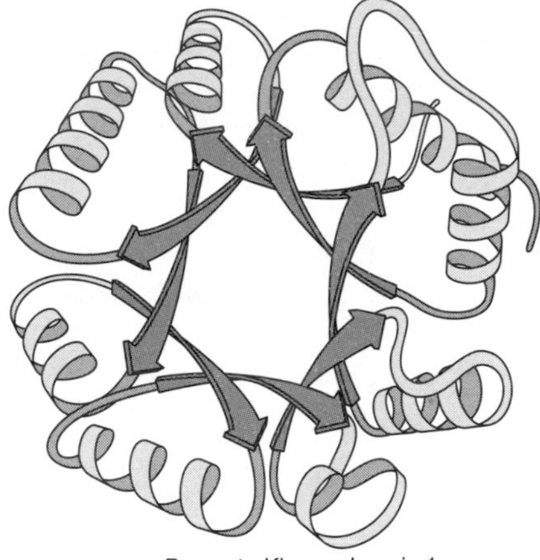

Pyruvate Kinase domain 1

FIGURE 2.47
Examples of α,β-folded domains in which β-structural strands form a β barrel in the center of the domain (see legend to Figure 2.46).
Redrawn with permission from J. S. Richardson, *Adv. Protein Chem.* 34:168, 1981.

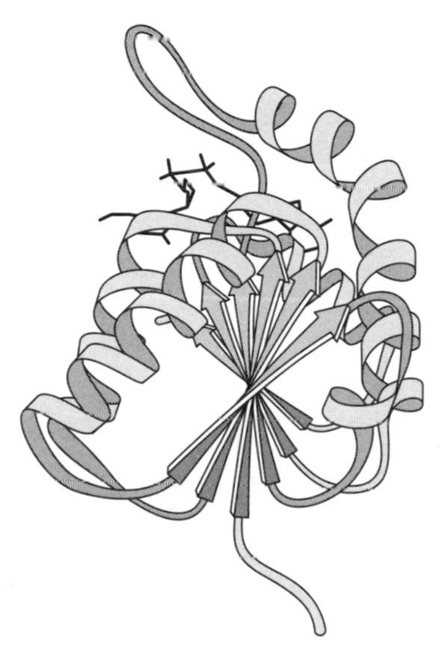

Lactate Dehydrogenase domain 1

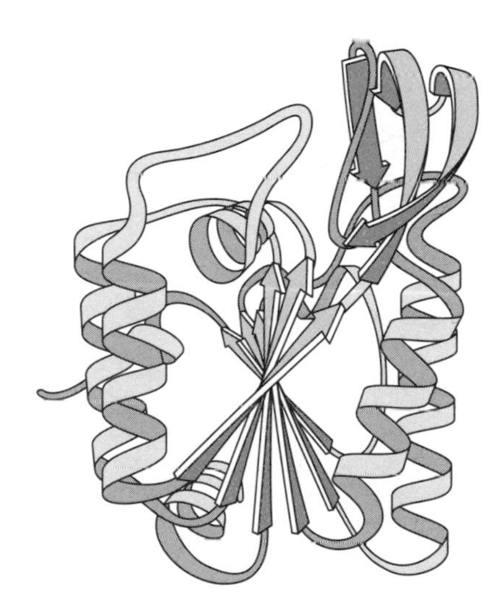

Phosphoglycerate Kinase domain 2

FIGURE 2.48
Examples of α,β-folded domains in which β-structure strands are in the form of a classical twisted β sheet (see legend to Figure 2.46).
Redrawn with permission from J. S. Richardson, *Adv. Protein Chem.* 34:168, 1981.

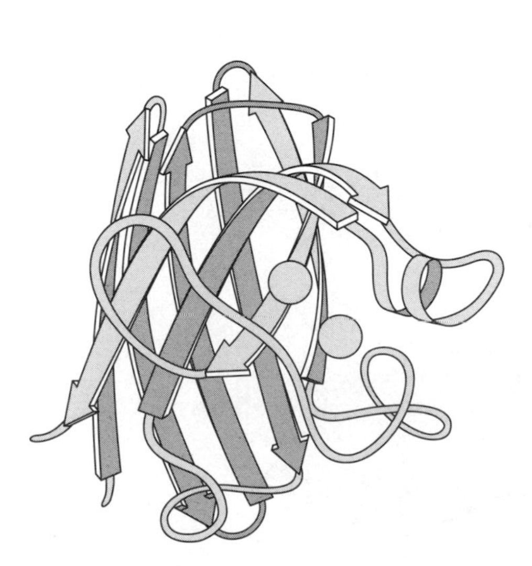

Cu_1Zn Superoxide Dismutase

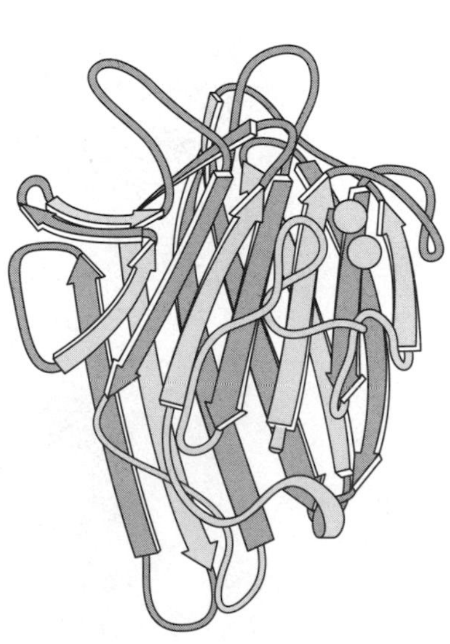

Concanavalin A

FIGURE 2.49
Examples of all β-folded domains (see legend to Figure 2.46).
Redrawn with permission from J. S. Richardson, *Adv. Protein Chem.* 34:168, 1981.

segments of antiparallel β structure assume a β-barrel pattern called a "jellyroll."

A fourth category of supersecondary structure folding patterns generally found for small cystine-rich or metal-rich proteins contain low amounts of regular secondary structure.

It should be noted that the protein structures used to define these classes have been observed by X-ray crystallographic analysis (Section 2.9) and are limited to globular proteins that are water soluble. Proteins that are not water soluble may differ significantly in their classification (see Sections 2.6).

A Quaternary Structure Occurs When Several Polypeptide Chains Form a Specific Noncovalent Association

The quaternary structure of a protein is the arrangement or conformation of polypeptide chain units in a multichain protein, which are required for biological function. The subunits in a quaternary structure must be in **noncovalent association.** Thus the enzyme α-chymotrypsin contains three polypeptide chains covalently joined together by interchain disulfide bonds into a single covalent unit. The protein myoglobin is composed of a single polypeptide chain and contains no quaternary structure. However, hemoglobin contains four polypeptide chains ($\alpha_2\beta_2$) held together *noncovalently* in a specific conformation as required for its function (see Chapter 3). Thus hemoglobin has a quaternary structure. The enzyme aspartate transcarbamylase (see Chapter 13) has a quaternary structure comprised of 12 polypeptide subunits. The poliovirus protein coat contains 60 polypeptide subunits, and the tobacco mosaic virus protein has 2120 polypeptide subunits held together noncovalently in a specific structural arrangement. The quaternary structure of a multidomain viral coat protein is shown in Figure 2.50.

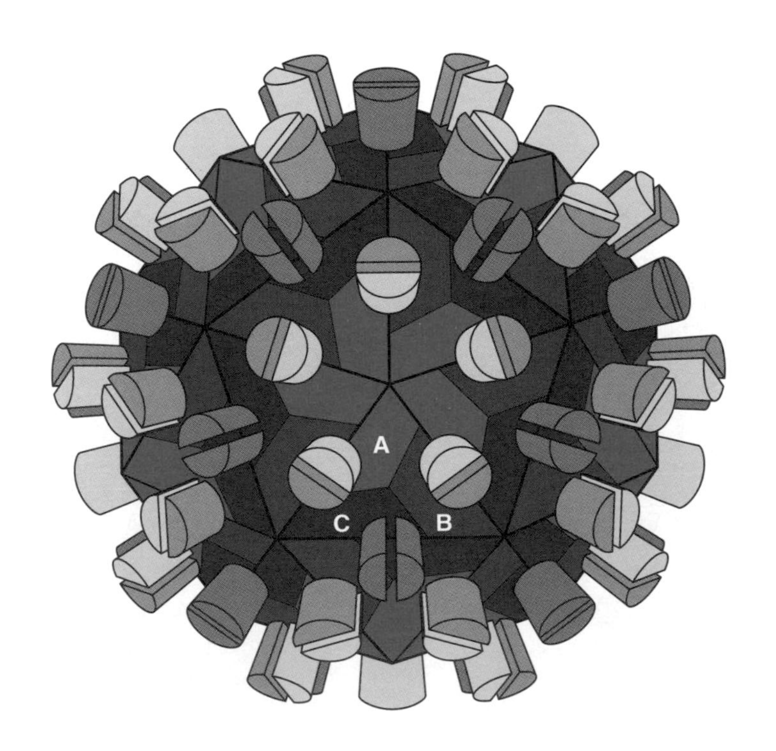

FIGURE 2.50
Cartoon representation of the quaternary structure of the tomato bushy stunt virus (TBSV) capsid.
Subunits labeled A, B, and C have an identical sequence but are present in three different conformations. There are 180 protein subunits in the capsid quaternary structure arranged in 60 ABC groups of three. The interface between protein subunits is shown by a heavy black line in diagram and the half-cylindrical structure represents a domain of each protein that protrudes from the viral surface.
Modified from S. C. Harrison, *Trends Biochem. Sci.* 9:345, 1984.

2.6 OTHER TYPES OF PROTEINS

The models for levels of protein structure are based on observations on globular, water soluble proteins. But this is only one group of the protein types. Other proteins are structural proteins (nonglobular) such as fibrous proteins. In addition, there are classes of proteins that have a heterogenous composition (lipoproteins and glycoproteins) that may or may not be soluble.

Fibrous Proteins Include Collagen, α-Keratin, and Tropomyosin

The characteristic of globular proteins are a spheroid shape, variable molecular weights, relatively high water solubility, and a variety of functional roles such as catalysts, transporters, and control proteins for the regulation of metabolic pathways and gene expressions. In contrast, the **fibrous proteins** characteristically contain higher amounts of regular secondary structure, a long cylindrical (rodlike) shape, a low solubility in water, and a structural rather than a dynamic role in the cell or organism. Examples of fibrous proteins are **collagen, α-keratin,** and **tropomyosin.**

Distribution of Collagen in Humans

Collagen is found in all tissues and organs where it provides the framework that gives the tissues their form and structural strength. The importance of collagen is shown by its high concentration in all organs; the percentage of collagen by weight for some representative human tissues and organs is 4% of the liver, 10% of lung, 12–24% of the aorta, 50% of cartilage, 64% of the cornea, 23% of whole cortical bone, and 74% of skin (see Clin. Corr. 2.4).

CLIN. CORR. **2.4**
SYMPTOMS OF DISEASES OF ABNORMAL COLLAGEN SYNTHESIS

Collagen is present in virtually all tissues and is the most abundant protein in the body. Certain organs depend heavily on normal collagen structure in order to function physiologically. Abnormal collagen synthesis or structure causes dysfunction of cardiovascular organs (aortic and arterial aneurysms and heart valve malfunction), bone (fragility and easy fracturing), skin (poor healing and unusual distensibility), joints (hypermobile joints, arthritis), and eyes (dislocation of the lens). Examples of diseases caused by abnormal collagen synthesis include Marfan's disease, Ehlers–Danlos syndrome, osteogenesis imperfecta, and scurvy.

These diseases may result from abnormal collagen genes, abnormal posttranslational modification of collagen, or deficiency of cofactors needed by the enzymes that carry out posttranslational modification of collagen.

Pyeritz, R. E. and McKusick, V. A. The Marfan syndrome: diagnosis and management. *N. Engl. J. Med.* 300:772, 1979, and Hollister, D., Byers, P. H., and Holbrook, K. A. Genetic disorders of collagen metabolism. *Adv. Hum. Genetics* 12:1, 1982.

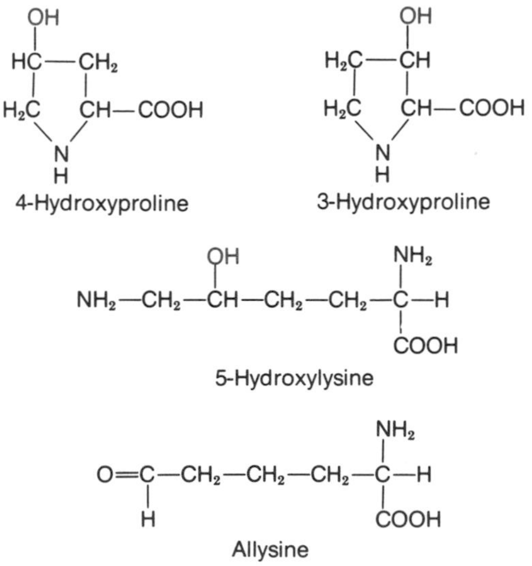

Amino Acid Composition of Collagen

The amino acid composition of collagen is quite different from that found in a typical globular protein. Comparisons of the amino acid composition of type I skin collagen with the globular proteins ribonuclease and hemoglobin are given in Table 2.10. Skin collagen is comparatively very high in glycine (33% of the amino acids), proline (13%), the derived amino acid 4-hydroxyproline (9%), and a second derived amino acid 5-hydroxylysine (0.6%) (Figure 2.51). Hydroxyproline, which is unique to collagens, is formed enzymatically from prolines incorporated in a collagen polypeptide chain. The enzymatic reaction that carries out the hydroxylation of proline requires the presence of ascorbic acid (vitamin C); thus humans with vitamin C deficiency (scurvy) have poor synthesis of new collagen. Most of the hydroxyprolines formed in a collagen polypeptide chain are substituted in the 4-position (γ carbon) of the proline structure, though a small amount of 3-hydroxyproline is also formed (Table 2.10).

Collagens are glycoproteins and a small amount of carbohydrate is found covalently joined to the derived amino acid, 5-hydroxylysine, making a *O*-glycosidic bond through the δ-carbon hydroxyl group. Enzyme catalyzed formation of the 5-hydroxylysine from lysine and the addition of the carbohydrate occur after polypeptide chain formation but prior to the folding of the collagen chains into their unique supercoiled structure.

4-Hydroxyproline

3-Hydroxyproline

$NH_2—CH_2—CH(OH)—CH_2—CH_2—C(NH_2)(COOH)—H$

5-Hydroxylysine

$O{=}C(H)—CH_2—CH_2—CH_2—C(NH_2)(COOH)—H$

Allysine

FIGURE 2.51
Derived amino acids found in collagen.
Carbohydrate is attached to 5-OH in 5-hydrosylysine by a type III glycosidic linkage (see Figure 2.60).

Amino Acid Sequence of Collagen

The molecular unit of mature collagen or **tropocollagen** is composed of three polypeptide chains. Various distinct collagen types exist, each with their own genes, which express their characteristic polypeptide chains. In

TABLE 2.10 Comparison of the Amino Acid Content of Human Skin Collagen (Type I) and Mature Elastin with That for Two Typical Globular Proteins[a]

Amino acid	Collagen (human skin)	Elastin (mammalian)	Ribonuclease (bovine)	Hemoglobin (human)
COMMON AMINO ACIDS		PERCENT OF TOTAL		
Ala	11	22	8	9
Arg	5	0.9	5	3
Asn			8	3
Asp	5	1	15	10
Cys	0	0	0	1
Glu	7	2	12	6
Gln			6	1
Gly	33	31	2	4
His	0.5	0.1	4	9
Ile	1	2	3	0
Leu	2	6	2	14
Lys	3	0.8	11	10
Met	0.6	0.2	4	1
Phe	1	3	4	7
Pro	13	11	4	5
Ser	4	1	11	4
Thr	2	1	9	5
Trp	2	1	9	2
Tyr	0.3	2	8	3
Val	2	12	8	10
DERIVED AMINO ACIDS				
Cystine	0	0	7	0
3-Hydroxyproline	0.1		0	0
4-Hydroxyproline	9	1	0	0
5-Hydroxylysine	0.6	0	0	0
Desmosine and isodesmosine	0	1	0	0

[a] Boxed numbers emphasize important differences in amino acid composition between the fibrous proteins (collagen and elastin) and typical globular proteins.

some collagen types, each of the three polypeptide chains have an identical amino acid sequence. In other collagens such as type I (Table 2.11), two of the chains are identical in sequence while the amino acid sequence of the third chain is slightly different. For type I collagen, the two identical chains are designated $\alpha 1$(I) chains and the third nonidentical chain, $\alpha 2$(I). In type V collagen all three polypeptide chains are of different amino acid sequences and are designated $\alpha 1$(V), $\alpha 2$(V), and $\alpha 3$(V). The different types of collagen are characterized by differences in their physical properties due to differences in their amino acid sequence and in their percent of carbohydrate. Table 12.11 describes some of the characteristics of collagens types I–VI; additional collagen types (designated up through type XII) have been reported.

The amino acid sequence of the polypeptide chains of collagens is most unusual. In large sequences of the polypeptide chain glycine is found to be every third residue, and proline or hydroxyproline also occur three residue apart in the same regions. In these segments of polypeptide chain with glycine every third amino acid and proline and hydroxyproline so abundant, the triplet amino acid sequences **Gly-Pro-Y** and **Gly-X-Hyp** (where X and Y are any of the amino acids) are shown to be reiterated in tamdem several hundred times. In type I collagen, these two triplet sequences will be reiterated over 100 times each, encompassing over 600 amino acids within a chain of approximately 1000 amino acids.

Structure of Collagen

Polypeptides that contain only the amino acid proline can be synthesized in the laboratory. These polyproline polypeptide chains will assume a

TABLE 2.11 Classification of Collagen Types

Type	Chain designations	Tissue found	Characteristics
I	$[\alpha1(I)]_2\alpha2(I)$	Bone, skin, tendons, scar tissue, heart valve, intestinal, and uterine wall	Low carbohydrate; <10 hydroxylysines per chain; 2 types of polypeptide chains
II	$[\alpha1(II)]_3$	Cartilage, vitreous	10% carbohydrate; >20 hydroxylysines per chain
III	$[\alpha1(III)]_3$	Blood vessels, newborn skin, scar tissue, intestinal, and uterine wall	Low carbohydrate; high hydroxyproline and Gly; contains Cys
IV	$[\alpha1(IV)]_3$ $[\alpha2(IV)]_3$	Basement membrane, lens capsule	High 3-hydroxyproline; >40 hydroxylysines per chain; low Ala and Arg; contains Cys; high carbohydrate (15%)
V	$[\alpha1(V)]_2\alpha2(V)$ $[\alpha1(V)]_3$ $\alpha1(V)\alpha2(V)\alpha3(V)$	Cell surfaces or exocytoskeleton; widely distributed in low amounts	High carbohydrate, relatively high glycine, and hydroxylysine
VI	—	Aortic intima, placenta, kidney, and skin in low amounts	Relatively large globular domains in telopeptide region; high Cys and Tyr; mol wt relatively low (~160,000); equimolar amounts of hydroxylysine and hydroxyproline

regular secondary structure in aqueous solution in which the chain is in a tightly twisted extended helix with three residues per turn of the helix ($n = 3$). This helix with all trans peptide bonds is designated the **polyproline type II helix** (Type I helices have all *cis*-proline peptide bonds, see Figure 2.11 for the difference between cis and trans peptide bonds). The polyproline helix has the same characteristics as the helix found for the polypeptide chains of collagen in which the primary structure contains a proline or hydroxyproline at approximately every third position.

Since the helix structure in collagen is the same as found in polyproline, the thermodynamic forces leading to the formation of the helix structure are due to the properties of proline. In proline, the angle contributed to the polypeptide chain is also part of a five-member ring side chain group. The five-member ring constrains the C—N bond to an angle that is compatible with the polyproline helix structure. In addition, the solvation properties of proline and hydroxyproline stabilize the novel polyproline helix conformation.

In the polyproline type II helix, the plane of each peptide bond is positioned perpendicular to the axis of the helix. In this geometry the peptide carbonyl groups are pointed in a direction where they form strong interchain hydrogen bonds to other chains of the collagen molecule. This is in contrast to the α helix, in which the plane containing the atoms of the peptide bonds are directed parallel to the axis of the helix and form intrachain hydrogen bonds with other peptide bonds in the same polypeptide chain rather than interchain hydrogen bonds as in the polyproline helix.

The three polypeptide chains of the collagen molecule, where each of the chains is in a polyproline type II helix conformation, are wound about

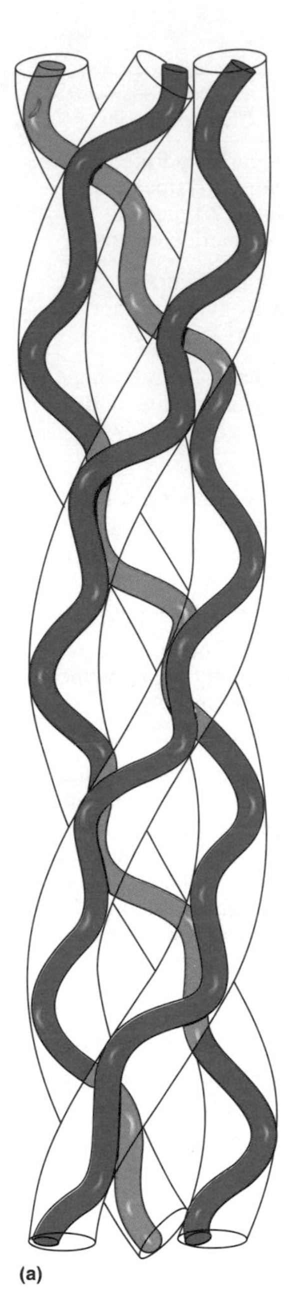

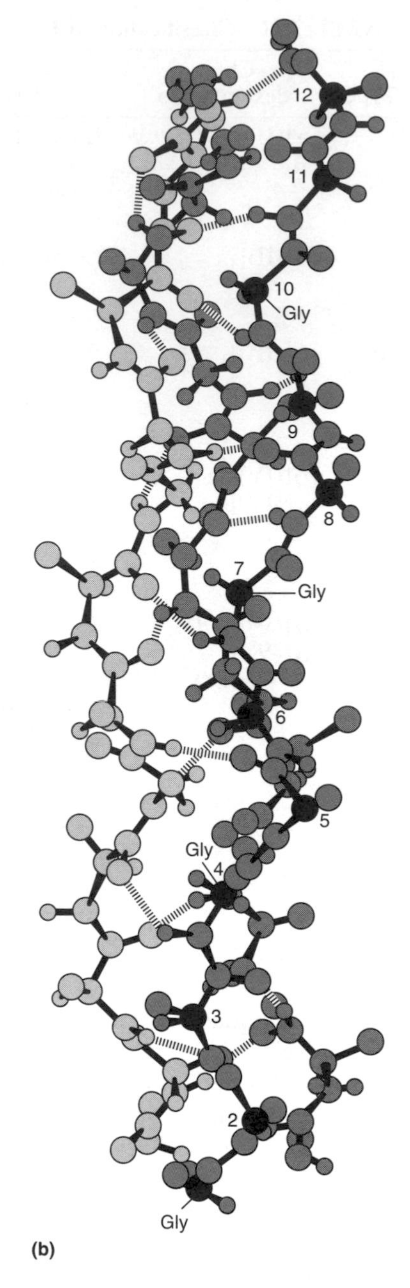

FIGURE 2.52
Diagram of collagen demonstrating the necessity for glycine in every third residue to allow the different chains to be in close proximity in the structure. All α-carbon atoms are numbered and proposed hydrogen bonds are shown by dashed lines.
(*a*) Ribbon model for supercoiled structure of collagen with each of the individual chains in a polyproline type II helix. (*b*) More detailed model of supercoiled conformation.
Redrawn with permission from R. E. Dickerson and I. Geis, *The Structure and Actions of Proteins,* Menlo Park, CA: W. A. Benjamin, Inc., 1969, pp. 41, 42.

each other in a defined way (Figure 2.52) to form a **superhelix structure.** The three chain superhelix has a characteristic rise (d) and pitch (p) as does a single chain helix.

The superhelix forms because glycines have a side chain of low steric bulk (R = H). As the polyproline type II helix has three residues per turn ($n = 3$) and glycine is at every third position, the glycines in each of the polypeptide chains will all be aligned along one side of the helix, forming an **apolar edge** of the polypeptide chain. The glycine edges among the three polypeptide chains can then noncovalently associate in a close arrangement, held together by hydrophobic interactions. A larger side chain group than glycine would impede the adjacent chains from coming together in this superhelix structure. The peptide bond NH and C═O are pointed in a plane perpendicular to the chain axis so that each makes an interchain hydrogen bond with one of the other two chains, stabilizing the structure of the superhelix (Figure 2.52).

In collagen molecules the superhelix conformation may propagate for long stretches of the sequence, which is especially true for type I collagen

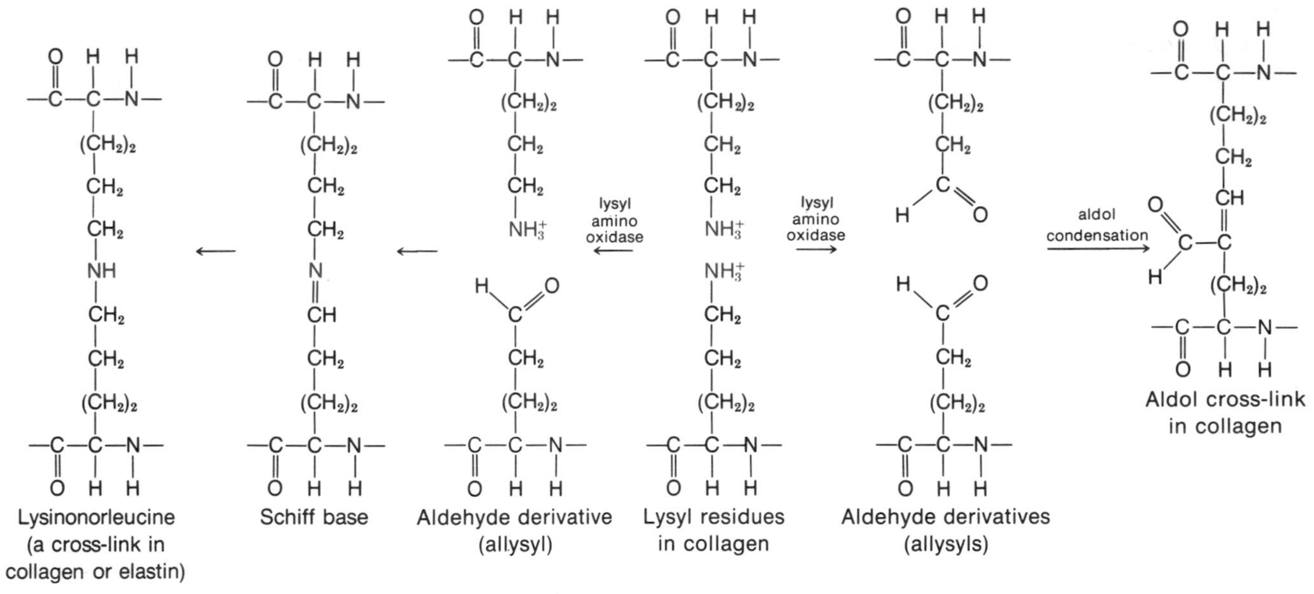

Lysinonorleucine (a cross-link in collagen or elastin)

Schiff base

Aldehyde derivative (allysyl)

lysyl amino oxidase

Lysyl residues in collagen

lysyl amino oxidase

Aldehyde derivatives (allysyls)

aldol condensation

Aldol cross-link in collagen

FIGURE 2.53
Covalent cross-links formed in collagen from allysine.

where only the COOH terminal and NH_2 terminal segments (known as the **telopeptides**) are not in a superhelical conformation. The type I collagen molecule has a length of 3000 Å and a width of only 15 Å, a very long cylindrical structure. In other collagen types, the superhelical regions may be periodically broken by regions of the polypeptide chain that fold into globular type domains.

Formation of Covalent Cross-Links in Collagen

An enzyme present extracellularly acts on secreted collagen molecules (see Chapter 17) to convert some of the ε-amino groups of lysine side chains to δ aldehydes (Figure 2.53). The resulting derived amino acid, containing the aldehydic R group, is known as **allysine.** The newly formed aldehyde side chain group in the polypeptide chain will spontaneously undergo nucleophilic addition reactions with nonmodified lysine ε-amino groups and also with the δ-carbon atoms of other allysine aldehydic groups to form covalent interchain bonds (Figure 2.53). These covalent linkages can be with side chain groups from chains without the superhelical structure or to adjacent collagen molecules in a collagen fibril. If the covalent cross-links are not formed because of a deficiency of the enzyme that converts lysine to allysine, lysine amino oxidase, the collagen fibrils are easily dissociated and rapidly degraded.

Allysines are also important for cross-link formation in another fibrous protein, elastin. **Elastin** does not contain repeating sequences of Gly-Pro-Y or Gly-X-Hyp and, therefore does not fold into either a polyproline helix or a superhelix. It appears to lack a regular secondary structure, but contains an unordered coiled structure in which amino acid residues in the structure are highly mobile. Individual residues in the chain apparently tumble freely in three dimensions over a time scale of 10^{-7}s. The highly mobile, kinetically free, though extensively cross-linked structure gives the protein the property of a rubberlike elasticity. Elastin, which is found in high concentrations in ligaments, lungs, walls of arteries, and skin, gives tissues and organs the capacity to stretch without tearing. It is classified with the fibrous protein due to its structural function and relative insolubility.

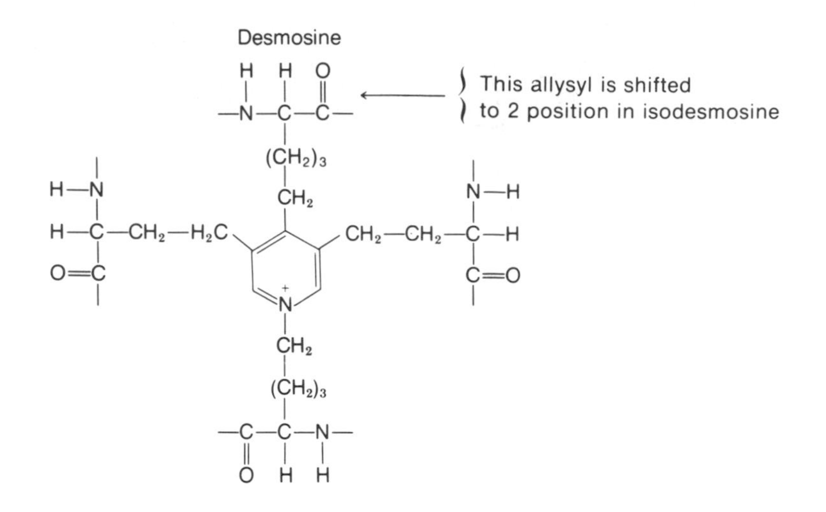

FIGURE 2.54
Desmosine and isodesmosine covalent cross-link in elastin.

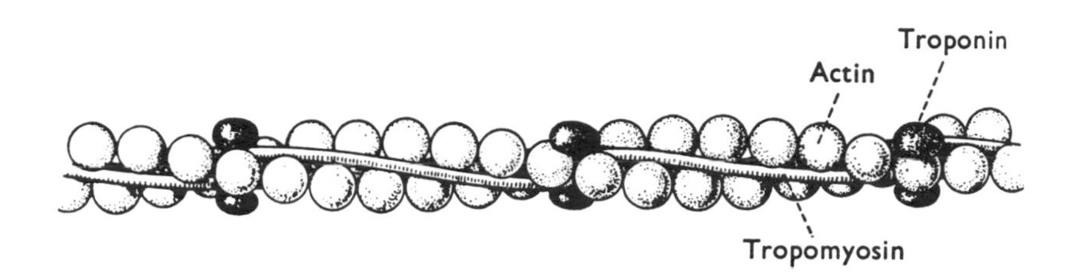

FIGURE 2.55
Structure of the thin filament of muscle with tropomyosin strands (containing two polypeptide chains) showing association to actin monomers and troponin.
Reprinted with permission from S. Ebashi, M. Endo, and I. Ohtsuki, *Q. Rev. Biophys.* 2:351, 1969.

An extracellular lysine amino oxidase acts on particular lysine side chains in elastin to convert lysines to allysines, particularly where lysines appear in the sequence -Lys-Ala-Ala-Lys- and -Lys-Ala-Ala-Ala-Lys-. Four lysines or allysines in these sequences from different regions of the polypeptide chains, react to form the heterocyclic structure of **desmosine** or **isodesmosine** that covalently cross-links the polypeptide chains in elastin fibers (Figure 2.54).

Alpha-Keratin and Tropomyocin

Alpha-keratin and tropomyosin are fibrous proteins in which each of the chains are in α-helical conformation. Alpha-keratin is found in the epidermal layer of skin, in nails, and in hair. Tropomyosin is a component of the thin filament in muscle tissue (Figure 2.55).

Analysis of the amino acid sequence in both these proteins shows repetitive segments of seven amino acids, of which the first and fourth are hydrophobic and the fifth and seventh are polar. The reiterated side chain properties in seven amino acid segments can be symbolically represented by $(\text{a-b-c-d-e-f-g})_i$, where a and d contain hydrophobic amino acid side chain groups, and e and g contain polar or ionized side chain groups. Since a seven amino acid segment represents approximately two complete turns of an α helix (n = 3.6), the apolar residues at a and d will align to form an apolar edge along one side of the α helix (Figure 2.56), similar to the way in which every third residue of glycine in a collagen helix (n = 3.0) forms an apolar edge. The apolar edge in α-keratin interacts with other polypeptide edges of α-keratin to form a superhelical structure containing two or three polypeptide chains. The polar edges interact with the water solvent on the outside to stabilize the multistrand superhelical structure. Similarly, two tropomyosin strands are wound around each other into a superhelical coil.

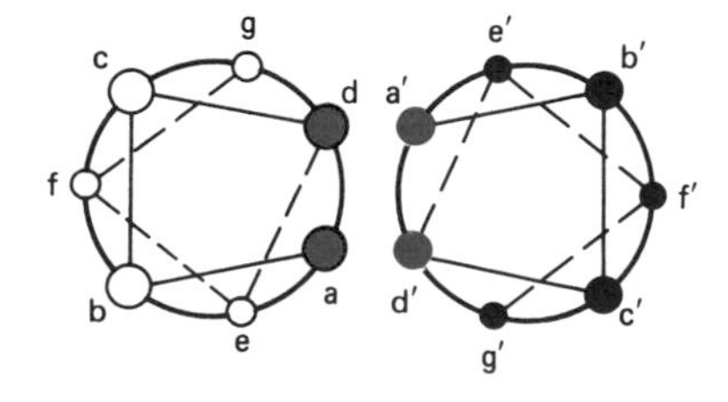

FIGURE 2.56
Interaction of an apolar edge of two chains in α-helical configuration as in α-keratin and in tropomyosin.
Interaction of apolar a-d′ and d-a′ residues of two α helices aligned parallel in an NH_2 terminal (top) to COOH terminal direction is presented.
Redrawn from A. D. McLachlan and M. Stewart, *J. Mol. Biol.* 98:293, 1975.

Thus collagen, α-keratin, and tropomyosin molecules are each composed of multistrand structures in which polypeptide chains with a highly regular secondary structure (polyproline type II helix in collagen, α helix in α-keratin and tropomyosin) are wound around each other in a supercoil conformation. In turn, the supercoiled molecules are aligned in connective fibrils that are stabilized by covalent cross-links. The amino acid sequences of the polypeptide chains are repetitive, generating edges on the cylindrical surfaces of each of the polypeptide chains that stabilize the hydrophobic interactions between the polypeptide chains and generate the supercoil conformation.

Lipoproteins Are a Complex of Lipids With Protein

Lipoproteins are multicomponent complexes of protein and lipids that form a distinct molecular aggregate with an approximate stoichiometry between the specific components within the complex. Each type of lipoprotein will have a characteristic molecular weight, size, chemical composition, density, and physiological role. The protein and lipid in the complex is held together by noncovalent forces.

Lipoproteins serve a wide variety of functions in blood, transporting lipids from tissue to tissue and participating in lipid metabolism (see Chapter 9). The blood lipoproteins are the most extensively characterized of the lipoproteins and changes in their relative amounts are predictive of a major human disease, atherosclerosis (see Clin. Corr. 2.5). Four classes of blood plasma lipoproteins exist in normal fasting humans (Table 2.12); in the postabsorptive period a fifth type of blood lipoprotein, **chylomicrons,** are also present. The lipoproteins are classified on the basis of their density, as determined by ultracentrifugation; they can also be identified by electrophoreses (Figure 2.57).

The purified protein components of a lipoprotein particle are called **apolipoproteins;** each type of lipoprotein has a characteristic apolipoprotein composition. The different apolipoproteins in a particular class of lipoprotein particles will most often be present in a set ratio. For example, the most prominent apolipoprotein in the **high density lipoprotein** (HDL) of plasma is **apolipoprotein A-I** (ApoA-I). Other apolipoproteins (ApoA-II–ApoA-IV and ApoD) are found in low concentrations in the HDL particles relative to ApoA-I (see Table 2.13). In **low density lipoproteins** (LDL) the prominent apolipoprotein protein is **ApoB-100,** which is also present in the **intermediate density lipoproteins** (IDL) and **very low density lipoprotein** (VLDL). The ApoC family of proteins are also present in high amounts in IDL and VLDL. Each of the apolipoprotein classes (A, B, etc.) are distinct (see Clin. Corr. 2.6). Proteins within a class will cross-react with antibodies made to another class. The molecular weights of the

TABLE 2.12 Hydrated Density Classes of Plasma Lipoproteins

Lipoprotein Fraction	Density (g mL^{-1})	Flotation rate (S_f) (Svedberg units)	Molecular weight	Particle diam. (Å)
HDL	1.063–1.210		HDL_2, 4×10^5	70–130
			HDL_3, 2×10^5	50–100
LDL (or LDL_2)	1.019–1.063	0–12	2×10^6	200–280
IDL (or LDL_1)	1.006–1.019	12–20	4.5×10^6	250
VLDL	0.95–1.006	20–400	5×10^6–10^7	250–750
Chylomicrons	<0.95	>400	10^9–10^{10}	10^3–10^4

SOURCE: Data from A. K. Soutar and N. B. Myant, in *Chemistry of Macromolecules,* IIB, R. E. Offord (Ed.), Baltimore, MD: University Park Press, 1979.

CLIN. CORR. 2.5 HYPERLIPIDEMIAS

Hyperlipidemias are disorders of the rates of synthesis or clearance of lipoproteins from the blood stream. Usually they are detected by measuring plasma triglyceride and cholesterol and are classified on the basis of which class of lipoproteins is elevated.

Type I hyperlipidemia is due to accumulation of chylomicrons. Two genetic forms are known: lipoprotein lipase deficiency and ApoCII deficiency. ApoCII is required by lipoprotein lipase for full activity. Patients with type I hyperlipidemia have exceedingly high plasma triglyceride levels (over 1000 mg dL^{-1}) and suffer from eruptive xanthomas (triglyceride deposits in the skin) and pancreatitis.

Type II hyperlipidemia is characterized by elevated LDL. Most cases are due to genetic defects in the synthesis, processing, or function of LDL receptor. Heterozygotes have elevated LDL levels, hence the trait is dominantly expressed. Homozygous patients have very high LDL levels and may suffer myocardial infarctions before age 20.

Type III hyperlipidemia is due to abnormalities of ApoE, which interfere with the uptake of chylomicron and VLDL remnants. Hypothyroidism can produce a very similar hyperlipidemia. These patients have an increased risk of atherosclerosis.

Type IV hyperlipidemia is the commonest abnormality. The VLDL levels are increased, often due to obesity, alcohol abuse, or diabetes. Familial forms are also known but the molecular defect is unknown.

Type V hyperlipidemia is, like type I, associated with high chylomicron triglyceride levels, pancreatitis, and eruptive xanthomas.

Hypercholesterolemia also occurs in certain types of liver disease in which biliary excretion of cholesterol is reduced. An abnormal lipoprotein called lipoprotein X accumulates. This disorder is not associated with increased cardiovascular disease from atherosclerosis.

Goldstein, J. L. and Brown, M. S. Familial hypercholesterolemia. *The Metabolic Basis of Inherited Disease, 5th ed.,* J. B. Stanbury, J. B. Wyngaarden, D. S. Frederickson, J. L. Goldstein, and M. S. Brown (Eds.). New York: McGraw-Hill, 1983.

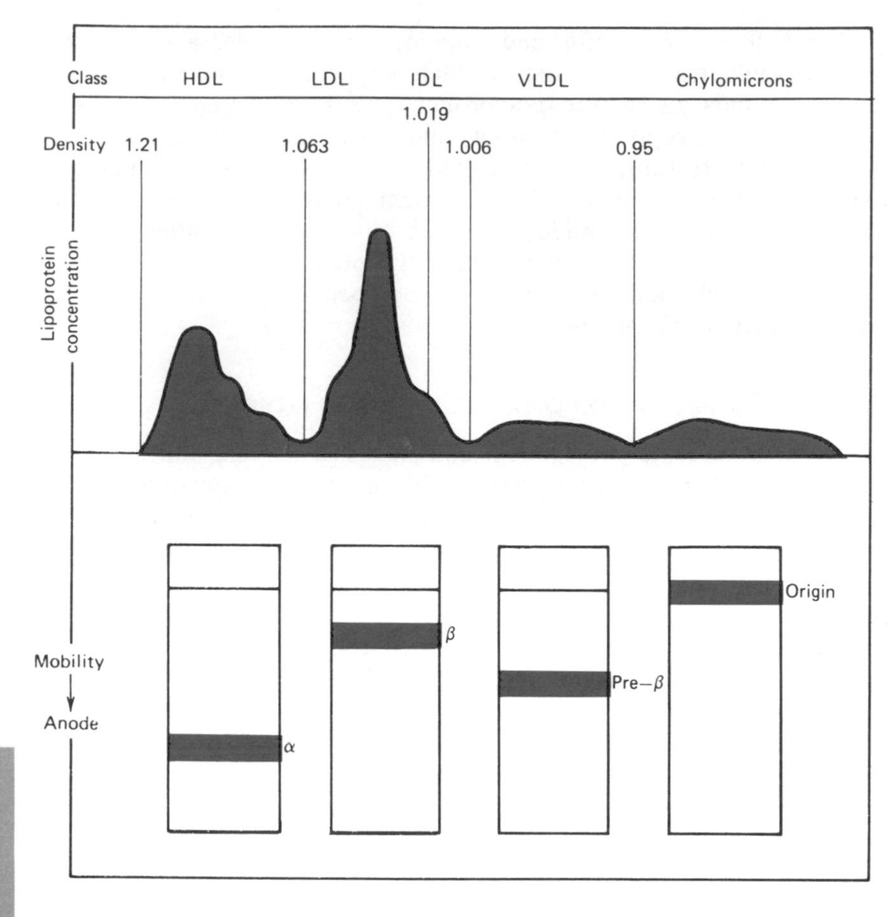

FIGURE 2.57
Correspondence of plasma lipoprotein density classes with electrophoresis mobility in a plasma electrophoresis.
In the upper diagram the ultracentrifugation schlieren pattern is shown. At the bottom, electrophoresis on a paper support shows the mobilities of major plasma lipoprotein classes with respect to α- and β-globulin electrophoresis bands.
Reprinted with permission from A. K. Soutar and N. B. Myant, in *Chemistry of Macromolecules,* IIB, R. E. Offord (Ed.). Baltimore, Md: University Park Press, 1979.

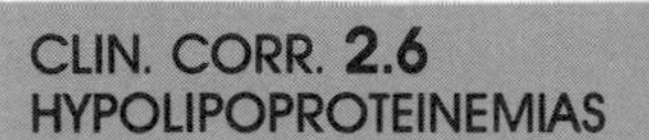

CLIN. CORR. 2.6 HYPOLIPOPROTEINEMIAS

A β-lipoproteinemia is a genetic disease that is characterized by an absence of chylomicrons, VLDL, and LDL due to an inability to synthesize apolipoprotein B-100.

These patients show accumulation of lipid droplets in small intestinal cells, malabsorption of fat, acanthocytosis (spiny shaped red cells), and neurological disease (retinitis pigmentosa, ataxia, and retardation).

Tangier disease, an α-lipoprotein deficiency, is a rare autosomal recessive disease in which the HDL is 1–5% of its normal value. The clinical features are due to the accumulation of cholesterol in the lymphoreticular system, which may lead to hepatomegaly and splenomegaly. In this disease the plasma cholesterol and phospholipids are greatly reduced.

Deficiency of the enzyme lecithin : cholesterol acyltransferase is a rare disease that results in the production of lipoprotein X (see Clin. Corr. 2.5). Also characteristic of this disease is the decrease in the α-lipoprotein and pre-β-lipoprotein bands, with the increase in the β-lipoprotein (lipoprotein X) in electrophoresis.

Herbert, P. N., Assmann, G., Gotto, A. M., and Frederickson, D. S. Familial lipoprotein deficiency: a *beta*-lipoproteinemia, hypo-*beta*-lipoproteinemia, and Tangier disease. *The Metabolic Basis of Inherited Disease, 5th ed.,* J. B. Stanbury, J. B. Wyngaarden, D. S. Frederickson, J. L. Goldstein, and M. S. Brown (Eds.), New York: McGraw-Hill, 1983.

apoproteins of the plasma lipoproteins vary from 6000 Da (ApoC-I) to 550,000 Da for ApoB-100. This later protein, which has been cloned, is one of the longest single chain polypeptides known (4536 amino acids).

A model for a VLDL particle is shown in Figure 2.58. Structural studies indicate that the inside of a particle contains neutral lipids such as cholesterol esters and triacylglycerols. On the outside of this inner core of neutral lipids, in a shell ~20-Å thick, reside the proteins and the charged amphoteric lipids such as unesterified cholesterol and the phosphatidylcholines (Chapter 10). The amphoteric lipids and proteins in the outer shell place their hydrophobic apolar parts towards the inside of the parti-

TABLE 2.13 Apolipoprotein of the Human Plasma Lipoproteins (Values in Percent of Total Protein Present)[a]

Apolipoprotein	HDL_2	HDL_3	LDL	IDL	VLDL	Chylomicrons
ApoA-I	85	70–75	Trace	0	0–3	0–3
ApoA-II	5	20	Trace	0	0–0.5	0–1.5
ApoD	0	1–2			0	1
ApoB	0–2	0	95–100	50–60	40–50	20–22
ApoC-I	1–2	1–2	0–5	<1	5	5–10
ApoC-II	1	1	0.5	2.5	10	15
ApoC-III	2–3	2–3	0–5	17	20–25	40
ApoE	Trace	0–5	0	15–20	5–10	5
ApoF	Trace	Trace				
ApoG	Trace	Trace				

SOURCE: Data from A. K. Soutar and N. B. Myant, in *Chemistry of Macromolecules,* IIB, R. E. Offord (Ed.), Baltimore, MD: University Park Press, 1979; G. M. Kostner, *Adv. Lipid Res.* **20,** 1 (1983).

[a] Values show variability from different laboratories.

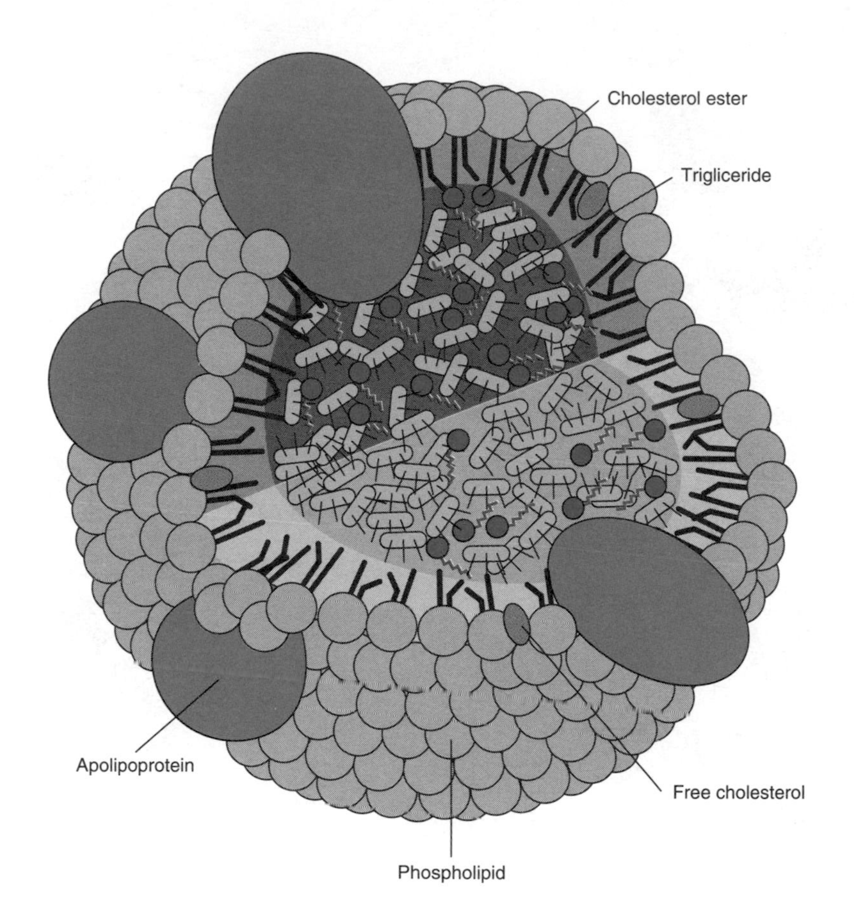

FIGURE 2.58
Generalized structure for the plasma lipoproteins.
Spherical particle consisting of a core of triglycerides and cholesterol esters with a shell ~20-Å thick of apolipoproteins, phospholipids, and unesterified cholesterol. Apolipoproteins are embedded with their hydrophobic edges oriented toward the core and their hydrophylic edges toward the outside.
Modified from R. B. Weinberg, Lipoprotein metabolism; hormonal regulation. In *Hospital Practice,* June 15, 224, 1987.

cle and their charged groups facing the outside where they interact with each other and with the water solvent. This model exists for all plasma lipoproteins, irrespective of their density classes and particle size. The smaller particles, such as the HDL, have a smaller diameter. As the diameter of a spherical particle decreases the molecules in the outer "crust" will make up a greater percentage of the molecules in the particle. Therefore HDL particles have a higher percentage of protein and amphoteric lipids (located on or near the surface) than the larger VLDL particles, which have a higher percentage of its lipid in the core volume and a lower percentage of its total molecules occupying the outer shell. Thus the HDL particles are 45% protein and 55% lipid, while the larger VLDL is only 10% protein and 90% lipid (see Table 2.14). The placement of the

TABLE 2.14 Chemical Composition of the Different Plasma Lipoprotein Classes

			Percent composition of lipid fraction			
Lipoprotein class	Total protein (%)	Total lipid (%)	Phospholipids	Esterified cholesterol	Unesterified cholesterol	Triacylglycerides
HDL_2[a]	40–45	55	35	12	4	5
HDL_3[a]	50–55	50	20–25	12	3–4	3
LDL	20–25	75–80	15–20	35–40	7–10	7–10
IDL	15–20	80–85	22	22	8	30
VLDL	5–10	90–95	15–20	10–15	5–10	50–65
Chylomicrons	1.5–2.5	97–99	7–9	3–5	1–3	84–89

SOURCE: Data from A. K. Soutar and N. B. Myant, in *Chemistry of Macromolecules,* IIB, R. E. Offord (Ed.), Baltimore, MD: University Park Press, 1979.

[a] Subclasses of HDL.

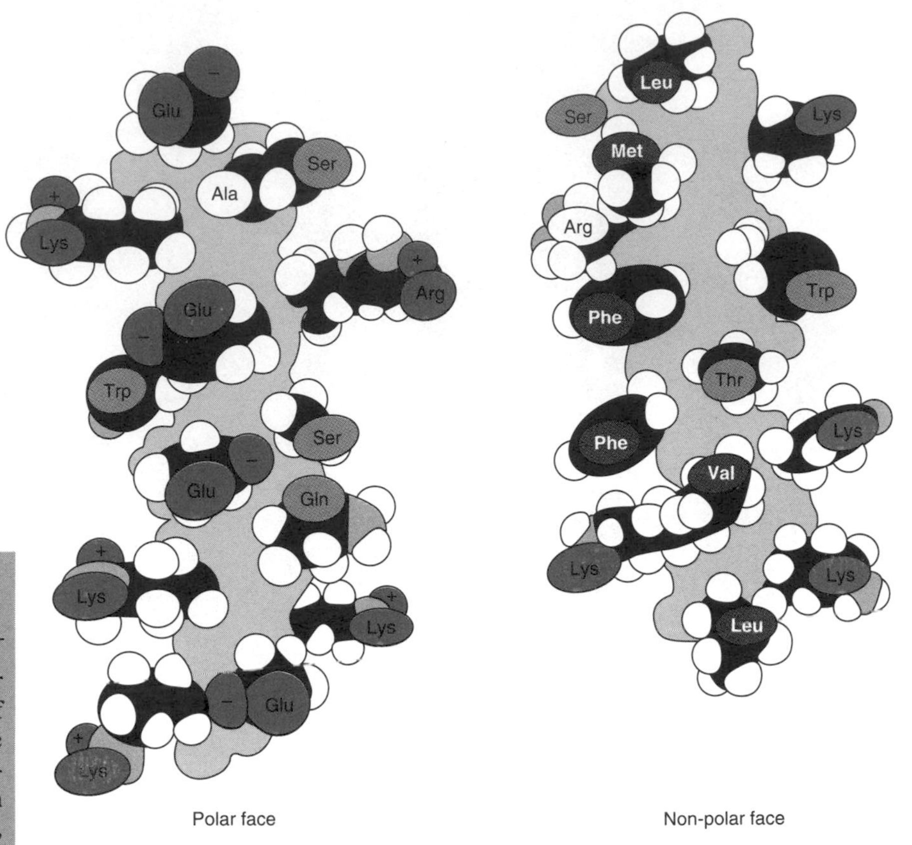

FIGURE 2.59
Illustration showing side chain groups of a helical segment of apolipoprotein C-1 between residues 32 and 53.
The polar face shows ionizable acid residues in the center and basic residues at the edge. On the other side of the helix, the hydrophobic residues form a nonpolar longitudinal face.
Redrawn with permission from J. T. Sparrow and A. M. Gotto, Jr., *CRC Crit. Rev. Biochem.* 13:87, 1983. Copyright © CRC Press, Inc., Boca Raton, FL.

protein in the outer shell of the VLDL particle is diagramatically shown in Figure 2.58.

The apoproteins of the plasma lipoproteins have a high amount of α-helical structure when associated with lipid. These helical regions have an amino acid sequence in which every third or fourth amino acid has an ionic side chain. Thus a polar edge is formed along the longitudinal axis of the helix that associates with the polar heads of phospholipids and the aqueous solvent on the outside. The other side of the helix contains hydrophobic side chain groups that associate with the nonpolar lipid core of the phospholipid particle. The α-helical structure for part of apolipoprotein C-I is shown in Figure 2.59.

Glycoproteins Contain Covalently Bound Carbohydrate

Many proteins contain covalently attached sugar molecules especially proteins secreted from cells and proteins localized on the outer surface of cells. Secreted glycoproteins include proteins that function as hormones, extracellular matrix proteins, proteins involved in blood coagulation (see Clin. Corr. 2.7), and as part of mucus secretion from epithelial cells. On the cell surface glycoproteins function as receptors to transmit signals of hormones or growth factors from the outside environment into the cell.

The percent of the carbohydrate composition of glycoproteins is variable. The IgG antibody molecules contain low amounts of carbohydrate (4%), whereas glycophorin, a glycoprotein of the human red blood cell membrane, is 60% carbohydrate by weight and the human gastric glycoprotein is 82% carbohydrate. The carbohydrate can be distributed evenly along the polypeptide chain of the protein or concentrated in defined regions. For example, in plasma membrane proteins, only that portion of

CLIN. CORR. 2.7 FUNCTIONS OF GLYCOPROTEINS

Glycoproteins participate in a large number of normal and disease-related functions of clinical relevance. For example, many of the proteins in the plasma membranes are glycoproteins. Such proteins are antigens, which determine the blood antigen system (A, B, O), and the histocompatibility and transplantation determinants of an individual. Immunoglobulin antigenic sites and viral and hormone receptor sites in cellular membranes are often glycoproteins. In addition, the carbohydrate portions of glycoproteins in membranes provide a surface code for cellular identification by other cells and for contact inhibition in the regulation of cell growth. As such, changes in the membrane glycoproteins can be correlated with tumorigenesis and malignant transformation in cancer.

Most of the important plasma proteins, except albumin, are glycoproteins. These plasma proteins include the blood-clotting proteins, the immunoglobulins, and many of the complement proteins. Some of the protein hormones, such as follicle-stimulating hormone (FSH) and thyroid-stimulating hormone (TSH), are glycoproteins. The important structural proteins collagen, laminin, and fibronectin contain carbohydrate. The proteins found in the mucus secretions are carbohydrate-containing proteins, where they perform a role in lubrication and in the protection of epithelial tissue. The antiviral protein interferon is a glycoprotein. Plasma glycoproteins that lose their terminal sialic acid are removed from the circulation by a hepatic asialoglycoprotein receptor.

Paulson, J. C. Glycoproteins: what are the sugar chains for?, *Trends Biochem. Sci.* 14:272, 1989.

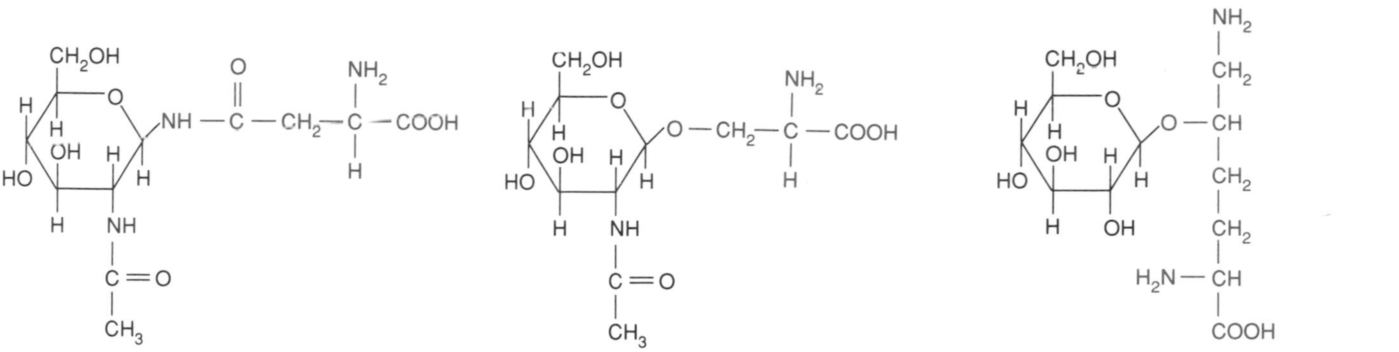

Type I *N*-Glycosyl linkage to asparagine

Type II *O*-Glycosyl linkage to serine

Type III *O*-Glycosyl linkage to 5-hydroxylysine

FIGURE 2.60
Examples of glycosidic linkages to amino acids in proteins.
Type I is an *N*-glycosidic linkage through an amide nitrogen of Asn; type II is an *O*-glycosidic linkage through the OH of Ser or Thr; and type III is an *O*-glycosidic linkage to the 5-OH of 5-hydroxylysine.

the polypeptide chain located on the outside of the cell will have carbohydrate covalently attached.

The carbohydrate attached at one or at multiple points along the polypeptide chain usually contains less than 15 sugar residues and in some cases only a single sugar residue. In comparing a glycoprotein with the same function from different animal species, they all have similar amino acid sequences but a variable carbohydrate structure. This heterogeneity in carbohydrate may occur even on the same protein within a single organism. For example, pancreatic ribonuclease, found in an A and a B form, have identical primary structure but differ in their carbohydrate composition.

Functional glycoproteins may also be found in different stages of "completion." The addition of complex carbohydrate units occurs in a series of individual enzyme-catalyzed reactions as the polypeptide chain is transported through the endoplasmic reticulum and golgi (see Chapter 17). As a consequence, "immature" glycoproteins are sometimes secreted with intermediate stages of carbohydrate additions.

Types of Carbohydrate–Protein Covalent Linkages

Different types of covalent linkages join the sugar moieties and protein in a glycoprotein. The most common types are the ***N*-glycosidic linkages** (type I linkages) formed between an asparagine side chain amide and a sugar, and the ***O*-glycosidic linkage** (type II linkage) made between a serine or threonine hydroxyl side chain and a sugar (Figure 2.60). In type I linkage the bond to the asparagine is made to asparagine side chains within the sequence ***Asn*-X-Thr.** Another linkage found in mammalian glycoproteins is the *O*-glycosidic bond to the 5-hydroxylysine residue (type III linkage) found in collagens and in the serum complement protein C1q. Other less common linkages include attachment to the hydroxyl group of the derived amino acid 4-hydroxyproline (type IV linkage), through the cysteine thiol side chain (type V linkage), and through the NH_2 terminal α-amino group of a polypeptide chain (type VI linkage). High concentrations of type VI linkages are spontaneously formed with red blood cell hemoglobin and blood glucose in uncontrolled diabetics. Assays of the concentration of glycosylated hemoglobin are used to follow changes in blood glucose concentration (see Clin. Corr. 2.8).

CLIN. CORR. 2.8
GLYCOSYLATED HEMOGLOBIN, HbA_{1c}

A glycosylated hemoglobin, designated HbA_{1c}, is formed spontaneously in the red blood cell by combination of the NH_2 terminal amino groups of the hemoglobin β chain and glucose. The aldehyde group of the glucose first forms a Schiff base with the NH_2 terminal amino group,

$$-\mathrm{N}{=}\underset{\mathrm{H}}{\mathrm{C}}-\underset{\mathrm{H}}{\overset{\mathrm{OH}}{\mathrm{C}}}-$$

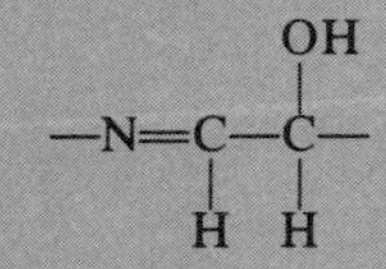

which then rearranges to a more stable amino ketone linkage,

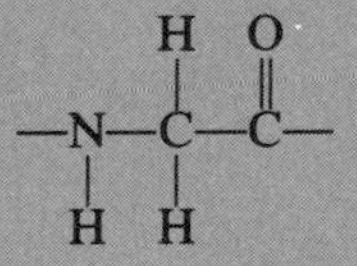

$$-\underset{\mathrm{H}}{\mathrm{N}}-\underset{\mathrm{H}}{\overset{\mathrm{H}}{\mathrm{C}}}-\overset{\mathrm{O}}{\overset{\|}{\mathrm{C}}}-$$

by a spontaneous (nonenzymatic) reaction known as the Amadori rearrangement. The concentration of HbA_{1c} is dependent on the concentration of glucose in the blood and the duration of hyperglycemia. In prolonged hyperglycemia the concentration may rise to 12% or more of the total hemoglobin. Patients with diabetes mellitus will have high concentrations of glucose and therefore high amounts of HbA_{1c}. The changes in the concentration of HbA_{1c} in diabetic patients can be used to follow the effectiveness of treatment for the diabetes.

Bunn, H. F. Evaluation of glycosylated hemoglobin in diabetic patients. *Diabetes* 30:613, 1980, and Brown, S. B. and Bowes, M. A. Glycosylated haemoglobins and their role in management of diabetes mellitus. *Biochem. Educ.* 13:2, 1985.

2.7 FOLDING OF PROTEINS FROM RANDOMIZED TO UNIQUE STRUCTURES: PROTEIN STABILITY

The ability of a primary structure to spontaneously fold to its native secondary and tertiary conformation, without any special information other than the existence of **noncovalent interactions,** is demonstrated by experiments in which proteins are denatured without the hydrolysis of peptide bonds. These proteins, on standing, will refold to their native conformation. A polypeptide sequence usually contains physical properties sufficient to promote protein folding to the unique conformation characteristic of the protein under the correct solvent conditions and in the presence of prosthetic groups that may be a part of its structure. Quaternary structures also assemble spontaneously, after the tertiary structure of the individual polypeptide subunits are formed.

It may appear surprising that a protein folds into a single unique conformation from all the possible a priori rotational conformations available around single bonds in the primary structure of a protein. For example, the α chain of hemoglobin contains 141 amino acids in which there are 4–9 single bonds per amino acid residue around which free rotation can occur. If each bond about which free rotation occurs has two or more stable rotamer conformations accessible to it, then there are a minimum of 4^{141}–9^{141} possible conformations for this chain. However, only a limited set of conformations is found for the α chain. The folded conformation of a protein is that conformation of the lowest Gibbs free energy accessible to the amino acid primary structure within a specified time frame. Thus the folding process is under both thermodynamic and kinetic control. A discrete number of pathways to the native structure of lowest energy conformation for the protein in its native solution environment exist.

Although knowledge of the process of de novo folding of a polypeptide chain is at present an unattainable goal, certain constraints are accepted. In the folding process short-range interactions initiate folding of small regions of secondary structure in the polypeptide strand. Short-range interactions are the noncovalent interactions that occur between a side chain (R) group and the polypeptide chain to which it is covalently attached. Particular side chain R groups have a propensity to promote the formation of α helices, β structure, and sharp turns or bends (β turns) in the polypeptide strand. The interaction of a particular side chain group with those of its nearest neighbors in the polypeptide sequence form a grouping that determines the secondary structure into which the section of the polypeptide strand folds. Sections of the polypeptide strand, called **initiation sites,** will thus spontaneously fold into small regions of secondary structure. **Medium- and long-range interactions** between different initiation sites then stabilize a folded tertiary structure for the polypeptide chain. Disulfide cystine bonds are formed between cysteines in the primary structure after the protein has folded correctly.

Since it is noncovalent forces that act on the primary structure to cause a protein to fold into a unique conformational structure and then stabilize the native structure against denaturation processes, it is of importance to understand the properties of these forces. Some of the important properties of these noncovalent forces are discussed in the section that follows.

TABLE 2.15 Bond Strength for Typical Bonds Found in Protein Structures

Bond type	Bond strength (kcal mol^{-1})
Covalent bonds	>50
Noncovalent bonds	0.6–7
Hydrophobic bond (i.e., two benzyl side chain groups of Phe)	2–3
Hydrogen bond	1–7
Ionic bond (low dielectric environment)	1–6
van der Waals	<1
Average energy of kinetic motion (37°C)	0.6

Noncovalent Forces Lead to Protein Folding and Contribute to a Protein's Stability

Noncovalent forces are weak bonding forces of bonding strength of 1–7 kcal mol^{-1} (4–29 kJ mol^{-1}). This may be compared to the strength of covalent bonds that have a bonding strength of at least 50 kcal mol^{-1}. The noncovalent bonding forces are just higher then the average kinetic energy of molecules at 37°C (0.6 kcal mol^{-1}) (Table 2.15). However, the large number of individually weak noncovalent contacts within a protein add up to a large energy factor that is a net thermodynamic force favoring protein folding.

Hydrophobic Interaction Forces

The most important of the noncovalent forces that will cause a randomized polypeptide conformation to lose rotational freedom and fold into its native structure are hydrophobic interaction forces. It is important to realize that the strength of a hydrophobic interaction is not due to a high intrinsic attraction between nonpolar groups, but rather to the properties of the water solvent in which the nonpolar groups are dissolved.

A nonpolar residue dissolved in water induces in the water solvent a solvation shell in which water molecules are highly ordered. When two nonpolar groups come together on the folding of a polypeptide chain, the surface area exposed to solvent is reduced and a part of the highly ordered water in the **solvation shell** is released to bulk solvent. Accordingly, the entropy of the water (i.e., net disorder of the water molecules in the system) is increased. The increase in entropy (disorder) is a thermodynamically favorable process, and is the driving force causing nonpolar moieties to come together in aqueous solvent. A favorable free energy change of ~2 kcal mol^{-1} for the association of two phenylalanine side chain groups in water is due to this favorable solvent entropy gain.

Calculations show that in transition from a randomized conformation into a regular secondary conformation such as an α helix or β structure, approximately one-third of the ordered water of solvation about the unfolded polypeptide is lost to bulk solvent. This is an approximate driving force favoring folding in a typical globular protein of 0.5–0.9 kcal mol^{-1} for each peptide residue. It is calculated that an additional one-third of the original solvation shell is lost when a protein already folded into a secondary structure then folds into a tertiary structure. The tertiary folding brings different segments of folded polypeptide chains into close proximity with the release of water of solvation between the polypeptide chains.

Hydrogen Bonds

A second important noncovalent force in proteins is hydrogen bonding. Hydrogen bonds are formed when a hydrogen atom covalently bonded to an electronegative atom is shared with a second electronegative atom. The atom to which the hydrogen atom is covalently bonded is designated the donor atom. The atom with which the hydrogen atom is shared is the hydrogen acceptor atom. Typical hydrogen bonds found in proteins are shown in Figure 2.61. It has been previously shown that α-helical and β-structure conformations are extensively hydrogen bonded.

The strength of a hydrogen bond is at a first approximation dependent on the distance between the donor and acceptor atoms. High bonding energies occur when the donor and acceptor atoms are between 2.7 and 3.1 Å apart. Of lesser importance to bonding strength than the distance requirement, but still of some importance, is the dependence of hydrogen-bond strength on geometry. Bonds of higher energy are geometrically colinear, with donor, hydrogen, and acceptor atoms lying in a straight line. The dielectric constant of the medium around the hydrogen bond may also be reflected in the bonding strength. Typical hydrogen bond strengths in proteins are 1–7 kcal mol^{-1}.

Although hydrogen bonds do contribute to the thermodynamic stability of a folded protein's conformation, their formation within a native protein structure may not be as major a driving force for folding as we might at first believe. This is because peptide groups and other hydrogen-bonding

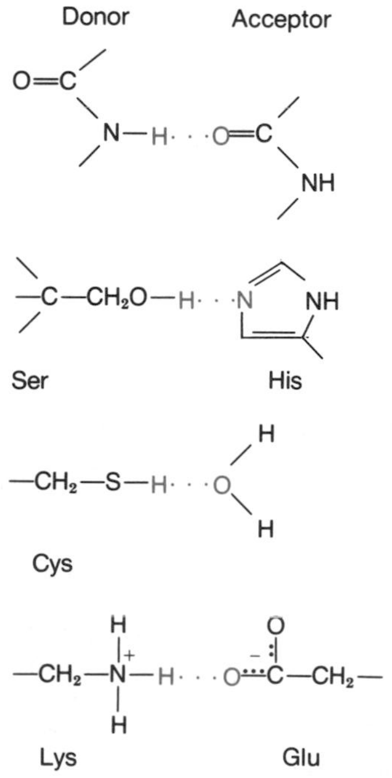

FIGURE 2.61
Some common hydrogen bonds found in proteins.

$$\Delta E_{el} \sim \frac{Z_A \cdot Z_B \cdot \varepsilon^2}{D \cdot r_{ab}}$$

FIGURE 2.62
Strength of electrostatic interactions.

$$E = -\frac{A}{r_{ab}^6} + \frac{B}{r_{ab}^{12}}$$

FIGURE 2.63
van der Waals forces.

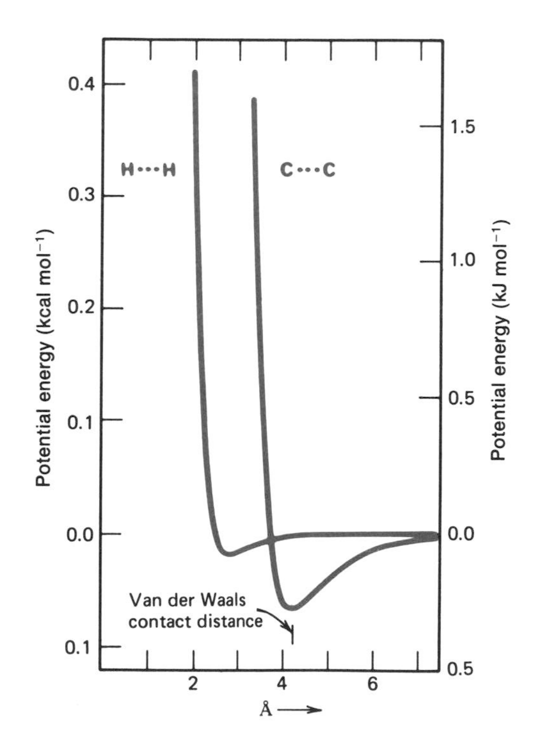

FIGURE 2.64
van der Waals–London dispersion interaction energies between two hydrogen atoms and two (tetrahedral) carbon atoms. Negative energies are favorable and positive energies unfavorable.
Redrawn from A. Fersht, *Enzyme Structure and Mechanism*, San Francisco: W. H. Freeman, 1977, p. 228.

groups in proteins form hydrogen bonds to the water solvent in the denatured state, and these bonds must be broken before the protein folds. The energy required to break the hydrogen bonds to water must be subtracted from the energy gained from the formation of the new hydrogen bonds between atoms in the folded protein in the calculation of the net thermodynamic contribution of hydrogen bonding to the folding.

Electrostatic Interactions

Electrostatic interactions between charged groups are of importance to particular protein structures and in the binding of charged ligands and substrates to proteins. Electrostatic forces can be repulsive or attractive depending on whether the interacting charges are of the same or opposite sign. The strength of an electrostatic force (ΔE_{el}) is directly dependent on the charge (Z) for each ion, and is inversely dependent on the **dielectric constant** (D) of the solvent and the distance between the charges (r_{ab}) (Figure 2.62).

Water has a high dielectric constant ($D = 80$), and ionic charge interactions in water are relatively weak in comparison to electrostatic interactions in the interior of a protein, where the dielectric constant ($D = 2$–40) is approximately a factor of 1 : 40 to 1 : 2 that of water. Consequently, the strength of an electrostatic interaction in the interior of a protein, where the dielectric constant is low, may be of significant energy. However, most charged groups of proteins are on the surface of the protein where they do not strongly interact with other charged groups from the protein due to the high dielectric constant of the water solvent, but are stabilized by hydrogen bonding and polar interactions to the water. These water interactions are the driving force leading to the placement of most ionic groups of a protein on the outside of the protein structure, where they can make energetically favorable contacts with the solvent.

Van der Waals–London Dispersion Forces

Van der Waals and London dispersion forces are a fourth type of weak noncovalent force of great importance in protein structure. This force has an attractive term (A) dependent on the 6th power of the distance between two interacting atoms (r_{ab}), and a repulsive term (B) dependent on the 12th power of r_{ab} (Figure 2.63). The A term contributes at its optimum distance an attractive force of less than 1 kcal mol^{-1} per atomic interaction. This attractive component is due to the induction of complementary partial charges or dipoles in the electron density of adjacent atoms when the electron orbitals of the two atoms approach to a close distance. The repulsive component (term B) of the van der Waals force predominates at closer distances than the attractive force when the electron orbitals of the adjacent atoms begin to overlap. This type of repulsion is commonly called steric hindrance.

The distance of maximum favorable interaction between two atoms is known as the **van der Waals contact distance,** which is equal to the sum of the van der Waals radii for the two atoms (Figure 2.64). Some van der Waals radii for atoms commonly found in proteins are given in Table 2.16. While a van der Waals–London dispersion interaction between any two atoms in a protein is usually less than 1 kcal mol^{-1}, the total number of these weak interactions in a protein molecule is in the thousands. Thus the sum of the attractive and repulsive van der Waals–London dispersion forces are extremely important to protein folding and stability.

The van der Waals contact distances of 2.8–4.1 Å are longer than hydrogen-bond distances of 2.6–3.1 Å, and at least twice as long as normal covalent bond distances of 1.0–1.6 Å between C, H, N, and O atoms.

TABLE 2.16 Covalent Bond Radii and van der Waals Radii for Selected Atoms

Atom	Covalent radius (Å)	van der Waals radius (Å)[a]
Carbon (tetrahedral)	0.77	2.0
Carbon (aromatic)	0.69 along=bond	1.70
	0.73 along—bond	
Carbon (amide)	0.72 to amide N	1.50
	0.67 to oxygen	
	0.75 to chain C	
Hydrogen	0.33	1.0
Oxygen (—O—)	0.66	1.35
Oxygen (=O)	0.57	1.35
Nitrogen (amide)	0.60 to amide C	1.45
	0.70 to hydrogen bond H	
	0.70 to chain C	
Sulfur, diagonal	1.04	1.70

SOURCE: *CRC Handbook of Biochemistry and Molecular Biology,* 3rd ed., Sect. D, Vol. II, G. D. Fasman (Ed.), 1976, p. 221.

[a] The van der Waals contact distance is the sum of the two van der Waals radii for the two atoms in proximity.

Inasmuch as the latter bonds are shorter than the van der Waals contact distance, a repulsive van der Waals force must be overcome in forming hydrogen bonds and covalent bonds between atoms. The energy to overcome van der Waals repulsive force must form a part of the energy of activation for hydrogen bond and covalent bond formation.

A special type of interaction (π electron–π electron) occurs when two aromatic rings approach each other with the plane of their aromatic rings overlapping (Figure 2.65). This type of interaction can result in a noncovalent attractive force of up to 6 kcal mol^{-1}. A significant number of π–π aromatic interactions occur in a typical folded protein, which contribute to the stability of the folded structure.

Denaturation of Proteins Leads to Loss of Native Structure

Denaturation occurs in a protein upon the loss of its native secondary, tertiary, and quaternary structures. In the denatured state the higher structural levels of the native conformation are randomized or scrambled. However, the primary (covalent) structure is not necessarily broken in denaturation.

The loss of protein function, such as a catalytic activity in an enzyme, may result from small modifications of the protein's native conformation that do not lead to a complete scrambling of the secondary, tertiary, and quaternary levels of conformation. Denaturation is thus correlated with the loss of a protein's function; but loss of a protein's function, which may be due to only a small conformational change, is not necessarily synonymous with denaturation.

Even though the conformational differences between denatured and native structure are substantial, the free energy difference between the denatured structure and the native structure of a protein may in some cases be as low as the free energy of a single noncovalent bond. Thus the loss of a single structurally essential hydrogen bond, or electrostatic or hydrophobic interaction can lead to denaturation of a folded structure. A change in the stability of a noncovalent bond leading to denaturation can in turn be caused by a change in pH, ionic strength, and temperature, which affect the strength of noncovalent bonds. The presence of prosthetic groups, cofactors, and the substrates of a protein may also affect

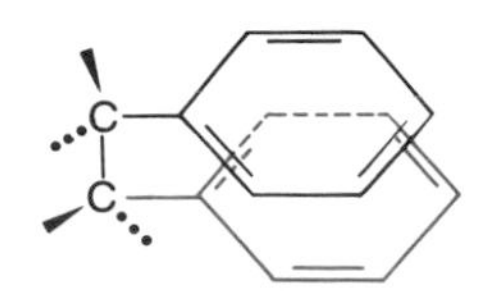

FIGURE 2.65
π Electron–π electron interactions between two aromatic rings.

$$\text{Amino acids} \xrightarrow{\text{rate of synthesis}} [\text{protein}] \xrightarrow{\text{rate of denaturation}} \text{protein digest}$$

FIGURE 2.66
Steady-state concentration of a protein is due to both its rate of synthesis and denaturation.

the stability of the native conformation. These later ligands, which may have a role in the function of the protein, often also act to stabilize the native conformation of a protein.

The statement that the breaking of a single noncovalent bond in a protein can cause denaturation may appear to conflict with the observation, discussed in Section 2.4, that the amino acid sequence for a protein can often be extensively varied without loss of the native structure and related functional ability of the protein. The key to the resolution of the apparent conflict, between the extensive variability of the amino acid sequence present in many proteins and the possible ease of denaturation, is the word "essential." Many noncovalent interactions in a protein are either not essential to the protein's overall thermodynamic stability or when mutated in a conservative manner (see Section 2.4), the new amino acid side chain can perform a similar structural role. However, the substitution or modification of an *essential* nonvariant amino acid residue of a protein that provides a critical noncovalent interaction will dramatically affect the stability of a native protein structure relative to a denatured conformation.

The concentration of a protein in vivo is under the influence of processes that both control the rate of the protein's synthesis and control the protein's rate of degradation (Figure 2.66). Therefore, an understanding of the processes that control protein degradation may be as important as an understanding of the process of protein synthesis. It is believed that the inherent denaturation rate for a protein, in many cases, is the rate-determining step in a protein's degradation. Enzymes and cellular organelles that participate in the digestion of proteins appear to "recognize" denatured protein conformations and digest these denatured conformations at faster rates than proteins in their native conformation.

In experimental situations, denaturation of a protein can often be achieved by addition of urea or detergents (sodium dodecyl sulfate or guanidine hydrochloride) that act to weaken hydrophobic bonding in proteins. Thus these reagents stabilize the denatured state and shift the equilibrium toward the denatured form of the protein. Addition of strong base, acid, or organic solvent, or heating to temperatures above 60°C are also common ways to denature a protein.

2.8 DYNAMIC ASPECTS OF PROTEIN STRUCTURE

The atomic coordinates obtained for a protein structure by X-ray diffraction actually represent time averaged coordinates rather than the exact location of an atom for any instant in time. The experimental evidence from NMR spectroscopy, fluorescence spectroscopy, and X-ray diffraction reveal that the atoms in a folded protein molecule do not exist in a single static position but rather have a **fluidlike dynamic motion.** In addition, while the X-ray structures show the interior of a protein so closely packed that the addition of a single methylene group to a typical side chain would appear to prevent the side chain from fitting into the structure, the X-ray structure also shows small "defects" in the packing of the folded structure indicating the existence of "holes" or small spaces in the

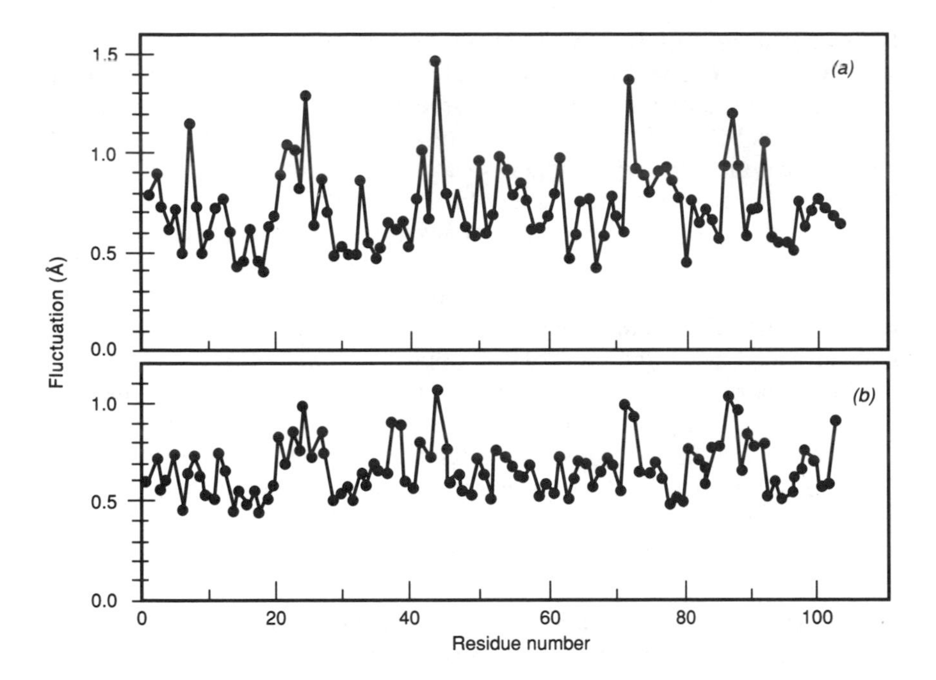

FIGURE 2.67
Calculated and experimentally observed fluctuation of each amino acid residue (averaged) in the primary structure of the protein cytochrome *c*.
(*a*) Molecular dynamic stimulation. (*b*) Based on experimental observation by X-ray crystallography (see Section 2.9). Cytochrome *c* has 103 amino acids.
Redrawn from M. Karplus and J. A. McCammon, *Annu. Rev. Biochem.* 53:263, 1983.

structure that will allow the protein space for flexibility. The concept that each atom in a protein structure is in constant motion such as molecules within a fluid, although constrained by its covalent bonds and secondary and tertiary structure, is an important aspect of protein structure.

Theoretical calculations describe the changes in conformation of regions of structure due to the summation of the movements of atoms in that region. The dynamic motion is computed based on the change of the energy components with time by solving Newton's equations of motion simultaneously for all the atoms of the protein and the atoms of the solvent that interact with the protein. The **energy functions** used in the equation include representations of covalent and noncovalent bonding energies due to electrostatic forces, hydrogen bonding, and van der Waals forces between atoms. The individual atoms are randomly assigned a velocity from a theoretical distribution and Newton's equations are used to "relax" the system at a given "temperature." Computations can be carried out to allow the system to reach an equilibrium temperature (*energy minimization*) or the motion of the system (*dynamic trajectory*) can be followed at a specific temperature. The calculation is computationally intensive, even when limited to less than several hundred picoseconds (ps) of protein dynamic time (1 ps = 10^{-12} s), and requires supercomputers. These calculations indicate that the average atom within a protein is oscillating over a distance of 0.7 Å on the picosecond time scale. Some atoms or groups of atoms will be moving smaller distance and others larger distances than this calculated average (Figure 2.67).

The small amplitude fluctuations are essential motions in proteins. The net movement of any segment of a polypeptide chain over time represents the sum of forces due to rapid, local jiggling and elastic movements of covalently attached groups of atoms. These movements within the closely packed interior of a protein molecule are large enough to allow the planar aromatic rings of buried tyrosines to flip in the interior of a protein. The occurrence of this flip is at least 10^4 times per s. Furthermore, the small amplitude fluctuations provide the "lubricant" for large motions in proteins such as domain motions and quaternary structure changes, like those observed in hemoglobin on O_2 binding (Chapter 3). In the case of the proteolytic digestive enzyme trypsin, a comparison has been made between the atomic oscillations in the active enzyme and its inactive precursor trypsinogen. The trypsinogen molecule has a higher conformational disorder, correlated with a higher degree of internal atomic mobility than trypsin. It is believed that the extent of atomic flexibility in the substrate binding region of the trypsinogen structure is a factor that decreases the ability of the protein to effectively bind and catalytically transform substrate to product. On conversion of trypsinogen to trypsin, the loss in atomic movements at the substrate binding site enhances the ability of the enzyme to act as a catalyst. This gives a functional significance to the relatively high degree of atomic motion for particular regions of the trypsinogen molecule in the control of the protein's activity.

The dynamic behavior of proteins is implicated in conformational changes induced by substrate, inhibitor, or drug binding to enzymes and receptors, generation of allosteric effects in hemoglobins, electron transfer in cytochromes, and in the assembly of supramolecular assemblies such as viruses.

2.9 METHODS FOR THE STUDY OF HIGHER LEVELS OF PROTEIN STRUCTURE AND ORGANIZATION

X-Ray Diffraction Techniques Are Used to Determine the Three-Dimensional Structure of Proteins

The most powerful method for determining the three-dimensional structure of molecules is **X-ray diffraction.** The approach requires formation of a protein crystal but in return affords the opportunity to "observe" macromolecules at near atomic resolution. Our present understanding of the detailed components of secondary, tertiary, and quaternary structure are thus derived from experiments performed in the crystalline state, which, however, correlate with other physical measurements that are based on proteins in solution, such as NMR spectroscopy (see below).

The generation of the crystal remains the most elusive and time consuming aspect of the technique. A crystal exhibits an infinite regular repeat along each of its physical growth directions by successive stable interactions between adjacent molecules. The crystal may include or exclude solvent components, and conformational variability among the different molecular units may occur to provide degrees of freedom necessary to generate a crystalline lattice in which the molecules serve as lattice points.

A significant factor in both the experimental and computational handling of protein crystals, in contrast with most small molecule crystals, stems from the contents of the crystalline material. Small molecules (molecular weight <2000 Da) typically crystallize with fewer than two solvent components, complexed in a stoichiometric manner within spacings along the growth axes. Thus the atoms of the molecules occupy greater than 90% of the volume of the crystal. Proteins typically exhibit molecular dimensions an order of magnitude greater than small molecules, and a packing of protein molecules into crystal lattice points generates a crystal with large "holes" or solvent channels. A protein crystal typically contains 40–60% solvent, and thus may be considered a concentrated solution rather than the hard crystalline solid associated with most small molecules. This attribute proves both beneficial and detrimental, as the presence of solvent and unoccupied volume in the crystal enables the infusion of inhibitors and substrates to the protein molecules in the "crystalline state" but also permits a dynamic flexibility within regions of the protein structure. The flexibility may be seen as "disorder" in the X-ray diffraction experiment. Disorder is used to describe the situation where the observed electron density can be fitted by more than a single local conformation. Two explanations for the apparent disorder exist and must be distinguished. The first one involves the presence of two or more rigid static molecular conformations, for example, rotation about a single bond in the side chain of lysine, which are present in a stoichiometric manner and may or may not be uniformly distributed among the molecules throughout the lattice. The second involves the actual dynamic range of motions exhibited by atoms or groups of atoms in localized regions of the molecule. An experimentally derived distinction can be made by lowering the environmental temperature of the crystal to a point where dynamic disorder is "frozen out," in contrast the **static disorder** is not temperature dependent and persists. Analysis of **dynamic disorder** by its temperature dependency using X-ray diffraction determinations is an important method for studying protein dynamics (see Figure 2.67). Dynamic disorder demonstrates movements within regions of a protein's structure that may have a functional role in the physiological action of the protein.

The first proteins crystallized were from materials that were readily available in large amounts and included hemoglobin, myoglobin, lysozyme, and the pancreatic protease chymotrypsin. Crystallization techniques have advanced so that crystals are now obtainable from much less abundant proteins. Interesting structures have been recently reported for proteins in which specific amino acid residues have been substituted, of antibody–antigen complexes, and of viral products such as the protease required for the infection of the human immunodeficiency virus (HIV) that causes acquired immunodeficiency syndrome (AIDS). Approximately 400 protein structures have now been solved by X-ray diffraction and the details are stored in a data base called the Protein Data Bank, which is readily available.

Diffraction of X-ray radiation by a crystal occurs with the use of incident radiation of a characteristic wavelength (e.g., copper, $K_\alpha = 1.54$ Å). The X-ray beam is diffracted by the electron cloud surrounding the atomic nuclei in the crystal, with an intensity proportional to the number of electrons around the nucleus. Thus the technique establishes the **electron distribution** of the molecule and infers the nuclear distribution. The positions of atomic nuclei can be determined by diffraction with neutron beam radiation, an interesting but very expensive technique as it requires a source of neutrons (nuclear reactor or particle accelerator). With the highest resolution now available for X-ray diffraction determinations of protein structures, the electron diffraction from C,N,O, and S atoms can be observed. The diffraction from hydrogen atoms is not observed due to the low density of electrons (a single electron) around a hydrogen nucleus. Detectors of the diffracted beam, typically photographic film or electronic area detectors, permit the recording of the amplitude (intensity) of the radiation diffracted in a defined orientation. However, the data do not give information about the phases of the radiation, which is essential to the solution of the protein structure. The determination of the **phase angles** for the diffracted radiation historically required the placement of heavy atoms (such as iodine, mercury, or lead) in the protein structure. Modern procedures, however, can often solve the phase problem without use of a heavy atom.

It is convenient to consider the analogy between X-ray crystallography and light microscopy to understand the processes involved in structure determination. In light microscopy, incident radiation is reflected by an object under study and the reflected beam is recondensed by the objective lens to form an image of the object. The analogy is appropriate to incident X-rays with the notable exception that no known material exists that is

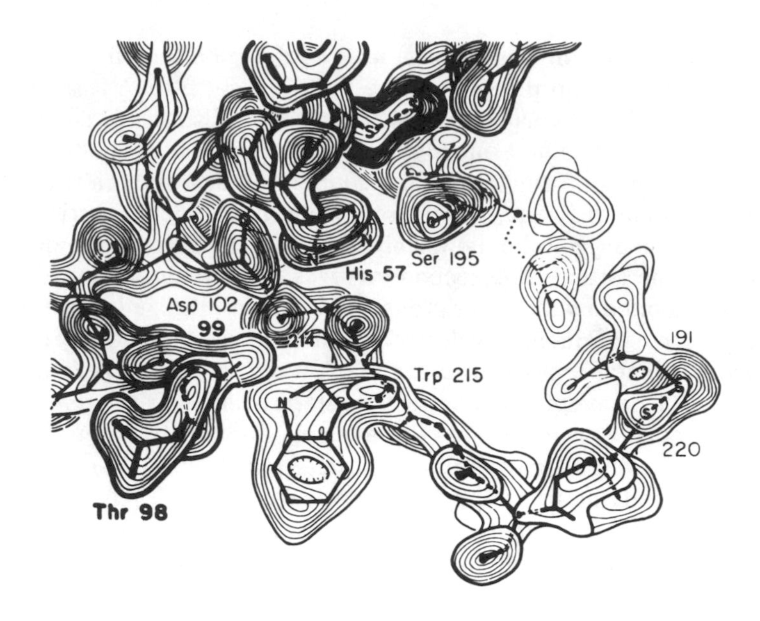

FIGURE 2.68
Electron-density map at 1.9-Å resolution of active site region of proenzyme (trypsinogen) form of trypsin.
The catalytically active residues in trypsin (Asp 102, His 57, and Ser 195) are superposed on the map.
Reprinted with permission from A. A. Kossiakoff, J. L. Chambers, L. M. Kay, and R. M. Stroud, *Biochemistry,* 16:654, 1977. Copyright © 1977, American Chemical Society.

capable of serving as an objective lense for X-ray radiation. To replace the objective lens, amplitude and phase angle measurements of the diffracted radiation are mathematically reconstructed by Fourier synthesis to yield an image of the diffracting material.

The Fourier synthesis generates a three-dimensional electron-density map of the diffracted object. Initially a few hundred reflections are obtained to construct a low-resolution electron-density map. For example, in one of the first protein crystallographic structures, 400 reflections were utilized to obtain a 6-Å map of the protein myoglobin. At this level of resolution it is possible to locate clearly the molecule within the unit cell of the crystal and study the overall packing of the subunits in a protein with a quaternary structure. A trace of the polypeptide chain of an individual protein molecule may be made with difficulty. However, utilizing the low resolution structure as a base, further reflections may be used to obtain higher resolution maps. For myoglobin, whereas 400 reflections were utilized to obtain the 6-Å map, 10,000 reflections were needed for a 2-Å map, and 17,000 reflections for an extremely high-resolution 1.4-Å map. A two-dimensional slice through a three-dimensional electron-den-

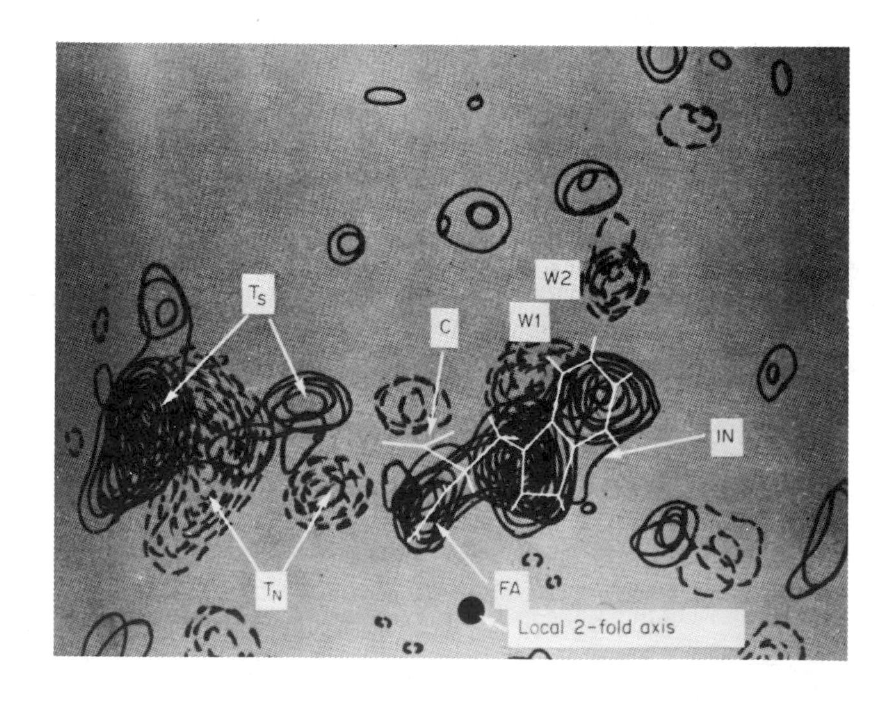

FIGURE 2.69
Difference electron-density map between the structures of the enzyme α-chymotrypsin with substrate analog (*N*-formyl-L-tryptophan) bound to the enzyme active site and native α-chymotrypsin, at 2.5-Å resolution.
The smooth contours (positive density) represent electron density present in the enzyme: substrate complex and not in the native enzyme; the dashed contours (negative density) represent density present in native enzyme and not in the enzyme: substrate complex of the enzyme. Density arising from *N*-formyl-L-tryptophan is at IN (indole) and FA (formylamido) with structure superimposed. Negative density at W1 and W2 is due to water molecules displaced by substrate binding. The negative density T_N is the position of the carboxy terminal of Tyr 146 in the native enzyme, which moves to position T_s in the enzyme: substrate complex.
Reprinted with permission from T. A. Steitz, R. Henderson, and D. M. Blow, *J. Mol. Biol.* 46:337, 1969. Copyrighted © by Academic Press, Inc. (London) Ltd.

sity map of the protein trypsinogen is shown in Figure 2.68. The known primary structure of the protein is fitted to the electron-density pattern (Figure 2.68). The process of aligning a protein's amino acid sequence to the electron-density pattern until the best fit is obtained is known as refinement.

In order to observe changes in a protein's conformation that may occur on the binding of an inhibitor, activator, or substrate molecule of a protein, a **difference electron-density map** may be computed. In this procedure, the diffraction pattern is obtained for the crystalline protein with the ligand bound and is substracted from the electron-density pattern obtained for the protein without the ligand bound. The resulting difference map shows the changes that occur on the binding of the ligand (Figure 2.69).

Whereas X-ray diffraction has provided extensive knowledge on protein structure, it should be emphasized that an X-ray derived structure provides incomplete evidence for a protein's mechanism of action. The X-ray determined structure is an average structure of a molecule in which atoms are normally undergoing rapid fluctuations in solution (see Section 2.8). In any one case, the average crystalline structure determined by X-ray diffraction may not be the active structure of a particular protein in solution. A second important consideration is that it currently takes at least a day to collect data in order to determine a structure. On this time scale, the structures of reactive enzyme–substrate complexes, intermediates, and transition states of enzyme proteins are not observed. Rather, these mechanistically important structures must be inferred from the static pictures of an *inactive* form of the protein or from complexes with inactive analogs of the normally reactive substrates of the protein.

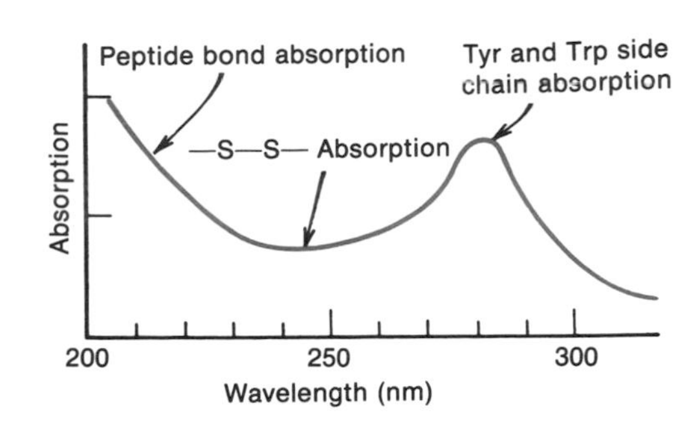

FIGURE 2.70
Ultraviolet absorption spectrum of the globular protein α-chymotrypsin.

Various Spectroscopic Methods Are Employed in Evaluating Protein Structure and Function

Ultraviolet Light Spectroscopy

The side chain groups of tyrosine, phenylalanine, tryptophan, and cystine, as well as the peptide bonds in proteins, can absorb ultraviolet (UV) light. The efficiency of light energy absorption for each of these different types of absorbing chromophores is related to a molar extinction coefficient (ε), which has a characteristic value for each type of chromophoric group.

A typical protein spectrum is shown in Figure 2.70. The absorbance between 260 and 300 nm is due to phenylalanine R groups, tyrosine R groups, and tryptophan R groups. The **molar extinction coefficients** for these chromophoric amino acids are plotted in Figure 2.71. When the

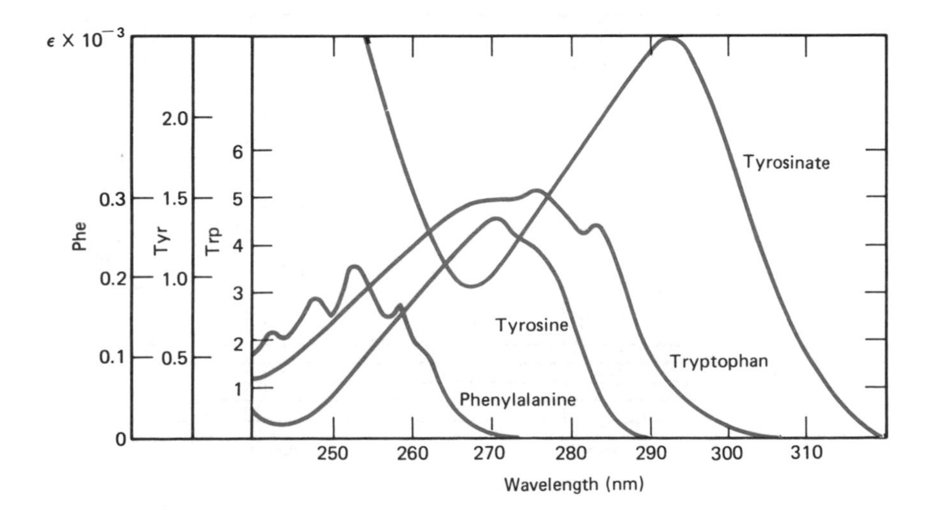

FIGURE 2.71
Ultraviolet absorption for the aromatic chromophores in Phe, Tyr, Trp, and tyrosinate.
Note differences in extinction coefficients on left axis for the different chromophores.
Redrawn from A. d'Albis and W. B. Gratzer, in *Companion to Biochemistry,* A. T. Bull, J. R. Lagmado, J. O. Thomas, and K. F. Tipton (Eds.). London: Longmans, 1974, p. 170.

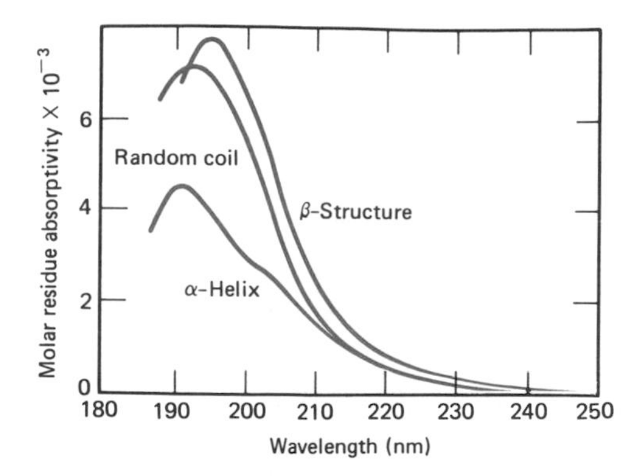

FIGURE 2.72
Ultraviolet absorption of the peptide bonds of a polypeptide chain in α-helical, random-coil, and antiparallel β-structure conformations.
Redrawn from A. d'Albis and W. B. Gratzer, in *Companion to Biochemistry,* A. T. Bull, J. R. Lagmado, J. O. Thomas, and K. F. Tipton (Eds.). London: Longmans, 1970, p. 175.

tyrosine side chain is ionized at high pH (tyrosine R group $pK_a' \simeq 10$), the absorbance for tyrosine is shifted to higher wavelength (red shifted) and its molar absorptivity is increased (Figure 2.71).

The peptide bond absorbs in the far-UV (180–230 nm). A peptide bond in a helix conformation interacts with the electrons of other peptide bonds above and below it in the spiral conformation to create an **exciton system** in which electrons are delocalized. The result is a shift of the absorption maximum from that of an isolated peptide bond to either a lower or higher wavelength (Figure 2.72).

Thus UV spectroscopy can be used to study changes in a protein's secondary and tertiary structure. As a protein is denatured (helix unfolded), differences are observed in the absorption characteristics of the peptide bonds between 180 and 230 nm due to the disruption of the exciton system. In addition, the absorption maximum for an aromatic chromophore appears at a lower wavelength in an aqueous environment than in a nonpolar environment.

The molar absorbancy of a chromophoric substrate or ligand will often change on binding to a protein. This change in the binding molecule's extinction coefficient can be used to measure its binding constant. Changes in chromophore extinction coefficients during enzyme catalysis of a chemical reaction can often be used to obtain the kinetic parameters for the reaction.

Fluorescence Spectroscopy

The energy of an excited electron produced by light absorption can be lost by a variety of mechanisms. Most commonly the excitation energy is dissipated as thermal energy in a collision process. In some chromophores the excitation energy is dissipated by fluorescence.

The fluorescent emission is always at a longer wavelength of light (lower energy) than the absorption wavelength of the fluorophore. This is because vibrational energy levels formed in the excited electron state during the excitation event are lost during the time it takes the fluorescent event to occur (Figure 2.73).

In the presence of a second chromophore that can absorb light energy at the wavelength of a fluorophore's emission, the fluorescence that is normally emitted may not be observed. Rather, the fluorescence energy can be transferred to the second molecule. The acceptor molecule, in turn, can then either emit its own characteristic fluorescence or lose its

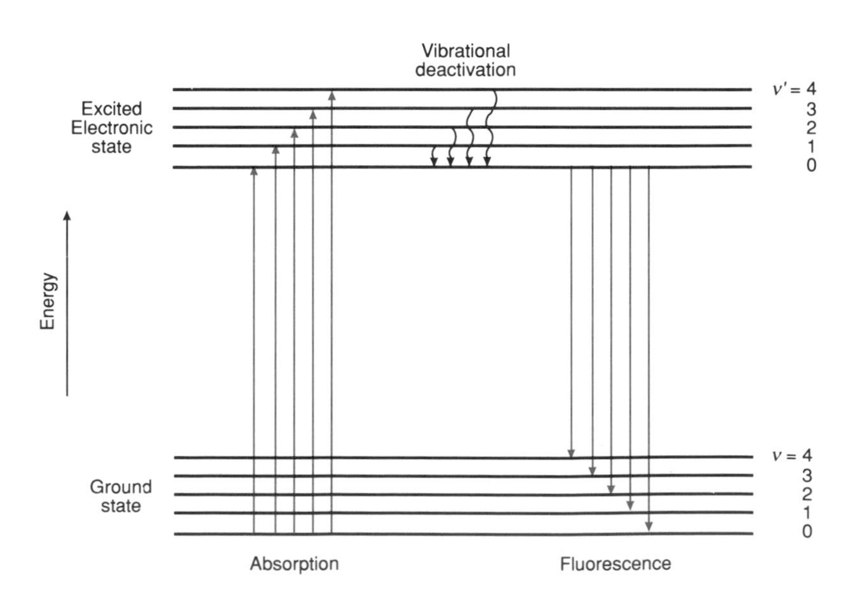

FIGURE 2.73
Absorption and fluorescence electronic transitions.
Excitation is from the zero vibrational level in the ground state to various higher vibrational levels in the excited state. Fluorescence is from the zero vibrational level in the excited electronic state to various vibrational levels in the ground state.
Redrawn from A. d'Albis and W. B. Gratzer, in *Companion to Biochemistry,* A. T. Bull, J. R. Lagmado, J. O. Thomas, and K. F. Tipton (Eds.). London: Longmans, 1970, p. 166.

excitation energy by an alternative process. If the acceptor molecule loses its excitation energy by a nonfluorescent process, it is acting as a **quencher** of the donor molecule's fluorescence. The efficiency of excitation transfer is dependent on the distance and the orientation between donor and acceptor molecules as well as the degree of overlap between the emission wavelengths of the donor molecule and the absorption wavelengths characteristic of the acceptor molecule.

The fluorescence emission spectra for phenylalanine, tyrosine, and tryptophan side chains are shown in Figure 2.74. A comparison of their emission and absorption spectra (Figure 2.74) show that the emission wavelengths for phenylalanine overlaps with the absorption wavelengths for tyrosine. In turn, the emission wavelengths for tyrosine overlap with the absorption wavelengths for tryptophan. Because of the overlap in emission and absorption wavelengths, primarily only the tryptophan fluorescence is observed in proteins that contain all three of these types of amino acids.

Excitation energy transfers occur over distances up to 80 Å, which are typical diameter distances in folded globular proteins. When a protein is denatured the distances between donor and acceptor groups become greater. The increased distance between donor and acceptor groups decreases the efficiency of energy transfer. Accordingly, an increase in the intrinsic fluorescence of the tyrosines and/or phenylalanines to the protein's emission spectrum will be observed with denaturation.

Since excitation transfer processes in proteins are distance- and orientation dependent, the fluorescence yield is dependent on the conformation of the protein. As such, fluorescence is a highly sensitive tool with which to study protein conformation and changes in a protein's conformation related to its function.

Some common prosthetic groups in enzyme proteins are fluorophores and changes in fluorescence yields from enzymes that contain these prosthetic moieties can be used to follow the chemical reactions catalyzed by the enzymes.

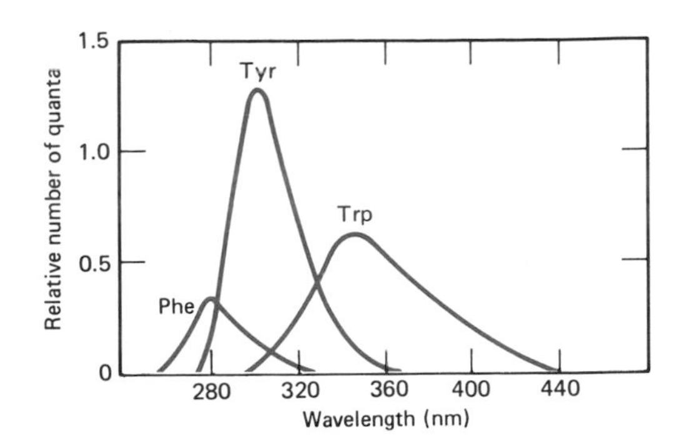

FIGURE 2.74
Characteristic fluorescence of aromatic groups in proteins.
Redrawn from A. d'Albis and W. B. Gratzer, in *Companion to Biochemistry,* A. T. Bull, J. R. Lagmado, J. O. Thomas, and K. F. Tipton (Eds.). London: Longmans, 1970, p. 478.

Optical Rotatory Dispersion and Circular Dichroism Spectroscopy

Optical rotation is caused by differences in the *refractive index* encountered by the clockwise and counterclockwise vector components of a beam of polarized light in a solution containing an asymmetric solute. **Circular dichroism** (CD) is caused by differences in the *light absorption* between the clockwise and counterclockwise component vectors of a beam of polarized light. In proteins the aromatic amino acids of asymmetric configuration give an optical rotation and CD. Also the polypeptide chains in regular helical conformation will form either a right-handed or left-handed direction spiral conformation. These two spiral conformations are not superimposable, and they generate a significant optical rotation and CD.

Circular dichroism spectra for different conformations of the polypeptide chain are shown in Figure 2.75. Because of the differences, circular dichroism is a sensitive assay for the amount and type of secondary structure in a protein.

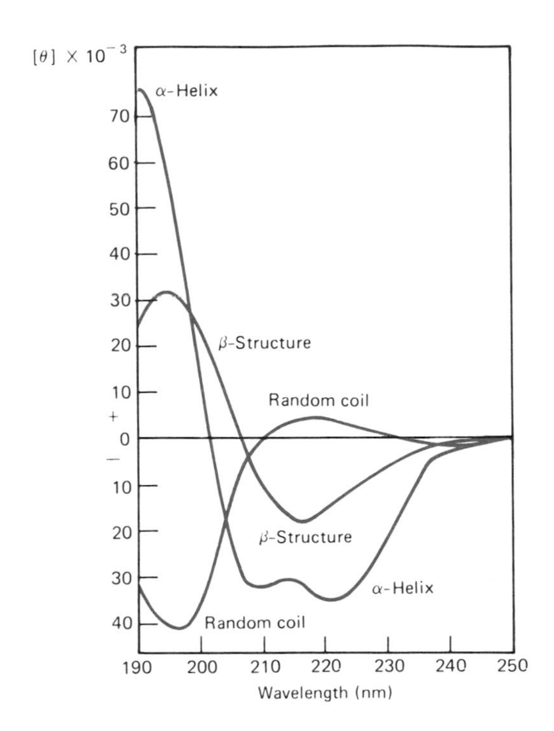

FIGURE 2.75
Circular dichroism spectra for polypeptide chains in α helical, β structure, and random-coil conformations.
Redrawn from A. d'Albis and W. B. Gratzer, in *Companion to Biochemistry,* A. T. Bull, J. R. Lagmado, J. O. Thomas, and K. F. Tipton (Eds.). London: Longmans, 1970, p. 190.

Nuclear Magnetic Resonance

Using the newer technique of **two-dimensional** (2D) **NMR** with powerful NMR spectrometers it is possible to obtain the solution conformation of small proteins. At this time it is confined to proteins of less than 100 amino acids, but one major advantage is that crystallization is not required. A high concentration of protein (~1-mM solution) is, however, required.

Conventional NMR techniques involve the use of radio frequency (rf) radiation to study the environment of atomic nuclei that are magnetic. The material under study is placed in a homogeneous magnetic field and subjected to a short pulse of a specific frequency. The requirement for magnetic nuclei is absolute and is based on an unpaired spin state in the nucleus. Thus the naturally abundant carbon (^{12}C) and oxygen (^{16}O) do not absorb, while ^{13}C and ^{17}O do absorb.

The actual absorption bands in an NMR spectrum are characterized by: (1) a position or *chemical shift* value, reported as the frequency difference between that observed for a specific absorption band and that for a standard reference material; (2) the *intensity* of the peak or integrated area, which is proportional to the total number of absorbing nuclei; (3) the half-height *peak width,* which reflects the degree of motion in solution of the absorbing species; and (4) the *coupling constant,* which measures the extent of direct interaction or influence of neighboring nuclei on the absorbing nuclei. These four measurements enable the determination of the identity and number of nearest neighbor groups that can affect the response of the absorbing species through bonded interactions. They give no information on through-space (nonbonded) interaction due to the three-dimensional structure of the protein. To determine through-space interaction and protein tertiary structure requires additional techniques. Recent enhancements to NMR, which are applicable to the determination of protein structure, include the ability to synthesize proteins containing isotopically enriched (e.g., containing ^{13}C or ^{15}N) amino acids, the development of paramagnetic shift reagents to study localized environments on paramagnetic resonances, such as the lanthanide ion reporting groups, the use of **nuclear Overhauser effects** (NOEs) to permit the determination of through-space nearest neighbor interactions, and the application of the two-dimensional technique.

The development of **two-dimensional proton** (^{1}H) **NMR** has proven the most significant in the evolution of the method for actual protein structure determination. The major difference in two-dimensional versus one-

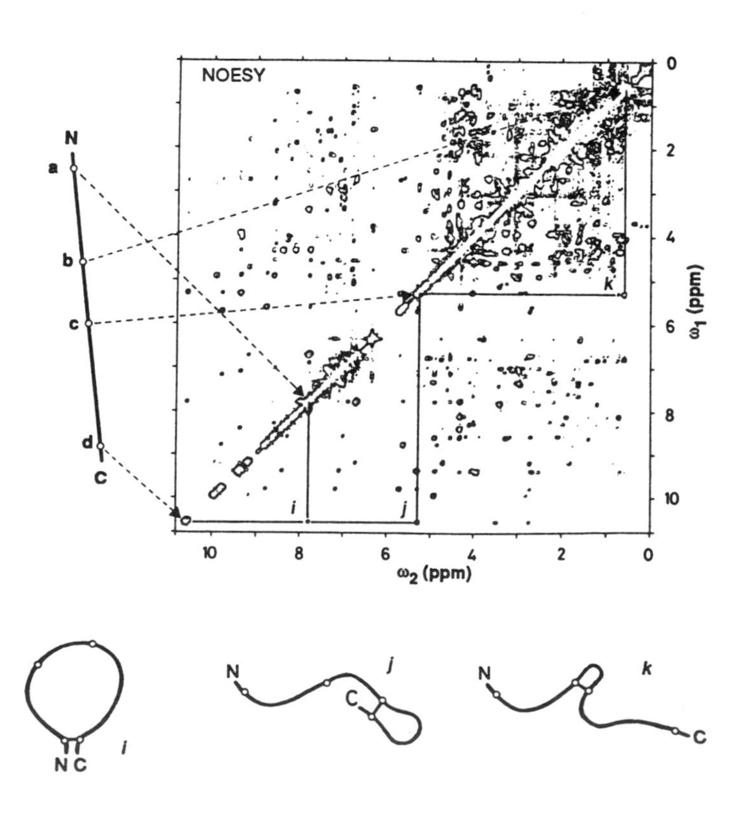

FIGURE 2.76
Two-dimensional NMR Spectrum of the protein pancreatic trypsin inhibitor.
The spectrum (nuclear Overhauser affect spectroscopy or NOESY) was taken at 500 MHz. A countour plot is shown with two frequency axes ω_1 and ω_2. The cross peaks are marked *i, j,* and *k* and are connected by horizontal and vertical lines. The straight line on the left represents an extended polypeptide chain with the NH_2 terminal identified as N and the COOH terminal identified as C, and four proton absorbancies (a–d) identified on the diagonal. By a simple geometric pattern, each cross peak establishes a correlation between peaks identified on the diagonal. Thus the spectrum shows that the following proton absorbances are within 5 Å of one another: a and d, c and d, and b and c. Below the spectrum three structures are schematically drawn representing these distance relationships.
Reprinted with permission from K. Würthrich, *Science* 243:45, 1989.

dimensional (1D) NMR is the addition of a second time delay pulse. The usefulness of the two-dimensional technique is in systems where spins of two or more nuclei are coupled, as occur in the complex protein structure. The data are collected in a two-dimensional array (Figure 2.76) to plot chemical shift versus chemical shift or chemical shift versus coupling constants. The plot of the data reveals the normal spectrum occurring along the diagonal of the two-dimensional plot (Figure 2.76). The technique first requires the identification in the spectrum of a proton absorbance from a particular position in the protein structure. Then through the combined approach of **NOESY** (nuclear Overhauser effect spectroscopy) and **COSY** (correlated spectroscopy) the assignment of ^1H NMR peaks to sequence related positions is achieved. In the two-dimensional plot, the peaks that occur as off-diagonal indicate the interaction between nuclei that occur through-space. A maximum distance of approximately 5 Å is the limit for which these through-space interactions can be observed.

Upon the generation of distance information for interresidue pairs through the protein structure, it is possible to generate three-dimensional protein conformations consistent with the spectra. The method used to accomplish this is termed *distance geometry* and is related to the process of forming a three-dimensional structure from a distance matrix. In this calculation, a distance matrix is constructed filled with ranges of distances (minimum and maximum) for as many interresidue interactions as may be measured. A number of possible structures are generated from the data consistent with the constraints imposed by the NMR spectra. Computational refinements of the initially calculated structures can be made to optimize covalent bond distances and bond angles. The method generates a family of structures, the variability showing either the imprecision of the technique or the dynamic "disorder" of the folded structure (see Figure 2.77). Such computations based on NMR experiments have yielded structures for proteins that do not significantly differ from the time-averaged structure observed with X-ray diffraction methods.

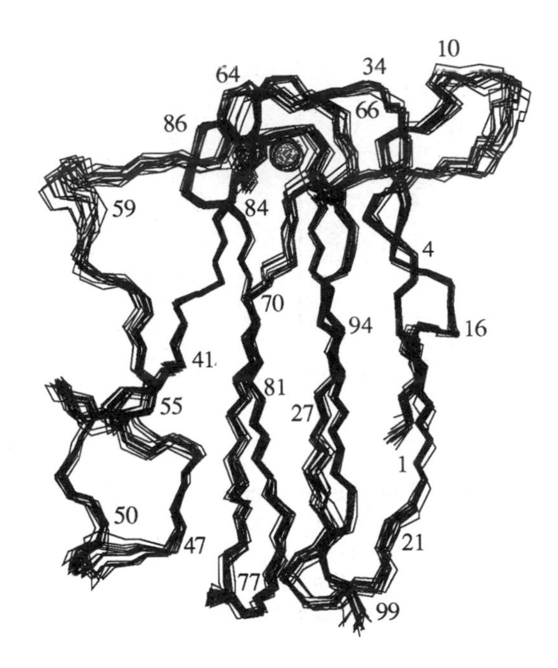

FIGURE 2.77
NMR Structure of the protein plastocyanin from the French bean.
The structures shows the superposition of eight structures of the polypeptide backbone for the protein, calculated from the constraints of the NMR spectrum.
J. M. Moore, C. A. Lepre, G. P. Gippert, W. J. Chazin, D. A. Case and P. E. Wright, *J. Mol. Biol.* 221:533, 1991. Figure generously supplied by P. E. Wright.

Proteins Can Be Separated and Characterized Based on Molecular Weight or Size

Ultracentrifugation: Definition of Svedberg Coefficient

A protein subjected to centrifugal force will move in the direction of the force at a velocity dependent on the protein's mass. The rate of movement can be measured with the appropriate optical system, and from the measured rate of movement the sedimentation coefficient (s) calculated in **Svedberg units** (units of 10^{-13} s). In the equation, which can be used to calculate a sedimentation coefficient for a molecule (Figure 2.78), γ is the measured velocity of protein movement, ω the angular velocity of the centrifuge rotor, and r the distance from the center of the tube in which the protein is placed to the center of rotation. Sedimentation coefficients between 1 and 200 Svedberg units (S) have been found for proteins (Table 2.17).

$$s = \frac{\gamma}{\omega^2 r}$$

FIGURE 2.78
Equation for calculation of the Svedberg coefficient.

Equations have been derived to relate the sedimentation coefficient to the molecular weight for a protein. One of the more simple equations is shown in Figure 2.79, in which R is the gas constant, T the temperature, s the sedimentation coefficient, D the diffusion coefficient of the protein, $\bar{v}$ the partial specific volume of the protein, and ρ the density of the solvent. The quantities D and $\bar{v}$ must be measured in independent experiments. In addition, *the equation assumes a spheroidal geometry for the protein.* Because the assumption of a spheroidal geometry may not be true for any particular case, and independent measurements of D and $\bar{v}$ are difficult,

$$\text{Molecular weight} = \frac{RT_s}{D(1 - \bar{v}\rho)}$$

FIGURE 2.79
An equation relating the Svedberg coefficient to molecular weight.

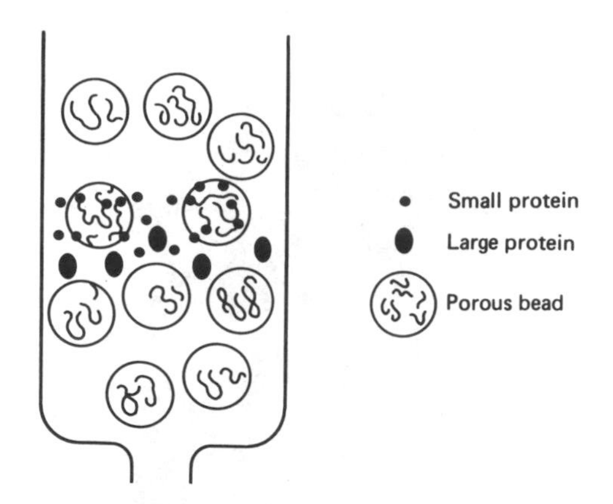

FIGURE 2.80
Molecular exclusion chromatography.
The small protein can enter the porous gel particles and will be retarded on the column with respect to the larger protein that cannot enter the porous gel particles.

TABLE 2.17 Svedberg Coefficients for Some Plasma Proteins of Different Molecular Weights

Protein	s_{20}, $\times 10^{-13}$ cm s^{-1} dyn^{-1} (*a*)	Molecular weight
Lysozyme	2.19	15,000–16,000
Albumin	4.6	69,000
Immunoglobulin G	6.6–7.2	153,000
Fibrinogen	7.63	341,000
C1q (factor of complement)	11.1	410,000
α_2-Macroglobulin	19.6	820,000
Immunoglobulin M	18–20	1,000,000
Factor VIII of blood coagulation	23.7	1,120,000

SOURCE: *CRC Handbook of Biochemistry and Molecular Biology*, 3rd ed., Sect. A, Vol. II, G. D. Fasman (Ed.), Boca Raton, FL, CRC Press, Inc., 1976, p. 242.

[a] s_{20}, $\times 10^{-13}$ is sedimentation coefficient in Svedberg units, referred to water at 20°C, and extrapolated to zero concentration of protein.

often only the sedimentation coefficient for a molecule is reported. The magnitude of the protein's sedimentation coefficient will give a relative value that can be used in a qualitative way to characterize a protein's molecular weight.

Molecular Exclusion Chromatography

A porous gel in the form of small insoluble beads is commonly used to separate proteins by size in column chromatography. Small protein molecules can penetrate the pores of the gel and will have a larger solvent volume through which to travel in the column than large proteins, which are sterically excluded from the pores. Accordingly, a protein mixture will be separated by size, the larger proteins eluted first, followed by the smaller proteins, which are retarded by their accessibility to a larger solvent volume (Figure 2.80).

As with ultracentrifugation, an assumption must be made as to the geometry of the unknown protein, and nonspheroid proteins will give anomalous molecular weights when compared to standard proteins of spherical conformation.

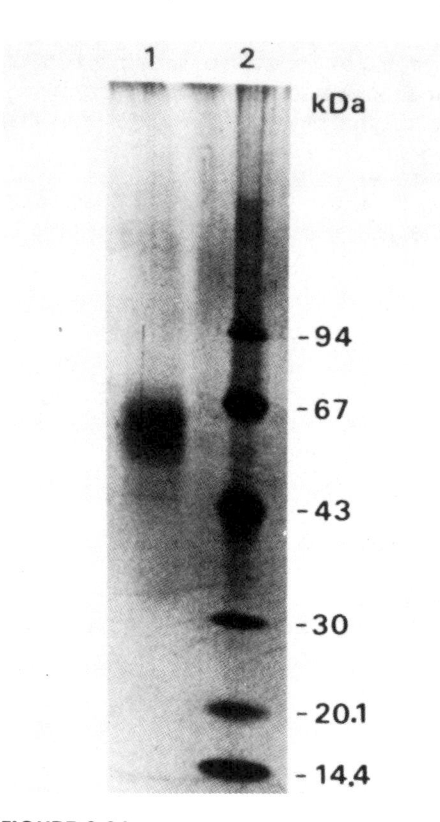

FIGURE 2.81
Example of SDS–polyacrylamide gel electrophoresis.
Molecular weight determination for the urokinase receptor protein isolated from the membrane of human lymphoma cells. Molecular weight standard proteins are shown in column 2 with membrane receptor protein in column 1. Its molecular weight is estimated between 55–60 kDa.
Reprinted with permission from N. Behrendt et al. *J. Biol. Chem.* 265:6453, 1990.

Polyacrylamide Gel Electrophoresis in the Presence of a Detergent

If a charged detergent is added to an electrophoresis buffer and a protein is electrophoresed on a sieving support, a separation of proteins occurs based on protein size but not charge. A detergent commonly used in protein electrophoresis based on size is sodium dodecyl sulfate (SDS). The dodecyl sulfates are amphiphilic C_{12} alkyl sulfate molecules that act to denature the protein and form a charged micelle about the denatured molecule. The inherent charge of the native protein is obliterated by the charged micelle layer of SDS, and each protein–SDS solubilized aggregate has an identical charge per unit volume due to the charge characteristics of the SDS micelle. The negatively charged micelle particles will move through an electrophoresis gel toward the anode (+ pole). A common gel for SDS electrophoresis is cross-linked polyacrylamide. In the migration toward the positive pole, polyacrylamide acts as a molecular sieve and the protein–micelle complexes are separated by size. As the proteins are denatured, artifacts caused by nonspheroid shapes of protein native structures will not be significant (Figure 2.81).

It should be realized that the detergents dissociate quaternary structure into its constituent subunits, and only the molecular weights of covalent subunits are determined by this method.

Chromatographic Techniques Can Separate Proteins on the Basis of Charge and Reactive Groups

In **high-performance liquid chromatography** (HPLC), a liquid solvent containing a mixture of components to be identified or purified is passed through a column densely packed with a small-diameter insoluble beadlike resin. In column chromatography, the smaller and more tightly packed the resin beads, the greater the resolution of the separation technique. In this technique, the resin is so tightly packed that in order to overcome the resistance the liquid must be pumped through the column at high pressure. Therefore, HPLC uses precise high-pressure pumps with metal plumbing and columns rather than glass and plastics as used in gravity chromatography. The resin beads can be coated with charged groupings to separate compounds by ion exchange or with hydrophobic groupings to retard nonpolar molecules passed through the resin. In hydrophobic chromatography nonpolar compounds can be eluted from the hydrophobic beads in aqueous eluents containing various percentages of an organic reagent. The higher the percentage of organic solvent in the eluent, the faster the nonpolar component is eluted from the hydrophobic resin. This latter type of chromatography over nonpolar resin beads is called reverse-phase HPLC (see Figure 2.28 for data showing separation of dansyl amino acids by reverse-phase HPLC). Alternatively, the beads can be porous in order to separate large molecules from smaller ones on the basis of size by molecular exclusion chromatography. The HPLC separations are carried out with extremely high resolution and reproducibility in column retention times.

Affinity Chromatography

Proteins have a high affinity for their substrates, prosthetic groups, membrane receptors, specific noncovalent inhibitors, and specific antibodies made against them. These high affinity compounds can be covalently attached to an insoluble resin and utilized to purify a protein in column chromatography. In a mixture of compounds eluted through the resin, the protein of interest will be selectively retarded.

General Approach to Protein Purification

A protein must be purified prior to a meaningful characterization of its chemical composition, structure, and function. As living cells contain thousands of genetically distinct proteins, the purification of a single protein from a mixture of cellular molecules may be difficult.

The first task in the purification of a protein is the development of a facile assay for the protein. A protein assay, whether it utilizes the rate of a substrate's transformation to a product, an antibody–antigen reaction, or a physiological response in an animal assay system, must in some way give a quantitative measurement of activity per unit weight for a particular sample containing the protein. This quantity is known as the sample's specific activity. The purpose of the purification is to increase a sample's specific activity to the value equal to the specific activity expected for the pure protein.

A typical protocol for purification of a soluble cellular protein may first involve the disruption of the cellular membrane, followed by a differential centrifugation in a density gradient to isolate the protein activity from subcellular particles and high molecular weight aggregates. A further purification step may utilize selective precipitation by addition of inorganic salts (salting out) or addition of miscible organic solvent to the solution containing the protein. Final purification will include a combination of techniques previously discussed, which include methods based on molecular charge (Section 2.3), molecular size, and affinity chromatography.

BIBLIOGRAPHY

Physical and Structural Properties of Proteins

Chothia, C. Principles that determine the structure of proteins. *Annu. Rev. Biochem.* 53:537, 1984.

Eisenberg, D. and McLachlan, A. D. Solvation energy in protein folding and binding. *Nature (London)* 319:199, 1986.

Fasman, G. D. Protein conformational prediction. *Trends Biochem. Sci.* 14:295, 1989.

Jennings, M. L. Topography of membrane proteins. *Annu. Rev. Biochem.* 58:999, 1989.

Kantrowitz, E. R. and Lipscomb, W. N. *Escherichia coli* aspartate transcarbamylase: the relation between structure and function. *Science* 241:669, 1988.

Richards, F. M. and Richmond, T. Solvents, interfaces and protein structure. *Molecular Interactions and Activity in Proteins,* Ciba Foundation Symposium 60 (new series), Amsterdam: Excerpta Medica, 1978, p. 23.

Richardson, J. S. The anatomy and taxonomy of protein structure. *Adv. Protein Chem.* 34:168, 1981.

Protein Folding

Anfinsen, C. B. and Scheraga, H. Experimental and theoretical aspects of protein folding. *Adv. Protein Chem.* 29:205, 1975.

Baldwin, R. L. How does protein folding get started? *Trends Biochem. Sci.* 14:291, 1989.

Fischer, G. and Schmid, F. X. The mechanism of protein folding. Implications of *in vitro* refolding models for *de novo* protein folding and translocation in the cell. *Biochem.* 29:2205, 1990.

Jaenicke, R. Protein folding: local structures, domains, subunits and assemblies. *Biochem.* 30:3147, 1991.

Rothman, J. E. Polypeptide chain binding proteins: catalysts of protein folding and related processes in cells. *Cell* 59:591, 1989.

Spolar, R. S., Ha, J.-H., and Record, M. T., Jr. Hydrophobic effect in protein folding and other noncovalent processes involving proteins. *Proc. Natl. Acad. Sci. U.S.A.* 86:8382, 1989.

Techniques for the Study of Proteins

Hunkapiller, M. W., Strickler, J. E., and Wilson K. J. Contemporary methodology for protein structure determination. *Science* 226:304, 1984.

Lakowicz, J. R. Fluorescence spectroscopy; principles and application to biological macromolecules. *Modern Physical Methods in Biochemistry, Part B,* A. Neuberger and L. L. M. Van Deenen (Eds.), 1988. Amsterdam: Elsevier, pp. 1–26.

Shively, J. E., Paxton, R. J., and Lee, T. D. Highlights of protein structural analysis. *Trends Biochem. Sci.* 14:246, 1989.

Wilson, K. J. Micro-level protein and peptide separations. *Trends Biochem. Sci.* 14:252, 1989.

NMR

Bax, A. Two-dimensional NMR and protein structure. *Annu. Rev. Biochem.* 58:223, 1989.

Markley, J. One and two dimensional NMR, spectroscopic investigations of the consequences of amino acid replacements in proteins. *Protein Engineering,* D. L. Oxender and C. F. Fox (Eds.), New York: Alan R. Liss, Inc., 1987, pp. 15–33.

Wright, P. E. What can two-dimensional NMR tell us about proteins? *Trends Biochem. Sci.* 14:255, 1989.

Wüthrich, K. Protein structure determination in solution by nuclear magnetic resonance spectroscopy. *Science* 243:45, 1989.

X-ray Crystallography

Eisenberg, D. and Hill, C. P. Protein cyrstallography: more surprises ahead. *Trends Biochem. Sci.* 14:260, 1989.

Johnson, L. N. Protein crystallography. *Modern Physical Methods in Biochemistry, Part A,* A. Neuberger and L. L. M. Van Deenen (Eds.), Amsterdam: Elsevier, 1985, pp. 347–415.

Dynamics in Folded Proteins

Joseph, D., Petsko, G. A., and Karplus, M. Anatomy of a conformational change: hinged lid motion of the triosephophate isomerase loop. *Science* 249:1425, 1990.

Karplus, M. and McCammon, J. A. Dynamics of proteins: elements and function. *Ann. Rev. Biochem.* 53:263, 1983.

Karplus, M. and McCammon, J. A. The dynamics of proteins. *Sci. Am.* 254:42, 1986.

Karplus, M. and Petsko, G. A. Molecular dynamics simulations in biology. *Nature* 347:631, 1990.

Levitt, M. and Sharon, R. Accurate simulation of protein dynamics in solution. *Proc. Natl. Acad. Sci. U.S.A.* 85:7557, 1988.

Glycoproteins

Paulson, J. C. Glycoproteins: what are the sugar chains for? *Trends Biochem. Sci.* 14:272, 1989.

Sharon, N. and Lis, H. Glycoproteins. *The Proteins.* 3rd ed., Vol. 5, H. Neurath and R. L. Hill (Eds.), New York: Academic Press, 1982, pp. 1–144.

Lipoproteins

Gotto, A. M., Jr. (Ed.), Plasma Lipoproteins. *New Comprehensive Biochemistry,* A. Neuberger and L. L. M. van Deenen (Eds.), Amsterdam: Elsevier, 1987.

Kostner, G. M. Apolipoproteins and lipoproteins of human plasma: Significance in health and in disease. *Adv. Lipid. Res.* 20:1, 1983.

Collagen

Kuivaniemi, H., Tromp, G., and Prockop, D. J. Mutations in collagen genes: causes of rare and some common diseases in humans. *FASEB J.* 5:2052, 1991.

Miller, E. J. Chemistry of collagens and their distributions. *Extracellular matrix biochemistry.* K. A. Piez and A. H. Reddi (Eds.), New York: Elsevier, 1984, pp. 41–81.

Piez, K. A. Molecular and aggregate structures of collagens. *Extracellular matrix biochemistry.* K. A. Piez and A. H. Reddi (Eds.), New York: Elsevier, 1984, pp. 1–39.

Shaw, L. M. and Olsen, B. R. Facit collagens: diverse molecular bridges in extracellular matrices. *Trends Biochem. Sci.* 16:191, 1991.

QUESTIONS

J. BAGGOTT AND C. N. ANGSTADT

*Q*uestion *T*ypes are described on page xxiii.

Refer to the following structure for Questions 1 and 2.

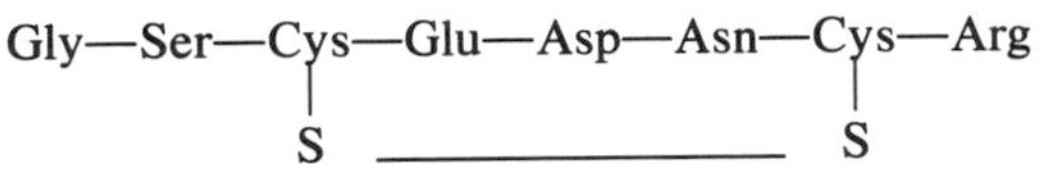

Gly—Ser—Cys—Glu—Asp—Asn—Cys—Arg
S ——————— S

1. (QT2) The peptide shown above:
   1. has arginine in position 1 of the sequence.
   2. contains a derived amino acid.
   3. is basic.
   4. consists primarily of amino acids that are either of intermediate polarity or charged at pH 7.

2. (QT1) The charge on the peptide shown above is about:
   A. −2 at pH > 13.5
   B. −1 at pH ~ 11.5
   C. +1 at pH ~ 6.5
   D. +2 at pH ~ 5.5
   E. 0 at pH ~ 4.5

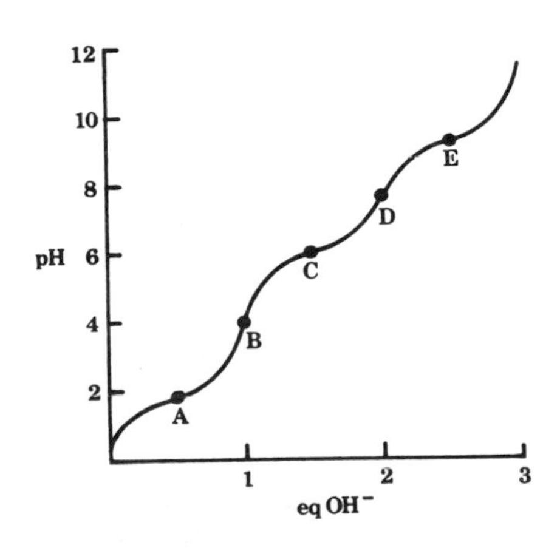

3. (QT2) The figure above shows the titration curve of one of the common amino acids. From this curve we can conclude:
   1. at point B the amino acid is zwitterionic.
   2. point D corresponds to the pK_a' of an ionizable group.
   3. the amino acid contains two carboxyl groups.
   4. at point E the amino acid has a net negative charge.
4. (QT2) Which of the following can be used for a quantitative determination of amino acids in general?
   1. Pauly's reagent
   2. Fluorescamine
   3. The Sakaguchi reaction
   4. Ninhydrin

5. (QT3) A. Freedom of rotation of the peptide bond in an α-helix.
   B. Freedom of rotation of the peptide bond in a "random" structure.

6. (QT2) Which of the following has quaternary structure?
   1. myoglobin
   2. insulin
   3. α-chymotrypsin
   4. hemoglobin

Refer to the drawing for Questions 7 and 8.

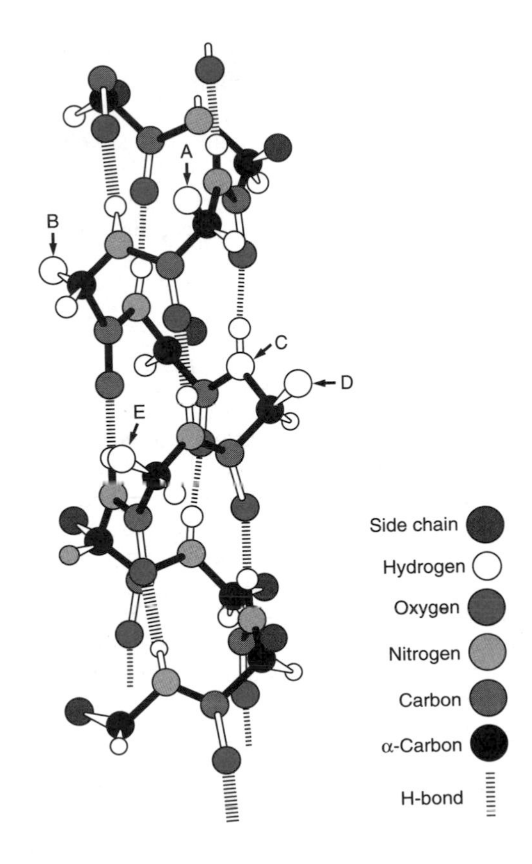

7. (QT2) When group E contains a negatively charged carboxyl function, the structure is destabilized by:
   1. aspartate at position D.
   2. glutamate at position B.
   3. alanine at position A.
   4. proline at position D.

8. (QT1) The properties of atom C are essential to which stabilizing force in the structure?
   A. hydrogen bonding
   B. steric effects
   C. ionic attraction
   D. disulfide bridge
   E. none of the above

9. (QT1) When protein subunits combine to form a quaternary structure, all of the following interactions may arise *except:*
   A. disulfide bond formation.
   B. hydrogen bonding.
   C. hydrophobic interaction.
   D. electrostatic interactions.
   E. van der Waals forces.

10. (QT3) A. Contribution of hydrophobic interactions to the thermodynamic stability of proteins.
    B. Contribution of hydrogen bonding to the thermodynamic stability of proteins.
11. (QT3) A. Number of X-ray reflections necessary for a low-resolution electron-density map.
    B. Number of X-ray reflections necessary for a high-resolution electron-density map.
12. (QT3) A. Molar absorptivity of tyrosine at pH 7.
    B. Molar absorptivity of tyrosine at pH 11.
13. (QT3) A. Wavelength of maximum absorption by a fluorophore.
    B. Wavelength of maximum emission by the same fluorophore.

14. (QT2) Proteins may be separated according to size by:
    1. isoelectric focusing.
    2. polyacrylamide gel electrophoresis in the presence of SDS.
    3. ion-exchange chromatography.
    4. molecular exclusion chromatography.

A. Primary structure
B. Secondary structure
C. Tertiary structure
D. Quaternary structure
E. Random conformation

15. (QT5) All ϕ angles are equal and all ψ angles are equal.
16. (QT5) May bring distant segments of a single polypeptide chain into close juxtaposition.
17. (QT5) Unaffected by binding of a charged detergent, such as SDS.
18. (QT5) Circular dichroism spectrum exhibits a maximum in the 210–220-nm region.
19. (QT5) Exemplified by the β structure (pleated sheet).

ANSWERS

1. C 2 and 4 true. 1: The convention is to write the N-terminal to the left. Numbering begins at the N-terminal, so glycine is in position 1 (p. 33). 2: Cystine, formed by joining two cysteine residues through a disulfide bridge, is a derived amino acid (p. 33). 3: The peptide contains two acidic amino acids, glutamate and aspartate, and only one basic amino acid, arginine, so it is acidic (p. 35). 4: Only cysteine is nonpolar; all the others are of intermediate polarity or are charged (p. 30).
2. E At pH 4.5 the peptide is in the following ionic state: The N-terminal amino group is +1, the side chain carboxyls of glutamate and aspartate each average about −0.5 (since this pH is about at their pK values), the side chain of arginine is +1, and its terminal carboxyl group is −1. The sum is zero (pp. 36).
3. D Only 4 true. The axes of this titration curve are reversed from the presentation in the text. The abscissa shows that three ionizable groups are present. The pK values, where the groups are 50% titrated, are at points A (pH ~ 2), C (pH ~ 6.5), and E (pH ~ 9.5). Histidine is the only common amino acid with these pK values. At point B, its net charge is $-1 + 1 + 1 = +1$. At point E, the net charge is $-1 + 0 + 0.5 = -0.5$ (pp. 36–37).
4. C 2 and 4 true. Pauly's reagent reacts only with histidine and tyrosine, and the Sakaguchi reaction is for arginine. 2 and 4 react with all amino acids (Table 2.7).
5. C The peptide bond has partial double bond character and does not rotate freely at physiological temperatures (p. 32).
6. D Only 4 true. Quaternary structure consists of a specific noncovalent association of subunits having their own tertiary structures (p. 49). Myoglobin is a single polypeptide chain (p. 60). Insulin (p. 48) and chymotrypsin (p. 60) are multichain proteins, with the chains joined by disulfide bridges.
7. C 2 and 4 true. Like charges in the third or fourth position in either direction from the designated position destabilize the helix due to charge repulsion. Thus aspartate at position D is harmless, whereas glutamate at position A or B would destabilize. Alanine has a small side chain. Proline destabilizes α-helix conformation and is usually not found in either α-helix or β-structure (p. 53).
8. A Atom C is an amide nitrogen. The attached hydrogen atom participates in hydrogen bonding (pp. 52–53). Hydrogen bonds contribute to the stability of the structure (p. 53).
9. A Quaternary structure is stabilized exclusively by noncovalent interactions. Disulfide bonds are covalent (p. 60).
10. A Hydrophobic interactions are a major contributor to the thermodynamic stability of proteins (p. 73).
11. B More reflections must be analyzed to achieve higher resolution (p. 80). In fact, the resolution is related to the third power of the number of reflections analyzed.
12. B The pK'_a of tyrosine's phenolic —OH group is about 10 (Table 2.4). In the dissociated state, tyrosine's molar absorptivity increases, and the wavelength of maximum absorption changes (p. 82).
13. B Fluorescence emission is always less energetic than the radiation required for excitation. This is because of vibrational loss of energy from the excited molecules. Longer wavelengths are less energetic (p. 82).
14. C 2 and 4 true. 2 and 4 separate on the basis of size (p. 86). 1 and 3 separate on the basis of charge (p. 39).
15. B This statement is a definition of secondary structure (p. 50).
16. C This is a consequence of folding into a compact structure (p. 57).
17. A SDS binding produces an extended conformation of a polypeptide chain due to charge repulsion, but no peptide bonds are broken (p. 86).
18. E Any protein containing α-helix or β-structure will exhibit a minimum in this range. Random coils exhibit a small maximum (p. 83, Figure 2.75).
19. B β structure is an important type of secondary structure (pp. 54–55).

3

Proteins II: Structure-Function Relationship of Protein Families

RICHARD M. SCHULTZ
MICHAEL N. LIEBMAN

3.1 OVERVIEW 92
3.2 ANTIBODY MOLECULES AND THE IMMUNOGLOBULIN SUPERFAMILY 93
Antibody (Immunoglobulin) Molecules Have a Tetrameric Structure 93
Immunoglobulins in a Single Class Contain Homologous Regions 95
There Are Two Antigen Binding Sites per Antibody Molecule 97
Repeating Amino Acid Sequence and Homologous Three-Dimensional Domains Occur Within an Antibody Structure 98
The Genetics of the Immunoglobulin Molecule Have Been Determined 100
The Immunoglobulin Fold Is a Tertiary Structure Found in a Large Family of Proteins With Different Functional Roles 102
3.3 PROTEINS WITH A COMMON CATALYTIC MECHANISM: THE SERINE PROTEASES 102
Proteolytic Enzymes Are Classified Based on Their Catalytic Mechanism 102
Serine Proteases Exhibit a Specificity for the Site of Peptide Bond Hydrolysis 105
Serine Proteases Are Synthesized in a Zymogen Form 108
There Are Specific Protein Inhibitors of Serine Proteases 109
Serine Proteases Have Similar Structural-Function Relationships 110
Amino Acid Sequence Homology Occurs in the Serine Protease Family 111
The Tertiary Structure of Serine Proteases Are Similar 112
3.4 DNA BINDING PROTEINS 116
Three Major Structural Motifs of DNA Binding Proteins 116
3.5 HEMOGLOBIN AND MYOGLOBIN 119
Human Hemoglobin Occurs in Several Forms 119
Myoglobin Is a Single Polypeptide Chain With One O_2 Binding Site 120
A Heme Prosthetic Group Is at the Site of O_2 Binding 120
X-Ray Crystallography Has Assisted in Defining the Structure of Hemoglobin and Myoglobin 122
Primary, Secondary, and Tertiary Structures of Myoglobin and the Individual Hemoglobin Chains 122
A Simple Equilibrium Defines O_2 Binding to Myoglobin 124
Binding of O_2 to Hemoglobin Involves Cooperativity Between the Hemoglobin Subunits 126
The Molecular Mechanism of Cooperativity in O_2 Binding 127
The Bohr Effect Involves Dissociation of a Proton on Binding of Oxygen 131
BIBLIOGRAPHY 132

QUESTIONS AND ANNOTATED ANSWERS 133
CLINICAL CORRELATIONS
3.1 The Complement Proteins 96
3.2 Functions of the Different Antibody Classes 96
3.3 Immunization 97
3.4 Fibrin Formation in Myocardial Infarct and the Action of Recombinant Tissue Plasminogen Activator (rt-PA) 104
3.5 Involvement of Serine Proteases in Tumor Cell Metastasis 105

3.1 OVERVIEW

In Chapter 2 we discussed the fundamentals of protein architecture, including the structural organization and physical properties of the amino acid constituents, the hierarchical organization of primary, secondary, supersecondary, tertiary and quaternary structure, and the energetic forces that both hold these molecules together and provide the flexibility observed in their dynamic motion. Computational and experimental tools were introduced which permit the analysis of high-resolution structural features and their conformational response to perturbations which may be a simple perturbation of the solution environment or aspects of their interactions with other molecules that define their biological function. The concept that structure and function are interrelated was introduced through the examples of conservation of primary structure with function, and the reoccurrence of secondary, supersecondary, tertiary, and quaternary structural patterns in molecules that may not share similar functional or evolutionary origin.

In this chapter we examine in depth the relationship between structure and function in four example **protein families:** immunoglobulins, serine proteases, DNA binding proteins, and hemoglobins. We pursue this study through the examination of the variation in amino acid sequence and structural organization, similarities and differences in structure and in biological function, and the gene organization of members of each of these four protein families. The significance of the structure–function relationship can be best appreciated through the observation of such variations within specific protein families. Information about these relationships contain important data concerning what sequence and structural elements are maintained through evolution to carry out homologous function within a family as well as functional diversity or specificity.

In the **immunoglobulin family,** we have examples of a multidomain architecture that supports recognition and binding to a series of foreign molecules that leads to their inactivation. Diversity among family members is the source of specific recognition and binding capabilities and appears to be of particular interest to gene regulation.

In the **serine proteases,** we have examples of a family of enzymes that appear to have diverged to carry out specific physiological functions, frequently highly organized within enzyme cascade processes. The inherent similarity in mechanism and structural fold remains a common link, although certain bacterial forms of serine proteases only show a functional similarity.

In the **DNA binding proteins** there exist multifamilies of proteins that bind to regulatory sites in DNA and regulate gene expression. These proteins contain unusual supersecondary structure motifs that allow them to selectively bind to regulatory sites of a specific gene, an amazing feat as the mammalian genome contains tens of thousands of unique genes.

In the **hemoglobins,** we have examples of a highly fine-tuned system that permits small substitutions, many of which have been studied as to their clinical implications. This family reveals the potential diversity of amino acid sequence substitutions that can be tolerated and where the protein can continue to function in an accepted physiological manner. Gene regulation can also be addressed as a developmental issue in this protein family.

3.2 ANTIBODY MOLECULES AND THE IMMUNOGLOBULIN SUPERFAMILY

Antibody molecules are immunoglobulins produced by an organism in response to the invasion of foreign compounds, such as proteins, carbohydrates, and nucleic acid polymers. The antibody molecule noncovalently associates to the foreign substance, initiating a process by which the foreign substance is eliminated from the organism.

Materials that elicit antibody production in an organism are called **antigens.** An antigen may contain multiple **antigenic determinants,** which are small regions of the antigen molecule that specifically elicit the production of antibody to which the antigen binds. In proteins, for example, an antigenic determinant may comprise only six or seven amino acids of the total protein.

A **hapten** is a small molecule that cannot alone elicit the production of antibodies specific to it. However, when covalently attached to a larger molecule it can act as an antigenic determinant and elicit antibody synthesis. Whereas the hapten molecule requires attachment to a larger molecule to elicit the synthesis of antibody, when detached from its carrier, the hapten will retain its ability to bind strongly to antibody.

It is estimated that each human can potentially produce about 1×10^8 different antibody structures. All the antibodies, however, have a similar basic structure. The elucidation of this structure has been accomplished to a great extent from studies of immunoglobulin primary structures. Recent studies with X-ray diffraction, show the three-dimensional structure of the antibody molecule alone and in complex with antigen.

Structural studies of proteins require pure homogeneous preparations. Such a sample of an antibody protein is extremely difficult to prepare from blood because of the wide diversity of antibody molecules present in any one organism. Homogeneous antibodies are now obtained by the monoclonal hybridoma technique in which mouse or rat myeloma cells are fused with mouse antibody producing B lymphocytes to produce immortalized hybridoma cells that produce a single antibody.

Antibody (Immunoglobulin) Molecules Have a Tetrameric Structure

Antibody molecules are composed of four polypeptide chains, comprised of two identical copies of each of two nonidentical polypeptide chains, chains L and H giving $(LH)_2$. In the most common immunoglobulin, IgG, the two H or **heavy chains** have approximately 440 amino acids (mol wt 50,000). The smaller L or **light chains** contain about one-half the number of amino acids of the H chain (mol wt 25,000). The four polypeptide chains are covalently interconnected by disulfide bonds (Figure 3.1).

In the other classes of immunoglobulins (see Table 3.1) the H chains have a slightly higher molecular weight than those of the IgG class. There is a small amount of carbohydrate (2–12%, depending on immunoglobulin class) attached to the H chain, and thus the antibodies are glycoproteins.

In the three-dimensional antibody structure, each H chain is associated with an L chain such that the NH_2 terminal ends of both chains are near each other. Since the L chain is one-half the size of the H chain, only the NH_2-terminal one-half of the H chain (~214 amino acids in the IgG class)

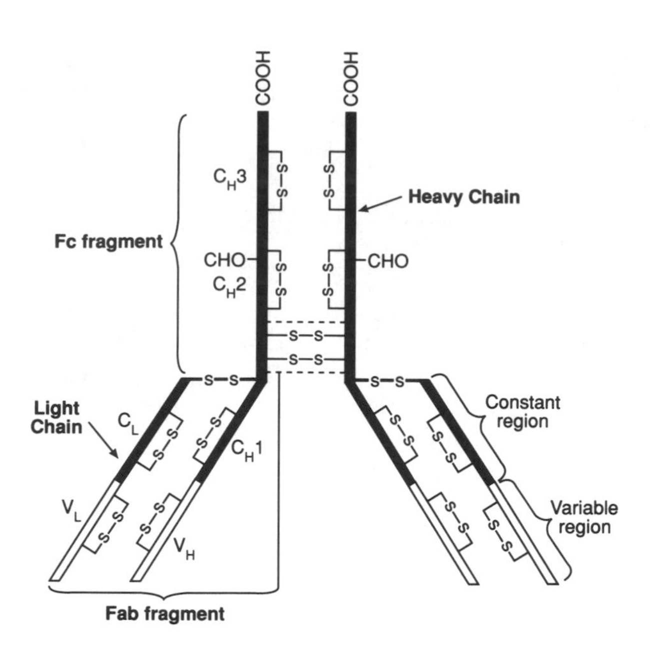

FIGURE 3.1
Linear representation of four-chain IgG antibody molecule.
Disulfide bonds link the heavy (H) chains, and the light (L) chains to the H chains. In addition, intrachain disulfide bonds exist within each of the domains. Domains of constant (C) region of H chain are C_H1, C_H2, and C_H3. The constant region of L chain is designated C_L, and variable (V) regions are V_H and V_L of H and L chains, respectively.
Based on figure by D. R. Burton, in *Molecular Genetics of Immunoglobulin*, F. Calabi and M. S. Neuberger (Eds.), 1987, Amsterdam: Elsevier, pp. 1–50.

is associated with the L chain. This is diagrammatically shown in Figures 3.1 and 3.2.

Constant and Variable Regions of Primary Structure

A comparison of the amino acid sequences of antibody molecules elicited toward different antigens shows that there are regions of exact homology and regions of high variability. In particular, the sequences of the NH_2-terminal one-half of the L chains and the NH_2-terminal one-quarter of the larger H chains are highly variable. These NH_2-terminal segments are designated the **variable** (V) **regions.** Within the V region certain segments are observed to be even more variable than other segments and are termed **"hypervariable" regions.** Three hypervariable regions of between 5 and 7 residues in the NH_2-terminal region of the L chain and three or four hypervariable regions of between 6 and 17 residues in the NH_2-terminal region of the H chain are commonly found. The hypervariable sequences are called the **complementary determining regions** (CDRs) as these are the sections of polypeptide that form the antigen binding site complementary to the topology of the antigen structure.

In contrast, a comparison of the amino acid sequences from different antibodies shows that the COOH-terminal three-quarters of the H chains and the COOH-terminal one-half of the L chains are mostly homologous in sequence. These regions of the polypeptide sequences are named the **constant regions** of primary structure and designated C_H and C_L.

TABLE 3.1 Immunoglobulin Classes

Classes of immunoglobulin	Approximate molecular weight	H Chain isotype	Carbohydrate by weight (%)	Concentration in serum (mg 100mL^{-1})
IgG	150,000	γ, 53,000	2–3	600–1800
IgA	170,000–720,000[a]	α, 64,000	7–12	90–420
IgD	160,000	δ, 58,000		0.3–40
IgE	190,000	ε, 75,000	10–12	0.01–0.10
IgM	950,000[a]	μ, 70,000	10–12	50–190

[a] Forms polymer structures of basic structural unit.

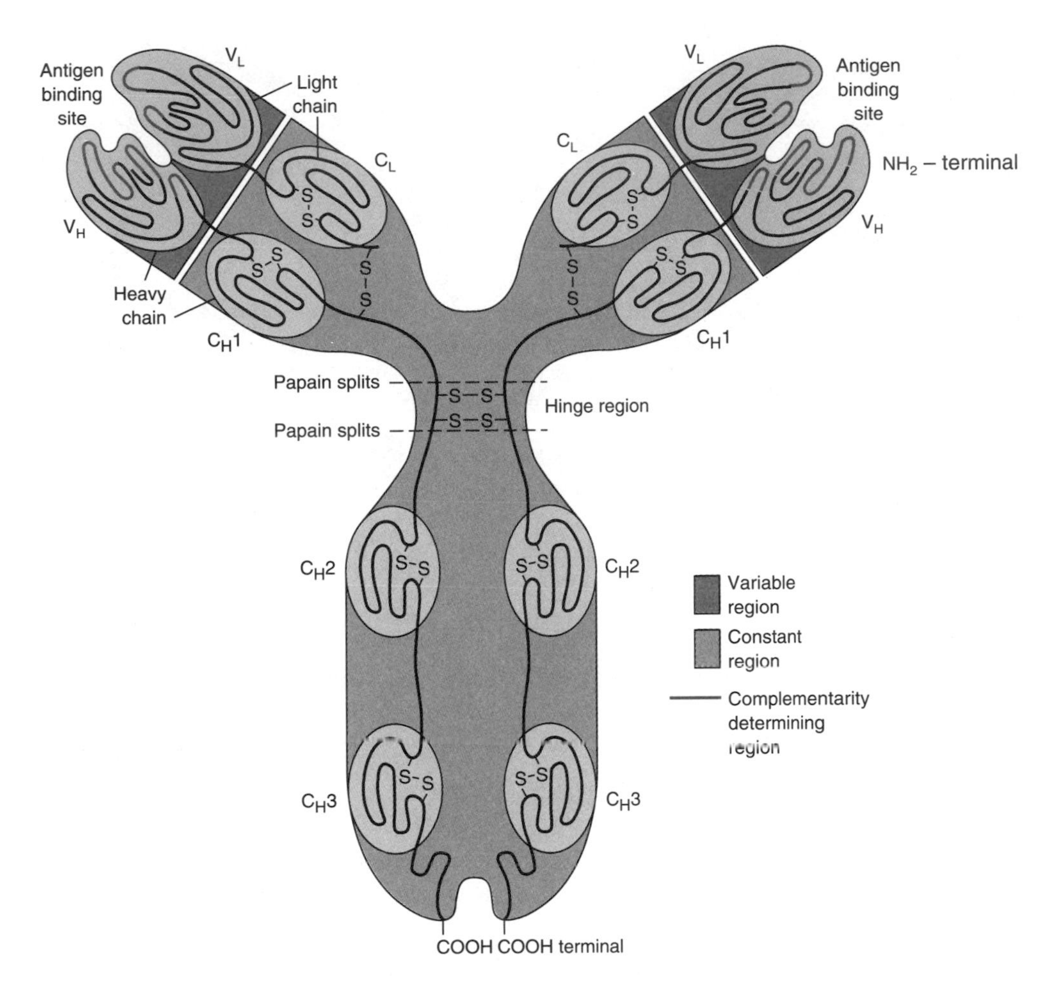

FIGURE 3.2
Diagrammatic structure for IgG.
The light chains (L) are divided into domains V_L (variable amino acid sequence) and C_L (constant amino acid sequence). The heavy chains (H) are divided into domains V_H (variable amino acid sequence) and C_H1, C_H2, and C_H3. The antigen binding sites are V_H–V_L. "Hinge" polypeptides interconnect the domains. The positions of inter- and intrachain cystine bonds are shown.
Based on figure in C. R. Cantor and P. R. Schimmel, *Biophysical Chemistry*, Part I, San Francisco: W. H. Freeman, 1980. Redrawn with permission of Mr. Irving Geis, N.Y.

It is the C_H regions that determine the antibody class. In addition, the C region provides for binding of complement proteins (Clin. Corr. 3.1) and the site necessary for antibodies to cross the placental membrane. The V regions determine the antigen specificity of the antibody molecule.

Immunoglobulins in a Single Class Contain Homologous Regions

The C_H regions within a particular class of immunoglobulins are homologous but differ significantly from the amino acid sequence of the C_H regions of other antibody classes. These differences are responsible for the different physical characteristics of the immunoglobulin classes.

In some cases the H chain sequence promotes polymerization of the antibody molecules. Thus $(LH)_2$ antibody units of the IgA class are sometimes found in covalently linked dimeric forms of the structure $[(LH)_2]_2$. Similarly, IgM antibodies are pentamers of the basic $(LH)_2$ covalent struc-

CLIN CORR. 3.1 THE COMPLEMENT PROTEINS

At least 11 distinct complement proteins exist in plasma. They are activated by IgG or IgM antibody binding to antigens on the outer cell membrane of invading bacterial cells, protozoa, or tumor cells. On initiation by the immunoglobulin binding event, the 11 complement proteins are sequentially activated and associate with the cell membrane causing a lysis of the membrane and death of the target cell.

Many of the complement proteins are precursors of proteolytic enzymes that are present in a nonactive form prior to activation. On their activation during the complementation process, they will in turn activate a succeeding protein of the pathway by the hydrolysis of a specific peptide bond in the second protein. The inactive forms of enzymes are referred to as proenzymes or zymogens (see p. 104). The activation of enzymes by specific proteolysis (i.e., hydrolysis of a specific peptide bond in its primary structure) is an important general method for activating extracellular enzymes. For example, the enzymes that catalyze blood clot formation, induce fibrinolysis of blood clots, and digest dietary proteins in the gut are all activated by a specific proteolysis catalyzed by a second enzyme (see pp. 967, 1075).

The classical complement reaction is initiated by the binding of IgG or IgM to cell surface antigens. On association to a cellular antigen the exposure of a complement binding site in the antibody's F_c region occurs and causes the binding of the C1 complement proteins, which are a protein complex composed of three individual proteins: C1q, C1r, and C1s. The C1r and C1s proteins undergo a conformational change and become active enzymes on association with the immunoglobulin on the cell surface. The activated C1 complex (C1a) catalytically hydrolyzes a peptide bond in complement proteins C2 and in C4, which then form a complex that also associates on the cell surface. The now active C2–C4 complex has a proteolytic activity that hydrolyzes a peptide bond in complement protein C3. Activated C3 protein binds to the cell surface, and the activated C2–C4–C3 complex activates protein C5. Activated protein C5 will associate with complement proteins C6, C7, C8, and six molecules of complement protein C9. This multiprotein complex binds to the cell surface and initiates membrane lysis.

The mechanism is a cascade type in which amplification of the trigger event occurs. In summary, activated C1 can activate a number of molecules of C4–C2–C3, and each activated C4–C2–C3 complex can in turn activate many molecules of C5 to C9.

The series of reactions in the classical complement pathway is summarized in the scheme below, where "a" and "b" designate the proteolytically modified proteins and a line above a protein indicates an enzyme activity.

$$\text{IgG or IgM} \xrightarrow{\text{C1q, C1r, C1s}} \overline{\text{C1a}} \xrightarrow{\text{C2,C4}}$$

$$\text{C4b.}\overline{\text{C2a}} \xrightarrow{\text{C3}}$$

$$\text{C4b.}\overline{\text{C2a}}\text{.C3b} \xrightarrow{\text{C5,C6,C7,C8,C9}}$$

$$\text{C5b.C6.C7.C8.6C9}$$

There is an "alternative pathway" for C3 complement activation, initiated by aggregates of IgA or by bacterial polysaccharide in the absence of immunoglobulin binding to cell membrane antigens. This alternative pathway involves the proteins properdin, C3 proactivator convertase, and C3 proactivator.

A major role of the complement systems is to generate opsonins. Opsonin is an old term for proteins that stimulate phagocytosis by neutrophils and macrophages. The major opsonin is C3b; macrophages have specific receptors for this protein. Patients with inherited deficiency of C3 are subject to repeated bacterial infections.

Schur, P. H. Inherited complement component abnormalities. *Annu. Rev. Med.* 37:333, 1986.

CLIN. CORR. 3.2 FUNCTIONS OF THE DIFFERENT ANTIBODY CLASSES

The IgA class of immunoglobulins are primarily found in the mucosal secretions—bronchial, nasal, and intestinal mucus secretions, tears, milk, and colostrum. As such, these immunoglobulins are the initial defense against invading viral and bacterial pathogens prior to their entry into plasma or other internal space.

The IgM class are primarily found in plasma. They are the first of the antibodies elicited in significant quantity on the introduction of a foreign antigen into a host's plasma. The IgM antibodies can promote phagocytosis of microorganisms by macrophage and polymorphonuclear leukocytes and are also potent activators of complement (see Clin. Corr. 3.1). The IgM can be found in many external secretions but at levels lower than those of IgA.

The IgG class is found in high concentration in plasma. Its response to foreign antigens takes a longer period of time than that of IgM. At its maximum concentration it is present in significantly higher concentrations than the IgM antibodies. Like IgM antibodies, IgG antibodies can promote phagocytosis by phagocytic cells in plasma and can activate complement.

The normal biological functions of the IgD and IgE classes of immunoglobulins are not known, however, it is clear that the IgE antibodies play an important role in allergic responses, which are a cause of anaphylactic shock, hay fever, and asthma.

Immunoglobulin deficiency usually causes increased susceptibility to infection. X-linked agammaglobulinemia and common variable immunodeficiency are two examples. The commonest disorder is selective IgA deficiency, which results in recurrent infections of the sinus and respiratory tract.

Rosen, F. S., Cooper, M. D., and Wedgewood, R. J. P. The primary immunodeficiencies. *N. Engl. J. Med.* 311:235 (Part I); 300 (Part II), 1984.

ture, giving it the formula $[(LH)_2]_5$. The different H chains designated γ, α, μ, δ, and ε are found in the IgG, IgA, IgM, IgD, and IgE classes, respectively (Table 3.1; Clin. Corr. 3.2).

Although IgG is the major immunoglobulin in plasma, the biosynthesis of a specific IgG antibody in significant concentrations after exposure to a new antigen takes about 10 days (Clin. Corr. 3.3). In the absence of an initially high concentration of IgG to a specific antigen, antibodies of the IgM class, which is synthesized at faster rates, will associate with the antigen and serve as the first line of defense against the foreign antigen until the large quantities of IgG are produced (Figure 3.3; Clin. Corr. 3.3).

Two types of L chain constant sequences are synthesized, either of which are found combined with the five classes of H chains. The chains are designated the *lambda* (λ) chain and the *kappa* (κ) chain.

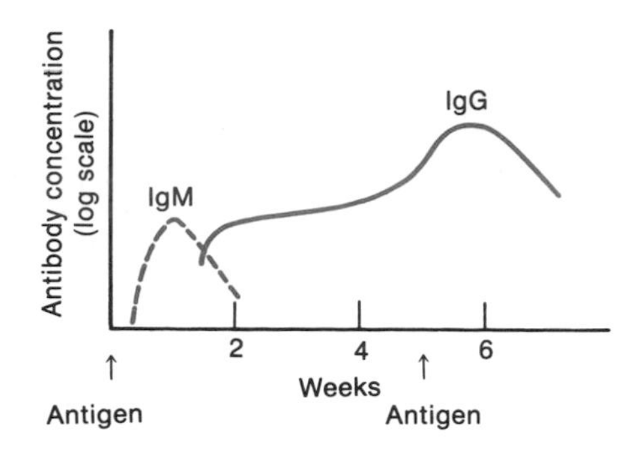

FIGURE 3.3
Time course of specific antibody IgM and IgG response to added antigen.
Based on a figure in L. Stryer, *Biochemistry*, San Francisco: W. H. Freeman, 1988, p. 890.

There Are Two Antigen Binding Sites per Antibody Molecule

The NH_2-terminal V regions of the L and H chain pairs comprise an antigen binding site. As the basic antibody structure contains two LH pairs, there are two antigen binding sites per antibody molecule. Clear evidence that an antigen binding site exists in the LH pair NH_2-terminal region was obtained by chemical techniques. In these experiments an antibody is hydrolyzed by the proteolytic enzyme papain at a single peptide bond located in the hinge peptide of each H chain (see Figures 3.2 and 3.4). On papain hydrolysis of this peptide bond, the antibody molecule is cleaved into three products, two of which are identical and comprise the NH_2-terminal segments of the H chain associated with the full L chain (Figure 3.4). *These NH_2-terminal H · L fragments can bind antigens with a similar affinity as does the whole antibody molecule.* The two NH_2-terminal fragments are designated the **Fab** (antigen binding) **fragments.** The second type of product from the papain hydrolysis is the COOH-terminal one-half of the H chains bound covalently in a single covalent fragment designated the **Fc** (crystallizable) **fragment.** The Fc fragment cannot bind antigen. It is thus clear that there are two antigen binding sites per antibody molecule present in the NH_2-terminal H · L fragments, which comprise the variable (V) amino acid sequences of each of the chains.

CLIN. CORR. 3.3
IMMUNIZATION

An immunizing vaccine can consist of killed bacterial cells, inactivated viruses, killed parasites, a nonvirulent form of live bacterium related to a virulent bacterium, or a denatured bacterial toxin. The introduction of such a preparation into a human can lead to protection against the virulent forms of the infection or toxic agents that contain the same antigen. The antigens in the nonvirulent material not only cause the differentiation of lymphoid cells into cells that produce antibody toward the foreign antigen but also cause the differentiation of some lymphoid cells into memory cells. Memory cells do not secrete antibody but place antibodies to the antigen into their outer membrane where they act as future sensors for the antigen. These memory cells are like a long-standing radar for the potentially virulent antigen. On the reintroduction of the antigen at a later time, the binding of the antigen to the cell surface antibody in the memory cells stimulates the memory cell to divide into antibody-producing cells as well as new memory cells. This mechanism reduces the time for antibody production that is required on introduction of an antigen and increases the concentration of antigen-specific antibody produced. It is the basis for the protection provided by immunization.

Recent additions to available vaccines, which are used for adults, include pneumococcal vaccine (to prevent pneumonia due to *Diplococcus pneumoniae*), hepatitis B vaccine, and influenza vaccine. The latter changes each year to account for antigenic variation in the influenza virus.

Center for Disease Control, Recommendations of the immunization practices advisory committee: adult immunization. *MMWR* (Morbidity and Mortality Weekly Report) Vol. 33 Supplement: 1S, 1984.

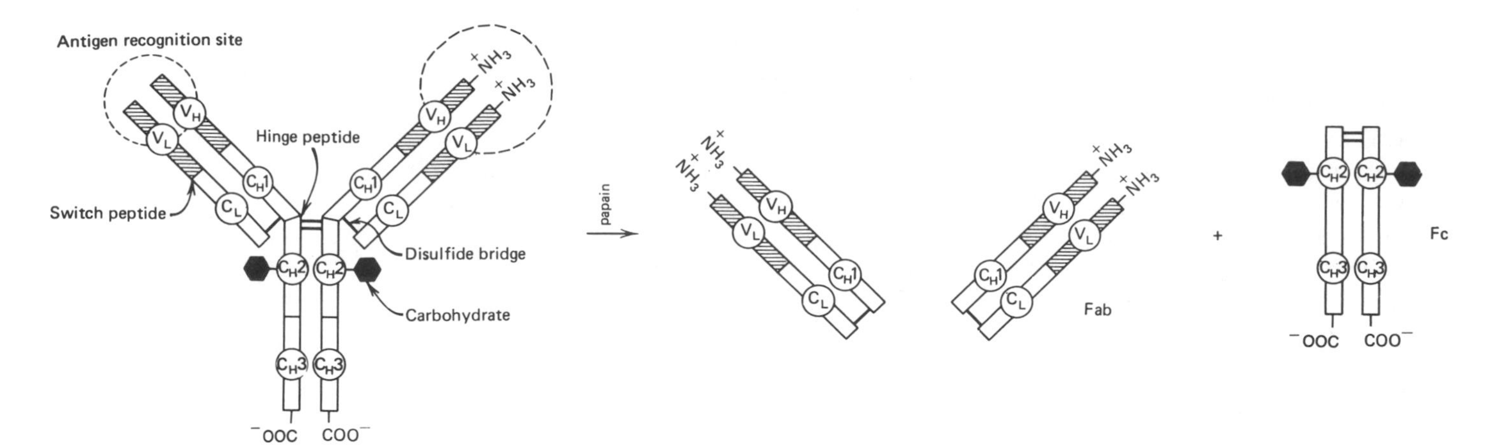

FIGURE 3.4
Hydrolysis of IgG into two Fab and one Fc fragments by papain, a proteolytic enzyme.

The valency of 2 for each antibody molecule allows each antibody to bind to two antigen molecules. This property of bivalency facilitates the agglutination and precipitation of antigen molecules by allowing the antibodies to form an interconnected matrix of antigens and antibodies.

The L chain can be dissociated from its H chain segment within the Fab fragment by oxidation of disulfide bonds, followed by chromatography. The dissociation of the L and H chains in this way eliminates antigen binding. Accordingly, each antigen binding site must be formed from components of both the L and H variable regions acting together.

In support of these conclusions, the X-ray crystallographic structure of an Fab fragment with a hapten associated has been obtained (Figure 3.5). The structure shows that the hypervariable sequences of the V regions are specifically utilized in forming the antigen binding site (Figure 3.5). The sequence of the hypervariable regions apparently give a unique three-dimensional conformation for each antibody that makes it specific to the antigenic determinant with which it associates.

The strength of association between antibody and antigen is due to noncovalent forces (see Chapter 2). The complementarity of the structures of the antigenic determinant and antigen binding site within the antibody results in extremely high equilibrium affinity constants, between 10^5 and 10^{10} M^{-1} (strength of 7–14 kcal mol^{-1}) for this noncovalent association.

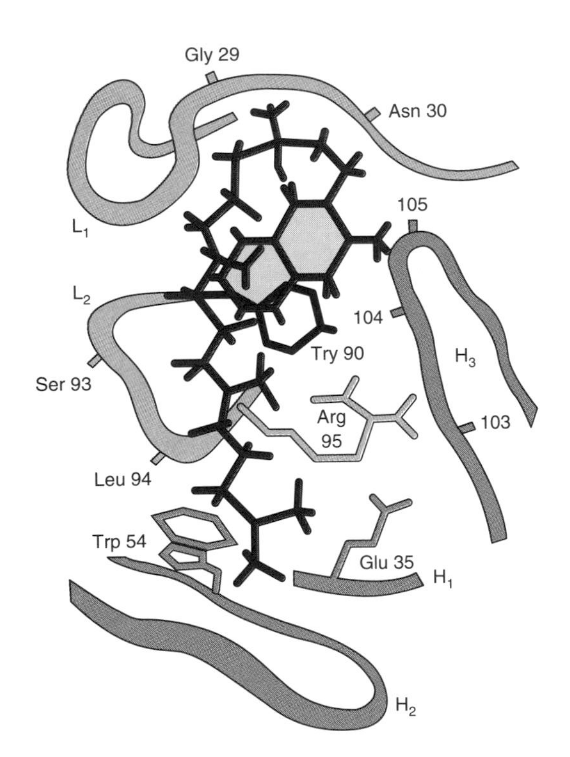

FIGURE 3.5
Structure of antigen (vitamin K_1-OH) bound to variable region of antibody.
Hypervariable regions designated L_1 and L_3 in light chains and H_1, H_2, and H_3 in heavy chains are shown in van der Waals contact with antigen.
Redrawn with permission from L. M. Amzel, R. J. Poljak, F. Saul, J. M. Varga, and F. F. Richards, *Proc. Natl. Acad. Sci. U.S.A.*, 71:1427, 1974.

Repeating Amino Acid Sequence and Homologous Three-Dimensional Domains Occur Within an Antibody Structure

We have discussed previously how the NH_2-terminal segments of both the L and H chains contain a variable (V) region, and the COOH-terminal segments of the L and H chains contain constant (C) regions of primary structure. The definition of these V and C regions resulted from analyses comparing different antibody amino acid sequences within the same antibody class. An even closer examination of primary structure can be made between different segments within each chain of a single antibody molecule. Such a comparison shows a repeating pattern of amino acid sequences within each of the chains. For the IgG class of L and H chains, this repetitive pattern is observed between segments of approximately 110 amino acids within both L and H chains. This homology is far from exact, but clearly a number of amino acids match identically following alignment of 110 amino acid segments. Other amino acids are matched in the sequence by having a similar nonpolar or polar side chain group. The repetition of the homologous sequence occurs four times along an immunoglobulin H chain and the chain is divided into one V region and three C regions (designated C_H1, C_H2, and C_H3) (see Figure 3.2). The L chain is divided into one V region and one C region. Each of these sequence repeats contains an intrachain cystine linking two cysteines within each sequence repeat.

Based on the presence of homologous regions in the amino acid sequence, it has been proposed that each of the segments have a similar tertiary conformation. X-ray diffraction studies confirm this proposal. The X-ray data dramatically show 110 amino acid domains in the H and L chains with a similar three-dimensional conformation as predicted from the study of the primary structures (Figure 3.6). The globular domains are formed from the winding of the polypeptide chain back and forth upon itself forming two β-barrel structures (Figure 3.6*a* and *b*) that run roughly parallel to each other. In three dimensions the two sheets form a structure that appears like a double blanket with a bulky hydrophobic core (Figure

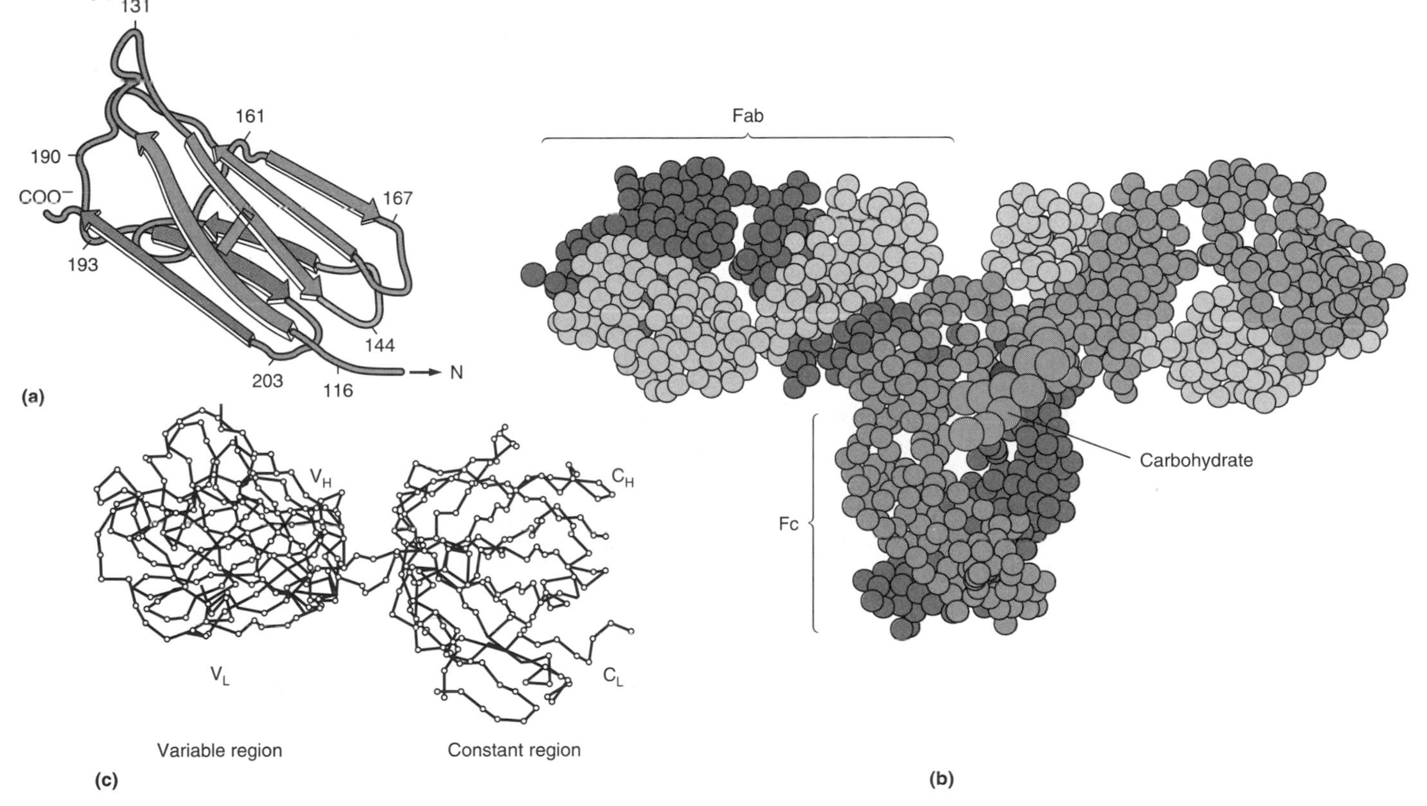

FIGURE 3.6
(*a*) Schematic diagram of folding of a C_L domain, showing β-pleated sheet structure. (*b*) The α carbon (○) structure of the Fab fragment of IgG KOL showing the V_L–V_H and C_H1–C_L domains interconnected by hinge polypeptide, (*c*) Space-filling model of an IgG molecule. One of the H chains is in white, and the other H chain is in dark gray. The two L chains are lightly shaded. The black spheres represent the carbohydrate attached to the protein.
Redrawn with permission. Figure *a* from A. B. Edmundson, K. R. Ely, E. E. Abola, M. Schiffer, and N. Pavagiotopoulos, Biochemistry 14:3953, 1975. Copyright © 1975 American Chemical Society. Figure *b* from R. Huber, J. Deisenhofer, P. M. Coleman, M. Matsushima, and W. Palm, in *The Immune System*. 27th Mosbach Colloquium, Berlin: Springer-Verlag, 1976, p. 26. Figure *c* from E. W. Silverston, M. A. Navia, and D. R. Davies, *Proc. Natl. Acad. Sci. U.S.A.* 74:5140, 1977.

3.6*b* and *c*). In the NH_2-terminal domain a shallow crevice is opened into the center of the hydrophobic core that binds to antigen. The hypervariable sequences are present in this NH_2-terminal domain crevice, standing out as different in conformation from the rest of the folded domain. Each of the hypervariable regions extends from and interconnects strands participating in the β-barrel structure. The hypervariable loops associate noncovalently in the three-dimensional structure of the V_L and V_H domains to form the binding site for the antigen (see Figure 3.5).

The Genetics of the Immunoglobulin Molecule Have Been Determined

The genes that code for the amino acid sequence of the human IgG L chains are located on different chromosomes than the H chain genes. In addition, in the synthesis of either an L or H chain the V and C regions are specified by separate genes. The V and C region genes of the human IgG H chain are located on chromosome 14, the genes that code for the *kappa*

```
Constant Region C1:
Cγ1: AlaSerThrLysGlyProSerValPheProLeuAlaProSerSerLysSerThrSerGlyGlyThrAlaAlaLeuGly
Cγ2                                          C     R           E  S
Cγ4

Cγ1  CysLeuValLysAspTyrPheProGluProValThrValSerTrpAsnSerGlyAlaLeuThrSerGlyValHisThr
Cγ2
Cγ4

Cγ1  PheProAlaValLeuGlnSerSerGlyLeuTyrSerLeuSerSerValValThrValProSerSerSerLeuGly
Cγ2                                                                     N  F
Cγ4

Cγ1  ThrGlnThrTyrIleCysAsnValAsnHisLysProSerAsnThrLysValAspLysLysVal
Cγ2              T          D                                  T
Cγ4     K                                                      R

Hinge Region H:
Cγ1  GluProLysSerCysAspLysThrHisThrCysProProCysPro
Cγ2      R     C     V  E  C  P  P         –  –  –
Cγ4      S     Y  G  P  P        S         –  –  –

Constant Region C2:
Cγ1  AlaPro    GluLeuLeuGlyGlyProSerValPheLeuPheProProLysProLysAspThrLeuMetIleSerArg
Cγ2            –  P  V  A
Cγ4        E   F     G

Cγ1  ThrProGluValThrCysValValValAspValSerHisGluAspProGluValLysPheAsnTrpTyrValAspGly
Cγ2                                                         Q
Cγ4                                      Q

Cγ1  ValGluValHisAsnAlaLysThrLysProArgGluGluGlnTyrAsnSerThrThrArgValValSerValLeuThr
Cγ2                                             F           F
Cγ4                                                         Y

Cγ1  ValLeuHisGlnAspTrpLeuAsnGlyLysGluTyrLysCysLysValSerAsnLysAlaLeuProAlaProIleGlu
Cγ2     V                                                       G
Cγ4                                                                      S  S

Cγ1  LysThrIleSerLysAlaLys
Cγ2                 T
Cγ4

Constant Region C3:
Cγ1  GlyGlnProArgGluProGlnValTyrThrLeuProProSerArgAspGluLeuThrLysAsnGlnValSerLeuThr
Cγ2                                                 E     M
Cγ4                                             Q

Cγ1  CysLeuValLysGlyPheTyrProSerAspIleAlaValGluTrpGluSerAsnGlyGlnProGluAsnAsnTyrLys
Cγ2
Cγ4

Cγ1  ThrThrProProValLeuAspSerAspGlySerPhePheLeuTyrSerLysLeuThrValAspLysSerArgTrpGln
Cγ2              M
Cγ4                                                    R

Cγ1  GlnGlyAsnValPheSerCysSerValMetHisGluAlaLeuHisAsnHisTyrThrGlnLysSerLeuSerLeuSer
Cγ2
Cγ4  E

Cγ1  ProGlyLysStop
Cγ2
Cγ4  L
```

FIGURE 3.7
Amino acid sequence of the constant regions of the IgG heavy chain γ_1, γ_2, and γ_4 genes.
Domains of constant domain C1, hinge region H, constant domain C2, and constant domain C3 are presented. The sequence for γ_1 is fully given and differences in γ_2 and γ_4 from the γ_1 sequence shown using single letter amino acid abbreviations. Dash line (–) indicates the absence of an amino acid in position correlated with γ_1, due to the desire to align the sequences in order to show the maximum homology.
Sequence of γ_1 chain from J. W. Ellison, B. J. Berson, and L. E. Hood, *Nucleic Acid Res.* 10:4071, 1982; and sequences of the γ_2 and γ_4 genes from J. Ellison and L. Hood, *Proc. Natl. Acad. Sci. U.S.A.* 79:1984, 1982.

(κ)-type L chains are on chromosome 2 and those for the *lambda* (λ)-type L chains are on chromosome 22. During the differentiation of individual antibody forming cells, a V region gene is assembled and comes together with a C region gene for the synthesis of a particular antibody chain and the whole antibody molecule (two L chains and two H chains) is formed in the endoplasmic reticulum after each of the polypeptide chains are synthesized.

There are actually four unique genes on human chromosome 14 that code for the C region of the H chain in the IgG antibody class. Each of these genes code for a complete constant region, thus containing all the amino acids of the H chain except for the V region sequence. The four genes are known as *gamma* (γ) genes, that is, γ_1, γ_2, γ_3, and γ_4. Figure 3.7 compares the amino acid sequences of three of the four γ-gene proteins. These gene products give rise to the **IgG isotypes** IgG1, IgG2, and IgG4. There is a 95% homology in amino acid sequence among the γ genes. Figure 3.8 shows the organization of one of these genes (γ_4) in the chromosome. The gene is composed of nucleotide bases that are translated into the amino acid sequence shown in Figure 3.7. The gene is read from the left (5′ end) to the right (3′ end), which translates into the corresponding NH_2- to COOH-terminal direction of the γ_4 polypeptide chain. Only certain regions of the gene are translated into protein. These translated or expressed regions are shown by large boxes and are known as **exons.** They are separated by nontranslatable or inserted nucleotide regions of the gene known as **introns.** There are four exon regions that code for the γ_4 polypeptide chain. Three of these exons each code for one of the constant domains: exon 1 = C_H1 domain, exon 3 = C_H2 domain, and exon 4 = C_H3 domain. The remaining exon (exon 2) codes for a connector region known as the hinge region (H; not to be confused in this context with the letter symbol for the heavy chain) located between domains C_H1 and C_H2, giving the heavy chain constant region polypeptide C_H1–H–C_H2–C_H3.

It is probable that a primordial gene coded for a single segment of approximately 110 amino acids, and multiple **exon gene duplications** resulted in the three repeating exon units within the same γ gene. Over time mutations slowly changed the sequence of one exon from the other so that the present correspondence in sequence is no longer exact. Each of the exons that code for an immunoglobulin domain give a similar domain length and immunoglobulin folding pattern stabilized by a cystine linkage. Later in evolution **gene duplications** led to the multiple genes (γ_1, γ_2, γ_3, γ_4) that code for the constant regions of the IgG class H chain and are found together in the human genome on chromosome 14.

Human γ_4 — C_H1 (294) — 390 — H (36) — 118 — C_H2 (330) — 97 — C_H3 (321) — UT (~130)

FIGURE 3.8
Domain organization of immunoglobulin heavy chain γ_4 gene.
Other genes have a similar exon–intron organization (see p. 712), with different exons coding for each of the immunoglobulin domains C_H1, C_H2, and C_H3. Also the hinge region between C_H1 and C_H2 is expressed as a separate exon. Exons are designated by solid blocks and introns by lines connecting the blocks. The number of nucleotide pairs in each exon and intron are given above or below the respective exon and intron. Open block on the right-hand end (3′ end) depicts region transcribed into the messenger RNA (mRNA) for the gene, but not translated into the protein product.
Redrawn from J. Ellison, J. Buxbaum, and L. Hood, *DNA* 1:11, 1981.

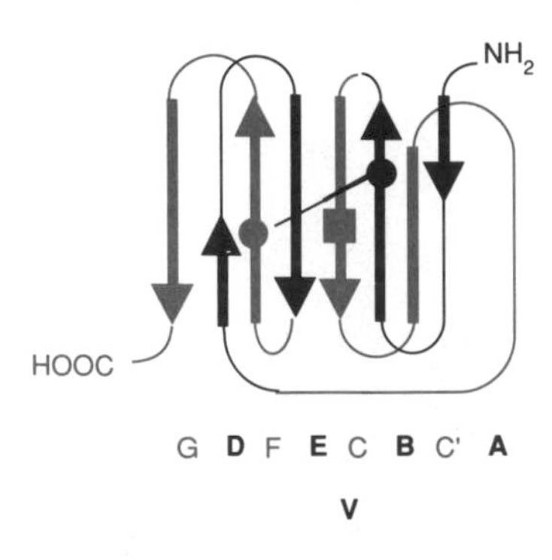

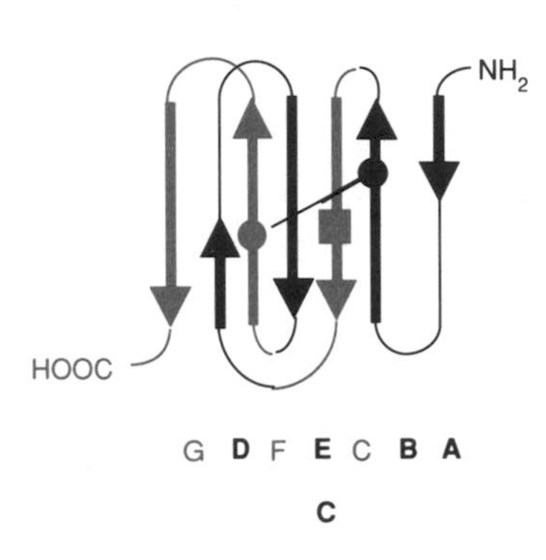

FIGURE 3.9
Immunoglobulin fold motif found in proteins of immunoglobulin superfamily.
Domain examples are those for IgG variable and constant regions. Thick arrows indicate β strands and thin line loops that interconnect the β strands. Circles indicate cysteines that form the intradomain disulfide bond. Squares show position of tryptophan residues that are an invariant component of the core of the immunoglobulin fold. Black letters indicate strands that form one plane of the sheet, while red letters form a parallel plane behind the first plane.
Based on a figure by F. Calabi, in *Molecular Genetics of Immunoglobulin,* F. Calabi and M. S. Neuberger (Eds.), Amsterdam: Elsevier, 1987, pp. 203–239.

The Immunoglobulin Fold Is a Tertiary Structure Found in a Large Family of Proteins With Different Functional Roles

Each of the 110 segment domains of the H and L chain folded into a supersecondary structure contain a unique grouping of β strands known as an immunoglobulin fold (Figures 3.6*a* and 3.9). As described above, individual exons in the genes for the IgG H chain code for an immunoglobulin fold. This three-dimensional folding pattern is more easily studied in two dimensions through a distance matrix (see p. 51) (Figure 3.10). Using this two-dimensional map of distances between different amino acid residues in the tertiary structure, both secondary and supersecondary structures are clearly shown. In addition, insertions and deletions to the amino acid sequence of a motif will modify the characteristic pattern by inserting or deleting strips in the matrix pattern. These maps can therefore be used for the comparison of different polypeptide domains. In the matrix, α helices are shown by a cross pattern along the major diagonal that broadens the diagonal (see Figure 3.33 for hemoglobin domain, which is almost all α helical). A parallel β sheet gives a series of close contacts on a diagonal line offset from the main diagonal but *parallel* to the main diagonal. Two strands of antiparallel β sheet give rise to a line of contact points *perpendicular* to the main diagonal.

The **immunoglobulin fold motif** or β barrel is found in a large number of nonimmunological proteins. These proteins exhibit widely different functions but based on their structural homology they can be grouped into a single **superfamily.** For example, a distance matrix map of the Class I major histocompatibility complex proteins that act as antigens for cytotoxic T cells, shows a tertiary structure like that of an immunoglobulin domain and thus places this protein into the immunoglobulin superfamily (Figure 3.11). Each of the diverse proteins in this superfamily, contain an immunoglobulin fold (Figure 3.9) for its domain structure, which is usually encoded by a single exon. As for the C_H, the motif consists of two stacked antiparallel β sheets that enclose an internal space filled mainly by hydrophobic amino acids. Two cysteines in the structure form a cystine that link the facing β sheets. It may be speculated that exon duplication during evolution led to the distribution of the structural motif in the functionally diverse protein superfamily, all containing one or more domain with an immunoglobulin domainlike structure.

3.3 PROTEINS WITH A COMMON CATALYTIC MECHANISM: THE SERINE PROTEASES

The serine proteases are a family of enzymes that utilize a uniquely activated serine residue in its substrate binding site to catalytically hydrolyze peptide bonds. This active site serine can be characterized by the irreversible reaction of its side chain hydroxyl group with diisopropylfluorophosphate (DFP) (Figure 3.12). Of all the serines in the protein, DFP only reacts with the catalytically active serine to form a phosphate ester.

Proteolytic Enzymes Are Classified Based on Their Catalytic Mechanism

Proteolytic enzymes are classified based on a knowledge of their catalytic mechanism. Other classes of proteases have an activated cysteine **(cysteine-proteases),** aspartate **(aspartate-proteases),** or metal ion **(metalloproteases).** The serine proteases can be identified by their stoichiometric

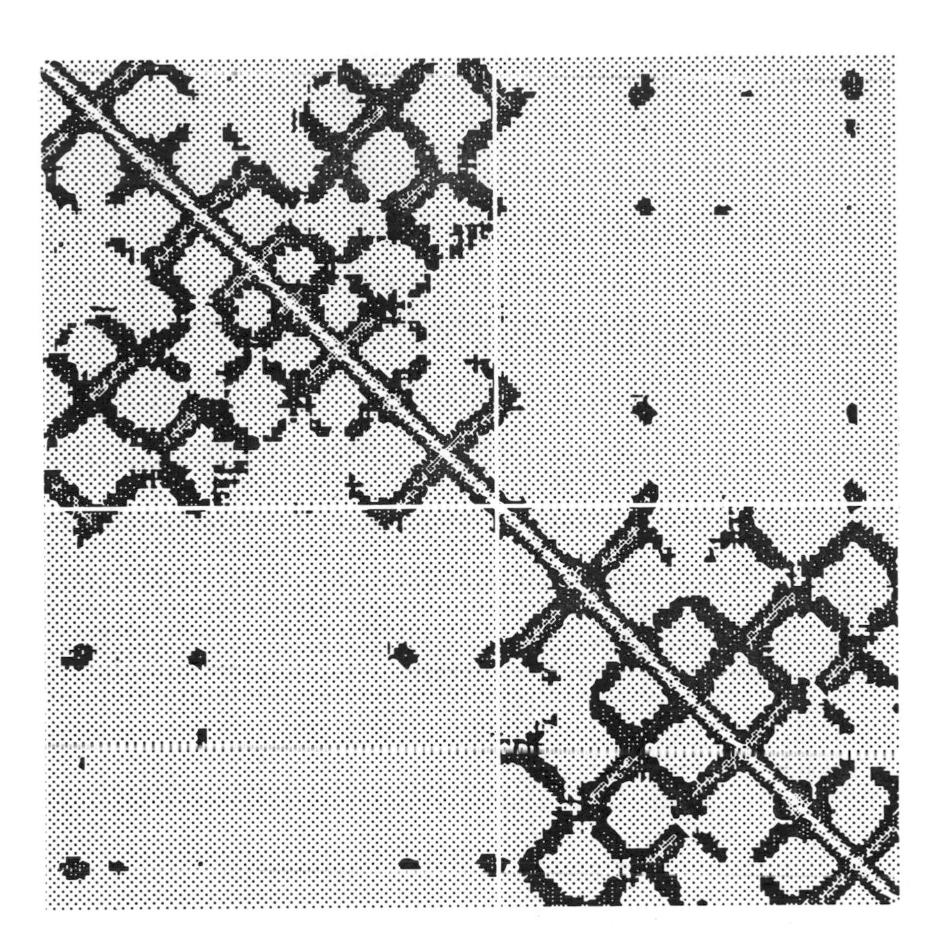

FIGURE 3.10
Interresidue distance matrix of the V_H and C_H1 domains of the antibody MOPc603 heavy chain, shaded to show distances of less than 15 Å
The V_H and C_H1 domains are separated by the dividing lines. The matrix patterns show the super secondary β barrels that comprise each immunoglobulin domain and that exhibits both parallel and antiparallel β structure.

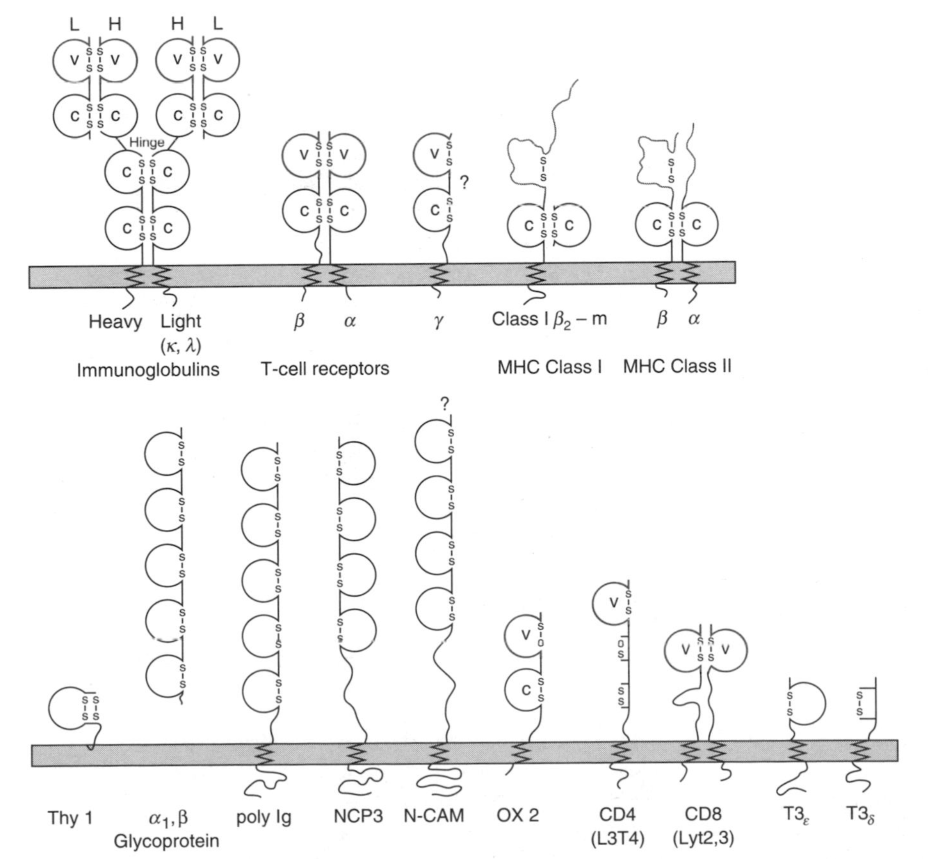

FIGURE 3.11
Diagrammatic representation of immunoglobulin domain structures from different proteins of the immunoglobulin gene superfamily.
The proteins presented include the heavy and light chains of immunoglobulins, T-cell receptors, major histocompatibility complex (HMC) class I and class II proteins, T-cell accessory proteins involved in class I (CD8) and class II (CD4) MHC recognition and possible ion channel formation, a receptor responsible for transporting certain classes of immunoglobulin across mucosal membranes (poly-Ig), beta(2)-microglobulin, which associates with class I molecules, a human plasma protein with unknown function (α/β-glycoprotein), two molecules of unknown function with a tissue distribution that includes both lymphocytes and neurons (Thy-1, OX-2), and two brain specific molecules, neuronal cell-adhesive molecule (N-CAM) and neurocytoplasmic protein 3 (NP3).
Reprinted with permission from T. Hunkapiller and L. Hood, *Nature (London)* 323:15, 1986.

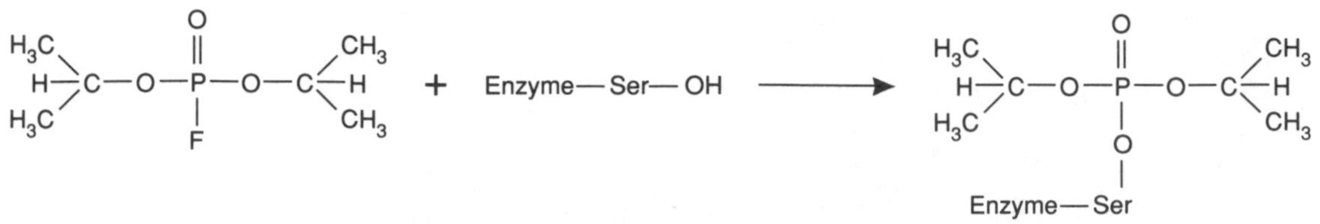

$$(H_3C)_2CH{-}O{-}P(=O)(F){-}O{-}CH(CH_3)_2 + \text{Enzyme—Ser—OH} \longrightarrow (H_3C)_2CH{-}O{-}P(=O)(O{-}\text{Ser—Enzyme}){-}O{-}CH(CH_3)_2$$

FIGURE 3.12
Reaction of diisopropylfluorophosphate (DFP) with the active site serine in a serine protease.

reaction with DFP which reacts only with proteases containing an active serine in the catalytic site. In addition, proteases that hydrolyze peptide bonds within a polypeptide chain are classified as **endopeptidases** and those that cleave the peptide bond of either the COOH- or NH_2-terminal amino acid are classified as **exopeptidases.**

The target of the serine proteases are specific peptide bonds in proteins and often their substrates are other serine proteases that are activated from an inactive precursor form, termed a **zymogen,** by the catalytic cleavage of a specific peptide bond in their structure. Serine proteases participate in carefully controlled physiological processes such as blood coagulation (Clin. Corr. 3.4), fibrinolysis, complement activation (see Clin. Corr. 3.1), fertilization, and hormone production (Table 3.2). The protein activations catalyzed by serine proteases are examples of "limited proteolysis" because the peptide hydrolysis is limited to the hydrolysis of only one or two particular peptide bonds of the hundreds of peptide bonds

CLIN. CORR. 3.4
FIBRIN FORMATION IN MYOCARDIAL INFARCT AND THE ACTION OF RECOMBINANT TISSUE PLASMINOGEN ACTIVATOR (rt-PA)

Coagulation is an example of an enzyme cascade process and contrasts with the cyclic pathways observed in intermediary metabolism. In the coagulation cascade multiple inactive serine protease enzymes (zymogens) are catalytically activated by other serine proteases in a step-wise manner (the coagulation pathway is described in Chapter 22). These multiple activation events generating catalytic products result in a dramatic amplification of the initial signal of the pathway. The end product of the coagulation pathway is a cross-linked fibrin clot. The zymogen forms of the serine protease components of coagulation include factor II (prothrombin), factor VII (proconvertin), factor IX (Christmas factor), factor X (Stuart factor), factor XI (plasma thromboplastin antecedent), and factor XII (Hageman factor). The protein factors within the coagulation pathway have a roman numeral designation given in order of their discovery and not from their order of action within the pathway. On activation of their zymogen form, the activated enzymes are noted with the suffix "a." Thus prothrombin is denoted in the scheme below as Factor II, and its activated enzyme, thrombin, is also known as Factor II_a. The active enzymes of the pathway are all proteases with an active site serine.

The main function of coagulation is to maintain the integrity of the closed, mammalian circulatory system after blood vessel injury. This process, however, can be dangerously activated in a myocardial infarction, causing a deficiency in the flow of blood to the heart muscle. Present statistics indicate that 1.5 million individuals suffer heart attacks each year, resulting in 600,000 deaths. Like the coagulation pathway that leads to fibrin clot formation, a fibrinolysis pathway exists in blood to degrade fibrin clots. This pathway also utilizes zymogen factors that are activated to serine proteases. The end reaction of the fibrinolysis pathway is the activation of the serine protease plasmin. Plasmin acts directly on fibrin to catalyze the degradation of the fibrin clot. Tissue plasminogen activator (t-PA) is one of the plasminogen activators that activates plasmin from plasminogen. The recombinant-t-PA (rt-PA) is produced by the gene cloning technology. Clinical studies show that the administration of rt-PA during a myocardial infarct significantly enhances recovery. Other plasminogens activators such as urokinase and streptokinase may also be of similar benefit. Currently, clinical trials are studying which of the plasminogen activators or combinations of plasminogen activators has the most beneficial effect and least risk for patients.

Topol, E. J. Recombinant tissue plasminogen activator: implications in therapy. *Semin. Hematol.* 26:25, 1989; and Bates, E. R. Reperfusion therapy in inferior myocardial infarction. *J. Am. College of Cardiology* 12:44A, 1988.

TABLE 3.2 Some Serine Proteases and Their Biochemical and Physiological Roles

Protease	Action	Possible disease due to deficiency or malfunction
Plasma kallikrein Factor XIIa Factor XIa Factor IXa Factor VIIa Factor Xa Factor IIa (thrombin) Activated protein C	Coagulation (Clin. Corr. 3.4)	Cerebral infarction (stroke), coronary infarction, thrombosis, bleeding disorders
Factor C$\bar{1}$r Factor C$\bar{1}$s Factor D Factor B C3 convertase	Complement (Clin. Corr. 3.1)	Inflammation, rheumatoid arthritis, autoimmune disease
Trypsin Chymotrypsin Elastase (pancreatic) Enterokinase	Digestion	Pancreatitis
Urokinase plasminogen activator Tissue plasminogen activator Plasmin	Fibrinolysis, cell migration, embryogenesis, menstruation	Clotting disorders, tumor metastasis (Clin. Corr. 3.4)
Tissue kallikreins	Hormone activation	
Acrosin	Fertilization	Infertility
α-Subunit nerve growth factor γ-Subunit nerve growth factor	Growth factor activation	
Granulocyte elastase Cathepsin G Mast cell chymases Mast cell tryptases	Extracellular protein and peptide degradation, mast cell function	Inflammation, allergic response

in a protein substrate. Under denaturing conditions, however, these same type of enzymes are capable of hydrolysis of multiple peptide bonds in a polypeptide chain leading to the digestion of simple peptides, proteins, and even self-digestion (autolysis). Several diseases such as emphysema, arthritis, thrombosis, cancer metastasis (Clin. Corr. 3.5), and some forms of hemophilia, are thought to result from the lack of regulation of serine protease activities.

Serine Proteases Exhibit a Specificity for the Site of Peptide Bond Hydrolysis

There are multiple aspects to the description of a serine protease's specificity for substrates. Each of the serine proteases exhibits preference for hydrolysis of peptide bonds adjacent to a particular class of amino acid. Thus the serine protease trypsin cleaves peptide bonds following basic amino acids such as arginine and lysine, and chymotrypsin cleaves peptide bonds following large hydrophobic amino acid residues such as tryptophan, phenylalanine, tyrosine, and leucine. The serine protease elastase cleaves peptide bonds following small hydrophobic residues such as alanine. As a result of these preferences, a serine protease may be called

CLIN. CORR. 3.5 INVOLVEMENT OF SERINE PROTEASES IN TUMOR CELL METASTASIS

The serine protease urokinase is believed to be required for the metastasis of cancer cells. Metastasis is the process in which a cancer cell leaves a primary tumor and migrates through the blood or lymph systems to a new tissue or organ, where it starts the growth of a secondary tumor. Increased synthesis of urokinase has been correlated with an increased ability to metastasize in many cancers. Urokinase activates plasmin from its zymogen plasminogen. Plasminogen is ubiquitously located in extracellular space and its activation to plasmin can cause the catalytic degradation of the proteins in the extracellular matrix through which the metastasizing tumor cells invade. Plasmin can also activate the precursor forms of the collagenases, known as procollagenase, thus promoting the degradation of the collagen in the basement membrane surrounding the capillaries and lymph system. This promotion of proteolytic degradative activity by the urokinase secreted by tumor cells allows the tumor cells to invade into its target tissue and form secondary tumor sites.

Dano, K., Andreasen, P. A., Grondahl-Hansen, J., Kristensen, P., Nielsen, L. S., and Skriver, L., Plasminogen activators, tissue degradation and cancer, *Adv. Cancer Res.* 44:139, 1985.

trypsin-like if it prefers to cleave peptide bonds of lysine and arginine, chymotrypsin-like if it prefers to cleave peptide bonds of aromatic amino acids, and elastase-like if it prefers to cleave peptide bonds of amino acids with small side chain groups like alanine. The selectivity of a specific protease for a certain type of amino acid only indicates its preference. Trypsin can also cleave peptide bonds following hydrophobic amino acids, but at a much slower rate than for the basic amino acids lysine or arginine. Thus specificity for hydrolysis of the peptide bond of a particular type of amino acid may not be absolute, but specificity may be more accurately described as a range of most likely targets. Each of the potential hydrolysis sites within a protein substrate is not equally susceptible. Thus trypsin hydrolyzes each of the multiple arginine peptide bonds in a particular protein substrate at different catalytic rates, and some may require a conformational change to become accessible. Some arginine bonds are thus more susceptible to hydrolysis by trypsin than other arginine bonds. Finally, not every protein, in its native folded conformation, is susceptible to limited proteolysis.

Detailed experimental studies of the specificity of serine proteases for a particular peptide bond have been performed with synthetic peptide substrates of fewer than 10 amino acids in length (Table 3.3). Because these

TABLE 3.3 Reactivity of α-Chymotrypsin and Elastase Towards Substrates of Various Structures

Structure	Variation in side chain group (chymotrypsin)	Relative reactivity[a]
Glycyl	H—	1
Leucyl	$(H_3C)_2CH-CH_2-$	1.6×10^4
Methionyl	$CH_3-S-CH_2-CH_2-$	2.4×10^4
Phenylalaninyl	$C_6H_5-CH_2-$	4.3×10^6
Hexahydrophenylalaninyl	(S)$-CH_2-$	8.2×10^6
Tyrosyl	$HO-C_6H_4-CH_2-$	3.7×10^7
Tryptophanyl	indolyl (N H)$-CH_2-$	4.3×10^7
Variation in chain length (elastase-hydrolysis of Ala *N* terminal amide)[b]		
Ac-Ala-NH_2		1
Ac-Pro-Ala-NH_2		1.4×10^1
Ac-Ala-Pro-Ala-NH_2		4.2×10^3
Ac-Pro-Ala-Pro-Ala-NH_2		4.4×10^5
Ac-Ala-Pro-Ala-Pro-Ala-NH_2		2.7×10^5

[a] Calculated from values of k_{cat}/K_m found for *N*-acetyl amino acid methyl esters in the chymotrypsin substrates.

[b] Calculated from values of k_{cat}/K_m in R. C. Thompson and E. R. Blout, *Biochemistry* 12:57, 1973.

peptide substrates are smaller than their natural targets, they interact only with the catalytic site (primary binding site S_1, see below) and are said to be **active site directed.** As outlined in the data of Table 3.3, the first descriptions of the specificity of the serine protease α-chymotrypsin towards amino acid side chain groups may be misleading in terms of its actual physicochemical basis. This specificity has been most frequently referred to as a preference for binding aromatic residues in S_1 with a shape that is flat and aromatic in nature. This identification is based on the characteristics of the naturally occurring amino acids of which those with the largest side chains tend to be aromatic and flat, such as phenylalanine, tyrosine, and tryptophan. Expansion of the set of substrates to nonamino acid side chains reveals that a nonaromatic hexahydrocyclohexyl side chain has a similar binding specificity to that of the side chains of tyrosine or phenylalanine. This suggests that the correct definition of the binding specificity for this serine protease is towards "large, hydrophobic" rather than aromatic side chains.

The studies with small substrates and inhibitors indicate that the site of hydrolysis is flanked by approximately four amino acid residues in both directions that can bind to the enzyme and impact on the reactivity of the peptide bond hydrolyzed. A simple nomenclature has been developed for the designation of the region around the scissile peptide bond in protease substrates. The two amino acids in the substrate that contribute the hydrolyzable bond are designated P_1—P_1'. Thus in trypsin-like substrates P_1 will be lysine or arginine and in chymotrypsin-like substrates P_1 will be a hydrophobic amino acid such as phenylalanine or tryptophan. The other interacting residues in the substrate are labeled P_4–P_2 on the NH_2-terminal side of P_1—P_1' and the COOH-terminal residues to the scissile bond are P_2'–P_4'. Thus the residues in the substrate that interact with the extended active site in the serine protease will be P_4—P_3—P_2—P_1—P_1'—P_2'—P_3'—P_4'. The complementary regions in the enzyme that bind the amino acid residues in the substrate are designated subsites $S_4 \cdots S_4'$ (Figure 3.13). It is the secondary interaction, outside of S_1—S_1', with the substrate that ultimately determines a protease's specificity toward a particular protein substrate. Thus the serine protease in coagulation, Factor Xa, only cleaves a particular arginine peptide bond in prothrombin, activating prothrombin to thrombin. It is the **secondary in-**

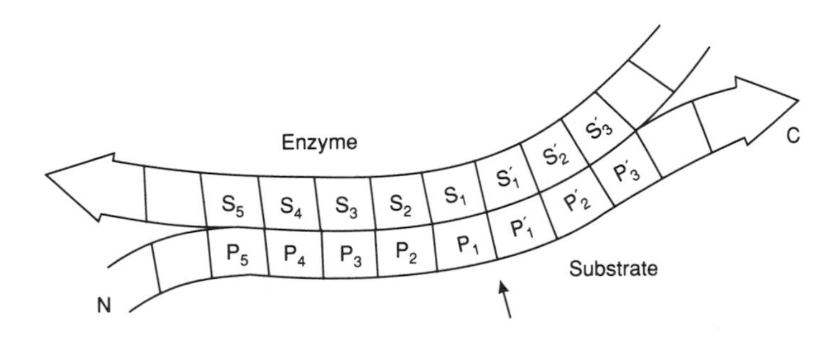

FIGURE 3.13
Schematic diagram of the binding of a polypeptide substrate to the binding site in a proteolytic enzyme.
$P_5 \cdots P_5'$ are amino acid residues in the substrate that are binding to subsites $S_5 \cdots S_5'$ in the enzyme with enzyme-catalyzed peptide hydrolysis occurring between P_1—P_1' (arrow). The NH_2-terminal direction of substrate polypeptide chain is indicated by N, and the COOH-terminal direction by C.
Redrawn from L. Polgar, in *Hydrolytic Enzymes*. A. Neuberger and K. Brocklehurst (Eds.), Amsterdam: Elsevier, 1987, p. 174.

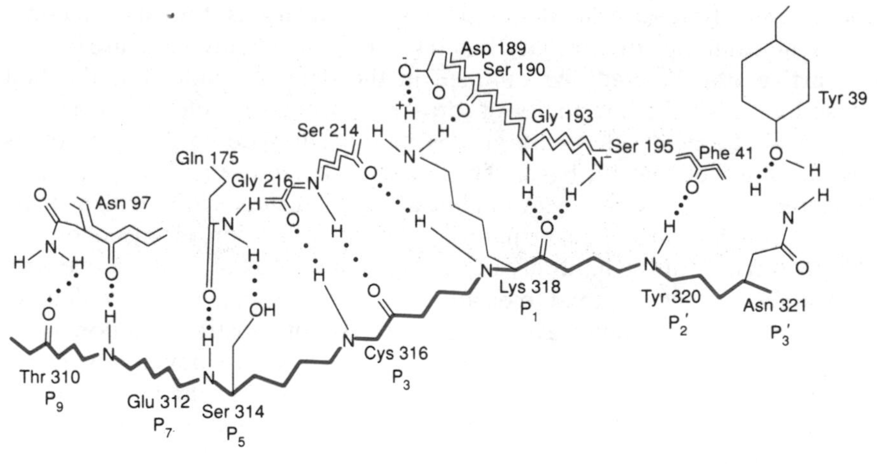

FIGURE 3.14
Schematic drawing of the binding of the protein protease inhibitor, pancreatic trypsin inhibitor, to trypsinogen based on X-ray diffraction data.
The binding site region of trypsinogen in the complex assumes a conformation like that of active trypsin. Note that the inhibitor has an extended conformation so that amino acids $P_9, P_7, P_5, P_3, P_1 \cdots P'_3$ interact with binding subsites $S_5 \cdots S'_3$. Potentially hydrolyzable bond in inhibitor is between P_1—P'_1.
Redrawn from Bolognesi, M., Gatti, B., Menegatti, E., Guarneri, M., Papamokos, E., and Huber, R. *J. Mol. Biol.* 162:839, 1983.

teraction that allows Factor Xa to recognize the particular arginine in the structure of prothrombin to be cleaved. The interaction of the substrate residues $P_4 \cdots P'_4$ with the enzyme binding subsites $S_4 \cdots S'_4$ are due to noncovalent interactions. The substrate interacts with the enzyme binding site to extend a β-sheet structure between the polypeptide chain in the enzyme and the polypeptide chain of the substrate, which places the scissile peptide bond of the substrate into S_1—S'_1 (Figure 3.14).

Serine Proteases Are Synthesized in a Zymogen Form

Serine proteases are synthesized in an inactive *zymogen* form, which requires limited proteolysis to produce the active enzyme. It is in this

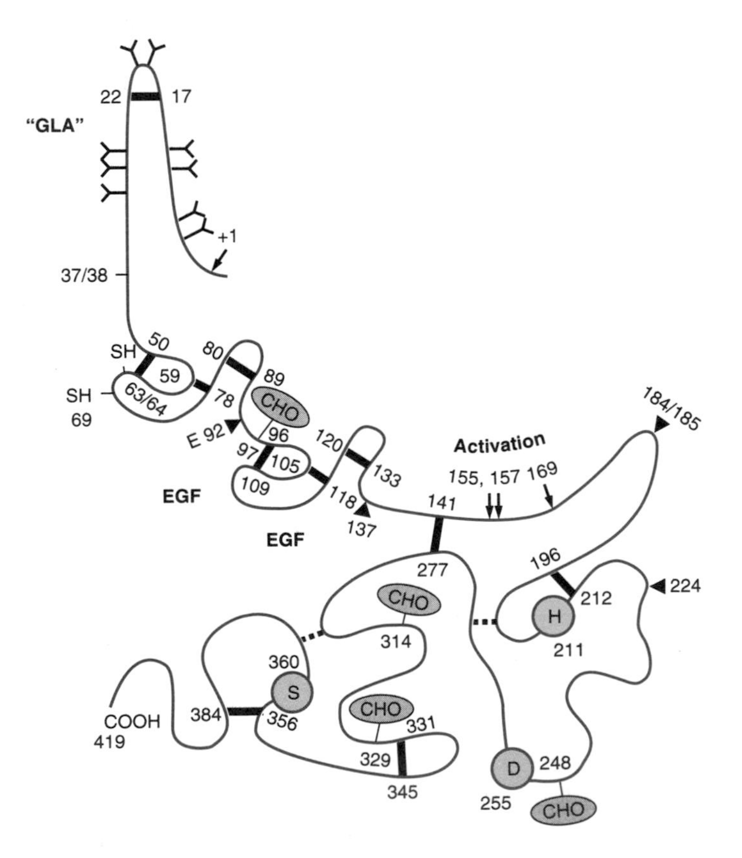

FIGURE 3.15
Schematic of domain structure for Protein C showing multidomain structure.
"GLA" refers to the γ-carboxy-glutamic residues (indicated by tree [Y] structures) in the NH_2-terminal domain, disulfide bridges by thick bars, and CHO indicates the positions where sugar residues are joined to the polypeptide chain. The proteolytic cleavage sites leading to catalytic activation are shown by arrows. The amino acid sequence is numbered from the NH_2-terminal end, and the catalytic site residues of serine, histidine, and aspartate are shown in the catalytic domains by the circled one letter abbreviations S, H, and D, respectively.
Redrawn from a figure in G. L. Long, *J. Cellular Biochem.* 33:185, 1987.

manner, for example, that the serine proteases for coagulation are synthesized in liver cells and are secreted into the blood for subsequent activation through limited proteolysis by other serine proteases when required during vascular injury. Zymogen forms are usually designated by the suffix -ogen following their enzyme name; the zymogen form of trypsin is termed trypsin*ogen* and chymotrypsin is chymotrypsin*ogen*. In some cases the zymogen form is referred to as a **proenzyme;** the zymogen form of thrombin is referred to as prothrombin.

A unique characteristic shared by several plasma serine proteases is the observation that the secreted zymogen form of the enzyme contains **multiple domains.** For example, the schematic diagram for the serine protease protein C, involved in a fibrinolysis pathway in blood, reveals that the zymogen has four distinct domains (Figure 3.15). The NH_2-terminal domain contains the derived amino acid, **γ-carboxyglutamic acid** (Figure 3.16), which is enzymatically formed by the carboxylation of glutamic acid residues in the sequence in a reaction dependent on vitamin K. The γ-carboxyglutamic acids in the NH_2-terminal domain chelate to calcium ions and form part of a binding site to membranes. Finally, the COOH-terminal segment contains the catalytic domain (serine protease domain). Activation of serine protease requires specific proteolysis just outside the catalytic domain (see Figure 3.15), and is controlled by the binding through the nine γ-carboxyglutamic acid residues at the NH_2-terminal end to a membrane.

There Are Specific Protein Inhibitors of Serine Proteases

Evolutionary selection of this enzyme family for participation in physiological processes requires a parallel evolution of controlling factors. Specific proteins have been identified that inhibit the activities of serine proteases after the proteases' physiological role has ended (Table 3.4). Thus coagulation is limited to the site of vascular injury and complementation activation leads to lysis only of cells exhibiting foreign antigens. The inability to control these protease systems, which may be caused by a deficiency of a specific protease inhibitor, can lead to undesirable conse-

TABLE 3.4 Some Human Proteins that Inhibit Serine Proteases

Inhibitor	Action
α_1-Proteinase inhibitor	Inhibits tissue proteases including neutrophil elastase, deficiency leads to pulmonary emphysema
α_1-Antichymotrypsin	Inhibits proteases of chymotrypsin-like specificity from neutrophils, basophils, and mast cells including cathepsin G and chymase
Inter-α-trypsin inhibitor	Inhibits broad range of serine protease activities in plasma
α_2-Antiplasmin	Inhibits plasmin
Antithrombin III	Inhibits thrombin and other coagulation proteases
C_1 Inhibitor	Inhibits complement reaction
α_2-Macroglobulin	General protease inhibitor
Protease nexin I	Inhibits thrombin, urokinase, and plasmin
Protease nexin II	Inhibits growth factor associated serine proteases, identical to NH_2-terminal domain of amyloid protein secreted in Alzheimer's disease
Plasminogen activator inhibitor I	Inhibits plasminogen activators
Plasminogen activator inhibitor II	Inhibits urokinase plasminogen activator

$$^{-}OOC-CH(COO^{-})-CH_2-CH(COO^{-})-\overset{+}{N}H_3$$

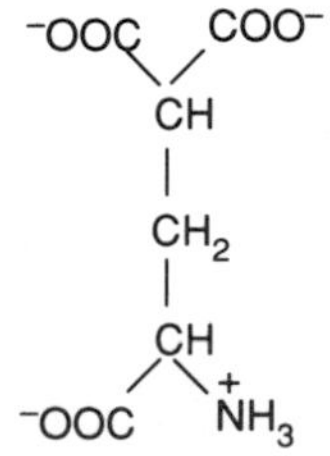

FIGURE 3.16
Structure of the derived amino acid γ-carboxyglutamic acid (abbreviation Gla), found in the NH_2-terminal domain of many clotting proteins.

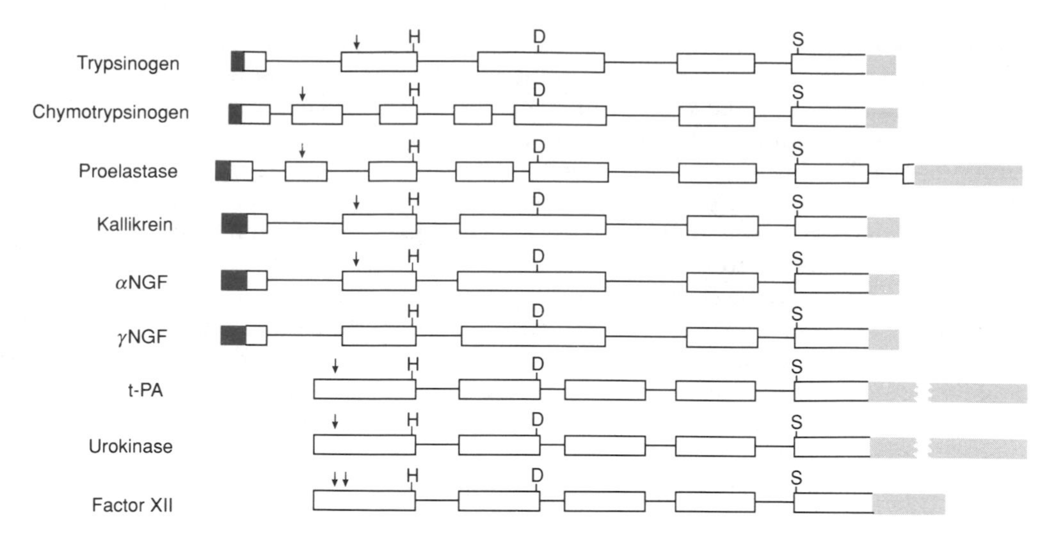

FIGURE 3.17
Organization of the exons and introns in the genes that code for serine proteases.
t-PA is tissue plasminogen activator, and NGF is nerve growth factor. Exons are shown by open boxes and introns by connecting lines. The position of the nucleotide codons for the active site serine, histidine, and aspartate are denoted by S, H, and D, respectively. Dark colored boxes, on left, show regions that code for an NH_2-terminal part of polypeptide chain (signal peptide) that is cleaved from the chain before protein is secreted from the cells. Light colored boxes, on right, represent a part of the gene sequence that is transcribed into messenger RNA (mRNA), but is not translated into protein. Arrows show codons for residues at which proteolytic activation of zymogen forms occur.
Based on a Figure in D. M. Irwin, K. A. Roberts, and R. T. MacGillivray, *J. Mol. Biol.* 200:31, 1988.

quences such as thrombi formation in myocardial infarction and stroke, or uncontrolled complement reactions in autoimmune disease. A class of natural inhibitors of serine proteases, termed **serpins** for *ser*ine *p*rotease *in*hibitors, appear to have evolved with related structures to control the activities of these proteases. This family of inhibitors is observed in animals where the proteases are also observed, but surprisingly it is also seen in plants where the proteases are not found.

Serine Proteases Have Similar Structure-Function Relationships

The understanding of the complex relationships between structure and physiological function in the serine proteases requires analysis of a number of intriguing observations. In summary: (a) Only one serine residue in the protein is catalytically active and participates in peptide bond hydrolysis. Thus bovine trypsin contains 34 serine residues with only one catalytically active or able to react with the inhibitor DFP (see Figure 3.12). (b) X-ray diffraction studies and amino acid sequence homology among the serine proteases demonstrates that there are two residues, a histidine and an aspartate, that are always associated with the activated serine in the catalytic site of the protein. Based on their position in the sequence of chymotrypsinogen, these three invariant active residues of serine proteases are named Ser 195, His 57, and Asp 102. This numbering based on their sequence number in chymotrypsinogen are used to identify these residues irrespective of their exact position in the primary structure of a particular serine protease. (c) The eucaryotic serine proteases exhibit a high amino acid sequence similarity and corresponding structural similarity with each other. (d) There exists an analogous organization of the genes that code for serine proteases (Figure 3.17). The exon–intron pattern of the gene structure for the proteases show that each of the catalytically essential amino acid residues [the Ser 195 (S), His 57 (H), and Asp 102 (D)] are on different exons. The catalytically essential histidine and serine are all almost immediately adjacent to their exon boundary. The active site of the enzyme is thus formed from amino acid side chains contained within three different exons. The homology in exon–intron organization exists not only for the many serine proteases in a particular organism but for the serine protease family of enzymes among the different eucaryotic species. The cross-species homology in serine protease gene structure supports the evolution of the serine proteases from a common primordial gene. (e) The catalytic unit of serine proteases exhibit two structural domains, of approximately equal size. The catalytic site is at the interface (crevice) between the two domains. (f) Serine proteases that function through a direct interaction with membranes typically have an additional domain to provide this function. (g) Natural protein substrates and inhibitors of serine proteases bind through an extended specificity site ($P_4 \cdots P'_4$) (see Figures 3.13 and 3.14). (h) Specificity with natural protein inhibitors is marked by extremely tight binding. The binding constant for trypsin to pancreatic trypsin inhibitor is in the order of 10^{13} M^{-1} reflecting a binding free energy of approximately 18 kcal mol^{-1}. (i) Natural protein inhibitors are usually poor substrates with strong inhibition by the inhibitor requiring hydrolysis of a peptide bond in the inhibitor by the protease. (j) The serine proteases are formed in inactive precursor zymogen forms to permit their production and transport in an inactive state to their sites of action. (k) Zymogen activation frequently involves a precursor serine protease. (l) Several serine proteases undergo **autolysis** or self-hydrolysis. Sometimes the self-reaction leads to specific peptide bond cleavage and activation of the catalytic activity. At other times autolysis leads to destruction of the catalytic activity of the protease.

Amino Acid Sequence Homology Occurs in the Serine Protease Family

Much of our early knowledge of the serine protease enzyme family came from trypsin and chymotrypsin purified from bovine materials obtained from a slaughterhouse. The natural tendency was to view this family of enzymes as serving primarily a function in protein digestion. This has yielded a nonintuitive form of nomenclature, which uses a sequence alignment against the amino acid sequence of chymotrypsin, to name and number residues in all the serine proteases. As mentioned previously, the catalytically essential residues are Ser 195, His 57, and Asp 102. Insertions and deletions of the amino acids in the primary structure of another

TABLE 3.5 Structural Superposition of Selected Serine Proteases and the Resultant Amino Acid Sequence Comparison

Comparison	Number of amino acids in sequence		Number of structurally equivalent residues	Number of structurally equivalent or chemically identical residues
	Protease 1	Protease 2		
Trypsin–elastase	223	240	188	81
Trypsin–chymotrypsin	223	241	185	93
Trypsin–mast cell protease	223	224	188	69
Trypsin–prekallikrein	223	232	194	84
Trypsin–*S. griseus* protease	223	180	121	25

serine protease, compared to the numbering of amino acid residues in the amino acid sequence of chymotrypsin, complicates the nomenclature of the second serine protease. Thus trypsin contains 22 less amino acids than chymotrypsin. The nomenclature forces trypsin in a part of its structure to have two residues sequentially linked but numbered Lys 204 and Leu 209 to account for the fewer amino acids in trypsin than in chymotrypsin. By contrast, this structural alignment of trypsin to chymotrypsin also forces trypsin to include two residues numbered 184 and two residues numbered 188. The data of Table 3.5 contrast the extent of structural superimposability with the comparison of the amino acid sequences that are brought into coincidence by the structural superposition. This table shows the total number of amino acids in each structure and then the number of amino acids that are statistically identical, by X-ray diffraction, in their topological position, even if they are chemically different amino acids. *Topologically equivalent amino acids have the same relationship in three-dimensional space to the point where they cannot be distinguished from one another.* Finally, the last column of the table gives the number of amino acids that are chemically identical from those amino acids that are structurally indistinguishable.

Serine protease family enzymes of similar structure are found in procaryotes as well. Thus the bacterial serine protease from *Strepyomyces griseus* and *Myxobacter* 450 are shown to have a structural and functional homology with trypsin. An apparently separate class of serine protease enzymes have been isolated, however, from bacteria that have no structural homology to the mammalian serine protease family. The serine protease subtilisin isolated from *Bacillus subtilis* hydrolyzes peptide bonds and contains an activated serine with a histidine and aspartate in its active site but the active site arises from structural regions of the protein that bear no sequence or structural homology with the mammalian serine proteases. This bacterial serine protease is an example of a genetically distinct class of serine proteases from those discussed above, and may provide evidence for the **convergent evolution** of an enzyme catalytic mechanism. Apparently a gene completely different than those that code for the eucaryotic serine proteases evolved the same catalytic mechanism for peptide bond hydrolysis, utilizing an active site serine activated by its interaction with a histidine and aspartate. The primary and tertiary structure that contains the catalytic site, however, is different from that of the trypsin- and chymotrypsin-like structure.

The Tertiary Structure of Serine Proteases Are Similar

As described previously, Ser 195 in chymotrypsin reacts with diisopropylfluorophosphate (DFP), which proceeds with a 1 : 1 enzyme : DFP stoichiometry and inhibits the enzymes ability to catalyze peptide bond hydrolysis. Determination of the three-dimensional structure of chymotrypsin reveals that the Ser 195 is situated within an internal pocket, with access to the solvent interface. In addition, the residues His 57 and Asp 102 are oriented in three-dimensional space so that they participate together with the Ser 195 in the catalytic mechanism of the enzyme (see Chapter 4). Furthermore, these three active site residues are invariant in all serine proteases and the sequences surrounding these residues in their primary structure are invariant among the known eukaryotic serine proteases (see Table 3.6).

Since the first structure determination of α-chymotrypsin, structure determinations by X-ray crystallography have been carried out on a wide variety of members of this class of proteins (see Table 3.7). Structural data is available for catalytically active enzyme forms, zymogens, the same enzyme in multiple species, enzyme–inhibitor complexes, a particu-

TABLE 3.6 Invariant Sequences Found Around the Catalytically Essential Serine (S) and Histidine (H)

Enzyme	Sequence (identical residues to chymotrypsin are in bold)
	RESIDUES AROUND CATALYTICALLY ESSENTIAL HISTIDINE
	* (above H)
Chymotrypsin A	F **H** **F** C **G** **G** **S** **L** **I** N **E** N **W** **V** **V** **T** **A** **A** **H** **C** G V T T S D
Trypsin	Y **H** **F** C **G** **G** **S** **L** **I** N S Q **W** **V** **V** S **A** **A** **H** **C** Y K S G I Q
Panc. elastase	A **H** T C **G** **G** T **L** **I** R Q **N** **W** **V** M **T** **A** **A** **H** **C** V D R E L T
Thrombin	E L L C **G** A **S** **L** **I** S D R **W** **V** L **T** **A** **A** **H** **C** L L Y P P W
Factor X	E G **F** C **G** **G** T I L **N** **E** F Y **V** L **T** **A** **A** **H** **C** L H Q A K R
Plasmin	M **H** **F** C **G** **G** T **L** **I** S P E **W** **V** L **T** **A** **A** **H** **C** L E K S P R
Pl. kallikrein	S F Q C **G** **G** V **L** V **N** P K **W** **V** L **T** **A** **A** **H** **C** K N D N Y E
Strep. trypsin	— — — C **G** **G** A **L** Y A Q D I **V** L **T** **A** **A** **H** **C** V S G S G N
Subtilisin	V G G A S F V A G E A Y N T D G N G **H** G T H V A G T
	RESIDUES AROUND CATALYTICALLY ESSENTIAL SERINE
	* (above S)
Chymotrypsin A	**C** **A** **G** — — — A S **G** **V** — — **S** **S** **C** **M** **G** **D** **S** **G** **G** **P** **L** **V**
Trypsin	**C** **A** **G** Y — — L E **G** G K — D **S** **C** Q **G** **D** **S** **G** **G** **P** V **V**
Panc. elastase	**C** **A** **G** — — — G N **G** **V** R — **S** G **C** Q **G** **D** **S** **G** **G** **P** **L** **H**
Thrombin	**C** **A** **G** Y K P G E **G** K R G D A **C** E **G** **D** **S** **G** **G** **P** F **V**
Factor X	**C** **A** **G** Y — — D T **Q** P E — D A **C** Q **G** **D** **S** **G** **G** **P** H **V**
Plasmin	**C** **A** **G** H — — L A **G** G T — D **S** **C** Q **G** **D** **S** **G** **G** **P** **L** **V**
Pl. kallikrein	**C** **A** **G** Y — — L P **G** G K D T **C** **M** **G** **D** **S** **G** **G** **P** **L** I
Strep. trypsin	**C** **A** **G** Y — P D T **G** G V — D T **C** Q **G** **D** **S** **G** **G** **P** M F
Subtilisin	A G V Y S T Y P T N T Y A T L N **G** T **S** M A S P H

SOURCE: A. J. Barrett, in *Proteinase Inhibitors*, A. J. Barrett and G. Salvesen (Eds.), Elsevier, Amsterdam: 1986, p. 7.

TABLE 3.7 Serine Protease Structures Determined by X-Ray Crystallography

Enzyme	Species source	Inhibitors present	Resolution (Å)
Chymotrypsin[a]	Bovine	Yes[c]	1.67[b]
Chymotrypsinogen	Bovine	No	2.5
Elastase	Porcine	Yes	2.5
Kallikrein	Porcine	Yes	2.05
Proteinase A	*S. griseus*	No	1.5
Proteinase B	*S. griseus*	Yes	1.8
Proteinase II	Rat	No	1.9
Trypsin[a]	Bovine	Yes[c]	1.4[b]
Trypsinogen[a]	Bovine	Yes[c]	1.65[b]

[a] Structure of this enzyme molecule independently determined by two or more investigators.
[b] Highest resolution for this molecule of the multiple determinations.
[c] Structure obtained with no inhibitor present (native structure) as well as with inhibitors. Inhibitors used include low molecular weight inhibitors (i.e., benzamidine, DFP, and tosyl) and protein inhibitors (i.e., bovine pancreatic trypsin inhibitor).

lar enzyme at different temperatures, and in different solvents. The most complete analysis is that of trypsin. Its X-ray diffraction structure has been determined at better than 1.7-Å resolution. At this resolution one can resolve atoms at a separation of 1.3 Å, which approaches the C=O separation of the carbonyl group. This resolution represents a limiting value, however, and is not uniform over the entire trypsin structure. Different regions of the molecule have a different tendency to be localized in space during the time course of the X-ray diffraction experiment and for some atoms in the structure their exact position cannot be as precisely defined as for others. The structural disorder is especially apparent in those residues on the surface of the molecule not in contact with neighboring molecules.

The X-ray structure clearly shows trypsin is globular in its overall shape and consists of two domains of approximately equal size (Figure 3.18). The two domains do not penetrate one another, although the NH_2-termini and COOH-termini of the full molecule extend over the surface of the alternative domains. The domain organization is most readily seen in Figures 3.18*b* and 3.19, which displays the contoured distance matrix for trypsin.

The secondary structure of trypsin has little α-helix, except in the COOH-terminal region of the molecule. The structure is predominantly β structure, with each of the domains in a "deformed" β-barrel. These barrels are not as regular as those observed in the immunoglobulin fold but are oriented such that one will come into approximate coincidence with the other. The similarity of the two domains is further seen by examination of Figure 3.19, which shows that domain A and B have an

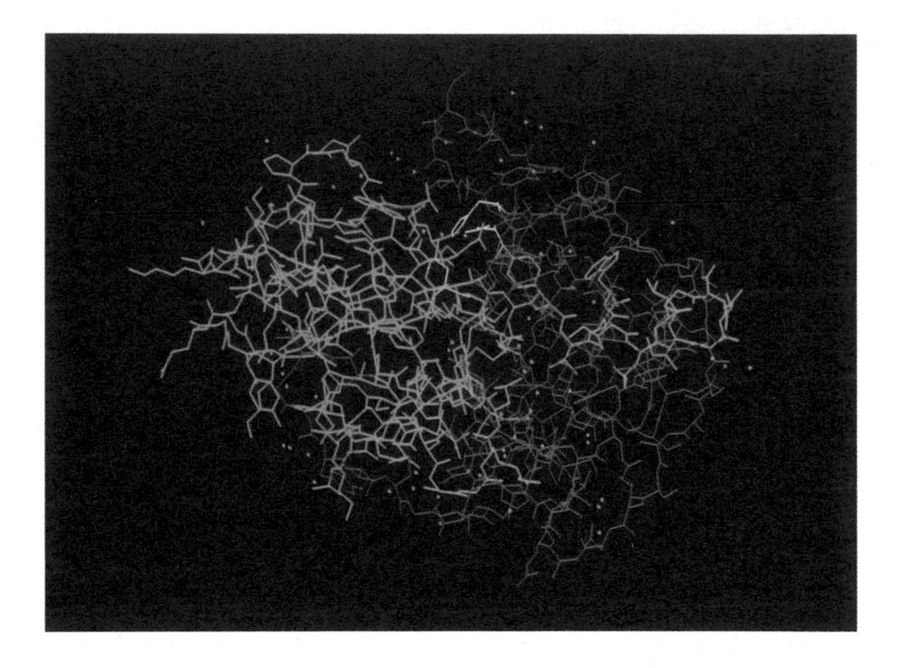

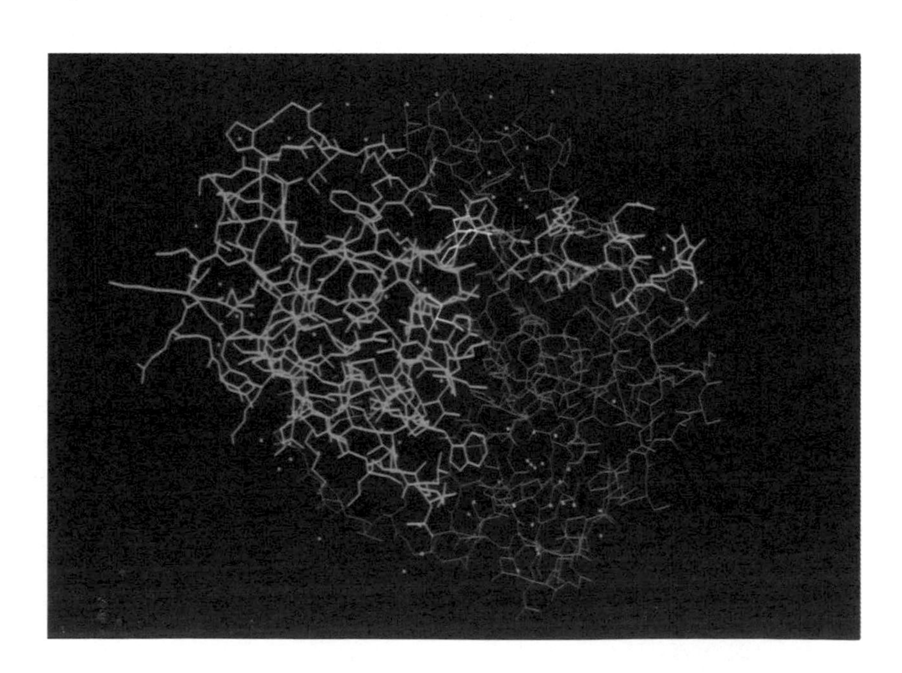

FIGURE 3.18
Two views of the structure of trypsin showing the tertiary structure of the two domains.
Active site serine, histidine, and aspartate indicated in yellow.

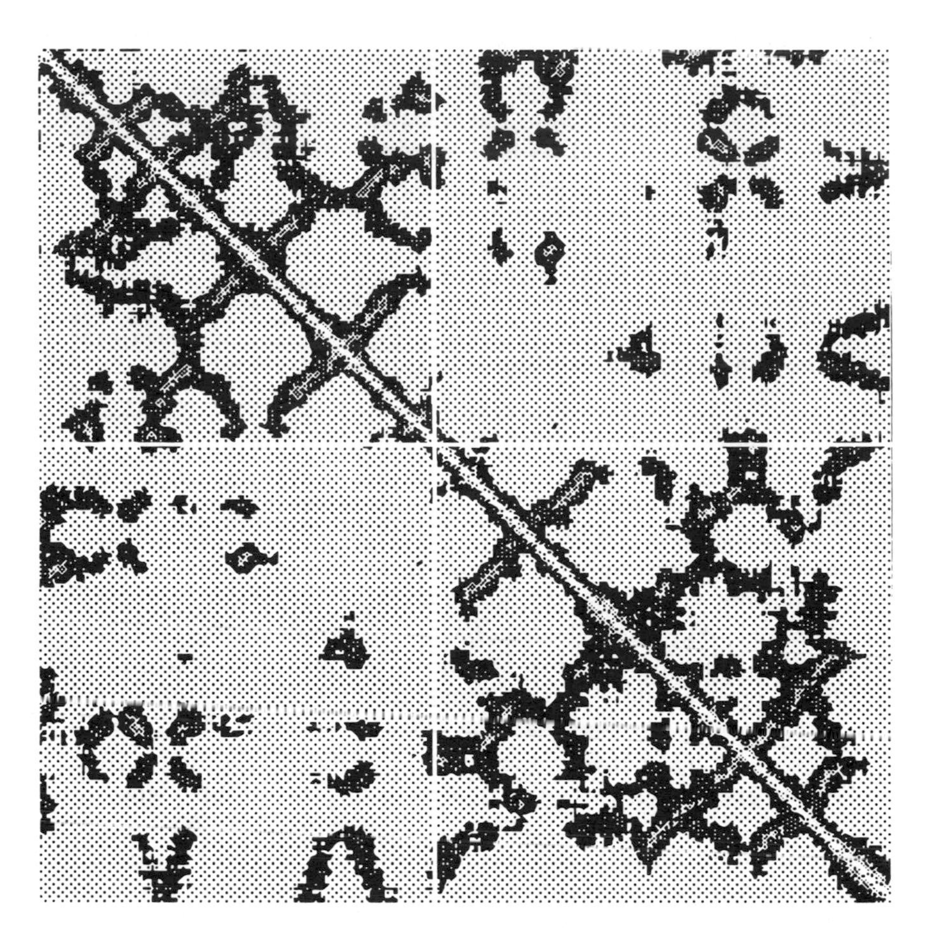

FIGURE 3.19
Interresidue distance matrix of trypsin showing shaded distances of less than 15 Å.
The two separate β-barrel domains of the trypsin molecule as well as the interdomain interactions are presented. Parallel β structure is shown by interresidue interactions parallel to the diagonal, and antiparallel β structure by interactions offset but perpendicular to the diagonal.

approximate palindromic-type (a sequence that reads the same backwards or forwards) structural organization. This palindrome results from the observation that the two domains have an analogous folding at the tertiary structural level when comparing the NH_2-terminus to residue 122 from one end of the molecule compared to the COOH-terminal residue backwards to residue 123 in the other one-half of the molecules.

The result of the two distorted β-barrel domain structures is that the loop regions, which protrude from the barrel ends, are almost symmetrically presented by each of the two folded domains. These loop structures combine to form a surface region of the enzyme that extends outward, above the catalytic site region of the S_1 substrate binding site.

An alignment of the three-dimensional structure can be performed on the serine proteases by a mathematical algorithm, which provides the greatest structural equivalence and allows for insertions and deletions of amino acids in a particular sequence. It is interesting to note in these structural alignments that the regions of difference appear to be localized to the loop regions, which extend from the β-barrel domains to form the surface region out from the catalytic site. The effect of altering the amino acids in these loops is to alter the **macromolecular binding specificity** for the protease. It is the structure of the loop in Factor Xa, for example, that allows this protease to specifically bind to prothrombin. Serpins interact with different proteases based on their affinity for the loop structures that generate their extended binding sites. Bacterial proteases related to the eucaryotic serine protease family contain the same two domains as do the eucaryotic proteases of the family but lack most of the loop structures. This is in line with the lack of a requirement of the bacterial proteases for complex interactions that the eucaryotic protease must carry out in multicellular organisms and the corresponding observation that bacterial proteases are not produced in an inactive zymogen form.

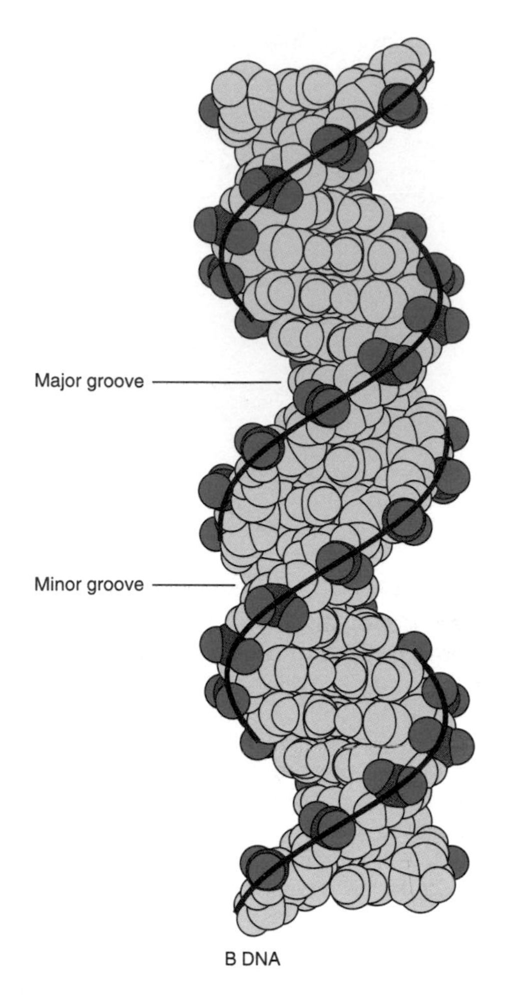

FIGURE 3.20
Space-filling model of DNA in B conformation showing major and minor groove.

Thus the serine protease family of proteins constitute a structurally related series of proteins with a similar utilization of a catalytically activated serine. During evolution, the basic two domain structure and the catalytically essential residues have been maintained, but the region of the secondary interactions (loop regions) have changed to give the different proteins of the family their different specificities towards protein substrates, activators, and inhibitors, characteristic of the important physiological function of the individual serine proteases.

3.4 DNA BINDING PROTEINS

Regulatory sites exist near genes in the DNA that regulate gene expression. These sites in DNA contain a nucleotide sequence that forms a binding site for regulatory proteins. The binding of the regulatory protein to its DNA sequence can promote either an activation or a repression of the rate of gene transcription into mRNA, the initial step in gene expression. Along the helical spiral of a DNA molecule in its most common form (B form) are two grooves, the major and minor grooves (Figure 3.20) (See Chapter 15). The major groove can nicely accommodate a small α-helix region of a protein, allowing noncovalent interactions (hydrogen bonding, ionic, and van der Waals) with the DNA. These noncovalent binding interactions formed between the protein and DNA give the protein an ability "to read" the nucleotide sequence and specifically bind to the regulatory site of a particular gene. This is a highly selective feat as the human genome has tens of thousands of genes, each with its own regulatory sequences. The specific DNA sequence that forms the protein binding site, known as the **binding element,** is usually less than 10 nucleotides long. The tertiary structure of the DNA in the binding site is also a factor in binding of regulatory proteins. Very little is currently known about how binding proteins regulate gene expressions, however, certain structural motifs in some of the DNA binding proteins are apparent.

Three Major Structural Motifs of DNA Binding Proteins

A type of structural motif found within many DNA binding proteins is the **helix-turn-helix** as presented in Figure 3.21. A current model for the association of this structural motif with the DNA places the short α-helix across the major groove of DNA, and the second α helix partly within the major groove, where it makes specific residue–base interactions with the DNA. As proteins containing this motif most often bind as dimers, there will be two helix-turn-helix motifs per active regulatory protein. The structures obtained for some of these proteins show the two helix-turn-helix motifs separated by about 36 Å in the dimer protruding from the surface of the folded protein. The two protrusions can orient at roughly the same angle to each other as the major grooves in DNA to which it binds. The two helix-turn-helix structures of the dimer protein may thus bind into adjacent major grooves, without significant perturbations in the protein structure (Figure 3.22).

The **zinc finger** motif is a second type of structure found in some DNA binding proteins. Zinc finger proteins contain repeating motifs of a Zn^{2+} atom bonded to four amino acid side chains as shown in Figure 3.23, in which the Zn^{2+} atom is bound to two cysteine and two histidine side chains. In some cases the histidines may be substituted by additional cysteines. The primary structure for the motif will thus contain two close cysteines separated from a second pair of Zn^{2+} liganding amino acids (histidine or cysteine) by about 12 amino acids (Figure 3.23). How do zinc finger structures bind to the DNA polymer? An X-ray structure of a

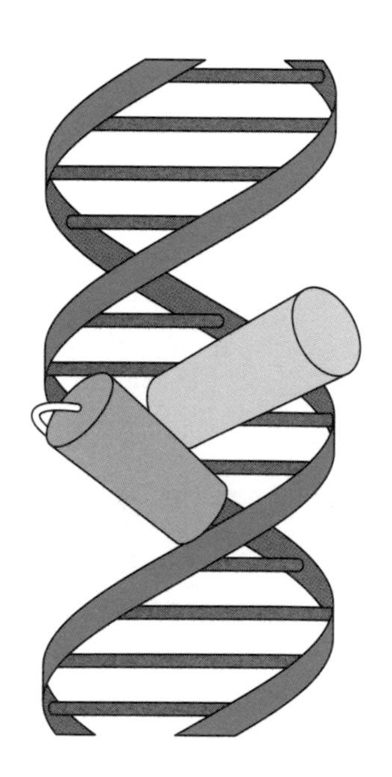

FIGURE 3.21
Binding of a helix-turn-helix motif into the major groove of B-DNA.
Redrawn from R. Schleif, *Science* 241:241, 1988.

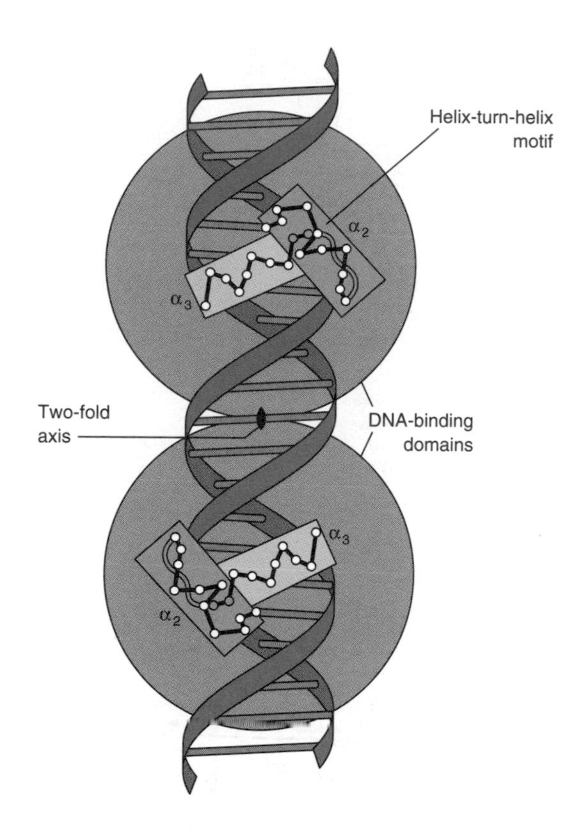

FIGURE 3.22
Association of a DNA binding protein (dimer) with helix-turn-helix motifs into adjacent major grooves of B-DNA.
Redrawn from R. G. Brennan and B. W. Matthews, *Trends Biochem. Sci.* 14:287 (Fig. 1, Part b), 1989.

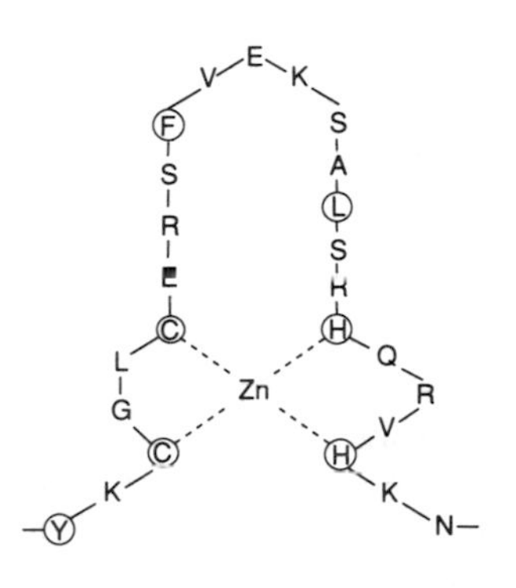

FIGURE 3.23
Primary sequence of a zinc finger motif found in the DNA binding protein *Xfin* from *Xenopus*.
Invariant and highly conserved amino acids in the structure are circled.
Redrawn from M. S. Lee, G. P. Gippert, K. V. Soman, D. A. Case, and P. E. Wright, *Science* 245:635, 1989.

protein containing a zinc finger has been obtained; the three-dimensional structure of one such zinc finger has been deduced by ^{1}H NMR (Figure 3.24). The motif contains an α-helix segment that can bind within the major groove at the regulatory site of the DNA and make specific interactions with the DNA sequence in the regulatory site.

A third structural motif found in some of the DNA binding regulatory proteins is the **leucine zipper.** Leucine zippers are formed from a region of α-helix containing at least 4 leucines, each leucine separated by 6 amino acids from one another (i.e., Leu-X_6-Leu-X_6-Leu-X_6-Leu, where X is any of the common amino acids). As there are 3.6 residues per turn of the α-

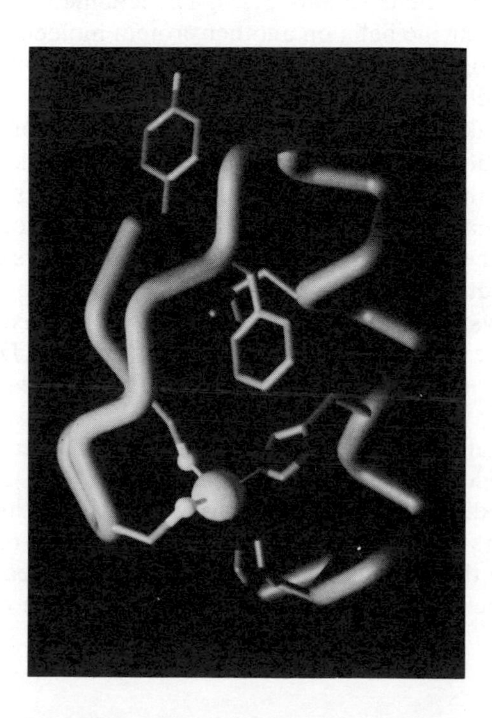

FIGURE 3.24
Three-dimensional structure obtained by ^{1}H NMR of zinc finger motif from the *Xenopus* protein *Xfin* (sequence shown in Figure 3.23).
Superposition of 37 possible structures derived from calculations based on the ^{1}H NMR. The NH_2-terminal is at the upper left and the COOH-terminal is at the bottom right. The zinc is the sphere at the bottom with the Cys residues to the left and the histidines to the right.
Photograph provided by Michael Pique and Peter E. Wright, Department of Molecular Biology, Research Institute of Scripps Clinic, La Jolla, California.

Leu
Leu
Leu
Leu
22
15
8
1
Ile
Glu
Arg
Asp
Phe
Leu
Val
Asn
4
5
Arg 7
Gln
Thr
Gly
2 Thr
Arg
Ser
Arg
3
6
Ser
Lys
Arg
Gly
Asp
Glu
Asp
Arg

(a)

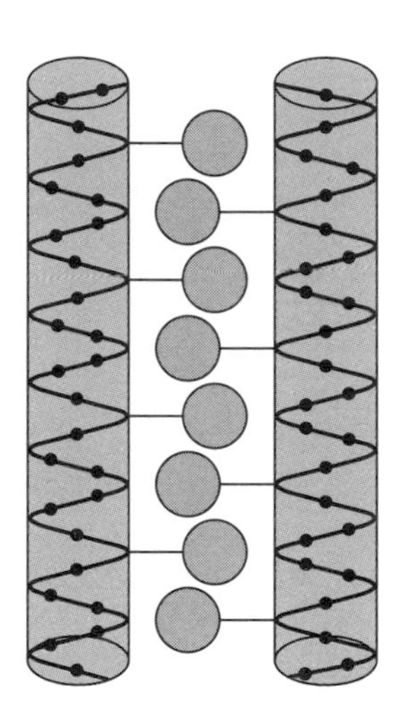

(b)

FIGURE 3.25
Leucine zipper of DNA binding proteins.
(*a*) Helical wheel analysis of the leucine zipper motif in DNA enhancer binding protein. The amino acid sequence in the wheel analysis is displayed end-to-end down the axis of a schematic α-helix structure. The leucines (L) are observed in alignment along one edge of the helix (residues 1, 8, 15, and 22 in the sequence). Amino acids are designated by their one letter abbreviations.
(*b*) Schematic diagram showing how the leucines between two α-helix leucine zipper motifs can interdigitate.
Redrawn from W. H. Landschulz, P. F. Johnson, and S. L. McKnight, *Science* 240:1759, 1988.

helix, the leucines align on one edge of the helix, with a leucine at every second turn of the helix (Figure 3.25). The leucine helix can interdigitate with a second leucine helix on another protein molecule, to "zipper" the two proteins together (Figure 3.26). The leucine zipper motif does not directly interact with the DNA, as do the zinc finger or helix-turn-helix motifs, but facilitates dimerization of two protein subunits. Mutations in the zipper motif show that if the dimmer is not formed by association of the protomers through the zipper, the protein will not strongly bind to the DNA. Just adjacent to the α-helix of the zipper motif in the primary structures of the DNA binding proteins, there appears a region of amino acids in the sequences containing large numbers of the basic amino acids, arginine and lysine. These evolutionary conserved basic residues are believed to be the region in the protein that binds to the DNA. The positive charges of the arginine and lysine side chains are drawn to the negatively charged DNA.

Many regulatory proteins with the leucine zipper motif have been shown to be oncogene products with names like *Myc, Jun,* and *Fos*. Fos forms a heterodimer with Jun through a leucine zipper interaction, and the Fos/Jun dimers bind to gene regulatory sites. If the regulatory proteins are mutated or these proteins, themselves, are produced in an unregulated manner, the cell can be transformed to a phenotype characteristic of a cancer cell.

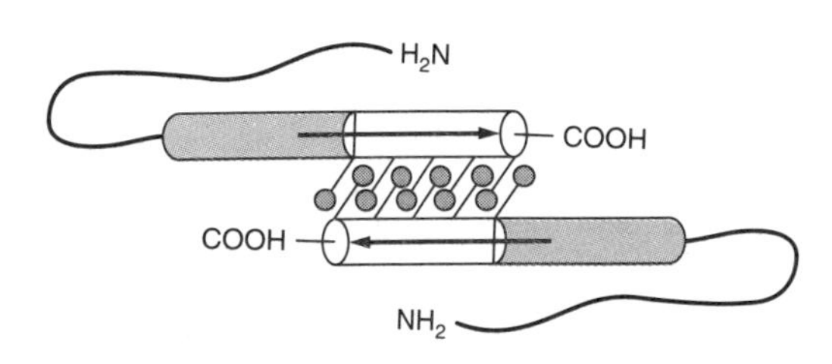

FIGURE 3.26
Schematic diagram of two proteins with leucine zippers in antiparallel association.
The DNA binding domains containing a high content of basic amino acids (arginines and lysines) are shown by the stripped boxes.
Redrawn from W. H. Landschulz, P. F. Johnson, and S. L. McKnight, *Science* 240:1759, 1988.

3.5 HEMOGLOBIN AND MYOGLOBIN

The hemoglobins are globular proteins, in high concentrations in red blood cells, that bind oxygen in the lungs and transport the oxygen in blood to the tissues and cells around the capillary beds of the vascular system. On returning to the lungs from the capillary beds, hemoglobins act to transport carbon dioxide and protons. In this section the structural and molecular aspects of hemoglobin and myoglobin are described. The physiological roles of these proteins are discussed in Chapter 25.

Human Hemoglobin Occurs in Several Forms

A hemoglobin molecule consists of four polypeptide chains of two different primary structures. In the common form of human adult hemoglobin, **HbA_1,** two chains of one kind are designated the α chains, and the second two chains of the same kind are designated the β chains. The polypeptide chain composition of HbA_1 is therefore $\alpha_2\beta_2$. The α-polypeptide chain contains 141 amino acids and the β-polypeptide chain contains 146 amino acids.

While HbA_1 is the major form of hemoglobin in the adult human, other forms of hemoglobin predominate in the blood of the human fetus and early embryo. The fetal form **(HbF)** contains two of the same α chains found in HbA_1, but their second kind of chain (γ chain) in the tetramer molecule differs in amino acid sequence from that of the β chain of adult HbA_1 (Table 3.8). Additional forms appear in the first months after conception (embryonic) in which the α chains are substituted for by *zeta* (ζ) chains of different amino acid sequence and the ε chains serve as the β chains. A minor form of adult hemoglobin, **HbA_2,** comprises about 2% of normal adult hemoglobin and is composed of two α chains and two chains designated *delta* (δ) (Table 3.8).

TABLE 3.8 Chains of Human Hemoglobin

Developmental stage	Symbol	Chain designations
Adult	HbA_1	$\alpha_2\beta_2$
Adult	HbA_2	$\alpha_2\delta_2$
Fetus	HbF	$\alpha_2\gamma_2$
Embryo	Hb Gower-1	$\zeta_2\varepsilon_2$
Embryo	Hb Portland	$\zeta_2\gamma_2$

The globin genes that code for the hemoglobin polypeptide chains are organized into two clusters. The human genes for the α-type polypeptide chains are located on chromosome 16 and the genes for the β-type polypeptide chains on chromosome 11. These genes are diagrammatically drawn in Figure 3.27. The dark boxes represent the globin genes that are expressed, and the clear boxes are **pseudogenes.** These later genes are whole but because of a deficiency in their structure, these genes are never expressed as polypeptide chains.

The α-like gene designated ζ is expressed in the early embryo (first 6-weeks post conception) and forms an α-like chain in hemoglobin molecules named Gower 1 ($\zeta_2\varepsilon_2$) and hemoglobin Portland ($\zeta_2\gamma_2$). The ζ chains

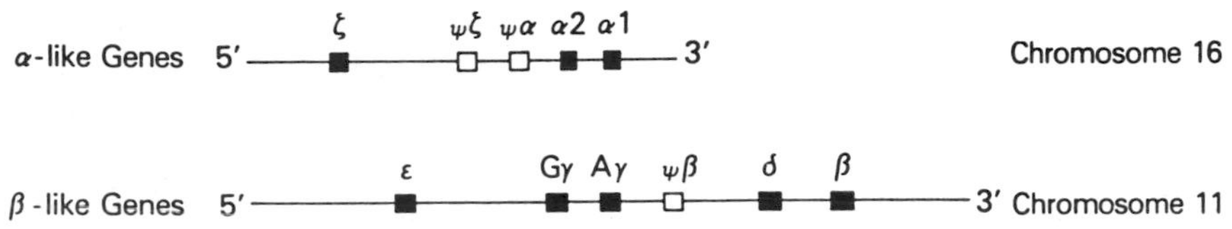

α-like Genes 5′ — ζ — $\psi\zeta$ $\psi\alpha$ $\alpha2$ $\alpha1$ — 3′ Chromosome 16

β-like Genes 5′ — ε — $G\gamma$ $A\gamma$ $\psi\beta$ δ β — 3′ Chromosome 11

FIGURE 3.27
The human globin gene clusters.
The α-like genes on chromosome 16 and the β-like genes on chromosome 11. Open boxes represent the location of pseudogenes. Pseudogenes code for a globin chain but are not expressed. The G_γ and A_γ are allotypes of the γ gene found in HbF.
Based on A. W. Nienhuis and T. Maniatis, in *The Molecular Basis of Blood Diseases*, G. Stamatoyannopoulos, A. W. Nienhuis, P. Leder, and P. W. Majerus (Eds.), Philadelphia: W. B. Saunders, 1987, p. 32.

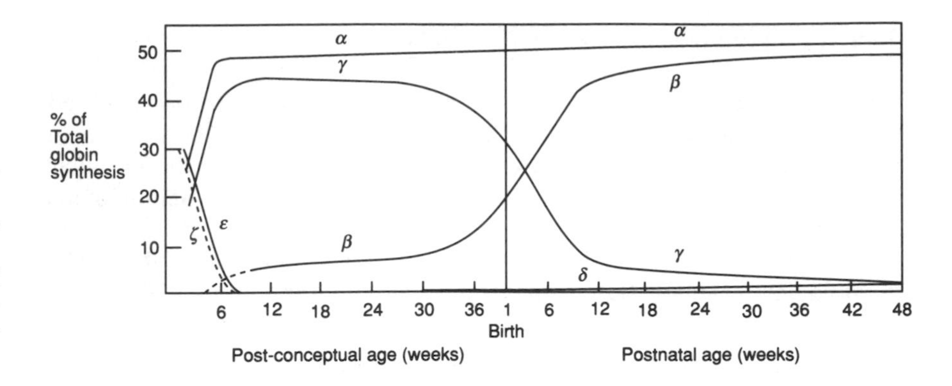

FIGURE 3.28
Changes in globulin chain production during development.
Based on a figure in A. W. Nienhuis and T. Maniatis, in *The Molecular Basis of Blood Diseases*, G. Stamatoyannopoulos, A. W. Nienhuis, P. Leder, and P. W. Majerus (Eds.), Philadelphia: W. B. Saunders, 1987, p. 68, in which reference of Weatherall, D. J. and Clegg, J. B., *The Thalassemia Syndrome*, 3rd ed., Oxford: Blackwell Scientific Publications, 1981, is acknowledged.

are then replaced by α chains in the fetus where the predominant hemoglobin molecule is hemoglobin F (HbF, $\alpha_2\gamma_2$). In the adult the predominant hemoglobin is HbA_1 ($\alpha_2\beta_2$) with minor amounts of HbA_2 ($\alpha_2\delta_2$) (Figure 3.28). There are two complete genes in the α-like gene cluster that code for the α chain (α_1 and α_2, Figure 3.27). The polypeptide chains produced by the α_1 and α_2 genes are identical.

In the β-like gene cluster on chromosome 11, there are five functional genes. One each of the ε, δ, and β chains. Two γ genes are functional and produce γ chains for HbF. The two γ gene products are identical with the exception of either a glycine (G_γ) or alanine (A_γ) at position 136 of the polypeptide chain.

Myoglobin Is a Single Polypeptide Chain With One O_2 Binding Site

Myoglobin (Mb) is an O_2 carrying protein that binds and releases O_2 with changes in the oxygen concentration in the cytoplasm of muscle cells. In contrast to hemoglobin, which has four polypeptide chains and four O_2 binding sites, myoglobin is composed of only a single polypeptide chain and a single O_2 binding site.

A comparison between myoglobin and hemoglobin is instructive in that myoglobin is a model for what occurs when a single protomer molecule acts alone without the interactions exhibited among the four O_2 binding sites in the more complex tetramer molecule of hemoglobin.

A Heme Prosthetic Group Is at the Site of O_2 Binding

The four polypeptide chains in hemoglobin and the single polypeptide chain of myoglobin each contain a heme prosthetic group. A **prosthetic group** is a nonpolypeptide moiety that forms a part of a protein in its native functional state. A protein without its prosthetic group is designated an **apoprotein.** A complete protein with its prosthetic group is a **holoprotein.**

The heme is a porphyrin molecule containing an iron atom in its center. The type of porphyrin found in hemoglobin and myoglobin (protoporphyrin IX) contains two propionic acid, two vinyl, and four methyl side chain groups attached to the pyrrole rings of the porphyrin structure (Figure 3.29). The iron atom is in the ferrous (2+ charge) oxidation state in functional hemoglobin and myoglobin.

The ferrous atom in the heme can form five or six ligand bonds, depending on whether or not O_2 is bound to the protein. Of the five or six bonds, four are to the pyrrole nitrogen atoms of the porphyrin. Since all the pyrrole rings of the porphyrin lie in a common plane, the four ligand bonds from the porphyrin to the iron atom at its center will also have a

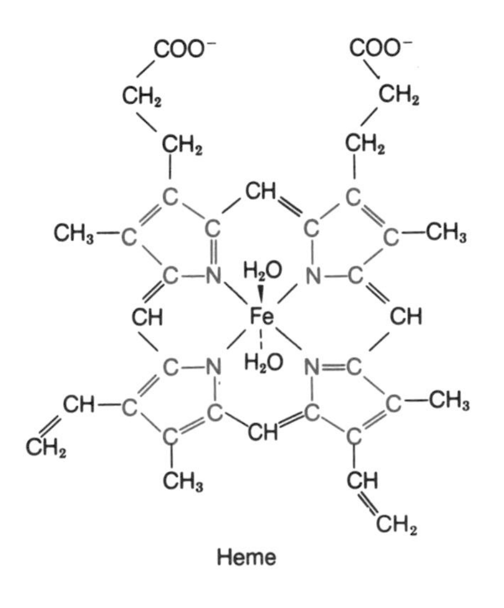

FIGURE 3.29
Structure of heme.

tendency to lie in the plane of the porphyrin ring. This is especially true for six-coordinate ferrous iron when O_2 is bound to Hb (oxyhemoglobin). (A later section describes how the five-coordinate bond ferrous atom of deoxyhemoglobin sits out of the plane of the porphyrin rings by about 0.6 Å.) The fifth and the potentially sixth ligand bonds to the ferrous atom of the heme are directed along an axis perpendicular to the plane of the porphyrin ring (Figure 3.30).

The fifth coordinate bond of the ferrous atom in each of the hemes is to a nitrogen of a histidine imidazole. This histidine is designated the **proximal histidine** in the hemoglobin and myoglobin structures (Figures 3.30 and 3.31).

In each of the polypeptide chains with O_2 bound, the O_2 molecule forms a sixth coordinate bond to the ferrous atom. In this bonded position the O_2 is placed between the ferrous atom to which it is liganded and a second histidine imidazole, designated the **distal histidine,** in hemoglobin and myoglobin structures. In deoxyhemoglobin, the sixth coordinate position (O_2 binding position) of the ferrous atom is unoccupied.

The porphyrin part of the heme is positioned within a hydrophobic pocket formed in each of the polypeptide chains. In the heme pocket X-ray diffraction studies show that approximately 80 interactions are provided by approximately 18 amino acids to the heme. Most of these noncovalent interactions are between apolar side chains of amino acids and the nonpolar regions of the porphyrin. As discussed in Chapter 2, the driving force for these interactions is the release of water of solvation on association of the hydrophobic heme with the apolar residues of the heme pocket of the protein. In myoglobin additional noncovalent interactions are made between the negatively charged propionate groups of the heme and positively charged arginine and histidine R groups of the protein. However, in hemoglobin chains a difference in the amino acid sequence in this region of the heme binding site leads to the stabilization of the porphyrin propionates by interaction with an uncharged histidine imidazole and with water molecules of the solvent toward the outer surface of the molecule.

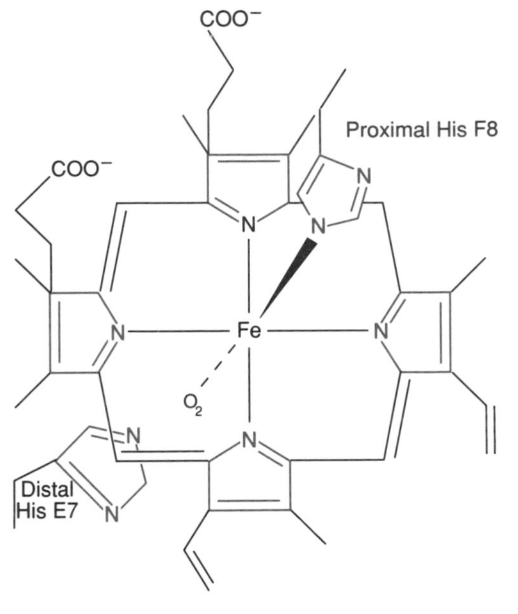

FIGURE 3.30
The ligand bonds to the ferrous atom in oxyhemoglobin.

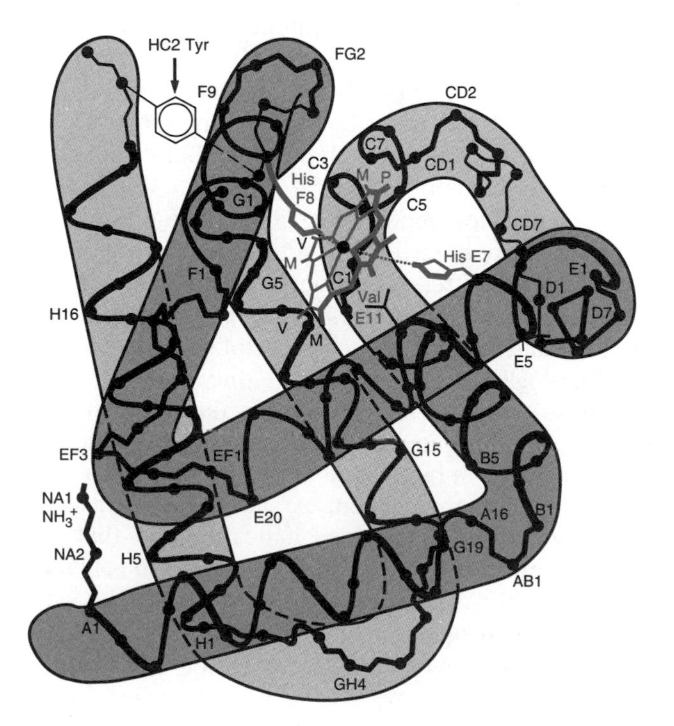

FIGURE 3.31
Secondary and tertiary structure characteristics of chains of hemoglobin.
The proximate His F8, distal His E7, and Val E11 side chains are shown. The other amino acids of the polypeptide chain are represented by their α-carbon positions only; the letters M, V, and P refer to the methyl, vinyl, and propionate side chains of the heme.
Redrawn with permission, from M. Perutz, *Br. Med. Bull.* 32:195, 1976.

X-Ray Crystallography Has Assisted in Defining the Structure of Hemoglobin and Myoglobin

Information on the structure of the deoxy and oxy forms of hemoglobin and myoglobin has primarily come from X-ray diffraction crystallography. In fact, sperm whale myoglobin was the first globular protein whose three-dimensional structure was determined by the technique of X-ray crystallography. The determination of the structure of myoglobin was soon followed by the X-ray structure of the more complex horse hemoglobin molecule. The mechanism of cooperative associations of O_2, discussed below, is based on model building from the X-ray structures of oxyhemoglobin, deoxyhemoglobin, and a variety of hemoglobin derivatives.

Primary, Secondary, and Tertiary Structures of Myoglobin and the Individual Hemoglobin Chains

The amino acid sequences for the polypeptide chain of myoglobin found in 23 different animal species have been determined. These myoglobins contain 153 amino acids in their polypeptide chains, of which 83 positions are shown to be invariant in the comparison of the amino acid sequences. Only 15 of these invariant residues in the myoglobin sequence are identical to the invariant residues of the currently sequenced mammalian hemoglobin chains. It should be noted, however, that the changes in residues at particular positions in the primary structure are, in the great majority of cases, varied in a conservative manner (Table 3.9). Since myoglobin is active as a monomer unit, many of its surface positions are formed to interact with water and prevent another molecule of myoglobin from associating. In contrast, the surface residues of the individual chains in hemoglobin are designed to provide hydrogen bonds and nonpolar contacts with other subunits in the hemoglobin quaternary structure. The proximal and distal histidines are, of course, preserved in the sequences of all the polypeptide chains. Other invariant residues are in the hydrophobic heme pocket and form essential nonpolar contacts with the heme that stabilizes the heme–protein complex. In addition, particular prolines in the sequence, which act to break some of the helical sections to allow the chain to fold back upon itself, are predominantly retained in most of the chains.

While there is a surprising variability in amino acid sequence among the different polypeptide chains, to a first approximation the secondary and tertiary structures of each of the polypeptide chains of hemoglobin and myoglobin appear to be almost identical (Figure 3.32).

The distance matrix for hemoglobin chains (Figure 3.33) reveals the high degree of similarity in tertiary structure among the different chains, as well as the specific differences due to insertions in the amino acid sequence. The significant differences in the physiological properties between the α, β, γ, and δ chains in the hemoglobins and the single polypeptide chains of myoglobin are, therefore, due to rather small specific changes in their structures. The similarity in tertiary structure, resulting from widely varied amino acid sequences, show that the same tertiary structure for a protein can be arrived at in many different ways.

The X-ray crystallographic structures show that each of the polypeptide chains are composed of multiple α-helical regions that are broken by turns of the polypeptide chain, allowing the protein to fold into a spheroidal shape (Figure 3.32). Approximately 70% of the residues in the protein participate in the α-helical secondary structure, which generates seven helical segments in the α chain and eight helical segments in the β chain.

TABLE 3.9 Amino Acid Sequences of Human Hemoglobin Chains and of Sperm Whale Myoglobin[a]

	NA 1	2	3	A 1	2	3	4	5	6	7	8	9	10	11	12	13	14	15	A 16	AB 1	B 1	2	3	4	5	6
MYOGLOBIN	val	.....	leu	ser	glu	gly	glu	trp	gln	leu	val	leu	his	val	trp	ala	lys	val	glu	ala	asp	val	ala	gly	his	gly
HEMOGLOBIN Horse α	val	.....	leu	ser	ala	ala	asp	lys	thr	asn	val	lys	ala	ala	trp	ser	lys	val	gly	gly	his	ala	gly	glu	tyr	gly
β	val	gln	leu	ser	gly	glu	glu	lys	ala	ala	val	leu	ala	leu	trp	asp	lys	val	asn	.....	.....	glu	glu	glu	val	gly
Human α	val	.....	leu	ser	pro	ala	asp	lys	thr	asn	val	lys	ala	ala	trp	gly	lys	val	gly	ala	his	ala	gly	glu	tyr	gly
β	val	his	leu	thr	pro	glu	glu	lys	ser	ala	val	thr	ala	leu	trp	gly	lys	val	asn	.....	.....	val	asp	glu	val	gly
γ	gly	his	phe	thr	glu	glu	asp	lys	ala	thr	ilu	thr	ser	leu	trp	gly	lys	val	asn	.....	.....	val	glu	asp	ala	gly
δ	val	his	leu	thr	pro	glu	glu	lys	thr	ala	val	asn	ala	leu	trp	gly	lys	val	asn	.....	.....	val	asp	ala	val	gly

	7	8	9	10	11	12	13	14	15	16	C 1	2	3	4	5	6	7	CD 1	2	3	4	5	6	7	8	D 1
MYOGLOBIN	gln	asp	ilu	leu	ilu	arg	leu	phe	lys	ser	his	pro	glu	thr	leu	glu	lys	phe	asp	arg	phe	lys	his	leu	lys	thr
HEMOGLOBIN Horse α	ala	glu	ala	leu	glu	arg	met	phe	leu	gly	phe	pro	thr	thr	lys	thr	tyr	phe	pro	his	phe	.....	asp	leu	ser	his
β	gly	glu	ala	leu	gly	arg	leu	leu	val	val	tyr	pro	trp	thr	gln	arg	phe	phe	asp	ser	phe	gly	asp	leu	ser	gly
Human α	ala	glu	ala	leu	glu	arg	met	phe	leu	ser	phe	pro	thr	thr	lys	thr	tyr	phe	pro	his	phe	.....	asp	leu	ser	his
β	gly	glu	ala	leu	gly	arg	leu	leu	val	val	tyr	pro	trp	thr	gln	arg	phe	phe	glu	ser	phe	gly	asp	leu	ser	thr
γ	gly	glu	thr	leu	gly	arg	leu	leu	val	val	tyr	pro	trp	thr	gln	arg	phe	phe	asp	ser	phe	gly	asn	leu	ser	ser
δ	gly	glu	ala	leu	gly	arg	leu	leu	val	val	tyr	pro	trp	thr	gln	arg	phe	phe	glu	ser	phe	gly	asp	leu	ser	ser

	2	3	4	5	6	7	E 1	2	3	4	5	6	7	8	9	10	11	12	13	14	E 15	16	17	18	19	20
MYOGLOBIN	glu	ala	glu	met	lys	ala	ser	glu	asp	leu	lys	lys	his	gly	val	thr	val	leu	thr	ala	leu	gly	ala	ilu	leu	lys
HEMOGLOBIN Horse α	.....	.....	.....	.....	.....	gly	ser	ala	gln	val	lys	ala	his	gly	lys	lys	val	ala	asp	gly	leu	thr	leu	ala	val	gly
β	pro	asp	ala	val	met	gly	asn	pro	lys	val	lys	ala	his	gly	lys	lys	val	leu	his	ser	phe	gly	glu	gly	val	his
Human α	.....	.....	.....	.....	.....	gly	ser	ala	gln	val	lys	gly	his	gly	lys	lys	val	ala	asp	ala	leu	thr	asn	ala	val	ala
β	pro	asp	ala	val	met	gly	asn	pro	lys	val	lys	ala	his	gly	lys	lys	val	leu	gly	ala	phe	ser	asp	gly	leu	ala
γ	ala	ser	ala	ilu	met	gly	asn	pro	lys	val	lys	ala	his	gly	lys	lys	val	leu	thr	ser	leu	gly	asp	ala	ilu	lys
δ	pro	asp	ala	val	met	gly	asn	pro	lys	val	lys	ala	his	gly	lys	lys	val	leu	gly	ala	phe	ser	asp	gly	leu	ala

	EF 1	2	3	4	5	6	7	8	F 1	2	3	4	F 5	6	7	8	9	FG 1	2	3	4	5	G 1	2	3	4
MYOGLOBIN	lys	lys	gly	his	his	glu	ala	glu	leu	lys	pro	leu	ala	gln	ser	his	ala	thr	lys	his	lys	ilu	pro	ilu	lys	tyr
HEMOGLOBIN Horse α	his	leu	asp	asp	leu	pro	gly	ala	leu	ser	asp	leu	ser	asn	leu	his	ala	his	lys	leu	arg	val	asp	pro	val	asn
β	his	leu	asp	asn	leu	lys	gly	thr	phe	ala	ala	leu	ser	glu	leu	his	cys	asp	lys	leu	his	val	asp	pro	glu	asn
Human α	his	val	asp	asp	met	pro	asn	ala	leu	ser	ala	leu	ser	asp	leu	his	ala	his	lys	leu	arg	val	asp	pro	val	asn
β	his	leu	asp	asn	leu	lys	gly	thr	phe	ala	thr	leu	ser	glu	leu	his	cys	asp	lys	leu	his	val	asp	pro	glu	asn
γ	his	leu	asp	asp	leu	lys	gly	thr	phe	ala	gln	leu	ser	glu	leu	his	cys	asp	lys	leu	his	val	asp	pro	glu	asn
δ	his	leu	asp	asn	leu	lys	gly	thr	phe	ser	gln	leu	ser	glu	leu	his	cys	asp	lys	leu	his	val	asp	pro	glu	asn

	5	6	7	8	G 9	10	11	12	13	14	15	16	17	18	19	GH 1	2	3	4	5	6	H 1	2	H 3	4	5
MYOGLOBIN	leu	glu	phe	ilu	ser	glu	ala	ilu	ilu	his	val	leu	his	ser	arg	his	pro	gly	asn	phe	gly	ala	asp	ala	gln	gly
HEMOGLOBIN Horse α	phe	lys	leu	leu	ser	his	cys	leu	leu	ser	thr	leu	ala	val	his	leu	pro	asn	asp	phe	thr	pro	ala	val	his	ala
β	phe	arg	leu	leu	gly	asn	val	leu	ala	leu	val	val	ala	arg	his	phe	gly	lys	asp	phe	thr	pro	glu	leu	gln	ala
Human α	phe	lys	leu	leu	ser	his	cys	leu	leu	val	thr	leu	ala	ala	his	leu	pro	ala	glu	phe	thr	pro	ala	val	his	ala
β	phe	arg	leu	leu	gly	asn	val	leu	val	cys	val	leu	ala	his	his	phe	gly	lys	glu	phe	thr	pro	pro	val	gln	ala
γ	phe	lys	leu	leu	gly	asn	val	leu	val	thr	val	leu	ala	ilu	his	phe	gly	lys	glu	phe	thr	pro	glu	val	gln	ala
δ	phe	arg	leu	leu	gly	asn	val	leu	val	cys	val	leu	ala	arg	asn	phe	gly	lys	glu	phe	thr	pro	gln	met	gln	ala

	6	7	8	9	10	11	12	13	14	15	16	17	18	19	20	H 21	22	23	2	HC 1	2	3	4	5
MYOGLOBIN	ala	met	asn	lys	ala	leu	glu	leu	phe	arg	lys	asp	ilu	ala	ala	lys	tyr	lys	glu	leu	gly	tyr	gln	gly
HEMOGLOBON Horse α	ser	leu	asp	lys	phe	leu	ser	ser	val	ser	thr	val	leu	thr	ser	lys	tyr	arg						
β	ser	tyr	gln	lys	val	val	ala	gly	val	ala	asn	ala	leu	ala	his	lys	tyr	his						
Human α	ser	leu	asp	lys	phe	leu	ala	ser	val	ser	thr	val	leu	thr	ser	lys	tyr	arg						
β	ala	tyr	gln	lys	val	val	ala	gly	val	ala	asn	ala	leu	ala	his	lys	tyr	his						
γ	ser	trp	gln	lys	met	val	thr	ala	val	ala	ser	ala	leu	ser	ser	arg	tyr	his						
δ	ala	tyr	gln	lys	val	val	ala	gly	val	ala	asn	ala	leu	ala	his	lys	tyr	his						

SOURCE: Based on diagram in R. E. Dickerson and I. Geis, *The Structure and Function of Proteins*, New York: Harper & Row, 1969, p. 52.

[a] Residues that are identical are enclosed in box. A, B, C, . . . designate different helices of tertiary structure (see text).

These latter eight helical regions are commonly lettered A–H, starting from the A helix at the NH_2-terminal end, and the interhelical regions designated as AB, BC, CD, . . . , GH, respectively. The nonhelical region that lies between the NH_2-terminal end and the A helix is designated the NA region; and the region between the COOH-terminal end and the H helix is designated the HC region (Figure 3.31). This naming system allows discussion of particular residues that have similar functional and structural roles in each of the hemoglobin and myoglobin molecules.

The similarity in structure can be observed in the distance matrix plots for the hemoglobin α and β chains (Figure 3.33). In these plots the α-helix segments appear along the diagonal as characteristic of α-helical secondary structure.

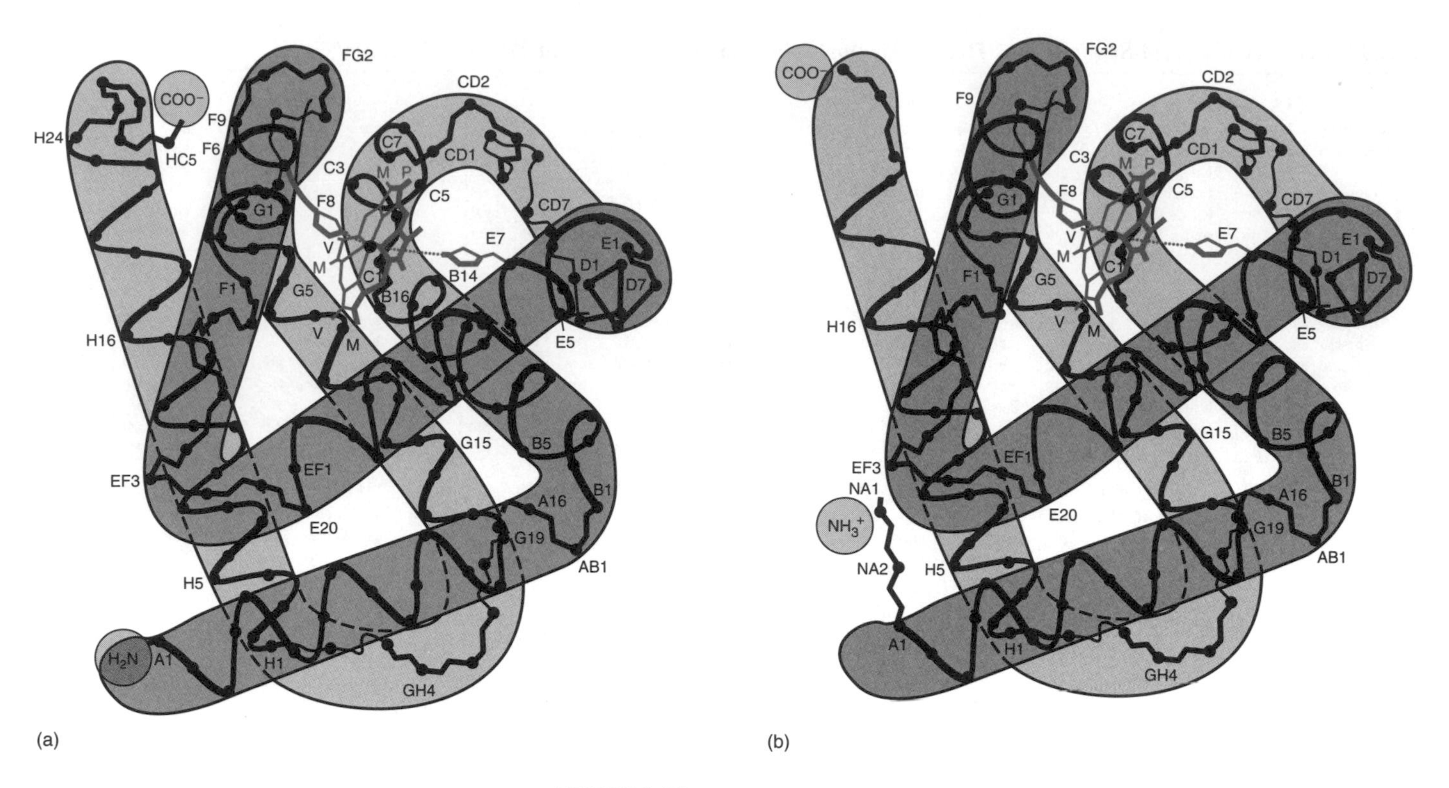

FIGURE 3.32
Comparison of the conformation of myoglobin (*a*) and β chain of HbA_1 (*b*).
The overall structures are very similar, except at the NH_2-terminal and COOH-terminal ends.
Redrawn with permission, from A. Fersht, *Enzyme Structure and Mechanism,* San Francisco, CA: W. H. Freeman, 1977, pp. 12 and 13.

A Simple Equilibrium Defines O_2 Binding to Myoglobin

Since myoglobin contains a single O_2 binding site per molecule, the association of oxygen to myoglobin is characterized by a simple equilibrium constant (Eqs. 3.1 and 3.2). In Eq. 3.2 [MbO_2] is the solution concentration of oxymyoglobin, [Mb] is the concentration of deoxymyoglobin, and [O_2] is the concentration of oxygen, expressed in units of moles per liter. The equilibrium constant, K_{eq}, will also have the units of moles per liter. As for any true equilibrium constant, the value of K_{eq} is dependent on pH, ionic strength, and temperature.

$$\text{Mb} + O_2 \overset{K_{eq}}{\rightleftharpoons} \text{MbO}_2 \tag{3.1}$$

$$K_{eq} = \frac{[\text{Mb}][O_2]}{[\text{MbO}_2]} \tag{3.2}$$

Since oxygen is a gas, it is more convenient to express O_2 concentration in terms of the pressure of oxygen in units of torr (1 torr is equal to the pressure of 1 mm Hg at 0°C and standard gravity). In Eq. 3.3 this transfer of units has been made, with P_{50} the equilibrium constant and pO_2 the concentration of oxygen, now expressed in units of torr.

$$P_{50} = \frac{[\text{Mb}] \cdot pO_2}{[\text{MbO}_2]} \tag{3.3}$$

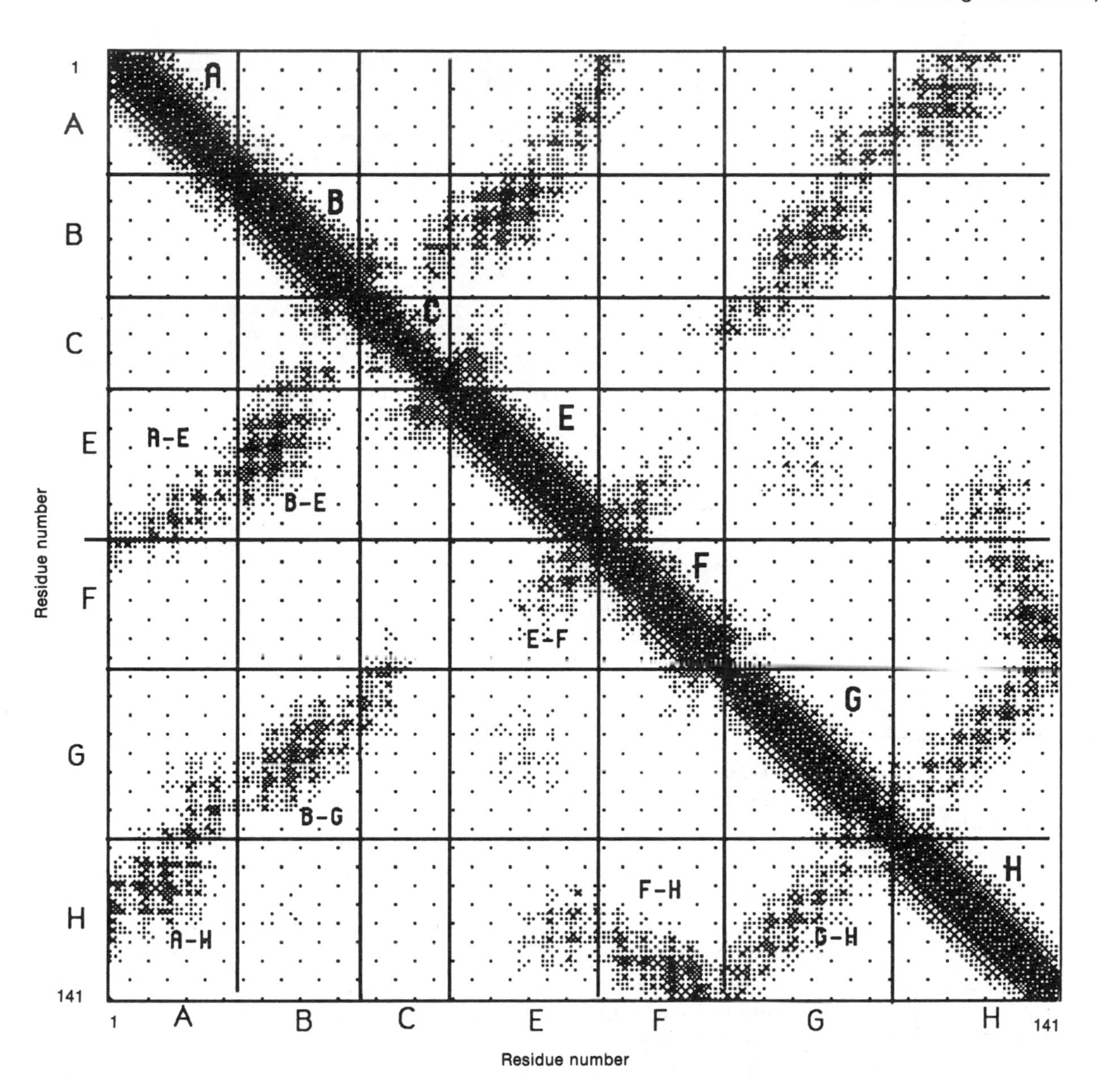

FIGURE 3.33
Interresidue distance matrix of deoxy-hemoglobin. The α-chain shows shaded distances of less than 15 Å.
The α-helical secondary structures are shown by the broadening of the diagonal of the distance matrix. The standard helix nomenclature (A–H) is noted for the different helices of the α chain. Independent helices and helix–helix interactions are apparent from the partitioning.

An **oxygen saturation curve** is used to characterize the properties of an oxygen binding protein. In this type of plot the fraction of oxygen binding sites in solution that contain oxygen (Y, Eq. 3.4) is plotted on the ordinate versus the pO_2 (oxygen concentration) on the abscissa. The Y value is simply defined for myoglobin by Eq. 3.5. Substitution into Eq. 3.5 of the value of [MbO_2] obtained from Eq. 3.3 and then dividing through by [Mb], results in Eq. 3.6, which shows the dependence of Y on the value of the equilibrium constant P_{50}, and the oxygen concentration. It is seen from Eqs. 3.3 and 3.6 that the value of P_{50} is equal to the oxygen concentration, pO_2, when $Y = 0.5$ (50% of the available sites occupied). Hence the designation of the equilibrium constant by the subscript 50.

$$Y = \frac{\text{number of binding sites occupied}}{\text{total number of binding sites in solution}} \tag{3.4}$$

$$Y = \frac{[MbO_2]}{[Mb] + [MbO_2]} \tag{3.5}$$

$$Y = \frac{pO_2}{P_{50} + pO_2} \tag{3.6}$$

A plot of Eq. 3.6 of Y versus pO_2 generates an oxygen saturation curve for myoglobin in the form of a rectangular hyperbola (Figure 3.34).

A simple algebraic manipulation of Eq. 3.6 leads to Eq. 3.7. Taking the logarithm of both sides of Eq. 3.7 results in Eq. 3.8, which is known as the **Hill equation.** A plot of log $(Y/1 - Y)$ versus log pO_2, according to Eq. 3.8, yields a straight line with a slope equal to 1 for myoglobin (Figure 3.35). This is called the Hill plot, and the slope (n_h) is referred to as the **Hill coefficient** (see Eq. 3.9).

$$\frac{Y}{1 - Y} = \frac{pO_2}{P_{50}} \tag{3.7}$$

$$\log \frac{Y}{1 - Y} = \log pO_2 - \log P_{50} \tag{3.8}$$

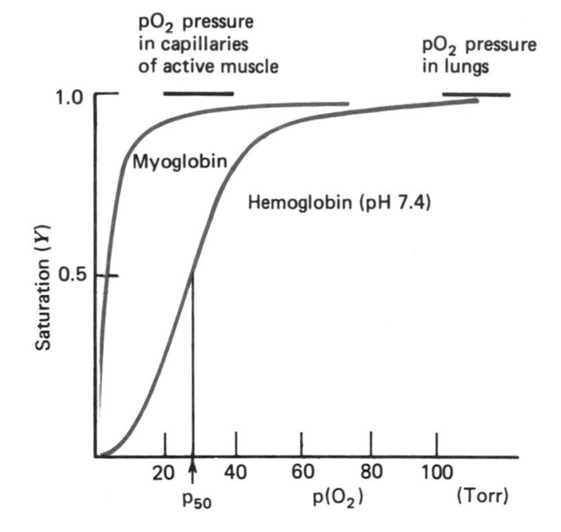

FIGURE 3.34
Oxygen binding curves for myoglobin and hemoglobin.

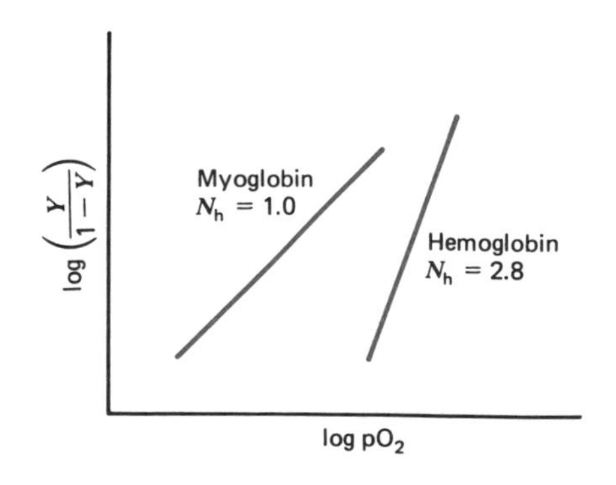

FIGURE 3.35
Hill plots for myoglobin and hemoglobin HbA$_1$.

Binding of O_2 to Hemoglobin Involves Cooperativity Between the Hemoglobin Subunits

Whereas myoglobin has a single O_2 binding site per molecule, hemoglobins contain a quaternary structure of four polypeptide chains, each with a heme binding site for O_2. The binding of the four O_2 molecules in hemoglobin is found to be **positively cooperative,** so that the binding of the first O_2 to deoxyhemoglobin facilitates the binding of O_2 to the other subunits in the molecule. Conversely, the dissociation of the first O_2 from fully oxygenated hemoglobin, $Hb(O_2)_4$, will make the dissociation of O_2 from the other subunits of the tetramer molecule easier.

Based on the cooperativity in oxygen association and dissociation, the oxygen saturation curve for hemoglobin differs from that previously derived for myoglobin. A plot of Y versus pO_2 for hemoglobin follows a *sigmoid* line, indicating cooperativity in oxygen association (Figure 3.34). A plot of the Hill Eq. 3.9 gives a value of the slope (n_h) equal to 2.8 (Figure 3.35).

$$\log \frac{Y}{1 - Y} = n_h \log pO_2 - \text{constant} \tag{3.9}$$

The meaning of the Hill coefficient to cooperative O_2 association can be quantitatively evaluated as presented in Table 3.10. A parameter known as the **cooperativity index,** R_x, is calculated, which shows the ratio of pO_2 required to change Y from a value of $Y = 0.1$ (10% of sites filled) to a value of $Y = 0.9$ (90% of sites filled) for designated Hill coefficient values found experimentally. In the case of myoglobin, $n_h = 1$, and an 81-fold change in oxygen concentration is required to change from $Y = 0.1$ to $Y = 0.9$. In hemoglobin, where positive cooperativity is observed, $n_h = 2.8$, and *only a 4.8-fold change in oxygen concentration* is required to change the fractional saturation from 0.1 to 0.9.

The Molecular Mechanism of Cooperativity in O_2 Binding

The X-ray diffraction data on deoxyhemoglobin show that the ferrous atoms sit out of the plane of their porphyrins by about 0.4–0.6 Å. This is thought to occur because of two factors. The electronic configuration of the five-coordinated ferrous atom in deoxyhemoglobin has a slightly larger radius than the distance from the center of the porphyrin to each of the pyrrole nitrogen atoms. Accordingly, the iron can be placed in the center of the porphyrin only with some distortion of the most stable porphyrin conformation. Probably a more important consideration is that if the iron atom sits in the plane of the porphyrin, the proximal His F8 imidazole will interact unfavorably with atoms of the porphyrin. The strength of this unfavorable steric interaction would, in part, be due to conformational constraints on the His F8 and the porphyrin in the deoxyhemoglobin conformation that energetically forces the approach of the His F8 toward the porphyrin to a particular path (Figure 3.36). These constraints will become less significant in the oxy conformation of hemoglobin.

The conformation with the iron atom out of the plane of the porphyrin is unstrained and energetically favored for the five-coordinate ferrous atom. When O_2 binds to the sixth coordinate position of the iron, however, this conformation becomes strained. A more energetically favorable conformation for the O_2 liganded iron is one in which the iron atom is within the plane of the porphyrin structure.

On the binding of O_2 to a ferrous atom into its sixth coordinate position, the favorable free energy of bond formation is used to overcome the repulsive interaction between the His F8 and porphyrin, and the ferrous atom moves into the plane of the porphyrin ring. This is the most thermodynamically stable position for the now six-bonded iron atom; one axial

TABLE 3.10 Relationship Between Hill Coefficient (n_h) and Cooperativity Index (R_x)

n_h	R_x	Observation
0.5	6560	Negative substrate cooperativity
0.6	1520	
0.7	533	
0.8	243	
0.9	132	
1.0	81.0	Noncooperativity
1.5	18.7	
2.0	9.0	
2.8	4.8	Positive substrate cooperativity
3.5	3.5	
6.0	2.1	
10.0	1.6	
20.0	1.3	

SOURCE: Based on Table 7.1 in A. Cornish-Bowden, *Principles of Enzyme Kinetics*, London and Boston: Butterworths Scientific Publishers, 1976.

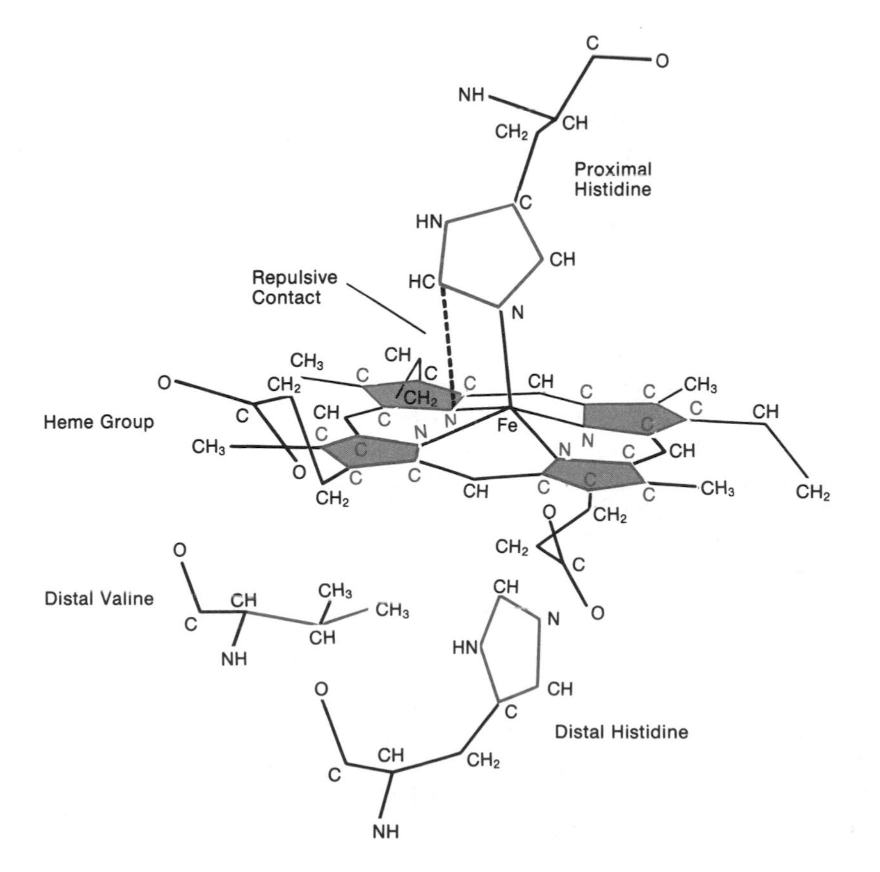

FIGURE 3.36
Steric hindrance between proximal histidine and prophyrin in deoxyhemoglobin.
Redrawn from M. Perutz, *Sci. Am.*, 239:92, 1978 by Scientific American, Inc.

ligand is on either side of the plane of the porphyrin ring, and the steric repulsion of one of the axial ligands with the porphyrin is balanced by the repulsion of the second axial ligand on the opposite side when the ferrous atom is in the center. If the iron atom is displaced from the center, the steric interactions of the two axial ligands with the porphyrin in the deoxy conformation are unbalanced, and the stability of the unbalanced structure will be lower than that of the equidistant conformation. Also, the radius of the iron atom with six ligands is reduced so that it can just fit into the center of the porphyrin without distortion of the porphyrin conformation.

Since the steric repulsion between the porphyrin and the His F8 must be overcome on O_2 association, the binding of the first O_2 to hemoglobin is characterized by a relatively low affinity constant. However, when an O_2 association does occur to the first heme in a deoxyhemoglobin molecule, the change in the iron atom position from above the plane of the porphyrin into the center of the porphyrin *triggers a conformational change in the whole hemoglobin molecule*. The change in conformation results in a greater affinity of O_2 to the other heme sites after the first O_2 has bound. This conformation change is thought to occur as described below.

The conformation of the deoxyhemoglobin appears to be stabilized by noncovalent interactions of the quaternary structure at the interface between α and β subunits in which the FG corner of one subunit noncovalently binds to the C helix of the adjacent subunit (Figure 3.37). In addition, ionic interactions stabilize the deoxy-quaternary conformation of the protein (Figure 3.38). These particular noncovalent interactions of the deoxy conformation are now *destabilized* on the binding of O_2 to one of the heme subunits of a deoxyhemoglobin molecule. The binding of O_2 pulls the Fe^{2+} atom into the porphyrin plane, which correlated with the movement of the His F8 toward the porphyrin moves the F helix of which the His F8 is a part. The movement of the F helix, in turn, moves the FG corner of its subunit, destabilizing the FG noncovalent interaction with the C helix of the adjacent subunit at an $\alpha_1\beta_2$ or $\alpha_2\beta_1$ subunit interface (Figures 3.37 and 3.39).

The FG to C intersubunit contacts are thought to act as a "switch," because they apparently can exist in two different arrangements with different modes of contact between the FG corner of one subunit and the C helix of the adjacent subunit. On the binding of O_2 to deoxyhemoglobin, the movement of the FG corner in the subunit to which the O_2 is bound forces the intersubunit contacts to switch to their alternative position. The switch in noncovalent interactions between the two positions involves a relative movement of FG and C in adjacent subunits of about 6 Å. In the second position of the "switch," *the tertiary conformation of the subunits participating in the FG to C intersubunit contact are less constrained and the adjacent subunit changes to a new tertiary conformation* (oxy conformation). This second conformation allows the His F8 to approach their porphyrins, on O_2 association, with a less significant steric repulsion than in the deoxy conformation of the hemoglobin subunits (Figure 3.39). An O_2 molecule can bind to the empty hemes in the less constrained oxy conformation more easily than to a subunit conformation held by the quaternary interactions in the deoxy conformation.

In addition, the Val E11 in the deoxy conformation of the β subunits is at the entrance to the O_2 binding site, where it sterically impedes O_2 association to the heme (see Figure 3.31). In the oxy conformation the heme in the β subunits appears to move approximately 1.5 Å further into the heme binding site of the protein, changing the geometric relationship of the O_2 binding site in the heme to the Val E11 side chain, so that the Val

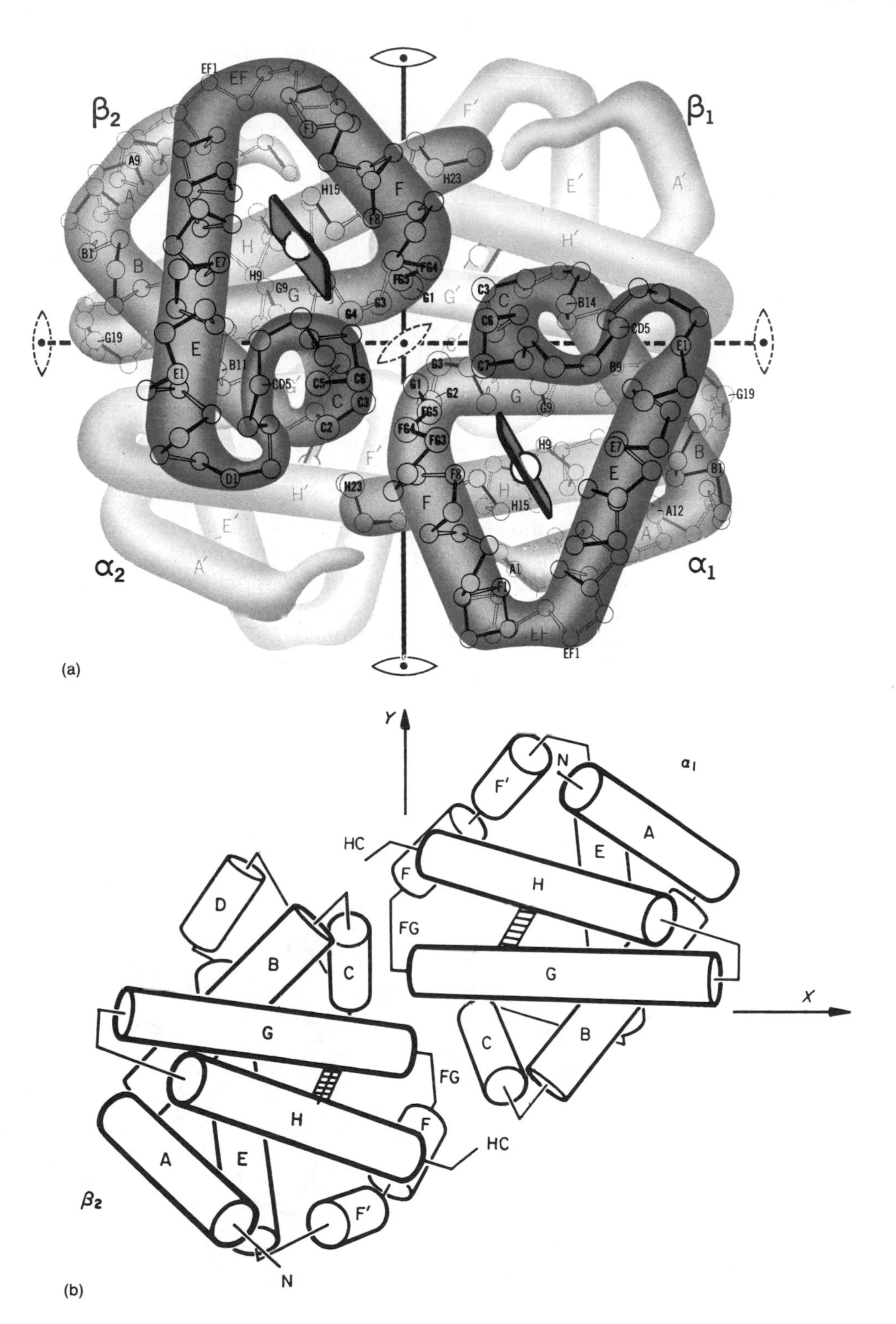

FIGURE 3.37
Quaternary structure of hemoglobin.
(*a*) The $\alpha_1\beta_2$ interface contacts between FG corners and C helix are shown. (*b*) Cylinder representation of α_1 and β_2 subunits in hemoglobin molecule showing α_1 and β_2 interface contacts between FG corner and C helix, viewed from opposite side of x–y plane from in (*a*).

(*a*) Figure reprinted with permission from R. E. Dickerson and I. Geis, *The Structure and Action of Proteins,* Menlo Park: W. A. Benjamin, Inc., 1969, p. 56.
(*b*) Figure reprinted with permission from J. Baldwin and C. Chothia, *J. Mol. Biol.* 129:175.

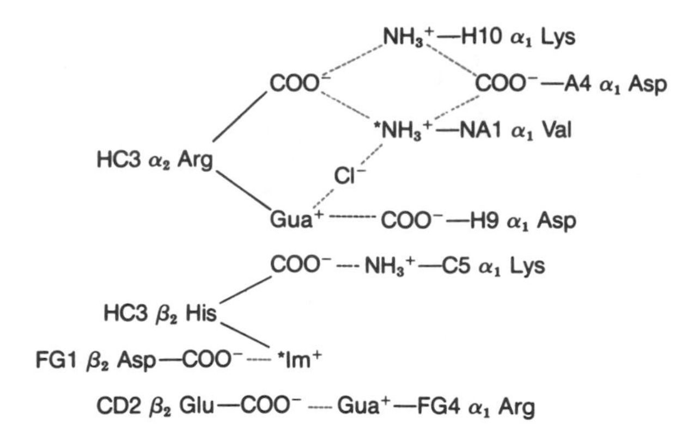

FIGURE 3.38
Salt bridges between subunits in deoxyhemoglobin.
Im^+ is imidazolium; Gua^+ is quanidinium; starred residues account for approximately 60% of the alkaline Bohr effect.
Redrawn from M. Perutz, *Br. Med. Bull.*, 32:195, 1976.

E11 no longer sterically interferes with O_2 binding. This appears to be an important additional factor that results in a higher affinity of O_2 to the oxy conformation of the β chain than to the deoxy conformation.

The deoxy conformation of hemoglobin is referred to as the "tense" or **T conformational state.** The oxyhemoglobin conformational form is referred to as the "relaxed" or **R conformational state.** On the binding of the initial oxygen to the heme subunits of the tetramer molecule, the molecular conformation is pushed from the T to the R conformational

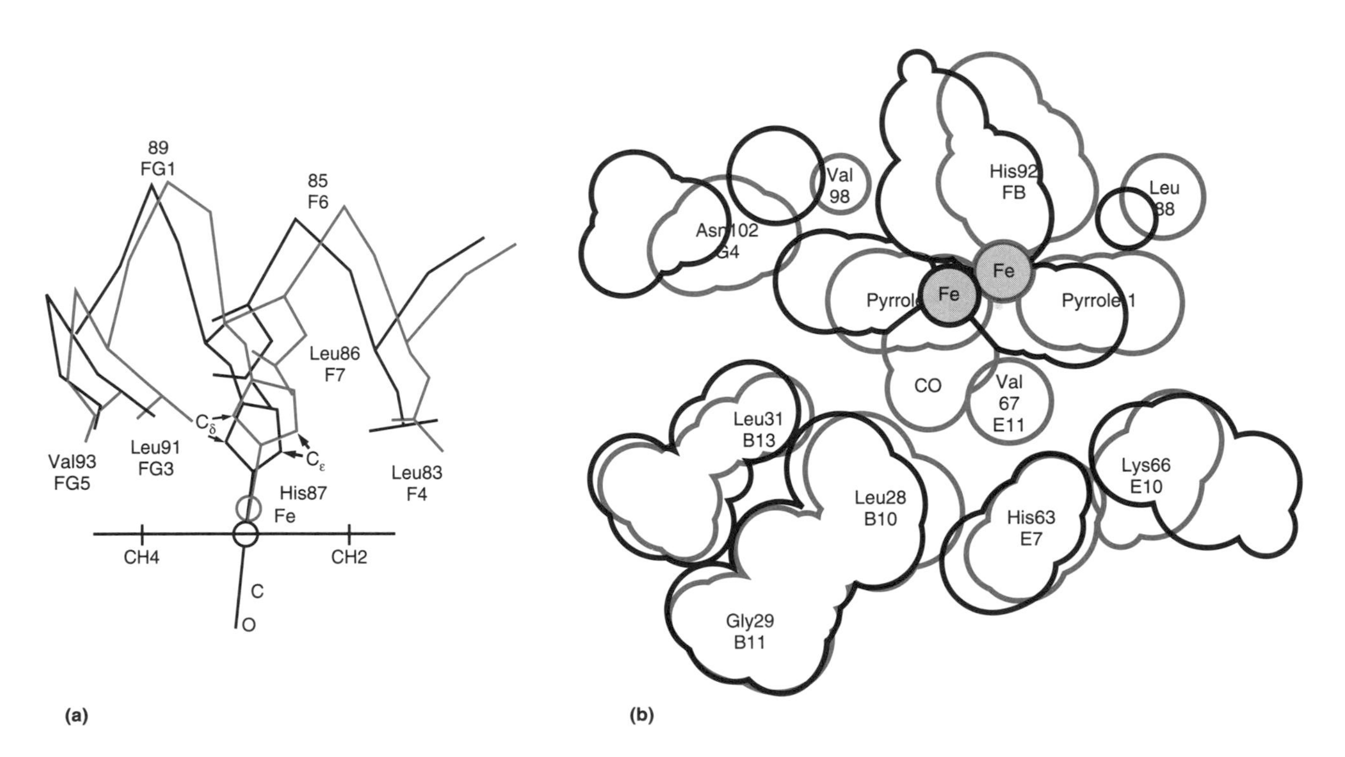

FIGURE 3.39
Stick and space-filling diagrams drawn by computer graphics showing movements of residues in heme environment on transition from deoxyhemoglobin to oxyhemoglobin.
(*a*) Black line outlines position of polypeptide chain and His F8 in carbon monoxyhemoglobin, which is a model for oxyhemoglobin. Red line outlines the same for deoxyhemoglobin. Position of iron atom shown by circle. Movements are for an α subunit. (*b*) Similar movements in a β subunit using space-filling diagram shown. Residue labels centered in density for the deoxyconformation.
Redrawn with permission from J. Baldwin and C. Chotia, *J. Mol. Biol.* 129:175. Copyright © 1979 by Academic Press Inc. (London) Ltd.

state. The affinity constant of O_2 is greater for the R state hemes than the T state by a factor of 150–300, depending on the solution conditions.

The Bohr Effect Involves Dissociation of a Proton on Binding of Oxygen

The equilibrium expression for oxygen association to hemoglobin includes a term that indicates the participation of hydrogen ion in the equilibrium.

$$\text{Hb} + 4\text{O}_2 \rightleftharpoons \text{Hb(O}_2)_4 + x\text{H}^+ \qquad \textbf{(3.10)}$$

Equation 3.10 shows that the R form is more acidic, and protons dissociate when the hemoglobin is transformed to the R form. The number of equivalents of protons that dissociate per mole of hemoglobin depends on the pH of the solution and the concentration of other factors that can bind to hemoglobin, such as Cl^- and diphosphoglycerate (see Chapter 25). At pH 7.4, the value of x may vary from about 1.8–2.8, depending on the solution conditions. This production of protons at alkaline pH (pH > 6), when deoxyhemoglobin is transformed to oxyhemoglobin, is known as the alkaline **Bohr effect.**

The protons are derived from the partial dissociation of acid residues with pK'_a values within ± 1.5 pH units of the solution pH, that change from a higher to lower pK'_a on the transformation of the T to R conformation of the hemoglobin. For example, the HC3 His 146(β) in the deoxy conformation is predominantly in its imidazolium form (positively charged acid form), which is stabilized by a favorable interaction with the negatively charged side chain of the FG1 Asp 94(β) (Figure 3.38). This ion-pair interaction makes it more difficult to remove the imidazolium proton, and thus raises the pK'_a of the imidazolium to a higher value than normally found for a free imidazolium ion in solution, where a stabilization by a proximate negatively charged group does not normally occur. However, on conversion of the protein to the R conformation, the strength of this ionic interaction is broken and the imidazolium assumes a lower pK_a. The decrease in the histidine's pK'_a at blood pH results in the conversion of some of the acid form of the histidine to its conjugate base (imidazole) form, with the dissociation of free protons that form a part of the Bohr effect. Breakage of this ion pair with release of protons accounts for 50% of the protons released on conversion to the R conformation. Other acid groups in the protein contribute the additional protons due to changes in their pK'_a to lower values on changing from the T to R conformation.

The Bohr effect may fit the definition of an allosteric mechanism. An **allosteric mechanism** is a common process in protein molecules in which substrate association is influenced by the binding of other molecules that are not direct substrates of the protein. In an allosteric process there must be separate binding sites on the protein for substrate (e.g., O_2 in the case of hemoglobin) and effector (inhibitor or activator) molecules that exert allosteric control. As the effector molecule's binding site by definition is distinctly separate from that of the substrate's binding site, the effector molecule acts to increase or decrease the affinity of the substrate at the substrate binding site by either causing a conformational change or stabilizing a particular conformation of the protein. With regard to the Bohr effect, it is evident that proton binding sites exist in the hemoglobin molecule to which the binding of protons, the effector "molecules," thermodynamically stabilizes and thus increases the concentration of the T form with respect to the R form. The binding of a proton to form the ion-pair site between the His 146(β) and Asp 94(β) is one such interaction, for example, that favors the T conformation. By increasing the ratio T/R on stabilizing the T conformation, the binding of protons to their effector sites is correlated with a poorer affinity of hemoglobin for oxygen.

The Bohr effect has important physiological consequences. Cells metabolizing at high rates, with high requirements for molecular oxygen, produce carbonic acid and lactic acid, which act to increase the hydrogen ion concentration in the cell's environment. As the increase in hydrogen ion concentration forces the equilibrium of Eq. 3.10 to the left, from the higher O_2 affinity conformation (R) to the lower affinity conformation (T), an increased amount of oxygen is dissociated from the hemoglobin molecule.

BIBLIOGRAPHY

Serine Proteases

Birk, Y. Proteinase inhibitors. A. Neuberger and K. Brocklehurst (Eds.), *Hydrolytic Enzymes,* Amsterdam: Elsevier, 1987, p. 257.

Dufton, M. J. Could domain movements be involved in the mechanism of trypsin-like serine proteases? *FEBS Lett.* 271:9, 1990.

Liebman, M. N. Structural organization in the serine proteases. *Enzyme* 36:115, 1986.

Neurath, H. Proteolytic processing and physiological regulation, *Trends Biochem. Sci.* 14:268, 1989.

Polgar, L. Structure and function of serine proteases. *Hydrolytic Enzymes,* A. Neuberger and K. Brocklehurst (Eds.), series in New Comprehensive Biochemistry, Vol. 16, Amsterdam: Elsevier, 1987, p. 159.

Zwaal, R. F. A. and Hemker, H. C. (Eds.). *Blood Coagulation,* series in New Comprehensive Biochemistry, Vol. 13, Amsterdam: Elsevier, 1986.

DNA Binding Proteins

Landschulz, W. H., Johnson, P. F., and McKnight, S. L. The leucine zipper: a hypothetical structure common to a new class of DNA binding proteins. *Science* 240:1759, 1988.

Lee, M. S., Gippert, G. P., Soman, K. V., Case, D. A., and Wright, P. E. Three-dimensional solution structure of a single zinc finger DNA-binding domain. *Science* 245:635, 1989.

O'Shea, E. K., Rutkowski, R., Stafford, W. F., III, and Kim, P. S. Preferential heterodimer formation by isolated leucine zippers from Fos and Jun. *Science* 245:646, 1989.

Pavletich, N. P. and Pabo, C. O. Zinc finger—DNA Recognition: Crystal structure of a Zif-268-DNA complex at 2.1 A. *Science* 252:809, 1991.

Schleif, R. DNA binding by proteins. *Science* 241:1182, 1988.

Immunoglobulins

Alzari, P. M., Lascombe, M.-B., and Poljak, R. J. Structure of antibodies, *Annu. Rev. Immunol.* 6:555, 1988.

Calabi, F. The immunoglobulin superfamily. F. Calabi and M. S. Neuberger (Eds.), *Molecular Genetics of Immunoglobulin,* 1987, Amsterdam: Elsevier, p. 203.

Chothia, C., Lesk, A. M., Tramontano, A., Levitt, M., Smith-Gill, S. J., Air, G., Sheriff, S., Padlan, E. A., Davies, D., Tulip, W. R., Colman, P. M., Spinelli, S., Alzari, P. M., and Poljak, R. J. Conformations of immunoglobulin hypervariable regions, *Nature (London)* 342:877, 1989.

Davies, D. R., Sheriff, S., and Padlan, E. A. Antibody–antigen complexes, *J. Biol. Chem.* 263:10541, 1988.

Maizels, N. To understand function, study structure, *Cell* 60, 887, 1990.

Nienhuis, A. W. and Maniatis, T. Structure and Expression of globin genes in erythroid cells. *The Molecular Basis of Blood Diseases,* G. Stamatoyannopoulos, A. W. Nienhuis, P. Leder, and P. W. Majerus (Eds.), Philadelphia: Saunders, 1987, p. 28.

Padlan, E. A., Silverton, E. W., Sheriff, S., Cohen, G. H., Smith-Gill, S. J., and Davies, D. R. Structure of an antibody–antigen complex: crystal structure of the HyHEL-10 Fab–lysozyme complex, *Proc. Natl. Acad. Sci. U.S.A.* 86, 5938, 1989.

Hemoglobin

Baldwin, J. and Chothia, C. Haemoglobin: the structural changes related to ligand binding and its allosteric mechanism, *J. Mol. Biol.* 129:175, 1979.

Busch, M. R., Mace, J. E., Ho, N. T., and Ho, C. Roles of the *beta*-146 histidyl residue in the molecular basis of the Bohr effect of hemoglobin: a protein nuclear magnetic resonance study. *Biochem.* 30:1865, 1991.

Dickerson, R. E. and Geis, I. *Hemoglobin: Structure, function, evolution and pathology,* Benjamin–Cummings, Menlo Park, CA: 1983

Ho, C. and Russu, I. M. How much do we know about the Bohr effect of hemoglobin? *Biochemistry* 26:6299, 1987.

Liddington, R., Derewenda, Z., Dodson, G., and Harris, D. Structure of the liganded T state of haemoglobin identifies the origin of cooperative oxygen binding, *Nature (London)* 331:725, 1988.

Mathews, A. J., Rohlfs, R. J., Olson, J. S., Tame, J., Renaud, J. P., and Nagai, K. The effects of E7 and E11 mutations on the kinetics of ligand binding to R state human hemoglobin, *J. Biol. Chem.* 264:16573, 1989.

Perutz, M. Hemoglobin structure and respiratory transport. *Sci. Am.* 239:92, 1978.

Perutz, M. F., Fermi, G., and Shih, T.-B. Structure of deoxy cowtown [His HC3(146)*beta* to Leu]: Origin of the alkaline Bohr effect, *Proc. Natl. Acad. Sci. U.S.A.* 81:4781, 1984.

QUESTIONS

J. BAGGOTT AND C. N. ANGSTADT

*Q*uestion *T*ypes are described on page xxiii.

1. (QT2) Haptens:
   1. can function as antigens.
   2. strongly bind to antibodies specific for them.
   3. may be macromolecules.
   4. can act as antigenic determinants.

   A. H chains of immunoglobulins C. Both
   B. L chains of immunoglobulins D. Neither
2. (QT4) C regions determine the class to which the antibody belongs.
3. (QT4) Contain variable regions and hypervariable regions.

4. (QT1) Study of the papain hydrolysis products of an antibody indicates:
   A. antibodies are bivalent.
   B. the products have decreased affinity for antigens.
   C. each antibody molecule is hydrolyzed into many small peptides.
   D. the hypervariable sequences are in the hinge region of the intact molecule.
   E. None of the above is true.

5. (QT2) In immunoglobulins:
   1. there are four polypeptide chains.
   2. there are two copies of each type of chain.
   3. all chains are linked by disulfide bonds.
   4. carbohydrate is covalently bound to the protein.

6. (QT1) Serine proteases:
   A. hydrolyze peptide bonds involving the carboxyl groups of serine residues.
   B. are characterized by having several active sites per molecule, each containing a serine residue.
   C. are inactivated by reacting with one molecule of diisopropylfluorophosphate per molecule of protein.
   D. are exopeptidases.
   E. are synthesized in an active form in eukaryotes.

7. (QT2) The active sites of all serine proteases contain which of the following amino acid residues?
   1. Aspartate
   2. γ-carboxyglutamate
   3. Histidine
   4. Arginine *or* lysine
8. (QT2) Natural inhibitors of the serine proteases:
   1. bind to a limited area of the surface of the protease.
   2. bind very tightly to the protease.
   3. form a covalent bond to an essential histidine residue of the protease.
   4. undergo peptide bond hydrolysis catalyzed by the protease.

   A. β barrel C. Both
   B. loop regions D. Neither
9. (QT4) Two domains, conserved in a wide variety of eukaryotic and prokaryotic serine proteases.
10. (QT4) Site of determinants of binding specificity in eukaryotic serine proteases.
11. (QT4) Site of interaction with serpins.

    A. Helix-turn-helix motif C. Both
    B. Leucine zipper D. Neither
12. (QT4) Associated with dimeric DNA-binding proteins.
13. (QT4) Direct interaction with double helical DNA.
14. (QT4) Two α helices.

15. (QT2) Hemoglobin and myoglobin *both:*
    1. bind heme in a hydrophobic pocket.
    2. are highly α helical.
    3. bind one molecule of heme per globin chain.
    4. consist of subunits designed to provide hydrogen bonds to and nonpolar interaction with other subunits.
16. (QT2) Hemoglobin, but not myoglobin, when it binds oxygen, exhibits:
    1. a sigmoid saturation curve.
    2. a Hill coefficient of 1.
    3. positive cooperativity.
    4. a cooperativity index of 81.

17. (QT1) All of the following are believed to contribute to the stability of the deoxy or T conformation of hemoglobin *except:*
    A. a salt bridge involving specific valyl and arginyl residues.
    B. the larger ionic radius of the six-coordinated ferrous ion as compared to the five-coordinated ion.
    C. steric interaction of His F8 with the porphyrin ring.
    D. interactions between the FG corner of one subunit and the C helix of the adjacent subunit.
    E. a valyl residue that tends to block O_2 from approaching the hemes of the β-chains.

18. (QT3)
    A. pK'_a of HC3 His 146 (β) in oxyhemoglobin.
    B. pK'_a of HC3 His 146 (β) in deoxyhemoglobin.

ANSWERS

1. C 2 and 4 true. Haptens are small molecules and cannot alone elicit antibody production; thus they are not antigens. They can act as antigenic determinants if covalently bound to a larger molecule, and free haptens may bind strongly to the antibodies thereby produced (p. 93).

2. A There are significant differences among the C regions of the H chains of the different antibody classes (p. 95).
3. C See p. 94. These regions form the antigen binding sites and differ among immunoglobulins of differing specificity.
4. A In these hydrolysis experiments, three fragments are produced: two identical Fab fragments, each of which binds antigen with an affinity similar to that of the whole antibody molecule, and one Fc fragment, which does not bind antigens (p. 97).
5. E 1, 2, 3, and 4 true. See p. 93. There are two copies of each of two types of polypeptide chains.
6. C This is the distinguishing characteristic of the serine proteases, and of the serine hydrolases in general. A: They have various specificities (p. 105). B: There is only one active site per molecule (p. 102). D: They are all endopeptidases (p. 104). E: In eukaryotes they are synthesized as inactive precursors, zymogens or proenzymes (p. 109).
7. B 1 and 3 true. See p. 110. 2: γ-carboxyglutamate is essential to some of the serine proteases, but it is not at the active site. 4: These are the substrate specificities of the trypsin-like proteases.
8. C 2 and 4 true. See p. 111. 1: The inhibitor binding site covers a broad area. 3: A covalent bond is formed to the serine residue.
9. A The β barrel domain is found in all serine proteases, but the loop regions may be lacking in the bacterial serine proteases that are related to the eucaryotic proteins (p. 115).
10. B See p. 116.
11. B See p. 116.
12. C The helix-turn-helix motif is frequently found in dimeric DNA-binding proteins (p. 116), and the very purpose of the leucine zipper is to effect dimerization of DNA-binding monomers (p. 118).
13. A With the helix-turn-helix, one α-helix lies in the major groove of DNA, and the other α helix lies across the first helix, in contact with the DNA (p. 116). The leucine zipper consists of two leucine-containing α helices that interact with each other, forming a head-to-head protein dimer (p. 118).
14. C See answer 13 above.
15. A 1, 2, and 3 true. 1: See p. 121. 2: See p. 122. 3: Hemoglobin has four chains and four oxygen binding sites, whereas myoglobin has one chain and one oxygen binding site. Each oxygen binding site is a heme (pp. 120–121). 4: Only hemoglobin is designed to form a quaternary structure; myoglobin is structured to interact with water and to prevent association with other myoglobin molecules (p. 122).
16. B 1 and 3 true. 1: See p. 126, Figure 3.34. 2 and 3: Myoglobin has a Hill coefficient of 1; hemoglobin's Hill coefficient of 2.8 indicates positive cooperativity (p. 126). 4: A cooperativity index of 81 indicates noncooperativity; hemoglobin's lower value of 4.8 reflects cooperative oxygen binding (p. 126).
17. B Six-coordinated ferrous ion has a smaller ionic radius than the five-coordinated species and just fits into the center of the porphyrin ring without distortion (p. 128).
18. B HC3 His 146 (β) is a major contributor to the Bohr effect. Thus its pK'_a will be lower (it will be a stronger acid) in oxyhemoglobin (p. 131).

4

Enzymes: Classification, Kinetics, and Control

J. LYNDAL YORK

4.1 GENERAL CONCEPTS 136
4.2 CLASSIFICATION OF ENZYMES 137
Class 1. Oxidoreductases 137
Class 2. Transferases 138
Class 3. Hydrolases 140
Class 4. Lyases 140
Class 5. Isomerases 140
Class 6. Ligases 140
4.3 KINETICS 141
Kinetics Studies the Rate of Change of Reactants to Products 141
Enzymes Show Saturation Kinetics 144
An Enzyme Catalyzes Both Forward and Reverse Directions of a Reversible Reaction 148
Multisubstrate Reactions Follow Either a Ping–Pong or Sequential Mechanism 149
4.4 COENZYMES: STRUCTURE AND FUNCTION 149
Adenosine Triphosphate May Be Either a Second Substrate or a Modulator of Activity 150
NAD and NADP Are Coenzymes of Niacin 151
FMN and FAD Are Coenzymes of Riboflavin 152
Metal Cofactors Have Various Functions 153
4.5 INHIBITION OF ENZYMES 157
Competitive Inhibition May Be Reversed by Increased Substrate 157
Noncompetitive Inhibitors Do Not Prevent Substrate From Binding 158
Irreversible Inhibition Involves Covalent Modification of an Enzyme Site 158
Many Drugs Are Enzyme Inhibitors 159
4.6 ALLOSTERIC CONTROL OF ENZYME ACTIVITY 162
Allosteric Effectors Bind at Sites Different From Substrate-Binding Sites 162
Allosteric Enzymes Exhibit Sigmoidal Kinetics 164
Cooperativity Explains the Interaction Between Ligand Sites in an Oligomeric Protein 165
Regulatory Subunits Modulate the Activity of Catalytic Subunits 167
4.7 ENZYME SPECIFICITY: THE ACTIVE SITE 167
Complementarity of Substrate and Enzyme Explains Substrate Specificity 169
4.8 MECHANISM OF CATALYSIS 171
Enzymes Decrease Activation Energy 172
Abzymes Are Antibodies With Catalytic Activity Which Have Been Artifically Synthesized 177
Environmental Parameters Influence Catalytic Activity 178
4.9 CLINICAL APPLICATIONS OF ENZYMES 179
Coupled Assays Utilize the Optical Properties of NAD, NADP or FAD 182
Clinical Analyzers May Use Immobilized Enzymes as Reagents 183
Enzyme-Linked Immunoassays Employ Enzymes as Indicators 183
Measurement of Isozymes Is Used Diagnostically 184

Some Enzymes Are Used as Therapeutic Agents 186
Enzymes Linked to Insoluble Matrices Are Used as Chemical Reactors 186
4.10 REGULATION OF ENZYME ACTIVITY 186
Compartmentation of Opposing Pathways Maximizes Cellular Economy 187
Enzyme Concentrations in Cells Are Controlled 187
Modulation of Enzyme Activity Is a Short-Term Regulation 188
Some Enzymes Are Regulated by Covalent Modification 189
4.11 AMPLIFICATION OF REGULATORY SIGNALS 190
The Phosphorylase Cascade Amplifies the Hormonal Signal 190
BIBLIOGRAPHY 191
QUESTIONS AND ANNOTATED ANSWERS 192
CLINICAL CORRELATIONS
4.1 A Case of Gout Demonstrates Two Phases in the Mechanism of Enzyme Action 146
4.2 The Physiological Effect of Changes in Enzyme K_m Value 147
4.3 Mutation of a Coenzyme Binding Site Results in Clinical Disease 149
4.4 A Case of Gout Demonstrates the Difference Between an Allosteric Site and the Substrate Binding Site 163
4.5 Thermal Lability of G6PD Results in Hemolytic Anemia 179
4.6 Alcohol Dehydrogenase Isoenzymes With Different pH Optima 180
4.7 Identification and Treatment of an Enzyme Deficiency 181
4.8 Ambiguity in the Assay of Mutated Enzymes 183

4.1 GENERAL CONCEPTS

Enzymes are proteins evolved by the cells of living organisms for the specific function of catalyzing chemical reactions. Enzymes increase the rate at which reactions approach equilibrium. *Rate* is defined as the change in the amount (moles, grams) of starting materials or products per unit time. The enzyme triggers the increased rate by acting as a catalyst. A true catalyst increases the rate of a chemical reaction, but is not itself changed in the process. The enzyme may become temporarily covalently bound to the molecule being transformed during intermediate stages of the reaction, but in the end the enzyme will be regenerated in its original form as the product is released.

There are two important characteristics of catalysts in general and enzymes in particular that should not be forgotten. The first is that the enzyme is not changed by entering into the reaction. The second is that the enzyme does not change the equilibrium constant of the reaction, it simply increases the rate at which the reaction approaches equilibrium. Therefore, a catalyst is responsible for increasing the rate but not changing the thermodynamic properties of the system with which it is interacting. In biological systems, a catalyst is necessary because at the temperature and pH of the human body reactions would not occur at a rate sufficient to support rapid muscular activity, nerve impulse generation, and all the other processes required to support life.

At this point, we need to define several terms before entering into a discussion of the mechanism of enzyme action. An **apoenzyme** is the protein part of the enzyme minus any cofactors or prosthetic groups that may be required for the enzyme to be functionally active. The apoenzyme is therefore catalytically inactive. Not all enzymes require cofactors or prosthetic groups to be active. The **cofactors** are those small organic or inorganic molecules that the enzyme requires for its activity. For example, lysine oxidase is a copper-requiring enzyme. Copper in this case is loosely bound but is required for the enzyme to be active. The **prosthetic group** is similar to the cofactor but is tightly bound to the apoenzyme. For example, in the cytochromes, the heme prosthetic group is very tightly bound and requires strong acids to disassociate it from the cytochrome. The addition of cofactor or prosthetic group to the apoprotein yields the **holoenzyme,** which is the active enzyme. The molecule the enzyme acts upon to form product is called the **substrate**. Since most reactions are reversible, the products of the forward reaction will become the substrates of the reverse reaction. Enzymes have a great deal of specificity. For example, glucose oxidase will oxidize glucose but not galactose. This specificity resides in a particular region on the enzyme surface called the **substrate binding site,** which is a particular arrangement of chemical groups on the enzyme surface that is specially formulated to bind a specific substrate. The substrate binding site may have integrated within it the **active site.** In some cases the active site may not be within the substrate binding site but may be contiguous to it in the primary sequence. In other cases the active site lies in distant regions of the primary sequence but is brought adjacent to the substrate binding site by folding of the tertiary structure. The active site contains the machinery, in the form of particular chemical groups, that is involved in catalyzing the reaction under consideration. The chemical groups involved in both binding of substrate and catalysis are often part of the side chains of the amino acids of the apoenzyme.

In some cases variant forms or isoenzymes (isozymes) are found. These isoenzymes are electrophoretically distinguishable, but they all catalyze the same chemical reaction.

In some enzymes there is another region of the molecule, the **allosteric site,** that is not at the active site or substrate binding site, but is somewhere else on the molecule. The allosteric site is the site where small molecules bind and effect a change in the active site or the substrate binding site. The binding of a specific small organic molecule at the allosteric site causes a change in the conformation of the enzyme, and that conformational change may cause the active site to become either more active or less active. It may cause the binding site to have a greater affinity for substrate, or it may actually cause the binding site to have less affinity for substrate. Such interactions are involved in the regulation of the activity of enzymes and are discussed in more detail on page 162.

4.2 CLASSIFICATION OF ENZYMES

The International Union of Biochemistry (IUB) has established a system whereby all enzymes are placed into one of six major classes. Each class is then subdivided into several subclasses, which are further subdivided. A number is assigned to each class, subclass, and sub-subclass so that an enzyme is assigned a four-digit number as well as a name. The fourth digit identifies a specific enzyme. For example, alcohol : NAD oxidoreductase is assigned the number 1.1.1.1. because it is an oxidoreductase, the electron donor is an alcohol and the acceptor is the coenzyme NAD^+. Notice that in naming an enzyme, the substrates are stated first, followed by the reaction type to which the ending *-ase* is affixed. The trivial name of the enzyme 1.1.1.1. is alcohol dehydrogenase. Many common names persist but are not very informative. For example, "aldolase" does not tell much about the substrates, although it does identify the reaction type. We will use the trivial names that are recognized by the IUB.

Each of the six major enzyme classes will be briefly described in the following subsections.

Class 1. Oxidoreductases

These enzymes catalyze oxidation–reduction reactions. For example, alcohol : NAD oxidoreductase catalyzes the oxidation of an alcohol to an aldehyde. This enzyme removes two electrons as two hydrogen atoms from the alcohol to yield an aldehyde, and in the process, the two electrons that were originally in the carbon–hydrogen bond of the alcohol are

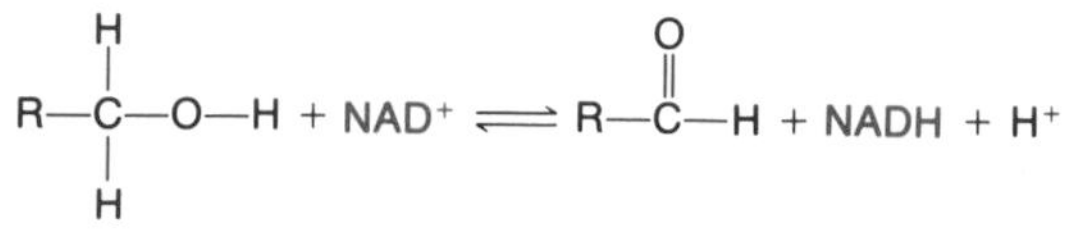

$$R{-}\underset{H}{\overset{H}{C}}{-}O{-}H + NAD^+ \rightleftharpoons R{-}\overset{O}{\overset{\|}{C}}{-}H + NADH + H^+$$

FIGURE 4.1
The alcohol dehydrogenase catalyzed oxidation of ethanol.

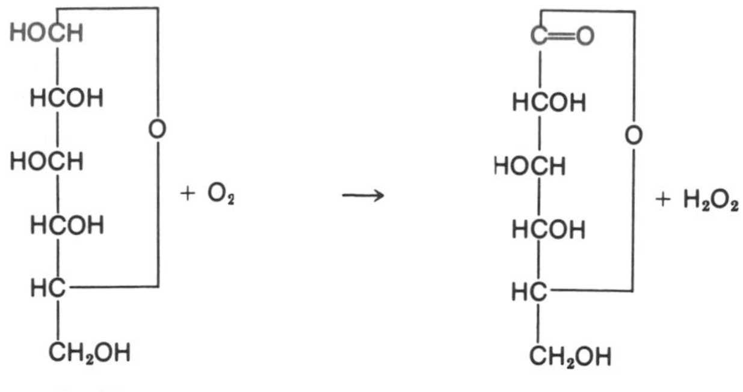

β-D-Glucose + O_2 → δ-Gluconolactone + H_2O_2

FIGURE 4.2
The glucose oxidase catalyzed oxidation of glucose.

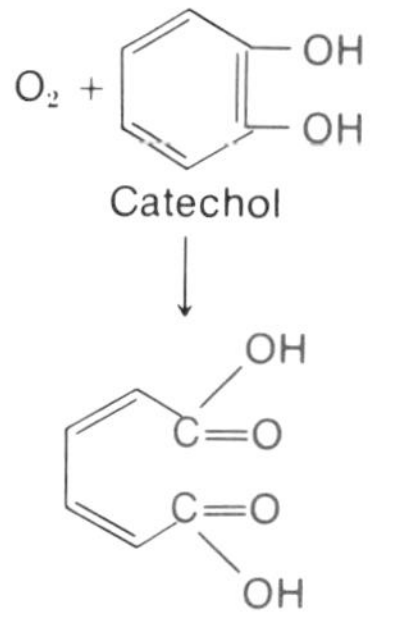

O_2 + Catechol → *cis,cis*-Muconic acid

FIGURE 4.3
Oxygenation of catechol.

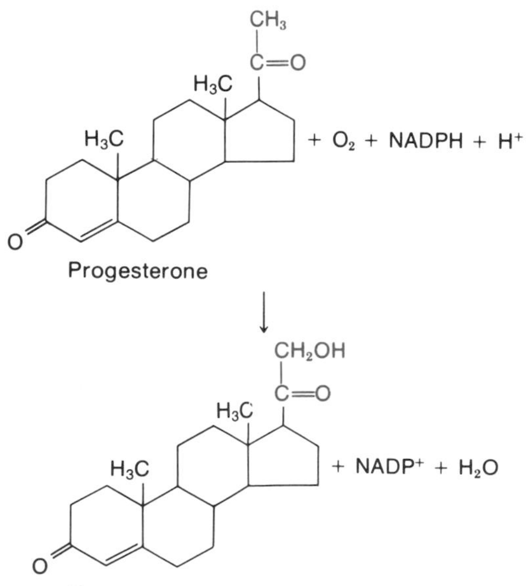

Progesterone + O_2 + NADPH + H^+ → Deoxycorticosterone + $NADP^+$ + H_2O

FIGURE 4.4
Hydroxylation of progesterone by oxygen.

transferred to the NAD^+, which then becomes reduced as shown in Figure 4.1. In addition to the alcohol and aldehyde functional groups, **dehydrogenases** also act on the following functional groups as electron donors: $—CH_2—CH_2—$, $—CH_2—NH_2$, $—CH{=}NH$, as well as the nucleotides NADH and NADPH.

Other major subclasses of the oxidoreductases are summarized as follows:

Oxidases transfer two electrons from the donor to oxygen, resulting usually in hydrogen peroxide (H_2O_2) formation. For example, glucose oxidase catalyzes the reaction shown in Figure 4.2. In the case of cytochrome oxidase, H_2O rather than H_2O_2 is the product.

Oxygenases catalyze the incorporation of oxygen into a substrate. In the case of the **dioxygenases,** both atoms of O_2 are incorporated in a single product, whereas with the **monoxygenases,** only a single atom of oxygen is incorporated as a hydroxyl group and the other oxygen in reduced to water by electrons from the substrate or from a second substrate that is not oxygenated. Catechol oxygenase catalyzes the dioxygenase reaction presented in Figure 4.3; the steroid hydroxylase illustrates the monoxygenase (mixed function oxygenase) reaction type as shown in Figure 4.4.

Peroxidases utilize H_2O_2 rather than oxygen as the oxidant. The NADH peroxidase catalyzes the reaction

$$NADH + H^+ + H_2O_2 \rightleftharpoons NAD^+ + 2H_2O$$

Catalase is unique in that H_2O_2 serves as both donor and acceptor. Catalase functions in the cell to detoxify H_2O_2:

$$H_2O_2 + H_2O_2 \rightleftharpoons O_2 + 2H_2O$$

Class 2. Transferases

These enzymes are involved in transferring functional groups between donors and acceptors. The amino, acyl, phosphate, one-carbon C_1, and glycosyl groups are the major moieties that are transferred.

Aminotransferases (*transaminases*) transfer the amino group from one amino acid to a keto acid acceptor, resulting in the formation of a new amino acid and a new keto acid (Figure 4.5).

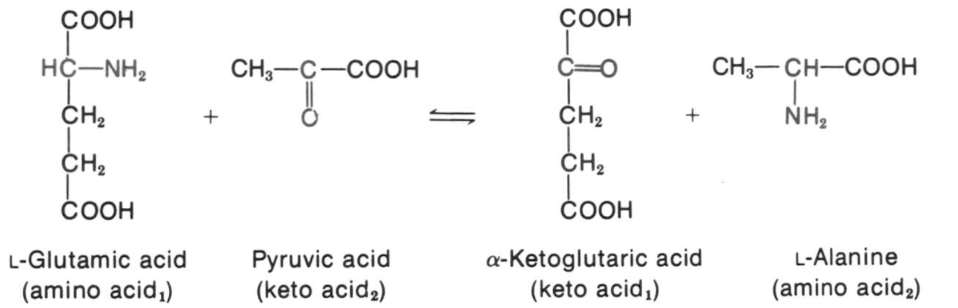

FIGURE 4.5
Examples of a reaction catalyzed by an aminotransferase.

Kinases are the phosphorylating enzymes that catalyze the transfer of the phosphoryl group from ATP or another nucleoside triphosphate, to alcohol or amino group acceptors. For example, glucokinase catalyzes the reaction:

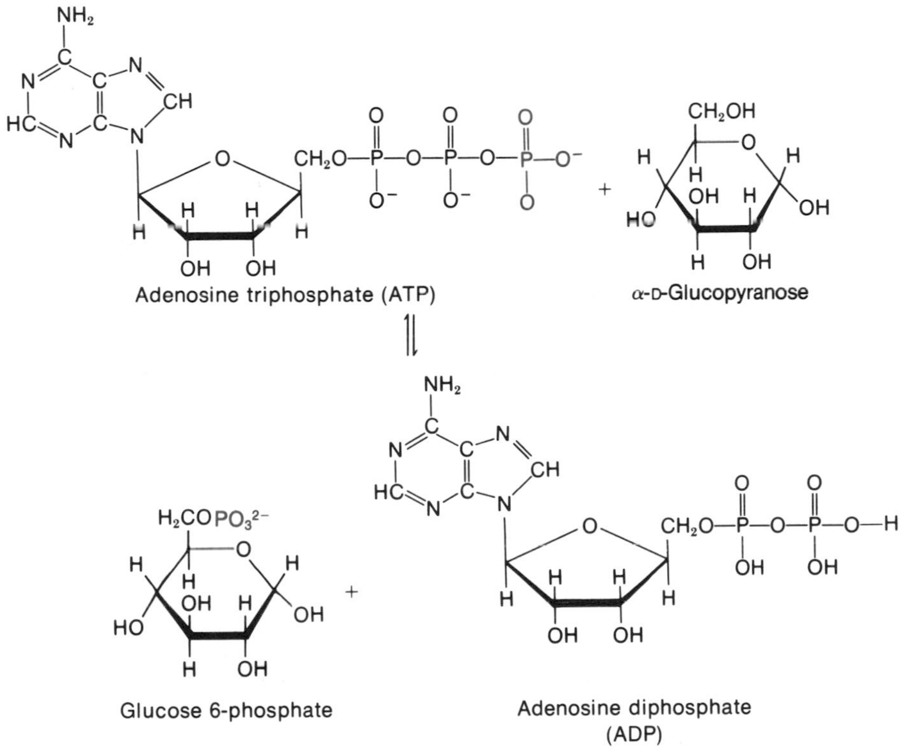

The synthesis of glycogen depends on **glucosyltransferases,** which catalyze the transfer of an activated glucosyl residue to a glycogen primer. The phosphoester bond in uridine diphosphoglucose is labile, which allows the glucose to be transferred to the growing end of the glycogen primer, that is,

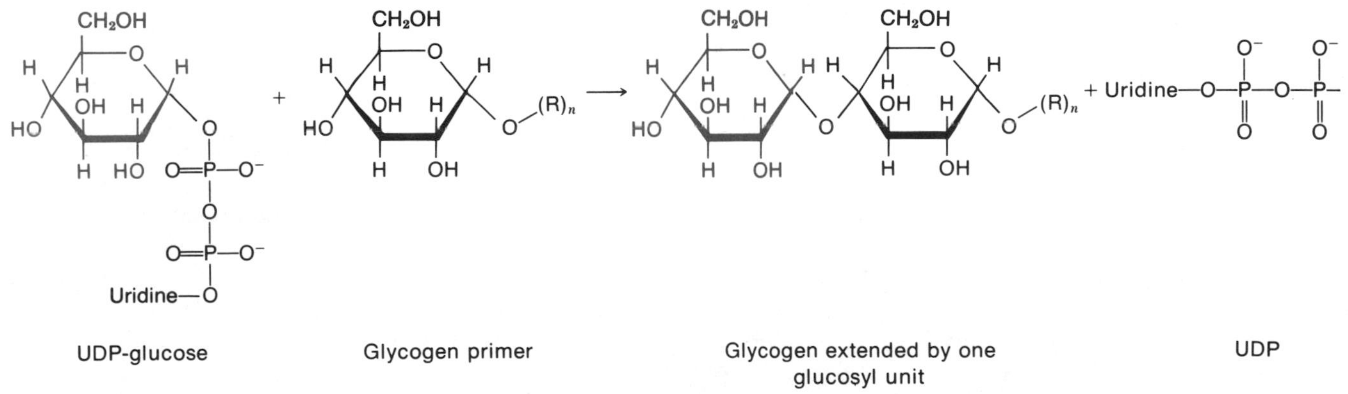

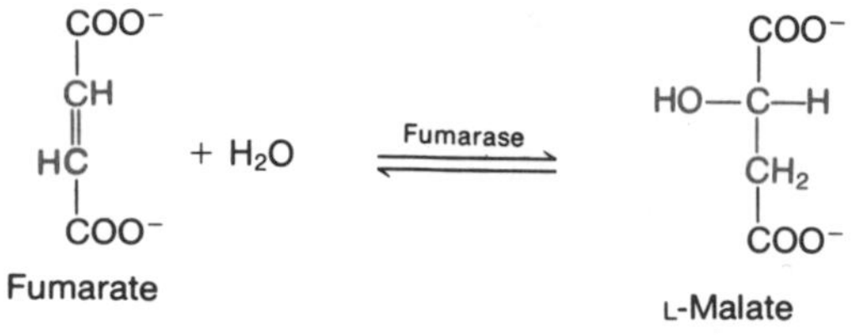

$$\text{Fumarate } (^{-}OOC-CH=CH-COO^-) + H_2O \xrightleftharpoons{\text{Fumarase}} \text{L-Malate } (^{-}OOC-CH_2-CH(OH)-COO^-)$$

FIGURE 4.6
The fumarase reaction.

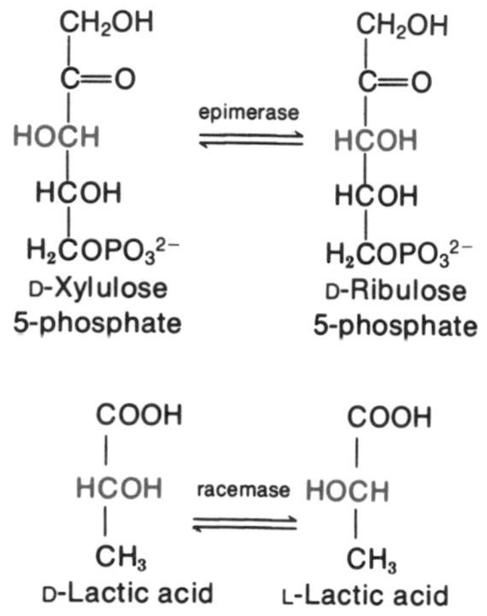

$$\text{D-Xylulose 5-phosphate} \xrightleftharpoons{\text{epimerase}} \text{D-Ribulose 5-phosphate}$$

$$\text{D-Lactic acid} \xrightleftharpoons{\text{racemase}} \text{L-Lactic acid}$$

FIGURE 4.7
Examples of reactions catalyzed by an epimerase and a racemase.

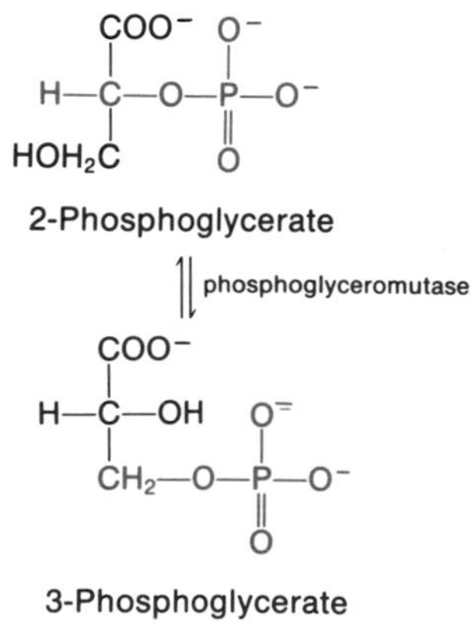

$$\text{2-Phosphoglycerate} \xrightleftharpoons{\text{phosphoglyceromutase}} \text{3-Phosphoglycerate}$$

FIGURE 4.8
Interconversion of the 2- and 3-phosphoglycerates.

It should be noted that although a polymer is synthesized, the reaction is not of the ligase type, which we discuss in Class 6 below, as is the formation of protein from activated amino acids.

Class 3. Hydrolases

This group of enzymes can be considered as a special class of the transferases in which the donor group is transferred to water. The generalized reaction involves the hydrolytic cleavage of C—O, C—N, O—P, and C—S bonds. The cleavage of the peptide bond is a good example of this reaction:

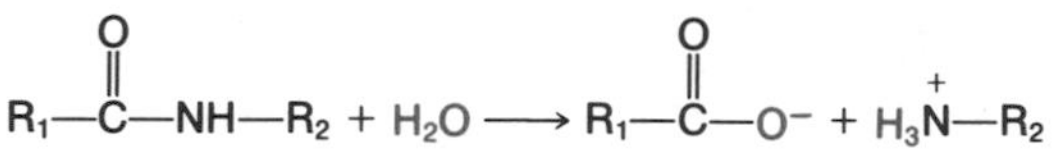

$$R_1-\overset{O}{\overset{\|}{C}}-NH-R_2 + H_2O \longrightarrow R_1-\overset{O}{\overset{\|}{C}}-O^- + H_3\overset{+}{N}-R_2$$

The proteolytic enzymes are a special class of hydrolases called **peptidases**.

Class 4. Lyases

Lyases are enzymes that add or remove the elements of water, ammonia, or carbon dioxide.

The **decarboxylases** remove the element of CO_2 from α- or β-keto acids or amino acids:

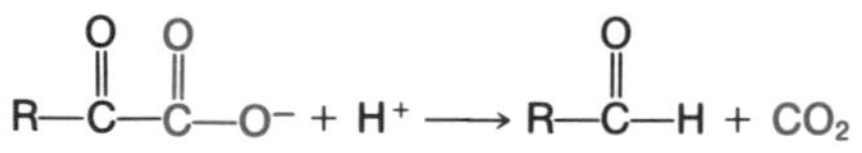

$$R-\overset{O}{\overset{\|}{C}}-\overset{O}{\overset{\|}{C}}-O^- + H^+ \longrightarrow R-\overset{O}{\overset{\|}{C}}-H + CO_2$$

The **dehydratases** remove the elements of H_2O in a dehydration reaction. Fumarase catalyzes the conversion of fumarate to malate (Figure 4.6).

Class 5. Isomerases

This is a very heterogeneous group of enzymes that catalyze isomerizations of several types. These include cis–trans and aldose–ketose interconversions. Isomerases that catalyze inversion at asymmetric carbon atoms are either **epimerases** or **racemases** (Figure 4.7). **Mutases** involve the intramolecular transfer of a group such as the phosphoryl. The transfer need not be direct but can involve a phosphorylated enzyme as an intermediate. An example is phosphoglycerate mutase, which catalyzes the conversion of 2-phosphoglycerate to 3-phosphoglycerate as shown in Figure 4.8.

Class 6. Ligases

Since to ligate means to bind, these enzymes are involved in synthetic reactions where two molecules are joined at the expense of an ATP "high-energy phosphate bond." The use of **synthetase** is reserved for this particular group of enzymes. The formation of amino acyl tRNAs, acyl coenzyme A, glutamine, and the addition of CO_2 to pyruvate, are reactions catalyzed by ligases. Pyruvate carboxylase is a good example of a ligase enzyme. The reaction is presented in Figure 4.9. The two substrates bicarbonate and pyruvate are ligated to form a four-carbon (C_4) keto acid.

The six enzyme classes and most of the important subclass members are compiled in Table 4.1. The accepted trivial names are used for members of the subclass.

TABLE 4.1 Summary of the Enzyme Classes and Major Subclasses

1. Oxidoreductases	2. Transferases
Dehydrogenases	Transaldolase
Oxidases	and transketolase
Reductases	Acyl-, methyl-,
Peroxidases	glucosyl-, and
Catalase	phosphoryl-transferase
Oxygenases	Kinases
Hydroxylases	Phosphomutases
3. Hydrolases	4. Lyases
Esterases	Decarboxylases
Glycosidases	Aldolases
Peptidases	Hydratases
Phosphatases	Dehydratases
Thiolases	Synthases
Phospholipases	Lyases
Amidases	
Deaminases	
Ribonucleases	
5. Isomerases	6. Ligases
Racemases	Synthetases
Epimerases	Carboxylases
Isomerases	
Mutases (not all)	

$$\text{ATP} + \text{HCO}_3^- + \underset{\text{Pyruvate}}{\text{CH}_3\text{-C(=O)-COO}^-} \xrightarrow{\text{biotin}} \underset{\text{Oxaloacetate}}{\text{HOOC-CH}_2\text{-C(=O)-COO}^-} + \text{ADP} + \text{P}_i$$

FIGURE 4.9
The reaction catalyzed by pyruvate carboxylase.

4.3 KINETICS

Kinetics Studies the Rate of Change of Reactants to Products

Since enzymes affect the rate of chemical reactions, it is important to understand basic chemical kinetics and how kinetic principles apply to enzyme-catalyzed reactions. *Kinetics* is a study of the rate of change of the initial state of reactants and products to the final state of reactants and products. The term *velocity* is often used rather than rate. Velocity is expressed in terms of change in the concentration of substrate or product per unit time, whereas rate refers to changes in total quantity (moles or grams) per unit time. Biochemists tend to use the two terms interchangeably.

The velocity of a reaction A → P is determined from a progress curve or velocity profile of a reaction. The progress curve can be determined by following the disappearance of reactants or the appearance of product at several different times. Such a curve is shown in Figure 4.10, where product appearance is plotted against time. The slope of tangents to the progress curve yields the instantaneous velocity at that point in time. The initial velocity represents an important parameter in the assay of enzyme concentration, as we learn later. Notice that the velocity constantly changes as the reaction proceeds to equilibrium, and becomes zero at equilibrium. Mathematically, the velocity is expressed as

$$\text{Velocity} = v = \frac{-d[\text{A}]}{dt} = \frac{d[\text{P}]}{dt} \qquad \textbf{(4.1)}$$

and represents the change in concentration of reactants or products per unit time.

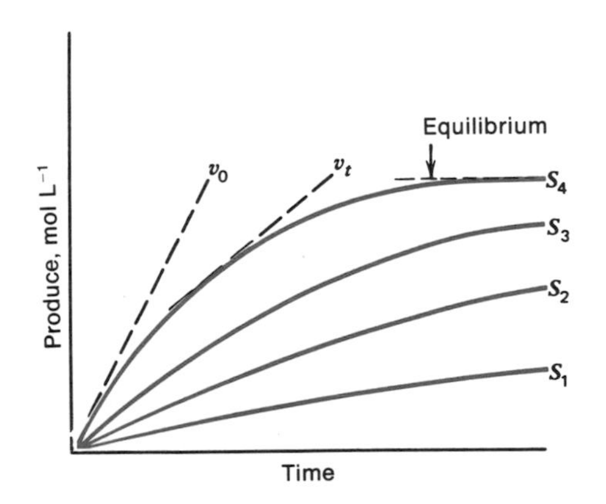

FIGURE 4.10
Progress curves for an enzyme-catalyzed reaction.
The initial velocity (v_0) of the reaction is determined from the slope of the progress curve at the beginning of the reaction. The initial velocity increases with increasing substrate concentration (S_1–S_4) but reaches a limiting value that is characteristic of each enzyme. The velocity at any time, *t*, is denoted as v_t.

The Rate Equation

Determination of the velocity of a reaction reveals nothing about the stoichiometry of the reactants and products or about the mechanism of the reaction. What is needed is an equation that relates the experimentally determined initial velocity to the concentration of reactants. Such a relation is the velocity or rate equation. In the case of the reaction $A \rightarrow P$, the velocity equation is

$$\frac{-d[A]}{dt} = v = k[A]^n \quad \textbf{(4.2)}$$

That is to say, the observed initial velocity will depend on the starting concentration of A to the nth power multiplied by a proportionality constant k. The latter is known as the **rate constant.** The exponent n is usually an integer from 1 to 3 that is required to satisfy the mathematical identity of the velocity expression.

Characterization of Reactions Based on Order

Another term that is useful in describing a reaction is the *order* of reaction. Empirically the order is determined as the sum of the exponents on each concentration term in the rate expression. In the case under discussion the reaction is **first order,** since the velocity depends on the concentration of A to the first power, $v = k[A]^1$. In a reaction such as $A + B \rightarrow C$, if the order with respect to A and B is 1, that is, $v = k[A]^1[B]^1$, overall the reaction is second order. It should be noted that the order of reaction is independent of the stoichiometry of the reaction, that is, if the reaction were third order, the rate expression could be either $v = k[A][B]^2$ or $v = k[A]^2[B]$, depending on the order in A and B. Since the velocity of the reaction is constantly changing as the reactant concentration changes, it is obvious that first-order reaction conditions would not be ideal for assaying an enzyme-catalyzed reaction because one would have two variables, the changing substrate concentration and the unknown enzyme concentration.

If the differential first-order rate expression Eq. 4.2 is integrated, one obtains

$$k_1 \cdot t = 2.3 \log \left(\frac{[A]}{[A] - [P]}\right) \quad \textbf{(4.3)}$$

where [A] is the initial reactant concentration and [P] is the concentration of product formed at time t. The first-order rate constant k_1 has the units of reciprocal time. If the data shown in Figure 4.10 were replotted as log [P] versus time for any one of the substrate concentrations, a straight line would be obtained whose slope is equal to $k_1/2.303$. The rate constant k_1 should not be confused with the rate or velocity of the reaction.

Many biological processes proceed under first-order conditions. The clearance of many drugs from the blood by peripheral tissues is a first-order process. A specialized form of the rate equation can be used in these cases. If we define $\boldsymbol{t_{1/2}}$ as the time required for the concentration of the reactants or the blood level of a drug to be reduced by one-half the initial value, then Eq. 4.3 reduces to

$$k_1 \cdot t_{1/2} = 2.3 \log \left(\frac{1}{1 - \frac{1}{2}}\right) = 2.3 \log 2 = 0.69 \quad \textbf{(4.4)}$$

or

$$t_{1/2} = \frac{0.69}{k_1} \quad \textbf{(4.5)}$$

Notice that $t_{1/2}$ is not one-half the time required for the reaction to be completed. The term $t_{1/2}$ is referred to as the *half-life* of the reaction.

Many **second-order** reactions that involve water or any one of the reactants in large excess can be treated as pseudo-first-order reactions. In the case of the hydrolysis of an ester,

$$\mathrm{R{-}\overset{\overset{\displaystyle O}{\|}}{C}{-}O{-}CH_3 + H_2O \rightleftharpoons R{-}\overset{\overset{\displaystyle O}{\|}}{C}{-}OH + CH_3OH}$$

the second-order rate expression is

$$\text{velocity} = v = k_2[\text{ester}]^1[\mathrm{H_2O}]^1 \tag{4.6}$$

but since water is in abundance (55.5 M) compared to the ester (10^{-3}–10^{-2}M), the system obeys the first-order rate law Eq. 4.2, and the reaction appears to proceed as if it were a first-order reaction. Those reactions in the cell that involve hydration, dehydration, or hydrolysis are pseudo-first order.

The rate expression for the **zero-order** reaction is $v = k_0$. Notice that there is no concentration term for reactants; therefore, the addition of more reactant does not augment the rate. The disappearance of reactant or the appearance of product proceeds at a constant velocity irrespective of reactant concentration. The units of the rate constant are concentration per unit time. Zero-order reaction conditions only occur in catalyzed reactions where the concentration of reactants is large enough to saturate all the catalytic sites. Under these conditions the catalyst is operating at maximum velocity, and all catalytic sites are filled; therefore, addition of more reactant cannot increase the rate.

Reversibility of Reactions

Although most chemical reactions are reversible, some directionality may be imposed on particular steps in a metabolic pathway through rapid removal of the end product by subsequent reactions in the pathway.

Many ligase reactions involving the nucleoside triphosphates result in release of pyrophosphate (PP_i). These reactions are rendered irreversible by the hydrolysis of the pyrophosphate to 2 mol of inorganic phosphate, P_i. Schematically,

$$\mathrm{A + B + ATP \longrightarrow A{-}B + AMP + PP_i}$$
$$\mathrm{PP_i + H_2O \longrightarrow 2P_i}$$

The conversion of the "high-energy" pyrophosphate to inorganic phosphate imposes irreversibility on the system by virtue of the thermodynamic stability of the products.

For those reactions that are reversible, the equilibrium constant for the reaction

$$\mathrm{A + B \rightleftharpoons C}$$

is

$$K_{eq} = \frac{[\mathrm{C}]}{[\mathrm{A}][\mathrm{B}]} \tag{4.7}$$

and can also be expressed in terms of the rate constants of the forward and reverse reactions:

$$\mathrm{A + B} \underset{k_2}{\overset{k_1}{\rightleftharpoons}} \mathrm{C}$$

where

$$\frac{k_1}{k_2} = K_{eq} \tag{4.8}$$

Equation 4.8 shows the relationship between thermodynamic and kinetic quantities. The term K_{eq} is a thermodynamic expression of the state of the system, while k_1 and k_2 are kinetic expressions that are related to the speed at which that state is reached.

Enzymes Show Saturation Kinetics

Terminology

Enzyme activity is usually expressed in units of micromoles (μmol) of substrate converted to product per minute under specified assay conditions. One **standard unit** of enzyme activity (U) is an amount of activity that catalyzes the transformation of 1 μmol/min. The **specific activity** of an enzyme preparation is defined as the number of enzyme units per milligram of protein (μmol min^{-1} mg of protein^{-1} or U/mg of protein). This expression, however, does not indicate whether the sample tested contains only the enzyme protein; during an enzyme purification the value will increase as contaminating protein is removed. The **catalytic constant,** or **turnover number,** for an enzyme is equal to the units of enzyme activity per mol of enzyme (μmol/min/mol of enzyme). In cases where the enzyme has more than one catalytic center, the catalytic constant is often given on the basis of the particle weight of the subunit rather than the molecular weight of the entire protein. The Commission on Enzyme Nomenclature of the International Union of Biochemistry has recommended that enzyme activity be expressed in units of moles per second, instead of micromoles per minute, to conform with the rate constants used in chemical kinetics. A new unit, the **katal** (abbreviated kat), is proposed where 1 kat denotes the conversion of 1 mol substrate per second. Activity can be expressed, however, as millikatals (mkat), microkatals (μkat), and so forth. The specific activity and catalytic constant can also be expressed in this unit of activity.

The catalytic constant or turnover number allows a direct comparison of relative catalytic ability between enzymes. For example, the constants for catalase and α-amylase are 5×10^6 and 1.9×10^4, respectively, indicating that catalase is ~2500 times more active than amylase.

The **maximum velocity** V_{max} is the velocity obtained under conditions of substrate saturation of the enzyme under a given set of conditions of pH, temperature, and ionic strength.

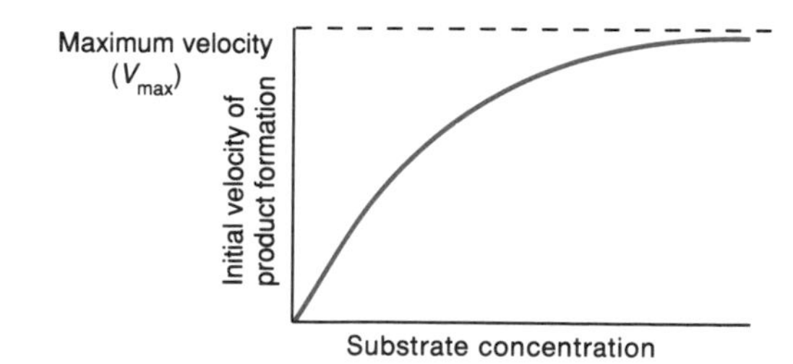

FIGURE 4.11
Plot of velocity versus substrate for an enzyme-catalyzed reaction.
Initial velocities are plotted against the substrate concentration at which they were determined. The curve is a rectangular hyperbola, which asymptotically approaches the maximum velocity possible with a given amount of enzyme.

Interaction of Enzyme and Substrate

The initial velocity of an enzyme-catalyzed reaction is dependent on the concentration of substrate (S) as shown in Figure 4.10. As the concentration of substrate is increased (S_1–S_4), the initial velocity increases until the enzyme is completely saturated with the substrate. If one plots the initial velocities obtained at given substrate concentrations (Figure 4.11), a rectangular hyperbola is obtained. The same type of curve will be obtained for the binding of oxygen to myoglobin as a function of increasing oxygen pressure. In general, the rectangular hyperbola will be obtained for any process that involves an interaction or binding of reactants or other substances at a specific but limited number of sites. The velocity of the reaction reaches a limiting maximum at the point at which all the available sites are saturated. The curve in Figure 4.11 is referred to as the *substrate saturation curve* of an enzyme-catalyzed reaction and reflects the fact that the enzyme has a specific binding site for the substrate. Obviously the enzyme (E) and substrate must interact in some way if the substrate is to be converted to products. Initially there is formation of a complex between the enzyme and substrate:

$$\mathrm{E} + \mathrm{S} \underset{k_2}{\overset{k_1}{\rightleftharpoons}} \mathrm{ES} \tag{4.9}$$

The rate constant for formation of this ES complex is defined as k_1, and the rate constant for disassociation of the ES complex is defined as k_2. So far, we have described only an equilibrium binding of enzyme and substrate. The actual chemical event in which bonds are made or broken occurs in the ES complex. The conversion of substrate to products (P) then occurs from the ES complex with a rate constant k_3. Therefore, Eq. 4.9 is transformed to

$$\mathrm{E} + \mathrm{S} \underset{k_2}{\overset{k_1}{\rightleftharpoons}} \mathrm{ES} \xrightarrow{k_3} \mathrm{E} + \mathrm{P} \tag{4.10}$$

Equation 4.10 is a general statement of the mechanism of enzyme action. The equilibrium between E and S can be expressed as an affinity constant, K_a, only if the rate of the chemical phase of the reaction, k_3, is small compared to k_2; then $K_a = k_1/k_2$. Earlier we used K_{eq} to describe chemical reactions. In enzymology the association or affinity constant K_a is preferred.

The **initial velocity** of an enzyme-catalyzed reaction is dependent on the amount of substrate present and on the enzyme concentration. Figure 4.12 shows progress curves for increasing concentrations of enzyme, where there is enough substrate initially to saturate the enzyme at all levels. The initial velocity doubles as the concentration of enzyme doubles. At the lower concentrations of enzyme, equilibrium is reached more slowly than at higher concentrations, but the final equilibrium position is the same.

From our discussion thus far, we can conclude that the velocity of an enzyme reaction is dependent on both substrate and enzyme concentration.

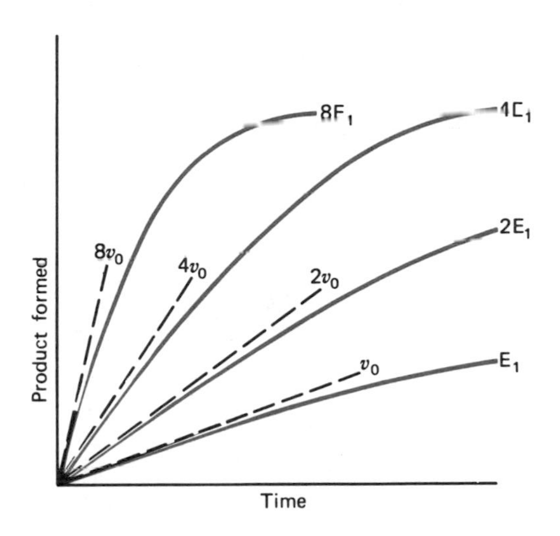

FIGURE 4.12
Progress curves at variable concentrations of enzyme and saturating levels of substrate.
The initial velocity (v_0) doubles as the enzyme concentration doubles. Since the substrate concentrations are the same, the final equilibrium concentrations of product will be identical in each case; however, equilibrium will be reached at a slower rate in those assays containing small amounts of enzyme.

Formulation of the Michaelis–Menten Equation

It should be recalled that in the discussion of chemical kinetics, rate equations were developed so that the velocity of the reaction could be expressed in terms of the substrate concentration. This philosophy also holds for enzyme-catalyzed reactions where the ultimate goal is to develop a relationship that will allow the velocity of a reaction to be correlated with the amount of enzyme. First, a rate equation must be developed that will relate the velocity of the reaction to the substrate concentration.

In the development of this rate equation, which is known as the **Michaelis–Menten** equation, three basic assumptions are made. The first is that the ES complex is in a steady state. That is, during the initial phases of the reaction, the concentration of the ES complex remains constant, even though many molecules of substrate are converted to products via the ES complex. The second assumption is that under saturating conditions all of the enzyme is converted to the ES complex, and none is free. This occurs when the substrate concentration is high. The third assumption is that if all the enzyme is in the ES complex, then the rate of formation of products will be the maximum rate possible, that is,

$$V_{max} = k_3[\mathrm{ES}] \tag{4.11}$$

CLIN. CORR. **4.1**
A CASE OF GOUT DEMONSTRATES TWO PHASES IN THE MECHANISM OF ENZYME ACTION

The partitioning of the Michaelis–Menten model of enzyme action into two phases: binding, followed by chemical modification of substrate, is illustrated by studies on a family with hyperuricemia and gout. The patient excreted three times the normal amount of uric acid per day and had markedly increased levels of 5′-phosphoribosyl-α-pyrophosphate (PRPP) in his red blood cells. Assays in vitro revealed that his red cell PRPP synthetase activity was increased threefold. The pH optimum and the K_m of the enzyme for ATP and ribose 5-phosphate were normal, but V_{max} was increased threefold! This increase was not because of an increase in the amount of enzyme; immunologic testing with a specific antibody to the enzyme revealed similar quantities of the enzyme protein in the normal and in the patient's red cells. This finding demonstrates that the binding of substrate as reflected by K_m and the subsequent chemical event in catalysis, which is reflected in V_{max}, are separate phases of the overall catalytic process. This situation holds only for those enzyme mechanisms in which $k_2 >> k_3$. It is distinctly unusual for mutant enzymes to be more active than the normal form.

Becker, M. A., Kostel, P. J., Meyer, L. J., and Seegmiller, J. E. Human phosphoribosylpyrophosphate synthetase: increased enzyme specific activity in a family with gout and excessive purine synthesis. *Proc. Natl. Acad. Sci. U.S.A.* 70:2749, 1973.

If one then writes the steady-state expression for the formation and breakdown of the ES complex as

$$K_m = \frac{k_2 + k_3}{k_1} \qquad \textbf{(4.12)}$$

then the rate expression can be obtained after suitable algebraic manipulation as

$$\text{Velocity} = v = \frac{V_{max} \cdot [S]}{K_m + [S]} \qquad \textbf{(4.13)}$$

The complete derivation of this equation is at the end of this section. The two constants in this rate equation, V_{max} and K_m, are unique to each enzyme under specific conditions of pH and temperature. For those enzymes in which $k_3 \ll k_2$, $\mathbf{K_m}$ becomes the reciprocal of the enzyme–substrate binding constant, that is,

$$K_m = \frac{1}{K_a}$$

and the V_{max} reflects the catalytic phase of the enzyme mechanism as suggested by Eq. 4.11. In other words, in this simple model the activity of the enzyme can be separated into two phases: binding of substrate followed by chemical modification of the substrate. This biphasic nature of enzyme mechanism is reinforced in the clinical example discussed in Clin. Corr. 4.1.

Significance of K_m

The concept of K_m may appear to have no physiological or clinical relevance. The truth is quite the contrary. As discussed in Section 4.9, all valid enzyme assays performed in the clinical laboratory are based on knowledge of the K_m values for each substrate.

In terms of the physiological control of glucose and phosphate metabolism, two hexokinases have evolved, one with a high K_m and one with a low K_m for glucose. Together, they contribute to maintaining steady-state levels of blood glucose and phosphate, as discussed in more detail on page 309.

In general K_m values are found to be near concentrations of substrate found in the cell. Perhaps enzymes have evolved substrate binding sites with affinities comparable to in vivo levels of their substrates. Occasionally, mutation of the enzyme binding site occurs, or a different form of an enzyme (isoenzyme) with an altered K_m is expressed. Either one of these events can result in an abnormal physiology. An interesting example given in Clin. Corr. 4.2 is the case of the expression of only the atypical form of aldehyde dehydrogenase in people of Asiatic origin.

Notice that if one allows the initial velocity, v_0, to be equal to $\frac{1}{2}V_{max}$ in Eq. 4.13, K_m will become equal to [S]:

$$\tfrac{1}{2}V_{max} = \frac{V_{max} \cdot [S]}{K_m + [S]}$$

$$K_m + [S] = \frac{2V_{max} \cdot [S]}{V_{max}}$$

$$K_m = [S]$$

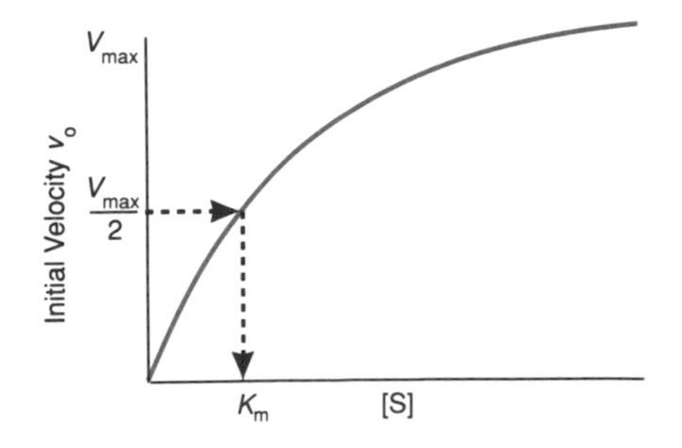

FIGURE 4.13
Graphic estimation of K_m for the v_0 versus [S] plot.
K_m is the substrate concentration at which the enzyme has half-maximal activity.

Therefore, from a substrate saturation curve the numerical value of the K_m can be derived by graphical analysis, as shown in Figure 4.13. In other words, the K_m is equal to the substrate concentration that will give one-half the maximum velocity.

Linear Form of the Michaelis–Menten Equation

In practice the determination of K_m from the substrate saturation curve is not very accurate, because V_{max} is approached asymptotically. If one takes the reciprocal of Eq. 4.13 and separates the variables into a format consistent with the equation of a straight line ($y = mx + b$), then

$$\frac{1}{v_0} = \frac{K_m}{V_{max}} \times \frac{1}{[S]} + \frac{1}{V_{max}}$$

A plot of the reciprocal of the initial velocity versus the reciprocal of the initial substrate concentration yields a line whose slope is K_m/V_{max} and whose y intercept is $1/V_{max}$. Such a plot is shown in Figure 4.14. It is often easier to obtain the K_m from the intercept on the x axis, which is $-1/K_m$.

This linear form of the Michaelis–Menten equation is often referred to as the **Lineweaver–Burk** or double reciprocal plot. Its advantage is that statistically significant values of K_m and V_{max} can be obtained directly with six to eight data points.

Derivation of the Michaelis–Menten Equation

The generalized statement of the mechanism of enzyme action is

$$E + S \underset{k_2}{\overset{k_1}{\rightleftharpoons}} ES \xrightarrow{k_3} E + P \qquad \textbf{(4.10)}$$

If we assume that the rate of formation of the ES complex is balanced by its rate of breakdown (the steady-state assumption), then we can write

$$v_{formation} = k_1[S]\,[E]$$

and

$$v_{breakdown} = k_2[ES] + k_3[ES] = [ES](k_2 + k_3)$$

If we set the rate of formation equal to the rate of breakdown, then

$$k_1[S]\,[E] = [ES](k_2 + k_3)$$

After dividing both sides of the equation by k_1, we have

$$[S]\,[E] = [ES]\left[\frac{k_2 + k_3}{k_1}\right] \qquad \textbf{(4.14)}$$

If we now define the ratio of the rate constants $(k_2 + k_3)/k_1$ as K_m, the Michaelis constant, and substitute it into Eq. 4.14, then

$$[S]\,[E] = [ES]K_m \qquad \textbf{(4.15)}$$

Since [E] is equal to the free enzyme, we must express its concentration in terms of the total enzyme added to the system minus any enzyme in the [ES] complex, that is,

$$[E] = ([E_t] - [ES])$$

CLIN. CORR. 4.2
THE PHYSIOLOGICAL EFFECT OF CHANGES IN ENZYME K_m VALUE

The unusual sensitivity of Asians to alcoholic beverages has a biochemical basis. In some Japanese and Chinese, much less ethanol is required to produce the vasodilation that results in facial flushing and tachycardia than is required to achieve the same effect in Europeans. The physiological effects arise from the acetaldehyde generated by liver alcohol dehydrogenase.

$$CH_3CH_2OH + NAD^+ \rightleftharpoons CH_3CHO + H^+ + NADH$$

The acetaldehyde formed is normally removed by a mitochondrial aldehyde dehydrogenase that converts it to acetate. In some individuals of Mongoloid origins, the normal form of the mitochondrial aldehyde dehydrogenase is missing. This form has a low K_m for acetaldehyde. In these people, only the cytosolic high K_m (lower affinity) form of the enzyme is present. This form of the enzyme allows a higher steady-state level of acetaldehyde in the blood after alcohol consumption, which accounts for the increased sensitivity to alcohol.

Crabb, D. W., Edenberg, H. J., Bosron, W. F., and Li, T.-K. Genotypes for aldehyde dehydrogenase deficiency and alcohol sensitivity: The ALDH22 allele is dominant. *J. Clin. Invest.* 83:314, 1989.

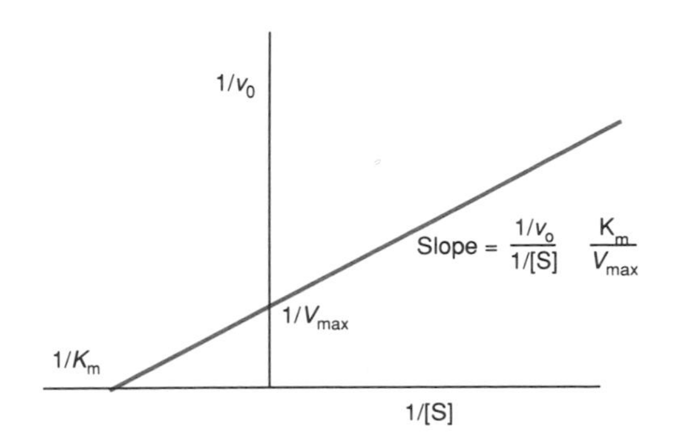

FIGURE 4.14
Determination of K_m and V_{max} from the Lineweaver–Burk double-reciprocal plot.
Plots of the reciprocal of the initial velocity versus the reciprocal of the substrate concentration used to determine the initial velocity yield a line whose x intercept is $-1/K_m$.

Upon substitution of the equivalent expression for [E] into Eq. 4.15 we have

$$[S]([E_t] - [ES]) = [ES]K_m$$

Dividing through by [S] yields

$$[E_t] - [ES] = \frac{[ES]K_m}{[S]}$$

and dividing through by [ES] yields

$$\frac{[E_t]}{[ES]} - 1 = \frac{K_m}{[S]} \quad \text{or} \quad \frac{[E_t]}{[ES]} = \frac{K_m}{[S]} + 1 = \frac{K_m + [S]}{[S]} \qquad \textbf{(4.16)}$$

We now need to obtain an alternative expression for $[E_t]/[ES]$, since [ES] cannot be measured easily, if at all. When the enzyme is saturated with substrate all the enzyme will be in the ES complex, and none will be free, $[E_t] = [ES]$, and the velocity observed will be the maximum possible; therefore, $V_{max} = k_3[E_t]$ [see Eq. 4.11.] When $[E_t]$ is not equal to [ES], $v = k_3[ES]$. From these two expressions we can obtain the ratio of $[E_t]/[ES]$, that is,

$$\frac{[E_t]}{[ES]} = \frac{V_{max}/k_3}{v/k_3} = \frac{V_{max}}{v} \qquad \textbf{(4.17)}$$

Substituting this value of $[E_t]/[ES]$ into Eq. 4.16 yields a form of the Michaelis–Menten equation:

$$\frac{V_{max}}{v} = \frac{K_m + [S]}{[S]}$$

or

$$v = \frac{V_{max}[S]}{K_m + [S]}$$

An Enzyme Catalyzes Both Forward and Reverse Directions of a Reversible Reaction

As has been indicated previously, enzymes do not alter the equilibrium constant of a reaction; consequently, in a reaction

$$S \underset{k_2}{\overset{k_1}{\rightleftharpoons}} P$$

the direction of flow of material, either in the forward direction or the reverse direction, will depend on the concentration of S relative to P and the equilibrium constant of the reaction. Since enzymes catalyze the forward as well as the reverse reaction, a problem may arise if the product has an affinity for the enzyme that is similar to that of the substrate. In this case the product can easily rebind to the active site of the enzyme and will compete with the substrate for that site. In such cases the product inhibits the reaction as the concentration of product increases. The Lineweaver–Burk plot will not be linear in those cases where the enzyme is susceptible to product inhibition. If the subsequent enzyme in the metabolic pathway removes the product rapidly, then product inhibition may not occur.

Product inhibition in a metabolic pathway provides a limited means of controlling or modulating the flux of substrates through the pathway. As the end product of the pathway increases, each intermediate will also increase via mass action. If one or more enzymes in the pathway are particularly sensitive to product inhibition, the output of the end product of the pathway will be suppressed.

Reversibility of a pathway or a particular enzyme-catalyzed reaction is dependent on the rate of product removal. If the end product is quickly removed, then the pathway may be physiologically unidirectional.

Multisubstrate Reactions Follow Either a Ping–Pong or Sequential Mechanism

Most enzymes utilize more than one substrate, or they act upon one substrate plus a coenzyme and generate one or more products. In any case, a K_m must be determined for each substrate and coenzyme involved in the reaction when establishing an enzyme assay.

Mechanistically, enzyme reactions are divided into two major categories, ping–pong or sequential. There are many variations on these major mechanisms.

The **ping–pong mechanism** can be diagrammatically outlined as follows:

$$E + A \longrightarrow EA \xrightarrow{\uparrow P_1} E' \xrightarrow{\downarrow B} E'B \longrightarrow P_2 + E$$

in which substrate A reacts with E to produce product P_1, which is released before the second substrate B will bind to the modified enzyme E′. The substrate B is then converted to product P_2 and the enzyme regenerated. A good example of this mechanism is the transaminase catalyzed reaction (p. 478) in which the α-amino group of amino $acid_1$ (A) is transferred to the enzyme and the newly formed keto $acid_1$ is released (P_1) followed by the binding of the acceptor keto $acid_2$ (B) and release of amino $acid_2$ (P_2). This reaction is schematically outlined in Figure 4.15.

In the **sequential mechanism,** if the two substrates A and B can bind in any order, it is a **random mechanism;** if the binding of A is required before B can be bound, then it is an **ordered mechanism.** In either case the reaction is bimolecular, that is, both A and B must be bound before reaction occurs. Examples of both these mechanisms can be found among the dehydrogenases in which the second substrate is the coenzyme (NAD^+, FAD, etc., p. 138). The release of products may or may not be ordered in either mechanism.

4.4 COENZYMES: STRUCTURE AND FUNCTION

Coenzymes function with the enzyme in the catalytic process. Often the **coenzyme** has an affinity for the enzyme that is similar to that of the substrate; consequently, the coenzyme can be considered a second substrate. In other cases, the coenzyme is covalently bound to the enzyme and functions at or near the active site in the catalytic event. There are other examples of enzymes where the role of the coenzyme falls between these two extremes.

Several, but not all, of the coenzymes are synthesized from the B vitamins. Vitamin B_6, pyridoxine, requires little modification to be transformed to the active coenzyme, pyridoxal phosphate (Section 26.5). Clin. Corr. 4.3 points out the importance of the coenzyme binding site and how alterations in this site cause metabolic dysfunction.

CLIN. CORR. 4.3
MUTATION OF A COENZYME BINDING SITE RESULTS IN CLINICAL DISEASE

Cystathioninuria is a genetic disease in which the enzyme γ-cystathionase is either deficient or inactive. Cystathionase catalyzes the reaction:

$$\text{Cystathionine} \longrightarrow \text{cysteine} + \alpha\text{-ketobutyrate}$$

Deficiency of the enzyme leads to accumulation of cystathionine in the plasma .

Since cystathionase is a pyridoxal phosphate-dependent enzyme, vitamin B_6 was administered to patients whose fibroblasts contained material that cross-reacts with antibody against cystathionase. Many responded to B_6 therapy with a fall in plasma levels of cystathionine. Such patients produced the apoenzyme. In one particular patient the activity of the enzyme was undetectable in fibroblast homogenates but increased to 31% of normal with the addition of 1 mM of pyridoxal phosphate to the assay mixture. It is thought that the K_m for pyridoxal phosphate binding to the enzyme was increased because of a mutation in the binding site. Activity is partially restored by increasing the concentration of coenzyme. Apparently these patients require a higher steady-state concentration of coenzyme to maintain any γ-cystathionase activity.

Pascal, T. A., Gaull, G. E., Beratis, N. G., Gillam, B. M., Tallan, H. H., and Hirschhorn, K. Vitamin B6-responsive and unresponsive cystathionuria: two variant molecular forms. *Science* 190:1209, 1975.

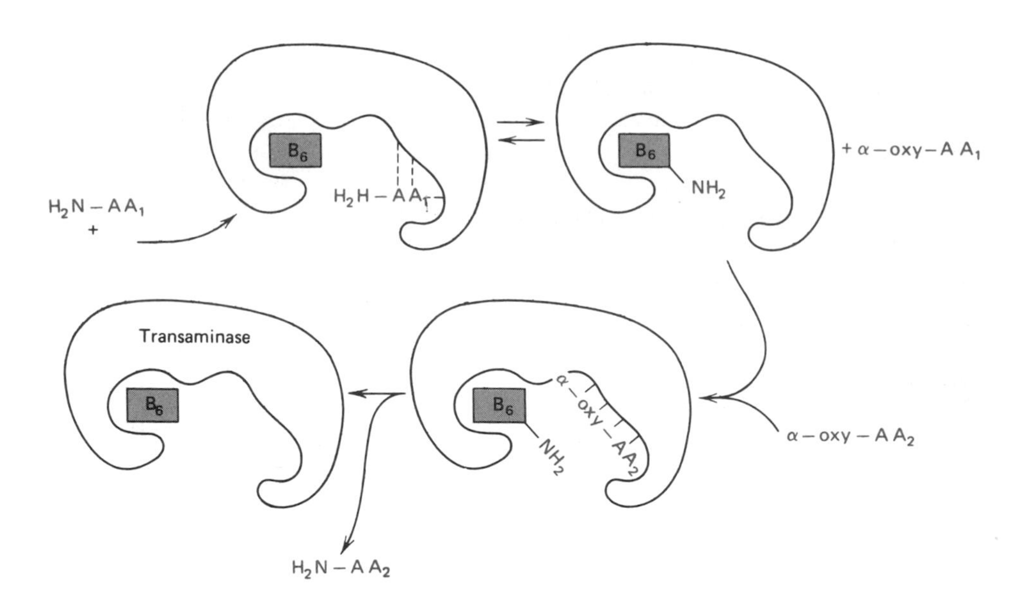

FIGURE 4.15
Schematic representation of the transaminase reaction mechanism—an example of a ping–pong mechanism.
Enzyme-bound vitamin B_6 coenzyme accepts the α-amino group from the first amino acid (AA_1), which is then released from the enzyme as an α-keto acid. The acceptor α-keto acid (AA_2) is then bound to the enzyme, and the bound amino group is transferred to it, forming a new amino acid, which is then released from the enzyme. The terms "oxy" and "keto" are used interchangeably.

In contrast to vitamin B_6, niacin requires major alteration by the mammalian cell before it is capable of acting as a coenzyme. This metabolic interconversion is outlined in Section 13.9.

The structure and function of the coenzymes of only two B vitamins, niacin and riboflavin, and of ATP will be discussed in this chapter. The structure and function of coenzyme A (CoA) (Section 13.9), thiamine (Section 26.5), biotin, and vitamin B_{12} are included in those chapters dealing with enzymes dependent on the given coenzyme for activity.

Adenosine Triphosphate May Be Either a Second Substrate or a Modulator of Activity

Adenosine triphosphate (ATP) often functions as a second substrate but can also serve as a cofactor in modulating the activity of specific enzymes. This compound is so pivotal that its structure and function will be introduced here. Adenosine triphosphate (Figure 4.16) can be synthesized de novo in all mammalian cells.

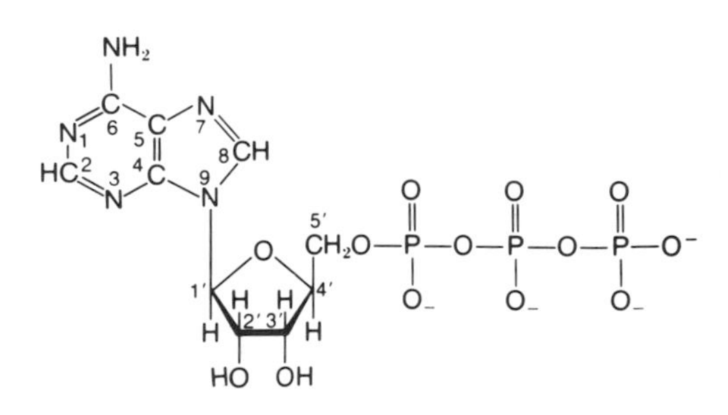

FIGURE 4.16
Adenosine triphosphate (ATP).

The nitrogenous heterocyclic ring is adenine. To the adenine is affixed a ribosyl 5′-triphosphate. The functional end of the molecule is the reactive triphosphate, which is shown in the state of ionization found in the cell. As a cosubstrate ATP is utilized by the kinases for the transfer of the terminal phosphate to various acceptors. A typical example is the glucokinase-catalyzed reaction:

$$\text{Glucose} + \text{ATP} \longrightarrow \text{glucose 6-phosphate} + \text{ADP}$$

ADP is adenosine diphosphate. The combination of adenine plus ribose is adenosine.

ATP has an additional role, other than cosubstrate: in a number of specific enzyme reactions it serves as a modulator of the activity of the enzyme. These particular enzymes have binding sites for ATP, occupancy of which changes the affinity or reactivity of the enzyme toward its substrates. Mechanistically, ATP is acting as an allosteric effector in these cases (p. 162).

NAD and NADP Are Coenzymes of Niacin

Niacin is pyridine-3-carboxylic acid. It is converted to two major coenzymes that are involved in the oxidoreductase class of enzymes. These coenzymes are **NAD (nicotinamide adenine dinucleotide)** and **NADP (nicotinamide adenine dinucleotide phosphate).** It is convenient to use the abbreviations NAD and NADP when referring to the coenzymes regardless of their state of oxidation or reduction, NAD^+ and $NADP^+$ to represent the oxidized forms, and NADH and NADPH to represent the reduced forms. There are dehydrogenases that function with NADP as coenzyme but not with NAD. The reverse is also true and some enzymes function with either coenzyme. Such an arrangement allows for specificity and control over dehydrogenases that reside in the same subcellular compartment.

Structurally, NAD is composed of adenosine and *N*-ribosyl-nicotinamide linked through a pyrophosphate linkage between the 5′-OH groups of the two ribosyl moieties (Figure 4.17). The NADP differs structurally from NAD in having an additional phosphate esterified to the 2′-OH group of the adenosine moiety.

Both coenzymes function as intermediates in the transfer of two electrons between an electron donor and an acceptor. The donor and acceptor need not be involved in the same metabolic pathway. In other words, the reduced form of these nucleotides acts as a common "pool" of electrons that arise from many oxidative reactions and can be used for various reductive reactions.

The adenine, ribose, and pyrophosphate components of NAD are involved in the binding of NAD to the enzyme. Enzymes requiring NADP have a specific cationic region in their NADP binding site that is positioned so as to form an ionic bond with the 2′-phosphate of NADP. This enhances binding of NADP in preference to NAD in these particular enzymes.

The nicotinamide portion of the molecule is involved in reversibly accepting and donating two electrons at a time. It is the active center of the coenzyme. In the oxidation of deuterated ethanol by alcohol dehydrogenase, NAD^+ accepts two electrons and the deuterium from the ethanol, and the other hydrogen is released as a proton (Figure 4.18).

The specific binding of NAD^+ to the enzyme surface confers a chemically recognizable "top side" and "bottom side" to the planar nicotinamide. The former is known as the **A** face and the latter the **B** face. In the case of alcohol dehydrogenase, the proton or the deuterium ion that serves as a tracer is added to the **A** face. Other dehydrogenases utilize the **B** face. The particular effect just described demonstrates how enzymes

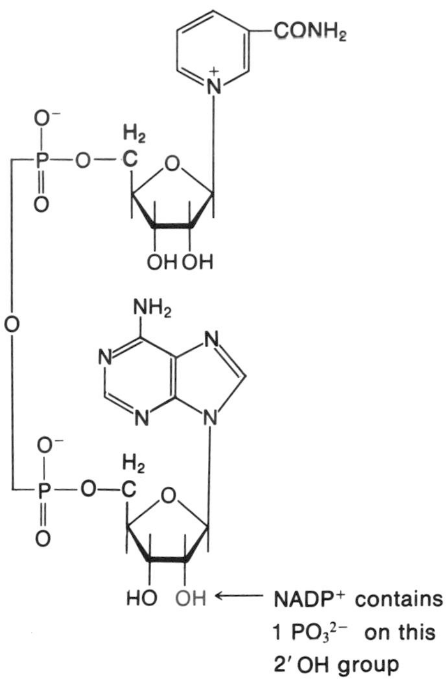

NADP$^+$ contains 1 PO_3^{2-} on this 2′ OH group

FIGURE 4.17
Nicotinamide adenine dinucleotide (NAD^+).

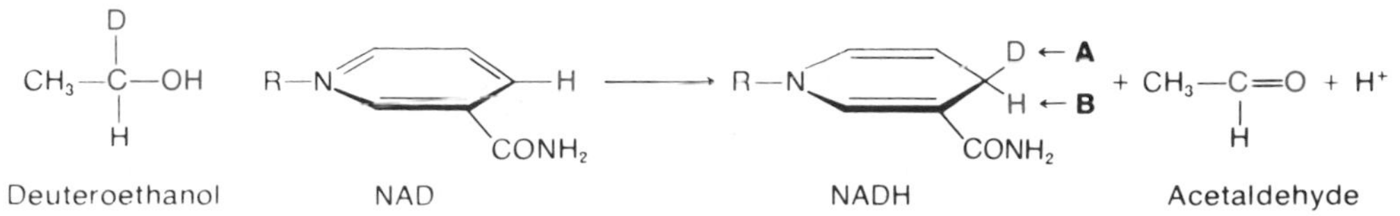

Deuteroethanol NAD NADH Acetaldehyde

FIGURE 4.18
Transfer of deuterium from deuterated ethanol to NAD^+.

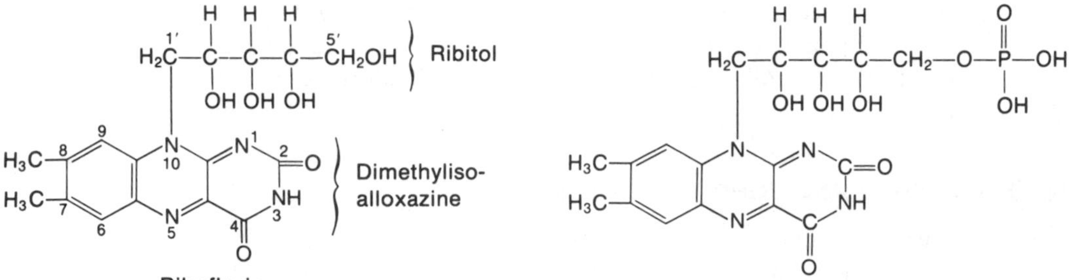

Riboflavin

Flavin mononucleotide

FIGURE 4.19
Riboflavin and flavin mononucleotide.

are able to induce stereospecificity into chemical reactions by virtue of the asymmetric binding of coenzymes and substrates.

FMN and FAD Are Coenzymes of Riboflavin

The two coenzyme forms of riboflavin are **FMN (flavin mononucleotide)** and **FAD (flavin adenine dinucleotide).** The vitamin riboflavin consists of the heterocyclic ring, isoalloxazine (flavin) connected through N-10 to the alcohol ribitol as shown in Figure 4.19.

The FMN has a phosphate esterified to the 5′-OH group of riboflavin. The FAD is structurally analogous to NAD in having adenosine linked through a pyrophosphate linkage to a heterocyclic ring, in this case riboflavin.

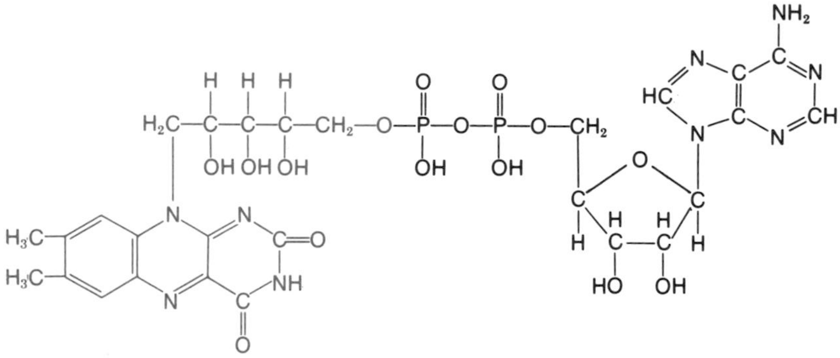

Flavin adenine dinucleotide (FAD)

Both FAD and FMN function in oxidation–reduction reactions by accepting and donating $2e^-$ through the isoalloxazine ring. A typical example of FAD participation in an enzyme reaction is the oxidation of succinate to fumarate by succinate dehydrogenase (p. 258).

In some cases, these coenzymes are $1e^-$ acceptors, which lead to flavin semiquinone formation (a free radical).

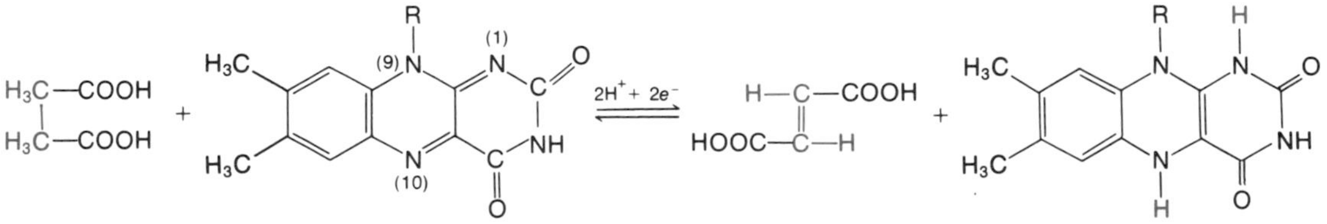

Succinic acid + FAD $\underset{2H^+ + 2e^-}{\rightleftharpoons}$ Fumaric acid + $FADH_2$

The flavin coenzymes have a tendency to be bound much tighter to their enzymes than the niacin coenzymes, and thus they may often function as prosthetic groups rather than as cofactors.

Metal Cofactors Have Various Functions

Metals are not coenzymes in the sense of FAD and NAD^+, but are required as cofactors in approximately two-thirds of all enzymes. There are two major areas in which metals participate in enzyme reactions—through their ability to act as *Lewis acids* and through various modes of *chelate* formation. *Chelates* are organometallic coordination complexes. A good example of a chelate is the complex between iron and porphyrin to form a heme (p. 1009).

Those metals that act as Lewis acid catalysts are found among the transition metals like Zn, Fe, Mn, and Cu, which have empty *d* electron orbitals that can act as electron sinks. The alkaline earth metals such as K and Na do not possess this ability.

A good example of a metal functioning as a Lewis acid is found in carbonic anhydrase. Carbonic anhydrase is a zinc enzyme (Enz) that catalyzes the reaction

$$CO_2 + H_2O \rightleftharpoons H_2CO_3$$

The first step can be visualized as the in situ generation of a proton and a hydroxyl group from water:

$$\text{Enz—Zn}^{2+} + \text{H—O—H} \rightleftharpoons \text{Enz—Zn}^{2+}\text{---}^{-}\text{O—H} + H^+$$

The proton and hydroxyl group are subsequently added to the carbon dioxide and carbonic acid is released. The reactions are presented in a stepwise fashion for clarity. Actually, the reactions may occur in a concerted fashion, that is, all at one time.

$$\text{Enz—Zn}^{2+}\text{---}^{-}\text{O—H} + H^+ + O{=}C{=}O \rightleftharpoons \text{Enz—Zn}^{2+}\text{---}^{-}\text{O(H)}\cdots O{=}C{=}O + H^+ \rightleftharpoons$$

$$\text{Enz—Zn}^{2+}\text{---O(H)—C(=O)—OH} \rightleftharpoons \text{Enz—Zn}^{2+} + H_2CO_3$$

Metals can also promote catalysis by either binding substrate and promoting electrophilic catalysis at the site of bond cleavage or by stabilizing intermediates in the reaction pathway. In the case of carboxypeptidase and thermolysin, zinc proteases with identical active sites, the zinc functions in the latter role although formerly it was thought that polarization of the carbonyl oxygen of the peptide bond was the function of the metal. This dual role of zinc is illustrated in the active site mechanism for carboxypeptidase as outlined in Figure 4.20.

The metal can also promote catalysis by binding substrate at the site of the bond cleavage. In carboxypeptidase, the carbonyl oxygen atom is chelated to the zinc. The resulting flow of electrons from the carbonyl carbon to the electropositive metal increases the susceptibility of the peptide bond to cleavage by nucleophiles such as water or carboxylate. This is schematically shown in Figure 4.20.

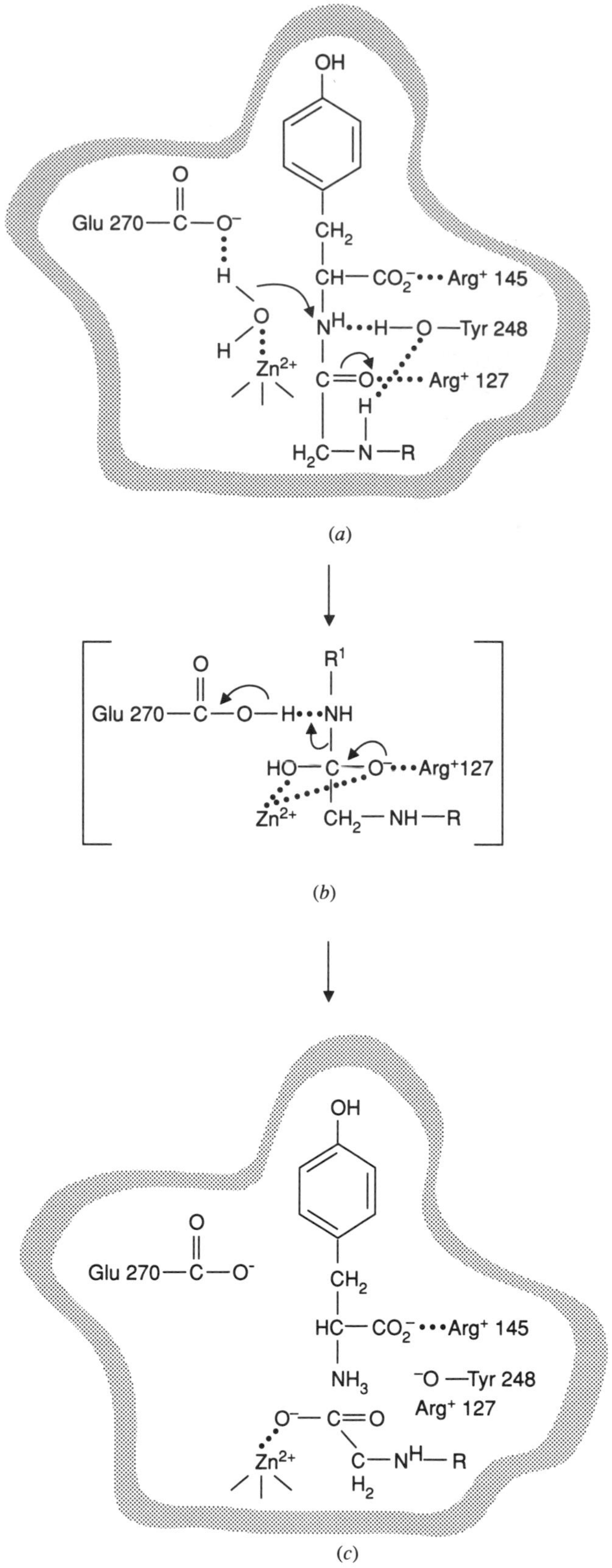

OH
O
Glu 270—C—O⁻
CH_2
H
CH—CO_2^-•••Arg^+ 145
O
NH•••H—O—Tyr 248
H
Zn^{2+}
C=O•••Arg^+ 127
H
H_2C—N—R

(a)

O
R^1
Glu 270—C—O—H•••NH
HO—C—O⁻•••Arg^+127
Zn^{2+}
CH_2—NH—R

(b)

OH
O
Glu 270—C—O⁻
CH_2
HC—CO_2^-•••Arg^+ 145
NH_3
⁻O—Tyr 248
Arg^+ 127
O⁻—C=O
Zn^{2+}
C—NH—R
H_2

(c)

FIGURE 4.20
The role of zine in the mechanism of reaction of carboxypeptidase A.
Enzyme bound zine generates a hydroxyl nucleophile from bound water, which then attacks the polarized carbonyl of the peptide bond (*a*). The zinc stabilizes the tetrahedral intermediate (*b*), which then collapses, as indicated by the arrows to yield products (*c*), which disassociate from the enzyme. Replacement of Tyr 248 with Phe has no effect on the enzymatic rate; demonstrating that this residue is not involved in catalysis contrary, to previous thinking.

Role of the Metal as a Structural Element

The functioning of a metal as a Lewis acid requires chelate formation. In addition, various modes of chelation occur between metal, enzyme, and substrate that are structural in nature, but in which no acid catalysis occurs.

In several of the kinases, creatine kinase being the best example, the true substrate is not ATP but Mg^{2+}–ATP (Figure 4.21).

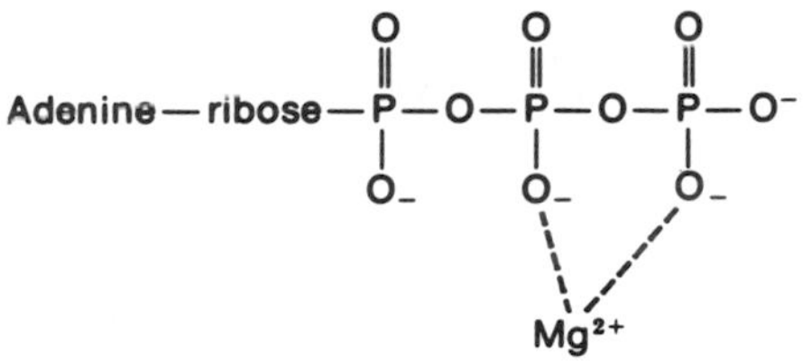

In this case, the magnesium does not interact directly with the enzyme. It may serve to neutralize the negative charge density on ATP and facilitate binding to the enzyme. Ternary complexes of this conformation are known as "substrate-bridged" complexes and can be schematically represented as Enz—S—M. A hypothetical scheme for the binding of Mg^{2+}–ATP and glucose in the active site of hexokinase is shown in Figure 4.22. All the kinases except muscle pyruvate kinase and phosphoenol pyruvate carboxykinase are substrate-bridged complexes.

In pyruvate kinase Mg^{2+} serves to chelate the ATP to the enzyme as shown in Figure 4.23. The absence of the metal cofactor results in failure of the ATP to bind to the enzyme. Enzymes of this class are "metal-bridged" ternary complexes, Enz—M—S. All metalloenzymes are of this type. **Metalloenzymes** are enzymes containing a tightly bound transition metal such as Zn^{2+} or Fe^{2+}. Several enzymes catalyzing enolization and elimination reactions are metal-bridge complexes.

In addition to the role of binding enzyme and substrate, metals may also bind directly to the enzyme to stabilize it in the active conformation or perhaps to induce the formation of a binding site or active site. Not only do the strongly chelated metals like Mn^{2+} play a role in this regard, but the weakly bound alkali metals (Na^+ or K^+) are also important. In pyruvate kinase, K^+ has been found to induce an initial conformation change, which is necessary, but not sufficient, for the ternary complex formation.

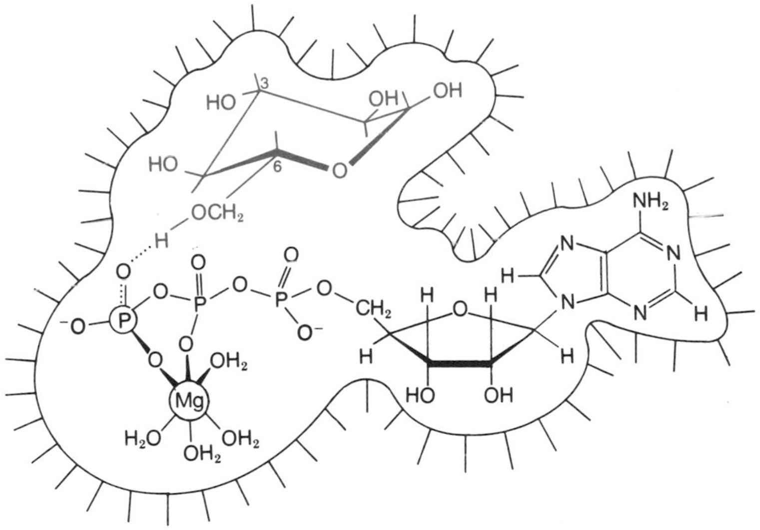

FIGURE 4.21
Mg^{2+}–ATP.

FIGURE 4.22
Model of the role of magnesium as a substrate-bridged complex in the active site of the kinases.
In hexokinase the terminal phosphate of ATP is transferred to glucose, yielding glucose 6-phosphate. Magnesium coordinates with the ATP to form the true substrate and in addition may labilize the terminal P—O bond of ATP to facilitate transfer of the phosphate to glucose. There are specific binding sites in the active site for glucose (upper left) as well as the adenine and ribose moieties of ATP.

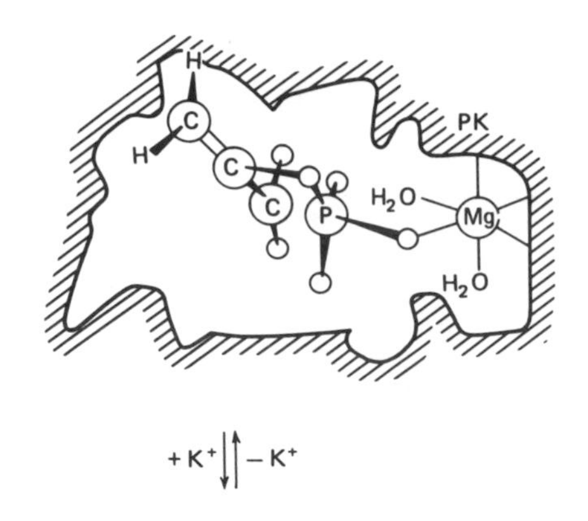

$+K^+ \updownarrow -K^+$

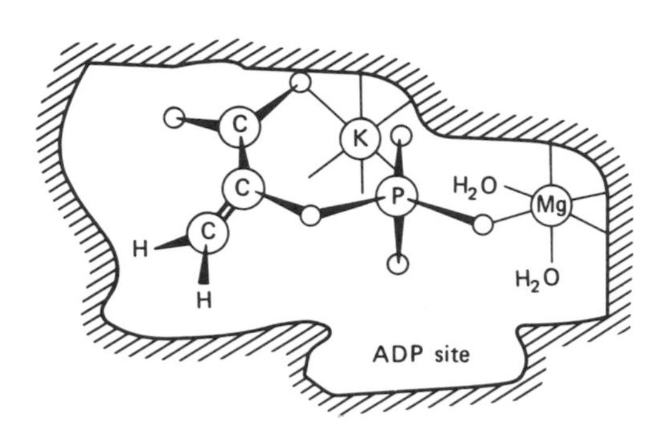

FIGURE 4.23
Model of the role of potassium ion in the active site of pyruvate kinase.
Pyruvate kinase catalyzes the reaction: phosphoenolpyruvate + ADP → ATP + pyruvate. Initial binding of K^+ induces conformational changes in the kinase, which result in increased affinity for phosphoenolpyruvate. In addition, K^+ orients the phosphoenolpyruvate in the correct position for transfer of its phosphate to the second substrate, which is ADP. Magnesium coordinates the substrate to the enzyme active site.
Modified, with permission, from A. S. Mildvan, *Annu. Rev. Biochem.* 43:365, 1974. © Annual Reviews, Inc.

Upon substrate binding, K^+ induces a second conformational change to the catalytically active ternary complex as indicated in Figure 4.23. In more general terms it is thought that Na^+ and K^+ stabilize the active conformation of the enzyme, but are passive from the catalytic standpoint.

Role of Metals in Oxidation and Reduction

The iron–sulfur enzymes, often referred to as **nonheme iron proteins,** are a unique class of metalloenzymes in which the active center consists of one or more clusters of sulfur-bridged iron chelates. These are of greater prominence in bacterial and plant systems than in mammalian cells. In mammalian systems succinate dehydrogenase (p. 259), NADH dehydrogenase, and adrenodoxin are good representatives of this group of proteins. The structure of the iron chelate in these nonheme iron proteins is represented in Figure 4.24.

In these proteins the bridging sulfide is released as H_2S on acidifying the enzyme. Cysteine thiol groups from the enzyme hold the bridged binuclear iron complex in the enzyme. These particular enzymes have reasonably low reducing potentials (E_0') and function in electron-transfer reactions. Adrenodoxin functions in the activation of oxygen in the steroid 11β- and 18-hydroxylases as a cosubstrate. It is not an enzyme.

The **cytochromes,** which are heme iron proteins, also function as cosubstrates for their respective reductases (p. 276). The iron in the hemes of the cytochromes undergoes reversible 1 e^- transfers. In addition, the heme is bound to the enzyme through coordination of an amino acid side chain to the iron of the heme. Thus, in the cytochromes the metal serves not only a structural role, but also participates in the chemical event.

The last role discussed here is the role of metals, specifically copper and iron, in activation of molecular oxygen. Copper is an active participant in several oxidase and hydroxylase enzymes. For example, **dopamine β-hydroxylase** catalyzes the introduction of one oxygen atom from O_2 into dopamine to form norepinephrine as shown in Figure 4.25. It is thought that the active form of the enzyme contains one atom of cuprous ion that reacts with oxygen to form an activated oxygen–copper complex. The copper–hydroperoxide shown in Figure 4.25 is thought to be converted to a copper(II)—O^- species that serves as the "active oxygen" in the hydroxylation of DOPA. In other metalloenzymes other species of "active oxygen" are generated and used for hydroxylation.

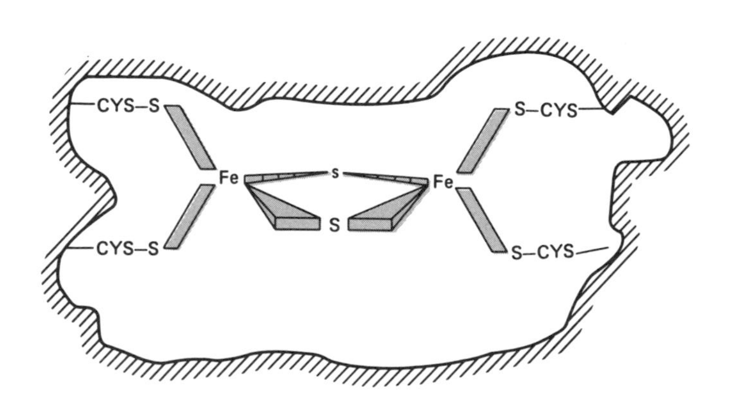

FIGURE 4.24
The iron binding site of adrenodoxin.
Two iron atoms are chelated to the protein via cysteine sulfhydryl groups. The two iron atoms are bridged by sulfides, which are released as hydrogen sulfide upon acidification of the protein. A formal valence state cannot be assigned to the iron atoms because they are magnetically coupled.
Redrawn, with permission, from W. Orme-Johnson in H. Sigel (Ed.). *Metal Ions in Biological Systems,* 7:129, 1979 Marcel Dekker, Inc.

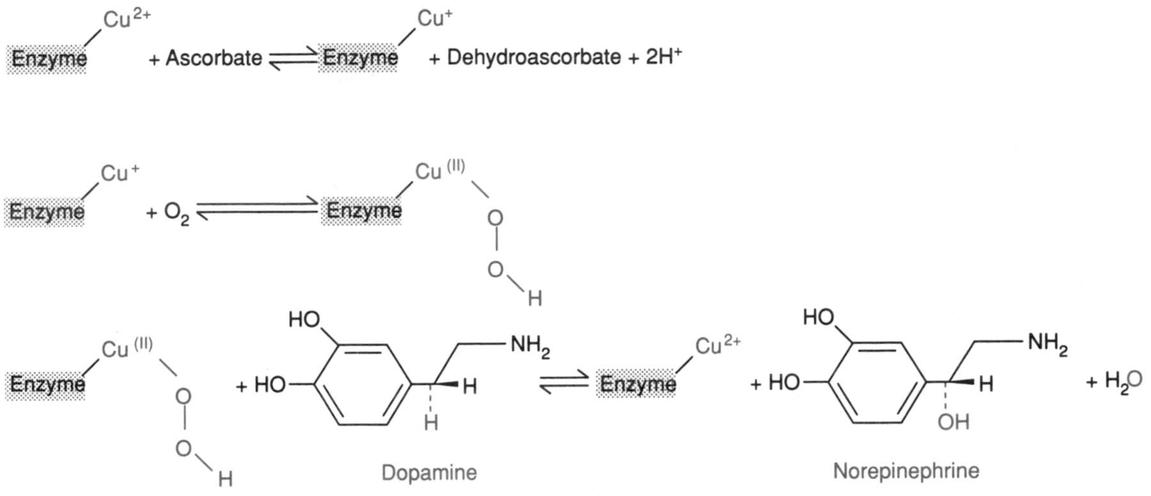

Enzyme–Cu^{2+} + Ascorbate ⇌ Enzyme–Cu^{+} + Dehydroascorbate + $2H^{+}$

Enzyme–Cu^{+} + O_2 ⇌ Enzyme–Cu(II)–O–O–H

Enzyme–Cu(II)–O–O–H + Dopamine ⇌ Enzyme–Cu^{2+} + Norepinephrine + H_2O

Dopamine

Norepinephrine

FIGURE 4.25
Role of copper in the activation of molecular oxygen by dopamine hydroxylase.
The normal cupric form of the enzyme is not reactive with oxygen but on reduction by the cosubstrate, ascorbate, generates a reactive enzyme–copper bound oxygen radical that subsequently reacts with dopamine to form norepinephrine and an inactive cupric enzyme.

4.5 INHIBITION OF ENZYMES

Mention was made earlier of product inhibition of enzyme activity and how an entire pathway could be controlled or modulated by this mechanism (p. 149). In addition to inhibition by the immediate product, products of other enzymes can also inhibit or even activate a particular enzyme. Much of current drug therapy is based on inhibition of specific enzymes with a substrate analog. It is important to discuss inhibition in more detail. Basically, there are three major classes of inhibitors: competitive, noncompetitive, and uncompetitive.

Competitive Inhibition May Be Reversed by Increased Substrate

Competitive inhibitors are defined as inhibitors whose action can be reversed by increasing amounts of substrate. Competitive inhibitors are usually enough like the substrate structurally that they bind at the substrate binding site and compete with the substrate for the enzyme. Once bound, the enzyme cannot convert the inhibitor to products. Increasing substrate concentrations will displace the reversibly bound inhibitor by the law of mass action. A competitive inhibitor need not be structurally related to the substrate.

In the succinate dehydrogenase reaction, malonate is structurally similar to succinate and is a competitive inhibitor (Figure 4.26).

Since the substrate and inhibitor are competing for the same site on the enzyme, the K_m for the substrate shows an apparent increase in the presence of inhibitor. This can be seen in a double-reciprocal plot as a shift in the x intercept ($-1/K_m$) and in the slope of the line (K_m/V_{max}). If we first establish the velocity at several levels of substrate and then repeat the experiment with a given but constant amount of inhibitor at various substrate levels, two different straight lines will be obtained as shown in Figure 4.27. As can be seen, the V_{max} does not change; hence the intercept on the y axis remains the same. In the presence of inhibitor, the x inter-

COO^-–CH_2–CH_2–COO^-
Succinate

COO^-–CH_2–COO^-
Malonate

FIGURE 4.26
Substrate and inhibitor of succinate dehydrogenase.

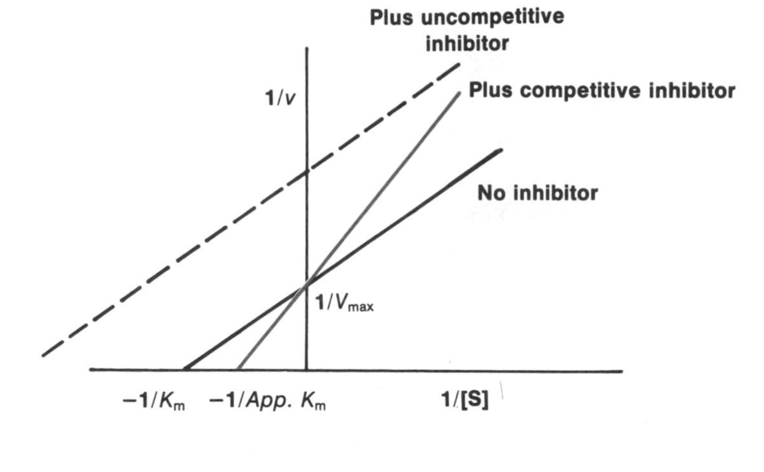

FIGURE 4.27
Double-reciprocal plots for competitive and uncompetitive inhibition.
A competitive inhibitor binds at the substrate binding site and effectively increases the K_m for the substrate. An uncompetitive inhibitor causes an equivalent shift in both V_{max} and K_m, resulting in a line parallel to that given by the uninhibited enzyme.

cept is no longer the negative reciprocal of the true K_m, but of an apparent value, $K_{m_{app}}$ where

$$K_{m_{app}} = K_m \cdot \left(1 + \frac{[I]}{K_I}\right)$$

Thus the inhibitor constant, K_I, can be determined from the concentration of inhibitor [I] used and the K_m, which was obtained from the x intercept of the line, obtained in the absence of inhibitor.

Noncompetitive Inhibitors Do Not Prevent Substrate From Binding

A **noncompetitive inhibitor** binds at a site other than the substrate binding site. The inhibition is not reversed by increasing concentrations of substrate. Both binary (EI) and ternary (EIS) complexes form, both of which are catalytically inactive and are therefore, dead-end complexes. The noncompetitive inhibitor behaves as though it were removing active enzyme from the solution, resulting in a decrease in V_{max}. This effect is seen graphically in the double-reciprocal plot (Figure 4.28), where K_m does not change but V_{max} does change. Inhibition can often be reversed by exhaustive dialysis of the inhibited enzyme provided that the inhibitor has not reacted covalently with the enzyme. This case is considered under the irreversible inhibitors and is discussed below.

The **uncompetitive inhibitor** binds only with the ES form of the enzyme in the case of a one-substrate enzyme. The result is an apparent equivalent change in K_m and V_{max}, which is reflected in the double-reciprocal plot as a line parallel to that of the uninhibited enzyme (Figure 4.27). In the case of multisubstrate enzymes the interpretation is complex and will not be considered further.

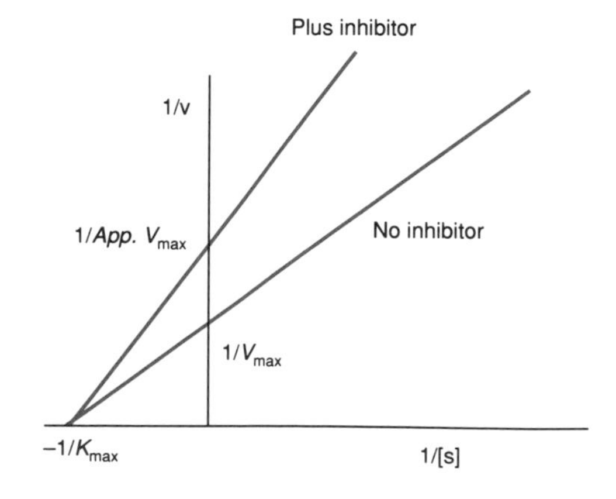

FIGURE 4.28
Double-reciprocal plot for an enzyme subject to reversible noncompetitive inhibition.
A noncompetitive inhibitor binds at a site other than the substrate binding site; therefore, the effective K_m does not change, but the apparent V_{max} decreases.

Irreversible Inhibition Involves Covalent Modification of an Enzyme Site

In cases of covalent modification of the binding site or the active site, inhibition will not be reversed by dialysis unless the linkage is chemically labile like that of an ester or thioester. The active site thiol in glyceraldehyde 3-phosphate dehydrogenase reacts with p-chloromercuribenzoate to form a mercuribenzoate adduct of the enzyme as shown in Figure 4.29. Such adducts are not reversed by dialysis or by addition of substrate. Double reciprocal plots show a characteristic pattern for noncompetitive inhibition (Figure 4.28).

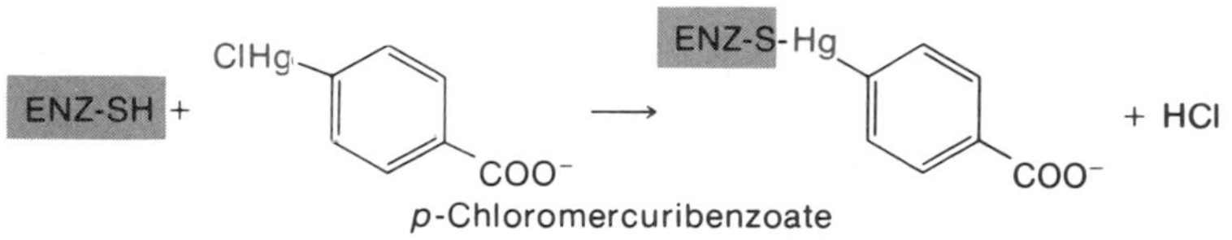

$$\text{ENZ-SH} + \text{ClHg-}C_6H_4\text{-}COO^- \longrightarrow \text{ENZ-S-Hg-}C_6H_4\text{-}COO^- + HCl$$

p-Chloromercuribenzoate

FIGURE 4.29
Enzyme inhibition by a covalent modification of an active center cysteine.

Many Drugs Are Enzyme Inhibitors

Most if not all of modern drug therapy is based on the concepts of enzyme inhibition that were covered in the previous section.

Drugs are designed with a view toward inhibiting a specific enzyme in a specific metabolic pathway. This application is most easily appreciated with the antiviral, antibacterial, and antitumor drugs, which are administered to the patient under conditions of limited toxicity. Such toxicity to the patient is often unavoidable because, with the exception of cell wall biosynthesis in bacteria, there are few critical metabolic pathways that are unique to tumors, viruses, or bacteria. Hence, drugs that will kill these organisms will often kill the host cells. The one characteristic that can be taken advantage of is the comparatively short generation time of the undesirable organisms. They are much more sensitive to antimetabolites and in particular those that inhibit enzymes involved in replication. **Antimetabolites** are compounds with some structural deviation from the natural substrate. In subsequent chapters, numerous examples of antimetabolites will be brought to your attention. Here we will present only a few examples that illustrate the concept.

Sulfa Drugs

Modern chemotherapy had its beginning in these compounds whose general formula is $R—SO_2—NHR'$. Sulfanilamide is the simplest member of the class and is an antibacterial agent because of its competition with *p*-aminobenzoic acid, which is required for bacterial growth. Structures of these compounds are shown in Figure 4.30.

It is now known that bacteria cannot absorb folic acid, a required vitamin, from the host, but must synthesize it. The synthesis of folate involves the series of reactions shown in Figure 4.31.

Since sulfanilamide is a structural analog of *p*-aminobenzoate, the enzyme dihydropteroate synthetase is tricked into making a dihydropteroate containing sulfanilamide that cannot be converted to folate. Thus the bacterium is starved of the required folate and cannot grow or divide. Since humans require folate from external sources, the sulfanilamide is not harmful at the doses that will kill bacteria.

Methotrexate

The biosynthesis of purines and pyrimidines, heterocyclic bases employed in the synthesis of RNA and DNA, requires folic acid, which serves as a coenzyme in the transfer of one-carbon units from various amino acid donors (p. 516).

Methotrexate (Figure 4.32) is a structural analog of folate. It has been used with great success in childhood leukemia. Its mechanism of action is based on competition with dihydrofolate for the dihydrofolate reductase.

$H_2N-C_6H_4-SO_2NH_2$ — Sulfanilamide

$H_2N-C_6H_4-COOH$ — *p*-Aminobenzoic Acid

FIGURE 4.30
Structure of *p*-aminobenzoic acid and sulfanilamide, a competitive inhibitor.

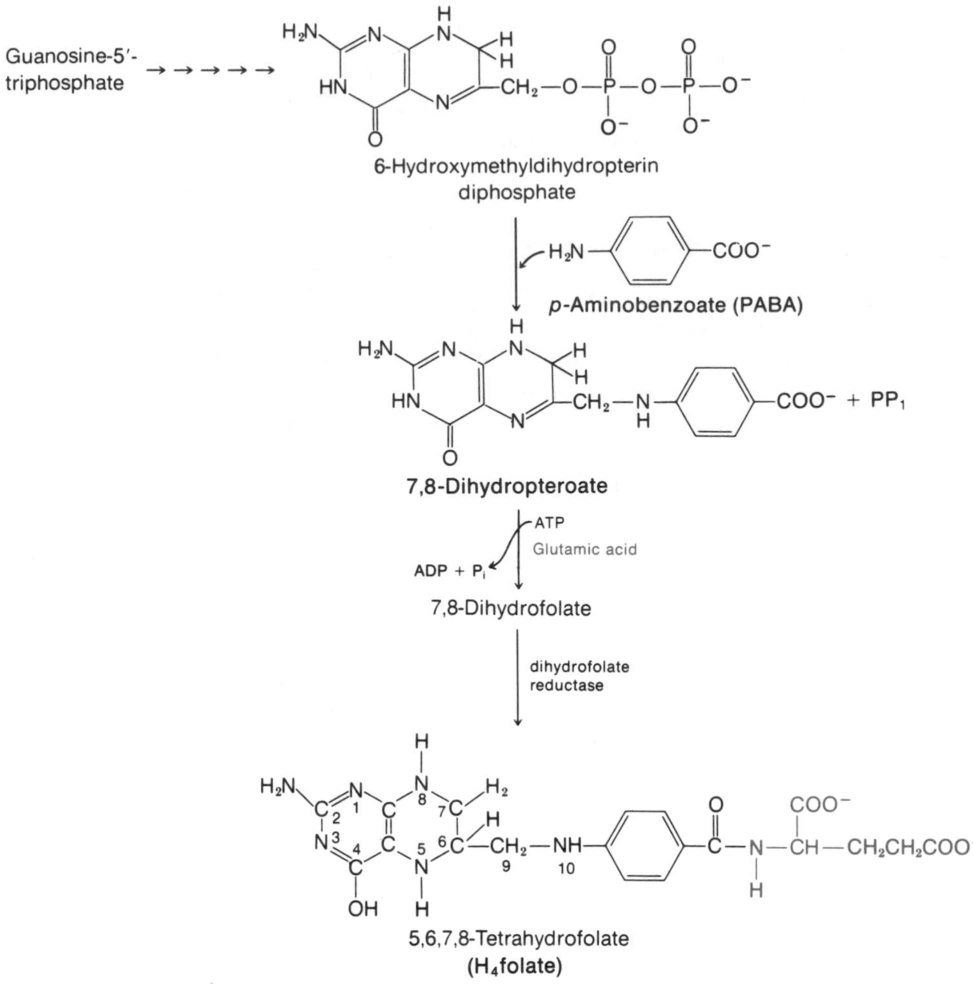

Guanosine-5′-triphosphate → → → → → 6-Hydroxymethyldihydropterin diphosphate

↓ $H_2N-C_6H_4-COO^-$ p-Aminobenzoate (PABA)

7,8-Dihydropteroate + PP_i

↓ ATP, Glutamic acid → ADP + P_i

7,8-Dihydrofolate

↓ dihydrofolate reductase

5,6,7,8-Tetrahydrofolate (H_4folate)

FIGURE 4.31
The route of synthesis of folate in bacteria.
p-Aminobenzoate (PABA) is a required cofactor for the synthesis of dihydrofolate. Sulfanilamide is a competitive inhibitor of PABA and is incorporated into a metabolically inactive, 7,8-dihydropteroylsulfonamide.

It binds 1000-fold better than the natural substrate and is a powerful competitive inhibitor of the enzyme. This being the case, the synthesis of thymidine monophosphate stops in the presence of methotrexate because of failure of the one-carbon metabolic system. Since cell division is dependent on thymidine monophosphate as well as the other nucleotides, the leukemia cell cannot multiply. One problem is that rapidly dividing human cells such as those in bone marrow are sensitive to the drug for the same reasons. Also, prolonged usage leads to amplification of the gene for dihydrofolate reductase, with increased levels of the enzyme and preferential growth of the resistant cells.

FIGURE 4.32
Methotrexate (4-amino-N^{10}-methyl folic acid).

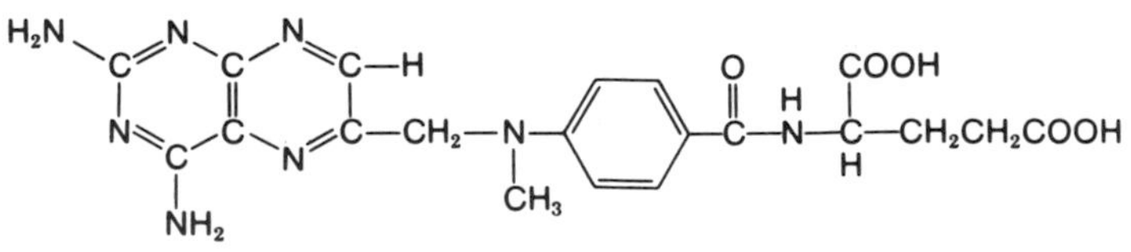

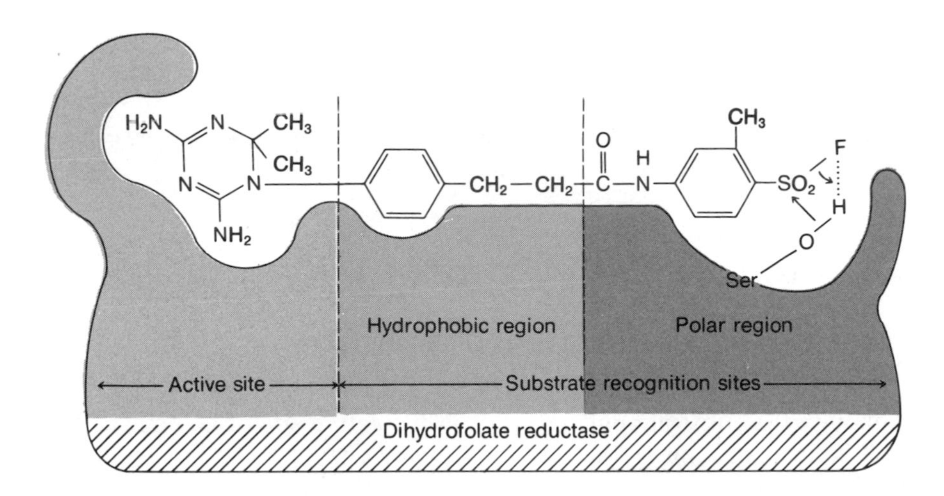

FIGURE 4.33
Suicide substrate inactivation of tetrahydrofolate reductase.
The suicide substrate, a substituted dihydrotriazine, structurally resembles dihydrofolate and binds specifically to the dihydrofolate site on dihydrofolate reductase. The triazine portion of the suicide substrate resembles the pterin moiety and therefore, binds to the active site. The ethylbenzene group binds to the hydrophobic site normally occupied by the *p*-aminobenzoyl group. The reactive end of the suicide substrate contains a reactive sulfonyl fluoride that forms a covalent linkage with a serine hydroxyl on the enzyme surface. Thus the suicide substrate irreversibly inhibits the enzyme by blocking access of dihydrofolate to the active site.

Nonclassical Antimetabolites

A nonclassical antimetabolite is a substrate for an enzyme, which upon action of the enzyme produces a product with a chemically reactive group that will form a covalent adduct with the enzyme at the active site. The addition causes irreversible inactivation of the enzyme. These inhibitors are referred to as suicide substrates and are very specific. In addition, some nonclassical inhibitors have been synthesized that have a high affinity for the active site and contain a reactive group which covalently reacts with an amino acid in the active center; in this case the enzyme does not catalyze the formation of the reactive group. For example, the compound shown in Figure 4.33 is an irreversible inhibitor of dihydrofolate reductase because the compound structurally resembles dihydrofolate and is specifically bound at the active site where the reactive benzylsulfonyl fluoride is positioned so as to react with a serine hydroxyl in the substrate binding site. Covalent binding of this analog of the substrate to the enzyme prevents binding of the normal substrate and leads to inhibition of the enzyme.

Other Antimetabolites

Two other analogs of the purines and pyrimidines will be mentioned in order to emphasize the structural similarity of chemotherapeutic agents to normal substrates.

Fluorouracil (Figure 4.34) is an analog of thymine in which the ring bound methyl is substituted by fluorine. The deoxynucleotide of this compound is an irreversible inhibitor of the enzyme thymidylate synthetase.

6-Mercaptopurine (Figure 4.34) is an analog of hypoxanthine and therefore of adenine and guanine, which is converted to the 6-mercaptopurine nucleotide by hypoxanthine-guanine phosphoribosyl transferase in cells.

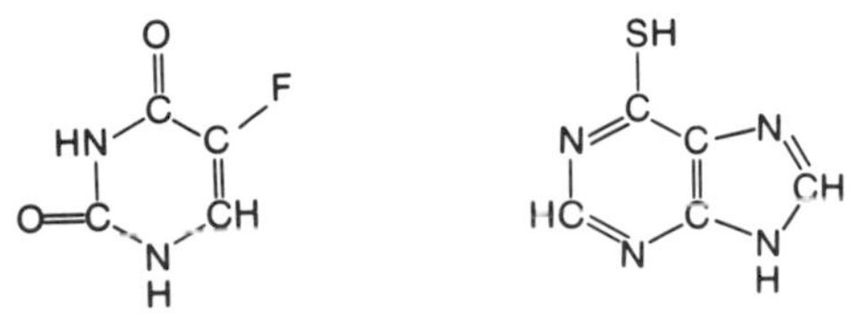

5-Fluorouracil 6-Mercaptopurine

FIGURE 4.34
Structures of two antimetabolites.

The nucleotide is a broad spectrum antimetabolite because of its competition in most reactions involving adenine and guanine nucleotides.

The antimetabolites discussed have been related to purine and pyrimidine metabolism. The general concepts developed here, however, can be applied to any enzyme or metabolic pathway.

4.6 ALLOSTERIC CONTROL OF ENZYME ACTIVITY

Allosteric Effectors Bind at Sites Different From Substrate-Binding Sites

Although the substrate binding site and the active site of an enzyme are well-defined structures, the activity of many enzymes can be modulated by ligands acting in ways other than as competitive or noncompetitive inhibitors. A **ligand** is any molecule that is bound to a macromolecule; the term is not limited to small organic molecules such as ATP, but is extended to low molecular weight proteins. Ligands can be activators, inhibitors, or even the substrates of enzymes. Those ligands that cause a change in enzymatic activity, but are unchanged as a result of enzyme action, are referred to as **effectors, modifiers,** or **modulators.** Most of the enzymes subject to modulation by ligands are rate-determining enzymes in metabolic pathways. In order to appreciate the mechanisms of control of metabolic pathways, the principles governing the allosteric and cooperative behavior of individual enzymes must be understood.

In addition to the substrate binding site and the active site, which we have previously discussed, those enzymes that respond to modulators have additional site(s) known as allosteric site(s). **Allosteric** is derived from the Greek root *allo,* meaning "the other"; hence the allosteric site is a unique region of the enzyme that is different from the substrate binding site, which is often the catalytic site. The existence of allosteric inhibitor or activator sites distinct from the substrate binding site is illustrated by the case of gout described in Clin. Corr. 4.4. The modulating ligands that bind at the allosteric site are known as the allosteric effectors or modulators. Binding of the allosteric effector causes an allosteric transition in the enzyme, that is, the conformation of the enzyme changes, so that the affinity for the substrate or other ligand changes. Positive (+) allosteric effectors increase the enzyme affinity for substrate or other ligand. The reverse is true for negative allosteric effectors. The allosteric site at which the positive effector binds is referred to as the activator site. The negative effector binds at an inhibitory site.

Allosteric enzymes are divided into two classes based on the effect of the allosteric effector on the K_m and V_{max}. In the K class the effector alters the K_m but not the V_{max}, whereas in the V class the effector alters the V_{max} but not the K_m. The K class enzymes give double-reciprocal plots like those given by competitive inhibitors (Figure 4.27) and V class enzymes give double-reciprocal plots like those of noncompetitive inhibitors (Figure 4.28). However, it is inappropriate to use the terms competitive and noncompetitive with allosteric enzyme systems because the mechanism of the effect of an allosteric inhibitor on a V or K enzyme is quite different from the mechanism of a simple competitive or noncompetitive inhibitor. For example, in the ***K* class** the inhibitor binds at an allosteric site, which then affects the affinity of the substrate binding site for the substrate, whereas in simple competitive inhibition the inhibitor competes with substrate for the substrate binding site. In the ***V* class** enzymes, positive and negative allosteric modifiers increase or decrease the rate of breakdown

CLIN. CORR. **4.4**
A CASE OF GOUT DEMONSTRATES THE DIFFERENCE BETWEEN AN ALLOSTERIC SITE AND THE SUBSTRATE BINDING SITE

The realization that allosteric inhibitory sites are separate from activator sites as well as from the substrate binding and the catalytic sites is illustrated by a study of a gouty patient whose red blood cell PRPP levels are increased. It was found that the patient's PRPP synthetase had a normal K_m, V_{max}, and sensitivity to activation by P_i. However, it displayed reduced sensitivity to the normal allosteric inhibitors AMP, ADP, GDP, and 2,3-diphosphoglycerate (2,3-DPG). The increased PRPP levels and hyperuricemia arose because endogenous end products of the pathway (ATP, GTP) were not able to control the activity of the synthase as is normally the case. Apparently a mutation had resulted in a change in the inhibitory site (I) or in the coupling mechanism between the I and the catalytic site (C). The cosubstrate binding sites (S) and the phosphate activator (A) site are shown for reference.

Sperling, O., Persky-Brosh, S., Boen, P., and DeVries, A. Human erythocyte phosphoribosylpyrophosphate synthetase mutationally altered in regulatory properties. *Biochem. Med.* 7:389, 1973.

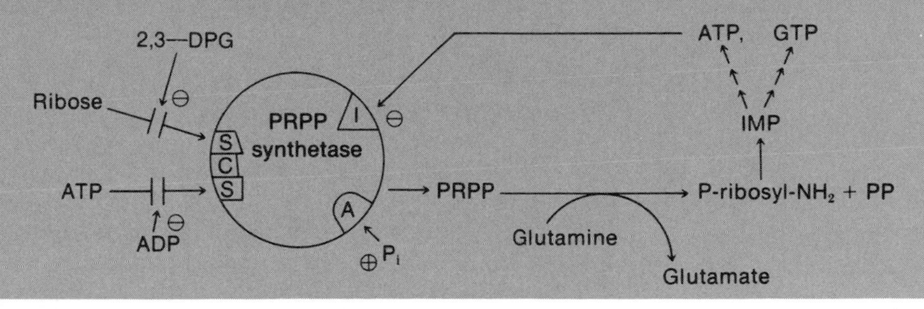

of the ES complex to products, that is, the catalytic rate constant, k_3, is affected and not the substrate binding constant. There are a few enzymes in which both K_m and V_{max} are affected.

In theory a monomeric enzyme can undergo an allosteric transition in response to a modulating ligand. In practice only two monomeric allosteric enzymes have been found, ribonucleoside diphosphate reductase and pyruvate-UDP-*N*-acetylglucosamine transferase. Most allosteric enzymes are oligomeric, that is, they consist of several subunits. The identical subunits are designated as *protomers*. Each protomer may consist of one or more polypeptide chains. As a consequence of the oligomeric nature of allosteric enzymes, binding of ligand to one protomer can affect the binding of ligands on other protomers in the oligomer. Such ligand effects are referred to as **homotropic interactions.** The transmission of the homotropic effects between protomers is one aspect of cooperativity, considered in detail in the section on Cooperativity. Substrate influencing substrate, activator influencing activator, or inhibitor influencing inhibitor binding are homotropic interactions. Homotropic interactions are almost always positive.

A **heterotropic interaction** is defined as the effect of one ligand on the binding of a different ligand. For example, the effect of an allosteric inhibitor on the binding of substrate or the effect of an allosteric inhibitor on the binding of an allosteric activator are heterotropic interactions. Heterotropic interactions can be either positive or negative and can occur in monomeric allosteric enzymes. Both heterotropic and homotropic effects, in an oligomeric enzyme, are mediated by cooperativity between subunits.

Based on the foregoing descriptions of allosteric enzymes, two models are pictured in Figure 4.35. In panel *a* a model for a monomeric enzyme is shown, and in panel *b* a model for an oligomeric enzyme consisting of two protomers is visualized. In both models heterotropic interactions can occur between the activator and substrate sites. In model *b*, homotropic

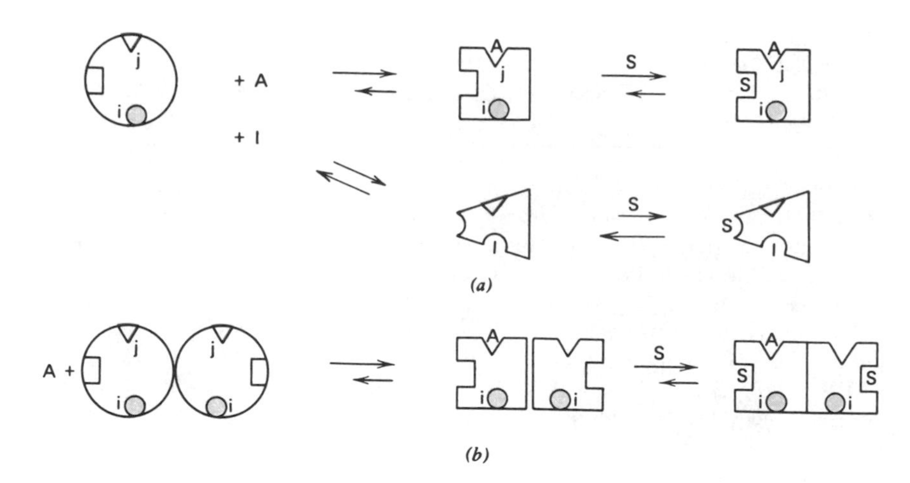

FIGURE 4.35
Models of allosteric enzyme systems.
(*a*) A model of a monomeric enzyme. Binding of a positive allosteric effector, A, to the activator site, j, induces a new conformation to the enzyme, one that has a greater affinity for the substrate. Binding of a negative allosteric effector to the inhibitor site, i, results in an enzyme conformation having a decreased affinity for substrate. (*b*) A model of a polymeric allosteric enzyme. Binding of the positive allosteric effector, A, at the j site causes an allosteric change in the conformation of the protomer to which the effector binds. This change in the conformation is transmitted to the second promoter through cooperative protomer–protomer interactions. The affinity for the substrate is increased in both protomers. A negative effector decreases the affinity for substrate of both protomers.

interactions can occur between the activator sites or between the substrate sites.

Allosteric Enzymes Exhibit Sigmoidal Kinetics

As a consequence of the interaction between the substrate site, the activator site, and the inhibitor site, a characteristic sigmoid or S-shaped curve, as shown in Figure 4.36 (curve A), is obtained in [S] versus v_0 plots of allosteric enzymes. Negative allosteric effectors move the curve toward higher substrate concentrations and enhance the sigmoidicity of the curve. If we use $\frac{1}{2}V_{max}$ as a guideline, it can be seen from Figure 4.36 that a higher concentration of substrate would be required to achieve $\frac{1}{2}V_{max}$ in the presence of a negative effector (curve C) than is required in the absence of negative effector (curve A). In the presence of a positive modulator (curve B), $\frac{1}{2}V_{max}$ can be reached at a lower substrate concentration

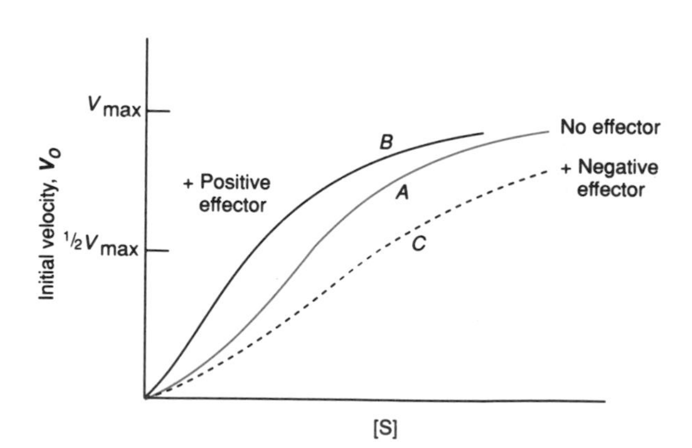

FIGURE 4.36
The kinetic profile of a *K* class allosteric enzyme.
The enzyme shows sigmoid S versus v_0 plots. Negative effectors shift the curve to the right resulting in an increase in K_m. Positive effectors shift the curve to the left and effectively lower the apparent K_m. The term V_{max} is not changed.

than is required in the absence of the positive modulator (curve A). Positive modulators shift the v_0 versus [S] plots toward the hyperbolic plots observed in Michaelis–Menten kinetics.

From the viewpoint of **metabolic control,** allosteric enzymes allow fine control of the activity of individual enzymes through small fluctuations in the level of substrate. Often the in vivo concentration of substrate corresponds with the sharply rising segment of the sigmoid [S] versus v_0 plot; consequently, large changes in enzyme activity are effected by small changes in substrate concentration (see Figure 4.36.). It is also possible to "turn the enzyme off" with small amounts of a negative allosteric effector by its shifting the apparent K_m to higher values, values that are far above the in vivo level of substrate. Notice that at a given in vivo concentration of substrate the initial velocity, v_0, is decreased in the presence of a negative effector (compare curves A and C).

Cooperativity Explains the Interaction Between Ligand Sites in an Oligomeric Protein

The realization that most allosteric enzymes are oligomeric and also have sigmoid [S] versus v_0 plots led to the concept of cooperativity to explain the interaction between ligand sites in oligomeric enzymes. **Cooperativity** is defined as the influence that the binding of ligand to one protomer has on the binding of ligand to a second protomer of an oligomeric protein. It should be emphasized that kinetic mechanisms other than cooperativity can also produce sigmoid v_0 versus [S] plots; consequently, sigmoidicity is not diagnostic of cooperativity in a v_0 versus [S] plot. We have previously discussed cooperativity in terms of the binding of oxygen to hemoglobin and will now expand the concept.

The relationship between allosterism and cooperativity has been confused in many recent texts. The conformational changes occurring in a given protomer in response to ligand binding at an allosteric site is an allosteric effect. Cooperativity generally involves a change in conformation of an effector-activated protomer, which in turn transforms an adjacent protomer into a new conformation with an altered affinity for the effector ligand or for a second ligand. The conformation change may be induced by an allosteric effector or it may be induced by substrate, as it is in the case of hemoglobin. In hemoglobin the oxygen binding site on each protomer corresponds to the substrate site on an enzyme rather than to an allosteric site. Therefore, the oxygen-induced conformational change in the hemoglobin protomers is technically not an allosteric effect, although some authors identify it as such. It is a homotropic cooperative interaction. Those who consider the oxygen-induced changes in hemoglobin to be "allosteric" are using the term in a much broader sense than the original definition allows; however, "allosteric" is now used by many to describe any ligand-induced change in the tertiary structure of a protomer.

It should be emphasized that one can have an allosteric effect in the absence of any cooperativity. For example, in alcohol dehydrogenase, conformational changes can be demonstrated in each of the protomers upon the addition of positive allosteric effectors, but the active site of each protomer is completely independent of the other and there is no cooperativity between protomers, that is, induced conformational changes in one protomer are not transmitted to adjacent protomers.

In an attempt to describe experimentally observed ligand saturation curves mathematically, several models of cooperativity have been proposed. The two most prominent models are the **concerted** and the **sequential-induced fit.**

Although the *concerted model* is rather restrictive, most of the nomenclature associated with allosterism and cooperativity arose from this model. The model proposes that the enzyme exists in only two states, the T (tense or taut) and the R (relaxed). The T and R states are in equilibrium. Activators and substrates favor the R state and shift the preexisting equilibrium toward the R state by the law of mass action. Inhibitors favor the T state. A conformational change in one protomer causes a corresponding change in all protomers. No hybrid states occur. The model is diagrammed in Figure 4.37*a*. Although the model accounts for the kinetic behavior of many enzymes, it cannot account for negative cooperativity.

The *sequential-induced fit* model proposes that ligand binding induces a conformational change in a protomer. A corresponding conformational change is then partially induced in an adjacent protomer contiguous with the protomer containing the bound ligand. The effect of ligand binding is sequentially transmitted through the oligomer, giving rise to an increasing or decreasing affinity for the ligand by contiguous protomers as suggested by the scheme in Figure 4.37*b*. In this model numerous hybrid states occur giving rise to the cooperativity and the sigmoid $[S]$ versus v_0 plots. Both positive and negative cooperativity can be accommodated by the model. A positive modulator induces a conformation in the protomer,

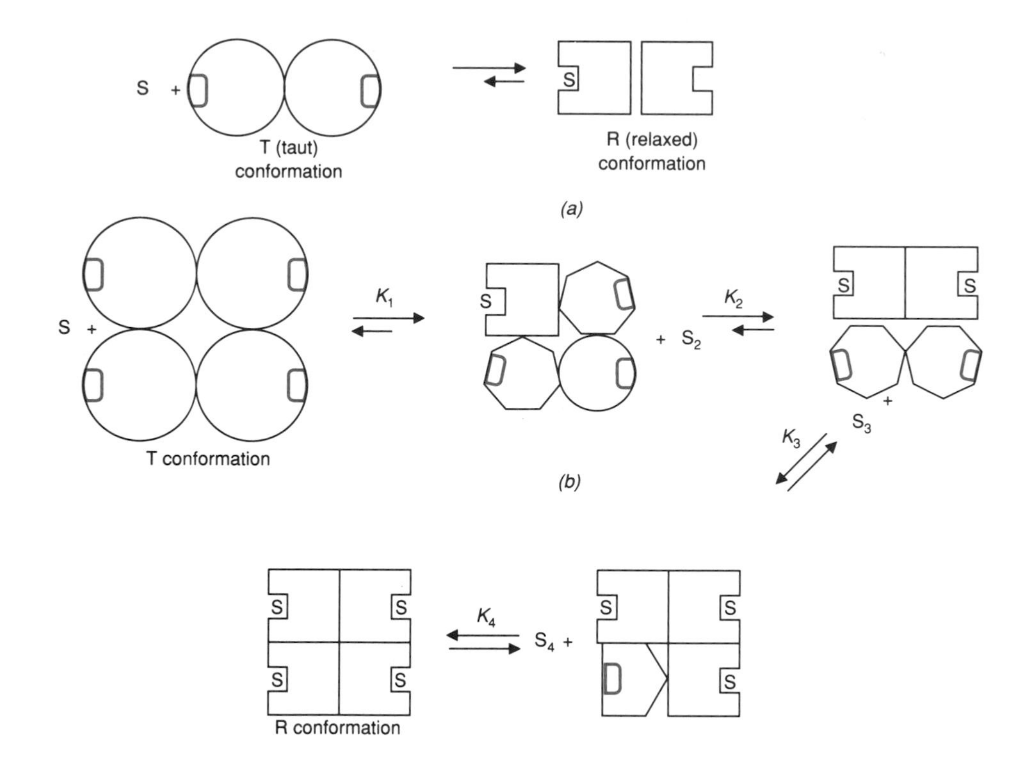

FIGURE 4.37
Models of cooperativity.
(*a*) The concerted model. The enzyme exists in only two states, the T (tense of taut) and the R (relaxed) conformation. Substrates and activators have a greater affinity for the R state and inhibitors for the T state. Ligands shift the equilibrium between the T and R states. (*b*) The sequential induced fit model. Binding of a ligand to any one subunit induces a conformational change in that subunit. This conformational change is transmitted partially to adjoining subunits through subunit–subunit interaction. Thus the effect of the first ligand bound is transmitted cooperatively and sequentially to the other subunits (protomers) in the oligomer resulting in a sequential increase or decrease in ligand affinity of the other protomers. The cooperativity may be either positive or negative, depending on the ligand.

which has an increased affinity for the substrate. A negative modulator induces a different conformation in the protomer, one that has a decreased affinity for substrate. Both effects are cooperatively transmitted to adjacent protomers. For the V class enzymes the same reasoning applies, but the effect is on the catalytic event (k_3) rather than on K_m.

Regulatory Subunits Modulate the Activity of Catalytic Subunits

In the foregoing discussion the allosteric site has been considered to reside on the same protomer as the catalytic site. Furthermore, all protomers were considered to be identical. There are several very important enzymes in which a distinct regulatory protomer exists. These **regulatory subunits** have no catalytic function per se, but their binding with the catalytic protomer modulates the activity of the catalytic subunit through an induced conformational change.

Ribonucleotide reductase converts ribonucleoside diphosphates to deoxyribonucleotides for DNA biosynthesis (p. 551). The active site is generated at the interface of two different subunits, B_1 and B_2. Both B_1 and B_2 are catalytically inactive by themselves. The allosteric sites reside on B_1; however, B_2 does not contain the active site in and of itself. The active site is generated at the interface between polymerized B_1 and B_2, as suggested in Figure 4.38, ATP is a positive effector, which binds to the ℓ allosteric site, and dATP is a negative effector, which competes with ATP at the ℓ site. This particular arrangement of subunits to form the catalytic site is not common.

Completely separate regulatory subunits are observed in aspartate transcarbamoylase in bacteria. This enzyme is involved in pyrimidine biosynthesis and catalyzes the transfer of the carbamoyl group from carbamoyl phosphate to the α-amino group of aspartate:

$$\text{Carbamoyl phosphate} + \text{aspartate} \longrightarrow \text{carbamoyl aspartate} + P_i$$

The enzyme is a complex consisting of two active catalytic (C) and three regulatory (R) subunits. The end product of the pyrimidine pathway, cytidine triphosphate (CTP), binds to the R subunits as a negative effector, whereas ATP functions as a positive effector.

4.7 ENZYME SPECIFICITY: THE ACTIVE SITE

Enzymes are the most specific catalysts known, both from the viewpoint of the substrate as well as the type of reaction performed on the substrate. Specificity inherently resides in the **substrate binding site,** which lies on the enzyme surface. The tertiary structure of the enzyme is folded in such a way as to create a region that has the correct molecular dimensions, the appropriate topology, and the optimal alignment of counterionic groups and hydrophobic regions to accommodate a specific substrate. The tolerances in the active site are so small that usually only one isomer of a diastereomeric pair will bind. For example, D-amino acid oxidase will bind only D-amino acids and not L-amino acids. Some enzymes show absolute specificity for substrate. Others have broader specificity and will accept several different analogs of a specific substrate. For example, hexokinase catalyzes the phosphorylation of glucose, mannose, fructose, glucosamine, and 2-deoxyglucose, but not all at the same rate. Glucokinase, on the other hand, is specific for glucose.

The specificity of the reaction catalyzed rests in the active site and the particular arrangement of amino acids that participate in the bond-making

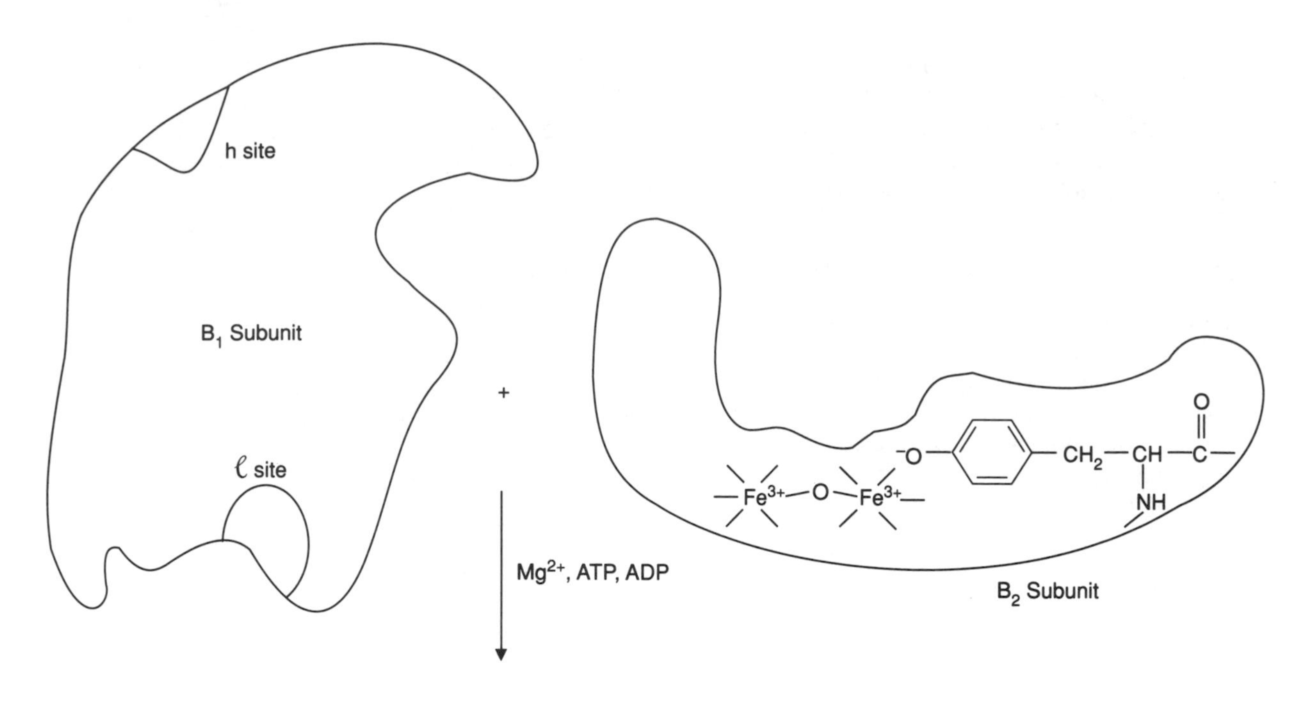

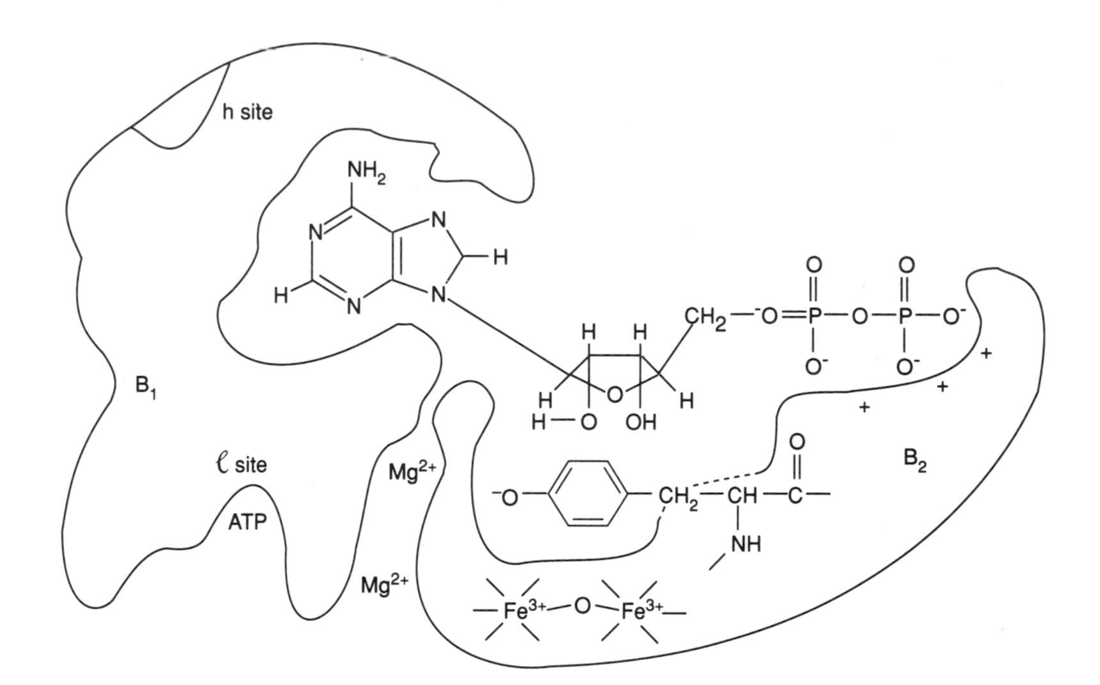

FIGURE 4.38
Model of an allosteric enzyme with a separate regulatory subunit.
The enzyme is ribonucleotide reductase. B_1 is the regulatory subunit. ATP binds at the "ℓ site" and is a positive effector. B_2 is the catalytic subunit but only when combined with the B_1 subunit. B_2 contains the iron cluster and tyrosyl radical, which is involved in reduction of the 2′-OH of the ribose of the ribonucleoside diphosphates, which in this example is ADP. dATP is a negative modifier that competes at the ℓ site for ATP. Other modifiers such as dTTP bind at the "*h*" site and are either positive or negative modulators depending on the substrate.

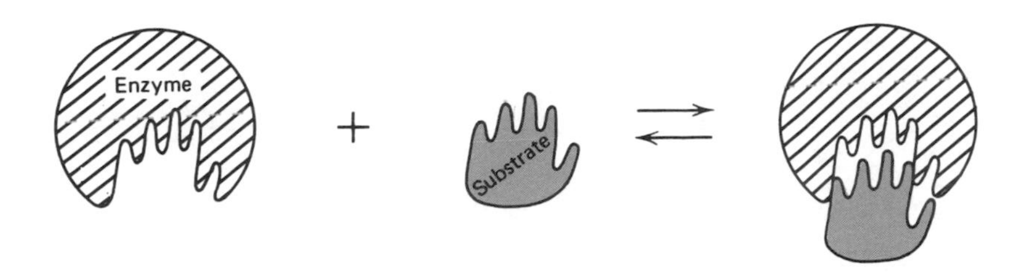

FIGURE 4.39
Lock-and-key model of the enzyme binding site.
The enzyme contains a negative impression of the molecular features of the substrate, thus allowing specificity of the enzyme for a particular substrate.

and bond-breaking phase of catalysis. The mechanism of catalysis is discussed in Section 4.8.

Complementarity of Substrate and Enzyme Explains Substrate Specificity

Various models have been proposed to explain the substrate specificity of enzymes. The first proposal was the "lock-and-key" model (Figure 4.39), in which a negative impression of the substrate is considered to exist on the enzyme surface. The substrate fits to this binding site just as a key fits into the proper lock or a hand into the proper sized glove. This model gives a rigid picture of the enzyme and cannot account for the effects of allosteric ligands.

A more flexible model of the binding site is the **induced fit** model. In this model, the binding site and certainly the active site are not fully formed. The essential elements of the binding site are present to the extent that the correct substrate can position itself properly in the nascent binding site. The interaction of the substrate with the enzyme induces a conformational change in the enzyme, resulting in the formation of a stronger binding site and the repositioning of the appropriate amino acids to form the active site. There is excellent X-ray evidence for the correctness of this model in the enzyme carboxypeptidase A. A diagram of the induced fit model is shown in Figure 4.40*a*.

The concept of *induced fit* combined with substrate strain accounts for more of the experimental observations concerning enzyme action than do other models. In this model (Figure 4.40*b*), the substrate is "strained" toward product formation as a result of an induced conformational transition of the enzyme. A good example of enzyme-induced substrate strain is that of lysozyme (Figure 4.41) where the conformation of the sugar residue "D" at which bond breaking occurs is strained from the stable chair to the unstable half-chair conformation upon binding. These conformations of glucose are shown in Figure 4.42.

The concept of substrate strain is useful in explaining the role of the enzyme in increasing the rate of reaction. This effect is considered in Section 4.8.

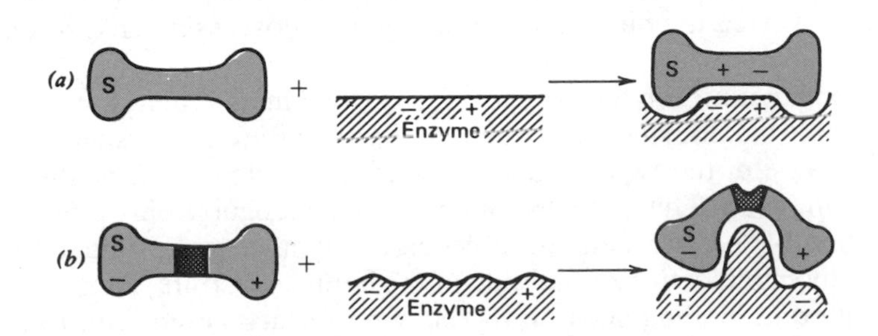

FIGURE 4.40
Models for induced fit and substrate strain.
(*a*) Approach of substrate to the enzyme induces the formation of the active site. (*b*) Substrate strain, induced by substrate binding to the enzyme, contorts normal bond angles and "activates" the substrate.
Reprinted, with permission, from D. Koshland *Annu. Rev. Biochem.* 37:374, 1968. Copyright © by Annual Reviews, Inc.

FIGURE 4.41
Hexasaccharide binding at the active site of lysozyme.
In the model substrate pictured, the ovals represent individual pyranose rings of the repeating units of the lysozyme substrate shown to the right in the figure. Ring D is strained by the enzyme to the half-chair conformation and hydrolysis occurs between the D and E rings. Six subsites on the enzyme bind substrate. Alternate sites are specific for acetamido groups (a) but are unable to accept the lactyl (P) side chains, which occur on the *N*-acetylmuramic acid residues. Thus the substrate can bind to the enzyme in only one orientation.
Reprinted, with permission from T. Imoto, et al. *The Enzymes,* 3rd, P. Boyer (Ed.) 7:713, 1972. Academic Press.

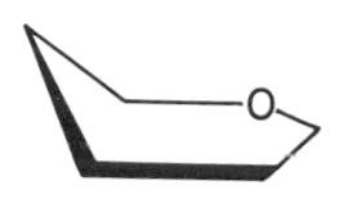

Half-chair conformation of the pyranose ring

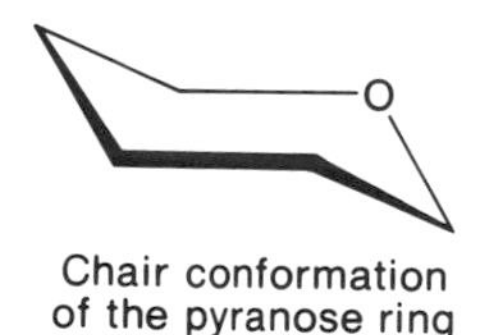

Chair conformation of the pyranose ring

FIGURE 4.42
Two possible conformations of glucose.

Asymmetry of the Binding Site

Not only are enzymes able to distinguish between isomers, but they are able to distinguish between two equivalent atoms in a symmetrical molecule. For example, the enzyme glycerol kinase is able to distinguish between configurations of H and OH on C-2 in the symmetric substrate glycerol, so that only the asymmetric product L-glycerol 3-phosphate is formed. These prochiral substrates have two identical substituents and two additional but dissimilar groups on the same carbon ($C_{aa'bd}$).

$$\begin{array}{ll} (a)\ CH_2OH & CH_2OH \\ \quad\ CHOH\ (b,d) \xrightarrow[ATP]{\text{glycerol kinase}} & HOCH \\ (a')\ CH_2OH & CH_2OPO_3^{2-} \\ \quad\ \text{Glycerol} & \text{L-Glycerol 3-phosphate} \end{array}$$

Prochiral substrates are substances that possess no optical activity but can be converted to chiral compounds, that is, possessing an asymmetric center.

The explanation for this enigma was forthcoming when it was recognized that if the enzyme binds the two dissimilar groups at specific sites, then only one of the two similar substituents is able to bind at the active site (Figure 4.43). Thus, the enzyme is able to recognize only one specific orientation of the symmetrical molecule. Asymmetry is produced in the product by modification of one side of the bound substrate. A minimum of three different binding sites on the enzyme surface is required to distinguish between identical groups on a prochiral substrate.

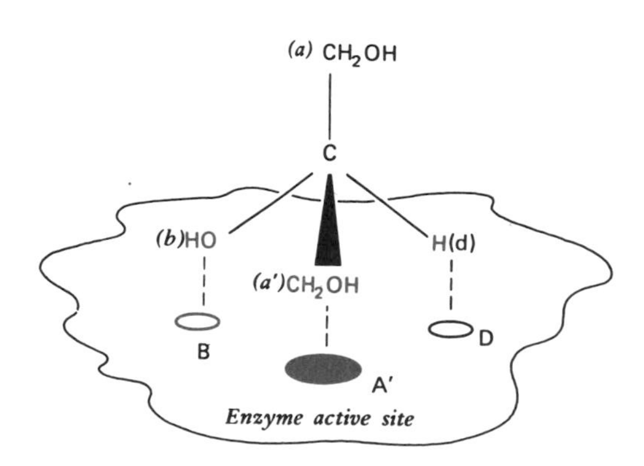

FIGURE 4.43
Three-point attachment of a symmetrical substrate to an asymmetric enzyme binding site.
Glycerol kinase by virtue of dissimilar binding sites for the —H and —OH group of glycerol binds only the a′ hydroxymethyl group to the active site. Thus, only one stereoisomer results from the kinase reaction, the L-glycerol 3-phosphate.

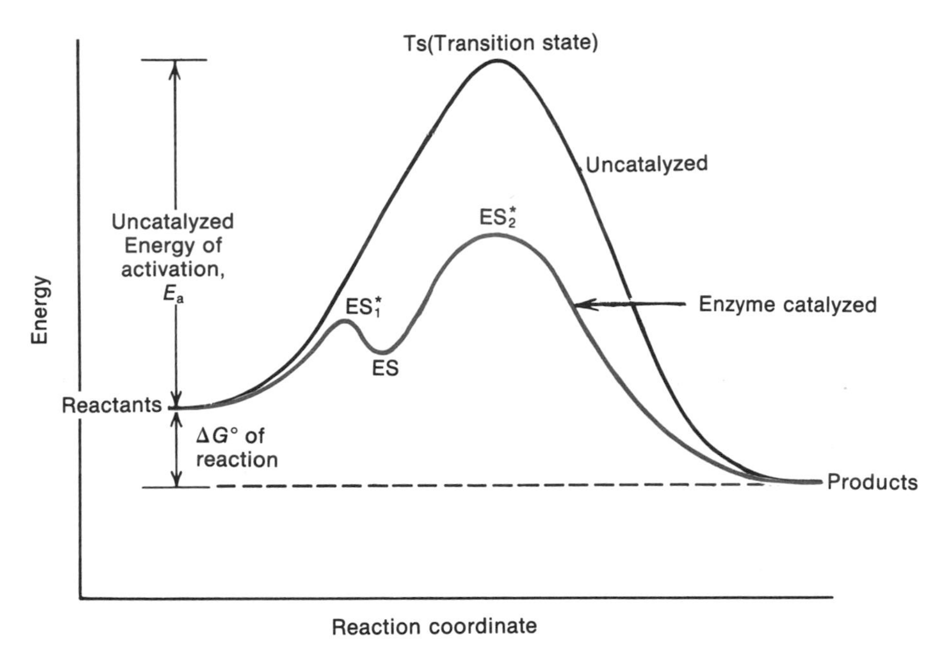

FIGURE 4.44
Energy diagrams for catalyzed versus noncatalyzed reactions.
The overall energy difference between reactants and products is the same in catalyzed and noncatalyzed reactions. The enzyme-catalyzed reaction proceeds at a faster rate because the energy of activation is lowered.

4.8 MECHANISM OF CATALYSIS

All chemical reactions have a potential energy barrier that must be overcome before reactants can be converted to products. In the gas phase the reactant molecules can be given enough kinetic energy by heating so that collisions result in product formation. The same is true with solutions. However, a well-controlled body temperature of 37°C does not allow temperature to be increased to accelerate the reaction, and 37°C is not warm enough to provide the reaction rates required for fast-moving species of animals. Enzymes employ other means of overcoming the barrier to reaction, and these will be discussed after some useful definitions are covered.

A comparison of the enzyme diagrams for catalyzed and noncatalyzed reactions is shown in Figure 4.44. The energy barrier represented by the uncatalyzed curve in Figure 4.44 is a measure of the **activation energy,** E_a. The reaction coordinate is simply the pathway in terms of bond stretching between reactants and products. At the apex of the energy barrier is the activated complex known as the **transition state,** Ts. The transition state represents the reactants in their activated state. In this state reactants are in an intermediate stage along the reaction pathway and cannot be identified as starting material or products. For example, in the hydrolysis of ethyl acetate:

$$CH_3-\overset{\overset{\large O}{\|}}{C}-O-CH_2-CH_3 \xrightarrow{H_2O} CH_3-CH_2-OH + CH_3-\overset{\overset{\large O}{\|}}{C}-OH$$

the Ts might look like

$$\left[CH_3-\underset{\underset{\overset{\large O}{\diagup\ \diagdown}\ \ }{\vdots}}{\overset{\overset{\large O^-}{|}}{C}}-O^{-}CH_2-CH_3 \right] \quad (\text{O bonded to H, H})$$

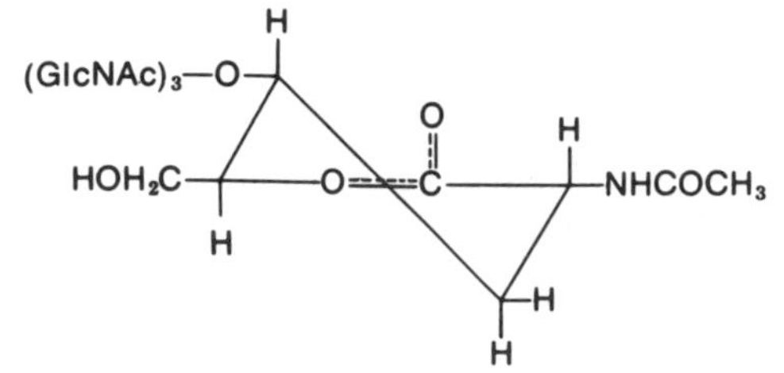

FIGURE 4.45
A transition state analog (tetra-*N*-acetylchitotetrose-δ-lactone) of ring D of the substrate for lysozyme.

The transition state complex can break down to products or go back to reactants. The Ts is not an intermediate and cannot be isolated!

Notice that in the case of the enzyme-catalyzed reaction (Figure 4.44) the energy of the reactants and products is no different than in the uncatalyzed reaction. Enzymes do not change the thermodynamics of the system but they do change the pathway for reaching the final state.

As noted on the energy diagram, there may be several plateaus or valleys on the energy contour for an enzyme reaction. At these points metastable intermediates exist. An important point is that each valley may be reached with the heat input available in a 37°C system. In other words, the enzyme allows the energy barrier to be scaled in increments. The Michaelis–Menten, ES, complex is not the transition state, but may be found in one of the valleys. This is the case because in the ES complex, the substrates are properly oriented and the substrate may be "strained," and therefore the bonds to be broken lie further along the reaction coordinate.

If our concepts of the transition state are correct, one would expect that compounds designed to closely resemble the transition state would bind much more tightly to the enzyme than does the natural substrate. This has proven to be the case. In such substrate analogs one finds affinities 10^2–10^5 times greater than those for substrate. These compounds are called **transition state analogs** and are potent enzyme inhibitors. Previously, lysozyme was discussed in terms of substrate strain, and mention was made of the conversion of sugar ring D from a chair to a strained half-chair conformation. Synthesis of a transition state analog in the form of the δ-lactone of tetra-*N*-acetylchitotetrose (Figure 4.45), which has a distorted half-chair conformation, followed by binding studies, showed that this transition state analog was bound 6×10^3 times better than the normal substrate.

Enzymes Decrease Activation Energy

Enzymes are able to enhance the rates of reaction by a factor of 10^9–10^{12} times that of the noncatalyzed reaction. Most of this rate enhancement can be accounted for by four processes: acid–base catalysis, substrate strain (transition state stabilization), covalent catalysis, and entropy effects.

Acid–Base Catalysis

Specific acids and bases are H^+ and OH^-, respectively. Free protons and hydroxide ions are not encountered in most enzyme reactions and then only in some metal-dependent enzymes (page 153).

General acids and bases are important in enzymology. A *general* acid or base is any substance that is weakly ionizable. In the physiological pH range, the protonated form of histidine is the most important general acid and its conjugate base an important general base. These structures are shown in Figure 4.46. Other acids are the thiol —SH, tyrosine —OH and the ε-amino group of lysine. Other bases are carboxylic acid anions and the conjugate bases of the general acids.

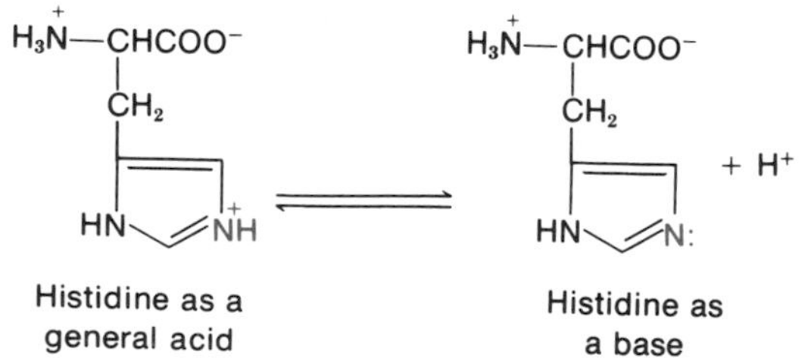

Histidine as a general acid

Histidine as a base

FIGURE 4.46
Acid and base forms of histidine.

Consideration of the mutarotation of glucose illustrates the principle of **acid–base catalysis.** If one dissolves pure α-D-glucose in water, over several days an equilibrium mixture of 34% α-D-glucose and 66% β-D-glucose forms. The conversion of the α anomer to the β anomer proceeds through intermediate formation of the open-chain aldehyde. Upon reforming the pyranose ring, there is no stereospecificity in the orientation of the resulting hydroxyl on C-1 (the anomeric carbon), and the ring can reclose with α-D-Glucopyranose

D-Glucose (open chain aldehyde form)

β-D-Glucopyranose

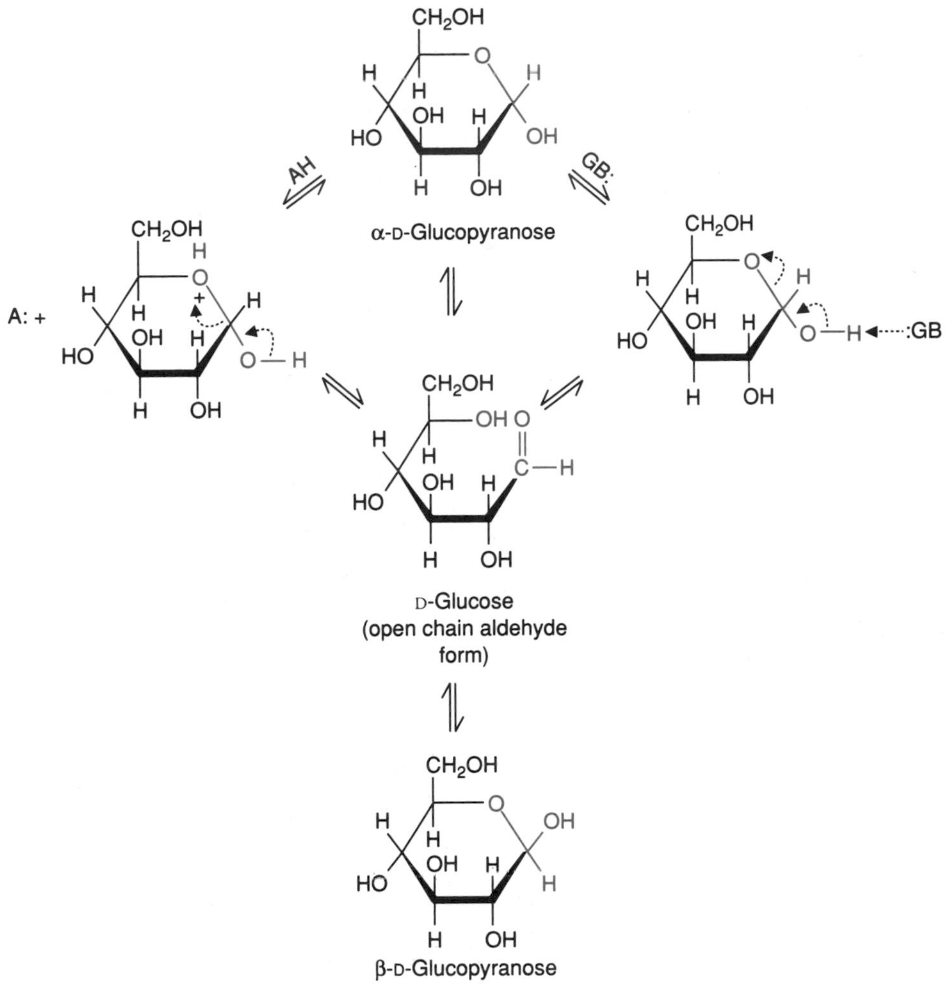

FIGURE 4.47
Mutarotation of glucose.
The center series of structures illustrate the uncatalyzed reaction. The structure on the left represents an intermediate in acid catalysis and the one on the right an intermediate in base catalysis. The open-chain form of glucose is a common intermediate in both the catalyzed and uncatalyzed pathways.

the hydroxyl group in either the α or β orientation. See center reaction sequence in Figure 4.47.

General acids increase the rate of mutarotation by protonating the pyranose oxygen, which labilizes the C—O bond to yield the aldehyde. The rate of mutarotation is increased 1000-fold over the uncatalyzed reaction. General bases increase the rate another 1000-fold over the acid rate by generating an alkoxide ion from the hydroxyl on C-1, which quickly opens the pyranose ring. These concepts are diagrammed in Figure 4.47. Both the acid and base catalyzed reactions yield the open chain form of glucose. Again, reclosure to the pyranose is not stereospecific, and a mixture of anomers is obtained.

These concepts of acid–base catalysis have been extended to the enzyme active site where specific amino acid functional groups are implicated as acid and base catalysts. Ribonuclease is a good example of the role of acid and base catalysis at the enzyme active site. Ribonuclease (RNase) cleaves the RNA chain at the 3′-phosphodiester linkage of pyrimidine nucleotides with an obligatory formation of a cyclic 2′,3′-

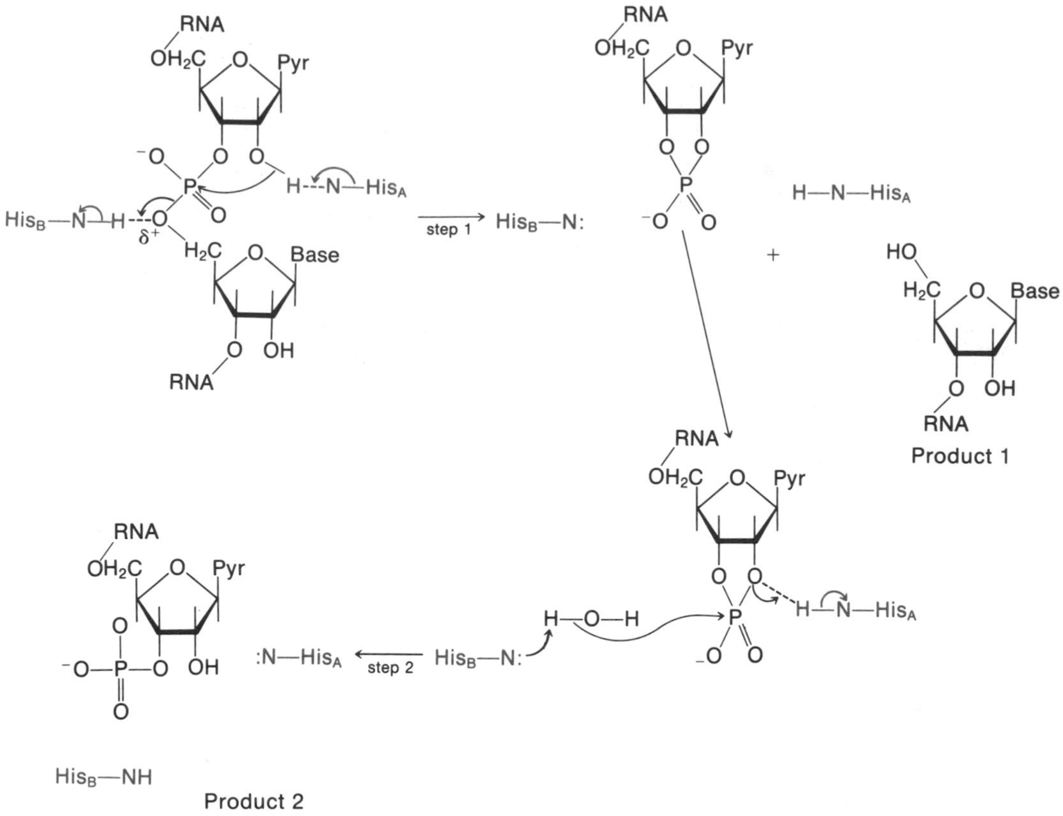

FIGURE 4.48
Role of acid and base catalysis in the active site of ribonuclease.
RNase cleaves the phosphodiester bond in a pyrimidine loci in RNA. Both HisA and HisB are histidine residues 12 and 119, respectively, at the ribonuclease active site that function as acid and base catalysts in enhancing the formation of an intermediate 2′,3′-cyclic phosphate and release of a shorter fragment of RNA (product 1). These same histidines then play a reverse role in the hydrolysis of the cyclic phosphate and release of the other fragment of RNA (product 2) that ends in a pyrimidine nucleoside 3′-phosphate. As a result of the formation of product 2, the active site of the enzyme is regenerated.

phosphoribose on a pyrimidine nucleotide as intermediate. In the mechanism outlined in Figure 4.48, His 119 acts as a general acid to protonate the phosphodiester bridge, whereas His 12 acts as a base in generating an alkoxide on the ribose-3′-hydroxyl. The latter then attacks the phosphorus, resulting in formation of the cyclic phosphate and breakage of the RNA chain at this locus. The cyclic phosphate is then cleaved in phase 2 by a reversal of the reactions in phase 1, but with water replacing the leaving group. The active site histidines revert to their original protonated state.

Substrate Strain

Our previous discussion of this topic related to induced fit of enzymes to substrate. It is also possible that binding of substrate to a preformed site on the enzyme induces strain into the substrate. Irrespective of the mechanism of strain induction, the important point is that the energy level of the substrate is raised, and the substrate is propelled toward the bonding found in the transition state.

A combination of **substrate strain** and acid–base catalysis is observed in the mechanism of lysozyme action (Figure 4.49). Ring D of the hexa-

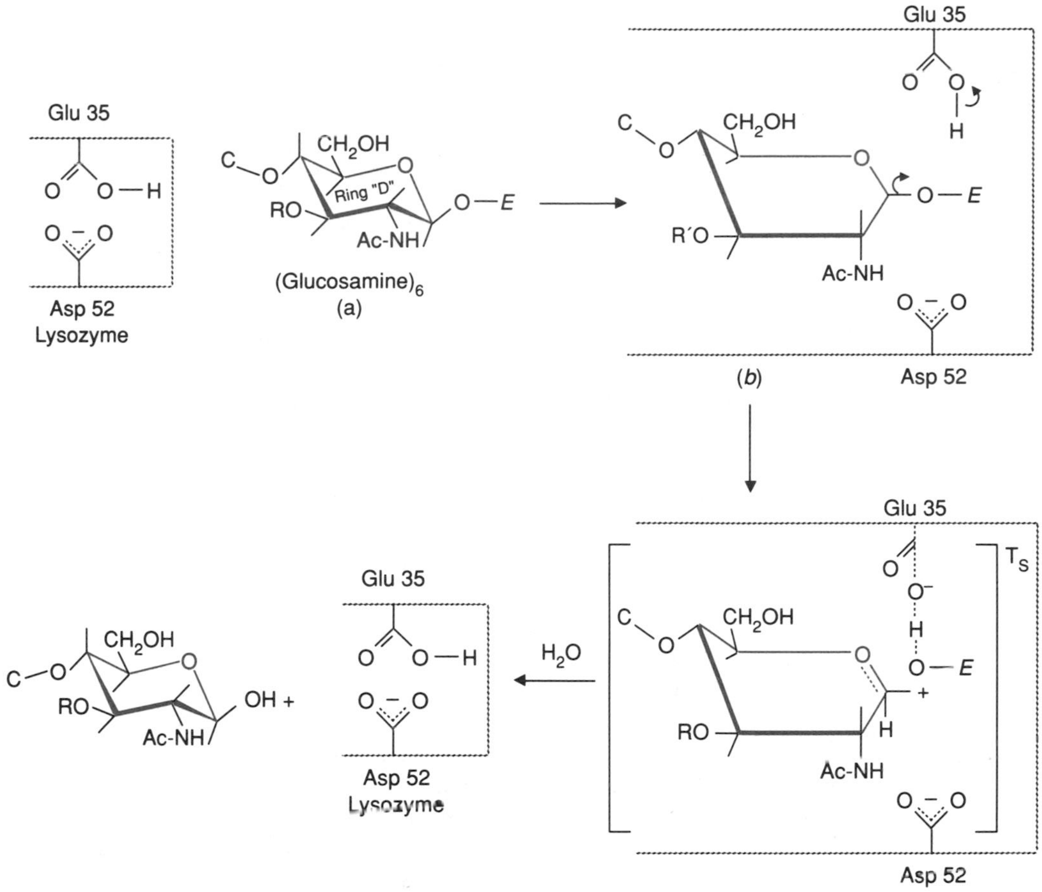

FIGURE 4.49
A mechanism for lysozyme action–substrate strain.
The binding of the stable chair (*a*) conformation of the substrate to the enzyme generates the strained half-chair conformation (*b*) in the ES complex. In the transition state, acid-catalyzed hydrolysis of the glycosidic linkage by an active site glutamic acid residue generates a carbonium ion on the D ring, which relieves the strain generated in the initial ES complex and results in collapse of the transition state to products.

saccharide substrate upon binding to the enzyme is strained to the half-chair conformation. General acid catalysis by active site glutamic acid promotes the unstable half-chair into the transition state. The carbonium ion formed in the transition state is stabilized by a negatively charged aspartate. Breakage of the glycosidic linkage between rings D and E relieves the strained transition state by allowing rings D and E to return to the stable chair conformation.

An alternative interpretation of substrate strain relates to the greater affinity of the enzyme for that small fraction of substrate molecules that exist in a transition state geometry (see Figure 4.45). Such binding stabilizes the transition state and causes a decrease in the energy required to promote substrate molecules into the transition state. Advocates of this view do not ascribe a role to the enzyme of physically distorting the substrate toward the transition state as is depicted in Figure 4.40*b*.

Covalent Catalysis

In **covalent catalysis,** the attack of a nucleophilic (negatively charged) or electrophilic (positively charged) group in the enzyme active site upon the substrate results in covalent binding of the substrate to the enzyme as an

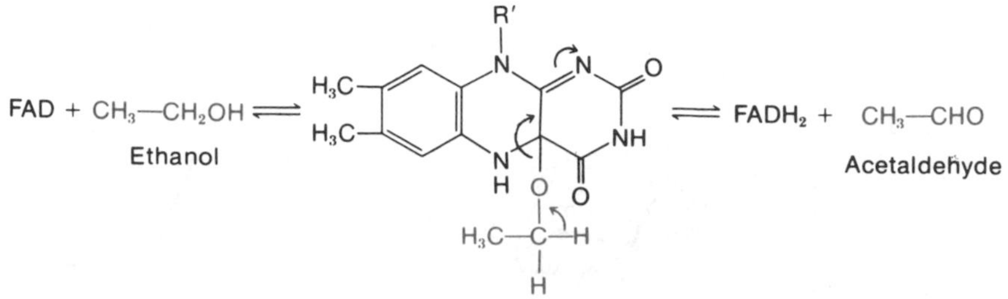

FIGURE 4.50
Proposed covalent substrate–coenzyme complex in aldehyde dehydrogenase.
The enzyme catalyzes a nucleophilic addition of the alcohol to the 4a position of the coenzyme FAD followed by collapse of this intermediate, as indicated by the arrows, to products.

FAD + CH_3—CH_2OH ⇌ [covalent FAD–substrate adduct] ⇌ $FADH_2$ + CH_3—CHO
Ethanol — Acetaldehyde

intermediate in the reaction sequence. Also, enzyme-bound coenzymes often form covalent bonds with the substrate. For example, in the transaminases, the amino acid substrate forms a Schiff's base with enzyme-bound pyridoxal phosphate (p. 478). Evidence is now accumulating that some oxidoreductases, utilizing FAD as coenzyme, form intermediate covalent adducts of substrate and FAD. For example, in the oxidation of alcohols the scheme shown in Figure 4.50 has been proposed.

In all cases of covalent catalysis, the enzyme- or coenzyme-bound substrate is more labile than the original substrate. The enzyme–substrate adduct represents one of the valleys on the energy profile (Figure 4.44).

The serine proteases, such as trypsin, chymotrypsin, and thrombin, are good representatives of the covalent catalytic mechanism. The name "serine protease" arises because serine is involved in the active site of all these enzymes. Acylated enzyme has been isolated in the case of chymotrypsin. Covalent catalysis is assisted by acid–base catalysis in these particular enzymes.

In chymotrypsin the attacking nucleophile is Ser 195. Since the serine hydroxyl is not dissociated at pH 7.4, a mechanism for ionizing it is required. This is achieved by a *charge relay* system consisting of a buried Asp 102 and solvent accessible His 57. At pH 7.4 Asp is ionized and can pull a proton from the His rendering it a much stronger base that can in turn enhance the removal of the serine hydroxyl as outlined in Figure 4.51*a*. The resulting serine alkoxide attacks the carbonyl carbon of the peptide bond, releasing the amino terminal end of the protein and forming an acylated enzyme (through Ser 195). The acylated enzyme is then cleaved by reversal of the reaction sequence in Figure 4.51*b*, but with water as the nucleophile rather than Ser 195.

The catalytic triad of Asp, His, and Ser has been conserved in the serine proteases. Questions concerning the role of Asp 102 as described above have been resolved by genetic engineering in which the Asp was changed to Asn. In the Asn mutant the carboxyl is not free to function as an acid or base catalyst consequently, Ser 195 becomes a million times less reactive.

Entropy Effect

Entropy is a thermodynamic term, S, which defines the extent of disorder in a system. At equilibrium, the entropy is maximal. For example, in solution two reactants A—□ and B—□ exist in many different orientations. The chances of A—□ and B—□ coming together with the correct geometric orientation and with enough energy to react is small at 37°C and in dilute solution. However, if an enzyme with two high-affinity binding sites for A—□ and B—□ is introduced into the dilute solution of these reactants, as suggested in Figure 4.52, A—□ and B—□ will be bound to the enzyme in the correct orientation for the reaction to occur. They will be bound with the correct stoichiometry, and the effective concentration of the reactants will be increased on the enzyme surface—all of which will contribute to an increased rate of reaction.

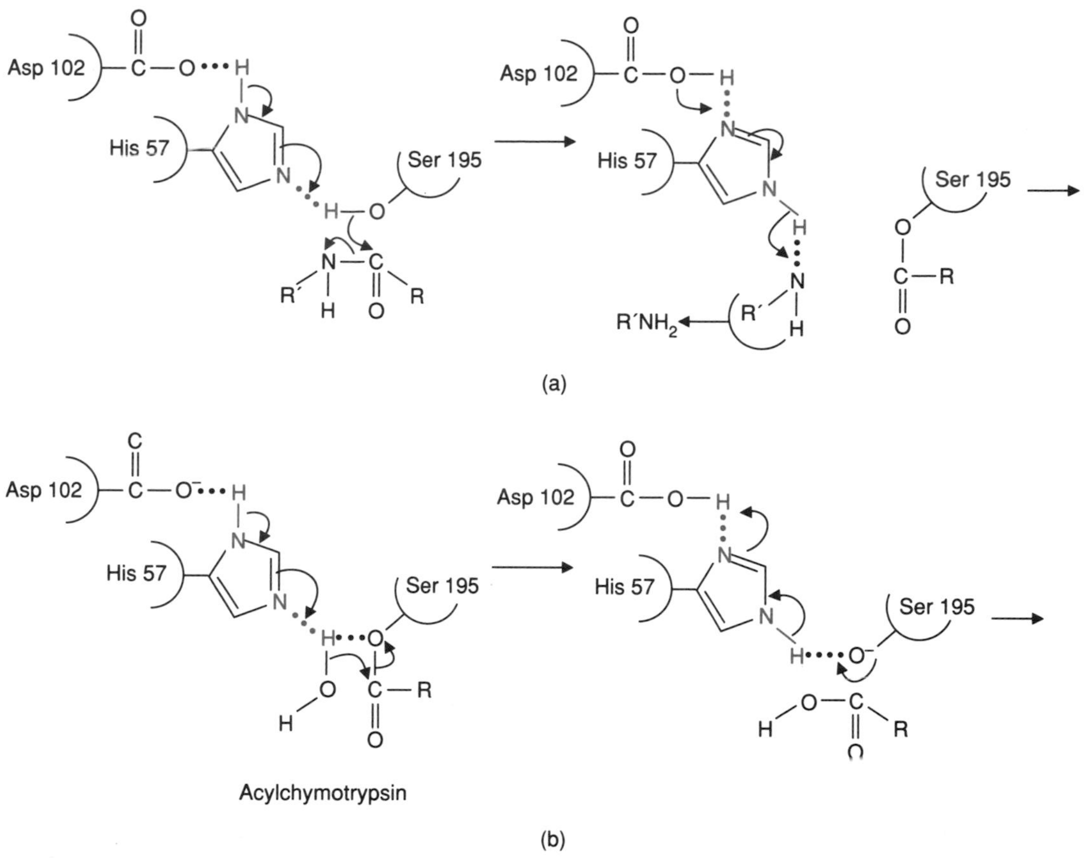

FIGURE 4.51
Covalent catalysis in the active site of chymotrypsin.
Through acid-catalyzed nucleophilic attack, the stable amide linkage of the peptide substrate is converted into an unstable acylated enzyme. The latter is hydrolyzed in the rate-determining step. The new amino-terminal peptide is released concomitant with formation of the acylated enzyme. Scheme (*a*) represents the acylation step and (*b*) the deacylation step.

Once correctly positioned on the enzyme surface, and as a result of binding, the substrates may be "strained" toward the transition state. At this point the substrates have been "set up" for acid–base and/or covalent catalysis. Thus, the proper orientation and the nearness of the substrate with respect to the catalytic groups, which has been dubbed the "proximity effect," contributes 10^3–10^4-fold to the rate enhancement observed with enzymes. It has been estimated that the decrease in entropy contributes a factor of 10^3 to the rate enhancement.

Abzymes Are Antibodies With Catalytic Activity Which Have Been Artificially Synthesized

If the principles discussed above for enzyme catalysis are correct, then one should be able to design an artificial enzyme. This feat has been accomplished by the use of several different approaches, but only the synthesis of antibodies that have catalytic activity will be considered in this discussion. These antibodies are referred to as **abzymes.**

The design of abzymes is based on two principles. The first principle is the ability of the immune system to recognize any arrangement of atoms in the foreign antigen and to make a binding site on the resulting immunoglobulin that is exquisitely suited to binding that antigen. The sec-

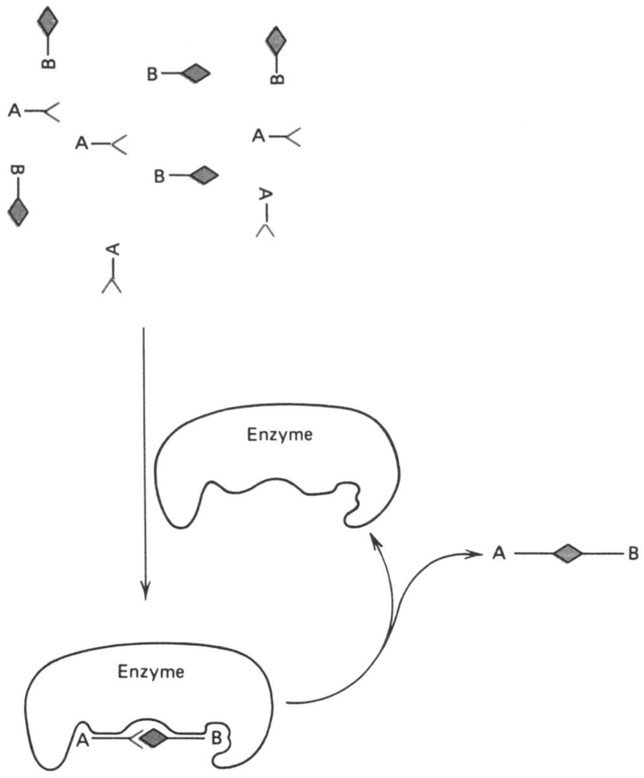

FIGURE 4.52
Diagrammatic representation of the entropy effect.
Substrates in dilute solution are concentrated and oriented on the enzyme surface so as to enhance the rate of reaction.

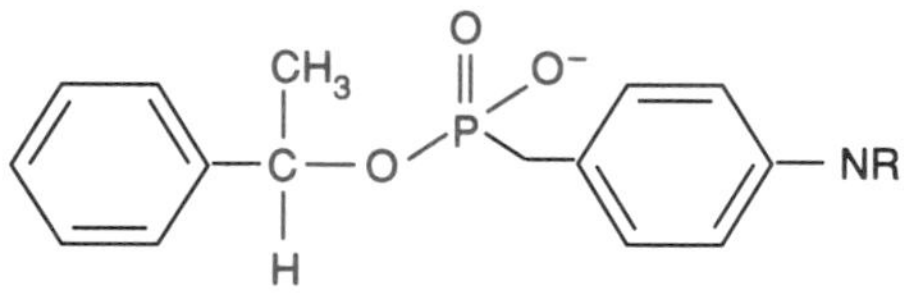

ond principle is that strong binding of transition statelike substrates reduces the energy barrier along the reaction pathway (see discussion on p. 174.)

In abzymes a transition state analog serves as the hapten. For a lipase abzyme, the racemic phosphonate shown in Figure 4.53*a* serves as a hapten. Two enantiomeric fatty acid ester substrates are shown in *b* and *c*. See page 171 for the transition state structure expected for ester hydrolysis. Among many antibodies produced by rabbits on challenge with the protein bound transition state analog *a*, one was found that hydrolyzed only the (*R*) isomer *b* and another only the (*S*) isomer. The antibodies, abzymes, enhanced the rate of hydrolysis of substrates *a* and *b* 10^3–10^5-fold above the background rate in a stereospecific manner. Acceleration of a million-fold, which is close to the enzymatic rate, has been achieved in another esterase-like system.

(*a*) Transition state analog

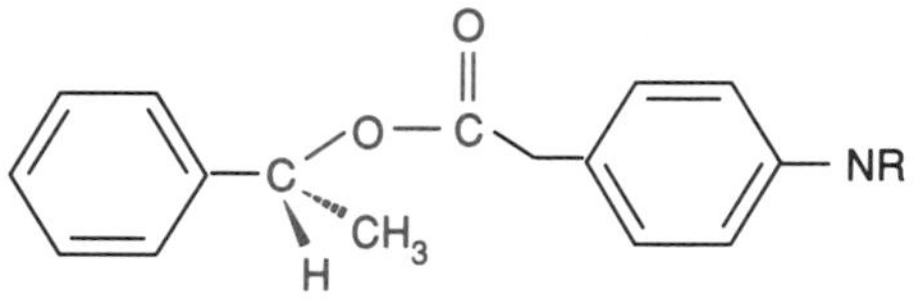

(*b*) Substrate—(*R*) isomer

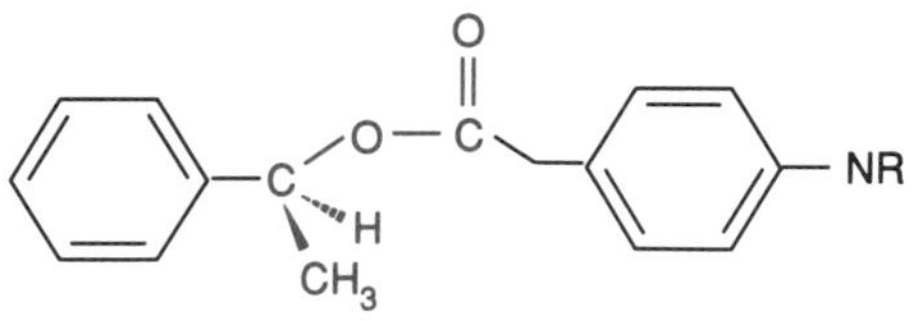

(*c*) Substrate—(*S*) isomer

FIGURE 4.53
Hapten and substrate for a catalytic antibody (abzyme).
The phosphanate (*a*) is the transition state analog used as the hapten to generate antibodies with lipase like catalytic activity. Specific abzymes can be generated that are specific for either the (*R*)-isomer (*b*) or the (*S*)-isomer (*c*) of methyl benzyl esters.

Environmental Parameters Influence Catalytic Activity

A number of external parameters, including pH, temperature, and salt concentration, affect the enzyme activity. These effects are probably not important in vivo, under normal conditions, but are very important in setting up enzyme assays in vitro to measure enzyme activity in a patient's plasma or tissue sample.

Temperature Dependence

Plots of velocity versus temperature for most enzymes reveal a bell-shaped curve with an optimum between 40 and 45°C for mammalian enzymes, as indicated in Figure 4.54. Above this temperature, heat denaturation of the enzyme occurs. Between 0 and 40°C, most enzymes show a twofold increase in activity for every 10°C rise in temperature. Under conditions of hypothermia most enzyme reactions are depressed, which

CLIN. CORR. 4.5
THERMAL LABILITY OF G6PD RESULTS IN HEMOLYTIC ANEMIA

Glucose-6-phosphate dehydrogenase (G6PD) is an important enzyme in the red cell for the maintenance of the red cell membrane. A deficiency or inactivity of this enzyme leads to a hemolytic anemia. In other cases, a variant enzyme is present that normally has sufficient activity to maintain the membrane, but under conditions of oxidative stress fails. A mutation of this enzyme leads to a protein with normal kinetic constants but a decreased thermal stability. This condition is especially critical to the red cell, since it is devoid of protein-synthesizing capacity and cannot renew enzymes as they denature. The end result is a greatly decreased lifetime for those red cells that have an unstable G6PD. These red cells are also susceptible to drug-induced hemolysis. See Clin. Corr. 8.1.

Beutler, E. Glucose-6-phosphate dehydrogenase deficiency. Stanbury J. B., Wynaarden J. B., and Frederickson D. S. (Eds.). *The Metabolic Basis of Inherited Disease*, 4th ed. McGraw-Hill: New York, pp. 1430, 1978.

accounts for the decreased oxygen demand of living organisms at low temperature. Mutation of an enzyme to a thermolabile form can have serious consequences as discussed in Clin. Corr. 4.5.

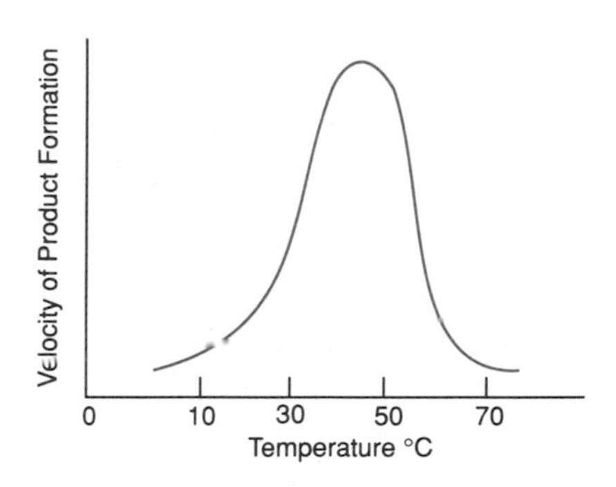

FIGURE 4.54
Temperature dependence of a typical mammalian enzyme.
To the left of the optimum the rate is low because the environmental temperature is too low to provide enough kinetic energy to overcome the energy of activation. To the right of the optimum, the enzyme is inactivated by heat denaturation.

pH Effects

Nearly all enzymes show a bell-shaped pH–velocity profile, but the maximum (pH optimum) varies greatly with different enzymes. Alkaline and acid phosphatases are both found in humans, but their pH optima are greatly different, as shown in Figure 4.55.

The bell-shaped curve and its position on the x axis are dependent on the particular ionized state of the substrate that will be optimally bound to the enzyme. This in turn is related to the ionization of specific amino acids that constitute the substrate binding site. In addition, those amino acids that are involved in catalyzing the reaction must be in the correct charge state to be functional in the catalytic event. For example, if aspartic acid is involved in catalyzing the reaction, the pH optimum may be in the region of 4.5 at which the α-carboxyl of aspartate ionizes, whereas if the ε-amino of lysine is the catalytic group, the pH optimum may be around pH 9.5, the pK_a of the ε-amino group. Studies of the pH dependence of enzymes are useful for suggesting which amino acid(s) may be operative in the catalytic event in the active site.

Clin. Corr. 4.6 points out the physiological effect of a mutation leading to a change in the pH optimum of a physiologically important enzyme. Such a mutated enzyme may function on the shoulder of the pH-rate profile, but not be optimally active, even under normal physiological conditions. Then when an abnormal condition such as alkalosis (observed in vomiting) or acidosis (observed in pneumonia and often in surgery) occurs, the enzyme activity may disappear because the pH is inappropriate. The point is that under normal conditions, the enzyme may be active enough to meet normal requirements, but under physiological stress in vivo environmental conditions may change so that the enzyme is less active and cannot meet its metabolic obligations.

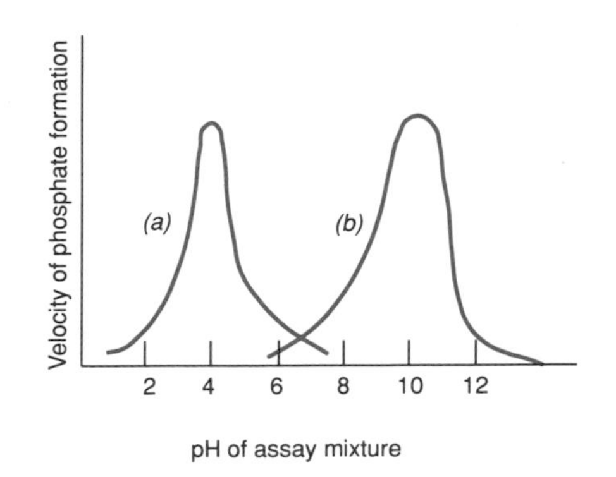

FIGURE 4.55
The pH dependence of (*a*) acid and (*b*) alkaline phosphatase reactions.
In each case the optimum represents the ideal ionic state for binding of enzyme and substrate and the correct ionic state for the amino acids involved in the catalytic event.

4.9 CLINICAL APPLICATIONS OF ENZYMES

The principles of enzymology outlined in previous sections find practical application in the clinical laboratory in the measurement of plasma or tissue enzyme activities and concentrations of substrates in the sick individual.

The rationale for measuring **plasma enzyme activities** is based on the premise that changes in activities reflect changes that have occurred in a

CLIN. CORR. 4.6 ALCOHOL DEHYDROGENASE ISOENZYMES WITH DIFFERENT pH OPTIMA

In addition to the change in aldehyde dehydrogenase isoenzyme composition in Asians (Clin. Corr. 4.2), different alcohol dehydrogenase isoenzymes are also observed.

Alcohol dehydrogenase (ADH) is encoded by three genes, which produce three different polypeptides: α, β, and γ. Three alleles are found for the β-gene. These alleles differ in a single nucleotide base, which causes substitutions for arginine. The substitutions are shown below:

	Residue 47	Residue 369
β_1	Arg	Arg
β_2	His	Arg
β_3	Arg	Cys

The β_3 form was detected by finding ADH activity in liver with a pH optimum near 7, compared with a normal optimum of 10 for β_1, and 8.5 for β_2. The shift in pH optimum occurs because the release of NADH from the enzyme active site determines the rate of the catalysis. Binding of NADH involves ionic bonds between the phosphates of the coenzyme and the arginines at positions 47 and 369. The high pH optimum for β_1 ADH is due to weakening of this interaction under alkaline conditions. Substitution of amino acids with lower pK_a values (histidine or cysteine) also weakens the interaction and lowers the pH optimum. Since the release of NADH is facilitated, the V_{max} of β_2 and β_3 are also higher than that for β_1 ADH.

Burnell, J. C., Carr, L. G., Dwulet, F. E., Edenberg, H. J., Li, T-K., and Bosron, W. F. The human β_3 alcohol dehydrogenase subunit differs from β_1 by a cys- for arg-369 substitution which decreased NAD(H) binding. *Biochem. Biophys. Res. Commun.* 146:1227, 1987.

specific tissue or organ. Plasma enzymes are of two types: one is present in the highest concentration, is specific to plasma, and has a functional role; the other is normally present at very low levels and plays no functional role in the plasma. The former includes the enzymes associated with blood coagulation (thrombin), fibrin dissolution (plasmin), and processing of chylomicrons (lipoprotein lipase).

In disease of tissues and organs, the nonplasma-specific enzymes are most important. Normally, the plasma levels of these enzymes are low to absent. An insult in the form of any disease process may cause changes in cell membrane permeability or increased cell death, resulting in release of intracellular enzymes into the plasma. In cases of permeability change, those enzymes of lower molecular weight will appear in the plasma first. The greater the concentration gradient between intra- and extracellular levels, the more rapidly the enzyme diffuses out. Cytoplasmic enzymes will appear in the plasma before mitochondrial enzymes, and of course the greater the quantity of tissue damaged, the greater the increase in the plasma level. The nonplasma-specific enzymes will be cleared from the plasma at varying rates, which depend on the stability of the enzyme and its susceptibility to the reticuloendothelial system.

In the diagnosis of specific organ involvement in a disease process it would be ideal if enzymes unique to each organ could be identified; however, this is unlikely, since the metabolism of various organs is very similar. Alcohol dehydrogenase of the liver and acid phosphatase of the prostate are useful for specific identification of disease in these organs. Other than these two examples, there are few enzymes that are tissue or organ specific. However, the ratio of various enzymes does vary from tissue to tissue. This fact, combined with a study of the kinetics of appearance and disappearance of particular enzymes in plasma, allows a diagnosis of specific organ involvement to be made. Figure 4.56 illustrates the time dependence of the plasma activities of enzymes released from the myocardium following a heart attack. Such profiles allow one to establish when the attack occurred and whether treatment is effective.

Clin. Corr. 4.7 demonstrates how diagnosis of a specific enzyme defect led to a rational clinical treatment that restored the patient to health.

Studies of the kinetics of appearance and disappearance of plasma en-

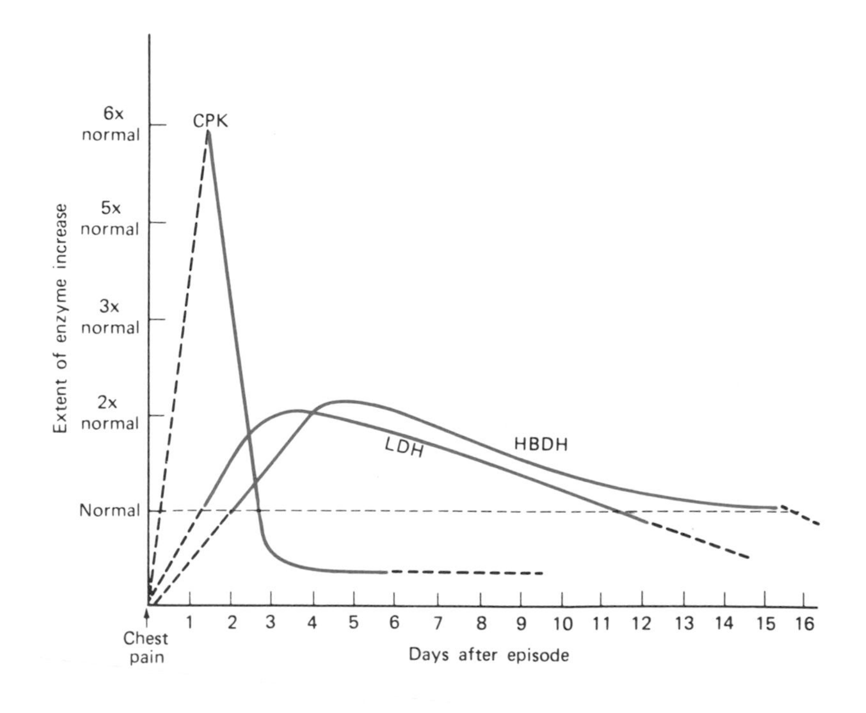

FIGURE 4.56
Kinetics of release of cardiac enzymes into serum following a myocardial infarction.
Creatine phosphokinase, CPK; lactic dehydrogenase, LDH; α-hydroxybutyric dehydrogenase, HBDH. Such kinetic profiles allow one to determine where the patient is with respect to the infarct and recovery. Note: CPK rises sharply but briefly; HBDH rises slowly but persists.
Reprinted, with permission, from *Diagnostic Enzymes*, E. L. Coodley, page 61. 1970, Philadelphia: Lea & Febiger.

CLIN. CORR. 4.7
IDENTIFICATION AND TREATMENT OF AN ENZYME DEFICIENCY

Enzyme deficiencies usually lead to increased accumulation of specific intermediary metabolites in plasma and hence in urine. The recognition of the intermediates that are accumulating is useful in pinpointing possible enzyme defects. After the enzyme deficiency is established, metabolites that normally occur in the pathway but are distal to the block may be supplied exogenously in order to overcome the metabolic effects of the enzyme deficiency. For example, in hereditary orotic aciduria there is a double enzyme deficiency in the pyrimidine biosynthetic pathway leading to accumulation of orotic acid.

Both orotate phosphoribosyltransferase and orotidine 5′-phosphate decarboxylase are deficient. The pyrimidine nucleotides, both dCTP and dTTP, are required for cell division, particularly in erythropoiesis. The patients are pale, weak, and fail to thrive. Administration of the missing pyrimidines as uridine or cytidine promotes growth, general well-being, and also decreases orotic acid excretion. The latter occurs because the TTP and CTP formed from the supplied nucleosides repress carbamoyl phosphate synthetase, the committed step, by feedback inhibition.

Smith, L. H., Jr. Hereditary orotic aciduria: pyrimidine auxotrophison in man. *Am. J. Med.* 38:1, 1965.

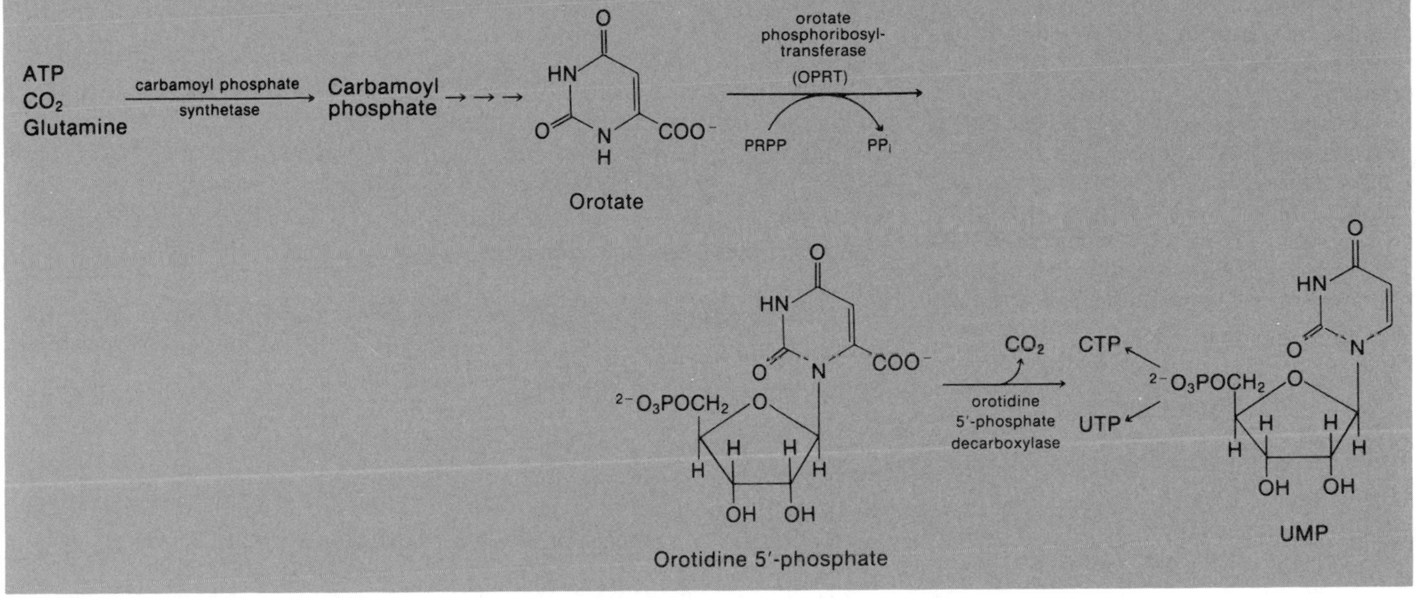

zymes require a valid enzyme assay. The establishment of a good assay is based on good temperature and pH control, as well as saturating levels of all substrates, cosubstrates, and cofactors. In order to accomplish the latter, the K_m must be known for those particular conditions of pH, ionic strength, and so on, that are to be used in the assay. You will recall that K_m is the substrate concentration at half-maximal velocity (V_{max}). To be assured that the system is saturated, the substrate concentration is generally increased 5- to 10-fold over the K_m. The value of saturation of the enzyme with substrate is that under these conditions the reaction is zero order. This fact is emphasized in Figure 4.57. Under zero-order conditions changes in velocity are proportional to enzyme concentration alone. Under first-order conditions in substrate, the velocity is dependent on both the substrate and the enzyme concentration.

Clin. Corr. 4.8 demonstrates the importance of determining if the assay conditions accurately reflect the amount of enzyme actually present. Clinical laboratory assay conditions are optimized routinely for the properties of the normal enzyme and may not reflect levels of mutated enzyme. The pH dependence and/or the K_m for substrate and cofactors may drastically change in a mutated enzyme.

Under optimal conditions a valid enzyme assay will reflect a linear dependence of velocity and amount of enzyme. This can be tested by

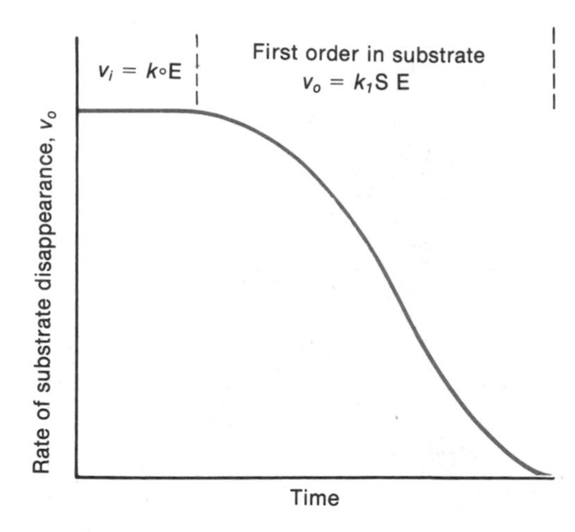

FIGURE 4.57
Relation of substrate concentration to order of the reaction.
When the enzyme is completely saturated, the kinetics are zero order with respect to substrate and are first order in enzyme, that is, the rate depends only on enzyme concentration. When the substrate level falls below saturating levels, the kinetics are first order in both substrate and enzyme and are therefore second order, that is, the observed rate is dependent on both enzyme and substrate.

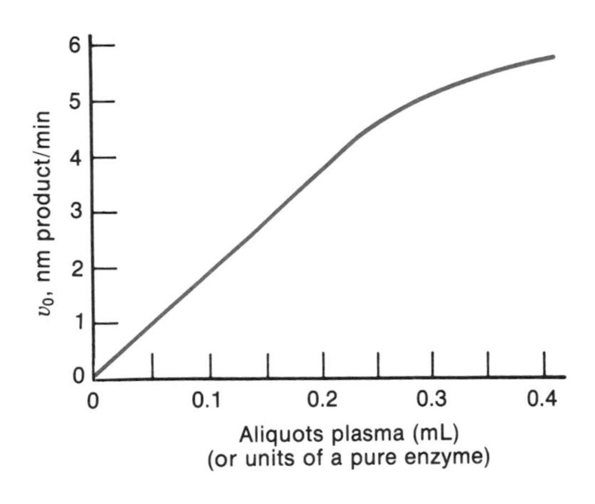

FIGURE 4.58
Assessing the validity of an enzyme assay.
The line shows what is to be expected for any reaction where the concentration of substrate is held constant and the aliquots of enzyme increased. In this particular example linearity between the initial velocity observed and the amount of enzyme, whether pure or in a plasma sample, is only observed up to 0.2 mL of plasma or 0.2 units of pure enzyme. If one were to measure the velocity with an aliquot of plasma greater than 0.2 mL, the actual amount of enzyme present in the sample would be underestimated.

determining if the velocity of the reaction doubles when the plasma sample size is doubled, while keeping the total volume of the assay constant, as demonstrated in Figure 4.58.

Coupled Assays Utilize the Optical Properties of NAD, NADP or FAD

Enzymes that employ the coenzymes NAD^+, $NADP^+$, and FAD are easy to measure because of the optical properties of NADH, NADPH, and FAD. The absorption spectra of NADH and FAD in the ultraviolet and visible light regions are shown in Figure 4.59. Oxidized FAD absorbs strongly at 450 nm, while NADH has maximal absorption at 340 nm. The concentration of both FAD and NADH is related to their absorption of light at the respective absorption maximum by the Beer–Lambert relation

$$A = \varepsilon \cdot c \cdot l$$

where l is the pathlength of the spectrophotometer cell in centimeters (usually 1 cm), ε is the absorbance of a molar solution of the substance being measured at a specific wavelength of light, A is the absorbance, and c is the concentration. Absorbance is the log of transmittance (I_0/I). The term ε is a constant that varies from substance to substance; its value can be found in a handbook of biochemistry. In an optically clear solution, the concentration c can be found after a determination of the absorbance A is made.

Many enzymes do not employ either NAD^+ or FAD but do generate products that can be utilized by a NAD^+- or FAD-linked enzyme. For example, glucokinase catalyzes the reaction

$$\text{Glucose} + \text{ATP} \rightleftharpoons \text{glucose 6-phosphate} + \text{ADP}$$

Both ADP and glucose 6-phosphate (G6P) are difficult to measure directly; however, the enzyme G6P dehydrogenase catalyzes the reaction,

$$\text{G6P} + \text{NADP}^+ \rightleftharpoons \text{6-phosphogluconolactone} + \text{NADPH} + \text{H}^+$$

Thus by adding an excess of the enzyme G6P dehydrogenase and NADP to the assay mixture, the velocity of production of G6P by gluco-

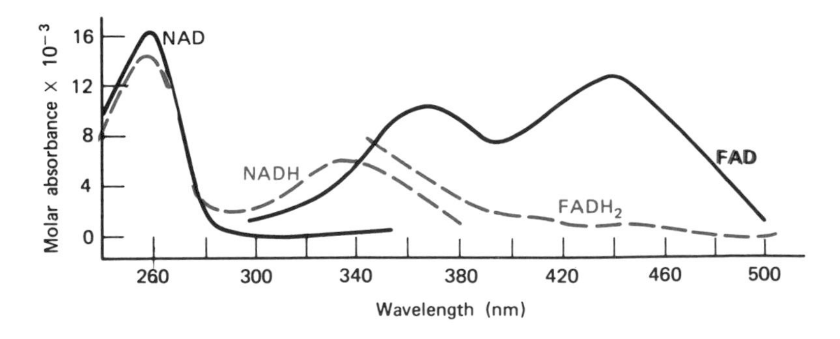

FIGURE 4.59
Absorption spectra of niacin and flavin coenzymes.
The reduced form of NAD (NADH) absorbs strongly at 340 nm. The oxidized form of flavin coenzymes absorbs strongly at 450 nm. thus, one can follow the rate of reduction of NAD^+ by observing the increase in the absorbance at 340 nm and the formation of $FADH_2$ by following the decrease in absorbance at 450 nm.

kinase is proportional to the rate of reduction of $NADP^+$, which can be measured directly in the spectrophotometer.

Clinical Analyzers May Use Immobilized Enzymes as Reagents

Enzymes are now being used as chemical reagents in desk-top clinical analyzers that can be used in the office or for screening purposes in shopping centers and malls. For example, screening tests for cholesterol and triglycerides can be completed in a few minutes using 10 μL of plasma. The active components are cholesterol oxidase for the cholesterol determination and lipase for the triglycerides. The enzymes are immobilized in a bilayer along with the necessary buffer salts, cofactors or cosubstrates, and indicator reagents. These ingredients are arranged in a multilayered vehicle about the size and thickness of a 35-mm slide. The plasma sample provides the substrate and water necessary to activate the system. In the case of cholesterol oxidase, hydrogen peroxide is a product that subsequently oxidizes a colorless dye to a colored product that can be measured by reflectance spectroscopy. Peroxidase is included in the film packet to catalyze the latter reaction.

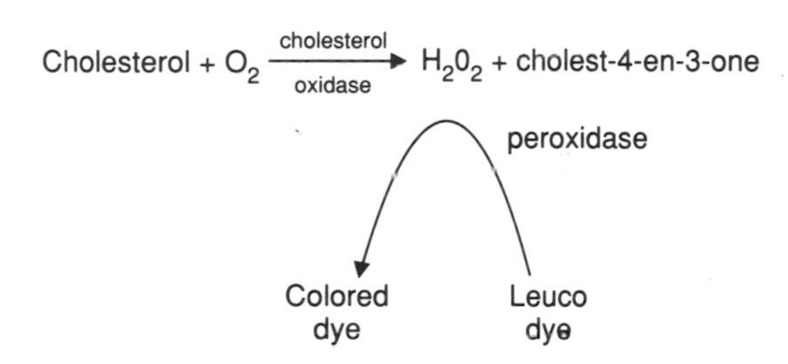

Each slide packet is constructed to measure a specific substance or enzyme and can be stored in the cold for use as needed. In many cases the slide packet contains several enzymes in a coupled assay system that eventually generates a reduced nucleotide or a colored dye for spectroscopic measurement purposes.

This technology has been made possible, in part, by the fact that the enzymes involved are stabilized when bound to immobilized matrices and are stored in the dry state or in the presence of a stabilizing solvent such as glycerol.

Enzyme-Linked Immunoassays Employ Enzymes as Indicators

Modern clinical chemistry has benefited greatly from the marriage of enzyme chemistry and immunology. Antibodies specific to a protein antigen are coupled to an indicator enzyme such as horseradish peroxidase to generate a very specific and sensitive assay system. After binding of the peroxidase-coupled antibody to the antigen, the peroxidase can be used to generate a colored product that is measurable and whose concentration is related to the amount of antigen in a sample. Because of the catalytic nature of the enzyme the system greatly amplifies the signal. This assay has been given the acronym **ELISA** for enzyme-linked immunosorbant assay.

The manner in which these principles can be utilized in an assay system can be demonstrated by an assay for the acquired immunodeficiency syndrome (AIDS) virus coat protein antigens. Antibodies are prepared in a rabbit to the AIDS virus coat proteins. In addition, a reporter antibody is prepared in a goat against rabbit IgG against the AIDS virus protein. To this goat antirabbit IgG is attached to the enzyme, horseradish perox-

CLIN. CORR. 4.8
AMBIGUITY IN THE ASSAY OF MUTATED ENZYMES

Structural gene mutations leading to the production of enzymes with changes in K_m are frequently observed. The K_m may be either increased or decreased, depending on the mutation. A case in point is a patient with hyperuricemia and gout, whose red blood cell hypoxanthine-guanine-phosphoribosyltransferase (HGPRT) showed little activity in assays in vitro. This enzyme is involved in the salvage of purine bases and catalyzes the reaction

$$\text{Hypoxanthine} + \text{PRPP} \longrightarrow \text{inosine monophosphate} + \text{PP}_i$$

where PRPP = phosphoribosyl pyrophosphate.

The absence of HGPRT activity results in a severe neurological disorder known as Lesch–Nyhan syndrome (p. 543), yet this patient did not have the clinical signs of this disorder. Immunological testing with a specific antibody to the enzyme revealed as much cross-reacting material in the patient's red blood cells as in normal controls. The enzyme was therefore being synthesized but was inactive in the assay in vitro. Additional experimentation revealed that by increasing the substrate concentration in the assay, full activity was measurable in the patient's red cell hemolysates. This anomaly is explained as a mutation in the substrate binding site of HGPRT, leading to an increased K_m. Neither the substrate concentration in the assay nor in the red blood cells was high enough to bind to the enzyme. This case reinforces the point that an accurate enzyme determination is dependent on zero-order kinetics, that is, the enzyme being saturated with substrate.

Sorenson, L. and Benke P. J. Biochemical evidence for a distinct type of primary gout. *Nature (London)* 213:1122, 1967.

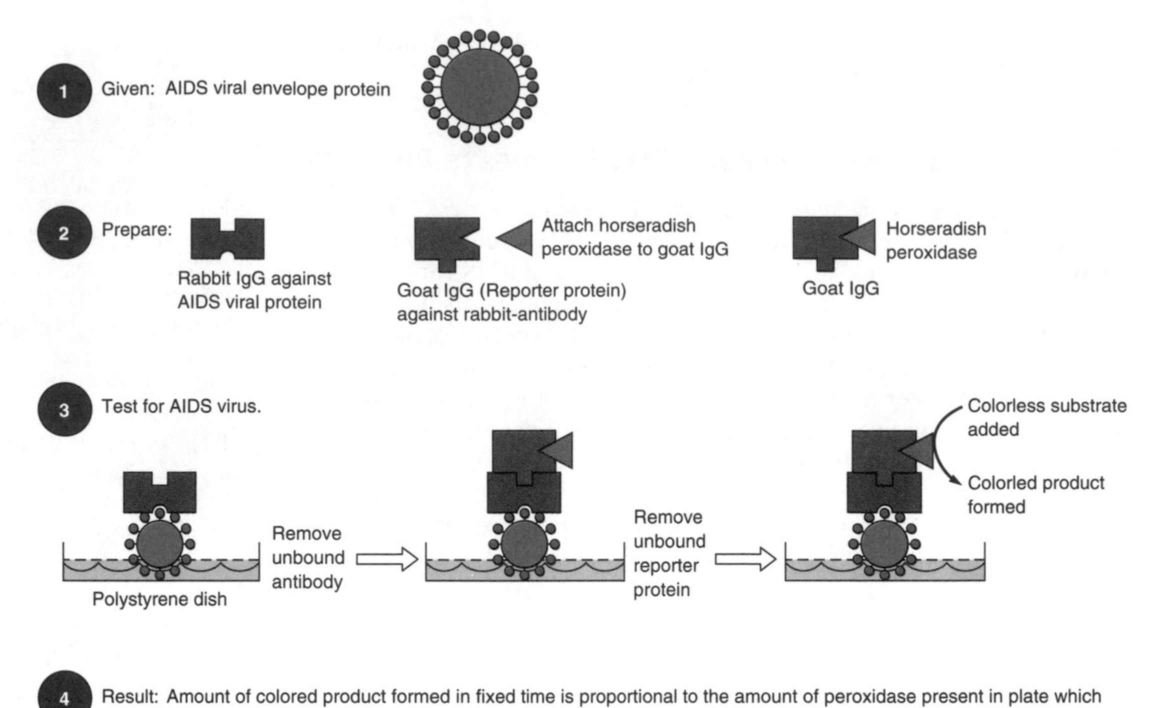

FIGURE 4.60
Schematic of the ELISA (enzyme-linked immunosorbant assay) system for detecting the AIDS virus envelope proteins.

idase. The test for the AIDS virus is performed by incubating patient serum in a polystyrene dish that binds the proteins in the serum sample. Any free protein binding sites on the dish are then covered by incubating with a nonspecific protein like bovine serum albumin. Next the rabbit IgG antibody against the AIDS virus protein is incubated in the dish during which time the IgG attaches to any AIDS viral coat proteins that are attached to the polystyrene dish. All unbound rabbit IgG is washed out with buffer. The goat antirabbit IgG-peroxidase is now placed in the dish resulting in binding of any rabbit IgG attached to the dish via the viral coat protein. Unattached antibody-peroxidase is then washed out. Peroxidase substrates are then added to the dish and the amount of color developed in a given time period is a measurement of the amount of AIDS viral coat protein present in a given volume of patient plasma when compared against a standard curve. This procedure is schematically diagrammed in Figure 4.60. Such amplified enzyme assays allow the measurement of remarkably small amounts of antigens.

Measurement of Isozymes Is Used Diagnostically

Isozymes (or *isoenzymes*) are enzymes that catalyze the same reaction but migrate differently on electrophoresis. Their physical properties may also be different, but not necessarily. The most common mechanism for the formation of isozymes involves the arrangement of subunits arising from two different genetic loci in different combinations to form the active polymeric enzyme. The isozymes that have been studied for clinical applications are lactate dehydrogenase, creatine kinase, and alkaline phosphatase.

Creatine kinase (p. 963) occurs as a dimer. There are two types of subunits, M (muscle type) and B (brain type). These designations arise

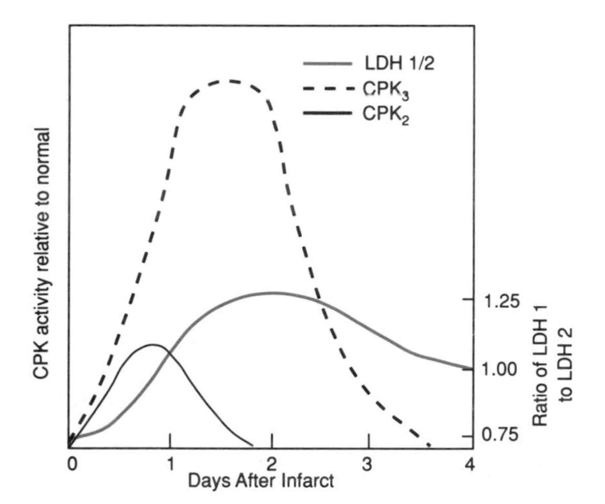

FIGURE 4.61
Characteristic changes in serum CPK and LDH isozymes following a myocardial infarction.
CPK_2 (MB) isozyme increases to a maximum within 1 day of the infarction. CPK_3 lags behind CPK_2 by about 1 day. The total LDH level increases more slowly. The increase of LDH_1 and LDH_2 within 12–24 h coupled with an increase in CPK_2 is diagnostic of myocardial infarction.

from the fact that in brain both subunits are electrophoretically of the same type and are arbitrarily given the designation B. In skeletal muscle the subunits are both of the M type. The isozyme containing both M and B type subunits (MB) is found only in the myocardium. Other tissues contain variable amounts of the MM and BB isozymes. The isozymes are numbered beginning with the species migrating the fastest to the anode thus, CPK_1 (BB), CPK_2 (MB), and CPK_3 (MM).

Lactate dehydrogenase is a tetrameric enzyme, but only two distinct subunits have been found: those designated H for heart (myocardium) and M for muscle. These two subunits are combined in five different ways. The lactate dehydrogenase isozymes, subunit compositions, and major location are as follows:

Type	Composition	Location
LDH_1	HHHH	Myocardium and RBC
LDH_2	HHHM	Myocardium and RBC
LDH_3	HHMM	Brain and kidney
LDH_4	HMMM	
LDH_5	MMMM	Liver and skeletal muscle

As an illustration of how measurement of amounts of isozymes and kinetic analysis of plasma enzyme activities are useful in medicine, activities of some CPK and LDH isozymes are plotted in Figure 4.61 as a function of time after infarction. After damage to heart tissue the cellular breakup releases CPK_2 into the blood within the first 6–18 h after an infarct, but LDH release lags behind the appearance of CPK by 1–2 days. Normally the activity of the LDH_2 isozyme is higher than that of LDH_1; however, in the case of infarction the activity of LDH_1 becomes greater than LDH_2, at about the time CPK_2 levels are back to base line (48–60 h). Figure 4.62 shows the fluctuations of all five LDH isozymes after an infarct. The increased ratio of LDH_2 and LDH_1 can be seen in the 24-h tracing. The LDH isozyme "switch" coupled with increased CPK_2 is diagnostic of myocardial infarct (MI) in virtually 100% of the cases. Increased activity of LDH_5 is an indicator of liver congestion. Thus secondary complications of heart failure can be monitored.

Formerly, plasma levels of the transaminases SGOT (serum glutamate-oxaloacetate transaminase also called aspartate transaminase or AST) and SGPT (serum glutamate-pyruvate transaminase also called alanine transaminase or ALT) were followed; however, these enzymes have much less specificity and predictive accuracy in diagnosing MI and liver disease. The rationale for assaying these two enzymes is that liver and

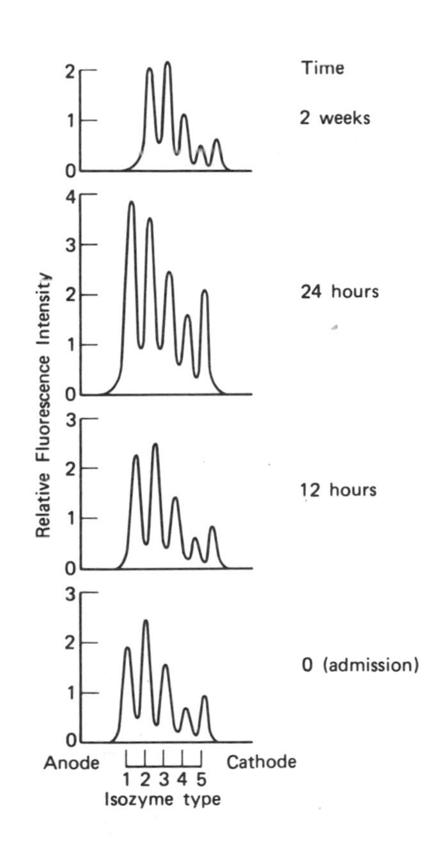

FIGURE 4.62
Tracings of densitometer scans of LDH isozymes at time intervals following a myocardial infarction.
As can be seen total LDH increases and LDH_1 becomes greater than LDH_2 between 12 and 24 h. Increases in LDH_5 is diagnostic of a secondary congestive liver involvement. After electrophoresis on agarose gels the LDH activity is assayed by measuring the fluorescence of the NADH formed in the LDH catalyzed reaction.
Courtesy of Dr. A. T. Gajda, Clinical Laboratories, The University of Arkansas for Medical Science.

heart contain high levels of both enzymes, but liver contains more GPT than GOT and the reverse is true in heart.

Some Enzymes Are Used as Therapeutic Agents

In a few cases enzymes have been used as drugs in the therapy of specific medical problems. Streptokinase is an enzyme mixture prepared from a streptococcus. It is useful in clearing blood clots that occur in myocardial infarcts and in the lower extremities. Streptokinase activates the fibrinolytic proenzyme plasminogen that is normally present in plasma. The activated enzyme is plasmin. Plasmin is a serine protease-like trypsin that attacks fibrin, cleaving it into several soluble components.

Another serine protease, human tissue plasminogen activator, t-PA, is being commercially produced by bioengineered *Escherichia coli* (*E. coli*) for use in dissolving blood clots in patients suffering myocardial infarction. t-PA also functions by activating the patient's plasminogen.

Asparaginase therapy is used for some types of adult leukemia. Tumor cells have a nutritional requirement for asparagine and must scavenge it from the host's plasma. By intravenous (i.v.) administration of asparaginase, the host's plasma level of asparagine is markedly depressed, which results in depressing the viability of the tumor.

Most enzymes do not have a long half-life in blood; consequently, unreasonably large amounts of enzyme are required to keep the therapeutic level up. Work is now in progress to enhance enzyme stability by coupling enzymes to solid matrices and implanting these materials in areas that are well perfused. In the future, enzyme replacement in individuals that are genetically deficient in a particular enzyme may be feasible.

Enzymes Linked to Insoluble Matrices Are Used as Chemical Reactors

Specific enzymes linked to insoluble matrices are being used in the pharmaceutical industry as highly specific chemical reactors. For example, immobilized β-galactosidase has been used to decrease the lactose content of milk so that lactose intolerant people can have access to milk and milk products. In the production of prednisolone, immobilized steroid 11-β-hydroxylase and a δ-1,2-dehydrogenase are used to convert a cheap precursor to prednisolone in a rapid, stereospecific, and economical manner.

4.10 REGULATION OF ENZYME ACTIVITY

Our discussion up to this point has centered upon the chemical and physical characteristics of individual enzymes, but physiologically we must be concerned with the integration of many enzymes into a metabolic pathway and the interrelationship of the products of one pathway with the metabolic activity of other pathways. For example, dietary glucose can be either converted to glycogen, fat, some nonessential amino acids, or oxidized to carbon dioxide. In each case, glucose is converted to a different end product through a specific metabolic pathway involving several enzymes each of which is unique to the type of reaction catalyzed. After eating, there is an abundance of glucose in the system, but it is not diverted in equal amounts to each of the end products mentioned. Rather, there are very tightly controlled homeostatic mechanisms, which work to maintain a constant blood glucose level, utilize the glucose needed for energy production, maintain the glycogen stores, and, if any excess glucose remains, convert it to fat. The point is that all metabolic pathways are not operating at maximum capacity at all times. In fact many pathways may be shut down during certain phases in the life cycle of a cell. If this were not the case, wild, uncontrolled and uneconomical growth of the cell would occur.

Control of metabolic regulation of a pathway occurs through modulation of the enzymatic activity of one or more key enzymes in the pathway. Although the overall catalytic efficiency of a metabolic pathway is dependent on the activity of all the individual enzymes in the pathway, the pathway can be controlled by one **rate-limiting** enzyme in the pathway. Usually this rate-controlling enzyme is the first enzyme that can be identified as unique to that particular pathway. The chemical reaction that is unique to a metabolic pathway is referred to as the **committed step.** For example, in the de novo synthesis of purines, the committed step is the reaction catalyzed by the 5-phosphoribosyl-α-pyrophosphate (PRPP) amidotransferase, which in this case is also the rate-limiting enzyme. The rate-limiting enzyme is not necessarily the enzyme associated with the committed step. The substrate of the amidotransferase, PRPP, is also used as substrate by the pyrimidine biosynthetic pathway; hence the enzyme PRPP synthase, which produces PRPP, does not catalyze the committed step in the biosynthesis of purines because it occurs before the branch point in the two pathways (see p. 563).

The activity of the enzyme associated with the committed step or with the rate-limiting enzyme can be regulated in a number of ways. First, the absolute amount of the enzyme can be regulated either by substrate or hormone stimulation of the de novo synthesis of more enzyme. Hormones can also suppress the de novo synthesis of enzyme. Second, the activity of the enzyme can be modulated by activators, inhibitors, and by covalent modification through mechanisms previously discussed. Finally, the activity of a pathway can be regulated by partitioning the pathway from its initial substrate and by controlling access of the substrate to the enzymes of the pathway. This is referred to as **compartmentation.** We will now consider each of these general mechanisms of control in more detail.

Compartmentation of Opposing Pathways Maximizes Cellular Economy

Generally anabolic and catabolic pathways are segregated into different organelles in order to maximize the cellular economy. There would be no point to the oxidation of fatty acids occurring at the same time and in the same compartment as biosynthesis of fatty acids. If such occurred, a futile cycle would exist. By maintaining fatty acid biosynthesis in the cytoplasm and oxidation in the mitochondria, control can be exerted by regulating transport of common intermediates across the mitochondrial membrane. For example, coenzyme A derivatives of fatty acids cannot diffuse across the mitochondrial membrane; fatty acids, however, are transported by a specific transport system. If the metabolic situation requires fatty acid biosynthesis rather than fatty acid oxidation, there could be hormonal or other control over the mitochondrial membrane fatty acid transport system such that it is depressed during fatty acid biosynthesis, but activated when fatty acid oxidation is required for cellular energy.

Table 4.2 contains a compilation of some of the important enzymes, metabolic pathways, and their intracellular distribution.

Enzyme Concentrations in Cells Are Controlled

As indicated earlier, the velocity of any reaction is dependent on the amount of enzyme present. Many rate-controlling enzymes are present in very low concentrations. More enzyme may be synthesized or existing

TABLE 4.2 Intracellular Location of Major Enzymes and Metabolic Pathways[a]

Cytoplasm	Glycolysis: hexose monophosphate pathway; glycogenesis and glycogenolysis; fatty acid synthesis; purine and pyrimidine catabolism; peptidases; aminotransferases; amino acyl synthetases
Mitochondria	Tricarboxylic acid cycle; fatty acid oxidation; amino acid oxidation; fatty acid elongation; urea synthesis; electron transport and coupled oxidative phosphorylation
Lysosomes	Lysozyme; acid phosphatase; hydrolases, including proteases, nucleases, glycosidases, arylsulfatases, lipases, phospholipases, and phosphatases
Endoplasmic reticulum (microsomes)	NADH- and NADPH-cytochrome *c* reductases; cytochrome b_5 and cytochrome P-450 related mixed function oxidases; glucose 6-phosphatase; nucleoside diphosphatase, esterase, β-glucuronidase, and glucuronyltransferase; protein synthetic pathways; phosphoglyceride and triacylglycerol synthesis; steroid synthesis and reduction
Golgi	Galactosyl- and glucosyltransferase; chondroitin sulfotransferase; 5′-nucleotidase; NADH-cytochrome *c* reductase; glucose 6-phosphatase
Peroxisomes	Urate oxidase; D-amino acid oxidase; α-hydroxy acid oxidase; catalase; long-chain fatty acid oxidation
Nucleus	DNA and RNA biosynthetic pathways

[a] NADH-cytochrome b_5 reductase has been found in endoplasmic reticulum, Golgi, outer mitochondrial membrane, and in the nuclear envelope. Several of the enzymes noted in the table are common to one or more of the membranous organelles.

rates of synthesis repressed through hormonally instituted activation of the mechanisms controlling gene expression. For example, insulin is an anabolic hormone that induces the synthesis of increased amounts of glucokinase, phosphofructokinase, pyruvate kinase, and glycogen synthetase, but represses synthesis of several key gluconeogenic enzymes. The detailed mechanism of these effects are not known in mammalian systems; however, the general concepts of the regulation of eucaryotic gene expression are discussed in Chapter 19.

In some instances substrate can repress the synthesis of enzyme. For example, glucose represses the de novo synthesis of pyruvate carboxykinase. This enzyme is the rate-limiting enzyme in the conversion of pyruvate to glucose. In other words, if there is plenty of glucose available there is no point in synthesizing glucose at the expense of amino acids, which are the alternative source of pyruvate; consequently, this pathway is repressed by the effect of its end product, glucose, on the synthesis of the carboxykinase enzyme. This effect of glucose may be mediated via insulin and is not direct feedback inhibition.

Many rate-controlling enzymes have relatively short half-lives, for example, that of pyruvate carboxykinase is 5 h. Teleologically this is reasonable because it provides a mechanism for effecting much larger fluctuations in the activity of a pathway than would be possible by inhibition or activation of existing levels of enzyme.

Modulation of Enzyme Activity Is a Short-Term Regulation

Regulation at the gene level is long term. Short-term regulation occurs through modification of the activity of existing levels of enzyme by means of mechanisms we will now consider.

During various phases of the cell cycle, specific metabolic pathways are turned on or off, depending on the special requirements of a given phase of the cell cycle for a particular product. For example, there is no point to continued production of deoxyribonucleotides at all times during a cell's life, but only during replicative phases; consequently, during nonreplicative phases, the concentration of deoxyribonucleotide builds up to such an extent that the ribonucleoside diphosphate reductase is inhibited by the end products of the pathway. This type of control is referred to as **feedback inhibition.** The inhibition may take the form of competitive inhibition or allosteric inhibition. In any case, the apparent K_m may be raised above the in vivo levels of substrate, and the reaction ceases or decreases in velocity.

In addition to feedback within the pathway, feedback on other pathways also occurs. This is referred to as **cross-regulation.** In cross-regulation a product of one pathway serves as an inhibitor or activator of an enzyme occurring early in another pathway as depicted in Figure 4.63. A good example, which will be considered in detail in Chapter 13, is the cross-regulation of the production of the four deoxyribonucleotides for DNA synthesis so that approximately equal amounts of each are produced.

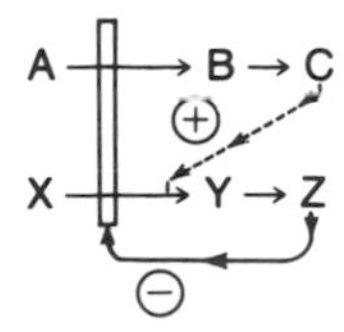

FIGURE 4.63
Model of feedback inhibition and cross-regulation.
The open bar indicates inhibition, and the broken line indicates activation. The product Z cross-regulates the production of C by its inhibitory effect on the enzyme responsible for the conversion of A to B in the A → B pathway. C in turn cross-regulates the production of Z. The product Z inhibits its own formation by feedback inhibition of the conversion of X to Y.

In addition to the inhibition and activation mechanisms just discussed, another very important mechanism of control of enzyme activity is that of reversible covalent modification, which will now be discussed in some detail.

Some Enzymes Are Regulated by Covalent Modification

The first example of this mechanism of regulation was glycogen phosphorylase, in which the interconvertible a (active) and b (inactive) forms were recognized to be phosphorylated and dephosphorylated proteins, respectively.

Other examples of reversible covalent modification include acetylation–deacetylation, adenylylation–deadenylylation, uridylylation–deuridylylation, and methylation–demethylation.

The phosphorylation–dephosphorylation scheme is most common and will be considered in detail. There are four different modes of phosphorylation based upon the cofactor requirement. These are the cAMP dependent, the non-cAMP-dependent, the calcium-dependent, and the double-stranded RNA-dependent protein kinases. The cAMP-dependent phosphorylation can be considered as characteristic of the other types of covalent modification. Details of the mechanisms will be covered in Chapter 7.

cAMP-Dependent Phosphorylation

The phosphorylation of an enzyme occurs as the end result of a cascade of reactions initiated by the binding of a hormone such as epinephrine to a specific extracellular membrane receptor on the target cell. Such binding activates the adenylate cyclase on the intracellular surface of the plasma membrane through an induced conformational change. **Adenylate cyclase** catalyzes the cyclization of ATP to cAMP (Figure 4.64), which then allosterically activates cAMP-dependent protein kinases.

The cAMP activates the protein kinase by combining with its regulatory (R) subunits, resulting in the release of active catalytic subunits (C) as outlined in Figure 4.65. Cyclic AMP is referred to as the "second messenger," since it transmits the signal from the hormone, the first messenger, to the intracellular protein kinases, which then effect the intracellular response to the hormone. This system demonstrates another mode of

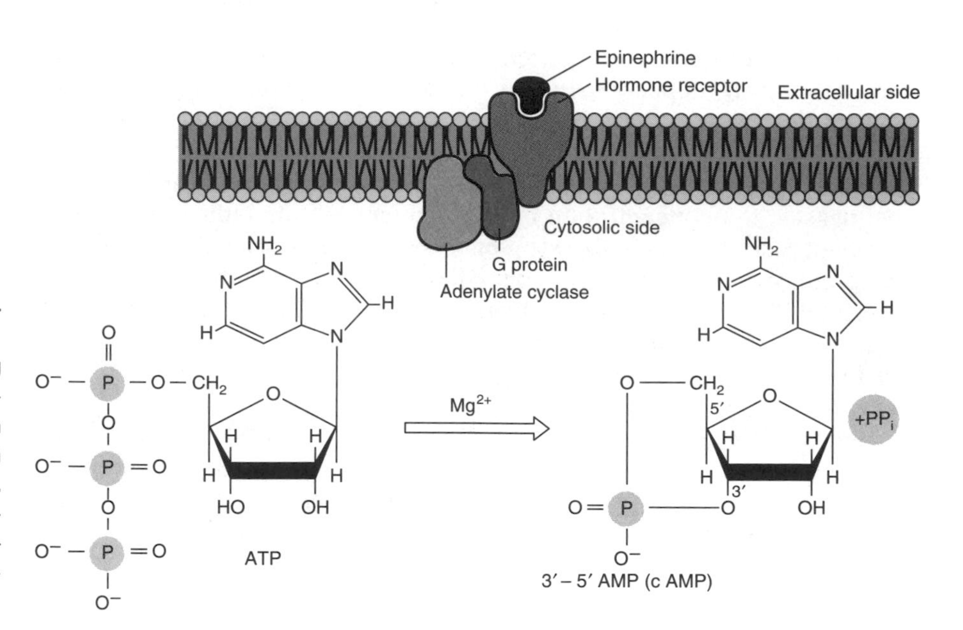

FIGURE 4.64
Epinephrine stimulation of adenylate cyclase.
Activation of the adenylate cyclase, residing on the intracellular side of the cell membrane, by extracellular epinephrine through the intermediate "G" proteins results in an increase in the intracellular levels of cAMP. The cAMP diffuses throughout the intracellular space and serves as a second messenger in activating the phosphorylation of various protein kinases.

enzyme regulation, that is, activation of catalytic subunits by binding of intracellular ligands to regulatory subunits of a multisubunit enzyme. The hormone does not enter the cell to produce its ultimate metabolic effect.

The activated cAMP-dependent protein kinases can phosphorylate various other inactive protein kinases and activate them. The latter are kinases that phosphorylate specific enzymes. This is the point of specificity in the hormone stimulation of adenylate cyclase. The phosphorylated form of a particular enzyme resulting from the protein kinase reaction may be either activated or inhibited (or rendered less active).

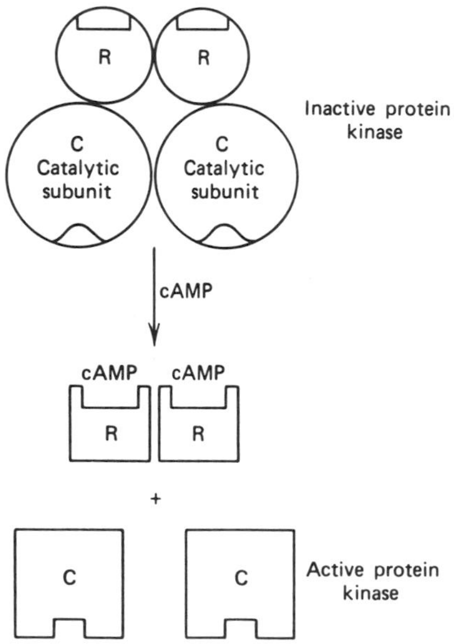

R R
C Catalytic subunit C Catalytic subunit
Inactive protein kinase
cAMP
cAMP cAMP
R R
+
C C
Active protein kinase

FIGURE 4.65
cAMP activation of protein kinase.
The enzyme is tetrameric with two catalytic and two regulatory subunits. Binding of cAMP to the regulatory subunits (R) results in dissociation of the complex and activation of the catalytic subunits (C). The active protein kinase is rather nonspecific with respect to the protein substrate.

4.11 AMPLIFICATION OF REGULATORY SIGNALS

In biological systems many of the signal molecules, such as hormones, that function in interorgan communication are very low in concentration but have tremendous effects on the target organ. The signal generated by the binding of very small amounts of hormone is multiplied many-fold inside the cell through a process of biological amplification. The mechanism of this **amplification** involves a cascade of reactions whereby the activation of the initial enzyme in the cascade activates a second proenzyme, and it in turn activates a third proenzyme, and so on. Since catalytic proteins are involved at each step in the cascade, the initial signal can be increased many times in terms of amounts of the final product generated.

The Phosphorylase Cascade Amplifies the Hormonal Signal

A beautiful example of amplification is the epinephrine-stimulated phosphorolysis of glycogen. The end product of the cascade is a phosphorylated form of phosphorylase that is the active enzyme involved in converting glycogen to glucose 1-phosphate (see p. 346 for details). Levels of epinephrine of the order of 10^{-10} mol/g of muscle will stimulate the formation of 25×10^{-6} mol of glucose 1-phosphate/min/g of muscle. This is an amplification factor of 250,000. If we look at the initial stages of the amplification, we find that the epinephrine raises the steady-state cAMP concentration only three- to fivefold. This may at first appear to be a rather small amplification, but we must remember that the hormone is binding to only a limited number of receptor sites on the cell surface, which only affects the activation of an equivalent number of adenylate cyclase molecules. Theoretical considerations have shown that the logarithm of the concentration of effector needed to activate 50% of the ultimate target enzymes is inversely proportional to the number of steps in the cascade. Since there are three steps between cAMP and the activation of phosphorylase, it has been calculated that only a 1% increase in the steady-state level of cAMP would produce a 50% activation of phosphorylase; consequently, three- to fivefold increase in cAMP is theoretically sufficient to activate all the phosphorylase.

In general the factors that limit the extent of amplification in a system involving a catalytic enzyme cascade are the relative amounts and the turnover numbers of each enzyme in the cascade. If each step in the cascade is bicyclic, phosphorylation–dephosphorylation, for example, then the ratio of the forward rate (activation) to the backward rate (deactivation) is important.

It is interesting that the concentrations of the enzymes in the phosphorylase cascade are of the ratio expected for an amplification system. In fast twitch muscles, the molar ratios of cAMP-dependent protein kinase, phosphorylase kinase, and phosphorylase are 1 : 10 : 240. In slow twitch muscles the ratios are 1 : 0.15 : 25. Thus the absolute concentration as well as the ratio of the enzymes in a cascade are altered to provide the chemical response appropriate to the functioning of a particular tissue.

BIBLIOGRAPHY

Blackburn, G. M., Kang, A. S., Kingsbury, G. A., and Burton, D. R. Review of Abzymes. *Biochem. J.* 262:381, 1989.

Boyer, P. D. (Ed.) *The enzymes,* 3rd ed. (28 vols.). New York: Academic, 1970–1987.

Dixon, M. and Webb, E. C. *Enzymes* (3rd ed.) New York: Academic Press, 1979.

Fersht, A. *Enzyme structure and mechanism* (2nd ed.) New York: Freeman, 1985.

Knowles, J. R. and Alberty, W. J. Evolution of enzyme function and the development of catalytic efficiency. *Biochemistry* 15:5631, 1976.

Meister, A. (Ed.) *Advances in enzymology*. New York: Wiley–Interscience, issued annually.

Neurath, H. Evolution of proteolytic enzymes. Science 234:350, 1984.

Richardson, C. C. (Ed.) *Annual reviews of biochemistry*. Palo Alto: Annual Reviews, Inc., issued annually.

Segel, I. H., *Enzyme kinetics*. New York: Wiley, 1975.

Westley, J. *Enzyme catalysis*. New York: Harper and Row, 1969.

Weber, G. (Ed.). *Advances in enzyme regulation,* Elmsford, N.Y.: Pergamon, issued annually.

QUESTIONS

J. BAGGOTT AND C. N. ANGSTADT

*Q*uestion *T*ypes are described on page xxiii.

1. (QT1) The reaction

$$HO-\underset{CH_2-COOH}{\overset{CH_2-COOH}{\underset{|}{\overset{|}{C}}}}-COOH + CoA\text{-}SH \rightleftharpoons$$

$$HOOC-CH_2-\overset{O}{\overset{\|}{C}}-COOH + CH_3-\overset{O}{\overset{\|}{C}}-S-CoA + H_2O$$

is catalyzed by:
A. an oxidoreductase.
B. a transferase.
C. a hydrolase.
D. a lyase.
E. a ligase.

2. (QT2) Although enzymic catalysis is reversible, a given reaction may appear irreversible:
1. if the products are thermodynamically far more stable than the reactants.
2. under initial velocity conditions.
3. if a product is rapidly removed from the system.
4. at high enzyme concentrations.

3. (QT2) Metal ions may:
1. serve as Lewis acids in enzymes.
2. participate in oxidation–reduction processes.
3. stabilize the active conformation of an enzyme.
4. form chelates with the substrate, with the chelate being the true substrate.

A. Metal cofactor
B. Nonmetal cofactor
C. Both
D. Neither

4. (QT4) Can serve as a $1e^-$ acceptor–donor.
5. (QT4) Enzyme can be generally expected to exhibit activity in the absence of the cofactor.

6. (QT2) Which of the following pairs can be distinguished on the basis of Michaelis–Menten inhibition patterns?
   1. Competitive–noncompetitive
   2. Competitive–irreversible
   3. Noncompetitive–uncompetitive
   4. Noncompetitive–irreversible

A. Michaelis–Menten kinetics
B. Allosteric kinetics with cooperative interaction
C. Both
D. Neither

7. (QT4) Lineweaver–Burk plot is useful for determining K_m and V_{max}.
8. (QT4) Inhibitor may increase the apparent K_m.
9. (QT4) Small changes in [S] produce the largest changes in v when [S] = K_m.

10. (QT2) Drugs that act as enzyme inhibitors:
    1. may function as competitive inhibitors.
    2. are clinically useful only when directed against an enzyme unique to a cell that is to be killed.
    3. may serve as irreversible inhibitors.
    4. must be harmless to the patient.
11. (QT2) Enzymes may be specific with respect to:
    1. chemical identity of the substrate.
    2. optical activity of product formed from a symmetrical substrate.
    3. type of reaction catalyzed.
    4. which of a pair of optical isomers will react.
12. (QT2) Which of the following necessarily result(s) in formation of an enzyme–substrate intermediate?
    1. Substrate strain
    2. Acid–base catalysis
    3. Entropy effects
    4. Covalent catalysis

13. (QT1) An enzyme with histidyl residues that participate in both general acid and general base catalysis would be most likely to have a pH-activity profile resembling:
    A. Curve A
    B. Curve B
    C. Curve C
    D. Curve D
    E. Curve E

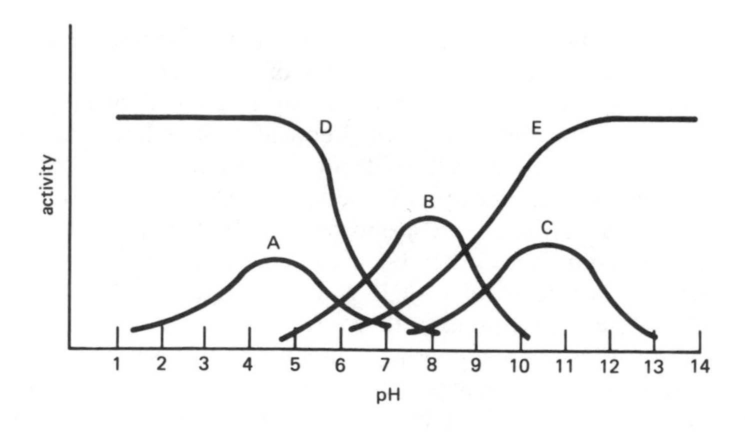

14. (QT1)

$$\begin{array}{c} \text{Cpd 3} \\ \Updownarrow \text{B} \\ \text{Cpd 1} \xrightarrow{A} \text{Cpd 2} \overset{C}{\rightleftharpoons} \text{Cpd 4} \xrightarrow{D} \text{Cpd 5} \overset{E}{\rightleftharpoons} \text{Cpd 6} \end{array}$$

In the reaction sequence above, the best point for controlling production of Compound 6 is reaction:
A. A
B. B
C. C
D. D
E. E

15. (QT2) If the plasma activity of an intracellular enzyme is abnormally high:
    1. the rate of removal of the enzyme from plasma may be depressed.
    2. tissue damage may have occurred.
    3. the enzyme may have been activated.
    4. determination of the isozyme distribution may yield useful information.
16. (QT2) Features of enzyme cascades include:
    1. circulating hormone binds to the cell membrane.
    2. participating enzymes may be activated, inactivated, or reactivated.
    3. different properties in different tissues.
    4. signal amplification.
17. (QT2) Types of physiological regulation of enzyme activity include:
    1. covalent modification.
    2. changes in rate of synthesis of the enzyme.
    3. allosteric activation.
    4. competitive inhibition.

ANSWERS

1. D This is an unusually complicated lyase reaction, since secondary reactions are involved. It is a lyase because it removes a group (the acetyl group) with formation of a double bond (the C=O bond of the C_4 product, oxalacetate). The common name of this enzyme is citrate synthase (p. 140).

2. A 1, 2, and 3 true. 1: Such a system is theoretically reversible, but it would be difficult to reverse it in practice (p. 143). 2: At the beginning of a reaction (initial velocity conditions), product concentration is so low that the reverse reaction is insignificant. 3: If a product is removed, it is no longer available for the

reverse reaction (p. 143). 4: Is false because enzymes merely catalyze reactions; they do not alter the equilibrium, no matter what their concentration (p. 136).

3. E All of the statements are true (p. 153).
4. C The most familiar example of a metal cofactor serving as a $1e^-$ acceptor–donor is any of the cytochromes (p. 157). Other examples exist. Flavins may also participate in $1e^-$ transfers (p. 152).
5. D Not all enzymes have cofactors, but when they do, the cofactor is generally essential for activity (p. 136).
6. A 1, 2, and 3 true. Competitive inhibitors increase the apparent K_m, noncompetitive and irreversible inhibitors both decrease V_{max}, and uncompetitive inhibitors decrease the apparent K_m (p. 158, Figures 4.27 and 4.28).
7. A The Lineweaver–Burk plot is a rearrangement of the Michaelis–Menten equation into a form that makes graphical evaluation of K_m and V_{max} easier. It is not applicable to systems that cannot be described by the Michaels–Menten equation (p. 147).
8. C Competitive inhibitors do this in systems described by Michaelis–Menten kinetics (p. 158, Figure 4.27), and negative allosteric effectors affect *K* class allosteric enzymes in this way (p. 162, p. 164, Figure 14.36).
9. B For Michaelis–Menten enzymes, the v versus [S] curve is steepest at the lowest [S]. For allosteric enzymes, the steepest part of the curve is in the neighborhood of the [S] that corresponds to $v = \frac{1}{2} V_{max}$, that is, K_m (p. 165).
10. B 1 and 3 true. Drugs may serve as competitive inhibitors, such as sulfanilamide (p. 159), or as irreversible inhibitors, such as fluorouracil (p. 161). Pathways unique to pathogenic bacteria, viruses, and so on, are rare, so drugs are often developed that are merely *less* harmful to the host than the target cell (because of differences in cell permeability, metabolic rate, etc.) (p. 159).
11. E 1, 2, 3, and 4 true. Enzymes are specific for the substrate and the type of reaction (p. 167). The asymmetry of the binding site generally permits only one of a pair of optical isomers to react, and only one optical isomer is generated when a symmetric substrate yields an asymmetric product (p. 170).
12. D Only 4 true. All enzyme-catalyzed reactions involve an enzyme–substrate complex. There is always at least one transition state involved, but only in covalent catalysis is a covalent bond between enzyme and a portion of the substrate involved (p. 176).
13. B A group must be in the correct ionization state to act catalytically. For a histidyl group to serve as a general acid and a general base (as it does in chymotrypsin), the pH must be compatable with *both* ionization states of histidine. Since the p*K* of the histidyl side chain is about 6.8, the maximum activity is likely to be near that pH. Chymotrypsin's pH optimum is in the 7–9 range (p. 176).
14. D Control of reaction A would control production of Cpds 3 and 6. Reaction B is not on the direct route. Reaction C is freely reversible, so it does not need to be controlled. Reaction D is irreversible; if it were not controlled, Cpd 5 might build up to toxic levels (p. 187).
15. E 1, 2, 3, and 4 true. Intracellular enzymes may appear in abnormal amounts when tissues are damaged. Different tissues have characteristic distributions of isozymes. Choices 1 and 3 are theoretically possible, but are less common (p. 180).
16. E 1, 2, 3, and 4 are true. 1: Epinephrine binds to extracellular receptors to activate adenylate cyclase (p. 189). 2: The forms of glycogen phosphorylase are interconvertible (p. 189). 3: Enzymes of the phosphorylase cascade are present in different ratios in different types of muscle (p. 191). 4: Amplification of a small signal by activating a sequence of enzymes (catalysts) is the purpose of a cascade (p. 190).
17. E 1, 2, 3, and 4 true. 1: Covalent modification includes zymogen activation and phospho–dephospho protein conversions (p. 189). 2: Enzyme levels may be controlled (p. 187). 3: Allosteric activation is common (p. 189). 4: End products of a reaction or reaction sequence may inhibit their own formation by competitive inhibition (p. 189).

5

Biological Membranes: Structure and Membrane Transport

THOMAS M. DEVLIN

5.1 OVERVIEW 196
5.2 CHEMICAL COMPOSITION OF MEMBRANES 197
Lipids Are a Major Component of Membranes 197
Phosphoglycerides Are the Most Abundant Lipid of Membranes 197
Sphingolipids Are Also Present in Membranes 201
Most Membranes Also Contain Cholesterol 204
The Lipid Composition Varies in Different Membranes 204
Membrane Proteins Are Classified Based on Their Ease of Removal 205
Carbohydrates of Membranes Are Present as Glycoproteins or Glycolipids 205
5.3 MICELLES AND LIPOSOMES 206
Lipids Form Vesicular Structures 206
Liposomes Have a Membrane Structure Similar to Biological Membranes 207
5.4 STRUCTURE OF BIOLOGICAL MEMBRANES 208
The Fluid Mosaic Model of Biological Membranes 208
Integral Membrane Proteins Are Immersed in the Lipid Bilayer 208
Peripheral Membrane Proteins Have Various Modes of Attachment 211
The Human Erythrocyte Is Ideal for Studying Membrane Structure 212
Lipids Are Distributed in an Asymmetric Manner in Membranes 212
Proteins and Lipids Diffuse in Membranes 213
5.5 MOVEMENT OF MOLECULES THROUGH MEMBRANES 215
Some Molecules Can Diffuse Through Membranes 215
Movement of Molecules Across Membranes Can Be Facilitated 216
Membrane Transport Systems Have Common Characteristics 217
There Are Four Common Steps in the Transport of Solute Molecules 218
Energetics of Membrane Transport Systems 219
5.6 CHANNELS AND PORES 220
Channels and Pores in Membranes Function Differently 220
Opening and Closing of Channels Are Controlled 221
Examples of Pores Are Gap Junctions and Nuclear Pores 222
5.7 PASSIVE MEDIATED TRANSPORT SYSTEMS 223
In Some Membranes Glucose Is Transported Down Its Chemical Gradient 224
Cl^- and HCO_3^- Are Transported by an Antiport Mechanism 224
Mitochondria Contain a Number of Passive Transport Systems 224
5.8 ACTIVE MEDIATED TRANSPORT SYSTEMS 225
Translocation of Na^+ and K^+ Is a Primary Active Transport System 226
All Plasma Membranes Contain a Na^+-K^+-Activated ATPase 226

Erythrocyte Ghosts Are Used to Study the Na^+-K^+ Transporter 227
Ca^{2+} Translocation Is Another Example of a Primary Active Transport System 228
Na^+-Dependent Transport of Glucose and Amino Acids Are Secondary Active Transport Systems 229
Group Translocation Involves the Chemical Modification of the Substrate Transported 230
Summary of Transport Systems 231
5.9 IONOPHORES 231
BIBLIOGRAPHY 234
QUESTIONS AND ANNOTATED ANSWERS 235
CLINICAL CORRELATIONS
5.1 Liposomes as Carriers of Drugs and Enzymes 208
5.2 Abnormalities of Cell Membrane Fluidity in Disease States 214
5.3 Diseases Due to Loss of Membrane Transport Systems 232

5.1 OVERVIEW

All biological membranes, whether from eucaryotic or procaryotic cells, have the same classes of chemical components, a similarity in structural organization, and a number of properties in common. There are major differences in the specific lipid, protein, and carbohydrate components but not in the physicochemical interaction of these molecules in the membrane. Membranes have a trilaminar appearance when viewed by electron microscopy (Figure 5.1), with two dark bands on each side of a light band. The overall width of various mammalian membranes is 7–10 nm but some membranes have significantly smaller widths. Differences are also observed in both the size and the density of membranes depending on the staining technique employed. Intracellular membranes are usually thinner than the plasma membrane. In addition, many membranes do not appear symmetrical, with the inner dense layer often thicker than the outer dense layer; as discussed below there is a chemical asymmetry of the membrane, with some components only on one or the other side. With the development of sophisticated techniques for preparation of tissue samples and staining, including negative staining and freeze fracturing, the surfaces of membranes have been viewed; at the molecular level the surfaces are not smooth but dotted with globular-shaped components protruding from the membrane, which have been identified as proteins.

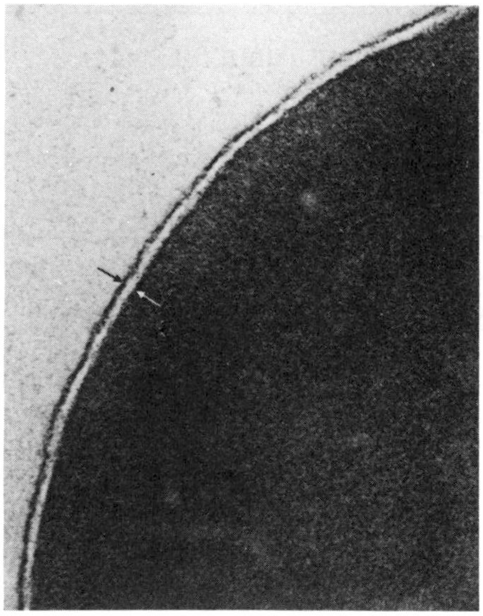

Membranes are very dynamic structures with a movement that permits the cell as well as subcellular structures to adjust their shape and to move. The chemical components of membranes, that is, lipids and protein, are ideally suited for the dynamic role of membranes. Membranes are visualized as essentially a semistructured, organized sea of lipid in a fluid state in which the various components are able to move. The lipid membrane in the fluid state is a nonaqueous compartment of the cell in which components can interact.

Cellular membranes control the composition of the space they enclose not only by their ability to exclude a variety of molecules but also because of the presence of selective transport systems permitting the movement of specific molecules from one side to the other. By controlling the translocation of substrates, cofactors, ions, and so on, from one compartment to another, membranes modulate the concentration of substances, thereby exerting an influence on metabolic pathways. The plasma membrane of eucaryotic cells also has a role in cell–cell recognition, maintenance of the shape of the cell, and in cell locomotion. The site of action of many hormones and metabolic regulators is on the plasma membrane (Chapter 20), where there are specific receptors, and the information to be imparted to the cell by the hormone or regulator is transmitted by the membrane

FIGURE 5.1
Electron micrograph of the erythrocyte plasma membrane showing the trilaminar appearance.
A clear space separates the two-electron dense lines. Electron microscopy has demonstrated that the inner dense line is frequently thicker than the outer line. Magnification about 150,000×.
Courtesy of Dr. J. D. Robertson, Duke U., Durham, NC.

component to the appropriate metabolic pathway by a series of intracellular intermediates, termed second messengers.

The discussion that follows is directed primarily to the chemistry and function of membranes of mammalian cells but the basic observations and activities described are applicable to all biological membranes.

5.2 CHEMICAL COMPOSITION OF MEMBRANES

Lipids and proteins are the two major components of all membranes but the amount of each varies greatly between different membranes (Figure 5.2). The percent of protein ranges from about 20% in the myelin sheath to over 70% in the inner membrane of the mitochondria. Intracellular membranes have a high percentage of protein because of the greater enzymatic activity of these membranes. Membranes also contain a small amount of various polysaccharides in the form of glycoprotein and glycolipid; there is no free carbohydrate in membranes.

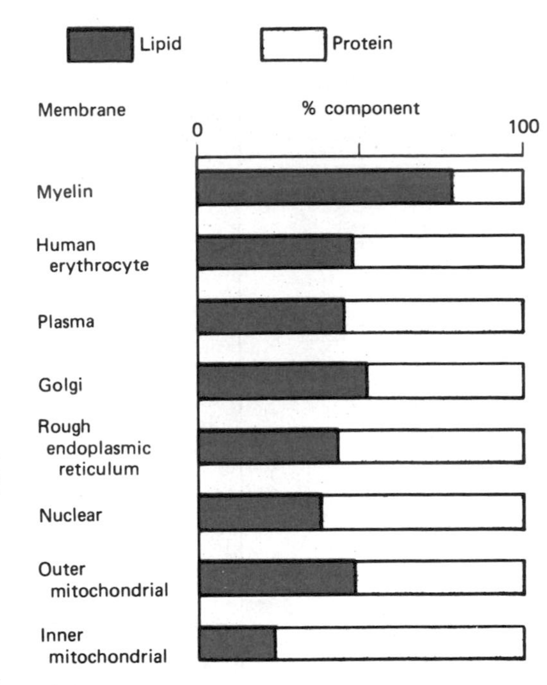

FIGURE 5.2
Representative values for the percentage of lipid and protein in various cellular membranes.
Values are for rat liver, except for the myelin and human erythrocyte plasma membrane. Values for liver from other species, including human, indicate a similar pattern.

Lipids Are a Major Component of Membranes

The three major lipid components of eucaryotic cell membranes are phosphoglycerides, sphingolipids, and cholesterol. The phosphoglycerides and sphingomyelin, a sphingolipid containing phosphate, are classified as **phospholipids**. In addition, bacteria and blue-green algae contain glycerolipids where a carbohydrate is attached directly to the glycerol. Individual cellular membranes also contain small quantities of other lipids, such as triacylglycerol and diol derivatives (see the appendix).

The percentage of each of the major classes varies significantly in different membranes and is presumably related to the specific roles of the individual membranes. This is discussed in more detail below.

Phosphoglycerides Are the Most Abundant Lipid of Membranes

Phosphoglycerides, also referred to as glycerophospholipids, have a glycerol molecule as the basic component to which phosphoric acid is esterified at the α carbon (Figure 5.3) and two long-chain fatty acids are esterified at the remaining carbon atoms (Figure 5.4). Even though glycerol does not contain an asymmetric carbon, the α-carbon atoms are not stereochemically identical. Esterification of a phosphate to an α carbon makes the molecule asymmetric. The naturally occurring phosphoglycerides are designated by the stereospecific numbering system (*sn*) as presented in Figure 5.3 and also discussed in Section 10.2.

Phosphatidic acid, 1,2-diacylglycerol 3-phosphate, is the parent compound of a series of phosphoglycerides, where different hydroxyl-containing compounds are also esterified to the phosphate groups. The major compounds attached by the phosphodiester bridge to glycerol are choline, ethanolamine, serine, glycerol, and inositol. These structures are presented in Figure 5.5. *Phosphatidylethanolamine* (also called ethanolamine phosphoglyceride and the trivial name *cephalin*) and *phosphatidylcholine* (choline phosphoglyceride or *lecithin*) are the two most common phosphoglycerides in membranes (Figure 5.6). *Phosphatidylglycerol phosphoglyceride* (Figure 5.7) (or diphosphatidylglycerol or *cardiolipin*) contains two phosphatidic acids linked by a glycerol and is found nearly exclusively in the inner membrane of mitochondria and in bacterial membranes.

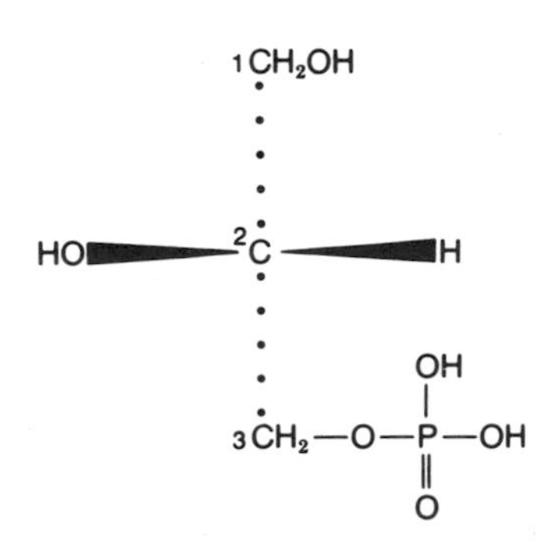

FIGURE 5.3
Stereochemical configuration of L-glycerol 3-phosphate (*sn*-glycerol 3-phosphate).
The H and OH attached to C-2 are above and C-1 and C-3 are below the plane of the page.

X–O–P(=O)(–O⁻)–O–$^{1}CH_2$–^{2}CH–$^{3}CH_2$ (polar head), with fatty acyl chains C=O–$(CH_2)_{14}$–CH_3 and C=O–$(CH_2)_7$–CH=CH–$(CH_2)_7$–CH_3 (hydrophobic tails)

Polar head

Hydrophobic tails

FIGURE 5.4
Structure of phosphoglyceride.
Long-chain fatty acids are esterified at C-1 and C-2 of the L-glycerol 3-phosphate. X can be a H (phosphatidic acid) or one of several alcohols presented in Figure 5.5.

Choline	$HO-CH_2-CH_2-N^+(CH_3)_3$
Ethanolamine	$HO-CH_2-CH_2-\overset{+}{N}H_3$
Serine	$HO-CH_2-CH(^+NH)-C(=O)-OH$
Glycerol	$HO-CH_2-CH(OH)-CH_2-OH$
Inositol	(hexahydroxycyclohexane ring: HO, OH, H, OH, H, H, OH, H, H, HO, H, OH)

FIGURE 5.5
Structures of the major alcohols esterified to phosphatidic acid to form the phosphoglycerides.

The hexahydroxy alcohol *inositol* is esterified to the phosphate in *phosphatidylinositol* (Figure 5.8); this compound should be differentiated from the class of lipids termed glycosylacylglycerols, which contain a sugar in glycosidic linkage with the 3-hydroxyl group of a diacylglycerol. 4–Phospho- and 4,5-bisphosphoinositol phosphoglycerides (Figure 5.8) are found in the plasma membrane and the latter is the source of two second messengers and the amplification of hormonal signals across the membrane (see p. 874).

Phosphoglycerides contain two fatty acyl groups esterified to carbon atoms 1 and 2 of the glycerol; some of the major fatty acids found in phosphoglycerides are presented in Table 5.1. A saturated fatty acid is

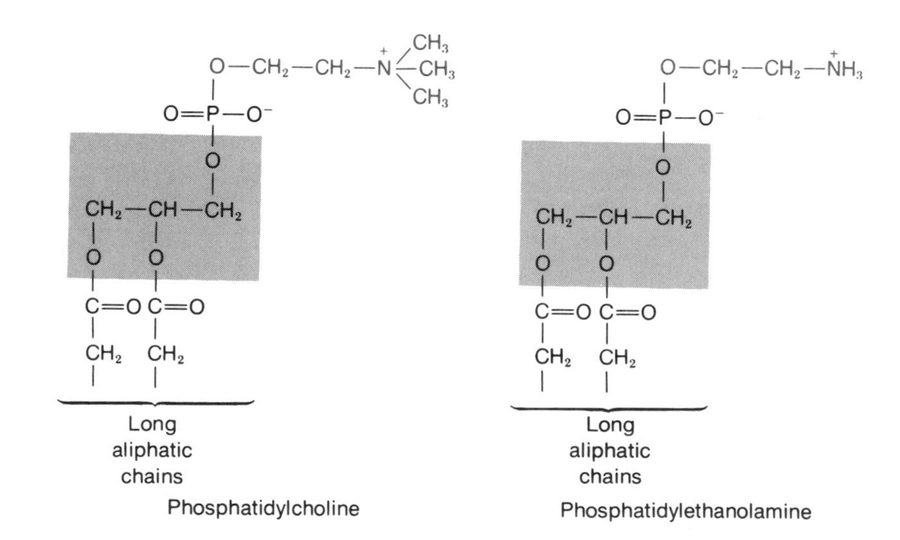

FIGURE 5.6
Structures of the two most common phosphoglycerides, phosphatidylcholine and phosphatidylethanolamine.

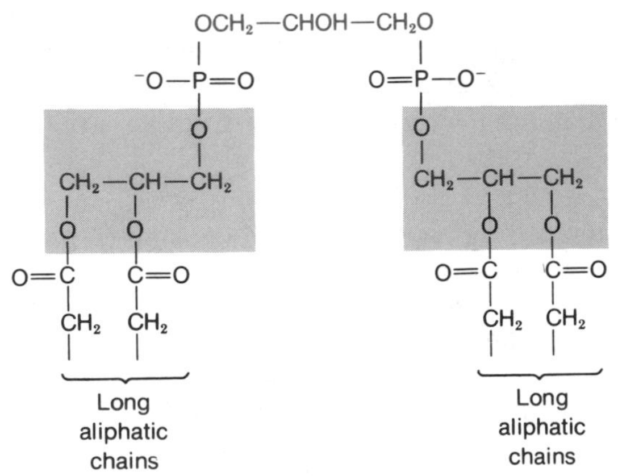

OCH$_2$—CHOH—CH$_2$O
$^-$O—P=O O=P—O$^-$
O O
CH$_2$—CH—CH$_2$ CH$_2$—CH—CH$_2$
O O O O
O=C C=O O=C C=O
CH$_2$ CH$_2$ CH$_2$ CH$_2$
Long aliphatic chains Long aliphatic chains

FIGURE 5.7
Phosphatidylglycerol phosphoglyceride (cardiolipin).

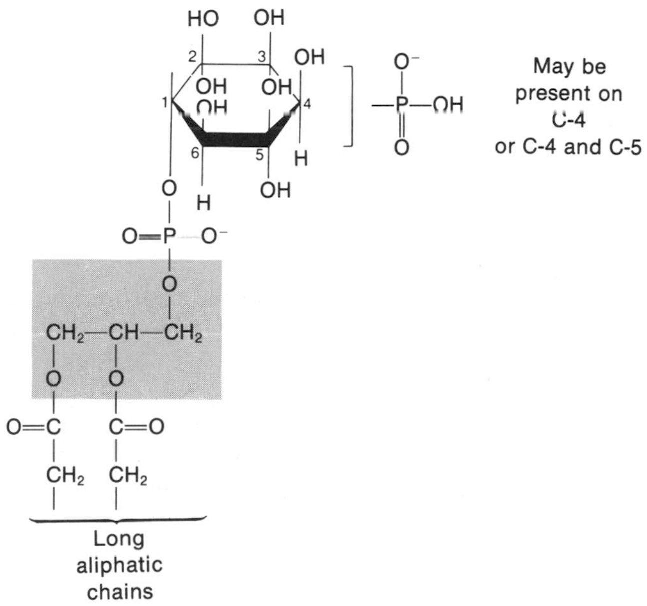

HO OH OH OH OH OH OH H H OH
O$^-$ —P—OH O
May be present on C-4 or C-4 and C-5
O O=P—O$^-$ O
CH$_2$—CH—CH$_2$
O O
O=C C=O
CH$_2$ CH$_2$
Long aliphatic chains

FIGURE 5.8
Phosphatidylinositol.
Phosphate groups are also found on C-4 or C-4 and C-5 of the inositol. The additional phosphate groups increase the charge on the polar head of this phosphoglyceride.

TABLE 5.1 Major Fatty Acids in Phosphoglycerides

Common name	Systematic name	Structural formula
Myristic acid	*n*-Tetradecanoic	$CH_3—(CH_2)_{12}—COOH$
Palmitic acid	*n*-Hexadecanoic	$CH_3—(CH_2)_{14}—COOH$
Palmitoleic acid	*cis*-9-Hexadecenoic	$CH_3—(CH_2)_5—CH{=}CH—(CH_2)_7—COOH$
Stearic acid	*n*-Octadecanoic	$CH_3—(CH_2)_{16}—COOH$
Oleic acid	*cis*-9-Octadecenoic acid	$CH_3—(CH_2)_7—CH{=}CH—(CH_2)_7—COOH$
Linoleic acid	*cis-cis*-9,12-Octadecadienoic	$CH_3—(CH_2)_3—(CH_2—CH{=}CH)_2—(CH_2)_7—COOH$
Linolenic acid	*cis,cis,cis*-9,12,15-Octadecatrienoic	$CH_3—(CH_2—CH{=}CH)_3—(CH_2)_7—COOH$
Arachidonic acid	*cis,cis,cis,cis*-5,8,11,14-Icosatetraenoic	$CH_3—(CH_2)_3—(CH_2—CH{=}CH)_4—(CH_2)_3—COOH$

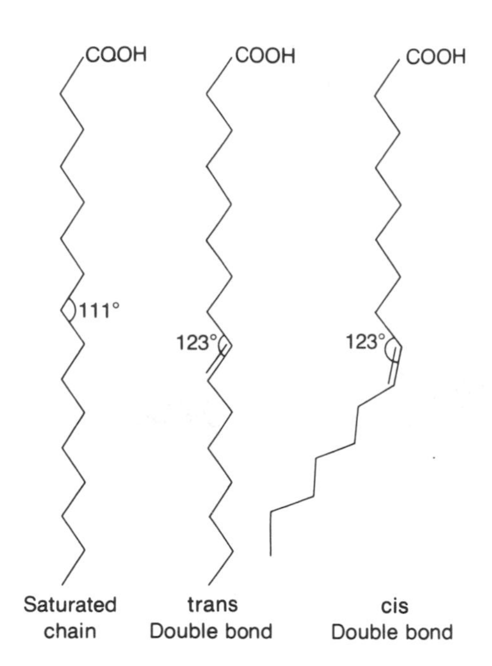

FIGURE 5.9
Conformation of fatty acyl groups in phospholipids.
The saturated and unsaturated fatty acids with trans double bonds are straight chains in their minimum energy conformation, whereas a chain with a cis double bond has a bend. The trans double bond is rare in naturally occurring fatty acids.

usually found on C-1 of the glycerol and an unsaturated fatty acid on C-2. The nomenclature for the different phosphoglycerides does not specify a specific compound because of the variety of possible fatty acid substitutions. Phosphatidylcholine usually contains palmitic or stearic in the *sn*-1 position and a C_{18} unsaturated fatty acid, oleic, linoleic, or linolenic, on the *sn*-2-carbon. Phosphatidylethanolamine also contains palmitic or oleic on *sn*-1 but one of the longer chain polyunsaturated fatty acids, that is, arachidonic, on the *sn*-2 position.

A saturated fatty acid is a straight chain, as is a fatty acid with an unsaturation in the trans position. The presence of a cis double bond, however, creates a kink in the hydrocarbon chain (Figure 5.9). A straight-chain diagram, as shown in Figures 5.4 and 5.9, does not adequately represent the chemical configuration of a long-chain fatty acid. Actually, there is a high degree of coiling of the hydrocarbon chain in a phosphoglyceride that is disrupted by the presence of a double bond. As described in Section 5.3, the presence of unsaturated fatty acids has a marked effect on the physicochemical state of the membrane.

Another group of phosphoglycerides are the **glycerol ether phospholipids** in which a long aliphatic chain is attached in ether linkage to the glycerol at the *sn*-1 position as presented in Figure 5.10. The ether phospholipids can contain an alkyl group (alkyl acylglycerophospholipid) or an α,β-unsaturated ether, termed a **plasmalogen.** The latter groups are more prevalent in membranes; plasmalogens containing ethanolamine (ethanolamine plasmalogen) and choline (choline plasmalogen) esterified to the phosphate are very abundant in nervous tissue and heart but not in liver. In human hearts more than 50% of the ethanolamine phosphoglycerides are plasmalogens.

The phosphoglycerides contain both a polar end, referred to as the head group, due to the charged phosphate and the substitutions on the phosphate, and a nonpolar tail due to the hydrophobic hydrocarbon chains of the fatty acyl groups. These polar lipids are **amphipathic,** that is, they contain both polar and nonpolar groups. The polar groups are charged at pH 7.0 with a negative charge due to the ionization of the phosphate group (pK ~2) and the charges from the groups esterified to the phosphate (Table 5.2). Choline and ethanolamine phosphoglycerides are zwitterions at pH 7.0, with both a negative charge from the phosphate and a positive charge on the nitrogen. Phosphatidylserine has two negative charges, one on the phosphate and one on the carboxyl group of serine, and a positive charge on the α-amino group of serine, with a net negative charge of 1 at pH 7.0. In contrast, the phosphoglycerides containing inositol and glycerol have only a single negative charge on the phosphate; the 4-phospho- and 4,5-bisphosphoinositol derivatives are very polar compounds with additional negative charges on the phosphate groups.

TABLE 5.2 Predominant Charge on Phosphoglycerides and Sphingomyelin at pH 7.0

Lipid	Phosphate group	Base	Net charge
Phosphatidylcholine	−1	+1	0
Phosphatidylethanolamine	−1	+1	0
Phosphatidylserine	−1	+1,−1	−1
Phosphatidylglycerol	−1	0	−1
Diphosphatidylglycerol (cardiolipin)	−2	0	−2
Phosphatidylinositol	−1	0	−1
Sphingomyelin	−1	+1	0

Every tissue and respective cellular membrane has a distinctive composition of phosphoglycerides. Not only are there differences in the classes of phosphoglycerides, but there are definite patterns in the fatty acid composition of the individual phosphoglycerides between tissues. There appears to be some degree of specificity for particular fatty acids in the individual tissues. There is a greater variability in the fatty acyl groups of different tissues in a single species than in the fatty acid composition of the same tissue in a variety of species. In addition, the fatty acid content of the phosphoglycerides can vary, depending on the physiological or pathophysiological state of the tissue.

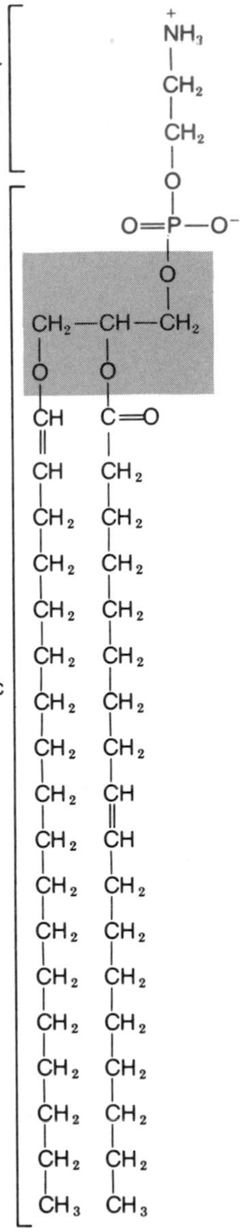

Polar head

Hydrophobic tails

FIGURE 5.10
Ethanolamine plasmalogen.
Note the ether linkage of the aliphatic chain on C-1 of glycerol.

Sphingolipids Are Also Present in Membranes

The amino alcohols sphingosine (D-4-sphingenine) and dihydrosphingosine (Figure 5.11) serve as the basis for another series of membrane lipids, the **sphingolipids.** On the amino group of sphingosine, a saturated or unsaturated long-chain fatty acyl group is present in amide linkage. This compound, termed a **ceramide** (Figure 5.12), with two nonpolar tails is similar in structure to the diacylglycerol portion of phosphoglycerides. Various substitutions are found on the hydroxyl group at position 1 of the ceramides. The sphingomyelin series has phosphorylcholine esterified to the 1-OH (Figure 5.13) and is the most abundant sphingolipid in mammalian tissues. The similarity of this structure to the choline phosphoglyceride is apparent, and they have many properties in common; note that the sphingomyelins are amphipathic compounds with a charged head group. It has been a common practice to classify the sphingomyelin series and the phosphoglycerides in one class of compounds, termed phospholipids. The sphingomyelin of myelin contains predominantly the longer

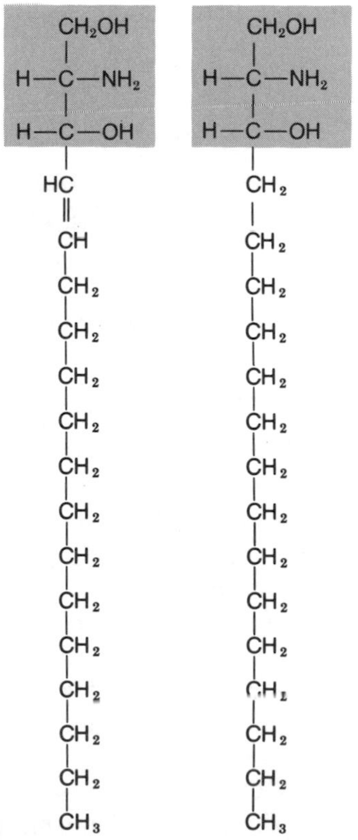

Sphingosine (D-4-sphingenine)   Dihydrosphingosine (D-sphinganine)

FIGURE 5.11
Structures of sphingosine and dihydrosphingosine.

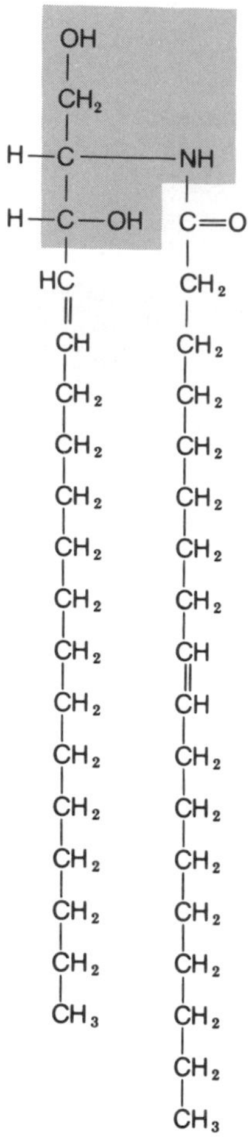

FIGURE 5.12
Structure of a ceramide.

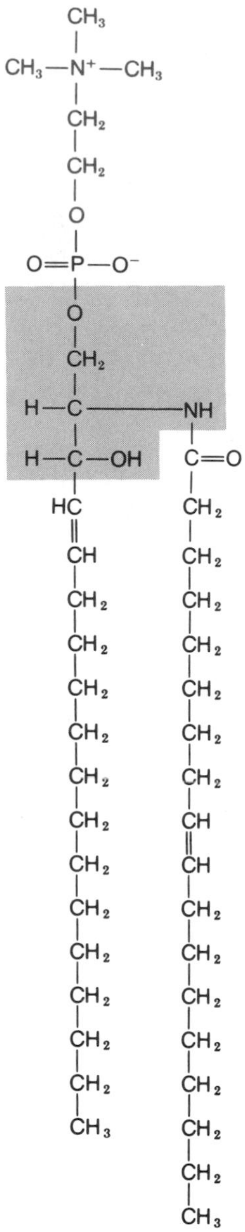

FIGURE 5.13
Structure of a choline containing sphingomyelin.

chain fatty acids, with carbon lengths of 24; as with phosphoglycerides, there is a specific fatty acid composition of the sphingomyelin, depending on the tissue.

The **glycosphingolipids** do not contain phosphate and have a sugar attached by a β-glycosidic linkage to the 1-OH group of the sphingosine in a ceramide. One subgroup is the **cerebrosides,** which contain either a glucose or galactose attached to a ceramide and are referred to as glucocerebrosides or galactocerebrosides, respectively (Figure 5.14). Cerebrosides are neutral compounds. Galactocerebrosides are found predominantly in brain and nervous tissue, whereas the small quantities of cerebrosides in nonneural tissues usually contain glucose. The specific galactocerebroside phrenosine contains a 2-OH C_{24} fatty acid. Galactocerebrosides may contain a sulfate group esterified on the 3 position of the sugar. They are called **sulfatides** (Figure 5.15). Cerebrosides and sulfatides usually contain very long-chain fatty acids with 22–26 carbon atoms.

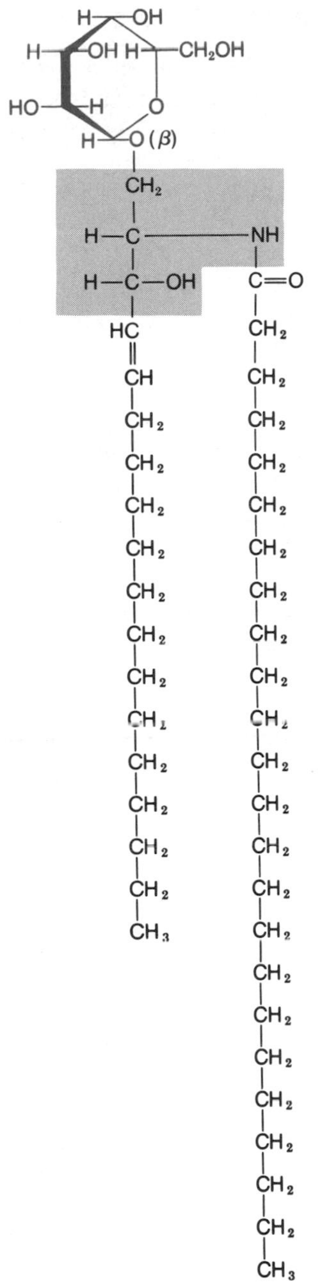

FIGURE 5.14
Structure of a galactocerebroside containing a C_{24} fatty acid.

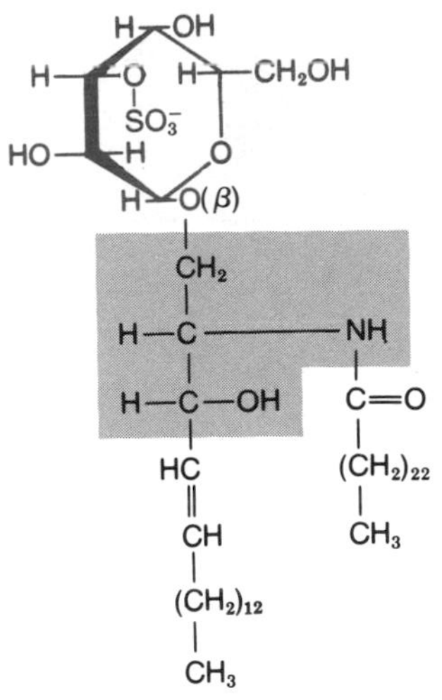

FIGURE 5.15
Structure of a sulfatide.

In place of monosaccharides, neutral glycosphingolipids often have 2 (dihexosides), 3 (trihexosides), or 4 (tetrahexosides) sugar residues attached to the 1-OH group of sphingosine. Diglucose, digalactose, *N*-acetylglucosamine, and *N*-acetyldigalactosamine are the usual sugars.

The most complex group of glycosphingolipids is the **gangliosides,** which contain oligosaccharide head groups with one or more residues of sialic acid; these are amphipathic compounds with a negative charge at pH 7.0. The gangliosides represent 5–8% of the total lipids in brain, and some 20 different types have been identified differing in the number and relative position of the hexose and sialic acid residues, which form the basis of their classification. A detailed description of the nomenclature and structures of the gangliosides is presented in Chapter 10.

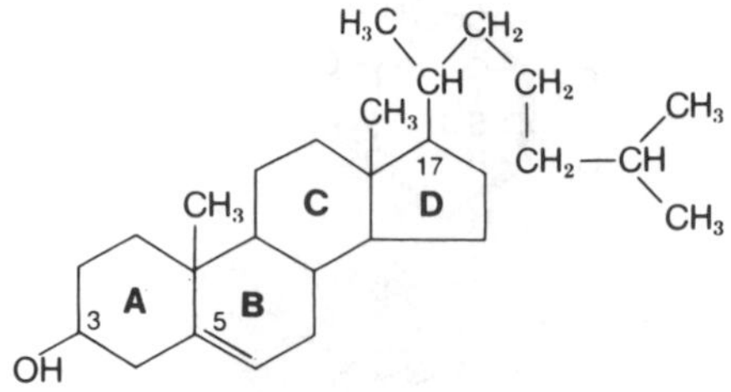

FIGURE 5.16
Structure of cholesterol.

Most Membranes Also Contain Cholesterol

The third major lipid present in membranes is **cholesterol.** As presented in Figure 5.16 cholesterol contains four fused rings, which makes it a rigid structure, a polar hydroxyl group at C-3, and an eight-member branched hydrocarbon chain attached to the D ring at position 17. Cholesterol is a compact hydrophobic molecule.

The Lipid Composition Varies in Different Membranes

There are large quantitative differences between the classes of lipids and individual lipids in various cell membranes as presented in Figure 5.17.

There is a resemblance among animal species in the lipid composition of the same intracellular membrane of cells in a specific tissue, such as liver mitochondria of rat and humans. The plasma membrane exhibits the greatest variation in percentage composition because the amount of cholesterol is affected by the nutritional state of the animal. Plasma membranes have the highest concentration of neutral lipids and sphingolipids; the myelin membranes of axons of neural tissue are rich in sphingolipids, with a high proportion of glycosphingolipids. Intracellular membranes primarily contain phosphoglycerides with little sphingolipids or cholesterol. When comparing intracellular structures, the membrane lipid composition of mitochondria, nuclei, and rough endoplasmic reticulum are similar, with the Golgi membrane being somewhere between the other intracellular membranes and the plasma membrane. As indicated previously, cardiolipin is found nearly exclusively in the inner mitochondrial membrane. The choline containing lipids, phosphatidylcholine and sphingomyelin, are predominant, with ethanolamine phosphoglyceride second. The constancy of composition of the various membranes indi-

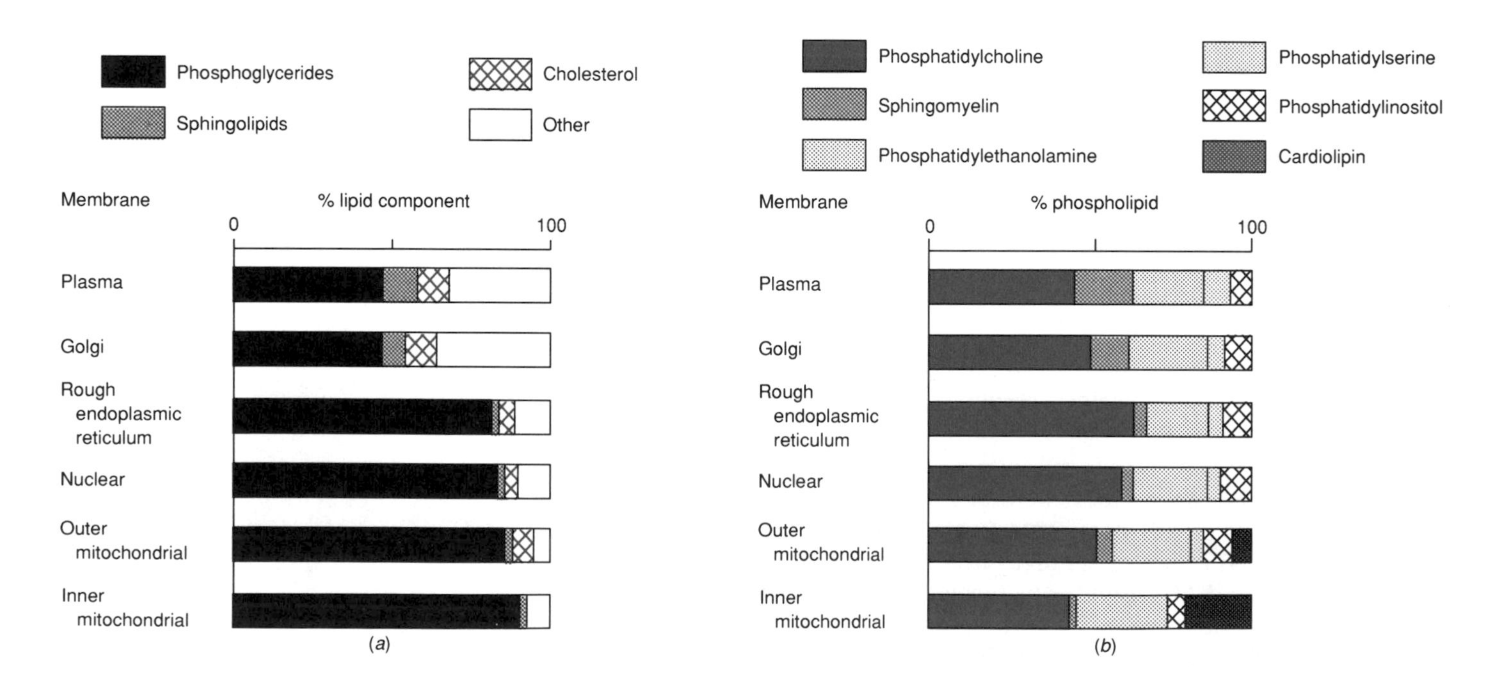

FIGURE 5.17
Lipid composition of cellular membranes isolated from rat liver.
(*a*) Amount of major lipid components as percentage of total lipid. The area labeled "Other" includes mono-, di-, and triacylglycerol, fatty acids, and cholesterol esters. (*b*) Phospholipid composition as a percentage of total phospholipid. Values from R. Harrison and G. G. Lunt, *Biological Membranes*. New York: Wiley, 1975.

cates the relationship between the lipids and the specific functions of the individual membranes.

Membrane Proteins Are Classified Based on Their Ease of Removal

Membrane proteins are classified on the basis of the ease of removal from isolated membrane fractions. **Peripheral (or extrinsic) proteins** are easily isolated by treatment of the membrane with salt solutions of low or high ionic strength, or extremes of pH, and the name is used to imply a physical location on the surface of the membrane. Peripheral proteins, many with specific enzymatic activity, are usually soluble in water and free of lipids. **Integral (or intrinsic) proteins** require rather drastic treatment, such as use of detergents or organic solvents, to be extracted from the membrane. They usually contain tightly bound lipid, which if removed leads to denaturation of the protein and loss of biological function. Removal of the integral protein leads to disruption of the membrane, whereas peripheral proteins can be removed with little or no change in the integrity of the membrane.

Of particular value in studying the chemistry and structure of integral proteins has been the use of sodium dodecyl sulfate (SDS), a detergent that dissociates the lipid–protein complex and solubilizes the protein permitting separation and analysis. The integral proteins studied have sequences of hydrophobic amino acids, which could create domains with a high degree of hydrophobicity in the tertiary structure of the protein. These hydrophobic regions of the protein interact with the hydrophobic hydrocarbons of the lipids stabilizing the protein–lipid complex.

A special class of integral proteins are the **proteolipids,** which are hydrophobic lipoproteins soluble in chloroform and methanol but insoluble in water. Proteolipids are present in many membranes but are particularly abundant in myelin, where they represent about 50% of the membrane protein component. An example is lipophilin, a major lipoprotein of brain myelin that contains over 65% hydrophobic amino acids and covalently bound fatty acids.

Another class of integral membrane proteins is the glycoproteins; plasma membrane of cells contain a number of different glycoproteins, each with its own unique carbohydrate content.

The complexity, variety, and interaction of membrane proteins with lipids are just being resolved. Many of the proteins are enzymes located within or on the cellular membranes. Membrane proteins also have a role in transmembrane movement of molecules and in many cells, such as neurons and erythrocytes, specific proteins have a structural role to maintain the integrity of the cell. Thus individual membrane proteins can have a catalytic, transport, structural, or recognition role, and it is not surprising to find a high protein content in a membrane being correlated with complexity and variety of function of the membrane.

Carbohydrates of Membranes Are Present as Glycoproteins or Glycolipids

Carbohydrates present in membranes are exclusively in the form of oligosaccharides covalently attached to proteins to form glycoproteins and to a lesser amount to lipids to form glycolipids. The sugars found in glycoproteins and glycolipids include glucose, galactose, mannose, fucose, *N*-acetylgalactosamine, *N*-acetylglucosamine, and sialic acid (see Figure 5.18 and the appendix for structures). Details of the structures of glycoproteins and glycolipids are given on pages 373 and 452, respectively. The

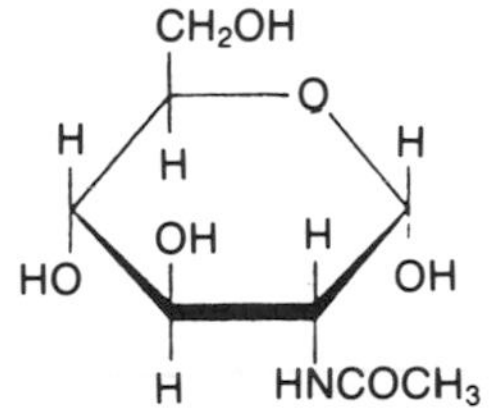

N-Acetyl-α-D-glucosamine

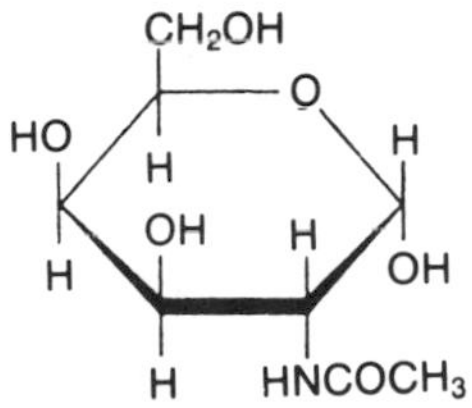

N-Acetyl-α-D-galactosamine

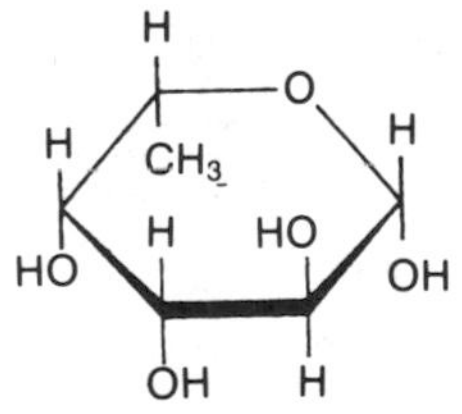

α-L-Fucose

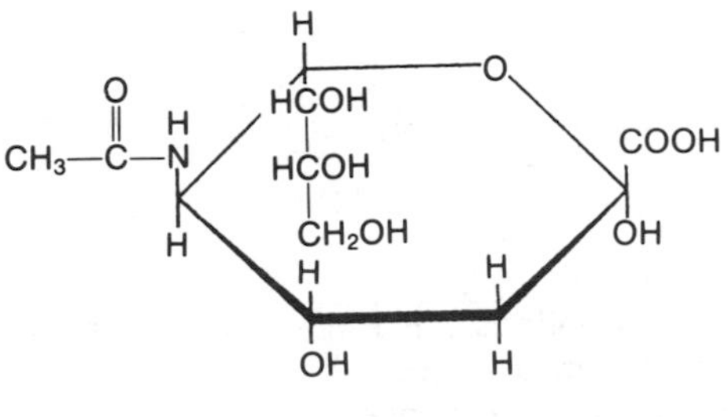

II

N-Acetyl-D-neuraminic acid

FIGURE 5.18
Structures of some membrane carbohydrates.

carbohydrate is on the exterior side of the plasma membrane or the luminal side of the endoplasmic reticulum. Roles for membrane carbohydrates include cell–cell recognition, adhesion, and receptor action.

5.3 MICELLES AND LIPOSOMES

Lipids Form Vesicular Structures

The basic structural characteristic of all membranes is derived from the physicochemical properties of the major lipid components, the phosphoglycerides and sphingolipids. These amphipathic compounds, with a hydrophilic head and a hydrophobic tail (Figure 5.19*a*), will interact with each other in an aqueous system. At an appropriate concentration, the lipid molecules will come together spontaneously to form spheres, termed **micelles.** The hydrophobic tails interact to exclude water and the charged polar head groups will interact with water and be on the outside of the sphere. This structure is shown in Figure 5.19*b*. The specific concentra-

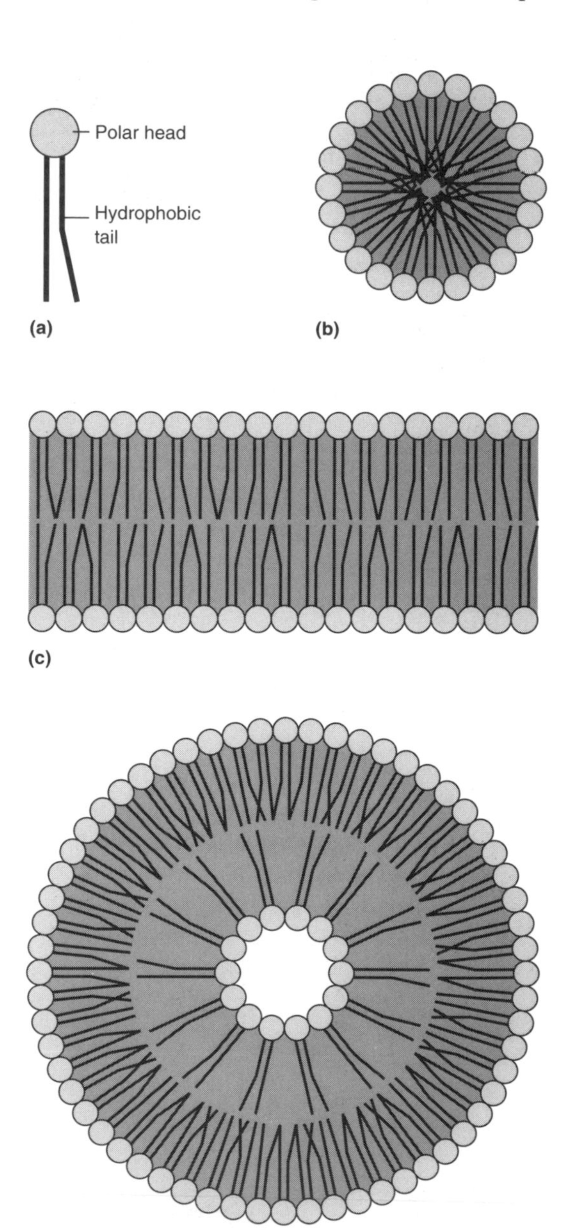

FIGURE 5.19
Representations of the interactions of phospholipids in an aqueous medium.
(*a*) Representation of an amphipathic lipid. (*b*) Cross-sectional view of the structure of a micelle. (*c*) Cross-sectional view of the structure of lipid bilayer. (*d*) Cross section of a liposome. Each structure has an inherent stability due to the hydrocarbon chains and the attraction of the polar head groups to water.

tion of lipid required for micelle formation is referred to as the **critical micelle concentration.** Micelles with a single lipid or a mixture of lipids can be made. The formation of the micelle depends also on the temperature of the system and, if a mixture of lipids are used, on the ratio of concentrations of the different lipids in the mixture. The micelle structure is very stable because of the hydrophobic interaction of the hydrocarbon chains and the attraction of the polar groups to water. As discussed in Section 26.6, micelles are important in the digestion of lipids.

Liposomes Have a Membrane Structure Similar to Biological Membranes

Depending on the conditions, amphipathic lipids such as the phosphoglycerides will interact to form a bimolecular leaf structure with two layers of lipid in which the polar head groups are at the interface between the aqueous medium and the lipid and the hydrophobic tails interact to form an environment that excludes water (Figure 5.19*c*). This bilayer conformation is the basic lipid structure of all biological membranes.

Lipid bilayers are extremely stable structures held together by noncovalent interactions of the hydrocarbon chains of the acyl groups and the ionic interactions of the charged head groups with water. Hydrophobic interactions of the hydrocarbon chains lead to the smallest possible area for water to be in contact with the chains, and water is essentially excluded from the interior of the bilayer. If disrupted, the bilayers have a tendency to self-seal because the hydrophobic groups will seek to establish a structure in which there is the least contact with water of the hydrocarbon chains, a condition that is most thermodynamically favorable. A lipid bilayer will close in on itself, forming a spherical vesicle separating the external space from an internal compartment. These vesicles are termed **liposomes.** Because the individual lipid–lipid interactions have low energies of activation, the lipids in a bilayer have a circumscribed mobility, breaking and forming interactions with surrounding molecules but not readily escaping from the lipid bilayer (Figure 5.19*d*). The ability of amphipathic lipids to self-assemble into bilayers is an important characteristic and is involved in the formation of cell membranes.

Individual phospholipid molecules can readily exchange places with neighboring molecules in a bilayer, which leads to rapid lateral diffusion in the plane of the membrane (see Figure 5.20). In addition, the fatty acyl chains can rotate around the carbon–carbon bonds; in fact, there is a greater degree of rotation nearer the methyl end, leading to greater motion at the center of the lipid bilayer. Individual lipid molecules cannot migrate readily from one monolayer to the other, a transverse movement, termed flip–flop, because of the thermodynamic constraints on movement of a charged molecule (the head group) through the lipophilic core. Thus the lipid bilayer has not only an inherent stability but also a fluidity in which individual molecules can move rapidly in their own monolayer but do not exchange with an adjoining monolayer. In artificial bilayer membranes composed of different lipids, the components will be randomly distributed.

Artificial membrane systems have been studied extensively as a means to determine the properties of biological membranes. A variety of techniques are available to prepare liposomes, using synthetic phospholipids and lipids extracted from natural membranes. Depending on the procedure, unilamellar vesicles and multilamellar vesicles (vesicles within vesicles) of various sizes (20 nm–1 μm in diameter) can be prepared. Figure 5.19*d* contains a representation of the structure of a liposome. The interior of the vesicle is an aqueous environment, and it is possible to prepare

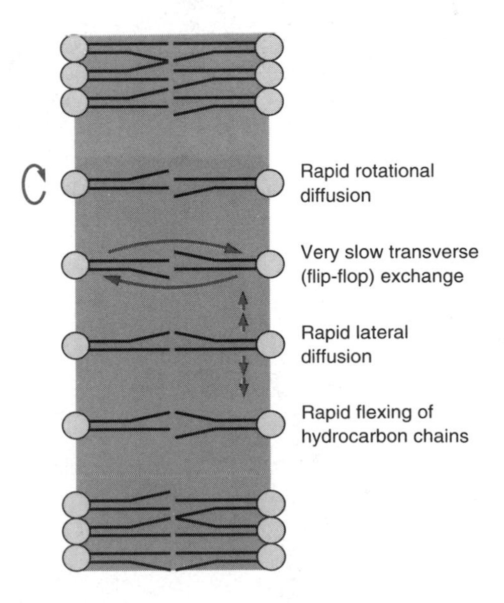

FIGURE 5.20
Mobility of lipid components in membranes.

liposomes with different substances entrapped. Thus the external and internal environments of the liposome can be manipulated and studies conducted on a variety of properties of these synthetic membranes, including their ability to exclude molecules, their interaction with various substances, and their stability under different conditions. Na^+, K^+, Cl^-, and most polar molecules do not readily diffuse across the lipid bilayer of liposomes, whereas the membrane presents no barrier to water. Lipid-soluble nonpolar substances such as triacylglycerol and undissociated organic acids readily diffuse into the membrane remaining in the hydrophobic environment of the hydrocarbon chains. Proteins, both synthetic and those isolated from cell membranes, have been incorporated into liposomes to mimic the natural membrane. Membrane-bound enzymes and proteins involved in translocating ions have been isolated from various tissues and incorporated into the membrane of liposomes for evaluation of the protein's function. With the liposome it is easier to manipulate the various parameters of the membrane system and, thus, study the catalytic activity free of possible interfering reactions that are present in the cell membrane. Liposomes are used in drug therapy (Clin. Corr. 5.1).

CLIN. CORR. 5.1 LIPOSOMES AS CARRIERS OF DRUGS AND ENZYMES

A major obstacle in the use of many drugs is the lack of tissue specificity in the action of the drug. Administration of drugs orally or intravenously leads to the drug acting on many tissues and not exclusively on a target organ. This may result in toxic side effects. An example is the commonly observed suppression of bone marrow cells by anticancer drugs. In addition, some drugs are metabolized rapidly and their period of effectiveness is relatively short. Liposomes have been prepared with drugs, enzymes and DNA encapsulated inside and used as carriers for these substances to target organs. Liposomes prepared from purified phospholipids and cholesterol are nontoxic and biodegradable. Alteration of surface charge enhances drug incorporation and release. Attempts have been made to prepare liposomes for interaction at a specific target organ. Antibiotic, antineoplastic, antimalarial, antiviral, antifungal, and antiinflammatory agents have been found to be effective when administered in liposomes. Some drugs have a longer period of effectiveness when administered encapsulated in liposomes. It is hoped, as our understanding of the structure and chemistry of cellular membranes increases, that it will be possible to prepare liposomes with a high degree of tissue specificity so that drugs and perhaps even enzyme replacement can be carried out with this technique.

Ranade, V. V. Drug delivery systems. 1. Site-specific drug delivery using liposomes as carriers. *J. Clin. Pharm.* 29:685, 1989; and Perez-Soler, R. Liposomes as carriers of antitumor agents: toward a clinical reality. *Cancer Treatment Rev.* 16:67, 1989.

5.4 STRUCTURE OF BIOLOGICAL MEMBRANES

The Fluid Mosaic Model of Biological Membranes

Based on evidence from physicochemical, biochemical, and electron microscopic investigations, knowledge of the structure of biological membranes has evolved. The basic structure of all biological membranes is a bimolecular leaf arrangement of lipids, as in liposomes, in which the amphoteric lipids and cholesterol are oriented so that the hydrophobic portions of the molecules interact, minimizing their contact with water or other polar groups, and the polar head groups of the lipids are at the interface with the aqueous environment. J. D. Davson and J. Danielli in 1935 proposed this basic structure in their model for a membrane; their proposal was later refined by J. D. Robertson. A major question with the earlier models was how to explain the interaction of membrane proteins with the lipid bilayer. In the early 1970s, S. J. Singer and G. L. Nicolson proposed the mosaic model for membranes in which some proteins are actually immersed in the lipid bilayer while others are loosely attached to the surface of the membrane. It was suggested that some proteins actually spanned the lipid bilayer being in contact with the aqueous environment on both sides. Figure 5.21 is a representation of a biological membrane as proposed by Singer and Nicolson. Results of extensive studies have confirmed the basic structure of this model that is now referred to as the **fluid mosaic model** to indicate the movement of both lipids and proteins in the membrane. The characteristics of the lipid bilayer explain many cellular membrane properties, including the fluidity, the flexibility that permits changes of shape and form, the ability to self-seal, and the impermeability of the membrane. The model continues to undergo modification and refinement; as an example, under certain conditions the lipids can assume structural variations other than the bimolecular leaf arrangement.

Integral Membrane Proteins Are Immersed in the Lipid Bilayer

An important difference between the Singer–Nicolson model and earlier models was the proposal that the lipid bilayer is discontinuous with proteins embedded in the hydrophobic portion of the bilayer. The develop-

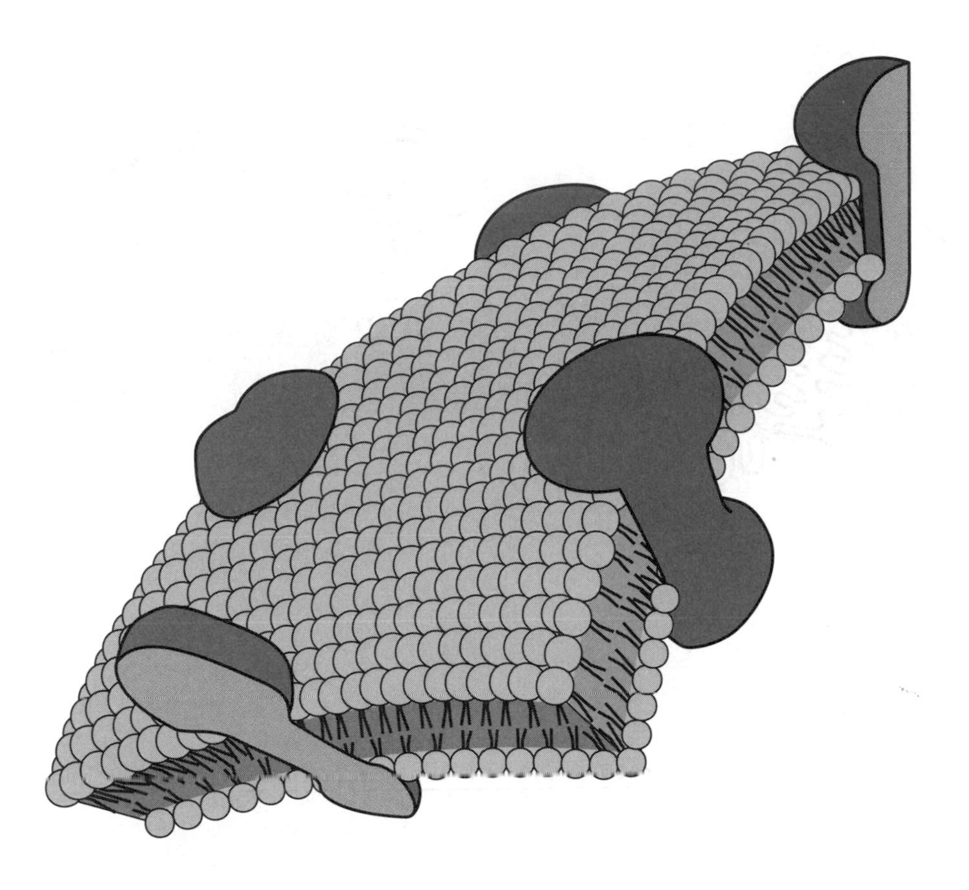

FIGURE 5.21
The fluid mosaic model of biological membranes.
The membrane consists of a fluid phospholipid bilayer with globular integral proteins penetrating the bilayer.

ment of techniques for isolation of integral membrane proteins, for determination of their primary structure, and for identification of specific functional domains in the protein has led to an understanding of the structural relationship between the hydrophobic lipid bilayer and membrane proteins. Figure 5.22 illustrates the various ways of attachment of proteins to a biological membrane. Some **integral membrane proteins** (see p. 205) span the membrane, whereas others may only be immersed partially in the lipid. Based on measurements of the hydrophobicity of the amino acid residues and partial proteolytic digestion of the proteins, those sequences of amino acids embedded in the membrane have been determined. Some proteins contain an α-helical structure consisting primarily of hydrophobic amino acids (such as leucine, isoleucine, valine, and phenylalanine), which is the transmembrane sequence. This is illustrated in Figure 5.22*a*. An example of such a protein is **glycophorin** that is present in the plasma membrane of human erythrocytes; amino acid residues 73–91, of the 131 total amino acids, are the transmembrane sequence and are predominantly hydrophobic. Glycophorin has three domains: a sequence exterior to the cell containing the amino terminal end, the transmembrane sequence, and a sequence extending into the cell with the carboxyl terminal end. In contrast, the amino acid chain of other transmembrane proteins loop back and forth across the membrane (Figure 5.22*b*). In some cases there are 12 loops snaking across the lipid bilayer. Often these multiple membrane spanning α helices are organized to form a

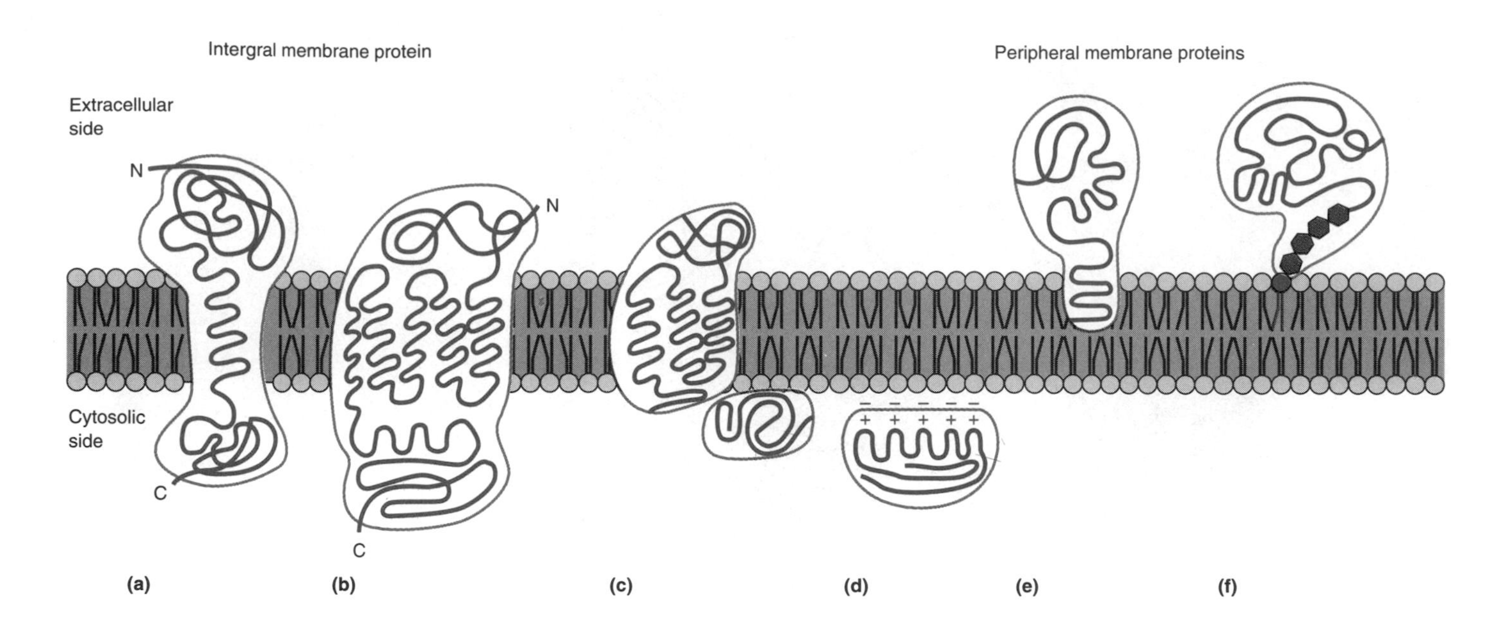

FIGURE 5.22
Interactions of membrane proteins with the lipid bilayer.
The diagram illustrates the multiple types of binding of proteins in or to the lipid bilayer. (*a*) A single transmembrane segment; (*b*) multiple transmembrane segments; (*c*) bound to an integral protein; (*d*) bound electrostatically to the lipid bilayer; (*e*) attached by a short terminal hydrophobic sequence of amino acids; and (*f*) attached by covalently bound lipid.

tubular structure; such is the case of the anion channel of human erythrocytes, which has 926 amino acids and is responsible for the exchange of Cl^- and HCO_3^- across the membrane (p. 224). Thus the secondary and tertiary structure of the proteins are critical in the topography of the protein in the membrane. In addition, some proteins form a quaternary structure in the membrane with multiple subunits.

As with other proteins, integral membrane proteins have specific domains, including those for ligand binding, catalytic activity, and attachment of carbohydrate or lipid. The anion channel of the erythrocyte can be separated into two major domains: a hydrophilic amino-terminal domain located on the cytosolic side of the membrane that contains binding sites for ankyrin, a protein that anchors the cytoskeleton and other cytosolic proteins, and a domain with 509 amino acids that traverses the membrane and mediates the exchange of Cl^- and HCO_3^-. Another example is glycophorin that contains 60% carbohydrate, all of which is attached to the protein domain on the extracellular side of the membrane. It is obvious that with such well-defined domains that integral membrane proteins will have a defined orientation in the membrane and their placement and interaction with the membrane is not random. This defined structural orientation demonstrates another important aspect of membrane structure; biological membranes are asymmetric, with each surface having specific characteristics. The orientation of proteins is fixed during the synthesis of the membrane; the bulkiness of the proteins, as well as thermodynamic restrictions, prevent transverse (flip–flop) movement.

The activity of many enzymes which are integral membrane proteins require the presence of the membrane lipid. As an example, D-β-hydroxybutyrate dehydrogenase, located in the inner mitochondrial membrane, has a requirement for phosphatidylcholine for its activity. Cholesterol has been implicated in the activity of various membrane ion pumps, including the Na^+, K^+- and Ca^{2+}-ATPase (see p. 226), and the acetylcholine receptor. Some of these modulating effects of lipids may be a reflection of a change in the ordering and fluidity of the membrane but the lipid may also have a direct influence on the activity.

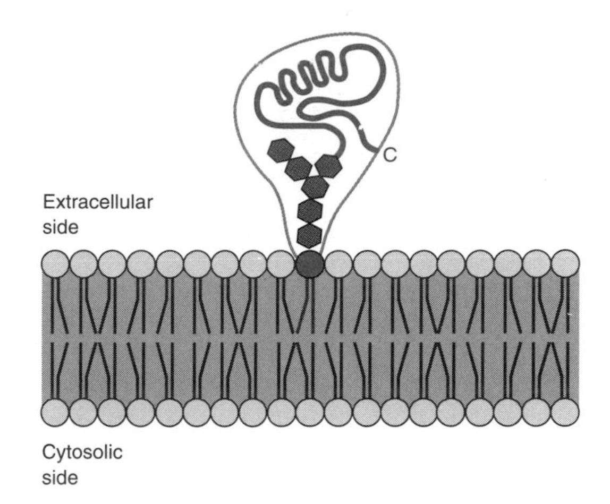

FIGURE 5.23
Attachment of a protein to a membrane by a glycosyl phosphatidylinositol anchor.

Peripheral Membrane Proteins Have Various Modes of Attachment

Peripheral membrane proteins are loosely attached to the membrane and if removed do not disrupt the lipid bilayer. Some apparently bind to integral membrane proteins, such as in the case of ankyrin binding to the anion channel protein in the erythrocyte (Figure 5.22*c*). In addition, negatively charged phospholipids of the membrane can interact with positively charged regions of proteins allowing electrostatic binding (Figure 5.22*d*). Some peripheral membrane proteins have sequences of hydrophobic amino acids at one end of the peptide chain that can serve as an anchor in the membrane lipid (Figure 5.22*e*); cytochrome b_5 is attached to the endoplasmic reticulum by such an anchor.

In the last decade several additional modes of attachment have been discovered involving fatty acid residues covalently linked to proteins. Phosphatidylinositol has been demonstrated to have a role in anchoring proteins to membranes; the proposed general structure of this form of anchor is presented as Figure 5.22*f* and in Figure 5.23. A glycan, consisting of ethanolamine, phosphate, mannose, mannose, mannose, and glycosamine is covalently bound to the carboxyl terminal of the protein. This glycan appears to have been conserved throughout evolution because it is found in a number of different species attached to carboxyl terminal amino acid residues of various membrane bound proteins. Additional carbohydrate can be attached to the last mannose. The glycosamine of the glycan is covalently bonded to phosphatidylinositol. The fatty acids of the phosphoglyceride are inserted into the lipid membrane, thus anchoring the protein. These molecules are now referred to as **glycosyl phosphatidylinositol** (GPI) **anchors.** A variety of proteins are attached in this manner including enzymes, antigens, and cell adhesion proteins; a partial list is presented in Table 5.3. The fatty acyl groups of phosphatidylinositol are apparently specific for different proteins. To date all of the proteins found to be attached by the GPI anchor are on the external surface of the plasma membrane. The significance of this form of anchoring has yet to be determined but it may be important for the specific localization of the protein on the membrane, control of the function of the protein, and controlled release of the protein from the membrane. A specific enzyme, phosphatidylinositol-specific phospholipase C catalyzes the hydrolysis of the phosphate–inositol bond leading to release of the protein.

In addition to the phosphatidylinositol anchor, myristic acid and palmitic acid have been found to be covalently linked to proteins and serve to anchor proteins in membranes by insertion of the acyl chain into the lipid bilayer (Figure 5.22*f*). Myristic acid (C_{14}) is attached directly by an amide linkage to an amino terminal glycine, and palmitic acid (C_{16}) is most often attached by a thioester linkage to cysteine or by a hydroxyester bond to serine or threonine.

Even though the model would suggest that proteins are randomly distributed throughout and on the membrane, evidence from a variety of sources supports a high degree of functional organization with definite restrictions on the localization of some proteins. As an example, proteins participating in electron transport in the inner membrane of mitochondria

TABLE 5.3 Proteins with a Glycosyl–Phosphatidylinositol Anchor

Alkaline phosphatase
5′-Nucleotidase
Acetylcholinesterase
Trehalase
Renal dipeptidase
Lipoprotein lipase
Carcinoembryoic antigen
Neural cell adhesion molecule
Scrapie prion protein
Oligodendrocyte–myelin protein

SOURCE: M. G. Low, Glycosyl–phosphatidylinositol: A versatile anchor for cell surface proteins. *FASEB J.* 3:1600, 1989.

function in consort and are organized into a functional unit both laterally and transversely in the membrane. The actual location of specific proteins on the surface of plasma membranes is also controlled. Cells lining the lumen of the kidney nephron have specific plasma membrane enzymes on the luminal surface but not on the contraluminal surface of the cell; the enzymes restricted to a particular region of the membrane are located to meet the specific functions of these cells. Thus there is a high degree of molecular organization of biological membranes that is not apparent from the diagrammatic models.

The Human Erythrocyte Is Ideal for Studying Membrane Structure

The structure of the plasma membrane of the human erythrocyte is under active investigation because of the ease with which the membrane can be purified from other cellular components. Figure 5.24 is a representation of the interaction of some of the many proteins in this membrane. Results from these studies have aided in our understanding of other membranes.

Lipids Are Distributed in an Asymmetric Manner in Membranes

In contrast to the random distribution of lipids between the outer and inner lipid monolayers of liposomes, *there is an asymmetric distribution of lipid components across biological membranes*. Each layer of the bilayer has a different composition with respect to individual phosphoglycerides and sphingolipids. An example is the asymmetric distribution of lipids in the human erythrocyte membrane, as presented in Figure 5.25. Sphingomyelin is predominantly in the outer layer, whereas phosphatidylethanolamine is predominantly in the inner lipid layer. In contrast, cholesterol is equally distributed on both sides of the plasma membrane.

The asymmetry of the lipids may be maintained by specific membrane proteins that promote the transverse movement of specific lipids from one side to the other. Recent studies suggest the involvement of metabolic energy in this process. Uncatalyzed transverse movement from one side to the other (i.e., flip–flop movement) of the phosphoglycerides and sphingolipids is slow and is measured in days or weeks. The slow rate of transverse movement is not unexpected, considering how unfavorable in thermodynamic terms it is to push or pull the hydrophilic polar head group of a phospholipid through the hydrophobic interior of a membrane

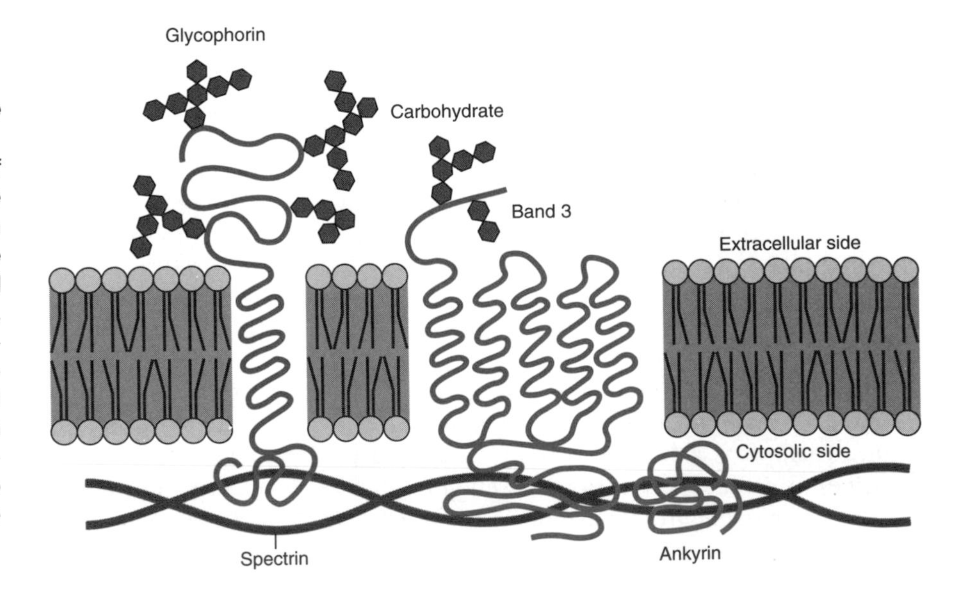

FIGURE 5.24
Schematic diagram of the erythrocyte membrane.
The diagram indicates the relationship of four membrane-associated proteins with the lipid bilayer. Glycophorin is a glycoprotein that contains 131 amino acids but whose function is unknown. Band 3, so designated because of its mobility in electrophoresis, contains over 900 amino acids and is involved in interacting with ankyrin and possibly in the facilitated diffusion of Cl^- and HCO_3^- (see Section 5.1). Ankyrin and spectrin are part of the cytoskeleton and are peripheral membrane proteins. Ankyrin binds to band 3 and spectrin is anchored to the membrane by ankyrin.

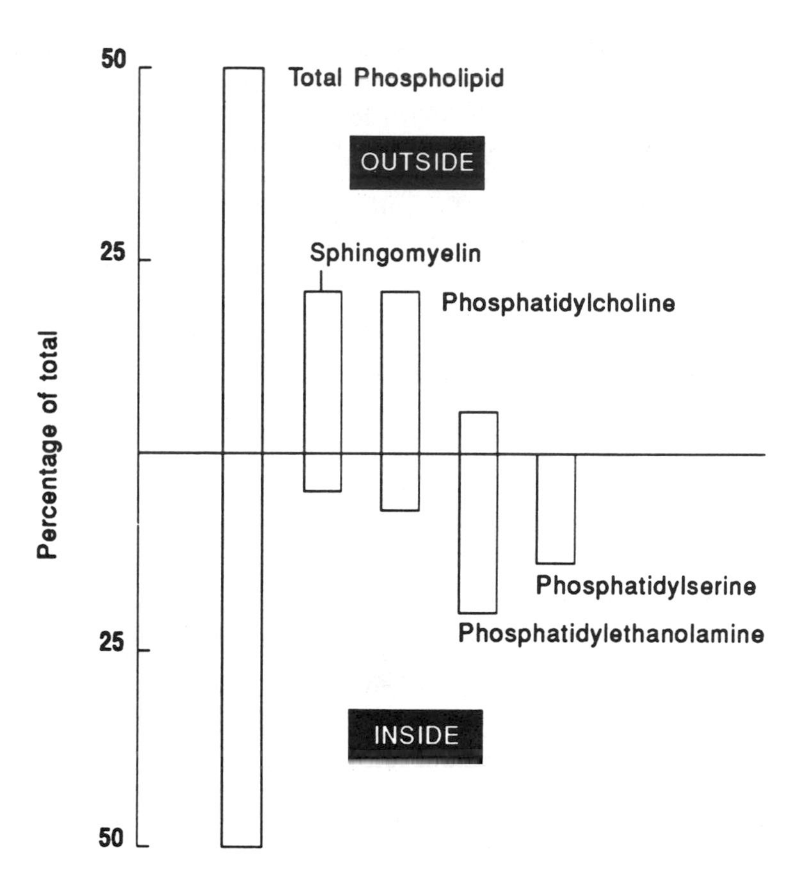

FIGURE 5.25
Distribution of phospholipids between inner and outer layers of the human erythrocyte membrane.
Values are percentage of each phospholipid in the membrane.
Redrawn from A. J. Verkeij, R. F. A. Zwaal, B. Roelotsen, P. Comfurius, D. Kastelijn, and L. L. M. Van Deenan. The asymmetric distribution of phospholipids in the human red cell membrane. *Biochim. Biophys. Acta* 323:178, 1973.

and then reorient the group on the opposite side. The asymmetry of lipids in the erythrocyte membrane is an example of how slow the transverse movement of membrane lipids is. The mature erythrocyte has a lifetime of about 120 days, during which time there is no new membrane synthesis or even significant repair. Even so, there appears to be little mixing of the phospholipids between the molecular layers. Individual lipids can exchange with lipids in the cell matrix, as well as with lipids of other membranes. Specific mechanisms to maintain both the composition and asymmetry of lipids in membranes apparently exist.

Proteins and Lipids Diffuse in Membranes

The interactions among the different lipids and between lipids and proteins are very complex and dynamic. There is a fluidity in the lipid portion of the membrane in which both the lipids and proteins move. The degree of fluidity is dependent on the temperature and composition of the membrane. At low temperatures, the lipids are in a gel–crystalline state, with the lipids restricted in their mobility. As the temperature is increased, there is a phase transition into a liquid–crystalline state, with an increase in fluidity. With liposomes prepared from a single pure phospholipid, the **phase transition temperature,** T_m, is rather precise; but with liposomes prepared from a mixture of lipids, the T_m becomes less precise because individual clusters of lipids may be in either the gel–crystalline or the liquid–crystalline state. The T_m is not precise for biological membranes because of their heterogeneous chemical composition. Interactions between lipids and proteins also lead to variations in the gel–liquid state throughout the membrane and differences in fluidity in different areas of the membrane.

The specific composition of the individual biological membranes leads to differences in fluidity. Phosphoglycerides containing short-chain fatty acids will increase the fluidity as does an increase in unsaturation of the fatty acyl groups. The cis-double bond in an unsaturated fatty acid of phospholipid leads to a kink in the hydrocarbon chain, preventing the tight packing of the chains, and creates pockets in the hydrophobic areas. It is assumed that these spaces, which will also be mobile due to the mobility of the hydrocarbon chains, are filled with water molecules and small ions. Cholesterol with its flat stiff ring structure reduces the coiling of the fatty acid chain and decreases fluidity. Consideration has been given to the potential clinical significance of high blood cholesterol on the fluidity of cell membranes (see Clin. Corr. 5.2.). The Ca^{2+} ion directly decreases the fluidity of a number of membranes because of its interaction with the negatively charged phospholipids, which reduces repulsion between the polar groups and increases the packing of lipid molecules. This ion also causes aggregation of lipids into clusters, which reduces membrane fluidity.

Fluidity at different levels within the membrane also varies. The hydrocarbon chains of the lipids have a motion, which produces a fluidity in the hydrophobic core. The central area of the bilayer is occupied by the ends of the hydrocarbon chains and is more fluid than the areas closer to the two surfaces, where there are more constraints due to the stiffer portions of the hydrocarbon chains. Cholesterol makes the membrane more rigid toward the periphery because it does not reach into the central core of the membrane.

Individual lipids and proteins can move rapidly in a lateral motion along the surface of the membrane. However, electrostatic interactions of polar head groups, hydrophobic interactions of cholesterol with selected phospholipids or glycolipids, and protein–lipid interactions all lead to constraints on the movement. Thus there may be lipid domains in which lipids move together, such as an island floating in a sea of lipid.

Movement of integral membrane proteins in the lipid environment has been demonstrated by the fusion of human and rat cells after antigenic membrane proteins on cells of each species were labeled with a different antibody marker. The markers permitted localization of the two different proteins on the membrane. Immediately following fusion of the cells, proteins on the membranes of the human and rat cells were segregated in different hemispheres of the new cell, but within 40 min the two groups of proteins were evenly distributed over the new cell membrane. Movement of protein is slower than that of lipids and may be restricted by other membrane proteins, matrix proteins, or cellular structural elements such as microtubules or microfilaments to which they may be attached.

Evidence is accumulating that the fluidity of cellular membranes can change in response to changes in diet or physiological state. Their content of fatty acid and cholesterol is modified by a variety of factors. In addition, pharmacological agents may have a direct effect on membrane fluidity. It is now considered that some of the actions of anesthetics, which induce sleep and muscular relaxation, may be due to their effect on membrane fluidity of specific cells. A number of structurally unrelated compounds induce anesthesia, but their common feature is lipid solubility. Anesthetics increase membrane fluidity in vitro.

Thus cellular membranes are in a constantly changing state, with not only movement of proteins and lipids laterally on the membrane but with molecules moving into and out of the membrane. The membrane creates a number of microenvironments, from the hydrophobic portion of the core of the membrane to the interface with the surrounding environments. It is difficult to express in words or pictures the very fluid and dynamic state, in that neither captures the time-dependent changes that occur in the structure of biological membranes. Figure 5.26 attempts to illustrate the structural and movement aspects of cellular membranes.

CLIN. CORR. 5.2
ABNORMALITIES OF CELL MEMBRANE FLUIDITY IN DISEASE STATES

Changes in membrane fluidity can control the activity of membrane-bound enzymes, membrane functions such as phagocytosis, and cell growth. A major factor in controlling plasma membrane fluidity is the concentration of cholesterol. Higher organisms and mammals have a significant concentration of cholesterol in their membranes, which presumably has a major role in controlling the fluidity of the lipid bilayer. With increasing cholesterol content membranes become less fluid on their outer surface but more fluid in the hydrophobic core. Individuals with spur cell anemia have an increased cholesterol content of the red cell membrane. This condition occurs in severe liver disease such as cirrhosis of the liver in alcoholics. The intoxicating effect of ethanol on the nervous system is probably due to modification of membrane fluidity and alteration of membrane receptors and ion channels. Erythrocytes have a spiny shape and are destroyed prematurely in the spleen. The cholesterol content is increased 25–65%, and the fluidity of the membrane is decreased. The erythrocyte membrane requires a high degree of fluidity for its function and any decrease would have serious effects on the cell's ability to pass through the capillaries. The increased plasma membrane cholesterol in other cells leads to an increase in intracellular membrane cholesterol, which also affects the fluidity of other cellular membranes. Individuals with abetalipoproteinemia have an increase in sphingomyelin content and a decrease in phosphatidylcholine, thus causing a decrease in fluidity. The ramifications of these changes in fluidity are still not understood, but it is presumed that, as techniques for the measurement and evaluation of cellular membrane fluidity improve, some of the pathological manifestations in disease states will be explained on the basis of changes in membrane structure and function.

Cooper, R. A. Abnormalities of cell membrane fluidity in the pathogenesis of disease. *N. Engl. J. Med.* 297:371, 1977.

5.5 MOVEMENT OF MOLECULES THROUGH MEMBRANES

The lipid nature of biological membranes severely restricts the type of molecules that will readily diffuse from one side to another. Substrates carrying a charge, whether inorganic ions or charged organic molecules, will not diffuse at a significant rate because of the attraction of these molecules to water molecules and the exclusion of the charged species by the hydrophobic environment of the lipid membrane. The diffusion rate of molecules, such as carbohydrates, amino acids, and inorganic ions, however, is not zero but may be too slow to accommodate a cell's requirements. Where there is a need to move a nondiffusible substance across a particular cellular membrane, specific mechanisms are available for its translocation.

The following discussion describes the mechanisms by which molecules cross various cellular membranes; examples of specific systems will be described for illustrative purposes, but throughout this book individual systems are described in the context of specific metabolic processes.

Some Molecules Can Diffuse Through Membranes

Movement through the membrane by **diffusion** involves three major steps: (1) the solute must leave the aqueous environment on one side and enter the membrane; (2) the solute must traverse the membrane; and (3) the solute must leave the membrane to enter a new environment on the opposite side (Figure 5.27). Each step involves an equilibrium of the solute between two states, and thermodynamic and kinetic constraints control the equilibrium established for the concentrations of the substance on the two sides of the membrane and the rate at which it can attain the equilibrium. For diffusion of a solute with strong interaction with water molecules, the solute must have the shell of water stripped away to enter the lipid milieu but regains it on leaving the membrane. For hydrophobic substances, the distribution between the aqueous phase and lipid membrane will depend on the degree of lipid solubility of the substance; very lipid-soluble materials will dissolve in the membrane.

The rate of diffusion of a solute is directly proportional to its lipid solubility and diffusion coefficient in lipids; the latter is a function of the size and shape of the substance. Uncharged lipophilic molecules, for example, fatty acids and steroids, diffuse relatively rapidly but water-

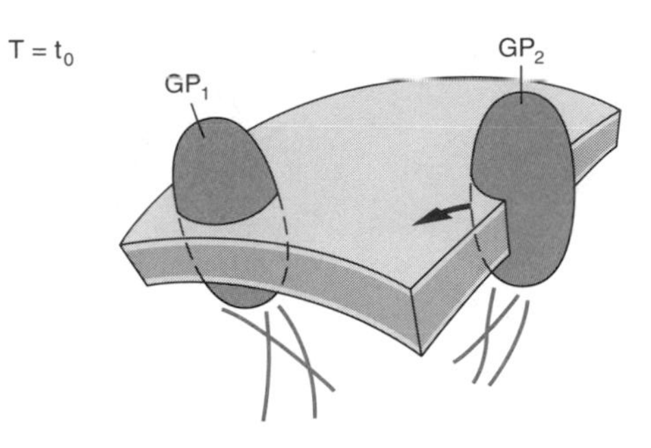

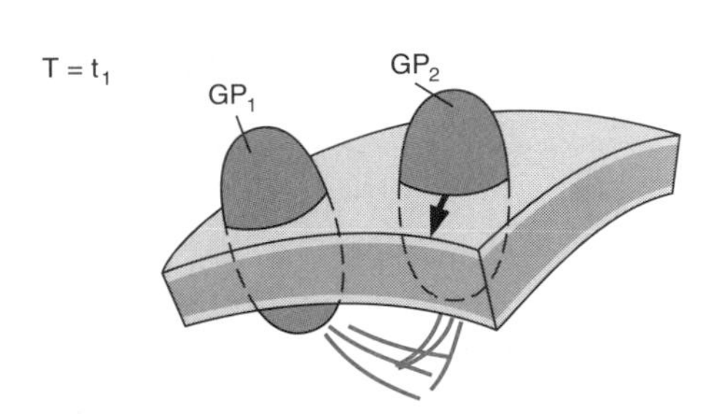

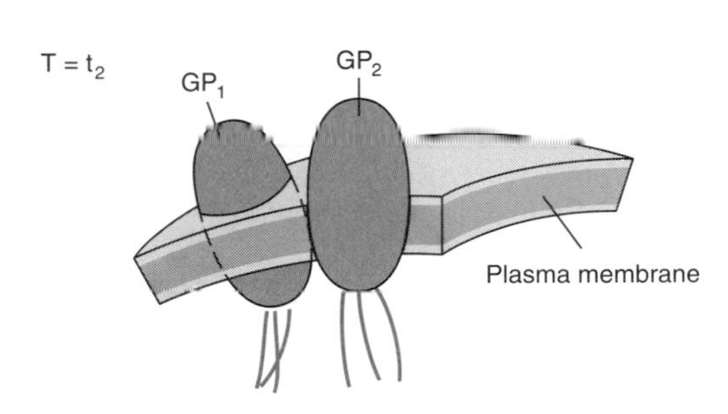

FIGURE 5.26
A modified version of the fluid mosaic model of biological membranes to indicate the mobility of membrane proteins.
t_0, t_1 and t_2 represent successive points in time. Some integral proteins (GP$_2$) are free to diffuse laterally in the plane of the membrane directed by the cytoskeletal components, whereas others (GP$_1$) may be restricted in their mobility.
Redrawn from J. L. Nicolson, *Int. Rev. Cytol.* 39:89, 1974.

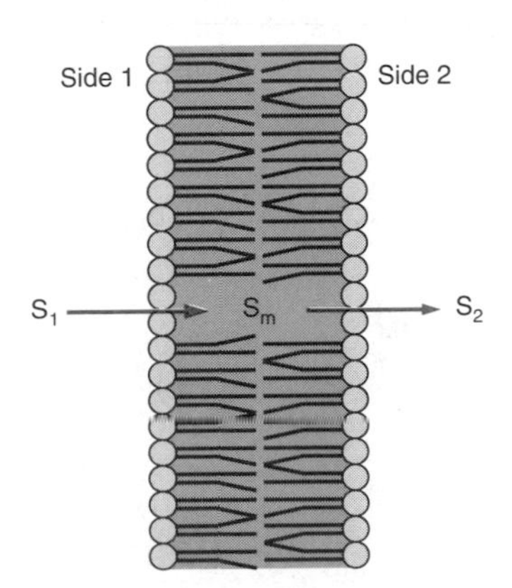

FIGURE 5.27
Diffusion of a solute molecule through a membrane.
S_1 and S_2 are solute on each side of membrane, and S_m solute in membrane.

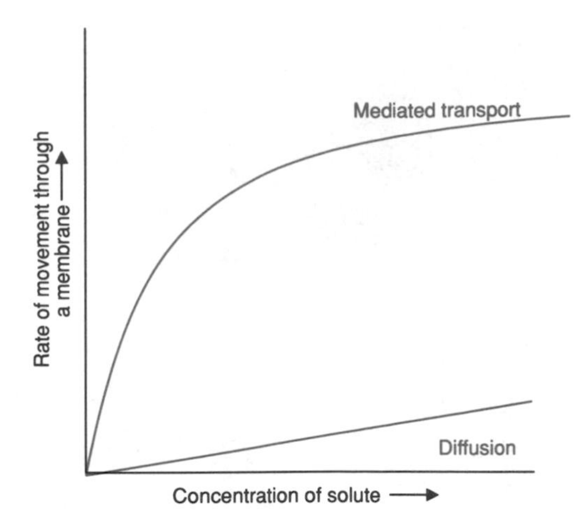

FIGURE 5.28
Kinetics of movement of a solute molecule through a membrane.
The initial rate of diffusion is directly proportional to the concentration of the solute. In mediated transport, the rate will reach a V_{max} when the carrier is saturated.

soluble substances, for example, sugars and inorganic ions, diffuse very slowly. Water diffuses readily through biological membranes; its movement apparently occurs via the gaps in the hydrophobic environment created by the random movement of the fatty acyl chains of the lipids. Water and other small polar molecules can move into these transitory spaces and equilibrate across the membrane from one gap to another.

The direction of movement of solutes by diffusion is always from a higher to a lower concentration and the rate is described by **Fick's first law of diffusion:**

$$J = -D\left(\frac{\delta c}{\delta x}\right)$$

where J = net amount of substance moved per time, D = the diffusion coefficient, and $\delta c/\delta x$ = the chemical gradient of the substance. As the concentration of solute on one side of the membrane is increased, there will be an increasing initial rate of diffusion as illustrated in Figure 5.28. A net movement of molecules from one compartment to another will continue until the concentration in each is at a chemical equilibrium. After equilibrium is attained, there will be a continued exchange of solute molecules from one side to another, but no net accumulation on one side can occur because this would recreate a concentration gradient.

Movement of Molecules Across Membranes Can Be Facilitated

Mechanisms for membrane translocation of various substances including sugars, amino acids, metabolic intermediates, inorganic ions, and even H^+ have been identified. The plasma membrane of both procaryotic and eucaryotic cells, as well as the membranes of subcellular organelles, contain transport systems that play an important role in the uptake of nutrients, the maintenance of ion concentrations, and control of metabolism. All of these systems, which involve intrinsic membrane proteins, are classified on the basis of their mechanism of translocation of the molecule across the membrane and the energetics of the system. A classification of the transport systems is presented in Table 5.4. Each of these will be discussed in more detail in subsequent sections but for now it is important to distinguish the three main types.

Membrane Channels

Membranes of most cells contain specific **channels,** in some cases referred to as **pores,** which permit the rapid movement of specific molecules or ions from one side of the membrane to the other. The tertiary and quater-

TABLE 5.4 Classification of Membrane Translocation Systems

Type	Class	Example
Channel	1. Voltage regulated	Na^+ channel
	2. Chemically regulated	Acetylcholine receptor
	3. Other	Pressure sensitive
Transporter	1. Passive mediated	Glucose transporter
	2. Active mediated	
	a. Primary-redox coupled	Respiratory chain linked
	Primary-ATPases	Na^+,K^+-ATPase
	b. Secondary	Na^+-dependent glucose transport
Group translocation		Amino acid translocation

nary structure of these intrinsic membrane proteins create an aqueous hole in the membrane that permits the diffusion of substances in both directions through the membrane. Like diffusion, the substances will move only in the direction of lower concentration, that is, down a concentration gradient. In contrast to transporters, the channel proteins do not bind the molecules or ions to be transported. The channels have some degree of specificity, however, based on the size and charge of the substance. Flow through the channel can be regulated by opening and shutting the passageway, like a gate on an enclosed grazing pasture.

Transporters

Transporters actually translocate the molecule or ion across the membrane by binding and physically moving the substance. In this case the translocating protein functions like an enzyme reacting with a substrate and the activity can be evaluated in the same kinetic terms as an enzyme catalyzed reaction. Transporters have specificity for the substance, frequently referred to as the substrate, defined reaction kinetics, and can be inhibited by both competitive and noncompetitive inhibitors. Some transporters move the molecules down their concentration gradient (referred to as passive), while others move the substrate against its concentration gradient (active) requiring the expenditure of some form of energy. With both channels and transporters the molecule is unchanged during the translocation across the membrane.

Group Translocation

Another mechanism for transport is referred to as **group translocation** and involves not only the movement of the substance across the membrane but also a chemical modification of the substance once inside the cell. This process is a major mechanism for the uptake of sugars by bacteria in which the sugar is transported and then phosphorylated before release into the cytoplasm of the cell.

A major difference between membrane channels and transporters is the rate of substrate translocation; for a channel, rates in the range of 10^7 ions s^{-1} are usual, whereas with a transporter the rate is in the range of 10^2–10^3 molecules s^{-1}. The activity of all translocation systems can be modulated, permitting cells and tissues to control the movement of substances across membranes. Drugs for specific channels and transporters have been developed to control these processes.

Membrane Transport Systems Have Common Characteristics

Membranes of all cells contain highly specific transporters for the movement of inorganic anions and cations (e.g., Na^+, K^+, Ca^{2+}, HPO_4^{2+}, Cl^-, and HCO_3^-), as well as uncharged and charged organic compounds (e.g., amino acids and sugars). Different cellular membranes have different transport mechanisms; as examples, the mitochondrial membrane has a mechanism to translocate adenine nucleotides that is not present in other cellular membranes. In contrast to channels and pores, the transport systems involve integral membrane proteins with a high degree of specificity for the substances transported. These proteins or protein complexes have been designated by a variety of names, including *transporter, translocase, translocator, and permease or termed transporter system, translocation mechanism, and mediated transport system* to name a few. For some, the term *pump* is applied, but this is not a very descriptive term for membrane transport systems. The designations above are used interchangeably, but for convenience we will use transporter in referring to the actual protein involved in the translocation.

TABLE 5.5 Characteristics of Membrane Transporters

Passive mediated	Active mediated
1. Saturation kinetics	1. Saturation kinetics
2. Specificity for solute transported	2. Specificity for solute transported
3. Can be inhibited	3. Can be inhibited
4. Solute moves down concentration gradient	4. Solute can move against concentration gradient
5. No expenditure of energy	5. Requires coupled input of energy

Membrane transporters have a number of characteristics in common. Each facilitates the movement of a molecule or molecules through the lipid bilayer at a rate which is significantly faster than can be accounted for by simple diffusion. If S_1 is the solute on side 1 and S_2 on side 2, then the transporter promotes an equilibrium to be established as follows:

$$[S_1] \rightleftharpoons [S_2]$$

where the brackets represent the concentration of solute. If the transporter (T) is included in the equilibrium the reaction is

$$[S_1] + T \rightleftharpoons [S\text{—}T] \rightleftharpoons [S_2] + T$$

If there is no energy input by the system, the concentration on both sides of the membrane will be equal at equilibrium, but if there is an expenditure of energy, a concentration gradient can be established, which will depend on the thermodynamic properties of the system. Note the similarity of the role of the transporter to that of an enzyme, which increases the rate of a chemical reaction but does not determine the final equilibrium.

Table 5.5 lists the major characteristics of membrane transport systems. Mediated transport systems like enzyme-catalyzed reactions demonstrate saturation kinetics; as the concentration of the substance to be translocated increases, the initial rate of transport increases but reaches a maximum when the substance saturates the protein transporter on the membrane. A plot of solute concentration versus initial rate of transport is hyperbolic, as presented in Figure 5.28. Simple diffusion does not demonstrate saturation kinetics. Constants such as V_{max} and K_m can be calculated for transporters. As with enzymes, transporters can theoretically catalyze movement of a solute in both directions across the membrane, but this, of course, depends on the $\Delta G'$ for the reaction.

Most transporters have a high degree of structural and stereospecificity for the substance transported. An example is the mediated transport system for D-glucose in the erythrocyte, where the K_m is 10 times larger for D-galactose than for D-glucose and for L-glucose it is 1000 times larger. The transporter has essentially no activity with D-fructose or disaccharides. Competitive and noncompetitive inhibitors have been found for many transporters. Structural analogs of the substrate inhibit competitively and reagents that react with specific groups on proteins are noncompetitive inhibitors.

$$\begin{aligned} &\text{Recognition:} \quad S_1 + T_1 \rightleftharpoons S\text{–}T_1 \\ &\text{Transport:} \quad S\text{–}T_1 \rightleftharpoons S\text{–}T_2 \\ &\text{Release:} \quad S\text{–}T_2 \rightleftharpoons T_2 + S_2 \\ &\text{Recovery:} \quad T_2 \rightleftharpoons T_1 \end{aligned}$$

FIGURE 5.29
Reactions involved in mediated transport across a biological membrane.
S_1 and S_2 are the solutes on side 1 and 2 of the membrane, respectively; T_1 and T_2 are the binding sites on the transporter on side 1 and side 2, respectively.

There Are Four Common Steps in the Transport of Solute Molecules

A key question is how does the transporter actually facilitate the movement of a molecule across a distance in space. We need to expand the equation above and consider four aspects of mediated transport, as present in Figure 5.29. These are (1) *recognition* by the transporter of the appropriate solute from a variety of solutes in the aqueous environment, (2) *translocation* of the solute across the membrane, (3) *release* of the solute by the transporter, and (4) *recovery* of the transporter to its original condition to accept another solute molecule.

The first step, *recognition* by the transporter of a specific substrate, is explained on the same basis as that described for recognition of a substrate by an enzyme. The presence of very specific binding sites on the protein permits the transporter to recognize the correct structure of the solute to be translocated.

The second step, *translocation,* is not understood. Models have been proposed based on studies on different transporters, but none has received universal support. A reasonable model (Figure 5.30) is one in which the protein transporter creates a channel between the environments on each side of the membrane with access through the channel being controlled by a gating mechanism in order to control which solutes can move into the channel. Transporters have receptor sites to which the solute attaches. After association of the solute and transporter a conformational change of the protein moves the solute molecule a short distance, perhaps only 2 or 3 Å, but into the new environment of the opposite side of the membrane. In this manner, it is not necessary for the transporter physically to move the molecule the whole distance across the membrane. Earlier suggestions for the translocation step included the possibility of a diffusible or rotating carrier, but both are improbable considering that transporters are integral membrane proteins that do not diffuse transversely.

Release, step 3, of the solute can occur readily based on simple equilibrium considerations if the concentration of solute is lower in the new compartment than on the initial side of binding. This does not require the dissociation constant of transporter–solute complex to change. For those transporters that move a solute against a concentration gradient release of the solute at the higher concentration occurs because the affinity for the solute by the transporter is decreased. A change in the conformation of the transporter decreases the affinity. In group translocation (p. 230) the solute is chemically altered while attached to the transporter and the new molecule has a lower affinity for the transporter.

Finally, step 4, the transporter must return to its original state that is *recovery*. If a conformational change has occurred, the transporter reverts to the original conformation.

The discussion above has centered on the movement of a single solute molecule by the transporter. Actually systems are known that move two molecules simultaneously in one direction (**symport** mechanisms), two molecules in opposite directions (**antiport** mechanism), as well as a single molecule in one direction (**uniport** mechanism) (Figure 5.31). Some transporters move charged substances such as K^+ in which there is no direct and simultaneous movement of an ion of the opposite charge. This creates a charge separation across the membrane, and the mechanism is termed **electrogenic**. If a counterion is moved to balance the charge, the mechanism is called neutral or electrically silent.

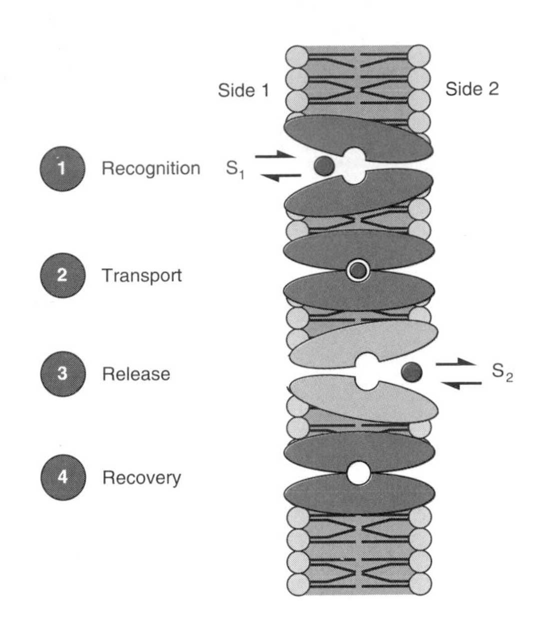

FIGURE 5.30
Model for a mediated transport system in a biological membrane.
The model is based on the concept of a gated channel in which conformational changes in the transporter move the bound solute a short distance but into the environment of the other side of the membrane. Once moved, the solute is released from the transporter.

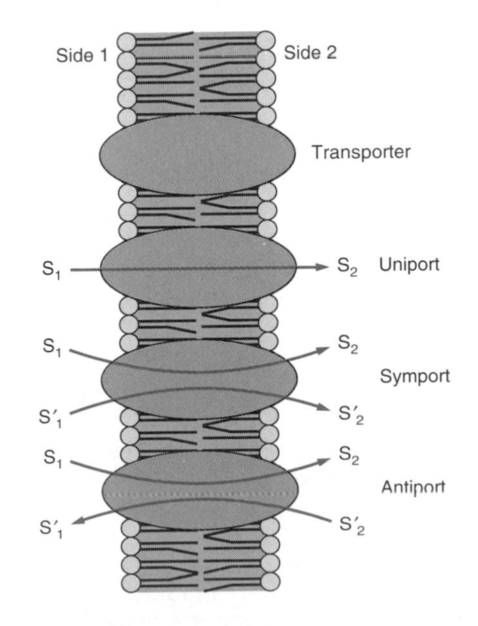

FIGURE 5.31
Uniport, symport, and antiport mechanisms for translocation of substances.
S and S′ represent different molecules.

Energetics of Membrane Transport Systems

The change in free energy when an uncharged molecule moves from a concentration of C_1 to a concentration of C_2 on the other side of a membrane is given by Eq. 5.1

$$\Delta G' = 2.3RT \log \left(\frac{C_2}{C_1}\right) \tag{5.1}$$

When $\Delta G'$ is negative, that is, there is release of free energy, the movement of solute will occur without the need for a driving force. When $\Delta G'$

is positive, as would be the case if C_2 is larger than C_1, then there needs to be an input of energy to drive the transport. For a charged molecule (e.g., Na^+) both the electrical potential and concentrations of solute are involved in calculating the change in free energy as in Eq. 5.2

$$\Delta G' = 2.3RT \log \left(\frac{C_2}{C_1}\right) + Z\mathcal{F}\Psi \quad \textbf{(5.2)}$$

where Z is the charge of the species moving, $\mathcal{F}$ is the Faraday (23.062 kcal V^{-1} mol^{-1}), and Ψ is the difference in electrical potential in volts across the membrane. The electrical component is the membrane potential and $\Delta G'$ is the electrochemical potential.

A passive transport system is one in which $\Delta G'$ is negative, that is, free energy is released, and the movement of solute occurs spontaneously. When $\Delta G'$ is positive, coupled input of energy from some source is required for movement of the solute and the process is called active transport. Several different forms of energy are available for driving active transport systems, including hydrolysis of adenosine triphosphate (ATP) to adenosine diphosphate (ADP), and the electrochemical gradient of the Na^+ ions or of H^+ ions across the membrane. In the first the chemical energy released on hydrolysis of a pyrophosphate bond drives the reaction, whereas in the latter the electrochemical gradient is dissipated to transport the solute.

Transport systems that can maintain very large concentration gradients are present in various membranes. An example is the plasma membrane transport system that maintains the Na^+ and K^+ gradients. One of the most striking examples of an active transport system is that present in the parietal cells of gastric glands, which are responsible for secretion of HCl into the lumen of the stomach (see Chapter 26). The pH of plasma is about 7.4 (4×10^{-8} M of H^+), and the luminal pH of the stomach can reach 0.8 (0.15 M of H^+). The cells transport H^+ against a concentration gradient of $1 \times 10^{6.6}$. Assuming there is no electrical component, the energy for H^+ secretion under these conditions can be calculated from Eq. 5.1 and is 9.1 · kcal mol^{-1} of HCl.

5.6 CHANNELS AND PORES

Channels and Pores in Membranes Function Differently

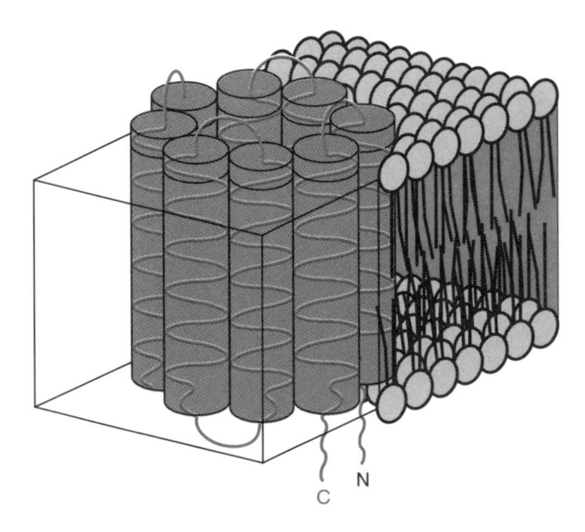

FIGURE 5.32
Arrangement of protein subunits or domains to form a membrane channel.

Membrane channels are differentiated from membrane pores on the basis of their degree of specificity for the molecules crossing the membrane. **Channels** are highly selective for specific inorganic cations and anions, whereas pores are not selective, permitting inorganic and organic molecules to pass through the membrane. The Na^+ channel of the plasma membrane of eucaryotic cells, for example, permits movement of Na^+ at a rate more than 10 times greater than that for K^+. This difference between channels and pores is due to differences in the size of the aqueous area created in the protein structure as well as the amino acid residues lining the channel area. Channels and pores are intrinsic membrane proteins and amino acid sequences in the proteins of many channels suggest the existence of structurally related super families of proteins in which similar amino acid sequences occur. A common motif is a structure formed by amphipatic α-helices of associated protein subunits or from domains within a single polypeptide chain creating a central aqueous space as pictured in Figure 5.32. An exception to the α-helical structure are the porins (see below) of Gram-negative bacteria, which have a β-

sheet structure lining the central pore. The opening and closing of membrane channels involves a conformational change in the channel protein.

Opening and Closing of Channels Are Controlled

As indicated in Table 5.4, the opening and closing of some channels can be controlled by changes in the **transmembrane potential.** These are referred to as **voltage gated channels.** In the case of the Na^+ channel, depolarization of the membrane leads to an opening of the channel. Voltage gated channels for Na^+, K^+, and Ca^{2+} are present in the plasma membrane of most cells and mitochondria have a voltage dependent channel for anions. Another mechanism to control the opening of channels is the binding by the channel protein of a specific agent, termed an **agonist.** Such is the case of the nicotinic-acetylcholine receptor, which on binding acetylcholine opens a channel permitted the flow of Na^+ into the cell. This mechanism is important to neuronal electrical signal transmission (see Chapter 22). Both forms of control for opening the channel are very fast, permitting bursts of ion flow through the membrane at rates of over 10^7 ions s^{-1}, which is near the diffusion rate of these ions in water. This, of course, would be expected considering that these channels are involved in nerve conduction and muscle contraction. A number of pharmacological agents which modulate these channels are used therapeutically.

The Sodium Channel

The voltage sensitive **sodium channel** mediates the rapid increase in intracellular Na^+ following depolarization of the plasma membrane in nerve and muscle cells. The channel consists of a single large glycopolypeptide with several smaller glycoproteins also involved. The genes for some of the Na^+ channels have been cloned and the amino acid sequence determined. It has been suggested that there are four repeat homology units, with four, six, or eight transmembrane α helices. One such model is presented in Figure 5.33*a*. A possible arrangement of these α-helices in the membrane as viewed down on a membrane is presented in Figure 5.33*b*. The channel size created by the protein, however, cannot totally explain the specificity for Na^+. How a change in membrane potential controls the opening and shutting of the channel is still unknown.

The Nicotinic-Acetylcholine Channel (nAChR)

The **nicotinic-acetylcholine channel,** also referred to as the acetylcholine receptor, is an example of a chemically regulated channel, where the binding of acetylcholine (Figure 5.34) opens the channel. The dual name is used to differentiate this receptor from other acetylcholine receptors, which function in a different manner. Acetylcholine is a neurotransmitter that is released at the neuromuscular junction by a neuron when electrically excited. The acetylcholine diffuses to the skeletal muscle membrane where it interacts with the acetylcholine receptor, opening the channel and allowing selective cations to move across the membrane (see Chapter 22). The change in transmembrane potential leds to a series of events culminating in muscle contraction. The nicotinic-acetylcholine receptor has been extensively purified and studied because of its ready availability; the channel consists of five polypeptide subunits, with 2α subunits and one each of β, γ, and δ; each α subunit is phosphorylated and glycosylated and two others contain covalently bound lipid. The channel opens when two acetylcholine molecules bind to the α subunits and cause a change in protein conformation; reclosure of the channel occurs within a millisecond due to the hydrolysis of acetylcholine to acetate and choline

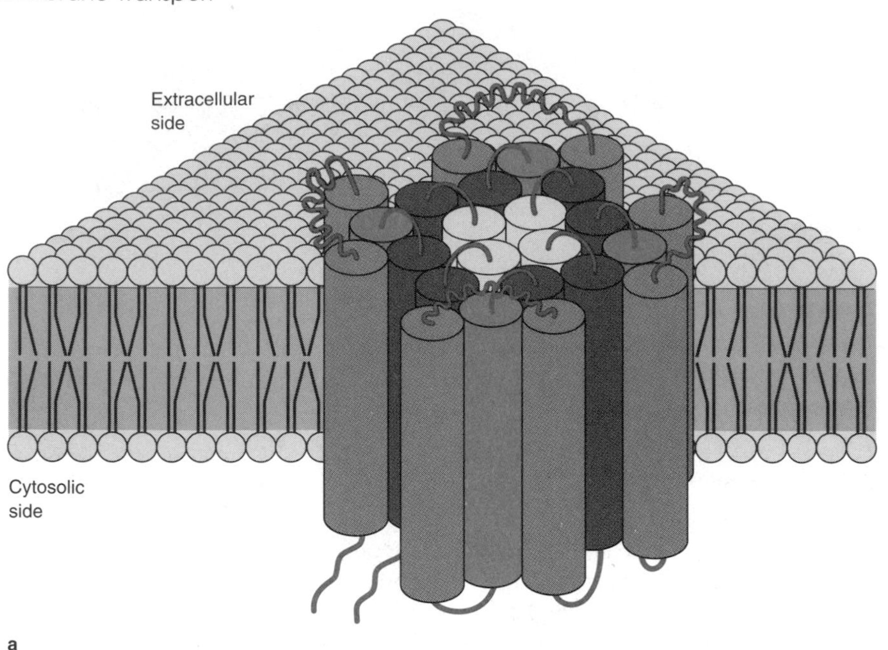

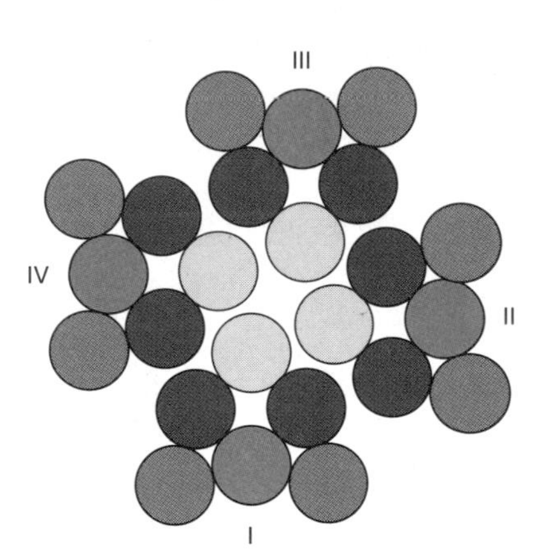

FIGURE 5.33
A possible model of the Na^+ channel.
(*a*) The single peptide consists of four repeating units with each unit folding into six transmembrane helices. (*b*) Proposed arrangement of the transmembrane sequence as viewed down on the membrane.
Redrawn from M. Noda, et al., *Nature (London)* 320:188, 1986.

and the release of the bound ligand. A desensitized state of the receptor has been reported that does not open when acetylcholine binds. In the open conformation, cations and small nonelectrolytes can flow through the channel but not anions; negatively charged amino acid residues in the channel are sufficient to repel negatively charge ions from passing.

The nicotinic-acetylcholine receptor is inhibited by a number of deadly neurotoxins including *d*-tubocurarine, the active ingredient of curare, and several toxins from snakes including α-bungarotoxin, erabutoxin, and cobratoxin, the latter from the cobra. Succinyl choline, a muscle relaxant, activates the channel leading to the depolarization of the membrane; succinyl choline is used in surgical procedures because its activity is reversible due to the rapid hydrolysis of the compound after cessation of administration.

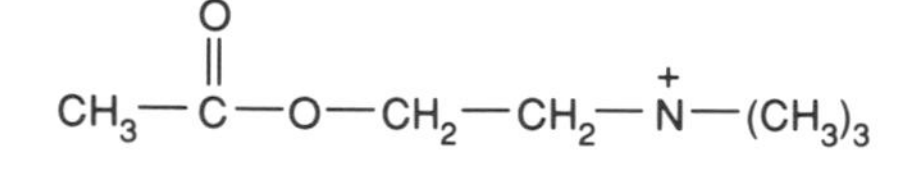

FIGURE 5.34
Structure of acetylcholine.

Examples of Pores Are Gap Junctions and Nuclear Pores

In contrast to channels, plasma membrane gap junctions and nuclear membrane pores are relatively large aqueous openings in the membrane created by specific proteins. **Gap junctions** are clusters of membrane channels that create aqueous connections between two cells; these chan-

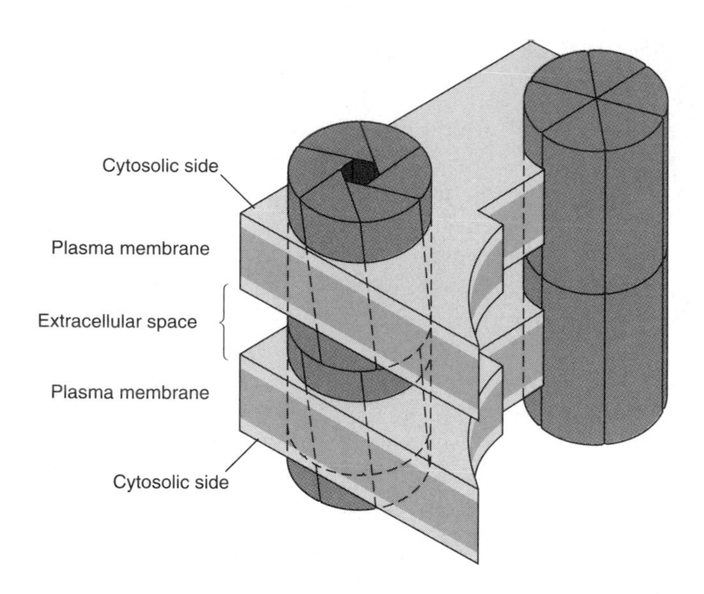

FIGURE 5.35
Model for a channel in the gap junction.

nels are lined by proteins spanning two plasma membranes and permit the exchange between cells of ions and metabolites but not large molecular weight compounds such as proteins. The diameter of the opening ranges from 12 to 20 Å. The channel is composed of an oligomer of a single polypeptide (32 kDa); the gap junction protein, referred to as connexin, is dissimilar from tissue to tissue. The channels, composed of 12 subunits, 6 from each cell, form a hexameric structure in each membrane as shown in Figure 5.35. Even though the channels are normally open, increases in cytosolic Ca^{2+}, a change in metabolism, a drop in transmembrane potential, or acidification of the cytosol mediate closure. When the channel is open the subunits appear to be slightly tilted but when closed they appear to be more nearly parallel to a perpendicular to the membrane, suggesting that the subunits slide over each other. The actual mechanism of opening and closing, however, is unknown.

Like gap junctions **nuclear pores** cover two membranes creating aqueous channels in the nuclear envelope. The pore is about 90 Å in diameter and thus permits the movement of large macromolecules. They are presumably lined with protein but the protein has not been isolated. The plasma membrane of Gram-negative bacteria also contain protein pores, termed porins. Over 40 different porins have been isolated and they range in size from 28 to 48 kDa. In contrast to most mammalian channels, these transmembrane segments are β sheets not α helices and exist in the membrane as trimers. Porins are water filled transmembrane channels and range in diameter from 6 to 23 Å with some degree of selectivity for inorganic ions; some, however, permit the uptake of sugars.

5.7 PASSIVE MEDIATED TRANSPORT SYSTEMS

Passive mediated transport, also referred to as **facilitated diffusion,** leads to the translocation of solutes through cell membranes without the expenditure of metabolic energy (see Table 5.5, p. 218). As with nonmediated diffusion the direction of flow is always from a higher to a lower concentration. The distinguishing differences between measurements of simple diffusion and passive-mediated transport are the demonstration of saturation kinetics, a structural specificity for the class of molecule moving across the membrane, and specific inhibition of solute movement.

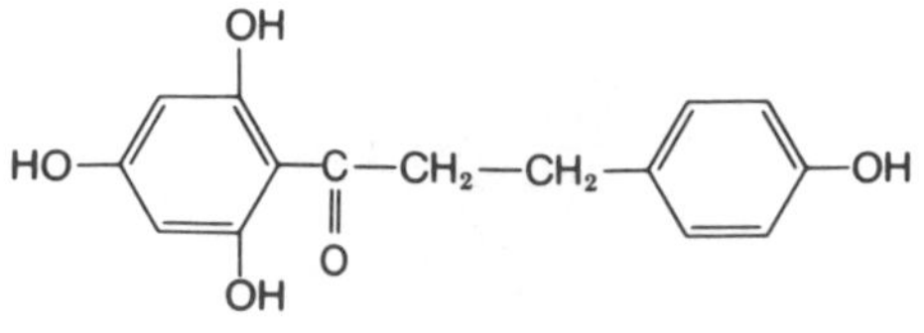

Phloretin

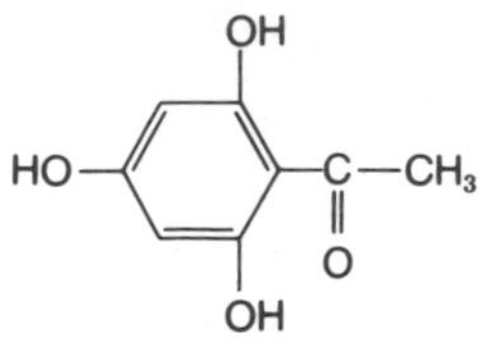

2,4,6-Trihydroxyacetophenone

FIGURE 5.36
Inhibitors of passive mediated transport of D-glucose in erythrocytes.

In Some Membranes Glucose Is Transported Down Its Chemical Gradient

The plasma membrane of many mammalian cells, but not all, has a passive mediated transport system for D-glucose. Most of our knowledge about this system is derived from studies of erythrocytes. The physiological direction of movement is into the cell because the extracellular level of glucose is about 5 mM and most cells metabolize glucose rapidly thus maintaining low intracellular concentrations. Transport is by a uniport mechanism, which demonstrates saturation kinetics and is inhibitable. The system is most active with D-glucose, but D-galactose, D-mannose, and D-arabinose, and several other D-sugars as well as glycerol can be translocated by the same transporter. The L-isomers are not transported. It has been proposed that the β-D-glucopyranose is transported with carrier interaction at the hydrogen atoms on at least C-1, C-3, and C-6 of the sugar. The affinity of the erythrocyte carrier for D-glucose is highest with a K_m of ~6.2 mM, whereas for the other sugars the K_m values are much higher. The carrier has a very low affinity for D-fructose, precluding a role in cellular uptake of fructose. A separate carrier for fructose has been proposed. With isolated erythrocytes, glucose will move either into or out of the erythrocyte, depending on the direction of the experimentally established concentration gradient, demonstrating the reversibility of the system.

Several sugar analogs as well as phoretin and 2,4,6-trihydroxyacetophenone (Figure 5.36) are competitive inhibitors.

Cl^- and HCO_3^- Are Transported by an Antiport Mechanism

Another passive mediated transporter is the **anion transport system** in erythrocytes that involves the movement of Cl^- and HCO_3^- by an antiport mechanism (Figure 5.37). The transporter is also referred to as band 3 because of its position in SDS polyacrylamide gel electrophoresis of erythrocyte membrane proteins. This transporter moves Cl^- in one direction and simultaneously a HCO_3^- in the opposite direction depending on the concentration gradients of the ions across the membrane. The transporter has an important role in adjusting the erythrocyte HCO_3^- concentration in arterial and venous blood (see Chapter 25).

Mitochondria Contain a Number of Passive Transport Systems

The inner mitochondrial membrane contains several antiport systems for the exchange of anions between the cytosol and mitochondrial matrix. These include: (1) a transporter for exchange of ADP and ATP; (2) a transporter for exchange of phosphate and OH^-; (3) a dicarboxylate carrier that catalyzes an exchange of malate for phosphate; and (4) a translocator for exchange of aspartate and glutamate (Figure 5.38). The relationship of these translocases and energy coupling are discussed in Section 6.5. In the absence of an input of energy these transporters will catalyze a passive exchange of metabolites down their concentration gradient to achieve a thermodynamic equilibrium of all intermediates. As an antiport mechanism, a concentration gradient of one compound can drive the movement of the other solute. In several cases, the transporter catalyzes the antiport movement of an equal number of charges on the substrate; in such movement the mitochondrial membrane potential influences the equilibrium and the anions can be moved against their concentration gradients. The ADP–ATP and the phosphate transporter, as well as an uncoupling protein which translocates H^+, have significant amino acid ho-

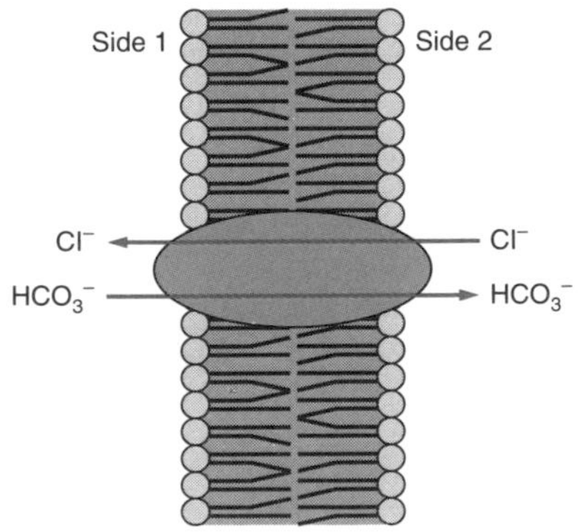

FIGURE 5.37
The passive anion antiport mechanism for movement of Cl^- and HCO_3^- across the erythrocyte plasma membrane.

mology and are presumably derived from a common ancestor by divergent evolution. It has been suggested that each subunit has six transmembrane α helices. The uncoupling protein, which is found in the mitochondria of brown adipose tissue, has been proposed to be involved in generation of heat.

The **ATP–ADP translocase** has been extensively studied; it is very specific for ATP and ADP and their deoxyribose derivatives, dATP and dADP, but does not transport adenosine monophosphate (AMP) and other nucleotides. The protein responsible for the translocation, a dimer containing two subunits of 33 kDa each, represents about 12% of the total protein in heart mitochondria. It is very hydrophobic and can exist in two conformations. Atractyloside and bongkrekic acid (Figure 5.39) are specific inhibitors, each apparently reacting with a different conformation of the protein. The mitochondrial membrane potential can drive the movement of the nucleotides by this translocator, but in the absence of the potential it will function as a passive mediated transporter.

It is sometimes difficult to differentiate passive mediated transport from simple diffusion, but specific inhibition is good evidence of a carrier; this has been the case for the anion carriers of mitochondria, which have been differentiated on the basis of specific inhibitors.

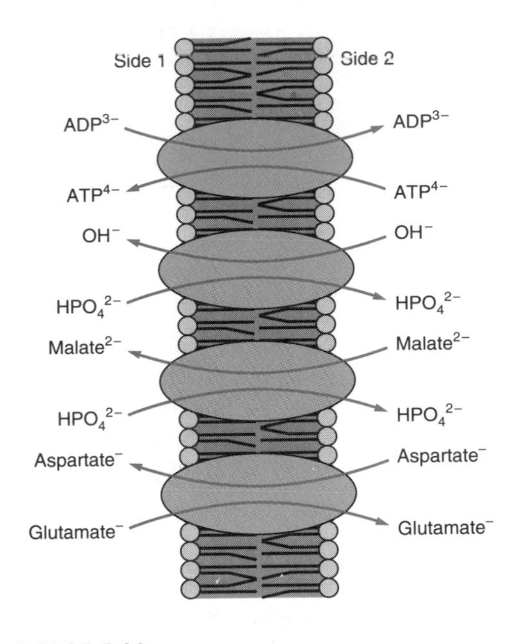

FIGURE 5.38
Repesentative anion transport systems in liver mitochondria.
Note that each is an antiport mechanism. Several other transport systems are known and are discussed in Chapter 6.

5.8 ACTIVE MEDIATED TRANSPORT SYSTEMS

Cellular membranes contain a number of mediated transporters that require the utilization of energy to translocate solutes. Active mediated transporters have the same three characteristics as passive transporters, that is, saturation kinetics, substrate specificity, and inhibitibility, but in addition these translocators require a coupled input of energy (see Table 5.5, p. 218). If the energy source is removed or inhibited, the transport

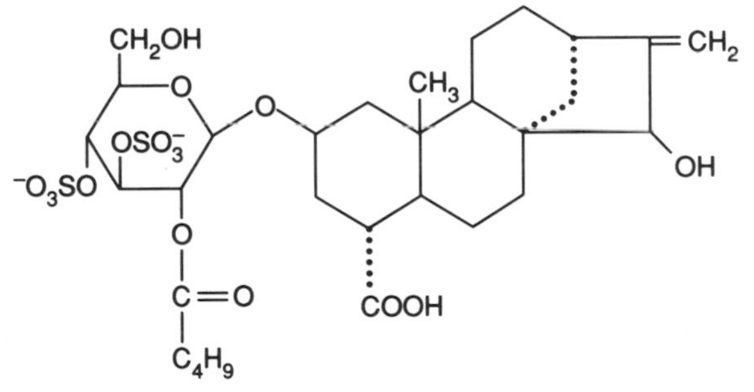

Atractyloside

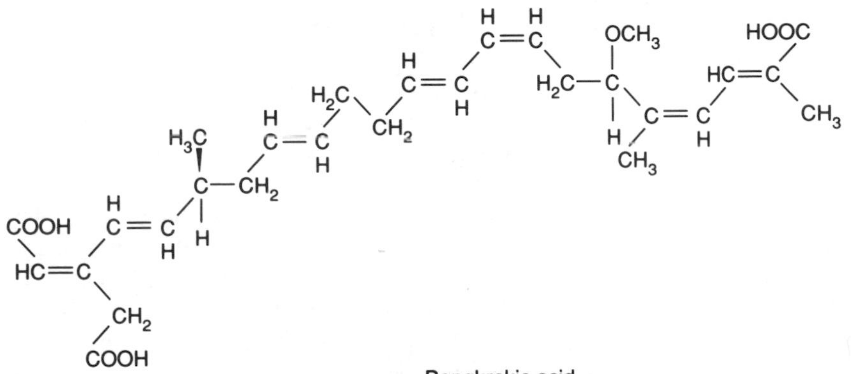

Bongkrekic acid

FIGURE 5.39
Structure of two inhibitors of the ATP–ADP transport system of liver mitochondria.

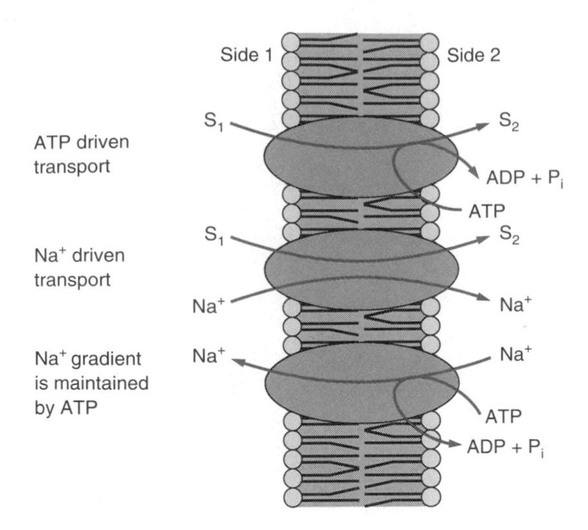

FIGURE 5.40
Involvement of metabolic energy (ATP) in active mediated transport systems.
The chemical energy released on the hydrolysis of ATP to ADP and inorganic phosphate is used to drive the active transport of various substances, including Na^+. The transmembrane concentration gradient of Na^+ is also used for the active transport of substances.

system will not function. These active transporters can be classified as either **primary transporters,** in that they require the direct utilization of ATP, or **secondary transporters** in which the transmembrane electrochemical gradient of Na^+ is utilized. A special case of primary active transport is the translocation of protons across the inner mitochondrial membrane; this mechanism is discussed in detail in Chapter 6, page 285. As indicated below in the discussion on the active mediated transport of glucose, which utilizes the transmembrane Na^+ gradient in a symport mechanism, metabolic energy in the form of ATP is required for maintenance of the Na^+ gradient but not directly for moving a glucose molecule; inhibition of ATP synthesis, however, leads to a dissipation of the Na^+ electrochemical gradient, which in turn decreases transport activity utilizing the gradient. This is visualized in Figure 5.40.

Translocation of Na^+ and K^+ Is a Primary Active Transport System

For years a major research effort has been directed toward an understanding of the cellular mechanism for maintenance of the Na^+ and K^+ gradients across the plasma membrane of cells. All mammalian cells contain a Na^+-K^+ antiporter, which utilizes the direct hydrolysis of ATP for movement of ions. Knowledge of this transporter has developed along two paths: (1) from studies of a membrane enzyme that catalyzes ATP hydrolysis and has a requirement for Na^+ and K^+ ions, and referred to as the **Na^+,K^+-ATPase,** and (2) from measurements of Na^+ and K^+ movements across intact plasma membranes by a protein referred to as a **Na^+,K^+-pump;** the two activities are catalyzed by the same protein.

All Plasma Membranes Contain a Na^+-K^+-Activated ATPase

All mammalian membranes catalyze the reaction

$$\text{ATP} \xrightarrow[\text{Mg}^{2+}]{\text{Na}^+ + \text{K}^+} \text{ADP} + \text{P}_i$$

The enzyme which is termed the Na^+,K^+-ATPase, has a requirement for both Na^+ and K^+ ions, as well as Mg^{2+}, required for ATP-requiring reactions. The level of the ATPase in plasma membranes correlates with the Na^+,K^+-transport activity; excitable tissue such as muscle and nerve have a high capacity of both the Na^+,K^+-ATPase and the Na^+,K^+ transport system as do cells actively involved in the movement of Na^+ ion such as those in the salivary gland and kidney cortex. The protein responsible for the Na^+,K^+ transporting ATPase activity is an oligomer containing two α subunits of about 110,000 Da each and two β subunits of about 55,000 Da each. The smallest subunits are glycoproteins, and the complex has the characteristics of a typical integral membrane protein. Figure 5.41 is a schematic diagram of the Na^+,K^+-transporter. The ATPase activity has a requirement for phospholipids indicating its close relationship to membrane function. During the hydrolysis of ATP, the larger subunit is cyclicly phosphorylated and dephosphorylated on a specific aspartic acid residue forming a β-aspartyl phosphate. Phosphorylation of the protein requires Na^+ and Mg^{2+} but not K^+, whereas dephosphorylation of the protein requires K^+ but not Na^+ or Mg^{2+}. The isolated enzyme has an absolute requirement for Na^+, but K^+ can be replaced with NH_4^+ or Rb^+. Two distinguishable conformations of the protein complex have been observed; this is the basis for referring to the ATPase as an **E_1–E_2 type transporter.** A possible sequence of reactions for the enzyme is presented in Figure 5.42.

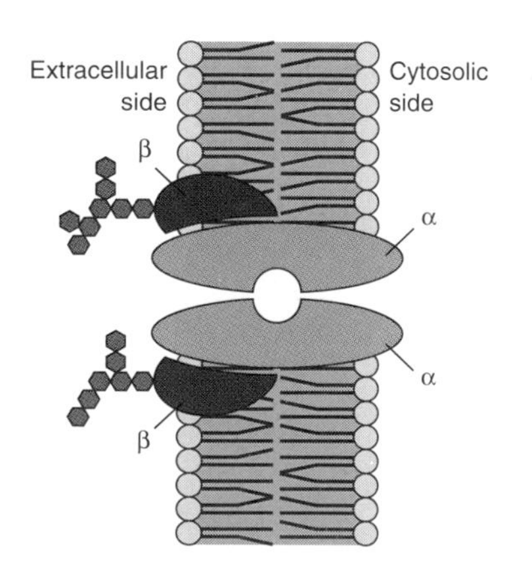

FIGURE 5.41
Schematic drawing of the Na^+,K^+-transporting ATPase of plasma membranes.

Of particular significance to its physiological role as a transporter, the enzyme is inhibited by a series of cardiotonic steroids. These pharmacological agents, which include digitalis, increase the force of contraction of heart muscle by altering the excitability of the tissue, which is a function of the Na^+,K^+ concentration across the membrane. **Ouabain** (Figure 5.43) is one of the most active Na^+,K^+-ATPase inhibitors of the series and its action has been studied extensively. The site of binding of ouabain is on the smaller subunit of the enzyme complex and at some distance from the ATP binding site on the larger monomer. An inhibitor of the transporter, which competes with ouabain binding, has been isolated from human serum and may be involved in the control of Na^+,K^+ transport.

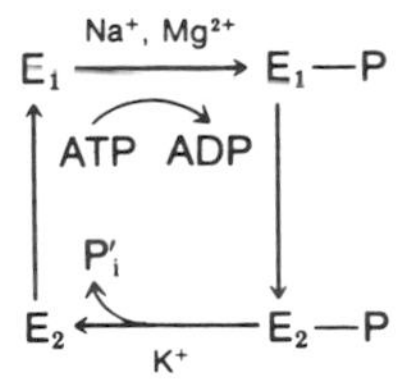

FIGURE 5.42
Proposed sequence of reactions and intermediates in hydrolysis of ATP by the Na^+,K^+-ATPase.
E_1 and E_2 are different conformations of the enzyme. Phosphorylation of the enzyme requires Na^+ and Mg^{2+} and dephosphorylation involves K^+.

Erythrocyte Ghosts Are Used to Study the Na^+-K^+ Transporter

Studies of the **Na^+,K^+ transporter** activity have been facilitated by use of **erythrocyte ghosts,** which are intact erythrocyte preparations free of hemoglobin. By carefully adjusting the tonicity of the medium, erythrocytes will swell with breaks in the phospholipid bilayer, permitting the leaking from the cell of cytosolic material, including hemoglobin. The cytosol can be replaced with a defined medium by readjusting the tonicity so that the membrane reseals, trapping the isolation medium inside. In this manner the intracellular ionic and substrate composition and even protein content can be altered. With erythrocyte ghosts the intra- and extracellular Na^+ and K^+ can be manipulated as well as ATP or inhibitor content. With such preparations it has been demonstrated that movement of Na^+ and K^+ is an antiport vectorial process, with Na^+ moving out and K^+ moving into the cell. The ATP binding site on the protein is on the inner surface of the membrane in that hydrolysis occurs only if ATP, Na^+, and Mg^{2+} are inside the cell. The K^+ ion is required externally for internal dephosphorylation of the protein. Ouabain inhibits the translocation of Na^+ and K^+ but only if it is present externally. The actual number of translocase molecules on an erythrocyte has been estimated by binding studies of radiolabeled ouabain. It is estimated that there are between 100 and 200 molecules per erythrocyte, but the number is significantly larger for other tissues.

ATP hydrolysis by the translocator occurs only if Na^+ and K^+ are translocated, demonstrating that the enzyme is not involved in dissipation of energy in a useless activity. For each ATP hydrolyzed three ions of Na^+ are moved out of the cell but only two ions of K^+ in, which leads to an increase in external positive charges. The electrogenic movement of Na^+ and K^+ is part of the mechanism for the maintenance of the transmembrane potential in tissues. Even though the energetics of the system dictate that it functions in only one direction, the translocator can be reversed by adjusting the Na^+,K^+ levels; a small net synthesis of ATP has been observed when transport is forced to run in the reverse direction. Obviously, under physiological conditions translocation does not occur in the opposite direction.

A hypothetical model for the movement of Na^+ and K^+ is presented in Figure 5.44. It is proposed that the protein goes through a series of conformational changes during which the Na^+ and K^+ are moved short distances. During the transition a change in the affinity of the binding protein for the cations can occur such that there is a decrease in affinity constants, resulting in the release of the cation into a milieu where the concentration is higher than that from which it was transported.

As an indication of the importance of this enzyme, it has been estimated that the Na^+,K^+-translocator uses 60–70% of the ATP synthesized by cells such as nerve and muscle, and may utilize about ~35% of all ATP generated in a resting individual.

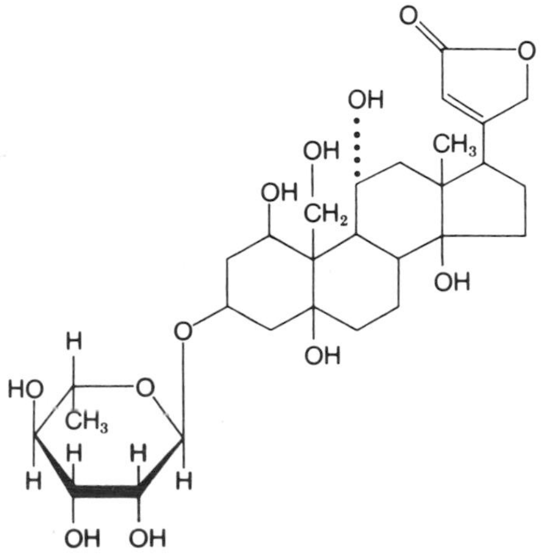

FIGURE 5.43
Structure of ouabain, a cardiotonic steroid, which is a potent inhibitor of the Na^+,K^+-ATPase and of active Na^+ and K^+ transport.

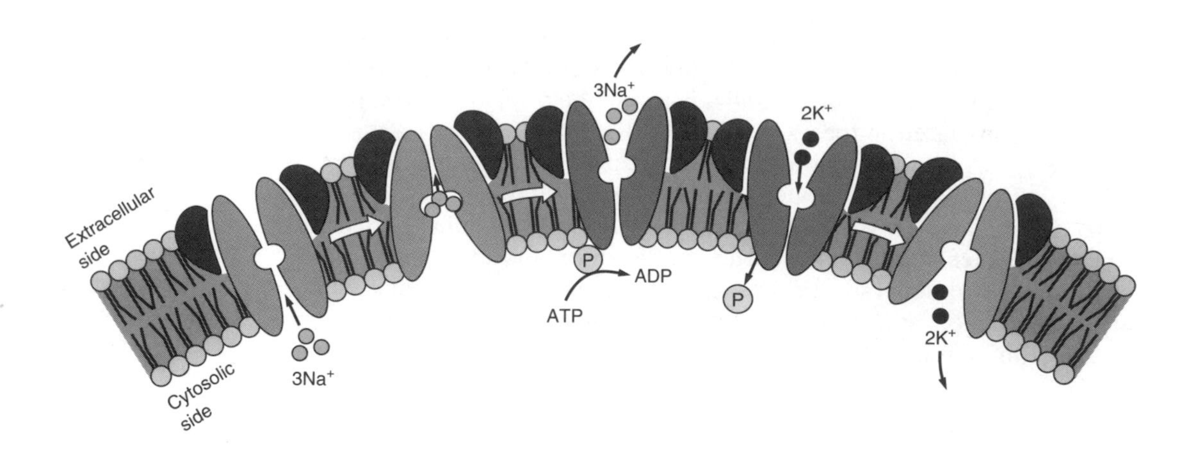

FIGURE 5.44
A hypothetical model for the translocation of Na^+ and K^+ across the plasma membrane by the Na^+,K^+-ATPase.

Ca^{2+} Translocation Is Another Example of a Primary Active Transport System

The Ca^{2+} ion is an important intracellular messenger regulating cellular processes as varied as muscle contraction and carbohydrate metabolism. The signal initiated by many hormones, the primary messenger to signal cells to alter their function, is transmitted in the cell by changes in cytosolic Ca^{2+}; for this reason Ca^{2+} is referred to as a **second messenger.** Cytosolic Ca^{2+} is in the range of 0.10 μM, over 10,000 times lower than extracellular Ca^{2+}. Intracellular Ca^{2+} concentrations can be increased rapidly by (1) the transient opening of Ca^{2+} channels in the plasma membrane, permitting the flow of Ca^{2+} down the large concentration gradient or (2) by release from the stores of Ca^{2+} in the endoplasmic or sarcoplasmic reticulum. In order to reestablish the low cytosolic levels, Ca^{2+} is actively transported out of the cell across the plasma membrane or into the endoplasmic or sarcoplasmic reticulum. With both membrane systems, a **Ca^{2+} transporter** of the E_1–E_2 type is involved in which ATP is hydrolyzed during the translocation. Thus, the transporter catalyzes a Ca^{2+} stimulated ATPase activity. As with the Na^+,K^+-transporter, both the Ca^{2+} translocation activity and ATPase activity have been studied extensively.

The **Ca^{2+}-ATPase** of the sarcoplasmic reticulum of muscle, which is involved in the contraction–relaxation cycles of muscle (see Chapter 2), has many properties similar to the Na^+,K^+-ATPase and represents 80% of the integral membrane protein of the sarcoplasmic reticulum and occupies one-third of the surface area. The protein has 10 membrane spanning helicies and is phosphorylated on an aspartyl residue during the Ca^{2+} translocation reaction. Two Ca^{2+} ions are translocated for each ATP hydrolyzed and the energetics are such that it can move Ca^{2+} against a very large concentration gradient.

The Ca^{2+} transporter of the plasma membrane has properties similar to the enzyme of the sarcoplasmic reticulum. In eucaryotic cells, the transporter is regulated by the cytosolic Ca^{2+} levels through a calcium binding protein termed **calmodulin** (CaM). As cellular Ca^{2+} levels rise, Ca^{2+} is bound to calmodulin, which has a dissociation constant of $\sim$1 μM. The Ca^{2+}–calmodulin complex binds to the Ca^{2+} transporter leading to an increased rate in Ca^{2+} transport. The rate is increased by lowering the K_m for Ca^{2+} of the transporter from about 20 to 0.5 μM. The increased activity reduces cytosolic Ca^{2+} to its normal resting level (~0.10 μM) at which concentration the Ca^{2+}–calmodulin complex dissociates and the activity of the Ca^{2+} transporter returns to a lower value. Thus the Ca^{2+}–calmodulin complex exerts fine control on the Ca^{2+} transporter. Calmodulin is one of several Ca^{2+} binding proteins, including parvalbumin and troponin C, all of which have very similar structures. The Ca^{2+}–calmodulin complex is also involved in control of other cellular processes, which are affected by Ca^{2+}. The protein (17 kDa) has the shape of a dumbbell with two globular ends connected by a seven turn α helix; there are four Ca^{2+} binding sites, two high affinity on one lobe and two low affinity on the other. It is believed that the binding of Ca^{2+} to the lower affinity binding sites causes a conformational change in the protein revealing a hydrophobic area that can interact with a protein that it controls. Each Ca^{2+} binding site consists of a **helix–loop–helix** structural motif (Figure 5.45) and Ca^{2+} is bound in the loop connecting the helicies. A similar structure is found in other Ca^{2+} binding proteins. The motif is referred to as the **EF hand,** based on studies with provabumin where the Ca^{2+} is bound between helices E and F of the protein.

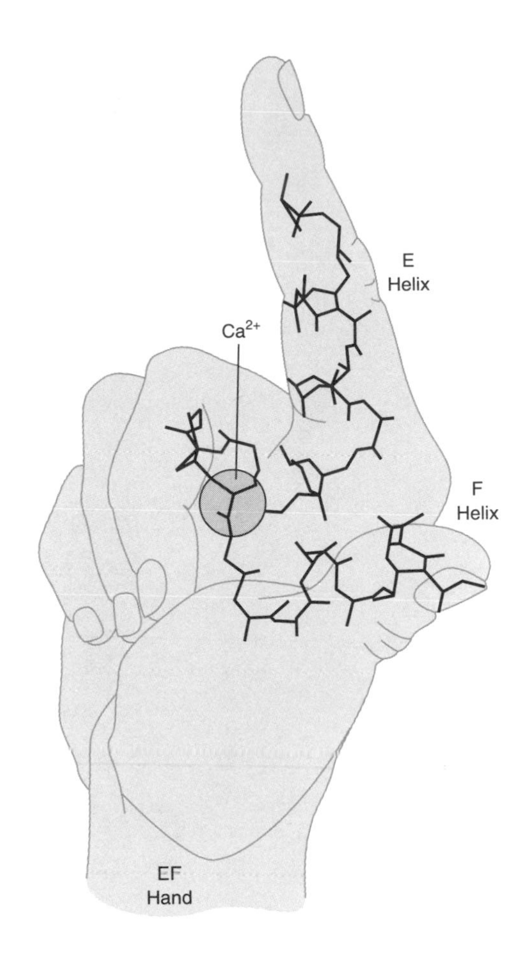

FIGURE 5.45
A binding site for Ca^{2+} in calmodulin. Calmodulin contains four Ca^{2+} binding sites, each with a helix–loop–helix motif. The Ca^{2+} ion is bound in the loop that connects two helices. This motif occurs in various Ca^{2+} binding proteins and is referred to as the EF hand.

Na^+-Dependent Transport of Glucose and Amino Acids Are Secondary Active Transport Systems

The mechanisms described above for the active transport of cations involve the direct hydrolysis of ATP as the driving force. *Cells have another energy source, the electrochemical gradient of Na^+ ion, which is utilized to move sugars, amino acids, and Ca^{2+} actively.* A symport translocation system involving the simultaneous movement of both a Na^+ ion and glucose in the same direction is present in the plasma membrane of cells of the kidney tubule and intestinal epithelium; other tissues may also contain similar transport systems. The general mechanism is presented in Figure 5.46. The diagram represents the transport of D-glucose driven by the movement of Na^+ ion down its concentration gradient. Note that in the transport of the sugar no hydrolysis of ATP occurs. There is an absolute requirement for Na^+, and in the process of translocation one Na^+ is moved with each glucose molecule. It can be considered that Na^+ is moving by passive facilitated transport down its electrochemical gradient. It is obligatory that the transporter also translocates a glucose with the Na^+ ion. In the transport the electrochemical gradient of Na^+ ion is dissipated and glucose can be translocated against its concentration gradient. Unless the Na^+ ion gradient is continuously regenerated, the movement of glucose will cease. The Na^+ gradient is maintained by the Na^+,K^+-transport system described above and also represented in Figure 5.46. Thus, metabolic energy in the form of ATP is indirectly involved in glucose transport because it is utilized to maintain the Na^+ ion gradient. Inhibition of ATP synthesis and a subsequent decrease in ATP will alter the Na^+ ion gradient and inhibit glucose uptake. Ouabain, the inhibitor of the Na^+,K^+-transporter, also inhibits uptake of glucose by preventing the cell from maintaining the Na^+,K^+ gradient. Each glucose molecule requires only one-third of an ATP to be translocated because three Na^+ ions are translocated for the hydrolysis of each ATP in the Na^+,K^+-transporting ATPase.

Amino acids are also translocated by the luminal epithelial cells of the intestines by Na^+-dependent pathways similar to the Na^+-dependent glucose transporter. At least four different translocators have been identified: (1) one for neutral amino acids such as alanine, valine, and leucine;

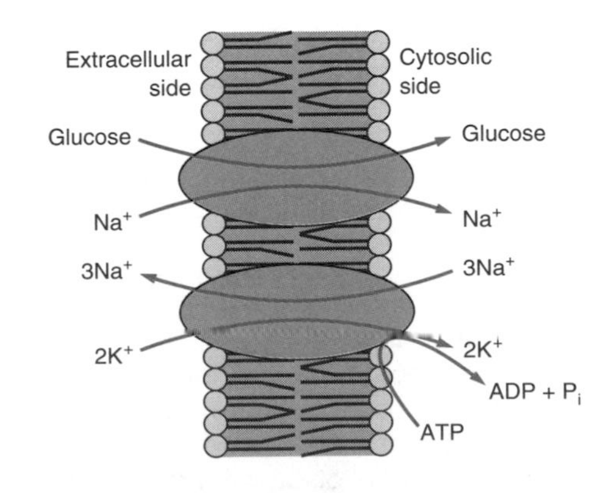

FIGURE 5.46
Na^+-dependent symport transport of glucose across the plasma membrane.

(2) one for basic amino acids, including lysine and arginine; (3) one for the acidic amino acids, aspartate and glutamate; and (4) one for the amino acids proline and glycine.

The Na^+ gradient is also utilized to drive the transport of other ions, including a symport mechanism in the small intestines for the uptake of Cl^- with Na^+ and an antiport mechanism for the excretion of Ca^{2+} out of the cell.

The chemical mechanism for the symport movement of molecules utilizing the Na^+ ion gradient involves a cooperative interaction of the Na^+ ion and the other molecule translocated on the protein. A conformational change of the protein occurs following association of the two ligands, which moves them the necessary distance to bring them into contact with the cytosolic environment. The dissociation of the Na^+ ion from the transporter because of the low Na^+ ion concentration inside the cell leads to a return of the protein to its original conformation, a decrease in the affinity for the other ligand, and a release of the ligand into the cytosol.

Group Translocation Involves the Chemical Modification of the Substrate Transported

As discussed previously, a major hurdle for any active transport system is the release of the transported molecule from the binding site after translocation. If the affinity of the transporter for the translocated molecule does not change there cannot be movement against a concentration gradient. In the active transport systems previously described it is believed that there is a change in the affinity for the substance by the transporter by a conformational change of the protein. An alternate mechanism for the release of the substrate is the chemical alteration of the molecule after translocation but before release from the transporter leading to a new compound with a lower affinity for the transporter. The **γ-glutamyl cycle** for the transport of amino acids across the plasma membrane of some tissues is an example where the substrate is altered during transport and released into the cell as a different molecule. The reactions of the transport mechanism are presented in Figure 5.47. The pathway involves the enzyme γ-glutamyltransferase, which is membrane bound and catalyzes a transpeptidation reaction, leading to the formation of a dipeptide involving the amino acid transported. The amino acid transported is the substrate to which the γ-glutamyl residue of glutathione (Figure 5.48) is transferred. The new dipeptide is not part of the chemical gradient for the amino acid across the membrane. The γ-glutamyl derivative is then hydrolyzed by a separate

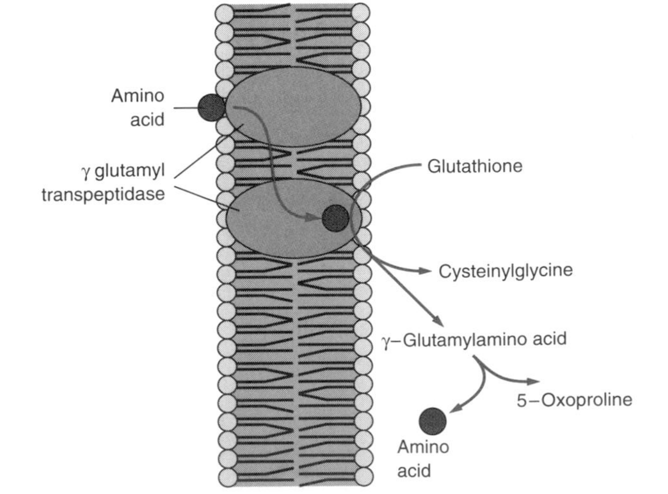

FIGURE 5.47
The γ-glutamyl cycle.
Represented are the key reactions involved in the group translocation of amino acids across liver cell plasma membranes. The continued uptake of amino acids requires the constant resynthesis of glutathione via a series of ATP requiring reactions described in Chapter 12, page 524.

enzyme, not on the membrane, leaving the free amino acid and oxoproline. The process is termed **group translocation.**

The pathway is active in many tissues but some doubt has been raised about its physiological significance in that individuals have been identified with a genetic absence of the γ-glutamyltransferase activity without any apparent difficulty in amino acid transport. It is possible, of course, that cells may have several alternate methods for the transport of amino acids and are not dependent on only one mechanism.

All the amino acids except proline can be transported by this group translocation process. The energy for transport comes from the hydrolysis of a peptide bond in glutathione. For the system to continue, glutathione must be resynthesized, which requires the expenditure of three ATP molecules. Thus for each amino acid translocated, three ATPs are required. Recall that the expenditure of only one-third of an ATP is required for each amino acid transported in the Na^+-dependent translocase system. This group translocation is an expensive energetic mechanism for the transport of amino acids.

A group translocation mechanism for the uptake of sugars is found in bacteria. This pathway involves the phosphorylation of the sugar, using phosphoenolpyruvate as the phosphate donor. The mechanism is referred to as the **phosphoenolpyruvate dependent phosphotransferase system** (PTS).

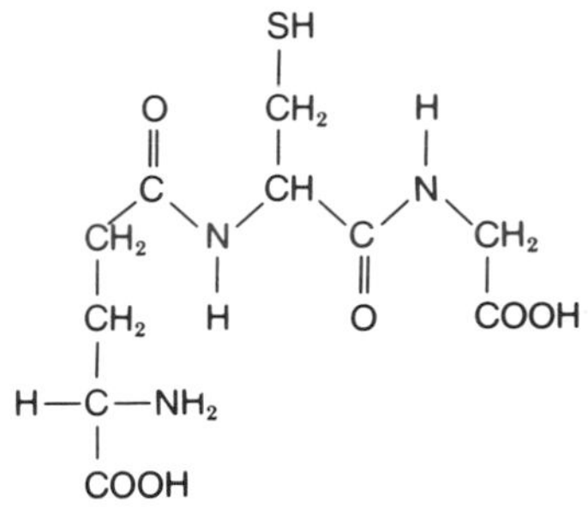

FIGURE 5.48
Glutathione (γ-glutamylcysteinylglycine).

Summary of Transport Systems

The foregoing has presented the major mechanisms for the movement of molecules across cellular membranes, particularly the plasma membrane. Mitochondria also contain several active transport mechanisms utilizing the pH gradient, that is, a hydrogen ion gradient, developed across the inner membrane. These will be presented in Chapter 6. Bacteria have a number of transport systems analogous to those observed in mammalian cells including passive mediated transporters, one involving a H^+ ion gradient and group translocation.

Table 5.6 summarizes some of the characteristics of the major transport systems found in mammalian cells (see Clin. Corr. 5.3).

5.9 IONOPHORES

An interesting class of antibiotics of bacterial origin has been discovered, which facilitates the movement of monovalent and divalent inorganic ions across biological and synthetic lipid membranes. These molecules are not large macromolecules such as proteins but are relatively small molecular weight compounds (up to several thousand daltons); the class of compounds are called **ionophores.** Ionophores are divided into two major groups: (1) mobile carriers are those ionophores which diffuse back and forth across the membrane carrying the ion from one side of the membrane to the other, and (2) ionophores which form a channel that transverses the membrane and through which ions can diffuse. With both types, ions are translocated by a passive-mediated transport mechanism. The ionophores which diffuse back and forth across the membrane are more affected by the changes in the fluidity of the membrane than those that form a channel. Some major ionophores are listed in Table 5.7.

Each ionophore has a definite ion specificity; **valinomycin,** whose structure is given in Figure 5.49, has an affinity for K^+ 1,000 times greater than that for Na^+, and the antibiotic A23187 (Figure 5.50) translocates Ca^{2+} 10 times more actively than Mg^{2+}. Several of the diffusion type ionophores

TABLE 5.6 Major Transport Systems in Mammalian Cells[a]

Substance transported	Mechanism of transport	Tissues
Sugars		
Glucose	Passive	Most tissues
	Active symport with Na^+	Small intestines and renal tubular cells
Fructose	Passive	Intestines and liver
Amino acids		
Amino acid specific transporters	Active symport with Na^+	Intestines, kidney, and liver
All amino acids except proline	Active group translocation	Liver
Specific amino acids	Passive	Small intestine
Other organic molecules		
Cholic acid, deoxycholic acid, and taurocholic acid	Active symport with Na^+	Intestines
Organic anions, e.g., malate, α-ketoglutarate, glutamate	Antiport with counterorganic anion	Mitochondria
ATP–ADP	Antiport transport of nucleotides; can be active transport	Mitochondria
Inorganic ions		
H^+	Active	Mitochondria
Na^+	Passive	Distal renal tubular cells
Na^+,H^+	Active antiport	Proximal renal tubular cells and small intestines
Na^+,K^+	Active ATP driven	Plasma membrane of all cells
Ca^{2+}	Active ATP driven	Plasma membrane and endoplasmic (sarcoplasmic) reticulum
Ca^{2+},Na^+	Active antiport	Most tissues
H^+,K^+	Active antiport	Parietal cells of gastric mucosa secreting H^+
Cl^-/HCO_3^-	Passive antiport	Erythrocytes and many other cells

[a] The transport systems are only indicative of the variety of transporters known; others responsible for a variety of substances have been proposed. Most systems have been studied in only a few tissues and their localization may be more extensive than indicated.

TABLE 5.7 Major Ionophores

Compound	Major cations transported	Action
Valinomycin	K^+ or Rb^+	Uniport, electrogenic
Nonactin	NH_4^+, K^+	Uniport, electrogenic
A 23187	$Ca^{2+}/2H^+$	Antiport, electroneutral
Nigericin	K^+/H^+	Antiport, electroneutral
Monensin	Na^+/H^+	Antiport, electroneutral
Gramicidin	H^+, Na^+, K^+, Rb^+	Forms channels
Alamethicin	K^+, Rb^+	Forms channels

CLIN. CORR. 5.3 DISEASES DUE TO LOSS OF MEMBRANE TRANSPORT SYSTEMS

A number of pathological conditions are due to an alteration in the transport systems for specific cellular components. Some of these are discussed in the appropriate sections describing the metabolism of the intermediates. Individuals have been observed with a decrease in glucose uptake from the intestinal tract due to a loss of the specific sodium-coupled glucose–galactose transporter. Fructose malabsorption syndromes have been observed, which are due to an alteration in the activity of the transport system for fructose. In Hartnup's disease there is a decrease in the transport of neutral amino acids in the epithelial cells of the intestine and renal tubules.

In cystinuria, the renal reabsorption of cystine and the basic amino acids lysine and arginine is abnormal, resulting in formation of cystine kidney stones. In hypophosphatemic, vitamin D resistant rickets, renal absorption of phosphate is abnormal. Little is known concerning possible changes of transport activities in tissues such as muscle, liver, and brain but it has been suggested that there may be a number of pathological states due to the loss of specific transport mechanisms.

Evans, L., Grasset, E., Heyman, M., Dumontier, A. M., Beau, J. P., and Desjeux, J. F. Congenital selective malabsorption of glucose and galactose. *J. Pediatr. Gastroenterol. Nutr.* 4:878, 1985.

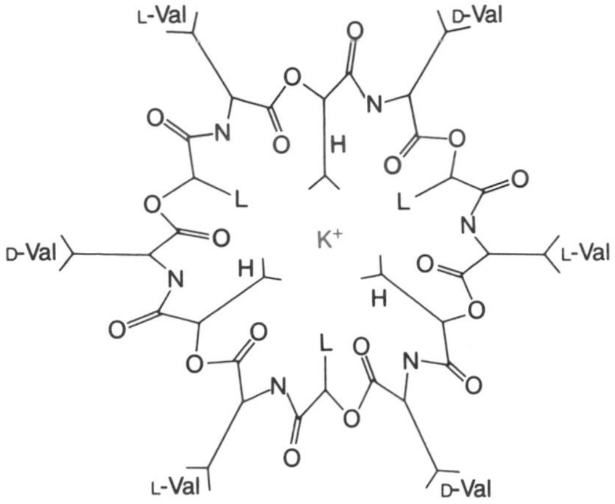

FIGURE 5.49
The structure of the valinomycin–K^+ complex.
Abbreviations: D-Val = D-valine; L-Val = L-valine; L = L-lactate; and H = D-hydroxyisovalerate.

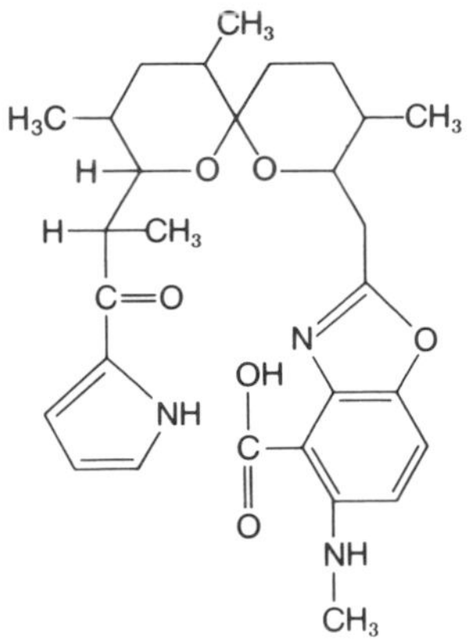

FIGURE 5.50
Structure of A23187, a Ca^{2+} ionophore.

have a common structural characteristic being cyclic structures, shaped like a doughnut. The metal ion is coordinated to several oxygen atoms in the core of the ionophore, and the periphery of the molecule consists of hydrophobic groups. The interaction of the ionophore leads to a chelation of the ion, stripping away its surrounding water shell and encompassing the ion by a hydrophobic shell. The ionophore–ion complex is soluble in the lipid membrane and freely diffuses across the membrane. Since the interaction of ion and ionophore is an equilibrium reaction, a steady state develops in the concentration of ions on both sides of the membrane. The specificity of the ionophore is due in part to the size of the pore into which the ion fits and to the attraction of the ionophore for the ion in competition with water molecules.

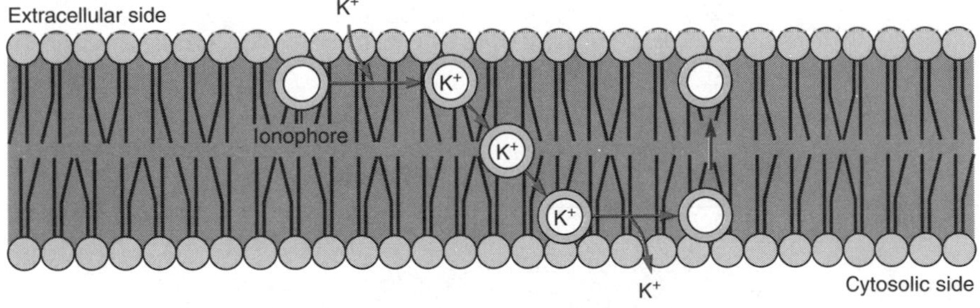

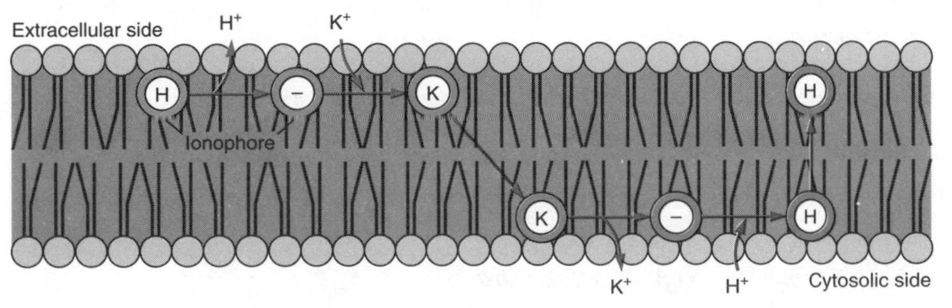

FIGURE 5.51
Proposed mechanism for the ionophoretic activities of valinomycin and nigericin.
(*a*) Transport by valinomycin. (*b*) Transport by nigericin. I represents the ionophore. The valinomycin–K^+ complex is positively charged and translocation of K^+ is electrogenic leading to the creation of a charge separation across the membrane. Nigericin translocates K^+ in exchange for a H^+ across the membrane and the mechanism is electrically neutral.
Diagram adopted from B. C. Pressman, *Annu. Rev. Biochem.* 45:501, 1976.

Valinomycin transports K^+ by a uniport mechanism and is thus electrogenic, that is, it can create an electrochemical gradient. It carries a positive charge in the form of the K^+ ion across the membrane. **Nigericin** functions as an antiporter having a free carboxyl group, which when dissociated can pick up a K^+ ion, leading to a neutral molecule. Thus on diffusion back through the membrane it transports a proton; the overall mechanism is electrically neutral, with a K^+ ion exchanging for a H^+ ion. These mechanisms are shown in Figure 5.51.

Gramicidin A is an example of an ionophore that creates a pore in the membrane. This type of ionophore has a lower degree of selectivity toward ions, in that the ions are essentially diffusing through a hole in the membrane. Two molecules of gramicidin A form a channel and the dimer is in constant equilibrium with the free monomer form. By the association and dissociation of the monomers in the membrane, channels can be formed and broken; the rate of interaction of two molecules of gramicidin A controls the rate of ion flux. The structure of the molecule suggests that polar peptide groups line the channel and hydrophobic groups are on the periphery of the channel interacting with the lipid membrane.

The antibiotic ionophores have been a valuable experimental tool in studies involving ion translocation in biological membranes and for the manipulation of the ionic compositions of cells. There have been reports that proteolipids, prostaglandins, and perhaps other lipids present in mammalian tissues may function as ionophores.

BIBLIOGRAPHY

General

Andreoli, T. E., Hoffman, J. F., and Fanestil, D. D. (Eds.). *Physiology of membrane disorders.* New York: Plenum, 1978.

Bittar, E. E. (Ed.). *Membrane structure and function,* Vols. 1–3. New York: Wiley, 1980.

Bonner, F. *Current topics in membranes and transport.* A series published regularly. New York: Academic.

Gennis, R. B. *Biomembranes: molecular structure and function.* New York: Springer-Verlag, 1989.

Houslay, M. D. and Stanley, K. K. *Dynamics of biological membranes.* New York: Wiley, 1982.

Shohet, S. B. and Lux, S. E. The erythrocyte membrane skeleton: Biochemistry and pathophysiology. *Hosp. Pract.* 19, 77–83 and 89–108, 1984.

Weissmann, G. and Claiborne, R. (Eds.). *Cell membranes,* New York: H. P. Publishing Co., 1975.

Membrane Structure

Devaux, P. F. Static and dynamic lipid asymmetry in cell membranes, *Biochem.* 30:1163, 1991.

Finean, J. B. and Michell, R. H. (Eds.). *Membrane structure.* Amsterdam: Elsevier, 1981.

Lipowsky, R. The conformation of membranes, *Nature* 349:475, 1991.

Low, M. and Saltiel, A. R. Structural and functional roles of glycosylphosphatidyl inositol in membranes. *Science* 239:268, 1988.

Marchesi, V. T., Furthmayr, H., and Tomita, M. The red cell membrane. *Annu. Rev. Biochem.* 45:667, 1976.

Martonosi, A. N. (Ed). *The enzymes of biological membranes; membrane structure and dynamics,* Vol. 1. New York: Plenum, 1985.

McMurchie, E. J. Dietary lipids and the regulation of membrane fluidity and function. *Physiological Regulation of Membrane Fluidity,* New York: Liss, 1988, p. 189.

Quinn, P. J., and Chapman, D. The dynamics of membrane structure. *CRC Crit. Rev. Biochem.* 8:1, 1980.

Singer, S. J. and Nicolson, G. L. The fluid mosaic model of the structure of cell membranes. *Science* 175:720, 1972.

Stubbs, C. D. Membrane fluidity: Structure and dynamics of membrane lipids, P. N. Campbell and R. D. Marshall (Eds.), *Essays in biochemistry,* Vol. 19, 1983, p. 1.

Wallach, D. F. H. *Membrane molecular biology of neoplastic cells.* Amsterdam: Elsevier, 1975.

Yeagle, P. L. Lipid regulation of cell membrane structure and function. *FASEB J.* 3:1833, 1989.

Transport Processes

Christensen, H. N. *Biological transport,* 2nd ed. Reading, MA: Benjamin, 1975.

Eisenman, G. and Dani, J. A. An introduction to molecular architecture and permeability of ion channels. *Ann. Rev. Biophys. Chem.* 16:205, 1987.

Martonosi, A. (Ed.). *The enzymes of biological membranes: membrane transport,* Vol. 3, New York: Plenum, 1985.

Miller, M., Park, N. K., and Hanover, J. A. Nuclear pore complex: structure function and regulation, *Physiol. Rev.* 71:909, 1991.

Skou, J. C. and Norby, D. G. (Eds.). *Na^+,K^+-ATPase: structure and kinetics.* New York: Academic, 1979.

Stein, W. D. *Transport and diffusion across cell membranes.* New York: Academic, 1986.

Wilson, D. B. Cellular transport mechanisms. *Annu. Rev. Biochem.* 47:933, 1978.

QUESTIONS

C. N. ANGSTADT AND J. BAGGOTT

*Q*uestion *T*ypes are described on page xxiii.

1. (QT2) Cell membranes typically:
   1. contain phospholipids.
   2. have both integral and peripheral proteins.
   3. have some cholesterol.
   4. contain free carbohydrate such as glucose.

   A. Phosphatidylcholines C. Both
   B. Sphingomyelins D. Neither
2. (QT4) Contain a fatty acid in an ether linkage
3. (QT4) Contain phosphate

   A. Cerebroside C. Both
   B. Ganglioside D. Neither
4. (QT4) Incorporate(s) an oligosaccharide containing sialic acid
5. (QT4) Belong(s) to the class of neutral glycosphingolipids

6. (QT2) According to the *fluid mosaic* model of a membrane:
   1. proteins may be embedded in the lipid bilayer.
   2. transverse movement (flip–flop) of a protein in the membrane is thermodynamically favorable.
   3. the transmembrane domain has largely hydrophobic amino acids.
   4. proteins are distributed symmetrically in the membrane.
7. (QT2) Characteristics of a mediated transport system include:
   1. the ability to bind an appropriate solute specifically.
   2. release of the transporter from the membrane following transport.
   3. a mechanism for translocating the solute from one side of the membrane to the other.
   4. release of the solute only if the concentration on the new side is lower than that on the original side.
8. (QT2) Membrane channels:
   1. have a large aqueous area in the protein structure so are not very selective.
   2. commonly contain amphipathic α helices.
   3. are opened or closed only as a result of a change in the transmembrane potential.
   4. may form clusters that create gap junctions between two cells.

   A. Passive mediated transport C. Both
   B. Active mediated transport D. Neither
9. (QT4) Require(s) a transporter that specifically binds a solute.
10. (QT4) Can transport a solute against its concentration gradient.

11. (QT1) The transport system that maintains the Na^+ and K^+ gradients across the plasma membrane of cells:
    A. involves an enzyme that is an ATPase.
    B. is a symport system.
    C. moves Na^+ either into or out of the cell.
    D. is an electrically neutral system.
    E. in the membrane, hydrolyzes ATP independently of the movement of Na^+ and K^+.

12. (QT2) A mediated transport system would be expected to:
    1. show a continuously increasing initial rate of transport with increasing substrate concentration.
    2. exhibit structural and/or stereospecificity for the substance transported.
    3. be slower than that of a simple diffusion system.
    4. establish a concentration gradient across the membrane if there is an expenditure of energy.

The answers to Questions 13 and 14 are based on the following figure:

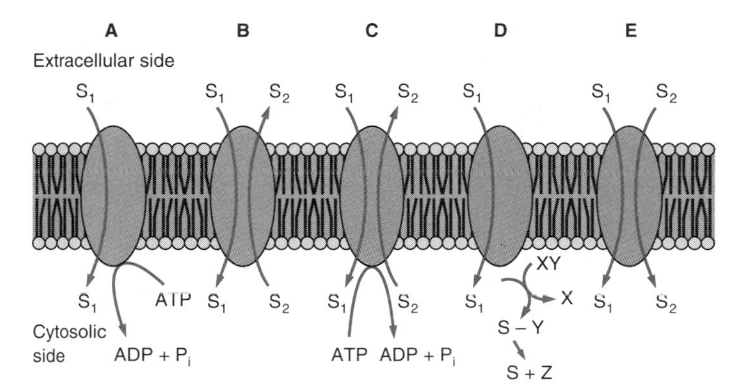

13. (QT5) Represents a passive mediated antiport system
14. (QT5) Could represent the Na^+-driven uptake of glucose

15. (QT1) The translocation of Ca^{2+} across a membrane:
    A. is a passive mediated transport.
    B. is an example of a symport system.
    C. involves the phosphorylation of a serine residue by ATP.
    D. may be regulated by the binding of a Ca^{2+}–calmodulin complex to the transporter.
    E. maintains $[Ca^{2+}]$ very much higher in the cell than in extracellular fluid.
16. (QT1) The group translocation type of transport system:
    A. does not require metabolic energy.
    B. involves the transport of two different solute molecules simultaneously.
    C. has been demonstrated for fatty acids.
    D. results in the alteration of the substrate molecule during the transport process.
    E. uses ATP to maintain a concentration gradient.

17. (QT2) An ionophore:
    1. may diffuse back and forth across a membrane.
    2. may form a channel across a membrane through which an ion may diffuse.
    3. may catalyze electrogenic mediated transport of an ion.
    4. requires the input of metabolic energy for mediated transport of an ion.

ANSWERS

1. A 1, 2, and 3 correct (Figures 5.2 and 5.18). 4: All carbohydrate in membranes is in the form of glycoproteins and glycolipids (p. 205).
2. D Phosphatidylcholines have ester linkages and a sphingomyelin binds a fatty acid in amide linkage (pp 197–201).
3. C A sphingomyelin is the only type of sphingolipid-containing phosphate (Figures 5.6 and 5.13).
4. B Cerebrosides have only a single sugar (p. 202).
5. A Cerebrosides are neutral; no phosphate; uncharged sugar. Gangliosides, by virtue of the presence of sialic acid, are acidic. *Note:* Sulfatides, which are acidic, are derived from cerebrosides but are not, themselves, classified as cerebrosides (p. 203).
6. B 1 and 3 correct (Figure 5.21). 2: Transverse motion of proteins is even less than that of lipids (p. 214). 3: Hydrophilic domains will be at either surface of the membrane (p. 209). 4: Both proteins and lipids are distributed asymmetrically (p. 212).
7. B 1 and 3 correct. 2: Recovery of the transporter to its original condition is one of the characteristics of mediated transport (p. 219). 4: Active transport, movement against a gradient, is also mediated transport.
8. C 2 and 4 correct. 1: This describes a pore; channels are quite specific. 3: Voltage gated channels, like that for Na^+, are controlled this way but others, like the nicotinic acetylcholine channel, are chemically regulated (p. 220).
9. C Specific binding by the transporter is a characteristic of mediated systems (p. 219).
10. B Transportation against a gradient requires the input of energy (p. 225).
11. A The Na^+-K^+-transporter is the Na^+-K^+-ATPase. It is an antiport, vectorial (Na^+ out), electrogenic ($3Na^+$, $2K^+$) system. ATP hydrolysis is not useless (p. 226).
12. C 2 and 4 correct. 1, 2: Mediated transport systems show saturation kinetics and substrate specificity (p. 218). 3: The purpose of the transporter is to aid the transport of water-soluble substances across the lipid membrane (p. 217). 4: This is a characteristic of active transport that is a mediated transport.
13. B The figure is a modified composite of Figures 5.31, 5.38, 5.40, and 5.46.
14. E All systems are mediated. A: An active uniport. B: A passive antiport; for example, Cl^--HCO_3^-. C: An *active* antiport; for example, Na^+,K^+-ATPase. D: A group translocation representing a change in S_1 during transport. E: A symport system; in this case, S_1 could be glucose and S_2 Na^+.
15. D This occurs with the eucaryotic plasma membrane (p. 228). A,B: Ca^{2+} translocation is an active uniport (p. 232). C: Like Na^+,K^+-ATPase, phosphorylation occurs on an aspartyl residue (p. 228). E: Extracellular [Ca^{2+}] is about 10,000 times higher (p. 228).
16. D In eucaryotic cells, amino acids are transported by group translocation in which they are converted to a γ-glutamyl amino acid during transport (Figure 5.47). A,E: It is an active system with the ATP used to resynthesize the intermediate, glutathione. B,C: The system transports a single amino acid at a time (p. 231).
17. A 1, 2, and 3 correct. 1, 2: These are the two major types of ionophores (p. 231). 3: Valinomycin transports K^+ by a uniport mechanism (p. 234). There are also antiport systems that are electroneutral. 4: Ionophores transport by passive mediated mechanisms (p. 231).

6

Bioenergetics and Oxidative Metabolism

MERLE S. OLSON

6.1 ENERGY-PRODUCING AND ENERGY-UTILIZING SYSTEMS 238
ATP Is the Link Between Energy-Producing and Energy-Utilizing Systems 238
6.2 THERMODYNAMIC RELATIONSHIPS AND ENERGY-RICH COMPONENTS 241
Free Energy Is the Energy Available for Useful Work 241
The Caloric Value of Dietary Substances 243
Compounds Are Classified on the Basis of the Energy Released on Hydrolysis of Specific Groups 243
Free Energy Changes Can Be Determined in Coupled Enzyme Reactions 245
High-Energy Bond Energies of Various Groups Can Be Transferred From One Compound to Another 246
6.3 SOURCES AND FATES OF ACETYL COENZYME A 247
Metabolic Sources and Fates of Pyruvate 248
Pyruvate Dehydrogenase Is a Multienzyme Complex 249
Pyruvate Dehydrogenase Is Strictly Regulated 251
Acetyl CoA Is Used by Several Different Pathways 252
6.4 THE TRICARBOXYLIC ACID CYCLE 253
The Reactions of the Tricarboxylic Acid Cycle 253
The Conversion of the Acetyl Group of Acetyl CoA to CO_2 and H_2O Conserves a Significant Portion of the Energy Available 260
The Activity of the Tricarboxylic Acid Cycle Is Carefully Regulated 260
6.5 STRUCTURE AND COMPARTMENTATION OF THE MITOCHONDRIAL MEMBRANES 261
The Inner and Outer Mitochondrial Membranes Have Different Compositions and Functions 263
Mitochondrial Inner Membranes Contain a Variety of Substrate Transport Systems 265
Substrate Shuttles Transport Reducing Equivalents Across the Mitochondrial Membrane 266
Acetyl Units Are Transported by Citrate 267
Transport of Adenine Nucleotides and Phosphates Are Controlled 267
Mitochondria Have a Specific Calcium Transport Mechanism 268
6.6 ELECTRON TRANSFER 270
Oxidation–Reduction Reactions 270
Free Energy Changes in Redox Reactions 272
Mitochondrial Electron Transport Is a Multicomponent System 272
The Mitochondrial Electron Transport Chain Is Located in the Inner Membrane in a Specific Sequence 277
Electron Transport Can Be Inhibited at Specific Sites 279
Electron Transport Is Reversible 281
Oxidative Phosphorylation Is Coupled to Electron Transport 282
6.7 OXIDATIVE PHOSPHORYLATION 283
The Chemical Coupling Hypothesis Proposed the Formation of a Specific High-Energy Intermediate 284

The Conformational Coupling Hypothesis Proposes That Electron Transport Causes a Change in Conformation of a Protein 285
The Chemiosmotic Coupling Hypothesis Involves the Generation of a Proton Gradient and Reversal of an ATP-Dependent Proton Pump 285
BIBLIOGRAPHY 287
QUESTIONS AND ANNOTATED ANSWERS 287
CLINICAL CORRELATIONS
6.1 Pyruvate Dehydrogenase Deficiency 252
6.2 Mitochondrial Myopathies 268
6.3 Cyanide Poisoning 282
6.4 Hypoxic Injury 284

6.1 ENERGY-PRODUCING AND ENERGY-UTILIZING SYSTEMS

Living cells are composed of a complex intricately regulated system of energy-producing and energy-utilizing chemical reactions. Metabolic reactions involved in energy generation sequentially break down ingested or stored macromolecular fuels such as carbohydrate, lipid, or protein in what are termed catabolic pathways. **Catabolic** reactions usually result in the conversion of large complex molecules to smaller molecules (ultimately CO_2 and H_2O), usually result in the production of storable or conservable energy, and often require the consumption of oxygen during this process. Such reactions are accelerated during periods of fuel deprivation or stress to an organism.

Energy-utilizing reactions are necessary to maintain, to reproduce, or to perform various necessary, and in many instances tissue-specific, cellular functions, for example, nerve impulse conduction and muscle contraction. Metabolic pathways in a cell, which are involved in the biosynthesis of various macromolecules, are termed anabolic pathways. **Anabolic** reactions usually result in the synthesis of large, complex molecules from smaller precursors and require the expenditure of energy. Such reactions are accelerated during periods of relative energy excess, during periods when there occurs a ready availability of precursor molecules or during periods of growth or regeneration of cellular material.

ATP Is the Link Between Energy-Producing and Energy-Utilizing Systems

The relationship between the energy-producing and energy-utilizing functions of the cell is illustrated in Figure 6.1. Energy may be derived from the oxidation of appropriate metabolic fuels such as carbohydrate, lipid, or protein. The proportion of each of these metabolic fuels that may be utilized as an energy source depends on the tissue and the dietary and hormonal state of the organism. For example, the mature erythrocyte and the adult brain in the fed state use only carbohydrate as a metabolic fuel, whereas the liver of a diabetic or fasted mammal metabolizes primarily lipid or fat. Energy may be consumed during the performance of various energy-linked (work) functions, some of which are indicated in Figure 6.1. Again the proportion of energy expended or utilized depends largely on the tissue and the physiological state of that tissue, for example, the liver and the pancreas are tissues primarily involved in biosynthetic and secretory work functions, whereas cardiac and skeletal muscle primarily are involved in converting metabolic energy into mechanical energy during the muscle contraction process.

The essential linkage between the energy-producing and the energy-utilizing pathways is maintained by the nucleoside triphosphate, **adenosine 5′-triphosphate** (ATP) (Figure 6.2). The ATP molecule is a purine

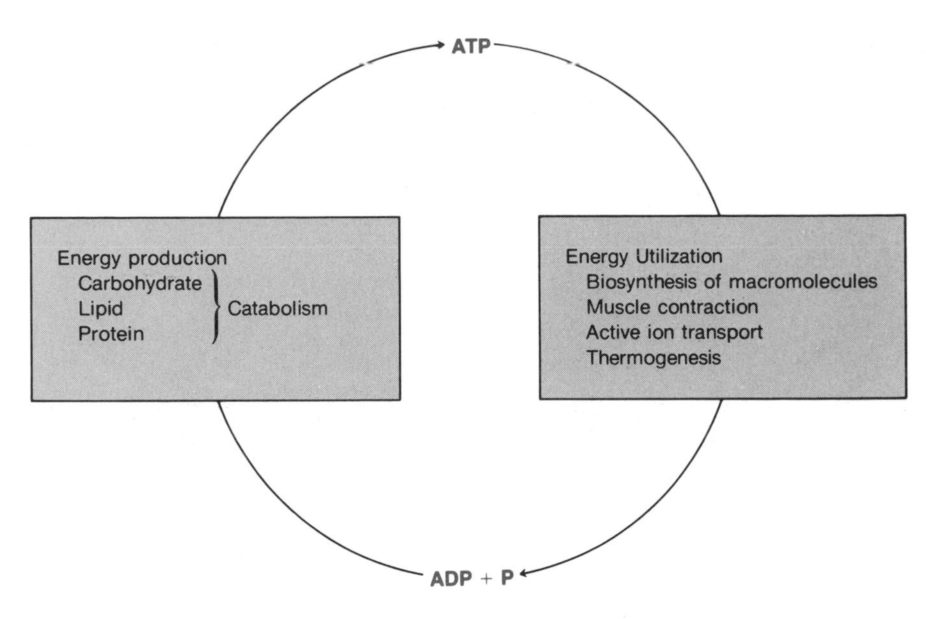

FIGURE 6.1
The relationship between energy production and energy utilization.

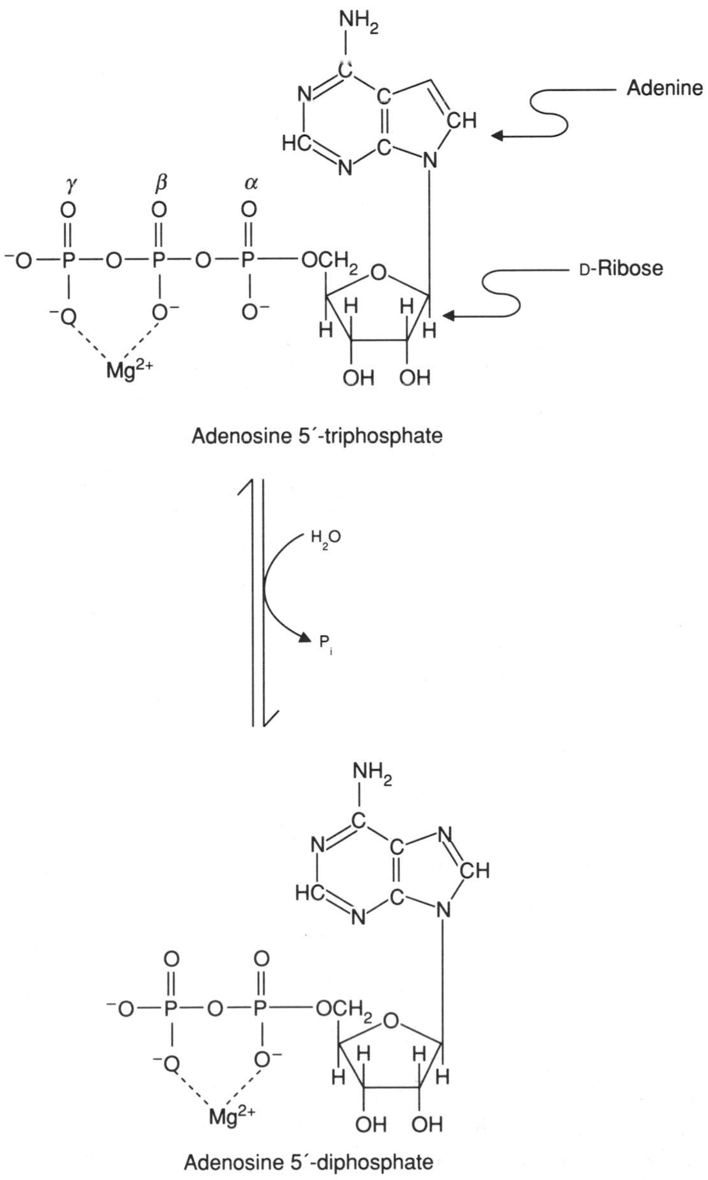

Adenine

D-Ribose

Mg^{2+}

Adenosine 5′-triphosphate

H_2O

P_i

Mg^{2+}

Adenosine 5′-diphosphate

Figure 6.2
Structure of ATP and ADP complexed with Mg^{2+}.

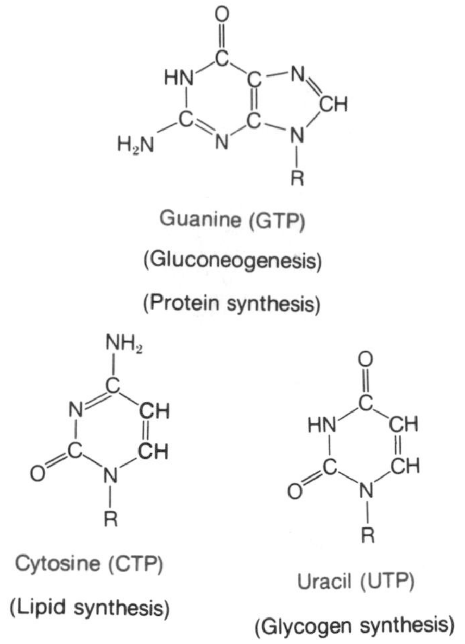

Guanine (GTP)

(Gluconeogenesis)

(Protein synthesis)

Cytosine (CTP)

(Lipid synthesis)

Uracil (UTP)

(Glycogen synthesis)

Figure 6.3
Structures of purine and pyrimidine bases involved in various biosynthetic pathways.

(adenine) nucleotide in which the adenine ring is attached in a glycosidic linkage to D-ribose. Three phosphoryl groups are esterified to the 5 position of the ribose moiety in what are termed phosphoanhydride bonds. The two terminal phosphate groups (i.e., β and γ), which are involved in the phosphoric acid anhydride bonding, are designated as *energy-rich* or *high-energy bonds*. Synthesizing ATP as a result of a catabolic process or consuming ATP in some type of energy-linked cellular function alternately involves the formation and either the hydrolysis or transfer of the terminal phosphate group of ATP. The physiological form of this nucleotide is chelated with a divalent metal cation such as magnesium. Adenosine diphosphate also can chelate magnesium, but the affinity of the metal cation for ADP is considerably less than for ATP. Although adenine nucleotides are the compounds mainly involved in the process of energy generation or conservation, various nucleoside triphosphates, including ATP are involved in transferring energy during biosynthetic processes. As indicated in Figure 6.3, the guanine nucleotide GTP serves as the source of energy input into the processes of gluconeogenesis and protein synthesis, whereas UTP (uracil) and CTP (cytosine) are utilized in glycogen synthesis and lipid synthesis, respectively. The energy in the terminal phosphate bonds of ATP may be transferred to the other nucleotides, using either the nucleoside diphosphate kinase or the nucleoside monophosphate kinase as illustrated in Figure 6.4. Two nucleoside diphosphates can be converted to a nucleoside triphosphate and a nucleoside monophosphate in various nucleoside monophosphate kinase reactions, such as the adenylate kinase reaction as indicated in Figure 6.5. A consequence of the action of these types of enzymes is that the terminal energy-rich phosphate bonds of ATP may be transferred to the appropriate nucleotides and utilized in a variety of biosynthetic processes.

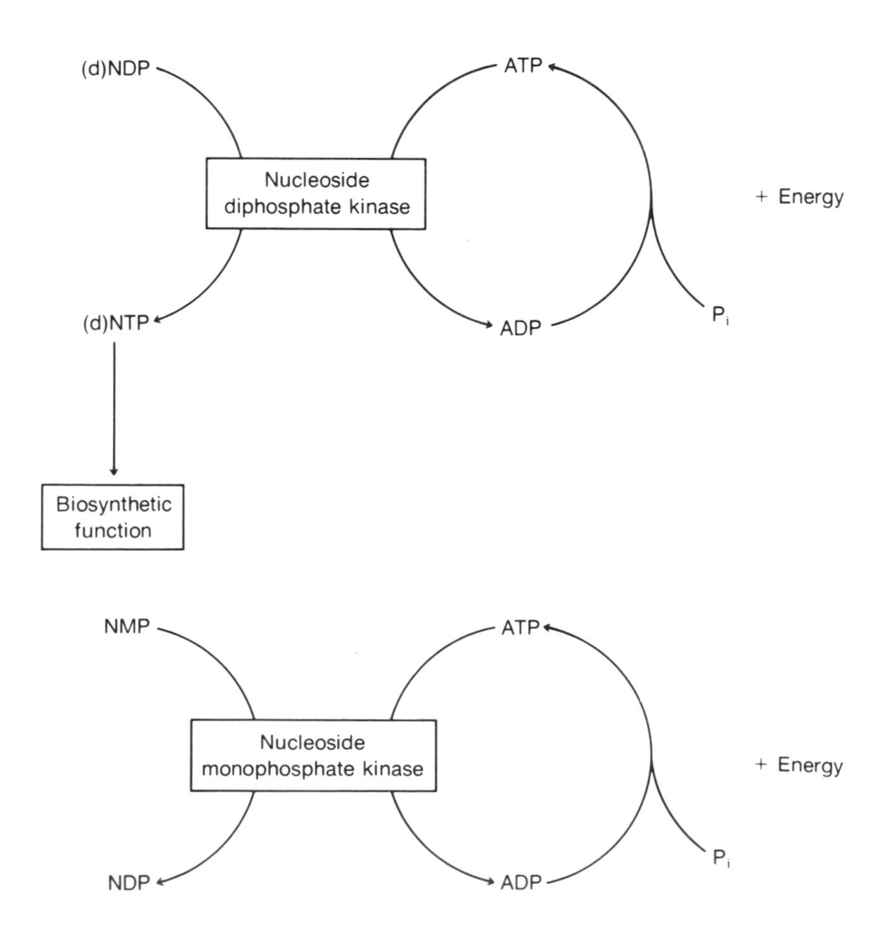

Figure 6.4
Nucleoside disphosphate kinase and nucleoside monophosphate kinase reactions.
N = any purine or pyrimidine base; (d) indicates a deoxyribonucleotide.

6.2 THERMODYNAMIC RELATIONSHIPS AND ENERGY-RICH COMPONENTS

Because living cells are capable of the interconversion of different forms of energy and may exchange energy with their surroundings, it is helpful to review certain laws or principles of thermodynamics. Knowledge of these principles will facilitate a perception of how energy-producing and energy-utilizing metabolic reactions are permitted to occur within the same cell and how an organism is able to accomplish various work functions.

The **first law of thermodynamics** indicates that energy can neither be created nor destroyed. This law of energy conservation stipulates that, although energy may be converted from one form to another, the total energy in a system must remain constant. For example, the chemical energy that is available in a metabolic fuel such as glucose may be converted in the process of glycolysis to another form of chemical energy, ATP.

In skeletal muscle chemical energy involved in the energy-rich phosphate bonds of ATP may be converted to mechanical energy during the process of muscle contraction. It has been demonstrated that the energy involved in an osmotic electropotential gradient of protons across the mitochondrial membrane may be converted to chemical energy using such a gradient to drive ATP synthesis.

In order to discuss the second law of thermodynamics the term entropy must be defined. **Entropy,** which is designated by the symbol S, is a measure or indicator of the degree of disorder or randomness in a system. In addition, entropy can be viewed as the energy in a system that is unavailable to perform useful work. All processes, whether chemical or biological, tend to progress toward a situation of maximum entropy. Equilibrium in a system will result when the randomness or disorder (entropy) is at a maximum. However, it is nearly impossible to quantitate entropy changes in systems that may be useful to study in biochemistry, and such systems are rarely at equilibrium. For the sake of simplicity and its inherent utility in these types of considerations, the quantity termed free energy is employed.

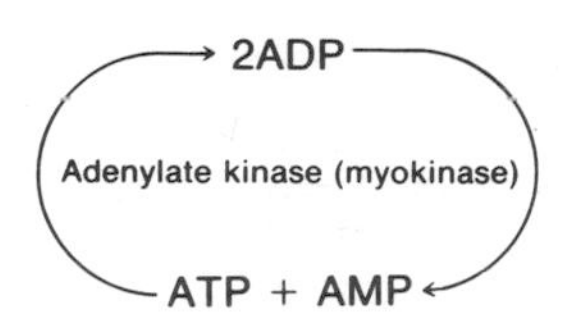

Figure 6.5
The adenylate kinase (myokinase) reaction.

Free Energy Is the Energy Available for Useful Work

The free energy (denoted by the letter G) of a system is that portion of the total energy in a system that is available for useful work and may be further defined by the equation

$$\Delta G = \Delta H - T\Delta S$$

In this expression for a system proceeding toward equilibrium at a constant temperature and pressure, ΔG is the change in free energy, ΔH is the change in enthalpy or the heat content, T is the absolute temperature, and ΔS is the change in entropy of the system. It can be deduced from this relationship that at equilibrium $\Delta G = 0$. Furthermore, any process that exhibits a negative free energy change proceeds to equilibrium since energy is given off, and is called an **exergonic reaction.** A process that exhibits a positive free energy change will not occur independently; energy from some other source must be applied to this process to allow it to proceed toward equilibrium, and this type of process is termed an **endergonic reaction.** It should be noted that *the change in free energy in a biochemical process is the same regardless of the pathway or mechanism employed to attain the final state.* Whereas the rate of a given reaction depends on the free energy of activation, *the magnitude of the ΔG is not related to the rate of the reaction*. The change in free energy for a chemical reaction is related to the **equilibrium constant** of that reaction. For example, an enzymatic reaction may be described as

$$A + B \rightleftharpoons C + D$$

And an expression for the equilibrium constant may be written as

$$K_{eq} = \frac{[C][D]}{[A][B]}$$

The free energy change (ΔG) at a constant temperature and pressure is defined as

$$\Delta G = \Delta G^\circ + RT \ln \left(\frac{[C][D]}{[A][B]}\right)$$

where ΔG is the free energy change; ΔG° is the standard free energy change, which is a constant for each individual chemical reaction; the reactants and products in the reaction are present at concentrations of 1.0 M; R is the gas constant, which is 1.987 cal mol^{-1} K^{-1} or 8.134 J mol^{-1} K^{-1}, depending on whether the resultant free energy change is expressed in calories (cal) or joules (J) per mole; and T is the absolute temperature in degrees Kelvin (K).

Because at equilibrium $\Delta G = 0$, the expression reduces to

$$\Delta G^\circ = -RT \ln K_{eq}$$

or

$$\Delta G^\circ = -2.3RT \log K_{eq}$$

Hence, if the equilibrium constant for a reaction can be determined, the standard free energy change (ΔG°) for that reaction also can be calculated. The relationship between ΔG° and K_{eq} is illustrated in Table 6.1. When the equilibrium constant of a reaction is less than unity, the reaction is endergonic, and ΔG° is positive. When the equilibrium constant is greater than 1, the reaction is exergonic, and ΔG° is negative.

During any consideration of the energy-producing and energy-utilizing metabolic pathways in cellular systems it is important to understand that the free energy changes characteristic of individual enzymatic reactions in an entire pathway are additive, for example,

$$A \longrightarrow B \longrightarrow C \longrightarrow D$$

$$\Delta G^\circ_{A\to D} = \Delta G^\circ_{A\to B} + \Delta G^\circ_{B\to C} + \Delta G^\circ_{C\to D}$$

Although any given enzymatic reaction in a sequence may have a characteristic positive free energy change, as long as the sum of all the free energy changes is negative, the pathway will proceed.

Another way of expressing this principle is that enzymatic reactions with positive free energy changes may be coupled to or driven by reactions with negative free energy changes associated with them. This is an important point because in a metabolic pathway such as the glycolytic pathway various individual reactions either have positive ΔG° values or ΔG° values that are close to 0. On the other hand, there are other reac-

TABLE 6.1 Tabulation of Values of K_{eq} and ΔG°

K_{eq}	ΔG° (kcal M^{-1})
10^{-4}	5.46
10^{-3}	4.09
10^{-2}	2.73
10^{-1}	1.36
1	0
10	−1.36
10^{2}	−2.73
10^{3}	−4.09
10^{4}	−5.46

tions that have large and negative ΔG° values, which drive the entire pathway. The crucial consideration is that the sum of the ΔG° values for the individual reactions in a pathway must be negative in order for such a metabolic sequence to be thermodynamically feasible. Also, it is important to remember that, as for all chemical reactions, individual enzymatic reactions in a metabolic pathway or the pathway as a whole would be facilitated if the concentrations of the reactants (substrates) of the reaction exceed the concentrations of the products of the reaction.

The Caloric Value of Dietary Substances

During the complete stepwise oxidation of glucose, one of the primary metabolic fuels in cellular systems, a large quantity of energy is available. The free energy released during the oxidation of glucose in a living functioning cell, is illustrated in the following equation:

$$C_6H_{12}O_6 + 6O_2 \longrightarrow 6CO_2 + 6H_2O \qquad \Delta G^\circ = -686{,}000 \text{ cal mol}^{-1}$$

When this process is performed under aerobic conditions in most types of cells, there exists a potential to conserve less than one half of this "available" energy in the form of ATP. The enzymatic machinery in cellular systems is capable of synthesizing 38 molecules of ATP during the complete oxidation of glucose. The ΔG° values for the oxidation of other metabolic fuels are listed in Table 6.2. Carbohydrates and proteins (amino acids) have a **caloric value** of 3–4 kcal g^{-1}, while lipid (i.e., the long-chain fatty acid palmitate or the triglyceride tripalmitin) exhibits a caloric value nearly three times greater. The reason that more energy can be derived from lipid than from carbohydrate or protein relates to the average oxidation state of the carbon atoms in these substances.

Carbon atoms in carbohydrates are considerably more oxidized (or less reduced) than those in lipids (see Figure 6.6). Hence during the sequential breakdown of these fuels nearly three times as many reducing equivalents (a reducing equivalent is defined as a proton plus an electron, i.e., $H^+ + e^-$) can be extracted from the lipid than from the carbohydrate. Reducing equivalents may be utilized for ATP synthesis in the mitochondrial energy transduction sequence.

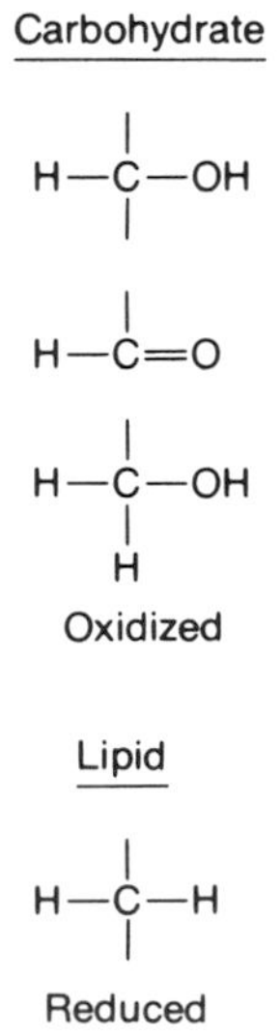

Compounds Are Classified on the Basis of the Energy Released on Hydrolysis of Specific Groups

The two terminal phosphoryl groups in the ATP molecule contain energy-rich or **high-energy bonds.** What this description is intended to convey is that the free energy of hydrolysis of an energy-rich phosphoanhydride bond is much greater than would be obtained for a simple phosphate ester. High-energy is not synonymous with stability of the bonding arrangement

TABLE 6.2 Free Energy Changes and Caloric Values Associated With the Total Metabolism of Various Metabolic Fuels

Compound	Molecular weight	ΔG° (kcal mol^{-1})	Caloric value (kcal g^{-1})
Glucose	180	−686	3.81
Lactate	90	−326	3.62
Palmitate	256	−2380	9.30
Tripalmitin	809	−7510	9.30
Glycine	75	−234	3.12

Carbohydrate

H—C—OH
H—C=O
H—C—OH
H

Oxidized

Lipid

H—C—H

Reduced

Figure 6.6
Oxidation states of typical carbon atoms in carbohydrates and lipids.

TABLE 6.3 Examples of Energy-Rich Compounds

Type of bond	ΔG° of hydrolysis (kcal mol^{-1})	Example
Phosphoric acid anhydrides	−7.3	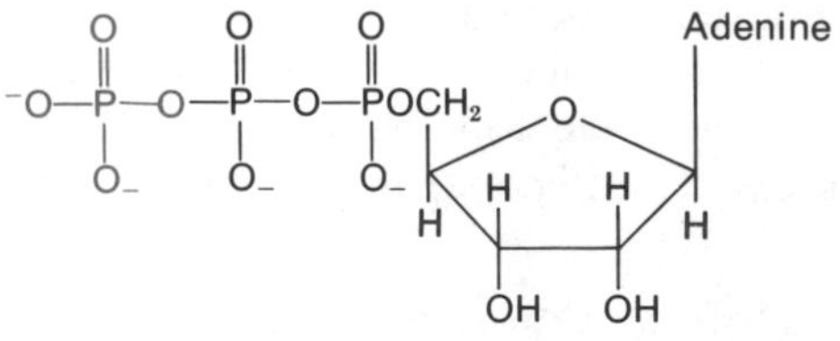 ATP
	−11.9	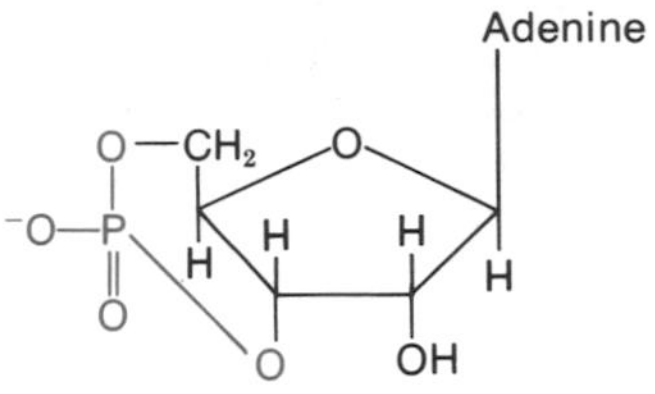 3′5′ cyclic AMP
Phosphoric-carboxylic acid anhydrides	−10.1	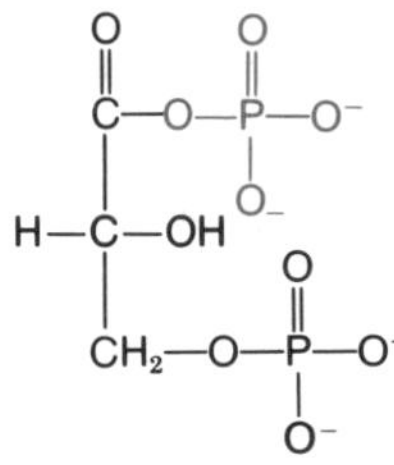 1,3-Bisphosphoglycerate
	−10.3	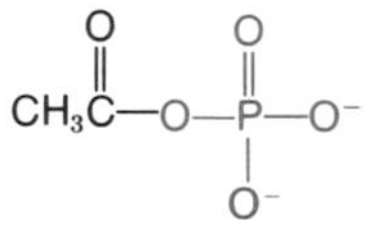 Acetyl phosphate
Phosphoguanidines	−10.3	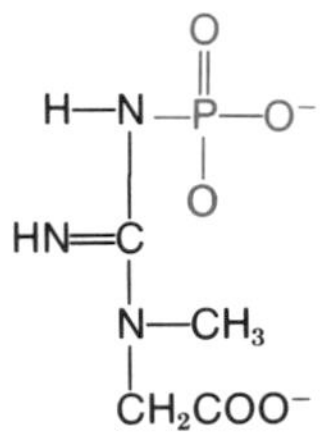 Creatine phosphate
Enol phosphates	−14.8	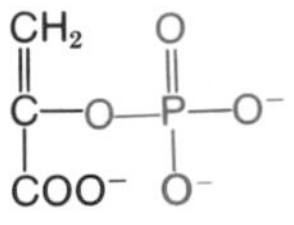 Phospho*enol*pyruvate
Thiolesters	−7.7	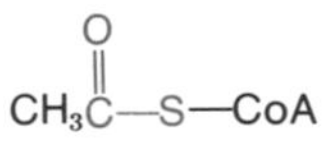 Acetyl CoA

in question, nor does high-energy refer to the energy required to break such bonds. The concept of the high-energy compound does imply that the products of the hydrolytic cleavage of the energy-rich bond are in more stable forms than the original compound. As a rule simple phosphate esters (low-energy compounds) exhibit negative ΔG° values of hydrolysis in the range 1–3 kcal mol^{-1}, whereas high-energy bonds have negative ΔG° values in the range 5–15 kcal mol^{-1}. Simple phosphate esters such as glucose 6-phosphate and glycerol 3-phosphate are examples of low-energy compounds. Table 6.3 lists various types of energy-rich compounds with approximate values for their ΔG° values of hydrolysis.

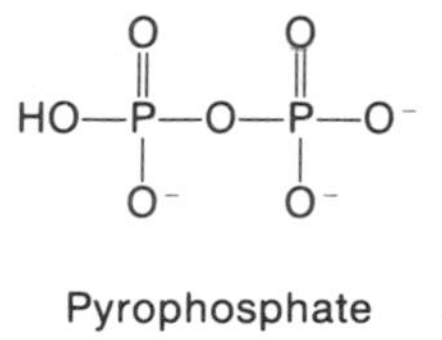

There are various reasons why certain compounds or bonding arrangements are energy rich. First, the products of the hydrolysis of an energy-rich bond may exist in more resonance forms than the precursor molecule. The more possible resonance forms in which a molecule can exist tend to stabilize that molecule. The resonance forms for inorganic phosphate (P_i) can be written as follows:

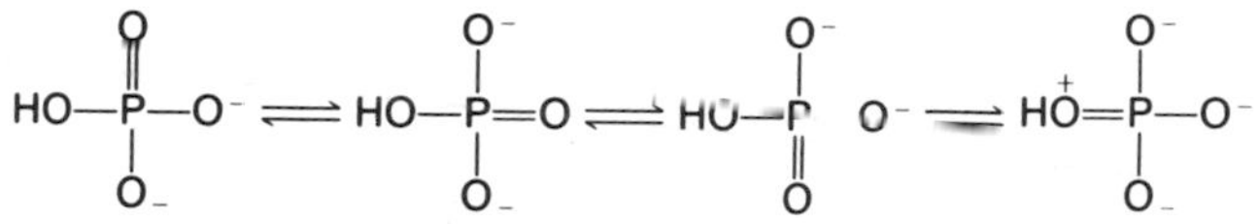

HO—P(=O)(O⁻)—O⁻ ⇌ HO—P(O⁻)(O⁻)=O ⇌ HO—P(O⁻)(=O)—O⁻ ⇌ HO⁺=P(O⁻)(O⁻)—O⁻

Fewer resonance forms may be written for ATP or a compound such as pyrophosphate (PP_i) (Figure 6.7) than for P_i.

Second, many high-energy bonding arrangements have groups of similar electrostatic charges located in close proximity to each other in such compounds. Because like charges tend to repulse one another, the hydrolysis of the energy-rich bond alleviates this situation and, again, lends stability to the products of hydrolysis. Third, the hydrolysis of certain energy-rich bonds results in the formation of an unstable compound, which may isomerize spontaneously to form a more stable compound. The hydrolysis of phosphoenolpyruvate is an example of this type of compound (Figure 6.8). The ΔG° value of the isomerization reaction is considerable, and the final product, in this case pyruvate, is much more stable. Finally, if a product of the hydrolysis of a high-energy bond is an undissociated acid, the dissociation of the proton from the acidic function and its subsequent buffering may contribute to the overall ΔG° of the hydrolytic reaction. In general, any property or process that lends stability to the products of hydrolysis of a compound tends to confer a high-energy character to that compound.

The high-energy character of 3′,5′-cyclic adenosine monophosphate (cAMP) has been attributed to the fact that the phosphoanhydride bonding character in this compound is strained as it bridges the 3′ and 5′ positions on the ribose. Furthermore, the energy-rich character of thiol ester compounds such as acetyl CoA or succinyl CoA results from the relatively acidic character of the thiol function. Hence, acetyl CoA is nearly equivalent to an anhydride bonding arrangement rather than a simple thioester.

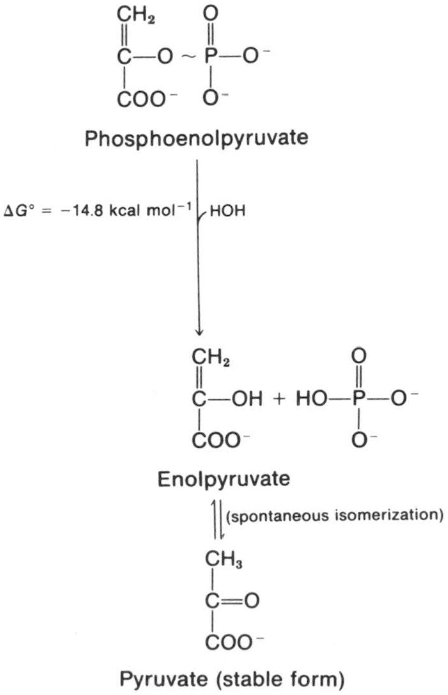

Free Energy Changes Can Be Determined in Coupled Enzyme Reactions

The ΔG° value of hydrolysis of the terminal phosphate of ATP is difficult to determine by simply utilizing the K_{eq} of the hydrolytic reaction because of the position of the equilibrium.

$$\text{ATP} + \text{HOH} \rightleftharpoons \text{ADP} + \text{P}_i + \text{H}^+$$

HO—P(=O)(O⁻)—O—P(=O)(O⁻)—O⁻

Pyrophosphate

Figure 6.7
Structure of pyrophosphate.

Phosphoenolpyruvate

$\Delta G^\circ = -14.8$ kcal mol^{-1} HOH

Enolpyruvate

(spontaneous isomerization)

Pyruvate (stable form)

Figure 6.8
Hydrolysis of phosphoenolpyruvate.

However, ΔG° of hydrolysis of ATP may be determined indirectly because of the additive nature of free energy changes. Hence, the free energy of hydrolysis of ATP can be determined by adding ΔG° of an ATP-utilizing reaction such as hexokinase to ΔG° of the reaction that cleaves the phosphate from the product of the hexokinase reaction, glucose 6–phosphate (G-6-P), as indicated below:

$$\text{Glucose} + \text{ATP} \xrightleftharpoons{\text{hexokinase}} \text{G-6-P} + \text{ADP} + \text{H}^+ \qquad \Delta G^\circ = -4.0 \text{ kcal mol}^{-1}$$

$$\text{G-6-P} + \text{HOH} \xrightleftharpoons{\text{glucose 6-phosphatase}} \text{glucose} + \text{P}_i \qquad \Delta G^\circ = -3.3 \text{ kcal mol}^{-1}$$

$$\Sigma \quad \text{ATP} + \text{HOH} \rightleftharpoons \text{ADP} + \text{P}_i + \text{H}^+ \qquad \Delta G^\circ = -7.3 \text{ kcal mol}^{-1}$$

Free energies of hydrolysis for other energy-rich compounds may be determined in a similar fashion.

High-Energy Bond Energies of Various Groups Can Be Transferred From One Compound to Another

Energy-rich compounds are capable of transferring various groups from the parent (donor) compound to an acceptor compound in a thermodynamically feasible fashion as long as an appropriate enzyme is present to facilitate the transfer. The energy-rich intermediates in the glycolytic pathway such as 1,3-bisphosphoglycerate and phosphoenolpyruvate can transfer their high-energy phosphate moieties to ATP in the phosphoglycerate kinase and pyruvate kinase reactions, respectively (Figure 6.9*a* and *b*). The ΔG° values of these two reactions are -4.5 and -7.5 kcal mol^{-1}, respectively, and hence the transfer of "high-energy" phosphate is thermodynamically possible, and ATP synthesis is the result. The ATP can transfer its terminal high-energy phosphoryl groups to form either compounds of relatively similar high-energy character (i.e., creatine phosphate in the creatine kinase reaction) or compounds that are of considerably lower energy, such as glucose 6-phosphate formed in the hexokinase reaction (Figure 6.9*c*).

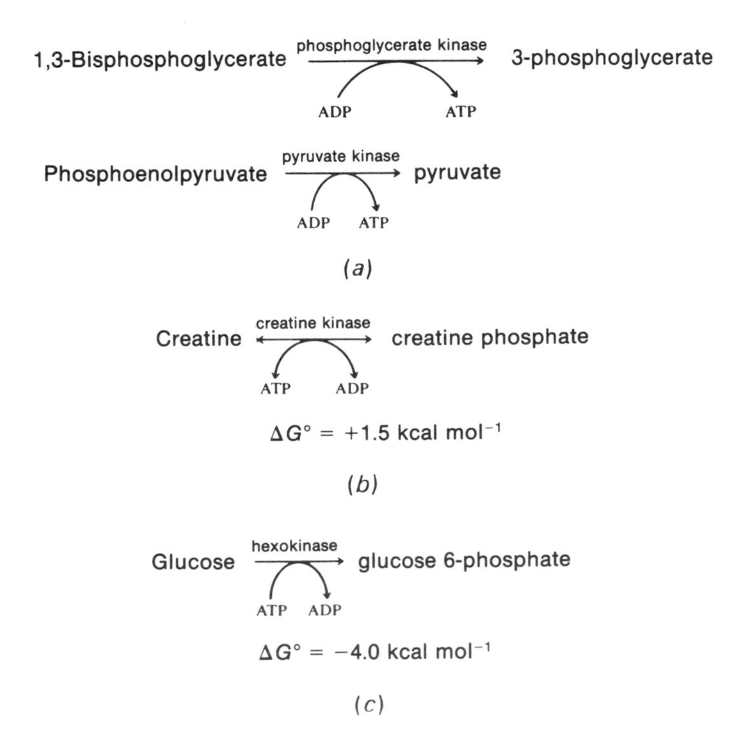

Figure 6.9
Examples of reactions involved in the transfer of "high-energy" phosphate.

The major point of this discussion is that phosphate, or for that matter other transferable groups, can be transferred from compounds that contain energy-rich bonding arrangements to compounds that have bonding characteristics of a lower energy in thermodynamically permissible enzymatic reactions. This principle is a major premise of the interaction between energy-producing and energy-utilizing metabolic pathways in living cells.

6.3 SOURCES AND FATES OF ACETYL COENZYME A

Most of the major energy-generating metabolic pathways of the cell eventually result in the production of the two-carbon unit **acetyl coenzyme A** (CoA). As illustrated in Figure 6.10, the catabolic breakdown of ingested or stored carbohydrate in the glycolytic pathway, long-chain fatty acids resulting from the lipolysis of triglycerides in the β-oxidation sequence, or certain amino acids resulting from proteolysis following transamination or deamination and subsequent oxidation, provide precursors for the formation of acetyl CoA.

The structure of acetyl CoA is shown in Figure 6.11. This complex coenzyme, abbreviated either as CoA or CoASH, is composed of β-mercaptoethylamine, the vitamin pantothenic acid and the adenine nucleotide, adenosine 3'-phosphate 5'-diphosphate. Coenzyme A exists as the reduced thiol and is involved in a variety of acyl group transfer reactions, where CoA alternately serves as the acceptor, then the donor, of the acyl function. Various metabolic pathways involve only acyl CoA derivatives, for example, β oxidation of fatty acids and branched-chain amino acid degradation. Specific information concerning the nutritional aspects of the vitamin pantothenic acid will be detailed in Chapter 28. Like many other nucleotide species, CoA derivatives are not freely transported across cellular membranes. This property has necessitated the evolution of certain transport or shuttle mechanisms by which various intermediates or groups can be transferred across the membranes of the cell. Such acyl transferase reactions for acetyl groups and long-chain acyl groups will be discussed in Chapter 9. Finally, as indicated previously, the thiolester linkage in acyl CoA derivatives is an energy-rich bond, and hence these compounds can serve as effective donors of acyl groups in

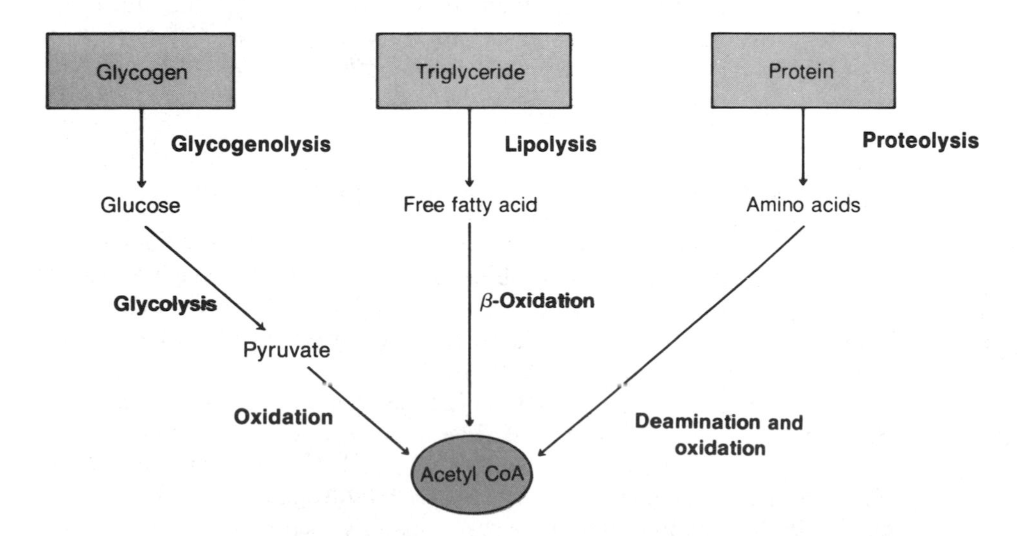

Figure 6.10
General precursors of acetyl CoA.

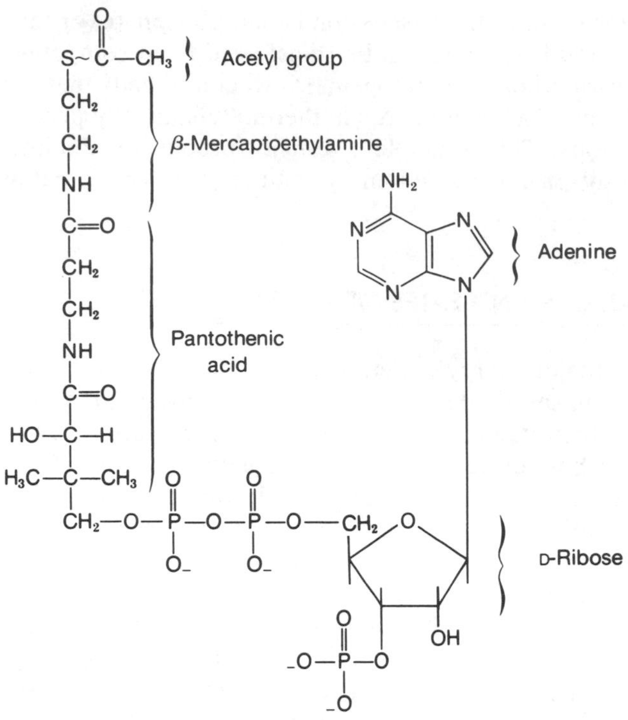

Acetyl group

β-Mercaptoethylamine

Adenine

Pantothenic acid

D-Ribose

Figure 6.11
Structure of acetyl CoA.

acyl transferase reactions. Also, in order to synthesize an acyl CoA derivative, such as in the acetate thiokinase reaction, a high-energy bond of ATP must be expended.

$$\text{Acetate} + \text{CoASH} + \text{ATP} \xrightarrow{\text{acetate kinase}} \text{acetyl CoA} + \text{AMP} + \text{PP}_i$$

As was mentioned above, the β oxidation of fatty acids is a primary source of acetyl CoA in many tissues. Whereas a more detailed description of the mobilization, transport, and oxidation of fatty acids is presented in Chapter 9, it is important to note that the products of the β-oxidation sequence are acetyl CoA and reducing equivalents (i.e., NADH + H^+). In certain tissues (e.g., cardiac muscle) and under somewhat special metabolic conditions in other tissues (e.g., in the brain of an individual during prolonged starvation) acetyl CoA for energy generation may be derived from the ketone bodies, acetoacetate and β-hydroxybutyrate.

Metabolic Sources and Fates of Pyruvate

During aerobic glycolysis (Chapter 7) glucose or other monosaccharides are converted to pyruvate, and hence in the presence of oxygen pyruvate is the end product per se of this cytosolic pathway. Also the degradation of amino acids such as alanine, serine, and cysteine results in the production of pyruvate (Chapter 12).

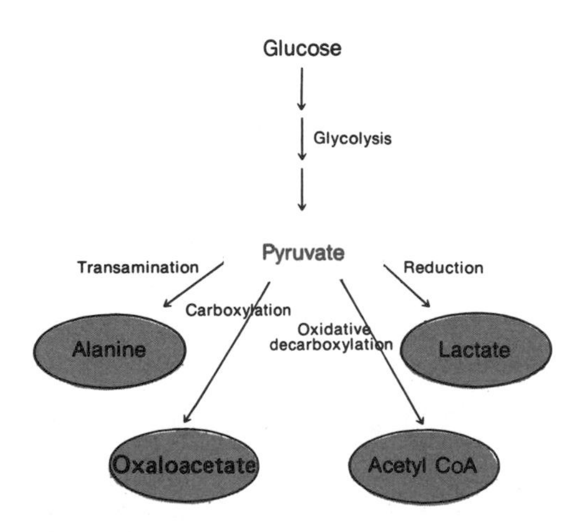

Figure 6.12
Metabolic fates of pyruvate.

Pyruvate has a variety of metabolic fates, depending on the tissue and the metabolic state of that tissue. The major types of reactions in which pyruvate participates are indicated in Figure 6.12. The oxidative decarboxylation of pyruvate in the pyruvate dehydrogenase reaction is discussed in this chapter; the other reactions in which pyruvate is involved are discussed in Chapter 7.

Pyruvate Dehydrogenase Is a Multienzyme Complex

Pyruvate is converted to acetyl CoA by the **pyruvate dehydrogenase** multienzyme complex.

$$\text{Pyruvate} + \text{NAD}^+ + \text{CoASH} \xrightarrow{\text{pyruvate dehydrogenase}}$$
$$\text{acetyl CoA} + CO_2 + \text{NADH} + \text{H}^+ \qquad \Delta G^\circ = -8 \text{ kcal mol}^{-1}$$

This enzyme is located exclusively in the mitochondrial compartment and is present in high concentrations in tissues such as cardiac muscle and kidney. Because of the large negative ΔG° of this reaction, under physiological conditions the pyruvate dehydrogenase reaction is essentially irreversible, and this fact is the primary reason that a net conversion of fatty acid carbon to carbohydrate cannot occur, for example,

$$\text{Acetyl CoA} \not\rightarrow \text{pyruvate}$$

Molecular weights of the multienzyme complex derived from kidney, heart, or liver range from 7 to 8.5×10^6. The mammalian pyruvate dehydrogenase enzyme complex consists of three different types of catalytic subunits:

Number of subunits/complex	Type	Molecular weight	Subunit structure
20 or 30[a]	Pyruvate dehydrogenase	154,000	$\alpha_2\beta_2$ Tetramer
60	Dihydrolipoyl transacetylase	52,000	Identical
6	Dihydrolipoyl dehydrogenase	110,000	α_2 Dimer

[a] Depending on source.

The structure of the pyruvate dehydrogenase complex derived from *E. coli* (particle weight, 4.6×10^6) is somewhat different from that of the mammalian enzyme. Electron micrographs of the bacterial enzyme complex (Figure 6.13) indicate that the transacetylase, which consists of 24 identical polypeptide chains (mol wt = 64,500), forms the cubelike core of the complex (white spheres in the model shown in Figure 6.11). Twelve pyruvate dehydrogenase dimers (black spheres; mol wt = 90,500) are distributed symmetrically on the 12 edges of the transacetylase cube. Six

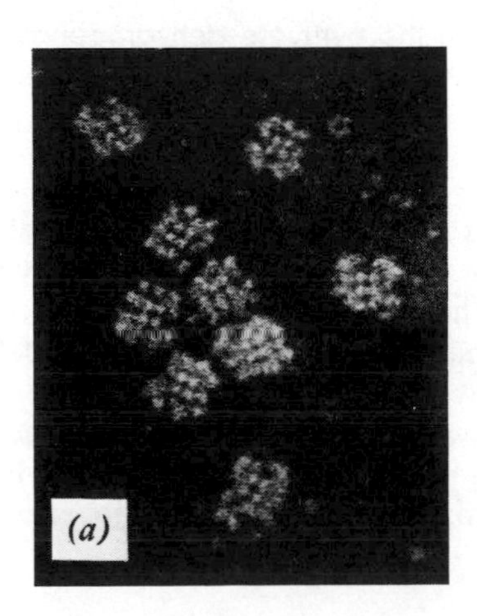

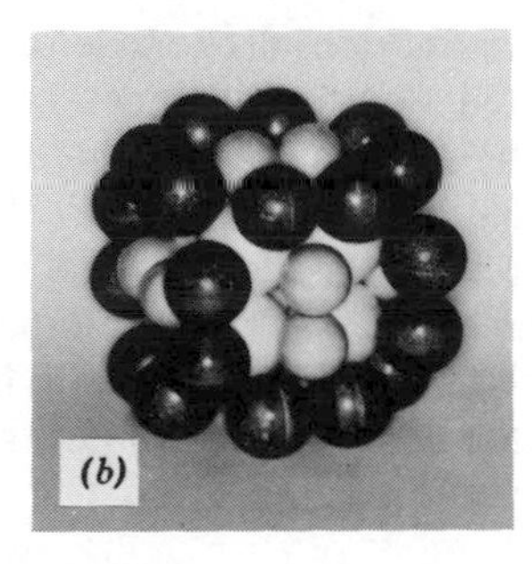

Figure 6.13
The pyruvate dehydrogenase complex from E. coli.
(*a*) Electron micrograph. (*b*) Molecular model. The enzyme complex was negatively stained with phosphotungstate (×200,000).
Courtesy of Dr. Lester J. Reed, University of Texas, Austin.

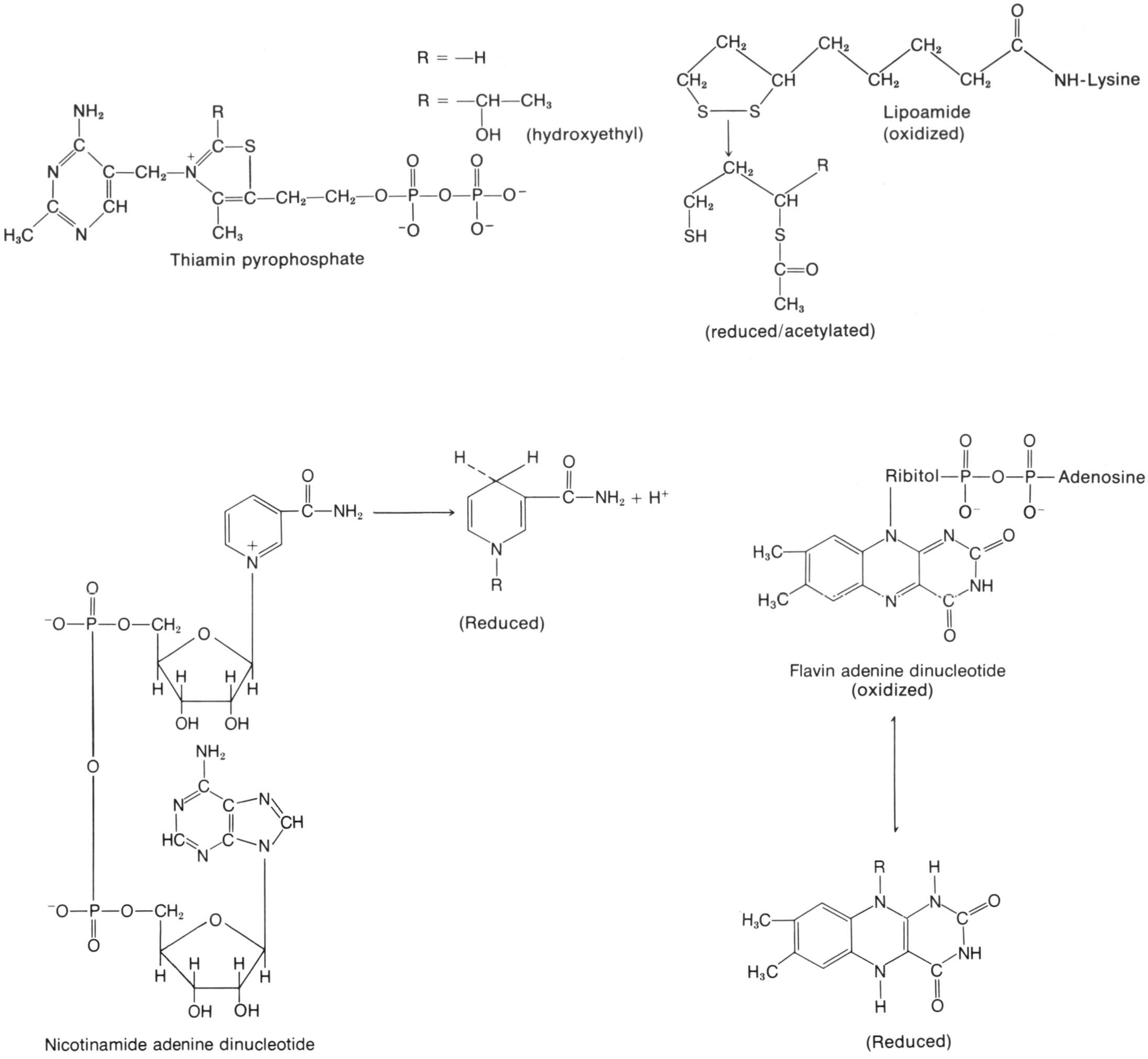

R = —H

R = —CH—CH₃ (hydroxyethyl)

Thiamin pyrophosphate

Lipoamide (oxidized)

(reduced/acetylated)

(Reduced)

Flavin adenine dinucleotide (oxidized)

(Reduced)

Nicotinamide adenine dinucleotide

Figure 6.14
Structures of the coenzymes involved in the pyruvate dehydrogenase reaction. See Figure 6.11 for structure of CoA.

dihydrolipoyl dehydrogenase dimers (gray spheres; mol wt = 56,000) are distributed on the six faces of the cube. Five different coenzymes or prosthetic groups are involved in the pyruvate dehydrogenase reaction (Table 6.4 and Figure 6.14). The mechanism of the pyruvate dehydrogenase reaction occurs as illustrated in Figure 6.15.

Because of the active participation of thiol groups in the catalytic mechanism of the enzyme, agents which either oxidize or complex with thiol groups are strong inhibitors of the enzyme complex. Arsenite is an example of such an inhibitor.

TABLE 6.4 Function of Coenzymes and Prosthetic Groups of the Pyruvate Dehydrogenase Reaction

Coenzyme or prosthetic group	Location	Function
Thiamin pyrophosphate	Bound to pyruvate dehydrogenase	Reacts with substrate, pyruvate
Lipoic acid	Covalently attached to a lysine residue on the dihydrolipoyl transacetylase	Accepts acetyl group from thiamine pyrophosphate
Coenzyme A	Free in solution	Accepts acetyl group from lipoamide group on the transacetylase
Flavin adenine dinucleotide (FAD)	Tightly bound to dihydrolipoyl dehydrogenase	Accepts reducing equivalents from reduced lipoamide group
Nicotinamide adenine dinucleotide	Free in solution	Terminal acceptor of reducing equivalents from the reduced flavoprotein

Pyruvate Dehydrogenase Is Strictly Regulated

Two types of regulation of the pyruvate dehydrogenase complex have been elucidated. First, it has been demonstrated that two of the products of the pyruvate dehydrogenase reaction, acetyl CoA and NADH, inhibit the complex in a competitive fashion. Second, the pyruvate dehydrogenase complex exists in two forms: (1) an active, dephosphorylated complex, and (2) an inactive, phosphorylated complex. The inactivation of the complex is accomplished by a Mg^{2+}–ATP-dependent protein kinase, which is tightly bound to the enzyme complex. The reactivation of the complex is accomplished by a phosphoprotein phosphatase, which dephosphorylates the complex in a Mg^{2+}- and Ca^{2+}-dependent reaction. Three separate serine residues on the α subunit of the pyruvate dehydro-

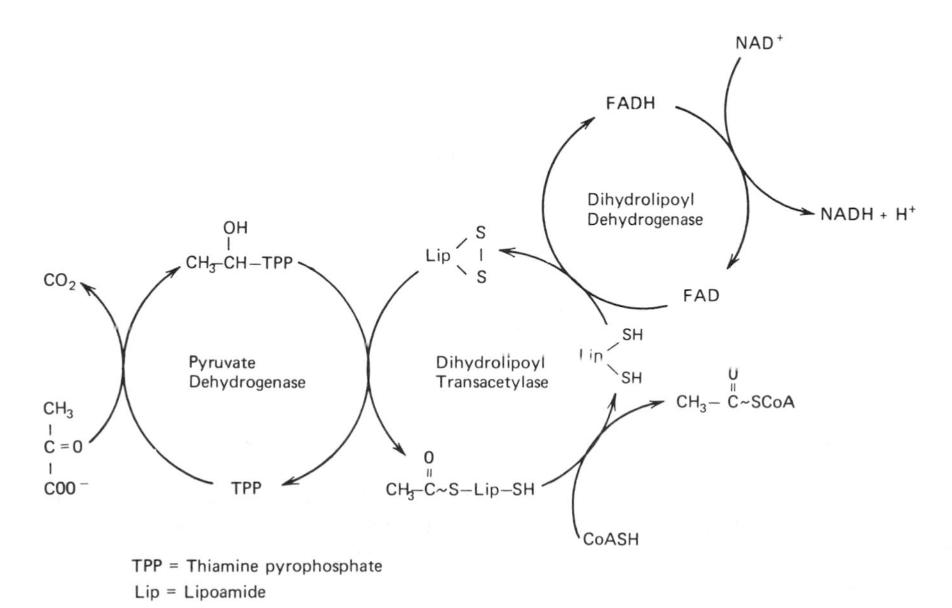

Figure 6.15
The mechanism of the pyruvate dehydrogenase reaction; the pyruvate dehydrogenase multienzyme complex.

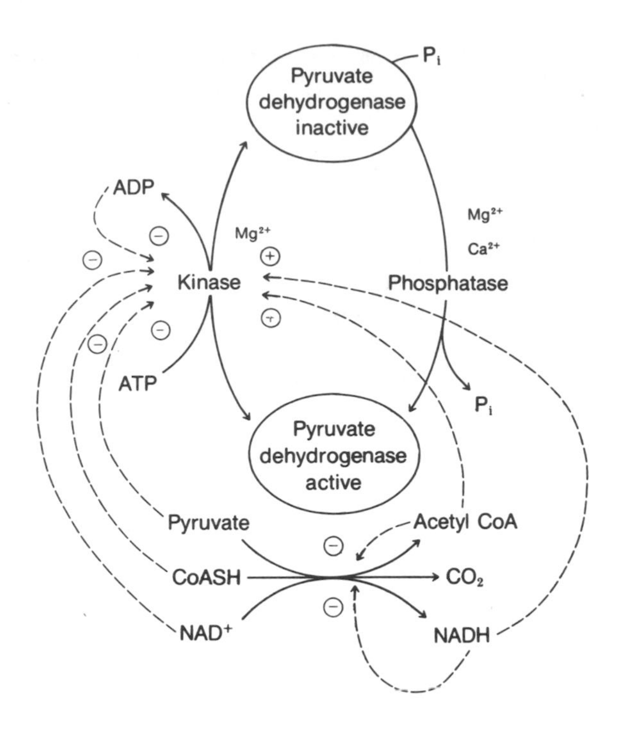

Figure 6.16
The regulation of the pyruvate dehydrogenase multienzyme complex.

genase are phosphorylated by the protein kinase, but the phosphorylation of only one of these sites is related to the activity of the complex. The differential regulation of the pyruvate dehydrogenase kinase and phosphatase is the key to the regulation of the pyruvate dehydrogenase complex. The essential features of this complex regulatory system are illustrated in Figure 6.16. Not only can acetyl CoA and NADH, the products of the pyruvate dehydrogenase reaction, inhibit the dephospho (active) form of the enzyme, but these two compounds stimulate the protein kinase reaction, leading to an interconversion of the complex to its inactive form. In addition, free CoASH and NAD^+ inhibit the protein kinase. Hence, with any increase of the mitochondrial $NADH/NAD^+$ or acetyl CoA/CoA ratio, such as during rapid β oxidation of fatty acids, pyruvate dehydrogenase will be inactivated by the kinase reaction. In addition, the substrate of the enzyme complex, pyruvate, is a potent inhibitor of the protein kinase, and therefore in the presence of elevated tissue pyruvate levels the kinase will be inhibited and the complex maximally active. Finally, it has been demonstrated that insulin administration can activate pyruvate dehydrogenase in adipose tissue, and catecholamines, such as epinephrine, can activate pyruvate dehydrogenase in cardiac tissue. The mechanisms of these hormonal effects are not well understood, but alterations of the intracellular distribution of calcium, such that the phosphoprotein phosphatase reaction is stimulated in the mitochondrial compartment, may be involved in these effects. These hormonal effects are not mediated directly by alterations in the tissue cAMP levels because the pyruvate dehydrogenase protein kinase and phosphatase are cAMP-independent or insensitive (see Clin. Corr. 6.1).

CLIN. CORR. 6.1 PYRUVATE DEHYDROGENASE DEFICIENCY

A variety of disorders in pyruvate metabolism has been detected in children. Some of these defects have been shown to involve deficiencies of the different component catalytic or regulatory subunits of the pyruvate dehydrogenase multienzyme complex. Children diagnosed with pyruvate dehydrogenase deficiency usually exhibit elevated serum levels of lactate, pyruvate, and alanine, which produce a chronic lactic acidosis. Such patients frequently exhibit severe neurological defects, and in most situations this type of enzymatic defect results in death. The diagnosis of pyruvate dehydrogenase deficiency is usually made by assaying the enzyme complex and/or its various enzymatic subunits in cultures of skin fibroblasts taken from the patient. In certain instances patients respond to dietary management in which a ketogenic diet is administered and carbohydrates are minimized. Patients in shock have lactic acidosis because decreased delivery of O_2 to tissues inhibits pyruvate dehydrogenase and increases anaerobic metabolism. This has been treated with dichloroacetate, an inhibitor of pyruvate dehydrogenase kinase and therefore an activator of the enzyme.

Sheu, K. F., Hu, C. C., and Utter, M. F. Pyruvate dehydrogenase complex activity in normal and deficient fibroblasts. *J. Clin. Invest.* 67:1463, 1981; and Stagpoole, P. W., Harmon, E. M., Curry, S. H., Baumgarten, T. G., and Misbin, R. I. Treatment of lactic acidosis with dichloroacetate. *N. Engl. J. Med.* 309:390, 1983.

Acetyl CoA Is Used by Several Different Pathways

The various fates of acetyl CoA generated in the mitochondrial compartment include: (a) complete oxidation of the acetyl group in the tricarboxylic acid cycle for energy generation; (b) in the liver, conversion of an excess of acetyl CoA into the ketone bodies, acetoacetate and β-hydroxybutyrate; and (c) transfer of the acetyl units to the cytosol with subse-

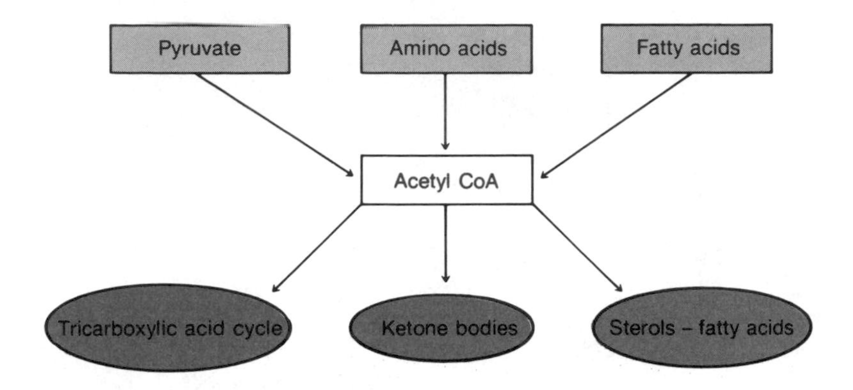

Figure 6.17
Sources and fates of acetyl CoA.

quent biosynthesis of such complex molecules as sterols (Chapter 10) and long-chain fatty acids (Chapter 9) (see Figure 6.17).

6.4 THE TRICARBOXYLIC ACID CYCLE

The primary metabolic fate of acetyl CoA produced in the various energy-generating catabolic pathways of most cells is its complete oxidation in a cyclic series of oxidative reactions termed the **tricarboxylic acid (TCA) cycle.** This metabolic cycle also is commonly referred to as the citric acid cycle or the **Krebs cycle** after Sir Hans Krebs who postulated the essential features of this pathway in 1937. Various investigators defined many of the enzymes and di- and tricarboxylic acid intermediates in this pathway, but it was Krebs who pieced together these components in his formulation of the "Krebs cycle." Although certain of the cycle enzymes are found in the cytosol, the primary location of enzymes of the TCA cycle is in the mitochondrion. This type of distribution is appropriate because the pyruvate dehydrogenase multienzyme complex and the fatty acid β-oxidation sequence, the two primary sources for generating acetyl CoA, are located in the mitochondrial compartment. Also, one of the primary functions of the TCA cycle is to generate reducing equivalents, which are utilized to generate energy, that is, ATP, in the electron transport–oxidative phosphorylation sequence, another process contained exclusively in the mitochondrion (see Figure 6.18). Mitochondrial energy transduction is discussed in Section 6.7.

The individual enzymatic reactions of the TCA cycle are illustrated in Figure 6.19. Figure 6.18 illustrates the essential process involved in the TCA cycle. The substrate or input into the cycle is the two-carbon unit acetyl CoA, and the products of a complete turn of the cycle are two CO_2 plus one high-energy phosphate bond (as GTP) and four reducing equivalents (i.e., three NADH and one $FADH_2$).

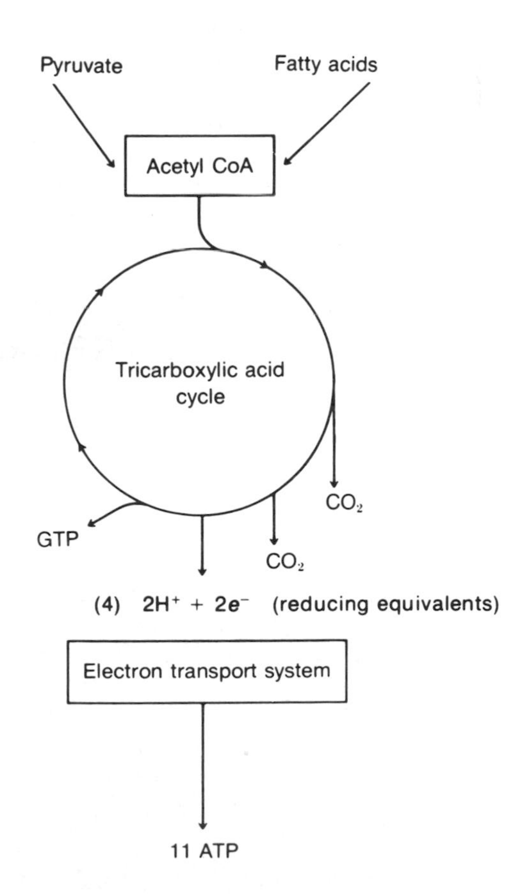

Figure 6.18
General description of mitochondrial ATP synthesis.

The Reactions of the Tricarboxylic Acid Cycle

The initial step of the cycle is catalyzed by the enzyme **citrate synthase.** This is a highly exergonic reaction and essentially commits acetyl groups toward citrate formation or oxidation in the Krebs cycle. As shown below the citrate synthase reaction involves the condensation of the acetyl moiety and the α-keto function of the dicarboxylic acid oxaloacetate. Citrate synthase is an enzyme with mol wt 100,000 and exists in the mitochondrial matrix.

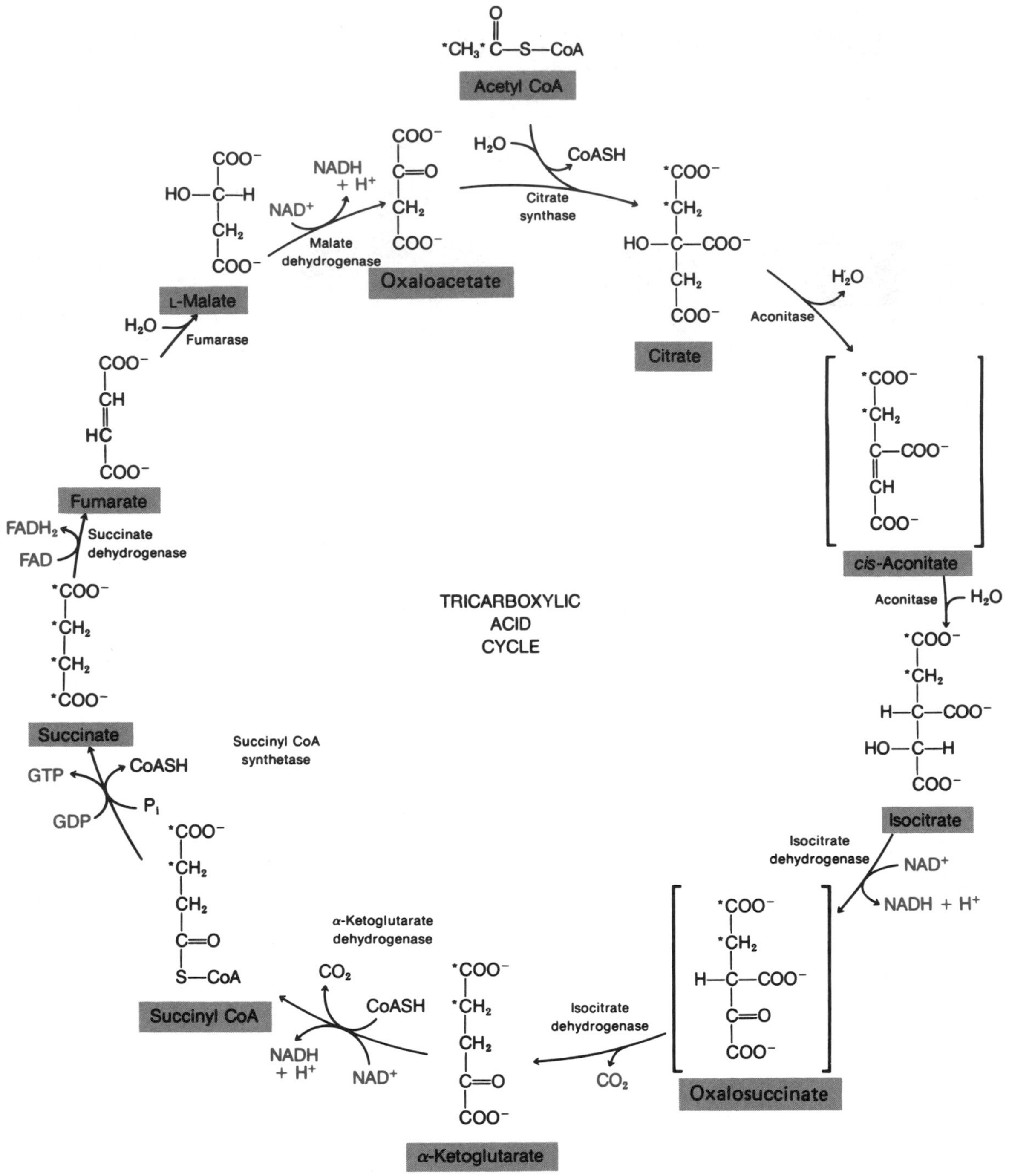

Figure 6.19
The tricarboxylic acid cycle.
The asterisked carbons indicate the fate of the carbons of the acetyl group.

$$\underset{\text{Acetyl CoA}}{\overset{*}{C}H_3\overset{*}{C}(=O)\text{—SCoA}} + \underset{\text{Oxaloacetate}}{O{=}C(\text{—COO}^-)CH_2COO^-} \longrightarrow \underset{\text{Citroyl-SCoA}}{\left[\begin{array}{c} O{=}\overset{*}{C}\text{—SCoA} \\ | \\ {}^{*}CH_2 \\ | \\ HO\text{—}C\text{—}COO^- \\ | \\ CH_2 \\ | \\ COO^- \end{array}\right]_{\text{Enzyme Bound}}} \xrightarrow{H_2O} \text{CoASH} + \underset{\text{Citrate}}{\begin{array}{c} {}^{*}COO^- \\ | \\ {}^{*}CH_2 \\ | \\ HO\text{—}C\text{—}COO^- \\ | \\ CH_2 \\ | \\ COO^- \end{array}}$$

CITRATE SYNTHASE

The equilibrium of this reaction is far toward citrate formation with a $\Delta G°$ near -9 kcal mol^{-1}. The citroyl-SCoA intermediate in this reaction is not released from the enzyme during the reaction and is thought to remain bound to the catalytic site on citrate synthase. It has been estimated that the citrate synthase reaction is considerably displaced from equilibrium under in situ conditions, which makes this step a primary candidate for regulatory modulation. It has been proposed that this enzyme is regulated (inhibited) by ATP, NADH, succinyl CoA, and long-chain acyl CoA derivatives on the basis of experiments performed with the purified enzyme, but none of these effects has been proven to be operative in intact metabolic systems under physiological conditions.

It is most probable that the primary regulator of the citrate synthase reaction is the availability of its two substrates, acetyl CoA and oxaloacetate. It is important to note the many important fates and effects of citrate in energy and biosynthetic metabolism. Figure 6.20 depicts the involvement of citrate as a regulatory effector of other metabolic pathways and as a source of carbon and reducing equivalents for various synthetic purposes (see Chapters 7 and 9 for further details).

Citrate synthase can react with monofluoroacetyl CoA to form monofluorocitrate, which is a potent inhibitor of the next step in the TCA cycle, the aconitase reaction. In fact, whether monofluorocitrate is synthesized in situ as a result of fluoroacetate poisoning or administered experimentally, a nearly complete block of TCA cycle activity is observed.

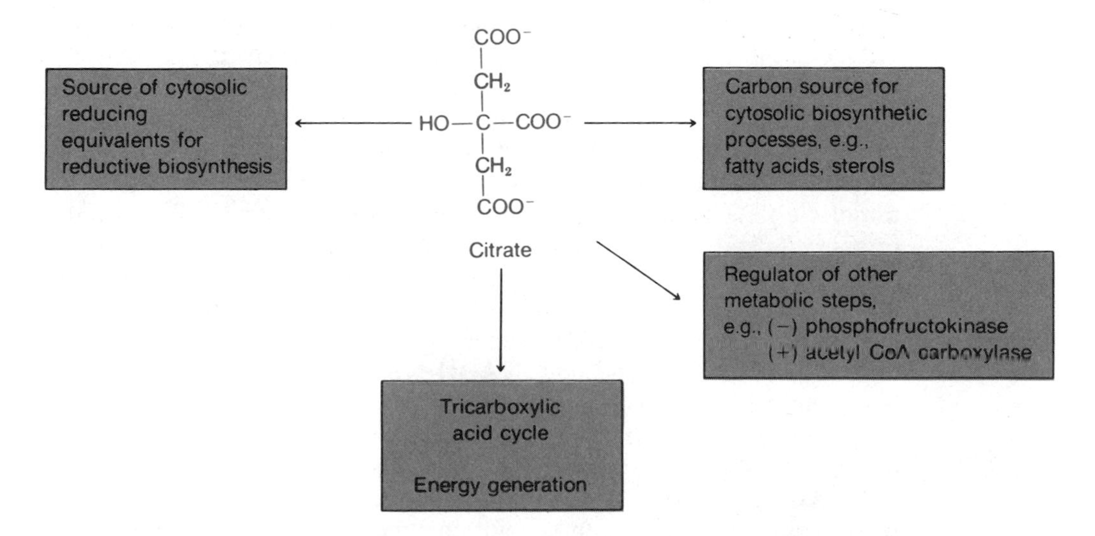

Figure 6.20
Fates and functions of citrate.

Citrate is converted to isocitrate in the **aconitase** reaction.

Citrate ⇌ (H_2O) [*cis*-Aconitate] ⇌ (H_2O) Isocitrate

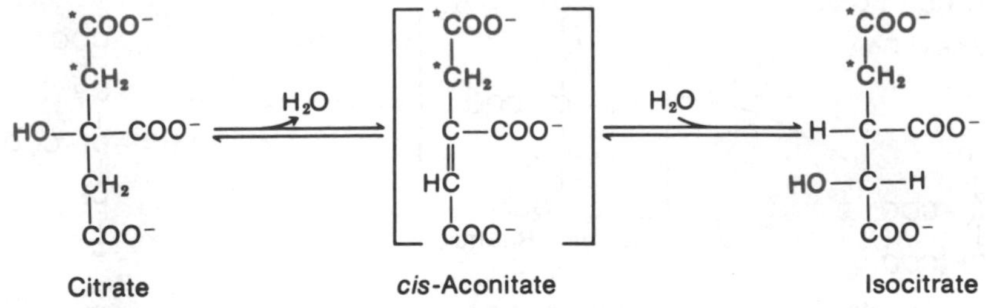

ACONITASE

This reaction involves the generation of an enzyme-bound intermediate, *cis*-aconitate. At equilibrium there exists 90% citrate, 3% *cis*-aconitate, and 7% isocitrate, hence the equilibrium of aconitase lies toward citrate formation. Although the aconitase reaction does not require cofactors, it requires ferrous (Fe^{2+}) iron in its catalytic mechanism. There is evidence that this Fe^{2+} may be involved in an iron–sulfur center, which is an essential component in the hydratase activity of aconitase.

Isocitrate dehydrogenase catalyzes the first dehydrogenase reaction in the TCA cycle. Isocitrate is converted to α-ketoglutarate in an oxidative decarboxylation reaction. In this step of the cycle the initial (of two) CO_2 is produced and the initial (of three) NADH + H^+ are generated. The isocitrate dehydrogenase involved in the TCA cycle in mitochondria from mammalian tissues requires NAD^+ as the oxidized acceptor of reducing equivalents.

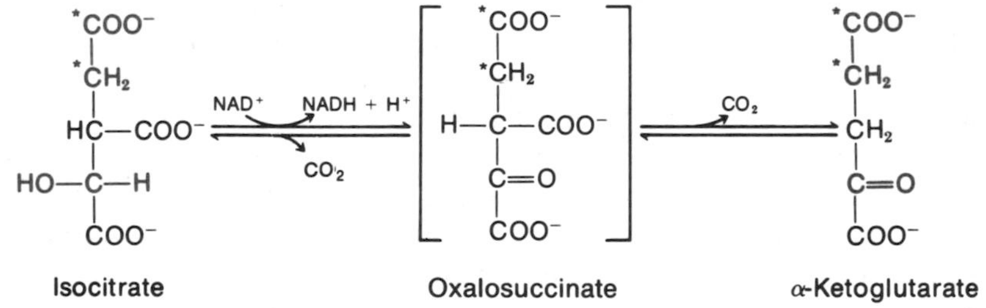

Isocitrate ⇌ (NAD^+ → NADH + H^+) [Oxalosuccinate] ⇌ (CO_2) α-Ketoglutarate

ISOCITRATE DEHYDROGENASE

Mitochondria also possess an isocitrate dehydrogenase that requires $NADP^+$ as the oxidized coenzyme. The $NADP^+$-linked enzyme may also be found in the cytosol, where it is probable that it is involved in providing reducing equivalents for cytosolic reductive processes. The equilibrium of this reaction lies strongly toward α-ketoglutarate formation with a ΔG° of nearly -5 kcal mol^{-1}. The NAD^+-linked isocitrate dehydrogenase has a mol wt 380,000 and consists of eight identical subunits. The reaction requires a divalent metal cation (e.g., Mn^{2+} or Mg^{2+}) in the decarboxylation of the β position of the oxalosuccinate. The NAD^+-linked isocitrate dehydrogenase is stimulated by ADP and in some cases AMP and is inhibited by ATP and NADH. Hence, under high-energy conditions (i.e., high ATP/ADP + P_i and high NADH/NAD^+ ratios) the NAD^+-linked isocitrate dehydrogenase of the TCA cycle is inhibited. During periods of low energy, on the other hand, the activity of this enzyme is stimulated in order to accelerate energy generation in the TCA cycle.

The conversion of α-ketoglutarate to succinyl CoA is catalyzed by the **α-ketoglutarate dehydrogenase** multienzyme complex. This enzyme complex is nearly identical to the pyruvate dehydrogenase complex in terms of the reactions catalyzed and some of its structural features. Again, thiamine pyrophosphate, lipoic acid, CoASH, FAD, and NAD^+ participate in the catalytic mechanism. The multienzyme complex consists of the α-ketoglutarate dehydrogenase, the dihydrolipoyl transuccinylase and the dihydrolipoyl dehydrogenase as the three catalytic subunits. The equilibrium of the α-ketoglutarate dehydrogenase reaction lies strongly toward succinyl CoA formation with a ΔG° of -8 kcal mol^{-1}. In this reaction the second molecule of CO_2 and the second reducing equivalent (i.e., NADH + H^+) of the TCA cycle are produced. The other product of this reaction, succinyl CoA is an example of an energy-rich thiolester compound similar to acetyl CoA. Unlike the pyruvate dehydrogenase complex, the α-ketoglutarate dehydrogenase is not regulated by a protein kinase mediated phosphorylation reaction. The nucleoside triphosphates, ATP and GTP, NADH, and succinyl CoA have been shown to inhibit this enzyme complex while Ca^{2+} has been shown to activate α-ketoglutarate dehydrogenase.

$$^{*}COO^- - {}^{*}CH_2 - CH_2 - C{=}O - COO^- \;(\alpha\text{-Ketoglutarate}) \xrightarrow[NAD^+ \;\to\; NADH + H^+]{CoASH \;\to\; CO_2} {}^{*}COO^- - {}^{*}CH_2 - CH_2 - C{=}O - S{-}CoA \;(\text{Succinyl CoA})$$

α-KETOGLUTARATE DEHYDROGENASE

It is at the level of α-ketoglutarate in the Krebs cycle where an intermediate may leave this oxidative pathway to be reductively aminated in the glutamate dehydrogenase reaction. This mitochondrial enzyme converts α-ketoglutarate to glutamate in the presence of NADH or NADPH and ammonia. Using various transamination reactions the amino group thus incorporated into glutamate is transferred to a variety of other amino acids. These enzymes and the relevance of the incorporation or release of ammonia into or from α-keto acids will be discussed further in Chapters 11 and 12.

$$\overset{+}{N}H_4 + COO^- - C{=}O - CH_2 - CH_2 - COO^- \;(\alpha\text{-Ketoglutarate}) \xrightleftharpoons{NAD(P)H \;\to\; NAD(P)^+} COO^- - H{-}C{-}\overset{+}{N}H_3 - CH_2 - CH_2 - COO^- \;(\text{Glutamate})$$

GLUTAMATE DEHYDROGENASE

$$COO^- - H{-}C{-}NH_3 - CH_2 - CH_2 - COO^- \;(\text{Glutamate}) + COO^- - C{=}O - CH_2 - COO^- \;(\text{Oxaloacetate}) \rightleftharpoons COO^- - H{-}C{-}\overset{+}{N}H_3 - CH_2 - COO^- \;(\text{Aspartate}) + COO^- - C{=}O - CH_2 - CH_2 - COO^- \;(\alpha\text{-Ketoglutarate})$$

GLUTAMATE-OXALOACETATE TRANSAMINASE

The energy-rich character of the thiolester linkage of succinyl CoA is conserved in a substrate-level phosphorylation reaction in the next step of the TCA cycle. The **succinyl CoA synthetase** or **succinate thiokinase** reaction converts succinyl CoA to succinate and in mammalian tissue results in the phosphorylation of GDP to GTP.

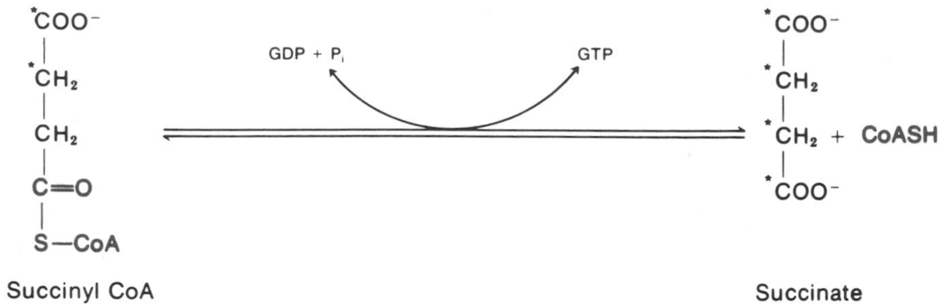

$$\begin{array}{c} {}^{*}COO^- \\ | \\ {}^{*}CH_2 \\ | \\ CH_2 \\ | \\ C{=}O \\ | \\ S{-}CoA \end{array} \xrightleftharpoons[]{GDP + P_i \quad GTP} \begin{array}{c} {}^{*}COO^- \\ | \\ {}^{*}CH_2 \\ | \\ {}^{*}CH_2 \\ | \\ {}^{*}COO^- \end{array} + CoASH$$

Succinyl CoA — Succinate

SUCCINYL CoA SYNTHETASE

This reaction is freely reversible with a $\Delta G^\circ = -0.7$ kcal mol^{-1} and the catalytic mechanism involves an enzyme–succinyl phosphate intermediate.

$$\text{Succinyl CoA} + P_i + \text{Enz} \rightleftharpoons \text{Enz—succinyl phosphate} + \text{CoASH}$$

$$\text{Enz—succinyl phosphate} \rightleftharpoons \text{Enz—phosphate} + \text{succinate}$$

$$\text{Enz—phosphate} + \text{GDP} \rightleftharpoons \text{Enz} + \text{GTP}$$

The enzyme is phosphorylated on the 3 position of a histidine residue during the succinyl CoA synthetase reaction. Hence, in this step of the TCA cycle, a high-energy bond is conserved as GTP. Because of the presence of the nucleoside diphosphate kinase discussed earlier in this chapter, the γ-phosphate of GTP may generate ATP by transfer to ADP.

Succinyl CoA represents a metabolic branch point in that intermediates may enter or exit the TCA cycle at this point (see Figure 6.21). Succinyl CoA may be formed either from α-ketoglutarate in the TCA cycle or from methylmalonyl CoA in the final steps of the breakdown of odd-chain length fatty acids or the branched-chain amino acids valine and isoleucine. The metabolic fates of succinyl CoA include its conversion to succinate in the succinyl CoA synthetase reaction of the Krebs cycle and its condensation with glycine to form δ-aminolevulinate in the δ-aminolevulinate synthetase reaction, which is the initial reaction in porphyrin biosynthesis (see Chapter 24).

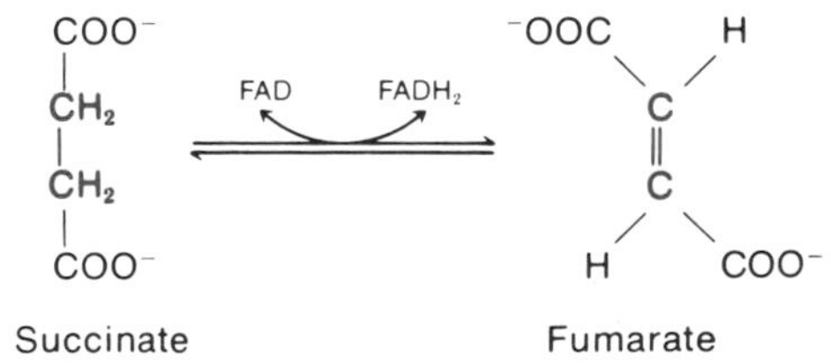

$$\begin{array}{c} COO^- \\ | \\ CH_2 \\ | \\ CH_2 \\ | \\ COO^- \end{array} \xrightleftharpoons[]{FAD \quad FADH_2} \begin{array}{c} ^-OOC \quad\; H \\ \backslash \;\; / \\ C \\ \| \\ C \\ / \;\; \backslash \\ H \quad\; COO^- \end{array}$$

Succinate — Fumarate

SUCCINATE DEHYDROGENASE

Succinate is oxidized to fumarate in the **succinate dehydrogenase** reaction of the Krebs cycle. Succinate dehydrogenase is tightly bound to the inner mitochondrial membrane and is composed of two subunits with mol wt 70,000 and 30,000. The 70,000 mol wt subunit contains the substrate binding site, the covalently bound (to a lysine residue) FAD, four non-

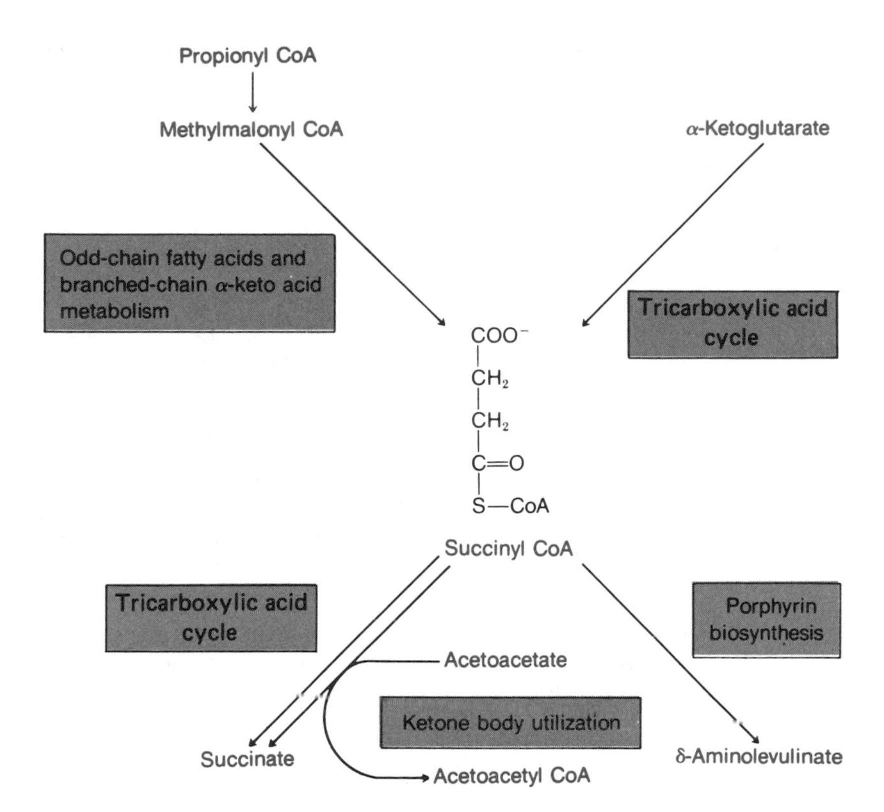

Figure 6.21
Sources and fates of succinyl CoA.

heme iron atoms, and four acid-labile sulfur atoms, whereas the 30,000 mol wt subunit contains four nonheme irons and four acid-labile sulfur atoms. It is thought that this enzyme is a typical example of an iron–sulfur protein in which the nonheme iron of succinate dehydrogenase undergoes valence changes (e.g., $Fe^{2+} \rightarrow Fe^{3+}$) during the removal of electrons and protons from succinate and the subsequent transfer of these reducing equivalents through the covalently bound FAD to the mitochondrial electron-transfer chain at the coenzyme Q–cytochrome b level.

Succinate dehydrogenase is strongly inhibited by malonate and oxaloacetate and is activated by ATP, P_i, and succinate. Malonate inhibits succinate dehydrogenase competitively with respect to succinate. This inhibitory characteristic of malonate is due to the very close structural similarity between malonate and the substrate succinate (Figure 6.22). Malonate is used experimentally as a very effective inhibitor of the Krebs cycle in complex metabolic systems. In fact, the ability of malonate to inhibit the cycle was used by Krebs as evidence for the cyclic nature of this oxidative metabolic pathway.

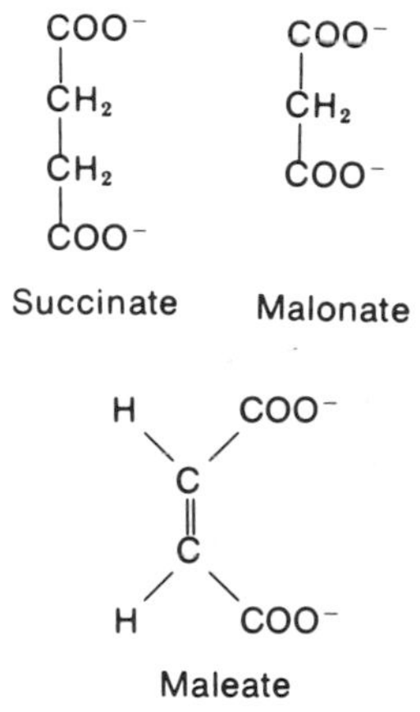

Fumarate is hydrated to form L-malate in the next step in the TCA cycle by the enzyme **fumarase.** Fumarase is a tetramer with a mol wt of 200,000 and is stereospecific for the trans form of the substrate (the cis form, maleate, is not a substrate; Figure 6.22), and the product of the fumarase reaction is only L-malate. The fumarase reaction is freely reversible under physiological conditions.

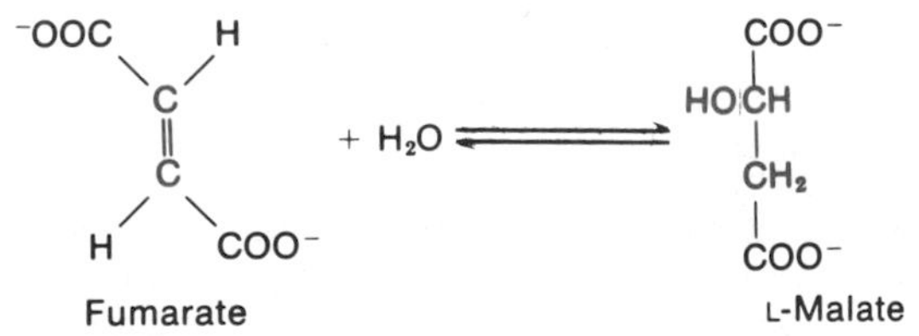

$$^{-}OOC{-}CH{=}CH{-}COO^{-}\ (\text{Fumarate}) + H_2O \rightleftharpoons {}^{-}OOC{-}CH(OH){-}CH_2{-}COO^{-}\ (\text{L-Malate})$$

FUMARASE

Succinate: $^{-}OOC{-}CH_2{-}CH_2{-}COO^{-}$

Malonate: $^{-}OOC{-}CH_2{-}COO^{-}$

Maleate: $(H)(^{-}OOC)C{=}C(H)(COO^{-})$ (cis)

Figure 6.22
Structures of succinate, a TCA cycle intermediate; malonate, a cycle inhibitor; and maleate, a compound not involved in the cycle.

The final reaction in the Krebs cycle is the **malate dehydrogenase** reaction in which the final (of three) reducing equivalents as NADH + H^+ are removed from the cycle intermediates.

$$\begin{array}{c} COO^- \\ | \\ HOCH \\ | \\ CH_2 \\ | \\ COO^- \end{array} \xrightleftharpoons[]{NAD^+ \quad NADH + H^+} \begin{array}{c} COO^- \\ | \\ C{=}O \\ | \\ CH_2 \\ | \\ COO^- \end{array}$$

L-Malate Oxaloacetate

MALATE DEHYDROGENASE

The equilibrium of the malate dehydrogenase reaction lies far toward L-malate formation, because $\Delta G^\circ = +7.0$ kcal mol^{-1}. Thus the malate dehydrogenase reaction is an endothermic reaction when considered in the forward direction of the Krebs cycle. However, the citrate synthase reaction and other reactions of the cycle pull malate dehydrogenase toward oxaloacetate formation by removing oxaloacetate. In addition, NADH produced in the various cycle NAD-linked dehydrogenases is oxidized rapidly to NAD^+ in the mitochondrial respiratory chain.

The Conversion of the Acetyl Group of Acetyl CoA to CO_2 and H_2O Conserves a Significant Portion of the Energy Available

In summary, the TCA cycle (Figure 6.18) serves as a terminal oxidative pathway for most metabolic fuels. Two-carbon moieties as acetyl CoA are taken into the cycle and are oxidized completely to CO_2 and H_2O. During this process 4 reducing equivalents (3 as NADH + H^+ and 1 as $FADH_2$) are produced, which are used subsequently for energy generation. As we discuss later in this chapter, oxidation of each NADH + H^+ results in the formation of 3 ATP molecules in the mitochondrial respiratory chain oxidative phosphorylation sequence, while the oxidation of the $FADH_2$ formed in the succinate dehydrogenase reaction yields 2 molecules of ATP. Also, a high-energy bond is formed in the succinyl CoA synthetase reaction. Hence, the net yield of ATP or its equivalent (i.e., GTP) for the complete oxidation of an acetyl group in the Krebs cycle is 12.

During the complete oxidation of glucose to CO_2 and H_2O there is a net formation of (a) 2 molecules of ATP per glucose in the conversion of glucose to 2 molecules of pyruvate; (b) 6 molecules of ATP per glucose as a result of the translocation and subsequent oxidation in the mitochondrial compartment of 2 molecules of NADH + H^+ formed in the glyceraldehyde 3-phosphate dehydrogenase reaction of the glycolytic pathway; and (c) 30 molecules of ATP per glucose from the oxidation of the 2 molecules of pyruvate in the pyruvate dehydrogenase reaction and subsequent conversion of 2 molecules of acetyl CoA to CO_2 and H_2O in the TCA cycle. Hence, the net ATP yield during the complete oxidation of glucose to $6CO_2 + 6H_2O$ is 38 molecules of ATP.

The Activity of the Tricarboxylic Acid Cycle Is Carefully Regulated

A variety of factors is involved in the regulation of the activity of the TCA cycle. First, the supply of acetyl units, whether they are derived from pyruvate (i.e., carbohydrate) or fatty acids, is a crucial factor in determining the rate of the Krebs cycle. Regulatory influences on the pyruvate

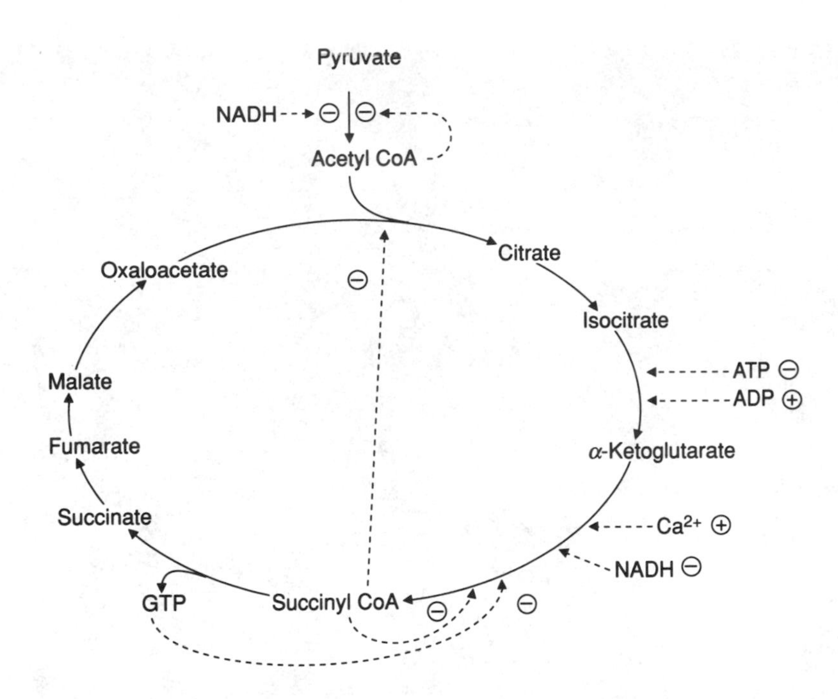

Figure 6.23
Representative examples of the regulatory interactions in the TCA cycle.

dehydrogenase complex have an important effect on the activity of the cycle. Likewise any control exerted on the processes of transport and β oxidation of fatty acids would be an effective determinant of the Krebs cycle activity.

Second, because the primary dehydrogenase reactions of the Krebs cycle are dependent on a continuous supply of both NAD^+ and FAD, their activities are very stringently controlled by the mitochondrial respiratory chain, which is responsible for oxidizing the NADH and $FADH_2$ produced as a result of substrate oxidation in the cycle. Because the activity of the respiratory chain is coupled obligatorily to the generation of ATP in the oxidative phosphorylation sequence of reactions, the activity of the Krebs cycle is very much dependent on a **respiratory control,** which is strongly affected by the availability of ADP + P_i and oxygen. Hence an inhibitory agent or metabolic condition which might interrupt the supply of oxygen, the continuous supply of ADP or the source of reducing equivalents (e.g., substrate for the cycle) would shut down cycle activity. This type of control of the cycle is generally referred to as the "coarse control" of the cycle. There are, of course, a variety of postulated effector-mediated regulatory interactions between various intermediates or nucleotides and the individual enzymes of the cycle, which may serve to exert a fine control on the activity of the cycle. Some illustrations of these interactions are shown in Figure 6.23 and have also been noted during the discussions of individual enzymes of the Krebs cycle. It must be stressed that the physiological relevance of many of these types of individual regulatory interactions has not been firmly established in intact metabolic systems.

6.5 STRUCTURE AND COMPARTMENTATION OF THE MITOCHONDRIAL MEMBRANES

Because the metabolic pathways for the oxidation of pyruvate, the end product of glycolysis, and fatty acids are located in mitochondria, a major portion of the energy-generating capacity of most cells resides in the mitochondrial compartment of the cell. The number of mitochondria in

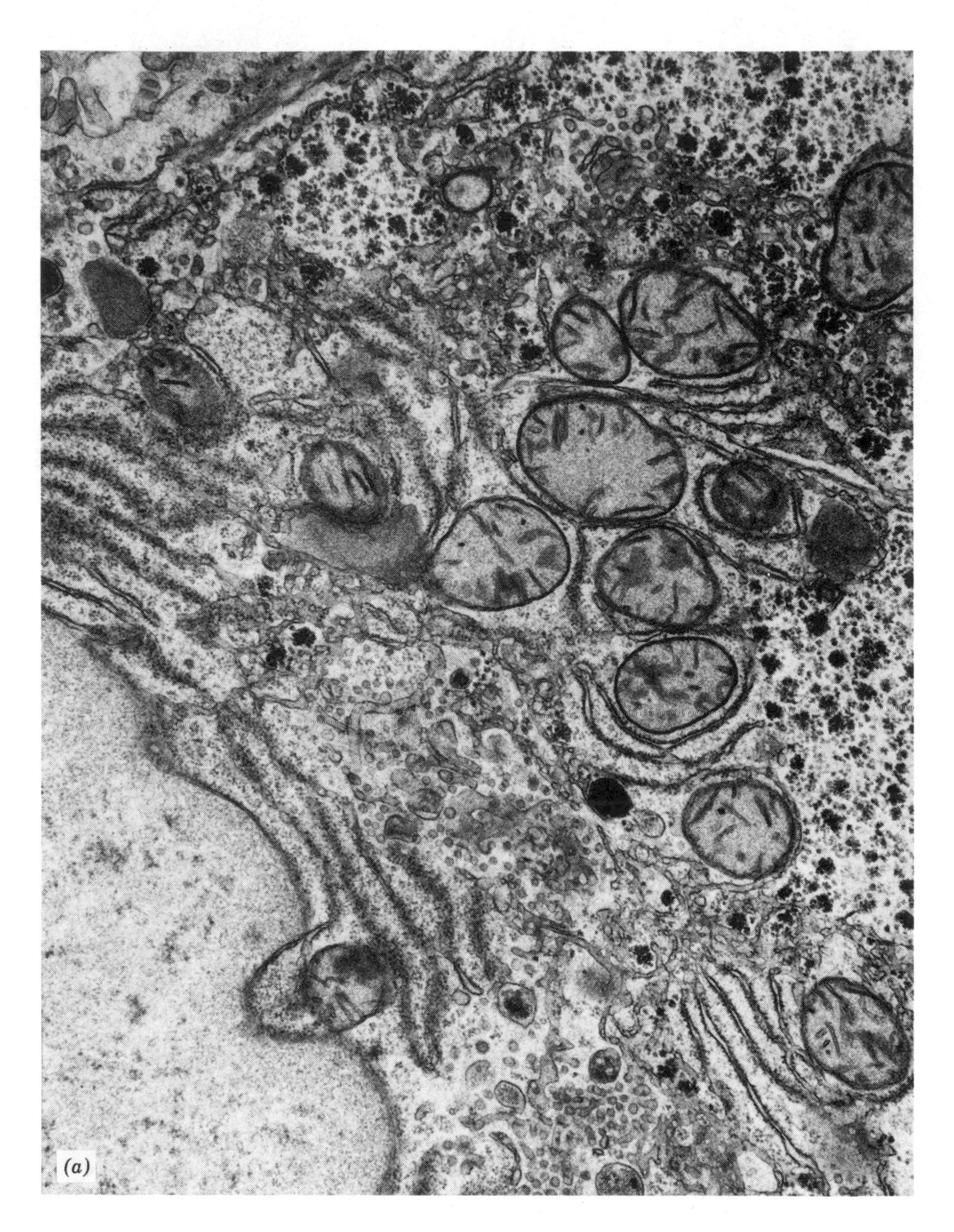

Figure 6.24 (a)
Electron micrographs of mitochondria in hepatocytes from rat liver.
Magnification 39,600×. Courtesy of Dr. W. B. Winborn, Department of Anatomy, The University of Texas Health Science Center at San Antonio, and the Electron Microscopy Laboratory, Department of Pathology, The University of Texas Health Science Center at San Antonio.

various tissues reflects the physiological function of the tissue and determines its capacity to perform aerobic metabolic functions. For example, the erythrocyte has no mitochondria and hence does not possess the capacity to generate energy using oxygen as a terminal electron acceptor. On the other hand, cardiac tissue is a highly aerobic tissue, and it has been estimated that about one-half of the cytoplasmic volume of cardiac cells is composed of mitochondria. The liver is another tissue that is highly dependent on aerobic metabolic processes for its various functions, and it has been estimated that mammalian hepatocytes contain between 800 and 2000 mitochondria per cell. Mitochondria exist in a variety of different shapes, depending on the cell type from which they are derived. As can be

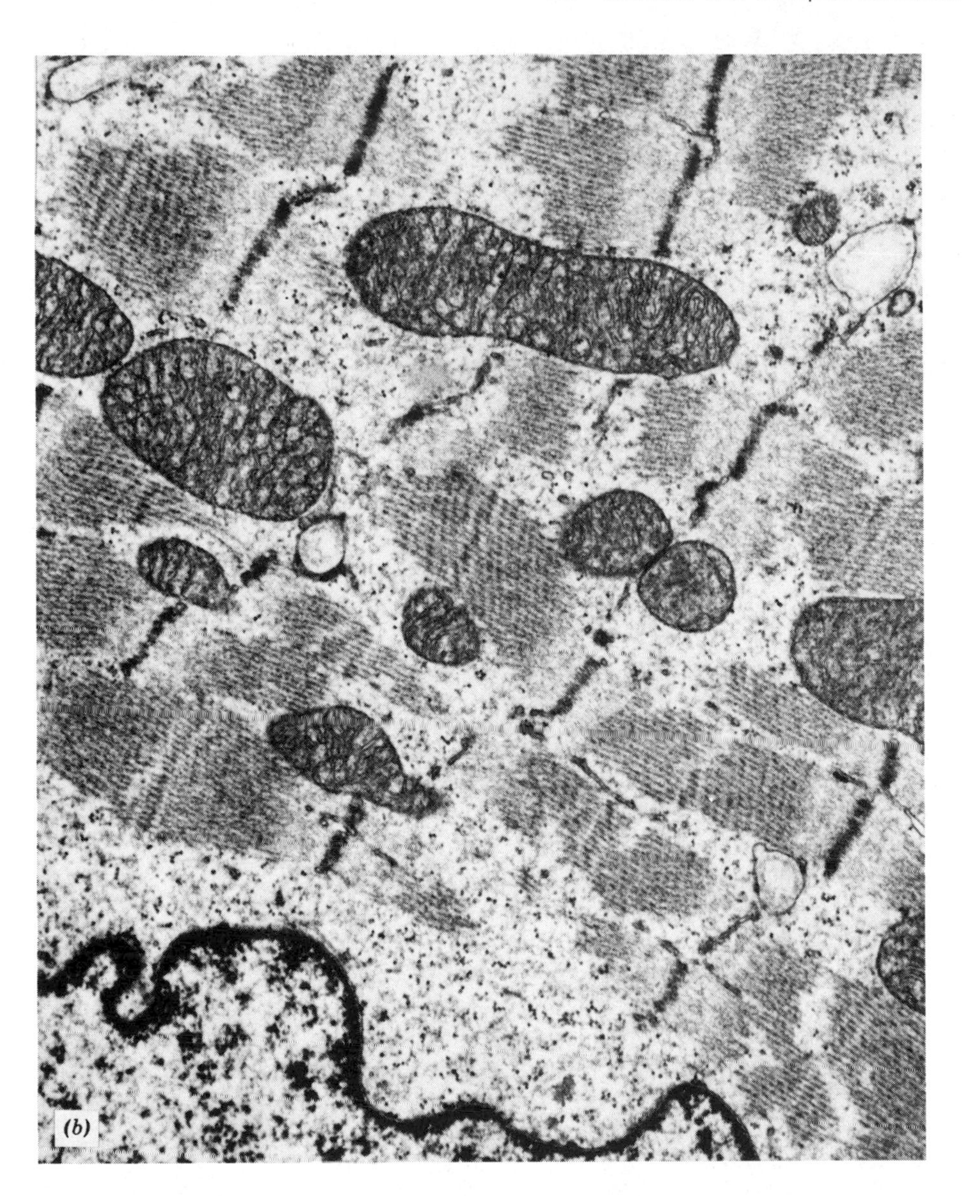

Figure 6.24 (b)
Electron micrograph of mitochondria in muscle fibers from rabbit heart.
Magnification 39,600×. Courtesy of Dr. W. B. Winborn, Department of Anatomy, The University of Texas Health Science Center at San Antonio, and the Electron Microscopy Laboratory, Department of Pathology, The University of Texas Health Science Center at San Antonio.

seen in Figure 6.24 mitochondria from liver are nearly spherical in shape, whereas those found in cardiac muscle are oblong or cylindrical.

The Inner and Outer Mitochondrial Membranes Have Different Compositions and Functions

Mitochondria are composed of two membranes, an outer membrane and a highly invaginated inner membrane (see Figure 6.25). The outer mitochondrial membrane is thought to be a rather simple membrane, which is composed of about 50% lipid and 50% protein, with relatively few enzymatic or transport functions. Table 6.5 defines some of the enzymatic components of the outer membrane.

The inner membrane is structurally and functionally much more complex than the outer membrane. Roughly 80% of the inner membrane is

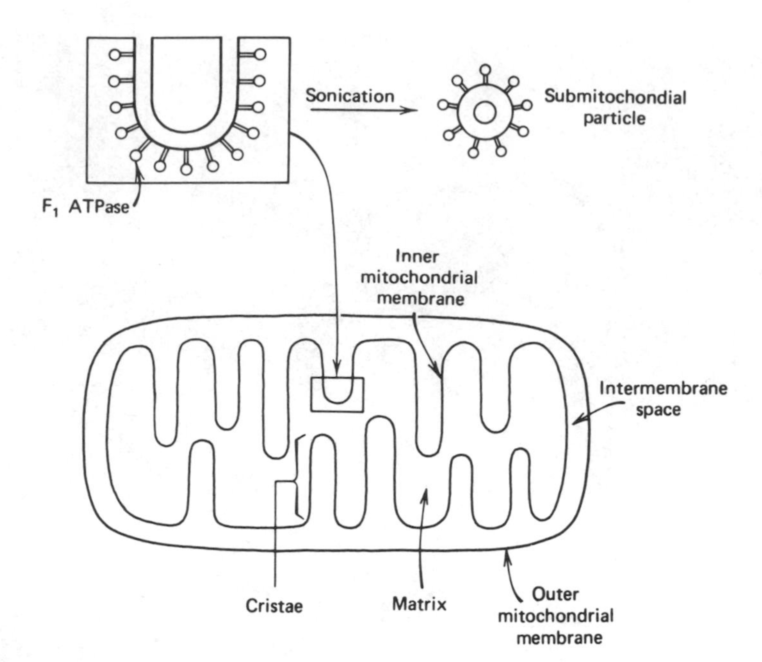

Figure 6.25
Diagram of the various submitochondrial compartments.

protein. The inner membrane contains most of the enzymes involved in electron transport and oxidative phosphorylation, various dehydrogenases, and several transport systems, which are involved in transferring substrates, metabolic intermediates, and adenine nucleotides between the cytosol and the mitochondrial matrix (Table 6.5).

Some of the enzymatic components associated with the inner mitochondrial membrane are only loosely associated with the membrane, whereas other enzymatic components are either tightly bound or are actual structural elements of the membrane. Hence, there is a wide variability in the extent to which physical (ultrasonic irradiation or freezing and thawing), chemical (organic solvent or detergent treatment), or enzymatic (protease or lipase) treatments remove, release, or inactivate the enzymes associated with the inner membrane.

TABLE 6.5 Enzymatic Composition of the Various Mitochondrial Subcompartments

Outer membrane	Intermembrane space	Inner membrane	Matrix
Monoamine oxidase	Adenylate kinase	Succinate dehydrogenase	Pyruvate dehydrogenase
Kynurenine hydroxylase	Nucleoside diphosphate kinase	F_1-ATPase	Citrate synthase
Nucleoside diphosphate kinase		NADH dehydrogenase	Isocitrate dehydrogenase
Phospholipase A		β-Hydroxybutyrate dehydrogenase	α-Ketoglutarate dehydrogenase
Fatty acyl CoA synthetases		Cytochromes b, c_1, c, a, a_3	Aconitase
NADH : cytochrome c reductase (rotenone-insensitive)		Carnitine : acyl CoA transferase	Fumarase
Choline phosphotransferase		Adenine nucleotide translocase	Succinyl CoA synthetase
		Mono-, di-, and tricarboxylate translocase	Malate dehydrogenase
		Glutamate–aspartate translocase	Fatty acid β-oxidation system
			Glutamate dehydrogenase
			Glutamate–oxaloacetate transaminase
			Ornithine transcarbamoylase

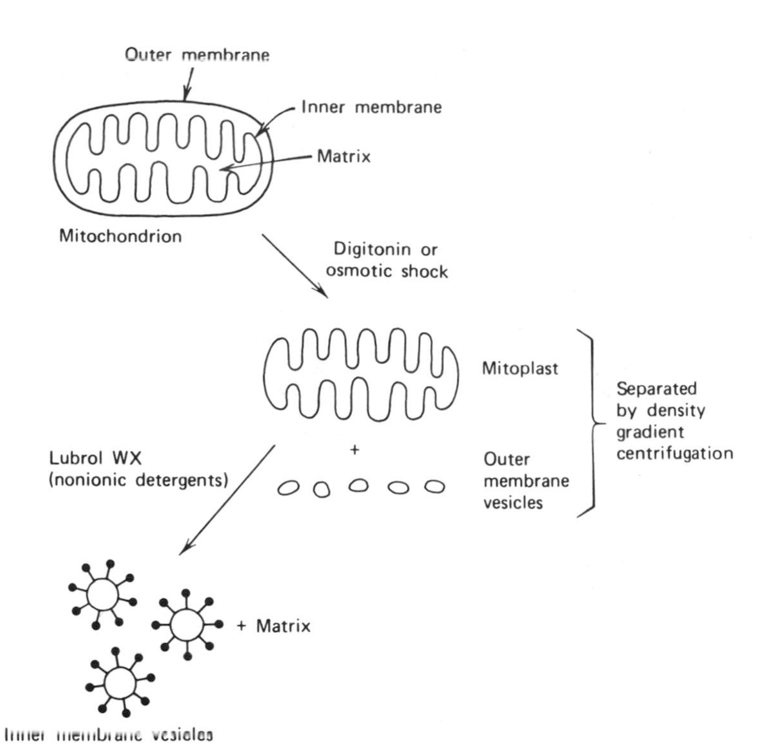

Figure 6.26
Separation of mitochondrial membranes.

Experimental procedures have been developed that allow separation of inner from outer mitochondrial membranes. As indicated in Figure 6.26, the outer membrane may be stripped off and isolated, using digitonin (a detergent), osmotic shock, or ultrasonic irradiation followed by density-gradient ultracentrifugation. The resulting inner membrane plus matrix fraction is referred to as a **mitoplast.** The mitochondrial matrix may be released from the mitoplast, by treatment with a nonionic detergent or vigorous sonication. Once the various subcompartments of the mitochondrion have been separated, analyses may be performed to determine the location of the various characteristic marker enzymes, some of which are listed in Table 6.5. Enzymatic markers have been used effectively to detect the presence of mitochondria or even a particular portion of mitochondria in membrane preparations of diverse derivation.

Mitochondrial Inner Membranes Contain a Variety of Substrate Transport Systems

Whereas the outer membrane presents little or no permeability barrier to substrate or nucleotide molecules of interest in energy metabolism, the inner membrane has very restricted limitations on the types of substrates, intermediates, and nucleotide species that may be transported across into the matrix compartment.

Figure 6.27 depicts various transport systems that have been described in mitochondria. Some of these transporters are well characterized, but others are not. The primary responsibility of these transport functions is to facilitate the selective movement of various substrates, intermediates, and nucleotides back and forth across the inner mitochondrial membrane from the cytosol to the mitochondrial matrix. By virtue of these transporters, various substrates and other molecules can be accumulated in the mitochondrial matrix since the transporters can facilitate the movement of the substrate against a concentration gradient. The importance of the

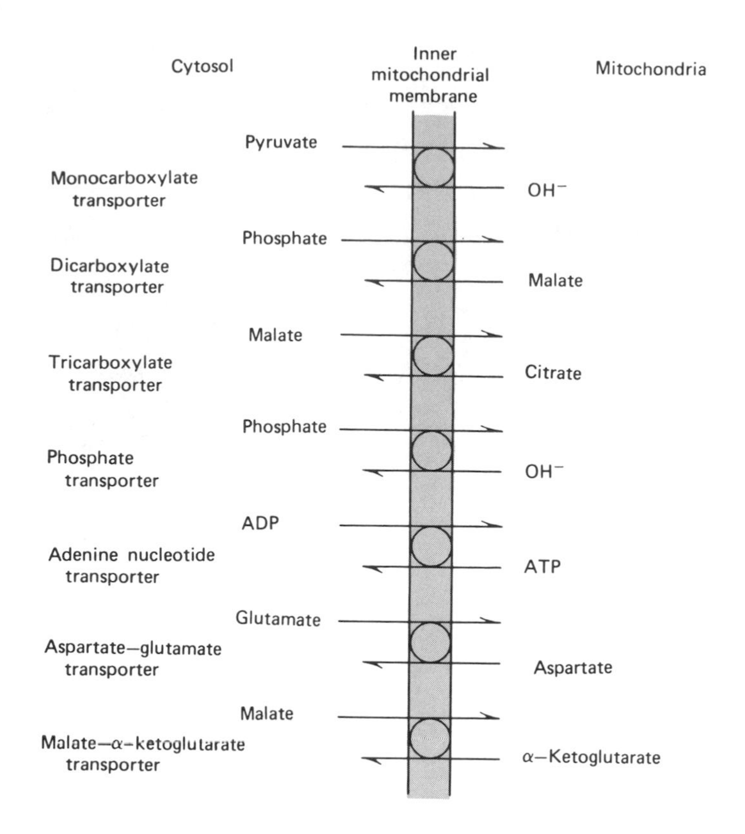

Figure 6.27
Mitochondrial metabolite transporters.

mitochondrial transporter systems derives from the involvement of the substances transported in a variety of mitochondrial metabolic processes.

Substrate Shuttles Transport Reducing Equivalents Across the Mitochondrial Membrane

The various nucleotides involved in cellular oxidation–reduction reactions (e.g., NAD^+, NADH, $NADP^+$, NADPH, FAD, and $FADH_2$) and CoA and its derivatives are not permeable to the inner mitochondrial membrane. Hence, for example, in order to transport reducing equivalents (e.g., protons and electrons) from the cytosol to the mitochondrial matrix or vice versa, **"substrate shuttle mechanisms"** involving the reciprocal transfer of reduced and oxidized members of various oxidation–reduction couples are used to accomplish the net transfer of reducing equivalents across the membrane. Two examples of how this transfer of reducing equivalents from the cytosol to the mitochondria occurs are shown in Figure 6.28. The **malate–aspartate shuttle** and the **α-glycerol phosphate shuttle** are employed in various tissues to translocate reducing equivalents from the cytosol, where they are generated, to the mitochondrial compartment, where they are oxidized to yield energy. The operation of such substrate shuttles requires that the appropriate enzymes are localized on the correct side of the membrane and that appropriate transporters or translocases are present on/in the membrane to shuttle the various intermediates. In this regard the operation of the malate–aspartate shuttle depends first on the fact that NADH, NAD^+, and oxaloacetate are not permeable to the inner mitochondrial membrane, second on the distribution of malate dehydrogenase and aspartate aminotransferase on both sides of the inner mitochondrial membrane, and third on the existence of membrane transporters, which allow the exchange of intramitochondrial aspartate for cytosolic glutamate and cytosolic malate for intramitochondrial α-ketoglutarate.

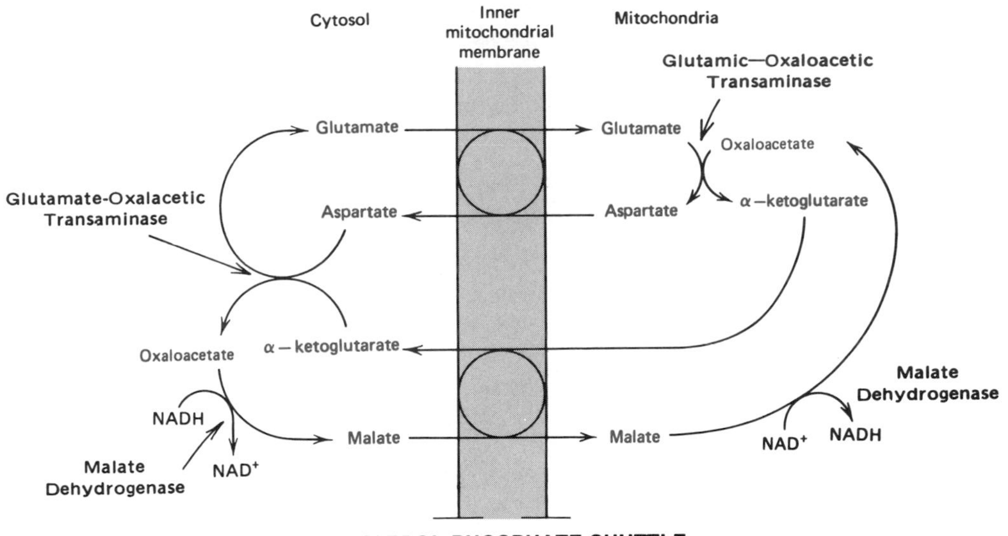

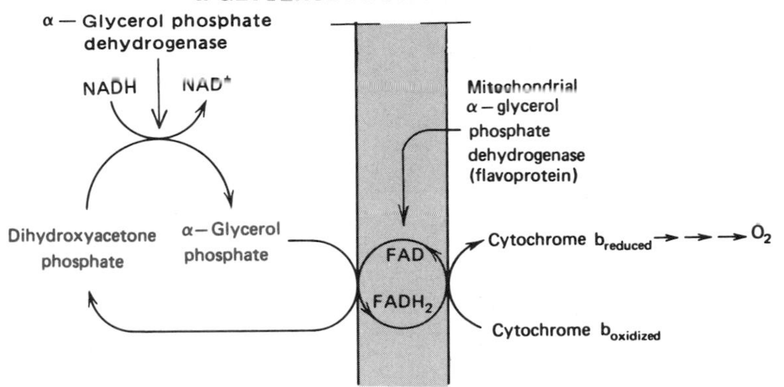

Figure 6.28
Transport shuttles for reducing equivalents.

Acetyl Units Are Transported by Citrate

Acetyl CoA is an impermeable substance but it can transfer the C_2 fragment (the acetyl group) from the mitochondrial compartment to the cytosol, where acetyl moieties are required for fatty acid or sterol biosynthesis, as illustrated in Figure 6.29.

Intramitochondrial acetyl CoA is converted to citrate in the citrate synthase reaction of the Krebs cycle. Subsequently the citrate is exported to the cytosol on the **tricarboxylate transporter** in exchange for a dicarboxylic acid such as malate. Cytosolic citrate may be cleaved to acetyl CoA and oxaloacetate at the expense of an ATP molecule in the ATP : citrate lyase reaction, which is discussed in Chapter 9. Substrate shuttle mechanisms are also involved in the movement of appropriate substrates and intermediates in both directions across the inner mitochondrial membrane in the liver during periods of active gluconeogenesis and ureogenesis (see Chapters 7 and 11).

Transport of Adenine Nucleotides and Phosphates Are Controlled

Adenine nucleotides are transported across the inner mitochondrial membrane by the very specific **adenine nucleotide translocator.** Nucleotide species such as the guanine, uridine, or cytosine nucleotides are neither

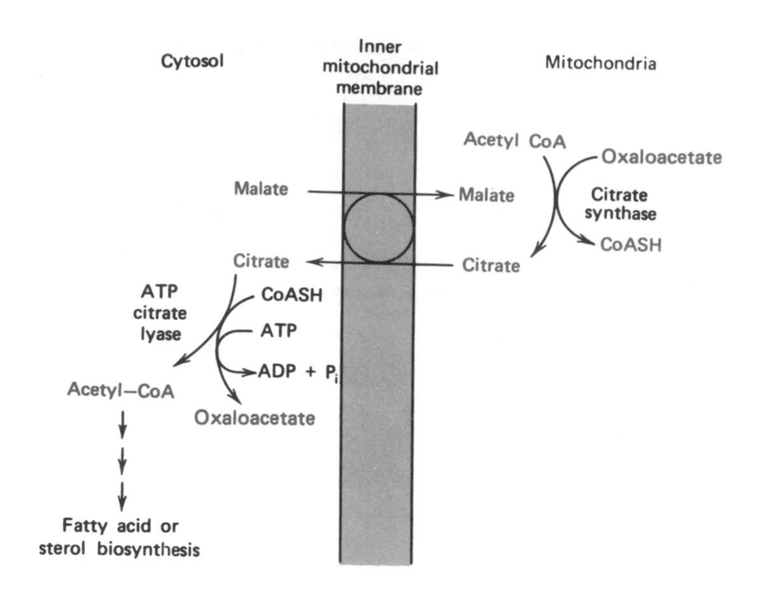

Figure 6.29
The export of intramitochondrially generated citrate to the cytosol to serve as a source of acetyl CoA for biosynthesis of fatty acids or sterols.

exchanged across the inner membrane on the adenine nucleotide specific translocator nor transported by a comparable carrier specific for non-adenine nucleotides. As indicated in Figure 6.30 cytosolic ADP, which is formed during energy-consuming reactions, is exchanged for mitochondrial ATP, which is generated in the process of oxidative phosphorylation. At pH 7 ADP has three negative charges and ATP has four, so that a 1 : 1 exchange of ADP : ATP would cause a charge imbalance across the membrane. Hence the ADP for ATP exchange across the mitochondrial membrane is an electrogenic process, which requires that in the end the charge imbalance must be compensated for by the movement of a proton or another charged species. An adenine nucleotide carrier has been isolated due to its capacity to bind very tightly to atractyloside, a specific inhibitor of the carrier. The carrier protein is a dimer with subunit mol wt 30,000. It is unlikely that the rate of transport of adenine nucleotides across the mitochondrial membrane is ever limiting to the overall process of mitochondrial ATP synthesis. Furthermore, it has been observed that low concentrations of long-chain fatty acyl CoA derivatives inhibit (i.e., $K_i = 1\ \mu M$) the transport of ATP and ADP in isolated liver mitochondria. However, experimental results performed under in vivo conditions in intact liver cells indicate that there occurs little, if any, inhibition of the adenine nucleotide transporter under metabolic conditions in which a large concentration of long-chain fatty acyl CoA accumulates.

There is also a transporter that transports cytosolic P_i into the mitochondrial matrix in exchange for negatively charged hydroxyl ions in an electroneutral exchange (see Figure 6.30). This **phosphate transport** may also be accomplished in a proton-compensated mechanism, for example, phosphate and protons are transported in a 1 : 1 ratio. Phosphate transport is strongly inhibited by the compound mersalyl and various mercurial reagents.

Mitochondria Have a Specific Calcium Transport Mechanism

Finally, mitochondria from most tissues possess a transport system capable of translocating calcium across the mitochondrial inner membrane. It is difficult to overestimate the importance of the distribution of cellular calcium pools in different cell functions, such as muscle contraction, neural transmission, and hormone action and secretion. Calcium exists in

CLIN. CORR. 6.2 MITOCHONDRIAL MYOPATHIES

Diseases that involve defects in various metabolic functions of muscle have been described. Clinically, patients with myopathies complain of weakness and cramping of the affected muscles; infants have difficulty feeding and crawling; severe fatigue results from minimal exertion; and there is usually evidence of muscle wasting. On the basis of electron microscopic examination and enzymatic characterization of muscle biopsy material, many myopathies have been found that have a primary lesion in mitochondrial function.

Deficiencies in mitochondrial transport functions (i.e., carnitine: palmitoyl CoA transferase) and in components of the mitochondrial electron transport chain (NADH dehydrogenase, cytochrome b, cytochrome a,a_3, or the mitochondrial ATPase) have been described. In many mitochondrial myopathies large paracrystalline inclusions have been found within the mitochondrial matrix (see figure). It is not known whether this crystalline material is inorganic or organic in composition. In certain mitochondrial myopathies electron transport is only loosely coupled to ATP production; in other cases these processes exhibit normal tight coupling. Because some of these disorders involve defects in enzymes encoded by mitochondrial genes, they have the unique pattern of inheritance from the mother, since all mitochondria are derived from mitochondria in the ovum.

Petty, R. K. H., Harding, A. E., and Morgan-Hughes, J. A. The clinical features of mitochondrial myopathy. *Brain* 109:915, 1986.

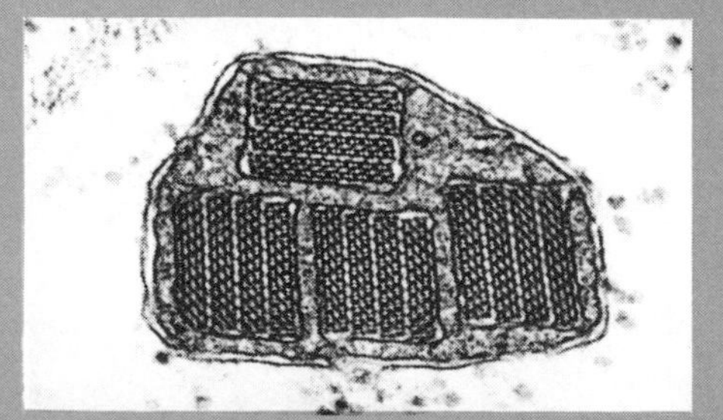

Example of paracrystalline inclusions in mitochondria from muscles of ocular myopathic patients.
Magnification 36,000×. Courtesy of Dr. D. N. London, Inst. of Neurology, Univ. of London.

distinct pools in the cell. The cytosol, mitochondria, endoplasmic reticulum, nuclei, and the Golgi membranes have their own pools of intracellular calcium. Some of the intracellular calcium is bound to nucleotides, metabolites, or membrane ligands, while a portion of the intracellular calcium is free in solution. A gradient of calcium exists from outside to inside a cell. Estimates of intracellular cytosolic calcium are in the range of 10^{-7} M, whereas extracellular calcium is at least four orders of magnitude greater. Total intramitochondrial calcium has been estimated to be $\sim 10^{-4}$ M but the free ionic calcium concentration in the mitochondrion is in the range of 10^{-7} M. Hence, processes involved in the alternate sequestering and release of an intracellular store of calcium can greatly influence intracellular calcium pools and various cell functions. Mitochondria have been known to accumulate rather large quantities of calcium at the expense of ATP hydrolysis, respiration, or an electrochemical gradient. Mitochondrial calcium transport is inhibited by low concentrations of lanthanides (trivalent metal cations) and by a compound called ruthenium red. Magnesium can compete with calcium for the carrier in certain types of mitochondria. The current view is that there is a specific carrier in the inner mitochondrial membrane, which is likely a glycoprotein (Figure 6.31). The mitochondrial calcium carrier exhibits saturation kinetics, has a high affinity for calcium, and is highly specific for calcium. Permeant counterions such as phosphate or acetate stimulate calcium transport and allow the metal cation to be retained by the mitochondria. Most interesting is the finding that certain hormones may affect intracellular calcium distribution (e.g., epinephrine or vasopressin) as part of the mechanism for the mediation of the hormone response. Various cytosolic protein kinases such as those involved in glycogen metabolism are calcium sensitive.

In summary, the inner mitochondrial membrane possesses a variety of transport systems that are involved in the movement of nucleotides, substrates, metabolites, and metal cations into and out of the mitochondrial compartment. An understanding of these transport functions is essential in order to understand complex cellular metabolic pathways and their regulation (see Clin. Corr. 6.2).

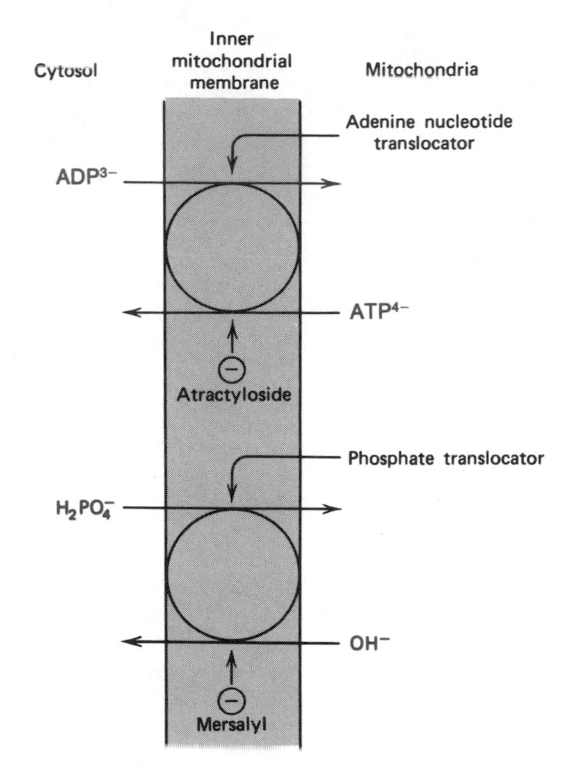

Figure 6.30
The adenine nucleotide and phosphate translocators.

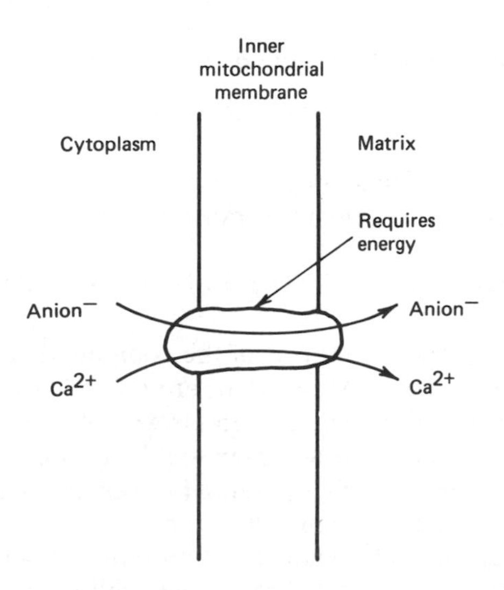

Figure 6.31
Mitochondrial calcium carrier.
The energy requirement can be met from ATP, ΔpH, or membrane potential.

6.6 ELECTRON TRANSFER

During the enzymatic reactions involved in glycolysis, fatty acid oxidation, and the TCA cycle, reducing equivalents are derived from the sequential breakdown of the initial metabolic fuel. In the case of glycolysis, NADH is produced in the glyceraldehyde 3-phosphate dehydrogenase reaction, and this reducing equivalent must be either reoxidized in the cytosol (e.g., by lactate dehydrogenase) or transported to the mitochondrial matrix via one of the substrate shuttle mechanisms in order to realize the maximum energy yield from the oxidation of glucose. In the case of fatty acid oxidation and in the TCA cycle, reducing equivalents as both NADH and $FADH_2$ are produced in the mitochondrial matrix. In order to transduce this reducing power into utilizable energy, mitochondria have a system of electron carriers in or associated with the inner mitochondrial membrane, which convert reducing equivalents in the presence of oxygen into utilizable energy by synthesizing ATP. This process is called **electron transport,** and, as will be seen later, NADH and $FADH_2$ oxidation in this process results in the production of 3 and 2 mol of ATP per mole of reducing equivalent transferred to oxygen, respectively.

Oxidation–Reduction Reactions

Prior to the presentation of a description of the many components and the mechanism of the electron transport sequence, it is important to discuss some basic information concerning oxidation–reduction reactions. The mitochondrial electron transport system is little more than a sequence of linked oxidation–reduction reactions, for example,

$$AH_2 + B \rightleftharpoons A + BH_2$$

Oxidation–reduction reactions occur when there is a transfer of electrons from a suitable electron donor (the reductant) to a suitable electron acceptor (the oxidant). In some oxidation–reduction reactions only electrons are transferred from the reductant to the oxidant (i.e., electron transfer between cytochromes),

$$\text{Cytochrome } c\ (Fe^{2+}) + \text{cytochrome } a\ (Fe^{3+}) \rightleftharpoons \text{cytochrome } c\ (Fe^{3+}) + \text{cytochrome } a\ (Fe^{2+})$$

whereas in other types of reactions, both electrons and protons (hydrogen atoms) are transferred (i.e., electron transfer between NADH and FAD).

$$NADH + H^+ + FAD \rightleftharpoons NAD^+ + FADH_2$$

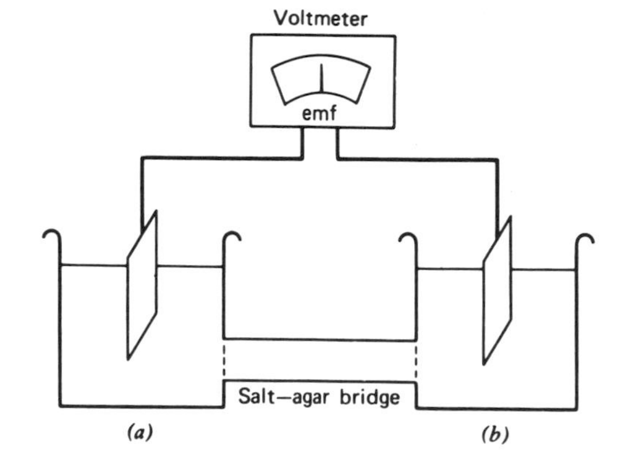

Figure 6.32
A system for determining oxidation–reduction potentials in half-cell reactions.
(*a*) Sample half-cell; Red:H + Ox $\rightleftharpoons$ Ox:H + Red, where Red = reductant and Ox = oxidant (both initially present at 1 M). (*b*) Reference half-cell; $2H^+ + 2e^- \rightleftharpoons H_2$, where emf = 0.0 V (1 M H^+ in solution and H_2 gas at 1 atm).

The oxidized and the reduced forms of the compounds or groups operating in oxidation–reduction type reactions are referred to as redox couples or pairs. The facility with which a given **electron donor** (reductant) gives up its electrons to an **electron acceptor** (oxidant) is expressed quantitatively as the **oxidation–reduction potential** of the system. An oxidation–reduction potential is measured in volts as an electromotive force (emf) of a half-cell made up of both members of an oxidation–reduction couple when compared to a standard reference half-cell (usually the hydrogen electrode reaction) (see Figure 6.32). The potential of the standard hydrogen reference electrode is set by convention at 0.0 V at pH 0.0. However, when this standard potential is corrected for pH 7.0 the reference elec-

TABLE 6.6 Standard Oxidation–Reduction Potentials for Various Biochemical Reactions

Oxidation–reduction system	Standard oxidation–reduction potential E_0' (V)
Acetate + $2H^+$ + $2e^-$ ⇌ acetaldehyde	−0.60
$2H^+$ + $2e^-$ ⇌ H_2	−0.42
Acetoacetate + $2H^+$ + $2e^-$ ⇌ β-hydroxybutyrate	−0.35
NAD^+ + $2H^+$ + $2e^-$ ⇌ NADH + H^+	−0.32
Acetaldehyde + $2H^+$ + $2e^-$ ⇌ ethanol	−0.20
Pyruvate + $2H^+$ + $2e^-$ ⇌ lactate	−0.19
Oxaloacetate + $2H^+$ + $2e^-$ ⇌ malate	−0.17
Coenzyme Q_{ox} + $2e^-$ ⇌ coenzyme Q_{red}	+0.10
Cytochrome *b* (Fe^{3+}) + e^- ⇌ cytochrome *b* (Fe^{2+})	+0.12
Cytochrome *c* (Fe^{3+}) + e^- ⇌ cytochrome *c* (Fe^{2+})	+0.22
Cytochrome *a* (Fe^{3+}) + e^- ⇌ cytochrome *a* (Fe^{2+})	+0.29
$\frac{1}{2}O_2$ + $2H^+$ + $2e^-$ ⇌ H_2O	+0.82

trode potential becomes −0.42 V. The oxidation–reduction potentials for a variety of important biochemical reactions have been determined and are tabulated in Table 6.6.

An important concept is indicated in this listing of oxidation–reduction potentials. The reductant of an oxidation–reduction pair with large negative oxidation–reduction potential will give up its electrons more readily than pairs with smaller negative or positive redox potentials. On the other hand, a strong oxidant (e.g., characterized by a large positive potential) has a very high affinity for electrons.

The **Nernst equation** characterizes the relationship between the standard oxidation–reduction potential of a particular redox pair (E_0'), the observed potential (E), and the ratio of the concentrations of the oxidant and reductant in the system:

$$E = E_0' + \frac{2.3RT}{n\mathcal{F}} \log \left(\frac{[\text{oxidant}]}{[\text{reductant}]}\right)$$

E is the observed potential with all concentrations at 1 M.

E_0' is the standard potential at pH 7.0.

R is the gas constant of 8.3 J deg^{-1} mol^{-1}.

T is the absolute temperature in degrees Kelvin (K).

n is the number of electrons being transferred.

$\mathcal{F}$ is the Faraday of 96,500 J V^{-1}.

When the observed potential is equal to the standard potential, a potential is defined that is referred to as the midpoint potential. At the midpoint potential the concentration of the oxidant is equal to that of the reductant. Knowing the standard oxidation–reduction potentials of a diverse variety of biochemical reactions allows one to predict the direction of electron flow or transfer when more than one redox pair is linked together by the appropriate enzyme that causes a reaction to occur. For example, as shown in Table 6.6 the NAD^+–NADH pair has a standard potential of −0.32 V, and the pyruvate/lactate pair possesses a potential of −0.19. This means that electrons will flow from the NAD^+–NADH system to the pyruvate–lactate system as long as the enzymatic component (lactate dehydrogenase) is present, for example,

$$\text{Pyruvate} + \text{NADH} + \text{H}^+ \rightleftharpoons \text{lactate} + \text{NAD}^+$$

Hence in the mitochondrial electron-transfer system electrons or reducing equivalents are being produced in NAD- and FAD-linked dehydrogenase reactions, which have standard potentials at or close to that of NAD^+–NADH and are passed through the electron-transfer chain, which has as its terminal acceptor the oxygen–water couple.

Free Energy Changes in Redox Reactions

Oxidation–reduction potential differences between two redox pairs are similar to free energy changes in a chemical reaction, in that both quantities depend on the concentration of the reactants and of the products of the reaction and the following relationship exists:

$$\Delta G^{\circ\prime} = -n\mathcal{F}\Delta E_0{}'$$

Using this expression the free energy change for electron-transfer reactions can be readily calculated if the potential difference between two oxidation–reduction pairs is known. Hence, for the mitochondrial electron transfer process in which electrons are transferred between the NAD^+–NADH couple ($E_0' = -0.32$ V) and the $\frac{1}{2}O_2$–H_2O couple ($E_o' = +0.82$ V) the free energy change for this process can be calculated:

$$\Delta G^{\circ} = -n\mathcal{F}\Delta E_0' = -2 \times 23.062 \times 1.14 \text{ V}$$

$$\Delta G^{\circ} = -52.6 \text{ kcal mol}^{-1}$$

where 23.062 is the Faraday in kcal V^{-1} and n is the number of electrons transferred; for example, in the case of NADH $\rightarrow O_2$, $n = 2$. Thus the free energy available from the potential span between NADH and oxygen in the electron-transfer chain is capable of generating more than enough energy to synthesize three molecules of ATP per two reducing equivalents or two electrons transported to oxygen. In addition, because of the negative sign of the free energy available in the process of mitochondrial electron transfer, this process is exergonic and will proceed provided that the necessary enzymatic components are present.

Mitochondrial Electron Transport Is a Multicomponent System

Before cataloging the mechanistic details of the mitochondrial electron transport chain it is necessary to describe the various components that participate in the transfer of electrons in this system. The major enzymes or proteins functioning as electron-transfer components involved in the mitochondrial electron-transfer system are as follows:

1. NAD^+-linked dehydrogenases
2. Flavin-linked dehydrogenases
3. Iron–sulfur proteins
4. Cytochromes

NAD-Linked Dehydrogenases

The initial stage in the mitochondrial electron transport sequence consists of the generation of reducing equivalents in the TCA cycle, the fatty acid β-oxidation sequence, and various other dehydrogenase reactions. The

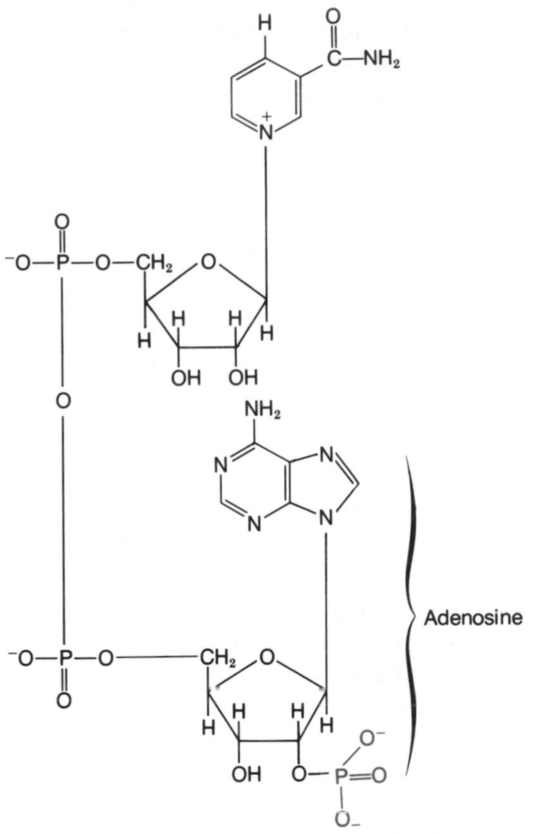

Figure 6.33
Structure of nicotinamide adenine dinucleotide phosphate: $NADP^+$.

NAD-linked dehydrogenase reactions of these pathways reduce NAD^+ to NADH while converting the reduced member of an oxidation–reduction couple to the oxidized form, for example, for the isocitrate dehydrogenase reaction

$$\text{Isocitrate} + NAD^+ \rightleftharpoons \alpha\text{-ketoglutarate} + CO_2 + NADH + H^+$$

Two nicotinamide nucleotides are involved in various metabolic reactions, NAD and NADP (Figure 6.33). Nicotinamide adenine dinucleotide phosphate has a phosphate esterified to the 2 position of the ribose in the adenosine portion of the dinucleotide. Each NAD(P)-linked dehydrogenase catalyzes a stereospecific transfer of the reducing equivalent from the substrate to the nucleotide:

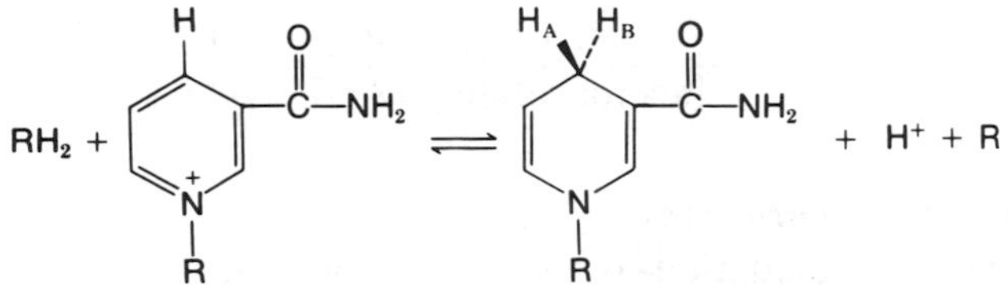

$$RH_2 + NAD^+ \rightleftharpoons NADH + H^+ + R$$

NAD(P)-linked dehydrogenases are either A specific or B specific in that the transfer of hydrogen occurs between either the oxidized or reduced metabolite and the A side (projecting out from the plane of the pyridine ring) or the B side (below the plane of the ring). Table 6.7 lists examples of the stereospecificity of NAD(P)-linked dehydrogenases. Once formed, NAD(P)H is released from the primary dehydrogenase and serves as the substrate for the mitochondrial electron transport system. The NADPH is not a substrate for the mitochondrial respiratory chain but is used in the

TABLE 6.7 The Stereospecificity of NAD(P)-Linked Dehydrogenases

NAD(P)-linked dehydrogenase	Specificity
Alcohol dehydrogenase	A
Malate dehydrogenase	A
Lactate dehydrogenase	A
Isocitrate dehydrogenase ($NADP^+$)	A
Hydroxyacyl CoA dehydrogenase	B
Glyceraldehyde 3-phosphate dehydrogenase	B
Glucose 6-phosphate dehydrogenase ($NADP^+$)	B

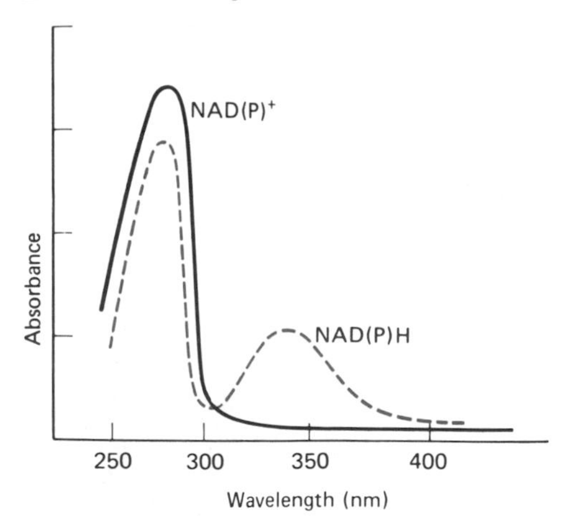

Figure 6.34
Absorbance properties of NAD+ and NADH.

reductive biosynthetic reactions of such processes as fatty acid and sterol synthesis. When NAD(P)$^+$ is converted to NAD(P)H, there is a characteristic change in the absorbant and fluorescent properties of these nucleotides, which occurs as a result of the reduction of NAD(P)$^+$. As seen in Figure 6.34, the reduced form of the nicotinamide nucleotide [NAD(P)H] has an absorbance maximum at 340 nm, not present in the oxidized NAD(P)$^+$ form. Furthermore, when the reduced form of the nicotinamide nucleotide is excited by light at 340 nm a fluorescence emission maximum is seen at 465 nm. These absorbant and fluorescent properties of the nicotinamide nucleotides have been employed extensively in developing assays for dehydrogenase reactions and have been utilized to monitor the oxidation–reduction state of a tissue or a preparation of intact mitochondria. With an appropriate spectrophotometer (e.g., a dual wavelength), capable of measuring small absorbancy changes in turbid cell or mitochondrial suspensions, the relative changes in the oxidized–reduced nicotinamide nucleotides may be determined as a function of the metabolic condition of the cell or subcellular suspension (e.g., changes in substrate, oxygen concentration, or on drug or hormone additions). This type of spectrophotometric technique and more sophisticated techniques, in which a light guide is used to direct a beam of excitation light to the surface of an intact organ or tissue, and another light guide is employed to pick up the reflected fluorescence emission at a longer wavelength, have been valuable tools in understanding the very complicated relationships that exist between the mitochondrial respiratory chain and the metabolic characteristics of various tissues.

Another effective method for monitoring the oxidation–reduction state of the cytosolic or the mitochondrial compartments is to measure the oxidized and reduced members of various **redox couples** in tissue extracts, in the bathing solution of a tissue or in the effluent perfusate of an isolated, perfused organ. Because lactate dehydrogenase is exclusively a cytosolic enzyme the pyruvate/lactate ratio in the tissue or organ perfusate should accurately reflect the cytosolic NAD$^+$/NADH ratio under a variety of metabolic conditions. In a like manner the β-hydroxybutyrate dehydrogenase is exclusively mitochondrial, and hence the ratio of acetoacetate/β-hydroxybutyrate should reflect the oxidation–reduction state of the mitochondrial NAD$^+$–NADH system. If the ratio of acetoacetate/β-hydroxybutyrate and the equilibrium constant for β-hydroxybutyrate dehydrogenase are known, the NAD$^+$/NADH ratio under any condition can be calculated:

$$\text{Acetoacetate} + \text{NADH} + \text{H}^+ \rightleftharpoons \beta\text{-hydroxybutyrate} + \text{NAD}^+$$

$$K_{eq} = \frac{[\beta\text{-hydroxybutyrate}][\text{NAD}^+]}{[\text{acetoacetate}][\text{NADH}][\text{H}^+]}$$

Flavin-Linked Dehydrogenases

The second type of oxidation–reduction reaction essential to a discussion of mitochondrial electron transport employs a flavin (e.g., derived from riboflavin) as the electron acceptor in the reaction. These reactions are catalyzed by a group of flavin-linked dehydrogenases. The two flavins commonly utilized in oxidation–reduction reactions are FAD (flavin adenine dinucleotide) and FMN (flavin mononucleotide) (Figure 6.35).

Among the flavin-containing enzymes that have been described, five play an essential role in energy metabolism in mammalian mitochondria (Table 6.8). In the discussion of the pyruvate and α-ketoglutarate dehydrogenase multienzyme complexes, the final reaction catalyzed by this complex involved the flavoprotein enzyme, dihydrolipoyl dehydrogenase, which accepts electrons via a bound FAD moiety from reduced lipoamide groups on the transacylase subunit and transfers these reducing equivalents to NAD^+. Also, in the TCA cycle, succinate dehydrogenase is a flavin-linked enzyme, which oxidizes succinate to fumarate and converts FAD to $FADH_2$. The first dehydrogenation reaction in β oxidation of fatty acids is catalyzed by acyl CoA dehydrogenase, another flavin-linked enzyme. Finally, oxidation of NADH in the mitochondrial respiratory chain is catalyzed by a FMN-containing enzyme, the NADH dehydrogenase, and the reducing equivalents are then transferred to another flavoprotein called the electron-transferring flavoprotein.

The flavins FAD and FMN either may be bound very tightly with noncovalent bonding (i.e., with dissociation constants in the range of 10^{-10} M) to their respective enzymes, as is the case with the NADH dehydrogenase, or they may be bound covalently to the enzyme (i.e., to a histidine residue), as is the case with succinate dehydrogenase. Flavoproteins may be classified into two groups: (1) the dehydrogenases in which the reduced flavin is reoxidized by electron carriers other than oxygen (e.g., coenzyme Q, and other flavins, or in vitro with chemicals such as ferricyanide, methylene blue, or phenazine methosulfate), and (2) the oxidases in which the flavin may be reoxidized using as the electron acceptor molecular oxygen, O_2, yielding H_2O_2 as the product. The H_2O_2 may then be broken down to water and oxygen by the enzyme catalase,

$$2H_2O_2 \xrightarrow{\text{catalase}} 2H_2O + O_2$$

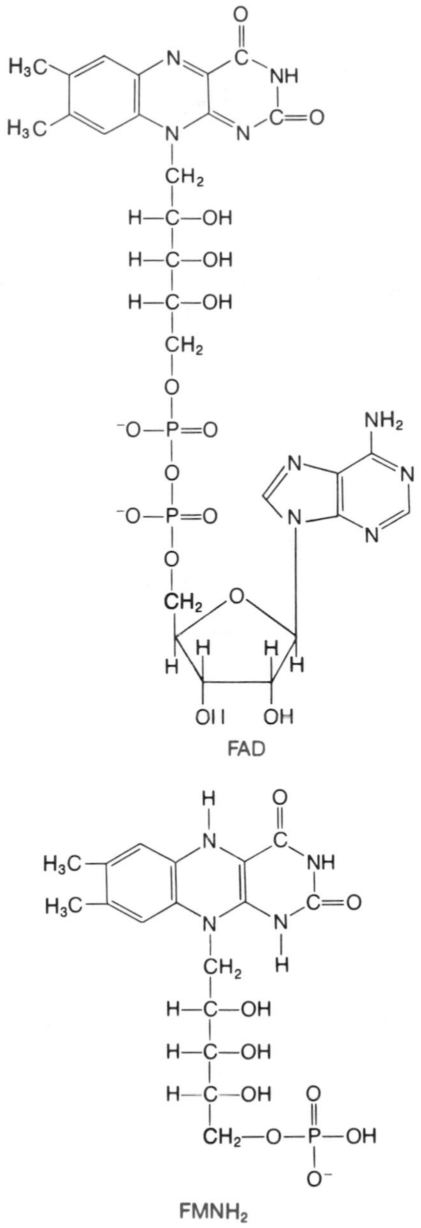

Iron–Sulfur Centers

A number of flavin-linked enzymes have **nonheme iron** (i.e., an **iron–sulfur center**) involved in the catalytic mechanism. In these enzymes the iron is converted from the oxidized (Fe^{3+}) form to the reduced (Fe^{2+}) form during the transfer of reducing equivalents on and off the flavin moiety. Both succinate dehydrogenase and NADH dehydrogenase contain iron–sulfur centers. The iron component of the iron–sulfur center is bound in various arrangements to cysteine residues in the protein and to acid-labile sulfur, for example, $Fe_4S_4Cys_4$; $Fe_2S_2Cys_4$; $Fe_1S_0Cys_4$ (Figure 6.36). Iron–sulfur proteins are found in abundance in all species from the simplest microorganism to the mammal. Certain flavin-linked enzymes (xanthine oxidase) have one or two molybdenum atoms associated

Figure 6.35
Structures of flavin adenine dinucleotide (FAD) and flavin mononucleotide (FMN).

TABLE 6.8 Various Flavin-Linked Dehydrogenases

Enzyme	Function	Flavin nucleotide
Succinate dehydrogenase	Tricarboxylic acid cycle	FAD
Dihydrolipoyl dehydrogenase	Component in pyruvate and α-ketoglutarate dehydrogenase complexes	FAD
NADH dehydrogenase	Electron transport chain	FMN
Electron-transferring flavoprotein	Electron transport chain	FAD
Acyl CoA dehydrogenase	Fatty acid β oxidation	FAD
D-Amino acid oxidase	Amino acid oxidation	FAD
Monoamine oxidase	Oxidation of monoamines	FAD

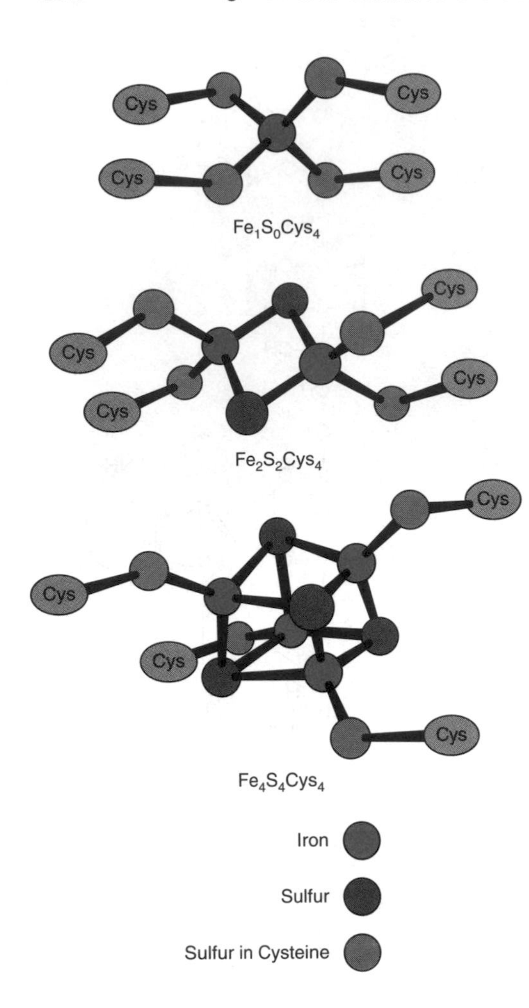

Figure 6.36
The structures of iron–sulfur centers.

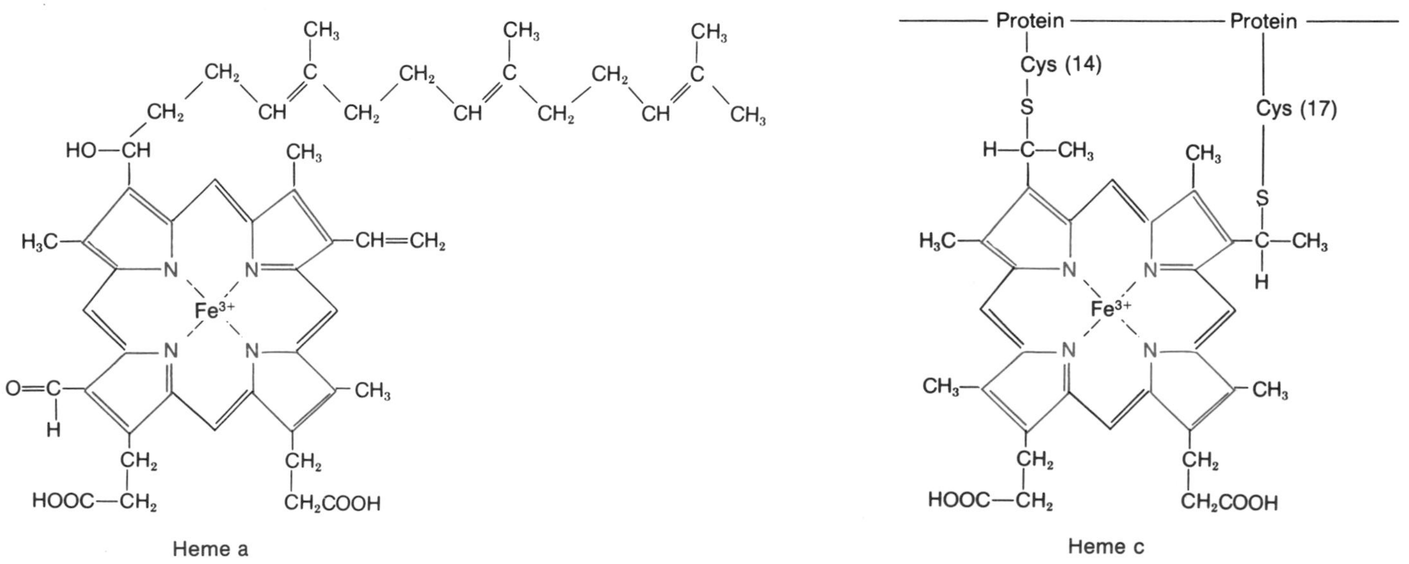

with their catalytic mechanism. The tightly bound molybdenum undergoes a valence change during the transfer of electrons $Mo^{6+} \longrightarrow Mo^{5+}$.

Cytochromes

Organisms that require oxygen (i.e., aerobic organisms) in their energy-generating functions possess various cytochromes that are involved in electron-transferring systems. **Cytochromes** are a class of proteins characterized by the presence of an iron-containing heme group covalently bound to the protein. Unlike the heme group in hemoglobin or myoglobin in which the heme iron remains in the Fe^{2+} state, the iron in the heme of a cytochrome alternately is oxidized (Fe^{3+}) or reduced (Fe^{2+}) as the cytochrome transfers electrons toward oxygen in the electron transport chain.

The cytochromes of mammalian mitochondria were designated as *a*, *b*, and *c* on the basis of the α band of their absorption spectrum and the type of heme group (see Figure 6.37). Cytochrome *c* is a small protein (104 amino acid residues) with mol wt = 13,000. Amino acid sequences of cytochrome *c* from a great many species have been described and show that 20 out of 104 amino acid residues are invariant. The iron of the heme group in cytochrome *c* is coordinated between the four nitrogen atoms of the tetrapyrrole structure of the porphyrin group, whereas the fifth and sixth coordination positions are occupied by the methionine residue at position 80 and histidine residue at position 18 of the protein (Figure 6.38). The fact that all six coordination positions are filled in most of the cytochromes prohibits oxygen from binding directly to the iron and prevents such respiratory inhibitors as cyanide, azide, and carbon monoxide from binding to most cytochromes. The notable exception is cytochrome a_3, which is involved in the terminal step in the mitochondrial electron transport chain. The heme group in cytochrome *c* is attached to the protein, not only by the fifth and sixth coordination positions of the heme iron, but also by the vinyl side chains of the protoporphyrin IX structure, from which both heme a and c are derived. These vinyl side chains are reduced by the addition of H—S across the vinyl group and the resulting sulfhydryl linkages attach the heme group to cysteine residues at positions 14 and 17 in cytochrome *c*. Hence, the heme group is covalently linked to the protein as well as being coordinated through the Fe^{2+} group in the heme. The three-dimensional structure of cytochrome *c* is shown in Figure 6.39.

Heme a

Heme c

Figure 6.37
Structures of heme a and heme c.

Coenzyme Q

Coenzyme Q, also called **ubiquinone,** is neither a nucleotide species nor a protein but a lipophilic electron carrier. Like the pyridine nucleotides and to a certain extent cytochrome *c*, coenzyme Q serves as a "mobile" electron transport component that operates between the various flavin-linked dehydrogenases, for example, NADH dehydrogenase, succinate dehydrogenase, or fatty acyl CoA dehydrogenase, and cytochrome *b* of the electron transport chain. As shown in Figure 6.40, the quinone portion of the coenzyme Q molecule is alternately oxidized and reduced by the addition of 2 reducing equivalents, that is 2 H^+, and 2 e^-. The number (n) of isoprene units in the side chain of coenzyme Q varies between 6 and 10, depending on the source of the coenzyme Q. The side chain renders the coenzyme Q lipid soluble and facilitates the accessibility of this electron carrier to the lipophilic portions of the inner mitochondrial membrane, where the enzymatic aspects of the mitochondrial electron-transfer chain are localized.

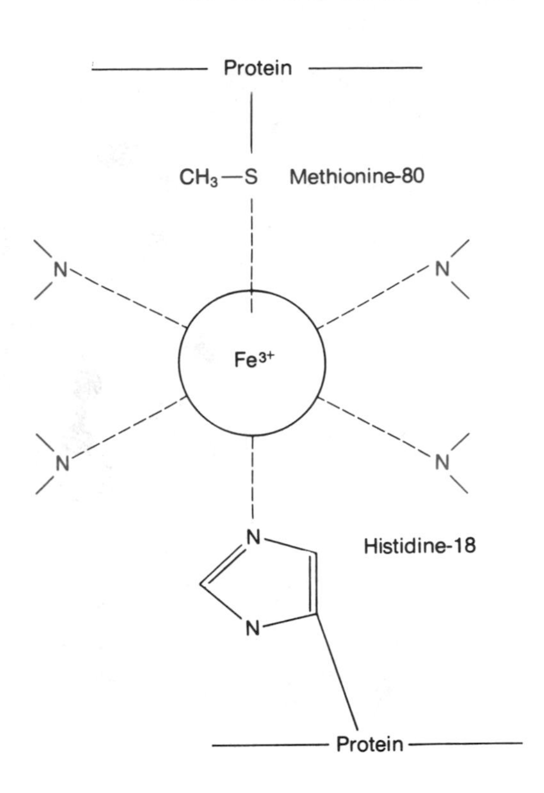

Figure 6.38
The six coordination positions of cytochrome *c*.

The Mitochondrial Electron Transport Chain Is Located in the Inner Membrane in a Specific Sequence

The various electron-transferring proteins and other carriers that comprise the mitochondrial electron-transfer chain are arranged in a sequential pattern in the inner mitochondrial membrane. Reducing equivalents are extracted from substrates in the TCA cycle, the fatty acid β-oxidation sequence, and indirectly from glycolysis, and passed sequentially through the electron transport chain to molecular oxygen. The current arrangement of the mitochondrial electron transport carriers is illustrated in Figure 6.41. Electrons or reducing equivalents are fed into the electron transport chain at the level of NADH or coenzyme Q from the primary NAD^+- and FAD-linked dehydrogenase reactions and are transported to molecular oxygen through the cytochrome chain. This electron transport system is set up so that the reduced member of one redox couple is oxidized by the oxidized member of the next component in the system:

$$\text{NADH} + \text{H}^+ + \text{FMN} \rightleftharpoons \text{FMNH}_2 + \text{NAD}^+$$

or

$$\text{Cytochrome } b\ (\text{Fe}^{2+}) + \text{cytochrome } c_1\ (\text{Fe}^{3+}) \rightleftharpoons \text{cytochrome } b\ (\text{Fe}^{3+}) + \text{cytochrome } c_1\ (\text{Fe}^{2+})$$

It should be noted that the electron-transfer reactions from NADH through coenzyme Q transfer 2 e^-, whereas the reactions between coenzyme Q and oxygen involving the various cytochromes are 1 e^- transfer reactions.

The various components of the respiratory chain have characteristic absorption spectra that can be visualized in suspensions of isolated mitochondria or submitochondrial particles using a dual beam spectrophotometer. The different absorption bands are shown in Figure 6.42. One of the light beams of the spectrophotometer was passed through a suspension of liver mitochondria, which was maintained under fully reduced conditions (e.g., substrate plus no oxygen), and the other beam was passed through an identical suspension in the presence of oxygen. Hence, the resulting spectrum is a difference spectrum of the reduced minus the oxidized states of the mitochondrial respiratory chain.

During the transfer of electrons from the NADH–NAD^+ couple ($E_0' =$ -0.32) to molecule oxygen ($E_0' = +0.82$) there occurs an oxidation–re-

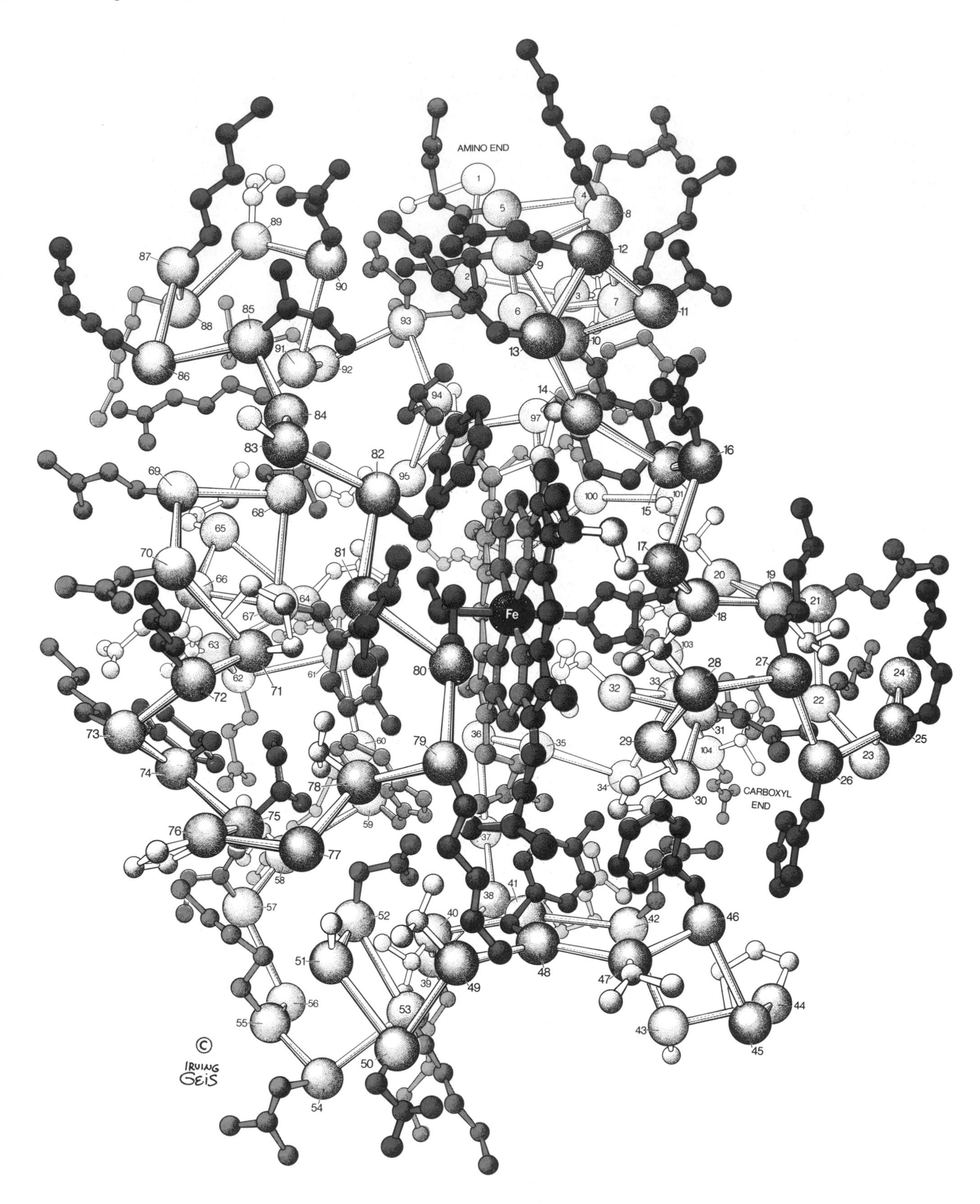

Figure 6.39
The three-dimensional structure of cytochrome *c*.
Copyright 1992 Irving Geis.

duction potential decrease of 1.14 V. As shown in Figure 6.43, this drop in potential occurs in discrete steps as reducing equivalents or electrons are passed between the different segments of the chain. There is at least a 0.3 V decrease in potential between each of the three coupling or phosphorylation sites. A potential drop of 0.3 V is more than sufficient to accommodate the synthesis of a high-energy phosphate bond such as occurs in ATP synthesis, for example,

$$\Delta E_0' = 0.3 \text{ V}$$

$$\Delta G^\circ = -n\mathcal{F}\Delta E_0'$$

$$\Delta G^\circ = -2 \times 23.062 \times 0.3$$

$$\Delta G^\circ = -13.8 \text{ kcal mol}^{-1}$$

Various components of the electron transport chain are located asymmetrically in the mitochondrial membrane. An example of this asymmetric localization is shown in Figure 6.44. Cytochrome *c* oxidase, which catalyzes the terminal step in the electron-transfer chain, consists of a dimer that spans the membrane between the matrix and the intermembrane space or the cytosol. Cytochrome *c* binds to the oxidase from the cytosolic side of the membrane, whereas oxygen binds from the matrix side of the membrane during the electron-transferring event.

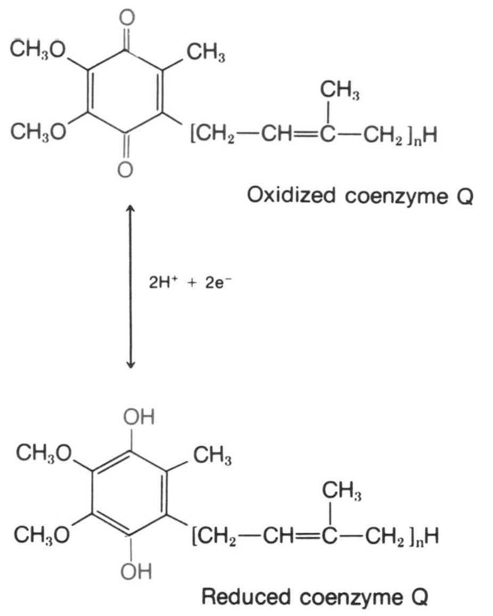

Figure 6.40
The oxidation–reduction of coenzyme Q.

Electron Transport Can Be Inhibited at Specific Sites

The illustration of the mitochondrial respiratory chain shown in Figure 6.41 indicates that a number of chemical compounds are capable of specifi-

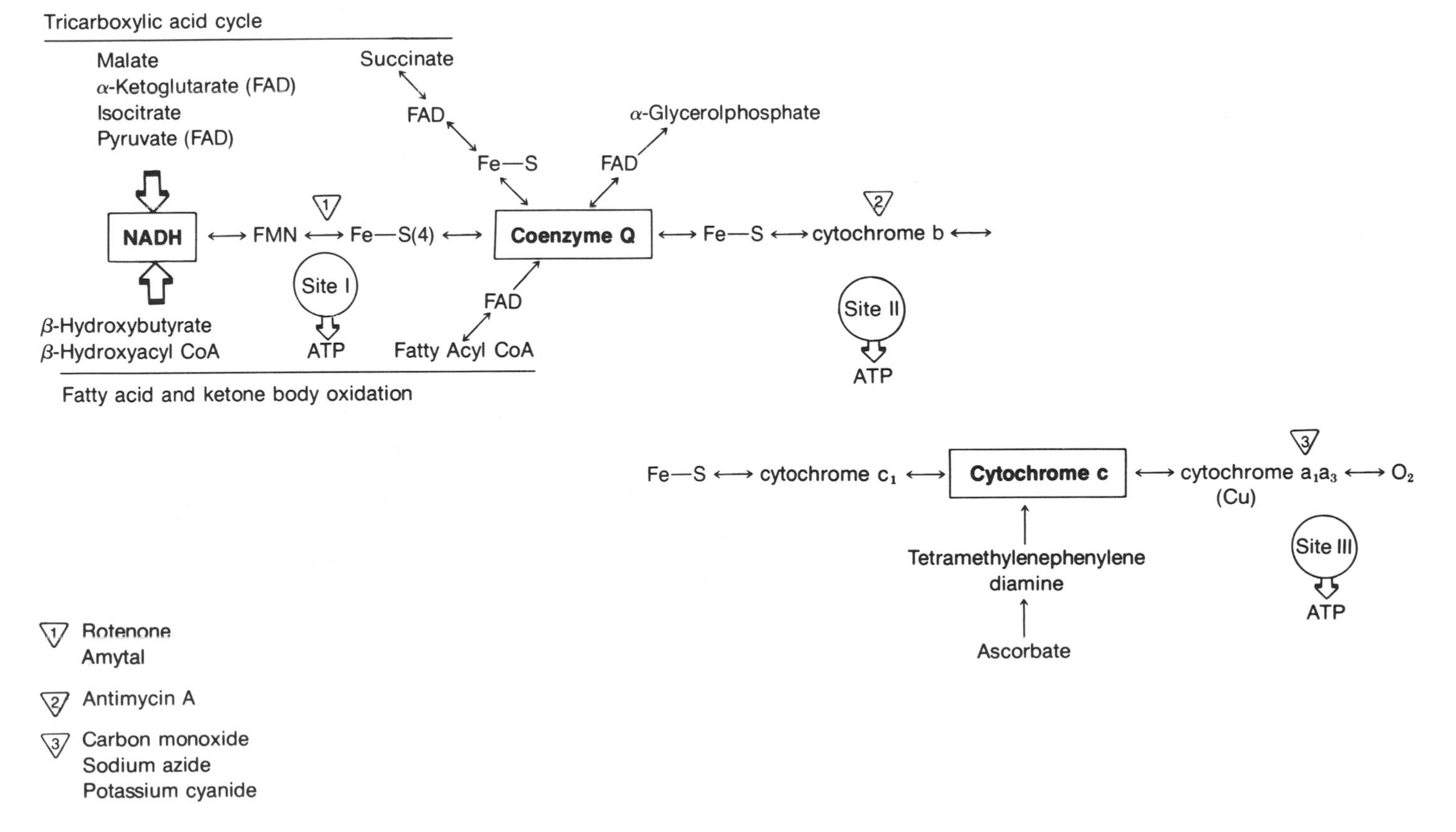

Figure 6.41
Mitochondrial electron transport chain.

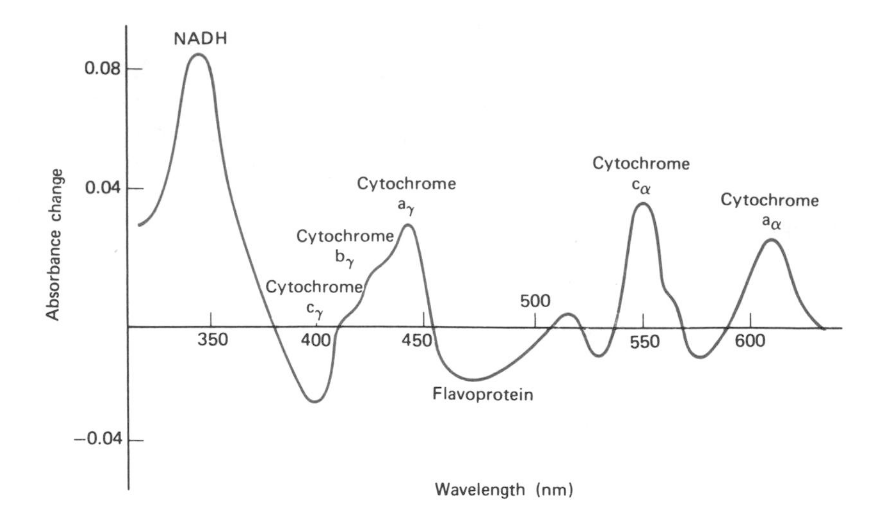

Figure 6.42
Difference spectra of liver mitochondrial suspension (oxidized–reduced).

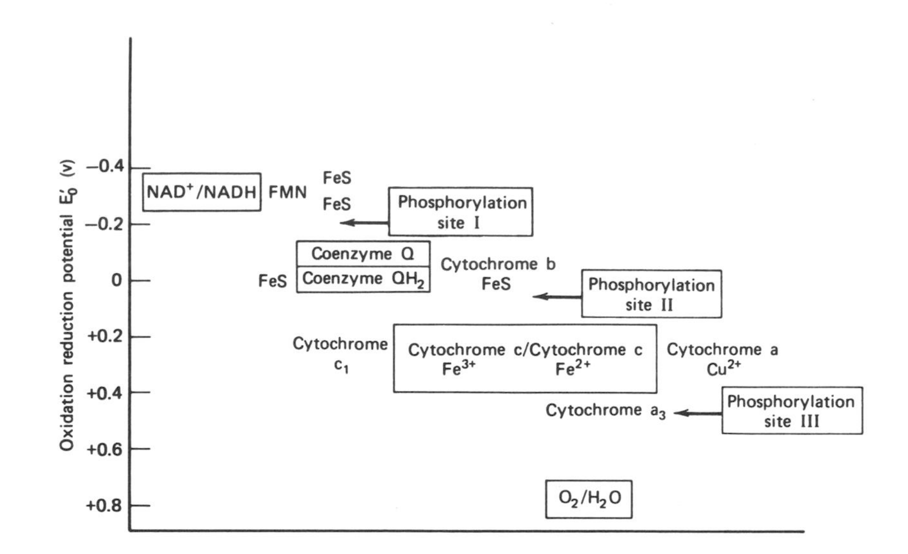

Figure 6.43
Oxidation–reduction potentials of the mitochondrial electron transport chain carriers.

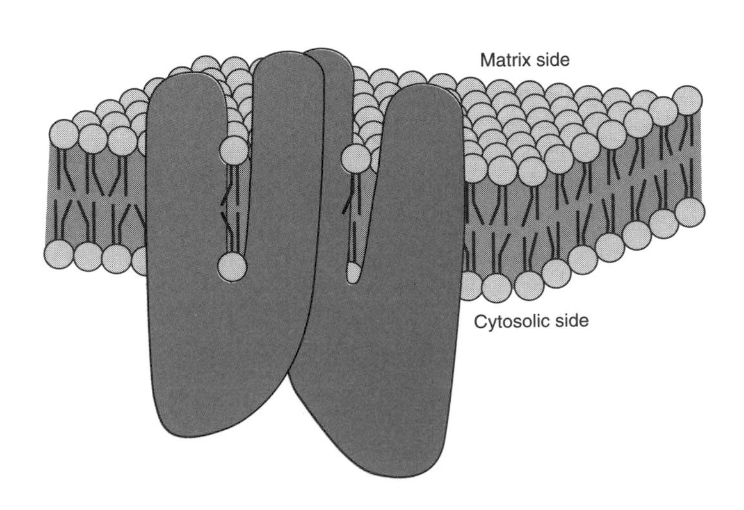

Figure 6.44
Model of cytochrome *c* oxidase dimer in the mitochondrial inner membrane.
Redrawn with permission from T. G. Frey, M. J. Costello, B. Karlsson, J. C. Haselgrove, and J. S. Leigh, *J. Mol. Biol.* 162:113, 1982.

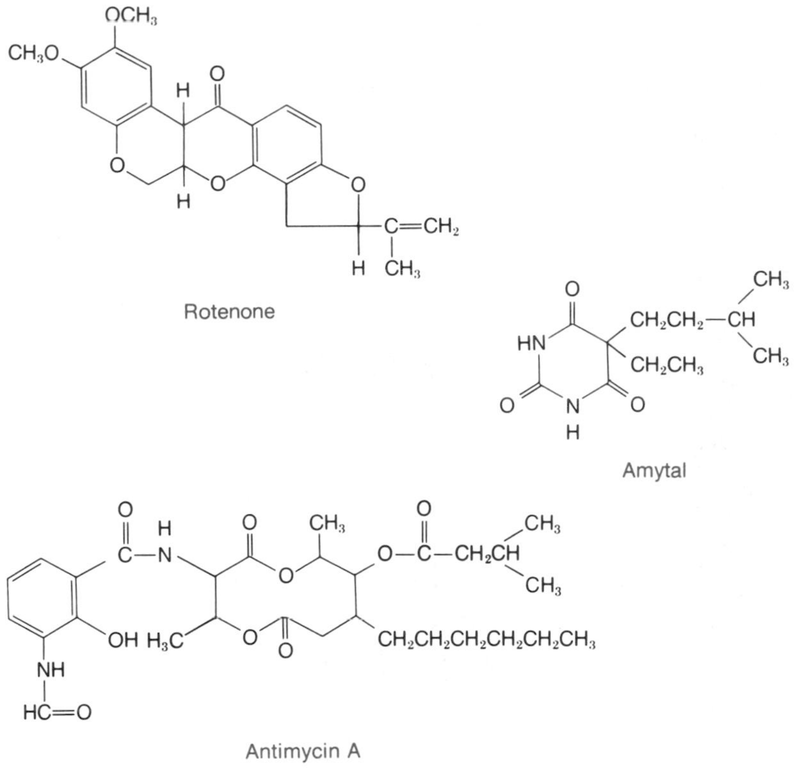

Rotenone

Amytal

Antimycin A

Figure 6.45
Structures of respiratory chain inhibitors.

cally inhibiting electron flow in the chain at different points. The fish poison **rotenone** (Figure 6.45) and the barbiturate **amytal** (Figure 6.45) inhibit the electron-transfer chain at the level of the flavoprotein, NADH dehydrogenase. Hence, electrons or reducing equivalents derived from NAD^+-linked dehydrogenases are not oxidized by a rotenone-inhibited respiratory chain, whereas those derived from flavin-linked dehydrogenases are freely oxidized. The antibiotic **antimycin A** (Figure 6.45) inhibits electron transfer at the level of cytochrome *b*, whereas the terminal step in the respiratory chain catalyzed by cytochrome oxidase is inhibited by **cyanide, azide,** or **carbon monoxide** (see Clin. Corr. 6.3.) Cyanide and azide combine with the oxidized heme iron (Fe^{3+}) in cytochromes *a* and a_3 and prevent the reduction of heme iron by electrons derived from reduced cytochrome *c*. Carbon monoxide binds to the reduced iron (Fe^{2+}) of cytochrome oxidase. Hence ingestion or injections of respiratory chain inhibitors leads to a blockage of electron transfer and impairment of the normal energy generating function of the mitochondrial electron transport chain, and if the exposure to such an inhibitor is prolonged, death of the organism would result.

Electron Transport Is Reversible

It should be pointed out that the various events in the mitochondrial electron transport system and the closely coupled reactions or processes in the oxidative phosphorylation sequence are reversible, provided an appropriate amount of energy is supplied to drive the system. In mitochondrial systems, reducing equivalents derived from succinate can be transferred to NADH with the concomitant hydrolysis of ATP (see Figure 6.46). Electron transport across the other two phosphorylation sites can be reversed in a similar fashion.

CLIN. CORR. **6.3**
CYANIDE POISONING

Inhalation of hydrogen cyanide gas or ingestion of potassium cyanide causes a rapid and extensive inhibition of the mitochondrial electron transport chain at the cytochrome oxidase step. Cyanide is one of the most potent and rapidly acting poisons known. Cyanide binds to the Fe^{3+} in the heme of the cytochrome a,a_3 component of the terminal step in the electron transport chain and prevents oxygen from reacting with cytochrome a,a_3. Mitochondrial respiration and energy production cease, and cell death occurs rapidly. Death due to cyanide poisoning occurs from tissue asphyxia, most notably of the central nervous system. If cyanide poisoning is diagnosed very rapidly, an individual who has been exposed to cyanide is given various nitrites that convert oxyhemoglobin to methemoglobin, which merely involves converting the Fe^{2+} of hemoglobin to Fe^{3+} in methemoglobin. Methemoglobin (Fe^{3+}) competes with cytochrome a,a_3 (Fe^{3+}) for cyanide, forming a methemoglobin–cyanide complex. Administration of thiosulfate causes the cyanide to react with the enzyme rhodanese forming the nontoxic compound thiocyanate.

Holland, M. A. and Kozlowski, L. M. Clinical features and management of cyanide poisoning. *Clin. Pharmacol.* 5:737, 1986.

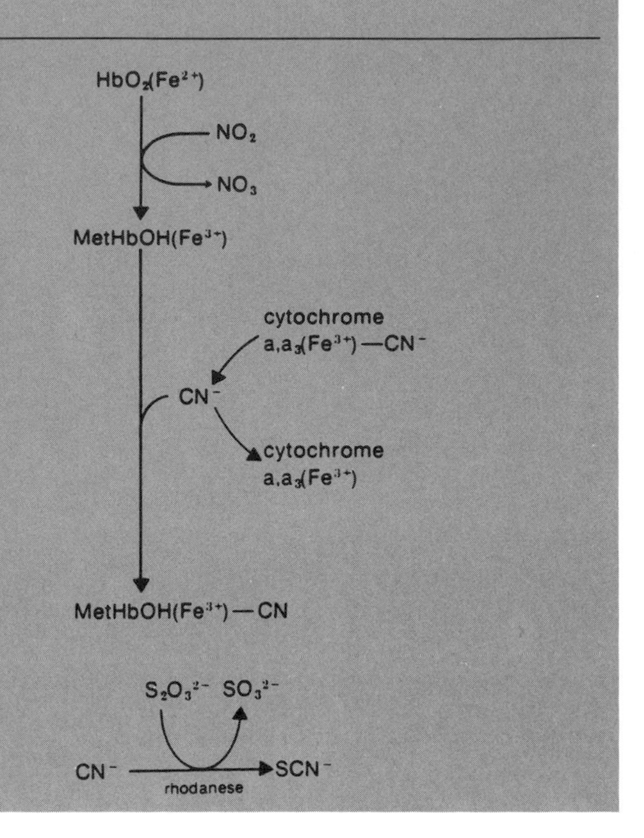

Oxidative Phosphorylation Is Coupled to Electron Transport

The obligatory tight coupling between the electron-transferring reactions and the reactions in oxidative phosphorylation can best be illustrated in the experiment shown in Figure 6.47. Mitochondrial electron transport

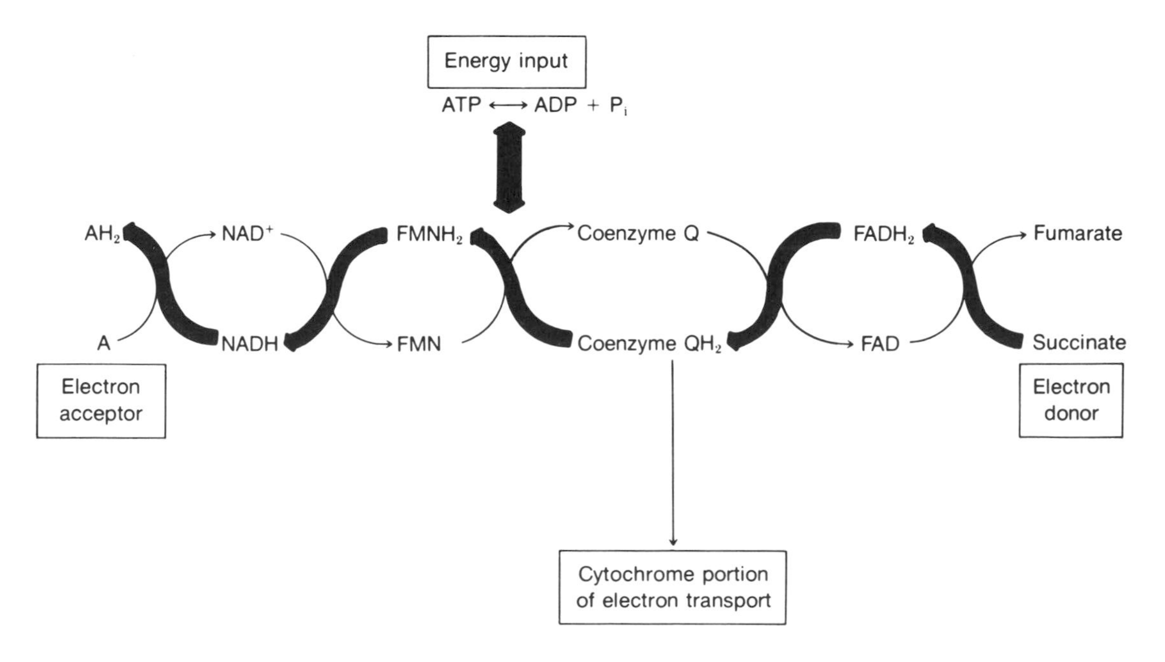

Figure 6.46
Reversal of mitochondrial electron transfer.

monitored by measuring the rate of oxygen consumption by a suspension of liver mitochondria can occur at a rapid rate only following the addition of an oxidizable substrate (the electron donor) and ADP (a phosphate acceptor) plus P_i. The "active" state in the presence of substrate and ADP has been designated state 3 and is a situation in which there occurs rapid electron transfer, oxygen consumption, and rapid synthesis of ATP. Following the conversion of all of the added ADP to ATP the rate of electron transfer subsides back to the rate observed prior to ADP addition. Hence, respiration is tightly coupled to ATP synthesis and this relationship has been termed respiratory control or phosphate acceptor control. The ratio of the active (state 3) to the resting (state 4) rates of respiration is referred to as the **respiratory control ratio** and is a measure of the "tightness" of the coupling between electron transfer and oxidative phosphorylation. Damaged mitochondrial preparations and preparations to which various uncoupling compounds (see below) have been added exhibit low respiratory control ratios, indicating that the integrity of the mitochondrial membrane is required for tight coupling.

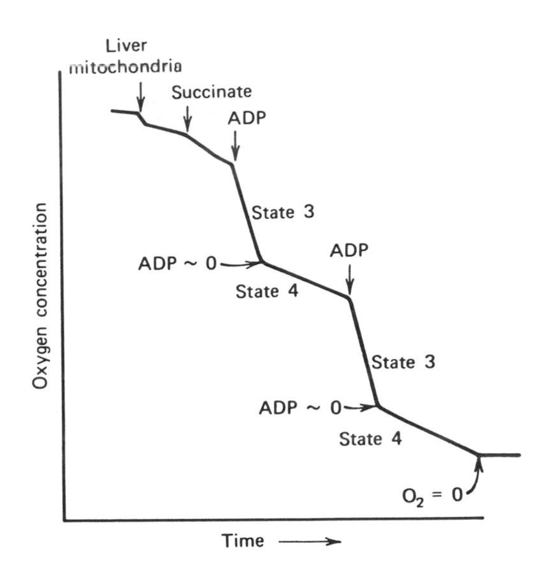

Figure 6.47
A demonstration of the coupling of electron transport to oxidative phosphorylation in a suspension of liver mitochondria.
State 3/state 4 = respiratory control ratio.

The effect of uncouplers and inhibitors of the electron transport–oxidative phosphorylation sequence is illustrated in Figure 6.48. Following the addition of ADP, which initiates a rapid state 3 rate of respiration, an inhibitor of the oxidative phosphorylation sequence (actually the mitochondrial ATPase), oligomycin, is added. **Oligomycin** stops ATP synthesis, and because the processes of electron transport and ATP synthesis are coupled tightly, respiration or electron transport is inhibited nearly completely. Following the inhibition of both oxygen consumption and ATP synthesis, the addition of an uncoupler of these two processes such as **2,4-dinitrophenol** or **carbonylcyanide-*p*-trifluoromethoxyphenylhydrazone** (FCCP),

2,4-Dinitrophenol

Carbonylcyanide-*p*-trifluoromethoxy phenylhydrozone (FCCP)

causes a rapid initiation of oxygen consumption. Because respiration or electron transport is now uncoupled from ATP synthesis, electron transport may continue but ATP synthesis may not occur.

It should be noted that regulation of the respiration rate of a tissue by the provision of a phosphate acceptor, ADP, is a normal physiological situation. For example, when a muscle is exercised, ATP is broken down to ADP and P_i, and creatine phosphate is converted to creatine as the high-energy phosphate bond is transferred to ATP in the creatine phosphokinase reaction. As ADP accumulates during the muscular activity, respiration or oxygen consumption is activated, and the energy generated in this fashion allows the ATP and creatine phosphate levels to be replenished (see Clin. Corr. 6.4).

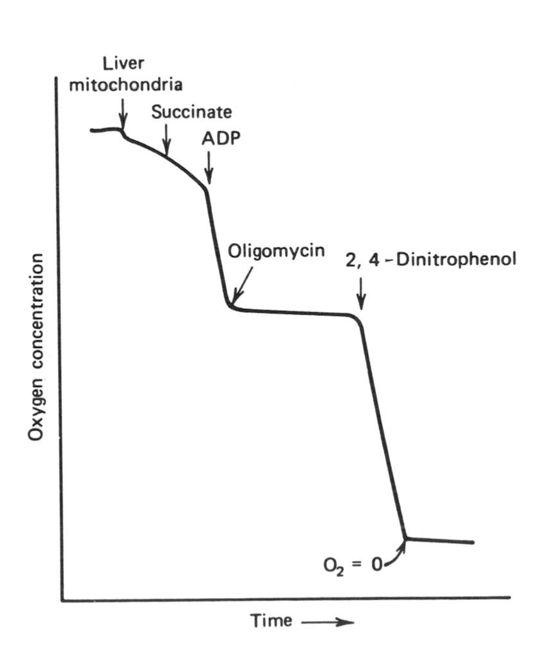

Figure 6.48
Inhibition and uncoupling of oxidative phosphorylation in liver mitochondria.

6.7 OXIDATIVE PHOSPHORYLATION

One of the most vexing problems that has confronted biochemists during the past three decades is the delineation of the mechanism of oxidative phosphorylation. Despite the countless years of experimental consideration that have been expended on this problem, it is only recently that consensus has been reached on many of the details of the mechanism by which energy derived from the passage of electrons sequentially along the electron transport chain is transduced into the chemical energy involved in the phosphoanhydride bonds of ATP. Many hypotheses for the mechanism of oxidative phosphorylation have been tested, and three general theories have emerged as reasonable proposals, and one of these theories is now widely accepted.

CLIN. CORR. 6.4
HYPOXIC INJURY

Acute hypoxic tissue injury has been studied in a variety of human tissues. The occlusion of one of the major coronary arteries during a myocardial infarction produces a large array of biochemical and physiological sequelae. When a tissue is deprived of its oxygen supply, the mitochondrial electron transport–oxidative phosphorylation sequence is inhibited, resulting in the decline of cellular levels of ATP and creatine phosphate. As cellular ATP levels diminish, anaerobic glycolysis is activated in an attempt to maintain normal cellular functions. Glycogen levels are rapidly depleted and lactic acid levels in the cytosol increase, reducing the intracellular pH. Hypoxic cells in such an energetic deficit begin to swell as they can no longer maintain their normal intracellular ionic environments. Mitochondria swell and begin to accumulate calcium, which may be deposited in the matrix compartment as calcium phosphate. The cell membranes of swollen cells become more permeable, leading to the leakage of various soluble enzymes, coenzymes, and other cell constituents from the cell. As the intracellular pH falls, damage occurs to lysosomal membranes, which release various hydrolytic proteases, lipases, glucosidases, and phosphatases into the cell. Such lysosomal enzymes begin an autolytic digestion of cellular components.

Cells that have been exposed to short periods of hypoxia can recover, without irreversible damage, upon reperfusion with an oxygen-containing medium. The exact point at which hypoxic cell damage becomes irreversible is not precisely known. This process is of great practical importance for transplantation of organs (heart, kidney and liver), which always undergo a period of hypoxia between the time they are removed from the donor and implanted into the recipient.

Kehrer, J. P. Concepts related to the study of reactive oxygen and cardiac reperfusion injury. *Free Rad. Res. Commun.* 5:305, 1986; and Granger, D. N. Role of xanthine oxidase and granulocytes in ischemia—reperfusion injury. *Am. J. Physiol.* 255:H1269, 1988.

The Chemical Coupling Hypothesis Proposed the Formation of a Specific High-Energy Intermediate

The **chemical-coupling hypothesis** for oxidative phosphorylation was developed in the early 1950s. This mechanism was based upon an analogy with the mechanism for substrate level phosphorylation observed in the glyceraldehyde 3-phosphate dehydrogenase reaction of the glycolytic pathway. In this reaction glyceraldehyde 3-phosphate is oxidized and a high-energy phosphoric–carboxylic acid anhydride bond is generated in the product of the reaction, 1,3-bisphosphoglycerate. An enzyme-bound high-energy intermediate is generated in this reaction, which is utilized to form the intermediate high-energy compound 1,3-bisphosphoglycerate and ultimately to form ATP in the next reaction in the glycolytic pathway, that of phosphoglycerate kinase (see Chapter 7). Another example of a substrate level phosphorylation reaction, which was defined in the 1960s, is the succinyl CoA synthetase reaction of the TCA cycle. In this reaction the high-energy character of succinyl CoA is converted to the phosphoric acid anhydride bond in GTP with the intermediate participation of a high-energy, phosphorylated histidine moiety on the enzyme. Originally it was thought (incorrectly) that this phosphohistidine was a high-energy intermediate in the oxidative phosphorylation sequence. Because of these types of substrate level phosphorylation reactions it was proposed that the mechanism of mitochondrial energy transduction involved a series of high-energy intermediates that were generated in the mitochondrial membrane as a consequence of electron transport:

$$AH_2 + B + I \longrightarrow A \sim I + BH_2$$

$$A \sim I + P_i \longrightarrow I \sim P_i + A$$

$$I \sim P_i + ADP \longrightarrow ATP + I$$

In this representation A and B are electron carriers, whereas I is a hypothetical ligand that participates in the formation initially of a high-energy compound with the respiratory carrier ($A \sim I$) and thereafter with P_i to form a phosphorylated high-energy intermediate ($I \sim P_i$). The "$\sim$" is used to indicate a bond which when hydrolyzed releases a large amount of energy, thus a "high-energy" bond. The phosphorylated high-energy intermediate is then utilized to form ATP in the ATP synthetase reaction. Uncouplers of oxidative phosphorylation were proposed to act by hydrolyzing the nonphosphorylated high-energy intermediate prior to the incorporation of phosphate into the system. Oligomycin was suggested to inhibit the incorporation of phosphate into ATP. The strongest argument for this type of mechanism was its basic simplicity, while its primary detraction is the fact that none of the proposed high-energy intermediates have ever been defined or isolated. Hence, it is believed that such intermediates do not actually exist.

The Conformational Coupling Hypothesis Proposes That Electron Transport Causes a Change in Conformation of a Protein

A second proposal for the mechanism of oxidative phosphorylation was the **conformational-coupling hypothesis.** This hypothesis has an analogy in the process of muscle contraction in which ATP hydrolysis is used to drive conformational changes in myosin head groups, which result in the disruption of cross-bridges to the actin thin filaments. The conformation coupling hypothesis suggests that a consequence of electron transport in the inner mitochondrial membrane is the induction of a conformational change in a membrane protein. ATP is synthesized by a mechanism that allows the membrane protein in its high-energy conformation to revert to its low-energy or random state, with the resultant formation of ATP from ADP and P_i. Hence, the high-energy state of the membrane protein is transduced into the bond energy of the γ-phosphate group of ATP (see Figure 6.49).

There are various experimental observations indicating that mitochondrial membrane proteins undergo conformational changes during the process of active electron transport. However, there is relatively little evidence demonstrating conclusively that such conformational changes are actually involved in the mechanism of ATP synthesis.

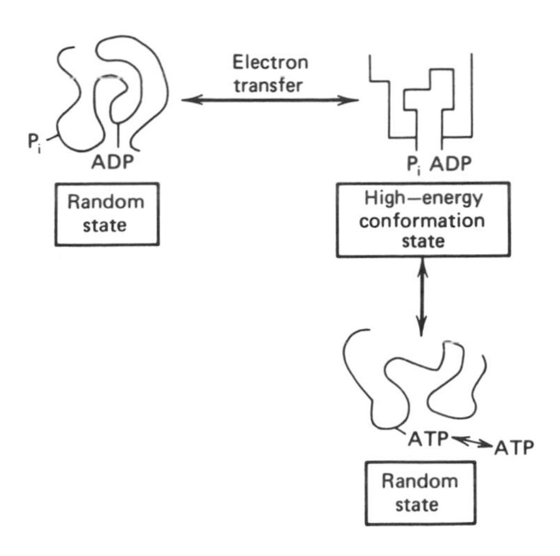

Figure 6.49
The conformational coupling hypothesis.

The Chemiosmotic Coupling Hypothesis Involves the Generation of a Proton Gradient and Reversal of an ATP-Dependent Proton Pump

Finally, the chemiosmotic-coupling hypothesis originally proposed by Peter Mitchell has gained widespread appreciation as a mechanism for energy transduction in mitochondria, as well as other biological systems. Mitchell's original proposition of the chemiosmotic theory of oxidative phosphorylation compared the energy-generating systems in biological membranes to a common storage battery. Just as energy may be stored in batteries because of the separation of positive and negative charges in the different components of the battery, energy may be generated as a consequence of the separation of charges in complex membranous systems. The **chemiosmotic hypothesis** (Figure 6.50) suggests that an electrochemical or proton gradient is established across the inner mitochondrial membrane during electron transport. This proton gradient is formed by pumping protons from the mitochondrial matrix side of the inner membrane to the cytosolic side of the membrane. Once there is a substantial electrochemical gradient established, the subsequent dissipation of the gradient is coupled to the synthesis of ATP by the mitochondrial ATPase. The chemiosmotic hypothesis requires that the electron transport carriers and the **F_1F_0-ATPase** are localized in such a fashion in the inner mitochondrial membrane that protons are pumped out of the matrix compartment during the electron transport phase of the process, and protons are pumped or allowed back through the membrane during the ATP synthetase aspect of the process.

Uncouplers, which are usually relatively lipophilic weak acids, act to dissipate the proton gradient by transporting protons through the membrane from the intermembrane space to the matrix, essentially short-circuiting the normal flow of protons through the ATP synthetic portion of the system. One of the strongest arguments supporting the chemiosmotic hypothesis is that ATPases can be purified, incorporated into artificial membrane vesicles, and are able to synthesize ATP when an electrochemical gradient is established across the membrane. In recent years a consid-

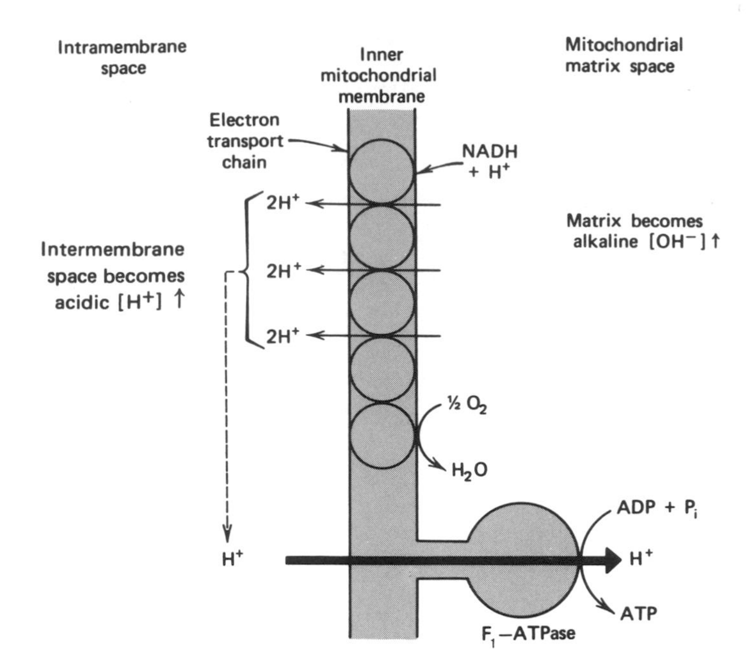

Figure 6.50
The chemiosmotic coupling hypothesis.

erable experimental effort has been expended to purify the various components of the mitochondrial ATPase. It has been determined that proton-translocating ATPases are present and may be purified from a variety of mammalian tissues, bacteria, and yeast. The ATPase is a multicomponent complex with a suggested molecular weight of 480,000–500,000 (Figure 6.51). These ATPases can be incorporated into artificial membranes and can catalyze ATP synthesis. The ATPase complex consists of a water-soluble portion called F_1 and a hydrophobic portion called F_0. The F_1 consists of five nonidentical subunits (α, β, γ, δ, and ε) with a subunit stoichiometry of $\alpha_3\beta_3\gamma\delta\varepsilon$ and a mol wt 350,000–380,000. Nucleotide binding sites of the enzyme have been localized on the α and β subunits. The γ subunit has been proposed to function as a gate to the proton translocating activity of the complex, while the δ subunit has been suggested to be necessary for the attachment of the F_1 to the membrane. The ε subunit has been proposed to be involved in regulating the F_1-ATPase. The F_0 portion of the ATPase consists of three or four nonidentical subunits and is an integral part of the membrane from which the ATPase is derived. When the purified F_0 portion of the ATPase is incorporated into an artificial membrane, it renders the membrane permeable to protons. In addition, the F_0 contains a subunit called the oligomycin-sensitivity-conferring protein which, as the name implies, causes the ATPase complex to show sensitivity to the inhibitory action of oligomycin.

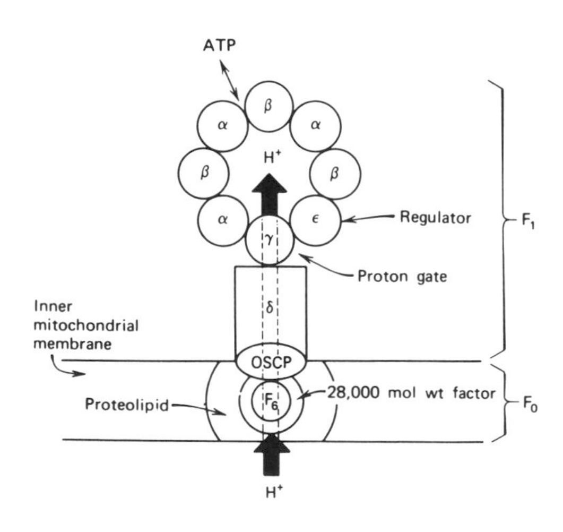

Figure 6.51
A model for the mitochondrial F_1F_0-ATPase.

While the chemiosmotic-coupling hypothesis for oxidative phosphorylation is widely accepted in principle as the mechanism for energy transduction in various biological systems, there are various questions that remain unanswered concerning the exact mechanism by which this important biochemical process occurs. For instance, what is the mechanism by which protons are pumped out of the mitochondrial matrix during electron transport? What is the stoichiometry of protons pumped per ATP synthesized? What is the mechanism by which protons are pumped back into the matrix "through" the F_1-ATPase? Is there a high-energy intermediate involved at some point in the ATP synthetic process?

BIBLIOGRAPHY

Energy-Producing and Energy-Utilizing Systems

Atkinson, D. E. *Cellular energy metabolism and its regulation.* New York: Academic, 1977.

Bock, R. M. Adenosine nucleotides and properties of pyrophosphate compounds. P. D. Boyer, H. Lardy, and K. Myrback (Eds.). *The enzymes,* 2nd ed., Vol. 2. New York: Academic, 1960, p. 3.

Lipmann, F. Metabolic generation and utilization of phosphate bound energy, *Adv. Enzymol.* 1:99, 1941.

Sources and Fates of Acetyl Coenzyme A

Denton, R. M. and Halestrap, A. Regulation of pyruvate metabolism in mammalian tissues. *Essays Biochem.* 15:37, 1979.

Reed, L. J. and Hackert, M. L. Structure–function relationships in dihydrolipoamide acyl transferases, *J. Biol. Chem.* 265:8971, 1990.

Roche, T. E. and Patel, M. S. α-Keto acid dehydrogenase complexes: organization, regulation and biomedical ramifications, *Ann. N. Y. Acad. Sci.* 573:1, 1989.

Weiland, O. H. The mammalian pyruvate dehydrogenase complex: structure and regulation, *Rev. Physiol. Biochem. Pharmacol.* 96:123, 1983.

The Tricarboxylic Acid Cycle

Hansford, R. G. Control of mitochondrial substrate oxidation. *Curr. Topics Bioenerg.* 10:217, 1980.

Krebs, H. A. The history of the tricarboxylic acid cycle, *Perspect. Biol. Med.* 14:154, 1970.

Kornberg, H. L. Tricarboxylic acid cycles, *BioEssays* 7:236 (1987).

Lowenstein, J. M. (Ed.). *Citric acid cycle: control and compartmentation.* New York: Marcel Dekker, 1969.

McCormack, J. G. and R. M. Denton. Ca^{2+} as a second messenger within mitochondria. *Trends Biochem. Sci.* 11:258, 1986.

Srere, P. M. The enzymology of the formation and breakdown of citrate. *Adv. Enzymol.* 43:57, 1975.

Williamson, J. R. and Copper, R. V. Regulation of the citric acid cycle in mammalian systems. *FEBS Lett.* 117 (Suppl.), K73, 1980.

Structure and Compartmentation on Mitochondrial Membranes

Ernster, L. (Ed.) *Bioenergetics.* Amsterdam: Elsevier, 1984.

Klingenberg, M. The ATP–ADP carrier in mitochondrial membranes, A. N. Martinosi (Ed.). *The enzymes of biological membranes.* New York: Plenum, 1976, p. 383.

LaNoue, K. F. and Schoolwerth, A. C. Metabolite transport in mitochondria. *Annu. Rev. Biochem.* 48:871, 1979.

Tzagoloff, A. *Mitochondria.* New York: Plenum, 1982.

Williamson, J. R. The role of anion transport in the regulation of metabolism, R. W. Hanson and M. A. Mehlman (Eds.). *Gluconeogenesis: its regulation in mammalian species.* New York: Wiley-Interscience, 1976, p. 165.

Yaffe, M. and G. Schatz. The future of mitochondrial research. *Trends Biochem. Sci.* 9:179, 1984.

Electron Transfer

Baltsheffsky, H. and Baltsheffsky, M. Electron transport phosphorylation. *Annu. Rev. Biochem.* 43:871, 1974.

Chance, B. The nature of electron transfer and energy coupling reactions, *FEBS Lett.* 23:3, 1972.

Hatefi, Y. The mitochondrial electron transport and oxidative phosphorylation system. *Annu. Rev. Biochem.* 54:1015, 1985.

Prince, R. C. The proton pump of cytochrome oxidase. *Trends Biochem. Sci.* 13:159, 1988.

Wikstrom, M., Krab, K., and Saraste, M. Proton-Translocating cytochrome complexes. *Annu. Rev. Biochem.* 50:623, 1981.

Oxidative Phosphorylation

Boyer, P. D., Chance, B., Ernster, L., Mitchell, P., Racker, E., and Slater, E. Oxidative phosphorylation and photophosphorylation. *Annu. Rev. Biochem.* 49:955, 1977.

Boyer, P. D. The unusual enzymology of ATP synthase. *Biochemistry* 26:8503, 1987.

Chernyak, B. V. and I. A. Kozlov. Regulation of H^+-ATPases in oxidative and photophosphorylation. *Trends Biochem. Sci.* 11:32, 1986.

Fillingame, R. H. The proton-translocating pumps of oxidative phosporylation. *Annu. Rev. Biochem.* 49:1079, 1980.

Mitchell, P. Vectorial chemistry and the molecular mechanism of chemiosmotic coupling: power transmission by proticity. *Biochem. Soc. Trans.* 4:398, 1976.

Mitchell, P. Keilin's respiratory chain concept and its chemiosmotic consequences. *Science* 206:1148, 1979.

Racker, E. From Pasteur to Mitchell: a hundred years of bioenergetics. *Fed. Proc.* 39:210, 1980.

QUESTIONS

J. BAGGOTT AND C. N. ANGSTADT

*Q*uestion *T*ypes are described on page xxiii.

1. (QT1) At 37°C, $-2.303RT = -1.42$ kcal mol^{-1}. For the reaction A $\rightleftharpoons$ B, if $\Delta G° = -7.1$ kcal mol^{-1}, what is the equilibrium ratio of B/A?
   A. 10,000,000/1
   B. 100,000/1
   C. 1000/1
   D. 1/1000
   E. 1/100,000
2. (QT1) A bond may be "high energy" for any of the following reasons *except:*
   A. products of its cleavage are more resonance stablized than the original compound.
   B. the bond is unusually stable, requiring a large energy input to cleave it.
   C. electrostatic repulsion is relieved when the bond is cleaved.
   D. a cleavage product may be unstable, tautomerizing to a more stable form.
   E. the bond may be strained.

3. (QT2) The active form of pyruvate dehydrogenase is favored by the influence of which of the following on pyruvate dehydrogenase kinase?
   1. Low NADH/NAD^+
   2. Low acetyl CoA/CoASH
   3. High [pyruvate]
   4. Low [Ca^{2+}]
4. (QT2) At which of the following enzyme-catalyzed steps of the tricarboxylic acid cycle do(es) net incorporation of the elements of water into an intermediate of the cycle occur?
   1. Citrate synthase
   2. Succinyl CoA synthase
   3. Fumarase
   4. Aconitase
5. (QT2) The freely reversible reactions of the tricarboxylic acid cycle include:
   1. the citrate synthase reaction.
   2. the isocitrate dehydrogenase reaction.
   3. the α-ketoglutarate dehydrogenase reaction.
   4. the succinyl CoA synthase reaction.
6. (QT2) Which of the following tricarboxylic acid cycle intermediates may be added or removed by other metabolic pathways?
   1. Oxalosuccinate
   2. α-ketoglutarate
   3. Isocitrate
   4. Succinyl CoA
7. (QT2) Regulation of tricarboxylic acid cycle activity in vivo may involve the concentration(s) of:
   1. acetyl CoA.
   2. ATP.
   3. ADP.
   4. oxygen.

8. (QT2) The mitochondrial membrane contains a transporter for:
   1. NADH.
   2. acetyl CoA.
   3. GTP.
   4. ATP.

9. (QT1) Which line of the accompanying table correctly describes the indicated properties of *both* the malate shuttle and the α-glycerophosphate shuttle?

Table for Question 9

Property	*Malate shuttle*	*α-Glycerophosphate shuttle*
A. Location	Inner mitochondrial membrane	Outer mitochondrial membrane
B. ATP generated per cytoplasmic NADH	3	2
C. Transporter	Malate dehydrogenase	α-Glycerophosphate dehydrogenasc
D. Species transported	Malate	α-Glycerophosphate
E. Matrix electron acceptor	Oxaloacetate	Cytochrome *b*

A. NAD C. Both
B. NADP D. Neither

10. (QT4) Sterospecific hydrogen transfer.
11. (QT4) Reduced form is the usual source of reducing equivalents for anabolic processes.
12. (QT4) Irradiation with light of 300 nm wavelength causes the reduced form to emit fluorescence at 465 nm.

13. (QT1) If rotenone is added to the mitochondrial electron transport chain:
    A. the P/O ratio of NADH is reduced from 3/1 to 2/1.
    B. the rate of NADH oxidation is diminished to 2/3 of its initial value.
    C. succinate oxidation remains normal.
    D. oxidative phosphorylation is uncoupled at site 1.
    E. electron flow is inhibited at site II.
14. (QT1) If cyanide is added to tightly coupled mitochondria that are actively oxidizing succinate:
    A. subsequent addition of 2,4-dinitrophenol will cause ATP hydroylsis.
    B. subsequent addition of 2,4-dinitrophenol will restore succinate oxidation.
    C. electron flow will cease, but ATP synthesis will continue.
    D. electron flow will cease, but ATP synthesis can be restored by subsequent addition of 2,4-dinitrophenol.
    E. subsequent addition of 2,4-dinitrophenol *and* the phosphorylation inhibitor, oligomycin, will cause ATP hydrolysis.

15. (QT2) The heme iron of which of the following is bound to the protein by only one coordination linkage?
    1. Cytochrome a
    2. Cytochrome a_3
    3. Cytochrome b
    4. Cytochrome P450

A. Substrate level phosphorylation C. Both
B. Oxidative phosphorylation D. Neither

16. (QT4) Occur(s) in mitochondria.
17. (QT4) High-energy intermediate compound has been found.
18. (QT4) ATP synthesis is linked to dissipation of a proton gradient.

ANSWERS

1. B $\Delta G° = -2.3RT \log K$. log 100,000 = 5. Substitution gives $\Delta G° = -7.1$ (p. 242).
2. B A "high-energy" bond is so designated because it has a high free energy of hydrolysis. This could arise for reasons A, C, D, or E. High-energy does not refer to a high energy of formation (bond stability) (p. 245).
3. A 1, 2, and 3 true. NADH and acetyl CoA activate pyruvate dehydrogenase kinase, thus inactivating pyruvate dehydrogenase. Pyruvate inhibits the kinase, favoring the active dehydrogenase. High Ca^{2+} favors the active dehydrogenase but by activating the phosphatase (p. 252, Figure 6.16).
4. B 1 and 3 true. 1 and 3 clearly incorporate water, whereas 4 merely removes and then adds water (p. 254, Figure 6.19). In 2, a thioester bond is cleaved, just as it is in 1. The mechanism, however, involves a molecule of inorganic phosphate, not water; the phosphate is ultimately incorporated into GTP (p. 258).
5. D Only 4 true. 1 is irreversible due to cleavage of the thioester link, a high-energy bond. In 2 and 3, CO_2 is released. In 4 there are high-energy compounds on both sides of the reaction, namely GTP and succinyl CoA (p. 258).
6. C 2 and 4 true. 2 can be formed from glutamate, and 4 can be formed from methylmalonyl CoA (p. 257, p. 258).
7. E 1, 2, 3, and 4 true. 1 is the substrate (p. 260). 2 inhibits isocitrate dehydrogenase, and 3 activates it (p. 261, Figure 6.23). The cycle requires oxygen to oxidize NADH and ADP to be converted to ATP (respiratory control) (p. 261).
8. D Only 4 true. Reducing equivalents from NADH are shuttled across the membrane, as is the acetyl group of acetyl CoA, but NADH and acetyl CoA themselves cannot cross (p. 267, Figures 6.28 and 6.29). Of the nucleotides, only ATP and ADP are transported. The translocator is inhibited by atractyloside (p. 268).
9. B A: Both shuttles operate across the inner membrane. C: Two transporters are used by the malate shuttle, the malate α-ketoglutarate antiporter and the aspartate–glutamate antiporter. D: α-Glycerophosphate is not translocated; only reducing equivalents are. E: Oxaloacetate is a reaction product. NAD^+ is the electron acceptor (p. 267, Figure 6.28).
10. C NAD- and NADP-linked dehydrogenases both exhibit specificity for the A or the B side of the pyridine ring (p. 273).
11. B NADPH is not a substrate for mitochondrial electron transport (p. 273).
12. D Fluorescence excitation of the reduced pyridine ring occurs in a wavelength range where it absorbs light, about 340 nm. 300 nm is an absorbance minimum (p. 274, Figure 6.34).
13. C Rotenone inhibits at the level of NADH dehydrogenase (site 1), preventing all electron flow and all ATP synthesis from NADH. Flavin-linked dehydrogenases feed in electrons below site 1 and are unaffected by site 1 inhibitors (p. 279, p. 281, Figure 6.41).
14. A Cyanide inhibits electron transport at site III, blocking electron flow throughout the system. In coupled mitochondria, ATP synthesis ceases too. Addition of an uncoupler permits the mitochondrial ATPase (which is normally driven in the synthetic direction) to operate, and it catalyzes the favorable ATP hydrolysis reaction unless it is inhibited by a phosphorylation inhibitor such as olgomycin (p. 283).
15. C 2 and 4 true. Fe^{2+} has six coordination positions. In heme, four are filled by the porphyrin ring. In cytochromes a and b, the other two are filled by the protein. But in cytochromes a_3 and P450, one position must be left vacant to provide an oxygen-binding site (p. 276).
16. C Substrate level phosphorylation is catalyzed by succinate thiokinase (p. 284).
17. A Enzyme-bound high-energy intermediates of substrate level phosphorylation have been found (p. 284), but none has been found for oxidative phosphorylation (p. 284).
18. B Current thinking is that the energy required for ATP synthesis comes from a proton gradient rather than from a chemical intermediate or a high-energy conformational state (p. 285).

7

Carbohydrate Metabolism I: Major Metabolic Pathways and Their Control

ROBERT A. HARRIS

7.1 OVERVIEW 292
7.2 GLYCOLYSIS 293
Glycolysis Occurs in All Human Cells 293
Glucose Is Metabolized Differently in Various Cells 295
7.3 THE GLYCOLYTIC PATHWAY 297
Glycolysis Occurs in Three Stages 297
Stage One Primes the Glucose Molecule 299
Stage Two Is Splitting of a Phosphorylated Intermediate 300
Stage Three Involves Oxidoreduction Reactions and the Synthesis of ATP 300
A Balance of Reduction of NAD^+ and Reoxidation of NADH Is Required: Role of Lactate Dehydrogenase 304
NADH Generated During Glycolysis Can Be Reoxidized via Substrate Shuttle Systems 304
The Shuttles Are Important in Other Oxidoreductive Pathways 306
The Two Shuttle Pathways Yield Different Amounts of ATP 307
Glycolysis Can Be Inhibited at Different Stages 308
7.4 REGULATION OF THE GLYCOLYTIC PATHWAY 309
Hexokinase and Glucokinase Have Different Properties 309
6-Phosphofructo-1-kinase Is the Major Regulatory Site 312
Crossover Theorem Explains Regulation of 6-Phosphofructo-1-kinase by ATP, AMP, and Inorganic Phosphate 313
Intracellular pH Can Regulate 6-Phosphofructo-1-kinase 316
Intracellular Citrate Levels Also Regulate 6-Phosphofructo-1-kinase 317
Hormonal Control of 6-Phosphofructo-1-kinase by cAMP and Fructose 2,6-bisphosphate 317
cAMP Activates a Protein Kinase 319
6-Phosphofructo-2-kinase and Fructose 2,6-bisphosphatase Are Part of a Bifunctional Enzyme Regulated by Phosphorylation-Dephosphorylation 321
Pyruvate Kinase Is Also a Regulated Enzyme of Glycolysis 322
7.5 GLUCONEOGENESIS 324
Glucose Synthesis Is Required for Survival of Humans and Other Animals 324
Substrate Cycles Between Tissues Are Required: The Cori and Alanine Cycles 324
Pathway of Glucose Synthesis From Lactate 326
Pyruvate Carboxylase and Phosphoenolpyruvate Carboxykinase 327
Gluconeogenesis Uses Some of the Glycolytic Enzymes but in the Reverse Direction 328
Glucose Is Synthesized From the Carbon Chain of Some Amino Acids 329
Glucose Can Be Synthesized From Odd-Chain Fatty Acids 332
Glucose Is Synthesized From Other Sugars 333
Gluconeogenesis Requires the Expenditure of ATP 334
Gluconeogenesis Has Several Sites of Regulation 334
Hormonal Control of Gluconeogenesis Is Critical for Homeostasis 335
Ethanol Ingestion Inhibits Gluconeogenesis 336

7.6 GLYCOGENOLYSIS AND GLYCOGENESIS 337
Glycogen, a Storage Form of Glucose, Is Required as a Ready Source of Energy 337
Glycogen Phosphorylase Catalyzes the First Step in Glycogen Hydrolysis 340
A Debranching Enzyme Is Required for Complete Hydrolysis of Glycogen 341
Synthesis of Glycogen Utilizes Some of the Enzymes Involved in Degradation but Also Some Unique Enzymes 343
Glycogen Synthase 343
Special Features of Glycogen Degradation and Synthesis: Glycogenin as a Primer 344
Glycogen Synthesis and Degradation Are Highly Regulated Pathways 346
BIBLIOGRAPHY 357
QUESTIONS AND ANNOTATED ANSWERS 357
CLINICAL CORRELATIONS
7.1 Alcohol and Barbiturates 307
7.2 Arsenic Poisoning 309
7.3 Fructose Intolerance 311
7.4 Diabetes Mellitus 312
7.5 Lactic Acidosis 316
7.6 Pickled Pigs and Malignant Hyperthermia 317
7.7 Angina Pectoris and Myocardial Infarction 318
7.8 Pyruvate Kinase Deficiency and Hemolytic Anemia 324
7.9 Hypoglycemia and Premature Infants 325
7.10 Hypoglycemia and Alcohol Intoxication 336
7.11 Glycogen Storage Diseases 340

7.1 OVERVIEW

The major pathways of carbohydrate metabolism either begin or end with glucose (Figure 7.1). This chapter will describe the utilization of glucose as a source of energy, the formation of glucose from noncarbohydrate precursors, the storage of glucose in the form of glycogen for later use, and the release of glucose from this storage form for use by cells. An understanding of the pathways and their regulation is necessary because of the important role played by glucose in the body. Glucose is the major form in which the carbohydrate absorbed from the intestinal tract is presented to cells of the body. Glucose is the only fuel used to any significant extent by a few specialized cells, and it is the major fuel used by the brain. Indeed, glucose is so important to these specialized cells and the brain that several of the major tissues of the body work together to ensure a continuous supply of this essential substrate. Glucose metabolism is defective in two very common metabolic diseases, obesity and diabetes, which in turn are contributing factors in the development of a number of major medical problems, including atherosclerosis, hypertension, small vessel disease, kidney disease, and blindness.

The discussion begins with glycolysis, a pathway used by all cells of the body to extract part of the chemical energy inherent in the glucose molecule. This pathway also converts glucose to pyruvate and thus sets the stage for the complete oxidation of glucose to CO_2 and H_2O. The de novo synthesis of glucose, that is, gluconeogenesis, is considered next. It is a function of the liver and kidneys and can be conveniently discussed following glycolysis because gluconeogenesis will seem, without careful examination of the pathway, to be simply the reverse of the glycolytic pathway. In contrast to glycolysis, which produces ATP, gluconeogenesis requires ATP and is therefore an energy-requiring process. The consequence is that only some of the enzyme-catalyzed steps can be common

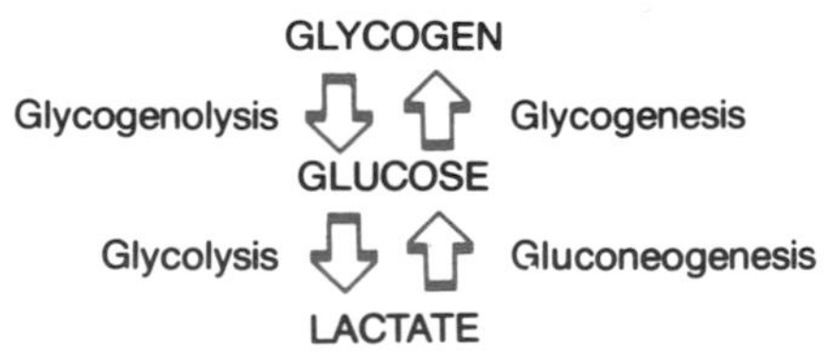

GLYCOGEN
Glycogenolysis ⇩ ⇧ Glycogenesis
GLUCOSE
Glycolysis ⇩ ⇧ Gluconeogenesis
LACTATE

FIGURE 7.1
Relationship of glucose to the major pathways of carbohydrate metabolism.

to both the glycolytic and gluconeogenic pathways. Indeed, additional enzyme-catalyzed steps and even mitochondria become involved to make the overall process of gluconeogenesis exergonic. Regulation of the rate-limiting and key enzyme-catalyzed steps will be stressed throughout this chapter. This will be particularly true for glycogen synthesis (glycogenesis) and glycogen degradation (glycogenolysis). Many cells store glycogen for the purpose of having glucose available for later use. The liver is less selfish, storing glycogen not for its own use, but rather for the maintenance of blood glucose levels to help ensure that other tissues of the body, especially the brain, have an adequate supply of this important substrate. Regulation of the synthesis and degradation of glycogen has been extensively studied and now serves as a model for our current understanding of how hormones work and how other metabolic pathways may be regulated. This subject will be emphasized because it contributes to our understanding of the diabetic condition, starvation, and how tissues of the body respond to stress, severe trauma, and injury. See the appendix for a discussion of the nomenclature and chemistry of the carbohydrates.

7.2 GLYCOLYSIS

Glycolysis Occurs in All Human Cells

The Embden–Meyerhof or **glycolytic pathway** represents an ancient process, possessed by all cells of the human body, in which anaerobic degradation of glucose to lactate occurs. This is one example of anaerobic fermentation, a term used to refer to pathways by which organisms extract chemical energy from high-energy fuels in the absence of molecular oxygen. For many tissues glycolysis represents an emergency energy-yielding pathway, capable of yielding 2 mol of ATP from 1 mol of glucose in the absence of molecular oxygen (Figure 7.2). This means that when the oxygen supply is shut off to a tissue, ATP levels can still be maintained by glycolysis for at least a short period of time. Many examples could be given, but the capacity to use glycolysis as a source of energy is particularly important to the human being at birth. With the exception of the brain, circulation of blood decreases to most parts of the body of the neonate during delivery. The brain is not normally deprived of oxygen during delivery, but other tissues must depend on glycolysis for their supply of ATP until circulation returns to normal and oxygen becomes available once again. This conserves oxygen for use by the brain, illustrating one of many mechanisms that have evolved to assure survival of brain tissue in times of stress. Glycolysis also sets the stage for aerobic oxidation of carbohydrate in cells. Oxygen is not necessary for glycolysis, and the presence of oxygen can indirectly suppress glycolysis, a phenomenon called the Pasteur effect, that is considered in a later section. Nevertheless, glycolysis can and does occur in cells with an abundant supply of molecular oxygen. Provided that the cells also contain mitochondria, the end product of glycolysis in the presence of oxygen becomes pyruvate rather than lactate. Pyruvate can then be completely oxidized to CO_2 and H_2O by enzymes housed within the mitochondria. The overall process of glycolysis *plus* the subsequent mitochondrial processing of pyruvate to CO_2 and H_2O has the following equation:

$$\underset{(C_6H_{12}O_6)}{\text{D-Glucose}} + 6O_2 + 38ADP^{3-} + 38P_i^{2-} + 38H^+ \longrightarrow 6CO_2 + 6H_2O + 38ATP^{4-}$$

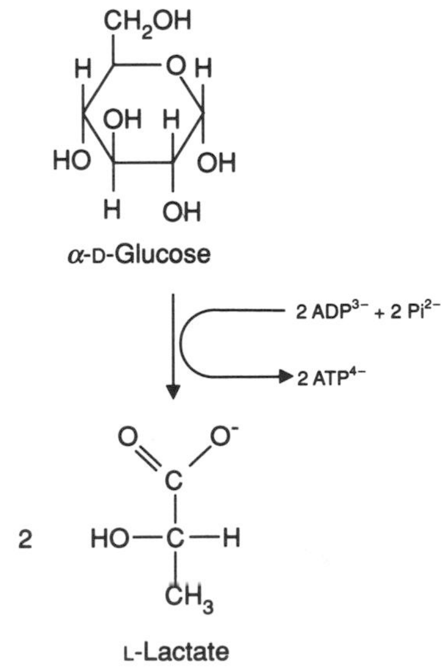

α-D-Glucose

$2\,ADP^{3-} + 2\,P_i^{2-}$ → $2\,ATP^{4-}$

2 L-Lactate

FIGURE 7.2
Overall balanced equation for the sum of the reactions of the glycolytic pathway.

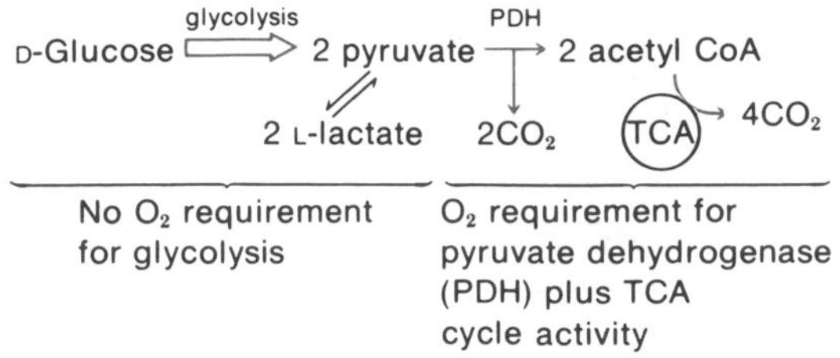

D-Glucose $\xrightarrow{\text{glycolysis}}$ 2 pyruvate $\xrightarrow{\text{PDH}}$ 2 acetyl CoA → TCA → $4CO_2$
2 pyruvate ⇌ 2 L-lactate; PDH → $2CO_2$

No O_2 requirement for glycolysis | O_2 requirement for pyruvate dehydrogenase (PDH) plus TCA cycle activity

FIGURE 7.3
Glycolysis is a preparatory pathway for aerobic metabolism of glucose.
TCA refers to the tricarboxylic acid cycle (see Figure 6.19).

Much more ATP is produced in the complete oxidation of glucose to CO_2 and H_2O than in the conversion of glucose to lactate. This has important consequences, which are considered in detail later. In order for glucose to be completely oxidized to CO_2 and H_2O, it must first be converted to pyruvate by glycolysis. This makes glycolysis a preparatory pathway for aerobic metabolism of glucose, as shown in Figure 7.3.

The importance of glycolysis as a preparatory pathway is best exemplified by the brain. This tissue has an absolute need for glucose and processes most of it via the glycolytic pathway. The pyruvate obtained is then oxidized completely to CO_2 and H_2O in brain mitochondria. Approximately 120 g of glucose is used by the adult human brain each day in order to meet its extraordinary need for ATP. The brain makes extensive use of glycolysis as a means of "preparing" the carbon of glucose for complete oxidation. In contrast, glycolysis with lactate as the end product, is the major mechanism of ATP production in a number of other tissues. Red blood cells (erythrocytes) lack mitochondria and therefore are unable to convert pyruvate to CO_2 and H_2O. The cornea, lens, and regions of the retina have a limited blood supply and also lack mitochondria (because mitochondria would absorb and scatter light) and likewise depend on glycolysis as the major mechanism for ATP production. Kidney medulla, testis, leukocytes, and white muscle fibers are almost totally dependent on glycolysis as a source of ATP, again because these tissues have relatively few mitochondria. Combined, the tissues that are dependent primarily on glycolysis for ATP production consume about 40 g of glucose per day in the normal human adult.

In any listing of the importance of glycolysis, it is impossible to ignore **alcoholic fermentation.** The overall balanced equation for the most common type is given in Figure 7.4. This pathway plays an important role in the making of "good brews," one of those things that make life worth living. This pathway is found in many yeast and certain bacteria but, somewhat surprisingly, the human body also produces a significant amount of ethanol. Most of the production, if not all, may be accounted for by microorganisms present in the intestinal tract. (There is a theory, however, that some of us have a better intestinal flora for the production of ethanol than others!) The pathway of alcoholic fermentation involves the same enzyme-catalyzed reactions as the glycolytic pathway, with the exception that lactate is not the end product. Rather than being reduced to lactate, pyruvate is decarboxylated to give acetaldehyde, which is reduced to ethanol to complete the pathway.

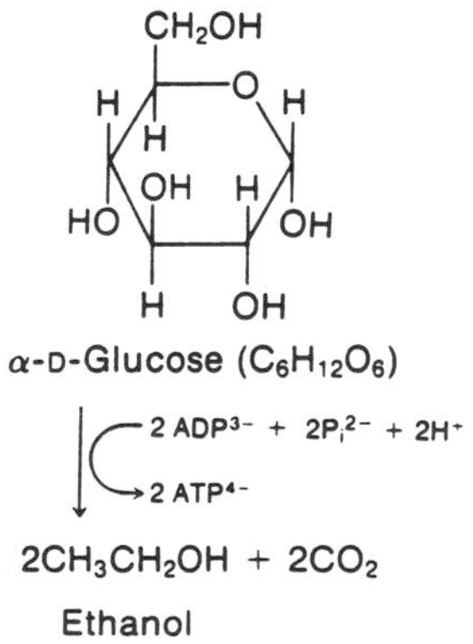

α-D-Glucose ($C_6H_{12}O_6$)

$$\xrightarrow[\;2\,ATP^{4-}]{2\,ADP^{3-} + 2P_i^{2-} + 2H^+}$$

$2CH_3CH_2OH + 2CO_2$
Ethanol

FIGURE 7.4
Overall balanced equation for ethanol production by alcoholic fermentation.

The major dietary sources of glucose are indicated in Chapter 26 in the discussion of the enzymes of digestion. Recall that **starch** is the storage form of glucose in plants and that it contains $\alpha[1 \rightarrow 4]$ glucosidic linkages along with $\alpha[1 \rightarrow 6]$ branches. **Glycogen** is the storage form of glucose in animal tissues and contains the same sort of glucosidic linkages and branches. It is important to distinguish between endogenous and exogenous sources of glucose. Exogenous refers to that which we eat and digest in the intestinal tract, whereas endogenous refers to that which is stored or synthesized in our tissues. Exogenous starch or glycogen is hydrolyzed within the lumen of the intestinal tract with the production of glucose, whereas stored glycogen endogenous to our tissues is converted to glucose or glucose 6-phosphate by enzymes present within the cells of these tissues. The disaccharides, which are important sources of glucose in our diet, include milk sugar (lactose) and grocery store sugar (sucrose). The hydrolysis of these sugars by enzymes of the brush border of the intestinal tract is discussed on page 1079. Glucose can be used as a source of energy to satisfy the needs of the cells of the intestinal tract. However, these cells are not designed to depend on glucose to any great extent, most of their energy requirement being met by glutamine catabolism. Most of the glucose passes through the cells of the intestinal tract into the blood, where it goes by way of the portal blood and the general circulation to be used by other cells of the body. The first major tissue to have an opportunity to remove it from the portal blood is the liver. When blood glucose is too high, the liver removes glucose from the blood by the glucose-consuming processes of glycogenesis and glycolysis. When blood glucose is too low, the liver supplies the blood with glucose by the glucose-producing processes of glycogenolysis and gluconeogenesis. The liver is also the first organ exposed to the blood flowing from the pancreas and therefore "sees" the highest concentrations of the hormones released from this endocrine tissue—glucagon and insulin. These are important hormonal regulators of blood glucose levels, in part because of their regulatory effects upon enzyme-catalyzed steps in the liver.

Glucose Is Metabolized Differently in Various Cells

After penetrating the plasma membrane by mediated transport, glucose is metabolized mainly by glycolysis in red blood cells (Figure 7.5*A*). Since red blood cells lack mitochondria, the end product of glycolysis is lactic acid, which is released from the cells back into the blood plasma. Glucose used by the **pentose phosphate pathway** (see Chapter 8) in red blood cells provides NADPH, necessary in these cells primarily to keep glutathione in the reduced state. Reduced glutathione, in turn, plays an important role in the destruction of organic peroxides and H_2O_2 by the reaction catalyzed by glutathione peroxidase (Figure 7.6). NADPH is required for the reduction of oxidized glutathione (GSSG) back to reduced glutathione (GSH) by glutathione reductase. Peroxides cause irreversible oxidative damage to membranes, DNA, and numerous other cellular components and must, therefore, be destroyed.

The brain, like red blood cells, takes up glucose by mediated transport in an insulin-independent manner (Figure 7.5*B*). Glycolysis in the brain yields pyruvate, which is then oxidized completely to CO_2 and H_2O, as discussed above. The pentose phosphate pathway is also quite active in these cells, generating part of the NADPH needed for reductive synthesis and the maintenance of glutathione in the reduced state. Muscle and heart cells readily utilize glucose (Figure 7.5*C*), and transport of glucose into both is dependent on the presence of insulin in the blood. Once taken up by these cells, glucose can be utilized by glycolysis to give pyruvate and lactate. Again pyruvate can be further utilized by the pyruvate dehydrogenase complex and the TCA cycle within the mitochondria to provide considerable energy in the form of ATP. Muscle and heart cells, in contrast to the other cells just considered, are capable of synthesizing significant quantities of glycogen. The synthesis and degradation of glycogen are important processes in these cells. Adipose tissue also accumulates glucose by an insulin-dependent mechanism (Figure 7.5*D*). Pyruvate, as in other cells, is generated by glycolysis and can be oxidized by the pyruvate dehydrogenase complex to give acetyl CoA within adipocytes.

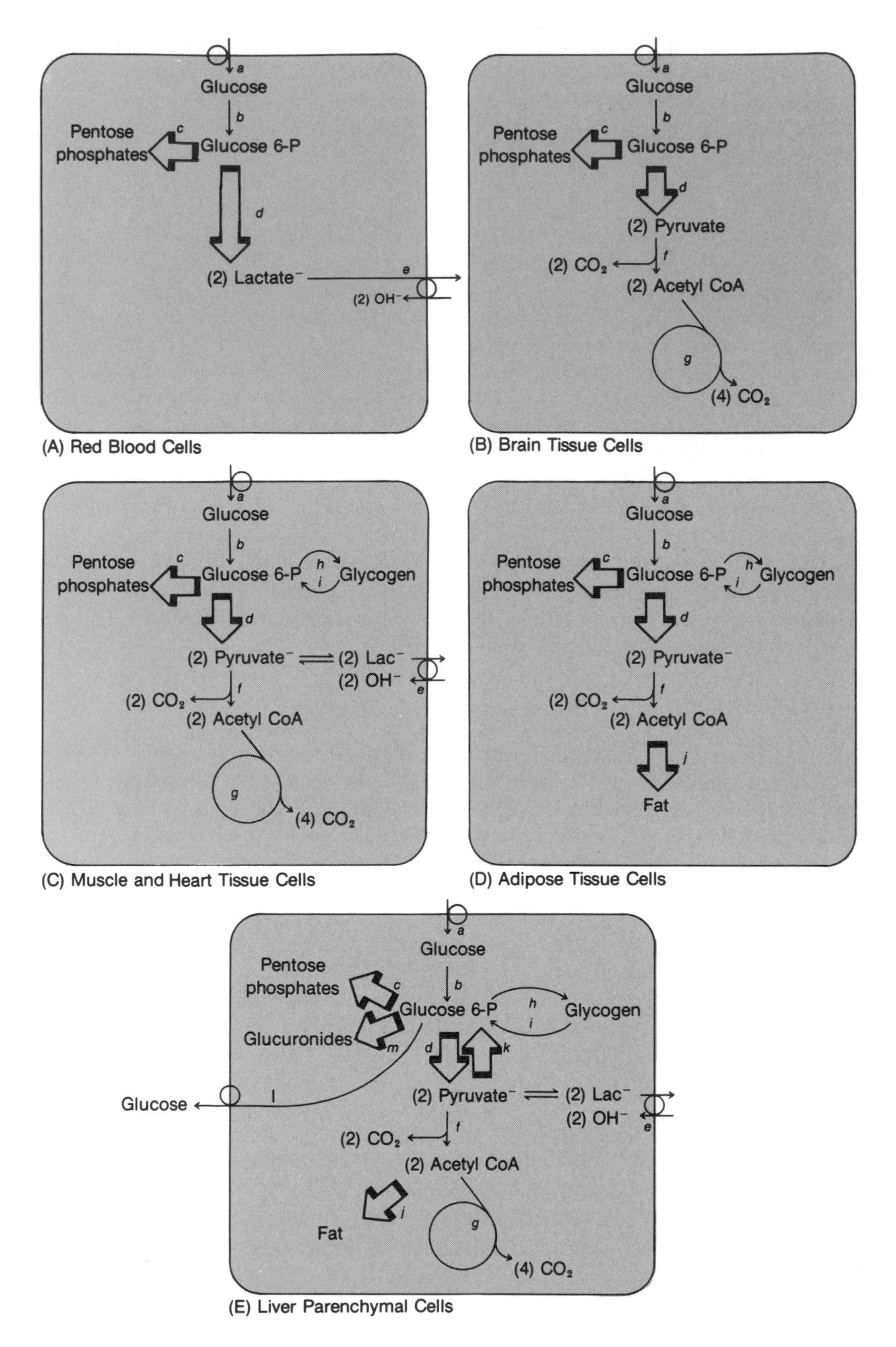

FIGURE 7.5

Overviews of the major ways in which glucose is metabolized within cells of selected tissues of the body.

(a) Glucose transport into the cell; (b) glucose phosphorylation by hexokinase; (c) the pentose phosphate pathway; (d) glycolysis; (e) lactic acid transport out of the cell; (f) pyruvate decarboxylation by pyruvate dehydrogenase; (g) TCA cycle; (h) glycogenesis; (i) glycogenolysis; (j) lipogenesis; (k) gluconeogenesis; (l) hydrolysis of glucose 6-phosphate and release of glucose from the cell into the blood; (m) formation of glucuronides (drug and bilirubin detoxification by conjugation) by the glucuronic acid pathway.

However, instead of being completely oxidized to CO_2 and H_2O for the production of ATP, the acetate moiety of acetyl CoA is used primarily for de novo fatty acid synthesis in this tissue. Generation of NADPH by the pentose phosphate pathway is an important process in adipose tissue, considerable quantities of NADPH being necessary for the reductive steps of fatty acid synthesis. Adipose tissue also has the capacity for glycogenesis and glycogenolysis, but these processes are much more limited in this tissue than in muscle and heart. The cells of the liver are involved in the greatest number of ways with glucose metabolism (Figure 7.5*E*). Uptake of glucose by the liver occurs independent of insulin by means of a high-capacity transport system. Glucose is used rather extensively by the pentose phosphate pathway for the production of NADPH, which is needed for reductive synthesis, maintenance of glutathione in the reduced state, and numerous reactions catalyzed by endoplasmic reticulum enzyme systems (see Chapter 6). A quantitatively less important but nevertheless vital function of the pentose phosphate pathway is the provision of ribose phosphate, required for the synthesis of nucleotides such as ATP and those found in DNA and RNA. Glucose is also used for glycogen synthesis, making glycogen storage an important feature of the liver. Glucose can also be used in the glucuronic acid pathway, important in drug and bilirubin detoxification (see Chapter 24). The liver also has significant capacity for glycolysis, the pyruvate produced being used as a source of acetyl CoA for complete oxidation by the TCA cycle and for the synthesis of fat by the process of de novo fatty acid synthesis. In contrast to the other tissues discussed above, the liver is unique in that it also has the capacity to convert three-carbon precursors, such as lactate, pyruvate, and alanine, into glucose by the process of gluconeogenesis. The glucose produced can then be used to meet the need for glucose of other cells of the body.

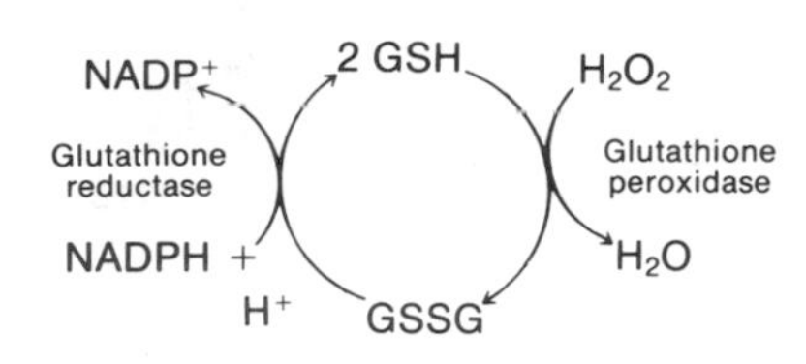

FIGURE 7.6
Destruction of H_2O_2 is dependent on reduction of glutathione by NADPH generated by the pentose phosphate pathway.

7.3 THE GLYCOLYTIC PATHWAY

Glucose is combustible and will burn in a test tube to yield heat and light but, of course, no ATP. Cells use some 30 steps to take glucose to CO_2 and H_2O, a seemingly inefficient process, since it can be done in a single step in a test tube. However, side reactions and some of the actual steps used by the cell to "burn" glucose to CO_2 and H_2O lead to the conservation of a significant amount of energy in the form of ATP. In other words, ATP is produced by the controlled "burning" of glucose in the cell, glycolysis representing only the first few steps, shown in Figure 7.7, in the overall process.

Glycolysis Occurs in Three Stages

Glycolysis can be conveniently pictured as occurring in three major stages (also see Figure 7.7):

Priming stage

$$\text{D-Glucose} + 2\text{ATP}^{4-} \longrightarrow \text{D-fructose 1,6-bisphosphate}^{4-} + 2\text{ADP}^{3-} + 2\text{H}^{+}$$

Splitting stage

$$\text{D-Fructose 1,6-bisphosphate}^{4-} \longrightarrow 2\ \text{D-glyceraldehyde 3-phosphate}^{2-}$$

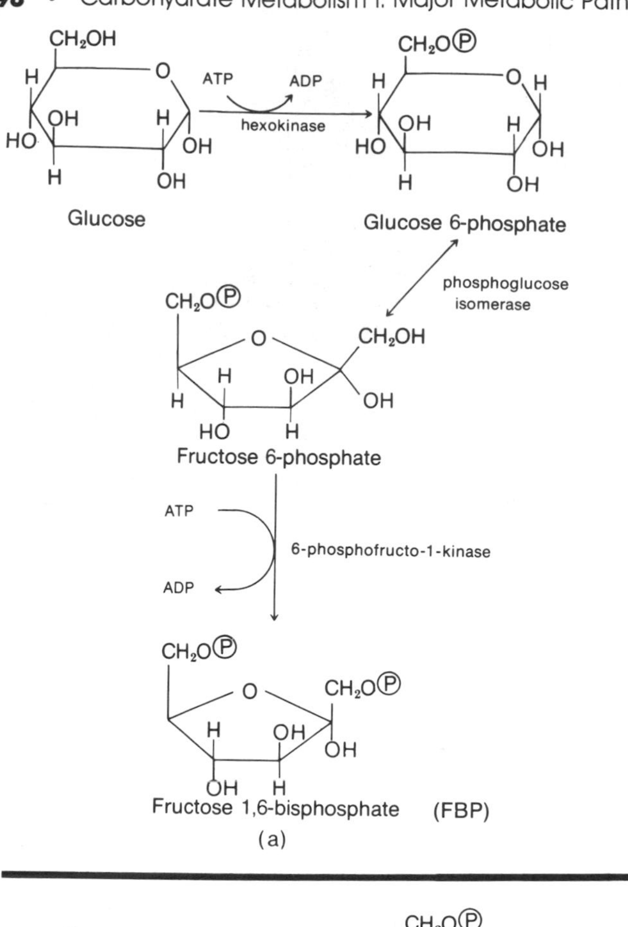

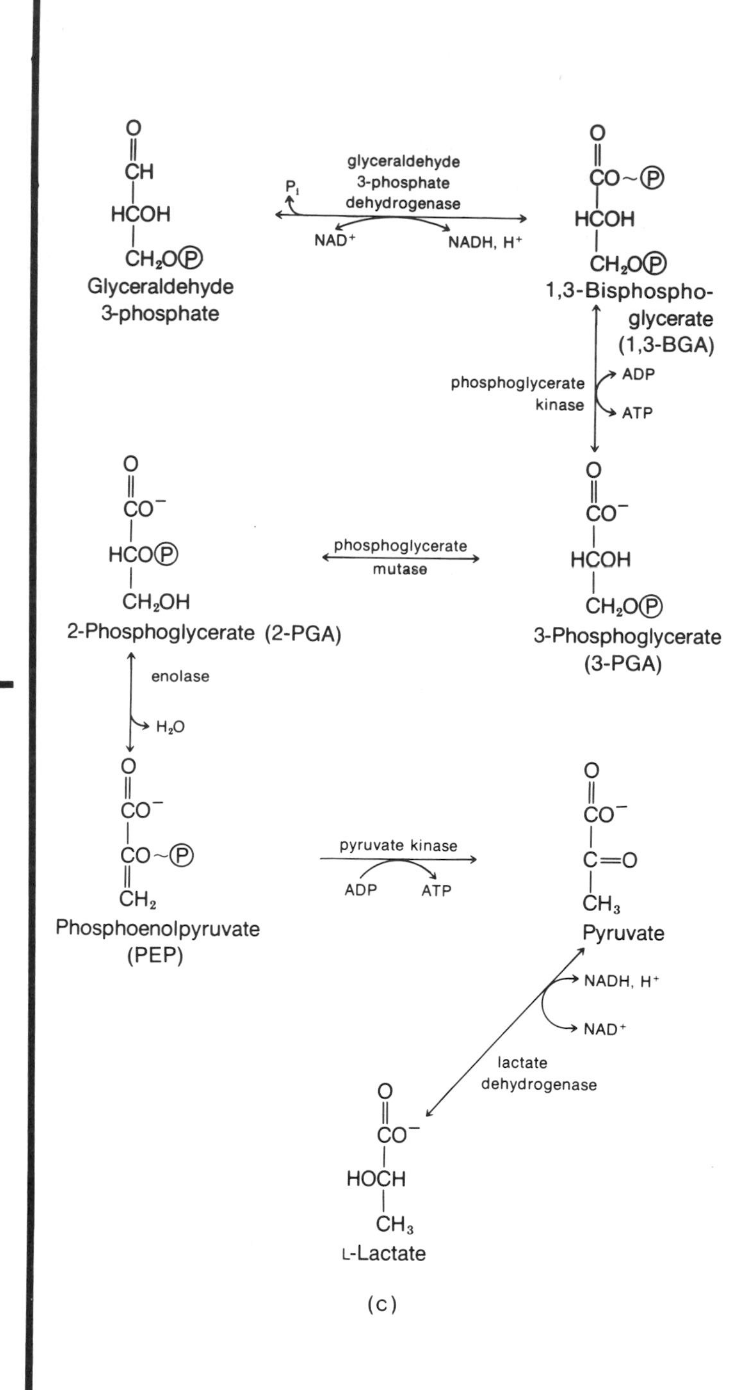

FIGURE 7.7
The glycolytic pathway, divided into its three stages.
The symbol Ⓟ refers to the phosphoryl group PO_3^{2-}; ~ indicates a high-energy phosphate bond. (a) Priming stage. (b) Splitting stage. (c) Oxidoreduction–phosphorylation stage.

Oxidoreduction–phosphorylation stage

$$2\ \text{D-Glyceraldehyde 3-phosphate}^{2-} + 4\text{ADP}^{3-} + 2\text{P}_i^{2-} + 2\text{H}^+ \longrightarrow 2\ \text{L-lactate}^- + 4\text{ATP}^{4-}$$

Sum

$$\text{D-Glucose} + 2\text{ADP}^{3-} + 2\text{P}_i^{2-} \longrightarrow 2\ \text{L-lactate}^- + 2\text{ATP}^{4-}$$

The priming stage involves the input of two molecules of ATP with the conversion of glucose into a molecule of fructose 1,6-biphosphate (ATP is "invested" in the priming stage of glycolysis). However, ATP beyond that invested is gained from the subsequent completion of the glycolytic process. The splitting stage "splits" the six-carbon molecule fructose 1,6-bisphosphate into two molecules of glyceraldehyde 3-phosphate. In the oxidoreduction–phosphorylation stage two molecules of glyceraldehyde 3-phosphate are converted into two molecules of lactate with the production of four molecules of ATP. The sum reaction for the overall process of glycolysis comes to the generation of two molecules of lactate and two molecules of ATP at the expense of one molecule of glucose.

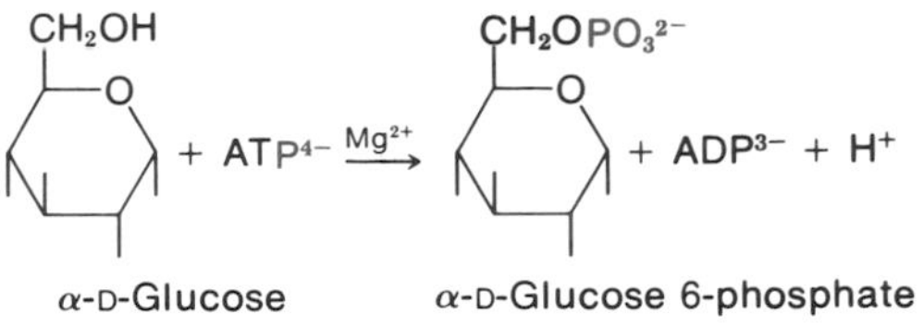

Stage One Primes the Glucose Molecule

Hexokinase catalyzes the first step of the glycolytic pathway (see Figure 7.7*a* and Step 1). Although this reaction consumes ATP, it gets glycolysis off to a good start by trapping glucose in the form of **glucose 6-phosphate (G6P)** within the cytosol of the cell where all of the glycolytic enzymes are located. Phosphate esters of charged, hydrophilic compounds do not readily penetrate cell membranes. The phosphorylation of glucose with ATP is a thermodynamically favorable reaction, requiring the use of one high-energy phosphate bond. It is an irreversible reaction under the conditions that exist in cells and represents, therefore, a way to synthesize G6P. However, it is *not,* by the reverse reaction, a way to synthesize ATP or to hydrolyze G6P to give glucose. Hydrolysis of G6P is accomplished by a completely different reaction, catalyzed by the enzyme **glucose 6-phosphatase:**

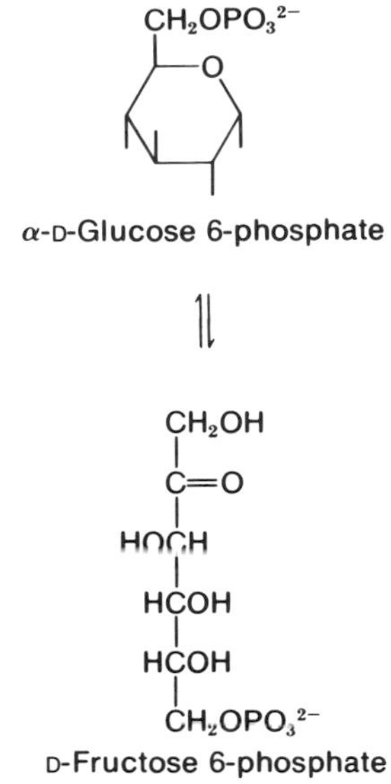

$$\text{Glucose 6-phosphate}^{2-} + \text{H}_2\text{O} \longrightarrow \text{glucose} + \text{P}_i^{2-}$$

This reaction is thermodynamically favorable in the direction written and cannot be used under conditions existing within biological cells for the synthesis of G6P from glucose. (A common mistake is to notice that ATP and ADP are involved in the reaction catalyzed by hexokinase but not to notice that they are *not* involved in the reaction catalyzed by glucose 6-phosphatase.) Glucose 6-phosphatase is an important enzyme in liver, functioning to produce free glucose from G6P in the last step of both gluconeogenesis and glycogenolysis, but it plays no role in the glycolytic pathway.

The next reaction is a readily reversible step of the glycolytic pathway, catalyzed by the enzyme **phosphoglucose isomerase** (Step 2). This step is not subject to regulation and, since it is readily reversible, functions in both glycolysis and gluconeogenesis.

6-Phosphofructo-1-kinase (also referred to as phosphofructokinase-1) catalyzes the next reaction of the glycolytic pathway, an ATP-dependent phosphorylation of **fructose 6-phosphate (F6P)** to give **fructose 1,6-bisphosphate (FBP)** (Step 3). This is a favorite enzyme of many students of

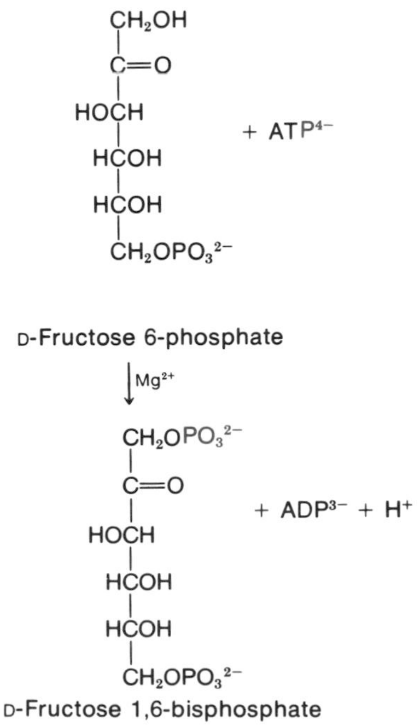

CH_2OH ... α-D-Glucose $+ ATP^{4-} \xrightarrow{Mg^{2+}}$ $CH_2OPO_3^{2-}$ ... α-D-Glucose 6-phosphate $+ ADP^{3-} + H^+$

Step 1

$CH_2OPO_3^{2-}$

α-D-Glucose 6-phosphate

$\rightleftharpoons$

CH_2OH, $C{=}O$, HOCH, HCOH, HCOH, $CH_2OPO_3^{2-}$

D-Fructose 6-phosphate

Step 2

CH_2OH, $C{=}O$, HOCH, HCOH, HCOH, $CH_2OPO_3^{2-}$ $+ ATP^{4-}$

D-Fructose 6-phosphate

$\downarrow Mg^{2+}$

$CH_2OPO_3^{2-}$, $C{=}O$, HOCH, HCOH, HCOH, $CH_2OPO_3^{2-}$ $+ ADP^{3-} + H^+$

D-Fructose 1,6-bisphosphate

Step 3

biochemistry, being subject to regulation by a score of effectors and considered the rate-limiting enzyme of the glycolytic pathway. The reaction is irreversible under intracellular conditions, that is, it represents a way to produce FBP but not a way to produce either ATP or F6P by the reverse reaction. This reaction utilizes the second ATP needed to "prime" glucose, thereby completing the first stage of glycolysis.

$CH_2OPO_3^{2-}$—C=O—HOCH—HCOH—HCOH—$CH_2OPO_3^{2-}$
D-Fructose 1,6-bisphosphate

$CH_2OPO_3^{2-}$—C=O—CH_2OH
Dihydroxyacetone phosphate

O=CH—HCOH—$CH_2OPO_3^{2-}$
D-Glyceraldehyde 3-phosphate

Step 4

Stage Two Is Splitting of a Phosphorylated Intermediate

Fructose 1,6-bisphosphate aldolase catalyzes the next step of the glycolytic pathway (see Figure 7.7*b*), cleaving fructose 1,6-bisphosphate into a molecule each of dihydroxyacetone phosphate and **glyceraldehyde 3-phosphate (GAP)** (Step 4). This is a reversible reaction, the enzyme being called aldolase because the overall reaction is a variant of an aldol cleavage in one direction and an aldol condensation in the other. **Triose phosphate isomerase** then catalyzes the reversible interconversion of dihydroxyacetone phosphate and GAP to complete the splitting stage of glycolysis (Step 5). With the transformation of **dihydroxyacetone phosphate (DHAP)** into GAP, the net conversion of one molecule of glucose into two molecules of GAP has been accomplished.

CH_2OH—C=O—$CH_2OPO_3^{2-}$
Dihydroxyacetone phosphate

$\rightleftharpoons$

O=CH—HCOH—$CH_2OPO_3^{2-}$
D-Glyceraldehyde 3-phosphate

Step 5

Stage Three Involves Oxidoreduction Reactions and the Synthesis of ATP

The first reaction of the last stage of glycolysis (Figure 7.7*c*) is catalyzed by the enzyme **glyceraldehyde 3-phosphate dehydrogenase** (Step 6). This reaction is of considerable interest, not so much because of the regulation of the enzyme involved, but rather because of what is accomplished in a single enzyme-catalyzed step. In this reaction an aldehyde (glyceraldehyde 3-phosphate) is oxidized to a carboxylic acid with the reduction of NAD^+ to NADH. Besides producing NADH, however, the reaction also produces a high-energy phosphate compound (1,3-bisphosphoglycerate), which is a mixed anhydride of a carboxylic acid and phosphoric acid. 1,3-Bisphosphoglycerate has a large negative free energy of hydrolysis, enabling it to participate in a subsequent reaction that yields ATP. The overall reaction catalyzed by glyceraldehyde 3-phosphate dehydrogenase can be visualized as the coupling of a very favorable exergonic reaction with a very unfavorable endergonic reaction on the surface of the enzyme. The exergonic reaction can be thought of as being composed of a half-reaction in which an aldehyde (glyceraldehyde 3-phosphate) is oxidized to a carboxylic acid (1,3-bisphosphoglycerate), which is then coupled with a half-reaction in which NAD^+ is reduced to NADH:

$$R\text{—}\overset{O}{\overset{\|}{C}}H + H_2O \longrightarrow R\text{—}\overset{O}{\overset{\|}{C}}OH + 2H^+ + 2e^-$$

$$NAD^+ + 2H^+ + 2e^- \longrightarrow NADH + H^+$$

The overall reaction (sum of the half-reactions) is quite exergonic, with the aldehyde being oxidized to a carboxylic acid and NAD^+ being reduced to NADH:

$$R\text{—}\overset{O}{\overset{\|}{C}}H + NAD^+ + H_2O \longrightarrow R\text{—}\overset{O}{\overset{\|}{C}}OH + NADH + H^+, \quad \Delta G^{\circ\prime} = -10.3 \text{ kcal mol}^{-1}$$

O=CH—HCOH—$CH_2OPO_3^{2-}$ + NAD^+ + P_i^{2-}
D-Glyceraldehyde 3-phosphate

$\rightleftharpoons$

O=$COPO_3^{2-}$—HCOH—$CH_2OPO_3^{2-}$ + NADH + H^+
1,3-Bisphospho-D-glycerate

Step 6

The endergonic component of the reaction corresponds to the formation of a mixed anhydride between the carboxylic acid and phosphoric acid:

$$R{-}\overset{O}{\overset{\|}{C}}OH + P_i^{2-} \longrightarrow R{-}\overset{O}{\overset{\|}{C}}{-}OPO_3^{2-} + H_2O, \quad \Delta G^{\circ\prime} = +11.8 \text{ kcal mol}^{-1}$$

The overall reaction involves coupling of the endergonic and exergonic components to give an overall standard free energy change of +1.5 kcal mol^{-1}.

$$\text{Sum: } R{-}\overset{O}{\overset{\|}{C}}H + NAD^+ + P_i^{2-} \longrightarrow$$

$$R{-}\overset{O}{\overset{\|}{C}}OPO_3^{2-} + NADH + H^+, \quad \Delta G^{\circ\prime} = +1.5 \text{ kcal mol}^{-1}$$

The reaction is freely reversible under intracellular conditions and is used in both the glycolytic and gluconeogenic pathways. The proposed mechanism for the enzyme-catalyzed reaction is shown in Figure 7.8. Glyceraldehyde 3-phosphate reacts with a sulfhydryl group of a cysteine residue of the enzyme to generate a **thiohemiacetal.** An internal oxidation–reduction reaction takes place on the surface of the enzyme in which the bound NAD$^+$ is reduced to NADH and the thiohemiacetal is oxidized to give a high-energy thiol ester. Exogenous NAD$^+$ then replaces the bound NADH and the high-energy thiol ester reacts with P_i to form the mixed anhydride and regenerate the free sulfhydryl group. The mixed anhydride then dissociates from the enzyme. It should be noted that, in contrast to the exergonic and endergonic components of the reactions discussed above, a carboxylic acid (RCOOH) is not considered to be an intermediate in the actual reaction mechanism. Instead, the enzyme uses the strategy of generating a high-energy thiol ester, which can be readily con-

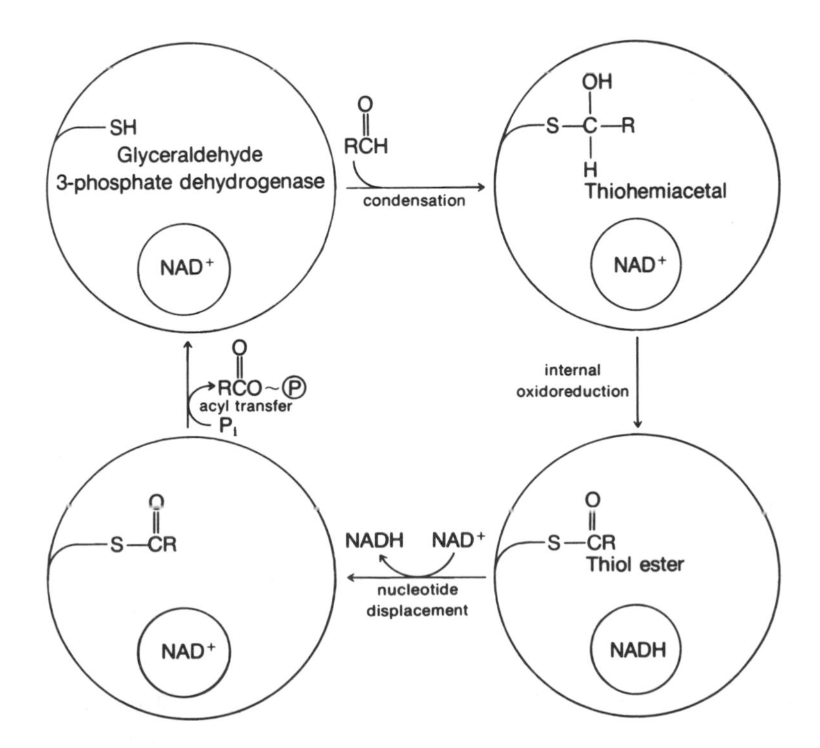

FIGURE 7.8
Mechanism of action of glyceraldehyde 3-phosphate dehydrogenase.
Large circle represents the enzyme; small circle, the binding site for NAD$^+$; $R\overset{O}{\overset{\|}{C}}H$, glyceraldehyde 3-phosphate; —SH, the sulfhydryl group of the cysteine residue located at the active site; and ~P, the high-energy phosphate bond of 1,3-bisphosphoglycerate.

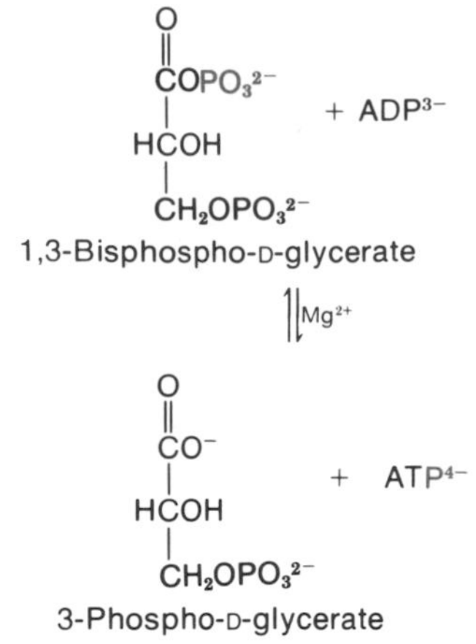

$$\begin{array}{l} O \\ \| \\ COPO_3^{2-} \\ | \\ HCOH \\ | \\ CH_2OPO_3^{2-} \end{array} + ADP^{3-}$$

1,3-Bisphospho-D-glycerate

$$\Big\downarrow\!\uparrow Mg^{2+}$$

$$\begin{array}{l} O \\ \| \\ CO^- \\ | \\ HCOH \\ | \\ CH_2OPO_3^{2-} \end{array} + ATP^{4-}$$

3-Phospho-D-glycerate

Step 7

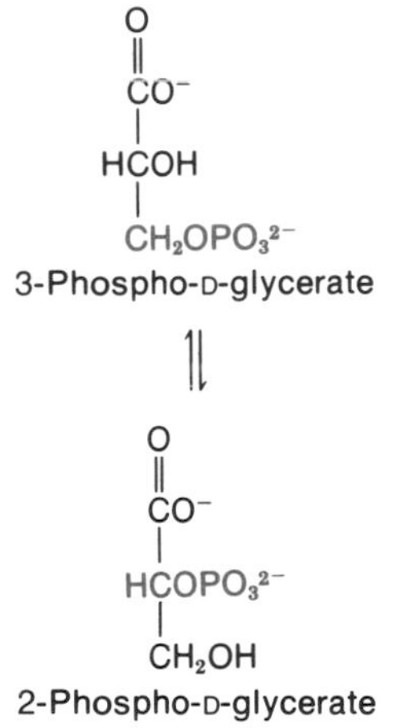

$$\begin{array}{l} O \\ \| \\ CO^- \\ | \\ HCOH \\ | \\ CH_2OPO_3^{2-} \end{array}$$

3-Phospho-D-glycerate

$$\rightleftharpoons$$

$$\begin{array}{l} O \\ \| \\ CO^- \\ | \\ HCOPO_3^{2-} \\ | \\ CH_2OH \end{array}$$

2-Phospho-D-glycerate

Step 8

verted into another high-energy compound, a mixed anhydride of carboxylic and phosphoric acids.

The reaction catalyzed by glyceraldehyde 3-phosphate dehydrogenase requires NAD^+ and produces NADH. Since the cytosol of cells has only a limited amount of NAD^+, it is imperative for continuous glycolytic activity that the NADH be converted back (turned over) to NAD^+. Without turnover of NADH, glycolysis will stop for want of NAD^+. The options that cells have for accomplishing the regeneration of NAD^+ are considered in detail later (see page 304).

The next reaction, catalyzed by the enzyme **phosphoglycerate kinase,** produces ATP from the high-energy compound 1,3-bisphosphoglycerate (Step 7 in Figure 7.7). This is the first site of ATP production in the glycolytic pathway. Since two molecules of ATP were "invested" for each glucose molecule in the priming stage [one at the hexokinase-catalyzed step (1) and one at the 6-phosphofructo-1-kinase-catalyzed step (3)], and since two molecules of 1,3-bisphosphoglycerate are produced from each glucose, all of the ATP "invested" in the priming stage is recovered in this step of glycolysis. Since ATP production occurs in the forward direction and ATP utilization in the reverse direction, it is somewhat surprising that the reaction is freely reversible and can be used in both the glycolytic and gluconeogenic pathways. The reaction provides a means for the generation of ATP in the glycolytic pathway but, when needed for glucose synthesis, can also be used in the reverse direction for the synthesis of 1,3-bisphosphoglycerate at the expense of ATP. The glyceraldehyde 3-phosphate dehydrogenase-phosphoglycerate kinase system is an example of substrate-level phosphorylation, a term used to refer to a process in which a substrate participates in an enzyme-catalyzed reaction that yields ATP or GTP. Substrate-level phosphorylation stands in contrast to oxidative phosphorylation in which electron transport by the respiratory chain of the mitochondrial inner membrane is used to provide the energy necessary for ATP synthesis (see Chapter 6). Note, however, that the combination of the reactions catalyzed by glyceraldehyde 3-phosphate dehydrogenase and phosphoglycerate kinase accomplishes the coupling of an oxidation (an aldehyde goes to a carboxylic acid) to a phosphorylation.

Phosphoglycerate mutase catalyzes the step in which 3-phosphoglycerate is converted to 2-phosphoglycerate (Step 8). This is a freely reversible reaction in which 2,3-bisphosphoglycerate functions as an obligatory intermediate at the active site of the enzyme (E):

$$\begin{array}{lrcl} & \text{E + 2,3-bisphosphoglycerate} & \rightleftharpoons & \text{E-phosphate + 2-phosphoglycerate} \\ & \text{E-phosphate + 3-phosphoglycerate} & \rightleftharpoons & \text{E + 2,3-bisphosphoglycerate} \\ \text{Sum:} & \text{3-Phosphoglycerate} & \rightleftharpoons & \text{2-phosphoglycerate} \end{array}$$

2,3-Bisphosphoglycerate is synthesized by a reaction catalyzed by another enzyme, 2,3-bisphosphoglycerate mutase:

$$\begin{array}{l} O \\ \| \\ COPO_3^{2-} \\ | \\ HCOH \\ | \\ CH_2OPO_3^{2-} \end{array} \rightleftharpoons \begin{array}{l} O \\ \| \\ CO^- \\ | \\ HCOPO_3^{2-} \\ | \\ CH_2OPO_3^{2-} \end{array} + H^+$$

1,3-Bisphospho-D-glycerate 2,3-Bisphospho-D-glycerate

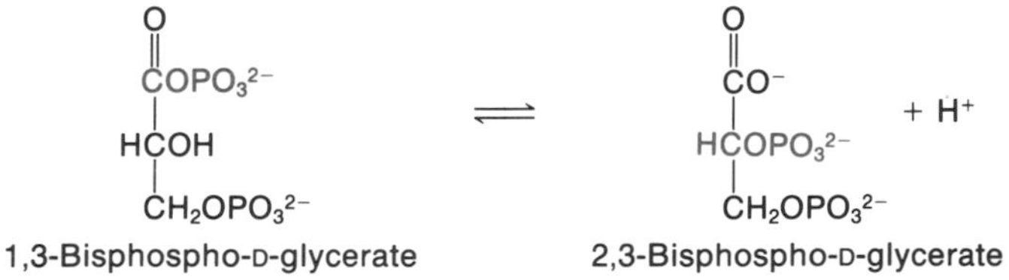

The mutase is unusual in that it is a **bifunctional enzyme,** serving also as a phosphatase that converts 2,3-bisphosphoglycerate to 3-phosphoglycerate and P_i. In the older literature 2,3-bisphosphoglycerate was abbrevi-

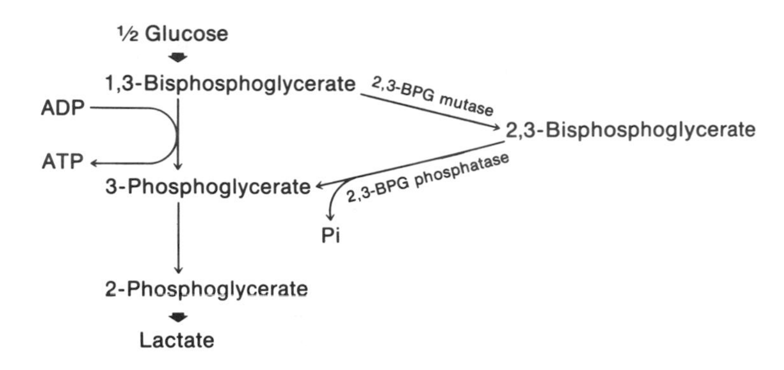

FIGURE 7.9
The 2,3-bisphosphoglycerate (2,3-BPG) shunt consists of reactions catalyzed by the bifunctional enzyme, 2,3-BPG mutase/phosphatase.

ated **DPG** for diphosphoglycerate. Most cells contain minute quantities of **2,3-bisphosphoglycerate** since it is only needed in catalytic amounts for the reaction catalyzed by phosphoglycerate mutase. Note that E-phosphate in the phosphoglycerate mutase reaction scheme cannot be generated without 2,3-bisphosphoglycerate. Red blood cells represent a special case, in which 2,3-bisphosphoglycerate accumulates to high concentrations and functions as a physiologically important allosteric effector of the association of oxygen with hemoglobin (see Chapter 25). From 15 to 25% of the glucose converted to lactate in red blood cells goes by way of the "BPG shunt" (Figure 7.9). Glucose catabolized by the shunt generates no net ATP since the reaction catalyzed by the phosphoglycerate kinase reaction is bypassed.

Enolase catalyzes the elimination of water from 2-phosphoglycerate to form **phosphoenolpyruvate (PEP)** in the next reaction (Step 9 in Figure 7.7). This is a remarkable reaction from the standpoint that a high-energy phosphate compound is generated from one of markedly lower energy level. The standard free energy change ($\Delta G^{\circ\prime}$) for the hydrolysis of phosphoenolpyruvate is -14.8 kcal mol^{-1}, a value strikingly greater than that of 2-phosphoglycerate (-4.2 kcal mol^{-1}). Although the reaction catalyzed by the enzyme is freely reversible, a large change in the distribution of energy occurs as a consequence of the action of enolase upon 2-phosphoglycerate. The free energy levels of PEP and 2-phosphoglycerate are not markedly different; however, the free energy levels of their products of hydrolysis (pyruvate and glycerate, respectively) are quite different. This accounts for the marked differences in the standard free energy of hydrolysis of these two compounds.

The next step of the glycolytic sequence is catalyzed by **pyruvate kinase** (Step 10). This enzyme accomplishes substrate level phosphorylation; that is, the synthesis of ATP with the conversion of the high-energy compound PEP into pyruvate. The reaction is not reversible under intracellular conditions. It constitutes a way to synthesize ATP, but in contrast to the reaction catalyzed by phosphoglycerate kinase is not reversible under conditions that exist in cells and is not a reaction that can be used for the synthesis of PEP when needed for glucose synthesis.

The last step of the glycolytic pathway is an oxidoreduction reaction catalyzed by **lactate dehydrogenase** (Step 11). Pyruvate is reduced in this reaction to give L-lactate, whereas NADH is oxidized to give NAD$^+$. This is a freely reversible reaction and the only one of the body in which L-lactate participates, that is, the only reaction that can result in L-lactate formation or L-lactate utilization. It should be noted that NADH gener-

$O{=}CO^-$ / $HCOPO_3^{2-}$ / CH_2OH
2-Phospho-D-glycerate
⇅
$O{=}CO^-$ / $C{-}OPO_3^{2-}$ ($={}CH_2$) $+ H_2O$
Phospho*enol*pyruvate

Step 9

$O{=}CO^-$ / $C{-}OPO_3^{2-}$ ($={}CH_2$) $+ ADP^{3-} + H$
Phosphoenolpyruvate
↓ $+ Mg^{2+}$
$O{=}CO^-$ / $C{=}O$ / CH_3 $+ ATP^{4-}$
Pyruvate

Step 10

$O{=}CO^-$ / $C{=}O$ / CH_3 $+ NADH + H^+$
Pyruvate
⇅
$O{=}CO^-$ / $HOCH$ / CH_3 $+ NAD^+$
L-Lactate

Step 11

ated by glyceraldehyde 3-phosphate dehydrogenase is converted back to NAD^+ by lactate dehydrogenase (see Figure 7.7), the major option used by cells under anaerobic conditions for the regeneration of cytosolic NAD^+.

A Balance of Reduction of NAD^+ and Reoxidation of NADH Is Required: Role of Lactate Dehydrogenase

An examination of the overall glycolytic pathway will show that there is a perfect coupling between the generation of NADH and its utilization (Figure 7.7). Two molecules of NADH are generated at the level of glyceraldehyde 3-phosphate dehydrogenase and two molecules of NADH are utilized at the level of lactate dehydrogenase in the overall conversion of one molecule of glucose into two molecules of lactate. The NAD^+, a soluble molecule present in the cytosol, is available in only limited amounts to participate in the glycolytic pathway. It is essential, therefore, that NAD^+ be regenerated from NADH for the glycolytic pathway to continue unabated. The NAD^+ reacts at the level of glyceraldehyde 3-phosphate dehydrogenase to produce NADH, which diffuses through the cytosol until it makes contact with lactate dehydrogenase, which, if pyruvate is also available, forms lactate with the regeneration of NAD^+. The overall reaction catalyzed by the combined actions of these two enzymes is the conversion of pyruvate, glyceraldehyde 3-phosphate, and P_i into lactate and 1,3-bisphosphoglycerate.

$$\text{D-Glyceraldehyde 3-phosphate} + NAD^+ + P_i \longrightarrow \text{1,3-bisphospho-D-glycerate} + NADH + H^+$$

$$\text{Pyruvate} + NADH + H^+ \longrightarrow \text{L-lactate} + NAD^+$$

Sum:

$$\text{D-Glyceraldehyde 3-phosphate} + \text{pyruvate} + P_i \longrightarrow \text{1,3-bisphosphoglycerate} + \text{L-lactate}$$

This perfect coupling of reducing equivalents in the glycolytic pathway only has to occur under conditions of anaerobiosis, or in cells that lack mitochondria. With the availability of oxygen and mitochondria, reducing equivalents in the form of NADH generated at the level of glyceraldehyde 3-phosphate dehydrogenase can be "shuttled" into the mitochondria for the synthesis of ATP. When this occurs, the end product of glycolysis becomes pyruvate. Two **shuttle systems** are known to exist for the transport of reducing equivalents from the cytosolic space to the mitochondrial matrix space (mitosol). The mitochondrial inner membrane is not permeable to NADH; therefore, NADH cannot penetrate directly across the mitochondrial inner membrane to gain access to the NADH dehydrogenase of the mitochondrial electron-transfer chain.

NADH Generated During Glycolysis Can Be Reoxidized via Substrate Shuttle Systems

The **glycerol phosphate shuttle** is shown in Figure 7.10*a*; the **malate–aspartate shuttle** in Figure 7.10*b*. All tissues that have mitochondria appear also to have the capability of "shuttling" reducing equivalents from the cytosol to the mitosol. The relative proportion of the activities of the two shuttles varies from tissue to tissue, with liver making greater use of the malate–aspartate shuttle, whereas some muscle cells may be more

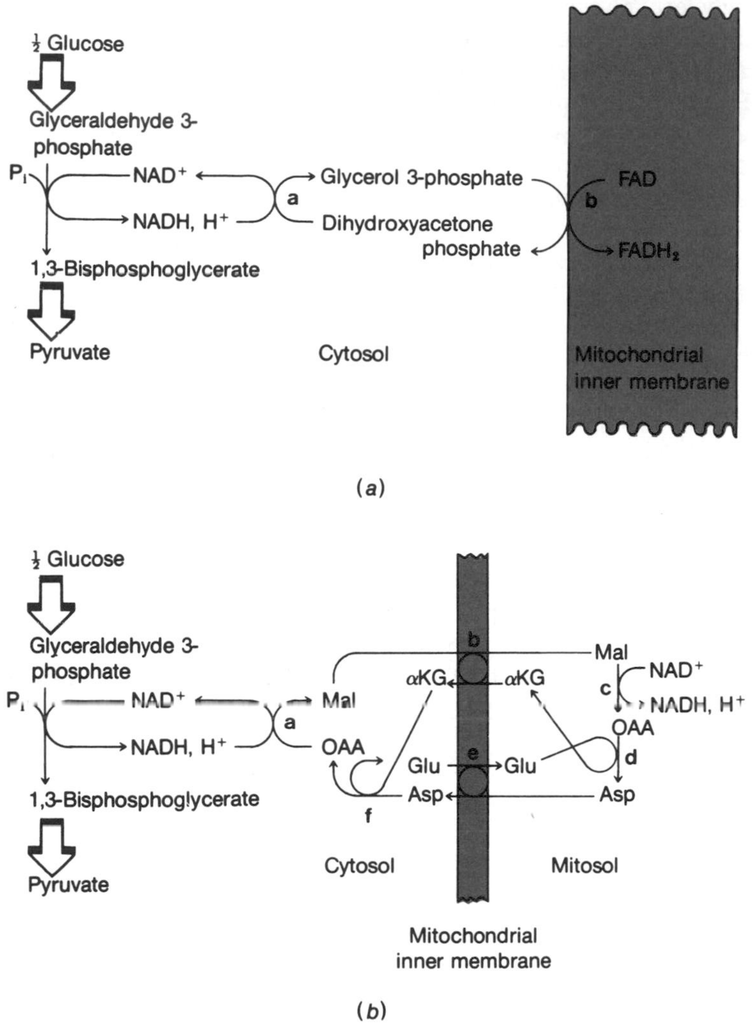

FIGURE 7.10
Shuttles for the transport of reducing equivalents from the cytosol to the mitochondrial electron-transfer chain.
(*a*) Glycerol phosphate shuttle; (*a*) cytosolic glycerol 3-phosphate dehydrogenase oxidizes NADH; (*b*) mitochondrial glycerol 3-phosphate dehydrogenase of the outer surface of the inner membrane reduces FAD. (*b*) Malate–aspartate shuttle: (*a*) cytosolic malate dehydrogenase reduces oxaloacetate (OAA) to malate; (*b*) dicarboxylic acid antiport of the mitochondrial inner membrane catalyzes electrically silent exchange of malate for α-ketoglutarate (α-KG); (*c*) mitochondrial malate dehydrogenase produces intramitochondrial NADH; (*d*) mitochondrial aspartate aminotransferase transaminates glutamate and oxaloacetate; (*e*) glutamate-aspartate antiport of the mitochondrial inner membrane catalyzes electrogenic exchange of glutamate for aspartate; (*f*) cytosolic aspartate aminotransferase transaminates aspartate and α-ketoglutarate.

dependent on the glycerol phosphate shuttle. The shuttle systems are irreversible, that is, they represent mechanisms for moving reducing equivalents into the mitosol, but not mechanisms for moving mitochondrial reducing equivalents into the cytosol.

The transport of aspartate out of the mitochondria in exchange for glutamate is the irreversible step in the malate–aspartate shuttle. The mitochondrial inner membrane has a large number of transport systems (see Chapter 6), but lacks one that is effective for oxaloacetate. For this reason oxaloacetate transaminates with glutamate to produce aspartate, which then exits irreversibly from the mitochondrion in exchange for glutamate. The aspartate entering the cytosol transaminates with α-ketoglutarate to give oxaloacetate and glutamate. The oxaloacetate accepts the reducing equivalents of NADH and becomes malate. Malate then penetrates the mitochondrial inner membrane, where it is oxidized by the mitochondrial malate dehydrogenase. This produces NADH within the mitosol and regenerates oxaloacetate to complete the cycle. The overall balanced equation for the sum of all the reactions of the malate–aspartate shuttle is simply

$$\mathrm{NADH_{cytosol}} + \mathrm{H^+_{cytosol}} + \mathrm{NAD^+_{mitosol}} \longrightarrow \mathrm{NAD^+_{cytosol}} + \mathrm{NADH_{mitosol}} + \mathrm{H^+_{mitosol}}$$

The glycerol phosphate shuttle is simpler, in the sense that fewer reactions are involved, but it should be noted that $FADH_2$ is generated as the end product within the mitochondrial inner membrane, rather than NADH within the mitosolic compartment. The irreversible step of the shuttle is catalyzed by the mitochondrial glycerol 3-phosphate dehydrogenase. The active site of this enzyme is exposed on the cytosolic surface of the mitochondrial inner membrane, making it unnecessary for glycerol 3-phosphate to penetrate completely into the mitosol for oxidation. The overall balanced equation for the sum of the reactions of the glycerol phosphate shuttle is

$$\text{NADH}_{\text{cytosol}} + \text{H}^+ + \text{FAD}_{\text{inner membrane}} \longrightarrow \text{NAD}^+_{\text{cytosol}} + \text{FADH}_{2,\text{ inner membrane}}$$

The Shuttles Are Important in Other Oxidoreductive Pathways

Alcohol Oxidation

The glycerol phosphate and malate–aspartate shuttles are important in using NADH generated in the cytosol in other ways as well. For example, the first step of alcohol (i.e., **ethanol**) metabolism is its oxidation to acetaldehyde with the production of NADH by the enzyme **alcohol dehydrogenase.**

$$\underset{\text{Ethanol}}{CH_3CH_2OH} + NAD^+ \longrightarrow \underset{\text{Acetaldehyde}}{CH_3\overset{O}{\overset{\|}{C}}H} + NADH + H^+$$

This enzyme is located almost exclusively in the cytosol of liver parenchymal cells. The acetaldehyde generated is able to traverse the mitochondrial inner membrane for oxidation by a mitosolic aldehyde dehydrogenase.

$$CH_3\overset{O}{\overset{\|}{C}}H + NAD^+ \longrightarrow CH_3\overset{O}{\overset{\|}{C}}O^- + NADH + 2H^+$$

The NADH generated by the last step can be used directly by the mitochondrial electron-transfer chain. However, the NADH generated by cytosolic alcohol dehydrogenase cannot be used directly, and must be oxidized back to NAD^+ by one of the shuttles. Thus, the capacity of human beings to oxidize alcohol is dependent on the ability of their liver to transport reducing equivalents from the cytosol to the mitosol by these shuttle systems.

Glucuronide Formation

Another situation in which the shuttles play an important role has to do with the formation of water-soluble **glucuronides** of bilirubin and various drugs. Conjugation with **glucuronic acid** occurs so that these compounds can be eliminated from the body in the aqueous media of urine and bile. In this process UDP-glucose (for structure, see page 367) is oxidized to UDP-glucuronic acid (structure on page 369).

$$\text{UDP-D-glucose} + 2\text{NAD}^+ + \text{H}_2\text{O} \longrightarrow \text{UDP-D-glucuronic acid} + 2\text{NADH} + 2\text{H}^+$$

In a reaction that occurs primarily in the liver, the "activated" glucuronic acid molecule is then transferred to a nonpolar, acceptor molecule, such as bilirubin or some compound (e.g., a drug) foreign to the body:

$$\text{UDP-D-glucuronic acid} + \text{R—OH} \longrightarrow \text{R—}O\text{—glucuronic acid} + \text{UDP}$$

Excess NADH generated by the first reaction has to be eliminated from the cytosol by the shuttles for this process to continue. Since ethanol oxidation and drug conjugation are properties of the liver, the two of them occurring together may overwhelm the combined capacity of the shuttles. A good thing to tell patients is not to mix the intake of pharmacologically active compounds and alcohol (see Clin. Corr. 7.1).

The Two Shuttle Pathways Yield Different Amounts of ATP

The mitosolic NADH formed as a consequence of malate–aspartate shuttle activity can be used in the presence of oxygen by the mitochondrial respiratory chain for the production of three molecules of ATP by oxidative phosphorylation:

$$NADH_{mitosol} + H^+ + \tfrac{1}{2}O_2 + 3ADP + 3P_i \longrightarrow NAD^+_{mitosol} + 3ATP + H_2O$$

In contrast, the $FADH_2$ obtained by glycerol phosphate shuttle activity yields only two ATP molecules:

$$FADH_{2\,inner\ membrane} + \tfrac{1}{2}O_2 + 2ADP + 2P_i \longrightarrow FAD_{inner\ membrane} + 2ATP + H_2O$$

Without the intervention of these shuttle systems, the conversion of one molecule of glucose to two molecules of lactate by glycolysis results in the *net* formation of two molecules of ATP. Two molecules of ATP are used in the priming stage to set glucose up so that it can be cleaved. However, subsequent steps then yield four molecules of ATP so that the overall net production of ATP by the glycolytic pathway is two molecules of ATP. Biological cells have only a limited amount of ADP and P_i. Flux through the glycolytic pathway is also dependent, therefore, on an adequate supply of these substrates. Consequently, the ATP generated has to be used, that is, turned over, in normal work-related processes in order for glycolysis to occur. The equation for the use of ATP for any work-related process is simply

$$ATP^{4-} \longrightarrow ADP^{3-} + P_i^{2-} + H^+ + \text{"work"}$$

When this equation is added to that given above for glycolysis, excluding the work accomplished, the overall balanced equation for the glycolytic process becomes

$$\text{D-Glucose} \longrightarrow 2\ \text{lactate}^- + 2H^+$$

If the ATP is not utilized for performance of work glycolysis will stop for want of ADP and/or P_i. Thus glycolytic activity is dependent on the turnover of ATP to ADP and P_i, just as it is dependent on the turnover of NADH to NAD^+.

CLIN. CORR. 7.1
ALCOHOL AND BARBITURATES

Acute alcohol intoxication causes increased sensitivity of an individual to the general depressant effects of barbiturates. Barbiturates and alcohol both interact with the γ-aminobutyrate (GABA)-activated chloride channel. Activation of the chloride channel inhibits neuronal firing, which may explain the depressant effects of both compounds. This drug combination is very dangerous and normal prescription doses of barbiturates have potentially lethal consequences in the presence of ethanol. In addition to the depressant effects of both ethanol and barbiturates on the central nervous system (CNS), ethanol inhibits the metabolism of barbiturates, thereby prolonging the time barbiturates remain effective in the body. Hydroxylation of barbiturates by the endoplasmic reticulum of the liver is inhibited by ethanol. This reaction, catalyzed by the NADPH-dependent cytochrome system, forms water-soluble derivatives of the barbiturates that are eliminated readily from the circulation by the kidneys. Blood levels of barbiturates remain high when ethanol is present, causing increased CNS depression.

Surprisingly, the alcoholic when sober is less sensitive to barbiturates. Chronic ethanol consumption apparently causes adaptive changes in the sensitivity of the CNS to barbiturates (cross-tolerance). It also results in the induction of the enzymes of liver endoplasmic reticulum involved in drug hydroxylation reactions. Consequently, the sober alcoholic is able to metabolize barbiturates more rapidly. This sets up the following scenario. A sober alcoholic has trouble falling asleep, even after taking several sleeping pills, because his liver has increased capacity to hydroxylate the barbiturate contained in the pills. In frustration he consumes more pills and then alcohol. Sleep results, but may be followed by respiratory depression and death because the alcoholic, although less sensitive to barbiturates when sober, remains sensitive to the synergistic effect of alcohol.

Misra, P. S., Lefevre, A., Ishii, H., Rubin, E., and Lieber, C. S. Increase of ethanol, meprobamate and pentobarbital metabolism after chronic ethanol administration in man and in rats. *Am. J. Med.* 51:346, 1971.

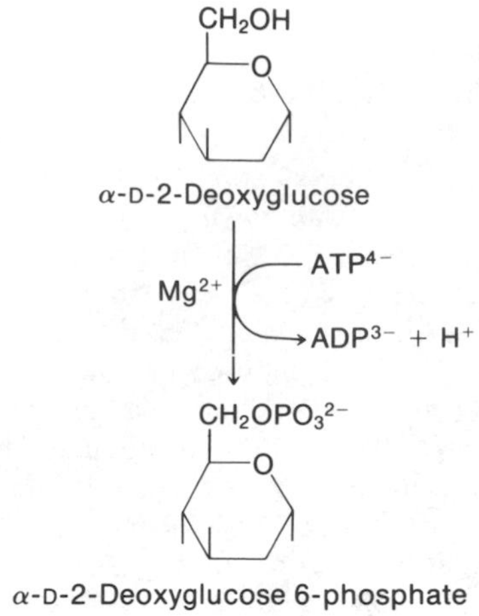

CH_2OH

α-D-2-Deoxyglucose

Mg^{2+}, ATP^{4-} → $ADP^{3-} + H^+$

$CH_2OPO_3^{2-}$

α-D-2-Deoxyglucose 6-phosphate

FIGURE 7.11
Hexokinase catalyzes the phosphorylation of 2-deoxyglucose.

$$\text{E—SH} + CH_3\text{—}Hg^+Cl^- \longrightarrow \text{E—S—Hg—}CH_3 + Cl^-$$

Glyceraldehyde 3-phosphate dehydrogenase; Methyl mercuric chloride; Inactive enzyme

$$\text{E—SH} + ICH_2CO_2^- \longrightarrow \text{E—S—}CH_2CO_2^- + H^+ + I^-$$

Iodoacetate; Inactive enzyme

FIGURE 7.12
Mechanism responsible for inactivation of glyceraldehyde 3-phosphate dehydrogenase by sulfhydryl reagents.

Glycolysis Can Be Inhibited at Different Stages

The best known inhibitors of the glycolytic pathway include 2-deoxyglucose, sulfhydryl reagents, and fluoride. **2-Deoxyglucose** causes inhibition of the reaction catalyzed by hexokinase. 2-Deoxyglucose serves as a substrate for this enzyme, being converted to the 6-phosphate ester by the reaction shown in Figure 7.11. Like glucose 6-phosphate, 2-deoxyglucose 6-phosphate is an effective inhibitor of the reaction catalyzed by hexokinase, but unlike glucose 6-phosphate, 2-deoxyglucose 6-phosphate will not function as a substrate for the reaction catalyzed by phosphoglucose isomerase. 2-Deoxyglucose 6-phosphate inhibition of hexokinase prevents glucose from being phosphorylated, resulting in the inhibition of glycolysis at the very first step.

Sulfhydryl reagents bring about an inhibition at the level of glyceraldehyde 3-phosphate dehydrogenase. As discussed above this enzyme has a cysteine residue at the active site, the sulfhydryl group of which reacts with glyceraldehyde 3-phosphate to give a thiohemiacetal. Sulfhydryl reagents are usually mercury-containing compounds or alkylating compounds, such as iodoacetate, which readily react with the sulhydryl group of glyceraldehyde 3-phosphate dehydrogenase to prevent the formation of the thiohemiacetal (see Figure 7.12).

Fluoride is a potent inhibitor of enolase. The Mg^{2+} ion and P_i are believed to form an ionic complex with fluoride ion, which is responsible for inhibition of the enzyme, apparently by interfering with the combination of the enzyme with its substrate (a Mg^{2+}–2-phosphoglycerate complex).

Arsenate has important effects on the glycolytic pathway and can be toxic. In the sense that it does not prevent flux through glycolysis, arsenate is not an inhibitor of the process. However, by causing arsenolysis at the step catalyzed by glyceraldehyde 3-phosphate dehydrogenase, arsenate prevents net synthesis of ATP by the pathway. Arsenate looks like P_i and is able to substitute for P_i in enzyme-catalyzed reactions. The result, in the case of glyceraldehyde 3-phosphate dehydrogenase, is the formation of a mixed anhydride of arsenic acid and the carboxyl group of 3-phosphoglycerate (Figure 7.13). 1-Arsenato 3-phosphoglycerate is unstable, undergoing spontaneous hydrolysis to give 3-phosphoglycerate and inorganic arsenate. Hence, glycolysis continues unabated in the presence of arsenate, but 1,3-bisphosphoglycerate is not formed, resulting in the loss of the capacity to synthesize ATP at the step catalyzed by phosphoglycerate kinase. The consequence is that net ATP synthesis does not occur when glycolysis is carried out in the presence of arsenate, the ATP invested in the priming stage being only balanced by the ATP generated in the pyruvate kinase step. This means that in the presence of arsenate, glycolysis does not generate the ATP required to meet the energy needs of a cell. This, along with the fact that arsenolysis also interferes with ATP formation by oxidative phosphorylation, makes arsenate a toxic compound (see Clin. Corr. 7.2).

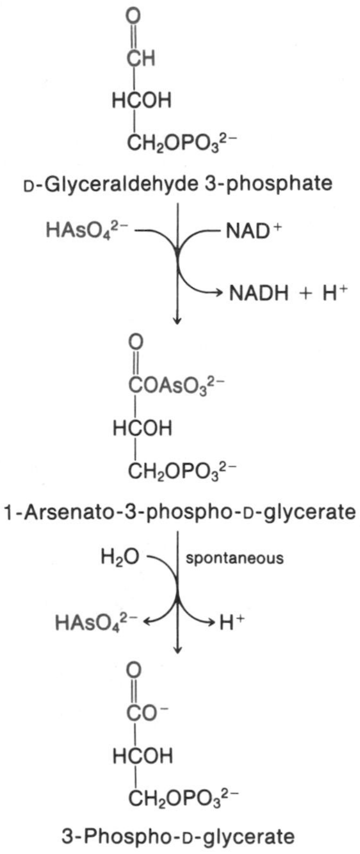

O=CH–HCOH–$CH_2OPO_3^{2-}$

D-Glyceraldehyde 3-phosphate

$HAsO_4^{2-}$, NAD^+ → $NADH + H^+$

O=$COAsO_3^{2-}$–HCOH–$CH_2OPO_3^{2-}$

1-Arsenato-3-phospho-D-glycerate

H_2O (spontaneous) → $HAsO_4^{2-}$ + H^+

O=CO^-–HCOH–$CH_2OPO_3^{2-}$

3-Phospho-D-glycerate

FIGURE 7.13
Arsenate uncouples oxidation from phosphorylation at the step catalyzed by glyceraldehyde 3-phosphate dehydrogenase.

7.4 REGULATION OF THE GLYCOLYTIC PATHWAY

Depending somewhat on the tissue under consideration, the regulatory enzymes of the glycolytic pathway are commonly considered to be hexokinase, 6-phosphofructo-1-kinase, and pyruvate kinase. A summary of the important regulatory features of these enzymes is presented in Figure 7.14. A **regulatory enzyme** is defined as an enzyme that is subject to control by either allosteric effectors or covalent modification. Both mechanisms are used by cells to control the most important of the regulatory enzymes. A regulatory enzyme can often be identified by determining whether the concentrations of the substrates and products within a cell indicate that the reaction catalyzed by the enzyme is close to equilibrium. An enzyme that is not subject to regulation will catalyze a **"near-equilibrium reaction"**, whereas a regulatory enzyme will catalyze a **"nonequilibrium reaction"** under intracellular conditions. This makes sense because flux through the step catalyzed by a regulatory enzyme is restricted because of controls imposed upon that enzyme. Whether an enzyme-catalyzed reaction is near equilibrium or nonequilibrium can be determined by comparing the established equilibrium constant for the reaction with the mass–action ratio as it exists within a cell. The equilibrium constant for the reaction A + B → C + D is defined as

$$K_{eq} = \frac{[C][D]}{[A][B]}$$

where the brackets indicate the concentrations at equilibrium. The mass–action ratio is calculated in a similar manner, except that the steady-state (ss) concentrations of reactants and products within the cell are used in the equation:

$$\textbf{Mass–action ratio} = \frac{[C]_{ss}[D]_{ss}}{[A]_{ss}[B]_{ss}}$$

If the mass–action ratio is approximately equal to the K_{eq}, the enzyme is said to be active enough to catalyze a near-equilibrium reaction and the enzyme is not considered subject to regulation. When the mass–action ratio is considerably different from the K_{eq}, the enzyme is said to catalyze a nonequilibrium reaction and usually will be found subject to regulation by one or more mechanisms. Mass–action ratios and equilibrium constants are compared for the glycolytic enzymes of liver in Table 7.1. The reactions catalyzed by glucokinase (liver isoenzyme of hexokinase), 6-phosphofructo-1-kinase, and pyruvate kinase in the intact liver are considered far enough from equilibrium to indicate that these enzymes are "regulatory" in this tissue.

CLIN. CORR. 7.2 ARSENIC POISONING

Most forms of arsenic are toxic, but the trivalent form (arsenite as AsO_2^-) is much more toxic than the pentavalent form (arsenate or $HAsO_4^{2-}$). Less ATP is produced whenever arsenate substitutes for P_i in biological reactions. Arsenate competes for P_i-binding sites on enzymes, resulting in the formation of arsenate esters that are unstable. Arsenite works by a completely different mechanism, involving the formation of a stable complex with enzyme bound lipoic acid:

(SH)(SH)–$(CH_2)_4C(=O)$—enzyme + AsO_2^- + H^+
↓
(S–As(OH)–S ring)–$(CH_2)_4C(=O)$—enzyme + H_2O

For the most part arsenic poisoning is explained by inhibition of those enzymes which require lipoic acid as a coenzyme. These include pyruvate dehydrogenase, α-ketoglutarate dehydrogenase, and branched-chain α-keto acid dehydrogenase. Chronic arsenic poisoning from well water contaminated with arsenical pesticides or through the efforts of a murderer is best diagnosed by determining the concentration of arsenic in the hair or fingernails of the victim. About 0.5 mg of arsenic would be found in a kilogram of hair from a normal individual. The hair of a person chronically exposed to arsenic could have 100 times as much.

Hindmarsh, J. T. and McCurdy, R. F. Clinical and environmental aspects of arsenic toxicity. *CRC Crit. Rev. Clin. Lab. Sci.* 23:315, 1986.

Hexokinase and Glucokinase Have Different Properties

Different isoenzymes of hexokinase are found in different tissues of the body. The hexokinase isoenzymes found in most tissues have a low K_m for glucose (<0.1 mM) relative to its concentration in blood (~5 mM) and are strongly inhibited by the product of the reaction, glucose 6-phosphate. The latter is an important regulatory feature because it prevents hexokinase from tying up all of the P_i of a cell in the form of phosphorylated hexoses (see Clin. Corr. 7.3). Thus the reaction catalyzed by hexokinase is not at equilibrium within cells that contain this enzyme because of the inhibition imposed by G6P. Liver parenchymal cells are unique in that

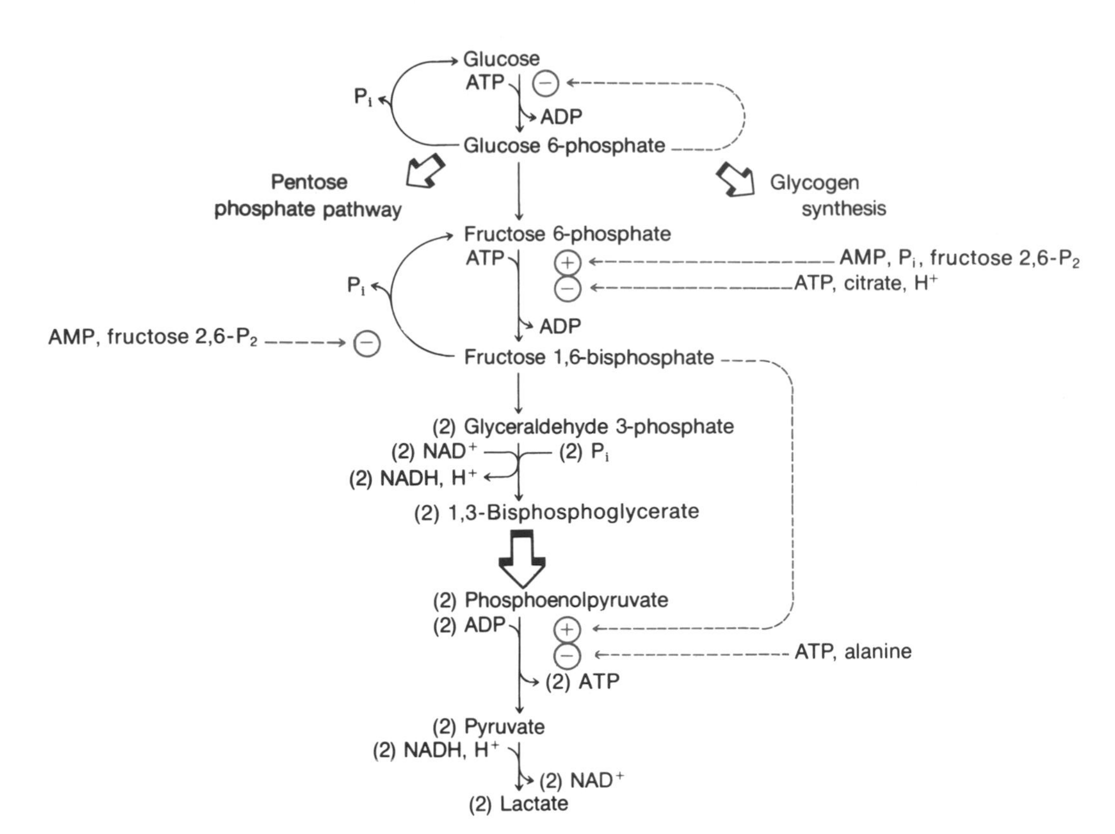

FIGURE 7.14
Important regulatory features of the glycolytic pathway.
Because of differences in isoenzyme distribution, not all tissues of the body have all of the regulatory mechanisms shown here.

TABLE 7.1 Apparent Equilibrium Constants and Mass–Action Ratios for the Reactions of Glycolysis and Gluconeogenesis in Liver

Reaction catalyzed by	Reaction in the pathway of: Glycolysis	Gluconeogenesis	Apparent equilibrium constant (K'_{eq})	Mass–action ratios	Considered near-equilibrium reaction?
Glucokinase	Yes	No	2×10^3	0.02	No
Glucose 6-phosphatase	No	Yes	850 M	120 M	No
Phosphoglucoisomerase	Yes	Yes	0.36	0.31	Yes
6-Phosphofructo-1-kinase	Yes	No	1×10^3	0.09	No
Fructose 1,6-bisphosphatase	No	Yes	530 M	19 M	No
Aldolase	Yes	Yes	13×10^{-5} M	12×10^{-7} M	Yes[a]
Glyceraldehyde 3-phosphate dehydrogenase + phosphoglycerate kinase	Yes	Yes	2×10^3 M^{-1}	0.6×10^3 M^{-1}	Yes
Phosphoglycerate mutase	Yes	Yes	0.1	0.1	Yes
Enolase	Yes	Yes	3.0	2.9	Yes
Pyruvate kinase	Yes	No	2×10^4	0.7	No
Pyruvate carboxylase + phosphoenolpyruvate carboxykinase	No	Yes	7.0 M	1×10^{-3} M	No

[a] Reaction catalyzed by aldolase appears to be out of equilibrium by two orders of magnitude. However, in vivo concentrations of fructose 1,6-bisphosphate and glyceraldehyde 3-phosphate are so low (μM concentration range) that significant enzyme binding of both metabolites is believed to occur. Although only the total concentration of any metabolite of a tissue can be measured, only that portion of the metabolite that is not bound should be used in the calculations of mass–action ratios. This is usually not possible, introducing uncertainty in the comparison of in vitro equilibrium constants to in vivo mass–action ratios.

they contain **glucokinase,** an isoenzyme of hexokinase with strikingly different kinetic properties from the other hexokinases. This isoenzyme catalyzes the same reaction, that is, an ATP-dependent phosphorylation of glucose, but has a much higher K_m for glucose and is not subject to product inhibition by G6P. It is, however, inhibited by fructose 6-phosphate and activated by fructose 1-phosphate. These effects depend on an inhibitory protein that inhibits by binding tightly to glucokinase. Fructose 6-phosphate promotes but fructose 1-phosphate inhibits binding of the inhibitory protein to glucokinase. The high K_m of glucokinase for glucose contributes to the capacity of the liver to "buffer" blood glucose levels. Glucose equilibrates readily across the plasma membrane of the liver, the concentration within the liver reflecting that of the blood. Since the K_m of glucokinase for glucose (~10 mM) is considerably greater than normal blood glucose concentrations (~5 mM), any increase in glucose concentration leads to a proportional increase in the rate of glucose phosphorylation by glucokinase (see Figure 7.15). Likewise, any decrease in glucose concentration leads to a proportional decrease in the rate of glucose phosphorylation. The result is that the liver uses glucose at a significant rate only when blood glucose levels are greatly elevated. This buffering effect of liver glucokinase on blood glucose levels would not occur if glucokinase had the low K_m for glucose characteristic of other hexokinases and was, therefore, completely saturated at physiological concentrations of glucose (see Figure 7.15). On the other hand, a low K_m form of hexokinase is a good choice for tissues such as the brain in that it allows phosphorylation of glucose even when blood and tissue glucose concentrations are dangerously low.

The reaction catalyzed by glucokinase is not at equilibrium under the intracellular conditions of liver cells (Table 7.1). Part of the explanation lies in the rate restriction imposed by the high K_m of glucokinase for glucose and part is due to the inhibitory protein mentioned above. Yet, another important factor is that the activity of glucokinase is opposed in liver by that of glucose 6-phosphatase. Like glucokinase, this enzyme has an unusually high K_m (3 mM) with respect to the normal intracellular concentration (~0.2 mM) of its primary substrate, glucose 6-phosphate. The result is that flux through this enzyme-catalyzed step is almost directly proportional to the intracellular concentration of glucose 6-phosphate. As shown in Figure 7.16, the combined action of glucokinase and glucose 6-phosphatase constitutes a futile cycle, that is, the sum of their reactions is simply the hydrolysis of ATP to give ADP and P_i without the performance of any work. It turns out that when blood glucose concentra-

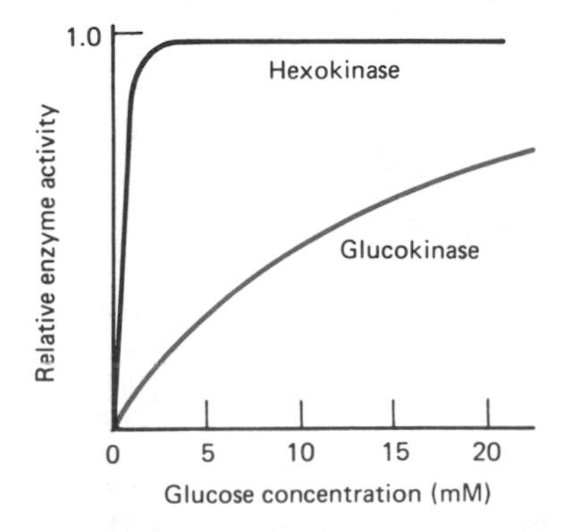

FIGURE 7.15
Comparison of the substrate saturation curves for hexokinase and glucokinase.

CLIN. CORR. 7.3
FRUCTOSE INTOLERANCE

Patients with hereditary fructose intolerance are deficient in the liver aldolase responsible for splitting fructose 1-phosphate into dihydroxyacetone phosphate and glyceraldehyde. Consumption of fructose by these patients results in the accumulation of fructose 1-phosphate and depletion of P_i and ATP in the liver. The reactions involved are those catalyzed by fructokinase and the enzymes of oxidative phosphorylation:

Fructose + ATP ⟶ fructose 1-phosphate + ADP

ADP + P_i + "energy provided by electron transfer chain" ⟶ ATP

Net

P_i + fructose ⟶ fructose 1-phosphate

Tying up P_i in the form of fructose 1 phosphate makes it impossible for liver mitochondria to generate ATP by oxidative phosphorylation. The ATP levels fall precipitously, making it also impossible for the liver to carry out its normal work functions. Damage results to the cells in large part because they are unable to maintain normal ion gradients by means of the ATP-dependent cation pumps. The cells swell and eventually lose their internal contents by osmotic lysis (see Clin. Corr. 6.4).

Although patients with fructose intolerance are particularly sensitive to fructose, humans in general have a limited capacity to handle this sugar. The capacity of the normal liver to phosphorylate fructose greatly exceeds its capacity to split fructose 1-phosphate. This means that fructose use by the liver is poorly controlled and that excessive fructose could deplete the liver of P_i and ATP. Fructose was actually tried briefly in hospitals as a substitute for glucose in patients being maintained by parenteral nutrition. The rationale was that fructose would be a better source of calories than glucose because fructose utilization is relatively independent of the insulin status of a patient. Delivery of large amounts of fructose by intravenous feeding was soon found to result in severe liver damage. Similar attempts have been made to substitute sorbitol and xylitol for glucose. These sugars also tend to deplete the liver of ATP and, like fructose, should not be used for parenteral nutrition.

Perheentupa, J., Raivio, K. O., and Nikkila, E. A. Hereditary fructose intolerance. *Acta Med. Scand.* *542* (Suppl.):65, 1972.

CLIN. CORR. 7.4 DIABETES MELLITUS

Diabetes mellitus is a chronic disease characterized by derangements in carbohydrate, fat, and protein metabolism. Two major types are recognized clinically—the juvenile-onset or insulin-dependent type (see Clin. Corr. 14.7) and the maturity-onset or insulin-independent type (see Clin. Corr. 14.9).

In patients who do not have fasting hyperglycemia, the oral glucose tolerance test can be used for the diagnosis of diabetes. It consists of determining the blood glucose level in the fasting state and at intervals of 30–60 min for 2 h or more after consuming a 100-g carbohydrate meal. In a normal individual blood glucose returns to normal levels within 2 h after ingestion of the carbohydrate meal. In the diabetic patient, blood glucose will reach a higher level and remain elevated for longer periods of time, depending on the severity of the disease. However, many factors may contribute to an abnormal glucose tolerance test. The patient must have consumed a high carbohydrate diet for the preceding 3 days, presumably to allow for induction of enzymes of glucose-utilizing pathways, for example, glucokinase, fatty acid synthase, and acetyl CoA carboxylase. In addition, almost any infection (even a cold) and less well-defined "stress," (presumably by effects on the sympathetic nervous system) can result in (transient) abnormalities of the glucose tolerance test. Because of problems with the glucose tolerance test, elevation of the fasting glucose level should probably be the sine qua non for the diagnosis of diabetes. Glucose uptake by cells of insulin-sensitive tissues, that is, muscle and adipose, is decreased in the diabetic state. Insulin is required for glucose uptake by these tissues, and the diabetic patient either lacks insulin or has developed "insulin resistance" in these tissues. Resistance to insulin is an abnormality of the insulin receptor or in subsequent steps mediating the metabolic effects of insulin. Parenchymal cells of the liver do not require insulin for glucose uptake. Without insulin, however, the liver has diminished enzymatic capacity to remove glucose from the blood. This is explained in part by decreased glucokinase activity plus the loss of insulin's action on key enzymes of glycogenesis and the glycolytic pathway.

National Diabetes Data Group. Classification and diagnosis of diabetes mellitus and other categories of glucose intolerance. *Diabetes 28:*1039, 1979.

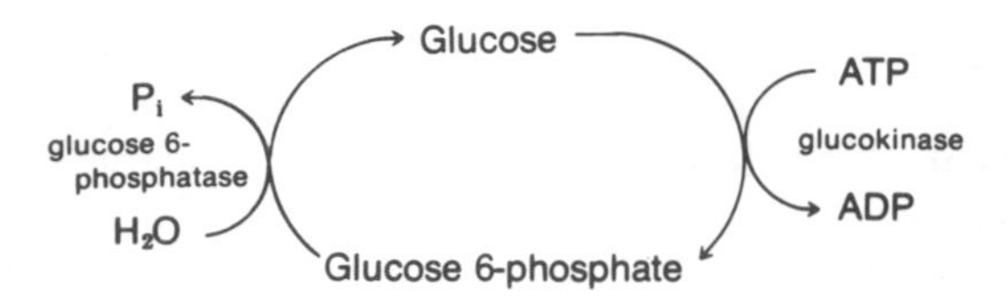

FIGURE 7.16
Phosphorylation of glucose followed by dephosphorylation constitutes a futile cycle in parenchymal cells of the liver.

tions are about 5 mM, the activity of glucokinase is almost exactly balanced by the opposing activity of glucose 6-phosphatase. The result is that no net flux occurs in either direction. This futile cycling between glucose and glucose 6-phosphate is wasteful of ATP but, combined with the process of gluconeogenesis, contributes significantly to the "buffering" action of the liver on blood glucose levels. Furthermore, it provides a mechanism for preventing glucokinase from tying up all of the P_i of the liver (see Clin. Corr. 7.3).

Fructose, a component of many vegetables, fruits, and sweeteners, promotes hepatic glucose utilization by an indirect mechanism. Fructose is converted in liver directly to fructose 1-phosphate (see Clin. Corr. 7.3) and, as mentioned above, fructose 1-phosphate activates glucokinase activity by promoting dissociation of an inhibitory protein. This may be a factor in the adverse effect, for example, hypertriglyceridemia, sometimes associated with excessive dietary fructose consumption.

Glucokinase is an **inducible enzyme.** This means that under various physiological conditions the *amount* of the enzyme either increases or decreases. Induction of the synthesis of an enzyme and the opposite—repression of the synthesis of an enzyme—are relatively slow control processes, usually requiring several hours before significant changes are realized. Insulin increases the amount of glucokinase by promoting transcription of the glucokinase gene. Thus, as long as insulin is also present, the amount of glucokinase in the liver tends to reflect how much glucose is being delivered to the liver via the portal vein. In other words, a person consuming large meals rich in carbohydrate will have greater amounts of glucokinase in the liver than one who is not. The liver in which glucokinase has been induced can make a greater contribution to the lowering of elevated blood glucose levels. The absence of insulin makes the liver of the diabetic patient deficient in glucokinase, in spite of high blood glucose levels, and this is one of the reasons why the liver of the diabetic has less blood glucose "buffering" action (see Clin. Corr. 7.4).

6-Phosphofructo-1-kinase Is the Major Regulatory Site

Much evidence suggests that 6-phosphofructo-1-kinase is the rate-limiting enzyme and most important regulatory site of glycolysis in most tissues. Usually we think of the first step of a pathway as the most logical choice for the rate-limiting step. Notice, however, that 6-phosphofructo-1-kinase catalyzes the first **committed step** of the glycolytic pathway. The phosphoglucose isomerase catalyzed reaction is reversible, and most cells can use glucose 6-phosphate for glycogen synthesis and in the pentose phosphate pathway. The reaction catalyzed by 6-phosphofructo-1-kinase commits the cell to the metabolism of glucose by glycolysis and is, therefore, a logical site for the step of the pathway that is rate limiting and subject to the greatest degree of regulation by allosteric effectors. Citrate, ATP, and hydrogen ions (low pH) are the most important negative allosteric effectors, whereas AMP, fructose 2,6-bisphosphate, and P_i are the most important positive allosteric effectors (Figure 7.14). Through their actions as strong inhibitors or activators of 6-phosphofructo-1-kinase, these compounds signal different rates of glycolysis in response to changes in (a) the energy state of the cell (ATP, AMP, and P_i), (b) the internal environment of the cell (hydrogen ions), (c) the availability of alternate fuels such as fatty acids and ketone bodies (citrate), and (d) the insulin/glucagon ratio in the blood (fructose 2,6-bisphosphate). Evidence for the physiological importance of these effectors comes in part from application of the crossover theorem to the glycolytic pathway.

Crossover Theorem Explains Regulation of 6-Phosphofructo-1-kinase by ATP, AMP, and Inorganic Phosphate

For the hypothetical pathway A → B → C → D → E → F · · · , the **crossover theorem** proposes that an inhibitor that partially inhibits the conversion of C to D will cause a "crossover" in the metabolite profile between C and D. This means that when the steady-state concentrations of the intermediates in the presence and absence of an inhibitor are compared, the concentrations of the intermediates before the site of inhibition should increase in response to the inhibitor, whereas those after the site should decrease. Crossover plots are constructed by setting the concentrations of all intermediates without some effector of the pathway equal to 100%. The concentrations of the intermediates observed in the presence of the effector are then expressed as percentages of these values. The expected result with a negative effector is shown in Figure 7.17*a*. The effect of returning the perfused rat heart from an anoxic condition to a well-oxygenated state is also shown (Figure 7.17*b*). This transition with the perfused rat heart is known to establish new steady-state concentrations of the glycolytic intermediates, the flux being much greater through the glycolytic pathway in the absence of oxygen. Under experimental conditions used, the perfused hearts consumed glucose at rates some 20 times greater in the absence than in the presence of oxygen. This example illustrates what is known as the **Pasteur effect,** defined as the inhibition of glucose utilization and lactate accumulation by the initiation of respiration (oxygen consumption). This is readily understandable on a thermodynamic basis, the complete oxidation of glucose to CO_2 and H_2O yielding much more ATP than anaerobic glycolysis:

Glycolysis: $\text{D-Glucose} + 2ADP^{3-} + 2P_i^{2-} \longrightarrow 2\ \text{L-lactate}^- + 2ATP^{4-}$

Complete Oxidation: $\text{D-Glucose} + 6O_2 + 38ADP^{3-} + 38P_i^{2-} + 38H^+ \longrightarrow 6CO_2 + 6H_2O + 38ATP^{4-}$

ATP is used by a cell only to meet its metabolic demand, that is, to provide the necessary energy for the work processes (metabolic demand) inherent to that cell. Since so much more ATP is produced from glucose in the presence of oxygen, much less glucose has to be consumed to meet the metabolic demand of the cell. The "crossover" at the conversion of fructose 6-phosphate to fructose 1,6-bisphosphate argues that oxygen imposes an inhibition at the level of 6-phosphofructo-1-kinase. This can be readily rationalized on the basis that ATP is a well-recognized inhibitor of 6-phosphofructo-1-kinase, and more ATP can be generated in the presence than in the absence of oxygen. However, ATP levels do not

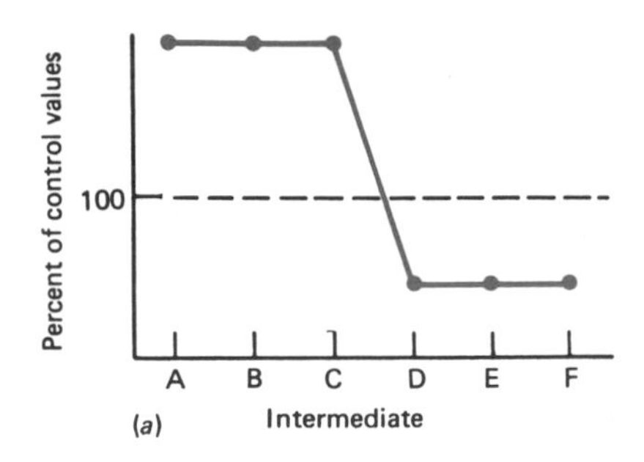

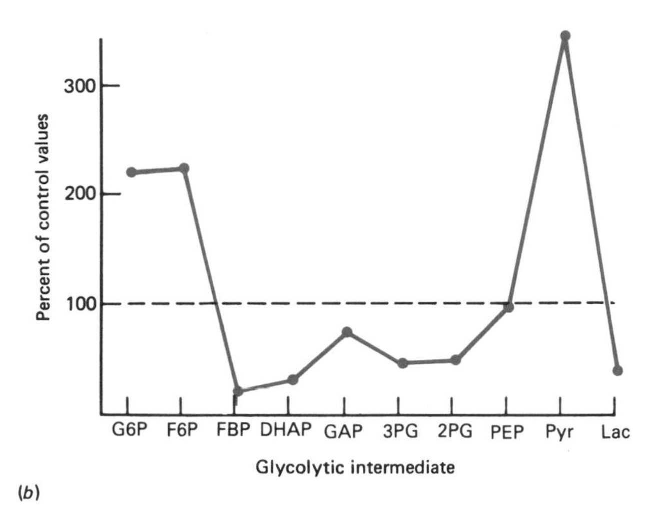

FIGURE 7.17
Crossover analysis is used to locate sites of regulation of a metabolic pathway.
(*a*) Theoretical effect of an inhibitor of the C to D step in the pathway of A → B ⇌ C → D ⇌ E → F. Steady-state concentrations of all intermediates of the pathway without the inhibitor present are arbitrarily set equal to 100%. Steady-state concentrations of all intermediates when the inhibitor is present are then expressed as percentages of the control values. (*b*) Effect of oxygen on the relative steady-state concentrations of the intermediates of the glycolytic pathway in the perfused rat heart. The changes in concentrations of metabolites of hearts perfused without oxygen caused by subsequent perfusion with oxygen (95% O_2 and 5% CO_2) are recorded as percentages of the anoxic values. Oxygen strongly inhibits glucose utilization and lactate production under such conditions. The dramatic increase in pyruvate concentration occurs as a consequence of greatly increased utilization of cytosolic NADH by the shuttle systems. Abbreviations: G6P, glucose 6-phosphate; F6P, fructose 6-phosphate; FBP, fructose 1,6-bisphosphate; DHAP, dihydroxyacetone phosphate; GAP, glyceraldehyde 3-phosphate; 3PG, 3-phosphoglycerate; 2PG, 2-phosphoglycerate; PEP, phosphoenolpyruvate; Pyr, pyruvate; and Lac, lactate.
Redrawn with permission from J. R. Williamson, *J. Biol. Chem.* 241:5026, 1966. © The American Society of Biological Chemists, Inc.

change greatly between these two conditions (in the experiment of Figure 7.17*b*, ATP increased from 4.7 μmol/g of wet weight in the absence of oxygen to 5.6 μmol/g of wet weight in the presence of oxygen). Since 6-phosphofructo-1-kinase is severely inhibited at concentrations of ATP (2.5–6 mM) normally present in cells, such a small difference in ATP concentration cannot account completely for the change in flux through 6-phosphofructo-1-kinase. However, much greater changes, percentage-wise, occur in the concentrations of AMP and P_i, both positive allosteric effectors of 6-phosphofructo-1-kinase. The changes that occur in the steady-state concentrations of AMP and P_i when oxygen is introduced into the system are exactly what might have been predicted, that is, the levels of both go down dramatically. These changes result in less 6-phosphofructo-1-kinase activity, greatly suppressed glycolytic activity, and account in large part for the Pasteur effect. The levels of AMP automatically go down in a cell when ATP levels increase. Although this is not intuitively obvious, the reason is simple. The sum of the adenine nucleotides in a cell, that is, ATP + ADP + AMP, is nearly constant under most physiological conditions, but the relative concentrations are such that the ATP concentration is always much greater than the AMP concentration. Furthermore, the adenine nucleotides are maintained in equilibrium in the cytosol through the action of **adenylate kinase** (also referred to as **myokinase**), which catalyzes the reaction 2ADP $\rightleftharpoons$ ATP + AMP. The equilibrium constant (K'_{eq}) for this reaction is given by

$$K'_{eq} = \frac{[\text{ATP}][\text{AMP}]}{[\text{ADP}]^2}$$

Since this reaction is "near equilibrium" under intracellular conditions, the concentration of AMP is given by

$$[\text{AMP}] = \frac{K'_{eq}[\text{ADP}]^2}{[\text{ATP}]}$$

Because intracellular [ATP] >> [ADP] >> [AMP], a small decrease in [ATP] causes a substantially greater percentage increase in [ADP]; and, since [AMP] is related to the square of [ADP], an even greater percentage increase in [AMP]. Because of this relationship, a small decrease in ATP concentration leads to a greater percent increase in [AMP] than in the percent decrease in [ATP]. This makes AMP an excellent signal of the energy status of the cell and allows it to function as an important allosteric effector of 6-phosphofructo-1-kinase activity. Furthermore, AMP influences in yet another way the effectiveness of the reaction catalyzed by 6-phosphofructo-1-kinase. An enzyme called **fructose 1,6-bisphosphatase** catalyzes an irreversible reaction, which opposes that of 6-phosphofructo-1-kinase:

$$\text{Fructose 1,6-bisphosphate} + H_2O \longrightarrow \text{fructose 6-phosphate} + P_i$$

This enzyme sits "cheek by jowl" with 6-phosphofructo-1-kinase in the cytosol of many cells. Together they catalyze a futile cycle (ATP $\rightarrow$ ADP + P_i + "heat"), and, at the very least, they decrease the "effectiveness" of one another. The AMP concentration is a perfect signal of the energy status of the cell—not only because AMP activates 6-phosphofructo-1-kinase but also because AMP *inhibits* fructose 1,6-bisphosphatase. The result is that a small decrease in ATP concentration triggers, via the increase in AMP concentration, a large increase in the net conversion of fructose 6-phosphate into fructose 1,6-bisphosphate. This increases the glycolytic flux by increasing the amount of substrate available for the splitting stage. In cells containing hexokinase, it also results in greater phosphorylation of glucose because a decrease in fructose 6-phosphate automatically causes a decrease in glucose 6-phosphate, which, in turn, results in less inhibition of hexokinase activity.

The decrease in lactate production in response to the onset of respiration is another feature of the Pasteur effect that can be readily explained. The most important factor is the decreased glycolytic flux caused by oxygen; however, secondary factors include competition between lactate dehydrogenase and the mitochondrial pyruvate dehydrogenase complex for pyruvate, as well as competition between lactate dehydrogenase and the shuttle systems for NADH. For the most part, lactate dehydrogenase loses the competition in the presence of oxygen.

Intracellular pH Can Regulate 6-Phosphofructo-1-kinase

It would be natural to suspect that lactate, as the end product of glycolysis, would inhibit the rate-limiting enzyme of the glycolytic pathway. It does not. However, hydrogen ions, the other glycolytic end product, do inhibit 6-phosphofructo-1-kinase. As shown in Figure 7.18, glycolysis in effect generates lactic acid, and the cell must dispose of it as such. This explains why excessive glycolysis in the body lowers blood pH and leads to an emergency medical situation termed *lactic acidosis* (see Clin. Corr. 7.5). Plasma membranes of cells contain a symport for lactate and hydrogen ions. Thus, lactic acid is released from the cell into the bloodstream. This ability to transport lactic acid out of the cell is a defense mechanism, preventing the pH from getting so low that everything becomes pickled (see Clin. Corr. 7.6). The sensitivity of 6-phosphofructo-1-kinase to hydrogen ions is also part of this mechanism. Hydrogen ions are able to shut off glycolysis, the process responsible for decreasing the pH. Note that transport of lactic acid out of a cell requires that the blood be available to the cell in order to carry this "end product" away. When blood flow to a group of cells is inadequate, for example, in heavy exercise of a skeletal muscle or an attack of angina pectoris in the case of the heart, hydrogen ions cannot escape from the cells fast enough. Yet, the need for ATP within such cells, because of the lack of oxygen, may partially override the inhibition of 6-phosphofructo-1-kinase by hydrogen ions. The unabated accumulation of hydrogen ions then results in pain, which, in the case of skeletal muscle, can be relieved by simply terminating the exercise. In the case of the heart, rest or pharmacologic agents that increase

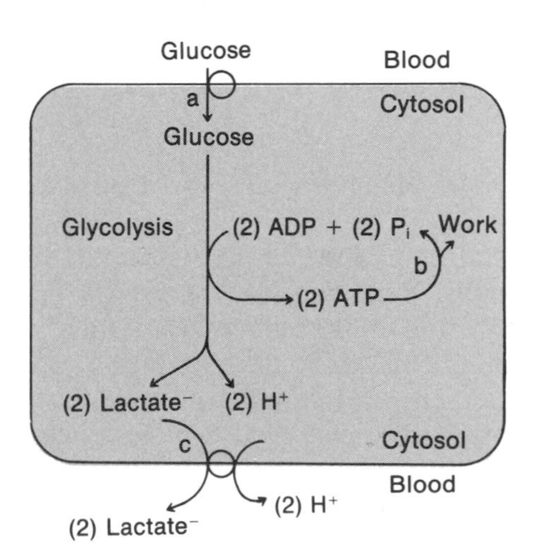

FIGURE 7.18
Unless lactate formed by glycolysis is released from the cell, the intracellular pH is decreased as a consequence of the accumulation of intracellular lactic acid.
The low pH decreases 6-phosphofructo-1-kinase activity so that further lactic acid production by glycolysis is shut off. (a) glucose transport into the cell; (b) all work performances that convert ATP back to ADP and P_i; (c) lactate-hydrogen ion symport (actual stoichiometry of one lactate$^-$ and one H^+ per symport).

CLIN. CORR. 7.5
LACTIC ACIDOSIS

This problem is characterized by elevated blood lactate levels, usually greater than 5 mM, along with decreased blood pH and bicarbonate concentrations. Lactic acidosis is the most commonly encountered form of metabolic acidosis and can be the consequence of overproduction of lactate, underutilization of lactate, or both. Lactate production is normally balanced by lactate utilization, with the result that lactate is usually not present in the blood at concentrations greater than 1.2 mM. All tissues of the body have the capacity to produce lactate by anaerobic glycolysis, but most tissues do not produce large quantities because much more ATP can be gained by the complete oxidation of the pyruvate produced by glycolysis. However, all tissues respond with an increase in lactate generation when oxygenation is inadequate. A decrease in ATP resulting from reduced oxidative phosphorylation allows the activity of 6-phosphofructo-1-kinase to increase. These tissues have to rely on anaerobic glycolysis for ATP production under such conditions and this results in lactic acid production. A good example is muscle exercise, which can deplete the tissue of oxygen and cause an overproduction of lactic acid. Tissue hypoxia occurs, however, in all forms of shock, during convulsions, and in diseases involving circulatory and pulmonary failure.

The major fate of lactate in the body is either complete combustion to CO_2 and H_2O or conversion back to glucose by the process of gluconeogenesis. Both require oxygen. Decreased oxygen availability, therefore, increases lactate production and decreases lactate utilization. The latter can also be decreased by liver diseases, ethanol, and a number of other drugs. Phenformin, a drug that was once used to treat the hyperglycemia of insulin-independent diabetes, was well-documented to induce lactic acidosis in certain patients.

Bicarbonate is usually administered in an attempt to control the acidosis associated with lactic acid accumulation. The key to successful treatment, however, is to find and eliminate the cause of the overproduction and/or underutilization of lactic acid and most often involves the restoration of circulation of oxygenated blood.

Kruse, J. A. and Carlson, R. W. Lactate metabolism. *Crit. Care Clin.* 3:725, 1985.

CLIN. CORR. 7.6
PICKLED PIGS AND MALIGNANT HYPERTHERMIA

In patients with malignant hyperthermia, a variety of agents, especially the widely used general anesthetic halothane, will produce a dramatic rise in body temperature, metabolic and respiratory acidosis, hyperkalemia, and muscle rigidity. This genetic abnormality occurs in about 1 in 15,000 children and 1 in 50,000–100,000 adults. It is dominantly inherited. Death often results the first time a susceptible person is anesthetized. Onset occurs within minutes of drug exposure and the hyperthermia must be recognized immediately. Packing the patient in ice is effective and should be accompanied by measures to combat acidosis. The drug dantrolene is also effective.

A phenomenon similar, if not identical, to malignant hyperthermia is known to occur in pigs. Pigs with this problem, called porcine stress syndrome, respond poorly to stress. This genetic disease usually manifests itself as the pig is being shipped to market. Pigs with the syndrome can be identified by exposure to halothane, which triggers the same response seen in patients with malignant hyperthermia. The meat of pigs that have died as a result of the syndrome is pale, watery, and of very low pH (i.e., nearly pickled).

Muscle is considered the site of the primary lesion in both malignant hyperthermia and porcine stress syndrome. In response to halothane the skeletal muscles become rigid and generate heat and lactic acid. Although much experimental work has been conducted, the biochemical basis for the increased heat production remains obscure. Heat produced by glycolytic activity and muscle contraction is not believed sufficient to explain the dramatic increase in body temperature. Uncontrolled futile cycling in which ATP hydrolysis is greatly accelerated has been suggested to be involved:

$$\text{ATP} + H_2O \longrightarrow \text{ADP} + P_i + \text{heat}$$

Indeed halothane has been shown to accelerate futile cycling at the level of 6-phosphofructo-1-kinase/fructose 1,6-bisphosphatase in muscles of pigs with porcine stress syndrome. Perhaps one of these regulatory enzymes will be found defective with respect to allosteric effector control in patients with malignant hyperthermia. There is also evidence that the sarcoplasmic reticulum of such patients may have abnormalities in a calcium channel and that the anesthetic triggers inappropriate release of Ca^{2+} from the sarcoplasmic reticulum. This could result in uncontrolled stimulation of a number of heat-producing processes, that is, myosin ATPase, glycogenolysis, glycolysis, and cyclic uptake and release of Ca^{2+} by mitochondria and sarcoplasmic reticulum.

Mickelson, J. R., Gallant, E. M., Litterer, L. A., Johnson, K. M., Rempel, W. E., and Louis, C. F. Abnormal sarcoplasmic reticulum ryanodine receptor in malignant hyperthermia. *J. Biol. Chem.* 263:9310, 1989; Mitchell, G., Heffron, J. J. A., and van Rensberg, A. J. J. A halothane-induced biochemical defect in muscle of normal and malignant hyperthermia-susceptible land-race pigs. *Anesth. Analg.* 59:250, 1980; Nelson, T. E. Sarcoplasmic reticulum function in malignant hyperthermia. *Cell Calcium* 9:257, 1988; and Symposium on Malignant Hyperthermia. *Br. J. Anesthesia* 60:253, 268, 274, 279, 287, 303, and 317, 1988.

blood flow or decrease the need for ATP within the myocytes may be effective (see Clin. Corr. 7.7).

Intracellular Citrate Levels Also Regulate 6-Phosphofructo-1-kinase

Many tissues prefer to use fatty acids and ketone bodies as oxidizable fuels in place of glucose. Most of these tissues have the capacity to use glucose but actually prefer to oxidize fatty acids and ketone bodies. This unselfish act helps preserve glucose for those tissues, such as brain, that are absolutely dependent on glucose as an energy source. The mechanism responsible for this preference is relatively simple. Oxidation of both fatty acids and ketone bodies elevates the levels of cytosolic citrate, which inhibits 6-phosphofructo-1-kinase. The result is decreased glucose utilization by the tissue when fatty acids or ketone bodies are available.

Hormonal Control of 6-Phosphofructo-1-kinase by cAMP and Fructose 2,6-bisphosphate

The structure of fructose 2,6-bisphosphate is given in Figure 7.19. This compound is probably present in all tissues, but its role in regulation of glycolysis is best understood for liver. Fructose 2,6-bisphosphate behaves like AMP in that it functions as a positive allosteric effector of 6-phosphofructo-1-kinase and as a negative allosteric effector of fructose

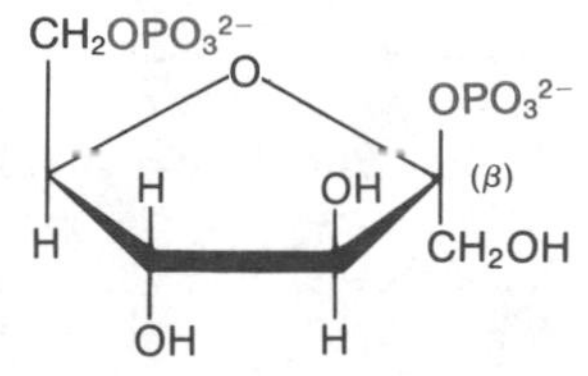

FIGURE 7.19
Structure of fructose 2,6-bisphosphate.

CLIN. CORR. 7.7 ANGINA PECTORIS AND MYOCARDIAL INFARCTION

Chest pain associated with reversible myocardial ischemia is termed angina pectoris (literally, strangling in the chest). The pain is the result of an imbalance between demand for and supply of blood flow to cardiac muscles and is most commonly caused by narrowing of the coronary arteries. The patient experiences a heavy squeezing pressure or ache substernally, often radiating to either the shoulder and arm or occasionally to the jaw or neck. Attacks occur with exertion, last from 1 to 15 min, and are relieved by rest. The coronary arteries involved are obstructed by atherosclerosis (i.e., lined with characteristic fatty deposits) or less commonly narrowed by spasm. Myocardial infarction occurs if the ischemia persists long enough to cause severe damage (necrosis) to the heart muscle. Commonly, a blood clot forms at the site of narrowing and completely obstructs the vessel. In myocardial infarction, tissue death occurs and the characteristic pain is longer lasting, and often more severe.

Nitroglycerin and other nitrates are frequently prescribed to relieve the pain caused by the myocardial ischemia of angina pectoris. These drugs can be used prophylactically, enabling patients to participate in activities that would otherwise precipitate an attack of angina. Nitroglycerin may work in part by causing dilation of the coronary arteries, improving oxygen delivery to the heart and washing out lactic acid. Probably more important is the effect of nitrates on the peripheral circulation. Nitrates relax smooth muscle, causing venodilation throughout the body. This reduces arterial pressure and allows blood to accumulate in the veins. The result is decreased return of blood to the heart, and a reduced volume of blood the heart has to pump, which reduces the energy requirement of the heart. In addition, the heart empties itself against less pressure, which also spares energy. The overall effect is a lowering of the oxygen requirement of the heart, bringing it in line with the oxygen supply via the diseased coronary arteries. Other useful agents are calcium channel blockers, which are coronary vasodilators, and β-adrenergic blockers. The β-blockers prevent the increase in myocardial oxygen consumption induced by sympathetic nervous system stimulation of the heart, as occurs with physical exertion.

The coronary artery bypass operation is

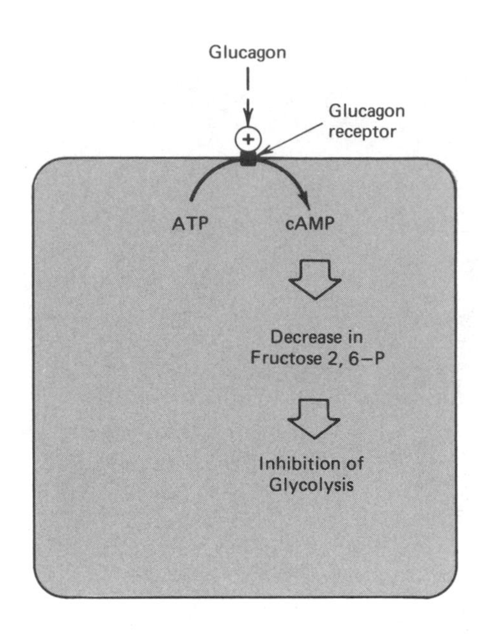

FIGURE 7.20
Overview of the mechanism responsible for glucagon inhibition of hepatic glycolysis.
The interaction of glucagon with its receptor (■) results in activation of adenylate cyclase and the production of cAMP.

1,6-bisphosphatase. Indeed, without the presence of this compound, glycolysis could not occur in the liver because 6-phosphofructo-1-kinase would have insufficient activity and fructose 1,6-bisphosphatase would have too much activity for net conversion of fructose 6-phosphate to fructose 1,6-bisphosphate.

Figure 7.20 gives a brief overview of the role of fructose-2,6-bisphosphate in hormonal control of hepatic glycolysis. Understanding this mechanism requires an appreciation of the role of **cAMP** (Figure 7.21) as the **"second messenger"** of hormone action. As discussed in more detail in Chapters 14 and 20, glucagon is released from the α cells of the pancreas and circulates in the blood until it comes in contact with glucagon receptors located on the outer surface of the liver plasma membrane (Figure 7.20). Binding of glucagon to these receptors is sensed by adenylate cy-

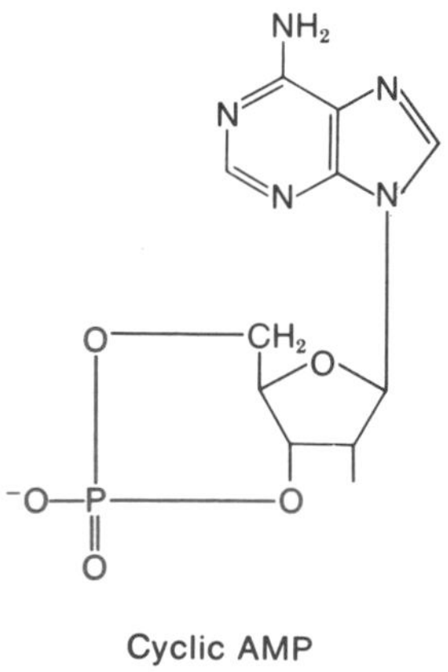

Cyclic AMP

FIGURE 7.21
Structure of cAMP.

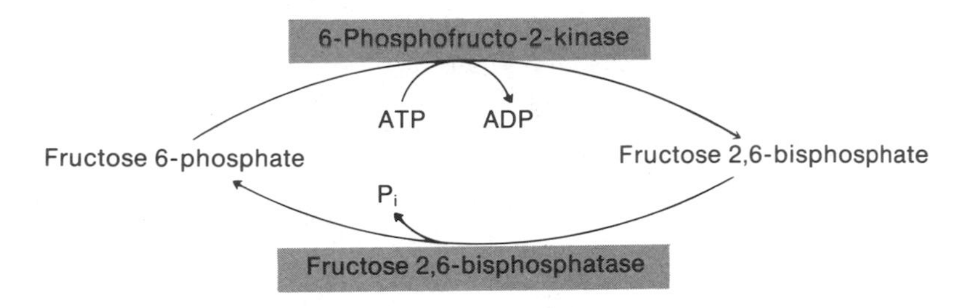

FIGURE 7.22
Reactions involved in the formation and degradation of fructose 2,6-bisphosphate.

used in severe cases of angina that cannot be controlled by medication. In this operation veins are removed from the leg and interposed between the aorta and coronary arteries of the heart. The purpose is to bypass the portion of the artery diseased by atherosclerosis and provide the affected tissue with a greater blood supply. Remarkable relief from angina can be achieved by this operation, with the patient being able to return to normal productive life in some cases.

Hugenholtz, P. G. Calcium antagonists for angina pectoris. *Ann. N.Y. Acad. Sci.* 522:565, 1988; and Silverman, K. J. and Grossman, M. Angina pectoris: natural history and strategies for evaluation and management. *N. Engl. J. Med.* 310:1712, 1984.

clase, an enzyme located on the inner surface of the plasma membrane, stimulating it to convert cytosolic ATP into cytosolic cAMP and PP_i. Cyclic AMP triggers a series of intracellular events, the details of which are discussed below, that result ultimately in a decrease in fructose 2,6-bisphosphate levels. A decrease in this compound makes 6-phosphofructo-1-kinase less effective but makes fructose 1,6-bisphosphatase more effective, thereby severely restricting flux from fructose 6-phosphate to fructose 1,6-bisphosphate in the glycolytic pathway.

Note that fructose 2,6-bisphosphate is not an intermediate of the glycolytic pathway. As shown in Figure 7.22, fructose 2,6-bisphosphate is produced from F6P by an enzyme called **6-phosphofructo-2-kinase** (also referred to as **phosphofructokinase-2**). We now have two "phosphofructokinases" to contend with, one producing an intermediate (FBP) of the glycolytic pathway and the other producing an important allosteric effector (fructose 2,6-bisphosphate) of the first enzyme. Before the discovery of fructose 2,6-bisphosphate, 6-phosphofructo-1-kinase could be called phosphofructokinase or even PFK among friends. Now we must carefully distinguish between these two very important enzymes.

Fructose 2,6-bisphosphate is destroyed in cells by being converted back to F6P by **fructose 2,6-bisphosphatase** (Figure 7.22). This is a simple hydrolysis, with no ATP or ADP being involved, in contrast to fructose 2,6-bisphosphate synthesis. An interesting fact is that a bifunctional enzyme carries out both the synthesis and degradation of fructose 2,6-bisphosphate. You may recall that a bifunctional enzyme is responsible for the synthesis and degradation of 2,3-bisphosphoglycerate (see page 302) and you will find in Chapter 9 that a multifunctional enzyme (fatty acid synthase) catalyzes numerous reactions during the process of fatty acid synthesis. Because of its bifunctional nature, the combined name of 6-phosphofructo-2-kinase/fructose 2,6-bisphosphatase is used to refer to the enzyme that makes and degrades fructose 2,6-bisphosphate. As mentioned above, cAMP is responsible for regulation of fructose 2,6-bisphosphate levels in the liver. How is this possible when the same enzyme carries out both the synthesis and degradation of the molecule? The answer is that a mechanism exists whereby cAMP is able to inactivate the kinase function and, at the same time, activate the phosphatase function of this bifunctional enzyme.

cAMP Activates a Protein Kinase

Cyclic AMP is an activator of an enzyme called **cAMP-dependent protein kinase.** This enzyme—in its inactive state—consists of two regulatory subunits of mol wt 85,000 each plus two catalytic subunits of mol wt 40,000 each. Cyclic AMP binds only to the regulatory subunits. Binding of

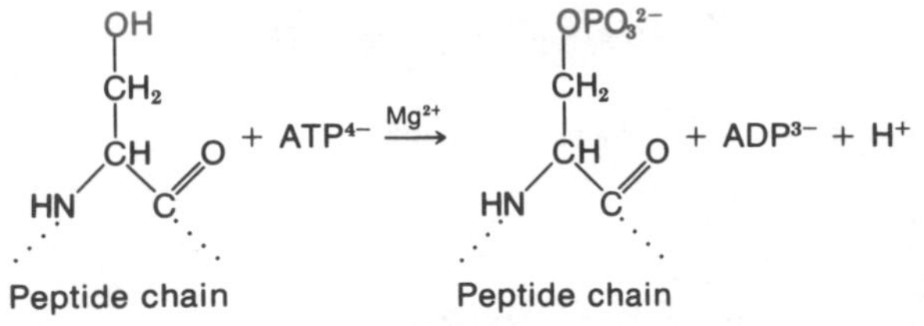

$$\text{HN–CH(CH}_2\text{OH)–C(=O) (peptide chain)} + ATP^{4-} \xrightarrow{Mg^{2+}} \text{HN–CH(CH}_2\text{OPO}_3^{2-}\text{)–C(=O) (peptide chain)} + ADP^{3-} + H^+$$

Peptide chain   Peptide chain

FIGURE 7.23
Enzymes subject to covalent modification are usually phosphorylated on specific serine residues.
Tyrosine and threonine residues are also important sites of covalent modification by phosphorylation.

cAMP causes conformational changes in the regulatory subunits, which sets the catalytic subunits free. The catalytic subunits are active only after being dissociated from the regulatory subunit by this action of cAMP. The liberated protein kinase then catalyzes the phosphorylation of specific serine residues of the polypeptide chains of several different enzymes (Figure 7.23).

Phosphorylation of an enzyme can conveniently be abbreviated as

$$\boxdot + ATP \longrightarrow \odot\text{–P} + ADP$$

where $\boxdot$ and $\odot$–P are used to indicate the dephosphorylated and phosphorylated enzymes, respectively. The circle and square symbols are used because phosphorylation of enzymes subject to regulation by covalent modification causes a change in their conformation, which affects the active site. It turns out that the change in conformation due to phosphorylation greatly increases the catalytic activity of some enzymes but greatly decreases the catalytic activity of others. It depends on the enzyme involved. A number of enzymes are subject to this type of regulation, called covalent modification. Regardless of whether phosphorylation or dephosphorylation activates the enzyme, the active form of the enzyme is called the "a" form and the inactive form the "b" form. Likewise, regardless of the effect of phosphorylation on catalytic activity, the action of cAMP-dependent protein kinase is always opposed by that of a **phosphoprotein phosphatase,** which catalyzes the reaction of

$$\odot\text{–P} + H_2O \longrightarrow \boxdot + P_i$$

Putting these together creates a cyclic control system (see Figure 7.24), such that the ratio of phosphorylated enzyme to dephosphorylated enzyme is a function of the relative activities of the cAMP-dependent protein kinase and the phosphoprotein phosphatase. If the kinase has greater activity than the phosphatase, more enzyme will be in the phosphorylated mode, and vice versa. Since the activity of an interconvertible enzyme (i.e., an enzyme subject to covalent modification) is determined by whether it is in the phosphorylated or dephosphorylated mode, the relative activities of the kinase and phosphatase determine the amount of a particular enzyme which is in the catalytically active state.

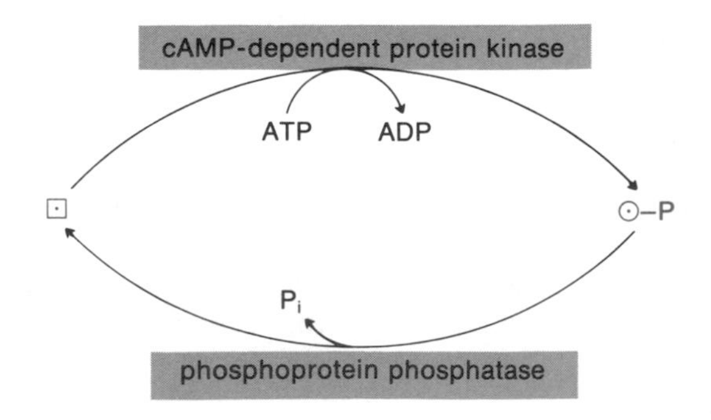

FIGURE 7.24
General model for the mechanism responsible for regulation of enzymes by phosphorylation–dephosphorylation.
The symbols $\boxdot$ and $\odot$–P indicate that different conformational and activity states of the enzyme are produced as a result of phosphorylation–dephosphorylation.

6-Phosphofructo-2-kinase and Fructose 2,6-bisphosphatase Are Part of a Bifunctional Enzyme Regulated by Phosphorylation-Dephosphorylation

As discussed above, most enzymes are either turned on or off by phosphorylation but, in the case of 6-phosphofructo-2-kinase and fructose 2,6-bisphosphatase, advantage is taken of the *bifunctional* nature of the enzyme. Phosphorylation causes inactivation of the active site responsible for synthesis of fructose 2,6-bisphosphate but activation of the active site responsible for hydrolysis of fructose 2,6-bisphosphate. Dephosphorylation of the enzyme has the opposite effects (Figure 7.25). A sensitive mechanism has evolved, therefore, to set the intracellular concentration of fructose 2,6-bisphosphate in response to changes in blood levels of glucagon (Figure 7.26). Increased levels of glucagon cause an increase in intracellular levels of cAMP. The second messenger activates cAMP-dependent protein kinase, which, in turn, phosphorylates 6-phosphofructo-2-kinase/fructose 2,6-bisphosphatase. The latter event inhibits fructose 2,6-bisphosphate synthesis and promotes its degradation. The resulting decrease in fructose 2,6-bisphosphate makes 6-phosphofructo-1-kinase less effective and fructose 1,6-bisphosphatase more effective. The overall result is inhibition of glycolysis at the level of the conversion of fructose 6-phosphate to fructose 1,6-bisphosphate. Decreased levels of glucagon in the blood result in less cAMP in the liver because adenylate cyclase is less active and the cAMP that had accumulated is converted to AMP by the action of cAMP phosphodiesterase (Figure 7.27). Loss of the cAMP signal results in inactivation of cAMP-dependent protein kinase and a corresponding decrease in the rate of phosphorylation of 6-phosphofructo-2-kinase/fructose 2,6-bisphosphatase by cAMP-dependent protein kinase. A phosphoprotein phosphatase removes phosphate from the bifunctional enzyme to produce active 6-phosphofructo-2-kinase and inactive fructose 2,6-bisphosphatase. Fructose 2,6-bisphosphate can now accumulate to a higher steady-state concentration and, by activating 6-phosphofructo-1-kinase and inhibiting fructose 1,6-bisphosphatase, greatly increase the rate of glycolysis. It should be apparent from this discussion that glucagon is an extracellular signal that stops the liver from using glucose, whereas fructose 2,6-bisphosphate is an intracellular signal that promotes glucose utilization by this tissuc.

The role of **insulin** in regulation of fructose 2,6-bisphosphate levels is not well understood. Although it is clear that this hormone opposes the action of glucagon, exactly how insulin works after binding to the plasma membrane remains to be established (see Chapter 20). One hypothesis is presented in Figure 7.27. The idea is that insulin binding may promote the formation of intracellular secondary mediators, much as **glucagon** promotes the formation of its intracellular messenger, cAMP. Obvious en-

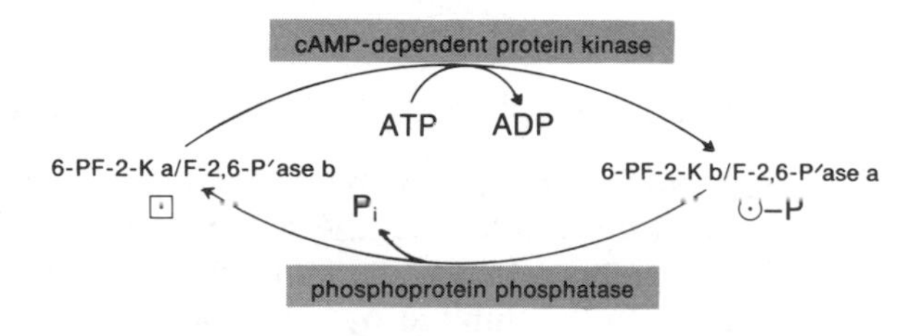

FIGURE 7.25
Mechanism responsible for covalent modification of the bifunctional enzyme 6-phosphofructo-2-kinase/fructose 2,6-bisphosphatase.
Name of the enzyme is abbreviated as 6-PF-2-K/F-2,6-P. Letters *a* and *b* indicate the active and inactive forms of the enzymes respectively.

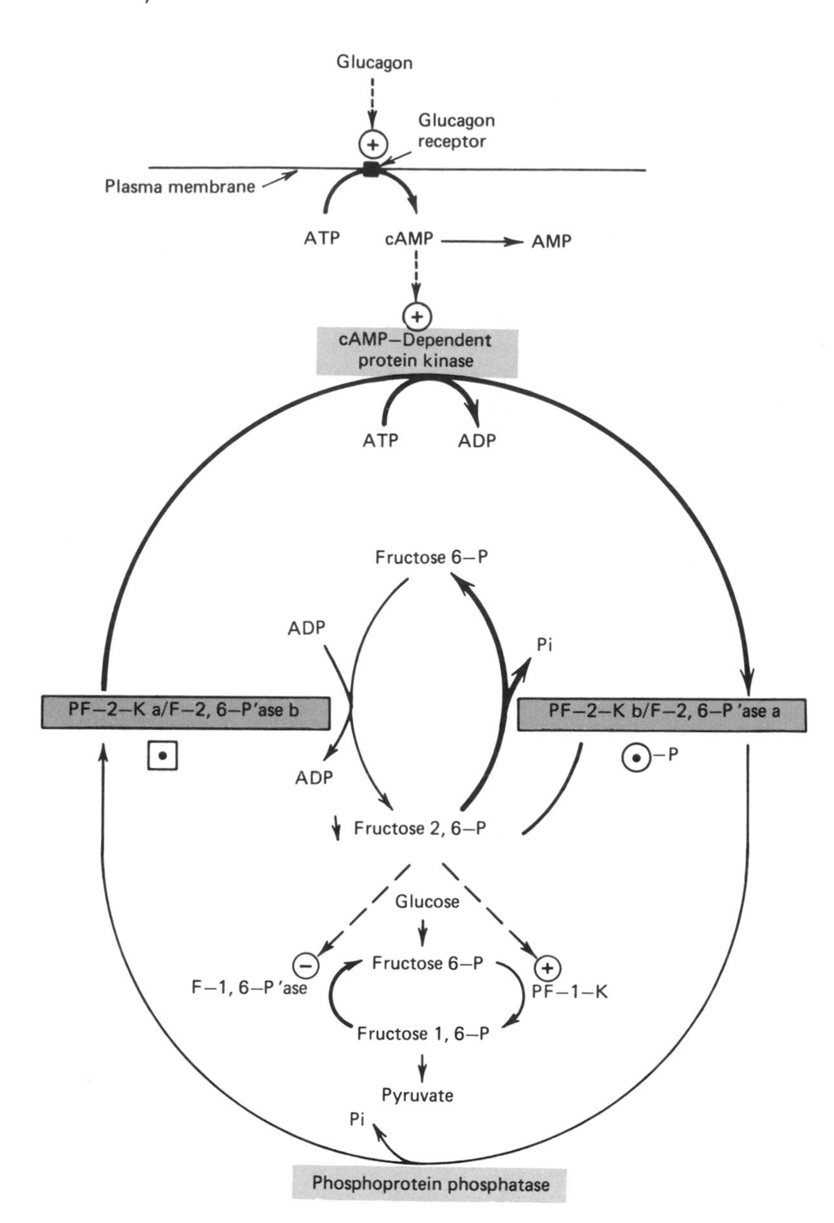

FIGURE 7.26
Mechanism responsible for glucagon inhibition of hepatic glycolysis via cAMP-mediated decrease in fructose 2,6-bisphosphate concentration.
The glucagon receptor (■) and adenylate cyclase are intrinsic components of the plasma membrane. The (+) and (−) symbols indicate activation and inhibition of the designated enzymes, respectively. The heavy arrows indicate the reactions that predominate in the presence of glucagon. The small arrow (↓) in front of fructose 2,6-bisphosphate indicates a decrease in concentration of this compound in response to glucagon.

zyme targets that secondary mediators might influence include cAMP phosphodiesterase, cAMP-dependent protein kinase, and phosphoprotein phosphatase (Figure 7.27). Regardless of the exact mechanism of action of insulin, glucagon and insulin clearly act in opposition to one another, and the insulin/glucagon ratio of the blood determines intracellular levels of fructose 2,6-bisphosphate and, therefore, the rate of glycolysis.

Pyruvate Kinase Is Also a Regulated Enzyme of Glycolysis

Pyruvate kinase is another regulatory enzyme of glycolysis (see Clin. Corr. 7.8). However, as with hexokinase, the reaction catalyzed by pyruvate kinase has to be considered a secondary site of regulation of glycolysis. This enzyme is drastically inhibited by physiological concentrations of ATP, so much so that its potential activity is never fully realized under physiological conditions. The isoenzyme found in liver is greatly activated by fructose 1,6-bisphosphate, thereby linking regulation of pyruvate kinase to what is happening at the level of 6-phosphofructo-1-

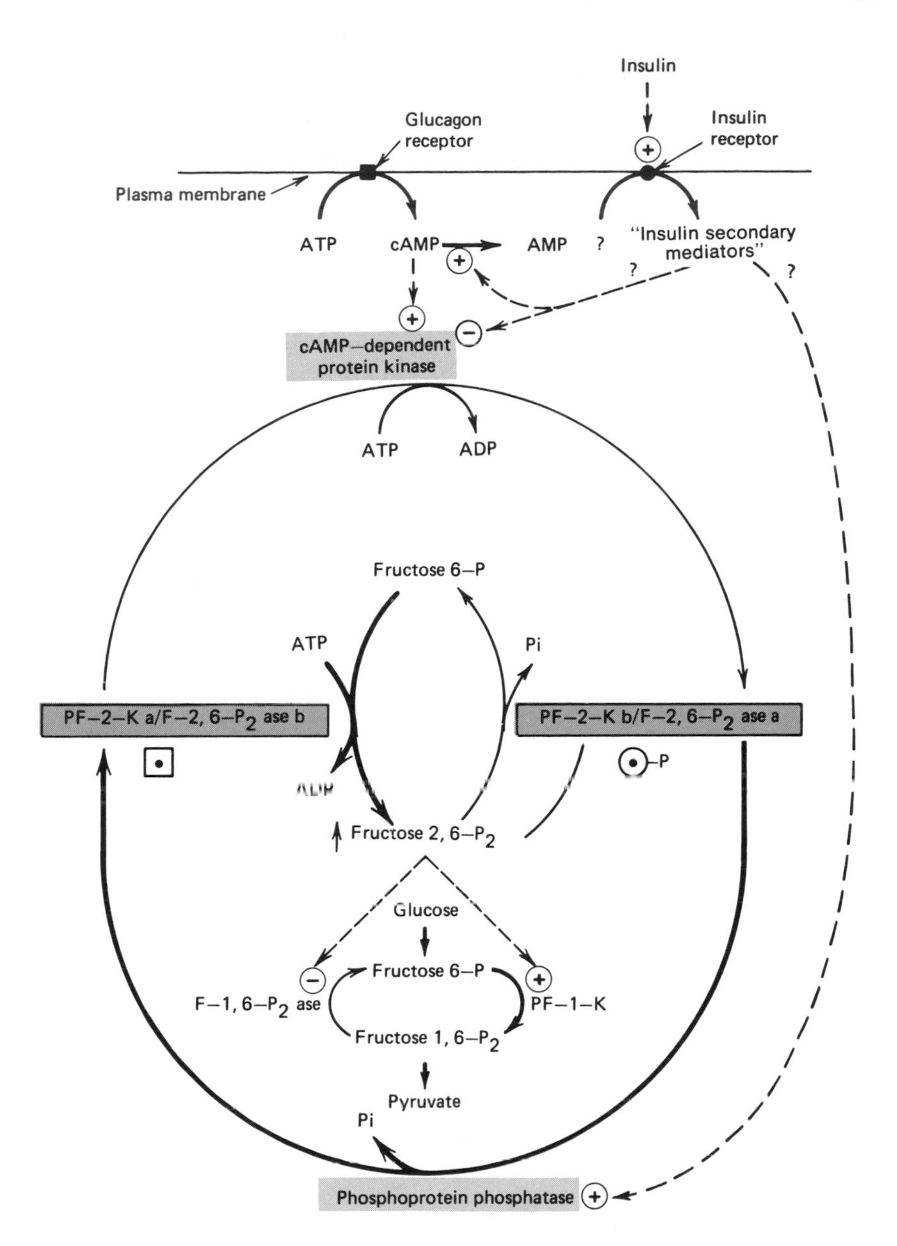

FIGURE 7.27
Mechanism responsible for accelerated rates of hepatic glycolysis when the concentration of glucagon is low and that of insulin is high in the blood.
See legend to Figure 7.26 for the meaning of symbols and heavy arrows. The insulin receptor (●) is an intrinsic component of the plasma membrane. The small arrow in front of fructose 2,6-bisphosphate indicates an increase in concentration of this compound in response to an increased insulin signal and decreased glucagon signal. The question marks indicate that the details of the mechanism of action of insulin are unknown at this time.

kinase. Thus, if conditions favor increased flux through 6-phosphofructo-1-kinase, the level of FBP increases and acts as a feed-forward activator of pyruvate kinase. The liver enzyme is also subject to covalent modification, being active in the dephosphorylated state and inactive in the phosphorylated state (Figure 7.28). Inactivation of pyruvate kinase by phosphorylation is a function of cAMP-dependent protein kinase in the liver. Glucagon inhibition of hepatic glycolysis and stimulation of hepatic gluco-

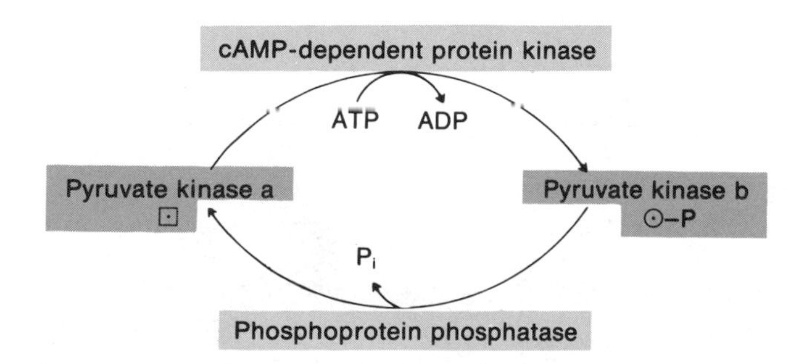

FIGURE 7.28
Glucagon acts via cAMP to cause the phosphorylation and inactivation of hepatic pyruvate kinase.

CLIN. CORR. 7.8
PYRUVATE KINASE DEFICIENCY AND HEMOLYTIC ANEMIA

Mature erythrocytes are absolutely dependent on glycolytic activity for ATP production. ATP is needed for the ion pumps, especially the Na^+-K^+-ATPase, which maintain the biconcave disk shape of erythrocytes, a characteristic that helps erythrocytes slip through the capillaries as they deliver oxygen to the tissues. Without ATP the cells swell and lyse. Anemia due to excessive erythrocyte destruction is referred to as hemolytic anemia. Pyruvate kinase deficiency is rare, but is by far the most common genetic defect of the glycolytic pathway known to cause hemolytic anemia. Most pyruvate kinase-deficient patients have 5–25% of normal red blood cell pyruvate kinase levels and flux through the glycolytic pathway is restricted severely, resulting in markedly lower ATP concentrations. The expected crossover of the glycolytic intermediates is observed, that is, those intermediates proximal to the pyruvate kinase-catalyzed step accumulate, whereas pyruvate and lactate concentrations decrease. Normal ATP levels are observed in reticulocytes of patients with this disease. Although deficient in pyruvate kinase, these "immature" red blood cells have mitochondria and can generate ATP by oxidative phosphorylation. Maturation of reticulocytes into red blood cells results in the loss of mitochondria and complete dependence on glycolysis for ATP production. Since glycolysis is defective, the mature cells are lost rapidly from the circulation. Anemia results because the cells cannot be replaced rapidly enough by erythropoiesis.

Valentine, W. N. The Stratton lecture: Hemolytic anemia and inborn errors of metabolism. *Blood* 54:549, 1979.

neogenesis are explained in part by the elevation of cAMP levels caused by this hormone. This aspect is explored more thoroughly in Section 7.5 (Gluconeogenesis).

Pyruvate kinase, like glucokinase, is induced to higher steady-state concentrations in the liver by the combination of high carbohydrate intake and high insulin levels. This increase in enzyme concentration is a major reason why the liver of the well-fed individual has much greater capacity for utilizing carbohydrate than a fasting or diabetic person (see Clin. Corr. 7.4).

7.5 GLUCONEOGENESIS

Glucose Synthesis Is Required for Survival of Humans and Other Animals

The net synthesis or formation of glucose from a large variety of noncarbohydrate substrates is termed **gluconeogenesis.** This includes the use of various amino acids, lactate, pyruvate, propionate and glycerol, as sources of carbon for the pathway (see Figure 7.29). Glucose is also synthesized from galactose and fructose. **Glycogenolysis,** that is, the formation of glucose or glucose 6-phosphate from glycogen, should be carefully differentiated from gluconeogenesis; glycogenolysis refers to

$$\text{Glycogen or (glucose)}_n \longrightarrow n \text{ molecules of glucose}$$

and thus does not correspond to de novo or new synthesis of glucose, the hallmark of the process of gluconeogenesis.

The capacity to synthesize glucose is crucial for the survival of humans and other animals. Blood glucose levels have to be maintained to support metabolism of those tissues that use glucose as their primary substrate (see Clin. Corr. 7.9). This includes brain, red blood cells, kidney medulla, lens and cornea of the eye, testis, and a number of other tissues. Gluconeogenesis enables the maintenance of blood glucose levels long after all dietary glucose has been absorbed and completely oxidized.

Substrate Cycles Between Tissues Are Required: The Cori and Alanine Cycles

Two important cycles between tissues are recognized in gluconeogenesis. The **Cori cycle** and the **alanine cycle,** given in Figure 7.30, consist of gluconeogenesis in the liver followed by transport of glucose to a peripheral tissue. The purpose of both is to provide a mechanism for continuously supplying glucose to tissues that are dependent on it as their primary energy source. The cycles are only functional, however, between the liver and tissues that do not completely oxidize glucose to CO_2 and H_2O. In order to participate in these cycles, the peripheral tissue must release either alanine or lactate as the end product of glycolysis. The type

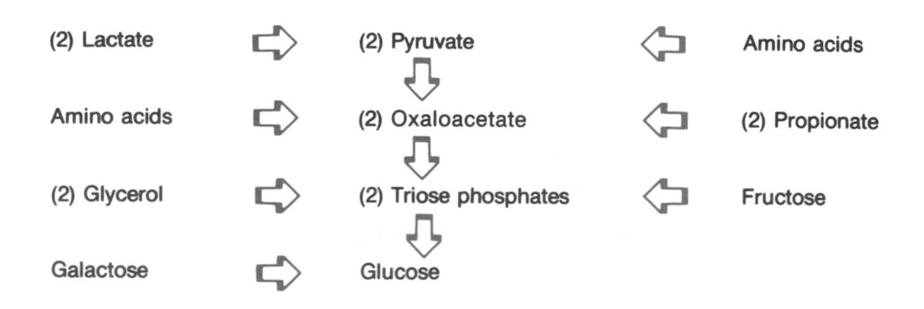

FIGURE 7.29
Abbreviated pathway of gluconeogenesis, illustrating the major substrate percursors for the process.

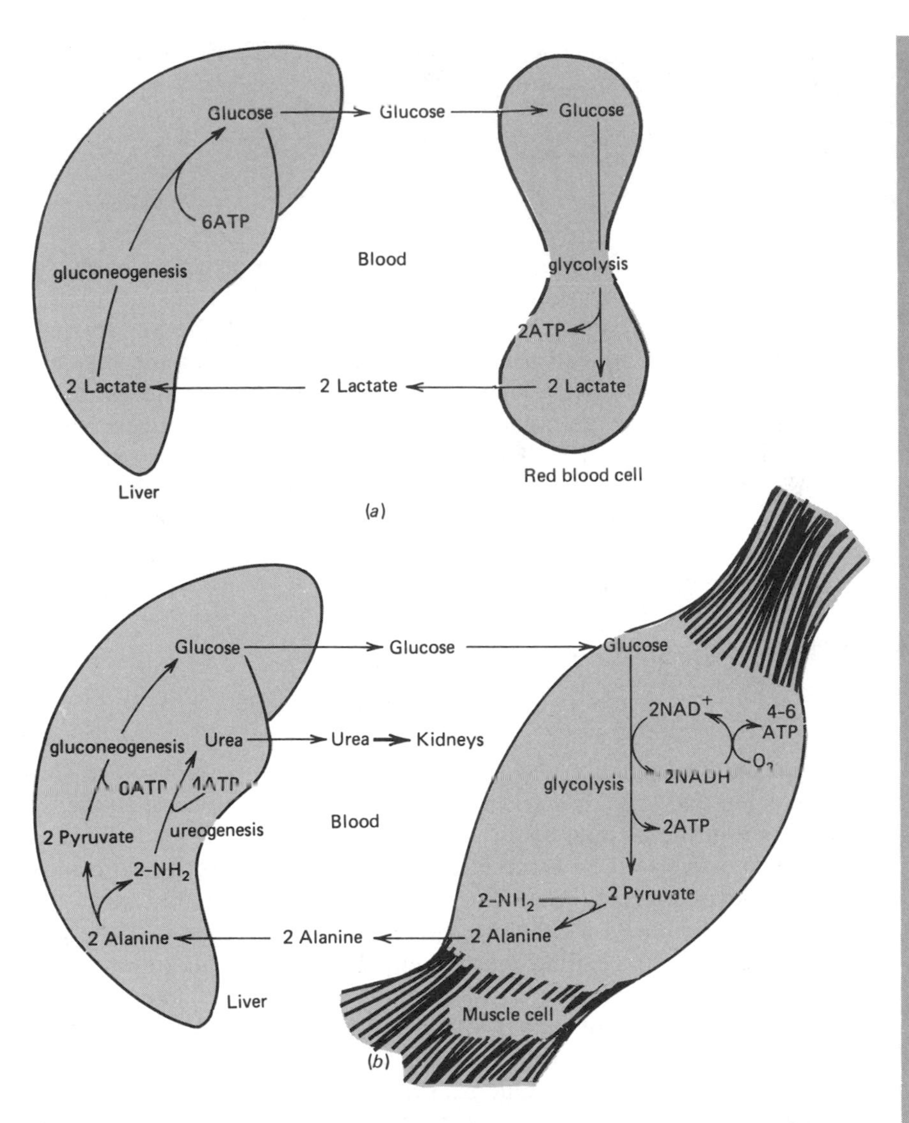

FIGURE 7.30
Relationship between gluconeogenesis in the liver and glycolysis in the rest of the body.
(*a*) Cori cycle. (*b*) Alanine cycle.

of recycled three-carbon intermediate is the major difference between the Cori cycle and the alanine cycle, carbon returning to the liver in the form of lactate in the Cori cycle but in the form of alanine in the alanine cycle. Another difference is that the NADH generated by glycolysis in the alanine cycle cannot be used to reduce pyruvate to lactate. In tissues that have mitochondria, the electrons of NADH can be transported into the mitochondria by the malate–aspartate shuttle or the glycerol phosphate shuttle for the synthesis of ATP by oxidative phosphorylation:

$$NADH + H^+ + \tfrac{1}{2}O_2 + 3ADP + 3P_i \longrightarrow NAD^+ + 3ATP$$

or

$$FADH_2 + \tfrac{1}{2}O_2 + 2ADP + 2P_i \longrightarrow FAD + 2ATP$$

The consequence is that six to eight molecules of ATP can be formed per glucose molecule in peripheral tissues that participate in the alanine cycle. This stands in contrast to the Cori cycle, in which only two molecules of

CLIN. CORR. 7.9 HYPOGLYCEMIA AND PREMATURE INFANTS

Premature and small-for-gestational-age neonates have a greater susceptibility to hypoglycemia than full-term, appropriate-for-gestational-age infants. Several factors appear to be involved. Children in general are more susceptible than adults to hypoglycemia, simply because they have larger brain/body weight ratios and the brain utilizes disproportionately greater amounts of glucose than the rest of the body. Newborn infants have a limited capacity for ketogenesis, apparently because the transport of long-chain fatty acids into liver mitochondria of the neonate is poorly developed. Since ketone body use by the brain is directly proportional to the circulating ketone body concentration, the neonate is unable to spare glucose to any significant extent by using ketone bodies. The consequence is that the neonate's brain is almost completely dependent on glucose obtained from liver glycogenolysis and gluconeogenesis.

The capacity for hepatic glucose synthesis from lactate and alanine is also limited in newborn infants. This is because the rate-limiting enzyme phosphoenolpyruvate carboxykinase is present in very low amounts during the first few hours after birth. Induction of this enzyme to the level required to prevent hypoglycemia during the stress of fasting requires several hours. Premature and small-for-gestational-age infants are believed to be more susceptible to hypoglycemia than normal infants because of smaller stores of liver glycogen. Fasting depletes their glycogen stores more rapidly, making these neonates more dependent on gluconeogenesis than normal infants.

Ballard, F. J. The development of gluconeogenesis in rat liver: controlling factors in the newborn. *Biochem. J.* 124:265, 1971.

ATP per molecule of glucose are produced. Inspection of Figure 7.30*a* will reveal that the overall stoichiometry for the Cori cycle is

$$6ATP_{liver} + 2(ADP + P_i)_{red\ blood\ cells} \longrightarrow 6(ADP + P_i)_{liver} + 2ATP_{red\ blood\ cells}$$

The six molecules of ATP are needed in the liver to provide the energy necessary for glucose synthesis. The alanine cycle also transfers the energy from liver to peripheral tissues and, because of the six to eight molecules of ATP produced per molecule of glucose, is an energetically more efficient cycle. However, as shown in Figure 7.30*b*, the participation of alanine in the cycle presents the liver with amino nitrogen, which must be disposed of as urea. In terms of ATP, urea synthesis is expensive (four ATP molecules per urea molecule). The concurrent need for urea synthesis results in more ATP being needed per glucose molecule synthesized in the liver. The overall stoichiometry for the alanine cycle, as presented in Figure 7.30*b*, is then

$$10ATP_{liver} + 6\text{–}8(ADP + P_i)_{muscle} + O_{2\ muscle} \longrightarrow 10(ADP + P_i)_{liver} + 6\text{–}8ATP_{muscle}$$

Note that the last equation makes the point that, in contrast to the Cori cycle, oxygen and mitochondria are required in the peripheral tissue for participation in the alanine cycle.

Liver was used as the example in Figure 7.30 because it is the most important gluconeogenic tissue. The kidneys, on a wet weight basis, have about the same capacity for the process. However, the liver is the largest organ in the body, exceeding the combined weight of the kidneys by a factor of 4, and thus contributes much more to the maintenance of blood glucose levels by gluconeogenesis. Certain muscle fibers may have the capacity for limited gluconeogenesis. Since the adult human has 18 times more muscle mass than liver, glucose synthesis in muscle may eventually be shown to be quantitatively important. However, muscle tissue lacks glucose 6-phosphatase, the enzyme that catalyzes the last step of the gluconeogenic pathway. Thus, any gluconeogenesis occurring in muscle should be pictured as taking place in order to help replenish glycogen stores in this tissue, rather than for the production of free glucose for the maintenance of blood sugar levels.

Pathway of Glucose Synthesis From Lactate

Gluconeogenesis from lactate is an ATP-requiring process with the overall equation of

$$2\ \text{L-Lactate}^- + 6ATP^{4-} \longrightarrow \text{glucose} + 6ADP^{3-} + 6P_i^{2-} + 4H^+$$

Many of the enzymes of the glycolytic pathway are common to the gluconeogenic pathway but, it is obvious from the overall equation for glycolysis,

$$\text{Glucose} + 2ADP^{3-} + 2P_i^{2-} \longrightarrow 2\ \text{L-lactate}^- + 2ATP^{4-}$$

that additional reactions have to be involved. Also, as pointed out in the discussion of glycolysis, certain steps of this pathway are irreversible under intracellular conditions and are replaced by irreversible steps of the gluconeogenic pathway. The reactions of gluconeogenesis from lactate

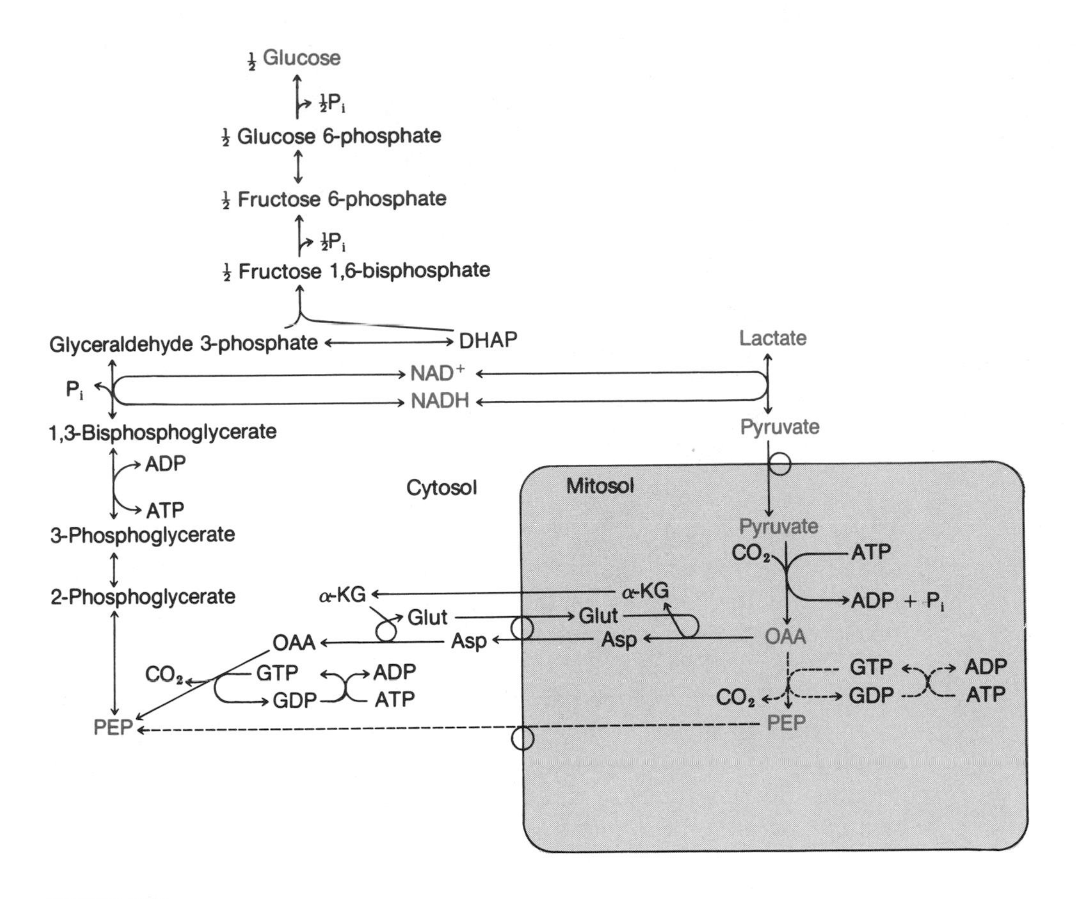

FIGURE 7.31
Pathway of gluconeogenesis from lactate.
The involvement of the mitochondrion in the process is indicated in the figure. Dashed arrows refer to an alternate route, which employs mitosolic phosphoenolpyruvate carboxykinase rather than the cytosolic isoenzyme. Abbreviations: OAA, oxaloacetate; α-KG, α-ketoglutarate; PEP, phosphoenolpyruvate; and DHAP, dihydroxyacetone phosphate.

are given in Figure 7.31. The initial step is the conversion of lactate to pyruvate by lactate dehydrogenase. NADH is generated and is also needed for a subsequent step in the pathway. Pyruvate cannot be converted to phosphoenolpyruvate (PEP) by reversing the step used in glycolysis because the reaction catalyzed by pyruvate kinase is irreversible under intracellular conditions. Pyruvate is converted into the high-energy phosphate compound PEP by the coupling of two reactions requiring high-energy phosphate compounds (an ATP and a GTP). The first is catalyzed by **pyruvate carboxylase** and the second by **phosphoenolpyruvate carboxykinase** (see Figure 7.32).

Pyruvate Carboxylase and Phosphoenolpyruvate Carboxykinase

Since the GTP required for the PEP carboxykinase catalyzed reaction is equivalent to an ATP through the action of nucleoside diphosphate kinase (GDP + ATP $\rightleftharpoons$ GTP + ADP), and since CO_2 and HCO_3^- readily equilibrate by the action of carbonic anhydrase ($CO_2 + H_2O \rightleftharpoons H_2CO_3 \rightleftharpoons H^+ + HCO_3^-$), the sum of these reactions is

$$\text{Pyruvate}^- + 2\text{ATP}^{4-} \longrightarrow \text{phosphoenolpyruvate}^{3-} + 2\text{ADP}^{3-} + 2\text{P}_i^{2-} + 4\text{H}^+$$

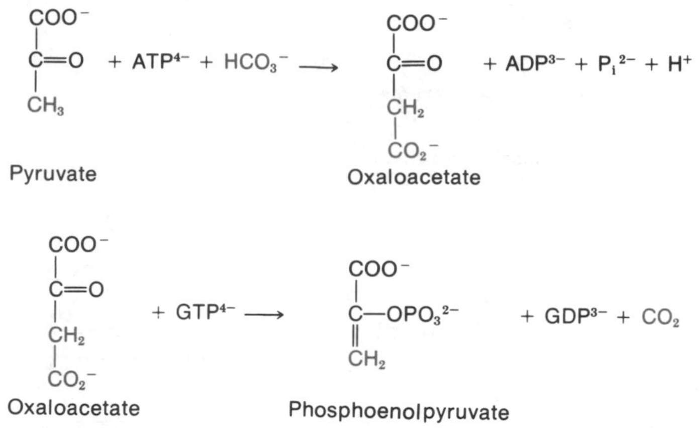

$$COO^- - C(=O) - CH_3 + ATP^{4-} + HCO_3^- \longrightarrow COO^- - C(=O) - CH_2 - CO_2^- + ADP^{3-} + P_i^{2-} + H^+$$

Pyruvate → Oxaloacetate

$$COO^- - C(=O) - CH_2 - CO_2^- + GTP^{4-} \longrightarrow COO^- - C(-OPO_3^{2-}) = CH_2 + GDP^{3-} + CO_2$$

Oxaloacetate → Phosphoenolpyruvate

FIGURE 7.32
Energy requiring steps involved in phosphoenolpyruvate formation from pyruvate. Reactions are catalyzed by pyruvate carboxylase and phosphoenolpyruvate carboxykinase, respectively.

Whereas the conversion of PEP to pyruvate by the enzyme pyruvate kinase yields the cell one molecule of ATP, the conversion of pyruvate into PEP by the combination of pyruvate carboxylase and PEP carboxykinase costs the cell two molecules of ATP.

As shown in Figure 7.31, the conversion of cytosolic pyruvate into cytosolic PEP requires the participation of the mitochondrion. Pyruvate carboxylase is housed within the mitochondrion, making these particles mandatory for glucose synthesis. There are two routes that oxaloacetate can then take to glucose—and both are important in human liver. This happens because PEP carboxykinase occurs in both the cytosolic and mitosolic compartments. The simplest pathway to follow is the one involving the mitochondrial PEP carboxykinase. In this case, oxaloacetate is simply converted within the mitochondrion into PEP, which then traverses the mitochondrial inner membrane in search of the rest of the enzymes of the gluconeogenic pathway. The second pathway would also be simple if oxaloacetate could traverse the mitochondrial inner membrane to reach the cytosolic PEP carboxykinase; however, as already discussed with respect to the malate–aspartate shuttle (Figure 7.10), oxaloacetate per se cannot escape from the mitochondrion. Thus, the trick is again used, as in the malate–aspartate shuttle (Figure 7.10*b*), of converting oxaloacetate into aspartate that traverses the mitochondrial inner membrane by way of the aspartate–glutamate antiport. Aspartate is converted back to oxaloacetate in the cytosol by transamination with α-ketoglutarate.

Gluconeogenesis Uses Some of the Glycolytic Enzymes but in the Reverse Direction

The steps from PEP to fructose 1,6-bisphosphate are already familiar, being just the reverse of steps of the glycolytic pathway. Note that the NADH generated by lactate dehydrogenase is utilized by the reaction

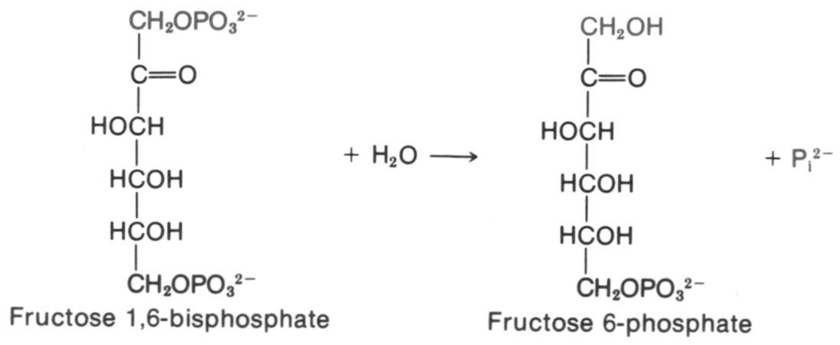

$$CH_2OPO_3^{2-} - C(=O) - HOCH - HCOH - HCOH - CH_2OPO_3^{2-} + H_2O \longrightarrow CH_2OH - C(=O) - HOCH - HCOH - HCOH - CH_2OPO_3^{2-} + P_i^{2-}$$

Fructose 1,6-bisphosphate → Fructose 6-phosphate

FIGURE 7.33
Reaction catalyzed by fructose 1,6-bisphosphatase.

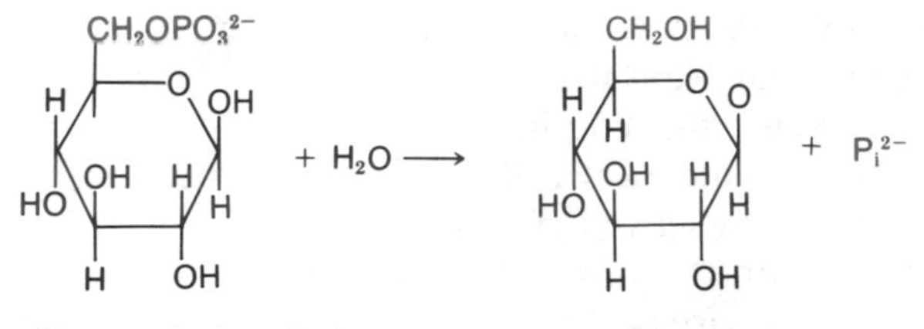

$$\alpha\text{-D-Glucose 6-phosphate} + H_2O \longrightarrow \alpha\text{-D-Glucose} + P_i^{2-}$$

FIGURE 7.34
Reaction catalyzed by glucose 6-phosphatase.

catalyzed by glyceraldehyde 3-phosphate dehydrogenase. This is tidy—nothing is left over and nothing extra is required.

6-Phosphofructo-1-kinase catalyzes an irreversible step in the glycolytic pathway and cannot be used for the conversion of FBP to fructose 6-phosphate. A way around this problem is offered by the enzyme **fructose 1,6-bisphosphatase,** which catalyzes the irreversible reaction shown in Figure 7.33. Note that ATP and ADP are not involved and that this reaction can be used to yield F6P, but since it is irreversible, cannot be used in glycolysis to yield FBP.

The reaction catalyzed by phosphoglucose isomerase is freely reversible and functions in both the glycolytic and gluconeogenic pathways. However, **glucose 6-phosphatase** has to be used instead of glucokinase for the last step. Glucose 6-phosphatase catalyzes an irreversible reaction under intracellular conditions (Figure 7.34). It should be noted again that nucleotides do not have a role in this reaction and that the function of this enzyme is to generate glucose, not to convert glucose into glucose 6-phosphate. Glucose 6-phosphatase is unique among the enzymes required for gluconeogenesis. It is a membrane-bound enzyme, housed within the endoplasmic reticulum, with its active site available for G6P hydrolysis on the cisternal surface of the tubules (see Figure 7.35). A translocase for G6P is required to move G6P from the cytosol to its site of hydrolysis within the **endoplasmic reticulum.** A genetic defect in either the translocase or the phosphatase interferes with gluconeogenesis and results in accumulation of glycogen in the liver. This will be discussed later in our consideration of glycogen metabolism (Section 7.6).

Glucose Is Synthesized From the Carbon Chain of Some Amino Acids

All amino acids except leucine and lysine can supply carbon for the net synthesis of glucose by gluconeogenesis. The details of the pathways of amino acid catabolism are covered in Chapter 12. For our purposes here, it is very important to note that if the catabolism of an amino acid can

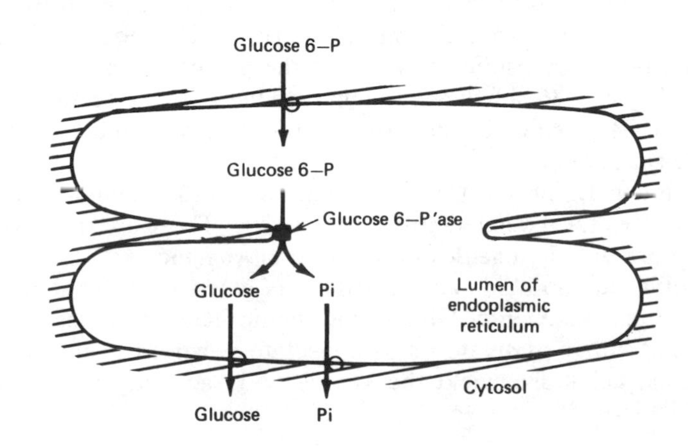

FIGURE 7.35
Glucose 6-phosphate is hydrolyzed by glucose 6-phosphatase (■) located on the cisternal surface of the endoplasmic reticulum.
Three transporters (○) are involved: one moves glucose 6-phosphate into the lumen, a second moves P_i back to the cytosol, and a third moves glucose back into the cytosol.

yield either net pyruvate or net oxaloacetate formation, then net glucose synthesis can occur from that amino acid. As shown in Figure 7.31 oxaloacetate is an intermediate in gluconeogenesis and pyruvate is readily converted to oxaloacetate by the action of pyruvate carboxylase. The abbreviated pathway, given in Figure 7.29, illustrates how amino acid catabolism fits with the process of gluconeogenesis. The catabolism of amino acids feeds carbon into the tricarboxylic cycle at more than one point. As long as net synthesis of a TCA cycle intermediate occurs as a consequence of the catabolism of a particular amino acid, net synthesis of oxaloacetate will follow. Reactions that "fill up" the TCA cycle with intermediates, that is, lead to the net synthesis of TCA cycle intermediates, are called **anaplerotic reactions.** Such reactions support gluconeogenesis because they provide for the net synthesis of oxaloacetate. The reactions catalyzed by pyruvate carboxylase and glutamate dehydrogenase are good examples of anaplerotic reactions (anaplerosis):

$$\text{Pyruvate}^- + \text{ATP}^{4-} + \text{HCO}_3^- \longrightarrow \text{oxaloacetate}^{2-} + \text{ADP}^{3-} + \text{P}_i^{2-} + \text{H}^+$$

$$\text{Glutamate}^- + \text{NAD(P)}^+ \longrightarrow \alpha\text{-ketoglutarate}^{2-} + \text{NAD(P)H} + \text{NH}_4^+ + \text{H}^+$$

On the other hand, the glutamate–oxaloacetate transaminase reaction is not an anaplerotic reaction,

$$\alpha\text{-Ketoglutarate} + \text{aspartate} \rightleftharpoons \text{glutamate} + \text{oxaloacetate}$$

because net synthesis of a TCA cycle intermediate is not accomplished (note the presence of an intermediate of the TCA cycle on both sides of the equation).

Gluconeogenesis from amino acids imposes an additional nitrogen load upon the liver, as pointed out in the description of the alanine cycle in Figure 7.30*b*. Since the liver has to convert this nitrogen into urea, there is a close relationship between urea synthesis and glucose synthesis from amino acids. This relationship is illustrated in Figure 7.36 for alanine, the most important gluconeogenic amino acid. Two alanine molecules are shown being transaminated to give two molecules of pyruvate, which enter the mitochondrion where each is used by a separate pathway, one yielding malate plus NH_4^+ and the other aspartate. The latter two nitrogen-containing compounds should be recognized as the primary substrates for the urea cycle. The aspartate leaves the mitochondrion and becomes part of the urea cycle after reacting with citrulline. The carbon of aspartate is released from the urea cycle in the form of fumarate, the latter then being converted to malate by cytosolic fumarase. Both this malate and the malate exiting from the mitochondria are converted to glucose by the action of the cytosolic enzymes of the gluconeogenic pathway. As shown in Figure 7.36, a balance is achieved between the reducing equivalents (NADH) generated and those required in both the cytosolic and mitosolic spaces.

TABLE 7.2 The Glucogenic and Ketogenic Amino Acids

Glucogenic	Ketogenic	Both
Glycine	Leucine	Threonine
Serine	Lysine	Isoleucine
Valine		Phenylalanine
Histidine		Tyrosine
Arginine		Tryptophan
Cysteine		
Proline		
Hydroxyproline		
Alanine		
Glutamate		
Glutamine		
Aspartate		
Asparagine		
Methionine		

Leucine and lysine are the only amino acids that cannot function as carbon sources for the net synthesis of glucose. These are the only amino acids that are only **ketogenic** and not also **glucogenic.** As shown in Table 7.2, all other amino acids are classified as glucogenic, or at least both glucogenic and ketogenic. Glucogenic amino acids give rise to the net synthesis of either pyruvate or oxaloacetate, whereas glucogenic–ketogenic amino acids also yield the ketone body acetoacetate, or at least

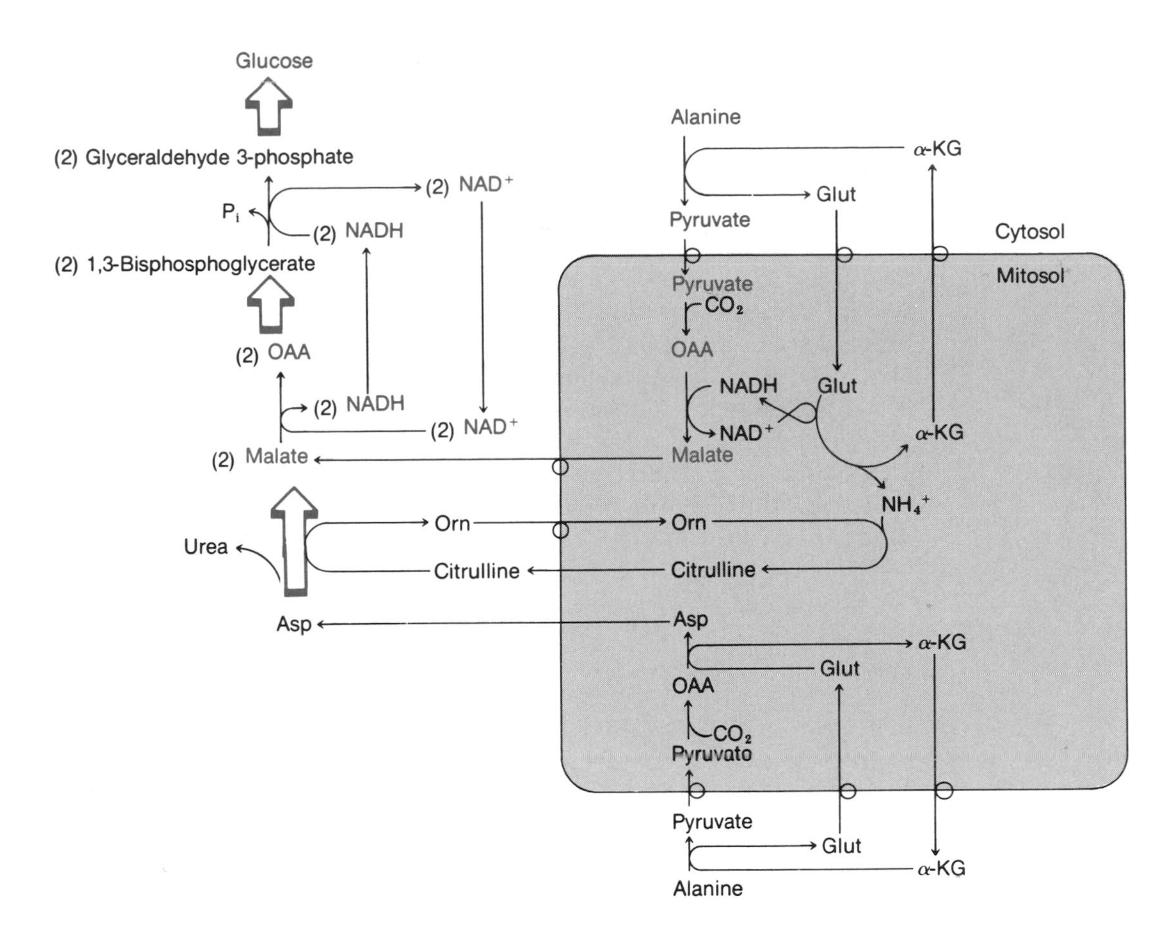

FIGURE 7.36
Pathway of gluconeogenesis from alanine and its relationship to urea synthesis.

acetyl CoA, which is readily converted into ketone bodies. Acetyl CoA is the end product of lysine metabolism, and acetoacetate and acetyl CoA are the end products of leucine metabolism. In the human and other animals, no pathway exists for converting acetoacetate or acetyl CoA into pyruvate or oxaloacetate. It may not be immediately obvious why acetyl CoA cannot be used for net synthesis of glucose but remember that the reaction catalyzed by pyruvate dehydrogenase complex is irreversible:

$$\text{Pyruvate} + \text{NAD}^+ + \text{CoASH} \longrightarrow \text{acetyl CoA} + \text{NADH} + \text{CO}_2$$

meaning this reaction cannot be used to synthesize pyruvate from acetyl CoA. It might be argued that oxaloacetate is generated from acetyl CoA by way of the TCA cycle:

$$\text{Acetyl CoA} \longrightarrow \text{citrate} \xrightarrow{\text{TCA}} 2\text{CO}_2 + \text{oxaloacetate}$$

However, this is a fallacious argument because oxaloacetate must react with acetyl CoA to give citrate by way of citrate synthase:

$$\text{Acetyl CoA} + \text{oxaloacetate} \longrightarrow \text{citrate} + \text{CoA}$$

$$\text{Citrate} \xrightarrow{\text{TCA}} 2\text{CO}_2 + \text{oxaloacetate}$$

$$\text{Sum:} \quad \text{Acetyl CoA} \longrightarrow 2\text{CO}_2 + \text{CoA}$$

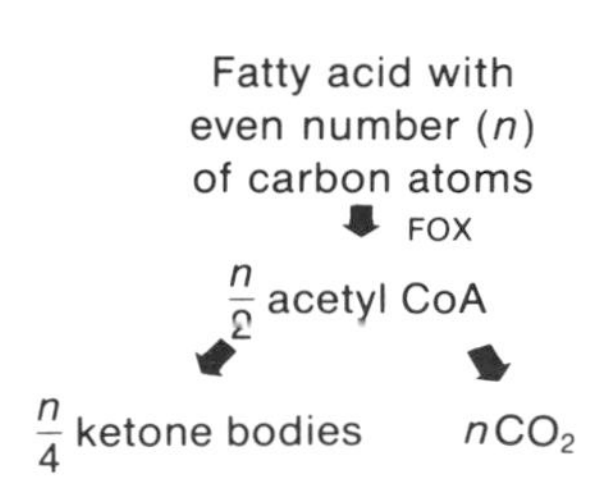

FIGURE 7.37
Overview of the catabolism of fatty acids to ketone bodies and CO_2.

The point is that, although students of biochemistry have tried every conceivable way in the laboratory and on examinations, it turns out to be impossible for animals to synthesize net oxaloacetate or glucose from acetyl CoA.

Glucose Can Be Synthesized From Odd-Chain Fatty Acids

This lack of an anaplerotic pathway from acetyl CoA also means that in general it is impossible to synthesize glucose from fatty acids. Most fatty acids found in the human body are of the straight-chain variety with an even number of carbon atoms. Their catabolism by fatty acid oxidation (FOX) followed by ketogenesis or complete oxidation to CO_2 can be abbreviated as given in Figure 7.37. Since acetyl CoA and other intermediates of even numbered fatty acid oxidation cannot be converted to oxaloacetate or any other intermediate of gluconeogenesis, it is impossible to synthesize glucose from fatty acids. An exception to this general rule applies to fatty acids with methyl branches (e.g., phytanic acid, obtained as a breakdown product of chlorophyll; see discussion of Refsum's disease, Clin. Corr. 9.4) and fatty acids with an odd number of carbon atoms. The catabolism of such compounds yields propionyl CoA:

$$\text{Fatty acid with an odd number } (n) \text{ of carbon atoms} \longrightarrow \frac{(n-3)}{2} \text{ acetyl CoA} + 1 \text{ propionyl CoA}$$

Propionate is a good precursor for gluconeogenesis, generating oxaloacetate by the anaplerotic pathway shown in Figure 7.38. Propionate is also produced in the catabolism of valine and isoleucine and the conversion of cholesterol into bile acids.

It is sometimes loosely stated that fat *cannot* be converted into carbohydrate (glucose) by the liver. In a sense this is certainly true, that is, fatty acid metabolism, with the exception of fatty acids with branched chains or an odd number of carbon atoms, cannot give rise to net synthesis of glucose. However, the term "fat" is usually used to refer to triacylglycerols, which are composed of three *O*-acyl groups combined with one glycerol molecule. Hydrolysis of this molecule of fat yields three fatty acids and glycerol, the latter compound being an excellent substrate for gluconeogenesis as shown in Figure 7.39. Phosphorylation of glycerol by glycerol kinase produces glycerol 3-phosphate, which can be converted

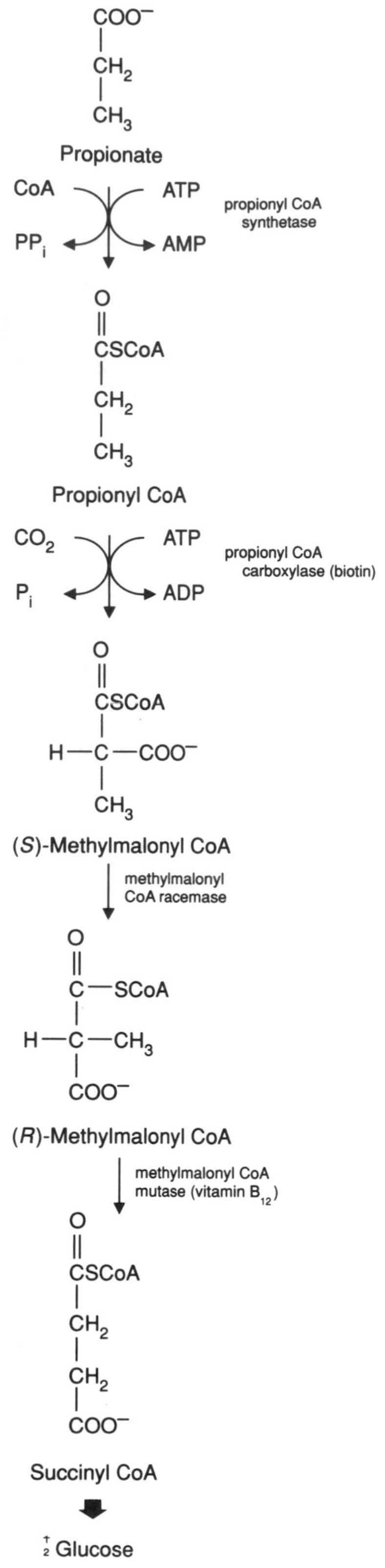

Propionate
CoA, ATP → PP$_i$, AMP (propionyl CoA synthetase)
Propionyl CoA
CO_2, ATP → P$_i$, ADP (propionyl CoA carboxylase (biotin))
(*S*)-Methylmalonyl CoA
methylmalonyl CoA racemase
(*R*)-Methylmalonyl CoA
methylmalonyl CoA mutase (vitamin B_{12})
Succinyl CoA
$\frac{1}{2}$ Glucose

FIGURE 7.38
Pathway of gluconeogenesis from propionate.
The large arrow refers to steps of the tricarboxylic acid cycle (see Figure 6.17) plus steps of lactate gluconeogenesis (see Figure 7.31).

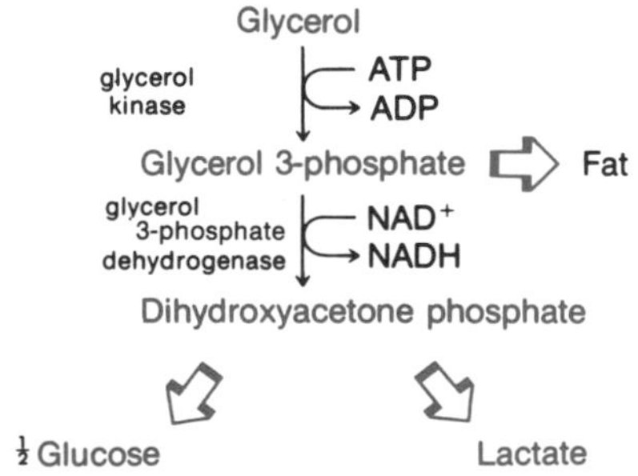

Glycerol
glycerol kinase: ATP → ADP
Glycerol 3-phosphate ⇨ Fat
glycerol 3-phosphate dehydrogenase: NAD^+ → NADH
Dihydroxyacetone phosphate
$\frac{1}{2}$ Glucose   Lactate

FIGURE 7.39
Pathway of gluconeogenesis from glycerol, along with competing pathways.
Large arrows indicate steps of the glycolytic and gluconeogenic pathways that have been given in detail in Figures 7.7 and 7.31, respectively. The large arrow pointing to fat refers to the synthesis of triacylglycerols and glycerophospholipids.

by **glycerol 3-phosphate dehydrogenase** into dihydroxyacetone phosphate, an intermediate of the gluconeogenic pathway (see Figure 7.31). As indicated in Figure 7.39, the last stage of glycolysis can compete with the gluconeogenic pathway and convert dihydroxyacetone phosphate into lactate (or into pyruvate for subsequent complete oxidation to CO_2 and H_2O).

Glucose Is Synthesized From Other Sugars

Fructose

Humans consume considerable quantities of **fructose** in the form of sucrose, and much of the fructose obtained by sucrose hydrolysis in the small bowel is converted into glucose in the liver. Like glucose, fructose is phosphorylated in the liver by a special ATP-linked kinase (Figure 7.40). Phosphorylation of fructose occurs in the 1 position to yield fructose 1-phosphate (see Clin. Corr. 7.3). A special aldolase then cleaves fructose 1-phosphate to give one molecule of dihydroxyacetone phosphate and one molecule of glyceraldehyde. The latter compound can be reduced to glycerol and used by the same pathway given for glycerol in the previous figure. The two molecules of dihydroxyacetone phosphate obtainable from one molecule of fructose can then be converted to glucose by enzymes of the gluconeogenic pathway or, alternatively, into pyruvate or lactate by the last stage of glycolysis. In analogy to glycolysis, the conversion of fructose into lactate is termed **fructolysis.**

Fructose is also generated in the human body for an interesting purpose. The major energy source of spermatozoa is fructose, formed from glucose by cells of the seminal vesicles by the pathway given in Figure 7.41. An NADPH-dependent reduction of glucose to sorbitol is followed by an NAD^+-dependent oxidation of sorbitol to fructose. Fructose is secreted from the seminal vesicles in a fluid that becomes part of the semen. Although the fructose concentration in human semen can exceed 10 mM, tissues that come in contact with semen utilize fructose poorly, allowing this substrate to be conserved to meet the energy demands of spermatozoa in their search for ova. Spermatozoa contain mitochondria and thus can metabolize fructose completely to CO_2 and H_2O by the combination of fructolysis and TCA cycle activity. The mitochondria of sperm are unique. They are the only mitochondria known to contain lactate dehydrogenase. In all other cells this enzyme is confined to the cytosol. This enables sperm mitochondria to oxidize lactate obtained by

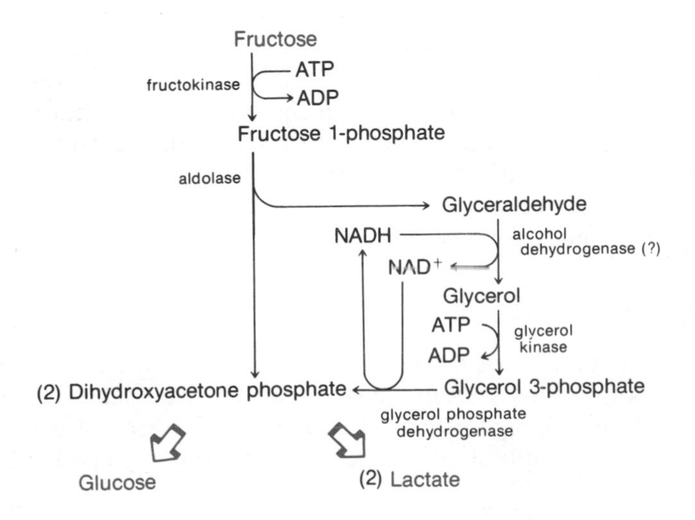

FIGURE 7.40
Pathway of glucose formation from fructose, along with the competing pathway of fructolysis.
Large arrows indicate steps of the glycolytic and gluconeogenic pathways that have been given in detail in Figures 7.7 and 7.31, respectively.

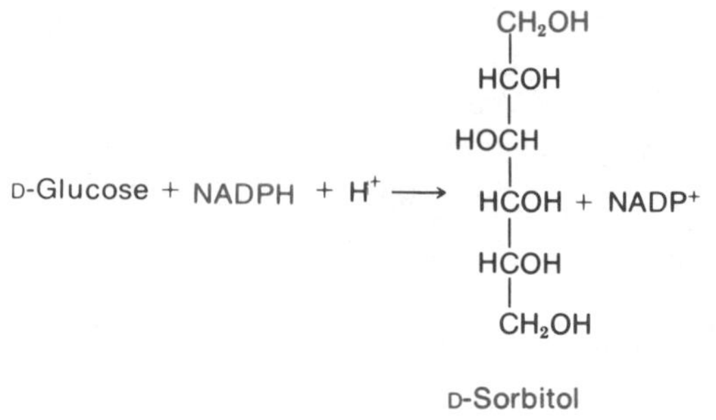

$$\text{D-Glucose} + \text{NADPH} + H^+ \longrightarrow \begin{array}{c} CH_2OH \\ | \\ HCOH \\ | \\ HOCH \\ | \\ HCOH \\ | \\ HCOH \\ | \\ CH_2OH \end{array} + \text{NADP}^+$$

D-Sorbitol

$$\text{D-Sorbitol} + \text{NAD}^+ \longrightarrow \text{D-fructose} + \text{NADH} + H^+$$

FIGURE 7.41
The pathway responsible for the formation of sorbitol and fructose from glucose.

fructolysis and makes shuttle systems for the transport of reducing equivalents into the mitosol unnecessary.

Galactose

Milk sugar or **lactose** constitutes an important source of **galactose** in the human diet. Glucose formation from galactose follows the pathway shown in Figure 7.42. The role of UDP-glucose as a recycling intermediate in the overall process of converting galactose into glucose should be noted. The absence of the enzyme **galactose 1-phosphate uridylyl transferase** accounts for most cases of galactosemia (see Clin. Corr. 8.3).

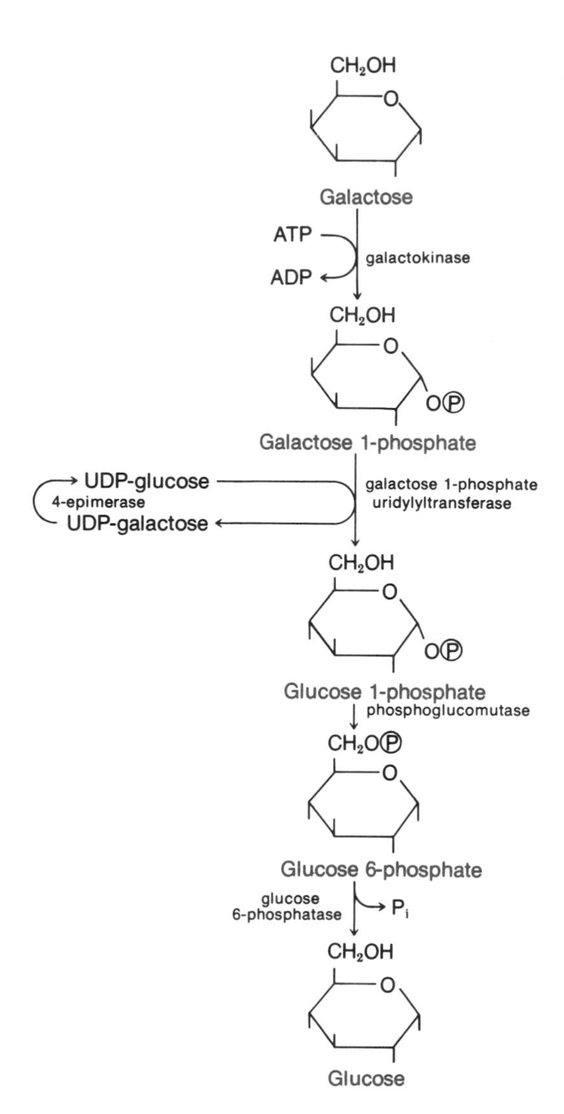

FIGURE 7.42
Pathway of glucose formation from galactose.

Mannose

Mannose is found in our diet, but in very limited quantities. Its pathway of metabolism is short and simple. It is first phosphorylated by hexokinase and then converted into fructose 6-phosphate by **mannose phosphate isomerase:**

$$\text{D-Mannose} + \text{ATP} \longrightarrow \text{D-mannose 6-phosphate} + \text{ADP}$$

$$\text{D-Mannose 6-phosphate} \rightleftharpoons \text{D-fructose 6-phosphate}$$

The latter compound can then be used in either the glycolytic pathway or the gluconeogenic pathway.

Gluconeogenesis Requires the Expenditure of ATP

The synthesis of glucose is costly in terms of ATP. At least six molecules of ATP are required for the synthesis of one molecule of glucose from two molecules of lactate. The ATP needed by the liver cell for glucose synthesis is provided in large part by fatty acid oxidation. Metabolic conditions under which the liver is required to synthesize glucose generally favor increased availability of fatty acids in the blood. These fatty acids are oxidized by liver mitochondria to ketone bodies with the concurrent production of large amounts of ATP. This ATP is used to support the energy requirements of gluconeogenesis, regardless of the substrate being used as the carbon source for the process.

Gluconeogenesis Has Several Sites of Regulation

The sites of regulation of the gluconeogenic pathway are apparent from the mass action ratios and equilibrium constants in Table 7.1, and are further indicated in Figure 7.43. Those enzymes that are used to "go around" the irreversible steps of glycolysis are primarily involved in regu-

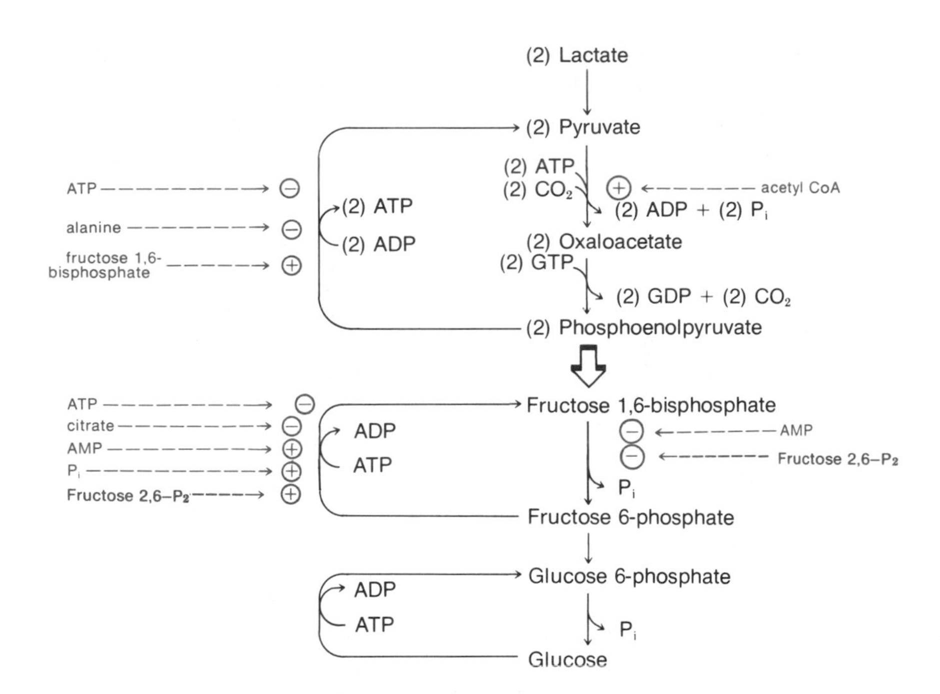

FIGURE 7.43
Important allosteric regulatory features of the gluconeogenic pathway.

lation of the pathway, that is, pyruvate carboxylase, phosphoenolpyruvate carboxykinase, fructose 1,6-bisphosphatase, and glucose 6-phosphatase. Considering the regulation of hepatic gluconeogenesis is almost the same as considering the regulation of hepatic glycolysis, which was discussed in some detail in earlier sections of this chapter. Inhibition of glycolysis at its chief regulatory sites, or repressing the synthesis of the enzymes involved at these sites (glucokinase and pyruvate kinase), greatly increases the effectiveness of the opposing gluconeogenic enzymes. Turning on gluconeogenesis is accomplished in large part, therefore, by shutting off glycolysis. Fatty acid oxidation does more than just supply ATP for the process. It actually promotes glucose synthesis by increasing the steady-state concentration of mitochondrial acetyl CoA, a positive allosteric effector of the mitochondrial enzyme pyruvate carboxylase. The increase in acetyl CoA and in pyruvate carboxylase activity results in a greater synthesis of citrate, a negative effector of 6-phosphofructo-1-kinase. A secondary effect of inhibition of 6-phosphofructo-1-kinase is a decrease in fructose 1,6-bisphosphate concentration, an activator of pyruvate kinase. This decreases the flux of PEP to pyruvate by pyruvate kinase, and increases the effectiveness of the combined efforts of pyruvate carboxylase and PEP carboxykinase in the conversion of pyruvate to PEP. An increase in ATP levels with the consequential decrease in AMP levels would favor gluconeogenesis by way of inhibition of 6-phosphofructo-1-kinase and pyruvate kinase and activation of fructose 1,6-bisphosphatase (see Figure 7.43 and the discussion of the regulation of glycolysis, page 309). A shortage of oxygen for respiration, a shortage of fatty acids for oxidation, or any inhibition or uncoupling of oxidative phosphorylation would be expected to cause the liver to turn from gluconeogenesis to glycolysis.

Hormonal Control of Gluconeogenesis Is Critical for Homeostasis

Hormonal control of gluconeogenesis is a matter of regulating the supply of fatty acids to the liver and, in addition, regulating the enzymes of both the glycolytic and gluconeogenic pathways. Glucagon increases plasma fatty acids by promoting lipolysis in adipose tissue, an action that is opposed by insulin. The greater availability of fatty acids results in more fatty acid oxidation by the liver which, as discussed above, promotes glucose synthesis. Insulin has the opposite effect. **Glucagon** and **insulin** also regulate gluconeogenesis by influencing the state of phosphorylation of hepatic enzymes subject to covalent modification. As discussed in detail previously, pyruvate kinase of the glycolytic pathway is active in the dephosphorylated mode and inactive in the phosphorylated mode (see Figure 7.14). Glucagon activates adenylate cyclase to produce cAMP, which activates cAMP-dependent protein kinase, which, in turn, phosphorylates and inactivates pyruvate kinase. Inactivation of this glycolytic enzyme stimulates the opposing pathway (gluconeogenesis) by blocking the futile conversion of PEP to pyruvate. Glucagon also stimulates gluconeogenesis at the conversion of fructose 1,6-bisphosphate to fructose 6-phosphate by decreasing the concentration of fructose 2,6-bisphosphate present in the liver. Recall (page 317) that fructose 2,6-bisphosphate is an allosteric activator of 6-phosphofructo-1-kinase and an allosteric inhibitor of fructose 1,6-bisphosphatase. Glucagon, again working via its second messenger cAMP, lowers fructose 2,6-bisphosphate levels by stimulating the phosphorylation of the bifunctional enzyme 6-phosphofructo-2-kinase/fructose 2,6-bisphosphatase. Phosphorylation of this enzyme inactivates the site (kinase moiety) that makes fructose 2,6-bisphosphate from F6P but activates the site (phosphatase moiety) that hydrolyzes fructose 2,6-bisphosphate back to F6P. The consequence of a glucagon-induced fall in fructose 2,6-bisphosphate levels is that 6-phosphofructo-1-kinase becomes less active while fructose 1,6-bisphosphatase become more active (Figure 7.43). The overall effect is an increased conversion of FBP to F6P and a corresponding increase in the rate of gluconeogenesis. Insulin has effects opposite to those of glucagon—but the exact mechanism of insulin action is not known.

Glucagon and insulin also have long-term effects upon the levels of hepatic enzymes involved in glycolysis and gluconeogenesis. A high glucagon/insulin ratio in the blood increases the capacity for gluconeogenesis and decreases the capacity for glycolysis in the liver. A low glucagon/insulin ratio has the opposite effects. In addition to the short-term or acute mechanisms discussed above, this is accomplished by induction and repression of the synthesis of key enzymes of the pathways. Thus the glucagon/insulin ratio increases when gluconeogenesis is needed. This serves to signal the induction within the liver of the synthesis of greater quantities of phosphoenolpyruvate carboxykinase, glucose 6-phosphatase, and various aminotransferases. The same signal causes the repression of the synthesis of glucokinase and pyruvate kinase. The opposite response occurs when glucose synthesis is not needed, that is, when a low glucagon/insulin ratio prevails because of maintenance of high blood glucose levels by glucose input from the gastrointestinal tract.

Ethanol Ingestion Inhibits Gluconeogenesis

Ethanol inhibits gluconeogenesis by the liver (see Clin. Corr. 7.10). Ethanol is oxidized primarily in the liver with the production of a large load of reducing equivalents that must be transported into the mitochondria by the malate–aspartate shuttle.

This excess NADH in the cytosol creates problems for liver gluconeogenesis because it forces the equilibrium of the lactate dehydrogenase- and malate dehydrogenase-catalyzed reactions in the directions of lactate

CLIN. CORR. 7.10 HYPOGLYCEMIA AND ALCOHOL INTOXICATION

Consumption of alcohol, especially by an undernourished person, can cause hypoglycemia. The same effect can result from drinking alcohol after strenuous exercise. In both cases the hypoglycemia results from the inhibitory effects of alcohol on hepatic gluconeogenesis and thus occurs under circumstances of hepatic glycogen depletion. The problem is caused by the NADH produced during the metabolism of alcohol. The liver simply cannot handle the reducing equivalents provided by ethanol oxidation fast enough to prevent metabolic derangements. The extra reducing equivalents block the conversion of lactate to glucose and promote the conversion of alanine into lactate, resulting in considerable lactate accumulation in the blood. Since lactate has no place to go, lactic acidosis (see Clin. Corr. 7.5) can develop, although it is usually mild.

Low doses of alcohol cause impaired motor and intellectual performance; high doses have a depressant effect that can lead to stupor and anesthesia. Low blood sugar can contribute to these undesirable effects of alcohol. What is more, a patient may be thought to be inebriated when in fact the patient is suffering from hypoglycemia that may lead to irreversible damage to the central nervous system. Children are highly dependent on gluconeogenesis while fasting, and accidental ingestion of alcohol by a child can produce severe hypoglycemia (see Clin. Corr. 7.9).

Krebs, H. A., Freedland, R. A., Hems, R., and Stubbs, M. Inhibition of hepatic gluconeogenesis by ethanol. *Biochem. J.* 112:117, 1969.

and malate formation, respectively:

$$\underset{\text{Ethanol}}{CH_3CH_2OH} + NAD^+ \longrightarrow \underset{\text{Acetaldehyde}}{CH_3\overset{\overset{O}{\|}}{C}H} + NADH + H^+$$

$$\text{Pyruvate} + NADH + H^+ \longrightarrow \text{lactate} + NAD^+$$

Sum: $\text{Ethanol} + \text{pyruvate} \longrightarrow \text{acetaldehyde} + \text{lactate}$

or

$$\text{Oxaloacetate} + NADH + H^+ \longrightarrow \text{malate} + NAD^+$$

Sum: $\text{Ethanol} + \text{oxaloacetate} \longrightarrow \text{acetaldehyde} + \text{malate}$

In the presence of ethanol there is no shortage of NADH for the gluconeogenic pathway at the level of glyceraldehyde 3-phosphate dehydrogenase; however, forcing the equilibrium of lactate dehydrogenase and malate dehydrogenase as shown above inhibits glucose synthesis because pyruvate and oxaloacetate are no longer available in sufficient concentrations for the reactions catalyzed by pyruvate carboxylase and PEP carboxykinase, respectively. The take home message is the following: Don't drink while synthesizing glucose!

7.6 GLYCOGENOLYSIS AND GLYCOGENESIS

Glycogen, a Storage Form of Glucose, Is Required as a Ready Source of Energy

Glycogenolysis refers to the intracellular breakdown of glycogen; **glycogenesis** to the intracellular synthesis of glycogen. We will be concerned here mainly with these processes in muscle and liver because of their greater quantitative importance in these tissues. However, these processes are of some importance in almost every tissue of the body.

The liver has tremendous capacity for storing glycogen. In the well-fed human the liver glycogen content can account for as much as 10% of the wet weight of this organ. Muscle stores less when expressed on the same basis—a maximum of only 1–2% of its wet weight. However, since the average person has more muscle than liver, there is about twice as much total muscle glycogen as liver glycogen.

Muscle and liver glycogen stores serve completely different roles. Muscle glycogen serves as a fuel reserve for the synthesis of ATP within that tissue, whereas liver glycogen functions as a glucose reserve for the maintenance of blood glucose concentrations. Liver glycogen levels vary greatly in response to the intake of food, accumulating to high levels shortly after a meal and then decreasing slowly as it is mobilized to help maintain a nearly constant blood glucose level (see Figure 7.44). Liver glycogen reserves in the human are called into play between meals and to an even greater extent during the nocturnal fast. In both humans and the rat, the store of glycogen in the liver lasts somewhere between 12 and 24 h during fasting, depending greatly, of course, on whether the individual under consideration is caged or running wild.

Glycogen in muscle is used within this tissue when needed as a source of ATP for increased muscular activity. Most of the glucose of the glycogen molecule is consumed within muscle cells without the formation of free glucose as an intermediate. However, because of a special feature of

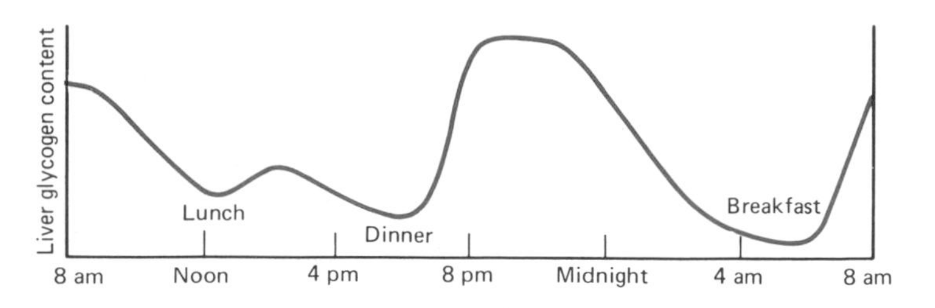

FIGURE 7.44
Variation of liver glycogen levels between meals and during the nocturnal fast.

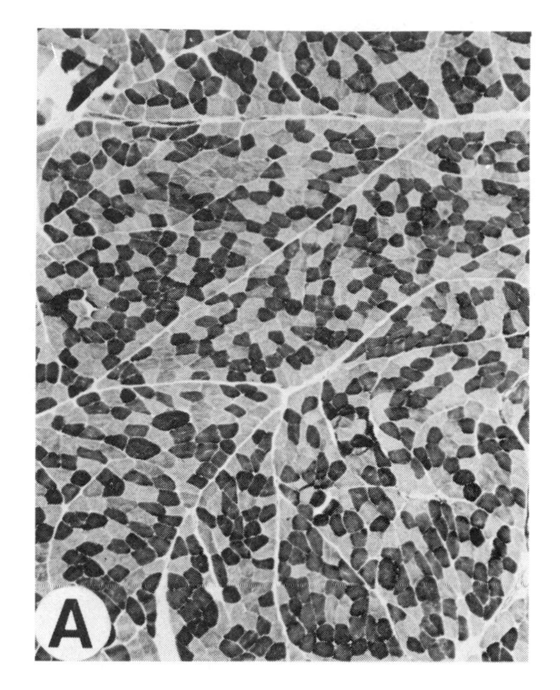

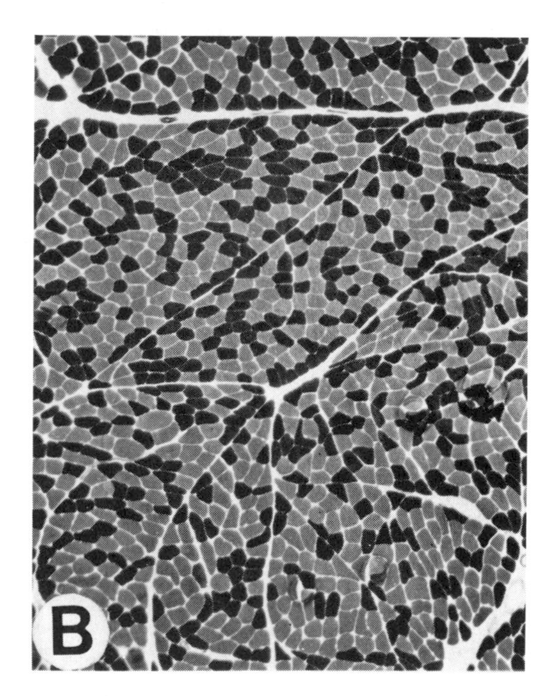

FIGURE 7.45
Cross section of human skeletal muscle showing red and white muscle fibers.
Sections were stained for NADH diaphorase activity in *a*; for ATPase activity in *b*. The red fibers are dark and the white fibers are light in *a*; vice versa in *b*.
Pictures generously provided by Dr. Michael H. Brooke of the Jerry Lewis Neuromuscular Research Center, St. Louis, Mo.

glycogen catabolism to be discussed below, about 8% of muscle glycogen is converted into free glucose within the tissue. Some of this glucose is released into the bloodstream, but most gets metabolized by the glycolytic pathway (Figure 7.5) in the muscle. Since muscle cells lack glucose 6-phosphatase, and since most of the free glucose formed during glycogen breakdown is further catabolized, muscle glycogen is not of quantitative importance in the maintenance of blood glucose levels. Muscle glycogen levels vary much less than liver glycogen levels in response to food intake. The processes of glycogenesis and glycogenolysis within the liver work to **"buffer" blood glucose levels,** but this is not an important role of these processes in muscle. Exercise of a muscle triggers mobilization of muscle glycogen for the formation of ATP. The yield of ATP and the fate of the carbon of glycogen depend on whether a "white" or "red" muscle is under consideration. Red muscle fibers are supplied with a rich blood flow, contain large amounts of myoglobin, and are packed with mitochondria. Glycogen mobilized within these cells is converted into pyruvate, which, because of the availability of O_2 and mitochondria, can be converted into CO_2 and H_2O. In contrast, white muscle fibers have a poorer blood supply and fewer mitochondria. Glycogenolysis within this tissue supplies substrate for glycolysis, with the end product being primarily lactate. White muscle fibers have enormous capacity for glycogenolysis and glycolysis, much more than red muscle fibers. Since their glycogen stores are limited, however, muscles of this type can only function at full capacity for relatively short periods of time. Breast muscle and the heart of chicken are good examples of white and red muscles, respectively. The heart has to beat continuously and has many mitochondria and a rich supply of blood via the coronary arteries. The heart stores glycogen to be used when a greater work load is imposed. The breast muscle of the chicken is not continuously carrying out work. Its important function is to enable the chicken to fly rapidly for short distances, as in fleeing from predators (or amorous roosters). Because glycogen can be mobilized so rapidly, these muscles are designed for maximal activity for a relatively short period of time. Although it was easy to point out readily recognizable white and red muscles in the chicken, most skeletal muscles of the human body are composed of a mixture of red and white fibers in order to provide for both rapid and sustained muscle activity. The distribution of white and red muscle fibers in cross sections of a human skeletal muscle can be readily shown by using special staining procedures (see Figure 7.45).

Glycogen granules are abundant in the liver of the well-fed animal but are virtually absent from the liver of the 24-h-fasted animal (Figure 7.46). Heavy exercise causes the same loss of glycogen granules in muscle fibers. These granules of glycogen correspond to clusters of glycogen molecules, the molecular weights of which can approach 2×10^7. Glyco-

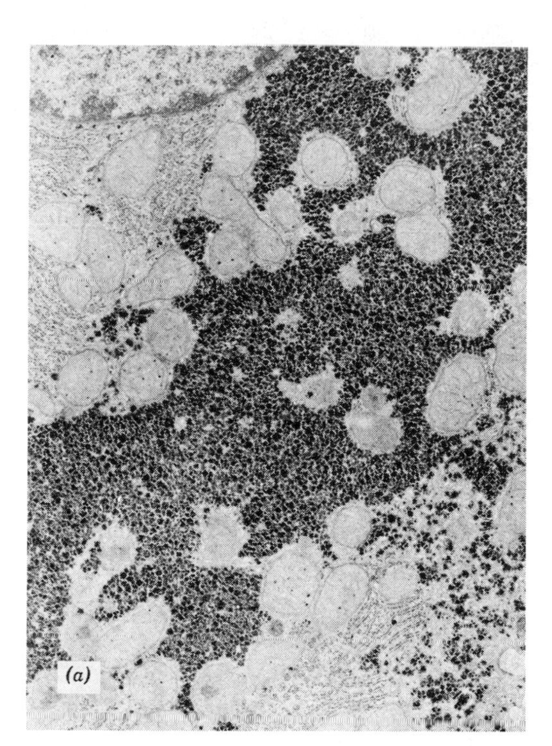

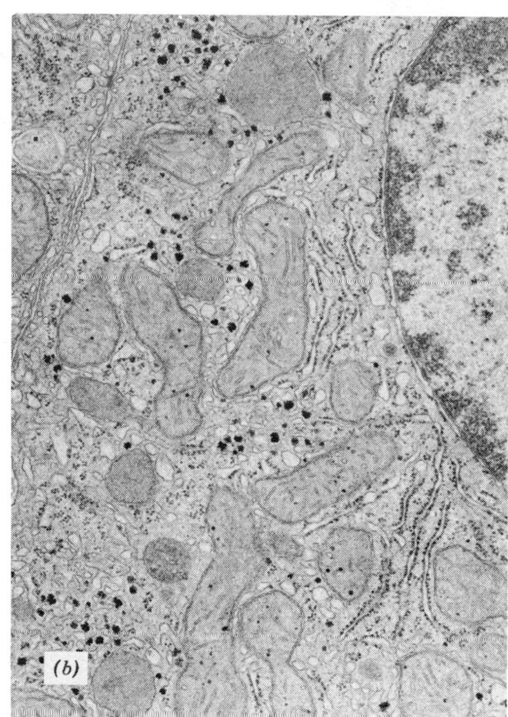

FIGURE 7.46
Electron micrographs showing glycogen granules (darkly stained material) in the liver of a well-fed rat (a) and the relative absence of such granules in the liver of a rat starved for 24 h (b).
Micrographs generously provided by Dr. Robert R. Cardell of the Department of Anatomy at the University of Cincinnati.

gen is composed entirely of glucosyl residues, the majority of which are linked together by $\alpha[1 \rightarrow 4]$ glucosidic linkages (Figure 7.47). Branches also occur in the glycogen molecule, however, because of frequent $\alpha[1 \rightarrow 6]$ glucosidic linkages (Figure 7.47). A limb of the glycogen "tree" (see Figure 7.48) is characterized by branches at every fourth glucosyl residue within the more central core of the molecule. These branches occur much less frequently in the outer regions of the molecule. An interesting question, which we shall attempt to answer below, is why this polymer is constructed by the cell with so many intricate branches and loose ends? Glycogen certainly stands in contrast to proteins and nucleic acids in this

CH_2OH CH_2OH O O ····O O O···· α [1 → 4] linkage (a)

CH_2OH O ······O O CH_2 O ······O O····· α [1 → 6] linkage (b)

FIGURE 7.47
Two types of linkage between glucose molecules are present in glycogen.

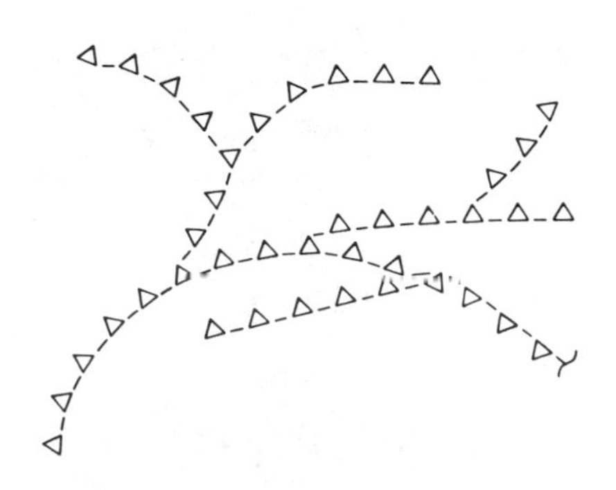

FIGURE 7.48
The branched structure of glycogen.

CLIN. CORR. **7.11** GLYCOGEN STORAGE DISEASES

There are a number of well-characterized glycogen storage diseases, all due to inherited defects of one or more of the enzymes involved in the synthesis and degradation of glycogen. The liver is usually the tissue most affected, but heart and muscle glycogen metabolism can also be defective.

VON GIERKE'S DISEASE

The most common glycogen storage disease, referred to as type I or von Gierke's disease, is caused by a deficiency of liver, intestinal mucosa, and kidney glucose 6-phosphatase. Thus, diagnosis by small bowel biopsy is possible. Patients with this disease can be further subclassified into those lacking the glucose 6-phosphatase enzyme per se (type Ia) and those lacking the glucose 6-phosphatase translocase (type Ib) (see Figure 7.35). A genetic abnormality in glucose 6-phosphate hydrolysis occurs in only about 1 person in 200,000 and is transmitted as an autosomal recessive trait. Clinical manifestations include fasting hypoglycemia, lactic acidema hyperlipidemia, and hyperuricemia with gouty arthritis. The fasting hypoglycemia is readily explained as a consequence of the glucose 6-phosphatase deficiency, the enzyme required to obtain glucose from liver glycogen and gluconeogenesis. The liver of these patients does release some glucose by the action of the glycogen debrancher enzyme. The lactic acidemia occurs because the liver cannot use lactate effectively for glucose synthesis. In addition, the liver inappropriately produces lactic acid in response to glucagon. This hormone should trigger glucose release without lactate production; however, the opposite occurs because of the lack of glucose 6-phosphatase. Hyperuricemia results from increased purine degradation in the liver; hyperlipidemia because of increased availability of lactic acid for lipogenesis and lipid mobilization from the adipose tissue caused by high glucagon levels in response to hypoglycemia. The manifestations of von Gierke's disease can be greatly diminished by providing carbohydrate throughout the day to prevent hypoglycemia. During sleep this can be done by infusion of carbohydrate into the gut by a naso-gastric tube.

Cori, G. T. and Cori, C. F. Glucose-6-phosphatase of the liver in glycogen storage disease. *J. Biol. Chem.* 199:661, 1952.

regard but, of course, it is a storage form of fuel and never has to catalyze a reaction nor convey information within a cell.

Glycogen Phosphorylase Catalyzes the First Step in Glycogen Hydrolysis

The first step of glycogen degradation is catalyzed by the enzyme **glycogen phosphorylase** (see Figure 7.49). This enzyme catalyzes the phosphorolysis of glycogen, a reaction in which the elements of P_i are used in the cleavage of an $\alpha[1 \rightarrow 4]$ glucosidic bond to yield glucose 1-phosphate. This always occurs at a terminal, nonreducing end of a glycogen molecule:

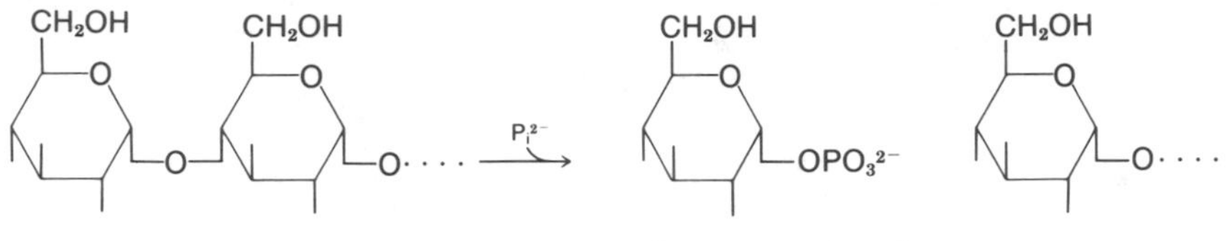

Glycogen (partial structure) $\xrightarrow{P_i^{2-}}$ α-D-Glucose 1-phosphate

The reaction catalyzed by glycogen phosphorylase should be distinguished from that catalyzed by α-amylase, the enzyme responsible for glycogen (and starch) degradation in the gut (see Chapter 26). α-Amylase acts by simple hydrolysis, using the elements of water rather than inorganic phosphate to cleave $\alpha[1 \rightarrow 4]$ glucosidic bonds. Since a molecule of glycogen may contain up to 100,000 glucose residues, its structure is usually abbreviated $(\text{glucose})_n$. The reaction catalyzed by the enzyme glycogen phosphorylase can then be written as

$$(\text{Glucose})_n + P_i^{2-} \longrightarrow (\text{glucose})_{n-1} + \alpha\text{-D-glucose 1-phosphate}^{2-}$$

The next step of glycogen degradation is catalyzed by **phosphoglucomutase:**

$$\text{Glucose 1-phosphate} \rightleftharpoons \text{glucose 6-phosphate}$$

This is a near-equilibrium reaction under intracellular conditions, allowing it to function in both glycogen degradation and synthesis. It has the

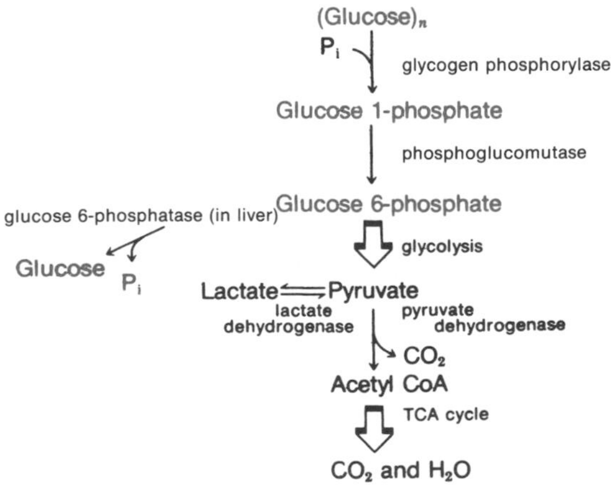

FIGURE 7.49
Glycogenolysis and the fate of glycogen degraded in liver versus its fate in peripheral tissues.

interesting feature of having a reaction mechanism analogous to that catalyzed by phosphoglyceromutase (page 302) in that a bisphosphate compound is an obligatory intermediate:

$$\text{E—P} + \text{glucose 1-phosphate} \rightleftharpoons \text{E} + \text{glucose 1,6-bisphosphate}$$

$$\text{E} + \text{glucose 1,6-bisphosphate} \rightleftharpoons \text{E—P} + \text{glucose 6-phosphate}$$

$$\text{Sum:} \quad \text{Glucose 1-phosphate} \rightleftharpoons \text{glucose 6-phosphate}$$

As with phosphoglyceromutase, a catalytic amount of the bisphosphate compound must be present for the reaction to occur. It is produced in small quantities for this specific purpose by an enzyme called **phosphoglucokinase:**

$$\text{Glucose 6-phosphate} + \text{ATP} \longrightarrow \text{glucose 1,6-bisphosphate} + \text{ADP}$$

The next enzyme involved in glycogenolysis depends on the tissue under consideration (see Figure 7.49). In liver the glucose 6-phosphate produced by glycogenolysis would be primarily hydrolyzed by **glucose 6-phosphatase** to give free glucose:

$$\text{Glucose 6-phosphate}^{2-} + H_2O \longrightarrow \text{glucose} + P_i^{2-}$$

Lack of this enzyme or of the translocase that transports G6P into the endoplasmic reticulum (see p. 329) results in type I **glycogen storage disease** (see Clin. Corr. 7.11). The overall balanced equation for the removal of one glucosyl residue from glycogen in the liver by glycogenolysis is then

$$(\text{Glucose})_n + H_2O \longrightarrow (\text{glucose})_{n-1} + \text{glucose}$$

In other words, glycogenolysis in the liver involves phosphorolysis but, because the phosphate ester is cleaved by a phosphatase, the overall reaction adds up to be hydrolysis of glycogen. It should be noted that no ATP is used or formed in the process of glycogenolysis.

In peripheral tissues the G6P generated by glycolysis would be used by the glycolytic pathway, which would lead primarily to the generation of lactate in white muscle fibers and primarily to the complete oxidation of the glucose to CO_2 and H_2O in red muscle fibers. Since no ATP had to be invested to produce the G6P obtained from glycogen, the overall equation for glycogenolysis followed by glycolysis is

$$(\text{Glucose})_n + 3\text{ADP}^{3-} + 3P_i^{2-} + H^+ \longrightarrow (\text{glucose})_{n-1} + 2\ \text{lactate}^{-1} + 3\text{ATP}^{4-}$$

A Debranching Enzyme Is Required for Complete Hydrolysis of Glycogen

The branches that exist in the glycogen molecule cause a complication that must be dealt with now. The first enzyme involved in glycogen degradation, glycogen phosphorylase, is specific for $\alpha[1 \rightarrow 4]$ glucosidic linkages. It stops attacking $\alpha[1 \rightarrow 4]$ glucosidic linkages four glucosyl residues from an $\alpha[1 \rightarrow 6]$ branch point. A glycogen molecule that has been degraded to the limit by phosphorylase is called **phosphorylase-limit dextrin.** The action within cells of a **debranching enzyme** is what allows glycogen phosphorylase to continue to degrade glycogen. The "debranching" enzyme is a *bifunctional* enzyme that catalyzes two reactions necessary for the debranching of glycogen. The first is a **4-α-D-glucanotransferase** activ-

POMPE'S DISEASE

Type II glycogen storage disease or Pompe's disease is caused by the absence of α-1,4-glucosidase (or acid maltase), an enzyme normally found in lysosomes. The absence of this enzyme leads to the accumulation of glycogen in virtually every tissue. This is somewhat surprising, but lysosomes take up glycogen granules and become defective with respect to other functions if they lack the capacity to destroy the granules. Because other synthetic and degradative pathways of glycogen metabolism are intact, metabolic derangements such as those in von Gierke's disease are not seen. The reason for extralysosomal glycogen accumulation is unknown. Massive cardiomegaly occurs and death results at an early age from heart failure.

Hers, H. G. α-Glucosidase deficiency in generalized glycogen storage disease (Pompe's disease). *Biochem. J.* 86:11, 1963.

CORI'S DISEASE

Also called type III glycogen storage disease, Cori's disease is caused by a deficiency of the glycogen debrancher enzyme. Glycogen accumulates because only the outer branches can be removed from the molecule by phosphorylase. Hepatomegaly occurs, but diminishes with age. The clinical manifestations are similar to but much milder than those seen in von Gierke's disease, because gluconeogenesis is unaffected, and hypoglycemia and its complications are less severe.

Van Hoff, F. and Hers, H. G. The subgroups of type III glycogenosis. *Eur. J. Biochem.* 2:265, 1967.

McARDLE'S DISEASE

Also called the type V glycogen storage disease, McArdle's disease is caused by an absence of muscle phosphorylase. Patients suffer from painful muscle cramps and are unable to perform strenuous exercise, presumably because muscle glycogen stores are not available to the exercising muscle. Thus, the normal increase in plasma lactate (released from the muscle) following exercise is absent. The muscles are probably damaged because of inadequate energy supply and glycogen accumulation. Release of muscle enzymes creatine phosphokinase and aldolase and of myoglobin is common; elevated levels of these substances in the blood suggests a muscle disorder.

McArdle, B. Myopathy due to a defect in muscle glycogen breakdown. *Clin. Sci.* 10:13, 1951.

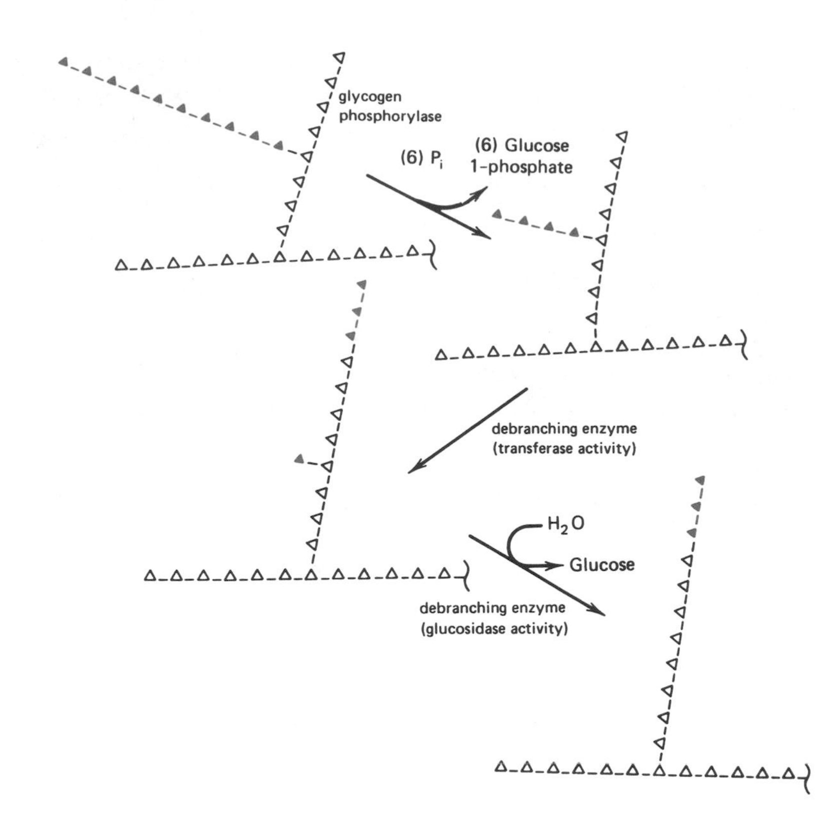

FIGURE 7.50
Action of the glycogen debranching enzyme.

ity in which a strand of three glucosyl residues is removed from a four glucosyl residue branch of the glycogen molecule (see Figure 7.50). The strand remains covalently attached to the enzyme until it can be transferred to a free 4-hydroxyl of a glucosyl residue at the end of the same or an adjacent glycogen molecule (see Figure 7.50). The result is a longer amylose chain with only one glucosyl residue remaining in $\alpha[1 \rightarrow 6]$ linkage. This linkage is broken hydrolytically by the other enzyme activity of the "debranching" enzyme, that is, its **amylo-α-[1,6]-glucosidase activity:**

CH$_2$OH — O — CH$_2$ ... Glycogen $\xrightarrow{H_2O}$ α-D-Glucose (CH$_2$OH) + Glycogen (CH$_2$OH)

The cooperative and repetitive action of phosphorylase and debranching enzyme results in complete phosphorolysis and/or hydrolysis of the glycogen molecule. Glycogen storage diseases result when either of these enzymes is defective. The average molecule of glycogen yields about 12 molecules of glucose 1-phosphate by the action of phosphorylase for every molecule of free glucose produced by the action of the debranching enzyme.

There is another pathway for glycogen degradation that is quantitatively not very important. Major problems result, however, when this pathway is defective in an individual. As discussed in Clin. Corr. 7.11, a glucosidase of lysosomes degrades glycogen, which has entered into these organelles during normal turnover of intracellular components.

Synthesis of Glycogen Utilizes Some of the Enzymes Involved in Degradation but Also Some Unique Enzymes

The pathway involved in glycogen synthesis is given in Figure 7.51. The first reaction is already familiar, being catalyzed by glucokinase in hepatic tissue and hexokinase in peripheral tissues:

$$\text{Glucose} + \text{ATP} \longrightarrow \text{glucose 6-phosphate} + \text{ADP}$$

The next enzyme involved, phosphoglucomutase, was discussed in relation to glycogen degradation, although this reversible reaction was written in the opposite direction:

$$\text{Glucose 6-phosphate} \longrightarrow \text{glucose 1-phosphate}$$

A unique reaction found in the next step, involves the formation of UDP-glucose by the action of **glucose 1-phosphate uridylyltransferase:**

$$\text{Glucose 1-phosphate} + \text{UTP} \longrightarrow \text{UDP-glucose} + \text{PP}_i$$

This reaction generates an ''activated'' glucosyl residue, which can be used to build the glycogen molecule. The formation of UDP-glucose is made energetically favorable and the reaction is irreversible by the subsequent hydrolysis of pyrophosphate by **pyrophosphatase:**

$$\text{PP}_i^{4-} + \text{H}_2\text{O} \longrightarrow 2\text{P}_i^{2-}$$

Glycogen Synthase

Glycogen synthase, utilizing glycogen and UDP-glucose as substrates, then catalyzes the transfer of the activated glucosyl moiety to the glycogen molecule so that a new glucosidic bond is formed between the hy-

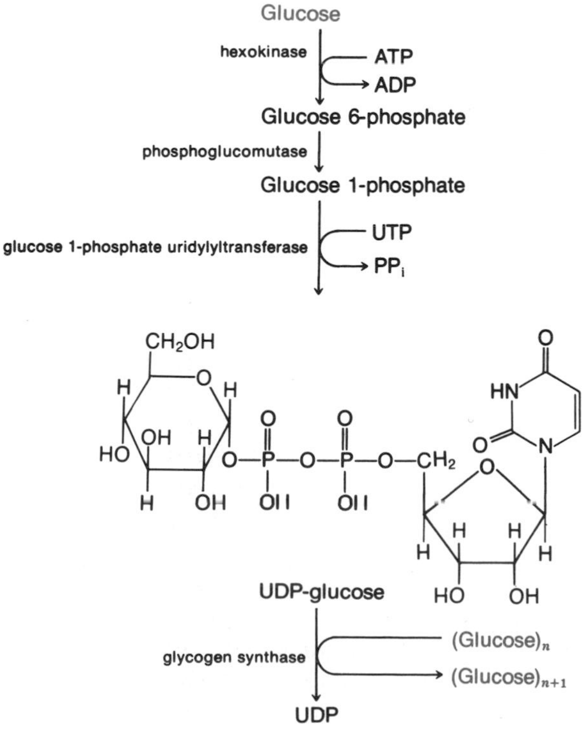

FIGURE 7.51
Pathway of glycogen synthesis.

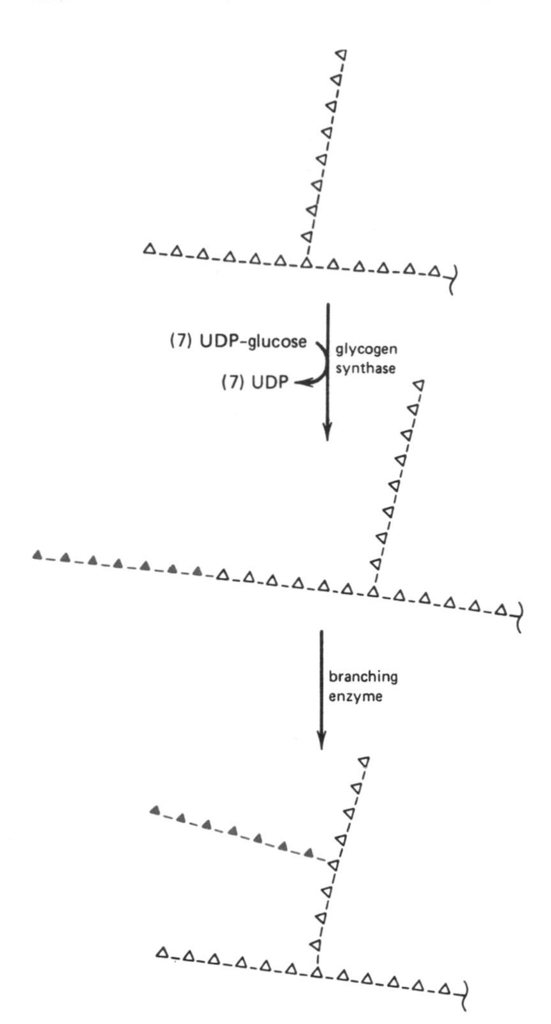

FIGURE 7.52
Action of the glycogen branching enzyme.

droxyl group of C-1 of the activated sugar and C-4 of a glucosyl residue of the growing glycogen chain. The reducing end of glucose (C-1 of glucose is an aldehyde that can reduce other compounds) is always added to a nonreducing end of the glycogen chain. Note that the glycogen molecule, regardless of its size, theoretically has only one free reducing end tucked away within the core. Also note that UDP, *not* UMP, is the product of the reaction catalyzed by glycogen synthase. UDP can be converted back to UTP by the action of nucleoside diphosphate kinase:

$$\text{UDP} + \text{ATP} \longrightarrow \text{UTP} + \text{ADP}$$

Glycogen synthase is very specific, that is, it will create chains of glucose molecules with $\alpha[1 \rightarrow 4]$ linkages but will not participate in the formation of $\alpha[1 \rightarrow 6]$ branches. Its action alone would only produce amylose, the straight-chain polymer of glucose with $\alpha[1 \rightarrow 4]$ linkages. Once an amylose chain of at least 11 residues has been formed, a **"branching" enzyme** comes into play. Its name is **1,4-α-glucan branching enzyme** because it removes a block of about 7 glucosyl residues from a growing chain and transfers it to another chain to produce an $\alpha[1 \rightarrow 6]$ linkage (see Figure 7.52). The last special feature to be mentioned is that the new branch has to be introduced at least four glucosyl residues from an adjacent branch point. Thus the creation of the highly branched structure of glycogen requires the concerted efforts of glycogen synthasc and thc branching enzyme. The overall balanced equation for glycogen synthesis by the pathway just outlined is

$$(\text{Glucose})_n + \text{glucose} + 2\text{ATP} \longrightarrow (\text{glucose})_{n+1} + 2\text{ADP} + 2\text{P}_i$$

As noted above, the combination of glycogenolysis and glycolysis yields only three molecules of ATP per glucosyl residue:

$$(\text{Glucose})_n + 3\text{ADP} + 3\text{P}_i \longrightarrow (\text{glucose})_{n-1} + 2\text{ lactate} + 3\text{ATP}$$

Thus the combination of glycogen synthesis plus glycogen degradation to lactate actually yields the cell only one ATP, that is, the sum of the last two equations is

$$\text{Glucose} + \text{ADP} + \text{P}_i \longrightarrow 2\text{ lactate} + \text{ATP}$$

It should be realized, however, that glycogen synthesis and degradation are normally carried out at different times in a cell. For example, white muscle fibers synthesize glycogen at rest when glucose is plentiful and ATP for muscle contraction is not needed. Glycogen is then used during periods of exertion. Although in such terms glycogen storage is not a very efficient process, it provides cells with a fuel reserve that can be quickly and efficiently mobilized.

Special Features of Glycogen Degradation and Synthesis: Glycogenin as a Primer

Since glycogen is such a good fuel reserve, it is obvious why we synthesize and store glycogen in liver and muscle. But why store glucose as glycogen? Why not store our excess glucose calories entirely as fat instead of glycogen? The answer is at least threefold: (1) we do store fat, some of us lots of it, but fat cannot be mobilized as rapidly in muscle as glycogen; (2) fat cannot be used as a source of energy in the absence of oxygen; and (3) fat cannot be converted to glucose by any pathway of the human body in order to maintain blood glucose levels for use by tissues such as the brain. Why not just pump glucose into cells and store it as free glucose until needed? Why waste so much ATP making a polymer out of glucose? The problem is that glucose is osmotically active. It would cost ATP to "pump" glucose into a cell, regardless of the mechanism, and glucose would have to reach concentrations of 400 mM in liver cells to match the **"glucose reserve"** provided by the usual liver glycogen levels. Unless balanced by the outward movement of some other osmotically active compound, the accumulation of such concentrations of glucose would cause the uptake of considerable water and the osmotic lysis of the cell. Assuming the molecular mass of a glycogen molecule is of the order of 10^7 Da, 400 mM of glucose is in effect stored at an intracellular glycogen concentration of 0.01 μM. Storage of glucose as glycogen, therefore, creates no osmotic pressure problem for the cell.

Another interesting feature about glycogen is that a **primer** is needed for its synthesis. No template is required, but like DNA synthesis, a primer is necessary. Glycogen itself is the usual primer, that is, glycogen synthesis usually takes place by the addition of glucosyl units to glycogen "core" molecules, which are almost invariably present in the cell. The outer regions of the glycogen molecule get removed and resynthesized much more rapidly than the inner core. Glycogen within a cell is frequently sheared by the combined actions of glycogen phosphorylase and debranching enzyme but is seldom ever obliterated before glycogen synthase and branching enzyme rebuild the molecule. This is a good time to point out why nature has evolved such an elaborate mechanism for creating and disposing of the branched structure of glycogen. In other words, why is glycogen a branched molecule with only one real beginning (the reducing end) and many branches terminating with nonreducing glucosyl units? The answer is that this gives numerous sites of attack for glycogen phosphorylase on a mature glycogen molecule and the same number of sites for glycogen synthase to add glucosyl units. If cells synthesized amylose, that is, an unbranched glucose polymer, there would only be one nonreducing end per molecule. The result would be that glycogen degradation and synthesis would surely be much slower processes. As it is, glycogen phosphorylase and glycogen synthase are usually found in tight association with glycogen granules in a cell, as though they exist in the branches of the glycogen tree with ready access to a multitude of nonreducing sugars at the ends of its limbs.

We digressed, however, from the problem of a need of a primer for glycogen synthesis. Perhaps as a consequence of the great number of nonreducing ends, glycogen synthase has a very low K_m for very large glycogen molecules. However, the K_m gets larger and larger as the glycogen molecule gets smaller and smaller. This phenomenon is so pronounced that it is clear that glucose, at its physiological concentration, could never function as a primer. This led to the notion that glycogen must be immortal, that is, some glycogen must be handed down from one cell generation to the next in order for glycogen to be synthesized. Although immortality is attractive, it is now thought that a protein called **glycogenin** functions as a primer for glycogen synthesis. Glycogenin is a self-glucosylating enzyme that uses UDP-glucose to link glucose to one of its own tyrosine residues (Figure 7.53). Glucosylated glycogenin then serves as a nucleus for the synthesis of glycogen. Alas, glycogen is probably not immortal.

If glycogen synthase becomes more efficient as the glycogen molecule gets bigger, how is synthesis of this ball of sugar attached to glycogenin curtailed? Fat cells have an almost unlimited capacity to pack away fat—but then fat cells do not have to do anything else. Muscle cells participate

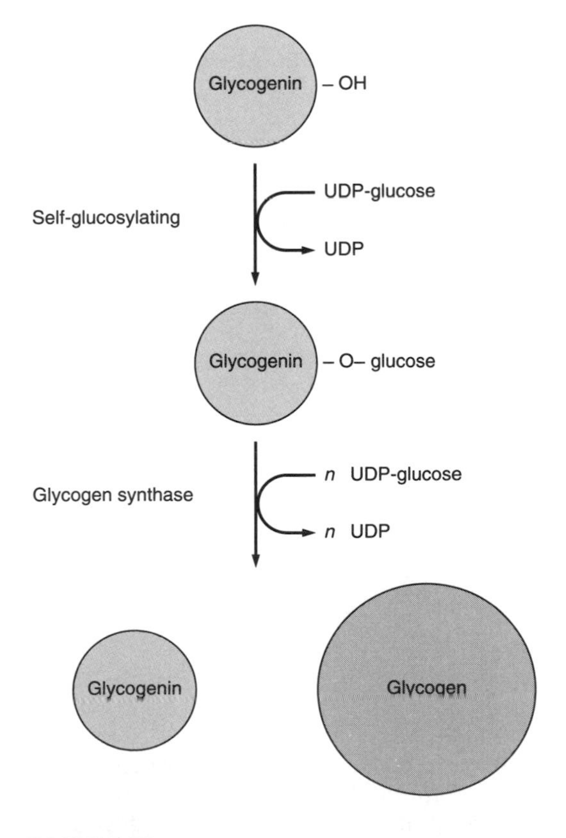

FIGURE 7.53
Glycogenin provides a primer for glycogen synthesis by glycogen synthase.

in mechanical activity and liver cells carry out many processes other than glycogen synthesis. Even in the face of excess glucose, there has to be a way to limit the intracellular accumulation of glycogen. It turns out that glycogen itself inhibits glycogen synthase by a mechanism discussed in a later section (p. 351).

Glycogen Synthesis and Degradation Are Highly Regulated Pathways

Regulatory Enzymes of the Pathways

Glycogen synthase and glycogen phosphorylase are the regulatory enzymes of glycogen synthesis and degradation, respectively. Both catalyze nonequilibrium reactions, and both are subject to control by allosteric effectors and covalent modification.

Regulation of Glycogen Phosphorylase

The mechanisms responsible for the regulation of glycogen phosphorylase are summarized in a rather formidable fashion in Figure 7.54. The enzyme is subject to allosteric activation by AMP and allosteric inhibition by glucose and ATP. Although these effectors are considered to be of physiological significance in the regulation of glycogen metabolism, effector control of phosphorylase is integrated with a very elaborate control by covalent modification. Phosphorylase exists in an "*a*" form, which is active, and a "*b*" form, which is inactive. These forms of the enzyme are interconverted by the actions of phosphorylase kinase and phosphoprotein phosphatase (Figure 7.54). A conformational change caused by phosphorylation transforms the enzyme into a more active catalytic state. **Phosphorylase *b*** has some catalytic activity and can be greatly activated by AMP. This allosteric effector has little activating effect, however, on the already active **phosphorylase *a*.** Hence the covalent modification mechanism can be bypassed by the allosteric mechanism or vice versa. Phosphorylase is composed of two identical 97,000-mol wt subunits. Serine 14, counting from the amino terminus, of both subunits can be

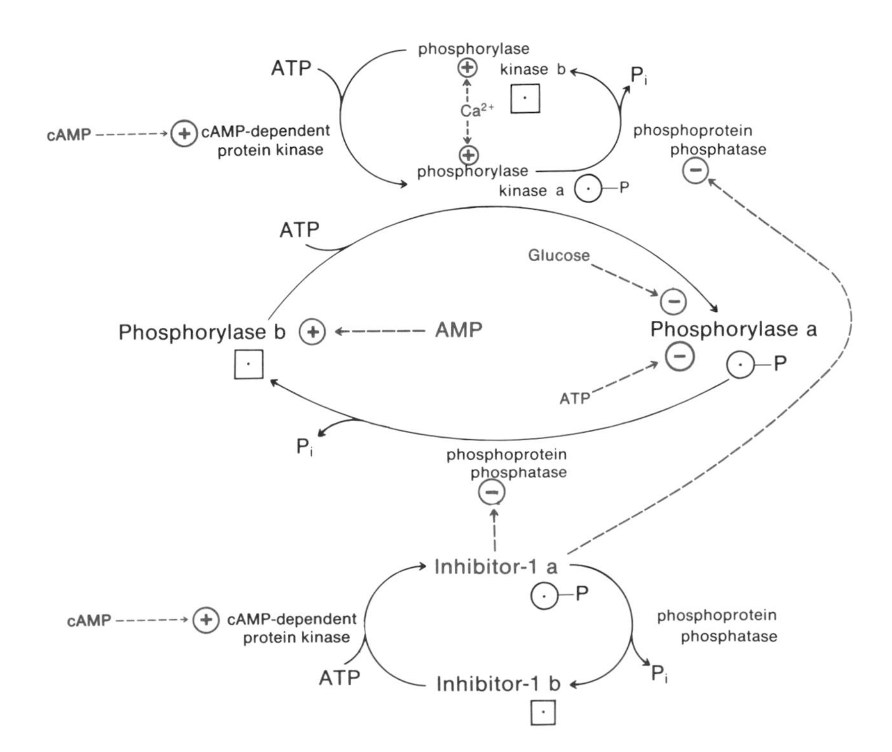

FIGURE 7.54
Regulation of glycogen phosphorylase by covalent modification.
Note that phosphorylation converts glycogen phosphorylase, phosphorylase kinase, and phosphatase inhibitor-1 from their inactive *b* forms to their active *a* forms.

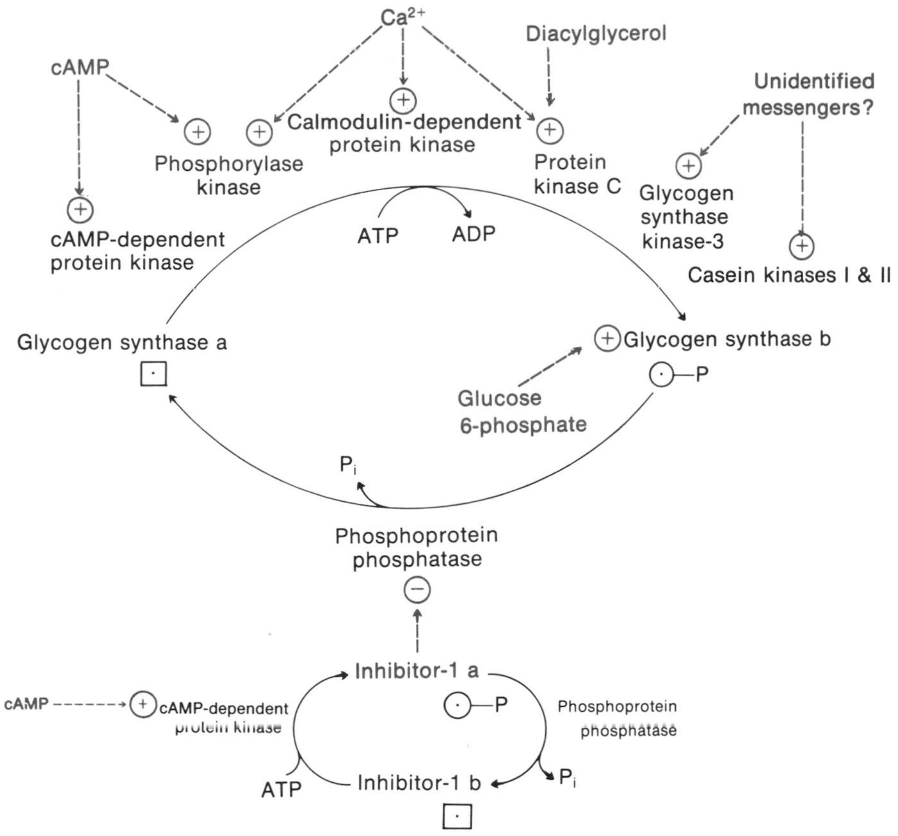

FIGURE 7.55
Regulation of glycogen synthase by covalent modification.
Note that phosphorylation converts glycogen synthase from its active *a* form to its inactive *b* form, whereas phosphorylation converts phosphatase inhibitor-1 from its inactive *b* form to its active *a* form.

phosphorylated by phosphorylase kinase. Thus, two ATP molecules are converted to two ADP molecules in the conversion of phosphorylase *b* to phosphorylase *a*. This detail is ignored for the sake of simplicity in our attempt (Figure 7.55) to present the overall mechanism involved in the regulation of this enzyme.

Phosphorylase kinase is responsible for the phosphorylation and activation of phosphorylase (Figure 7.54). Moreover, phosphorylase kinase itself is also subject to regulation by a cyclic phosphorylation–dephosphorylation mechanism. Cyclic AMP-dependent protein kinase is responsible for phosphorylation and activation of phosphorylase kinase; phosphoprotein phosphatase in turn is responsible for dephosphorylation and inactivation of phosphorylase kinase. Phosphorylase kinase is a large enzyme complex (1.3 million Da), composed of four subunits with four molecules of each subunit in the complex ($\alpha_4\beta_4\gamma_4\delta_4$). Catalytic activity resides with the γ subunit; α, β, and δ subunits exert regulatory control. The α and β subunits are phosphorylated in the transition from the inactive *b* form to the active *a* form of the enzyme. Cyclic AMP-dependent protein kinase does not interact with phosphorylase directly—this protein kinase can only exert an effect on phosphorylase via its ability to phosphorylate phosphorylase kinase. Thus, a bicyclic system is required for the activation of phosphorylase in response to cAMP-mediated signals.

The δ subunit of phosphorylase kinase also plays a regulatory role. The δ subunit is a Ca^{2+}-binding regulatory protein, called **calmodulin.** Calmodulin is not unique to phosphorylase kinase. It is found in cells as the free molecule and is also bound to other enzyme complexes. Calmodulin functions as a Ca^{2+} receptor in the cell, responding to changes in intracellular Ca^{2+} concentration and affecting the relative activities of a number of enzyme systems. The binding of Ca^{2+} to the calmodulin subunit of phosphorylase kinase changes the conformation of the complex, making the enzyme more active with respect to the phosphorylation of phosphorylase. Note in Figure 7.54 that Ca^{2+} is indicated as an activator of both phosphorylase kinase *a* and phosphorylase kinase *b*. This means that maximum activation of phosphorylase kinase requires both the phosphorylation of specific serine residues of the enzyme and the interaction of Ca^{2+} with the calmodulin subunit of the enzyme. This is one mechanism by which Ca^{2+} functions as an important "second messenger" of hormone action, as will be discussed in detail below.

It is obvious that activation of phosphorylase kinase by phosphorylation and Ca^{2+} will have a substantial effect on the activity of glycogen phosphorylase. It is equally obvious that turning off **phosphoprotein phosphatase** could achieve the same thing. But ultimate control for the activation of phosphorylase would involve the simultaneous turning off of phosphoprotein phosphatase and turning on of phosphorylase kinase, and vice versa, for the inactivation of the enzyme. Since phosphoprotein phosphatase also acts on phosphorylase kinase, turning off phosphoprotein phosphatase would also achieve greater activation of phosphorylase kinase. Such a mechanism exists for this sort of reciprocal relationship, with just a slight twist to make it interesting. Cells of many tissues contain a protein called inhibitor-1 that inhibits phosphoprotein phosphatase. Best studied in muscle, inhibitor-1 is subject to covalent modification by cAMP-dependent protein kinase and phosphoprotein phosphatase, as shown in Figure 7.54. Only the *a* form (phosphorylated form) of **phosphatase inhibitor-1** will inhibit phosphoprotein phosphatase. It should be noted that cAMP, by completely indirect mechanisms, causes the activation of phosphorylase kinase and the inhibition of phosphoprotein phosphatase—making it possible theoretically to activate glycogen phosphorylase completely (see Figure 7.54).

It is interesting that the phosphorylated form of phosphatase inhibitor-1 is converted back to its dephosphorylated form by the enzyme it inhibits—the phosphoprotein phosphatase! However, the inhibitor cleverly does not inhibit its own dephosphorylation, just the dephosphorylation of phosphorylase kinase *a* and phosphorylase *a*. In contrast to many other interconvertible enzymes that become phosphorylated on serine residues, inhibitor-1 becomes phosphorylated on a threonine residue. This may account in part for the reason why it can inhibit the action of phosphoprotein phosphatase against other phosphorylated enzymes and yet serve itself as a substrate for phosphoprotein phosphatase.

There is a good reason for the existence of the bicyclic control system for the phosphorylation of phosphorylase plus the additional control on its dephosphorylation. This provides a tremendous **amplification mechanism.** Think about it with relation to Figure 7.54. One molecule of epinephrine or one molecule of glucagon can cause, by the activation of adenylate cyclase, the formation of many molecules of cAMP. Cyclic AMP can then activate cAMP-dependent protein kinase, which, in turn, can cause the activation of many molecules of phosphatase inhibitor-1 as well as the activation of many molecules of phosphorylase kinase. In turn, phosphorylase kinase can cause the phosphorylation of many molecules of glycogen phosphorylase, which in turn can cause the phosphorolysis of many glucosidic bonds of glycogen. This mechanism provides an elaborate amplification system in which the signal provided by just a few molecules of hormone can be amplified into production of an enormous number of glucose 1-phosphate molecules. If each step represents, for argument's sake, an amplification factor of 100, then a total of four steps would result in an amplification of 100 million! This system is so rapid, in large part because of the amplification system, that all of the stored glycogen of white muscle fibers could be completely mobilized within just a few seconds.

Regulation of Glycogen Synthase

Glycogen synthase has to be active for glycogen synthesis and inactive during glycogen degradation. The combination of the reactions catalyzed by glycogen synthase, glycogen phosphorylase, glucose 1-phosphate uridylyltransferase, and nucleoside diphosphate kinase adds up to a futile cycle with the overall equation: ATP $\rightarrow$ ADP + P_i. Hence glycogen synthase needs to be turned off when glycogen phosphorylase is turned on, and vice versa.

The allosteric mechanism of glucose 6-phosphate activation of glycogen synthase might be of physiological significance under some circumstances. However, as with glycogen phosphorylase, this mode of control is integrated with regulation by covalent modification (see Figure 7.55). Glycogen synthase is known to exist in two forms. One is designated the **D form** because this form of the enzyme is dependent on the presence of G6P for activity. The other is designated the **I form** because this form of the enzyme is independent, that is it is active in the absence of G6P. These are old names for the two forms of the enzyme, used before the enzyme was established to be subject to covalent modification. The D form corresponds to the *b* or inactive form of the enzyme, the I form to the *a* or active form of the enzyme. Phosphorylation of this enzyme can be catalyzed by several different kinases, which, in turn, are regulated by several different second messengers of hormone action, including cAMP, Ca^{2+}, diacylglycerol, and perhaps some yet to be identified compounds (Figure 7.55). Each of the protein kinases shown in Figure 7.55 is capable of catalyzing the phosphorylation and contributing to the inactivation of glycogen synthase. Although glycogen synthase is a simple tetramer (α_4) with only one subunit type of mol wt 85,000, this subunit can be phosphorylated on at least nine different serine residues! At last count, eight different protein kinases have been identified that could phosphorylate glycogen synthase at one or more specific sites. This stands in striking contrast to glycogen phosphorylase, which is phosphorylated only by phosphorylase kinase and only at one specific site. Also in contrast to glycogen phosphorylase, the phosphorylation of glycogen synthase results in inactivation rather than activation of the enzyme. Likewise, phosphorylation of phosphatase inhibitor-1 by cAMP-dependent protein kinase results in inhibition of phosphoprotein phosphatase, which prevents reactivation of glycogen synthase. Since cAMP plays such an important role in the regulation of glycogen phosphorylase (Figure 7.54), it should be immediately appreciated that cAMP is an extremely important intracellular signal for reciprocally controlling glycogen synthase and glycogen phosphorylase. An increase in intracellular cAMP signals the activation of glycogen phosphorylase by the two different mechanisms, as shown in Figure 7.54, and signals the inactivation of glycogen synthase by the same two mechanisms, as shown in Figure 7.55. Note, however, that the regulation of glycogen synthase by cAMP is not bicyclic. Cyclic AMP-dependent protein kinase directly phosphorylates glycogen synthase, bypassing the need for phosphorylase kinase. Exactly why cAMP-dependent protein kinase directly phosphorylates glycogen synthase, thereby eliminating bicyclic control of glycogen synthase and losing some of the amplification factor important in the regulation of glycogen phosphorylase, is not known. This is particularly surprising in view of the fact that phosphorylase kinase is one of the enzymes capable of phosphorylating glycogen synthase (Figure 7.55). This may be important because phosphorylase kinase, in contrast to cAMP-dependent protein kinase, is sensitive to regulation by Ca^{2+}. Thus, both cAMP and Ca^{2+} influence the phosphorylation state and, therefore, the activity state of the two regulatory enzymes of glycogen metabolism. Furthermore, two cAMP-independent, Ca^{2+}-activated protein kinases have been identified that also may have physiological significance. One of these has been named calmodulin-dependent protein kinase and the other Ca^{2+}- and phospholipid-dependent protein kinase (protein kinase C). Both enzymes phosphorylate glycogen synthase, but neither enzyme can use glycogen phosphorylase as substrate. **Protein kinase C** requires phospholipid, diacylglycerol, and Ca^{2+} for full activity. There is considerable interest in protein kinase C because tumor-promoting agents called **phorbol esters** have been found to mimic diacylglycerol as activators of this enzyme. Thus, diacylglycerol is an important ''second messenger'' of hormone action, acting via protein kinase C to regulate numerous cellular processes.

Glycogen synthase is also phosphorylated by glycogen synthase kinase-3, casein kinase I, and casein kinase II. These kinases are not subject to regulation by either cAMP or Ca^{2+}. It is likely, however, that special regulatory mechanisms exist to regulate these kinases. Herein may lie solutions to unsolved problems such as the mechanism of action of insulin and other hormones. Of considerable current interest is the finding that phosphorylation by casein kinase II is required for the phosphorylation of glycogen synthase by glycogen synthase kinase-3. This is an example of **synergistic phosphorylation,** meaning phosphorylation of a protein is dependent on the action of a second kinase. Also referred to as **hierarchal phosphorylation** because phosphates are added in an exact order (Figure 7.56), phosphorylation of glycogen synthase by casein kinase II on one specific serine residue creates a phosphorylation site recognized by glycogen synthase kinase-3. Phosphorylation of the latter site creates another recognition site for glycogen synthase kinase-3 and this continues until appropriately spaced serine residues are no longer available and glycogen synthase has been phosphorylated five times (Figure 7.56). Hierarchal

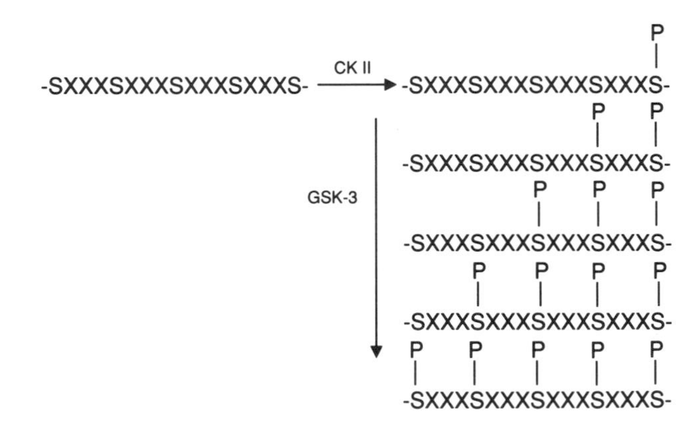

FIGURE 7.56
Sequential model for synergistic phosphorylation of glycogen synthase by casein kinase II and glycogen synthase kinase-3.
A portion of the amino acid sequence of glycogen synthase is shown; S denotes a serine residue, X refers to any other amino acid. Casein kinase II (CK II) is responsible for the first phosphorylation and generates a recognition site —SXXX-S(P)— for glycogen synthase kinase-3 (GSK-3). The GSK-3 then introduces phosphates sequentially at additional sites. A new GSK-3 recognition site —SXXXS(P)— is created with each phosphorylation.
Redrawn with modification from C. J. Fiol, A. M. Mahrenholz, Y. Wang, R. W. Roeske, and P. J. Roach. Formation of protein kinase recognition sites by covalent modification of the substrate. *J. Biol. Chem.* 262:14042, 1987.

phosphorylation is likely involved in the control of other multiply phosphorylated proteins and may prove of great importance in the regulation of numerous cellular processes.

Regulation of Phosphoprotein Phosphatases

There are mechanisms other than those shown in Figures 7.54 and 7.55, which are believed to be important in the regulation of phosphoprotein phosphatases. There may be a limited number of catalytically active phosphatase subunits (two major ones have been studied in detail) but these subunits associate with a number of different regulatory subunits to create complexes subject to unique mechanisms of regulation. As an example, a **glycogen-binding protein** (named G subunit) has been identified that binds both glycogen and a catalytically active phosphatase subunit (Figure 7.57). This association makes the phosphatase 10 times more active toward glycogen synthase and glycogen phosphorylase. Phosphorylation of the G subunit by cAMP-dependent protein kinase results in release of the catalytic subunit, which is then less active and also subject to inhibition by phosphorylated inhibitor-1. Another mechanism of current interest involves a catalytic subunit that is inactive because of association with a protein called **inhibitor-2** (Figure 7.58). In this case, phosphorylation of inhibitor-2 results in activation of the phosphatase catalytic subunit. In another example, a catalytic subunit associates with calmodulin and binding of Ca^{2+} to calmodulin regulates phosphatase activity. Thus, the phosphatase story is a complex one and will continue to evolve as more work is done on the regulatory features of these enzymes. An important point to appreciate is that the phosphatases, like the kinases, are subject to intricate regulatory mechanisms and that their relative activities set the activity of an enzyme subject to covalent modification.

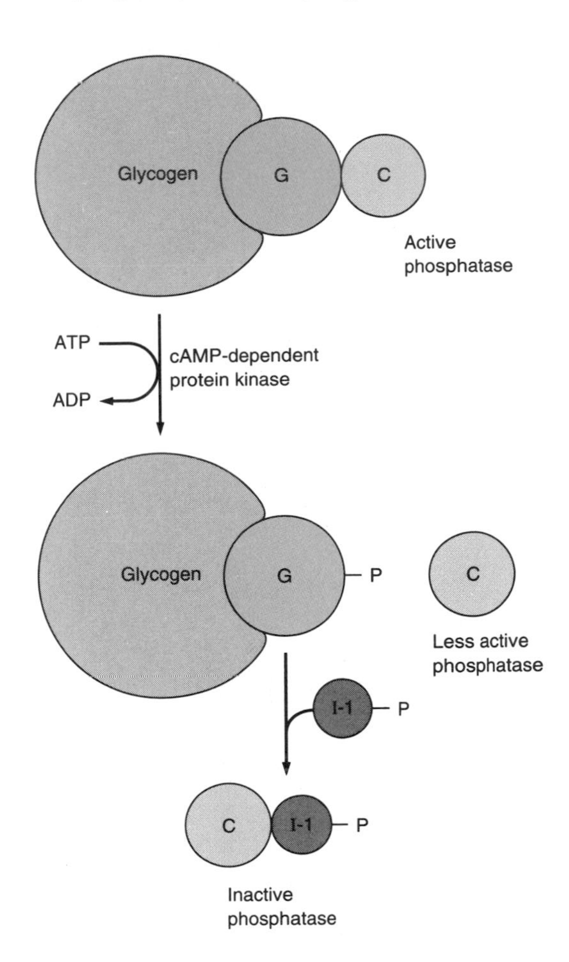

FIGURE 7.57
Mechanism for regulation of a phosphatase that binds to glycogen.
The G subunit binds directly to glycogen; the C (catalytic subunit) binds via the G subunit; phosphorylated I-1 (inhibitor 1) binds released C.

Effector Control of Glycogen Metabolism

Certain muscles under anaerobic conditions have been shown to mobilize their glycogen stores rapidly without marked conversion of phosphorylase *b* into phosphorylase *a* or glycogen synthase *a* into glycogen synthase *b*. Presumably this is accomplished by effector control in which ATP levels decrease, causing less inhibition of phosphorylase; glucose 6-phosphate levels decrease, causing less activation of glycogen synthase; and AMP levels increase, causing activation of phosphorylase. This enables the muscle to keep working, for at least a short period of time, by using the ATP produced by glycolysis of the glucose 6-phosphate obtained from glycogen.

Proof that effector control can operate has also been obtained in studies of a special strain of mice that are deficient in muscle phosphorylase kinase. Phosphorylase *b* in the muscle of such mice cannot be converted into phosphorylase *a*. Nevertheless, heavy exercise of these mice results in depletion of muscle glycogen, presumably because of stimulation of phosphorylase *b* by effectors.

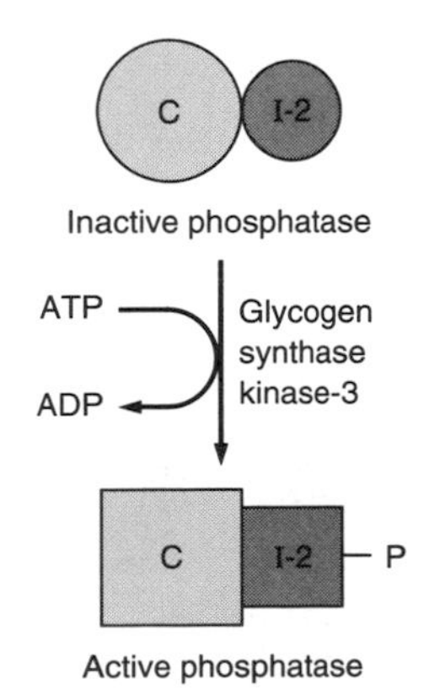

FIGURE 7.58
Mechanism for regulation of a phosphatase that binds inhibitor-2.
Inhibitor-2 (I-2) binds and inhibits the catalytic subunit (C). Phosphorylation by glycogen synthase kinase-3 results in conformational changes in the proteins and activation of the complex.

Negative Feedback Control of Glycogen Synthesis by Glycogen

As mentioned previously (see p. 345), glycogen is able to exert feedback control over its own formation. The portion of glycogen synthase in the active *a* form decreases as glycogen accumulates in a particular tissue. The mechanism is not well understood, but glycogen may make the *a* form of glycogen synthase a better substrate for one or more of the protein kinases, or, alternatively, glycogen may inhibit dephosphorylation of glycogen synthase *b* by phosphoprotein phosphatase. Either of these mechanisms would account for the shift in the steady state in favor of glycogen synthase *b* that occurs in response to glycogen accumulation.

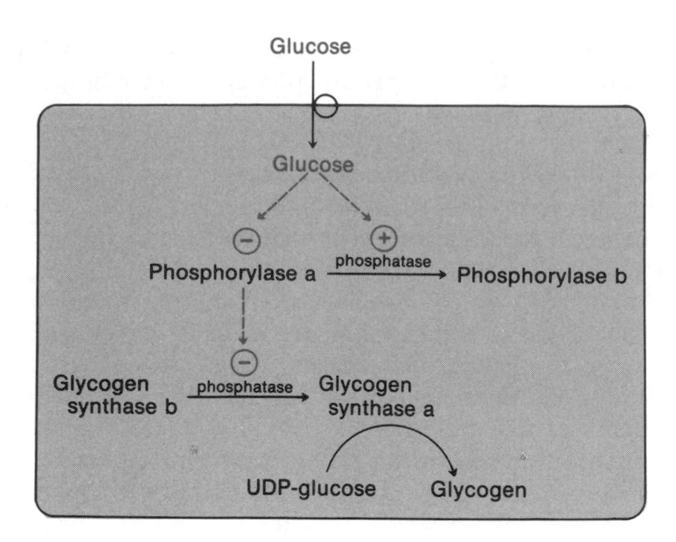

FIGURE 7.59
Overview of the mechanism responsible for glucose stimulation of glycogen synthesis in the liver.

Phosphorylase a Functions as a "Glucose Receptor" in the Liver

Consumption of a carbohydrate-containing meal results in an increase in blood and liver glucose, which, in turn, signals an increase in glycogen synthesis in the latter tissue. The mechanism involves glucose stimulation of insulin release from the pancreas and subsequent effects of this hormone on hepatic glycogen phosphorylase and glycogen synthase. However, hormone-independent mechanisms also appear to be important in the liver (Figure 7.59). Direct inhibition of phosphorylase *a* by glucose is probably of importance. Moreover, binding of glucose to phosphorylase makes the *a* form of phosphorylase a better substrate for dephosphorylation by phosphoprotein phosphatase. This has led to the hypothesis that phosphorylase *a* can function as a glucose receptor in the liver. Binding of glucose to phosphorylase *a* promotes the inactivation of phosphorylase *a*, with the overall result being inhibition of glycogen degradation by glucose. This **"negative feedback"** control of glycogenolysis by glucose would not necessarily promote glycogen synthesis. However, there also is evidence that phosphorylase *a* is an inhibitor of the dephosphorylation of glycogen synthase *b* by phosphoprotein phosphatase. This inhibition is lost once phosphorylase *a* has been converted to phosphorylase *b* (Figure 7.59). In other words, phosphoprotein phosphatase can turn its attention to glycogen synthase *b* only following dephosphorylation of phosphorylase *a*. Thus, as a result of the interaction of glucose with phosphorylase *a*, phosphorylase becomes inactivated, glycogen synthase becomes activated, and glycogen is synthesized rather than degraded in the liver. Phosphorylase *a* can serve this function of "glucose receptor" in liver because the concentration of glucose in liver always reflects the blood concentration of glucose. This is not true for extrahepatic tissues. Liver cells have a very high-capacity transport system for glucose and a high K_m enzyme for glucose phosphorylation (glucokinase). Cells of extrahepatic tissues as a general rule have low-capacity transport systems for glucose and low K_m enzymes for glucose phosphorylation (hexokinase) that maintain intracellular glucose at concentrations too low for phosphorylase *a* to function as a "glucose receptor."

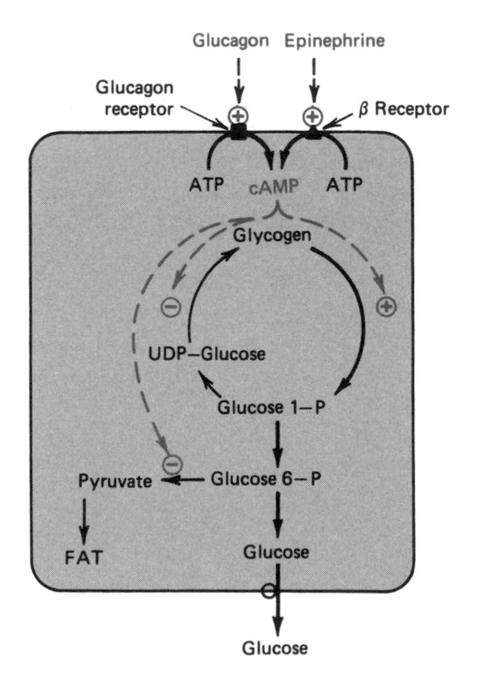

FIGURE 7.60
Cyclic AMP mediates the stimulation of glycogenolysis in liver by glucagon and β agonists.
The glucagon receptor (■), β-adrenergic receptor (▲), glucose transporter (○) are intrinsic components of the plasma membrane. Membrane bound adenylate cyclase converts ATP to cAMP.

Glucagon Stimulates Glycogen Degradation in the Liver

Glucagon is released from α cells of the pancreas in response to low glucose levels in the blood. One of glucagon's primary jobs during periods of low food intake (fasting or starvation) is to mobilize liver glycogen, that is, stimulate glycogenolysis, in order to ensure that adequate blood glucose is available to meet the needs of glucose-dependent tissues. Glucagon circulates in the blood until it interacts with glucagon receptors such as those located on the plasma membrane of liver cells (Figure 7.60). Binding of glucagon to these receptors activates adenylate cyclase and triggers the cascades that result in activation of glycogen phosphorylase and inactivation of glycogen synthase by the mechanisms given in Figures 7.54 and 7.55, respectively. Glucagon also inhibits the use of glucose by glycolysis at the level of 6-phosphofructo-1-kinase and pyruvate kinase by the mechanisms given in Figures 7.26 and 7.28, respectively. The net result of these effects of glucagon, all mediated by the second messenger cAMP and covalent modification, is a very rapid increase in blood glucose levels. Hyperglycemia might be expected but does not occur because less glucagon is released from the pancreas as blood glucose levels increase.

Epinephrine Stimulates Glycogen Degradation in the Liver

Epinephrine is released into the blood from the chromaffin cells of the adrenal medulla in response to stress. This hormone is our "fright, flight or fight" hormone, preparing the body for either combat or escape.

Epinephrine mobilizes liver glycogen by at least three different mechanisms, but the physiologically most important mechanism is not clearly established for humans. One mechanism involves epinephrine stimulation of glucagon release from the α cells of the pancreas. Glucagon then travels by way of the blood to mobilize liver glycogen as discussed above. Epinephrine can also interact directly with receptors in the plasma membrane of the liver cells to activate adenylate cyclase (Figure 7.60). The resulting increase in cAMP has the same effect as that caused by glucagon. The binding site for epinephrine on the plasma membrane, which is in communication with adenylase cyclase, is called the **β-adrenergic receptor.** The plasma membrane of liver cells also has another binding site for epinephrine, called the **α-adrenergic receptor.** Interaction of epinephrine with α-adrenergic receptors leads to the formation of **inositol 1,4,5-trisphosphate (IP_3)** and diacylglycerol (Figure 7.61). These compounds are second messengers, produced in the plasma membrane by the action of a **phospholipase C** on phosphatidylinositol 4,5-bisphosphate by the reaction shown in Figure 7.62. Inositol 1,4,5-trisphosphate stimulates the release of Ca^{2+} from the endoplasmic reticulum (Figure 7.61). As previously discussed for Figure 7.54, the increase in Ca^{2+} activates phosphorylase kinase, which in turn activates glycogen phosphorylase. Likewise, as previously discussed for Figure 7.55, Ca^{2+}-mediated activation of phos-

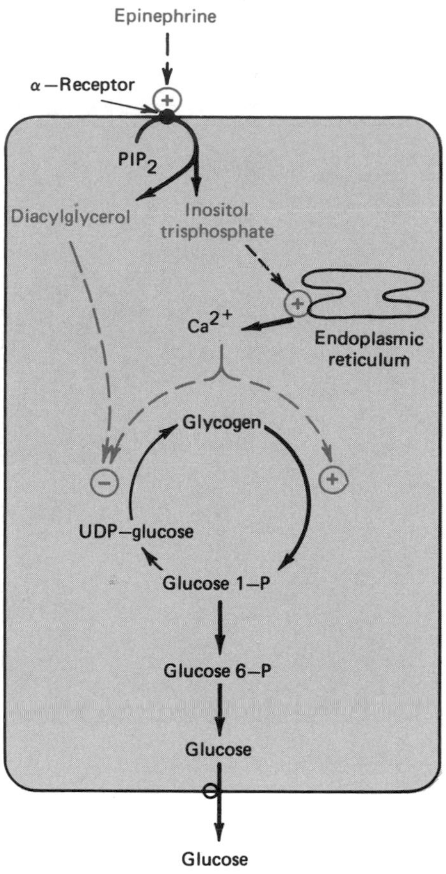

FIGURE 7.61
Inositol trisphosphate and Ca^{2+} mediate the stimulation of glycogenolysis in liver by α agonists.
The α-adrenergic receptor (●) and glucose transporter (○) are intrinsic components of the plasma membrane. Although not indicated, phosphatidylinositol 4,5-bisphosphate (PIP_2) is also a component of the plasma membrane.

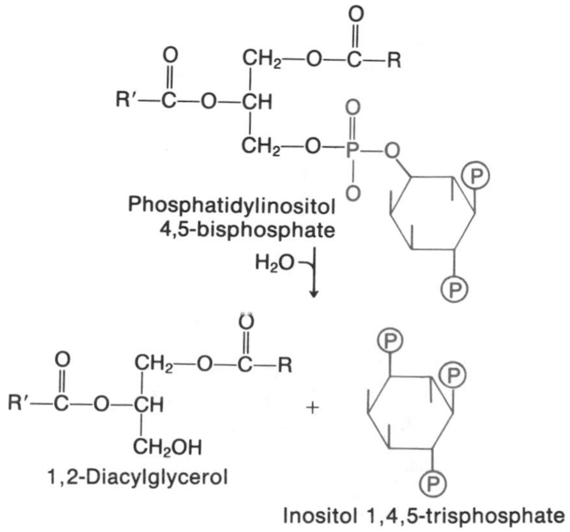

FIGURE 7.62
Phospholipase C cleaves phosphatidylinositol 4,5-bisphosphate to produce 1,2-diacylglycerol and inositol 1,4,5-trisphosphate.
$R-\overset{O}{\overset{\|}{C}}-$ refers to long-chain acyl groups of the molecules. Ⓟ refers to phosphate (PO_4^{2-}) groups.

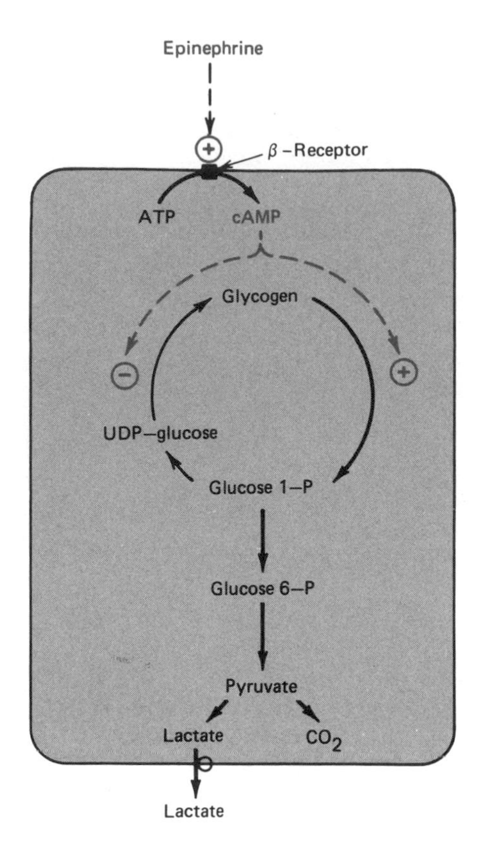

FIGURE 7.63
Cyclic AMP mediates the stimulation of glycogenolysis in muscle by β agonists.
The β-adrenergic receptor (■) is an intrinsic component of the plasma membrane.

phorylase kinase, calmodulin-dependent protein kinase, and protein kinase C, as well as diacylglycerol-mediated activation of protein kinase C, may all be important for inactivation of glycogen synthase.

The consequences of all of the mechanisms described for epinephrine are the same—increased release of glucose into the blood from the glycogen stored in the liver. This makes more blood glucose available to tissues that are called upon to meet the challenge of the stressful situation that triggered the release of epinephrine from the adrenal medulla.

Epinephrine Stimulates Glycogen Degradation in Skeletal Muscle

Epinephrine also stimulates glycogen degradation in skeletal muscle. This tissue lacks glucagon receptors but has β-adrenergic receptors. Cyclic AMP, produced in response to epinephrine stimulation of adenylate cyclase via β-adrenergic receptors (Figure 7.63), signals the concurrent activation of glycogen phosphorylase and inactivation of glycogen synthase by the mechanisms given previously in Figures 7.54 and 7.55, respectively. This does not lead, however, to glucose release into the blood from this tissue. In contrast to liver, skeletal muscle lacks glucose 6-phosphatase, and in this tissue cAMP does not inhibit glycolysis. Thus, the role of epinephrine on glycogen metabolism in skeletal muscle is to make more

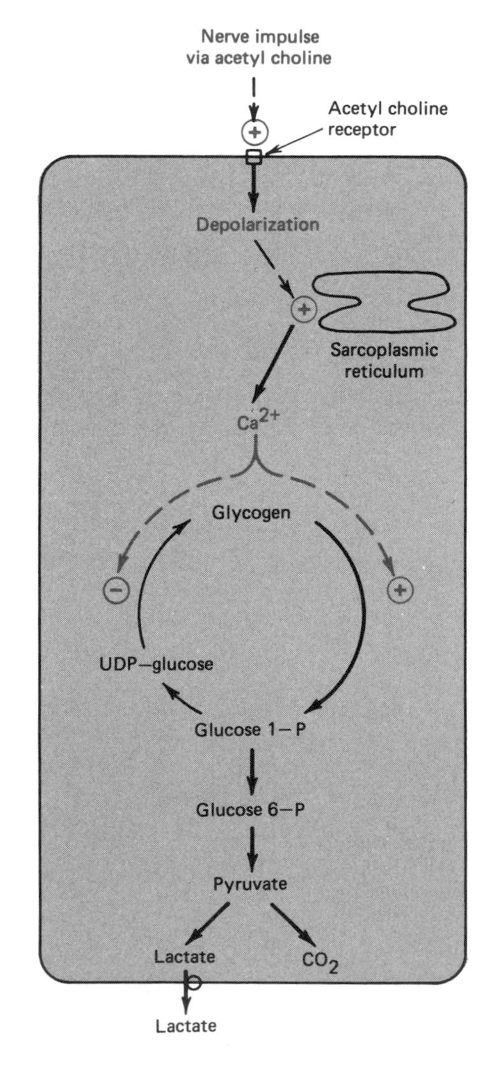

FIGURE 7.64
Ca^{2+} mediates the stimulation of glycogenolysis in muscle by nervous excitation.

substrate (glucose 6-phosphate) available for glycolysis. ATP generated by glycolysis can then be used to meet the metabolic demand imposed upon skeletal muscle by the stress that triggered epinephrine release.

Neural Control of Glycogen Degradation in Skeletal Muscle

Nervous excitation of muscle activity is mediated via changes in intracellular Ca^{2+} concentrations (Figure 7.64). The nerve impulse causes membrane depolarization which, in turn, causes **Ca^{2+}** release from the sarcoplasmic reticulum into the sarcoplasm of muscle cells. This release of Ca^{2+} triggers muscle contraction, whereas reaccumulation of Ca^{2+} by the sarcoplasmic reticulum causes relaxation. The same change in Ca^{2+} concentration effective in causing muscle contraction (from 10^{-8} to 10^{-6} M) also greatly affects the activity of phosphorylase kinase. As Ca^{2+} concentrations increase there is more muscle activity and a greater need for ATP. The activation of phosphorylase kinase by Ca^{2+} leads to the subsequent activation of glycogen phosphorylase and perhaps the inactivation of glycogen synthase. The result is that more glycogen is converted to glucose 6-phosphate so that more ATP can be produced to meet the greater energy demand of muscle contraction.

Synergistic Effects of Epinephrine and Neural Signals on Glycogen Degradation in Skeletal Muscle

As has been discussed for Figures 7.54 and 7.55, phosphorylase kinase can be activated both by phosphorylation of the α and β subunits and by interaction of Ca^{2+} with the δ subunit. These two different mechanisms for phosphorylase kinase activation provide independent routes to the activation of phosphorylase by epinephrine (Figure 7.63) and neural signals (Figure 7.64). However, maximum activation of phosphorylase kinase requires phosphorylation of the enzyme plus the binding of Ca^{2+} to the enzyme. This provides a mechanism useful to the skeletal muscle cell. For example, if a muscle is not signaled to contract in response to a particular stressful situation, that muscle will require no additional burst of ATP production from glycolysis and there will have been no reason for epinephrine to have triggered degradation of glycogen. Epinephrine serves to make the muscle more sensitive to changes in intracellular Ca^{2+} concentration. The release of epinephrine into the blood in response to stress triggers the cascade that causes the conversion of phosphorylase kinase *b* to its *a* form in skeletal muscle. However, this will only partially activate phosphorylase because phosphorylase kinase a is not fully active without Ca^{2+}. If nervous stimulation triggers work to be done in the muscle, intracellular Ca^{2+} levels will increase, phosphorylase kinase *a* will be maximally activated, phosphorylase will be maximally converted to its *a* form, and appropriate amounts of glycogen will be mobilized to meet the metabolic demand of the stimulated tissue.

Insulin Stimulates Glycogen Synthesis in Muscle and Liver

An increase in blood glucose signals the release of insulin from β cells of the pancreas. Insulin circulates in the blood, serving as a first messenger to inform several tissues that excess glucose is present. Insulin receptors, located on the plasma membranes of insulin-responsive cells, respond to insulin binding by producing secondary mediators of insulin action that promote glucose use within these tissues. The pancreas responds to a decrease in blood glucose with less release of insulin but greater release of glucagon. These hormones have opposite effects on glucose utilization by the liver, thereby establishing the pancreas as a fine-tuning device that prevents dangerous fluctuations in blood glucose levels.

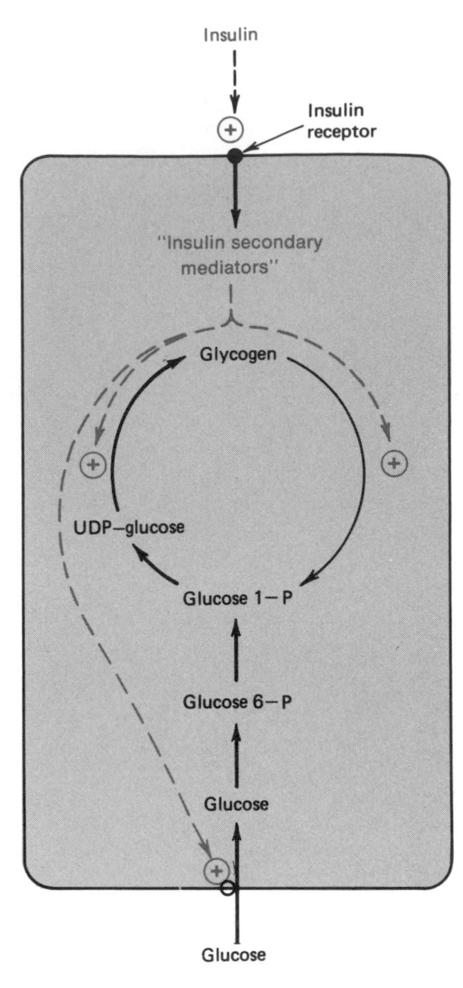

FIGURE 7.65
Insulin acts via secondary mediators of its action to promote glycogen synthesis in muscle.

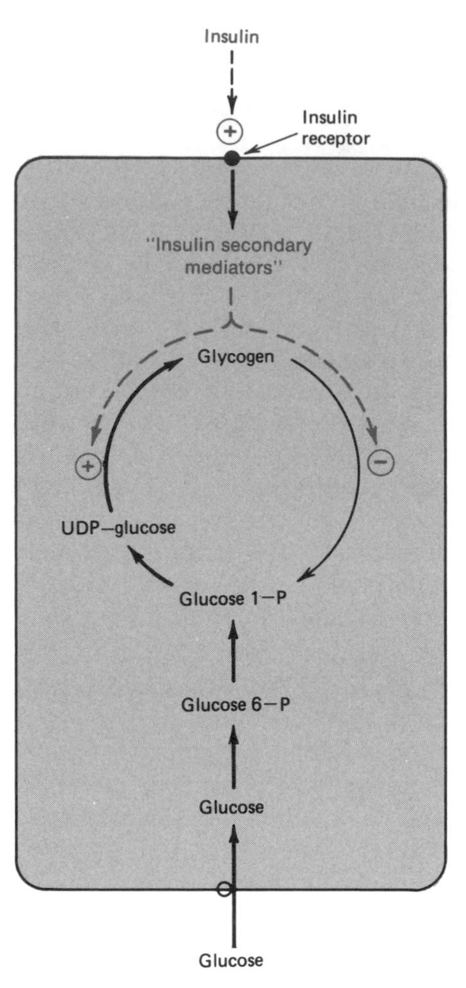

FIGURE 7.66
Insulin acts via secondary mediators of its action to promote glycogen synthesis in liver.

Insulin increases glucose utilization rates in part by promoting glycogenesis and inhibiting glycogenolysis in muscle (Figure 7.65) and liver (Figure 7.66). Insulin stimulation of glucose transport at the plasma membrane is important for these effects in muscle but not liver. Hepatocytes have a high-capacity, insulin-insensitive transport system, whereas muscle cells are equipped with a low capacity system that requires insulin for maximum rates of glucose uptake. Insulin stimulates muscle (and adipose tissue) glucose transport by signaling an increase in the number of functional glucose transporters associated with the plasma membrane. This is accomplished by promoting the translocation of glucose transporters from an intracellular pool to the plasma membrane (see Chapter 20). Insulin further promotes glycogen accumulation in both tissues by activating glycogen synthase. The secondary mediators of insulin action have effects opposite to those of cAMP. Recall from Figures 7.54 and 7.55 that cAMP promotes phosphorylation of glycogen phosphorylase and glycogen synthase by activating protein kinases and inhibiting phosphoprotein phosphatase. Thus, insulin promotes dephosphorylation of these enzymes by inhibiting protein kinases and activating phosphoprotein phosphatase. The mechanism by which **insulin receptor** activation results in modulation of enzyme activity and changes in gene expression is further discussed in Chapter 20.

BIBLIOGRAPHY

Arion, W. J., Lange, A. J., Walls, H. E., and Ballas, L. M. Evidence for the participation of independent translocases for phosphate and glucose 6-phosphate in the microsomal glucose 6-phosphatase system. *J. Biol. Chem.* 255:10396, 1980.

Berridge, M. J. Review article: Inositol trisphosphate and diacylglycerol as second messengers. *Biochem. J.* 220:345, 1984.

Brooke, M. H. and Kaiser, K. K. The use and abuse of muscle histochemistry. *Ann. N.Y. Acad. Sci.* 228:121, 1974.

Claus, T. H., et al. The role of fructose 2,6-bisphosphate in the regulation of carbohydrate metabolism. *Curr. Topics Cell. Regul.* 23:57, 1984.

DeFronzo, R. A. and Ferrannini, E. Regulation of hepatic glucose metabolism in humans. *Diabetes/Metabolism Rev.* 3:415, 1987.

DePaoli-Roach, A. A. Regulatory components of type 1 protein phosphatases. *Adv. Prot. Phosphatases* 5:479, 1989.

Exton, J. H. Mechanisms of hormonal regulation of hepatic glucose metabolism. *Diabetes/Metabolism Reviews* 3:163, 1987.

Geelen, M. J. H., Harris, R. A., Beynen, A. C., and McCune, S. A. Short-term hormonal control of hepatic lipogenesis. *Diabetes* 29:1006, 1980.

Greene, H. L., Slonin, A. E., and Burr, I. M. Type I glycogen storage disease: A metabolic basis for advances in treatment, L. A. Barness (Ed.), *Advances in pediatrics,* Vol. 26, Chicago: Year Book Publishers, 1979, p. 63.

Hallfrisch, J. Metabolic effects of dietary fructose. *FASEB J.* 4:2652, 1990.

Hanson, R. W. and Mehlman, M. A. (Eds.). *Gluconeogenesis, its regulation in mammalian species.* New York: Wiley, 1976.

Houslay, M. D. The search for a molecular mechanism for the action of insulin. *Biochem. Educ.* 12:49, 1984.

Ingebritsen, T. S. and Cohen, P. Protein phosphatases: Properties and role in cellular regulation. *Science* 221:331, 1983.

Isselbacher, K. J., Adams, R. D., Braundwald, E., Petersdorf, R. B., and Wilson, J. D. (Eds.). *Harrison's principles of internal medicine,* 9th ed. New York: McGraw-Hill, 1980.

Lieber, C. S. The metabolism of alcohol. *Sci. Am.* 234:25, 1976.

Lomako, J., Lomako, W. M., and Whelan, W. J. A self-glucosylating protein is the primer for rabbit muscle glycogen biosynthesis. *FASEB J.* 2:3097, 1988.

Metzler, D. E. *Biochemistry, the chemical reactions of living cells,* New York, Academic, 1977.

Newsholme, E. A. and Leech, A. R., *Biochemistry for the medical sciences.* New York: Wiley, 1983.

Newsholme, E. A. and Start, C. *Regulation in metabolism.* New York: Wiley, 1973.

Nishizuka, Y. Protein kinases in signal transduction. *Trends Biochem. Sci.* 9:163, 1984.

Pilkis, S. J. and El-Maghrabi, M. R. Hormonal regulation of hepatic gluconeogenesis and glycolysis. *Annu. Rev. Biochem.* 57:755, 1988.

Roach, P. J. Hormonal control of glycogen metabolism, H. Rupp (Ed.), *Regulation of heart function: Basic concepts and clinical applications.* New York: Thieme-Stratton Inc., 1985.

Roach, P. J. Principles of the regulation of enzyme activity, L. Goldstein and D. M. Prescott (Eds.), *Cell biology, a comprehensive treatise,* Vol. IV. New York: Academic, 1980.

Scriver, C. R., Beaudet, A. L., Sly, W. S., and Valle, D. (Eds.), *The metabolic basis of inherited disease,* 5th ed. New York: McGraw-Hill, 1988.

Stanley, C. A., Anday, E. K., Baker, L., and Delivoria-Papadopolous, M. Metabolic fuel and hormone responses to fasting in newborn infants. *Pediatrics* 64:613, 1979.

van de Werve, G. and Jeanrenaud, B. Liver glycogen metabolism: an overview. *Diabetes Metabolism Rev.* 3:47, 1987.

Van Schaftingen, E. A protein from rat liver confers to glucokinase the property of being antagonistically regulated by fructose 6-phosphate and fructose 1-phosphate. *Eur. J. Biochem.* 179:179, 1989.

QUESTIONS

J. BAGGOTT AND C. N. ANGSTADT

*Q*uestion *T*ypes are described on page xxiii.

1. (QT1) In glycolysis ATP synthesis is catalyzed by
   A. hexokinase.
   B. 6-phosphofructo-1-kinase.
   C. glyceraldehyde-3-phosphate dehydrogenase.
   D. phosphoglycerate kinase.
   E. none of the above.

2. (QT2) NAD^+ can be regenerated in the cytoplasm if NADH reacts with:
   1. pyruvate.
   2. dihydroxyacetone phosphate
   3. oxaloacetate
   4. the flavin bound to NADH dehydrogenase

A. Glucokinase C. Both
B. Hexokinase D. Neither

3. (QT4) K_m is well above normal blood glucose concentrations.
4. (QT4) Found in muscle.
5. (QT4) Inhibited by G6P.

6. (QT1) 6-Phosphofructo-1-kinase can be inhibited by all of the following *except:*
   A. ATP at high concentrations.
   B. citrate.
   C. AMP.
   D. low pH.

7. (QT2) Which of the following supports gluconeogenesis:
   1. α-ketoglutarate + aspartate $\rightarrow$ glutamate + oxaloacetate
   2. pyruvate + ATP + HCO_3^- $\rightarrow$ oxaloacetate + ADP + P_i + H^+
   3. acetyl CoA + oxaloacetate + H_2O $\rightarrow$ citrate + CoA
   4. glutamate + NAD^+ $\rightarrow$ α-ketoglutarate + NADH + NH_4^+

A. Cori cycle C. Both
B. Alanine cycle D. Neither

8. (QT4) Involves only tissues with aerobic metabolism (i.e., mitochondria and O_2).
9. (QT4) Three-carbon compounds arising from glycolysis are converted to glucose at the expense of energy from fatty acid oxidation.

10. (QT2) The uncontrolled production of NADH from NAD^+ during ethanol metabolism blocks gluconeogenesis from:
    1. pyruvate.
    2. oxaloacetate.
    3. glycerol.
    4. galactose.
11. (QT2) Gluconeogenic enzymes include:
    1. pyruvate carboxylase.
    2. fructose-1,6-bisphosphatase.
    3. phosphoenolpyruvate carboxykinase.
    4. phosphoglucomutase.

A. Glucose production from pyruvate C. Both
B. Glucose production from glycogen D. Neither

12. (QT4) Consumes ATP.
13. (QT4) Glucose 6-phosphatase is involved.

A. Synthesis of glycogen C. Both
B. Breakdown of glycogen D. Neither

14. (QT4) Phosphoglucomutase.
15. (QT4) Glucose 1-phosphate uridylyltransferase.
16. (QT4) Chain of glucosyl residues is transferred.
17. (QT4) Stimulated by epinephrine.

18. (QT2) Phosphorylation activates:
    1. glycogen phosphorylase.
    2. inhibitor-1.
    3. phosphorylase kinase.
    4. protein kinase.
19. (QT2) AMP activates:
    1. 6-phosphofructo-1-kinase.
    2. protein kinase.
    3. glycogen phosphorylase.
    4. hexokinase.

ANSWERS

1. D A and B use ATP; both catalyze irreversible reactions. C synthesizes 1,3-bisphosphoglycerate. D synthesizes ATP in the forward direction; the reaction is reversible (p. 302).
2. A 1, 2, and 3 true. 1 may be converted to lactate. 2 and 3 are the cytoplasmic acceptors for shuttle systems. 4 is mitochondrial (p. 304).
3. A Blood glucose is ~5 mM. K_m of glucokinase is ~10 mM. K_m of hexokinase is <0.1 mM (p. 311).
4. B Hexokinases are widely distributed. Glucokinase is hepatic (p. 309).
5. B Glucose 6-phosphate inhibition is an important control of hexokinase (p. 309).
6. C AMP is an allosteric regulator that relieves inhibition by ATP (p. 313).
7. C 2 and 4 true. 1 : α-Ketoglutarate and oxaloacetate both give rise to glucose; interconversion of one to the other accomplishes nothing (p. 330). 3: Citrate ultimately gives rise to oxaloacetate, losing two carbon atoms in the process; again nothing is gained (p. 331). 2 is on the direct route of conversion of pyruvate to glucose. 4 converts an amino acid to a compound that is converted to oxaloacetate (p. 330). *Net* synthesis of a TCA intermediate is required to support gluconeogenesis.
8. B If alanine is the end product of glycolysis, NADH must be reoxidized aerobically (p. 325, Figure 7.30*b*).
9. C Lactate or alanine is transported to the liver for glucose synthesis. The liver's major energy source is fatty acid (p. 334).
10. A 1, 2, and 3 true. High $NADH/NAD^+$ converts pyruvate to lactate and oxaloacetate to malate (p. 336). It prevents oxidation of glycerol 3-phosphate to dihydroxyacetone phosphate (p. 332, Figure 7.39). Gluconeogenesis from galactose is not affected by the redox state of the cell (p. 334, Figure 7.42).
11. A 1, 2, and 3, along with glucose 6-phosphatase, are the gluconeogenic enzymes (p. 327). 4 is on the pathway between glucose and glycogen (p. 340), which is not part of gluconeogenesis (p. 324).
12. A ATP is required for gluconeogenesis but not for glycogenolysis (p. 341).
13. C Both processes produce glucose 6-phosphate, which must be hydrolyzed (p. 341). Some free glucose is, however, produced by the debranching enzyme (p. 342).
14. C The enzyme interconverts glucose 6-phosphate and glucose 1-phosphate (p. 341).
15. A The enzyme synthesizes UDP-glucose (p. 343).
16. C Synthesis involves "branching" enzyme (p. 344), and breakdown requires "debranching" enzyme (p. 342).
17. B Epinephrine activates glycogen phosphorylase and inactivates glycogen synthase via Ca^{2+} or the cAMP-regulated cascade (pp. 353–354, Figures 7.61 and 7.63).
18. A 1, 2, and 3 are activated by phosphorylation (p. 346, Figure 7.54). Protein kinase is not a phospho–dephospho enzyme (p. 319).
19. B 1 and 3 true. 1 (p. 313) and 3 (p. 346) are allosterically activated by AMP. 2 is activated by cAMP (p. 347). 4 is controlled by glucose 6-phosphate.

8

Carbohydrate Metabolism II: Special Pathways

NANCY B. SCHWARTZ

8.1 OVERVIEW 360
8.2 PENTOSE PHOSPHATE PATHWAY 361
The Pentose Phosphate Pathway Has Two Phases 361
Glucose 6-Phosphate Is Oxidized and Decarboxylated to a Pentose Phosphate 361
Interconversions of Pentose Phosphates Lead to Glycolytic Intermediates 362
Glucose 6-Phosphate Can Be Completely Oxidized to CO_2 364
The Pentose Phosphate Pathway Produces NADPH 365
8.3 SUGAR INTERCONVERSIONS AND NUCLEOTIDE SUGAR FORMATION 365
Isomerization and Phosphorylation Are Common Reactions for Interconverting Carbohydrates 365
Nucleotide-Linked Sugars Are Intermediates in Many Sugar Transformations 366
Epimerization Interconverts Glucose and Galactose 367
Glucuronic Acid Is Formed by Oxidation of UDP-Glucose 369
Decarboxylation, Oxidoreduction, and Transamination of Sugars Produce Necessary Products 370
Sialic Acids Are Derived From N-Acetylglucosamine 371
8.4 BIOSYNTHESIS OF COMPLEX CARBOHYDRATES 372
8.5 GLYCOPROTEINS 373
Glycoproteins Contain Variable Amounts of Carbohydrate 374
Carbohydrates Are Covalently Linked to Glycoproteins by *N*- or *O*- Glycosyl Bonds 375
Synthesis of N-Linked Glycoproteins Involves Dolichol Phosphate 376
8.6 PROTEOGLYCANS 378
Hyaluronate Is a Copolymer of *N*-Acetylglucosamine and Glucuronic Acid 378
Chondroitin Sulfates Are the Most Abundant Glycosaminoglycans 379
Dermatan Sulfate Contains L-Iduronic Acid 380
Heparin and Heparan Sulfate Differ From Other Glycosaminoglycans 380
Keratan Sulfate Is Very Heterogeneous 380
Biosynthesis of Chondroitin Sulfate Is Typical of Glycosaminoglycan Formation 381
BIBLIOGRAPHY 384
QUESTIONS AND ANNOTATED ANSWERS 384
CLINICAL CORRELATIONS
8.1 Glucose-6-Phosphate Dehydrogenase: Genetic Deficiency or Presence of Genetic Variants in Erythrocytes 362
8.2 Essential Fructosuria and Fructose Intolerance: Deficiency of Fructokinase and Fructose 1-Phosphate Aldolase 366
8.3 Galactosemia: Inability to Transform Galactose Into Glucose 368
8.4 Pentosuria: Deficiency of Xylitol Dehydrogenase 368
8.5 Glucuronic Acid: Physiological Significance of Glucuronide Formation 369
8.6 Blood Group Substances 373
8.7 Aspartylglycosylaminuria: Absence of 4-L-Aspartylglycosamine Amidohydrolase 374
8.8 Heparin as an Anticoagulant 375
8.9 Mucopolysaccharidoses 383

8.1 OVERVIEW

In addition to the catabolism of glucose for the specific purpose of energy production in the form of ATP, several other pathways involving sugar metabolism exist in cells. One, the **pentose phosphate pathway,** is particularly important in animal cells. As will be discussed, this pathway does not operate instead of glycolysis and the tricarboxylic acid cycle, but rather it functions side by side with them for production of reducing power and pentose intermediates. Thus, the metabolic significance of this pathway, known also as the **hexose monophosphate shunt** or the **6-phosphogluconate pathway,** is not to obtain energy from the oxidation of glucose in animal tissues. In fact, starting with glucose 6-phosphate, no ATP is generated, nor is any required. The pentose phosphate pathway is rather a multifunctional pathway whose primary purpose is to generate reducing power in the form of NADPH. It has previously been mentioned that the fundamental distinction between NADH and NADPH in most biochemical reactions is that NADH is oxidized by the respiratory chain to produce ATP, whereas NADPH serves as a hydrogen and electron donor in reductive biosynthetic reactions. The enzymes involved in this pathway are located in the cytosol, indicating that the oxidation that occurs is not dependent on mitochondria or the tricarboxylic acid cycle. Another important function is to convert hexoses into pentoses, particularly ribose 5-phosphate. This C_5 sugar or its derivatives are components of ATP, CoA, NAD, FAD, RNA, and DNA. The pentose phosphate pathway also catalyzes the interconversion of C_3, C_4, C_6, and C_7 sugars, some of which can enter the glycolytic sequence. In order to fulfill another function, which will not be discussed further, the pentose phosphate pathway may be modified to participate in the formation of glucose from CO_2 in photosynthesis.

There are specific pathways for synthesis and degradation of monosaccharides, oligosaccharides, and polysaccharides (other than glycogen), and a profusion of chemical interconversions, whereby one sugar can be changed into another. All of the monosaccharides, and the oligo- and polysaccharides synthesized from the monosaccharides, can originate from glucose. The interconversion reactions by which one sugar is changed into another may occur directly or at the level of nucleotide-linked sugars. In addition to their important role in sugar transformation, nucleotide sugars are the obligatory activated form for saccharide synthesis. Monosaccharides are also often found as components of more complex macromolecules like oligo- and polysaccharides, glycoproteins, glycolipids, and proteoglycans. Certain oligosaccharides, covalently linked to a protein or lipid, form the major structural components of bacterial cell walls, and include peptidoglycan, teichoic acids, and lipopolysaccharides. Chitin, a linear homopolymer of *N*-acetylglucosamine residues, is the predominant organic structural component of the exoshells of the invertebrates, as well as most fungi, many algae, and some yeast. Cellulose, an unbranched polymer of glucose residues, is the major structural component in the plant kingdom. In higher animals some of the complex carbohydrate molecules are also predominantly structural elements found in ground substance filling the extracellular space in tissues and as components of cell membranes. Increasingly though, more dynamic functions for these complex macromolecules, such as recognition markers and as determinants of biological specificity, are being discovered. The discussion of complex carbohydrates in this chapter is limited to the chemistry and biology of those complex carbohydrates found in animal tissues and fluids. See the appendix, for a discussion of the nomenclature and chemistry of the carbohydrates.

8.2 PENTOSE PHOSPHATE PATHWAY

The Pentose Phosphate Pathway Has Two Phases

The oxidative pentose phosphate pathway provides a means for cutting the carbon chain of a sugar molecule one carbon at a time. However, in contrast to glycolysis and the tricarboxylic acid cycle, the operation of this pathway does not occur as a consecutive set of reactions leading directly from glucose 6-phosphate (G6P) to six molecules of CO_2. For simplification, the overall pathway can be visualized as occurring in two stages. In the first, hexose is decarboxylated to pentose. The two oxidation reactions that lead to formation of NADPH also occur in this stage. The pathway may continue further, and, by a series of transformations, six molecules of pentose may undergo rearrangements to yield five molecules of hexose. It is these various transformations that give the pentose phosphate pathway its characteristic complexity. To understand this pathway, it is necessary to examine each reaction individually.

Glucose 6-Phosphate Is Oxidized and Decarboxylated to a Pentose Phosphate

The first reaction of the pentose phosphate pathway (Figure 8.1) is the enzymatic dehydrogenation of G6P at C-1 to form 6-phosphoglucono-δ-

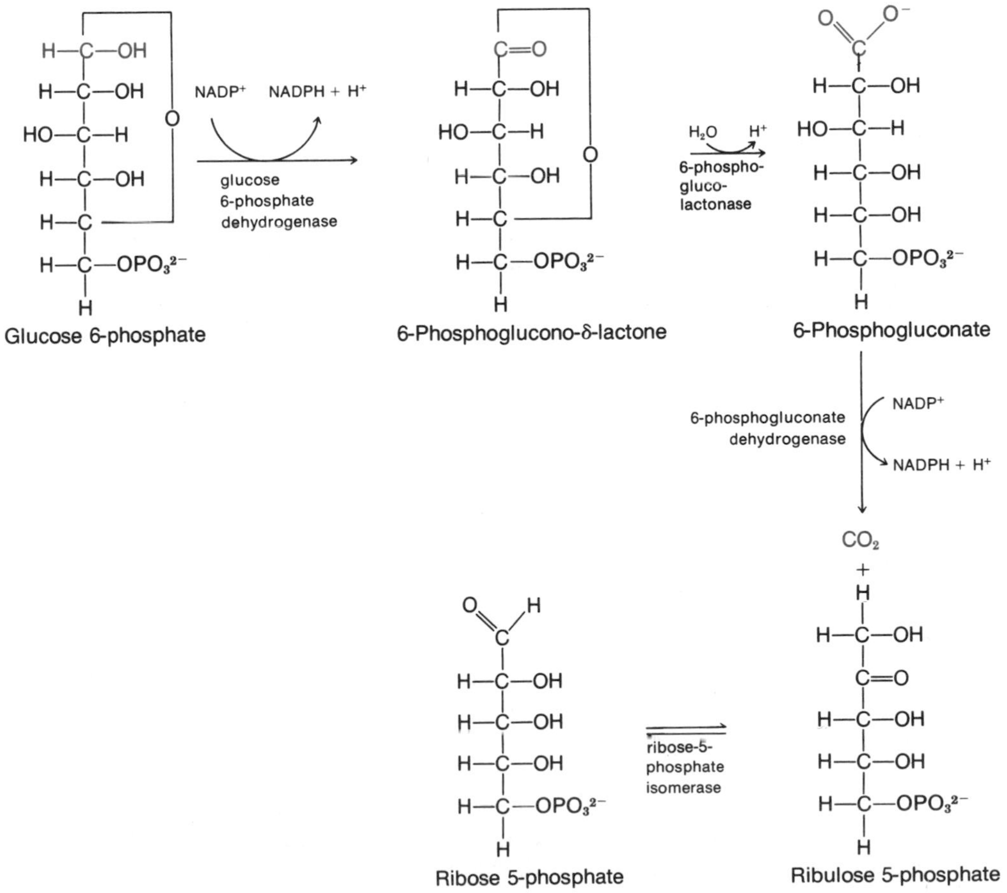

Glucose 6-phosphate
$NADP^+$ NADPH + H^+
glucose 6-phosphate dehydrogenase
6-Phosphoglucono-δ-lactone
H_2O H^+
6-phospho-gluco-lactonase
6-Phosphogluconate
6-phosphogluconate dehydrogenase
$NADP^+$
NADPH + H^+
CO_2 +
Ribulose 5-phosphate
ribose-5-phosphate isomerase
Ribose 5-phosphate

FIGURE 8.1
Formation of pentose phosphate.

CLIN. CORR. 8.1
GLUCOSE 6-PHOSPHATE DEHYDROGENASE: GENETIC DEFICIENCY OR PRESENCE OF GENETIC VARIANTS IN ERYTHROCYTES

When certain seemingly harmless drugs, such as antimalarials or sulfa antibiotics, are administered to susceptible patients, an acute hemolytic anemia may result. Susceptibility to drug-induced hemolytic disease may be due to a deficiency of G6P dehydrogenase activity in the erythrocyte, and was one of the early indications that genetic deficiencies of this enzyme exist. This enzyme, which catalyzes the oxidation of G6P to 6-phosphogluconate and the reduction of $NADP^+$, is particularly important, since the pentose phosphate pathway is the major pathway of NADPH production in the red cell. For example, red cells with the relatively mild A-type of glucose 6-phosphate dehydrogenase deficiency are able to oxidize glucose at a normal rate when the demand for NADPH is normal. However, if the rate of NADPH oxidation is increased, the enzyme-deficient cells are unable to increase the rate of glucose oxidation and carbon dioxide production adequately. In addition, cells lacking glucose 6-phosphate dehydrogenase do not reduce enough $NADP^+$ to maintain glutathione in its reduced state. Reduced glutathione is necessary for the integrity of the erythrocyte membrane, thus rendering enzyme-deficient red cells more susceptible to hemolysis by a wide range of compounds. Therefore, the basic abnormality in G6P deficiency is the formation of mature red blood cells that have diminished glucose 6-phosphate dehydrogenase activity. This enzymatic deficiency, which is usually undetected until administration of certain drugs, illustrates the interplay of heredity and environment on the production of disease. Enzyme absence is only one of several abnormalities that can affect enzyme activity, and others have been detected independent of drug administration. Increasingly, these variants can be distinguished from one another by clinical, biochemical, and molecular differences (see Clin. Corr. 4.5).

lactone and NADPH. The enzyme catalyzing this reaction is **G6P dehydrogenase,** the first enzyme found to be specific for $NADP^+$. Special interest in this enzyme stems from the severe anemia that may result from the absence of G6P dehydrogenase in erythrocytes or from the presence of one of several genetic variants of the enzyme (Clin. Corr. 8.1). Formation of the intermediate product of this reaction, a lactone, is freely reversible. Although the lactone is unstable and hydrolyzes spontaneously, a specific **gluconolactonase** causes a more rapid ring opening and ensures that the reaction goes to completion. The overall equilibrium of these two reactions lies far in the direction of NADPH maintaining a high $[NADPH]/[NADP^+]$ ratio within cells. A second dehydrogenation and decarboxylation is catalyzed by **6-phosphogluconate dehydrogenase,** a Mg^{2+}-dependent enzyme. The pentose phosphate, ribulose 5-phosphate, and a second molecule of NADPH are produced. The final step in the synthesis of ribose 5-phosphate is the isomerization of ribulose 5-phosphate by **ribose isomerase.** Like a similar reaction in the glycolytic pathway, this ketose–aldose isomerization proceeds through an enediol intermediate.

These first reactions resulting in decarboxylation may be considered to be the most important, since all the oxidation reactions leading to production of NADPH occur in this early part of the pathway. Under certain metabolic conditions, the pentose phosphate pathway may end at this point, with the utilization of NADPH for reductive biosynthetic reactions and ribose 5-phosphate as a precursor for nucleotide synthesis. The overall equation may be written as

$$\text{Glucose 6-Phosphate} + 2NADP^+ + H_2O \longrightarrow \text{ribose 5-phosphate} + 2NADPH + 2H^+ + CO_2$$

Interconversions of Pentose Phosphates Lead to Glycolytic Intermediates

In certain cells more NADPH is needed for reductive biosynthesis than ribose 5-phosphate for incorporation into nucleotides. A sugar rearrangement system (Figure 8.2), which leads to the formation of triose, tetrose, hexose, and heptose from the pentoses exists, thus creating a disposal mechanism for ribose 5-phosphate, as well as providing a reversible link between the pentose phosphate pathway and glycolysis. This occurs because certain of these sugars are common intermediates of both pathways. In order to allow the interconversions, another pentose phosphate, xylulose 5-phosphate, must first be formed through isomerization of ribulose 5-phosphate by the action of **phosphopentose epimerase.** As a consequence, these three pentose phosphates exist as an equilibrium mixture, and can then undergo further transformations catalyzed by transketolase and transaldolase. Both enzymes catalyze chain cleavage and transfer reactions involving the same group of substrates.

Transketolase, an enzyme involving thiamine pyrophosphate (TPP) and Mg^{2+}, transfers a C_2 unit from xylulose 5-phosphate to ribose 5-phosphate, producing the C_7 sugar sedoheptulose and glyceraldehyde 3-phosphate, an intermediate of glycolysis. The essential feature of this reaction involves the transfer of a C_2 group, designated "active glycolaldehyde," from a suitable donor ketose to an acceptor aldose (another example will be encountered later). A further transfer reaction, catalyzed by **transaldolase,** results in the recovery of the first hexose phosphate. In this reaction a C_3 unit (dihydroxyacetone) from sedoheptulose 7-phosphate is transferred to glyceraldehyde 3-phosphate, forming the tetrose, erythrose 4-phosphate, and fructose 6-phosphate, another intermediate of glycolysis.

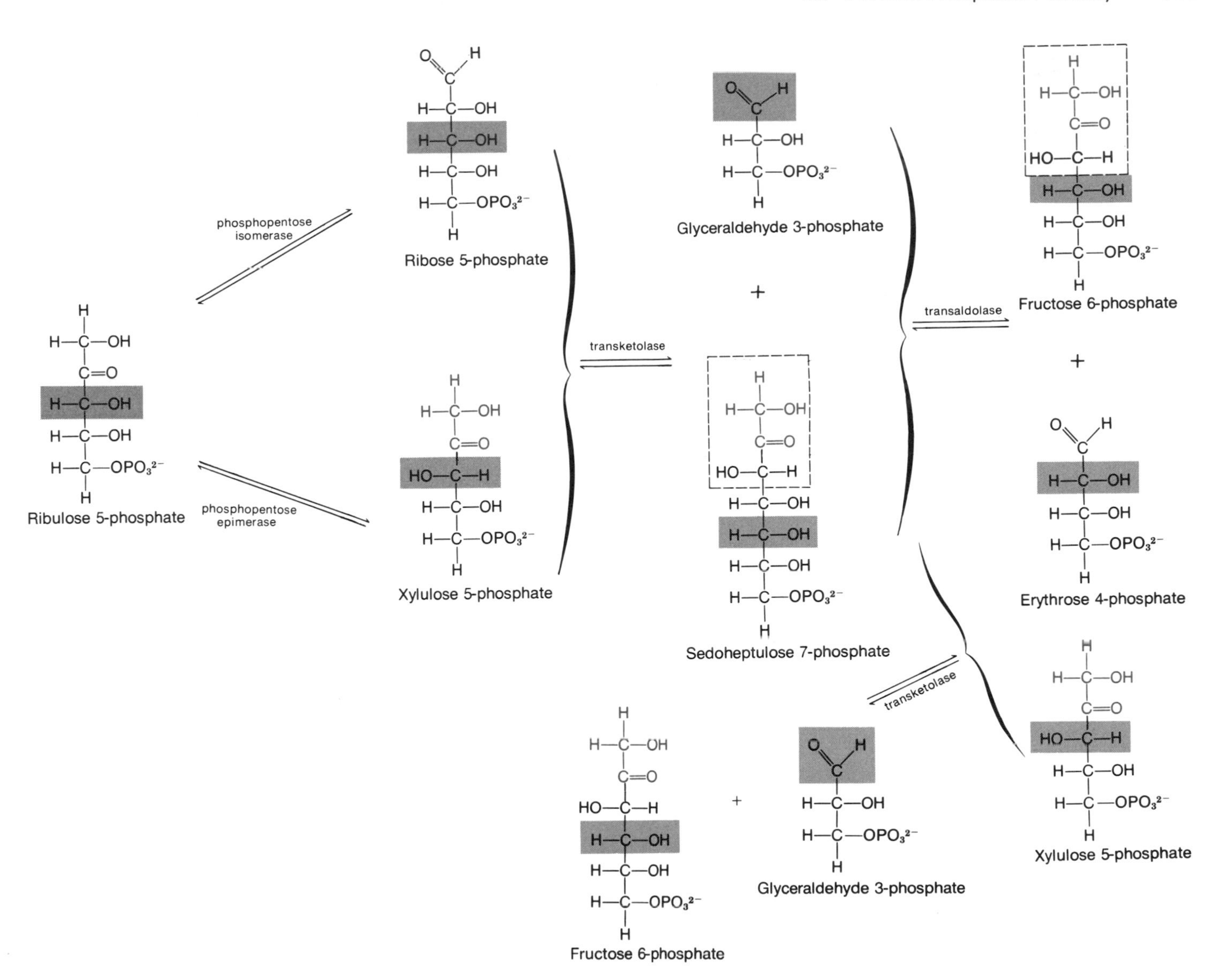

FIGURE 8.2
Interconversions of pentose phosphates.

This reaction proceeds by a mechanism similar to that of fructose 1,6-bisphosphate aldolase (Section 7.3), except that transaldolase is unable to react with or form free dihydroxyacetone or its phosphate.

In a third reaction, specific to this pathway, transketolase catalyzes the synthesis of fructose 6-phosphate and glyceraldehyde 3-phosphate from erythrose 4-phosphate and a second molecule of xylulose 5-phosphate. In this case, the C_2 unit is transferred from xylulose 5-phosphate to an acceptor C_4 sugar, now forming two glycolytic intermediates. The sum of these reactions is

$$\text{2 Xylulose 5-phosphate} + \text{ribose 5-phosphate} \rightleftharpoons \text{2 fructose 6-phosphate} + \text{glyceraldehyde 3-phosphate}$$

Since xylulose 5-phosphate is derived from ribose 5-phosphate, the net reaction starting from ribose 5-phosphate is

$$\text{3 Ribose 5-phosphate} \rightleftharpoons \text{2 fructose 6-phosphate} + \text{glyceraldehyde 3-phosphate}$$

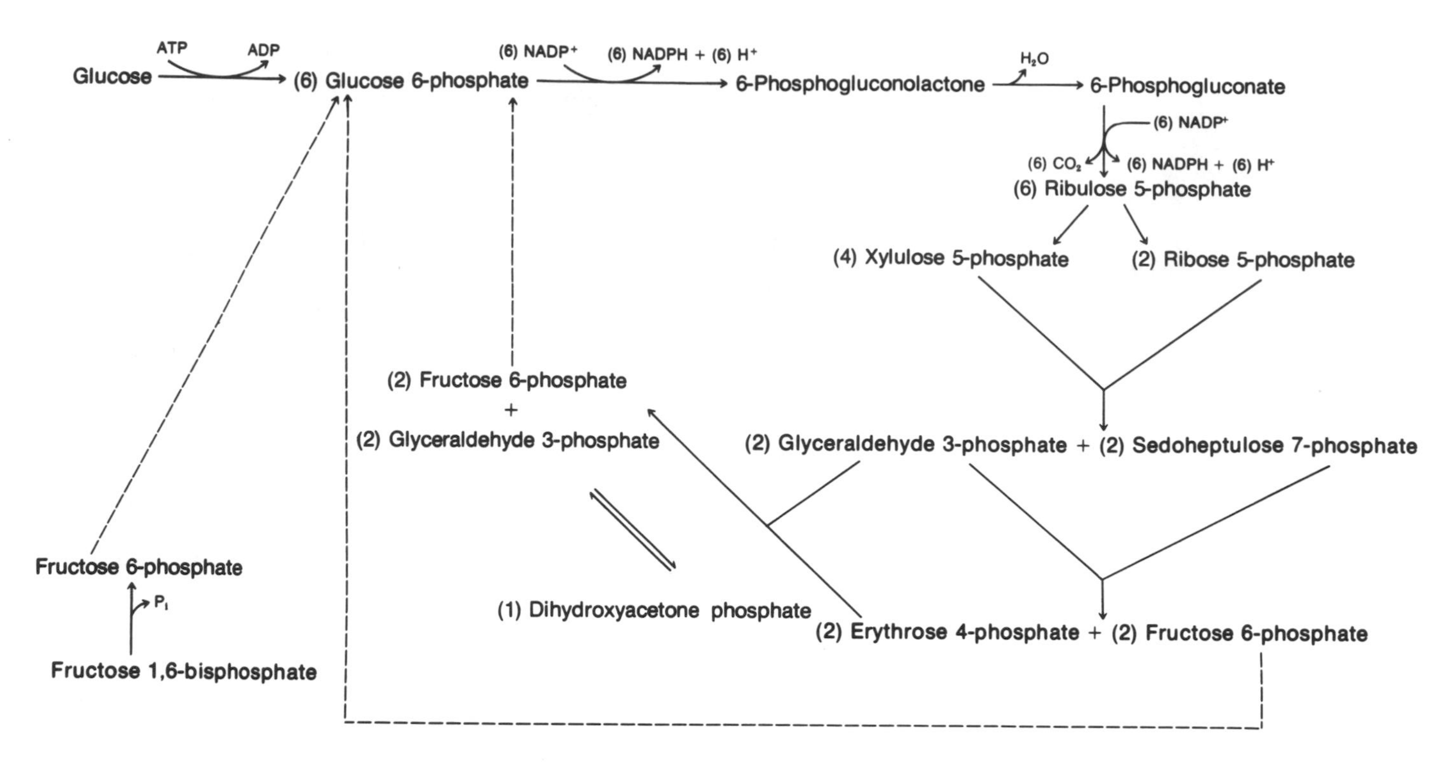

FIGURE 8.3
Pentose phosphate pathway.

Therefore, excess ribose 5-phosphate, whether it arises from the initial oxidation of G6P or from the degradative metabolism of nucleic acids, is effectively scavenged by conversion to intermediates that can enter the carbon flow of glycolysis.

Glucose 6-Phosphate Can Be Completely Oxidized to CO_2

In certain cells and tissues, like lactating mammary gland, a pathway for complete oxidation of G6P to CO_2, with concomitant reduction of $NADP^+$ to NADPH, also exists (Figure 8.3). By a complex sequence of reactions, the ribulose 5-phosphate produced by the pentose phosphate pathway is recycled into G6P by transketolase, transaldolase, and certain enzymes of the gluconeogenic pathway. Hexose continually enters this system, and CO_2 evolves as the only carbon compound. A balanced equation for this process would involve the oxidation of six molecules of G6P to six molecules of ribulose 5-phosphate and six molecules of CO_2. This represents essentially the first part of the pentose phosphate pathway and results in transfer of 12 pairs of electrons to $NADP^+$, the requisite amount for total oxidation of one glucose to six CO_2. The remaining six molecules of ribulose 5-phosphate are then rearranged by the pathway described above to regenerate five molecules of G6P. The overall equation can be written as

$$\text{6 Glucose 6-phosphate} + 12\text{NADP}^+ + 7\text{H}_2\text{O} \longrightarrow$$
$$\text{5 glucose 6-phosphate} + 6\text{CO}_2 + 12\text{NADPH} + 12\text{H}^+ + \text{P}_i$$

The net reaction is therefore

$$\text{Glucose 6-phosphate} + 12\text{NADP}^+ + 7\text{H}_2\text{O} \longrightarrow$$
$$6\text{CO}_2 + 12\text{NADPH} + 12\text{H}^+ + \text{P}_i$$

The Pentose Phosphate Pathway Produces NADPH

The pentose phosphate pathway serves several purposes, including a mechanism for synthesis and degradation of sugars other than hexoses, particularly pentoses necessary for nucleotides and nucleic acids, and other glycolytic intermediates. Most important, however, is the ability to synthesize NADPH, which plays a unique role in biosynthetic reactions. The direction of flow and path taken by G6P after entry into the pentose phosphate pathway is determined largely by the needs of the cell for NADPH or sugar intermediates. The situation in which more NADPH than ribose 5-phosphate is required has already been examined and results in a continuation of the pathway, leading to complete oxidation of G6P to CO_2 and resynthesis of G6P from ribulose 5-phosphate. Alternatively, if more ribose 5-phosphate than NADPH is required, G6P is converted to fructose 6-phosphate and glyceraldehyde 3-phosphate by the glycolytic pathway. Two molecules of fructose 6-phosphate and one molecule of glyceraldehyde 3-phosphate are then converted into three molecules of ribose 5-phosphate by a reversal of the transaldolase and transketolase reactions.

The distribution of the pentose phosphate pathway in the tissues of the body is consistent with its functions. As previously mentioned, it is present in erythrocytes for production of NADPH, which in turn is used to generate reduced **glutathione,** essential for maintenance of normal red cell structure. It is also active in tissues such as liver, mammary gland, testis, and adrenal cortex, which are active sites of fatty acid or steroid synthesis, processes that require the reducing power of NADPH. In contrast, in mammalian striated muscle, where little fatty acid or steroid synthesis occurs, there is no direct oxidation of glucose 6-phosphate through the pentose phosphate pathway. Rather, all catabolism proceeds via glycolysis and the TCA cycle. In some other tissues like liver, 20–30% of the CO_2 may arise from the pentose phosphate pathway, and the balance between glycolysis and the pentose phosphate pathway depends on the metabolic requirements of the cell.

8.3 SUGAR INTERCONVERSIONS AND NUCLEOTIDE SUGAR FORMATION

In preceding discussions, the general principles of carbohydrate metabolism, specifically those involving glucose, were considered. Now we shall examine certain aspects of the metabolism of other monosaccharides, oligosaccharides, and polysaccharides. Most of the monosaccharides found in biological compounds derive from glucose. The most common reactions for sugar transformations in mammalian systems are summarized in Figure 8.4.

Isomerization and Phosphorylation Are Common Reactions for Interconverting Carbohydrates

The formation of some saccharides may occur directly, starting from glucose via modification reactions, such as the conversion of G6P to fructose 6-phosphate by phosphoglucose isomerase in the glycolytic pathway. A similar aldose–ketose isomerization catalyzed by **phosphomannose isomerase** results in synthesis of mannose 6-phosphate.

Internal transfer of a phosphate group on the same sugar molecule from one hydroxyl group to another is a modification that has been previously described. Glucose 1-phosphate, resulting from enzymatic phosphorolysis of glycogen, is converted to G6P for entry into the glycolytic pathway

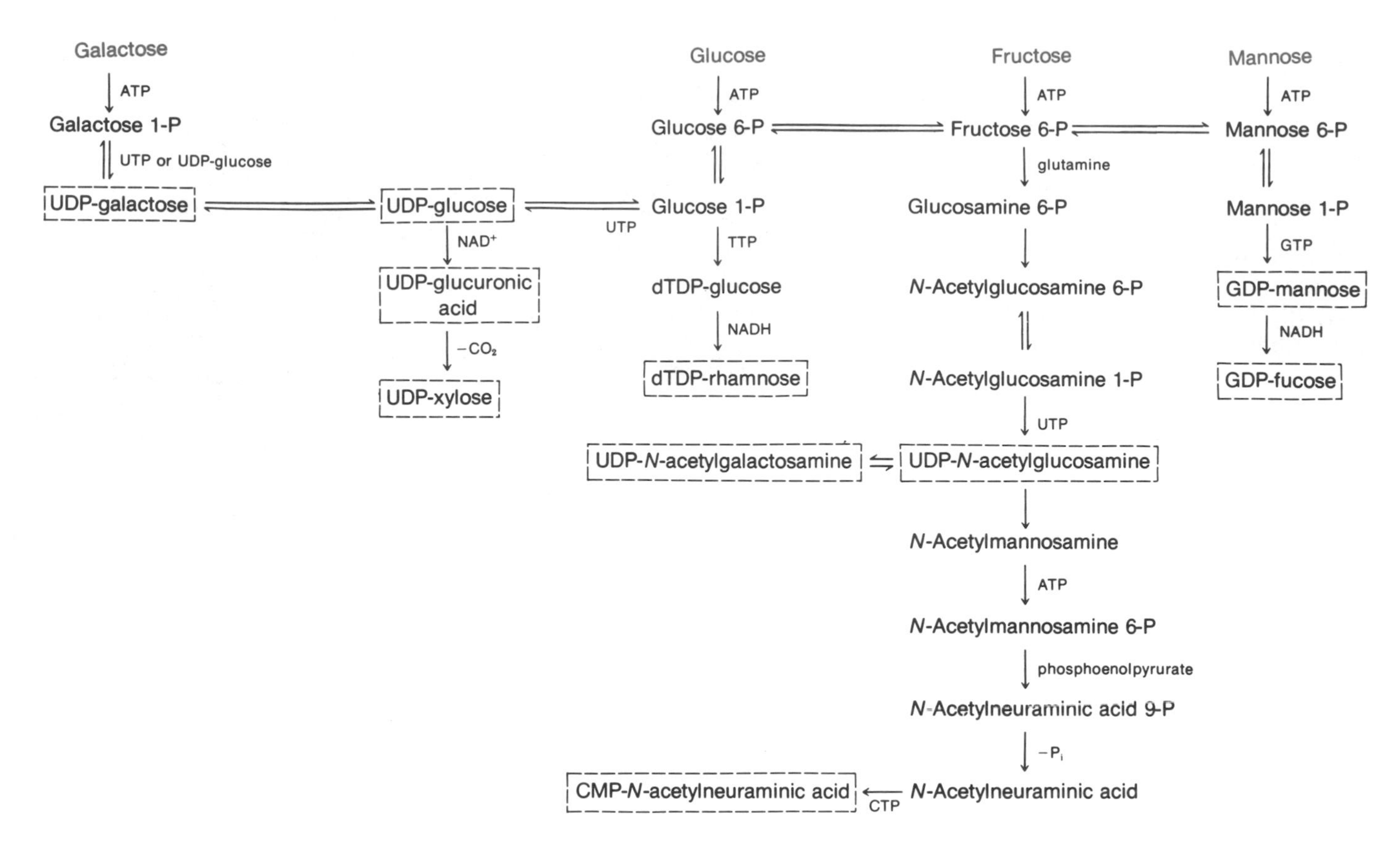

FIGURE 8.4
Pathways of formation of nucleotide-sugars and interconversions of some hexoses.

by phosphoglucomutase. Galactose may also be phosphorylated directly to galactose 1-phosphate by a **galactokinase** and mannose to mannose 6-phosphate by a **mannokinase;** the latter may equilibrate with fructose 6-phosphate as discussed above. Similarly, free fructose, an important dietary constituent, may be phosphorylated in the liver to fructose 1-phosphate by a special **fructokinase.** However, no mutase exists to interconvert fructose 1-phosphate and fructose 6-phosphate, nor can phosphofructokinase synthesize fructose 1,6-bisphosphate from fructose 1-phosphate. Rather, a **fructose 1-phosphate aldolase** cleaves fructose 1-phosphate to dihydroxyacetone phosphate (DHAP), which enters the glycolytic pathway directly, and glyceraldehyde, which must first be reduced to glycerol, phosphorylated, and then reoxidized to DHAP. Lack of this aldolase may lead to fructose intolerance (see Clin. Corr. 8.2).

Nucleotide-Linked Sugars Are Intermediates in Many Sugar Transformations

Most other sugar transformation reactions require the prior conversion into **nucleotide-linked sugars.** Nucleotides are phosphoric acid esters of nucleosides; nucleosides are pentosyl derivatives of a purine or pyrimidine base. Formation of nucleoside diphosphate (NDP) sugar involves the reaction of hexose 1-phosphate and nucleoside triphosphate (NTP), catalyzed by a **pyrophosphorylase.** These reactions are readily reversible. However, in vivo pyrophosphate is rapidly hydrolyzed irreversibly by pyrophosphatase, thereby driving the synthesis of nucleotide sugars.

CLIN. CORR. 8.2
ESSENTIAL FRUCTOSURIA AND FRUCTOSE INTOLERANCE: DEFICIENCY OF FRUCTOKINASE AND FRUCTOSE 1-PHOSPHATE ALDOLASE

Fructose may account for 30–60% of the total carbohydrate intake of mammals. It is predominantly metabolized by a specific fructose pathway. The first enzyme in this pathway, fructokinase, is deficient in essential fructosuria. This disorder is a benign asymptomatic metabolic anomaly, which appears to be inherited as an autosomal recessive. Biochemically, following intake of fructose, blood levels and urinary fructose are unusually high; however, 90% of fructose is eventually metabolized. In contrast, hereditary fructose intolerance is characterized by severe hypoglycemia after ingestion of fructose. Prolonged ingestion in young children may lead to death. In this disorder fructose 1-phosphate aldolase is deficient, and fructose 1-phosphate accumulates intracellularly (see Clin. Corr. 7.3).

Steinitz, H. and Mizrohy, O. Essential fructosuria and hereditary fructose intolerance. *N. Eng. J. Med.* 280:222, 1969.

These reactions are summarized as follows:

$$
\begin{array}{r}
\text{NTP} + \text{sugar 1-phosphate} \rightleftharpoons \text{NDP-sugar} + \text{PP}_i \\
\text{PP}_i + \text{H}_2\text{O} \longrightarrow 2\text{P}_i \\
\hline
\text{NTP} + \text{sugar 1-phosphate} + \text{H}_2\text{O} \longrightarrow \text{NDP-sugar} + 2\text{P}_i
\end{array}
$$

A common nucleotide sugar involved in synthesis of glycogen and presumably certain glycoproteins is **UDP-glucose.** It is synthesized from glucose 1-phosphate and UTP in a reaction catalyzed by **UDP-glucose pyrophosphorylase.** Pyrophosphate derives from the terminal two phosphoryl groups of UTP.

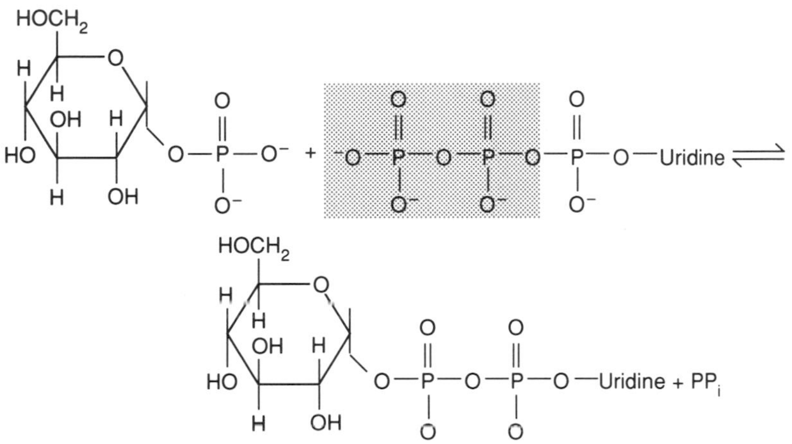

Nucleoside diphosphate-sugars contain two phosphoryl bonds with a large negative ΔG of hydrolysis, which contribute to the energized character of these compounds as glycosyl donors in further transformation and transfer reactions, as well as conferring specificity on the enzymes catalyzing these reactions. This is discussed in greater detail in Section 8.4. Uridine diphosphate usually serves as the glycosyl carrier in higher animals; however, ADP, GDP, and CMP have also been shown to act as carriers in other reactions. As previously mentioned, many of the sugar transformation reactions occur only at the level of nucleotide sugars. Some of these modification reactions involving nucleotide sugars include epimerization, oxidation, decarboxylation, reduction, and rearrangement.

Epimerization Interconverts Glucose and Galactose

One of the most common types of reaction in carbohydrate metabolism is epimerization. For example, the reversible conversion of glucose to galactose in animals occurs by epimerization of UDP-glucose to UDP-galactose, by **UDP-glucose epimerase.** The UDP-galactose that participates in the epimerization reaction is an important intermediate in the metabolism of free galactose, which is derived from the hydrolysis of lactose in the intestinal tract. Galactose can first be phosphorylated by galactokinase and ATP to yield galactose 1-phosphate:

$$\text{Galactose} + \text{ATP} \rightleftharpoons \text{galactose 1-phosphate} + \text{ADP}$$

An enzyme, **galactose 1-phosphate uridylyltransferase,** transforms galactose 1-phosphate into UDP-galactose by displacing glucose 1-phosphate from UDP-glucose:

UDP-glucose + galactose 1-phosphate ⇌

UDP-galactose + glucose 1-phosphate

A hereditary disorder, galactosemia, results from the absence of this uridylyltransferase (Clin. Corr. 8.3).

An NAD^+-dependent **galactose 4-epimerase** converts an equatorial hydroxyl group at C-4 in glucose to an axial one in galactose in the epimerization of these two sugars. Most likely, NAD^+ accepts the hydrogen atom at C-4, and a 4-keto intermediate is formed. Inversion of the hydroxyl group then occurs when NADH transfers its hydrogen to the other side of C-4.

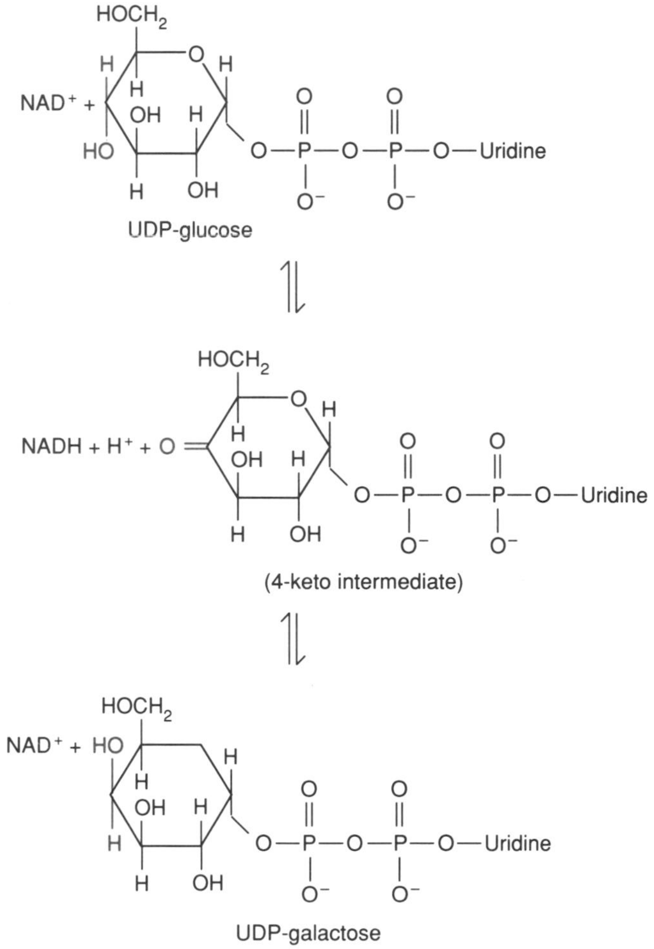

NAD^+ + UDP-glucose ⇌ $NADH + H^+$ + (4-keto intermediate) ⇌ NAD^+ + UDP-galactose

A combination of these reactions allows an efficient transformation of galactose derived from the diet into glucose 1-phosphate, which can then be further metabolized by previously described pathways. Alternatively, the 4-epimerase can operate in the reverse direction when UDP-galactose is needed for biosynthesis. Other 4-epimerases are known such as those that convert UDP-*N*-acetylglucosamine to UDP-*N*-acetylgalactosamine and UDP-xylose to UDP-arabinose. Presumably these operate by a mechanism similar to the UDP-galactose 4-epimerase.

More recent work has shown that epimerization reactions are not exclusively restricted to nucleotide-linked sugars but may also occur at the polymer level. Thus D-mannuronic acid is epimerized to L-glucuronic acid after incorporation into alginic acid (a polyglycosyluronate com-

CLIN. CORR. 8.3 GALACTOSEMIA: INABILITY TO TRANSFORM GALACTOSE INTO GLUCOSE

The reactions of galactose are of particular interest because in humans they are subject to genetic defects resulting in the hereditary disorder galactosemia. When the defect is present, individuals are unable to metabolize the galactose derived from lactose (milk sugar) to glucose metabolites, often with resultant cataract formation, growth failure, mental retardation, or eventual death from liver damage. The genetic disturbance is expressed as a cellular deficiency of either galactokinase, causing a relatively mild disorder characterized by early cataract formation, or of galactose 1-phosphate uridylyltransferase, resulting in severe disease.

Galactose can be reduced to galactitol in a reaction similar to the reduction of glucose to sorbitol. Galactitol is the initiator of cataract formation in the lens and may play a role in the central nervous system damage.

Segal, S., Blair, A., and Roth, H. The metabolism of galactose by patients with congenital galactosemia. *Am. J. Med.* 83:62, 1965.

CLIN. CORR. 8.4 PENTOSURIA: DEFICIENCY OF XYLITOL DEHYDROGENASE

The glucuronic acid oxidation pathway presumably is not essential to human carbohydrate metabolism, since individuals in whom the pathway is blocked suffer no ill effects. A metabolic variation, called idiopathic pentosuria, results from reduced activity of NADP-linked L-xylulose reductase, the enzyme that catalyzes the reduction of xylulose to xylitol. Hence, affected individuals excrete large amounts of pentose into the urine especially following intake of glucuronic acid.

Wang, Y. M. and van Eys, J. The enzymatic defect in essential pentosuria. *N. Eng. J. Med.* 282:892, 1970.

pound produced by seaweed and certain bacteria), and D-glucuronic acid is epimerized to L-iduronic acid after incorporation into heparin and dermatan sulfate (see Section 8.6).

Glucuronic Acid Is Formed by Oxidation of UDP-Glucose

Oxidation and reduction interconversions also result in formation of many additional sugars. One of the most important is formation of **glucuronic acid,** which serves as a precursor of **L-ascorbic acid** in those animals that synthesize vitamin C, but in humans is converted to L-xylulose, the ketopentose excreted in essential pentosuria (Clin. Corr. 8.4), and participates in physiological processes of detoxification by production of glucuronide conjugates (Clin. Corr. 8.5). Glucuronic acid is formed by oxidation of UDP-glucose catalyzed by **UDP-glucose dehydrogenase.**

UDP-glucose + 2NAD$^+$ ⟶

UDP-glucuronic acid + 2NADH + 2H$^+$

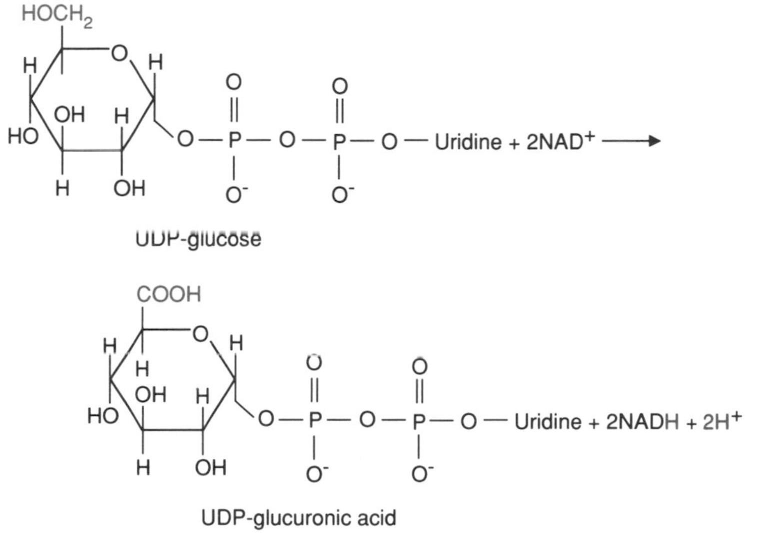

The UDP-glucose dehydrogenase-catalyzed reaction is important in the overall pathway for conversion of glucose to glucuronic acid, and most likely follows the scheme outlined in Figure 8.5. The UDP-glucuronic acid may then be epimerized to UDP-galacturonic acid. In a similar manner, GDP-mannose is oxidized to GDP-mannuronic acid, which may undergo epimerization to GDP-guluronic acid.

Following its formation, free glucuronic acid is further metabolized by reduction with NADPH to L-gulonic acid (Figure 8.6).Gulonic acid is converted by a two-step process through L-gulonolactone to L-ascorbic acid (vitamin C) in plants and most higher animals. Humans, other primates, and the guinea pig lack the enzyme that converts L-gulonolactone to L-ascorbic acid and therefore must satisfy their needs for ascorbic acid by its ingestion. Gulonic acid may also be oxidized to 3-ketogulonic acid and decarboxylated to L-xylulose. L-Xylulose is in turn converted by reduction to xylitol, reoxidized to D-xylulose and phosphorylated with ATP and an appropriate kinase to xylulose 5-phosphate. The latter compound may then reenter the pentose phosphate pathway described previously. This complex catabolic pathway for glucuronic acid represents another shunt pathway for oxidation of glucose. It should be noted that, in contrast to other pathways of carbohydrate metabolism in which only phosphate esters participate, these reactions also involve free sugars or sugar acids. Evidence suggests that this pathway operates in adipose tissue, and its activity may be increased in tissue from starved or diabetic animals; however, the regulation and extent to which these reactions proceed has not been adequately evaluated.

CLIN. CORR. 8.5 GLUCURONIC ACID: PHYSIOLOGICAL SIGNIFICANCE OF GLUCURONIDE FORMATION

The biological significance of glucuronic acid extends to its ability to be conjugated with certain endogenous and exogenous substances, forming a group of compounds collectively termed glucuronides in a reaction catalyzed by UDP-glucuronyltransferase. Conjugation of a compound with glucuronic acid produces a strongly acidic compound that is more water soluble at physiological pH than its precursor and therefore may alter the metabolism, transport, or excretion properties. Glucuronide formation is important in several processes, including drug detoxification, steroid excretion, and bilirubin metabolism. Bilirubin is the major metabolic breakdown product of heme, the prosthetic group of hemoglobin. The central step in excretion of bilirubin is conjugation with glucuronic acid by UDP-glucuronyltransferase. The development of this conjugating mechanism occurs gradually and may take several days to 2 weeks after birth to be fully active in humans. So called "physiological jaundice of the newborn" results in most cases from the inability of the neonatal liver to form bilirubin glucuronide at a rate comparable to that of bilirubin production. A defect in glucuronide synthesis has been found in a mutant strain of Wistar ("Gunn") rats, due to a deficiency of UDP-glucuronyltransferase and results in hereditary hyperbilirubinemia. In humans a similar defect is found in congenital familial nonhemolytic jaundice (Crigler–Najjar syndrome). Patients with this condition are also unable to conjugate foreign compounds efficiently with glucuronic acid.

Crigler, J. F. and Najjar, V. A. Congenital familial non-hemolytic jaundice with kernicterus. *Pediatrics* 10:169, 1952; Gunn, C. H. Hereditary acholuric jaundice in a new mutant strain of rats. *J. Hered.* 29:137, 1938; and Ostrow, J. D. (Ed.). Bile Pigments and Jaundice. New York: Marcel Dekker, 1986.

Glucose
↓
Glucose 6-phosphate
↓
Glucose 1-phosphate
↓
UDP-Glucose
↓
UDP-Glucuronic acid
↓
D-Glucuronic acid 1-phosphate
↓
D-Glucuronic acid

FIGURE 8.5
Biosynthesis of D-glucuronic acid.

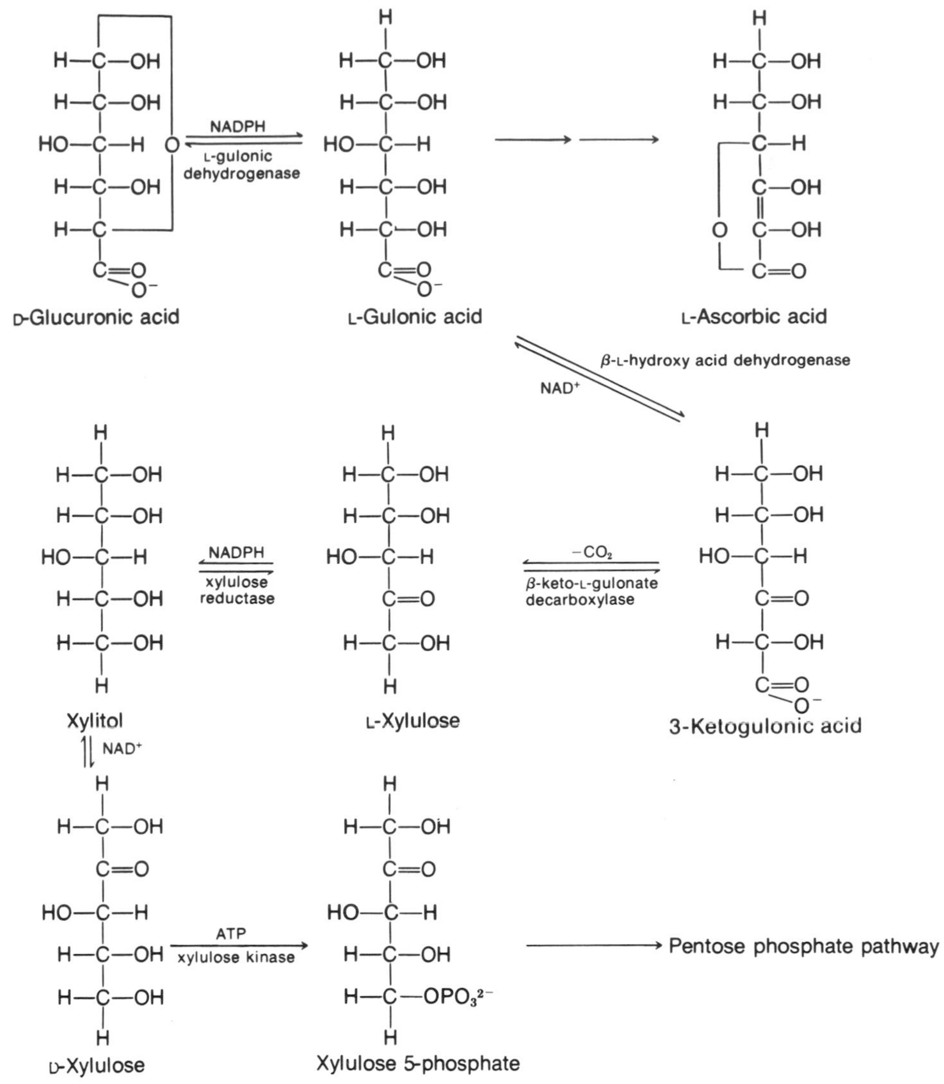

D-Glucuronic acid
NADPH
L-gulonic dehydrogenase
L-Gulonic acid
L-Ascorbic acid
NAD+
β-L-hydroxy acid dehydrogenase
3-Ketogulonic acid
$-CO_2$
β-keto-L-gulonate decarboxylase
L-Xylulose
NADPH
xylulose reductase
Xylitol
NAD+
D-Xylulose
ATP
xylulose kinase
Xylulose 5-phosphate
Pentose phosphate pathway

FIGURE 8.6
Glucuronic acid oxidation pathway.

Decarboxylation, Oxidoreduction, and Transamination of Sugars Produce Necessary Products

Decarboxylation, which is an important mechanism for degrading sugars one carbon atom at a time, has been previously encountered in the major metabolic pathways. The only known decarboxylation of a nucleotide sugar is the conversion of UDP-glucuronic acid to **UDP-xylose,** necessary for synthesis of proteoglycans (Section 8.6). UDP-xylose is a potent inhibitor of UDP-glucose dehydrogenase, which oxidizes UDP-glucose to UDP-glucuronic acid (Figure 8.4). Thus the level of these nucleotide sugar precursors is regulated by this sensitive feedback mechanism.

Deoxyhexoses and dideoxyhexoses are also synthesized, while the sugars are attached to nucleoside diphosphates, by a multistep process. For example, L-rhamnose is synthesized from glucose by a series of oxidation–reduction reactions starting with dTDP-glucose and yielding dTDP-rhamnose, catalyzed by oxidoreductases: Presumably, similar reactions account for the synthesis of GDP-fucose from GDP-mannose, as well as for various dideoxyhexoses.

Formation of **amino sugars,** major components of human oligo- and polysaccharides and often found as constituents of antibiotics, occurs by **transamidation.** For example, synthesis of glucosamine 6-phosphate occurs by the reaction of fructose 6-phosphate with glutamine.

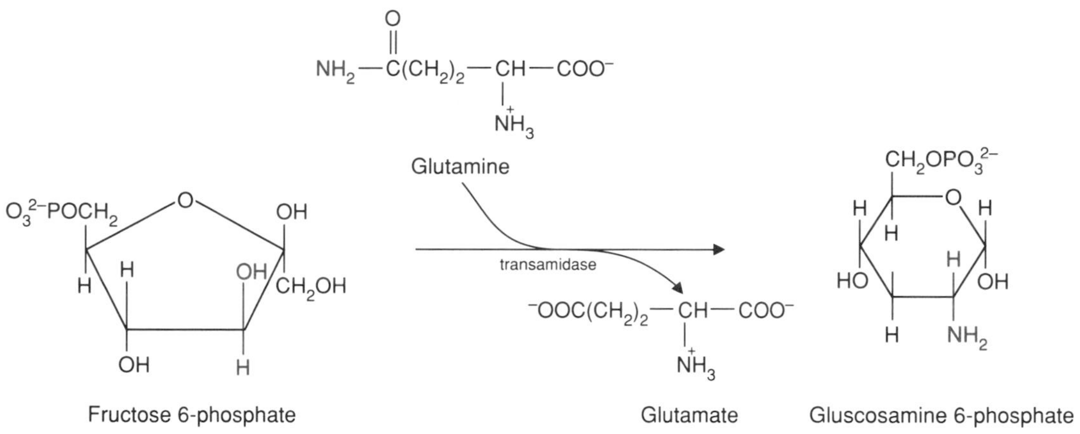

$NH_2—C(=O)(CH_2)_2—CH(NH_3^+)—COO^-$

Glutamine

transamidase

$^-OOC(CH_2)_2—CH(NH_3^+)—COO^-$

Fructose 6-phosphate Glutamate Gluscosamine 6-phosphate

Glucosamine 6-phosphate may then be *N*-acetylated, forming *N*-acetylglucosamine 6-phosphate, followed by isomerization to *N*-acetylglucosamine 1-phosphate. This latter sugar is converted to UDP-*N*-acetylglucosamine by reactions similar to those of UDP-glucose synthesis.

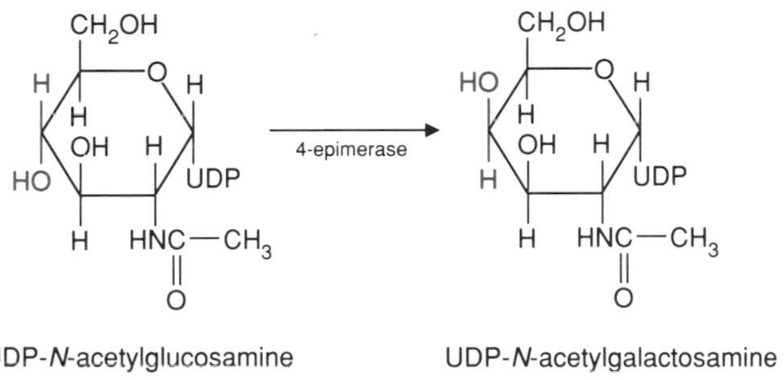

4-epimerase

UDP-*N*-acetylglucosamine UDP-*N*-acetylgalactosamine

UDP-*N*-acetylglucosamine, a precursor of glycoprotein synthesis, may be epimerized to UDP-*N*-acetylgalactosamine, necessary for proteoglycan synthesis. This first reaction in hexosamine synthesis, the fructose 6-phosphate-glutamine transamidase reaction, is under negative feedback control by UDP-*N*-acetylglucosamine; thus, the synthesis of both nucleotide sugars is regulated (Figure 8.4). This regulation is meaningful in certain tissues such as skin, in which this pathway may involve up to 20% of the glucose flux.

Sialic Acids Are Derived From N-Acetylglucosamine

Another product derived from UDP-*N*-acetylglucosamine is **acetylneuraminic acid,** one of a family of C_9 sugars, called **sialic acids** (Figure 8.7). The first reaction in this complex pathway involves epimerization of UDP-*N*-acetylglucosamine by a 2-epimerase to *N*-acetylmannosamine, concomitant with elimination of UDP. Since the monosaccharide product is no longer bound to nucleotide, this epimerization is clearly different from those previously encountered. Most likely, this 2-epimerase reaction proceeds by a trans elimination of UDP, with formation of the unsaturated intermediate, 2-acetamidoglucal. In mammalian tissues *N*-acetylmannosamine is phosphorylated with ATP to *N*-acetylmannosamine 6-phosphate, which then condenses with phosphoenolpyruvate to form *N*-acetylneuraminic acid 9-phosphate. This product is cleaved by a phosphatase and activated by CTP to form the CMP derivative, CMP-*N*-

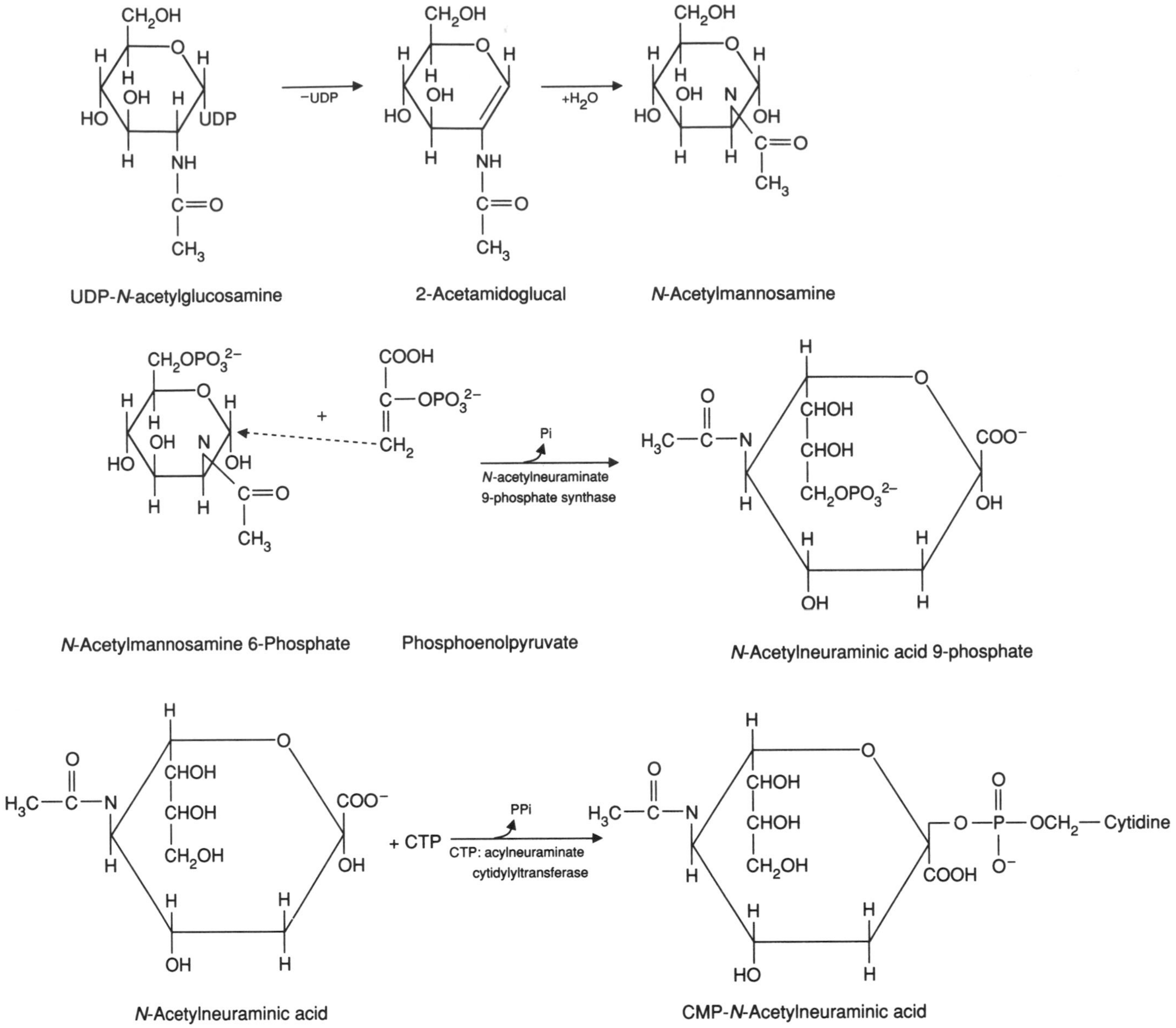

UDP-*N*-acetylglucosamine → (−UDP) → 2-Acetamidoglucal → ($+H_2O$) → *N*-Acetylmannosamine

N-Acetylmannosamine 6-Phosphate + Phosphoenolpyruvate → (*N*-acetylneuraminate 9-phosphate synthase; Pi) → *N*-Acetylneuraminic acid 9-phosphate

N-Acetylneuraminic acid + CTP → (CTP: acylneuraminate cytidylyltransferase; PPi) → CMP-*N*-Acetylneuraminic acid

FIGURE 8.7
Biosynthesis of CMP-*N*-acetylneuraminic acid.

acetylneuraminic acid. This is an unusual nucleotide sugar containing only one phosphate group, and is formed by a reaction that is irreversible. *N*-Acetylneuraminic acid is a precursor of other sialic acid derivatives, some of which evolve by modification of *N*-acetyl to *N*-glycolyl or *O*-acetyl after incorporation into glycoprotein.

8.4 BIOSYNTHESIS OF COMPLEX CARBOHYDRATES

In complex carbohydrate-containing molecules, sugars are linked to other sugars by glycosidic bonds, which are formed by specific **glycosyltransferases.** Energy is required for synthesis of a glycosidic bond and is made available through the use of nucleotide sugars as donor substrates. A glycosyltransferase reaction proceeds by donation of the glycosyl unit from the nucleotide derivative to the nonreducing end of an acceptor sugar. The nature of the bond formed is

$$\underset{\text{(donor)}}{\text{Nucleoside diphosphate—glycose}} + \underset{\text{(acceptor)}}{\text{glycose}_2} \xrightarrow{\text{glycosyltransferase}}$$

$$\underset{\text{(glycoside)}}{\text{glycosyl}_1\text{—}O\text{—glycose}_2} + \text{nucleoside diphosphate}$$

determined by the specificity of an individual glycosyltransferase, which is unique for the sugar acceptor, the sugar transferred, and the linkage formed. Thus polysaccharide synthesis is controlled by a nontemplate mechanism (see Chapter 15) in which genes code for specific glycosyltransferases.

At least 40 different glycosidic bonds have been identified in mammalian oligosaccharides and about 15 additional ones in connective tissue polysaccharides. The number of possible linkages is even greater and arises both from the diversity of monosaccharides covalently bonded and from the formation of both α and β linkages, with each of the available hydroxyl groups on the acceptor saccharide. The large and diverse number of molecules that can be generated suggests that oligosaccharides have the potential for great informational content. In fact, it is known that the specificity of many biological molecules is determined by the nature of the composite sugar residues. For example, the specificity of the major blood types is determined by sugars (see Clin. Corr. 8.6). *N*-Acetylgalactosamine is the immunodeterminant of blood type A and galactose of blood type B. Removal of *N*-acetylgalactosamine from type A erythrocytes, or of galactose from type B erythrocytes, will convert both to type O erythrocytes. Increasingly, other examples of sugars as determinants of specificity for cell surface receptor and lectin interactions, targeting of cells to certain tissues, and survival or clearance from the circulation of certain molecules, are being recognized.

All the glycosidic bonds that have been identified in biological compounds are degraded by specific hydrolytic enzymes, **glycosidases.** In addition to being valuable tools for the structural elucidation of oligosaccharides, recent interest in this class of enzymes stems arises because many genetic diseases of complex carbohydrate metabolism result from defects in glycosidases (see Clin. Corrs. 8.7 and 8.8).

8.5 GLYCOPROTEINS

Glycoproteins have been restrictively defined as conjugated proteins containing as a prosthetic group one or more saccharides lacking a serial repeat unit and bound covalently to a peptide chain. This definition excludes proteoglycans, which are discussed in Section 8.6.

The functions of glycoproteins in the human are currently of great interest. Glycoproteins in cell membranes apparently have an important role in the group behavior of cells and other important biological functions of the membrane. Glycoproteins form a major part of the mucus that is secreted by epithelial cells, where they perform an important role in lubrication and in the protection of tissues lining the body's ducts. Many other proteins secreted from cells into extracellular fluids are glycoproteins. These proteins include hormone proteins found in blood, such as follicle stimulating hormone, luteinizing hormone, and chorionic gonadotropin; and plasma proteins such as the orosomucoids, ceruloplasmin, plasminogen, prothrombin, and the immunoglobulins (Clin. Corr. 2.7).

CLIN. CORR. 8.6 BLOOD GROUP SUBSTANCES

The surface of the human erythrocyte is covered with a complex mosaic of specific antigenic determinants, many of which are saccharides. There are about 100 blood group determinants, belonging to 21 independent human blood group systems. The most widely studied are the antigenic determinants of the ABO blood group system and the closely related Lewis system. From the study of these systems, a definite correlation was established between gene activity as it relates to specific glycosyltransferase synthesis and oligosaccharide structure. The genetic variation is achieved through specific glycosyltransferases responsible for synthesis of the heterosaccharide determinants. For example, the *H* gene codes for a fucosyltransferase, which adds fucose to a peripheral galactose in the heterosaccharide precursor. The *A*, *B*, and *O* genes are located on chromosome 9. The *A* gene encodes an *N*-acetylgalactosamine glycosyltransferase, the *B* gene encodes a galactosyltransferase, and the *O* gene encodes an inactive enzyme. The sugars are added to the *H*-specific oligosaccharide. The Lewis (*Le*) gene codes for another fucosyltransferase, which adds fucose to a peripheral *N*-acetylglucosamine residue in the precursor. Absence of the *H* gene gives rise to the Le^a specific determinant, whereas in the absence of both the *H* and *Le* genes, the interaction product responsible for the Le^b specificity is found. The elucidation of the structures of these oligosaccharide determinants represents a milestone in carbohydrate chemistry. This knowledge is essential to blood transfusion practices and for legal and historical purposes. For example, tissue dust containing complex carbohydrates has been used in serological analysis to establish the blood group of Tutankhamen and his probable ancestral background.

Watkins, W. M. Blood group substances. *Science* 152:172, 1966.

CLIN. CORR. 8.7
ASPARTYLGLYCOSYLAMINURIA: ABSENCE OF 4-L-ASPARTYLGLYCOSAMINE AMIDOHYDROLASE

A group of human inborn errors of metabolism involving storage of glycolipids, glycopeptides, mucopolysaccharides, and oligosaccharides exists. These diseases are caused by defects in glycosidase activity, which prevents the catabolism of oligosaccharides. The disorders involve accumulation in tissues and urine of compounds derived from incomplete degradation of the oligosaccharides, and may be accompanied by skeletal abnormalities, hepato-spleno-megaly, cataracts or mental retardation. One disorder resulting from a defect in catabolism of asparagine-*N*-acetylglucosamine-linked oligosaccharides is aspartylglycosylaminuria. A deficiency in the enzyme 4-L-aspartylglycosylamine amidohydrolase allows the accumulation of aspartylglucosamine-linked structures. (See accompanying table.)

Other disorders have been described involving accumulation of oligosaccharides derived from both glycoproteins and glycolipids, which may share common oligosaccharide structures (see Table). Examples of genetic diseases include mannosidosis (α-mannosidase), the G_{M2} gangliosidosis variant O (Sandhoff–Jatzkewitz disease; β-*N*-acetylhexosaminidases A and B), and G_{M1} gangliosidosis (β-galactosidase). Mucolipidosis II ("I-Cell Disease") is a generalized degradative disorder resulting from a deficiency of UDP-GlcNAc: lysosomal enzyme precursor GlcNAc phosphotransferase, which attacks Man-6-PO_4 (see also Chapter 10).

Sewell, A. C. Urinary oligosaccharide excretion in disorders of glycolipid, glycoprotein, and glycogen metabolism: A review of screening for differential diagnosis. *Eur. J. Pediatr.* 134:183, 1980.

Enzymic Defects in Degradation of Asn-GlcNAc Type Glycoproteins[a]

Disease	Deficient enzyme[b]
Aspartylglycosylaminuria	4-L-Aspartylglycosylamine amidohydrolase (1)
β-Mannosidosis	β-Mannosidase (7)
α-Mannosidosis	α-Mannosidase (3)
G_{M2} Gangliosidosis variant O (Sandhoff–Jatzkewitz disease)	β-*N*-Acetylhexosaminidases (A and B) (4)
G_{M1} Gangliosidosis	β-Galactosidase (5)
Mucolipidosis I (Sialidosis)	Sialidase (6)
Fucosidosis	α-Fucosidase (8)

[a] A typical Asn-GlcNAc oligosaccharide structure.

Dermatan sulfate: —IdUA $\xrightarrow[\alpha]{(2)}$ GalNAc $\xrightarrow[\beta]{(4)}$ GlcUA $\xrightarrow[\beta]{(5)}$ GalNAc $\xrightarrow[\beta]{}$; IdUA –(1)– OSO_3H; GalNAc –(3)– OSO_3H; GalNAc – OSO_3H

Heparan sulfate: —IdUA $\xrightarrow[\alpha]{(2)}$ GlcN $\xrightarrow[\alpha]{(7)}$ GlcUA $\xrightarrow[\beta]{(5)}$ GlcNAc $\xrightarrow[\alpha]{(9)}$; IdUA –(1)– OSO_3H; GlcN –(6)– OSO_3H; GlcNAc –(8)– OSO_3H

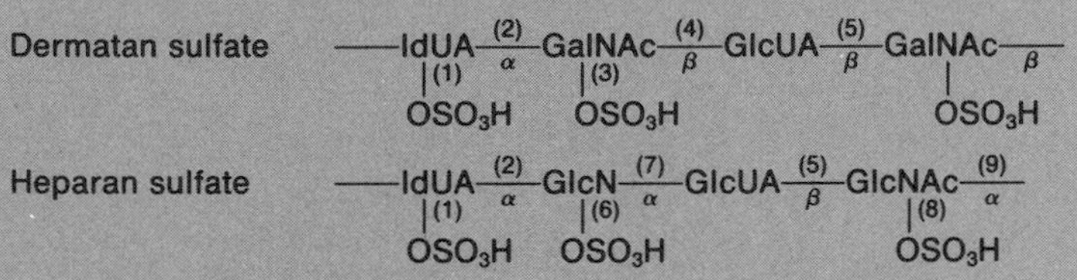

[b] The numbers in parentheses refer to the enzymes that hydrolyze those bonds.

Glycoproteins Contain Variable Amounts of Carbohydrate

The percent of carbohydrate within the glycoproteins is highly variable. Some glycoproteins such as IgG contain low amounts of carbohydrate (4%), while a human red cell membrane glycoprotein, glycophorin, has been found to contain 60% carbohydrate. Human ovarian cyst glycoprotein is composed of 70% carbohydrate, and human gastric glycoprotein is 82% carbohydrate. The carbohydrate can be distributed fairly evenly along the polypeptide chain of the protein component or concentrated into defined regions of the polypeptide chain. For example, in human glycophorin A the carbohydrate is found in the NH_2-terminal one-half of the polypeptide chain that lies on the outside of the cellular membrane.

The carbohydrate attached at one or at multiple points along a polypeptide chain usually contains less than 12–15 sugar residues. In some cases the carbohydrate component consists of only a single sugar moiety, as in the submaxillary gland glycoprotein (single *N*-acetyl-α-D-galactosaminyl residue) and in some types of mammalian collagens (single α-D-galactosyl residue). In general, glycoproteins contain sugar residues in the D form, except for L-fucose, L-arabinose, and L-iduronic acid. A glycoprotein from different animal species often has an identical primary structure in the protein component, but a variable carbohydrate component. This heterogeneity of a given protein may even be true within a single organism. For example, pancreatic ribonuclease is found in an A and a B form. The two forms have an identical amino acid sequence and a similar kinetic specificity toward substrates, but differ significantly in their carbohydrate composition.

Carbohydrates Are Covalently Linked to Glycoproteins by *N*- or *O*- Glycosyl Bonds

At present, the structures of a limited number of oligosaccharide components have been completely elucidated. The microheterogeneity of glycoproteins, arising from incomplete synthesis or partial degradation, makes structural analyses extremely difficult. However, certain generalities about the structure of glycoproteins have emerged. The covalent linkage of sugars to the peptide chain is a central part of glycoprotein structure, and only a limited number of bonds are found (see Chapter 2). The three major types of glycopeptide bonds, as shown in Figure 8.8 and Figure

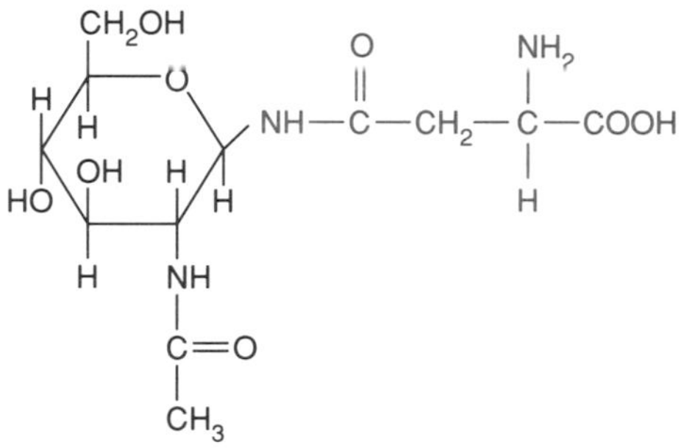

Type I *N*-Glycosyl linkage to asparagine

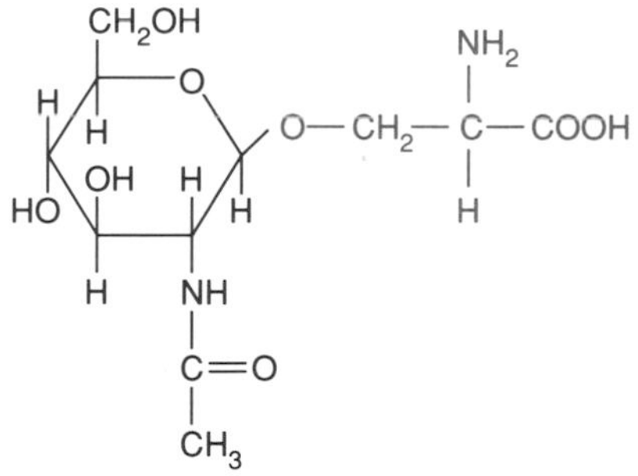

Type II *O*-Glycosyl linkage to serine

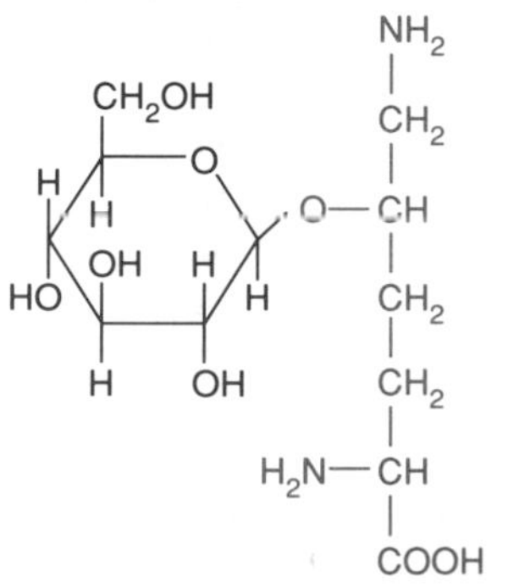

Type III *O*-Glycosyl linkage to 5-hydroxylysine

FIGURE 8.8
Structure of three major glycopeptide bonds.

CLIN. CORR. 8.8
HEPARIN AS AN ANTICOAGULANT

Heparin is a naturally occurring sulfated polysaccharide that is used to reduce the clotting tendency of patients. Both in vivo and in vitro heparin prevents the activation of clotting factors, but does not act directly on the clotting factors. Rather, the anticoagulant activity of heparin is brought about by the binding interaction of heparin with an inhibitor of the coagulation process. Presumably, heparin binding induces a conformational change in the inhibitor that generates a complementary interaction between the inhibitor and the activated coagulation factor, thereby preventing the factor from participating in the coagulation process. The inhibitor that interacts with heparin is antithrombin III, a plasma protein inhibitor of serine proteases. In the absence of heparin, antithrombin III slowly (10–30 min) combines with several clotting factors, yielding complexes devoid of proteolytic activity. In the presence of heparin, inactive complexes are formed within a few seconds. Antithrombin III contains an arginine residue that combines with the active site serine of factors Xa and IXa; thus, the inhibition is stoichiometric. Heterozygous antithrombin III deficiency results in an increased risk of thrombosis in the veins and resistance to the action of heparin.

Rosenberg, R. D. and Rosenberg, J. S. Natural anticoagulant mechanisms. *J. Clin. Invest.* 74:1, 1984.

2.60, are *N*-glycosyl to asparagine (Asn), *O*-glycosyl to serine (Ser), or threonine (Thr) and *O*-glycosyl to 5-hydroxylysine. The latter linkage, representing the carbohydrate side chains of either a single galactose or the disaccharide glucosylgalactose covalently bonded to hydroxylysine, is generally confined to the collagens. The other two linkages occur in a wide variety of glycoproteins. Of the three major types, only the *O*-glycosidic linkage to serine or threonine is labile to alkali cleavage. By this procedure two types of oligosaccharides (simple and complex) are released. Examination of the simple class from porcine submaxillary mucins reveals some general structural features. A core structure exists, consisting of galactose (Gal) linked $\beta(1 \rightarrow 3)$ to *N*-acetylgalactosamine (GalNAc) *O*-glycosidically linked to serine or threonine residues. Residues of L-fucose (Fuc), sialic acid (NeuAc), and another *N*-acetylgalactosamine are found at the nonreducing periphery of this class of glycopeptides. The general structure of this type of glycopeptide is as follows:

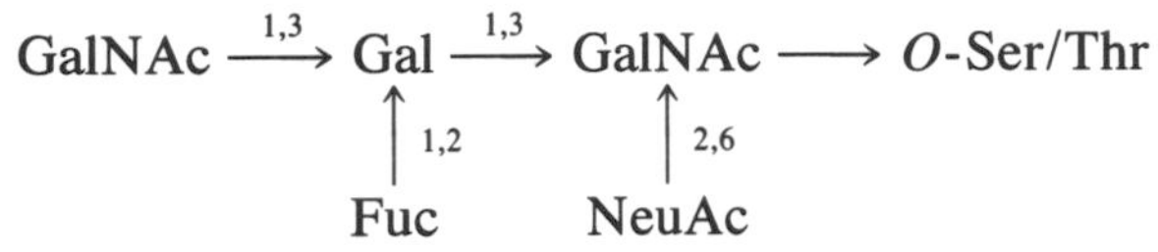

$$\begin{array}{ccccccc} \text{GalNAc} & \xrightarrow{1,3} & \text{Gal} & \xrightarrow{1,3} & \text{GalNAc} & \longrightarrow & O\text{-Ser/Thr} \\ & & \uparrow {\scriptstyle 1,2} & & \uparrow {\scriptstyle 2,6} & & \\ & & \text{Fuc} & & \text{NeuAc} & & \end{array}$$

More complex heterosaccharides are also linked to peptides via serine or threonine and are exemplified by the blood group substances. The study of these determinants has shown how complex and variable these structures are, as well as how the oligosaccharides of cell surfaces are assembled and how that assembly pattern is genetically determined. An example of how oligosaccharide structures on the surface of red blood cells determine blood group specificity is presented in Clin. Corr. 8.6.

Certain common structural features of the oligosaccharide *N*-glycosidically linked to asparagine have also emerged. These glycoproteins commonly contain a core structure consisting of mannose (Man) residues linked to *N*-acetylglucosamine (GlcNAc) in the following structure:

$$(\text{Man})_n \xrightarrow[1,4]{} \text{Man} \xrightarrow[1,4]{} \text{GlcNAc} \xrightarrow[1,4]{} \text{GlcNAc} \longrightarrow \text{Asn}$$

Synthesis of N-Linked Glycoproteins Involves Dolichol Phosphate

In contrast to the synthesis of *O*-glycosidically linked glycoproteins, which involves the sequential action of a series of glycosyltransferases, the synthesis of *N*-glycosidically linked peptides, involves a somewhat different and more complex mechanism (Figure 8.9). The common core is preassembled as a lipid-linked oligosaccharide prior to incorporation into the polypeptide. Similar assembly processes for synthesis of precursor units followed by transfer en bloc have been reported in bacterial cell wall synthesis, but are uncommon in mammalian heterosaccharide synthesis. During synthesis, the oligosaccharide intermediates are bound to derivatives of **dolichol phosphate.**

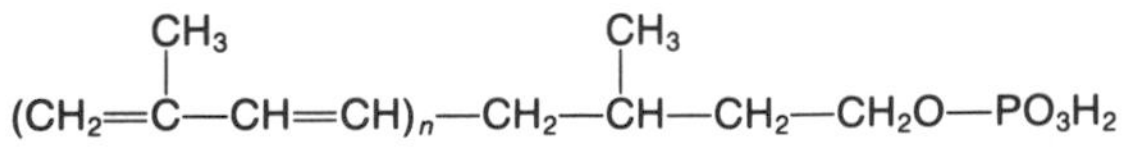

$$(CH_2{=}\overset{\overset{\large CH_3}{|}}{C}{-}CH{=}CH)_n{-}CH_2{-}\overset{\overset{\large CH_3}{|}}{C}H{-}CH_2{-}CH_2O{-}PO_3H_2$$

Dolichol phosphate

Dolichols are a class of polyprenols (C_{80}–C_{100}) containing 16–20 isoprene units, in which the final isoprene unit is saturated. These lipids participate in two types of reactions in core oligosaccharide synthesis. The first reac-

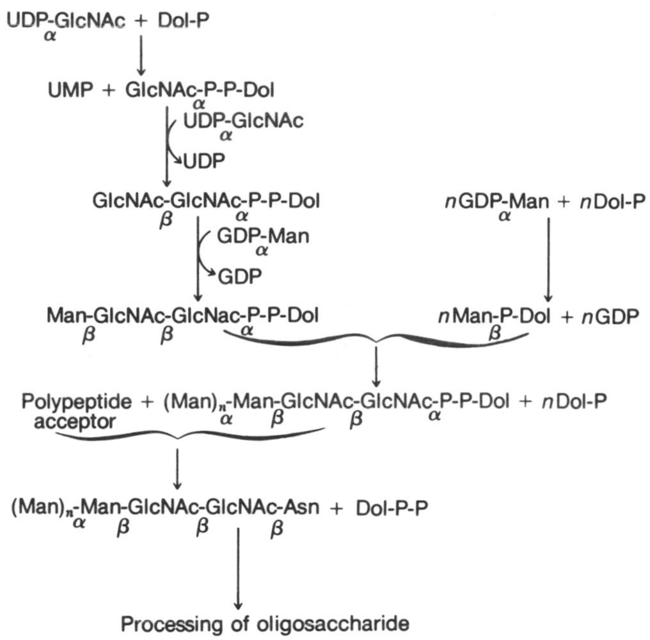

UDP-GlcNAc + Dol-P
↓
UMP + GlcNAc-P-P-Dol
↓ UDP-GlcNAc → UDP
GlcNAc-GlcNAc-P-P-Dol
↓ GDP-Man → GDP
Man-GlcNAc-GlcNac-P-P-Dol

nGDP-Man + nDol-P
↓
nMan-P-Dol + nGDP

↓
Polypeptide acceptor + $(\text{Man})_n$-Man-GlcNAc-GlcNAc-P-P-Dol + nDol-P
↓
$(\text{Man})_n$-Man-GlcNAc-GlcNAc-Asn + Dol-P-P
↓
Processing of oligosaccharide

FIGURE 8.9
Biosynthesis of the oligosaccharide core in asparagine-*N*-acetylgalactosamine-linked glycoproteins.
Abbreviation: Dol, dolichol.

tion involves formation of ***N*-acetylglucosaminylpyrophosphoryldolichol** with release of UMP from the respective nucleotide sugars. The second *N*-acetylglucosamine and mannose transferase reactions proceed by sugar transfer from the nucleotide without formation of intermediates. The subsequent addition of a mannose unit also occurs via a dolichol-linked mechanism. In the final step, the oligosaccharide is transferred from the dolichol pyrophosphate to an asparagine residue in the polypeptide chain.

After synthesis of the specific core region, the oligosaccharide chains are completed by the action of glycosyltransferases without further participation of lipid intermediates. Recent evidence suggests that extensive "processing," involving the addition and subsequent removal of certain glycosyl residues, occurs during the course of synthesis of asparagine-*N*-acetylglucosamine-linked glycoproteins. The pathways of "oligosaccharide processing" have been elucidated using certain enveloped viruses (VSV and Sindbis) and some secretory glycoprotein systems. It is thought that oligosaccharides destined to become simple or oligomannoside-type glycoproteins undergo more limited processing, whereas the more complex glycoproteins undergo more extensive processing and elongation. It is not yet understood why some glycoproteins become one type or another, but it may be of evolutionary significance.

Initiation of glycosylation by addition of the core saccharides may occur while the nascent peptide is still bound to ribosomes or soon after completion. Processing and elongation reactions then take place as the peptide moves through the rough endoplasmic reticulum to the Golgi. Once elongation is complete by addition of external sugars within the Golgi apparatus, the glycoprotein migrates toward the plasma membrane within a vesicle. The membrane of the transport vesicle fuses with the plasma membrane, and secretory glycoproteins are extruded, while internal membrane glycoproteins remain part of the plasma membrane.

Just as the synthesis of oligosaccharides requires specific glycosyltransferases, the degradation requires specific glycosidases. **Exoglycosidases** remove sugars sequentially from the nonreducing end, exposing the substrate for the subsequent glycosidase. The absence of a particular glycosidase prevents the action of the next enzyme, resulting in cessation of catabolism and accumulation of the product (Clin. Corr. 8.7). There is also evidence for the presence of **endoglycosidases** with broader specificity. However, the sequence of action of endo- and exoglycosidases in the catabolism of glycoproteins is not well understood. The primary degradation process occurs in lysosomes, but there are also specific microsomal glycosidases involved in the processing of glycoproteins during synthesis.

8.6 PROTEOGLYCANS

In addition to glycoproteins, which usually contain proportionally less carbohydrate than protein by weight, there is another class of complex macromolecules, which may contain as much as 95% or more carbohydrate. The properties of these compounds resemble polysaccharides more than proteins. To distinguish these compounds from other glycoproteins, they are referred to as **proteoglycans** and their carbohydrate chains as glycosaminoglycans. An older name, **mucopolysaccharides,** is still in use, especially in reference to the group of storage diseases, mucopolysaccharidoses, which result from an inability to degrade these molecules (see Clin. Corr. 8.9).

The proteoglycans are high molecular weight polyanionic substances consisting of many different glycosaminoglycan chains linked covalently to a protein core. Although six distinct classes of glycosaminoglycans are now recognized, certain features are common to all classes. The long heteropolysaccharide chains are made up largely of disaccharide repeating units, in which one sugar is a hexosamine and the other a uronic acid. Other common constituents of glycosaminoglycans are sulfate groups, linked by ester bonds to certain monosaccharides or by amide bonds to the amino group of glucosamine. An exception, hyaluronate, is not sulfated, and as yet has not been shown to exist covalently attached to protein. The carboxyl and sulfate groups contribute to the nature of glycosaminoglycans as highly charged polyanions. Both their electrical charge and macromolecular structure aid in their biological role as lubricants and support elements in connective tissue. The proteoglycans form solutions with high viscosity and elasticity by absorbing large volumes of water. This allows them to act in stabilizing and supporting fibrous and cellular elements of tissues, as well as contributing to the maintenance of water and salt balance in the body.

Hyaluronate Is a Copolymer of *N*-Acetylglucosamine and Glucuronic Acid

Among the glycosaminoglycans, **hyaluronate** is very different from the other five types. As previously mentioned, it is unsulfated, is not covalently complexed with protein, and is the only glycosaminoglycan not limited to animal tissue, but is also produced by bacteria. It is nevertheless classified as a glycosaminoglycan because of its structural similarity to these other polymers. It consists solely of repeating disaccharide units of *N*-acetylglucosamine and glucuronic acid (Figure 8.10). Although hyaluronate has the least complex chemical structure of all the glycosaminoglycans, the chains may reach molecular weights of 10^5–10^7. The large molecular weight, polyelectrolyte character, and large volume of water it occupies in solution all contribute to the properties of hyaluronate as a lubricant and shock absorbant. Hence, it is found predominantly in synovial fluid, vitreous humor, and umbilical cord.

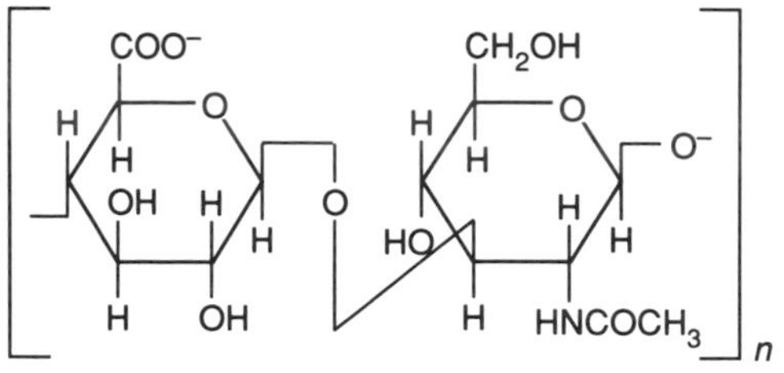

Repeat unit of hyaluronic acid

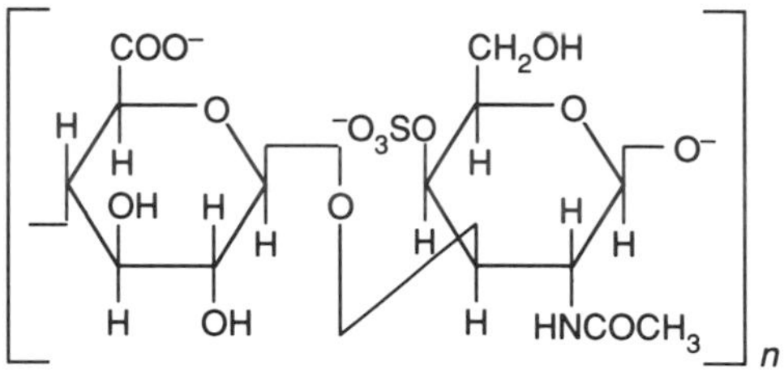

Repeat unit of chondroitin 4-sulfate

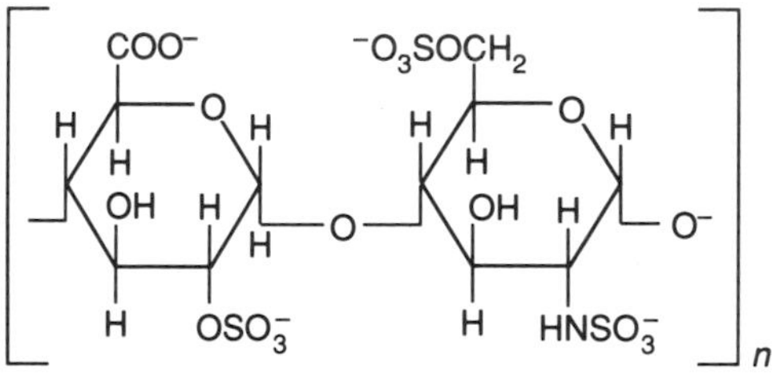

Repeat unit of heparin

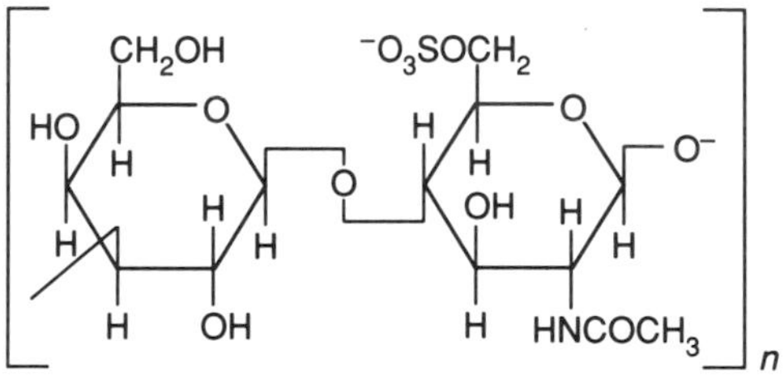

Repeat unit of keratan sulfate

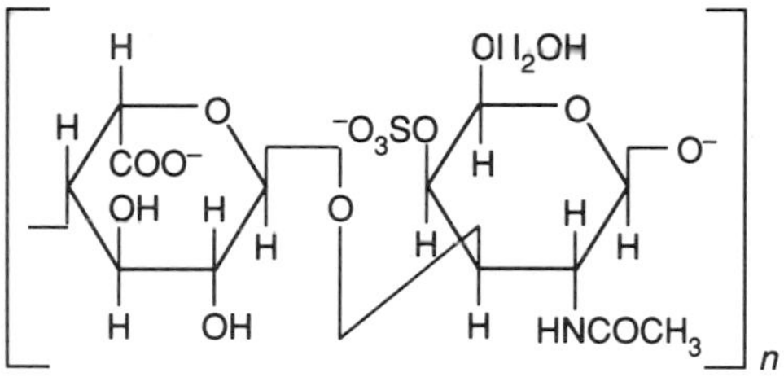

Repeat unit of dermatan sulfate

FIGURE 8.10
Major repeat units of glycosaminoglycan chains.

Chondroitin Sulfates Are the Most Abundant Glycosaminoglycans

The most abundant glycosaminoglycans in the body are the **chondroitin sulfates.** The individual polysaccharide chains are attached to specific serine residues in a protein core of variable molecular weight through a tetrasaccharide linkage region.

$$\text{GlcUA} \xrightarrow[1,3]{} \text{Gal} \xrightarrow[1,3]{} \text{Gal} \xrightarrow[1,4]{} \text{Xyl} \longrightarrow O\text{-Ser}$$

The characteristic repeating disaccharide units of *N*-acetylgalactosamine and glucuronic acid are covalently attached to this linkage region (Figure 8.10). The disaccharides may be sulfated in either the 4 or 6 position of *N*-acetylgalactosamine. Each polysaccharide chain contains between 30 and 50 such disaccharide units, corresponding to molecular weights of 15,000–25,000. An average chondroitin sulfate proteoglycan molecule has approximately 100 chondroitin sulfate chains attached to the protein core, giving rise to a molecular weight of 1.5–2 × 10^6. Proteoglycan preparations are, however, extremely heterogeneous, differing in length of protein core, degree of substitution, and distribution of polysaccharide chains, length of chondroitin sulfate chains, and degree of sulfation.

Chondroitin sulfate proteoglycans have also been shown to aggregate noncovalently with hyaluronate, forming much larger structures. They appear to exist in vivo in this aggregated form in the ground substance of cartilage, and have also been isolated from tendons, ligaments, and aorta.

Dermatan Sulfate Contains L-Iduronic Acid

Dermatan sulfate differs from chondroitin 4- and 6-sulfates in that its predominant uronic acid is L-iduronic acid, although D-glucuronic acid is also present in variable amounts. The glycosidic linkages are the same in position and configuration as in the chondroitin sulfates. with average polysaccharide chains of molecular weights of $2–5 \times 10^4$. The physiological function of dermatan sulfate is poorly understood, Unlike the chondroitin sulfates, dermatan sulfate is antithrombic like heparin, but in contrast to heparin, it shows only minimal whole blood anticoagulant and blood lipid-clearing activities. As a connective tissue macromolecule, dermatan sulfate is found in skin, blood vessels, and heart valves.

Heparin and Heparan Sulfate Differ From Other Glycosaminoglycans

Heparin differs from other glycosaminoglycans in a number of important respects. Glucosamine and D-glucuronic acid or L-iduronic acid form the characteristic disaccharide repeat unit, as in dermatan sulfate (Figure 8.10). In contrast to the glycosaminoglycans in ground substance, heparin contains α-glycosidic linkages. Almost all glucosamine residues are bound in sulfamide linkages, but a small number of glucosamine residues are *N*-acetylated. The sulfate content of heparin, although variable, approaches 2.5 sulfate residues per disaccharide unit in preparations with the highest biological activity. In addition to *N*-sulfate and *O*-sulfate on C-6 of glucosamine, heparin may also contain sulfate on C_3 of the hexosamine and C-2 of the uronic acid. Unlike the other glycosaminoglycans, which are predominantly extracellular components, heparin is an intracellular component of mast cells. Heparin is known as an anticoagulant and lipid-clearing agent; however, the natural physiological role of this polysaccharide still remains unclear (see Clin. Corr. 8.8 on p. 375).

Heparan sulfate contains a similar disaccharide repeat unit but has more *N*-acetyl groups, fewer *N*-sulfate groups, and a lower degree of *O*-sulfate groups. Heparan sulfate appears to be extracellular in distribution and has been isolated from blood vessel walls, amyloid, and brain. Recently, it has been shown to be an integral and ubiquitous component of the cell surface.

Keratan Sulfate Is Very Heterogeneous

More than any of the other glycosaminoglycans, **keratan sulfate** is characterized by molecular heterogeneity. This polysaccharide is composed principally of a repeating disaccharide unit of *N*-acetylglucosamine and galactose, with no uronic acid in the molecule (Figure 8.10). Sulfate content is variable, with ester sulfate present on C-6 of both galactose and hexosamine. Two types of keratan sulfate have been distinguished, which differ in their overall carbohydrate content and tissue distribution. Both contain as additional monosaccharides, mannose, fucose, sialic acid, and *N*-acetylgalactosamine. Keratan sulfate I, isolated from cornea, is linked to protein by an *N*-acetylglucosamine–asparaginyl bond, typical of glycoproteins. Keratan sulfate II, isolated from cartilage, is attached to protein through N-acetylgalactosamine in O-glycosidic linkage to either serine or threonine. Skeletal keratan sulfates are often found covalently attached to the same core protein as are the chondroitin sulfate chains.

Biosynthesis of Chondroitin Sulfate Is Typical of Glycosaminoglycan Formation

Most of the problems in studies on the biosynthesis of proteoglycans are similar to those previously encountered in the formation of other glycoproteins. The polysaccharide chains are assembled by the sequential action of a series of glycosyltransferases, which catalyze the transfer of a monosaccharide from a nucleotide sugar to an appropriate acceptor, either the nonreducing end of another sugar or a polypeptide. Since the biosynthesis of the chondroitin sulfates is most thoroughly understood, this pathway will be discussed as the prototype for glycosaminoglycan formation (Figure 8.11).

In analogy with the mechanisms established for glycoprotein biosynthesis, the formation of the core protein of the chondroitin sulfate proteoglycan is the first step in this process. However, whether the initiation of polysaccharide chains precedes completion of the peptide chain and release from the ribosomes has not been determined. The polysaccharide chains are assembled by six different glycosyltransferases. Strict substrate specificity is required for completion of the unique tetrasaccharide linkage region. Polymerization then results from the concerted action of two glycosyltransferases, an **N-acetylgalactosaminyltransferase** and a **glucuronosyltransferase,** which alternately add the two monosaccharides, forming the characteristic repeating disaccharide units. Sulfation of N-acetylgalactosamine residues in either the 4 or 6 position apparently occurs along with chain elongation. The sulfate donor in these reactions,

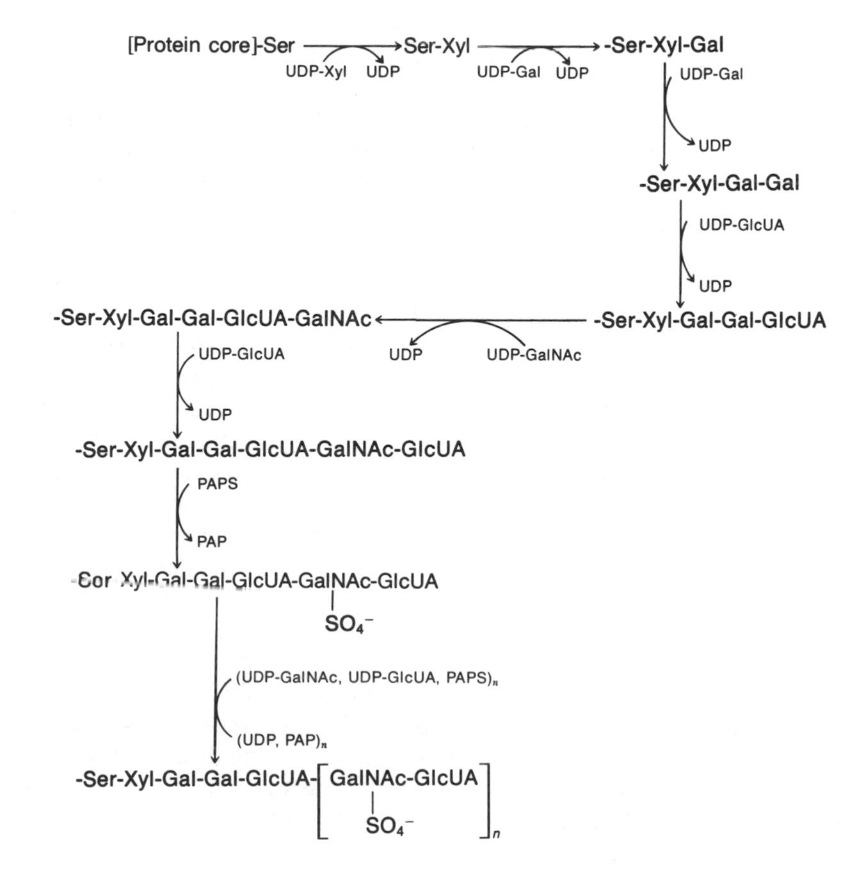

FIGURE 8.11
Synthesis of chondroitin sulfate proteoglycan.
Abbreviations: Xyl, xylose; Gal, galactose; GlcUA, glucuronic acid; GalNAc, N-acetylgalactosamine; PAPS, phosphoadenosine phosphosulfate.

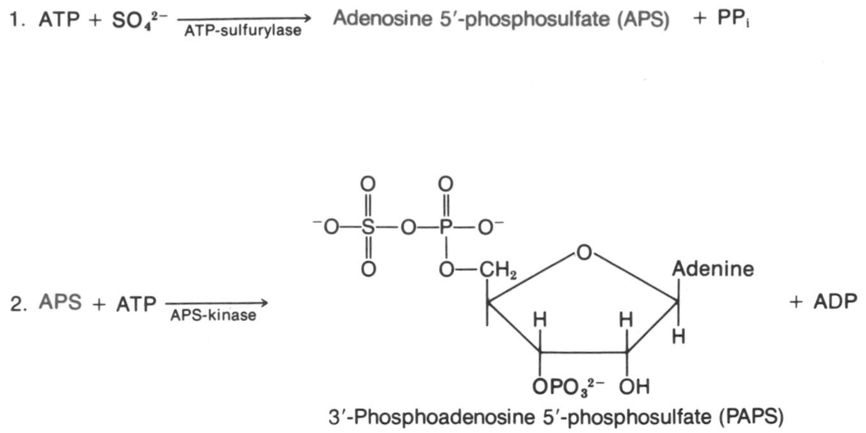

1. ATP + SO_4^{2-} $\xrightarrow{\text{ATP-sulfurylase}}$ Adenosine 5′-phosphosulfate (APS) + PP_i

2. APS + ATP $\xrightarrow{\text{APS-kinase}}$ 3′-Phosphoadenosine 5′-phosphosulfate (PAPS) + ADP

$^{-}O-S(=O)_2-O-P(=O)(O^{-})-O-CH_2$ (ribose: H, H, H; Adenine; OPO_3^{2-}, OH)

FIGURE 8.12
Biosynthesis of PAPS.

as in other biological systems, is **3′-phosphoadenosine 5′-phosphosulfate (PAPS)**, which is formed from ATP and sulfate in two steps (Figure 8.12).

Synthesis of the other glycosaminoglycans requires additional transferases specific for the sugars and linkages found in these molecules. Completion of these glycosaminoglycans often involves modifications in addition to *O*-sulfation, including epimerization, acetylation, and *N*-sulfation. Interestingly, the epimerization of D-glucuronic acid to L-iduronic acid occurs after incorporation into the polymer chain and is coupled with the process of sulfation.

It still remains unclear what fundamental central mechanisms determine the quantity as well as the qualitative nature of the proteoglycans synthesized. It appears that different proteoglycans are synthesized because of the presence and strict substrate specificity of the enzymes and the formation of specific acceptor proteins in a cell. Two mechanisms by which the levels of nucleotide sugars may be regulated have previously been described. The first specific reaction in hexosamine synthesis, the fructose 6-phosphate-glutamine transamidase reaction (Figure 8.4), is subject to feedback inhibition by UDP-*N*-acetylglucosamine, which is in equilibrium with UDP-*N*-acetylgalactosamine. Hence, synthesis of both proteoglycans and glycoproteins is regulated by the same mechanism. More specific to proteoglycan synthesis, the levels of UDP-xylose and UDP-glucuronic acid are stringently controlled by the inhibition by UDP-xylose of the UDP-glucose dehydrogenase conversion of UDP-glucose to UDP-glucuronic acid (Figure 8.4). Since xylose is the first sugar added during synthesis of chondroitin sulfate, dermatan sulfate, heparin, and heparan sulfate, the earliest effect of decreased core protein synthesis would be accumulation of UDP-xylose. This sensitive regulatory mechanism may be responsible for maintaining a balance between synthesis of the protein and the polysaccharide moieties of these complex macromolecules.

Proteoglycans, like glycoproteins, are presumably degraded by the sequential action of proteases and glycosidases, as well as deacetylases and sulfatases. Much of the information about metabolism and degradation of proteoglycans has been derived from the study of **mucopolysaccharidoses** (Clin. Corr. 8.9). This group of human genetic disorders is characterized by accumulation in tissues and excretion in urine of oligosaccharide products derived from incomplete breakdown of the proteoglycans, due to a deficiency of one or more lysosomal hydrolases. In the diseases for which the biochemical defect has been identified, it has been shown that a product accumulates with a nonreducing terminus that would have been the substrate for the deficient enzyme.

CLIN. CORR. 8.9 MUCOPOLYSACCHARIDOSES

A group of human genetic disorders characterized by excessive accumulation and excretion of the oligosaccharides of proteoglycans exists, collectively called mucopolysaccharidoses. These disorders result from a deficiency of one or more lysosomal hydrolases responsible for the degradation of dermatan and/or heparan sulfate. The enzymes lacking in specific mucopolysaccharidoses that have been identified are presented in the accompanying table. Presumably, for complete sequential degradation of the glycosaminoglycans, additional enzymes are required to hydrolyze bonds (4) *N*-acetyl β-galactosaminidase, (7) α-glucosaminidase, and (8) *N*-acetylglucosamine sulfatase, for which deficiency diseases remain unknown.

Although the chemical basis for this group of disorders is similar, their mode of inheritance as well as clinical manifestations may vary. Hurler's syndrome and Sanfilippo's syndrome are transmitted as autosomal recessives, whereas Hunter's disease is sex-linked. Both Hurler's syndrome and Hunter's disease are characterized by skeletal abnormalities and mental retardation, which in severe cases may result in early death. In contrast, in the Sanfilippo syndrome, the physical defects are relatively mild, while the mental retardation is severe. Collectively, the incidence for all mucopolysaccharidoses is 1 per 30,000 births.

In addition to those listed in the table, some others are less well understood. Morquio syndrome involves impaired degradation of keratan sulfate, and two types have been identified, type A due to defiency of *N*-acetylgalactosamine 6-sulfatase and type B due to deficiency of β-galactosidase. Multiple sulfatase deficiency involves the deficiency of at least six sulfatases; and "I-cell" disease involves a marked decrease in several hydrolase enzymes.

These disorders are amenable to prenatal diagnosis, since the pattern of metabolism by affected cells obtained from amniotic fluid is strikingly different from normal.

McKusick, V. and Neufeld, E. F. The mucopolysaccharide storage diseases. Stansbury, J. B., Wyngaarden, J. B., Frederickson, D. S., Goldstein, J. L., and Brown, M. S. (Eds.). The Metabolic Basis of Inherited Disease, 5th ed., New York: McGraw-Hill, 1983, p. 751.

Enzyme Defects in the Mucopolysaccharidoses

Disease	Accumulated products[a]	Deficient enzyme[b]
Hunter	Heparan sulfate Dermatan sulfate	Iduronate sulfatase (1)
Hurler + Scheie	Heparan sulfate Dermatan sulfate	α-L-Iduronidase (2)
Maroteaux–Lamy	Dermatan sulfate	*N*-Acetylgalactosamine (3) sulfatase
Mucolipidosis VII	Heparan sulfate Dermatan sulfate	β-Glucuronidase (5)
Sanfilippo A	Heparan sulfate	Heparan sulfamidase (6)
Sanfilippo B	Heparan sulfate	*N*-Acetylglucosaminidase (9)
Sanfilippo D	Heparin sulfate	*N*-Acetylglucosamine 6-sulfatase (8)

[a] Structures of dermatan sulfate and heparan sulfate.

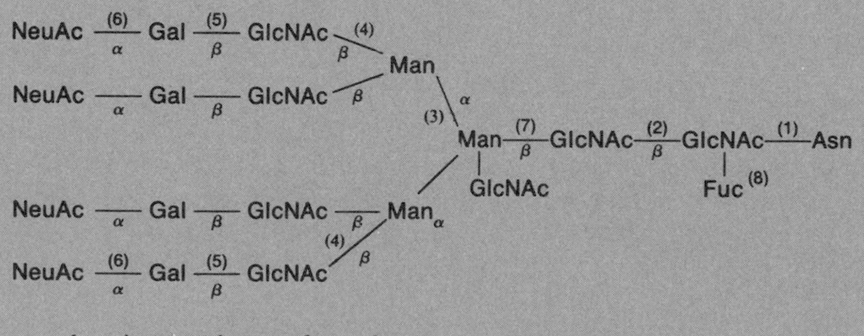

NeuAc —(6) α— Gal —(5) β— GlcNAc —(4) β— Man
NeuAc — α — Gal — β — GlcNAc — β — Man
Man —α (3)— Man —(7) β— GlcNAc —(2) β— GlcNAc —(1)— Asn
Man | GlcNAc; GlcNAc | Fuc (8)
NeuAc — α — Gal — β — GlcNAc — β — Man α
NeuAc —(6) α— Gal —(5) β— GlcNAc —(4) β— Man

[b] The numbers in parentheses refer to the enzymes that hydrolyze those bonds.

BIBLIOGRAPHY

Dutton, G. J. (Ed.). *Glucuronic acid, free and combined*. New York: Academic, 1966.

Ginsburg, V. and Robbins, P. (Eds.). *Biology of carbohydrates*. New York: Wiley, 1984.

Horecker, B. L. *Pentose metabolism in bacteria*. New York: Wiley, 1962.

Hughes, R. C. *Glycoproteins*. London: Chapman & Hall, 1983.

Kornfeld, R. and Kornfeld, S. Assembly of Asn-linked Oligosaccharides. *Annu. Rev. Biochem.* 54:631, 1985.

Lennarz, W. J. (Ed.). *The biochemistry of glycoproteins*. New York: Plenum, 1980.

Rademacker, T. W., Parck, R. B., and Dwek, R. A. Glycobiology, *Annu. Rev. Biochem.* 57:285, 1988.

Schwartz, N. B. and Smalheiser, N. Biosynthesis of glycosaminoglycans and proteoglycans, Neurobiology of Glycoconjugates. New York: Plenum, 1989, p. 151.

Scriver, C. P., Beaudet, A. L., Sly, W. S., and Vallee, D. (Eds.) The metabolic basis of inherited disease, 6th ed. New York: McGraw-Hill, 1989.

Sharon, N. *Complex carbohydrates—their chemistry, biosynthesis and functions*. Reading, MA: Addison-Wesley, 1975.

QUESTIONS

C. N. ANGSTADT AND J. BAGGOTT

*Q*uestion *T*ypes are described on page xxiii.

1. (QT1) [NADPH]/[$NADP^+$] is maintained at a high level in cells primarily by:
   A. lactate dehydrogenase.
   B. the combined actions of glucose 6-phosphate dehydrogenase and gluconolactonase.
   C. the action of the electron transport chain.
   D. shuttle mechanisms such as the α-glycerophosphate dehydrogenase shuttle.
   E. the combined actions of transketolase and transaldolase.
2. (QT1) Transketolase:
   A. transfers a C_2 fragment to an aldehyde acceptor.
   B. transfers a C_3 ketone-containing fragment to an acceptor.
   C. converts the ketose sugar ribulose 5-phosphate to ribose 5-phosphate.
   D. is part of the irreversible oxidative phase of the pentose phosphate pathway.
   E. converts two C_5 sugar phosphates to fructose 6-phosphate and erythrose 4-phosphate.
3. (QT1) If a cell requires more NADPH than ribose 5-phosphate:
   A. only the first phase of the pentose phosphate pathway would occur.
   B. glycolytic intermediates would flow into the reversible phase of the pentose phosphate pathway.
   C. there would be sugar interconversions but no net release of carbons from glucose 6-phosphate.
   D. the equivalent of the carbon atoms of glucose 6-phosphate would be released as $6CO_2$.
   E. only part of this need could be met by the pentose pathway, and the rest would have to be supplied by another pathway.

4. (QT2) All of the following interconversions of monosaccharides (or derivatives) require a nucleotide-linked sugar intermediate except:
   1. galactose 1-phosphate to glucose 1-phosphate
   2. glucose 6-phosphate to mannose 6-phosphate
   3. glucose to glucuronic acid
   4. D-glucuronic acid to L-iduronic acid
5. (QT1) Fructose:
   A. unlike glucose, cannot be catabolized by the glycolytic pathway.
   B. in the liver, enters directly into glycolysis as fructose 6-phosphate.
   C. must be isomerized to glucose before it can be metabolized.
   D. is converted to a UDP-linked form and then epimerized to UDP-glucose.
   E. catabolism in liver uses fructokinase and a specific aldolase that recognizes fructose 1-phosphate.
6. (QT1) Galactosemia:
   A. is a genetic deficiency of a uridylyltransferase that exchanges galactose 1-phosphate for glucose on UDP-glucose.
   B. results from a deficiency of an epimerase.
   C. is not apparent at birth but symptoms develop in later life.
   D. is an inability to form galactose 1-phosphate.
   E. would be expected to interfere with the use of fructose as well as galactose because the deficient enzyme is common to the metabolism of both sugars.

7. (QT2) Glucuronic acid:
   1. enhances the water solubility of compounds to which it is conjugated.
   2. as a UDP derivative can be decarboxylated to a component used in proteoglycan synthesis.
   3. is a precursor of ascorbic acid in many mammalian species but not in humans.
   4. formation from glucose is under feedback control by a UDP-linked intermediate.
8. (QT2) The conversion of fructose 6-phosphate to glucosamine 6-phosphate:
   1. is a transamination reaction with glutamate as the nitrogen donor.
   2. is under feedback inhibition by UDP-*N*-acetylglucosamine.
   3. requires that fructose 6-phosphate first be linked to a nucleotide.
   4. is a first step in the formation of *N*-acetylated amine sugars.

9. (QT2) *N*-Acetylneuraminic acid:
   1. is a sialic acid.
   2. is activated by conversion to CMP-*N*-acetylneuraminic acid.
   3. is derived from UDP-*N*-acetylglucosamine.
   4. formation includes a condensation of *N*-acetylmannosamine 6-phosphate with phosphoenolpyruvate.
10. (QT2) Roles for the complex carbohydrate moiety of glycoproteins include:
   1. determinant of blood type.
   2. cell surface receptor specificity.
   3. determinant of the rate of clearance from the circulation of certain molecules.
   4. template for the synthesis of glycosaminoglycans.

11. (QT3) A. Percentage of carbohydrate by weight in glycoproteins
    B. Percentage of carbohydrate by weight in proteoglycans

12. (QT2) In glycoproteins, the carbohydrate portion may be linked to the protein by:
   1. an *N*-glycosyl bond to asparagine.
   2. an *O*-glycosyl bond to lysine.
   3. an *O*-glycosyl bond to serine.
   4. noncovalent bonds.

A. *O*-linked glycoproteins C. Both
B. *N*-linked glycoproteins D. Neither

13. (QT4) Core structure is assembled on dolichol phosphate before transfer to the protein.
14. (QT4) Core structure typically contains fucose and sialic acid.
15. (QT4) Processing and elongation of the core structure occurs in the rough endoplasmic reticulum and Golgi apparatus.

16. (QT2) Glycosaminoglycans:
   1. are the carbohydrate portion of proteoglycans.
   2. contain large segments of a repeating unit typically consisting of a hexosamine and a uronic acid.
   3. are polyanions.
   4. exist as at least six different classes.

A. Hyaluronate
B. Chondroitin sulfate
C. Dermatan sulfate
D. Heparin
E. Keratan sulfate

17. (QT5) Differs from other glycosaminoglycans in being predominantly intracellular rather than extracellular.
18. (QT5) Only glycosaminoglycan not covalently linked to protein.
19. (QT5) Only glycosaminoglycan that is not sulfated.

20. (QT2) Proteoglycan:
   1. specificity is determined, in part, by the action of glycosyltransferases.
   2. synthesis is regulated, in part, by UDP-xylose inhibition of the conversion of UDP-glucose to UDP-glucuronic acid.
   3. synthesis involves sulfation of carbohydrate residues by PAPS.
   4. degradation is catalyzed in the cytosol by nonspecific glycosidases.

ANSWERS

1. B Although the glucose 6-phosphate dehydrogenase reaction, specific for NADP, is reversible, hydrolysis of the lactone assures that the overall equilibrium lies far in the direction of NADPH. A, C, D: These all use NAD, not NADP. E: These enzymes are part of the pentose phosphate pathway but catalyze freely reversible reactions that do not involve NADP (p. 361).
2. A Both reactions catalyzed by transketolase are of this type. B, E: Describe transaldolase. C. Describes an isomerase. D. Transketolase is part of the reversible phase of the pentose phosphate pathway that also allows glycolytic intermediates to be converted to pentose sugars, if necessary (p. 362).
3. D A, C, D, E: Glucose 6-phosphate yields ribose 5-phosphate + CO_2 in the oxidative phase. If this is multiplied by six, the six ribose 5-phosphates can be rearranged to five glucose 6-phosphates by the second, reversible phase. B: If more ribose 5-phosphate than NADPH were required, the flow would be in this direction to supply the needed pentoses (p. 365).
4. C 2, 4 correct. 1: Occurs via an epimerase at the UDP-galactose level. 2: The glucose and mannose phosphates are both in equilibrium with fructose 6-phosphate by phosphohexose isomerases. 3: This oxidation of glucose is catalyzed by UDP-glucose dehydrogenase. 4: Although most epimerizations involve nucleotide-linked sugars, this one does not, occurring after the glucuronic acid is incorporated into heparin or dermatan sulfate (p. 366).
5. E A, C, E: Fructokinase produces fructose 1-phosphate. Since this cannot be converted to fructose 1,6-bisphosphate, a specific aldolase cleaves it to dihydroxyacetone phosphate and glyceraldehyde. The first product is a glycolytic intermediate; the second requires modification to enter glycolysis. D: Glucose and fructose are not epimers (p. 366).
6. A B: The epimerase is normal. C: Galactose is an important sugar for infants. E: Fructose metabolism does not use the uridylyltransferase that is deficient in galactosemia (p. 367).
7. E All four correct. 1: Enhancing water solubility is a major physiological role for glucuronic acid, for example, bilirubin metabolism. 2, 4: Decarboxylation of UDP-glucuronic acid gives UDP-xylose, which is a potent inhibitor of the oxidation of UDP-glucose to

the acid. 3: The reduction of D-glucuronic acid to L-gulonic acid leads to ascorbate as well as xylulose 5-phosphate for the pentose phosphate pathway (p. 369, Figure 8.6).

8. C 2, 4 correct. 1, 3: This conversion is a transamidation of the amide nitrogen of glutamine and does not involve nucleotide intermediates. 2, 4: Glucosamine 6-phosphate is acetylated. UDP-*N*-acetylglucosamine is formed, and the UDP derivative can be epimerized to the galactose derivative. It also is a feedback inhibitor of the transamidase reaction, thus controlling formation of the nucleotide sugars (p. 371).
9. E All four correct. 2, 4: This complex pathway begins with an unusual epimerization of UDP-*N*-acetylglucosamine to free *N*-acetylmannosamine that is phosphorylated and then condensed with phosphoenolpyruvate. 3: Unlike most sugars, the activated form of *N*-acetylneuraminic acid is a CMP rather than a UDP derivative (p. 371).
10. A 1, 2, 3 correct. Because of the diversity possible with oligosaccharides, they play a significant role in determining the specificity of many biological molecules. 4: The synthesis of complex carbohydrates is not template directed but determined by the specificity of individual enzymes (p. 373).
11. B The term "proteoglycans" is reserved for species that contain at least 95% carbohydrate (p. 378).
12. B 1, 3 correct. 2: *O*-linkage occurs to hydroxylysine in collagen not to lysine. 4: Carbohydrates are covalently linked to protein (p. 376).
13. B Synthesis of *O*-linked glycoproteins involves the sequential addition to the *N*-acetylgalactosamine linked to serine or threonine (p. 376).
14. A Core also contains galactose and *N*-acetylgalactosamine; core structure of *N*-linked carbohydrates contains mannose and *N*-acetylglucosamine (p. 376).
15. C Both types of sugar require processing of the core structure (p. 377).
16. E All four correct. 1: These are quite different from the carbohydrate of glycoproteins. 2: This is a major distinction from glycoproteins, which, by definition, do not have a serial repeating unit. 3: The anionic character contributed by carboxyl and sulfate (another common feature) groups is important to the biological function (p. 378).
17. D (p. 379).
18. A Classified as a glycosaminoglycan because of its structural similarity to the others (p. 378).
19. A Both heparin and heparan sulfate are sulfated (p. 378).
20. A 1, 2, 3 correct. 1: Strict substrate specificity of the enzymes is important in determining the type and quantity of proteoglycans synthesized. Formation of specific protein acceptors for the carbohydrate is also important. 2: Both xylose and glucuronic acid levels are controlled by this; xylose is the first sugar added in the synthesis of four of the six types. 3: This is necessary for the formation of all proteoglycans (hyaluronic acid is not part of a proteoglycan). 4: Degradation is lysosomal; deficiencies of one or more lysosomal hydrolases leads to accumulation of proteoglycans in the mucopolysaccharidoses (p. 381).

9

Lipid Metabolism I: Utilization and Storage of Energy in Lipid Form

J. DENIS McGARRY

9.1 OVERVIEW 388
9.2 THE CHEMICAL NATURE OF FATTY ACIDS AND ACYLGLYCEROLS 389
Fatty Acids Are Alkyl Chains Terminating in a Carboxyl Group 389
Nomenclature of Fatty Acids 390
Most Fatty Acids in Humans Occur as Triacylglycerols 390
The Hydrophobic Nature of Lipids Is Important for Their Biological Function 391
9.3 SOURCES OF FATTY ACIDS 392
Most Fatty Acids Are Supplied in the Diet 392
Palmitate Is Synthesized From Acetyl CoA 392
Formation of Malonyl CoA Is the Commitment Step of Fatty Acid Synthesis 393
Reaction Sequence for the Synthesis of Palmitic Acid 393
Mammalian Fatty Acid Synthase Is a Multienzyme Complex 395
Stoichiometry of Fatty Acid Biosynthesis 397
Acetyl CoA Must Be Transported From Mitochondria to the Cytosol for Palmitate Synthesis 397
Palmitate Is the Precursor of Other Fatty Acids 398
Fatty Acid Synthase Can Produce Fatty Acids Other Than Palmitate 400
Fatty Acyl CoAs May Be Reduced to Fatty Alcohols 401
9.4 STORAGE OF FATTY ACIDS AS TRIACYLGLYCEROLS 401
Triacylglycerols Are Synthesized From Fatty Acyl CoAs and Glycerol-3-Phosphate in Most Tissues 401
Mobilization of Triacylglycerols Requires Hydrolysis 403
9.5 METHODS OF INTERORGAN TRANSPORT OF FATTY ACIDS AND THEIR PRIMARY PRODUCTS 404
Lipid-Based Energy Is Transported in Blood in Different Forms 404
Lipases Must Hydrolyze Blood Triacylglycerols for Their Fatty Acids to Be Transferred to Tissues 406
9.6 UTILIZATION OF FATTY ACIDS FOR ENERGY PRODUCTION 407
β-Oxidation of Straight-Chain Saturated Fatty Acids Is the Major Energy-Producing Process 407
The Energy-Yield From β-Oxidation of Fatty Acids 410
Comparison of the β-Oxidation Scheme With Palmitate Biosynthesis 411
Some Fatty Acids Require Modifications of β-Oxidation for Their Metabolism 411
Ketone Bodies Are Formed From Acetyl CoA 414
Peroxisomal Oxidation of Fatty Acids Serves Many Functions 417

9.7 CHARACTERISTICS, METABOLISM, AND FUNCTIONAL ROLE OF POLYUNSATURATED FATTY ACIDS 417
Some Polyunsaturated Fatty Acids Can Be Synthesized or Modified 418
Polyunsaturated Fatty Acids Are Susceptible to Autooxidation 419
BIBLIOGRAPHY 420
QUESTIONS AND ANNOTATED ANSWERS 420

CLINICAL CORRELATIONS

9.1 Obesity 404
9.2 Genetic Abnormalities in Lipid-Energy Transport 405
9.3 Genetic Deficiencies in Carnitine or Carnitine Palmitoyl Transferase 409
9.4 Genetic Deficiencies in the Acyl CoA Dehydrogenases 410
9.5 Refsum's Disease 414
9.6 Diabetic Ketoacidosis 417

9.1 OVERVIEW

As the human body builds and renews its structures, obtains and stores energy, and performs its various functions, there are numerous circumstances in which it is essential to use molecules or parts of molecules that do not associate with water. This property of being nonpolar and hydrophobic is largely supplied by the substances classed as lipids. Most of these are molecules that contain or are derived from fatty acids. In the early stages of biochemical research these substances were not studied as intensively as other body constituents, largely because the techniques for studying aqueous systems were easier to develop. This benign neglect led to early assumptions that the lipids were relatively inert and their metabolism was of lesser importance than that of carbohydrates, for instance.

As the methodology for investigation of lipid metabolism developed, however, it soon became evident that fatty acids and their derivatives had at least two major roles in the human body. On the one hand, the oxidation of fatty acids was shown to be a major means of metabolic energy production, and it became clear that their storage in the form of triacylglycerols was more efficient and quantitatively more important than storage of carbohydrates as glycogen. On the other hand, as details of the chemistry of biological structures were elucidated, hydrophobic structures were found to be largely composed of fatty acids and their derivatives. Thus the major separation of cells and subcellular structures into separate aqueous compartments is accomplished by the use of membranes whose hydrophobic characteristics are largely supplied by the fatty acid moieties of complex lipids. These latter compounds contain constituents other than fatty acids and glycerol. They frequently have significant covalently bound hydrophilic moieties, notably carbohydrates in the glycolipids and organic phosphate esters in the phospholipids.

In addition to these two major functions of lipids, energy production and structure building, there are several other quantitatively less important roles, which are nonetheless of great functional significance. These include the use of the surface active properties of some complex lipids for specific functions, such as maintenance of lung alveolar integrity and solubilization of nonpolar substances in body fluids. In addition, several classes of lipids, the steroid hormones and the prostaglandins, have highly potent and specific physiological roles in control of metabolic processes. The interrelationships of some of the processes involved in lipid metabolism are outlined in Figure 9.1.

Since the metabolism of fatty acids and triacylglycerols is so crucial to proper functioning of the human body, imbalances and deficiencies in these processes lead to significant pathological processes, and disease states related to fatty acid and triacylglycerol metabolism include some of the major clinical problems to be encountered by physicians, for instance, ketoacidosis, obesity, and abnormalities in transport of lipids in blood. In

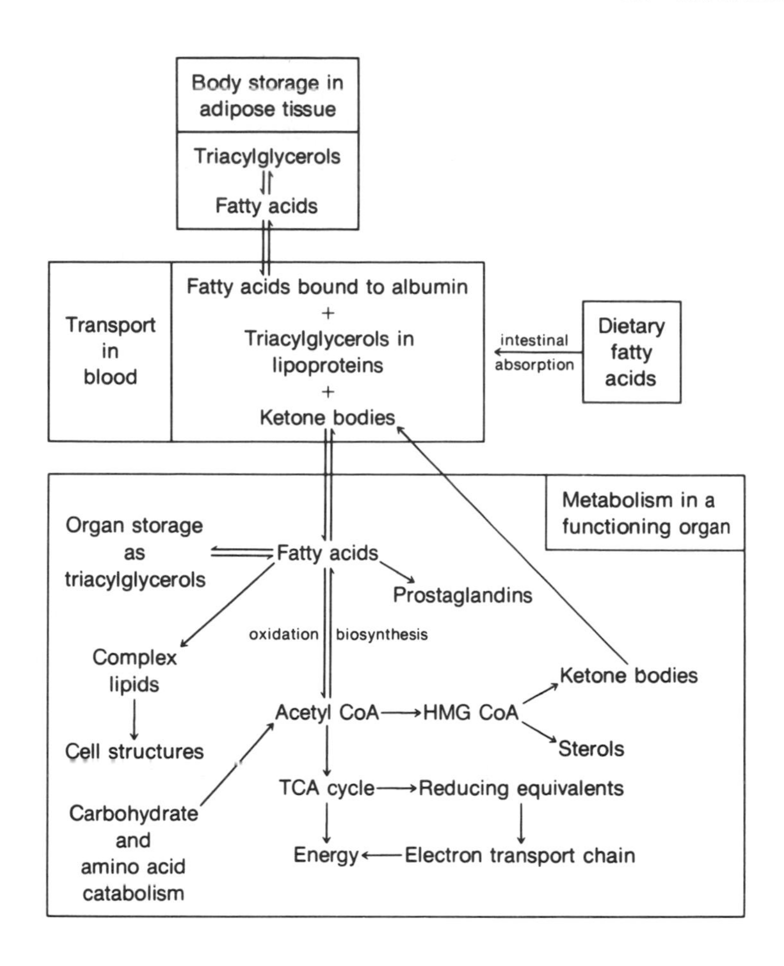

FIGURE 9.1
The metabolic interrelationships of fatty acids in the human body.

addition, some unique deficiencies have been found, such as Refsum's disease and familial hypercholesterolemia, which have helped to elucidate some pathways in lipid metabolism.

In this chapter we will be primarily concerned with the structure and metabolism of the fatty acids themselves and of their major storage form, the triacylglycerols. After a discussion of the structures of the more important fatty acids in the human body we describe how they are supplied to the human metabolic machinery from the diet or by biosynthesis. Since their storage as triacylglycerols is a major process we next discuss how this storage is accomplished and how the fatty acids themselves are mobilized and transported throughout the body to sites where they are needed. The central process of energy production from fatty acids is then discussed, and finally we introduce some concepts about the role and metabolism of polyunsaturated fatty acids.

See the appendix, for a discussion of the nomenclature and chemistry of lipids and Chapter 26 for a presentation of the digestion and absorption of lipids.

9.2 THE CHEMICAL NATURE OF FATTY ACIDS AND ACYLGLYCEROLS

Fatty Acids Are Alkyl Chains Terminating in a Carboxyl Group

Fatty acids consist of an alkyl chain with a terminal carboxyl group, and the simplest configuration is a completely saturated straight chain. The basic formula is $CH_3—(CH_2)_n—COOH$. The fatty acids of importance for humans have relatively simple configurations, although fatty acids in

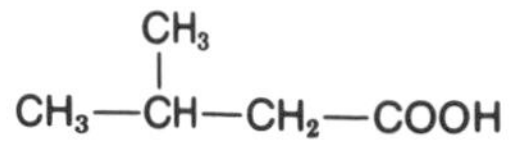

$$CH_3—\underset{|}{\overset{CH_3}{}}CH—CH_2—COOH$$

FIGURE 9.2
Isovaleric acid.

some organisms are occasionally quite complex, containing cyclopropane rings or extensive branching. Unsaturation occurs commonly in human fatty acids, with up to six double bonds per chain, and the bonds are almost always of the cis configuration. If there is more than one double bond per molecule, these bonds are always separated by a methylene ($—CH_2—$) group. The most common fatty acids in biological systems have an even number of carbon atoms, although some organisms do synthesize those with an odd number of carbon atoms. Human beings can use the latter for energy and incorporate them into complex lipids to a minimal degree.

A few fatty acids with an α-OH group are produced and used structurally by humans. However, more oxidized forms are normally produced only as metabolic intermediates during energy production or for specific physiological activity in the case of prostaglandins and thromboxanes. Higher animals, including humans, also produce relatively simple **branched-chain acids,** the branching being limited to methyl groups along the chain at one or more positions. These are apparently produced to contribute specific physical properties to some secretions and structures. For instance, large amounts of branched-chain fatty acids, particularly isovaleric acid (Figure 9.2), occur in the lipids of echo-locating structures in marine mammals. The elucidation of the role of these lipids in sound focusing should be fascinating.

The bulk of the fatty acids in the human body have C_{16}, C_{18}, or C_{20} atoms, but there are several with longer chains that occur principally in the lipids of the nervous system. These include nervonic acid, and a C_{22} acid with six double bonds (Figure 9.3).

$$CH_3—(CH_2)_7—CH{=}CH—(CH_2)_{13}—COOH$$

Nervonic acid

$$CH_3—(CH_2—CH{=}CH)_6—(CH_2)_2—COOH$$

All-*cis*-4,7,10,13,16,19-docosahexaenoic acid

FIGURE 9.3
Long-chain fatty acids.

Nomenclature of Fatty Acids

The most abundant fatty acids have common names that have been accepted for use in the official nomenclature. Examples are given in Table 9.1 along with the official systematic names. The approved abbreviations consist of the number of carbon atoms followed, after a colon, by the number of double bonds. Carbon atoms are numbered with the carboxyl carbon as number 1, and the double bond locations are designated by the number of the carbon atom on the carboxyl side of it. These designations of double bonds are in parentheses after the rest of the symbol. See Table 9.1 for examples.

Most Fatty Acids in Humans Occur as Triacylglycerols

Fatty acids occur primarily as esters of glycerol, as shown in Figure 9.4, when they are stored for future utilization. Compounds with one **(monoacylglycerols)** or two **(diacylglycerols)** acids esterified are present only

TABLE 9.1 Fatty Acids of Importance to Humans

Numerical symbol	Structure	Trivial name	Systematic name
16:0	$CH_3—(CH_2)_{14}—COOH$	Palmitic	Hexadecanoic
16:1(9)	$CH_3—(CH_2)_5—CH{=}CH—(CH_2)_7—COOH$	Palmitoleic	*cis*-9-Hexadecenoic
18:0	$CH_3—(CH_2)_{16}—COOH$	Stearic	Octadecanoic
18:1(9)	$CH_3—(CH_2)_7—CH{=}CH—(CH_2)_7—COOH$	Oleic	*cis*-9-Octadecenoic
18:2(9,12)	$CH_3—(CH_2)_3—(CH_2—CH{=}CH)_2—(CH_2)_7—COOH$	Linoleic	*cis,cis*-9,12-Octadecadienoic
18:3(9,12,15)	$CH_3—(CH_2—CH{=}CH)_3—(CH_2)_7—COOH$	Linolenic	*cis,cis,cis*-9,12,15-Octadecatrienoic
20:4(5,8,11,14)	$CH_3—(CH_2)_3—(CH_2—CH{=}CH)_4—(CH_2)_3—COOH$	Arachidonic	*cis,cis,cis,cis*-5,8,11,14-Icosatetraenoic

in relatively minor amounts and occur largely as metabolic intermediates in the biosynthesis and degradation of glycerol-containing lipids. The bulk of the fatty acids in the human body exist as **triacylglycerols,** in which all three hydroxyl groups on the glycerol are esterified with a fatty acid. Historically, these compounds have been termed *neutral fats* or *triglycerides,* and these terms are still in common usage. However, there are other types of "neutral fats" in the body, and the term "triglyceride" is chemically incorrect and should no longer be used. The same can be said for the terms "monoglyceride" and "diglyceride."

The distribution of various fatty acids in the different positions of the glycerol moiety of triacylglycerols in the body at any given time is the result of a number of factors, some of which are not completely understood. Suffice it to say that the fatty acid pattern varies with the time, diet, and anatomical location of the triacylglycerol. Compounds with the same fatty acid in all three positions are rare and the usual case is for a complex mixture.

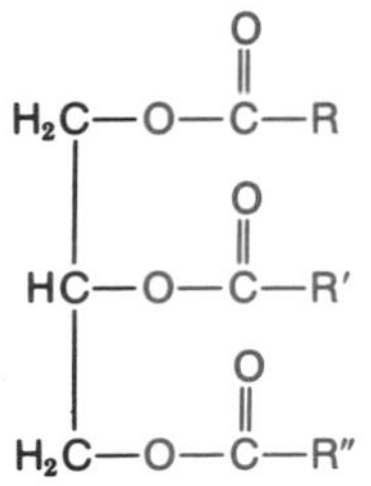

Triacylglycerol

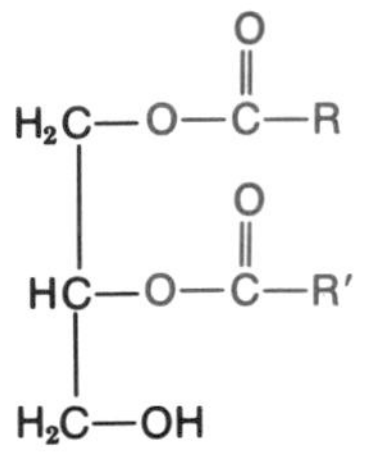

Diacylglycerol

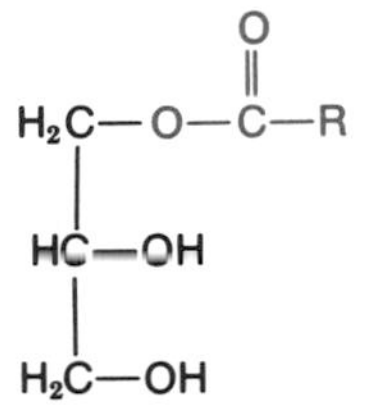

1-Monoacylglycerol

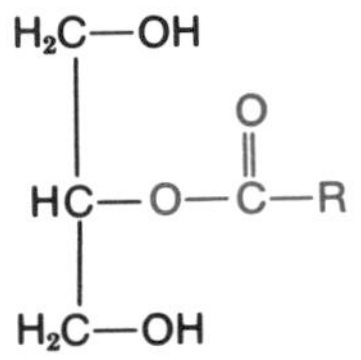

2-Monoacylglycerol

FIGURE 9.4
Acylglycerols.

The Hydrophobic Nature of Lipids Is Important for Their Biological Function

Certainly one of the most prominent and significant properties of fatty acids and triacylglycerols is their lack of affinity for water. The long hydrocarbon chains have negligible possibility for hydrogen bonding, and the acids, whether unesterified or in a complex lipid, have a much greater tendency to associate with each other or with other hydrophobic structures, such as sterols and the hydrophobic side chains of amino acids, than they do with water or polar organic compounds. It has been calculated that the van der Waals London forces between closely packed, relatively long-chain fatty acid moieties in lipids can approach the strength of a covalent bond. This hydrophobic character is essential for construction of complex biological structures and the separation of aqueous compartments as described in Chapter 5. It is also essential for use in biological surface active molecules, as in the intestinal tract.

Of major significance is the fact that the hydrophobic nature of triacylglycerols and their relatively reduced state make them efficient compounds for storing energy. Three points deserve emphasis. First, on a weight basis pure triacylglycerols yield nearly two and one-half times the amount of ATP on complete oxidation that pure glycogen does. Second, the triacylglycerols can be stored as pure lipid without associated water, whereas glycogen is quite hydrophilic and binds about twice its weight of water when stored in tissues. Thus the equivalent amount of metabolically recoverable energy stored as hydrated glycogen would weigh about four times as much as if it were stored as triacylglycerols. Third, the average 70 kg person stores about 100 g of carbohydrate as liver glycogen and 250 g as muscle glycogen. This represents about 1400 kcal of available energy, barely enough to sustain bodily functions for 24 h of fasting. By contrast, a normal complement of fat stores will provide sufficient energy to allow several weeks of survival during total food deprivation.

The bulk of the fatty acids in the lipids of the human body are either saturated or contain only one double bond. Consequently, although they are readily catabolized by appropriate enzymes and cofactors, they are fairly inert chemically. This is an added advantage of their use for energy storage. However, the smaller amounts of the more highly unsaturated fatty acids in the tissues are much more susceptible to oxidation. Some possible biological consequences of this oxidation are discussed later in Section 9.7.

9.3 SOURCES OF FATTY ACIDS

Both diet and biosynthesis supply the fatty acids needed by the human body for energy and for construction of hydrophobic parts of biomolecules. Excess amounts of protein and carbohydrate obtained in the diet are readily converted to fatty acids and stored as triacylglycerols.

Most Fatty Acids Are Supplied in the Diet

A great proportion of the fatty acids utilized by humans is supplied in their diet. Various animal and vegetable lipids are ingested, hydrolyzed at least partially by digestive enzymes, and absorbed through the intestinal wall to be distributed through the body, first in the lymphatic system and then in the bloodstream. These processes are extensively discussed in Chapter 26. To some extent, then, dietary supply governs the composition of the fatty acids in the body lipids. On the other hand, metabolic processes in the tissues of the normal human body can modify the dietary fatty acids, and/or those that are synthesized in these tissues, to produce almost all the various structures that are needed. For this reason, with one exception, the actual composition of the fatty acids supplied in the diet is relatively unimportant. This one exception involves the need for appropriate proportions of the relatively highly unsaturated fatty acids and particularly relates to the fact that many higher mammals, including humans, are unable to produce fatty acids with double bonds very far toward the methyl end of the molecule, either during de novo synthesis or by modification of dietary acids. Despite this inability, certain **polyunsaturated acids** with double bonds within the last seven linkages toward the methyl end are essential for some specific functions. Although all the reasons for this need are not yet elucidated, certainly one is that some of these acids are precursors of prostaglandins, highly active oxidation products (see Chapter 10).

In humans a dietary precursor is essential for two series of fatty acids. These are the **linoleic series** and the **linolenic series** (Figure 9.5).

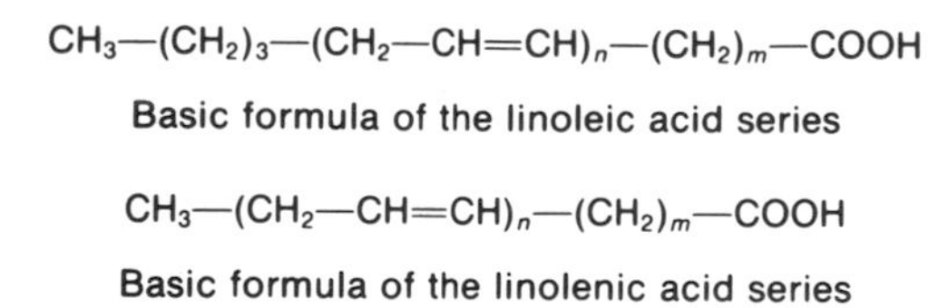

FIGURE 9.5
Linoleic and linolenic acid series.

Palmitate Is Synthesized From Acetyl CoA

Besides dietary supply, the second major source of fatty acids for humans is their biosynthesis from small-molecule intermediates, which can be derived from metabolic breakdown of sugars, of some amino acids, and of other fatty acids. In a majority of instances the saturated, straight-chain C_{16} acid, **palmitic acid,** is first synthesized, and all other fatty acids are made by modification of palmitic acid. Acetyl CoA is the direct source of all carbon atoms for this synthesis, and the fatty acids are made by sequential addition of two-carbon units to the activated carboxyl end of the growing chain. In mammalian systems the sequence of reactions is carried out by **fatty acid synthase.**

Fatty acid synthase is a fascinating enzyme complex that is still being intensively studied. In bacteria it is a complex of several proteins, whereas in mammalian cells it is a single multifunctional protein. For the most part its function is to form palmitate, but in some circumstances this pathway can be altered to produce other short-chain fatty acids. Some of the details of these modifications are discussed in later paragraphs, but first we will outline the basic scheme for synthesis of palmitate.

Either acetyl CoA or butyryl CoA is the priming unit for fatty acid synthesis, and the methyl end of these primers becomes the methyl end of palmitate. The addition of the rest of the two-carbon units requires further activation of the methyl carbon of acetyl CoA by carboxylation to

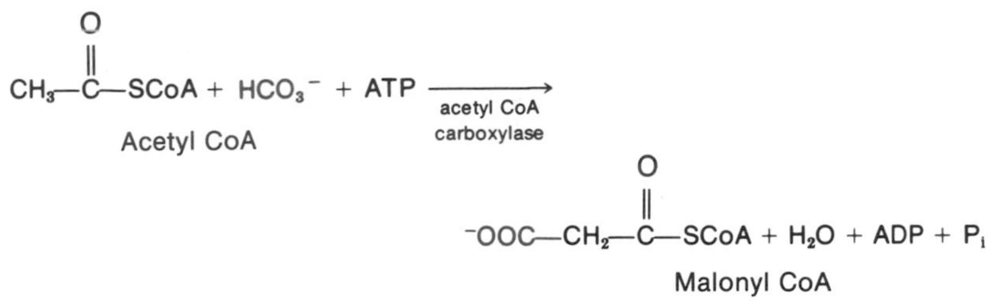

$$CH_3-\overset{\overset{\large O}{\|}}{C}-SCoA + HCO_3^- + ATP \xrightarrow[\text{carboxylase}]{\text{acetyl CoA}} {}^-OOC-CH_2-\overset{\overset{\large O}{\|}}{C}-SCoA + H_2O + ADP + P_i$$

Acetyl CoA → Malonyl CoA

FIGURE 9.6
Acetyl CoA carboxylase reaction.

malonyl CoA. However, the CO_2 added in this process is lost when the condensation of malonyl CoA to the growing chain occurs, so the carbon atoms in the palmitate chain originate only from the acetyl CoA.

Formation of Malonyl CoA Is the Commitment Step of Fatty Acid Synthesis

The metabolic process that commits acetyl CoA to fatty acid synthesis is its carboxylation to **malonyl CoA** by the enzyme **acetyl CoA carboxylase** (Figure 9.6). This reaction is similar in a number of ways to the carboxylation of pyruvate, which starts the process of gluconeogenesis. The reaction requires energy from ATP and uses dissolved bicarbonate as the source of CO_2. As in the case of pyruvate carboxylase, the first step in this reaction is the formation of activated CO_2 on the biotin moiety of the acetyl CoA carboxylase using the energy from ATP. This is then transferred to the acetyl CoA.

Acetyl CoA carboxylase catalyzes the committed step in the process of fatty acid synthesis and is thus an essential control point. The enzyme can be isolated in an inactive protomeric state, and these protomers aggregate to active polymers upon addition of citrate in vitro. In vitro studies also demonstrated that palmitoyl CoA inhibits the active enzyme. The action of these two effectors is very logical; increased synthesis of fatty acids to store energy being desirable when citrate is in high concentration, and decreased synthesis being necessary if high levels of the product accumulate. However, the degree to which these regulatory mechanisms actually operate in vivo is unknown.

Acetyl CoA carboxylase is also controlled by a cAMP-mediated phosphorylation–dephosphorylation mechanism in which the phosphorylated enzyme is less active than the dephosphorylated one. There is evidence suggesting that the phosphorylation is promoted by glucagon and that the presence of the active form is fostered by insulin. These effects of hormone-mediated phosphorylation are probably separate from the allosteric effects of citrate and palmitoyl CoA (see Table 9.2).

In longer-term effects the rate of synthesis of acetyl CoA carboxylase is regulated. More enzyme is produced by animals on high-carbohydrate or fat-free diets, and fasting or high-fat diets decrease the rate of enzyme synthesis.

Reaction Sequence for the Synthesis of Palmitic Acid

The first step catalyzed by the fatty acid synthase is the transacylation of the primer molecule, either acetyl CoA or butyryl CoA, to a 4′-phosphopantetheine moiety on a protein constituent of the enzyme complex in bacteria. This protein is **acyl carrier protein (ACP),** and its phosphopantetheine unit is identical with that in CoA. The mammalian enzyme also contains a **phosphopantetheine** unit. Six or seven two-carbon units are then added sequentially to the enzyme complex until the palmitate mole-

TABLE 9.2 Regulation of Fatty Acid Synthesis

Enzyme		Regulatory agent	Effect
		PALMITATE BIOSYNTHESIS	
Acetyl CoA carboxylase	Short term	Citrate	Allosteric activation
		C_{16}–C_{18} acyl CoAs	Allosteric inhibition
		Insulin	Stimulation
		Glucagon	Inhibition
		cAMP-mediated phosphorylation	Inhibition
		Dephosphorylation	Stimulation
	Long term	High-carbohydrate diet	Stimulation by increased enzyme synthesis
		Fat-free diet	Stimulation by increased enzyme synthesis
		High-fat diet	Inhibition by decreased enzyme synthesis
		Fasting	Inhibition by decreased enzyme synthesis
		Glucagon	Inhibition by decreased enzyme synthesis
Fatty acid synthase		Phosphorylated sugars	Allosteric activation
		High carbohydrate diet	Stimulation by increased enzyme synthesis
		Fat-free diet	Stimulation by increased enzyme synthesis
		High-fat diet	Inhibition by decreased enzyme synthesis
		Fasting	Inhibition by decreased enzyme synthesis
		Glucagon	Inhibition by decreased enzyme synthesis
		BIOSYNTHESIS OF FATTY ACIDS OTHER THAN PALMITATE	
Fatty acid synthase		High ratio of $\frac{\text{methylmalonyl CoA}}{\text{malonyl CoA}}$	Increased synthesis of methylated fatty acids
		Thioesterase cofactor	Termination of synthesis with short-chain product
Stearyl CoA desaturase		Various hormones	Stimulation of unsaturated fatty acid synthesis by increased enzyme synthesis
		Dietary polyunsaturated fatty acids	Decreased activity

cule is completed. After each addition of a two-carbon unit a series of reductive steps takes place. The reaction sequence starting with an acetyl CoA primer and leading to butyryl-ACP is as presented in Figure 9.7.

The next round of synthesis is initiated by transfer of the fatty acid chain from the 4′-phosphopantetheine moiety of ACP to the functional SH group of ***β*-ketoacyl-ACP synthase** (analogous to Reaction 3a). This liberates the SH group of ACP for acceptance of a second malonyl unit from malonyl CoA (Reaction 2) and allows Reactions 3b–6 to generate hexanoyl-ACP. The process is repeated five more times at which point

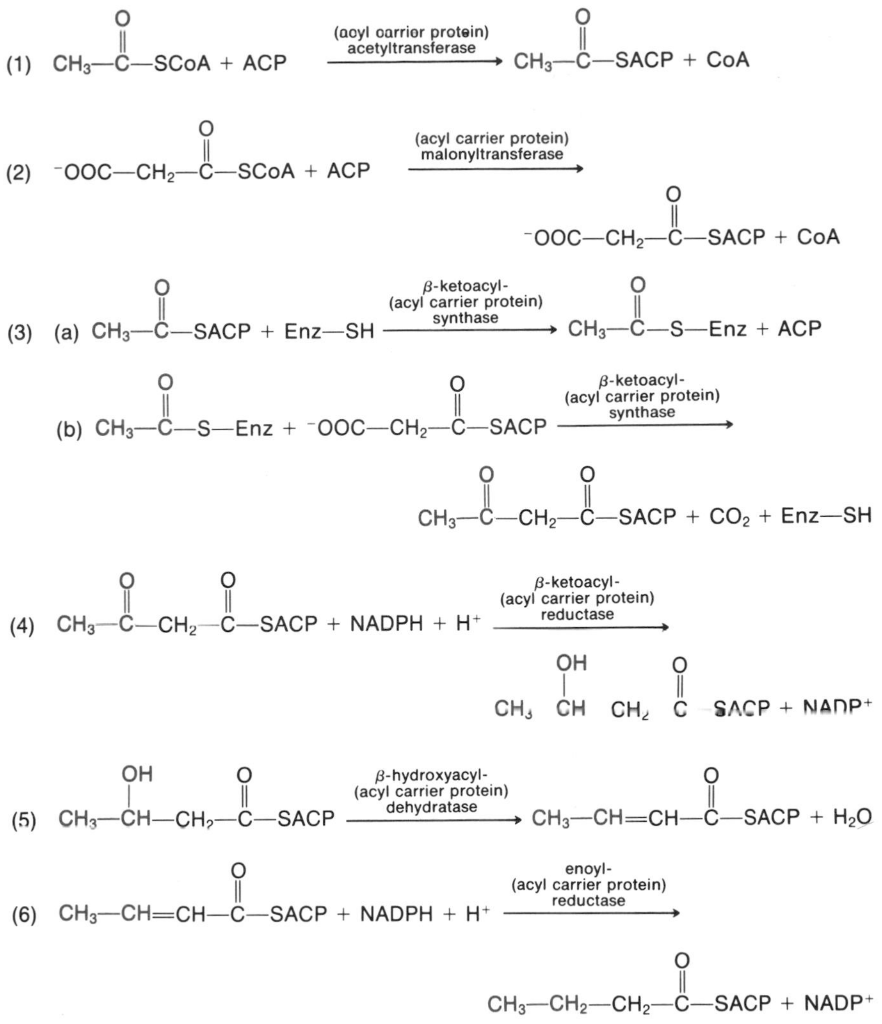

$$\text{(1)}\quad CH_3-\overset{O}{\overset{\|}{C}}-SCoA + ACP \xrightarrow[\text{acetyltransferase}]{\text{(acyl carrier protein)}} CH_3-\overset{O}{\overset{\|}{C}}-SACP + CoA$$

$$\text{(2)}\quad {}^{-}OOC-CH_2-\overset{O}{\overset{\|}{C}}-SCoA + ACP \xrightarrow[\text{malonyltransferase}]{\text{(acyl carrier protein)}} {}^{-}OOC-CH_2-\overset{O}{\overset{\|}{C}}-SACP + CoA$$

$$\text{(3) (a)}\quad CH_3-\overset{O}{\overset{\|}{C}}-SACP + Enz-SH \xrightarrow[\text{synthase}]{\beta\text{-ketoacyl-(acyl carrier protein)}} CH_3-\overset{O}{\overset{\|}{C}}-S-Enz + ACP$$

$$\text{(b)}\quad CH_3-\overset{O}{\overset{\|}{C}}-S-Enz + {}^{-}OOC-CH_2-\overset{O}{\overset{\|}{C}}-SACP \xrightarrow[\text{synthase}]{\beta\text{-ketoacyl-(acyl carrier protein)}} CH_3-\overset{O}{\overset{\|}{C}}-CH_2-\overset{O}{\overset{\|}{C}}-SACP + CO_2 + Enz-SH$$

$$\text{(4)}\quad CH_3-\overset{O}{\overset{\|}{C}}-CH_2-\overset{O}{\overset{\|}{C}}-SACP + NADPH + H^+ \xrightarrow[\text{reductase}]{\beta\text{-ketoacyl-(acyl carrier protein)}} CH_3-\overset{OH}{\overset{|}{C}H}-CH_2-\overset{O}{\overset{\|}{C}}-SACP + NADP^+$$

$$\text{(5)}\quad CH_3-\overset{OH}{\overset{|}{C}H}-CH_2-\overset{O}{\overset{\|}{C}}-SACP \xrightarrow[\text{dehydratase}]{\beta\text{-hydroxyacyl-(acyl carrier protein)}} CH_3-CH{=}CH-\overset{O}{\overset{\|}{C}}-SACP + H_2O$$

$$\text{(6)}\quad CH_3-CH{=}CH-\overset{O}{\overset{\|}{C}}-SACP + NADPH + H^+ \xrightarrow[\text{reductase}]{\text{enoyl-(acyl carrier protein)}} CH_3-CH_2-CH_2-\overset{O}{\overset{\|}{C}}-SACP + NADP^+$$

FIGURE 9.7
Reactions catalyzed by fatty acid synthase.

palmitoyl-ACP is acted upon by a **thioesterase** with the production of free palmitic acid (Figure 9.8). Note that at this stage the sulfhydryl groups of ACP and β-ketoacyl-ACP synthase are both free so that another cycle of fatty acid synthesis can begin.

Mammalian Fatty Acid Synthase Is a Multienzyme Complex

The reaction sequence given above is fairly well established as the basic pattern for fatty acid biosynthesis in living systems. However, the details of the reaction mechanisms are still far from clear and may vary among species. The enzyme complex termed fatty acid synthase catalyzes all these reactions, but its structure and properties vary considerably. The enzymes in *E. coli* are dissociable, and the reaction sequence was worked out with that organism. This sequence has been confirmed in mammalian systems, but the enzyme complex itself has not been dissociated. Some investigators postulate that the mammalian synthase is composed of two possibly identical subunits, each of which is a multienzyme polypeptide. Even among mammalian species and tissues there are certainly variations.

Despite the gaps in present knowledge, it appears likely that the growing fatty acid chain is continually bound to the enzyme complex and is sequentially transferred between the 4′-phosphopantetheine group of

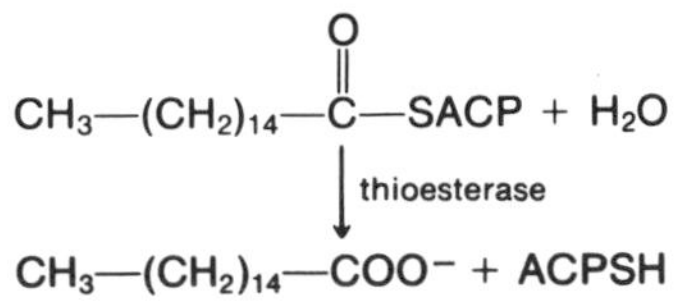

$$CH_3-(CH_2)_{14}-\overset{O}{\overset{\|}{C}}-SACP + H_2O \xrightarrow{\text{thioesterase}} CH_3-(CH_2)_{14}-COO^- + ACPSH$$

FIGURE 9.8
Release of palmitic acid from fatty acid synthase.

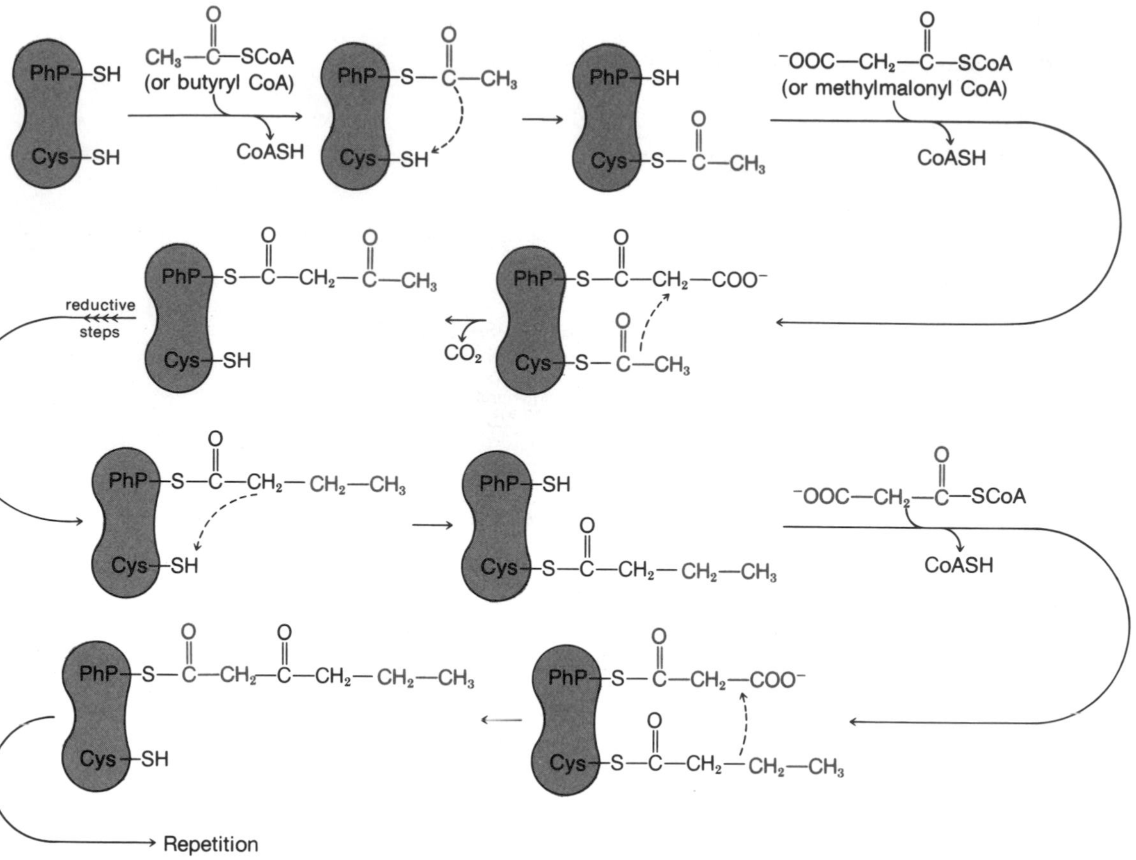

PhP = 4′-phosphopantetheine on acyl carrier protein
Cys = enzyme protein cysteine

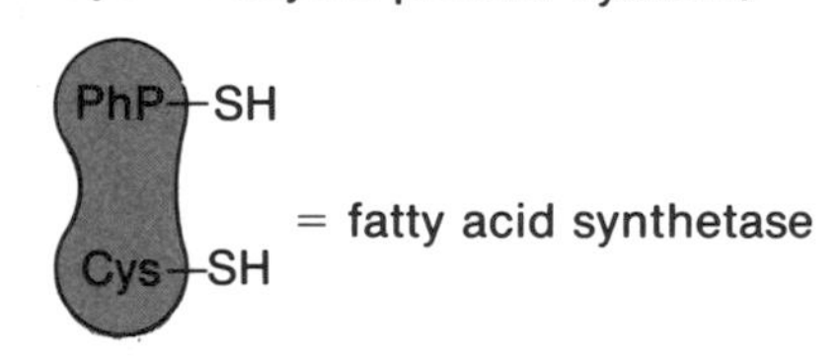

FIGURE 9.9
Proposed mechanism of elongation reactions taking place on mammalian fatty acid synthase.

ACP and the sulfhydryl group of a cysteine residue on β-ketoacyl-ACP synthase during the condensation reaction (Reaction 3, Figure 9.7) (see Figure 9.9). It is also probable that an intermediate acylation to an enzyme serine takes place when acyl CoA units add to the enzyme-bound ACP in the transacylase reactions.

There is suggestive evidence that some short-term regulation of fatty acid production is carried out by control of the activity of fatty acid synthase, but this is yet to be established firmly. Regulation of palmitate biosynthesis through fatty acid synthase probably occurs primarily by controlling the rate of synthesis and degradation of the enzyme. The agents and conditions that do this are given in Table 9.2. They are logical in terms of balancing an efficient utilization of the various biological energy substrates.

Stoichiometry of Fatty Acid Biosynthesis

If acetyl CoA is the primer for palmitate biosynthesis, the overall reaction is

$$CH_3{-}\overset{\overset{\displaystyle O}{\|}}{C}{-}SCoA + 7\,{}^{-}OOC{-}CH_2{-}\overset{\overset{\displaystyle O}{\|}}{C}{-}SCoA + 14NADPH + 14H^+ \longrightarrow$$
$$CH_3{-}(CH_2)_{14}{-}COO^- + 7CO_2 + 14NADP^+ + 8CoASH + 6H_2O$$

To calculate the energy needed for the overall conversion of acetyl CoA to palmitate, we must add the ATP used in formation of malonyl CoA:

$$7CH_3{-}\overset{\overset{\displaystyle O}{\|}}{C}{-}SCoA + 7CO_2 + 7ATP \longrightarrow 7\ OOC{-}CH_2{-}\overset{\overset{\displaystyle O}{\|}}{C}{-}SCoA + 7ADP + 7P_i$$

Then the overall stoichiometry for conversion of acetyl CoA to palmitate is

$$8CH_3{-}\overset{\overset{\displaystyle O}{\|}}{C}{-}SCoA + 7ATP + 14NADPH + 14H^+ \longrightarrow$$
$$CH_3{-}(CH_2)_{14}{-}\overset{\overset{\displaystyle O}{\|}}{C}{-}O \ + 8CoASH + 7ADP + 7P_i + 6H_2O + 14NADP^+$$

Acetyl CoA Must Be Transported From Mitochondria to the Cytosol for Palmitate Synthesis

Fatty acid synthase and acetyl CoA carboxylase are found primarily in the cytosol, and palmitate biosynthesis occurs largely in that subcellular compartment. However, mammalian tissues must use special processes to ensure an adequate supply of acetyl CoA and NADPH for this synthesis.

Specifically, the major source of acetyl CoA is the pyruvate dehydrogenase reaction inside the mitochondria. Since mitochondria are not readily permeable to acetyl CoA, a bypass mechanism moves it to the cytosol for palmitate biosynthesis. This mechanism, outlined in Figure 9.10, takes advantage of the facts that citrate does exchange freely from the mitochondria to the cytosol and that an enzyme exists in the cytosol to convert citrate to acetyl CoA and oxaloacetate. When there is an excess of citrate for the TCA cycle, citrate will pass into the cytosol and supply acetyl CoA for fatty acid biosynthesis. **Citrate cleavage enzyme** catalyzes the cleavage which requires 1 mol of ATP:

$$\text{Citrate} + ATP + CoA \xrightarrow[\text{enzyme}]{\text{citrate cleavage}} \text{acetyl CoA} + ADP + P_i + \text{oxaloacetate}$$

This mechanism has other advantages because CO_2 and NADPH for synthesis of palmitate can be produced from excess cytoplasmic oxaloacetate. As shown in Figure 9.10, the process produces NADPH from NADH, which was formed during glycolysis, by the sequential action of **NAD-linked malate dehydrogenase** and **NADP-linked malic enzyme** (malate : NADP oxidoreductase–decarboxylating). The products are pyruvate and CO_2. The cycle is completed by return of the pyruvate to the mitochondrion where it can be carboxylated to regenerate oxaloacetate, as has been described in the process of gluconeogenesis.

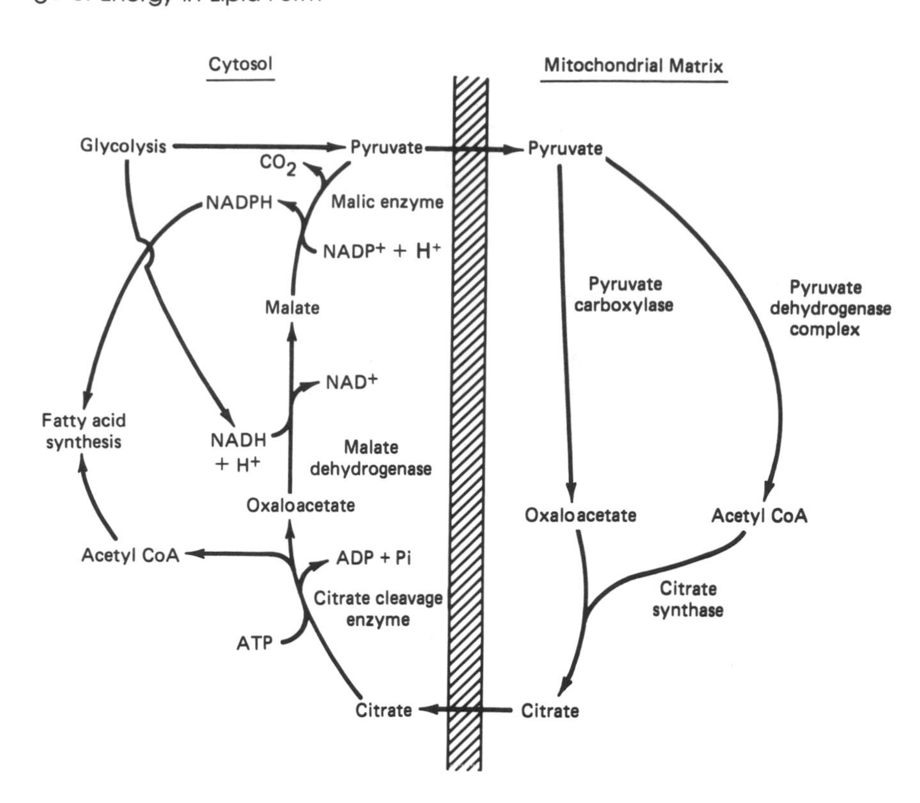

FIGURE 9.10
Mechanism for transfer of acetyl CoA from mitochondria to cytosol for fatty acid biosynthesis.

In sum, 1 NADH can be converted to NADPH for each acetyl CoA transferred from mitochondria to the cytosol, each transfer requiring 1 ATP. The transfer of the 8 acetyl CoA used for each molecule of palmitate can thus supply 8NADPH. Since palmitate biosynthesis requires 14NADPH mol^{-1}, the other 6 must be supplied from the cytosolic pentose phosphate pathway. This stoichiometry is, of course, hypothetical. The in vivo relationships are complicated by the fact that the transport of citrate and the other di- and tricarboxylic acids across the inner mitochondrial membrane takes place by several one-for-one exchange mechanisms. The actual flow rates are probably controlled by a composite of the concentration gradients of several of these exchange systems.

Palmitate Is the Precursor of Other Fatty Acids

The human body can synthesize all of the fatty acids it needs except for the essential, polyunsaturated fatty acids. These syntheses involve a variety of enzyme systems in a number of locations, and the palmitic acid produced by fatty acid synthase is modified by three processes: elongation, desaturation, and hydroxylation. In this section we discuss the process of elongation, the initial mechanism of desaturation, and the hydroxylation of brain fatty acids that are destined for incorporation into sphingolipids in nerve tissue. An α-oxidation process, involved in fatty acid degradation, and the more elaborate desaturation schemes producing polyunsaturated fatty acids are outlined in later sections.

Elongation Reactions

In mammalian systems **elongation** of fatty acids can occur either in the endoplasmic reticulum or in mitochondria, and the processes are slightly different in these two loci. In the endoplasmic reticulum the sequence of reactions is similar to that which occurs in the cytosolic fatty acid synthase in that the source of two-carbon units is again malonyl CoA, and

NADPH provides the reducing power. The preferred substrate for elongation in most cases is palmitoyl CoA, but in contrast to the system for de novo fatty acid synthesis, the intermediates in subsequent reactions are CoA esters, suggesting that the process is carried out by separate enzymes rather than a complex of the fatty acid synthase type. It appears that in most tissues this elongation system in the endoplasmic reticulum almost exclusively converts palmitate to stearate. However, brain contains one or more additional elongation systems, which synthesize the longer chain acids (up to C_{24}) that are needed for the brain lipids. These other systems also use malonyl CoA as substrate.

The elongation system in mitochondria is different from that in the endoplasmic reticulum in that acetyl CoA is the source of the added two-carbon units and both NADH and NADPH serve as reducing agents (Figure 9.11). Note that this system operates by simple reversal of the opposing pathway of fatty acid β oxidation (see Section 9.6) with the exception that NADPH-linked enoyl-CoA reductase (last step of elongation) replaces FAD-linked acyl CoA dehydrogenase (first step in β oxidation). The process has little activity with acyl CoA substrates of C_{16} atoms or longer, suggesting that it serves primarily in the elongation of shorter chain species.

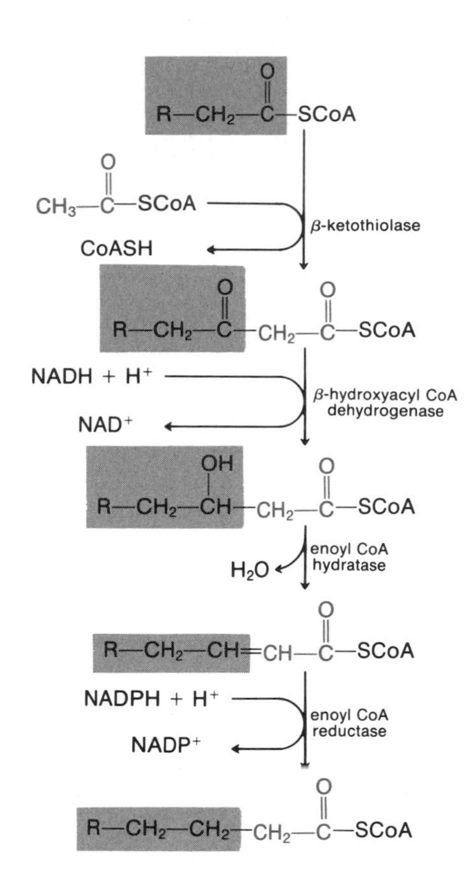

FIGURE 9.11
Mitochondrial elongation of fatty acids.

Formation of Monoenoic Acids by Stearoyl CoA Desaturase

In higher animals **desaturation** of fatty acids occurs in the endoplasmic reticulum, and the oxidizing system used to introduce cis double bonds is significantly different from the main fatty acid oxidation process in mitochondria. The systems in endoplasmic reticulum have sometimes been termed "mixed function oxidases" because the terminal enzymes simultaneously oxidize two substrates. In the case of fatty acid desaturation one of these substrates is NADPH and the other is the fatty acid. The electrons from NADPH are transferred through a specific flavoprotein reductase and a cytochrome to "active" oxygen so that it will then oxidize the fatty acid. Although the complete mechanism is not worked out, this latter step may involve a hydroxylation. The three components of the system are the desaturase enzyme, cytochrome b_5, and NADPH-cytochrome b_5 reductase. The overall reaction is

$$\text{R—CH}_2\text{—CH}_2\text{—(CH}_2)_7\text{—COOH} + \text{NADPH} + \text{H}^+ + \text{O}_2 \longrightarrow \text{R—CH=CH—(CH}_2)_7\text{—COOH} + \text{NADP}^+ + 2\text{H}_2\text{O}$$

As noted before, the enzyme specificity is such that the R group must contain at least six carbon atoms. The two main products in most organs are palmitoleic and oleic acids.

The control mechanisms that govern the conversion of the palmitate product of fatty acid synthase to unsaturated fatty acids are largely unexplored. One of the most important considerations is the control of the proportions of the unsaturated fatty acids available for a proper maintenance of physical state of stored triacylglycerols and membrane phospholipids. A critical committed step in the formation of unsaturated fatty acids from palmitate is the introduction of the first double bond by **stearoyl CoA desaturase.** The activity of this enzyme and its synthesis are controlled by both dietary and hormonal mechanisms. Increasing the amounts of polyunsaturated fatty acids in the diet of experimental animals decreases the activity of stearoyl CoA desaturase in liver, and insulin, triiodothyronine, and hydrocortisone cause its induction.

Formation of Hydroxy Fatty Acids in Nerve Tissue

There are apparently two different processes that produce **α-hydroxy fatty acids** in higher animals. One occurs in the mitochondria of many tissues and acts on relatively short-chain fatty acids. This is discussed in Section 9.6. The second process has so far been demonstrated only in tissues of the nervous system where it produces long-chain fatty acids with an hydroxyl group on C-2. These are needed for the structure of some myelin lipids. The specific case of α hydroxylation of lignoceric acid to cerebronic acid has been studied. These enzymes preferentially use C_{22} and C_{24} fatty acids and show characteristics of the "mixed function oxidase" systems, requiring molecular oxygen and reduced NAD^+ or $NADP^+$. This synthesis may be closely coordinated with the biosynthesis of the sphingolipids, which contain the hydroxylated fatty acids.

Fatty Acid Synthase Can Produce Fatty Acids Other Than Palmitate

The schemes outlined in previous sections, which utilize palmitate synthesized by fatty acid synthase and modify it by further enzymatic action, account for the great bulk of fatty acid biosynthesis in the human body, particularly that involved in energy storage. However, there are a number of special instances where smaller amounts of different fatty acids are needed for specific structural or functional purposes, and these acids are produced by modifications of the process carried out by fatty acid synthase. Two examples are the production of fatty acids shorter than palmitate in mammary glands and the synthesis of branched-chain fatty acids in certain secretory glands.

Recent work has shown that milk produced by many animals contains varying amounts of fatty acids with shorter chain lengths than palmitate. The amounts produced by the mammary gland apparently vary with species and especially with the physiological state of the animal. This is probably true of humans, although most investigations have been carried out with rats, rabbits, and various ruminants. The same fatty acid synthase that produces palmitate synthesizes the shorter chain acids when the linkage of the growing chain with the acyl carrier protein is split before the full C_{16} chain is completed. This hydrolysis is caused by soluble **thioesterases** whose activity is under hormonal control.

As noted in an earlier section, there are relatively few branched-chain fatty acids in higher animals, and, until recently, their metabolism has been studied mostly in primitive species such as *Mycobacteria,* where they are present in greater variety and amount. It is now known that simple branched-chain fatty acids are synthesized by tissues of higher animals for specific purposes, such as the production of waxes in sebaceous glands and avian preen glands and the elaboration of structures in the echo-locating systems of porpoises.

The majority of branched-chain fatty acids in higher animals are simply methylated derivatives of the saturated, straight-chain acids, and they are synthesized by fatty acid synthase. When **methylmalonyl CoA** is used as a substrate instead of malonyl CoA, a methyl side chain is inserted in the fatty acid, and the reaction is as follows:

$$CH_3{-}(CH_2)_n{-}\overset{\displaystyle O}{\overset{\|}{C}}{-}SACP + HOOC{-}\overset{\displaystyle CH_3}{\overset{|}{C}H}{-}\overset{\displaystyle O}{\overset{\|}{C}}{-}SCoA \longrightarrow$$

$$CH_3{-}(CH_2)_n{-}\overset{\displaystyle O}{\overset{\|}{C}}{-}\overset{\displaystyle CH_3}{\overset{|}{C}H}{-}\overset{\displaystyle O}{\overset{\|}{C}}{-}SACP + CO_2 + CoA$$

The regular reduction steps then follow. Apparently these reactions occur in many tissues normally at a rate several orders of magnitude slower than the utilization of malonyl CoA to produce palmitate. However, it has been suggested that the proportion of branched-chain fatty acids synthesized is largely governed by the relative availability of the two precursors, and an increase in branching can occur by decreasing the ratio of malonyl CoA to methylmalonyl CoA. A malonyl CoA decarboxylase capable of causing this decrease occurs in many tissues. It has also been suggested that increased levels of methylmalonyl CoA in pathological situations, such as vitamin B_{12} deficiency, can lead to excessive production of branched-chain fatty acids.

$$CH_3-(CH_2)_n-CH_2OH$$

FIGURE 9.12
Fatty alcohol.

Fatty Acyl CoAs May Be Reduced to Fatty Alcohols

As will be discussed in Chapter 10, many phospholipids contain fatty acid chain moieties in ether linkage rather than ester linkage. The biosynthetic precursors of these ether-linked chains are fatty alcohols (Figure 9.12) rather than fatty acids. These alcohols are formed in higher animals by a two-step, NADPH-linked reduction of fatty acyl CoAs in the endoplasmic reticulum. In organs that produce relatively large amounts of ether-containing lipids, the concurrent production of fatty acids and fatty alcohols is probably closely coordinated.

9.4 STORAGE OF FATTY ACIDS AS TRIACYLGLYCEROLS

Most tissues in the human body can convert fatty acids to triacylglycerols by a common sequence of reactions, but liver and adipose tissue carry out this process to the greatest extent. The latter organ is a specialized connective tissue, which is designed for the synthesis, storage, and hydrolysis of triacylglycerols, and this is the main mechanism that the human body has for relatively long-term energy storage. We are concerned here with white adipose tissue as opposed to brown adipose tissue, which occurs in much lesser amounts and has other specialized functions. The triacylglycerols are stored as liquid droplets in the cytoplasm, but this is not "dead storage," since they turn over with an average half-life of only a few days. Thus, in a homeostatic situation there is continuous synthesis and breakdown of triacylglycerols in adipose tissue. Some storage also occurs in skeletal and cardiac muscle, but this is only for local consumption.

Triacylglycerol synthesis in the liver is used primarily for production of blood lipoproteins, although the products can serve as energy sources for other liver functions. The required fatty acids may come from the diet, from adipose tissue via blood transport, or from liver biosynthesis. The acetyl CoA for biosynthesis is principally derived from glucose catabolism.

Triacylglycerols Are Synthesized From Fatty Acyl CoAs and Glycerol-3-Phosphate in Most Tissues

Triacylglycerols are synthesized in most tissues from activated fatty acids and a phosphorylated C_3 product of glucose catabolism (see Figure 9.13). The latter can be either glycerol 3-phosphate or dihydroxyacetone phosphate. Glycerol phosphate is formed either by reduction of dihydroxyacetone phosphate produced in glycolysis or by phosphorylation of glycerol.

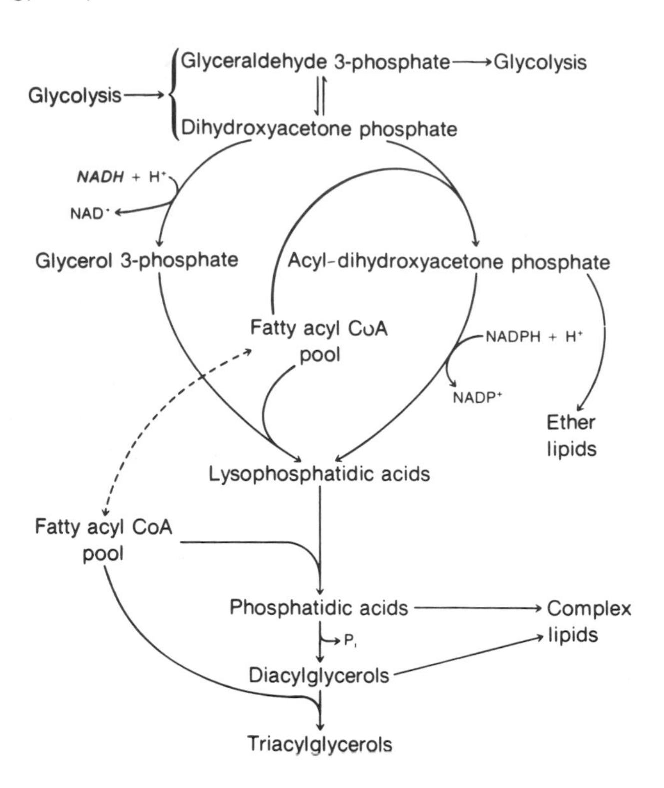

FIGURE 9.13
Alternative pathways for biosynthesis of triacylglycerols from dihydroxyacetone phosphate.

CH_2OH
HO—CH
$CH_2—O—PO_3^{2-}$
Glycerol 3-phosphate

acyltransferase: R—C(=O)—SCoA → CoASH

$CH_2—O—C(=O)—R$
HO—CH
$CH_2—O—PO_3^{2-}$
Lysophosphatidic acid

acyltransferase: R′—C(=O)—SCoA → CoASH

$CH_2—O—C(=O)—R$
R′—C(=O)—O—CH
$CH_2—O—PO_3^{2-}$
Phosphatidic acid

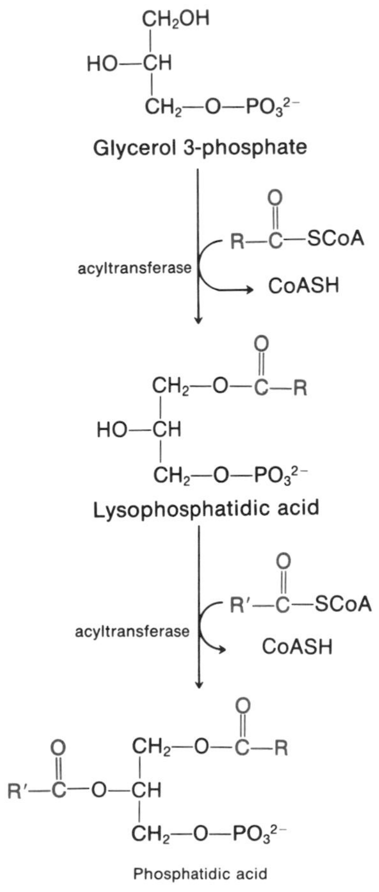

FIGURE 9.14
Synthesis of phosphatidic acid from glycerol 3-phosphate.

It is important to note that there is little or no **glycerol kinase** in white adipose tissue, so in that particular organ glycerol phosphate must be supplied from glycolytic intermediates. The fatty acids are activated by conversion to their CoA esters in the following reaction:

$$R—\overset{O}{\overset{\|}{C}}—O^- + ATP + CoASH \xrightarrow{\text{acyl CoA synthetase}} R—\overset{O}{\overset{\|}{C}}—SCoA + AMP + PP_i + H_2O$$

This is a two-step reaction with an acyl adenylate as intermediate and is driven by hydrolysis of the pyrophosphate to P_i.

The synthesis of triacylglycerols from the phosphorylated C_3 fragments involves formation of **phosphatidic acid,** which is a key intermediate in the synthesis of other lipids as well (see Chapter 10). This may be formed by two sequential acylations of glycerol 3-phosphate, as shown in Figure 9.14. Alternatively, dihydroxyacetone phosphate may be directly acylated at C-1 followed by reduction at C-2. The resultant lysophosphatidic acid can then be further esterified, as illustrated in Figure 9.15. If phosphatidic acid from either of these routes is to be used for synthesis of triacylglycerols, the phosphate group is next hydrolyzed by **phosphatidate phosphatase** to yield diacylglycerols. The latter are then acylated to triacylglycerols, as shown in Figure 9.16.

There is at least one tissue, the intestinal mucosa, in which the synthesis of triacylglycerols does not require formation of phosphatidic acid as described above. A major product of intestinal digestion of lipids is 2-monoacylglycerols, which are absorbed as such into the mucosa cells. An enzyme in these cells catalyzes the acylation of these monoacylglycerols with acyl CoA to form 1,2-diacylglycerols, which then can be further acylated as shown above.

The degree of specificity of the acylation reactions in all the steps above is still quite controversial. Analysis of fatty acid patterns in triacylglycerols from various human tissues shows that the distribution of different

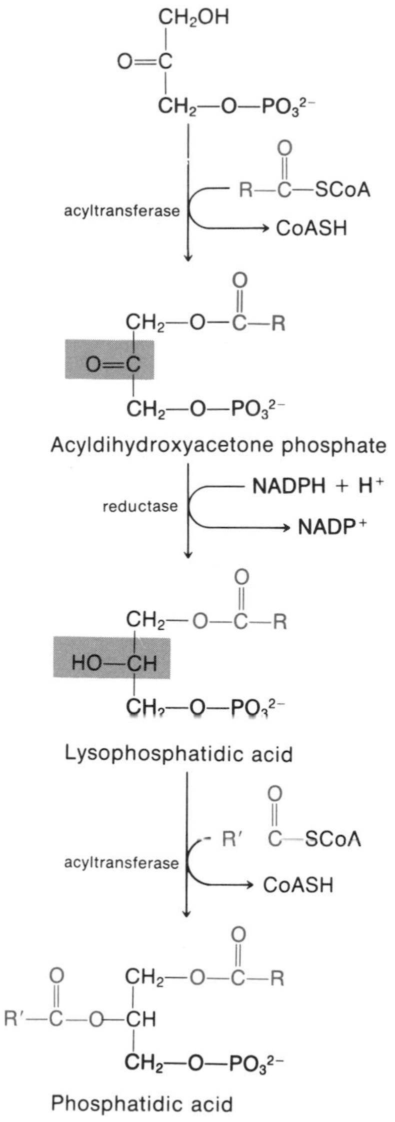

FIGURE 9.15
Synthesis of phosphatidic acid from dihydroxyacetone phosphate.

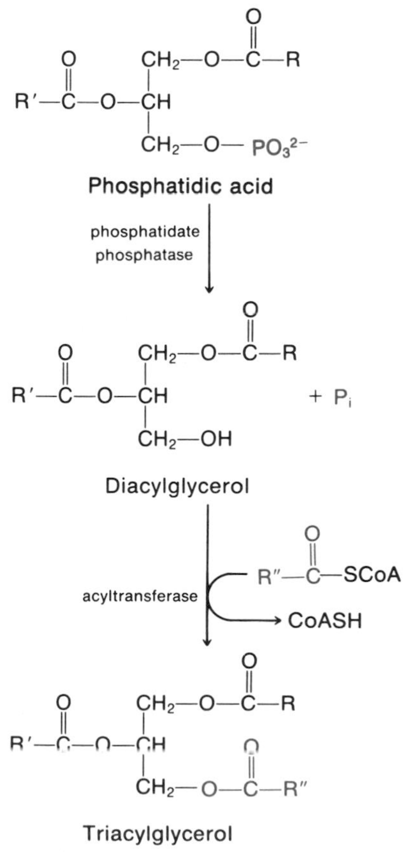

FIGURE 9.16
Synthesis of triacylglycerol from phosphatidic acid.

acids on the three positions of glycerol is neither random nor absolutely specific. The patterns in different tissues show some characteristic tendencies. Palmitic acid tends to be concentrated in position 1 and oleic acid in positions 2 and 3 of human adipose tissue triacylglycerols. (Position 3 is the one from which phosphate was removed in hydrolysis of phosphatidic acid.) The two main factors that determine the localization of a given fatty acid to a given position on glycerol are the specificity of the acyltransferase involved and the relative availability of the different fatty acids in the fatty acyl CoA pool. Other factors are probably involved also, but their relative importance is yet to be determined.

Mobilization of Triacylglycerols Requires Hydrolysis

The first step in recovering stored fatty acids for energy production is the hydrolysis of triacylglycerols. A variety of lipases catalyze this reaction, the sequence of hydrolysis from the three positions on glycerol depending on the specificities of the particular lipases involved.

CLIN. CORR. 9.1
OBESITY

The terms obesity and overweight refer to excess in body weight relative to height. Their definitions are arbitrary and are based upon actuarial estimates of ideal body weight (IBW), that is body weight associated with the lowest morbidity and mortality. Relative weight is body weight relative to IBW: overweight is defined as relative weight up to 20% above normal and obesity is relative weight over 20% above IBW. The body mass index (BMI) is well correlated with measures of body fat and is defined as weight (kilograms) divided by height^2 (M^2). Overweight is defined as a BMI of 25–30 kg per M^2 and obesity as BMI > 30 kg per M^2. Skinfold thickness also is a measure of body fat stores.

The cause of most cases of obesity is not known. Endocrine diseases such as hypothyroidism or Cushing's disease (overproduction of corticosteroids) are rare causes. Genetic factors interact with environmental factors: 80% of children of two obese parents will be obese, while only 14% of children of normal weight parents are obese. The major mechanism of weight gain is consumption of calories in excess of daily energy requirements, but the normal processes controlling food intake are not very well understood. Rarely, tumors of the hypothalamus result in pathological overeating (hyperphagia). However, a specific defect in most cases of obesity has not been demonstrated.

The treatment of obesity revolves about dietary restriction, increased physical activity, and behavior modification. The real problem is to modify the patients' eating patterns long term, and even in those who lose weight, regain of the weight is very common. Currently no pharmacological agents are effective in promoting long-term weight control. Surgery to limit the size of the gastric reservoir can be considered for patients over 100% above IBW. Medical complications of obesity include a two- to threefold increase in hypertension, gallstones, and diabetes, and fivefold increase in risk of endometrial carcinoma. Obese patients have decreased plasma antithrombin III levels, which predisposes them to venous thrombosis (see Clin. Corr. 8.8).

Bray, G. A. Complications of obesity. *Ann. Int. Med.* 103:1052, 1985; and Bray, G. A. Obesity: an endocrine perspective. *Endocrinology* Vol. 3, 2nd ed., L. J. DeGroot (Ed.). Philadelphia: Saunders, 1989, p. 2303.

$$\mathrm{R'{-}C({=}O){-}O{-}CH(CH_2{-}O{-}C({=}O){-}R)(CH_2{-}O{-}C({=}O){-}R'')} + 3H_2O \xrightarrow{\text{lipases}} \mathrm{HO{-}CH(CH_2{-}OH)_2} + \mathrm{R{-}C({=}O){-}OH} + \mathrm{R'{-}C({=}O){-}OH} + \mathrm{R''{-}C({=}O){-}OH}$$

The lipases in adipose tissue are, of course, the key enzymes for release of the major energy stores. The lipase that removes the first fatty acid is a carefully controlled enzyme, which is sensitive to a variety of circulating hormones. This control of triacylgylcerol hydrolysis must be balanced with the process of triacylglycerol synthesis described in the previous section to assure adequate energy stores and avoid obesity (see Clin. Corr. 9.1). The fatty acids and glycerol produced by the adipose tissue lipases are released to the circulating blood, where the fatty acids are bound by serum albumin and transported to tissues for use. The glycerol returns to the liver, where it is converted to dihydroxyacetone phosphate and enters the glycolytic or gluconeogenic pathways.

9.5 METHODS OF INTERORGAN TRANSPORT OF FATTY ACIDS AND THEIR PRIMARY PRODUCTS

The energy available in fatty acids needs to be distributed throughout the body from the site of fatty acid absorption, biosynthesis, or storage to the functioning tissues that consume them. This transport is closely integrated with the transport of other lipids, especially cholesterol. Since these transport systems appear intimately involved in the pathological processes leading to atherosclerosis, they are being intensively studied, but many important questions are still unanswered.

The human body uses three types of substances as vehicles to transport lipid-based energy: (a) chylomicrons and other plasma lipoproteins in which triacylglycerols are carried in protein-coated lipid droplets, the latter also containing other lipids; (b) fatty acids bound to serum albumin; and (c) the so-called "ketone bodies," acetoacetate and β-hydroxybutyrate. These three vehicles are used in varying proportions to carry the energy in the bloodstream via three routes. One is transport of dietary fatty acids as chylomicrons throughout the body from the intestine after absorption. Another is the transport of lipid-based energy processed by or synthesized in the liver and distributed either to adipose tissue for storage or to other tissues for utilization; in this case they use "ketone bodies" and plasma lipoproteins other than chylomicrons. Finally, there is transport of energy released from storage in adipose tissue to the rest of the body in the form of fatty acids that are bound to serum albumin.

Lipid-Based Energy Is Transported in Blood in Different Forms

The proportions of energy being transported in any one of the modes outlined above varies considerably with metabolic and physiological state. At any one time, the largest amount of lipid in blood is in the form of triacylglycerols in the various lipoproteins. However, the fatty acids bound to albumin are utilized and replaced very rapidly so the total energy transport for a given period of time by this mode may be very significant.

Plasma Lipoproteins Carry Triacylglycerols

The plasma **lipoproteins** are synthesized both in the intestine and in the liver and are a heterogeneous group of lipid–protein complexes composed of various types of lipids and apoproteins (see p. 67 for detailed discussion of structure). The two most important categories for delivery of lipid-based energy are the **chylomicrons** and the **very low density lipoprotein** (VLDL), since they contain relatively large amounts of triacylglycerols. Chylomicrons are formed in the intestine and function in the absorption and transport of dietary fat. The exact precursor–product relationships between the other types of plasma lipoproteins are yet to be completely defined, as are the roles of the various protein components. It seems clear, however, that the liver synthesizes VLDL and that the fatty acids from the triacylglycerols are taken up by adipose tissue and other tissues. In the process the VLDL are converted to **low density lipoprotein** (LDL). The role, if any, of **high density lipoprotein** (HDL) in transport of lipid-based energy is yet to be clarified. All of these lipoproteins are integrally involved in transport of other lipids, especially cholesterol. The lipid components can interchange to some extent between different classes of lipoprotein, and some of the apoproteins probably have functional roles in modifying enzyme activity during exchange of lipids between plasma lipoproteins and tissues. Other apoproteins serve as specific recognition sites for cell surface receptors. Such interaction constitutes the first step in receptor-mediated endocytosis of certain lipoproteins. Studies of rare genetic abnormalities have been helpful in elucidating the roles of some of these apoproteins (see Clin. Corr. 9.2).

Fatty Acids Are Bound to Serum Albumin

Serum albumin acts as a carrier for a number of substances in the blood, some of the most important being fatty acids. These acids are of course, water insoluble in themselves, but when they are released into the plasma

CLIN. CORR. **9.2**
GENETIC ABNORMALITIES IN LIPID-ENERGY TRANSPORT

Diseases that affect the transport of lipid-based energy frequently result in abnormally high plasma triacylglycerols, cholesterol, or both. They are classified as hyperlipidemias. Some of them are genetically transmitted, and presumably they result from the alteration or lack of one or more proteins involved in the production or processing of plasma lipids. The nature and function of all of these proteins is yet to be determined, so the elucidation of exact causes of the pathology in most of these diseases is still in the early stages. However, in several cases a specific protein abnormality has been associated with altered lipid transport in the patients' plasma.

In the extremely rare disease analbuminemia, there is an almost complete lack of serum albumin. In a rat strain with analbuminemia, a 7 base-pair deletion in an intron of the albumin gene results in the inability to process the nuclear mRNA for albumin. Despite the many functions of this protein, the symptoms of the disease are surprisingly mild. Lack of serum albumin effectively eliminates the transport of fatty acids unless they are esterified in acylglycerols or complex lipids. However, since patients with analbuminemia do have elevated plasma triacylglycerol levels, presumably the deficiency in lipid-based energy transport caused by the absence of albumin to carry fatty acids is filled by increased use of plasma lipoproteins to carry triacylglycerols.

A more serious genetic defect is the absence of lipoprotein lipase. The major problem here is the inability to process chylomicrons after a fatty meal. Pathological fat deposits occur in the skin (eruptive xanthomas) and the patients typically suffer from pancreatitis. If patients are put on a low-fat diet they respond reasonably well.

Another rare but more severe disease, abetalipoproteinemia, is caused by defective synthesis of apoprotein B, an essential component in the formation of chylomicrons and VLDL. Under these circumstances the major pathway for transporting lipid-based energy from the diet to the body is unavailable. Chylomicrons, VLDL, and LDL are absent from the plasma and fat absorption is deficient or nonexistent. There are other serious symptoms, including neuropathy and red cell deformities whose etiology is less clear.

Esumi, H., Takahashi, Y., Sato, S., Nagase, S., and Sugimura, T. A seven base pair deletion in an intron of the albumin gene of analbuminemic rats. *Proc. Natl. Aca. Sci.* U.S. 80:95, 1983; Dammacco, F., Miglietta, A., D'Addabbo, A., Fratello A., Moschetta, R., and Bonomo, L. Analbuminemia: Report of a case and review of the literature. *Vox Sang.* 39:153, 1980.

$$CH_3-\overset{\overset{O}{\|}}{C}-CH_2-\overset{\overset{O}{\|}}{C}-OH \qquad CH_3-\overset{\overset{OH}{|}}{CH}-CH_2-\overset{\overset{O}{\|}}{C}-OH$$

Acetoacetic acid   β-Hydroxybutyric acid

FIGURE 9.17
Structures of ketone bodies.

during triacylglycerol hydrolysis they are quickly bound to albumin. This protein has a number of binding sites for fatty acid, two of them bind with high affinity. At any one time the proportion of sites on albumin actually loaded with fatty acids is far from complete, but the rate of turnover is high, so binding by this mechanism constitutes a major route of energy transfer.

Ketone Bodies Are a Lipid-Based Energy Supply

The third mode of transport of lipid-based energy-yielding molecules is in the form of small water-soluble molecules, **acetoacetate** and **β-hydroxybutyrate** (Figure 9.17), which are produced primarily by the liver during the oxidation of fatty acids. The reactions involved in their formation and utilization will be discussed in a later section of this chapter. Under certain conditions, these substances can reach excessive concentrations in blood, leading to ketosis and acidosis. Spontaneous decarboxylation of acetoacetate to **acetone** also occurs, which is detectable when acetoacetate concentrations are high. This led early investigators to call the group of soluble products **"ketone bodies."** In fact, β-hydroxybutyrate and acetoacetate are continually produced by the liver and, to a lesser extent, by the kidney. Skeletal and cardiac muscle then utilize them to produce ATP. Nervous tissue, which obtains almost all of its energy from glucose if it is available, is unable to take up and utilize the fatty acids bound to albumin for energy production. However, it can use β-hydroxybutyrate when glucose supplies are insufficient.

Lipases Must Hydrolyze Blood Triacylglycerols for Their Fatty Acids to Be Transferred to Tissues

Lipid-based energy distributed as fatty acids bound to albumin or as "ketone bodies" is readily taken up by various tissues for oxidation and production of ATP. However, the energy in fatty acids stored or circulated as triacylglycerols is not directly available, but rather the latter compounds must be enzymatically hydrolyzed to release the fatty acids and glycerol. There are two types of lipases involved in this hydrolysis: (1) **lipoprotein lipases,** which hydrolyze triacylglycerols in the plasma lipoproteins, and (2) so-called **"hormone-sensitive triacylglycerol lipase,"** which initiates hydrolysis of triacylglycerols in adipose tissue and the release of fatty acids and glycerol into the plasma.

Lipoprotein lipases are located on the surface of the endothelial cells of capillaries and possibly of adjoining tissue cells. They hydrolyze fatty acids from the 1 and/or 3 position of tri- and diacylglycerols when the latter are present in VLDL or chylomicrons. One of the lipoprotein apoproteins must be present to activate the process. The fatty acids that are released are either bound to serum albumin or taken up by the tissue. The monoacylglycerol products may either pass into the cells or be further hydrolyzed by serum **monoacylglycerol hydrolase.**

A completely distinct type of lipase controls the mobilization of fatty acids from the triacylglycerols stored in the adipose tissue. One of them is

TABLE 9.3 Regulation of Triacylglycerol Metabolism

Enzyme	Regulatory agent	Effect
	TRIACYLGLYCEROL MOBILIZATION	
"Hormone-sensitive" lipase	"Lipolytic hormones," e.g., epinephrine, glucagon, ACTH, etc.	Stimulation by cAMP-mediated phosphorylation of relatively inactive enzyme
	Insulin	Inhibition
	Prostaglandins	Inhibition
Lipoprotein lipase	Lipoprotein apoprotein C-II	Activation
	Insulin	Activation
	TRIACYLGLYCEROL BIOSYNTHESIS	
Phosphatidate phosphatase	Steroid hormones	Stimulation by increased enzyme synthesis

hormonally controlled by a cAMP-mediated mechanism. There are a number of lipase activities in the tissue, but the enzyme attacking triacylglycerols initiates the process. Two other lipases then rapidly complete the hydrolysis of mono- and diacylglycerols, releasing fatty acids to the plasma where they are bound to serum albumin (see Table 9.3).

9.6 UTILIZATION OF FATTY ACIDS FOR ENERGY PRODUCTION

The fatty acids that arrive at the surface of tissues are taken up by the cells and can be used for energy production. This process occurs primarily inside the mitochondria and is intimately integrated with the processes of energy production from other sources. The energy-rich intermediates produced from fatty acids are the same as those obtained from sugars, that is, NADH and $FADH_2$, and the final stages of the oxidation process are exactly the same as for carbohydrates, that is, the metabolism of acetyl CoA by the TCA cycle and the production of ATP in the mitochondrial electron transport system.

The degree of utilization of fatty acids for energy production varies considerably from tissue to tissue and depends to a significant degree on the metabolic status of the body, whether it is fed or fasted, exercising, and so on. For instance, nervous tissue apparently oxidizes fatty acids to a minimal degree if at all, but cardiac and skeletal muscle depend heavily on fatty acids as a major energy source. During prolonged fasting most tissues are able to use fatty acids or ketone bodies for their energy requirements.

β-Oxidation of Straight-Chain Saturated Fatty Acids Is the Major Energy-Producing Process

For the most part, fatty acids are oxidized by a mechanism that is similar to, but not identical with, a reversal of the process of palmitate synthesis described earlier in this chapter. That is, two-carbon fragments are removed sequentially from the carboxyl end of the acid after steps of **dehydrogenation, hydration,** and **oxidation** to form a β-keto acid, which is split by **thiolysis**. These processes take place while the acid is activated in a thioester linkage to the 4′-phosphopantetheine of CoA.

Fatty Acids Are Activated by Conversion to Fatty Acyl CoA

The first step in oxidation of a fatty acid must therefore be its activation to a fatty acyl CoA. This is the same reaction described for synthesis of triacylglycerols in Section 9.4 and occurs in the endoplasmic reticulum or the outer mitochondrial membrane.

Fatty acids occurring inside the mitochondria can also be activated to a limited extent. This process is analogous to the extramitochondrial one, except that it is dependent on energy from guanine nucleotides instead of adenine nucleotides. The physiological significance of this mitochondrial process is not yet clear.

Carnitine Carries Acyl Groups Across the Mitochondrial Membrane

Since most of the fatty acyl CoAs are formed outside the mitochondria while the oxidizing machinery is inside the inner membrane, which is impermeable to CoA and its derivatives, the cell has a major logistical problem. An efficient shuttle system overcomes this problem by using **carnitine** as the carrier of acyl groups across the membrane. The steps involved are outlined in Figure 9.18. There are enzymes on both sides of the barrier that transfer the fatty acyl group between CoA and carnitine according to the equation:

$$CH_3{-}(CH_2)_n{-}\overset{O}{\overset{\|}{C}}{-}SCoA + HO{-}\underset{}{CH}(CH_2{-}\overset{+}{N}{-}(CH_3)_3){-}CH_2{-}COOH \underset{\text{carnitine palmitoyl transferase}}{\rightleftharpoons} CH_3{-}(CH_2)_n{-}\overset{O}{\overset{\|}{C}}{-}O{-}CH(CH_2{-}\overset{+}{N}{-}(CH_3)_3){-}CH_2{-}COOH + CoASH$$

On the outer surface, the acyl group is transferred to carnitine in a reaction catalyzed by **carnitine palmitoyltransferase I** (CPT I). The acyl carnitine exchanges across the inner mitochondrial membrane with free carnitine. The latter becomes available on the inner surface when the fatty acyl group is transferred back to CoA under the influence of **carnitine palmi-**

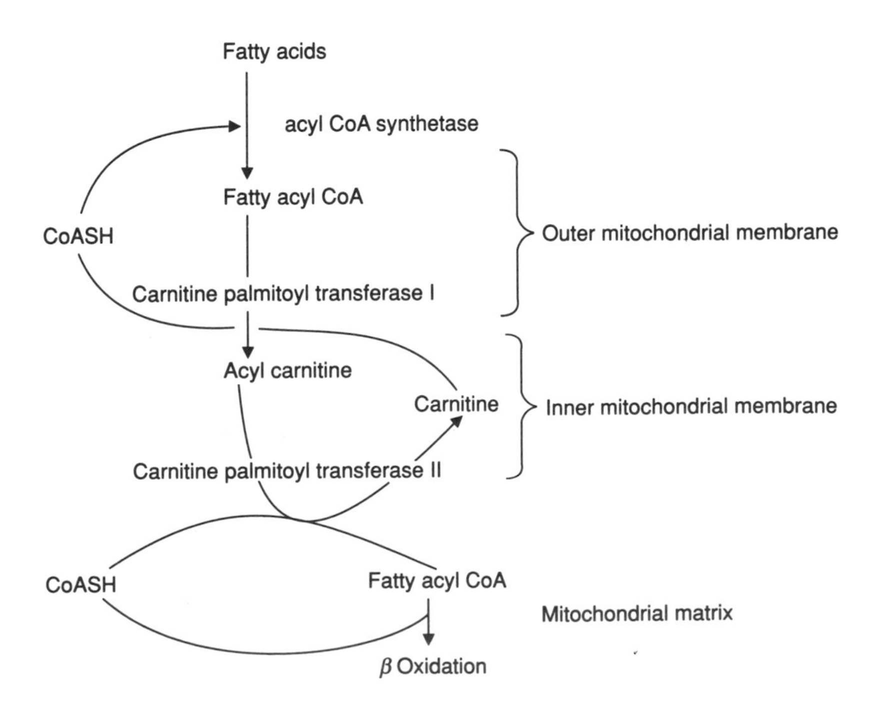

FIGURE 9.18
Mechanism for transfer of fatty acids from the cytosol through the mitochondrial membranes for oxidation.

CLIN. CORR. 9.3
GENETIC DEFICIENCIES IN CARNITINE OR CARNITINE PALMITOYL TRANSFERASE

There are several known diseases that are caused by genetic abnormalities in the carnitine system for transporting fatty acids across the inner mitochondrial membrane. They cause deficiencies either in the level of carnitine or in the functioning of the carnitine palmitoyl transferase enzymes. In both cases it is difficult to correlate the degree of the deficiency with the severity of the symptoms.

The clinical symptoms of carnitine deficiency, in particular, seem to vary greatly. They can range from mild, recurrent muscle cramping to severe weakness and death. The deficiency has been classified into two categories; that which is a generalized systemic carnitine deficiency and that which appears to be limited to deficiency only in muscles. The etiology of both disorders remains obscure. Some patients with systemic carnitine deficiency have an abnormality in renal reabsorption of carnitine. The myopathic form may result from impaired transport of carnitine into myocytes. Pathological accumulation of triacylglycerols in muscles usually occurs, since fatty acids are inefficiently transported into mitochondria for oxidation. The degree of muscle wasting is variable and results in unpredictable blood carnitine levels, since carnitine can be released into the blood as a result of muscle tissue breakdown. Carnitine therapy has proved effective in some cases, as has the replacement of normal dietary fat by triacylglycerols containing medium chain length fatty acids whose oxidation is carnitine independent. Secondary carnitine deficiency occurs in some inherited organic acidurias. The metabolic intermediates are CoA esters; they form acylcarnitine derivatives and deplete the body of carnitine.

The deficiency of carnitine palmitoyl transferase is usually a less serious situation and the symptoms are limited to recurrent muscle pain and myoglobinuria following prolonged strenuous exercise. Muscle mitochondria from patients with this disease have been studied, and the data show that the deficiency can occur at the level of carnitine palmitoyltransferase I or carnitine palmitoyltransferase II or both. Strangely, triacylglycerols do not accumulate in muscles in this disease. Inherited defects in the acyl CoA dehydrogenases can lead to serious disease (see Clin. Corr. 9.4).

Bank, W. J., Banilla, E., Capuzzi, D. M., and Rowland, C. P. A disorder of muscle lipid metabolism and myoglobinuria. Absence of carnitine palmityl transferase. *N. Engl. J. Med.* 292:443, 1975; Engel, A. G. and Angelini, C. Carnitine deficiency of human skeletal muscle with associated lipid storage myopathy; A new syndrome. *Science* 179:899, 1973; and Stanley, C. A. New genetic defects in mitochondrial fatty acid oxidation. *Adv. Pediatr.* 34:59, 1987.

toyltransferase II (CPT II). This process functions primarily in the mitochondrial transport of fatty acyl CoAs with chain lengths of C_{12}–C_{18}, and genetic abnormalities in the system lead to muscle pathology (see Clin. Corr. 9.3). By contrast, shorter chain fatty acids can cross the inner mitochondrial membrane directly and become activated to their CoA derivatives in the matrix compartment; that is, their oxidation is carnitine independent.

β-Oxidation Is a Sequence of Four Reactions

Once the fatty acyl groups have been transferred back to CoA at the inner surface of the inner mitochondrial membrane they can be oxidized by a group of **acyl CoA dehydrogenases** in this membrane that remove hydrogen atoms and form enoyl CoA with a trans double bond between C-2 and C-3 atoms. The several dehydrogenases with different specificities for chain length of the acyl CoA oxidized are flavoproteins (see Clin. Corr. 9.4). The reaction is

$$CH_3—(CH_2)_n—CH_2—CH_2—\overset{\overset{\displaystyle O}{\|}}{C}—SCoA + \text{FAD-protein} \xrightarrow[\text{acyl CoA dehydrogenase}]{} CH_3—(CH_2)_n—CH{=}CH—\overset{\overset{\displaystyle O}{\|}}{C}—SCoA + \text{FADH}_2\text{-protein}$$

As is the case in the TCA cycle, the enzyme-bound FADH transfers electrons through several other flavoproteins to ubiquinone in the electron transport scheme and only two ATP molecules can be obtained for each double bond formed.

CLIN. CORR. 9.4
GENETIC DEFICIENCIES IN THE ACYL CoA DEHYDROGENASES

The acyl CoA dehydrogenase deficiencies represent a recently discovered group of inherited defects that impair the β oxidation of fatty acids at different stages of the chain shortening process. The affected enzyme may be the long-chain acyl CoA dehydrogenase (LCAD), the medium-chain acyl CoA dehydrogenase (MCAD), or the short-chain acyl CoA dehydrogenase (SCAD) whose substrate specificities are for acyl CoA chains of greater than C_{12}, C_6–C_{12}, and C_4–C_6, respectively. The three conditions are inherited in autosomal recessive fashion and share many of the same clinical features. The best characterized is MCAD deficiency which, though first recognized as late as 1982, is now thought to be one of the most common of all inborn errors of metabolism.

Medium-chain acyl CoA dehydrogenase deficiency usually manifests itself within the first 2 years of life after a fasting period of 12 h or more. Typical symptoms include vomiting, lethargy, and frequently coma, are accompanied by hypoketotic hypoglycemia and dicarboxylic aciduria. The absence of starvation ketosis is accounted for by the block in hepatic fatty acid oxidation, which also causes a slowdown of gluconeogenesis. This, coupled with impaired fatty acid oxidation in muscle, which promotes glucose utilization, leads to profound hypoglycemia. Accumulation of medium-chain acyl CoAs in tissues forces their metabolism through alternative pathways including ω oxidation and transesterification to glycine or carnitine. Excessive urinary excretion of the reaction products (medium-chain dicarboxylic acids together with medium-chain esters of glycine and carnitine) provide diagnostic clues.

Most patients with this disorder do well simply by avoiding prolonged periods of starvation, which is consistent with the fact that the metabolic complications of MCAD deficiency are seen only when body tissues become heavily dependent on fatty acids as a source of energy (e.g., with carbohydrate deprivation). In retrospect it now seems likely that many cases previously diagnosed loosely as "Reye-like syndrome" or "sudden infant death syndrome" were in fact due to MCAD deficiency.

Rinaldo, P. et al. Medium chain acyl CoA dehydrogenase deficiency. *N. Eng. J. Med.* 319:1308, 1988.

The second step in β oxidation is hydration of the trans double bond to a 3-L-hydroxyacyl CoA.

$$\underset{\text{trans-}\Delta^2\text{-Enoyl CoA}}{CH_3{-}(CH_2)_n{-}CH{=}CH{-}\overset{O}{\overset{\|}{C}}{-}SCoA} + H_2O \xrightarrow[\text{enoyl CoA hydratase}]{} \underset{\text{3-L-hydroxyacyl CoA}}{CH_3{-}(CH_2)_n{-}\overset{OH}{\overset{|}{C}H}{-}CH_2{-}\overset{O}{\overset{\|}{C}}{-}SCoA}$$

This reaction is stereospecific, in that the L isomer is the product when the trans double bond is hydrated. The stereospecificity of the oxidative pathway is governed by the next enzyme, which is specific for the L isomer as its substrate.

$$CH_3{-}(CH_2)_n{-}\overset{OH}{\overset{|}{C}H}{-}CH_2{-}\overset{O}{\overset{\|}{C}}{-}SCoA + NAD^+ \xrightarrow[\text{3-L-hydroxyacyl CoA dehydrogenase}]{} \underset{\beta\text{-ketoacyl CoA}}{CH_3{-}(CH_2)_n{-}\overset{O}{\overset{\|}{C}}{-}CH_2{-}\overset{O}{\overset{\|}{C}}{-}SCoA} + NADH + H^+$$

The final step is the cleavage of the two-carbon fragment by a thiolase, which, like the preceding two enzymes, has relatively broad specificity with regard to chain length of the acyl group being oxidized.

$$CH_3{-}(CH_2)_n{-}\overset{O}{\overset{\|}{C}}{-}CH_2{-}\overset{O}{\overset{\|}{C}}{-}SCoA + CoASH \xrightarrow[\beta\text{-ketothiolase}]{} \underset{\text{Fatty acyl CoA}}{CH_3{-}(CH_2)_n{-}\overset{O}{\overset{\|}{C}}{-}SCoA} + CH_3{-}\overset{O}{\overset{\|}{C}}{-}SCoA$$

In the overall process then, an acetyl CoA is produced and the acyl CoA product is ready for the next round of oxidation starting with acyl CoA dehydrogenase.

As yet it has been impossible to show conclusively that any of the enzymes in the β-oxidation scheme are control points, although under rather rigid in vitro conditions some apparently have slower maximum rates of reaction than others. It is generally assumed that control is exerted by the availability of substrates and cofactors and by the rate of processing of the acetyl CoA product by the TCA cycle. One way in which substrate availability is controlled is by regulation of the shuttle mechanism that transports fatty acids into the mitochondria, a phenomenon of central importance in the regulation of hepatic ketone body production (see p. 408).

The Energy-Yield From β-Oxidation of Fatty Acids

Each set of oxidations resulting in production of a two-carbon fragment yields, in addition to the acetyl CoA, one reduced flavoprotein and one NADH. In the oxidation of palmitoyl CoA seven such cleavages take place, and in the last cleavage two acetyl CoA molecules are formed. The products of β oxidation of palmitate are thus eight acetyl CoAs, seven reduced flavoproteins, and seven NADH.

Each of the reduced flavoproteins can yield 2ATP and each NADH can yield 3 when processed through the electron transport chain, so the reduced nucleotides yield 35ATP per palmitoyl CoA. As described earlier in Chapter 6, the oxidation of each acetyl CoA through the TCA cycle yields 12ATP, so the 8C-2 fragments from a palmitate molecule produce 96ATP. However, 2ATP equivalents (1ATP going to 1AMP) were used to activate palmitate to palmitoyl CoA. Therefore, each palmitic acid entering the cell from the action of lipoprotein lipase or from its combination with serum albumin can yield 129ATP mol^{-1} by complete oxidation.

Comparison of the β-Oxidation Scheme With Palmitate Biosynthesis

In living metabolic systems the reactions in a catabolic pathway are sometimes quite similar to those in a reversal of the corresponding anabolic pathway, but there are significant differences that provide for separate control of the two schemes. This is true of the palmitate biosynthetic scheme and the scheme for β oxidation of fatty acids. The critical differences between these two pathways are outlined in Table 9.4. This comparison illustrates some basic mechanisms for separation of metabolic pathways. These include separation by subcellular compartmentation (β-oxidation occurring inside the mitochondria and palmitate biosynthesis in the cytosol), and use of different cofactors (NADPH in biosynthesis; FAD and NAD^+ in oxidation).

Some Fatty Acids Require Modifications of β-Oxidation for Their Metabolism

The β-oxidation scheme described in the previous sections accounts for the bulk of energy production from fatty acids in the human body. However, it is clear that these reactions must be supplemented by a few other

TABLE 9.4 Comparison of Schemes for Biosynthesis and β Oxidation of Palmitate

Parameter	Biosynthesis	β Oxidation
Subcellular localization	Primarily cytosolic	Primarily mitochondrial
Phosphopantetheine containing active carrier	Acyl carrier protein	Coenzyme A
Nature of small carbon fragment added or removed	C-1 and C-2 atoms of malonyl CoA after initial priming	Acetyl CoA
Nature of oxidation–reduction coenzyme	NADPH	FAD when saturated chain dehydrogenated, NAD^+ when hydroxy acid dehydrogenated
Stereochemical configuration of β-hydroxy intermediates	D-β-Hydroxy	L-β-Hydroxy
Energy equivalents yielded or utilized in interconversion of palmitate ↔ acetyl CoA	7ATP + 14NADPH = 49ATP equiv	$7FADH_2$ + 7NADH − 2ATP = 33ATP equiv

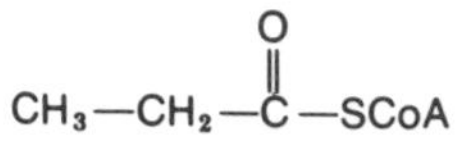

$$CH_3—CH_2—\overset{\overset{\displaystyle O}{\|}}{C}—SCoA$$

FIGURE 9.19
Propionyl CoA.

mechanisms so that all fatty acids that are ingested can be oxidized. The principal modifications are those required to oxidize odd-chain fatty acids and unsaturated fatty acids, and those that catalyze α and ω oxidation. α Oxidation occurs at C-2 instead of C-3 as occurs in the β-oxidation scheme. ω Oxidation occurs at the methyl end of the fatty acid molecule. Partial oxidation of fatty acids with cyclopropane ring structures probably occurs in humans, but the mechanisms are not worked out.

Propionyl CoA Is a Product of Oxidation of Odd-Chain Fatty Acids

The oxidation of fatty acids with an odd number of carbon atoms proceeds exactly as described above, but the final product is a molecule of propionyl CoA (Figure 9.19). In order that this compound can be further oxidized, it undergoes carboxylation, molecular rearrangement, and conversion to succinyl CoA. These reactions are identical with those described in Chapter 12 for the metabolism of propionyl CoA when it is formed as a product of the metabolic breakdown of some amino acids.

Oxidation of Unsaturated Fatty Acids Requires Additional Enzymes

The many unsaturated fatty acids in the diet are readily available for the production of energy by the human body. However, in several respects the structures encountered in these dietary acids may differ from those required by the specificity of the enzymes in the β-oxidation pathway.

One problem is that in the β oxidation of unsaturated fatty acids the sequential excision of C_2 fragments can generate an acyl CoA intermediate with a double bond between C-3 and C-4 atoms instead of between C-2 and C-3 atoms as required for the enoyl CoA hydratase reaction. If so, the cis bond between C-3 and C-4 atoms is converted into a trans bond between C-2 and C-3 atoms (Figure 9.20) by an auxiliary enzyme, enoyl CoA isomerase. The regular process can then proceed.

A second problem occurs if the cis double bond of the acyl CoA intermediate resides between C-4 and C-5 atoms. In this case the action of acyl CoA dehydrogenase gives rise to a *trans*-2, *cis*-4-enoyl CoA. This is acted upon by a recently discovered enzyme, 2, 4-dienoyl CoA reductase which, using reducing equivalents from NADPH, produces a *trans*-3-enoyl CoA. This will serve as a substrate for enoyl CoA isomerase producing the *trans*-2-enoyl CoA needed for the next round of β oxidation.

The oxidation of linoleoyl CoA, outlined in Figure 9.21, illustrates both of these points.

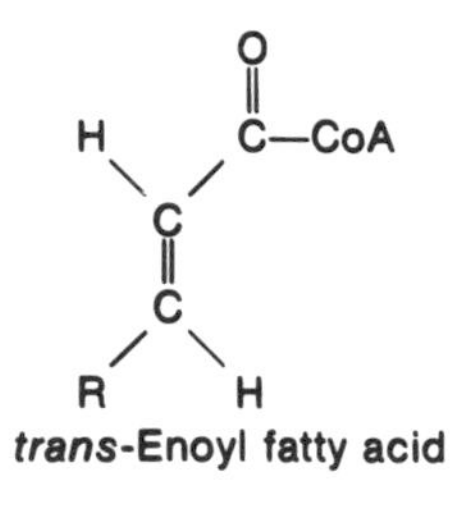

O
H C—CoA
C
C
H R
cis-Enoyl fatty acid

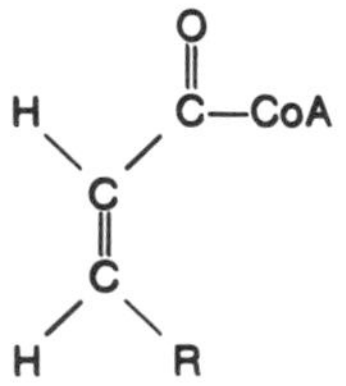

FIGURE 9.20
Geometric isomers of fatty acids.

Some Fatty Acids Undergo α-Oxidation

As noted in the earlier discussion of fatty acid biosynthesis, there are several mechanisms for hydroxylation of fatty acids. The one discussed previously is for α hydroxylation of the long-chain acids needed for the synthesis of sphingolipids. In addition, there are systems in other tissues that hydroxylate the α carbon of shorter chain acids in order to start their oxidation. The sequence is as follows:

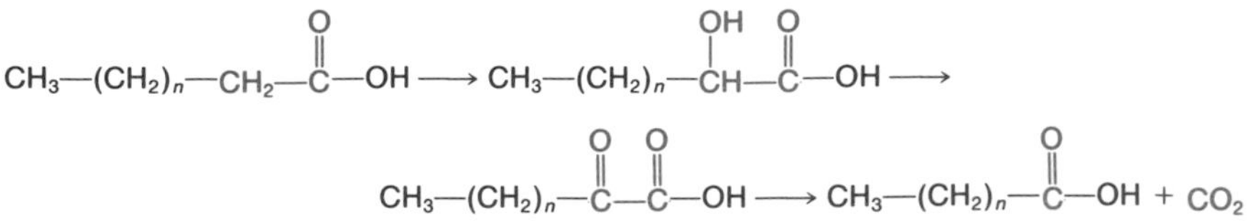

$$CH_3—(CH_2)_n—CH_2—\overset{\overset{\displaystyle O}{\|}}{C}—OH \longrightarrow CH_3—(CH_2)_n—\overset{\overset{\displaystyle OH}{|}}{C}H—\overset{\overset{\displaystyle O}{\|}}{C}—OH \longrightarrow$$

$$CH_3—(CH_2)_n—\overset{\overset{\displaystyle O}{\|}}{C}—\overset{\overset{\displaystyle O}{\|}}{C}—OH \longrightarrow CH_3—(CH_2)_n—\overset{\overset{\displaystyle O}{\|}}{C}—OH + CO_2$$

These hydroxylations probably occur in the endoplasmic reticulum and mitochondria and involve the "mixed function oxidase" type of mechanism discussed previously, because they require molecular oxygen, reduced nicotinamide nucleotides and specific cytochromes. These reac-

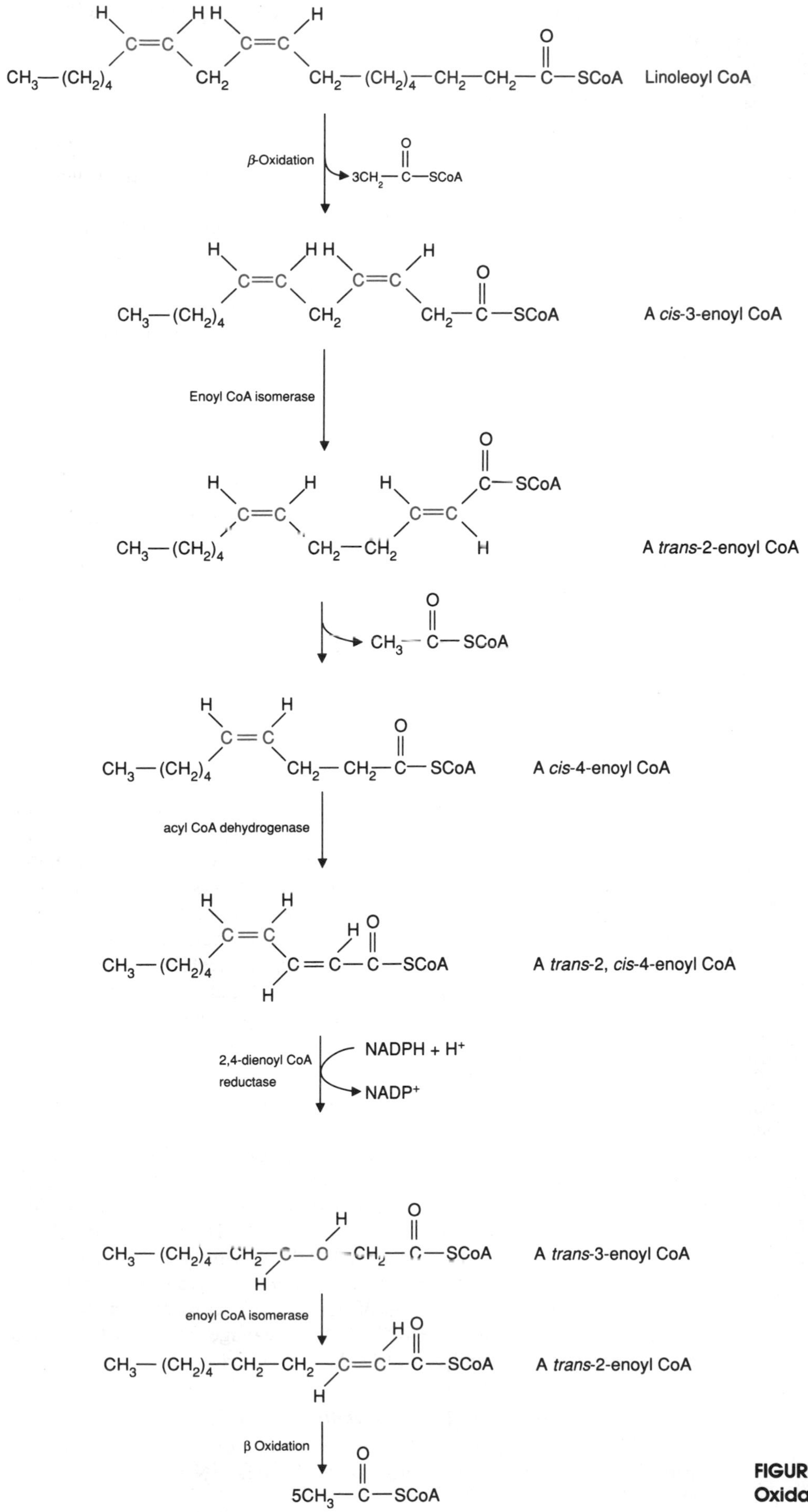

$CH_3-(CH_2)_4-CH=CH-CH_2-CH=CH-CH_2-(CH_2)_4-CH_2-CH_2-\overset{O}{\overset{\|}{C}}-SCoA$ Linoleoyl CoA

β-Oxidation → $3CH_3-\overset{O}{\overset{\|}{C}}-SCoA$

$CH_3-(CH_2)_4-CH=CH-CH_2-CH=CH-CH_2-\overset{O}{\overset{\|}{C}}-SCoA$ A *cis*-3-enoyl CoA

Enoyl CoA isomerase

$CH_3-(CH_2)_4-CH=CH-CH_2-CH_2-CH=CH-\overset{O}{\overset{\|}{C}}-SCoA$ A *trans*-2-enoyl CoA

→ $CH_3-\overset{O}{\overset{\|}{C}}-SCoA$

$CH_3-(CH_2)_4-CH=CH-CH_2-CH_2-\overset{O}{\overset{\|}{C}}-SCoA$ A *cis*-4-enoyl CoA

acyl CoA dehydrogenase

$CH_3-(CH_2)_4-CH=CH-CH=CH-\overset{O}{\overset{\|}{C}}-SCoA$ A *trans*-2, *cis*-4-enoyl CoA

2,4-dienoyl CoA reductase
$NADPH + H^+$ → $NADP^+$

$CH_3-(CH_2)_4-CH_2-CH=CH-CH_2-\overset{O}{\overset{\|}{C}}-SCoA$ A *trans*-3-enoyl CoA

enoyl CoA isomerase

$CH_3-(CH_2)_4-CH_2-CH_2-CH=CH-\overset{O}{\overset{\|}{C}}-SCoA$ A *trans*-2-enoyl CoA

β Oxidation

$5CH_3-\overset{O}{\overset{\|}{C}}-SCoA$

FIGURE 9.21
Oxidation of linoleoyl CoA.

tions are particularly important in oxidation of methylated fatty acids (see Clin. Corr. 9.5).

ω Oxidation Gives Rise to a Dicarboxylic Acid

Another minor pathway for fatty acid oxidation also involves hydroxylation and occurs in the endoplasmic reticulum of many tissues. In this case the hydroxylation takes place on the methyl carbon at the other end of the molecule from the carboxyl group or on the carbon next to the methyl end. It also uses the "mixed function oxidase" type of reaction requiring cytochrome P450, O_2, and NADPH as well as the necessary enzymes (See Chapter 23). The hydroxylated fatty acid can be further oxidized to a dicarboxylic acid via the sequential action of cytosolic alcohol and aldehyde dehydrogenases. The process occurs primarily with medium chain fatty acids. The overall reactions are

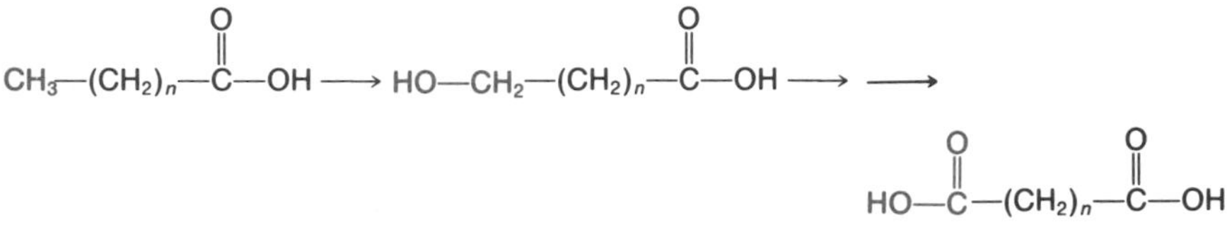

$$CH_3\text{—}(CH_2)_n\text{—}\overset{\overset{\displaystyle O}{\|}}{C}\text{—}OH \longrightarrow HO\text{—}CH_2\text{—}(CH_2)_n\text{—}\overset{\overset{\displaystyle O}{\|}}{C}\text{—}OH \longrightarrow \longrightarrow HO\text{—}\overset{\overset{\displaystyle O}{\|}}{C}\text{—}(CH_2)_n\text{—}\overset{\overset{\displaystyle O}{\|}}{C}\text{—}OH$$

The dicarboxylic acid so formed can be activated at either end of the molecule to form a CoA ester, which in turn can undergo β oxidation to produce shorter chain dicarboxylic acids such as adipic (C_6) and succinic (C_4) acids. This process appears to occur primarily in peroxisomes (see p. 21).

Ketone Bodies Are Formed From Acetyl CoA

As noted previously, the so-called ketone bodies, which are the most water-soluble form of lipid-based energy, consist mainly of acetoacetic acid and β-hydroxybutyric acid. The latter is a reduction product of the former. β-Hydroxybutyryl CoA and acetoacetyl CoA are intermediates near the end of the β-oxidation sequence, and it was initially presumed that enzymatic removal of CoA from these compounds was the main route for production of the free acids. However, more definitive studies indicated that β oxidation proceeds completely to acetyl CoA production without accumulation of any intermediates, and that acetoacetate and β-hydroxybutyrate are formed subsequently from acetyl CoA by a separate mechanism.

HMG CoA Is an Intermediate in the Synthesis of Acetoacetate From Acetyl CoA

The primary site for the formation of ketone bodies is the liver, with lesser activity occurring in the kidney. The entire process takes place within the mitochondrial matrix and begins with the condensation of two acetyl CoA molecules to form acetoacetyl CoA (Figure 9.22). The enzyme involved, **β-ketothiolase,** is probably an isozyme of that which catalyzes the reverse reaction as the last step of β oxidation. Acetoacetyl CoA then condenses with another molecule of acetyl CoA to form **β-hydroxy-β-methylglutaryl coenzyme A** (HMG CoA). Cleavage of HMG CoA then yields acetoacetic acid and acetyl CoA.

Acetoacetate Forms Both β-Hydroxybutyrate and Acetone

In mitochondria a proportion of acetoacetate is reduced to β-hydroxybutyrate depending on the intramitochondrial $NADH/NAD^+$ ratio.

CLIN. CORR. 9.5
REFSUM'S DISEASE

Although the use of the α-oxidation scheme is a relatively minor one in terms of total energy production, it is significant in the metabolism of dietary fatty acids that are methylated. A principal example of these is phytanic acid,

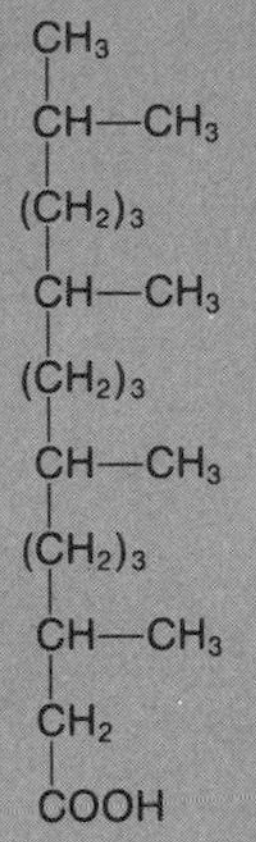

$$\begin{array}{l} CH_3 \\ | \\ CH\text{—}CH_3 \\ | \\ (CH_2)_3 \\ | \\ CH\text{—}CH_3 \\ | \\ (CH_2)_3 \\ | \\ CH\text{—}CH_3 \\ | \\ (CH_2)_3 \\ | \\ CH\text{—}CH_3 \\ | \\ CH_2 \\ | \\ COOH \end{array}$$

Phytanic acid

a metabolic product of phytol, which occurs as a constituent of chlorophyll. Phytanic acid is a significant constituent of milk lipids and animal fats, and normally it is metabolized by an initial α hydroxylation followed by dehydrogenation and decarboxylation. β Oxidation cannot occur initially because of the presence of the 3-methyl group, but it can proceed after the decarboxylation. The whole reaction produces three molecules of propionyl CoA, three molecules of acetyl CoA, and one molecule of isobutyryl CoA.

In a rare genetic disease called Refsum's disease, the patients lack the α-hydroxylating enzyme and accumulate large quantities of phytanic acid in their tissues and serum. This leads to serious neurological problems such as retinitis pigmentosa, peripheral neuropathy, cerebellar ataxia, and nerve deafness. The restriction of dietary dairy products and meat products from ruminants results in lowering of plasma phytanic acid and regression of neurologic symptoms.

Herndon, J. H., Steinberg, D., and Uhlendorf, B. W. Refsum's disease: defective oxidation of phytanic acid in tissue cultures derived from homozygotes and heterozygotes. *N. Engl. J. Med.* 281:1034, 1969.

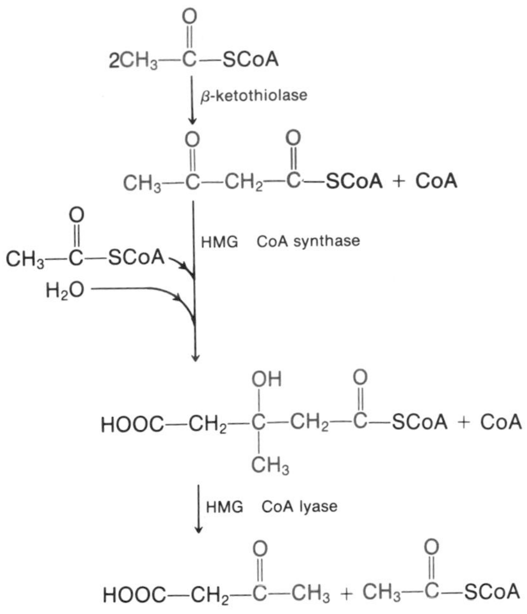

$$2CH_3-\overset{O}{\overset{\|}{C}}-SCoA \xrightarrow{\beta\text{-ketothiolase}} CH_3-\overset{O}{\overset{\|}{C}}-CH_2-\overset{O}{\overset{\|}{C}}-SCoA + CoA$$

$$CH_3-\overset{O}{\overset{\|}{C}}-CH_2-\overset{O}{\overset{\|}{C}}-SCoA + CH_3-\overset{O}{\overset{\|}{C}}-SCoA + H_2O \xrightarrow{\text{HMG CoA synthase}} HOOC-CH_2-\overset{OH}{\underset{CH_3}{C}}-CH_2-\overset{O}{\overset{\|}{C}}-SCoA + CoA$$

$$HOOC-CH_2-\overset{OH}{\underset{CH_3}{C}}-CH_2-\overset{O}{\overset{\|}{C}}-SCoA \xrightarrow{\text{HMG CoA lyase}} HOOC-CH_2-\overset{O}{\overset{\|}{C}}-CH_3 + CH_3-\overset{O}{\overset{\|}{C}}-SCoA$$

FIGURE 9.22
Pathway of acetoacetate formation.

$$CH_3-\overset{O}{\overset{\|}{C}}-CH_2-\overset{O}{\overset{\|}{C}}-O^- + NADH + H^+ \underset{\beta\text{-hydroxybutyrate dehydrogenase}}{\rightleftharpoons} CH_3-\overset{OH}{\overset{|}{C}H}-CH_2-\overset{O}{\overset{\|}{C}}-O^- + NAD^+$$

Note that the product of this reaction is D-β-hydroxybutyrate, whereas β-hydroxybutyryl CoA formed during the course of β oxidation is of the L configuration.

β-Hydroxybutyrate dehydrogenase is tightly associated with the inner mitochondrial membrane and, because of its high activity in liver, the concentrations of substrates and products of the reaction are maintained close to equilibrium. Thus, the ratio of β-hydroxybutyrate to acetoacetate in blood leaving the liver can be taken as a reflection of the mitochondrial NADH/NAD^+ ratio.

A certain amount of acetoacetate is continually undergoing slow, spontaneous nonenzymatic decarboxylation to acetone.

$$CH_3-\overset{O}{\overset{\|}{C}}-CH_2-\overset{O}{\overset{\|}{C}}-O^- + H^+ \longrightarrow CH_3-\overset{O}{\overset{\|}{C}}-CH_3 + CO_2$$

Under normal conditions acetone formation is negligible, but when pathological accumulations of acetoacetate occur, as for example in severe diabetic ketoacidosis (see Clin. Corr. 9.6), the amount of acetone in blood can be sufficient to cause it to be detectable in a patient's breath.

As seen from Figure 9.23, the pathway leading from acetyl CoA to HMG CoA also operates in the cytosolic space of the liver cell (indeed, this applies to essentially all tissues of the body). However, in this compartment HMG CoA lyase is absent and the HMG CoA formed is used for the purposes of cholesterol biosynthesis (see Chapter 10). What distinguishes liver from nonhepatic tissues is its high complement of intramitochondrial **HMG CoA synthase,** thus providing an enzymological basis for the primacy of this organ in ketone body production.

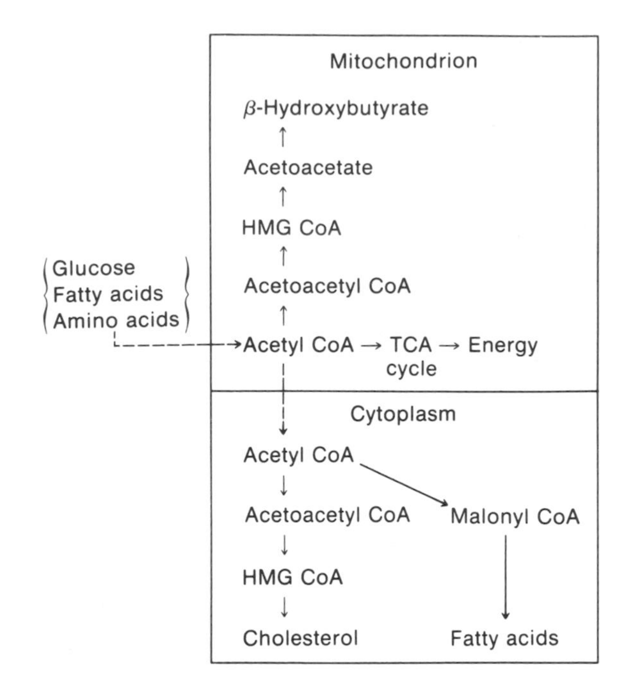

FIGURE 9.23
Interrelationships of ketone bodies with lipid, carbohydrate, and amino acid metabolism in liver.

Acetoacetate + succinyl CoA

⇅ acetoacetate : succinyl CoA CoA transferase

Acetoacetyl CoA + succinate

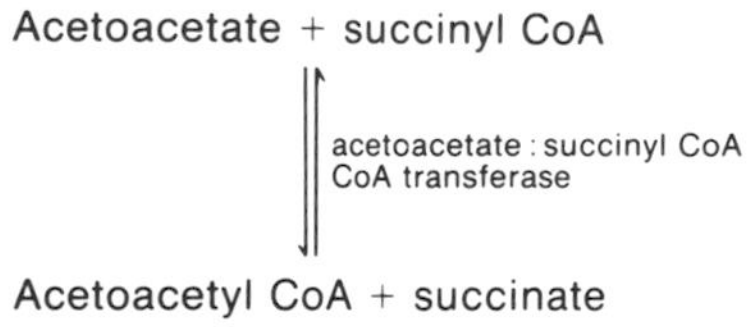

FIGURE 9.24
Initial step in the utilization of acetoacetate by nonhepatic tissues.

$$CH_3—(CH_2)_n—CH_2—CH_2—\overset{O}{\overset{\|}{C}}—SCoA$$

flavoprotein → flavoprotein-H_2; O_2 → H_2O_2

$$CH_3—(CH_2)_n—CH{=}CH—\overset{O}{\overset{\|}{C}}—SCoA$$

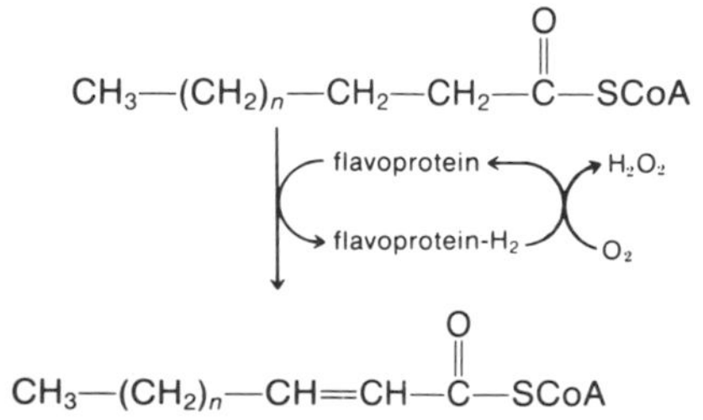

FIGURE 9.25
Initial step in peroxisomal fatty acid oxidation.

Utilization of Ketone Bodies by Non-Hepatic Tissues Requires Formation of Acetoacetyl CoA

Acetoacetate and β-hydroxybutyrate produced by the liver serve as excellent fuels of respiration for a variety of nonhepatic tissues, such as cardiac and skeletal muscle, particularly when glucose is in short supply (starvation) or inefficiently used (insulin deficiency). But since under these conditions the same tissues can readily use free fatty acids (whose blood concentration rises as insulin levels fall) as a source of energy, a nagging question for many years was why the liver should produce ketone bodies in the first place. The answer emerged in the late 1960s with the recognition that during prolonged starvation in humans the ketone bodies replace glucose as the major fuel of respiration for the central nervous system, which has a low capacity for fatty acid oxidation. Also noteworthy is the fact that during the neonatal period of development, acetoacetate and β-hydroxybutyrate serve as important precursors for cerebral lipid synthesis.

The mechanism for use of ketone bodies requires that acetoacetate first be reactivated to its CoA derivative. This is accomplished by a mitochondrial enzyme present in most nonhepatic tissues (but absent from liver) that uses succinyl CoA as the source of the coenzyme. The reaction is depicted in Figure 9.24. Through the action of β-ketothiolase, acetoacetyl CoA is then converted into acetyl CoA, which in turn enters the TCA cycle with the production of energy. Mitochondrial β-hydroxybutyrate dehydrogenase reconverts β-hydroxybutyrate into acetoacetate as the concentration of the latter is decreased.

Starvation and Certain Pathological Conditions Lead to Ketosis

Under normal feeding conditions, the hepatic production of acetoacetate and β-hydroxybutyrate is minimal and the concentration of these compounds in the blood is very low (<0.2 mM). However, with food deprivation ketone body synthesis is greatly accelerated, and the circulating level of acetoacetate plus β-hydroxybutyrate may rise to the region of 3–5 mM. This is a normal response of the body to a shortage of carbohydrate and, as alluded to above, subserves a number of crucial roles. In the early stages of fasting, use of the ketone bodies by heart and skeletal muscle conserves glucose for support of the central nervous system. With more prolonged starvation, the increased blood concentration of acetoacetate and β-hydroxybutyrate ensures their efficient uptake by the brain, thereby further sparing glucose consumption.

In contrast to the **physiological ketosis of starvation,** certain pathological conditions, most notably **diabetic ketoacidosis** (Clin. Corr. 9.6), are characterized by excessive accumulation of ketone bodies in the blood (up to 20 mM). The hormonal and biochemical factors operative in the overall control of hepatic ketone body production are discussed in detail in Chapter 14.

Peroxisomal Oxidation of Fatty Acids Serves Many Functions

Although the bulk of cellular fatty acid oxidation occurs in the mitochondria it has recently become clear that a significant fraction also takes place in the peroxisomes of liver, kidney, and other tissues. **Peroxisomes** are a class of subcellular organelles with distinctive morphological and chemical characteristics. Their initial distinguishing property was a high content of the enzyme catalase and it has been suggested that peroxisomes may function in a protective role against oxygen toxicity. Several lines of evidence suggest that they are also involved in lipid catabolism. First, the analogous structures in plants, glyoxysomes, are capable of oxidizing fatty acids. Second, a number of drugs used clinically to decrease triacylglycerol levels in patients cause a marked increase in histologically detectable peroxisomes. Third, liver peroxisomes, isolated by differential centrifugation, have been shown to oxidize fatty acids and to contain most of the enzymes needed for the β-oxidation process.

The mammalian peroxisomal fatty acid oxidation scheme, which is similar to that in plant glyoxysomes, differs from the mitochondrial β-oxidation system in three important respects: first, the initial dehydrogenation is accomplished by a cyanide-insensitive oxidase system, as shown in Figure 9.25. The H_2O_2 is then eliminated by catalase, and the remaining steps are the same as in the mitochondrial system. Second, there is evidence that the peroxisomal and mitochondrial enzymes are slightly different and that the specificity in peroxisomes is for somewhat longer chain length. Third, although rat liver mitochondria will oxidize a molecule of palmitoyl CoA to eight molecules of acetyl CoA, the β-oxidation system in peroxisomes from the same organ will not proceed beyond the stage of octanoyl CoA (C_8). The possibility is thus raised that one function of peroxisomes is to shorten the chains of relatively long-chain fatty acids to a point at which β oxidation can be completed in mitochondria.

Other peroxisomal events include the chain shortening of dicarboxylic acids, as noted earlier, as well as steps involved in the conversion of cholesterol into bile acids and in the formation of ether lipids. Given these diverse metabolic roles it is not surprising that the congenital absence of functional peroxisomes, an inherited defect known as Zellweger syndrome, has such devastating effects (see Clin. Corr. 1.3).

9.7 CHARACTERISTICS, METABOLISM, AND FUNCTIONAL ROLE OF POLYUNSATURATED FATTY ACIDS

In recent years there has been considerable renewed interest in elucidating the specific physiological roles of the **polyunsaturated fatty acids** at the biochemical level. This is due to some extent to the results of initial studies which suggested that a diet in which the proportion of polyunsaturated to saturated fatty acids was relatively high could help to lower blood cholesterol levels in some patients. The relationship between these diet

CLIN. CORR. 9.6 DIABETIC KETOACIDOSIS

Diabetic ketoacidosis (DKA) is a common illness among patients with insulin-dependent diabetes mellitus. Although mortality rates have declined, they are still in the range of 6–10%. The condition is triggered by severe insulin deficiency coupled with glucagon excess and is frequently accompanied by concomitant elevation of other stress hormones, such as epinephrine, norepinephrine, cortisol, and growth hormone. The major metabolic derangements are marked hyperglycemia, excessive ketonemia, and ketonuria. Blood concentrations of acetoacetic plus β-hydroxybutyric acids as high as 20 mM are not uncommon. Because these are relatively strong acids (pK ~ 3.5), the situation results in life-threatening metabolic acidosis.

The massive accumulation of ketone bodies in the blood in DKA stems from a greatly accelerated hepatic production rate such that the capacity of nonhepatic tissues to use them is exceeded. In biochemical terms the initiating events are identical with those operative in the development of starvation ketosis; that is, increased glucagon/insulin ratio → elevation of liver [cAMP] → decreased [malonyl CoA] → deinhibition of CPT I → activation of fatty acid oxidation and ketone production (see text for details). However, in contrast to physiological ketosis, where insulin secretion from the pancreatic β cells limits free fatty acid (FFA) availability to the liver, this restraining mechanism is absent in the diabetic individual. As a result, plasma FFA concentrations can reach levels as high as 3–4 mM, which drive hepatic ketone production at maximal rates.

Correction of DKA requires rapid treatment that will be dictated by the severity of the metabolic abnormalities and the associated tissue water and electrolyte imbalance. Insulin is essential. It lowers the plasma glucagon level, antagonizes the catabolic effects of glucagon on the liver, inhibits the flow of ketogenic and gluconeogenic substrates (FFA and amino acids) from the periphery, and stimulates glucose uptake in target tissues.

Foster, J. D. and McGarry, J. D. Metabolic derangements and treatment of diabetic ketoacidosis. *N. Engl. J. Med.* 309:159, 1983; and Foster, D. W. and McGarry, J. D. Acute complications of diabetes: ketoacidosis, hyperosmolar coma, lactic acidosis. Endocrinology Vol. 2, 2nd ed., L. J. DeGroot (Ed.), Philadelphia, Saunders 1989, p. 1439.

modifications and the development of atherosclerosis, if any, is not simple, but the initial reports did tend to spur interest in the polyunsaturated fatty acids.

It is important to emphasize that the essential fatty acids mentioned in Section 9.3 are polyunsaturated but that most of the individual polyunsaturated fatty acids need not be specifically supplied in the diet. All fatty acids with three or more double bonds, without regard to their position in the chain, are polyunsaturated and most of these can be synthesized by human tissues. The essential acids refer only to the linoleic and the linolenic series, which have double bonds near the methyl end of the chain, and which cannot be synthesized by humans. The degree to which these acids really are essential for humans is still to be determined, although clearly reproducible deficiency states can be produced in rats and some other animals by carefully controlled diets. The need for the linoleic acid series is clearer since the discovery of the prostaglandins, which are derived from arachidonic acid, one of the linoleic series. The need for the linolenic acid series is very obscure, although the C_{22} hexaenoic acid derivative of it shown in Section 9.2 is concentrated in some membranes of nerve and retina. A deficiency syndrome for linolenate has yet to be produced in any animal except rainbow trout, but this is possibly due to extremely efficient mechanisms for conservation of linolenate and its derivatives in the body.

A major role of all polyunsaturated fatty acids seems to be to produce the proper fluidity in biological membranes. As described in Chapter 10, the various phospholipids have variable amounts of polyunsaturated fatty acids as constituents, and it has been conclusively demonstrated that lower organisms can alter the fatty acid patterns in their membrane phospholipids to maintain proper fluidity under changing conditions, such as temperature alterations. This can be done by increasing the proportion of fatty acids with a few double bonds in them or by increasing the degree of unsaturation of the fatty acids.

Some Polyunsaturated Fatty Acids Can Be Synthesized or Modified

The human body can synthesize a variety of polyunsaturated fatty acids by the elongation and desaturation reactions described in Section 9.3. The **stearoyl CoA desaturase** introduces an initial double bond between C-9 and C-10 atoms in a saturated fatty acid, and then double bonds can be introduced just beyond C-4, C-5, or C-6 atoms. Desaturation at C-8 probably occurs also in some tissues. The positions of these desaturations are shown in Figure 9.26. The relative specificities of the various enzymes are still to be elucidated completely, but it seems likely that elongation and desaturation can occur in either order. The conversion of linolenic acid to all *cis*-4, 7, 10, 13, 16, 19-docosahexaenoic acid in brain is a specific example of such a sequence.

$CH_3—(CH_2—CH{=}CH)_3—CH_2—CH_2—CH_2—CH_2—CH_2—CH_2—CH_2—COOH$

Linolenic acid

$\downarrow$ "Δ^6-desaturase"

$CH_3—(CH_2—CH{=}CH)_3—CH_2—CH{=}CH—CH_2—CH_2—CH_2—CH_2—COOH$

$\downarrow$ elongation

$CH_3—(CH_2—CH{=}CH)_3—CH_2—CH{=}CH—CH_2—CH_2—CH_2—CH_2—CH_2—CH_2—COOH$

$\downarrow$ "Δ^5-desaturase"

$CH_3—(CH_2—CH{=}CH)_3—CH_2—CH{=}CH—CH_2—CH{=}CH—CH_2—CH_2—CH_2—COOH$

$\downarrow$ elongation

$CH_3—(CH_2—CH{=}CH)_3—CH_2—CH{=}CH—CH_2—CH{=}CH—CH_2—CH_2—CH_2—CH_2—CH_2—COOH$

$\downarrow$ "Δ^4-desaturase"

$CH_3—(CH_2—CH{=}CH)_3—CH_2—CH{=}CH—CH_2—CH{=}CH—CH_2—CH{=}CH—CH_2—CH_2—COOH$

All-*cis*-4,7,10,13,16,19-docosahexaenoic acid

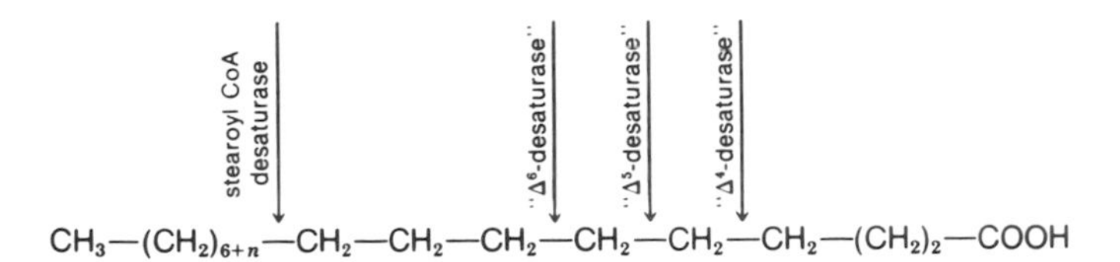

FIGURE 9.26
Positions in the fatty acid chain where desaturation can occur in the human body.
There must always be at least six single bonds in the chain toward the methyl end of the molecule just beyond the bond being desaturated.

The polyunsaturated fatty acids, particularly arachidonic acid, are the precursors of the highly active prostaglandins and thromboxanes. A number of different classes of prostaglandins are formed depending on the precursor fatty acid and the sequence of various oxidations that convert the acids to the active compounds. A detailed discussion of these substances and their formation is given in Chapter 10.

Polyunsaturated Fatty Acids Are Susceptible to Autooxidation

Polyunsaturated fatty acids in living systems have a significant potential for autooxidation, a process that may have important physiological and/or pathological consequences. This is the set of reactions that causes rancidity in fats and the curing of linseed oil in paints.

The basis behind the process is that the methylene carbon between any two double bonds in the polyunsaturated fatty acids is quite susceptible to hydrogen abstraction and free radical formation. Once this abstraction occurs the reactions can take place in any sequence, and many of the reactive breakdown products can contribute to further oxidation. Oxygen can attach to acids which that hydrogen has been abstracted, forming **free radicals** that can then react with another lipid molecule, leading to abstraction of hydrogen from the second molecule. The products of this reaction are a **lipid hydroperoxide** in the first molecule and a new free radical in the molecule attacked. The lipid hydroperoxide molecules break up, forming dialdehydes, the most prominent being malondialdehyde. This product can cause cross-linking between various types of molecules, such linkages leading to cytotoxicity, mutagenicity, membrane breakdown, and enzyme modification. Malondialdehyde also polymerizes with itself and other tissue breakdown products, forming an insoluble pigment, lipofuscin, which accumulates in some aging tissues.

Autooxidation May Be Initiated by External Agents

A number of external agents can initiate autooxidation in vitro. The extent to which they can lead to such reactions in vivo in humans is undetermined. Various types of radiation, including sunlight, and environmental pollutants such as oxides of nitrogen and carbon tetrachloride are examples of such external agents. The detoxification mechanisms for CCl_4 in the liver use cytochrome P450 and generate transient free radicals. The latter can initiate lipid autooxidation and lead to carcinogenesis. Metabolism of the herbicide paraquat, sometimes used for marijuana control, produces superoxide anions, which can also initiate fatty acid autooxidation.

It is quite possible that autooxidation can be initiated without the need for an external agent. Theoretically at least, the enzymes involved in various oxidative processes can produce singlet oxygen and transient partial reduction products of oxygen (superoxide anion, H_2O_2, and the hydroxyl radical), any or all of which could potentially lead to lipid free radicals and/or lipid hydroperoxides. For instance, under proper circumstances rat liver microsomes cause extensive formation of lipid peroxides in vitro, presumably initiated by an enzyme-bound reactive form of iron. A number of enzymes such as xanthine oxidase, superoxide dismutase, and lipoxidases can initiate lipid peroxide formation in vitro.

Possible Protective Mechanisms In Vivo

Recent evidence suggests that in special circumstances the partial reduction products of oxygen, which can potentially initiate lipid autooxidation, may actually be produced for beneficial purposes, for example, by leukocytes in killing bacteria. However, under most conditions the human body utilizes potent mechanisms to ensure against accumulation of these substances. Three types of enzymes, the **catalases, peroxidases,** and **superoxide dismutases,** seem primarily designed to destroy them rapidly and keep tissue levels of their substrates negligible. An additional safeguard is the presence of scavenging molecules in the body, which interact with any free radicals produced, but do not in turn produce self-propagating chain reactions. The principal one present in humans is α-tocopherol (vitamin E). Evidence for its protective role in vivo, however, is purely circumstantial.

BIBLIOGRAPHY

Brown, M. S. and Goldstein, J. L. The hyperlipoproteinemias and other disorders of lipid metabolism. *Harrison's Principles of Internal Medicine,* 11th ed. Braunwald et al. (Eds.). McGraw-Hill, 1987, p. 1650.

Foster, D. W. and McGarry, J. D., The metabolic derangements and treatment of diabetic ketoacidosis. *N. Engl. J. Med.* 309:159, 1983.

Gurr, M. I. and James, A. T. *Lipid biochemistry, an introduction,* 3rd ed. London: Chapman and Hall, 1980.

IUPAC–IUB Commission on Biochemical Nomenclature. The nomenclature of lipids. *Biochem. J.* 171:21, 1978.

McGarry, J. D. and Foster, D. W., Regulation of hepatic fatty acid oxidation and ketone body production. *Annu. Rev. Biochem.* 49:395, 1980.

Nilsson-Ehle, P., Garfinkel, A. S., and Schotz, M. C. Lipolytic enzymes and plasma lipoprotein metabolism. *Annu. Rev. Biochem.* 49:667, 1980.

Robinson, A. M. and Williamson, D. H., Physiological roles of ketone bodies as substrates and signals in mammalian tissues. *Physiol. Rev.* 60:143, 1980.

Scow, R. O., Blanchette-Mackie, E. J., and Smith, L. C. Transport of lipid across capillary endothelium. *Fed. Proc.* 39:2610, 1980.

Stanley, C. A. New genetic defects in mitochondrial fatty acid oxidation. *Adv. Pediatr.* 34:59, 1987.

Volpe, J. J. Lipid metabolism: fatty acid and cholesterol biosynthesis. *Adv. Mod. Nutr.* 37, 1978.

Wakil, S. J., Stoops, J. K., and Joshi, V. C., Fatty acid synthesis and its regulation. *Annu. Rev. Biochem.* 52:537, 1983.

QUESTIONS

C. N. ANGSTADT AND J. BAGGOTT

*Q*uestion *T*ypes are described on page xxiii.

1. (QT2) Fatty acids occurring in humans most commonly:
   1. are straight chain but may have some methyl branches.
   2. have double bonds present in trans configuration.
   3. contain an even number of carbon atoms.
   4. do not contain more than 16 carbon atoms.

2. (QT1) Triacylglycerols:
   A. would be expected to be good emulsifying agents.
   B. yield about the same amount of ATP on complete oxidation as would an equivalent weight of glycogen.
   C. are stored as hydrated molecules.
   D. in the average individual, represent sufficient energy to sustain life for several weeks.
   E. are generally negatively charged molecules at physiological pH.

3. (QT1) In humans, fatty acids:
   A. can be synthesized from excess dietary carbohydrate or protein.
   B. are not required at all in the diet.
   C. containing double bonds cannot be synthesized.
   D. must be supplied entirely by the diet.
   E. other than palmitate, must be supplied in the diet.

4. (QT2) Acetyl CoA carboxylase:
   1. undergoes protomer–polymer interconversion during its physiological regulation.
   2. requires biotin.
   3. is inhibited by cAMP-mediated phosphorylation.
   4. content in a cell responds to changes in fat content in the diet.

5. (QT2) In the synthesis of palmitate:
   1. the addition of malonyl CoA to fatty acid synthase elongates the growing chain by three carbon atoms.
   2. a β-keto residue on the 4′-phosphopantetheine moiety is reduced to a saturated residue by NADPH.
   3. palmitoyl CoA is released from the synthase.
   4. transfer of the growing chain from ACP to another —SH must precede the addition of the next malonyl CoA.
6. (QT2) Citrate stimulates fatty acid synthesis by:
   1. allosterically activating acetyl CoA carboxylase.
   2. providing a mechanism to transport acetyl CoA from the mitochondria to the cytosol.
   3. participating in a pathway that ultimately produces CO_2 and NADPH in the cytosol.
   4. participating in the production of ATP.

A. Fatty acid elongation in mitochondria C. Both
B. Fatty acid elongation in cytosol D. Neither

7. (QT4) Malonyl CoA is the source of carbon atoms.
8. (QT4) Preferred substrates are fatty acyl CoAs shorter than 16 carbon atoms.

9. (QT1) Fatty acid synthase:
   A. synthesizes only palmitate.
   B. yields an unsaturated fatty acid by skipping a reductive step.
   C. produces hydroxy fatty acids in nerve tissue.
   D. can stop with the release of a fatty alcohol instead of an acid.
   E. can produce a branched-chain fatty acid if methyl malonyl CoA is used as a substrate.

10. (QT2) Which of the following events is/are usually involved in the synthesis of triacylglycerols in adipose tissue?
    1. Addition of a fatty acyl CoA to a diacylglycerol
    2. Addition of a fatty acyl CoA to a lysophosphatide
    3. Hydrolysis of phosphatidic acid by a phosphatase
    4. Glycerol kinase reaction
11. (QT2) Plasma lipoproteins:
    1. are not the only carriers of lipid-based energy in the blood.
    2. usually have a nonpolar core containing triacylglycerols and cholesterol esters.
    3. do not generally include free (unesterified) fatty acids.
    4. include chylomicrons generated in the intestine.

12. (QT1) Lipoprotein lipase:
    A. is an intracellular enzyme.
    B. is stimulated by cAMP-mediated phosphorylation.
    C. functions to mobilize stored triacylglycerols from adipose tissue.
    D. is stimulated by one of the apoproteins present in VLDL.
    E. readily hydrolyzes three fatty acids from a triacylglycerol.
13. (QT1) A deficiency of carnitine might be expected to interfere with:
    A. β oxidation.
    B. ketone body formation from acetyl CoA.
    C. palmitate synthesis.
    D. mobilization of stored triacylglycerols from adipose tissue.
    E. uptake of fatty acids into cells from the blood.

14. (QT2) β Oxidation of fatty acids:
    1. has the potential to generate ATP even if acetyl CoA is not subsequently oxidized.
    2. is controlled primarily by allosteric effectors.
    3. can use odd-chain and unsaturated fatty acids as substrates.
    4. uses $NADP^+$.

15. (QT1) Ketone bodies:
    A. are formed by removal of CoA from the corresponding intermediate of β oxidation.
    B. are synthesized from cytoplasmic β-hydroxy-β-methyl glutaryl coenzyme A (HMGCoA).
    C. are excellent energy substrates for liver.
    D. include both β-hydroxybutyrate and acetoacetate, the ratio reflecting the intramitochondrial $NADH/NAD^+$ ratio in liver.
    E. form when β oxidation is interrupted.

16. (QT2) The high glucagon/insulin ratio seen in starvation:
    1. promotes mobilization of fatty acids from adipose stores.
    2. stimulates β oxidation by inhibiting the production of malonyl CoA.
    3. leads to increased concentrations of ketone bodies in the blood.
    4. produces a condition that results in an increased utilization of ketone bodies by the brain.
17. (QT2) Polyunsaturated fatty acids:
    1. cannot be synthesized by humans.
    2. are important in determining fluidity of membranes.
    3. have no known functions other than as membrane components.
    4. are quite susceptible to autooxidation.

ANSWERS

1. B 1, 3 correct. 2: Most naturally occurring double bonds are cis, an important factor in β oxidation of unsaturated fatty acids. 4: C-18 and C-20 fatty acids are very common (p. 390).
2. D A, C, E. Triacylglycerols are neutral, hydrophobic molecules with no hydrophilic portion and, therefore, are not emulsifying agents and are stored anhydrously. B: Their more reduced state, compared to carbohydrates, makes them more energy-rich (p. 391).
3. A It is important to realize that triacylglycerol is the ultimate storage form of excess dietary intake. B–E: We can synthesize most fatty acids, including those with double bonds, except for the essential fatty acids, linoleic and linolenic (p. 392).
4. E All four correct. 1: Acetyl CoA carboxylase shifts between its protomeric (inactive) and polymeric (active) forms under the influence of a variety of regulatory factors. 3: Since cAMP increases at times when energy is *needed,* it is consistent that a process that uses energy would be inhibited. 4: Long-term control is related to enzyme synthesis and responds appropriately to dietary changes (Table 9.2, p. 393).
5. C 2, 4 correct. 1: Splitting CO_2 from malonyl CoA is the driving force for the condensation reaction so the chain grows two carbon atoms at a time. 3: Palmitate is released as the free acid; the conversion to the CoA ester is by a different enzyme (p. 395). 4: It is important to realize that only ACP binds the incoming malonyl CoA so it must be freed before another addition can be made (p. 394).
6. A 1, 2, 3, correct. 1: Table 9.2. 2: Acetyl CoA is generated primarily in mitochondria but does not cross the membrane readily. 3: Oxaloacetate generated by citrate cleavage enzyme, when converted to malate, yields CO_2 and NADPH by the malic enzyme (Figure 9.10). 4: Citrate *consumes* ATP when acted upon by citrate cleavage enzyme (p. 397).
7. B The cytosolic system is very similar to fatty acid synthase except that the enzymes are not part of a multienzyme complex (p. 399).
8. A The role of mitochondrial fatty acid elongation seems to be to elongate short-chain fatty acids; the cytoplasmic system is most active with palmitate (p. 399).
9. E E. This is much slower than reaction with malonyl CoA, but it is significant. A: In certain tissues, for example, mammary glands, shorter-chain products are formed. B–D: These products are all formed by other processes. Reactions proceeding on a multienzyme complex generally do not "stop" at intermediate steps (p. 400).
10. A 1, 2, 3 correct. 4: Does not occur to any significant extent in adipose tissue. The sequential addition of fatty acyl CoAs to glycerol 3-phosphate forms lysophosphatidic acid, then phosphatidic acid whose phosphate is removed before the addition of the third fatty acyl residue (p. 402).
11. E All four correct. 1, 3: Fatty acids bound to serum albumin and ketone bodies are other sources. 2: All lipoproteins (Section 9.5) have this same general structure, a nonpolar core surrounded by a more polar shell.
12. D A–C: These are characteristics of hormone-sensitive lipase. E: It generally requires more than one lipase to hydrolyze all of the fatty acids (p. 403, Table 9.3).
13. A Carnitine functions in transport of fatty acyl CoA esters formed in cytosol into the mitochondria (p. 408).
14. B 1, 3 correct. 1, 4: It is important to realize that β oxidation, itself, generates $FADH_2$ and NADH, which can be reoxidized to generate ATP. 2: Carnitine transport to provide the substrate and reoxidation of reduced cofactors control β oxidation. 3: β Oxidation is a general process requiring only minor modifications to oxidize nearly any fatty acid in the cell (p. 409, Table 9.4).
15. D A, E: β Oxidation proceeds to completion; ketone bodies are formed by a separate process. B, C: Ketone bodies are formed, but not used, in liver mitochondria; cytosolic HMG CoA is a precursor of cholesterol (p. 414).
16. E All four correct. High glucagon/insulin ratio results in cAMP-mediated phosphorylations that activate hormone-sensitive lipase and inhibit acetyl CoA carboxylase. Both of these, as well as other events, promote ketone body formation by greatly increasing acetyl CoA production in mitochondria, thereby assuring efficient uptake and utilization by brain (p. 416).
17. C 2, 4 correct. 1: Humans can introduce additional double bonds between C-9 and the carboxyl end but not toward the methyl end of the molecule. 2, 3: Maintaining proper membrane fluidity is a major role; some polyunsaturated fatty acids are precursors of prostaglandins and thromboxanes. 4: They can form free radicals that initiate a sequence of events (p. 418).

10

Lipid Metabolism II: Pathways of Metabolism of Special Lipids

ROBERT H. GLEW

10.1 OVERVIEW 424
10.2 PHOSPHOLIPIDS 424
Phospholipids Contain 1,2-Diacylglycerol and a Base Connected by a Phosphodiester Bridge 425
Phospholipids in Membranes Serve a Variety of Roles 427
Biosynthesis of Phospholipids 431
10.3 CHOLESTEROL 438
Cholesterol, an Alicyclic Compound, Is Widely Distributed in Free and Esterified Forms 438
Cholesterol Is a Membrane Component and the Precursor of Bile Salts and Steroid Hormones 439
Cholesterol Is Synthesized From Acetyl CoA 439
Cholesterol Biosynthesis Is Carefully Regulated 443
Plasma Cholesterol Is in a Dynamic State 446
Cholesterol Is Excreted Primarily as Bile Acids 447
Vitamin D Is Synthesized From an Intermediate of the Cholesterol Biosynthetic Pathway 448
10.4 SPHINGOLIPIDS 449
Biosynthesis of Sphingosine 449
Ceramides Are Fatty Acid Amide Derivatives of Sphingosine 450
Sphingomyelin Is the Only Sphingolipid Containing Phosphorus 451
Glycosphingolipids Usually Have a Galactose or Glucose Unit 452
The Sphingolipidoses Are Lysosomal Storage Diseases With Defects in the Catabolic Pathway for Sphingolipids 456
10.5 PROSTAGLANDINS AND THROMBOXANES 461
Prostaglandins and Thromboxanes Are Derivatives of Twenty-Carbon, Monocarboxylic Acids 461
Synthesis of Prostaglandins Involves a Cyclooxygenase 462
Prostaglandins Exhibit a Multitude of Physiological Effects 465
10.6 LIPOXYGENASE AND THE OXY-EICOSATETRAENOIC ACIDS 466
Monohydroperoxyeicosatetraenoic Acids Are the Products of Lipoxygenase Action 467
Leukotrienes and Hydroxyeicosatetraenoic Acids Are the Hormones Derived From HPETEs 468
Leukotrienes and HETEs Affect Several Physiological Processes 469
BIBLIOGRAPHY 470
QUESTIONS AND ANNOTATED ANSWERS 471
CLINICAL CORRELATIONS
10.1 Respiratory Distress Syndrome 428
10.2 Treatment of Hypercholesterolemia 445
10.3 Atherosclerosis 446
10.4 Diagnosis of Gaucher's Disease in an Adult 461

10.1 OVERVIEW

Lipid is a general term that describes substances that are relatively water-insoluble and extractable by nonpolar solvents. The complex lipids of humans fall into one of two broad categories: the nonpolar lipids, such as triacylglycerols and cholesterol esters, and the polar lipids, which are amphipathic in that they contain both a hydrophobic domain and a hydrophilic region in the same molecule. This chapter will discuss the two major subdivisions of the polar lipids, the *phospholipids* and the *sphingolipids*. The hydrophobic and hydrophilic domains are bridged by a glycerol moiety in the case of glycerophospholipids and by sphingosine in the case of sphingomyelin and the glycosphingolipids. In terms of location, triacylglycerol is confined largely to storage sites in adipose tissue, whereas the polar lipids occur primarily in cellular membranes. Membranes generally contain ~40% of their dry weight as lipid and 60% as protein.

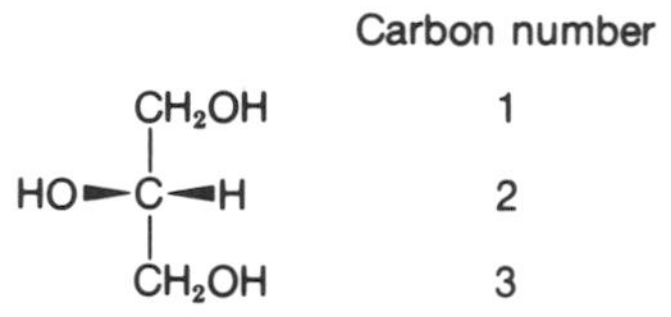

The processes of cell–cell recognition, phagocytosis, contact inhibition, and rejection of transplanted tissues and organs are all phenomena of medical significance that involve highly specific recognition sites on the surface of the plasma membrane. The synthesis of these complex glycosphingolipids that appear to play a role in these important biological events will be described.

The glycolipids are worthy of study because the ABO antigenic determinants of the blood groups are primarily glycolipid in nature. In addition, various sphingolipids are the storage substances that accumulate in the liver, spleen, kidney, or nervous tissue of persons suffering from certain genetic disorders called sphingolipidoses. In order to understand the basis of these enzyme-deficiency states, a knowledge of the relevant chemical structures involved is required.

A very important lipid is *cholesterol*. This chapter describes the pathway of cholesterol biosynthesis and its regulation and shows how cholesterol functions as a precursor to the bile salts and steroid hormones. Also described is the role of high density lipoprotein (HDL) and lecithin : cholesterol acyltransferase (LCAT) in the management of plasma cholesterol.

Finally, the metabolism and function of two pharmacologically powerful classes of hormones derived from arachidonic acid, namely, the prostaglandins and the leukotrienes will be discussed. See the Appendix, for a discussion of the nomenclature and chemistry of lipids.

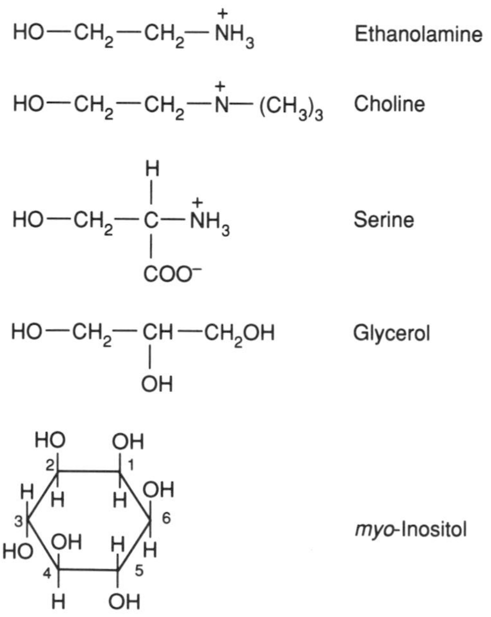

	Carbon number
CH_2OH	1
HO—C—H	2
CH_2OH	3

FIGURE 10.1
Stereospecific numbering of glycerol.

$HO—CH_2—CH_2—\overset{+}{N}H_3$	Ethanolamine
$HO—CH_2—CH_2—\overset{+}{N}—(CH_3)_3$	Choline
$HO—CH_2—CH(COO^-)—\overset{+}{N}H_3$	Serine
$HO—CH_2—CH(OH)—CH_2OH$	Glycerol
(ring structure)	*myo*-Inositol

FIGURE 10.2
Structure of some common polar head groups of phospholipids.

10.2 PHOSPHOLIPIDS

The two principal classes of acylglycerolipids are the **triacylglycerols** and **glycerophospholipids.** They are referred to as glycerolipids because the core of these compounds is provided by the C_3 polyol, glycerol.

The two primary alcohol groups of glycerol are not stereochemically identical and in the case of the phospholipids, it is usually the same hydroxyl group that is esterified to the phosphate residue. The stereospecific numbering system is the best way to designate the different hydroxyl groups. In this system, when the structure of glycerol is drawn in the Fischer projection with the C-2 hydroxyl group projecting to the left of the page, the carbon atoms are numbered as shown in Figure 10.1. When the stereospecific numbering (*sn*) system is employed, the prefix *sn*- is used before the name of the compound. Glycerophospholipids usually contain an *sn*-glycerol 3-phosphate moiety. Although each contains the glycerol moiety as a fundamental structural element, the neutral triacylglycerols and the charged, ionic phospholipids have very different physical properties and functions.

Phospholipids Contain 1,2-Diacylglycerol and a Base Connected by a Phosphodiester Bridge

The phospholipids are polar, ionic lipids composed of 1,2-diacylglycerol and a phosphodiester bridge that links the glycerol backbone to some base, usually a nitrogenous one, such as choline, serine, or ethanolamine (Figures 10.2 and 10.3). The most abundant phospholipids in human tissues are **phosphatidylcholine** (also called lecithin), **phosphatidylethanolamine,** and **phosphatidylserine** (Figure 10.4). Note that C-2 of the phospholipids represents an asymmetric center (Figure 10.3). At physiologic pH, phosphatidylcholine and phosphatidylethanolamine have no net charge and exist as dipolar zwitterions, whereas phosphatidylserine has a net charge of -1, causing it to be an acidic phospholipid. Phosphatidylethanolamine (PE) is related to phosphatidylcholine in that trimethylation of PE produces lecithin. Most phospholipids contain more than one kind of fatty acid per molecule, so that a given class of phospholipids from any tissue actually represents a family of molecular species. Phosphatidylcholine (PC) contains mostly palmitic acid (16 : 0) or stearic acid (18 : 0) in the *sn*-1 position and primarily the unsaturated C_{18} fatty acids oleic, linoleic, or linolenic in the *sn*-2 position. Phosphatidylethanolamine has the same saturated fatty acids as PC at the *sn*-1 position but contains more of the

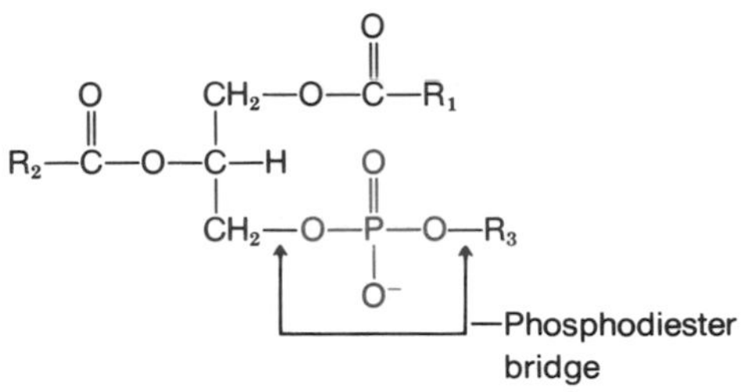

Phosphodiester bridge

FIGURE 10.3
Generalized structure of a phospholipid where R_1 and R_2 represent the aliphatic chains of fatty acids, and R_3 represents a polar head group.

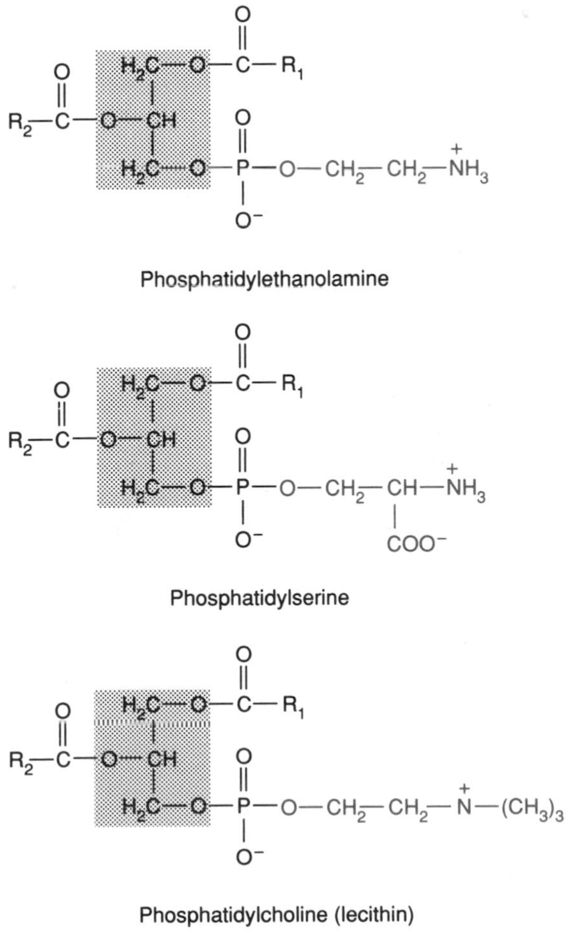

Phosphatidylethanolamine

Phosphatidylserine

Phosphatidylcholine (lecithin)

FIGURE 10.4
Structures of some common phospholipids.

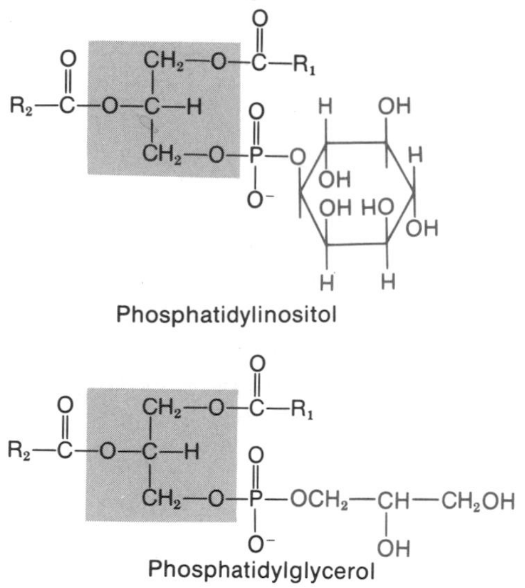

Phosphatidylinositol

Phosphatidylglycerol

FIGURE 10.5
Structures of phosphatidylglycerol and phosphatidylinositol.

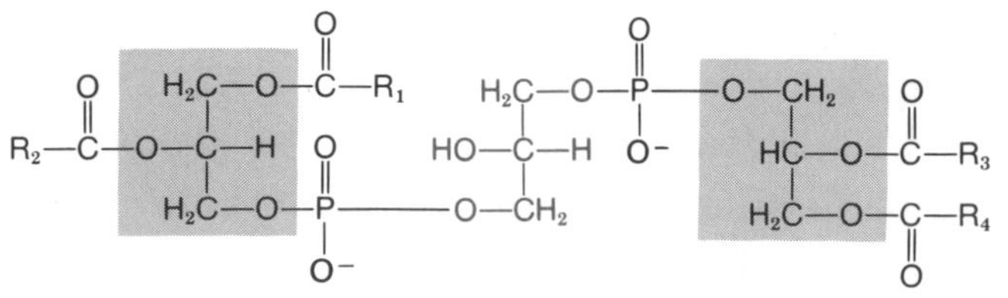

FIGURE 10.6
Structure of cardiolipin.

long-chain polyunsaturated fatty acids—namely, 18 : 2, 20 : 4, and 22 : 6—at the *sn*-2 position.

Phosphatidylinositol is an acidic phospholipid that occurs in mammalian membranes (Figure 10.5). Phosphatidylinositol is rather unusual because it often contains almost exclusively stearic acid in the *sn*-1 position and arachidonic acid (20 : 4) in the *sn*-2 position.

Another phospholipid comprised of a polyol polar head group is **phosphatidylglycerol.** Phosphatidylglycerol (Figure 10.5) occurs in relatively large amounts in mitochondrial membranes and pulmonary surfactant and is a precursor of cardiolipin. Phosphatidylglycerol and phosphatidylinositol both carry a formal charge of −1 at neutral pH and are therefore acidic lipids.

Cardiolipin, a very acidic (charge, −2) phospholipid, is composed of two molecules of phosphatidic acid linked together covalently through a molecule of glycerol. Cardiolipin is found primarily in the inner membrane of mitochondria and in bacterial membranes (Figure 10.6).

The phospholipids that have been discussed so far contain only *O*-acyl residues attached to glycerol. *O*-(1-Alkenyl) substituents occur at C-1 of the *sn*-glycerol moiety of phosphoglycerides in combination with an *O*-acyl residue esterified to the C-2 position; compounds in this class are known as **plasmalogens** (Figure 10.7). Relatively large amounts of ethanolamine plasmalogen (also called plasmenylethanolamine) occur in myelin with lesser amounts in heart muscle where choline plasmalogen is abundant.

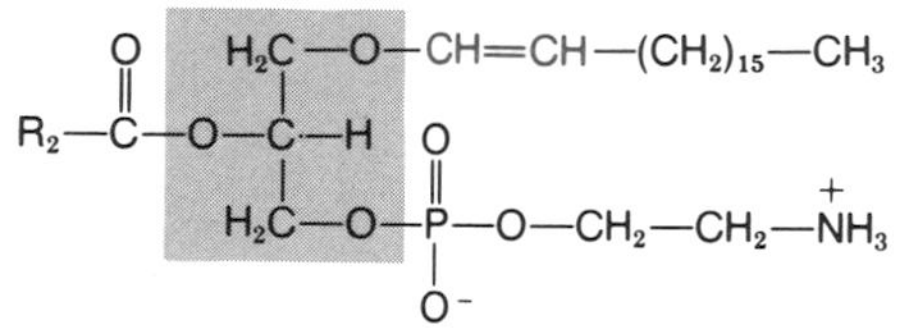

FIGURE 10.7
Structure of ethanolamine plasmalogen.

An unusual plasmalogen called **"platelet activating factor"** (PAF) (Figure 10.8) is a major mediator of hypersensitivity, acute inflammatory reactions and anaphylactic shock. In hypersensitive individuals, cells of the polymorphonuclear (PMN) leukocyte family (basophils, neutrophils, and eosinophils) macrophages, and monocytes are coated with IgE molecules that are specific for a particular antigen (e.g., ragweed pollen and bee venom). Subsequent reexposure to the antigen and formation of antigen–IgE complexes on the surface of the aforementioned inflammatory cells provokes the synthesis and release of PAF. Platelet activating factor is a choline plasmalogen containing an acetyl residue instead of a long-chain fatty acid in position 2 of the glycerol moiety. Platelet activating

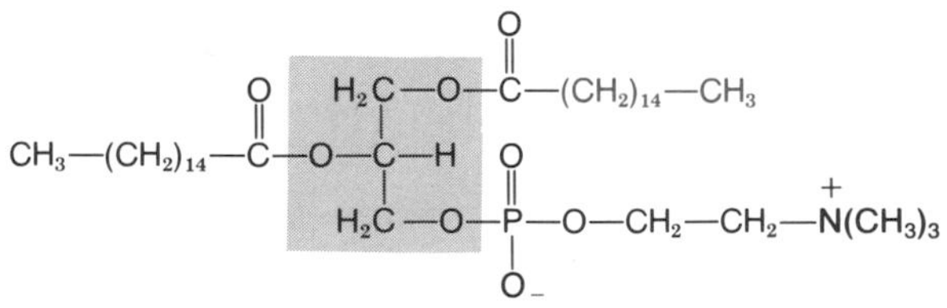

FIGURE 10.8
Structure of platelet activating factor (PAF).

factor is not stored; it is synthesized and released when PMNs are stimulated. Platelet aggregation, cardiovascular and pulmonary changes, edema, hypotension, and PMN cell chemotaxis are affected by PAF.

Phospholipids in Membranes Serve a Variety of Roles

Although present in body fluids such as plasma and bile, the phospholipids are found in highest concentration in the various cellular membranes where they perform many different functions, such as to serve as structural components of membranes. Nearly one half of the mass of the erythrocyte membrane is comprised of various phospholipids (see Chapter 5).

Phospholipids also play a role in activating certain enzymes. β-Hydroxybutyrate dehydrogenase, an enzyme imbedded in the inner membrane of mitochondria (see p. 415), has an absolute requirement for phosphatidylcholine; phosphatidylserine and phosphatidylethanolamine cannot substitute for phosphatidylcholine in activating the enzyme.

Dipalmitoyllecithin Is Necessary for Normal Lung Function

Normal lung function depends on a constant supply of an unusual phospholipid called **dipalmitoyllecithin** in which the lecithin molecule contains palmitic acid (16:0) residues in both the *sn*-1 and *sn*-2 positions. More than 80% of the phospholipid in the extracellular liquid layer that lines alveoli of normal lungs is contributed by dipalmitoyllecithin. This particular phospholipid—called surfactant–is produced by type II epithelial cells and prevents atelectasis at the end of the expiration phase of breathing (Figure 10.9). This lipid decreases the surface tension of the aqueous surface layer of the lung. Lecithin molecules that do not contain two residues of palmitic acid are not effective in lowering the surface tension of the fluid layer lining alveoli. Surfactant also contains phosphatidylglycerol, phosphatidylinositol, and 18- and 36-kDa proteins (designated surfactant protein), which contribute significantly to the surface tension lowering property of pulmonary surfactant.

During the third trimester—before the 28th week of gestation—the fetal lung synthesizes primarily sphingomyelin. Normally, at this time, glycogen that has been stored in epithelial type II cells is converted to fatty acids and then to dipalmitoyllecithin. During lung maturation there is a good correlation between the increase in lamellar inclusion bodies that represent the intracellular pulmonary surfactant (phosphatidylcholine) storage organelles, called lamellar bodies, and the simultaneous decrease in glycogen content of type II pneumocytes. At the 24th week of gestation the type II granular pneumocytes appear in the alveolar epithelium, and within a few days they produce their typical osmiophilic lamellar inclusion bodies. The number of type II cells increases until the 32nd week at which time surface active agent appears in the lung and amniotic fluid.

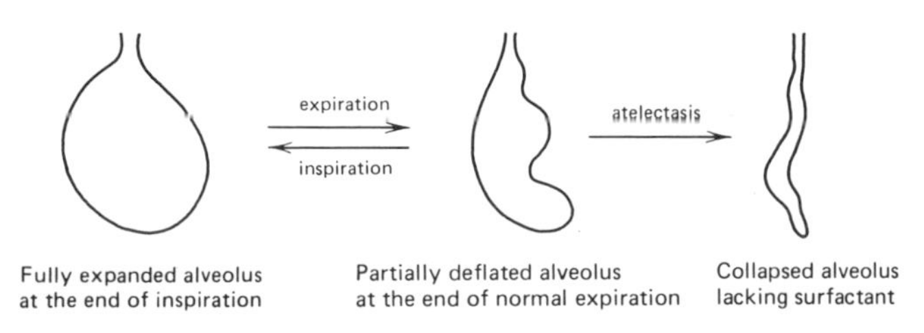

FIGURE 10.9
Role of surfactant in preventing atelectasis.

Surface tension decreases when the inclusion bodies increase in the type II cells. In the few weeks before term one can perform screening tests on amniotic fluid to detect newborns that are at risk for respiratory distress syndrome (RDS) (Clin. Corr. 10.1). These tests are useful in timing elective deliveries, in applying vigorous preventive therapy to the newborn infant, and to determine if the mother should be treated with a glucocorticoid drug to accelerate maturation of the fetal lung. Dexamethasone therapy has also been used in neonates with chronic lung disease (bronchopulmonary dysplasia); however, while such corticosteroid therapy may be effective in some cases in improving lung function, in others it causes periventricular abnormalities in the brain.

Respiratory failure due to an insufficiency in surfactant can also occur in adults whose type II cells or surfactant-producing pneumocytes have been destroyed as an adverse side effect of the use of immunosuppressive medications or chemotherapeutic drugs.

The detergent properties of the phospholipids, especially phosphatidylcholine, play an important role in bile where they function to solubilize cholesterol. An impairment in phospholipid production and secretion into bile can result in the formation of cholesterol and bile pigment gallstones.

Phosphatidylinositol and phosphatidylcholine also serve as donors of arachidonic acid for the synthesis of prostaglandins, thromboxanes, leukotrienes and related compounds.

Inositides Play a Role in Signal Transduction

Inositol-containing phospholipids (inositides) play a central role in signal transduction systems; the most important of these lipids is **phosphatidylinositol 4,5-bisphosphate** (PIP_2) (Figure 10.10). When certain ligands bind to their respective receptors on the plasma membrane of mammalian cells (see Chapter 20), PIP_2 localized to the inner leaflet of the membrane becomes a substrate for a receptor-dependent phosphoinositidase C (PIC), which hydrolyzes it into two intracellular signals (Figure 10.11): **Inositol 1,4,5-trisphosphate** (IP_3), which triggers the release of Ca^{2+} from special vesicles of the endoplasmic reticulum and 1,2-diacylglycerol, which stimulates the activity of protein kinase C. The regulatory functions of these products of the PIC reaction are discussed in Chapter 20.

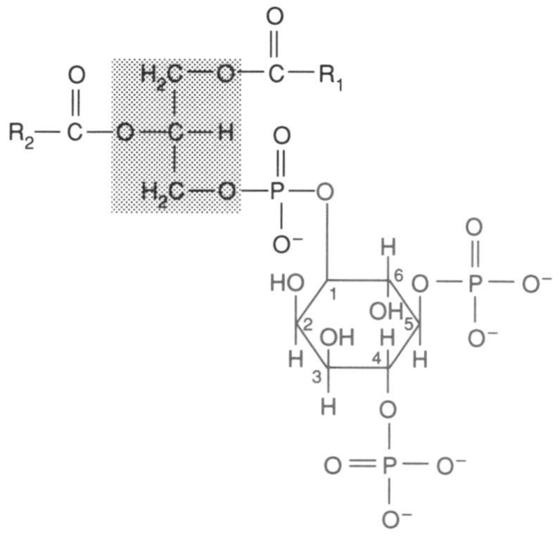

FIGURE 10.10
Structure of phosphatidylinositol 4,5-bisphosphate (PIP_2 or PtdIns $(4,5)P_2$).

CLIN. CORR. 10.1
RESPIRATORY DISTRESS SYNDROME

The respiratory distress syndrome (RDS) is a major cause of neonatal morbidity and mortality in many countries. It accounts for approximately 15–20% of all neonatal deaths in Western countries and somewhat less in the developing countries. The disease affects only premature babies and its incidence varies directly with the degree of prematurity. Premature babies develop RDS because of immaturity of their lungs, resulting from a deficiency of pulmonary surfactant. The maturity of the fetal lung can be predicted antenatally by measuring the lecithin/sphingomyelin (L/S) ratio in the amniotic fluid. The mean L/S ratio in normal pregnancies increases gradually with gestation until about 31 or 32 weeks when the slope rises sharply. The ratio of 2.0 that is characteristic of the term infant at birth is achieved at the gestational age of about 34 weeks. In terms of predicting pulmonary maturity, the critical L/S ratio is 2.0 or greater. The risk of developing RDS when the L/S ratio is less than 2.0 has been worked out: for an L/S ratio of 1.5–1.9, the risk is approximately 40%, and for a ratio less than 1.5 the calculated risk of developing RDS is about 75%.

Although the L/S ratio in amniotic fluid is still widely used to predict the risk of RDS, the results are unreliable if the amniotic fluid specimen has been contaminated by blood or meconium obtained during a complicated pregnancy.

In recent years the determination of saturated palmitoylphosphatidylcholine (SPC), phosphatidylglycerol, and phosphatidylinositol have been found to be additional predictors of the risk of RDS. Exogenous surfactant replacement therapy using surfactant from human and animal lungs is effective in the prevention and treatment of RDS.

Merritt, T. A. et al., Prophylactic treatment of very premature infants with human surfactant. *N. Engl. J. Med.* 315:785, 1986; and Simon, N. V., Williams, G. H., Fairbrother, P. F., Elser, R. C., and Perkins, R. P. Prediction of fetal lung maturity by amniotic fluid fluorescence polarization, L/S ratio, and phosphatidylglycerol. *Obstet. Gynecol.* 57:295, 1981.

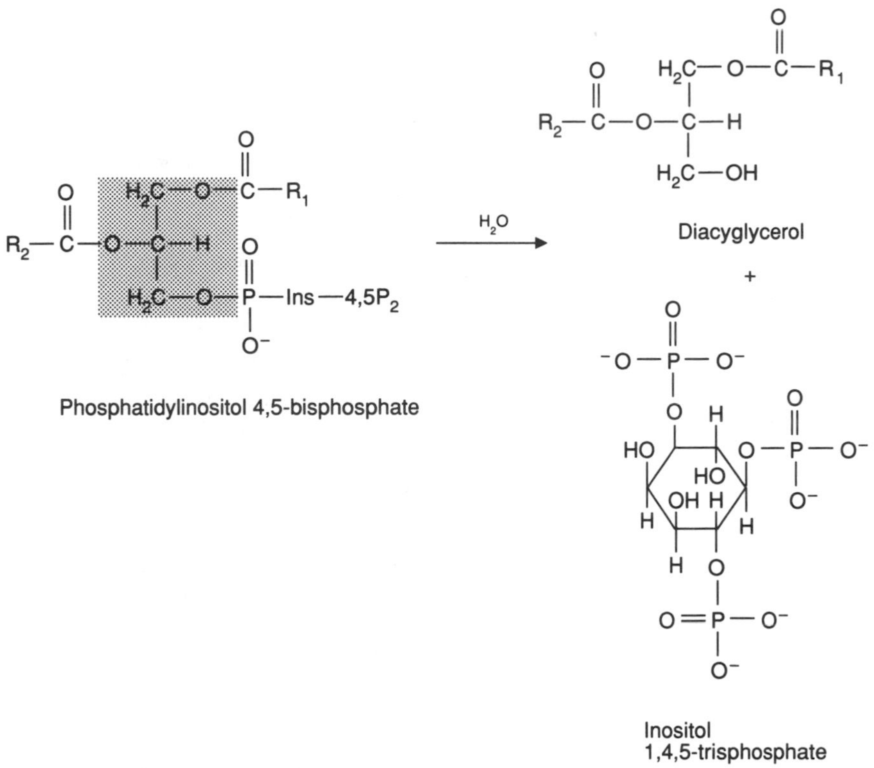

FIGURE 10.11
The generation of 1,2-diacylglycerol and inositol 1,4,5-trisphosphate by the action of phosphoinositidase C on phosphatidylinositol 4,5-bisphosphate.

The complex pathways of inositol phosphate metabolism serve three roles: (1) removal and inactivation of the potent intracellular signal, IP_3, (2) conservation of inositol, and (3) synthesis of polyphosphates such as inositol pentakisphosphate ($InsP_5$) and inositol hexakisphosphate ($InsP_6$) whose functions have not been determined.

Inositol 1,4,5-trisphosphate is metabolized by two enzymes: first a 5-phosphomonoesterase that converts IP_3 to inositol 1,4-bisphosphate [$Ins(1,4)P_2$] and second a 3-kinase that catalyzes the formation of inositol 1,3,4,5-tetraphosphate. A family of phosphatases in turn convert $Ins(1,4)P_2$ to myo-inositol (Figure 10.12). Inositol is eventually reincorporated into the phospholipid pool.

Phosphatidylinositol Serves to Anchor Glycoproteins to the Plasma Membrane

In addition to its role as a structural component of membranes and as a reservoir of arachidonic acid for prostaglandin and leukotriene synthesis (p. 461), phosphatidylinositol also serves as an anchor to tether certain glycoproteins to the external surface of the plasma membrane of cells. In trypanosome parasites (e.g., *Trypanosoma brucei,* which causes sleeping sickness), the external surface of the plasma membrane is coated with a protein called variable surface glycoprotein (VSG) that is linked to the membrane through a glycophospholipid anchor, specifically phosphatidylinositol (Figure 10.13). The salient structural features of the protein–lipid linkage region are (a) the diacylglyceride (DAG) moiety of phosphatidylinositol is integrated into the outer leaflet of the lipid bilayer of the plasma membrane; (b) the inositol residue is linked to the DAG moiety through a phosphodiester bond; (c) inositol is bonded to an amino sugar (glucosamine), which contains a free, nonacetylated amino group; (d) the presence of a mannose-rich glycan domain; and (e) a phosphoethanolamine residue linked to the carboxy terminus of the protein. Some other proteins

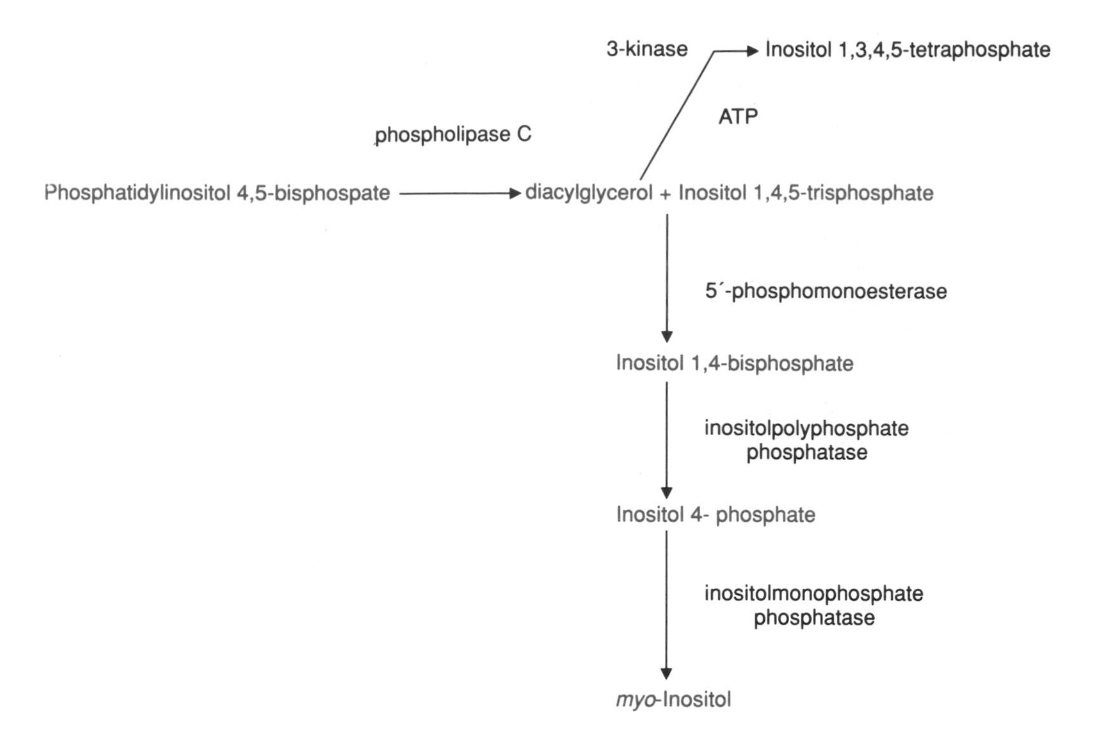

FIGURE 10.12
Pathways for the removal of intracellular inositol 1,4,5-trisphosphate.

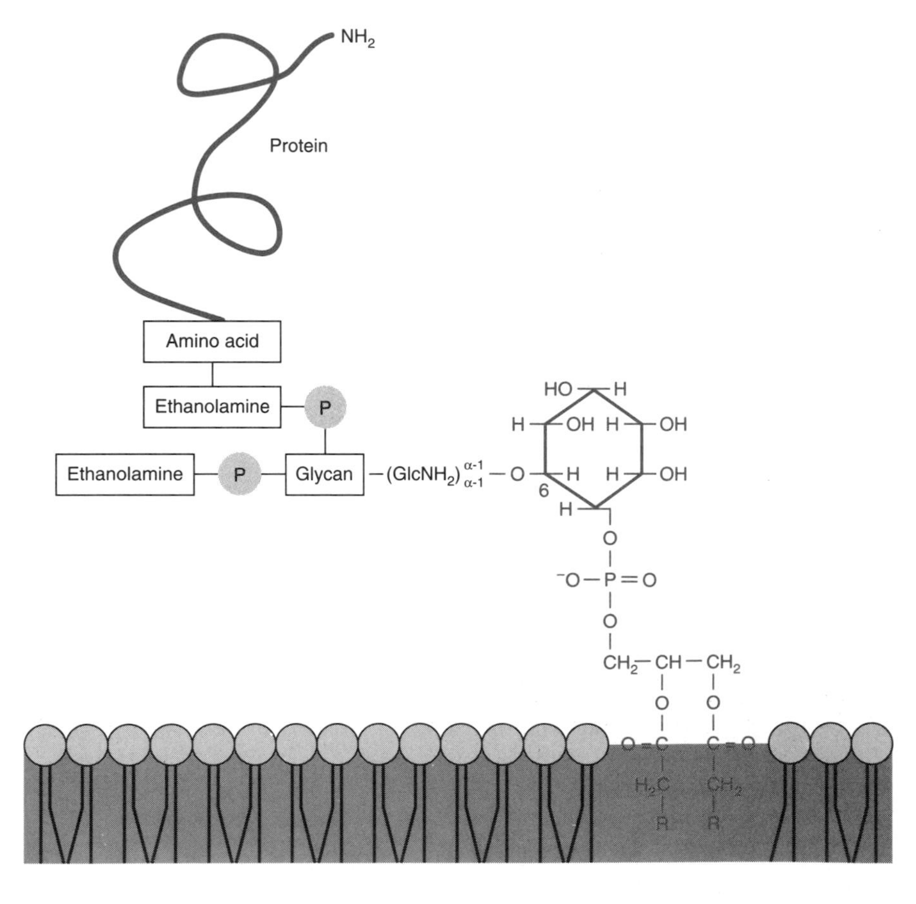

FIGURE 10.13
Structure of a typical phosphatidylinositol membrane protein anchor; $GlcNH_2$ = glucosamine.

that are attached to the external surface of the plasma membrane include acetylcholine esterase, alkaline phosphatase and 5′-nucleotidase.

The glycosyl-phosphatidylinositol (GPI) anchor serves several functions. First, it confers upon the protein to which it is attached an unrestricted and lateral mobility within the lipid bilayer, thereby allowing the protein to move about rapidly on the surface of the plasma membrane. Second, the presence of phospholipase C-type activity on the cell surface permits the shedding of the phosphatidylinositol-anchored protein. As an example, this provides trypanosomes with a means for discarding surface antigens, thus changing their coat and escaping antibodies of the host's immune system. Third, the action of phospholipase C on the phosphatidylinositol anchor releases diacylglyceride, a second messenger that can activate protein kinase C (see p. 875), thereby providing a means for transducing signals across the plasma membrane.

Biosynthesis of Phospholipids

Phosphatidic Acid Is Synthesized From α-Glycerophosphate and Fatty Acyl CoA

L-α-Phosphatidic acid (commonly called phosphatidic acid) and *sn*-1,2-diglyceride (1,2-diacyl-*sn*-glycerol) are common intermediates in the pathways of phospholipid and triacylglycerol biosynthesis (Figure 10.14) and both pathways share some of the same enzymes (see Chapter 9). Essentially all cells are capable of synthesizing phospholipids to some degree

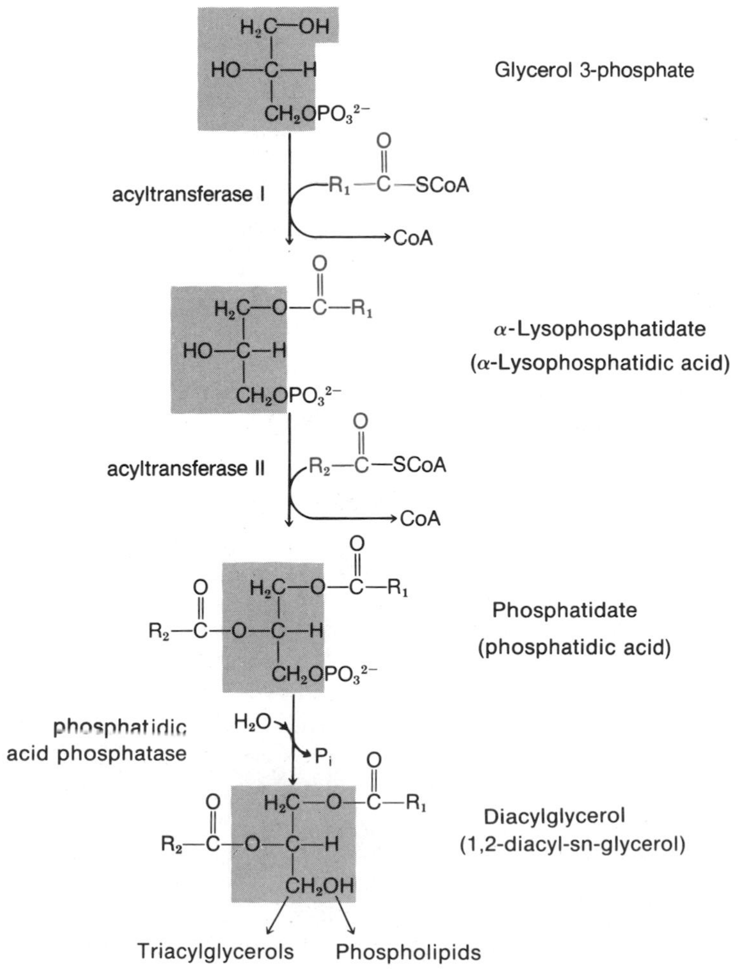

Glycerol 3-phosphate

acyltransferase I

α-Lysophosphatidate
(α-Lysophosphatidic acid)

acyltransferase II

Phosphatidate
(phosphatidic acid)

phosphatidic acid phosphatase

Diacylglycerol
(1,2-diacyl-sn-glycerol)

Triacylglycerols   Phospholipids

FIGURE 10.14
Phosphatidic acid biosynthesis from glycerol 3-phosphate and the role of phosphatidic acid phosphatase in the synthesis of phospholipids and triacylglycerols.

(except mature erythrocytes), whereas triacylglycerol biosynthesis occurs only in the liver, adipose tissue, and the intestine. In most tissues, the pathway for phosphatidic acid synthesis begins with α-glycerophosphate (*sn*-glycerol 3-phosphate), and there are two sources of this triose phosphate. The most general source of α-glycerophosphate, particularly in adipose tissue, is from reduction of the glycolytic intermediate, dihydroxyacetone phosphate, in the reaction catalyzed by α-glycerophosphate dehydrogenase:

$$\text{Dihydroxyacetone phosphate} + \text{NADH} + \text{H}^+ \rightleftharpoons \text{glycerol 3-phosphate} + \text{NAD}^+$$

A few specialized tissues, including the liver, kidney, and intestine, derive α-glycerophosphate by means of the glycerol kinase reaction:

$$\text{Glycerol} + \text{ATP} \xrightarrow{\text{Mg}^{2+}} \text{glycerol 3-phosphate} + \text{ADP}$$

The next two steps in phosphatidic acid biosynthesis involve stepwise transfer of long-chain fatty acyl groups from fatty acyl CoA. The first acyltransferase (I) is called **glycerol phosphate : acyltransferase** and attaches predominantly saturated fatty acids and oleic acid to the *sn*-1 position to produce 1-acylglycerol phosphate or α-lysophosphatidic acid. The second enzyme (II), **1-acylglycerol phosphate : acyltransferase,** catalyzes the acylation of the *sn*-2 position, usually with an unsaturated fatty acid (Figure 10.14). In both cases the donor of acyl groups is the CoA thioester derivative of the long-chain fatty acids.

The specificity of the two acyltransferases does not always match the fatty acid asymmetry that we find in the phospholipids of a particular cell. Remodeling reactions discussed below function to modify the fatty acid composition at the C-1 and C-2 positions of the glycerol phosphate backbone.

Cytoplasmic phosphatidic acid phosphatase (also called phosphatidic acid phosphohydrolase) hydrolyzes phosphatidic acid (1,2-diacylglycerophosphate) that is generated on the endoplasmic reticulum, thereby yielding 1,2-diacyl-*sn*-glycerol that serves as the branch point in triacylglycerol and phospholipid synthesis (Figure 10.14).

Phosphatidic acid can also be formed by a second pathway that begins with DHAP. The DHAP pathway is usually an alternative supportive route used by some tissues to produce phosphatidic acid (see Chapter 9).

Specific Phospholipids Are Synthesized by Addition of a Base to Phosphatidic Acid

The major pathway for the biosynthesis of phosphatidylcholine (lecithin) involves the sequential conversion of choline to phosphocholine, CDP-choline and phosphatidylcholine. In this pathway, the phosphocholine polar head group is activated using CTP, according to the following reactions. Free choline is first phosphorylated by ATP in a reaction catalyzed by choline kinase (Figure 10.15). Choline is a dietary requirement for most mammals. Phosphocholine in turn is converted to CDP-choline at the expense of CTP in the reaction catalyzed by **phosphocholine cytidylyltransferase.** Note that inorganic pyrophosphate (PP_i) is a product of this reaction resulting from attack by the phosphoryl residue of phosphocholine on the internal α-phosphorus atom of CTP. The high-energy pyrophosphoryl bond in CDP-choline is very unstable and reactive such that the phosphocholine moiety can be transferred readily to the nucleophilic center provided by the OH group at position 3 of 1,2-diacylglycerol

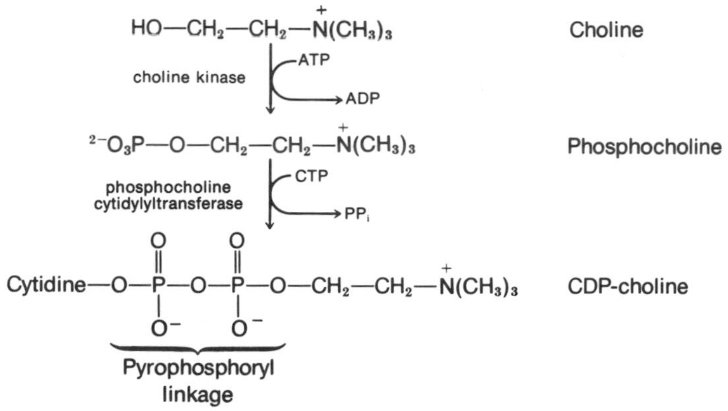

$$HO{-}CH_2{-}CH_2{-}\overset{+}{N}(CH_3)_3$$ Choline

choline kinase (ATP → ADP)

$$^{2-}O_3P{-}O{-}CH_2{-}CH_2{-}\overset{+}{N}(CH_3)_3$$ Phosphocholine

phosphocholine cytidylyltransferase (CTP → PP_i)

Cytidine—O—P(=O)(O⁻)—O—P(=O)(O⁻)—O—CH_2—CH_2—$\overset{+}{N}(CH_3)_3$ CDP-choline

Pyrophosphoryl linkage

FIGURE 10.15
The biosynthesis of CDP-choline from choline.

in the reaction catalyzed by choline phosphotransferase shown in Figure 10.16. This is the principal pathway for the synthesis of dipalmitoyllecithin in the lung.

The rate-limiting step for phosphatidylcholine biosynthesis is the cytidylyltransferase reaction that forms CDP-choline (Figure 10.15). This enzyme is regulated by a novel mechanism that involves exchange of the enzyme between the cytosol and the endoplasmic reticulum. The cytosolic form of cytidylyltransferase is inactive and appears to function as a reservoir of enzyme; binding of the enzyme to the membrane results in activation. Translocation of cytidylyltransferase from the cytosol to the endoplasmic reticulum is regulated by cAMP and fatty acids. Reversible phosphorylation in which a cAMP-dependent kinase phosphorylates the enzyme causes it to be released from the membrane, rendering it inactive. Subsequent dephosphorylation will cause the cytidylyltransferase to rebind to the membrane and become active. Fatty acyl CoAs activate the enzyme by promoting its binding to the endoplasmic reticulum.

In liver only, phosphatidylcholine can also be formed by repeated methylation of the phospholipid phosphatidylethanolamine. **Phosphatidylethanolamine *N*-methyltransferase,** an enzyme of the endoplasmic reticulum, catalyzes the transfer of methyl groups—one at a time—from ***S*-adenosylmethionine (AdoMet)** to phosphatidylethanolamine to produce phosphatidylcholine (Figure 10.17). It is not known if one or more enzymes are involved in the conversion of phosphatidylethanolamine to phosphatidylcholine.

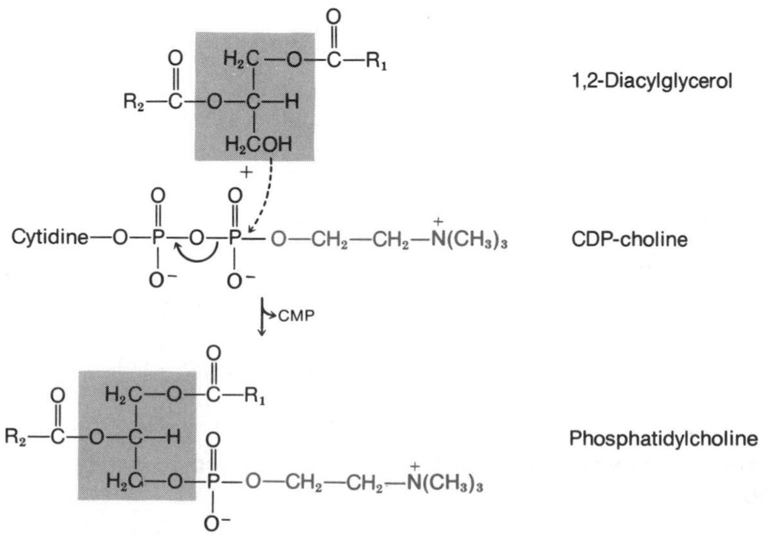

H_2C—O—C(=O)—R_1; R_2—C(=O)—O—C—H; H_2COH 1,2-Diacylglycerol

\+

Cytidine—O—P(=O)(O⁻)—O—P(=O)(O⁻)—O—CH_2—CH_2—$\overset{+}{N}(CH_3)_3$ CDP-choline

→ CMP

H_2C—O—C(=O)—R_1; R_2—C(=O)—O—C—H; H_2C—O—P(=O)(O⁻)—O—CH_2—CH_2—$\overset{+}{N}(CH_3)_3$ Phosphatidylcholine

FIGURE 10.16
The choline phosphotransferase reaction.

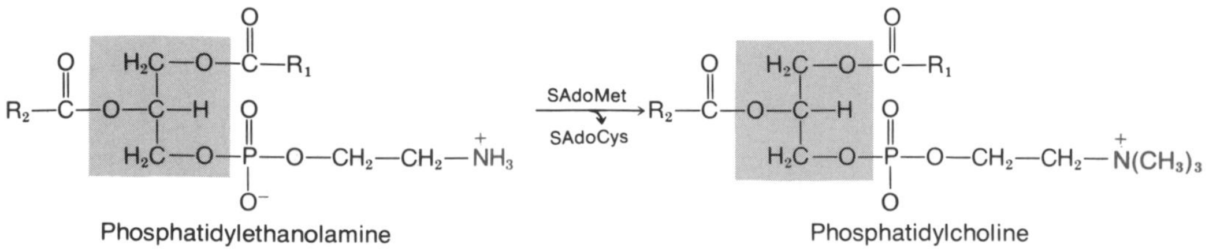

FIGURE 10.17
Biosynthesis of phosphatidylcholine from phosphatidylethanolamine and *S*-adenosylmethionine (AdoMet); *S*-adenosylhomocysteine (AdoCys).

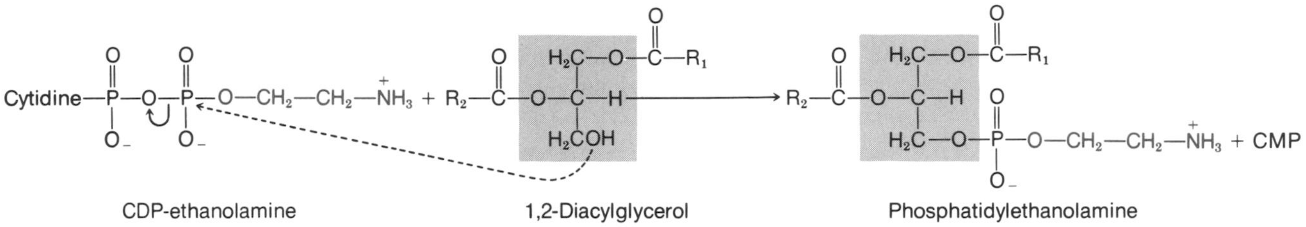

FIGURE 10.18
Biosynthesis of phosphatidylethanolamine from CDP-ethanolamine and diacylglycerol; the reaction is catalyzed by ethanolamine phosphotransferase.

The primary pathway for phosphatidylethanolamine synthesis in liver and brain involves **ethanolamine phosphotransferase** of the endoplasmic reticulum that catalyzes the reaction shown in Figure 10.18. This enzyme is particularly abundant in liver. CDP-ethanolamine is formed through the reaction catalyzed by **ethanolamine kinase:**

$$\text{Ethanolamine} + \text{ATP} \xrightarrow{Mg^{2+}} \text{phosphoethanolamine} + \text{ADP}$$

and the **phosphoethanolamine cytidylyltransferase** reaction:

$$\text{Phosphoethanolamine} + \text{CTP} \xrightarrow{Mg^{2+}} \text{CDP-ethanolamine} + \text{PP}_i$$

Liver mitochondria can also generate phosphatidylethanolamine by decarboxylation of phosphatidylserine; however, this is thought to represent only a minor pathway in phosphatidylethanolamine synthesis (Figure 10.19).

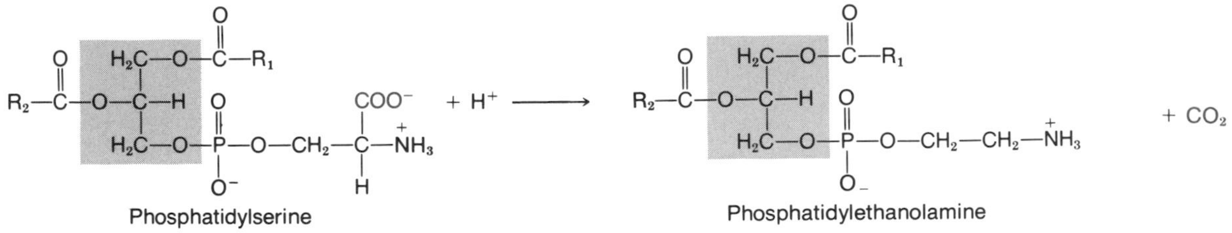

FIGURE 10.19
Formation of phosphatidylethanolamine by the decarboxylation of phosphatidylserine.

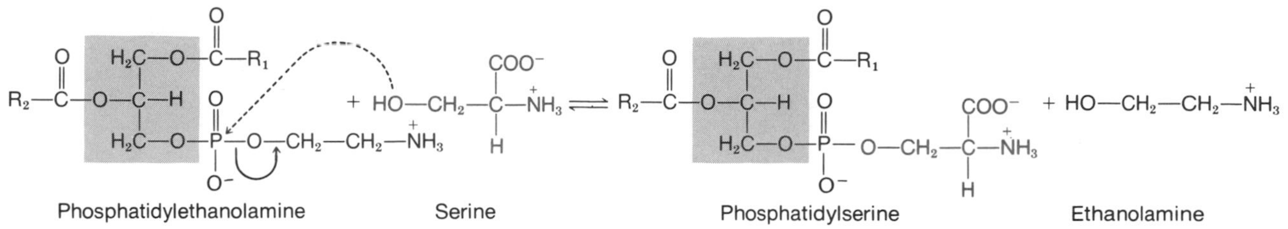

FIGURE 10.20
Biosynthesis of phosphatidylserine from serine and phosphatidylethanolamine by "base exchange."

The major source of phosphatidylserine in mammalian tissues is provided by the "base-exchange" reaction shown in Figure 10.20 in which the polar head group of phosphatidylethanolamine is exchanged for the amino acid serine; since there is no net change in the number or kinds of bonds, this reaction is reversible and has no requirement for ATP or any other high-energy compound. The reaction is initiated by attack on the phosphodiester bond of phosphatidylethanolamine by the hydroxyl group of serine.

Phosphatidylinositol is made via CDP-diacylglycerol and free myoinositol (Figure 10.21) in a reaction catalyzed by **phosphatidylinositol synthase** another enzyme of the endoplasmic reticulum.

The Asymmetric Distribution of Fatty Acids in Phospholipids Is Due to Remodeling Reactions

Two phospholipases, phospholipase A_1 and phospholipase A_2, occur in many tissues, and play a role in the formation of specific phospholipid structures containing the appropriate fatty acids in the *sn*-1 and *sn*-2 positions. Most of the fatty acyl CoA transferases and phospholipid synthesizing enzymes discussed above lack the specificity required to account for the asymmetric position or distribution of fatty acids found in many tissue phospholipids. The fatty acids that are found in the *sn*-1 and *sn*-2 positions of the various phospholipids are often not the same ones that were transferred to the glycerol backbone in the initial acyl transferase reactions of the phospholipid biosynthetic pathways. Phospholipases A_1 and A_2 catalyze the reactions indicated in Figure 10.22 where X represents the polar head group of a phospholipid. The products of the action of phospholipases A_1 and A_2 are called lysophosphatides.

If it becomes necessary for a cell to remove some undesired fatty acid, such as stearic acid from the *sn*-2 position of phosphatidylcholine, and

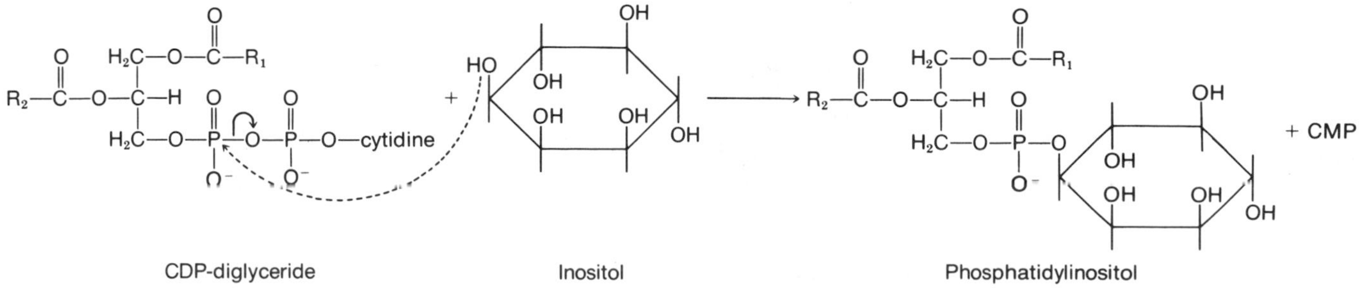

FIGURE 10.21
Biosynthesis of phosphatidylinositol.

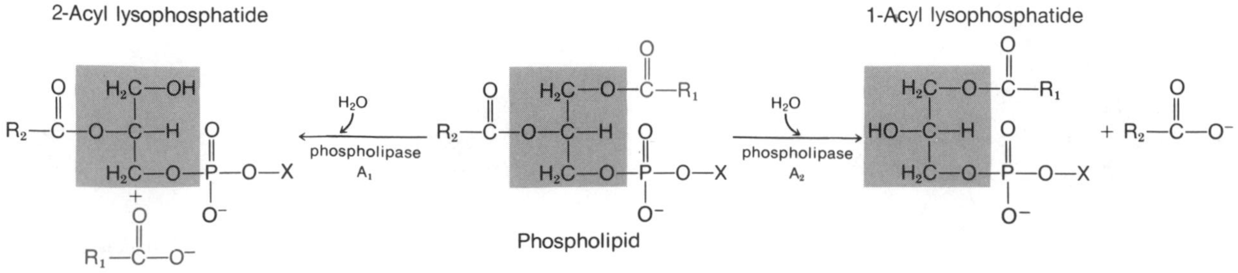

FIGURE 10.22
Reactions catalyzed by phospholipase A_1 and phospholipase A_2.

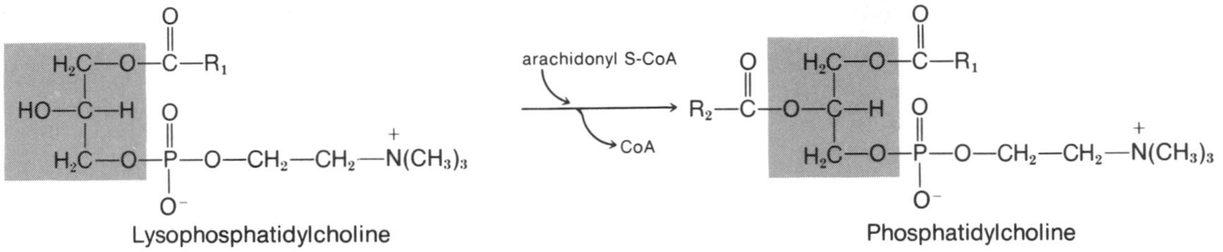

FIGURE 10.23
Synthesis of phosphatidylcholine by reacylation of lysophosphatidylcholine where $R_2{-}\overset{\overset{\displaystyle O}{\|}}{C}{-}O{-}$ represents arachidonic acid. This reaction is catalyzed by acyl-CoA:1-acylglycerol-3-phosphocholine *O*-acyltransferase.

replace it by a more unsaturated one like arachidonic acid, then this can be accomplished by the action of phospholipase A_2 followed by a reacylation step. The insertion of arachidonic acid into the 2 position of *sn*-2-lysophosphatidylcholine can then be accomplished by one of two means; either by direct acylation from arachidonyl CoA involving arachidonic

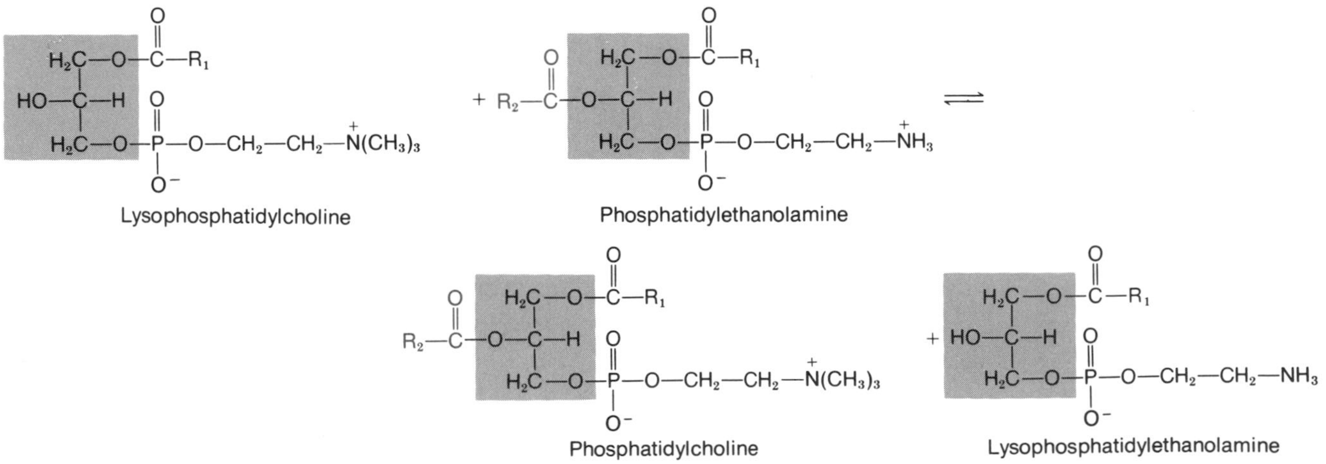

FIGURE 10.24
Formation of phosphatidylcholine by lysolecithin exchange, where $R_2{-}\overset{\overset{\displaystyle O}{\|}}{C}{-}O{-}$ represents arachidonic acid.

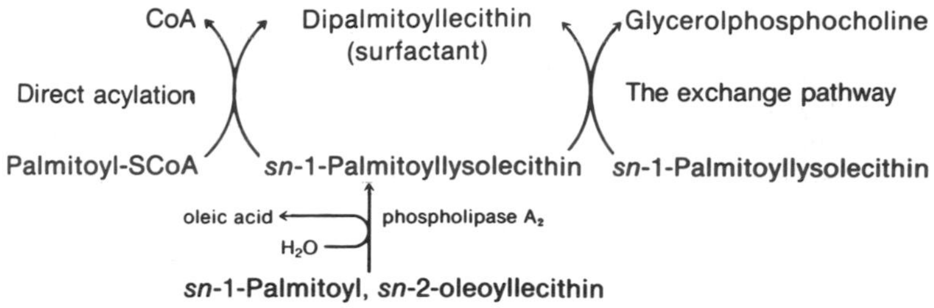

FIGURE 10.25
Two pathways for the biosynthesis of dipalmitoyllecithin from *sn*-1 palmitoyllysolecithin.

acid specific acyl CoA transacylase (Figure 10.23) or from some other arachidonic acid-containing phospholipid by an exchange-type reaction (Figure 10.24) catalyzed by **lysolecithin : lecithin acyltransferase** (LLAT) (Figure 10.24). There is no change in either the number or nature of the bonds involved in products and reactants, thus ATP is not required for this acylation reaction. Reacylation of lysophosphatidylcholine is the major route for the remodeling of phosphatidylcholine.

Lysophospholipids, particularly *sn*-1-lysophosphatidylcholine, can also serve as sources of fatty acid in the remodeling reactions. The remodeling reactions that are involved in the synthesis of dipalmitoyllecithin (surfactant) from 1-palmitoyl-2-oleoylphosphatidylcholine are presented in Figure 10.25. Note that *sn*-1-palmitoyl lysolecithin is the source of palmitic acid in the acyltransferase exchange reaction.

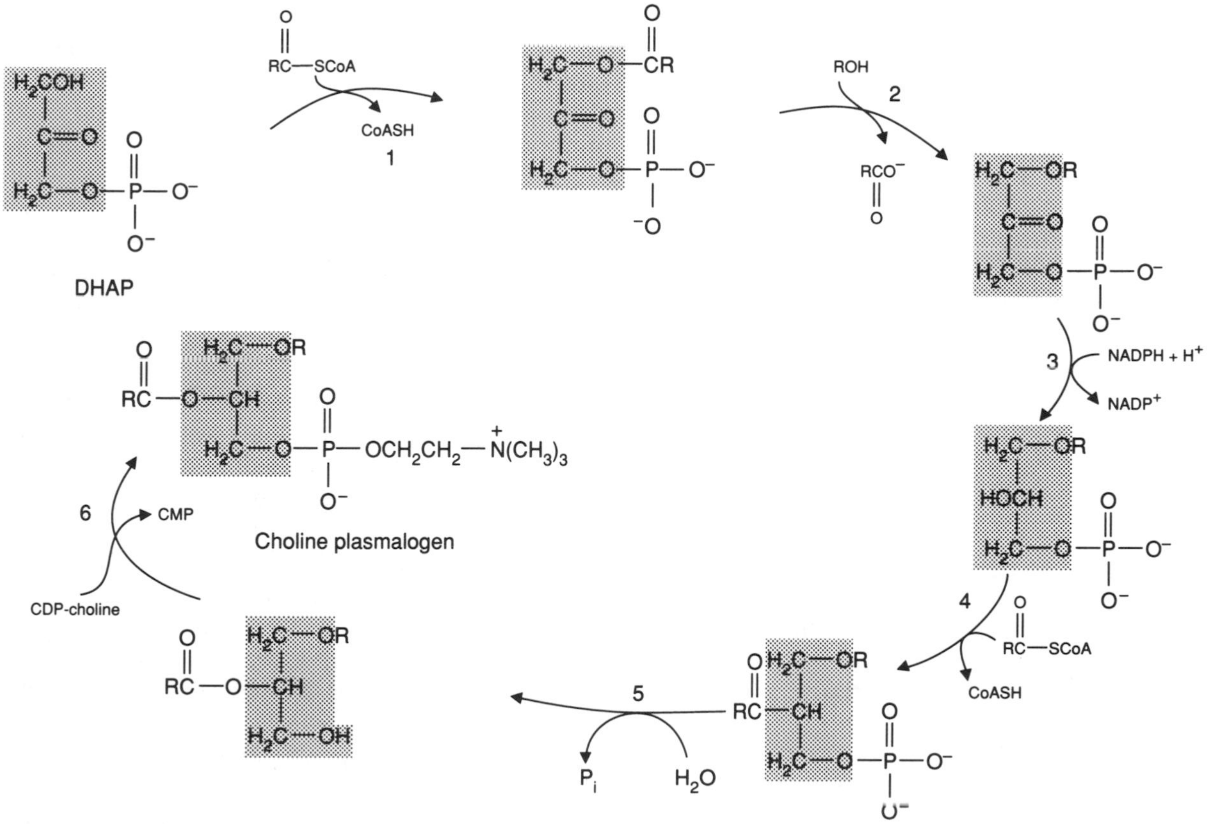

FIGURE 10.26
The pathway of choline plasmalogen biosynthesis from DHAP.
1, acyl CoA: dihydroxyacetone-phosphate acyltransferase; 2, alkyldihydroxyacetone-phosphate synthase; 3 NADPH: alkyldihydroxyacetone-phosphate oxidoreductase; 4, acyl CoA: 1-alkyl-2-lyso-*sn*-glycero-3-phosphate acyltransferase; 5, 1-alkyl-2-acyl-*sn*-glycerol-3-phosphate phosphohydrolase; 6, CDP-choline: 1-alkyl-2-acyl-*sn*-glycerol cholinephosphotransferase.

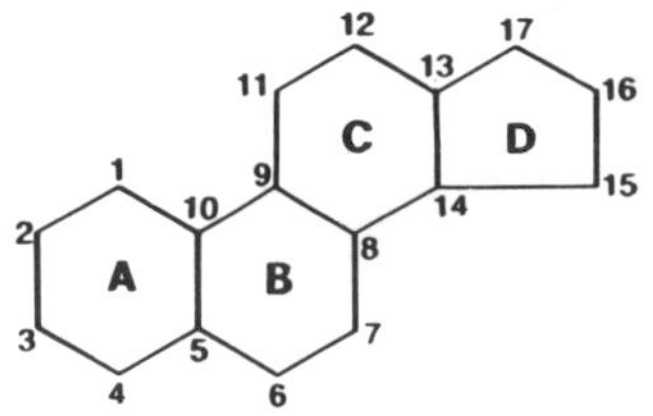

FIGURE 10.27
The cyclopentanophenanthrene ring.

Plasmalogens Are Synthesized From Fatty Alcohols

The ether glycerolipids are synthesized from DHAP, long-chain fatty acids and long-chain fatty alcohols; the reactions involved in their biosynthesis are summarized in Figure 10.26. Acyldihydroxyacetone phosphate is formed by acyl CoA: dihydroxyacetone phosphate acyltransferase (enzyme I) acting on dihydroxyacetone phosphate and long-chain fatty acyl CoA. The ether bond is introduced by alkyldihydroxyacetone phosphate synthase (Figure 10.26, enzyme II), which exchanges the 1-*O*-acyl group of acyldihydroxyacetone phosphate with a long-chain fatty alcohol. The synthase occurs in peroxisomes. Plasmalogen synthesis is completed by the transfer of a long-chain fatty acid from its respective CoA donor to the *sn*-2 position of 1-alkyl-2-lyso-*sn*-glycero-3-phosphate (Figure 10.26, Reaction IV). Patients with Zellweger's disease lack peroxisomes and cannot synthesize adequate amounts of plasmalogen.

10.3 CHOLESTEROL

Cholesterol, an Alicyclic Compound, Is Widely Distributed in Free and Esterified Forms

Cholesterol is an alicyclic compound whose structure includes (a) the perhydrocyclopentanophenanthrene nucleus with its four fused rings, (b) a single hydroxyl group at C-3, (c) an unsaturated center between C-5 and C-6 atoms, (d) an eight-membered branched hydrocarbon chain attached to the D ring at position 17, and (e) a methyl group (designated C-19) attached at position 10 and another methyl group (designated C-18) attached at position 13 (see Figures 10.27 and 10.28).

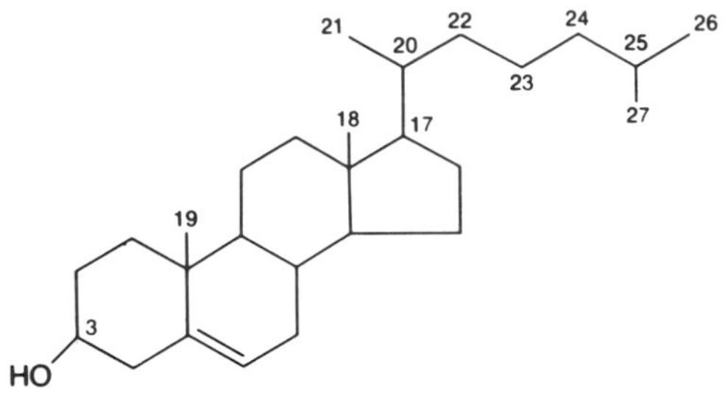

FIGURE 10.28
Structure of cholesterol (5-cholesten-3β-ol).

In terms of physical properties, cholesterol is a lipid with very low solubility in water; at 25°C, the limit of solubility is approximately 0.2 mg/100 mL, or 4.7 μM. The actual concentration of cholesterol in plasma of healthy people is usually 150–200 mg dL^{-1}: on a milligram basis, this value is almost twice the normal concentration of blood glucose. The very high solubility of cholesterol in blood is due to the presence of proteins called plasma lipoproteins (mainly LDL and VLDL) that have the ability to bind and thereby solubilize large amounts of cholesterol (see p. 67).

Actually, only about 30% of the total circulating cholesterol occurs free as such; approximately 70% of the cholesterol in plasma lipoproteins exists in the form of cholesterol esters where some long-chain fatty acid, usually linoleic acid, is attached by an ester bond to the OH group on C-3 of the A ring. The presence of the long-chain fatty acid residue enhances the hydrophobicity of cholesterol (Figure 10.29). Cholesterol is a ubiquitous and essential component of mammalian cell membranes.

Cholesterol is also abundant in bile where the normal concentration is 390 mg dL^{-1}. In contrast to the finding of predominantly cholesterol esters in plasma, only 4% of the cholesterol in bile is esterified to a long-

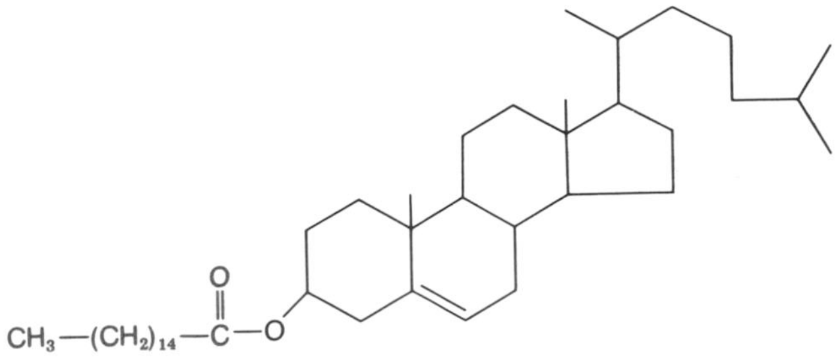

FIGURE 10.29
Structure of cholesterol (palmitoyl-) ester.

chain fatty acid. Bile does not contain appreciable amounts of any of the lipoproteins and the solubilization of free cholesterol is achieved in part by the detergent property of phospholipids present in bile that are produced in the liver. A chronic disturbance in phospholipid metabolism in the liver can result in the deposition of cholesterol-rich gallstones. Bile salts, which are derivatives of cholesterol, also aid in keeping cholesterol in solution in bile. Cholesterol also appears to protect the membranes of the gallbladder from the potentially irritating or harmful effects of bile salts.

In the clinical laboratory total cholesterol is estimated by the Liebermann–Burchard reaction. The proportions of free and esterified cholesterol can be determined by gas–liquid chromatography or reverse phase high-pressure liquid chromatography (HPLC).

Cholesterol Is a Membrane Component and the Precursor of Bile Salts and Steroid Hormones

Cholesterol, which can be derived from the diet or manufactured de novo in virtually all the cells of humans, plays a number of important roles. It is the major sterol in humans and a component of virtually all plasma and intracellular membranes. Cholesterol is especially abundant in the myelinated structures of the brain and central nervous system but is present in small amounts in the inner membrane of the mitochondrion. In contrast to the situation in plasma, most of the cholesterol in cellular membranes occurs in the free, unesterified form.

Cholesterol is the immediate precursor of the **bile acids** that are synthesized in the liver and that function to facilitate the absorption of dietary triacylglycerols and fat-soluble vitamins (Chapter 26). It is important to realize that the ring structure of cholesterol cannot be metabolized to CO_2 and water in humans. The route of excretion of cholesterol is by way of the liver and gallbladder through the intestine in the form of bile acids.

Another physiological role of cholesterol is as the precursor of the various steroid hormones (Chapter 21). Progesterone is the C_{21} keto steroid sex hormone secreted by the corpus luteum of the ovary. The metabolically powerful corticosteroids of the adrenal cortex are derived from cholesterol; these include deoxycorticosterone, corticosterone, cortisol, and cortisone. The mineralocorticoid aldosterone is derived from cholesterol in the zona glomerulosa tissue of the cortex of the adrenal gland. Cholesterol is also the precursor of the female steroid hormones, the estrogens (e.g., estradiol) in the ovary and of the male steroids (e.g., testosterone) in the testes.

Although all of the steroid hormones are structurally related to and biochemically derived from cholesterol, they have widely different physiological properties that relate to spermatogenesis, pregnancy, lactation and parturition, mineral balance, and energy (amino acids, carbohydrate and fat) metabolism.

The hydrocarbon skeleton of cholesterol is also found in the plant sterols, for example ergosterol, a precursor to vitamin D (Figure 10.30). Ergosterol is converted in the skin by ultraviolet irradiation to vitamin D_2. Vitamin D_2 is involved in calcium and phosphorus metabolism (Chapter 28).

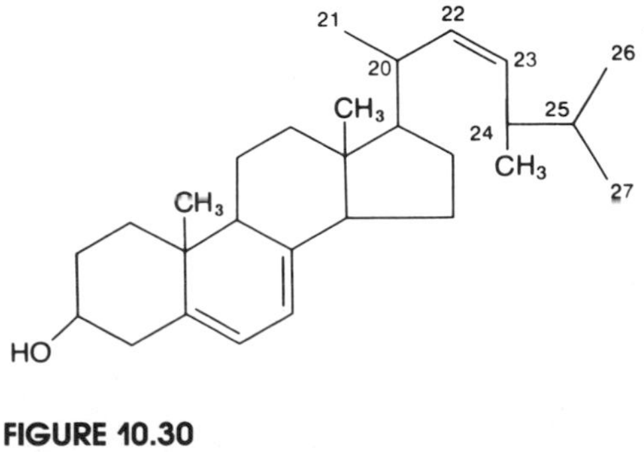

FIGURE 10.30
Structure of ergosterol.

Cholesterol Is Synthesized From Acetyl CoA

Although de novo biosynthesis of cholesterol occurs in virtually all cells, this capacity is greatest in liver, intestine, adrenal cortex, and reproductive tissues, including ovaries, testes, and placenta. From an inspection of its structure it is apparent that cholesterol biosynthesis will require a source of carbon atoms and considerable reducing power to generate the numerous carbon–hydrogen and carbon–carbon bonds. All of the carbon atoms of cholesterol are derived from acetate. Reducing power in the form of NADPH is provided mainly by enzymes of the hexose monophosphate shunt, specifically, glucose 6-phosphate dehydrogenase and 6-phosphogluconate dehydrogenase (p. 361). The pathway of cholesterol synthesis occurs in the cytoplasm and is driven in large part by the hydrolysis of the high-energy thioester bonds of acetyl CoA and the high-energy phosphoanhydride bonds of ATP.

Mevalonic Acid Is a Key Intermediate

The first compound unique to the pathway of cholesterol biosynthesis is mevalonic acid, which is derived from acetyl CoA. Acetyl CoA can be obtained from several sources: (a) the β oxidation of long-chain fatty acids (Chapter 9); (b) the oxidation of ketogenic amino acids such as leucine and isoleucine (Chapter 12); and (c) the pyruvate dehydrogenase reaction. In addition, free acetate can be activated to its thioester derivative at the expense of ATP by the enzyme **acetokinase,** which is also referred to as acetate thiokinase:

$$\text{ATP} + CH_3COO^- + \text{CoASH} \longrightarrow CH_3{-}\overset{\overset{\displaystyle O}{\|}}{C}{-}\text{SCoA} + \text{AMP} + PP_i$$

The first two steps in the pathway of cholesterol synthesis are shared by the pathway that also produces ketone bodies (Chapter 9). Two molecules of acetyl CoA condense to form acetoacetyl CoA in a reaction catalyzed by acetoacetyl CoA thiolase (acetyl CoA : acetyl CoA acetyltransferase):

$$CH_3{-}\overset{\overset{\displaystyle O}{\|}}{C}{-}\text{SCoA} + CH_3{-}\overset{\overset{\displaystyle O}{\|}}{C}{-}\text{SCoA} \longrightarrow CH_3{-}\overset{\overset{\displaystyle O}{\|}}{C}{-}CH_2{-}\overset{\overset{\displaystyle O}{\|}}{C}{-}\text{SCoA} + \text{CoA}$$

Note that the formation of the carbon–carbon bond in acetoacetyl CoA in this reaction is favored energetically by the cleavage of a thioester bond and the generation of free coenzyme A.

The next step introduces a third molecule of acetyl CoA into the cholesterol pathway and forms the branched-chain compound **3-hydroxy-3-methylglutaryl CoA** (HMG CoA) (Figure 10.31). This condensation reaction is catalyzed by **HMG CoA synthase** (3-hydroxy-3-methylglutaryl CoA : acetoacetyl CoA lyase). Liver parenchymal cells contain two isoenzyme forms of HMG CoA synthase; one is found in the cytosol and is involved in cholesterol synthesis, while the other has a mitochondrial location and functions in the pathway that forms ketone bodies (Chapter 9). In the HMG CoA synthase reaction, an aldol condensation occurs between the methyl carbon of acetyl CoA and the β-carbonyl group of acetoacetyl CoA with the simultaneous hydrolysis of the thioester bond of

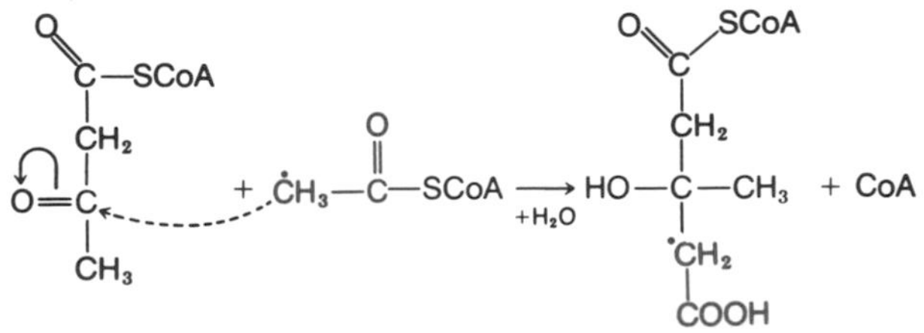

$$\underset{\text{Acetoacetyl CoA}}{O{=}C(CH_3){-}CH_2{-}C({=}O){-}\text{SCoA}} + \underset{\text{Acetyl CoA}}{CH_3{-}\overset{\overset{\displaystyle O}{\|}}{C}{-}\text{SCoA}} \xrightarrow[+H_2O]{} \underset{\text{HMG CoA}}{HO{-}C(CH_3)(CH_2{-}C(O)\text{SCoA})(CH_2{-}COOH)} + \text{CoA}$$

FIGURE 10.31
The HMG CoA synthase reaction.

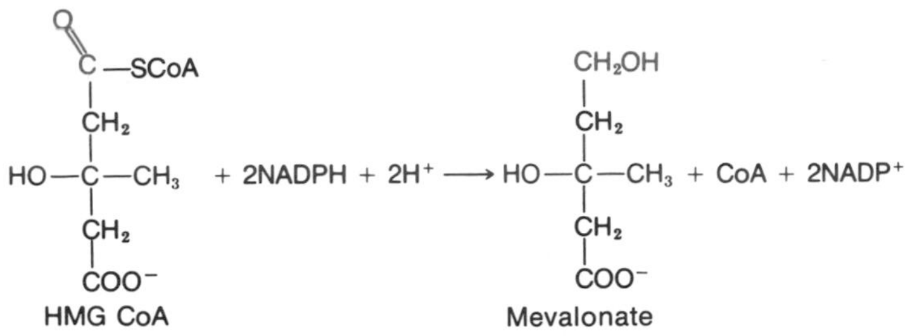

HMG CoA + 2NADPH + $2H^+$ ⟶ Mevalonate + CoA + $2NADP^+$

FIGURE 10.32
The HMG CoA reductase reaction.

acetyl CoA. Note that the thioester bond in the original acetoacetyl CoA substrate molecule remains intact.

HMG CoA can also be formed from the oxidative degradation of the branched-chain amino acid leucine, which proceeds through the intermediates 3-methylcrotonyl CoA and 3-methylglutaconyl CoA (Chapter 12).

The step that produces the unique compound mevalonic acid from HMG CoA is catalyzed by the important microsomal enzyme **HMG CoA reductase** (mevalonate : $NADP^+$ oxidoreductase) that has an absolute requirement for NADPH as the reductant (Figure 10.32). Note that this reductive step (a) consumes two molecules of NADPH from the pentose phosphate pathway, (b) results in the hydrolysis of the thioester bond of HMG CoA, and (c) generates a primary alcohol residue in mevalonate. This reduction reaction is irreversible and produces (*R*)-(+) mevalonate, which contains six carbon atoms. HMG CoA reductase catalyzes the rate-limiting reaction in the pathway of cholesterol biosynthesis. HMG CoA reductase is an intrinsic membrane protein of the endoplasmic reticulum whose carboxyl terminus extends into the cytoplasm and carries the enzyme's active site. Phosphorylation regulates the HMG CoA reductase activity of the cell in two ways: (1) diminishes its catalytic activity (V_{max}), and (2) enhances the rate of its degradation by increasing its susceptibility to proteolytic attack. Increased amounts of intracellular cholesterol stimulate phosphorylation of HMG CoA reductase.

Mevalonic Acid Is a Precursor of Farnesyl Pyrophosphate

The various reactions involved in the conversion of mevalonate to farnesyl pyrophosphate are described below and summarized in Figure 10.33.

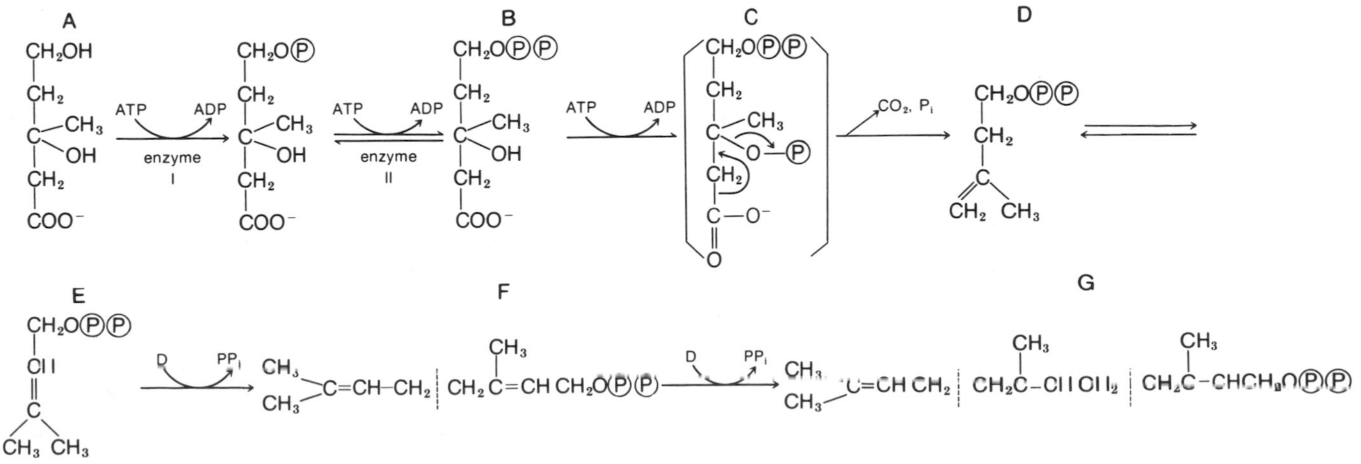

FIGURE 10.33
Formation of farnesyl-PP (F) from mevalonate (A).
The dotted lines divide the molecules into isoprenoid-derived units. D is 3-isopentenyl pyrophosphate.

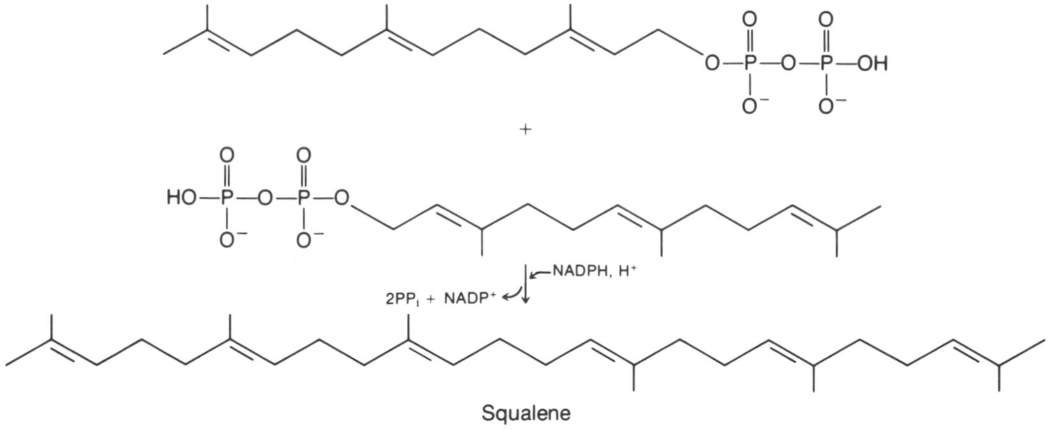

O–P–O–P–OH
+
HO–P–O–P–O
NADPH, H⁺
2PPᵢ + NADP⁺
Squalene

FIGURE 10.34
Formation of squalene from two molecules of farnesyl pyrophosphate.

The stepwise transfer of the terminal γ-phosphate group from two molecules of ATP to mevalonate **(A)** to form 5-pyrophosphomevalonate **(B)** are catalyzed by mevalonate kinase (enzyme I) and phosphomevalonate kinase (enzyme II). The next step affccts the decarboxylation of 5-pyrophosphomevalonate and generates Δ^3-isopentenyl pyrophosphate **(D)**; this reaction is catalyzed by pyrophosphomevalonate decarboxylase. In this ATP-dependent reaction in which ADP, P_i, and CO_2 are produced it is thought that decarboxylation–dehydration proceeds by way of the triphosphate intermediate, 3-phosphomevalonate 5-pyrophosphate **(C)**. Next, isopentenyl pyrophosphate is converted to its allylic isomer 3,3-dimethylallyl pyrophosphate **(E)** in a reversible reaction catalyzed by isopentenyl pyrophosphate isomerase. The condensation of 3,3-dimethylallyl pyrophosphate **(E)** and 3-isopentenyl pyrophosphate **(D)** generates geranyl pyrophosphate **(F)**.

The stepwise condensation of three C_5 isopentenyl units to form the C_{15} unit farnesyl pyrophosphate **(G)** is catalyzed by one enzyme, a cytoplasmic prenyl transferase called geranyl transferase.

Cholesterol Is Formed from Farnesyl Pyrophosphate via Squalene

The last steps in cholesterol biosynthesis involve the "head-to-head" fusion of two molecules of farnesyl pyrophosphate to form **squalene** and finally the cyclization of squalene to yield cholesterol. The reaction that produces the C_{30} molecule of squalene from two C_{15} farnesyl pyrophosphate moieties (Figure 10.34) is unlike the previous carbon–carbon bond-forming reactions in the pathway (Figure 10.33).

In this reaction catalyzed by the enzyme **squalene synthase** present in the endoplasmic reticulum, two pyrophosphate groups are released, with loss of a hydrogen atom from one molecule of farnesyl pyrophosphate and replacement by a hydrogen from NADPH. Several different intermediates probably occur between farnesyl pyrophosphate and squalene. By rotation about carbon–carbon single bonds, the conformation of squalene indicated in Figure 10.35 can be obtained. Note the similarity of the overall shape of the compound to cholesterol. Observe also that squalene is devoid of oxygen atoms.

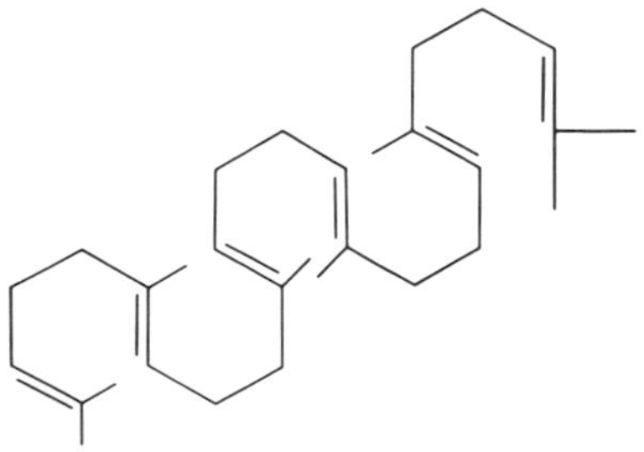

FIGURE 10.35
Structure of squalene, C_{30}.

Cholesterol biosynthesis from squalene proceeds through the intermediate **lanosterol,** which contains the fused tetracyclic ring system and a C_8 side chain:

$$\text{Squalene} \longrightarrow \text{squalene 2,3-epoxide} \longrightarrow \text{lanosterol}$$

The many carbon–carbon bonds that are formed during the cyclization of squalene are generated in a concerted fashion as indicated in Figure 10.36. Note that the OH group of lanosterol projects above the plane of the A ring; this is referred to as the β orientation. Groups that extend down below the ring in a trans relationship to the OH group are designated as α (alpha) by a dotted line. During this reaction sequence an OH group is added to C-3, two methyl groups undergo shifts, and a proton is eliminated. The oxygen atom is derived from molecular oxygen. The reaction is catalyzed by a microsomal enzyme system called **squalene oxidocyclase** that appears to be composed of at least two enzymes, squalene epoxidase or monoxygenase and a cyclase (lanosterol cyclase).

The cyclization process is initiated by epoxide formation between what will become C-2 and C-3 of cholesterol, the epoxide being formed at the expense of NADPH:

$$\text{Squalene} + O_2 + \text{NADPH} + H^+ \longrightarrow \text{squalene 2,3-epoxide} + H_2O + \text{NADP}^+$$

This reaction is catalyzed by the monoxygenase or epoxidase component of squalene oxidocyclase. Hydroxylation at C-3 by way of the epoxide intermediate triggers the cyclization of squalene to form lanosterol as shown in Figure 10.36. In the process of cyclization, two hydrogen atoms and two methyl groups migrate to neighboring positions.

The transformation of lanosterol to cholesterol involves many poorly understood steps and a number of different enzymes. These steps include (a) removal of the methyl group at C-14, (b) removal of the two methyl groups at C-4, (c) migration of the double bond from C-8 to C-5, and (d) reduction of the double bond between C-24 and C-25 in the side chain (see Figure 10.37).

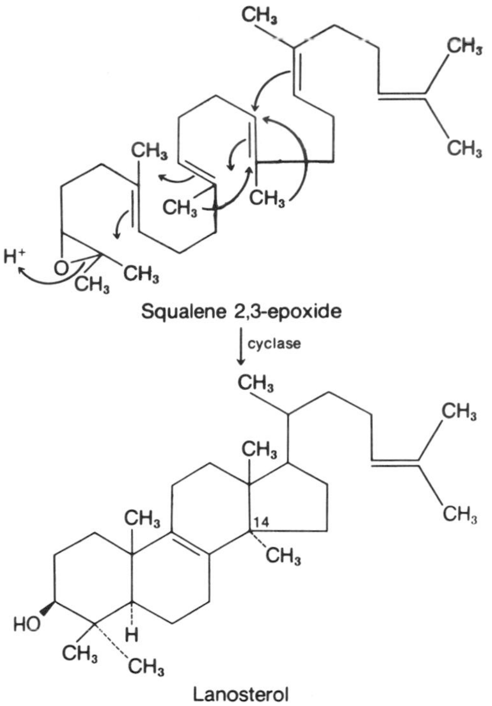

FIGURE 10.36
Conversion of squalene 2,3-epoxide to lanosterol.

Cholesterol Biosynthesis Is Carefully Regulated

The cholesterol pool of the body is derived from two sources: absorption of dietary cholesterol and biosynthesis de novo, primarily in the liver and the intestine. When the amount of dietary cholesterol is reduced, cholesterol synthesis is increased in the liver and intestine to satisfy the needs of other tissues and organs. Cholesterol synthesized de novo is transported from the liver and intestine to peripheral tissues in the form of lipoproteins. These two tissues are the only ones that can manufacture **apolipoprotein B,** the protein component of the cholesterol transport proteins LDL and VLDL. Most of the apolipoprotein B is secreted into the circulation as VLDL, which is converted into LDL by removal of triacylglycerol and the apolipoprotein C components, probably in peripheral tissues and the liver. In contrast, when the quantity of dietary cholesterol increases, cholesterol synthesis in the liver and intestine is almost totally suppressed. Thus, the rate of de novo cholesterol synthesis is inversely related to the amount of dietary cholesterol taken up by the body.

The primary site for control of cholesterol biosynthesis is HMG CoA reductase, which catalyzes the step that produces mevalonic acid. This is the committed step and the rate-limiting reaction in the pathway of cholesterol biosynthesis (Figure 10.38). Cholesterol effects feedback inhibition of its own synthesis by inhibiting the activity of preexisting HMG CoA reductase and also by promoting rapid inactivation of the enzyme by mechanisms that remain to be elucidated. The dietary cholesterol that suppresses HMG CoA reductase activity and cholesterol synthesis emerges from the intestine in the form of chylomicrons.

In a normal healthy adult on a low cholesterol diet about 1300 mg of cholesterol is returned to the liver each day for disposal. This cholesterol

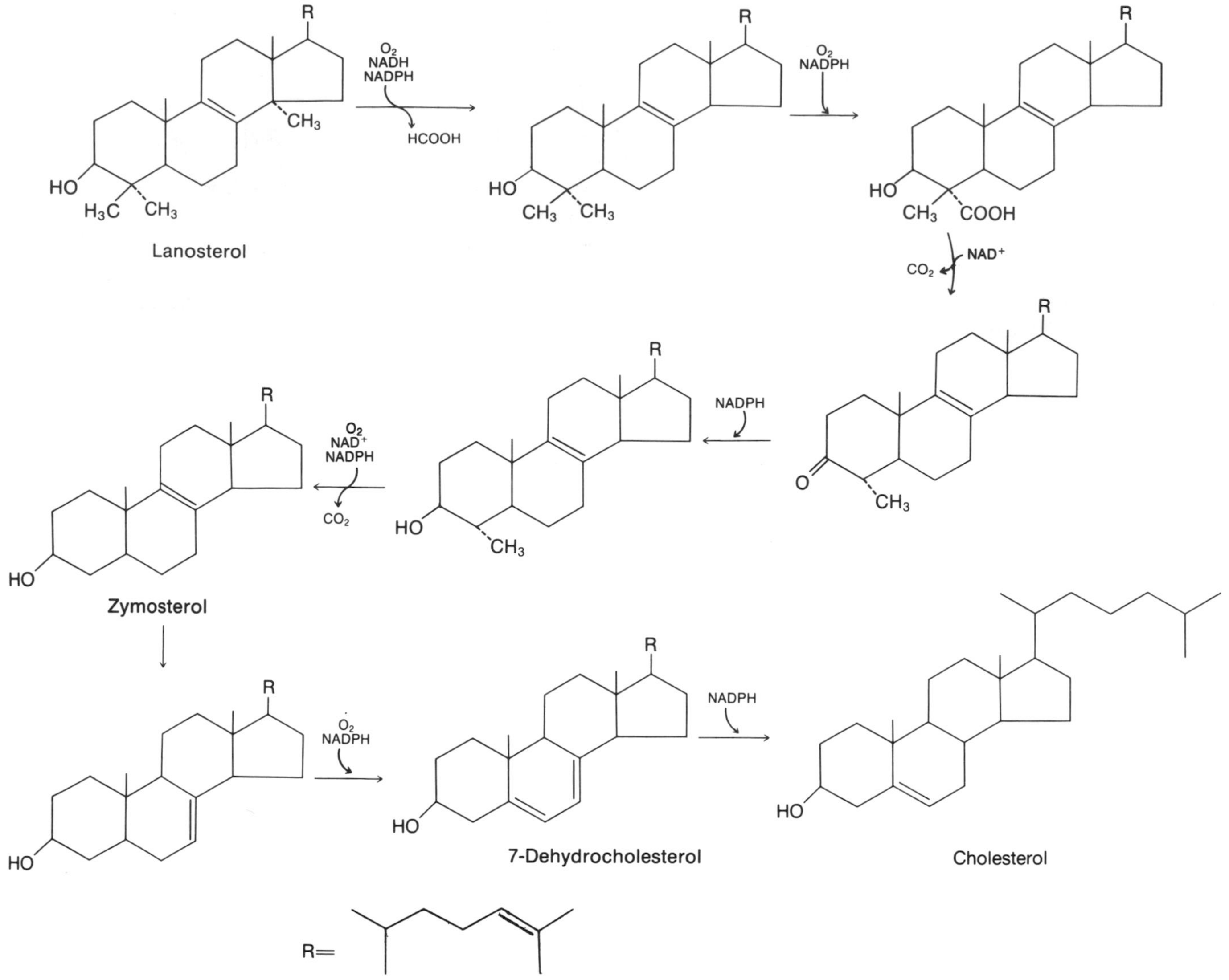

FIGURE 10.37
Conversion of lanosterol to cholesterol.

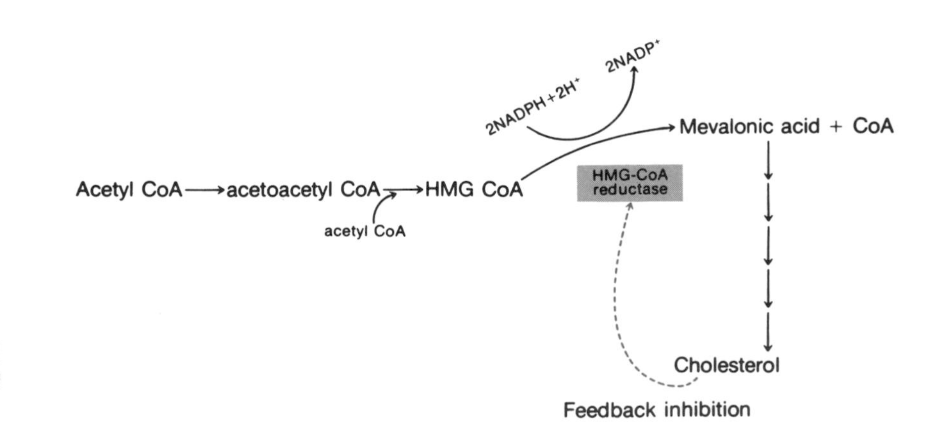

FIGURE 10.38
Summary of the pathway of cholesterol synthesis indicating feedback inhibition of HMG CoA reductase by cholesterol.

comes from (a) cholesterol reabsorbed from the gut by means of the enterohepatic circulation and (b) HDL that carries cholesterol to the liver from peripheral tissues. The liver disposes of cholesterol in one of three ways: (1) excretion in bile as free cholesterol and after conversion to bile salts; each day, about 250 mg of bile salts and 550 mg of cholesterol are lost from the enterohepatic circulation; (2) esterification and storage in the liver as cholesterol esters; and (3) incorporation into lipoproteins (VLDL and LDL) and secretion into the circulation. On a low-cholesterol diet, the liver will synthesize ~800 mg of cholesterol per day to replace bile salts and cholesterol lost from the enterohepatic circulation in the feces.

The mechanism of suppression of cholesterol biosynthesis by LDL-bound cholesterol involves specific **LDL receptors** that project from the surface of human cells. The first step of the regulatory mechanism involves the binding of the lipoprotein LDL to these LDL receptors, thereby extracting the LDL particles from the blood. The binding reaction is characterized by its saturability, high affinity, and high degree of specificity. The receptor recognizes only LDL and VLDL, the two plasma lipoproteins that contain apolipoprotein B. Once LDL binds to the LDL receptor at sites on the plasma membrane that contain pits coated with a protein called **clathrin,** the cholesterol-charged lipoprotein is endocytosed in the form of clathrin-coated vesicles. Intracellularly, the coated vesicle loses its clathrin and becomes an endosome. This process is termed **receptor-mediated endocytosis.** The next step involves the fusion of the endosome with a lysosome that contains numerous hydrolytic enzymes, including proteases and cholesterol esterase. Eventually the LDL receptor separates from LDL and returns to the cell surface. Inside the lysosome the cholesterol ester component of LDL is hydrolyzed by lysosomal cholesterol esterase to produce free cholesterol and a molecule of long-chain fatty acid. Free cholesterol then diffuses into the cytoplasm where, by some unknown mechanism, it inhibits the activity of HMG CoA reductase and suppresses the synthesis of HMG CoA reductase enzyme. There is evidence that cholesterol acts at the level of DNA and protein synthesis to decrease the rate of synthesis of HMG CoA reductase. At the same time, **fatty acyl CoA : cholesterol acyltransferase (ACAT)** in the endoplasmic reticulum is activated by cholesterol, thereby promoting the formation of cholesterol esters, principally cholesterol oleate. The accumulation of intracellular cholesterol eventually inhibits the replenishment of LDL receptors on the cell surface, a phenomenon called **down regulation,** thereby blocking further uptake and accumulation of cholesterol by the cell.

The LDL receptor is a single chain glycoprotein that spans the plasma membrane once; its carboxyl terminus is on the cytoplasmic face of the membrane and the amino terminus, which contains the LDL binding site, extends into the extracellular space. **Apoprotein B-100** (Apo B-100), which is the protein component of LDL, and **apoprotein E,** which is present in IDL (intermediate density lipoproteins) and some forms of HDL, are the two proteins through which particular lipoproteins bind to the LDL receptor.

The correlation between high levels of blood cholesterol, particularly LDL cholesterol, and heart attacks and strokes have lead to the development of dietary and therapeutic approaches to lower blood cholesterol (Clin. Corr. 10.2).

Patients with *familial* (genetic) *hypercholesterolemia* suffer from accelerated *atherosclerosis* (Clin. Corr. 10.3) and have a defect in this regulatory system. In most of these cases, there is a lack of functional LDL receptors on the cell surface because the mutant alleles produce little or no LDL receptor protein; these patients are referred to as receptor nega-

CLIN. CORR. 10.2 TREATMENT OF HYPERCHOLESTEROLEMIA

Many authorities recommend screening asymptomatic individuals by measuring plasma cholesterol. A level less than 200 mg% is considered desirable, and a level over 240 mg% requires lipoprotein analysis, especially determination of LDL cholesterol. Reduction of LDL cholesterol depends on dietary restriction of cholesterol to less than 300 mg day^{-1}, of calories to attain ideal body weight, and of total fat intake to less than 30% of total calories. Approximately two-thirds of the fat should be *mono-* or polyunsaturated. The second line of therapy is with drugs. Cholestyramine and colestipol are bile salt-binding drugs that promote excretion of bile salts in the stool. This in turn increases the rate of hepatic bile salt synthesis and of LDL uptake by the liver. A second drug that has recently become available is lovastatin, an inhibitor of HMG-CoA reductase. Since HMG-CoA reductase is limiting for cholesterol synthesis, lovastatin decreases endogenous synthesis of cholesterol and stimulates uptake and LDL via the LDL receptor. The combination of lovastatin and cholestyramine is sometimes used for severe hyperlipidemia.

Goodman, D. S., et al., Evaluation and treatment of high blood cholesterol in adults. *Arch. Int. Med.* 148:36, 1988.

tive. In others the LDL receptor is synthesized and transported normally to the cell surface; an amino acid substitution or other alteration in the protein's primary structure, however, adversely affects the LDL binding region of the receptor. As a result, there is no binding of LDL to the cell, cholesterol is not transferred into the cell, cholesterol synthesis is not inhibited, and the cholesterol content of the blood increases. Another LDL-deficient group of hypercholesterolemic patients is able to synthesize the LDL receptor but has a defect in the transport mechanism that delivers the glycoprotein to its proper location on the plasma membrane. And finally, there is another subclass of genetically determined hypercholesterolemics whose LDL receptors have a defect in the cytoplasmic carboxyl terminus; they populate their cell surfaces with LDL receptors normally but are unable to internalize the LDL–LDL receptor complex due to an inability to cluster this complex in coated pits.

In specialized tissues such as the adrenal gland and ovary, the cholesterol derived from LDL serves as a precursor to the steroid hormones made by these organs, such as cortisol and estradiol, respectively. In the liver, cholesterol extracted from LDL and HDL is converted into bile salts that function in intestinal fat digestion.

CLIN. CORR. 10.3
ATHEROSCLEROSIS

Atherosclerosis is a disorder of the arterial wall characterized by accumulation of cholesteryl esters in cells derived from the monocyte-macrophage line, smooth muscle cell proliferation, and fibrosis. The earliest abnormality is the migration of blood monocytes to the subendothelium of the artery. Once there, they differentiate into macrophages. These cells accumulate cholesteryl esters derived from plasma LDL. Why these cells do not normally regulate cellular cholesterol stores is not completely understood. Some of the LDL may be taken up via pathways distinct from the classical LDL receptor pathway. For instance, receptors that mediate uptake of acetylated LDL or LDL complexed with dextran sulfate have been described and these are not regulated by cellular cholesterol content. Distortion of the subendothelium leads to platelet aggregation on the arterial surface, and release of platelet-derived mitogens such as platelet-derived growth factor (PDGF). This in turn is thought to stimulate smooth muscle cell growth. Death of the foam cells results in the accumulation of a cellular lipid that can stimulate fibrosis. The resulting atherosclerotic plaque narrows the blood vessel and serves as the site of thrombus formation, which precipitates myocardial infarction (heart attack).

Ross, R. The pathogenesis of atherosclerosis—an update. *N. Eng. J. Med.* 314:488, 1986.

Plasma Cholesterol Is in a Dynamic State

Plasma cholesterol is in a dynamic state, entering the blood complexed with lipoproteins that keep the lipid in solution and leaving the blood as tissues remove cholesterol from these lipoproteins. Cholesterol occurs in plasma lipoproteins in two forms: as free cholesterol and esterified to some long-chain fatty acid. From 70 to 75% of plasma cholesterol is esterified to long-chain fatty acids. It is the free, unesterified form of cholesterol that exchanges readily between different lipoproteins and the plasma membranes of cells.

High density lipoproteins and the enzyme **lecithin : cholesterol acyltransferase** (LCAT) play important roles in the transport and elimination of cholesterol from the body. LCAT catalyzes the freely reversible reaction shown in Figure 10.39, which transfers the fatty acid in the *sn*-2 position of phosphatidylcholine to the 3-hydroxyl of cholesterol. LCAT is a plasma enzyme produced mainly by the liver. The actual substrate for LCAT is cholesterol contained in HDL. The LCAT–HDL system functions to protect cells, especially their plasma membranes, from the damaging effects of excessive amounts of free cholesterol. Cholesterol ester generated in the LCAT reaction diffuses into the core of the HDL particle where it is then transported from the tissues and plasma to the liver, the latter being the only organ capable of metabolizing and excreting cholesterol. Thus, by this mechanism, referred to as the *reverse transport of*

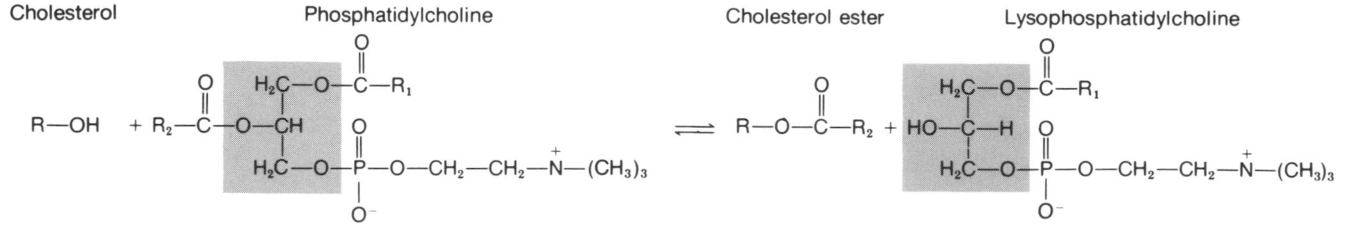

Cholesterol + Phosphatidylcholine ⇌ Cholesterol ester + Lysophosphatidylcholine

$$\mathrm{R{-}OH} + \mathrm{R_2{-}\overset{O}{\overset{\|}{C}}{-}O{-}CH(CH_2{-}O{-}\overset{O}{\overset{\|}{C}}{-}R_1)(CH_2{-}O{-}\overset{O}{\overset{\|}{P}}(O^-){-}O{-}CH_2{-}CH_2{-}\overset{+}{N}{-}(CH_3)_3)} \rightleftharpoons \mathrm{R{-}O{-}\overset{O}{\overset{\|}{C}}{-}R_2} + \mathrm{HO{-}CH(CH_2{-}O{-}\overset{O}{\overset{\|}{C}}{-}R_1)(CH_2{-}O{-}\overset{O}{\overset{\|}{P}}(O^-){-}O{-}CH_2{-}CH_2{-}\overset{+}{N}{-}(CH_3)_3)}$$

FIGURE 10.39
The lecithin : cholesterol acyltransferase (LCAT) reaction where R—OH = cholesterol.

cholesterol, LCAT acting on HDL provides a vehicle for transporting cholesterol from peripheral tissues to the liver.

Cholesterol Is Excreted Primarily as Bile Acids

The bile acids are the end products of cholesterol metabolism. Primary bile acids are those that are synthesized in hepatocytes directly from cholesterol. The most abundant bile acids in humans are derivatives of cholanic acid (Figure 10.40), that is **cholic acid** and **chenodeoxycholic acid** (Figure 10.41). The primary bile acids are composed of 24 carbon atoms, made in liver parenchymal cells, contain two or three OH groups, and have a side chain that ends in a carboxyl group that is ionized at pH 7.0 (hence the name bile salt). The carboxyl group of the primary bile acids is often conjugated via an amide bond to either glycine (NH_2—CH_2—COOH) or taurine (NH_2—CH_2—CH_2—SO_3H) to form **glycocholic** or **taurocholic acid,** respectively. The structure of glycocholic acid is shown in Figure 10.42.

When the primary bile acids undergo further chemical reactions by microorganisms in the gut, they give rise to secondary bile acids that also possess 24 carbon atoms. Examples of secondary bile acids are deoxycholic acid and lithocholic acid, which are derived from cholic acid and chenodeoxycholic acid, respectively, by the removal of one OH group (Figure 10.41).

The changes in the cholesterol molecule that occur during its transformation into bile acids include (a) epimerization of the 3β-OH group; (b) reduction of the Δ^5 double bond; (c) introduction of OH groups at C-7 (chenodeoxycholic acid) or at C-7 and C-12 (cholic acid); and (d) conversion of the C-27 side chain into a C-24 carboxylic acid by elimination of a propyl equivalent.

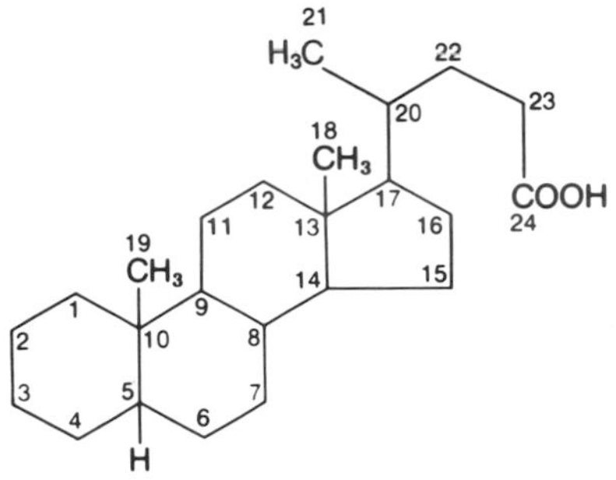

FIGURE 10.40
Structure of cholanic acid.

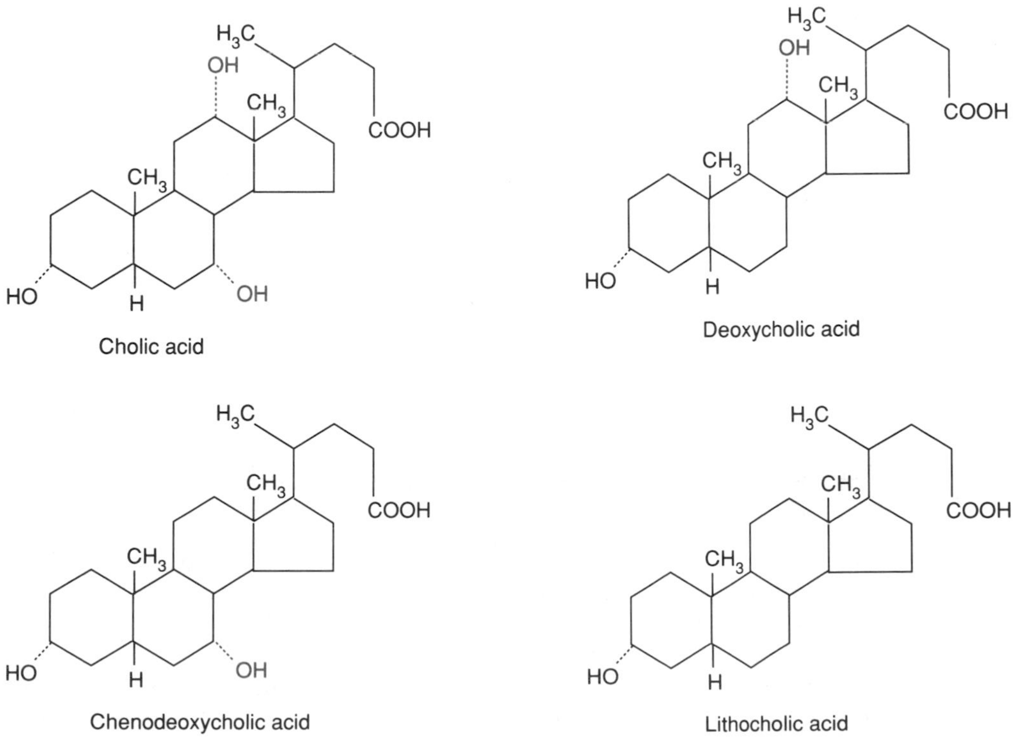

Cholic acid

Deoxycholic acid

Chenodeoxycholic acid

Lithocholic acid

FIGURE 10.41
Structures of some common bile acids.

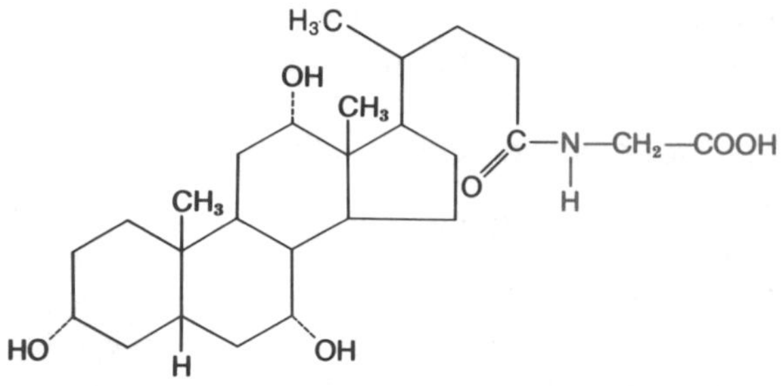

FIGURE 10.42
Structure of glycocholic acid, a conjugated bile acid.

Bile acids formed in liver parenchymal cells are secreted into the bile canaliculi, which are specialized channels formed by adjacent hepatocytes. Bile canaliculi unite with bile ductules, which in turn come together to form bile ducts. The bile acids are carried to the gallbladder for storage and ultimately to the small intestine where they are excreted. The capacity of the liver to produce bile acids is insufficient to meet the physiological demands, so the body relies upon an efficient enterohepatic circulation that carries the bile acids from the intestine back to the liver several times each day. The primary bile acids, after removal of the glycine or taurine residue in the gut, are reabsorbed by an active transport process from the intestine, primarily in the ileum, and returned to the liver by way of the portal vein. Bile acids that are not reabsorbed are acted upon by bacteria in the gut and converted into secondary bile acids: a portion of secondary bile acids, primarily deoxycholic acid and lithocholic acid, are reabsorbed passively in the colon and returned to the liver where they are secreted into the gallbladder. Hepatic synthesis normally produces 0.2–0.6 g of bile acids per day to replace those lost in the feces. The gallbladder pool of bile acids is 2–4 g. Because the enterohepatic circulation recycles 6–12 times each day, the total amount of bile acids absorbed per day from the intestine corresponds to 12–32 g.

In terms of function the bile acids are significant in medicine for several reasons:

1. They represent the only significant way in which cholesterol can be excreted. The carbon skeleton of cholesterol is not oxidized to CO_2 and H_2O in human beings but is excreted in bile as free cholesterol and as bile acids.
2. They prevent the precipitation of cholesterol out of solution in the gallbladder. Bile acids and phospholipids function to solubilize cholesterol in bile.
3. They act as emulsifying agents to prepare dietary triacyglycerols for attack by pancreatic lipase. Bile acids may also play a direct role in activating pancreatic lipase (see Chapter 26).
4. They facilitate the absorption of fat-soluble vitamins, particularly vitamin D, from the intestine.

Vitamin D Is Synthesized From an Intermediate of the Cholesterol Biosynthetic Pathway

Another function of the cholesterol biosynthetic pathway is to provide substrate for the photochemical production of **vitamin D_3** in the skin. The metabolism and function of vitamin D_3 are discussed in Chapter 28. Vitamin D_3 is a secosteroid in which the 9,10 carbon bond of the B ring of cholesterol has undergone fission (Figure 10.43). The most important supply of vitamin D_3 is that manufactured in the skin. 7-Dehydrocholesterol is an intermediate in the pathway of cholesterol biosynthesis and is converted in the skin to previtamin D_3 by irradiation with UV rays of the sun (285–310 nm). Previtamin D_3 is biologically inert and labile and converted thermally and slowly (~36 h) to the double bond isomer by a nonenzy-

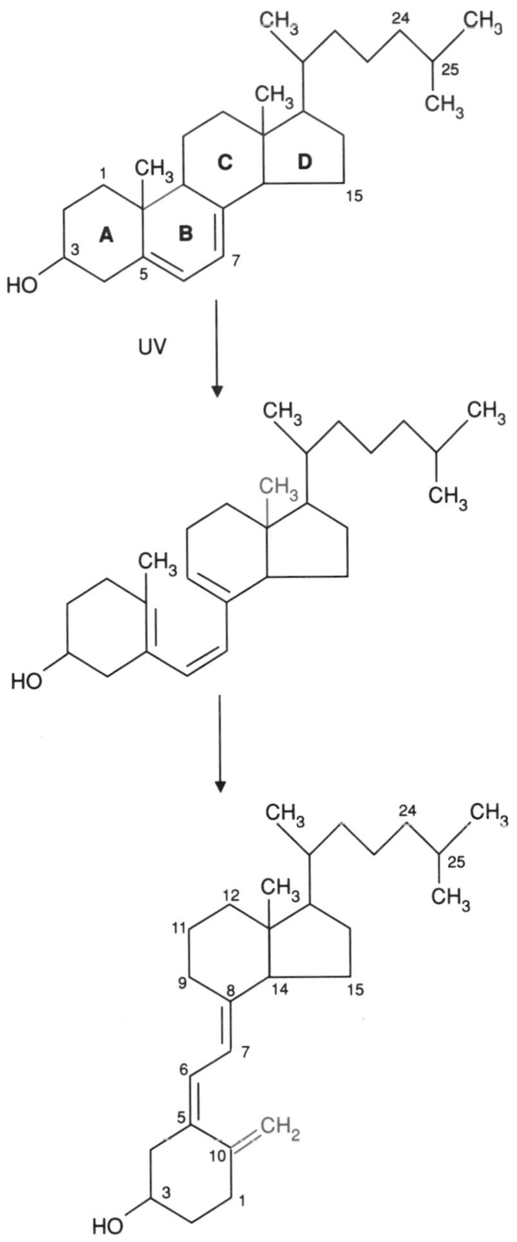

7-Dehydrocholesterol

UV

Previtamin D_3

Vitamin D_3 (cholecalciferol)

FIGURE 10.43
The photochemical conversion of 7-dehydrocholesterol to vitamin D_3 (cholecalciferol).

matic reaction to the biologically active vitamin, **cholecalciferol** (vitamin D_3). As little as 10-min exposure each day of the hands and face to sunlight will satisfy the body's need for vitamin D. Photochemical action on the plant sterol ergosterol also provides a dietary precursor to a compound designated vitamin D_2 (calciferol) that can satisfy the vitamin D requirement.

10.4 SPHINGOLIPIDS

Biosynthesis of Sphingosine

Sphingolipids are complex lipids whose core structure is provided by the long-chain amino alcohol **sphingosine** (Figure 10.44). Another common name for sphingosine is 4-sphingenine and the formal name for sphingosine is *trans*-1,3-dihydroxy-2-amino-4-octadecene. Sphingosine possesses two asymmetric carbon atoms (positions C-2 and C-3); of the four

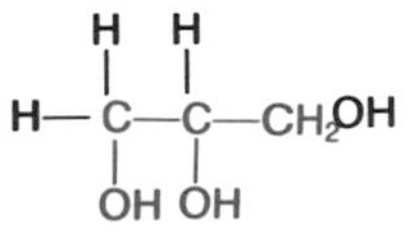

Glycerol

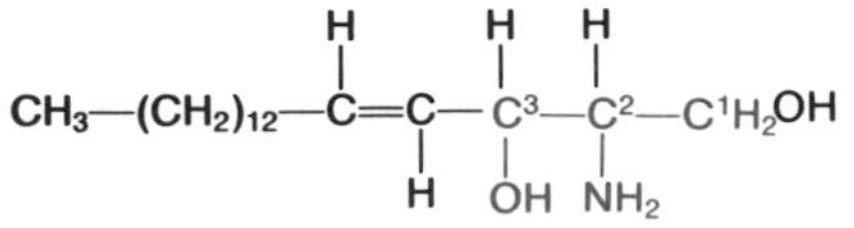

Sphingosine

FIGURE 10.44
Comparison of the structures of glycerol and sphingosine (*trans*-1,3,dihydroxy-2-amino-4-octadecene).

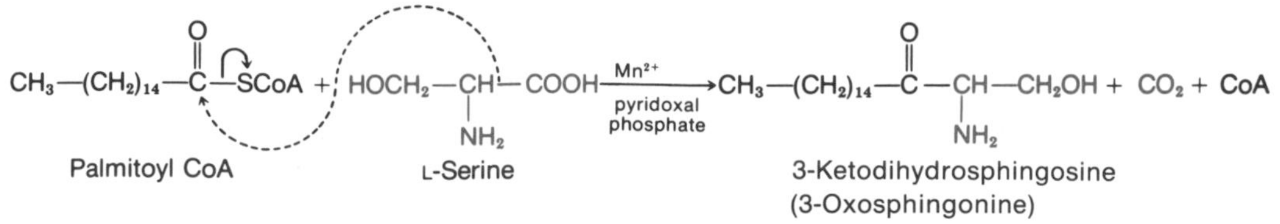

$$CH_3{-}(CH_2)_{14}{-}\overset{O}{\overset{\|}{C}}{-}SCoA + HOCH_2{-}\underset{NH_2}{\underset{|}{CH}}{-}COOH \xrightarrow[\text{pyridoxal phosphate}]{Mn^{2+}} CH_3{-}(CH_2)_{14}{-}\overset{O}{\overset{\|}{C}}{-}\underset{NH_2}{\underset{|}{CH}}{-}CH_2OH + CO_2 + CoA$$

Palmitoyl CoA L-Serine 3-Ketodihydrosphingosine (3-Oxosphingonine)

FIGURE 10.45
Formation of 3-ketodihydrosphingosine from serine and palmitoyl CoA.

possible optical isomers, naturally occurring sphingosine is of the D-erythro form. The Δ^4 carbon–carbon double bond of sphingosine has the trans configuration and the primary alcohol group at C-1 is a nucleophilic center that forms covalent bonds with sugars to form sphingomyelin. The amino group at C-2 always bears a long-chain fatty acid (usually C_{20}–C_{26}) in amide linkage. The secondary alcohol at C-3 is always free. It is useful to appreciate the structural similarity of a part of the sphingosine molecule to the glycerol moiety of the acyl glycerols. When one views the structure of sphingosine from another perspective, the similarity between carbon atoms 1, 2, and 3 of sphingosine and glycerol becomes apparent (Figure 10.44). Note that both glycerol and sphingosine have nucleophilic groups, hydroxyl or amino, at positions C-1, C-2, and C-3.

The sphingolipids occur in blood and nearly all of the tissues of humans. The highest concentrations of sphingolipids are found in the white matter of the central nervous system. Various sphingolipids are components of the plasma membrane of practically all cells.

Sphingosine is synthesized by way of **sphinganine** (dihydrosphingosine) in two steps from the precursors L-serine and palmitoyl CoA. Serine is the source of C-1, C-2, and the amino group of sphingosine, while palmitic acid provides the remaining carbon atoms. The condensation of serine with palmitoyl CoA is catalyzed by a pyridoxal phosphate-dependent enzyme serine palmitoyltransferase and the driving force for the reaction is provided by both the cleavage of the reactive, high-energy, thioester bond of palmitoyl CoA and the release of CO_2 from serine (Figure 10.45). The next step involves the reduction of the carbonyl group in 3-ketodihydrosphingosine with reducing equivalents being derived from NADPH to produce sphinganine (Figure 10.46). The insertion of the double bond into sphinganine to produce sphingosine occurs at the level of ceramide (see below).

Ceramides Are Fatty Acid Amide Derivatives of Sphingosine

Sphingosine as such with its free amino group does not occur naturally. The fundamental building block or core structure of the natural sphingolipids is **ceramide** which is a long-chain fatty acid amide derivative of sphingosine. The long-chain fatty acid is attached to the 2-amino group of sphingosine through an amide bond (Figure 10.47). Most often the acyl group is behenic acid, a saturated C_{22} fatty acid, but other long-chain acyl

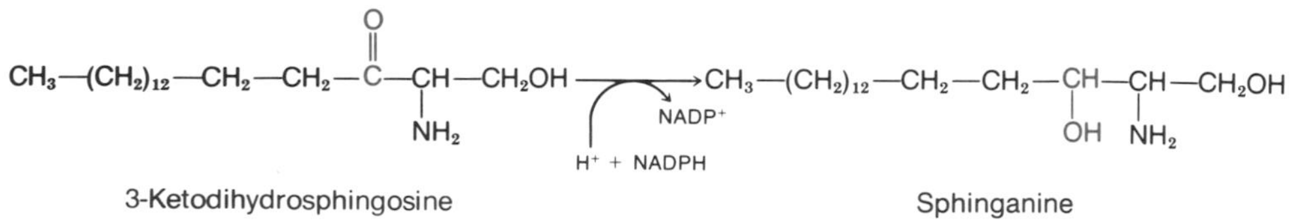

$$CH_3{-}(CH_2)_{12}{-}CH_2{-}CH_2{-}\overset{O}{\overset{\|}{C}}{-}\underset{NH_2}{\underset{|}{CH}}{-}CH_2OH \xrightarrow[H^+ + NADPH]{NADP^+} CH_3{-}(CH_2)_{12}{-}CH_2{-}CH_2{-}\underset{OH}{\underset{|}{CH}}{-}\underset{NH_2}{\underset{|}{CH}}{-}CH_2OH$$

3-Ketodihydrosphingosine Sphinganine

FIGURE 10.46
Conversion of 3-ketodihydrosphingosine to sphinganine.

groups can be used. There are two long-chain hydrocarbon domains in the ceramide molecule; these hydrophobic regions are responsible for the lipoidal character of the sphingolipids.

Ceramide is synthesized from dihydrosphingosine and a molecule of long-chain fatty acyl CoA by a microsomal enzyme with dihydroceramide as an intermediate that is then oxidized by dehydrogenation at C-4 and C-5 (Figure 10.48). Free ceramide is not a component of membrane lipids but rather is an intermediate in the biosynthesis and catabolism of glycosphingolipids and sphingomyelin. The structures of the prominent sphingolipids of humans are presented in Figure 10.49 in diagrammatic form.

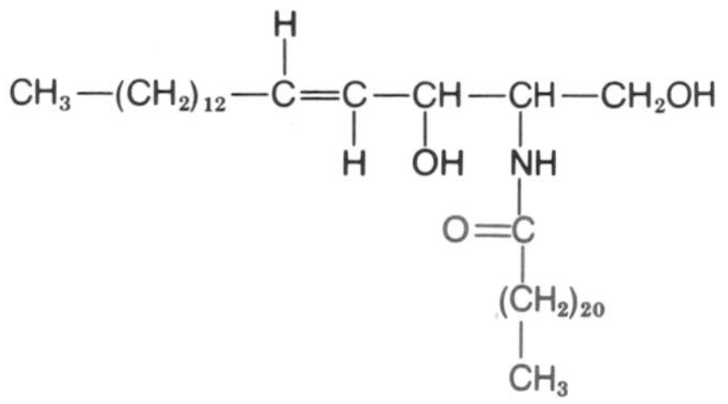

$CH_3—(CH_2)_{12}—CH{=}CH—CH(OH)—CH(NH—CO—(CH_2)_{20}—CH_3)—CH_2OH$

FIGURE 10.47
Structure of a ceramide (*N*-acylsphingosine).

Sphingomyelin Is the Only Sphingolipid Containing Phosphorus

Sphingomyelin, one of the principal structural lipids of the membranes of nervous tissue, is the only sphingolipid that is a phospholipid. In sphingomyelin the primary alcohol group at C-1 of sphingosine is esterified to choline through a phosphodiester bridge of the kind that occurs in the acyl glycerophospholipids and the amino group of sphingosine is attached to a long-chain fatty acid by means of an amide bond. Sphingomyelin is therefore a ceramide phosphocholine. It contains one negative and one positive charge so that it is neutral at physiological pH (Figure 10.50).

The most common fatty acids in sphingomyelin are palmitic (C_{16}), stearic (C_{18}), lignoceric (a C_{24}, saturated fatty acid), and nervonic acid [24 : 1, $CH_3—(CH_2)_7CH{=}CH—(CH_2)_{13}—COOH$]. The sphingomyelin of myelin contains predominantly longer chain fatty acids, mainly lignoceric and nervonic, whereas that of gray matter contains largely stearic acid. Excessive accumulations of sphingomyelin occur in Niemann–Pick disease.

Sphingomyelin Is Synthesized From a Ceramide and Phosphatidylcholine

The conversion of ceramide to sphingomyelin involves the transfer of a phosphorylcholine moiety not from CDP–choline as was suspected for

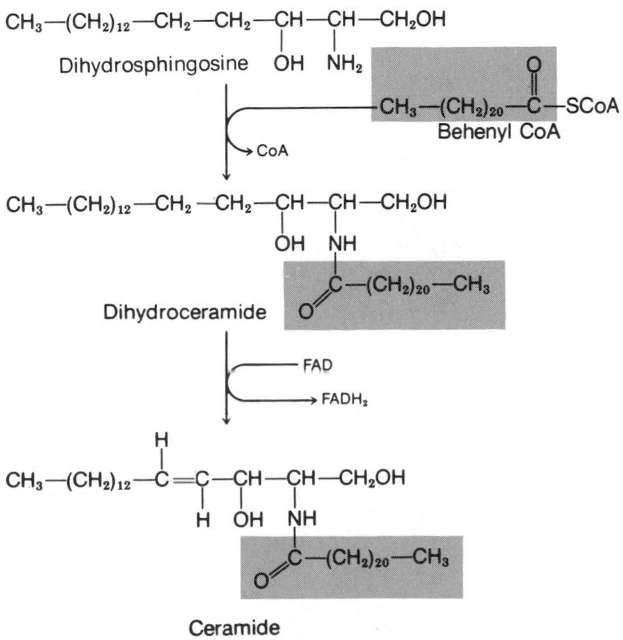

Dihydrosphingosine: $CH_3—(CH_2)_{12}—CH_2—CH_2—CH(OH)—CH(NH_2)—CH_2OH$

+ Behenyl CoA: $CH_3—(CH_2)_{20}—CO—SCoA$ → CoA

Dihydroceramide: $CH_3—(CH_2)_{12}—CH_2—CH_2—CH(OH)—CH(NH—CO—(CH_2)_{20}—CH_3)—CH_2OH$

FAD → $FADH_2$

Ceramide: $CH_3—(CH_2)_{12}—CH{=}CH—CH(OH)—CH(NH—CO—(CH_2)_{20}—CH_3)—CH_2OH$

FIGURE 10.48
Formation of ceramide from dihydrosphingosine.

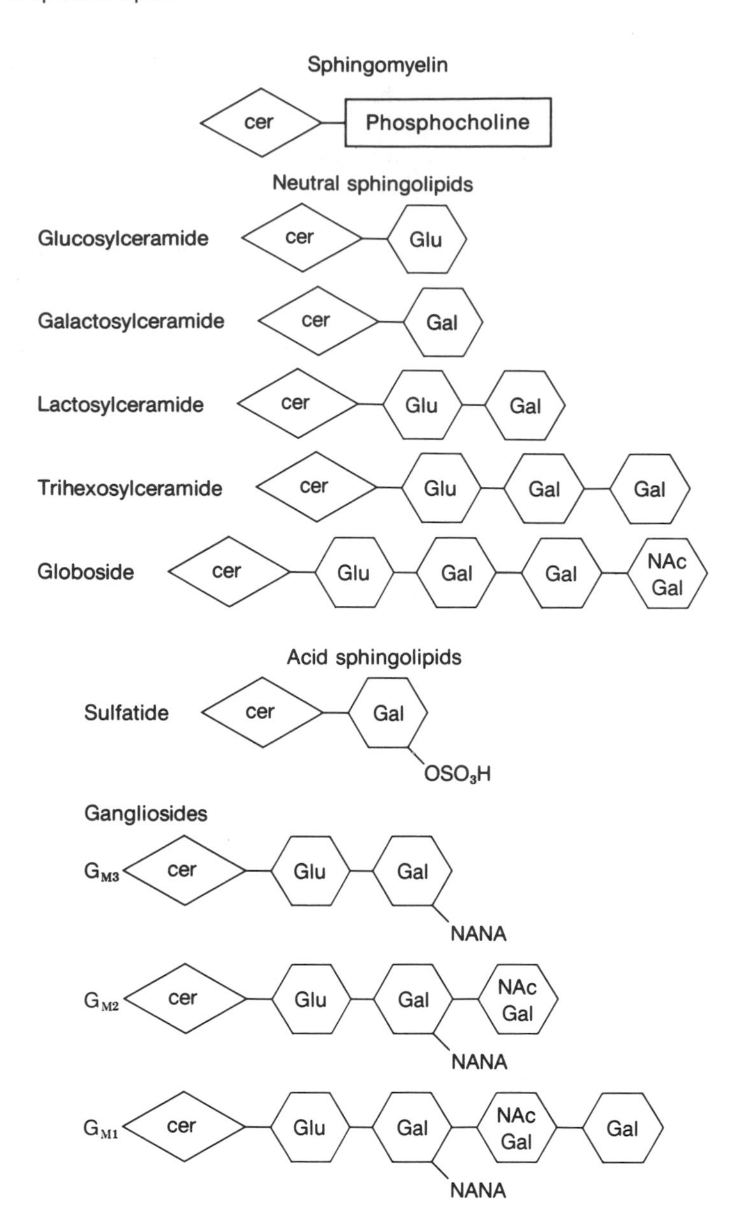

FIGURE 10.49
Structures of some common sphingolipids in diagrammatic form.
Cer = ceramide, Glu = glucose, Gal = galactose, NAcGal = *N*-acetylgalactosamine, and NANA = *N*-acetylneuraminic acid (sialic acid).

many years, but from phosphatidylcholine (lecithin) instead; this reaction is catalyzed by **sphingomyelin synthase** (Figure 10.51).

Glycosphingolipids Usually Have a Galactose or Glucose Unit

The principal glycosphingolipid classes are cerebrosides, sulfatides, globosides, and gangliosides. In the glycolipid class of compounds the polar head group is attached to sphingosine via the glycosidic linkage of a sugar molecule rather than a phosphate ester bond, as is the case in the phospholipids.

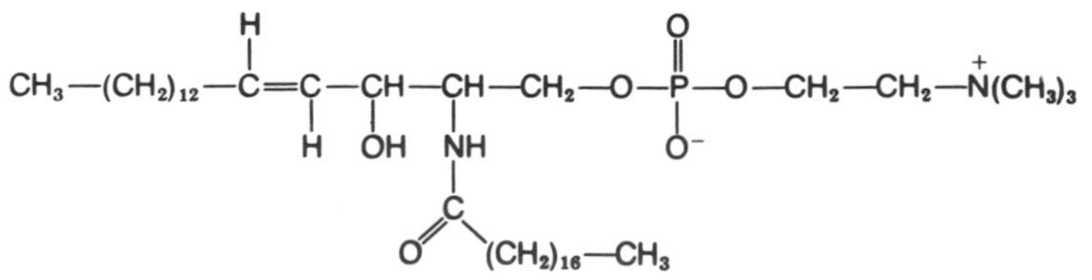

$$CH_3\text{—}(CH_2)_{12}\text{—}CH\text{=}CH\text{—}CH(OH)\text{—}CH(NH\text{—}C(\text{=}O)\text{—}(CH_2)_{16}\text{—}CH_3)\text{—}CH_2\text{—}O\text{—}P(\text{=}O)(O^-)\text{—}O\text{—}CH_2\text{—}CH_2\text{—}\overset{+}{N}(CH_3)_3$$

FIGURE 10.50
Structure of sphingomyelin.

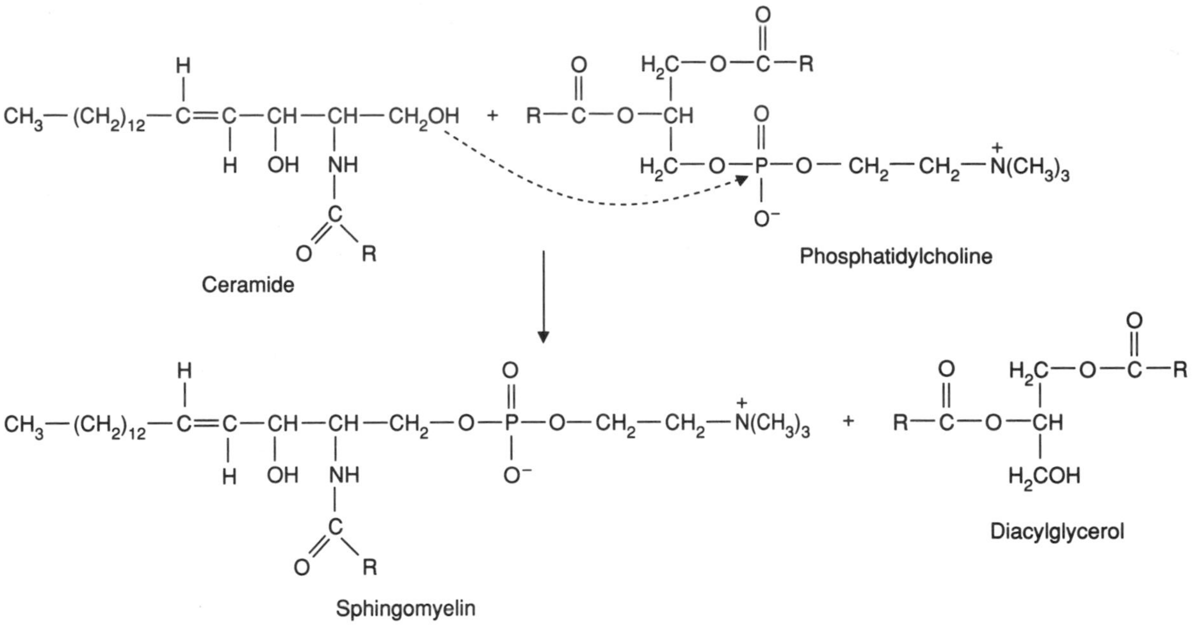

FIGURE 10.51
Sphingomyelin synthesis from ceramide and phosphatidylcholine.

Cerebrosides Are Glycosylceramides

The cerebrosides are a group of ceramide monohexosides. The two most common cerebrosides of 1-β-glycosylceramides are **galactocerebroside** and **glucocerebroside.** Unless specified otherwise, the term cerebroside usually refers to galactocerebroside, also called "galactolipid." In Figure 10.52 note that the monosaccharide units are attached at C-1 of the sugar moiety to the C-1 position of ceramide, and the anomeric configuration of the glycosidic bond between ceramide and hexose in both galactocerebroside and glucocerebroside is β. The largest amount of galactocerebroside in healthy individuals is found in the brain. Moderately increased amounts of galactocerebroside accumulate in the white matter in Krabbe's disease, also called globoid leukodystrophy, due to a deficiency in the lysosomal enzyme galactocerebrosidase.

Glucocerebroside (glucosylceramide) is not normally a structural component of membranes and is an intermediate in the synthesis and degradation of more complex glycosphingolipids (see Figure 10.53). However, 100-fold increases in the glucocerebroside content of spleen and liver

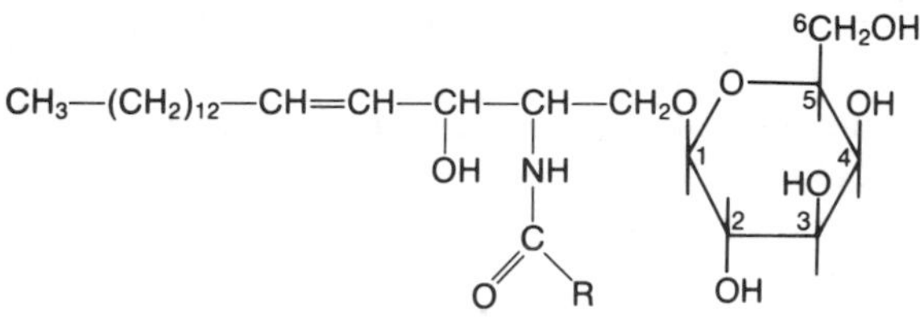

FIGURE 10.52
Structure of galactocerebroside (galactolipid).

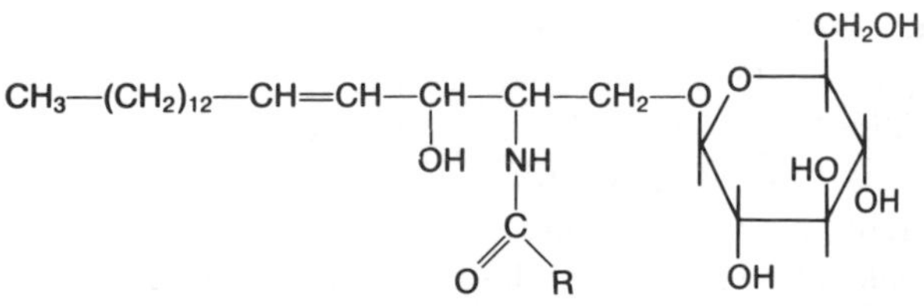

FIGURE 10.53
Structure of glucocerebroside.

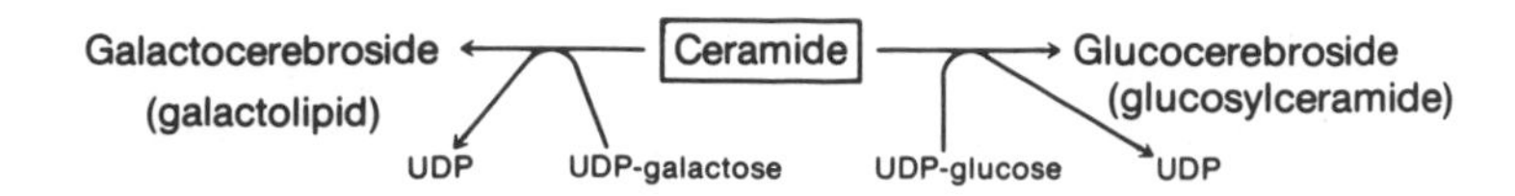

FIGURE 10.54
Synthesis of galacto- and glucocerebroside.

occur in the genetic lipid storage disorder called Gaucher's disease, which results from a deficiency of lysosomal glucocerebrosidase.

Galactocerebroside and glucocerebroside are synthesized from ceramide and the activated nucleotide sugars UDP-galactose and UDP-glucose, respectively. The enzymes that catalyze these reactions, glucosyl and galactosyl transferases, are associated with the endoplasmic reticulum (Figure 10.54). Alternatively, in some tissues, the synthesis of glucocerebroside (glucosylceramide) proceeds by way of glucosylation of sphingosine in a reaction catalyzed by glucosyltransferase:

$$\text{Sphingosine} + \text{UDP-glucose} \longrightarrow \text{glucosylsphingosine} + \text{UDP}$$

followed by fatty acylation:

$$\text{Glucosylsphingosine} + \text{stearoyl CoA} \longrightarrow \text{glucocerebroside} + \text{CoASH}$$

A Sulfatide Is a Sulfuric Acid Ester of Galactocerebroside

Sulfatide, or **sulfogalactocerebroside** as it is sometimes called, is a sulfuric acid ester of galactocerebroside. Galactocerebroside 3-sulfate is the major sulfolipid in brain and accounts for approximately 15% of the lipids of white matter (see Figure 10.55).

Galactocerebroside sulfate is synthesized from galactocerebroside and 3′-phosphoadenosine 5′-phosphosulfate (PAPS) in a reaction catalyzed by sulfotransferase:

$$\text{Galactocerebroside} + \text{PAPS} \longrightarrow \text{PAP} + \text{galactocerebroside 3-sulfate}$$

The structure of PAPS, sometimes referred to as "activated sulfate," is indicated in Figure 10.56. Large quantities of sulfatide accumulate in the tissues of the central nervous system in metachromatic leukodystrophy due to a deficiency in a specific sulfatase.

Globosides Are Ceramide Oligosaccharides

This family of compounds represents cerebrosides that contain two or more sugar residues, usually galactose, glucose, or *N*-acetylgalactosamine. The ceramide oligosaccharides are neutral compounds and contain no free amino groups.

Lactosyl ceramide is a component of the erythrocyte membrane (Figure 10.57).

Another prominent globoside is **ceramide trihexoside** or ceramide galactosyllactoside: ceramide-β-glc(4 $\leftarrow$ 1)-β-gal-(4 $\leftarrow$ 1)-α-gal. Note that the terminal galactose residue of this globoside has the α-anomeric configuration. Ceramide trihexoside accumulates in the kidneys of patients with Fabry's disease who are deficient in lysosomal α-galactosidase A activity.

Gangliosides Contain Sialic Acid

The name ganglioside was adopted for the class of sialic acid-containing glycosphingolipids that are highly concentrated in the ganglion cells of the central nervous system, particularly in the nerve endings. The central nervous system is unique among human tissues because more than one half of the sialic acid is in ceramide–lipid bound form, with the remainder

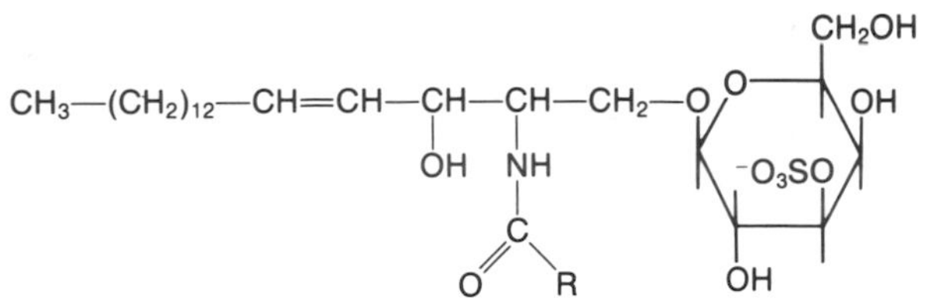

FIGURE 10.55
Structure of galactocerebroside sulfate (sulfolipid).

of the sialic acid occurring in the oligosaccharides of glycoproteins. Lesser amounts of gangliosides are contained in the surface membranes of the cells of most extraneural tissues where they account for less than 10% of the total sialic acid.

Neuraminic acid (abbreviated Neu) is present in gangliosides, glycoproteins, and mucins. The amino group of neuraminic acid occurs most often as the *N*-acetyl derivative, and the resulting structure is called *N*-acetylneuraminic acid or sialic acid, commonly abbreviated NANA (see Figure 10.58).

The OH group on C-2 occurs most often in the α-anomeric configuration and the linkage between NANA and the oligosaccharide ceramide always involves the OH group on position 2 of *N*-acetylneuraminic acid.

The structures of some of the common gangliosides are indicated in Table 10.1. The principal gangliosides in brain are G_{M1}, G_{D1a}, G_{D1b}, and G_{T1b}. Nearly all of the gangliosides of humans are derived from the family of compounds originating with glucosylceramide.

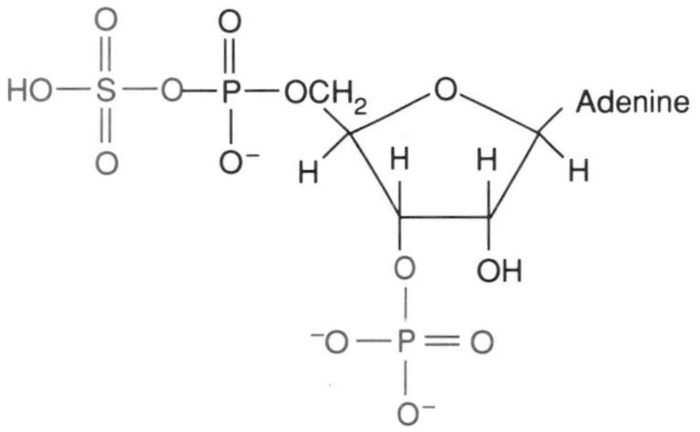

With regard to the nomenclature of the sialoglycosylsphingolipids, the letter G refers to the name ganglioside. The subscripts M, D, T, and Q indicate mono-, di-, tri-, and quatra(tetra)-sialic acid-containing gangliosides. The numerical subscripts 1, 2, and 3 designate the carbohydrate sequence that is attached to ceramide as indicated as follows: 1, Gal-GalNAc-Gal-Glc-ceramide; 2, GalNAc-Gal-Glc-ceramide; and 3, Gal-Glc-ceramide. Consider the nomenclature of the Tay–Sachs ganglioside; the designation G_{M2} denotes the ganglioside structure shown in Table 10.1.

A specific ganglioside on intestinal mucosal cells mediates the action of cholera toxin. Cholera toxin, a protein of mol wt 84,000, is secreted by the pathogen *Vibrio cholerae*. The toxin stimulates the secretion of chloride ions into the gut lumen, resulting in the severe diarrhea characteristic of cholera. Two kinds of subunits, A and B, comprise the cholera toxin; there is one copy of the A subunit (28,000 Da) and five copies of the B subunit (~11,000 Da each). After binding to the cell surface membrane through a domain on the B subunit, the active subunit A passes into the cell, where it activates adenylate cyclase on the inner surface of the membrane. The cAMP that is generated then stimulates chloride ion transport and produces diarrhea. The choleragenoid domain, as the B

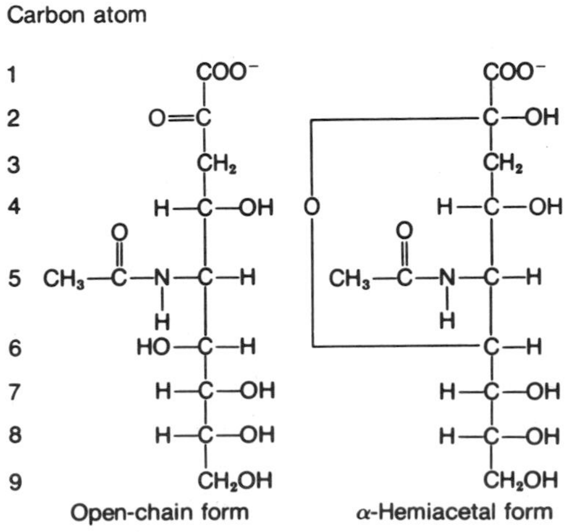

FIGURE 10.56
Structure of PAPS (3′-phosphoadenosine-5′-phosphosulfate).

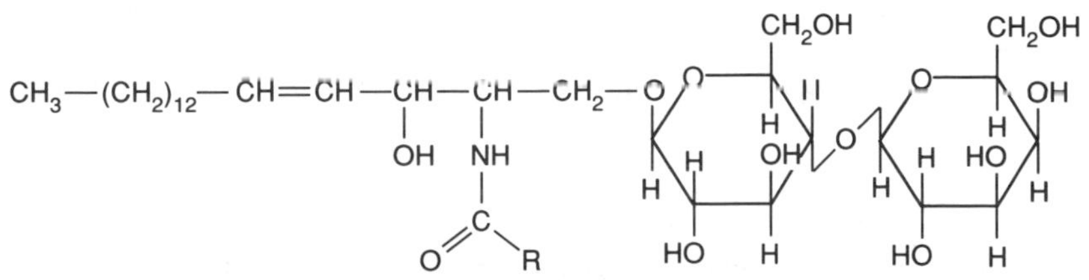

FIGURE 10.57
Structure of ceramide-β-glc-(4 $\leftarrow$ 1)-β-gal (lactosylceramide).

FIGURE 10.58
Structure of *N*-acetylneuraminic acid (NANA).

TABLE 10.1 The Structures of Some Common Gangliosides

Code name	Chemical structure
G_{M3}	Galβ → 4Glcβ → Cer 3 ↑ αNANA
G_{M2}	GalNAcβ → 4Galβ → 4Glcβ → Cer 3 ↑ αNANA
G_{M1}	Galβ → 3GalNAcβ → 4Galβ → 4Glcβ → Cer 3 ↑ αNANA
G_{D1a}	Galβ → 3GalNAcβ → 4Galβ → 4Glcβ → Cer 3 3 ↑ ↑ αNANA αNANA
G_{D1b}	Galβ → 3GalNAcβ → 4Galβ → 4Glcβ → Cer 3 ↑ αNANA8 ← αNANA
G_{T1a}	Galβ → 3GalNAcβ → 4Galβ → 4Glcβ → Cer 3 3 ↑ ↑ αNANA8 ← αNANA αNANA
G_{T1b}	Galβ → 3GalNAcβ → 4Galβ → 4Glcβ → Cer 3 3 ↑ ↑ αNANA αNANA8 ← αNANA
G_{Q1b}	Galβ → 3GalNAcβ → 4Galβ → 4Glcβ → Cer 3 3 ↑ ↑ αNANA8 ← αNANA αNANA8 ← αNANA

subunits are called, binds to the **ganglioside G_{M1}** that has the structure shown in Table 10.1.

Gangliosides are also thought to be receptors for other toxic agents, such as tetanus toxin, and certain viruses, such as the influenza viruses. There is also speculation that gangliosides may play an informational role in cell–cell interactions by providing specific recognition determinants on the surface of cells. There are several lipid storage disorders that involve the accumulation of sialic acid-containing glycosphingolipids. The two most common gangliosidoses involve the storage of the gangliosides G_{M1} (G_{M1} gangliosidosis) and G_{M2} (Tay–Sachs disease). The G_{M1} gangliosidosis is an autosomal recessive metabolic disease characterized by impaired psychomotor function, mental retardation, hepatosplenomegaly, and death within the first few years of life. The massive cerebral and visceral accumulation of G_{M1} ganglioside is due to a profound deficiency of β-galactosidase.

The Sphingolipidoses Are Lysosomal Storage Diseases With Defects in the Catabolic Pathway for Sphingolipids

The various sphingolipids are normally degraded within lysosomes of phagocytic cells, particularly the histiocytes or macrophages of the reticuloendothelial system located primarily in the liver, spleen, and bone marrow. Degradation of the sphingolipids by visceral organs begins with the engulfment of the membranes of white cells and erythrocytes that are rich in lactosylceramide (Cer-Glc-Gal) and hematoside (Cer-Glc-Gal-NANA). In the brain, the majority of the cerebroside-type lipids are gangliosides. Particularly during the neonatal period, ganglioside turnover in the central nervous system is extensive so that glycosphingolipids are rapidly being broken down and resynthesized. The pathway of sphingolipid catabolism is summarized in Figure 10.59. Note that among the various sphingolipids that comprise this pathway, there occurs a sulfate ester (in sulfolipid or sulfogalactolipid); *N*-acetylneuraminic acid groups (in the gangliosides); an α-linked galactose residue (in ceramide trihexoside); several β-galactosides (in galactocerebroside and the ganglioside G_{M1});

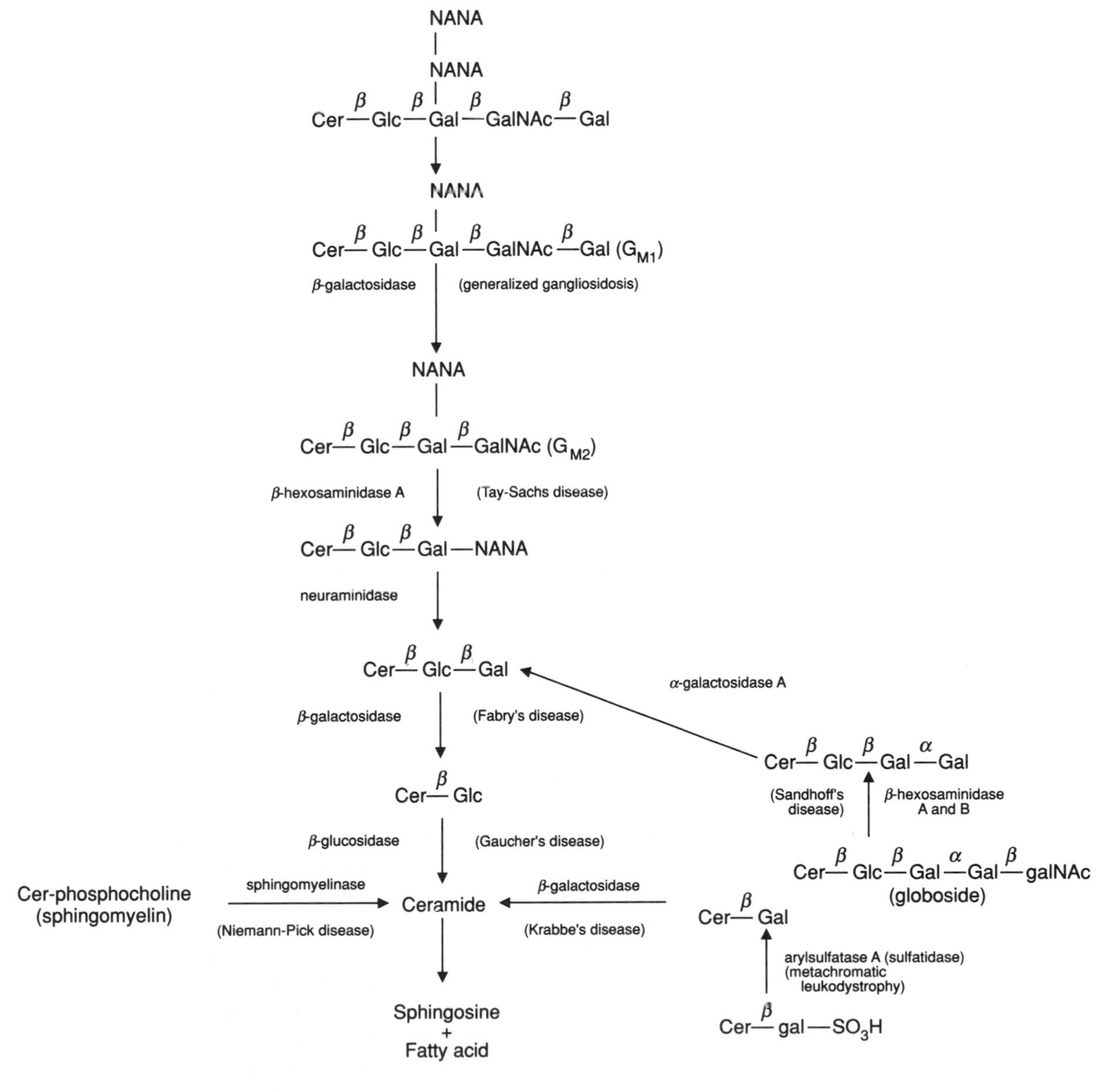

FIGURE 10.59
Summary of the pathways for the catabolism of sphingolipids by lysosomal enzymes.
The genetically determined enzyme deficiency diseases are indicated in the parentheses.

the ganglioside G_{M2}, which terminates in a β-linked *N*-acetylgalactosamine unit; and glucocerebroside, which is composed of a single glucose residue attached to ceramide through a β linkage. The phosphodiester bond in sphingomyelin is broken to produce ceramide, which is turn is converted in sphingosine by the cleavage of an amide bond to a long-chain fatty acid. This overall pathway of sphingolipid catabolism is composed of a series of enzymes that cleave specific bonds in the compounds including α- and β-galactosidases, a β-glucosidase, a neuraminidase, hexosaminidase, a sphingomyelin-specific phosphodiesterase (sphingomyelinase), a sulfate esterase (sulfatase), and a ceramide-specific amidase (ceramidase). The important features of the sphingolipid catabolic pathway are: (a) all the reactions take place within the lysosome; that is, the enzymes of the pathway are contained in lysosomes; (b) the enzymes are hydrolases; therefore, one of the substrates in each reaction is water; (c) the pH optimum of each of the hydrolases is in the acid range, pH 3.5–5.5; (d) most of the enzymes are relatively stable and occur as isoenzymes. For example, **hexosaminidase** occurs in two forms: hexosaminidase A (HexA) and hexosaminidase B (HexB); (e) the hydrolases of the sphingolipid pathway are glycoprotein in character and often occur firmly bound to the lysosomal membrane; and (f) the pathway is composed of a series of compounds, which differ by only one sugar molecule, a sulfate group, or a fatty acid residue. The substrates are converted to products by the sequential, stepwise removal of constituents such as sugars and sulfate, by hydrolytic, irreversible reactions.

In most cases, the pathway of sphingolipid catabolism functions smoothly, and all of the various complex glycosphingolipids and sphingomyelin are degraded to the level of their basic building blocks, namely, sugars, sulfate, fatty acid, phosphocholine, and sphingosine. However, when the activity of one of the hydrolytic enzymes in the pathway is markedly reduced in the tissues of a person due to a genetic, inborn error, then the substrate for the defective or missing enzyme accumulates and is deposited within the lysosomes of the tissue responsible for the catabolism of that lipid. For most of the reactions in Figure 10.59, patients have been identified who lack the enzyme that normally catalyzes that reaction. These disorders, called sphingolipidoses, are summarized in Table 10.2.

One can generalize about some of the common features of lipid storage diseases: (a) usually only a single sphingolipid accumulates in the involved organs; (b) the ceramide portion is shared by the various storage lipids; (c) the rate of biosynthesis of the accumulating lipid is normal; (d) a catabolic enzyme is missing in each of these disorders; and (e) the extent of the enzyme deficiency is the same in all tissues.

The diagnosis of a given sphingolipidosis can be made from a biopsy of the involved organ, usually bone marrow, liver, or brain, or on morphologic grounds on the basis of the highly characteristic appearance of the storage lipid within lysosomes. Analysis of the enzymes is used to confirm the diagnosis of a particular lipid storage disease. Of great practical value is the fact that, for most of the diseases, peripheral leukocytes, cultured skin fibroblasts and chorionic villus express the relevant enzyme deficiency and can be used as a source of enzyme for diagnostic purposes. In some cases (e.g., Tay–Sachs disease) serum, and even tears, have been used as a source of enzyme for the diagnosis of a lipid storage disorder. The sphingolipid storage diseases, for the most part, are recessive in terms of their hereditary mode of transmission, with the disease occurring only in homozygotes with a defect in both chromosomes. Enzyme assays can be used to identify carriers or heterozygotes. Let us consider some representative examples of the use of enzyme assays for diagnostic purposes.

TABLE 10.2 Sphingolipid Storage Diseases of Humans

Disorder	Principal signs and symptoms	Principal storage substance	Enzyme deficiency
1. Tay–Sachs disease	Mental retardation, blindness, cherry red spot on macula, death between second and third year	Ganglioside G_{M2}	Hexosaminidase A
2. Gaucher's disease	Liver and spleen enlargement, erosion of long bones and pelvis, mental retardation in infantile form only	Glucocerebroside	Glucocerebrosidase
3. Fabry's disease	Skin rash, kidney failure, pains in lower extremities	Ceramide trihexoside	α-Galactosidase A
4. Niemann–Pick disease	Liver and spleen enlargement, mental retardation	Sphingomyelin	Sphingomyelinase
5. Globoid leukodystrophy (Krabbe's disease)	Mental retardation, absence of myelin	Galactocerebroside	Galactocerebrosidase
6. Metachromatic leukodystrophy	Mental retardation, nerves stain yellowish brown with cresyl violet dye (metachromasia)	Sulfatide	Arylsulfatase A
7. Generalized gangliosidosis	Mental retardation, liver enlargement, skeletal involvement	Ganglioside G_{M1}	G_{M1} ganglioside: β-galactosidase
8. Sandhoff–Jatzkewitz disease	Same as 1; disease has more rapidly progressing course	G_{M2} ganglioside, globoside	Hexosaminidase A and B
9. Fucosidosis	Cerebral degeneration, muscle spasticity, thick skin	Pentahexosylfucoglycolipid	α-L-Fucosidase

In Niemann–Pick disease, the deficient enzyme is *sphingomyelinase,* which normally catalyzes the reaction shown in Figure 10.60. Sphingomyelin, radiolabeled in the methyl groups of choline with carbon-14, provides a useful substrate for determining sphingomyelinase activity. Extracts of white blood cells from healthy, appropriate controls, will hydrolyze the labeled substrate and produce the water-soluble product, [^{14}C]-phosphocholine. Extraction of the final incubation medium with an organic solvent such as chloroform will result in radioactivity in the up-

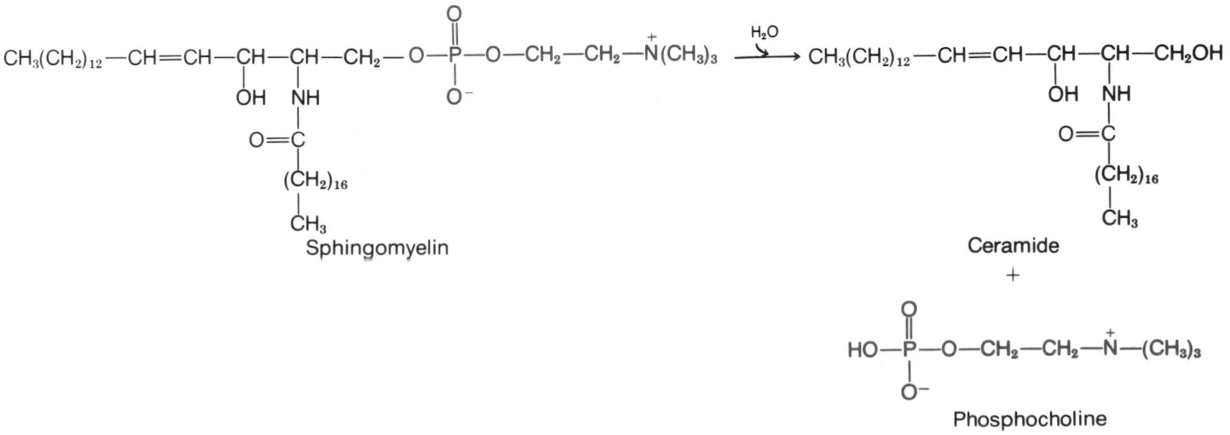

$$CH_3(CH_2)_{12}{-}CH{=}CH{-}CH(OH){-}CH(NH{-}C({=}O){-}(CH_2)_{16}{-}CH_3){-}CH_2{-}O{-}P({=}O)(O^-){-}O{-}CH_2{-}CH_2{-}\overset{+}{N}(CH_3)_3 \xrightarrow{H_2O}$$

Sphingomyelin

$$CH_3(CH_2)_{12}{-}CH{=}CH{-}CH(OH){-}CH(NH{-}C({=}O){-}(CH_2)_{16}{-}CH_3){-}CH_2OH$$

Ceramide

+

$$HO{-}P({=}O)(O^-){-}O{-}CH_2{-}CH_2{-}\overset{+}{N}{-}(CH_3)_3$$

Phosphocholine

FIGURE 10.60
The sphingomyelinase reaction.

per, aqueous phase; the unused, lipid-like substrate sphingomyelin will be found in the chloroform phase. On the other hand, if the white blood cells were derived from a patient with Niemann–Pick disease, then after incubation with labeled substrate and extraction with chloroform, little or no radioactivity (i.e., phosphocholine) would be found in the aqueous phase and the diagnosis of sphingomyelinase deficiency or Niemann–Pick disease would be confirmed.

Many of the lysosomal hydrolases have broad or versatile substrate specificity and hydrolyze not only natural sphingolipid substrate but also nonphysiologic synthetic substrates that can be measured colorimetrically and fluorometrically.

Another disease that can be diagnosed by use of an artificial substrate is Tay–Sachs disease. Tay–Sachs disease is the most common form of **G_{M2} gangliosidosis.** In this fatal disorder the ganglion cells of the cerebral cortex are swollen and the lysosomes are engorged with the acidic lipid, G_{M2} ganglioside. This results in a loss of ganglion cells, proliferations of glial cells, and demyelination of peripheral nerves. The pathognomonic finding is a cherry red spot on the macula caused by swelling and necrosis of ganglion cells in the eye. In Tay–Sachs disease, the commercially available artificial substrate 4-methylumbelliferyl-β-*N*-acetylglucosamine is used to confirm the diagnosis. The compound is recognized and hydrolyzed by hexosaminidase A, the deficient lysosomal hydrolase, to produce the intensely fluorescent product 4-methylumbelliferone (Figure 10.61). Unfortunately, the diagnosis may be confused by the presence of hexosaminidase B in tissue extracts and body fluids. This enzyme is not deficient in the Tay–Sachs patient and will hydrolyze the test substrate, thereby confusing the interpretation of results. The problem is usually resolved by taking advantage of the relative heat lability of hexosaminidase A and the heat stability of hexosaminidase B. The tissue extract or serum specimen to be tested is first heated at 55°C for 1 h and then assayed for hexosaminidase activity. The amount of heat-labile activity is a measure of hexosaminidase A, and this value is used in making the diagnosis of Tay–Sachs disease.

Enzyme assays of serum or extracts of tissues, peripheral leukocytes and fibroblasts have proven useful in heterozygote detection. Once carriers of a lipid storage disease have been identified, or if there has been a previously affected child in a family, the pregnancies at risk for these diseases can be monitored. All nine of the lipid storage disorders are transmitted as recessive genetic abnormalities. In all but one the allele is carried on an autosomal chromosome. Fabry's disease is linked to the X chromosome. In all of these conditions statistically one of four pregnancies will be homozygous (or hemizygous in Fabry's disease), two fetuses

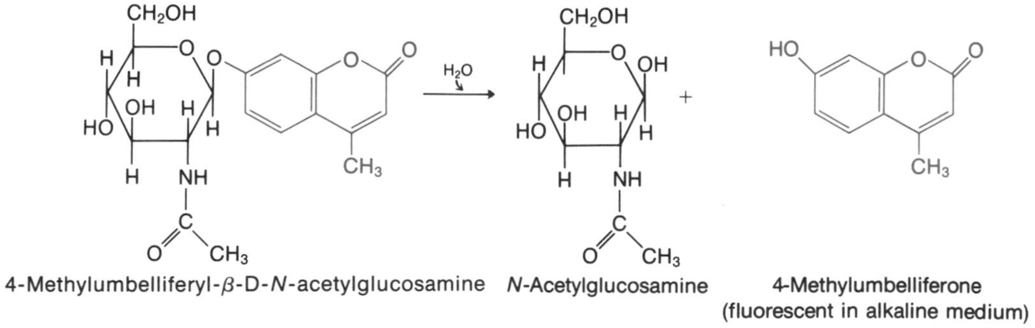

4-Methylumbelliferyl-β-D-*N*-acetylglucosamine $\xrightarrow{H_2O}$ *N*-Acetylglucosamine + 4-Methylumbelliferone (fluorescent in alkaline medium)

FIGURE 10.61
The β-hexosaminidase reaction.

will be carriers, and one will not be involved at all. The enzyme assay procedures have been used to detect affected fetuses and carriers in utero, using cultured fibroblasts obtained by amniocentesis as a source of enzyme.

There is no therapy for the sphingolipidoses; the role of medicine at the present time is prevention through genetic counseling based upon enzymologic assays of the type discussed above. A discussion of the diagnosis of Gaucher's disease is presented in Clin. Corr. 10.4.

10.5 PROSTAGLANDINS AND THROMBOXANES

Prostaglandins and Thromboxanes Are Derivatives of Twenty-Carbon, Monocarboxylic Acids

In mammalian cells there are two major pathways of arachidonic acid metabolism that produce important mediators of cellular and bodily functions: the *cyclooxygenase* and the *lipoxygenase* pathways. The substrate for both pathways is unesterified arachidonic acid. The cyclooxygenase pathway leads to a series of compounds including prostaglandins and thromboxanes. The prostaglandins were discovered through their effects on smooth muscle, specifically their ability to promote the contraction of intestinal and uterine muscle and the lowering of blood pressure. Although the complexity of their structures and the diversity of their sometimes conflicting functions often create a sense of frustration, the potent pharmacological effects of the prostaglandins have afforded them an important place in human biology and medicine. With the exception of the red blood cell, the prostaglandins are produced and released by nearly all mammalian cells and tissues; they are not confined to specialized cells as insulin is to the pancreas. Furthermore, unlike most other hormones, the prostaglandins are not stored in cells but instead are synthesized and released immediately.

There are three major classes of primary prostaglandins, the A, E, and F series. The structures of the more common prostaglandins A, E, and F are shown in Figure 10.62. They are all related to prostanoic acid (Figure 10.63). Note that the prostaglandins contain a multiplicity of functional groups; for example, PGE_2 contains a carboxyl group, a β-hydroxyketone, a secondary alkylic alcohol and two carbon–carbon double bonds. The three classes (A, E, and F) are distinguished on the basis of the functional groups about the cyclopentane ring: the E type is a β-hydroxyketone, the F series are 1,3-diols, and those in the A series are α,β-unsaturated ketones. The subscript numerals 1, 2, or 3, refer to the number of double bonds in the side chains. The subscript "α" refers to the configuration of the C-9 OH group: an α-hydroxyl group projects "down" from the plane of the ring.

The most important dietary precursor of the prostaglandins is linoleic acid (18:2), which is an essential fatty acid. In adults linoleic acid is ingested daily in amounts of about 10 g. Only a very minor part of this total intake is converted by elongation and desaturation in the liver to arachidonic acid (eicosatetraenoic acid) and to some extent also to dihomo-γ-linoleic acid. Since the total daily excretion of prostaglandins and their metabolites is only about 1 mg, it is clear that the formation of prostaglandins is a quantitatively unimportant pathway in the overall metabolism of fatty acids. At the same time, however, the metabolism of prostaglandins is completely dependent on a regular and constant supply of linoleic acid. When the diet is deficient in linoleic acid, there is decreased production of prostaglandins.

CLIN. CORR. 10.4
DIAGNOSIS OF GAUCHER'S DISEASE IN AN ADULT

Gaucher's disease is an inherited disease of lipid catabolism that results in deposition of glucocerebroside in macrophages of the reticuloendothelial system. Because of the large numbers of macrophages in the spleen, bone marrow and liver, hepatomegaly, splenomegaly and its sequelae (thrombocytopenia or anemia) and bone pain are the most common signs and symptoms of the disease.

Gaucher's disease results from a deficiency of glucocerebrosidase. Although this enzyme deficiency is inherited, different clinical patterns are observed. Some patients suffer severe neurologic deficits as infants, while others do not exhibit symptoms until adulthood. The diagnosis can be made by assaying leukocytes or fibroblasts for their ability to hydrolyze the β-glycosidic bond of artificial substrates (β-glucosidase activity) or of glucocerebroside (glucocerebrosidase activity). Gaucher's disease has been treated with regular infusions of purified glucocerebrosidase but the long-term efficacy of the therapy is not yet known.

Brady, R. O., Kanfer, J. N., Bradley, R. M., and Shapiro, D. Demonstration of a deficiency of glucocerebroside-cleaving enzyme in Gaucher's disease. *J. Clin. Invest.* 45:1112, 1966.

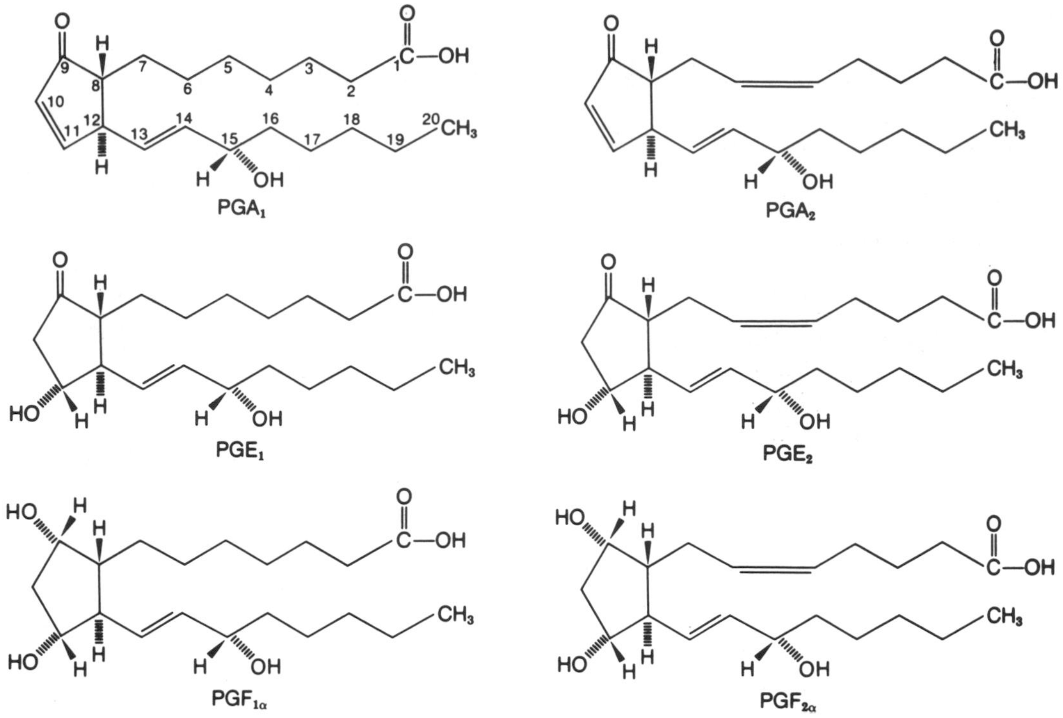

FIGURE 10.62
Structures of the major prostaglandins.

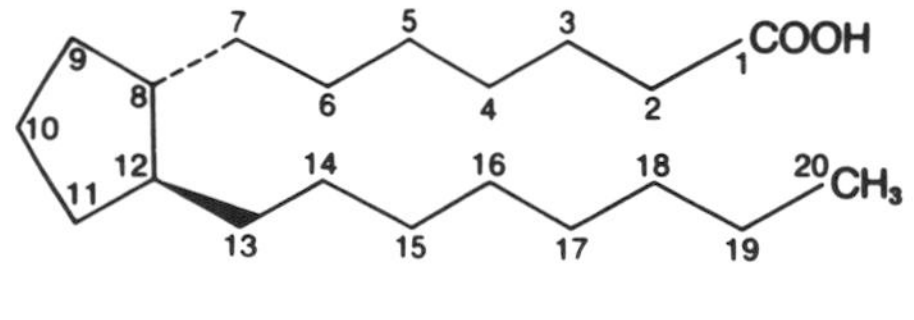

FIGURE 10.63
Structure of prostanoic acid.

Synthesis of Prostaglandins Involves a Cyclooxygenase

The immediate precursors to the prostaglandins are C_{20} polyunsaturated fatty acids containing 3, 4, and 5 carbon–carbon double bonds. Since **arachidonic acid** and most of its metabolites contain 20 carbon atoms, they are referred to as **eicosanoids.** During their transformation into various prostaglandins they are cyclized and take up oxygen. Dihomo-γ-linoleic acid (C_{20}-Δ8,11,14) is the precursor to PGE$_1$ and PGF$_{1\alpha}$; arachidonic acid (C_{20}-Δ5,8,11,14) is the precursor to PGE$_2$ and PGF$_{2\alpha}$; and eicosopentaenoic acid (C_{20}-Δ5,8,11,14,17) is the precursor to PGE$_3$ and PGF$_{3\alpha}$ (See Figure 10.64).

Compounds of the 2-series derived from arachidonic acid are the principal prostaglandins in humans and are of the greatest significance biologically. Thus, one should focus attention primarily on the metabolism of arachidonic acid.

The central enzyme system in prostaglandin biosynthesis is the **prostaglandin synthase complex,** which catalyzes the oxidative cyclization of polyunsaturated fatty acids. The major pathway of prostaglandin biosynthesis will be illustrated, using arachidonic acid as substrate. Arachidonic acid is derived from membrane phospholipids by the action of the hydrolase phospholipase A$_2$. This esterolytic cleavage step is important because it is the rate-limiting step in prostaglandin synthesis and because certain agents that stimulate prostaglandin production act by stimulating the activity of phospholipase A$_2$. Cholesterol esters containing arachidonic acid may also serve as a source of arachidonic acid substrate.

The first step catalyzed by the **cyclooxygenase** component of the prostaglandin synthase complex involves the cyclization of C-9–C-12 of arachidonic acid to form the cyclic 9, 11-endoperoxide 15-hydroperoxide, PGG$_2$, and the reaction requires two molecules of oxygen as shown in Figure 10.65. The PGG$_2$ is then converted to prostaglandin H$_2$ (PGH$_2$) by

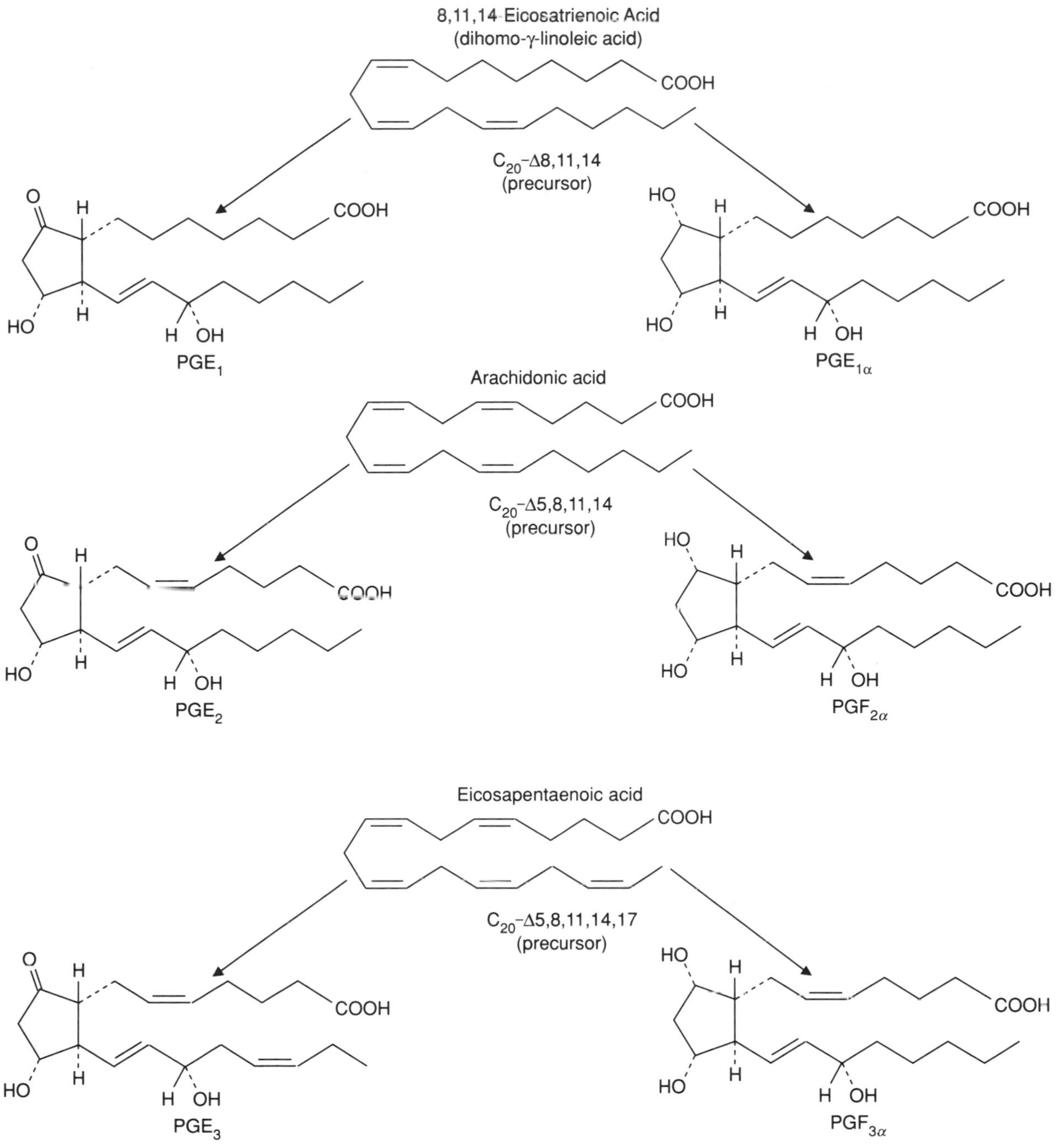

8,11,14-Eicosatrienoic Acid
(dihomo-γ-linoleic acid)
COOH
C_{20}–Δ8,11,14
(precursor)
COOH
COOH
HO H OH
PGE$_1$
PGE$_{1\alpha}$

Arachidonic acid
COOH
C_{20}–Δ5,8,11,14
(precursor)
COOH
COOH
PGE$_2$
PGF$_{2\alpha}$

Eicosapentaenoic acid
COOH
C_{20}–Δ5,8,11,14,17
(precursor)
COOH
COOH
PGE$_3$
PGF$_{3\alpha}$

FIGURE 10.64
Synthesis of E and F prostaglandins from fatty acid precursors.

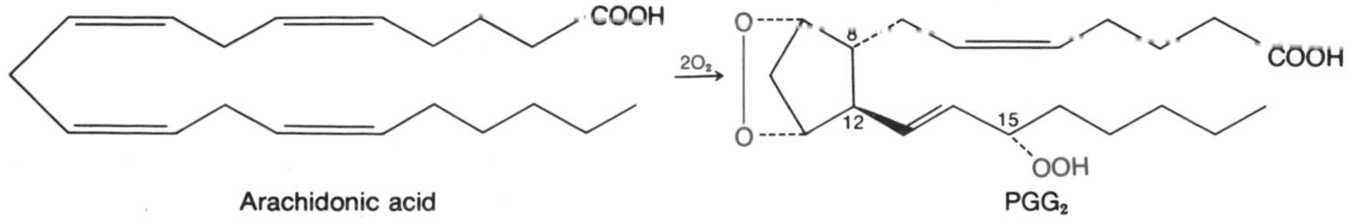

COOH
$2O_2$
8
12
15
OOH
COOH
Arachidonic acid
PGG$_2$

FIGURE 10.65
The cyclooxygenase reaction.

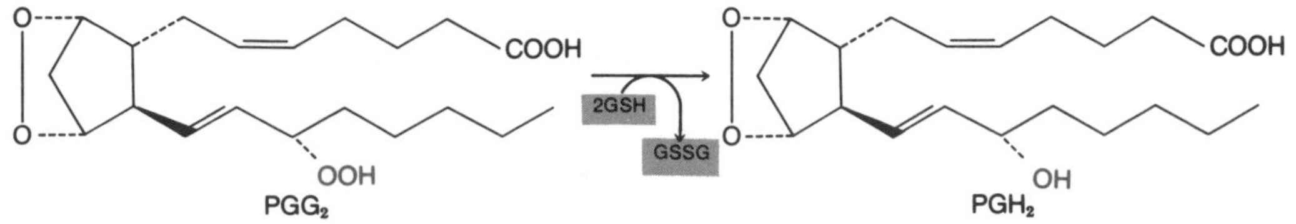

FIGURE 10.66
Conversion of PGG_2 to PGH_2; the PG hydroperoxidase (PGH synthase) reaction.

a reduced glutathione (GSH)-dependent peroxidase (PG hydroperoxidase) that catalyzes the reaction shown in Figure 10.66. The details of the additional steps leading to the individual prostaglandins remain to be elucidated. The reactions that cyclize polyunsaturated fatty acids are found in the membranes of the endoplasmic reticulum. The major pathways of prostaglandin biosynthesis are summarized in Figure 10.67. The formation of the primary prostaglandins of the D, E, and F series and of thromboxanes or prostacyclin (PGI_2) is mediated by different specific enzymes, whose presence varies depending upon the cell type and tissue. This results in a degree of tissue specificity as to the type and quantity of prostaglandin produced. Thus, in the kidney and spleen PGE_2 and $PGF_{2\alpha}$ are the major prostaglandins produced. In contrast, blood vessels produce mostly PGI_2. In the heart PGE_2, $PGF_{2\alpha}$, and PGI_2 are formed in about equal amounts. Thromboxane A_2 (TXA_2) is the main prostaglandin endoperoxide formed in platelets.

The prostaglandins have a very short half-life. Soon after release they are rapidly taken up by cells and inactivated. The lungs appear to play an important role in inactivating prostaglandins.

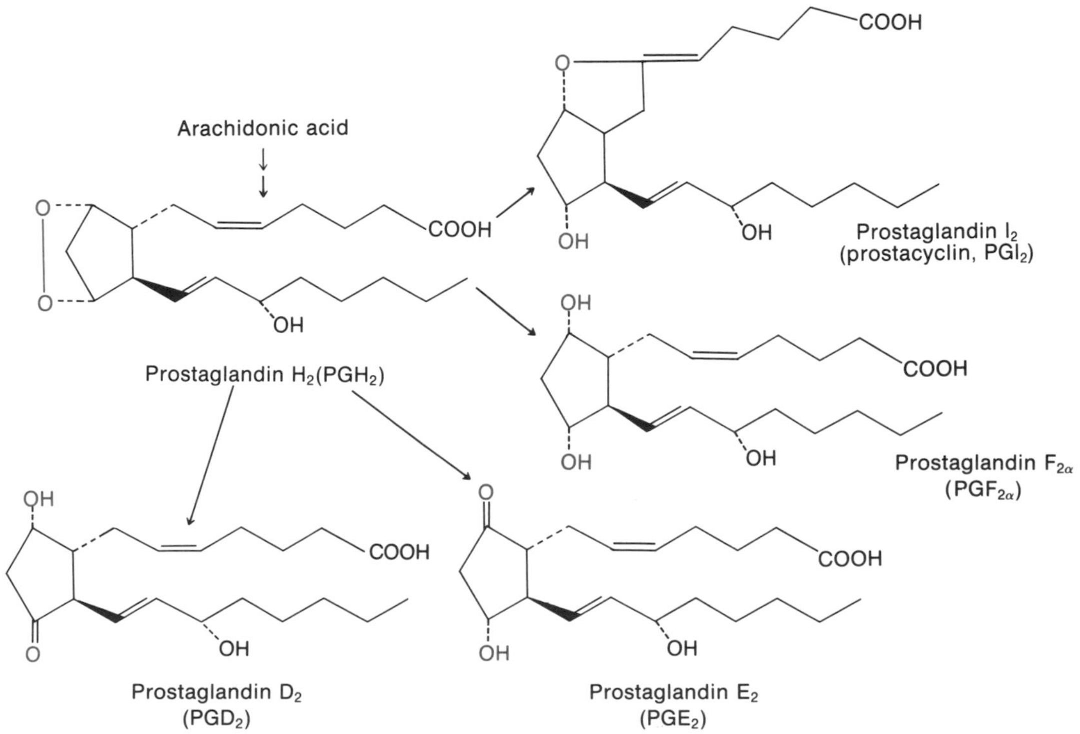

FIGURE 10.67
Major routes of prostaglandin biosynthesis.

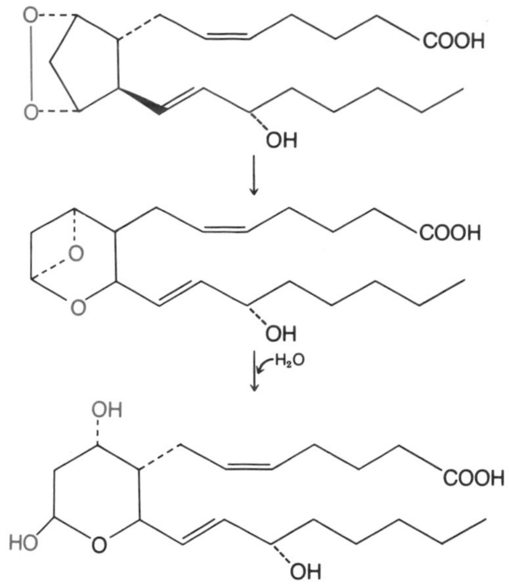

PGH_2

TXA_2

TXB_2

FIGURE 10.68
Synthesis of TXB_2 from PGH_2.

Thromboxanes, mentioned above, are the highly active metabolites of the PGG_2- and PGH_2-type prostaglandin endoperoxides that have the cyclopentane ring replaced by a six-membered oxygen-containing (oxane) ring. The term thromboxane is derived from the fact that these compounds have a thrombus-forming potential. **Thromboxane A synthase,** present in the endoplasmic reticulum, is abundant in lung and platelets and catalyzes the conversion of endoperoxide PGH_2 to TXA_2. The half-life of TXA_2 is very short in water ($t_{1/2}$, 1 min) as the compound is transformed rapidly into biologically inactive thromboxane B_2 (TXB_2) by the reaction shown in Figure 10.68.

Prostaglandin Production Is Inhibited by Both Steroidal and Non-Steroidal Anti-Inflammatory Agents

Clinically there are two types of drugs that affect prostaglandin metabolism and are therapeutically useful. First, there are the nonsteroidal, antiinflammatory agents such as aspirin (acetylsalicylic acid), indomethacin, and phenylbutazone, which block prostaglandin production by irreversibly inhibiting the enzyme cyclooxygenase. In the case of aspirin, inhibition occurs presumably by acetylation of the enzyme. These drugs are not without their undesirable side effects; aplastic anemia can result from phenylbutazone therapy. The second group, the steroidal antiinflammatory drugs like hydrocortisone, prednisone, and betamethasone, appear to act by blocking prostaglandin release by inhibiting phospholipase A_2 activity so as to interfere with mobilization of arachidonic acid, the substrate for cyclooxygenase (see Figure 10.69).

The factors that govern the biosynthesis of prostaglandins are poorly understood, but, in general, prostaglandin release seems to be triggered following hormonal or neural excitation or after muscular activity. For example, histamine stimulates an increase in the prostaglandin concentration in gastric perfusates. Also, prostaglandins are released during labor and after cellular injury (e.g., platelets exposed to thrombin, lungs irritated by dust).

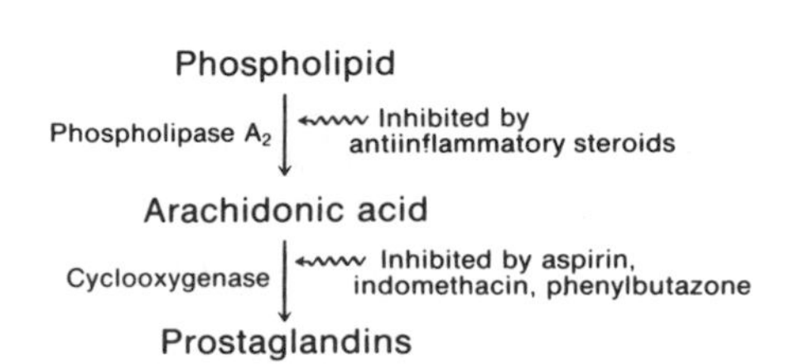

FIGURE 10.69
Site of action of inhibitors of prostaglandin synthesis.

Prostaglandins Exhibit a Multitude of Physiological Effects

Inflammation: Prostaglandins appear to be one of the natural mediators of inflammation. Inflammatory reactions most often involve the joints (rheumatoid arthritis), skin (psoriasis), and eyes, and inflammation of these sites is frequently treated with corticosteroids that inhibit prostaglandin synthesis. Administration of the prostaglandins PGE_2 and PGE_1 induce the signs of inflammation that include redness and heat (due to arteriolar vasodilation), and swelling and edema resulting from increased capillary permeability.

Pain and fever: PGE_2 in amounts that alone do not cause pain, prior to administration of the autocoids, histamine and bradykinin, enhance both the intensity and the duration of pain caused by these two agents. It is thought that pyrogen activates the prostaglandin biosynthetic pathway, resulting in the release of PGE_2 in the region of the hypothalamus where body temperature is regulated. Aspirin, which is an antipyretic drug, acts by inhibiting cyclooxygenase.

Reproduction: The prostaglandins have been used extensively as drugs in the reproductive area. Both PGE_2 and PGF_2 have been used to induce parturition and for the termination of an unwanted pregnancy. There is also evidence that the PGE series of prostaglandins may play some role in infertility in males.

Gastric secretion and peptic ulcer: Synthetic prostaglandins have proven to be very effective in inhibiting gastric acid secretion in patients with peptic ulcers. The inhibitory effect of PGE compounds appears to be due to inhibition of cAMP formation in gastric mucosal cells. Prostaglandins also accelerate the healing of gastric ulcers.

Regulation of blood pressure: Prostaglandins play an important role in controlling blood vessel tone and arterial pressure. The vasodilator prostaglandins, PGE, PGA, and PGI_2, lower systemic arterial pressure, thereby increasing local blood flow and decreasing peripheral resistance. TXA_2 causes contraction of vascular smooth muscle and glomerular mesangium. There is hope that the prostaglandins may eventually prove useful in the treatment of hypertension.

Ductus arteriosus and congenital heart disease: PGE_2 functions in the fetus to maintain the patency of the ductus arteriosus prior to birth. There are two clinical applications of prostaglandin biochemistry in this area. First, if the ductus remains open after birth, closure can be hastened by administration of the cyclooxygenase inhibitor indomethacin. In other situations it may be desirable to keep the ductus open. For example, in the case of infants born with congenital abnormalities where the defect can be corrected surgically, infusion of prostaglandins will maintain blood flow through the ductus over this interim period.

Platelet aggregation and thrombosis: Certain prostaglandins, especially PGI_2, inhibit platelet aggregation, whereas PGE_2 and TXA_2 promote this clotting process. TXA_2 is produced by platelets and accounts for the spontaneous aggregation that occurs when platelets contact some foreign surface, collagen or thrombin. Endothelial cells lining blood vessels release PGI_2 and may account for the lack of adherence of platelets to the healthy blood vessel wall.

10.6 LIPOXYGENASE AND THE OXY-EICOSATETRAENOIC ACIDS

In addition to cyclooxygenase, which directs polyunsaturated fatty acids into the prostaglandin pathway, there exists another equally important arachidonic acid oxygenating enzyme, called **lipoxygenase,** which is a dioxygenase. Actually, there is a family of lipoxygenases that differs in the position of the double bond on the arachidonic acid molecule at which oxygen attack initially occurs (e.g., positions 5, 11, or 15).

Monohydroperoxyeicosatetraenoic Acids Are the Products of Lipoxygenase Action

The products of the lipoxygenase reaction, which arises by addition of hydroperoxy groups to arachidonic acid, are designated **monohydroperoxy–eicosatetraenoic acids** (HPETEs). Figure 10.70 shows the conversion of arachidonic acid to the three major HPETEs. Thus, in contrast to the cyclooxygenase component of prostaglandin endoperoxide synthetase, which catalyzes the bis-dioxygenation of unsaturated fatty acids to endoperoxides, lipoxygenases catalyze the monodioxygenation of unsaturated fatty acids to allylic hydroperoxides. Hydroperoxy substitution of arachidonic acid by lipoxygenases in humans may occur at position 5, 12, or 15. 5-HPETE is the major lipoxygenase product in basophils, polymorphonuclear (PMN) leukocytes, macrophages, mast cells, and any organ undergoing an inflammatory response; 12-HPETE predominates in platelets, pancreatic endocrine islet cells, vascular smooth muscle, and glomerular cells; 15-HPETE is the principal lipoxygenase product in reticulocytes, eosinophils, T-lymphocytes and tracheal epithelial cells. The 5-, 12-, and 15- lipoxygenases occur mainly in the cytoplasm. Specific stimuli or signals determine which type of lipoxygenase product a given type of cell produces. The oxygenated carbon atom in the HPETEs is an asymmetric one and there are two possible stereoisomers of the hydroperoxy acid, namely, (*R*) or (*S*). All three of the major HPETEs are of the (*S*) configuration.

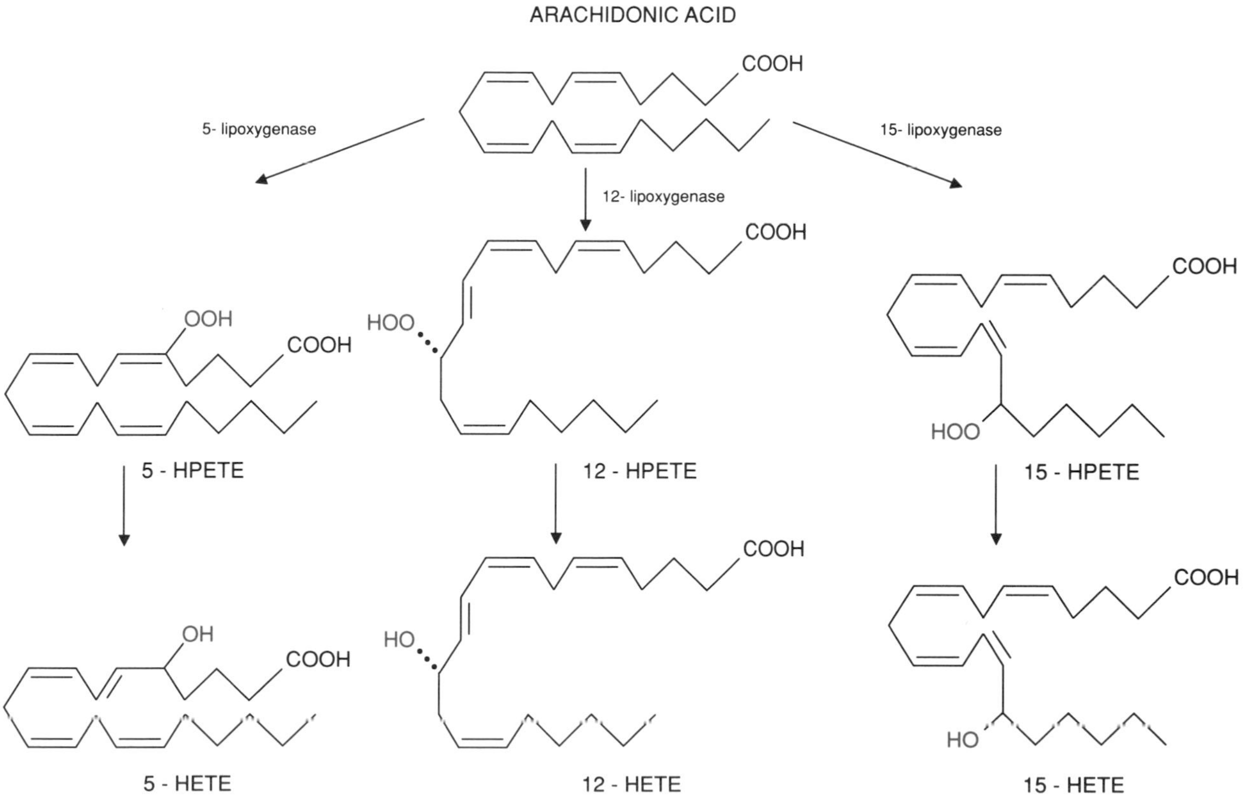

FIGURE 10.70
The lipoxygenase reaction and the role of 5-hydroperoxyeicosatetraenoic acids (HPETEs) as precursors of hydroxyeicosatetraenoic acids (HETEs).

Leukotrienes and Hydroxyeicosatetraenoic Acids Are the Hormones Derived From HPETEs

The HPETE-hydroperoxides themselves are not hormones, but instead are highly reactive, unstable intermediates that are converted either to the analogous alcohol (hydroxy fatty acid) by reduction of the peroxide moiety or to leukotrienes. The term leukotriene applies to lipoxygenase products that contain at least three conjugated double bonds. Figure 10.71 shows how 5-HPETE rearranges to the epoxide **leukotriene A_4** (LTA_4), which is then converted to LTB_4 or LTC_4, thus emphasizing that this particular HPETE occurs at an important branch point in the lipoxygenase pathway.

The peroxidative reduction of 5-HPETE to the stable **5-hydroxyeicosatetraenoic acid** (5-HETE) is illustrated in Figure 10.70. Note that the carbon–carbon double bonds in 5-HETE occur at positions 6, 8, 11, and 14, that these double bonds are unconjugated, and that the geometry of the double bonds is trans, cis, cis, cis, respectively. Two other common forms of HETE are 12- and 15-HETE. The HPETEs are reduced either spontaneously, or by the action of peroxidases, to the corresponding HETEs.

Leukotrienes are derived from the unstable precursor 5-HPETE by a reaction catalyzed by **LTA_4 synthase** that generates an epoxide called LTA_4. In the leukotriene series, the subscript indicates the number of carbon–carbon double bonds. Thus, while double bond rearrangement may occur, the number of double bonds in the leukotriene product is the

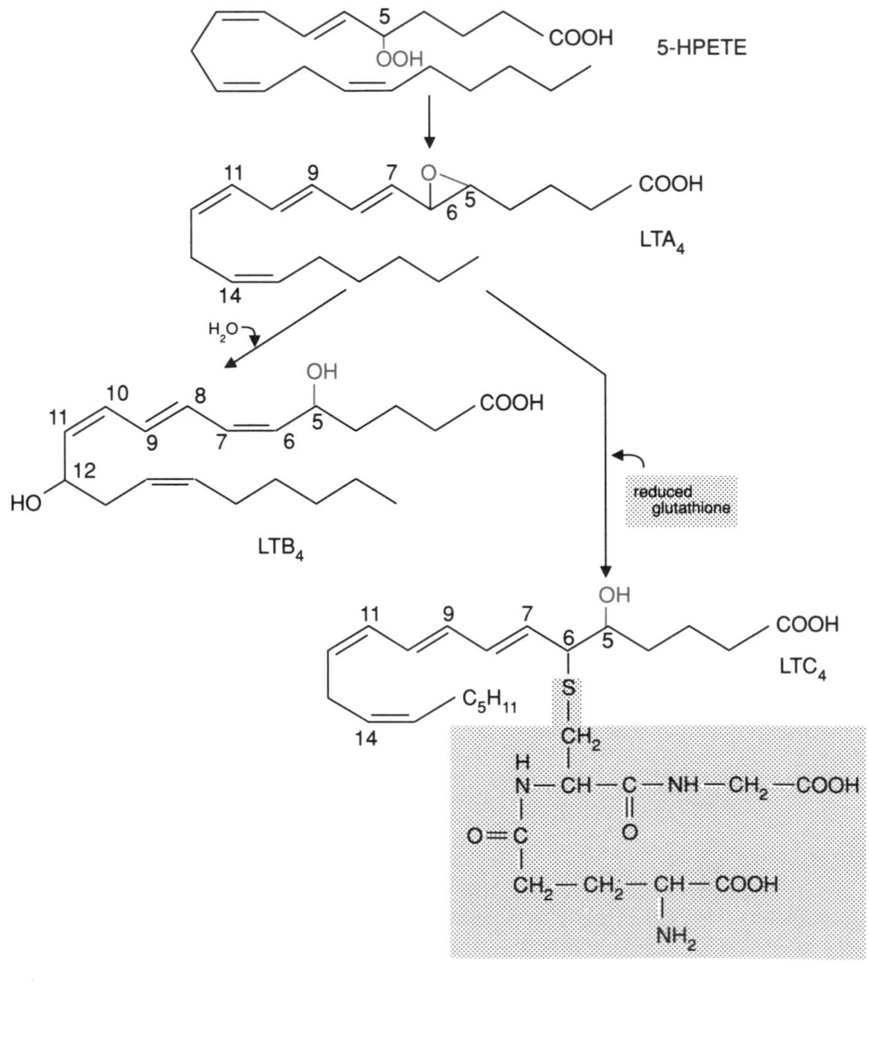

FIGURE 10.71
Conversion of 5-HPETE to LTB_4 and LTC_4 through the intermediate of LTA_4.

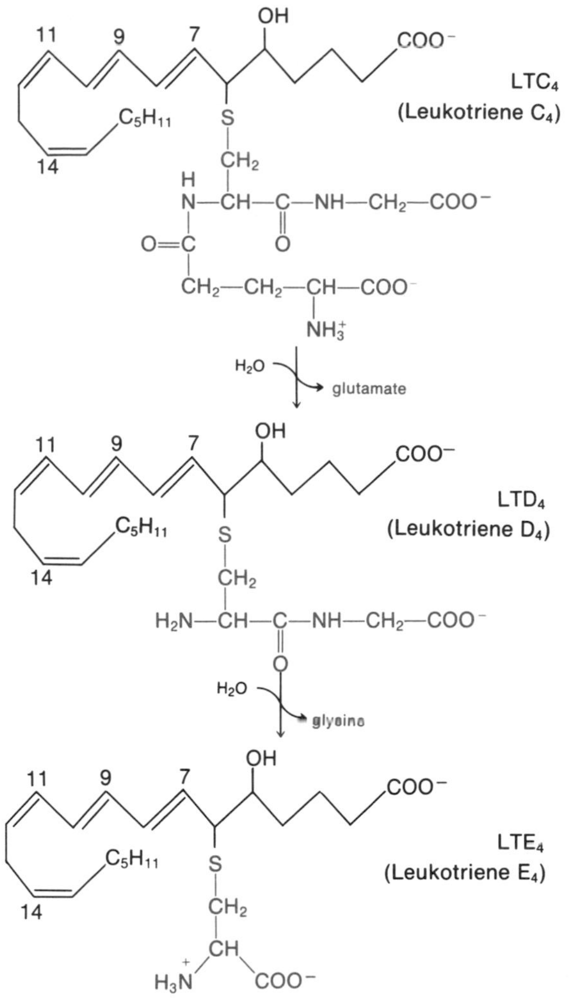

FIGURE 10.72
Conversion of LTC_4 to LTD_4 and LTE_4.

same as in the original arachidonic acid-lipoxygenase substrate. As indicated in Figure 10.71, LTA_4 occurs at a branch point; it can be converted either to 5, 12-dihydroxyeicosatetraenoic acid (designated leukotriene B_4 or LTB_4) or to the leukotrienes LTC_4 and LTD_4.

Conversion of 5-HPETE to the diol LTB_4, shown in Figure 10.71, is catalyzed by a cytosolic enzyme, **LTB_4 synthase** (LTA_4 hydrolase), in a reaction initiated by the addition of water to the double bond between C-11 and C-12 atoms.

The diversion of LTA_4 to leukotrienes LTC_4, LTD_4, and LTE_4 requires the participation of reduced glutathione that opens the epoxide ring in LTA_4 to produce LTC_4 (Figure 10.71). Sequential removal of glutamic acid and glycine residues by specific peptidases yields the leukotrienes LTD_4 and LTE_4 (Figure 10.72). The subscript 4 denotes the total number of carbon–carbon double bonds.

Leukotrienes and HETEs Affect Several Physiological Processes

The leukotrienes persist for as long as 4 h in the body, but little is known about the mechanisms that degrade or eliminate them. The biological actions of the thionyl peptides LTC_4, and LTD_4, and LTE_4 comprise what has been referred to for decades as the *slow-reacting substance of anaphylaxis* (SRS-A). They cause slowly evolving but protracted contraction of smooth muscles in the airways and gastrointestinal tract. The

LTC_4 is rapidly converted to LTD_4 and then slowly converted to LTE_4. These conversions are catalyzed by enzymes in plasma.

In general, the HETEs (especially 5-HETE) and LTB_4 are involved mainly in regulating neutrophil and eosinophil function: They mediate chemotaxis, stimulate adenylate cyclase, and induce PMNs to degranulate and release lysosomal hydrolytic enzymes. In contrast, the leukotrienes LTC_4 and LTD_4 are humoral agents that promote smooth muscle contraction, constriction of pulmonary airways, trachea and intestine, and changes in capillary permeability (edema). The HETEs appear to exert their effects by being incorporated into the phospholipids of target cells. It is thought that the presence of fatty acyl chains containing a polar OH group may disturb the packing of lipids and thus the structure and function of the membrane.

The monohydroxyeicosatetraenoic acids that comprise the lipoxygenase pathway are potent mediators of processes involved in allergy (hypersensitivity) and inflammation, secretion (e.g., insulin), cell movement, cell growth, and calcium fluxes. The initial allergic event, namely, the binding of IgE antibody to receptors on the surface of the mast cell, causes the release of substances, including leukotrienes, that are referred to as mediators of immediate hypersensitivity. Lipoxygenase products are usually produced within minutes after the stimulus. The leukotrienes LTC_4, LTD_4, and LTE_4 are much more potent than histamine in contracting nonvascular smooth muscles of bronchi and intestine. Leukotriene LTD_4 increases the permeability of the microvasculature. The mono-HETEs and LTB_4 stimulate migration (chemotaxis) of eosinophils and neutrophils, making them the principal mediators of PMN-leukocyte infiltration in inflammatory reactions.

Eicosatrienoic acids (e.g., dihomo-γ-linolenic acid) and eicosapentaenoic acid (Fig. 10.64) also serve as lipoxygenase substrates. The content of these C_{20} fatty acids with three and five carbon–carbon double bonds in tissues is less than that of arachidonic acid, but special diets can increase their levels. The lipoxygenase products of these tri- and pentaeicosanoids are usually less active than LTA_4 or LTB_4. It remains to be determined if fish oil diets rich in eicosapentaenoic acid are useful in the treatment of allergic and autoimmune diseases.

Although a potential fertile ground for the application of pharmacologic agents, to date, therapeutic use of lipoxygenase inhibitors has been prevented by their toxicity or lack of specificity.

BIBLIOGRAPHY

Phospholipid Metabolism

Downes, C. P. The cellular functions of *myo*-inositol. *Biochem. Soc. Trans.* 17:259, 1989.

Low, M. G. Biochemistry of the glycosyl-phosphatidyl inositol membrane protein anchors. *Biochem. J.* 244:1, 1987.

Raetz, C. R. H. Molecular genetics of membrane phospholipid synthesis. *Am. Rev. Genet.* 20:253, 1986.

Shears, S. B. Metabolism of the inositol phosphates produced upon receptor activation. *Biochem. J.* 260:313, 1989.

Vance, D. E. and Vance, J. E. Biochemistry of Lipids and Membranes. Menlo Park: Benjamin/Cummings, 1985.

Cholesterol Synthesis

Brown, M. S. and Goldstein, J. L. A receptor-mediated pathway for cholesterol homeostasis. *Science* 232:68, 1986.

Goldstein, J. L. and Brown, M. S. Regulation of the mevalonate pathway. *Nature* 343:425, 1990.

Gordon, D. J. and Rifkind, B. M. High-density lipoprotein: the clinical implications of recent studies. *N. Engl. J. Med.* 321:1311, 1989.

Sphingolipids and the Sphingolipidoses

Grabowski, G. A., Gatt, S., and Horowitze, M. Acid β-glucosidase: enzymology and molecular biology of Gaucher disease. *Crit. Rev. Biochem. Mol. Biol.* 25:385, 1990.

Robinson, D. Shedding light on lysosomes—applications of fluorescence techniques to cell biology and diagnosis of lysosomal disorders. Biochem. Soc. Trans. 16:11, 1988.

Tsuji, S., Choudary, P. V., Martin, B. M., Stubblefield, B. K., Mayor, J. A., Barranger, J. A., and Ginns, E. I. A mutation in the human glucocerebrosidase gene in neuronopathic Gaucher's disease. *N. Engl. J. Med.* 316:570, 1987.

Lung Surfactant

Caminici, S. P. and Young, S. The pulmonary surfactant system. *Hosp. Pract.* 26:87, 1991.

Konishi, M., Fujiwara, T., Naito, T., Tokeuchi, Y., Ogawa, Y., Inukai, K., Fujimura, M., Nakamura, H., and Hashimoto, T. Surfactant replacement therapy in neonatal respiratory distress syndrome. *Eur. J. Pediatr.* 147:20, 1988.

Prostaglandins, Thromboxanes, and Leukotrienes

Fitzpatrick, F. A. and Murphy, R. C. Cytochrome P-450 metabolism of arachidonic acid: formation and biological actions of "epoxygenase"-derived eicosanoids. *Pharmacol. Rev.* 40:229, 1989.

Parker, C. W. Lipid mediators produced through the lipoxygenase pathway. *Annu. Rev. Immunol.* 5:65, 1987.

Yamamoto, S. Mammalian lipoxygenases: molecular and catalytic properties. *Prostaglandins, Leukotrienes and Essential Fatty Acids* 35:219, 1989.

Bile Acids

Angelin, B. and Einarsson, K. Bile acids and lipoprotein metabolism. *Bile Acids and Atherosclerosis,* S. M. Grundy (Ed.), Raven, New York: 1986, p. 41.

Bjorkhem, I. Mechanism of bile acid biosynthesis in mammalian liver. *New Comprehensive Biochemistry,* H. Danielsson and J. Sjovall (Eds.), Elsevier, Amsterdam: 1985, pp. 231–278.

QUESTIONS

C. N. ANGSTADT AND J. BAGGOTT

*Q*uestion *T*ypes are described on page xxiii.

A.

$H_2C-O-C(=O)-(CH_2)_{14}CH_3$
$CH_3(CH_2)_{16}-C(=O)-O-CH$
$H_2C-O-C(=O)-(CH_2)_{14}CH_3$

B.

$H_2C-O-C(=O)-(CH_2)_7CH{=}CH-(CH_2)_5CH_3$
$CH_3(CH_2)_3(CH_2-CH{=}CH)_4-(CH_2)_3-C(=O)-O-CH$
$H_2C-O-C(=O)-(CH_2)_{14}CH_3$

C.

$H_2C-O-CH{=}CH(CH_2)_{15}CH_3$
$R-C(=O)-O-CH$
$H_2C-O-P(=O)(O^-)-O-(CH_2)_2-\overset{+}{N}H_3$

D.

$H_2C-O-C(=O)-R_1$
$R_2-C(=O)-O-CH$
$H_2C-O-P(=O)(O^-)-O-(CH_2)_2-\overset{+}{N}(CH_3)_3$

E.

$H_2C-O-C(=O)-R_1 \quad CH_2-O-P(=O)(O^-)-O-CH_2$
$R_2C(=O)-O-CH \quad HO-C-H \quad HC-O-C(=O)-R_3$
$H_2C-O-P(=O)(O^-)-O-CH_2 \quad R_4C(=O)-O-CH_2$

1. (QT5) A plasmalogen.
2. (QT5) A cardiolipin.
3. (QT5) An acylglycerol that would likely be liquid at room temperature.

4. (QT2) Roles of various phospholipids include:
   1. a surfactant function in the lung.
   2. activation of certain membrane enzymes.
   3. signal transduction.
   4. cell–cell recognition.

5. (QT1) Which of the following represents a correct group of enzymes involved in phosphatidylcholine synthesis in ADIPOSE tissue?
   A. choline phosphotransferase, glycerol kinase, phosphatidic acid phosphatase.
   B. choline phosphotransferase, glycerol phosphate:acyl transferase, phosphatidylethanolamine N-methyltransferase.
   C. glycerol phosphate:acyl transferase, α-glycerolphosphate dehydrogenase, phosphatidic acid phosphatase.
   D. glycerol phosphate:acyl transferase, α-glycerolphosphate dehydrogenase, glycerol kinase.
   E. α-glycerolphosphate dehydrogenase, glycerol kinase, phosphatidic acid phosphatase.

6. (QT2) CDP-X (where X is the appropriate alcohol) reacts with 1,2-diacylglycerol in the primary synthetic pathway for:
   1. phosphatidylethanolamine.
   2. phosphatidylinositol.
   3. phosphatidylcholine.
   4. phosphatidylserine.

7. (QT1) Phospholipases A_1 and A_2:
   A. have no role in phospholipid synthesis.
   B. are responsible for the initial insertion of fatty acids in *sn*-1 and *sn*-2 positions during synthesis.
   C. are responsible for base exchange in the interconversion of phosphatidylethanolamine and phosphatidylserine.
   D. hydrolyze a phosphatidic acid to a diglyceride.
   E. remove a fatty acid in an *sn*-1 or *sn*-2 position so it can be replaced by another in phospholipid synthesis.

8. (QT2) In the biosynthesis of cholesterol:
   1. 3-hydroxy-3-methyl glutaryl CoA (HMG CoA) is synthesized by cytosolic HMG CoA synthase.
   2. HMG CoA reductase catalyzes the rate-limiting step.
   3. the conversion of mevalonic acid to farnesyl pyrophosphate requires more than 1 mol of ATP/mol of mevalonic acid.
   4. the conversion of squalene to lanosterol is initiated by formation of an epoxide.

9. (QT2) The cholesterol present in LDL (low-density lipoproteins):
   1. binds to a cell receptor and diffuses across the cell membrane.
   2. when it enters a cell, suppresses the cell's cholesterol synthesis by inhibiting HMG CoA reductase.
   3. once in the cell is converted to cholesterol esters by LCAT (lecithin-cholesterol acyl transferase).
   4. once it has accumulated in the cell, inhibits the replenishment of LDL receptors.

10. (QT1) Primary bile acids:
    A. are any bile acids that are found in the intestinal tract.
    B. are any bile acids reabsorbed from the intestinal tract.
    C. are synthesized in the intestinal tract by bacteria.
    D. are synthesized in hepatocytes directly from cholesterol.
    E. are converted to secondary bile acids by conjugation with glycine or taurine.

11. (QT2) A ganglioside may contain which of the following?
    1. One or more sialic acids
    2. Glucose or galactose
    3. A ceramide structure
    4. Phosphate

12. (QT1) Sphingomyelins differ from the other sphingolipids in that they are:
    A. not based on a ceramide core.
    B. acidic rather than neutral at physiological pH.
    C. the only types containing *N*-acetylneuraminic acid.
    D. the only types that are phospholipids.
    E. not amphipathic.

13. (QT2) The degradation of sphingolipids:
    1. occurs by hydrolytic enzymes contained in lysosomes.
    2. is a sequential, stepwise removal of constituents.
    3. is inhibited in the types of diseases known as sphingolipidoses (lysosomal storage diseases).
    4. is catalyzed by enzymes that are specific for a type of linkage rather than for a particular compound.

14. (QT2) Structural features that are common to all prostaglandins include:
    1. 20 carbon atoms.
    2. an internal ring structure.
    3. at least one double bond.
    4. a peroxide group at C-15.

15. (QT2) The prostaglandin synthase complex:
    1. contains both a cyclooxygenase and a peroxidase component.
    2. is inhibited by antiinflammatory steroids.
    3. produces PGH_2.
    4. uses as substrate the pool of free arachidonic acid in the cell.

16. (QT1) Thromboxane A_2:
    A. is a long-lived prostaglandin.
    B. is an inactive metabolite of PGE_2.
    C. is the major prostaglandin produced in all cells.
    D. does not contain a ring structure.
    E. is synthesized from the intermediate PGH_2.

17. (QT2) Hydroperoxy eicosatetraenoic acids (HPETEs):
    1. are derived from arachidonic acid by a lipoxygenase reaction.
    2. are mediators of hypersensitivity reactions.
    3. are intermediates in the formation of leukotrienes.
    4. are relatively stable compounds (persist for as long as 4 h).

ANSWERS

1. C Only one with an ether instead of an ester link at *sn*-1. D is a phosphatidylcholine (p. 426).
2. E Two phosphatidic acids connected by glycerol (p. 426).
3. B Note the two unsaturated fatty acids. A: With all saturated fatty acids, would likely be solid at room temperature.
4. A 1, 2, 3 correct. 1: Especially dipalmitoyllecithin (p. 427). 2: For example, β-hydroxybutyrate dehydrogenase (p. 427). 3: Especially phosphatidylinositol (p. 428). 4: This function appears to be associated with complex glycosphingolipids (p. 456).
5. C A, D, E: Glycerol kinase is not present in adipose tissue, which must rely on the α-glycerolphosphate dehydrogenase. This is a liver process only (p. 432).
6. B 1, 3 correct. 2: Phosphatidylinositol is formed from CDP-diglyceride reacting with myoinositol (Fig. 10.21). 4: This is formed by "base exchange" (Fig. 10.20).
7. E Phospholipases A_1 and A_2, as their names imply, hydrolyze a fatty acid from a phospholipid and so are part of phospholipid degradation. They are also important in synthesis, however, in assuring the asymmetric distribution of fatty acids that occurs in phospholipids (p. 435).
8. E All four correct (p. 440). Remember that cholesterol biosynthesis is cytosolic; mitochondrial biosynthesis of HMG CoA leads to ketone body formation.
9. C 2, 4 correct. 1: The LDL binds to the cell receptor and is endocytosed and then degraded in lysosomes to release cholesterol. 3: LCAT is a plasma enzyme; the tissue enzyme that forms cholesterol esters is ACAT (p. 445).
10. D The intestinal tract contains a mixture of primary and secondary bile acids, both of which can be reabsorbed. Secondary bile acids are formed by bacteria in the intestine by chemical reactions, such as the removal of the C-7 OH group (p. 447).
11. A 1, 2, 3 correct. The glycosphingolipids do not contain phosphate. Ceramide is the base structure from which the glycosphingolipids are formed (p. 452).
12. D Sphingomyelins are not glycosphingolipids. They are formed from ceramides, are amphipathic, and are neutral. C is the definition of gangliosides (p. 451).
13. E All four correct. 4: Many of the sphingolipids share the same types of bonds, for example, a β-galactosidic bond, and one enzyme, for example, β-galactosidase, will hydrolyze it whenever it occurs (p. 457, Figure 10.59).
14. A 1, 2, 3 correct. This is so whether or not you include thromboxane A_2 as a prostaglandin (in TXA_2 the ring is 6-membered rather than 5). 4 is true only of the intermediate of synthesis, PGG_2 (Figures 10.64–10.68).
15. B 1, 3 correct. 2: Antiinflammatory steroids inhibit the release of the precursor fatty acid by phospholipase A_2. 4: Arachidonic acid is not free in the cell but is part of the membrane phospholipids (p. 462–465).
16. E TXA_2 is very active, has a very short half-life, contains a six-membered ring, and is the main prostaglandin in platelets but not all tissues (p. 464).
17. B 1, 3 correct. 2–4: HPETEs, themselves, are not hormones but highly unstable intermediates that are converted to either the HETEs (mediators of hypersensitivity) or leukotrienes (p. 467).

11

Amino Acid Metabolism I: General Pathways

ALAN H. MEHLER

11.1 OVERVIEW 475
11.2 GENERAL REACTIONS OF AMINO ACIDS 477
Transamination Is the Transfer of the Amino Group to an Acceptor 477
Glutamate Dehydrogenase Catalyzes the Metabolism or Production of Amino Acids 478
Amino Acid Oxidases Catalyze the Production of Ammonia and Hydrogen Peroxide 480
11.3 REACTIONS OF AMMONIA 480
11.4 THE UREA CYCLE 481
Carbamoyl Phosphate Formation Is a Prerequisite for the Urea Cycle 482
Citrulline Is Formed From Carbamoyl Phosphate and Ornithine 483
Aspartate Provides the Additional Nitrogen to Form Argininosuccinate 483
Arginine Is Synthesized in the Urea Cycle 483
Arginase Forms Urea and Regenerates Ornithine 484
11.5 REGULATION OF THE UREA CYCLE 485
11.6 ALTERNATIVE REACTIONS FOR ELIMINATION OF EXCESS NITROGEN 488
BIBLIOGRAPHY 488
QUESTIONS AND ANNOTATED ANSWERS 489
CLINICAL CORRELATIONS
11.1 Hyperammonemia and Hepatic Coma 483
11.2 Deficiencies of Urea Cycle Enzymes 487

11.1 OVERVIEW

Amino acids are a group of molecules of major importance that contain a common chemical structure R—CH(NH_3^+)COO^-. As written, this structure represents two stereoisomers because of the asymmetry of the α-carbon atom; except where specifically noted, all of the statements in this chapter apply to the L isomers. The primary amino acids share the function of being polymerized with each other to form proteins, and some of these compounds are interconverted by metabolic reactions. Each naturally occurring amino acid, however, is a unique compound with individ-

TABLE 11.1 Dietary Requirements of Amino Acids

Essential	Nonessential
Arginine[a]	Alanine
Histidine	Aspartate
Isoleucine	Cysteine
Leucine	Glutamate
Lysine	Glycine
Methionine[b]	Proline
Phenylalanine[c]	Serine
Threonine	Tyrosine
Tryptophan	
Valine	

[a] Arginine is synthesized by mammalian tissues, but the rate is not sufficient to meet the need during growth.
[b] Methionine is required in large amounts to produce cysteine if the latter is not supplied adequately by the diet.
[c] Phenylalanine is needed in larger amounts to form tyrosine if the latter is not supplied adequately by the diet.

ual biological functions and metabolism. In this chapter, the metabolic processes that affect amino acids collectively will be described; the metabolism of the individual compounds will be presented in Chapter 12.

Amino acids occur mainly as constituents of proteins, which make up most of the dry weight of the human body. Many of the amino acids incorporated into human proteins are the compounds absorbed by the intestine from digests of dietary protein.

The composition of dietary protein, however, does not correspond to the needs of any individual for the synthesis of protein and numerous essential metabolites derived from amino acids. Therefore, metabolic pathways exist to derive energy from excess exogenous amino acids and to synthesize certain amino acids from other dietary constituents, carbohydrates, and fats, which can supply carbon, hydrogen, and oxygen atoms. However, the nitrogen is contributed almost entirely by ingested amino acids.

Feeding experiments have shown that hydrolysates of proteins serve as well as the intact polymers. Subsequent studies with artificial mixtures of pure amino acids led to the grouping of these compounds as **essential** or **nonessential** (Table 11.1). That several amino acids needed for protein can be produced by human metabolism is the basis for listing those amino acids as nonessential. It must be understood that a deficiency of these amino acids in the diet can be overcome only when sufficient amounts of precursors are available to permit them to be formed as needed. With one exception, those amino acids listed as essential cannot be synthesized by the human body because our cells lack the necessary biosynthetic enzymes. A net synthesis of arginine apparently can meet the needs of adult humans, but normal growth requires an exogenous supply of this amino acid. In some of the enzyme deficiency diseases to be discussed in Chapter 12, needs for other amino acids, tyrosine and cysteine, must be met by the diet.

The ordinary diets of developed countries contain more than adequate amounts of both essential and nonessential amino acids. Therefore, the categories are of practical importance only in disease, when specific supplements are administered, or in designing diets for people in certain areas. Even when syntheses are not apparently necessary to meet nutritional needs, the reactions described in Chapter 12 constantly redistribute the chemical elements of the metabolic systems involved and, thus, are of great theoretical relevance in interpreting studies on human and animal metabolism.

The amino acids listed as essential or nonessential are only a small fraction of those that occur naturally. These lists include only those amino acids that must be present wherever proteins are synthesized. The 18 compounds are activated and incorporated into proteins (Chapter 17). In addition, two derivatives, glutamine from glutamic acid and asparagine from aspartic acid, are synthesized for subsequent incorporation into proteins, increasing the list of primary building blocks for proteins to 20. Asparagine survives digestion and is absorbed from the intestine like other amino acids, whereas glutamine is mainly hydrolyzed to glutamate. All of the many other amino acids found in specific proteins are derivatives made by so-called posttranslational modifications of one or more of the 20 primary building blocks after they have been assembled into polypeptide chains; none of these compounds, such as hydroxyproline, is required in the diet.

Since the amino acids comprise a complex group of compounds, the biochemistry of these substances can be considered from several points of view. The nutritional requirements and the demands of protein synthesis have already been mentioned. In Chapter 12, the metabolism of each

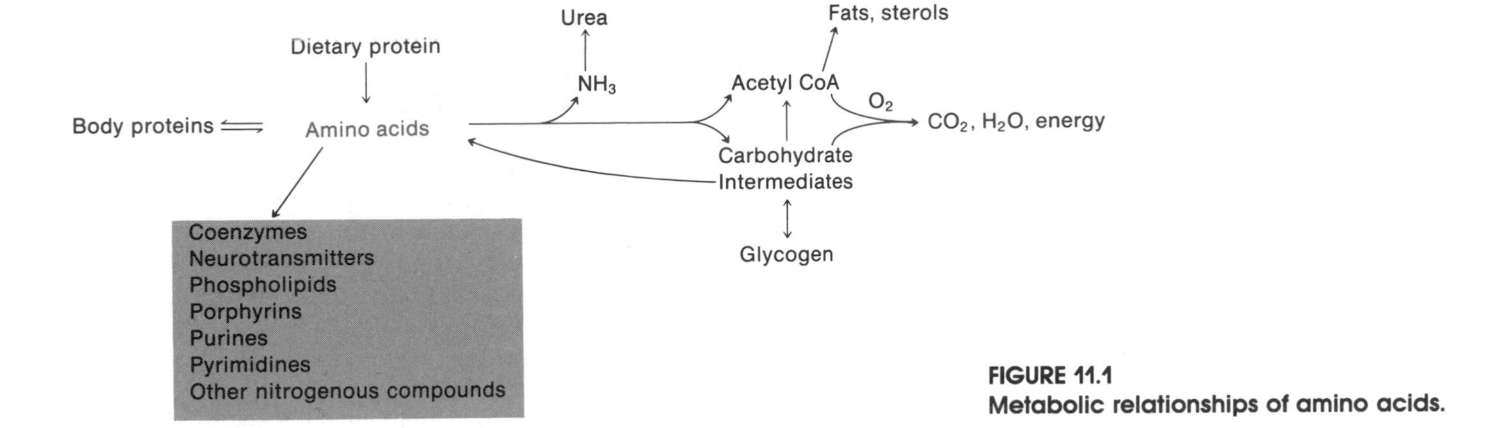

FIGURE 11.1
Metabolic relationships of amino acids.

amino acid will be considered from the point of view of degradation of the portion not needed for protein synthesis and also from the overlapping point of view of use of amino acids for the biosynthesis of other essential cellular materials. Another way of looking at amino acid metabolism is from the point of view of **nitrogen balance.**

The amount of nitrogen in each individual is regulated to stay relatively constant except during growth, when the amount must be increased in proportion to growth. There is no storage form for nitrogen reserves; only a small part of this nitrogen exists in the form of free amino acids or other compounds that can be used for synthesis of amino acids. Therefore, for maintenance of optimum body structure, an adequate supply of amino acids must be eaten frequently. If for any reason the supply of protein in the diet is insufficient, the need to synthesize specific proteins for vital physiological functions results in a redistribution of amino acids among proteins. For example, hemoglobin is degraded to the extent of almost 1% a day as red blood cells die, and under normal circumstances the degradation is balanced by resynthesis. In a deficiency of amino acids, relatively less hemoglobin is synthesized because a degree of anemia is more tolerable than a deficiency of certain other proteins.

Amino acids are substrates for many essential biosynthetic processes as shown in the outline of Figure 11.1; these will be discussed in Chapter 12. However, normal diets include a large excess of amino acids over the amount needed for synthesis of proteins or other cell constituents, and most excess amino acids are degraded to products that are either oxidized for energy or stored as fat and glycogen. In each case the nitrogen is liberated as ammonia. Some ammonia is reused in the synthesis of amino acids, some is used in other biosynthetic reactions, some is excreted in the urine, but the largest part is converted to urea, which is excreted by the kidneys. Urea synthesis occurs mainly in the liver, which is also the site of most of the biosynthesis of nonessential amino acids and a large part of the degradation of all amino acids.

11.2 GENERAL REACTIONS OF AMINO ACIDS

Transamination Is the Transfer of the Amino Group to an Acceptor

Quantitatively, the most important reaction of amino acid metabolism is the transfer of the amino group to an α-keto acid. This reaction, **transamination,** is involved in the synthesis of nonessential amino acids, in the

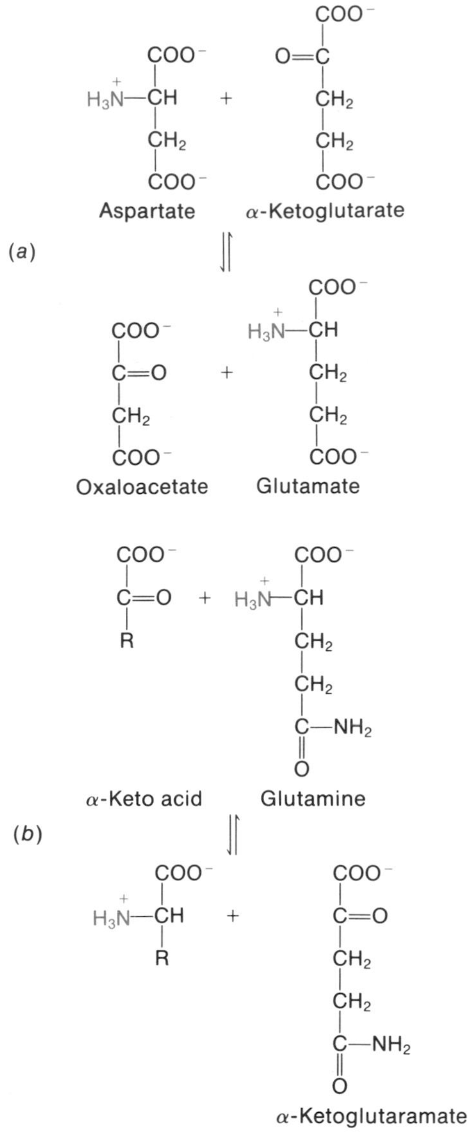

(a) Aspartate + α-Ketoglutarate ⇌ Oxaloacetate + Glutamate

(b) α-Keto acid + Glutamine ⇌ $H_3\overset{+}{N}$—CH(R)—COO^- + α-Ketoglutaramate

FIGURE 11.2
Transamination reactions.

degradation of most amino acids, and in the exchange of amino groups; transamination has been demonstrated in vivo for all of the primary amino acids except lysine and threonine. The best studied enzyme of the family of transaminases (also called aminotransferases) is the **aspartate transaminase** of the cytosol; a similar but different enzyme for the reaction has been found in mitochondria. Most transaminases specifically require α-ketoglutarate or glutamate as one of the reacting pair; the specificities of various transaminases for the other substrate (amino or keto acid) may be strict or broad. In each case an amino acid and its corresponding keto acid are equilibrated with α-ketoglutarate and glutamate as shown in Figure 11.2*a*, and the equilibrium constant is near 1. The actual direction taken thus depends on the concentrations of the four reactants, which are determined by the other cellular processes that produce or consume these compounds. A few transaminases use glutamine instead of glutamate as the amino donor and produce α-ketoglutaramate (Figure 11.2*b*). One of the first discoveries that resulted from the introduction of ^{15}N as a tracer in biochemistry by Schoenheimer and his students was that amino groups of many amino acids of proteins are rapidly labeled in vivo when the isotope is administered in a single compound. This led to the concept of **"dynamic equilibrium,"** meaning a steady state in which not only the proteins are constantly degraded and resynthesized but also the atoms of the amino acids are exchanged. Transaminases are responsible for an active redistribution of amino groups among amino acids in vivo.

All transaminases require a coenzyme, **pyridoxal phosphate** (Figure 11.3). The coenzyme is bound to the protein by ionic forces involving the pyridine ring and the phosphate group. In addition, the aldehyde group forms a Schiff's base, an aldimine, with an ε-amino group of a lysine residue. A series of steps illustrated in Figure 11.4 takes advantage of the versatility of the bound coenzyme. In the reaction with an appropriate amino acid, the carbon of the aldimine is transferred to the amino group of the substrate. The ability of the pyridine ring to serve as an electron sink facilitates the dissociation of the α-hydrogen of the amino acid and permits a tautomerization in which the double bond of the aldimine shifts to form a ketimine; this is hydrolyzed to liberate the keto acid and the enzyme is left bearing pyridoxamine as its coenzyme. A similar series of reactions with the other keto acid results in the formation of a new amino acid while the enzyme reverts to its original state.

Pyridoxal phosphate is used as the coenzyme of a large number of enzymes that catalyze many different kinds of reactions. In each case, it is likely that the chemical mechanism is initiated by the same sort of aldimine formation used by transaminases, which is followed by an electronic rearrangement, but the different enzymes are able to labilize bonds other than that of the α hydrogen.

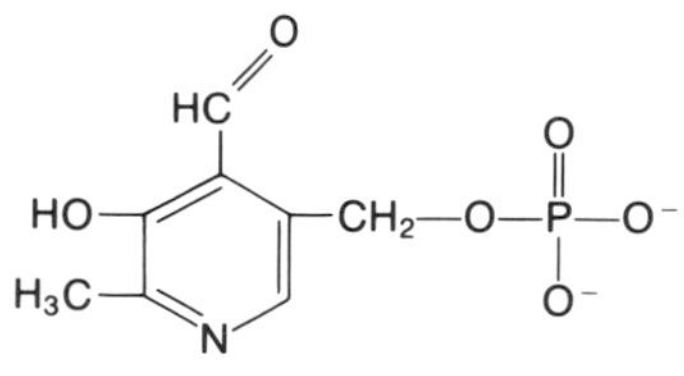

FIGURE 11.3
Pyridoxal phosphate.

Glutamate Dehydrogenase Catalyzes the Metabolism or Production of Amino Acids

The transfer of amino groups from one carbon chain to another is an efficient way to maintain proper ratios of the various amino acids while individual compounds are being synthesized or degraded. In addition to this means of redistribution, a mechanism is needed to dispose of surplus amino groups when ingestion of excess protein results in a surfeit of amino acids that must be metabolized and, in some circumstances, a mechanism is needed to use ammonia to increase the total amino acid concentration. The major mechanism for net use or production of amino acids is the reaction catalyzed by **glutamate dehydrogenase.**

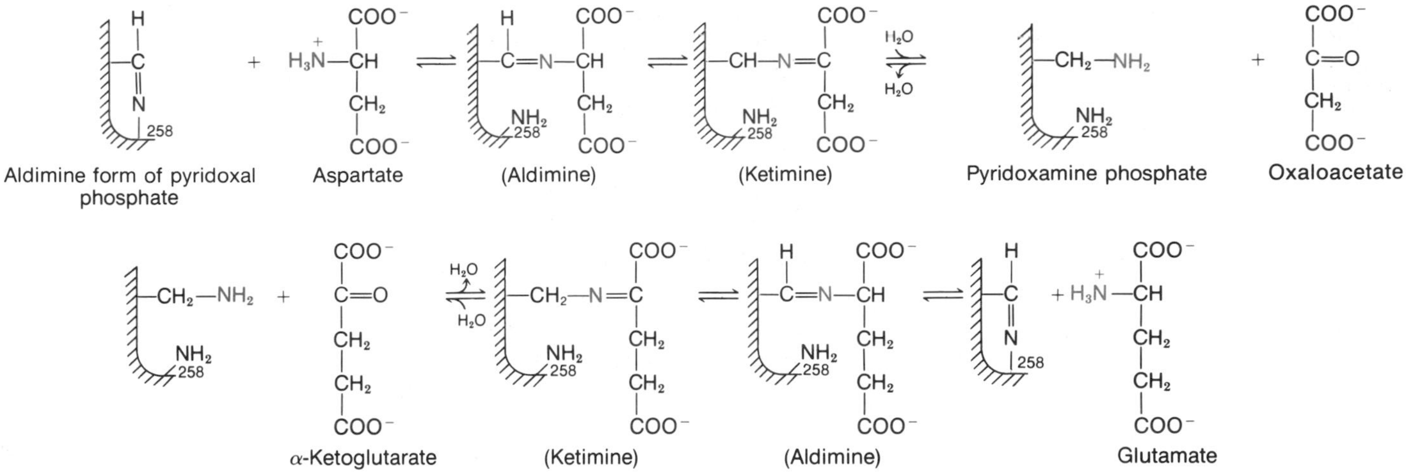

Aldimine form of pyridoxal phosphate + Aspartate ⇌ (Aldimine) ⇌ (Ketimine) ⇌ Pyridoxamine phosphate + Oxaloacetate

Pyridoxamine phosphate + α-Ketoglutarate ⇌ (Ketimine) ⇌ (Aldimine) ⇌ Aldimine form of pyridoxal phosphate + Glutamate

FIGURE 11.4
Mechanism of enzymatic transamination represented in schematic form.

$$\text{Glutamate} + \text{NAD}^+\text{(P)} \rightleftharpoons \alpha\text{-ketoglutarate} + \text{ammonia} + \text{NAD(P)H} + \text{H}^+$$

In vitro glutamate dehydrogenase uses either NAD^+ or $NADP^+$. At neutral pH, the equilibrium of the reaction lies to the side of glutamate synthesis. It should be noted that the equilibrium constant is a value derived from a theoretical consideration of a hypothetical situation with arbitrarily selected concentrations of reagents. Under physiological conditions in which other reactions compete for the products of the reaction (especially NADH oxidase and α-ketoglutarate dehydrogenase, Chapter 6, and ammonia conversion to urea, p. 480), an efficient oxidation of glutamate can be catalyzed. It should be noted that the equilibrium constant for this reaction is more complicated than most, even when the H^+ term is eliminated, because the number of products is greater than the number of reactants:

$$K_{eq} = \frac{[\alpha\text{-ketoglutarate}][\text{NAD(P)H}][\text{NH}_3]}{[\text{glutamate}][\text{NAD}^+\text{(P)}]}$$

The efficient removal of any one of the products can shift the equilibrium concentrations of the other reactants; lowering the concentration of a second product has a compounding effect. The facts that most of the coenzyme in vivo is in the oxidized form, that ammonia is removed from cells to give a very low steady-state concentration, and that intracellular glutamate occurs at more than 10^{-3}M concentrations create conditions that favor oxidative deamination of glutamate. Like many reactions of biochemistry, however, the concentrations are poised so that glutamate can be formed when excess ammonia is produced. A major source of ammonia in the liver is from bacterial metabolism in the intestine and transport through the portal system.

Glutamate dehydrogenase from bovine liver, the best studied representative of this ubiquitous enzyme family, is a complex enzyme composed of six identical subunits of M_r 56,000. It is subject to extensive allosteric control by a diverse group of substances; of these, GTP and ATP inhibit, whereas GDP and ADP activate. The less active form of the enzyme demonstrates a weak alanine dehydrogenase activity. These properties suggest a teleological rationale; when amino acids are needed for energy

TABLE 11.2 Relative Rates of Oxidation by Representative Amino Acid Oxidases Acting on the Appropriate Optical Isomer of a Series of Amino Acids

	L-Amino acid oxidase[a]	D-Amino acid oxidase[a]
Alanine	3.5	100
Arginine	0	0
Aspartate	0	0
Cysteine	0	0
Glutamate	0	0
Glycine	0	15
Histidine		20
Isoleucine	9	90
Leucine	100	20
Lysine	0	0
Methionine	87.5	110
Phenylalanine	32	90
Proline	77	300
Serine	0	50
Threonine	0	30
Tryptophan	35.7	45
Valine	5.3	65

[a] The L-amino acid oxidase was purified from rat kidney and the D-amino acid oxidase was from sheep kidney.

production, the activity of the dehydrogenase is greatest, and when nucleoside triphosphate levels are high, the activity is decreased. Inhibition by several steroid hormones and by thyroxine has been described but not established as having physiological significance.

Amino Acid Oxidases Catalyze the Production of Ammonia and Hydrogen Peroxide

For many years two enzymes have been known to oxidize D- and L-amino acids, respectively, to the corresponding α-keto acids and ammonia with molecular oxygen as the electron acceptor. In these reactions the oxygen is reduced to H_2O_2. The chemical properties of these enzymes, purified from liver and kidney, have been extensively studied, but the biological functions remain obscure. D-Amino acid oxidase uses FAD as its coenzyme and occurs in amounts sufficient to metabolize large quantities of substrate. However, D-amino acids are of limited natural occurrence in mammals. **L-Amino acid oxidase** has a much lower activity and contains FMN. These enzymes do not attack some amino acids with ionizable side chains but otherwise display broad specificity as shown in Table 11.2. L-Amino acid oxidase can also attack α-hydroxy acids, and D-amino acid oxidase oxidizes glycine and sarcosine (*N*-methyl glycine). In vitro these flavoproteins can substitute many oxidants, such as methylene blue, for molecular oxygen, and the reaction can be seen to proceed in steps:

$$\text{Amino acid} + \text{flavoprotein} + H_2O \longrightarrow \alpha\text{-keto acid} + NH_3 + \text{flavoprotein } H_2$$

$$\text{flavoprotein } H_2 + X \longrightarrow \text{flavoprotein} + XH_2$$

When the oxidant is O_2, the product of its reduction is H_2O_2; in liver and kidney this potentially destructive material is decomposed by a very active enzyme, catalase, which catalyzes the reaction

$$2H_2O_2 \longrightarrow 2H_2O + O_2$$

Superficially the overall reaction catalyzed by the amino acid oxidases appears to resemble the complex of reactions initiated by glutamate dehydrogenase and followed by the mitochondrial oxidation of NADH. An important difference is that in the latter case the oxidation is coupled with oxidative phosphorylation and the generation of ATP.

11.3 REACTIONS OF AMMONIA

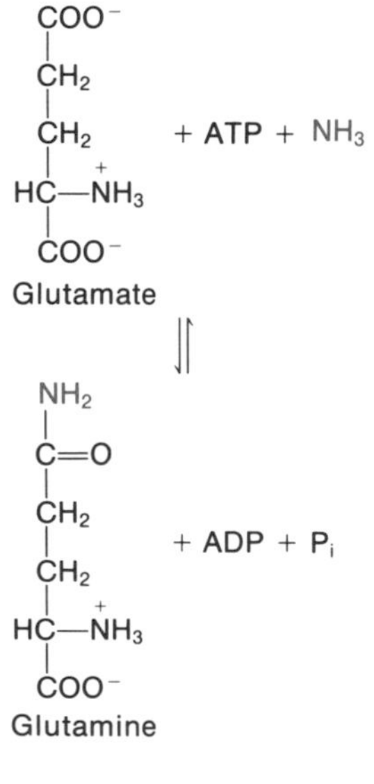

In addition to forming the α-amino groups of amino acids, ammonia is incorporated into several other metabolites. The mechanisms of formation of the purine ring, the amino group of the pyrimidine cytosine, and amino sugars all involve the intermediate formation of glutamine. As seen in Figure 11.5, **glutamine** is synthesized from glutamate and ammonia by the enzyme **glutamine synthetase.** This amino acid amide is an important compound because it is one of the 20 primary amino acids used for protein synthesis. In addition, in the kidney, glutamine supplies the bulk of the ammonia excreted. Thus the ammonia is transported in the blood as a nontoxic, nonionized amide that is used in regulation of urinary pH by release of ammonia in a simple hydrolysis of the amide group catalyzed by the enzyme **glutaminase** (Figure 11.6) followed by the removal of the amino group by glutamate dehydrogenase. Blood levels of glutamine nor-

COO^-–CH_2–CH_2–HC–$\overset{+}{N}H_3$–COO^- (Glutamate) + ATP + NH_3 ⇌ NH_2–C=O–CH_2–CH_2–HC–$\overset{+}{N}H_3$–COO^- (Glutamine) + ADP + P_i

FIGURE 11.5
Reaction catalyzed by glutamine synthetase.

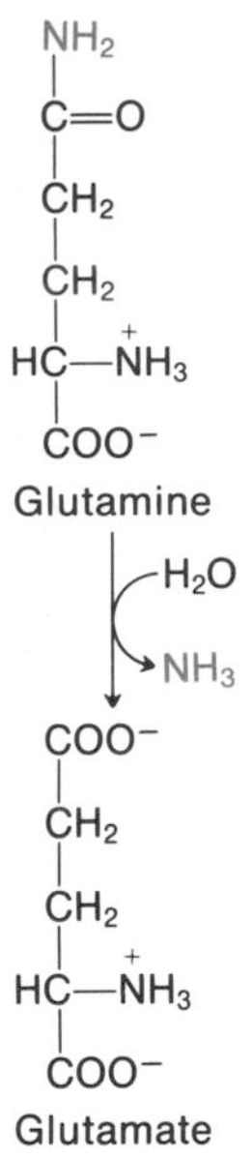

FIGURE 11.6
Reaction catalyzed by glutaminase.

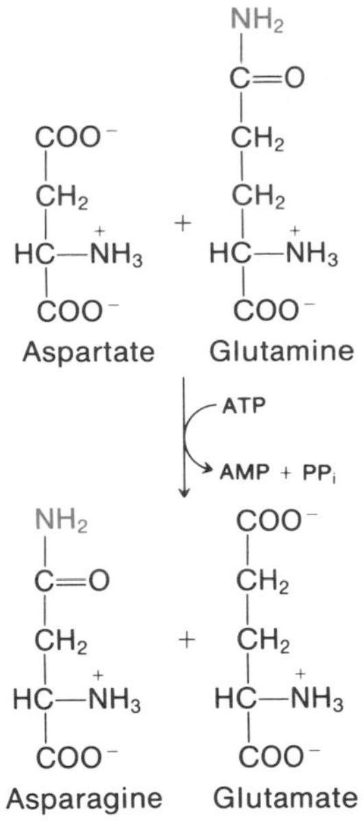

FIGURE 11.7
Synthesis of asparagine.

mally exceed those of any other amino acid. Ammonia released by degradation of amino acids in the liver is made available as glutamine to cells in other organs for the synthesis of purines, pyrimidines, and other compounds.

Glutamine also donates the amide group to form its lower homolog, asparagine, from aspartate. Although there appears to be no need for energy to drive a reaction in which the substrates and products have similar groups, that is, a carboxyl group and a carboxamide, the synthesis of asparagine requires ATP as shown in Figure 11.7. The ATP is needed to activate the β-carboxyl group of aspartate. In general it is not possible to alter carboxyl groups under biological conditions until they have been activated by reaction with a nucleoside triphosphate.

Asparagine is not known to have any function in mammals other than incorporation into proteins. In general animal cells appear to be able to synthesize enough asparagine for their own needs. Some rapidly dividing leukemic cells have little or no ability to produce asparagine and depend on asparagine taken from blood to synthesize their proteins; this is the basis for the therapeutic use of asparaginase (Figure 11.8) to treat leukemic patients, depriving the neoplastic cells of asparagine by lowering the serum concentration drastically.

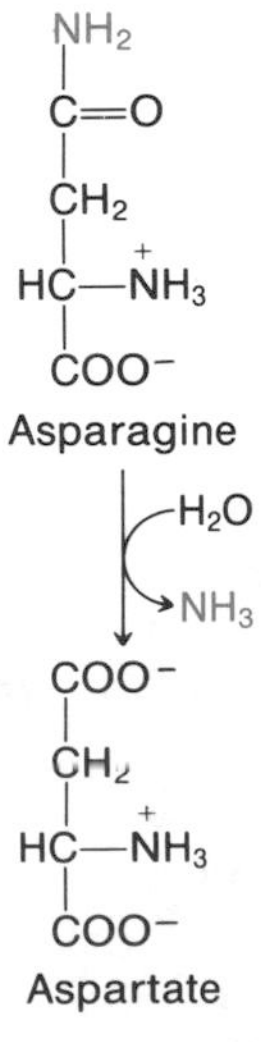

FIGURE 11.8
Reaction catalyzed by asparaginase.

11.4 THE UREA CYCLE

Normal adults are in nitrogen balance; that is, the amount of nitrogen ingested is balanced by the excretion of an equivalent amount (Chapter 27). About 80% of the excreted nitrogen is in the form of urea (Figure 11.9). The discovery in 1828 of a laboratory synthesis of this simple compound was momentous in the history of science because it demonstrated for the first time that compounds characteristic of living organisms can also be made by the methods of chemistry. The elucidation of the relatively complex method by which our bodies produce urea had a similar impact on scientific thought in developing the concept of metabolic cy-

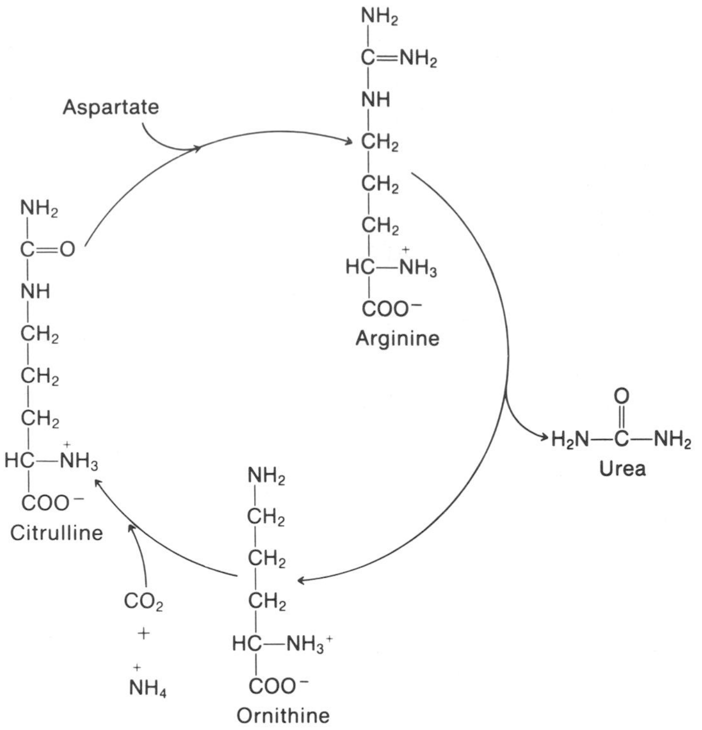

FIGURE 11.9
Concept of the urea cycle.

cles. The **urea cycle** (also the Krebs, Krebs–Henseleit, or **ornithine cycle**) is a true cycle in which the carrier molecule, ornithine, is regenerated with the same atoms in its skeleton after the formation of each molecule of urea.

Amino acids are metabolized extensively throughout the body, but most urea synthesis occurs in the liver. The importance of the liver in nitrogen metabolism is further illustrated in Clin. Corr. 11.1. The concept of the urea cycle was derived from observations on urea synthesis in liver slices. Only a few compounds were found to cause a major increase in the rate of urea synthesis and were effective in catalytic amounts; that is, the extra synthesis of urea exceeded the amounts added of arginine, ornithine, or citrulline (Figure 11.9). The explanation of the stimulation by these three compounds on the basis of "paper chemistry" has been confirmed and elaborated by the identification and characterization of each of the enzymes involved. Thus, the urea cycle is established as a process, not merely a theory.

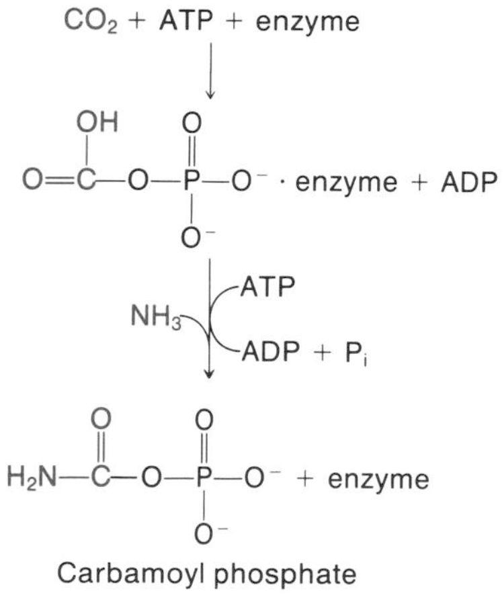

Carbamoyl Phosphate Formation Is a Prerequisite for the Urea Cycle

The formation of citrulline from ornithine requires the addition of NH_3 and CO_2. These are joined in a preliminary reaction in which a phosphate group from ATP is also used to produce a relatively stable compound, carbamoyl phosphate.

In the reaction of **carbamoyl phosphate synthetase,** one molecule of ATP activates CO_2 to form enzyme-bound carboxy phosphate, which reacts with ammonia to form carbamate. A second molecule of ATP reacts with the enzyme-bound intermediate in a kinase reaction to form carbamoyl phosphate, shown in Figure 11.10. Actually, two distinct carbamoyl phosphate synthetases are found in mammalian livers. The one that participates in the urea cycle, carbamoyl phosphate synthetase I, is

FIGURE 11.10
Reactions catalyzed by carbamoyl phosphate synthetase.
The enclosed atoms in this and subsequent figures are destined to be incorporated into urea.

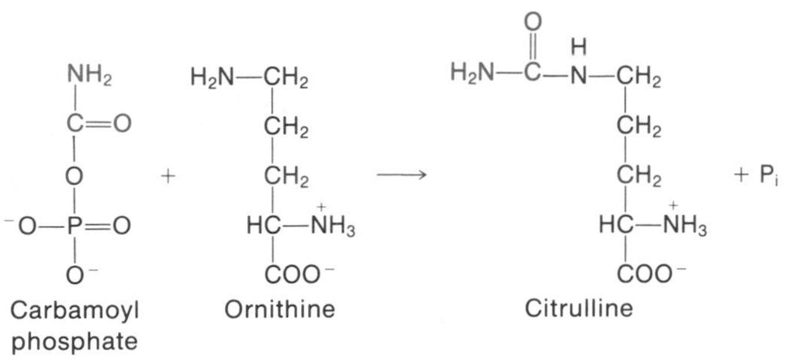

Carbamoyl phosphate + Ornithine ⟶ Citrulline + P_i

FIGURE 11.11
Reaction catalyzed by ornithine transcarbamoylase.

part of the mitochondrial matrix and is inactive in the absence of an *N*-acylated glutamate as an allosteric activator. The physiological activator is *N*-acetylglutamate. A cytosolic carbamoyl phosphate synthetase II participates in the biosynthesis of pyrimidines (Chapter 13) and does not require an acylated glutamate for activity.

Citrulline Is Formed From Carbamoyl Phosphate and Ornithine

Citrulline is formed by the transfer of the carbamoyl group from its phosphoric acid anhydride to the δ-amino group of ornithine (Figure 11.11). The enzyme **ornithine transcarbamoylase** is also mitochondrial. Citrulline diffuses from mitochondria to the cytosol, where the rest of the urea cycle occurs.

Aspartate Provides the Additional Nitrogen to Form Argininosuccinate

In two steps citrulline is converted to arginine, which has a similar structure but with a nitrogen in place of the ureido oxygen. The new nitrogen atom comes from aspartate. The transfer involves a condensation, as shown in Figure 11.12, to form argininosuccinate. Since the ureido group is very stable, the condensation requires activation by ATP. The reaction probably proceeds in steps analogous to the formation of active derivatives of fatty acids (Chapter 9) and amino acids (Chapter 17). In each case an intermediate compound containing AMP is formed with the release of PP_i. The activated molecule is then transferred to its acceptor without dissociating from the enzyme. The reaction is reversible, but no partial reactions have been observed, leading to the conclusion that everything is bound firmly to the enzyme until the reaction is complete. The reaction proceeds strongly in the direction of synthesis of argininosuccinate because of the hydrolysis of the PP_i.

Arginine Is Synthesized in the Urea Cycle

The formation of arginine is catalyzed by **argininosuccinate lyase** (Figure 11.13). The reaction catalyzed by this enzyme is an elimination; the nitrogen of what was originally aspartate is eliminated along with a proton from the C_4 dicarboxylic compound, leaving a double bond. The unsaturated product is fumarate, a TCA cycle intermediate (Chapter 6). The nitrogen atom that is eliminated becomes one of two equivalent atoms of the guanidinium group of arginine.

CLIN. CORR. 11.1 HYPERAMMONEMIA AND HEPATIC COMA

An increase of ammonia in blood over the normal concentration range of 30–60 μM, hyperammonemia, may cause coma. Loss of consciousness in general may be a consequence of ATP depletion. Ammonia crosses the blood–brain barrier and might shift the glutamic dehydrogenase equilibrium to an extent that α-ketoglutarate becomes limiting for the TCA cycle. Synthesis of glutamine might also contribute to a decrease in the ATP level.

Hyperammonemia is usually caused by inability of the patient to form urea at a sufficient rate to maintain tolerable levels of ammonia. Inherited deficiency of individual urea cycle enzymes has been implicated in many cases. A delay in development of urea cycle enzymes has been observed to cause a transient hyperammonemia in newborns. Ammonia is not cleared from the blood efficiently in cases in which the normal blood flow through the liver is reduced. Cirrhosis of the liver can result in the development of collateral circulation of blood from the portal system to the inferior vena cava, bypassing the liver. In some cases this type of bypass is deliberately produced surgically as a portal–caval shunt.

The reactions described in Section 11.6 are useful in eliminating nitrogen as glycine and glutamine. Since anaplerotic reactions can supply α-ketoglutarate to consume two equivalents of ammonia in the formation of glutamine and since glycine can also be produced from ammonia by reactions described in Chapter 12, careful administration of benzoate and/or phenylacetate can compensate for deficiencies in the urea cycle in preventing ammonia toxicity. An additional treatment for this purpose is substitution in the diet of α-keto analogs of some amino acids to reduce the total intake of nitrogen and to divert amino groups from an excretory fate to the production of essential amino acids for protein synthesis.

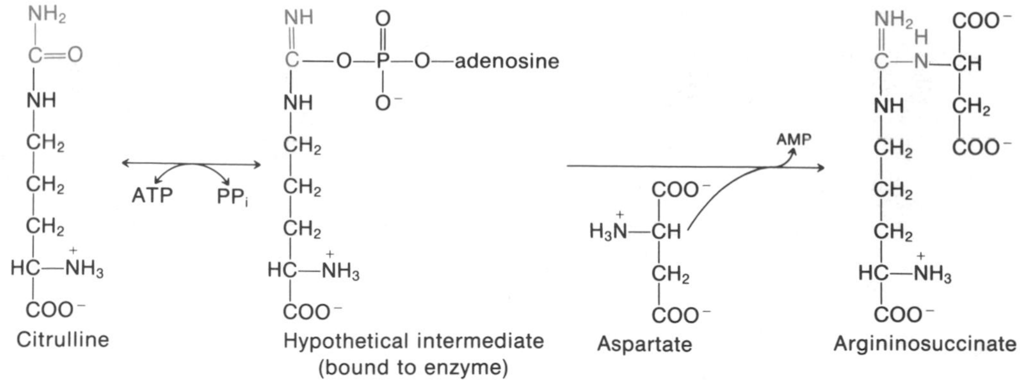

FIGURE 11.12
Reactions catalyzed by argininosuccinate synthetase.

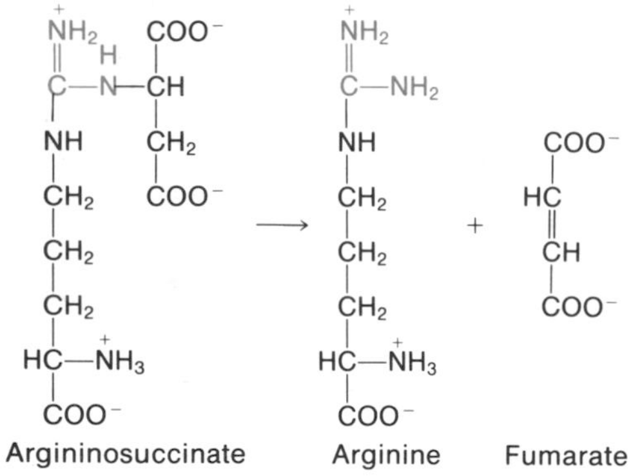

FIGURE 11.13
Formation of arginine by argininosuccinate lyase.

Arginase Forms Urea and Regenerates Ornithine

The urea cycle is completed by the hydrolysis of arginine to ornithine and urea, a reaction catalyzed by **arginase** (Figure 11.14). Arginase is a tetrameric protein that depends on easily dissociable Mn^{2+} for its activity. This enzyme is found in brain, kidney, and other organs in which the production of ornithine from dietary arginine may have more physiological importance than formation of urea. The ornithine that is released in this reaction in the urea cycle is the identical molecule that was used to initiate the series of reactions that resulted in the synthesis of urea and, after transport into mitochondria, reacts with carbamoyl phosphate to start another cycle. The urea diffuses into the blood from which it is cleared by the kidneys.

The biological formation of urea is much more complicated than the chemical synthesis, which can be as simple as heating ammonium cyanate, NH_4CNO. The urea cycle uses the energy of ATP instead of high temperature and, equally important, it uses the ornithine structure as a handle for the enzymes that build the urea molecule and enables each step to be integrated in a smooth process that maintains the optimum physiological concentration of ammonia. The outline in Figure 11.15 shows the essential simplicity of the cycle in which complex enzymes participate.

Urea has no role in animal metabolism other than as an end product that is excreted. An enzyme, **urease,** present in plants and microorganisms hydrolyzes urea to CO_2 and two equivalents of ammonia. A portion of the urea synthesized by the urea cycle diffuses into the intestine and is degraded by the urease of intestinal bacteria to produce ammonia that is

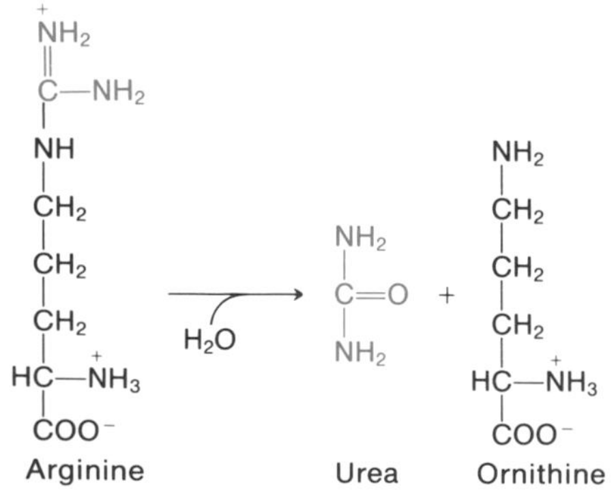

FIGURE 11.14
Hydrolysis of arginine by arginase.

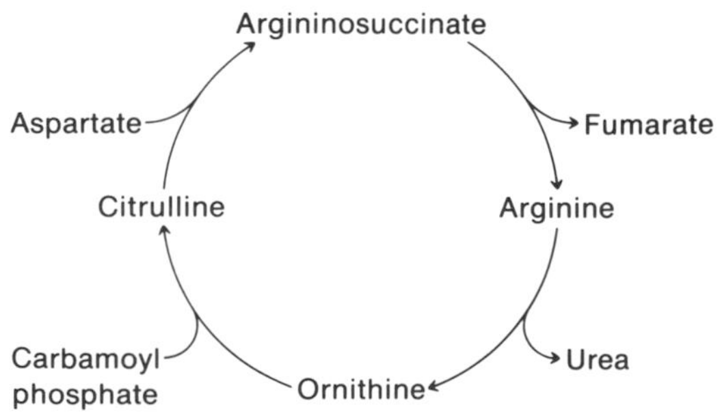

FIGURE 11.15
Outline of the urea cycle.

recycled to the liver. This is of significance only in cases of kidney failure, when a large amount of urea may pass through the intestine. Urease from the jackbean is renowned in biochemistry as the first enzyme to be shown to form crystals.

The urea cycle is considered as a mechanism for elimination of excess ammonia as urea. The mechanism, however, requires that one of the nitrogen atoms be donated by aspartate. In the synthesis of carbamoyl phosphate two equivalents of ATP lose their terminal phosphate groups, and another ATP is degraded to AMP and pyrophosphate in the synthesis of argininosuccinate. The overall chemical balance of the biosynthesis of urea is as follows:

$$\begin{array}{lrl} & NH_3 + CO_2 + 2ATP & \longrightarrow \text{carbamoyl phosphate} + 2ADP + P_i \\ & \text{Carbamoyl phosphate} + \text{ornithine} & \longrightarrow \text{citrulline} + P_i \\ & \text{Citrulline} + ATP + \text{aspartate} & \longrightarrow \text{argininosuccinate} + AMP + PP_i \\ & \text{Argininosuccinate} & \longrightarrow \text{arginine} + \text{fumarate} \\ & \text{Arginine} & \longrightarrow \text{urea} + \text{ornithine} \\ \hline \text{Sum:} & 2NH_3 + CO_2 + 3ATP & \longrightarrow \text{urea} + 2ADP + AMP + PP_i + 2P_i \end{array}$$

The reaction has a large negative free energy derived largely from the cleavage of the pyrophosphate bonds of ATP and the increased number of ionized groups in the split products (15 in the products compared with 12 in the three molecules of ATP). Therefore, the cycle is not reversible. In effect, four equivalents of ATP are required for each turn of the cycle since two phospho anhydride bonds must be used to regenerate ATP from AMP.

The essential aspartate is generated almost entirely by transamination of oxaloacetate by glutamate, which is thereby converted to α-ketoglutarate. Since oxaloacetate and α-ketoglutarate are both intermediates in the TCA cycle, the two cycles discovered by Krebs are metabolically linked (Figure 11.16 illustrates this relationship). The fumarate produced in the urea cycle is also a TCA cycle intermediate. A theoretical sequence of reactions can regenerate the aspartate from these intermediates and ammonia:

$$\begin{array}{lrl} & \alpha\text{-Ketoglutarate} + NH_3 + NADH & \longrightarrow \text{glutamate} + NAD^+ + H_2O \\ & \text{Fumarate} + H_2O & \longrightarrow \text{malate} \\ & \text{Malate} + NAD^+ & \longrightarrow \text{oxaloacetate} + NADH \\ & \text{Oxaloacetate} + \text{glutamate} & \longrightarrow \text{aspartate} + \alpha\text{-ketoglutarate} \\ \hline \text{Sum:} & \text{Fumarate} + NH_3 & \longrightarrow \text{aspartate} \end{array}$$

In living cells the dicarboxylic acids are equilibrated with TCA cycle intermediates and pass rapidly between the cytosol and mitochondria.

11.5 REGULATION OF THE UREA CYCLE

The urea cycle operates only to eliminate excess **ammonia.** The excess comes mainly from ingested amino acids not used promptly for the synthesis of proteins; only a few nitrogenous compounds comprising a small percentage of nitrogen turnover, such as purines, pyrimidines, creatine, and nicotinic acid, are not degraded to ammonia. The general nutritional status does not influence the degradation of excess amino acids; whether the carbon skeletons from amino acids are used immediately to produce ATP or are stored as fat and glycogen, the amino groups are released as ammonia, which must be eliminated. In nitrogen balance the amount of urea produced represents "metabolic wear and tear," a poorly under-

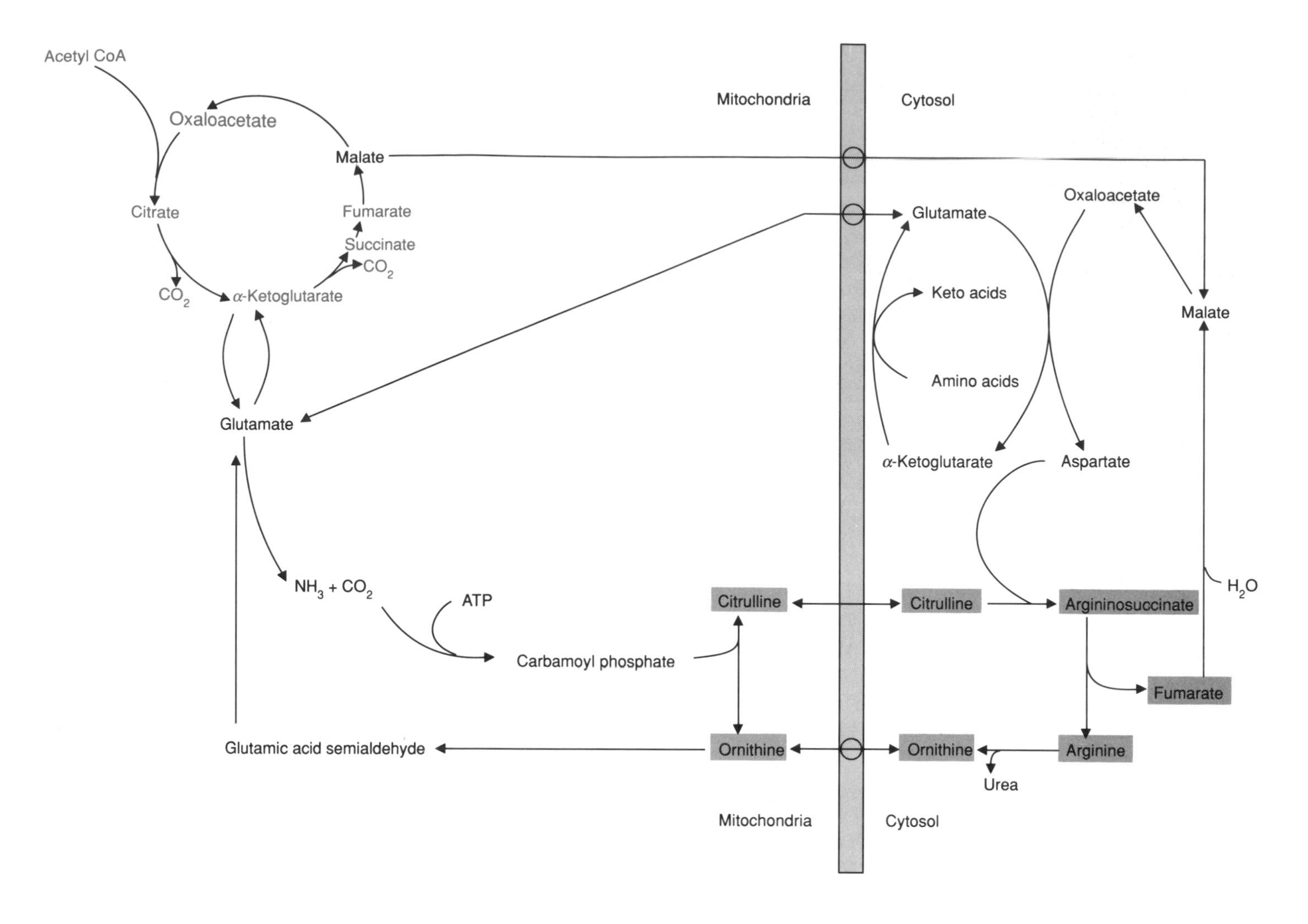

FIGURE 11.16
Interactions between the urea and TCA cycles.

$$CH_3-\overset{O}{\overset{\|}{C}}-SCoA + H_3\overset{+}{N}-CH(COO^-)-CH_2-CH_2-COO^-$$

Acetyl CoA Glutamate

$$\downarrow$$

$$CoA + CH_3-\overset{O}{\overset{\|}{C}}-\overset{H}{N}-CH(COO^-)-CH_2-CH_2-COO^-$$

N-Acetylglutamate

FIGURE 11.17
Reaction catalyzed by acetylglutamate synthetase.

stood combination of cell death and regeneration and protein turnover with partial, but never complete, reuse of amino acids. These metabolic processes persist even during protein deficiency in a high-fat, high-carbohydrate diet so that urea production never ceases. The rate, however, is precisely regulated to the need to eliminate potentially toxic ammonia. Interference with this process results in serious diseases, described in Clin. Corr. 11.2.

Gross control of the urea cycle is accomplished by altering the amounts of the enzymes of the cycle. The levels of the individual enzymes in laboratory animals have been seen to change 10–20-fold in response to extreme changes in diet, and similar changes presumably occur in humans. During extensive starvation, when muscle proteins are broken down to amino acids to be oxidized for energy, the importance of removing ammonia is accompanied by elevated levels of urea cycle enzymes in the liver.

Fine control of the urea cycle appears to be applied primarily at the synthesis of carbamoyl phosphate. Carbamoyl phosphate synthetase is inactive in the absence of its allosteric activator, ***N*-acetylglutamate.** This activator is synthesized by a liver enzyme from acetyl CoA and glutamate (Figure 11.17) and is hydrolyzed by a specific deacylase. The steady-state concentration of *N*-acetylglutamate is determined by the concentrations

CLIN CORR. 11.2
DEFICIENCIES OF UREA CYCLE ENZYMES

Even though the normal elimination of nitrogen and the prevention of ammonia toxicity require the function of the urea cycle, infants born with total deficiency of one or more enzymes of the cycle survive at least for several days. Many of the enzyme deficiencies that have been identified are partial and are reported as so much percent of normal activity. Analogy with mutations in other enzymes suggests that the enzymes are likely to have altered K_m values in many cases, rather than altered V_{max}, but there is as yet little information on characterizing the human mutations.

Cases are known of deficiencies of each of the urea cycle enzymes. Interruption of the cycle at each point affects nitrogen metabolism differently because some of the intermediates can diffuse from hepatocytes, accumulate in the blood, and pass into the urine. Therefore, the symptoms, prognosis, and treatment differ for the various enzyme deficiencies. In general, these are severe diseases with high incidences of mental retardation, seizures, coma, and early death.

N-ACETYLGLUTAMATE SYNTHETASE DEFICIENCY

The importance of an activator for carbamoyl phosphate synthetase was demonstrated by the finding of hyperammonemia and general hyperaminoacidemia in a newborn whose liver contained no detectable ability to synthesize *N*-acetylglutamate. In this case normal nitrogen metabolism was maintained by a low protein diet and the administration of carbamoyl glutamate, an analog of *N*-acetylglutamate that is also an activator of carbamoyl phosphate synthetase.

CARBAMOYL PHOSPHATE SYNTHETASE DEFICIENCY

Hyperammonemia has been observed in infants with 0–50% of the normal level of carbamoyl phosphate synthetase synthesis in their livers. Treatment with benzoate and phenylacetate has been effective in maintaining such infants. Low protein diets supplemented with arginine have been attempted on the hypothesis that activation of *N*-acetylglutamate synthesis by arginine would indirectly stimulate the low level of carbamoyl phosphate synthetase to provide enough of its product to sustain the urea cycle. In general this deficiency is associated with mental retardation, which may be an indirect consequence of the periods of uncontrolled hyperammonemia.

ORNITHINE TRANSCARBAMOYLASE DEFICIENCY

The most common deficiency disease involving urea cycle enzymes is lack of ornithine transcarbamoylase. Early death can be prevented by removal of excess ammonia and prevention of further accumulation by the same sort of diet effective with carbamoyl phosphate synthetase deficiency. The occasional finding of normal development in treated patients supports the idea that the mental retardation usually associated with this deficiency is caused by the excess ammonia before adequate therapy.

Genetic analysis of ornithine transcarbamoylase deficiency indicates that the gene is located on the X chromosome. Therefore, males generally are more seriously affected than females, who are often asymptomatic as heterozygotes. In addition to ammonia and amino acids appearing in the blood in increased amounts, orotic acid also increases, presumably because carbamoyl phosphate that cannot be used to form citrulline diffuses into the cytosol, where it condenses with aspartate, ultimately forming orotate (Chapter 13). This suggests that normal orotate synthesis is limited by the cytosolic carbamoyl phosphate synthetase II and that the production of excess carbamoyl phosphate intended for urea synthesis bypasses the normal control mechanism.

ARGININOSUCCINATE SYNTHETASE DEFICIENCY

The inability to continue the urea cycle by condensing citrulline with aspartate results in accumulation of citrulline in the blood and excretion in the urine (citrullinemia). In some cases a majority of the nitrogen excreted may be as citrulline. In addition to restricted nitrogen intake, therapy for this disease requires specific supplementation with arginine for protein synthesis and for the formation of creatine (Chapter 12) and ornithine, which also has essential functions other than its role in urea synthesis (Chapter 12).

ARGININOSUCCINATE LYASE DEFICIENCY

Impaired ability to split argininosuccinate to form arginine resembles argininosuccinate synthetase deficiency in that the substrate, in this case argininosuccinate, is excreted in large amounts. For undetermined reasons, the severity of symptoms in this disease varies greatly so that it is hard to evaluate the effect of therapy. Nevertheless, dietary restriction of nitrogen, alternative excretion of nitrogen by administration of benzoate and phenylacetate, and supplementation of arginine appear to be useful.

ARGINASE DEFICIENCY

Arginase deficiency is a rare disease that causes many abnormalities in the development and function of the central nervous system. In this condition arginine accumulates and is excreted. In addition precursors of arginine and products of arginine metabolism (Chapter 12) may also be excreted. Unexpectedly, some urea is also excreted; this has been attributed to a second type of arginase found in the kidney. Feeding low nitrogen diets including essential amino acids but excluding arginine, or in some cases using the keto analog in place of essential amino acids, has been used successfully; the addition of sodium benzoate has also been effective.

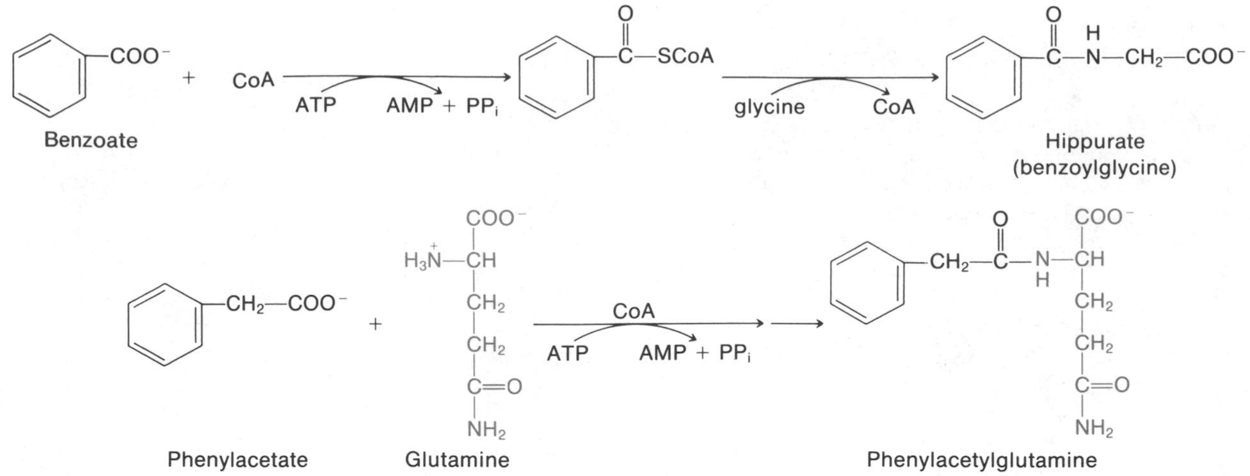

FIGURE 11.18
Detoxification reactions used as alternatives to the urea cycle.

of the substrates, acetyl CoA and glutamate, and by the concentration of arginine, which is an activator of ***N*-acetylglutamate synthetase.**

An additional control of the urea cycle is effected by the concentration of the intermediates of the cycle. Although the cycle is theoretically perfect in that the ornithine is regenerated completely in each turn, in real life there are reactions that convert the intermediates to other products. A normal diet supplies enough arginine to maintain the necessary concentration of ornithine; an enzyme system in the intestinal mucosa synthesizes ornithine from glutamate (Chapter 12).

11.6 ALTERNATIVE REACTIONS FOR ELIMINATION OF EXCESS NITROGEN

Many organic acids are excreted as conjugates with amino acids. A prominent reaction used as a liver function test is the conversion of benzoate to hippurate, shown in Figure 11.18. This reaction forms an activated carboxyl group with CoA as in the case of fatty acid activation; the carboxyl group is subsequently transferred efficiently to glycine to form hippurate, which is excreted in the urine. A related reaction of the higher homolog of benzoic acid, phenylacetic acid, also requires CoA but does not involve an accumulation of a thioester intermediate; instead the carboxyl group is derivitized with the α-amino group of glutamine. These reactions are not prominent in normal metabolism but can be important when aromatic acids are administered.

BIBLIOGRAPHY

General

Meister, A. *Biochemistry of the amino acids,* 2nd ed., New York: Academic, 1965.

Pyridoxal Phosphate

Dolphin, D., Poulson, R., and Avramovic, O. (Eds.). *Vitamin B_6 Pyridoxal Phosphate.* New York: Wiley, 1986.

Glutamate Dehydrogenase

Fisher, H. F. Glutamate Dehydrogenase. *Methods Enzymol.* 113:16, 1985.

Asparaginase

Ellem, K. A. O., Fabrizio, A. M., and Jackson, L. The dependence of DNA and RNA synthesis on protein synthesis in asparaginase-treated lymphoma cells. *Cancer Res.* 30:515, 1970.

Urea Cycle

Grisolia, S., Baguena, R., and Mayor, F. *The urea cycle*. New York: Wiley, 1976.

Holmes, F. L. Hans Krebs and the discovery of the ornithine cycle. *Fed. Proc.* 39:216, 1980.

Alternative Reactions for Excretion of Nitrogen

Brusilow, S. W., et al. Treatment of episodic hyperammonemia in children with inborn errors of urea synthesis. *N. Eng. J. Med.* 310:1630, 1984.

QUESTIONS

C. N. ANGSTADT AND J. BAGGOTT

*Q*uestion *T*ypes are described on page xxiii.

1. (QT1) Amino acids considered nonessential for humans are
   A. those not incorporated into protein.
   B. not necessary in the diet if sufficient amounts of precursors are present.
   C. the same for adults as for children.
   D. the ones made in specific proteins by post-translational modifications.
   E. generally not provided by the ordinary diet.

2. (QT2) Aminotransferases:
   1. usually require α-ketoglutaramate or glutamine as one of the reacting pair.
   2. require pyridoxal phosphate as an essential cofactor for the reaction.
   3. catalyze reactions that result in a net use or production of amino acids.
   4. catalyze freely reversible reactions.
3. (QT2) The production of ammonia in the reaction catalyzed by glutamate dehydrogenase:
   1. requires the participation of NAD^+ or $NADP^+$.
   2. does not proceed through a Schiff's base intermediate.
   3. is favored by removal of the products of the reaction.
   4. may be reversed to consume ammonia if it is present in excess.
4. (QT2) The net synthesis of aspartate using ammonia as the source of nitrogen would be expected to involve:
   1. glutamate dehydrogenase.
   2. a transamination reaction.
   3. oxaloacetate.
   4. L-amino acid oxidase.

5. (QT1) L-Amino acid oxidase:
   A. catalyzes an oxidation coupled to the production of ATP.
   B. is present in large amounts in normal cells.
   C. in vivo, catalyzes a reaction producing H_2O_2.
   D. uses pyridoxal phosphate as its coenzyme.
   E. transfers the amino group of an amino acid to an acceptor molecule.

6. (QT2) Glutamine:
   1. amide nitrogen represents a nontoxic transport form of ammonia.
   2. is a major source of ammonia for urinary excretion.
   3. is used in the synthesis of asparagine in humans.
   4. is an intermediate in the synthesis of purines and pyrimidines.

   A. Aspartate
   B. Arginine
   C. Carbamoyl phosphate
   D. Citrulline
   E. Ornithine
7. (QT5) Urea is formed by a hydrolysis of this compound.
8. (QT5) As part of the urea cycle, synthesized in the cytosol but used in the mitochondria.
9. (QT5) May be synthesized in both mitochondria and cytosol by different isozymes.

10. (QT2) In the formation of urea from ammonia by the urea cycle:
    1. aspartate supplies one of the nitrogens found in urea.
    2. part of the large negative free energy change of the process may be attributed to the hydrolysis of pyrophosphate.
    3. fumarate is produced.
    4. genetic deficiency of any one of the enzymes can lead to hyperammonemia.
11. (QT2) Urea production, as a process to eliminate ammonia:
    1. is not necessary during periods of starvation.
    2. fluctuates as amounts of urea cycle enzymes change in response to changing diets.
    3. because it is cyclic, is unaffected by changing concentrations of intermediates.
    4. has its primary control at the level of carbamoyl phosphate synthesis.

12. (QT1) Carbamoyl phosphate synthetase I:
    A. is a flavoprotein.
    B. is controlled primarily by feedback inhibition.
    C. is unresponsive to changes in arginine.
    D. requires acetyl glutamate as an allosteric effector.
    E. requires ATP as an allosteric effector.
13. (QT1) Asparagine:
    A. is formed in proteins after the protein has been synthesized.
    B. is an essential amino acid in humans.
    C. is an intermediate in the urea cycle.
    D. is a primary source of ammonia in the kidney.
    E. from external sources is a requirement for certain abnormal cells.
14. (QT1) Pyridoxal phosphate:
    A. is covalently bound to its enzyme.
    B. facilitates labilization of bonds by acting as an electron-attracting entity.
    C. forms a Schiff's base with amino acids.
    D. all of the above.
    E. none of the above.

15. (QT1) If glutamate labeled with ^{14}C in the α-carbon atom and ^{15}N is included in the diet, the ratio of ^{14}C to ^{15}N in aspartate derived from the reactions of the citric acid cycle and transamination will most likely be:
    A. identical to the ratio in the glutamate fed.
    B. greater than the ratio in the glutamate fed.
    C. lower than the ratio in the glutamate fed.
    D. depending on circumstances, higher or lower than the ratio in the glutamate fed.

16. (QT2) Hippurate:
    1. excretion represents a mechanism for elimination of excess nitrogen.
    2. is a conjugate of glycine and benzoic acid.
    3. formation involves conversion of benzoic acid to a CoA derivative.
    4. excretion is the primary mechanism for disposing of excess nitrogen from glycine.

ANSWERS

1. B A: All of the 20 common amino acids are incorporated into protein. B, E: Although most of our supply of nonessential amino acids comes from the diet, we can make them if necessary, given the precursors. C: Arginine is not believed to be required for adults (Section 11.1).
2. C 2, 4 correct. 1: Most mammalian aminotransferases use glutamate or α-ketoglutarate. 3: One amino acid is converted into another amino acid; there is neither net gain nor net loss (Section 11.2).
3. E All four correct. 1, 2: The cofactor is a pyridine nucleotide, and the enzyme recognizes both NAD^+ and $NADP^+$ (Note: Production of NH_4^+ requires the oxidized form). 3, 4: The reaction is freely reversible and is usually pulled in the direction of ammonia production by removal of products, but it can easily reverse to meet changing cellular conditions (p. 479).
4. A 1, 2, 3 correct. 1: Glutamate dehydrogenase is necessary to convert the ammonia to an organic compound, glutamate. 2, 3: A transamination from glutamate to oxaloacetate produces aspartate. 4: The L-amino acid oxidase reaction is essentially irreversible in the direction of producing ammonia (pp. 478–480).
5. C A, D: The reduced flavoprotein is reoxidized directly by O_2, not by the electron transport system. B: It is present in rather small amounts. E: Free ammonia is produced (p. 480).
6. E All four correct. It is in the form of the amide nitrogen of glutamine that much of amino acid nitrogen is made available in a nontoxic form (p. 480).
7. B This is the direct production of urea (p. 484).
8. E Ornithine transcarbamoylase, which utilizes ornithine, is mitochondrial; ornithine is regenerated in the cytosol by arginase (Section 11.4).
9. C Carbamoyl phosphate synthetase I, used in the urea cycle, is mitochondrial; CPSII is cytoplasmic and leads to pyrimidines (p. 482).
10. E All four correct. 1, 2, 3: One of the nitrogen atoms is supplied as aspartate, with its carbon atoms being released as fumarate. This reaction is physiologically irreversible because of the hydrolysis of pyrophosphate. 4: Since this is the main pathway for disposal of ammonia, any defect leads to hyperammonemia (p. 485).
11. C 2, 4 correct. 1, 2: Regardless of the fate of the amino acid carbon chain, it is necessary to remove ammonia so this process never ceases although its rate may fluctuate depending on how many amino acid carbon chains are being produced. 3: This would be true only if the cycle were completely closed but, in fact, intermediates, especially ornithine, can be diverted to other processes. 4: The primary control in humans is on CPSI (p. 486).
12. D B, C, D: The primary control is by the allosteric effector, *N*-acetylglutamate. Synthesis of the effector, and therefore activity of CPSI, is increased in the presence of arginine. E: ATP is a substrate (p. 486).
13. E This is the basis for the therapeutic use of asparaginase in leukemia. A, B: Amides are formed from their corresponding acids before incorporation into protein. C, D: In mammals, asparagine has no known function other than protein synthesis (p. 481).
14. D A, C: Pyridoxal phosphate forms a Schiff's base (covalent bond) with first an ε-amino of lysine in its enzyme and then with an amino acid substrate. B: The ability of the pyridine ring to attract electrons is the basis of its action (p. 478).
15. B ^{15}N is diluted, without an accompanying dilution of the carbon label, by the actions of glutamate dehydrogenase and other transaminases than that using aspartate. Therefore, the oxaloacetate formed from the labeled carbon receives nitrogen with relatively less ^{15}N than originally present. Since glutamate is the most efficient precursor of oxaloacetate, it is unlikely that transamination would transfer nitrogen from glutamate more rapidly than this substrate is oxidized (p. 477).
16. A 1, 2, 3, correct. 1: Some organic acids can be excreted as amino acid conjugates, thus excreting that amino acid nitrogen. This is advantageous in hyperammonemia. (p. 478, Clin. Corr. 11.2) 2, 3: See Figure 11.18. 4: Excess nitrogen from glycine, like any other amino acid, would normally be excreted as urea (p. 481).

12

Amino Acid Metabolism II: Metabolism of the Individual Amino Acids

ALAN H. MEHLER

12.1 OVERVIEW 492
12.2 GLUCOGENIC AND KETOGENIC AMINO ACIDS 492
12.3 METABOLISM OF INDIVIDUAL AMINO ACIDS 493
Alanine Is in Equilibrium With Pyruvate via Transamination 493
Branched-Chain Amino Acids: Valine, Leucine, and Isoleucine Have Similar Catabolism 493
Hydroxyamino Acids, Serine and Threonine, and Glycine 496
Glutamate Is Converted to and Formed From Proline, Ornithine, and Arginine 498
Methionine and Cysteine Are Sources of Sulfur 501
Phenylalanine Is Irreversibly Oxidized to Tyrosine 506
Aspartate Has Several Roles 509
Tryptophan Catabolism in Liver Produces Nicotinic Acid 510
Histidine Catabolism Produces a One-Carbon Group 512
Lysine's Amino Group Does Not Equilibrate With the Amino Pool 513
Carnitine Is Synthesized From Trimethyllysine Released by Proteolysis 514
12.4 FOLIC ACID AND ONE-CARBON METABOLISM 516
12.5 NITROGENOUS DERIVATIVES OF AMINO ACIDS 518
Creatine to Phosphocreatine to Creatinine 518
Choline Is a Highly Methylated Compound 519
Amines and Polyamines Are Metabolized by Oxidation and Methylation 520
12.6 GLUTATHIONE 522
Glutathione Functions as a Reductant 522
Glutathione-S-Transferases Function in Detoxication Reactions 523
The γ-Glutamyl Cycle Functions in Amino Acid Transport 524
Control of Glutathione Levels May Have Physiological Significance 525
BIBLIOGRAPHY 525
QUESTIONS AND ANNOTATED ANSWERS 526
CLINICAL CORRELATIONS
12.1 Diseases of Metabolism of Branched-Chain Amino Acids 492
12.2 Diseases of Propionate and Methylmalonate Metabolism 496
12.3 Nonketotic Hyperglycinemia 498
12.4 Diseases of Sulfur Amino Acids 503
12.5 Phenylketonuria 506
12.6 Disorders of Tyrosine Metabolism 508
12.7 Parkinson's Disease 509
12.8 Histidinemia 512
12.9 Diseases Involving Lysine and Ornithine 515
12.10 Folic Acid Deficiency 517
12.11 Diseases of Folate Metabolism 518
12.12 Clinical Problems Related to Glutathione Deficiency 523

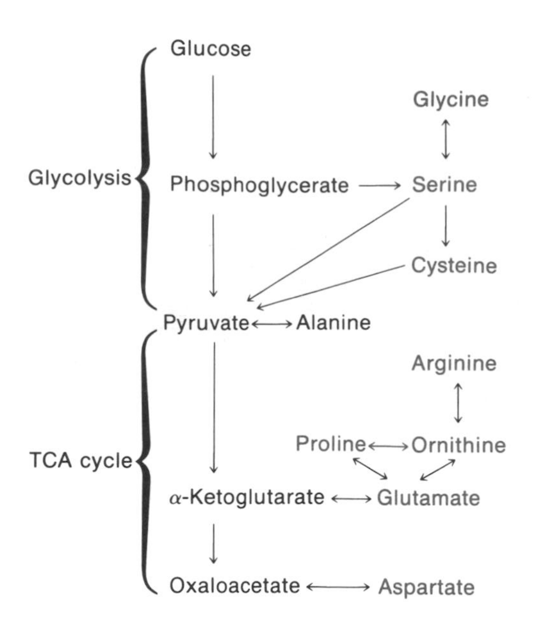

FIGURE 12.1
Interconversion of amino acids and intermediates of carbohydrate metabolism.

$CH_3-CH(OH)-CH_2-COO^-$
β-Hydroxybutyrate

NAD
NADH + H^+

$CH_3-C(=O)-CH_2-COO^-$
Acetoacetate

CO_2

$CH_3-C(=O)-CH_3$
Acetone

FIGURE 12.2
Ketone bodies.
The primary product of fatty acid metabolism, acetoacetate, gives rise to β-hydroxybutyrate and acetone when it accumulates during ketosis.

12.1 OVERVIEW

The metabolism of amino acids is a major part of all intermediary metabolism. Since transamination reactions are freely reversible, the nonessential amino acids alanine, aspartate, and glutamate, which feed into glycolytic and TCA cycle reactions, are synthesized, when needed, from glycolytic and TCA cycle intermediates. Proline and arginine (via ornithine) participate indirectly in these processes through glutamate; serine and glycine are interconverted and related to glycolysis in both anabolic and catabolic reactions. These relationships are outlined in Figure 12.1.

Each of the amino acids is metabolized by an individual pathway, although some steps may be shared, notably the transamination and oxidation of the branched-chain amino acids valine, leucine, and isoleucine. The subsequent steps even in this group of similar molecules are completely different for each compound.

An important aspect of amino acid metabolism is the dependence on coenzymes derived from the vitamins *folic acid,* B_{12}, and *biotin* for certain of the reactions in addition to the more widely used *pyridoxal phosphate* and nicotinamide nucleotides. Some of these reactions and their abnormally accumulated substrates are sensitive indicators of deficiency diseases involving both enzymes and coenzymes.

Individual amino acids serve not only to form other amino acids but also as precursors for many other essential groups of diverse compounds. These include the porphyrins, phospholipids, catecholamines and other hormones and neurotransmitters, a large number of methylated compounds, the sulfate esterified to both endogenous and foreign molecules, the nicotinic acid of the pyridine nucleotides, purines, pyrimidines, the pigment melanin, the phosphagen creatine, carnitine, the polyamines spermine and spermidine, and many other compounds, some of whose functions are not yet known. This long list is not intended to be discouraging but is to emphasize the importance of each of the amino acids in intermediary metabolism. An understanding of the interrelationships, controls, and physiological significance of the amino acids requires that each be recognized as an important metabolite in its own right. The amount of information presented in outlining this area might appear to be excessive for the nonspecialist to assimilate, but familiarity with the broad scope of amino acid metabolism is a precondition for seeking precise information from references when problems are encountered involving this part of biochemistry.

12.2 GLUCOGENIC AND KETOGENIC AMINO ACIDS

Before the detailed metabolism of the amino acids was studied by isolation of enzymes and identification of intermediates, overall relationships of amino acids to the processes of fat and carbohydrate metabolism were established through physiological experiments. By measuring variables such as blood glucose, liver glycogen, and circulating ketone bodies (Figure 12.2), it was seen that some amino acids increase the total amount of glucose in animals, others increase the fat or ketone bodies associated with fatty acid oxidation, and some do both. This evidence led to a classification of amino acids as glucogenic, ketogenic, or both (p. 330). The reaction pathways described in the following sections provide explanations for the fates of each of the amino acids derived from proteins. In each case, the reactions are specific for the L isomer of the amino acid. Although D-amino acids are not found in normal diets, when these are administered to animals in some cases they are metabolized like the

CLIN. CORR. 12.1
DISEASES OF METABOLISM OF BRANCHED-CHAIN AMINO ACIDS

Metabolic deficiencies in the catabolism of the branched-chain amino acids are not common but several abnormalities have been de-

L-isomers after oxidation to the common α-keto analog by D-amino acid oxidase. The keto acids may be transaminated to form the natural isomers or may be further metabolized by the pathways described in this chapter.

12.3 METABOLISM OF INDIVIDUAL AMINO ACIDS

Alanine Is in Equilibrium With Pyruvate via Transamination

The only known functions for alanine are its incorporation into proteins and participation in transamination. The ability of glutamate dehydrogenase to use alanine or pyruvate in place of glutamate or α-ketoglutarate is of doubtful significance in mammalian metabolism. A large amount of nitrogen is transported from muscle and other peripheral tissues to the liver. This physiological process uses pyruvate produced by glycolysis to accept nitrogen from other amino acids in the formation of alanine, which is converted back to pyruvate in the liver where it participates in gluconeogenesis. Since the liver supplies glucose to other tissues, pyruvate and alanine constitute a shuttle mechanism for carrying nitrogen to be reutilized or converted to urea.

Branched-Chain Amino Acids: Valine, Leucine, and Isoleucine Have Similar Catabolism

Valine, leucine, and isoleucine bear nonpolar side chains, each with a methyl group branch. They are essential in the diet of all higher animals, which totally lack the relevant biosynthetic enzymes. The physical and chemical similarities are emphasized by the finding of homologous proteins in various organisms in which these branched-chain compounds replace each other in certain positions without greatly altering the functional properties of the proteins (conservative substitution, Chapter 2). The biosynthetic pathways in plants and microorganisms involve some common steps, and the initial reactions in mammalian catabolism are at some steps carried out by the same enzymes in the case of all three of these amino acids. Subsequent steps, however, are entirely different and yield different products. Thus, valine is glucogenic, leucine is ketogenic, and isoleucine is both.

Where the branched-chain amino acids are present in excess over what is needed for protein synthesis, they are transaminated with α-ketoglutarate to form the corresponding branched-chain α-keto acids. This first step in degradation can be deficient, as seen in Clin. Corr. 12.1. The deaminated products can be considered as higher homologs of pyruvate, and they are oxidized by a complex of enzymes very similar to those that oxidize pyruvate and α-ketoglutarate. The oxidation products include CO_2, NADH that feeds electrons into the electron transport system, and the branched-chain acyl CoAs. These analogs of fatty acyl CoA are oxidized by specific dehydrogenases as if they were unbranched to form the corresponding α,β unsaturated compounds. These reactions are illustrated in Figure 12.3. At this point, the products derived from valine and isoleucine follow one pathway (Figure 12.4), whereas the leucine pathway (Figure 12.5) is quite different.

The unsaturated compounds derived from valine (methylacrylyl CoA) and isoleucine (tiglyl CoA) are hydrated like the unbranched fatty acyl CoA thioesters to form β-hydroxyisobutyryl CoA and α-methyl-β-hydroxyisobutyryl CoA, respectively. The C_4 product from valine at this point is hydrolyzed from the CoA, then the OH group is oxidized by a dehydrogenase to an aldehyde, methylmalonic semialdehyde. A new

scribed that have given insights into the physiology of these metabolites. In general these diseases are detected because of acidosis in newborns or young children and correct identification of the primary cause is necessary for appropriate therapy and prognosis.

Very rare instances have been reported of hypervalinemia and hyperleucine isoleucinemia. These cases do not provide sufficient information to understand the basic pathology or to design an effective therapy. It has been suggested that the two types of disease might indicate the existence of a specific transaminase for valine and one for leucine and isoleucine. Alternatively, rare mutations could alter the specificity of a single enzyme to restrict its specificity.

The most common cause of abnormality in the metabolism of branched-chain amino acids is deficiency of the branched-chain keto acid dehydrogenase complex. There are several variations according to the severity of the deficiency and the enzyme component involved, but all patients with this disease excrete the α-keto acids and corresponding hydroxyacids and other side products; an unidentified excretory product is responsible for a characteristic odor that gives the name maple syrup urine disease to the group. A few cases respond to high doses of thiamine. A large percentage of the cases show serious mental retardation, ketoacidosis, and short life span, but treatment with diets to reduce the ketoacidemia seems to be effective in some cases. A few cases have been reported of diseases caused by deficiency of enzymes in later reactions of branched-chain amino acids. These include a block at the oxidation of isovaleryl CoA and accumulation of isovalerate, β-methylcrotonyl CoA carboxylase deficiency as judged from excretion of β-methylcrotonylglycine and β-hydroxyisovalerate (an abnormal product of hydration of β-methylcrotonate), deficiency of β-hydroxy-β-methylglutaryl CoA lyase and deficiency of the β-ketothiolase that splits α-methylacetoacetyl CoA (with no defect in acetoacetate cleavage). In the latter condition, development is normal and symptoms appear to be related only to episodes of ketoacidosis.

Yeaman, S. J. The mammalian 2-oxoacid dehydrogenases: a complex family. *Trends Biochem. Sci.* 11:293, 1986; Zhang, B., Edenberg, H. J., Crabb, D. W., and Harris, R. A. Evidence for both a regulatory and structural mutation in a family with Maple Syrup Urine Disease. *J. Clin. Invest.* 83:1425, 1989.

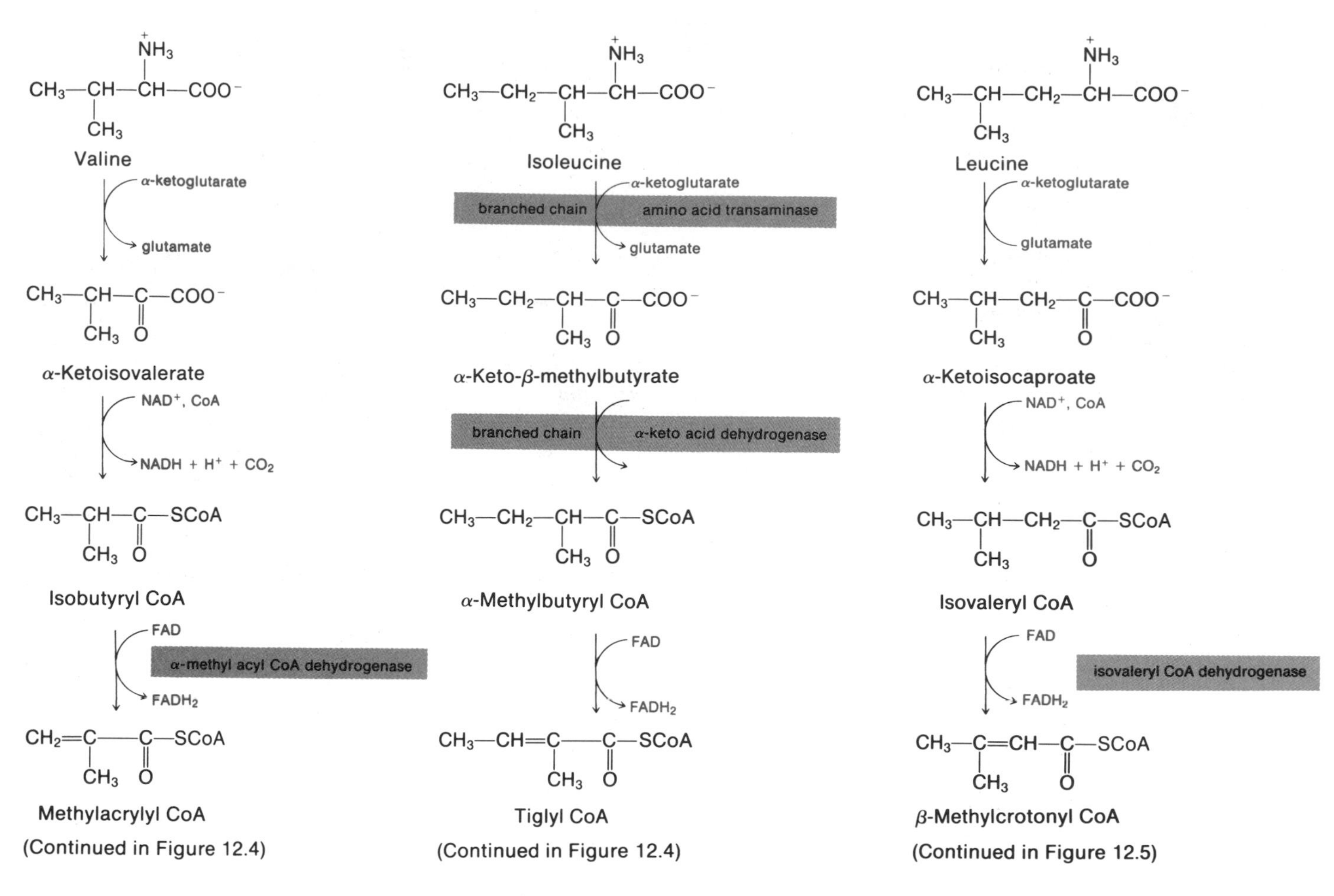

FIGURE 12.3
Common reactions in the degradation of branched-chain amino acids.

thioester is formed when the aldehyde is oxidized by an NAD^+ requiring dehydrogenase and during the oxidation the original carboxyl group is lost as CO_2. This complex reaction results in the formation of propionyl CoA. The C_5 intermediate from isoleucine is oxidized as its CoA derivative to α-methylacetoacetyl CoA, which is cleaved by thiolase to form two thioesters, acetyl CoA and propionyl CoA.

The acetyl CoA formed from isoleucine enters the acetyl CoA pool and can be used for any of the functions of this compound, that is, oxidation in the TCA cycle, fatty acid synthesis, acetylations, and so on. **Propionyl CoA** from isoleucine and valine for the most part is not oxidized further as a fatty acid but is the substrate for a novel reaction sequence, which is also used in the metabolism of the propionyl CoA produced from odd-chain fatty acid oxidation.

Propionyl CoA is carboxylated by a biotin-containing enzyme; the reaction is driven by coupling with ATP breakdown and the product is D-methylmalonyl CoA. This and subsequent reactions are shown in Figure 12.6. Further metabolism requires racemization to the L isomer by an enzyme that labilizes the α hydrogen so that an equilibrium mixture of racemic (equal amounts of D and L) methylmalonyl CoA is formed. Only the L isomer is a substrate for the **methylmalonyl CoA mutase,** an enzyme with a vitamin B_{12} coenzyme, that shifts the carboxy–thioester group to the methyl carbon atom to form the straight-chain succinyl CoA. This compound is metabolized as if it were produced during the operation of the TCA cycle. Deficiency of any of the three enzymes that together

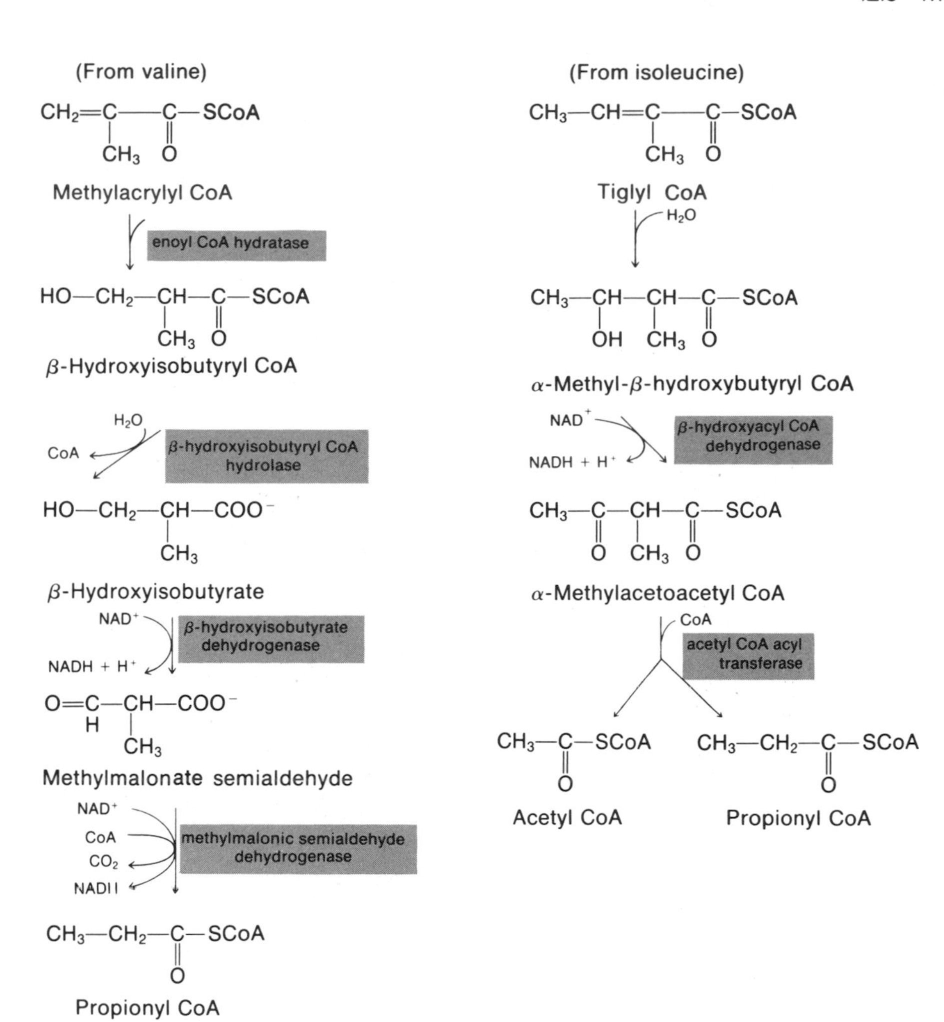

FIGURE 12.4
Terminal reactions in degradation of valine and isoleucine.

convert propionate to succinyl CoA causes disease, as discussed in Clin. Corr. 12.2. The three carbon atoms derived from valine and isoleucine may be oxidized completely to CO_2, but they can be incorporated into carbohydrate by oxidation of succinate to oxaloacetate and the reactions of gluconeogenesis (Chapter 7).

The unsaturated β-methylcrotonyl CoA produced from leucine is carboxylated by another specific biotin-containing enzyme. An unusual aci-

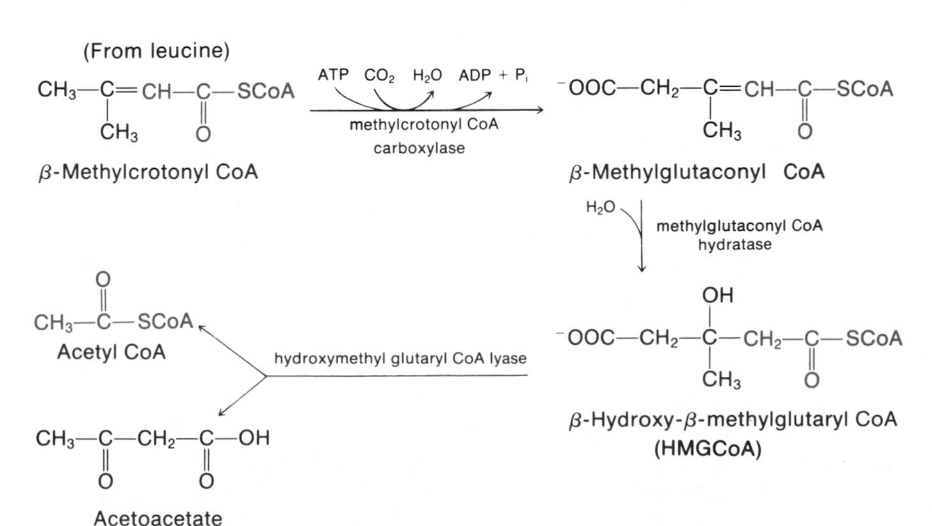

FIGURE 12.5
Terminal reactions of leucine degradation.

CLIN. CORR. 12.2
DISEASES OF PROPIONATE AND METHYLMALONATE METABOLISM

The three enzymes shown in Figure 12.6 have all been implicated in the production of ketoacidosis. Propionate is formed in the degradation of valine, isoleucine, methionine, threonine, the side chain of cholesterol, and odd-chain fatty acids; therefore, each of these could be implicated as a source of the accumulated acid but quantitatively the amino acids appear to be the main precursors since decreasing or eliminating protein from the diet has an immediate effect in minimizing acidosis.

A defect in propionyl CoA carboxylase results in accumulation of propionate, and secondarily propionylglycine, methylcitrate (formed by propionyl CoA condensing with oxaloacetate as an abnormal reaction of citrate synthetase), and tiglate. Increased propionyl CoA is diverted to alternative pathways, oxidation like long-chain fatty acids to β-hydroxypropionate, and incorporation of the C_3 chain into fatty acids in place of an acetyl group to form odd-chain fatty acids. The extent of these reactions is very limited. In one case administration of large amounts of biotin was reported to produce beneficial effects. This suggests that more than one defect results in decreased propionyl CoA carboxylase.

Children have been found who suffer from acidosis caused by high levels of methylmalonate, which is normally not detectable in blood. Analyses of enzymes from liver taken at autopsy or from cultured fibroblasts have established that some cases were due to deficiency of methylmalonyl CoA mutase. One such group was unable to convert methylmalonyl CoA to succinyl CoA under any conditions, but extracts from cells of another group of cases carried out the conversion when the coenzyme adenosylcobalamin was added. Clearly, patients with a structural defect in the enzyme cannot metabolize methylmalonate, but some patients with defects in handling vitamin B_{12} do respond to massive doses of the vitamin. Other cases of methylmalonic aciduria suffer from a more fundamental inability to use vitamin B_{12} that results in deficiency in methylcobalamin, the coenzyme of methionine salvage, as well as adenosylcobalamin deficiency.

Mahoney, M. J. and Bick, D. Recent advances in the inherited methylmalonic acidemias. *Acta Ped. Scand.* 76:689, 1987.

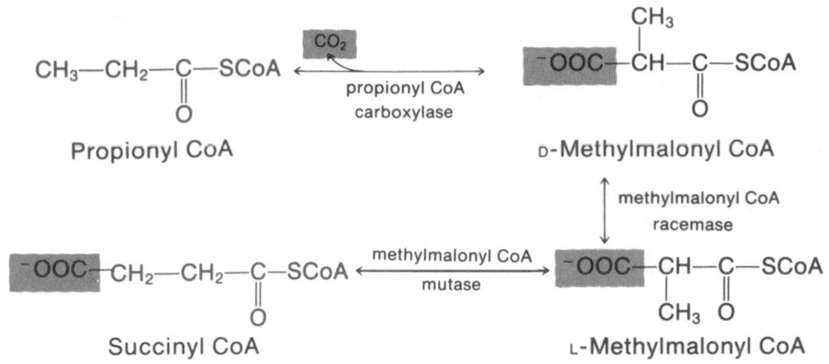

Propionyl CoA ⇌ (CO_2; propionyl CoA carboxylase) D-Methylmalonyl CoA ⇌ (methylmalonyl CoA racemase) L-Methylmalonyl CoA ⇌ (methylmalonyl CoA mutase) Succinyl CoA

FIGURE 12.6
Interconversion of propionyl CoA, methylmalonyl CoA, and succinyl CoA.

dosis is caused by hydrolysis of the thioester to free β-methylcrotonate when the carboxylase is deficient (Clin. Corr. 12.1). In the normal pathway hydration of the double bond forms β-hydroxy-β-methylglutaryl CoA. This compound is identical to an intermediate in the biosynthesis of sterols (Chapter 10), but the fate of the material derived from leucine is cleavage by a lyase to acetoacetate and acetyl CoA; it cannot be converted to mevalonate because it is formed in mitochondria and does not mix with the sterol precursor in the cytosol. Thus, in contrast to the metabolites from valine and isoleucine that can form carbohydrate, both products of leucine degradation are characteristic of fatty acid oxidation.

Hydroxyamino Acids, Serine and Threonine, and Glycine

The synthesis of serine from glycolytic intermediates starts with the oxidation of 3-phosphoglycerate to 3-phosphopyruvate, which is converted to a 3-phosphoserine by transamination (Figure 12.7*a*). The final step is hydrolysis by a specific phosphatase, whose specificity is emphasized by its inhibition by serine. As will be described below, serine is freely interconvertible with glycine so that alternative routes are available to produce this amino acid. In contrast, threonine is an essential amino acid.

Under conditions favoring gluconeogenesis there is an increase in enzymes that convert serine to 3-phosphoglycerate in a "nonphosphorylated pathway" (Figure 12.7*b*) that resembles the biosynthetic pathway but does not participate in formation of the amino acid. The transamination of serine forms 3-hydroxypyruvate, which is reduced by D-glycerate dehydrogenase. D-Glycerate kinase then produces the gluconeogenic intermediate 3-phosphoglycerate.

When serine is used to produce energy, the principal pathway is conversion to pyruvate by a pyridoxal phosphate requiring enzyme, **serine dehydratase,** that eliminates water and NH_3 from the amino acid (Figure 12.8). A similar reaction forms α-ketobutyrate from threonine. Oxidation of α-ketobutyrate, probably by the pyruvate dehydrogenase complex, produces propionyl CoA.

The formation of glycine from serine is a reversible reaction in which a pyridoxal phosphate enzyme, **serine hydroxymethyl transferase,** uses another coenzyme, tetrahydrofolic acid (H_4folate), as a substrate. In the transfer reaction shown in Figure 12.9, C-3 of serine, which is at the oxidation level of formaldehyde, is transferred to the acceptor where it bridges the nitrogen atoms at positions 5 and 10. This product, N^5,N^{10}-methylene H_4folate, can transfer the 1-carbon element back to glycine, which is thus in equilibrium with serine. It will be shown later that the methylene group is part of a pool of 1-carbon fragments at several levels

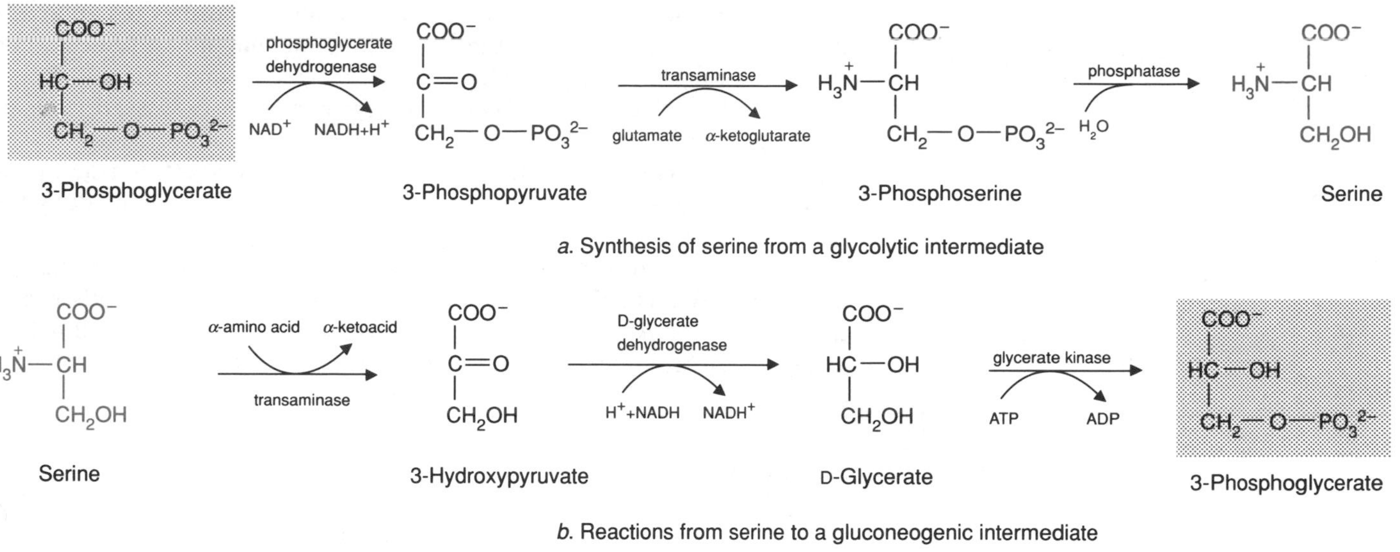

a. Synthesis of serine from a glycolytic intermediate

b. Reactions from serine to a gluconeogenic intermediate

FIGURE 12.7
Pathways for synthesis of serine and for gluconeogenesis from serine.

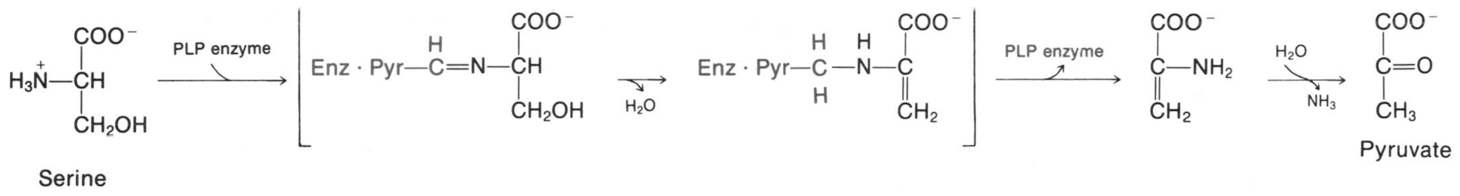

FIGURE 12.8
Reaction of serine dehydratase, a pyridoxal phosphate (PLP) requiring enzyme.

of oxidation. A significant amount of serine is used in the formation of phospholipids.

Glycine is oxidized by two distinct enzyme mechanisms. D-Amino acid oxidase attacks glycine efficiently to form glyoxalate. This product can be converted back to glycine by transamination but is also oxidized further to oxalate (Figure 12.10). Oxalate is of major importance in the formation of renal calculi, some of which are mainly precipitates of calcium oxalate. The α carbon of glycine can also enter the 1-carbon pool through the action of an enzyme that contains pyridoxal phosphate and uses both NAD^+ and H_4folate (see Figures 12.34–12.36). The products are NH_3, CO_2, NADH, and 5,10-methylene H_4folate. Since the reaction is reversible, it provides a mechanism for the synthesis of glycine and also explains why the α- and β-carbon atoms of isotopically labeled serine and the α-carbon atom of glycine are rapidly randomized in vivo. The metabolic importance of this reaction is seen in Clin. Corr. 12.3.

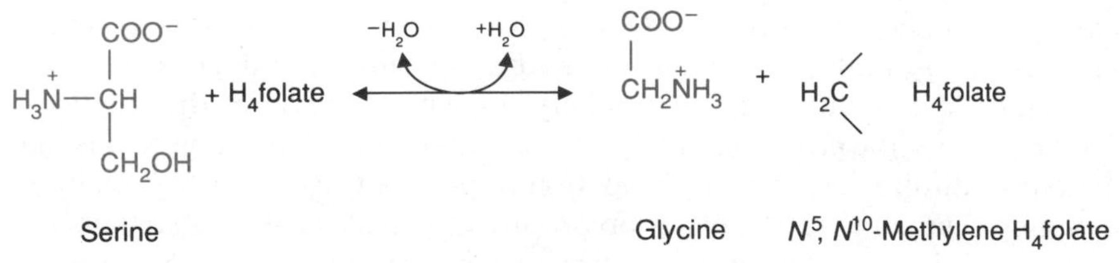

FIGURE 12.9
Interconversion of serine and glycine.

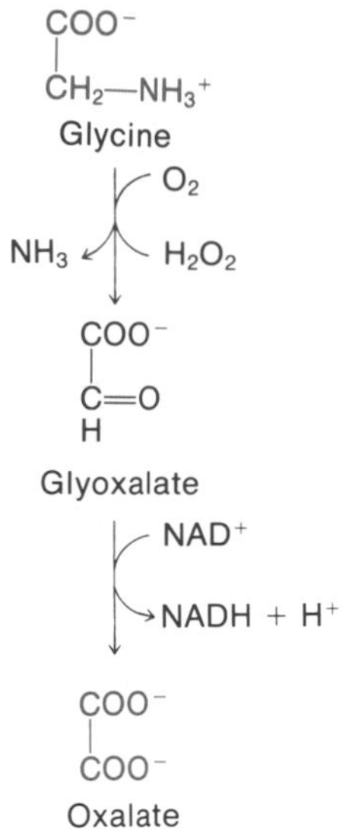

COO^-
|
$CH_2{-}NH_3^+$
Glycine
O_2 →
NH_3 ← → H_2O_2
COO^-
|
$C{=}O$
H
Glyoxalate
NAD^+ →
→ $NADH + H^+$
COO^-
|
COO^-
Oxalate

FIGURE 12.10
Oxidation of glycine.

Glycine is incorporated intact as a constituent of purines (Chapter 13). It also contributes to the synthesis of δ-aminolevulinate en route to porphyrin formation (Chapter 24). Creatine, discussed in Section 12.5, is a substituted glycine. Other substituted glycines to be discussed later are sarcosine and betaine.

The major pathway of threonine degradation is presented in Figure 12.11 and involves oxidation of the alcohol group by a specific dehydrogenase to form α-amino-β-ketobutyrate. This unstable intermediate can be decarboxylated to aminoacetone, which is oxidized by an amine oxidase to methylglyoxal, a precursor of lactate and pyruvate. Another reaction of α-amino-β-ketobutyrate is thiolysis by CoA to yield acetyl CoA and glycine. Although this reaction is reversible in vitro, the physiological reaction appears to proceed only in the direction of threonine catabolism.

Selenocysteine

An amino acid containing the trace element **selenium** in place of the sulfur of cysteine has been identified as a component of a very few enzymes. In mammals it is an essential component of glutathione peroxidase. Unlike other modified amino acids, selenocysteine is not synthesized as a posttranslational modification of a protein. It is made from serine only when esterified to a specific isoacceptor tRNA that contains an anticodon complementary to the stop codon UGA. In mRNAs containing appropriate signals UGA (see Chapter 17) does not act to terminate peptide chain elongation but acts as the codon for selenocysteine.

Glutamate Is Converted to and Formed From Proline, Ornithine, and Arginine

In Chapter 11 a major role was described for glutamate in the incorporation of ammonia into amino acids, transferring amino groups, and eliminating nitrogen. Glutamate is also a key intermediate in the metabolism of other amino acids that share with it a common carbon skeleton. The nonessential amino acids proline and arginine are built directly from the C_5 chain of glutamate.

Proline

The cyclization of glutamate to a five-membered ring at neutral pH requires prior modification of the γ-carboxyl group. The pyrrolidine ring of proline is more reduced than the direct cyclization product, 5-oxoproline (pyrrolidone carboxylate, p. 524). The pathway brackets the ring closure with reduction steps. Glutamate is activated in a reaction with ATP; the putative γ-carboxyl phosphate does not dissociate from the protein but is reduced to the corresponding aldehyde by NADH. The resulting glutamic semialdehyde is an equilibrium mixture composed of only a trace of an open chain and mainly the cyclic Schiff base, Δ^1-pyrroline-5-carboxylate. A second reduction by NADPH saturates the ring and gives a stable product, proline, as seen in Figure 12.12.

The derivatives of proline found in collagen and a few other proteins, **3-** and **4-hydroxyprolines**, are formed by mixed-function oxygenases from proline residues incorporated in the polypeptide chains (Chapter 17). During growth and development there is extensive turnover of the structural proteins, and both proline and the hydroxyprolines are degraded.

Catabolism of proline superficially appears to reverse the synthetic pathway, but the conversion to glutamate involves different enzymes that catalyze different reactions. Oxidation of proline is carried out by a mitochondrial enzyme (probably a flavoprotein by analogy with a similar bacterial activity) that transfers electrons to the electron transport system in

CLIN. CORR. 12.3 NONKETOTIC HYPERGLYCINEMIA

Nonketotic hyperglycinemia is characterized by severe mental deficiency, and many patients do not survive infancy. The name of this very serious disease is meant to distinguish it from ketoacidosis in abnormalities of branched-chain amino acid metabolism in which, for unknown reasons, the glycine level in the blood is also elevated. Deficiency of glycine cleavage enzyme has been demonstrated in homogenates of liver from several patients, and isotopic studies in vivo have confirmed that this enzyme is not active in the patients. The glycine cleavage enzyme consists of four protein subunits. Inherited abnormalities have been found in three of these. The severity of this disease suggests that the glycine cleavage enzyme pathway is of major importance in the catabolism of glycine. Glycine is a major inhibitory neurotransmitter, which probably explains some of the neurological complications of the disease. Vigorous measures to reduce the glycine levels [infusion of 5-formyltetrahydrofolate, administration of sodium benzoate (see Clin. Corr. 11.1), and exchange transfusion] fail to alter the course of the disease.

Nyhan, W. L. Metabolism of glycine in the normal individual and in patients with non-ketotic hyperglycinemia. *J. Inherited Metab. Dis.* 5:105, 1982.

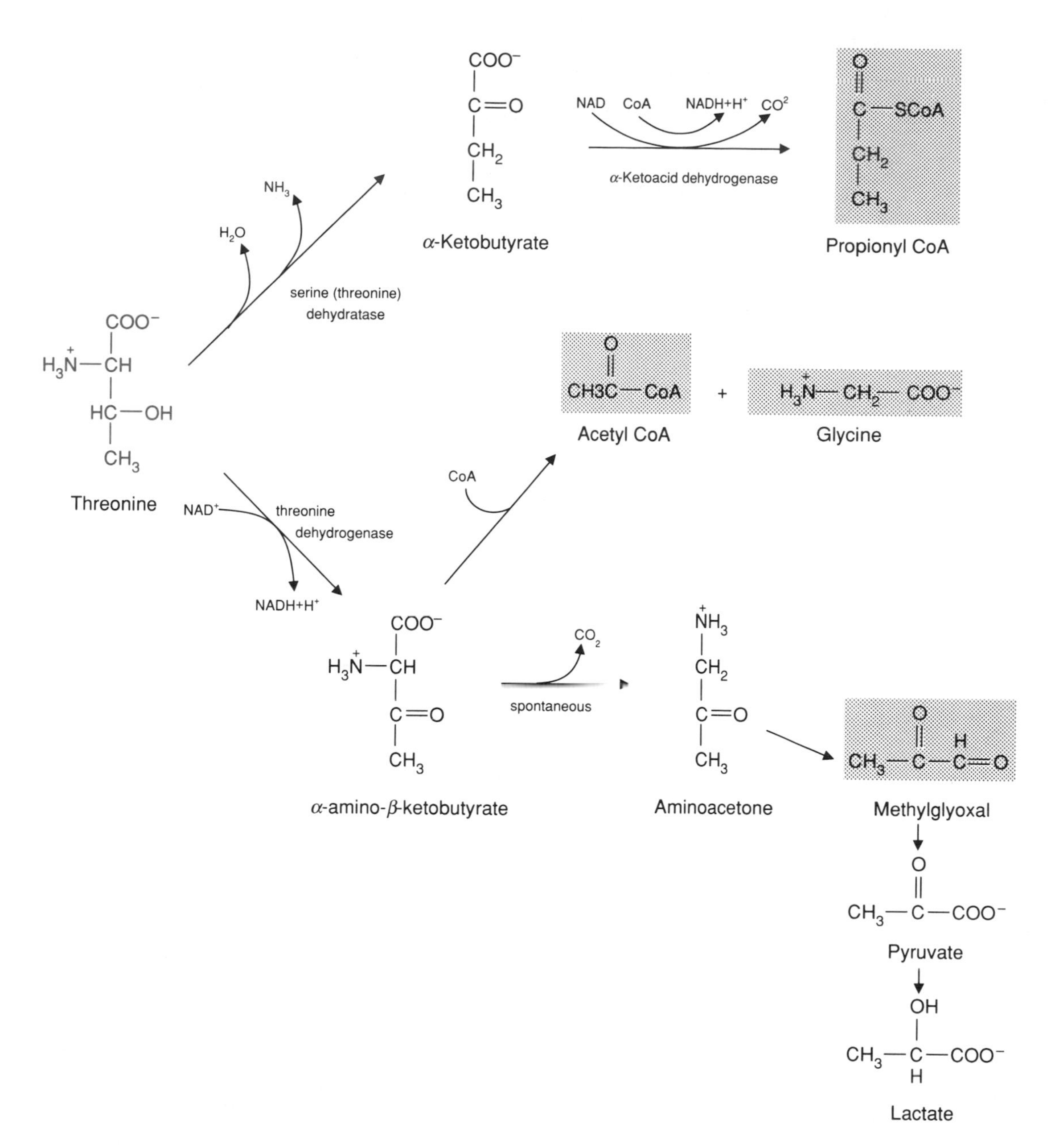

FIGURE 12.11
Outline of threonine metabolism.

forming Δ^1-pyrroline 5-carboxylate. The further oxidation of pyrroline carboxylate is a dehydrogenation by an enzyme nonspecific for NAD^+ and $NADP^+$ to form glutamate with no involvement of phosphate. Like all oxidations of aldehydes to free carboxyl groups, this reaction is effectively irreversible. 4-Hydroxyproline is oxidized by an oxidase similar to that for proline but not the same. The second oxidation, however, is catalyzed by the same dehydrogenase involved in proline oxidation. The product, 4-hydroxy-2-ketoglutarate, is split by an aldolase to pyruvate and glyoxylate. The structures of these compounds are shown in Figure 12.13. Little is known about the oxidation of 3-hydroxyproline.

Ornithine

Δ^1-Pyrroline-5-carboxylate is in equilibrium with the open-chain aldehyde that participates in a transamination reaction to form ornithine. In mammals the conversion of glutamate to ornithine has been demonstrated

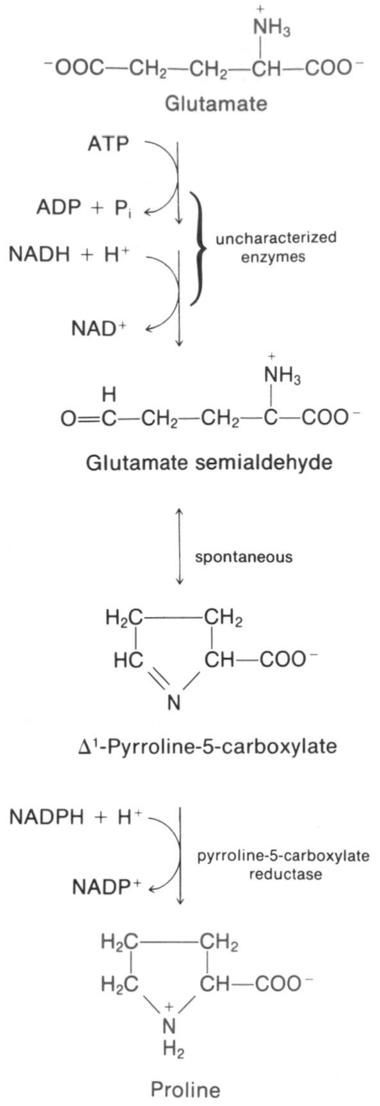

FIGURE 12.12
Synthesis of proline from glutamate.

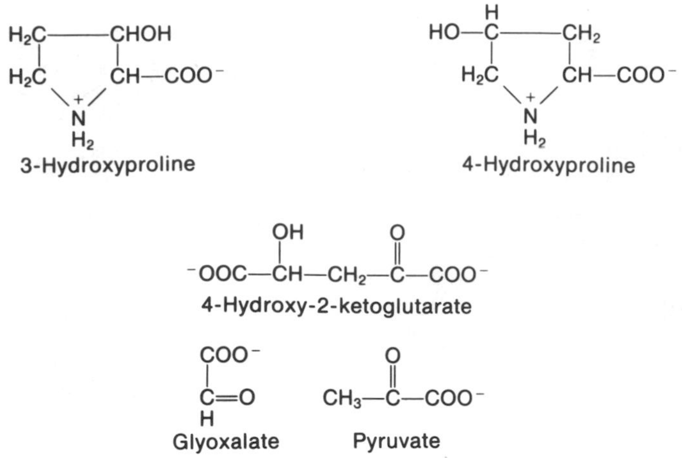

FIGURE 12.13
Hydroxyprolines and metabolic products.

only in intestinal mucosa, not in liver (Figure 12.14). The existence of a common intermediate makes it possible for ornithine and proline to be interconverted without involving the formation of glutamate.

In addition to its participation in urea cycle reactions, arginine is the donor in a **transamidinase** reaction. The physiological acceptor is glycine, which is converted to guanidinoacetate. This compound is an intermediate in the synthesis of creatine (p. 518).

Ornithine is prominent in metabolism primarily in relationship to arginine, as a precursor and product in the urea cycle, or simply as a product when arginine is degraded by arginase in tissues other than liver or when the transamidinase reaction produces guanidinoacetate. Since most of the arginine ingested is not required for protein synthesis, a large amount of ornithine is produced and started on a degradative pathway by ornithine transaminase (Clin. Corr. 12.9). Another very important function for ornithine is the formation of putrescine by decarboxylation (Figure 12.14).

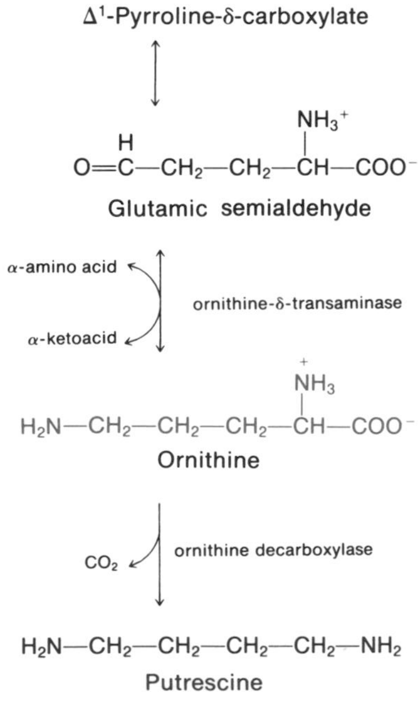

FIGURE 12.14
Formation and decarboxylation of ornithine.

The putrescine is required for the synthesis of polyamines, spermidine, and spermine. The exact function of polyamines is still uncertain, but it is clear that initiation of cell division depends on ornithine decarboxylase.

Arginine

In addition to its role in the urea cycle and in transamidination, arginine is the source of nitric oxide, NO, which is produced by vascular endothelium as a muscle relaxing agent, by cytotoxic macrophages when activated and by neutrophiles as a platelet inhibitory factor. The remainder of arginine becomes citrulline. Different mixed-function oxygenases carry out this synthesis in various cells; the enzyme in macrophages behaves like a single protein that requires tetrahydrobiopterin, whereas the enzyme purified from rat cerebellum uses NADPH and also requires calmodulin.

γ-Aminobutyrate

Removal of the α-carboxyl group of glutamate leaves γ-aminobutyrate (GABA). The glutamate decarboxylase reaction is prominent in the brain, where GABA is an important neurotransmitter. Gamma-aminobutyrate is converted to succinic semialdehyde by transamination and then oxidized to succinate (Figure 12.15).

$^{-}OOC-CH_2-CH_2-CH(NH_3^+)-COO^-$
Glutamate

↓ (CO_2)

$^{-}OOC-CH_2-CH_2-CH_2-NH_3^+$
γ-Aminobutyrate

↓ (α-ketoglutarate → glutamate)

$OOC-CH_2-CH_2-CHO$
Succinic semialdehyde

↓ (NAD → NADH + H⁺)

$OOC-CH_2-CH_2-COO^-$
Succinate

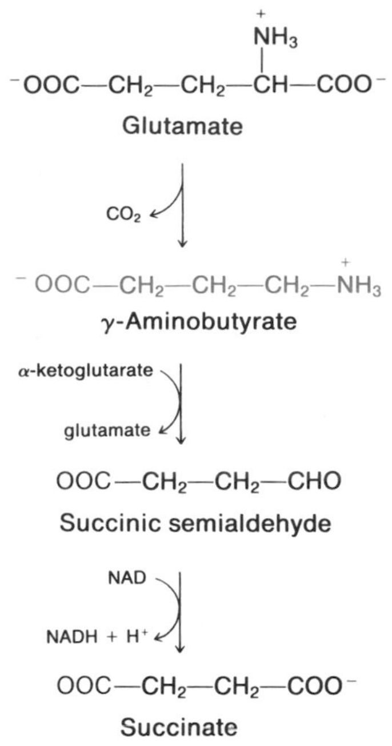

FIGURE 12.15
Metabolism of γ-aminobutyrate (GABA).

Methionine and Cysteine Are Sources of Sulfur

In the human adult methionine can serve as the sole source of sulfur, but on a normal diet a larger amount of sulfur is ingested as the more abundant cysteine. The synthesis of cysteine requires that the thioether of methionine be made available as a sulfhydryl group. The mechanism of exposing the sulfur is the same that is used to make methionine the principal source of methyl groups that are transferred to many diverse acceptors in reactions of great physiological importance and also to enable the side chain of methionine to be used for other biosynthetic reactions.

S-*Adenosylmethionine*

The activation of methionine, described in Figure 12.16, is a unique reaction in that the sulfur of the thioether becomes a sulfonium atom by the addition of a third carbon atom, the C-5′ of the ribose of ATP. In this reaction all of the phosphates of ATP are eliminated, one as P_i and the others as PP_i. The product, ***S*-adenosylmethionine** (Ado Met), is reactive because of the positive charge that reverts to a neutral thioether when any of its three substituents is lost. Enzymatic reactions are known that specify transfer to appropriate acceptors of either the methyl group, a C_3- or C_4-carbon fragment from the rest of the methionine structure, or the adenosyl group, in each case leaving the other two substituents in the thioether.

The most prominent reactions of *S*-adenosylmethionine are methyl transferases. These are for the most part presented as reactions of the methyl acceptors. In each case, as illustrated in Figure 12.17 with nicotinamide, the methyl donor becomes *S*-adenosylhomocysteine. Hydrolysis of the latter yields adenosine and **homocysteine.** There are three principal routes of metabolism of homocysteine; the distribution of homocysteine among these pathways is determined by the physiological needs of the organism. If cysteine is needed, the homocysteine condenses with serine to form **cystathionine** in a reaction catalyzed by an enzyme with pyridoxal phosphate as a coenzyme. Another pyridoxal phosphate enzyme, cystathionase, cleaves this thioester to cysteine, α-ketobutyrate, and ammo-

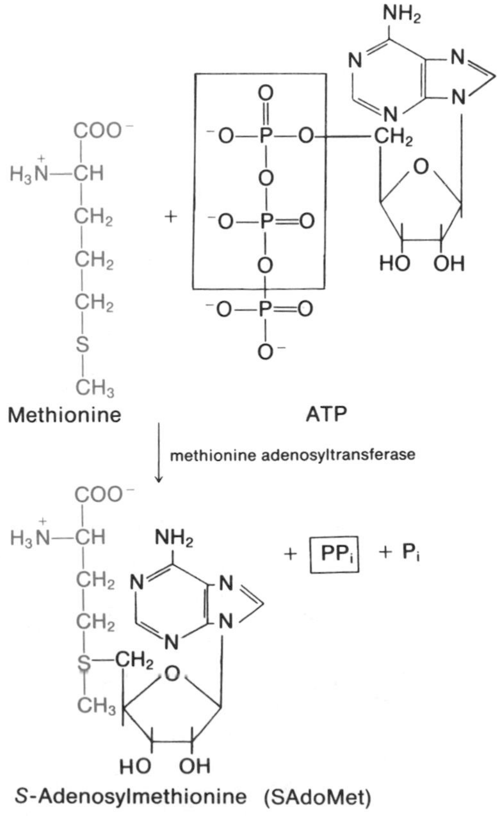

Methionine + ATP
↓ methionine adenosyltransferase
S-Adenosylmethionine (SAdoMet) + PP_i + P_i

FIGURE 12.16
Synthesis of *S*-adenosylmethionine.

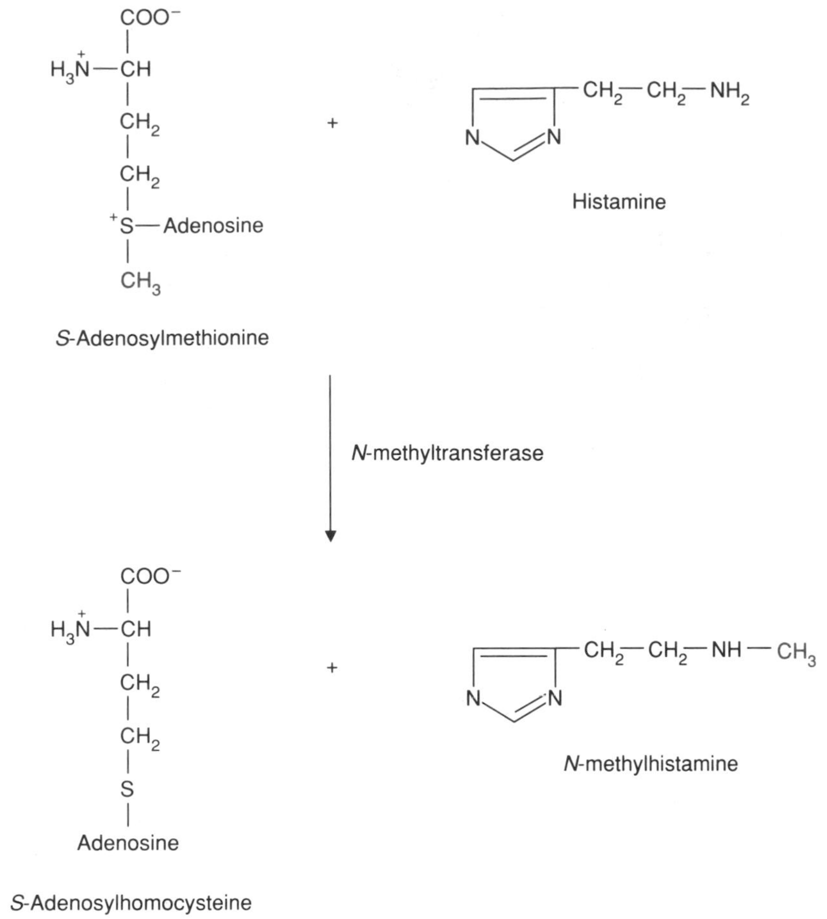

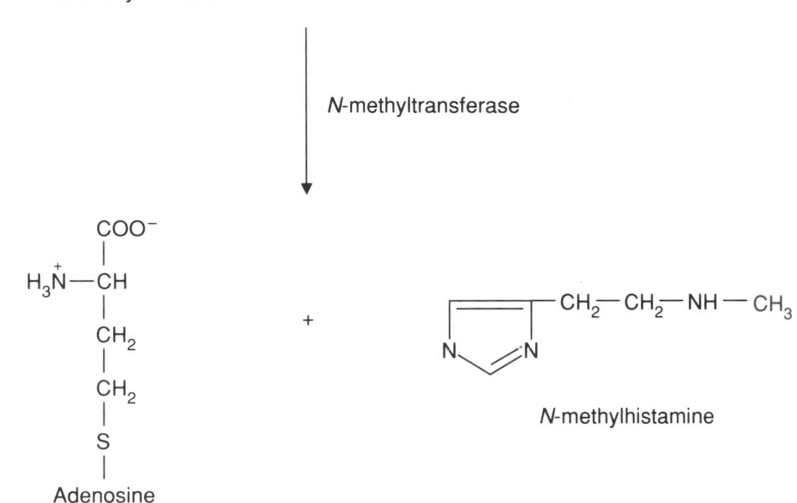

FIGURE 12.17
A representative methylation reaction.
N-Methylhistamine is formed in one of the alternative pathways of inactivation of histamine, a potent physiological effector of smooth muscles. Histamine is also inactivated by oxidation of the primary amino group to an aldehyde by diamine oxidase (histaminase).

nia (Figure 12.18). The sum of cystathionine synthase and cystathionase reactions is transsulfuration. If, however, methionine is in short supply, homocysteine is remethylated by $\mathbf{N}^5$-methyl H_4folate or betaine. When both sulfur amino acids are present in adequate amounts, the same enzyme, cystathionase, appears to be responsible for the activity called homocysteine desulfhydrase that hydrolyzes homocysteine to α-ketobutyrate, NH_3, and H_2S. *S*-Adenosylmethionine activates cystathionine synthase and thus regulates the competing pathways of homocysteine metabolism. Defects of enzymes involved in the metabolism of cystathionine result in accumulation of methionine, homocysteine, or cystathionine (Clin. Corr. 12.4).

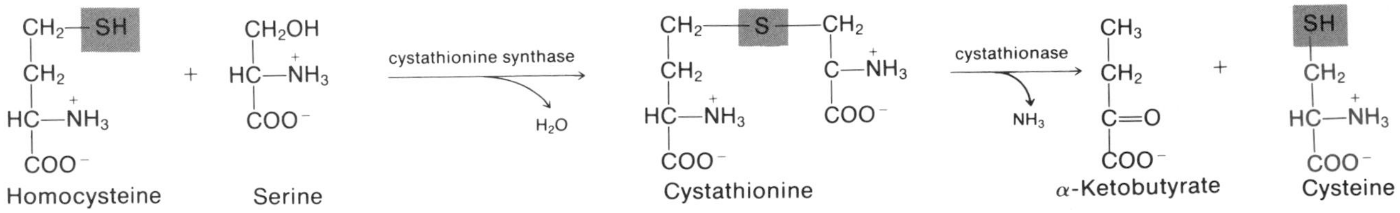

Homocysteine + Serine —(cystathionine synthase; H_2O)→ Cystathionine —(cystathionase; NH_3)→ α-Ketobutyrate + Cysteine

FIGURE 12.18
Biosynthesis of cysteine.
The enzyme cystathionine synthase is sometimes named β-cystathionine synthase, and cystathionase is sometimes designated γ-cystathionase. The Greek letters are used to distinguish these enzymes from similar activities in bacteria that participate in a reversal of this pathway, making homocysteine (and methionine) from cysteine.

CLIN. CORR. 12.4
DISEASES OF SULFUR AMINO ACIDS

TRANSSULFURATION DEFECTS

Congenital deficiency of any of three enzymes involved in transsulfuration results in accumulation of sulfur-containing amino acids. Hypermethioninemia has been attributed to a deficiency of methionine adenosyltransferase. A total lack of this enzyme would eliminate most methylation reactions and has never been found. The partial deficiencies are probably manifestations of K_m mutants; that is, the patients have enzymes that require higher than normal concentrations of methionine for saturation so that a steady state is established in which methionine at a high concentration permits a rate of formation of *S*-adenosylmethionine adequate for the various functions illustrated in Figure 12.19. The accumulation of methionine in these patients has no other consequence so the condition is benign. This is of significance in making a differential diagnosis since hypermethioninemia also occurs in cystathionine synthase deficiency (see below), in severe liver disease, and in tyrosinemia.

Deficiency of cystathionine synthase causes homocysteine to accumulate, and this causes even higher levels of methionine to be formed by remethylation, hence it is a cause of homocystinuria. Many minor products of these amino acids are formed and excreted as a result of these accumulations. No mechanism has been established to explain why the accumulation of homocysteine should lead to pathological changes, but the deficiency of the synthase is associated with many abnormalities. Homocysteine may react with and block lysyl aldehyde groups on collagen and interfere with cross-linking (Chapter 2). The lens of the eye frequently is dislocated some time after age 3 years, and other ocular abnormalities are often seen. Osteoporosis and other skeletal abnormalities develop during childhood. Mental retardation is often the first indication of this deficiency. Thromboembolism and vascular occlusion may occur at any age. Rational attempts to treat this biochemical lesion are complex because of the variety of symptoms. Restriction of methionine intake and feeding of betaine (or its precursor, choline) have been found to lower the homocystine level. In a number of cases significant improvement has been obtained by feeding pyridoxine (vitamin B_6), but other cases are nonresponsive to the vitamin. This suggests that the deficiency can be caused by more than one type of gene mutation; one type may affect the K_m for pyridoxal phosphate and others may alter the K_m for other substrates, V_{max}, or the amount of enzyme. Other causes of homocystinuria are due to abnormalities of the remethylation of homocysteine to methionine. This can result from deficiency of N^5-methyl H_4folate homocysteine methyltransferase or its cofactor methylcobalamin. The latter in turn may result from malabsorption of cobalamin or inability to convert it to methylcobalamin.

In contrast to the severe manifestations of cystathionine synthase deficiency, lack of cystathionase does not seem to cause any clinical abnormalities other than accumulation of cystathionine and excretion of this compound in the urine. It is worth noting that the first discovery of cystathioninuria was in a mental patient, and subsequent searches for additional cases tended to concentrate on the mentally retarded. Even so, a majority of reported cases are mentally normal and the retarded cases include cases with endocrine disorders, phenylketonuria, and explanations other than cystathioninuria. A large majority of reported cases are responsive to pyridoxine. The amount of cysteine synthesized in the various deficiencies of enzymes of transsulfuration is not known, but supplementation is not necessary except when a low methionine diet is used in cases of hypermethioninemia.

DISEASES INVOLVING CYSTINE

Two clinical conditions involve the disulfide cystine. Cystinuria is a disease of defective membrane transport of cystine and the basic amino acids that results in increased renal excretion of these compounds. Since extracellular sulfhydryl compounds are quickly oxidized to disulfides, cysteine in the blood and urine exists as cystine. The low solubility of cystine results in the formation of calculi, the serious feature of this disease. Treatment is limited to attempts to remove stones or to prevent precipitation by drinking large amounts of water, or drugs to make soluble derivatives of cystine.

A much more serious disease is cystinosis in which cystine accumulates in lysosomes. The mechanism of normal transport is unknown. The stored cystine forms crystals in many cells, with most serious loss of function of the kidneys, usually causing renal failure within 10 years.

Four cases have been reported in which a mixed disulfide of cysteine and β-mercaptolactate is excreted. Two of the reports describe severely retarded individuals, and two concern individuals with normal development. It is accepted that the abnormal metabolite is derived from β-mercaptopyruvate, but why this intermediate accumulates to be reduced and excreted together with cysteine remains a mystery.

Seashore, M. R., Durant, J. L., and Rosenberg, L. E. Studies on the mechanisms of pyridoxine responsive homocystinuria. *Pediatr. Res.* 6:187, 1972; Mudd, S. H. The natural history of homocystinuria due to cystathione β synthase deficiency. *Am. J. Hum. Genet.* 37:1, 1985; and Frimpter, G. W. Cystathionuria: nature of the defect. *Science* 149:1095, 1965.

S-Adenosylmethionine can be decarboxylated by a specific enzyme. The decarboxylated product, *S*-adenosylmethylthiopropylamine, is the donor of the C_3 fragment that is used in the formation of **polyamines.** The acceptor of the propylamino group is the diamine **putrescine** produced by decarboxylation of ornithine and the first condensation product is spermidine. A second transfer of a propylamino residue produces the symmetrical polyamine, spermine. **Spermine** is a highly cationic molecule that

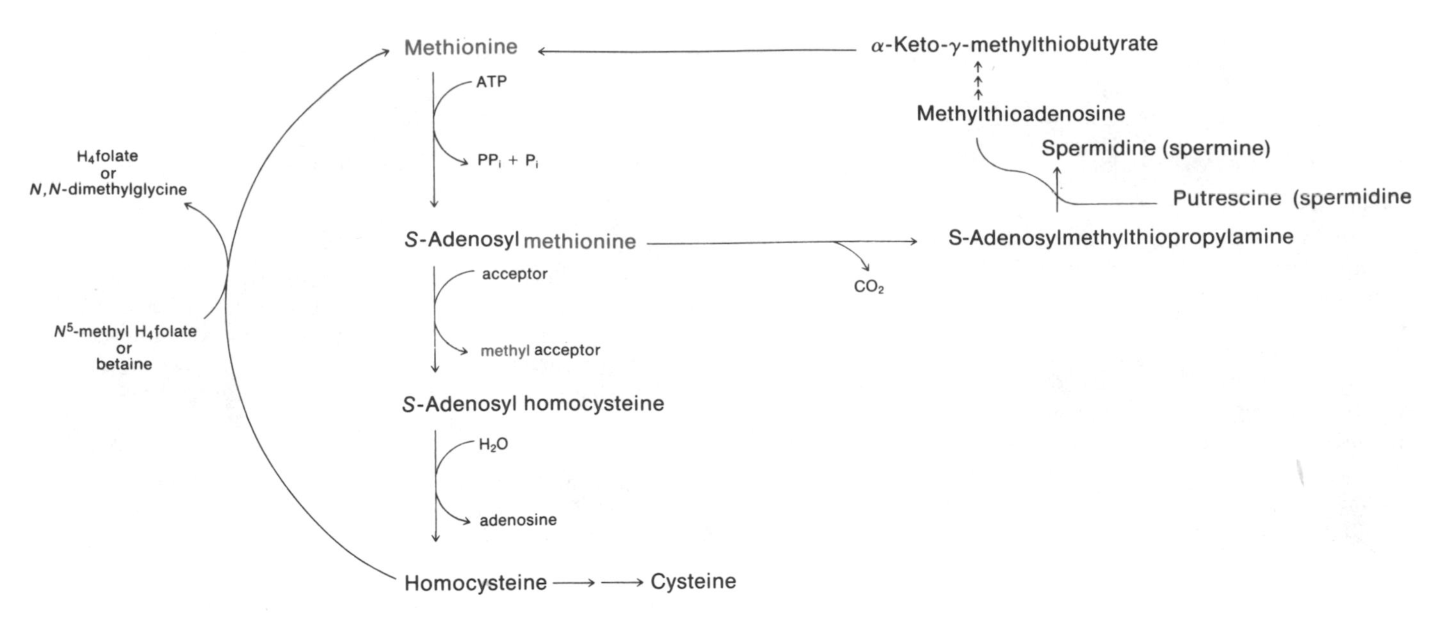

FIGURE 12.19
Uses and resynthesis of methionine.

binds tightly to nucleic acids and other polyanions. It has been found to stimulate a variety of reactions in vitro and appears to be essential for the function of a topoisomerase (Chapter 17). This may explain the early increase of ornithine decarboxylase activity in cell division, but the precise biological functions of the polyamines remain to be determined.

Although methionine is an essential amino acid, it is resynthesized to a large extent by salvage pathways. N^5,N^{10}-Methylene H_4folate produced from serine or glycine or by reduction of other 1-carbon derivatives of H_4-folate can be reduced by NADH to N^5-methyl H_4folate. The only known reaction to use N^5-methyl H_4folate is the methyltransferase with homocysteine as the methyl acceptor to regenerate methionine. The reaction requires *cobalamin,* the coenzyme form of vitamin B_{12}. The salvage reaction in which betaine is the methyl donor appears to be a simpler reaction.

The sulfur-containing product of polyamine synthesis, methylthioadenosine, is not wasted. The steps in the salvage pathway have not yet all been identified, but a phosphorylase is known to replace the adenine with a phosphate. Isotopic evidence shows that the sulfur, the methyl carbon, and the ribose chains are all incorporated into methionine, presumably through formation of α-keto-γ-methylthiobutyrate, which is transaminated to regenerate methionine. It should be noted that this process, pictured in Figure 12.19, was initiated by activation of methionine and cannot give a net synthesis of the essential amino acid.

Obviously, a transaminase that makes methionine can also degrade it via the same α-keto-γ-methylthiobutyrate. A degradative sequence that starts with this transamination is probably responsible for the formation of toxic mercaptans, including methanethiol and H_2S. Normally the amounts of these compounds are negligible, but they become significant in methionine toxicity, caused by excessive ingestion or decreased transsulfuration during liver disease.

Cysteine

Cysteine is a major source of sulfur for the body. Sulfur occurs at several oxidation levels, from sulfide to sulfate. Reports of a direct desulfhydration of cysteine, analogous to dehydration of serine, has been claimed, but at this time the best evidence indicates that H_2S (or S^{2-}) is derived indirectly from cysteine in a series of reactions catalyzed by **cystathionase.** Very little of this toxic material is produced in normal metabolism. The sulfur of cysteine is made available indirectly for detoxification reactions and for incorporation into proteins such as ferridoxin. A sulfur donor produced by cystathionase is thiocysteine.

$$\text{Cystine} + H_2O \longrightarrow \text{pyruvate} + NH_3 + HS{-}S{-}CH_2{-}CH(\overset{+}{N}H_3){-}COO^- \quad \text{(Thiocysteine)}$$

Rhodanese catalyzes the transfer of sulfur from thiocysteine (and other sulfane donors, such as thiosulfate) to many acceptors; the best known of these is cyanide, which is converted to the much less toxic thiocyanate.

$$\text{Thiocysteine} + CN^- \longrightarrow \text{cysteine} + SCN^-$$

Glutamate aminotransferase and other aminotransferases are able to use cysteine as an amino donor. The resulting 3-mercaptopyruvate is the substrate of a sulfur transferase. In the absence of an acceptor elemental sulfur is formed, but with acceptors such as sulfite and cyanide, the transfer reaction shown in Figure 12.20 predominates.

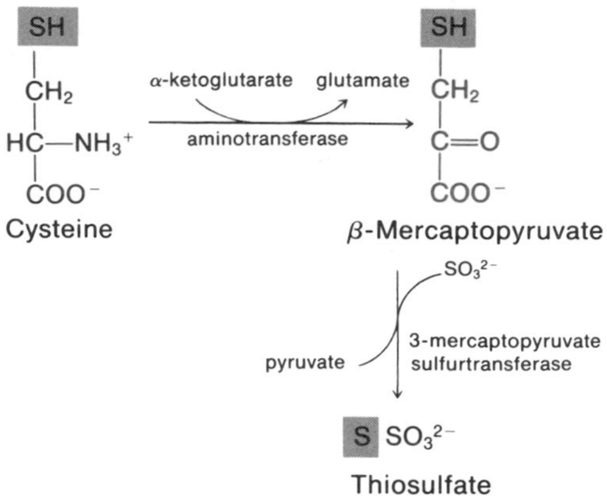

FIGURE 12.20
Formation and use of β-mercaptopyruvate.

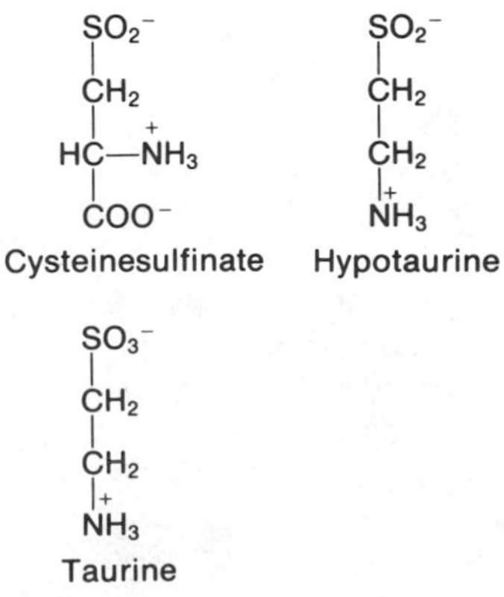

A major pathway of cysteine metabolism is oxidation by an oxygenase to **cysteinesulfinate.** The enzyme has the requirements of a mixed-function oxygenase, Fe^{2+} and a reduced nicotinamide nucleotide, but nevertheless inserts two atoms of oxygen from O_2 into the product. Cysteinesulfinate and compounds derived from it react spontaneously to give several products, so it has been difficult to evaluate the biochemical pathways using this intermediate. At this time it seems likely that two reactions account for most of its metabolism. Decarboxylation yields hypotaurine, which is rapidly oxidized to taurine shown in Figure 12.21. An alternative pathway of cysteinesulfinate metabolism starts with transamination, which should give β-sulfinylpyruvate, but only sulfite and pyruvate are found. Sulfite oxidase, an enzyme containing molybdenum and cytochrome b_5, catalyzes the main production of sulfate from bisulfite.

$$HSO_3^- + O_2 + H_2O \xrightarrow{\text{sulfite oxidase}} SO_4^{2-} + H_2O_2 + H^+$$

3′-Phosphoadenosine-5′-phosphosulfate

Although much of the sulfate is excreted, some is also used as a component of many structural elements; some proteins, polysaccharides, and lipids are sulfated. In addition, many compounds are excreted as sulfate esters. Metabolites of the steroidal sex hormones and exogenous phenolic compounds are prominent among the urinary sulfates. Each of these materials is made by transfer of sulfate from a nucleotide carrier, 3′-phosphoadenosine-5′-phosphosulfate (PAPS), which is synthesized in two steps:

$$SO_4^{2-} + \text{ATP} \xrightarrow{\text{ATP sulfurase, } Mg^{2+}} \text{adenosine-5'-phosphosulfate (AMPS)} + PP_i$$

$$\text{Adenosine-5'-phosphosulfate} + \text{ATP} \xrightarrow[Mg^{2+}]{\text{AMPS phosphokinase}} \text{3'-phosphoadenosine-5'-phosphosulfate} + \text{ADP}$$

Sulfate incorporated into this nucleotide is also transferred to some unknown acceptor to form **taurine**. The physiological role of taurine is still conjectural, but the high concentrations found in the brain suggest a

FIGURE 12.21
Taurine and its precursors.

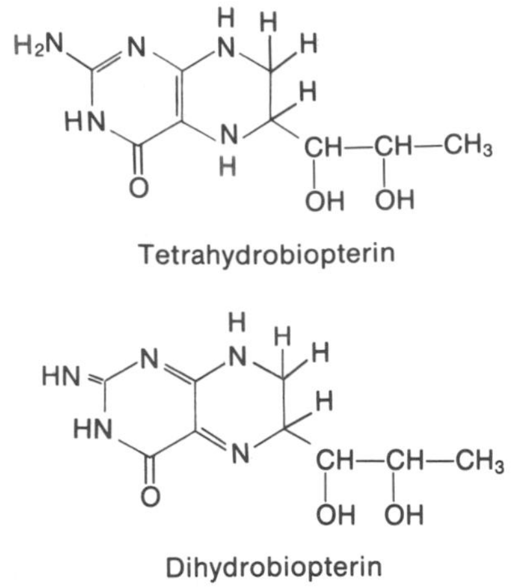

Tetrahydrobiopterin

Dihydrobiopterin

FIGURE 12.22
Coenzyme of aromatic amino acid hydroxylation.
The quinonoid form is produced in the oxygenation reaction and is reduced to the coenzyme by a dehydrogenase using NADH.

more important function than its use by the liver for conjugation with bile acids. An alternative route for taurine synthesis starts with cysteamine, but the source of this precursor appears to be limited to the breakdown of CoA; a direct decarboxylation of cysteine has not been observed. Children maintained for long periods on parenteral nutrition have been found to have decreased plasma levels of taurine and to produce abnormal electroretinograms; these conditions were corrected by addition of taurine to the solutions that were administered intravenously. It is not yet known at which stages in human development or under which clinical conditions dietary taurine is necessary.

The vitamins thiamine, biotin, and lipoic acid contain sulfur but do not contribute measurably to the pool of sulfur metabolites.

Phenylalanine Is Irreversibly Oxidized to Tyrosine

Under normal circumstances almost all of the **phenylalanine** that is degraded undergoes only a single metabolic conversion—hydroxylation at position 4 to form **tyrosine**. The hydroxylation is a mixed-function oxygenation that uses as cosubstrate a specific coenzyme, **tetrahydrobiopterin** (Figure 12.22). This relative of folic acid serves as the coenzyme for several hydroxylases. When the hydroxylation reaction is deficient (Clin. Corr. 12.5), excess phenylalanine accumulates in the blood and equilibrates with its transamination product, phenylpyruvate. This compound is in part excreted and also is converted to phenyllactate and phenylacetate by reduction and oxidation, respectively; these products are also excreted in the urine (Figure 12.23).

Tyrosine is degraded by a major pathway in terms of the quantity of amino acid metabolized, but the so-called minor pathways have great physiological importance. The major pathway, shown in Figure 12.24, occurs in the liver, where it begins with transamination to *p*-hydroxyphenylpyruvate. The aminotransferase is a prominent "inducible" enzyme, which is increased manyfold in animals treated with both tyrosine and glucocorticoid hormones. The oxidation of *p*-hydroxyphenylpyruvate is very complex. A single enzyme adds the atoms of O_2 to the α-carbon atom of the side chain and to the ring carbon to which the side chain is attached. The enzyme contains Fe^{2+} and is protected by ascorbic acid, but a biological role of vitamin C has not been established in this system. The oxidation of the carbonyl group to a carboxyl is accompanied by decarboxylation of the original acid group. To accommodate the new

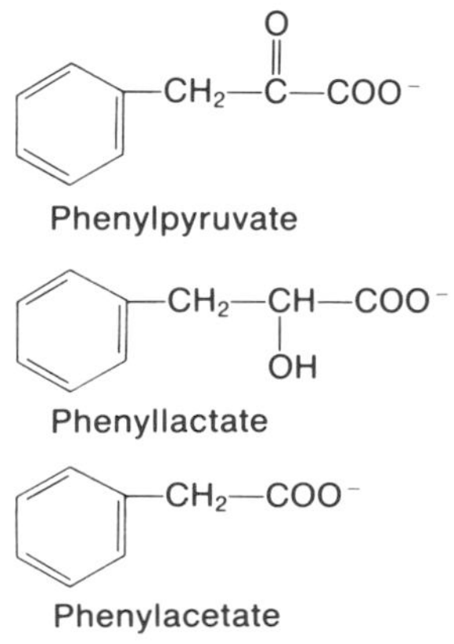

CLIN. CORR. 12.5 PHENYLKETONURIA

Phenylketonuria (PKU) is the most common disease caused by a deficiency of an enzyme of amino acid metabolism. The name comes from the excretion of phenylpyruvic acid (a phenylketone) in the urine. An oxidation product of phenylpyruvate, phenylacetate, is also excreted and gives the urine a "mousey" odor. Phenylpyruvate is detected as a green color produced with ferric chloride, but other compounds also give similar colors. Therefore, the routine screening legally required in many states is an estimation of phenylalanine in the blood; over 20 mg per 100 mL is considered positive. The requirement for this laboratory analysis to be performed on all infants is a reflection of the high incidence of the genetic deficiencies that interfere with the hydroxylation of phenylalanine and the fact that mental retardation associated with the condition can be prevented by dietary control.

Classical PKU is a deficiency of phenylalanine hydroxylase. Genetic analysis shows that the deficiency is caused by an autosomal recessive gene; thus, the enzyme is produced normally by genes on two chromosomes, one from each parent, and the presence of one defective gene results in a heterozygous individual whose liver produces less than normal amounts of the enzyme. Six different mutations have been identified. When the normal pathway of phenylalanine metabolism is

Phenylpyruvate

Phenyllactate

Phenylacetate

FIGURE 12.23
Minor products of phenylalanine metabolism.

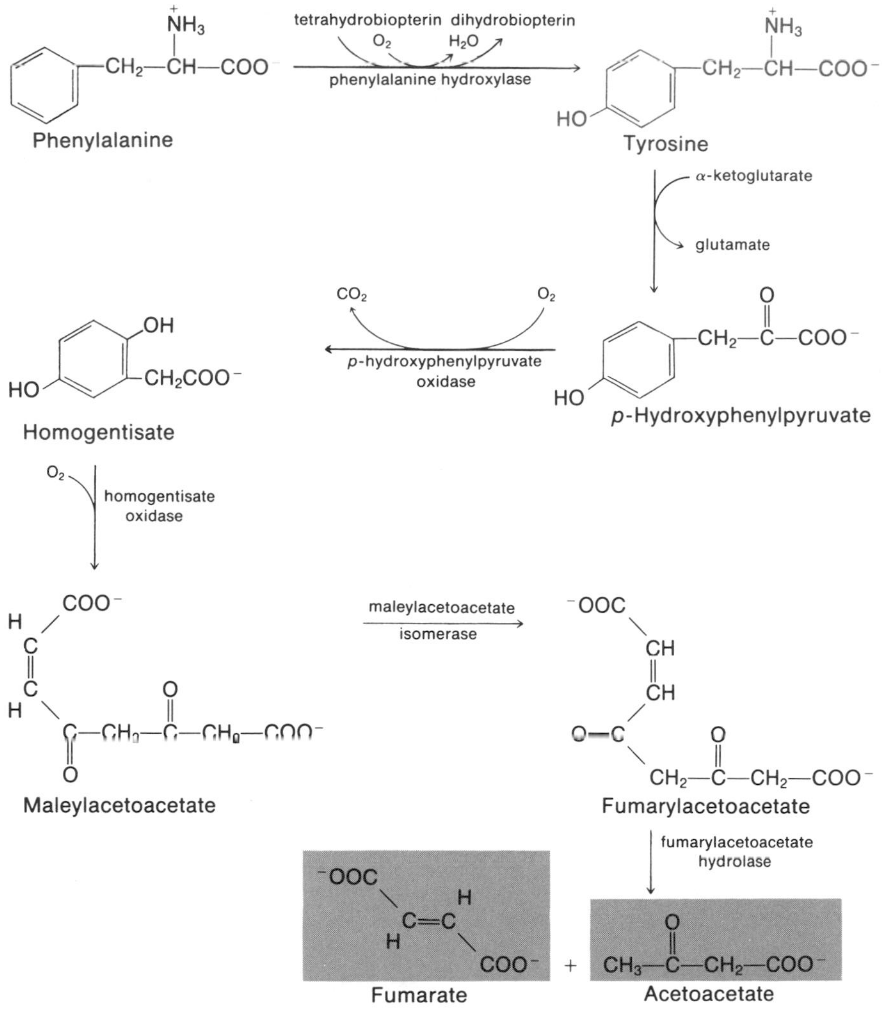

FIGURE 12.24
Synthesis and degradation of tyrosine.

OH group in the ring, the shortened side chain is shifted to an adjacent carbon atom. The product is **homogentisate.** Another iron-containing oxygenase cleaves the aromatic ring by adding oxygen atoms to C-1 and C-2 atoms; the *cis*-configuration of the double bond in the open-chain compound is the basis for the designation maleylacetoacetate. Further degradation requires the isomerization of the double bond to the trans configuration of fumarylacetoacetate; the isomerase appears to have a specific requirement for glutathione. Hydrolysis yields fumarate (a TCA cycle intermediate) and acetoacetate (a ketone body). Phenylalanine and tyrosine are, therefore, both glycogenic and ketogenic. Several genetic diseases are associated with this series of reactions (Clin. Corr. 12.6).

Several prominent physiological agents are produced by a sequence of reactions (Figure 12.25) initiated by hydroxylation of tyrosine to **3,4-dihydroxyphenylalanine (DOPA).** The hydroxylase is very similar to the enzyme that forms tyrosine from phenylalanine; Fe^{2+} is essential for the activity and the cosubstrate is tetrahydrobiopterin. The decarboxylation of DOPA by a pyridoxal phosphate enzyme yields the corresponding amine, which is formed in specific regions of the brain where it functions as a neurotransmitter. Decreased production of **dopamine** is the cause of Parkinson's disease (Clin. Corr. 12.7). In the adrenal medulla, dopamine is also made from tyrosine and is further hydroxylated on the β-carbon

blocked by this deficiency (in most cases there is no enzyme detected by either catalytic or immunochemical methods), the concentration of phenylalanine increases in all body fluids and "minor" products accumulate and are excreted. In some cases there are severe neurological symptoms and very low IQ. These are generally attributed to toxic effects of phenylalanine, possibly on the transport and metabolism of other aromatic amino acids in the brain. It should be noted that there is great variation in the symptoms of children with this enzyme deficiency. Another characteristic of the condition is light color of skin, eyes, and internal tissues where melanin is normally accumulated. This suggests that tyrosine deficiency is also characteristic; that is, dietary tyrosine is not sufficient to compensate for lack of endogenous production. The conventional treatment is to feed infants a synthetic diet low in phenylalanine for about 4–5 years and to restrict proteins in the diet for several more years or for life.

The diagnosis of phenylketonuria is complicated by the increase of phenylalanine in the blood (phenylalaninemia) by several mechanisms. About 3% of infants with high levels of phenylalanine have normal hydroxylase but are defective in either the synthesis or reduction of the essential cofactor, biopterin. Since this cofactor is also necessary for the synthesis of neurotransmitters, central nervous functions are more seriously affected and treatment at this time includes administration of 5-hydroxytryptophan and DOPA to overcome the deficiency of precursors of serotonin and catecholamines.

Brewster, T. G., et al. Dihydropteridine reductase deficiency associated with severe neurologic disease and mild hyperphenylalaninemia. *Pediatrics* 63:94, 1979; Scriver, C. R. and Clow, L. L. Phenylketonuria: epitome of human biochemical genetics. *N. Engl. J. Med.* 303:1336, 1980; and Woo, S. L. C. Molecular basis and population genetics of phenylketonuria. *Biochem.* 28:1, 1989.

CLIN. CORR. 12.6
DISORDERS OF TYROSINE METABOLISM

TYROSINEMIAS

The absence or deficiency of cytosolic tyrosine transaminase (tyrosine aminotransferase, TAT) is responsible for the accumulation and excretion of tyrosine and metabolites including *N*-acetyltyrosine, *p*-hydroxyphenylpyruvate, *p*-hydroxyphenyllactate, *p*-hydroxyphenylacetate, and tyramine. Since *p*-hydroxyphenylpyruvate, the presumed precursor of some of the other products, is also the product of the transaminase, it is likely that these products come from mitochondrial transaminases or oxidases in extrahepatic tissues. The disease is characterized by eye and skin lesions, and most but not all of the cases reported have been mentally retarded. This condition is called oculocutaneous or type II tyrosinemia.

Type I, hepatorenal tyrosinemia, is a more serious disease involving liver failure, renal tubular dysfunction, rickets, and polyneuropathy in addition to excretion of tyrosine, other amino acids, and other metabolites. All of this appears to be caused by a deficiency of fumarylacetoacetate hydrolase. For unknown reasons there is an increased risk of hepatoma development in children with tyrosinemia.

Deficiency of *p*-hydroxylphenylpyruvate oxidase is believed to be responsible for neonatal tyrosinemia, which is usually a temporary condition and in some cases responds to ascorbic acid, given on the hypothesis that this compound protects the enzyme from substrate inhibition.

The very different consequences of deficiencies at various points in tyrosine metabolism show the necessity of analyzing all of the factors that might be relevant and avoiding a simplistic explanation. Interruption of the pathway at homogentisate oxidation causes this compound to accumulate without any other metabolic effects. Homogentisate is not an intrinsically reactive compound, and it does not alter any enzyme before it is excreted. In contrast, in the absence of its hydrolase, fumarylacetoacetate accumulation causes maleylacetoacetate to accumulate as well and this is chemically reactive, especially combining with sulfhydryl compounds. Another toxic compound, succinylacetone, has been suggested as a secondary product of fumarylacetoacetate metabolism that might be responsible for some of the biochemical lesions in this disease.

ALBINISM

Skin and hair color are controlled by an unknown number of genetic loci in humans; in mice 147 genes have been identified in color determination. It is not surprising, therefore, that skin color exists in infinite variations and also that formation of pigment can be interfered with in many ways. Many conditions have been described in which the skin has little or no pigment, but the chemical basis is not established for any except classical albinism. In this condition the enzyme tyrosinase is deficient and melanin is not formed. Lack of pigment in the skin makes albinos sensitive to sunlight, which may cause carcinoma of the skin in addition to burns; lack of pigment in the eyes causes photophobia. Lack of eye pigment does not imply impaired eyesight; a description of albinos among American Indians in 1699 indicated that their vision at night was superior to normal.

ALCAPTONURIA

The first condition to be identified as an "inborn error of metabolism" was alcaptonuria. People deficient in homogentisate oxidase excrete almost all ingested tyrosine as homogentisic acid in their urine. This hydroquinone is colorless, but on standing it autooxidizes to the corresponding quinone, which polymerizes to form an intensely dark color. Concern about the dark urine is the only consequence of this condition early in life. Homogentisate is slowly oxidized to pigments that are deposited in bones, connective tissue, and other organs. This generalized pigmentation is called ochronosis because of the ochre color seen in the light microscope. Pigment deposition is thought to be responsible for the arthritis that develops in many alcaptonuric individuals, especially in males.

The analysis of alcaptonuria by Archibald Garrod who first indicated its genetic basis as an autosomal recessive deficiency condition includes an unusual historical description of the condition. This is of great value in appreciating the iatrogenic suffering that can be inflicted by physicians who act on the basis of inadequate information and false assumptions.

Fellman, J. H., Varbellinghen, P. J., Jones, R. T. and Koler, R. D. Soluble and mitochondrial tyrosine aminotransferase. Relationship to human tyrosinuria. *Biochemistry* 8:615, 1969; and Kvittingen, E. A. Hereditary tyrosinemia type I.—An overview. *Scand. J. Clin. Lab. Invest.* 46:27, 1986.

atom of the side chain by a copper-containing enzyme located in chromaffin granules. Some of the product, *norepinephrine,* is stored in secretory granules until the cells are stimulated to release the hormone into the blood. The larger part of the norepinephrine is methylated by a relatively nonspecific **phenylethanolamine *N*-methyltransferase** to become *epinephrine,* which is also stored in the chromaffin granules. Epinephrine and norepinephrine are hormones known collectively as catecholamines. Norepinephrine is stored in vesicles in the termini of axons to be released for synaptic transmission of nerve impulses. They affect the physiological functions and metabolism of most organs very rapidly, at least in some cases by stimulating the synthesis of cAMP.

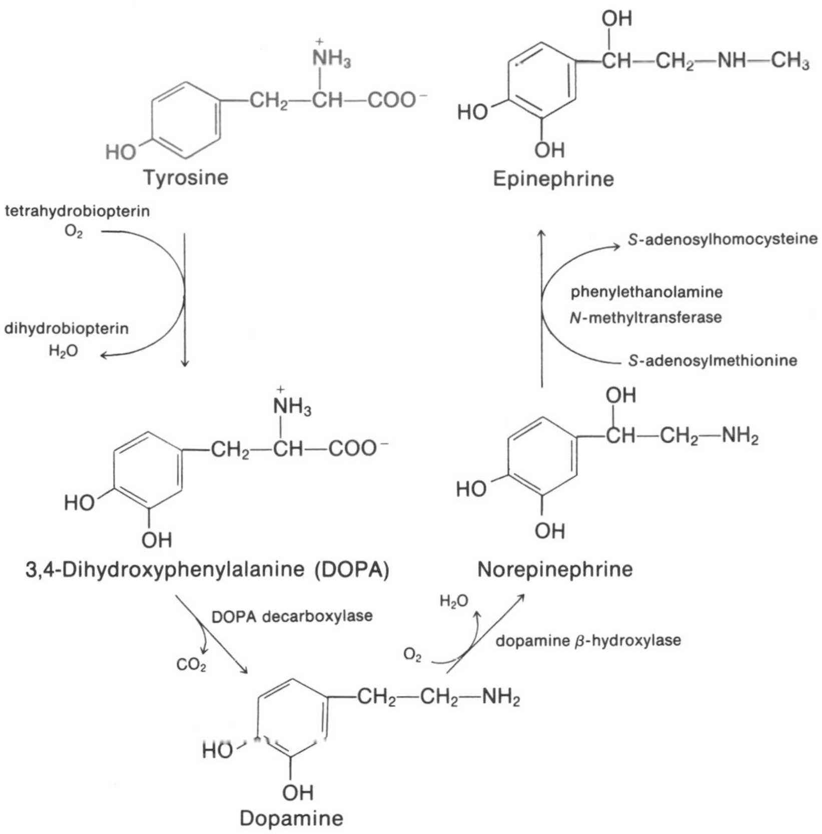

Tyrosine
tetrahydrobiopterin
O_2
dihydrobiopterin
H_2O
3,4-Dihydroxyphenylalanine (DOPA)
DOPA decarboxylase
CO_2
Dopamine
O_2
H_2O
dopamine β-hydroxylase
Norepinephrine
S-adenosylmethionine
phenylethanolamine *N*-methyltransferase
S-adenosylhomocysteine
Epinephrine

FIGURE 12.25
Synthesis of the catecholamines.

An oxidation that closely resembles that carried out by tyrosine hydroxylase is carried out by a copper-containing enzyme, tyrosinase. This is an oxygenation that does not use a cofactor but that uses the presumed product of the hydroxylation, DOPA, as the hydrogen donor in a mixed-function oxygenation so that the material that accumulates is dopaquinone (Figure 12.26). This is a very reactive molecule that cyclizes and condenses to form **melanin.** Melanin is a family of high molecular weight polymers that may include cysteine; melanin granules are very insoluble and very dark in color. Melanin is concentrated in special cells called melanocytes in the skin and also in the choroid plexus, the retina and ciliary body of the eye and in the substantia nigra of the brain. Inability to form melanin is the basis of albinism (Clin. Corr. 12.6).

Thyroglobulin, the precursor of the hormone thyroxin, contains iodinated aromatic amino acids derived from tyrosine. These are produced by posttranslational modification of the protein in the thyroid gland.

Aspartate Has Several Roles

Aspartate is a metabolically active compound because of its interconversion with the C_4 dicarboxylic acids of the TCA cycle after transamination. It is important as a precursor of asparagine, a primary amino acid in protein synthesis, as is aspartate itself. The other reactions of aspartate are as amino donors in urea synthesis (p. 483) and in purine synthesis (pp. 538 and 540) and as a precursor of pyrimidine rings through the formation of carbamoyl aspartate in the cytosol (p. 548).

CLIN. CORR. 12.7 PARKINSON'S DISEASE

Usually in people over the age of 60 years but occasionally in much younger people, tremors may develop that gradually interfere with motor function of various muscle groups. This condition is named Parkinson's disease for the physician who described "shaking palsy" in 1817. The primary cause has not been identified, and there may be more than one etiological agent. The defect, however, is established as degeneration of cells in certain small areas of the brain, substantia nigra and locus coeruleus. Cells in these nuclei produce dopamine as a neurotransmitter, and the amount of dopamine released decreases as a function of the number of surviving cells. A dramatic cause of Parkinsonism occurred in drug addicts using a derivative of pyridine (methylphenyltetrahydropyridine, MPTP). The drug appears to be directly toxic to the dopamine producing cells of the *substantia nigra*. Symptomatic relief, often dramatic, is obtained by increasing the availability of DOPA, the precursor of dopamine. A number of clinical problems were recognized when DOPA (L-DOPA, levoDOPA) became available for treating the large number of people suffering from Parkinson's disease. Side effects included nausea, vomiting, hypotension, cardiac arrhythmias and varied central nervous symptoms. These were explained as effects of dopamine produced outside the central nervous system. Administration of analogs of DOPA that inhibit DOPA decarboxylase and are unable to cross the blood–brain barrier has been effective both in decreasing the side effects and increasing the effectiveness of the DOPA. It should be emphasized that interactions of the many neurotransmitters are very complex, that degeneration of cells continues after treatment, and that elucidation of the major biochemical abnormality has not yet led to complete control of Parkinson's disease. Very recent attempts have been made to treat Parkinson's disease by transplanting adrenal medullary tissue into the brain. The adrenal tissue synthesizes dopamine and improves the movement disorder.

Calne, D. B. and Langston, J. W. Aetiology of Parkinson's disease. *Lancet* 2:1457, 1983; and Cell and tissue transplantation into the adult brain. *Ann. N.Y. Acad. Sci.* 495, 1987.

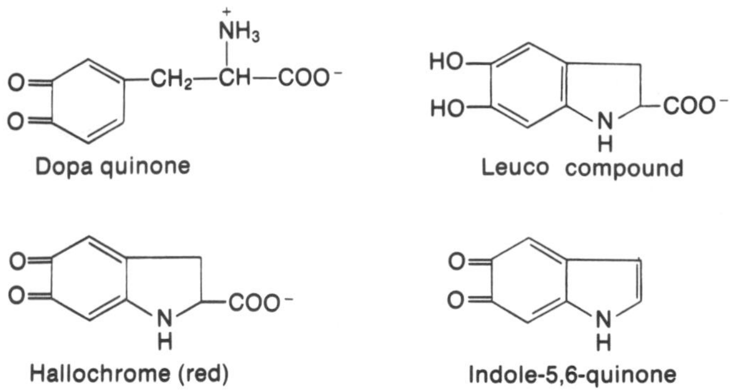

Dopa quinone

Leuco compound

Hallochrome (red)

Indole-5,6-quinone

FIGURE 12.26
Some intermediates in the formation of melanin by tyrosinase.

Tryptophan Catabolism in Liver Produces Nicotinic Acid

The metabolism of tryptophan does not resemble that of any other metabolite. Its principal degradative pathway occurs in the liver and leads to the formation of **nicotinic acid,** usually classified as a vitamin, and many byproducts that accumulate under normal circumstances. Tryptophan metabolism also leads to a physiologically active amine.

The initial reaction in the main pathway (Figure 12.27) is an oxygenation that opens the five-membered ring of the indole nucleus of tryptophan to form **formylkynurenine.** The enzyme, **tryptophan oxygenase** (tryptophan pyrrolase, tryptophan dioxygenase), contains an iron porphyrin and is one of the most studied inducible mammalian enzymes. Administration of large amounts of tryptophan causes an increase of more than a factor of 10 in the level of tryptophan oxygenase by protecting the enzyme against degradation. High levels of the adrenal glucocorticoid hormones (cortisone, dihydrocortisone, and synthetic analogs) cause similar increases by stimulating enzyme synthesis.

The oxidation and cleavage of the ring occur without the formation of a detectable intermediate. *N*-Formylkynurenine is hydrolyzed by a constitutive enzyme that acts on a variety of aromatic formamides but most rapidly on its natural substrate. The formate is handled as a C_1 fragment by the H_4folate system (p. 516). Kynurenine is a branch point; the main pathway continues with a mixed-function oxygenase that includes FAD and uses NADH or NADPH as the cosubstrate in the synthesis of 3-hydroxykynurenine. One side branch splits off the bulk of the side chain as alanine through the action of the pyridoxal phosphate enzyme **kynureninase,** and another side branch removes the α-amino group by transamination (also using pyridoxal phosphate), but the expected keto group forms a Schiff base with the aromatic amine to form the stable aromatic compound kynurenate. **Kynurenate** has recently been found to be produced in brain, where it antagonizes the effects of excitatory amino acids.

3-Hydroxykynurenine is also a branch point. The main pathway now uses the kynureninase that caused a branch earlier to remove alanine but to produce 3-hydroxyanthranilate. The branch is again caused by transamination, which also results in a quinoline ring by cyclization of the presumed carbonyl group with the amine, to form xanthurenic acid. The quinoline compounds are significant components of normal urine and are responsible for part of its yellow color.

A third oxygenase cleaves 3-hydroxyanthranilate to an unstable intermediate, 2-amino-3-carboxymuconic 6-semialdehyde. This enzyme uses Fe^{2+}. In the absence of a competing enzyme, the unstable intermediate cyclizes to a Schiff base in a first-order reaction, probably limited in rate by the isomerization of the double bond to bring the amino group near the

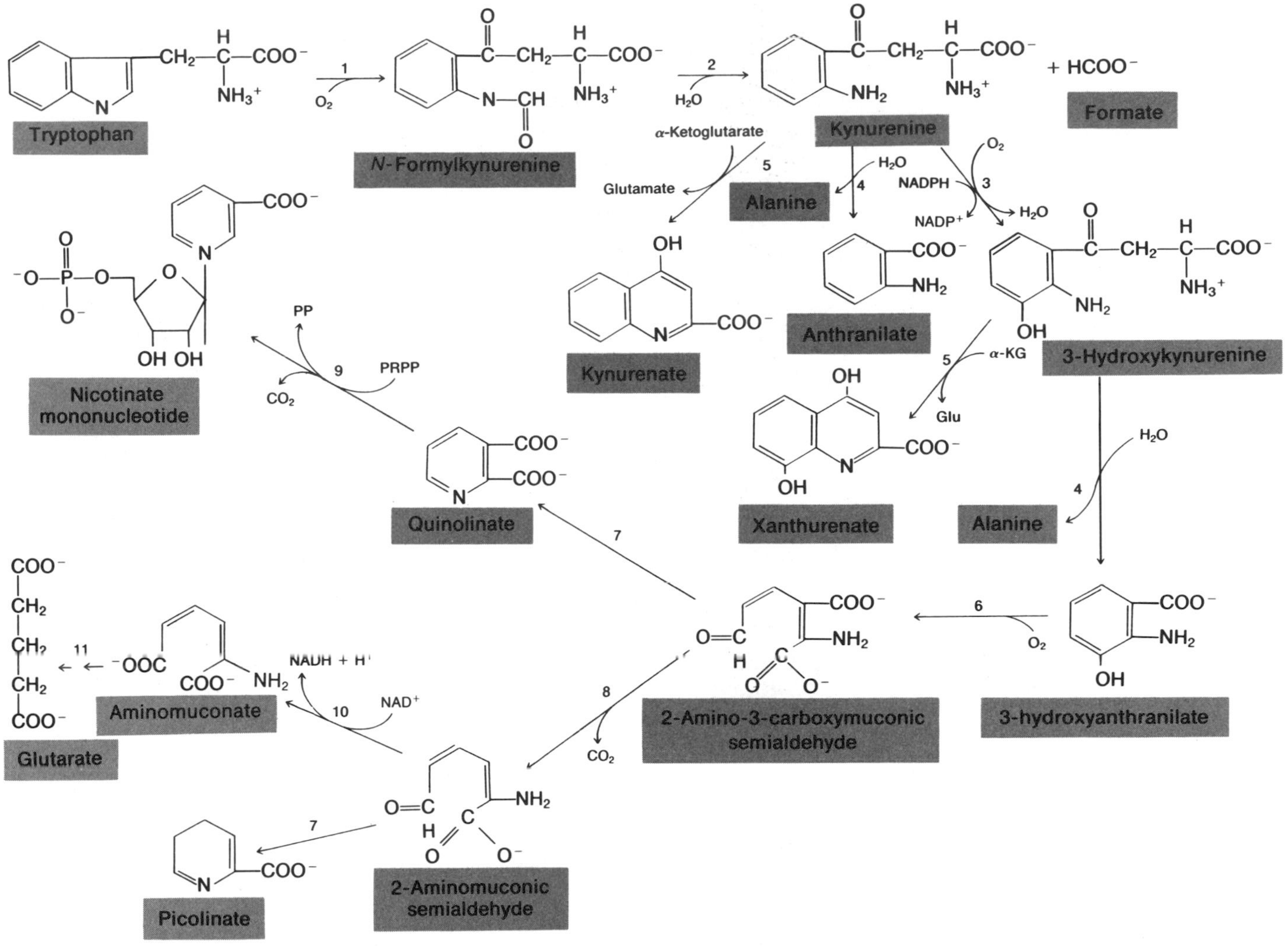

FIGURE 12.27
Metabolism of tryptophan.
The enzymes indicated by bold numbers are (1) tryptophan oxygenase; (2) kynurenine formamidase; (3) kynurenine hydroxylase; (4) kynureninase; (5) transaminase; (6) 3-hydroxyanthranilate oxidase; (7) spontaneous nonenzymatic reaction; (8) picolinate carboxylase; (9) quinolinate phosphoribosyl transferase; (10) aldehyde dehydrogenase; (11) complex series of reactions including reduction and deamination to α-ketoadipate and further metabolism as described for lysine (p. 514). The compounds named in boxes are end products or familiar metabolites of energy metabolism.

aldehyde. The product is a pyridine dicarboxylate, quinolinate. For unknown reasons in some species, an enzyme, picolinic carboxylase, exists that competes with the formation of quinolinate, and in other species, for equally mysterious reasons, this enzyme appears in animals that develop diabetes. It decarboxylates the intermediate to one that cyclizes more rapidly than its precursor to form picolinate. Most of the decarboxylated material is caught by a dehydrogenase, however, that converts the aldehyde to an acid and leads through α-ketoadipate and glutaryl CoA to acetoacetyl CoA (see lysine metabolism, p. 513). Thus, to the extent that tryptophan contributes to energy metabolism it is ketogenic in forming acetoacetyl CoA as well as glucogenic because of the alanine produced in the early steps.

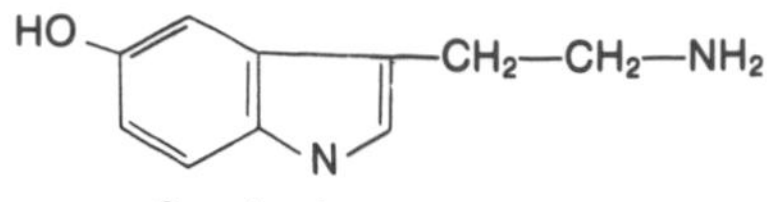

HO—(indole ring)—CH_2—CH_2—NH_2

Serotonin

FIGURE 12.28
Structure of serotonin.

Nicotinic acid is not formed in mammals as a free compound but as a product of a concerted reaction in which quinolinate is decarboxylated in the act of forming a nucleotide, nicotinate mononucleotide, as described in Chapter 13. At this point it should be noted that this biosynthetic pathway is unusual in containing a nonenzymatic step, the formation of quinolinate, which may be regulated by diversion of an intermediate to an oxidative degradation by a completely enzymatic route.

Serotonin (Figure 12.28) is an important neurotransmitter, but it probably has additional physiological functions since it is found in many organs outside the central nervous system, especially in mast cells and platelets, and it causes contraction of smooth muscle in arterioles and bronchioles. This amine may act as a transmitter in the gastrointestinal tract to evoke release of peptide hormones. The formation of serotonin is very similar to the synthesis of the catecholamines. Tryptophan is hydroxylated at C-5 by a mixed-function oxygenase that uses tetrahydrobiopterin as a cosubstrate. The resulting 5-hydroxytryptophan is decarboxylated by a pyridoxal phosphate enzyme to give serotonin, 5-hydroxytryptamine.

Histidine Catabolism Produces a One-Carbon Group

The catabolism of histidine is of greater importance in 1-carbon metabolism than in energy production. The principal pathway of histidine degradation leads to glutamate by a series of reactions of different types but without oxidation. An obligate step requires H_4folate. Minor pathways in terms of the percentage of histidine consumed are also of physiological importance.

The first step in histidine degradation is the elimination of ammonia with the formation of a double bond. This kind of reaction is known to occur with other amino acids in bacteria and plants, but the **histidase** reaction is unique in higher animals although elimination reactions do occur in the transfer of amino groups after condensation of aspartate in the synthesis of arginine and purines. A disease associated with deficiency of histidase is described in Clin. Corr. 12.8. The unsaturated urocanate is hydrated to 4-imidazolone-5-propionate. Histidase and urocanase have been well characterized only as bacterial enzymes that catalyze the same reactions that occur in liver. The bacterial enzymes have no organic cofactors but contain altered amino acids, α-ketobutyrate at the former *N*-terminus of urocanase and a tentatively identified dehydroalanine in histidase; each of these participates in the catalytic mechanism. Hydrolysis of imidazolonepropionate opens the five-membered ring to form *N*-formiminoglutamate. The only reaction in which this compound participates is a transfer of the formimino group, a 1-carbon fragment bearing a nitrogen atom, to N-5 of H_4folate, leaving glutamate. The medical significance of this reaction is illustrated in Clin. Corr. 12.11. These reactions are illustrated in Figure 12.29.

The decarboxylation of histidine to give the corresponding amine, **histamine** (Figure 12.30), occurs in various organs, especially in mast cells. Histamine is secreted in the physiological control of such diverse functions as production of acid by the gastric mucosa and the dilation and constriction of specific blood vessels. Excess reaction to histamine causes the symptoms of asthma and various allergic reactions.

Two unusual peptides containing histidine, **carnosine** and **anserine,** are found in muscle. They are both derivatives of β-alanine. This compound is produced as a degradation product of pyrimidines (Chapter 13) but can be synthesized from malonic semialdehyde, which is derived from propionate. Transfer to the α-amino group of histidine of a β-alanyl residue from an anhydride with AMP, with no carrier such as CoA or tRNA, forms

CLIN. CORR. 12.8
HISTIDINEMIA

Histidinemia, an elevated level of histidine in the blood, was first detected as a false positive in the screening of the urine of infants with ferric chloride. The green color was given by an analog of phenylpyruvate, imidazolepyruvate, formed by transamination of histidine. The accumulation of histidine is due to a deficiency of histidase. A convenient assay for this enzyme uses skin, which produces urocanate as a constituent of sweat; urocanase and the other enzymes of histidine catabolism in liver are not found in skin. Thus, histidase deficiency can be confirmed by assay for the enzyme in skin biopsies and the absence of urocanate in sweat.

The incidence of the disorder is high—1 in 10,000 newborns screened. Most reported cases of histidinemia have shown normal mental development. Restriction of dietary histidine normalizes the biochemical abnormalities but is not generally required.

Scriver, C. R. and Levy, H. L. Histidinemia: Reconciling retrospective and prospective findings. *J. Inherited Metab. Dis.* 6:51, 1983.

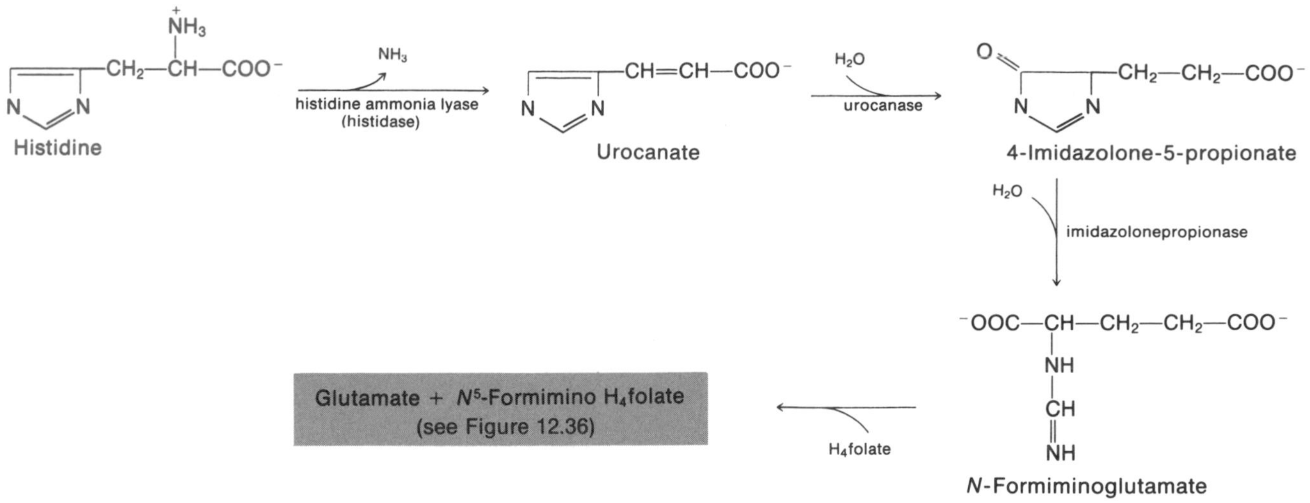

FIGURE 12.29
Degradation of histidine.

carnosine; a similar reaction with 1-methylhistidine forms anserine. The methylation of carnosine by *S*-adenosylmethionine has been demonstrated in vitro but is of doubtful significance in vivo. The functions of these peptides in muscle is conjectural; a role has been proposed for carnosine in olfactory function. The origin of methyl histidine is unknown, but its occurrence in certain proteins suggests that it is formed in a posttranslational modification of histidine in a polypeptide, and the free amino acid is liberated when the protein is degraded.

Lysine's Amino Group Does Not Equilibrate With the Amino Pool

Lysine is one of two essential amino acids whose α-amino group does not equilibrate with the body pool of amino groups; the other is threonine. The amino group of lysine is transferred to other amino acids, but the reverse does not occur. The α-keto acid corresponding to lysine, α-keto-ε-aminocaproic acid, cannot replace lysine in the diet; this compound cyclizes by Schiff base formation and exists mainly as Δ^1-piperideine-2-carboxylic acid. Derivatives of lysine in which the ε-amino group is blocked, such as ε-*N*-acetyllysine and ε-*N*-methyllysine, are converted to lysine in vivo by hydrolytic or oxidative enzymes found in the liver and kidney.

Most degradation of lysine occurs by a unique pathway shown in Figure 12.31 in which a secondary amine is formed between the ε-amino group and the carbonyl group of α-ketoglutarate. The product, saccharopine, is formed by an enzyme that reduces the hypothetical Schiff base with NADPH. Normally saccharopine does not accumulate but is oxidized by another dehydrogenase that splits the linkage on the other side of the bridge nitrogen. The sum of the reduction–oxidation reactions is effectively a transamination yielding glutamate and α-aminoadipic semialdehyde. The latter, of course, can also form a Schiff base but with the double bond on the side of the nitrogen atom away from the carboxylate. This compound can be oxidized by another dehydrogenase to become α-aminoadipate. A conventional transamination converts this higher homologue of glutamate to the corresponding α-ketoadipate. In a reaction analogous to the oxidation of α-ketoglutarate to succinyl CoA, glutaryl CoA is formed. Another oxidation introduces a double bond, forming glutaconyl CoA, which is decarboxylated to crotonyl CoA. This unsatu-

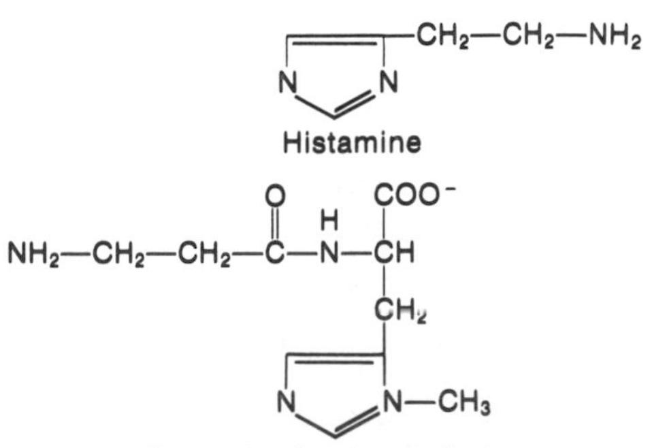

FIGURE 12.30
Some derivatives of histidine.

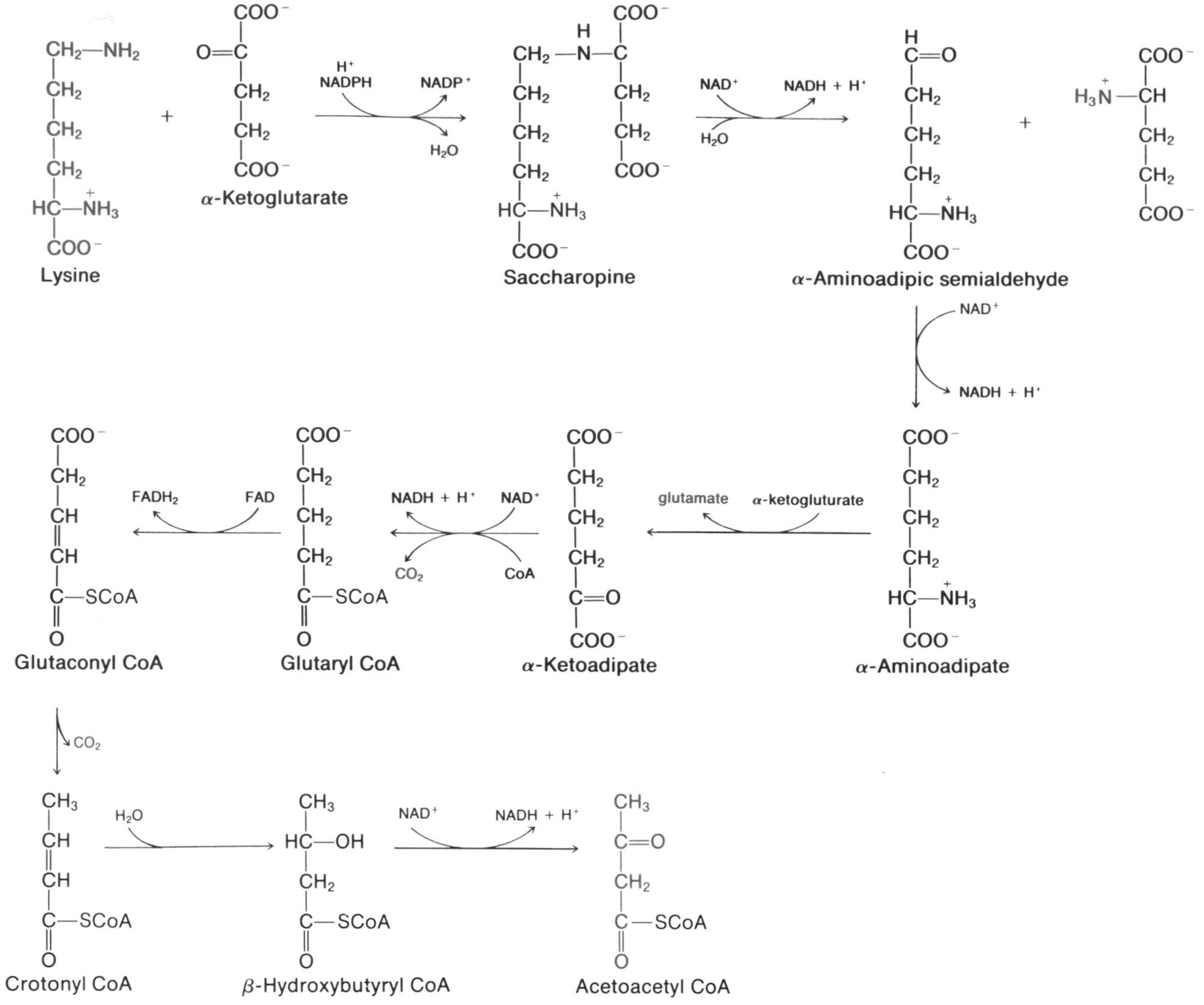

FIGURE 12.31
Principal pathway of lysine degradation.

rated fatty acyl CoA is an intermediate in the normal oxidation of fatty acids, and subsequent reactions of the material derived from lysine are those of fatty acid oxidation leading to acetoacetyl CoA.

Although lysine does not participate in transamination, the α-amino group is removed by a dehydrogenase and Δ^1-piperideine-2-carboxylate is formed. This reaction is of minor importance in that it cannot compensate for deficiency in the saccharopine pathway. The ring is saturated by reduction with a nicotinamide nucleotide to form pipecolate, shown in Figure 12.32. Following the reduction, an oxidation produces an unsaturation on the other side of the nitrogen to reform the lysine passing through the saccharopine pathway at α-aminoadipate semialdehyde. Abnormalities associated with lysine metabolism are discussed in Clin. Corr. 12.9.

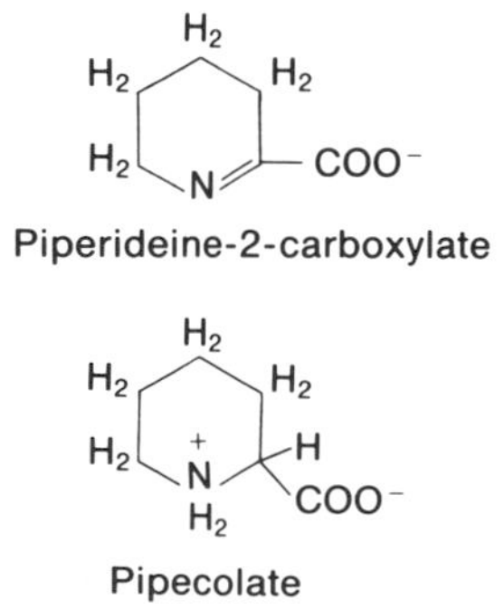

FIGURE 12.32
Minor products of lysine metabolism.

Carnitine Is Synthesized From Trimethyllysine Released by Proteolysis

For reasons that are not understood, some of the ε-amino groups of lysine residues in many proteins are methylated to **mono-, di-, or trimethyllysine**

CLIN. CORR. 12.9
DISEASES INVOLVING LYSINE AND ORNITHINE

LYSINE

Two metabolic disorders of lysine are recognized. The enzyme α-amino adipic semialdehyde synthase is deficient in a small number of patients. They excrete lysine in the urine as well as smaller amounts of saccharopine. This has led to the discovery that this enzyme has two activities: lysine-α-ketoglutarate reductase and saccharopine dehydrogenase activities. Single enzymes with multiple enzymatic activities are also found in the pathway of pyrimidine synthesis. It is currently thought that the hyperlysinemia is benign.

A more serious condition is familial lysinuric protein intolerance. The underlying defect is failure to transport dibasic amino acids at the intestinal cell and renal tubular epithelium. Plasmic lysine, arginine, and ornithine are decreased to one-third or one-half of normal. The patients develop marked hyperammonemia after a meal containing protein. This is thought to result from a deficiency of urea cycle intermediates ornithine and arginine in the hepatocyte. This limits the maximal capacity of the cycle. Consistent with this view, oral supplementation with citrulline prevents the hyperammonemia. Other clinical features are thin hair, muscle wasting, and osteoporosis. These are thought to reflect protein malnutrition due to lysine and arginine deficiency.

ORNITHINE

Elevated ornithine levels are generally due to a deficiency of the aminotransferase that enables ornithine to be oxidized through glutamate and the TCA cycle. A well-defined clinical entity, gyrate atrophy of the choroid and retina, characterized by progressive loss of vision leading to blindness by the fourth decade, is caused by a deficiency of the mitochondrial pyridoxine-dependent enzyme ornithine transaminase (ornithine-δ-aminotransferase). The mechanism of the specific pathological changes in the eye is not known. Progression of the disease may be slowed by restriction of arginine intake and/or administration of pyridoxine, which reduces the level of ornithine in body fluids. A rare disorder, hyperornithinemia, hyperammonemia, and homocitrullinuria (HHH) syndrome, appears to result from defective transport of ornithine into the mitochondria. Hyperammonemia results from intramitochondrial ornithine deficiency, and homocitrulline is thought to be formed by transcarbomylation of lysine.

Markovitz, P. J. and Chuang, D. T. The bifunctional bovine aminoadipic semialdehyde synthase in lysine degradation: Separation of reductase and dehydrogenase domain by limited proteolysis. *J. Biol. Chem.* 262:9353, 1987; Rajantic, J., Simell, O., and Perheentupa, J. Lysinuric protein intolerance. Basolateral transport defect in renal tubuli. *J. Clin. Invest.* 67:1078, 1981; O'Donnell, J. J., Sandman, R. P., and Martin, S. R. Gyrate atrophy of the retina: inborn error of L-ornithine: 2-oxoacid amino transferase. *Science* 200:200, 1978.

by an N-methyl transferase that uses S-adenosylmethionine as the methyl donor. Speculation that the methylation plays a regulatory role in the function of such proteins as histones is supported by the observation that the methyl groups are turned over rather rapidly by oxidation to formaldehyde. Some lysyl residues in proteins are also acetylated. Probably the same enzyme that demethylates lysyl residues in proteins removes the methyl groups from free mono- and dimethyllysine, which are produced when the methylated proteins are degraded in normal protein turnover. The trimethyllysine that is liberated by proteolysis has an unusual metabolic role as the precursor of carnitine. The biosynthesis of carnitine in mammals is absolutely dependent on the formation of trimethyllysine residues in proteins. Free lysine is not methylated, and the intermediates in carnitine formation are derived exclusively from trimethyllysine.

Trimethyllysine is oxidized by a specific oxygenase that requires as cosubstrate α-ketoglutarate. The resulting β-hydroxytrimethyllysine is a substrate for serine hydroxymethyl transferase and is converted to γ-butyrobetaine aldehyde and glycine. The aldehyde is oxidized by a specific NAD-requiring dehydrogenase to form γ-butyrobetaine. Subsequent hydroxylation at the β position to form carnitine is catalyzed by an enzyme with the same requirements as for the earlier hydroxylation of trimethyllysine and that may be the same enzyme. These reactions are illustrated in Figure 12.33. The role of carnitine in fatty acid transport across the inner mitochondrial membrane is described in Chapter 9.

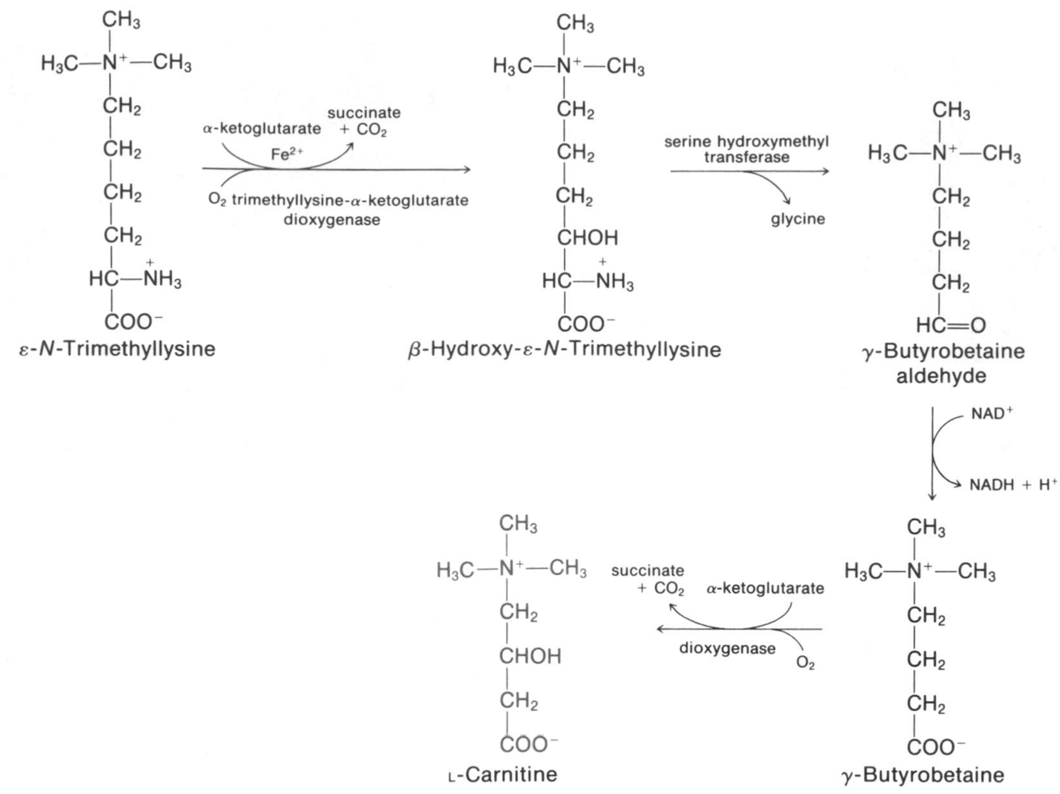

ε-N-Trimethyllysine
α-ketoglutarate
succinate + CO_2
Fe^{2+}
O_2 trimethyllysine-α-ketoglutarate dioxygenase
β-Hydroxy-ε-N-Trimethyllysine
serine hydroxymethyl transferase
glycine
γ-Butyrobetaine aldehyde
NAD^+
$NADH + H^+$
γ-Butyrobetaine
α-ketoglutarate
O_2
dioxygenase
succinate + CO_2
L-Carnitine

FIGURE 12.33
Biosynthesis of carnitine.

12.4 FOLIC ACID AND ONE-CARBON METABOLISM

One-carbon metabolism is a term applied to reactions in which a chemical group built around a single carbon atom is transferred from one compound to another. The groups that correspond to derivatives of methanol, formaldehyde, and formate are transferred mainly by enzymes that use the cofactor H_4folate. The most reduced 1-carbon compound, methane, is inert in animal metabolism, and the most oxidized molecule, carbon dioxide, is produced by a variety of decarboxylases and is used in a similarly diverse group of reactions not related to the rest of 1-carbon metabolism.

The coenzyme **H_4folate** is derived from the vitamin folic acid (Figure 12.34) that is absorbed from the intestine (Clin. Corr. 12.10). The vitamin is reduced in two steps catalyzed by the same enzyme, named for the

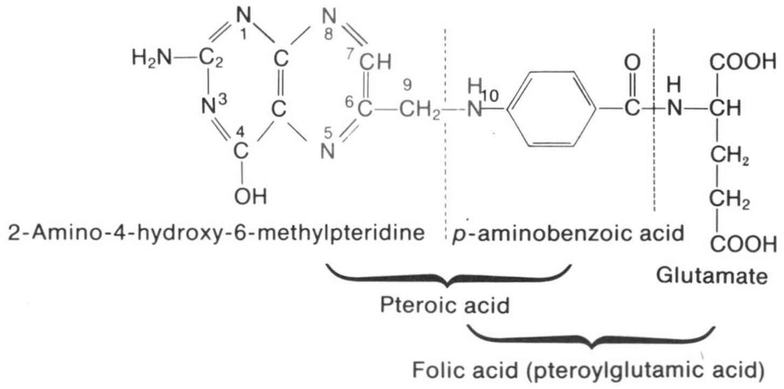

2-Amino-4-hydroxy-6-methylpteridine
p-aminobenzoic acid
Glutamate
Pteroic acid
Folic acid (pteroylglutamic acid)

FIGURE 12.34
Components of folic acid.

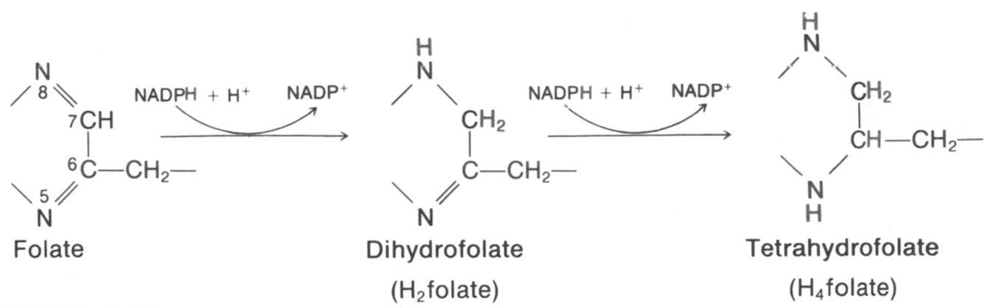

Folate → Dihydrofolate (H_2folate) → Tetrahydrofolate (H_4folate)

(NADPH + H⁺ → NADP⁺ at each step)

FIGURE 12.35
Reduction of folate by dihydrofolate reductase.

second step **dihydrofolate reductase** (Figure 12.35). Deficiency of this enzyme and other enzymes of 1-carbon metabolism are discussed in Clin. Corr. 12.11. Actually, there are several species of H_4folate with up to seven glutamate residues in γ-amide linkage. Until now no functional differences have been attributed to these species, and attention will be focused on the pteroic acid portion of the molecule where interactions with substrates occur.

One-carbon elements at the oxidation levels of formaldehyde and formate are combined with H_4folate to produce the structures illustrated in Figure 12.36. Free formaldehyde adds spontaneously to H_4folate and forms a bridge between N-5 and N-10, N^5,N^{10}-methylene H_4folate. The major donor of the 1-carbon bridge of N^5,N^{10}-methylene H_4folate is serine in the formation of glycine. Oxidation of glycine also forms this bridge compound.

N^5,N^{10}-Methylene H_4folate can be both oxidized and reduced by dehydrogenases using pyridine nucleotide coenzymes. Reduction breaks the bond to N-10 and forms N^5-methyl H_4folate. Oxidation, in contrast, pro-

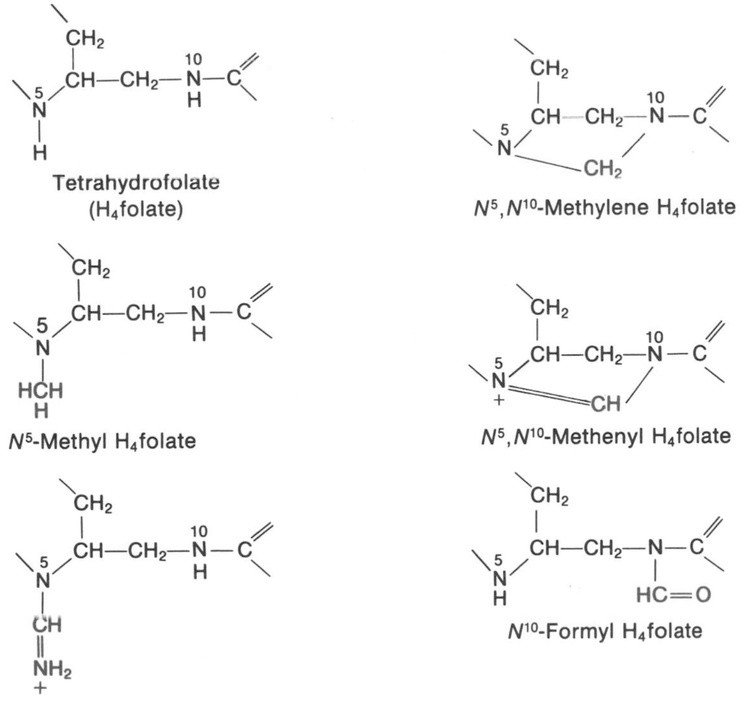

Tetrahydrofolate (H_4folate)

N^5,N^{10}-Methylene H_4folate

N^5-Methyl H_4folate

N^5,N^{10}-Methenyl H_4folate

N^5-Formimino H_4folate

N^{10}-Formyl H_4folate

FIGURE 12.36
Active center of H_4-folate and its C_1 derivatives.

CLIN. CORR. 12.10
FOLIC ACID DEFICIENCY

The 100–200 μg of folic acid required daily by an average adult are easily obtained from conventional Western diets. Deficiency of folic acid, however, is not uncommon. It may result from limited diets, especially when food is cooked at high temperatures for long periods, which destroys the vitamin. Intestinal diseases, notably coeliac disease, often are characterized by folic acid deficiency caused by malabsorption.

In contrast to pernicious anemia, a fairly common disease in adults caused by decreased production of gastric intrinsic factor needed for absorption of vitamin B_{12} (see Chapter 28), specific inability to absorb folate is rare. Folate deficiency usually is seen in newborns. Both conditions are characterized by megaloblastic anemia. Of the few cases studied, some were responsive to large doses of oral folate but one required parenteral administration, suggesting a carrier-mediated process for absorption. Besides the anemia, mental and other central nervous symptoms are seen in patients with folate deficiency, and all respond to continuous therapy although permanent damage appears to be caused by delayed or inadequate treatment.

A classical experiment was carried out by a physician, apparently serving as his own experimental subject, to study the human requirements for folic acid. His diet consisted only of foods boiled repeatedly to extract the water-soluble vitamins to which vitamins (and minerals) were added, omitting folic acid. Symptoms attributable to folate deficiency did not appear for 7 weeks, altered appearance of blood cells and formiminoglutamate excretion were seen only at 13 weeks, and serious symptoms (irritability, forgetfulness, and macrocytic anemia) appeared only after 4 months. The neurological symptoms were alleviated within 2 days after folic acid was added to the diet; the blood picture became normal more slowly.

The occurrence of folic acid in essentially all natural foods makes deficiency difficult, and apparently a normal person accumulates more than adequate reserves of this vitamin. For pregnant women the situation is very different. The needs of the fetus for normal growth and development include constant, uninterrupted supplies of coenzymes (in addition to amino acids and other cell constituents). Recently, folate deficiency has been implicated in spina bifida.

Herbert, V. Experimental nutritional folate deficiency in man. *Trans. Assoc. Am. Phys.* 75:307, 1962.

duces N^5,N^{10}-methenyl H_4folate, which is reversibly hydrolyzed to N^{10}-formyl H_4folate. Two other mechanisms are of importance in forming bound 1-carbon units at this level of oxidation: (1) free formate is activated by ATP and forms 10-formyl H_4folate and (2) the formimino group attached to glutamate in the catabolism of histidine is transferred to N-5 and in a second step the carbon forms the cyclic N^5,N^{10}-methenyl compound while eliminating ammonia. Both the transfer reaction (formiminotransferase) and the elimination reaction (cyclodeaminase) are associated with a single enzyme.

Both the methenyl compound and N^{10}-formyl H_4folate are 1-carbon donors in specific reactions. This is seen in the synthesis of purines (Chapter 13) in which the former compound is the specific donor of C-8 and the latter the donor of C-2. The methyl group associated with H_4folate is used only in the resynthesis of methionine from homocysteine. The methyl group of methionine is the source of most other methyl groups in the body through transfer reactions of *S*-adenosylmethionine.

The methyl group of thymine is formed by transfer of the 1-carbon unit of N^5,N^{10}-methylenetetrahydrofolate to deoxyuridine 5-phosphate in a unique reaction in which the coenzyme reduces the 1-carbon group at the same time that it joins the pyrimidine ring. The coenzyme is liberated as H_2folate (Chapter 13) and must be reduced to the tetrahydro state in order to continue to function in 1-carbon metabolism. Inhibition of dihydrofolate reductase by folic acid analogs (methotrexate and amethopterin) is the basis for the cytotoxic action of these agents that are used widely in the chemotherapy of leukemias and other neoplasms.

12.5 NITROGENOUS DERIVATIVES OF AMINO ACIDS

Creatine to Phosphocreatine to Creatinine

The energy for driving most biochemical reactions is derived from the hydrolysis of ATP. Both ATP and ADP are powerful effectors of many enzymes. Therefore, the concentrations of adenine nucleotides must be carefully regulated and cannot be increased to provide a reservoir of energy. To permit bursts of activity, a large amount of phosphate is stored as phosphocreatine, which has about the same free energy of hydrolysis

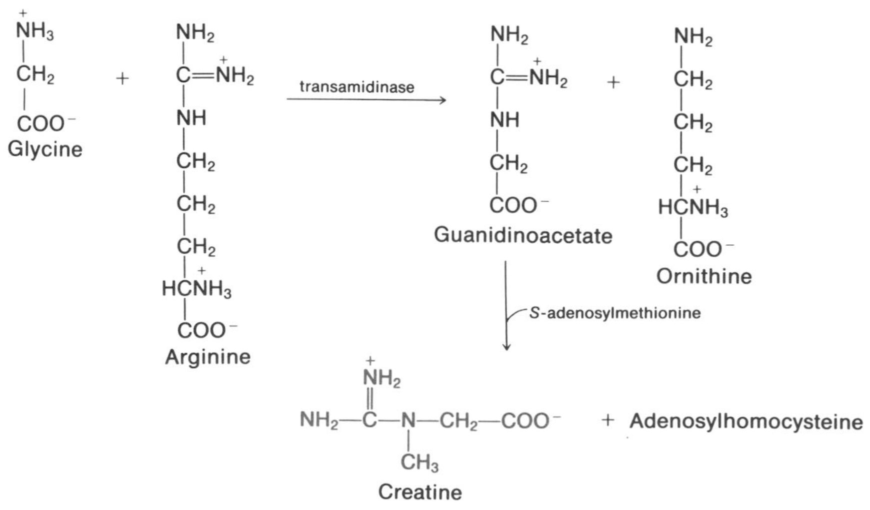

CLIN. CORR. 12.11
DISEASES OF FOLATE METABOLISM

DIHYDROFOLATE REDUCTASE DEFICIENCY

Although folic acid derivatives in all tissues that have been examined are mainly reduced to the level of H_4folate, these compounds are easily oxidized so that some significant fraction of the absorbed vitamin must be reduced in order to function as a coenzyme. In a few cases symptoms of folate deficiency were found to be due to a deficiency of H_2folate reductase (which catalyzes both steps of the conversion of folate to H_4folate). An effective treatment for such cases is parenteral administration of N^5-formyl H_4folate, the most stable of the reduced folates.

METHYLENE TETRAHYDROFOLATE REDUCTASE DEFICIENCY

In the few cases of central nervous system abnormality attributed to a deficiency of methylene H_4folate reductase, an accompanying biochemical abnormality is homocystinuria. A younger sibling of a patient was found to have homocystinuria, the same reductase level as the patient (~18% of normal), but no symptoms. It is not known whether symptoms develop late or whether some people function adequately without this enzyme activity. The lowered enzyme activity decreases the amount of N^5-methyl H_4folate that is formed so that the source of methyl groups for the salvage of homocysteine is limiting. Administering large amounts of folic acid, betaine, and methionine reverses the biochemical abnormalities and, in at least one case, the neurological disorder.

FORMIMINOTRANSFERASE DEFICIENCIES

A series of patients with widely divergent presentations has shown deficiencies in the transfer of the formimino group from formiminoglutamate (FIGLU) to H_4folate. The patients excrete varying amounts of FIGLU; some respond to administration of large doses of folate, but others do not. Histidine loading generally increases the excretion of FIGLU, and a low histidine diet decreases the excretion. No clear mechanism is known, whereby a deficiency of formiminotransferase should cause pathological changes. Speculation on possible mechanisms includes distortion of the normal distribution of 1-carbon forms of folate and hypothetical effects of accumulation of FIGLU. It is not entirely clear whether this deficiency really causes a disease state.

$$\underset{\text{Glycine}}{{}^{+}NH_3{-}CH_2{-}COO^-} + \underset{\text{Arginine}}{NH_2{-}C({=}\overset{+}{N}H_2){-}NH{-}CH_2{-}CH_2{-}CH_2{-}HC\overset{+}{N}H_3{-}COO^-} \xrightarrow{\text{transamidinase}} \underset{\text{Guanidinoacetate}}{NH_2{-}C({=}\overset{+}{N}H_2){-}NH{-}CH_2{-}COO^-} + \underset{\text{Ornithine}}{NH_2{-}CH_2{-}CH_2{-}CH_2{-}HC\overset{+}{N}H_3{-}COO^-}$$

$$\text{Guanidinoacetate} \xrightarrow{S\text{-adenosylmethionine}} \underset{\text{Creatine}}{NH_2{-}C({=}\overset{+}{N}H_2){-}N(CH_3){-}CH_2{-}COO^-} + \text{Adenosylhomocysteine}$$

FIGURE 12.37
Synthesis of creatine.

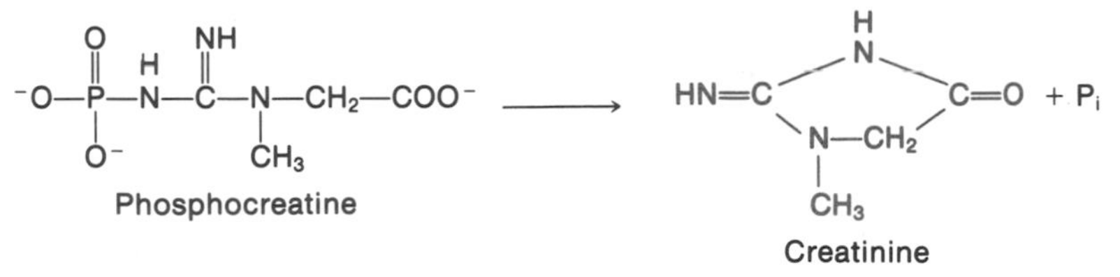

Phosphocreatine → Creatinine + P_i

FIGURE 12.38
Spontaneous reaction forming creatinine.

as ATP; **creatine kinase** catalyzes the transfer of phosphate from ATP to form creatine phosphate during periods of rest and regenerates ATP when it is needed.

Creatine is synthesized from three amino acids: glycine, arginine, and methionine (Figure 12.37). The transfer of the amidine group from arginine to glycine produces guanidinoacetate and ornithine. Creatine is formed by addition of a methyl group from *S*-adenosylmethionine. The formation of creatine accounts for more utilization of *S*-adenosylmethionine than all other transfers of methyl groups combined.

The reactivity of the phosphoguanidine group is responsible for a nonenzymatic cyclization of creatine in which the carboxyl group displaces the phosphate (Figure 12.38). The cyclic compound is called **creatinine.** Since the amount of creatine phosphate is roughly proportional to the muscle mass of the body, the spontaneous formation of creatinine is characteristic of each individual and proceeds at a constant rate from day to day. All of the creatinine produced is excreted in the urine, where it is the most accurate measure of the time during which the urine was produced. Thus, measurement of creatinine in 24-h urine samples is carried out to ensure the accuracy of the collection.

TETRAHYDROFOLATE METHYLTRANSFERASE DEFICIENCY

A single patient has been described who suffered from severe central nervous system maldevelopment and hematological signs of folate deficiency, all attributable to a deficiency of H_4folate methyltransferase. Other patients with similar deficiencies have been found to lack vitamin B_{12}, but treatment with cobalamin did not help this patient although the anemia responded partially to folic acid therapy. The favored explanation for these findings is the "methyl trap hypothesis," which proposes that conversion of folate to methyl H_4folate is an irreversible blind alley when the transferase is less active than normal. Since the patient had 30% of the normal methyltransferase activity, this was adequate to prevent accumulation of homocysteine, but it is possible that the activity in vivo depends on higher than normal levels of methyl H_4folate. The hypothesis proposes that additional vitamin is needed to maintain other folate-dependent reactions by replacing the coenzyme lost by the drain into methyl H_4folate. A similar mechanism is thought to be responsible for anemia in vitamin B_{12} deficiency.

Rosenblatt, D. S. Inherited disorders of folate transport and metabolism. *The Metabolic Basis of Inherited Disease*, 6th ed. C. R. Scriver, A. L. Beaudet, W. S. Sly, and D. Valle (Eds.). New York: McGraw-Hill, 1989.

Choline Is a Highly Methylated Compound

Choline (Figure 12.39) is a highly methylated compound that plays a central role in nerve conduction and also is a component of the phospholipid lecithin. Isotopic evidence demonstrated long ago that the C_2 skeleton of choline is derived from serine, but the details of the conversion are not well established. A decarboxylation of the serine in a phosphotidylserine is generally considered to initiate a minor pathway and exogenous ethanolamine is rapidly assimilated. However, little is known about the formation of free ethanolamine or the relative amounts of methyl transfer to ethanolamine and its derivatives. It is clear that three methyl

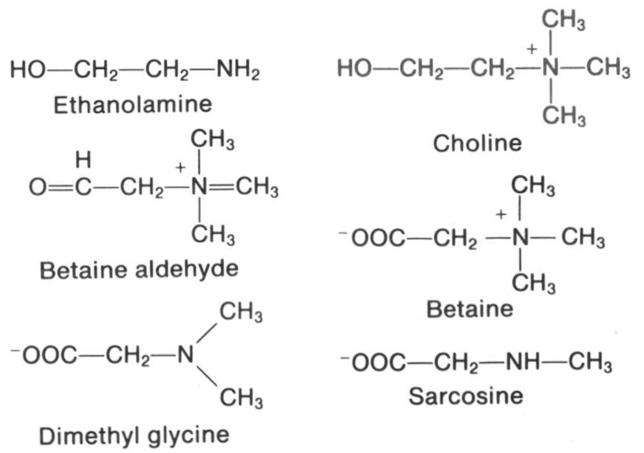

$HO—CH_2—CH_2—NH_2$ Ethanolamine

$HO—CH_2—CH_2—N^+(CH_3)_3$ Choline

$O=CH—CH_2—N^+(CH_3)_3$ Betaine aldehyde

$^-OOC—CH_2—N^+(CH_3)_3$ Betaine

$^-OOC—CH_2—N(CH_3)_2$ Dimethyl glycine

$^-OOC—CH_2—NH—CH_3$ Sarcosine

FIGURE 12.39
Choline and related compounds.

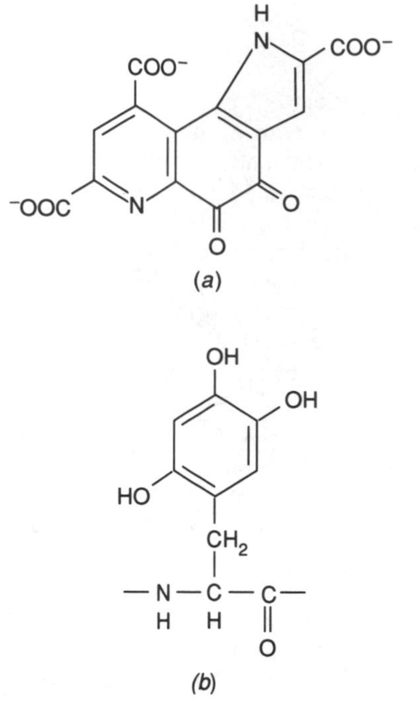

groups are transferred from *S*-adenosylmethionine to make the quaternary ammonium group of choline.

Choline is acetylated by acetyl CoA in nervous tissue to form the neurotransmitter **acetylcholine.** Cholinesterase hydrolyzes the ester to acetate and choline. Choline is oxidized in a succession of reactions to the corresponding aldehyde and carboxylic acid, named betaine aldehyde and betaine, respectively. **Betaine** is an effective methyl donor and participates in a minor variant of the salvage pathway to reform methionine from homocysteine. Only one of the three methyl groups can be transferred; the other methyl groups are removed as formaldehyde by oxidation by flavoproteins, giving *N*-methylglycine (sarcosine) then glycine.

FIGURE 12.40
Pyrroloquinoline quinone (PQQ) (*a*) and trihydroxyphenylalanine (TOPA) (*b*).
The biosynthesis of this recently discovered cofactor has not been reported.

Amines and Polyamines Are Metabolized by Oxidation and Methylation

As described earlier, decarboxylation of certain amino acids produces the corresponding amines, and some of these are combined to form polyamines. The amines that act as extracellular effectors, such as neurotransmitters, must be metabolized quickly to limit responses and allow other messages to be effective. In the nervous system a physiological mechanism recovers a large part of the amines released from nerve endings by reuptake back into the cells that produced them. The part that is not taken up is inactivated by two mechanisms, oxidation of the amino group or methylation of the phenolic OH groups.

In most cells an enzyme located in the outer membrane of mitochondria oxidizes amines to the corresponding aldehydes as follows:

$$RCH_2NH_2 + O_2 + H_2O \longrightarrow RCHO + NH_3 + H_2O_2$$

This is one type of **monoamine oxidase.** It contains covalently bound FAD and uses only O_2 as an electron acceptor. In addition to primary amines, secondary and tertiary amines with methyl groups added to the nitrogen also serve as substrates. In the serum a very different monoamine oxidase

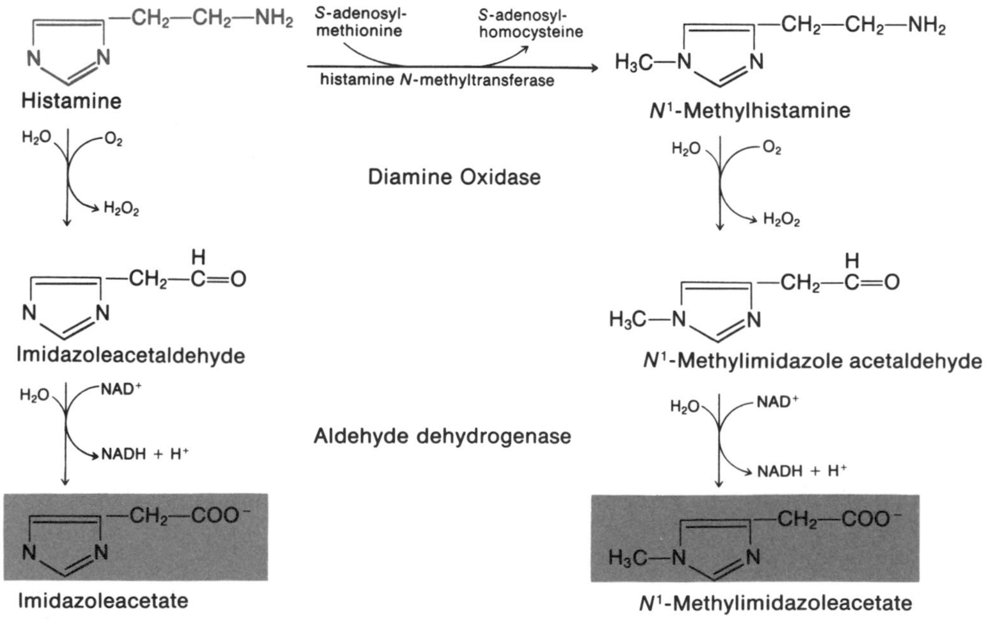

FIGURE 12.41
Metabolism of histamine.

oxidizes only primary amines but with little specificity; in some mammalian species the so-called monoamine oxidase also attacks typical diamine oxidase substrates, including histamine and spermine. The serum enzymes contain copper and a reactive carbonyl group. The carbonyl was first attributed to pyridoxal phosphate, then to another recently discovered cofactor necessary for normal mammalian development but whose precise function is not yet established, pyrroloquinoline quinone (PQQ) (Figure 12.40*a*). However, the active group in the enzyme has been identified as a modified amino acid that is part of the primary structure, 6-hydroxydopa or trihydroxyphenylalanine (TOPA), shown in Figure 12.40*b*. **Diamine oxidase** has been purified from kidney and also shown to contain copper and a cofactor that remains to be identified (the purified enzyme has a pink color, similar to some of the serum monoamine oxidases). It is sometimes referred to as *histaminase* for its only known physiological substrate.

Methylation at N-1 of the imidazole ring by an enzyme specific for histamine inactivates most of the histamine released; the methylhistamine produced is also a substrate of diamine oxidase. For the most part the aldehydes produced by amine oxidases are oxidized by aldehyde dehydrogenases to the corresponding carboxyl compounds, but a small part is sometimes found as the alcohol, presumably the result of reduction of the aldehyde by a pyridine nucleotide and alcohol dehydrogenase. Thus major products of histamine metabolism are methylhistamine, imidazoleacetate, and methylimidazoleacetate (Figure 12.41). In addition, some histamine is acetylated (Figure 12.42). An unusual derivative was isolated from the urine of animals given large doses of histamine; the imidazoleacetate was conjugated with ribose. The nucleoside is formed in a reaction with 5-phosphoribosyl-1-pyrophosphate catalyzed by a liver enzyme that liberates both ortho- and pyrophosphate.

Methylation of norepinephrine and similar compounds is catalyzed by **catechol-*O*-methyl transferase** (COMT). This enzyme uses *S*-adenosylmethionine as the donor for a great variety of substrates. The presence of methyl groups does not affect the action of amine oxidases, and COMT does not require an amine in the substrate, so that the major degradative

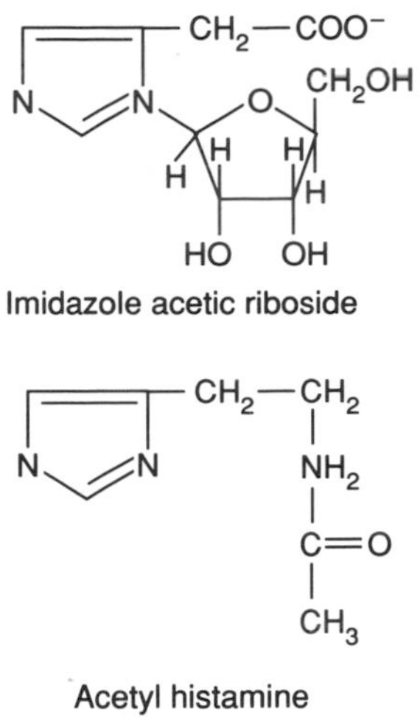

Imidazole acetic riboside

Acetyl histamine

FIGURE 12.42
Additional metabolites of histamine.

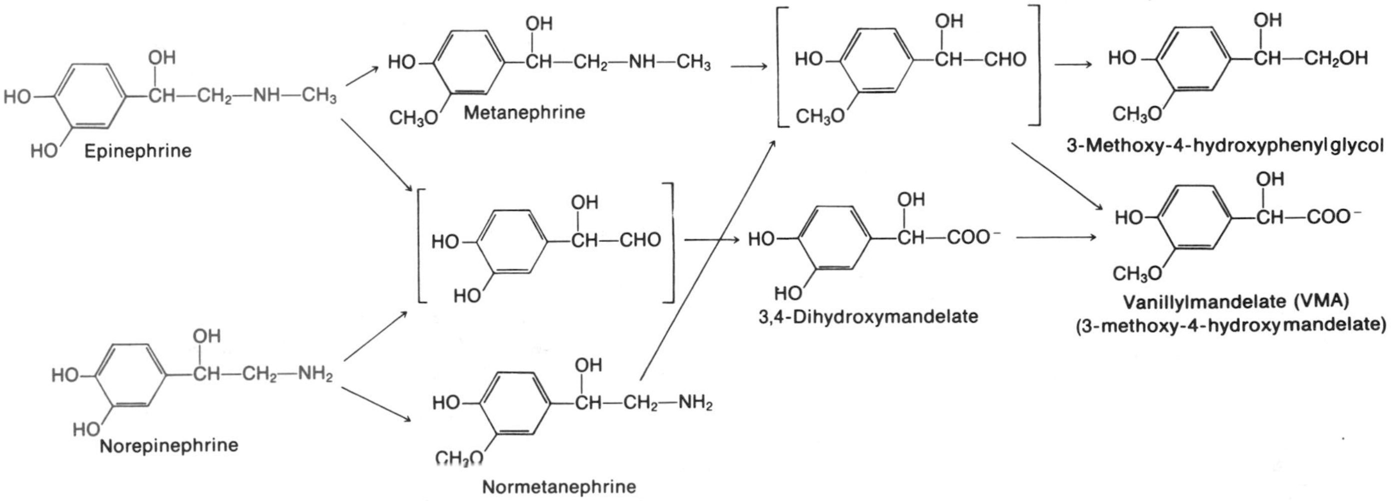

Epinephrine
Metanephrine
3-Methoxy-4-hydroxyphenylglycol
3,4-Dihydroxymandelate
Vanillylmandelate (VMA) (3-methoxy-4-hydroxymandelate)
Norepinephrine
Normetanephrine

FIGURE 12.43
Metabolism of the catecholamines.
Each of the compounds named is excreted in the urine. Epinephrine, norepinephrine, and their methyl derivatives are excreted mainly as conjugates with sulfate or glucuronic acid.

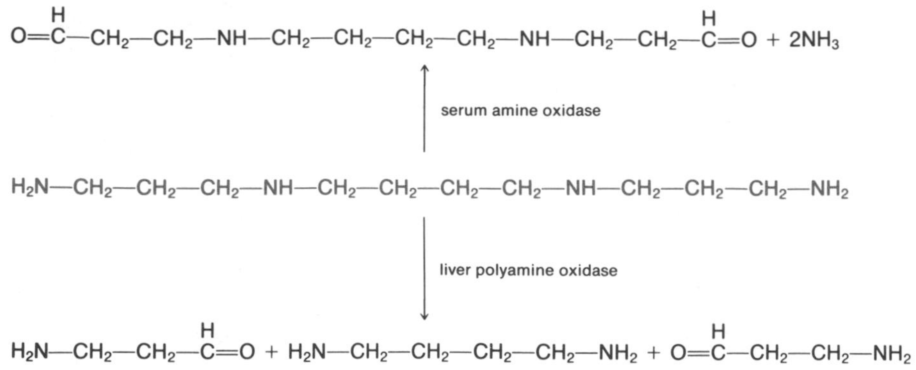

$$O{=}CH{-}CH_2{-}CH_2{-}NH{-}CH_2{-}CH_2{-}CH_2{-}CH_2{-}NH{-}CH_2{-}CH_2{-}CH{=}O + 2NH_3$$

↑ serum amine oxidase

$$H_2N{-}CH_2{-}CH_2{-}CH_2{-}NH{-}CH_2{-}CH_2{-}CH_2{-}CH_2{-}NH{-}CH_2{-}CH_2{-}CH_2{-}NH_2$$

↓ liver polyamine oxidase

$$H_2N{-}CH_2{-}CH_2{-}CH{=}O + H_2N{-}CH_2{-}CH_2{-}CH_2{-}CH_2{-}NH_2 + O{=}CH{-}CH_2{-}CH_2{-}NH_2$$

FIGURE 12.44
Oxidation of spermine.

pathways of the catechol amines converge to give vanillylmandelic acid as the major product. In addition to the other products shown in Figure 12.43, sulfate esters and glucuronides of the catechols are also excreted.

Spermine and spermidine are substrates for the serum amine oxidase. Both primary amino groups of spermine are converted to aldehydes, but only the propionylamine of spermidine is oxidized. In liver a flavoprotein oxidizes spermine internally to produce 3-aminopropionaldehyde and spermidine, which is oxidized similarly to form putrescine. The oxidative reactions are illustrated in Figure 12.44. Liver also contains two types of acetylase, nuclear enzymes that form N^1 and N^8-acetylspermidine and an inducible cytosolic enzyme that forms exclusively N^1-acetylspermidine. Both enzymes also acetylate spermine, but only the nuclear enzymes acetylate the basic histone proteins. N^1-Acetylspermidine and N^1-acetyl spermine are substrates for the liver polyamine oxidase and also for a hydrolytic deacylase. The reason for acetylating and deacetylating polyamines is unknown.

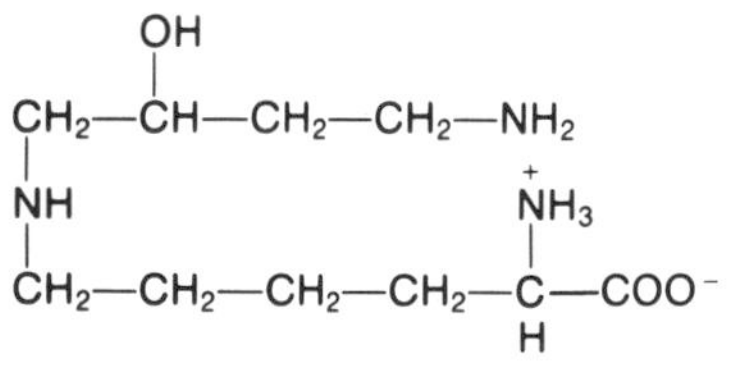

A unique reaction produces a new amino acid by transfer of the butylamino residue of spermidine to a single lysine residue in a specific protein, the initiation factor of protein synthesis elF-4D. After the amino-butyl group is transferred, it is hydroxylated by an iron-dependent enzyme, presumably a mixed-function oxygenase. The modified amino acid, N^1-(4-amino-2-hydroxybutyl)-lysine, is named hypusine (Figure 12.45).

OH
CH₂—CH—CH₂—CH₂—NH₂
NH
CH₂—CH₂—CH₂—CH₂—C(H)(NH₃⁺)—COO⁻

FIGURE 12.45
Structure of hypusine.

12.6 GLUTATHIONE

Glutathione Functions as a Reductant

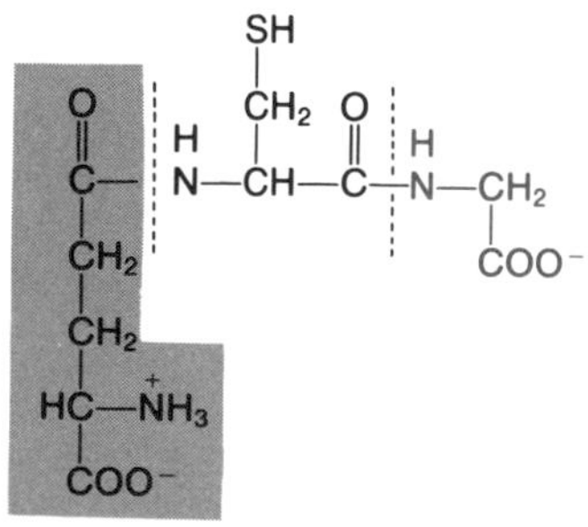

The tripeptide glutathione, **γ-glutamylcysteinylglycine** (Figure 12.46), is a major constituent of cells and serves several independent functions. Recent insight into the metabolism of glutathione suggests several potential clinical applications, discussed in Clin. Corr. 12.12. One group of functions is based on the reducing powers of the sulfhydryl group, which is the reason the accepted abbreviation of glutathione is GSH. Since the primary sulfhydryl compound synthesized (cysteine) is toxic in high concentrations, possibly because of the effects of degradation products, chelation of metals, or other properties of this reactive molecule, a derivative that is better tolerated is used as the reservoir of reducing power. This is graphically illustrated in red blood cells, in which GSH is present in high concentration and serves to reduce methemoglobin back to hemoglobin. The primary reductants are the pentose phosphate shunt reactions that

SH
CH₂ O H
O=C—NH—CH—C—N—CH₂—COO⁻
CH₂
CH₂
HC—NH₃⁺
COO⁻

FIGURE 12.46
Structure of glutathione (γ-glutamylcysteinylglycine).

produce NADPH, which is used by **glutathione reductase** to reduce the disulfide of oxidized glutathione (GSSG):

$$2GSH + 2MetHb \longrightarrow GSSG + 2Hb + 2H^+$$

$$NADPH + H^+ + GSSG \longrightarrow NADP^+ + 2GSH$$

The steady state within cells generally maintains a ratio of about 100 : 1 of GSH/GSSG.

Toxic amounts of peroxides and free radicals are produced in vivo by irradiation and other mechanisms; these are scavenged by **glutathione peroxidase,** a selenium-containing enzyme:

$$2GSH + H_2O_2 \longrightarrow GSSG + 2H_2O$$

The sulfhydryl group of GSH also participates in a disulfide interchange reaction that rearranges disulfide bonds in proteins until the thermodynamically most stable structure is formed. The sulfhydryl group is also involved in the cofactor function of GSH with several enzymes, including glyoxalase, maleylacetoacetate isomerase, and prostaglandin PGE_2 synthetase. An important physiological addition of glutathione occurs in the synthesis of leukotrienes (Chapter 10).

Glutathione-*S*-Transferases Function in Detoxication Reactions

Glutathione is a substrate for a group of enzymes, **glutathione *S*-transferases,** that add its sulfur to a great variety of acceptor molecules. Xenobiotic acceptors include halogenated and nitro compounds, organophosphates (including insecticides), allylic compounds, epoxides (including metabolites of carcinogens), and more. The conjugation is usually considered to be a detoxication and the conjugate is degraded by the enzymes of the γ-glutamyl cycle, first losing the γ-glutamyl residue by transpeptidation, then the glycyl residue by hydrolysis. The resulting cysteinyl derivative is acetylated with acetyl CoA to become a mercapturic acid, which is excreted in the urine, along with intermediate compounds of this pathway (Figure 12.47). The fates of the glutathione conjugates vary greatly as a function of the chemical structure of the acceptors.

$$RX + GSH \longrightarrow RSG + HX$$

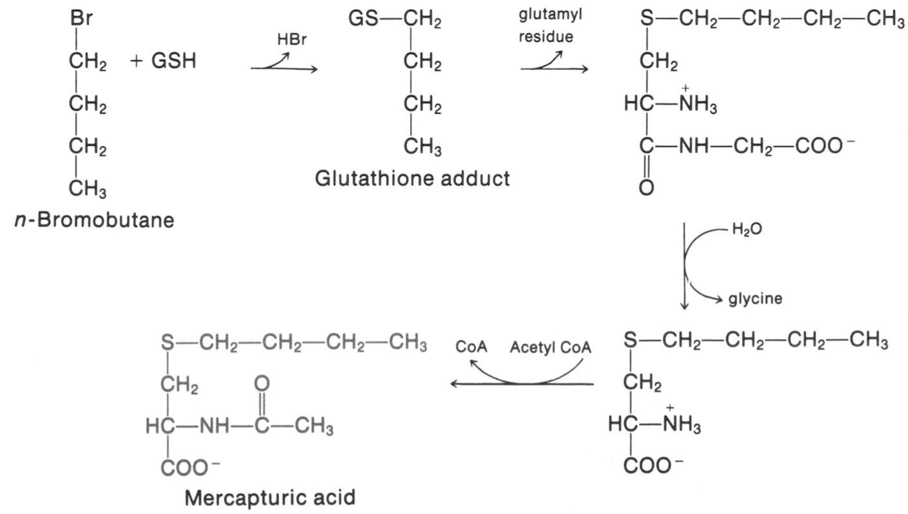

$Br-CH_2-CH_2-CH_2-CH_3$ + GSH $\xrightarrow{HBr}$ $GS-CH_2-CH_2-CH_2-CH_3$ $\xrightarrow{\text{glutamyl residue}}$ $S(-CH_2-CH_2-CH_2-CH_3)-CH_2-HC(-\overset{+}{N}H_3)-C(=O)-NH-CH_2-COO^-$

n-Bromobutane; Glutathione adduct

$\xrightarrow{H_2O \to \text{glycine}}$ $S(-CH_2-CH_2-CH_2-CH_3)-CH_2-HC(-\overset{+}{N}H_3)-COO^-$

$\xrightarrow{\text{Acetyl CoA} \to \text{CoA}}$ $S(-CH_2-CH_2-CH_2-CH_3)-CH_2-HC(-NH-C(=O)-CH_3)-COO^-$

Mercapturic acid

FIGURE 12.47
Formation of a mercapturic acid.

CLIN. CORR. 12.12
CLINICAL PROBLEMS RELATED TO GLUTATHIONE DEFICIENCY

DEFICIENCY OF ENZYMES OF γ-GLUTAMYL CYCLE

Decreased ability to form glutathione has been associated with hemolytic disease and neurological and mental abnormalities. Two different deficiencies result in excretion of 5-oxoproline. Deficiency of 5-oxoprolinase is a benign condition that does not cause acidosis or lack of glutamate. A lack of glutathione synthetase, however, causes much more 5-oxoproline to be accumulated so that acidosis affects the central nervous system. In this disease large amounts of γ-glutamylcysteine are formed. This dipeptide is a good substrate for γ-glutamylcyclotransferase and cannot be used appreciably to make GSH, so 5-oxoproline is formed at almost the rate of formation of γ-glutamylcysteine. Experimental approaches to therapy include giving inhibitors for γ-glutamyltranspeptidase to maintain higher intracellular concentrations of GSH and use of an inhibitor of γ-glutamylcysteine synthetase, buthionine sulfoximine, a structural analog of γ-glutamate derivatives.

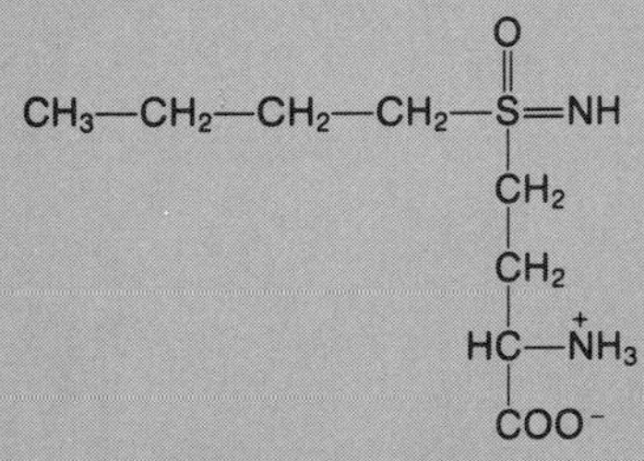

$CH_3-CH_2-CH_2-CH_2-S(=O)(=NH)-CH_2-CH_2-HC(-\overset{+}{N}H_3)-COO^-$

Buthionine sulfoximine

Deficiency of γ-glutamylcysteine synthetase results in low levels of glutathione. For many years patients with this disease show no abnormality except hemolytic anemia, but later in life serious neurologic defects develop rapidly. In contrast, two patients deficient in γ-glutamyltranspeptidase were mentally deficient and excreted glutathione.

Spielberg, S. P., et al. Biochemical heterogeneity in glutathione synthetase deficiency. *J. Clin. Invest.* 61:1417, 1978; and Konrad, P. N., Richards, F., Valentine, W. N., and Paglia, D. γ-glutamyl cysteine synthetase deficiency. *N. Engl. J. Med.* 286:557, 1972.

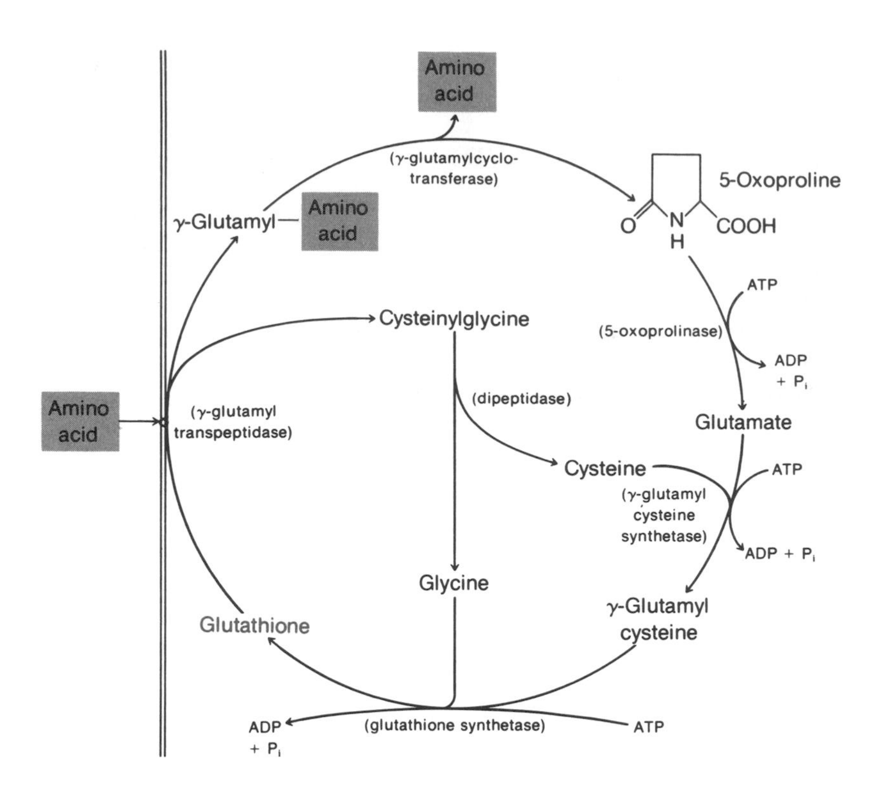

FIGURE 12.48
The γ-glutamyl cycle.

Glutathione is synthesized from free amino acids in two steps: first the γ-carboxyl group of glutamate is activated by ATP and forms an amide with the amino group of cysteine and second a similar activation of the carboxyl group of the cysteinyl residue of the dipeptide permits condensation with glycine. These reactions comprise the lower one-half of the cycle pictured in Figure 12.48.

The γ-Glutamyl Cycle Functions in Amino Acid Transport

Although glutathione is present in high concentration relative to free amino acids in the cytosol, isotopic evidence shows that it is rapidly turned over; that is, it is constantly being broken down and resynthesized. An explanation of this metabolic activity is the **γ-glutamyl cycle** (Figure 12.48), a mechanism used by some cells to transport amino acids across the cell membrane. The enzyme **γ-glutamyl transpeptidase** is associated with cell membranes and forms isopeptides of glutamate and various free amino acids, releasing cysteinyl glycine. Both peptides are brought into the cytosol, where the cysteinyl glycine is converted to free amino acids as a result of action of a dipeptidase. The γ-glutamyl peptide is split by a specific enzyme, cyclotransferase, which releases the other amino acid while converting the glutamate residue to a ring compound 5-oxoproline (also called pyrrolidone carboxylic acid). To open this ring at neutral pH requires energy; an enzyme 5-oxoprolinase couples hydrolysis of ATP with hydrolysis of the cyclic amide to form glutamate, ADP, and P_i. The sum of the reactions is the transport of an amino acid at the expense of three molecules of ATP. This appears to be a needless extravagance but may be necessary for rapid, high capacity transport in the kidney and other organs for certain amino acids, especially cysteine and glutamine. The physiological significance of the cycle is not clear, but the high level of activity, even when the system is incomplete (Clin. Corr. 12.12) suggests an important function.

Control of Glutathione Levels May Have Physiological Significance

Animal experiments offer possibilities for both increasing and decreasing cellular GSH, which might be useful in the future for treating different conditions. Two ways to increase GSH are administration of the intermediate γ-glutamylcysteine and giving glutathione esters. Increased intracellular cysteine has been accomplished by administration of 2-oxothiazolidine-4-carboxylate, an analog of 5-oxoproline that is a good substrate for 5-oxoprolinase. Hydrolysis of the analog should form 5-carboxycysteine, which would be expected to decompose spontaneously to cysteine and CO_2. Depletion of glutathione is brought about by inhibitors of γ-glutamyl cycle enzymes. Buthionine sulfoximine acts as an analog of glutamate and binds to γ-glutamylcysteine synthetase but does not inhibit glutamine synthetase as analogs with shorter alkyl groups do (e.g., ethionine sulfoxime).

Why should depletion of GSH be desirable? Again, in animal studies the administration of buthionine sulfoximine has been shown to increase the sensitivity of tumors to radiation. In some situations chemotherapy might also be made more effective by lowering GSH. Certain parasites, including trypanosomes, contain marginal levels of GSH and, therefore, they may be more sensitive to inhibitors of GSH synthesis than human cells. Some drugs are inactivated by conjugation with GSH and may be more effective when the GSH concentration is lowered. The role of GSH is removal of H_2O_2 suggests that catalase-deficient organisms may become less virulent when GSH is decreased. Other reactions of GSH including metabolism of prostaglandins and leukotrienes may also be modulated for therapeutic purposes by adjusting the GSH level. Besides possible use in γ-glutamylcysteine synthetase deficiency, increasing GSH might be useful in protecting normal cells against radiation, free radicals, and other toxic materials.

BIBLIOGRAPHY

General

Meister, A. *Biochemistry of the amino acids,* 2nd ed. New York: Academic, 1965.

Branched-Chain Amino Acids

Odessey, R. *Problems and potential of branched-chain amino acids in physiology and medicine.* Amsterdam: Elsevier, 1986.

Serine

Snell, K. The duality of pathways for serine biosynthesis is a fallacy. *Trends Biochem. Sci.* 11:241, 1986.

Nitric Oxide

Bredt, D. S. and Snyder, S. H. Isolation of nitric oxide synthetase, a calmodulin-requiring enzyme. *Proc. Natl. Acad. Sci. U.S.A.* 87:682, 1990.

Tayeh, M. A. and Marletta, M. A. Macrophage oxidation of L-arginine to nitric oxide, nitrite and nitrate. *J. Biol. Chem.* 264:19654, 1989.

Sulfur Amino Acids

Cooper, A. J. L. Biochemistry of the sulfur-containing amino acids. *Annu. Rev. Biochem.* 52:187, 1983.

Furfine, E. S. and Abeles, R. H. Intermediates in the conversion of 5′-*S*-methylthioadenosine to methionine in Klebsiella pneumoniae. *J. Biol. Chem.* 263:9598, 1988.

Lee, B. J., Worland, P. J., Davis, J. N., Stadtman, T. C., and Hatfield, D. L. Identification of a selenocysteyl–$tRNA^{Ser}$ in mammalian cells that recognizes the nonsense codon, UGA. *J. Biol. Chem.* 264:9724, 1989.

Stepanuk, M. H. Metabolism of sulfur-containing amino acids. *Annu. Rev. Nutr.* 6:179, 1986.

Wright, C. E., Tallan, H. H., Lin, Y. Y., and Gaull, G. E. Taurine: biological update. *Annu. Rev. Biochem.* 55:427, 1986.

Polyamines

Tabor, C. W. and Tabor, H. Polyamines. *Annu. Rev. Biochem.* 53:749, 1984.

Folates and Pterins

Blakley, R. L. and Benkovic, S. J. *Folate and pterins.* New York: Wiley, Vol. 1, 1984; Vol. 2, 1985.

Pyrroloquinoline Quinone

Killgore, J., et al. Nutritional importance of pyrroloquinoline quinone. *Science* 245:850, 1989.

Carnitine

Bieber, L. L. Carnitine. *Annu. Rev. Biochem.* 57:261, 1988.

Glutathione

Beutler, E. Nutritional and metabolic aspects of glutathione. *Annu. Rev. Nutr.* 9:287, 1989.

Meister, A. and Anderson, M. E. Glutathione. *Annu. Rev. Biochem.* 52:711, 1976.

Disorders of Amino Acid Metabolism

Kaiser-Kupfer, M. I., Fujikawa, L., Kuwabara, T., Jain, S., and Jahe, W. A. Removal of corneal crystals by topical cysteamine in nephropathic cystinosis. *N. Eng. J. Med.* 316:775, 1983.

Kaufman, S. Regulation of the activity of hepatic phenylalanine hydroxylase. *Adv. Enzyme Regul.* 25:37, 1986.

Rosenberg, L. E. and Scriver, C. R. Disorders of amino acid metabolism. *Metabolic control and disease,* 8th ed. P. K. Bondy and L. E. Rosenberg (Eds.). Philadelphia: Saunders, 1980.

Scriver, C. R., Beaudet, A. L., Sly, W. S., and Valle, D. *The metabolic basis of inherited disease.* New York: McGraw-Hill, 6th ed. 1989.

Wellner, D. and Meister, A. A survey of inborn errors of amino acidmetabolism and transport. *Annu. Rev. Biochem.* 50:911, 1980.

QUESTIONS

C. N. ANGSTADT AND J. BAGGOTT

*Q*uestion *T*ypes are described on page xxiii.

1. (QT1) Pyruvate and alanine are components of a shuttle that involves:
   A. hepatic and renal gluconeogenesis.
   B. hepatic gluconeogenesis and transport of muscle nitrogen to liver as alanine.
   C. transport of alanine to muscle to supply pyruvate.
   D. the production of alanine for use in protein synthesis in most peripheral tissues.
   E. transport of alanine between cytosol and mitochondria of liver.

2. (QT2) The branched-chain amino acids:
   1. are essential in the diet.
   2. differ in that one is glucogenic, one is ketogenic, and one is classified as both.
   3. are catabolized in a manner that bears a resemblance to β oxidation of fatty acids.
   4. are oxidized by a dehydrogenase complex to branched-chain acyl CoAs one carbon shorter than the parent compound.

A. Serine　　C. Both
B. Threonine　　D. Neither

3. (QT4) In equilibrium with glycine, via a reaction catalyzed by a hydroxymethyltransferase.
4. (QT4) Can be synthesized from an intermediate of glycolysis.

5. (QT2) Glycine:
   1. is oxidized to glyoxylate by D-amino acid oxidase.
   2. if labeled in its α carbon can lead to a serine labeled in its α and/or β carbon atoms.
   3. is a contributor to the pool of C_1 H_4folate compounds.
   4. apart from its catabolism through serine or incorporation into proteins has no other known functions.

A. Proline　　C. Both
B. Ornithine　　D. Neither

6. (QT4) May be formed from or converted to glutamic semialdehyde or its cyclic Schiff base, Δ^1-pyrroline-5-carboxylate.
7. (QT4) Play(s) a major role in the urea cycle.

8. (QT2) Which of the following is/are a product of decarboxylation of an amino acid.
   1. Guanidinoacetate
   2. Putrescine
   3. Spermidine
   4. γ-Aminobutyrate (GABA)
9. (QT2) *S*-Adenosylmethionine:
   1. contains a positively charged sulfur (sulfonium) that facilitates the transfer of substituents to suitable acceptors.
   2. yields homocysteine when used as a methyl donor.
   3. participates in the formation of the polyamine, spermine.
   4. generates H_2S by transsulfuration.

10. (QT1) In humans, sulfur of cysteine may participate in all of the following *except* the:
    A. conversion of cyanide to less toxic thiocyanate.
    B. formation of thiosulfate.
    C. formation of urinary sulfur products.
    D. donation of the sulfur for methionine formation.
    E. formation of PAPS.

11. (QT2) An inability to generate tetrahydrobiopterin might be expected to:
    1. inhibit the normal degradative pathway of phenylalanine.
    2. lead to albinism.
    3. reduce formation of the neurotransmitter, serotonin.
    4. reduce the body's ability to transfer 1-carbon fragments.

12. (QT1) Both tyrosine aminotransferase and tryptophan oxygenase are enzymes that can be induced by adrenal glucocorticoids. This is reasonable because:
    A. tyrosine and tryptophan are precursors of physiologic amines.

B. glucocorticoids work by inducing enzymes.
C. tryptophan is the precursor of nicotinic acid needed for NAD^+ synthesis.
D. tyrosine is the precursor of catecholamines in the adrenal gland.
E. these two enzymes initiate the major catabolic pathways in the liver of tyrosine and tryptophan.

13. (QT2) Histidine:
    1. unlike most amino acids, is not converted to an α-keto acid when the amino group is removed.
    2. is a contributor to the tetrahydrofolate 1-carbon pool.
    3. decarboxylation produces a physiologically active amine.
    4. forms a peptide with β-alanine.
14. (QT2) Lysine as a nutrient:
    1. may be replaced by its α-keto acid analogue.
    2. produces acetoacetyl CoA in its catabolic pathway.
    3. is methylated by *S*-adenosylmethionine.
    4. is the only one of the common amino acids that is a precursor of carnitine.
15. (QT2) In folic acid-dependent 1-carbon metabolism:
    1. the formation of the methyl group of thymine involves a reduction in the process of transfer of the 1-carbon group.
    2. the major donor of the 1-carbon group of N^5,N^{10}-methylene H_4folate is serine.
    3. carbons at different oxidation levels may be interconverted by suitable enzymes.
    4. the only acceptor for the methyl form is homocysteine.
16. (QT2) In the catabolism and excretion of physiologically active amines:
    1. mitochondrial monoamine oxidase oxidizes primary and methylated amines to aldehydes.
    2. methylation may occur.
    3. conjugation with sulfate and/or glucuronate may occur.
    4. opening of a ring, if present, is common.

17. (QT1) Glutathione does all of the following *except* to:
    A. participate in the transport of amino acids across some cell membranes.
    B. scavenge peroxides and free radicals.
    C. form sulfur conjugates for detoxication of compounds.
    D. convert hemoglobin to methemoglobin.
    E. act as a cofactor for some enzymes.

ANSWERS

1. B Peripheral tissues, especially muscle, collect amino group nitrogen that is eventually transferred to pyruvate by transamination to produce alanine. Alanine, in the liver, is converted back to pyruvate and then to glucose. The glucose is transported back to muscle to once again produce pyruvate through glycolysis (p. 493).
2. E All four correct. 2–4: Although their catabolism is similar, the end products are different because of the differences in the branching. After transamination, the α-keto acids are oxidized by a dehydrogenase complex in a fashion similar to pyruvate dehydrogenase. The similarity to β oxidation comes in steps like oxidation to an α,β unsaturated CoA, hydration of the double bond, and oxidation of an hydroxyl to a carbonyl (p. 493).
3. A This is an important source and utilization of N^5, N^{10}-methylene H_4folate. B: Threonine is an essential amino acid and, therefore, not synthesized (p. 496).
4. A Serine can be synthesized from either 2- or 3-phosphoglycerate, but threonine is an essential amino acid (p. 496).
5. A 1, 2, 3 correct. 1: Glycine is neither a D- nor an L-amino acid but is a substrate for this enzyme. 2, 3: Serine and glycine are interconvertible via N^5,N^{10}-methylene H_4folate, and glycine, itself, also produces this compound, resulting in a randomization of isotopic label among these carbons. 4: Glycine is a precursor of purines, creatine, and heme (p. 497).
6. C Ornithine and proline are interconvertible because they both give rise to and are formed from glutamic semialdehyde. They both enter the TCA cycle as glutamate for the same reason (p. 498).
7. B Ornithine is both a substrate and product of the urea cycle leading to or being formed from arginine (p. 500).
8. C 2, 4 correct. 1: Product of transamidination of arginine. 2, 3: Decarboxylation of ornithine produces putrescine, which is a precursor of the *polyamine* spermidine. 4: GABA is the decarboxylation product of glutamate (p. 501).
9. A 1, 2, 3 correct. 1, 2: The reactive, positively charged sulfur reverts to a neutral thioether when the methyl group is transferred to an acceptor. The product, *S*-adenosylhomocysteine, is hydrolyzed to homocysteine. 3: Decarboxylation of *S*-adenosylmethionine generates the C_3 amine fragment that is transferred to putrescine. 4: Transsulfuration refers to the combined action of cystathionine synthase and cystathionase transferring methionine's sulfur to serine to yield cysteine (p. 501).
10. D A, B: Transamination to β-mercaptopyruvate with subsequent formation of thiosulfate and/or conversion of cystine to thiocysteine allows transfer of the sulfur to detoxify cyanide. C, E: SO_4^{2-}, the most oxidized form of sulfur found physiologically, is either excreted or activated as PAPS for use in detoxifying phenolic compounds or in biosynthesis. D: Methionine is the source of sulfur for cysteine (via homocysteine), but the reverse is not true in humans (p. 505).
11. B 1, 3 correct. 1, 3: Tetrahydrobiopterin is a necessary cofactor for phenylalanine, tyrosine, and tryptophan hydroxylases. The first catalyzes the major pathway of phenylalanine catabolism, and the third leads to the formation of serotonin. Catecholamine formation, catalyzed by the second enzyme, would also be

deficient. 2: Albinism stems from a deficiency of tyrosinase, which, while giving the same product as tyrosine hydroxylase, is not a tetrahydrobiopterin-requiring enzyme. 4: One-carbon fragments are transferred from either *S*-adenosylmethionine or the H_4folate 1-carbon pool (p. 506).

12. E Although all of the statements are true, only E offers a suitable rationale. Tyrosine and tryptophan both yield a glucogenic fragment (fumarate and alanine) upon catabolism. Glucocorticoids are secreted in response to low blood glucose or stress (p. 506–510).
13. E All four correct. 1: Elimination of ammonia from histidine leaves a double bond (urocanate) unlike both transamination and oxidative deamination reactions. 2: A portion of the ring is released as formimino H_4folate. 3, 4: Histamine; carnosine (p. 512).
14. C 2, 4 correct. 1: Lysine does not participate in transamination (L- amino acid oxidase reaction is not reversible), probably in part because the α-keto acid exists as a cyclic Schiff's base. 3, 4: Free lysine is not methylated, but lysyl residues in a protein are methylated in a posttranslational modification. Intermediates of carnitine synthesis are derived from trimethyllysine liberated by proteolysis (p. 513).
15. E All four correct. 1: 5,10-methylene H_4folate is reduced to a methyl in the process of transfer to dUMP, releasing the carrier as dihydrofolate. 3: This is probably the most important point to remember about the H_4folate C_1 pool. 4: This reaction requires vitamin B_{12}. Once converted to methionine, this methyl is now available for transmethylation reactions in general (p. 516, Clin. Corr. 12.11).
16. A 1, 2, 3 correct. 1–3: Oxidation to aldehydes and/or methylation are extremely common. Frequently the products are conjugated to enhance solubility in urine. 4: Modification, rather than cleavage, of aromatic or imidazole rings is the norm (p. 520).
17. D Most of the functions of glutathione listed are dependent upon the sulfhydryl group (—SH). A major role of glutathione in red blood cells is reduction of methemoglobin. Glutathione reductase helps to maintain the ratio of GSH : GSSG at about 100 : 1 (p. 522).

13

Purine and Pyrimidine Nucleotide Metabolism

JOSEPH G. CORY

13.1 OVERVIEW 530
13.2 METABOLIC FUNCTIONS OF NUCLEOTIDES 530
Cells Have Different Distributions of Nucleotides 532
13.3 CHEMISTRY OF NUCLEOTIDES 532
Properties of Nucleotides 533
13.4 METABOLISM OF PURINE NUCLEOTIDES 536
Purine Nucleotides Are Synthesized by a Stepwise Building of the Ring to Form IMP 536
IMP Is the Common Precursor of AMP and GMP 539
Purine Nucleotide Synthesis Is Regulated at the Commitment Step and Branch Points 540
Purine Bases Can Be Salvaged to Reform Nucleotides 542
Interconversion Maintains an Appropriate Balance of Purine Nucleotides 542
Purine Degradation in Humans Produces Uric Acid 544
13.5 METABOLISM OF PYRIMIDINE NUCLEOTIDES 547
Pyrimidines Are Synthesized From Glutamine, Aspartate, and CO_2 547
Pyrimidine Nucleotide Synthesis Is Regulated at the Level of Carbamoyl Phosphate Synthetase II 550
Pyrimidines Can Be Salvaged to Re-Form Nucleotides 550
13.6 DEOXYRIBONUCLEOTIDE FORMATION 551
Deoxyribonucleotides Are Formed by Reduction of Ribonucleoside Diphosphates 551
Deoxythymidylate Synthesis Requires N^5,N^{10}-Methylene H_4Folate 552
Interconversion of Deoxyribopyrimidine Nucleotides Maintains an Appropriate Balance 553
Pyrimidine Nucleotides Are Degraded to β-Amino Acids 553
13.7 NUCLEOSIDE AND NUCLEOTIDE KINASES 555
13.8 NUCLEOTIDE METABOLIZING ENZYMES AS A FUNCTION OF THE CELL CYCLE AND RATE OF CELL DIVISION 557
The Cell Cycle Has Distinct Phases 557
Enzymes of Purine and Pyrimidine Nucleotide Synthesis Are Elevated During the S Phase 558
13.9 NUCLEOTIDE COENZYME SYNTHESIS 559
NAD Is Synthesized From Tryptophan, Nicotinate, and Nicotinamide 559
NADP Is Produced by Phosphorylation of NAD 561
NAD Is Degraded to Nicotinamide 561
FAD Synthesis Requires Exogenous Riboflavin 562
Coenzyme A Synthesis Requires Exogenous Pantothenic Acid 562
13.10 SYNTHESIS AND UTILIZATION OF 5-PHOSPHORIBOSYL 1-PYROPHOSPHATE 563
13.11 COMPOUNDS THAT INTERFERE WITH CELLULAR PURINE AND PYRIMIDINE NUCLEOTIDE METABOLISM: USE AS CHEMOTHERAPEUTIC AGENTS 566
Antimetabolites Are Structural Analogs of the Bases or Nucleosides 566

Antifolates Inhibit the Formation of Tetrahydrofolate 567
Glutamine Antagonists Inhibit Glutamine Utilization as a Nitrogen Donor 569
Other Agents Inhibit Cell Growth 569
13.12 PURINE AND PYRIMIDINE ANALOGS AS ANTIVIRAL AGENTS 569
13.13 BIOCHEMICAL BASIS FOR THE DEVELOPMENT OF DRUG RESISTANCE 570
BIBLIOGRAPHY 571
QUESTIONS AND ANNOTATED ANSWERS 572
CLINICAL CORRELATIONS
13.1 Lesch-Nyhan Syndrome 543
13.2 Immunodeficiency Diseases Associated With Defects in Purine Nucleotide Metabolism 544
13.3 Gout 546
13.4 Orotic Aciduria 548

13.1 OVERVIEW

Purine and pyrimidine nucleotides are critically important metabolites that participate in many cellular functions. These functions range from serving as the monomeric precursors of the nucleic acids, to serving as energy stores, effectors, group transfer agents, and mediators of hormone action. The nucleotides are formed in the cell de novo from amino acids, ribose, formate, and CO_2. The de novo pathways for the synthesis of the nucleotides require a relatively high input of energy. To compensate for this, most cells have very efficient "salvage" pathways by which the preformed purine or pyrimidine bases can be reutilized.

Because of the manner in which nucleotides are synthesized and "salvaged," the purines and pyrimidines occur primarily as nucleotides in the cell. The concentrations of free bases or free nucleosides under normal conditions are exceedingly small. The levels of nucleotides in the cell are very finely regulated by a series of allosterically controlled enzymes in the pathway. Nucleotides are the regulators of these reactions.

Ribonucleotides (at the diphosphate level) serve as the precursors of the deoxyribonucleotides. While the concentrations of ribonucleotides in the cell are in the millimolar range, the concentrations of deoxyribonucleotides are in the micromolar range. The replication of DNA requires that there be sufficient quantities of the deoxyribonucleoside triphosphates. To facilitate this, the activities of several enzymes required for deoxyribonucleotide metabolism increase during the late-G_1 phase of the cell cycle prior to DNA replication, which takes place during the S phase of the cell cycle.

There are several diseases or syndromes that result from defects in the metabolic pathways for the synthesis of nucleotides either de novo or by salvage or for the degradation of the nucleotides. These include gout, the Lesch–Nyhan syndrome, orotic aciduria, and immunodeficiency diseases. Since nucleotides are obligatory for DNA and RNA synthesis in dividing cells, the metabolic pathways involving the synthesis of nucleotides have been the sites at which many antitumor agents have been directed.

It should be kept in mind during the reading of this chapter that the metabolism discussed has been limited exclusively to mammalian cells. In certain instances there are major differences between the bacterial and mammalian cells in nucleotide synthesis, degradation, and regulation.

13.2 METABOLIC FUNCTIONS OF NUCLEOTIDES

All types of cells (mammalian, bacterial, and plant) contain a wide variety of nucleotides and their derivatives. Some of these nucleotides occur in relatively high concentrations (millimolar range) in the cells. The reason for the large number of nucleotides and their derivatives in the cell is that they are involved in many metabolic processes that must be carried out for normal cellular growth and function.

These functions include the following:

1. *Role in Energy Metabolism:* As we have already seen, ATP is the main form of chemical energy available to the cell. Quantitatively, ATP is generated in cells by oxidative phosphorylation and substrate-level phosphorylation. The ATP is utilized to drive metabolic reactions, as a phosphorylating agent, and is involved in such processes as muscle contraction, active transport, and maintenance of cell membrane integrity. As a phosphorylating agent, ATP serves as the phosphate donor for the generation of the other nucleoside 5′-triphosphates (e.g., GTP, UTP, and CTP).
2. *Monomeric Units of Nucleic Acids:* The nucleic acids, DNA and RNA, are composed of monomeric units of the nucleotides. In the reactions in which the nucleic acids are synthesized, the nucleoside 5′-triphosphates are the substrates and are linked in the polymer through 3′,5′-phosphodiester bonds with the release of pyrophosphate.
3. *Physiological Mediators:* More recently recognized functions of nucleotides and their derivatives involve those in which the nucleotides or nucleosides serve as mediators of key metabolic processes. The role of cAMP as a "second messenger" in epinephrine- and glucagon-mediated control of glycogenolysis and glycogenesis has already been discussed. The importance of cGMP as a mediator of cellular events has also been recognized. ADP has been shown to be very critical for normal platelet aggregation and hence blood coagulation. Adenosine has been shown to cause dilation of coronary blood vessels and therefore may be important in the regulation of coronary blood flow. The nucleotide GTP is required for functions such as the capping of mRNA, signal transduction through the GTP-binding proteins, and microtubule formation.
4. *Components of Coenzymes:* Coenzymes such as NAD, NADP, FAD, and CoA are critically important constituents of cells and are involved in many metabolic pathways.
5. *Activated Intermediates:* The nucleotides also serve as carriers of "activated" intermediates required for a variety of reactions. A compound such as UDP-glucose is a key intermediate in the synthesis of glycogen and glycoproteins. The compounds GDP-mannose, GDP-fucose, UDP-galactose, and CMP-sialic acid are all key intermediates in reactions in which sugar moieties are transferred for the synthesis of glycoproteins. CTP is utilized to generate CDP-choline, CDP-ethanolamine, and CDP-diacylglycerols, which are involved in phospholipid metabolism. Other activated intermediates include *S*-adenosylmethionine (SAM) and 3′-phosphoadenosine 5′-phosphosulfate (PAPS). *S*-Adenosylmethionine is a methyl donor in reactions involving the methylation of the sugar and base moieties of RNA and DNA and in the formation of compounds such as phosphatidylcholine from phosphatidylethanolamine, carnitine from lysine, and so on. *S*-Adenosylmethionine also provides the aminopropyl groups for the synthesis of spermine from ornithine. The intermediate PAPS is used as the sulfate donor to generate sulfated biomolecules such as the proteoglycans and the sulfatides.
6. *Allosteric Effectors:* Many of the regulated steps of the metabolic pathways are controlled by the intracellular concentrations of nucleotides.

Many examples have already been discussed in previous chapters, and the roles of nucleotides in the regulation of mammalian nucleotide metabolism will be discussed in this chapter.

Cells Have Different Distributions of Nucleotides

The principal form of purine and pyrimidine compounds found in cells is the 5′-nucleotide derivative. In normally functioning cells, the nucleotide of highest concentration is ATP. Depending on the cell type, the concentrations of the nucleotides vary greatly. For example, in the red cell the adenine nucleotides far exceed the other nucleotides, which are barely detectable. In liver cells and other tissues a complete spectrum of the mono-, di-, and triphosphates are found along with UDP-glucose, UDP-glucuronic acid, NAD^+, NADH, and so on. The presence of the free bases, nucleosides, or 2′- and 3′-nucleotides in the acid-soluble fraction of the cell represents degradation products of either the endogenous or exogenous nucleotides or nucleic acids. The presence of the so-called minor bases (see p. 688) is due to the degradation of nucleic acids.

The ribonucleotide concentration in the cell is in the millimolar range while the concentration of deoxyribonucleotides in the cell is in the micromolar range. As a specific example, the ATP concentration in Ehrlich tumor cells is 3600 pmol per 10^6 cells, while the dATP concentration in these cells is only 4 pmol per 10^6 cells. The deoxyribonucleotide levels, however, are subject to major fluctuations during the cell cycle, in contrast to the ribonucleotide levels, which remain relatively constant.

In normal cells the total concentrations of the nucleotides are fixed within rather narrow limits, although the concentration of the individual components can vary. That is, the total concentration of adenine nucleotides (AMP, ADP, and ATP) is constant, although there is a variation in the ratio of ATP to AMP + ADP, depending on the energy state of the cells. The basis for this "fixed concentration" is that the synthesis of nucleotides is one of the most finely regulated pathways occurring in the cell.

13.3 CHEMISTRY OF NUCLEOTIDES

Quantitatively, the major purine derivatives found in the cell are those of adenine and guanine. Other purine bases encountered are hypoxanthine and xanthine (Figure 13.1). Nucleoside derivatives of these molecules will contain either ribose or 2-deoxyribose linked to the purine ring through a

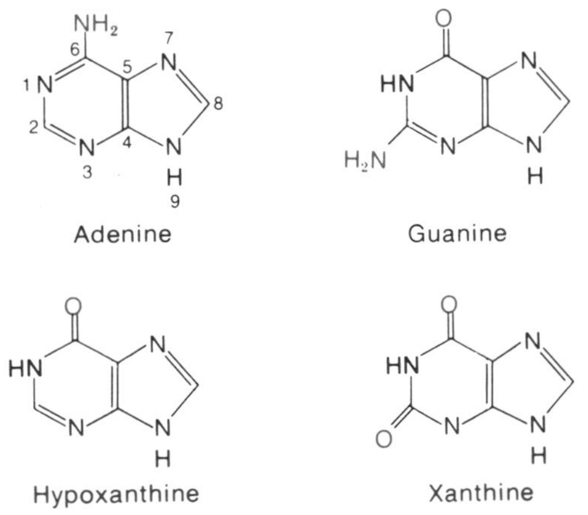

Adenine Guanine

Hypoxanthine Xanthine

FIGURE 13.1
Purine bases.

β-*N*-glycosidic bond at N-9. Ribonucleosides contain ribose, while deoxyribonucleosides contain deoxyribose as the sugar moiety (Figure 13.2). Nucleotides are phosphate esters of the purine nucleosides (Figure 13.3). 3′-Nucleotides such as adenosine 3′-monophosphate (3′-AMP) may occur in cells as a result of nucleic acid degradation.

In normally functioning cells, the tri- and diphosphates of the nucleosides are found to a greater extent than the monophosphates, nucleosides, or free bases.

The pyrimidine nucleotides found in highest concentrations in the cell are those containing uracil, cytosine, and thymine. The structures of the bases are presented in Figure 13.4. Uracil and cytosine nucleotides are the major pyrimidine components of RNA, whereas cytosine and thymine nucleotides are the major pyrimidine components of DNA. As with purine derivatives, the pyrimidine nucleosides or nucleotides contain either ribose or 2-deoxyribose. The sugar moiety is linked to the pyrimidine in a β-*N*-glycosidic bond at N-1. The nucleosides of the pyrimidines are uridine, cytidine, and thymidine (Figure 13.5). The phosphate esters of the pyrimidine nucleosides are UMP, CMP, and TMP. In the cell the major pyrimidine derivatives found are the tri- and diphosphates (Figure 13.6).

The symbols and abbreviations for the bases, nucleosides, and nucleotides are summarized in Table 13.1. See the appendix for a summary of the nomenclature and chemistry of the purines and pyrimidines.

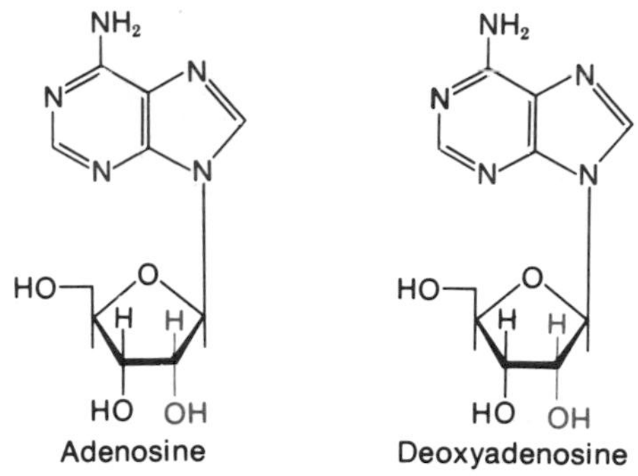

Properties of Nucleotides

Cellular components containing either the purine or pyrimidine bases can be easily detected because of the strong absorption of UV light by these

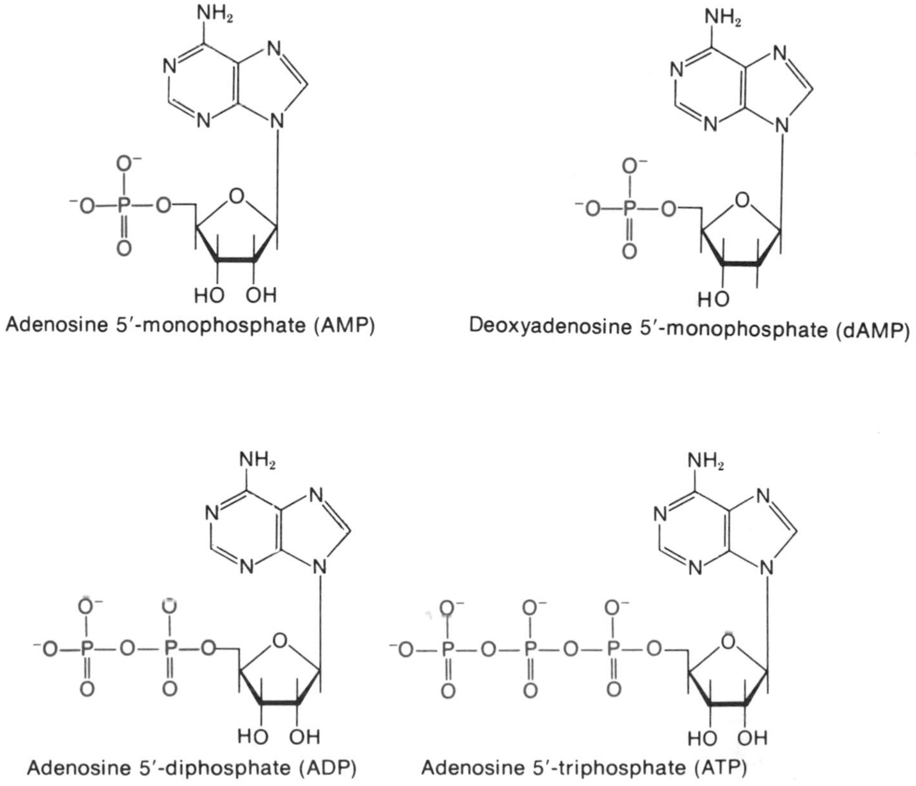

Adenosine

Deoxyadenosine

FIGURE 13.2
Adenosine and deoxyadenosine.

Adenosine 5′-monophosphate (AMP)

Deoxyadenosine 5′-monophosphate (dAMP)

Adenosine 5′-diphosphate (ADP)

Adenosine 5′-triphosphate (ATP)

FIGURE 13.3
Adenosine nucleotides.

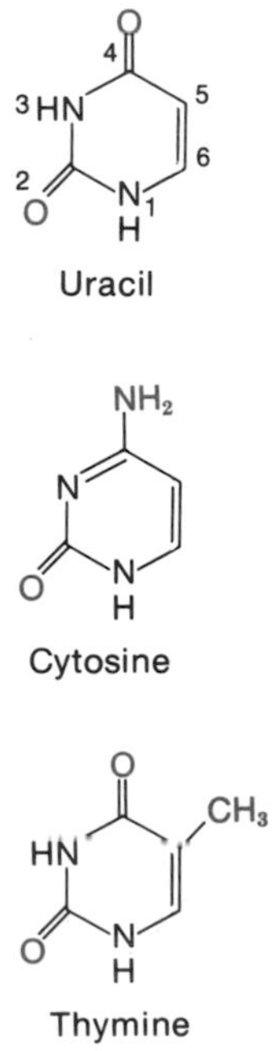

Uracil

Cytosine

Thymine

FIGURE 13.4
Pyrimidine bases.

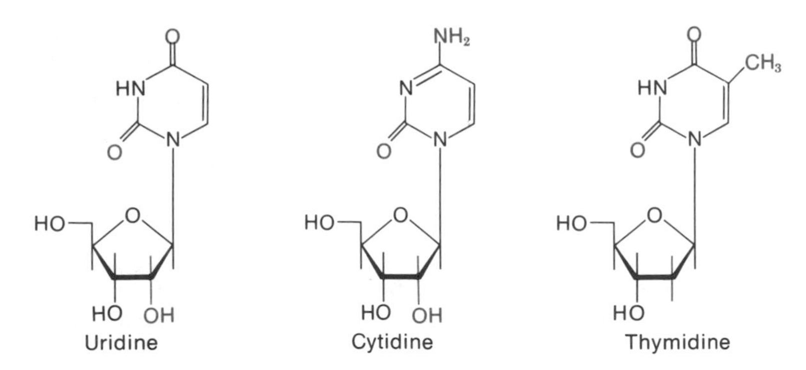

FIGURE 13.5
Pyrimidine nucleosides.

Uridine 5′-monophosphate (UMP)

Uridine 5′-diphosphate (UDP)

Uridine 5′-triphosphate (UTP)

FIGURE 13.6
Uracil nucleotides.

compounds. The purine bases, nucleosides, and nucleotides have stronger absorptions than the pyrimidines and their derivatives. The molar extinction coefficients (a measure of the light absorption at a specific wavelength of a compound) and λ_{max} for these are given in Table 13.2. The wavelength of light at which maximum absorption occurs varies with the particular base component, but in most cases the UV maximum is close to 260 nm. The UV spectrum for each of the nucleoside or nucleotide derivatives responds differently to changes in pH. The strong UV absorptions and the differences due to the specific structure of the base moiety provide the basis for sensitive methods in assaying these compounds both qualitatively and quantitatively. For example, the deamination of cyto-

TABLE 13.1 Symbols for Bases, Nucleosides, and Nucleotides

Compound	Abbreviations
Adenine	Ade
Cytosine	Cyt
Guanine	Gua
Thymine	Thy
Uracil	Ura
Adenosine	Ado
Cytidine	Cyd
Guanosine	Guo
Thymidine (2′-deoxythymidine)	dThd
Uridine	Urd
Adenosine 5′-mono-, di-, and triphosphate	AMP, ADP, and ATP
Cytidine 5′-mono-, di-, and triphosphate	CMP, CDP, and CTP
Guanosine 5′-mono-, di-, and triphosphate	GMP, GDP, and GTP
Thymidine 5′-mono-, di-, and triphosphate	TMP, TDP, and TTP
Uridine 5′-mono-, di-, triphosphate	UMP, UDP, and UTP

sine nucleosides or nucleotides to the corresponding uracil derivatives causes a marked shift in λ_{max} from 271 to 262 nm, which is easily determined. Because of the high molar extinction coefficients of the purine and pyrimidine bases and their high concentrations in the nucleic acids, a solution of RNA or DNA at a concentration of 1 mg mL^{-1} would have an absorbance at 260 nm of about 20, whereas a typical protein at a concentration of 1 mg mL^{-1} would have an absorbance at 280 nm of about 1. Consequently, the nucleic acids are easily detected at low concentrations.

The *N*-glycosidic bond of the purine and pyrimidine nucleosides and nucleotides is stable to alkali. However, the stability of this bond to acid hydrolysis differs markedly. The *N*-glycosidic bond of purine nucleosides and nucleotides is easily hydrolyzed by dilute acid at elevated temperatures (e.g., 60°C) to yield the free purine base and the sugar or sugar phosphate. On the other hand, the *N*-glycosidic bond of uracil, cytosine, and thymine nucleosides and nucleotides is very stable to acid treatment. Strong conditions, such as perchloric acid (60%) and 100°C, will cause the release of the free pyrimidine but with the complete destruction of the sugar moiety. The *N*-glycosidic bond of dihydrouracil nucleoside and dihydrouracil nucleotide is labile to mild acid treatment.

Because of the highly polar phosphate group, the purine and pyrimidine nucleotides are considerably more soluble in aqueous solutions than are their nucleosides and free bases. In general, the nucleosides are more soluble than the free purine or pyrimidine bases.

The purine and pyrimidine bases and their nucleoside and nucleotide derivatives can be easily separated by a variety of techniques. These methods include: paper chromatography; thin-layer chromatography (TLC), utilizing plates with cellulose or ion-exchange resins; electrophoresis; and ion-exchange column chromatography. The most recent advance in the separation of the purine and pyrimidine components involves the use of high-performance liquid chromatography (HPLC); nanomole quantities of these components are easily separated and detected in a brief period of time. Figure 13.7 shows the separation by HPLC of a mixture containing AMP, ADP, ATP, GMP, GDP, GTP, UMP, UDP, UTP, CMP, CDP, and CTP in 25 min. The development of this instrument and the high-resolution columns (anion and cation exchange and reverse phase) has allowed rapid determination of nucleoside and nucleotide pools under a variety of cellular conditions.

TABLE 13.2 Spectrophotometric Constants for Purine and Pyrimidine Nucleosides

Nucleoside	Molar extinction coefficient × 10^{-3}	λ_{max} pH 7
Adenosine	15.4	259
Guanosine	13.7	253
Cytidine	8.9	271
Uridine	10.0	262
Thymidine	10.0	262

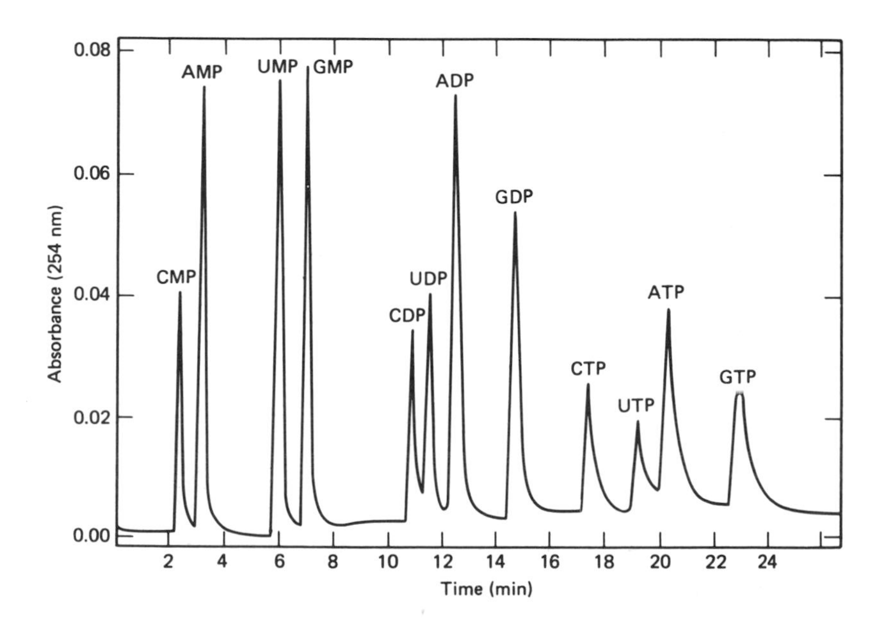

FIGURE 13.7
Separation of nucleotides by HPLC.

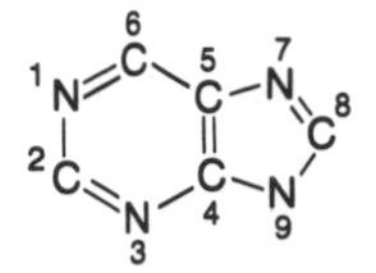

N-3 and N-9 from amide N of glutamine
C-4, C-5, and N-7 from glycine
C-2 and C-8 from formate via H_4folate
C-6 from CO_2
N-1 from aspartate

FIGURE 13.8
Sources of carbon and nitrogen atoms in the purine ring.

13.4 METABOLISM OF PURINE NUCLEOTIDES

The purine ring is synthesized de novo in mammalian cells utilizing amino acids as carbon and nitrogen donors, and formate and CO_2 as carbon donors.

The numbering system for the purine ring is shown in Figure 13.8 with the sources for the various carbon and nitrogen atoms indicated. From the sources of carbon and nitrogen that make up the purine ring it is very evident that amino acids play an important role in nucleotide metabolism. It has been shown that glutamine and aspartate levels can influence the rate of purine nucleotide synthesis in tumor cells. Many of the reactions that are required for the de novo synthesis utilize the hydrolysis of ATP. The overall set of reactions leading to the synthesis of a purine nucleotide is, therefore, expensive in terms of ATP required.

Purine Nucleotides Are Synthesized by a Stepwise Building of the Ring to Form IMP

All of the enzymes involved with purine nucleotide synthesis and degradation are found in the cytosol of the cell, but not all cells are capable of de novo purine nucleotide synthesis.

A series of 10 reactions leads to the de novo synthesis of **inosine 5′-monophosphate** (IMP) which serves as the precursor for adenosine 5′-monophosphate (AMP) and guanosine 5′-monophosphate (GMP) and consequently is not found to any extent in the cell under normal conditions. The reactions leading to the de novo synthesis of purine nucleotides are as follows:

1. *Formation of N-Glycosidic Bond* (N-9 of purine ring and C-1 of ribose)

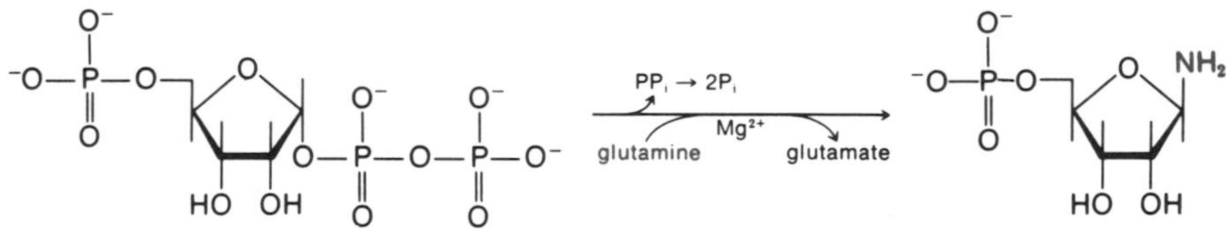

$PP_i \rightarrow 2P_i$; Mg^{2+}; glutamine → glutamate

5-Phosphoribosyl-1-pyrophosphate (PRPP) → 5-Phosphoribosylamine

This reaction is catalyzed by the enzyme **PRPP amidotransferase,** which is the **committed step** in this pathway and, as we will see later, the major regulated step.

In the formation of 5-phosphoribosylamine from PRPP there is inversion at C-1, giving rise to the β configuration of the *N*-glycosidic bond in purine nucleotides.

2. *Addition of Glycine* (C-4, C-5, and N-7 of purine ring)

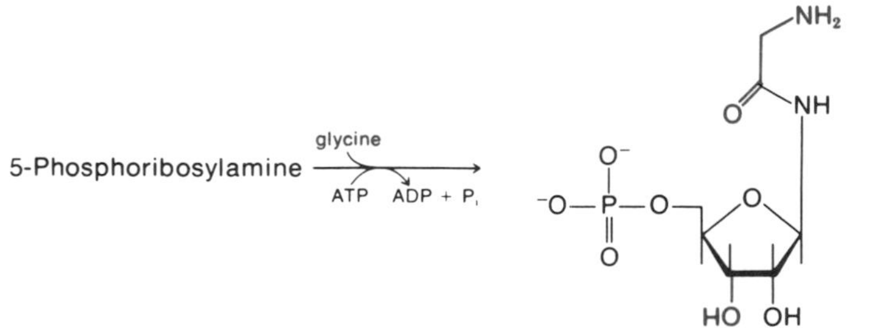

5-Phosphoribosylamine → (glycine; ATP → ADP + P_i) 5′-phosphoribosylglycinamide

This reaction is catalyzed by **phosphoribosylglycinamide synthetase** and requires ATP as the high-energy source to drive the reaction.

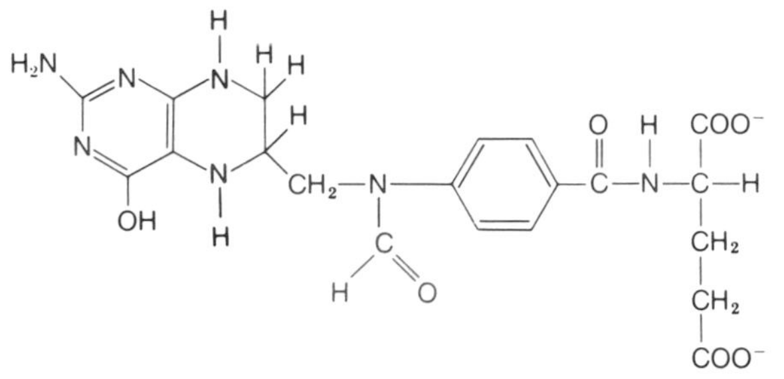

FIGURE 13.9
N^{10}-Formyl H_4folate.

3. *Introduction of Formyl Group via H_4-folate* (Figure 13.9) (C-8 of purine ring).

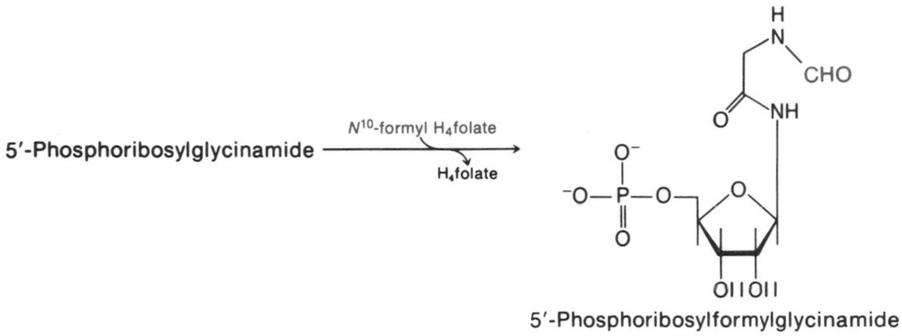

5′-Phosphoribosylglycinamide → (N^{10}-formyl H_4folate → H_4folate) → 5′-Phosphoribosylformylglycinamide

This reaction is catalyzed by **phosphoribosylglycinamide formyltransferase,** which requires N^{10}-formyl H_4folate as the C-1 donor.

4. *Addition of Nitrogen from Glutamine* (N-3 of purine ring)

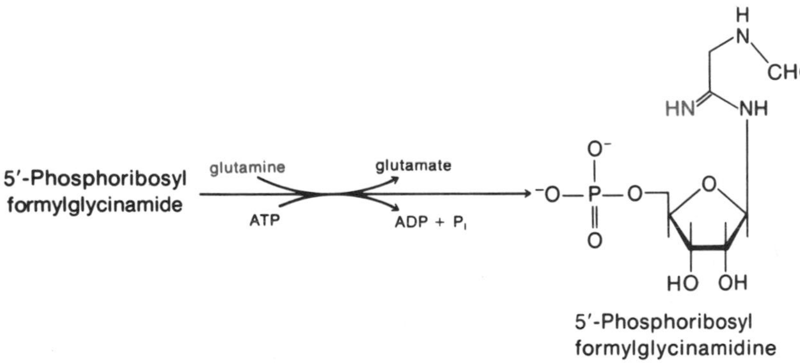

5′-Phosphoribosyl formylglycinamide → (glutamine, ATP → glutamate, ADP + P_i) → 5′-Phosphoribosyl formylglycinamidine

This reaction is catalyzed by **phosphoribosylformylglycinamide synthetase.** ATP hydrolysis provides the energy for this reaction.

5. *Ring Closure to Form Imidazole Ring*
The enzyme for this step is **phosphoribosylaminoimidazole synthetase,** which again requires ATP hydrolysis to close the ring.

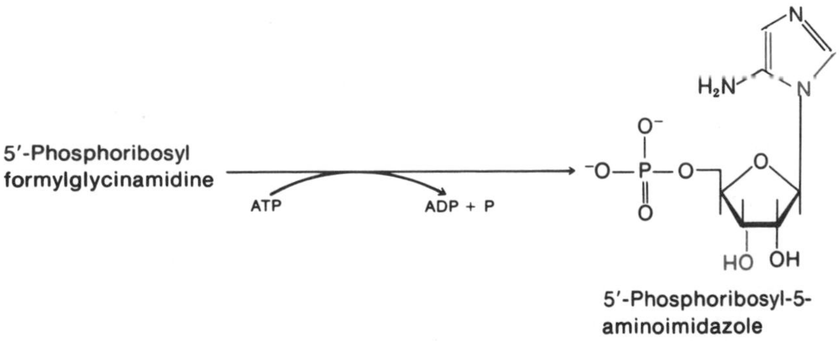

5′-Phosphoribosyl formylglycinamidine → (ATP → ADP + P) → 5′-Phosphoribosyl-5-aminoimidazole

6. *Addition of CO_2* (C-6 of purine ring)

5′-Phosphoribosyl-5-aminoimidazole —(CO_2)→ 5′-Phosphoribosyl 5-aminoimidazole-4-carboxylic acid

The enzyme that catalyzes this reaction is **phosphoribosylaminoimidazole carboxylase.** This reaction is not inhibited by avidin, indicating that this enzyme is not a biotin-requiring carboxylase.

7. *Addition of NH_2 Group from Aspartate* (N-1 of purine ring)

5′-Phosphoribosyl 5-aminoimidazole-4-carboxylic acid —(aspartate; ATP → ADP + P_i)→ 5′-Phosphoribosyl 5-aminoimidazole-4-N-succinocarboxamide

The enzyme catalyzing this reaction is **phosphoribosylaminoimidazolesuccinocarboxamide synthetase.** ATP is required for the formation of the amide bond between the amino group of aspartate and the carboxylate group of the imidazole derivative.

8. *Cleavage of N—C Bond of Aspartate*
Adenylosuccinase is the enzyme that cleaves the N—C bond, in effect, to transfer the amino group to the imidazole derivative. Adenylosuccinase is the same enzyme that is used in the conversion of adenylosuccinate to AMP and fumarate later in the pathway.

5′-Phosphoribosyl-5-aminoimidazole-4-N-succinocarboxamide —(→ fumarate)→ 5′-Phosphoribosyl-5-aminoimidazole-4-carboxamide

9. *Introduction of Formate via H_4folate* (Figure 13.9) (C-2 of purine ring)

5′-Phosphoribosyl-5-aminoimidazole-4-N-succinocarboxamide → (N^{10}-formyl-H_4folate → H_4folate) → 5′-Phosphoribosyl-5-formamidoimidazole-4-carboxamide

The enzyme catalyzing this reaction is phosphoribosylaminoimidazole carboxamide formyltransferase. N^{10}-formyl H_4folate is the cosubstrate in this reaction and is the C_1 donor.

10. *Ring Closure to Form IMP*

5′-Phosphoribosyl 5-formamidoimidazole-4-carboxamide → (H_2O) → Inosine 5′-monophosphate (IMP)

In the final reaction of the de novo synthesis of the purine ribonucleotide, IMP is formed from phosphoribosylformamidoimidazole carboxamide by **inosinicase.**

It has been found that the enzyme activities that catalyze Steps 3–5 (phosphoribosylglycinamide formyltransferase, phosphoribosylformylglycinamide synthetase, and phosphoribosylaminoimidazole synthetase) of the de novo pathway are encoded by the same genetic locus and reside on a **trifunctional protein.**

IMP Is the Common Precursor of AMP and GMP

As shown in Figure 13.10, IMP serves as the common precursor for GMP and AMP synthesis. From these reactions it is clear that the conversion of IMP to AMP and GMP does not occur randomly. The conversion of IMP to GMP requires ATP as the energy source, while the conversion of IMP to AMP requires GTP as the energy source. Therefore, when there is sufficient ATP in the cell, IMP will be converted to GMP, and conversely, when there is sufficient GTP in the cell IMP will be converted to AMP.

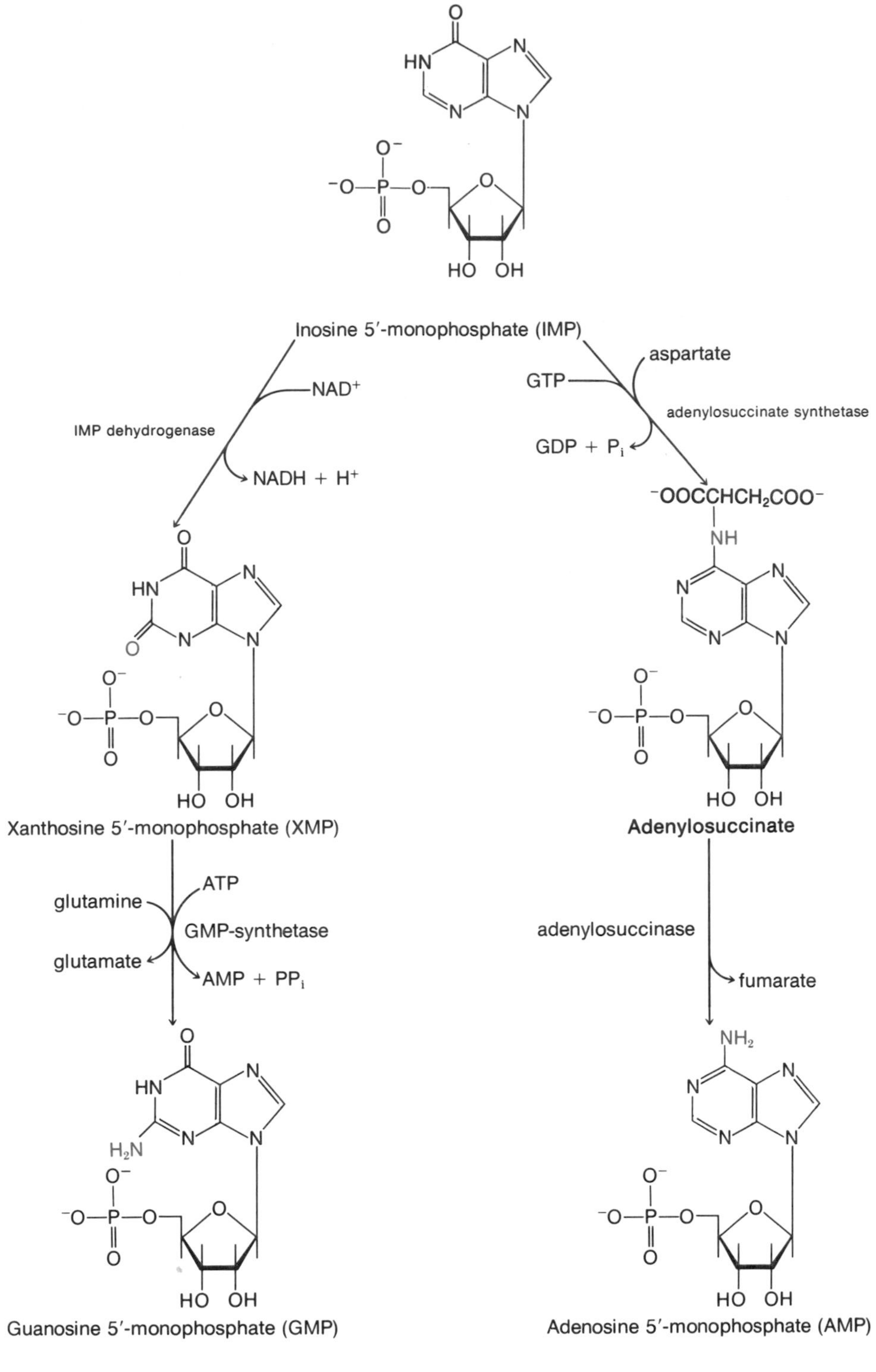

Inosine 5′-monophosphate (IMP)

IMP dehydrogenase

NAD^+

$NADH + H^+$

GTP

aspartate

adenylosuccinate synthetase

$GDP + P_i$

$^-OOCCHCH_2COO^-$

Xanthosine 5′-monophosphate (XMP)

Adenylosuccinate

glutamine

ATP

GMP-synthetase

glutamate

$AMP + PP_i$

adenylosuccinase

fumarate

Guanosine 5′-monophosphate (GMP)

Adenosine 5′-monophosphate (AMP)

FIGURE 13.10
Formation of AMP and GMP from IMP branch point.

Purine Nucleotide Synthesis Is Regulated at the Commitment Step and Branch Points

The committed step of a metabolic pathway is frequently the site of regulation. Such is the case in the de novo synthesis of purine ribonucleotides where the formation of 5-phosphoribosylamine from glutamine and 5-

phosphoribosyl-1-pyrophosphate, catalyzed by PRPP amidotransferase, is the committed step for purine nucleotide synthesis. This reaction is strongly regulated by IMP, GMP, and AMP. It has been found that PRPP amidotransferase from human placenta exists in two forms with molecular weights of 133,000 and 270,000. The enzyme activity is correlated with the smaller form. In the presence of 5′-nucleotides the small active form is converted to the large inactive form, whereas PRPP causes the large form to shift to the active form. This can be viewed as presented on the right.

The placental enzyme appears to have at least two effector binding sites. One site specifically binds the oxypurine nucleotides (IMP and GMP), while the other site binds the aminopurine nucleotide (AMP). The simultaneous binding of an oxypurine nucleotide and an aminopurine nucleotide results in a *synergistic inhibition* of the enzyme.

The enzyme PRPP amidotransferase displays hyperbolic kinetics (Figure 13.11) with respect to glutamine. With respect to the second substrate, PRPP, the enzyme shows sigmoidal kinetics. The presence of nucleotides further exaggerates the sigmoidal nature of the v (velocity) versus [PRPP] plot. The intracellular concentrations of glutamine are approximately equal to the K_m of the enzyme for glutamine. On the other hand, the PRPP levels in the cell, which do fluctuate, can be 10–100 times less than the K_m of the enzyme for PRPP. Consequently, the PRPP concentrations in the cell play an important role in controlling de novo purine nucleotide synthesis.

Between the formation of 5-phosphoribosylamine and IMP, there are no known regulated steps. However, there is regulation at the branch point of IMP to GMP and IMP to AMP. The two enzymes that utilize IMP at this branch point, **IMP dehydrogenase** and **adenylosuccinate synthetase,** have similar K_m values for IMP. Adenosine monophosphate is a competitive inhibitor (with respect to IMP) of adenylosuccinate synthetase, while GMP is a competitive inhibitor of IMP dehydrogenase. Two levels of control are therefore in effect at the IMP branch point. GTP serves as an energy source for the adenylosuccinate synthetase reaction, while AMP is a competitive inhibitor of this step; and ATP serves as the energy source in the conversion of XMP to GMP, while GMP acts as an inhibitor of XMP formation.

The regulation of purine nucleotide synthesis is summarized in Figure 13.12.

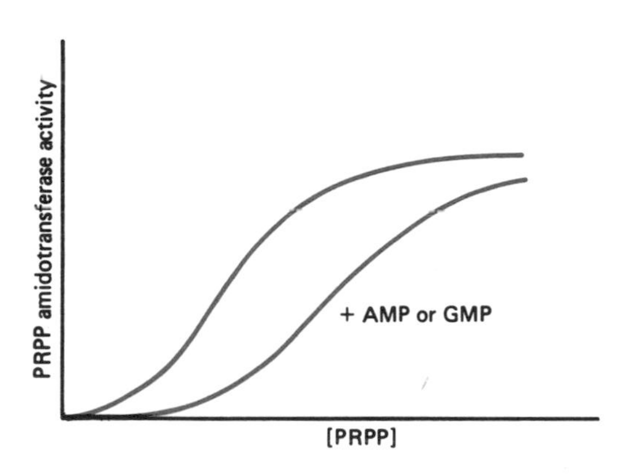

FIGURE 13.11
PRPP-amidotransferase activity as a function of glutamine or PRPP concentrations.

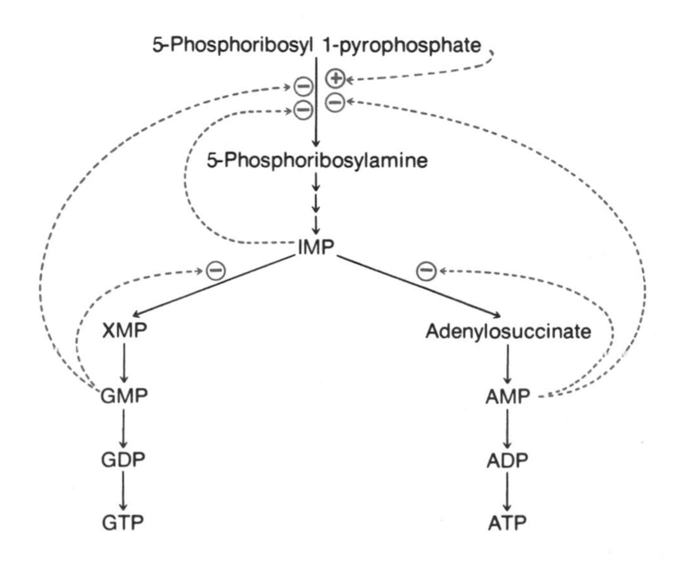

FIGURE 13.12
Regulation of purine nucleotide synthesis.
The dashed arrows indicate the regulated steps (+, activation; −, inhibition).

Purine Bases Can Be Salvaged to Reform Nucleotides

The efficiency of cellular metabolism under normal conditions is expressed by the presence of the so-called *"salvage pathways"* for the purine bases. In the metabolic pathway for the de novo synthesis of the purine nucleotides, a great deal of energy in the form of ATP is required to synthesize the purine nucleotides. In the "salvage" pathways the preformed bases (from exogenous sources or from the turnover of nucleic acids) can be reutilized, resulting in a considerable energy saving for the cell. There are two distinct enzymes involved. The reactions are as follows:

$$\text{Guanine} + \text{PRPP} \longrightarrow \text{GMP} + \text{PP}_i$$

$$\text{Hypoxanthine} + \text{PRPP} \longrightarrow \text{IMP} + \text{PP}_i$$

The enzyme that catalyzes both these reactions is **hypoxanthine-guanine phosphoribosyltransferase (HGPRTase)** and requires Mg^{2+}. The HGPRTase is regulated by the presence of IMP or GMP. IMP and GMP are competitive inhibitors with respect to PRPP in the HGPRTase reaction. The competitive nature of the inhibition implies that high concentrations of PRPP can overcome the regulation of this metabolic step.

$$\text{Adenine} + \text{PRPP} \longrightarrow \text{AMP} + \text{PP}_i$$

The enzyme that catalyzes this reaction is **adenine phosphoribosyltransferase (APRTase)** and also requires Mg^{2+}. The product of the reaction catalyzed by APRTase is AMP, which is an inhibitor of this reaction.

The source of adenine utilized in the APRTase reaction appears to be mainly from the synthesis of polyamines. For each molecule of spermine synthesized, two molecules of 5′-methylthioadenosine are generated. 5′-Methylthioadenosine is then degraded to 5-methylthioribose-1-phosphate and adenine via the 5′-methylthioadenosine phosphorylase-catalyzed reaction. The adenine base is salvaged through the APRTase reaction, whereas the carbon skeleton of 5-methylthioribose-1-phosphate is utilized in a reaction sequence that regenerates methionine.

The generation of AMP and GMP through these phosphoribosyltransferase reactions effectively shuts off the de novo pathway at the PRPP amidotransferase step. First, PRPP is consumed, decreasing the rate of formation of 5-phosphoribosylamine and second, AMP and GMP serve as feedback inhibitors at this step.

These reactions are important not only because they conserve energy, but also because they permit cells such as erythrocytes to form nucleotides from the bases. The erythrocyte, for example, does not have PRPP amidotransferase and hence cannot synthesize 5-phosphoribosylamine, the first unique metabolite in the pathway of purine nucleotide synthesis. As a consequence, the red cell must depend on the purine phosphoribosyltransferases to replenish the nucleotide pools.

The importance of these reactions is further demonstrated in the clinical situation in which the HGPRTase activity is markedly depressed. Such a deficiency results in the Lesch–Nyhan syndrome (Clin. Corr. 13.1), which is characterized clinically by hyperuricemia, mental retardation, and self-mutilation.

Interconversion Maintains an Appropriate Balance of Purine Nucleotides

Along with the very fine control exhibited in the cell for the de novo synthesis of purine nucleotides, there are enzymes that can be used to

CLIN. CORR. 13.1
LESCH–NYHAN SYNDROME

The Lesch–Nyhan syndrome is characterized clinically by hyperuricemia, excessive uric acid production, and neurological problems, which may include spasticity, mental retardation, and self-mutilation. This disorder is associated with a very severe or complete deficiency of HGPRTase activity. The gene for HGPRTase is on the X chromosome, hence the deficiency is virtually limited to males. In a study of the available patients, it was observed that if the HGPRTase activity was less than 2% of normal, mental retardation was present, and if the activity was less than 0.2% of normal, the self-mutilation aspect was expressed. This defect leads not only to the overproduction and excretion of uric acid, but also to increased excretion of hypoxanthine and xanthine.

There appear to be several variant forms of HGPRTase-deficient patients. In one form there appeared to be a complete lack of HGPRTase activity, and yet by titration with antibodies prepared against the purified enzyme from normal erythrocytes, the Lesch–Nyhan patient had an equivalent amount of immunoprecipitable protein. In another form, the enzyme appeared to be unstable. "Young" red cells separated from the "old" red cells by density centrifugation had much higher (although still markedly reduced compared to normal) HGPRTase activity than did the "old" red cells. In still another case, a child was studied who had all the clinical manifestations of the Lesch–Nyhan syndrome, but had normal levels of HGPRTase activity when determined in the laboratory. Further kinetic analysis showed that this variant was a "K_m mutant." When assayed under saturating levels of PRPP, the level of activity was comparable to normal. When assayed at the concentration of PRPP, which would approximate the intracellular concentration of PRPP, the activity was diminished to the range found in Lesch–Nyhan patients.

As discussed previously, the role of HGPRTase is to catalyze reactions in which hypoxanthine and guanine are converted to nucleotides. The hyperuricemia and excessive uric acid production that occur in patients with the Lesch–Nyhan syndrome are easily explained by the lack of HGPRTase activity. The hypoxanthine and guanine are not salvaged leading to increased intracellular pools of PRPP and decreased levels of IMP or GMP. Both of these factors promote the de novo synthesis of purine nucleotides without regard for the proper regulation of this pathway.

It is not understood why a severe defect in this salvage pathway leads to the neurological problems. The adenine phosphoribosyltransferase activity in these patients is normal or in fact elevated. With this salvage enzyme, presumably the cellular needs for purine nucleotides could be met via the pathway,

$$\text{Adenine} \xrightarrow{\text{PRPP}} \text{AMP} \longrightarrow \text{IMP} \longrightarrow \text{GMP}$$

if the cell's de novo pathway were not functioning. The normal tissue distribution of HGPRTase activity perhaps could explain the neurological symptoms. The brain (frontal lobe, basal ganglia, and cerebellum) has 10–20 times the enzyme activity found in liver, spleen, or kidney and from 4 to 8 times that found in the erythrocytes. Individuals who have primary gout with excessive uric acid formation and hyperuricemia do not display the neurological problems. It is argued that on this basis the products of purine degradation (hypoxanthine, xanthine, and uric acid) cannot be toxic to the central nervous system (CNS). However, it is possible that these metabolites are toxic to the developing CNS or that the lack of the enzyme leads to an imbalance in the concentrations of the purine nucleotides at critical times during development.

If IMP dehydrogenase activity in the brain were extremely low, the lack of HGPRTase could lead to decreased levels of intracellular GTP due to decreased salvage of guanine. Since GTP serves as a precursor of tetrabiopterin, a required cofactor in the biosynthesis of neurotransmitters, and is required in other functions such as signal transduction via G-proteins and protein synthesis, low levels of GTP during development could be the triggering factor in the observed neurological manifestations.

Treatment of Lesch–Nyhan patients with allopurinol will decrease the amount of uric acid formed, relieving some of the problems caused by sodium urate deposits. However, since the Lesch–Nyhan patient has a marked reduction in HGPRTase activity, hypoxanthine and guanine are not salvaged, PRPP is not consumed, and consequently the de novo synthesis of purine nucleotides is not shut down. There is no known treatment for the neurological problems. These patients usually die from kidney failure, resulting from the high sodium urate deposits.

Lesch, M. and Nyhan, W. L. A familial disorder of uric acid metabolism and central nervous system function. *Am. J. Med.* 36:561, 1964; Wilson, J. M., and Kelley, W. N. Molecular basis of hypoxanthine guanine phosphoribosyl transferase deficiency in a patient with the Lesch–Nyhan syndrome. *J. Clin. Invest.* 71:1331, 1983.

balance the levels of guanine and adenine nucleotides. As discussed earlier, IMP can be converted by one pathway to GMP and by a different pathway to AMP. There is no known direct pathway for the conversion of GMP to AMP or AMP to GMP. However, these purine nucleotides can be redistributed to meet the cellular needs through the conversion of GMP and AMP back to IMP. These reactions are carried out by separate enzymes, each under separate controls. The pathways are summarized in Figure 13.13. The reductive deamination of GMP to IMP by **GMP reductase** is activated by GTP and inhibited by XMP. The XMP is a competitive

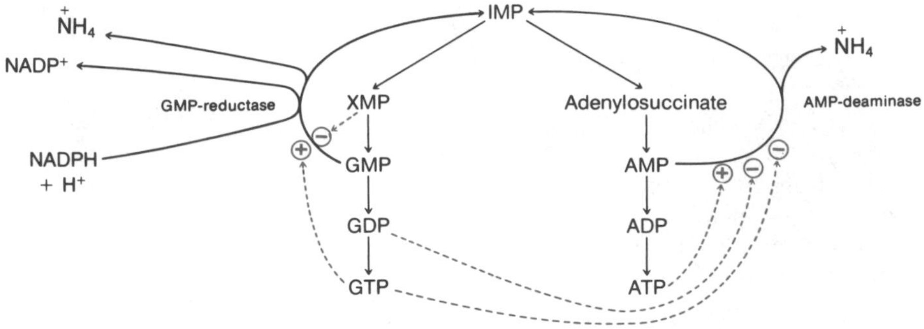

FIGURE 13.13
Interconversion of purine nucleotides.
The solid arrows represent enzyme-catalyzed reactions. The dashed lines represent steps of regulation (+, activation; −, inhibition).

inhibitor of human GMP reductase, having a $K_i = \sim 0.2\ \mu M$. Because of this low K_i, the concentration of XMP in the cell could influence the conversion of GMP to IMP. On the other hand, GTP is a nonessential activator of GMP reductase and serves to lower the K_m of the enzyme with respect to GMP and to increase the V_{max}.

The activity of **AMP deaminase** (5′-AMP aminohydrolase), which specifically catalyzes the deamination of AMP to yield IMP, is activated by K^+, and ATP and is inhibited by P_i, GDP, and GTP. In the absence of K^+ ions the v versus [AMP] curve is sigmoidal. The presence of K^+ ions is not required for maximum activity; rather K^+ acts as a positive allosteric effector to reduce the apparent K_m for AMP.

Purine Degradation in Humans Produces Uric Acid

The purine nucleotides, nucleosides, and bases funnel through a common pathway for the degradation of these biomolecules. The end product of purine degradation in humans is uric acid. The catabolic pathways are as shown in Figure 13.14. The enzymes involved in the degradation of the nucleic acids and the nucleotides and nucleosides vary in specificity. The nucleases show specificity toward either RNA or DNA and also toward the bases and position of cleavage of the 3′,5′-phosphodiester bonds. The nucleotidases range from those with relatively high specificity, such as 5′-AMP nucleotidase, to those with broad specificity, such as the acid and alkaline phosphatases, which will hydrolyze any of the 3′- or 5′-nucleotides. AMP deaminase is specific for AMP. **Adenosine deaminase** is much less specific, since not only adenosine, but also 2′-deoxyadenosine and many other 6-aminopurine nucleosides are deaminated by this enzyme.

A special comment should be made about **purine nucleoside phosphorylase.** As indicated, the reaction is readily reversible.

$$\text{Inosine} + P_i \rightleftharpoons \text{hypoxanthine} + \text{ribose 1-P}$$

or

$$\text{Guanosine} + P_i \rightleftharpoons \text{guanine} + \text{ribose 1-P}$$

or

$$\text{Xanthosine} + P_i \rightleftharpoons \text{xanthine} + \text{ribose 1-P}$$

CLIN. CORR. 13.2 IMMUNODEFICIENCY DISEASES ASSOCIATED WITH DEFECTS IN PURINE NUCLEOTIDE METABOLISM

Recently, two different immunodeficiency diseases have been recognized that are associated with deficiencies in the enzymes adenosine deaminase and purine nucleoside phosphorylase. Referring to Figure 13.14, it is seen that these two enzymes are involved in the degradation of purine nucleosides.

The deficiency in adenosine deaminase is associated with a severe combined immunodeficiency involving T-cell and usually B-cell dysfunction. Adenosine deaminase deficiency is not associated with the overproduction of purine nucleotides. The mechanism by which the lack of adenosine deaminase interferes with immune function is not completely understood. It has been shown, how-

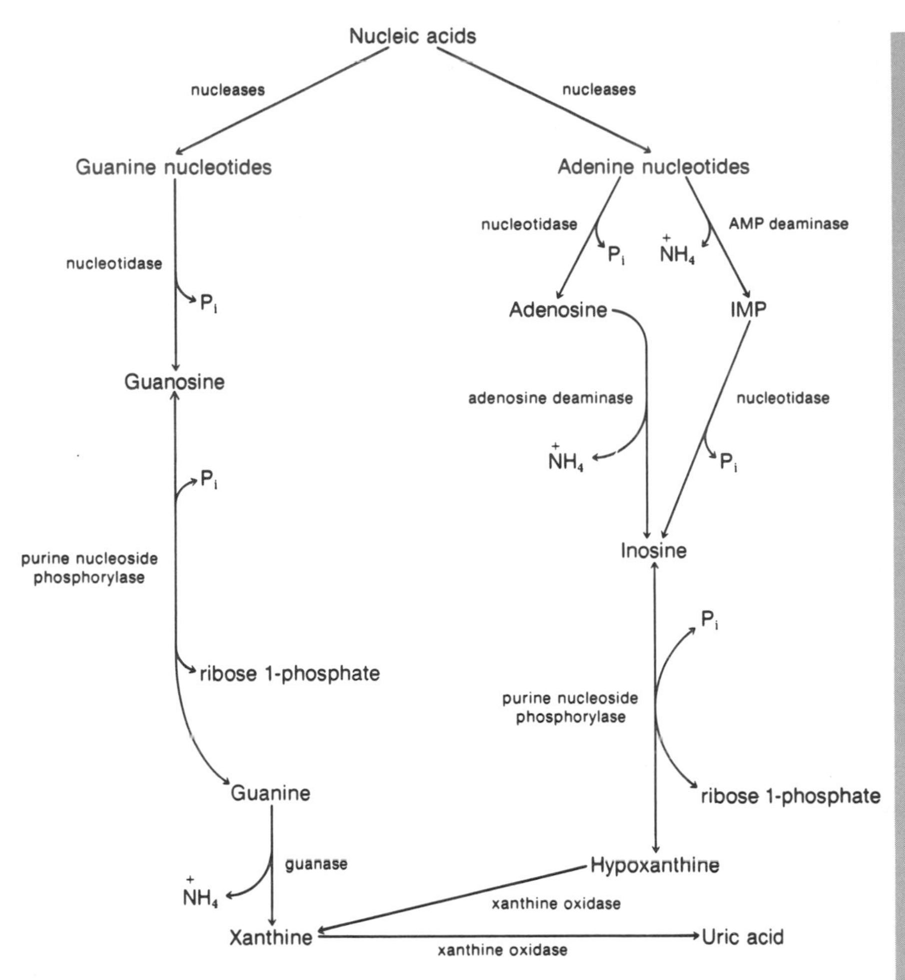

FIGURE 13.14
Degradation of purine nucleotides.

Deoxyinosine and deoxyguanosine are also excellent substrates for purine nucleoside phosphorylase. While the equilibrium for the reactions catalyzed by purine nucleoside phosphorylase favors nucleoside synthesis, it would appear that in the cell the concentrations of the free purine and ribose 1-phosphate are too low to support nucleoside synthesis under most conditions. The main function of purine nucleoside phosphorylase in the cells seems to be its role in purine nucleoside degradation. This conclusion is supported by the conditions observed in the cases where a deficiency of purine nucleoside phosphorylase has been detected. Under these conditions there is a large buildup of the substrates (inosine, guanosine, deoxyinosine, and deoxyguanosine) for purine nucleoside phosphorylase with a corresponding decrease in uric acid formation.

These enzymes may therefore be regarded as part of the degradative pathway. On the other hand, since the reaction is readily reversible, it may serve as part of the salvage pathway under certain metabolic conditions.

Deficiencies in two of the enzymes of this degradative pathway (adenosine deaminase and purine nucleoside phosphorylase) have been correlated with two disease states in humans. Adenosine deaminase deficiency has been associated with a severe combined immunodeficiency, and purine nucleoside phosphorylase deficiency is associated with a defective T-cell immunity and a normal B-cell immunity (Clin. Corr. 13.2).

ever, that in a patient with adenosine deaminase deficiency there is an extremely large buildup of deoxyadenosine triphosphate in the erythrocytes examined. In fact, the dATP concentration exceeded the ATP concentration in these cells. It is therefore thought that the failure of the cells to metabolize deoxyadenosine to deoxyinosine for further conversion to hypoxanthine leads to the increased levels of dATP. The dATP is known to be a very effective inhibitor of ribonucleotide reductase and consequently of DNA synthesis (cell replication). It is thought that this is the explanation for the deficiencies in the immune system since activation of an immune response requires proliferation of lymphocytes. Other suggestions have included the proposals that the elevated adenosine is toxic to the cells by virtue of its ability to increase the intracellular concentrations of cAMP or due to the inhibition of *S*-adenosyl homocysteine hydrolase, leading to increased intracellular levels of *S*-adenosyl-L-homocysteine.

The deficiency in purine nucleoside phosphorylase is associated with an impairment of T-cell function with no apparent effects on B-cell function. There is no overproduction of purine nucleotides associated with this deficiency. However, there is a marked *decrease* in uric acid formation with the corresponding increased levels of the purine nucleoside phosphorylase substrates, guanosine, deoxyguanosine, inosine, and deoxyinosine. When these various nucleosides were incubated with normal T lymphocytes in culture, deoxyguanosine was found to be the most toxic. In addition, it was found that dGTP was the major nucleotide that accumulated in the red cells from patients with purine nucleoside phosphorylase deficiency. It is suggested that dGTP, which acts as an inhibitor of CDP reductase, is the actual agent that is toxic to the development of normal T cells.

In both of these enzyme deficiencies, it is not entirely clear how these defects lead to the immune problems. However, it is clear that defects in enzymes that have been casually considered in metabolic pathways in the past, reveal their importance to normal metabolism when they are absent or severely decreased.

Akeson, A. L., et al. Mutations in the human adenosine deaminase gene that affect protein structure and RNA splicing. *Proc. Natl. Acad. Sci.* U.S.A. 84:5947, 1987; Gudas, L. J., Fannis, V. I., Clift, S. M., Ammann, A. J., Stael, G. E. J., and Martin, D. W. Characterization of mutant subunits of human purine nucleoside phosphorylase. *J. Biol. Chem.* 253:8916, 1978.

Formation of Uric Acid

Xanthine oxidase is an enzyme that contains FAD, Fe(III), and Mo(VI). In the reaction, molecular oxygen is a substrate with H_2O_2 being generated as a product. Uric acid is the end product of purine

Hypoxanthine —xanthine oxidase (O_2 → H_2O_2)→ Xanthine —xanthine oxidase (O_2 → H_2O_2)→ Uric acid

nucleotide catabolism and is excreted in the urine. There are clinical disorders in which the serum level of uric acid is markedly elevated (hyperuricemia) and which can lead to the deposit of sodium urate crystals. This condition is known as gout (Clin. Corr. 13.3). Allopurinol is an inhibitor of xanthine oxidase; this is discussed in detail as a clinical drug in Clin. Corr. 13.3.

CLIN. CORR. 13.3
GOUT

Primary gout is characterized by elevated uric acid levels in blood due to a variety of metabolic abnormalities that lead to the overproduction of purine nucleotides via the de novo pathway. With the overproduction of uric acid, the levels of uric acid in the serum and urine are elevated, and there are deposits of sodium urate crystals in the joints of extremities. The metabolic bases for the increased production of purine nucleotides which in turn are manifested by increased uric acid levels have been identified in several situations. Many, if not all, of the clinical symptoms associated with the overproduction of uric acid arise because uric acid is not very soluble. Formation of sodium urate crystals leads not only to the joint problems but also to renal disease. Hyperuricemia resulting from the overproduction of uric acid via the de novo pathway as opposed to hyperuricemia resulting from renal damage or increased cell death (e.g., radiation therapy) can be relatively easily distinguished. The feeding of ^{15}N-glycine to an "overproducer" will result in a marked ^{15}N enrichment of the N-7 of uric acid isolated from the urine or serum of these patients, whereas there would be little ^{15}N enrichment in the uric acid from individuals with renal problems or with increased nucleic acid degradation.

Various studies directed at determining the molecular basis in primary gout for the overproduction of purine nucleotides have uncovered a diverse group of metabolic defects. While the primary metabolic defects that have been determined may be seemingly unrelated to the de novo synthesis of purine nucleotides, a common feature of these defects evolves.

The defects described in human beings include the following:

1. *PRPP-Synthetase:* Mutant forms of PRPP-synthetase have been detected, which are not subject to allosteric regulation by P_i or to feedback inhibition by GDP and ADP. Under these conditions, the intracellular concentration of PRPP is elevated, leading to increased formation of 5-phosphoribosylamine.
2. *Partial HGPRTase Deficiency:* A characteristic of this deficiency is the overproduction of purine nucleotides. The basis for this appears to be twofold. First, the lack of HGPRTase activity decreases the amount of hypoxanthine or guanine that can be "salvaged." Consequently, the level of PRPP is increased because PRPP is not consumed via the salvage enzyme. The increased PRPP levels lead to increased PRPP amidotransferase activity. Second, the lack of salvage of hypoxanthine or guanine leads to decreased levels of IMP and GMP, which in turn act as feedback regulators of the PRPP amidotransferase step.

Recently, the basis of a HGPRTase deficiency has been defined at the molecular level as being due to a single base alteration (adenine to guanine transition); this base change resulted in the gene product (HGPRTase) having glycine substituted for aspartic acid at position 201. This altered HGPRTase has elevated K_m values for hypoxanthine and PRPP, which were much in excess of the physiological concentrations of these substrates. Consequently, the intracellular mutant HGPRTase was active but essentially nonfunctional due to the lack of substrates at the appropriate concentrations.

In both of these conditions, the common feature of the defect that leads to the overproduction of purine nucleotides is that the intracellular concentration of PRPP is elevated. These defined defects fully support the conclusion that the PRPP amidotransferase step is the rate-controlling step in purine nucleotide synthesis.

Further support comes from clinical cases of secondary gout (a consequence of another metabolic defect). A deficiency in glucose 6-phosphatase (glycogen storage disease, type I; von Gierke's disease) leads to increased purine nucleotide synthesis de novo. The lack of conversion of glucose 6-phosphate to glucose leads to increased hexose monophosphate shunt activity. The increased utilization of glucose 6-phosphate via the shunt results in increased ribose 5-phosphate levels and consequently increased PRPP levels.

These latter two examples show quite clearly that a defect in one pathway can cause major problems in a metabolic pathway that is seemingly unrelated and points to the critical nature of the interrelationships among various pathways.

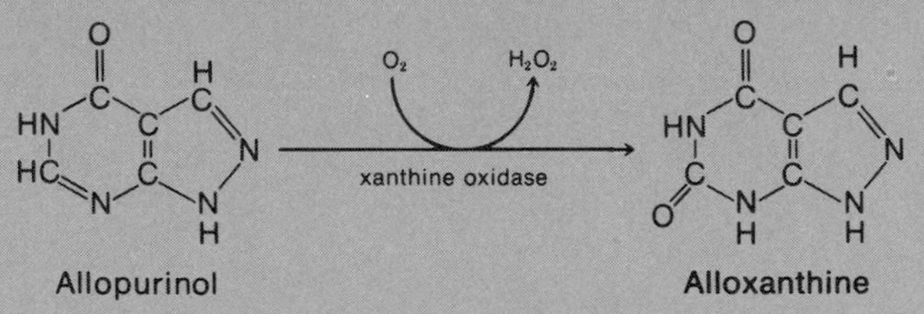

Allopurinol $\xrightarrow[\text{xanthine oxidase}]{O_2 \quad H_2O_2}$ Alloxanthine

Probably the major treatment for primary gout involves the use of the drug allopurinol. The overall effect of allopurinol treatment is to lower the uric acid levels in vivo. It is generally reported that allopurinol is an inhibitor of xanthine oxidase. However, allopurinol is oxidized by xanthine oxidase to alloxanthine and this product (alloxanthine) binds tightly to the reduced form of xanthine oxidase. The dissociation constant for the binding of alloxanthine to reduced xanthine oxidase is about 0.5 nM, which makes alloxanthine a very effective inhibitor of xanthine oxidase.

The inhibition of xanthine oxidase by alloxanthine decreases the formation of uric acid, while increasing the levels of hypoxanthine and xanthine excreted. This benefits the patient, since the amount of the purine degradative products will be distributed among three compounds instead of just uric acid. Hypoxanthine and xanthine are more soluble than uric acid, so that the total amount (hypoxanthine, xanthine, and uric acid) that will be soluble is increased by allopurinol treatment.

In "overproducers" who do not have a deficiency in HGPRTase, allopurinol treatment not only lowers the formation of uric acid with an increase in the excretion of hypoxanthine and xanthine, but also decreases the overall production of purine nucleotides via the de novo pathway. The metabolic basis for this appears to be due to the increased salvage of hypoxanthine and xanthine, which requires the consumption of PRPP and the subsequent formation of IMP and XMP, which can block de novo synthesis at the PRPP amidotransferase step.

Becker, M. A., Kostel, P. J., Meyer, L. J., and Seegmiller, J. E. Human phosphoribosyl pyrophosphate synthetase: Increased enzyme specific activity in a family with gout and excessive purine synthesis. *Proc. Natl. Acad. Sci. U.S.A.* 70:2749, 1973; Kelley, W. N., Rosenbloom, F. M., Henderson, J. F., and Seegmiller, J. E. A specific enzyme defect in gout associated with overproduction of uric acid. *Proc. Natl. Acad. Sci. U.S.A.* 57:1735, 1967.

13.5 METABOLISM OF PYRIMIDINE NUCLEOTIDES

Pyrimidines Are Synthesized From Glutamine, Aspartate, and CO_2

The pyrimidine ring is synthesized de novo in mammalian cells utilizing amino acids as carbon and nitrogen donors and CO_2 as a carbon donor.

The numbering system for the pyrimidine ring is shown in Figure 13.15 along with the sources for the various carbon and nitrogen atoms indicated.

As in the case of the de novo synthesis of purine nucleotides, amino acids also play an important role in the de novo synthesis of pyrimidine nucleotides. The reactions leading to the synthesis of pyrimidine nucleotides in mammalian cells are as follows:

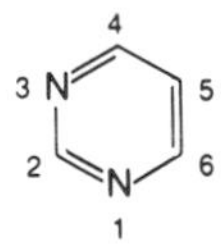

N-1, C-4, C-5, and C-6 from aspartate
C-2 from CO_2
N-3 from amide N of glutamine

FIGURE 13.15
Sources of carbon and nitrogen atoms in pyrimidines.

1. *Formation of Carbamoyl Phosphate* (C-2 and N-3 of pyrimidine ring)

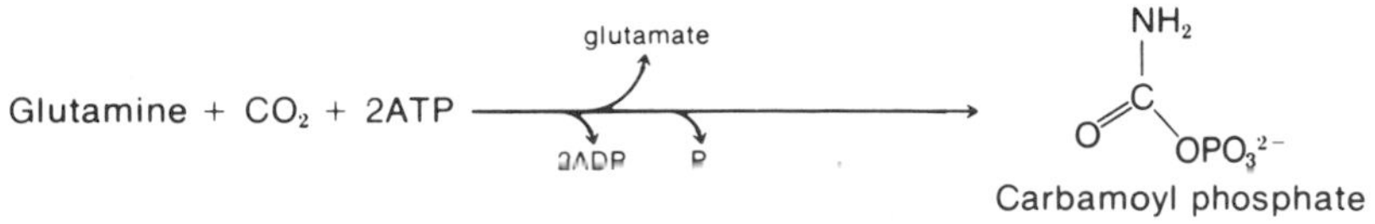

$$\text{Glutamine} + CO_2 + 2ATP \xrightarrow[\text{2ADP} \quad P]{\text{glutamate}} H_2N{-}C(=O){-}OPO_3^{2-}$$

Carbamoyl phosphate

The enzyme that catalyzes this reaction is cytosolic **carbamoyl phosphate synthetase II.** It is distinct from carbamoyl phosphate synthetase I, a mitochondrial enzyme involved in the urea cycle.

2. *Addition of Aspartate* (N-1, C-4, C-5, and C-6 of pyrimidine ring)

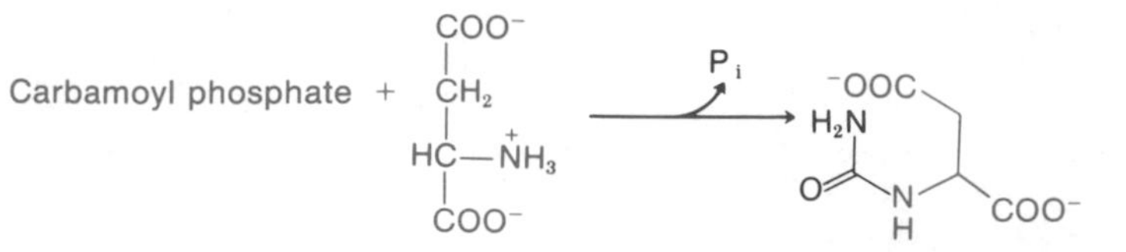

Carbamoyl phosphate + Aspartate → (P_i) N-Carbamoylaspartate

This reaction is catalyzed by the enzyme **aspartate carbamoyl transferase.** The mammalian enzyme is not allosterically regulated, although this could be considered to be the committed step for pyrimidine synthesis.

3. *Ring Closure to Form Pyrimidine Ring*
This reaction is catalyzed by the enzyme **dihydroorotase.**

N-Carbamoyl aspartate → Dihydroorotate

4. *Oxidation of Dihydroorotate*

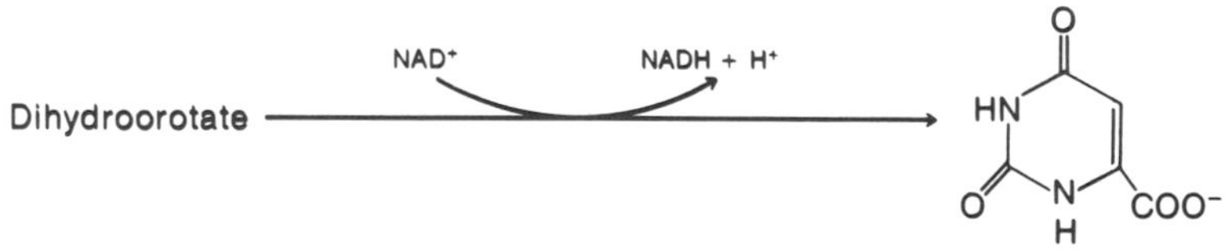

Dihydroorotate → (NAD^+ → $NADH + H^+$) Orotate

Dihydroorotate dehydrogenase, a flavoprotein, catalyzes the reaction in which orotic acid is formed.

5. *Addition of the Ribose 5-Phosphate* (formation of *N*-riboside bond)
The enzyme catalyzing the reaction in which the first pyrimidine nucleotide is formed is **orotate phosphoribosyltransferase.** PRPP is the ribose-5-phosphate donor.

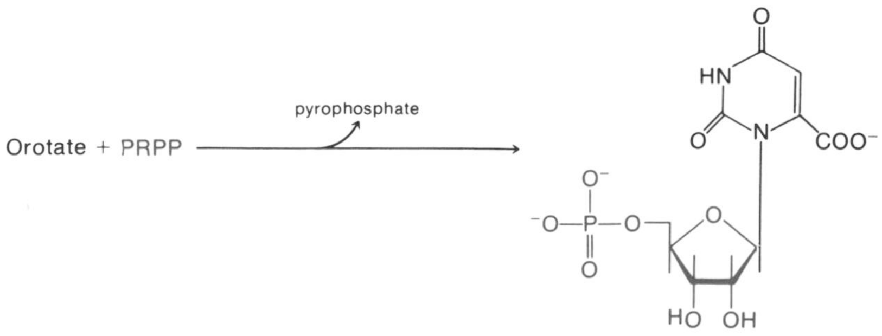

Orotate + PRPP → (pyrophosphate) Orotidine 5′-mono phosphate (OMP)

CLIN. CORR. 13.4
OROTIC ACIDURIA

In a clinical condition called hereditary orotic aciduria, characterized by retarded growth and severe anemia, high levels of orotic acid are excreted. The biochemical basis for this increased production of orotic acid is the absence of either or both of the enzymes orotate phosphoribosyltransferase and OMP-decarboxylase. While this is a relatively rare disease, the understanding of the metabolic basis for it has led not only to a successful treatment, but also has confirmed the site of regulation of pyrimidine nucleotide synthesis.

When these patients are fed either cytidine or uridine there is not only a marked improvement in the hematologic manifestation but also a decrease in orotic acid formation. The biochemical basis for this can be explained as follows. Uridine or cytidine, after conversion to the nucleotide by the cell, bypasses the block at orotate phosphoribosyltransferase/OMP-decarboxylase to provide rapidly growing cells, such as the erythropoietic cells, with the pyrimidine nucleotides required for RNA and DNA synthesis. In addition, the intracellular formation of UTP from these nucleosides acts as a feedback inhibitor of carbamoyl phosphate synthetase II to shut down the synthesis of orotic acid.

Haggard, M. E. and Lockhart, L. H. Megaloblastic anemia and orotic aciduria. A hereditary disorder of pyrimidine metabolism responsive to uridine. *Am. J. Dis. Child.* 113:733, 1967; Winkler, J. K. and Suttle, D. P. Analysis of UMP synthase gene and mRNA structure in hereditary orotic aciduria fibroblasts. *Am. J. Hum. Genet.* 43:86, 1988.

6. *Decarboxylation of OMP*

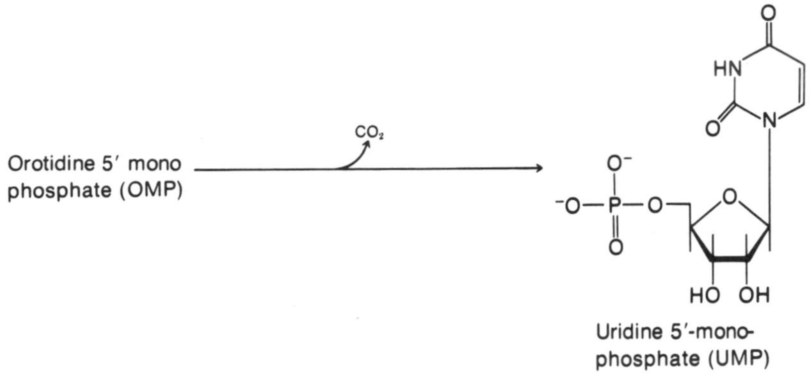

Orotidine 5′ mono phosphate (OMP) → (CO_2) → Uridine 5′-monophosphate (UMP)

OMP-decarboxylase catalyzes this reaction. The absence of both or either of these last two enzyme activities leads to a condition termed orotic aciduria (Clin. Corr. 13.4).

Although the de novo pathway for UMP synthesis requires six enzyme activities for the six steps, the activities are found on only three gene products. Carbamoyl phosphate synthetase, aspartate carbamoyl transferase, and dihydroorotase activities (abbreviated pyr 1-3 or CAD) are present on the same polypeptide chain (200,000 Da); dihydroorotate dehydrogenase activity is on a separate protein; and orotate phosphoribosyltransferase and OMP-decarboxylase activities (pyr 5,6) are on the same polypeptide (51,000 Da). The multifunctional enzymes (pyr 1-3 and pyr 5,6) are found in the cytosol, whereas dihydroorotate dehydrogenase is a mitochondrial enzyme. As a result of the channeling of intermediates through these enzyme systems, essentially none of the metabolites between the first step and the last step is found in the intracellular pool of the cells.

By these reactions the pyrimidine nucleotide UMP is synthesized. The formation of cytidine nucleotides proceeds from the uridine nucleotide but at the triphosphate level rather than at the monophosphate level. This reaction for the synthesis of this pyrimidine is as follows:

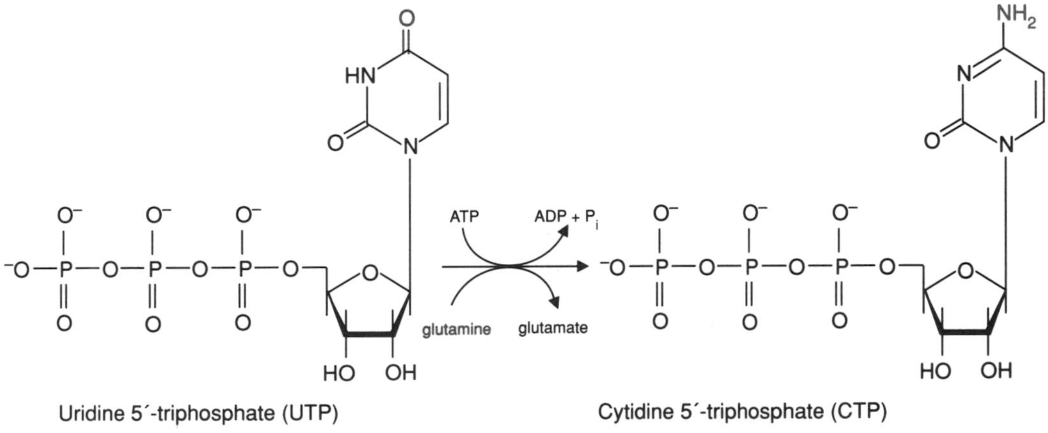

Uridine 5′-triphosphate (UTP) + ATP + glutamine → Cytidine 5′-triphosphate (CTP) + ADP + P_i + glutamate

CTP synthetase catalyzes this reaction. The enzyme does not have an absolute requirement for GTP, but concentrations as low as 0.2 mM GTP stimulate CTP synthetase activity 5- to 10-fold.

As indicated, carbamoyl phosphate is synthesized in the cytosol by a specific carbamoyl phosphate synthetase II. This is the only source of carbamoyl phosphate for pyrimidine synthesis in extrahepatic tissue. As is discussed in more detail in the section on regulation of pyrimidine nucleotide synthesis, carbamoyl phosphate synthetase II is the regulated enzyme of the pyrimidine nucleotide de novo pathway. It is quite clear, however, that in liver, carbamoyl phosphate synthetase I can provide cytosolic carbamoyl phosphate. Under stressed physiological conditions in which there is excessive ammonia, the liver, through carbamoyl phosphate synthetase I (a component of the urea cycle) can utilize this pathway to detoxify the NH_3 by forming carbamoyl phosphate in the mitochondria. This carbamoyl phosphate passes into the cytosol and becomes a substrate for aspartate carbamoyl transferase. High levels of orotic acid have been observed to be excreted as a result of ammonia toxicity in humans.

The pathway for de novo synthesis of pyrimidine nucleotides differs in two major respects (on a comparative basis) from the de novo pathway for purine nucleotides. First, in purine nucleotide synthesis the *N*-glycosidic bond is formed in the first committed step of the pathway. In the pyrimidine pathway, the pyrimidine ring is formed first, and then the sugar phosphate is added. Second, all the enzymes of the purine nucleotide pathway are in the cytosol. For pyrimidine synthesis one of the enzymes, dihydroorotate dehydrogenase, is found in the mitochondria, whereas the other five activities (present on only two proteins) are located in the cytosol.

Pyrimidine Nucleotide Synthesis Is Regulated at the Level of Carbamoyl Phosphate Synthetase II

Unlike in bacterial cells, the regulation of mammalian pyrimidine nucleotide synthesis does not occur at the aspartate carbamoyl transferase step. The regulation of the pyrimidine nucleotide synthesis in mammalian cells occurs at the level of carbamoyl phosphate synthetase II (the cytosolic enzyme), which is inhibited by UTP.

The next level of regulation of pyrimidine nucleotide synthesis is at the level of OMP-decarboxylase. UMP and, to a lesser extent, CMP are inhibitors of OMP-decarboxylase but not orotate phosphoribosyltransferase. Since these two enzymes activities are located on the same polypeptide, however, orotate phosphoribosyltransferase does not function. Under the conditions in which OMP-decarboxylase is inhibited, orotate, *not* OMP accumulates.

In a clinical condition called orotic aciduria excessive amounts of orotic acid are produced. This is caused by decreases in the activity of orotate phosphoribosyltransferase or OMP decarboxylase (Clin. Corr. 13.4).

Another possible regulatory site is the CTP synthetase step. This enzyme shows a hyperbolic curve for a plot of velocity (v) versus [UTP]. However, in the presence of CTP, the plot of v versus [UTP] curve becomes sigmoidal. In this way the activity of CTP synthetase is depressed, preventing all of the UTP from being converted to CTP.

Pyrimidines Can Be Salvaged to Re-Form Nucleotides

Pyrimidines can be "salvaged" by conversion to the nucleotide level by reactions involving **pyrimidine phosphoribosyltransferase.**

The general reaction is

$$\text{Pyrimidine} + \text{PRPP} \longrightarrow \text{pyrimidine nucleoside monophosphate} + \text{pyrophosphate}$$

The enzyme from human erythrocytes has been purified and can utilize orotate, uracil, and thymine as substrates. Cytosine is not a substrate.

It should be mentioned that as uracil becomes available to the cell, competing reactions can occur. Uracil can be degraded to β-alanine or can be salvaged. Normal liver, when presented with uracil, will readily degrade it, whereas regenerating liver would convert the uracil to UMP. This is the result of the availability of PRPP, the enzyme levels, and in general the metabolic state of the animal.

The **pyrimidine nucleoside kinases** can also be thought of as "salvage" enzymes. The net effect of the kinase reaction is to divert the pyrimidine nucleoside from the degradative pathway to the pyrimidine nucleotide level for cellular utilization.

It should be noted that the enzyme activities of the salvage pathways for purine and pyrimidine bases and nucleosides are in excess over that available for the de novo synthesis of purine and pyrimidine nucleotides.

13.6 DEOXYRIBONUCLEOTIDE FORMATION

As indicated earlier in this chapter, the concentrations of deoxyribonucleotides are extremely low in the "resting" cell. Only at the time of DNA replication (S phase) does the deoxyribonucleotide pool increase to support the required DNA synthesis.

Deoxyribonucleotides Are Formed by Reduction of Ribonucleoside Diphosphates

Deoxyribonucleotides are formed by the direct reduction of the 2′ position of the corresponding ribonucleotides. The reaction is strongly regulated not only by allosteric effectors (both activators and inhibitors) but also by drastic changes in the level of the enzyme catalyzing the formation of deoxyribonucleotides. This reaction occurs at the level of the nucleoside diphosphates. The general reaction can be summarized as follows:

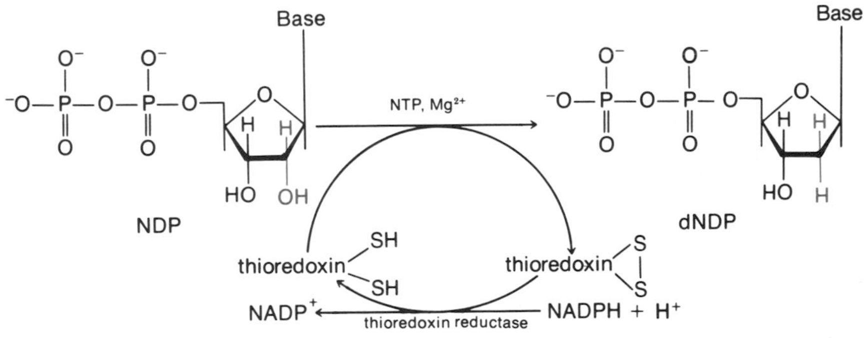

NDP → (NTP, Mg^{2+}) → dNDP
thioredoxin(SH)(SH) → thioredoxin(S—S)
$NADP^+$ ← thioredoxin reductase ← NADPH + H^+

The enzyme catalyzing the formation of 2′-deoxyribonucleotides is **nucleoside diphosphate reductase** (ribonucleotide reductase). The reduction of a specific nucleoside diphosphate (NDP) requires a specific nucleoside triphosphate (NTP) as a positive effector of the enzyme. This reaction is also subject to regulation by other NTPs, which can serve as negative effectors. The specificity of the effectors for the various substrates is summarized in Table 13.3.

The mammalian ribonucleotide reductase enzyme consists of two nonidentical subunits, neither of which alone has enzymatic activity. One of the subunits contains the effector-binding sites, while the other subunit contains the nonheme iron and a tyrosyl free radical. The two subunits

TABLE 13.3 Effectors of Ribonucleotide Reductase Activity

Substrate	Major positive effector	Major negative effector
CDP	ATP	dATP, dGTP, dTTP
UDP	ATP	dATP, dGTP
ADP	dGTP	dATP, ATP
GDP	dTTP	dATP

make up the active site of the enzyme. These subunits are encoded by separate genes, which are differentially expressed during the cell cycle transit. For the mammalian system it has not been completely resolved whether there is one enzyme for all four substrates, or whether there are separate enzymes or binding sites for each of the substrates.

Deoxy ATP is a potent inhibitor of the reduction of all four NDP substrates. This fact provides the biochemical basis for the toxicity of deoxyadenosine for a variety of mammalian cells.

Thioredoxin is a small molecular weight protein (12,000 Da), which is oxidized during the reduction of the 2′-OH group of the ribose moiety. To complete the catalytic cycle, reduced thioredoxin is regenerated by thioredoxin reductase (a flavoprotein) and NADPH.

The importance of ribonucleotide reductase for DNA replication cannot be overemphasized. Ribonucleotide reductase is uniquely responsible for catalyzing the reactions in which the deoxynucleoside triphosphates are generated for DNA replication. It appears that in controlling the deoxyribonucleotide levels in the cell, at least two approaches are utilized by the cell. These are (1) the actual concentration of reductase in the cells and (2) the very strict allosteric regulation of enzyme activity by the NTP.

Deoxythymidylate Synthesis Requires N^5,N^{10}-Methylene H_4Folate

Deoxythymidylate is formed in a unique reaction. The enzyme **thymidylate synthase** catalyzes the reaction in which a one carbon unit is not only transferred but also reduced to a methyl group (Figure 13.16). In the process, N^5,N^{10}-methylene H_4folate (Figure 13.17) acts as both a one-carbon transfer carrier and as a reducing agent. In the reaction H_2folate is generated.

The dUMP for this reaction can arise from two different pathways. In one reaction dCMP is deaminated directly to dUMP by dCMP deaminase, whereas in the other pathway UDP is reduced to dUDP, which is then

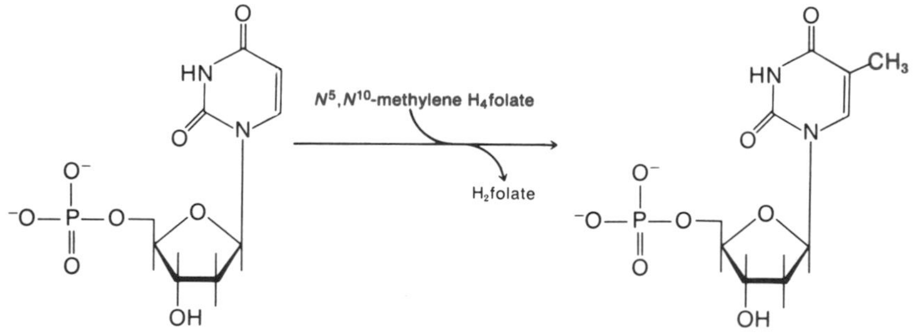

FIGURE 13.16
Synthesis of deoxythymidine nucleotide.

Deoxyuridine 5′-monophosphate (dUMP)

Thymidine 5′-monophosphate (TMP)

converted to dUMP (Figure 13.18). From labeling studies it appears that for most cells the major source of dUMP is the step in which dCMP is deaminated to dUMP. The dCMP deaminase activity in the mammalian cell far exceeds the level of UDP reductase activity.

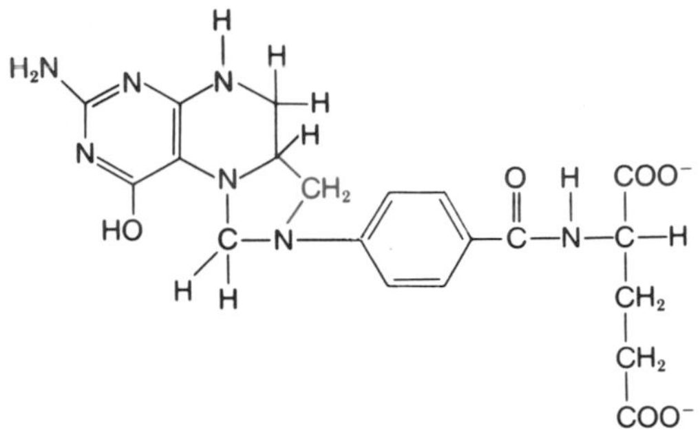

FIGURE 13.17
N^5,N^{10}-Methylene H_4folate.

Interconversion of Deoxyribopyrimidine Nucleotides Maintains an Appropriate Balance

Because of the cell's critical need for the deoxyribonucleotides to support DNA synthesis and hence cell replication, a series of enzymes is present that is responsible for the interconversion of the deoxyribopyrimidine nucleotides. These interconversions are summarized in Figure 13.18. The enzymes catalyzing these interconversions are strongly regulated. The control of dCDP and dUDP formation through the ribonucleotide reductase reaction has already been discussed. The **dCMP deaminase** is activated by dCTP and inhibited by dTTP. The inhibition by dTTP can be overcome by increasing concentrations of dCTP. A v versus [dCMP] curve is sigmoidal for dCMP deaminase. The presence of dCTP shifts the curve to a hyperbolic activity curve. The dTTP is also an inhibitor of thymidine kinase, an enzyme, which is important in "salvaging" thymidine and deoxyuridine for the cell. It is seen that dTTP serves as a negative effector of several conversions, and this is the basis for the toxicity of high concentrations of thymidine for mammalian cells.

Pyrimidine Nucleotides Are Degraded to β-Amino Acids

The turnover of nucleic acids results in the release of pyrimidine nucleotides. These nucleotides are also in a steady state, and there is constant synthesis and degradation. The degradation of pyrimidine nucleotides follows the pathways shown in Figure 13.19. In these degradative pathways, the nucleotides are first converted to the nucleosides and then to the free base uracil or thymine. The conversion of the pyrimidine nucleotides to nucleosides is catalyzed by various nonspecific phosphatases. **Deoxycytidylate deaminase** has preference for dCMP but can utilize CMP as substrate. Cytidine and deoxycytidine are deaminated to uridine and deoxyuridine, respectively, by the nucleoside deaminase. Uridine phos-

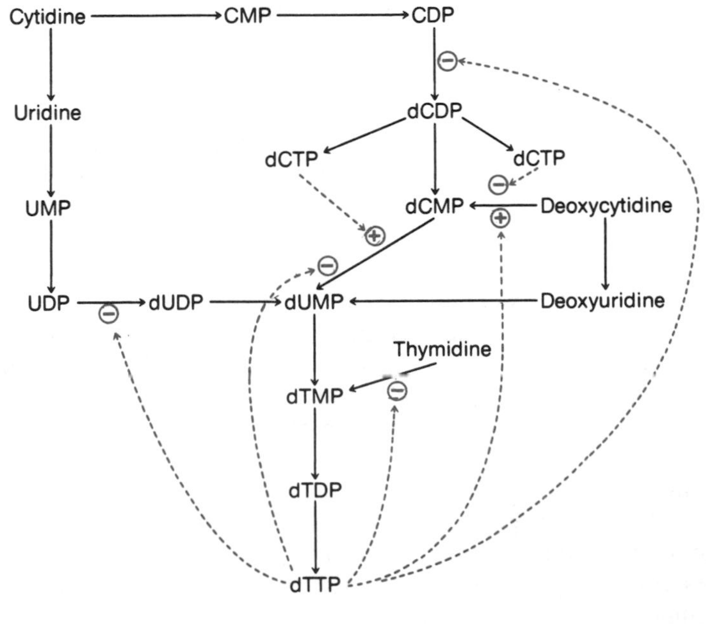

FIGURE 13.18
Pyrimidine interconversions.
The solid arrows represent enzyme-catalyzed reactions. The dashed lines represent the regulation that certain nucleotides have on the various steps (+, activation; −, inhibition).

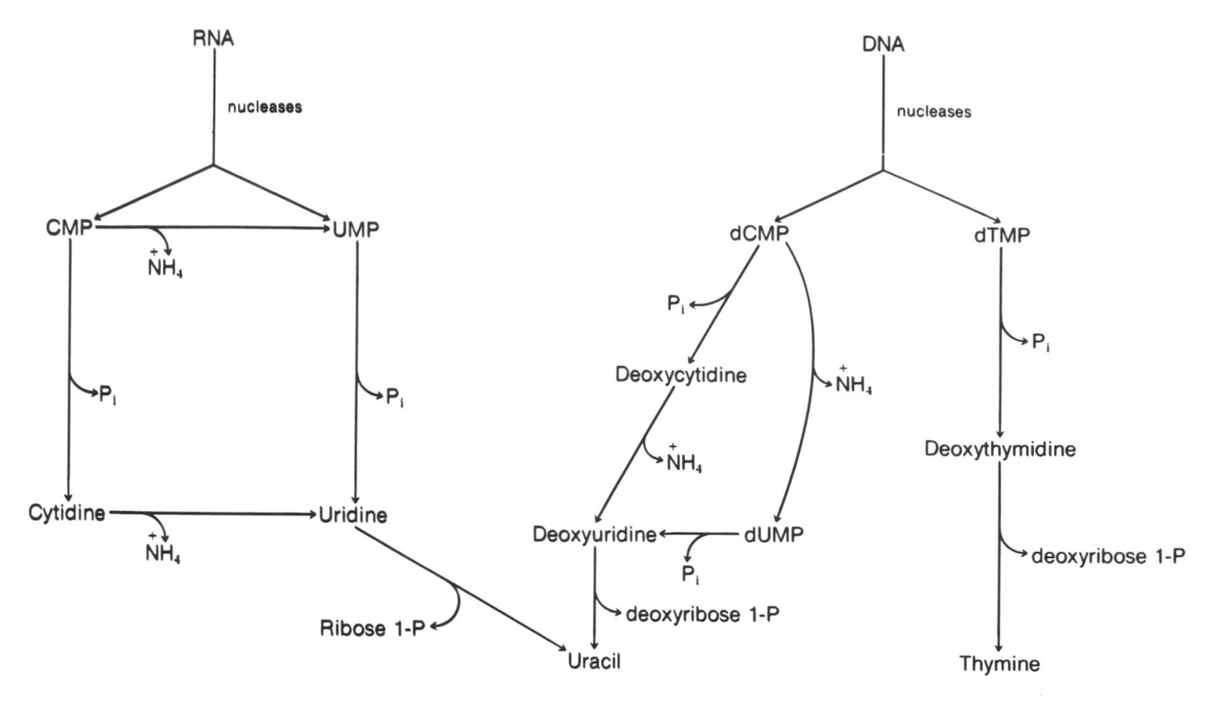

FIGURE 13.19
Pathways for the degradation of pyrimidine nucleotides.

phorylase catalyzes the phosphorolysis not only of uridine, but also of deoxyuridine and deoxythymidine.

It is important to note that mammalian cells have a very specific **dUTPase** in high concentration, which catalyzes the reaction:

$$\text{dUTP} \longrightarrow \text{dUMP} + \text{Pyrophosphate}$$

It is critical for normal DNA replication that dUTP not be present in the nucleotide pool. As is shown in Chapter 15, dUTP can very effectively replace dTTP as a substrate in the DNA polymerase reaction.

The K_m of dUTP for DNA polymerase is approximately 10 μM, whereas the K_m of dUTP for dUTPase is only 1 μM. Consequently, the K_m strongly favors the dUTPase reaction. Therefore, dUTPase, serves to prevent misincorporation of dUMP into the DNA, which would have other major consequences.

Uracil and thymine are then further degraded by analogous reactions, although the final products are different as shown in Figure 13.20. Uracil is degraded to β-alanine, NH_4^+, and CO_2. None of these is unique to uracil degradation, and consequently the turnover of cytosine or uracil nucleotides cannot be estimated from the end products of this pathway. Thymine degradation proceeds to β-aminoisobutyric acid, NH_4^+, and CO_2. β-Aminoisobutyric acid is excreted in the urine of humans and originates exclusively from the degradation of thymine. Increased levels of β-isoaminobutyric acid are excreted in cancer patients undergoing chemotherapy or radiation therapy in which large numbers of cells are killed and the DNA is degraded. It is therefore possible to estimate the turnover of DNA or thymidine nucleotides by the measurement of β-aminoisobutyric acid production.

The three enzymes (dihydropyrimidine dehydrogenase, dihydropyrimidinase, and ureidopropionase) required to catalyze the degradation of uracil or thymine to their respective products are separate proteins. These enzymes, however, appear to utilize uracil or thymine and their intermediates equally well as substrates.

Uracil → (dihydropyrimidine dehydrogenase; $NADPH + H^+ \rightarrow NADP^+$) → Dihydrouracil → (dihydropyrimidinase) → β-Ureidopropionate → (ureidopropionase; CO_2, NH_3) → $^+NH_3CH_2CH_2COO^-$ β-Alanine

Thymine → (dihydropyrimidine dehydrogenase; $NADPH + H^+ \rightarrow NADP^+$) → Dihydrothymine → (dihydropyrimidinase) → β-Ureidoisobutyrate → (ureidopropionase; CO_2, NH_3) → $^+H_3NCH_2CH(CH_3)COO^-$ β-Aminoisobutyrate

FIGURE 13.20
Degradation of uracil and thymine.

13.7 NUCLEOSIDE AND NUCLEOTIDE KINASES

As shown in the pathways for the de novo synthesis of purine and pyrimidine nucleotides, the nucleotide is synthesized as the monophosphate. However, most, if not all, reactions in which the nucleotides function require that these nucleotides be at the di- or triphosphate level.

There are specific kinases to "salvage" nucleosides to nucleotides and to convert the nucleoside monophosphates to the di- and triphosphates. Examples of these are as follows:

1. *Uridine–Cytidine Kinase*

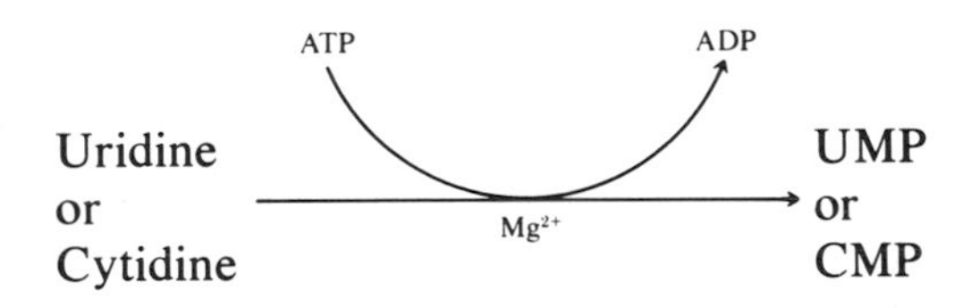

The enzyme is specific for uridine or cytidine as substrates. Both UTP and CTP are inhibitors of these reactions.

2. *Deoxycytidine Kinase*

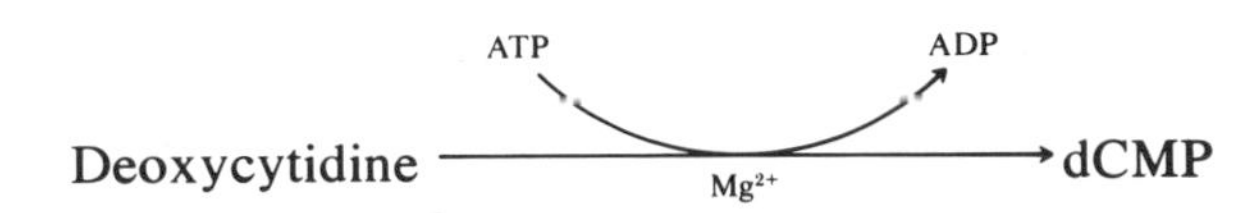

Deoxycytidine is the preferred substrate for this kinase. Cytidine, uridine, and thymidine are not substrates. However, deoxyadenosine and deoxyguanosine, although poorer substrates, are phosphorylated by this enzyme. The dCTP is a potent inhibitor of this reaction while dTTP will reverse the inhibition caused by dCTP.

3. *Pyrimidine Nucleoside Monophosphate Kinase*

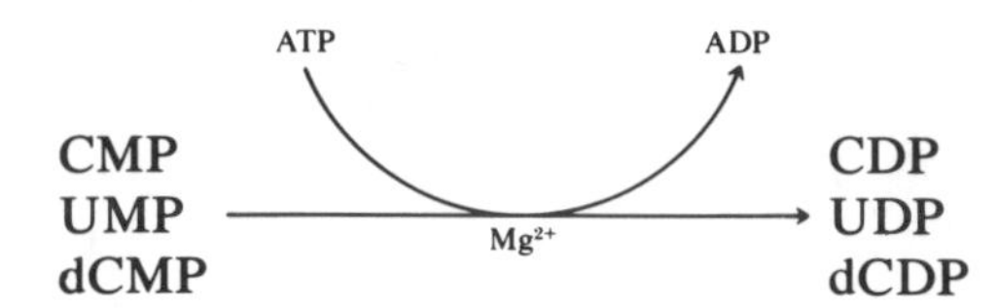

This enzyme shows specificity for the substrates CMP, UMP, and dCMP. The dUMP is not a substrate.

4. *Thymidine Kinase*

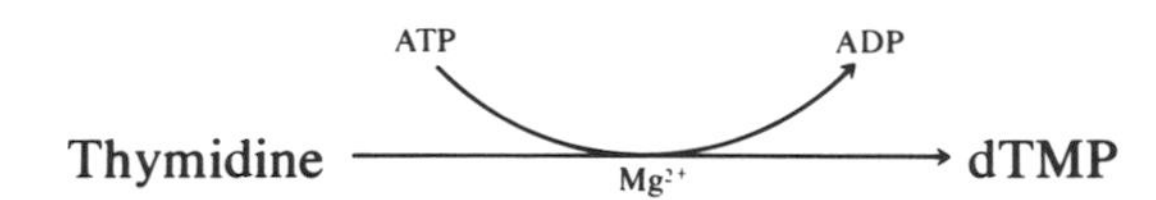

This kinase is specific for thymidine and deoxyuridine as substrates. This enzyme is elevated in rapidly growing tissues.

5. *Thymidylate Kinase*

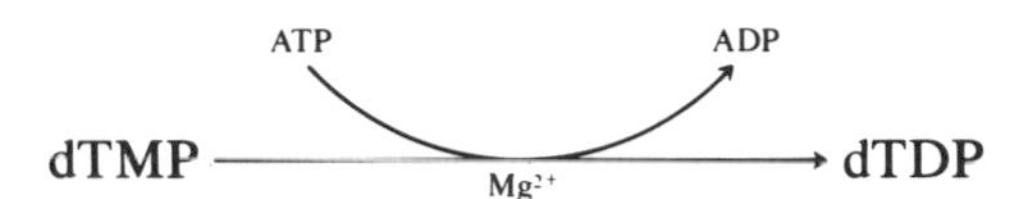

This enzyme is specific for dTMP.

6. *Adenosine Kinase*

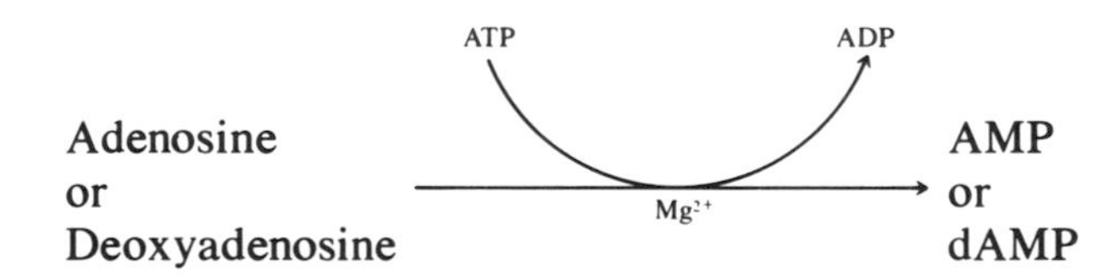

This enzyme is specific for adenosine or deoxyadenosine. Inosine is not a substrate for this enzyme.

7. *AMP Kinase*

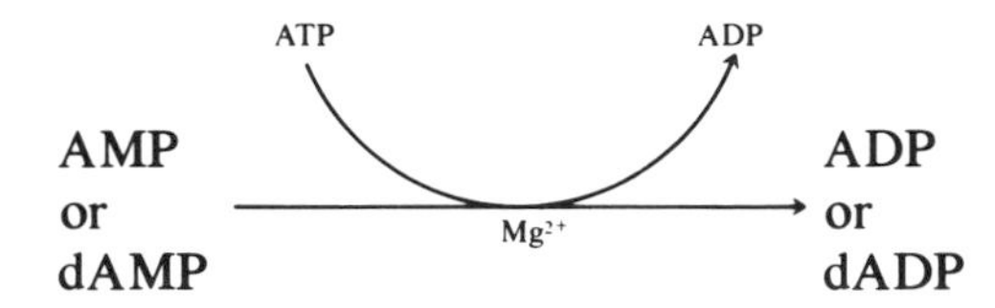

AMP kinase shows specificity for AMP. Although dAMP can be utilized as substrate, AMP is phosphorylated at a rate 10 times higher than that of dAMP.

8. *GMP Kinase*

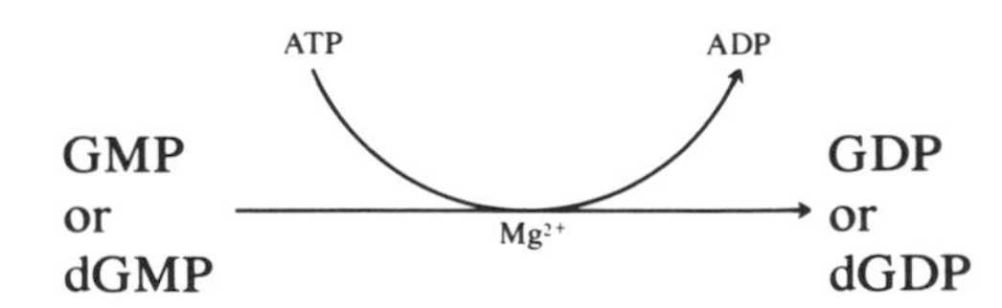

GMP kinase is distinct from AMP kinase showing specificity for GMP and dGMP.

9. *Nucleoside Diphosphokinase*

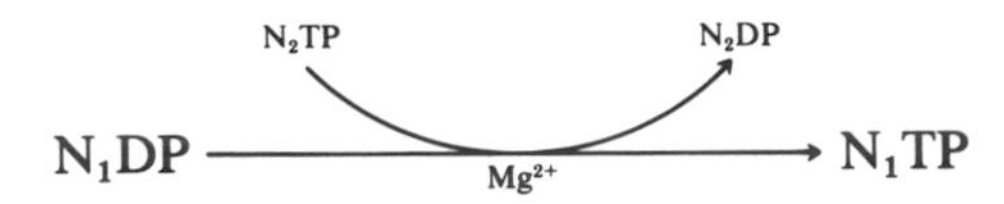

Mammalian cells contain an enzyme, nucleoside diphosphokinase, which is not specific for either the phosphate donor or phosphate acceptor in terms of either the purine or pyrimidine base or the sugar moiety. Since in most cells the concentration of ATP is the highest of the triphosphates and most easily regenerated via glycolysis or oxidative phosphorylation, ATP is probably the major phosphate donor for these reactions in intact cells.

13.8 NUCLEOTIDE METABOLIZING ENZYMES AS A FUNCTION OF THE CELL CYCLE AND RATE OF CELL DIVISION

The Cell Cycle Has Distinct Phases

For normal cell division to occur, essentially all components of the cell must double. The events that lead from the formation of a daughter cell as a result of mitosis, to the completion of the processes required for its own division into two daughter cells are described by the term cell cycle. The periods of the cell cycle have been termed mitosis (M), gap 1 (G_1), synthesis (S), and gap 2 (G_2). The total period of the cell cycle will vary with the particular cell type. Some cells will enter G_0; the cells are viable and metabolically functional but are in the nonproliferative or quiescent state. In most mammalian cells the periods of M, S, and G_2 are relatively constant, while G_1 varies widely causing cells to have long or short doubling times. The cell cycle is represented in Figure 13.21. In preparation for DNA replication during S phase of the cell cycle, there is considerable synthesis of enzymes involved in nucleotide metabolism during the G_1 phase, especially during late G_1/early S. Both RNA and protein synthesis occur continuously, although at varying rates during G_1, S, and G_2 phases. The DNA replication occurs only during the S phase. The cell cycle distribution of a population of cells can be determined by several techniques. Probably, the method in greatest use currently is that of flow cytometry. In this procedure, the DNA of the cells is stained with a fluorescent dye and the DNA content of each cell is measured to establish a histogram. The cell cycle data for a rapidly proliferating mouse leukemia L1210 cell line is shown in Figure 13.22. The G_0/G_1 population has the diploid content of DNA (2N). The G_0 cells cannot be distinguished from the G_1 cells by DNA staining and analysis by flow cytometry. The G_2/M

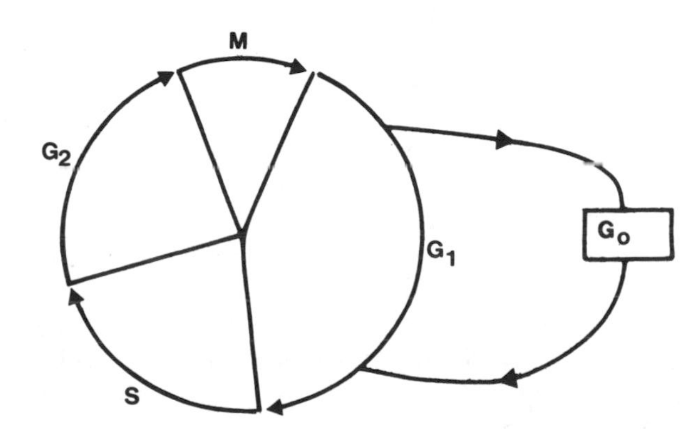

FIGURE 13.21
Diagrammatic representation of the cell cycle.
For a mammalian cell with a doubling time of 24 h, G_1 would last ~12 h; S, 7 h; G_2, 4h; and M, 1 h. Cells would enter the G_0 state if they became quiescent or nonproliferative.

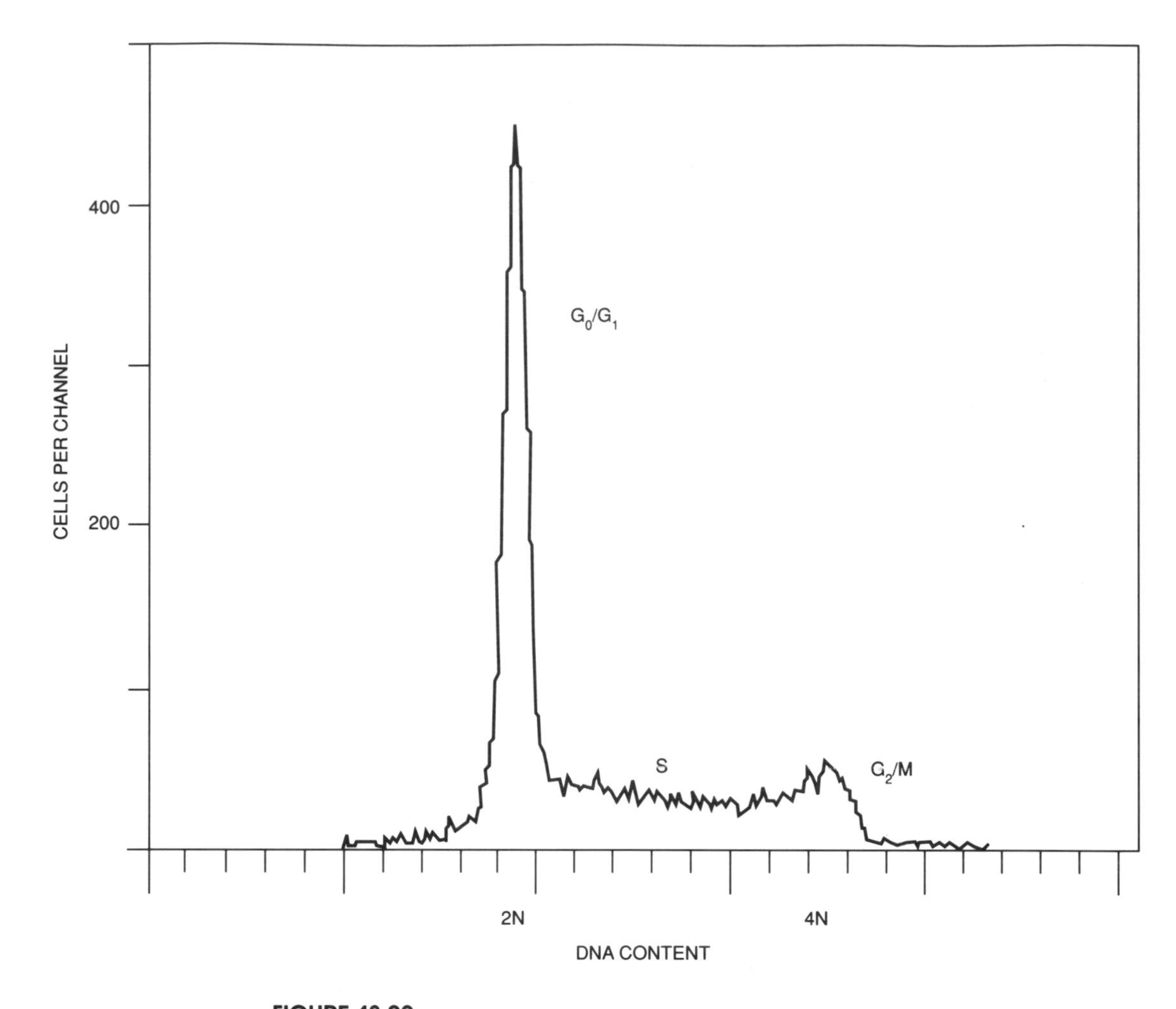

FIGURE 13.22
Analysis of cell cycle phases by flow cytometry.
The cell cycle distribution of the leukemia L1210 cells in log phase of growth was G_0/G_1, 45%; S, 45%; and G_2/M, 9%.

population has twice the normal amount of DNA (4N), since the DNA has been replicated but cell division has not yet been completed. The S phase is the period of DNA replication and consequently the DNA content varies between 2N and 4N.

Enzymes of Purine and Pyrimidine Nucleotide Synthesis Are Elevated During the S Phase

The strict regulation of nucleotide synthesis requires that certain mechanisms must be available to the cell to meet the requirements for the ribonucleotides and deoxyribonucleotide precursors at the time of increased RNA synthesis and DNA replication. To meet these needs, the cell responds by increasing the levels of specific enzymes involved with nucleotide formation during very specific periods of the cell cycle.

The enzymes involved in purine nucleotide synthesis and interconversions that are elevated during the S phase of the cell cycle are PRPP amidotransferase and IMP dehydrogenase. Adenylosuccinate synthetase and adenylosuccinase do not appear to increase.

The enzymes involved in pyrimidine nucleotide synthesis that are elevated during the S phase of the cell cycle include aspartate carbamoyltransferase, dihydroorotase, dihydroorotate dehydrogenase, orotate phosphoribosyltransferase, and CTP synthetase.

Many of the enzymes involved in the synthesis and interconversions of deoxyribonucleotides are also elevated during the S phase of the cell cycle. Included in these enzymes are ribonucleotide reductase, thymidine kinase, dCMP deaminase, thymidylate synthase, and TMP kinase. The importance to DNA replication of the increased levels of enzyme activities during late G_1/early S phase is worthy of further discussion with a specific example.

As discussed previously, the deoxyribonucleotide pool is extremely small in "resting" cells (less than 1 μM). As a result of the increase in ribonucleotide reductase the levels of deoxyribonucleotides reach levels of 10–20 μM during DNA synthesis. However, this concentration would sustain DNA synthesis for only minutes, while complete DNA replication would require hours. Consequently, the levels of ribonucleotide reductase activity not only must increase but must be sustained during S phase in order to provide the necessary substrates for DNA synthesis.

If one looks at a population of cells as a whole (i.e., tissue) rather than as individual cells going through the cell cycle, it is observed that rapidly growing tissues such as regenerating liver, embryonic tissues, intestinal mucosal cells, and erythropoietic cells are geared toward DNA replication and RNA synthesis. These tissues will show elevated levels of those key enzymes involved with purine and pyrimidine nucleotide synthesis and interconversions and complementary decreases in the levels of the enzymes that catalyze reactions in which these precursors are degraded. Of course these changes reflect the proportion of the cells in that tissue that are in S phase.

An understanding has evolved of the biochemical changes that occur to satisfy the proliferative life-style of tumor cells. It has been determined that gene expression has been altered to result not only in quantitative changes in enzyme levels but also qualitative changes (isozyme shifts). As a result of careful experimental study, utilizing a series of liver, colon, and kidney tumors of varying growth rates, it has been possible to categorize these biochemical changes as (1) transformation-linked (meaning that all tumors regardless of growth rate show certain increased and certain decreased enzyme levels); (2) progression-linked (alterations that correlate with the growth rate of the tumor); and (3) coincidental alterations (not connected to the malignant state). As very limited examples, the levels of ribonucleotide reductase, thymidylate synthase, and IMP dehydrogenase increase as a function of the tumor growth rate. PRPP amidotransferase, UDP kinase, and uridine kinase are examples of enzymes whose activity is increased in all tumors, whether they are slow-growing or the most rapidly growing tumors.

It is important to point out that, while certain of the enzymes are increased in both fast-growing normal tissue (e.g., embryonic and regenerating) and tumors, the total quantitative and qualitative patterns for normal and tumor tissue can easily be distinguished.

13.9 NUCLEOTIDE COENZYME SYNTHESIS

The coenzymes NAD, FAD, and CoA are synthesized in mammalian cells provided that there is a suitable source of the vitamin component (niacin, riboflavin, and pantothenic acid).

NAD Is Synthesized From Tryptophan, Nicotinate, and Nicotinamide

NAD is synthesized in mammalian cells by at least three different pathways shown in Figure 13.23.

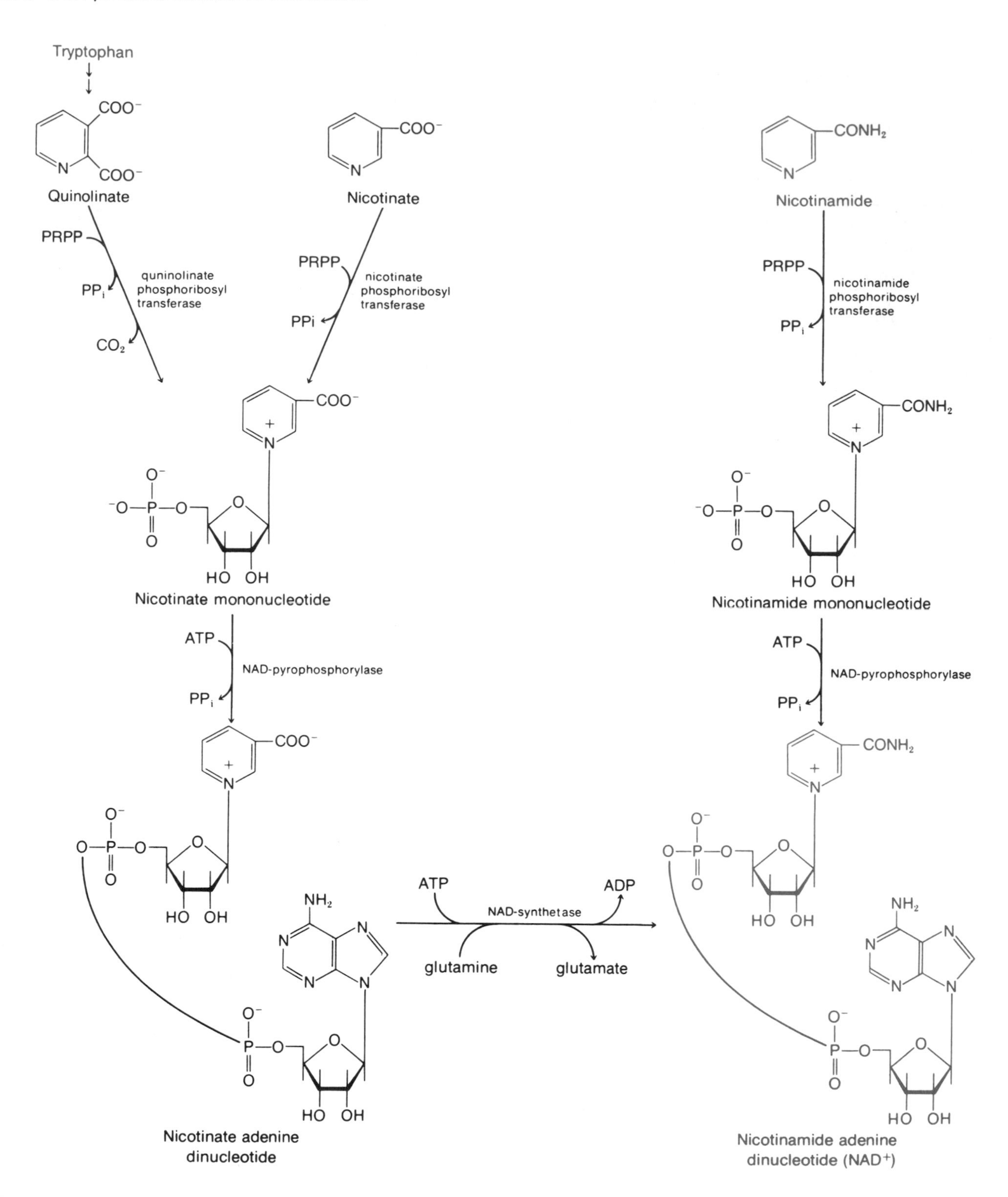

FIGURE 13.23
Pathways for NAD$^+$ synthesis.

When the exogenous source of tryptophan is in excess of the amount required for protein synthesis, tryptophan can be metabolized to quinolinic acid. **Quinolinic acid** is utilized in a reaction with PRPP to form nicotinate mononucleotide. The enzyme catalyzing this reaction, quinolinate phosphoribosyltransferase, is found in liver, kidney and brain. Therefore, this pathway is specific for these tissues.

Nicotinate reacts with PRPP to form nicotinate mononucleotide. The enzyme catalyzing this reaction is nicotinate phosphoribosyltransferase and is widely distributed in various tissues. Nicotinate mononucleotide reacts with ATP to yield nicotinate adenine dinucleotide. The enzyme catalyzing this reaction is NAD-pyrophosphorylase and is widely distributed in various tissues. Nicotinate adenine dinucleotide reacts with glutamine with the hydrolysis of ATP to yield nicotinamide adenine dinucleotide (NAD). The enzyme **NAD-synthetase** catalyzes this reaction.

Nicotinamide reacts with PRPP to give nicotinamide mononucleotide. The enzyme that catalyzes this reaction is nicotinamide phosphoribosyltransferase. The enzyme is specific for nicotinamide and is entirely distinct from nicotinate phosphoribosyltransferase. Nicotinamide mononucleotide reacts with ATP to yield NAD^+. The enzyme that catalyzes this reaction is the same enzyme that catalyzes the reaction between nicotinate mononucleotide and ATP.

In nucleated cells **NAD-pyrophosphorylase** is located exclusively in the nucleus of the cell. However, the erythrocyte is entirely capable of synthesizing NAD^+ from nicotinate or nicotinamide.

The intracellular concentration of pyridine nucleotides is maintained at a constant level, implying a pathway that is tightly regulated. **Nicotinamide phosphoribosyltransferase** appears to be the regulated enzyme in NAD^+ synthesis. NMN, NAD^+, $NADP^+$, and NADPH are strong inhibitors of nicotinamide mononucleotide synthesis. ATP stimulates this reaction, although it is not an absolute requirement; ATP lowers the K_m for PRPP 10-fold and nicotinamide 100-fold, while increasing the V_{max}. The resulting K_m for nicotinamide approaches the intracellular concentration of this compound. Nicotinate phosphoribosyltransferase does not appear to be regulated by the end products of the pathway utilizing nicotinate as the substrate.

NAD^+ is synthesized also by the mitochondria. However, the relative importance of the mitochondrial pathway to the pathways just described is not known.

There is considerable turnover of NAD^+ in cells. The NAD^+ is consumed by two distinct reactions. In one reaction **NAD-glycohydrolase** catalyzes the conversion of NAD^+ to nicotinamide and adenosine diphosphoribose. This enzyme is located in the endoplasmic reticulum. In the second, **poly(ADP-ribose) synthetase** catalyzes the polymerization of the ADP-ribose moiety of NAD^+ onto nuclear proteins with the release of nicotinamide. This enzyme is found exclusively in the nucleus and the level of activity is highest during G_2 and lowest during S phase of the cell cycle. Roles for poly(ADP) ribosylation in cell regulation and DNA repair have been proposed.

Since there is no known enzymatic route for the conversion of nicotinate directly to nicotinamide, the NAD-glycohydrolase reaction can provide nicotinamide to extrahepatic tissues for NAD^+ synthesis. In some tissues, nicotinate is a more efficient precursor of NAD^+ than is nicotinamide.

NADP Is Produced by Phosphorylation of NAD

NAD^+ is the immediate precursor of nicotinamide adenine dinucleotide phosphate ($NADP^+$). The reaction is catalyzed by **NAD-kinase** as presented in Figure 13.24. The NAD-kinase is found in the cytosol of the cell and NADPH, the reduced form, is a negative effector of this reaction.

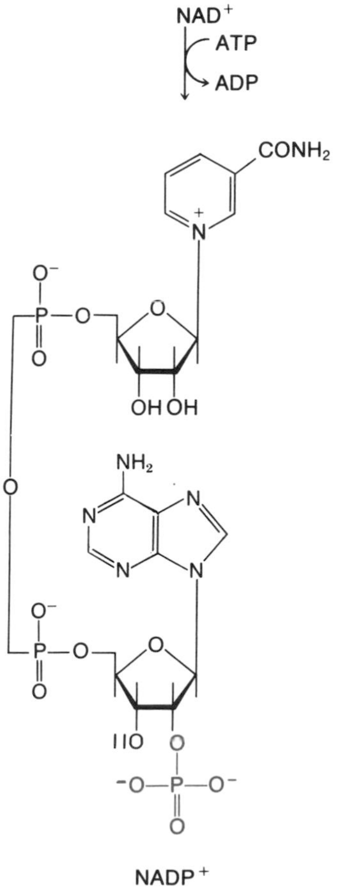

FIGURE 13.24
Synthesis of $NADP^+$.

NAD Is Degraded to Nicotinamide

NAD can be degraded by two pathways, although the product is nicotinamide by either pathway as shown in Figure 13.25. Nicotinamide is ex-

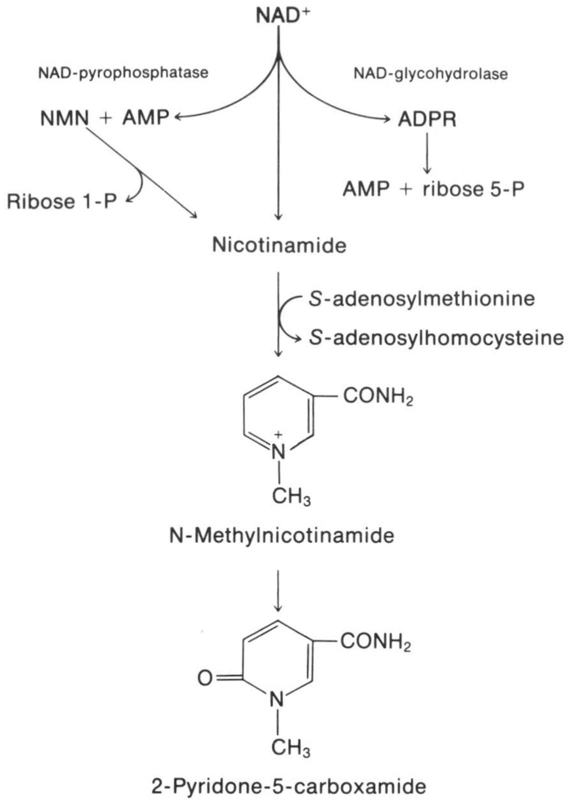

FIGURE 13.25
Degradation of NAD⁺.

creted in the urine of human beings as the *N*-methyl derivatives of nicotinamide and 2-pyridone-5-carboxamide. An excess of these products gives urine a bright yellow fluorescent color. It should be noted that the excretion of excess *N*-methylnicotinamide requires the consumption of *S*-adenosylmethionine.

FAD Synthesis Requires Exogenous Riboflavin

FAD is synthesized in a two-step reaction. Riboflavin is required from an exogenous source because human beings cannot synthesize the isoalloxazine moiety. Riboflavin is phosphorylated by ATP in a reaction catalyzed by **riboflavin kinase** (flavokinase) to give riboflavin phosphate. The Mg^{2+} ions are the preferred divalent cation required for this kinase. GTP will partially replace ATP as the phosphorylating agent. Flavokinase is found in the cytosol of the cells of a variety of tissues such as liver, kidney, brain, spleen, and heart. Many references are made to this compound as flavin mononucleotide (FMN), although this is not a true nucleotide. The FMN then reacts with ATP to yield FAD in a reaction catalyzed by FAD pyrophosphorylase. This enzyme shows an absolute requirement for ATP and Mg^{2+} ions are required. The FAD-pyrophosphorylase activity has been reported to be located mainly in mitochondria. The pathway for FAD synthesis is summarized in Figure 13.26.

Coenzyme A Synthesis Requires Exogenous Pantothenic Acid

Coenzyme A is synthesized in humans by a series of reactions with an absolute requirement for an exogenous source of pantothenic acid. The pathway for the synthesis of CoA in mammalian cells is shown in Figure 13.27. Pantothenic acid is phosphorylated by ATP to give 4-phosphopantothenic acid. In the next reaction, cysteine is added to provide the SH group, which will ultimately be the "business end" of CoA. The α-car-

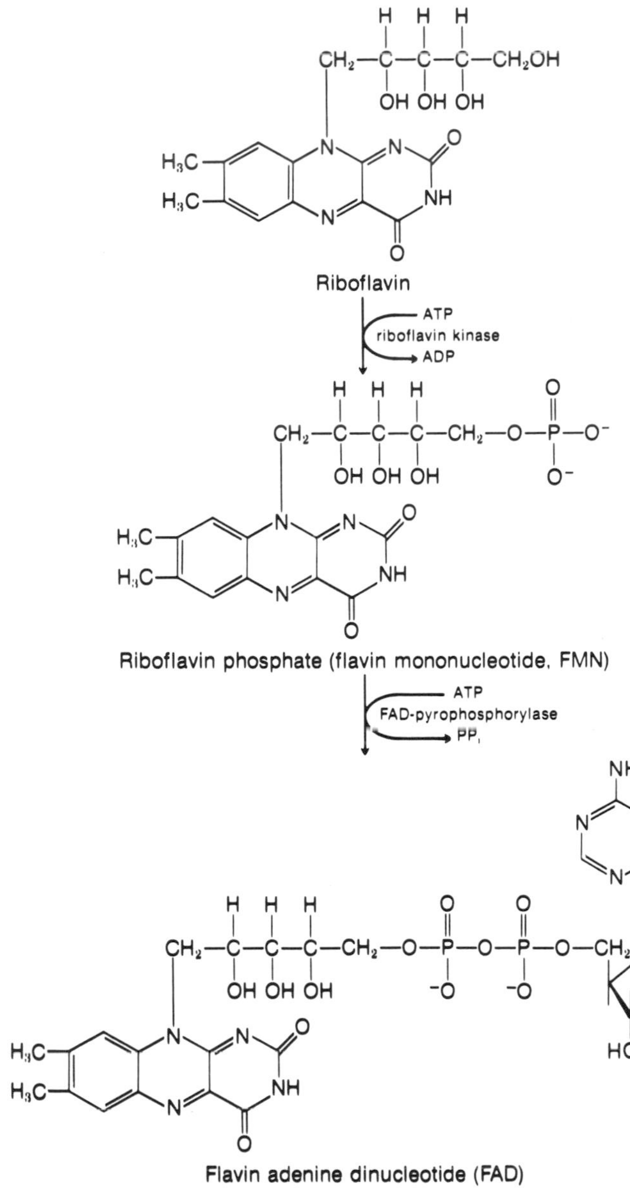

Riboflavin

ATP
riboflavin kinase
ADP

Riboflavin phosphate (flavin mononucleotide, FMN)

ATP
FAD-pyrophosphorylase
PP_i

Flavin adenine dinucleotide (FAD)

FIGURE 13.26
Synthesis of FAD.

boxyl group of cysteine is then removed from 4-phosphopantothenoyl-L-cysteine to yield 4-phosphopantotheine. In a pyrophosphorylase reaction, ATP is then added to give dephosphoCoA. The dephosphoCoA is then phosphorylated at the 3′ position of the adenosine moiety to give CoA. The enzymes, **dephosphoCoA pyrophosphorylase** and **dephospho-CoA kinase,** appear to exist in nature as a bifunctional enzyme complex. The enzymes copurify, with the ratio of the two activities remaining constant through many steps.

This pathway appears to be regulated at the **phosphopantothenoyl-cysteine decarboxylase** step. The product of the reaction catalyzed by this enzyme, 4-phosphopantotheine, is a relatively strong competitive inhibitor of this reaction.

13.10 SYNTHESIS AND UTILIZATION OF 5-PHOSPHORIBOSYL 1-PYROPHOSPHATE

The intracellular concentration of 5-phosphoribosyl 1-pyrophosphate (PRPP) plays an important role in regulating several important pathways.

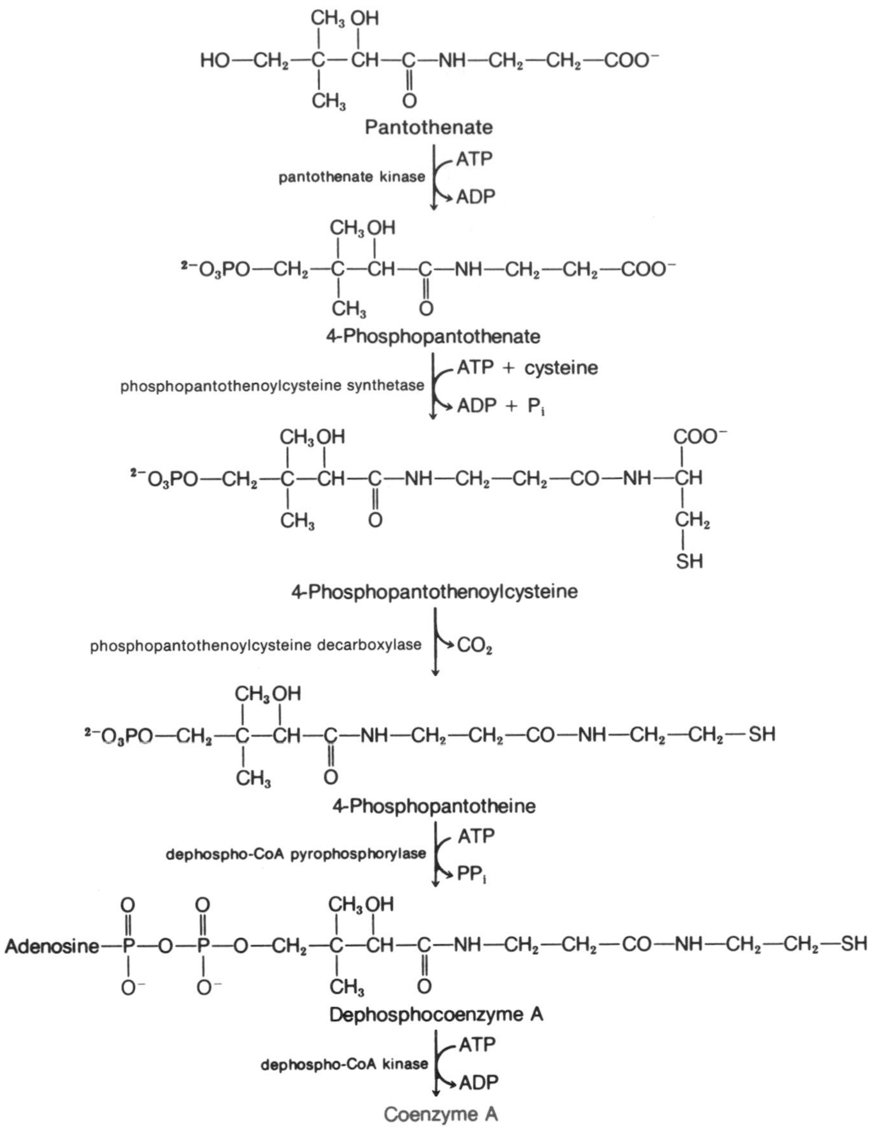

$HO{-}CH_2{-}C(CH_3)_2{-}CH(OH){-}C(=O){-}NH{-}CH_2{-}CH_2{-}COO^-$

Pantothenate

pantothenate kinase: ATP → ADP

$^{2-}O_3PO{-}CH_2{-}C(CH_3)_2{-}CH(OH){-}C(=O){-}NH{-}CH_2{-}CH_2{-}COO^-$

4-Phosphopantothenate

phosphopantothenoylcysteine synthetase: ATP + cysteine → ADP + P_i

$^{2-}O_3PO{-}CH_2{-}C(CH_3)_2{-}CH(OH){-}C(=O){-}NH{-}CH_2{-}CH_2{-}CO{-}NH{-}CH(COO^-){-}CH_2{-}SH$

4-Phosphopantothenoylcysteine

phosphopantothenoylcysteine decarboxylase: → CO_2

$^{2-}O_3PO{-}CH_2{-}C(CH_3)_2{-}CH(OH){-}C(=O){-}NH{-}CH_2{-}CH_2{-}CO{-}NH{-}CH_2{-}CH_2{-}SH$

4-Phosphopantotheine

dephospho-CoA pyrophosphorylase: ATP → PP_i

$Adenosine{-}P(=O)(O^-){-}O{-}P(=O)(O^-){-}O{-}CH_2{-}C(CH_3)_2{-}CH(OH){-}C(=O){-}NH{-}CH_2{-}CH_2{-}CO{-}NH{-}CH_2{-}CH_2{-}SH$

Dephosphocoenzyme A

dephospho-CoA kinase: ATP → ADP

Coenzyme A

FIGURE 13.27
Synthesis of CoA.

The synthesis and utilization of PRPP by the cell will determine the steady-state concentration of PRPP and hence the metabolic pathways that compete for PRPP.

PRPP is synthesized in the cell in the reaction catalyzed by **5-phosphoribose pyrophosphokinase (PRPP synthetase)** utilizing α-ribose 5-phosphate and ATP. The reaction requires Mg^{2+} ions (Figure 13.28). The ribose 5-phosphate used in this reaction is generated from glucose 6-phosphate metabolism via the hexose monophosphate shunt or from ribose 1-phosphate (generated by phosphorolysis of nucleotides) via a phosphoribomutase reaction.

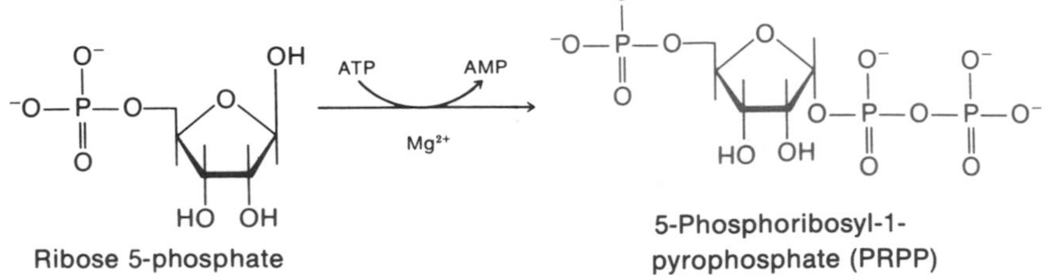

Ribose 5-phosphate → 5-Phosphoribosyl-1-pyrophosphate (PRPP)

ATP → AMP; Mg^{2+}

FIGURE 13.28
Synthesis of PRPP.

As expected for such a critical reaction, the formation of PRPP is regulated. The enzyme has an absolute requirement for P_i ions. The v versus $[P_i]$ curve for PRPP-synthetase is sigmoidal and at the concentration of P_i normally found in the cell, the activity is markedly depressed because of the sigmoidal rather than hyperbolic curve.

The importance of the sigmoidal curve relative to P_i in regulating PRPP formation, and hence the overproduction of uric acid, was shown in a "gouty" individual who had a marked increase in uric acid formation. Analysis of the patient's red cells for PRPP-synthetase revealed that this patient had a mutant form of the enzyme, which showed a hyperbolic v versus $[P_i]$ curve. At the intracellular concentration of P_i, the production of PRPP was greatly elevated. Presumably the increased levels of PRPP in other tissues lead to the overproduction of purine nucleotides and consequently of uric acid.

The levels of PRPP synthetase are elevated in cells undergoing rapid cell division and decrease to basal levels in cells that have reached confluence. The activity of PRPP synthetase is inhibited by nucleoside di- and triphosphates. ADP is the most potent inhibitor of PRPP synthetase activity in human placenta, rat liver, and mouse tumor cells. The ADP is a competitive inhibitor with respect to ATP. Its inhibition constant (K_i) is less than the intracellular concentration of ADP, indicating that it can serve as a physiological effector of PRPP synthetase. 2,3-Bisphosphoglycerate is also an inhibitor of PRPP synthetase, and this is of importance in red cell metabolism dealing with the salvage pathways.

Factors that lead to increased flux of glucose 6-phosphate through the hexose monophosphate shunt pathway can result in increased intracellular levels of PRPP. Pyrroline-5-carboxylate (an intermediate in the interconversions of the amino acids, ornithine, glutamate, and proline) stimulates the hexose monophosphate shunt via the generation of $NADP^+$ in the pyrroline-5-carboxylate reductase-catalyzed reaction. This can lead to elevated intracellular concentrations of PRPP.

PRPP formed in the cells is a required substrate for many key metabolic reactions depending on the cell type. The reactions and the pathways in which PRPP is utilized are as follows:

1. De novo purine nucleotide synthesis

   a. PRPP + glutamine $\longrightarrow$ 5-phosphoribosylamine + glutamate + PP_i

2. "Salvage" of purine bases

   a. PRPP + hypoxanthine (guanine) $\longrightarrow$ IMP (GMP) + PP_i

   b. PRPP + adenine $\longrightarrow$ AMP + PP_i

3. De novo pyrimidine nucleotide synthesis

   a. PRPP + orotate $\longrightarrow$ OMP + PP_i

4. "Salvage" of pyrimidine bases

   a. PRPP + uracil $\longrightarrow$ UMP + PP_i

5. NAD^+ synthesis

   a. PRPP + nicotinate $\longrightarrow$ nicotinate mononucleotide + PP_i

   b. PRPP + nicotinamide $\longrightarrow$ nicotinamide mononucleotide

   c. PRPP + quinolinate $\longrightarrow$ nicotinate mononucleotide + PP_i

In the red cell, the major reactions in which PRPP is consumed are reactions 2a and 2b and 5a and 5b. In a rapidly growing tumor cell, PRPP would be consumed by all five pathways. The direction in which PRPP would be consumed would depend on several factors, including the relative K_m values of the competing enzymes for PRPP, the availability of the second substrate, and the concentration of the effector for the particular reaction. For example, if the concentration of AMP in the cell were high, adenine phosphoribosyltransferase would be inhibited and PRPP could be utilized by hypoxanthine-guanine phosphoribosyltransferase to generate IMP or GMP provided that hypoxanthine or guanine were present in the cell and IMP and GMP concentrations were low.

13.11 COMPOUNDS THAT INTERFERE WITH CELLULAR PURINE AND PYRIMIDINE NUCLEOTIDE METABOLISM: USE AS CHEMOTHERAPEUTIC AGENTS

As has been discussed, the de novo synthesis of purine and pyrimidine nucleotides is critical to normal cell replication, maintenance, and function. The regulation of these pathways has also been shown to be important since disease states arise from defects in these steps.

Many compounds have been synthesized or isolated as natural products from plants, bacteria, or fungi that are relatively specific inhibitors of enzymes involved with nucleotide synthesis or interconversions. These drugs, useful in therapy in many clinical problems, have been classified as antimetabolites, antifolates, glutamine antagonists, and other compounds.

Antimetabolites Are Structural Analogs of the Bases or Nucleosides

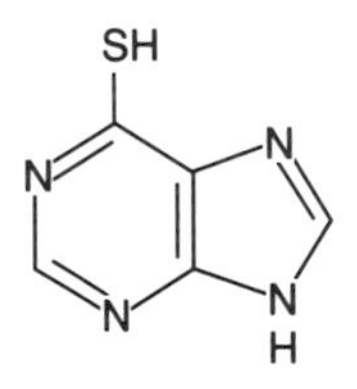

6-Mercaptopurine

5-Fluorouracil

Cytosine arabinoside

FIGURE 13.29
Structures of antimetabolites.

Antimetabolites, generally, are structural analogs of the purine and pyrimidine bases or nucleosides that interfere with rather specific metabolic sites. In 1988 the Nobel Prize in medicine was shared by Dr. George Hitchings, Dr. Gertrude Elion, and Dr. James Black. Dr. Hitchings and Dr. Elion were cited for their research on purine antimetabolites that led to agents used successfully in the treatment of several human diseases. These drugs include: *6-mercaptopurine* and *6-thioguanine* for the treatment of acute leukemia; *azathioprine* for immunosuppression in patients with organ transplants; *allopurinol* for the treatment of gout and hyperuricemia; and *acyclovir* for the treatment of herpesvirus infection. The detailed understanding of purine nucleotide metabolism greatly aided the development of these compounds as useful drugs in the treatment of human disease.

Only a few of these will be discussed to show: (a) the importance of the de novo pathways to normal cell metabolism; (b) that the regulation of these pathways occurs in vivo; (c) the concept of the requirement for metabolic activation of the drugs; and (d) that the inactivation of these compounds can greatly influence their usefulness.

6-Mercaptopurine (*6-MP*) (Figure 13.29) is a useful antitumor drug in humans. The cytotoxic activity of this agent is related to the formation of 6-mercaptopurine ribonucleotide by the tumor cell. Utilizing PRPP and HGPRTase, 6-MP is converted to its nucleotide form.

6-Mercaptopurine ribonucleoside 5′-monophosphate accumulates in the cell and serves as a negative effector of PRPP-amidotransferase, the committed step in the de novo pathway. This nucleotide also acts as an inhibitor of the conversion of IMP to GMP at the IMP-dehydrogenase step and IMP to AMP at the adenylosuccinate synthetase step. Since 6-mercaptopurine is a substrate for xanthine oxidase and is oxidized to

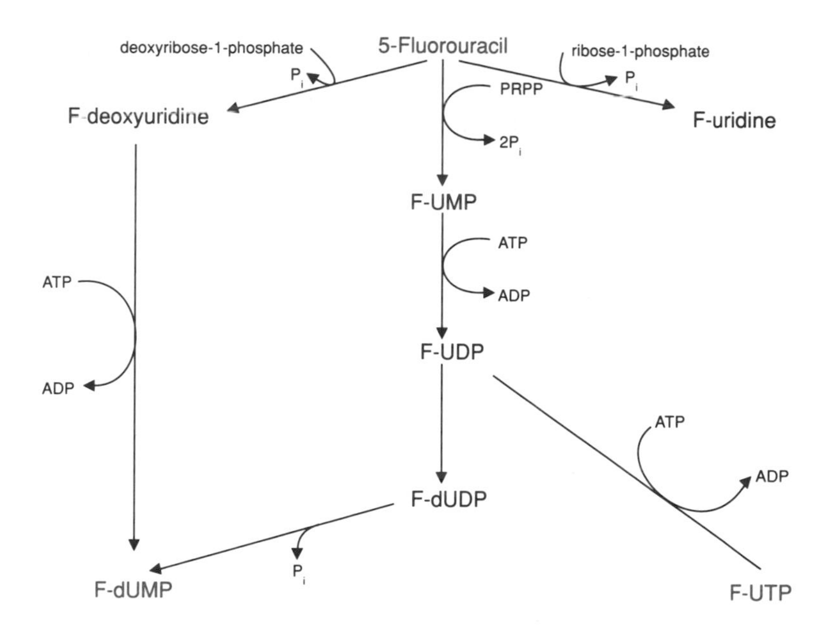

FIGURE 13.30
Activation of 5-fluorouracil to 5-fluorouridine 5′-triphosphate and 5-fluorodeoxyuridine 5′-monophosphate by cellular enzymes.

6-thiouric acid, allopurinol can be administered to inhibit the degradation of 6-MP and potentiate the antitumor properties of 6-MP.

5-Fluorouracil (*FUra*) (Figure 13.29) is a pyrimidine analog that has no biological activity as a cytotoxic drug. 5-Fluorouracil must be activated to the nucleotide level to exert its antitumor properties. The active metabolites of FUra are fluorodeoxyuridylate (FdUMP) and FUTP (Figure 13.30). Fluorodeoxyuridylate is a potent and specific inhibitor of thymidylate synthase. In the presence of thymidylate synthase, H_4folate and FdUMP, a ternary complex is formed that results in the covalent binding of FdUMP to thymidylate synthase. Fluorodeoxyuridylate, thus, causes what amounts to a "thymineless death" for the cells. The second metabolite, FUTP, is incorporated into various RNA species, which inhibits the maturation of 45S precursor rRNA into the 28S and 18S RNA species and alters the splicing of pre-mRNA into functional mRNA.

Cytosine arabinoside (*AraC*) (Figure 13.29) is used in the treatment of several forms of human cancer. The AraC must be metabolized by cellular enzymes to cytosine arabinoside 5′-triphosphate (araCTP) to exert its cytotoxic effects (Figure 13.31). AraCTP competes with dCTP in the DNA polymerase reaction; araCMP is incorporated into DNA and stops the synthesis of the growing DNA strand. Clinically, the efficacy of araC as an antileukemic drug correlates with the concentration of araCTP in the tumor cell, which determines the level of araCMP incorporated into DNA. The formation of araCMP via deoxycytidine kinase appears to be the rate-limiting step in the activation to araCTP.

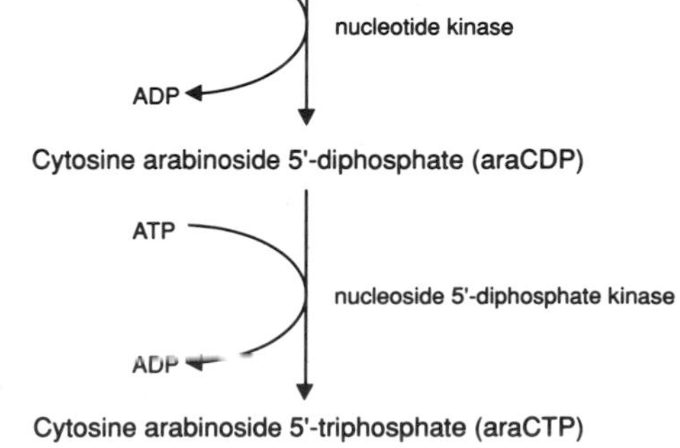

Cytosine arabinoside (araC)
ATP → ADP (deoxycytidine kinase)
Cytosine arabinoside 5′-monophosphate (araCMP)
ATP → ADP (nucleotide kinase)
Cytosine arabinoside 5′-diphosphate (araCDP)
ATP → ADP (nucleoside 5′-diphosphate kinase)
Cytosine arabinoside 5′-triphosphate (araCTP)

FIGURE 13.31
Activation of cytosine arabinoside to cytosine arabinoside 5′-triphosphate by cellular enzymes.

Antifolates Inhibit the Formation of Tetrahydrofolate

Antifolates are compounds that interfere with the formation of H_4folate from H_2folate or folate by inhibition of H_2folate reductase. **Methotrexate** (MTX) is an example of an antifolate that is currently in use as an antitumor agent in the treatment of human cancers. Methotrexate is a close structural analog of folic acid. The comparison of the two structures is seen in Figure 13.32. The differences are at the C-4 where an amino group replaces an OH group and at N-10 where a methyl group replaces a hydrogen atom. **Folylpolyglutamate synthetase,** which catalyzes the addition of 4 to 6 glutamic acid residues to H_4folate, also utilizes MTX as a substrate to form polyglutamylated-MTX. The mode of action of MTX is

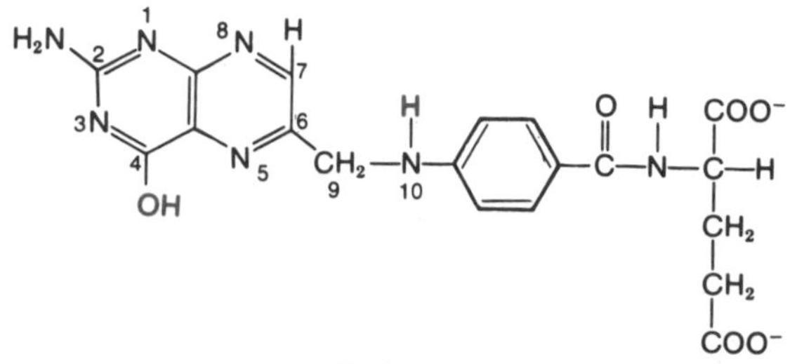

Folic acid

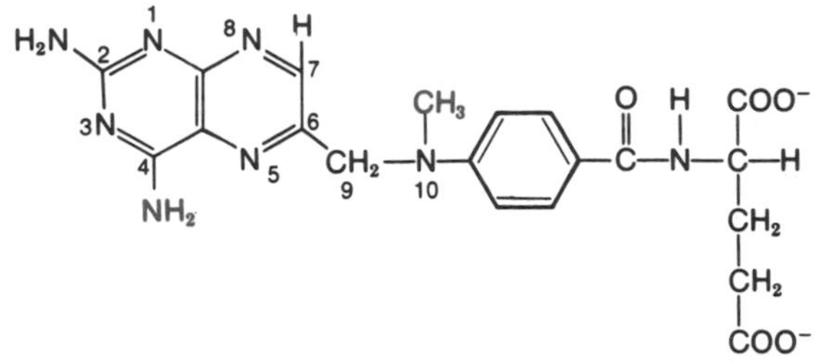

Methotrexate

FIGURE 13.32
Comparison of the structures of folic acid and methotrexate.

specific; it inhibits H_2folate reductase with a K_i in the range of 0.1 nM. The reactions inhibited are

$$\text{Folate} \xrightarrow[\text{MTX}]{\uparrow} H_2\text{folate} \xrightarrow[\text{MTX}]{\uparrow} H_4\text{folate}$$

Tumor cells in culture die when treated with MTX. The cytotoxic effects of MTX can be overcome by the addition of thymidine and hypoxanthine to the culture medium. The reversal by these metabolites indicates that the direct effect of MTX on cells leads to the depletion of thymidine and purine nucleotides. Figure 13.33 schematically shows the relationship between H_4folate, de novo purine nucleotide synthesis and dTMP formation.

In the treatment of human leukemias, normal cells can be rescued from the toxic effects of "high dose MTX" by N^5-formyl-H_4folate (leucovorin). This has led to increased clinical efficacy.

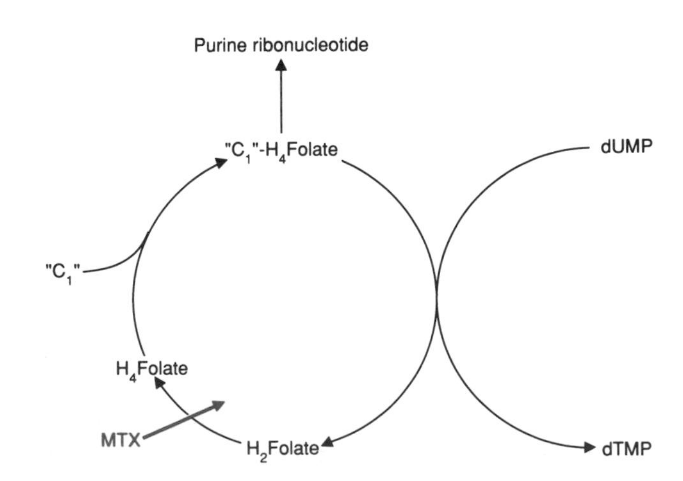

FIGURE 13.33
Relationship between the inhibition of H_2folate reductase and purine nucleotide and thymidine 5′-monophosphate syntheses.

Glutamine Antagonists Inhibit Glutamine Utilization as a Nitrogen Donor

Many reactions occur in mammalian cells in which glutamine serves as an amino donor. These amidation reactions are very critical for the synthesis of the purine ring de novo (N-3 and N-9), in the conversion of IMP to GMP, in the conversion of UTP to CTP and in the conversion of nicotinate adenine dinucleotide to NAD^+.

Compounds that inhibit these reactions are referred to as glutamine antagonists. **Azaserine** (*O*-diazoacetyl-L-serine) and **6-diazo-5-oxo-L-norleucine** (DON) (Figure 13.34), which were first isolated from cultures of *Streptomyces,* are very effective inhibitors of glutamine utilization. Since azaserine and DON inactivate the enzymes irreversibly, glutamine will not reverse the effects of these two drugs. It would require that many types of metabolites, such as guanine, cytidine, hypoxanthine (or adenine), and nicotinamide, would have to be utilized to overcome the sites blocked by these glutamine antagonists. As expected from the multiple sites of inhibition at key steps, the glutamine antagonists are extremely toxic.

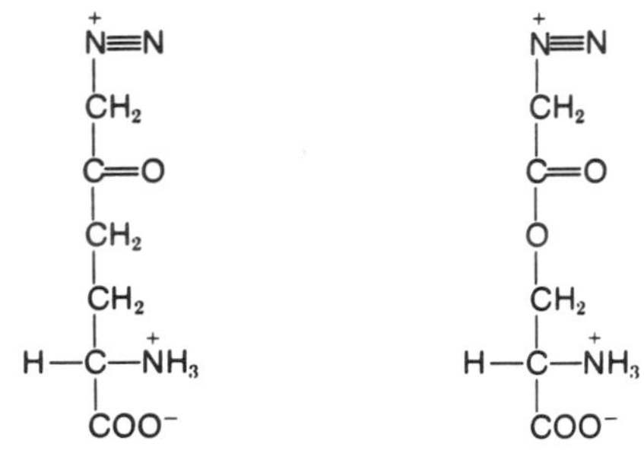

6-Diazo-5-oxo-L-norleucine

Azaserine

FIGURE 13.34
Structures of glutamine antagonists.

Other Agents Inhibit Cell Growth

Tumor cells treated with **hydroxyurea** (Figure 13.35) show a specific inhibition of DNA synthesis with little or no inhibition of RNA or protein synthesis. Hydroxyurea is an inhibitor of ribonucleotide reductase, blocking the reduction of all four nucleoside 5′-diphosphate substrates (CDP, UDP, GDP, and ADP) to the corresponding 2′-deoxyribonucleoside 5′-diphosphates. Toxicity to this drug results from the depletion of the 2′-deoxyribonucleoside 5′-triphosphates required for DNA replication. Although hydroxyurea is specific for the inhibition of ribonucleotide reductase, its clinical use is limited because of its rapid rate of clearance and the high drug concentration required for the effective inhibition of this enzyme.

Tiazofurin (Figure 13.35) is converted by cellular enzymes to the NAD^+ analog, tiazofurin adenine dinucleotide (TAD). Tiazofurin adenine dinucleotide inhibits IMP dehydrogenase, the rate-limiting enzyme in GTP synthesis, with a K_i of 0.1 μM. As a result of IMP dehydrogenase inhibition, the cellular concentration of GTP is markedly depressed in tiazofurin-treated cells.

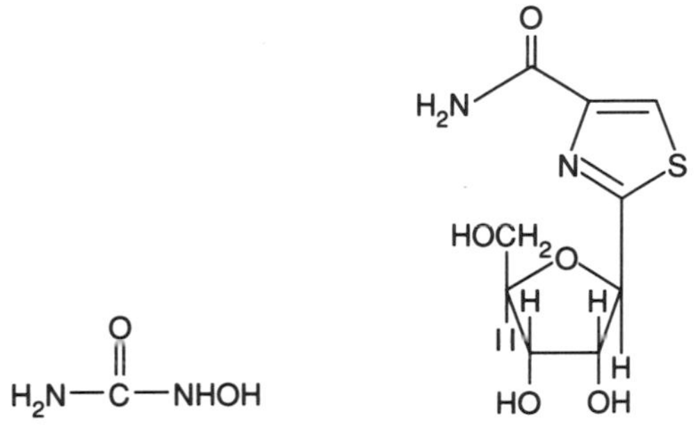

The drugs described above are only a few of the many compounds in clinical use, but these serve as examples in which the knowledge of basic biochemical pathways and mechanisms lead to the generation of effective drugs. Another important point regarding many of the antimetabolites used as drugs is that they must be activated to the nucleotide level by cellular enzymes to exert their cytotoxic effects.

Hydroxyurea

Tiazofurin

FIGURE 13.35
Structures of other inhibitors.

13.12 PURINE AND PYRIMIDINE ANALOGS AS ANTIVIRAL AGENTS

Herpesvirus (HSV) and human immunodeficiency virus (HIV) infections (AIDS) have become the plague of mankind in recent years. Two antimetabolites have been identified that can be used in the control–treatment (but, not cure) of HSV and HIV infections. These drugs (Figure 13.36), **acyclovir** (acycloguanosine), a purine analog, and **3′-azido-3′-deoxythymidine (AZT),** a pyrimidine analog, require metabolism to phosphorylated compounds to yield the active drug. These two demonstrate the specific nature of drug activation in human cells. Acycloguanosine is activated to

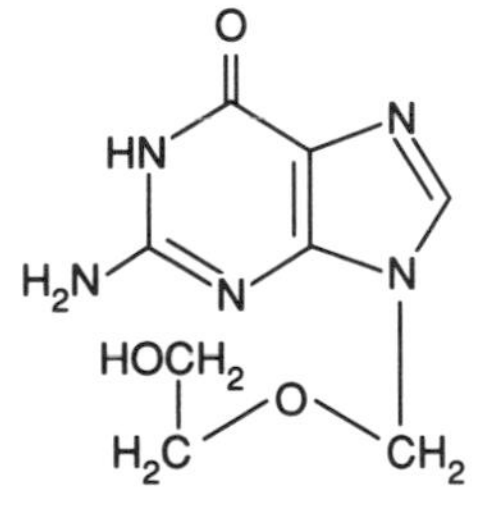

acycloguanosine

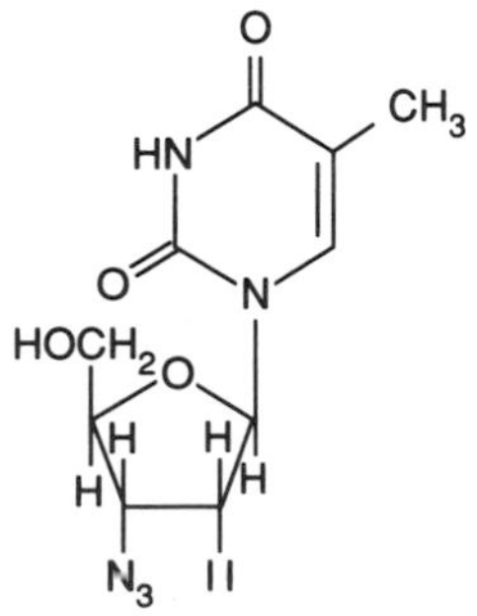

AZT

FIGURE 13.36
Structures of antiviral agents, 9-(2-hydroxymethylmethyl)guanine (acylcloguanosine) and 3′-deoxy-3′-azidothymidine (AZT).

the monophosphate by a specific HSV-thymidine kinase, encoded by the HSV genome, which can catalyze the phosphorylation of acycloguanosine. The host cellular thymidine kinase cannot utilize acyclovir as a substrate. The acycloguanosine monophosphate is then phosphorylated by the cellular enzymes to the di- and triphosphate forms. Acycloguanosine triphosphate serves as a substrate for the HSV-specific DNA polymerase and is incorporated into the growing viral DNA chain causing chain termination. The specificity of acycloguanosine and its high therapeutic index resides, therefore, in the fact that only HSV-infected cells can form the acycloguanosine monophosphate.

AZT is phosphorylated, on the other hand, by cellular kinases to AZT-triphosphate. The AZT-triphosphate blocks HIV replication by inhibiting HIV-DNA polymerase (an RNA-dependent polymerase). The selectivity of AZT for HIV-infected cells versus uninfected cells occurs because the DNA polymerase from HIV is at least 100-fold more sensitive to AZT-triphosphate than is host cell DNA-dependent DNA polymerase.

These two antiviral agents demonstrate the diversity of responses required for selectivity. In one case, enzyme activity encoded by the viral genome is mandatory for the activation of the drug (acycloguanosine); in the second example, although cellular enzymes activate the drug (AZT), the viral gene product (HIV-DNA polymerase) is the selective target.

13.13 BIOCHEMICAL BASIS FOR THE DEVELOPMENT OF DRUG RESISTANCE

The failure of chemotherapy in the treatment of human cancer is often related to the development of drug-resistant tumor cell populations. Tumors represent a very heterogeneous population of cells and in many instances drug-resistant cells are already present. Resistant cells are enriched as the drug-sensitive cells are killed by the cytotoxic drug. In other cases, drug treatment causes genetic alterations that result in drug-resistant cells. Resistance to drugs can fall into the categories of "specific drug resistance" or "multidrug resistance" (pleiotropic drug resistance).

An example of specific drug resistance is the resistance to methotrexate observed in tumor cells in culture and confirmed in clinically derived tumor samples. The biochemical bases for resistance to methotrexate can be multiple and include: (a) decreased uptake of methotrexate due to the impaired transporter of N^5-formyl-H_4folate and N^5-methyl-H_4folate; (b) amplification of the H_2folate reductase gene, which results in an increase in the target enzyme, H_2folate reductase; (c) altered H_2folate reductase gene whose gene product, H_2folate reductase, binds methotrexate less tightly; and (d) decreased activity of folylpolyglutamate synthetase. The net effect of any of the mechanisms of resistance is to decrease the ability of methotrexate to inhibit H_2folate reductase effectively in tumor cells at drug concentrations that minimize host toxicity.

Several mechanisms have been shown to account for the resistance to cystosine arabinoside in experimental systems. As discussed previously cytosine arabinoside must be converted to araCTP, the active metabolite. The rate-limiting step is the formation of araCMP in a reaction catalyzed by deoxycytidine kinase. A competing reaction for araC is that catalyzed by cytidine deaminase leading to the formation of uracil arabinoside, an inactive compound. A tumor cell population in which cytidine deaminase activity is elevated or deoxycytidine kinase activity is reduced would show resistance to araC. Other araC-resistant tumor cells have been shown to be able to generate sufficient intracellular araCTP; however, in these cells the half-life of araCTP is markedly reduced compared to the sensitive cells. This results in decreased incorporation of araC into DNA, which is the cytotoxic event.

In multiple drug resistance, a tumor cell population arises that is cross-resistant to a host of functionally unrelated antitumor agents. These compounds include drugs such as the vinca alkaloids, adriamycin, actinomycin D, and etoposide. Interesting points regarding the agents that fall into this multiple drug resistance category are that all of these drugs are natural products or derived from natural products and they are not chemically related in structure; they have different mechanisms of action as antitumor agents; and they appear to act on some nuclear event. Multiple drug resistant cells, regardless of the inducing drug, have elevated levels of a membrane glycoprotein (P 170) and when treated with drug do not accumulate cytotoxic levels of the drug. A cell that is resistant to vincristine will likewise not accumulate actinomycin D or any of the other agents in this category.

Drug resistance can be generated to just about any of the drugs that have been studied and the propensity for the development of resistance in tumor cells to specific antitumor drugs is quite high.

BIBLIOGRAPHY

Baliga, B. S. and Borek, E. Metabolism of thymine in tumor tissue: The origins of β-aminoisobutyric acid *Adv. Enzyme Regul.* 13:27, 1975.

Becker, M. A. and Kim, M. Regulation of purine synthesis *de novo* in human fibroblasts by purine nucleotides and phosphoribosylpyrophosphate. *J. Biol. Chem.* 262:14531, 1987.

Cory, J. G. Role of ribonucleotide reductase in cell division. *Inhibitors of Ribonucleoside Diphosphate Reductase Activity,* International Encyclopedia of Pharmacology and Therapeutics. J. G. Cory and A. H. Cory (Eds.). New York: Pergamon, p. 1, 1989.

Elion, G. B. The purine path to chemotherapy. *Science* 244:41, 1989.

Henderson, J. F. and Patterson, A. R. P. *Nucleotide metabolism: An introduction.* New York: Academic, 1973.

Jones, M. E. Pyrimidine nucleotide biosynthesis in animals: Genes, enzymes, and regulation of UMP biosynthesis. *Annu. Rev. Biochem.* 49:253, 1980.

Mitsuya, H. and Broder, S. Strategies for antiviral therapy in AIDS. *Nature (London)* 325:773, 1987.

Schimke, R. T. Gene amplification, drug resistance, and cancer. *Cancer Res.* 44:1735, 1984.

Scriver, C. R., Beaudet, A. L., Sly, W. S., and Valle, D. (Eds.). *The Metabolic Basis of Inherited Disease,* 6th ed. Vol. 1, Chapters 37–43, New York: McGraw-Hill, 1989.

Traut, T. W. and Loechel, S. Pyrimidine catabolism: Individual characterization of the three sequential enzymes with a new assay. *Biochemistry* 23:2533, 1984.

Tremblay, G. C., Crandall, D. E., Knott, C. E., and Alfant, M. Orotic acid biosynthesis in rat liver: Studies on the source of carbomylphosphate. *Arch. Biochem. Biophys.* 178:264, 1977.

Weber, G. Biochemical strategy of cancer cells and the design of chemotherapy: G. H. A. Clowes Memorial Lecture. *Cancer Res.* 43:3466, 1983.

Weber, G., Jayaram, H. N., Pillwein, K., Natsumeda, Y., Reardon, M. A., and Zhen, Y.-S. Salvage pathways as targets of chemotherapy. *Adv. Enzyme Regul.* 26:335, 1987.

Wyngaarden, J. B. Regulation of purine biosynthesis and turnover. *Adv. Enzyme Regul.* 14:25, 1976.

QUESTIONS

C. N. ANGSTADT AND J. BAGGOTT

Question Types are described on page xxiii.

1. (QT1) Nucleotides serve all of the following roles *except:*
   A. monomeric units of nucleic acids.
   B. physiological mediators.
   C. sources of chemical energy.
   D. structural components of membranes.
   E. structural components of coenzymes.

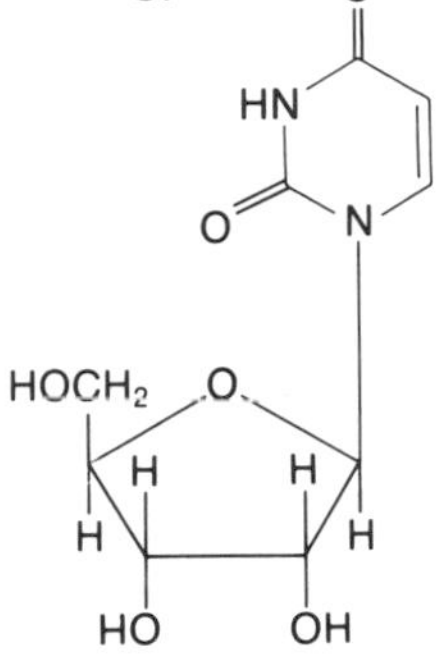

A. 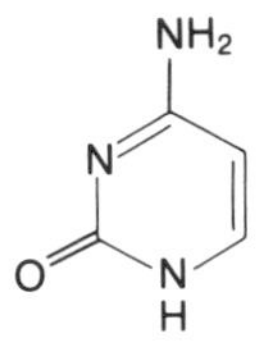 B. C. D. 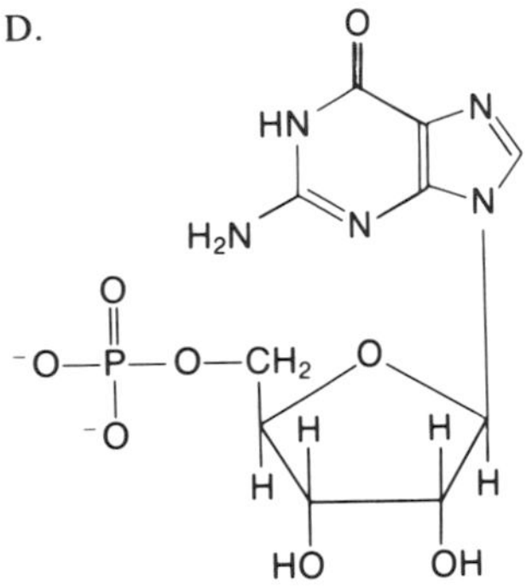 E.

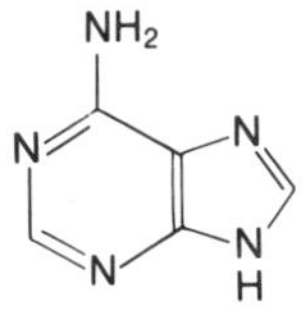

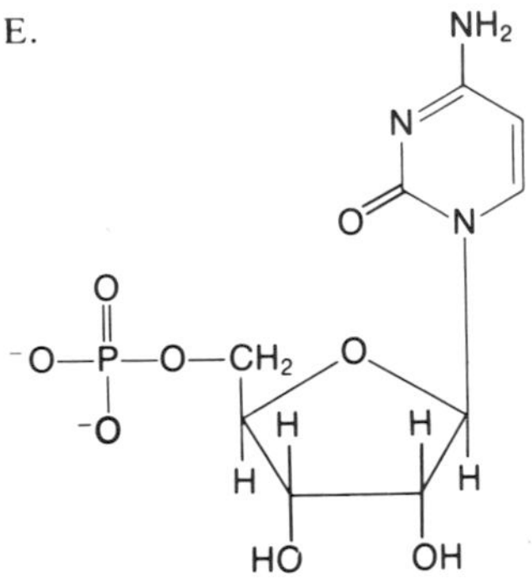

2. (QT5) Adenine
3. (QT5) A pyrimidine nucleoside
4. (QT5) CMP

A. De novo synthesis of purine nucleotides
B. De novo synthesis of pyrimidine nucleotides
C. Both
D. Neither

5. (QT4) A source of nitrogen is the amide nitrogen of glutamine.
6. (QT4) PRPP is a substrate of the rate-limiting step.
7. (QT4) The two nucleotides found in RNA are formed in a branched pathway from a common intermediate.
8. (QT4) A free base is formed in the process.
9. (QT2) Which of the following are aspects of the overall regulation of de novo purine nucleotide synthesis?
   1. AMP, GMP, and IMP cause a shift of PRPP amido transferase from a small form to a large form.
   2. PRPP levels in the cell can be severalfold less than the K_m of PRPP amidotransferase for PRPP.
   3. GMP is a competitive inhibitor of IMP dehydrogenase.
   4. UMP is a competitive inhibitor of OMP-decarboxylase.
10. (QT2) The type of enzyme known as a phosphoribosyltransferase is involved in:
    1. salvage of pyrimidine bases.
    2. the de novo synthesis of pyrimidine nucleotides.
    3. salvage of purine bases.
    4. the de novo synthesis of purine nucleotides.
11. (QT1) Uric acid is
    A. formed from xanthine in the presence of O_2.
    B. a degradation product of cytidine.
    C. deficient in the condition known as gout.
    D. a competitive inhibitor of xanthine oxidase.
    E. oxidized, in humans, before it is excreted in urine.
12. (QT2) In nucleic acid degradation:
    1. there are nucleases that are specific for either DNA or RNA.
    2. nucleotidases convert nucleotides to nucleosides.
    3. the conversion of a nucleoside to a free base is an example of a phosphorolysis.
    4. because of the presence of deaminases, hypoxanthine rather than adenine is formed.
13. (QT1) Deoxyribonucleotides:
    A. cannot be synthesized so they must be supplied preformed in the diet.
    B. are synthesized de novo using dPRPP.
    C. are synthesized from ribonucleotides by an enzyme system involving thioredoxin.
    D. are synthesized from ribonucleotides by nucleotide kinases.
    E. can be formed only by salvaging free bases.
14. (QT1) β-Aminoisobutyrate:
    A. is an intermediate in the degradation of both uracil and thymine.
    B. in the urine can be used to estimate the turnover of DNA.
    C. arises from uracil by cleavage of the pyrimidine ring.
    D. is in equilibrium with β-alanine.
    E. is the end product common to the degradation of both uracil and thymine.
15. (QT2) If a cell were unable to synthesize PRPP, which of the following processes would be likely to be *directly* impaired?
    1. FAD synthesis
    2. NAD synthesis
    3. Coenzyme A synthesis
    4. OMP synthesis

16. (QT2) Sources of the nicotinamide portion of NAD include:
    1. *N*-Methylnicotinamide.
    2. The vitamin riboflavin.
    3. PRPP.
    4. tryptophan.

17. (QT2) Which of the following antitumor agents work by impairing de novo purine synthesis?
    1. Azaserine (glutamine antagonist)
    2. 5-Fluorouracil (antimetabolite)
    3. Methotrexate (antifolate)
    4. Hydroxyurea

ANSWERS

1. D Both cAMP and cGMP are physiological mediators. NAD, FAD, and CoA all contain AMP as part of their structures (p. 531).
2. B Adenine is the free purine. (A is a pyrimidine).
3. C A nucleoside contains a base plus sugar but no phosphate.
4. E CMP is a pyrimidine nucleotide. (D is a purine nucleotide) (p. 532).
5. C Nitrogen atoms 3 and 9 of purine nucleotides (p. 536) and N-3 of pyrimidine nucleotides (p. 547) are supplied by glutamine.
6. A The rate-limiting step of purine nucleotide synthesis is the amido transfer between glutamine and PRPP (p. 536). PRPP is used in pyrimidine nucleotide synthesis but only after a pyrimidine has been formed (p. 548).
7. A GMP and AMP are both formed from the first purine nucleotide, IMP, in a branched pathway (Figure 13.13). The pyrimidine nucleotides UMP and CTP are formed in a sequential pathway from orotic acid (p. 549).
8. B There is no free purine base at any point of the pathway, but in pyrimidine nucleotide synthesis, orotic acid is formed. It is only then that the sugar phosphate is added (p. 548).
9. A 1, 2, 3 correct. 1 is the mechanism of inhibition since the large form of the enzyme is inactive (p. 541). 2: PRPP amidotransferase shows sigmoidal kinetics with respect to PRPP so large shifts in concentration of PRPP have the potential for altering velocity (p. 541). 3 plays a major role in controlling the branched pathway of IMP to GMP or AMP (p. 541). 4: OMP is a pyrimidine nucleotide so it would not be expected to be part of the regulation of purine nucleotide synthesis.
10. A 1, 2, 3 correct. Phosphoribosyltransferases are important salvage enzymes for both purines and pyrimidines (pp. 542, 550) and are also part of the synthesis of pyrimidines since OPRT catalyzes the conversion of orotate to OMP (p. 548). In purine nucleotide synthesis, though, the purine ring is built up stepwise on ribose-5-phosphate and not transferred to it (p. 536).
11. A The xanthine oxidase reaction produces uric acid. B, E: Uric acid is an end product of purines, not pyrimidines. C: Gout is characterized by excess uric acid (p. 546).
12. E All four correct. 1: They can also show specificity toward the bases and positions of cleavage. 2: A straight hydrolysis. 3: The product is ribose-1-phosphate rather than the free sugar. 4: AMP deaminase and adenosine deaminase remove the 6-NH_2 as NH_3. The IMP or inosine formed is eventually converted to hypoxanthine (Figure 13.17).
13. C Deoxyribonucleotides are synthesized from the ribonucleoside diphosphates by nucleoside diphosphate reductase that uses thioredoxin as the direct hydrogen-electron donor (p. 551). A, B, E: There is a synthetic mechanism as just described but it is not a de novo pathway. D: Nucleotide kinases are enzymes that add phosphate to a base or nucleotide.
14. B This compound originates exclusively from thymine, which is found primarily in DNA. A, E: It is an end product of degradation but only of thymine. C, D: β-Alanine arises from cleavage of the uracil ring.
15. C 2, 4 correct. PRPP is a substrate in both of these processes (p. 560). Both FAD and coenzyme A contain AMP, which is supplied as ATP (p. 562).
16. D Only 4 correct. 1: *N*-Methylnicotinamide is the excretory form of nicotinamide. 2: Nicotinamide comes from the vitamin nicotinic acid (FAD from riboflavin). 3: PRPP supplies the ribose-phosphate not the nicotinamide. 4: Tryptophan can be metabolized, through quinolinic acid, to nicotinate mononucleotide and eventually NAD (p. 559).
17. B 1, 3 correct. 1: Glutamine is the source of N-3 and N-9 for the purine ring. 2: 5-Fluorouracil is a pyrimidine analog not a purine analog. 3: Antifolates reduce the concentration of H_4folate compounds that are necessary for two steps of purine synthesis. 4: Hydroxyurea inhibits the reduction of ribonucleotides to deoxyribonucleotides so is not involved in de novo purine synthesis (p. 566).

14

Metabolic Interrelationships

ROBERT A. HARRIS
DAVID W. CRABB

14.1 OVERVIEW 576
14.2 STARVE–FEED CYCLE 577
In the Well-Fed State the Diet Supplies the Energy Requirements 577
In the Early Fasting State Hepatic Glycogenolysis Is an Important Source of Blood Glucose 579
The Fasting State Requires Gluconeogenesis From Amino Acids and Glycerol 580
In the Early Refed State, Fat Is Metabolized Normally, but Normal Glucose Metabolism Is Slowly Reestablished 582
The Insulin to Glucogen Ratio Is Critical for Caloric Homeostasis 583
Energy Requirements and Reserves 584
Glucose Homeostasis Has Five Phases 585
14.3 MECHANISMS INVOLVED IN SWITCHING THE METABOLISM OF THE LIVER BETWEEN THE WELL-FED STATE AND THE STARVED STATE 586
Substrate Availability Controls Many Metabolic Pathways 586
Negative and Positive Allosteric Effectors Regulate Many Pathways 586
Covalent Modification Regulates Key Enzymes 589
Changes in Levels of Key Enzymes Are a Longer Term Adaptive Mechanism 591
14.4 METABOLIC INTERRELATIONSHIPS OF TISSUES IN VARIOUS NUTRITIONAL AND HORMONAL STATES 593
Obesity Occurs When Stored Fuel Is Not Completely Consumed During Fasting 593
Aerobic and Anaerobic Exercise Use Different Fuels 593
Changes in Pregnancy Are Related to Fetal Requirements and Hormonal Changes 595
Lactation Requires the Synthesis of Lactose, Triacylglycerol, and Protein 596
Insulin-Dependent Diabetes Mellitus 597
Noninsulin-Dependent Diabetes Mellitus 597
Stress and Injury Lead to Metabolic Changes 598
Liver Disease Causes Major Metabolic Derangements 598
In Renal Disease Nitrogenous Wastes Accumulate 600
Ethanol Is Oxidized in the Liver Altering the NAD^+ to NADH Ratio 600
In Acid-Base Balance, Glutamine Has an Important Role 602
BIBLIOGRAPHY 603
QUESTIONS AND ANNOTATED ANSWERS 604
CLINICAL CORRELATIONS
14.1 Obesity 576
14.2 Protein Malnutrition 577
14.3 Starvation 577
14.4 Reye's Syndrome 582
14.5 Hyperglycemic, Hyperosmolar Coma 584
14.6 Hyperglycemia and Protein Glycosylation 585
14.7 Insulin-Dependent Diabetes Mellitus 597
14.8 Complications of Diabetes and the Polyol Pathway 598
14.9 Noninsulin-Dependent Diabetes Mellitus 600
14.10 Cancer Cachexia 602

CLIN. CORR. 14.1 OBESITY

Obesity is the most common nutritional problem in the United States and other affluent countries of the world. It causes a reduction in life span and is a risk factor in the development of diabetes mellitus, hypertension, endometrial carcinoma osteoarthritis, gallstones, and cardiovascular diseases. Obesity is easy to explain—an obese person has eaten more than he/she required. The accumulation of massive amounts of body fat is not otherwise possible. Appreciation of this fact, however, is of little or no consolation to the obese person. For unknown reasons, the neural control of caloric intake to balance energy expenditure is abnormal. Rarely, obesity is secondary to a correctable disorder. Hypothyroidism causes puffiness and some weight gain but is hardly ever the cause of obesity. Cushing's syndrome, the result of increased levels of glucocorticoids, causes fat deposition in the face and trunk, with wasting of the limbs, and glucose intolerance. These effects are due to increased protein breakdown in muscle and conversion of the amino acids to glucose and fat. Less commonly, tumors, vascular accidents, or maldevelopment of the nervous system hunger control centers in the hypothalamus cause obesity. The most common "gland" problem causing obesity is overactivity of the salivary glands, that is, overeating.

In the most common type of obesity, the number of adipocytes of the body does not increase, they just get large as they become engorged with triacylglycerols. If obesity develops before puberty, however, an increase in the number of adipocytes can also occur. In the latter case, both hyperplasia (increase in cell number) and hypertrophy (increase in cell size) are contributing factors to the magnitude of the obesity.

The only effective treatment of obesity is reduction in the ingestion, absorption, or use of calories. Practically speaking, this means dieting. Unfortunately, the body compensates for decreased energy intake with reduced formation of triiodothyronine and a corresponding decrease in the basal metabolic rate. Thus, there is a biochemical basis for the universal complaint that it is far easier to gain than to lose weight. Furthermore, about 95% of people who are able to lose a significant amount of weight regain it within 1 year.

Bray, G. D. Effect of caloric restriction on energy expenditure in obese patients. *Lancet* 2:397, 1969; and Bray, G. D. The overweight patient. *Adv. Int. Med.* 21:267, 1976.

14.1 OVERVIEW

In this chapter the interdependence of the metabolic processes of the major tissues of the body will be stressed. Not all of the major metabolic pathways and processes of the body operate in every tissue at any given time. Given the nutritional and hormonal status of a patient, we need to be able to say qualitatively which of the major metabolic pathways of the body are functional and how these pathways relate to one another.

The metabolic processes with which we are now concerned are glycogenesis, glycogenolysis, gluconeogenesis, glycolysis, fatty acid synthesis, fatty acid oxidation, citric acid cycle activity, ketogenesis, amino acid oxidation, protein synthesis, proteolysis, and urea synthesis. It is important to know: (a) which tissues are most active in these various processes, (b) when these processes are most or least active, and (c) how these processes are controlled and coordinated in different metabolic states.

The best way to gain an understanding of the relationships of the major metabolic pathways to one another is to become familiar with the changes in metabolism that occur during the **starve–feed cycle.** As shown in Figure 14.1, the starve–feed cycle allows a variable fuel consumption to meet a variable metabolic demand. Starve is a poor choice of words; fast is what we mean, but the phrase "fast–feed" brings to mind lunch at McDonald's rather than fasting followed by feeding. Feed refers to the intake of meals (the variable fuel input) after which we store the fuel (in the form of glycogen and fat) to be used to meet our metabolic demand while we fast. Note the participation of an ATP cycle within the starve–feed cycle (Figure 14.1). Adenosine triphosphate functions as the energy-transferring agent in the starve–feed cycle, being like money to the cell.

Humans have the capacity to consume food at a rate some 100 times greater than their basal caloric requirements. This allows us to survive from meal to meal without nibbling continuously between meals. We thus store the calories as glycogen and fat and utilize them as needed. Unfortunately, an almost unlimited capacity to consume food is matched by an almost unlimited capacity to store it as fat. Obesity is the consequence of excess food consumption and is common in affluent countries (Clin. Corr. 14.1), whereas other forms of malnutrition are more prevalent in developing countries (Clin. Corr. 14.2 and 14.3). Every day represents a series of starve–feed cycles. It balances out quite nicely for most, that is, fuel consumption equals fuel utilization. The regulation of food consumption is complex and not well understood. The tight control needed is indicated by the calculation that eating two extra pats of butter (~100 cal) per day over caloric expenditures results in a 10 lb weight gain per year. A weight

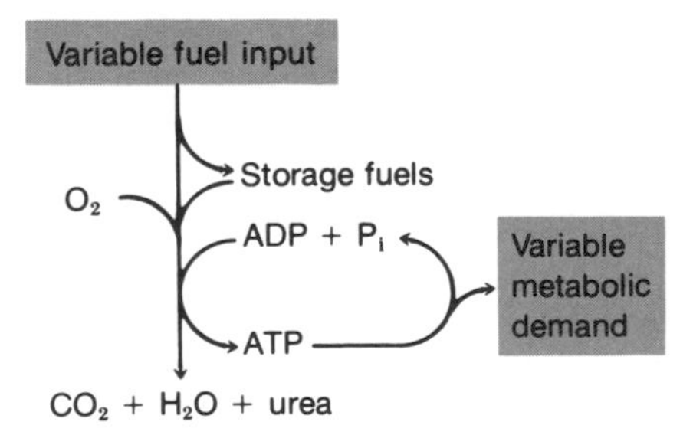

FIGURE 14.1
Humans are able to use a variable fuel input to meet a variable metabolic demand.

CLIN. CORR. 14.2
PROTEIN MALNUTRITION

Protein malnutrition is the most important and widespread nutritional problem among young children in the world today. The clinical syndrome, called kwashiorkor, occurs mainly in children 1–3 years of age and is precipitated by weaning an infant from breast milk onto a starchy, protein-poor diet. The name originated in Ghana, meaning "the sickness of the older child when the next baby is born." It is a consequence of feeding the child a diet adequate in calories but deficient in protein. It may become clinically manifest when protein requirements are increased by infection, for example, malaria, helminth infestation, or gastroenteritis. The syndrome is characterized by poor growth, low plasma protein levels, muscle wasting, edema, diarrhea, and increased susceptibility to infection. The presence of subcutaneous fat clearly differentiates it from simple starvation. The maintenance of fat stores is due to the high carbohydrate intake and resulting high insulin levels. In fact, the high insulin level interferes with the adaptations described for starvation. Specifically, fat is not mobilized as an energy source, ketogenesis does not take place, and there is no transfer of amino acids from the skeletal muscle to the visceral organs, that is, the liver, kidneys, heart, and immune cells. The lack of dietary amino acids results in diminished protein synthesis in all tissues. The liver becomes enlarged and infiltrated with fat, reflecting the need for hepatic protein synthesis for the formation and release of lipoproteins. In addition, protein malnutrition impairs the function of the gut, resulting in malabsorption of calories, protein, and vitamins, which accelerates the disease. The consequences of the disease include a permanent stunting of physical growth and poor intellectual and psychological development.

Bistrian, B. R., Blackburn, G. L., Vitale, J., and Cochran, D. Prevalence of malnutrition in general medical patients. *J. Am. Med. Assoc.* 235:1567, 1976; and Chase, H. P., Kumar, V., Caldwell, R. T., and O'Brien, D.: Kwashiorkor in the United States. *Pediatrics* 66:972, 1980.

gain of 10 lb may not sound excessive, but multiplied by 10 years it equals obesity!

14.2 STARVE–FEED CYCLE

In the Well-Fed State the Diet Supplies the Energy Requirements

Figure 14.2 shows the fate of glucose, amino acids, and fat obtained from food. Note the different route by which fat enters the bloodstream. Glucose and amino acids pass directly into the blood from the intestinal epithelial cells and are presented to the liver by way of the portal vein. Fat, contained in chylomicrons, is secreted by the intestinal epithelial cells into lymphatics, which drain the intestine. The lymphatics lead to the thoracic duct, which, by way of the subclavian vein, delivers chylomicrons to the blood at a site of rapid blood flow. The latter provides for rapid distribution of the chylomicrons and prevents coalescence of the fat particles.

Liver is the first tissue to have the opportunity to use dietary glucose. Glucose can be converted into glycogen by glycogenesis, into pyruvate and lactate by glycolysis, or can be used in the pentose phosphate pathway for the generation of NADPH for synthetic processes. Pyruvate can be oxidized to acetyl CoA, which, in turn, can be converted into fat or oxidized to CO_2 and water by the TCA cycle. Some of the glucose coming from the intestine escapes the liver and circulates to other tissues. The brain is almost solely dependent on glucose for the production of ATP. Other major users of glucose include red blood cells, which can only convert glucose to lactate and pyruvate, and the adipose tissue, which converts it into fat. Muscle also has the capacity to use glucose, converting it to glycogen or using it in the glycolytic and the TCA cycle pathways. A number of tissues produce lactate and pyruvate from circulating glucose by glycolysis. Lactate and pyruvate generated in peripheral tissues are taken up by the liver and converted to fat by the process of lipogene-

CLIN. CORR. 14.3
STARVATION

Starvation, or an overall deficit in food intake, including both calories and protein, leads to the development of a syndrome known as marasmus. Marasmus is a word of Greek origin meaning "to waste." Although not restricted to any age group, it is most common in children under 1 year of age. In developing countries early weaning of infants from breast milk is a common cause of marasmus. This may result from pregnancies in rapid succession, the desire of the mother to return to work, or switching to artificial formulas. Although the latter usually provides complete nutrition for a growing infant, poor parents tend to dilute it with water to make it last longer. This practice leads to insufficient intake of calories. Likewise, diarrhea and malabsorption can develop if safe water and sterile procedures are not used.

In contrast to kwashiorkor (see Clin. Corr. 14.2), subcutaneous fat, hepatomegaly, and fatty liver are absent in marasmus because fat is mobilized as an energy source and muscle temporarily provides amino acids to the liver for the synthesis of glucose and hepatic proteins. Low insulin levels allow the liver to oxidize fatty acids and to produce ketone bodies for other tissues. Ultimately, energy and protein reserves are exhausted, and the child starves to death. The immediate cause of death is often pneumonia, which occurs because the child is too weak to cough.

Waterlow, J. C., Childhood malnutrition—the global problem. *Proc. Nutr. Soc.* 38:1, 1979.

sis. In the very well-fed state, the liver uses glucose and does not engage in gluconeogenesis. Thus, the Cori cycle, which involves conversion of glucose to lactate in the peripheral tissues followed by conversion of lactate back to glucose in the liver, is interrupted in the well-fed state.

Dietary protein is hydrolyzed in the intestine, the cells of which use some amino acids as an energy source. Most dietary amino acids are transported into the portal blood, but the intestine metabolizes aspartate, asparagine, glutamate, and glutamine, and releases alanine, lactate, citrulline, and proline into the portal blood.

Liver then has the opportunity to remove absorbed amino acids from the blood (Figure 14.2). The liver lets most of each amino acid pass through, unless the concentration of the amino acid is unusually high. This is especially important for the essential amino acids, needed by all tissues of the body for protein synthesis. The liver can catabolize amino acids, but the K_m values for amino acids of many of the enzymes involved is high, allowing the amino acids to be present in excess before significant catabolism can occur. In contrast, the tRNA-charging enzymes have much lower K_m values for amino acids. This ensures that as long as all the amino acids are present, protein synthesis can occur as needed for growth and protein turnover. Excess amino acids can be oxidized completely to CO_2 and water, or the intermediates generated can be used as substrates

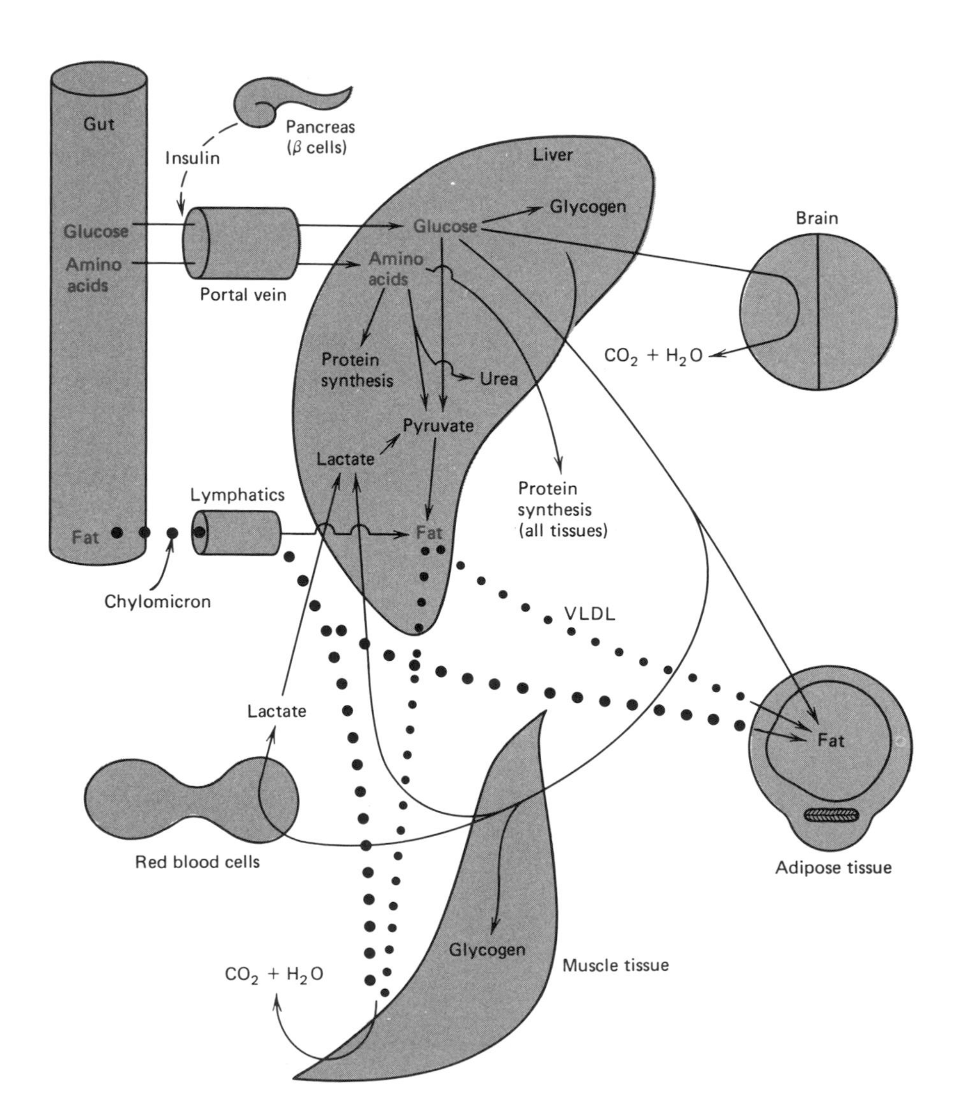

FIGURE 14.2
Disposition of glucose, amino acids, and fat by various tissues in the well-fed state.

for lipogenesis with the amino nitrogen converted to urea. Amino acids that escape the liver can be used for protein synthesis or for energy in other tissues. Skeletal muscle and heart muscle have a high capacity for transamination of amino acids and oxidation of the resulting α-keto acids to CO_2 and water. The **branched-chain amino acids** (leucine, isoleucine, and valine) are handled in an interesting manner. The liver has low capacity for transamination of these amino acids but considerable capacity for oxidative decarboxylation of their corresponding α-keto acids. On the other hand, skeletal muscle has considerable capacity for transamination but is relatively deficient in the enzymes responsible for subsequent catabolism. As a consequence, much of the transamination occurs in skeletal muscle, the α-keto acids escape into the blood, and the liver oxidizes the α-keto acid. Branched-chain amino acids are a major source of nitrogen for the production of alanine and glutamine in muscle. **Citrulline** produced in the gut is metabolized by the kidney to arginine. The liver can then use this arginine to generate urea and ornithine. This extra ornithine increases the liver's capacity for urea synthesis after a high protein meal.

When considering fat delivery to the tissues, we must differentiate between endogenous and exogenous fat (Figure 14.2). Glucose, lactate, pyruvate, and amino acids can be used to support hepatic lipogenesis. The fat formed from these substrates is released from the liver in the form of VLDL. Dietary fat is delivered to the bloodstream as chylomicrons. Both chylomicrons and VLDL circulate in the blood until they are acted upon by an extracellular enzyme attached to the endothelial cells of the capillaries. This enzyme, lipoprotein lipase, is particularly abundant in the capillaries in adipose tissue. It acts on both the VLDL and chylomicrons, liberating fatty acids by hydrolysis of the triacylglycerols. The fatty acids are then taken up by the adipocytes, reesterified with glycerol 3-phosphate to form triacylglycerols, and stored as large fat droplets within these cells. The glycerol 3-phosphate needed for triacylglycerol formation in adipose tissue is generated from glucose, using the first one-half of the glycolytic pathway to generate dihydroxyacetone phosphate, which is reduced to glycerol 3-phosphate by glycerol 3-phosphate dehydrogenase.

The β cells of the pancreas are very responsive to the influx of glucose and amino acids in the fed state. The β cells release insulin during and after eating, which is essential for the metabolism of these nutrients by liver, muscle, and adipose tissue. The role of insulin in the starve–feed cycle is discussed in more detail in Section 14.3.

In the Early Fasting State Hepatic Glycogenolysis Is an Important Source of Blood Glucose

Figure 14.3 shows what happens in early fasting after fuel stops coming in from the gut. Hepatic glycogenolysis is very important for the maintenance of blood glucose during this period. Lipogenesis is curtailed, and the lactate, pyruvate, and amino acids that were used by that pathway are diverted into the formation of glucose. The Cori cycle becomes an important pathway for maintaining blood glucose levels; glucose is produced from lactate by the liver and is then converted back to lactate by glycolysis in peripheral tissues such as red blood cells. The alanine cycle, in which carbon returns to the liver in the form of alanine rather than lactate, also becomes important. The catabolism of amino acids for energy is greatly diminished in the early fasting condition because less is available from the gut.

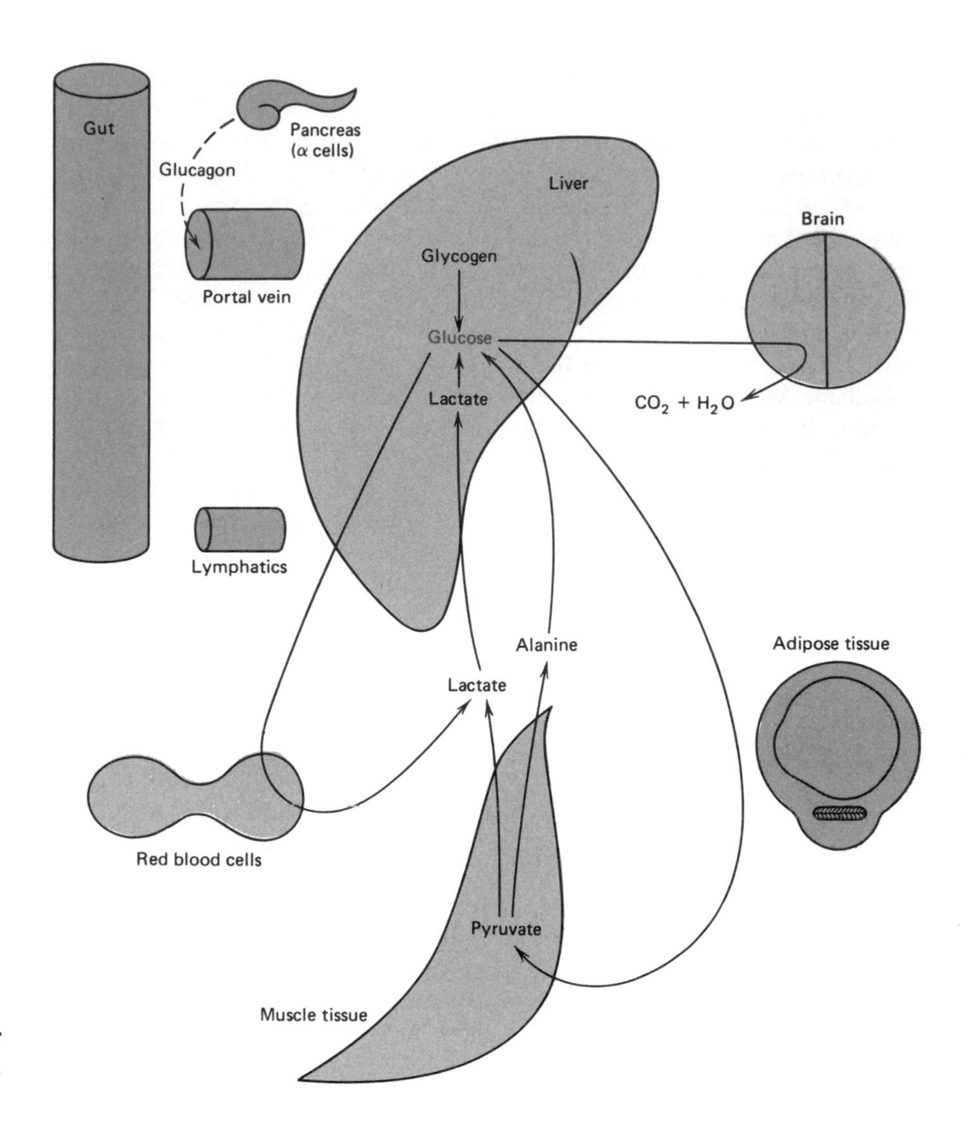

FIGURE 14.3
Metabolic interrelationships of the major tissues of the body in the early fasting state.

The Fasting State Requires Gluconeogenesis From Amino Acids and Glycerol

Figure 14.4 shows what happens in the fasting state. No fuel enters from the gut and little glycogen is left in the liver. Tissues that use glucose are then dependent on hepatic gluconeogenesis, primarily from lactate, glycerol, and alanine. The Cori and the alanine cycles described above play important roles; however, these cycles do not provide carbon for net synthesis of glucose. Glucose formed from lactate and alanine by the liver merely replaces that which was converted to lactate and alanine by the peripheral tissues. The brain oxidizes glucose completely to CO_2 and water. Hence net glucose synthesis from some source of carbon is mandatory in fasting. Fatty acids cannot be used for the synthesis of glucose, because acetyl CoA obtained by fatty acid catabolism cannot be converted to C_3 intermediates of gluconeogenesis. Glycerol, a byproduct of lipolysis in adipose tissue, is an important substrate for glucose synthesis in the fasted state. However, protein, especially from skeletal muscle, supplies most of the carbon needed for net glucose synthesis. Proteins are hydrolyzed within muscle cells (proteolysis) to produce amino acids. Most of the amino acids are not released but are partially metabolized within the muscle cell. Only three amino acids—alanine, glutamine, and glycine—are released in large amounts. The others are metabolized to give intermediates (pyruvate and α-ketoglutarate), which can yield

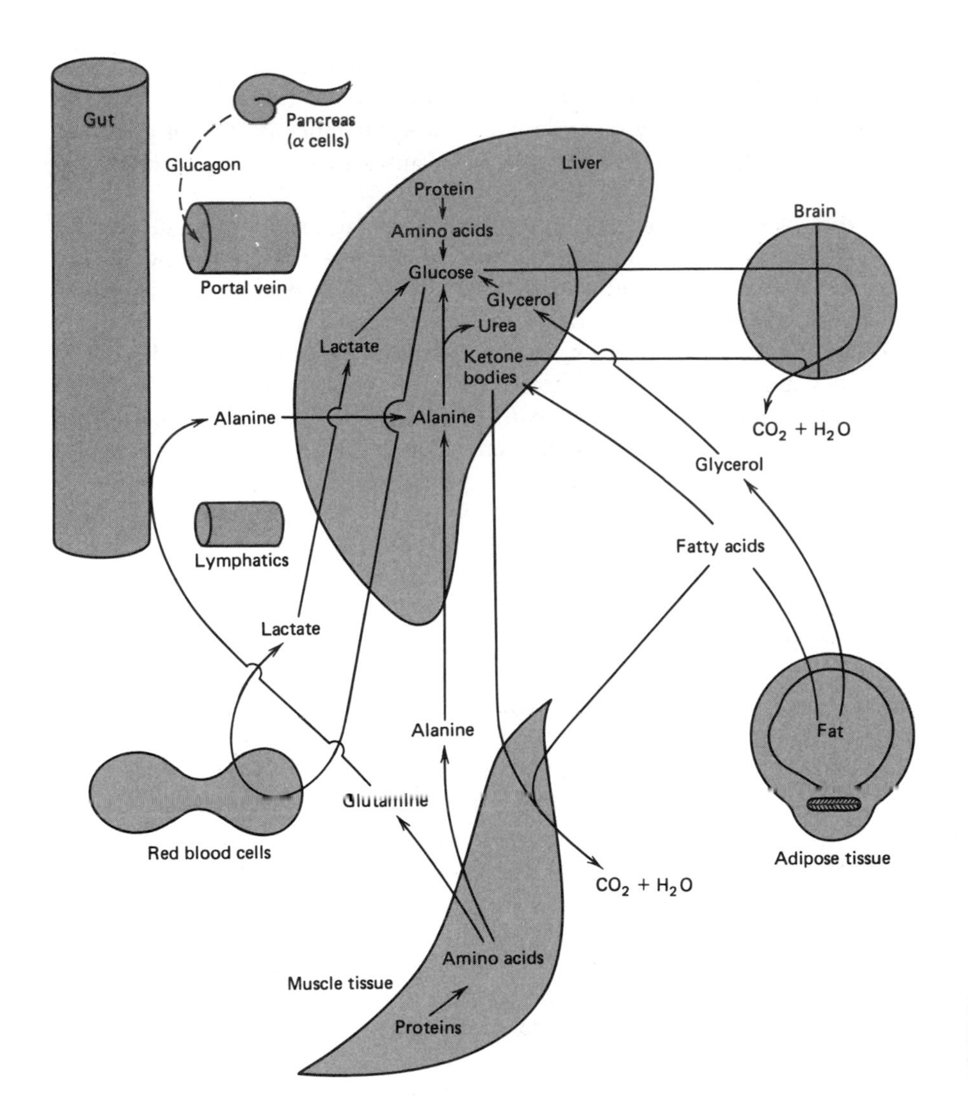

FIGURE 14.4
Metabolic interrelationships of the major tissues of the body in the fasting state.

alanine and glutamine. These amino acids are then released into the blood, from which they can be removed by the liver or kidney for net glucose formation. Alanine is quantitatively the most important gluconeogenic substrate. Kidney and muscle oxidize fatty acids in preference to glucose. Muscle also releases branched-chain α-keto acids to the liver, which synthesizes glucose from the keto acid of valine, ketone bodies from the keto acid of leucine, and both glucose and ketone bodies from the keto acid of isoleucine. There is evidence that much of the glutamine released from muscle is converted into alanine by the intestinal epithelium. Glutamine is partially oxidized in these cells to supply energy, and the carbon and amino groups left over are released back into the bloodstream in part as alanine and NH_4^+. This pathway probably involves the formation of oxaloacetate from glutamine via the TCA cycle and the conversion of oxaloacetate to phosphoenolpyruvate, and phosphoenolpyruvate to pyruvate. Direct oxidative decarboxylation of malate to pyruvate is also possible. Pyruvate is then transaminated to alanine. Glycine released from muscle is transformed in part to serine by the kidneys. Serine is subsequently converted into glucose by the liver or kidney.

The synthesis of glucose in the liver during fasting is closely linked to the synthesis of urea. Most amino acids can give up the amino nitrogen by transamination with α-ketoglutarate, forming glutamate and a new α-keto acid, which can be utilized for glucose synthesis. Glutamate provides both nitrogen compounds required for urea synthesis, ammonia from oxidative deamination by glutamate dehydrogenase, and aspartate from transamination of oxaloacetate by aspartate aminotransferase. An additional important source of ammonia is the gut mucosa, which converts glutamine to alanine and ammonia. The gut also releases precursors of ornithine such as citrulline, as described above.

Adipose tissue is also very important in the fasting state. Because of the low insulin : glucagon ratio during fasting, lipolysis is greatly activated. This raises the blood level of fatty acids, which are used in preference to glucose by many tissues. In heart and muscle, the oxidation of fatty acids inhibits glycolysis. The brain, on the other hand, does not oxidize fatty acids because fatty acids cannot cross the blood–brain barrier. In liver, fatty acid oxidation provides most of the ATP needed for gluconeogenesis. Very little of the acetyl CoA generated by β oxidation in the liver is oxidized completely. The acetyl CoA is converted into ketone bodies by liver mitochondria under these conditions. The ketone bodies (acetoacetate and β-hydroxybutyrate) are released into the blood and are a source of energy for many tissues. Like fatty acids, ketone bodies are preferred by many tissues over glucose. In contrast to fatty acids, ketone bodies penetrate the blood–brain barrier. Once their blood concentration is high enough, ketone bodies function as an alternative fuel for the brain. They are unable, however, to replace the need for glucose by the brain completely. Ketone bodies also suppress proteolysis in skeletal muscle and thereby decrease muscle wasting during starvation. As long as ketone body levels are maintained at a high level by hepatic fatty acid oxidation, there is less need for glucose, less need for gluconeogenic amino acids, and less need for using up precious muscle tissue by proteolysis.

The interrelationships between liver, muscle, and adipose tissue in supplying glucose for the brain are shown in Figure 14.4. The liver synthesizes the glucose, the muscle supplies the substrate (alanine), and the adipose tissue supplies the ATP (via fatty acids for oxidation) needed for hepatic gluconeogenesis. These relationships are disrupted in Reye's syndrome (Clin. Corr. 14.4) and by alcohol (Clin. Corr. 7.10). This interaction is dependent on a low insulin/glucagon ratio. Glucose levels are lower in the fasting condition, reducing the secretion of insulin but favoring the release of glucagon from the pancreas. In addition, fasting reduces the formation of triiodothyronine, the active form of thyroid hormone, from thyroxine. This reduces the daily basal energy requirements by as much as 25%.

CLIN. CORR. 14.4
REYE'S SYNDROME

Reye's syndrome is a devastating illness of children that tends to follow viral infections. It is characterized by evidence of brain dysfunction and edema (irritability, lethargy, and coma) and liver dysfunction (elevated plasma free fatty acids, fatty liver, hypoglycemia, hyperammonemia, and accumulation of short-chain organic acids). In some respects, it appears that hepatic mitochondria are specifically damaged, which impairs β oxidation and synthesis of carbamoyl phosphate and ornithine (for ammonia detoxification) and oxaloacetate (for gluconeogenesis). On the other hand, the accumulation of organic acids has suggested that the oxidation of these compounds is defective and that the CoA esters of some of these acids may inhibit specific enzymes, such as carbamoyl phosphate synthetase I, pyruvate dehydrogenase, pyruvate carboxylase, and the adenine nucleotide transporter, all present in mitochondria. The issue has not yet been resolved. Aspirin (acetylsalicylate) may contribute to the development of Reye's syndrome, and parents have been urged in a national campaign not to give aspirin to children with fever. The therapy for Reye's syndrome consists of measures to reduce brain edema (administration of hypertonic solutions of mannitol, hyperventilation, and occasionally pentobarbital-induced coma) and the provision of glucose intravenously. Glucose administration prevents hypoglycemia and elicits a rise in insulin levels that may (a) inhibit lipolysis in adipose cells and (b) reduce proteolysis in muscles and the release of amino acids, which (c) reduces the deamination of amino acids to ammonia. For unclear reasons, the incidence of Reye's syndrome has fallen dramatically.

Reye, R. D. K., Morgan, G., and Baval, J. Encephalopathy and fatty degeneration of the viscera, a disease entity in childhood. *Lancet* 2:749, 1963; and Snodgrass, P. J. and DeLong, G. R., Urea cycle enzyme deficiencies and an increased nitrogen load producing hyperammonemia in Reye's syndrome. *N. Engl. J. Med:* 294:855, 1976.

In the Early Refed State, Fat Is Metabolized Normally, but Normal Glucose Metabolism Is Slowly Reestablished

Figure 14.5 shows what happens soon after fuel is absorbed from the gut. Fat is metabolized as described above for the well-fed state. In contrast, glucose is poorly extracted by the liver and serves as a poor precursor of hepatic glycogen synthesis during this period of the starve–feed cycle. Glycolysis by the liver is also slowly established, so that the liver remains in the gluconeogenic mode for a few hours after feeding. Rather than providing blood glucose, however, hepatic gluconeogenesis provides glucose 6-phosphate for glycogenesis. This means that liver glycogen is not repleted after a fast by direct synthesis from blood glucose. Rather, the incoming glucose is catabolized in peripheral tissues to lactate, which is converted in the liver to glycogen. Gluconeogenesis from specific amino acids entering from the gut may also play an important role in reestablishing normal liver glycogen levels. After the maintenance of the well-fed state for a few hours, the metabolic interrelationships of Figure 14.2 become established. The rate of gluconeogenesis declines, glycolysis be-

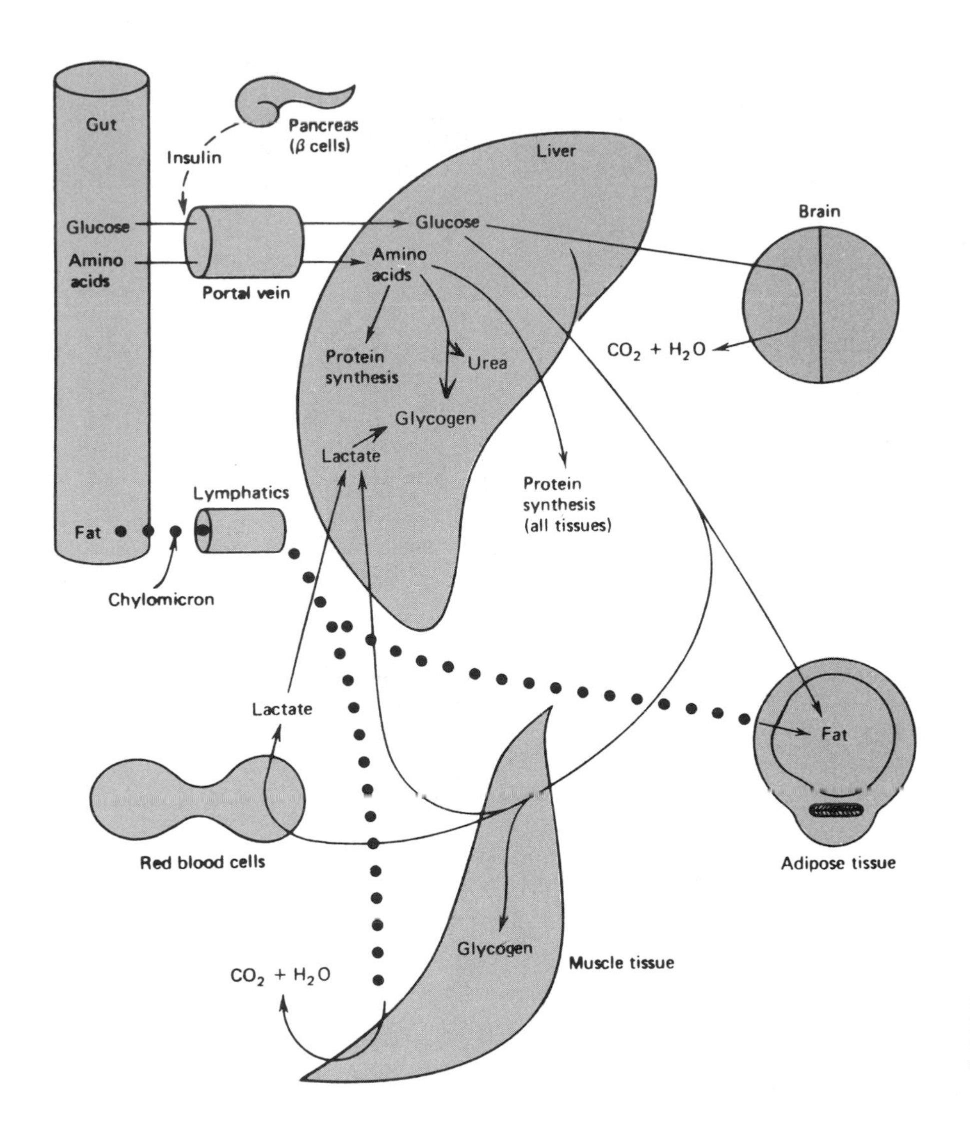

FIGURE 14.5
Metabolic interrelationships of the major tissues of the body in the early refed state.

comes the predominant means of glucose disposal in the liver, and liver glycogen is sustained by direct synthesis from glucose.

The Insulin to Glucogen Ratio Is Critical for Caloric Homeostasis

The tissues of the body work together to maintain a constant availability of fuels in the blood. This is termed **caloric homeostasis,** which, as illustrated in Table 14.1, means that regardless of whether a person is in the well-fed state, fasting, or starving to death, the blood level of fuels that supply an equivalent amount of ATP when metabolized does not fall below certain limits. The changes in insulin/glucagon ratio, discussed above and shown in Table 14.1, are crucial to the proper maintenance of caloric homeostasis. Note that blood glucose levels are controlled within very tight limits, whereas fatty acid concentrations in the blood can vary by an order of magnitude and ketone bodies by two orders of magnitude. Glucose is maintained within tight limits because of the absolute need of the brain for this substrate. If the blood glucose level falls too low (<1.5 mM), coma results from lack of ATP production, and death will follow shortly unless the situation can be rapidly corrected. On the other hand, hyperglycemia must be avoided because of the risk of hyperosmolar, hyperglycemic coma (see Clin. Corr. 14.5). Hyperglycemia also leads to

CLIN. CORR. 14.5
HYPERGLYCEMIC, HYPEROSMOLAR COMA

In contrast to young type I diabetic patients who can become ill with diabetic ketoacidosis within a day or so, older type II diabetic patients sometimes develop a condition called hyperglycemic, hyperosmolar coma. This is particularly common in the elderly. Hyperglycemia, perhaps worsened by failure to take insulin or hypoglycemic drugs, an infection, or a coincidental medical problem such as a heart attack, leads to urinary losses of water, glucose, and electrolytes (sodium, chloride, and potassium). This osmotic diuresis reduces the circulating blood volume, a stress that in turn worsens insulin resistance and hyperglycemia. In addition, elderly patients may be less able to sense thirst or to obtain fluids. Over the course of several days these patients can become extremely hyperglycemic (glucose >1000 mg dL^{-1}), dehydrated, and ultimately comatose. Ketoacidosis does not develop in these patients possibly because free fatty acids are not always elevated or because adequate insulin concentrations exist in the portal blood to inhibit ketogenesis (although it is not high enough to inhibit gluconeogenesis). Therapy is aimed at restoring water and electrolyte balance and correcting the hyperglycemia with insulin. The mortality of this syndrome is considerably higher than that of diabetic ketoacidosis.

Arieff, A. I. and Carroll, H. J. Nonketotic hyperosmolar coma with hyperglycemia. Clinical features, pathophysiology, renal function, acid–base balance, plasma-cerebrospinal fluid equilibria, and the effects of therapy in 37 cases. *Medicine* 51:73, 1972.

TABLE 14.1 Substrate and Hormone Levels in Blood of Well-Fed, Fasting, and Starving Humans[a]

Hormone or substrate (units)	Very well fed	Postabsorptive 12 h	Fasted 3 days	Starved 5 weeks
Insulin (μU mL^{-1})	40	15	8	6
Glucagon (pg ml^{-1})	80	100	150	120
Insulin : glucagon ratio (μU pg^{-1})	0.50	0.15	0.05	0.05
Glucose (mM)	6.1	4.8	3.8	3.6
Fatty acids (mM)	0.14	0.6	1.2	1.4
Acetoacetate (mM)	0.04	0.05	0.4	1.3
β-Hydroxybutyrate (mM)	0.03	0.10	1.4	6.0
Lactate (mM)	2.5	0.7	0.7	0.6
Pyruvate (mM)	0.25	0.06	0.04	0.03
Alanine (mM)	0.8	0.3	0.3	0.1
ATP equivalents (mM)	313	290	380	537

SOURCE: N. B. Ruderman, T. T. Aoki, and G. F. Cahill, Jr. Gluconeogenesis and its disorders in man, in R. W. Hanson and M. A. Mehlman (Eds.), *Gluconeogenesis, its regulation in mammalian species*. New York: Wiley, 1976, p. 515.

[a] Data are for normal weight subjects except for the 5-week starvation values, which are from obese subjects undergoing therapeutic starvation. ATP equivalents were calculated on the basis of the ATP yield expected on complete oxidation of each substrate to CO_2 and H_2O: 38 molecules of ATP for each molecule of glucose; 144 for the average fatty acid (oleate); 23 for acetoacetate; 26 for β-hydroxybutyrate; 18 for lactate; 15 for pyruvate; and 13 (corrected for urea formation) for alanine.

loss of glucose in the urine and to the glycosylation of a number of proteins, the latter being postulated to be a more insidious complication of prolonged high concentrations of glucose (see Clin. Corr. 14.6).

Energy Requirements and Reserves

The average person leading a sedentary life, such as a portly professor of biochemistry, consumes daily about 200 g of carbohydrate, 70 g of protein, 60 g of fat, and, during the academic year, 100 g of ethanol and an occasional graduate student. This meets a daily energy requirement of 1600–2400 kcal. As shown in Table 14.2, the **energy reserves** of an average-sized person are considerable. We tend to emphasize the details of glycogen metabolism, and the ability to mobilize glycogen rapidly is indeed very important. Table 14.2 demonstrates, however, that our glycogen reserves are minuscule with respect to our fat reserves. The fat stores of obese subjects can weigh as much as 80 kg, adding another 585,000 kcal to their energy reserves. Protein is listed in Table 14.2 as an energy reserve, and, in the sense that it can be used to provide amino acids for oxidation, protein is an energy source. On the other hand, protein is not inert like stored fat and glycogen. Proteins make up the muscles that allow

TABLE 14.2 The Energy Reserves of Humans[a]

Stored fuel	Tissue	Fuel reserves (g)	Fuel reserves (kcal)
Glycogen	Liver	70	280
Glycogen	Muscle	120	480
Glucose	Body fluids	20	80
Fat	Adipose	15,000	135,000
Protein	Muscle	6,000	24,000

[a] The data are for a normal subject weighing 70 kg. Carbohydrate contains 4 kcal g^{-1}; fat 9 kcal g^{-1}; protein 4 kcal g^{-1}.

us to move and breathe and the enzymes that carry out metabolism. Hence it is not as dispensable as fat and glycogen and is given up by the body more reluctantly.

Glucose Homeostasis Has Five Phases

Figure 14.6 comes from the work of Cahill and his colleagues with obese patients undergoing long-term **starvation** for therapeutic purposes. It illustrates the effects of starvation on those processes that are used by the body to maintain caloric homeostasis. For convenience, the time period involved has been divided into five phases. Phase I is the well-fed state, in which glucose is provided by dietary carbohydrate. Once this supply is exhausted, hepatic glycogenolysis maintains blood glucose levels during phase II. As this supply of glucose starts to dwindle, hepatic gluconeogenesis from lactate, glycerol, and alanine becomes increasingly important until, in phase III, it is the major source of blood glucose. Note that all of these changes occur within just 20 or so hours of fasting, depending of course on how well fed the individual was prior to the fast, how much hepatic glycogen was present, and the sort of physical activity occurring during the fast. Several days of fasting move one into phase IV, where the dependence on gluconeogenesis actually decreases. As discussed above, ketone bodies have accumulated to concentrations that are high enough for them to enter the brain and meet some of the energy needs of this tissue. Because of a decrease in liver size, renal gluconeogenesis becomes significant relative to hepatic gluconeogenesis in this phase. Phase V occurs after very prolonged starvation of extremely obese individuals. It is characterized by even less dependence of the body on gluconeogenesis, the energy needs of almost every tissue being met to an even greater extent by either fatty acid or ketone body oxidation.

As long as ketone body concentrations are high, proteolysis will be somewhat restricted, and conservation of muscle proteins and enzymes

CLIN. CORR. **14.6**
HYPERGLYCEMIA AND PROTEIN GLYCOSYLATION

Glycosylation of enzymes is known to cause changes in their activity, solubility and susceptibility to degradation. In the case of hemoglobin A, glycosylation occurs by a nonenzymatic reaction between glucose and the amino-terminal valine of the β chain. A Schiff base forms between glucose and valine, followed by a rearrangement of the molecule to give a 1-deoxyfructose molecule attached to the valine. The reaction is favored by high glucose levels and the resulting protein, called hemoglobin A_{1c}, is a good index of how high a person's average blood–glucose concentration has been over the previous several weeks. The concentration of this protein increases substantially in the red blood cells of an uncontrolled diabetic.

It has been proposed that increased glycosylation of proteins resulting from hyperglycemia may contribute to the medical complications caused by diabetes, for example, coronary heart disease, retinopathy, nephropathy, and neuropathy. Collagen, fibrin, and antithrombin III (an inhibitor of blood coagulation) can become glycosylated and undergo alterations in physical and enzymatic properties. It is possible that these changes favor the accelerated blood vessel damage that occurs in patients with diabetes. Likewise, increased glycosylation of the lens protein α-crystallin may contribute to the development of diabetic cataracts.

Cerami, A., Stevens, V. J., and Monnier, V. M. Role of nonenzymatic glycosylation in the development of the sequelae of diabetes mellitus. *Metabolism* 28 (Suppl. 1):431, 1979.

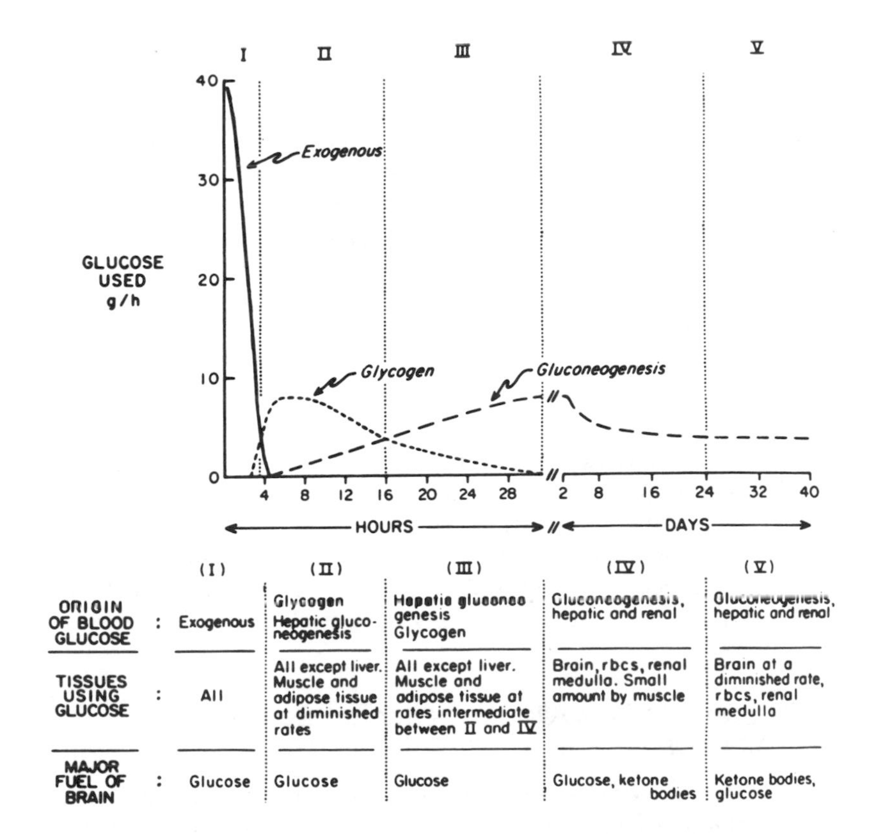

	(I)	(II)	(III)	(IV)	(V)
ORIGIN OF BLOOD GLUCOSE	Exogenous	Glycogen Hepatic gluconeogenesis	Hepatic gluconeogenesis Glycogen	Gluconeogenesis, hepatic and renal	Gluconeogenesis, hepatic and renal
TISSUES USING GLUCOSE	All	All except liver. Muscle and adipose tissue at diminished rates	All except liver. Muscle and adipose tissue at rates intermediate between II and IV	Brain, rbcs, renal medulla. Small amount by muscle	Brain at a diminished rate, rbcs, renal medulla
MAJOR FUEL OF BRAIN	Glucose	Glucose	Glucose	Glucose, ketone bodies	Ketone bodies, glucose

FIGURE 14.6
The five phases of glucose homeostasis in humans.
Reprinted with permission from N. B. Ruderman, T. T. Aoki, and G. F. Cahill, Jr. Gluconeogenesis and its disorders in man, in R. W. Hanson, and M. A. Mehlman (Eds.), *Gluconeogenesis, its regulation in mammalian species*. New York: Wiley, 1976, p. 515.

will occur. This continues until practically all of the fat is gone as a consequence of starvation. After all of it is gone, the body has to use muscle protein. Before it is gone—you are gone (see Clin. Corr. 14.3).

14.3 MECHANISMS INVOLVED IN SWITCHING THE METABOLISM OF THE LIVER BETWEEN THE WELL-FED STATE AND THE STARVED STATE

The liver of the well-fed person is actively engaged in processes that favor the synthesis of glycogen and fat; such a liver is glycogenic, glycolytic, lipogenic, and cholesterogenic. The liver of the fasting person is quite a different organ; it is glycogenolytic, gluconeogenic, ketogenic, and proteolytic. The strategy is to store calories when food is available, but then to be able to mobilize these stores when the rest of the body is in need. The liver is switched between these metabolic extremes by a variety of regulatory mechanisms: substrate supply, allosteric effectors, covalent modification, and induction-repression of enzymes.

Substrate Availability Controls Many Metabolic Pathways

Because of the other, more sophisticated levels of control, the importance of substrate supply is often ignored. However, the concentration of fatty acids in the blood of the portal vein is clearly a major determinant of the rate of ketogenesis. Excess fat is not synthesized and stored unless one consumes excessive amounts of substrates that can be used for lipogenesis. Glucose synthesis by the liver is also restricted by the rate at which gluconeogenic substrates flow to the liver. Delivery of excess amino acids to the liver of the diabetic, because of the accelerated and uncontrolled proteolysis that occurs in this metabolic condition, increases the rate of gluconeogenesis and exacerbates the hyperglycemia characteristic of diabetes. On the other hand, failure to supply the liver adequately with glucogenic substrate explains some types of hypoglycemia, such as that observed during pregnancy or advanced starvation.

Another pathway regulated by substrate supply is urea synthesis. Amino acid metabolism in the intestines provides a substantial fraction of the ammonia used by the liver for urea production. The intestines also releases citrulline and proline, metabolic precursors of ornithine. The citrulline is metabolized to arginine in the kidney, and the arginine is taken up by the liver. Conversion of arginine to ornithine provides the liver with a larger ornithine pool and increased capacity for urea synthesis after a high protein meal. Your pet cat could be used (literally) to illustrate this particular interaction between substrate supplies. Cats vomit, become comatose, and may even die from ammonia intoxication when given a single protein meal deficient in arginine. Arginine is needed in the cat to replenish ornithine levels in the liver.

From the examples provided above we can conclude that substrate supply is a major determinant of the rate at which virtually every metabolic process of the body operates. However, variations in substrate supply are not sufficient to account for the tremendous changes in metabolism that must occur in the starve–feed cycle. As discussed below, regulation by allosteric effectors plays an important role in the different states.

Negative and Positive Allosteric Effectors Regulate Many Pathways

Figures 14.7 and 14.8 summarize the effects of negative and positive allosteric effectors believed to be important in the well-fed and starved

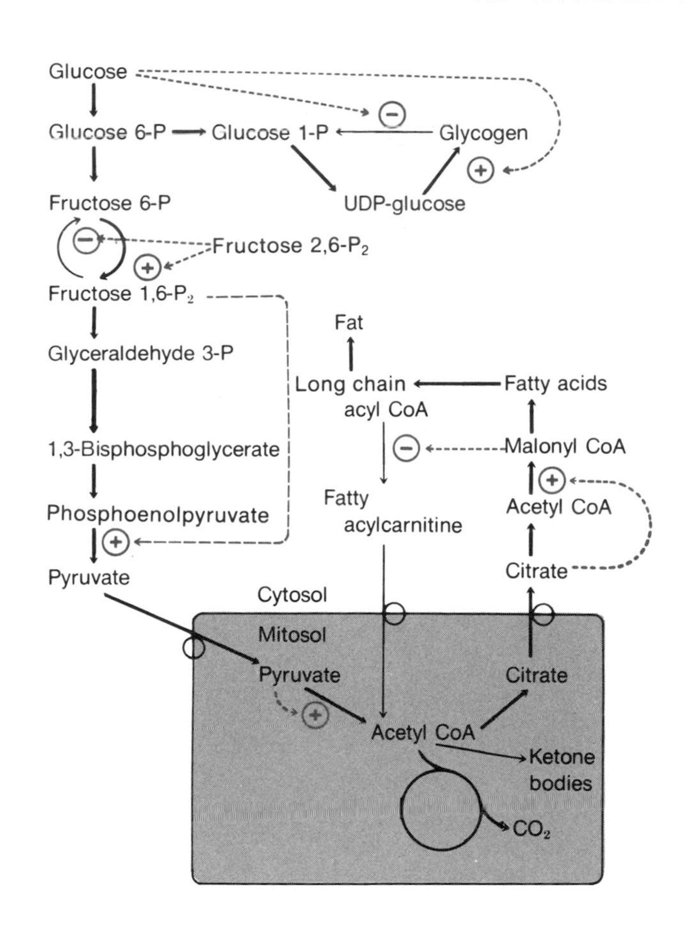

FIGURE 14.7
Control of hepatic metabolism in the well-fed state by allosteric effectors.

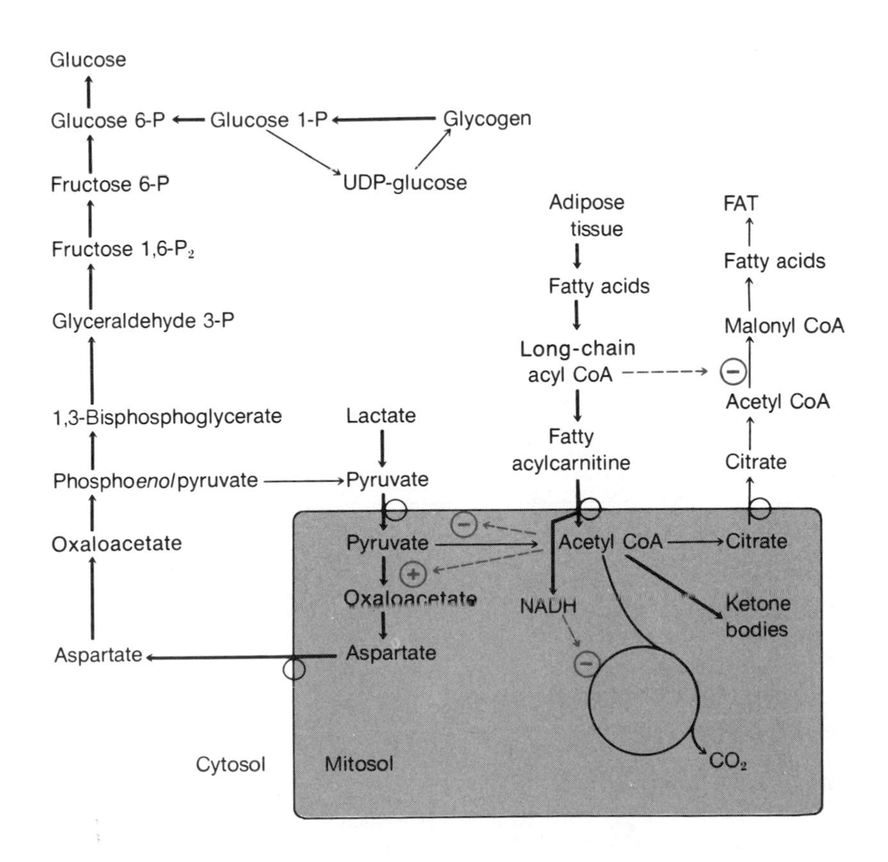

FIGURE 14.8
Control of hepatic metabolism in the fasting state by allosteric effectors.

states, respectively. As shown in Figure 14.7, glucose inactivates glycogen phosphorylase and activates glycogen synthase (indirectly; see Chapter 7, p. 352), thereby preventing degradation and promoting synthesis of glycogen; fructose 2,6-bisphosphate stimulates 6-phosphofructo-1-kinase and inhibits fructose 1,6-bisphosphatase, thereby stimulating glycolysis and inhibiting gluconeogenesis; fructose 1,6-bisphosphate activates pyruvate kinase, thereby stimulating glycolysis; pyruvate activates pyruvate dehydrogenase (indirectly by inhibition of pyruvate dehydrogenase kinase; see Chapter 6, p. 251) and citrate activates acetyl CoA carboxylase, thereby stimulating fatty acid synthesis; and malonyl CoA inhibits carnitine palmitoyltransferase I, thereby inhibiting fatty acid oxidation.

As shown in Figure 14.8, acetyl CoA stimulates gluconeogenesis in the fasted state by activating pyruvate carboxylase (direct allosteric effect) and inhibiting pyruvate dehydrogenase (direct allosteric effect and also via stimulation of pyruvate dehydrogenase kinase; see Chapter 7, p. 335); long-chain acyl CoA esters inhibit acetyl CoA carboxylase, which lowers the level of malonyl CoA and permits greater carnitine palmitoyltransferase I activity and fatty acid oxidation rates; NADH produced by fatty acid oxidation inhibits TCA cycle activity.

Although not shown in Figure 14.8, cAMP is an important allosteric effector. Its concentration in the liver is increased in the starved state. Cyclic AMP is a positive effector of cAMP-dependent protein kinase, which, in turn, is responsible for changing the kinetic properties of several regulatory enzymes by covalent modification, as summarized next.

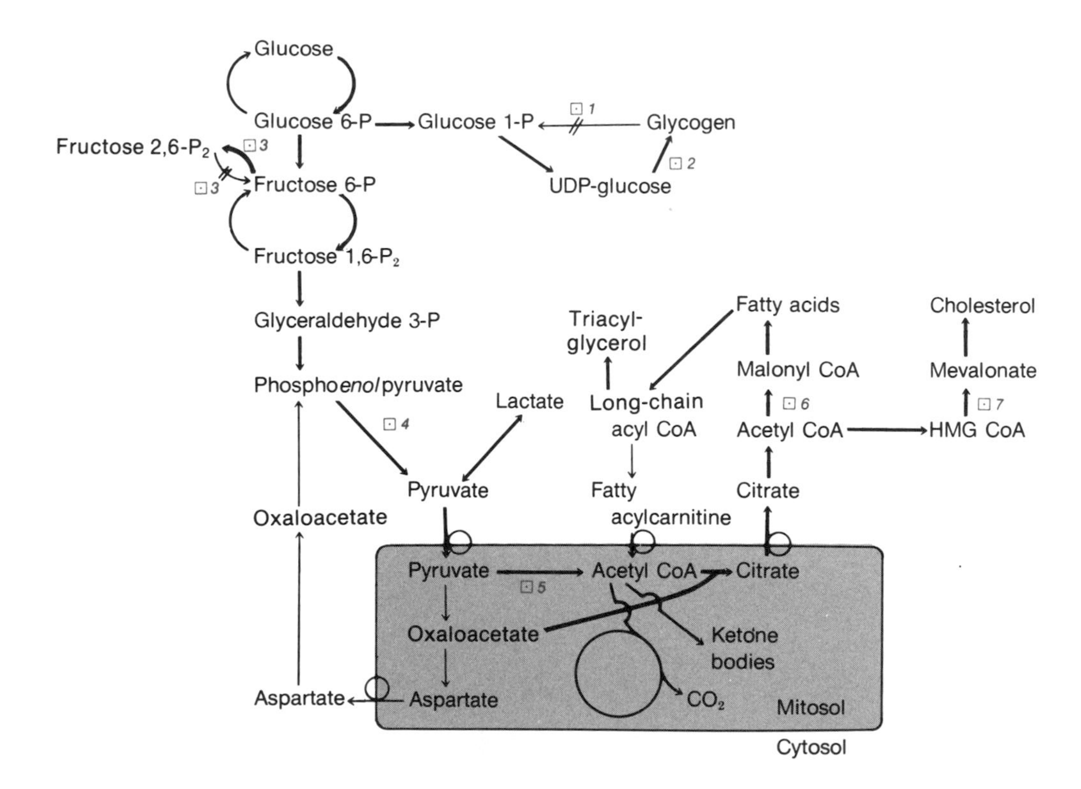

FIGURE 14.9
Activity and state of phosphorylation of the enzymes subject to covalent modification in the lipogenic liver.
The dephosphorylated mode is indicated by the symbol ⊡. The interconvertible enzymes numbered are 1, glycogen phosphorylase; 2, glycogen synthase; 3, 6-phosphofructo-2-kinase/fructose-2,6-bisphosphatase (bifunctional enzyme); 4, pyruvate kinase, 5, pyruvate dehydrogenase; 6, acetyl CoA carboxylase; and 7, β-hydroxy-β-methylglutaryl CoA reductase.

Covalent Modification Regulates Key Enzymes

Figures 14.9 and 14.10 point out the interconvertible enzymes that play important roles in switching the liver between the well-fed and starved states. The regulation of enzymes by covalent modification has been discussed in Chapter 7. Recall that ⊡- and ⊙-P represent interconvertible forms of an enzyme in the nonphosphorylated and phosphorylated states, respectively.

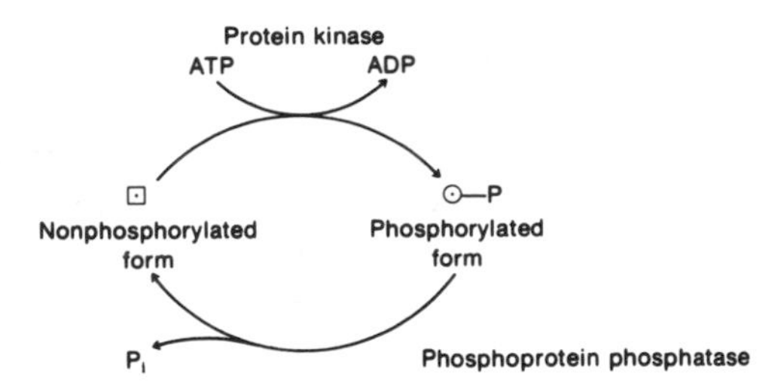

The important points are as follows: (a) enzymes subject to covalent modification undergo phosphorylation on one or more serine residues by a protein kinase; (b) the phosphorylated enzyme can be returned to the dephosphorylated state by phosphoprotein phosphatase; (c) hormonal regulation via protein kinases is better understood than hormonal regulation via phosphoprotein phosphatases; (d) phosphorylation of the enzyme changes its conformation and its catalytic activity; (e) some enzymes are

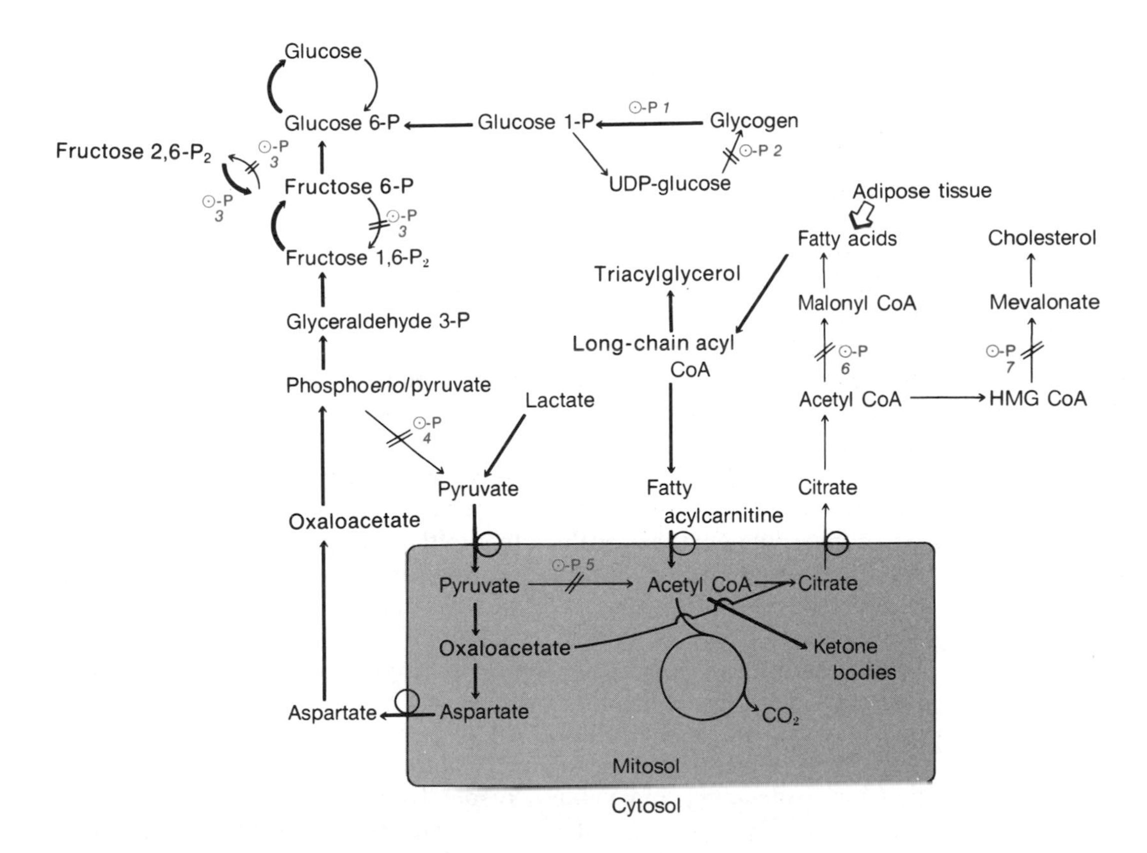

FIGURE 14.10
Activity and state of phosphorylation of the enzymes subject to covalent modification in the glucogenic liver.
The phosphorylated mode is indicated by the symbol ⊙—P. The numbers refer to the same enzymes as in Figure 14.9.

active only in the dephosphorylated state, others only in the phosphorylated state; (f) cAMP is the messenger that signals the phosphorylation of many, but not all, of the enzymes subject to covalent modification; (g) cAMP acts by activating cAMP-dependent protein kinase; (h) cAMP also indirectly promotes phosphorylation of interconvertible enzymes by signaling inactivation of phosphoprotein phosphatase by poorly understood mechanisms; (i) glucagon increases cAMP levels in the liver by activating adenylate cyclase; and (j) insulin, by mechanisms not completely defined (see Chapter 20, p. 890), opposes the action of glucagon and cAMP and thereby promotes dephosphorylation of the interconvertible enzymes.

As shown in Figure 14.9, hepatic enzymes currently believed subject to covalent modification are in the dephosphorylated mode in the liver of the well-fed animal. Although not shown in the figure, phosphorylase kinase is also in the dephosphorylated mode in the well-fed state. Insulin/glucagon ratios are high in the blood, and the cAMP levels are low in the liver. This results in a low activity for the cAMP-dependent protein kinase and, by poorly understood mechanisms, a high activity for phosphoprotein phosphatase. However, not all interconvertible enzymes are subject to phosphorylation by cAMP-dependent protein kinase. At this writing, it is clear that glycogen synthase, glycogen phosphorylase (via phosphorylase kinase), 6-phosphofructo-2-kinase, fructose-2,6-bisphosphatase, and pyruvate kinase are subject to regulation by the cAMP-dependent protein kinase. Whether there is a link to cAMP-dependent protein kinase for the other enzymes is a question of current research interest. Only three of the interconvertible enzymes, glycogen phosphorylase, phosphorylase kinase, and fructose-2,6-bisphosphatase, are inactive in the dephosphorylated mode. All of the other identified interconvertible enzymes (glycogen synthase, 6-phosphofructo-2-kinase, pyruvate kinase, pyruvate dehydrogenase, acetyl CoA carboxylase, and β-hydroxy-β-methylglutaryl CoA reductase) are active. Glycogenesis, glycolysis, and lipogenesis are greatly favored as a result of placing the interconvertible enzymes in the dephosphorylated mode. On the other hand, the opposing pathways, glycogenolysis, gluconeogenesis, and ketogenesis, are inhibited.

As shown in Figure 14.10, the hepatic enzymes subject to covalent modification are in the phosphorylated mode in the liver of the fasting animal. Insulin/glucagon ratios are low in the blood, and hepatic cAMP levels are high. This results in activation of cAMP-dependent protein kinase and inactivation of phosphoprotein phosphatase. The net effect is a much greater degree of phosphorylation of the interconvertible enzymes than in the well-fed state. In the starved state, three of the interconvertible enzymes—glycogen phosphorylase, phosphorylase kinase, and fructose-2,6-bisphosphatase—are in the active catalytic state. All the other interconvertible enzymes are inactive in the phosphorylated mode. As a result, glycogenesis, glycolysis and lipogenesis are shut down almost completely, and glycogenolysis, gluconeogenesis and ketogenesis predominate.

Two additional hepatic enzymes, phenylalanine hydroxylase and branched-chain α-keto acid dehydrogenase, are also controlled by phosphorylation/dephosphorylation. These enzymes catalyze rate-limiting steps in the disposal of phenylalanine and the branched-chain amino acids (leucine, isoleucine, and valine), respectively. These enzymes are not included in Figures 14.9 and 14.10 because of special features of their control by covalent modification. Phenylalanine hydroxylase, a cytosolic enzyme, is active in the phosphorylated state, and phosphorylation is stimulated by glucagon via cAMP-dependent protein kinase. Branched-chain α-keto acid dehydrogenase, a mitochondrial enzyme, is active in the dephosphorylated state, and its activity is regulated by branched-chain α-

keto acid dehydrogenase kinase and a phosphoprotein phosphatase. Phenylalanine acts as a positive allosteric effector for the phosphorylation and activation of phenylalanine hydroxylase by cAMP-dependent protein kinase. Branched-chain α-keto acids activate branched-chain α-keto acid dehydrogenase indirectly by inhibiting branched-chain α-keto acid dehydrogenase kinase. Covalent modification of these enzymes provides a very sensitive means for control of the degradation of phenylalanine and the branched-chain amino acids. The clinical experience with phenylketonuria (see Clin. Corr. 12.5) and maple syrup urine disease (see Clin. Corr. 12.1) emphasizes the importance of regulating low blood and tissue levels of these amino acids. On the other hand, the enzymes must be controlled to prevent depletion of the body stores of these essential amino acids. Therefore, the tissue requirements for these amino acids supersedes the phase of the starve–feed cycle in establishing the phosphorylation and activity state of these interconvertible enzymes.

Adipose tissue responds almost as dramatically as liver to the starve–feed cycle because it also contains enzymes subject to covalent modification. Pyruvate kinase, pyruvate dehydrogenase, acetyl CoA carboxylase, and hormone-sensitive lipase (not found in liver) are all in the dephosphorylated mode in the adipose tissue of the well-fed person. As in liver, the first three enzymes are active in the dephosphorylated mode. Hormone-sensitive lipase is inactive in the dephosphorylated mode. The high insulin/glucagon ratio, the low tissue cAMP concentration, and perhaps a high insulin "messenger" level are believed to be important determinants of the phosphorylation state of the interconvertible enzymes of adipocytes. Lipogenesis within adipose tissue is favored in the well-fed state. During fasting, adipocytes quickly shut down lipogenesis and activate lipolysis. This is accomplished in large part by the phosphorylation of the enzymes described above, a consequence of the decrease in the insulin/glucagon ratio induced by fasting and the resulting increase in cAMP levels and protein kinase activity. In this manner, adipose tissue is transformed from a fat storage tissue into a source of fatty acids for other tissues.

Regulation of the metabolic processes of liver and adipose tissue by hormonal effects upon the interconvertible enzymes is clearly of great importance in the starve–feed cycle. Such effects are probably important in muscle and kidney as well, but less is known about the starve–feed cycle in these tissues. This type of control is like allosteric effectors and substrate supply, a short-term regulatory mechanism, operating on a minute-to-minute basis. Adaptive changes in enzyme activities due to changes in the absolute amounts of key enzymes of a tissue are also subject to hormonal and nutritional factors but require several hours to come into effect.

Changes in Levels of Key Enzymes Are a Longer Term Adaptive Mechanism

The adaptive change in enzyme levels is a mechanism of regulation involving changes in the rate of synthesis or degradation of key enzymes of metabolism. Whereas allosteric effectors and covalent modification affect either the K_m or V_{max} of an enzyme, this mode of regulation involves the actual quantity of an enzyme in the tissue. In other words, because of the influence of hormonal and nutritional factors on its turnover, there are more or fewer enzyme molecules present in the tissue. For example, when a person is maintained in a well-fed or overfed condition, the liver improves its capacity to synthesize fat. To be sure, this can be explained in part by increased substrate supply, as well as appropriate changes in

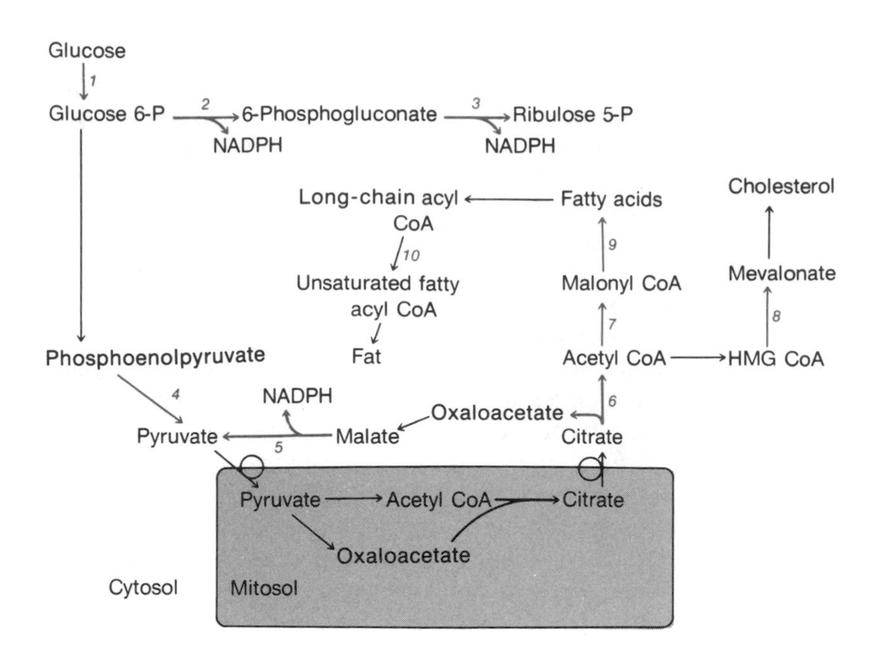

FIGURE 14.11
Enzymes induced in the liver of the well-fed individual.
The inducible enzymes are numbered: 1, glucokinase; 2, glucose 6-phosphate dehydrogenase; 3, 6-phosphogluconate dehydrogenase; 4, pyruvate kinase; 5, malic enzyme; 6, citrate cleavage enzyme; 7, acetyl CoA carboxylase; 8, β-hydroxy-β-methylglutaryl CoA reductase; 9, fatty acid synthase; and 10, Δ^9-desaturase.

allosteric effectors (Figure 14.7) and the conversion of the interconvertible enzymes into the dephosphorylated mode (Figure 14.9). This is not the entire story, however, because the liver also has more molecules of those enzymes that play a key role in fat synthesis (see Figure 14.11). A whole battery of enzymes is induced, including glucokinase and pyruvate kinase for faster rates of glycolysis; glucose 6-phosphate dehydrogenase, 6-phosphogluconate dehydrogenase, and malic enzyme to provide greater quantities of NADPH for reductive synthesis; and citrate cleavage enzyme, acetyl CoA carboxylase, fatty acid synthase, and Δ^9-desaturase for more rapid rates of fatty acid synthesis. All of these enzymes are present at higher levels in the well-fed state, possibly in response to the increased insulin/glucagon ratios. While these enzymes are induced, there is a decrease in the enzymes that favor glucose synthesis. Phosphoenolpyruvate carboxykinase, glucose 6-phosphatase, and some aminotransferases are decreased in amount; that is, their synthesis is reduced or degradation increased in response to increased circulating glucose and insulin.

If a person fasts for several hours, the enzyme pattern characteristic of the liver changes dramatically (Figure 14.12). The enzymes involved in lipogenesis decrease in quantity, possibly because their synthesis is decreased or degradation of these proteins is increased. At the same time a

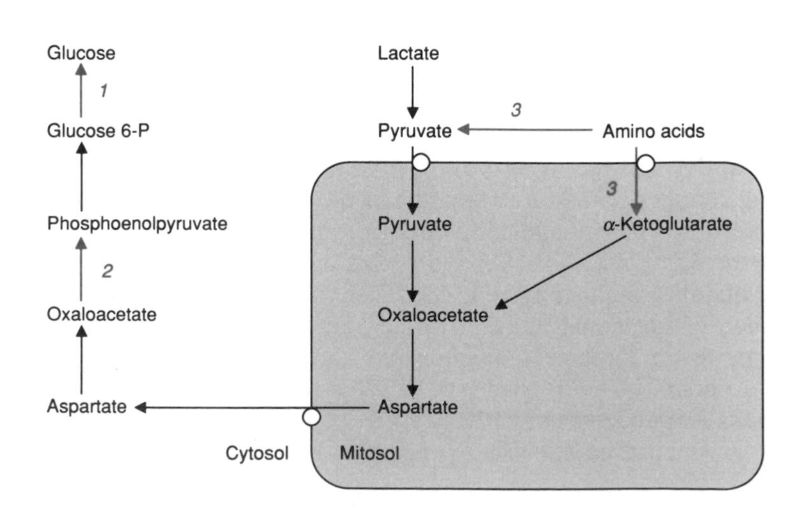

FIGURE 14.12
Enzymes induced in the liver of an individual during fasting.
The inducible enzymes are numbered: 1, glucose 6-phosphatase; 2, phosphoenol-pyruvate carboxykinase; and 3, various aminotransferases.

number of enzymes favoring gluconeogenesis are induced (Figure 14.12), making the liver much more effective in synthesizing glucose. In addition, the enzymes of the urea cycle are induced, possibly by the presence of higher blood glucagon levels. This permits the disposal of nitrogen, as urea, from the alanine used in gluconeogenesis. For several of these enzymes, increased rates of transcription of their genes is mediated by the interaction of a transcription factor(s) with a cAMP response element (CRE) in the promoter of the gene.

These adaptive changes are clearly important in the starve–feed cycle, greatly affecting the capacity of the liver for its various metabolic processes. Although often overlooked, the adaptive changes also influence the effectiveness of the short-term regulatory mechanisms. For example, long-term starvation or uncontrolled diabetes greatly decreases the level of acetyl CoA carboxylase. Taking away long-chain acyl CoA esters that inhibit this enzyme, increasing the level of citrate that activates this enzyme, or creating conditions that activate this interconvertible enzyme by dephosphorylation will not have any effect when the enzyme is virtually absent. Another example is afforded by the glucose intolerance of starvation. A chronically starved person, because of the absence of the key enzymes needed for glucose metabolism, cannot effectively utilize a sudden load of glucose. A glucose load, however, will set into motion the induction of the required enzymes and the reestablishment of short-term regulatory mechanisms.

14.4 METABOLIC INTERRELATIONSHIPS OF TISSUES IN VARIOUS NUTRITIONAL AND HORMONAL STATES

Many of the changes that occur in various nutritional and hormonal states are just variations on the starve–feed cycle and are completely predictable from what we have learned about the cycle. Some examples are given in Figure 14.13. Others are so obvious that a diagram is not necessary; for example, in rapid growth of a child, amino acids are directed away from catabolism and into protein synthesis. On the other hand, the changes that occur in some physiologically important situations are rather subtle and poorly understood. An example of the latter is aging, which seems to lead to a decreased "sensitivity" of the major tissues of the body to hormones. The important consequence is a decreased ability of the tissues to respond normally during the feed–starve cycle. Whether this is a contributing factor to or a consequence of the aging process is not known.

Obesity Occurs When Stored Fuel Is Not Completely Consumed During Fasting

Figure 14.13*a* illustrates the metabolic interrelationships prevailing much of the time in an obese person. Most of the body fat of the human is either provided by the diet or synthesized in the liver and transported to the adipose tissue for storage. Obesity is caused by a person staying in such a well-fed state that stored fuel (particularly fat) does not get used up during the fasting phase of the cycle. The body then has no option other than to accumulate fat.

Aerobic and Anaerobic Exercise Use Different Fuels

It is important to differentiate between two distinct types of exercise—aerobic and anaerobic. Aerobic exercise is exemplified by long-distance running, anaerobic exercise by sprinting or weight lifting. During anaero-

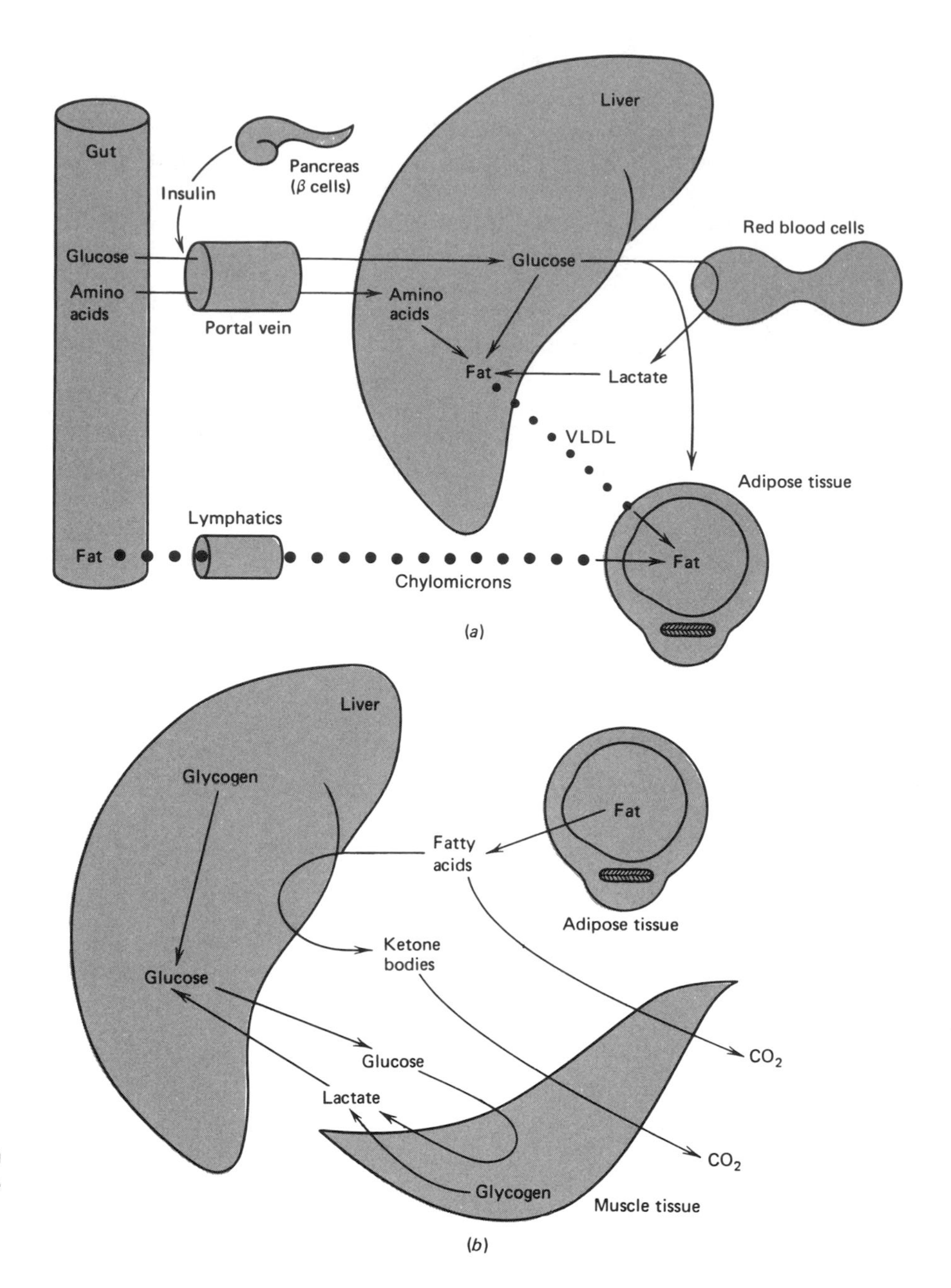

FIGURE 14.13
Metabolic interrelationships of tissues in various nutritional, hormonal, and disease states.
(*a*) Obesity. (*b*) Exercise.

bic exercise there is really very little interorgan cooperation. The muscle largely relies upon its own stored glycogen and phosphocreatine. Phosphocreatine serves as a source of high-energy phosphate bonds for ATP synthesis until glycogenolysis and glycolysis are stimulated. The blood vessels within these muscles are compressed during peak contraction, thus these cells are isolated from the rest of the body. Aerobic exercise is metabolically more interesting (Figure 14.13*b*). The body in the well-fed state does not store enough glucose and glycogen to provide the energy needed for running long distances. It is also known that the respiratory quotient, the ratio of carbon dioxide exhaled to oxygen consumed, falls during distance running. This indicates the progressive switch to using free fatty acids during the race. Apparently, lipolysis gradually increases as glucose stores are exhausted, and, as in the fasted state, the muscle will oxidize fatty acids in preference to glucose as the former become available. Unlike what happens in fasting, there is little increase in blood ketone body concentration. This may simply reflect a balance between hepatic ketone body synthesis and muscle ketone body oxidation.

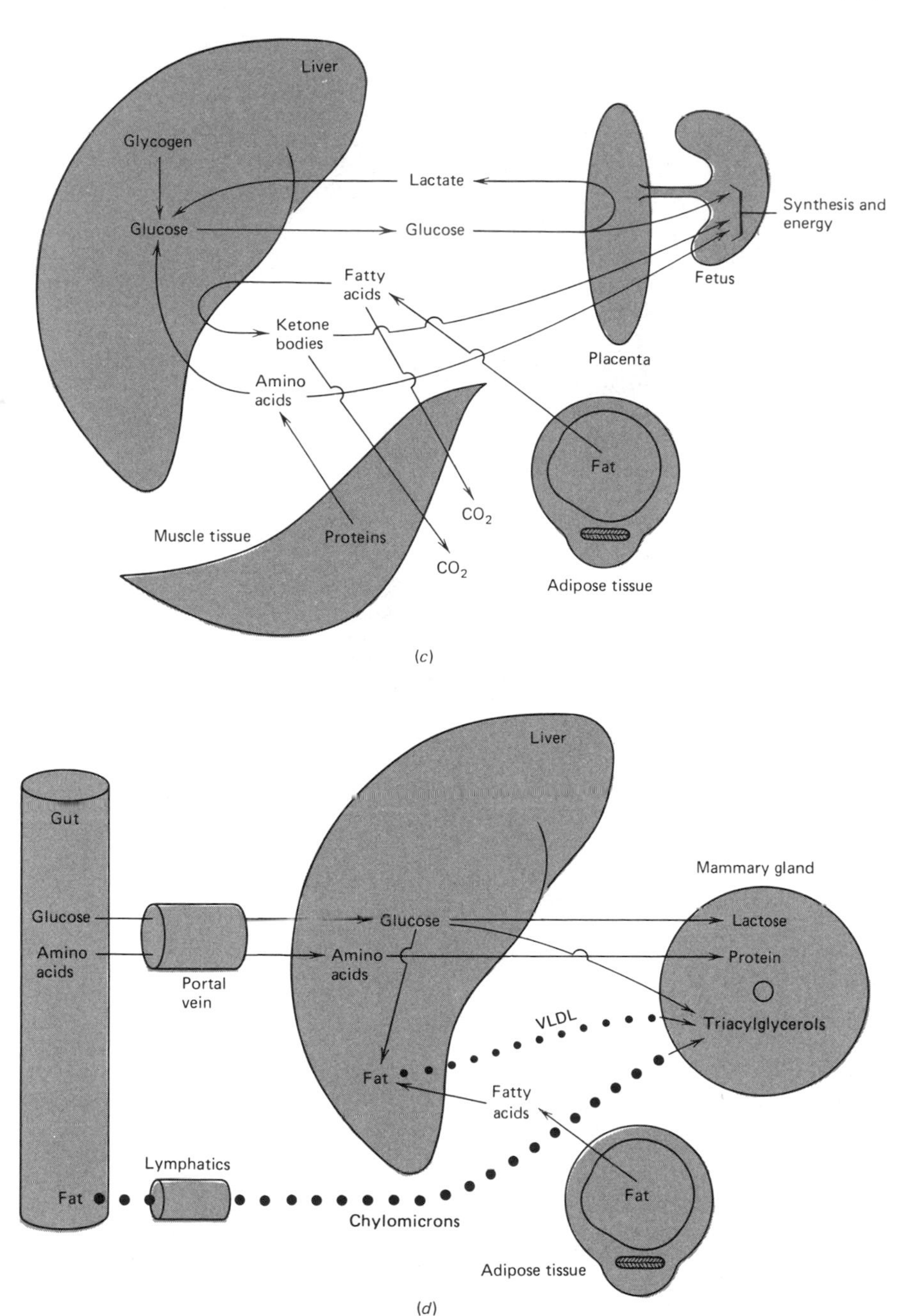

FIGURE 14.13 (Continued)
(*c*) Pregnancy. (*d*) Lactation.

Changes in Pregnancy Are Related to Fetal Requirements and Hormonal Changes

The fetus can be considered as another nutrient-requiring tissue (Figure 14.13*c*). It mainly uses glucose for energy, but it may also use amino acids, lactate, and ketone bodies. Fatty acids do not cross the placenta to the fetus, but maternal LDL cholesterol is an important precursor of placental steroids. During pregnancy, the starve–feed cycle is perturbed. The placenta secretes a polypeptide hormone, placental lactogen, and two steroid hormones, estradiol and progesterone. Placental lactogen stimulates lipolysis in adipose tissue, and the steroid hormones induce an insulin-resistant state. In the postprandial state, pregnant women enter the starved state more rapidly than do nonpregnant women. This results from increased consumption of glucose and amino acids by the fetus. Plasma glucose, amino acids, and insulin levels fall rapidly, and glucagon levels rise and stimulate lipolysis and ketogenesis. The consumption of glucose and amino acids by the fetus may be great enough to cause maternal

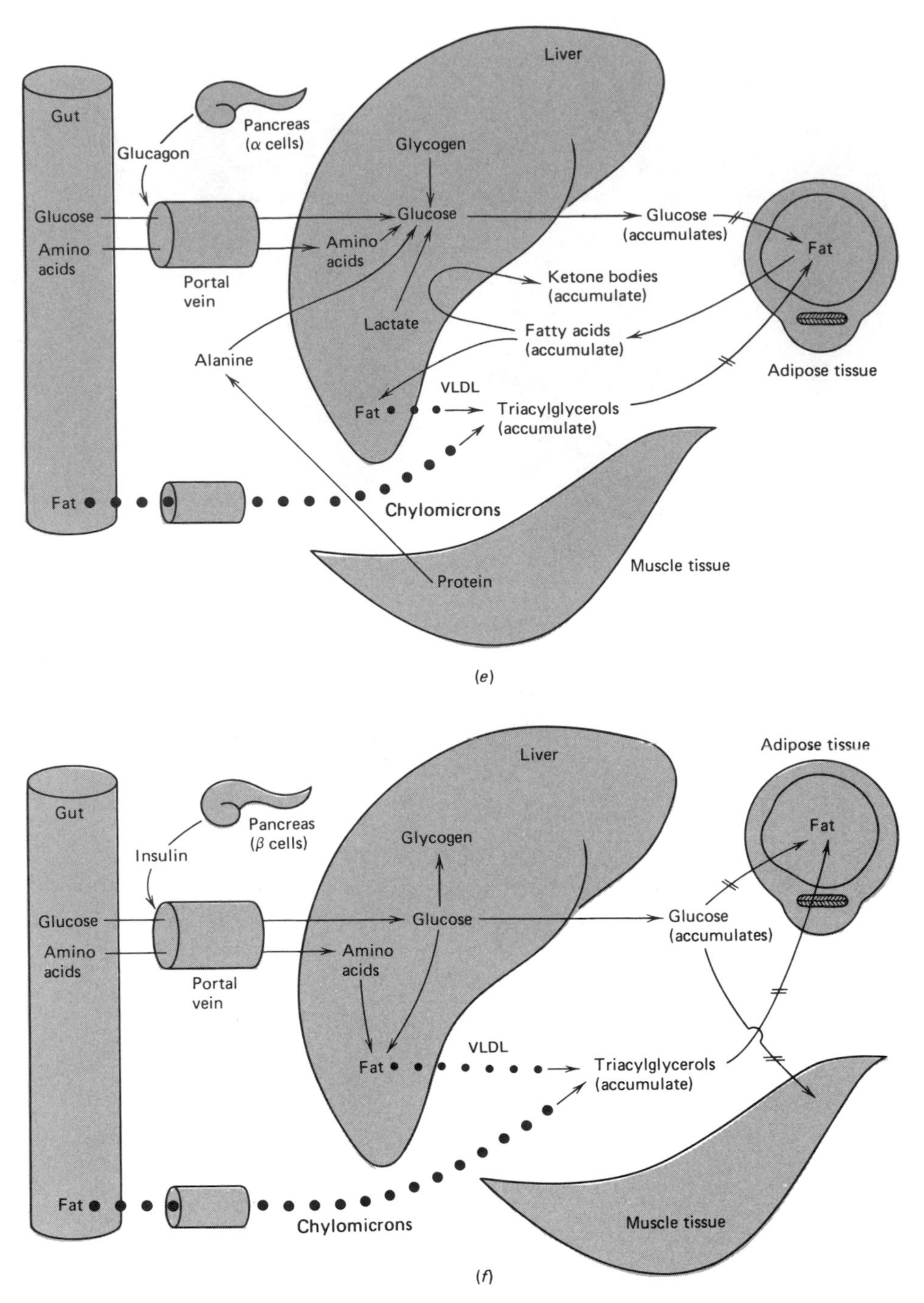

FIGURE 14.13 (Continued)
(*e*) Insulin-dependent diabetes mellitus. (*f*) Noninsulin-dependent diabetes mellitus.

hypoglycemia. On the other hand, in the fed state pregnant women have increased levels of insulin and glucose and demonstrate resistance to exogenous insulin. These swings of plasma hormones and fuels are even more exaggerated in diabetic women and make control of blood glucose difficult in these patients.

Lactation Requires the Synthesis of Lactose, Triacylglycerol, and Protein

In late pregnancy placental hormones induce lipoprotein lipase in the mammary gland and promote the development of milk-secreting cells and ducts. During lactation (see Figure 14.13*d*) the breast utilizes glucose for lactose and triacylglycerol synthesis, as well as its major energy source. Amino acids are taken up for protein synthesis, and chylomicrons and VLDL are utilized as sources of fatty acids for triacylglycerol synthesis. If these compounds are not supplied by the diet, proteolysis, gluconeogenesis, and lipolysis must supply them, resulting eventually in maternal malnutrition and poor quality milk. The lactating breast also secretes a hormone with some similarity to parathyroid hormone (see Chapter 20). This hormone probably is important for the absorption of calcium and phosphorous from the gut and bone.

Insulin-Dependent Diabetes Mellitus

Figure 14.13*e* shows the metabolic interrelationships that exist in insulin-dependent diabetes mellitus (see Clin. Corrs. 14.7 and 14.8). Because of defective β-cell production of insulin, blood levels of insulin remain low in spite of elevated blood glucose levels. Even when dietary glucose is being delivered from the gut, the insulin/glucagon ratio cannot increase, and the liver remains gluconeogenic and ketogenic. Since it is impossible to switch to the processes of glycolysis, glycogenesis, and lipogenesis, the liver cannot properly buffer blood glucose levels. Indeed, since hepatic gluconeogenesis is continuous, the liver contributes to hyperglycemia in the well-fed state. The failure of some tissues, especially muscle, to take up glucose in the absence of insulin contributes further to the hyperglycemia. Accelerated gluconeogenesis, fueled by substrate made available by body protein degradation, maintains the hyperglycemia even in the starved state.

It may seem an enigma that hypertriglyceridemia is characteristic of this condition, since fatty acid synthesis is greatly diminished in the diabetic state. However, the low insulin/glucagon ratio results in uncontrolled rates of lipolysis in the adipose tissue. This increases blood levels of fatty acids and results in accelerated ketone body production by the liver. If the ketone bodies are not used as rapidly as they are formed, a dangerous condition, known as ketoacidosis, develops due to the accumulation of ketone bodies and hydrogen ions. Not all of the fatty acid taken up by the liver can be handled by the pathway of fatty acid oxidation and ketogenesis. The excess is esterified and directed into VLDL synthesis. Hypertriglyceridemia results because VLDL is synthesized and released by the liver more rapidly than these particles can be cleared from the blood by lipoprotein lipase. The quantity of this enzyme is dependent on a high insulin/glucagon ratio. The defect in lipoprotein lipase also results in hyperchylomicronemia, since lipoprotein lipase is also required for chylomicron catabolism in adipose tissue.

The most important thing to remember about the diabetic state is that every tissue continues to play the catabolic role that it was designed to play in starvation, in spite of delivery of adequate or even excess fuel from the gut. The consequence is that metabolism becomes stuck in the starve phase of the starve–feed cycle, with life-threatening consequences.

CLIN. CORR. 14.7
INSULIN-DEPENDENT DIABETES MELLITUS

Insulin-dependent diabetes mellitus was once called juvenile-onset diabetes because it usually appears in childhood or in the teens, but it is not limited to these patients. Insulin is either absent or nearly absent in this disease because of defective or absent β cells in the pancreas. Untreated, it is characterized by hyperglycemia, hyperlipoproteinemia (chylomicrons and VLDL), and episodes of severe ketoacidosis. Far from being a disease of defects in carbohydrate metabolism alone, diabetes causes abnormalities in fat and protein metabolism in such patients as well. The hyperglycemia results in part from the inability of the insulin-dependent tissues to take up plasma glucose and in part by accelerated hepatic gluconeogenesis from amino acids derived from muscle protein. The ketoacidosis results from increased lipolysis in the adipose tissue and accelerated fatty acid oxidation in the liver. Hyperchylomicronemia is the result of low lipoprotein lipase activity in adipose tissue capillaries, an enzyme dependent on insulin for its synthesis. Although insulin does not cure the diabetes, its use markedly alters the clinical course of the disease. The injected insulin promotes glucose uptake by tissues and inhibits gluconeogenesis, lipolysis, and proteolysis. The life span of the treated diabetic is still decreased, perhaps because it remains impossible to maintain perfect control of metabolism by repeated injections of insulin.

National Diabetes Data Group. Classification and diagnosis of diabetes mellitus and other categories of glucose intolerance. *Diabetes* 28:1039, 1977.

Noninsulin-Dependent Diabetes Mellitus

Figure 14.13*f* shows the metabolic interrelationships characteristic of a person suffering from noninsulin-dependent diabetes. In contrast to insulin-dependent diabetes discussed above, insulin is not absent in noninsulin-dependent diabetes (see Clin. Corr. 14.9). Indeed high levels of insulin may be observed in this form of diabetes, and the problem is primarily insulin resistance rather than lack of insulin. Insulin resistance is a poorly understood phenomenon in which the tissues fail to respond to insulin. The number or affinity of insulin receptors is reduced in some patients; others have normal insulin binding, but abnormal postreceptor responses, such as the activation of glucose transport.

Obesity causes some degree of insulin resistance. Indeed, the majority of patients with noninsulin-dependent diabetes mellitus are obese. Their insulin levels, which may be high, are not as high as those of a nondiabetic but similarly obese person. Hence, this form of diabetes is also a form of β-cell failure, and exogenous insulin will reduce the hyperglycemia. Hyperglycemia results mainly because of poor uptake of glucose by peripheral tissues, especially muscle. In contrast to insulin-dependent diabetes, ketoacidosis does not develop because uncontrolled lipolysis in the adipose tissue is not a feature of this disease. On the other hand, hypertriglyceridemia is characteristic of noninsulin-dependent diabetes but usually results from an increase in VLDL without hyperchylomicronemia. This is most likely explained by rapid rates of de novo hepatic synthesis of fatty acids and VLDL rather than increased delivery of fatty acids from the adipose tissue.

CLIN. CORR. 14.8 COMPLICATIONS OF DIABETES AND THE POLYOL PATHWAY

Diabetes is complicated by several disorders that may share a common pathogenesis. The lens, peripheral nerve, renal papillae, Schwann cells, glomerulus, and possibly retinal capillaries contain two enzymes that constitute the polyol pathway (the term polyol refers to polyhydroxy sugars). The first is aldose reductase, an NADPH-requiring enzyme. It reduces glucose to form sorbitol. Sorbitol is further metabolized by sorbitol dehydrogenase, an NAD^+-requiring enzyme that oxidizes sorbitol to fructose. Aldose reductase has a high K_m for glucose; therefore this pathway is only quantitatively important during hyperglycemia. It is known that in diabetic animals the sorbitol content of lens, nerve, and glomerulus is elevated. Sorbitol accumulation may damage these tissues by causing them to swell. There are now inhibitors of the reductase that prevent the accumulation of sorbitol in these tissues. These may prove useful in treatment of the symptoms of nerve injury in humans. Another metabolic disturbance in nerve is a reduction in *myo*-inositol content. Inositol is a component of the membrane phospholipid phosphatidyl-inositol and of the second messenger inositol trisphosphate. By unknown mechanisms, aldose reductase inhibitors also increase the level of inositol in nerve.

Gabbay, K. H. Hyperglycemia, polyol metabolism, and the complications of diabetes mellitus. *Ann. Rev. Med.* 26:521, 1975.

Stress and Injury Lead to Metabolic Changes

Stress includes injury, surgery, renal failure, burns, and infections (Figure 14.13*g*). Characteristically, blood cortisol, glucagon, catecholamines, and growth hormone levels are increased. The patient is resistant to insulin. The basal metabolic rate and blood glucose and free fatty acid levels are elevated. However, ketogenesis is not accelerated as in fasting, and perhaps as a result of this, the body is less able to preserve protein stores in muscle. It can be very difficult to reverse this protein breakdown, although now it is common to replace amino acids, glucose, and fat by infusing solutions of these nutrients intravenously. It has been proposed that the negative nitrogen balance of injured or infected patients is mediated by monocyte and lymphocyte proteins, such as **interleukin 1, interleukin 6,** and **tumor necrosis factor (TNF)** (see Clin. Corr. 14.10). These proteins are endogenous pyrogens, that is, they cause fever. Interleukin 1 activates proteolysis in skeletal muscle. Interleukin 6 stimulates the synthesis of a number of hepatic proteins called acute phase reactants by the liver. Acute phase reactants include fibrinogen, complement proteins, some clotting factors, and α_2 macroglobulin, which are presumed to play a role in defense against injury and infection. Tumor necrosis factor suppresses adipocyte fat synthesis, prevents uptake of circulating fat by inhibiting lipoprotein lipase, and probably stimulates lipolysis. It is responsible for much of the wasting seen in chronic infections.

Liver Disease Causes Major Metabolic Derangements

Since the liver is central to the body's metabolic interrelationships, advanced liver disease can be associated with major metabolic derangements (Figure 14.13*h*). The most important abnormalities are those in the metabolism of amino acids. The liver is the only organ capable of urea synthesis. In patients with cirrhosis, the liver is unable to convert ammonia into urea rapidly enough, and the blood ammonia rises. Part of this problem is due to abnormalities of blood flow in the cirrhotic liver, which interferes with the intercellular glutamine cycle (see p. 602). Ammonia arises from several enzyme reactions, such as glutaminase, glutamate dehydrogenase, and adenosine deaminase, during the metabolism of amino acids by the intestines and liver, and from the intestinal lumen, where bacteria can split urea into ammonia and carbon dioxide. Ammonia is very toxic to the central nervous system and is a major reason for the coma that sometimes occurs in patients in liver failure.

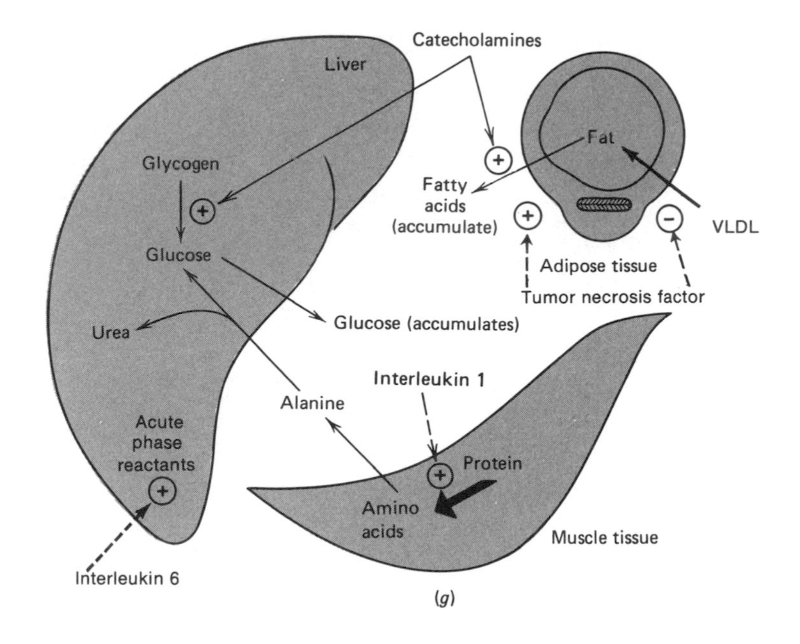

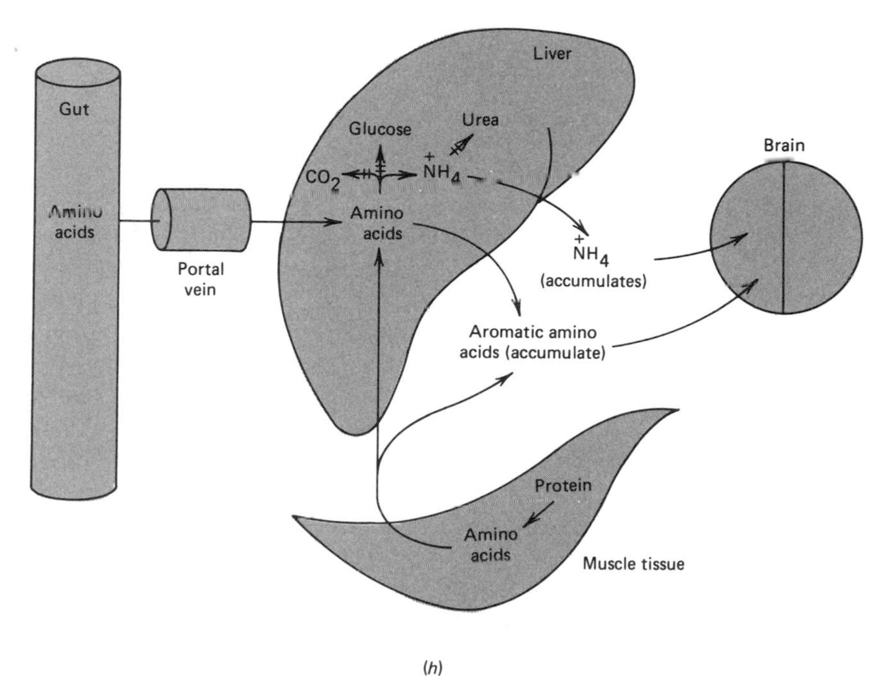

FIGURE 14.13 (Continued) (*g*) Stress. (*h*) Liver disease.

In advanced liver disease, aromatic amino acids accumulate in the blood to higher levels than branched-chain amino acids, apparently because of defective hepatic catabolism of the aromatic amino acids. This is important because aromatic amino acids and branched-chain amino acids are transported into the brain by the same carrier system. An elevated ratio of aromatic amino acids to branched-chain amino acids in liver disease results in increased brain uptake of aromatic amino acids. Increased synthesis of neurotransmitters such as serotonin in the brain as a consequence of increased availability of aromatic amino acids has been suggested to be responsible for some of the neurological abnormalities characteristic of liver disease. The liver is also a major source of insulin-like growth factor I (IGF-1). Cirrhotics suffer muscle wasting because of deficient IGF-1 synthesis. Finally, in outright liver failure, patients sometimes die of hypoglycemia because the liver is unable to maintain the blood glucose level by gluconeogenesis.

CLIN. CORR. 14.9 NONINSULIN-DEPENDENT DIABETES MELLITUS

Noninsulin-dependent diabetes mellitus, which accounts for 80–90% of the diagnosed cases of diabetes, is also called maturity-onset diabetes to differentiate it from insulin-dependent, juvenile diabetes. It usually occurs in middle-aged obese people. Insulin is present at normal to elevated levels in this form of the disease. The defect in these patients may be at the level of the insulin receptors located on the plasma membranes of insulin-responsive cells, that is, hepatocytes, adipocytes, and muscle cells. Noninsulin-dependent diabetes is characterized by hyperglycemia, often with hypertriglyceridemia. The ketoacidosis characteristic of the insulin-dependent disease is not observed. Increased levels of VLDL are probably the result of increased hepatic triacylglycerol synthesis stimulated by hyperglycemia and hyperinsulinemia. Obesity often precedes the development of insulin-independent diabetes and appears to be the major contributing factor. Obese patients are usually hyperinsulinemic. An inverse relationship between insulin levels and the number of insulin receptors has been established. The higher the basal level of insulin, the fewer receptors present on the plasma membranes. In addition, there are defects within insulin-responsive cells at sites beyond the receptor. An example is the ability of insulin to recruit glucose transporters from intracellular sites to the plasma membrane. As a consequence, insulin levels remain high, but glucose levels are poorly controlled because of the lack of normal responsiveness to insulin. Although the insulin level is high, it is not as high as in a person who is obese but not diabetic. In other words, there is a relative deficiency in the insulin supply from the β cells. Diet alone can control the disease in the obese diabetic. If the patient can be motivated to lose weight, insulin receptors will increase in number, and the postreceptor abnormalities will improve, which will increase both tissue sensitivity to insulin and glucose tolerance. The noninsulin-dependent diabetic tends not to develop ketoacidosis but nevertheless develops many of the same complications as the insulin-dependent diabetic, that is, nerve, eye, kidney, and coronary artery disease.

Olefsky, J. M. and Kolterman, O. G. Mechanisms of insulin resistance in obesity and non-insulin dependent (type II) diabetes. *Am. J. Med.* 70:151, 1981.

In Renal Disease Nitrogenous Wastes Accumulate

Nitrogenous wastes, including urea and creatinine, accumulate in patients with renal failure (Figure 14.13*i*). This accumulation is worsened by high dietary protein intake or accelerated proteolysis. The fact that gut bacteria can split urea into ammonia and that the liver can use ammonia and α-keto acids to form nonessential amino acids has been used to control the level of nitrogen wastes in renal patients. The patients are given a diet high in carbohydrate calories, and the amino acid intake is limited as much as possible to essential amino acids. Under these circumstances, the liver synthesizes nonessential amino acids from TCA cycle intermediates. This type of diet therapy may extend the time before the renal failure patient requires dialysis.

Ethanol Is Oxidized in the Liver Altering the NAD⁺ to NADH Ratio

The liver is primarily responsible for the first two steps of the ethanol catabolism:

$$\underset{\text{Ethanol}}{CH_3CH_2OH} + NAD^+ \longrightarrow \underset{\text{Acetaldehyde}}{CH_3CHO} + NADH + H^+$$

$$\underset{\text{Acetaldehyde}}{CH_3CHO} + NAD^+ \longrightarrow \underset{\text{Acetate}}{CH_3COO^-} + NADH + H^+$$

The first step, catalyzed by alcohol dehydrogenase, generates NADH and occurs in the cytosol; the second step, catalyzed by aldehyde dehydrogenase, also generates NADH but occurs in the mitochondrial matrix space. The intake of ethanol demands that the liver dispose of the NADH generated by these reactions by the only pathway it has available—the mitochondrial electron-transfer chain. Intake of even moderate amounts of ethanol generates too much NADH, resulting in the inhibition of important processes that require NAD^+, for example, gluconeogenesis and fatty acid oxidation (Figure 14.13*j*). Thus, fasting hypoglycemia and the accumulation of hepatic triacylglycerols (fatty liver) are consequences of alcohol ingestion. Lactate, formed in muscle and a number of other tissues, accumulates as a consequence of inhibition of lactate gluconeogenesis and can contribute to the development of a metabolic acidosis.

Liver mitochondria have a limited capacity to oxidize acetate to CO_2, because the activation of acetate to acetyl CoA requires GTP, which is the product of the succinyl-CoA synthetase reaction. The TCA cycle, and therefore GTP synthesis, are inhibited by high NADH levels during ethanol oxidation. Much of the acetate made from ethanol escapes to the blood where its concentration can approach 2 mM. Although heart is among the most active of the tissues for the disposal of acetate, virtually every cell with mitochondria can oxidize it to CO_2 and H_2O by way of the TCA cycle.

Acetaldehyde, the intermediate in the formation of acetate from ethanol, can also escape from the liver. The amount of this compound that accumulates in the blood is modest ($<20\ \mu M$) relative to that of acetate. However, acetaldehyde is a reactive compound that readily forms covalent bonds with functional groups of biologically important compounds. Formation of acetaldehyde adducts with proteins in tissues and blood of intoxicated animals has been demonstrated. Compromise of enzyme function as a consequence may occur but this remains to be established. Such

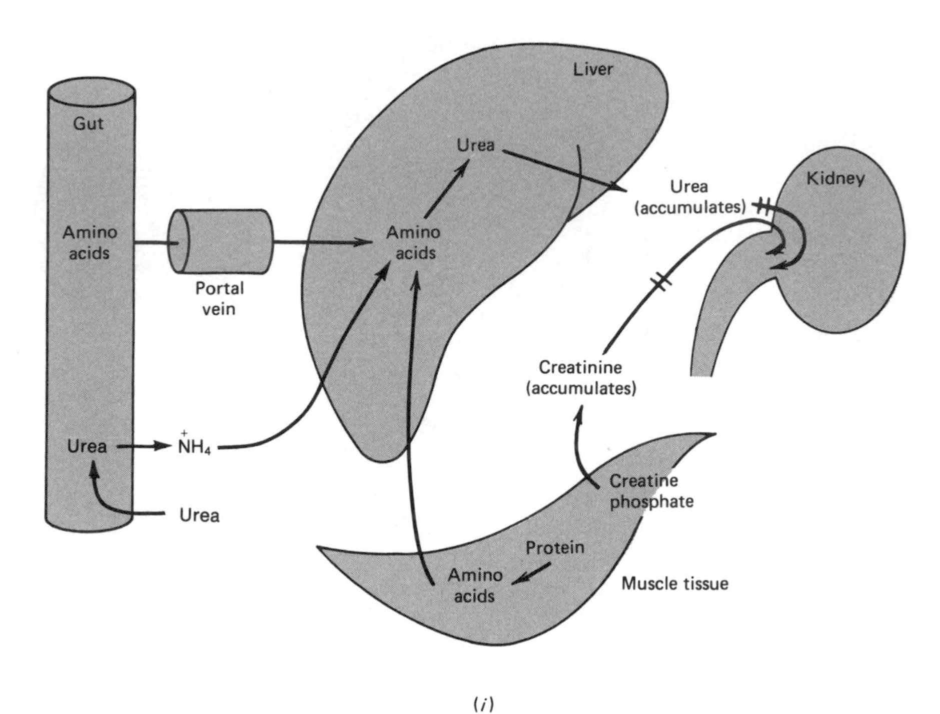

FIGURE 14.13 (Continued)
(*i*) Kidney failure.

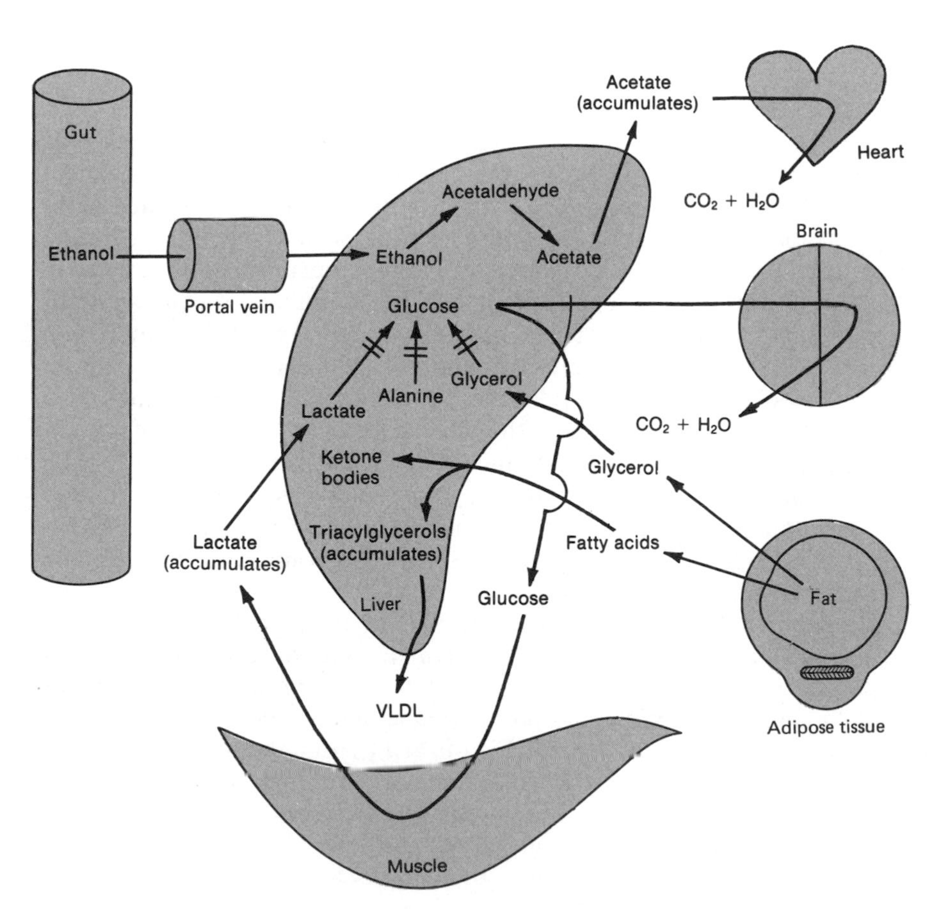

FIGURE 14.13 (Continued)
(*j*) Ethanol ingestion.

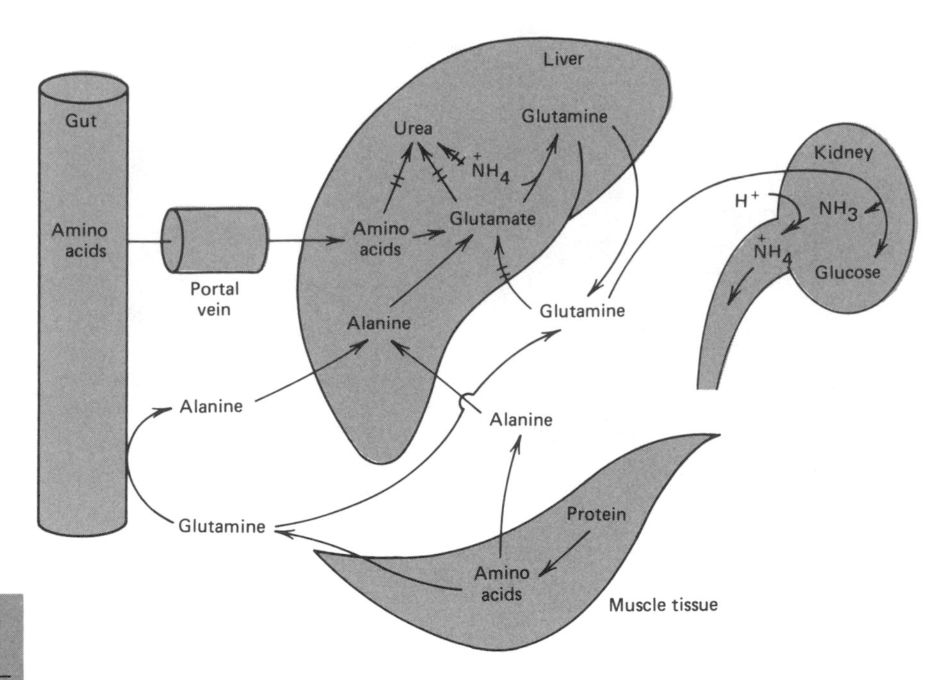

FIGURE 14.13 (Continued)
(*k*) Acidosis.

adducts may provide a marker for past drinking activity of an individual, just as hemoglobin A_{1c} has proved useful as an index of blood glucose control in the diabetic patient.

In Acid-Base Balance, Glutamine Has an Important Role

The regulation of acid–base balance, like that of nitrogen excretion, is shared by the liver and kidney. Metabolism of proteins generates excess hydrogen ions, which must be excreted by the kidney. The kidney helps regulate blood pH by excreting hydrogen ions, which is necessary for the reabsorption of bicarbonate and the titration of phosphate and ammonia in the tubular filtrate (see Chapter 25, p. 1057). Glutamine is the precursor of renal ammonia production. In chronic metabolic acidosis (see Figure 14.13*k*), the activities of renal glutaminase, glutamate dehydrogenase, phosphoenolpyruvate carboxykinase, and mitochondrial glutamine transporter increase and correlate with increased urinary excretion of ammonium ions and increased renal gluconeogenesis from amino acids. The liver participates in this process by synthesizing less urea, which makes more glutamine available for the kidney. In alkalosis, urea synthesis increases in the liver, and gluconeogenesis and ammonium ion excretion by the kidney decrease.

An intercellular glutamine cycle enables the liver to play a central role in the regulation of blood pH. The liver is composed of two types of hepatocytes involved in glutamine metabolism: periportal hepatocytes near the hepatic arteriole and portal venule and perivenous hepatocytes located near the central venule (Figure 14.14). Blood enters the liver by the hepatic artery and portal vein, and leaves the liver by way of the central vein. Glutaminase and the urea cycle enzymes are concentrated in the periportal hepatocytes while glutamine synthetase is found exclusively in perivenous hepatocytes. In alkalosis and at normal blood pH,

CLIN. CORR. 14.10
CANCER CACHEXIA

Unexplained weight loss may be a sign of malignancy, and weight loss is common in advanced cancer. Decreased appetite and food intake contribute to but do not entirely account for the weight loss. The weight loss is largely from skeletal muscle and adipose tissue, with relative sparing of visceral protein (i.e., liver, kidney, and heart). Although tumors commonly exhibit high rates of glycolysis and release lactate, the energy requirement of the tumor probably does not explain weight loss, because weight loss can occur with even small tumors. In addition, the presence of another energy-requiring growth, the fetus in a pregnant woman, does not normally lead to weight loss. Several endocrine abnormalities have been recognized in cancer patients. They tend to be insulin resistant, have higher cortisol levels, and have a higher basal metabolic rate compared with controls matched for weight loss. Two other phenomena may contribute to the metabolic disturbances. Some tumors synthesize and secrete biologically active peptides such as ACTH, nerve growth factor, and insulin-like growth factors, which could modify the endocrine regulation of energy metabolism. It is also possible that the host response to a tumor, by analogy to chronic infection, includes release of interleukin-1, interleukin 6, and tumor necrosis factor (TNF) by cells of the immune system. Tumor necrosis factor is also called cachexin because it produces wasting. These peptides stimulate fever, proteolysis, and lipolysis.

Beutler, B. and Cerami, A. Tumor necrosis, cachexia, shock, and inflammation: a common mediator. *Annu. Rev. Biochem.* 57:1505, 1988.

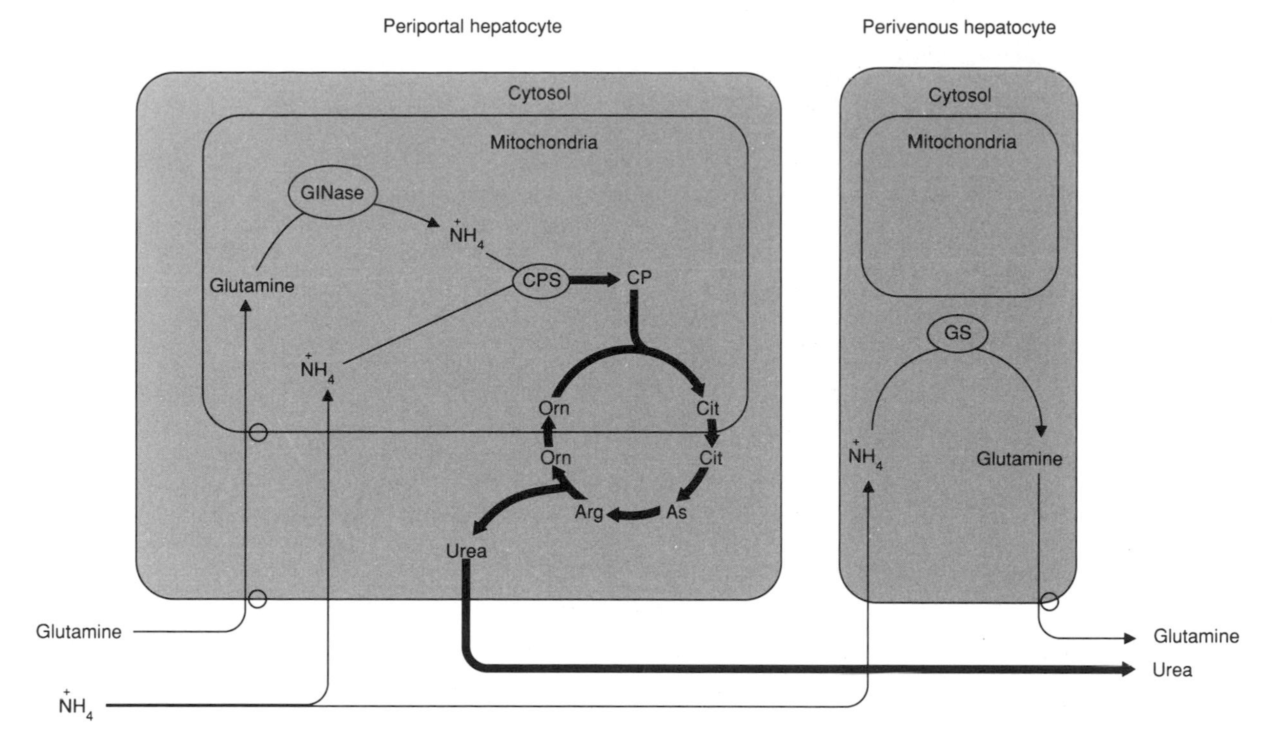

FIGURE 14.14
Intercellular glutamine cycle of the liver.
Abbreviations: GlNase = Glutaminase; GS = Glutamine synthetase; CPS = carbamoyl phosphate synthetase I; CP = carbamoyl phosphate; Cit = citrulline; AS = argininosuccinate; Arg = arginine; Orn = ornithine.
Redrawn from Häussinger, D. Glutamine metabolism in the liver: overview and current concepts. *Metabolism* 38 (Suppl. 1): 14, 1989.

glutamine enters the periportal cells and is hydrolyzed to contribute ammonium ion for urea synthesis. The bulk of glutamine and ammonium nitrogen entering the liver leaves the liver as urea. The perivenous cellular location of the glutamine synthetase is important because some ammonium ion escapes conversion to urea and this enzyme traps much of this potentially toxic compound in the form of glutamine. Thus, instead of the ammonium ion, glutamine is released from the liver and it circulates back to the liver where it can reenter the glutamine cycle in the periportal hepatocytes. Thus, in liver, both the donation of ammonium ion by glutamine for urea synthesis and the synthesis of glutamine are important in maintaining low blood ammonium levels. In acidosis, glutaminase of the periportal hepatocytes is less active and much of the blood glutamine escapes hydrolysis in the liver. Likewise, carbamoyl phosphate synthetase of the periportal hepatocytes is less active in acidosis, permitting perivenous cells to convert more ammonium ion to glutamine, which is then available for metabolism by the kidney to yield hydrogen ions that need to be eliminated in the urine.

BIBLIOGRAPHY

Brownlee, M., Vlassara, H., and Cerami, A. Nonenzymatic glycosylation and the pathogenesis of diabetes complications. *Ann. Intern. Med.* 101:527, 1984.

Cahill, G. F., Jr. Diabetes mellitus: a brief overview. *Johns Hopkins Med. J.* 143:155, 1978.

Cerami, A. and Koenig, R. H. Hemoglobin A_{Ic} as a model for the development of the sequelae of diabetes mellitus. *Trends Biochem. Sci.* 3:73, 1978.

Cohen, P. and Cohen, P. T. W. Protein phosphatases come of age. A review. *J. Biol. Chem.* 264:21435, 1989.

Crabb, D. W. and Lumeng, L. Metabolism of alcohol and the pathophysiology of alcoholic liver disease, G. Gitnick (Ed.). *Principles and practice of gastroenterology and hepatology*. New York: Elsevier, 1988, p. 1163.

Denton, R. M. and Pogson, C. I. *Metabolic regulation*. New York: Wiley, 1976.

Foster, D. W. Banting lecture 1984. From glycogen to ketones and back. *Diabetes* 33:1188, 1984.

Foster, D. W. Diabetes mellitus. Scriver, C. R., Beaudet, A. L., Sly, W. S., and Vallee, D. (Eds). *The metabolic basis of inherited disease*. New York: McGraw-Hill, 1989.

Fulop, M. Alcoholism, ketoacidosis, and lactic acidosis. *Diabetes/Metabolism Rev*. 5:365, 1989.

Geelen, M. J. H., Harris, R. A., Beynen, A. C., and McCune, S. A. Short-term hormonal control of hepatic lipogenesis. *Diabetes* 29:1006, 1980.

Gibson, D. M. and Parker, R. A. Control of HMG CoA reductase by reversible phosphorylation. E. G. Krebs (Ed.). *The enzymes* (series): *Enzyme control by phosphorylation*. New York: Academic, 1987, p. 179.

Goldberg, A. L., Baracos, V., Rodemann, P., Waxman, L., and Dinarello, C. Control of protein degradation in muscle by prostaglandins, calcium, and leukocytic pyrogen (interleukin 1). *Fed. Proc*. 43:1301, 1984.

Häussinger, D. Glutamine metabolism in the liver: Overview and current concepts. *Metabolism* 38 (Suppl. 1): 14, 1989.

Hers, H. G. and Hue, L. Gluconeogenesis and related aspects of glycolysis. *Annu. Rev. Biochem*. 52:617, 1983.

Ingebritsen, T. S. and Cohen, P. Protein phosphatases: properties and role in cellular regulation. *Science* 221:331, 1983.

Katz, J. and McGarry, J. D. The glucose paradox. Is glucose a substrate for liver metabolism? *J. Clin. Invest*. 74:1901, 1984.

Krebs, H. A. Some aspects of the regulation of fuel supply in omnivorous animals. *Adv. Enzyme Regul*. 10:387, 1972.

Krebs, H. A., Williamson, D. H., Bates, M. W., Page, M. A., and Hawkins, R. A. The role of ketone bodies in caloric homeostasis. *Adv. Enzyme Regul*. 9:387, 1971.

Kurkland, I. J. and Pilkis, S. J. Indirect and direct routes of hepatic glycogen synthesis. *FASEB J*. 3:2277, 1989.

Larner, J. *Intermediary metabolism and its regulation*. Englewood Cliffs, NJ: Prentice-Hall, 1971.

McGarry, J. D., Woeltje, K. F., Kuwajima, M., and Foster, D. W. Regulation of ketogenesis and the renaissance of carnitine palmitoyltransferase. *Diabetes/Metabolism Rev*. 5:271, 1989.

Newsholme, E. A. and Leech, A. R., *Biochemistry for the medical sciences*. New York: Wiley, 1983.

Newsholme, E. A. and Start, C. *Regulation in metabolism*. New York: Wiley, 1973.

Nosadini, R., Avogaro, A., Doria, A., Fioretto, P., Trevisan, R., and Morocutti, A. Ketone body metabolism: A physiological and clinical overview. *Diabetes Metab. Rev*. 5:299, 1989.

Pedersen, O. The impact of obesity on the pathogenesis of non-insulin-dependent diabetes mellitus: A review of current hypotheses. *Diabetes/Metabolism Rev*. 5:495, 1989.

Pedersen, O. and Beck-Nielsen, H. Insulin resistance and insulin-dependent diabetes mellitus. *Diabetes Care* 10:516, 1987.

Pilkis, S. J., Ed-Maghrabi, M. R. and Claus, T. H. Hormonal regulation of hepatic gluconeogenesis and glycolysis. *Annu. Rev. Biochem*. 57:755, 1988.

Sugden, M. C., Holness, M. J., and Palmer, T. N. Fuel selection and carbon flux during the starved-to-fed transition. A review article. *Biochem. J*. 263:313, 1989.

Toth, B., Bollen, M., and Stalmans, W. Acute regulation of hepatic phosphatases by glucagon, insulin, and glucose. *J. Biol. Chem*. 263:14061, 1988.

QUESTIONS

C. N. ANGSTADT AND J. BAGGOTT

*Q*uestion *T*ypes are described on page xxiii.

A. Well-fed state
B. Early fasting state
C. Fasting state
D. Early refed state

1. (QT5) Hepatic glycogenolysis is a primary source of blood glucose during this period.
2. (QT5) Ketone bodies supply a significant portion of the brain's fuel.

3. (QT1) The fact that the K_m of aminotransferases for amino acids is much higher than that of aminoacyl-tRNA synthetases means that:
   A. at low amino acid concentrations, protein synthesis will take precedence over amino acid catabolism.
   B. the liver cannot accumulate amino acids.
   C. amino acids will undergo transamination as rapidly as they are delivered to the liver.
   D. any amino acids in excess of immediate needs for energy must be converted to protein.
   E. amino acids can be catabolized only if they are present in the diet.
4. (QT1) Branched-chain amino acids:
   A. are normally completely catabolized by muscle to CO_2 and H_2O.
   B. can be catabolized by liver but not muscle.
   C. are the main dietary amino acids metabolized by intestine.
   D. are in high concentration in blood following the breakdown of muscle protein.
   E. are a major source of nitrogen for alanine and glutamine produced in muscle.
5. (QT1) The largest energy reserve (in terms of kilocalories) in humans is:
   A. blood glucose.
   B. liver glycogen.
   C. muscle glycogen.
   D. adipose tissue triacylglycerol.
   E. muscle protein.

Use the accompanying figure (representing hours postfeeding) to answer Questions 6 and 7.

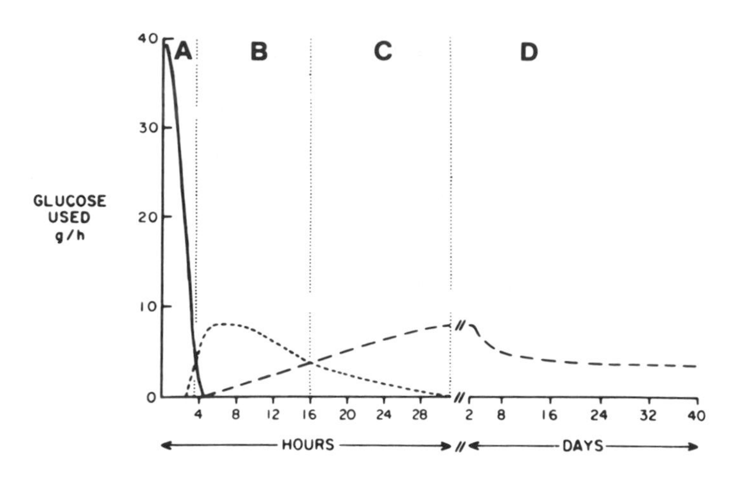

6. (QT5) The origin of most of the glucose in the period represented by _______ is liver glycogen.
7. (QT5) Ketone bodies are a significant fuel for brain as well as muscle in the period _______.

8. (QT2) Which of the following may represent control of a metabolic process by substrate availability?
   1. Increased urea synthesis after a high protein meal
   2. Rate of ketogenesis
   3. Hypoglycemia of advanced starvation
   4. Response of glycolysis to fructose 2,6-bisphosphate
9. (QT2) Which of the following would favor gluconeogenesis in the fasted state?
   1. Fructose 1,6-bisphosphate stimulation of pyruvate kinase
   2. Long-chain acyl CoA ester inhibition of acetyl CoA carboxylase
   3. Malonyl CoA inhibition of carnitine palmitoyltransferase I
   4. Acetyl CoA activation of pyruvate carboxylase
10. (QT2) Conversion of a nonphosphorylated enzyme to a phosphorylated one:
    1. usually changes its activity.
    2. may be catalyzed by a cAMP-dependent protein kinase.
    3. is favored when phosphoprotein phosphatase is phosphorylated.
    4. is more likely to occur in the well-fed than in the fasted state.

11. (QT1) Adipose tissue responds to low insulin : glucagon by:
    A. dephosphorylating the interconvertible enzymes.
    B. stimulating the deposition of fat.
    C. increasing the amount of pyruvate kinase.
    D. stimulating hormone-sensitive lipase.
    E. stimulating phenylalanine hydroxylase.

12. (QT2) Changing the level of enzyme activity by changing the number of enzyme molecules:
    1. is considerably slower than allosteric or covalent modification methods.
    2. may involve enzyme induction.
    3. may override the effectiveness of allosteric control.
    4. may be caused by hormonal influences or by changing the nutritional state.

13. (QT1) Muscle metabolism during exercise:
    A. is the same in both aerobic and anaerobic exercise.
    B. shifts from primarily glucose to primarily fatty acids as fuel during aerobic exercise.
    C. uses largely glycogen and phosphocreatine in the aerobic state.
    D. causes a sharp rise in blood ketone body concentration.
    E. uses only phosphocreatine in the anaerobic state.

A. Insulin-dependent diabetes mellitus — C. Both
B. Noninsulin-dependent diabetes mellitus — D. Neither

14. (QT4) Hepatic gluconeogenesis occurs in both the well-fed and fasted states.
15. (QT4) Hypertriglyceridemia and ketoacidosis are commonly present in the untreated state.

16. (QT2) The elevated liver concentration of NADH produced by ingestion of ethanol:
    1. is restricted to the mitochondria.
    2. may lead to an acidosis by inhibiting gluconeogenesis from lactate.
    3. leads to "fatty liver" by stimulating fatty acid synthesis.
    4. inhibits the oxidation of acetate to CO_2.

17. (QT3) A. Activity of glutaminase and carbamoyl phosphate synthetase in periportal hepatocytes in alkalosis.
    B. Activity of glutaminase and carbamoyl phosphate synthetase in periportal hepatocytes in acidosis.

ANSWERS

1. B The response of glycogenolysis to fasting is rapid, and during this period there is still glycogen present. In fasting, the glycogen is depleted and in the other two states, glycogenesis would occur (p. 579).
2. C If ketone body concentration in blood is high, ketone bodies can cross the blood–brain barrier and they are a good fuel. High ketone body concentrations do not occur in the other states (p. 582).
3. A A high K_m means that a reaction will proceed slowly at low concentration, whereas a low K_m means the reaction can be rapid under the same circumstances. Protein synthesis requires only that all amino acids be present. Unless amino acids are in high enough concentration, the liver does not catabolize them (p. 578).
4. E A, B: Muscle has high levels of the aminotransferases

for branched-chain amino acids, whereas liver has high levels of enzymes for the catabolism of the branched-chain α-keto acids. C: Intestine metabolizes several dietary amino acids but not these. D, E: When branched-chain amino acids are derived from muscle protein, transamination transfers the nitrogen to alanine or glutamine, which are transported to the liver and kidney (p. 579).

5. D A: Blood glucose must be maintained but is a relatively minor reserve. B, C: Glycogen is a rapidly mobilizable reserve of energy but not a large one. E: Protein can be used for energy, but that is not its primary role. D: The caloric content of adipose tissue fat is more than 5 times as great as that of muscle protein and almost 200 times as great as that of the combined carbohydrates (Table 14.2).
6. B Glycogen is the most rapidly mobilized source of glucose in fasting but lasts for only a little more than a day. A represents dietary glucose.
7. D The decreased rate of gluconeogenesis indicates that the brain is using ketone bodies and muscle proteolysis is restricted, conserving body protein (Figure 14.6).
8. A 1, 2, 3 correct. 1: After a high protein meal, the intestine produces ammonia and precursors of ornithine for urea synthesis. 2: Ketogenesis is dependent on the availability of fatty acids. 3: This represents lack of gluconeogenic substrates. 4: Fructose 2,6-bisphosphate is an allosteric effector (activates the kinase and inhibits the phosphatase) of the enzyme controlling glycolysis (p. 586).
9. C 2, 4 correct. 1: Stimulation of pyruvate kinase stimulates glycolysis, opposing gluconeogenesis. 2, 3: Decreased activity of acetyl CoA carboxylase results in decreased synthesis of malonyl CoA and greater transport of fatty acids into mitochondria for β oxidation, a necessary source of energy for gluconeogenesis. 4: Pyruvate carboxylase is a key gluconeogenic enzyme (p. 588).
10. A 1, 2, 3 correct. 1: Some enzymes are active when phosphorylated; for others the reverse is true. 2: This is the most common, though not only, mechanism of phosphorylation. 3: Phosphorylation inhibits phosphoprotein phosphatase. 4: In the well-fed state, insulin : glucagon is high and cAMP levels are low (p. 589).
11. D A: Low insulin/glucagon means high cAMP and, thus, high activity of cAMP-dependent protein kinase and protein phosphorylation. B, D: Phosphorylation activates hormone-sensitive lipase to mobilize fat. C: cAMP works by stimulating covalent modification of enzymes. E: This is a liver enzyme (p. 591).
12. E All four correct. 1: Adaptive changes are examples of long-term control. 2, 4: Both hormonal and nutritional effects are involved in inducing certain enzymes and/or altering their rate of degradation. 3: If there is little or no enzyme because of adaptive changes, allosteric control is irrelevant. This is important to keep in mind in refeeding a starved person (p. 591).
13. B A: Anaerobic muscle uses glucose almost exclusively; aerobic muscle uses fatty acids and ketone bodies. B: This is indicated by the drop in the respiratory quotient. D: Ketone bodies are good aerobic substrates so the blood concentration does not increase greatly. E: Phosphocreatine is only a short-term source of ATP (p. 593).
14. A Because the defect is an inability of the β cells of the pancreas to produce insulin, the insulin/glucagon ratio is always low and gluconeogenesis is stimulated. Insulin levels may be high in noninsulin-dependent diabetes, and hyperglycemia is primarily because of poor uptake by peripheral tissues (p. 599).
15. A Hypertriglyceridemia is present in both types, although for different reasons, but ketoacidosis is common only in the insulin-dependent type, again because the low insulin/glucagon ratio results in excessive lipolysis of adipose tissue (p. 599).
16. C 2, 4 correct. 1: The oxidation of ethanol, which also produces NADH, is cytosolic. Acetaldehyde oxidation is mitochondrial. 2: Failure to oxidize lactate to pyruvate because of the unfavorable NAD^+/NADH leads to lactate accumulation. 3: "Fatty liver" is a consequence of inhibition of fatty acid oxidation by high NADH. 4: Acetate activation requires GTP from the TCA cycle. Why is this cycle inhibited (p. 601)?
17. A A: This is part of the mechanism by which liver participates in regulation of blood pH and ammonia concentration. High activity of these enzymes leads to urea synthesis. B: Low activity of these enzymes results in glutamine, not urea, leaving liver for kidney to participate in acid excretion (p. 602).

15

DNA: The Replicative Process and Repair

STELIOS AKTIPIS

15.1 BIOLOGICAL PROPERTIES OF DNA 608
Overview 608
DNA Can Transform Cells 608
DNA's Informational Capacity Is Enormous 609
15.2 STRUCTURE OF DNA 609
Nucleotides Joined by Phosphodiester Bonds Form Polynucleotides 610
Nucleases Hydrolyze Phosphodiester Bonds 612
Periodicity Leads to Secondary Structures of DNA 613
Many Factors Stabilize the DNA Structure 619
15.3 TYPES OF DNA STRUCTURE 627
Size of DNA Is Highly Variable 627
DNA May Be Either Linear or Circular 630
Topology of Circular DNA Is That of a Superhelix 631
Nucleoproteins of Eucaryotes Contain Histones and Nonhistone Proteins 637
Nucleoproteins of Procaryotes Have Similarities to Those of Eucaryotes 641
15.4 DNA STRUCTURE AND FUNCTION 641
The Uniqueness of Each DNA Is Its Nucleotide Sequence 641
Most Procaryotic DNA Codes for Specific Proteins 644
Only a Small Percentage of Eucaryotic DNA Codes for Structural Genes 644
15.5 FORMATION OF THE PHOSPHODIESTER BOND IN VIVO 649
DNA-Dependent DNA Polymerase 649
15.6 MUTATION AND REPAIR OF DNA 653
Mutations Are Stable Changes in DNA Structure 653
Mutagens May Be Chemical Compounds or Radiation 654
DNA Is Repaired Rather Than Degraded 657
15.7 DNA REPLICATION 662
Complementary Strands Are Basic to the Mechanism of Replication 662
E. coli Provides the Basic Model for the Replication of DNA 669
Eucaryotic DNA Replication 673
15.8 DNA RECOMBINATION 675
Recombination May Be General or Site-Specific 675
15.9 SEQUENCING OF NUCLEOTIDES IN DNA 677
Restriction Maps Give the Sequence of Segments of DNA 677
BIBLIOGRAPHY 677
QUESTIONS AND ANNOTATED ANSWERS 678
CLINICAL CORRELATIONS
15.1 Diagnostic Use of Probes in Medicine 626
15.2 Mutations and the Etiology of Cancer 658
15.3 Diseases of DNA Repair and Susceptibility to Cancer 662
15.4 The Nucleotide Sequence of the Human Genome 677

15.1 BIOLOGICAL PROPERTIES OF DNA

Overview

This chapter reviews the chemical structure of DNA and examines the relationship between this structure and the biochemical function of DNA. Within this context the process of DNA replication and repair is detailed. The remaining processes through which DNA regulates the expression of biological information (i.e., transcription and translation) are the subjects of Chapters 16 and 17.

One of the striking aspects of natural order is the sense of unity that exists between the members of successive generations in each species. It is apparent that an almost totally stable bank of information must always be preserved and passed from one generation to the next if individual species are to maintain their identities relatively unchanged over millions of years. It is now well established that this bank of genetic information takes the form of a stable macromolecule, deoxyribonucleic acid (DNA), which serves as the carrier of genetic information in both procaryotes and eucaryotes. Deoxyribonucleic acid exhibits a rare purity of function by being one of the few macromolecules known to perform, with only minor exceptions, the same basic functions across species barriers.

It is apparent that the properties of cells are to a large extent determined by their constituent proteins. Many proteins serve as indispensable structural components of the cell. Other proteins, such as enzymes and certain hormones, are functional in character and determine most of the biochemical properties of the cell. As a result, the factors that control *which proteins* a cell may synthesize, at *what quantities,* and with *which sequence* are the same factors that primarily determine the function as well as the destiny of every living cell.

It is now well recognized that DNA is the macromolecule that ultimately controls every aspect of cellular function, primarily through protein synthesis. The DNA exercises this control as suggested by the sequence

$$\circlearrowright\text{DNA} \longrightarrow \text{RNA} \longrightarrow \text{protein}$$

The flow of biological information is clearly from one class of nucleic acid to another, from DNA to RNA, with only minor exceptions, and from there to protein. In order for this transfer of information to occur faithfully, each preceding macromolecule serves as a structure-specifying template for the synthesis of the subsequent member in the sequence.

In addition to regulating cellular expression, DNA plays an exclusive role in heredity. This role is suggested by a circular arrow engulfing DNA, which depicts DNA as a **replicon,** a molecule that can undergo self-replication. The significance of *replication* is far reaching. It permits DNA to make copies of itself as a cell divides. These copies are bestowed to the daughter cells, which can thus inherit each and every property and characteristic of the original cell.

First, the important message to be retained is that DNA ultimately determines the properties of a living cell by *regulating the expression of biological information,* primarily by the control of protein synthesis. Second, but not less importantly, it should be clear that DNA transfers biological information from one generation to the next, that is, it is essential for the *transmittance of genetic information.*

DNA Can Transform Cells

The above principles, universally accepted today, were rejected outright not long ago. In fact, prior to the 1950s the general view was that nucleic acids were substances of somewhat limited cellular importance. The first convincing suggestion that DNA is the genetic material was made during the mid-1940s. The experiment involved the **transformation** of one type of pneumococcus, surrounded by the presence of a polysaccharide capsule and referred to as the S form because of its property of forming colonies with smooth-looking cellular perimeters, to a mutant without capsule, called the R form, which forms colonies with rough-looking outlines. These two forms are genetically distinct and cannot interconvert spontaneously. The transformation experiment demonstrated that a pure extract of DNA from the S form, when incorporated into the R form of pneumococcus, conveyed to the R form the specific property of synthesizing the characteristic polysaccharide capsule. Furthermore, the bacteria transformed from the R form to the S form maintained the property of synthesizing the capsule over succeeding generations. It was thus demonstrated that DNA was the *transforming agent,* as well as the material responsible for *transmitting* genetic information from one generation to the next. Almost three-quarters of a century had to elapse from the time nucleic acids were discovered until their important biological role was generally recognized.

DNA's Informational Capacity Is Enormous

One of the striking characteristics of DNA is that it is able to encode an enormous quantity of biological information. An undifferentiated mammalian fetal cell contains only a few picograms (10^{-12} g) of DNA. Yet, this minute amount of material is sufficient to direct the synthesis of an enormous number of distinct proteins that will determine the form and biochemical behavior of a large variety of differentiated tissues in the adult animal.

The compactness with which such information is stored in DNA is unique. Even the sophisticated memory elements of contemporary computers would appear pitifully inadequate by comparison. How does DNA achieve such a supreme coding effectiveness? The answers must obviously be sought in the nature of its chemical structure. It turns out that this structure is not only consistent with the unique efficiency of DNA as a "memory bank," but also provides the basis for understanding how DNA eventually "translates" this information into proteins.

15.2 STRUCTURE OF DNA

Structurally DNA is a **polynucleotide.** A formal analogy between polynucleotides and proteins may therefore be perceived. Polynucleotides are the products of *nucleotide* condensation, just as proteins are produced by the polymerization of *amino acids.* This similarity of structures is an important element that facilitates the transfer of genetic information between these two distinct classes of macromolecules. The structure of nucleotides and their constituent purine and pyrimidine bases are examined in Chapter 13.

The base composition of DNA varies considerably among species, particularly procaryotes, which have a range of 25–75% in adenine–thymine content. This range narrows with evolution, reaching limiting values of about 45–53% in mammals.

In addition to the four common bases, adenine, guanine, thymine, and cytosine, which occur in DNA from all sources, DNA isolated from many plant and animal tissues (e.g., wheat germ and thymus gland) contains small amounts of the base 5-methylcytosine. Methylated derivatives of the bases are also present in all DNA molecules examined to date. In

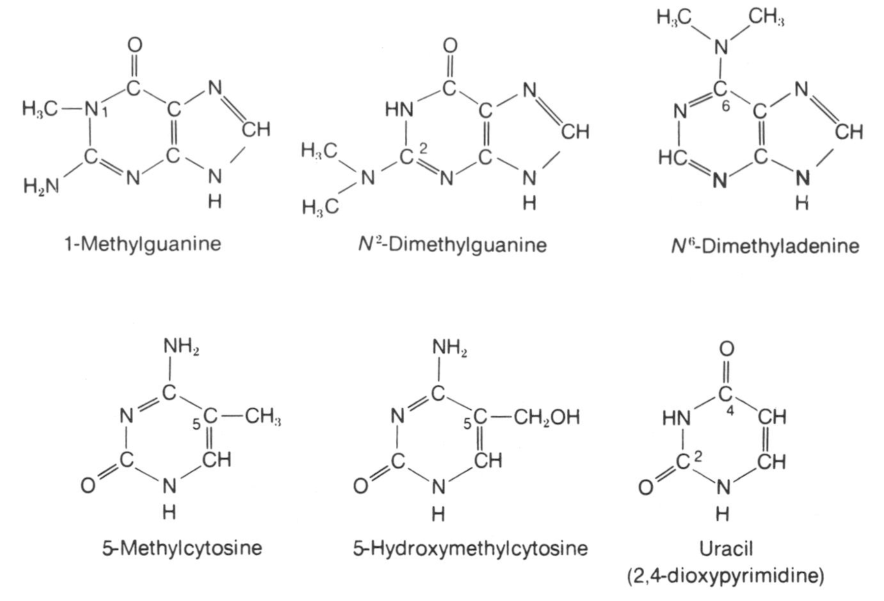

1-Methylguanine

N^2-Dimethylguanine

N^6-Dimethyladenine

5-Methylcytosine

5-Hydroxymethylcytosine

Uracil
(2,4-dioxypyrimidine)

FIGURE 15.1
Structures of some less common bases occurring in DNA.

addition, the DNA of certain bacteriophages (the T-even coliphages) contain 5-hydroxymethylcytosine in place of cytosine, and this derivative occurs in a glucosylated form. Even uracil, a base constituent of RNA, has been found in certain *Bacillus subtilis* phages, instead of thymine. The structures of some of these bases are shown in Figure 15.1.

Nucleotides Joined by Phosphodiester Bonds Form Polynucleotides

Polynucleotides are formed by the joining of nucleotides by phosphodiester bonds. The phosphodiester bond is the formal analog of the peptide bond in proteins. It serves to join, as a result of the esterification of two of the three OH groups of phosphoric acid, two adjoining nucleotide residues. Two free OH groups are present in deoxyribose on the C-3′ and C-5′ atoms. Therefore, these are the only OH groups that can participate in the formation of a phosphodiester bond. Indeed, it turns out that the nucleotide residues in DNA polynucleotides are jointed together by 3′, 5′-phosphodiester bonds, as shown in Figure 15.2.

In some instances polynucleotides are linear polymers. The last nucleotide residue at each of the opposite ends of the polynucleotide chain serve as the two terminals of the chain. It is apparent that these terminals are not structurally equivalent, since one of the nucleotides must terminate at a 3′-OH group and the other at a 5′-OH group. These ends of the polynucleotides are referred to as the 3′ and the 5′ termini, and they may be viewed as corresponding to the amino and carboxyl termini in proteins. Polyneucleotides also exist as cyclic structures, which contain no free terminals. Esterification between the 3′-OH terminus of a polynucleotide with its own 5′-phosphate terminus can produce a cyclic polynucleotide.

In this discussion long polymers of nucleotides joined by phosphodiester bonds are referred to as polynucleotides, in accordance with the prevailing nomenclature. A distinct name, **oligonucleotide,** is reserved for shorter nucleotide-containing polymers. According to formal rules of nomenclature, however, polynucleotides must be named by using roots derived from the names of the corresponding nucleotides, and using the ending *ylyl*. For example, the polynucleotide segment in Figure 15.2, in which the 5′ terminal is on the left of each nucleotide residue, should be

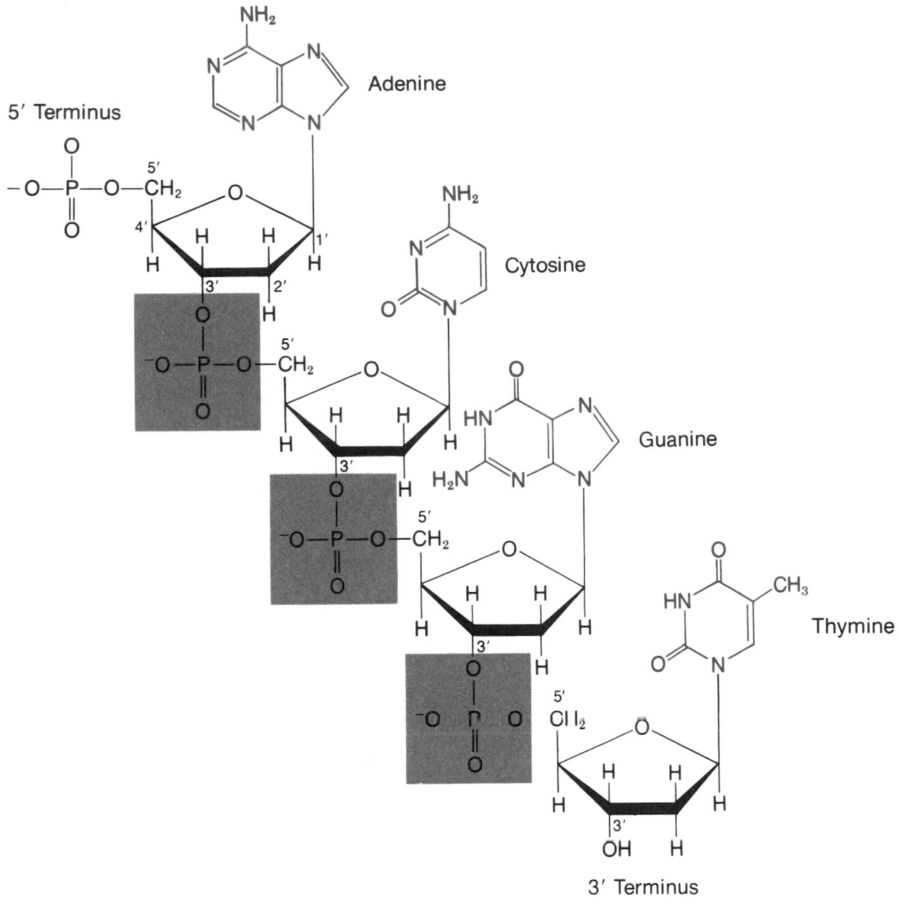

FIGURE 15.2
Structure of a DNA polynucleotide segment.
The example shown in this figure is a tetranucleotide, that is, an oligonucleotide consisting of four monomeric units. Although an exact polymerization size for this change in name does not exist, as a general rule a polymer containing less than 30–40 nucleotides is referred to as an oligonucleotide.

named from left to right as

· · ·deoxyaden*ylyl*, deoxycytid*ylyl*, deoxyguan*ylyl*,
deoxythymid*ylyl*· · ·

It is apparent, however, that the result of this approach is so cumbersome that abbreviations are generally preferred. For example, the oligonucleotide shown in Figure 15.2 is usually referred to as dAdCdGdT, and a polynucleotide containing only one kind of nucleotide, for example, dA, may be written as poly(dA). Oligo- and polynucleotide structures are also written out in shorthand, as shown in Figure 15.3. In every instance the sequence is written starting on the left with the nucleotide of the 5′ terminus.

DNA is made of polynucleotides, and it is the specific sequence of bases along a polynucleotide chain that determines the biological properties of the polymer. Although the structure of the nucleic acid building blocks, the bases, had been correctly known for many years, the polymeric structure initially proposed for DNA turned out to be one of the classical errors in the history of biochemistry. Experimental data obtained from what appears to have been partially degraded samples of DNA, and several other misconceptions, led to the erroneous conclusion that DNA consisted of repeating tetranucleotide units. Each tetranucleotide supposedly contained equimolar quantities of the four common bases. These impressions persisted to some degree until the late 1940s and early 1950s, when they were clearly shown to be in error. In the interim, however, these misconceptions were responsible for setting back the acceptance of the concept that the DNA of chromosomes carried genetic

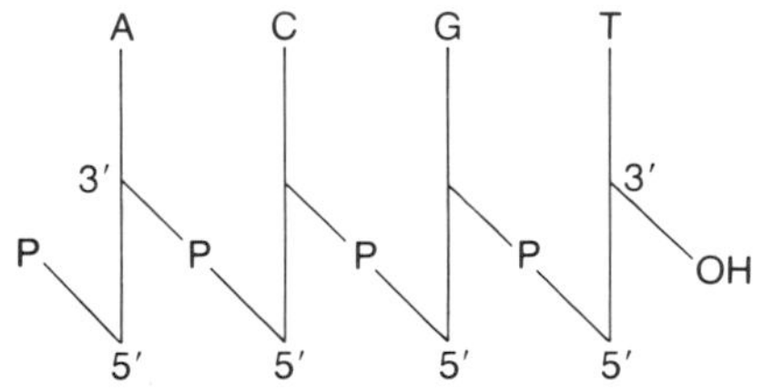

FIGURE 15.3
Shorthand form for structure of oligonucleotides.
The convention used in writing the structure of an oligo- or polynucleotide is a perpendicular bar representing the deoxyribose moiety, with the 5′-OH position of the sugar located at the bottom of the bar and the 3′-OH at a midway position. Bars joining the 3′ and 5′ positions represent the 3′,5′-phosphodiester bond, and the P on the left side of the perpendicular bar represents a 5′-phosphate ester. A 3′-phosphate ester is represented by placing the phosphate group on the right side of the bar. The base is indicated by its initial.

information. The monotonous structure of repeating tetranucleotides appeared incapable of having the versatility to encode for the enormous number of messages necessary to convey hereditary traits. Instead proteins, which can be ordered in an almost unlimited number of amino acid sequences, were favored as the most suitable candidates for a hereditary function. The transformation experiment carried out in the mid-1940s, and the subsequent finding that DNA consists of polynucleotide rather than tetranucleotide chains, were responsible for the general acceptance of the hereditary role of DNA that followed.

Nucleases Hydrolyze Phosphodiester Bonds

The nature of the linkage between nucleotides to form polynucleotides was elucidated primarily by the use of exonucleases, which are enzymes that hydrolyze these polymers in a selective manner. **Exonucleases** cleave the last nucleotide residue in either of the two terminals of an oligonucleotide. Oligonucleotides can thus be degraded by the stepwise removal of individual nucleotides or small oligonucleotides from either the 5′ or the 3′ terminus. Nucleases sever the bonds in one of two nonequivalent positions indicated in Figure 15.4 as proximal (p) or distal (d) to the base, which occupies the 3′ position of the bond. For example, the treatment of an oligodeoxyribonucleotide with venom diesterase, an enzyme obtained from snake venom, yields deoxyribonucleoside 5′-phosphates. In contrast, treatment with a diesterase isolated from animal spleen produces deoxyribonucleoside 3′-phosphates.

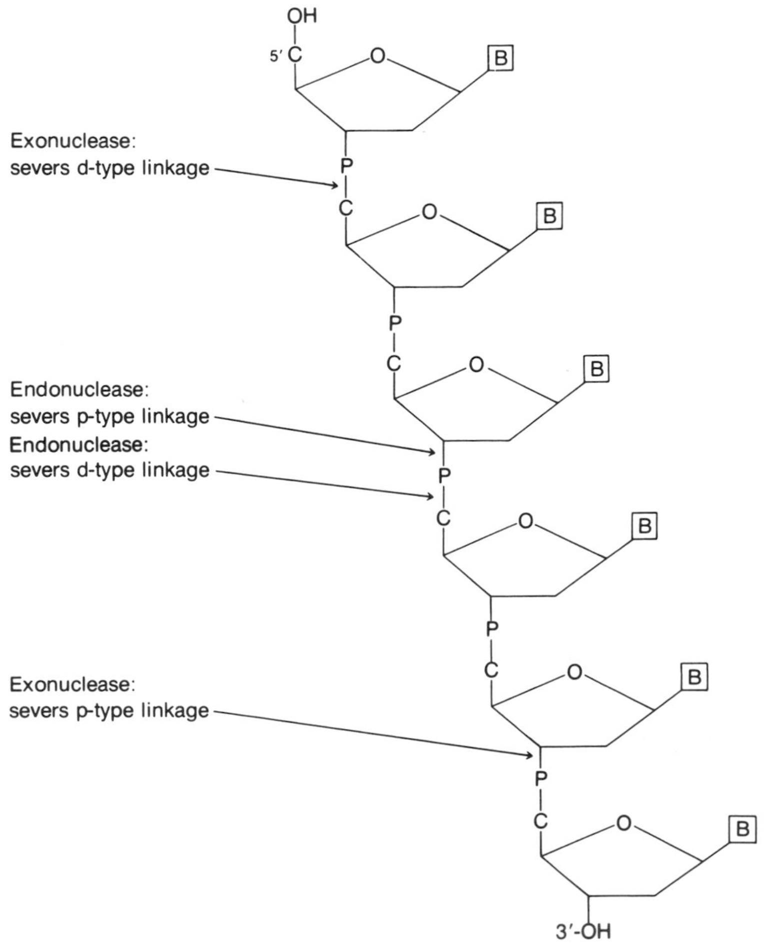

FIGURE 15.4
Nucleases of various specificities.
Exonucleases remove nucleotide residues from either of the terminals of a polynucleotide, depending on their specificity. Endonucleases hydrolyze interior phosphodiester bonds. Both endo- and exonucleases hydrolyze either d- or p-type linkages, as illustrated in this figure (see text for explanation of d- and p-type linkages).

TABLE 15.1 Specificities of Various Types of Nucleases

Enzyme	Substrate	Specificity[a]
EXONUCLEASES		
Snake venom phosphodiesterase	DNA or RNA single-stranded only	Cleaves all type p linkages, starting with a free 3′-OH group and moving toward the 5′ terminal; releases nucleoside 5′-phosphates; has no base specificity
Bovine spleen phosphodiesterase	DNA or RNA single-stranded only	Cleaves all type d linkages, starting at the free 5′-OH and proceeding to the 3′ terminal; releases nucleoside 3′-phosphates; has no base specificity
ENDONUCLEASES		
Bovine pancreas deoxyribonuclease (DNase I)	DNA single- or double-stranded	Cleaves all type p linkages but prefers those between purine and pyrimidine bases
Calf thymus deoxyribonuclease (DNase II)	DNA single- or double-stranded	Cleaves all type d linkages randomly

[a] See text for explanation of d and p type linkages.

It should be noted that other nucleases, which cleave phosphodiester bonds located in the interior of polynucleotides and are designated as **endonucleases,** behave similarly in this respect. For instance, DNase I cleaves only p linkages, while DNase II cleaves d linkages. The points of cleavage along an oligonucleotide chain are indicated by arrows in Figure 15.4. Some endonucleases have been particularly useful in the development of early methodologies for sequencing of RNA polynucleotides. More recently other endonucleases, known as restriction endonucleases, have provided the basis for the development of recombinant DNA techniques.

Many nucleases do not exhibit any specificity with respect to the base adjacent to the linkage that is hydrolyzed. Certain nucleases, however, act more discriminately next to specific types of bases or even specific individual bases. Restriction nucleases act only on sequences of bases specifically recognized by each restriction enzyme. Nucleases also exhibit specificities with respect to the overall structure of polynucleotides. For instance, some nucleases act on either single- or double-stranded polynucleotides, whereas others discriminate between these two types of structures. In addition, some nucleases exclusively designated as **phosphodiesterases** will act on either DNA or RNA, whereas other nucleases will limit their activity to only one type of polynucleotide. The nucleases listed in Table 15.1 illustrate some of the diverse properties of these enzymes.

Periodicity Leads to Secondary Structures of DNA

As has been emphasized previously, the polypetide chains of protein are often arranged in space in a manner that leads to the formation of *periodic structures*. For instance, in the α helix each residue is related to the next by a translation of 1.5 Å along the helix axis and a rotation of 100°. This arrangement places 3.6 amino acid residues in each complete turn of the

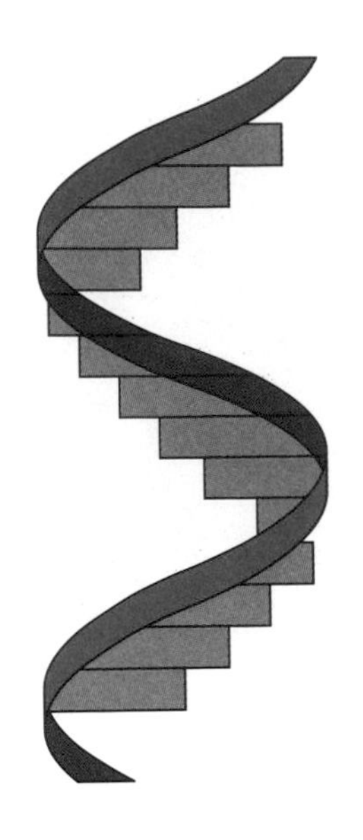

FIGURE 15.5
Conformation of a hypothetical, perfectly helical, single-stranded polynucleotide.
The helical band represents the phosphate backbone of the polynucleotide. The bases are shown in a side view as solid blocks in tight contact with their neighbors, above and below each base. The surfaces of the rings are in contact with each other and are not visible to the observer in the perspective from which the figure was drawn.

polypeptide helix. The property of periodicity is also encountered with polynucleotides, which usually occur in the form of helices.

Such preponderance of helical conformations among macromolecules is not surprising. The formation of helices tends to accommodate the effects of intramolecular forces, which in a helix can be distributed at regular intervals. The precise geometry of the polynucleotide helices varies, but the helical structure invariably results from the stacking of bases along the helix axis. In many instances stacking produces helices in which the bases are more or less perpendicularly oriented along the helix and touch one another. This arrangement, which obviously leaves no free space between two successive neighboring bases, is illustrated in Figure 15.5. Such stacked single-stranded helices, however, are not commonly encountered in nature. Rather, as it will become apparent from the subsequent discussion, polynucleotide helices tend to associate with one another to form double helices.

Forces That Determine Polynucleotide Conformation

The hydrophobic properties of the bases are, to a large extent, responsible for forcing polynucleotides to adopt helical conformations. Examination of molecular models of the bases reveals that the edges of the rings contain polar groups (i.e., amino and OH group residues) that are able to interact with other polar groups or surrounding water molecules. The faces of the rings, however, are unable to participate in such interactions and tend to avoid any contact with water. Instead they tend to interact with one another, producing the stacked conformation. The stability of this arrangement is further reinforced by an interchange between the electrons that circulate in the π orbitals located above and below the plane of each ring.

Clearly then, single-stranded polynucleotide helices are stabilized by both hydrophobic as well as stacking interactions involving the π orbitals of the bases. The stability of the helical structures is also influenced by the potential repulsion among the charged phosphate residues of the polynucleotide backbone. These repulsive forces introduce a certain degree of rigidity to the structure of the polynucleotide. Under physiological conditions, that is, at neutral pH and relatively high concentrations of salts, the charges on the phosphate residues are partially shielded by the cations present, and the structure can be viewed as a fairly flexible coil. Under more extreme conditions the stacking of the bases is disrupted and the helix collapses. A collapsed helix is commonly described as a random coil. A conversion between a stacked helix and a random-coil conformation is depicted in Figure 15.6.

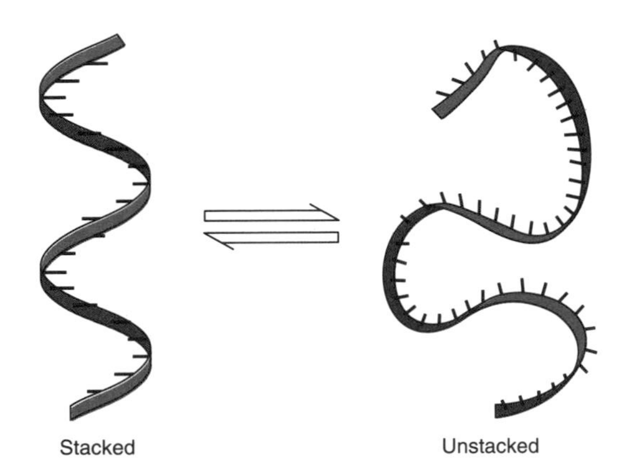

FIGURE 15.6
Stacked and unstacked conformations of a polynucleotide.
Stacking of the bases decreases the flexibility of a polynucleotide and tends to produce a more extended, often helical, structure.

DNA Double Helix

Although certain forms of cellular DNA exist as single-stranded structures, the most widespread DNA structure is the double helix. The double helix can be visualized as resulting from the interwinding around a common axis of two right-handed helical polynucleotide strands. The two strands achieve contact through hydrogen bonds, which are formed at the hydrophilic edges of their bases. These bonds extend between purine residues in one strand and pyrimidine residues in the other, so that the two types of resulting pairs are always adenine–thymine and guanine–cytosine. A direct consequence of these hydrogen-bonding specificities is that double-stranded DNA contains equal amounts of purines and pyrimidines. Examination of space-filling models clearly indicates the structural compatibility of these bases in forming linear hydrogen bonds.

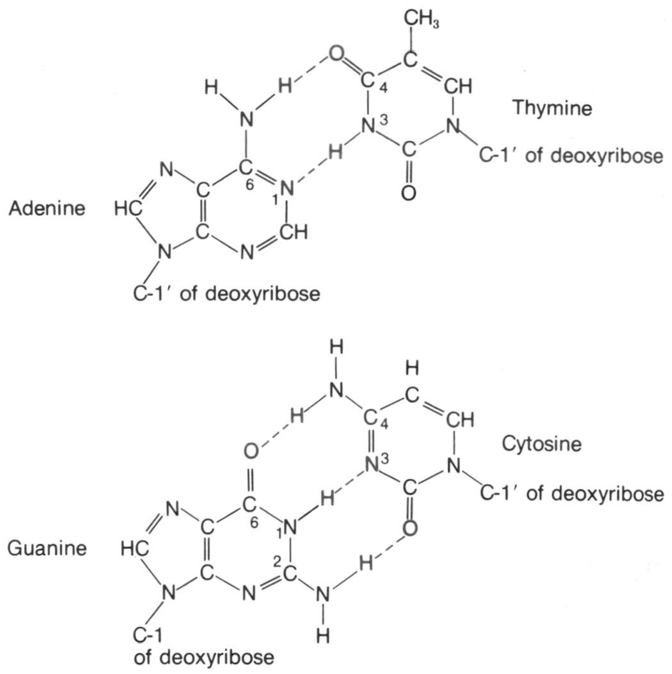

FIGURE 15.7
Formation of hydrogen bonds between complementary bases in double-stranded DNA.
The interaction between polynucleotide strands is a highly selective process. The property of complementarity depends not only on the geometric factors that allow the proper fitting between the complementary bases of the two strands, but also on the electronic specificity of interaction between complementary bases. Thus specificity of interaction between purines and pyrimidines has also been noted both in solution and in the crystal form, and it is expressed in terms of strong hydrogen bonding between monomers of adenine and uracil or monomers of guanine and cytosine. In double-stranded DNA adenine interacts instead with thymine, which is a structural analog of uracil.

This relationship between bases in the double helix is described as **complementarity.** The bases are complementary because every base of one strand is matched by a complementary hydrogen-bonding base on the other strand. For instance, for each adenine projecting toward the common axis of the double helix, a thymine must be projected from the opposite chain so as to fill exactly the space between the strands by hydrogen bonding with adenine. Neither cytosine nor guanine *fits precisely in the available space in a manner that allows the formation of hydrogen bonds across strands.* These hydrogen-bonding specificities, illustrated in Figure 15.7 ensure that the entire base sequence of one strand is complementary to that of the other strand.

The conventional double helix exists in various geometries designed as **DNA A, B, and C.** The formation of these different conformations depends on the base composition of the DNA and on physical conditions. These forms, however, share certain common characteristics. Specifically, the phosphate backbone is always located on the outside of the helix. Also, because the diesters of phosphoric acid are fully ionized at neutral pH, the exterior of the helix is negatively charged. The bases are well packed in the interior of the helix, where their faces are protected from contact with water. In this environment the strength of the hydrogen bonds that connect the bases can be maximized. The interwinding of the two polynucleotide strands produces a structure having two deep helical grooves that separate the winding phosphate backbone ridge.

However, the precise geometry of the double helix varies among the different forms. The original X-ray data obtained with highly oriented DNA fibers suggested the occurrence of a form, later designated as B, which appears to be the one commonly found in solution and in vivo (Figure 15.8). A characteristic of this form is that one of its grooves is wider than the other, and it is referred to as the *major groove* to distinguish it from the *minor,* or second, *groove.*

The disparity in the width between these two grooves results from the characteristic geometry of base pairs (bp). The glycosidic bonds between the sugars and the bases of each base pair are not arranged directly

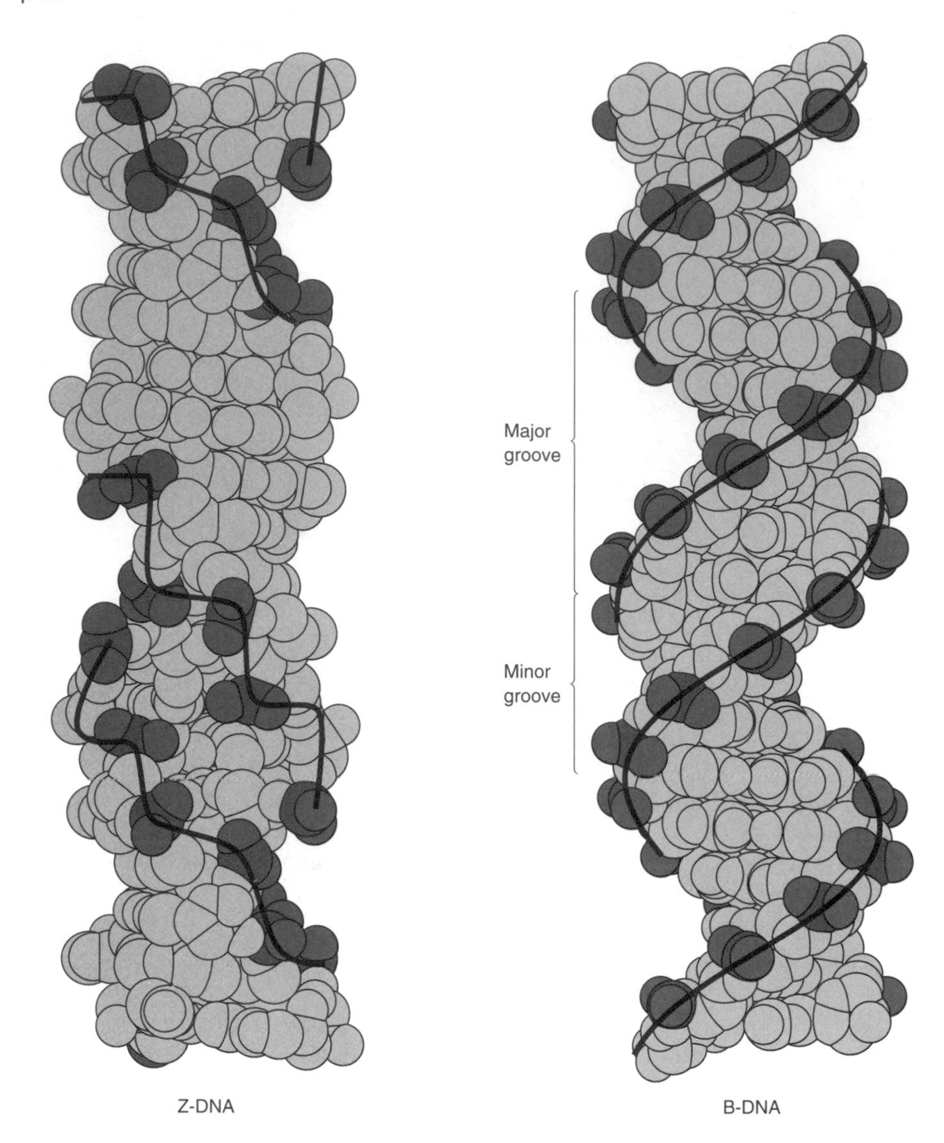

FIGURE 15.8
Space-filling molecular models of B- and Z-DNA.
Watson and Crick were the first to postulate a double-stranded model for the structure of DNA. The double helix is still referred to as the Watson and Crick model, although this structure has been substantially refined since it was proposed. B-DNA may be the most typical form of DNA occurring in the cell. Z-DNA may be present in the cell as small stretches, consisting of alternating purines and pyrimidines, incorporated between long stretches of B-DNA. The zigzag nature of the Z-DNA backbone is illustrated by the heavy lines that connect phosphate residues along the chain.
Redrawn based on figure from A. Rich, *J. Biomol. Struct. Dynamics* 1:1, 1983.

opposite to one another. Instead the edge of the helix, that is more than 180° from glycosidic bond to glycosidic bond, is the edge that forms part of the major groove. Clearly, the opposite edge corresponds to the minor groove. The nucleotide sequence of the polynucleotides can be discerned without dissociating the double helix by looking inside these grooves. As each of the four bases has its own orientation with respect to the rest of the helix, each base always shows the same atoms through the grooves. For instance, the C-6, N-7, and C-8 of the purine rings and the C-4, C-5, and C-6 of the pyrimidine rings line up in the major groove. The minor groove is paved with the C-2 and N-3 of the purine and the C-2 of the pyrimidine rings. Forms A and C differ from B in the pitch of the base pairs relative to the helix axis as well as in other geometric parameters of the double helix, as shown in Figure 15.9. Under conditions of low salt concentration and humidity, the thin B-DNA double helix shifts to a conformation characterized by a thicker helix. In this conformation the nucleotides move towards the major edge of each base pair, generating the A-DNA, which has a narrower and deeper major groove and a wider and shallower minor groove than B-DNA. The parameters for these different DNA conformations, listed in Table 15.2, have been determined on oriented fibers of DNA by X-ray diffraction methods. While such methods provide very accurate information about molecular geometry and dimensions of crystalline samples, they give only the average dimensions

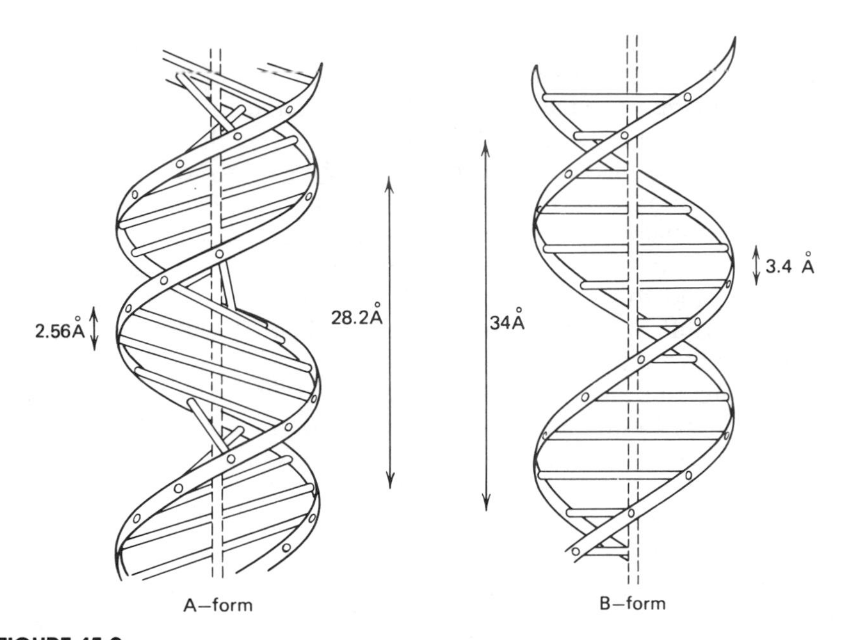

FIGURE 15.9
The various geometries of the DNA double helix.
Depending on conditions, the double helix can acquire various forms of distinct geometries. In the B form of DNA the centers of the bases are about 3.4 Å apart and produce a complete turn of a helix with a pitch of 34 Å. Such an arrangement results in a complete turn of the helix for every 10 base pairs. The diameter of the helix is 20 Å. Form C (not shown) is very similar to the B structure, with a pitch of 33 Å and 9 base pairs per turn. Form A, which is obtained from form B when the relative humidity of the fiber is reduced to 75%, differs from B in that the base pairs are not perpendicular to the helical axis but are tilted. This tilt results in a pitch of 28.2 Å and a shortening of the helix by the packing of 11 pairs per helical turn.
Redrawn from W. Guschelbauer, *Nucleic acid structure*. Berlin: Springer-Verlag, 1976.

for the monomeric units present in a noncrystalline macromolecule. Therefore, these parameters are listed as such and the listing does not imply that the same geometry characterizes each and every base pair in DNA. Rather, depending on base sequence, considerable local variation in the conformation of individual nucleotides may occur. Such variations may be important in the regulation of gene expression, since they influence the extent of DNA binding with various types of regulatory proteins.

A new form of DNA was discovered, which has geometric characteristics radically different from those of the conventional forms. In this DNA, termed **Z-DNA,** the polynucleotide phosphodiester backbone assumes a ''zigzag'' arrangement rather than the smooth conformation that characterizes other double-stranded forms. The Z-DNA structure, a longer and much thinner structure than B-DNA, which completes one turn in 12 bp rather than the 10 bp participating in a B-DNA turn, forms a single groove

TABLE 15.2 Nucleic Acid Helix Parameters

Family	A		B		Z
Environment	Crystal	Fiber	Crystal	Fiber	Crystal
Helix sense	Right	Right	Right	Right	Left
Sugar ring conformation (pucker)	C3′-*endo*	C3′-*endo*	Variable	C2′-*endo*	Alternating
Base pairs per turn	10.7	11	9.7	10	12
Rise per base pair (Å)	2.3	2.6	3.3	3.4	3.7

as opposed to the two grooves that characterize B-DNA. Therefore, the conformation of Z-DNA may be viewed as the result of the major groove of B-DNA having "popped out" in order to form the outer convex surface of Z-DNA. This change places the stacked bases on the outer part of Z-DNA rather than in their conventional positions in the interior of the double helix. Another highly unusual property of the Z structure is that it consists of left-handed rather than right-handed helices, which characterize the conventional forms. These major structural differences between the B-DNA and the Z-DNA, which are illustrated in Figure 15.8, are partly the result of different conformations in the nucleotide residues between the two forms.

Specifically, in B- and A-DNA the sugars and the bases are arranged in an extended conformation referred to as the *anti* conformation. In contrast, in Z-DNA some nucleotides rotate into the *syn* conformation, which places the sugar and the base on the same side of the glycosidic bond. The DNA sequences that consist of alternating GC nucleotides are the most prone to acquire the Z conformation, which places the glycosidic bond of each G in *syn*, with C residues maintaining the anti conformation. The zigzag arrangement of the phosphate backbone reflects the sudden turns of the backbone, as it follows the alternating arrangement of *syn* and *anti* geometries.

The biological function of Z-DNA is not known with certainty. Some evidence exists suggesting that Z-DNA influences gene expression and regulation. Apparently, Z-DNA is incorporated in small stretches, normally containing approximately one to two dozen nucleotide residues, in regions of the gene that regulate transcriptional activities. These stretches consist of alternating purines and pyrimidines in the sequence, which is a condition favoring the formation of the Z conformation. The Z form of DNA is stabilized by the presence of cations or polyamines and by methylation of either guanine residues in the C-8 and N-7 positions or cytosine residues in the C-5 position.

An important structural characteristic of all double-stranded DNA is that its strands are **antiparallel.** Polynucleotides are asymmetric structures with an intrinsic sense of polarity built into them. As it may be concluded from inspection of Figure 15.10, the two strands are aligned in opposite directions that is, if two adjacent bases in the same strand, for example, thymine and cytosine, are connected in the $5' \rightarrow 3'$ direction, their complementary bases adenine and guanine will be linked in the $3' \rightarrow 5'$ direction (directions are defined by linking the 3′ and 5′ positions within the same nucleotide). This antiparallel alignment produces a stable association between strands to the exclusion of the alternate parallel arrangement. Just as peptide geometries and the formation of α helices determine the overall preferred conformation of proteins, the formation of hydrogen bonds between complementary bases on antiparallel polynucleotide strands leads to the formation of the double helix.

The double-stranded structure for DNA was proposed in 1953. The proposal was partly based on the results of previously available X-ray diffraction studies, which suggested that the structures of DNAs from various sources exhibited remarkable similarities. These studies also suggested that DNA had a helical structure containing two or more polynucleotides. An additional piece of evidence of central importance to the proposal was the clarification of the quantitative base composition of DNA, which was obtained independently in 1950. These results indicated the existence of molar equivalence between purines and pyrimidines, which turned out to be the essential observation suggesting the existence of complementarity between the two strands.

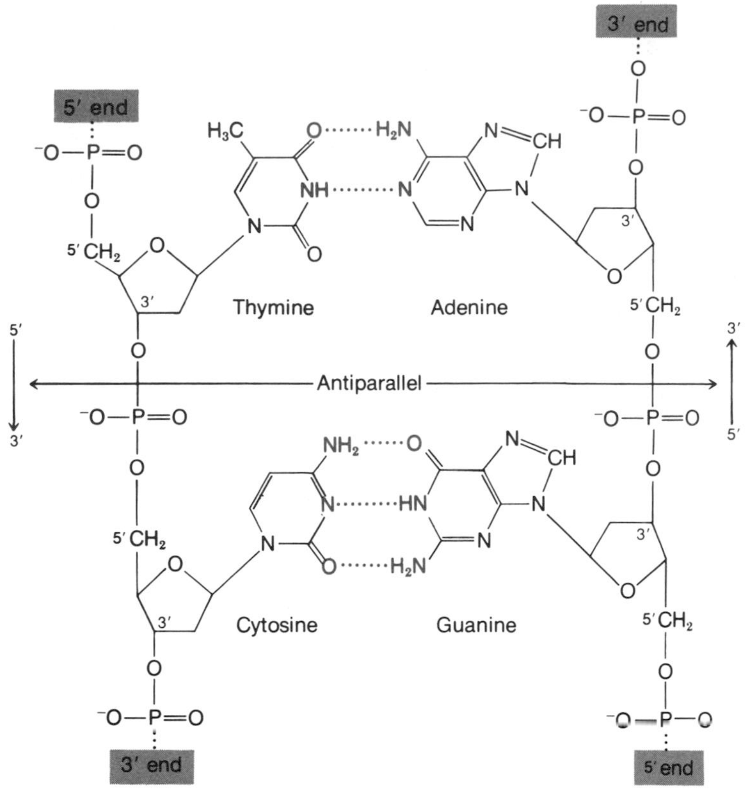

FIGURE 15.10
Antiparallel nature of the DNA strands.
The strands of a double-stranded DNA are arranged in such a manner that, as the complementary bases pair with one another, the two strands are aligned with opposite polarities, that is, the conventional assignment of the 5′ → 3′ direction to each of the strands suggests opposite directions. It should be noted that the geometry of the helices does not prevent a parallel alignment, but such an arrangement is not found in DNA.

Many Factors Stabilize the DNA Structure

The same factors that stabilize single-stranded polynucleotide helices, hydrophobic and stacking forces, are also instrumental in stabilizing the double helix. The separation between the hydrophobic core of the stacked bases and the hydrophilic exterior of the charged sugar-phosphate groups is even more striking in the double helix than with single-stranded helices. This arrangement, which produces substantial stabilization of the double-stranded structures over single-stranded conformations, explains the preponderance of the former. The stacking tendency of single-stranded polynucleotides may be viewed as resulting from a tendency of the bases to avoid contact with water. The double-stranded helix is by far a more favorable arrangement, as it permits the phosphate backbone to be highly solvated by water while the bases are essentially removed from the aqueous environment.

The stacking of the bases gives rises to additional stabilization, that results from the attractive forces generated among atoms that are optimally situated relative to one another within a molecular structure. Two atoms at optimum distance, meaning close but not too close to one another, are subjected to attractive forces known as van der Waals forces. Stacked DNA bases are stabilized by these relatively weak but additive forces.

Additional stabilization of the double helix results from its extensive network of cooperative hydrogen bonding. Although this bonding per se makes only a relatively minor contribution to the free energy of stabilization of the double helix, the physiological importance of hydrogen bonds should not be underestimated. By contrast to hydrophobic forces, hydrogen bonds are highly directional and for this reason are able to provide a discriminatory function for choosing between correct and incorrect base pairs. In addition, because of their directionality, hydrogen bonds tend to

TABLE 15.3 Effects of Various Reagents on the Stability of the Double Helix[a]

Reagent	Adenine solubility $\times 10^{-3}$ (in 1 M reagent)	Molarity producing 50% denaturation
Ethylurea	22.5	0.60
Propionamide	22.5	0.62
Ethanol	17.7	1.2
Urea	17.7	1.0
Methanol	15.9	3.5
Formamide	15.4	1.9

SOURCE: Data from L. Levine, J. Gordon, and W. P. Jencks, *Biochemistry* 2:168, 1963.

[a] The destabilizing effect of the reagents listed below on the double helix is independent of the ability of these reagents to break hydrogen bonds. Rather, the destabilizing effect is determined by the solubility of adenine. Similar results would be expected if the solubility of the other bases were examined.

orient the bases in a way that favors stacking. Therefore although hydrogen bonds make a minor contribution to the total energy of stabilization, their contribution is essential for the stability of the double helix.

In the past, the relative importance of hydrogen bonding and hydrophobic forces in stabilizing the double helix was not always appreciated. However, studies on the effect of various reagents on the stability of the double helix have suggested that the destabilizing effect of a reagent is not related to the ability of the reagent to break hydrogen bonds. Rather, the stability of the double helix is determined by the solubility of the free bases in the reagent, the stability decreasing as the solubility increases. Some of these findings, summarized in Table 15.3, emphasize the importance of hydrophobic forces in maintaining the structure of double stranded DNA.

Ionic forces also have an effect on the stability and the conformation of the double helix. At physiological pH the electrostatic *intrastrand* repulsion between negatively charged phosphates forces the double helix into a relatively rigid rodlike conformation. In addition, the repulsion between phosphate groups located on opposite strands tends to separate the complementary strands. In distilled water, DNA strands will separate at room temperature; near the physiological salt concentration, cations, particularly Mg^{2+}, (in addition to other charged groups, for example, the basic side chains of proteins) shield the phosphate groups and decrease repulsive forces. Therefore, the flexibility of the double helix is partially restored and its stability is enhanced.

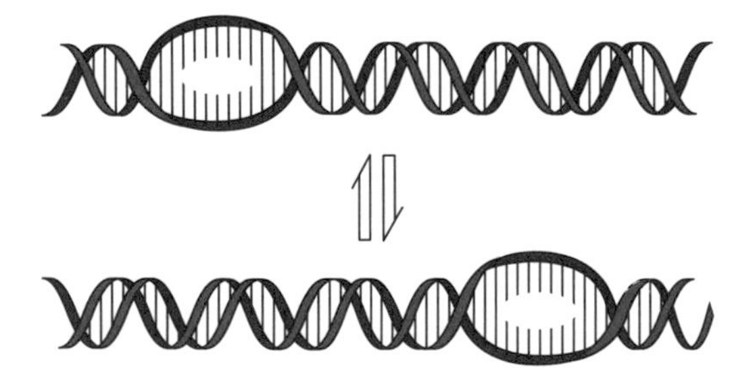

FIGURE 15.11
"Zipper" model for the DNA double helix. DNA contains short sections of open-strandedness that can "move" up and down the helix.

Denaturation

The double helix is stabilized by about 1 kcal per base pair. Therefore a relatively minor perturbation can produce disruption in double strandedness, provided that only a short section of the DNA is involved. As soon as the relatively few base pairs have separated, they close up again and release free energy, and then the adjacent base pairs unwind. In this manner minor disruptions of double strandedness can be propagated along the length of the double helix. Therefore, at any particular moment the large majority of the bases of the double helix remain hydrogen bonded, but all bases can pass through the single-stranded state, a few at a time. This dynamic state of the double helix is characterized by the movement of an "open-stranded" portion up and down the length of the helix,

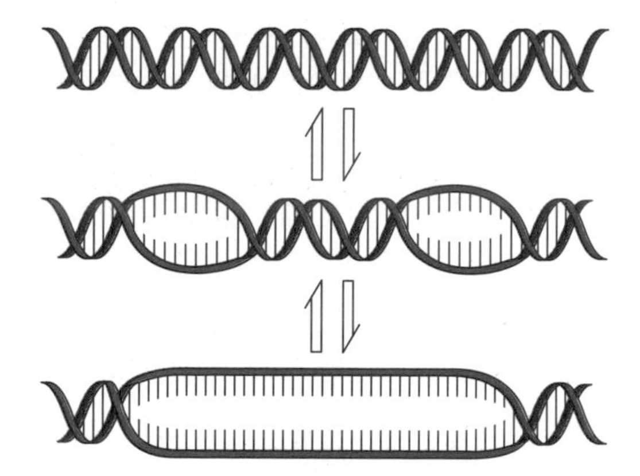

FIGURE 15.12
Structure of double-stranded DNA at increasing temperatures.
Disruptions of the double-stranded structure appear first in regions of relatively high adenine-thymine content. The size of these "bubbles" increases with increasing temperatures, leading to extensive disruptions in the structure of the double helix at elevated temperatures.

as indicated in Figure 15.11. The "dynamic" nature of this structure is an essential prerequisite for the biological function of DNA and especially the process of DNA synthesis.

Furthermore, the strands of DNA can be completely separated by increasing the temperature in solution. At relatively low temperatures a few base pairs will be disrupted, creating one or more "open-stranded bubbles." These "bubbles" form initially in sections that contain relatively higher proportions of adenine and thymine pairs. Adenine-thymine pairs are bound by two hydrogen bonds and are therefore less stable than guanine-cytosine pairs, which contain three such bonds per pair. As the temperature is raised, the size of the "bubbles" increases and eventually the thermal motion of the polynucleotides overcomes the forces that stabilize the double helix. This transformation is depicted in Figure 15.12. At even higher temperatures the strands can separate physically and acquire a random-coil conformation, as shown in Figure 15.13, the process is most appropriately described as a **helix-to-coil transition,** but it is commonly referred to as **denaturation.** The process of denaturation is accompanied by a number of physical changes, including a buoyant density increase, a reduction in viscosity, a change in the ability to rotate polarized light and changes in absorbancy.

Changes in absorbancy are frequently used for following experimentally the process of denaturation. DNA absorbs in the UV region due to the heterocyclic aromatic nature of its purine and pyrimidine constituents. Although each base has a unique absorption spectrum, all bases exhibit maxima at or near 260 nm. This property is responsible for the absorption of DNA at 260 nm. However, this absorbancy is almost 40% lower than that expected from adding up the absorbancy of each of the base components of DNA. This property of DNA, referred to as **hypochromic effect,** results from the close stacking of the bases along the DNA helices. In this special arrangement interactions between the electrons of neighboring bases produce a decrease in absorbancy. However, as the ordered structure of the double helix is disrupted at increasing temperatures, stacking interactions are gradually decreased. Therefore, a totally disordered polynucleotide, a random coil, eventually approaches an absorbance not very different from the sum of the absorbancy of its purine and pyrimidine constituents.

Slow heating of double-stranded DNA in solution is accompanied by a gradual change in absorbancy as the strands separate. However, since the interactions between the two strands are cooperative, the transition from double-stranded to random-coil conformation occurs over a narrow range of temperatures, as indicated in Figure 15.14. Before the rise of the *melting curve,* DNA is double stranded. In the rising section of the curve an increasing number of base pairs is interrupted as the temperature rises. Strand separation occurs at a critical temperature corresponding to the

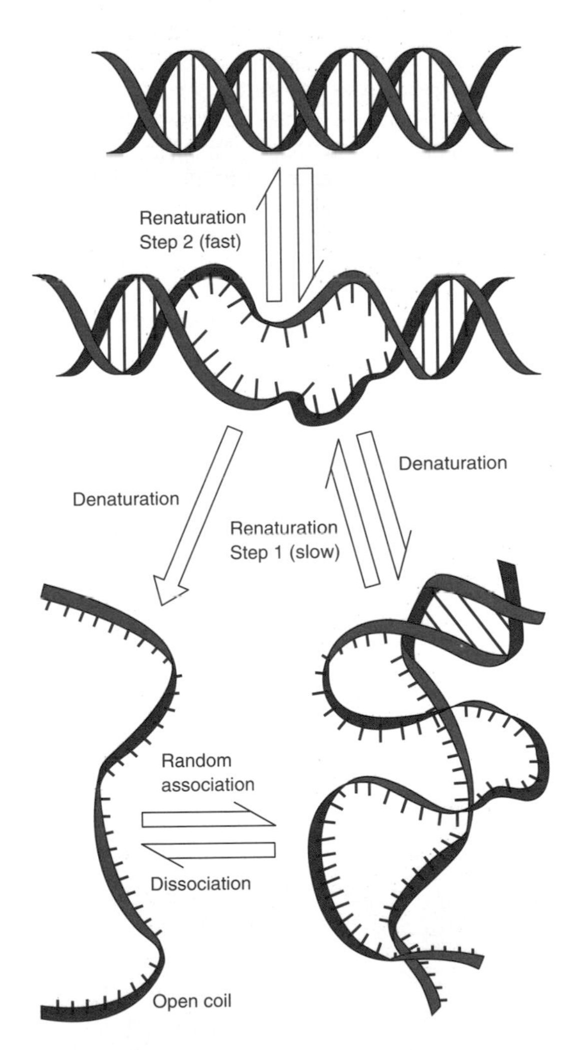

FIGURE 15.13
Denaturation of DNA.
At high temperatures the double-stranded structure of DNA is completely disrupted, with the eventual separation of the strands and the formation of single-stranded open coils. Denaturation also occurs at extreme pH ranges or at extreme ionic strengths.

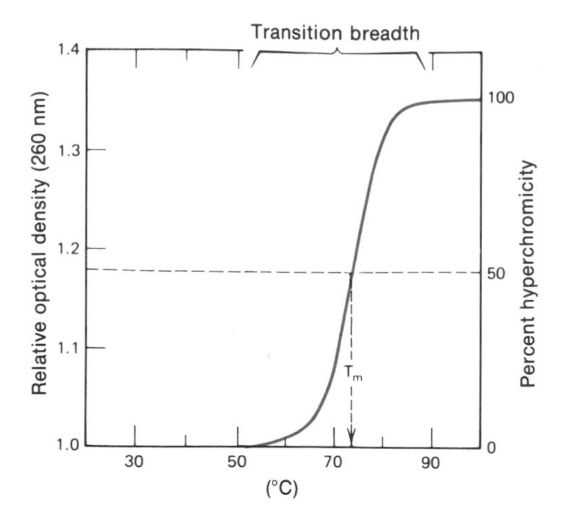

FIGURE 15.14
Temperature–optical density profile for DNA.
When DNA is heated, the optical density increases with rising temperature. A graph in which optical density versus temperature is plotted is called a "melting curve." Relative optical density is the ratio of the optical density at the temperature indicated to that at 25°C. The temperature at which one-half of the maximum optical density is reached is the midpoint temperature (T_m).
Redrawn from D. Freifelder, *The DNA molecule: structure and properties*. San Francisco: W. H. Freeman, 1978.

upper plateau of the curve. However, if the temperature is decreased before the complete separation of the strands, the native structure is completely restored.

The *midpoint temperature,* T_m, of this process, under standard conditions of concentration and ionic strength, is characteristic of the base content of each DNA. The higher the guanine-cytosine content, the higher the transition temperature between the double-stranded helix and the single strands. This difference in T_m values is attributed to the increased stability of guanine-cytosine pairs, as a result of the three hydrogen bonds that connect them in DNA, in contrast to only two hydrogen bonds that connect adenine and thymine pairs.

Rapid cooling of a heated DNA solution normally produces denatured DNA, a structure that results from the reformation of some hydrogen bonds either between the separate strands or between different sections of the same strand. The latter must contain complementary base sequences. By and large denatured DNA is a disordered structure containing substantial amounts of random-coil and single-stranded regions.

DNA can also be denatured at a pH above 11.3 as the charge on several substituents on the rings of the bases is changed preventing these groups from participating in hydrogen bonding. Alkaline denaturation is often used as an experimental tool in preference to heat denaturation to prevent breakage of phosphodiester bonds that can occur to some degree at high temperatures. Denaturation can also be induced at low ionic strengths, because of enhanced interstrand repulsion between negatively charged phosphates, as well as by various denaturing reagents, that is, compounds that weaken or break hydrogen bonds. A complete denaturation curve similar to that shown in Figure 15.14 can be obtained at a relatively low constant temperature, for instance room temperature, by variation of the concentration of an added denaturant.

Renaturation

Complementary DNA strands, separated by denaturation, can reform a double helix if appropriately treated by a process referred to as **renaturation** or **reannealing.**

If denaturation is not complete, that is, even if only a few bases remain hydrogen bonded between the two strands, the helix-to-coil transition is rapidly reversible. Annealing, however, is possible even after the complementary strands have been completely separated, except that under these conditions the renaturation process depends on the meeting of complementary DNA strands in an exact manner that can lead to the reformation of the original structure, and it is therefore a slow, concentration-dependent process. As a rule, maintaining DNA at temperatures 10–15°C below its T_m under conditions of moderate ionic strength (~0.15 M), provides the maximum opportunity for renaturation. At lower salt concentrations, the charged phosphate groups repel one another and prevent the strands from associating. As renaturation begins, some of the hydrogen bonds formed are extended between short tracts of polynucleotides that might have been distant in the original native structure. Short sequences, consisting, for example, of four to six base pairs, are reiterated many times within every DNA strand. Furthermore, eucaryotic DNA contains a large number of much longer nucleotide sequences reiterated many times within each genome. Such sequences provide sites for initial base pairing, which produces a partially hydrogen-bonded double helix. These randomly base paired structures are short-lived because the bases that surround the short complementary segments cannot pair and lead to the formation of a stable fully hydrogen-bonded structure. However, once the correct bases begin to pair by chance, the double helix over the entire

DNA molecule is rapidly reformed. Clearly then renaturation is a two-step process. The first step, which determines the rate of association, involves the chance meeting of two complementary sequences on different strands and it is, therefore, a second-order reaction. The rate of renaturation is thus proportional to the product of the concentrations of the two homologous dissociated strands and is expressed as $dt/dc = -kc^2$, where k is the rate constant for the association. Integration of this equation gives $C/C_0 = 1/(1 + kC_0t)$, where C is the concentration of single-stranded DNA expressed as moles of nucleotides per liter at time t, and C_0 is the concentration of DNA at time zero. A plot of C/C_0 (which is proportional to DNA that is single stranded or of the DNA fraction that is reassociated) versus C_0t can be constructed, Figure 15.15, and a $C_0t_{\frac{1}{2}}$ (C_0T-a-half) value, which corresponds to $C/C_0 = 0.5$ can be determined. The **C_0t-a-half value** is proportional to the complexity of the genome, which is equal to the molecular weight of the genome provided that the genome consists of unique nucleotide sequences.

For example, both the complexity and the molecular weight of a hypothetical genome consisting of three unique nucleotide sequences that may be represented as N_1, N_2, and N_3 is equal to the sum $N_1 + N_2 + N_3$. However, in eucaryotic genomes, which contain both unique as well as reiterated sequences, the complexity of the genome is significantly lower than the molecular weight. If, for instance, a eucaryotic genome contains

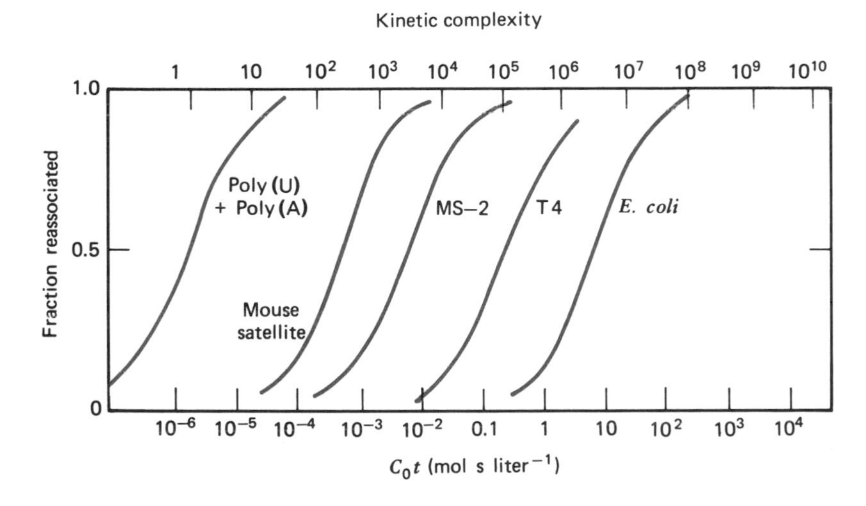

FIGURE 15.15
Reassociation kinetics for DNA isolated from various sources.
Each DNA is first fragmented to segments of approximately 400 nucleotides. The denatured segments are subsequently then allowed to renature. The fraction of each polynucleotide reassociated, calculated from changes in hypochromicity, is plotted against the total concentration of nucleotides multiplied by the renaturation time (C_0t). The top scale shows the kinetic complexity of each DNA sample. Whenever a DNA contains reiterated sequences, these sequences are present in the fragments at higher concentrations than they would have been if a unique sequence had been fragmented. As a result, renaturation of fragments, obtained from DNAs containing reiterated sequences, proceeds more rapidly the higher the degree of repetition. This is exemplified by the rates of renaturation of fragments obtained from the synthetic double-stranded polynucleotide poly(A)–poly(U) and mouse satellite DNA, a DNA that contains many repeated sequences. For a homogeneous DNA, which contains a distribution of different extents of reiterated sequences, kinetic complexity can be defined as the minimum length of DNA needed to contain a whole single copy of the reiterated sequence.
Based on figure in R. J. Britten and D. E. Kohne, *Science* 161:529, 1968.

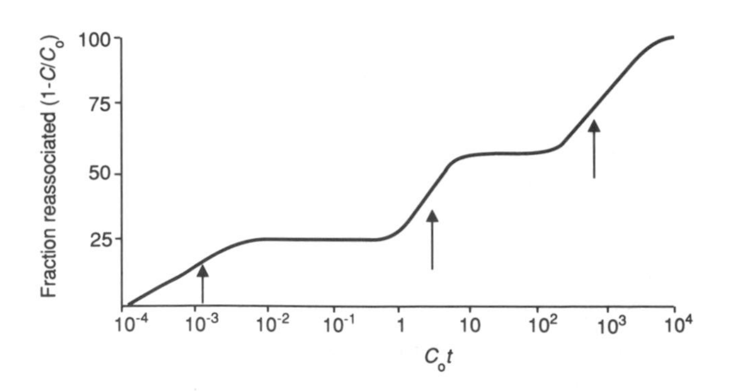

	Fast Component	Intermediate Component	Slow Component
Percent of genome	25	30	45
$C_ot_{1/2}$	0.0013	1.9	630
Complexity, bp	340	6.0×10^5	3.0×10^8
Repetition frequency	500,000	350	1

FIGURE 15.16
Reassociation kinetics of eucaryotic DNA. This idealized C_0t-a-half plot represents a eucaryotic DNA that consists of three distinct components with three different C_0t values. The percentage to which each one of these components is present in the DNA can be read from the ordinate (fraction reassociated axis) of the figure. Repetition frequencies and complexities are calculated based on the principles discussed in the text. In practice, the experimental separation of different DNA components is not as pronounced and their identification not as clear-cut as shown in this hypothetical example.

10^5 copies of sequence N_1, 10^3 copies of sequence N_2, and 1 copy of sequence N_1, the complexity will still be $N_1 + N_2 + N_3$ but the molecular weight will be equal to $10N_3 + 10N_2 + N_1$. Thus complexity may be defined as the minimum length of DNA that contains a *single complete copy* of all the single and the reiterated sequences that are found in the genome.

The C_0t curves of eucaryotic genomes with reiterated DNA segments show several kinetic components, each representing those parts of the genome that have similar reiteration frequencies (Figure 15.16). Highly reiterated sequences, that is, those present in many copies, will reassociate the fastest; unique sequence the slowest. The repetition frequency of sequences within a kinetic component can be calculated from the ratio of the association rate constant of the component to the rate constant for the corresponding single copy component. *Thus, C_0t curves provide information on genome complexity, on the number of repetitive classes, and on the proportion of the total genome represented by those classes.*

Hybridization

The self-association of complementary polynucleotide strands has also provided the basis for the development of the technique of **hybridization.** This technique depends on the association between any two polynucleotide chains, which may be of the same or of different length, provided that a relationship of base complementarity exists between these chains. Hybridization can take place not only between DNA chains but also between complementary RNA chains as well as DNA–RNA combinations.

Appropriate techniques have been developed for measuring the *maximum amount of polynucleotide* that can be hybridized as well as the *rates of hybridization*. These techniques are among the most indispensable basic tools of contemporary molecular biology and are also specifically used for the following: (a) determining whether or not a certain sequence occurs more than once in the DNA of a particular organism; (b) demonstrating a *genetic* or *evolutionary relatedness* between different organisms; (c) determining the number of genes transcribed in a particular mRNA (clearly DNA-RNA hybridizations are needed for accomplishing the last goal), and (d) determining the location of any given DNA sequence by annealing with a complementary polynucleotide that is appropriately tagged for easy detection of the hybrid. Such a polynucleotide is commonly referred to as a probe.

As the first step, DNA to be tested for hybridization is denatured. The resulting single strands are immobilized by binding to a suitable polymer, which is then used to pack a chromatography column. The DNA formed in the presence of labeled precursors, usually tritiated thymidine, is allowed to run through the column that contains the bound, unlabeled DNA. The rate at which radioactivity is retained by the column obviously equals the rate of annealing between complementary strands.

As discussed in the preceding section, measurements of such rates have established that the DNA of eucaryotic cells contain a given nucleotide sequence reiterated a number of times. The principle of this determination is simple. For a DNA of a given size the rate of annealing depends on the frequency with which two complementary segments can collide with each other. Therefore, the larger the number of reiterated sequences in a given DNA, the greater is the chance that a particular collision will result in the formation of annealed polynucleotides. On this basis the extent to which annealing takes place within a unit of time can be used to determine the number of reiterated sequences in the DNA.

Determinations of the maximum amount of DNA that can be hybridized have been used to establish homologies between the DNA of different species. This is possible because the base sequences of the DNA in each organism are unique for this organism. Therefore, the annealed helices represent the same unique sequences of DNA even if the individual annealed strands originate from different cells. On this basis annealing can be used to compare the degree to which DNAs isolated from different species are related to one another. Consequently, the observed homologies serve as indexes of *evolutionary relatedness* and have been particularly useful for defining *phylogenies* in procaryotes. "Hybridization" studies between DNA and RNA have, in addition, provided very useful information about the biological role of DNA, particularly the mechanism of transcription.

In recent years hybridization techniques using membrane filters, usually made of nitrocellulose, have found increasing application (see Chapter 18, p. 781). In general, hybridization can be quantitated by either measuring *the amount of hybrid in equilibrium or the rate of hybrid formation* under conditions in which one nucleic acid is present in large excess. The approach used for the latter determination is analogous to the C_0t procedure and when it is used for DNA-RNA hybridization and RNA is present in excess it is referred to as the R_0t method, or the D_0t method when DNA is in excess.

A variant of filter hybridization, known as the *Southern transfer,* can be used for identifying the location of specific genes (see Chapter 18, p. 782). Since a gene sequence represents a very small percentage of total DNA, the gene must be separated from the remaining DNA and the DNA detected by using appropriate probes.

Probes

Probes are short oligonucleotides, of either RNA or DNA, that are complementary to specific sequences of interest in genomic DNA. Probes are frequently synthesized by chemical means. The use of chemically synthesized probes may appear to be limited by the degree to which the desired genomic nucleotide sequence is known but in fact this approach has much

CLIN. CORR. 15.1 DIAGNOSTIC USE OF PROBES IN MEDICINE

Use of DNA probes is becoming increasingly important in the laboratory diagnosis of many diseases. For example, simple probes are very useful for the identification of mutant alleles responsible for genetic diseases, with the limitation that the mutations must be stable and few in number. Prominent examples of diseases in this category are sickle-cell anemia, hemoglobin C disease, and phenylketonuria. Sickle-cell anemia and hemoglobin C disease are the result of a single-base change in the genes coding for β-globin. By using three different probes, corresponding to the sequence of the normal and the two mutated hemoglobins, the presence of mutated β-globin genes can be detected. Similarly, a probe can identify a mutation in the phenylalanine 4-monoxygenase gene that is responsible for phenylketonuria.

wider applicability. As an example, if the protein product of a gene is known, the nucleotide sequence of the desired gene can be approximated by using a mixture of different synthetic oligonucleotides. These oligonucleotides represent the various alternate mRNA sequences that, because of the degeneracy of the code, can encode for the same protein. One of these oligonucleotide sequences is therefore complementary to the desired gene. In those instances when the gene of interest is transcribed to mRNA molecules that are *abundant* and *easily purified,* as, for example, in the case of histones and hemoglobin, mRNA can be used. As a rule, probes need to be at least 15 nucleotides long because shorter sequences may occur randomly along genomic DNA. To achieve easy detection probes are typically labeled by the incorporation of ^{32}P or are obtained by the use of biotin containing nucleotides that are incorporated into the probe and serve as fluorescent labels. Probes show great promise for the definitive and rapid diagnosis of genetic disorders, infectious disease and cancer as described briefly in Clin. Corr. 15.1.

Heteroduplexes

The principle of hybridization has also served as the basis for the development of a technique that has permitted the construction of precise physical maps of DNA genes. This technique depends on the direct visualization under the electron microscope of single-stranded loops in the structures of artificially formed double-stranded DNA molecules known as *heteroduplexes*. The principle of this technique is simple. Heteroduplexes are constructed by hybridization of two complementary DNA strands. One of these strands, however, is selected on the basis that, as the result of a known mutation, it misses the gene being mapped. As is apparent from Figure 15.17, the complementary strands of the heteroduplex pair perfectly throughout the length of the molecule, with one important exception. Across from the position of the missing gene in the mutant

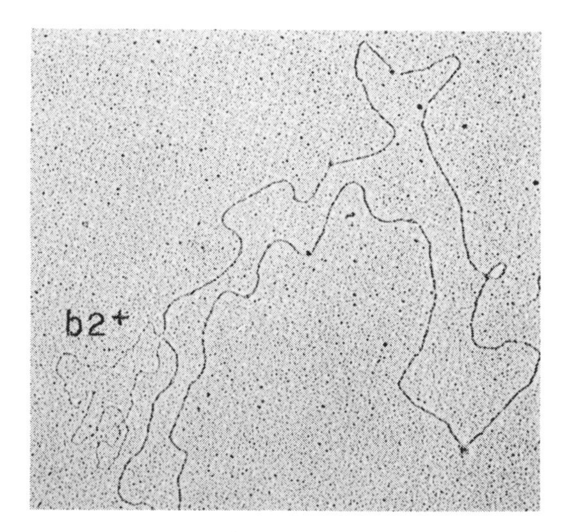

FIGURE 15.17
Heteroduplex formation in bacteriophage λ.
Electron micrograph of a heteroduplex DNA molecule constructed from complementary strands of bacteriophage λ and a bacteriophage λ deletion mutant (bacteriophage λβ2). In λβ2 a segment of DNA has been deleted, producing, at the site of deletion, a loop labeled $b2^+$.
Reprinted with permission from B. C. Westmoreland, W. Szybalski, and H. Ris, *Science* 163:1343, 1969.

strand the complementary strand forms a clearly visible loop. The position of the loop identifies the location of the deleted gene.

15.3 TYPES OF DNA STRUCTURE

The subject of DNA structure has been treated so far as though DNA were a "generic" substance, that is, only the essential features common to all DNAs have been presented. In fact, the specific structural features of DNA vary, depending on the origin and the function of each DNA molecule. Molecules of DNA differ in size, conformation, and topology.

Size of DNA Is Highly Variable

The size of DNA varies from a few thousand base pairs for the DNA of the small viruses, to millions for the chromosomal DNA of bacteria, and to billions for the chromosomal DNA of animals. Several types of expressions are commonly used to describe DNA size, including number of base pairs, molecular weights, the length of the strands, and even the actual weight of DNA. The units used in these expressions, however, can be easily interconverted, taking into account that a 1-million-mol-wt DNA contains approximately 1500 bp, which comprise a macromolecular segment of 0.5 nm length. Also, since DNA is a macromolecule, DNA weight can be converted to molecular weight by division with the average molecular weight of a DNA nucleotide pair.

As is apparent from Table 15.4, the amount of DNA per cell increases as the complexity of the cellular function increases. It should be noted that although mammalian cells contain some of the highest amounts of DNA per cell, some amphibian, fish, and plant cells may contain even higher amounts. In fact, lung fish cells contain more than 40 times the amount of DNA in human cells, but such extraordinary amounts of DNA reflect a reiteration of nucleotide sequences within the DNA macromolecule and do not represent an actual increase in the size of DNA in terms of unique sequences. But aside from these minor irregularities, the size of the DNA of higher cells is very large indeed. The DNA contained within a single human cell, if it were stretched end to end, would be about 1 m long. This suggests that the polynucleotides are exquisitely packed in order to fit within the minute dimensions of the cell nucleus.

Because of their extraordinary length, relative to the total mass, DNA molecules are extremely sensitive to shearing forces that develop during ordinary laboratory manipulations. Even careful pipetting may shear a DNA molecule. In addition, during the process of isolation it is difficult to prevent with absolute confidence the disruption of some phosphodiester bonds by contaminating endonucleases (nicking). For these reasons the

TABLE 15.4 The DNA Cell Content of Some Species

Type of cell	Organism	DNA per cell (pg)[a]
Phage	T4	2.4×10^{-4}
Bacterium	*E. coli*	4.4×10^{-3}
Fungi	*N. crassa*	1.7×10^{-2}
Avian erythrocyte	Chicken	2.5
Mammalian leukocyte	Human	3.4

SOURCE: B. Lewin, *Gene expression,* Vol. 2, 2nd ed. New York: Wiley, 1980, p. 958.

[a] pg = picograms.

precise size of DNA, especially that of the higher species, could not be determined until special handling techniques were developed, both for the isolation of DNA and the measurement of its molecular weight.

Techniques for Determining DNA Size

In any event, devising suitable methods for the measurement of the molecular size of DNA has been a scientific challenge. The classical methods for determining size in proteins, such as light scattering, sedimentation diffusion, sedimentation equilibrium, or osmometry proved to be unsuitable for measuring the molecular weight of even relatively small DNAs. For instance, because of the great mass of DNA the sedimentation coefficient of the macromolecule is so high that centrifuges could not be run slowly enough to yield useful data with existing methodology. Instead custom-tailored methods had to be devised. Equilibrium centrifugation in a density gradient (usually a concentrated cesium chloride solution), electron microscopy, and electrophoresis in agarose gels are among the principle methods providing reliable information about the molecular weights of various DNAs. Electron microscopy provides a measure of the length of DNA strands. Molecular weights can be calculated from known values of the mass per unit length. The DNA can be visualized under the electron microscope if it is first coated with protein and a metal film. Determination of molecular weights by electrophoresis depends on the molecular-sieving effect of porous agarose gels. Over a limited range of molecular weights the mobility of DNA is directly proportional to the logarithm of the molecule's weight. The range of the method is further extended by appropriate adjustments in the density of the agarose gels, which leads to changes in mobility.

In order to determine the molecular weight of DNA by **equilibrium centrifugation** a small portion of a DNA solution to be analyzed is layered on top of a gradient in a centrifuge tube. Upon centrifugation, the molecules of DNA sediment to equilibrium through the gradient. Under these conditions a homogeneous high molecular weight DNA will form a Gaussian band centered at a position in the gradient that corresponds to the density of the macromolecule. Molecules with different densities are resolved into a series of bands that sediment independently of one another, as shown in Figure 15.18. A relationship can be demonstrated between the width of the bands at equilibrium and the molecular weights, permitting the determination of accurate molecular weights.

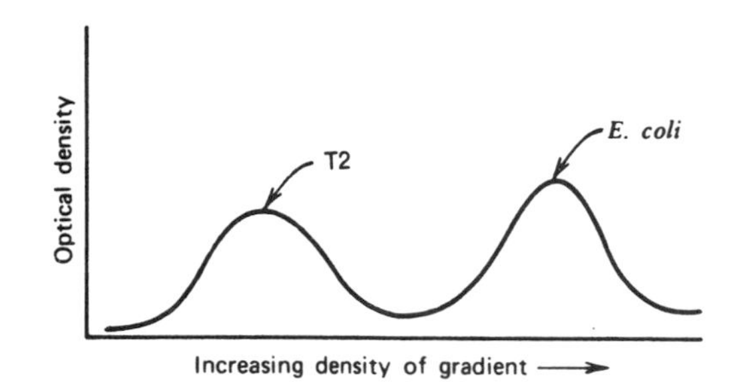

FIGURE 15.18
Equilibrium gradient centrifugation of DNA.
The DNA macromolecules travel into the increasingly dense regions of the gradient driven by centrifugal forces. The macromolecules equilibrate as soon as they reach an area of the gradient of density equal to their own. For example, bacteriophage T2 DNA and *E. coli* DNA can be resolved into two distinct bands. The width of the bands at equilibrium is related to the molecular weight of DNA.

A biochemical method based on the labeling of the terminals of a macromolecule has been used successfully for determining molecular weights in proteins. In this case the DNA is treated with the enzyme alkaline phosphatase, which converts the 5′-phosphate nucleotide terminals of double-stranded DNA to the corresponding OH groups. These terminals are then esterified, using [γ-^{32}P]ATP with the enzyme polynucleotide kinase. The free 5′ terminus of each polynucleotide chain becomes labeled as shown in Figure 15.19. The labeled DNA is then ana-

3′-HO———P-5′ / 5′-P———OH-3′ —(alkaline phosphatase)→ 3′-HO———OH-5′ / 5′-HO———OH-3′ —(polynucleotide kinase, [γ-^{32}P]ATP)→ 3′-HO———^{32}P-5′ / 5′-^{32}P———OH-3′

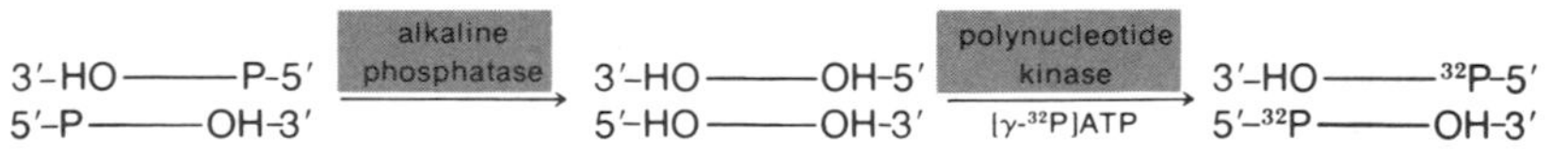

FIGURE 15.19
End-group labeling procedure.
The 5′ terminals on the opposite ends of DNA are labeled with ^{32}P by treatment with alkaline phosphatase and esterification of the resulting 5′-hydroxyl groups with [γ-^{32}P]ATP.

lyzed by zonal centrifugation and detected from both its absorbancy at 260 nm and ^{32}P, counting as indicated in Figure 15.20. The molecular weight is calculated from the ratio of the amount of ^{32}P to the absorbancy, both measured at the coinciding peaks of the bands.

The above methods have permitted the determination of DNA molecular weights with an accuracy of at least 10%, but the usefulness of each method is limited within certain molecular weight ranges. Electrophoresis is most suitable for molecular weights in the range between 1.5×10^5 and 1.5×10^7. This range can be extended upward to 2×10^8 by electron microscopy. The most versatile method, however, is equilibrium centrifugation, the range of which extends approximately between 2×10^5 and 10^9. The high range of the method is limited because of the effect of shear forces on larger molecules. Therefore, because even bacterial DNAs often have molecular weights in excess of 10^9, it is apparent that none of the above methods can be used for very large DNAs. For DNA molecules of mol wt 10^{10} a specifically designed low shear viscometric method, described in Figure 15.21 has been developed. This method, known as **viscoelastic retardation,** is based on mildly stretching long DNA molecules by hydrodynamic shear forces. Once these forces are removed, the DNA molecules can relax back to their normal unstressed conformation. The relaxation time is related to, and can be used to determine with accuracy, the molecular weight of DNA molecules of the size found in eucaryotic chromosomes.

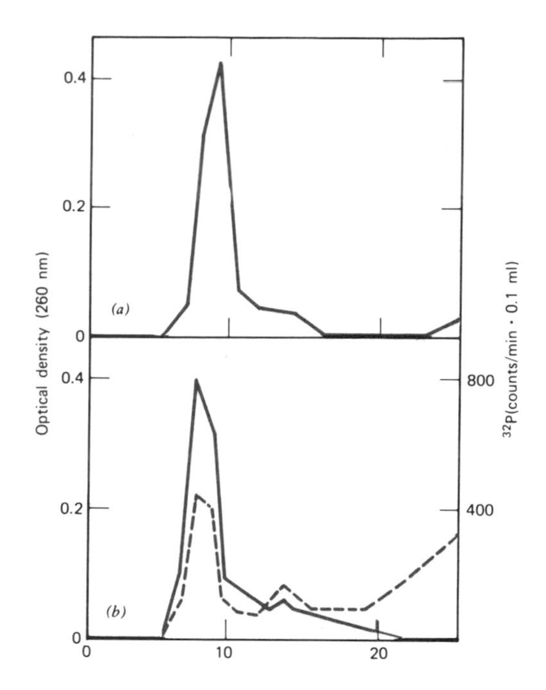

FIGURE 15.20
Zonal centrifugation profiles of denatured T7 DNA treated by the end-group labeling procedure.
Sedimentation is from right to left. (*a*) Untreated DNA. (*b*) DNA treated by the end-group labeling procedure. Zonal centrifugation is performed on a sucrose density gradient and should be distinguished from density gradient centrifugation. The latter is an equilibrium centrifugation with the macromolecules reaching equilibrium at regions within the tube at which their density equals the density of the environment. With zonal centrifugation the macromolecules move continuously until they reach the bottom of the tube or until the centrifuge is stopped. The molecular weight is calculated from the ratio of the amount of ^{32}P (dotted line) to the optical density (solid line) at the peak of the curve.
Redrawn from C. C. Richardson, *J. Mol. Biol.* 15:49, 1966.

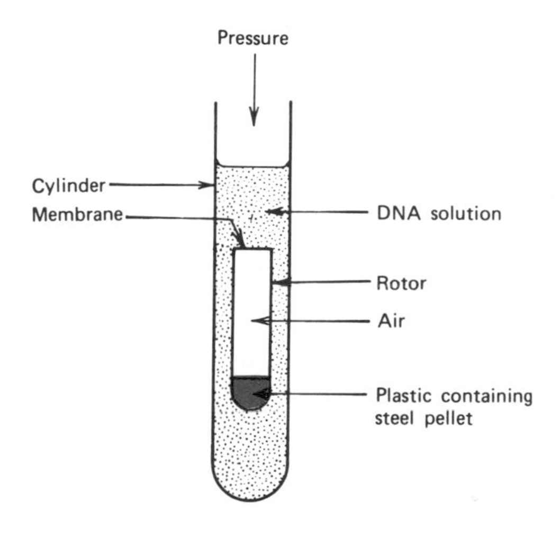

FIGURE 15.21
Viscoelastic retardation technique for the measurement of the molecular weight of large DNA molecules.
An appropriately constructed cylinder holds the solution to be measured. A free-floating rotor, which can maintain a suspended position by adjusting the pressure exerted on the surface of the solution, is inserted into the tube. The rotor is magnetically rotated, but once the magnetic field is removed the tube does not come to a complete stop. Instead, after slowing down it reverses direction before coming to a final stop. This reversal can be used to measure the relaxation time of DNA. In practice, cells are lysed in the measuring tube so as to avoid breakage of DNA caused by a transfer of DNA solution. Incubation with detergent at 65°C inactivates nucleases and insures the intactness of the resulting DNA strands. Proteins bound to DNA are removed by addition of the proteolytic enzyme pronase. The technique measures the size of the largest DNA molecule in the mixture rather than the average molecular weight of the molecules present.
Redrawn from D. Freifelder, *The DNA molecule: structure and properties*. San Francisco: W. H. Freeman, 1978. Copyright © 1978.

DNA May Be Either Linear or Circular

The DNA of several small viruses occurs in the form of typical linear double-stranded helices of equal size. In addition, certain DNAs have naturally occurring interior single-stranded breaks. The breaks found in natural bacteriophage molecules result mostly from broken phosphodiester bonds, although occasionally a deoxyribonucleoside may be missing. The DNA of coliphage T5 consists of one intact strand and a complementary strand, which is really four different well-defined complementary fragments ordered perfectly along the intact strand. A similar regularity in the points of strand breaks is noted with a few other DNAs, for example, *Pseudomonas aeruginosa* phage B3, but generally interior breaks seem to be randomly distributed along the strands. The overall structure of the double helix is maintained because the breaks that occur in one strand are generally in different locations from breaks in the complementary strand.

Double-Stranded Circles

Most naturally occurring DNA molecules exist in circular form. In some instances **circular DNA** exists even as interlocked circles. Provided that suitable precautions are taken to avoid shearing the DNA, the circular form can be isolated intact and observed by electron microscopy. The circular structure results from the circularization of a linear DNA by formation of a phosphodiester bond between the 3′ and 5′ terminals of a linear polynucleotide.

The circular nature of DNA of the small phage ϕX174 was first suspected from studies that showed no polynucleotide ends were available for reactions with exonucleases. Sedimentation studies also revealed that endonuclease cleavage yielded one rather than two polynucleotides. These suspicions were later confirmed by observation with electron microscopy.

During the early 1960s, after workable methods for avoiding the shear of large molecules were developed, the circular nature of the DNA chromosome of *E. coli* was demonstrated by the use of autoradiography techniques. Soon it became apparent that many other DNAs (e.g., those of mitochondria, chloroplasts, bacterial plasmids, and mammalian viruses) also existed as closed circles. Obviously, the strands of a circular DNA cannot be irreversibly separated because they exist as intertwined closed circles. The absence of 3′ or 5′ termini apparently provides an evolutionary advantage because it endows the circular DNA with complete resistance toward exonucleases, which act by hydrolyzing the phosphodiester bond of terminal nucleotides only. Thus circularity may be a protective mechanism against cellular exonucleases, which insures the longevity of DNA.

The DNA of some bacteriophages exists in a linear double-stranded form, which has the tendency to circularize when it enters the host cell. The linear DNA form of bacteriophage λ of *E. coli,* for instance, has single-stranded 5′ terminals of 20 nucleotides each. These terminals have complementary sequences, so that an **open circle** structure can be formed when the linear λ molecule acquires a circular shape, which allows the overlap of these complementary sequences. Subsequently, the enzyme **DNA ligase,** which forms phosphodiester bonds between properly aligned polynucleotides, joins the 3′- and 5′-terminal residues of each strand and transforms the DNA into a covalently **closed circle,** as illustrated in Figure 15.22.

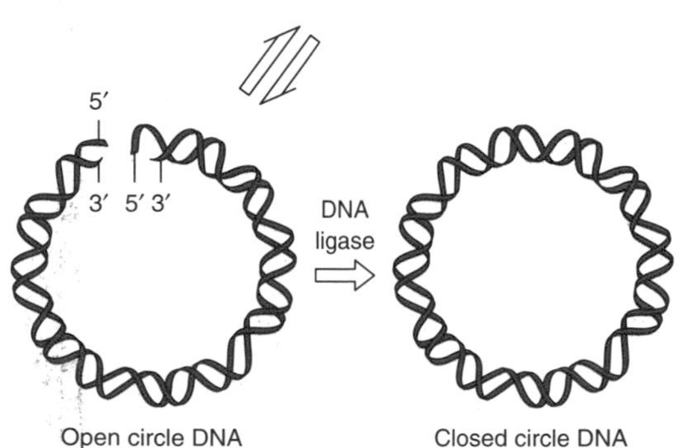

FIGURE 15.22
Circularization of λ DNA.
The DNA of bacteriophage λ exists in both a linear and a circular form, which are interconvertible. The circularization of λ DNA is possible because of the complementary nature of the single-stranded 5′ terminals of the linear form.

Single-Stranded DNA

With the exception of a few small bacteriophages (e.g., ϕX174 and G4) that can acquire a single-stranded form, most circular as well as linear

DNAs exist as double-stranded helices. The single-stranded nature of the nonreplicative form of ϕX174 DNA was first suspected in the 1950s when it was discovered that the base composition of this DNA did not conform to the base *equivalence* rules, that is, for this DNA A $\neq$ T and G $\neq$ C. The single-stranded nature of this structure was also confirmed by the observation that the amino groups of the bases reacted rapidly with formaldehyde, which indicated that the bases were exposed. Furthermore, electron micrographs of ϕX174 indicated that single-stranded DNA appears more "kinky" and less thick than the double-stranded form. It may be noted that the discovery of the single-stranded circular form of ϕX174 actually preceded the identification of the replicative double-stranded form, which has a normal complementary base composition.

Topology of Circular DNA Is That of a Superhelix

The double-stranded circular DNAs, with few apparent exceptions, possess an intriguing topological characteristic. The circular structure contains twists, which are referred to as **supercoils** and can actually be visualized by electron microscopy. In order to understand the origin of these twists, it may be helpful to consider two possible approaches by which, in principle, linear DNA can be converted to a circular molecule. Circular DNA may be formed by bringing together, and joining by a phosphodiester bond, the free terminals of linear DNA. If no other manipulations are used, the resulting circular DNA will be **relaxed;** that is, the circular molecule will have the thermodynamically favored structure of the linear double helix (B-DNA), which accommodates one complete turn of the helix for a unit of length of polynucleotide consisting of approximately 10 base pairs. However, if before sealing the circle, one DNA terminus is held steady while the other terminus is rotated one or more full turns in a direction that unwinds the double helix, the resulting structure will be strained. This strained structure, which is characterized by a deficiency of turns, is known as negative **superhelical DNA.** Twisting in the opposite direction produces positive superhelical DNA (Figure 15.23). Torsional strain increases the standard free energy of DNA by about 10 kcal mol^{-1} per each supercoil that is introduced into the structure. The strain produced by this deficit of turns can be accommodated by the disruption of hydrogen bonds and the opening of the double helix over a small region of the macromolecular structure. The resulting structure may be viewed as

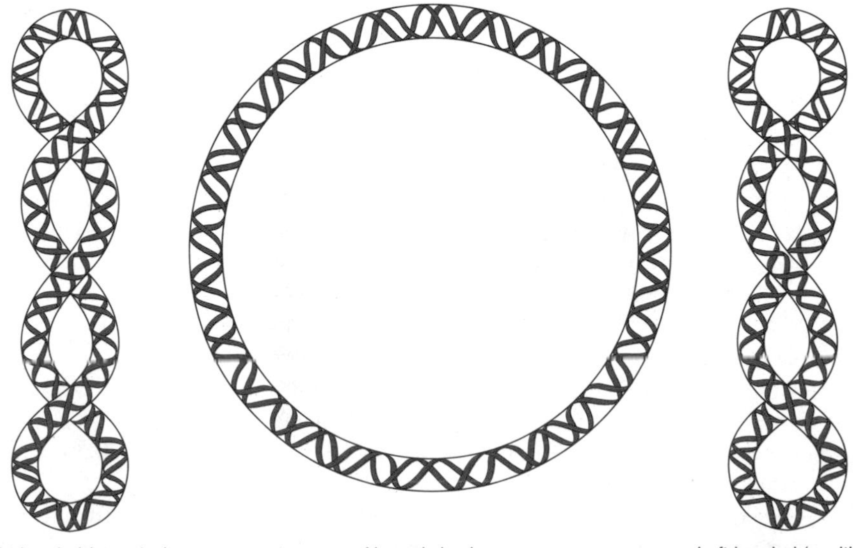

FIGURE 15.23
Relaxed and supercoiled DNA.
Relaxed DNA can be converted to either right- or left-handed superhelical DNA. Right-handed DNA (negatively supercoiled DNA) is the form that is normally present in cells. Left-handed DNA may also be transiently generated as DNA is subjected to enzymatically catalyzed transformations (replication, recombination, etc.) The distinctly different patterns of folding for right- and left-handed DNA are apparent in this representation of the two types of superhelices.
Redrawn from J. Darnell, H. Lodish, and D. Baltimore, *Molecular cell biology*. New York: W. H. Freeman, 1986.

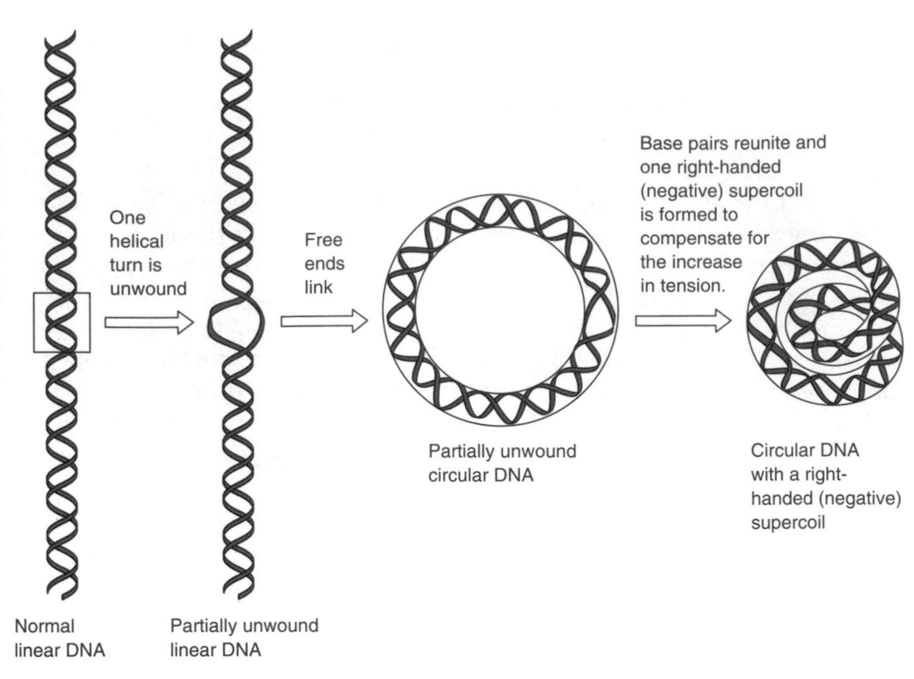

FIGURE 15.24
Right-handed (negative) DNA supercoiling. Right-handed supercoils (negatively supercoiled DNA) are formed if relaxed DNA is partially unwound. Unwinding may lead to a disruption of hydrogen bonds or alternatively produce negative supercoils. The negative supercoils are formed to compensate for the increase in tension that is generated when disrupted base pairs are reformed.
Redrawn from J. Darnell, H. Lodish, and D. Baltimore, *Molecular cell biology*. New York: W. H. Freeman, 1986.

consisting of a small-stranded loop along with regions of regularly spaced relaxed double helical turns. If, however, hydrogen bonds are not disrupted, the circular DNA will twist in a direction opposite to the one in which it was rotated in order to relieve the strain induced by the unwinding. Thus the rotational strain that was introduced before the circularization of DNA can also be accommodated by the formation of tertiary structures with visible *supercoils* (Figure 15.24). These two representations of the negative superhelix should not be viewed as two distinct types of superhelices, but rather as two manifestations of the same underlying phenomenon. In general a dynamically imposed compromise, determined by the environment and the status of circular DNA, is reached between hydrogen bond disruption and supertwisting. In practice this means that supercoiled DNA consists of *twisted structures* with *enhanced tendency* to contain regions with disrupted hydrogen bonding (bubbles).

In a circular DNA that is initially relaxed, the transient strand unwinding would tend to introduce compensating supertwists. However, if DNA is superhelical to begin with, the density of the superhelix will obviously tend to fluctuate with the ''breathing'' of the helix. All naturally occurring DNA molecules contain a deficit of helical turns; that is, they exist as *negative superhelices* with a superhelical density that remains remarkably constant among different DNAs. Normally one negative twist is found for every 20 turns of the helix.

If before converting a linear DNA to the corresponding circular structure one of the terminals of the linear polynucleotides is rotated in the direction of *overwinding* rather than *unwinding* the double helix, the resulting DNA will contain *positive* superhelices. The DNA cannot accommodate the added strain by overtwisting. While negatively superhelical DNA can accommodate unwinding stress either by unwinding (accompanied by the interruption of hydrogen bonds) or by the formation of negative superhelices, the *only available option for overwound DNA* is to accommodate the stress by acquiring positive superhelices. *Positive* supercoils can be experimentally produced by specialized enzymes, the topoisomerases, and may be present in vivo transiently.

The notion of superhelicity is often difficult to grasp fully without examining an appropriate physical model. In the absence of a more suitable alternative, you might attempt to twist two pieces of *fully extended* thick rope past the point at which considerable resistance develops. At that point the rope would represent a positive supercoil, which, in order to be accommodated without undue strain, must be allowed to escape the fully extended conformation and acquire the form of a compact coil. This model also highlights the concept that superhelicity is inseparably associated with the existence of a closed or restricted topological domain. Superhelicity in this example will be preserved only for as long as both hands grasp the rope firmly so as to maintain a closed topological system. Once this closed system is interrupted, the superhelix can unwind and acquire a relaxed form.

Geometric Description of Superhelical DNA

The conformations acquired by the interlocking rings of a closed circular complex can be formally characterized by three parameters: the linking number α, the number of helical turns β, and the number of supercoils or tertiary turns τ. These parameters are related by the equation $\alpha = \beta + \tau$. The nature of β and τ is self-explanatory. When interlocked rings are viewed with one ring held in a plane, the linking number α may be defined as the number of times one ring passes through the other. As is apparent in Figure 15.25, α can also be determined by counting the number of times the two rings appear to cross each other and dividing this number by 2. This is so because for each turn of the helix of a closed complex the second strand must pass through the circle formed by the first twice when viewed perpendicular to the helix axis.

Two important conclusions can be reached from consideration of these definitions and from examination of Figure 15.26. First, it is apparent that for every *relaxed* DNA the linking number and the number of helical turns are identical. However, as will be apparent shortly, the reverse is not true. Second, DNAs with a specific linking number can acquire various different arrangements in space. In the case of superhelical DNAs different types of supercoils may be formed. However, *all conformations with the same linking number α are interconvertible without the need of breaking any covalent bonds*. Therefore, the linking number is a *constant* for any covalently closed circular DNA.

The various forms of supercoiled DNAs can be described using the α, β, and τ numbers. The mental exercise shown in Figure 15.26 illustrates how these numbers apply. It should be recalled that the turns of the typical double helix are right handed. Therefore, if a hypothetical linear DNA duplex that is 10 turns long ($\alpha = 10$ and $\beta = 10$) is unwound by, say, one turn, the resulting structure will have the following characteristics: $\alpha = 9$ and $\beta = 9$. A *potentially* equivalent structure can be formed if instead the ends of the same hypothetical DNA are secured so that they cannot rotate and the molecule is looped in a counterclockwise manner. Since in this case untwisting is not permitted to occur, the number of helical turns remains unchanged, that is, $\beta = 10$. However, as a result of the "looping" operations, the linking number is now reduced by 1, that is, $\alpha = 9$. The structure resulting from this deliberate introduction of a loop is visibly superhelical. Furthermore, application of the equation that relates the values of α, β, and τ indicates that τ must be equal to -1, that is, the structure is a *negative* superhelix with *one* superhelical turn.

The two structures described above, $\alpha = 9$, $\beta = 9$, $\tau = 0$ and $\alpha = 9$, $\beta = 10$, $\tau = -1$ obviously have the same linking number and are therefore interconvertible without the disruption of any phosphodiester bonds. The potential equivalence of these two types of structure becomes more apparent when the ends of the polynucleotides in each structure are joined into a circle without the strands being allowed to rotate. Circularization

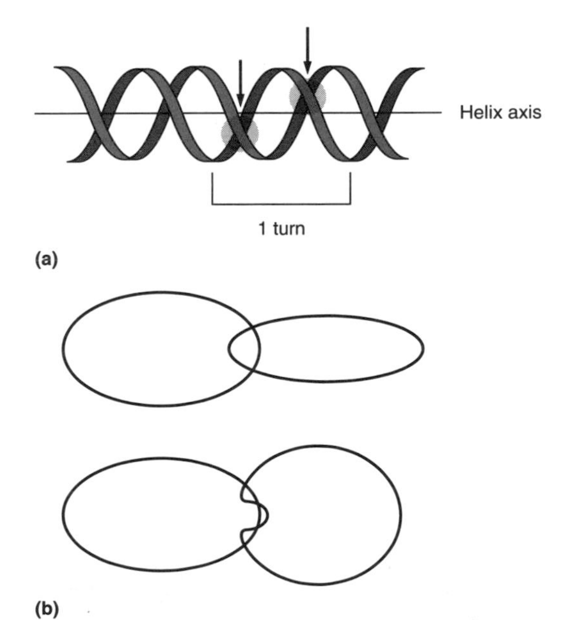

FIGURE 15.25
Determination of the linking number α in superhelical DNA.
(*a*) Side view of a schematic representation of the double helix. Note that the strands cross twice for each turn of the helix. (*b*) DNA circles interwound once and twice. Note that each pair of crossings is equivalent to one interwind.

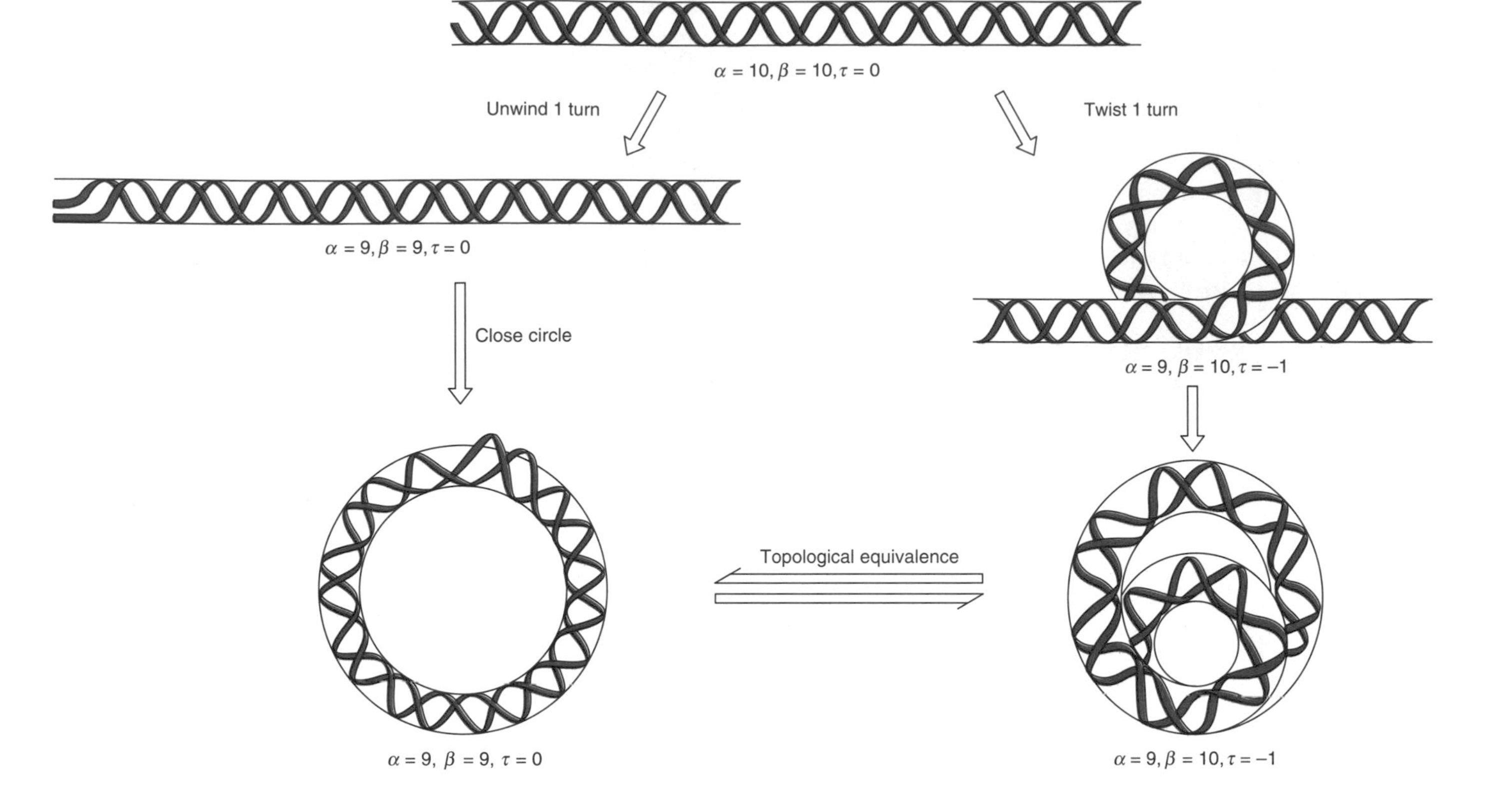

FIGURE 15.26
Various types of DNA superhelices.
An accurate representation of superhelical DNA structures can be made, using the number of helical turns β and the number of supercoils or tertiary turns τ along with a third parameter referred to as the linking number α as defined in the text. The figure shows ways of introducing one supercoil into a DNA segment of 10 duplex turns and the parameters of the resulting superhelices.
Redrawn with permission from C. R. Cantor and P. R. Schimmel, *Biophysical chemistry,* Part III. San Francisco: W. H. Freeman, 1980. Copyright © 1980.

produces an underwound circular structure and a doughnut-shaped superhelical arrangement referred to as a **toroidal turn,** which are freely interconvertible. A third equivalent structure, called an **interwound** turn, shown in Figure 15.27, can be produced by unfolding a toroidal turn along an axis which is distinct from the supercoil axis.

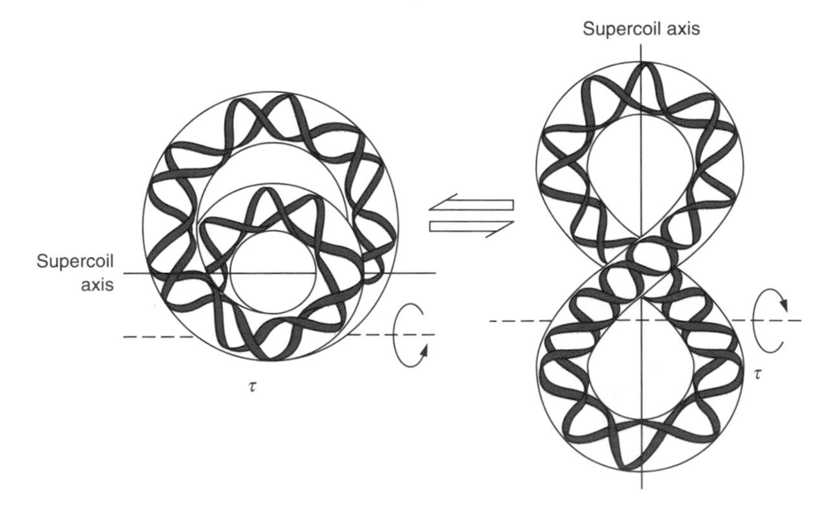

FIGURE 15.27
Equilibrium between two equivalent supercoiled forms of DNA.
The forms shown are freely interconvertible by unfolding the doughnut-shaped toroidal form along an axis parallel to the supercoil axis or by folding the 8-shaped interwound form along an axis perpendicular to the supercoil axis. The two forms have the same α, β, and τ numbers.
Redrawn with permission from C. R. Cantor and P. R. Schimmel, *Biophysical chemistry,* Part III. San Francisco: W. H. Freeman, 1980. Copyright © 1980.

In summary, if the termini of a linear DNA molecule are covalently attached, a "relaxed" covalent circle results. However, if one end of the double helix is maintained in a fixed and stationary position while the other end is rotated in either direction prior to closing the circle, the resulting structure will twist in the opposite direction so as to generate a superhelical structure. For each additional complete turn of the helix, the circle will acquire one more superhelical twist in the opposite direction of the rotation in order to relieve the intensifying strain. As a result, topologically equivalent structures, such as those shown in Figure 15.26 will be created. A real superhelical DNA must exist as an equilibrium among these forms and many other intermediate arrangements in space that have the same linking number.

Although the closed circular form of DNA is an ideal candidate for acquiring a superhelical structure, any segment of double-stranded DNA that is in some way immobilized at both of its terminals qualifies for superhelicity. This property therefore is not the exclusive province of circular DNA. Rather, any appropriately anchored DNA molecule can acquire a superhelical conformation.

The DNA of animal cells, for instance, normally associated with nuclear proteins, falls into this category. Because of the fragility and the large size of this DNA, it has been difficult to establish whether it generally consists of a single circular piece, although this may be the case. However, even in the absence of a circular structure, animal DNA can acquire a superhelical form because its association with nuclear proteins creates numerous closed topological domains. In addition, most bacterial phages, animal viruses, bacterial plasmids, and cell organelles, such as mitochondria and chloroplasts, contain superhelical DNA. The existence of negative superhelicity appears to be an important factor, promoting the packaging of DNA within the confines of the cell because supercoils generate compact structures. For instance, while the length of DNA in each human chromosome is of the order of centimeters, the condensed mitotic chromosomes that contain this DNA are only a few nanometers long. Negative superhelicity may also be instrumental in facilitating the process of localized DNA strand separation during the process of DNA synthesis.

Topoisomerases

Although much remains to be learned as to how superhelices are generated, specific enzymes known as **topoisomerases** appear to regulate the formation of superhelices. Topoisomerases act by catalyzing the concerted breakage and rejoining of DNA strands, which produces a DNA that is more or less superhelical than the original DNA. Topoisomerases are classified into type I, which break only one strand, and type II, which break both strands of DNA simultaneously. Topoisomerases I act by making a transient *single-strand break* in a supercoiled DNA duplex resulting in *relaxation of the supercoiled* DNA (Figure 15.28). In the past these enzymes were referred to as the nicking–closing enzymes. Topoisomerase I isolated from *E. coli* is also known as the **omega protein** (ω protein). Topoisomerase II acts by binding to a DNA molecule in a manner that generates two supercoiled loops, as shown in Step 1 of Figure 15.29. Since one of these loops is positive and the other negative, the overall linking number of the DNA remains unchanged. In subsequent steps, however, the enzyme *nicks both strands* and passes one DNA segment through this break before resealing it. This manipulation inverts the sign of the positive supercoil, resulting in the introduction of *two negative supercoils* in each catalytic step. This reaction occurs at the expense of ATP; that is, topoisomerases II are ATPases.

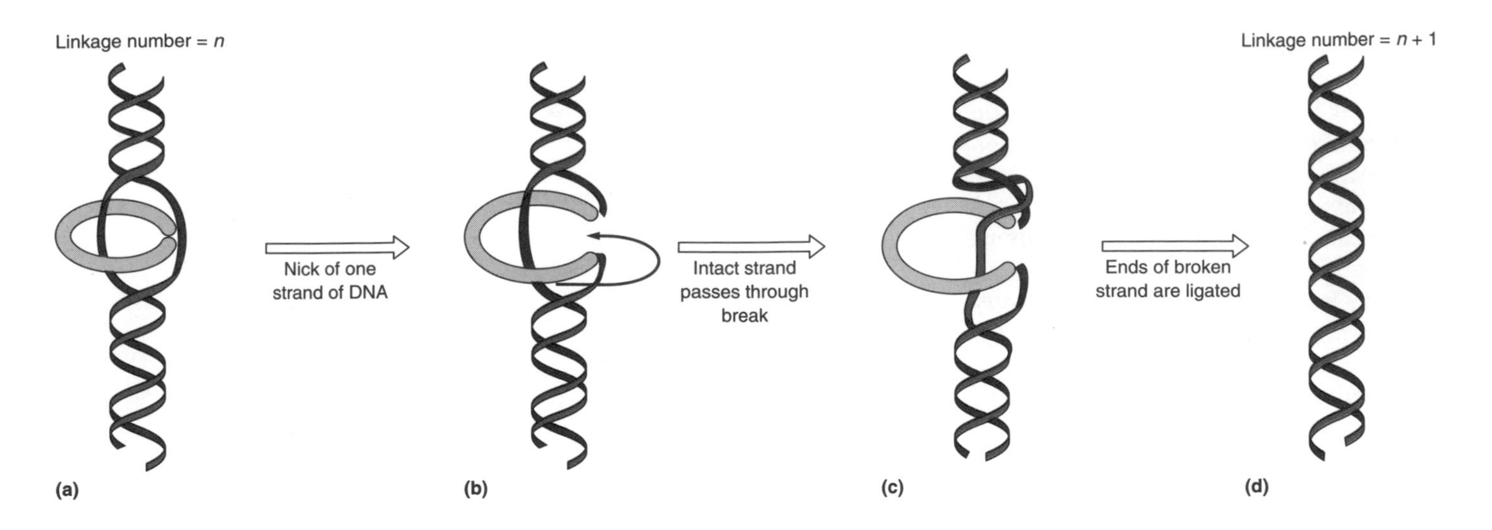

FIGURE 15.28
The mechanism of action of topoisomerases I.
Topoisomerases I can (*a*) relax DNA by first binding to it and locally separating the complementary polynucleotide strands. (*b*) Subsequently one of the strands is nicked, (*c*) the enzyme binds to the newly generated termini and prevents these termini from rotating freely. (*d*) In the last step the enzyme pulls the intact strand through the gap generated by the nick, closes the gap by restoring the phosphodiester bond, and gives rise to a relaxed structure.
Redrawn from F. Dean et al., *Cold Spring Harbor Symp. Quant. Biol.* 47:773, 1982.

Information about the mechanism of this reaction has been obtained by the use of nonhydrolyzable structural analogs of ATP. These analogs are capable of adding a single inversion (which generates two negative supercoils) to DNA but are unable to further catalyze the reaction. Apparently, the energy released by ATP hydrolysis is used for restoring topoisomerase II conformation, after the enzyme has catalyzed the formation of

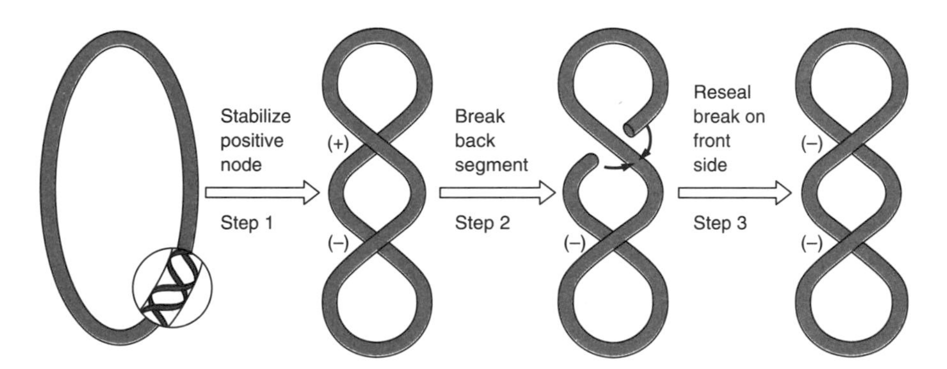

FIGURE 15.29
Mechanism of action of gyrase.
Gyrase, and other type II topoisomerases, can change the linking number of DNA by binding to a DNA molecule and passing one DNA segment through a reversible break formed at a different segment of the same DNA molecule. The mechanism of action of gyrase is illustrated above using as an example the conversion of a relaxed DNA molecule to a molecule that contains first two supercoils, one positive and one negative (Step 1). Passage of a DNA segment through the positive supercoil shown on the rightmost part of the figure (Step 3) changes the linking number, producing a molecule that contains two negative supercoils.
Redrawn with permission from P. O. Brown and N. R. Cozzarelli, *Science* 206:1081, 1979. Copyright © 1979 by the American Association for the Advancement of Science.

1-mol equivalent of product. The return of the enzyme in its original conformation, which is accompanied by hydrolysis of ATP, is, however, necessary for the participation of the enzyme in the next cycle of catalysis. The final substep of the reaction, that is, the addition of a single inversion, is inhibited by the antibiotic nalidixic acid. Another antibiotic novobiocin interferes with the second ATP-dependent substep of the reaction. Derivatives of nalidixic acid are used clinically in the treatment of infections caused by bacteria resistant to other more commonly used antibiotics.

Obviously, reversal of this reaction removes two supercoils in each step. Therefore, topoisomerases II can either *add* or *remove* supercoils. During the reaction the enzyme remains bound to DNA by forming a covalent bond extended between a tyrosyl residue and a phosphoryl group at the incision site. This enzyme–polynucleotide bond conserves the energy of the interrupted phosphodiester bond for the subsequent repair of the nick. The cleavage sites do not consist of unique nucleotide sequences, although certain sequences are preferentially found at cleavage sites. Many topoisomerases of type II have been purified. One of these, **Gyrase,** which has been isolated from *E. coli,* has been studied quite extensively. It is a tetrameric protein consisting of two A subunits and two B subunits, and it adds negative supercoils to DNA at a rate of about 100 per min.

Separation of superhelical DNA from the corresponding relaxed or linear forms can be achieved by gel electrophoresis or by equilibrium centrifugation. With the latter method separation is achieved because the density of supercoiled DNA differs from that of the relaxed forms.

Nucleoproteins of Eucaryotes Contain Histones and Nonhistone Proteins

The DNA in eucaryotes is associated in the cell with various types of protein known as **nucleoprotein.** The main protein constituents of nucleoproteins are a class of highly basic proteins known as histones. **Histones,** regardless of their source, consist of five distinct types of polypeptides of different size and composition as listed in Table 15.5. The most "conserved" histones are the H4 and H3, which differ very little even between

TABLE 15.5 The Structure of the Five Types of Histones[a]

Name	Structure[b]	Residues	Molecular weight
H4	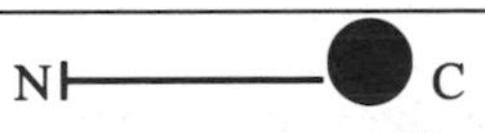 N C	102	11,300
H3	 N C	135	15,300
H2A	N C	129	14,000
H2B	 N C	125	13,800
H1	 N C	~216	~21,000

SOURCE: From D. E. Olins and A. L. Olins, *Am. Sci.* 66:704, 1978.

[a] Histones, which are highly basic polypeptides, are often classified as lysine-rich (H1), slightly lysine-rich (H2A and H2B), and arginine-rich (H3 and H4). Many of the basic amino acids are clustered on amino-terminal tails (i.e., the first 30–40 amino acids on the N side). The nonbasic mid- and carboxyl-terminal portions of the histones (C side) form globular structures that appear to be the sites of interaction between histones in the nucleohistones. The basic tail, on the other hand, interacts with DNA. The H1 class is almost twice as large and more basic than the other histones. Its globular region is nearer the N terminal.

[b] Scale ⊢⊣ 10 amino acids.

extremely diverse species; histone H4 from peas and cows are very similar, differing only by two amino acids, although these species diverged more than a billion years ago. The H2A and H2B histones are less highly conserved, but still exhibit substantial evolutionary stability, especially within their nonbasic portions. The H1 histones are quite distinct from the inner histones. They are larger, more basic, and by far the most tissue-specific and species-specific histones. As a result of their unusually high content of the basic amino acids lysine and arginine, histones are highly polycationic and interact with the polyanionic phosphate backbone of DNA so as to produce uncharged nucleoproteins. All five histones are characterized by a central nonpolar polypeptide domain, which under appropriate conditions of ionic strength tends to form a globular structure, and N-terminal and C-terminal regions that contain most of the basic amino acids. The basic N-terminal regions of histones H2A, H2B, H3, and H4 are the major, but not the exclusive, sites of interaction with DNA. The nonpolar domains and the C-terminal regions are involved in interactions both between histones and between DNA and histones.

In addition to histones, a heterogenous group of proteins with high species, and even organ, specificity is present in nucleoproteins. These proteins, which are grouped together under the somewhat unimaginative name of **nonhistone proteins,** consist of several hundred different proteins, most of which are present in trace amounts.

Nucleosomes and Polynucleosomes

Nucleoproteins interacting with DNA have a *periodic* structure in which an elementary unit known as **nucleosome** is regularly repeated. Each nucleosome consists of a DNA segment associated with a histone cluster and nonhistone proteins. Each DNA–histone cluster is a disk-shaped structure about 10 nm in diameter and 6 nm in height composed of two molecules each of H2A, H2B, H3, and H4 histones. The clusters are organized as tetramers consisting of $(H3)_2$ $(H4)_2$, with an H2A–H2B dimer stacked on each face in the disk. The DNA is wrapped around the octamer as a negative toroidal superhelix. For the purpose of describing the structure of nucleoproteins two distinct structures can be distinguished—the nucleosome core and the **chromatosome** as presented in Figure 15.30. The chromatosome constitutes the basic structural element of nucleoproteins.

The next level of organization of nucleoprotein is the **polynucleosome,** which consists of numerous nucleosomes joined by "linker" DNA the size of which differs between cell types. Usually the nucleosome core is used as the elementary unit for describing the polynucleosome, in which case linker DNA size varies anywhere from about 20–80 bp. (Linker sequences would of course be proportionally smaller if the chromatosome were to be used as the elementary unit for the polynucleosome.) Since in addition to the linker sequence 145 bp are wrapped around the nucleosome core, the polynucleosome has a minimum *nucleosome repeat frequency* of 165 bp. As a rule, repeat frequencies appear to be relatively long in transcriptionally inactive cells and are found to depend on both the organism and the organ from which the cell is isolated. For example, chick erythrocytes have a repeat frequency of 212 bp. Active cells, such as yeast cells that have a frequency of 165 bp, generally have shorter linker sequences. Although nucleosomes are periodically positioned along the polynucleosome, their distribution may not be random with respect to the base sequence of DNA.

Chromosomal Structure

Almost all the DNA of differentiated cells is present as **chromatin.** Chromatin, which fills the entire nucleus of resting cells, becomes highly con-

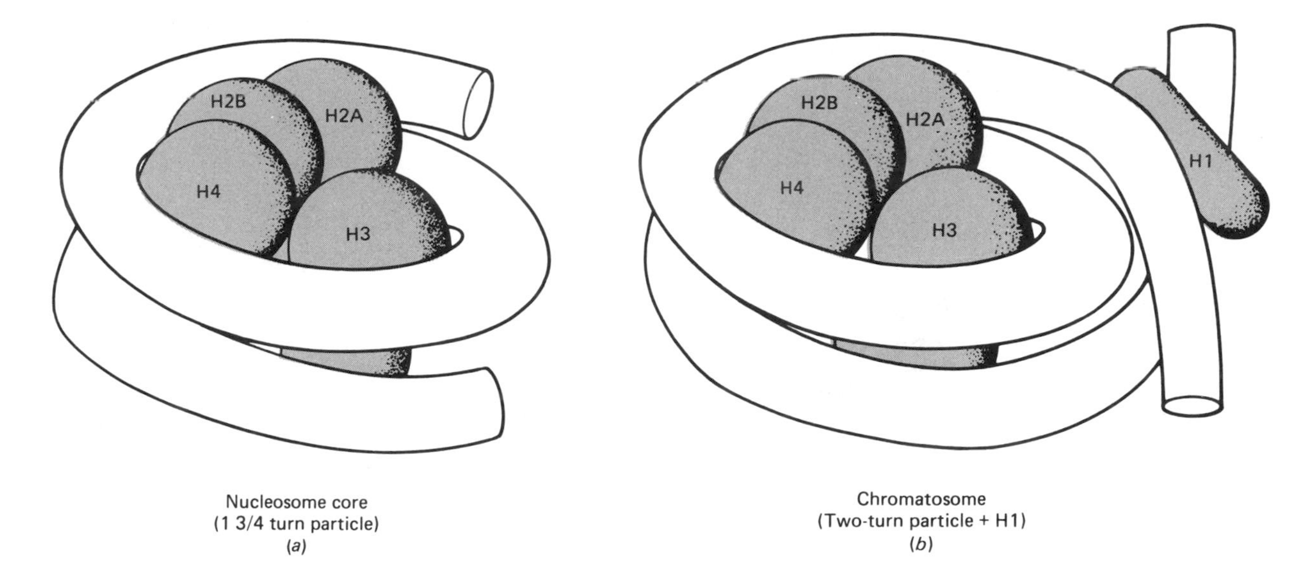

FIGURE 15.30
Postulated structures for the nucleosome core and the chromatosome.
The nucleosome core consists of 146 base pairs of DNA corresponding to $1\frac{3}{4}$ superhelical turns wound around a histone octamer. The chromatosome (two-turn particle) consists of 166 base pairs of DNA (two superhelical turns). The H1 subunit is retained by this particle and may be associated with it, as shown. Nucleosome particles containing less than 166 base pairs do not bind the H1 subunit.

densed into distinct *chromosomes* as soon as the process of DNA replication is completed. Chromosomes consist of fibers about 300 Å wide, which at low ionic strengths can be dissociated into 100-Å wide fibers known as **nucleofilaments** (Figure 15.31).

A small amount of eucaryotic DNA is located in the mitochondria and the chloroplasts of plant cells. This DNA, which occurs in the form of small superhelices, is generally free of protein.

Nucleosomes are universally present among eucaryotic organisms and appear to be the first level of chromosomal organization beyond the DNA helix. There is little doubt that the nucleosome has a definite packaging function for chromosomal DNA.

Histones H3 and H4, as well as the central regions of H2A and H2B, apparently participate in interactions essential for maintaining chromatin structure and function. In the H1 sequence, variations are also confined within the basic N terminal and the very basic C terminal of the histone.

Nucleosomes have a definite packaging function for chromosomal DNA. The DNA from a human cell, which has a length of the order of 1 m, must be condensed so that it can fit within a nucleus with a diameter of approximately 10 μm. In order for this DNA to fit within the confines of the nucleus, it must be made compact by various types of sequential folding that are stabilized by histones and other proteins. In addition to the nucleofilament and the supercoiled or solenoid structures, the eucaryotic chromosome can become further condensed by the formation of 600-Å knoblike structures. These various consecutive levels of chromosomal organization are depicted in Figure 15.32. For the last stage of packaging, the supercoiled DNA is organized into separate loops, each 40–80 thousand bp long, which merge in the center of the chromosome within a protein-rich region referred to as the *scaffold*.

During the first level of condensation, when a DNA molecule is organized as a nucleosome, the length of DNA is reduced by a factor of 50. During the subsequent steps of condensation further reduction in DNA

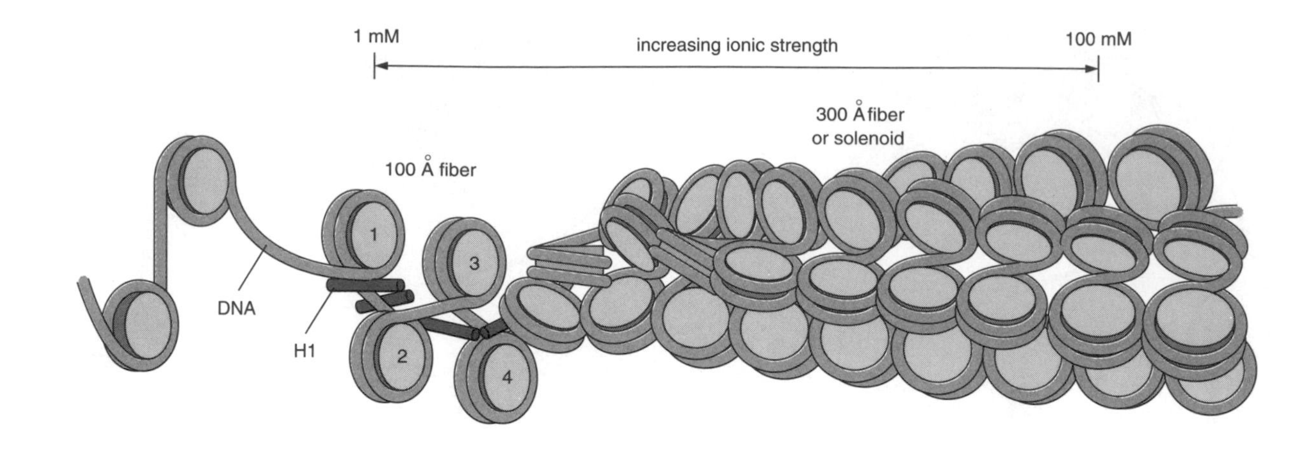

FIGURE 15.31
Nucleofilament structure.
Nucleofilament has the "string of beads" appearance, which corresponds to an extended polynucleosome chain. H1 histone is attached to the "linker" regions between nucleosomes, but in the resulting structure H1 molecules, associated to adjacent nucleosomes, are located close to one another. Furthermore, at higher salt concentrations polynucleosomes can be transformed into the higher order structure of the 300-Å fiber. It has been proposed that at higher ionic strengths the nucleofilament forms a very compact helical structure or a helical solenoid, as illustrated in the upper part of the figure. H1 histones appear to interact strongly with one another in this structure. In fact the organization of the 100-Å nucleofilament into the 300-Å coil or solenoid requires, and may be dependent on, the presence of H1.
Adapted from R. D. Kornberg and A. Klug, *The nucleosome.*

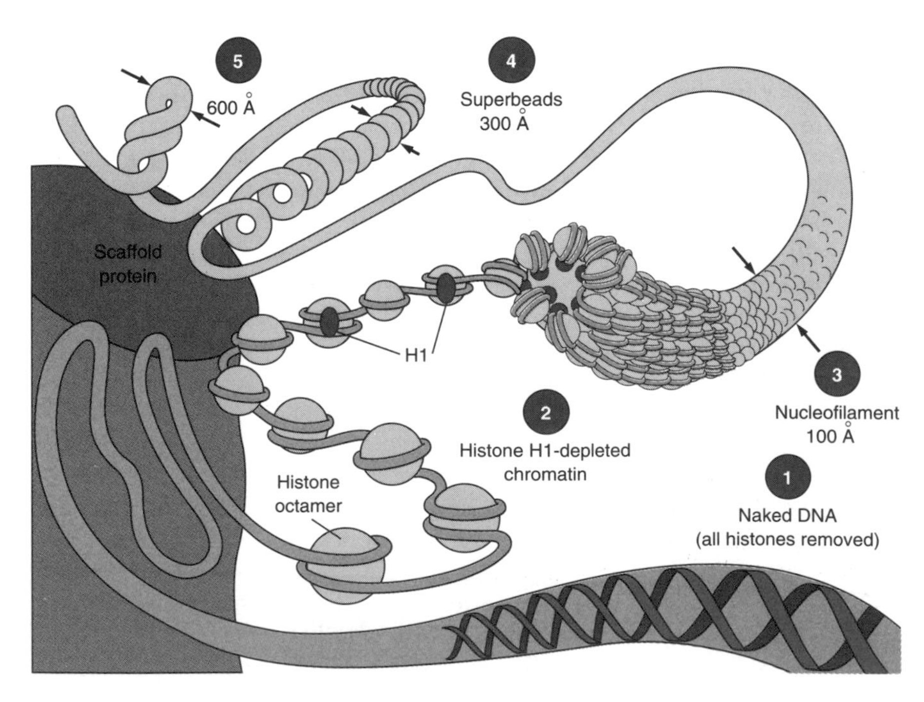

FIGURE 15.32
Various levels of organization of chromatin in the cell.
The "string of beads" structure of nucleofilament (Section 3) is generated as a result of histone interaction with histone-free DNA (Section 1). Superbeads (Section 4) are formed by condensation of the nucleofilament at physiological ionic strengths. The final packaging of chromatin is in the form of the 600-Å knoblike structures depicted in Section 5. Chromatin at all levels of its organization remains associated with a cellular protein scaffold.

length occurs. Eventually, at the level of condensed chromatin, which is found in chromosomes prior to cell division, DNA must be packaged with a remarkably high 7000-fold reduction in its length.

The function of histones in DNA packaging may be regulated by various in vivo reactions such as methylation, acetylation, and phosphorylation. For example, the OH group of the N-terminal serine residue in histone H4 is phosphorylated in a reaction catalyzed by a kinase. Also, four lysine residues near the N terminus of this histone undergo reversible acylation by two different enzymes, a **histone acetylase** and a **histone deacylase.** These reactions change the charge of the N-terminal region of histone H4 in the form that is both phosphorylated and acetylated, from +5 to −2. Such changes in polarity alter the affinity of histones for DNA, as well as other histones and proteins, and may thus have an important role in the regulation of DNA dependent processes. For instance maximum phosphorylation is observed before mitosis when chromosomes are most compact.

Much remains to be done before arriving at an understanding of the control of eucaryotic transcription and replication, but it is becoming increasingly apparent that both histone and nonhistone proteins are involved in these processes. While the dissociation of histones from chromosomal DNA may be a prerequisite for transcription, nonhistone proteins may provide more finely tuned transcription controls. However, the exact manner in which nonhistone proteins interact with DNA and regulate gene expression remains to be elucidated. Finally, nonhistone proteins may control gene expression during differentiation and development and may serve as sites for the binding of hormones and other regulatory molecules.

Viral DNA is almost always complexed with protein, where the function of the protein is generally one of "packaging." In essence the protein protects the DNA from mechanical damage or digestion by endonucleases by providing housing for the DNA within the tertiary structure of the protein.

Nucleoproteins of Procaryotes Have Similarities to Those of Eucaryotes

In procaryotic cells, DNA is generally present as a double-stranded circular supercoil, which is in part associated with the inner side of the plasma membrane. An abundant histone-like protein found in *E. coli* may be responsible for the presence of "beaded" chromatin fibers in procaryotes, which were long thought to lack histones. Histones together with various cations, polyamines (such as spermine, spermidine, putrescine and cadaverine), RNA, and nonhistone proteins apparently account for the organization of bacterial chromosomes. For example, the chromosome of *E. coli* exists as a highly compact structure, known as **nucleoid,** which consists of a single supercoiled DNA molecule organized into 45 loops merging into a scaffold rich in protein and RNA (Figure 15.33). In procaryotic scaffolds, the loops are maintained by interactions between DNA and RNA rather than DNA–protein interactions only, as is the case with eucaryotes.

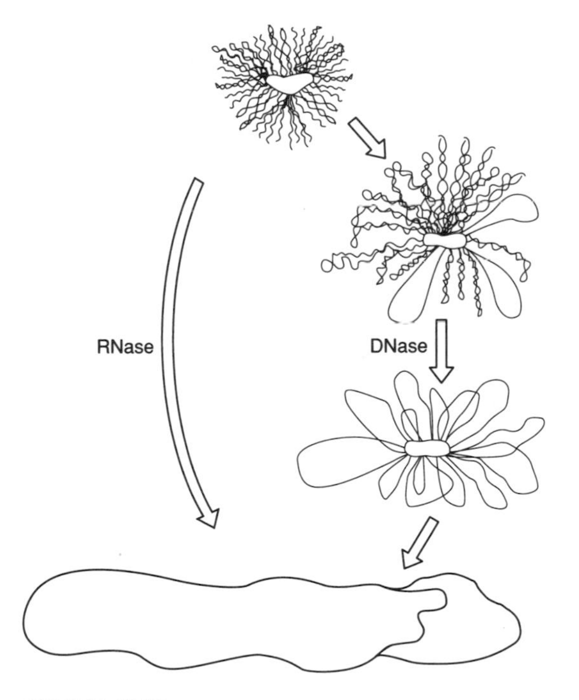

FIGURE 15.33
A schematic depiction of the folded chromosome of *E. coli.*
This chromosome contains about 50 loops of supercoiled DNA organized by a central RNA scaffold. DNase relaxes the structure progressively by opening individual loops, one at a time. RNase completely unfolds the chromosome in a single step.
Redrawn from A. Worcel and E. Burgi, *J. Mol. Biol.* 71:127, 1972.

15.4 DNA STRUCTURE AND FUNCTION

The Uniqueness of Each DNA Is its Nucleotide Sequence

Overall *base composition* characterizes DNA only in a very general manner. Yet, information on composition is often useful for DNA characterization. For determining this parameter, the nucleic acid is first hydrolyzed to its nucleotide components either by chemical or enzymatic means. The resulting nucleotides are separated usually by ion-exchange chromatography, and their amounts are determined spectrophotometrically. Alternatively, the composition can be indirectly estimated by equilibrium centrifugation of the DNA in a density gradient or by measurement of the melting temperature under standard conditions. In both of these techniques, the guanine–cytosine content of DNA influences in a quantitative manner the buoyant density and the thermal stability of the macromolecule, respectively.

A more specific property, which characterizes any DNA in a unique way, is its nucleotide *sequence*. Clearly the uniqueness of each DNA does not rest on its base composition but rather in the sequential arrangement of its individual bases. The direct determination of nucleotide sequences in DNA remained, until the 1980's, an intimidating undertaking. This has been the case, in spite of the fact that the amino acid sequences of proteins and the nucleotide sequences of certain small RNAs such as tRNAs, have been accessible for many years through the application of effective but tedious methods of digestion of these macromolecules by appropriate enzymes. These enzymes, which sever macromolecules at specific sequences, can be chosen so that they yield fragments with overlapping sequences from which overall sequences can be gathered. Such approaches could not be used for sequencing DNA, partly because even the smallest DNA molecules are very large compared to proteins or small RNAs. In addition, enzymes that cleave DNA next to either a specific base or a specific sequence were not available until recently.

Restriction Endonucleases and Palindromes

The discovery of *restriction endonucleases,* which cleave DNA chains in a specific sequence-dependent manner, has made possible the sectioning of large DNA molecules into small segments amenable to sequencing.

These highly specific bacterial enzymes act in vivo by making two cuts, one in each strand of double-stranded DNA of an invading phage, generating 3′-OH and 5′-P termini. This initial fragmentation exposes phage DNA to eventual degradation by bacterial exonucleases.

The terminology for these endonucleases originates from the bacterial sources from which they are isolated. Many hundreds of restriction endonucleases have been isolated in pure form and the list of new restriction enzymes is growing daily. With few exceptions, these enzymes have been found to recognize sequences of four to six nucleotides long. These sequences, known as *palindromes,* are characterized by local symmetry as illustrated by the examples listed in Table 15.6. The order of the bases is the same or nearly the same, when the two strands of the palindrome are read in opposite directions. For example, in the case of the restriction enzyme EcoR1, isolated from *E. coli,* the order of the bases is GAATTC when read from the 5′ terminus of either of the strands.

Restriction enzymes are classified into three categories: Enzymes of types I and III make cuts in the vicinity of the recognition site in a unpredictable manner; type II enzymes cleave specifically DNA within the recognition sequence. The cuts, made by type II enzymes, are indicated in Table 15.6 by arrows. Examples of the products generated by these various specificities are shown in Figure 15.34.

These enzymes recognize specific sequences that occur along large DNAs with relatively low frequencies, and, thus, fragment DNA very selectively. For example, a typical bacterial DNA, which may contain about 3 million bp, will be broken down only into a few hundred fragments. A small virus or plasmid may have few or no cutting sites at all for

TABLE 15.6 Examples of Sites of Cleavage of DNA by Restriction Enzymes of Various Specificities[a]

Enzyme	Microorganism	Specific sequence	Number of cleavage sites for 2 commonly used substrates	
			ϕX174	pBR 322
EcoR1	*E. coli*	-G↓AATT-C- -C-TTAA↓G-	25	9
Hae III	*Haemophilus aegyptus*	-GG ∣ CC- -CC ↓GG-	11	22
Hpa II	*Haemophilus parainfluenzae*	-C↓CG-G- -G-GC↑C-	5	26
Hind III	*Haemophilus influenzae* Rd.	-A↓AGCT-T- -T-TCGA↑A-	0	1

[a] Cleavage takes place within palindromes. The cleavage sites are indicated by arrows.

a particular restriction endonuclease. The practical significance of this selectivity of restriction enzymes is that a particular enzyme generates a unique family of fragments for any given DNA molecule. This unique fragmentation pattern is called a **restriction digest.**

The availability of restriction enzymes for sectioning large DNA sequences and the development of new gel electrophoresis techniques for separating DNA segments that differ from one another by only a single nucleotide has made the determination of sequences a simple matter. These sequencing techniques are described in Chapter 18, p. 781.

Early attempts to determine DNA sequences were limited to small DNA regions, which could be easily sectioned off from the remaining DNA. Sequences that bind selectively with various functional proteins, for example, RNA polymerase and the repressor proteins, have been among the first to be determined. As a rule these sections can be separated from the remaining DNA by nuclease digestion of the complexes formed between DNA and the respective proteins. The protein protects the DNA section over which it is bound from the action of nuclease, and the protected DNA is recovered after digestion by dissociation of the protein. These studies indicated that many functional proteins and enzymes interact with DNA over regions of palindromic sequence (Figure 15.35).

EcoRI	5′.... G↓AATTC....3′ 3′.... CTTAA↑G....5′	⟶	5′....G3′ 3′....CTTAA5′	+	5′AATTC....3′ 3′G....5′
PstI	5′....CTGCA↓G....3′ 3′....G↑ACGTC....5′	⟶	5′....CTGCA3′ 3′....G5′	+	5′G....3′ 3′ACGTC....5′
HaeIII	5′....GG↓CC....3′ 3′....CC↑GG....5′	⟶	5′....GG3′ 3′....CC5′	+	5′CC....3′ 3′GG....5′

FIGURE 15.34
Types of products generated by type II restriction endonucleases.
Enzymes exemplified by EcoRI and PstI nick on both sides of the center of symmetry of the palindrome generating single-stranded stubs. Commonly used enzymes generate 5′ ends, although some produce stubs with 3′ stubs as shown for PstI. Other restriction nucleases cut across the center of symmetry of the recognition sequence, producing flush or blunt ends as exemplified by HaeIII.

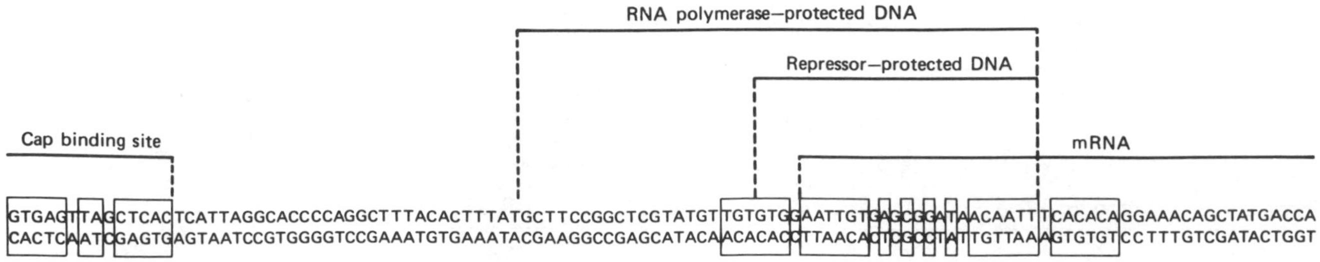

RNA polymerase–protected DNA

Repressor–protected DNA

Cap binding site

mRNA

GTGAGTTAGCTCACTCATTAGGCACCCCAGGCTTTACACTTTATGCTTCCGGCTCGTATGTTGTGTGGAATTGTGAGCGGATAACAATTTCACACAGGAAACAGCTATGACCA
CACTCAATCGAGTGAGTAATCCGTGGGGTCCGAAATGTGAAATACGAAGGCCGAGCATACAACACACCTTAACACTCGCCTATTGTTAAAGTGTGTCCTTTGTCGATACTGGT

FIGURE 15.35
The nucleotide sequence of part of the DNA segment that controls the synthesis of the enzyme β-galactosidase in *E. coli* (the lac operon).
The binding regions of the cap protein, which acts as an activator of transcription, and of the lac repressor protein, an inhibitor of transcription, are indicated. Also shown is the region of RNA polymerase interaction. The presence of two palindromic sequences is indicated by boxes.
Redrawn from C. R. Cantor and P. R. Schimmel, *Biophysical chemistry,* Part I. San Francisco: W. H. Freeman, 1980. Copyright © 1980.

Palindromes in DNA also serve as recognition sites for **methylases** that modify the host DNA by introducing methyl groups into two bases of the palindrome. Once methylated, these palindromes cannot be recognized by the corresponding restriction enzymes, and the DNA of the host is protected from cleavage.

The new sequencing methods have made possible the determination of the complete nucleotide sequences of the DNA of many small viruses containing thousands of nucleotide residues. The effectiveness of the new methods is such that sequencing the DNA of even higher cells is now becoming a routine undertaking.

Most Procaryotic DNA Codes for Specific Proteins

In procaryotes a large percentage of total chromosomal DNA codes for specific proteins. In certain bacteriophages the primary structure of DNA reveals that **structural genes,** nucleotide sequences coding for protein, do not always have distinct physical locations in DNA. Rather they frequently overlap with one another, as illustrated by the partial sequence of bacteriophage ϕX174 shown in Figure 15.36. It is believed that this type of overlap provides for the efficient and economic utilization of the limited DNA present in small procaryotes. This arrangement of genes may also be a factor in controlling the sequence in which genes are expressed.

Only a Small Percentage of Eucaryotic DNA Codes for Structural Genes

A typical mammalian DNA, with 20 times as many genes as that of *E. coli,* contains 500 times more DNA than *E. coli.* Clearly then, the DNA content of a mammalian cell is much too high on the assumption that it consists mostly of structural genes, along with some sequences used to control gene expression, as is the case with procaryotes. For example, only 10% of the DNA of the human cell may suffice for coding for the approximately 50 thousand genes that are probably present in the human genome. Determination of the complete nucleotide sequences of whole eucaryotic DNA could shed some light on the function of this excess DNA. This is an impractical task. However, the sequencing of large sections of eucaryotic DNA, for instance, structural genes and their sur-

```
                    (PROTEIN A) ...................Glu   Ser   Lys   Asn   Tyr   Leu   Asp   Lys   Ala   Gly   Ile   Thr   Thr
                       (ORIGIN OF PROTEIN K) Met   Ser   Arg   Lys   Ile   Ile   Leu   Ile   Lys   Gln   Glu   Leu   Leu   Leu

(NUCLEOTIDE SEQUENCE)...............A T G A G T C G A A A A A T T A T C T T G A T A A A G C A G G A A T T A C T A C T
                                    51                  61                  71                  81                  91

Ala   Cys   Leu   Arg   Ile   Lys   Ser   Lys   Trp   Thr   Ala   Gly   Gly   Lys (TERMINUS OF PROTEIN A)
  Leu   Val   Tyr   Glu   Leu   Asn   Arg   Ser   Gly   Leu   Leu   Ala   Glu   Asn   Glu   Lys   Ile   Arg   Pro   Ile
                                                              (ORIGIN OF PROTEIN C) Met   Arg   Lys   Phe   Asp   Leu   Ser
G C T T G T T T A C G A A T T A A A T C G G A G T G G A C T G C T G G C G G A A A A T G A G A A A A T C G A C C T A T
                  101                 111                 121                 131                 141                 151

  Leu   Ala   Gln   Leu   Glu   Lys   Leu   Leu   Leu   Cys   Asp   Leu   Ser   Pro   Ser   Thr   Asn   Asp   Ser   Val
    Leu   Arg   Ser   Ser   Arg   Ser   Ser   Tyr   Phe   Ala   Thr   Phe   Arg   His   Gln   Leu   Thr   Ile   Leu   Ser
C C T T G C G C A G C T G A C G A A G C T C T T A C T T G C G A C C T T T C G C C A T C A A C T A A C G A T T C T G T
                  161                 171                 181                 191                 201                 211

  Lys   Asn (TERMINUS OF PROTEIN K)
    Lys   Thr..........(PROTEIN C CONTINUES)
C A A A A A C T..............
```

FIGURE 15.36

Partial nucleotide sequences of contiguous and overlapping genes of bacteriophage ϕX174.

The complete nucleotide sequence of ϕX174 is known. Only the sequence starting with nucleotide 51 and continuing to nucleotide 219 is shown in this figure. This sequence codes for the complete amino acid sequence of one of the proteins of ϕX174, protein K. A part of the same sequence, nucleotide 51 to nucleotide 133, codes for part of the nucleotide sequence of another protein, protein A (the remaining part of protein A is coded by a sequence, not shown, extending on the left beyond nucleotide 51). The remaining part of the sequence coding for protein K, which starts with nucleotide 133, also codes for part of a third protein, protein C. Similar overlaps are noted between other genes of ϕX174; for instance the sequence coding for a fourth protein, protein B, extends on the left of nucleotide 51.

Adapted with permission from M. Smith, *Am. Sci.* 67:61, 1979. Journal of Sigma Xi, The Scientific Research Society.

rounding regions, has now become a relatively easy undertaking. As a result, sequence data of eucaryotic DNA segments are accumulating rapidly. These data indicate that, in contrast to procaryotes, eucaryotic genes not only do not overlap, but with few exceptions (e.g., the genes of histones and the majority of tRNA genes) are interrupted by intervening nucleotide sequences, **introns,** as shown in Figure 15.37. The nucleotide sequences in the gene which are expressed either in the final RNA product or as a protein are termed **exons** (see p. 714).

As a rule the sequence and the size of introns vary greatly among species, but generally these intervening segments may be 5–10 times longer than the sum of the length of the parts of the structural genes they separate. Some genes are interrupted only once, whereas others are highly fragmented. For instance, the conalbumin gene of the chicken, which codes for a major protein in egg white, may be divided by introns in as many as 17 distinct sections. The possibility exists that these sequences play a role in the control of gene expression. The suggestion has been made that genes subdivided by introns could, on an evolutionary scale of time, be more easily shuffled to produce new gene combinations than genes that are put together as one piece.

Repeated Sequences

Until recently, the nucleotide sequences of eucaryotic DNA had been extensively studied by reassociation techniques. The more recent application of direct sequencing methods on DNA fragments, obtained by restriction endonuclease digestion, further extended the scope of these studies. As a result, in addition to obtaining the sequences of specific

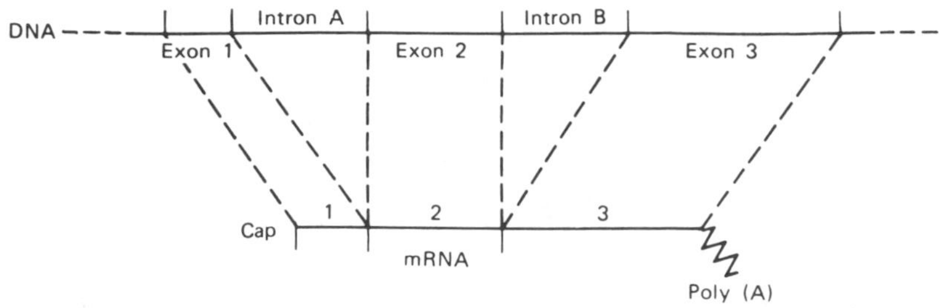

FIGURE 15.37
Schematic presentation of a eucaryotic gene.
The nucleotide sequences of eucaryotic genes are frequently separated by polynucleotide segments that are not present in mRNA and therefore are not translated to protein. These segments are referred to as introns. The gene is thus separated into noncontiguous segments called exons. The top horizontal line in the figure represents a part of the DNA genome of a eucaryote: the bottom line represents the mRNA produced by it. In this hypothetical example the DNA consists of two introns and three exons. The intron sequences are transcribed as hnRNA (precursor mRNA) but are not present in mature mRNA.
Redrawn from F. Crick, *Science* 204:264, 1979. Copyright © 1979 by the American Association for the Advancement of Science.

DNA sections, a good understanding of the complex characteristics of the primary structures of eucaryotic DNAs is now beginning to evolve.

As distinct from procaryotes, the DNA of eucaryotes contains multiple copies of certain nucleotide sequences that are repeated anywhere from a few times, for certain coding genes, to millions of times per genome for certain simple, relatively short, sequences. In addition to evidence obtained from sequencing data and DNA reassociation studies, the repetition of certain types of DNA sequences can be observed directly by electron microscopy, as in the cases of rRNA genes undergoing transcription.

Based on the number of times a sequence is repeated, three classes of sequences have been distinguished—**single copy, moderately reiterated, and highly reiterated.** These classes are defined experimentally from their rates of reassociation. Reassociation rates have also been used to define a fourth class of DNA, **inverted repeats.**

It may be recalled that the genome size of procaryotic DNA can be determined by fragmenting the DNA, denaturing the fragments, and then allowing them to reassociate and form double-stranded molecules (Figure 15.15). The kinetics of reassociation obey a single second-order equation, indicating that all the sequences in the procaryotic genomes occur as single copies. When a mouse DNA was first studied by this method, unexpected results were obtained, which lead to the realization that eucaryotic DNAs contain reiterated sequences. A priori, it was assumed that since mammalian genes are about three orders of magnitude larger than *E. coli* genes, the rates of reassociation of denatured mammalian DNA would be exceedingly slow. Instead it turned out that a fraction of the mouse DNA, the highly repetitive fraction reassociated far more rapidly than even the DNAs of small viruses (Figure 15.15). This is reasonable, since the probability that a fragment will encounter a complementary fragment leading to reassociation is proportional to the number of similar sequences repeated in the original DNA prior to fragmentation. The more reiterated the sequence, the more rapid the reassociation. Consequently, the reassociation kinetics of eucaryotic DNAs provided the first evidence for four classes or sequences. The inverted repeat and the highly repetitive sequences reassociate extremely rapidly. The unique sequences reassociate slowly, and the moderately reiterated renature at intermediate rates.

Most of the highly reiterated sequences have a distinct base composition from that of the remaining DNA. These sequences can be isolated from the total genomic DNA by shearing the DNA into segments of a few hundred nucleotides each and separating the fragments by density gradient centrifugation. These fragments are termed **satellite DNA** because after centrifugation they appear as satellites of the band of bulk DNA. For example, the highly reiterated DNA sequence in the rat consisting of the repeated sequence 5′-GCACAC-3′ can be separated as satellite DNA. Other highly reiterated sequences, however, cannot be isolated by centrifugation, although they can be identified by virtue of their property of rapid reannealing. Some of the highly reiterated sequences can also be isolated by digestion of total DNA with restriction endonucleases that cleave at specific sites within the reiterated sequence.

The exact boundaries separating the various types of reiterated DNAs do not appear to have been strictly defined. Keeping this limitation in mind, the following distinctions among reiterated eucaryotic DNA types may be used.

Single Copy DNA

About one-half of the human genome is made up of unique nucleotide sequences, but only a small fraction of it codes for specific proteins. A part of the remaining DNA is devoted to **pseudogenes;** that is, tracts of DNA that have significant nucleotide homology to a functional gene but that contain mutations that prevent gene expression. These genes, which may be present in a frequency as high as one pseudogene for every four functional genes, significantly increase the size of eucaryotic genomes without contributing to their *expressible* genetic content. Additional DNA sequences are committed to serve as introns and as regions flanking genes. For instance, the genes producing the ε, β, γ, and δ chains of hemoglobin are located in a cluster and are separated by a noncoding sequence 400 bp long interspersed among these genes. The function of the remaining single copy DNA remains unclear.

Moderately Reiterated DNA

In this class of DNA, we may include copies of identical or closely related sequences that are reiterated anywhere from a few to several hundred times. These sequences are relatively long, varying between a few hundred to many thousand nucleotides before the same polynucleotide sequence is repeated. Normally single copy and moderately reiterated sequences are present on the chromosome in an orderly pattern known as the **interspersion pattern,** which consists of alternating blocks of single copy DNA and moderately reiterated DNA. For example, in the human genome sequences 600 nucleotides long, present in about 100 copies per genome, are tandemly reiterated over about 6% of the genome. Another 52% of the genome consists of reiterated sequences interspersed with single copy sequences about 2250 nucleotides long. The remainder of the genome consists of unique sequences sparsely interspersed with repeated sequences. With the exception of some insects that have a distinctly different interspersion patterns, the human chromosome is typical of the observed pattern.

The short period of the interspersion pattern implicates the interspersed reiterated sequences in the control of transcription of the structural genes present in DNA. A role for the moderately reiterated DNA in controlling the transcription of the adjacent structural genes is plausible because the

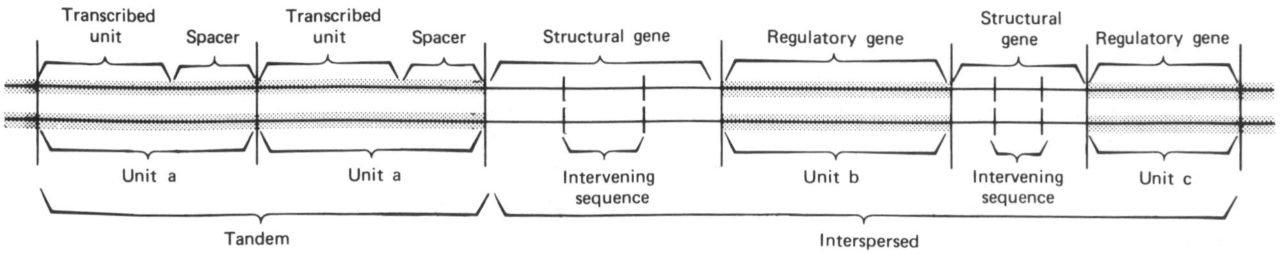

FIGURE 15.38
Structural genes with intervening sequences of the "interspersed" and "tandem array" types.
A hypothetical segment of eucaryotic DNA may be visualized as consisting of nonrepetitive sequences (indicated by a thin line) as well as moderately repetitive sequences (indicated by a shaded line). The latter can be of the interspersed or the tandem type. The interspersed sequences are separated by structural genes. The structural genes themselves are frequently interrupted by intervening sequences. The terms unit *a*, unit *b*, and unit *c* refer to distinct nucleotide sequences. The illustration does not imply that the various DNA elements shown are contiguous.

large majority of the structural gene sequences occur adjacent to reiterated sequences. Other moderately reiterated sequences are present as **segregated tandem arrays.** These two distinct types of arrangements of the moderately reiterated sequences appear to relate to different functions for these sequences. Tandem arrays are used for the synthesis of products that must be rapidly generated in numerous copies, such as ribosomal RNA and certain proteins of specialized function. For example, in sea urchin oocyte histone, genes are amplified so that sufficient amounts of histone are available for DNA packaging during the rapid cycles of DNA replication that follow fertilization. The genes for the five histones are arranged in tandemly repeated clusters, with each histone gene separated from its neighbor in the cluster by **spacers** that vary from about 400 to 900 nucleotides in length. These spacers are AT-rich and can therefore be separated as satellite DNA from the GC-rich DNA of the histone genes.

The arrangement of structural genes, with their introns, and the moderately repetitive segments of DNA of the interspersed and tandem array types is illustrated in Figure 15.38. In this hypothetical segment of eucaryotic DNA the single copy and moderately repetitive sequences are indicated, which together normally account for more than 80% of the total nucleotide content of the genome.

Highly Reiterated DNA

The major part of the remaining DNA consists of sequences constructed by the repetition, many thousand times, of a nucleotide sequence that is typically shorter than 20 nucleotides. Because of the manner in which they are constructed, highly reiterated DNAs are also referred to as **simple sequence DNA.** You should draw a distinction between the terms "reiterated" and "repetitive" in describing a DNA sequence. The term **reiterated** is used to describe a unique DNA sequence, usually several hundred nucleotides long, present in multiple copies in a genome. An individual DNA sequence is termed **repetitive** if a certain, usually short, nucleotide sequence is repeated many times over the DNA sequence. Simple sequences are typically present in the DNA of most, if not all, eucaryotes. In some eucaryotes only one major type of simple sequence may be present, as, for example, in the rat in which the sequence 5′-GCA-CAC-3′ is repeated every six bases. In other eucaryotes several simple sequences are repeated up to 1 million times. Some considerably longer repeat units for simple sequence DNA have also been identified. For instance, in the genome of the African green monkey a 172-bp segment has been found to be highly repeated. Determination of the nucleotide sequence of the highly repetitive DNA reveals that there are few sequence repetitions within the 172-base segment. The repeated units consist of a set of closely related but variant sequences. Because of its characteristic composition, simple sequence DNA can often be isolated as satellite DNA. The function of simple sequence DNA is not known, but this DNA appears to be concentrated in the centromers of chromosomes, and since it is not transcribed, a structural role in the organization of the eucaryotic chromosome is proposed.

Inverted Repeat DNA

Short inverted repeats, each consisting of no more than six nucleotides, such as the palindromic sequence GAATTC, occur by chance about once for every 3000 nucleotides. Such short repeats cannot form a stable "hairpin" structure that can be formed by longer palindromic sequences. *Inverted repeat sequences* that are long enough to form stable "hairpins" are not likely to occur by chance, and therefore they should be classified as a separate class of eucaryotic sequences. Experimentally they can be easily detected and quantitated by virtue of their extremely rapid rates of reassociation. In human DNA, about 2 million inverted repeats are present, with an average length of about 200 bp, although inverted sequences longer than 1000 base pairs have been detected. Some of these repeats may be separated by a spacer sequence that is not part of the inverted repeat. Most inverted repeat sequences are repeated 1000 or more times per cell.

15.5 FORMATION OF THE PHOSPHODIESTER BOND IN VIVO

The processes of enzymatic repair of certain randomly introduced changes in the chemical structure of the DNA bases and the process of DNA replication is discussed in this section.

The repair of DNA and particularly DNA replication are very complex processes. Although key similarities in the mechanisms of DNA replication and repair are discernible among different organisms, a considerable amount of diversity exists in terms of individual detail. This diversity further complicates any attempt to present a simplified and universally applicable model for each of these two processes. To resolve this difficulty the basic mechanistic elements of the substeps of each process are first described and subsequently integrated, using as an example the *E. coli* replication system but differences between *E. coli* and eucaryotic systems are pointed out.

DNA-Dependent DNA Polymerase

The common denominator between the processes of DNA replication and repair is the enzymatically catalyzed synthesis of DNA polynucleotide segments, which can be assembled with preexisting polynucleotides, leading to products of repair or replication. The synthesis of these polynucleotide segments is catalyzed by the enzyme DNA-dependent DNA polymerase, which in the case of *E. coli* has been isolated in three distinct forms, the polymerases I, II, and III listed in Table 15.7. The DNA polymerases are characterized by a $3' \rightarrow 5'$ exonuclease activity in addition to synthetic activities. Polymerases I and III are also $5' \rightarrow 3'$ exonu-

TABLE 15.7 DNA Polymerase I, II, and III of *E. coli*

	Polymerase		
Properties	I	II	III
Molecular weight	110,000	120,000	180,000
Molecules per cell	400	100	10
Polymerization activity (turnover number)	1,000	50	15,000
Exonuclease activity 3′ → 5′	Active	Active	Active
Exonuclease activity 5′→ 3′	Active	Inactive	Active

cleases. The involvement of all these enzymatic activities in the processes of repair and replication will be apparent shortly.

The synthetic activity of DNA polymerase can be described by referring to Figure 15.39 in which two complementary DNA strands of unequal length are shown. This conformation, in which the shorter strand has a free 3′ terminus, is essential for the function of DNA polymerase. The enzyme catalyzes the addition of free 5′-deoxynucleoside triphosphates to the 3′ terminus of the short strand, the **primer.** The term *primer* applies to the initial terminus of a molecule, in this instance the 3′-polynucleotide end, onto which additional monomeric units can be added stepwise to yield the final product. The free portion of the longer complementary strand is used as a **template** to direct the condensation of selected 5′-deoxynucleotides onto the growing primer. In the present context the term *template* refers to a single strand of nucleic acid, which provides the specific information necessary for the synthesis of a complementary strand. DNA polymerase requires both a primer and a template in order to function. The primer provides a site for the polymerization to begin, and the template provides the information that determines the precise nucleotide sequence of the new polymer.

As seen from Table 15.7, polymerase III catalyzes the elongation of a primer with a much higher degree of efficiency than polymerase I. The enhanced catalytic efficiency of polymerase III is partially attributable to the higher processivity of this enzyme. After a polymerase has added a nucleotide residue on the 3′-OH terminus of the primer, it may dissociate from the primer and bind at random to another partially completed polynucleotide chain or it may remain bound to the original template until many subsequent residues are added to it. Enzymes that tend to remain bound to their substrates through many rounds of polymerization are said to be **processive.** Polymerase I is not processive in that it tends to dissociate from the template after incorporating only a few nucleotides. Although processivity per se does not determine the catalytic rate of an enzyme, it is apparent that an enzyme with high catalytic activity, such as polymerase III, can achieve its optimal catalytic rate only if it is also highly processive.

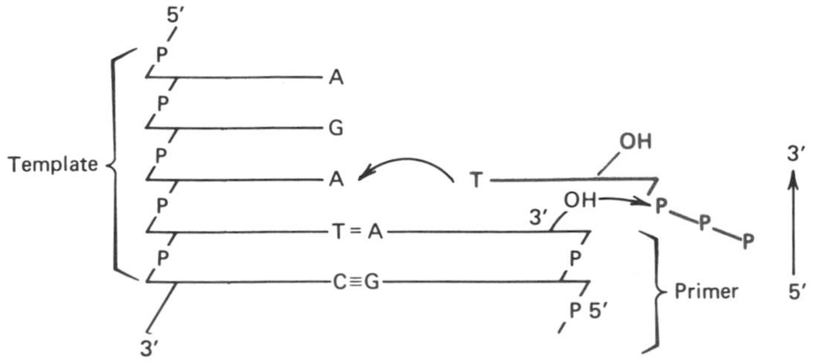

FIGURE 15.39
Synthetic activity of DNA polymerase.
DNA polymerase catalyzes the polymerization of nucleotides in the 5′ → 3′ direction. A phosphodiester bond is formed between a free 3′-hydroxyl group of the strand undergoing elongation (the primer) and an incoming deoxynucleoside 5′-triphosphate. Pyrophosphate is eliminated.
Redrawn from A. Kornberg, *Science* 163:1410, 1969. Copyright © 1969 by the American Association for the Advancement of Science.

The DNA polymerase-catalyzed reaction permits the selection of 5′-deoxyribonucleoside triphosphates, one at a time, with a base complementary to that present in the corresponding position of the template. The specificity of the polymerase reaction with respect to the template is vested in the strong association of each of the bases of the template with their normal complementary partners present in the cell as free 5′-deoxyribonucleotides. Strong binding between complementary bases is apparently achieved because the bases become confined within custom-fitted cages created by appropriate hydrophobic regions of the DNA polymerase. As a result the reading of the template is extremely accurate. Errors, however, occur in a rather indirect way, with a frequency of 1 in about 10^4–10^5 nucleotide additions, and originate from the presence of rare tautomeric forms of the four DNA bases (Figure 15.40) that occur transiently in DNA in ratios of 1 part to 10^4 or 10^5. These tautomeric forms mispair with the abundant keto forms as shown in Figure 15.41, and foster the incorporation into DNA of inappropriate bases.

The high fidelity of DNA polymerase thus depends on the existence of a **"proofreading"** mechanism that allows for the removal of these errone-

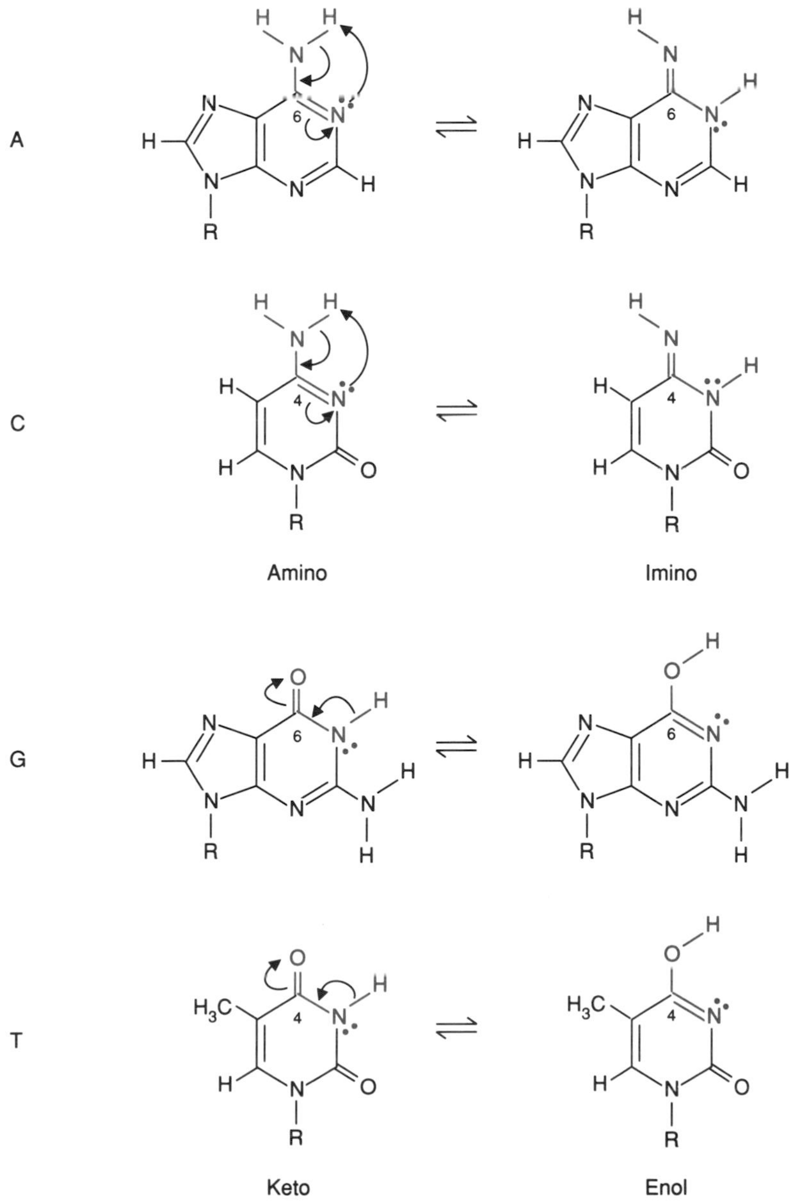

A

C

Amino

Imino

G

T

Keto

Enol

FIGURE 15.40
Tautomeric forms of the DNA bases.
Tautomeric forms of the bases can be generated by the shifting of protons. Under physiological conditions the equilibria of these tautomerizations give rise to only very minute amounts of the enol and the imino tautomers of the bases.

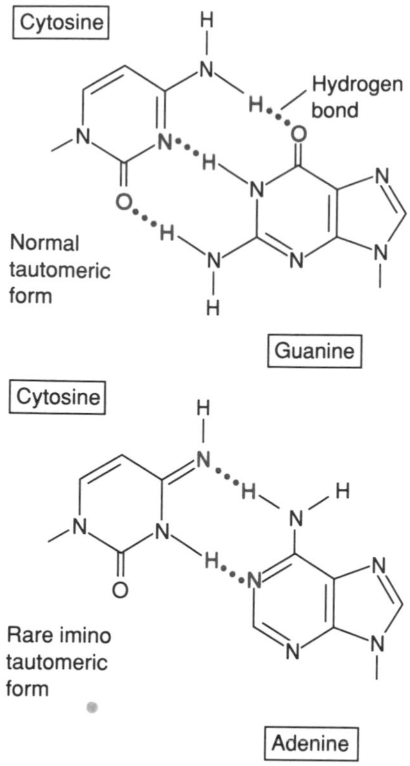

FIGURE 15.41
DNA base pairing with tautomeric forms. The imino form of cytosine does not form a hydrogen-bonded pair with the normal partner of cytosine, guanine (above). Instead cytosine mispairs with adenine (below).

ously introduced bases through the activation of the 3′ → 5′ exonuclease activity of the polymerase. Specifically, if a 5′-deoxyribonucleotide that is not complementary with the corresponding base on the template is erroneously condensed with the primer, the enzyme can temporarily reverse its synthetic activity and hydrolyze the phosphodiester bond formed between the primer and the erroneous base.

This correcting exonuclease activity is very accurate in that it will fail to remove no more than 1 in 10^3 improperly incorporated nucleotides, that is, it has an error rate of 10^{-3}. The operation of these two specific sequential reactions produces an overall error rate in the order of at least 10^{-6}, which makes polymerase III an enzyme of high fidelity. Yet, the cellular requirements for accuracy in the process of DNA replication are extremely high, with an allowable upper limit of 1 error in about 10^9 bp replications. This extremely high level of fidelity is maintained by further **postreplication repair** that removes replication errors missed by the proofreading activity of the polymerase. The operation of this mismatched repair system is described on page 657.

DNA polymerase III, which is the most complex of the three polymerases of *E. coli*, consists of 13 protein subunits. Three of these subunits, α, ε, and θ, constitute a core that exhibits partial enzymic activity. As shown in Table 15.8 enzymic activities have been assigned to two of the core subunits and seven other subunits appear to function as enhancers of processivity. The polymerases have well-defined selectivities also in a different sense. Only the 3′ terminus of a strand can be used for priming. Therefore the enzyme can elongate a strand only in the 5′ → 3′ direction, as indicated in Figure 15.39. The 5′ terminus of the strand is rejected as a primer because the polymerase is unable to elongate a polynucleotide in the opposite 3′ → 5′ direction.

The necessity to maintain high fidelity in replication is probably the reason why DNA Polymerase III synthesizes only in the 5′ → 3′ direction. If polynucleotide chains could be elongated in the 3′ → 5′ direction the hypothetical growing 5′ terminus, rather than the incoming nucleotide, would have to carry the triphosphate, which provides the free energy

TABLE 15.8 Subunits of DNA Polymerase III Holoenzyme

Subunit	Molecular weight		Activity
α	132,000	core	Polymerase
ε	27,000	core	Proofreading
θ	10,000	core	Unknown
β	37,000		Increasing processivity
τ	71,000		Increasing processivity
γ	52,000		Increasing processivity
δ	35,000		Increasing processivity
δ'	33,000		Increasing processivity
χ	15,000		Increasing processivity
ψ	12,000		Increasing processivity

SOURCE: From S. Maki and A. Kornberg, DNA polymerase III holoenzyme of *Escherichia coli*. Distinctive processive polymerase from purified subunits. *J. Biol. Chem.* 263:6561, 1988.

for the polymerizing reaction. Under these circumstances the 5′-chain terminus created by the removal of a mismatched base would be unsuitable for further elongation by the synthetic activity of the polymerase.

15.6 MUTATION AND REPAIR OF DNA

Mutations Are Stable Changes in DNA Structure

One of the fundamental requirements for a structure that serves as a permanent depository of genetic information is extreme stability. Such stability is essential, at least in terms of those characteristics of the structure that code for the genetic information. Therefore a prerequisite for the structure of DNA is extreme stability in its base content and in its sequence, in which hereditary information is encoded. Yet, the structure of the DNA bases is not totally exempt from gradual change. Normally, changes occur infrequently and then affect very few bases, but nevertheless they do take place. Chemical- or irradiation-induced reactions may modify the structure of some bases or may disrupt phosphodiester bonds and sever the strands. Errors may also occur during the processes of replication and strand recombination, leading to the incorporation of one or more erroneous bases into a new strand. In almost every instance, however, a few cycles of DNA replication are required before a modification in the structure of a base can lead to irreversible damage, that is, DNA polymerase must use the polynucleotide initially damaged as a template for the synthesis of a complementary strand for the initial change to become permanent. As Figure 15.42 suggests, use of the damaged strand as template extends the damage from a change of a single base to a change of a complete base pair and subsequent replication perpetuates the change.

Since the properties of cells and of the organisms constructed from them ultimately depend on the DNA sequences of their genes, irreversible alterations in a few DNA base pairs can cause substantial changes in the corresponding organism. These changes, referred to as **mutations,** may be hidden or visible, that is, phenotypically silent or expressed. Therefore, a *mutation* may be defined as a stable change in the DNA structure of a gene, which may be expressed as a phenotypic change in the corresponding organism. Mutations may be classified, depending on their origin, into two categories: **base substitutions** and **frame shift** mutations. Base substitutions include **transitions,** substitutions of one purine–pyrimidine pair by another, and **transversions,** substitutions of a purine–pyrimidine pair by a pyrimidine–purine pair. Frame shift mutations, which are the most radical, are the result of either the insertion of a new base pair or the deletion of a base pair or a block of base pairs from the DNA base sequence of the gene. These changes are illustrated in Figure 15.43.

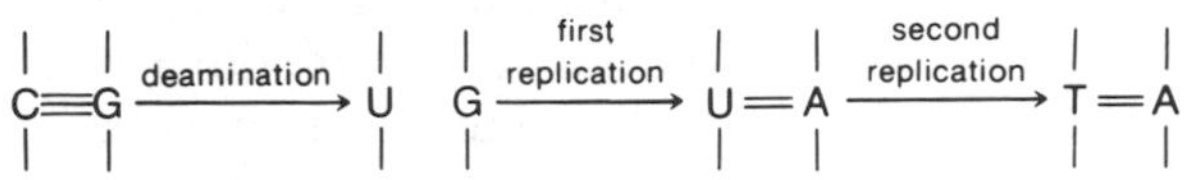

$$\mathrm{C{\equiv}G} \xrightarrow{\text{deamination}} \mathrm{U \quad G} \xrightarrow[\text{replication}]{\text{first}} \mathrm{U{=}A} \xrightarrow[\text{replication}]{\text{second}} \mathrm{T{=}A}$$

FIGURE 15.42
Mutation perpetuated by replication.
Mutations introduced on a DNA strand, such as the replacement of a cytosine residue by a uracil residue resulting from deamination of cytosine, extend to both strands when the damaged strand is used as a template during replication. In the first round of replication uracil selects adenine as the complementary base. In the second round of replication uracil is replaced by thymine. Similar events occur when the other bases are altered.

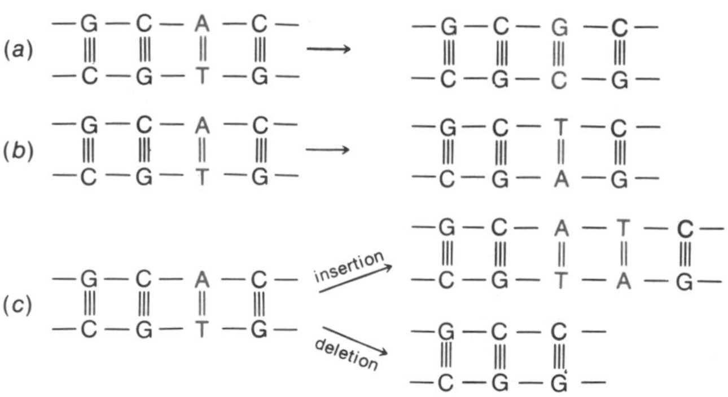

(a) —G—C—A—C— / —C—G—T—G— ⟶ —G—C—G—C— / —C—G—C—G—

(b) —G—C—A—C— / —C—G—T—G— ⟶ —G—C—T—C— / —C—G—A—G—

(c) —G—C—A—C— / —C—G—T—G— insertion ⟶ —G—C—A—T—C— / —C—G—T—A—G—; deletion ⟶ —G—C—C— / —C—G—G—

FIGURE 15.43
Mutations.
Mutations can be classified as transitions, transversions, and frame shift. Bases undergoing mutation are shown in boxes. (*a*) Transitions: A purine–pyrimidine base pair is replaced by another. This mutation occurs spontaneously, possibly as a result of adenine enolization or can be induced chemically by such compounds as 5-bromouracil or nitrous acid. (*b*) Transversions: A purine–pyrimidine base pair is replaced by a pyrimidine–purine pair. This mutation occurs spontaneously and is common in humans. About one-half the mutations in hemoglobin are of this type. (*c*) Frame shift: This mutation results from insertion or deletion of a base pair. Insertions can be caused by mutagens such as acridines, proflavin and ethidium bromide. Deletions are caused by deaminating agents. Alteration of bases by these agents prevents pairing.

Mutagens May Be Chemical Compounds or Radiation

A more systematic coverage of the subject of mutations, especially with respect to the expression of a mutation as a change in the product of the corresponding gene, must await the detailed description of the processes of replication, transcription, and translation. In this section, the factors that cause mutations are listed, and a few examples of structural DNA changes brought about by these factors are given. Irradiation and certain chemical compounds are recognized as the main **mutagens.** Some rare events of incorporation of erroneous bases by DNA polymerase can also lead to mutations.

Chemical Modification of the Bases

The bases in DNA are sensitive to the action of numerous chemicals. Among them are nitrous acid (HNO_2), hydroxylamine (NH_2OH), and various alkylating agents such as dimethyl sulfate and *N*-methyl-*N'*-nitro-*N*-nitrosoguanidine. Chemical modifications of bases, brought about by these reagents are shown in Figure 15.44. The conversion of guanine to xanthine by nitrous acid has no effect on the hydrogen-bonding properties of this base. The new base, xanthine, can pair with cytosine, the normal partner of guanine. However, the conversion of either adenine to hypoxanthine or the change from cytosine to uracil disrupts the normal hydrogen bonding of the double helix. This is because neither hypoxanthine nor uracil can form complementary pairs with the base present in the initial double helix (Figure 15.45). Subsequent replication of the DNA extends and perpetuates these base changes. Alkylating agents may affect both the structure of the bases as well as disrupt phosphodiester bonds so as to lead to the fragmentation of the strands. In addition, certain alkylating agents can interact covalently with both strands, creating interstrand bridges.

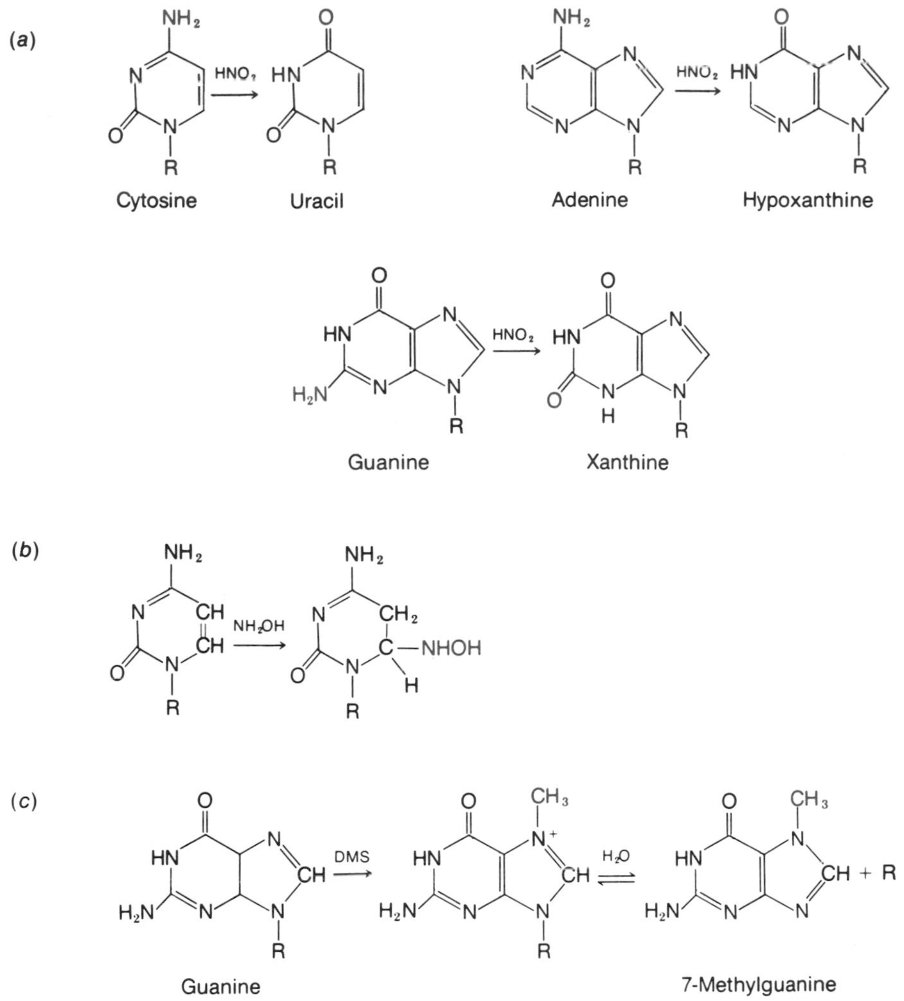

(a)

Cytosine $\xrightarrow{HNO_2}$ Uracil

Adenine $\xrightarrow{HNO_2}$ Hypoxanthine

Guanine $\xrightarrow{HNO_2}$ Xanthine

(b)

$\xrightarrow{NH_2OH}$

(c)

Guanine $\xrightarrow{DMS}$ $\underset{}{\overset{H_2O}{\rightleftharpoons}}$ 7-Methylguanine + R

FIGURE 15.44
Reactions of various mutagens.
(*a*) Deamination by nitrous acid (HNO_2) converts cytosine to uracil, adenine to hypoxanthine, and guanine to xanthine. (*b*) Reaction of bases with hydroxylamine (NH_2OH) as illustrated by the action of this reagent on cytosine. (*c*) Alkylations of guanine by dimethyl sulfate (DMS). The formation of a quaternary nitrogen destabilizes the deoxyriboside bond and releases deoxyribose. Among the effective agents for methylation of the bases are certain nitrosoguanidines such as *N*-methyl-*N'*-nitro-*N*-nitrosoguanidine.

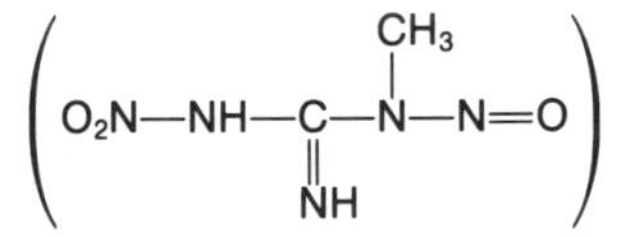

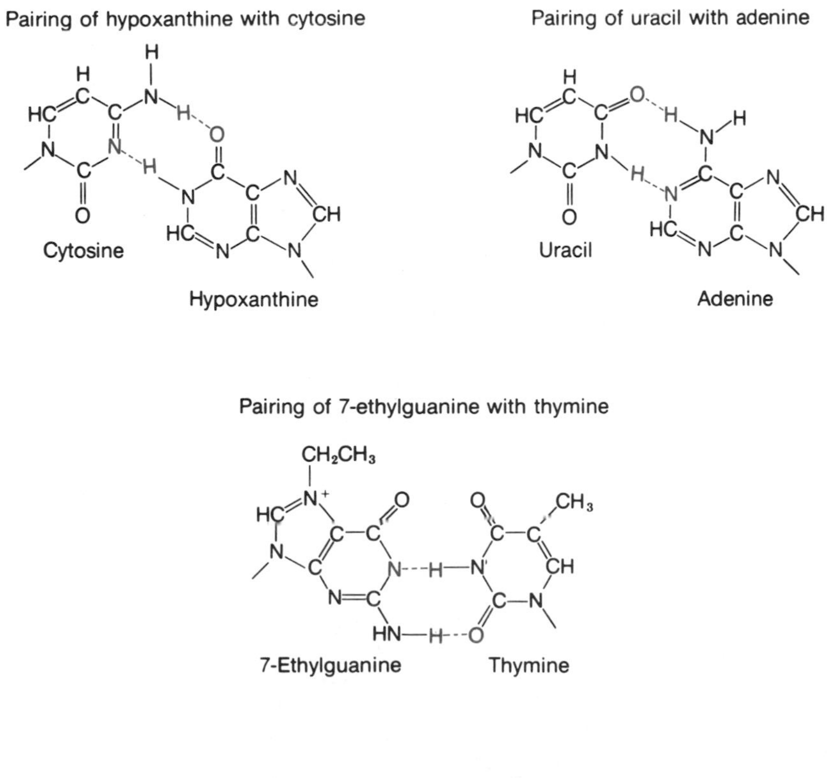

$$\left(O_2N{-}NH{-}\underset{\underset{NH}{\|}}{C}{-}\overset{\overset{CH_3}{|}}{N}{-}N{=}O \right)$$

Pairing of hypoxanthine with cytosine

Cytosine — Hypoxanthine

Pairing of uracil with adenine

Uracil — Adenine

Pairing of 7-ethylguanine with thymine

7-Ethylguanine — Thymine

FIGURE 15.45
Chemical modifications that alter the hydrogen-bonding properties of the bases.
Hypoxanthine, obtained by deamination of adenine, has different hydrogen-bonding properties from adenine, for example, it pairs with cytosine. Similarly, uracil obtained from cytosine, has a different hydrogen-bonding specificity than cytosine and pairs with adenine. Alkylation of guanine modifies the hydrogen-bonding properties of the base.

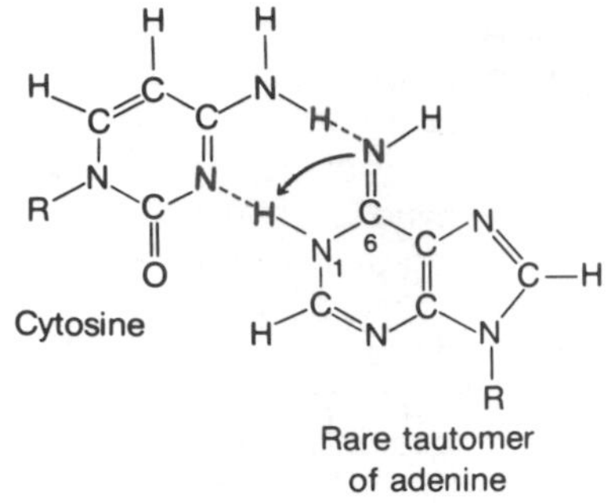

The DNA also undergoes changes as a result of various physical and chemical perturbations in the absence of reactive chemicals. For instance, adenine undergoes hydrolytic deamination to hypoxanthine, and cytosine is spontaneously deaminated to uracil. In addition, purine bases are lost from DNA because of thermal disruption of *N*-glycosyl bonds with deoxyribose.

Radiation Damage

Both UV as well as X-ray irradiation are generally very effective means of producing mutations. The bases normally exist as keto or amino forms in equilibrium, with only minor amounts of the enol or the imino structures. Radiation energy absorbed by the bases tends to shift the equilibrium to the minor forms. The minor forms, however, cannot pair with the normal partners of the bases. For example, the enol form of adenine pairs with cytosine instead of thymine, the normal adenine partner. This atypical base pairing is shown in Figure 15.46. It has been suggested that the existence of increased amounts of the enol forms of the bases at the moment of replication increases the frequency of mutations of the newly synthesized DNA strand because the enol forms select new bases that pair with them rather than the normal hydrogen-bonding partners of the more predominant keto forms.

Exposure of DNA to high-energy radiation (X-rays or γ rays) may also bring about direct modifications in the structure of the bases. Intermediates produced by electron expulsion can be rearranged, leading to the opening of the heterocyclic rings of the bases and the disruption of phosphodiester bonds. In the presence of oxygen additional reactions take place, yielding a variety of oxidation products.

Irradiation by UV light primarily affects the pyrimidines. Activation of the ethylene bond of these bases frequently leads to a photochemical dimerization of two adjacent pyrimidines, as shown in Figure 15.47. Thymine residues are particularly susceptible to this reaction, although cytosine dimers and thymine–cytosine combinations are also produced.

DNA Polymerase Errors

When the appropriate deoxyribonucleotides are available, DNA-dependent DNA polymerases function with a high degree of fidelity. Some

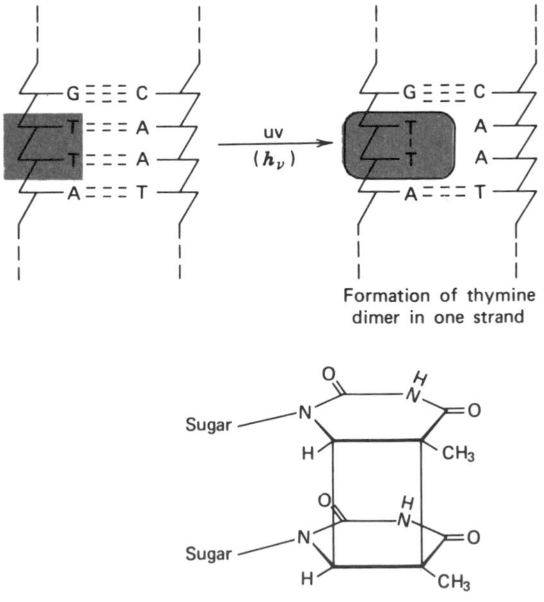

Cytosine

Rare tautomer of adenine

FIGURE 15.46
Base complementarity properties of the minor tautomeric form of adenine.
The hydrogen-bonding properties of adenine are fundamentally changed when adenine acquires the minor tautomeric form. The pairing with cystosine, shown in this figure, is very atypical of the normal properties of this base.

uv ($h\nu$)

Formation of thymine dimer in one strand

Sugar

Sugar

FIGURE 15.47
Dimerization of adjacent pyrimidines in irradiated DNA.
A residue of thymine that is activated by the absorption of UV light can react with a second neighboring thymine and form a thymine dimer.

mutations do occur during DNA replication, but these changes are limited by the high synthetic fidelity of DNA polymerase and the "proofreading" exonuclease properties of this enzyme. The fidelity of DNA replication is further enhanced, at least in procaryotic systems, by an excision repair process known as the **mismatched repair system.** This system recognizes and corrects mismatches in newly replicated DNA by detecting distortions on the outside of the helix that are produced from poor fit between paired noncomplementary bases. Clearly, accurate correction of mismatched bases requires that the mismatched repair system discriminate between preexisting and newly synthesized DNA strands. Such discrimination is feasible because certain adenine residues in DNA (which are part of a reoccurring GATC sequence) are subject to methylation that occurs posttranscriptionally, but not without some delay. Mismatched proofreading is carried out by a multienzyme complex that excises mismatched nucleotides only from newly synthesized strands. The complex identifies these nucleotides by searching for unmethylated adenine residues in the GATC sequences of each strand. The mechanism of excision repair is described in the subsequent section.

In spite of their high degree of fidelity in catalyzing DNA replication, DNA polymerases are unable to distinguish between the normal deoxyribonucleoside triphosphate substrates and other nucleotides with very similar structures. On the basis of the tendency of DNA polymerase to accept, in place of the normal substrates, structural analogs of the common bases, certain mutations can be introduced into DNA by design. For instance, 2-aminopurine, incorporated instead of adenine into a newly synthesized DNA strand, can associate with cytosine and produce an A–T → G–C transition. A somewhat more complex example is provided by 5-bromouracil. This base, in the form of the corresponding deoxynucleoside triphosphate, can be incorporated into a strand in place of thymine. However, the equilibrium between the enol and the keto forms for these two bases allows for the formation of a somewhat higher proportion of the enol form in 5-bromouracil than in thymine. This occurs presumably because of the higher electronegative nature of the bromine atom in comparison to the corresponding methyl group in thymine. Because the enol form of 5-bromouracil pairs with guanine, as shown in Figure 15.48 the substitution of thymine by bromouracil produces an A–T → G–C transition.

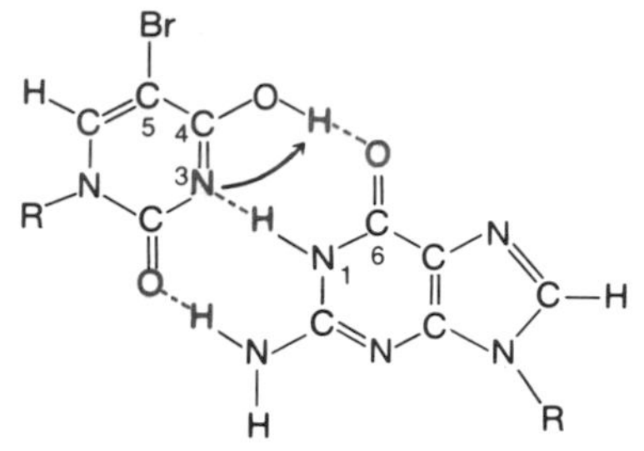

Minor tautomer of 5-bromouracil

Guanine

FIGURE 15.48
Hydrogen-bonding properties of the minor enol form of 5-bromouracil.
The enol form of 5-bromouracil, an analog of thymine, pairs with guanine instead of adenine, the normal partner of thymine.

Stretching of the Double Helix

Certain organic compounds, which are characterized by planar aromatic ring structures of appropriate size and geometry, can be inserted between base pairs in double-stranded DNA. This process is referred to as **intercalation.** During intercalation neighboring base pairs in DNA are separated to allow for the insertion of the intercalating ring system, causing an elongation of the double helix by stretching. The continuity of the base sequences in DNA is disrupted, and the reading of the bases by the DNA polymerase produces a new strand, with an additional base inserted near the site of intercalation. The resulting mutation is referred to as a *frame shift. Acridines, ethidium bromide,* and other intercalators are known to be effective *frame-shift mutagens* (Figure 15.49).

Clin. Corr. 15.2 discusses mutations and the etiology of cancer.

DNA Is Repaired Rather Than Degraded

DNA has the distinction of being the only macromolecule that is repaired rather than degraded. This is so perhaps because changes in the informational content of the genetic bank, which is encoded in the form of nucleo-

CLIN. CORR. 15.2
MUTATIONS AND THE ETIOLOGY OF CANCER

Considerable progress in our understanding of the etiology of cancer has been made in recent years by our ever-increasing realization that long-term exposure to certain chemicals leads to various forms of cancer. Some experts are now suggesting that the great majority of cancers are in fact triggered by environmental factors.

Carcinogenic (cancer-causing) compounds are not only introduced into the environment by the increasing use of new chemicals in industrial applications but are also present in the form of natural products. For instance, the *aflatoxins*, produced by certain molds, and *benz[a]anthracene*, present in cigarette smoke and charcoal broiled foods, are carcinogenic. Some carcinogens act directly, while others, such as benz[*a*]anthracene, must undergo prior hydroxylation by arylhydroxylases, present mainly in the liver, before their carcinogenic potential can be expressed.

Benz[*a*]anthracene

↓

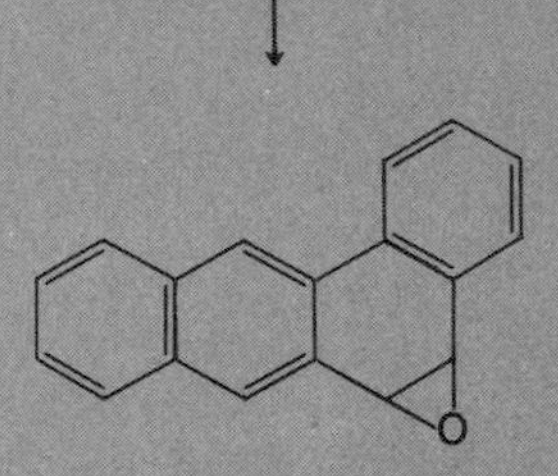

5,6-Epoxide (carcinogenic)

The reactivity of many carcinogenic compounds toward guanine residues results in modification of the guanine structure, usually by alkylation at the N-7 position, as well as in breaks of the phosphodiester bond, events that, upon replication, lead to permanent mutations. Chemicals that produce mutations generally turn out to be carcinogenic and vice versa.

The vulnerability of DNA toward alkylating agents, and other chemicals as well, underscores the concerns expressed today by many scientists about the ever-increasing exposure of our environment to new chemicals. What is particularly distressing is that the carcinogenic potential of new chemicals released into the environment cannot be predicted with confidence even when they appear chemically innocuous toward DNA.

Until recently, tests for carcinogenicity, that is, the ability of a substance to cause cancer, required the use of large numbers of experimental animals to which high doses of the suspected carcinogen were administered over a long period of time. Such tests, which are time consuming as well as expensive, are the only approach still available for testing carcinogenicity directly. Recently, however, a much simpler and much more inexpensive indirect test for carcinogenicity was developed. This test is based on the premise that carcinogenicity and mutagenicity are essentially manifestations of the same underlying phenomenon, the structural modification of DNA. The test measures the rate of mutation that bacteria undergo when exposed to chemicals suspected to be carcinogens. The test is referred to as the Ames test.

A major criticism advanced against the test is that the assumption of an equivalence between mutagenicity and carcinogenicity is not always valid. Because of the unusually large economic implication of labeling a chemical with widespread use as a potential carcinogen, the scrutiny that is often exercised in assessing the reliability of applicable tests for labeling a chemical as a carcinogen is understandable. Yet, certain exceptions notwithstanding, the great majority of chemicals tested has reinforced the view that a good correlation exists between the tendency of a chemical to produce bacterial mutations and animal cancer. Furthermore, even the direct and very costly tests for carcinogenicity, in which large numbers of animals are used, have not completely escaped criticism. The reliability of the test has been questioned because of the relatively large doses of chemicals employed in these tests, doses essential for shortening the long-term chemical exposure of the animals to a practically manageable period of time.

In addition, the necessity of projecting data from animals, usually rodents, to humans has often been used as an argument against the validity of the test.

The enzymes that activate carcinogens are often members of the cytochrome P450 family (Chapter 23). Cytochrome P450's can be induced by noncarcinogenic compounds such as ethanol; hence, alcohol can increase the potential risk of cancer development after exposure to carcinogens.

Ames, B., Durston, W. E., Yamasaki, E., and Lee, F. D. Carcinogens are mutagens: a simple test system combining liver homogenates for activation and bacteria for detection. *Proc. Natl. Acad. Sci. U.S.A.* 70:2281, 1973.

tide sequences in DNA, are essential for effective evolutionary response to environmental change. Fewer than 1 out of 1000 accidental changes result in mutations. The rest are corrected through various processes of DNA repair. Mutation rates can be estimated from the frequency with which new mutants arise in populations, such as fruit flies, or in specific proteins in cells growing in tissue culture. These experiments provide estimates of mutation rates of 1 base pair change per 10^9 base pairs for each cell generation. On this basis, for an average-sized protein, which

TABLE 15.9 DNA Lesions That Require Repair

DNA lesion	Cause
Missing base	Acid and heat remove purines (~10^4 purines per day per cell in mammals)
Altered base	Ionizing radiation; alkylating agents
Incorrect base	Spontaneous deaminations: C → U, A → hypoxanthine
Deletion–insertion	Intercalating agents (e.g., acridine dyes)
Cyclobutyl dimer	UV irradiation
Strand breaks	Ionizing radiation; chemicals (bleomycin)
Cross-linking of strands	Psoralin derivatives (light-activated); mitomycin C (antibiotic)

SOURCE: A. Kornberg, *DNA replication*. San Francisco: W. H. Freeman, 1980, p. 608.

contains about 1000 coding base pairs, a mutation may occur once in 10^6 cell generations.

DNA repair is a high priority process for maintaining cellular function. Germ cells must be protected against high rates of mutation, for preserving the evolution of the species and somatic mutation must be controlled in order to avoid uncontrolled cell growth and disease. Unchecked accumulation of damage would lead to accumulation of nonfunctional proteins and gradual loss of cellular function or the deregulated growth characteristic of malignant cells. Commonly encountered DNA lesions are listed in Table 15.9.

DNA damage is dealt with by a variety of different repair mechanisms. Two principal mechanisms that lead to the restoration of DNA structure are presented here; those that directly reverse the DNA damage and those that lead to the replacement of the damaged DNA section. The latter process, known as *excision repair,* depends on the existence of two complementary DNA strands to the extent that, upon accidental damage of the DNA sequence of one strand, the complementary strand is available to provide the necessary information for accurate repairs. Excision repair is catalyzed by a variety of enzymatic systems that are tailored to different types of damage.

Excision Repair

Excision repair mechanisms are characterized by four distinct sequential steps: Incision, excision, resynthesis, and ligation. Incision is the recognition step in the repair process and is individualized for the specific type of damage present. It is also the rate-controlling step in the process. During the subsequent excision step the damaged DNA section is excised leaving a gap in the DNA strand. In the resynthesis step the gap is filled by DNA polymerase I. This repair enzyme functions in a manner generally similar to the synthetic enzyme DNA polymerase III in that it catalyzes the stepwise addition of nucleotide triphosphates on a 3′-OH primer, which is generated by the preceding incision step. Polymerase I, however, differs from polymerase III in that it is less processive than the synthetic enzyme. Thus, polymerase I tends to dissociate from the 3′-OH primer as soon as it has incorporated 10–12 nucleotides into the growing DNA chain. At this stage the gap is reduced to the size of a single phosphodiester bond. Because of the combined synthetic–nucleolytic action of polymerase I, the nick can move along the strand undergoing repair until it is

NH_2

H_2N $^+N-C_2H_5\ Br^-$

Ethidium bromide

(a)

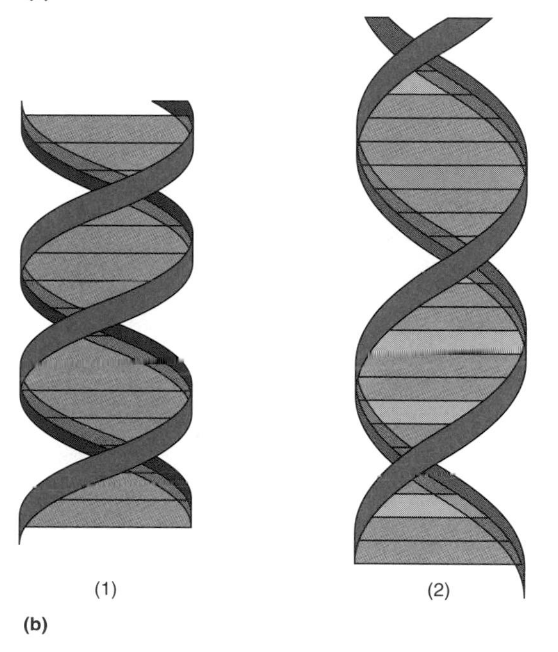

FIGURE 15.49
Intercalation between base pairs of the double helix.
The insertion of the planar ring system of intercalators (*a*) between two adjacent base pairs requires the stretching of the double helix (*b*). During replication this stretching apparently changes the frame used by DNA polymerase for reading the sequence of nucleotides. Consequently, newly synthesized DNA is frame shifted. (*b*-1) The original DNA helix; (*b*-2) the helix with intercalative binding of ligands.
Redrawn based on figure in S. J. Lippard, *Acct. Chem. Res.* 11:211, 1978. Copyright © 1978 by the American Chemical Society.

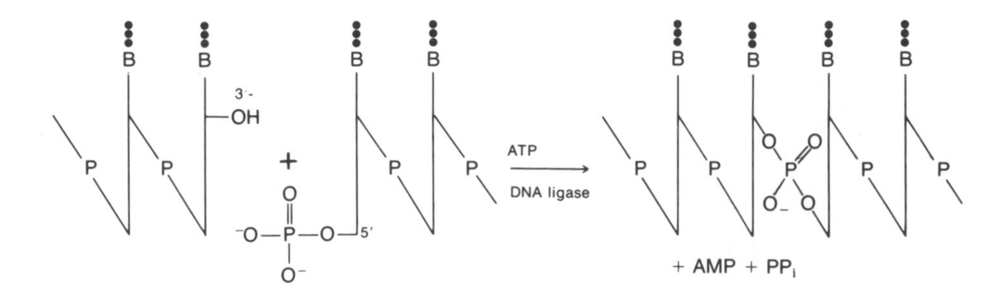

FIGURE 15.50
The action of DNA ligase.
The enzyme catalyzes the joining of polynucleotide strands that are part of a double-stranded DNA. A single phosphodiester bond is formed between the 3′-OH and the 5′-P ends of the two strands. In *E. coli* cells the energy for the formation of the bond is derived from the cleavage of the pyrophosphate bond of NAD^+. In eucaryotic cells and bacteriophage-infected cells energy is provided by the hydrolysis of the α,β-pyrophosphate bond of ATP.

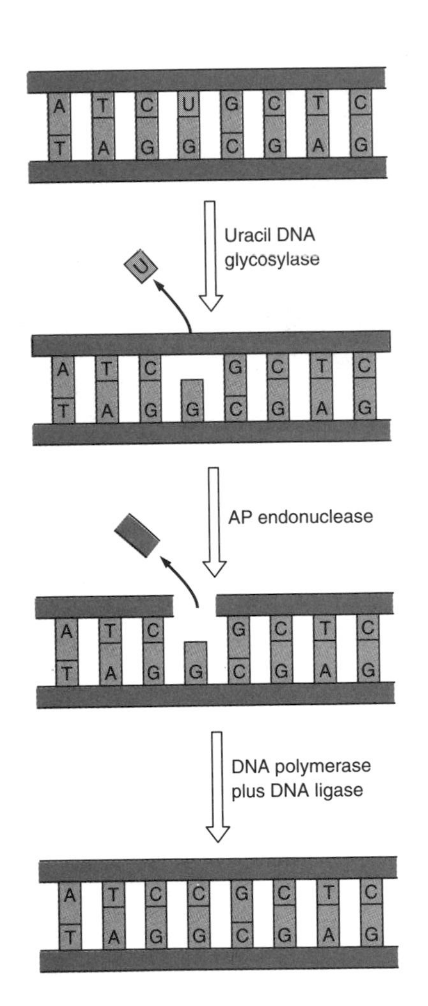

FIGURE 15.51
Uracil DNA glycosylase repair of DNA.
Uracil DNA glycosylase removes uracil, formed for example by accidental spontaneous deamination, by cutting the glycosidic bond, leaving DNA with a missing base. The AP endonuclease subsequently cuts out the sugar–phosphate remnant. The repair is completed by the action of DNA polymerase and ligase.

finally bridged during the ligation step by the action of DNA ligase (Figure 15.50).

The exact nature of the incision step depends on the type of damage that is being repaired. For example, depurination, which leaves behind a deoxyribose residue with its purine stripped away, is catalyzed by the enzyme **apurinic–apyrimidinic (AP) endonuclease.** This enzyme nicks the phosphodiester backbone at the depurinized site and excises the sugar phosphate residue, prior to the restoration of the damaged strand by the action of DNA polymerase I and ligase. Depyrimidination does not occur to an appreciable extent because the pyrimidine–glycoside bond is much more stable than the purine–glycoside bond. As discussed earlier exposure to various chemicals can lead to modification of purines, including methylation and ring opening. Ring opening may also result from exposure to ionizing radiation. In addition, the amino groups of cytosine, adenine, and guanine are susceptible to elimination. These deaminated bases are hydrolytically removed by enzymes referred to as **DNA glycosylases,** as illustrated using the removal of deaminated cytosine (i.e., uracil) by the enzyme uracil DNA glycosylase in Figure 15.51. This enzyme first removes the damaged cytosine producing a deoxyribose residue with the base missing. This structure is subsequently acted upon by AP endonuclease, with the resulting single-stranded gap filled and sealed.

A third type of incision is used for DNA that is damaged in a way that produces a ''bulky'' lesion, which occurs when DNA interacts with polycyclic aromatic hydrocarbons, such as benzo(a)pyrenes and dialyklbenzathracenes or the UV light-induced dimerization of adjacent pyrimidines. These lesions are recognized by a multienzymatic complex by virtue of their bulk rather than the presence of a specific base that has been modified. Once the lesion has been located, an endonuclease activity cleaves the modified strand on both sides of the distortion and the entire lesion is excised. Excision repair of this type occurs in *E. coli* for the removal of pyrimidine dimers (Figure 15.52). The repair in initiated by the recognition of the distortion of the DNA by an endonuclease system consisting of the products of three *E. coli* genes *UvrA*, *UvrB*, and *UvrC*. A tetramer consisting of two *UvrA* and two *UvrB* proteins, which is formed on DNA during a series of preincision steps, ''melts'' the DNA locally at the expense of ATP and locates the bulky lesion. The complex is subsequently subjected to incision at both sides of the bulky lesion by

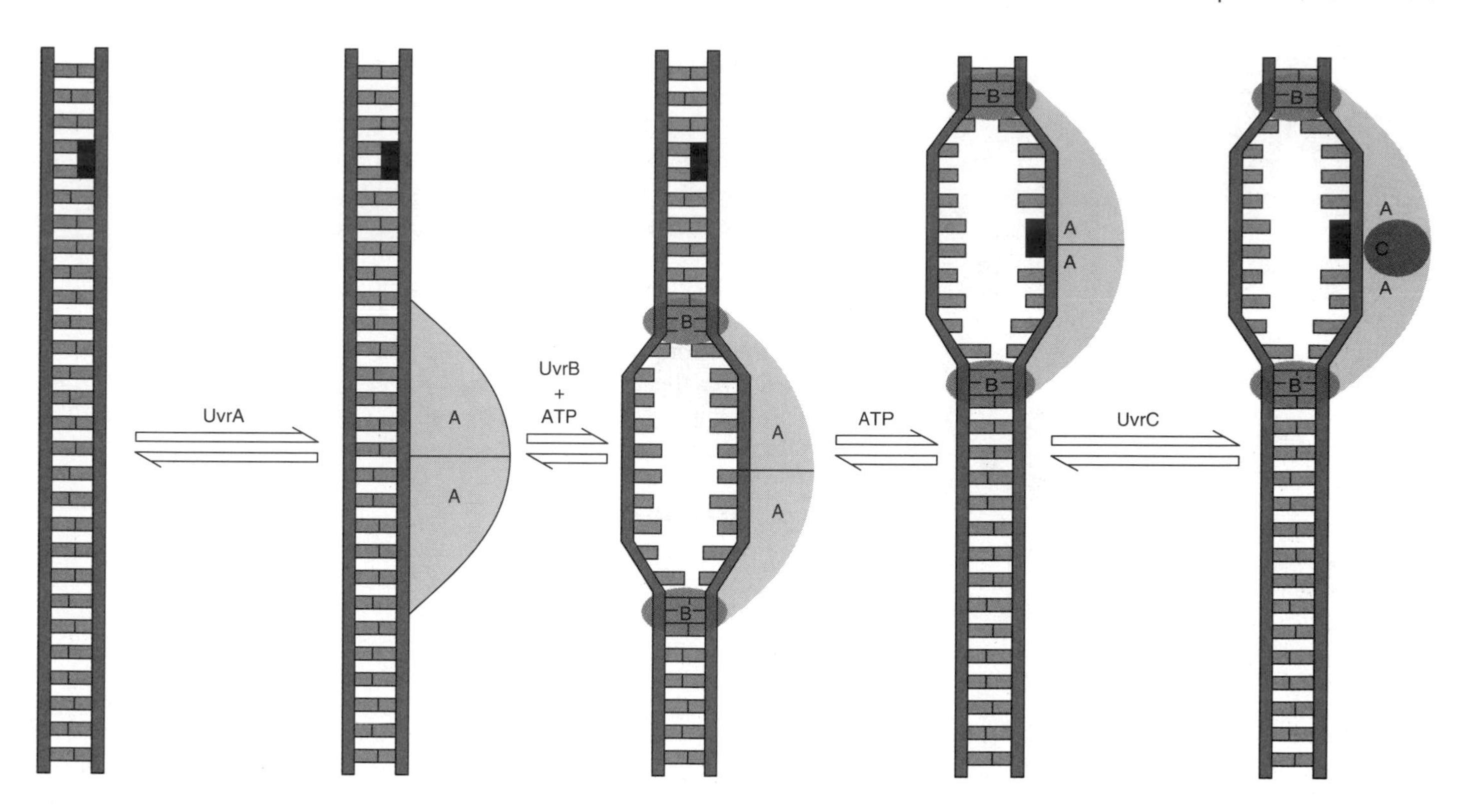

FIGURE 15.52
The preincision model for DNA excision repair.
According to this model a dimer of the product of UvrA initiates ATP dependent topological unwinding of DNA. The UvrA dimer subsequently associates with the UvrB product giving a UvrAB complex. The UvrAB complex catalyzes its own translocation along the DNA at the expense of ATP, until it reaches the damaged site. Incision at this site is catalyzed by the product of the UvrC gene.

UvrC protein and releases an oligonucleotide consisting of 12–13 residues that includes the pyrimidine dimer. For the remainder of the repair *E. coli* makes use of the protein product of a fourth gene, *UvrD,* which by virtue of its helicase activity unwinds and releases the oligonucleotide that was excised by *UvrC* protein. The repair is completed by polymerase I and ligase.

A second less complicated endonucleolytic mechanism for the excision of pyrimidine dimers has been found to operate in the UV-resistant organism *M. luteus* and in T7 phage infected *E. coli.* This process, illustrated in Figure 15.53, depends on the action of a *N-glycosylase* that acts on the pyrimidine dimer by nicking the *N*-glycoside bond that connects one of the dimeric pyrimidines to its polynucleotide strand. The DNA must subsequently be nicked on both sides of the nucleotide residue that carries the dimeric pyrimidine by *AP endonuclease* so that the dimeric residue can be removed. The glycosylase-AP endonuclease activity is present in a single protein induced by UV irradiation. The single nucleotidyl gap that results from the removal of the dimer can then be filled by DNA polymerase I.

Excision repair can also remove cross-links between complementary DNA strands, such as those introduced by the mustards and drugs used in cancer therapy (i.e., mitomycin D and platinum complexes). In such cases error-free repair is not possible if the cross-link extends across directly opposing bases. Clin. Corr. 15.3 discusses defects in DNA repair that are associated with human disease.

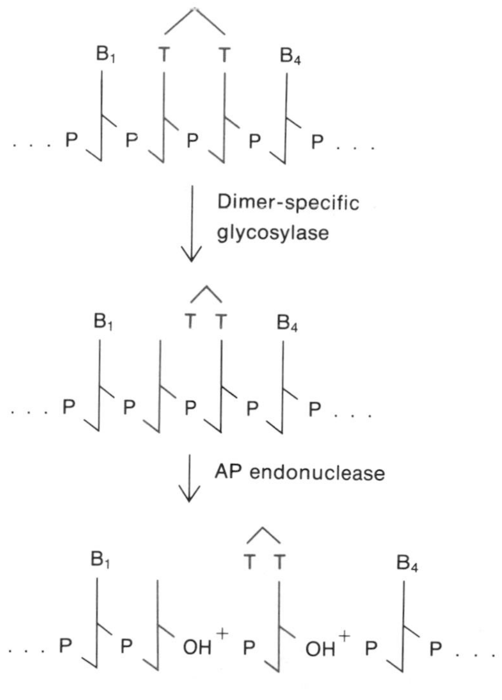

B_1 T T B_4
. . . P P P P P . . .
Dimer-specific glycosylase
B_1 T T B_4
. . . P P P P P . . .
AP endonuclease
B_1 T T B_4
. . . P P OH^+ P OH^+ P P . . .

FIGURE 15.53
DNA repair by the combined action of *N*-glycosylase and AP endonuclease.
Repair is initiated by a cleavage of an *N*-glycosidic bond connecting the thymine dimer with ribose by a dimer-specific glycosylase, followed by cleavage of phosphodiester bonds on both sides of the dimeric thymidine residue by AP endonuclease.

CLIN. CORR. 15.3 DISEASES OF DNA REPAIR AND SUSCEPTIBILITY TO CANCER

Defects in DNA repair may lead to a number of diseases including *xeroderma pigmentosum, ataxia telangiectasia, hereditary retinoblastoma* and *Fanconi's anemia. Xeroderma pigmentosum* is the best understood of these conditions. Those suffering from this affliction are particularly sensitive to the effects of sunlight. Exposure to the sun creates severe skin reactions that range initially from excessive freckling and skin ulceration to the eventual development of skin cancers. Some forms of the disease are also accompanied by various neurological abnormalities.

Normal mammalian cells can carry out excision repair of DNA. In contrast, skin cells from *xeroderma pigmentosum* patients are unable to repair the DNA damage produced by UV light. This defect originates from a reduced effectiveness of the first step of the repair mechanism, that is, the step at which an endonuclease nicks the DNA undergoing repair near a pyrimidine dimer. It has been suggested, based on indirect evidence, that in *xeroderma pigmentosum* patients this enzyme is defective. There are at least nine complementation groups of *xeroderma pigmentosum* patients. At present, the exact genetic basis for defective DNA repair is unknown.

One particularly intriguing aspect of diseases related to defective DNA repair is their possible relationship to carcinogenesis. Although these diseases are rare autosomal recessive conditions, the carriers of the defective genes are relatively common; carriers of *xeroderma pigmentosum* gene account for as much as 1% of the general population. An interesting finding is that not only are those who are afflicted by *xeroderma pigmentosum* susceptible to cancer but the carriers of the gene have a higher incidence of skin cancer. These and other similar findings suggest the presence of some subtle defect in DNA repair among the carriers of these genes. Further research on the predisposition of these carriers to cancer might provide some clues regarding the mechanism of carcinogenesis for some types of cancer.

Cleaver, J. E. *Xeroderma pigmentosum:* a human disease in which an initial stage of DNA repair is defective. Proc. Natl. Acad. Sci. U.S.A. 63:428, 1969.

Mechanisms That Reverse Damage

In addition to the two mechanisms of repair that respond to the presence of pyrimidine dimers described in the preceding section, the formation of dimers can be directly reversed by the action of light that leads to the regeneration of the base monomers. This photoreversal is catalyzed by **deoxyribodipyrimidine photolyase** that disrupts the covalent bonds that hold together the pyrimidine molecules in the dimer. Photolyases are activated by light in the range of 300–600 nm.

Another example of direct reversal of damage is the removal of a methyl or ethyl group from the 6 position of the enol form of a guanine residue in DNA; these alkyl groups can be removed, and the normal structure of guanine reestablished, by the action of a specific protein that accepts alkyl groups and in the process itself becomes alkylated.

Error-Prone (SOS) Repair

Error-prone repair is a process rather than a true repair mechanism. It is activated in response to an emergency, generated from severe DNA damage. The signal that activates this process, at least in the case of *E. coli,* is a block in DNA replication that results from damage in the double helix. More than a dozen different genes that produce proteins that function in DNA repair are activated. SOS repair allows DNA replication to take place in a stop-gap manner until the damage is permanently repaired by another mechanism. A daughter strand is synthesized by skipping over the damaged base. After synthesis the daughter strand is found to be missing a base that would be normally present across from the damaged base. The missing base is added *postreplicatively* after the parental strand, and with the damaged base still present, is separated from the daughter strand.

Another distinct, and perhaps unexpected, outcome of the induction of proteins involved in SOS repair, is the increase noted in the mutation rate of DNA. This increase in mutations results from an increase in the number of errors made in copying DNA sequences. Higher mutation rates certainly do not contribute positively to the repair process as such, but in some instances they are presumably advantageous for long-term survival, to the degree that they increase genetic diversity.

Eucaryotic DNA Repair

The process of DNA repair can be described in greater detail in bacteria than in eucaryotes, partly because in bacteria the genetic systems coding for and regulating these processes are better understood and the enzymes are easier to isolate and study.

Mammalian repair enzymes catalyzing relatively uncomplicated steps have been found to act in a manner analogous to their bacterial counterparts. Indirect evidence suggests that even in complicated mammalian repair systems, the analogy with known procaryotic repair mechanisms may be strong.

15.7 DNA REPLICATION

Complementary Strands Are Basic to the Mechanism of Replication

From the moment the double-stranded structure of DNA was proposed, it was apparent that this structure could serve as the basis of a mechanism for DNA replication. The complementary structure of the strands was immediately perceived as a characteristic, which, in principle, permitted each one of the strands to serve as a template for the synthesis of a new strand identical to the other strand, as suggested in Figure 15.54. A number of pathways could be easily visualized that lead to the synthesis of two new double-stranded helices, identical to one another and to the maternal double helix.

In the more than thirty years since the double helix was proposed, the correctness of this overall scheme of replication has been solidly established. Even bacteriophages, which contain single-stranded instead of double-stranded DNA, have been shown to convert their DNA to a double-stranded form before replication. Our expanding knowledge of the character of DNA replication has also revealed that the simplicity of the basic scheme conceals, in fact, a rather complex set of more intricate substeps. A multiplicity of enzymes and protein factors participate in the process of replication. Before the synthesis of a DNA molecule can be brought to successful completion, the enzymes involved in replication must deal with a variety of topological problems.

Complexities originate partly because the DNA-dependent DNA polymerase can synthesize new strands by operating *along the 5′ → 3′ direction only,* and therefore it is unable to elongate the two *antiparallel* strands of the helix in the *same macroscopic direction*. Also, the replication of DNA cannot proceed unless the complementary strands are separated at an early stage of the synthesis. Separation requires the commitment of energy for disrupting the thermodynamically favorable double-helical arrangement and the unwinding of a highly twisted double helix at extremely rapid rates. As if these difficulties were not enough, double-stranded DNA is normally a topologically closed domain which, unless properly modified, will not tolerate strand unwinding to any appreciable degree. Obviously, these multiple difficulties are enzymatically resolved before the replication of DNA can take place.

In order to ease the complexity of describing the process of DNA replication, various distinct elements of the overall process will be described first. The presentation of a unique model with universal applicability is not feasible because variations in the mode of replication occur among different species.

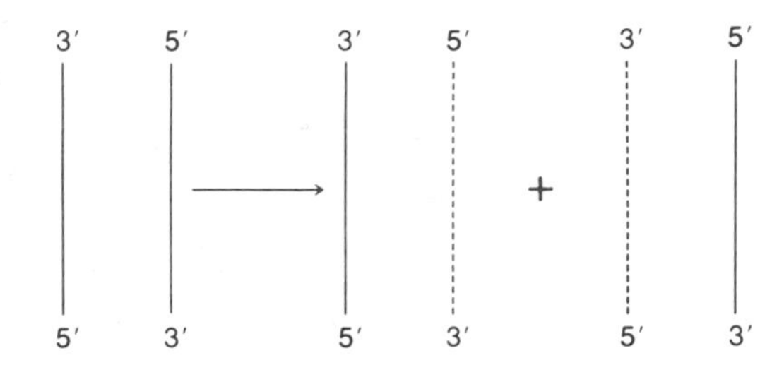

FIGURE 15.54
Each DNA strand serves as template for the synthesis of a new complementary strand. Replication of DNA proceeds by a mechanism in which a new DNA strand (indicated by a dashed line) is synthesized that matches each original strand (shown by solid lines).

Replication Is Semiconservative

The concept that DNA strands are separable and that new strands, complementary to the preexisting strands, can be assembled from free nucleotides on each separate strand, is not new. In a macroscopic sense three possibilities by which information transfer could take place during replication were initially visualized as indicated in Figure 15.55. **Conservative** replication could in principle yield a product consisting of a double helix

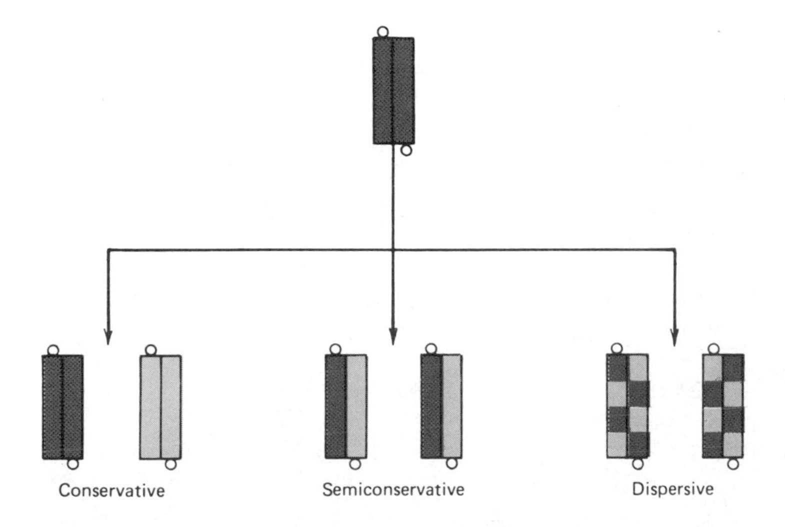

FIGURE 15.55
Three possible types of DNA replication. Replication has been shown to occur exclusively according to the semiconservative model, that is, after each round of replication one of the parental strands is maintained intact, and it combines with one newly synthesized complementary strand.

of the original two strands and a daughter DNA consisting of completely newly synthesized chains. A second possibility, labeled **dispersive,** would have resulted if the nucleotides of the parental DNA were randomly scattered along the strands of the newly synthesized DNA. The synthesis of DNA eventually proved to be a **semiconservative** process. After each round of replication, the structure of paternal DNA is found to preserve one of its own original strands combined with a newly synthesized complementary polynucleotide.

The semiconservative nature of replication was elegantly suggested by a classic experiment that allowed the physical separation and identification of the paternal and the newly synthesized strands. For this experiment *E. coli* was grown in a medium containing [^{15}N]ammonium chloride as the exclusive source of nitrogen. Several cell divisions were allowed to occur during which the naturally occurring ^{14}N in the DNA of *E. coli* was, for all practical purposes, replaced by the heavier ^{15}N isotope. The ^{14}N-containing nutrient was then added, and cells were removed at appropriate intervals. The DNA of these cells was extracted, and the ratios of ^{14}N to ^{15}N content were determined by equilibrium density gradient centrifugation. The separation between ^{14}N and ^{15}N DNA was achieved based on the lower density of DNA, which contained the lighter isotope. In subsequent experiments, the newly synthesized DNA was thermally denatured and the individual strands were completely separated. The results, shown in Figure 15.56, demonstrated that daughter DNA molecules consisted of two strands with different densities, corresponding to the densities of single-stranded polynucleotides containing exclusively ^{14}N or ^{15}N.

Clearly, the synthetic activity of DNA polymerase makes it possible for the enzyme to synthesize new complementary DNA strands by using in turn each parental DNA strand as template.

A Primer Is Required

The semiconservative nature of DNA replication requires that each strand serve as a DNA polymerase template for the synthesis of a new complementary strand. The polymerase that catalyzes the elongation of DNA polynucleotides is polymerase III (Table 15.7), as distinguished from polymerase I, which is primarily a repair enzyme. Polymerase III is

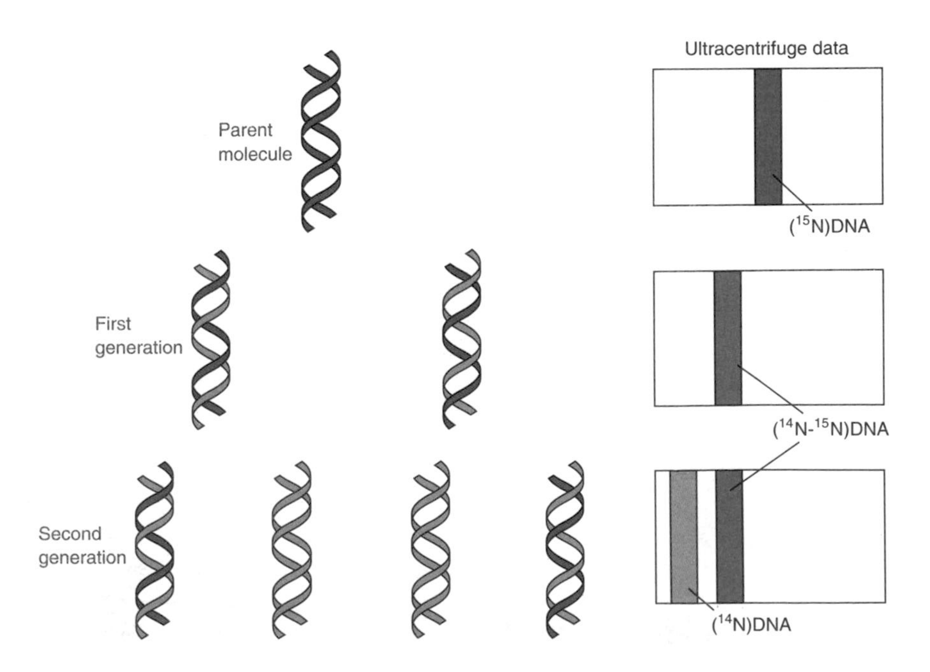

FIGURE 15.56
Semiconservative replication of DNA.
Schematic representation of the experiment of Meselson and Stahl that demonstrated semiconservtive replication of DNA. This model of replication requires that, if the parent molecule (shown in black) contains ^{15}N, each of the molecules produced during the first generation contain ^{15}N in one strand and ^{14}N in the other. Furthermore, in the second generation two molecules must contain only ^{14}N, and two molecules must contain equal amounts of ^{14}N and ^{15}N. The results of separating DNA molecules from successive generations, shown on the right are consistent with this model.

TABLE 15.10 RNA Primers

Replicating system	RNA oligonucleotide[a]
Bacteriophage T4	pppAC $(N)_3$
Bacteriophage T7	pppACCA
	pppACCC
Mouse polyoma virus	pppA $(N)_9$
	pppG $(N)_9$
Lymphoblastoid cells	pppA $(N)_8$
	pppG $(N)_8$

[a] N = any ribonucleotide. The primer lengths for the mouse polyoma virus and the animal cells are averages.

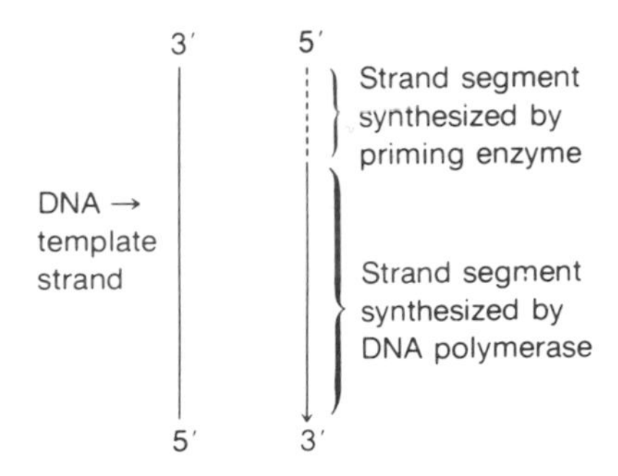

FIGURE 15.57
Synthesis of primer for DNA replication.
The primer (dashed line) is synthesized by specific enzymes. The existence of a primer permits new DNA (solid line) to be synthesized after which the primer is excised.

an ATP-dependent enzyme, unable to assemble the first few nucleotides of a new strand and needs a primer that varies in size between a few nucleotides in procaryotes (1–5) to about 10 nucleotides in animal cells. With few exceptions, the primer is an oligonucleotide synthesized by other enzymes, as indicated in Figure 15.57.

Primers are formed by enzymes known as **primases.** In some bacterial systems and phages, the priming enzyme has activity characteristic of an RNA polymerase because the ribonucleotides condense to form the primer. In other systems the primase does not discriminate between 5′-ribonucleotides and 5′-deoxyribonucleotides. Some enzymes that catalyze the synthesis of primers act exclusively as primases while others contain additional enzymatic activities, as is the case with phage T7 primase that also contains helicase activity. Helicase functions in the separation of the two DNA strands prior to replication. Once the primers have been synthesized, the DNA polymerase can move in and take over the process of synthesis. It is not clear what signal causes a switchover from primase to DNA polymerase, although it has been suggested that a ribonuclease (RNaseH) is involved. The RNaseH activity appears to be specifically directed at RNA that is hydrogen bonded to a DNA strand.

Although primers are almost invariably short RNA or RNA-like segments (Table 15.10) RNA priming is not used universally. In the "rolling circle" replication mechanism of DNA a 3′-OH primer is generated by endonuclease digestion of parental DNA, and with parvoviruses a 3′-OH primer is generated by the folding back of an existing 3′ terminus. A single deoxyribonucleotide can serve as primer in adenovirus. Such a nucleotide, with its 3′-OH terminus free, is attached to the end of a template strand through a virus-encoded specific protein (Figure 15.58).

FIGURE 15.58
An unusual primer used in the replication of adenovirus DNA.
This primer consists of a single nucleotide attached, by its 5′-terminal phosphate, to the serine residue of a protein. Adenovirus DNA is synthesized by the extension of the 3′ terminus of this nucleotide.

Both Strands of DNA Serve as Templates Concurrently

In the preceding section, the events leading to the synthesis of DNA by DNA polymerase were examined and attention was directed to one of the two parental DNA strands used as template. In fact, synthetic events occur at both strands almost concurrently. This would appear to generate some problems of geometry. Specifically, if a single initiation site is considered, and if the synthesis is assumed to continue until each template is completely copied, the result of the synthesis would be the creation of *two* new double-stranded molecules. Examination of Figure 15.59 indicates that at least in the case of linear double-stranded DNA, none of these two hypothetical DNA molecules would be identical to the parental DNA.

Such an outcome is not in agreement with the actual course of DNA replication. The discrepancy can be accounted for because it is recognized that the microscopic synthesis of the new strands does not proceed uninterrupted. In fact, the synthesis occurs in a discontinuous fashion and

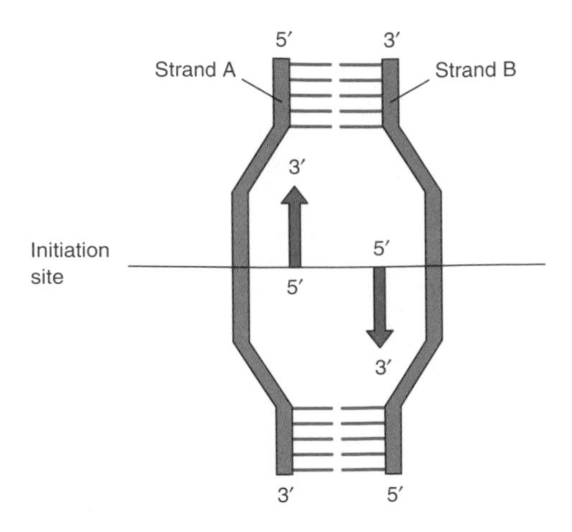

FIGURE 15.59
Both DNA strands serve as templates for DNA synthesis.
Each DNA must serve as a template for DNA synthesis. The new DNA can be synthesized only in the 5′ → 3′ direction. For these reasons if only a single initiation site were considered, the result of the synthesis would be the formation of two new nonidentical double-stranded DNA molecules (one above and one below the initiation origin). Also, the upper part of strand A and the lower part of strand B could not have been used as templates. More than a single initiation site is involved.

in a manner that permits the assembly of the synthesized polynucleotide portions into appropriate complete DNA strands.

Synthesis Is Discontinuous

Examination of the overall process of DNA synthesis should now be expanded past the immediate vicinity of initiation and encompass a larger section of DNA. The attention of the reader should be focused on only one of the two parts of DNA that would be generated if the macromolecule were divided at the site of chain initiation, as indicated in Figure 15.60. In most instances, the synthesis is bidirectional, which means that the synthetic events occurring at the part of the molecule indicated by solid lines are of the same general nature as those occurring on the other site with dashed lines.

A prerequisite for the semiconservative mechanism of replication is that the two complementary strands of DNA gradually separate as the synthesis of new strands takes place. The mechanics of this separation is addressed later, but it may be apparent that as a result of separating the strands at an interior position, two topologically equivalent forks are created at the point of diversion of the two strands.

The observation to be emphasized presently is that DNA polymerase acts in a discontinuous manner, that is, along each DNA molecule there are numerous points at which primers are formed. How these points are selected is not known in most cases; in the case of bacteriophage T7, primosomes appear to recognize two tetranucleotide sequences, TGGT and GGGT. These signals are recognized by a group of polypeptides known as **prepriming proteins.** Once a site for primer initiation has been recognized, **single-strand binding proteins (SSB),** which preferentially interact with single-stranded polynucleotides, are displaced and the primase lays down a primer. After promoting primer initiation at one point, prepriming proteins move along the template strand in order to synthesize the adjacent primer. At each one of these locations, DNA polymerase III makes use of the assembled primers for the synthesis of DNA. When DNA polymerase reaches the end of the single-stranded template, it comes upon a primer annealed to the template. The polymerase can overcome this hurdle by sliding over the intervening double-stranded DNA–RNA hybrid and resuming replication at the 3′ end of this new primer.

The segments built by DNA polymerase upon each primer, which are known as **precursor (Okazaki) fragments** or **nascent DNA,** vary in size

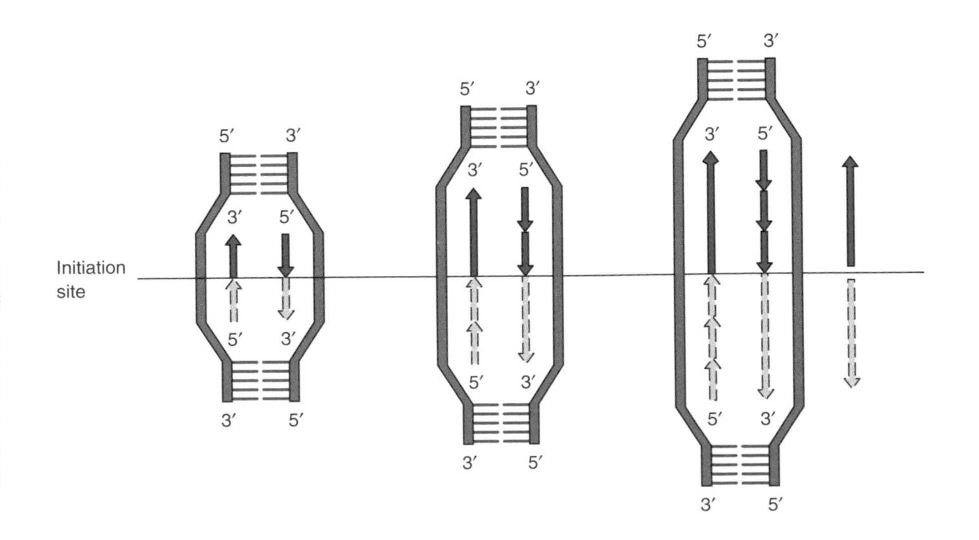

FIGURE 15.60
Discontinuous synthesis of DNA.
In this figure emphasis is placed on the synthetic events occurring at only one side of the initiation site (solid line). The two complementary strands of DNA separate as the discontinuous synthesis of small DNA segments takes place on both strands. After excision of the primers, the excised parts are repaired, and the segments are jointed together. Although the segments are clearly synthesized in opposite directions on the two strands, the overall macroscopic impression is that the DNA grows in the single direction suggested by the solid arrow on the right.

from 100 to 200 deoxyribonucleotides in the case of eucaryotes to about 10 times as long in the case of bacteria. Once the small segments of the new DNA strands are synthesized on both strands of a fork (upper part of Figure 15.60), the fork opens up further, and the same process of synthesis is repeated. Shortly after synthesis, the primer portions of the Okazaki fragments are excised by the $5' \rightarrow 3'$ exonuclease activity of DNA polymerase I, which serves both as an exonuclease as well as a repair enzyme.

This discontinuous mechanism compensates for the inability of DNA polymerase to synthesize strands in the $3' \rightarrow 5'$ direction. By synthesizing portions of DNA strands only in the $5' \rightarrow 3'$ direction on both antiparallel strands of the parental DNA, the polymerase is able to produce *the illusion that both strands are concurrently elongated in the same macroscopic direction*. In Figure 15.60 this direction is indicated by a large solid arrow. It should be noted that the first strand synthesized, often referred to as the **leading strand,** is synthesized *continuously*. It is the second strand, the **lagging strand,** *that must be synthesized discontinuously*.

Macroscopic Synthesis Is as a Rule Bidirectional

Examination of Figure 15.60 indicates that at the site of initiation of DNA synthesis two identical forks are created. Therefore, two possibilities exist for the synthesis of DNA: the process may occur at only one fork and proceed in a single direction, as shown by the thick solid arrow, or alternatively it may occur at both forks and in both directions away from the starting point. The events occurring in the forks located below the starting line are simply a mirror image repetition of what occurs in the fork that is located above the line. **Bidirectional replication** is the exclusive mechanism of DNA synthesis in animal cells. There are exceptions to the rule of bidirectionality in a small number of phages and plasmids that replicate unidirectionally. In the case of a small linear chromosome (e.g., bacteriophage λ) each fork moves along, synthesizing new DNA, until the end of the chromosome is reached. In a circular chromosome (e.g., *E. coli*) the two forks proceed in opposite directions until they meet at a predetermined site on the other side of the chromosome, as depicted in Figure 15.61. As the two forks meet, a new copy of the parental DNA is completed and released. The average rate at which each fork moves during replication is of the order of 60,000 bases per minute at 37°C. Upon completion, new DNA is released by the action of topoisomerase II as presented in Figure 15.62.

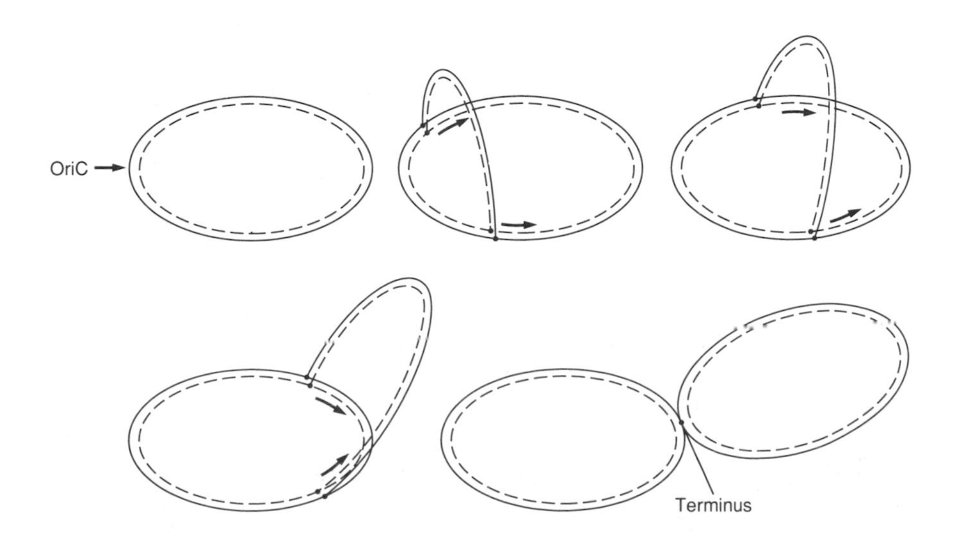

FIGURE 15.61
Bidirectional replication of a circular chromosome.
Replication starts at a fixed origin and proceeds at a constant rate in opposite directions until the two replication forks meet. Newly synthesized strands are indicated by dashed lines. After DNA synthesis is completed the two newly synthesized circular DNA molecules are separated by the action of topoisomerases.

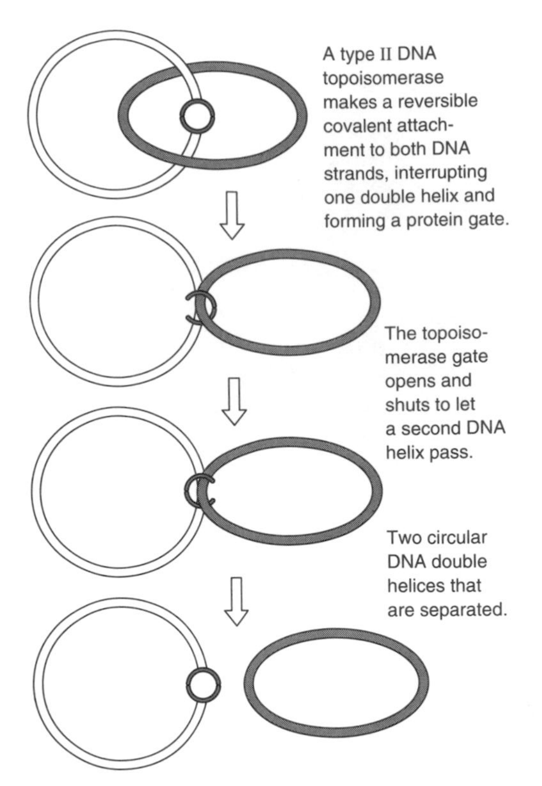

FIGURE 15.62
The function of topoisomerases II in separating interlocked DNA double helixes.
Topoisomerase II attaches to both strands of DNA through reversible covalent bonds, thus forming an interrupted double helix with a topoisomerase "gate." A second DNA helix can pass through the portal using an "open and shut the gate" mechanism and leading to two separated DNA molecules. After the separation of the two molecules the topoisomerase dissociates from the DNA.

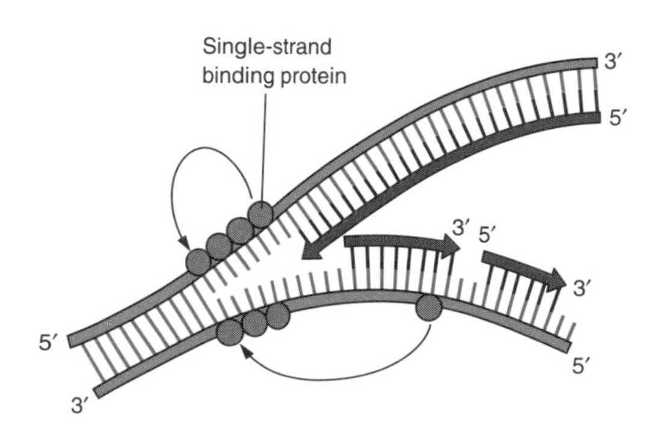

FIGURE 15.63
Function of SSB protein in replication.
Single-strand binding proteins operate at regions of single strandedness including the replication fork. The protein may follow the movement of the fork using the hypothetical scheme shown in this figure.
Redrawn with permission from A. Kornberg, *DNA replication*. San Francisco: W. H. Freeman, 1982.

Strands Must Unwind and Separate

Separation of the strands of the parental DNA prior to the synthesis of new strands is a requirement because the bases of each template must be made accessible to the complementary deoxyribonucleotides from which the new strands are constructed. The overall process of separation consists of a number of enzymatically catalyzed, coordinated steps, including the local unwinding of the helix, and the nicking and rejoining of the strands necessary for the continuation of the unwinding process. Once the strands are unwound, they must be kept separate so that they can operate freely as templates.

Helicase. The fact that most DNAs are circular supertwisted molecules facilitates the unwinding of the helix. They contain a net deficit of helical turns in comparison to the corresponding "relaxed" molecules, which introduces into the double helix a tendency toward partial unwinding. This tendency is even more pronounced in regions richer in A-T pairs, which are intrinsically less stable than G-C pairs. Therefore A-T-rich regions are susceptible to local melting, that is, unwinding and partial separation of the strands of the double helix.

The cell resorts to the services of specialized proteins to accomplish the rapid orderly unwinding of the strands. These proteins separate DNA strands in advance of the moving replication fork; for the *E. coli* system they are referred to as **helicase II** and **rep protein.** Helicases move unidirectionally along one or the other strand of the DNA and separate the strands in advance of replication. They destabilize the interaction between complementary base pairs at the expense of ATP.

Binding Proteins. Once the strands have been separated, the single-stranded regions are stabilized by specific proteins, the SSB proteins in both procaryotic and eucaryotic cells. The DNA single strands are covered by the binding proteins because of the high stability of the complexes formed between these proteins and single-stranded DNA regions. As the helicase moves in advance of the replication fork, binding proteins go on and off the DNA, with protein molecules that are displaced from one site reassociating with another (Figure 15.63). Binding proteins do not consume ATP and do not exhibit any enzymatic activities. Their role is only to keep the strands apart long enough for the priming process to occur.

Topoisomerases. Even after the local unwinding and separation of the strands is achieved, other practical problems must still be solved in order for the replication to proceed unimpeded. For the *E. coli* DNA it may be calculated that the parental double helix must unwind at a rate of about 6000 turns per minute. These high rates would generate serious difficulties if strands were to separate over an appreciable length of DNA. The large free energy requirements of bringing about the unwinding of large regions of DNA can, however, be reduced to manageable levels by the nicking of one or both of the DNA strands near the replicating fork. Since the fork is a moving entity, the nicking must be visualized as a reversible cut-and-rejoin process, which moves along with the fork.

Nicking is indispensable for a topological reason as well. Unwinding at one of the two forks requires that the parental double helix rotates in the opposite direction to that necessary for the unwinding of the opposite fork. Furthermore, even if only one of the forks is considered and the other is ignored, the topological problem still remains unresolved. In the absence of a nick as the unwinding at one of the forks would progress, an increasing number of positive supercoils would have to be introduced into the double helix. Once the limit of the helix to accommodate the supercoils is reached, the unwinding and the replication would have to stop.

The above topological restraints can be overcome if DNA is maintained during replication in the *negative superhelical* form. This form could serve as a "sink" for the positive supercoils that can potentially be generated during replication. In *E. coli, gyrase,* a topoisomerase type II, induces the formation of negative supercoils at the expense of ATP. Topoisomerases type I, on the other hand, tend to relax supercoiled DNA. The in vivo superhelicity of DNA may be negatively regulated through a balance between topoisomerases of types I and II; that is, a diminishment of topoisomerase II activity may bring about a decrease in the amount of negative superhelicity that can be created, whereas an inhibition of topoisomerase I activity may increase it. During replication the linking number between the parental strands decreases from a large value at the beginning of replication to zero at the end of a complete round of DNA synthesis.

E. coli Provides the Basic Model for the Replication of DNA

Our understanding of the mechanism of DNA replication is still far from complete. Extensive studies in *E. coli* and its phages have permitted the proposal of a replication model that depends on the action of a large number of proteins listed in Table 15.11. With the specific exceptions noted in the following two sections, this model may also be viewed as a basic scheme for DNA replication in other cells.

Synthesis of DNA begins at a specific site of the chromosome referred to as the replication origin, which in *E. coli* is referred to as *OriC* (Figure 15.64). The *OriC* consists of a sequence of 245 base pairs that contains four sites (nucleotide 9-mers) at which *dnaA* can bind and initiate the stepwise assembly of all the proteins and enzymes necessary to carry out replication (Figure 15.65). This final assembly is called a **replisome.**

The formation of a replisome begins with the binding of the dnaA molecule at each one of the four sites present at the replication origin. Several additional dnaA molecules are subsequently added via a cooperative binding process and in a manner that promotes the opening of the DNA strands in an AT-rich region adjacent to the origin. Finally dnaA, with the aid of dnaC, adds dnaB in the complex which, by virtue of its helicase

TABLE 15.11 The Components of the Replisome

Protein	Function
SSB	Single-strand binding
Protein i (dnaT) Protein n Protein n′ Protein n″	Primosome assembly and function
dnaG	Primase (Primer synthesis)
Pol III holoenzyme	Processive chain elongation
Pol I	Gap filling, and primer excision
Ligase	Ligation
Gyrase	Supercoiling
gyrA	
gyrB	
rep	Helicase
Helicase II	
dnaB	Helicase
dnaA dnaC	Origin of replication

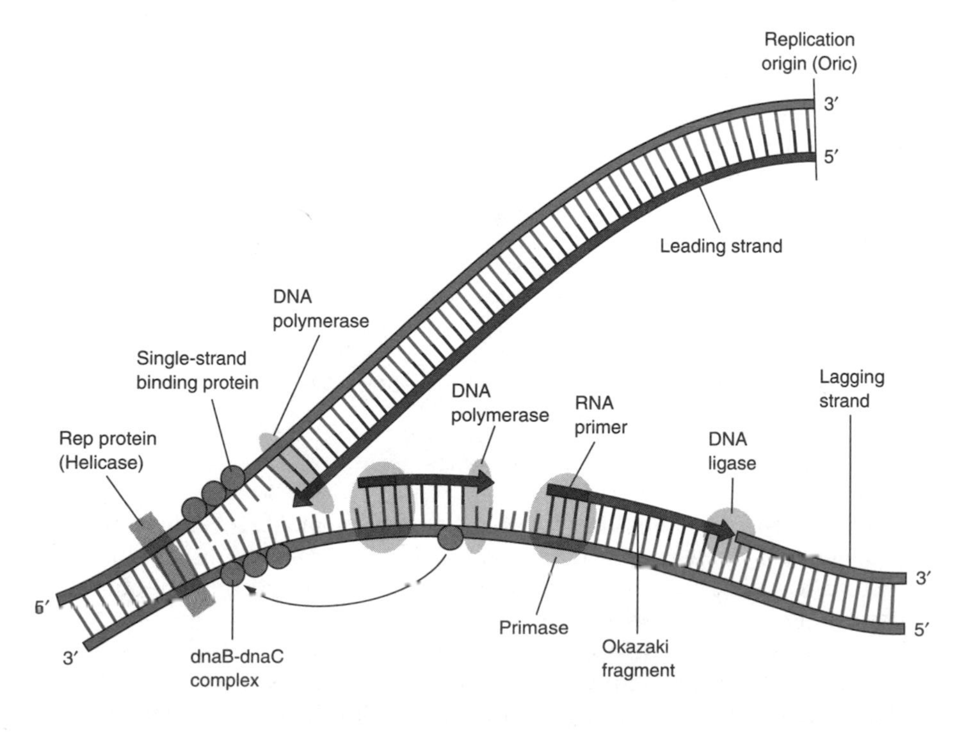

FIGURE 15.64
Model for DNA replication in *E. coli*.
In this figure the initial stages of replication are depicted. The primers are subsequently removed from the newly synthesized segments of DNA at the lagging strand, and the segments are joined. Since replication is normally bidirectional, similar events take place concurrently at the other side of the initiation origin.

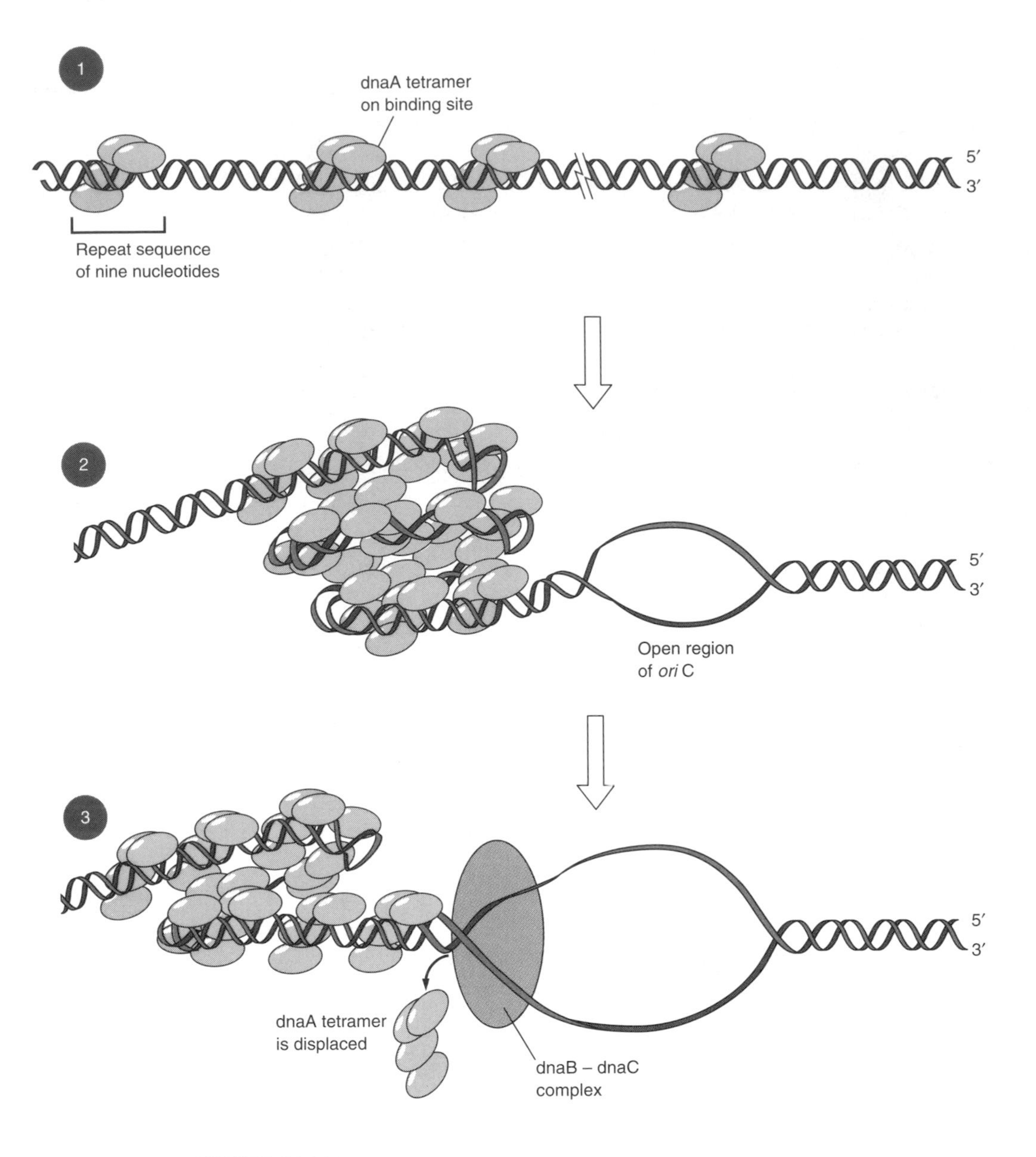

FIGURE 15.65
Model for the initiation of replication in *E. coli.*
Step 1: Initiation of replication begins with the binding of dnaA molecules to four sites consisting of nine-nucleotide long sequences each. These sequences are present at the origin of replication in *E. coli* (oriC). Step 2: The DNA bound dnaA molecules subsequently coalesce and are joined by additional dnaA molecules to form a nucleosome-like DNA–protein complex, which is characterized by nearby "melting" of the double helix. Step 3: The resulting opening of the strands allows a dnaB–dnaC complex to become attached to DNA so that the helicase activity of dnaB can unwind further the DNA. The unwinding is accompanied by a displacement of dnaA molecules.

activity, creates an initiation "bubble" consisting of a few hundred nucleotide pairs. The energy for the formation of this "bubble" is provided by ATP in a reaction catalyzed by topoisomerase II, and the "bubble" is stabilized by interaction with SSB molecules.

At this point the origin of replication is ready for the synthesis of an RNA primer and the subsequent initiation of DNA synthesis, which begins with the formation of a **prepriming complex.** The prepriming assembly consists of the *dnaB–dnaC* complex to which four other proteins (polypeptides n, n′, n″, and i) have been added. Addition of primase converts the prepriming complex to a **primosome** (Figure 15.66). The primosome interacts with a template, at each one of the two forks generated by the formation of a "bubble," and begins the synthesis of RNA primers on the two leading strands. The assembly of the *replisome* is completed by the addition to the primosome of DNA polymerase III and rep proteins. The replisome is an "operative" protein assembly that can bring about the efficient synthesis of DNA.

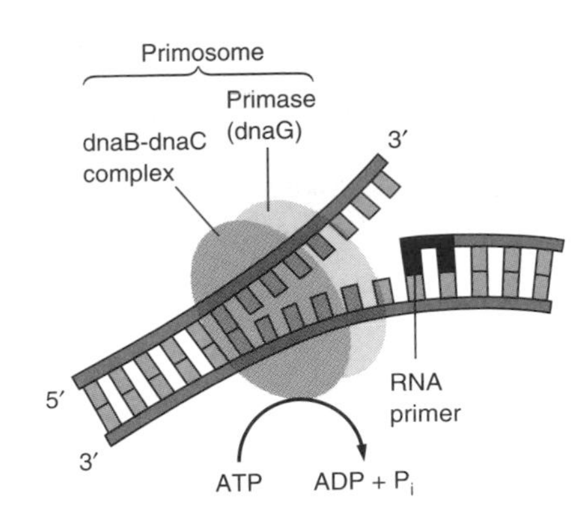

FIGURE 15.66
The primosome of *E. coli.*
The primosome is formed by the binding of primase, together with a complex of the dnaB and dnaC proteins, at specific sequences of DNA that serve as the sites for the formation of RNA primers. Additional factors, described as the N proteins, are the specific primosomal components that are responsible for placing the primosome at the appropriate sequences. In effect the primosome "searches" the DNA for these sequences at the expense of energy generated by the hydrolysis of ATP. Once the correct destination of the primosome is reached, RNA primer synthesis is initiated.

Initiation of the leading DNA strand at the replication origin by the primosome is a process distinct and perhaps more complex than the subsequent initiation of synthesis of precursor (Okazaki) fragments that make up the lagging strand. Okazaki fragment synthesis is initiated by primase at sites selected by the prepriming proteins.

The initiation of the leading strand does not present the cell with serious topological problems because of the negative superhelicity initially present in the circular DNA. For the continuation of the synthesis, however, the presence of helicase activity, which in *E. coli* originates from the two enzymes designated as *helicase II* and *rep protein,* is essential. These enzymes unwind and separate the strands in each of the two forks created by the initiation event. As the helicases move in advance of each fork, two single-stranded regions are generated on parental DNA. These regions are immediately covered by single-stranded binding protein that keeps the fork open and allows DNA polymerase III to take over the elongation of primers. A short time after initiation at the leading strand a signal, uncovered on the template for the initiation of the lagging strand by the movement of helicase, leads to the binding of primase. Primase bound to the lagging strand template moves in the same direction as helicase (a direction that is opposite to that of chain elongation) to set the stage for the discontinuous synthesis of the lagging strand. Primase, the action of which is triggered by the prepriming proteins, synthesizes a brief complementary segment of the strand. This segment serves as a primer for covalent extension of the strand synthesized by DNA polymerase III and for the formation of Okazaki fragments. DNA polymerase III appears to be organized as a dimer consisting of a pair of complexes, each characterized by a different distribution of polymerase subunits. These two complexes appear to be endowed with distinct properties, one tailored for the continuous synthesis of the leading strand and the other for the discontinuous synthesis of the lagging strand.

This polymerase assembly, which appears to combine primase activity with nonidentical twin active sites for polynucleotide synthesis, allows for concurrent replication to occur on both strands. In this scheme, looping of the lagging strand template by 180° brings it to the same orientation as the leading strand template (Figure 15.67). Thus, a primer synthesized at the lagging strand is drawn past it. When a nascent (Okazaki) fragment reaches the 5′ end of the previously synthesized Okazaki fragment, the lagging strand template is released and unlooped.

Removal of the primer portions at the 5′ end of the Okazaki fragments by DNA polymerase I, repair by the same enzyme, and joining of the repaired fragments by DNA ligase produces intact DNA strands. Termination of the synthesis occurs near the center of a 270-kb region across from *oriC*. It is postulated that upon completion of the synthesis the newly synthesized DNA is untangled from the parental DNA by the action of topoisomerase II.

Considerable variations in this basic replication scheme are apparent in both procaryotic and eucaryotic replication, some of which are examined below.

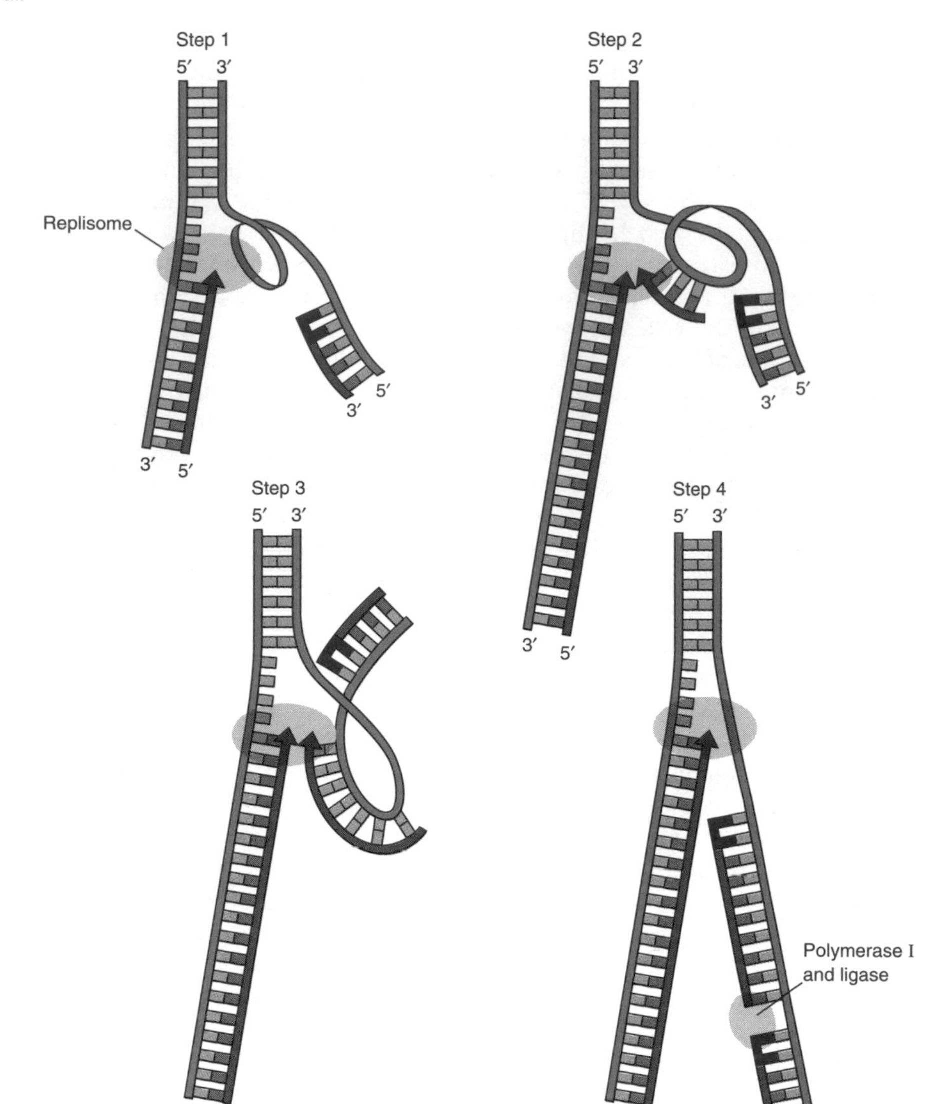

FIGURE 15.67
Model for the simultaneous synthesis of both leading and lagging DNA strands by DNA polymerase.
Two molecules of DNA polymerase operating in concert, and in the same rather than the opposite direction, may be participating in the simultaneous synthesis of DNA on both strands. In this model the replisome consists of a DNA polymerase dimer associated with the primosome and helicases. The primer made by the primosome is extended by the replisome as the lagging-strand template is looped through it. The primer continues to be extended until the previously completed Okazaki fragment is reached, at which point the loop is relaxed. The stretch of unpaired lagging strand template then loops back again to participate in the formation of the next Okazaki fragment.
Redrawn based on figure in A. Kornberg, *DNA replication,* 1982 supplement. San Francisco: W. H. Freeman, 1982.

Rolling Circle Model for Replication

DNA synthesis directed by the circular DNA of mitochondria and in some instances of bacteria and viruses occurs in a manner that initially gives rise to linear daughter DNA molecules that contain the base sequence of parental DNA repeated numerous times. These repeated linear DNAs, which are known as **concatemers,** are important intermediates in phage production, are essential for the process of bacterial mating, and may be involved in the process of gene amplification.

The synthesis of concatemer DNA occurs by a mechanism known as **rolling circle replication.** The essential aspects of continuous replication of the leading strand and discontinuous replication of lagging strand described above is present in the rolling circle model. The replication of DNA by the rolling circle mechanism, shown in Figure 15.68, is different from the basic scheme of DNA replication described above in two important ways. The initiation of the synthesis of the leading strand does not make use of an RNA primer; instead the initiation depends on the nicking of one strand by a phage-encoded endonuclease that generates 3′-OH and a 5′-P termini. The synthesis of the leading strand can thus occur by elongation of the 3′-OH terminus at a fork created under the influence of helicase and single-stranded binding protein. As the leading strand is elongated by the action of polymerase III, the parental template for the synthesis of the lagging strand is displaced and begins to replicate in the usual manner, that is, via Okazaki fragments.

A second characteristic that distinguishes the rolling circle model from conventional replication is that the circular template, used for the synthesis of the leading strand, does not dissociate from the complementary strand during the synthesis. Instead the replication of the leading strand goes on beyond the length of circle-generating linear concatemeric DNA. Appropriately sized DNA molecules are subsequently generated from concatemers by specific endonuclease cleavage.

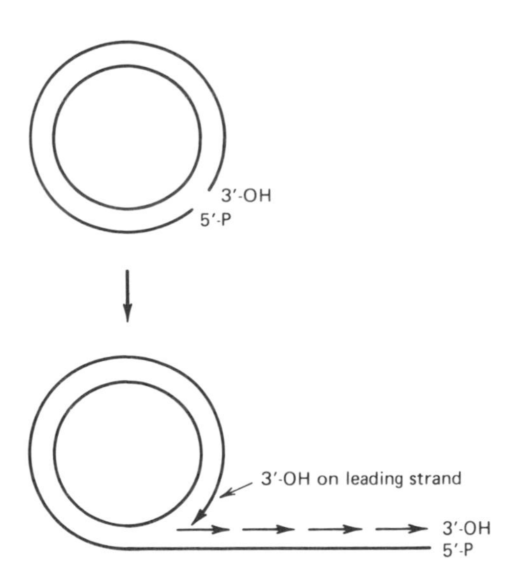

FIGURE 15.68
Replication by the rolling circle mechanism.
Synthesis of the leading strand occurs by elongation of the 3′-OH terminus generated by endonuclease cleavage of the DNA. As the leading strand is synthesized, the parental template directs the synthesis of the lagging strand in the form of precursor fragments.

Eucaryotic DNA Replication

The DNA synthesis in eucaryotes appears to be a process that is fundamentally similar to that occurring in procaryotes, both in terms of the elementary steps involved as well the enzymes and proteins that catalyze these steps. The formation of a replication fork, primer synthesis, Okazaki segment involvement, primer removal, and gap bridging between newly synthesized DNA segments, all parallel the corresponding steps that occur in procaryotes. Some of the enzymes catalyzing eucaryotic DNA synthesis, the eucaryotic DNA polymerases in particular, are now available in pure form for study.

Eucaryotic Polymerases

Eucaryotic cells have been found to contain four different DNA polymerizing enzymes designated as DNA polymerase α, β, γ, and δ. The cellular roles of these enzymes have not yet been firmly established. Polymerase α, which is exclusively found in the nucleus, appears to be the central enzyme in eucaryotic DNA replication. This is suggested by the observation that drugs that specifically inhibit polymerase α, but have no effect on polymerases β and γ, prevent DNA synthesis in growing cells. Polymerase δ is also associated with DNA replication. It has a similar size and organization to polymerase α and may in fact be associated with it. Polymerse β appears to be a repair enzyme and polymerase γ may be responsible for mitochondrial DNA synthesis.

Given the high fidelity with which these enzymes replicate DNA, which matches or exceeds the fidelity of procaryotic polymerases, it is clear that eucaryotic polymerases must also be endowed with an editing exonuclease function analogous to that present in their procaryotic counterparts. Until very recently it had not been possible to demonstrate the existence of a eucaryotic $3' \rightarrow 5'$ exonuclease (proofreading) activity, partially because of the sensitivity of these enzymes to proteolysis during isolation. Polymerase α is composed of four subunits, the largest of which exhibits a combined polymerase and $3' \rightarrow 5'$ exonuclease activity. The two smaller subunits provide primase activity. The fourth subunit endows polymerase α with unusual properties in that, when associated to the other three subunits, it prevents the enzyme from functioning as a nuclease. Polymerase δ, in contrast, has exonuclease activity.

The properties of the eucaryotic polymerases are summarized in Table 15.12. Because of the size and complexity of organization of eucaryotic

TABLE 15.12 Properties of Eucaryotic DNA Polymerases

Enzyme	α	β	γ	δ
Polymerization: $5' \rightarrow 3'$	+	+	+	+
Exonuclease action: $3' \rightarrow 5'$	+	−	+	+
Sensitivity to aphidicolin	+	−	−	+

genomes, other enzymes involved in replication remain elusive or poorly characterized, but their existence is inferred by comparison with the corresponding enzymes that have been isolated from procaryotic systems. Beyond these overall similarities, DNA replication in eucaryotes is also a process with distinctive characteristics.

Eucaryotic Replication Process

Replication of the eucaryotic chromosome must allow for additional complexities, including the organization of the DNA into chromosomes and the very large size of the eucaryotic chromosomes. In eucaryotes the rates of fork, and therefore polymerase, movement do not exceed 30,000 base pairs per minute, which is considerably slower than the rates observed for *E. coli*. Based on the higher DNA content of animal cells and the lower activities of DNA polymerases in comparison to bacteria, the replication cycle of eucaryotic cells could be expected to take as long as a month to complete. In fact, however, the replication cycle is completed within hours, because compensating factors are in operation. Eucaryotic cells contain a large number of DNA polymerase molecules (in excess of 20,000) as compared to no more than a few dozen molecules found in each *E. coli* cell. In addition, DNA polymerase molecules initiate bidirectional synthesis, not in one, but at several initiation points along the chromosome. Eucaryotic initiation appears to resemble that of *E. coli*. In both systems, proteins with helicase activity open the double-stranded template, in preparation for priming and replication by appropriate enzymes. The relative abundance of such initiating proteins may control the initiation process.

The DNA segments between two initiation points are termed *replicons*. Therefore, for a DNA molecule that contains 1000 replicons, replication may proceed simultaneously at as many as 2000 forks. At each of these forks, strands are being replicated as Okazaki fragments. Okazaki segments synthesized in the lagging strand in eucaryotes consists on the average of about 150 nucleotides in length, as opposed to 1000–2000 nucleotides that assemble in procaryotes, which may reflect eucaryotic nucleosome spacing along the template. The presence of nucleosomes may also explain why DNA polymerase molecules move on eucaryotic templates at a rate about one order of magnitude slower relative to their procaryotic counterparts. Replicons vary considerably in size and may extend across as many as 40,000 nucleotide pairs. Each mammalian chromosome may use for replication as many as several hundred replicons of different sizes as indicated in Figure 15.69.

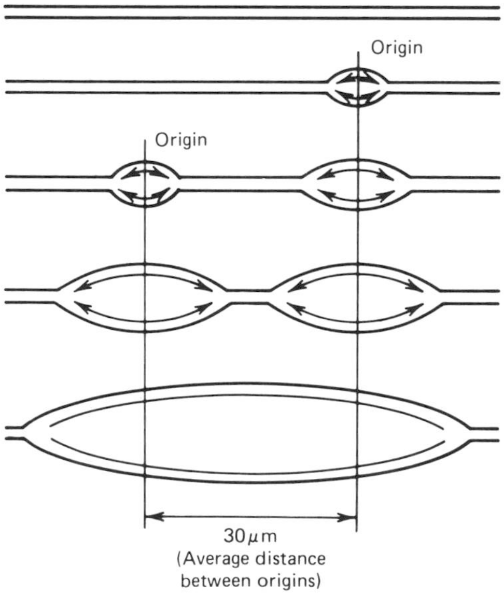

FIGURE 15.69
Replication of mammalian DNA.
Mammalian DNA replicates by using a very large number of replicating forks simultaneously. This mechanism serves to accelerate the process of replication, which in mammalian systems is limited by rates of fork movement that are considerably slower than those characteristic of procaryotes.
Redrawn from J. A. Huberman and A. D. Riggs, *J. Mol. Biol.* 32:327, 1968.

In eucaryotes DNA is present in packaged form as chromatin. Therefore DNA replication is sandwiched between two additional steps, namely, a carefully ordered and incomplete dissociation of the chromatin and, post replicatively, the reassociation of DNA with the histone octamers to form nucleosomes. The synthesis of new histones occurs simultaneously with DNA replication and the reassociation of newly synthesized nucleosomes contain only newly synthesized histones, indicating that the parental histone octamers are conserved into the constituent histones. Furthermore, the histone octamers appear not to dissociate from DNA completely as indicated by the observations that the newly synthesized histone octamers are associated exclusively with one of the two daughter strands rather than becoming distributed between both daughter strands (Figure 15.70).

Another area of difference between procaryotic and eucaryotic cells is the distinct manner in which the processes of DNA synthesis and cell division are coordinated in these two systems. Specifically in rapidly growing procaryotes, DNA is replicated through much of the cell cycle and cell division occurs as soon as DNA synthesis has ceased. In contrast, eucaryotic DNA synthesis (and histone synthesis) is confined to only one part of the cell cycle, specifically the synthetic (S) phase of the interphase. This phase is preceded and followed by two periods during which DNA is not synthesized (gap periods G_1 and G_2). Cell division occurs at a different time within the interphase, referred to as the mitotic (M) period.

(*a*) Cooperative

(*b*) Randomly dispersive

(*c*) Half-nucleosome formation

FIGURE 15.70
Distribution of nucleosomes at a replication fork.
Three possible mechanisms for the distribution of newly synthesized nucleosomes are depicted. The actual mechanism appears to be cooperative. (○, parental nucleosome; ●, newly synthesized nucleosomes).
Reproduced with permission from H. Weintraub, *Cell* 9:419, 1976. Copyright © 1976 by MIT Press.

15.8 DNA RECOMBINATION

DNA recombination is a general phenomenon during which two "parental" DNA molecules are spliced together, giving rise to a new DNA that contains genetic information from both parental strands. Recombination underlies many essential biological processes, including the crossing over between eucaryotic chromosomes during meiosis and the events that lead to exchange of genetic material between related DNAs. The extent to which DNA from different organisms "mixes" in nature is strictly controlled with combinations occurring only between "suitable" closely related DNA molecules. The most common example of DNA "mixing" is the **integration** of phage or plasmid DNA into the corresponding bacterial hosts. Another form of DNA exchange occurs as certain phages or animal viruses incorporate small segments of DNA of the host cell and transfer it to a recipient cell upon infection, a process termed **transduction.**

Recombination May Be General or Site-Specific

One type of recombination, known as **general recombination,** occurs between homologous DNA regions, that is, regions that are largely or completely complementary in their nucleotide sequences. However, a second type of recombination that results in the splicing of the parental DNA strands at specific sites, referred to as **site-specific recombination,** takes place at sites characterized by limited complementary. An example of site-specific recombination is the integration of bacteriophage λ that occurs at a predetermined site on the bacterial chromosome and is characterized by the alignment of the integrated phage in a specific orientation within the *E. coli* chromosome. Bacteriophage λ and the *E. coli* chromosome have distinct recombining sites and both sites have a common sequence of 15 nucleotides. Such short sequences of homology are, as a rule, characteristic of site specific recombination. Four sites have been identified as phage attachment sites, one of which is in the homologous region that binds a phage protein with type I topoisomerase activity known as **integrase.** Integrase produces a staggered cleavage seven base pairs apart within the homologous region and catalyzes the exchange of strands at the position of the cut. The integration reaction is formally reversible but is precisely controlled with its forward and reverse steps separately regulated.

Site-specific recombinations appear to be widespread in nature and lead to rearrangements in the DNA of many genes that are referred to as **transpositions.** Transposable elements of DNA, or **jumping genes,** are capable of movement from one chromosome to another or to a different site within the same chromosome. Transposable elements, which are also known as **transposons,** can modify gene expression and therefore introduce a substantial element of genetic flexibility into eucaryotic genomes. This flexibility challenges the traditional view that associates a specific gene with a particular chromosome.

Recombination that can occur at any complementary location along the length of the combining strands is referred to as **general** (or **homologous**) *recombination*. This process requires, in addition to extensive homology, that one of the parental DNA molecules be at least partially single stranded and that a free end exists at some location in either one of the parental DNA molecules. The example shown in Figure 15.71 illustrates the recombination between the single-stranded form of ϕX174 viral DNA (replicative form) and a linear DNA duplex. General recombination is catalyzed by a protein, recA, which pairs the single-stranded DNA with the DNA duplex by placing the two molecules in homologous register in a process referred to as **synapsis.** The recA also catalyzes the migration of the branch formed during synapsis that leads to the transfer of one of the strands of the duplex onto the circular DNA to form an heteroduplex. Both reactions require ATP, but ATP hydrolysis is essential only for the formation of the heteroduplex by the process of **branch migration.** The recA protein controls the process of general recombination via its *ATPase* and **recombinase** activities that catalyze the binding of the duplex DNA and the unwinding of the double strands. The recA protein has no topoisomerase activity, which explains why a free polynucleotide end must be present for recombination to occur. General recombination between two double-stranded DNAs may occur as indicated in Figure 15.72. The protein recA also exhibits a highly specific *protease* activity, activated by the presence of *unpaired* DNA strands, directed at specific regulatory proteins. Digestion of such regulatory molecules vests recA with unique properties for the coordinate regulation of a number of cellular functions that occur when DNA damage, or the interruption of DNA replication,

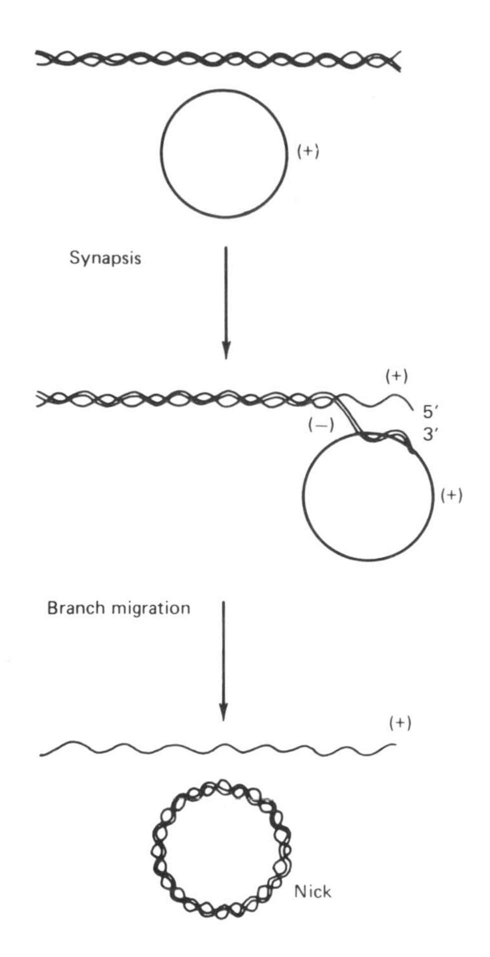

FIGURE 15.71
Homologous recombination.
In this example, a circular ϕX174 single strand is assimilated into a double-stranded (replicative form) ϕX174 DNA that is present in linear form. The process is initiated by a synapsis at one of the terminals of the linear duplex followed by branch migration. Both reactions require single-strand binding proteins, recA protein, and ATP.
Redrawn from A. Kornberg, *DNA replication*. 1982 supplement. San Francisco: W. H. Freeman, 1982.

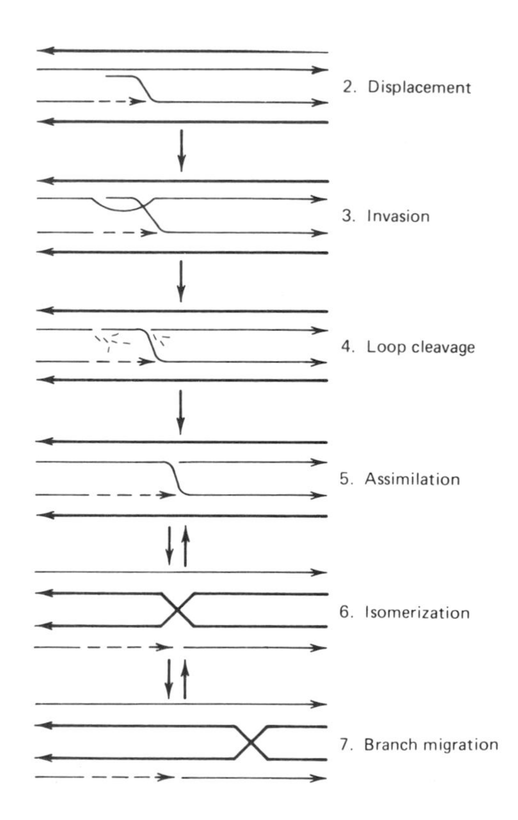

FIGURE 15.72
A model for general recombination between two duplex DNAs.
Recombination between two double-stranded DNA molecules may occur via the multistep process shown above. Dashed lines represent newly synthesized DNA.
Redrawn from M. Meselson and C. Radding, *Proc. Natl. Acad. Sci. U.S.A.* 72:358, 1975.

leads to the production of single-stranded DNA segments. An example of such a process is the postreplication repair of DNA damaged by UV light or other mutagens.

15.9 SEQUENCING OF NUCLEOTIDES IN DNA

Restriction Maps Give the Sequence of Segments of DNA

The sequences of a number of genes and adjoining DNA segments have been determined for bacteria, viruses, plants and humans. The determination of the sequence of a large DNA molecule begins by cutting the DNA into pieces of a more manageable size with appropriate restriction endonucleases. **Restriction digests** permit the construction of a characteristic **restriction map** for each DNA. One frequently used protocol depends on the generation of partial restriction digests of end-labeled DNA. Partial digests are obtained by setting the conditions so that the restriction endonuclease will not recognize all sites in every DNA molecule but will instead produce a digest that includes a collection of partial fragments. Double-stranded DNA is end labeled by treatment with alkaline phosphatase, which removes the phosphate residue at the 5′ end, then γ labeled ^{32}P ATP and a polynucleotide kinase, which catalyzes the incorporation of the ^{32}P label into the two 5′ termini of the DNA strands (see Figure 15.19). Alternately, the ^{32}P label can be introduced at the 3′ termini by the incorporation of ^{32}P labeled deoxyribonucleotide triphosphates using DNA polymerase. End labeling allows for each fragment to be identified on an electrophoresis gel, which separates the fragments by size. The details of this procedure are presented on page 771. Thus, with a series of different site cuts the fragments can be mapped directly relative to the labeled end.

Restriction maps are used for the characterization of various DNAs and for the ordering of smaller DNA fragments within a particular DNA sequence. Such ordering is essential before the nucleotide sequence of large DNA molecules can be elucidated. In recent years several effective methods have been developed for the rapid sequencing of relatively large polydeoxyribonucleotides. These methods are impressively accurate. Digests obtained using different restriction enzymes produce segments with overlapping lengths of nucleotide sequences. These sequences allow the sequencing of the DNA by use of multiple sets of independent sequence data. In addition, the accuracy of the sequencing methods are increased by sequencing the complementary strand. These procedures can also be used for the sequencing of RNA molecules by prior conversion of the polyribonucleotide sequence of RNA to complementary polydeoxyribonucleotides by use of reverse transcriptase.

Contemporary sequencing methods, described on page 770, allow the determination of the sequence of DNAs of any length. Sequences up to 500 nucleotides can be determined in a single automated operation. Clin. Corr. 15.4 discusses the application of these procedures for obtaining the sequence of the human genome.

CLIN. CORR. 15.4 THE NUCLEOTIDE SEQUENCE OF THE HUMAN GENOME

The human genome is the sum total of the approximately 100,000 different genes that determine the genetic characteristics of every individual human being. It is estimated that the human genome consist of about 3 billion base-paired nucleotides, assembled in the form of 23 pairs of chromosomes. The discovery of restriction endonucleases and the development of effective *physical mapping* procedures for DNA, combined with the increasing rapidity of contemporary nucleotide sequencing methods, have recently provided strong impetus for the very ambitious scientific undertaking of determining the nucleotide sequence of the entire human genome.

Extensive *physical mapping* of the human genome has already been completed. In addition, *genetic mapping* seeks to locate the over 500 known genetic markers on the chromosomes. It is estimated that complete sequencing of the genome would take more than a decade, provided that sufficient resources are committed to this project. However, because of the routine nature of determining the countless nucleotide sequences involved, many scientists have questioned the wisdom of diverting the massive effort required to sequence the human genome from other areas of more creative scientific endeavor. Proponents point out the great potential benefits of determining the imprint that controls the genetic properties of the human cell at the highest possible level of resolution, the nucleotide sequence. Searching for the imprint of human disease at the level of nucleotide sequences may eventually permit us to understand all disease states at the genomic level. Although this goal may remain elusive for the foreseeable future, except for a few diseases that are currently understood in terms of the mutations that cause them, the determination of the sequence of the human genome appears to be one of the prerequisites for understanding human disease at the molecular level. There is little doubt that the sequencing of the human genome will present us with many new challenges and opportunities in medicine.

BIBLIOGRAPHY

Books

Adams, R. L. P., Knowles, J. T., and Leader, D. P., *The Biochemistry of the nucleic acids,* 10th ed., New York: Chapman and Hall, 1986.

Bradbury, M. E., Maclean, N., and Mathews, H., *DNA chromatin and chromosomes*. New York: Chapman and Hall, 1981.

Cantor, C. R. and Schimmel, P. R., *Biophysical chemistry, part I: The conformation of biological macromolecules*. San Francisco: Freeman, 1980.

Cantor, C. R. and Schimmel, P. R., *Biophysical chemistry, part II: The behavior of biological macromolecules*. San Francisco: Freeman, 1980.

Freifelder, D., *Molecular biology:* 2nd ed., Boston, Jones and Barlett, 1987.

Friedberg, E. C., *DNA repair*. New York: Freeman, 1985.

Friedberg, E. C. and Hanawalt, P. C. (Eds.). *Mechanisms and consequences of DNA damage processing*. UCLA Symposia on Molecular and Cellular Biology, 1988.

Gait, M. J. and Blackburn, B. *Nucleic Acids in Chemistry and Biology*. New York: IRL Press, 1989.

Kahl, G. (Ed.). *Architecture of eucaryotic genes*. New York: VCH Publishers, 1988.

Kornberg, A., *DNA replication,* 3rd ed. New York: Wiley, 1987.

Lewin, B., *Genes III,* 3rd ed. New York: Wiley, 1987.

Low, K. B. (Ed.). *The recombination of genetic material*. San Diego: Academic, 1988.

Mainwaring, W. I. P., Paris, J. H., Pickering, J. D., and Mann, N. H., *Nucleic acid biochemistry and molecular biology*. London: Blackwell Scientific Publications, 1982.

Richardson, C. C. and Lehman, I. R. *Molecular Mechanisms in DNA Replication and Recombination*. New York: Wiley, 1990.

Saenger, W., *Principles of nucleic acid structure*. New York: Springer-Verlag, 1988.

Van Holde, K. E., *Chromatin*. New York: Springer-Verlag, 1989.

Watson, J. D., Hopkins, N. H., Roberts, J. W., Steitz, J. A., and Weiner, A. M., *Molecular biology of the gene,* 4th ed. Menlo Park, CA: Benjamin-Cummings, 1987.

Woods, R. A., *Biochemical genetics*. New York: Chapman and Hall, 1988.

Cleaver, J. E. and Karentz, D. DNA repair in man. Regulation by a multigene family and association with human disease. *Bioassays* 6:122, 1987.

Cox, M. M. and Lehman, I. R. Enzymes of general recombination. *Annu. Rev. Biochem.* 56:229, 1987.

Davidson, J. N. and Cohn, W. E. (Eds.). Progress in nucleic acid research and molecular biology. New York: Academic (a series published regularly).

DuBridge, R. B. and Carolos, M. P. Molecular approaches to the study of gene mutation in human cells. *Trends Genet.* 3:293, 1987.

Graig, N. The mechanism of conservative site-specific recombination. *Ann. Rev. Genet.* 22:77, 1988.

Grossman, L., Caron, P. R., Mazur, S. J., and Oti, E. Y. Repair of DNA-containing pyrimidine dimers. *FASEB J.* 2:2696, 1988.

Kogan, S. C. An improved method for prenatal diagnosis of genetic diseases by analysis of amplified DNA sequences. *N. Engl. J. Med.* 317:985, 1987.

Kornberg, A. DNA replication. *J. Biol. Chem.* 263:1, 1988.

Landy, A. Dynamic, structural and regulating aspects of site-specific recombination. *Annu. Rev. Biochem.* 58:913, 1989.

Lehman, I. R. and Kaguni, L. S. DNA polymerase. *J. Biol. Chem.* 264:4265, 1989.

Martin, J. B. Molecular genetics: Applications to the clinical neurosciences. *Science* 238:765, 1987.

Matson, S. W. and Kaiser-Rogers, K. A. DNA helicases. *Annu. Rev. Biochem.* 58:519, 1988.

Modrich, P. Methyl-directed DNA mismatch correction. *J. Biol. Chem.* 264:6597, 1989.

Sancar, A. and Sancar, G. B. DNA repair enzymes. *Annu. Rev. Biochem.* 57:29, 1988.

Tompkins, L. S., Nucleic acid probes in Infectious Diseases. *Curr. Clin. Top. Infect. Dis.* 10:174, 1989.

Reviews and Specialized Papers

Bramhill, D. and Kornberg, A. Duplex opening by dnaA proteins at novel sequences in initiation of replication at the origin of the *E. coli* chromosome. *Cell* 52:743, 1988.

QUESTIONS

C. N. ANGSTADT AND J. BAGGOTT

*Q*uestion *T*ypes are described on page xxiii.

1. (QT1) A polynucleotide is a polymer in which:
   A. the two ends are structurally equivalent.
   B. the monomeric units are joined together by phosphodiester bonds.
   C. there are at least 20 different kinds of monomers that can be used.
   D. the monomeric units are not subject to hydrolysis.
   E. purine and pyrimidine bases are the repeating units.
2. (QT1) The *best* definition of an endonuclease is an enzyme that hydrolyzes:
   A. a nucleotide from only the 3′ end of an oligonucleotide.
   B. a nucleotide from either terminal of an oligonucleotide.
   C. a phosphodiester bond located in the interior of a polynucleotide.
   D. a bond only in a specific sequence of nucleotides.
   E. a bond that is distal (d) to the base that occupies the 5′ position of the bond.

3. (QT2) Which of the following tends to favor a helical conformation of a single polynucleotide chain?
   1. Hydrophobic interactions of the rings of the purine and pyrimidine bases, which exclude water
   2. Interchange of electrons in the π orbitals of the purine and pyrimidine bases
   3. Charge–charge repulsion of phosphate residues of the polynucleotide backbone
   4. Hydrogen bonding between appropriate purine–pyrimidine pairs

Use the accompanying figure to answer the next two questions.

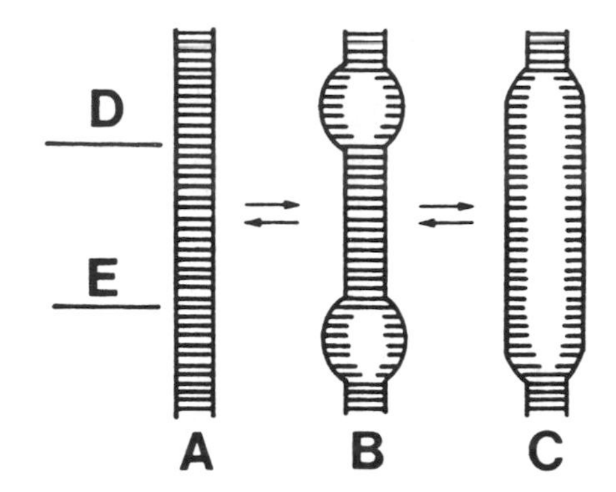

4. (QT5) A, B, and C represent conformations at different temperatures. Which one represents the highest temperature?
5. (QT5) Which section, D or E, has the higher content of guanine and cytosine?
   A. Annealing
   B. Viscoelastic retardation
   C. Equilibrium centrifugation
6. (QT5) A technique for determining the molecular weight of very large ($>10^9$ Da) DNA
7. (QT5) A technique involved in locating a specific gene on DNA with a probe.

8. (QT2) The superhelices that form in double-stranded circular DNA:
   1. may have fewer turns of the helix per unit length than does a linear double helix.
   2. are associated with a restricted topological domain.
   3. may exist in multiple conformations that are interconvertible without breaking covalent bonds.
   4. may be either formed or relaxed by enzymes called topoisomerases.

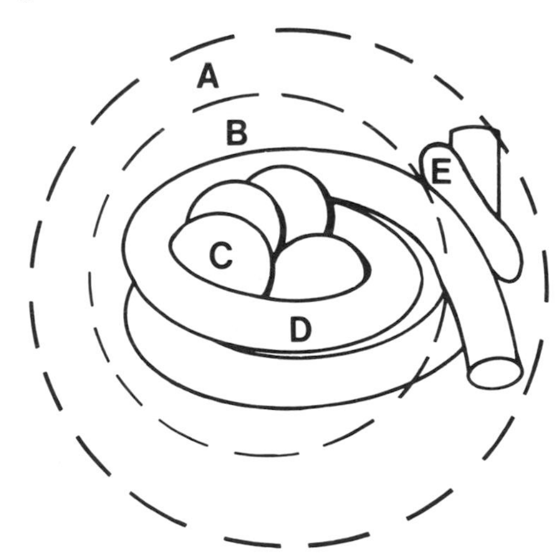

9. (QT5) A chromatosome
10. (QT5) DNA
11. (QT5) H1 class of histones

12. (QT1) A palindrome is a sequence of nucleotides in DNA that:
    A. is highly reiterated.
    B. is part of the introns of eucaryotic genes.
    C. is a structural gene.
    D. has local symmetry and may serve as a recognition site for various proteins.
    E. has the information necessary to confer antibiotic resistance in bacteria.

13. (QT2) Which of the following would result in a frame-shift mutation?
    1. Insertion of a new base pair in the DNA
    2. Substitution of a purine–pyrimidine pair by a pyrimidine–purine pair
    3. Intercalation of ethidium bromide into the nucleotide chain
    4. Deamination of cytosine to uracil

    A. DNA replication　　C. Both
    B. DNA excision repair　　D. Neither
14. (QT4) Both strands of DNA serve as templates concurrently.
15. (QT4) Require(s) the activity of DNA ligase.

16. (QT1) Replication:
    A. is semiconservative.
    B. requires only proteins with DNA polymerase activity.
    C. uses $5' \rightarrow 3'$ polymerase activity to synthesize one strand and $3' \rightarrow 5'$ polymerase activity to synthesize the complementary strand.
    D. requires a primer in eucaryotes but not in procaryotes.
    E. must begin with an incision step.

17. (QT2) In eucaryotic DNA replication:
    1. only one replisome forms because there is a single origin of replication.
    2. Okazaki fragment synthesis is initiated by a primase.
    3. helicase dissociates from DNA as soon as the initiation bubble forms.
    4. at least one DNA polymerase has a $3' \rightarrow 5'$ exonuclease activity.

ANSWERS

1. B The structure of a polynucleotide possesses an intrinsic sense of direction that does not depend on whether a 3′-OH or 5′-OH terminal is esterified. C, E: There are only four different monomers, and the repeating unit is the base monophosphate (p. 610).
2. C Both A and B describe exonucleases. D does refer to an endonuclease but only to a specific type, a restriction endonuclease, and is therefore not a definition of the general type. E: Both endo- and exonucleases show specificity toward the bond hydrolyzed and so this is not a definition of an endonuclease (p. 613).
3. A 1, 2, 3 correct. The exclusion of water by stacking of the bases is a strong stabilizing force that is enhanced by the interaction of π orbital electrons. The repulsive forces of the phosphate groups confer a certain rigidity to the structure. 4: This is very important in holding two different polynucleotide chains together, but it is unlikely that the proper positioning would occur within a single chain (p. 614).
4. C The figure represents the process of denaturation with
5. E the extent of disruption increasing as temperature increases. Since a guanine–cytosine pair has three hydrogen bonds and an adenine–thymine pair only has two, higher temperatures are required to disrupt regions high in G–C (Figure 15.12).
6. B This is a special low shear method. Equilibrium centrifugation (C) is also a method for determining molecular weight, but it is limited to a molecular weight of 10^9 Da or less because of the effects of shear forces on large molecules (p. 629).
7. A The probe is a labeled polynucleotide with a sequence complementary to the gene of interest. Annealing of the two permits location of the gene (p. 625).
8. E All four correct. 1 describes a negative superhelix. There may also be more turns per unit length in a positive superhelix. 2: Once a closed system is interrupted, a superhelix can unwind. 3: All conformations with the same linking number α are interconvertible without breaking covalent bonds. 4: Topoisomerase I (omega protein) from *E. coli* relaxes and gyrase (topoisomerase II) can introduce or remove superhelices (depending on the conditions) (p. 631).
9. A The chromatosome, the basic structural element of
10. D nucleoprotein, contains the nucleosome core with
11. E associated H_1 histones. B: The nucleosome core is a discrete particle consisting of an octamer of specific histones with a segment of DNA wrapped around it. D: The strand depicted represents DNA; the circles, histones. E: The H1 class of histones is bound to the spacer regions between nucleosomes. C: Represents one of the histones (H_{2A}, H_{2B}, H_3, or H_4), which are part of the nucleosomes (Figure 15.30).
12. D A palindrome, by definition, reads the same forward and backward. Short palindromic segments of DNA are recognized by a variety of proteins such as restriction endonucleases and CAP-binding protein. A is not likely since it would be incompatible with specific recognition. B is possible but has not been shown. C is not correct since genes are thousands of base pairs in length, whereas palindromes are short segments. E also would not be likely because palindromes are too short (p. 642).
13. B 1, 3 correct. Since the bases are read in groups of three, insertion of an additional base would shift the reading frame (p. 653). Intercalation stretches the DNA so when DNA is replicated an additional base is inserted near the intercalation site (p. 657). 2 and 4 are both examples of base substitution type of mutations.
14. A In excision repair, the damaged segment of a strand is removed so both strands are not available (pp. 659–665).
15. C In both cases, ligase is needed to seal the newly synthesized strand to the end of the strand already in place (pp. 660–661).
16. A B,D: Replication requires a primer, usually synthesized by a primase. Ligases, helicases, and other proteins are required as well. C: Replication involves Okazaki fragments because synthesis occurs only in the 5′ → 3′ direction. E: Incision is the recognition step for DNA repair (p. 663).
17. C 2,4 correct. 1: There are multiple initiation sites. The DNA segments between two initiation points are called replicons (p. 674). 2: Primase synthesizes the RNA primer required by DNA polymerases (p. 671). 3: Helicase activity is also necessary for the continuation of synthesis, that is, the opening of the forks (p. 668). 4: Polymerase α shows this activity that provides proofreading during synthesis (p. 673).

16

RNA: Structure, Transcription, and Posttranscriptional Modification

FRANCIS J. SCHMIDT

16.1 OVERVIEW 681
16.2 STRUCTURE OF RNA 683
RNA Is a Polymer of Ribonucleoside 5′-Monophosphates 683
Secondary Structure of RNA Involves Intramolecular Base Pairing 684
RNA Can Have a Tertiary Structure 685
16.3 TYPES OF RNA 686
Transfer RNA Is Involved in Activating Amino Acids and Recognition of Codons in mRNA 687
Ribosomal RNA Is Part of the Protein Synthesis Apparatus 688
Messenger RNAs Carry the Information for the Primary Structure of Proteins 692
RNA in Ribonucleoprotein Particles 694
Mitochondria Contain Unique RNA Species 694
16.4 MECHANISMS OF TRANSCRIPTION 695
The Initial Process of RNA Synthesis Is Transcription 695
The Template for RNA Synthesis Is DNA 696
RNA Polymerase Catalyzes the Transcription Process 696
The Steps of Transcription in Procaryotes Have Been Determined 698
Transcription in Eucaryotes Involves Additional Molecular Events 703
16.5 POSTTRANSCRIPTIONAL PROCESSING 708
Ribozymes Are Enzymes Whose RNA Component Has Catalytic Activity 708
Transfer RNA Is Modified by Cleavage, Additions, and Base Modification 710
Messenger RNA Processing Requires Maintenance of the Coding Sequence 712
16.6 NUCLEASES AND RNA TURNOVER 717
BIBLIOGRAPHY 719
QUESTIONS AND ANNOTATED ANSWERS 720
CLINICAL CORRELATIONS
16.1 Staphylococcal Resistance to Erythromycin 692
16.2 Antibiotics and Toxins Targeting RNA Polymerase 699
16.3 Involvement of Transcriptional Factors in Carcinogenesis 707
16.4 Thalassemia Due to Defects in Messenger RNA Synthesis 714
16.5 Autoimmunity in Connective Tissue Disease 717

16.1 OVERVIEW

The primary information store of a cell is its genetic complement, that is, its DNA. The DNA information is exactly analogous to the master copies of a computer program or any data base: It is the source of cellular information and therefore must be kept as error-free as possible. Chapter 15 has detailed some of the elaborate mechanisms that are employed to keep DNA information intact from one cell generation to the next. This chapter describes another type of information transfer that helps to ensure the integrity of genomic information. Just as a careful computer programmer makes working copies of a program or data set, the cell makes macromolecular copies of the information in DNA. Macromolecules that mediate the transfer of information must reflect the sequence of the purines and pyrimidines of the DNA. These macromolecules, called ribonucleic acids (RNAs) are linear polymers of ribonucleoside monophosphates. The process by which RNA copies of selected DNA copies are made is termed *transcription*. The primary role of RNA within the cell is its involvement in protein synthesis, that is, *translation*.

The overall process of information transfer in the cell is therefore given by the so-called **central dogma** of molecular biology:

$$\text{DNA} \xrightarrow{\text{transcription}} \text{RNA} \xrightarrow{\text{translation}} \text{protein}$$

RNA information is occasionally **reverse transcribed** into DNA, a process important in the life cycle of infectious retroviruses such as the human immunodeficiency virus (HIV), which causes the acquired immunodeficiency syndrome (AIDS). Reverse translation of protein sequence into nucleic acid sequence information, however, does not occur in nature.

The RNA molecules are classified according to the roles they play in the information transfer processes (Table 16.1). In eucaryotes these processes are spatially separated; transcription occurs in the nucleus and translation in the cytoplasmic portions of the cell. **Messenger RNAs (mRNA)** serve as templates for the synthesis of protein; they carry information from the DNA to the cellular protein synthetic machinery. Here a number of other RNA species contribute to the synthesis of the peptide bond.

The molecules that transfer specific amino acids from soluble amino acid pools to ribosomes, and ensure the alignment of these amino acids in the proper sequence prior to peptide bond formation, are **transfer RNAs (tRNA).** All the tRNA molecules are approximately the same size and shape. The assembly site, or factory, for peptide synthesis involves ribosomes, complex subcellular particles containing at least three different RNA molecules called **ribosomal RNAs (rRNA),** and 70–80 ribosomal proteins.

Protein synthesis requires a close interdependent relationship between mRNA, the informational template, tRNA, the amino acid adaptor molecule, and rRNA, part of the synthetic machinery. In order for protein synthesis to occur at the correct time in a cell's life, the synthesis of mRNA, tRNA, and rRNA must be coordinated with the cell's response to the intra- and extracellular environments.

All cellular RNA is synthesized on a DNA template and reflects a portion of the DNA base sequence. Therefore, all RNA is associated with DNA at some time. The RNA involved in protein synthesis functions in the cytoplasm outside the nucleus, while some RNAs remain in the nucleus, where they have structural and regulatory roles. The RNA synthesized in the mitochondria remains there and is involved in mitochondrial protein synthesis.

Although DNA is the more prevalent genetic store of information, RNA can also carry genetic information in the sequence of the bases and serves as the genome in several viruses. However, RNA is not normally found as the genome in eucaryotic or procaryotic cells. Genomic RNA is found in

TABLE 16.1 Characteristics of Cellular RNAs

Type of RNA	Abbreviation	Function	Size and sedimentation coefficient	Site of synthesis	Structural features
Messenger RNA Cytoplasmic	mRNA	Transfer of genetic information from nucleus to cytoplasm, or from gene to ribosome	Depends on size of protein 1000–10,000 nucleotides	Nucleoplasm	Blocked 5′ end; poly(A) tail on 3′ end; nontranslated sequences before and after coding regions; few base pairs and methylations
Mitochondrial	mt mRNA		9S–40S	Mitochondria	
Transfer RNA Cytoplasmic	tRNA	Transfer of amino acids to mRNA ribosome complex and correct sequence insertion	65–110 nucleotides 4S	Nucleoplasm	Highly base paired; many modified nucleotides; common specific structure
Mitochondrial	mt tRNA		3.2S–4S	Mitochondria	
Ribosomal RNA Cytoplasmic	rRNA	Structural framework for ribosomes	28S, 5400 nucleotides 18S, 2100 nucleotides 5.8S, 158 nucleotides 5S, 120 nucleotides	Nucleolus Nucleolus Nucleolus Nucleoplasm	5.8S and 5S highly base paired; 28S and 18S have some base paired regions and some methylated nucleotides
Mitochondrial	mt rRNA		16S, 1650 nucleotides 12S, 1100 nucleotides	Mitochondria	
Heterogeneous nuclear RNA	hnRNA	Some are precursors to mRNA and other RNAs	Extremely variable 30S–100S	Nucleoplasm	mRNA precursors may have blocked 5′ ends and 3′-poly(A) tails; many have base paired loops
Small nuclear RNA	snRNA	Structural and regulatory RNAs in chromatin	100–300 nucleotides	Nucleoplasm	
Small cytoplasmic RNA [7S(L) RNA]	scRNA	Selection of proteins for export	129 nucleotides	Cytosol and rough endoplasmic reticulum	Associated with protein as part of signal recognition particle

the RNA tumor viruses and the other small RNA viruses, such as poliovirus and reovirus.

16.2 STRUCTURE OF RNA

RNA Is a Polymer of Ribonucleoside 5′-Monophosphates

Chemically, RNA is very similar to DNA. Although RNA is one of the more stable components within a cell, it is not as stable as DNA. Some RNAs, such as mRNA, are synthesized, used, and degraded, whereas others, such as rRNA, are not turned over rapidly.

The RNA is an unbranched linear polymer in which the monomeric subunits are the ribonucleoside 5′-monophosphates. The purines found in RNA are *adenine* and *guanine;* the pyrimidines are *cytosine* and *uracil.* Except for uracil, which replaces thymine, these are the same bases found in DNA.

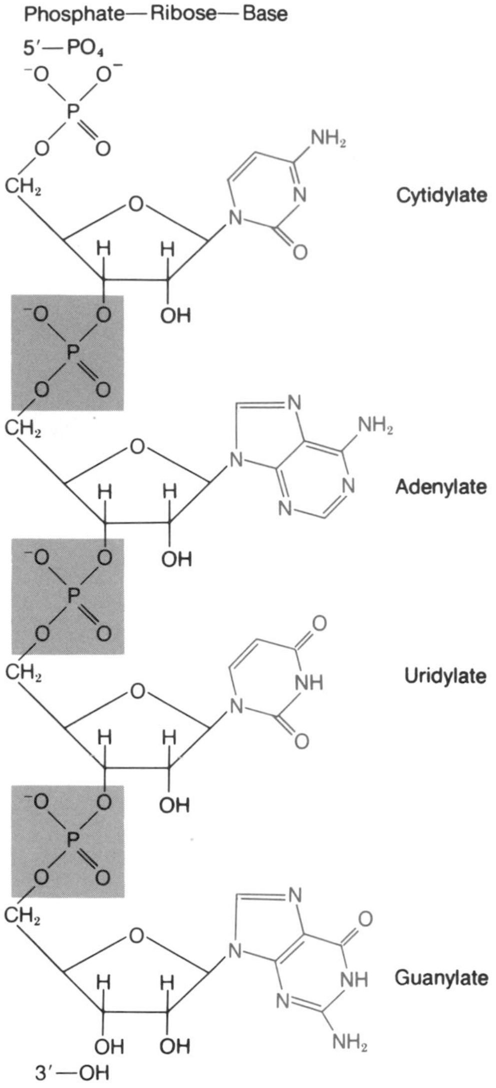

FIGURE 16.1
The structure of the 3′,5′-phosphodiester bonds between ribonucleotides forming a single strand of RNA.
The phosphate joins the 3′-OH group of one ribose with the 5′-OH group of the next ribose. This linkage produces a polyribonucleotide having a sugar–phosphate "backbone." The purine and pyrimidine bases extend away from the axis of the backbone and may pair with complementary bases to form double helical base paired regions.

Although A, C, G, and U nucleotides are incorporated into RNA during its transcription, other **modified nucleotides** are synthesized after transcription. As can be seen in Table 16.2 (pp. 688–689), there are a number of nucleotide modifications, some of them quite complex. Modified nucleotides are especially characteristic of stable RNA species (i.e., tRNA and rRNA); however, some methylated nucleotides are also present in eucaryotic mRNA. For the most part, the functions of the modified nucleotides in RNA have not been identified. Where known, the function of nucleotide modification seems to involve "fine tuning" rather an indispensable role in the cell.

The monomers are connected by single phosphate groups linking the 3′-C of one ribose with the 5′-C of the next ribose. The internucleotide link, a 3′,5′-phosphodiester, forms a chain or backbone from which the bases extend (Figure 16.1). The length of natural RNA molecules in eucaryotic cells varies from approximately 65 nucleotides to more than 6000 nucleotides. The sequences of the bases are complementary to the base sequences of specific portions of only *one strand of DNA*. Thus, unlike the base composition of DNA, molar ratios of A + U and G + C in RNA are not equal. All cellular RNA so far examined is linear and single stranded, but double-stranded RNA is present in some viruses.

Secondary Structure of RNA Involves Intramolecular Base Pairing

The RNA, being single stranded rather than double stranded as is DNA, does not usually form a double helix. Rather, the structure in an RNA molecule arises from **intramolecular base pairing.** In solutions of low ionic strength the molecules appear as extended polyelectrolyte chains with no base pairs. Shifting the molecules to solutions of higher ionic strength causes the RNA to contract as intramolecular hydrogen bonds form. Increasing the temperature of an RNA solution denatures the secondary and tertiary structure of the molecule as hydrogen bonding and base stacking are disrupted. These changes can be monitored by measuring absorption of UV light at 260 nm, much as in studies of DNA structure. Considerable helical structure exists in RNA even in the absence of extensive base pairing, for example, in the portions of an RNA that do not form intramolecular Watson–Crick base pairs. This helical structure is due to the strong base-stacking forces between A, G, and C residues. Base stacking is more important than simple hydrogen bonding in determining inter- and intramolecular interactions. These forces act to restrict the possible conformations of an RNA molecule (Figure 16.2). Base stacking is a result of the van der Waals forces between the π electron clouds above and below the unsaturated rings of the purines and pyrimidines and the hydrophobic nature of the bases. As with DNA, the distance restriction of the phosphodiester–ribosyl backbone and the near-perpendicular angle of the β-glycosidic bond do not permit the bases to stack directly over each other. Therefore each succeeding base is offset by about 35°, forming RNA helical structures with 10 or 11 nucleotides per turn in a double helix in comparison to 10 in DNA. In single-stranded helical RNA, each succeeding base is offset by 60°. The result is a turn with 6 nucleotides.

A single strand of RNA may often have double helical regions formed by hydrogen bonding between complementary base sequences located within the molecule. These intermolecular duplex structures, often called **"hairpins,"** may or may not have large unpaired loops at the end. There are considerable variations in the fine structural details of the "hairpin" structures. These variables include the length of base paired regions and the size and number of unpaired loops (Figure 16.3). Transfer RNAs have a large proportion of their bases involved in these helical structures and are excellent examples of base stacking and hydrogen bonding in a single-stranded molecule (Figure 16.4*a*). The anticodon region in tRNA is an unpaired, base-stacked loop of seven bases. The partial helix caused by base stacking in this loop binds, by specific base pairing, to a complementary codon in mRNA so that translation (peptide bond formation) can occur. Within the tRNA molecule itself about 60% of the bases are paired in four double helical regions called stems. In addition, the unpaired regions have the capability to form base pairs with free bases in the same or other looped regions, thereby contributing to the molecule's tertiary structure. The role and extent of base pairing in each type of RNA is described in the following sections.

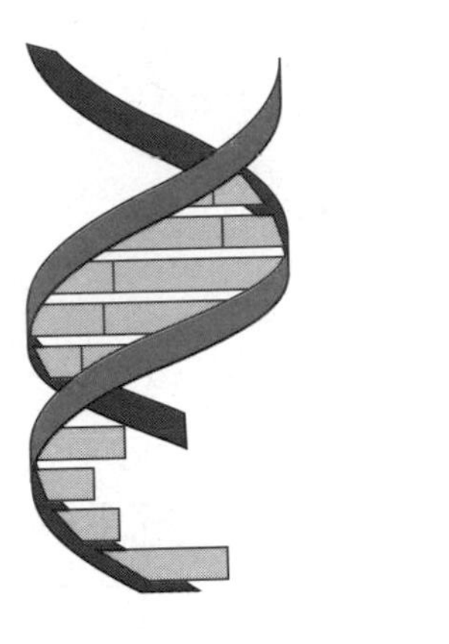

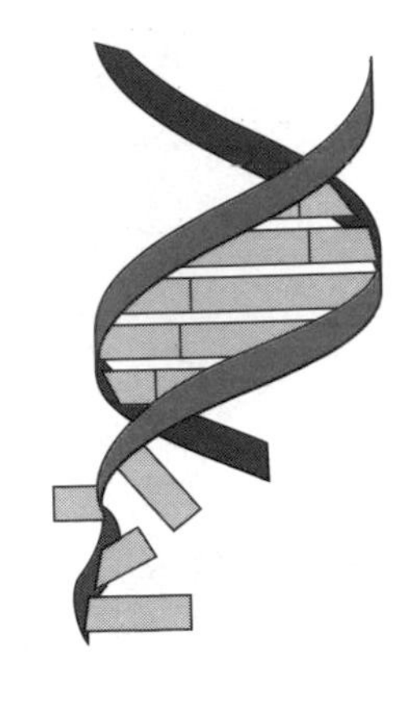

(a) (b)

FIGURE 16.2
Models indicating a helical structure due to (*a*) base stacking in the —CCA_{OH} terminus of tRNA and (*b*) the lack of an ordered helix when no stacking occurs in this non-base paired region.
Redrawn from M. Sprinzl and F. Cramer, *Prog. Nucl. Res. Mol. Biol.* 22:9, 1979.

RNA Can Have a Tertiary Structure

The actual functioning structures of RNA molecules are more complex than the base-stacked and hydrogen-bonded helices mentioned above. The RNAs in vivo are dynamic molecules that undergo changes in conformation during synthesis, processing, and functioning. Proteins associated with RNA molecules often lend stability to the RNA structure; in fact, it is perhaps more correct to think of RNA–protein complexes rather than naked RNA molecules as functioning components of the cell.

In addition to the secondary, base paired structure, RNA molecules also exhibit a large number of hydrogen bonds that are important in the tertiary structure of the molecule. The structure of tRNA provides a

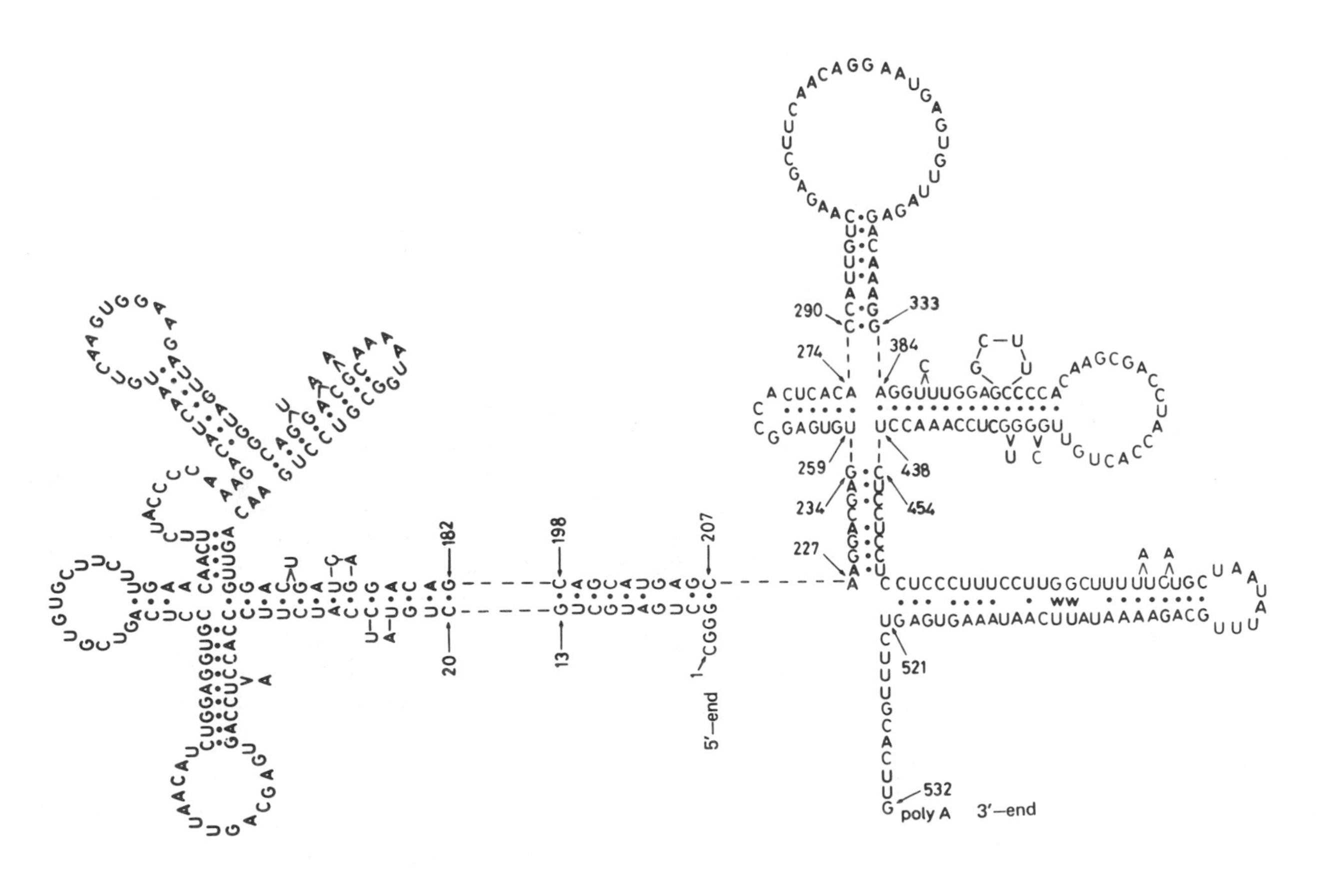

FIGURE 16.3
Proposed base pairing regions in the mRNA for mouse immunoglobulin light chain.
Base paired structures shown have free energies of at least −5 kcal. Note the variance in loop size and length of paired regions.
P. H. Hamlyn, G. G. Browniee, C. C. Cheng, M. J. Gait, and C. Milstein, *Cell,* 15:1067, 1978. Reproduced with permission.

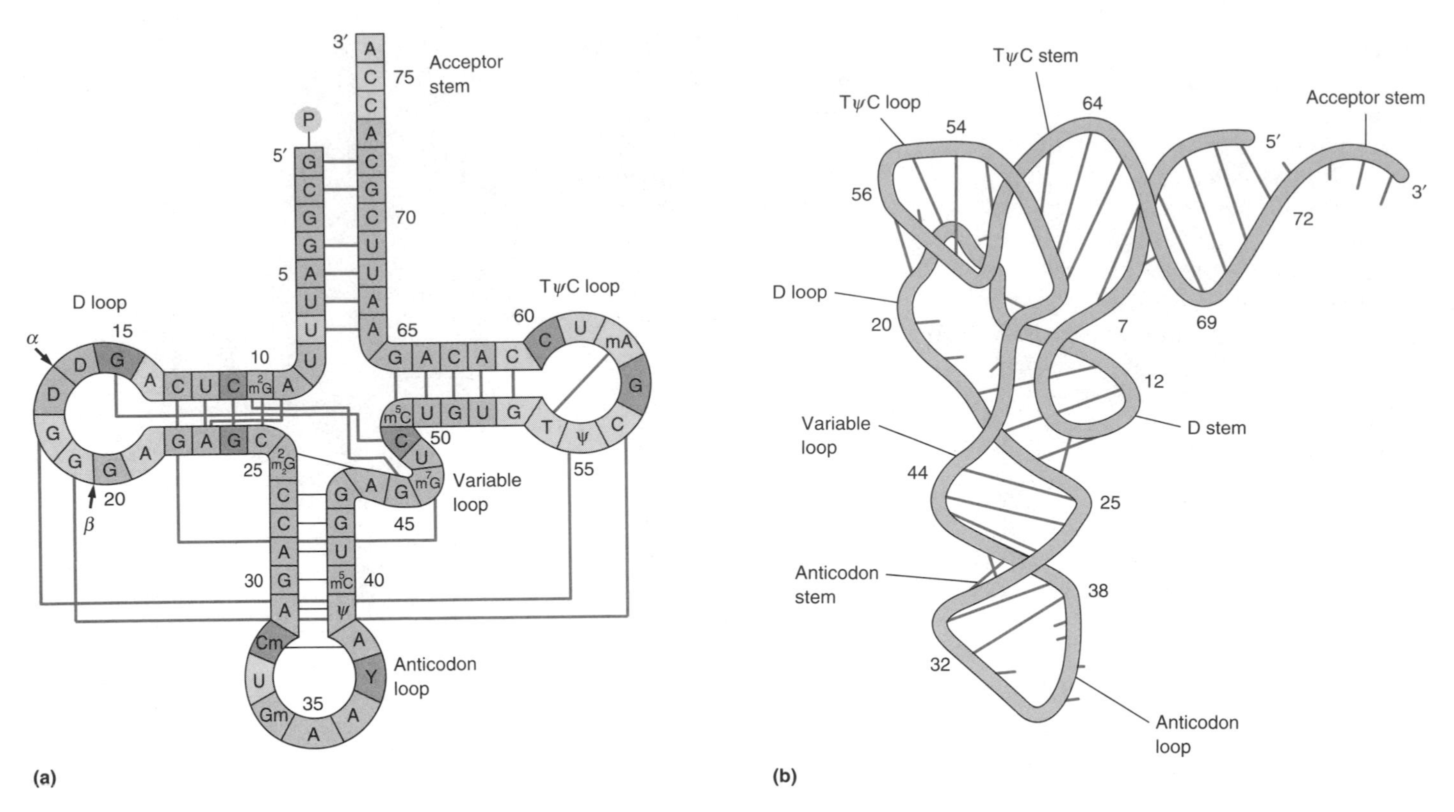

FIGURE 16.4
(*a*) Cloverleaf diagram of the two-dimensional structure and nucleotide sequence of yeast tRNA^{Phe}. Solid lines connecting circled nucleotides indicate hydrogen-bonded bases. Solid squares indicate constant nucleotides; dashed squares indicate a constant purine or pyrimidine. Insertion of nucleotides in the D loop occurs at positions α and β for different tRNAs. (*b*) Tertiary folding of the cloverleaf structure in (*a*). Hydrogen bonds are indicated by cross rungs.
Redrawn with permission from G. J. Quigley and A. Rich, *Science* 194:797, 1976. Copyright © 1976 by the American Association for the Advancement of Science.

number of examples. Transfer RNA in solution is folded in a compact "L-shaped" conformation (Figure 16.4*b*). The arms and loops are folded in specific conformations held in position not only by traditional Watson–Crick base pairing as found in DNA, but also base interactions involving more than two nucleotides. There are interactions between bases and the phosphodiester backbone and between bases with particular regions of the sugars, especially the 2′-OH group. The folding of the tRNA molecules apparently occurs during transcription. Transfer RNA is the only biologically functional RNA molecule for which a three-dimensional structure is known. The first structural determination was accomplished by X-ray crystallography of yeast tRNA specific for phenylalanine.

16.3 TYPES OF RNA

RNA molecules in the cell can be classified by a number of different schemes, including their function, stability, and cytological localization. Traditionally, RNA species have been classified as transfer, ribosomal, and messenger RNAs; however, more recently it has been recognized that RNA molecules perform or facilitate a variety of different functions in a cell.

Transfer RNA Is Involved in Activating Amino Acids and Recognition of Codons in mRNA

The **tRNAs** comprise about 15% of the total cellular RNA. Transfer RNA has two functions. First, tRNA molecules **activate amino acids** for protein synthesis so that formation of peptide bonds is energetically favored. The activated amino acid is transported to the polyribosome where it is transferred to the growing peptide chain (hence its name). The second function of tRNA is the recognition of the codons in mRNA so that the correct amino acid is incorporated into the growing peptide chain. These two functions are reflected in the fact that tRNAs have two primary active sites, the *3'-OH terminus* **—CCA_{OH},** to which specific amino acids are attached enzymatically, and the **anticodon triplet,** which recognizes mRNA codons. These two active sites together are responsible for the adapter function of tRNA; that is, the conversion of information encoded in a nucleic acid (DNA or mRNA) sequence into protein sequence during translation.

Each tRNA can transfer only a single amino acid. Although there are only 20 amino acids used in protein synthesis, there are at least 56 different species of tRNAs in any given cell, with each tRNA recognizing different anticodon triplets. (Mitochondria synthesize a much smaller number of tRNAs.) There is often more than one tRNA for a given amino acid and these tRNAs are called isoacceptors. A tRNA that accepts phenylalanine would be written as $tRNA^{Phe}$, whereas one accepting tyrosine would be written $tRNA^{Tyr}$.

Transfer RNAs are relatively small for nucleic acids, ranging in length from 65 to 110 nucleotides. This corresponds to a molecular weight range of about 22,000–37,000. The sedimentation constant for tRNAs as a group is 4S and the term 4S RNA is also used to designate tRNA. The sequences of all tRNA molecules (nearly 1000 are known) can be arranged into a common secondary structure that has the appearance of a **cloverleaf.** The cloverleaf structure is determined by complementary Watson–Crick base pairs forming three stem and loop structures. The anticodon triplet sequence is at one "leaf" of the cloverleaf while the C—C—A acceptor stem is at the "stem" (see Figure 16.4). This arrangement where the two active sites of a tRNA are spatially separated is preserved in the tertiary structure of $tRNA^{Phe}$ shown in Figure 16.4. Additional, non-Watson–Crick, hydrogen bonds fold the structure into the L-shaped molecule in solution.

From the nucleotide sequence and structure of the $tRNA^{Phe}$ shown in Figure 16.4, it is clear that tRNAs have several modified nucleotides as well as a high proportion of bases involved in secondary conformations, helices, and tertiary folding. Some of the modified nucleotides found in tRNA are listed in Table 16.2 and their positions in tRNAs indicated in Figure 16.5. The modified nucleotides affect tRNA structure and stability but are not essential for the formation or maintenance of tertiary conformation. The modifications do not appear to have a role in general aminoacylation, or "charging," of tRNAs, but may be involved in regulating specific and nonspecific recognition of enzymes and proteins and also specify interactions between tRNA and ribosomes.

Many structural features are common to all tRNA molecules (see Table 16.3). Seven base pairs are always in the amino acid acceptor stem, which, in functioning molecules, is terminated with the nucleotide triplet —CCA_{OH}. This —CCA_{OH} triplet is not base paired. The dihydrouracil or "D" stem has three or four base pairs, while the anticodon and —TΨC— stems have five base pairs each. Both the anticodon loop and —TΨC— loop have seven nucleotides. Differences in the number of nucleotides in

TABLE 16.2 Some Modified Nucleosides Found in RNA

Purine derivatives
Nucleosides with a methylated base
1-Methyladenosine (m^1A)
2-Methyladenosine (M^2A)
7-Methylguanosine (m^7G)
1-Methylguanosine (m^1G)
N^6-Methyladenosine (m^6A)
N^6,N^6-Dimethyladenosine ($m_2{}^6A$)
N^2-Methylguanosine (m^2G)
N^2,N^2-Dimethylguanosine ($m_2{}^2G$)
2′-*O*-Methylated derivatives
2′-*O*-Methyladenosine (Am)
2′-*O*-Methylguanosine (Gm)
Deaminated derivatives
Inosine (I)
1-Methylinosine (m^1I)
Adenosine derivatives with an isopentenyl group
N^6-(Δ^2-Isopentenyl)adenosine (i^6A)
N^6-(Δ^2-Isopentenyl)-2-methylthioadenosine (ms^2i^6A)
N^6-(4-Hydroxy-3-methylbut-2-enyl)-2-methylthioadenosine
Other nucleosides
N-[9-(β-D-Ribofuranosyl)purin-6-ylcarbamoyl]threonine (t^6A)
N-[9-(β-D-Ribofuranosyl)purin-6-yl-*N*-methylcarbamoyl]threonine (mt^6A)
7-4, 5-*cis*-dihydroxy-1-cyclopenten-3-ylaminomethyl-7-deaza-quanosine (Q), queuosine

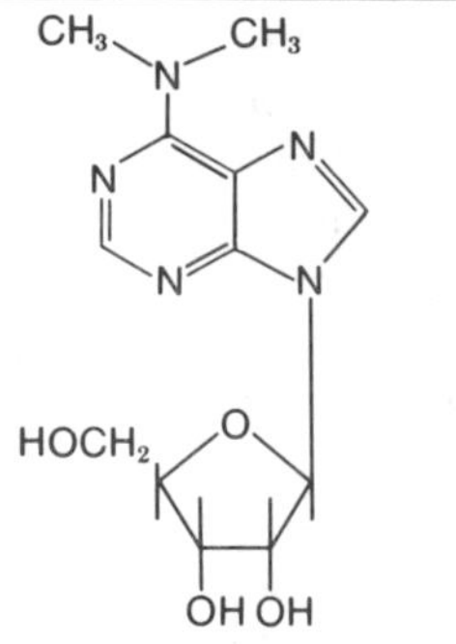

N^6,N^6-Dimethyladenosine

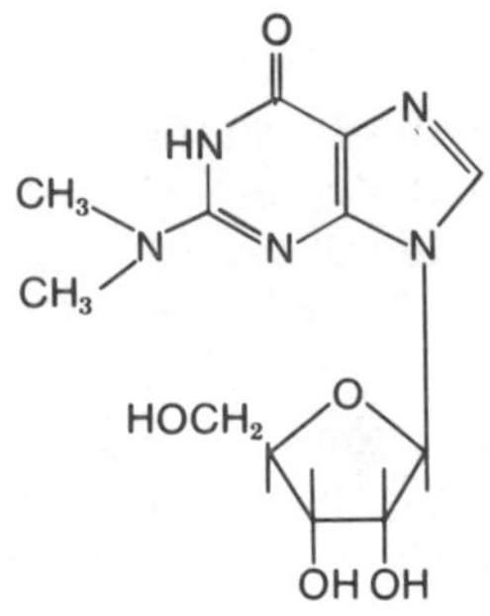

N^2,N^2-Dimethylguanosine

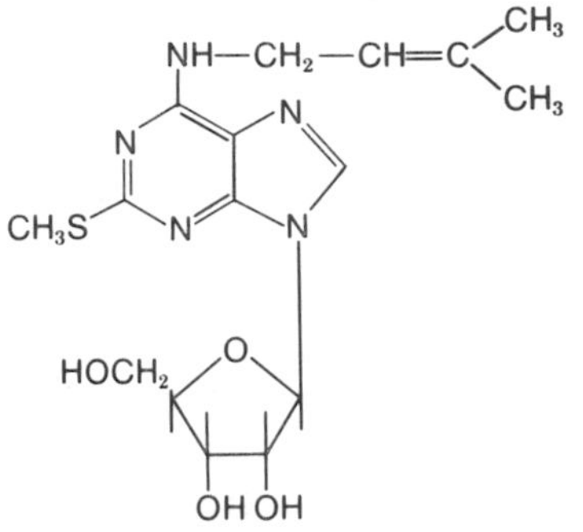

N^6-(Δ^2-Isopentenyl)-2-methylthioadenosine

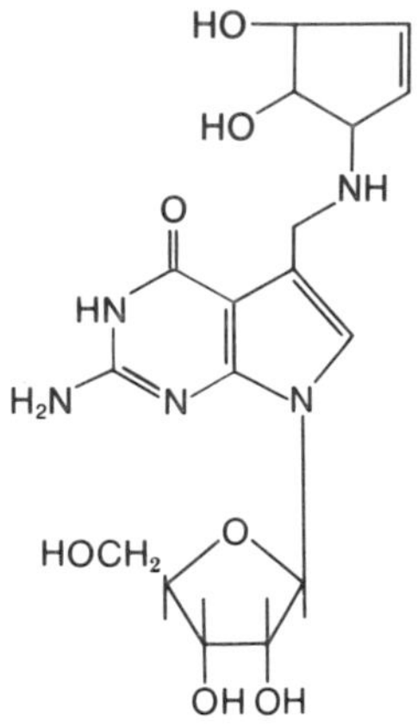

Queuosine

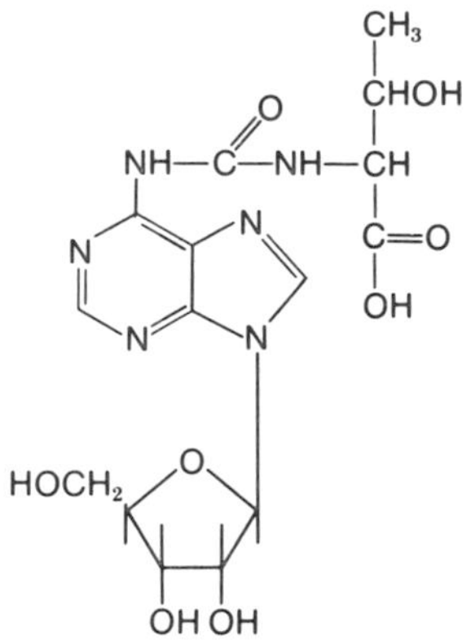

N-[9-(β-D-Ribofuranosyl)purin-6-yl carbamoyl]-L-Threonine

different tRNAs are accounted for by the variable loop. Thus 80% of tRNAs have small variable loops of 4–5 nucleotides, while the others have larger loops with 13–21 nucleotides. Some nucleotides are constant in all tRNAs (see Figure 16.4*a*).

Ribosomal RNA Is Part of the Protein Synthesis Apparatus

Protein synthesis takes place on ribosomes. These complex assemblies are composed in eucaryotes of four RNA molecules, representing about two-thirds of the particle mass, and 70–80 proteins. The smaller subunit, the **40S particle,** contains one **18S RNA** and 55% of the proteins. The larger subunit, the **60S particle,** contains the remaining rRNAs and pro-

TABLE 16.2 (*Continued*)

Pyrimidine derivatives
Nucleosides with a methylated base
Thymine riboside (T)
5-Methylcytidine (m^5C)
3-Methylcytidine (m^3C)
3-Methyluridine (m^3U)
2′-*O*-Methylated derivatives
2′-*O*-Methyluridine (Um)
2′-*O*-Methylcytidine (Cm)
2′-*O*-Methylpseudouridine (Ψm)
2′-*O*-Methylthymine riboside (Tm)
Sulfur-containing nucleosides
4-Thiouridine (s^4U)
5-Carboxymethyl-2-thiouridine methyl ester (cm^5s^2U)
5-Methylaminomethyl-2-thiouridine (mnm^5s^2U)
5-Methyl-2-thiouridine (2-thiothymidine, s^2T)
2-Thiocytidine (s^2C)
Other nucleosides
Pseudouridine (Ψ)
5,6-Dihydrouridine (D)
4-Acetylcytidine (ac^4C)
Uridine-5-oxyacetic acid (V)

*N*⁴-Acetylcytidine

5-Methyluridine

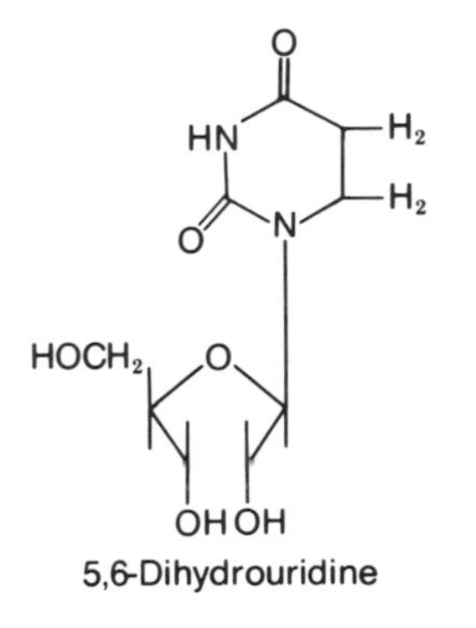

5,6-Dihydrouridine

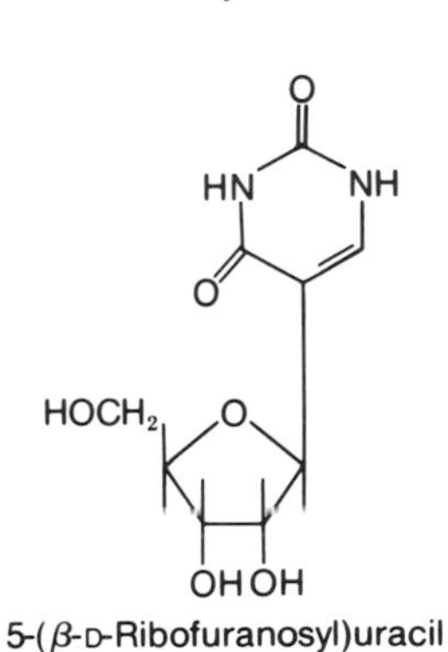

5-(β-D-Ribofuranosyl)uracil (Pseudouridine)

2-Thio-5-carboxymethyluridine methyl ester

teins. The three rRNAs in the large subunit are the **28S,** the **5.8S,** and the **5S rRNA.** The total assembly is called the **80S ribosome.**

The rRNAs account for 80% of the total cellular RNA and are metabolically stable. This stability, required for repeated functioning of the ribosome, is enhanced by close association with the ribosomal proteins. The 28S, 18S, and 5.8S rRNAs are synthesized in the nucleolar region of the nucleus. The 5S rRNA is not transcribed in the nucleolus but rather from separate genes within the nucleoplasm. Processing of the rRNAs (see Section 16.5) includes cleavage to the functional size, limited formation of internal base pairing via hydrogen bonds, modification of particular nucleotides, and association with ribosomal proteins to form a stable tertiary conformation.

The 5S rRNA is 68% base paired, with helical regions formed between proximal as well as distal internal complementary sequences (see Figure 16.6). The nucleotide sequences and proposed conformations for 5S rRNA have been highly conserved throughout the evolutionary scale. The

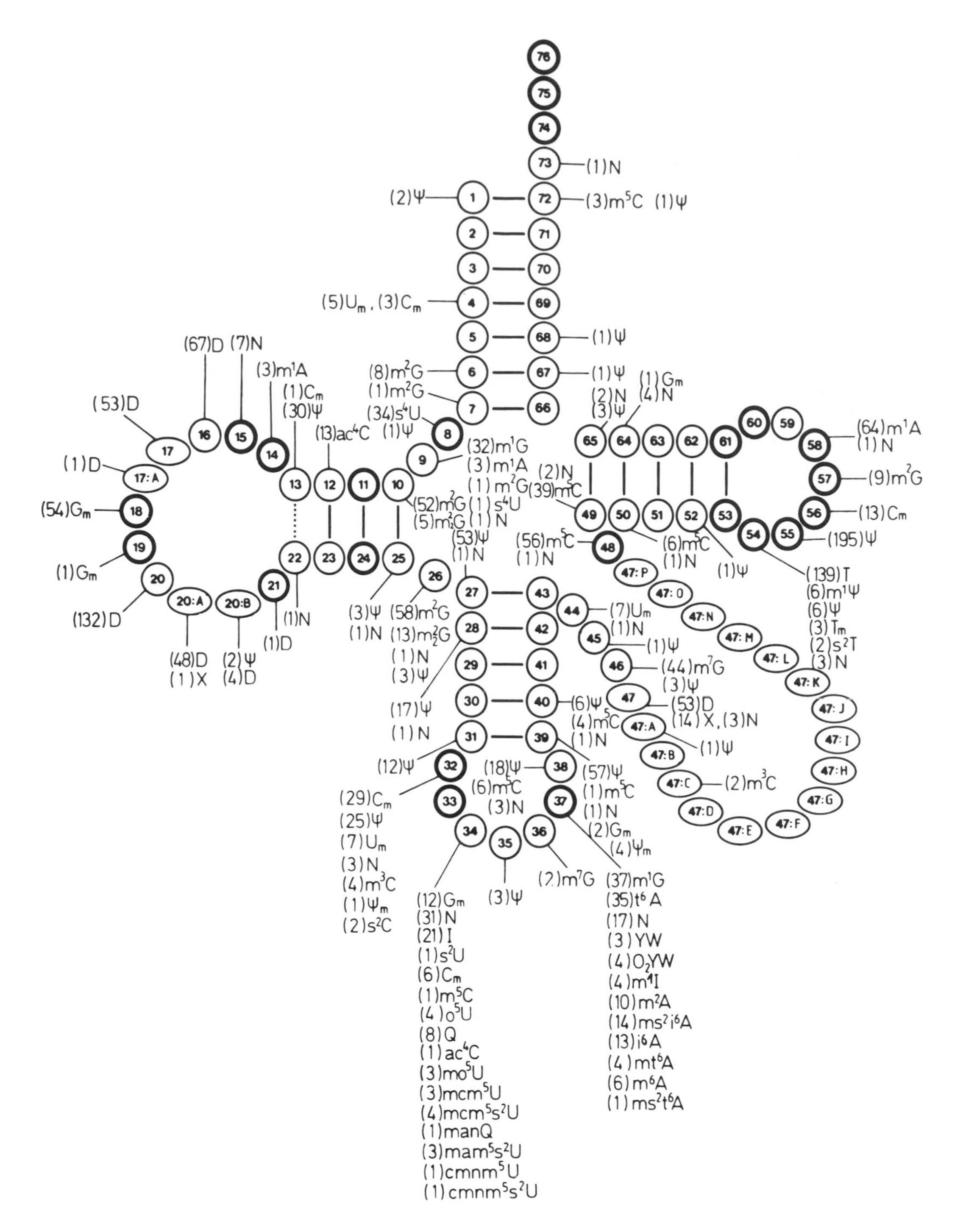

FIGURE 16.5
Cloverleaf secondary structure of tRNA showing positions and prevalence of modified nucleotides.
The figure depicts the cloverleaf secondary structure of tRNA, the standardized numbering of the nucleotide sequences, and those nucleotides that are always present (thick-edged circles) or commonly present (thin-edged circles) in the structure. Ovals represent nucleotides that are not present in each sequence. The location and prevalence of each of the modified nucleotides in the known sequences of 211 different tRNAs is shown as the frequency of appearance at a particular location in parentheses preceding the nucleotide abbreviation. Table 16.2 lists the full names and abbreviations of the modified nucleotides.
Reprinted with permission from M. Sprinzl and D. H. Gauss in P. F. Agris and R. A. Kopper (Eds.). *The modified nucleosides of transfer RNA II*. New York: Alan R. Liss, 1983.

TABLE 16.3 Characteristics of Regions in tRNA

Region	Number of nucleotides	Comments
Amino acid helix	14 (7 bp)	Region where base mispairing occurs frequently; $G \cdot U$ is common; —CCA_{OH} is added posttranscriptionally
Dihydrouracil stem	6 or 8 (3 or 4 bp)	First and last bp are usually $C \cdot G$
Dihydrouracil loop (loop I)	7–10	Region exhibits considerable variation
Anticodon stem	10 (5 bp)	Second base pair from anticodon loop is usually $C \cdot G$;
Anticodon loop (loop II)	7	from 5′ end, the 3rd, 4th, and 5th bases are the anticodon; 5′ side of anticodon is always a pyrimidine; 3′ side is usually a modified purine
Variable arm (loop III)	3–21	Extremely variable in structure and often lacks a helical stem; the arm probably forms hydrogen-bonded stem region (3–7 bp)
TΨC stem	10 (5 bp)	Base pair adjacent to TΨC loop is $C \cdot G$
TΨC loop (loop IV)	7	All tRNAs contain the sequence T-Ψ-C-purine at the same location in the loop; the purine is usually guanine

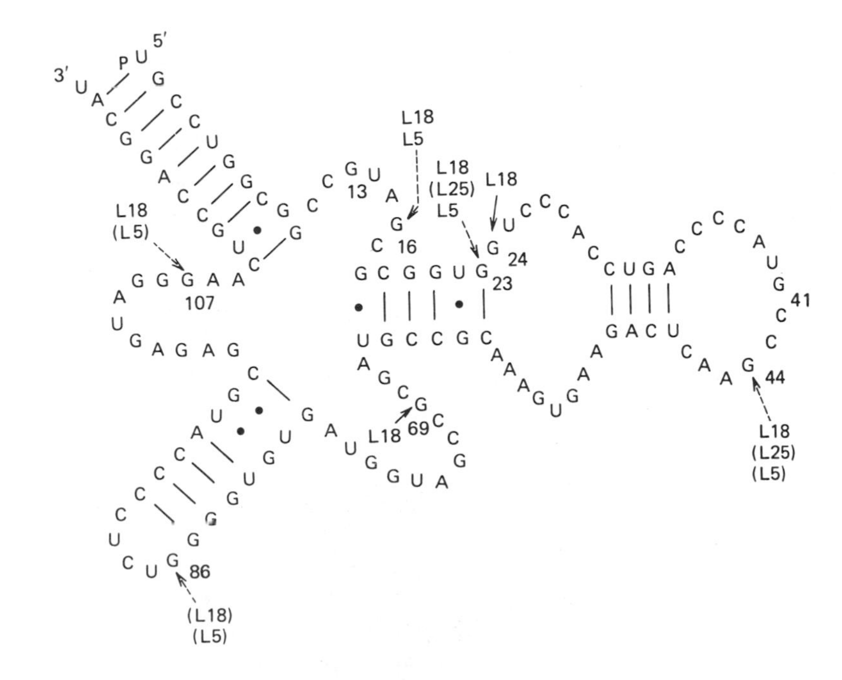

FIGURE 16.6
Secondary, base paired, structure proposed for 5S rRNA.
Arrows indicate regions protected by proteins in the large ribosomal subunit.
Combined information from G. E. Fox and C. R. Woese, *Nature* (*London*) 256:505, 1975, and R. A. Garrett and P. N. Gray.

CLIN. CORR. 16.1
STAPHYLOCOCCAL RESISTANCE TO ERYTHROMYCIN

Bacteria exposed to antibiotics in a clinical or agricultural setting often develop resistance to the drugs. This resistance can arise from a mutation in the target cell's DNA, which gives rise to resistant descendants. An alternative and clinically more serious mode of resistance arises when plasmids coding for antibiotic resistance proliferate through the bacterial population. These plasmids may carry multiple resistance determinants and render several antibiotics useless at the same time.

Erythromycin inhibits protein synthesis by binding to the large ribosomal subunit. *Staphylococcus aureus* can become resistant to erythromycin and similar antibiotics as a result of a plasmid-borne RNA methylase enzyme that converts a single adenosine in 23S rRNA to N^6-dimethyladenosine. Since the same ribosomal site binds lincomycin and clindamycin this plasmid causes cross-resistance to these antibiotics as well. Synthesis of the methylase enzyme is induced by erythromycin.

The microorganism that produces an antibiotic must also be immune to it or else it would be inhibited by its own toxic product. The organism that produces erythromycin, *Streptomyces erythreus,* itself possesses an rRNA methylase that acts at the same ribosomal site as the one from *S. aureus*.

Which came first? It is likely that many of the resistance genes in target organisms evolved from those of producer organisms. In several cases, DNA sequences from resistance genes of the same specificity are conserved between producer and target organisms. We may therefore look on plasmid-borne antibiotic resistance as a case of "natural genetic engineering," whereby DNA from one organism (the *Streptomyces* producer) is appropriated and expressed in another (the *Staphylococcus* target).

Cundliffe, E. How antibiotic-producing microorganisms avoid suicide. *Ann. Rev. Microbiol.* 43:207, 1989.

length, some sequences, and most helical regions are the same for *E. coli* and human 5S rRNA. A specific function for 5S rRNA has not been described, although it is apparently required, in a structural role, for protein synthesis. A lack of 5S rRNA in a ribosome or cleavage at specific locations in the 5S rRNA nucleotide chain render the ribosome inactive.

The 5.8S rRNA, with 158 nucleotides, is closely associated, by hydrogen bonds, with the 28S rRNA. The 5.8S rRNA has considerable internal base pairing, while the 5′ and 3′ ends are free to interact with the 28S rRNA. Like the 5S rRNA, the 5.8S rRNA nucleotide sequence has been conserved during evolution, but the 5.8S rRNA is not found in procaryotes.

The larger rRNAs contain most of the altered nucleotides found in rRNA. These are primarily methylations on the 2′ position of the ribose, giving 2′-*O*-methylribose. At least one methylation of rRNA has been directly related to bacterial antibiotic resistance in a pathogenic species (Clin. Corr. 16.1). There are a small number of N^6-dimethyladenines present in the 18S rRNA. The 28S rRNA has about 45 methyl groups and the 18S rRNA has 30 methyl groups, which may be involved in processing of the 45S precursor molecule (see Table 16.4).

In addition to the base pairing within each 18S and 28S rRNA there is evidence for a base sequence in mRNAs, which can base pair with the rRNA of the smaller subunit, forming a translation complex. The hinging mechanism between the two ribosomal subunits, which enables translocation and mRNA movement, is thought to involve protein–protein interactions and base pairing between the 18S and 28S rRNAs.

It is apparent from the above discussion that rRNA molecules are more than macromolecular scaffolds for enzymatic proteins. The exact extent to which rRNA participates in protein biosynthetic reactions is the subject of current investigation.

Messenger RNAs Carry the Information for the Primary Structure of Proteins

The **mRNAs** are the direct carriers of genetic information from the nucleus to the cytoplasmic ribosomes. Each eucaryotic mRNA contains information for only one polypeptide chain, and therefore these mRNAs have been designated **monocistronic,** whereas in procaryotes mRNA species can encode more than one protein in a **polycistronic** molecule. A cell's phenotype and functional state are related directly to the cytoplasmic mRNA content.

In the cytoplasm mRNAs have relatively short life spans, which in part are determined by the cell's particular needs at any given time. Some mRNAs are known to be synthesized and stored in an inactive or dormant state in the cytoplasm, ready for a quick protein synthetic response. An example of this is the unfertilized egg of the African clawed toad, *Xenopus laevis*. Immediately upon fertilization the egg undergoes rapid protein synthesis, indicating the presence of preformed mRNA and ribosomes.

Eucaryotic mRNAs have unique structural features not found in rRNA or tRNA (see Figure 16.7). These features aid in proper mRNA functioning. Since the information within mRNA lies in the linear sequence of the nucleotides, the integrity of this sequence is extremely important. Any loss or change of nucleotides could alter the protein being translated. The translation of mRNA on the ribosomes must also begin and end at specific sequences. Structurally, starting from the 5′ terminus, there is an inverted methylated base attached via **5′-phosphate-5′-phosphate bonds** rather than the usual internucleotide 3′,5′-phosphodiester linkages between adjacent riboses. This structure, called a **"cap,"** is a guanosine 5′-triphos-

TABLE 16.4 Ribosomal RNA Characteristics

Sedimentation value[a]	Molecular weight	G + C (%)	Methylations	Number of gene copies
Eucaryotes				
p45S (nucleolar)	4.3×10^6	70	Mostly 2′-*O*-methylribose All retained in 28S and 18S	500–1000
p41S	3.1×10^6	~70		
p32S	2.1×10^6	70		
p20S	0.95×10^6	<70		
m28S	1.7×10^6	65	9–13/1000 NT	
m18S	0.7×10^6	58	15–19/1000 NT + $m_2{}^6A$	
m5.8S	5.1×10^4	~65	?	
m5S (nuclear)	3.9×10^4	65	None	Several hundred
Mitochondrial				
16S	5.4×10^5		Very low	1
12S	3.5×10^5		Very low	1
Procaryotes				
23S	1.1×10^6		11/1000 NT	5–6
16S	0.6×10^6		17/1000 Nt	5–6
5S	4×10^4	65	None	~7

[a] p, precursor form of the RNA; m, mature form of the RNA.

phate methylated at the number 7 nitrogen of the ring ($m^7G^{5'}ppp$). The cap is attached to the first transcribed nucleotide, usually a purine, methylated on the 2′-OH of the ribose (see Figure 16.8). The cap is followed by a nontranslated or **"leader"** sequence to the 5′ side of the coding region. Following the leader sequence is the initiation sequence or codon, most often AUG, and then the translatable message or coding region of the molecule. At the end of the coding sequence is a termination sequence signaling termination of polypeptide formation and release from the ribosome. A second nontranslated or "trailer" sequence follows, terminated by a string of adenylic acids, called a **poly(A) tail,** which makes up the 3′ terminus of the mRNA. This poly(A) section may vary from 20 to 200 nucleotides.

The 5′ cap structure blocks the action of RNA exonucleases and phosphatases, which could attack the 5′ terminus of the message. The cap also has a positive effect on the initiation of message translation. In the initiation of translation of a mRNA, the cap structure is recognized by a single ribosomal protein, an initiation factor (see Chapter 17). Several methylated nucleotides also occur in the internal portions of some mRNAs in addition to those on the cap and adjacent nucleotides. The majority of the internal methylations are m^6-adenosines with some m^5-cytidines. The role for the poly(A) tail has not yet been fully established. Most evidence suggests that the poly(A) sequence is correlated with the stability of the mRNA molecule; for example, *histone mRNA* molecules lack a poly(A)

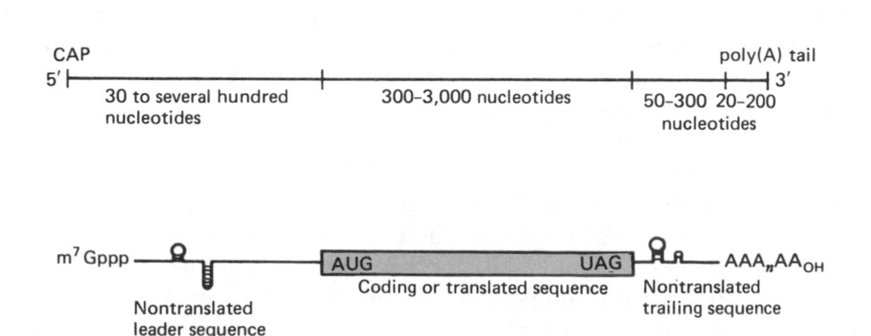

FIGURE 16.7
General structure for a eucaryotic mRNA. There is a "blocked" 5′ terminus, cap, followed by the nontranslated leader containing a promoter sequence. The coding region usually begins with the initiator codon AUG and continues to the translation termination sequence UAG, UAA, or UGA. This is followed by the nontranslated trailer and a poly(A) tail on the 3′ end.

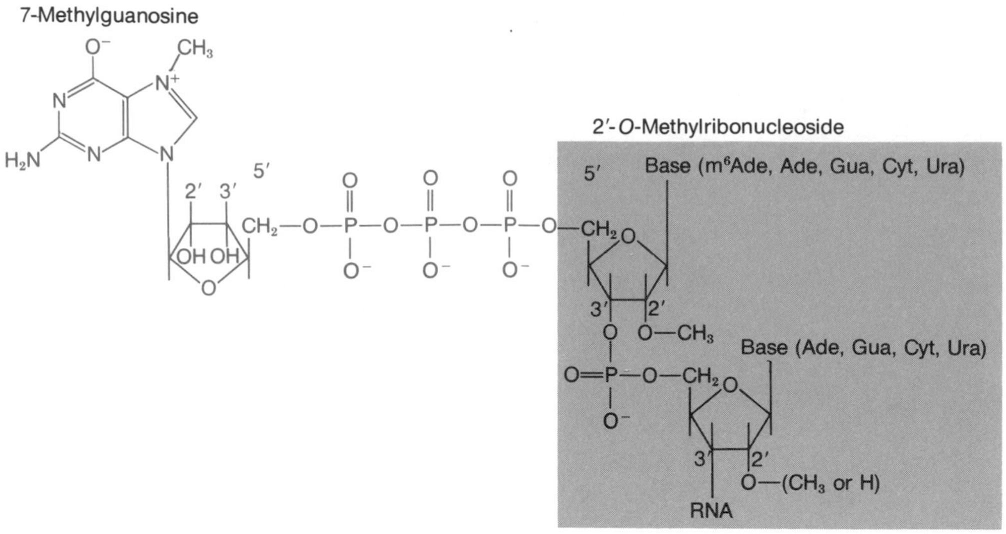

FIGURE 16.8
Diagram of the "cap" structure or blocked 5′ terminus in mRNA.
The 7-methylguanosine is inverted to form a 5′-phosphate to 5′-phosphate linkage with the first nucleotide of the mRNA. This nucleotide is often a methylated purine.

tail and are also present in the cell only transiently, during S phase of the cell cycle.

RNA in Ribonucleoprotein Particles

Besides tRNA, rRNA, and mRNA, small, stable RNA species can be found in the nucleus, cytoplasm, and mitochondria. These small RNA species function as ribonucleoprotein particles (RNPs), with one or more protein subunits attached. Different RNP species, both cytoplasmic and nuclear, have been implicated in a variety of cellular functions. These functions include RNA trimming, splicing, transport, and control of translation, as well as recognition of proteins due to be exported. The actual roles of these species, where known, are described more fully in the discussion of specific metabolic events.

Mitochondria Contain Unique RNA Species

Mitochondria (mt) have their own protein-synthesizing mechanisms, including ribosomes, tRNAs, and mRNAs. The mt rRNAs, 12S and 16S, are transcribed from the mitochondrial DNA (mt DNA) as are at least 19 specific tRNAs and some mRNAs. Note that there are fewer mt tRNAs than procaryotic or cytoplasmic tRNA species; there is only one mt tRNA species per amino acid. The mt RNAs account for 4% of the total cellular RNA. They are transcribed by a mitochondrial-specific RNA polymerase and are processed from a pair of mt RNA precursors. Each precursor is an exact copy of the entire mitochondrial genome, complementary to either the heavy (H) or light (L) strand of mt DNA. Genes for 12tRNAs are located on the heavy mt DNA strand and 7 on the light strand. Some of the mRNAs have eucaryotic characteristics, such as 3′-poly(A) tails. A large degree of coordination exists between the nuclear and mitochondrial genomes. Most of the aminoacylating enzymes for the mt tRNAs and most of the mitochondrial ribosomal proteins are specified by nuclear genes, translated in the cytoplasm and transported into the mitochondria. Furthermore, at least some of the modified bases in mt tRNA species are synthesized by enzymes encoded in nuclear DNA.

16.4 MECHANISMS OF TRANSCRIPTION

The Initial Process of RNA Synthesis Is Transcription

The process by which RNA chains are made from DNA templates is called **transcription.** All known transcription reactions take the following form:

$$\text{DNA template} + n(\text{NTP}) \longrightarrow \text{pppN(pN)}_{n-1} + (n-1)\text{PP}_i + \text{DNA template}$$

Enzymes that catalyze this reaction are designated RNA polymerases; it is important to recognize that they are absolutely template dependent. In contrast to DNA polymerases, however, RNA polymerases do not require a primer molecule. The energetics favoring the RNA polymerase reaction are twofold: First, the 5′ α-nucleotide phosphate of the ribonucleoside triphosphate is converted from a phosphate anhydride to a phosphodiester bond with a change in free energy ($\Delta G'$) of approximately 3 kcal mol^{-1} under standard conditions. Second, the released pyrophosphate, PP_i, can be cleaved to two phosphates by other enzymatic systems. This latter reaction means that [PP_i] is low and phosphodiester bond formation is favored relative to standard conditions (see Chapter 6 for a fuller discussion of metabolic coupling).

Since a DNA template is required for RNA synthesis, eucaryotic transcription takes place in the cell nucleus or mitochondrial matrix. Within the nucleus, the *nucleolus* is the site of rRNA synthesis, whereas mRNA and tRNA are synthesized in the nucleoplasm. Procaryotic transcription is accomplished on the cell's DNA, which is located in a relatively small region of the cell. In the case of procaryotic plasmids, the DNA template need not be associated with the chromosome.

Structural changes in DNA occur during its transcription. In the polytene chromosomes of *Drosophila,* transcriptionally active genes are visualized in the light microscope as puffs distinct from the condensed, inactive chromatin. Furthermore, the nucleosome patterns of active genes are disrupted so that active chromatin is more accessible to, for example, DNase attack. In procaryotes, the DNA double helix is transiently opened (unwound) as the transcription complex proceeds down the DNA.

These openings and unwindings are a manifestation of a topological necessity: If the RNA chain were copied off DNA without this unwinding, the transcription complex and growing end of the RNA chain would have to wind around the double helix once every 10 base pairs as they travel from the beginning of the gene to its end. Such a process would wrap the newly synthesized RNA chain around the DNA double helix; the problem then would be to unwind it before the RNA could be exported to the cytoplasm. Local opening and unwinding of the DNA solves this problem before it occurs by allowing transcription to proceed on a single face of the DNA. In addition, the opening of DNA base pairs during transcription allows Watson–Crick base pairing between template DNA and the newly synthesized RNA.

The process of transcription is divided into three parts: **Initiation** refers to the recognition of an active gene starting point by RNA polymerase and the beginning of the bond formation process. **Elongation** is the actual synthesis of the RNA chain, and is followed by chain **termination** and **release.**

The Template for RNA Synthesis Is DNA

Each cycle of transcription begins and ends with the recognition of certain sites in the DNA template. The DNA sequencing of a large number of transcription start regions, called **promoters,** has shown that certain sequences occur with great regularity. These sequences are called **conserved** or **consensus sequences.**

An example of the identification of a consensus sequence is shown in Figure 16.9. The sequences of a large number of promoter sequences were determined. Then a computer program was used to count how often each nucleotide occurred at each position. *Consensus nucleotides* are those that occur more often than would be predicted by random chance. This sort of comparative analysis has been applied to nucleic acid sequences involved in a wide variety of genetic functions. Defining a consensus sequence is usually the first step in studying a molecular genetic reaction. Similar considerations demonstrate that termination occurs at specific consensus-type sequences. In addition, sites within a transcript may allow premature termination of transcription. These sites can act as molecular switches affecting the continuation of synthesis of an RNA molecule.

Consensus sequences near the transcription start are found for both procaryotic and some eucaryotic promoters. In addition, eucaryotic transcription has been shown in some cases to be affected by **internal promoter** elements and other sequences called **enhancers.** Enhancers are gene-specific sequences that positively affect transcription. The most extensively studied is a twofold repeat of a 72 base pair sequence in the simian virus 40 (SV40) chromosome. The remarkable observation is that the SV40 and other enhancers function at various positions and in either orientation in the SV40 DNA. In contrast to, for example, a site in which RNA polymerase initially binds DNA, enhancer sequences can stimulate transcription whether they are located at the beginning or the end of a gene. The enhancer sequence must be on the same DNA strand as the transcribed gene (genetically in a position cis) but can function in either orientation. Cellular proteins are known that specifically bind enhancers. The most likely hypothesis is that eucaryotic enhancers serve to bring about a structural change in the DNA template, allowing transcription to occur.

RNA Polymerase Catalyzes the Transcription Process

The RNA polymerases all synthesize RNA in the $5' \rightarrow 3'$ direction using a DNA template; in this respect, they are similar to template-dependent DNA polymerases discussed in Chapter 15. Unlike DNA polymerases, however, RNA polymerases are not primer dependent. They are capable of initiating polymerization de novo at a promoter sequence. Cellular RNA polymerases in both procaryotic and eucaryotic organisms are large multisubunit enzymes whose mechanisms are only partially understood despite over 30 years of intense effort.

The most intensely studied procaryotic RNA polymerase is that from *E. coli,* which consists of five subunits having an aggregate molecular weight of over 500,000 (Table 16.5). Two α subunits, one β subunit and one β' subunit, constitute the **core enzyme,** which is capable of faithful transcription but not of specific (i.e., correctly initiated) RNA synthesis. The addition of a fifth protein subunit, designated σ, results in the **holoen-**

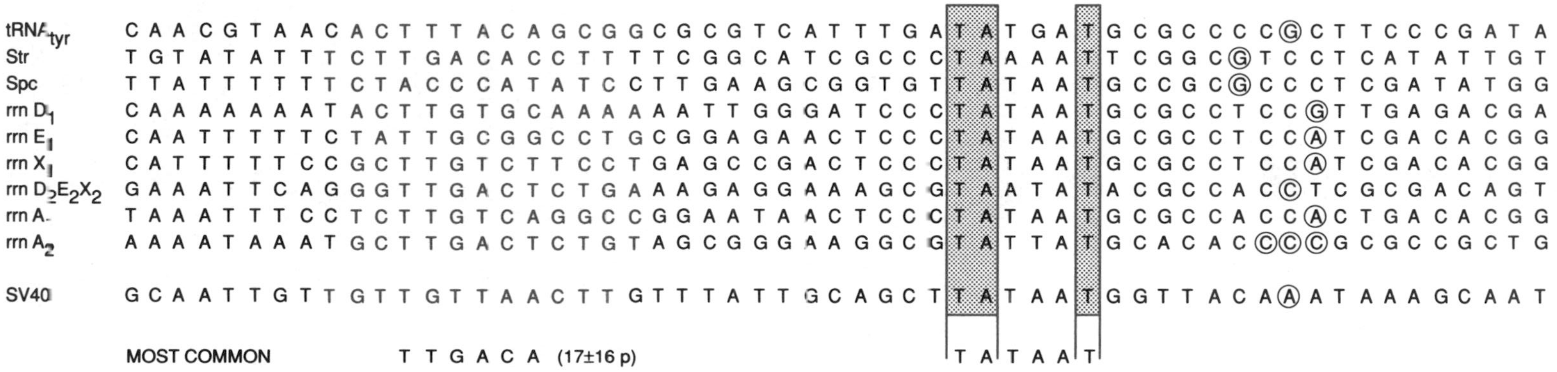

```
tRNA tyr     C A A C G T A A C A C T T T A C A G C G G C G C G T C A T T T G A T A T G A T G C G C C C G C T T C C C G A T A
Str          T G T A T A T T T C T T G A C A C C T T T T C G G C A T C G C C C T A A A A T T C G G C G T C C T C A T A T T G T
Spc          T T A T T T T T T C T A C C C A T A T C C T T G A A G C G G T G T T A T A A T G C C G C G C C C T C G A T A T G G
rrn D1       C A A A A A A A T A C T T G T G C A A A A A A T T G G G A T C C C T A T A A T G C G C C T C C G T T G A G A C G A
rrn E1       C A A T T T T T C T A T T G C G G C C T G C G G A G A A C T C C C T A T A A T G C G C C T C C A T C G A C A C G G
rrn X1       C A T T T T T C C G C T T G T C T T C C T G A G C C G A C T C C C T A T A A T G C G C C T C C A T C G A C A C G G
rrn D2E2X2   G A A A T T C A G G G T T G A C T C T G A A A G A G G A A A G C G T A A T A T A C G C C A C C T C G C G A C A G T
rrn A1       T A A A T T T C C T C T T G T C A G G C C G G A A T A A C T C C C T A T A A T G C G C C A C C A C T G A C A C G G
rrn A2       A A A A T A A A T G C T T G A C T C T G T A G C G G G A A G G C G T A T T A T G C A C A C C C C G C G C C G C T G

SV40         G C A A T T G T T G T T G T T A A C T T G T T T A T T G C A G C T T A T A A T G G T T A C A A A T A A A G C A A T

             MOST COMMON         T T G A C A  (17±16 p)                   T A T A A T
```

FIGURE 16.9

Determination of a consensus sequence for procaryotic promoters.

A portion of the data set used for the identification of the consensus sequence for *E. coli* promoter activity. Note that none of the individual promoters has the entire consensus sequence.

Modified from M. Rosenberg and D. Court, *Ann. Rev. Genet.* 13:319, 1979.

TABLE 16.5 Comparative Properties of Some RNA Polymerases

	Nuclear				
	I (A)	II (B)	III (C)	Mitochondrial	*E. coli*
High mol wt subunits[a]	195–197	240–214	155	65	160 (β')
	117–126	140	138		150 (β)
Low mol wt subunits	61–51	41–34	89		86 (σ)
	49–44	29–25	70		40 (α)
	29–25	27–20	53		10 (ω)
	19–16.5	19.5	49		
		19	41		
		16.5	32		
			29		
			19		
Variable forms	2–3 types	3–4 types	2–4 types	1	1
Specialization	Nucleolar; rRNA	mRNA Viral RNA	tRNA 5S rRNA	All mtRNA	None
Inhibition by α-amanitin	Insensitive (>1 mg mL^{-1})	Very sensitive 10^{-9}–10^{-8} M	Sensitive 10^{-5}–10^{-4} M	Insensitive, but sensitive to rifampicin	Rifampicin-sensitive

[a] Molecular weight × 10^{-3}.

zyme that is capable of specific RNA synthesis in vitro and in vivo. The logical conclusion, that σ is involved in the specific recognition of promoters, has been borne out by a variety of biochemical studies and is discussed below. Furthermore, specific σ factors can recognize different classes of genes. This has been shown to be the case in several systems. In *E. coli* a specific σ factor recognizes a class of promoters for genes that are turned on as a result of heat shock. In *B. subtilis* specific σ factors are made that recognize genes turned on during sporulation. Some bacteriophage synthesize σ factors that allow the appropriation of the cell's RNA polymerase for transcription of the viral DNA.

The common procaryotic RNA polymerases are inhibited by the antibiotic *rifampicin* (used in treating tuberculosis), which binds to the β subunit (Clin. Corr. 16.2). Eucaryotic nuclear RNA polymerases are inhibited differentially by the compound α-amanitin, which is synthesized by the poisonous mushroom *Amanita phalloides*. A particular concentration of amanitin can be used to inhibit synthesis of a class of RNA in vitro or in vivo. Using such experiments, three nuclear RNA polymerase classes can be distinguished. Very low concentrations of α-amanitin inhibit the synthesis of mRNA and some small nuclear RNAs (snRNAs); higher concentrations inhibit the synthesis of tRNA and other snRNAs, whereas rRNA synthesis is not inhibited at these concentrations of drug. Messenger RNA synthesis is the function of **RNA polymerase II.** Synthesis of transfer RNA, 5sRNA, and some snRNAs are carried out by **RNA polymerase III.** Ribosomal RNA genes are transcribed by **RNA polymerase I,** which is concentrated in the nucleolus. (The numbers refer to the order of elution of the enzymes from a chromatography column.) Each enzyme is highly complex structurally (Table 16.5), and functions have not been established for individual enzyme subunits.

In addition, a mitochondrial RNA polymerase is responsible for the synthesis of this organelle's mRNA, tRNA, and rRNA species. This enzyme, like bacterial RNA polymerase, is inhibited by rifampicin.

The Steps of Transcription in Procaryotes Have Been Determined

Transcription is a **strand-selective process;** most double helical DNA is transcribed in only one direction. This is illustrated as follows:

DNA: 5′ ———————— 3′

3′ ———————— 5′

RNA: 5′ ⟶ 3′ · · ·

Note that one of the two DNA strands serves as template for RNA synthesis. The template strand is sometimes called the *sense* strand because it is complementary to the RNA transcript. Conventionally, the sense strand is usually the "bottom" strand of a double-stranded DNA as written. The other strand, the "top" strand in a DNA sequence, has the same direction as the transcript when read in the 5′ → 3′ direction; this strand is sometimes (confusingly) called the antisense strand. When only a single DNA sequence is given in this book, the *antisense* strand is represented. Its sequence can be easily interpreted into the RNA transcript of a gene by simply substituting U (uridine) for T (thymidine) bases. Procaryotic transcription begins with the specific binding of RNA polymerase to a gene's promoter (Figures 16.9 and 16.10). RNA polymerase holoenzyme binds to one face of the DNA extending some 45 or so bp upstream and 10 bp downstream from the RNA initiation site. Two short oligonucleotide sequences in this region are highly conserved. One sequence that is located about 10 bp upstream from the transcription start is the consensus sequence (sometimes called a **Pribnow box**):

T*A*TAAT*

The positions marked with an asterisk are the most conserved; indeed, the last T residue is *always* found in *E. coli* promoters.

A second consensus sequence is located upstream from the Pribnow or "−10" box. This "−35 sequence" is centered about 35 bp upstream from the transcription start and its consensus sequence is

T*T*G*ACA

where the nucleotides with asterisks are most conserved. The spacing between the "−35 and −10 sequences" is crucial with 17 bp being highly conserved. Note that the nature of the promoter sequence imparts strand selectivity to the process of transcription. The consensus elements, TTGACA and TATAAT, are asymmetrical, that is, they do not have the same sequence if the complementary sequence is read. For an exercise, compare the palindromic recognition sequence of the Eco RI restriction enzyme, GAATTC, with the consensus sequence of the Pribnow box, TATAAT. Write down each sequence and its complement and read the sequences in the 5′ → 3′ direction. The nonpalindromic nature of the promoter consensus sequence means that a given promoter sequence directs transcription in only a single direction, as shown in Figure 16.11.

Consensus sequences are useful and striking, but what difference do they make to a gene? Measurements of RNA polymerase binding affinity and initiation efficiency to various promoter sequences have shown that the most active promoters fit the consensus sequences most closely. Statistical measurements of promoter homology conform closely to the measured "strength" of a promoter; that is, its kinetic ability to initiate transcription with purified RNA polymerase.

Other conserved elements in the promoter region are observed by statistical analysis: bases flanking the "−35 and −10 sequences," bases near the transcription start, and bases located near the −16 position are weakly conserved. In some of these weakly conserved regions, RNA polymerase may require that a particular nucleotide *not* be present or that local variations in DNA helical structure be present.

CLIN. CORR. 16.2
ANTIBIOTICS AND TOXINS TARGETING RNA POLYMERASE

RNA polymerase is obviously an essential enzyme for life since transcription is the first step of gene expression. No RNA polymerase means no enzymes (or anything else). Two natural products point out this principle; in both cases inhibition of RNA polymerase leads to death of the organism.

The "death cap" or "destroying angel" mushroom, *Amanita phalloides,* is highly poisonous and still causes several deaths each year despite widespread warnings to amateur mushroom hunters (it is reputed to taste delicious, incidentally). The most lethal toxin, α-amanitin, inhibits the largest subunit of eucaryotic RNA polymerase II, resulting in the inhibition of mRNA synthesis. The course of the poisoning is twofold: initial relatively mild gastrointestinal symptoms are followed about 48 h later by massive liver failure as essential mRNAs and their proteins are degraded but not replace by newly synthesized molecules. The only therapy is supportive including liver transplantation; but this is clearly a desperate measure of unproven efficacy.

More benign (at least from the point of view of our own species) is the action of the antibiotic rifampicin to inhibit the RNA polymerases of a variety of bacteria, most notably in the treatment of tuberculosis. *Mycobacterium tuberculosis,* the causative agent of tuberculosis, is insensitive to many commonly used antibiotics but it is sensitive to rifampicin, the product of a soil streptomycete. Since mammalian RNA polymerase is so different from the procaryotic variety, inhibition of the latter enzyme is possible without great toxicity to the host. This consideration implies a good therapeutic index for the drug, that is, the ability to treat a disease without causing undue harm to the patient. Together with improved public health measures, antibiotic therapy with rifampicin and isoniazid (an antimetabolite) has greatly reduced the morbidity due to tuberculosis in industrialized countries. Unfortunately the disease is still endemic in impoverished populations in the United States and in other countries. Furthermore, in increasing numbers immunocompromised individuals, especially AIDS patients, have active tuberculosis.

Mitchel, D. H. Amanita Mushroom Poisoning. *Ann. Rev. Med.* 31:51, 1980; and Mandell, G. L. and Sande, M. A., A. G. Gilman, L. S. Goodman, T. W. Rall, and F. Murad (Eds.). *The Pharmacological Basis of Therapeutics* 7th ed., New York: Macmillan, 1985, pp. 1199–1204.

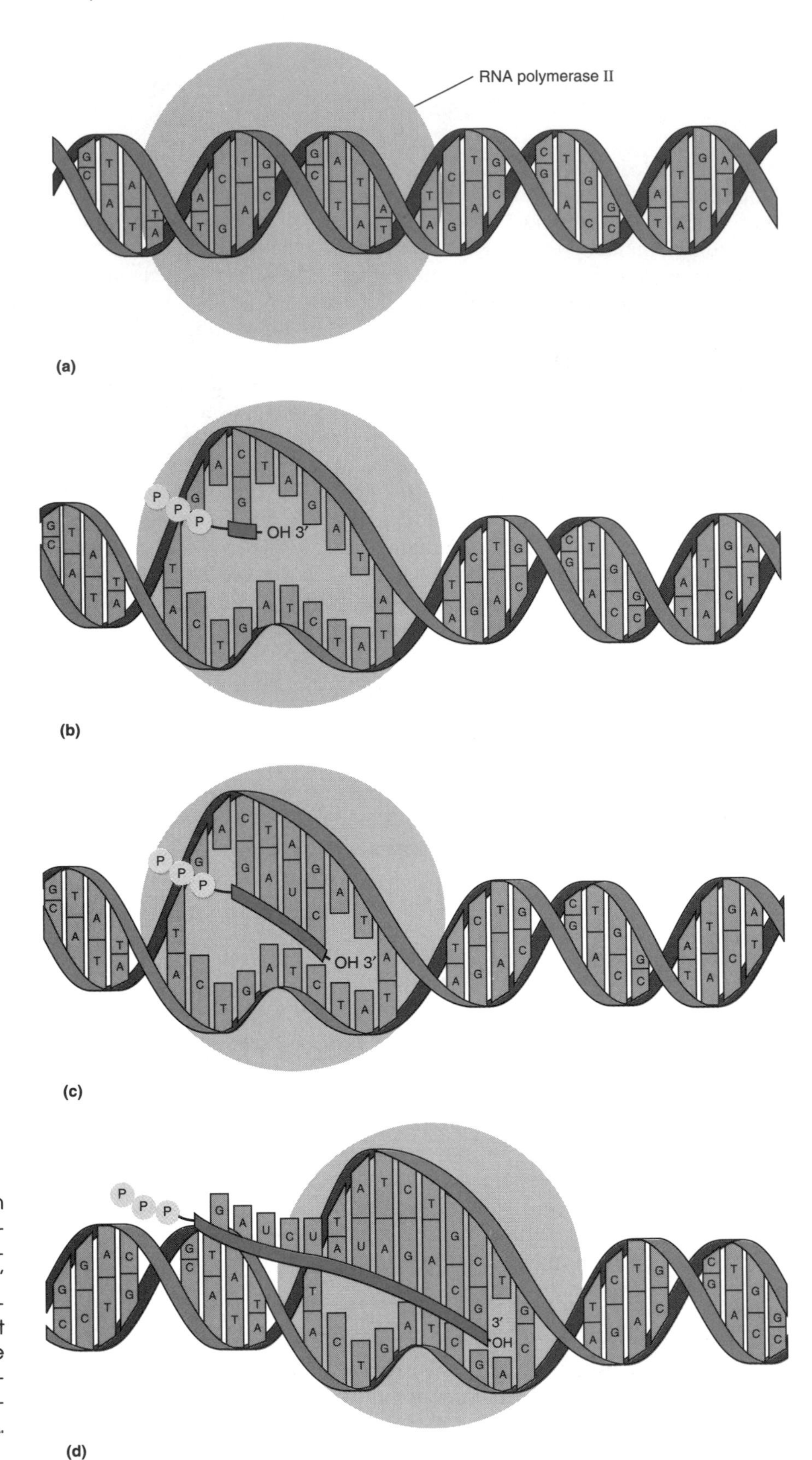

FIGURE 16.10
Early events in procaryotic transcription.
(*a*) Recognition: RNA polymerase with "sigma" factor binds to a DNA promoter region in a "closed" conformation. (*b*) Initiation: The complex is converted to an "open" conformation and the first nucleoside triphosphate aligns with the DNA. (*c*) The first phosphodiester bond is formed and the "sigma" factor released. (*d*) Elongation: Synthesis of nascent RNA proceeds with movement of the RNA polymerase along the DNA. The double helix reforms.

Promoters for *E. coli* heat shock genes have different consensus sequences at the "−35" and "−10" homologies. This is consistent with their being recognized by a different σ factor.

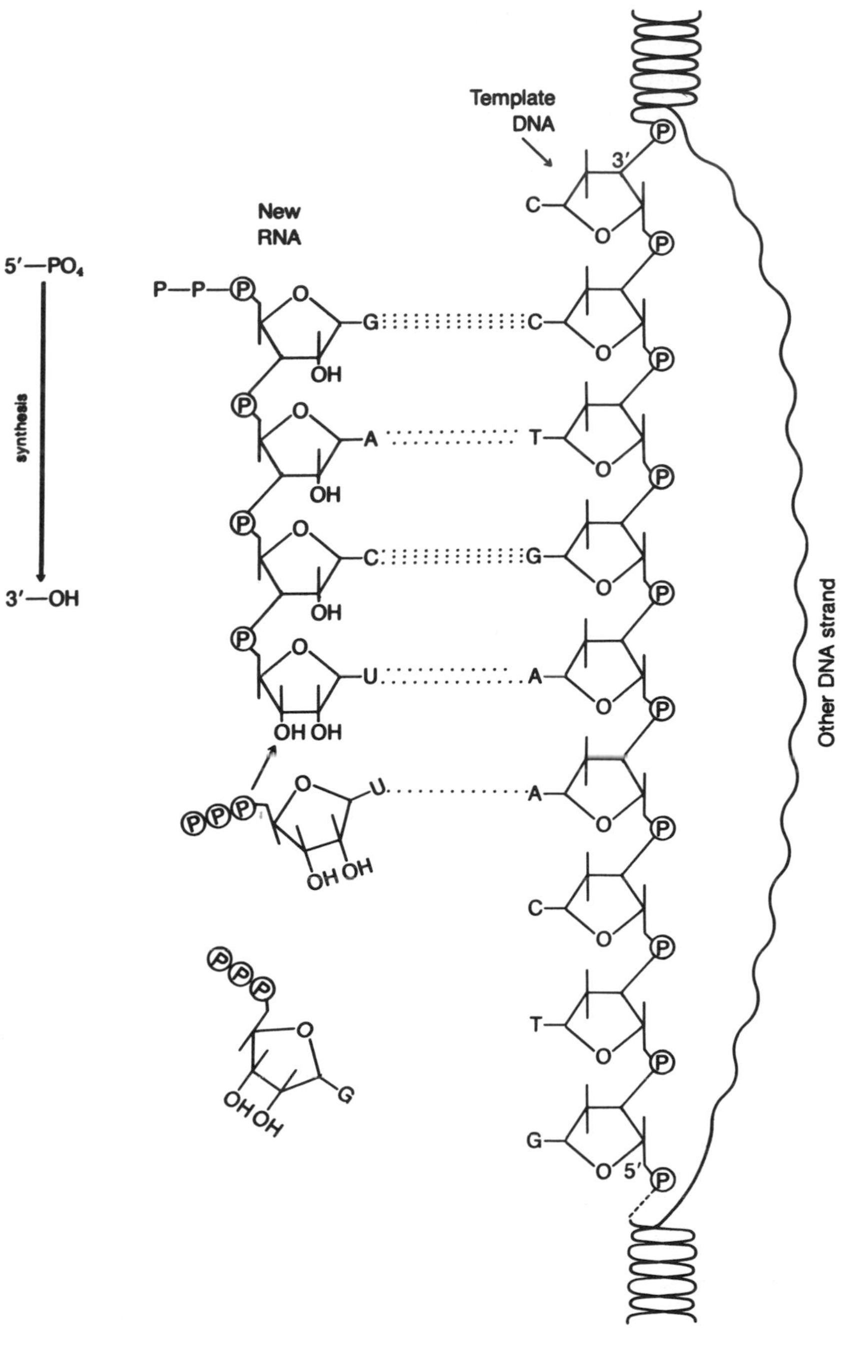

FIGURE 16.11
Biosynthesis of RNA showing asymmetry in transcription.
Nucleoside 5′-triphosphates align with complementary bases on one DNA strand, the template. RNA polymerase catalyzes the formation of the 3′,5′-phosphodiester links by attaching the 5′-phosphate of the incoming nucleotide to the 3′-OH group of the growing nascent RNA releasing P_i. The new RNA is synthesized from its 5′ end toward the 3′ end.

An RNA transcript usually starts with a purine riboside triphosphate; that is, pppG · · · or pppA · · · , but pyrimidine starts are also known (Figures 16.10 and 16.11). The position of an RNA chain initiation differs slightly between various promoters, usually occurring from five to eight base pairs downstream from the invariant T of the Pribnow box.

Initiation

Two kinetically distinct steps are required for RNA polymerase to initiate an RNA chain. In the first step, RNA polymerase holoenzyme binds electrostatically to the promoter DNA to form a **"closed complex."** In the second step, the holoenzyme forms a more tightly bound **"open complex,"** which is characterized by a local opening of about 10 bp of the DNA double helix. Note that the consensus Pribnow box is A-T rich; it therefore can serve as the initiation of this local unwinding. As discussed in Chapter 15, unwinding of a 10 base pair stretch of the DNA double helix is topologically equivalent to relaxation of a single negative supercoil. As might be predicted from this observation, the activity of some promoters has been shown to depend on the superhelical state of the DNA template; some promoters are more active on highly supercoiled DNA while others are more active when the superhelical density of the template is lower. The unwound DNA binds the initiating triphosphate and then forms the first phosphodiester bond. The enzyme translocates to the next position (this is the rifampicin-inhibited step) and continues synthesis. At or a short time after the initial bond formation, σ factor is released and the enzyme is considered to be in an *elongation* mode.

Elongation

RNA polymerase continues the binding–bond formation–translocation cycle at a rate of about 40 nucleotides per second. This rate is only an average, however, and there are many examples known for which RNA polymerase pauses or slows down at particular sequences, usually inverted repeats (palindrome sequence of nucleotides). As will be discussed below, these pauses can bring about termination.

As RNA polymerase continues down the double helix, it continues to separate the two strands of the DNA template. As seen in Fig. 16.11, this process allows the template (sense) strand of the DNA to base pair with the growing RNA chain. Thus a single mechanism of information transfer (Watson–Crick base pairing) is able to serve several processes: DNA replication, DNA repair, and transcription of genetic information into RNA. (As will be seen in chapter 17, base pairing is essential for translation as well.) The process of unwinding and restoring the DNA double helix is aided by DNA topoisomerases I and II, which are a part of the transcription complex.

Changes in the transcription complex during the elongation phase can affect subsequent termination events. These changes depend on the binding of another cellular protein (nusA protein) to core RNA polymerase. Failure to bind sometimes results in an increased frequency of termination and, consequently, a reduced level of gene expression.

Termination

The RNA polymerase complex recognizes the *ends* of genes as well (Figure 16.12). Transcription termination can occur in either of two modes, depending on whether or not it is dependent on the protein factor **rho.** Terminators are thus classified as rho-independent or rho-dependent.

Rho-independent terminators are better characterized (Figure 16.13). A consensus-type sequence is involved here: a G-C rich **palindrome** (inverted repeat) precedes a sequence of 6–7 U residues in the RNA chain.

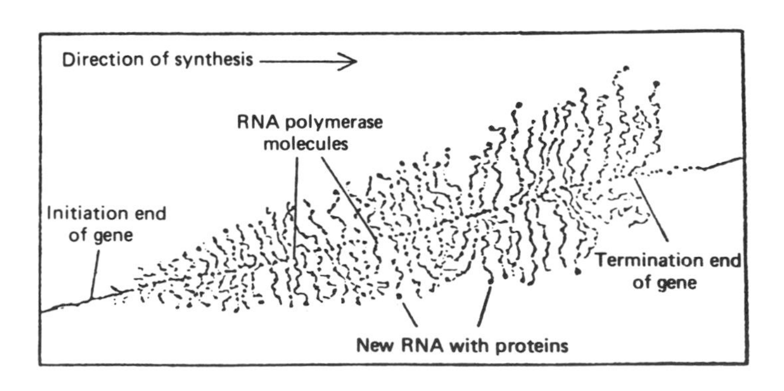

FIGURE 16.12
Simultaneous transcription of a gene by many RNA polymerases, depicting the increasing length of nascent RNA molecules.
Courtesy of Dr. O. L. Miller, University of Virginia. Reproduced with permission from O. L. Miller and B. R. Beatty, *J. Cell Physiol.* 74:225, 1969.

As a result the RNA chain forms a stem and loop structure preceding the U sequence. The secondary structure of the stem and loop is crucial for termination; base change mutations in the stem and loop that disrupt pairing also reduce termination. Furthermore, the most efficient terminators are the most G-C rich and therefore most stable. The terminator stem and loop stabilize procaryotic mRNA against nucleolytic degradation.

Rho-dependent terminators are less well characterized, and the biochemical mechanism of rho action is still unclear. Rho factor does possess an RNA-dependent ATPase activity that is required for termination. Further, rho does seem to act at polymerase pausing sites in the DNA sequence. This has suggested that rho-dependent termination occurs as a result of competition between elongation and termination.

It is also important to remember that procaryotic ribosomes usually attach to the nascent mRNA while it is still being transcribed. This coupling between transcription and translation is important in gene control by *attenuation,* which is discussed in Chapter 19.

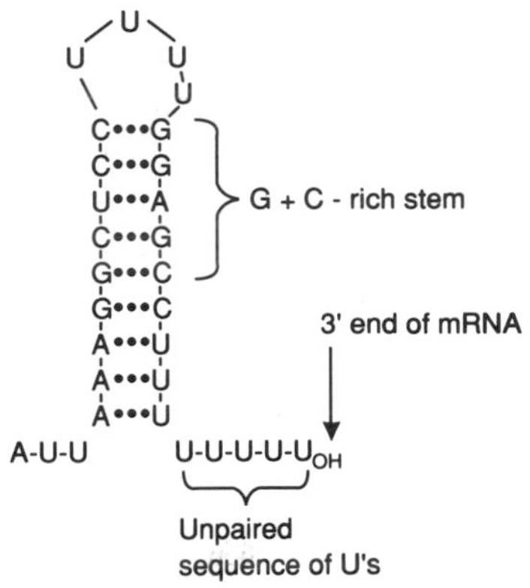

FIGURE 16.13
The stem-loop structure of the RNA transcript that determines ρ-independent transcriptional termination.
Note the two components of the structure: the G + C-rich stem and loop, followed by a sequence of U residues.

Transcription in Eucaryotes Involves Additional Molecular Events

Eucaryotic transcription is considerably more complex than the process in procaryotes. While the information specifying a promoter is still carried in a DNA sequence, the sequence of a promoter and an RNA polymerase enzyme to recognize it are only some of the components determining whether transcription occurs. Several other molecular events are required for eucaryotic transcription initiation. First, chromatin containing the promoter sequence must be spatially accessible to the transcription machinery. Second, protein transcription factors distinct from RNA polymerase must bind to sequences in the promoter region for a gene to be active. Third, other sequences located some distance away from the promoter affect transcription; these sequences are termed *enhancers* and they, too, bind protein factors to stimulate transcription. Finally, recall the eucaryotic RNA polymerase consists of three distinct enzyme forms, each specific form capable of transcribing only a single class of cellular RNA. Compare this number of events with transcription in procaryotes that requires, in its simplest case, only an appropriate sequence of DNA, RNA polymerase holoenzyme and nucleoside triphosphate substrates.

The Nature of Active Chromatin

The structural organization of eucaryotic chromosomes was discussed in Chapter 15. Chromatin is organized into nucleosomes whether or not it is capable of being transcribed. Beyond the similarity of nucleosome structure, some differences do exist. First, an active gene has a generally "looser" configuration than does transcriptionally inactive chromatin. This difference is most striking in the promoter sequences, parts of which are not organized into nucleosomes at all (Figure 16.14). The lack of nucleosomes is manifested experimentally by the enhanced sensitivity of promoter sequences to external reagents that cleave DNA, such as the enzyme DNase I. This enhanced accessibility of promoter sequences, termed **DNase I hypersensitivity,** ensures that transcriptional factors will be able to bind to appropriate regulatory sequences. Second, although the transcribed parts of a gene may be organized into nucleosomes, the nucleosomes are less tightly bound than those in an inactive gene. The overall pattern, therefore, is one of a partially unfolded chromatin structure being necessary but not sufficient for transcription. This suggests that specific proteins bind to these regions to activate or inactivate transcription.

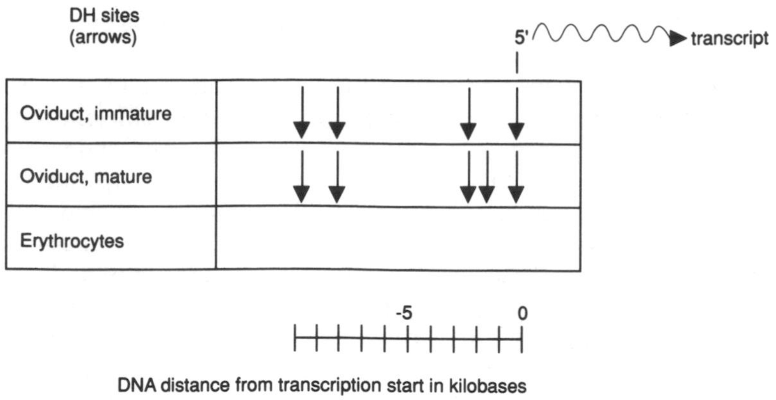

FIGURE 16.14
DNase-hypersensitive (DN) sites upstream of the promoter for the chick lysozyme gene, a typical eucaryotic transcriptional unit.
Hypersensitive sites, that is, sequences around the lysozyme gene where nucleosomes are not bound to the DNA, are indicated by arrows. Note that some hypersensitive sites are found in the lysozyme promoter whether the oviduct is synthesizing or not synthesizing lysozyme; the synthesis of lysozyme is accompanied by the opening up of a new hypersensitive site in mature oviduct. In contrast, no hypersensitive sites are present in nucleated erythrocytes that never synthesize lysozyme.
Adapted from S. C. R. Elgin, *J. Biol. Chem.* 263:19258, 1988.

Enhancers

Enhancer sequences were first discovered during recombinant DNA studies of transcription in vivo where it was observed that the maximum expression of a eucaryotic gene requires sequences other than those in the promoter. The presence of these sequences increased (enhanced) the expression of a gene about 100-fold, hence the name. Further experiments showed that enhancers share several properties. They function only when located on the same DNA molecule (chromosome) as the promoter whose activity they affect. They can function when located in either the 5′ or 3′ direction relative to the gene whose transcription they enhance. These sequences can enhance transcription of a gene when located some distance away from the affected promoter. This distance can be as much as 1000 bp away from the affected promoter. Protein factors bind to enhancer DNA; these protein factors are necessary for enhancer function.

Transcription of Ribosomal RNA Genes

Recall that the rRNA genes are located in a specialized nuclear structure, the nucleolus. There are several hundred copies of each rRNA gene in a eucaryotic cell. These sequences transcribed into rRNA are tandemly repeated in the DNA contained in a specific region of one chromosome, the **nucleolar organizer.** Each repeat unit of ribosomal DNA contains a copy of each RNA sequence separated by **nontranscribed spacer** regions. Figure 16.15 is a diagram of this arrangement. Each repeat unit is transcribed as a unit, yielding a primary transcript containing one copy each of the 28S, 5.8S, and 18S sequences. The primary transcript is then processed by ribonucleases and modifying enzymes to the three mature molecules (see Section 16.5). Termination of transcription occurs within the nontranscribed spacer region before RNA polymerase reaches the promoter of the next repeat unit.

The promoter recognized by RNA polymerase I is located within the nontranscribed spacer region of the gene cluster from about positions −40

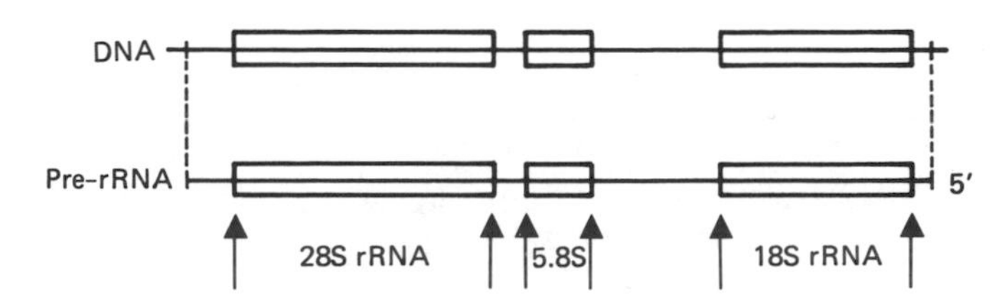

FIGURE 16.15
Structure of a rRNA transcription unit.
Ribosomal RNA genes are arranged with many copies one after another. Each copy is transcribed separately and each transcript is processed into three separate RNA species. Promoter and enhancer sequences are located in the nontranscribed regions of the tandemly repeated sequences.

to +10. A transcription factor binds to the promoter and thereby directs RNA polymerase recognition of the promoter sequence. In addition, an enhancer element is located about 250 bp upstream from the promoter in human ribosomal DNA. The size of the nontranscribed spacer varies considerably from one organism to the next, as does the position of the enhancer element.

Transcription of rRNA can be very rapid; this reflects the fact that synthesis of ribosomes is rate-limiting for cell growth. There is some evidence that phosphorylation of RNA polymerase I may activate transcription of rRNA, for example, during embryonic growth or liver regeneration.

Transcription by RNA Polymerase II

The RNA polymerase II is responsible for the synthesis of mRNA in the nucleus. Since the transcription of a gene is the first step in gene expression, much effort has gone into delineating the molecular events that make transcription occur. Three common themes have emerged from research on a large number of genes (Figure 16.16). (1) The DNA sequences controlling transcription are complex; a single gene may be controlled by as many as six or eight DNA sequence elements in addition to the promoter (RNA polymerase binding region) itself. The controlling sequence elements function in combination to give a finely tuned pattern of control. (2) The effect of the controlling sequences on transcription is mediated by the binding of protein molecules to each sequence element. These proteins, called factors, recognize the nucleotide sequence of the appropriate controlling sequence element. (3) Transcription factors when bound to the DNA sequence elements interact with each other and with RNA polymerase to activate transcription. The DNA binding and activation activities of the factors are carried on separate domains of the protein molecules.

Promoters for mRNA Synthesis

Accurate transcription by eucaryotic RNA polymerase II usually requires the presence of two consensus sequences upstream from the mRNA start site. The first and most prominent of these, sometimes called the **TATA box,** has the sequence

$$\text{TATA}^{\text{A}}_{\text{T}}\text{A}^{\text{A}}_{\text{T}}$$

These nucleotides are highly conserved, much more so than in the *E. coli*

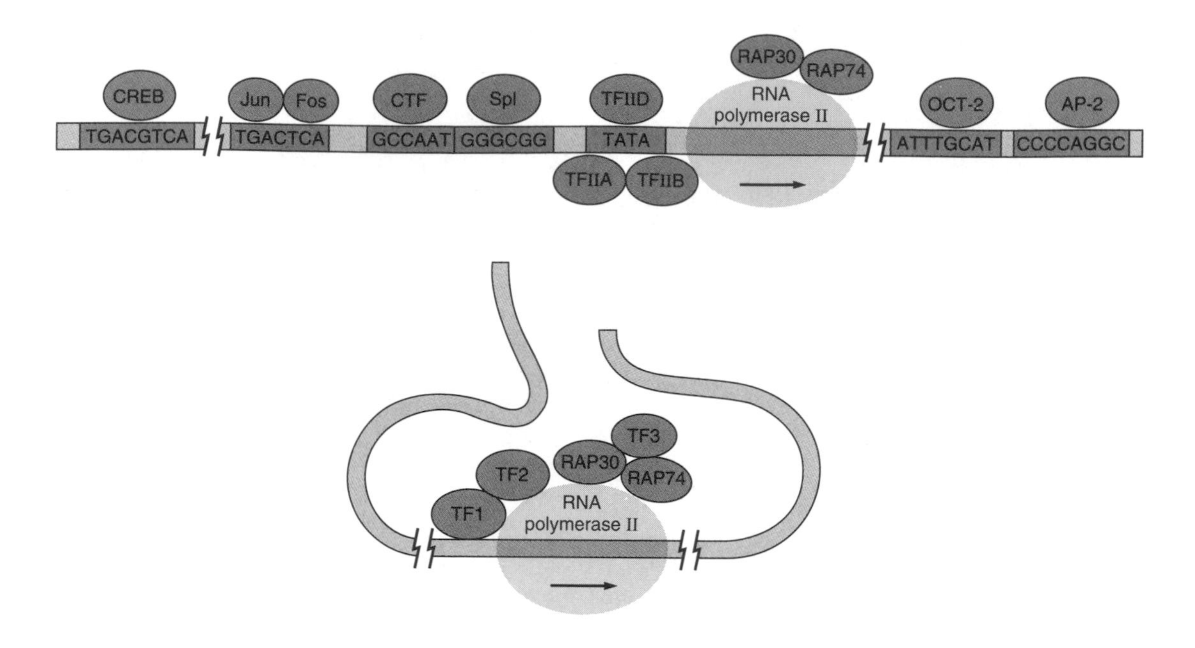

FIGURE 16.16
Interaction of transcription factors with promoters.
A large number of transcriptional factors interact with eucaryotic promoter regions. (*a*) A hypothetical array of factors that interact with specific DNA sequences near the promoter. This includes a factor, TFIID, which binds to the TATA box and the Jun and Fos proteins, which are protooncogenes (Clin. Corr. 16.3). The figure is not meant to imply that all of the DNA binding factors bind to the promoter simultaneously. (*b*) One way in which the DNA binding factors are hypothesized to bind to each other and to RNA polymerase. Although this model is not completely proven, it is known that proteins that bind to distant DNA sequences make protein–protein contacts with each other.
Reprinted with permission from P. J. Mitchell and R. Tjian, *Science* 245:371, 1989.

"−10" consensus sequence previously described. The TATA box is centered about 25 bp upstream from the transcription unit. Experiments in which it was deleted suggest that it is required for efficient transcription, although weak promoters may lack it entirely.

A second region of homology is located further upstream, in which the **CAAT** box sequence

$$\mathbf{GG^{T}_{C}CAATCT}$$

is found. This sequence is not as highly conserved as the TATA box, and some active promoters may not possess it.

Both the CAAT and TATA boxes require the binding of specific transcription factors to function. The current model for the activation of genes in this manner is shown in Figure 16.17. Note how protein factors bind not only to their recognition sequences but also to each other and to RNA polymerase, itself a very large and complex enzyme. Despite the complexities of the detailed interactions, the three principles elaborated above account for the known mechanisms of all Class II transcription factors. Mutated forms of several of these transcription factors function as nuclear oncogenes (Clin. Corr. 16.3).

Transcription by RNA Polymerase III

The themes elaborated above for the transcription of Class I and II promoters also hold for the transcription of 5 S RNA and tRNA by RNA

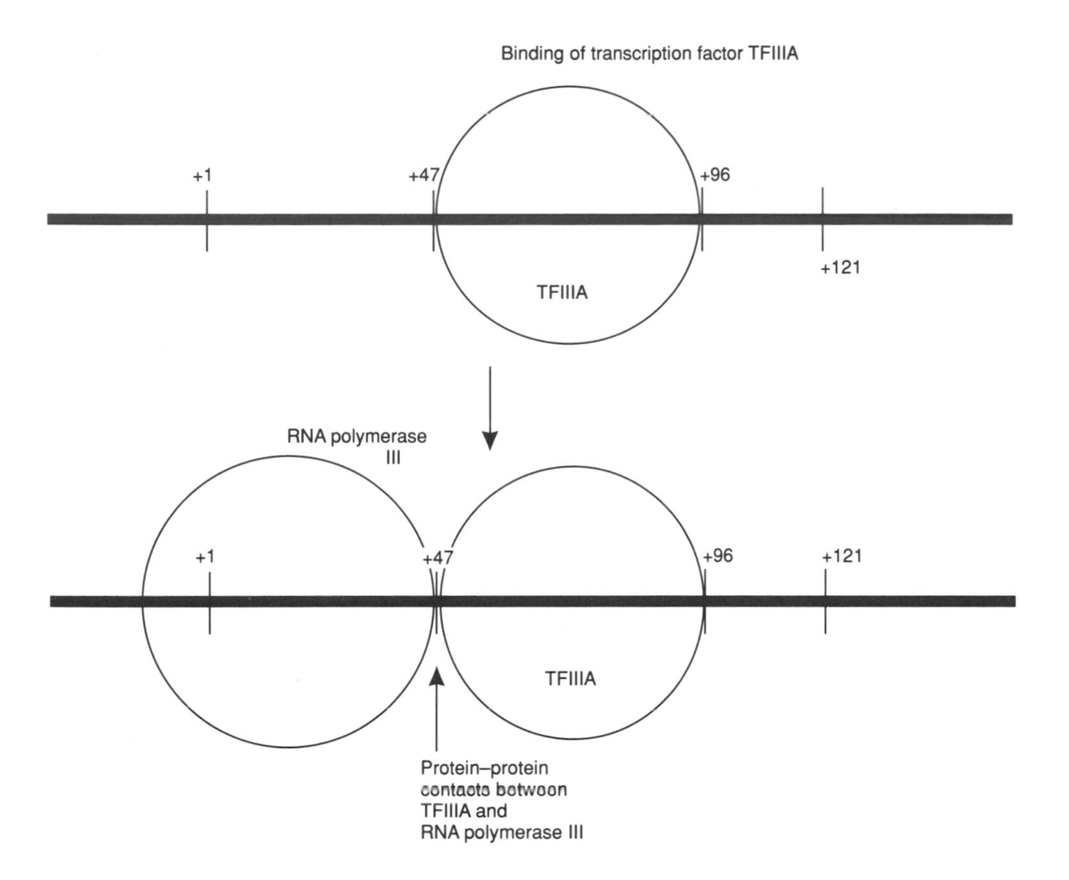

FIGURE 16.17
Transcription factor for a Class III eucaryotic gene.
The transcription factor TFIIIA binds to a sequence located within the *Xenopus* gene for 5S rRNA. The RNA polymerase III then binds to the factor and initiates transcription of the 5S sequence. No specific sequence in the DNA is required other than the factor binding sequence.

CLIN. CORR. 16.3 INVOLVEMENT OF TRANSCRIPTIONAL FACTORS IN CARCINOGENESIS

The conversion of a normally well-regulated cell into a cancerous one requires a number of independent steps whose end result is a mutated cell capable of uncontrolled growth and metastasis. Insights into this process have come about by recombinant DNA studies of the genes whose mutated or overexpressed products contribute to carcinogenesis. These genes are termed *oncogenes*. Oncogenes were first identified as products of DNA and RNA tumor viruses but, somewhat surprisingly, normal cells have copies of these genes as well. The normal, nonmutated cellular analogs of oncogenes are termed protooncogenes. The products of protooncogenes are components of the many pathways that regulate growth and differentiation of a normal cell; mutation into an oncogenic form involves a change that makes the regulatory product less responsive to normal control.

Some protooncogenic products are involved in the transduction of hormonal signals or the recognition of cellular growth factors and act cytoplasmically. A second class of protooncogenes has a nuclear site of action; their gene products are often associated with the transcriptional apparatus where they are synthesized in response to growth stimuli. It is easy to visualise how the overproduction or permanent activation of such a positive transcription factor could aid the transformation of a cell to malignancy: genes normally transcribed at a low or controlled level would be overexpressed by such a deranged control mechanism.

A more subtle genetic effect predisposing to cancer is exemplified by the product of the human gene responsible for heredity retinoblastoma. This tumor primarily affects young children; those patients who are successfully treated for the primary tumor are often at increased risk for other types of cancer in later life. Tumor cells have lost both copies of the *Rb* gene and are thereby predisposed to malignancy. This means that the wild-type *Rb* gene acts as a *suppressor* of tumorigenesis; therefore the *Rb* gene itself is sometimes called an antioncogene. Biochemical experiments suggest that the product of the *Rb* gene (a protein of M_r = 105,000 termed p105) is a negative regulator of transcription. Inactivation of this negative regulatory molecule apparently contributes to the multistep process of carcinogenesis.

Weinberg, R. A. Oncogenes, Antioncogenes, and the Molecular Basis of Multistep Carcinogenesis. *Cancer Res.* 49:3713, 1989.

polymerase III. Transcription factors bind to DNA and direct the action of RNA polymerase. One unusual feature of RNA polymerase III action in the transcription of 5S RNA is the location of the factor-binding sequence; it can be located within the DNA sequence encoding the RNA. The DNA in the region that would normally be thought of as a promoter, that is, the sequence immediately 5′ to the transcribed region of the gene, has no specific sequence and can be substituted by other sequences without any material effect on transcription. Figure 16.17 shows a diagram of this unusual sequence arrangement. In other cases, for example, tRNA transcription, the factor-binding sequence is located more conventionally at the 5′ region of the gene preceding the transcribed sequences.

16.5 POSTTRANSCRIPTIONAL PROCESSING

The immediate product of transcription is a **precursor RNA** molecule, the **primary transcript,** which is modified subsequently to a mature functional molecule. Primary transcripts have longer nucleoside sequences than the final RNA, have no modifications on the bases or sugars, and thus contain only A, G, C and U residues. The primary transcript is copied from a linear segment of DNA, a **transcriptional unit,** between specific initiation and termination sites. Transcriptional units may contain information in one or more forms: (a) information that is contiguous or without interruption; (b) information that is discontinuous, having sequences coding for a single protein or RNAs that are interrupted by unwanted nucleotide stretches; and (c) information that is in a tandem or repeated form in which information for multiple molecules is linked together and at some later stage requires separation. A gene, therefore, may not necessarily be colinear with the nucleotide or amino acid sequence of the final gene product.

The summation of all the enzymatic reactions leading to mature functional RNA molecules from primary transcripts is called **RNA processing.** Processing involves a variety of events, which include base modifications, sugar modifications, pyrimidine ring rearrangements, formation of helices and tertiary conformations, additions to the 5′ terminus, additions to the 3′ terminus, specific exonucleolytic cleavages, specific endonucleolytic cleavages, complex cleavages with splicing of pieces, and formation of RNA–protein complexes. The number, type, and order of processing events is different for each group of RNA and often varies for each specific type within the groups.

Ribozymes Are Enzymes Whose RNA Component Has Catalytic Activity

Some of the enzymes carrying out RNA processing reactions are **ribonucleoprotein particles.** In some cases the RNA component of the ribonucleoprotein particle is the active subunit of the enzyme. Enzymes whose RNA subunits carry out catalytic reactions are called **ribozymes.** There are four classes of ribozyme. Three of these RNA species carry out self-processing reactions while the fourth, ribonuclease P (RNase P), is a true catalyst.

Introns, noncoding sequences present within the coding portion of RNA, in rRNA from a ciliated protozoan, *Tetrahymena thermophila,* are removed from the rRNA precursor by a multistep reaction (Figure 16.18). A guanosine nucleoside or nucleotide reacts with the intron–exon phosphodiester linkage to displace the donor exon from the intron. This reaction is promoted by the folded intron itself. The free donor exon then

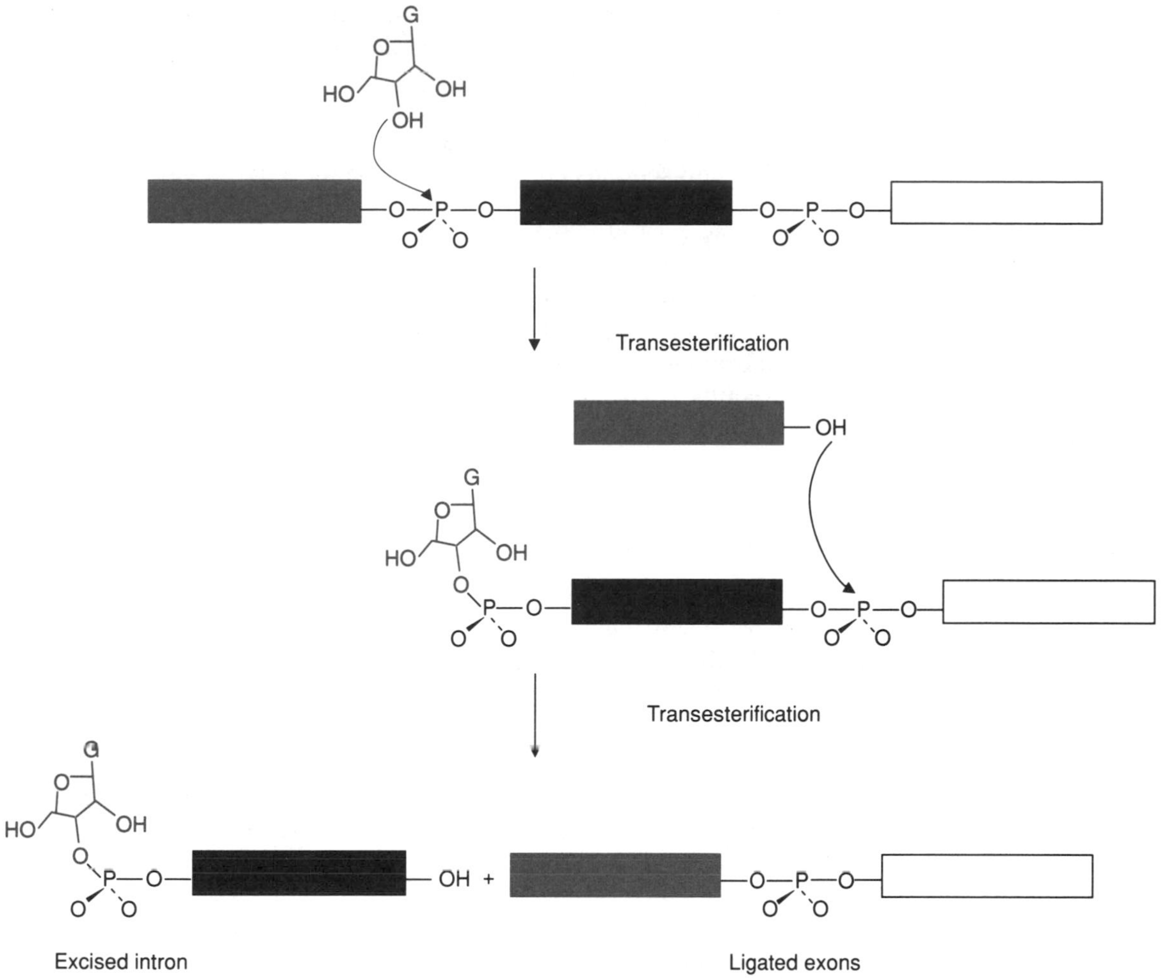

Transesterification

Transesterification

Excised intron

Ligated exons

FIGURE 16.18
Mechanism of self-splicing of the rRNA precursor of *Tetrahymena.*
The two exons of the rRNA are denoted by white and shaded boxes, respectively. Catalytic functions reside in the intron, which is black. This splicing function requires an added guanosine nucleoside or nucleotide.
Reproduced from T. R. Cech, *J. Am. Med. Assoc.* 260:308, 1988.

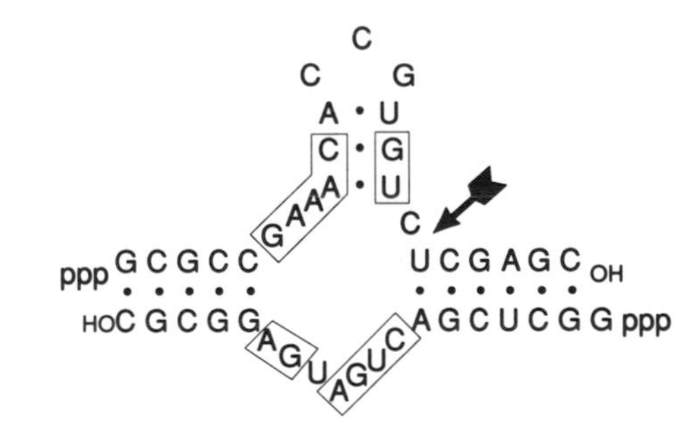

FIGURE 16.19
The "hammerhead" structure of a self-cleaving viral RNA.
This artificial molecule is formed by the base pairing of two separate RNAs. The cleavage of the RNA sequence at the site indicated by the *arrow* in the top strand requires its base pairing with the sequence at the bottom of the molecule. The boxed nucleotides are a consensus sequence found in self-cleaving RNA viral RNAs.
Redrawn from J. R. Sampson, F. X. Sullivan, L. S. Behlen, A. B. DiRenzo, and O. C. Uhlenbeck, *Cold Spring Harbor Symposium on Quantitative Biology* 52:267, 1987.

similarly attacks the intron–exon phosphodiester bond at the acceptor end of the intron. Introns of this type have been found in a variety of genes in fungal mitochondria and in the bacteriophage T4. Although these introns are not true enzymes as they do not turn over, they can be made to carry out catalytic reactions under specialized conditions. A second group of self-splicing introns are found in the mitochondrial RNA precursors of yeasts and other fungi. The self-splicing of these introns proceeds through a lariat intermediate similar to the lariat intermediate in the splicing of nuclear mRNA precursors. A third self-cleaving RNA structure is found in the genomic RNAs of several plant viruses. A folded RNA structure called a **hammerhead** self-cleaves during the generation of single genomic RNA molecules from large multimeric precursors. This structure is also found in a small satellite virus, hepatitis delta virus, that is implicated in severe cases of human infectious hepatitis. Ribonuclease P acts as a true RNA enzyme in the cell and in the test tube, carrying out a cleavage reaction on many molecules of tRNA precorsor substrates. In all of these events the structure of the RNA is essential for intramolecular or enzyme catalysis. The hammerhead structure is presented in Figure 16.19; other ribozymes have more complex secondary and tertiary structures that are required for catalysis.

The discovery of RNA catalysis has greatly altered our concepts of biochemical evolution and the range of allowable cellular chemistry. First of all, RNA can serve as both a catalyst and a carrier of genetic information. This has raised the possibility that the earliest living organisms were based entirely on RNA and that DNA and proteins evolved later. This model is sometimes referred to as the "RNA world." Second, we know that many viruses, including human pathogens, use RNA genetic information; some of these RNAs have been shown to be catalytic. Third, many of the information processing events in protein synthesis and mRNA splicing require RNA components. The possibility exists that these RNAs may also be fulfilling a catalytic function.

Transfer RNA Is Modified by Cleavage, Additions, and Base Modification

The proper recognition of mRNA by the tRNAs and the proper aminoacylation of the specific tRNAs are important and need to be rigorously controlled. Much of the specificity that results in overall tRNA function resides in the many nucleotide modifications (see Figure 16.5).

Cleavage

The product of transcription of a tRNA gene contains extra nucleotide sequences both 5′ and 3′ to the mature tRNA sequence. In some cases these primary transcripts contain introns in the anticodon region of the tRNA as well. The reactions occur in a closely defined but not necessarily rigid temporal order. First, the primary transcript is trimmed in a relatively nonspecific manner to yield a precursor molecule with relatively short 5′ and 3′ extensions. Then ribonuclease P, a ribozyme (see above), removes the 5′ extension by endonucleolytic cleavage. The 3′ end is trimmed exonucleolytically, followed by synthesis of the C—C—A_{OH} terminus. Synthesis of the modified nucleotides occurs in any order relative to the nucleolytic trimming. Intron removal is dictated by the secondary structure of the precursor (see Figure 16.20) and is carried out by a soluble, two-component enzyme system; one enzyme removes the intron and the other reseals the nucleotide chain.

Additions

Each functional tRNA has at its 3′ terminus the sequence —pCpCpA_{OH}. In most instances this sequence is added sequentially by nucleotidyltransferase. Cells grown in the presence of actinomycin D, an antibiotic that blocks transcription, still add —CCA quickly to presynthesized tRNAs. Nucleotidyltransferase prefers ATP and CTP as substrates and always incorporates them into tRNA at a ratio of 2C/1A. The —CCA_{OH} ends are found on both cytoplasmic and mitochondrial tRNAs.

Modified Nucleosides

Transfer RNA nucleotides are the most highly modified of all nucleic acids (Figure 16.5). More than 60 different modifications to the bases and ribose, requiring well over 100 different enzymatic reactions, have been found in tRNA. Many are simple, one-step methylations, but others involve multistep synthesis. Two derivatives, pseudouridine and queuosine (7–4, 5-*cis*-dihydroxy-1-cyclopenten-3-ylamino methyl-7-deazaguanosine) (see Table 16.2) actually require severing of the β-glycosidic bond to

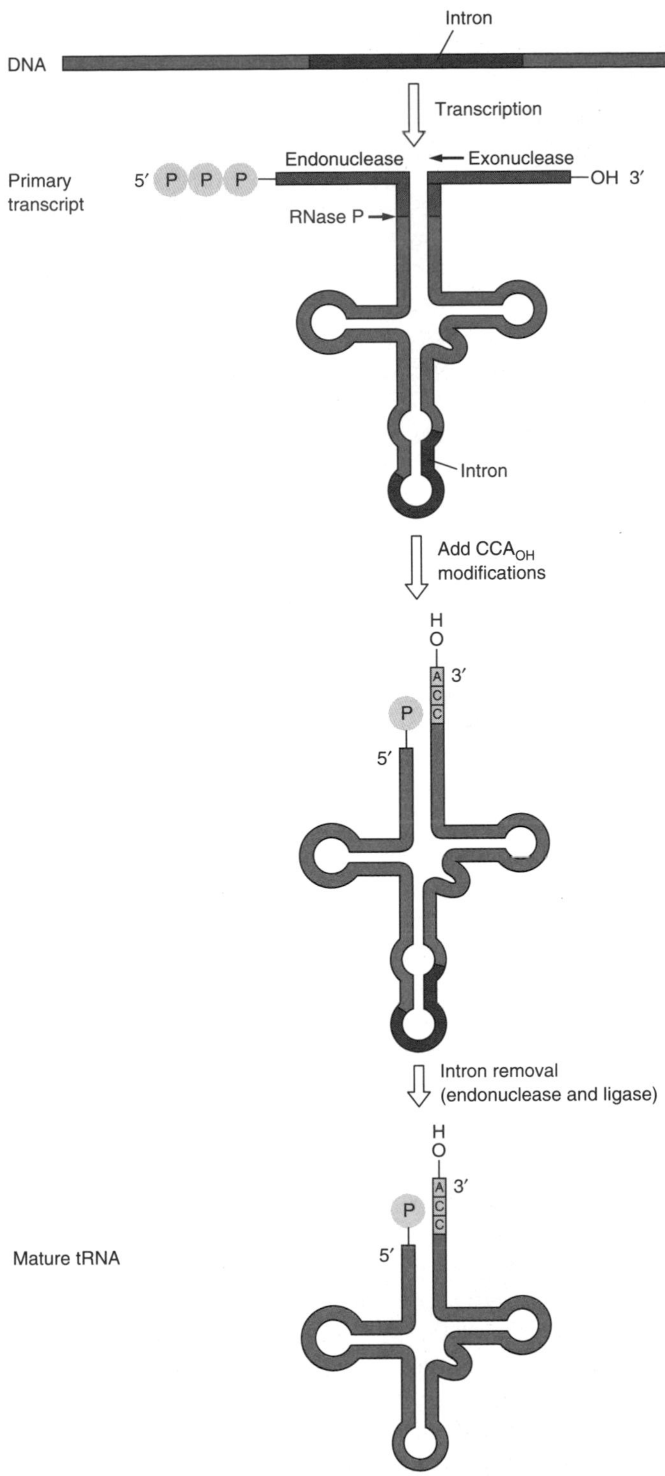

FIGURE 16.20
Scheme for processing a eucaryotic tRNA. The primary transcript is cleaved by RNase P and a 3′-exonuclease and the terminal C—C—A_{OH} is synthesized by tRNA nucleotidyltransferase before the intron is removed, if necessary.

the transcribed base. One enzyme or set of enzymes produces a single site-specific modification in more than one species of tRNA molecule. Separate enzymes or sets of enzymes produce the same modifications at more than one location in tRNA. In other words, most modification enzymes are site- or nucleotide sequence specific, not tRNA specific. All modifications are synthesized posttranscriptionally. Most are completed before the tRNA precursors have been cleaved to mature tRNA size. One noted exception is the methylation of ribose, which occurs late in processing.

Some modifications are found at extremely high frequencies in certain nucleotide sequences or locations within the tRNA structure. Ribothymidine (5-methyluridine) and pseudouridine (ψ) are almost always found together in the constant sequence GTψC of the hairpin loop proximal to the 3′ terminus, although ψ is also found elsewhere in tRNAs. Dihydrouridine is found in the loop nearest the 5′ terminus, and 7-methylguanosine is often found in the small or "extra" loop. The consistency of certain modifications hint at their possible functions in the tRNA structure.

Although first described over 20 years ago, the functions of the modifications are only just becoming evident through the use of a combination of biochemical, genetic, and biophysical techniques. Analyses of modified nucleotide function in tRNA has been limited to those steps in protein synthesis that can be assayed: aminoacylation, binding to initiation and elongation factors, ribosome binding, codon recognition and fidelity, for instance. There is direct evidence that modifications at the first position of the anticodon (the wobble base position) and immediately adjacent to the 3′ end of the anticodon are important for correct and effective recognition of the codon in mRNA.

Cleavage of the 45S rRNA occurs by sequential and specific endonucleolytic attack, yielding discrete intermediate RNAs. Both the RNase P and RNase III activities for this type of cleavage have been detected in human cells. Base compositions have been determined for the 45S, 32S, 28S, and 18S RNAs. Whereas the 45S RNA has a G + C content near 70%, the 28S and 18S rRNAs have G + C contents similar to the cellular average of 56–65%. This indicates preferential degradation of G—C rich regions (78% G + C). The unwanted regions may be bounded by methylated sequences serving as cleavage signals (see Table 16.4). Further stabilization of the rRNAs occurs by formation of base paired helical hairpin loops. Many of the base paired regions are retained in 28S rRNA. The remainder are degraded, but may first serve as processing signals for cleavage or ribosomal protein binding.

The locations of the 28S and 18S rRNAs within the 45S precursor rRNA have been determined by nucleotide sequence analysis, oligonucleotide mapping, and hybridization competition experiments, using the intermediate RNAs, 45S RNA, and nucleolar DNA. The 28S rRNA originates from the 3′ end of the 45S molecule, while the 18S rRNA is processed from the middle of the 5′ one-half of the 45S precursor.

Processing of rRNA in procaryotes also involves cleavage of high molecular weight precursors to smaller precursor and mature molecules (see Figure 16.21). Some of the bases are modified by methylation on the ring nitrogens of the bases rather than the ribose and by the formation of pseudouridine (only 0.06–0.3 mol%). The *E. coli* genome has approximately seven rRNA transcriptional units dispersed throughout the DNA. Each contains at least one 16S, one 23S, and one 5S rRNA or tRNA sequence. Processing of the rRNA is coupled directly to transcription, so that cleavage of a large precursor primary transcript, 30S, occurs at double helical regions yielding precursor 16S, precursor 23S, precursor 5S, and precursor tRNAs. These precursors are slightly larger than the functional molecules and only require trimming for maturation.

Messenger RNA Processing Requires Maintenance of the Coding Sequence

Most eucaryotic mRNAs have distinctive structural features, which are not a consequence of RNA polymerase action. These features, added in the nucleus, include a 3′-terminal poly(A) tail, methylated internal nucleotides, and a cap or blocked methylated 5′ terminus. Cytoplasmic mRNAs are shorter than their nuclear primary transcripts, which can contain additional terminal and internal sequences. The noncoding se-

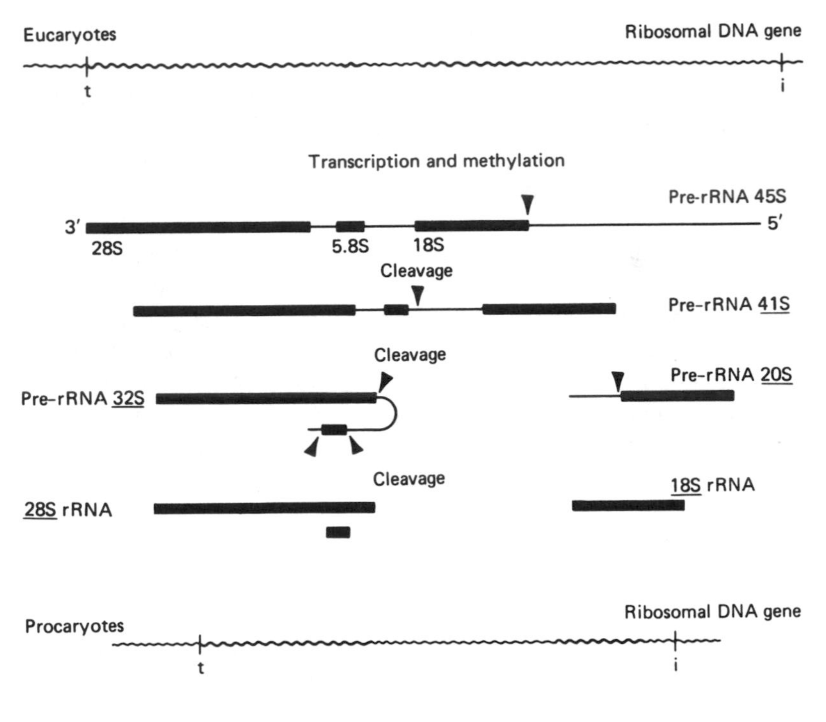

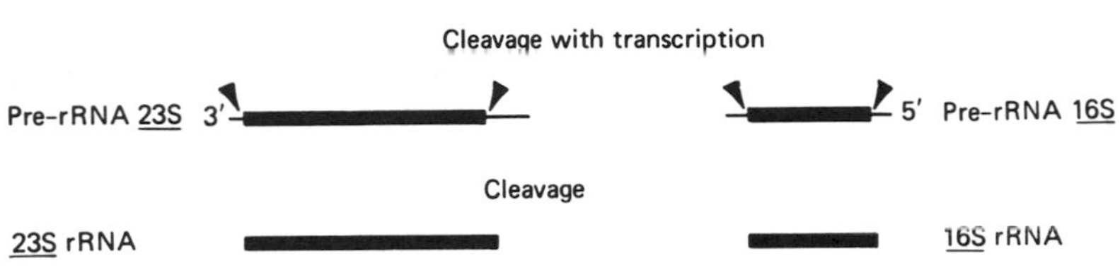

FIGURE 16.21
Schemes for transcription and processing of rRNAs.
Redrawn from R. Perry, *Annu. Rev. Biochem.* 45:611, 1976. Copyright © 1976 by Annual Reviews, Inc.

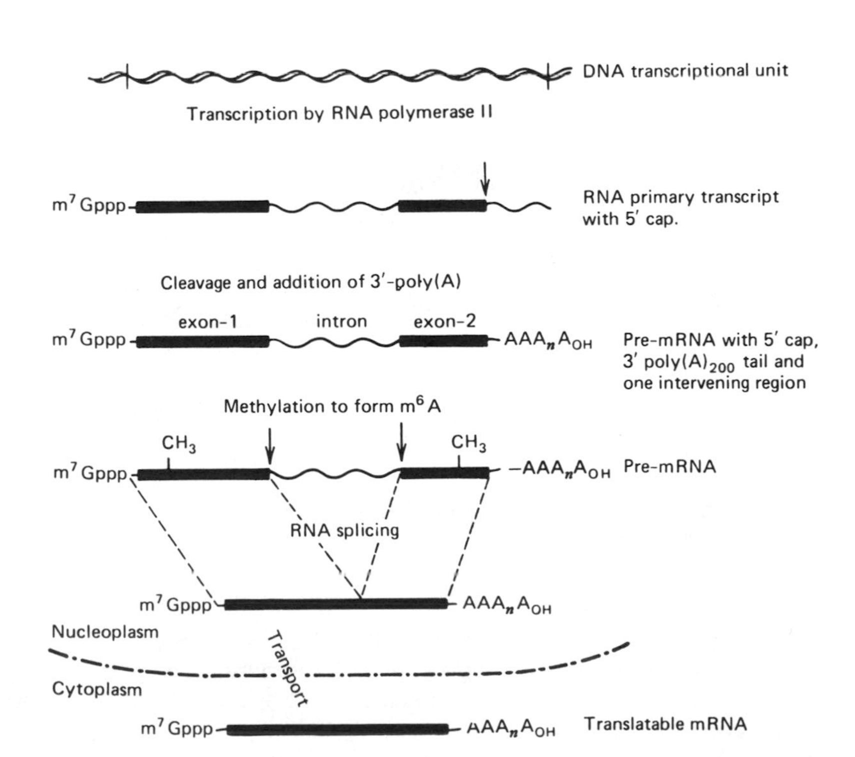

FIGURE 16.22
Scheme for processing mRNA.
The points for initiation and termination of transcription are indicated on the DNA. Arrows indicate cleavage points. The many proteins associated with the RNA and tertiary conformations are not shown.

CLIN. CORR. 16.4
THALASSEMIA DUE TO DEFECTS IN MESSENGER RNA SYNTHESIS

The thalassemias are genetic defects in the coordinated synthesis of α- and β-globin peptide chains; a deficiency of β chains is termed β-thalassemia while a deficiency of α chains is termed α-thalassemia. Patients suffering from either of these conditions present with anemia at about 6 months of age as Hb F synthesis ceases and Hb A synthesis would normally begin. The severity of symptoms leads to the classification of the disease into either thalassemia major, where a severe deficiency of globin synthesis occurs, or thalassemia minor, presenting a less severe imbalance. Occasionally an intermediate form is seen. Therapy for thalassemia major involves frequent transfusions, leading to a risk of complications from iron overload. Unless chelation therapy is successful, the deposition of iron in peripheral tissues can lead to death before adulthood. Carriers of the disease usually have thalassemia minor, involving a slight degree of anemia. Ethnographically, the disease is common in persons of Mediterranean, Arabian, and East Asian descent. As is also the case for sickle-cell anemia (HbS) and glucose 6-phosphate dehydrogenase deficiency, the fragility of the carriers' erythrocytes affords some protection from malaria. Maps of the regions where one or another of these diseases is frequent in the native population superimpose over the areas of the world where malaria is endemic.

Alpha thalassemia is usually due to a genetic deletion. These deletions occur because the α-globin genes are repeated; unequal crossing over between adjacent α alleles apparently has led to the loss of one or more loci. In contrast, β-thalassemia, which is more rare, can result from a wide variety of mutations. Known events include mutations leading to frame shifts in the β-globin coding sequence, as well as mutations leading to premature termination of peptide synthesis. Many β-thalassemias result from mutations affecting the biosynthesis of β-globin mRNA. Genetic defects are known that affect the promoter of the gene, leading to inefficient transcription. Other mutations result in aberrant processing of the nascent transcript, either in splicing out the two introns from the transcript or in polyadenylation of the mRNA precursor. Examples where the molecular defect illustrates a general principle of mRNA synthesis are discussed in the text.

Orkin, S. H. Disorders of hemoglobin synthesis: The thalassemias. G. Stamatoyannopoulis, A. W. Nienhuis, P. Leder, and P. W. Majerus (Eds.). *The*

quences present within the coding portion of pre-mRNA molecules, but not present in cytoplasmic mRNAs, are called **intervening sequences** or **introns.** The **expressed** or retained sequences are called **exons.** The general pattern for mRNA processing is depicted in Figure 16.22.

Processing of eucaryotic mRNA precursors involves a number of molecular reactions all of which must be carried out with exact fidelity. This principle is most clear in the removal of introns from an mRNA transcript. An extra nucleotide in the mRNA coding sequence will cause the reading frame of that message to be shifted and the resulting protein will almost certainly be nonfunctional. Indeed, mutations in the β-globin gene that interfere with intron removal are a major cause of the genetic disease β-thalassemia (Clin. Corr. 16.4). The task for the cell becomes even more daunting when seen in the light of the structure of some important human genes that consist of over 90% intron sequences. These complex reactions are accomplished by multicomponent enzyme systems that act in the nucleus; only after these reactions are completed is the mRNA exported to the cytoplasm where it interacts with ribosomes to initiate translation.

Blocking of the 5′ Terminus and Poly(A) Synthesis

Addition of the cap structures occurs during transcription by RNA polymerase II (Figure 16.22). As the transcription complex moves along the DNA, the capping enzyme complex modifies the 5′ end of the nascent mRNA. This is the only known eucaryotic mRNA processing event that is known to occur *cotranscriptionally,* that is, while RNA polymerase is still transcribing the downstream portions of the gene.

After initiation and cap synthesis, RNA polymerase continues transcribing the gene until a **polyadenylation signal sequence** is reached (Figure 16.23). This sequence, which has the consensus AAUAAA, appears in the mature RNA but usually does not form part of the coding region of the message. Rather, it functions as the signal for the cleavage of the nascent mRNA precursor about 20 or so nucleotides downstream. The poly(A) sequence is added by a soluble polymerase to the free 3′ end that results from this cleavage. Note that polyadenylation does not require a

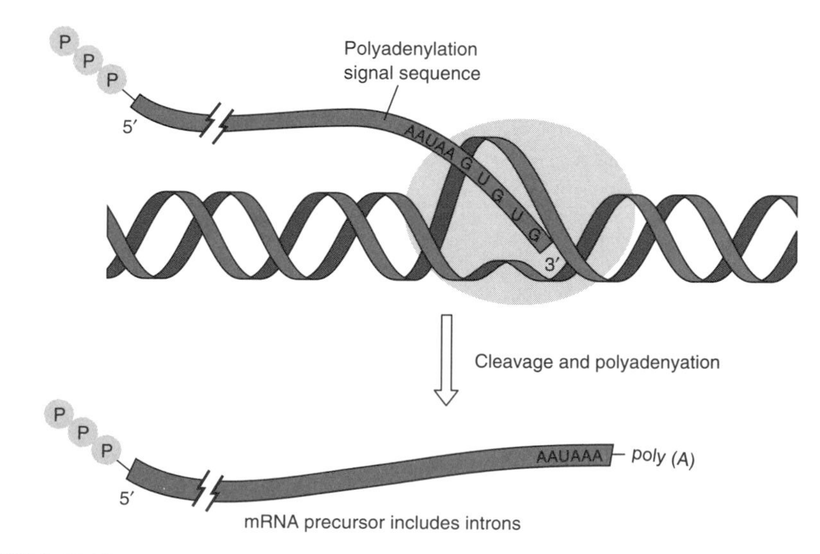

FIGURE 16.23
Cleavage and polyadenylation of eucaryotic mRNA precursors.
The 3′ termini of eucaryotic mRNA species are derived by processing. The sequence AAUAAA in the mRNA specifies the cleavage of the mRNA precursor. The free 3′-OH end of the mRNA is a primer for poly(A) synthesis.
Adapted from N. J. Proudfoot, *Trends Biochem. Sci.* 14:105, 1989.

template. Somewhat paradoxically, RNA polymerase II continues transcription for as many as 1000 nucleotides beyond the point at which the transcript is released from chromatin. The nucleotides incorporated into RNA by this process are apparently turned over and never appear in any cytoplasmic RNA species.

Molecular Basis of Blood Diseases. Philadelphia; Saunders, 1987; and Weatherall, D. J., Clegg, J. B., Higgs, D. R., and Wood, W. G. The hemoglobinopathies. C. R. Scriver, A. L. Beaudet, W. S. Sly, and D. Valle (Eds.). *The Metabolic Basis of Inherited Disease*, 6th ed. New York: McGraw-Hill, 1989.

Removal of Introns From mRNA Precursors

As the RNA is extruded from the RNA polymerase complex, it is rapidly bound by **small nuclear ribonucleoproteins, snRNPs** (snurps). Each of these snRNP species is composed of a single uridine-rich RNA and several proteins. While not all of these particles have been assigned a function, the U1 snRNP carries out an essential function in intron removal by recognizing the 5′ splice site. Recall that all introns begin with a GU sequence and end with AG; these are termed the donor and acceptor intron–exon junctions, respectively. Not all GU or AG sequences are spliced out of RNA, however. How does the cell know which GU sequences are in introns (and therefore must be removed) and which are destined to remain in mature mRNA? This discrimination is accomplished by the formation of base pairs between the U1 RNA and the sequence of the mRNA precursor preceding and following the GU dinucleotide sequence. See Figure 16.24 for an illustration of this process. Another snRNP containing U2 RNA recognizes important sequences at the 3′ end of the intron, the acceptor. Still other snRNP species then bind to the RNA precursor, forming a large complex termed a **spliceosome** (by analogy with the large ribonucleoprotein assembly involved in protein synthesis, the ribosome). The spliceosome then uses ATP energy to carry out the accurate removal of the intron. Selection of the 3′ splice site is carried out in a multistep process. First, the nucleotide bond is broken, leaving a free 3′-OH group at the end of the first exon and a 5′ phosphate on the donor guanosine of the intron. This pG is then used to form an unusual linkage with an adenosine nucleotide in the intron. The 5′ phosphate of the donor guanosine is ligated to the 2′-OH group of this adenosine to form a branched or *lariat* RNA structure, as shown in Figure 16.25. Notice how this branched structure is composed entirely of intron sequences. After the lariat is formed, the final reactions of intron removal occur about 20 or so nucleotides downstream from the branch: The phosphodiester bond immediatley following the AG is cleaved and the two exon sequences are ligated together. In genes with a large number of introns, the introns are removed roughly in sequence from the 5′ to the 3′ end of the mRNA precursor; however, this is not a hard and fast rule as

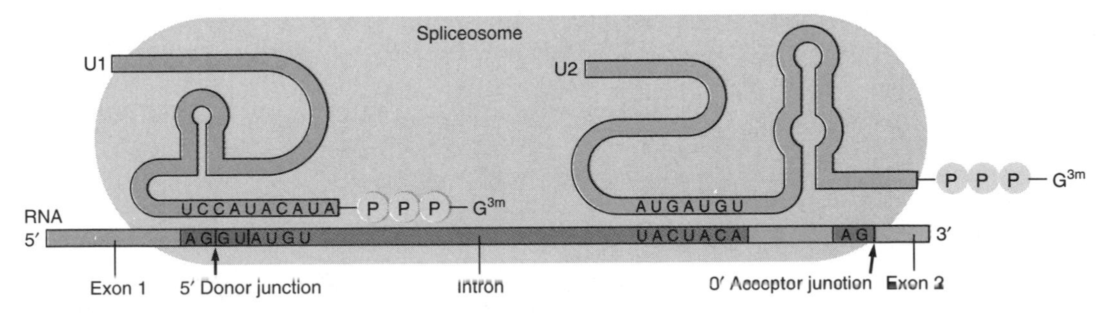

FIGURE 16.24
Mechanism of splice junction recognition.
The recognition of the 5′ splice junction involves base pairing between the intron–exon junction and the U1 RNA snRNP. This base pairing targets the intron for removal.
Adapted from P. A. Sharp, *J. Am. Med. Assoc.* 260:3035, 1988.

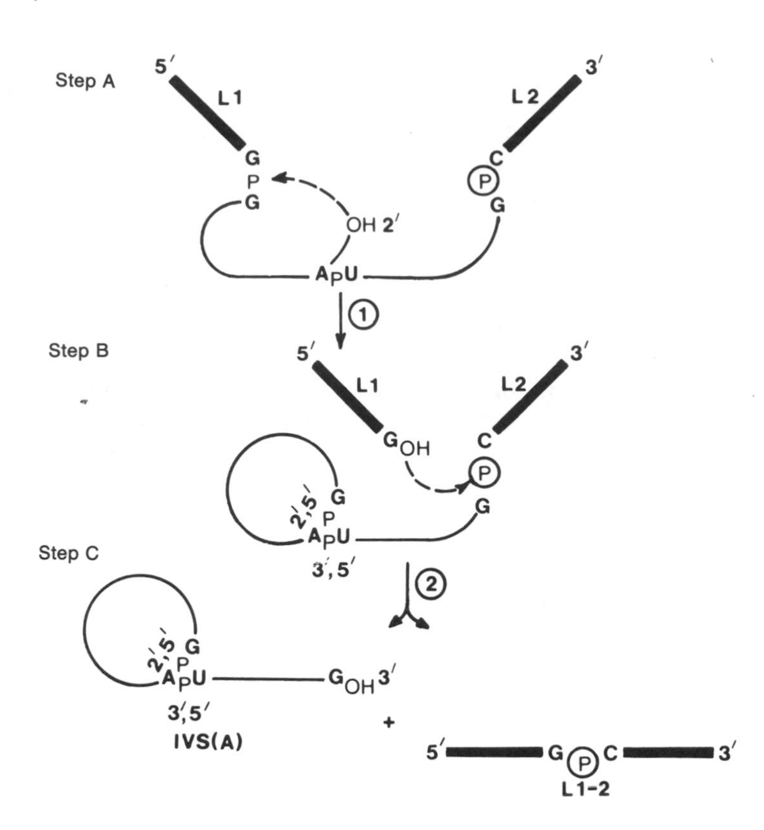

FIGURE 16.25
Proposed scheme for mRNA splicing to include the lariat structure.
A messenger RNA is depicted with two exons (L1 and L2, heavy lines) and intervening intron (light line). A 2′-OH group of the intron sequence reacts with the 5′-phosphate of the intron's 5′-terminal nucleotide producing a 2′–5′ linkage and the lariat structure. Simultaneously, the L1-exon–intron phosphate ester bond is broken leaving a 3′-OH terminus on this exon free to react with the 5′-phosphate of the L2 exon, displacing the intron and creating the spliced mRNA.

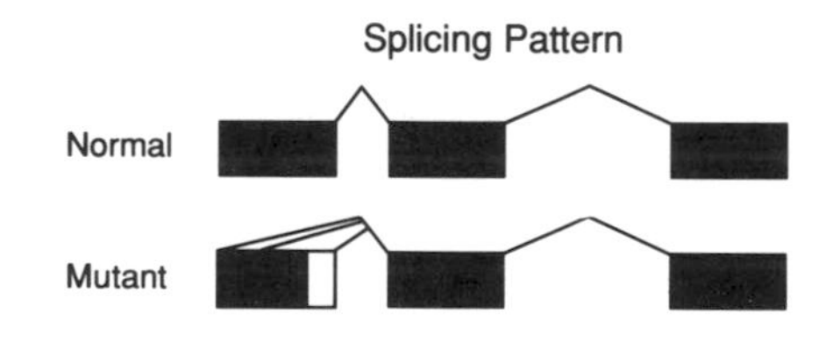

FIGURE 16.26
Nucleotide change at an intron–exon junction of the human β-globin gene, which leads to aberrant splicing and β-thalassemia.
This figure shows the splicing pattern of a mutated transcript containing a change of G-U to A-U at the first two nucleotides of the first intron. Loss of this invariant sequence means that the correct splice junction cannot be used; therefore transcript sequences that base pair with the U1 snRNA less well than the correct sequence jucntion are used as splice donors. The diagonal lines indicate the portions spliced together in mutant transcripts. Note that some of the mutant mRNA precursor molecules are spliced so that portions of the first intron (denoted as white boxes) appear in the processed product. In other instances the donor junction lies within the first exon and portions of the first exon are deleted. In no case is wild-type globin mRNA produced.
Adapted from S. H. Orkin, in G. Stamatoyannopoulis, *et al.* (Eds.). *The molecular basis of blood diseases*. Philadelphia: Saunders, 1987.

there is no singly preferred order for removal. The end result of all this processing is a fully functional coding mRNA, all introns removed and ready to direct synthesis of a protein.

Mutations in Splice Signals as a Cause of Human Disease

Splicing is an intricate process dependent on many molecular events. If these events are not carried out with precision, functional mRNA is not produced. This principle is illustrated in the human thalassemias, which affect the balanced synthesis of α- and β-globin chains (see Clin. Corr. 16.4). Some of the mutations leading to β-thalassemia interfere with the splicing of β-globin mRNA precursors. For example, we know that all intron sequences begin with the dinucleotide G-T (G-U in the mRNA precursor). A mutation of the G in this sequence to an A means that the splicing machinery will no longer recognize this as an intron donor and will "pass by" the correct exon–intron junction. This could lead to one of two results: either the intron will not be spliced out at all or the splicing machinery will attempt to use another donor that does not match the consensus as well as the normal (wild-type) junction. As a result, extra sequences that would normally be spliced out will appear in the processed mRNA. Alternatively, sequences could be deleted from the mRNA product (Figure 16.26). Functional globin protein will be made in reduced amounts if at all and the anemia characteristic of the disease will result.

Alternate Splicing Can Lead to Multiple Proteins Being Made From a Single DNA Coding Sequence

The existence of intron sequences has prompted many questions about the reason for their existence. The paradox is this: Introns must be removed accurately in order for the mRNA to accurately code for a protein. As we have seen in the case of the thalassemias, a single base change in a transcript can drastically interfere with splicing and cause a serious disease. Furthermore, the presence of introns in a gene means that the overall sequence of DNA is much larger than is required to encode the sequence of amino acids in its protein product. This is potentially harmful because the large gene is a target for more mutagenic events than is a small gene. Indeed, common human genetic diseases like Duchenne muscular dystrophy occur in genes that encompass millions of base pairs of DNA information. Why has nature not removed introns completely over the long timescale of eucaryotic evolution? Although there is not a clear answer to questions of this type, introns and RNA processing do have beneficial effects in at least a few cases.

The tropomyosin proteins are essential components of the contractile apparatus in a number of tissues (see Chapter 22, p. 961); furthermore, each contractile cell type contains a specific tropomyosin that differs from that in other cell types, for example, those found in striated versus smooth muscle. This diversity does not arise from a multigene family as does the diversity of globin molecules. Rather, a single gene is transcribed into a primary transcript that is then differentially processed as is diagrammed in Figure 16.27. All cells containing tropomyosin make the same primary transcript but each cell type processes this transcript in a characteristic fashion. The resulting mRNA species then encode the tropomyosins characteristic of each cell type. About 40 examples are well documented of tissue-specific splicing. Thus the existence of introns supplies the human organism with still another method of gene control.

16.6 NUCLEASES AND RNA TURNOVER

The different roles of RNA and DNA in genetic expression are reflected in their metabolic fates. A cell's information store (DNA) must be preserved; the results of this biological necessity are seen in the myriad DNA repair and editing systems in the nucleus. Thus, although individual stretches of nucleotides in DNA may turn over, the molecule as a whole is metabolically inert when not replicating. The RNA molecules, on the other hand, are individually dispensable; they can be replaced by newly synthesized species of the same specificity. It is therefore no surprise that RNA repair systems are not known. Instead, defective RNAs are removed from the cell by being degraded to nucleotides, which then are repolymerized into new RNA species.

This principle is clearest for mRNAs, which are classified as unstable. However, even the so-called stable RNAs turn over; for example, the half-life of tRNA species in liver is on the order of 5 days. A fairly long half-life for a eucaryotic mRNA would be 30 h.

This removal of RNAs from the cytoplasm is accomplished by cellular ribonucleases. Messenger RNAs are at least initially degraded in the cytoplasm. The rates vary for different RNAs, raising the possibility of control by differential degradation.

Two examples of the role of RNA stability in gene control illustrate how the stability of mRNA can influence gene expression. Tubulin is the major component of the microtubules found in many cell types, for example, as part of cilia. When there is an excess of tubulin in the cell, the free protein binds to and promotes the degradation of tubulin mRNA, thereby reducing tubulin synthesis. A second example is provided by *Herpes simplex* viruses (HSV), the agent causing cold sores and some genital infections. An early event in the establishment of HSV infection is the ability of the virus to destabilize all the cellular mRNA molecules, thereby reducing the competition for free ribosomes. Thus, the viral proteins are more efficiently translated.

CLIN. CORR. 16.5 AUTOIMMUNITY IN CONNECTIVE TISSUE DISEASE

Humoral antibodies in the sera of patients with various connective tissue diseases recognize cellular RNA–protein complexes. Patients with systemic lupus erythematosus exhibit a serum antibody activity designated Sm, and those with mixed connective tissue disease exhibit an antibody designated RNP. Each antibody recognizes a distinct site on the same RNA–protein complex, U1 RNP, that now has been implicated in mRNA processing in mammalian cells. The U1–RNP complex contains U1 RNA, a 165 nucleotide sequence highly conserved among eucaryotes, that at its 5′ terminus includes a sequence complementary to intron–exon splice junctions. Addition of antibody to in vitro splicing assays inhibits splicing presumably by removal of the U1 RNP from the reaction. Sera from patients with other connective tissue diseases recognize different nuclear antigens, nucleolar proteins, and the chromosome centromere, for example. Sera of patients with myositis have been shown to recognize cytoplasmic antigens such as aminoacyl-tRNA synthetases. Although humoral antibodies have been reported to enter live cells that have F_c receptors, there is no evidence of such as part of the autoimmune disease mechanism. The immunogens of these diseases remain unknown.

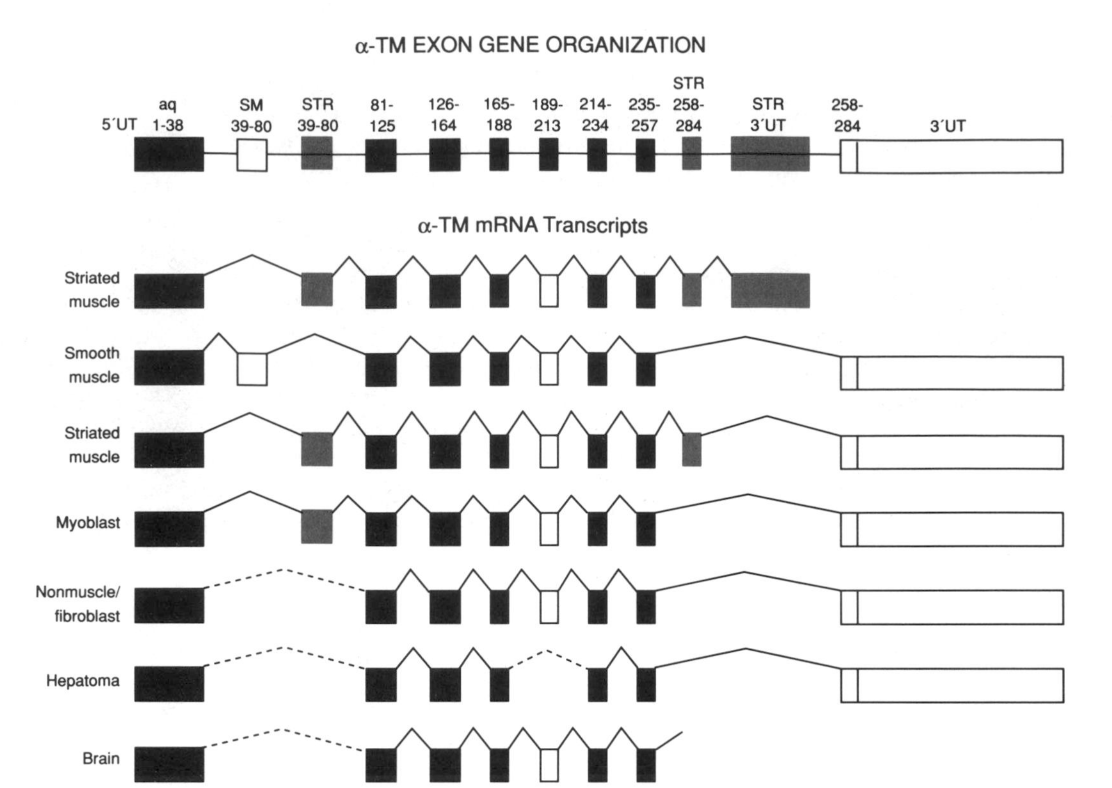

FIGURE 16.27
Alternate splicing of tropomyosin gene transcripts results in a family of tissue-specific tropomyosin proteins.
Redrawn from R. E. Breitbart, A. Andreadis, and B. Nadal-Ginard, *Annu. Rev. Biochem.* 56:467, 1986.

Lysosomes also contain large numbers of nucleases, and RNAs transported there could be degraded as well. The four common nucleotides resulting from RNA degradation are recycled to triphosphates and used again. Nucleosides containing modified bases do not recycle since the modifications are accomplished after the RNA is transcribed. The modified nucleosides released by degradation of tRNA or rRNA are ultimately excreted in the urine. The excretion is enhanced in human diseases exhibiting rapid cellular metabolism, division, and turnover, including cancers.

Nucleases are of several types and specificities (Table 16.6). The most useful distinction is between **exonucleases,** which degrade RNA from either the 5′ or 3′ end, and **endonucleases,** which cleave phosphodiester bonds within a molecule. The products of RNase action contain either 3′ or 5′ terminal phosphates, and both endo- and exonucleases can be further characterized by the position (5′ or 3′) at which the monophosphate created by the cleavage is located.

The structure of RNA also affects nuclease action. Most degradative enzymes are less efficient on highly ordered RNA structure. Thus, tRNAs are preferentially cleaved in unpaired regions of the sequence. Limited digestion of other RNAs leaves base paired regions uncleaved by the enzyme. On the other hand, many RNases involved in maturation of RNA require a defined three-dimensional structure for enzyme activity. These enzymes are discussed more fully above in the consideration of RNA processing pathways.

TABLE 16.6 Characteristics of Ribonucleases

Nuclease	Specificity	Products
Endonucleases yielding 3′ phosphates		from 5′-pApGpCpGpUpU$_{OH}$-3′
Pancreatic RNase	After pyrimidines	pApGpCp + GpUp + U$_{OH}$
T_1	After guanines	pApGp + CpGp + UpU$_{OH}$
U_1	After guanines	pApGp + CpGp + UpU$_{OH}$
T_2	After adenines	pAp + GpCpGpUpU$_{OH}$
Rat liver RNase-1	All phosphodiester bonds	pAp + 2Gp + Cp + Up + U$_{OH}$
Rat liver RNase-2	Between adjacent pyrimidines	pApGpCpGpUp + U$_{OH}$
E. coli RNase III	Double helical structures	Cleaves rRNA precursors
Endonucleases yielding 5′ phosphates		
Rat liver alkaline RNase I	Nonspecific	pA + 2pG + pC + pU + pU$_{OH}$
E. coli RNase P also in mammalian cells	Precursor tRNA	Mature tRNA + fragment
E. coli RNase H also in mammalian cells	RNA of DNA–RNA hybrid	Nucleoside 5′-monophosphates + DNA
Exonucleases		
E. coli RNase II also in mammalian cell nuclei	Single strands, 3′ ⟶ 5′	Nucleoside 5′-monophosphates, may trim RNA precursors after endonucleases
E. coli RNase V	Single strands, 5′ ⟶ 3′ precursor mRNAs	Nucleoside 5′-monophosphates
E. coli oligoribonuclease	Short oligoribonucleotides	Nucleoside 5′-monophosphates
RNase H from RNA tumor virus	RNA of DNA · RNA hybrid 5′ ⟶ 3′ 3′ ⟶ 5′	Nucleoside monophosphates
Polynucleotide phosphorylase	Single-strand RNA 3′ ⟶ 5′	Nucleoside 5′-diphosphates
Nonspecific nucleases		
Micrococcal endonuclease	Single-strand RNA or DNA	Nucleoside 3′-monophosphates
Nuclease S1	Single-strand RNA or DNA	5′-Phosphate oligonucleotides
Venom phosphodiesterase	Exonuclease 3′ ⟶ 5′ Blocked by 3′-phosphate end	Nucleoside 5′-monophosphates
Spleen phosphodiesterase	Exonuclease 5′ ⟶ 3′	Nucleoside 3′-monophosphates

BIBLIOGRAPHY

Bradshaw, R. A., *ed*. Transcription. Special issue of *Trends Biochem. Sci.* 16, November, 1991.

Breitbart, R. E., Andreadis, A., and Nadal-Ginard, B. Alternative splicing: A ubiquitous mechanism for the generation of multiple protein isoforms from single genes. *Annu. Rev. Biochem.* 56:467, 1987.

Cech, T. R. Ribozymes and their medical implications. *J. Am. Med. Assoc.* 260:3030, 1988.

Davies, K. E. and Read, A. P. *Molecular Basis of Inherited Disease*. Oxford: IRL Press, 1988.

Elgin, S. C. R. The formation and function of DNase I hypersensitive sites in the process of gene activation. *J. Biol. Chem.* 263:19259, 1988.

Kozak, M. A profusion of controls. *J. Cell Biol.* 107:1, 1988.

Manley, J. L. Polyadenylation of mRNA precursors. *Biochim. Biophys. Acta* 950:1, 1988.

Mitchell, P. J. and Tjian, R. Transcriptional regulation in mammalian cells by sequence-specific DNA binding proteins. *Science* 245:371, 1989.

Noller, H. F. Ribosomal RNA and translation. *Annu. Rev. Biochem.* 60:191, 1991.

Orkin, S. H. Disorders of hemoglobin synthesis: The thalassemias. G. Stamatoyannopoulis, A. W. Nienhuis, P. Leder, and P. W. Majerus (Eds.). *The Molecular Basis of Blood Diseases*. Philadelphia: Saunders, 1987, pp. 106–126.

Pace, N. R., and Smith, D. Ribonuclease P: Function and Variation. *J. Biol. Chem.* 262:3587, 1990.

Proudfoot, N. J. How RNA polymerase II terminates transcription in higher eukaryotes. *Trends Biochem. Sci.* 14:105, 1989.

Rosenberg, M. and Court, D. Regulatory sequences involved in the promotion and termination of RNA transcription. *Ann. Rev. Genet.* 12:319, 1979.

Ross, J. The turnover of messenger RNA. *Sci. Am.* 48, April 1989.

Sharp, P. A. RNA splicing and genes. *J. Am. Med. Assoc.* 260:3035, 1988.

Sollner-Webb, B. and Tower, J. Transcription of cloned eukaryotic ribosomal RNA genes. *Annu. Rev. Biochem.* 55:801, 1986.

Weinberg, R. A. Oncogenes, antioncogenes and the molecular basis of multistep carcinogenesis. *Cancer Res.* 49:3713, 1989.

QUESTIONS

C. N. ANGSTADT AND J. BAGGOTT

*Q*uestion *T*ypes are described on page xxiii.

1. (QT2) RNA:
   1. contains modified purine and pyrimidine bases that are formed posttranscriptionally.
   2. is usually single stranded in mammals.
   3. structures exhibit base stacking and hydrogen-bonded base pairing.
   4. usually contains about 65–100 nucleotides.

   A. HnRNA
   B. mRNA
   C. rRNA
   D. snRNA
   E. tRNA
2. (QT5) Has the highest percentage of modified bases of any RNA.
3. (QT5) Stable RNA representing the largest percentage by weight of cellular RNA.
4. (QT5) Contains both a 7-methylguanosine triphosphate cap and a polyadenylate segment.

5. (QT2) In eucaryotic transcription:
   1. RNA polymerase does not require a template.
   2. different kinds of RNA are synthesized in different parts of the nucleus.
   3. consensus sequences are the only known promoter elements.
   4. phosphodiester bond formation is favored, in part, because it is accomplished by pyrophosphate hydrolysis.

6. (QT1) An enhancer:
   A. is a consensus sequence in DNA located where RNA polymerase first binds.
   B. may be located in various places in different genes.
   C. may be on either strand of DNA in the region of the gene.
   D. functions by binding RNA polymerase.
   E. stimulates transcription in both procaryotes and eucaryotes.
7. (QT1) The sigma (σ) subunit of procaryotic RNA polymerase:
   A. is part of the core enzyme.
   B. binds the antibiotic rifampicin.
   C. is inhibited by α-amanitin.
   D. must be present for transcription to occur.
   E. specifically recognizes promoter sites.

Use this schematic representation of a procaryotic gene to answer questions 8–10. Numbers refer to positions of base pairs relative to the beginning of transcription.

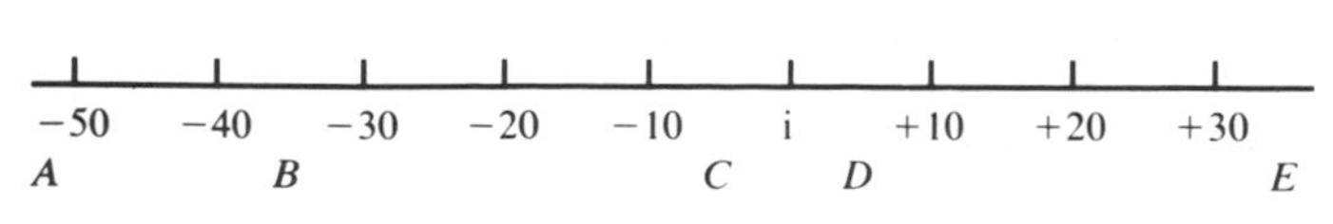

8. (QT5) Sigma (σ) factor might be released from RNA polymerase.
9. (QT5) An "open complex" should form in this region.
10. (QT5) Events beyond this region should be catalyzed by core enzyme.

11. (QT1) Termination of a procaryotic transcript:
    A. is a random process.
    B. requires the presence of the rho subunit of the holoenzyme.
    C. does not require rho factor if the end of the gene contains a G-C rich palindrome.
    D. is most efficient if there is an A-T rich segment at the end of the gene.
    E. requires an ATPase in addition to rho factor.

12. (QT2) Eucaryotic transcription:
    1. is independent of the presence of consensus sequences upstream from the start of transcription.
    2. may involve a promoter located within the region transcribed rather than upstream.
    3. requires a separate promoter region for each of the three ribosomal RNAs transcribed.
    4. often involves an alteration in chromatin structure.
13. (QT2) The primary transcript in eucaryotes:
    1. is usually longer than the functional RNA.
    2. may contain nucleotide sequences that are not present in functional RNA.
    3. will contain no modified bases.
    4. could contain information for more than one RNA molecule.

14. (QT1) Ribozymes:
    A. are any ribonucleoprotein particles.
    B. are enzymes whose catalytic function resides in RNA subunits.
    C. carry out self-processing reactions but cannot be considered true catalysts.
    D. bind to the mRNA precursor to recognize the 5′-splice site for intron removal.
    E. function only in the processing of mRNA.

15. (QT2) The processing of transfer RNA involves:
    1. cleavage of extra bases from both the 3′ and 5′ ends.
    2. nucleotide sequence-specific methylation of bases.
    3. addition of the sequence CCA by a nucleotidyl transferase.
    4. addition of a methylated guanosine at the 5′ end.

16. (QT1) Cleavage and splicing:
    A. are features of ribosomal RNA processing.
    B. cause sequences that are widely separated in a DNA molecule to be placed next to each other.
    C. remove noninformational sequences occurring anywhere within a primary transcript.
    D. are usually the first events in mRNA processing.
    E. are catalyzed by enzymes that recognize and remove specific introns.

17. (QT2) In the cellular degradation of RNA:
    1. any of the nucleotides released may be recycled.
    2. regions of extensive base pairing are more susceptible to cleavage.
    3. endonucleases may cleave the molecule starting at either the 5′ or 3′ end.
    4. the products are nucleotides with a phosphate at either the 3′- or 5′-OH group.

ANSWERS

1. A 1, 2, 3 correct. 1: Only the four bases A, G, U, and C are incorporated during transcription. 2, 3: Although single stranded, RNA exhibits considerable secondary and tertiary structure. 4: Only tRNA would be this small; sizes can range to more than 6000 nucleotides (Section 16.2).
2. E Modified bases seem to be very important in the three-dimensional structure of tRNA (Fig. 16.5).
3. C Stability of rRNA is necessary for repeated functioning of ribosomes.
4. B These are important additions during processing that yield a functional eucaryotic mRNA (Section 16.3, Table 16.1).
5. C 2, 4 correct. 1, 2: Transcription is directed by the genetic code, generating rRNA precursors in the nucleolus and mRNA and tRNA precursors in nucleoplasm. 3: Eucaryotic transcription may have internal promoter regions as well as enhancers. 4: This is an important mechanism for driving reactions (p. 695).
6. B B, C: Enhancer sequences seem to work whether they are at the beginning or end of the gene, but they must be on the same DNA strand as the transcribed gene. D: They seem to function by structurally altering the template (p. 696).
7. E A, D, E: Sigma factor is required for correct initiation and dissociates from the core enzyme after the first bonds have been formed. Core enzyme can transcribe but cannot correctly initiate transcription. B, C: Rifampicin binds to the β subunit, and α-amanitin is an inhibitor of eucaryotic polymerases (p. 696).
8. D Sigma factor is released when, or a short time after, the initial bond is formed.
9. C The high A-T content of the Pribnow box is believed to facilitate initial unwinding.
10. E Elongation, which requires only the core enzyme, is well underway in this region (p. 702).
11. C C, D: Rho-independent termination involves secondary structure, which is stabilized by high G-C content. A, B, E: There is a rho-dependent as well as a rho-independent process. Rho is a separate protein from RNA polymerase and appears to possess ATPase activity (p. 702).
12. C 2, 4 correct. 1: RNA polymerase II activity involves the TATA and CAAT boxes. 2: RNA polymerase III uses an internal promoter. 3: RNA polymerase I produces *one* transcript, which is later processed to yield three rRNAs. 4: This is a major difference between pro- and eucaryotic transcription (p. 703).
13. E All four correct. Modification of bases, cleavage, and splicing are all important events in posttranscriptional processing to form functional molecules (p. 708).
14. B A,B: Ribozymes are a very specific type of particle. C: One of the four classes, RNase P, catalyzes a cleavage reaction. D: This is the function of one of the snRNPs, several of which binding to mRNA result in a spliceosome. E: Ribozymes have been implicated in the processing of ribosomal and tRNAs (p. 708).
15. A 1, 2, 3 correct. 1: The primary transcript is longer than the functional molecule. 2: The same modifications, catalyzed by a certain (set of) enzyme(s), occurs at more than one location. 3: This is a posttranscriptional modification. 4: Capping is a feature of mRNA (p. 710).
16. C A: Cleavage occurs, but splicing does not. B: This occurs by DNA recombination. D: Splicing occurs after other events. E: Specificity of cleavage is related to specific sequences at the intron–exon junctions, not to the sequence of the intron itself (p. 712).
17. D Only 4 correct. 1: Modified bases cannot be recycled. 2: Although some enzymes of maturation may require an ordered structure, degradative enzymes are less efficient on an ordered structure. 3: An endonuclease cleaves an interior phosphodiester bond (p. 718).

17

Protein Synthesis: Translation and Posttranslational Modifications

DOHN GLITZ

17.1 OVERVIEW 724
17.2 COMPONENTS OF THE TRANSLATIONAL APPARATUS 724
Messenger RNA Is the Carrier of Information 724
Ribosomes Are the Workbench of Protein Biosynthesis 725
Transfer RNA Acts as the Bilingual Translator Molecule 727
The Genetic Code Uses a Four-Letter Alphabet of Nucleotides 729
Codon–Anticodon Interactions Permit the Reading of Messenger RNA 730
Aminoacylation of Transfer RNA Activates Amino Acids for Protein Synthesis 733
17.3 PROTEIN BIOSYNTHESIS 736
Translation Is Directional and Colinear With mRNA 736
Initiation of Protein Synthesis Is a Complex Process 737
Elongation Is a Step-Wise Formation of Peptide Bonds 739
Termination of Polypeptide Synthesis Requires a Stop Codon 742
Protein Synthesis in Mitochondria Differs From Cytoplasmic Synthesis 743
Some Antibiotics and Toxins Inhibit Protein Biosynthesis 744
17.4 PROTEIN MATURATION: MODIFICATION, SECRETION, AND TARGETTING TO ORGANELLES 746
Proteins for Export Follow the Secretory Pathway 746
Glycosylation of Proteins Occurs in the Endoplasmic Reticulum and Golgi Apparatus 748
Sorting of Proteins Targeted for Lysosomes Occurs in the Secretory Pathway 752
17.5 ORGANELLE BIOGENESIS AND TARGETTING 753
Import of Proteins by Mitochondria Requires Specific Signals 753
Targetting to Other Organelles Also Requires Specific Signals 754
17.6 FURTHER POSTTRANSLATIONAL PROTEIN MODIFICATIONS 755
Insulin Biosynthesis Involves Proteolysis 755
Proteolysis Leads to Zymogen Activation 755
Amino Acids Can Be Modified After Incorporation Into Proteins 756
Collagen Biosynthesis Requires Posttranslational Modifications 757
17.7 PROTEIN DEGRADATION AND TURNOVER 761
Intracellular Digestion of Some Proteins Occurs in Lysosomes 762
Ubiquitin Is a Marker in ATP-Dependent Proteolysis 762
BIBLIOGRAPHY 763
QUESTIONS AND ANNOTATED ANSWERS 764
CLINICAL CORRELATIONS
17.1 Missense Mutation: Hemoglobin 732
17.2 Disorders of Terminator Codons 732
17.3 Thalassemia 733
17.4 I-Cell Disease 753
17.5 Familial Hyperproinsulinemia 755
17.6 Defects in Collagen Synthesis 760

17.1 OVERVIEW

Protein biosynthesis is also called translation because it involves translation of information from the 4-letter language and structure of nucleic acids into the 20-letter language and structure of proteins. Different cells vary widely in their needs and abilities to synthesize proteins. At one extreme are terminally differentiated mature red blood cells, which have a finite life span of about 120 days, no nuclei, and cannot divide. Red blood cells also cannot synthesize proteins because they lack the major components of the biosynthetic apparatus. Many other cells must maintain levels of various enzymes or structural proteins and so, to remain viable, carry out a limited amount of protein synthesis. Cells that are growing and dividing are able to synthesize much larger amounts of protein needed for these processes. Finally, some cells synthesize proteins for export as well as for their own use. For example, liver cells synthesize large amounts of enzymes involved in the multiple metabolic pathways within the liver, and they also synthesize quantities of exported protein including serum albumin, the major protein of blood plasma or serum. Organs such as the liver are particularly rich in the organelles and macromolecules that are needed to translate genetic information into the structures of functional proteins. In essence the translation process involves the efficient and accurate organization of the RNA message and appropriate aminoacyl-tRNA translator molecules on the surface of ribosomes. Polypeptides are formed by the sequential addition of amino acids, arranged in a specific order in response to the information carried in the linear sequence of nucleotides in the mRNA. The completed protein is often then matured or processed by a variety of modifications. These modifications modulate its activity or function and may help target it to a specific intracellular location or for secretion from the cell. These complex processes are carried out with considerable speed and extreme precision. Finally, when the protein is no longer needed or becomes nonfunctional, it is degraded so that its amino acids can be recycled into new proteins.

17.2 COMPONENTS OF THE TRANSLATIONAL APPARATUS

Messenger RNA Is the Carrier of Information

The genetic information of the cell is stored and transmitted in the nucleotide sequences of DNA. Expression of this information requires its selective transcription into molecules of mRNA that carry specific and precise messages from the nuclear "data bank" to the cytoplasmic sites of protein synthesis. In eucaryotes the messengers (mRNA) are synthesized in the nucleus, often as significantly larger precursor molecules classified as heterogenous nuclear RNA (hnRNA). Mature mRNA in the cytoplasm has several identifying characteristics. In eucaryotes, it is almost always monocistronic, encoding a single polypeptide. The 5′ end is capped with a specific structure involving 7-methylguanosine linked through a 5′-triphosphate bridge to the 5′ end of the messenger sequence (see p. 694). A 5′-nontranslated region, which, depending on the specific message, may be quite short or as much as a few hundred nucleotides in length, separates the cap from the **translational initiation signal,** an **AUG** codon. Usually, but not always, this is the first AUG sequence encountered as the message is read 5′ → 3′. Informational sequences that specify a single polypeptide sequence follow the initiation signal in a continuous fashion. The informational sequence continues until a specific translational termination signal is reached. This is followed by a 3′-untranslated sequence, usually of about 100 nucleotides length, before the mRNA is terminated by a further 100–200 nucleotide long polyadenylate tail.

Procaryotic mRNA differs from eucaryotic mRNA in a few details. The 5′ terminus is not capped, but it retains a terminal triphosphate from initiation of its synthesis by RNA polymerase. Most messengers are polycistronic, encoding several polypeptides, and including more than one initiation AUG sequence. In each case a ribosome-positioning sequence is located about 10 nucleotides upstream of a valid AUG initiation signal. An untranslated sequence follows the last coding sequence, but there is no polyadenylate tail.

Ribosomes Are the Workbench of Protein Biosynthesis

Proteins are assembled on the surface of ribosomes. All ribosomes are made up of two dissimilar subunits, each of which contains both RNA and proteins. With one exception, each protein is present in a single copy per ribosomal subunit and each RNA species is found in a single copy per subunit. The composition of major ribosome types is shown in Table 17.1.

Ribosome architecture has been highly conserved in evolution. The similarities between ribosomes and subunits of different sources are more obvious than the differences; functional roles for each subunit are well defined. More detailed information on ribosome structure and its relationship to function has come from a variety of techniques. For example, the location of many ribosomal proteins on each subunit has been determined by electron microscopy of subunits that are complexed to antibodies directed against a single ribosomal protein. Further structural information has been obtained from chemical cross-linking of proteins, which identifies near neighbors within the structure, and from neutron diffraction measurements, which quantitate the separation distances of protein pairs. Sequence comparisons and chemical or immunological probes provide structural information about the rRNA. Correlation of structural data with functional measurements in protein synthesis have allowed the development of models, such as that in Figure 17.1, that link ribosome morphology to the several different functions of ribosomes in translation.

These experiments are made easier because procaryotic ribosomes can **self-assemble**—that is, the native structures can be reconstituted from

TABLE 17.1 Ribosome Classification and Composition

Ribosome source	Monomer size	Subunits	
		Small	Large
Eucaryotes:			
Cytoplasm	80S	40S: 34 proteins 18S RNA	60S: 50 proteins 28S, 5.8S, 5S RNAs
Mitochondria			
Animals	55S–60S	30S–35S: 12S RNA	40–45S: 16S RNA
		70–100 proteins	
Higher plants	77S–80S	40S: 19S RNA	60S: 25S, 5S RNAs
		70–75 proteins	
Chloroplasts	70S	30S: 20–24 proteins 16S RNA	50S: 34–38 proteins 23S, 5S, 4.5S RNAs
Procaryotes:			
E. coli	70S	30S: 21 proteins 16S RNA	50S: 34 proteins 23S, 5S RNAs

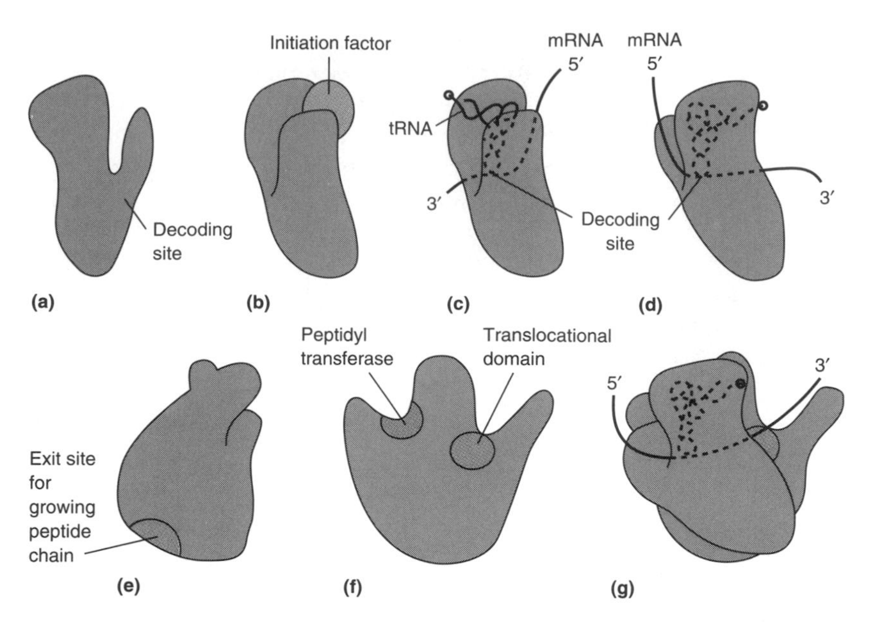

FIGURE 17.1
Ribosome structure and functional sites.
Different orientations of the small subunit are shown in the top row. In *a* the subunit is seen from the side and the platform is seen as a segment of the particle that projects from the body and is separated from the body by a well-defined crevice. Codon–anticodon interactions occur deep in the crevice, at the decoding site. In *b* the subunit has been rotated about 90° so that the platform is facing the reader and the binding site of eIF3 is shown. In *c* the orientation of mRNA and tRNA is depicted, with their interaction in the decoding site obscured by the platform (*c*) or, with the subunit rotated a further 180°, by the body of the subunit (*d*). Large subunits are depicted in the lower row of drawings. In *e* the subunit is shown from one side; the exit site is where newly synthesized protein emerges from the subunit. This area of the subunit is in contact with membranes in the "bound" ribosomes of the rough endoplasmic reticulum. In *f* the subunit is rotated so that the flat surface in *e* now faces the reader. The site of peptide bond formation, the peptidyl transferase center, is distant from the exit site; the growing peptide must pass through a groove or tunnel in the ribosome to reach the exit site. In *g* the relative orientation of the subunits is shown. Note how tRNA bound by the small subunit is oriented so that the aminoacyl acceptor end is near the peptidyl transferase and the translocational domain (at which EF1 and EF2 are bound) is near the decoding region and the area in which messenger enters the complex. These drawings are based primarily on electron microscopy. Progress in crystallographic determination of ribosome structure may soon allow construction of a more detailed and complete model of the ribosome.

mixtures of purified individual proteins plus the appropriate RNA. Reconstitution of subunits from mixtures in which a single component is omitted or modified can show if, for example, a given protein is required for assembly of the subunit or for some specific subunit function. An assembly map for large ribosomal subunits of *E. coli* is shown in Figure 17.2. Total reconstitution of subunits from eucaryotes has not yet been achieved but it is clear that most of the general conclusions about how ribosomes function, although determined using bacterial ribosomes, are fully applicable to eucaryotic systems.

Ribosomes in eucaryotic cells are organized in two additional ways. First, it is rare for mRNA to be associated with a single ribosome; instead, several ribosomes are normally involved in simultaneously translating a single mRNA molecule. These messenger-linked **polysomes** can be visualized by electron microscopy (Figure 17.3). Second, some ribosomes are found **free** in the cytoplasm of the cell, but many are **bound** to membranes

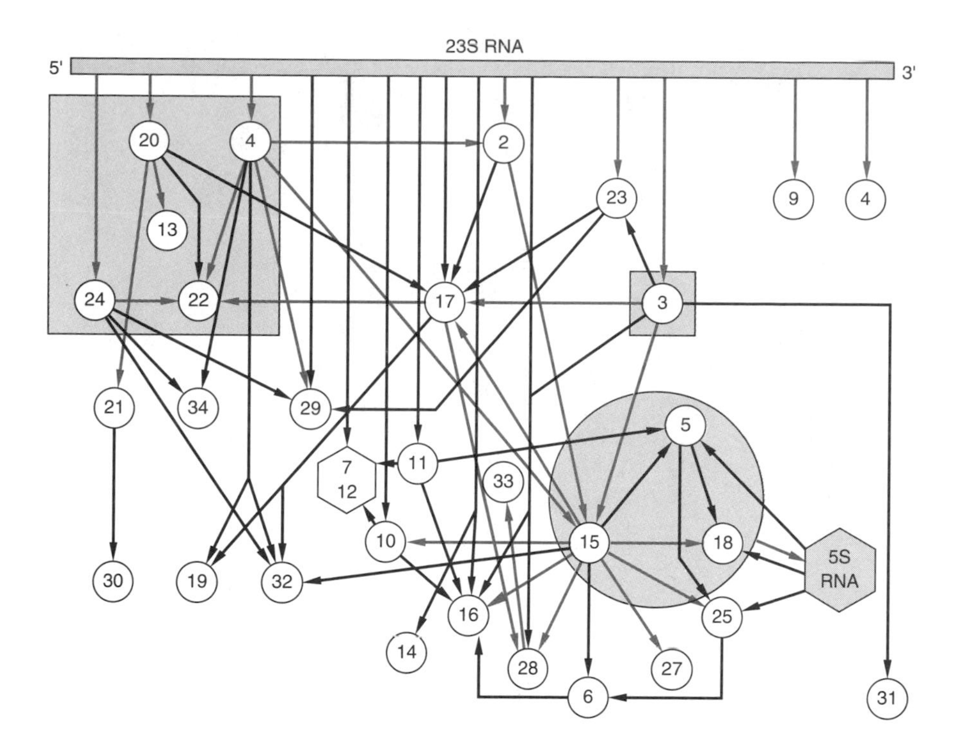

FIGURE 17.2
Assembly map of the large ribosomal subunit of *E. coli.*
The heavy bar at the top represents the 23S rRNA, and the individual ribosomal proteins are identified by numbers in circles. Heavy arrows from RNA to protein indicate that the protein binds directly and strongly to RNA, while lighter arrows indicate a less strong interaction. Similarly, heavy arrows between proteins show a strong binding dependence and light arrows show a lesser dependency. For example, protein L4 binds RNA strongly; it then strongly stimulates binding of proteins L2, L22, and L29. Protein L2 in turn stimulates binding of proteins L5 and L15. Proteins L5, L15, and L18 are essential for binding 5S RNA. Proteins within the larger gray boxes are required for a conformational transition that occurs during the assembly. The diagram shows both the orderly progression of the assembly process and the interdependence of the components and their specific reactions with other components during the assembly of the subunit.
Adapted from M. Herold and K. Nierhaus, *J. Biol. Chem.* 262:8826, 1987. A similar assembly map for the small subunit was elucidated earlier. (M. Nomura, *Cold Spring Harbor Symp. Quant. Biol.* 52:653, 1987.)

of the rough endoplasmic reticulum. In cellular homogenates the bound ribosomes are found in the fraction known as microsomes from which the ribosomes are released by detergents that disrupt the membranes. In general, these membrane-bound ribosomes are synthesizing proteins that will be secreted from the cell or sequestered in cellular membranes or lysosomes. In contrast, free ribosomes synthesize primarily proteins that will remain within the cell cytoplasm or will be targetted to the cell nucleus, mitochondria, or other organelles.

Transfer RNA Acts as the Bilingual Translator Molecule

All tRNA molecules have several common structural characteristics including the 3′-terminal CCA sequence to which amino acids are bound, a highly conserved cloverleaf secondary structure, and an L-shaped three-dimensional structure (see p. 686). But each of the more than 50 molecular species also has a unique nucleotide sequence that gives it individual characteristics, which allow it to be highly specific in its interactions with mRNA and with the enzymes that couple only one amino acid to it.

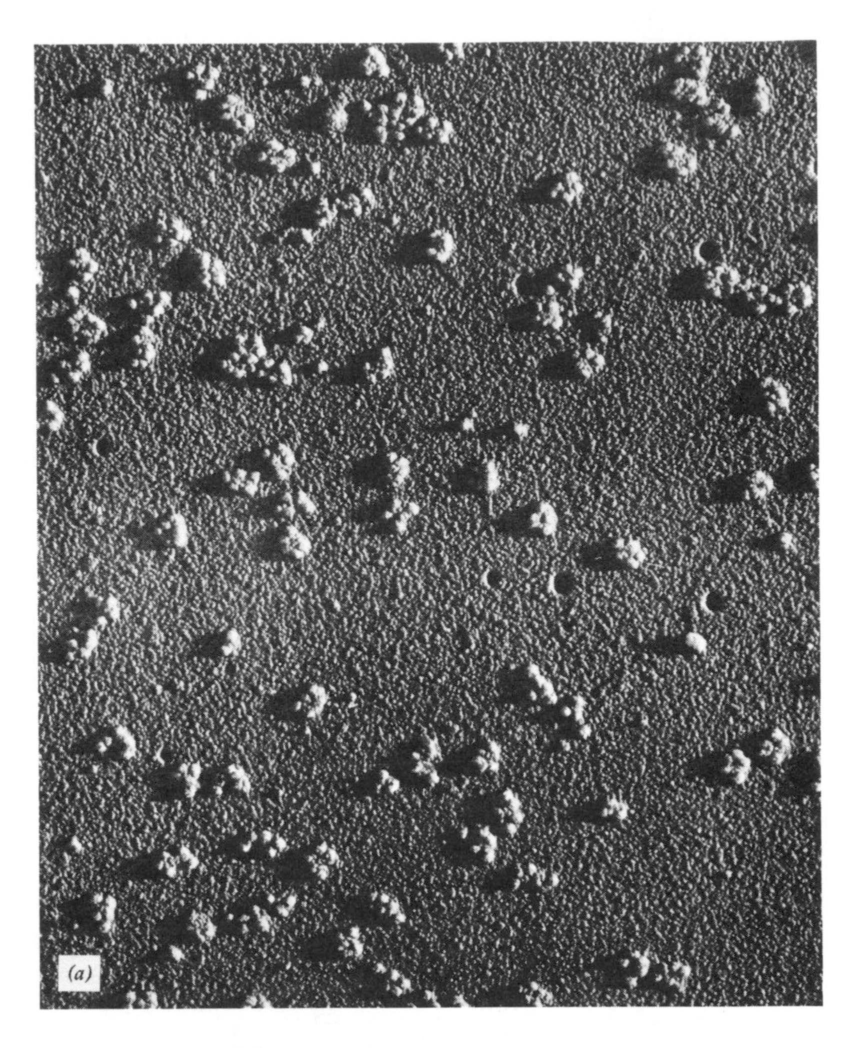

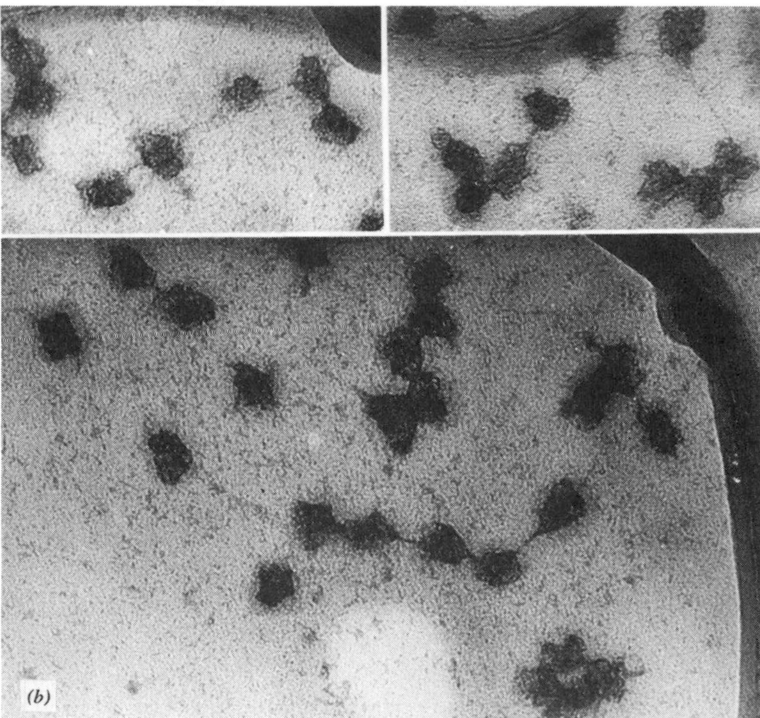

FIGURE 17.3
(*a*) Reticulocyte polyribosomes shadowed with platinum as shown above appear as clusters of three to six ribosomes, a number consistent with the size of mRNA for a globin chain. Further magnification after uranyl acetate staining as shown in (*b*) reveals one extraordinarily clear five-ribosome polysome with part of the mRNA visible.
Courtesy of Dr. Alex Rich, MIT.

The Genetic Code Uses a Four-Letter Alphabet of Nucleotides

Information in the cell is stored in the form of a linear sequence of nucleotides in DNA, in a manner that is analogous to the linear sequence of letters of the alphabet in the language you are now reading. The genetic language of DNA is very simple; it is written in a **4-letter alphabet** corresponding to the two purines, A and G (adenine and guanine) and the two pyrimidines, C and T (cytosine and thymine). In mRNA the information is encoded in a similar 4-letter alphabet in which U (uracil) replaces T. The language of RNA can be seen as a dialect of the genetic language of DNA. The genetic information is expressed in the form of proteins that derive their properties from their linear sequence of amino acids. Thus, in the process of protein biosynthesis the 4-letter language of nucleic acids is translated into the 20-letter language of proteins. Implicit in the analogy to language is the directional specificity of these sequences. By convention, nucleic acid sequences are written and read in a $5' \rightarrow 3'$ direction, and protein sequences are written and read from the amino terminus to the carboxy terminus. Later we will see that these directions in mRNA and protein correspond in both their reading and biosynthetic senses.

Codons in mRNA Are Three-Letter Words

A 1 : 1 correspondence of nucleotides to amino acids would only permit mRNA to encode four amino acids, while a 2 : 1 correspondence could possibly encode $4^2 = 16$ amino acids. Either ratio is insufficient since 20 amino acids occur in proteins. Moreover, a two letter code would leave no possibility for specific information to encode start signals (like a capital letter to begin a sentence) or stop signals (like a period at the end of a sentence). A code of three-letter words, which is the actual genetic code, is more than sufficient since it includes $4^3 = 64$ permutations or words. These three base words are called **codons** and they are customarily shown in the form of Table 17.2. This table shows that only two amino acids are designated by single codons: methionine as AUG and tryptophan as

TABLE 17.2 The Genetic Code[a]

5′ Base	U		C		A		G		3′ Base
U	UUU	Phe	UCU	Ser	UAU	Tyr	UGU	Cys	U
	UUC		UCC		UAC		UGC		C
	UUA	Leu	UCA		UAA	Stop	UGA	Stop	A
	UUG		UCG		UAG		UGG	Trp	G
C	CUU	Leu	CCU	Pro	CAU	His	CGU	Arg	U
	CUC		CCC		CAC		CGC		C
	CUA		CCA		CAA	Gln	CGA		A
	CUG		CCG		CAG		CGG		G
A	AUU	Ile	ACU	Thr	AAU	Asn	AGU	Ser	U
	AUC		ACC		AAC		AGC		C
	AUA		ACA		AAA	Lys	AGA	Arg	A
	AUG	Met	ACG		AAG		AGG		G
G	GUU	Val	GCU	Ala	GAU	Asp	GGU	Gly	U
	GUC		GCC		GAC		GGC		C
	GUA		GCA		GAA	Glu	GGA		A
	GUG		GCG		GAG		GGG		G

[a] The genetic code comprises 64 codons, which are permutations of four bases taken in threes. Note the importance of sequence: three bases, each used once per triplet codon, give six permutations: ACG, AGC, GAC, GCA, CAG, and CGA, for threonine, serine, aspartate, alanine, glutamine, and arginine, respectively.

TABLE 17.3 Nonuniversal Codon Usage in Mitochondria

Codon	Usual code	Mitochondrial code
UGA	Termination	Trp
AUA	Ile	Met
CUA	Leu	Thr

TABLE 17.4 Wobble Base Pairing Rules

5′ Codon base	3′ Anticodon bases possible
A	U or I
C	G or I
G	C or U
U	A or G or I

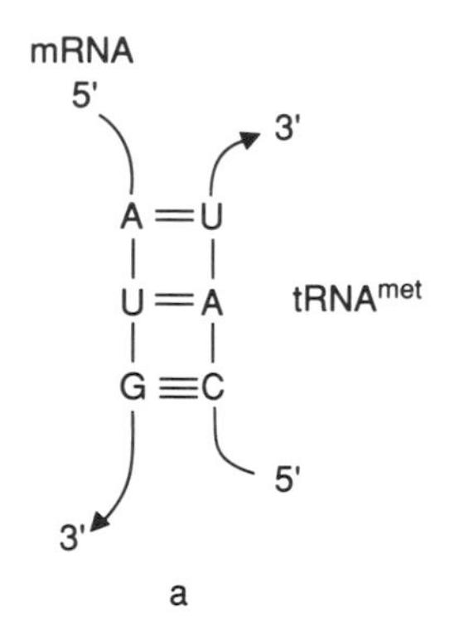

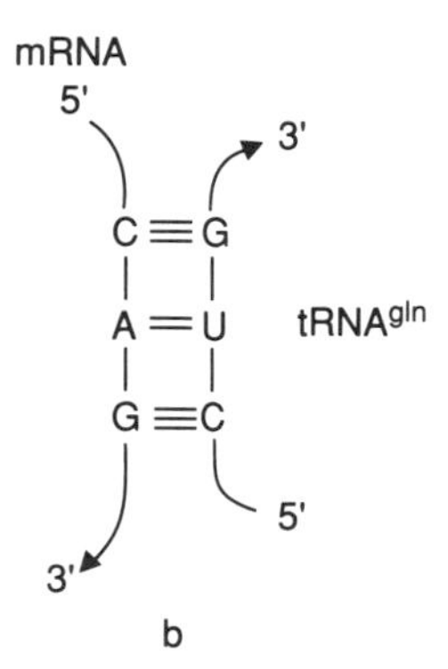

FIGURE 17.4
Codon–anticodon interactions.
Shown are interactions between the AUG (methionine) codon and its CAU anticodon (*a*) and the CAG (glutamine) codon and its CUG anticodon (*b*). Note that these interactions involve pairing of the mRNA with tRNA species in an antiparallel fashion.

UGG. Other amino acids are designated by two, three, four, or six codons. Multiple codons for a single amino acid are characteristic of degeneracy in the code. The code is not only **degenerate,** but also **universal,** or nearly so. This means that the same code words are used in all living organisms, procaryotic and eucaryotic. The exception to the universality of the code occurs in mitochondria. For reasons that are not clear a few codons specify different information in mitochondria than in the cytoplasm of the same organism. These differences are shown in Table 17.3.

Punctuation

Four of the 64 codons function partly or totally as punctuation, signaling the start and stop of protein synthesis. The **start signal, AUG,** also specifies the amino acid methionine. An AUG at an appropriate site and within an acceptable sequence in the mRNA signifies methionine as the initial, amino terminal amino acid. The AUG codons elsewhere in the messenger sequence simply specify methionine residues within the protein. Three codons, **UAG, UAA, and UGA,** are **stop signals** that specify no amino acid. For this reason they are also known as **nonsense codons.**

Codon–Anticodon Interactions Permit the Reading of Messenger RNA

The translation of the codons of mRNA involves the direct interaction of codons with complementary **anticodon sequences** of tRNA. The anticodons are three base sequences in tRNA as discussed in Chapter 16. The anticodon base pairs with a complementary codon sequence of mRNA in an antiparallel manner as shown in Figure 17.4. In the tRNA molecule the anticodon is distant from the amino acid acceptor stem. This is clear whether considering the two-dimensional cloverleaf secondary structure or the L-shaped three-dimensional structure of all tRNA molecules shown in Chapter 16, p. 686. In each instance the anticodon and amino acid residue are at opposite extremes of the molecule. This structure permits the tRNA molecule to both conceptually and physically bridge the gap between the nucleotide sequence of the ribosome-bound mRNA and the peptidyl transferase center of the ribosome at which the encoded polypeptide is assembled.

Since there are 61 codons that designate an amino acid, it might appear necessary to have exactly this number of different tRNA species. This is not the observed fact, and variances from standard base pairing are common in codon–anticodon interactions. Many tRNA molecules can recognize and bind more than one codon, many amino acids can be carried by more than one tRNA species, and several codons are read by more than one tRNA (but always one carrying the correct amino acid). Much of this complexity of interactions is explained by the **"wobble" hypothesis,** which permits base pairing between the third position of the codon and the first position of the anticodon to be less stringent than ordinary Watson–Crick pairing. The nucleotides are proposed to literally wobble, altering their relative geometry and allowing base pairs to form between G and U residues. Thus the first two positions of the anticodon predominate in tRNA selection and the third **(degenerate)** position is less important. A second major factor in determining codon–anticodon interactions resides in the presence of modified nucleotides at or beside the first position of the anticodon in many tRNA species. A frequent anticodon nucleotide is inosinic acid (I), the nucleotide of hypoxanthine. Wobble rules allow I to base pair with U, C, or A. The allowed wobble pairs are shown in Table 17.4.

If the wobble rules are followed, the 61 nonpunctuation codons could be read by as few as 31tRNA molecules. Again, this is too great a simplification. Most cells have many more species of tRNA, typically 50 or more. It seems likely that some codons are read more efficiently by one anticodon than a second even though both are capable. In human mitochondria from HeLa cells only 23tRNA species are found; some condons must be translated using only the first two positions. Finally, not all codons are used to an equal extent, and some are used very rarely. Examination of many mRNA sequences has allowed construction of "codon usage" tables that show different organisms preferentially use different code words to generate similar polypeptide sequences.

"Breaking" the Genetic Code

The genetic code of Table 17.2 was determined before biochemists had developed methods to sequence natural mRNA. These code-breaking experiments provided insight into how proteins are synthesized. Important experiments used simple artificial mRNAs or chemically synthesized trinucleotide codons.

The enzyme **polynucleotide phosphorylase** catalyzes the readily reversible reaction:

$$x\text{NDP} \rightleftharpoons \text{polynucleotide } (\text{pN}_x) + x\text{P}_i$$

where NDP is any nucleoside 5′-diphosphate or a mixture of two or more. If the nucleoside diphosphate is UDP, a polymer of U, designated poly(U), is formed. Under nonphysiological conditions with high levels of Mg^{2+} protein synthesis can occur in vitro without the specific initiation components normally required. With poly(U) as the mRNA, the "protein" synthesized is polyphenylalanine. Similarly, poly(A) encodes polylysine and poly(C) specifies polyproline. More complex messengers with random sequence copolymers, such as one containing only U and C produce polypeptides that contain not only proline and phenylalanine as predicted but also serine (from UCU and UCC) and leucine (from CUU and CUC). Because of degeneracy in the code and the complexity of the products, many of these experiments were difficult to interpret, and so use was made of synthetic messengers of defined sequence, transcribed from simple DNA sequences by RNA polymerase. Thus poly(AU), transcribed from a repeating poly(dAT), produces only a repeating copolymer of Ile-Tyr, Ile-Tyr read from successive triplets AUA UAU AUA UAU AUA, and so on. A synthetic poly(CUG), transcribed from a repeating CAG template strand, has possible codons CUG for Leu, UGC for Cys, and GCU for Ala, each repeating itself once the **reading frame** has been selected. Since selection of the initiation codon is random in these in vitro experiments, three different homopolypeptides are produced: polyleucine, polycysteine, and polyalanine. A perfect poly(CUCG) produces a single polypeptide with the internal sequence (Leu-Ala-Arg-Ser) whatever the initiation point. These relationships are summarized in Table 17.5; they show the codons to be triplets that are read in exact sequence, without overlaps or omissions of any base. An additional group of experiments used chemically synthesized trinucleotide codons as minimal messages. No proteins could be made, but the binding of only one amino acid (conjugated to an appropriate tRNA) by the ribosome was stimulated by a given codon. It was thus possible to decipher the meaning of each possible codon, and to identify termination codons that do not encode an amino acid. All of these conclusions were later verified by the determination of mRNA sequences.

TABLE 17.5 Polypeptide Products of Synthetic mRNAs[a]

mRNA	Codon sequence	Products
—$(AU)_n$—	— AUA UAU AUA UAU —	—$(Ile\text{-}Tyr)_{n/3}$—
—$(CUG)_n$—	— CUG CUG CUG CUG —	—Leu_n—
	— UGC UGC UGC UGC —	—Cys_n—
	— GCU GCU GCU GCU —	—Ala_n—
—$(CUCG)_n$—	CUC GCU CGC UCG	—$(Leu\text{-}Ala\text{-}Arg\text{-}Ser)_{n/3}$—

[a] The horizontal brackets accent the reading frame.

Mutations

An understanding of the genetic code and the way in which it is read provides a basis for understanding the nature of mutations. The term mutation simply means a change in a gene. **Point mutations** involve a change in a single base pair in the DNA and thus a single base in the mRNA. In some instances this change will occur in the third position of a degenerate codon and result in no change in the amino acid that is specified (e.g., UCC to UCA still codes for serine); such mutations are silent and would only be detected by gene sequence determination. They are commonly seen where genes for similar proteins, for example, hemoglobins from different species, are compared. The polypeptide sequences may be very similar or identical, while the majority of differences in the DNA sequence correspond to degenerate codon position. **Missense mutations** result from a base change that results in the incorporation of a different amino acid in the encoded protein (see Clin. Corr. 17.1). Point mutations that involve formation or modification of a termination codon change the length of the protein formed. Formation of a termination codon from one that encodes an amino acid (see Clin. Corr. 17.2) is often called a **nonsense mutation;** it results in premature termination and a shortened protein. Mutation of a natural termination codon to create one that encodes an amino acid allows the message to be "read through" until another stop codon is encountered. The result is a larger than normal protein. This phenomenon is the basis of several disorders as presented in Table 17.6 and Clin. Corr. 17.3.

Insertion or deletion of a single nucleotide within the coding region of a gene results in a **frameshift mutation.** The reading frame is altered and subsequent codons are read in the new context until a termination codon is reached. Table 17.7 illustrates this phenomenon with the mutant hemoglobin Wayne. The significance of reading frame selection is underscored by a phenomenon in some viruses in which a single segment of DNA

CLIN. CORR. 17.1 MISSENSE MUTATION: HEMOGLOBIN

Clinically the most important missense mutation known is the change from A to U in either the GAA or GAG codon for glutamate to give a GUA or GUG codon for valine in the sixth position of the β chain for hemoglobin. An estimated 1 of 10 American blacks are carriers of this mutation, which in its homozygous state is the basis for sickle-cell disease, the most common of all hemoglobinopathies (see Clin. Corr. 2.3 for the effects of this substitution). The second most common hemoglobinopathy is hemoglobin C disease, in which a change from G to A in either the GAA or GAG codon for glutamate results in an AAA or AAG codon for lysine in the sixth position of the β chain. Methods for diagnosis of these and other genetic disorders are discussed in Clin. Corr. 15.2.

CLIN. CORR. 17.2 DISORDERS OF TERMINATOR CODONS

In the disorder of hemoglobin McKees Rocks the UAU or UAC codon normally designating tyrosine in position 145 of the β chain has mutated to the terminator codon UAA or UAG. The result is a shortening of the β chain from its normal 146 residues to 144 residues, a change that gives the hemoglobin molecule an unusually high oxygen affinity. An attempt to compensate for decreased oxygen delivery by increased red blood cell production gives the polycythemic phenotype (see Clin. Corr. 23.2). Another human illness resulting from a terminator mutation is a variety of β-thalassemia. The thalassemias comprise a group of disorders characterized at the molecular level by an inbalance in the stoichiometry of globin synthesis. In β^0-thalassemia no β-globin is synthesized. As a result α-globin, unable to associate with β-globin to form hemoglobin, accumulates and precipitates in erythroid cells. The precipitation damages cell membranes, causing hemolytic anemia and interfering with erythropoiesis. One variety of β^0-thalassemia, which is common in Southeast Asia, results from a terminator mutation at codon 17 of the β-globin. The normal codon AAG designates a lysyl residue at β-17 but becomes the stop codon UAG in this variety of β^0-thalassemia. In contrast to the situation with hemoglobin McKees Rocks, in which the terminator mu-

TABLE 17.6 "Read Through" Mutation in Termination Codons Produce Abnormally Long α-Globin Chains

Hemoglobin	α-Codon 142	Amino acid 142	α-Globin length (residues)
A	UAA		141
Constant Spring	CAA	Glutamine	172
Icaria	AAA	Lysine	172
Seal Rock	GAA	Glutamate	172
Koya Dora	UCA	Serine	172

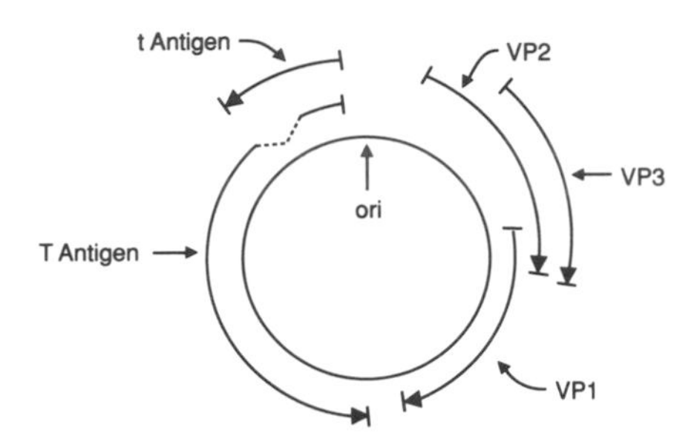

FIGURE 17.5
Map of the genome of simian virus 40 (SV40).
The DNA of SV40 is a double-stranded circle of slightly more than 5000 base pairs that encodes all of the information needed by the virus for its survival and replication within a host cell. It is an example of extremely efficient use of the information-coding potential of a small genome. Proteins VP1, VP2, and VP3 are structural proteins of the virus; VP2 and VP3 are translated from different initiation points to the same carboxyl terminus. The protein VP1 is translated in a different reading frame so that its amino-terminal section overlaps the *VP2* and *VP3* genes but its amino acid sequence in the overlapping segment is different from that of *VP2* and *VP3*. Two additional proteins, the large T and small t tumor antigens, which promote transformation of infected cells, have identical amino-terminal sequences. The carboxyl terminal segment of the small t protein is encoded by a segment of mRNA that is spliced out of the large T message, and the carboxyl-terminal sequence of large T is encoded by DNA that follows the termination of small t. This occurs through differential processing of a common mRNA precursor. The single site of origin of DNA replication (ori) is outside of all coding regions of the genome.

encodes different polypeptides that are translated using different reading frames. This may occur because the small size of the viruses physically limits the amount of DNA that can be packaged in them. One example is bacteriophage ϕX174 whose circular 5386 nucleotide genome is too small to separately encode all 9 viral proteins. A similar situation occurs in the tumor-causing simian virus SV40, as shown in Figure 17.5.

Aminoacylation of Transfer RNA Activates Amino Acids for Protein Synthesis

In order to be incorporated into proteins, amino acids first must be "activated" by being linked to their appropriate tRNA carriers. In every instance this is a two-step process that requires energy in the form of ATP and is catalyzed by one of a family of **aminoacyl-tRNA synthetases** each of

tation occurs late in the β-globin message, the terminator mutation occurs so early in the mRNA of β^0-thalassemia that no useful partial β-globin sequence can be synthesized, and β-globin is absent. In addition, β-globin mRNA levels are depressed, probably because premature termination of translation leads to instability of the mRNA.

Winslow, R. M. et al. Hemoglobin McKees Rocks (α_2 β_3 ^{145}Tyr leads to Term). A human nonsense mutation leading to a shortened β chain. *J. Clin. Invest.* 57:772, 1976. Chang, J. C. and Kan, Y. W.: β-Thalassemia: a nonsense mutation in man. *Proc. Natl. Acad. Sci. U.S.A.* 76:2886, 1979.

CLIN. CORR. 17.3 THALASSEMIA

The disorders summarized in Table 17.6 are forms of α-thalassemia, because the abnormally long α-globin molecules, which replace normal α-globin, are present only in small amounts. The small amounts of α-globin result either from decreased rate of synthesis or more likely, from increased rate of breakdown. As can be seen, the normal stop codon, UAA, for α-globin mutates to any of four sense codons with resultant placement of four different amino acids at position 142. Normal α-globin is only 141 residues in length, but the four abnormal α-globins are 172 residues in length, presumably because of a terminator codon in position 173. Other elongated globin chains can result from frame-shift mutations or insertions.

Weatherall, D. J. and Clegg, J. B. The α-chain termination mutants and their relationship to the α-thalassemias. *Philos. Trans. R. Soc. London* 271:411, 1975.

TABLE 17.7 A Frameshift Mutation Results in Production of Abnormal Hemoglobin Wayne[a]

Position	137	138	139	140	141	142	143	144	145	146	147
Normal α-globin amino acid sequence	- Thr	- Ser	- Lys	- Tyr	- Arg						
Normal α-globin codon sequence	- ACP	- UC(U)	- AAA	- UAC	- CGU	- [UAA]	- GCU	- GGA	- GCC	- UCG	- GUA
Wayne α-globin codon sequence	- ACP	- UCA	- AAU	- ACC	- GUU	- AAG	- CUG	- GAG	- CCU	- CGG	- [UAG]
Wayne α-globin amino acid sequence	- Thr	- Ser	- Asn	- Thr	- Val	- Lys	- Leu	- Glu	- Pro	- Arg	

[a] The base deletion causing the frameshift is encircled. The stop codons are boxed.
P = A, G, U, or C.

which is specific for a single amino acid and its appropriate tRNA species. The reactions are normally written as follows:

$$\mathrm{H{-}\overset{\displaystyle H}{\overset{|}{N}}{-}\overset{\displaystyle R}{\overset{|}{CH}}{-}\overset{\displaystyle O}{\overset{\|}{C}}{-}OH} + \mathrm{ATP} + \mathrm{E} \rightleftharpoons \left[\mathrm{H{-}\overset{\displaystyle H}{\overset{|}{N}}{-}\overset{\displaystyle R}{\overset{|}{CH}}{-}\overset{\displaystyle O}{\overset{\|}{C}}} \sim \mathrm{AMP\cdot E}\right] + \mathrm{PP_i} \quad \textbf{(1)}$$

$$\left[\mathrm{H{-}\overset{\displaystyle H}{\overset{|}{N}}{-}\overset{\displaystyle R}{\overset{|}{CH}}{-}\overset{\displaystyle O}{\overset{\|}{C}}} \sim \mathrm{AMP\cdot E}\right] + \mathrm{tRNA} \rightleftharpoons \mathrm{H{-}\overset{\displaystyle H}{\overset{|}{N}}{-}\overset{\displaystyle R}{\overset{|}{CH}}{-}\overset{\displaystyle O}{\overset{\|}{C}}{-}tRNA} + \mathrm{AMP} + \mathrm{E} \quad \textbf{(2)}$$

Sum:

$$\mathrm{H{-}\overset{\displaystyle H}{\overset{|}{N}}{-}\overset{\displaystyle R}{\overset{|}{CH}}{-}\overset{\displaystyle O}{\overset{\|}{C}}{-}OH} + \mathrm{ATP} + \mathrm{tRNA} \rightleftharpoons \mathrm{H{-}\overset{\displaystyle H}{\overset{|}{N}}{-}\overset{\displaystyle R}{\overset{|}{CH}}{-}\overset{\displaystyle O}{\overset{\|}{C}}{-}tRNA} + \mathrm{AMP} + \mathrm{PP_i}$$

The brackets surrounding the aminoacylAMP–enzyme complex indicate that it is a transient intermediate that is not usually encountered except on the enzyme surface. The ''squiggle'' (~) linkage of the amino acid to the AMP designates this aminoacyladenylate as a high-energy intermediate since it is a mixed acid anhydride with carboxyl and phosphoryl components. The aminoacyl ester linkage in tRNA is lower in energy than the aminoacyladenylate, but still ''activated'' relative to the free amino acid. The reactions are written to show their reversibility. In reality, pyrophosphatases rapidly cleave the pyrophosphate that is released in Equation 1 and the equilibrium is very strongly shifted toward formation of aminoacyl-tRNA. From the viewpoint of precision in protein biosynthesis, the amino acid, which had only its side chain (R group) to distinguish it, is now coupled to a much larger, more complex, and more easily recognized carrier.

Specificity and Fidelity of the Aminoacylation Reactions

Cells contain 20 different aminoacyl-tRNA synthetases, each specific for one amino acid, and at most a small family of carrier tRNAs for that amino acid. In translation, codon–anticodon interactions define the amino acid to be incorporated. If an incorrect amino acid is carried by the tRNA, it will be incorporated into the protein. Cysteinyl-tRNA can be reduced in vitro with Raney nickel to form alanyl-tRNA. Synthesis of hemoglobin with this tRNA results in vitro in the incorporation of alanine at positions normally occupied by cysteine. Correct selection of the tRNA and amino acids by the synthetases is necessary to avoid such mistakes. Accuracy of these enzymes is central to the fidelity of protein synthesis.

The aminoacyl-tRNA synthetases that catalyze amino acid activation reactions share a common mechanism and in the cell many of the synthetases are physically associated with one another. Nevertheless they are a diverse group of proteins that may contain one, two, or four identical subunits or pairs of dissimilar subunits. Detailed study of these enzymes indicates that separate structural domains are involved in aminoacyl-adenylate formation, tRNA recognition and, if it occurs, subunit interactions. Binding of tRNA may occur on a single subunit or the site may bridge two subunits. In spite of their structural diversity, each enzyme is capable of almost error-free formation of correct aminoacyl–tRNA complexes.

Selection and incorporation of the correct amino acid requires considerable discrimination on the part of some synthetases. While some amino acids may be easily recognized by their bulk (e.g., tryptophan), or lack of bulk (glycine), or by positive or negative charges on the side chains (e.g., arginine, lysine, aspartate, and glutamate), others appear much more difficult to discriminate. Recognition of valine rather than isoleucine by the valyl-tRNA synthetase or the selective reaction of isoleucine with its synthetase may be the most difficult since the two side chains differ by only a single methylene group. Two selective steps are involved in this phenomenon. First, the amino acid recognition and activation site of each enzyme has a great specificity that is characteristic of many enzymes. Mis-acylation is relatively rare, but it does occur. Second, an additional **"proofreading"** site exists on the enzyme and it preferentially recognizes and hydrolyzes mis-acylated tRNA. This is not simply the reversal of aminoacyl-tRNA formation but a separate hydrolytic reaction that does not utilize AMP. The two-step mechanism results in mis-acylation rates of about 1 in 3000. It is termed a **double sieve** mechanism to indicate that the size of the two sites accounts for their selectivity. The activation site of the valyl-tRNA synthetase is small enough to largely exclude isoleucine, but the larger hydrolytic site accommodates isoleucine and destroys it if it is mistakenly coupled. The isoleucyl-tRNA synthetase must have an activation site large enough to accommodate isoleucine and so may be less able to discriminate against valine, but its smaller hydrolytic site effectively fits only the valyl ester and hydrolyzes it. Proofreading is done by many synthetases and is analogous to the editing or proofreading of misincorporated nucleotides due to the $3' \rightarrow 5'$ exonuclease activity of DNA polymerases (Chapter 15).

Each synthetase must also correctly recognize one to several tRNA species that correctly serve to carry its amino acid, while also rejecting incorrect or improper tRNA species. Given the relative size and complexity of tRNA molecules this should be simpler than selection of a single amino acid. But the process is complicated by the overall conformational similarities of all tRNAs and the structural diversity of the aminoacyl-tRNA synthetases. The most obviously unique element of each tRNA is its anticodon, and so it appears a logical element of tRNA recognition by the synthetase. In some cases, for example, methionyl tRNA, this seems to be true; changing the anticodon also alters synthetase recognition. In other instances it is at least partly true. But in many cases the anticodon cannot be the primary determinant of synthetase recognition of tRNA. Evidence for this conclusion comes from the existence of **suppressor mutations,** that is, those that "suppress" the expression of classes of chain termination or nonsense mutations. For example, a point mutation in a glutamine (CAG) codon to form a termination (UAG) codon would normally cause the encoded protein to be terminated at that point. A second suppressor mutation in the anticodon of a tyrosine tRNA, in which the normal GUA anticodon is changed to CUA, allows the termination codon to be "read through," and a more nearly normal protein in which one glutamine residue is replaced by tyrosine can be made. Correct aminoacylation of the mutant tyrosine tRNA shows that in this case the anticodon is not the determinant of synthetase specificity but it does not tell us how specificity is determined. In one instance, alanine tRNA, the primary recognition characteristic is a G-U base pair in the acceptor stem; even if no other changes in the alanine tRNA occur, any variation at this position, even to U-G, destroys its acceptor ability with alanine tRNA synthetase, while incorporation of a G-U base pair in the same position of a cysteine tRNA makes it a good alanine acceptor. Other tRNAs look more complex, with important residues at several positions in their structures. It thus appears that the recognition characteristics may be very different for different synthetase-tRNA systems and the mechanisms for most are not yet known.

17.3 PROTEIN BIOSYNTHESIS

Translation Is Directional and Colinear With mRNA

In the English language words are read from left to right and not right to left. Similarly, messenger RNA is read in a defined fashion and proteins are also synthesized directionally. The conventions under which both RNA and proteins are written correspond to the directions in which they are read and synthesized, that is, $5' \rightarrow 3'$ for RNA sequences and from the amino terminal residue to the carboxy terminus for proteins. This was demonstrated by following the incorporation of radioactive amino acids into specific sites of a protein as a function of time. Figure 17.6 illustrates such an experiment in which reticulocytes were allowed to synthesize hemoglobin at a rate that was reduced by lowering the temperature. Only full length, completed globin chains were isolated and analyzed. The completed chains that incorporated radioactive amino acids during the shortest exposures to the radioactive precursors were near to being finished at the time of the pulse, and were found to have radioactive amino acids only in the carboxy-terminal segment (peptides f and g). Longer pulses with labeled amino acids resulted in central segments (peptides c, d, and e) of the protein also showing radioactivity, and only the longest pulse time, still corresponding to less than the time needed to synthesize a full-length polypeptide, showed radioactivity approaching the amino-terminal segments, peptides A and B. Hence, *the temporal sequence of protein synthesis is from the amino to the carboxy end.*

The existence of stable, isolatable polysomes and the directional, sequential nature of protein synthesis imply that ribosomes remain bound to a given mRNA molecule until the message is fully read, and that there is movement of each ribosome along the length of the mRNA. These conclusions are supported by comparisons of mRNA sequences with the se-

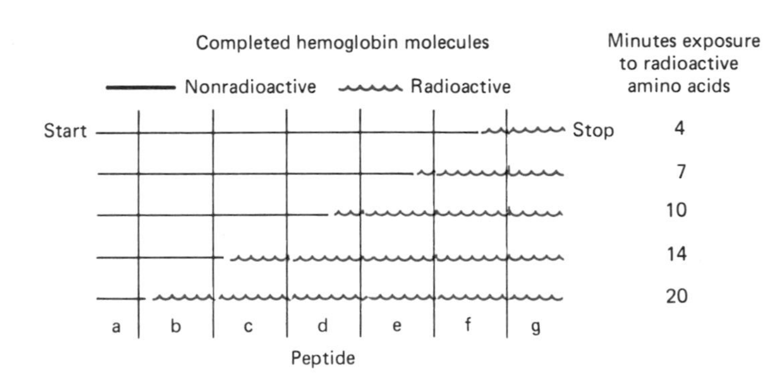

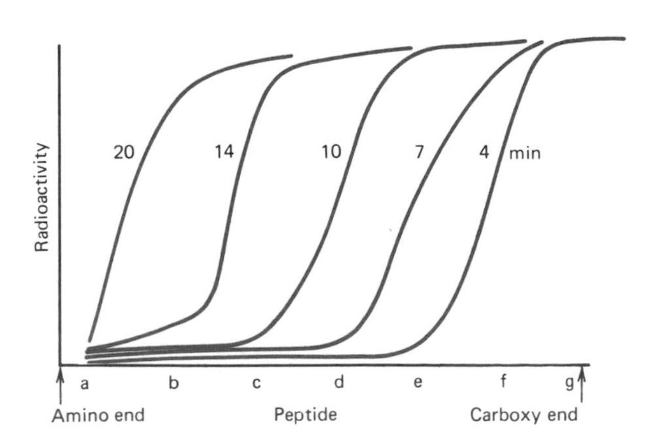

FIGURE 17.6
Hemoglobin synthesis starts at the amino end and stops at the carboxy end.
See text for explanation.

quences of the proteins they encode. There is a perfect, colinear, gap-free correspondence of the mRNA coding sequence and the sequence of the polypeptide that is synthesized. In fact, it is common to deduce the sequences of proteins solely from the nucleotide sequences of mRNAs or the DNA of genes encoding the proteins, although these deduced sequences may differ from the genuine protein because of posttranslational modifications.

Initiation of Protein Synthesis Is a Complex Process

A good novel can be analyzed in terms of its beginning, its development or middle section, and its satisfactory ending. Protein biosynthesis can be described in a similar conceptual and mechanical framework. We will separately consider the initiation of the process, elongation, during which the great bulk of the protein is formed, and termination of the synthesis and release of the finished polypeptide. We will then examine the posttranslational modifications that the protein may undergo.

Initiation requires the bringing together of the small (40S) ribosomal subunit, the amino terminal amino acid as its tRNA complex, and the mRNA, all in an appropriate orientation. This is followed by association of the large (60S) subunit to form a completed initiation complex on an 80S ribosome. The process is shown in Figure 17.7; it requires considerable precision, and is promoted by a complex group of proteins known as **initiation factors.** (The terminology reflects an initial uncertainty as to the chemical nature of these proteins.) These initiation factors participate only in the initiation process. Many of them bind to ribosomes transiently during initiation steps; they are not ribosomal structural proteins. The number of eucaryotic initiation factors is large and the specific functions of some remain unclear; procaryotic protein synthesis provides a useful and less complex model for comparison.

As a first step, ***e*ucaryotic *i*nitiation *f*actor 2 (eIF2)** binds to GTP and one species of methionyl tRNA, designated **Met–tRNA$_i^{met}$,** to form a ternary complex. The subscript i refers to this specific tRNA involved only in initiation. No other aminoacyl-tRNA, including Met–tRNA$_m^{met}$, which participates only in elongation step of protein synthesis, can replace the initiation-specific Met–tRNA$_i^{met}$ in this step. Recognition of the initiator tRNA by eIF2 is in some ways analogous to the recognition of a tRNA species by its cognate aminoacyl transferase. It is not yet possible to define the recognition element completely, although a comparison of initiator tRNA sequences from several species show conservation of base pairs in the anticodon and acceptor stems and a unique conformation of the anticodon loop. Procaryotes also utilize a specific initiator tRNA whose methionine is modified by formylation of its amino group (Figure 17.8). Only **fMet–tRNA$_i^{met}$** is recognized by procaryotic IF2.

The second step in the initiation process requires the participation of 40S ribosomal subunits, which are found in association with a very complex protein **eIF3,** also called a ribosome dissociation or antiassociation factor. The protein eIF3 consists of 7–10 different polypeptides and has a molecular weight of 500,000–700,000. In electron micrographs, eIF3 can be seen bound to the 40S subunit on the surface that contacts the larger 60S subunit, thus physically blocking association of 40S and 60S subunits. It is not yet clear if eIF3 alone also causes dissociation of 80S ribosomes into subunits upon termination of protein synthesis, or whether additional protein factors such as **eIF4c** or **eIF6** must also be involved.

An "entry" complex that includes eIF2–Met–tRNA$_i^{met}$–GTP, eIF3 40S, eIF4c, and perhaps additional protein factors can now form. This entry complex is able to bind mRNA and thus generate a preinitiation

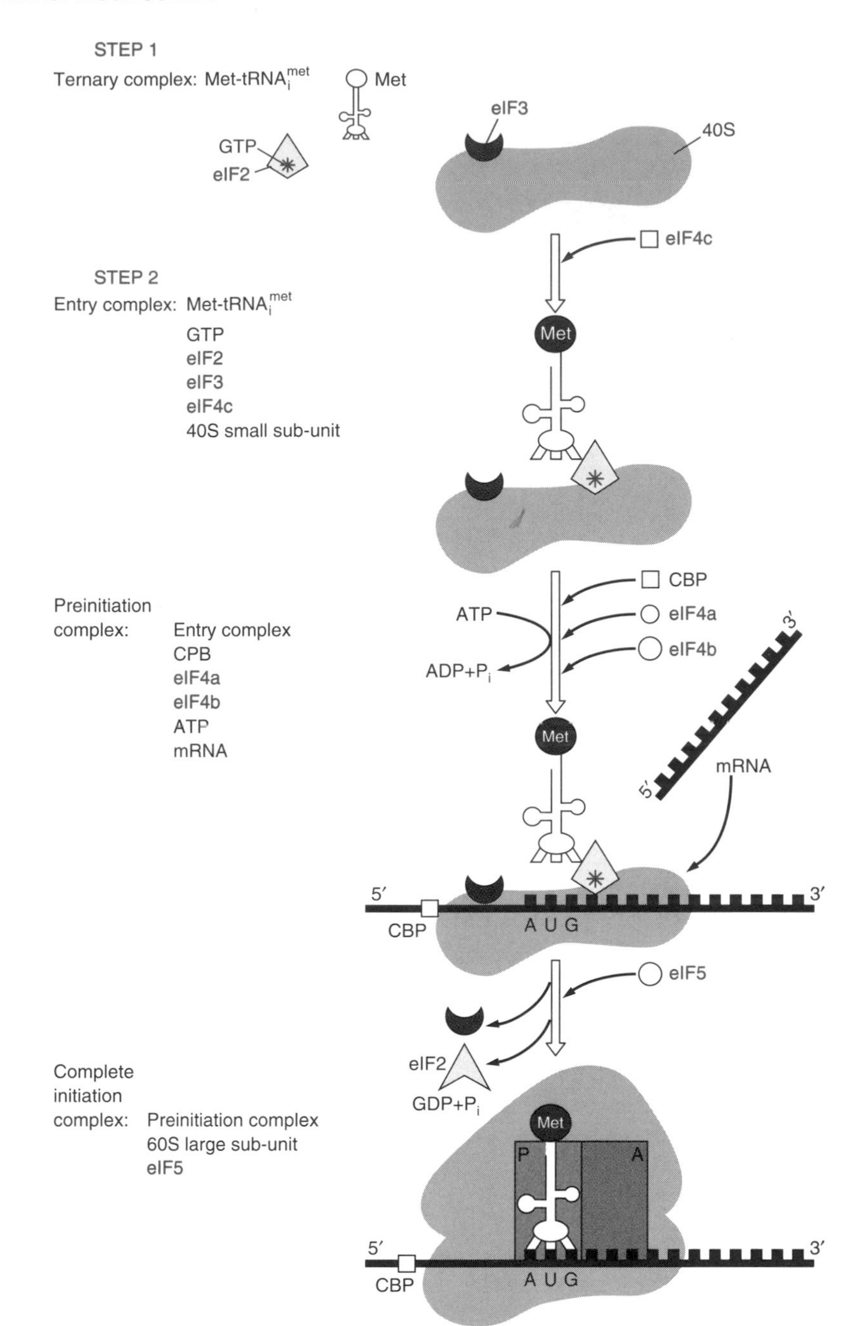

FIGURE 17.7
Initiation of translation in eucaryotes.
See the text for a description. The figure shows initiation on a naked mRNA molecule; additional subunits will later complex with the same message to form polysomes. Several additional initiation factors are involved in the formation of the initiation complex and are bound at times during the process, but are omitted from the figure for clarity. The different shape of eIF2 in complexes with GTP and GDP indicates that a conformational change in the protein occurs upon hydrolysis of the triphosphate.

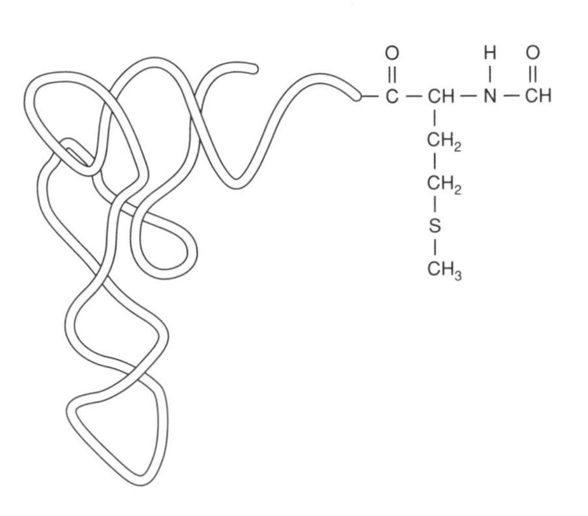

FIGURE 17.8
Formylmethionyl-tRNA$_i^{met}$.
The formylmethionyl residue is very greatly exaggerated in size relative to the RNA in this drawing.

complex but once again several protein factors play a role in this process. Proteins **eIF4a** and **eIF4b** and a **cap binding protein** (CBP) help place the message on the 40S subunit with the capped 5′ end correctly oriented. The ATP hydrolysis is required in this process, for opening, that is melting, the secondary structure of the message, for "scanning" the messenger to locate the initiation AUG sequence, and to place the AUG sequence in the correct site on the 40S subunit. Scanning of the messenger from its 5′-capped end may involve the passage of more than 100 nucleotides along the 40S until an AUG is encountered. Normally the first AUG sequence in the message is used to initiate translation, but sometimes the surrounding nucleotide sequence "context" is not appropriate for initiation and an AUG later in the message is selected for initiation. The final preinitiation complex thus includes the 40S subunit, the eIF2–Met–$tRNA_i^{met}$–GTP ternary complex, correctly oriented mRNA, and several protein factors.

Formation of the complete initiation complex now can proceed with the involvement of 60S subunits and an additional factor, eIF5. Protein **eIF5** first interacts with the preinitiation complex; GTP is hydrolyzed to GDP and P_i, and eIF2–GDP and eIF3 are released from the subunit. The 40S Met–$tRNA_i^{met}$–mRNA complex can now interact with a 60S subunit to generate an 80S ribosome with the message and initiator tRNA correctly positioned on the ribosome.

Procaryotes accomplish the same end using many fewer nonribosomal factors and a slightly different order of interaction. Procaryotic 30S subunits, as complexes with a much less complicated IF3, first bind mRNA. Correct orientation of the messenger with an AUG initiation codon in the proper site relies in part on base pairing between a pyrimidine-rich sequence of eight nucleotides in 16S rRNA and a purine-rich **"Shine–Dalgarno"** sequence (named for its discoverers) about 10 nucleotides upstream from the initiator signal. Complementarity between the rRNA and the message-positioning sequence of an mRNA is often not perfect and may include several mismatches. As a first approximation, the better the complementary pairing, the more efficient initiation at that AUG will be. It is interesting that eucaryotes do not utilize this mRNA–rRNA base pairing mechanism, but instead use many protein factors to accomplish a correctly positioned mRNA. After procaryotic mRNA is bound by the 30S subunit, a ternary complex of **IF2,** fMet–$tRNA_i^{met}$, and GTP is bound. A third initiation factor, **IF1,** also participates in formation of the preinitiation complex. The 50S subunit is now bound; in the process, GTP is hydrolyzed to GDP and P_i, and the initiation factors are released.

Elongation Is a Step-Wise Formation of Peptide Bonds

Protein synthesis now occurs by the stepwise elongation of the polypeptide chain. At each step the ribosomal **peptidyl transferase** transfers the growing peptide (or in the first step the initiating methionine residue) from its carrier tRNA to the α-amino group of the amino acid residue of the aminoacyl tRNA specified by the next codon of the messenger. Once again the efficiency and fidelity of this process is enhanced by the participation of nonribosomal proteins known as elongation factors. These factors utilize the energy released by GTP hydrolysis to ensure selection of the proper aminoacyl-tRNA species and to move the message and associated tRNAs through the decoding region of the ribosome. The elongation process is illustrated in Figure 17.9.

At a given instant, up to three different tRNA molecules may be bound at specific sites on the ribosome. The initiating methionyl-tRNA has been placed in position so that the methionyl residue may be transferred to the

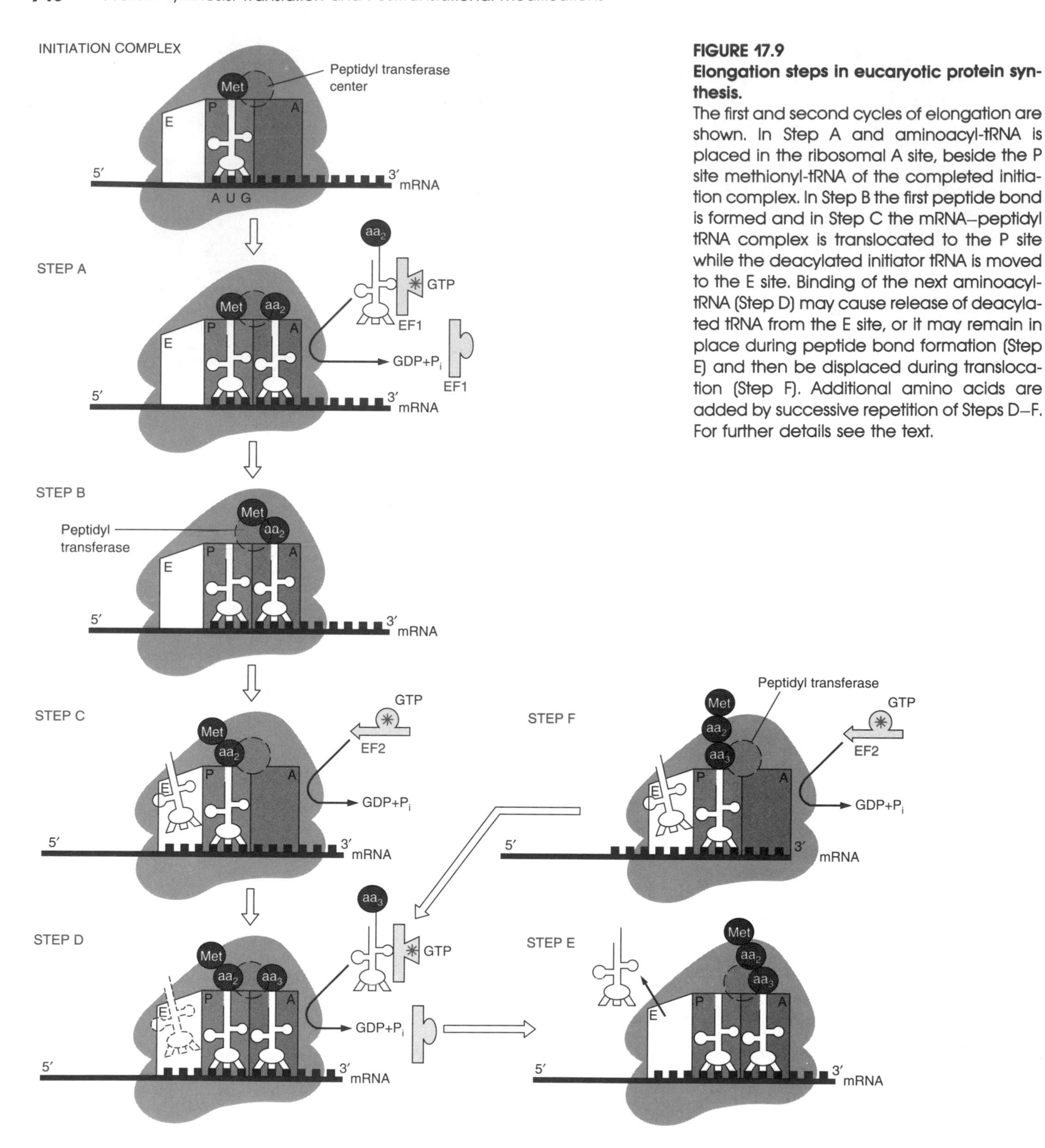

FIGURE 17.9
Elongation steps in eucaryotic protein synthesis.
The first and second cycles of elongation are shown. In Step A and aminoacyl-tRNA is placed in the ribosomal A site, beside the P site methionyl-tRNA of the completed initiation complex. In Step B the first peptide bond is formed and in Step C the mRNA–peptidyl tRNA complex is translocated to the P site while the deacylated initiator tRNA is moved to the E site. Binding of the next aminoacyl-tRNA (Step D) may cause release of deacylated tRNA from the E site, or it may remain in place during peptide bond formation (Step E) and then be displaced during translocation (Step F). Additional amino acids are added by successive repetition of Steps D–F. For further details see the text.

free amino group of the incoming aminoacyl-tRNA; it thus occupies the donor site, also called the **peptidyl site** or **P site** of the ribosome. The aminoacyl-tRNA specified by the next codon of the message is bound at the acceptor site, also called the **aminoacyl site** or **A site** of the ribosome. Selection of the correct aminoacyl-tRNA is enhanced by **elongation factor 1 (EF1);** a component of EF1, $EF1_{\alpha}$, first forms a ternary complex with aminoacyl-tRNA and GTP. The $EF1_{\alpha}$–aminoacyl–tRNA–GTP complex is bound by the ribosome and if codon–anticodon interactions are correct the aminoacyl-tRNA is placed at the A site, GTP is hydrolyzed to GDP and P_i, and the $EF1_{\alpha}$–GDP complex dissociates. At this point the initiating methionyl-tRNA and the incoming aminoacyl-tRNA are juxtaposed on the ribosome. Their anticodons are paired with successive codons of the message in the decoding region of the small subunit, and their amino acids are beside one another at the peptidyl transferase site of the large subunit. Peptide bond formation can now occur.

Peptidyl transferase catalyzes the attack of the amino group of the aminoacyl-tRNA on the carbonyl carbon of the methionyl-tRNA. The result is transfer of the methionine to the amino group of the aminoacyl-tRNA in the A site. This reaction does not require an external energy source such as ATP or GTP. Instead, the energy of the methionyl (or peptidyl) ester linkage to the tRNA is sufficient to drive the reaction in the direction of peptide bond formation (recall that ATP is used in the biosynthesis of each aminoacyl-tRNA and that these reactions are reversible).

The mRNA and the dipeptidyl-tRNA at the A site must now be repositioned to permit another addition cycle to begin. This is accomplished by means of **elongation factor 2 (EF2),** also called translocase. Elongation factor 2 moves the messenger and dipeptidyl-tRNA, in codon–anticodon register, from the A site to the P site. In the process a molecule of GTP is hydrolyzed to GDP plus P_i, thus providing energy for the movement. As the dipeptidyl-tRNA is moved to the P site, the deacylated donor (methionine) tRNA is moved to the third tRNA binding region, the exit site or E site on the ribosome. The ribosome is now ready to enter a new cycle. The next aminoacyl-tRNA specified by the message is delivered by EF1 to the A site and the deacylated tRNA in the E site is probably released. Now peptide transfer can again occur. Successive cycles of EF1 mediated binding of aminoacyl-tRNA, peptide bond formation, and EF2 mediated translocation result in the stepwise elongation of the polypeptide toward its eventual carboxyl terminus. Note that regardless of the length of the polypeptide chain, peptide bond formation always occurs through the attack of the α-amino group of the incoming aminoacyl-tRNA on the peptide carboxyl-tRNA linkage, hence their geometric arrangement at the peptidyl transferase site remains constant.

In order to study and understand the mechanism of peptidyl transferase, biochemists have tried to dissociate it from the large subunit or identify it as a specific ribosomal protein. These attempts have all failed. In the best understood case, the bacterium *E. coli,* peptidyl transferase activity requires several different large subunit proteins and the rRNA. Omission or significant modification of any of these components causes total loss of peptidyl transferase activity, while other proteins can be deleted with little or no effect. The discovery of catalytic RNA molecules (Chapter 16) has led to the speculation that the ribosome has evolved from an ancestral RNA particle in which peptide bond formation was catalyzed by the RNA; this is, however, unproven, but it illustrates the complexity of the peptidyl transferase center and the reaction it catalyzes.

The role of GTP in the action of EF1 and EF2 most probably relates to conformational changes in these proteins. Both $EF1_{\alpha}$ and EF2 bind ribosomes tightly as GTP complexes. The GDP complexes dissociate from the

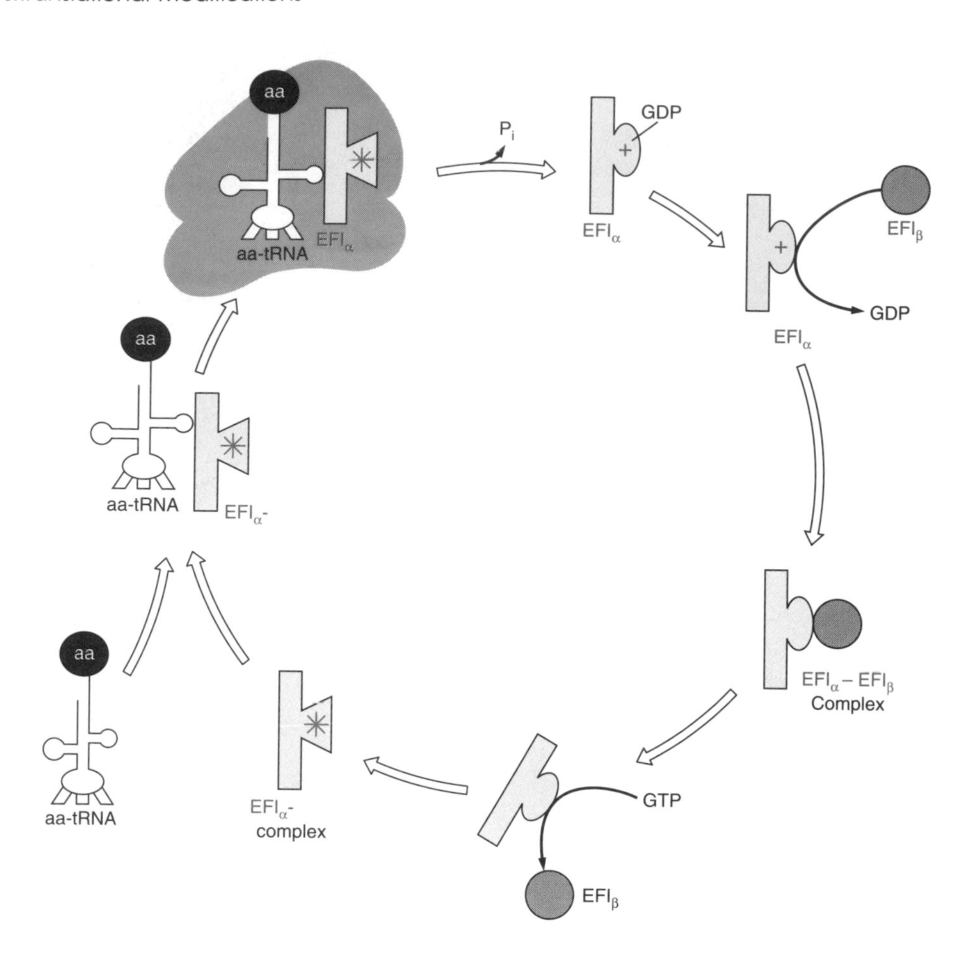

FIGURE 17.10
EF1 in the elongation cycle.
The $EF1_\alpha$–GTP–aminoacyl–tRNA complex (left center) transfers the aminoacyl-tRNA to the ribosome with the concomitant hydrolysis of GTP and a change in the conformation of $EF1_\alpha$ that reduces its affinity for tRNA and the ribosome. The GDP is then displaced from $EF1_\alpha$ by $EF1_\beta$, resulting in the complex at the lower right. Binding of GTP then displaces $EF1_\beta$ and allows binding of an aminoacyl-tRNA by $EF1_\alpha$ in its higher affinity conformation. In procaryotes a similar cycle exists; EF-Tu functions as the carrier of aminoacyl-tRNA and EF-Ts serves to displace GDP and regenerate the functional form of EF-Tu.

ribosome more easily because the protein conformation is changed. Viewed another way, GTP stabilizes a protein conformation that confers upon $EF1_\alpha$ high affinity toward aminoacyl-tRNA and the ribosome, while GDP stabilizes a conformation with lower affinity for aminoacyl-tRNA and ribosome, thus allowing tRNA delivery and factor dissociation. In the case of $EF1_\alpha$, restoration of the higher affinity GTP-associated conformation requires participation of an additional protein, $EF1_\beta$ (Figure 17.10). This protein displaces GDP from $EF1_\alpha$, forming an $EF1_\alpha$–$EF1_\beta$ complex. The GTP then displaces $EF1_\beta$, forming an $EF1_\alpha$–GTP complex, which can successively bind an aminoacyl-tRNA and then a ribosome. Procaryotes use a similar mechanism in which EF-Tu binds GTP and aminoacyl-tRNA and EF-Ts displaces GDP and thus helps recycle the carrier molecule. Procaryotes also utilize a GTP-dependent translocase, equivalent to EF2 but called **EF-G** or **G factor.**

Termination of Polypeptide Synthesis Requires a Stop Codon

A chain-terminating UAG, UAA, or UGA codon in the ribosomal A site does not promote the binding of any tRNA species. Instead, another nonribosomal protein, **release factor (RF)** binds to the ribosome as an RF–GTP complex (Figure 17.11). The peptide–tRNA ester linkage is then cleaved through the action of peptidyl transferase, acting in this instance as a hydrolase. The completed protein is thus released from its carrier tRNA and from the ribosome. Dissociation of the RF from the ribosome requires hydrolysis of the GTP molecule, again probably resulting in a change in RF conformation. This also frees the ribosome to be dissociated into subunits and then reenter the protein synthetic cycle at the initiation stage. In procaryotes three release factors, **RF1, RF2,** and **RF3** carry out

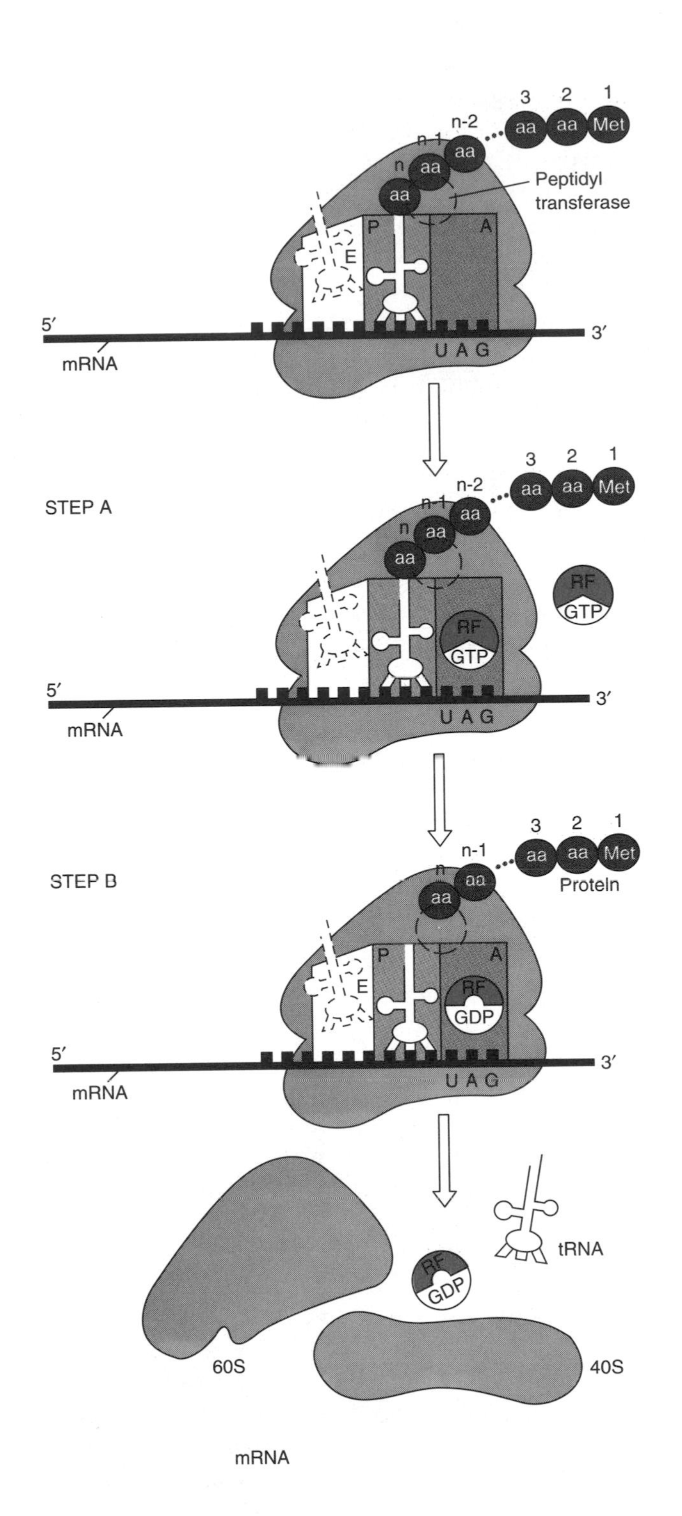

FIGURE 17.11
Termination of protein biosynthesis.
When a termination codon (UAG, UAA, or UGA) occupies the messenger portion of the ribosomal A site, binding of a GTP-release factor complex occurs (Step A). Peptidyl transferase now functions as a hydrolase; in Step B the protein is released by hydrolysis of the ester bond linking it to tRNA, and GTP is hydrolyzed to GDP and P_i. The complex now dissociates and the components can enter additional rounds of protein synthesis.

the termination function. The factor RF1 acts in response to UAG or UAA codons, RF2 acts in response to UGA or UAA codons, and RF3 stimulates the action of RF1 and RF2.

Protein Synthesis in Mitochondria Differs From Cytoplasmic Synthesis

Many of the characteristics of mitochondria suggest that they are evolutionary descendants of aerobic procaryotes that may have invaded and set up a symbiotic relationship within a eucaryotic host cell. Vestiges of their possible previous independence are retained, including their own circular DNA genome of about 16,500 base pairs. Most mitochondrial proteins are synthesized in the cytoplasm and are encoded in nuclear DNA, but some proteins, several tRNA species, and mitochondrion-specific rRNA are encoded within the mitochondrial genome. Mitochondria have an independent protein synthesis apparatus that includes RNA polymerase, aminoacyl-tRNA synthetases, tRNAs, and characteristic ribosomes.

Although the overall course of protein biosynthesis in mitochondria is like that previously described, some aspects are significantly different. The synthetic components—tRNAs, aminoacyl-tRNA synthetases, and ribosomes—are unique to the mitochondrion (Table 17.1). The number of tRNA species is small and the genetic code is slightly different from the universal code described previously (Table 17.3). Mitochondrial ribosomes are smaller and the rRNAs are shorter than those of either the eucaryotic cytoplasm or of procaryotes. As in procaryotes, *N*-formylmethionyl-tRNA is used to initiate protein synthesis. A specific initiator, Met-$tRNA_i^{met}$, is modified by a transformylase that uses N^{10}-formyl H_4-folate as a donor to produce fMet-$tRNA_i^{met}$. A significant question is how the cell coordinates protein synthesis within the mitochondrion with the cytoplasmic synthesis of proteins destined for import into the mitochondria.

Some Antibiotics and Toxins Inhibit Protein Biosynthesis

Protein biosynthesis is obviously central to the continuing existence and reproduction of cells. An aggressive competitor can gain a biological advantage by interfering in the ability of other organisms to synthesize proteins. Many antibiotics and toxins function in this manner, often interacting with very specific translational steps or components. These inhibitors have been of value in elucidating normal protein synthesis; some show selectivity for procaryotic rather than eucaryotic protein synthesis and are extremely useful in clinical practice. Some examples of antibiotic action are listed in Table 17.8.

The antibiotic **streptomycin** physically binds to the small (30S) subunit of procaryotic ribosomes. This antibiotic interferes with the initiation of protein synthesis and also causes misreading of mRNA information. The effects of streptomycin are mediated by a single ribosomal component, protein S12 of the small subunit, and although streptomycin does not directly bind S12, mutations in protein S12 can confer resistance to or even dependence on streptomycin. Protein S12 is known to be involved in binding of tRNA, and thus streptomycin may function by altering the interactions of tRNA with the ribosome and message. Other aminoglycoside antibiotics, such as the **neomycins** or **gentamicins,** also interact with the small ribosomal subunit, but at sites that differ from that for streptomycin. They also cause mistranslation of messages. Another aminoglycoside, **kasugamycin,** binds small subunits and inhibits the initiation of trans-

TABLE 17.8 Some Inhibitors of Protein Biosynthesis

Inhibitor	Processes affected	Site of action
Streptomycin	Initiation, elongation	Procaryotes: 30S subunit
Neomycins	Translation	Procaryotes: multiple sites
Tetracyclines	Aminoacyl–tRNA binding	30S or 40S subunits
Puromycin	Peptide transfer	70S or 80S ribosomes
Erythromycin	Translocation	Procaryotes: 50S subunit
Fusidic Acid	Translocation	Procaryotes: EF-G
Cycloheximide	Elongation	Eucaryotes: 80S ribosomes
Ricin	Multiple	Eucaryotes: 60S subunit

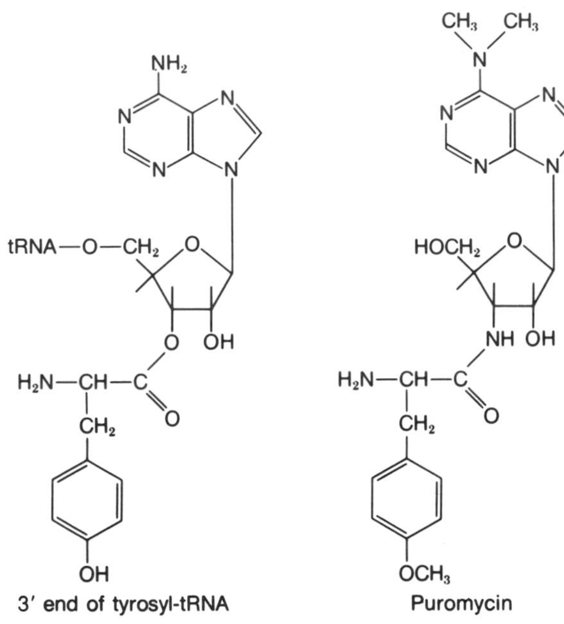

3′ end of tyrosyl-tRNA

Puromycin

FIGURE 17.12
Puromycin (right) interferes with protein synthesis by functioning as an analog of aminoacyl-tRNA, here tyrosyl-tRNA (left) in the peptidyltransferase reaction.

lation. Resistance to kasugamycin results from the absence of base methylation that normally occurs on two adjacent adenosine residues of small subunit RNA. **Tetracyclines** also bind directly to ribosomes, and function by interfering in aminoacyl-tRNA binding. Hence, several mechanisms of interfering in subunit–tRNA interactions are utilized by different antibiotics.

Other antibiotics function by interfering in peptide bond formation. **Puromycin** (Figure 17.12) resembles an aminoacyl-tRNA; it binds to the ribosomal A site and serves as an acceptor in the peptidyl transferase reaction. However, it cannot be translocated or serve as a peptide donor since its aminoacyl derivative is not in an ester linkage to the nucleoside. Thus puromycin prematurely terminates protein synthesis, leading to release of peptidyl-puromycin. Chloramphenicol directly inhibits peptidyl transferase upon binding at the transferase center; no transfer occurs, and peptidyl tRNA remains associated with the ribosome.

The translocation step is another potential target. **Erythromycin,** a macrolide antibiotic, interferes with translocation by procaryotic ribosomes. Eucaryotic translocation is inhibited by a protein toxin produced by *Corynebacterium diphtheria* **(diphtheria toxin)** through a mechanism in which the toxin binds at the cell membrane and a subunit enters the cytoplasm and catalyzes the **ADP-ribosylation** and inactivation of EF2, as represented in the following reaction.

$$\underset{\text{(active)}}{\text{EF2}} + \text{NAD}^+ \rightleftharpoons \underset{\text{(inactive)}}{\text{ADP-ribosyl EF2}} + \text{nicotinamide} + \text{H}^+$$

Attachment of the ADP-ribose moiety to EF2 is at a specific posttranslationally modified histidine residue known as diphthamide. (Posttranslational events will be discussed in the next section.)

A final group of toxins function by directly attacking the rRNA. The plant toxin **ricin** (from castor beans) and several related toxins are *N*-glycosidases that cleave a single adenine from the large subunit RNA backbone. The ribosome is inactivated by this apparently (to us) minor damage. A fungal toxin, **α-sarcin,** cleaves large subunit RNA at a single site and similarly inactivates the ribosome. Some strains of *E. coli* produce extracellular toxins that affect other bacteria. One of these, colicin

E3, is a ribonuclease that cleaves 16S RNA at a single site near the message-binding sequence; it thus inactivates the subunit and halts protein synthesis in competitors of the colicin-producing cell.

17.4 PROTEIN MATURATION: MODIFICATION, SECRETION, AND TARGETTING TO ORGANELLES

Some proteins emerge from the ribosome and are ready to function immediately. Many proteins, however, undergo a variety of posttranslational alterations. These modifications may result in their activation to a functional form, their physical placement in a specific subcellular compartment, or their secretion from the cell. It is important to remember that the information that determines the posttranslational fate of a protein resides in the primary structure of the protein. That is, the amino acid sequence and conformation of the polypeptide determine whether a protein will serve as a substrate for a modifying enzyme and/or identify it for direction to a subcellular or extracellular location.

Proteins for Export Follow the Secretory Pathway

Proteins destined for export are synthesized on the membrane-bound ribosomes of the rough endoplasmic reticulum (Figure 17.13). Their synthesis, however, begins on free cytoplasmic ribosomes, because the ribo-

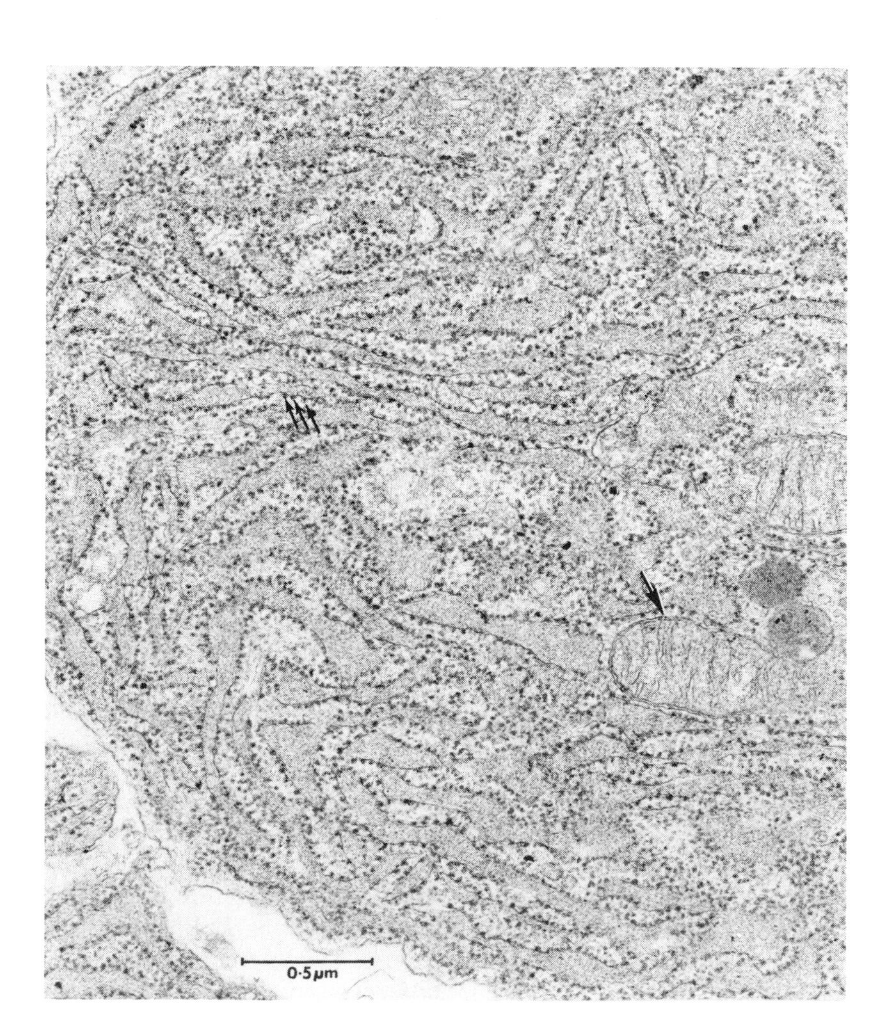

FIGURE 17.13
Rough endoplasmic reticulum of a plasma cell.
The three parallel arrows indicate three ribosomes among the many attached to the extensive membranes. The single arrow indicates a mitochondrion for comparison.
Courtesy of Dr. U. Jarlfors, University of Miami.

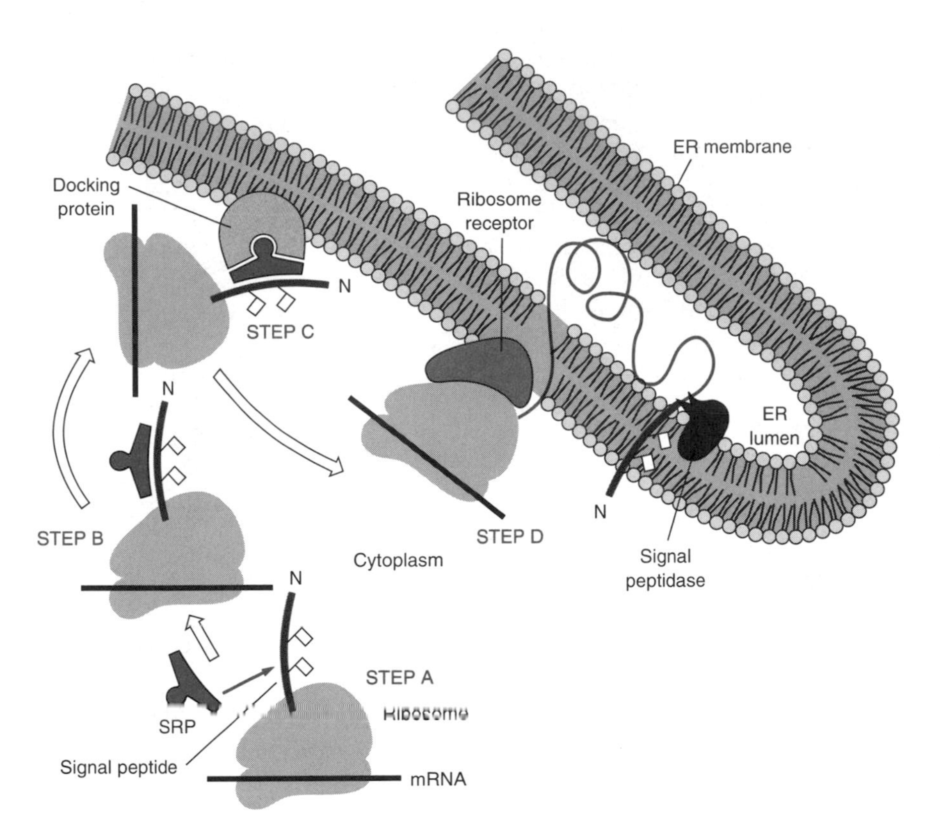

FIGURE 17.14
The secretory pathway: signal peptide recognition.
At Step A a hydrophobic signal peptide emerges from the exit site of a free ribosome in the cytoplasm of the cell. A signal recognition particle (SRP) recognizes and binds to the peptide and peptide elongation is temporarily halted (Step B). The ribosome moves to the ER membrane where the docking protein binds to the SRP (Step C). In Step D the ribosome is transferred to a ribosome receptor, protein biosynthesis is resumed and newly synthesized protein is extruded through the membrane into the lumen of the endoplasmic reticulum.

some has no means of identifying the protein it is about to synthesize. Initiation and elongation begin in the usual manner. Proteins of the secretory pathway include in their sequences a **hydrophobic signal peptide.** The sequences of amino acids of the signal peptide are almost always at or near the amino terminus of the protein, hence it emerges from the ribosome early in the process of synthesis. Signal peptides can be 15–30 amino acids long and always include a hydrophobic core of at least 8 nonpolar amino acids. There is no specific sequence for the signal peptide, but it is always rich in very hydrophobic amino acids.

As the signal peptide emerges from the ribosome it is bound by a cytoplasmic **signal recognition particle (SRP)** (see Figure 17.14). The SRP is made up of six different proteins plus one small (7S) RNA molecule. Binding of the SRP temporarily halts synthesis of the protein. Instead the ribosome moves to the endoplasmic reticulum where the SRP recognizes and binds to a **docking protein** that is localized at the cytoplasmic surface of the ER membrane. The ribosome is then transferred to a ribosome receptor on the membrane and the hydrophobic signal sequence, probably complexed by a signal sequence receptor protein, is inserted into the membrane, further anchoring the ribosome to the ER membrane. Both the SRP and the docking protein are no longer bound by the ribosome and can again function to direct other ribosomes to the ER. Meanwhile the translational block that was caused by SRP binding is relieved, and translation may again proceed. Translation and extrusion into or through the membrane are now coupled; even very hydrophilic and ionic segments can be directed through the hydrophobic ER membrane into the lumen of the endoplasmic reticulum. Other integral membrane proteins may participate in this process by forming a less hydrophobic pore or channel through which the growing polypeptide can pass.

The completed export-destined protein will be localized within the lumen of the ER but probably anchored to the membrane by its signal peptide. It can be released by the action of **signal peptidase,** an integral membrane protein localized at the luminal surface of the ER. Signal peptidase recognizes a cleavage site on the protein and frees it from the membrane-bound signal peptide. The protein can now fold into a stable three-dimensional conformation, disulfide bonds can form, and various components of multisubunit proteins may assemble. Other steps that may be necessary for maturation include proteolytic processing and glycosylation. These steps occur within the lumen of the ER and during the transit of the protein through the Golgi apparatus and secretory vesicles.

Glycosylation of Proteins Occurs in the Endoplasmic Reticulum and Golgi Apparatus

Glycosylation of proteins to form glycoproteins (see p. 373) is important for two distinct reasons. Glycosylation alters the physical properties of proteins, changing their solubility, stability, and physical bulk. Equally important, the carbohydrate groups of glycoproteins serve as recognition signals that are central to aspects of protein targetting and for cellular recognition of proteins and other cells. There is no glycosylation template or coding molecule. Sites and types of glycosylation are determined by the presence on the protein surface of appropriate amino acids and sequences and by the availability of enzymes and substrates to carry out the glycosylation reactions. Glycosylation can involve addition of a few carbohydrate residues or the formation of large branched oligosaccharide chains.

Glycosylation of proteins is carried out by a variety of **glycosyltransferases.** Up to 100 different enzymes may be involved but all carry out a similar basic reaction in which a sugar is transferred from an activated donor substrate to an appropriate receptor, usually another sugar residue that is part of an oligosaccharide "under construction." General classes of glycosyltransferases are summarized in Table 17.9. Individual enzymes show three kinds of specificity: for the monosaccharide that is transferred, for the structure and sequence of the acceptor molecule, and for the site and configuration of the anomeric linkage formed.

One major class of glycoproteins have sugars that are bound through either serine or threonine OH groups. Such ***O*-linked glycosylation** occurs only after the protein has been transported to the Golgi apparatus (see below). These *O*-linked carbohydrates always involve an *N*-acetylgalactosamine attachment to a serine or threonine residue of the protein. There is no defined amino acid sequence in which the serine or threonine residue must occur but only those residues whose side chains appear on the

TABLE 17.9 Glycosyltransferases in Eukaryotic Cells

Sugar transferred	Abbreviation	Donors	Glycosyltransferase
Mannose	Man	GDP-Man dolichol-Man	Mannosyltransferase
Galactose	Gal	UDP-Gal	Galactosyltransferase
Glucose	Glc	UDP-Glc dolichol-Glc	Glucosyltransferase
Fucose	Fuc	GDP-Fuc	Fucosyltransferase
N-Acetylgalactosamine	GalNAc	UDP-GalNac	*N*-acetylgalactosaminyltransferase
N-Acetylglucosamine	GlcNAc	UDP-GlcNAc	*N*-acetylglucosaminyltransferase
N-Acetylneuraminic acid (or sialic acid	NANA or NeuNAc SA	CMP-NANA CMP-SA	*N*-Acetylneuraminyltransferase Sialyltransferase)

surface of the folded peptide and in an appropriate environment serve as acceptors for the GalNAc-transferase.

Sequential addition of sugar residues to the attached GalNAc acceptor follows, and the structures synthesized depend on the nature of the specific glycosyltransferases in a given cell. If an acceptor is a substrate for more than one transferase, the quantity of each transferase controls the effective competition between them. Competition may lead to formation of some oligosaccharides that are not acceptors for any glycosyl transferase present, hence no further growth of the chain occurs. Other structures generated in the competition may be excellent acceptors that continue to grow until completed by one of a number of nonacceptor termination sequences. These processes may thus lead to many different kinds of oligosaccharide structures on otherwise identical polypeptides, so it is common to find considerable heterogeneity in the oligosaccharide components of glycoproteins.

The second class of glycoproteins have sugars that are linked through the amide nitrogen of asparagine residues, referred to as ***N*-linked glycosylation.** Biosynthesis of these *N*-linked oligosaccharides occurs in both the lumen of the ER and after transport of the protein to the Golgi apparatus. A specific glycosylation sequence, Asn-X-Thr (or Ser) in which X may be any amino acid except proline or aspartic acid, is required for *N*-glycosylation. Not all Asn-X-Thr/Ser sequences are glycosylated because some may be unavailable due to protein conformation.

Biosynthesis of the *N*-linked oligosaccharides begins with the synthesis of a lipid-linked intermediate (Figure 17.15). **Dolichol phosphate** at the cytoplasmic surface of the ER membrane serves as a glycosyl acceptor of *N*-acetylglucosamine. This reaction is inhibited by the antibiotic **tunicamycin,** which prevents *N*-glycosylation; this antibiotic has been valuable in elucidating this pathway. The GlcNAc-pyrophosphoryl dolichol is a substrate for stepwise glycosylation and formation of a branched $(Man)_5(GlcNAc)_2$-pyrophosphoryl dolichol on the cytoplasmic side of the membrane. After this intermediate is reoriented to the luminal surface of the membrane, four additional mannose and then three glucose residues are sequentially added to generate the completed intermediate. This complete oligosaccharide is then transferred from its dolichol carrier to a polypeptide as it emerges from the ER membrane. Thus *N*-glycosylation occurs cotranslationally, that is, at the same time as the protein is being synthesized, while *O*-glycosylation is posttranslational and so involves only fully folded polypeptides.

Processing of the oligosaccharide by glycosidases involves removal of some of the sugar residues from the newly transferred structure. Within the ER lumen the glucose residues, which were required for transfer of the *N*-linked oligosaccharide from the dolichol carrier, are sequentially removed, as is one mannose residue. These alterations mark the glycoprotein for transport to the Golgi apparatus where further trimming by glycosidases may occur. Additional sugars are also added by the same families of glycosyltransferases that generated the *O*-linked oligosaccharides. The resulting *N*-linked oligosaccharides are structurally related but diverse; all have a common core region ($GlcNAc_2Man_3$) linked to asparagine and originating from the dolichol-linked intermediate. Two classes of *N*-linked oligosaccharides are distinguishable. The high mannose type (Figure 17.16) includes mannose residues in a variety of linkages; it shows less processing from the dolichol-linked intermediate. The complex type is more highly processed, has a larger variety of sugars and linkages, and is as diverse as the *O*-linked group of oligosaccharides. Examples of mature oligosaccharides of each type are shown in Figure 17.17.

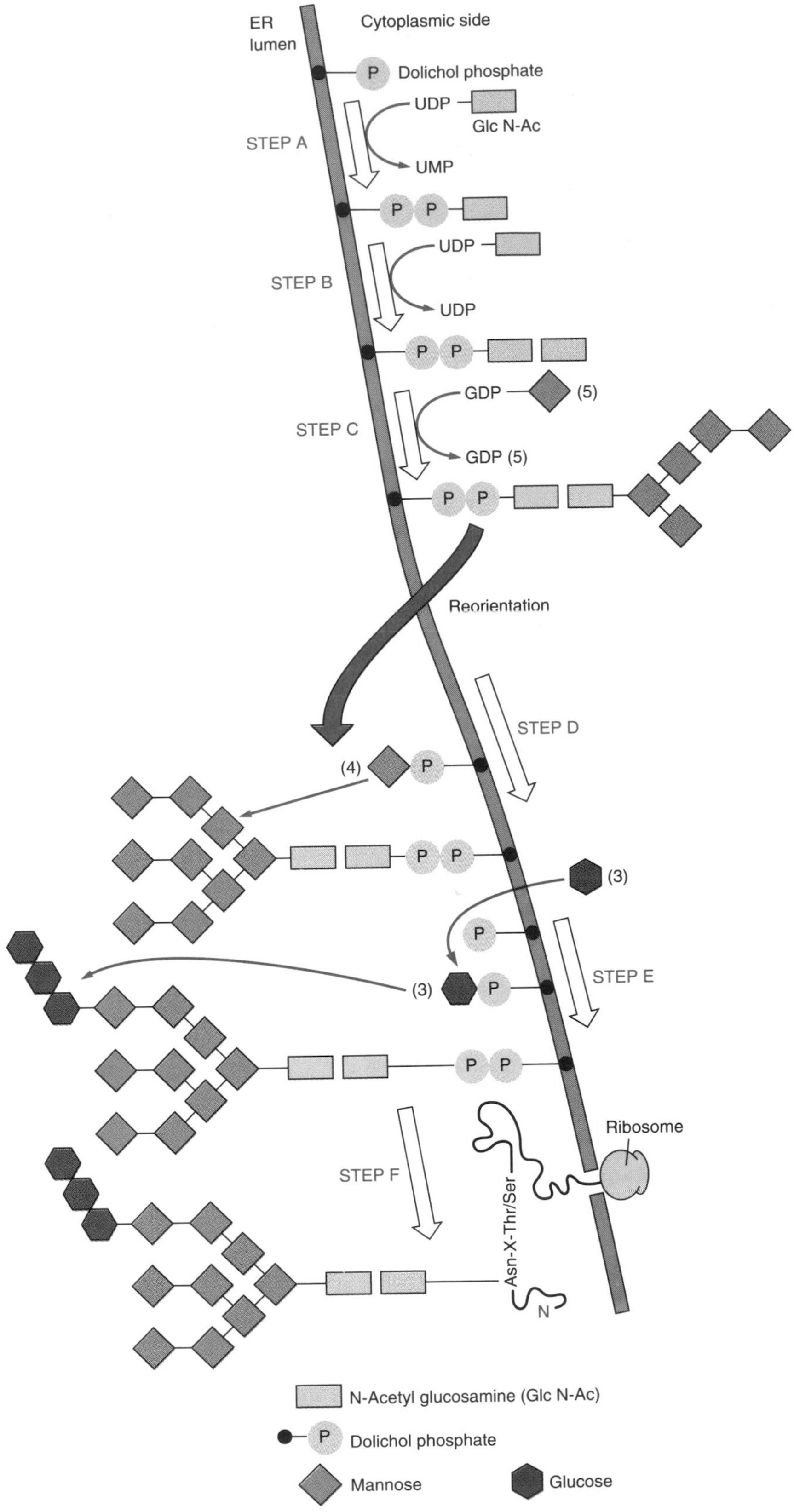

FIGURE 17.15
Biosynthesis of *N*-linked oligosaccharides at the membrane of the endoplasmic reticulum.
Synthesis is initiated on the cytoplasmic face of the ER membrane by transfer of *N*-acetyl glucosamine phosphate to a dolichol acceptor (Step A) followed by formation of the first glycosidic bond upon transfer of a second residue of *N*-acetyl glucosamine (Step B). Five residues of mannose are then added sequentially (Steps C) from a GDP mannose carrier. At this stage the lipid-linked oligosaccharide is reoriented to the luminal face of the membrane, and additional mannose (Steps D) and glucose (Steps E) residues are transferred from dolichol-linked intermediates. The dolichol-sugars are generated from cytoplasmic nucleoside diphosphate sugars. The completed oligosaccharide is finally transferred to a protein that is in the process of being synthesized at the membrane surface; the signal peptide may have already been cleaved at this point.

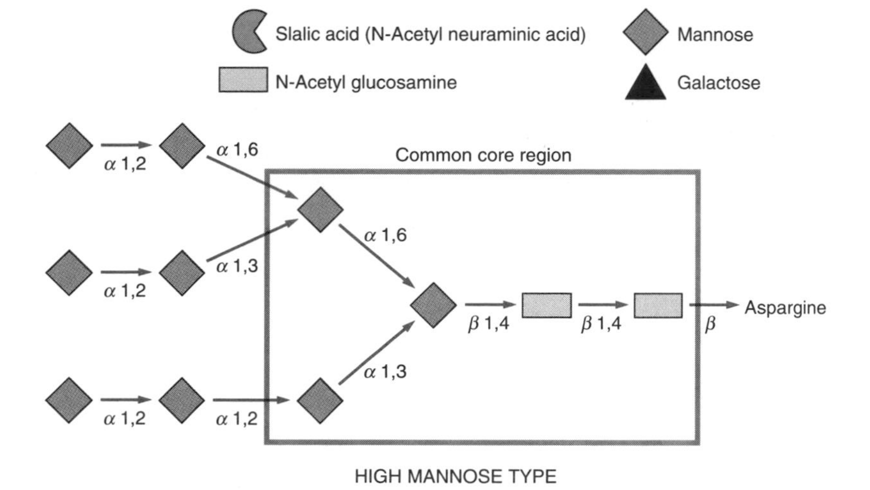

Common core region

α 2,3 β 1,4 β 1,2 α 1,6

β 1,4 β 1,4 β Aspargine

α 2,6 β 1,4 β 1,2 α 1,3

COMPLEX TYPE

FIGURE 17.16
Structure of *N*-linked oligosaccharides.
The basic structures of both types of *N*-linked oligosaccharides are shown. In each case the structure is derived from that of the initial dolichol-linked oligosaccharide through the action of glycosidases and glycosyltransferases. Note the variety of glycosidic linkages involved in these structures.

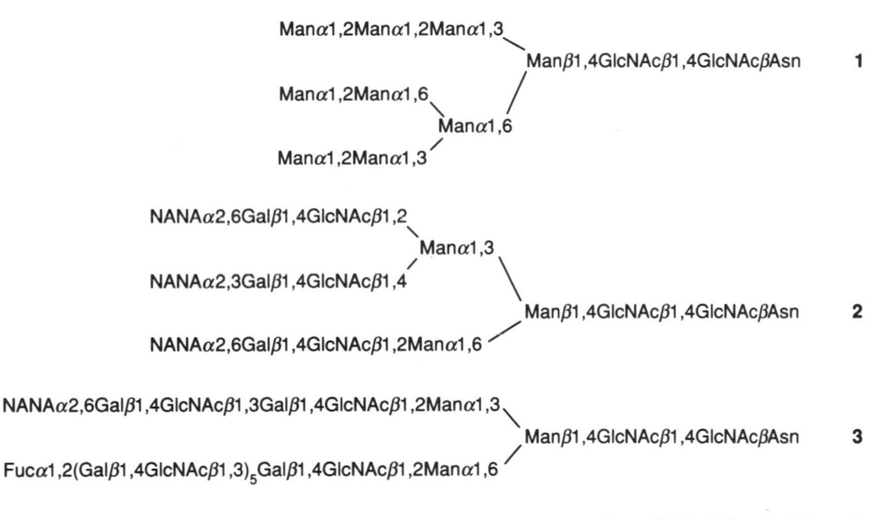

Nanaα2,6GalNAcαSer/Thr **4**

Fucα1,2Galβ1,3GlcNAcβ1,3GalNAcαSer/Thr **5**

Nanaα2,3Galβ1,3
GalNAcαSer/Thr **6**
Nanaα2,6

Nanaα2,3Galβ1,4
GalNAcβ1,3GalNAcαSer/Thr **7**
Fucα1,3

GlcNAcα1,4Galβ1,4GlcNAcβ1,3
Galβ1,3
Fucα1,2Galβ1,4GlcNAcβ1,6
GalNAcαSer/Thr **8**
Fucα1,2Galβ1,4GlcNAcβ1,6

FIGURE 17.17
Examples of oligosaccharide structure.
Structures 1–3 are typical *N*-linked oligosaccharides of the high mannose (1) and complex types (2,3); note the common core structure from the protein asparagine residue through the first branch point. Structures 4–8 are common *O*-linked oligosaccharides that may be quite simple or highly complex. Note that although the core structure (GalNAc-Ser/Thr) is unlike that of *N*-linked oligosaccharides, the termini can be quite similar (e.g., structures 2 and 6, 3, and 7). Abbreviations: Man = mannose; Gal = galactose; Fuc = fucose; GlcNAc = *N*-acetylglucosamine; GalNAc = *N*-acetylgalactosamine; NANA = *N*-acetyl neuraminic acid (sialic acid).
Adapted from J. Paulson, *Trends Biochem. Sci.* 14:272, 1989.

Sorting of Proteins Targeted for Lysosomes Occurs in the Secretory Pathway

Protein transport from the ER to the Golgi apparatus occurs through carrier vesicles that bud from the ER. The process requires energy in the form of GTP; in the presence of inhibitors of oxidative phosphorylation proteins accumulate in the ER and vesicles. Sorting of proteins for their ultimate destinations occurs in conjunction with their glycosylation and proteolytic trimming during passage through the cis, medial, and trans elements of the Golgi apparatus.

The best understood of the sorting processes involves **targetting** to lysosomes of specific glycoproteins. In the cis Golgi some element of structure, but not simply a linear amino acid sequence, allows lysosomal

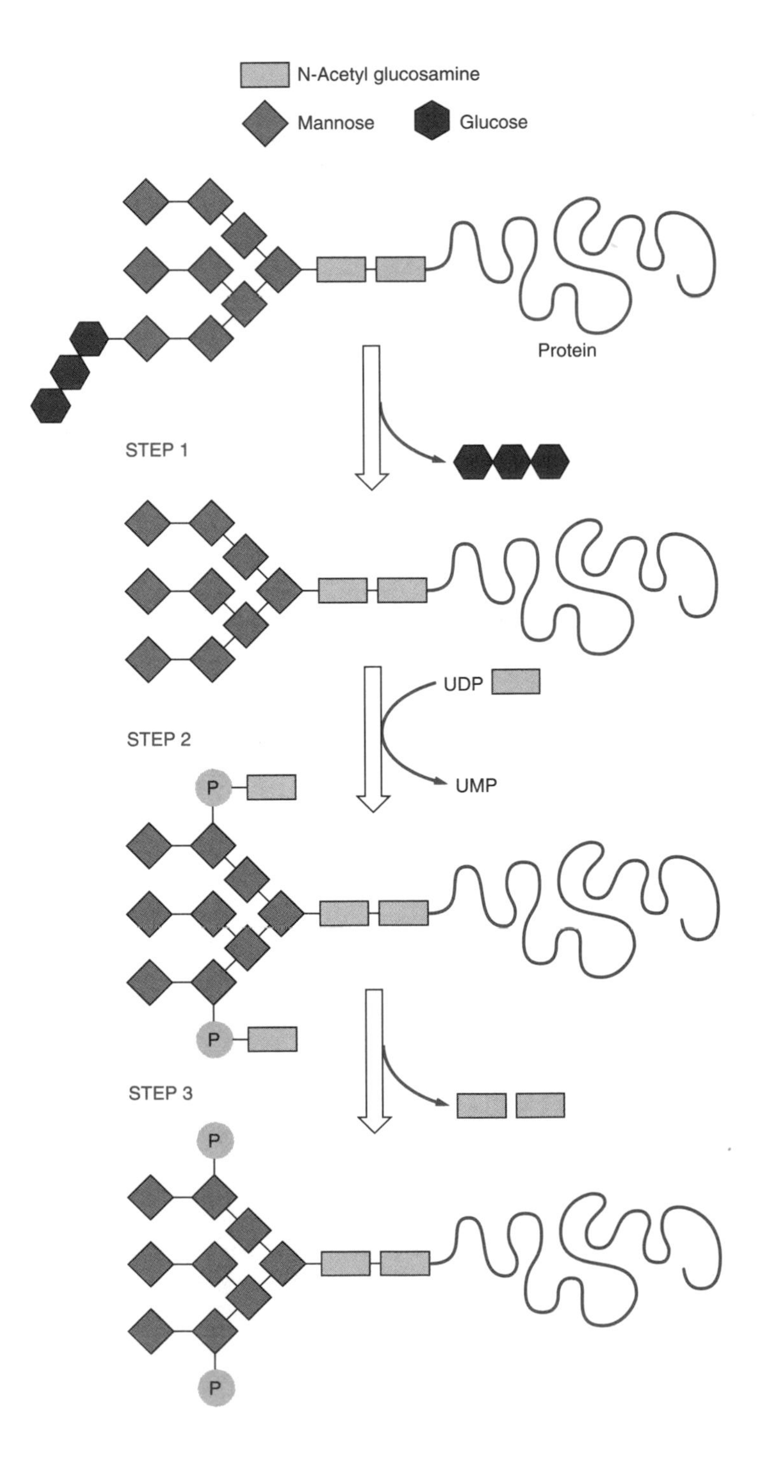

FIGURE 17.18
Targetting of enzymes to lysosomes.
N-linked glycoprotein is released from the ER membrane upon completion of its synthesis, and during its transport to and through the Golgi apparatus the oligosaccharide is modified by glycosidases that remove the glucose residues (Step 1). Some mannose residues may also be removed at this time. An element of protein structure is then recognized by a glycosyl transferase that transfers one or sometimes two *N*-acetyl glucosamine phosphate residues to the oligosaccharide (Step 2). A glycosidase removes the *N*-acetyl glucosamine residue(s), leaving one or two mannose 6-phosphate residues on the oligosaccharide (Step 3). The protein is then recognized by a mannose 6-phosphate receptor and directed to the lysosomes.
Adapted from R. Kornfeld and S. Kornfeld, *Annu. Rev. Biochem.* 54:631, 1985.

proteins to be recognized by a glycosyltransferase that attaches *N*-acetylglucosamine phosphate (GlcNAc-P) to high mannose type oligosaccharides. A glycosidase then removes the GlcNAc, leaving an oligosaccharide that includes mannose 6-phosphate (Figure 17.18). The mannose 6-phosphate is recognized by a specific receptor protein that is responsible for the compartmentation and vesicular transport of these proteins to lysosomes. Other oligosaccharide chains on the proteins may be further processed and glycosylated to form complex type structures, but the mannose 6-phosphate will still determine the lysosomal destination of these proteins. Patients with I-cell disease lack the GlcNAc-P glycosyltransferase and cannot correctly mark lysosomal enzymes for their destination. Thus the enzymes are secreted from the cell (Clin. Corr. 17.4).

A few other sorting signals are becoming understood. Polypeptide-specific glycosylation and sulfation of some glycoprotein hormones in the pituitary mediates their sorting into storage granules. Polysialic acid modification of a neural cell adhesion protein appears to be both specific to the protein and regulated developmentally. Many other sorting signals still must be deciphered in order to understand how the Golgi apparatus directs proteins to its own subcompartments, various storage and secretory granules, and specific elements of the plasma membrane.

CLIN. CORR. 17.4
I-CELL DISEASE

I-cell disease (mucolipidosis II) and pseudo-Hurler polydystrophy (mucolipidosis III) are related diseases that arise from defects in lysosomal enzyme targeting because of a deficiency in the enzyme that transfers *N*-acetylglucosamine phosphate to the oligosaccharides of proteins destined for the lysosome. Fibroblasts from affected individuals show dense *inclusion bodies* (hence I-cells) and are defective in multiple lysosomal enzymes, which are instead found secreted into the medium. Patients also have abnormally high levels of lysosomal enzymes in their sera and body fluids. The disease is characterized by severe psychomotor retardation, many skeletal abnormalities, coarse facial features, and restricted joint movement. Symptoms are usually observable at birth and progress until death, usually by age 8. Pseudo-Hurler polydystrophy is a much milder form of the disease. Onset is usually delayed until the age of 2–4 years, the disease progresses more slowly, and patients survive into adulthood. Prenatal diagnosis of both diseases is possible, but there is as yet no definitive treatment.

Kornfeld, S., *J. Clin. Invest.* 77:1, 1986 for a review of lysosomal enzyme trafficking, or Nolan, C. M. and Sly, W. S., I-Cell Disease and Pseudo-Hurler Polydystrophy, C. R. Scriver, A. L. Beaudet, W. S. Sly, and D. Valle (Eds.), *The Metabolic Basis of Inherited Disease,* 6th ed. New York: McGraw-Hall, 1989, pp. 1589–1601, for a comprehensive review of these diseases.

17.5 ORGANELLE BIOGENESIS AND TARGETTING

The secretory pathway directs proteins to the lysosomes and destinations either at the plasma membrane or totally external to the cell. Proteins of the ER and Golgi apparatus are targetted through partial use of the total pathway. For example, localization of proteins on either side of or spanning the ER membrane can utilize the signal recognition particle in slightly different ways (Figure 17.19). If the signal sequence is downstream from the amino terminus of the protein the amino end may not be inserted into the membrane but remains on the cytoplasmic surface. Internal hydrophobic anchoring sequences within a protein can allow a significant portion of the sequence to either remain on the cytoplasmic surface or to be retained, anchored on the luminal surface of the ER membrane. Multiple anchoring sequences in a single polypeptide can cause it to span the membrane several times and thus be largely buried in it. In the absence of *N*-linked glycosylation, proteins may remain in the lumen of the ER. Specific targetting amino acid sequences also lead to retention in the ER lumen.

Import of Proteins by Mitochondria Requires Specific Signals

Mitochondria provide a particularly complex targeting problem since specific proteins are localized to the mitochondrial matrix, the inner or outer membranes, or the intermembrane space. All but a few of these proteins are synthesized in the cell cytoplasm on free ribosomes and imported posttranslationally into the mitochondrion. Most are synthesized as larger preproteins; N-terminal presequences serve to mark the protein not only for the mitochondrion but also for a specific subcompartment. The **mitochondrial targetting signal** is not fully defined but it is rich in basic and hydroxyl amino acids and normally lacks acidic residues. It is recognized by a mitochondrial receptor, probably with the aid of a cytoplasmic **import factor.** Proteins are translocated across both membranes and into the mitochondrial matrix in an energy dependent reaction that appears to require unfolding of the protein. Passage occurs at adhesion sites where

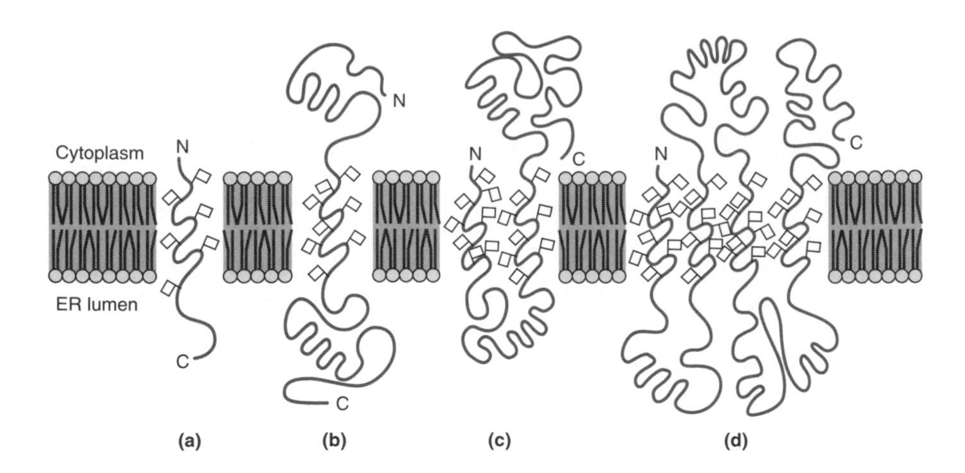

FIGURE 17.19
Topology of proteins at membranes of the endoplasmic reticulum. Proteins are shown in several orientations with respect to the membrane. In Example (a) the protein is anchored to the luminal surface of the membrane by an uncleaved signal peptide. In Example (b) the signal sequence is not near the N terminus; a domain of the protein was synthesized before emergence of the signal peptide. Insertion of the internal signal sequence, followed by completion of translation, resulted in a protein with a cytoplasmic N′-terminal domain, a membrane-spanning central segment, and a C-terminal domain in the ER lumen. Example (c) shows a protein with the opposite orientation: An N-terminal signal sequence, which might also have been cleaved by signal peptidase, resulted in extrusion of a segment of protein into the ER lumen. A second hydrophobic anchoring sequence remained membrane associated and prevented passage of the rest of the protein through the membrane, thus allowing formation of a C-terminal cytoplasmic domain. In Example (d), several internal signal and anchoring sequences allow various segments of the protein to be oriented on each side of the membrane.

the inner and outer mitochondrial membranes come close together. Proteolysis removes the mitochondrial targetting sequence but may leave other sequences that help sort the protein within the mitochondrion. For example, a clipped precursor of cytochrome b_2 is translocated back across the inner membrane in response to a localization signal sequence. Further proteolysis frees the protein to the intermembrane space.

Other mechanisms of mitochondrial targetting also exist. For example, cytochrome *c* is not formed as a larger precursor but it includes a localization signal within its structure. The apoprotein (without heme) binds at the outer membrane and is passed into the intermembrane space. There it acquires its covalently bound heme and undergoes a conformational change that prevents its return to the cytoplasm.

Targetting to Other Organelles Also Requires Specific Signals

Nuclei must import many proteins involved in their own structure as well as for DNA replication, transcription, and ribosome biogenesis. Nuclei have rather large pores that permit the passage of small proteins, but most nuclear proteins appear to be targetted by specific nuclear localization signals that are very rich in lysine and arginine residues. Histone localization and nucleosome assembly is also mediated by a protein called **nucleoplasmin.** Other nuclear proteins may be retained in the nucleus by forming complexes within the organelle.

Peroxisomes contain a limited array of enzymes. One targetting signal appears to be a simple carboxy-terminal tripeptide, Ser-Lys-Leu. Many more targetting signals must exist but have yet to be determined.

17.6 FURTHER POSTTRANSLATIONAL PROTEIN MODIFICATIONS

Several additional maturation events may modify newly synthesized polypeptides to help generate the final, functional form of the protein. Many of these events are specialized to one or a few known instances. In this section we will consider a few examples.

Insulin Biosynthesis Involves Proteolysis

Proteolysis of proteins is a common maturation step. Amino-terminal, carboxy-terminal, or internal sequences can be removed from the protein as part of its localization, activation, or stabilization. Proteolysis in both the ER and Golgi apparatus helps to mature the peptide hormone insulin (Figure 17.20). A **preproinsulin** molecule is encoded by the mRNA and inserted into the lumen of the ER. Signal peptidase cleaves off the signal peptide to generate **proinsulin.** The protein also folds to allow formation of the correct disulfide linkages and the proinsulin is transported to the Golgi apparatus. There it is packaged into secretory granules in which an internal connecting peptide, termed **C peptide,** is removed by proteolytic cleavage, and functional insulin is secreted. In familial hyperproinsulinemia a different situation exists (Clin. Corr. 17.5).

This pathway for insulin biosynthesis has clear advantages over synthesis of two separate polypeptides. First, it ensures production of exactly equal amounts of mature A and B chains and so does not require coordination of two translational activities. Second, proinsulin folds into a three-dimensional structure in which the cysteine residues are well-placed for correct disulfide bond formation. Proinsulin can be reduced and denatured, but under gentle renaturation will refold correctly to form functional proinsulin. Reduction and denaturation of insulin allows only inefficient renaturation, and many random incorrect disulfide linkages are formed. Correct formation of insulin from independently synthesized chains would probably have required evolution of a helper protein or molecular chaperone such as the immunoglobulin heavy chain binding protein that helps assemble immunoglobulins.

Proteolysis Leads to Zymogen Activation

Precursor protein cleavage is a common mechanism of enzyme activation. The digestive proteases are classic examples of this phenomenon

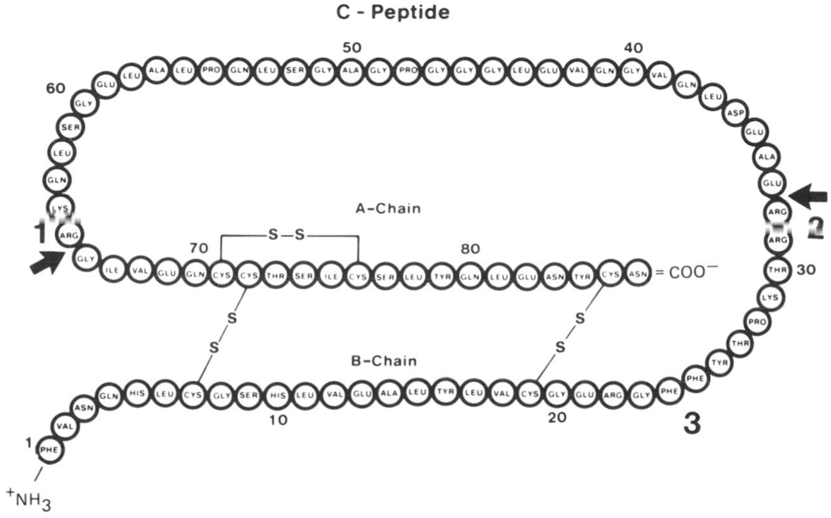

CLIN. CORR. 17.5 FAMILIAL HYPERPROINSULINEMIA

Familial hyperproinsulinemia, an autosomal dominant condition, results in approximately equal amounts of insulin and proinsulin being released into the circulation. Although affected individuals have high levels of proinsulin in the blood, they are apparently normal in terms of glucose metabolism, being neither diabetic nor hypoglycemic. The defect was originally thought to be the result of a deficiency of one of the proteases that process proinsulin. Three enzymes process proinsulin: endopeptidases that cleave the Arg^{31}-Arg^{32} and Lys^{64}-Arg^{65} peptide bands and a carboxypeptidase. Dominant inheritance is not easily reconciled with this complex processing mechanism. In one family it is now known that a point mutation ($His^{10} \rightarrow Asp^{10}$) causes the hyperproinsulinemia, but how this mutation interferes with processing is not known.

Chan, S. J., Seino, S., Gruppuso, P. A., Schwartz, R., and Steiner, D. F. A mutation in the β chain coding region is associated with impaired proinsulin conversion in a family with hyperproinsulinemia. *Proc. Natl. Acad. Sci., U.S.A.*, 84:2194, 1987.

FIGURE 17.20
Human proinsulin.
After cleavage at the two sites indicated by arrows the arginine residues 31, 32, and 65 and the lysine residue 64 are removed to give insulin and C-peptide.
Redrawn from G. I. Bell, W. F. Swain, R. Pictet, B. Cordell, H. M. Goodman, and W. J. Rutter, *Nature (London)* 282:525, 1979.

(see p. 1074). Inactive **zymogen** precursors are sequestered in storage granules until needed, and activated by proteolysis upon secretion. Thus trypsinogen is cleaved to give an amino terminal peptide plus trypsin and chymotrypsinogen is cleaved to form chymotrypsin and two peptides.

Amino Acids Can Be Modified After Incorporation Into Proteins

Only 20 amino acids are encoded genetically and incorporated during translation. Posttranslational modification of proteins, however, leads to formation of hundreds of different amino acid derivatives in mature proteins. The amounts of these modified amino acids may be small, but they sometimes play a major functional role in their proteins. Examples are listed in Table 17.10.

Protein amino termini are frequently modified. Procaryotic protein synthesis is initiated using *N*-formylmethionine; in many cases either the formyl group or the methionine is later removed, but the resulting amino terminus is sometimes then modified by, for example, acetylation or myristoylation. Amino-terminal glutamine residues spontaneously cyclize; one possible result is the stabilization of the protein (see Section 17.6 on protein degradation).

Disulfide bond formation, a posttranslational modification, is catalyzed by a **disulfide isomerase** and the cystine-containing cross-linked protein is conformationally stabilized. Disulfide formation can prevent unfolding of proteins and their passage across membranes, so it becomes a means of localization. As seen in the case of insulin, disulfide bonds hold the two chains together and are necessary for biological function.

Methylation of lysine ε-amino groups occurs at some positions in histone proteins and may modulate their interactions with DNA. A fraction

TABLE 17.10 Modified Amino Acids in Proteins[a]

Amino acid	Modifications found
Arginine	*N*-methylation, ADP-ribosylation
Lysine	*N*-acetylation, *N*-methylation, oxidation, hydroxylation, cross-linking
Histidine	Methylation, diphthamide formation, phosphorylation
Glutamic acid	γ-Carboxyglutamate formation, methylation
Aspartic acid	Phosphorylation, methylation, racemization
Asparagine	Glycosylation, deamidation
Glutamine	Deamidation, cross-linking, pyroglutamate formation
Serine	Phosphorylation, glycosylation, acetylation
Threonine	Phosphorylation, glycosylation
Tyrosine	Iodination, phosphorylation, sulfation, flavin linkage, nucleotide linkage
Phenylalanine	Hydroxylation
Proline	Hydroxylation, glycosylation
Cysteine	Cystine formation, selenocysteine formation, heme linkage, myristoylation
Amino terminus	Formylation, acetylation, pyroglutamate formation
Carboxyl terminus	Phosphatyidylinositol derivitization

SOURCE: Wold, F., Posttranslational protein modifications: Perspectives and prospectives, in B. C. Johnson (Ed.), *Posttranslational Covalent Modification of Proteins*. New York; Academic, 1983, pp. 1–12.

[a] The listing is not comprehensive. Note that no derivatives of glycine, alanine, leucine, isoleucine, valine, and tryptophan have been identified in proteins.

of the H2A histone protein is also modified through isopeptide linkage of a small protein, ubiquitin, from its C-terminal glycine to a lysine ε-amino group of the histone. A role in DNA interactions is postulated.

Serine and threonine OH groups are sites of phosphorylation by a variety of ATP-requiring protein kinases. These modifications are reversible through the action of protein phosphatases. The classic example of regulatory phosphorylation of a serine residue is glycogen phosphorylase, which is modified by a specific phosphorylase kinase (see p. 347). Tyrosine kinase activity is a property of many growth factors that stimulate the proliferation of specific cell types. Oncogenes, responsible in part for the proliferation of tumor cells, often show strong homology with normal growth factors and also possess tyrosine kinase activity. Dozens of other examples exist; together the kinases and phosphatases regulate the activity of many proteins and processes.

The ADP-ribosylation of EF2 at a modified histidine residue represents a doubling of posttranslation modifications. First, the EF2 histidine is modified to generate the diphthamide derivative (Figure 17.21) of the functional protein. This is a very specific and unusual set of reactions that are probably not absolutely required since yeast mutants that cannot make diphthamide survive. The ADP-ribosylation of the diphthamide by diphtheria toxin then inhibits the activity of EF2. In other instances, ADP-ribosylation is a reversible modification of proteins.

Formation of γ-carboxyglutamate from glutamic acid residues occurs in several blood clotting proteins including prothrombin and factors VII, IX, and X. The γ-carboxyglutamate residues chelate calcium ion which is required for normal blood clotting (see p. 967). In each case the modification requires vitamin K and can be blocked by coumarin derivatives which antagonize vitamin K. As a result the rate of coagulation, quantitated as an increased prothrombin time, is greatly increased.

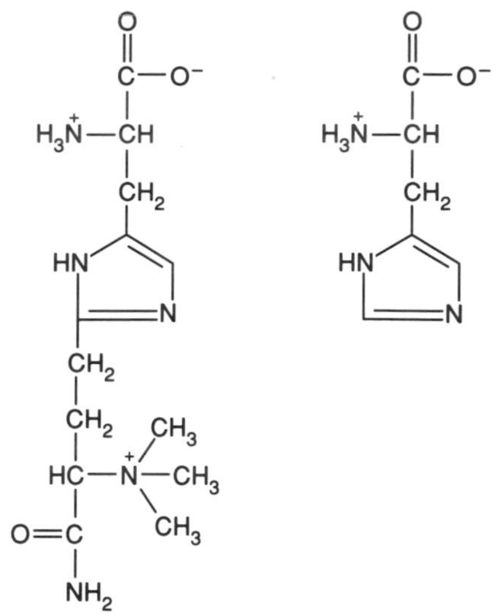

FIGURE 17.21
Diphthamide (left) is a posttranslational modification of a specific residue of histidine (right) in EF2.

Collagen Biosynthesis Requires Posttranslational Modifications

Collagen is the most abundant protein, or family of closely related proteins, in the human body. It is a fibrous protein that provides the structural framework for our tissues and organs. Collagen is a secretory protein that undergoes a wide variety of important posttranslational modifications, many of which are directly related to its function. Genetic defects involving posttranslational modification systems result in serious diseases. Collagen is thus an excellent example of the importance of posttranslational events in protein function.

Each collagen polypeptide, designated an α chain, exists as a left-handed helical polypeptide (not an α helix) about 1000 residues long in which every third residue is glycine (Figure 17.22). A collagen molecule includes three α chains intertwined in a right-handed **collagen helix** in which the glycine residues occupy the center of the structure. Given an approximate formula $(\text{Gly-X-Y})_{333}$, about one-third of the X positions are occupied by proline and a similar number of Y positions are 4-hydroxyproline, a posttranslationally modified form of proline. These proline residues impart considerable rigidity to the structure. Different tissues have different species of collagen, designated Types I, II, III, IV, V, and so on. For example, Type I collagen, made up of two chains of α1, type I, plus one chain of α2, type I (designated $[\alpha 1(\text{I})]_2\alpha 2(\text{I})$ in the usual shorthand), is found in skin, arteries, bone, and tendons. Type III collagen, made up of three identical α1, type III chains (hence $[\alpha 1(\text{III})]_3$) is found in skin, arteries, and the uterus and type II collagen, $[\alpha 1(\text{II})]_3$ is found in cartilage. At least 10 collagen types are now known, and the number may grow. Genes for various α chains are located on several chromosomes and the

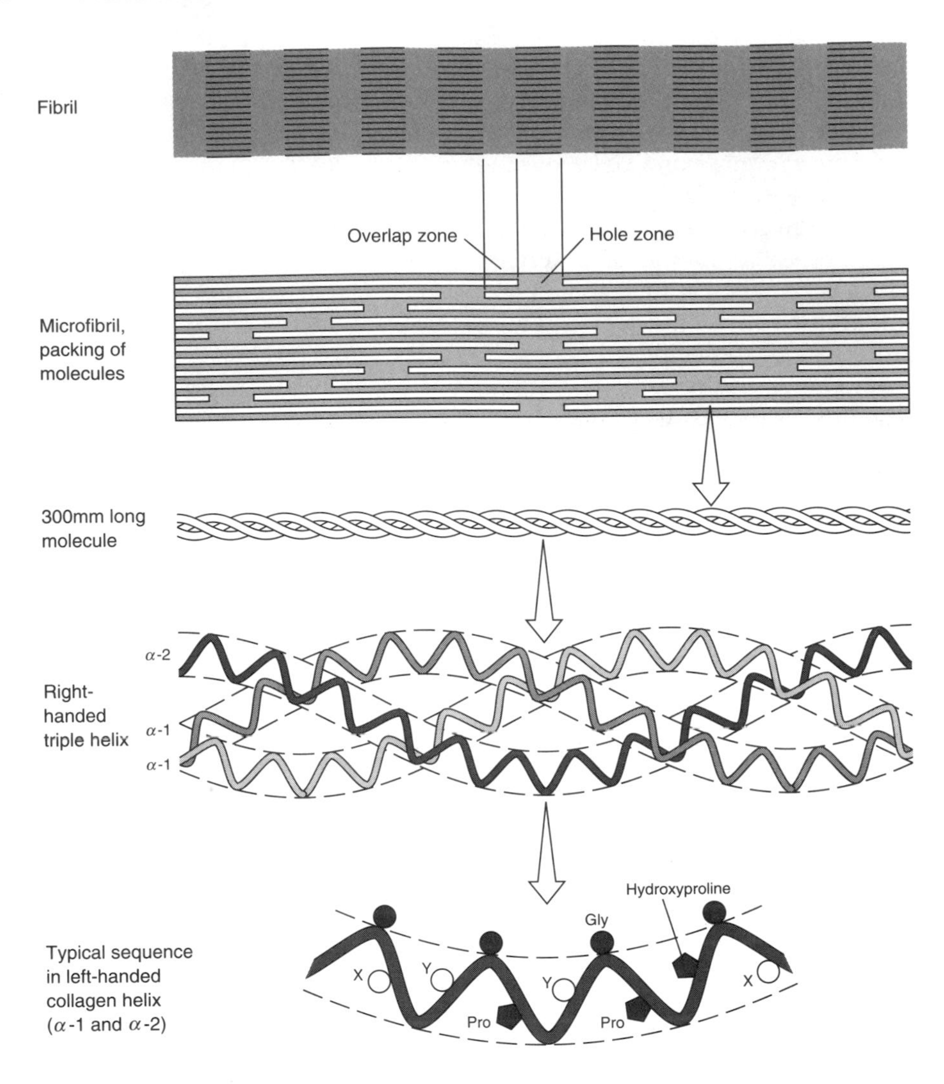

FIGURE 17.22
Collagen structure, illustrating the regularity of primary sequence, the left-handed α helix, the right-handed triple helix, the 300-nm molecule, and the organization of molecules in a typical fibril, within which the collagen molecules are cross-linked.

amino acid sequences they encode differ, but their overall structural similarity suggests a common evolutionary origin.

Procollagen Formation: Events in the Endoplasmic Reticulum and Golgi Apparatus

Synthesis of collagen α chains is initiated in the cytoplasm and amino terminal sequences are recognized by signal recognition particles. Then, for example, the prepro $\alpha 1(I)$ and prepro $\alpha 2(I)$ chains are extruded into the lumen of the ER and the signal peptides are cleaved. Hydroxylation of proline and lysine residues occurs cotranslationally, before assembly of a triple helix. **Prolyl 4-hydroxylase** requires an -X-Pro-Gly- sequence (hence 4-hydroxyproline is found only at Y positions in the -Gly-X-Y- sequence of α chains). Also present in the ER is a **prolyl 3-hydroxylase,** which modifies a smaller number of proline residues in the sequence and a **lysyl hydroxylase** that modifies some of the Y position lysine residues. The extent of modification depends on the specific α-chain type. These hydroxylases require Fe^{2+} and ascorbic acid. Proline hydroxylation makes the final collagen thermally stable, and lysine hydroxylation provides sites for interchain cross-linking and for glycosylation by specific glycosyltransferases of the ER. Asparagine residues are also glycosylated at this point, eventually leading to high mannose type oligosaccharides.

Assembly of the triple helix occurs only after the polypeptide chains have been completed. Carboxy-terminal globular proprotein domains fold and disulfide bonds are formed. Interaction of these domains initiates winding of the triple helix from the carboxyl end toward the amino terminus. The completed triple helix has globular proprotein domains at each end. It is then transported to the Golgi apparatus where the oligosaccharides are processed and matured, and in some cases tyrosine residues are modified by sulfation and some serines are phosphorylated. The completed procollagen is then released from the cell via secretory vesicles.

Collagen Maturation

Conversion of procollagen to collagen occurs extracellularly. The amino-terminal and carboxyl-terminal propeptides are cleaved by separate proteases that may also be type specific. Concurrently, the triple helices assemble into fibrils; other connective tissue proteins are also involved at this stage. The collagen is then stabilized by extensive cross-linking (Figure 17.23). The enzyme lysyl oxidase converts some lysine or hydroxylysine residues to the reactive aldehydes, allysine, or hydroxyallysine. These residues condense with lysine or hydroxylysine residues in adjacent chains to form cross-links. Further and less-well-characterized reactions can involve other residues including histidines and can link three α chains.

Defects at many of these steps are known. Some of the best characterized are listed in Table 17.11 and described in Clin. Corr. 17.6.

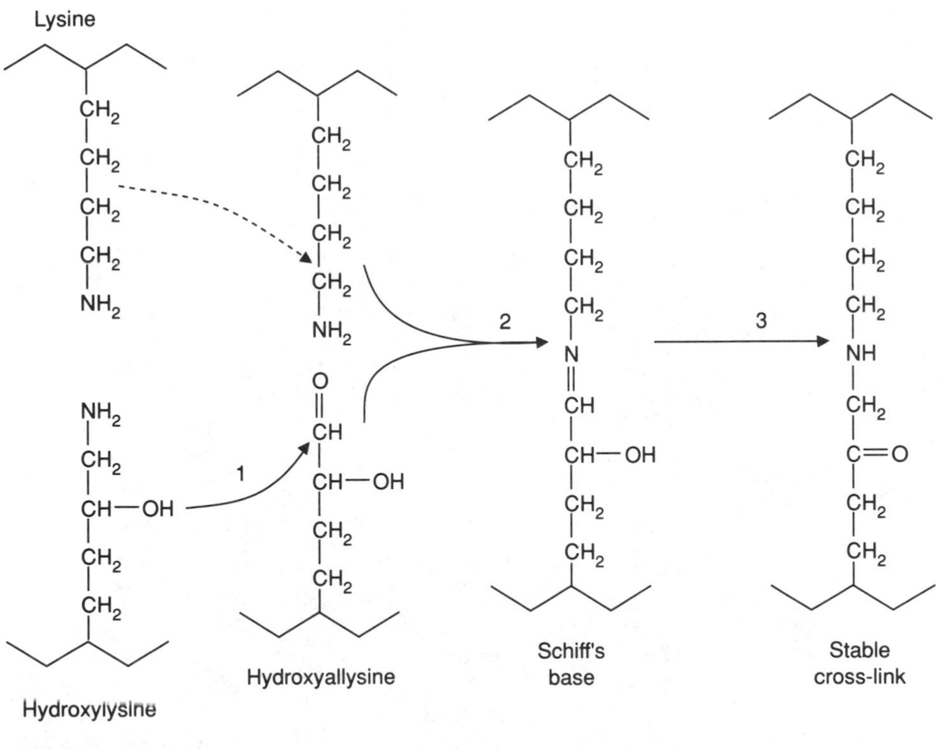

Lysine

Hydroxylysine

Hydroxyallysine

Schiff's base

Stable cross-link

FIGURE 17.23
Cross-linking of collagen chains.
In Step 1 a residue of hydroxylysine is converted to hydroxyallysine through the action of lysyl oxidase; the same enzyme catalyzes the oxidative deamination of lysine to form allysine. In Step 2 the aldehyde condenses with no enzyme involvement with a lysine ε-amino group from a different chain, forming a Schiff's base addition product in a highly reversible reaction. Step 3 involves the nonenzymic rearrangement of this product to form a chemically stable cross-link. In addition, less well-characterized reactions result in further cross-links between chains.

TABLE 17.11 Some Disorders of Collagen Structure

Disorder	Collagen defect	Clinical manifestations
Ehlers–Danlos IV	Decrease in type III	Arterial, intestinal, or uterine rupture; thin, easily bruised skin
Osteogenesis imperfecta	Decrease in type I	Blue sclerae, multiple fractures, and bone deformities
Scurvy	Decreased hydroxyproline	Poor wound healing, deficient growth; increased capillary fragility
Ehlers–Danlos VI	Decreased hydroxylysine	Hyperextensible skin and joints, poor wound healing, musculo-skeletal deformities
Ehlers–Danlos VII	Amino terminal propeptide present	Hyperextensible, easily bruised skin; hip dislocations
Ehlers–Danlos V and cutis laxa	Decreased cross-linking	Skin and joint hyperextensibility
Marfan syndrome	Decreased cross-linking	Skeletal deformities, aortic aneurism, valvular heart disease, ectopia lentis

CLIN. CORR. 17.6
DEFECTS IN COLLAGEN SYNTHESIS

EHLERS–DANLOS SYNDROME, TYPE IV

The Ehlers–Danlos syndrome is the name for a group of at least 10 disorders that are clinically, genetically, and biochemically distinguishable but that all share manifestations of structural weaknesses in connective tissue. The usual problems are with fragility and hyperextensibility of skin and hypermobility of the joints. The weaknesses result from defects in collagen structure. For example, type IV Ehlers–Danlos syndrome is caused by defects in type III collagen, which is particularly important in skin, arteries, and hollow organs. Manifestations may be severe, with arterial rupture, intestinal perforation, rupture of the uterus during pregnancy or labor, and easy bruisability of thin, translucent skin. The basic defect in several variants of type IV Ehlers–Danlos appears to be one of several changes in the primary structure of proα1(III) chains that slow synthesis, delay secretion, or increase rate of degradation, thereby preventing normal formation of type III collagen fibrils. From instance, in some patients, type III collagen is synthesized at a normal rate but accumulates in the rough ER. Insertions or deletions within one pro α, (III) allele could interfere with formation of the triple helical collagen, and explain the autosomal dominant inheritance of the diseases.

Superti-Furga, A., Gugler, E., Gitzelmann, R., and Steinmann, B. Ehlers–Danlos Syndrome type IV: A multi-exam deletion in one of the two COL 3A1 alleles affecting structure, stability, and processing of type III procollagen. *J. Biol. Chem.* 263:6226, 1988.

OSTEOGENESIS IMPERFECTA

Osteogenesis imperfecta is the name for a group of at least four clinically, genetically, and biochemically distinguishable disorders all characterized by multiple fractures with resultant bone deformities. Several variants result from mutations producing shortened α(I) chains. In the clearest example a deletion mutation causes absence of 84 amino acids in the α1(I) chain. The shortened α1(I) chains are synthesized, because the mutation leaves the reading frame in register. The short α1(I) chains associate with normal α1(I) and α2(I) chains, thereby preventing normal triple helix formation, with resultant degradation of all of the chains, a phenomenon aptly named "protein suicide." Three-fourths of all of the collagen molecules formed have at least one short (defective) α1(I) chain, an amplification of the effect of a heterozygous gene defect. Other forms of osteogenesis imperfecta result from point mutations that substitute another amino acid for one of the glycines. Since glycine has to fit into the interior of the triple helix, these substitutions destabilize the helix.

Barsh, G. S., Roush, C. L., Bonadio, J., Byers, P. H., and Gelinas, R. E. Intron mediated recombination causes an α (I) collagen deletion in a lethal form of osteogenesis imperfecta. *Proc. Natl. Acad. Sci. U.S.A.*, 82:2870, 1985.

SCURVY AND HYDROXYPROLINE SYNTHESIS

Scurvy results from dietary deficiency of ascorbic acid. Most animals can synthesize ascorbic acid from glucose, but humans have lost this enzymatic mechanism. Ascorbic acid deficiency causes decreased hydroxyproline synthesis because prolyl hydroxylase requires ascorbic acid. The hydroxyproline provides additional hydrogen-bonding atoms that stabilize the collagen helix. Collagen containing insufficient hydroxyproline loses temperature stability and is less stable than normal collagen at body temperature. The resultant clinical manifestations are distinctive and understandable: suppression of the orderly growth process of bone in children, poor wound healing, and increased capillary fragility with resultant hemorrhage, particularly in the skin. Severe ascorbic acid deficiency leads secondarily to a decreased rate of procollagen synthesis.

Crandon, J. H., Lund, C. C., and Dill, D. B. Experimental human scurvy. *N. Engl. J. Med.* 223:353, 1940.

DEFICIENCY OF LYSYLHYDROXYLASE

In type VI Ehlers–Danlos syndrome lysylhydroxylase is deficient. As a result collagen with decreased hydroxylysine content is synthesized, and subsequent cross-linking of collagen fibrils is less stable. Some cross-linking between lysine and allysine occurs but these are not as stable and do not mature as readily as do hydroxylysine containing cross-links. In addition, carbohydrates add to the hydroxylysine residues but the function of this carbohydrate is unknown. The clinical features include marked hyperextensibility of the skin and joints, poor wound healing, and musculoskeletal deformities. Some patients with this form of Ehlers–Danlos syndrome have a mutant form of lysylhydroxylase with a higher Michaelis constant for ascorbic acid than the normal enzyme. Accordingly, they respond to high doses of ascorbic acid.

Pinnell, S. R., Krane, S. M., Kenzora, J. E., and Glimcher, M. J. A heritable disorder of connective tissue: Hydroxylysine-deficient collagen disease. *N. Engl. J. Med.* 286:1013, 1972.

EHLERS–DANLOS SYNDROME, TYPE VII

In Ehlers–Danlos syndrome, type VII, the skin bruises easily and is hyperextensible, but the major manifestations are dislocations of major joints, such as the hips and knees. The laxity of ligaments is caused by incomplete removal of the amino-terminal propeptide. In one variant the deficiency is in procollagen aminopeptidase. The deficiency occurs in the autosomal recessive disease called dermatosparaxis of cattle, sheep, and cats, in which skin fragility is so extreme as to be lethal. In another variant the proα2(I) chain lacks amino acids at the cleavage site because of skipping of one exon in the gene. This prevents its normal cleavage by procollagen aminopeptidase. A similar defect in a pro α, (I) gene has also been observed.

Cole, W. G., Chan, W., Chambers, G. W., Walker, I. D., and Bateman, J. F. Deletion of 24 amino acids from the pro α, (I) chain of type I procollagen in a patient with the Ehlers–Danlos syndrome type VII. *J. Biol. Chem.* 261:5496, 1986.

CUTIS LAXA

In type V Ehlers–Danlos syndrome and in some forms of cutis laxa there is thought to be a deficiency in lysyl oxidase with consequent cross-linking defects in both collagen and elastin. In cutis laxa the defect is manifested in loose skin, which appears excessively wrinkled, hangs in folds, and lacks the elastic qualities of the hyperextensible skin in Ehlers–Danlos syndrome. In type V Ehlers–Danlos syndrome the skin hyperextensibility is accompanied by joint hypermobility, but the latter is limited to the digits. Copper-deficient animals have deficient cross-linking of elastin and collagen, apparently because of the requirement for cuprous ion by lysyl oxidase. In Menke's steely hair syndrome, there is a defect in intracellular copper transport that results in low activity of lysyl oxidase. A woman taking high doses of the copper-chelating drug, D-penicillamine, gave birth to an infant with an acquired Ehlers–Danlos-like syndrome, which subsequently cleared. Side effects of D-penicillamine therapy include poor wound healing and hyperextensible skin.

Peltonen, L., et al. Alterations of copper and collagen metabolism in the Menkes syndrome and a new subtype of Ehlers–Danlos syndrome. *Biochemistry* 22:6156, 1983.

MARFAN SYNDROME

The Marfan syndrome is an autosomal dominant disorder (probably a group of disorders) characterized by skeletal, eye, and heart manifestations and often ending with cardiovascular catastrophe, especially dissection of the aorta. The organ and tissue distribution of defects have pointed to type I collagen as the basis for this disorder. Many studies have indicated a defect in cross-linking, but no defect has ever been found in lysyl oxidase, the enzyme essential for cross-linking, in Marfan patients. The explanation must lie in a structural mutation. That has been found in one patient whose α2(I) chains have an insert of about 20 amino acid residues near the carboxy terminal. The insert alters the register of the mutant α2(I) chain relative to the normal chains, thereby interfering with normal cross-linking. The molecular basis for Marfan syndrome, however, remains unknown in most families. The treatment of Marfan syndrome often involves surgical replacement of the aortic valve and aortic arch when there is early evidence of dilation of the aorta.

Pyeritz, R. E. and McKusick, V. A. Boric defects in the Marfan syndrome. *N. Engl. J. Med.* 305:1011, 1981.

17.7 PROTEIN DEGRADATION AND TURNOVER

Proteins have finite lifetimes. They are subject to environmental damage such as oxidation, biochemical accidents such as proteolysis or conformational denaturation, and other irreversible modifications or distortions. Equally important, cells need to change their protein complements in order to respond to different needs and situations. Specific proteins exhibit very different lifetimes. At one extreme cells of the lens of the eye are not replaced and their proteins are not recycled. Hemoglobin in red blood cells lasts the life of the red cell—about 120 days. Other proteins have shorter life times measured in periods of a few days, hours, and even minutes. As an example, some of the blood clotting proteins survive for only a few days, thus hemophiliacs are only protected by transfusions or injections of required factors for a short period. Diabetics require insulin injections regularly since the hormone is not recycled. Metabolic enzymes vary quantitatively depending on need—for example, urea cycle enzyme levels change in response to diet.

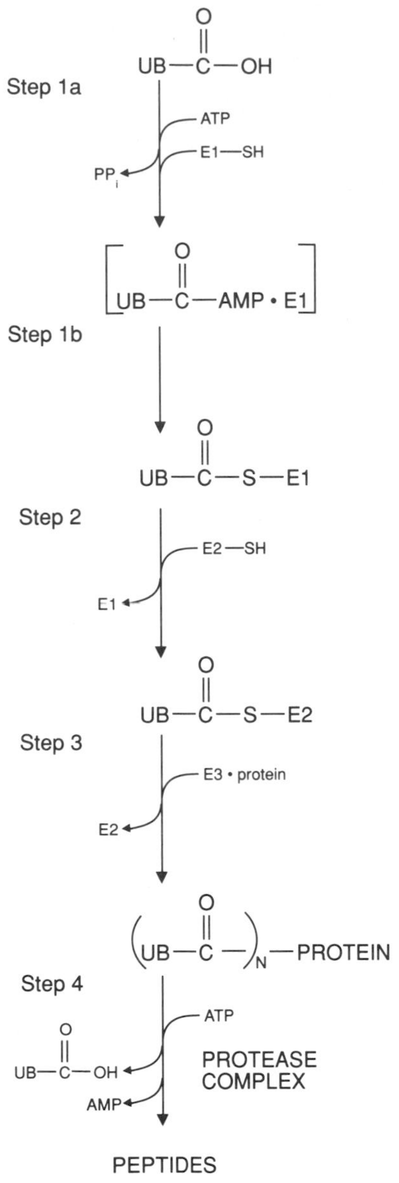

FIGURE 17.24
ATP and ubiquitin-dependent protein degradation.
Ubiquitin is first activated in a two-step reaction involving formation of a transient mixed anhydride of AMP and the carboxy terminus of ubiquitin (Step 1A), followed by generation of a thioester with enzyme E1 (Step 1B). Enzyme E2 can now form a thioester with ubiquitin (Step 2) and serve as a donor in the E3-catalyzed transfer of ubiquitin to a targetted protein (Step 3). Several ubiquitin molecules are usually attached to different lysine residues of a targetted protein at this stage. The ubiquitinylated protein is now degraded by an ATP dependent multicomponent protease (Step 4); the ubiquitin is not degraded, and can reenter the process at Step 1.

Most of the amino acids produced by protein degradation are recycled to synthesize new proteins. Some of the degradation products will also be excreted. Thus we should think of protein turnover as well as simple degradation. In either case, proteolysis first reduces the proteins in question to peptides and eventually amino acids. Several proteolytic systems exist to accomplish this end.

Intracellular Digestion of Some Proteins Occurs in Lysosomes

Digestive proteases such as pepsin, trypsin, chymotrypsin, and elastase are involved with the hydrolysis of dietary protein and do not play a role in protein turnover within an organism (see Chapter 26). Intracellular digestion of proteins from the extracellular environment occurs within **lysosomes.** Materials from outside the cell that are impermeable to the plasma membrane are imported by the processes of endocytosis. This could be through simple pinocytosis; large particles, molecular aggregates, or the other molecules present within the extracellular fluid are ingested by engulfing the material. Macrophages injest bacteria and dead cells by this mechanism. Receptor-mediated endocytosis uses different receptors on the cell surface to bind specific molecules. Endocytosis occurs at pits in the cell surface that are internally coated with the multisubunit protein **clathrin.** Uptake involves invagination of the plasma membrane and the receptors to form intracellular coated vesicles. One fate of such vesicles is fusion with a lysosome and degradation of the contents. Some intracellular protein turnover may also occur within lysosomes and under some conditions significant amounts of cellular material can be mobilized via lysosomes. For example, serum starvation of fibroblasts or starvation of rats leads to the lysosomal degradation of a subpopulation of cellular proteins. Recognition of a specific peptide sequence is involved, indicating that the lifetime of a protein is ultimately encoded in its sequence. This concept will be more apparent in the next section on ubiquitin-dependent proteolysis.

Although lysosomal degradation of cellular proteins does occur, it is almost surely not the primary route of protein turnover. Other proteolytic systems exist, but their quantitative importance is still unclear. Calcium-dependent proteases, also called **calpains,** are present in most cells. Both activators and inhibitors of these enzymes are also present, and so calpains are logical candidates for enzymes involved in protein turnover. However, an important role in these processes is not yet established. Golgi and ER proteases clearly degrade peptide fragments that arise during maturation of proteins in the secretory pathway. They could be involved in turnover of ER proteins. It is also likely that other uncharacterized mechanisms exist in both the cytoplasm and in the mitochondria.

Ubiquitin Is a Marker in ATP-Dependent Proteolysis

One clearly described proteolytic pathway requires energy in the form of ATP and involves the participation of a highly conserved 76 amino acid protein known as **ubiquitin.** Since enzymatic proteolysis readily proceeds without any outside energy source, the ATP requirement is only partly understood. It is speculated that it is needed to help select proteins for degradation and denature them to make them proteolytically susceptible, but this is not proven. Ubiquitin appears to function here as a marker that identifies proteins for degradation. Ubiquitin also has other roles; as an example, linkage of ubiquitin to histones H2A and H2B is unrelated to turnover since the proteins are stable but modification may effect chromatin structure or transcription.

The ubiquitin-dependent proteolytic cycle is shown in Figure 17.24. Ubiquitin is "activated" by an enzyme termed E1 to form a thioester; ATP is required in this reaction and a transient AMP–ubiquitin complex is involved. The ubiquitin is then passed to a second protein, enzyme E2, and finally via one of a group of E3 enzymes it is coupled to a targetted protein. Linkage of ubiquitin is through isopeptide bonds between ε-amino groups of lysine residues of the protein and the carboxyl-terminal glycine residues of ubiquitin. Several ubiquitin molecules may be attached to the protein and the α-amino group may be modified as well. An ATP-dependent protease then degrades the tagged protein and frees the ubiquitin for further rounds of degradation.

How are proteins selected for this process? A major determinant is simply the identity of the amino-terminal amino acid residue. Otherwise identical β-galactosidase proteins with different amino-terminal amino acids are degraded at widely differing rates. Some residues, including methionine and glycine, are "stabilizing" since they confer long half-lives on the protein. Others, including lysine, aspartic acid, and tryptophan, are "destabilizing" since they lead to rapid protein turnover. Some amino terminal residues are modified in the cell to alter their characteristics. Glutamine and asparagine are deamidated to become destabilizing. Others serve as acceptors for a destabilizing residue from an aminoacyl-tRNA. Selectivity occurs at the E3 enzyme level, since different E3 species recognize large hydrophobic residues, basic residues, or small uncharged residues at the amino terminus. This ***N*-end rule** is largely conserved; similar stability patterns are seen in yeast and reticulocytes, although details differ and may vary physiologically. Moreover, analysis of protein sequences shows that the rule applies throughout nature. Long-lived intracellular proteins have stabilizing amino acids at their N termini in species from humans to *E. coli,* while secreted or extracellular proteins show a much wider variety of terminal residues.

Few biochemists believe that the *N*-end rule is the complete solution to questions of protein stability. Damaged or mutant proteins are rapidly degraded through the ubiquitin pathway. Internal sequences and conformation are also likely to be important. Destabilizing **PEST sequences** (rich in proline, glutamic acid, serine, and threonine, and named on the basis of the one-letter amino acid codes) have been identified in several short-lived proteins. It is expected that additional degradative systems will be discovered. Ubiquitin is not found in procaryotes, so a different tagging mechanism must exist in these organisms. The complex *E. coli* protease La and similar enzymes in other microorganisms (and in mitochondria) require ATP hydrolysis for their action. Thus protein degradation may turn out to be as complex and important a problem as protein biosynthesis.

BIBLIOGRAPHY

Ribosomes and Transfer RNA

Dang, C. V. and Dang, C. V. Multienzyme complexes of aminoacyl-tRNA synthetases: an essence of being eukaryotic. *Biochem. J.* 239:249, 1986.

Fersht, A. and Dingwall, C. Evidence for the double-sieve mechanism in protein synthesis. Steric exclusion of isoleucine by valyl-tRNA. *Biochemistry* 18:2627, 1979.

Freist, W. Mechanisms of aminoacyl-tRNA synthetases: a critical consideration of recent results. *Biochemistry* 28:6787, 1989.

Herold, M. and Nierhaus, K. H. Incorporation of six additional proteins to complete the assembly map of the 50S subunit from *Escherichia coli* ribosomes. *J. Biol. Chem.* 262:8826, 1987.

Hill, W., Dahlberg, A., Garrett, R., Moore, P., Schlessinger, D., and Warner, J. (Eds.), *The Ribosome: Structure, Function and Evolution.* Washington D.C.: American Society for Microbiology, 1990.

Hou, Y.-M., Francklyn, C., and Schimmel, P. Molecular dissection of a transfer RNA and the basis for its identity. *Trends Biochem. Sci.* 14:233, 1989.

Noller, H. F. Ribosomal RNA and translation. *Annu. Rev. Biochem.* 60:191, 1991.

Nomura, M. The role of RNA and protein in ribosome function: a review of early reconstitution studies and prospects for future studies. *Cold Spring Harbor Symp. Quant. Biol.* 52:653, 1987.

Rould, M., Perona, J., and Steitz, T. Structural basis of anticodon loop recognition by glutaminyl-tRNA synthetase. *Nature* 352:213, 1991.

Protein Biosynthesis

Barrell, B. et al. Different pattern of codon recognition by mammalian mitochondrial tRNAs. *Proc. Natl. Acad. Sci. U.S.A.* 77:3164, 1980.

Gnirke, A., Geigenmuller, U., Rheinberger, H., and Nierhaus, K. The allosteric three-site model for the ribosomal elongation cycle. *J. Biol. Chem.* 264:7291, 1989.

Moldave, K. Eukaryotic protein synthesis. *Annu. Rev. Biochem.* 54:1109, 1985.

Ziff, E. B. Transcription and RNA processing by the DNA tumour viruses. *Nature* (*London*) 287:491, 1980.

Protein Targetting and Posttranslational Modification

Borst, P. Peroxisome biogenesis revisited. *Biochim. Biophys. Acta* 1008:1, 1989.

Dahms, N. M., Lobel, P., and Kornfeld, S. Mannose 6-phosphate receptors and lysosomal enzyme targeting. *J. Biol. Chem.* 264:12115, 1989.

Ellis, R. J. and Hemmingsen, S. M. Molecular chaperones: proteins essential for the biogenesis of some macromolecular structures. *Trends Biochem. Sci.* 14:339, 1989.

Hirschberg, C. B. and Snider, M. D. Topography of glycosylation in the rough endoplasmic reticulum and Golgi apparatus. *Annu. Rev. Biochem.* 56:63, 1987.

Jennings, M. L. Topography of membrane proteins. *Annu. Rev. Biochem.* 58:999, 1989.

Kornfeld, R. and Kornfeld, S. Assembly of asparagine-linked oligosaccharides. *Annu. Rev. Biochem.* 54:631, 1985.

Kuhn, K. The classical collagens, R. Mayne and R. E. Burgeson (Eds.), *Structure and Function of Collagen Types*. Orlando, FL: Academic, 1987, pp. 1–42.

Paulson, J. C. Glycoproteins: What are the sugar chains for? *Trends Biochem. Sci.* 14:272, 1989.

Paulson, J. C. and Colley, K. J. Glycosyltransferases: Structure, localization, and control of cell type-specific glycosylation. *J. Biol. Chem.* 264:17615, 1989.

Pfanner, N. and Neupert, W. The mitochondrial protein import apparatus. *Annu. Rev. Biochem.* 59:331, 1990.

Pfeffer, S. R. and Rothman, J. E. Biosynthetic protein transport and sorting by the endoplasmic reticulum and Golgi. *Annu. Rev. Biochem.* 56:829, 1987.

Roise, D. and Schatz, G. Mitochondrial presequences. *J. Biol. Chem.* 263:4509, 1988.

Rucker, R. B. and Wold, F. Cofactors in and as post translational protein modifications. *FASEB J.* 2:2252, 1988.

Silver, P. and Goodson, H. Nuclear protein transport. *Crit. Rev. Biochem. Mol. Biol.* 24:419, 1989.

Towler, D., Gordon, J., Adams, S., and Glaser, L. The biology and enzymology of eukaryotic protein acylation. *Annu. Rev. Biochem.* 57:69, 1988.

Wold, F. Posttranslational protein modifications: Perspectives and prospectives, B. C. Johnson (Ed.), *Posttranslational Covalent Modifications of Proteins*. New York: Academic, 1983, pp. 1–17.

Protein Turnover

Dice, J. F. Molecular determinants of protein half-lives in eukaryotic cells. *FASEB J.* 1:349, 1987.

Gonda, D. K. et al. Universality and structure of the N-end rule. *J. Biol. Chem.* 264:16700, 1989.

Hershko, A. Ubiquitin-mediated protein degradation. *J. Biol. Chem.* 263:15237, 1988.

Rechsteiner, M. Natural substrates of the ubiquitin proteolytic pathway. *Trends Biochem. Sci.* 66:615, 1991.

Rogers, S., Wells, R., and Rechsteiner, M. Amino acid sequences common to rapidly degraded proteins: the PEST hypothesis. *Science* 234:364, 1986.

Schlessinger, M. and Hershko, A. (Eds.). *The Ubiquitin System*. Cold Spring Harbor, NY: Cold Spring Harbor Laboratory, 1988.

QUESTIONS

J. BAGGOTT AND C. N. ANGSTADT

*Q*uestion *T*ypes are described on page xxiii.

1. (QT1) Degeneracy of the genetic code denotes the existence of:
   A. multiple codons for a single amino acid.
   B. codons consisting of only two bases.
   C. base triplets that do not code for any amino acid.
   D. different protein synthesis systems in which a given triplet codes for different amino acids.
   E. codons that include one or more of the ''unusual'' bases.

2. (QT2) Deletion of a single base from a coding sequence of mRNA may result in a polypeptide product with:
   1. a sequence of amino acids that differs from the sequence found in the normal polypeptide.
   2. more amino acids.
   3. fewer amino acids.
   4. a single amino acid replaced by another.

3. (QT1) During initiation of protein synthesis:
   A. methionyl-tRNA appears at the A site of the 80S initiation complex.
   B. eIF-3 and the 40S ribosomal subunit participate in forming the entry complex.
   C. eIF-2 is phosphorylated by GTP.
   D. the same methionyl-tRNA is used as is used during elongation.
   E. a complex consisting of mRNA, the 60S ribosomal subunit, and certain initiation factors is formed.

4. (QT2) Requirements for protein synthesis include:
   1. mRNA.
   2. ribosomes.
   3. GTP.
   4. 20 different amino acids in the form of aminoacyl-tRNAs.
5. (QT2) During the elongation stage of eucaryotic protein synthesis:
   1. the incoming aminoacyl-tRNA binds to the A site.
   2. a new peptide bond is synthesized by peptidyltransferase in a GTP-requiring reaction.
   3. the peptide, still bound to a tRNA molecule, is translocated to a different site on the ribosome.
   4. streptomycin can cause premature release of the incomplete peptide.

6. (QT1) Diphtheria toxin and *Pseudomonas* toxin differ from all antibiotic inhibitors of protein synthesis discussed in this chapter in that they:
   A. act catalytically.
   B. release incomplete polypeptide chains from the ribosome.
   C. inhibit translocase
   D. prevent release factor from recognizing termination signals.
   E. have no clinical usefulness.

Protein synthesis in:
A. Mitochondria
B. Eucaryotic free polysomes
C. Both
D. Neither

7. (QT4) UGA is a stop signal.
8. (QT4) AUG codes for initiation and for internal methionyl residues.
9. (QT4) Synthesis of proteins for export.

10. (QT2) Posttranslational modification of polypeptides can include:
    1. removal of a signal peptide.
    2. removal of one or more terminal amino acid residues.
    3. removal of a peptide from an internal region.
    4. disulfide bond formation.
11. (QT2) 4-hydroxylation of specific prolyl residues during collagen synthesis requires:
    1. Fe^{2+}.
    2. a specific amino acid sequence.
    3. ascorbic acid.
    4. succinate.

12. (QT1) In the formation of an aminoacyl-tRNA:
    A. ADP and P_i are products of the reaction.
    B. aminoacyl adenylate appears in solution as a free intermediate.
    C. the aminoacyl-tRNA synthetase is believed to recognize and hydrolyze incorrect aminoacyl-tRNA's it may have produced.
    D. there is a separate aminoacyl-tRNA synthetase for every amino acid appearing in the final, functional protein.
    E. there is a separate aminoacyl-tRNA synthetase for every tRNA species.

13. (QT2) During collagen synthesis, events that occur extracellularly include:
    1. disulfide bond formation.
    2. modification of lysyl residues.
    3. modification of prolyl residues.
    4. hydrolysis of N-terminal and C-terminal peptides from the triple helical procollagen.
14. (QT2) In the functions of ubiquitin
    1. ATP is required for activation of ubiquitin.
    2. linkage of a protein to ubiquitin does not always mark it for degradation.
    3. the identity of the N-terminal amino acid is a major determinant of selection for degradation.
    4. ATP is required by the protease that degrades the tagged protein.

A. export from the cell
B. lysosomes
C. mitochondria
D. nucleus
E. peroxisomes

Match each of the following numbered markers with the appropriate lettered target site.

15. (QT5) Lysine and arginine amino acid residues
16. (QT5) Mannose 6-phosphate
17. (QT5) Basic and hydroxyl amino acid residues
18. (QT5) Ser-Lys-Leu

ANSWERS

1. A A is the definition of degeneracy (p. 730). B and E are not known to occur, although sometimes tRNA reads only the first two bases of a triplet (wobble), and sometimes unusual bases occur in anticodons (p. 730). C denotes the stop (nonsense) codons (p. 730). D is a deviation from universality of the code, as found in mitochondria (p. 730).
2. A 1, 2, and 3 true. Deletion of a single base causes a frameshift mutation (p. 732). The frameshift would destroy the original stop codon; another one would be generated before or after the original location. In contrast, replacement of one base by another would cause replacement of one amino acid (missense mutation), unless a stop codon is thereby generated (p. 732).
3. B A: methionyl-tRNA$_f^{met}$ appears at the P site. C: phosphorylation of eIF-2 inhibits initiation. D: methionyl-tRNA$_m^{met}$ is used internally. E: mRNA associates first with the 40S subunit (pp. 737).
4. E 1–4 true. Absence of a required aminoacyl-tRNA stops protein synthesis.
5. B 1 and 3 true. 2: Peptide bond formation requires no energy source other than the aminoacyl-tRNA (p. 741). 4: Streptomycin inhibits formation of the

procaryotic 70S initiation complex (analogous to the eucaryotic 80S complex) and causes misreading of the genetic code when the initiation complex is already formed (p. 744).

6. A This toxin catalyzes the formation of an ADP ribosyl derivative of translocase, which irreversibly inactivates the translocase (p. 745). Erythromycin inhibits translocation stoichiometrically, not catalytically (p. 745). The toxin is, of course clinically useless, but so is puromycin (p. 745).
7. B UGA is a stop signal in all systems except mitochondria, in which it codes for tryptophan (p. 730).
8. C See p. 739.
9. D Proteins synthesized for export are synthesized on ribosomes bound to the endoplasmic reticulum (p. 746).
10. E 1–4 true. 1: See p. 748. 2: See p. 754. 3: The C peptide of insulin is a good example. 4: See p. 756.
11. A 1, 2, and 3 required (p. 758).
12. C A and B: ATP and the amino acid react to form an enzyme-bound aminoacyl adenylate; PP_i is released into the medium (p. 734). C: Bonds between a tRNA and an incorrect smaller amino acid may form but are rapidly hydrolyzed (p. 735). D: Some amino acids, such as hydroxyproline and hydroxylysine, arise by co- or posttranslational modification (p. 758). E: An aminoacyl-tRNA synthetase may recognize any of several tRNA's specific for a given amino acid (p. 734).
13. C 2 and 4 are extracellular (see p. 759). Some modification of lysyl residues also occurs intracellularly (p. 758).
14. E 1–4 true (see p. 762). 2: Linkage to histones does not result in their degradation.
15. D (see p. 754).
16. B (see p. 753).
17. C (see p. 753).
18. E (see p. 754). This tripeptide must occur at the carboxy terminal.

18

Recombinant DNA and Biotechnology

GERALD SOSLAU

18.1 OVERVIEW 768

18.2 RESTRICTION ENDONUCLEASES AND RESTRICTION MAPS 769

Restriction Endonucleases Permit the Selective Hydrolysis of DNA to Generate Restriction Maps 769

Restriction Maps Permit the Routine Preparation of Defined Segments of DNA 770

18.3 DNA SEQUENCING 770

Chemical Cleavage Method: Maxam–Gilbert Procedure 771

Interrupted Enzymatic Method: Sanger Procedure 772

18.4 RECOMBINANT DNA AND CLONING 773

DNA From Different Sources Can Be Ligated to Form a New DNA Species: Recombinant DNA 773

Recombinant DNA Vectors Can Be Produced in Significant Quantities by Cloning 775

DNA Can Be Inserted Into Vector DNA in a Specific Direction: Directional Cloning 776

Bacteria Can Be Transformed With Recombinant DNA 777

It Is Necessary to Be Able to Select Transformed Bacteria 777

Recombinant DNA Molecules in a Gene Library 777

18.5 SELECTION OF SPECIFIC CLONED DNA IN LIBRARIES 778

Loss of Antibiotic Resistance Is Used to Select Transformed Bacteria 778

Alpha-Complementation for Selecting Bacteria Carrying Recombinant Plasmids 780

18.6 TECHNIQUES FOR DETECTION AND IDENTIFICATION OF NUCLEIC ACIDS 781

Nucleic Acids Can Serve as Probes for Specific DNA or RNA Sequences 781

Southern Blot Technique Is Useful for Identifying DNA Fragments 782

18.7 COMPLEMENTARY DNA AND COMPLEMENTARY DNA LIBRARIES 784

Messenger RNA Is Used as a Template for DNA Synthesis Using Reverse Transcriptase 784

A Desired mRNA in a Sample Can Be Enriched by Separation Techniques 784

Complementary DNA Synthesis 785

18.8 BACTERIOPHAGE, COSMID, AND YEAST CLONING VECTORS 785

Bacteriophage as Cloning Vectors 785

Screening Bacteriophage Libraries 786

Cloning DNA Fragments Into Cosmid and Yeast Artificial Chromosome Vectors 786

18.9 TECHNIQUES TO FURTHER ANALYZE LONG STRETCHES OF DNA 788

Subcloning Permits Definition of Large Segments of DNA 788

Chromosome Walking Is a Technique to Define Gene Arrangement in Long Stretches of DNA 788

18.10 EXPRESSION VECTORS AND FUSION PROTEINS 788

Foreign Genes Can Be Expressed in Bacteria Allowing Synthesis of Their Encoded Proteins 790

18.11 EXPRESSION VECTORS IN EUCARYOTIC CELLS 790

DNA Elements Required for Expression of Vectors in Mammalian Cells 792

Transfected Eucaryotic Cells Can Be Selected by Utilizing Mutant Cells Which Require Specific Nutrients 792

Foreign Genes Can Be Expressed in Eucaryotic Cells by Utilizing Virus Transformed Cells 793
18.12 SITE-DIRECTED MUTAGENESIS 793
The Role of Flanking Regions in DNA Can Be Evaluated by Deletion and Insertion Mutations 793
Site-Directed Mutagenesis of a Single Nucleotide 794
18.13 POLYMERASE CHAIN REACTION 797
18.14 APPLICATIONS OF RECOMBINANT DNA TECHNOLOGIES 798
Antisense Nucleic Acids Hold Promise as Research Tools and in Therapy 799
Transgenic Animals 801
Recombinant DNA in Agriculture Will Have Significant Commercial Impact 801
18.15 CONCLUDING REMARKS 802
BIBLIOGRAPHY 802
QUESTIONS AND ANNOTATED ANSWERS 802
CLINICAL CORRELATIONS
18.1 Restriction Mapping and Evolution 771
18.2 Restriction Fragment Length Polymorphisms Determine the Clonal Origin of Tumors 783
18.3 Site Directed Mutagenesis of HSV I gD 795
18.4 Polymerase Chain Reaction and Screening for Human Immunodeficiency Virus 799
18.5 Transgenic Animal Models 800

18.1 OVERVIEW

By 1970, the stage was set for modern molecular biology based upon the studies of numerous scientists in the prior 30 years. In those three decades we advanced from ignorance of what biochemical entity orchestrated the replication of life forms with such fidelity to a state where sequencing and manipulating the expression of genes would be feasible. The relentless march towards a full understanding of gene regulation under normal and pathological conditions has moved with increasing rapidity since the 1970s. Deoxyribonucleic acid, composed of only four different bases covalently linked by a sugar–phosphate backbone, is deceptively complex. Complexity is conferred on the DNA molecule by the nonrandom sequence of its bases, multiple conformations that exist in equilibrium in the biological environment, and specific proteins that recognize and associate with selected regions. By the late 1960s the biochemical knowledge of the cellular processes and its macromolecular components had established several facts required for the surge forward. It was clear that gene expression was highly regulated. Enzymes involved in DNA replication and RNA transcription had been purified and their function in the synthetic process defined. The genetic code had been broken. Genetic maps of procaryotic chromosomes had been established based upon gene linkage studies with thousands of different mutants. Finally, RNA species could be purified, enzymatically hydrolyzed into discrete pieces, and laboriously sequenced. It was evident that further progress in the understanding of gene regulation would require techniques to selectively cut DNA into homogeneous pieces. Even small, highly purified viral DNA genomes were too complex to decipher. The thought of tackling the human genome with more than 3×10^9 base pairs was all the more onerous.

The identification, purification, and characterization of restriction endonucleases that faithfully hydrolyze DNA molecules at specific sequences permitted the development of recombinant DNA methodologies. The concomitant development of DNA sequencing opened the previously tightly locked molecular biology gates to the secrets held within the organization of the diverse biological genomes. Genes could finally be sequenced, but perhaps more importantly so could the flanking regions that regulate their expression. Sequencing regulatory regions of numerous genes defined consensus sequences such as those found in promoters, enhancers, and many binding sites for regulatory proteins (discussed in Chapter 19). Each gene contains an upstream promoter where a DNA-dependent RNA polymerase binds prior to the initiation of transcription. While some DNA regulatory sites lie just upstream of the transcription initiation site other regulatory regions are hundreds to thousands of bases removed and still others are downstream.

This chapter presents many of the sophisticated techniques, developed in the past 20 years, that allow for the dissection of complex genomes into defined fragments with the complete analysis of the nucleotide sequence and function of these DNA regions. The modification and manipulation of genes, that is, genetic engineering, facilitates the introduction and expression of genes in both procaryotic and eucaryotic cells. Proteins for experimental and clinical uses are readily produced by these procedures and it is anticipated that in the not too distant future these methods will allow for the treatment of genetic diseases with gene replacement therapy. Current and potential uses of recombinant DNA technologies are also described.

18.2 RESTRICTION ENDONUCLEASES AND RESTRICTION MAPS

Restriction Endonucleases Permit the Selective Hydrolysis of DNA to Generate Restriction Maps

Nature possesses a diverse set of tools, the **restriction endonucleases,** capable of selectively dissecting DNA molecules of all sizes and origin into smaller fragments. These enzymes appear to confer some protection on bacteria against invading viruses, bacteriophage. The DNA sequence normally recognized by the restriction endonuclease may be protected from cleavage in the host cell by methylation of bases within the palindrome while the unmethylated viral DNA is recognized as foreign and is hydrolyzed. The set of enzymes was discovered in the early 1970s and their potential in manipulating DNA was recognized immediately. Numerous Type II restriction endonucleases, with differing sequence specificities, have been identified and purified; many are now commercially available (see p. 642) for a discussion of the restriction endonuclease activities).

Restriction endonucleases permit the construction of a new type of genetic map, the **restriction map,** in which the site of enzyme cleavage within the DNA is identified. Purified DNA species that contain restriction endonuclease sequences can be, and in many cases has been, subjected to restriction endonuclease cleavage. By regulating the time of exposure of the purified DNA molecules to restriction endonuclease cleavage one can generate a population of DNA fragments that are partially to fully hydrolyzed. The separation of these enzyme generated fragments by agarose gel electrophoresis allows for the construction of restriction maps; an example of this procedure with circular DNA is presented in Figure 18.1. Analysis of a DNA completely hydrolyzed by a restriction endonuclease establishes how many sites the restriction endonuclease recognizes within the molecule and what size fragments are generated. The size distribution of composite fragments generated by the partial enzymatic cleavage of the DNA molecules demonstrates linkage of all potential fragments. The sequential use of different restriction endonucleases has permitted a detailed restriction map of numerous circular DNA species including bacterial plasmids, viruses, and mitochondrial DNA. The method is also equally amenable to linear DNA fragments that have been purified to homogeneity.

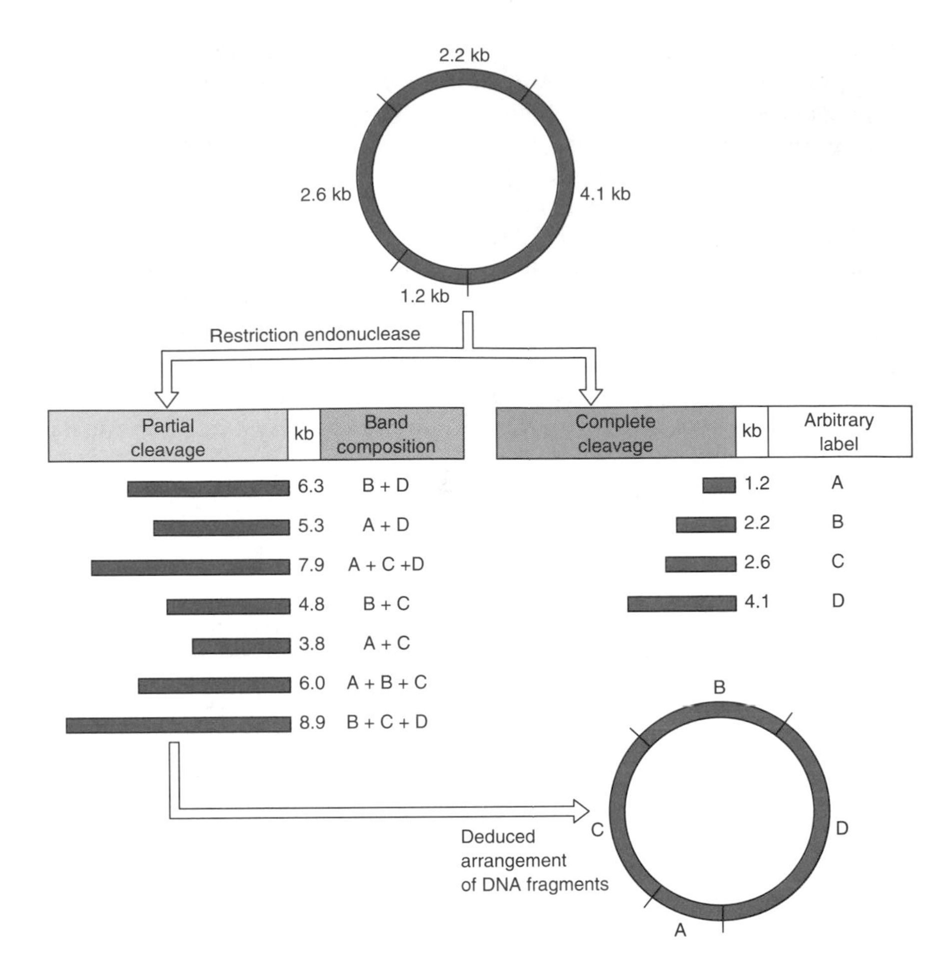

FIGURE 18.1
Restriction endonuclease mapping of DNA. Purified DNA is subjected to restriction endonuclease digestion for varying times which generates partially to fully cleaved DNA fragments. The DNA fragments are separated by agarose gel electrophoresis and stained with ethidium bromide. The DNA bands are visualized with a U.V. light source and photographed. The size of the DNA fragments is determined by the relative migration through the gel as compared to co-electrophoresed DNA standards. The relative arrangement of each fragment within the DNA molecule can be deduced from the size of the incompletely hydrolyzed fragments.

Restriction Maps Permit the Routine Preparation of Defined Segments of DNA

Restriction maps themselves may yield little information as to the genes or regulatory elements within the various DNA fragments. They have been used to demonstrate sequence diversity of organelle DNA, such as mitochondrial DNA, within species (see Clin. Corr. 18.1). Restriction maps can also be used to detect deletion mutations where a defined DNA fragment from the parental strain migrates as a smaller fragment in the mutated strain. Most importantly, the enzymatic microscissors used to generate restriction maps cut DNA into defined homogeneous fragments that can be readily purified. These maps are crucial for cloning and for sequencing genes and their flanking DNA regions.

18.3 DNA SEQUENCING

In order to determine the complexities of regulation of gene expression and to seek the basis for genetic diseases, it was obvious in the early years of study of gene structure that techniques were necessary to determine the exact sequence of bases in DNA. Early approaches at sequencing were very laborious and could only sequence relatively short segments of DNA. In the late 1970s two different sequencing techniques were developed, one by A. Maxam and W. Gilbert, the chemical cleavage approach, and the other by F. Sanger, the enzymatic approach. Both procedures may employ the labeling of a terminal nucleotide, followed by the separation and detection of generated oligonucleotides. The following section describes these two widely used procedures that along with the capability of automated sequencing have permitted rapid and reproducible results.

Chemical Cleavage Method: Maxam–Gilbert Procedure

The requirements for this procedure include the following steps: (1) labeling of the terminal nucleotide, (2) selective hydrolysis of the phosphodiester bond for each nucleotide separately to produce fragments with 1, 2, 3, or more bases, (3) quantitative separation of the hydrolyzed fragments, and (4) a qualitative determination of the label added in Step 1. The following describes one approach of the **Maxam–Gilbert procedure.** The overall approach is presented in Figure 18.2.

One end of each strand of DNA can be selectively radiolabeled with ^{32}P. This is accomplished when a purified double helix DNA fragment contains restriction endonuclease sites on either side of the region to be sequenced. Hydrolysis of the DNA with two different restriction endonucleases then results in different staggered ends, each with a different base in the first position of the single-stranded region. Labeling the 3′ end of each strand is accomplished with the addition of the next nucleotide as directed by the corresponding base sequence on the complementary DNA strand. A fragment of *E. coli* DNA polymerase I, termed the **Klenow fragment,** will catalyze this reaction. The Klenow fragment, produced by partial proteolysis of the polymerase holoenzyme, lacks 5′ → 3′ exonuclease activity but retains the 3′ → 5′ exonuclease and polymerase activity. Each strand can, therefore, be selectively labeled in separate experiments. The complementary unlabeled strand will not be detectable when analyzing the sequence of the labeled strand.

The hydrolysis of the labeled DNA into different lengths is accomplished by first selectively destroying one or two bases of the four nucleotides. The procedure used exposes the phosphodiester bond connecting adjoining bases and permits selective cleavage of the DNA at the altered base. In separate chemical treatments, samples of the labeled DNA are treated to alter the purines [(1) guanine and (2) adenine plus guanine] and the pyrimidines [(3) cytosine and (4) cytosine plus thymine] without disrupting the sugar–phosphate backbone; a method is not currently available to specifically alter adenine or thymine. Conditions for base modification are selected such that only one or a few bases are destroyed randomly within any one molecule. The four separate DNA samples are then reacted with piperidine, which chemically breaks the sugar–phosphate backbone at sites where a base has been destroyed, generating fragments of different size. Since labeling is specific at the end while the chemical alteration of the base is random and not total, some of the fragments will be end labeled. For example, wherever a cytosine residue had been randomly destroyed in the appropriate reaction tube a break will be introduced into the DNA fragment. The series of chemically generated, end-labeled DNA fragments from each of the four tubes are electrophoresed through a polyacrylamide gel. Bases destroyed near the end-labeled nucleotide will generate fragments that migrate faster through the gel, as low molecular weight species, while fragments derived from bases destroyed more distant from the end will migrate through the gel more slowly as higher molecular weight molecules. The gel is then exposed to X-ray film, which detects the [^{32}P], and the radioactively labeled bands within the gel can be visualized. The sequence can be read manually or by automated methods directly from the X-ray autoradiograph beginning at the bottom (smaller fragments), toward the top of the film (larger frag-

CLIN. CORR. **18.1**
RESTRICTION MAPPING AND EVOLUTION

Evolutionary studies of species have in the past depended solely on anatomical changes observed in fossil records along with carbon dating. More recently, these studies are being supported by the molecular analysis of the sequence and size of selected genes or whole DNA molecules. Evolutionary alterations of a selected DNA molecule from different species can be rapidly assessed by restriction endonuclease mapping. The generation of restriction endonuclease maps requires a pure preparation of DNA. Mammalian mitochondria contain a covalently closed circular DNA molecule of approximately 16,000 base pairs that can be rapidly purified from cells. The mitochondrial DNA (mtDNA) can be employed directly for the study of evolutionary changes in DNA without the need of cloning a specific gene.

Mitochondrial DNA molecules have been purified from four higher primate cells. The Guinea baboon, rhesus macaque, guenon, and human mtDNAs were cleaved with 11 different restriction endonucleases and a restriction map constructed for each species. The maps were all aligned relative to the direction and the nucleotide site where DNA replication is initiated. A comparison of shared and altered restriction endonuclease sites allowed for the calculation of the degree of divergence in nucleotide sequence between species. It was found that the rate of base substitution (calculated from the degree of divergence versus the time of divergence) has been about 10-fold greater than changes in the nuclear genome. This high rate of mutation of the readily purified mtDNA molecule makes it an excellent model to study evolutionary relationships between species.

Brown, W. M., George, M., Jr., and Wilson, A. C. Rapid evolution of animal mitochondrial DNA. *Proc. Natl. Acad. Sci.* U.S.A. 76:1967, 1979.

ments). Sequencing the complementary strand checks the correctness of the sequence.

Interrupted Enzymatic Method: Sanger Procedure

The **Sanger procedure** of DNA sequencing, developed by F. Sanger, is based on the random termination of a DNA chain during enzymatic synthesis. The technique is possible because the dideoxynucleotide analog of each of the four normal nucleotides (Figure 18.3) can be incorporated into

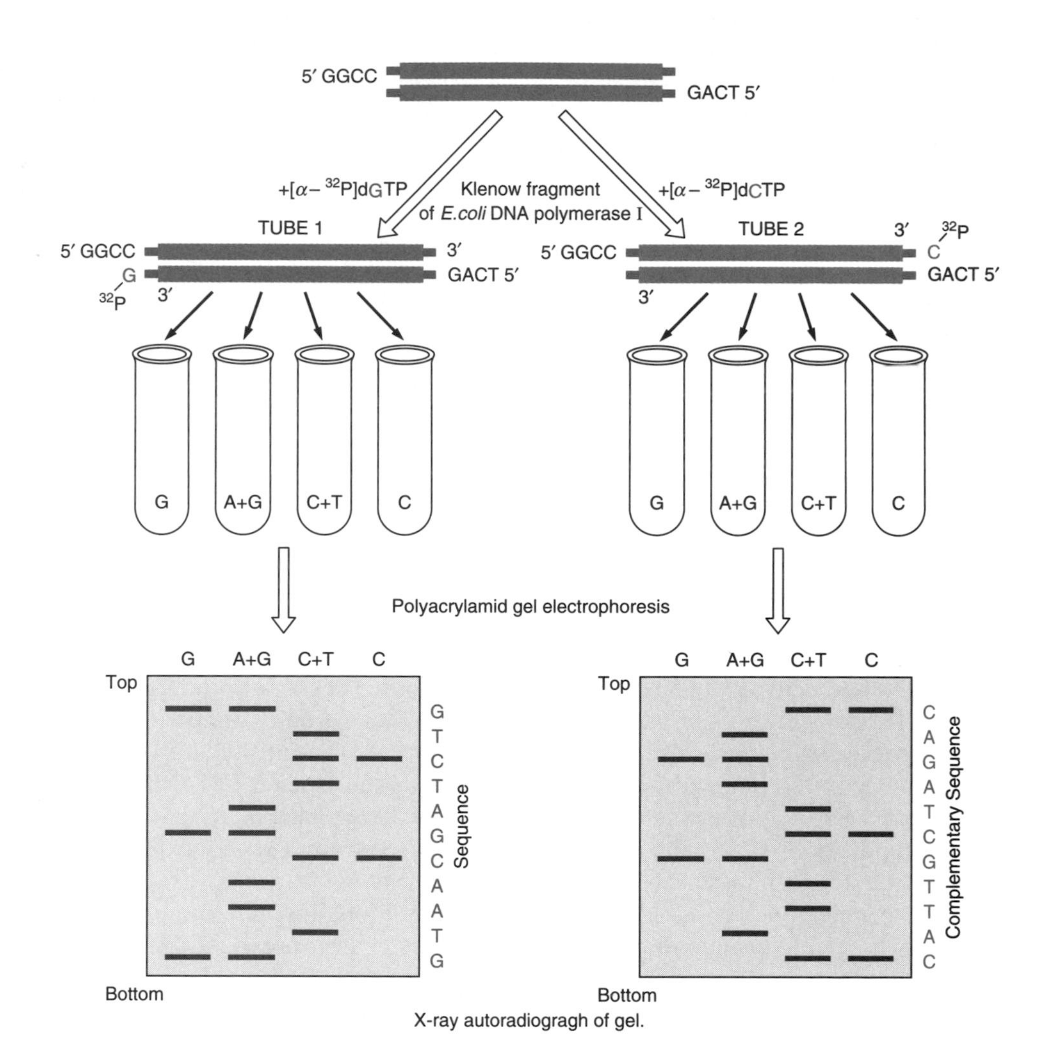

FIGURE 18.2
The Maxam–Gilbert chemical method to sequence DNA.
The double-stranded DNA fragment to be sequenced is obtained by restriction endonuclease cleavage and purified. Both strands are sequenced by selectively labelling the ends of each DNA strand. One strand of DNA is end-labelled with ^{32}P-dGTP in reaction tube 1 while the other is end-labelled with ^{32}P-dCTP in reaction tube 2. The end-labelled DNA is then subdivided into four fractions where the different bases are chemically destroyed at random positions within the single-stranded DNA molecule. The less selective chemical destruction of adenine simultaneously destroys G and the destruction of thymidine destroys the C bases. The single-stranded DNA is cleaved at the sites of the destroyed bases. This generates end-labelled fragments of all possible lengths corresponding to the distance from the end to the sites of base destruction. The labelled DNA fragments are separated according to size by electrophoresis. The DNA sequence can then be determined from the electrophoretic patterns detected on autoradiograms.

a growing DNA chain by DNA polymerase. The ribose of the **dideoxynucleotide (ddNTP)** has the OH group at both the 2′ and 3′ positions replaced with a proton, whereas dNTP has only a single OH group replaced by a proton at the 2′ position. Thus, the ddNTP incorporated into the growing chain is unable to form a phosphodiester bond with another dNTP because the 3′ position of the ribose does not contain a OH group. The growing DNA molecule can be terminated at random points, from the first nucleotide incorporated to the last, by including in the reaction system both the normal nucleotide and the ddNTP (e.g., dATP and ddATP) at concentrations such that the two nucleotides compete for incorporation.

Identification of the DNA fragments requires the labeling of the 5′ end of the DNA molecules or the incorporation of labeled nucleotides during synthesis. The technique, outlined in Figure 18.4, is best conducted with pure single-stranded DNA; however, denatured double-stranded DNA has also been used. Today, the DNA to be sequenced is frequently isolated from a recombinant single-stranded bacteriophage (Section 18.8) where a region flanking the DNA of interest contains a sequence that is complementary to a universal primer. The primer can be labeled with either ^{32}P or ^{35}S nucleotide. Primer extension is accomplished with one of several different available DNA polymerases; one with great versatility is a genetically engineered form of the bacteriophage T 7 DNA polymerase. The reaction mixture, composed of the target DNA, labeled primer and all four deoxynucleotide triphosphates, is divided into four tubes, each containing a different dideoxynucleotide triphosphate. The ddNTPs are randomly incorporated during the enzymatic synthesis of DNA and cause termination of the chain. Since the ddNTP is present in the reaction tube at a low level, relative to the corresponding dNTP, termination of DNA synthesis occurs randomly at all possible complementary sites to the DNA template. This yields a population of DNA molecules of varying sizes, labeled at the 5′ end, that can be separated by polyacrylamide gel electrophoresis. The labeled species are detected by X-ray autoradiography and the sequence read.

Initially, this method had some distinct drawbacks as compared to the Maxam–Gilbert method, including the requirement of a single-stranded DNA template, the production of a specific complementary oligonucleotide primer, and the need for a relatively pure preparation of the Klenow fragment of *E. coli* DNA polymerase I. These difficulties have been overcome and modifications have simplified the approach. The Sanger method can rapidly sequence as many as 400 bases while the Maxam–Gilbert method is limited to about 250 bases.

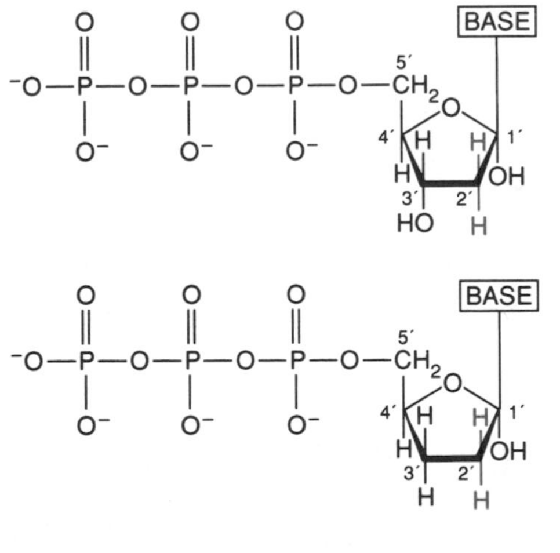

FIGURE 18.3
Structure of deoxynucleotide triphosphate and dideoxynucleotide triphosphate.
The 3′-OH group is lacking on the ribose component of the dideoxynucleotide triphosphate (ddNTP). This molecule can be incorporated into a growing DNA molecule through a phosphodiester bond with its 5′-phosphates. Once incorporated the ddNTP blocks further synthesis of the DNA molecule since it lacks the 3′-OH acceptor group for an incoming nucleotide.

18.4 RECOMBINANT DNA AND CLONING

DNA From Different Sources Can Be Ligated to Form a New DNA Species: Recombinant DNA

The ability to selectively hydrolyze a population of DNA molecules with a battery of restriction endonucleases led to the development of an important technique of joining together two different DNA molecules termed **recombinant DNA.** This technique combined with the various techniques for replication, separation, and identification permits the production of large quantities of purified DNA fragments. The combined techniques, referred to as recombinant DNA technologies, allows the removal of a piece of DNA out of a larger complex molecule, such as the genome of a virus or human, and amplification of the DNA fragment. Recombinant

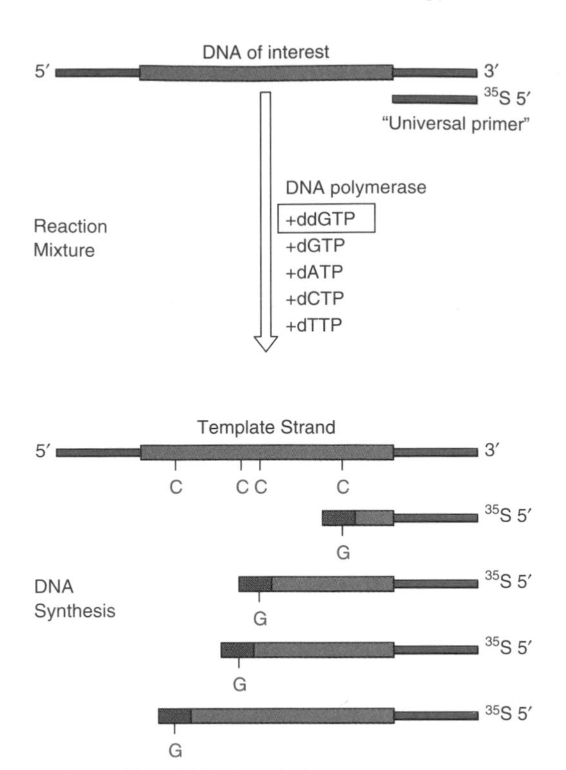

(a) Recombinant M 13 Bacteriophage

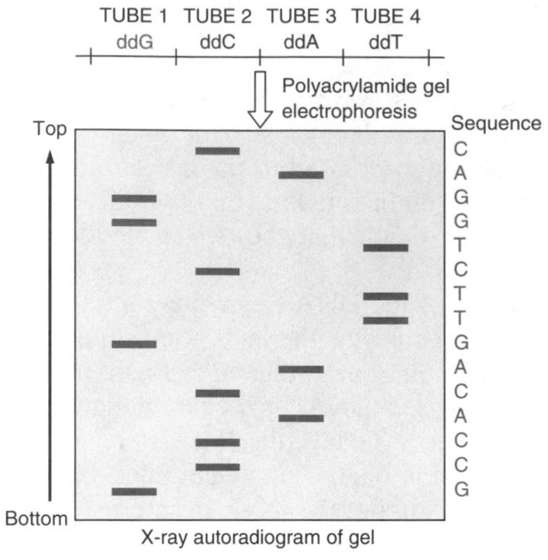

(b) Polyacrylamide Gel Electrophoresis of Reaction Mixture

FIGURE 18.4
The Sanger dideoxynucleotide triphosphate method to sequence DNA.
The DNA region of interest is inserted into a bacteriophage DNA molecule. Replicating bacteriophage produce a single-stranded recombinant DNA molecule that is readily purified. The known sequence of the bacteriophage DNA downstream of the DNA insert serves as a hybridization site for an end-labelled oligomer with a complementary sequence, a universal primer. Extension of this primer is catalyzed with a DNA polymerase in the presence of all four deoxynucleotide triphosphates plus one dideoxynucleotide triphosphate, for example ddGTP. Synthesis stops whenever a dideoxynucleotide triphosphate is incorporated into the growing molecule. Note that the dideoxynucleotide competes for incorporation with the deoxynucleotide. This generates end-labelled DNA fragments of all possible lengths that are separated by electrophoresis. The DNA sequence can then be determined from the electrophoretic patterns.

DNAs have been prepared with DNA fragments from bacteria combined with fragments from humans, viruses with viruses, and so on.

The ability to join two different pieces of DNA together at specific sites within the molecules is achieved with two enzymes, a restriction endonuclease and a DNA ligase. There are a number of different restriction endonucleases, varying in their nucleotide sequence specificity, that can be used (Section 18.2). Some hydrolyze the two strands of DNA in a staggered fashion producing "sticky or cohesive" ends (Figure 18.5) while others cut both strands symmetrically producing a blunt end. A specific restriction enzyme cuts DNA at exactly the same nucleotide sequence site regardless of the source of the DNA (bacteria, plant, mammal, etc.). A DNA molecule may have one, several, hundreds, thousands, or no recognition sites for a particular restriction endonuclease. The staggered cut results in a fragmented DNA molecule with ends that are single stranded. When different DNA fragments generated by the same restriction endonuclease are mixed, their single-stranded ends can hybridize, that is, anneal together. In the presence of DNA ligase the two fragments are connected covalently producing a recombinant DNA molecule.

The DNA fragments produced from restriction endonucleases that form blunt ends can also be ligated but with much lower efficiency. The efficiency can be increased by enzymatically adding a poly(dA) tail to one species of DNA and a poly(dT) tail to the ends of the second species of DNA. The DNA fragments with complementary tails can be annealed and ligated in the same manner as fragments with restriction enzyme generated cohesive ends.

Recombinant DNA Vectors Can Be Produced in Significant Quantities by Cloning

The synthesis of a recombinant DNA opened the way for the production of significant quantities of interesting DNA fragments. By incorporating a recombinant DNA into a cellular system that allows replication of the recombinant DNA, amplification of the DNA of interest can be achieved. A carrier DNA, termed a **cloning vector,** is employed. Bacterial plasmids are ideally suited as recombinant DNA vectors.

Many bacteria contain a single circular chromosome of approximately 4 million base pairs and minicircular DNA molecules called **plasmids.** Plasmids are usually composed of only a few thousand base pairs and are rarely associated with the large chromosomal molecule. Genes within the plasmid have various functions; one of the most useful is the ability to confer antibiotic resistance to the bacterium, an attribute useful in selecting specific colonies of the bacteria. Plasmids replicate independently of the replication of the main bacterial chromosome. One type of plasmid, the **relaxed-control** plasmids, may be present in tens to hundreds of copies per bacterium, and replication is dependent solely on host enzymes that have long half-lives. Therefore, replication of **"relaxed" plasmids** can occur in the presence of a protein synthesis inhibitor. Bacteria can accumulate several thousand plasmid copies per cell under these conditions. Other plasmid types are subjected to **stringent control** and their replication is dependent on the continued synthesis of plasmid-encoded proteins. These plasmids replicate at about the same rate as the large bacterial chromosome, and only a low number of copies occur per cell. The former plasmid type is routinely used for recombinant studies.

The first achievement of a practical recombinant DNA molecule that could be cloned involved as a vector the *E. coli* **plasmid pSC101,** which contains a single EcoRI restriction endonuclease site and a gene that encodes for a protein that confers antibiotic resistance to the bacteria. This plasmid contains an origin of replication and associated DNA regulatory sequences that are referred to as a **replicon.** This vector, however, suffers from a number of limiting factors. The single restriction endonuclease site limits the DNA fragments that can be cloned and the one antibiotic-resistance selectable marker reduces the convenience in selection; in addition it replicates poorly.

Early studies used naturally occurring plasmids, but in the last decade, plasmid vectors with broad versatilities have been constructed using recombinant DNA technology. The desirable features of a plasmid vector include: a relatively low molecular weight (3–5 kb) to accommodate larger fragments; several different restriction endonuclease sites useful in cloning a variety of restriction enzyme generated fragments; multiple selectable markers to aid in selecting bacteria with recombinant DNA molecules; and a high rate of replication. The first plasmid constructed (Figure 18.6) to satisfy these requirements was **pBR322** and this plasmid has been

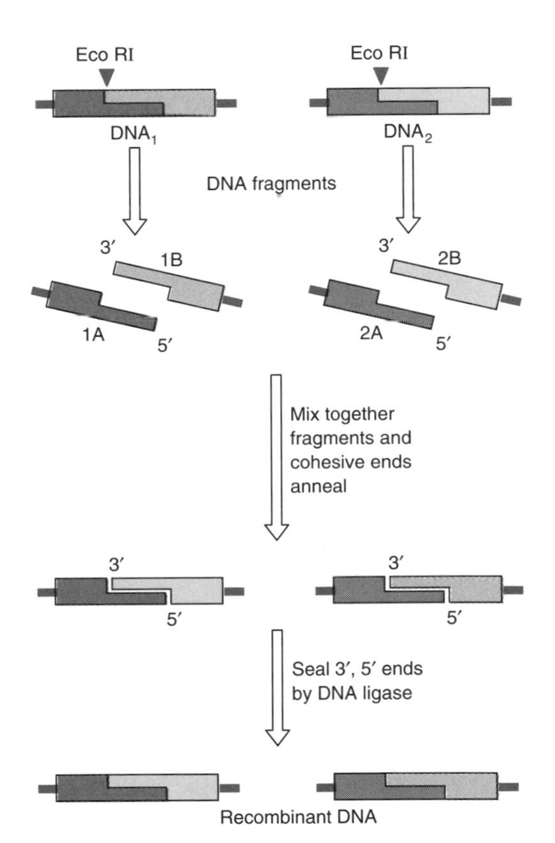

FIGURE 18.5
The formation of recombinant DNA from restriction endonuclease generated fragments containing cohesive ends.
Many restriction endonucleases hydrolyze DNA in a staggered fashion yielding fragments with single-stranded regions at their 5′ and 3′ ends. DNA fragments generated from different molecules with the same restriction endonuclease have complementary single-stranded ends that can be annealed and covalently linked together with a DNA ligase. All different combinations are possible in a mixture. When two DNA fragments of different origin combine it results in a recombinant DNA molecule.

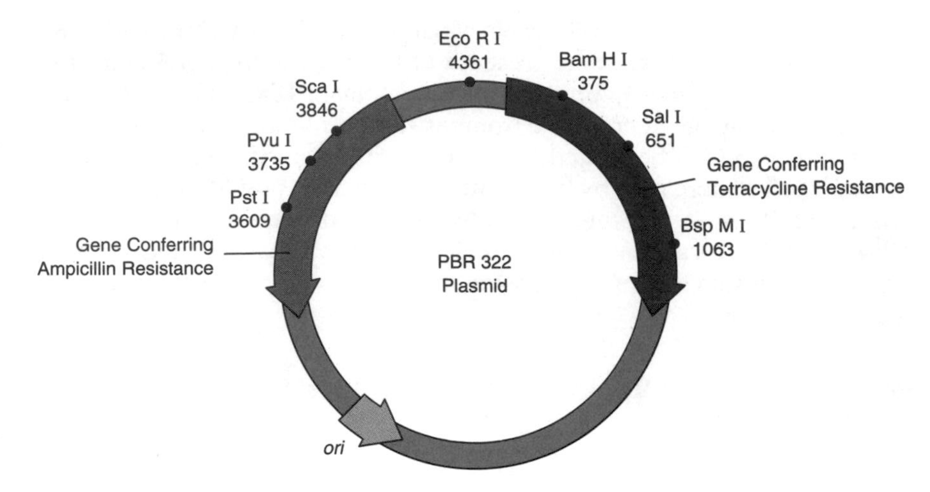

FIGURE 18.6
The pBR322 plasmid constructed in the laboratory to contain features that facilitate cloning foreign DNA fragments.
By convention the numbering of the nucleotides begins with the first T in the unique EcoRI recognition sequence (GAATTC) and the positions on the map refer to the 5′-base of the various restriction endonuclease recognition sequences. Only a few of the unique restriction sites within the antibiotic resistance genes and none of the numerous sites where an enzyme cuts more than once within the plasmid are shown.

used for the subsequent generation of newer vectors in use today. Most of the currently employed vectors contain an inserted sequence of DNA termed **polylinkers, restriction site banks** or **polycloning sites,** which is a DNA region consisting of a cluster of numerous restriction endonuclease sites unique in the plasmid.

DNA Can Be Inserted Into Vector DNA in a Specific Direction: Directional Cloning

Directional cloning reduces the number of variable "recombinants" and enhances the probability of selection of the desired recombinant. The insertion of a piece of foreign DNA, with a defined polarity, into a plasmid vector in the absence of the plasmid resealing itself can be accomplished by employing two restriction endonucleases to cleave the plasmids (Fig-

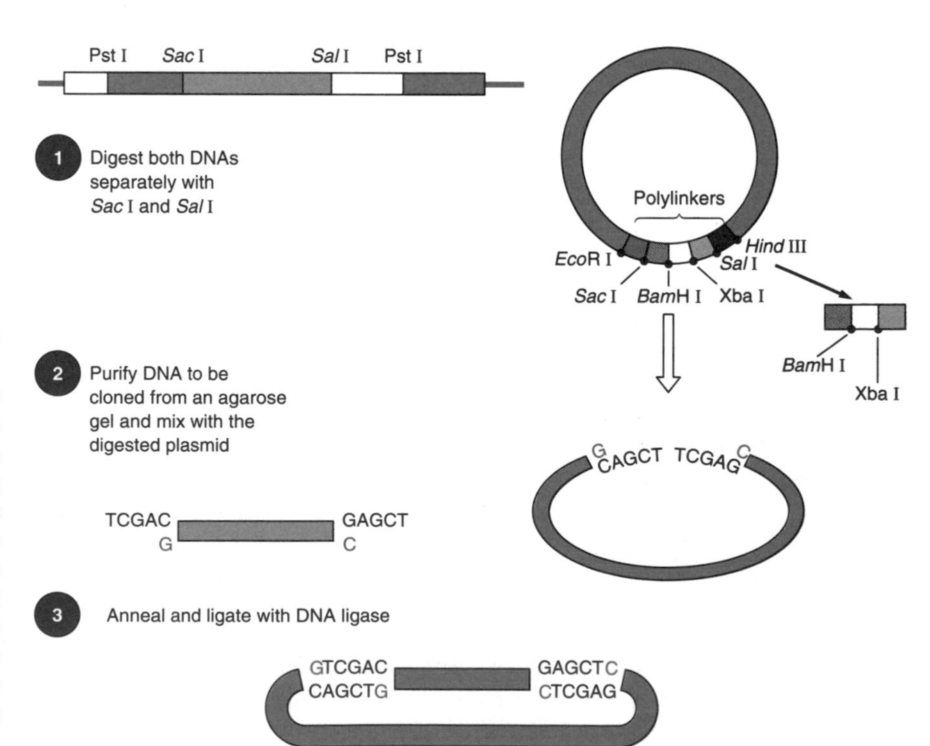

FIGURE 18.7
Directional cloning of foreign DNA into vectors with a specified orientation.
The insertion of a foreign DNA fragment into a vector with a specified orientation requires two different annealing sequences at each end of the fragment and the corresponding complementary sequence at the two ends generated in the vector. A polylinker with numerous unique restriction endonuclease sites within the vector facilitates directional cloning. Knowledge of the restriction map for the DNA of interest allows for the selection of appropriate restriction endonucleases to generate specific DNA fragments that can be cloned in a vector.

ure 18.7); vectors with polylinkers are ideally suited for this purpose. The use of two enzymes yields DNA fragments and linearized plasmids with different "sticky" ends. Under these conditions the plasmid is unable to reanneal with itself. In addition, the foreign DNA can be inserted into the vector in only one orientation. This is an extremely important point when one clones a potentially functional gene downstream from the promoter–regulatory elements in expression vectors (see p. 786).

Bacteria Can Be Transformed With Recombinant DNA

The process of artificially introducing DNA into bacteria is referred to as **transformation.** This is accomplished by briefly exposing the cells to divalent cations making them transiently permeable to small DNA molecules. Recombinant plasmid molecules, containing foreign DNA can be introduced into bacteria where it would replicate normally.

It Is Necessary to Be Able to Select Transformed Bacteria

Once the plasmid has been introduced into the bacterium, both the plasmid and the cells can replicate. Methods are available to select those bacteria that carry the recombinant DNA molecules. In the recombinant process some bacteria might not be transformed or may be transformed with a vector not carrying foreign DNA; in preparing the vector some reanneal without inclusion of the DNA of interest. In some experimental conditions one can generate restriction endonuclease DNA fragments that can be readily purified for recombinant studies. Such fragments can be generated from small, highly purified DNA species, for example, some DNA viruses. More typically, however, a single restriction endonuclease will generate hundreds to hundreds of thousands of DNA fragments depending on the size and complexity of DNA being studied. Individual fragments cannot be isolated from these samples to be individually incorporated into the plasmid. Methods have, therefore, been developed to select those bacteria containing the desired DNA.

Even after surmounting these problems other major obstacles remain. Restriction endonucleases do not necessarily hydrolyze DNA into fragments containing intact genes. If by chance the fragment contains an entire gene it would probably not contain the required flanking regulatory sequences, such as the promoter region. Furthermore, if the foreign gene is of mammalian origin, its regulatory sequences would not be recognized by the bacterial synthetic machinery. The primary gene transcript (mRNA) can also contain introns that cannot be processed by the bacteria. The following sections describe techniques that have been developed to address some of these difficulties.

Recombinant DNA Molecules in a Gene Library

When a complex mixture of thousands of different genes, arranged on different chromosomes, as in the human genome, is subjected to hydrolysis with a single restriction endonuclease, thousands of DNA fragments are generated. These DNA fragments are annealed with a plasmid vector that has been cleaved to a linear molecule with the same restriction endonuclease. By adjusting the ratio of plasmid to foreign DNA the probability of joining at least one copy of each DNA fragment within a cyclized recombinant-plasmid DNA approaches one. Usually, only one out of the multiple of DNA fragments is inserted into each plasmid vector. Bacteria are transformed with the recombinant molecules such that only one plasmid is taken up by a single bacterium. Each recombinant molecule can now be replicated within the bacterium and the bacterium will give rise to progeny, each carrying multiple copies of the recombinant DNA. The total population of bacteria now contain fragments of DNA that may represent the entire human genome. This is termed a **gene library.** As in any library containing thousands of volumes, a selection system must be available to retrieve the book or gene of interest.

Today, plasmids are most commonly employed to clone DNA fragments generated from molecules of limited size and complexity, such as viruses and subcloning large DNA fragments previously cloned in other vectors. Genomic DNA fragments are usually cloned from other vectors capable of carrying larger foreign DNA fragments than plasmids (see Section 18.8).

18.5 SELECTION OF SPECIFIC CLONED DNA IN LIBRARIES

Loss of Antibiotic Resistance Is Used to Select Transformed Bacteria

When a single transformed bacterium carrying a recombinant molecule multiplies its progeny are all genetically the same. If the transformed bacteria carries a recombinant DNA, all progeny will carry copies of the same recombinant plasmid. The foreign DNA has been amplified and is derived from a single cloned DNA fragment. The problem is how to identify the one colony containing the desired plasmid in a field of thousands to millions of different bacterial colonies. As stated earlier, plasmids often carry genes that confer antibiotic resistance to transformed cells. The laboratory derived plasmid construct pBR322 and its descendants carry two genes that confer antibiotic resistance. Within these antibiotic resistant genes are DNA sequences sensitive to restriction endonuclease. When a fragment of foreign DNA is inserted into a restriction site within the gene for antibiotic resistance, the gene becomes nonfunctional. Bacteria carrying this recombinant plasmid will be sensitive to the antibiotic (Figure 18.8). The second antibiotic resistance gene within the plasmid, however, remains intact and the bacteria will be resistant to this antibiotic. This technique of **insertional inactivation** of plasmid gene products affords a method to select bacteria that carry recombinant plasmids.

pBR322 contains genes that confer resistance to ampicillin (amp^r) and tetracycline (tet^r). A gene library with cellular DNA fragments inserted within the tet^r gene can be selected and screened in two stages (Figure 18.8). First, the bacteria are grown in an ampicillin containing growth medium. Bacteria that are not transformed by a plasmid (they lack a normal or recombinant plasmid) during the construction of the gene library, will not grow in the presence of the antibiotic, thus, eliminating this population of bacteria. This, however, does not indicate which of the remaining viable bacteria carry a recombinant plasmid vector versus a plasmid with no DNA insert. The second step is to identify bacteria carrying recombinant vectors with nonfunctional tet^r genes, that are, therefore, sensitive to tetracycline.

Bacteria insensitive to ampicillin are plated and grown on agar plates containing ampicillin (Figure 18.8). Replica plates can be made by touching the colonies on the original agar plate with a filter and then touching additional sterile plates with the filter. All the plates will contain portions of each original colony at identifiable positions on the plates. The replica plate can contain tetracycline, which will not support the growth of bacteria harboring recombinant plasmids with their tet^r gene disrupted. Comparison of replica plates with and without tetracycline will indicate which

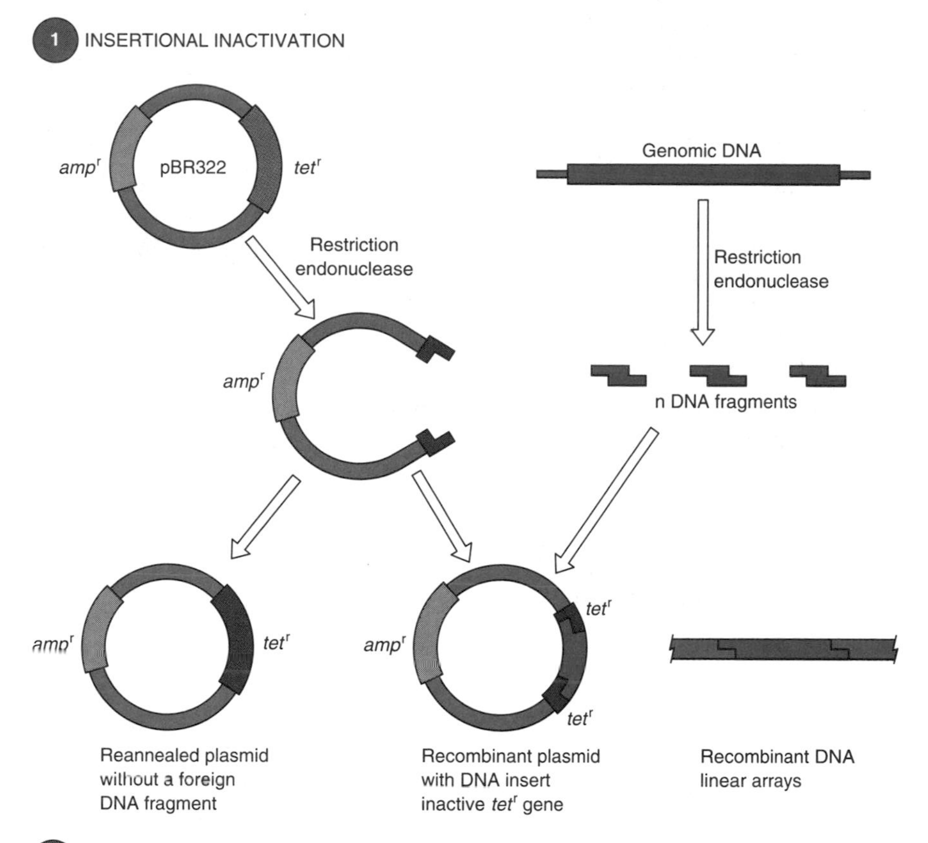

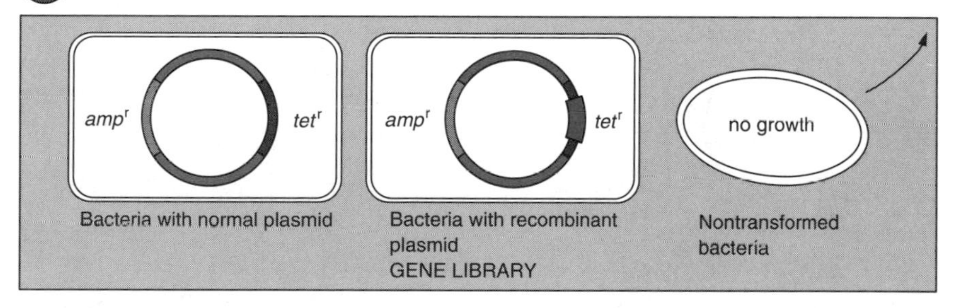

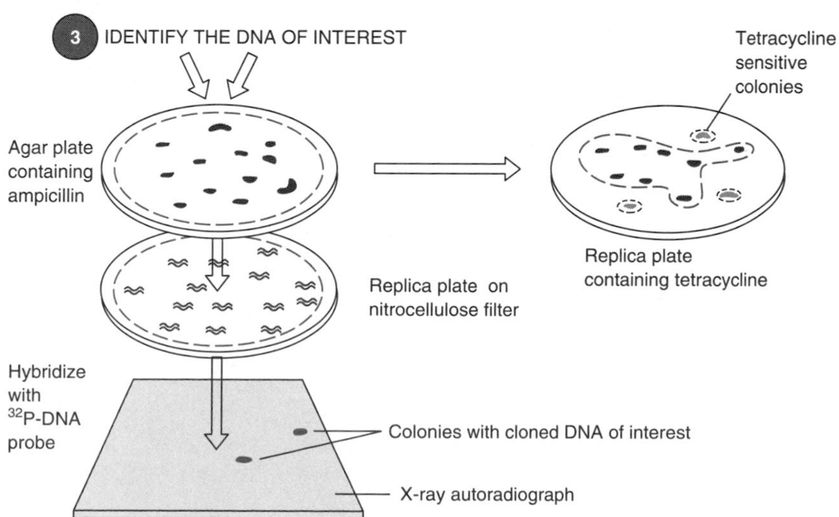

FIGURE 18.8
Insertional inactivation of recombinant plasmids and detection of transformed bacteria carrying a cloned DNA of interest. When the insertion of a foreign DNA fragment into a vector disrupts a functional gene sequence the resulting recombinant DNA will not express the gene. The gene that codes for antibiotic resistance to tetracycline (*Tet*^{r}) is destroyed by DNA insertion while the ampicillin resistance gene (*AMP*^{r}) remains functional. The destruction of one antibiotic resistance gene and the retention of a second antibiotic resistance gene allows for the detection of bacterial colonies carrying the foreign DNA of interest within the replicating recombinant vector.

colonies on the original ampicillin plate contain recombinant plasmids. Thus, individual colonies containing the recombinant DNA can be selected, grown in culture, and analyzed.

Either DNA or RNA probes (see pp. 625 and 781) can be utilized to identify the DNA of interest. Ampicillin resistant bacterial colonies on agar can be replica plated onto a nitrocellulose filter and adhering cells from each colony can be lysed with NaOH (Figure 18.8). The DNA within the lysed bacteria is also denatured by the NaOH and becomes firmly bound to the filter. A labeled DNA or RNA probe that is complementary to the DNA of interest can be hybridized to the nitrocellulose bound DNA. The filter is exposed to X-ray autoradiography. Any colony carrying the cloned DNA of interest will appear as a developed signal on the X-ray film. These spots would then correspond to the colony on the original agar plate that can then be grown in a large scale culture for further manipulation.

These cloned and amplified DNA fragments usually do not contain a complete gene and are not expressed. The DNA inserts, however, can be readily purified for sequencing or used as probes to detect genes within a mixture of genomic DNA, transcription levels of mRNA, and pathological conditions via clinical diagnostic tests.

Alpha-Complementation for Selecting Bacteria Carrying Recombinant Plasmids

Other selection techniques are available to identify bacteria carrying recombinant DNA molecules. Vectors have been constructed (the pUC

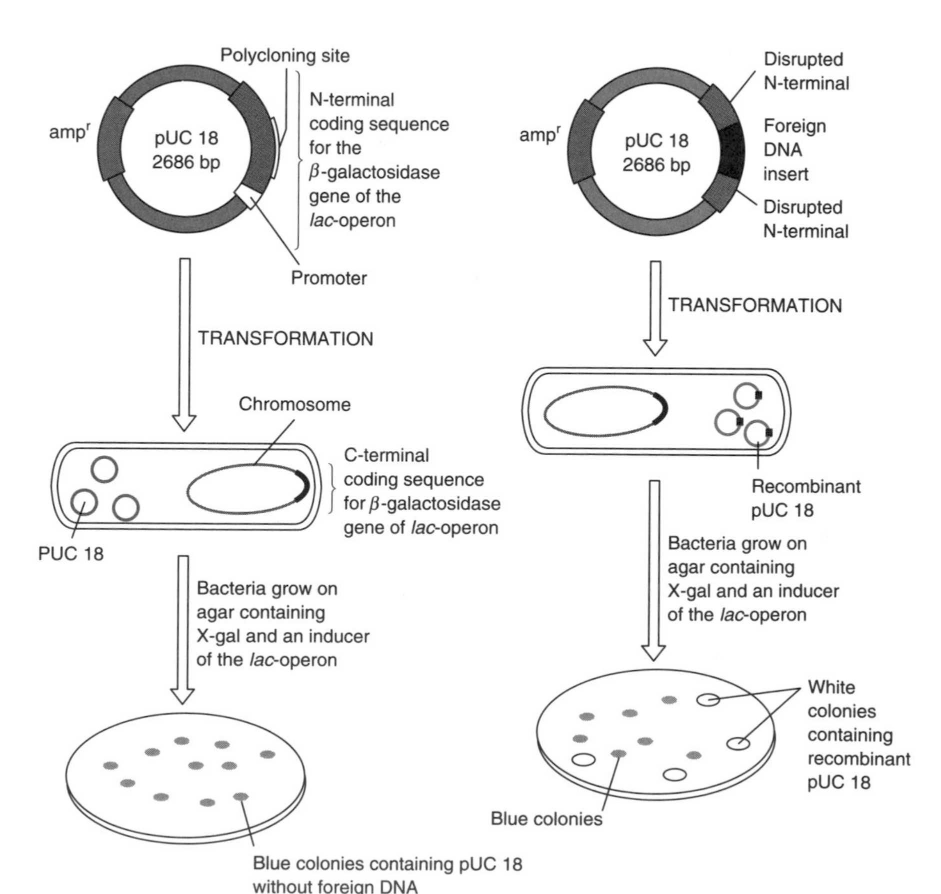

FIGURE 18.9
Alpha-complementation for the detection of transformed bacteria.
A vector has been constructed (pUC 18) which expresses the N-terminal coding sequence for the enzyme β-galactosidase of the lac operon. Bacterial mutants coding for the C-terminal portion of β-galactosidase are transformed with pUC 18. These transformed bacteria grown in the presence of a special substrate for the intact enzyme (X-gal) result in blue colonies because they contain the enzyme to react with substrate. The functional N-terminal and C-terminal coding sequences for the gene complement each other to yield a functional enzyme. If, however, a foreign DNA fragment insert disrupts the pUC 18 N-terminal coding sequence for β-galactosidase bacteria transformed with this recombinant molecule will not produce a functional enzyme. Bacterial colonies carrying these recombinant vectors can then be visually detected as white colonies.

series) such that selected bacteria transformed with these vectors carrying foreign DNA inserts, can be identified visually (Figure 18.9). The pUC plasmids contain the regulatory sequences and a portion of the 5′-end coding sequence (N-terminal 146 amino acids) for the β-galactosidase gene (*lac Z* gene) of the *lac* operon (Chapter 19, p. 808). The translated N-terminal 146 amino acid fragment of β-galactosidase is an inactive polypeptide. Mutant *E. coli,* that code for the missing inactive carboxy-terminal portion of β-galactosidase, can be transformed with the pUC plasmids. The translation of the host cell and plasmid portions of the β-galactosidase in response to an inducer, isopropylthio-β-D-galactoside, complement each other yielding an active enzyme. The process is referred to as α-complementation. When these transformed bacteria are grown in the presence of a chromogenic substrate (5-bromo-4-chloro-3-indolyl-β-D-galactoside [X-gal]) for β-galactosidase they form blue colonies. If, however, a foreign DNA fragment is inserted into the base sequences for the N-terminal portion of β-galactosidase, the active enzyme cannot be formed. Bacteria, transformed with these recombinant plasmids and grown on X-gal will yield white colonies and can be selected visually from the nontransformed blue colonies.

18.6 TECHNIQUES FOR DETECTION AND IDENTIFICATION OF NUCLEIC ACIDS

Nucleic Acids Can Serve as Probes for Specific DNA or RNA Sequences

The selection of bacteria harboring recombinant DNAs of interest or the analysis of mRNAs expressed in a cell or identification of the presence of DNA sequences within a genome require sensitive and specific detection methods. The use of DNA and RNA **probes** meets these requirements. These probes contain nucleotide sequences complementary to the target nucleic acid and will, thus, hybridize with the nucleic acid of interest. The degree of complementarity of the probe with the DNA under investigation will determine the tightness of binding of the probe. The probe does not need to contain the entire complementary sequence of the DNA. The probe, RNA or DNA, is labeled, usually with ^{32}P but nonradioactive labels are also employed. Enzyme substrates can be coupled to nucleotides which when incorporated into the nucleic acid can be detected by an enzyme catalyzed reaction.

Labeled probes can be produced by **nick translation** of double-stranded DNA. Nick translation (Figure 18.10) involves the random enzymatic hydrolysis of a phosphodiester bond in the backbone of one strand of DNA by DNase I; the enzymatic breaks in the DNA backbone are referred to as nicks. A second enzyme, *E. coli* DNA polymerase I, with its 5′ → 3′ exonucleolytic activity and its DNA polymerase activity, creates single-strand gaps by hydrolyzing nucleotides from the 5′ side of the nick and then filling in the gaps with its polymerase activity. The polymerase reaction is usually carried out in the presence of one α-^{32}P-labeled deoxynucleotide triphosphate and three unlabeled deoxynucleotide triphosphates. The DNA employed in this method is usually purified and is derived from cloned DNA, viral DNA, or cDNA.

Labeled RNA probes can be produced that have advantages over the DNA probes. For one, relatively large amounts of RNA can be transcribed from a template, which may be available in very limited quantities. A double-stranded DNA probe must be denatured prior to hybridiza-

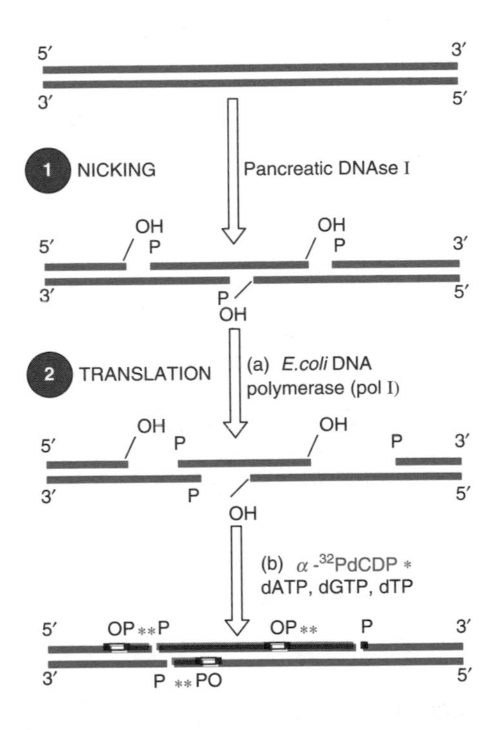

FIGURE 18.10
Nick translation to label DNA probes.
Purified DNA molecules can be radioactively labeled and used to detect, by hybridization, the presence of complementary RNA or DNA in experimental samples. (1) Nicking step: introduces random single-stranded breaks in the DNA; (2) Translation step: (*a*) E. coli DNA polymerase (pol I) has both 5′ → 3′ exonucleolytic activity that hydrolyzes nucleotides from the 5′ end of the nick; (*b*) pol I simultaneously fills back in the single-stranded gap using the 3′ end as a primer with radioactively labeled nucleotides.

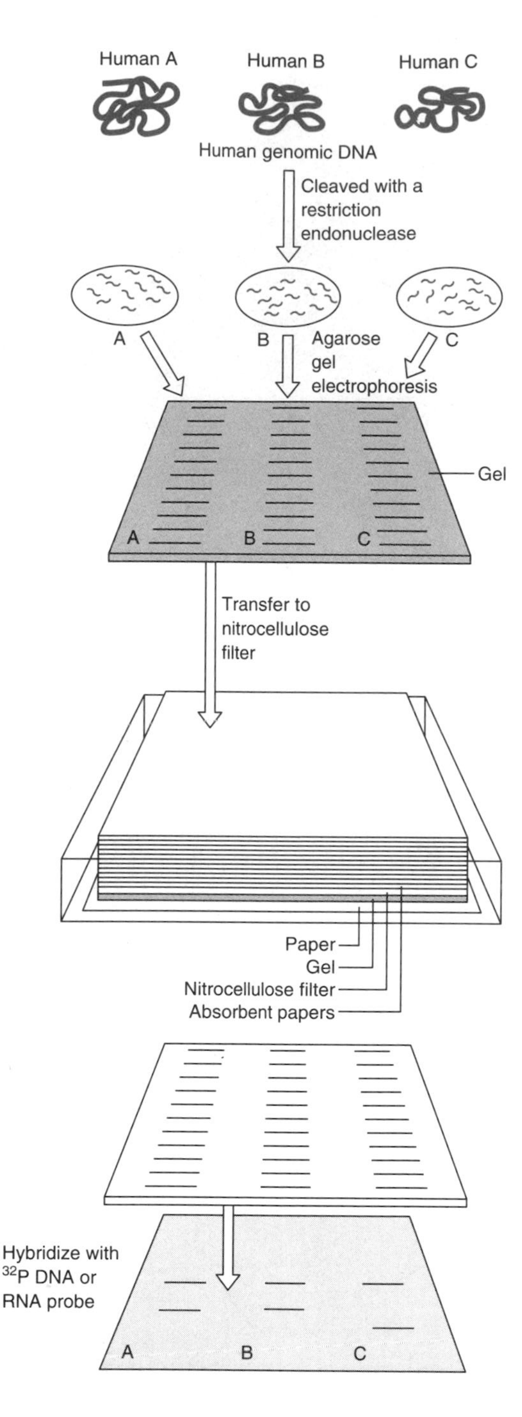

FIGURE 18.11
Southern blot to transfer DNA from agarose gels to nitrocellulose.
The transfer of DNA to nitrocellulose, as single-stranded molecules, allows for the detection of specific DNA sequences within a complex mixture of DNA. Hybridization with nick translated labelled probes can demonstrate if a DNA sequence of interest is present in the same or different regions of the genome.

tion with the target DNA and rehybridization with itself competes for hybridization with the DNA of interest. No similar competition occurs with the single-stranded RNA probes that hybridize with complementary DNA or RNA molecules. The RNA probe synthesis requires DNA as a template. In order to transcribe the template DNA it must be covalently linked to an upstream promoter that can be recognized by a DNA dependent RNA polymerase. Vectors have been constructed that are well suited for this technique.

A labeled DNA or RNA probe that is complementary to the DNA or RNA of interest can be hybridized to nitrocellulose bound nucleic acids and identified by the detection of the labeled probe. The nucleic acids of interest can be transferred to nitrocellulose from bacterial colonies grown on agar or from agarose gels where the nucleic acid species have been electrophoretically separated by size.

Southern Blot Technique Is Useful for Identifying DNA Fragments

A technique to transfer DNA species separated by agarose gel electrophoresis to a filter for further analysis was developed in the 1970s, and has become an indispensable tool in molecular biology. The method, developed by E. M. Southern, is referred to as the **Southern blot technique** (Figure 18.11). A DNA mixture of discrete restriction endonuclease generated fragments from any source and complexity can be separated according to size by electrophoresis through an agarose gel. The DNA is denatured by soaking the gel in alkali. The gel is then placed on absorbent paper and a nitrocellulose filter placed directly on top of the gel. Several layers of absorbent paper are placed on top of the nitrocellulose filter. The absorbent paper under the gel is kept wet with a concentrated salt solution that by capillary action is pulled up through the gel, the nitrocellulose, and into the absorbent paper layers above. The DNA is eluted from the gel by the upward movement of the high salt solution onto the nitrocellulose filter directly above, where it becomes bound. The position of the DNA bound to the nitrocellulose filter is exactly that which was present in the agarose gel. In its single-stranded membrane-bound form, the DNA can be analyzed with labeled probes that will hybridize to specific bands with complementary sequences.

The Southern blot technique has proven to be an invaluable tool in analytical procedures for the detection of the presence and the determination of the number of copies of particular sequences in complex genomic DNA, confirming DNA cloning results and the demonstration of polymorphic DNA arrangements of the human genome that correspond to pathological states. An example of the use of Southern blots is shown in Figure 18.11. Here whole human genomic DNA, isolated from three individuals, is digested with a restriction endonuclease generating thousands of fragments. These fragments are distributed throughout the agarose gel according to size in an electric field. The DNA is transferred (blotted) to a nitrocellulose filter and hybridized with a ^{32}P-labeled DNA or RNA probe that represents a portion of a gene of interest. The probe detects two bands in all three individuals indicating that the gene of interest is cleaved at one site within its sequence. Individuals A and B present a normal pattern while patient C has one normal band and one lower molecular weight band. This is an example of altered DNA within different individuals of a single species, **restriction fragment length polymorphism** (RFLP), and implies a deletion in a segment of the gene that may be associated with a pathological state. The gene from this patient can be cloned, sequenced, and fully analyzed to characterize the altered nature of the DNA (see Clin. Corr. 18.2).

CLIN. CORR. 18.2
RESTRICTION FRAGMENT LENGTH POLYMORPHISMS DETERMINE THE CLONAL ORIGIN OF TUMORS

It is generally assumed that most tumors are monoclonal in origin, that is, a rare event alters a single somatic cell genome in such a fashion that the cells grow abnormally into a tumor mass with all-daughter cells carrying the identically altered genome. Proof that a tumor is of monoclonal origin versus polyclonal in origin can help to distinguish hyperplasia (increased production and growth of *normal* cells) from neoplasia (growth of new or *tumor* cells). The detection of restriction fragment length polymorphisms (RFLPs) of Southern blotted DNA samples allows one to define the clonal origin of human tumors. If tumor cells were collectively derived from different parental cells they should contain a mixture of DNA markers characteristic of each cell of origin. However, an identical DNA marker in all tumor cells would indicate a monoclonal origin. The analysis is limited to females where one can take advantage of the fact that each cell carries only one active X chromosome of either paternal or maternal origin with the second X chromosome being inactivated. Activation occurs randomly during embryogenesis and is faithfully maintained in all-daughter cells with one-half the cells carrying an activated maternal X chromosome and the other one-half an activated paternal X chromosome.

Analysis of the clonal nature of a human tumor depends on the fact that activation of an X chromosome involves changes in the methylation of selected cytosine (C) residues within the DNA molecule. Several restriction endonucleases, such as Hha I, which cleaves DNA at GCGC sites, will not cleave DNA at their recognition sequences if a C is methylated within this site. Therefore, the methylated state (activated vs. inactivated) of the X chromosome can be probed with restriction endonucleases. Furthermore, the paternal X chromosome can be distinguished from the maternal X chromosome in a significant number of individuals based upon differences in the electrophoretic migration of restriction endonuclease generated fragments derived from selected regions of the chromosome. These DNA fragments are identified on a Southern blot by hybridization with a DNA probe that is complementary to this region of the X chromosome. An X-linked gene that is amenable to these studies is the hypoxanthine phosphoribosyltransferase (*HPRT*) gene. The *HPRT* gene consistently has two Bam HI restriction endonuclease sites (B_1 and B_3 in figure attached) within it, however, in many individuals a third site (B_2) is also present.

The presence of site B_2 in only one parental X chromosome *HPRT* allows for the detection of restriction enzyme-generated polymorphisms. Therefore, a female cell may carry one X chromosome with the *HPRT* gene possessing 2Bam HI sites (results in a single detectable DNA fragment of 24 kb) or 3Bam HI sites (results in a single detectable DNA fragment of 12 kb). This figure depicts the expected results for the analysis of tumor cell DNA to determine its monoclonal or polyclonal origin. As expected, three human tumors examined by this method were shown to be of monoclonal origin.

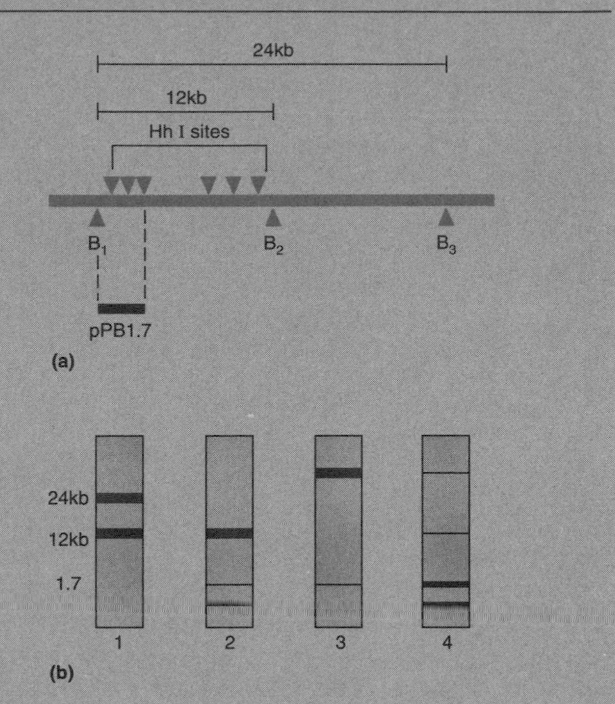

The analysis of genomic DNA to determine the clonal origin of tumors.
(*a*) The X chromosome-linked *HPRT* gene contains two invariant Bam HI restriction endonuclease sites (B_1 and B_3) while in some individuals a third site, B_2 is also present. The *HPRT* gene also contains several Hha I sites, however, all of these sites, except H_1, are usually methylated in the active X chromosome. Therefore, only the H_1 site would be available for cleavage by Hha I in the active X chromosome. A cloned, labelled probe, pPB1.7, can be employed to determine which form of the *HPRT* gene is present in a tumor and if it is present on an active X chromosome. (*b*) Restriction endonuclease patterns predicted for monoclonal versus polyclonal tumors are: (1) Cleaved with Bam HI alone. 24 Kb fragment derived from HPRT gene containing only B_1 and B_3 sites and 12 Kb fragment derived from *HPRT* gene containing extra B_2 site. Pattern characteristic for heterozygous individual; (2) Cleaved with Bam HI plus Hha I. Monoclonal tumor with the 12 Kb derived from an active X chromosome (methylated); (3) Cleaved with Bam HI plus Hha I. Monoclonal tumor with the 24 Kb derived from an active X chromosome (methylated); (4) Cleaved with Bam HI plus Hha I. Polyclonal tumor. All tumors studied displayed patterns characterized by Lane 2 or Lane 3.

Vogelstein, B., Fearon, E. R., Hamilton, S. R., and Feinberg, A. B. Use of restriction fragment length polymorphism to determine the clonal origin of tumors. *Science* 227:642, 1985.

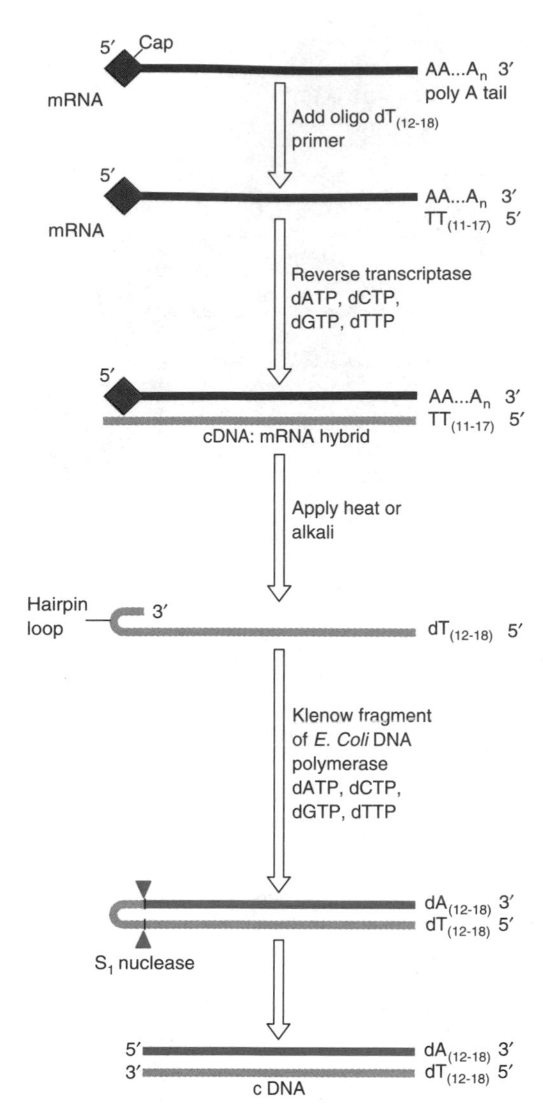

FIGURE 18.12
Synthesis of cDNA from mRNA.
The 3′ poly(A) tail of mRNA is hybridized with an oligomer of dT (oligo(dT)$_{12-18}$) that serves as a primer for reverse transcriptase which catalyzes the synthesis of the complementary DNA (cDNA) strand in the presence of all four deoxynucleotide triphosphates (dNTPs). The resulting cDNA: mRNA hybrid is separated into single-stranded cDNA by melting with heat or hydrolyzing the mRNA with alkali. The 3′ end of the cDNA molecule forms a hairpin loop that serves as a primer for the synthesis of the second DNA strand catalyzed by the Klenow fragment of *E. coli* DNA polymerase. The single-stranded unpaired DNA loop is hydrolyzed by S$_1$ nuclease to yield a double-stranded DNA molecule.

Other techniques that employ the principles of Southern blot are the transfer of RNA (northern blots) and of proteins (western blots) to nitrocellulose filters or nylon membranes.

18.7 COMPLEMENTARY DNA AND COMPLEMENTARY DNA LIBRARIES

The insertion of specific functional eucaryotic genes into vectors that can be expressed in a procaryotic cell is a highly sought goal. Such a system could produce large amounts of "genetically engineered" proteins with significant medical, agricultural, and experimental potential. Several potential replacement hormones and enzymes are currently produced by these methods, including insulin, erythropoietin, interleukins, interferons, and tissue plasminogen activator. Unfortunately it is impossible, except in rare instances, to clone functional genes from genomic DNA. One reason for this is that most genes within the mammalian genome yield transcripts that include introns that must be spliced out of the primary mRNA transcript. The procaryotic system cannot splice out the introns to yield functional mRNA transcripts. This problem can be circumvented by synthesizing **complementary DNA (cDNA)** from functional eucaryotic mRNA.

Messenger RNA Is Used as a Template for DNA Synthesis Using Reverse Transcriptase

Messenger RNA can be reverse transcribed to cDNA and the cDNA inserted into a vector for amplification, identification, and expression. Mammalian cells normally contain 10,000–30,000 different species of mRNA molecules at any time during the cell cycle. In some cases, however, a specific mRNA species may approach 90% of the total mRNA, such as mRNA for globin in reticulocytes. Many mRNAs are normally present at only a few (1–14) copies per cell. A cDNA library can be constructed from the total cellular mRNA but if only a few copies per cell of mRNA of interest are present, the cDNA may be very difficult to identify. Methods have been developed to enrich the population of mRNAs or their corresponding cDNAs. This permits the size or the number of different cDNA species within the cDNA library that must be screened to be reduced and greatly enhances the probability of identifying the clone of interest.

A Desired mRNA in a Sample Can Be Enriched by Separation Techniques

Messenger RNA can be separated into molecular weight ranges by gel electrophoresis or centrifugation. Utilization of mRNA in a specific molecular weight range will enrich an mRNA of interest severalfold. Knowledge of the molecular weight of the protein encoded by the gene of interest gives a clue to the approximate size of the mRNA transcript or its cDNA; variability in the predicted size, however, will arise from differences in the length of the untranslated regions of the mRNAs.

The enrichment of a specific mRNA molecule can also be accomplished by immunological procedures, but requires the availability of antibodies against the protein encoded by the gene of interest. Antibodies added to an in vitro protein synthesis mixture will react with the growing polypeptide chain associated with the polysome and precipitate it. The mRNA can be purified from the immunoprecipitated polysomal fraction.

Complementary DNA Synthesis

The isolated mRNA mixture is used as a template to synthesize a complementary strand of DNA using RNA-dependent DNA polymerase, reverse transcriptase (Figure 18.12). A primer is required for the reaction; advantage is taken of the poly(A) tail at the 3′ terminus of eucaryotic mRNA. An oligo(dT) with 12–18 bases is employed as the primer that will hybridize with the poly(A) sequence. After cDNA synthesis, the hybrid is denatured or the mRNA hydrolyzed in alkali in order to obtain the single-stranded cDNA. The 3′ termini of single-stranded cDNAs form a hairpin loop that serves as a primer for the synthesis of the second strand of the cDNA. Either the Klenow fragment or a reverse transcriptase can be used for this step. The resulting double-stranded cDNA contains a single-stranded loop that is selectively recognized and digested by S_1 nuclease. The ends of the cDNA must be modified prior to cloning in a vector. One method involves incubating blunt-ended cDNA molecules with linker molecules and an enzyme, bacteriophage T_4 DNA ligase, that catalyzes the ligation of blunt-ended molecules (Figure 18.13). The synthetic linker molecules contain restriction endonuclease sites that can now be hydrolyzed with the appropriate enzyme for insertion of the cDNA into a compatibly cut vector.

Bacteriophage DNA (see p. 786) has proven to be the most convenient and efficient vector to create cDNA libraries because they can be readily amplified and stored indefinitely. Two bacteriophage vectors, **λgt10** and **λgt11,** and their newer constructs, have been employed to produce cDNA libraries. The cDNA libraries in λgt10 can be screened only with labeled nucleic acid probes, whereas, those in λgt11, an expression vector, can also be screened with antibody for the production of the protein or antigen of interest.

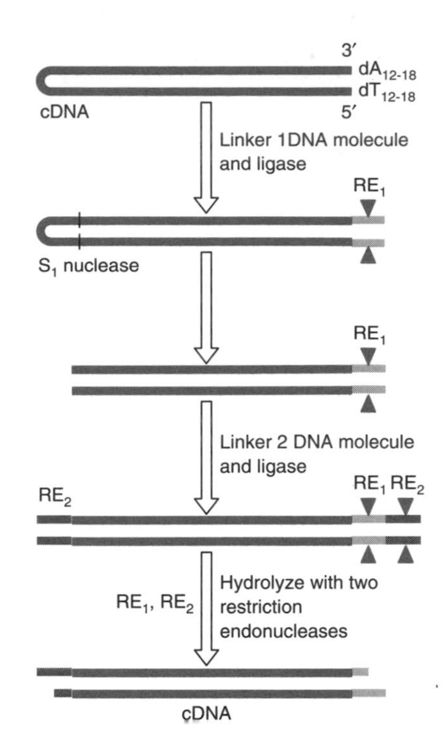

FIGURE 18.13
Modification of cDNA for cloning.
The procedure begins with double-stranded DNA containing a hairpin loop. A linker DNA containing a restriction endonuclease site (RE_1) is added to the free end of the cDNA by blunt end ligation. The single-stranded hairpin loop is next hydrolyzed with S_1 nuclease. A second linker with a different restriction endonuclease site within (RE_2) is blunt end ligated to the newly created free cDNA. The second linker will probably bind to both ends but will not interfere with the first restriction endonuclease site. The modified DNA is hydrolyzed with the two restriction endonucleases and can be inserted into a plasmid or bacteriophage DNA by directional cloning.

18.8 BACTERIOPHAGE, COSMID, AND YEAST CLONING VECTORS

The detection of noncoding sequences in most eucaryotic genes and distant regulatory regions flanking the genes necessitated new cloning strategies to package larger DNA fragments than could be cloned in plasmids. Plasmids can accommodate foreign DNA inserts with a maximum length in the range of 5–10 kb (kilobases). Portions of recombinant DNA fragments larger than this are randomly deleted during replication of the plasmid within the bacterium. Thus, alternate vectors have been developed.

Bacteriophage as Cloning Vectors

Bacteriophage λ (λ phage), viruses that infect and replicate in bacteria, are ideal vectors for DNA inserts of approximately 15-kb lengths. The λ phage selectively infects bacteria and can replicate by either a lytic or nonlytic (lysogenic) pathway. The λ phage contains a self-complementary 12 base single-stranded tail (cohesive termini) at both ends of its 50 kb double-stranded DNA molecule. Upon infection of the bacteria the cohesive termini (cos sites) of a single λ phage DNA molecule self-anneals and the ends are covalently linked with the host cell DNA ligase. The circular DNA molecule serves as a template for transcription and replication. Lambda phage, with restriction endonuclease generated fragments representing a cell's whole genomic DNA inserted into it, are used to infect bacteria. Recombinant bacteriophage, released from the lysed cells, are collected and constitute a genomic library in bacteriophage λ. The phage library can be screened more rapidly than a plasmid library due to the increased size of the DNA inserts.

Numerous bacteriophage λ vectors have been constructed for different cloning strategies. For the sake of simplicity only a generic λ phage vector will be described here. Cloning large fragments of DNA in bacteriophage λ takes advantage of the fact that a 15–25-kb segment of the phage DNA can be replaced without impairing its replication in *E. coli* (Figure 18.14). Packaging of phage DNA into the virus particle is constrained by its total length, which must be approximately 50 kb. The linear phage λ DNA can be digested with specific restriction endonucleases that generate small terminal fragments with their **cos sites** (arms), that are separated from the larger intervening fragments. Cellular genomic DNA is partially digested with the appropriate restriction enzymes to permit annealing and ligation with the phage arms. Genomic DNA is not enzymatically hydrolyzed completely in order to randomly generate fragments that can be properly packaged into phage particles. The DNA fragments that are smaller or larger than 15–25 kb can hybridize with the cos arms but are excluded from being packaged into infectious bacteriophage particles. All of the information required for phage infection and replication in bacteria is carried within the cos arms. The recombinant phage DNA is mixed with λ phage proteins in vitro, which assemble into infectious virions. The infectious recombinant bacteriophage λ are then propagated in an appropriate *E. coli* strain to yield a λ phage library. Many different *E. coli* strains have been genetically altered to sustain replication of specific recombinant virions.

Screening Bacteriophage Libraries

The bacteriophage library can be screened by plating the virus on a continuous layer of bacteria (a bacterial lawn) grown on agar plates (Figure 18.15). Individual phage will infect, replicate, and lyse one cell. The progeny virions will then infect and subsequently lyse bacteria immediately adjacent to the site of the first infected cell creating a clear region or plaque in the opaque bacterial field. Phage, within each plaque, can be picked up on a nitrocellulose filter (as for replica plating) and the DNA fixed to the filter with NaOH. The location of cloned DNA fragments of interest is determined by hybridizing the filter-bound DNA with a labeled DNA or RNA probe and identifying its position by autoradiography. The bacteriophage in the plaque corresponding to the labeled filter-bound hybrid are picked up and amplified in bacteria for further analysis.

Complementary DNA libraries in bacteriophage are also constructed with the phage *cos* arms. If the cDNA is recombined with phage DNA that permits expression of the gene, such as λgt 11, then plaques can be screened immunologically with antibodies specific for the antigen of interest.

Cloning DNA Fragments Into Cosmid and Yeast Artificial Chromosome Vectors

Even though bacteriophage λ are the most commonly used vectors to construct genomic DNA libraries today, the length of many genes exceed the maximum size of the DNA that can be inserted between the phage arms. A **cosmid vector** can accommodate foreign DNA inserts of approximately 45 kb. **Yeast artificial chromosomes (YAC)** have recently been developed to clone DNA fragments of 200–500-kb lengths. While cosmid and yeast artificial chromosome vectors are difficult to work with, their libraries permit the cloning of large genes with their flanking regulatory sequences, as well as families of genes or contiguous genes.

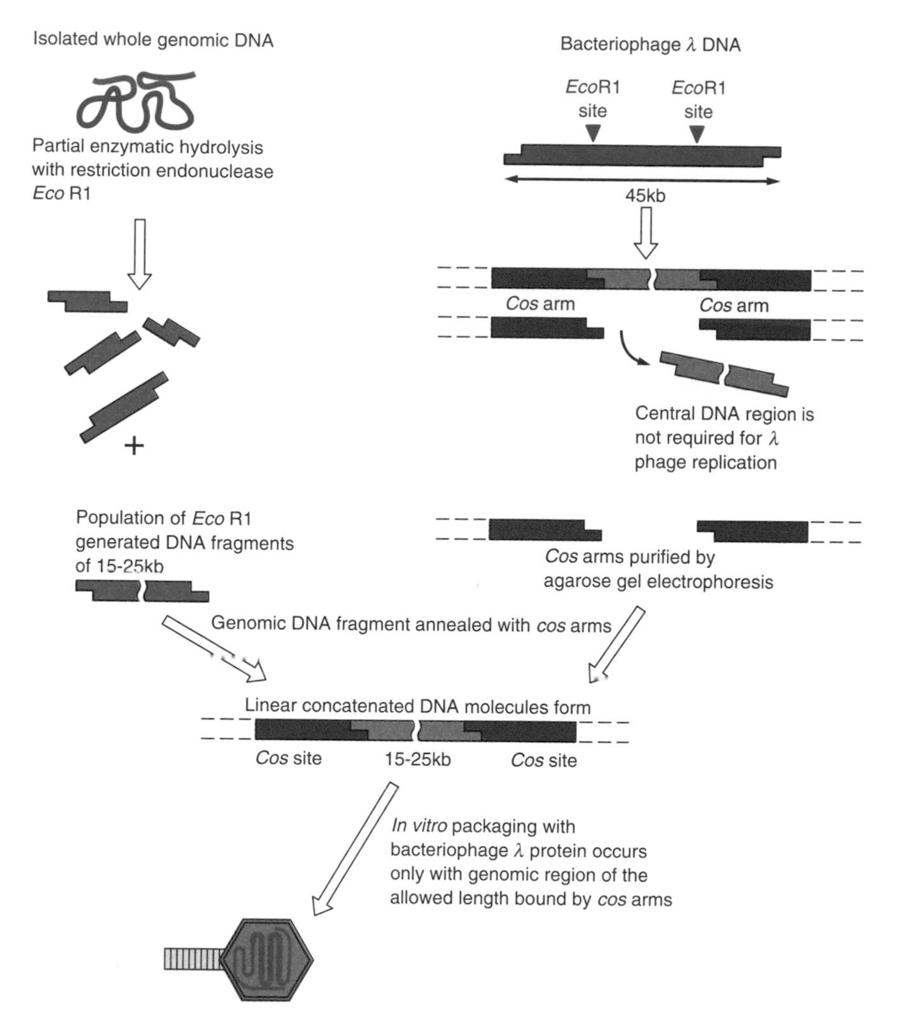

FIGURE 18.14
Cloning genomic DNA in bacteriophage λ.
Whole genomic DNA is incompletely digested with a restriction endonuclease (e.g. *EcoR1*). This results in DNA of random size fragments with single-stranded sticky ends. DNA fragments, Cos arms, are generated with the same restriction endonuclease from bacteriophage λDNA. The purified Cos arm fragments carry sequence signals required for packaging DNA into a bacteriophage virion. The genomic fragments are mixed with the Cos arms, annealed and ligated, forming linear concatenated DNA arrays. The in vitro packaging with bacteriophage λ proteins occurs only with genomic DNA fragments of allowed lengths (15–25 Kb) bounded by Cos arms.

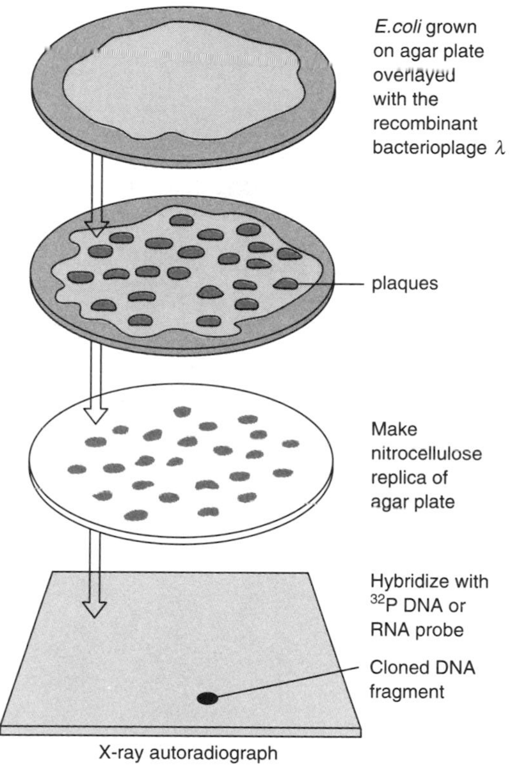

FIGURE 18.15
Screening genomic libraries in bacteriophage λ.
An agar plate with competent *E. coli* are grown to confluence and then overlayed with the recombinant bacteriophage. Plaques develop where bacteria are infected and subsequently lysed by the phage λ. Replicas of the plate can be made by touching the plate with nitrocellulose. The DNA is denatured and fixed to the nitrocellulose with NaOH. The fixed DNA is hybridized with a ^{32}P-labelled probe and exposed to X-ray film. The autoradiograph identifies the plaque(s) with recombinant DNA of interest.

Cosmid vectors are a cross between plasmid and bacteriophage vectors. Cosmids contain an antibiotic resistance gene for selection of recombinant DNA molecules, an origin of replication for propagation in bacteria, and a cos site for packaging of recombinant molecules in bacteriophage particles. The bacteriophages with recombinant cosmid DNA can infect *E. coli* and inject its DNA into the cell. Cosmid vectors contain only approximately 5 kb of the 50-kb bacteriophage DNA and, therefore, cannot direct its replication and assembly of new infectious phage particles. Instead, the recombinant cosmid DNA circularizes and replicates as a large plasmid. Bacterial colonies with recombinants of interest can be selected and amplified by methods similar to those described for plasmids.

Standard cloning procedures along with some novel methods are employed to construct YAC. Very large foreign DNA fragments are joined to yeast DNA sequences, one that functions as telomeres (distal extremity of a chromosome arm), and another that functions as a centromere and as an origin of replication. The recombinant YAC DNA is introduced into the yeast by transformation. The YAC constructs have been designed in such a way that yeast transformed with recombinant chromosomes grow as visually distinguishable colonies that facilitates the selection and analysis of cloned DNA fragments.

18.9 TECHNIQUES TO FURTHER ANALYZE LONG STRETCHES OF DNA

Subcloning Permits Definition of Large Segments of DNA

The complete analysis of the functional elements in a cloned DNA fragment requires sequencing the entire molecule. Current techniques can sequence 200–400 bases in a DNA fragment, yet cloned DNA inserts are frequently much larger. Restriction maps of the initial DNA clone are essential for cleaving the DNA into smaller pieces to be recloned, or **subcloned** for further analysis. The sequences of each of the small subcloned DNA fragments can be determined. Overlapping regions of the subcloned DNA properly align and confirm the entire sequence of the original DNA clone.

Chromosome Walking Is a Technique to Define Gene Arrangement in Long Stretches of DNA

Knowledge of how genes and their regulatory elements are arranged in a chromosome should lead to an understanding of how sets of genes may be coordinately regulated. Currently, it is difficult if not impossible to clone DNA fragments large enough to identify contiguous genes. The combination of several techniques allows for the analysis of very long stretches of DNA (50–100 kb). The method, **chromosome walking,** is possible because bacteriophage λ or cosmid libraries contain partially cleaved genomic DNA cut at specific restriction endonuclease sites. The cloned fragments will contain overlapping sequences with other cloned fragments. Overlapping regions can be identified by restriction mapping, subcloning, screening λ phage or cosmid libraries, and sequencing procedures.

The overall procedure of chromosome walking is shown in Figure 18.16. Initially the λ phage library is screened for a sequence of interest with a DNA or RNA probe. The cloned DNA is restriction mapped and a small segment is subcloned in a plasmid, amplified, purified, and labeled by nick translation. This labeled probe is then used to rescreen the λ phage library for complementary sequences, which are then cloned. The newly identified overlapping cloned DNA is then treated in the same fashion as the initial DNA clone to search for other overlapping sequences. Caution must be taken that the subcloned DNA does not contain a sequence common to the large numbers of repeating DNA sequences in higher eucaryotic genomes. If a subcloned DNA probe contains a repeat sequence it would hybridize to numerous bacteriophage plaques preventing the identification of a specific overlapping clone.

18.10 EXPRESSION VECTORS AND FUSION PROTEINS

The recombinant DNA methodology described to this point has dealt primarily with screening, amplifying, and purifying cloned DNA species.

An important goal of recombinant DNA studies, as stated earlier, is to have a foreign gene expressed in bacteria with the product in a biologically active form. Sequencing the DNA of many bacterial genes and their flanking regions have identified the spatial arrangement of regulatory sequences required for expression of genes. A promoter and other regulatory elements upstream of the gene are required to transcribe a gene (Chapter 19, Section 3). An mRNA transcript of a recombinant eucaryotic gene, however, is not translated in a bacterial system because it lacks the bacterial recognition sequence, the Shine–Dalgarno sequence, required to properly orient it with a functional bacterial ribosome. Vectors that facilitate the functional transcription of DNA inserts, termed **expression vectors,** have been constructed such that a foreign gene can be inserted into the vector downstream of a regulated promoter but within a bacterial

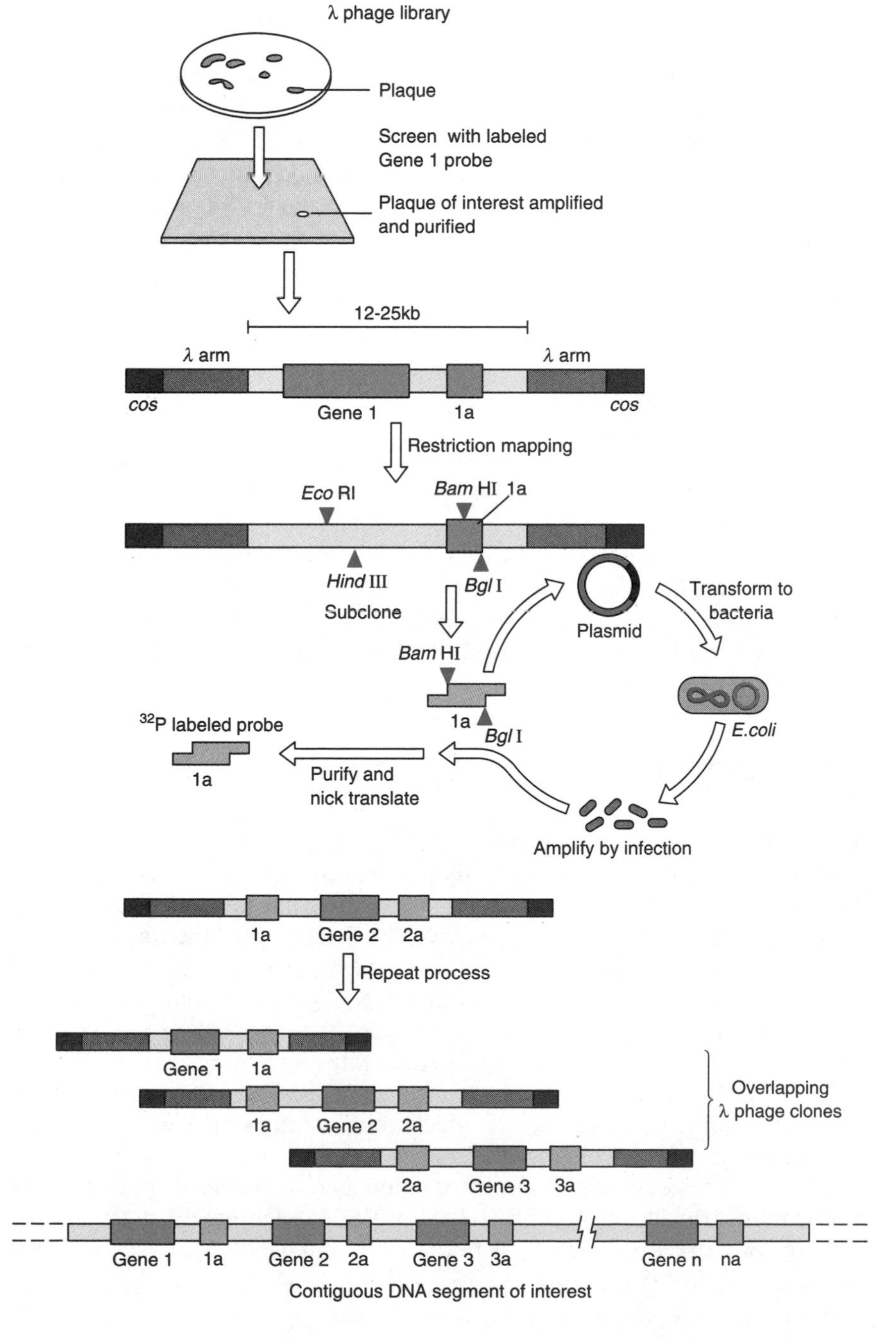

FIGURE 18.16
Chromosome walking to analyze contiguous DNA segments in a genome.
Initially, a DNA fragment is labeled by nick translation to screen a library for recombinant λ phage carrying a gene of interest. The amplified DNA is mapped with a battery of restriction endonucleases to select a new region (1a) within the original cloned DNA that can be recloned (subcloned). The subcloned DNA(1a) is used to identify other DNA fragments within the original library that would overlap the initially amplified DNA region. The process can be repeated many times to identify contiguous DNA regions upstream and downstream of the initial DNA (gene 1) of interest.

gene, commonly the *lacZ* gene. The mRNA transcript of the recombinant DNA contains the *lacZ* Shine–Dalgarno sequence, codons for a portion of the 3′ end of the *lacZ* gene protein, followed by the codons of the complete foreign gene of interest. The protein product is a **fusion protein** containing a few N-terminal amino acids of the *lacZ* gene protein and the complete amino acid sequence of the foreign gene product.

Foreign Genes Can Be Expressed in Bacteria Allowing Synthesis of Their Encoded Proteins

Many plasmid and bacteriophage vectors have been constructed to permit the expression of eucaryotic genes in bacterial cells. Rapidly replicating bacteria can serve as a biological factory to produce large amounts of specific proteins which have research, clinical, and commercial value. As an example, human protein hormones can be produced by recombinant technologies which serve as replacement or supplemental hormones in patients with aberrant or missing hormone production. Figure 18.17 depicts a generalized plasmid vector for the expression of a mammalian gene. Recall that the inserted foreign gene must be in the form of cDNA from its corresponding mRNA since the bacterial system cannot remove the introns of the primary mRNA transcript. The DNA must be inserted in register with the codons of the 3′-terminal codons of the bacterial protein when creating a fusion protein. That is, insertion must occur after a triplet codon of the bacterial protein and at the beginning of a triplet codon of the eucaryotic gene protein to ensure proper translation. Finally, the foreign gene must be inserted in the proper orientation relative to the promoter to yield a functional transcript. This can be achieved by directional cloning.

Eucaryotic proteins synthesized within bacteria are often unstable and are degraded by intracellular proteases. Fusion protein products, however, are usually stable. The fusion protein amino acids encoded by the procaryotic genome may be cleaved from the purified protein of interest by enzymatic or chemical procedures. An alternative cloning strategy to circumvent the intracellular instability of some proteins is to produce a foreign protein that is secreted. This requires cloning the foreign gene in a vector such that the fusion protein synthesized contains a signal peptide recognized by the bacterial signal peptidase that properly processes the protein for secretion.

18.11 EXPRESSION VECTORS IN EUCARYOTIC CELLS

Mammalian genetic diseases result from missing or defective intracellular proteins. To utilize recombinant techniques to treat these diseases, vectors have to be constructed that can be incorporated into mammalian cells. In addition, these vectors have to be selective for the tissue or cells containing the aberrant protein. Numerous vectors have been developed that permit the expression of foreign DNA genes in mammalian cells grown in tissue culture. These vectors have been used extensively for elucidation of the posttranslational processing and synthesis of proteins in cultured eucaryotic cells. Unfortunately, the goal to selectively express genes in specific tissues or at specific developmental stages within a whole animal has met with very limited success.

Several types of expression vectors have been developed that allow the replication, transcription, and translation of foreign genes in eucaryotic cells grown in vitro, including both RNA and DNA viral vectors that

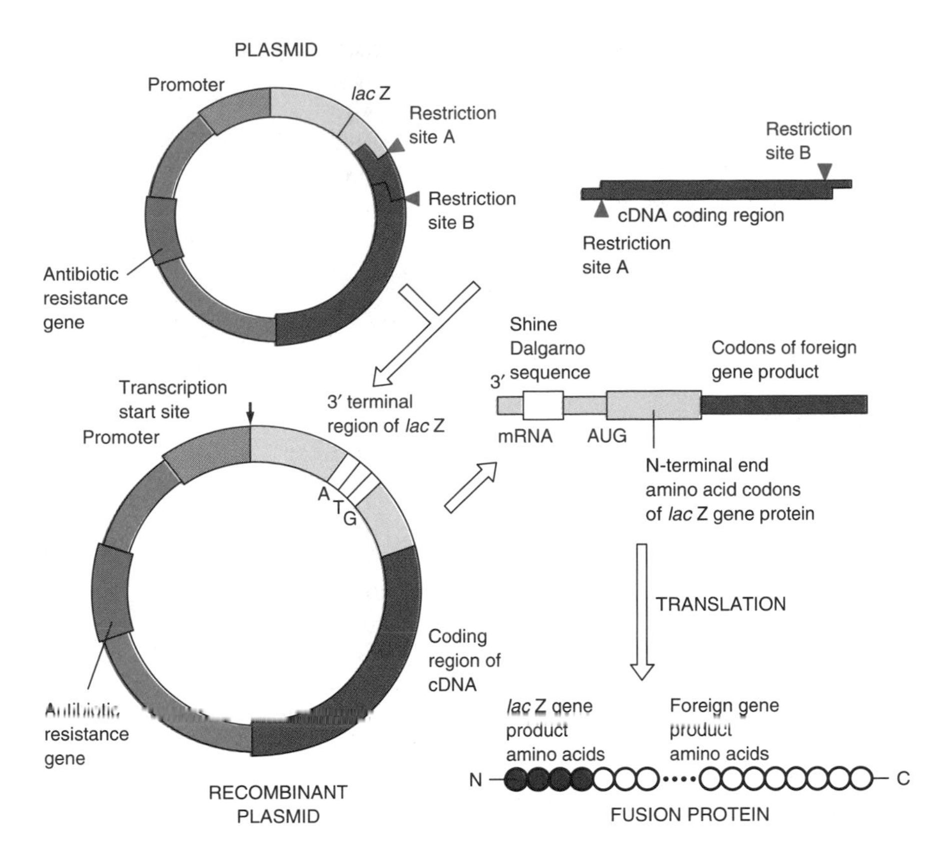

FIGURE 18.17
Construction of a bacterial expression vector.
A cDNA coding region of a protein of interest is inserted downstream of bacterial regulatory sequences (promoter (P)) for the lacZ gene, the coding sequence for the mRNA Shine-Dalgarno sequence, the AUG codon and a few codons for the N-terminal amino acids of the LacZ gene protein. The mRNA produced from this expression vector will, therefore, direct the synthesis of a foreign protein in the bacterium with a few of its N-terminal amino acids of bacterial protein origin (a fusion protein).

contain a foreign DNA insert. These viral vectors are able to infect and then replicate in a host cell. Experimentally constructed vectors that contain essential DNA elements, usually derived from a viral genome, permits expression of foreign gene inserts. **Shuttle vectors** are vectors that contain both bacterial and eucaryotic replication signals thus permitting replication of the vector in both bacteria and mammalian cells. The primary advantage of a shuttle vector is that it allows a gene to be cloned and purified in large quantities from a bacterial system and then the same recombinant vector can be expressed in a mammalian cell. Some expression vectors become integrated into the host cell genome while others remain as extrachromosomal entities (episomal) with stable expression of their recombinant gene in the daughter cells. Other expression vectors remain as episomal DNA permitting only transient expression of their foreign gene prior to cell death.

Foreign DNA, such as viral expression vectors, must be introduced into the cultured eucaryotic cells by **transfection,** a process that is analogous to transformation of DNA into bacterial cells. Several different methods are employed to transfect DNA into eucaryotic cells. The most commonly employed transfection methods involve the formation of a complex of DNA with calcium phosphate or diethylaminoethyl (DEAE)-dextran, which is then taken up by the cell by endocytosis. The DNA is subsequently transferred from the cytoplasm to the nucleus, where it is replicated and expressed. The details of the mechanism of transfection are not known. Both methods are employed to establish transiently expressed vectors while the calcium phosphate procedure is also used for permanently expressed foreign genes. Typically, 10–20% of the cells in culture can be transfected by these procedures.

DNA Elements Required for Expression of Vectors in Mammalian Cells

Expression of recombinant genes in mammalian cells requires the presence of DNA controlling elements within the vector that are not necessary in the bacterial system. To be expressed in a eucaryotic cell the cloned gene is inserted in the vector in the proper orientation relative to control elements, including a promoter, polyadenylation signals and an enhancer sequence. Expression may be improved by the inclusion of an intron. Some, or all of these DNA elements may be present in the recombinant gene if whole genomic DNA is used for cloning. A particular cloned fragment generated by restriction endonuclease cleavage, however, may not contain the required controlling elements. A cDNA would not possess these required DNA elements. It is necessary, therefore, that the expression vector to be used in mammalian cells be constructed such that it contains all of the required controlling elements.

An expression vector can be constructed by inserting the required DNA controlling elements into the vector by recombinant technologies. Enhancer and promoter elements, engineered into an expression vector, should be recognized by a broad spectrum of cells in culture for the greatest applicability of the vector. Controlling elements derived from viruses with a broad host range are used for this purpose and are usually derived from the **papovavirus, simian virus 40(SV40), Rous sarcoma virus,** or the **human cytomegalovirus.**

The vector must also replicate to increase the number of copies within each cell or to maintain copies in daughter cells. The vector, therefore, is constructed to contain DNA sequences that promote its replication in the eucaryotic cell. This DNA region is usually derived from a virus and is referred to as the origin of replication (ori). Specific protein factors, encoded by genes engineered into the vector or previously introduced into the host genome, recognize and interact with the ori sequences to initiate DNA replication.

Transfected Eucaryotic Cells Can Be Selected by Utilizing Mutant Cells Which Require Specific Nutrients

It is important to have a means of selectively growing the transfected cells since they often represent only 10–20% of the cell population. As was the case for the bacterial plasmid, a gene can be incorporated into the vector that encodes an enzyme that confers resistance to a drug or confers selective growth capability to the cells carrying the vector. Constructing vectors that express both a selectable marker and a foreign gene is difficult. **Cotransfection** circumvents this problem. Two different vectors are efficiently taken up by those cells capable of being transfected. In most cases greater than 90% of transfected cells carry both vectors, one with the selectable marker and the second carrying the gene of interest.

Two of the more commonly employed selectable markers are the thymidine kinase (*tk*) and the H_2folate reductase gene. The *tk* gene product, thymidine kinase, is expressed in most mammalian cells and participates in the salvage pathway for thymidine. Several mutant cell lines have been isolated that lack a functional thymidine kinase gene (tk^-) and in growth medium containing hypoxanthine, aminopterin, and thymidine these cells will not survive. Only those tk^- mutant cells cotransfected with a vector carrying a *tk* gene, usually of *Herpes simplex* virus origin, will grow in the medium. In most instances, these cells have been cotransfected with the gene of interest.

The dihydrofolate reductase gene (*dhfr*) is required to maintain cellular concentrations of H_4folate for nucleotide biosynthesis (see Chapter 13). Cells lacking this enzyme will only survive in media containing thymidine, glycine, and purines. Mutant cells (*dhfr*$^-$), which are transfected with the *dhfr* gene, therefore, can be selectively grown in a medium lacking these supplements. Expressing foreign genes in mutant cells, cotransfected with selectable markers, is limited to cell types that can be isolated with the required gene defect. Normal cells, however, transfected with a vector carrying the *dhfr* gene, are also resistant to methotrexate, an inhibitor of dihydrofolate reductase and these cells can be selected for by growth in methotrexate.

Another approach for selecting nonmutated cells involves the use of a bacterial gene coding for aminoglycoside 3′-phosphotransferase (APH) for cotransfection. Cells expressing APH are resistant to aminoglycoside antibiotics such as neomycin and kanamycin, which inhibits protein synthesis in both procaryotes and eucaryotes. Vectors carrying an *APH* gene, therefore, can be used as a selectable marker in both bacterial and mammalian cells.

Foreign Genes Can Be Expressed in Eucaryotic Cells by Utilizing Virus Transformed Cells

Figure 18.18 depicts the transient expression of a transfected gene in COS cells, one of the more commonly used systems to express foreign eucaryotic genes. The COS cells are permanently cultured simian cells, transformed with an origin-defective SV40 genome. The defective viral genome has integrated into the host cell genome and constantly expresses viral proteins. Infectious viruses which are normally lytic to infected cells are not produced because the viral origin of replication is defective. The SV40 proteins expressed by the transformed COS cell will recognize and interact with a normal SV40 ori carried in a vector transfected into these cells. These SV40 proteins will, therefore, promote the repeated replication of the vector. The transfected vector containing both an SV40 ori and a gene of interest may reach a copy number in excess of 10^5 molecules/cell. Transfected COS cells die after 3–4 days possibly due to a toxic overload of the episomal vector DNA.

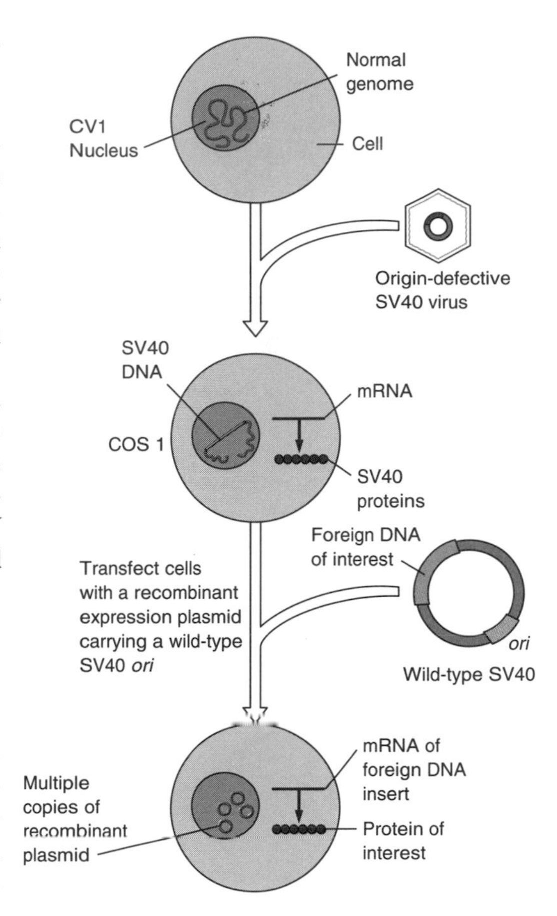

FIGURE 18.18
Expression of foreign genes in the eucaryotic COS cell.
CV1, an established tissue culture cell line of simian origin, can be infected and will support the lytic replication of the simian DNA virus, SV40. Cells are infected with an origin (*ori*)-defective mutant of SV40 whose DNA permanently integrate into the host CV-1 cell genome. The defective viral DNA continuously codes for proteins that can associate with a normal SV40 *ori* to regulate replication. Due to its defective ori, the integrated viral DNA will not produce viruses. The SV40 proteins synthesized in the permanently altered CV-1 cell line, COS-1, can, however, induce the replication of recombinant plasmids carrying a wild-type SV40 *ori* to a high copy number (as high as 10^5 molecules per cell). The foreign protein synthesized in the transfected cells may be detected immunologically or enzymatically.

18.12 SITE-DIRECTED MUTAGENESIS

By altering or mutating selected regions or single nucleotides within cloned DNA and using the technologies described above it is possible to define the role of DNA sequences in gene regulation and amino acid sequences in protein function. **Site-directed mutagenesis** is the controlled alteration of selected regions of a DNA molecule. It may involve the insertion or deletion of selected DNA sequences or the replacement of a specific nucleotide with a different base. A variety of chemical methods mutate DNA in vitro and in vivo, but these usually occur at random sites within the DNA molecule.

The Role of Flanking Regions in DNA Can Be Evaluated by Deletion and Insertion Mutations

Site-directed mutagenesis can be carried out in various regions of a DNA sequence including the gene itself or the flanking regions. Figure 18.19 depicts a simple deletion mutation strategy where the sequence of interest

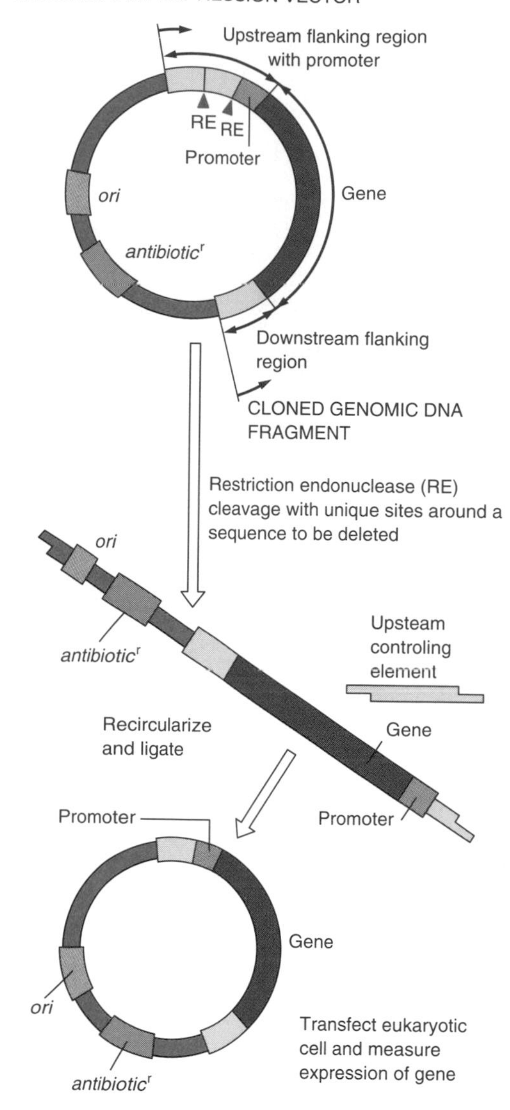

FIGURE 18.19
Use of expression vectors to study DNA regulatory sequences.
The gene of interest along with upstream and/or downstream DNA flanking regions are inserted and cloned in an expression vector and the base-line expression of the gene in an appropriate cell determined. Defined regions of potential regulatory sequences can be removed by restriction endonuclease cleavage and the truncated recombinant DNA vector recircularized, ligated and transfected into an appropriate host cell. The level of gene expression in the absence of the potential regulator is determined and compared to controls to ascertain the regulatory role of the deleted flanking DNA sequence.

is selectively cleaved with restriction endonucleases, specific sequences removed, and the altered recombinant vector recircularized with DNA ligase. The role of the deleted sequence can be determined by comparing the level of expression (translation) of the gene product, measured immunologically or enzymatically, to the unaltered recombinant expression vector. A similar technique is used to insert new sequences at the site of cleavage. The deletion of a DNA sequence within the flanking region of a cloned gene can help to define its regulatory role in gene expression. The presence or absence of a regulatory sequence may not be sufficient to evaluate its role in controlling expression. The spatial arrangement of regulatory elements to one another, to the gene, and to its promoter may be important in the regulation of gene expression (see Chapter 19). Therefore, several different strategies have been developed to define the regulatory elements flanking a variety of genes.

The analysis of potential regulatory sequences is conveniently conducted by inserting the sequence of interest upstream of a reporter gene in an expression vector. A reporter gene, usually of procaryotic origin, encodes for a gene product that can be readily distinguished from proteins normally present in the nontransfected cell and for which there is a convenient and rapid assay. The most widely used reporter gene is the chloramphenicol acetyltransferase (*CAT*) gene of bacteria. The gene product catalyzes the acetylation and inactivation of chloramphenicol, a protein synthesis inhibitor of procaryotic cells. The ability of a regulatory element to enhance or suppress expression of the *CAT* gene can be determined by assaying the level of acetylation of chloramphenicol in extracts prepared from transfected cells. The regulatory element can be mutated prior to insertion into the vector carrying the reporter gene to determine its spatial and sequence requirements as a regulator of gene expression.

A common difficulty encountered in the analysis of regulatory elements is the lack of restriction endonuclease sites at useful positions within the cloned DNA. **Deletion mutations** can be made, in the absence of appropriately positioned restriction endonuclease sites, by linearizing cloned DNA with a restriction endonuclease downstream of the potential regulatory sequence of interest. The DNA can then be systematically truncated with an exonuclease which hydrolyzes nucleotides from the free end of both strands of the linearized DNA. Increasing times of digestion generates smaller DNA fragments. Figure 18.20 demonstrates how larger deletion mutations (yielding smaller fragments) can be tested for functional activity. The enzymatic hydrolysis of the double strand of DNA occurs at both ends of the linearized recombinant vector destroying the original restriction endonuclease site (RE_2). A unique restriction endonuclease site is reestablished to recircularize the truncated DNA molecule for further manipulations to evaluate the function of the deleted sequence. This is accomplished by ligating the blunt ends with a linker DNA, a synthetic oligonucleotide containing one or more restriction endonuclease sites. The ligated linkers are cut with the appropriate enzyme permitting recircularization and ligation of the DNA.

Site-Directed Mutagenesis of a Single Nucleotide

The previously discussed procedures can elucidate the functional role of small to large DNA sequences. Frequently, however, one wants to evaluate the role of a single nucleotide at selected sites within the DNA molecule. A single base change permits evaluation of the role of specific amino acids in a protein (see Clin. Corr. 18.3). This method also allows one to create or destroy a restriction endonuclease site at specific locations within a DNA sequence.

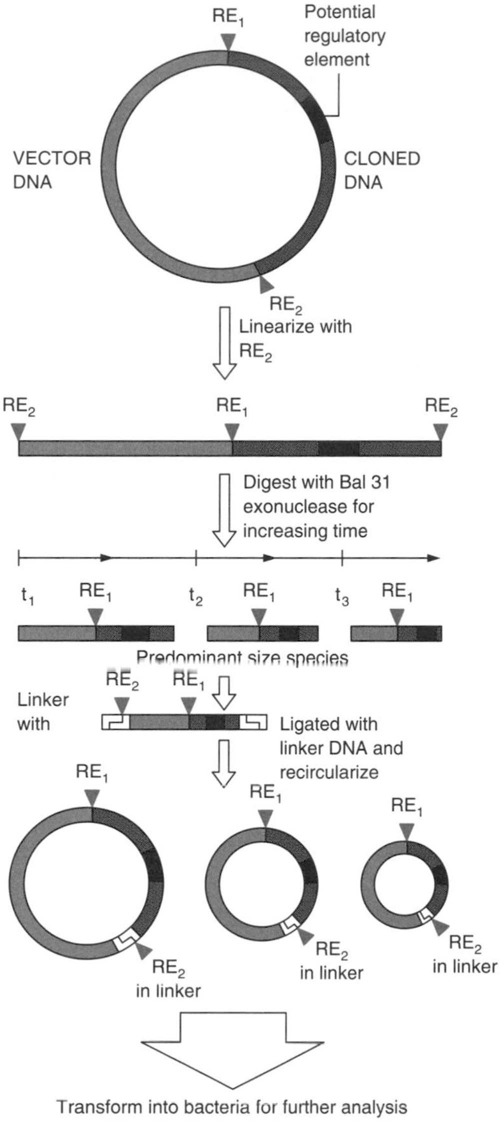

FIGURE 18.20
Enzymatic modification of potential DNA regulatory sequences.
A purified recombinant DNA molecule with a suspected gene regulatory element within flanking DNA regions is cleaved with a restriction endonuclease (RE_2). The linearized recombinant DNA is digested for varying time periods with the exonuclease, Bal31, reducing the size of the DNA flanking the potential regulatory element. The resulting recombinant DNA molecules of varying reduced sizes have small DNA oligomers (linkers) containing a restriction endonuclease sequence for RE_2 ligated to their ends. The linker modified DNA is hydrolyzed with RE_2 creating complementary single-stranded sticky ends that permit recircularization of the recombinant vectors. The potential regulatory element, bounded by various reduced-sized flanking DNA sequences, can be amplified, purified, sequenced and inserted upstream of a competent gene in an expression vector. Modification of the expression of the gene in an appropriate transfected cell can then be monitored to evaluate the role of the potential regulatory element placed at varying distances from the gene.

CLIN. CORR. 18.3
SITE DIRECTED MUTAGENESIS OF HSV I gD

The potential structural and functional role of a carbohydrate moiety, covalently linked to a protein, can be studied by site-directed mutagenesis. The gene that codes for a glycoprotein whose asparagine residue(s) is normally glycosylated (*N*-linked) must first be cloned. The *Herpes simplex* virus, type I (HSV I) glycoprotein D (gD) may contain as many as three *N*-linked carbohydrate groups. The envelop bound HSV I gD appears to play a central role in virus absorption and penetration. The role the carbohydrate groups may play in these processes is poorly understood.

The *HSV I gD* gene has been cloned and modified by site-directed mutagenesis to alter codons for the asparagine residue at the three potential glycosylation sites. These mutated genes, cloned within an expression vector, were transfected into eukaryotic cells (COS 1), where the gD protein was transiently expressed. The mutated *HSV I gD,* lacking one, or, all of its normal carbohydrate groups, can be analyzed with a variety of available monoclonal anti-gD antibodies to determine if immunological epitopes (specific sites on a protein recognized by an antibody) have been altered. Altered epitopes would indicate that the missing carbohydrate moiety was directly associated with the normal recognition site or played a role in the protein's native conformation. An altered protein conformation can impact on immunogenicity (e.g., for vaccines) and protein processing (movement of the protein from the endoplasmic reticulum, where it is synthesized, to the membrane where it normally is bound). Mutation at two of the glycosylation sites altered the native conformation of the protein such that it was less reactive with selected monoclonal antibodies. Alteration at a third site had no apparent affect on protein structure and loss of the carbohydrate chain at all three sites did not prevent normal processing of the protein.

Sodora, D. L., Cohen, G. H., and Eisenberg, R. J. Influence of asparagine-linked oligosaccharides on antigenicity, processing, and cell surface expression of Herpes Simplex Virus Type I glycoprotein D. *J. Virol.* 63:5184, 1989.

The site-directed mutagenesis of a specific nucleotide is a multistep process that begins with cloning the normal type gene in a bacteriophage (Figure 18.21). The M13 series of recombinant bacteriophage vectors are commonly employed for these studies. The **M13** is a filamentous bacteriophage that specifically infects male *E. coli* that express sex pili encoded for by a plasmid (F factor). The M13 bacteriophage contains DNA in a single-stranded or replicative form [the (+) strand], which is replicated to double-stranded DNA within an infected cell. The double-stranded form of the DNA is isolated from infected cells and used for cloning the gene to be mutated. The plaques of interest can be visually identified by α-complementation (see p. 780).

The M13 bacteriophage carrying the cloned gene of interest is used to infect susceptible *E. coli*. The progeny bacteriophage are released into the growth medium and contain single-stranded DNA. An oligonucleotide (18–30 nucleotides long) is synthesized that is complementary to a region of interest *except* for the nucleotide to be mutated. This oligomer, with one mismatched base, will hybridize to the single-stranded gene cloned in the M13 molecule and serves as a primer. Primer extension is accomplished with the bacteriophage T_4 DNA polymerase and the resulting double-stranded DNA can be transformed into susceptible *E. coli* where

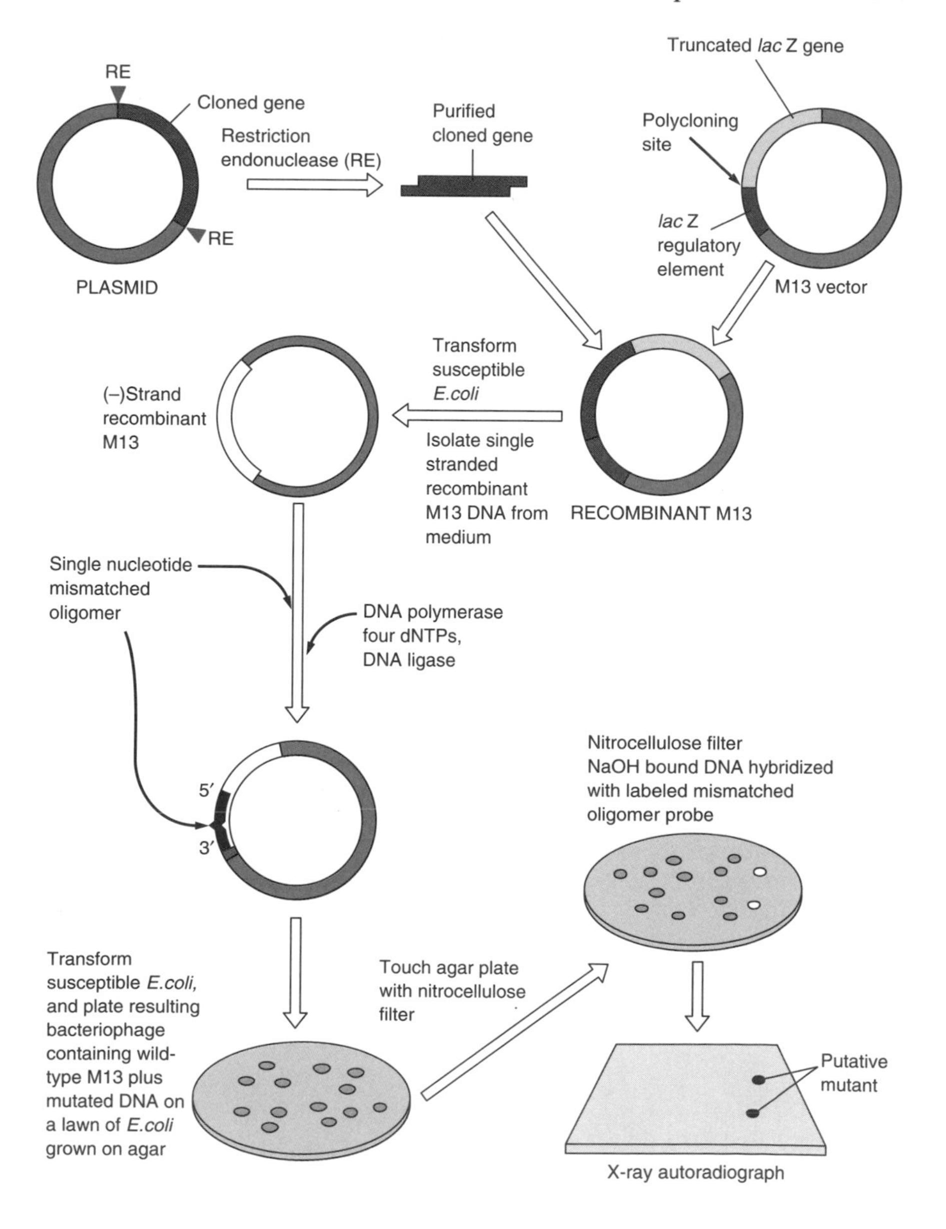

FIGURE 18.21
Site-directed mutagenesis of a single nucleotide and detection of the mutated DNA.
The figure is a simplified overview of the method. This process involves the insertion of an amplified pure DNA fragment into a modified bacteriophage vector, M13. Susceptible E. coli, transformed with the recombinant M13 DNA, synthesize the (+) strand DNA packaged within the viiron bacteriophage proteins. The bacteriophage are isolated from the growth medium and the single-stranded recombinant M13 DNA purified. The recombinant M13 DNA serves as a template for DNA replication in the presence of DNA polymerase, deoxynucleotide triphosphates (dNTP), DNA ligase and a special primer. The DNA primer (mismatched oligomer) is synthesized to be exactly complementary to a region of the DNA (gene) of interest except for the one base intended to be altered (mutated). The newly synthesized M13 DNA, therefore, contains a specifically mutated base which when reintroduced into susceptible *E. coli* will be faithfully replicated. The transformed *E. coli* are grown on agar plates with replicas of the resulting colonies picked up on a nitrocellulose filter. DNA associated with each colony is denatured and fixed to the filter with NaOH and the filter bound DNA hybridized with a ^{32}P-labelled mismatched DNA oligomer probe. The putative mutants are then identified by exposing the filter to X-ray film.

the mutated DNA strand serves as a template to replicate new (+) strands now carrying the mutated nucleotide.

The bacteriophage plaques, containing the mutated DNA, are screened by hybridizing with a labeled probe of the original oligonucleotide. By adjusting the wash temperature of the hybridized probe only the perfectly matched hybrid will remain complexed while the wild-type DNA–oligomer with mismatched nucleotide will dissociate. The M13 carrying the mutated gene is then replicated in bacteria, the DNA purified, and the mutated region of the gene sequenced to confirm the identity of the mutation. Many modifications have been developed to improve the efficiency of site-directed mutagenesis of a single nucleotide including a method to selectively replicate the mutated strand [the (−) strand]. The M13 bacteriophage, replicated in a mutant *E. coli,* incorporates some uracil residues into its DNA in place of thymine due to a metabolic defect in the synthesis of dTTP from dUTP and the lack of an enzyme that normally removes uracil residues from DNA. The purified single-stranded M13 uracil-containing DNA [the (+) strand] is hybridized with a complementary oligomer containing a mismatched base at the nucleotide to be mutated. The oligomer serves as the primer for DNA replication in vitro with the template (+) strand containing uracils and the new (−) strand containing thymines. When this double-stranded M13 DNA is transformed into a wild-type *E. coli* the uracil-containing strand is destroyed and mutated (−) strand serves as the template for the progeny bacteriophage, most of which will carry the mutation of interest.

18.13 POLYMERASE CHAIN REACTION

The rapid production of large quantities of a specific DNA sequence took a leap forward with the development of the **polymerase chain reaction (PCR).** The PCR requires two nucleotide oligomers that hybridize to the complementary DNA strands in a region of interest. The oligomers serve as primers for a DNA polymerase that extends each strand. Repeated cycling of the PCR yields large amounts of each DNA molecule of interest in a matter of hours as opposed to days and weeks associated with cloning techniques.

The PCR amplification of a specific DNA sequence can be accomplished with a purified DNA sample or a small region within a complex mixture of DNA. The principles of the reaction are shown in Figure 18.22. The nucleotide sequence of the DNA to be amplified must be known or it must be cloned in a vector where the sequence of the flanking DNA has been established. Two oligonucleotides are synthesized that hybridize to each strand of DNA to be amplified and flank the region of interest. The oligomers, hybridized to the denatured template, serve as primers for a DNA polymerase that extends each primer along its template. The product is a double-stranded DNA molecule and the reaction is completed when all of the template molecules have been copied. In order to initiate a new round of replication the sample is heated to melt the double-stranded DNA and, in the presence of excess oligonucleotide primers, cooled to permit hybridization of the single-stranded template with free oligomers. A new cycle of DNA replication will initiate in the presence of DNA polymerase and all four dNTPs. Heating to about 95°C as required for melting DNA inactivates most DNA polymerases, but a heat stable polymerase, termed Taq DNA polymerase isolated from Thermus aquaticus, is now employed, obviating the need to add fresh polymerse after each cycle. This has permitted the automation of PCR with each DNA molecule capable of being amplified one millionfold.

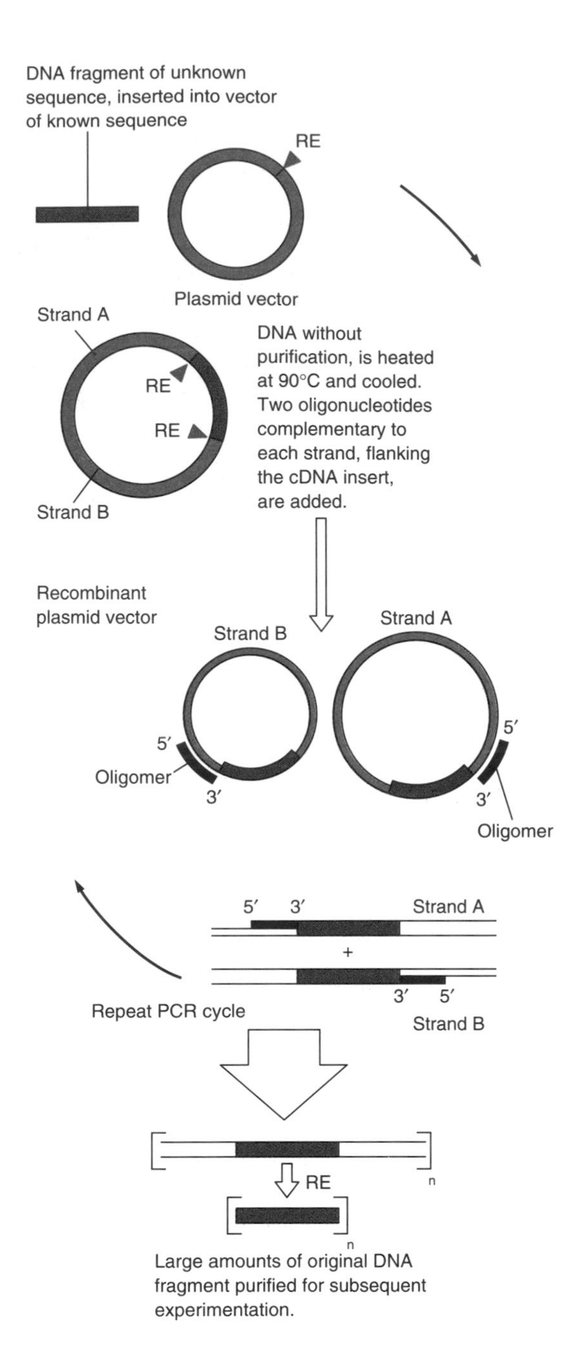

FIGURE 18.22
The polymerase chain reaction (PCR).
A DNA fragment of unknown sequence is inserted into a vector of known sequence by normal recombinant methodologies. The recombinant DNA of interest does not need to be purified from all other contaminating DNA species. The DNA is heated to 90°C to dissociate the double strands and cooled in the presence of excess amounts of two different complementary oligomers that hybridize to the known vector DNA sequences flanking the foreign DNA insert. Only recombinant single-stranded DNA species will serve as templates for DNA replication yielding double-stranded DNA fragments of the foreign DNA bounded by the oligomer DNA sequences. The heating-replication cycle is repeated many times to rapidly produce greatly amplified amounts of the original foreign DNA fragment. The DNA fragment of interest can be purified from the polymerase chain reaction mixture by cleaving it with the original restriction endonuclease (RE), electrophoresing the DNA mixture through an agarose gel and eluting the band of interest from the gel.

The polymerase chain reaction has many applications including gene diagnosis, forensic investigations, where only a drop of dried blood or a single hair is available, and evolutionary studies with preserved biological material. The use of PCR for screening for human immunodeficiency virus is presented in Clin. Corr. 18.4. Polymerase chain reaction can also be employed for site-directed mutagenesis. Strategies have been readily developed to incorporate a mismatched base into one of the oligonucleotides that primes the polymerase chain reaction.

18.14 APPLICATIONS OF RECOMBINANT DNA TECHNOLOGIES

The practical uses of recombinant DNA methods in biological systems is limited only by one's imagination. Recombinant DNA methods are applicable to numerous biological disciplines including agriculture, studies of evolution, forensic biology, and clinical medicine. Genetic engineering can introduce new or altered proteins into crops (e.g., corn), such that they contain amino acids essential to humans but often lacking in plant proteins. Toxins that are lethal to specific insects but harmless to humans can be introduced into crops to protect plants without the use of environmentally destructive pesticides. The DNA isolated from cells in the amniotic fluid of a pregnant woman can be analyzed for the presence or absence of genetic defects in the fetus. Minuscule quantities of DNA can be isolated from biological samples that have been preserved in ancient tar pits or frozen tundra, amplified and sequenced for evolutionary studies at the molecular level. The DNA from a single hair, a drop of blood, or sperm from a rape victim can be isolated, amplified, and mapped to aid in identifying felons. Current technologies in conjunction with future methods should permit the selective introduction of genes into cells with defective or absent genes. Developing methodologies are also likely to become available to introduce nucleic acid sequences into cells to selectively turn off the expression of detrimental genes.

Antisense Nucleic Acids Hold Promise as Research Tools and in Therapy

The cellular function of genes and their encoded products have been studied in a variety of ways. Recently, a new tool, antisense nucleic acids, has been introduced to study the intracellular expression and function of specific proteins. Natural and synthetic **antisense nucleic acids** that are complementary to mRNAs will hybridize in situ within the cell, and inactivate the mRNA and thus block translation. The introduction of antisense nucleic acids into cells has opened new avenues to explore how proteins, whose expression has been selectively repressed in a cell, function within that cell. This method also holds great promise in the control of diseased processes such as viral infections. Antisense technology, along with site-directed mutagenesis are part of a new discipline termed reverse genetics. **Reverse genetics** selectively modifies a gene to evaluate its function, as opposed to classical genetics, which depends on the isolation and analysis of cells carrying random mutations that can be identified. A second use of the term reverse genetics refers to the mapping and ultimate cloning of a human gene associated with a disease where no prior knowledge of the molecular agents causing the disease exist. The use of the term "reverse genetics" in this latter case is likely to be modified.

Antisense RNA can be introduced into a cell by employing common cloning techniques. Figure 18.23 demonstrates one method. A gene of interest is cloned in an expression vector in the wrong orientation. That is, the sense or coding strand that is normally inserted into the expression vector downstream of a promoter is intentionally inserted in the opposite direction. This now places the complementary or antisense strand of the DNA under the control of the promoter with expression or transcription yielding antisense RNA. Transfection of cells with the antisense expression vector introduces antisense RNA that is capable of hybridizing with normal cellular mRNA. The mRNA–antisense RNA complex is not translated due to a number of reasons such as its inability to bind to ribosomes, blockage of normal processing, and rapid enzymatic degradation.

DNA oligonucleotides have also been synthesized that are complementary to the known sequences of mRNAs of selected genes. Introduction of specific DNA oligomers to cells in culture have inhibited viral infections including infections by the human immunodeficiency virus (HIV). It is conceivable that one day bone marrow cells will be removed from AIDS patients and antisense HIV virus nucleic acids introduced into their cells in culture. These "protected" cells can then be reintroduced into the

CLIN. CORR. 18.4 POLYMERASE CHAIN REACTION AND SCREENING FOR HUMAN IMMUNODEFICIENCY VIRUS

The use of the polymerase chain reaction (PCR) to amplify minute quantities of DNA has revolutionized the ability to detect and analyze DNA species. With PCR it is possible to synthesize sufficient DNA for analysis. The detection and identification of the human immunodeficiency virus (HIV) by conventional methods, such as Southern blot–DNA hybridization and antigen analysis, is labor intensive, expensive, and has low sensitivity. An infected individual, with no sign of AIDS (acquired immune deficiency syndrome), may test false negative for HIV by these procedures. Early detection of HIV infections in these individuals is crucial to initiate treatment and/or monitor the progression of their disease. In addition, a sensitive method is required to be certain that contributed blood from donors does not contain HIV. The PCR amplification of potential HIV DNA sequences within DNA isolated from an individual's white blood cells permits the identification of viral infections prior to the presence of antibodies, the so-called seronegative state. Current methodologies are too costly to apply this testing to large scale screening of donor blood samples. Polymerase chain reaction can also be used to increase the sensitivity to detect and characterize DNA sequences of any other human infectious pathogen.

Kwok, S. and Sninsky, J. J. Application of PCR to the detection of human infectious diseases, in PCR technology. H. A. Erlich (Ed.). New York: Stockton Press, 1989, p. 235.

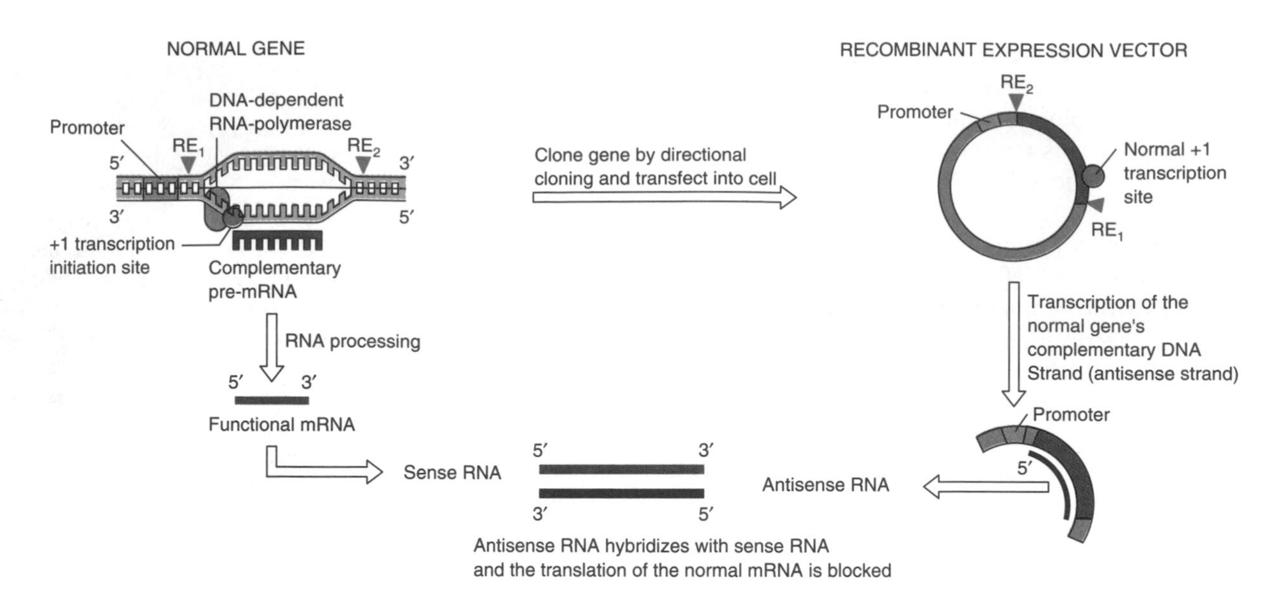

FIGURE 18.23
Production of antisense RNA.
A gene, or a portion of it, is inserted into a vector by directional cloning downstream of a promoter and in the reversed orientation normally found in the cell of origin. Transfection of this recombinant DNA into the parental cell carrying the normal gene results in the transcription of RNA (antisense RNA) from the cloned reversed polarity DNA along with a normal cellular mRNA (sense RNA) transcript. The two antiparallel complementary RNAs can hybridize within the cell resulting in blocked expression (translation) of the normal mRNA transcript.

AIDS patient's bone marrow (autologous bone marrow transplantation) and replace those cells normally destroyed by the virus. Experimental progress is also being made with antisense nucleic acids that can regulate the expression of oncogenes, genes involved in the cancer-forming process. Harnessing antisense technologies holds great promise for the treatment of human diseases.

CLIN. CORR. 18.5
TRANSGENIC ANIMAL MODELS

Transgenic animal model systems hold promise for future methodologies to correct genetic diseases early in fetal development. These animals are currently used to study the regulation of expression and function of specific gene products in a whole animal and in the near future may create new breeds of commercially valuable animals. Transgenic mice have been developed from fertilized mouse eggs with rat growth hormone (GH) genes microinjected into their male pronuclei. The rat GH gene DNA, fused to the mouse metallothionein-I (MT-I) promoter region, was purified from the plasmid in which it had been cloned. Approximately 600 copies of the promoter–gene complex were introduced into each egg, which was then inserted into the reproductive tract of a foster mother mouse. The resulting transgenic mouse was shown to carry the rat GH gene within its chromosome by hybridizing a labeled DNA probe to mouse DNA that had been purified from a slice of the tail, restriction endonuclease digested, electrophoresed, and Southern blotted. The diet of the animals was supplemented with $ZnSO_4$ at 33 days postparturition. The $ZnSO_4$ presumably activates the mouse MT-I promoter to initiate transcription of the rat GH gene. The continuous overexpression of rat GH in some transgenic animals produced mice nearly twice the size of litter mates that did not carry the rat GH gene. A transgenic mouse transmitted the rat GH gene to one-half of its offspring, indicating that the gene stably integrated into the germ cell chromosome and that new breeds of animals can be created.

Palmiter, R. D., et al. Dramatic growth of mice that develop from eggs microinjected with metallothionein-growth hormone fusion genes. *Nature* (*London*) 300:611, 1982.

Transgenic Animals

Recombinant DNA methods, described to this point, allow one to produce large amounts of foreign gene products in bacteria and individual cells. These methods also facilitate the evaluation of the role of a specific gene product in cell structure or function. In order to investigate the role of a selected gene product in the growth and development of a whole animal the gene must be introduced into the fertilized egg. Foreign genes can be inserted into the chromosome of a fertilized egg. Animals that develop from a fertilized egg with a foreign gene insert carry that gene in every cell and are referred to as **transgenic animals.**

The most commonly employed method to create transgenic animals is outlined in Figure 18.24. The gene of interest is usually a cloned recombinant DNA molecule, that includes its own promoter or cloned in a construct with a different promoter that can be selectively regulated. Multiple copies of the foreign gene are microinjected into the pronucleus of the fertilized egg. The foreign DNA inserts randomly within the chromosomal DNA. If the insert disrupts a critical cellular gene the embryo will die. Usually, nonlethal mutagenic events result from the insertion of the foreign DNA into the chromosome.

Transgenic animals are currently being used to study several different aspects of the foreign gene, including the analysis of DNA regulatory elements, expression of proteins during differentiation, tissue specificity, and the potential role of oncogene products on growth, differentiation and the induction of tumorigenesis. Eventually, it is expected that these and related technologies will allow for methods to replace defective genes in the developing embryo (see Clin. Corr. 18.5).

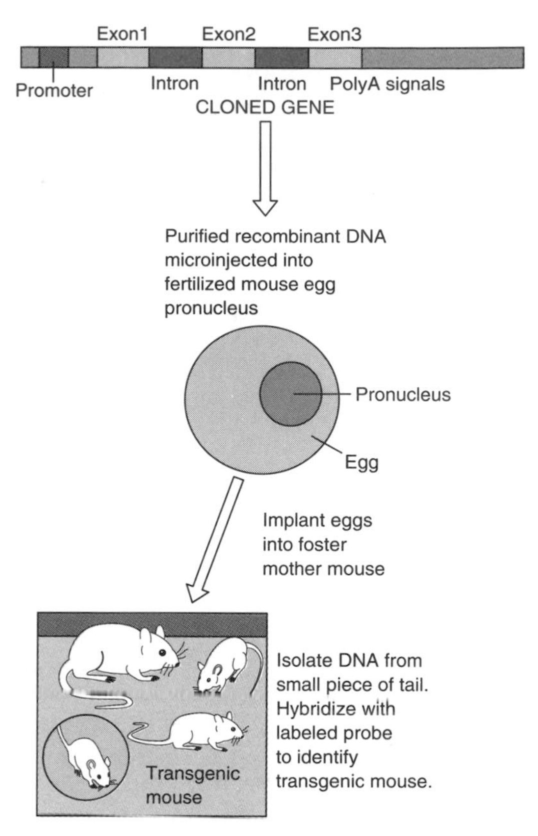

FIGURE 18.24
Production of transgenic animals.
Cloned, amplified, and purified functional genes are microinjected into several fertilized mouse egg pronuclei in vitro. The eggs are implanted into a foster mother. DNA is isolated from a small piece of each offspring pup's tail and hybridized with a labelled probe to identify animals carrying the foreign gene (transgenic mouse). The transgenic mice can be mated to establish a new strain of mice. Cell lines can also be established from tissues of transgenic mice to study gene regulation and the structure/function of the foreign gene product.

Recombinant DNA in Agriculture Will Have Significant Commercial Impact

Perhaps the greatest gain to all humanity would be the practical use of recombinant technologies to improve our agricultural crops. Genes must be identified and isolated that code for properties that include higher crop yield, rapid plant growth, resistance to adverse conditions such as arid or cold periods and plant size. New genes, not common to plants, may be engineered into plants that confer resistance to insects, fungi, or bacteria. Finally, genes encoding existing structural proteins can be modified to contain essential amino acids not normally present in the plant, without modifying the protein function. The potential to produce plants with new genetic properties depends on the ability to introduce genes into plant cells that can differentiate into whole plants.

New genetic information carried in crown gall plasmids can be introduced into plants infected with soil bacteria known as agrobacteria. Agrobacteria naturally contain a crown gall or Ti (tumor-inducing) plasmid whose genes integrate into an infected cell's chromosome. The plasmid genes direct the host plant cell to produce new amino acid species that are required for bacterial growth. A crown gall, or tumor mass of undifferentiated plant cells develops at the site of bacterial infection. New genes can be engineered into the Ti plasmid, and the recombinant plasmid introduced into plant cells upon infection with the Agrobacteria. Transformed plant cells can then be grown in culture and under proper conditions can be induced to redifferentiate into whole plants. Every cell would contain the new genetic information and would represent a transgenic plant.

Some limitations in producing plants with improved genetic properties must be overcome before significant advances in our world food supply can be realized. Clearly, proper genes must yet be identified and isolated for desired characteristics. Also, important crops such as corn and wheat cannot be transformed by Ti plasmids, therefore, other vectors must be identified.

18.15 CONCLUDING REMARKS

The old cliche, so close and yet so far away, seems appropriate for our current juncture in molecular biology. The entire human genome will likely be sequenced in the next decade or so. Genetic diseases now identified and to be identified should eventually be curable by gene replacement therapy if and when we surmount the technical roadblocks. If one looks at the enormous advances made in molecular biology in just the past two decades it is reasonable to believe the "when" will not be that far off.

BIBLIOGRAPHY

Brown, W. M., George, M. Jr., and Wilson, A. C. Rapid evolution of animal mitochondrial DNA. *Proc. Natl. Acad. Sci. U.S.A.* 76:1967, 1979.

Erlich, H. A. (Ed.). *PCR technology. Principles and applications for DNA amplification.* New York: Stockton Press, 1989.

Jaenisch, R. Transgenic animals. *Science* 240:1468, 1988.

Kunkel, T. A. Rapid and efficient site-specific mutagenesis without phenotypic selection. *Proc. Natl. Acad. Sci. U.S.A.* 82:488, 1985.

Maxam, A. M. and Gilbert, W. A new method of sequencing DNA. *Proc. Natl. Acad. Sci. U.S.A.* 74:560, 1977.

Palmiter, R. D., et al. Dramatic growth of mice that develop from eggs microinjected with metallothionein-growth hormone fusion genes. *Nature (London)* 300:611, 1982.

Rigby, P. W. J., Dieckmann, M., Rhodes, C., and Berg, P. Labelled deoxyribonucleic acid to high specific activity *in vitro* by nick translation with DNA polymerase I. *J. Mol. Biol.* 113:237, 1977.

Sambrook, J., Fritsch, E. F., and Maniatis, T. *Molecular Cloning. A laboratory manual.* 2nd ed. New York: Cold Spring Harbor Laboratory Press, 1989.

Sanger, F., Nicklen, S., and Coulson, A. R. DNA sequencing with chain-terminating inhibitors. *Proc. Natl. Acad. Sci. U.S.A.* 74:5463, 1977.

Southern, E. M. Detection of specific sequences among DNA fragments separated by gel electrophoresis. *J. Mol. Biol.* 98:503, 1975.

Vogelstein, B., Fearon, E. R., Hamilton, S. R., and Feinberg, A. P. Use of restriction fragment length polymorphism to determine the clonal origin of human tumors. *Science* 227:642, 1985.

Watson, J. D., Tooze, J., and Kurtz, D. T. *Recombinant DNA a Short Course.* Scientific American Books, San Francisco, Freeman, 1983.

Weintraub, H. M. Antisense RNA and DNA. *Sci. Am.* 262:40, 1990.

QUESTIONS

J. BAGGOTT AND C. N. ANGSTADT

*Q*uestion *T*ypes are described on page xxiii.

1. (QT2) In the Maxam–Gilbert method of DNA sequencing:
   1. cleavage of the DNA backbone occurs randomly at only some of the sites where the base had been destroyed.
   2. all nucleotides produced during cleavage of the DNA backbone are detected by radioautography.
   3. electrophoretic separation of DNA fragments is due to differences in both size and charge.
   4. the sequence of bands in the four lanes of the autoradiogram contains the base sequence information.
2. (QT2) The Sanger and Maxam–Gilbert methods of DNA sequencing differ in that:
   1. the Maxam–Gilbert method involves labeling the 5′ end, while the Sanger method requires labeling the 3′ end of the DNA.
   2. only the Maxam–Gilbert method involves electrophoresing a mixture of fragments of different sizes.
   3. only the Maxam–Gilbert method uses radioautography to detect fragments in which one of the termini is radioactively labeled.
   4. in the Maxam–Gilbert method, a complete DNA chain is cleaved, while in the Sanger method, synthesis of the chain is interrupted at different points.

   A. Recombinant DNA
   B. Plasmid
   C. Vector
3. (QT5) A molecule in which DNA sequences from more than one source have been joined.
4. (QT5) A small (a few thousand bases) circular DNA molecule found in bacteria.
5. (QT5) A species of DNA that is capable of carrying a DNA fragment which is to be cloned.

6. (QT2) In the selection of colonies of bacteria that carry cloned DNA in plasmids such as pBR322 that contain two antibiotic resistance genes:
   1. one antibiotic resistance gene is nonfunctional in the desired bacterial colonies.

2. Untransformed bacteria are antibiotic resistant.
3. the colonies of interest may be killed by antibiotic treatment.
4. radiolabeled DNA or RNA probes play a role.

A. DNA probes C. Both
B. RNA probes D. Neither

7. (QT4) DNA is a template for synthesis.
8. (QT4) Must be denatured before use.

9. (QT2) Which of the following pairs of vectors and DNA insert sizes is/are correct?
   1. plasmids 5–10 kb
   2. cosmids 15 kb
   3. yeast 200–500 kb
   4. Bacteriophage λ 45 kb
10. (QT2) Expression of a eucaryotic gene in procaryotes requires:
   1. a SD sequence in mRNA.
   2. absence of introns.
   3. regulatory elements upstream of the gene.
   4. a fusion protein.

A. Antisense nucleic acid
B. Polymerase chain reaction
C. Site-directed mutagenesis
D. Shuttle vector
E. Transfecton

11. (QT5) Contains both bacterial and eucaryotic replication signals.
12. (QT5) Complementary to mRNA, and will hybridize to it, blocking translation.
13. (QT5) Can rapidly produce large quantities of a specific DNA.
14. (QT5) Oligomer with one mismatched base is used as a primer.
15. (QT5) A process that introduces foreign DNA into a eucaryotic genome.

ANSWERS

1. D Only 4 correct. 1: Cleavage occurs at *all* such sites. Limited destruction of the bases is random (p. 771). 2: Only the nucleotides that contain the labeled 5′ terminal are detected. Other nucleotides are produced, but are not detected by the method, and do not contribute information to the analysis (p. 771). 3: Although charge is, of course, required in order to produce movement of a particle in a field, the separation of these fragments is not due to charge differences, but to size differences, with the smallest fragments migrating farthest (p. 771). 4: The relative positions of G are given by the bands in the lane corresponding to the destruction of G; of A by the bands in the AG lane that are *not* duplicated in the G lane; of C, by the bands in the C lane; of T, by the bands in the CT lane that are *not* duplicated in the C lane.
2. D Only 4 correct. 1: Both involve a labeled 5′ end. 2: Both methods do this. 3: They both use radioautography to detect fragments in which one of the termini is radioactively labeled. 4: (pp. 771–773).
3. A (see p. 773).
4. B (see p. 775).
5. C (see p. 775).
6. B 1, 3 correct. 1: The foreign DNA is inserted into one antibiotic resistance gene, thus destroying it (p. 778). 2: Resistance is due to the plasmids (p. 778). 3: But samples of these colonies had been saved (on replicate plates) so they can be grown (Figure 18.8, p. 779). 4: (pp. 779–780).
7. C DNA Polymerase I removes DNA, replacing it with radiolabeled DNA; RNA is made from a DNA template (p. 781).
8. A Double-stranded DNA must be denatured so one strand can bind to the DNA of interest (p. 781–782).
9. B 1, 3 correct. Bacteriophage λ will accept a 15-kb insert (p. 785), and a cosmid will accept a 45-kb insert (p. 786).
10. A 1, 2, 3 correct. 1: The SD sequence is necessary for the bacterial ribosome to recognize the mRNA. 2: Bacteria do not have the intracellular machinery to remove introns from mRNA. 3: Appropriate regulatory elements are necessary to allow the DNA to be transcribed. 4: A fusion protein may be a product of the reaction (p. 790).
11. D (see p. 791).
12. A (see p. 799).
13. B (see p. 797).
14. C (see p. 794).
15. E (see p. 791).

19

Regulation of Gene Expression

JOHN E. DONELSON

19.1 OVERVIEW 806
19.2 THE UNIT OF TRANSCRIPTION IN BACTERIA: THE OPERON 807
19.3 THE LACTOSE OPERON OF *E. COLI* 808
The Repressor of the Lactose Operon Is a Diffusible Protein 809
The Operator Sequence of the Lactose Operon Is Contiguous on the DNA With a Promoter and Three Structural Genes 811
The Promoter Sequence of the Lactose Operon Contains Recognition Sites for RNA Polymerase and a Regulator Protein 813
The Catabolite Activator Protein Binds at a Site on the Lactose Promoter 813
19.4 THE TRYPTOPHAN OPERON OF *E. COLI* 815
The Tryptophan Operon Is Controlled by a Repressor Protein 815
The Tryptophan Operon Has a Second Control Site: The Attenuation Site 817
Transcription Attenuation Is a Mechanism of Control in Operons for Amino Acid Biosynthesis 820
19.5 OTHER BACTERIAL OPERONS 821
Synthesis of Ribosomal Proteins Is Regulated in a Coordinated Manner 822
The Stringent Response Controls the Synthesis of rRNAs and tRNAs 823
19.6 BACTERIAL TRANSPOSONS 824
Transposons Are Mobile Segments of DNA 824
The *Tn3* Transposon Contains Three Structural Genes 825
19.7 INVERSION OF GENES IN *SALMONELLA* 827
19.8 ORGANIZATION OF GENES IN MAMMALIAN DNAs 828
Eucaryotic DNA Does Not Correlate With the Complexity or Function of an Organism 828
Eucaryotic Genes Usually Contain Intervening Sequences (Introns) 829
19.9 REPETITIVE DNA SEQUENCES IN EUCARYOTES 829
The Importance of the Highly Repetitive Sequences Is Unknown 829
A Variety of Repeating Units Are Defined as Moderately Repetitive Sequences 830
19.10 GENES FOR GLOBIN PROTEINS 833
Recombinant DNA Technology Has Been Used to Clone Genes for the Globin Molecules 833
Sickle-Cell Anemia Is Due to a Single Base Pair Change 836
Thalassemias Are Caused by Mutations in the Genes for the α or β Chains of Globin 836
19.11 GENES FOR HUMAN GROWTH HORMONE-LIKE PROTEINS 837
19.12 GENES IN THE MITOCHONDRION 839
19.13 BACTERIAL EXPRESSION OF FOREIGN GENES 840
Bacteria Can Synthesize Human Insulin 840
Bacteria Can Synthesize Human Growth Hormone 842
19.14 INTRODUCTION OF THE RAT GROWTH HORMONE GENE INTO MICE 842
BIBLIOGRAPHY 844
QUESTIONS AND ANNOTATED ANSWERS 845

CLINICAL CORRELATIONS

19.1 Transmissible Multiple Drug Resistances 825
19.2 Prenatal Diagnosis of Sickle-Cell Anemia 836
19.3 Prenatal Diagnosis of Thalassemia 837
19.4 Leber's Hereditary Optic Neuropathy (LHON) 840

19.1 OVERVIEW

To survive, a living cell must be able to respond to changes in its environment. One of many ways in which cells adjust to these changes is to alter the expression of specific genes, which, in turn, affects the number of the corresponding protein molecules in the cell. This chapter will focus on some of the molecular mechanisms that determine when a given gene will be expressed and to what extent. The attempt to understand how the expression of genes is regulated is one of the most active areas of biochemical research today.

It makes sense for a cell to vary the amount of a given gene product that is available under different conditions. For example, the bacterium *Escherichia coli* (*E. coli*) contains genes for about 3000 different proteins, but it does not need to synthesize all of these proteins at the same time. Therefore, it regulates the number of molecules of these proteins that are made. The classic illustration of this phenomenon, as we will see, is the regulation of the number of β-galactosidase molecules in the cell. This enzyme converts the disaccharide lactose to the monosaccharides, glucose and galactose. When *E. coli* is growing in a medium containing glucose as the carbon source, β-galactosidase is not required and only about five molecules of the enzyme are present in the cell. When lactose is the sole carbon source, however, 5000 or more molecules of β-galactosidase occur in the cell. Clearly, the bacteria respond to the need to metabolize lactose by increasing the synthesis of β-galactosidase molecules. If lactose is removed from the medium, the synthesis of this enzyme stops as rapidly as it began.

The complexity of eucaryotic cells means that they have even more extensive mechanisms of gene regulation than do procaryotic cells. The differentiated cells of higher organisms have a much more complicated physical structure and often a more specialized biological function that is determined, again, by the expression of their genes. For example, insulin is synthesized in specific cells of the pancreas and not in kidney cells even though the nuclei of all cells of the body contain the insulin genes. Molecular regulatory mechanisms facilitate the expression of insulin in the pancreas and prevent its synthesis in the kidney and other cells of the body. In addition, during development of the organism the appearance or disappearance of proteins in specific cell types are tightly controlled with respect to the timing and sequence of the developmental events.

As expected from the differences in complexities, far more is understood about the regulation of genes in procaryotes than in eucaryotes. However, studies on the control of gene expression in procaryotes often provide exciting new ideas that can be tested in eucaryotic systems. And, sometimes, discoveries about eucaryotic gene structure and regulation alter the interpretation of data on the control of procaryotic genes.

In this chapter several of the best studied examples of gene regulation in bacteria will be discussed, followed by some illustrations of the organization and regulation of related genes in the human genome. Finally, the use of recombinant DNA techniques to express some human genes of clinical interest will be presented.

19.2 THE UNIT OF TRANSCRIPTION IN BACTERIA: THE OPERON

The single *E. coli* chromosome is a circular double-stranded DNA molecule of about 4 million base pairs. Most of the approximately 3000 *E. coli* genes are not distributed randomly throughout this DNA; instead, the genes that code for the enzymes of a specific metabolic pathway are clustered in one region of the DNA. In addition, genes for associated structural proteins, such as the 70 or so proteins that comprise the ribosome, are frequently adjacent to one another. The members of a set of clustered genes are usually coordinately controlled; they are transcribed together to form a **"polycistronic" mRNA** species that contains the coding sequences for several proteins. The term **operon** is used to describe the complete regulatory unit of a set of clustered genes. An operon includes the adjacent **structural genes** that code for the related enzymes or associated proteins, a **regulatory gene** or genes that code for regulator protein(s), and **control elements** that are sites on the DNA near the structural genes at which the regulator proteins act. Figure 19.1 shows a partial genetic map of the *E. coli* chromosome that gives the locations of the structural genes of some of the different operons.

When transcription of the structural genes of an operon increases in response to the presence of a specific substrate in the medium, the effect is known as **induction.** The increase in transcription of the β-galactosidase gene when lactose is the sole carbon source is an example of induc-

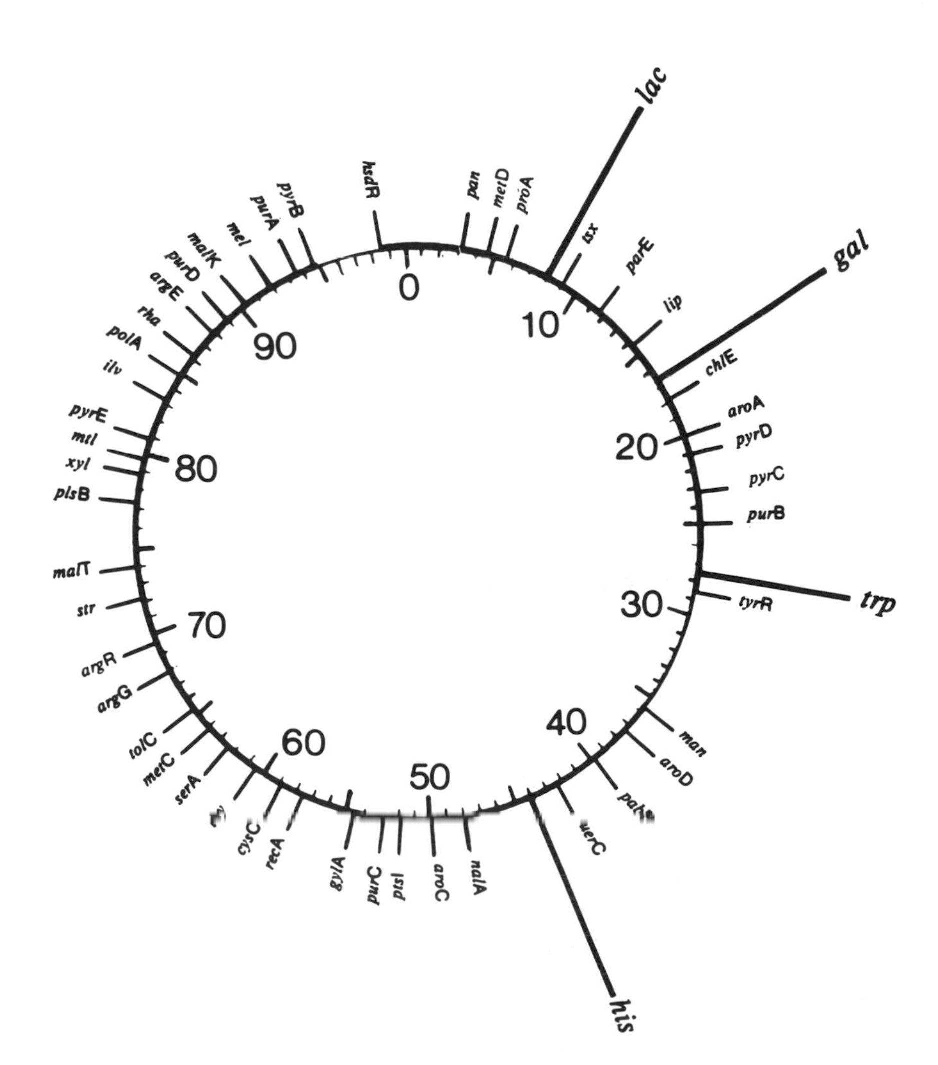

FIGURE 19.1
Partial genetic map of *E. coli*.
The locations of only a few of the genes identified and mapped in *E. coli* are shown here. Four operons discussed in this chapter are indicated.
Reproduced with permission from G. S. Stent and R. Calendar, *Molecular Genetics, An Introductory Narrative*. W. H. Freeman, San Francisco: 1978, p. 289; modified from B. J. Bachmann, K. B. Low, and A. L. Taylor, *Bacteriol. Rev.* 40:116, 1976.

tion. Bacteria can also respond to nutritional changes by quickly turning off the synthesis of enzymes that are no longer needed. As will be described below, *E. coli* synthesizes the amino acid tryptophan as the end product of a specific biosynthetic pathway. However, if tryptophan is supplied in the medium, the bacteria do not need to make it themselves, and the synthesis of the enzymes for this metabolic pathway is stopped. This process is called **repression.** It permits the bacteria to avoid using their energy for making unnecessary and even harmful proteins.

Induction and repression are manifestations of the same phenomenon. In one case the bacterium changes its enzyme composition so that it can utilize a specific substrate in the medium; in the other it reduces the number of enzyme molecules so that it does not overproduce a specific metabolic product. The signal for both types of regulation is the small molecule that is a substrate for the metabolic pathway or a product of the pathway, respectively. These small molecules are called **inducers** when they stimulate induction and **corepressors** when they cause repression to occur.

Section 19.3 will describe in detail the lactose operon, the best studied example of a set of inducible genes. Section 19.4 will present the tryptophan operon, an example of a repressible operon. The remaining sections will briefly describe some other operons as well as some gene systems in which physical movement of the genes themselves within the DNA (i.e., gene rearrangements) plays a role in their regulation.

19.3 THE LACTOSE OPERON OF *E. COLI*

The lactose operon contains three adjacent structural genes as shown in Figure 19.2. *LacZ* codes for the enzyme β-galactosidase, which is composed of four identical subunits of 1021 amino acids. *LacY* codes for a permease, which is a 275-amino acid protein that occurs in the cell membrane and participates in the transport of sugars, including lactose, across the membrane. The third gene, *lacA*, codes for β-galactoside transacetylase, a 275-amino acid enzyme that transfers an acetyl group from acetyl CoA to β-galactosides. Interestingly, of these three proteins, only β-

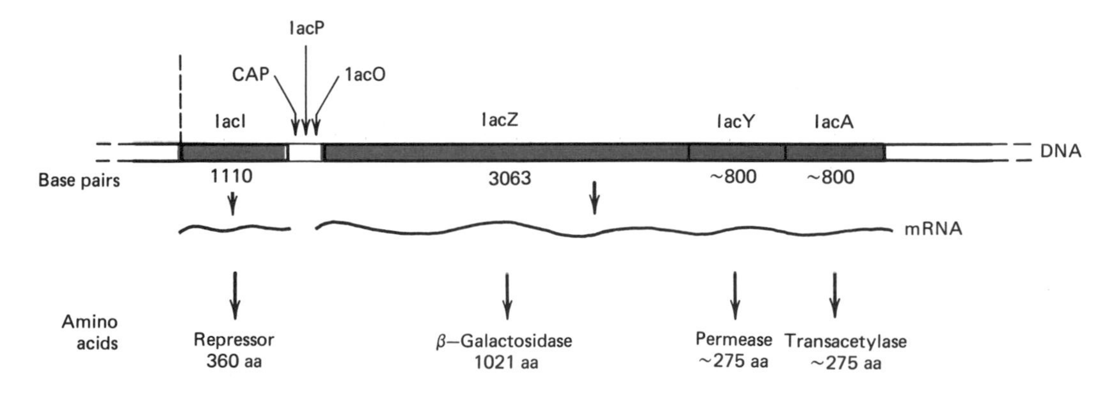

FIGURE 19.2
Lactose operon of *E. coli*.
The lactose operon is composed of the *lacI* gene, which codes for a repressor, the control elements of *CAP*, *lacP*, and *lacO;* and three structural genes, *lacZ*, *lacY*, and *lacA*, which code for β-galactosidase, a permease and a transacetylase, respectively. The *lacI* gene is transcribed from its own promotor. The three structural genes are transcribed from the promotor, *lacP*, to form a polycistronic mRNA from which the three proteins are translated.

galactosidase actually participates in a known metabolic pathway. However, the permease is clearly important in the utilization of lactose since it is involved in transporting lactose into the cell. The significance of the acetylation reaction has long been a mystery, but it may be associated with the detoxification and excretion reactions of nonmetabolized analogs of β-galactosides.

Mutations in *lacZ* or *lacY* that destroy the function of the β-galactosidase or the permease prevent the cells from cleaving lactose or acquiring it from the medium, respectively. Mutations in *lacA* that destroy the acetylase activity do not seem to have an identifiable effect on cell growth and division. Perhaps there are other related enzymes in the cell that serve as backups for this enzyme, or perhaps it has a specific unknown function that is required only under certain conditions.

A single mRNA species containing the coding sequences of all three structural genes is transcribed from a promoter that occurs just upstream from the *lacZ* gene. The induction of these three genes occurs during the initiation of their transcription. Without the inducer, transcription of the gene cluster occurs only at a very low level. In the presence of the inducer, transcription begins at the promoter, called *lacP,* and goes through all three genes to a transcription terminator located slightly beyond the end of *lacA*. Therefore, the genes are **coordinately expressed;** either all three are transcribed in unison or none is transcribed.

The presence of the three coding sequences on the same mRNA molecule suggests that the relative amounts of the three proteins are always the same under varying conditions of induction. An inducer that causes a high rate of transcription will result in a high level of all three proteins; an inducer that stimulates only a little transcription of the operon will result in a low level of the proteins. The **inducer** can be thought of as a molecular switch that influences synthesis of the single mRNA species for all three genes. The number of molecules of each protein in the cell may be different, but this does not reflect differences in transcription; it reflects differences in translation rates of the coding sequences or in degradation of the proteins themselves.

The lactose mRNA is very unstable; it is degraded with a half-life of about 3 min. Therefore, expression of the operon can be altered very quickly. Transcription ceases as soon as the inducer is no longer present, the existing mRNA molecules disappear within a few minutes, and the cell stops making the proteins.

The Repressor of the Lactose Operon Is a Diffusible Protein

The regulatory gene of the lactose operon, *lacI,* was originally identified by mutations that mapped outside the *lacZYA* region but affected the transcription of all three of these structural genes. *LacI* is now known to code for a protein whose only function is to control the transcription initiation of the three *lac* structural genes. This regulator protein is called the **lac repressor.** The *lacI* gene is located just in front of the controlling elements for the *lacZYA* gene cluster. However, it is not obligatory that a regulatory gene be physically close to the gene cluster that it regulates. In some of the other operons it is not. Transcription of *lacI* is not regulated; instead, this single gene is always transcribed from its own promoter at a low rate that is relatively independent of the cell's status. Therefore, the affinity of the *lacI* promoter for RNA polymerase seems to be the only factor involved in its transcription initiation.

The *lac* repressor is initially synthesized as monomers of 360 amino acids that associate to form a tetramer, the active form of the repressor. Usually there are about 10 tetramers per cell. The repressor has a strong affinity for a specific DNA sequence that lies between *lacP* and the start of *lacZ*. This sequence is called the **operator** and is designated *lacO*. The operator overlaps the promoter somewhat so that the presence of the repressor bound to the operator physically prevents RNA polymerase from binding to the promoter and initiating transcription.

In addition to recognizing and binding to the *lac* operator DNA sequence, the repressor also has a strong affinity for the inducer molecules of the *lac* operon. Each monomer has a binding site for an inducer molecule. The binding of inducer to the monomers causes an allosteric change in the repressor that greatly lowers its affinity for the operator sequence (Figure 19.3). In other words, when inducer molecules are bound to their sites on the repressor, a conformational change in the repressor occurs that alters the binding site for the operator. The result is that the repressor no longer binds to the operator so that RNA polymerase, in turn, can begin transcription from the promoter. A repressor molecule that is already bound to the operator when the inducer becomes available can still bind to the inducer so that the repressor–inducer complex immediately disassociates from the operator.

The study of the lactose operon has been greatly facilitated by the discovery that some small molecules fortuitously serve as inducers but are not actually metabolized by β-galactosidase. Isopropylthiogalactoside (IPTG) is onc of several thiogalactosides with this property. They are called **gratuitous inducers.** They bind to the inducer sites on the repressor molecule causing the conformational change but are not cleaved by the induced β-galactosidase. Therefore, they affect the system without themselves being altered (metabolized) by it. If it were not possible to manipulate the system with these gratuitous inducers experimentally, it would have been much more difficult to reach our current understanding of the lactose operon in particular and bacterial gene regulation in general.

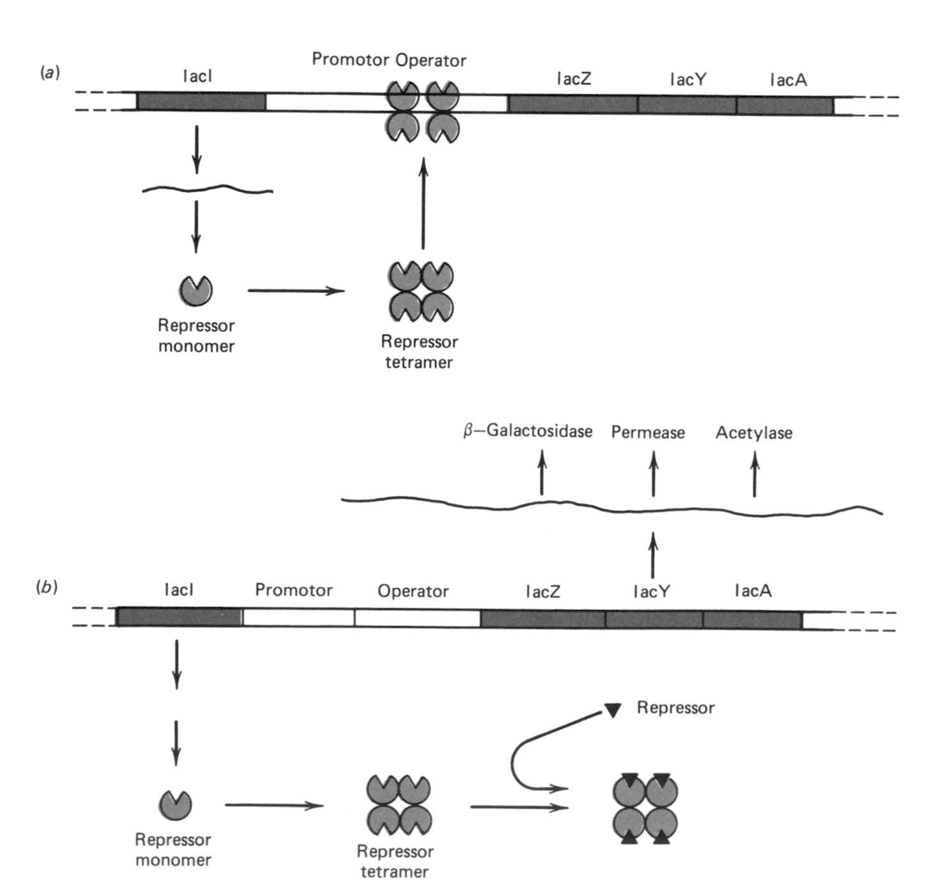

FIGURE 19.3
Control of the *lac* operon.
(*a*) The repressor tetramer binds to the operator and prevents transcription of the structural genes. (*b*) The inducer binds to the repressor tetramer, which prevents the repressor from binding to the operator. Transcription of the three structural genes can occur from the promotor.

The product of the *lacI* gene, the repressor protein, acts in trans; that is, it is a diffusible product that moves through the cell to its site of action. Therefore, mutations in the *lacI* gene can exert an effect on the expression of other genes located far away or even on genes located on different DNA molecules. *LacI* mutations can be of several types. One class of mutations changes or deletes amino acids of the repressor that are located in the binding site for the inducer. These changes interfere with the interaction between the inducer and the repressor but do not affect the affinity of the repressor for the operator. Therefore, the repressor is always bound to the operator, even in the presence of inducer, and the *lacZYA* genes are never transcribed above a very low basal level. Another class of *lacI* mutations changes the amino acids in the operator binding site of the repressor. Most of these mutations lessen the affinity of the repressor for the operator. This means that the repressor does not bind to the operator and the *lacZYA* genes are always being transcribed. These mutations are called **repressor-constitutive mutations** because the *lac* genes are permanently turned on. Interestingly, a few rare *lacI* mutants actually increase the affinity of the repressor for the operator over that of wild-type repressor. In these cases inducer molecules can still bind to the repressor, but they are less effective in releasing the repressor from the operator.

The repressor-constitutive mutants illustrate the features of a negative control system. An active repressor, in the absence of an inducer, shuts off the expression of the *lac* structural genes. An inactive repressor results in the constitutive, unregulated, expression of these genes. It is possible, using the recombinant DNA techniques described in Chapter 18, to introduce into constitutive *lacI* mutant cells a recombinant plasmid containing the wild-type *lacI* gene (but not the rest of the *lac* operon). Therefore, these cells have one wild-type and one mutant *lacI* gene and will synthesize both active and inactive repressor molecules. Under these conditions, normal wild-type regulation of the lactose operon occurs. In genetic terms, the wild-type induction is dominant over the mutant constitutivity. This property is the main feature of a negative control system.

The Operator Sequence of the Lactose Operon Is Contiguous on the DNA With a Promoter and Three Structural Genes

The known control elements in front of the structural genes of the lactose operon are the operator and the **promoter.** The operator was originally identified, like the *lacI* gene, by mutations that affected the transcription of the *lacZYA* region. Some of these mutations also result in the constitutive synthesis of *lac* mRNA, that is, they are **operator-constitutive mutations.** In these cases the operator DNA sequence has undergone one or more base pair changes so that the repressor no longer binds as tightly to the sequence. Thus, the repressor is less effective in preventing RNA polymerase from initiating transcription.

In contrast to mutations in the *lacI* gene that affect the diffusible repressor, mutations in the operator do not affect a diffusible product. They exert their influence on the transcription of only the three *lac* genes that lie immediately downstream of the operator on the same DNA molecule. This means that if a second *lac* operon is introduced into a bacterium on a recombinant plasmid, the operator of one operon does not influence action on the other operon. Therefore, an operon with a wild-type operator will be repressed under the usual conditions, whereas in the same bacterium a second operon that has an operator-constitutive mutation will be continuously transcribed.

Operator mutations are frequently referred to as **cis-dominant** to emphasize that these mutations affect only adjacent genes on the same DNA molecule and are not influenced by the presence in the cell of other copies of the unmutated sequence. Cis-dominant mutations occur in DNA sequences that are recognized by proteins rather than sequences that *code* for the diffusible proteins. **Trans-dominant** mutations occur in genes that specify the diffusible products. Therefore, cis-dominant mutations also occur in promoter and transcription termination sequences, whereas trans-dominant mutations occur in the genes for the subunit proteins of RNA polymerase, the ribosomes, and so on.

Figure 19.4 shows the sequence of both the *lac* operator and promoter. The operator sequence has an axis of dyad symmetry. The sequence of the upper strand on the left side of the operator is nearly identical to the lower strand on the right side; only three differences occur between these inverted DNA repeats. This symmetry in the DNA recognition sequence reflects symmetry in the tetrameric repressor. It probably facilitates the tight binding of the subunits of the repressor to the operator, although this has not been definitively demonstrated. A common feature of many protein binding or recognition sites on double-stranded DNA, including most recognition sites for restriction enzymes, is a dyad symmetry in the nucleotide sequence.

The 30 bp that constitute the *lac* operator are an extremely small fraction of the total *E. coli* genome of 4 million bp and occupy an even smaller fraction of the total volume of the cell. Therefore, it would seem that the approximately 10 tetrameric repressors in a cell might have trouble finding the *lac* operator if they just randomly diffuse about the cell. Although this remains a puzzling consideration, there are factors that confine the repressor to a much smaller space than the entire volume of the cell. First, it probably helps that the repressor gene is very close to the *lac* operator. This means that the repressor does not have far to diffuse if its translation begins before its mRNA is fully synthesized. Second, and more importantly, the repressor possesses a low general affinity for all DNA sequences. When the inducer binds to the repressor, its affinity for the operator is reduced about a 1000-fold, but its low affinity for random DNA sequences is unaltered. Therefore, all of the *lac* repressors of the cell probably spend the majority of the time in loose association with the DNA. As the binding of the inducer releases a repressor molecule from the operator, it quickly reassociates with another nearby region of the DNA. Therefore, induction redistributes the repressor on the DNA rather than generates freely diffusing repressor molecules. This confines the repressor to a smaller volume within the cell.

Another question is how does lactose enter a *lac* repressed cell in the first place if the *lacY* gene product, the permease, is repressed yet is required for lactose transport across the cell membrane? The answer is that even in the fully repressed state, there is a very low basal level of transcription of the *lac* operon that provides five or six molecules of the

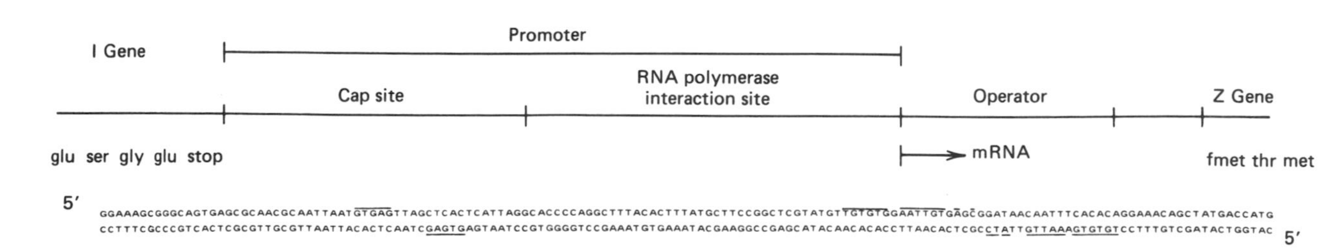

FIGURE 19.4
Nucleotide sequence of the control elements of the lactose operon.
The end of the *I* gene (coding for the lactose repressor) and the beginning of the *Z* gene (coding for β-galactosidase) are also shown. Lines above and below the sequence indicate symmetrical sequences within the CAP site and the operator.

permease per cell. Perhaps this is just enough to get a few molecules of lactose inside the cell and begin the process.

An even more curious observation is that, in fact, lactose is not the natural inducer of the lactose operon as we would expect. When the repressor is isolated from fully induced cells, the small molecule bound to each repressor monomer is *allolactose,* not lactose. Allolactose, like lactose, is composed of galactose and glucose, but the linkage between the two sugars is different. It turns out that a side reaction of β-galactosidase (which normally breaks down lactose to galactose and glucose) converts these two products to allolactose. Therefore, it appears that a few molecules of lactose are taken up and converted by β-galactosidase to allolactose, which then binds to the repressor and induces the operon. Further confirmation that lactose itself is not the real inducer comes from experiments indicating that lactose binding to the purified repressor slightly increases the repressor's affinity for the operator. Therefore, in the induced state a small amount of allolactose must be present in the cell to overcome this ''antiinducer'' effect of the lactose substrate.

The Promoter Sequence of the Lactose Operon Contains Recognition Sites for RNA Polymerase and a Regulator Protein

Immediately in front of the *lac* operator sequence is the promoter sequence. This sequence contains the recognition sites for two different proteins, RNA polymerase and the **CAP binding protein** (Figure 19.4). The site at which RNA polymerase interacts with the DNA to initiate transcription has been identified using several different genetic and biochemical approaches. Point mutations that occur in this region frequently affect the affinity to which RNA polymerase will bind the DNA. Deletions (or insertions) that extend into this region also dramatically affect the binding of RNA polymerase to the DNA. The end points of the sequence to which RNA polymerase binds have been identified by DNase protection experiments. Purified RNA polymerase was bound to the *lac* promoter region cloned in a bacteriophage DNA or a plasmid, and this protein–DNA complex was digested with DNase I. The DNA segment protected from degradation by DNase was recovered and its sequence determined. The ends of this protected segment varied slightly with different DNA molecules but corresponded closely to the boundaries of the RNA polymerase interaction site shown in Figure 19.4.

The sequence of the RNA polymerase interaction site is not composed of symmetrical elements similar to those described for the operator sequence. This is not surprising since RNA polymerase must associate with the DNA in an assymmetrical fashion for RNA synthesis to be initiated in only one direction from the binding site. However, that portion of the promoter sequence that is recognized by the CAP binding protein does contain some symmetry. A DNA–protein interaction at this region enhances transcription of the *lac* operon as described in the next section.

The Catabolite Activator Protein Binds at a Site on the Lactose Promoter

Escherichia coli prefers to use glucose instead of other sugars as a carbon source. For example, if the concentrations of glucose and lactose in the medium are the same, the bacteria will selectively metabolize the glucose and not utilize the lactose. This phenomenon is illustrated in Figure 19.5, which shows that the appearance of β-galactosidase, the *lacZ* product, is delayed until all of the glucose in the medium is depleted. Only then can lactose be used as the carbon source. This delay indicates that glucose interferes with the induction of the lactose operon. This effect is called

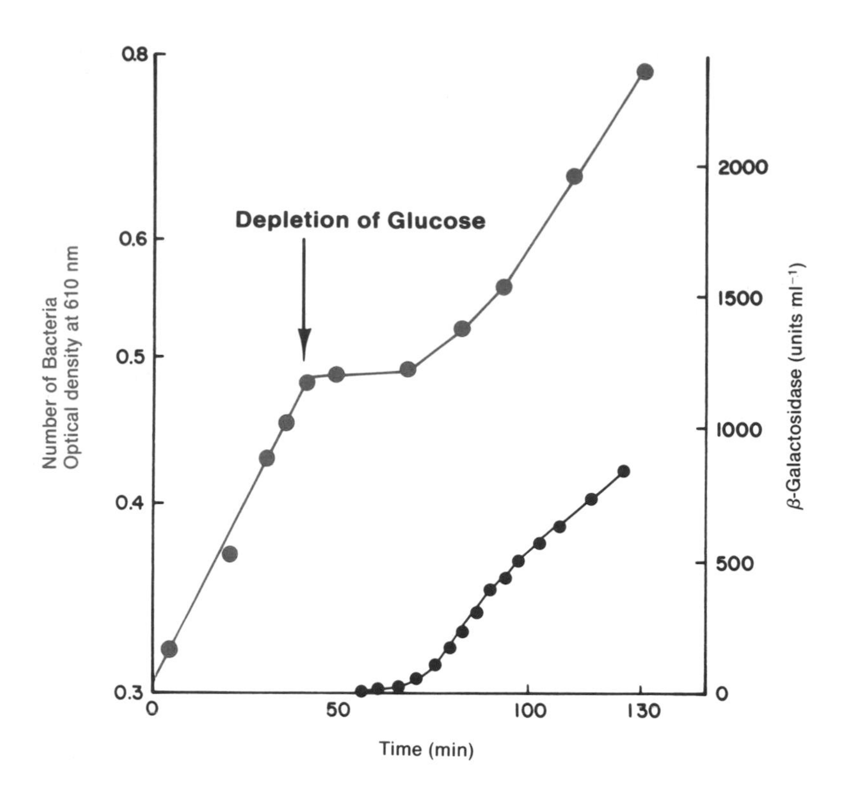

FIGURE 19.5
Lack of synthesis of β-galactosidase in *E. coli* when glucose is present.
The bacteria are growing in a medium containing initially 0.4 mg mL^{-1} of glucose and 2 mg mL^{-1} lactose. The left-hand ordinate indicates the cell density of the growing culture, an indicator of the number of bacterial cells. The right-hand ordinate indicates the units of β-galactosidase per milliliter. Note that the appearance of β-galactosidase is delayed until the glucose is depleted.
Redrawn from W. Epstein, S. Naono, and F. Gros, *Biochem. Biophys. Res. Commun.* 24:588, 1966.

catabolite repression because it occurs during the catabolism of glucose and may be due to a catabolite of glucose rather than glucose itself. An identical effect is exerted on a number of other inducible operons, including the arabinose and galactose operons, which code for enzymes involved in the utilization of various substances as energy sources. It probably is a general coordinating system for turning off synthesis of unwanted enzymes whenever the preferred substrate, glucose, is present.

Catabolite repression begins in the cell when glucose lowers the concentration of intracellular cyclic AMP (cAMP). The exact mechanism by

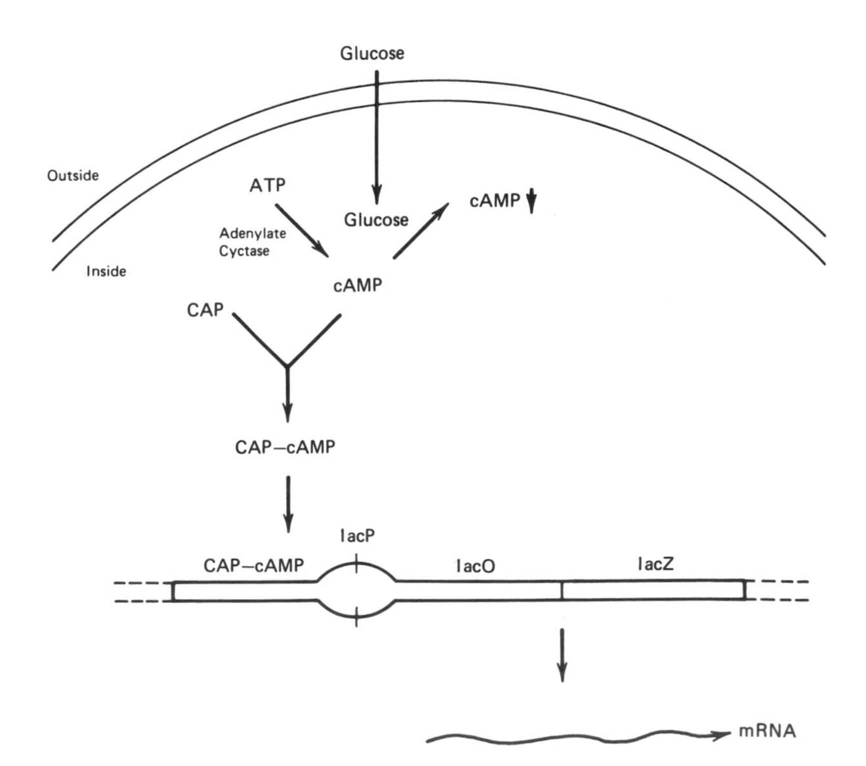

FIGURE 19.6
Control of *lacP* by cAMP.
A CAP–cAMP complex binds to the CAP site and enhances transcription at *lacP*. Catabolite repression occurs when glucose lowers the intracellular concentration of cAMP. This reduces the amount of the CAP–cAMP complex and decreases transcription from *lacP* and from the promotors of several other operons.

which this reduction in the cAMP level is accomplished is not known. Perhaps glucose influences either the rate of synthesis or degradation of cAMP. At any rate cAMP can bind to another regulatory protein, which has not been discussed yet, called **CAP** (for **catabolite *a*ctivator *p*rotein**) or CRP (for *c*AMP *r*eceptor *p*rotein). CAP is an allosteric protein, and when it is combined with cAMP, it is capable of binding to the CAP regulatory site that is at the promoter of the *lac* (and other) operons. The CAP–cAMP complex exerts positive control on the transcription of these operons. Its binding to the CAP site on the DNA facilitates the binding of RNA polymerase to the promoter (Figure 19.6). Alternatively, if the CAP site is not occupied, RNA polymerase has more difficulty binding to the promoter, and transcription of the operon occurs much less efficiently. Therefore, when glucose is present, the cAMP level drops, the CAP–cAMP complex does not form, and the positive influence on RNA polymerase does not occur. Conversely, if glucose is absent, the cAMP level is high, a CAP–cAMP complex binds to the CAP site, and transcription is enhanced.

19.4 THE TRYPTOPHAN OPERON OF *E. COLI*

Tryptophan is essential for bacterial growth; it is needed for the synthesis of all proteins that contain tryptophan. Therefore, if tryptophan is not supplied in sufficient quantity by the medium, the cell must make it. In contrast, lactose is not absolutely required for the cell's growth; many other sugars can substitute for it, and, in fact, as we saw in the previous section, the bacterium prefers to use some of these other sugars for the carbon source. As a result, synthesis of the tryptophan biosynthetic enzymes is regulated differently than synthesis of the proteins coded for in the lactose operon.

The Tryptophan Operon Is Controlled by a Repressor Protein

In *E. coli* tryptophan is synthesized from chorismic acid in a five-step pathway that is catalyzed by three different enzymes as shown in Figure 19.7. The **tryptophan operon** contains the five structural genes that code for these three enzymes (two of which have two different subunits). Upstream from this gene cluster is a promoter where transcription begins and an operator to which binds a repressor protein that is coded for by the unlinked *trpR* gene. Transcription of the lactose operon is generally "turned off" unless it is *induced* by the small molecule inducer. The tryptophan operon, on the other hand, is always "turned on" unless it is *repressed* by the presence of a small molecule **corepressor** (a term used to distinguish it from the repressor protein). Hence, the *lac* operon is inducible, whereas the *trp* operon is repressible. When the tryptophan operon is being actively transcribed, it is said to be **derepressed;** that is, the *trp* repressor is not preventing RNA polymerase from binding. This is mechanistically the same as an induced lactose operon in which the *lac* repressor is not interfering with RNA polymerase.

The biosynthetic pathway for tryptophan synthesis is regulated by mechanisms that affect both the synthesis and the activity of the enzymes that catalyze the pathway. For example, anthranilate synthetase, which catalyzes the first step of the pathway, is coded for by the *trpE* and *trpD* genes of the *trp* operon. The number of molecules of this enzyme that is present in the cell is determined by the transcriptional regulation of the *trp* operon. However, the catalytic activity of the existing molecules of the enzyme is regulated by feedback inhibition. This is a common short-term means of regulating the first committed step in a metabolic pathway. In

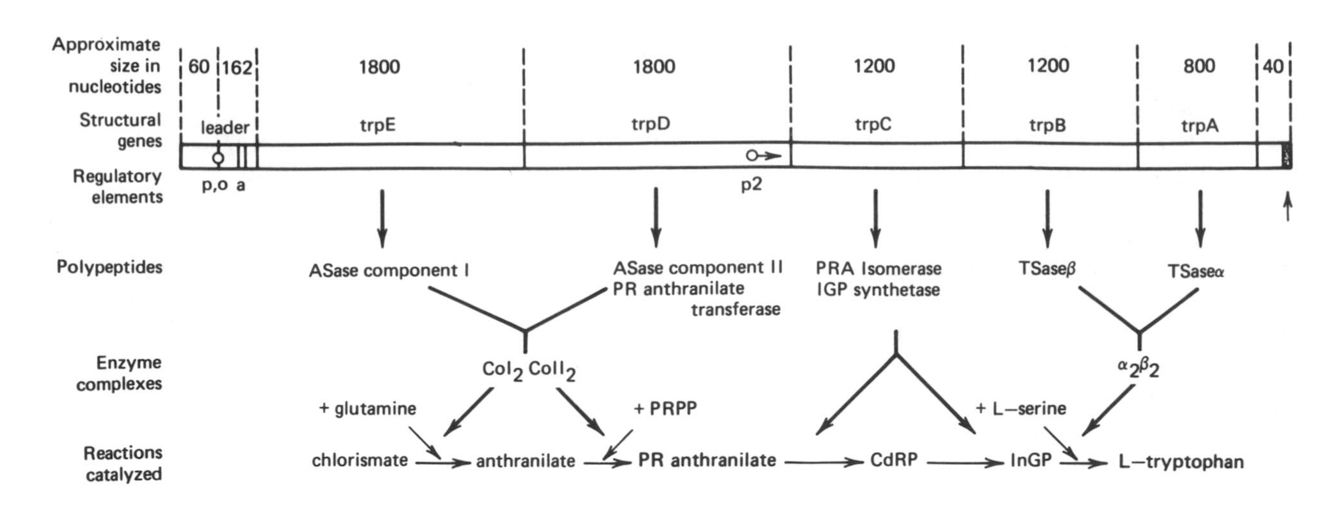

FIGURE 19.7
Genes of the tryptophan operon of *E. coli.*
Regulatory elements are the primary promoter (*trpP*), the operator (*trpO*), the attenuator (*trp a*), the secondary internal promoter (*trpP2*), and the terminator (*trp t*). Sites of mRNA initiation are given by ϕ, sites of termination by ■. CoI_2 and $CoII_2$ signify components I and II, respectively, of the anthranilate synthetase complex; PR-anthranilate is *N*-5′ phosphoribosyl-anthranilate; CdRP is 1-(*o*-carboxy-phenylamino)-l-deoxyribulose-5-phosphate; InGP is indole-3-glycerol phosphate; and PRPP is 5-phosphoribosyl-1-pyrophosphate.
Redrawn from T. Platt, The tryptophan operon, J. H. Miller and W. Reznikoff (Eds.). *The operon*. New York: Cold Spring Harbor, 1978, p. 263.

this case, tryptophan, the end product of the pathway, can bind to an allosteric site on the anthranilate synthetase and interfere with its catalytic activity at another site. Therefore, as the concentration of tryptophan builds up in the cell, it begins to bind to anthranilate synthetase and immediately decreases its activity on the substrate, chorismic acid. In addition, as we shall see, tryptophan also acts as a corepressor to shut down the synthesis of new enzyme molecules from the *trp* operon. Therefore, feedback inhibition is a short-term control that has an immediate effect on the pathway, whereas repression takes a little longer but has the more permanent effect of reducing the number of enzyme molecules.

The *trp* repressor is a tetramer of four identical subunits of about 100 amino acids each. Under normal conditions about 20 molecules of the repressor tetramer are present in the cell. The repressor by itself does not bind to the *trp* operator. It must be complexed with tryptophan in order to bind to the operator and therefore acts in vivo only in the presence of tryptophan. This is exactly the opposite of the *lac* repressor, which binds to its operator only in the absence of its small molecule inducer.

Interestingly, the *trp* repressor also regulates transcription of *trpR*, its own gene. As the *trp* repressor accumulates in the cell, the repressor–tryptophan complex binds to a region upstream of this gene turning off its transcription and maintaining the equilibrium of 20 repressors per cell. In addition, the repressor–tryptophan complex represses transcription of still another gene, *aroH*. This gene is not linked to any of the other genes of the *trp* operon. However, it codes for one of three enzymes that catalyze the first steps in the common pathway of aromatic amino acid biosynthesis. Therefore, the *trp* repressor influences the level of other amino acids besides tryptophan. The genes for the other two enzymes, *aroF* and *aroG* are controlled by other regulator molecules.

Another difference from the *lac* operon is that the *trp* operator occurs entirely within the *trp* promoter rather than adjacent to it, as shown in

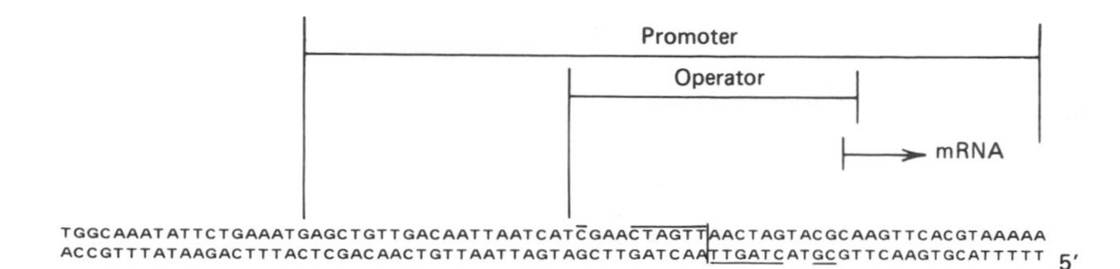

FIGURE 19.8
Nucleotide sequence of the control elements of the tryptophan operon.
Lines above and below the sequence indicate symmetrical sequences within the operator.

Figure 19.8. The operator sequence is a region of dyad symmetry, and the mechanism of preventing transcription is the same as in the *lac* operon. Binding of the repressor–corepressor complex to the operator physically blocks the binding of RNA polymerase to the promoter.

Repression results in only about a 70-fold decrease in the rate of transcription initiation at the *trp* promoter. (In contrast, the basal level of the *lac* gene products is about 1000-fold lower than the induced level.) However, the *trp* operon contains additional regulatory elements that impose further control on the extent of its transcription. One of these additional control sites is a secondary promoter, designated *trpP2,* which is located within the coding sequence of the *trpD* gene (shown in Figure 19.7). This promoter is not regulated by the *trp* repressor. Transcription from it occurs constitutively at a relatively low rate and is terminated at the same location as transcription from the regulated promoter for the whole operon, *trpP*. The resulting transcription product from *trpP2* is an mRNA that contains the coding sequences for *trpCBA,* the last three genes of the operon. Therefore, two polycistronic mRNAs are derived from the *trp* operon, one containing the coding sequences of all five structural genes and one possessing only the last three genes. Under conditions of maximum repression the basal level of mRNA coding sequence for the last three genes is about five times higher than the basal mRNA level for the first two genes.

The reason for the need of a second internal promoter is not clear, but there are several possibilities. Perhaps the best alternative comes from the observation that three of the five proteins do not contain tryptophan; only the *trpB* and *trpC* genes contain the single codon that specifies tryptophan. Therefore, under extreme tryptophan starvation, these two proteins would not be synthesized, which would prevent the pathway from being activated. However, since both of these genes lie downstream of the unregulated second promoter, their protein products will always be present at the basal level necessary to maintain the pathway.

The Tryptophan Operon Has a Second Control Site: The Site for Attenuation

Another important control element of the *trp* operon that is not present in the *lac* operon is the **attenuator** site (Figure 19.9). It lies within the 162 nucleotides between the start of transcription from *trpP* and the initiator codon of the *trpE* gene. Its existence was first deduced by the identification of mutations that mapped in this region and increased transcription of all five structural genes. Within the 162 nucleotides, called the **leader sequence,** are 14 adjacent codons that begin with a methionine codon and are followed by an in-phase termination codon. These codons are preceded by a canonical ribosome binding site and could potentially specify a 14 residue leader peptide. This peptide has never been detected in bacterial cells, perhaps because it is degraded very rapidly. It has been shown that the ribosome binding site does function properly when its corresponding DNA sequence is ligated upstream of a structural gene using recombinant DNA techniques.

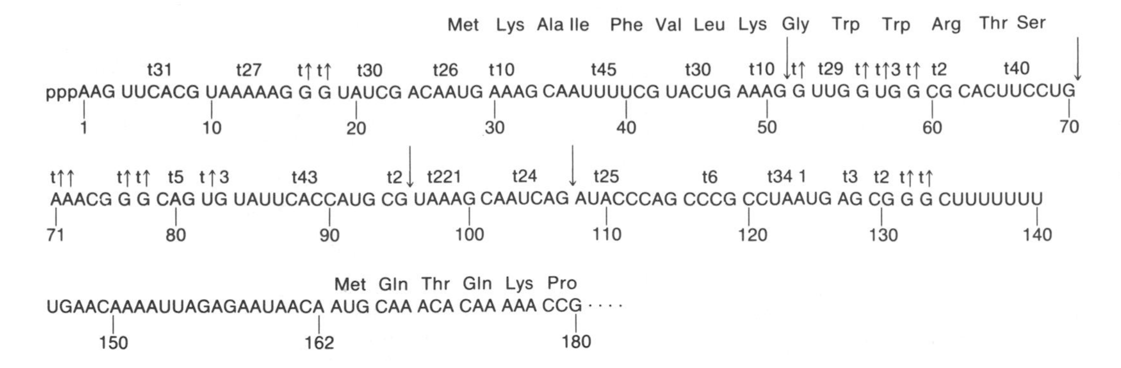

FIGURE 19.9
Nucleotide sequence of leader RNA from the *trp* operon.
The 14 amino acids of the putative leader peptide are indicated over their codons.
Redrawn with permission from D. L. Oxender, G. Zurawski, and C. Yanofsky, *Proc. Natl. Acad. Sci. U.S.A.* 76:5524, 1979.

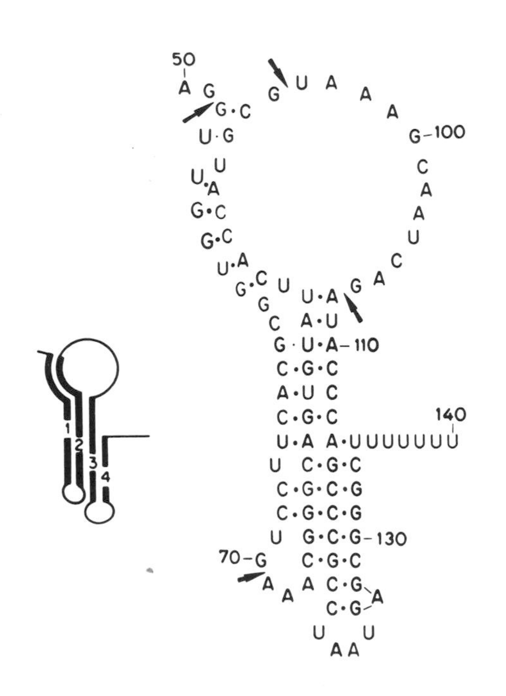

FIGURE 19.10
Schematic diagram showing the proposed secondary structures in *trp* leader RNA from *E. coli*.
Four regions can base pair to form three stem and loop structures. These are shown as 1–2, 2–3, and 3–4.
Reproduced with permission from D. L. Oxender, G. Zurawski, and C. Yanofsky, *Proc. Natl. Acad. Sci. U.S.A.* 76:5524, 1979.

The attenuator region provides RNA polymerase with a second chance to stop transcription if the *trp* enzymes are not needed by the cell. In the presence of tryptophan, it acts like a rho-independent transcription termination site to produce a short 140 nucleotide transcript. In the absence of tryptophan, it has no effect on transcription, and the entire polycistronic mRNA of the five structural genes is synthesized. Therefore, at both the operator and the attenuator, tryptophan exerts the same general influence. At the operator it participates in repressing transcription, and at the attenuator it participates in stopping transcription by those RNA polymerases that have escaped repression. It has been estimated that attenuation has about a 10-fold effect on transcription of the *trp* structural genes. When multiplied by the 70-fold effect of derepression at the operator, about a 700-fold range exists in the level at which the *trp* operon can be transcribed.

The molecular mechanism by which transcription is terminated at the attenuator site is a marvelous example of the cooperative interaction between bacterial transcription and translation to achieve the desired levels of a given mRNA. The first hints that ribosomes were involved in the mechanism of attenuation came from the observation that mutations in the gene for *trp*-tRNA synthetase (the enzyme that charges the tRNA with tryptophan) or the gene for an enzyme that modifies some bases in the tRNA prevents attenuation. Therefore, a functional *trp*-tRNA must participate in the process.

The leader peptide (Figure 19.9) of 14 residues contains two adjacent tryptophans in positions 10 and 11. This is unusual because tryptophan is a relatively rare amino acid in *E. coli*. It also provides a clue about the involvement of *trp*-tRNA in attenuation. If the tryptophan in the cell is low, the amount of charged *trp*-tRNA will also be low and the ribosomes may be unable to translate through the two *trp* codons of the leader peptide region. Therefore, they will stall at this place in the leader RNA sequence.

It turns out that the RNA sequence of the attenuator region can adopt several possible secondary structures (Figure 19.10). The position of the ribosome within the leader peptide-coding sequence determines the secondary structure that will form. This secondary structure, in turn, is recognized (or sensed) by the RNA polymerase that has just transcribed through the attenuator coding region and is now located a small distance downstream. The RNA secondary structure that forms when a ribosome is not stalled at the *trp* codons is a termination signal for the RNA polymerase. Under these conditions the cell does not need to make tryptophan, and transcription stops after the synthesis of a 140 nucleotide transcript, which is quickly degraded. On the other hand, the secondary structure that results when the ribosomes *are* stalled at the *trp* codons is not recognized as a termination signal, and the RNA polymerase continues on into the *trpE* gene. Figure 19.11 shows these different secondary structures in detail.

In the first structure in Figure 19.11, region 1 base pairs with region 2, and region 3 pairs with region 4. The two *trp* codons, UGG, occur at the beginning of the base paired region 1. The pairing between regions 3 and 4 results in a hairpin loop because of base pairing between the G and C residues. This is followed by eight U residues, a common feature of sequences that occur at sites of transcription termination. As the leader RNA sequence is being synthesized in the presence of tryptophan, it is likely that a loop between regions 1 and 2 will form first so that the region 3 and 4 loop will then occur and be recognized as a signal for termination by the RNA polymerase.

A different structure occurs if region 1 is prevented from base pairing with region 2 (Figure 19.11*b*). Under these circumstances, region 2 has the potential to base pair with region 3. This region 2 and 3 hairpin ties up the complementary sequence to region 4, which now must remain single

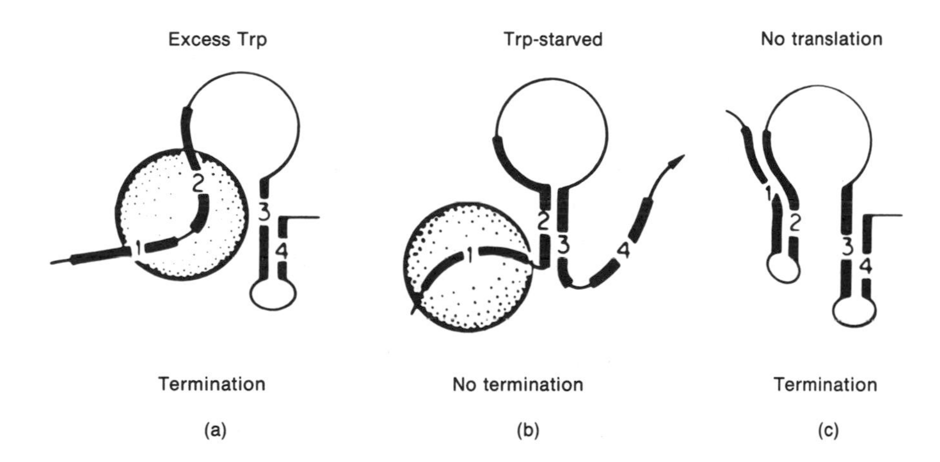

FIGURE 19.11
Schematic diagram showing the model for attenuation in the *trp* operon of *E. coli.*
(*a*) Under conditions of excess tryptophan, the ribosome (the shaded circle) translating the newly transcribed leader RNA will synthesize the complete leader peptide. During this synthesis the ribosome will mask regions 1 and 2 of the RNA and prevent the formation of stem and loop 1–2 or 2–3. Stem and loop 3–4 will be free to form and signal the RNA polymerase molecule (not shown) to terminate transcription. (*b*) Under conditions of tryptophan starvation, charged tryptophan-tRNA will be limiting, and the ribosome will stall at the adjacent *trp* codons in the leader peptide coding region. Because only region 1 is masked, stem and loop 2–3 will form, excluding the formation of stem and loop 3–4, which is required as the signal for transcription termination. Therefore RNA polymerase will continue transcription into the structural genes. (*c*) Under conditions in which the leader peptide is not translated, stem and loop 1–2 will form preventing the formation of stem and loop 2–3, and thereby permit the formation of stem and loop 3–4. This will signal transcription termination.
Reproduced with permission from D. L. Oxender, G. Zurawski, and C. Yanofsky, *Proc. Natl. Acad. Sci. U.S.A.* 76:5524, 1979.

stranded. Therefore, the region 3 and 4 hairpin loop that serves as the termination signal does not form, and the RNA polymerase continues on with its transcription. Thus, for transcription to proceed past the attenuator, region 1 must be prevented from pairing with region 2. This is accomplished if the ribosome stalls in region 1 due to an insufficient amount of charged *trp*-tRNA for translation of the leader peptide to continue beyond this point (Figure 19.11). When this happens, region 1 is bound within the ribosome and cannot pair with region 2. Since regions 2 and 3 are synthesized before region 4, they, in turn, will base pair before region 4 appears in the newly transcribed RNA. Therefore, region 4 remains single stranded, the termination hairpin does not form, and RNA polymerase continues transcription into the structural genes.

The tryptophan codons occur in region 1 right at the beginning of the hairpin between regions 1 and 2. This means that if the ribosome happens to stall at an earlier codon in the leader sequence, it will have little effect on attenuation. For example, starvation for lysine, valine, or glycine would be expected to reduce the amount of the corresponding charged tRNA and stall the ribosome at that codon, but a deficiency in these amino acids has no effect on transcription of the *trp* operon. An exception is arginine whose codon occurs immediately after the tryptophan codons. Starving for arginine does attenuate transcription termination somewhat, probably because of ribosome stalling at this codon, but to less of an extent than a deficiency in tryptophan.

Cis-acting mutations in the attenuator region support the alternate hairpin model. Most of these mutations result in increased transcription because they disrupt base pairing in the double-stranded portion of the termination hairpin and render it less stable. Some mutations, however, increase termination at the attenuator. One of these interferes with base pairing between regions 2 and 3, allowing region 3 to be available for pairing with region 4 even when region 1 is bound to a stalled ribosome. Another mutation occurs in the AUG initiator codon for the leader peptide so that the ribosome cannot begin its synthesis.

Transcription Attenuation Is a Mechanism of Control in Operons for Amino Acid Biosynthesis

Attenuation is a common phenomenon in bacterial gene expression; it occurs in at least six other operons that code for enzymes catalyzing amino acid biosynthetic pathways. Figure 19.12 shows the corresponding leader peptide sequences specified by each of these operons. In each

Operon	Leader peptide sequence	Regulatory amino acids
his	Met-Thr-Arg-Val-Gln-Phe-Lys-His-His-His-His-His-His-His-Pro-Asp	His
pheA	Met-Lys-His-Ile-Pro-Phe-Phe-Phe-Ala-Phe-Phe-Phe-Thr-Phe-Pro	Phe
thr	Met-Lys-Arg-Ile-Ser-Thr-Thr-Ile-Thr-Thr-Thr-Ile-Thr-Ile-Thr-Thr-Gly-Asn-Gly-Ala-Gly	Thr Ile
leu	Met-Ser-His-Ile-Val-Arg-Phe-Thr-Gly-Leu-Leu-Leu-Leu-Asn-Ala-Phe-Ile-Val-Arg-Gly-Arg-Pro-Val-Gly-Gly-Ile-Gln-His	Leu
ilv	Met-Thr-Ala-Leu-Leu-Arg-Val-Ile-Ser-Leu-Val-Val-Ile-Ser-Val-Val-Val-Ile-Ile-Ile-Pro-Pro-Cys-Gly-Ala-Ala-Leu-Gly-Arg-Gly-Lys-Ala	Leu, Val, Ile

FIGURE 19.12
Leader peptide sequences specified by biosynthetic operons of *E. coli*.
All contain multiple copies of the amino acid(s) synthesized by the enzymes coded for by the operon.

case, the leader peptide contains several codons for the amino acid end product of the pathway. The most extreme case is the 16 residue leader peptide of the histidine operon that contains seven contiguous histidines. Starvation for histidine results in a decrease in the amount of charged *his*-tRNA and a dramatic increase in transcription of the *his* operon. As with the *trp* operon, this effect is diminished by mutations that interfere with the level of charged *his*-tRNA. Furthermore, the nucleotide sequence of the attenuator region suggests that ribosome stalling at the histidine codons also influences the formation of alternate hairpin loops, one of which resembles a termination hairpin followed by several U residues. In contrast to the *trp* operon, transcription of the *his* operon is regulated entirely by attenuation; it does not possess an operator that is recognized by a repressor protein. Instead, the ribosome acts rather like a positive regulator protein, similar to the cAMP–CAP complex discussed with the *lac* operon. If the ribosome is bound to (i.e., stalled at) the attenuator site, then transcription of the downstream structural genes is enhanced. If the ribosome is not bound, then transcription of these genes is greatly reduced.

Transcription of some of the other operons shown in Figure 19.12 can be attenuated by more than one amino acid. For example, the *thr* operon is attenuated by either threonine or isoleucine; the *ilv* operon is attenuated by leucine, valine, or isoleucine. This effect can be explained in each case by stalling of the ribosome at the corresponding codon, which, in turn, interferes with the formation of a termination hairpin. Although it is not proven, it is possible that with some of the longer leader peptides stalling at more than one codon is necessary to achieve maximal transcription through the attenuation region.

19.5 OTHER BACTERIAL OPERONS

Many other bacterial operons have been studied and found to possess the same general regulatory mechanisms as the *lac, trp,* and *his* operons as discussed in Section 19.4. However, each operon has evolved its own distinctive quirks. For example, the **galactose operon** contains three structural genes coding for enzymes that metabolize galactose to glucose 1-phosphate. This operon is inducible. A repressor prevents transcription of the gene cluster unless galactose is present. Galactose binds to the repressor causing it to dislodge from the operator, which, in turn, allows transcription to begin. The cAMP–CAP complex greatly enhances this transcription. However, this operon has two different promoters and a different order to the operator, CAP binding site, and the promoters as shown in Figure 19.13. The operator occurs first in the sequence followed

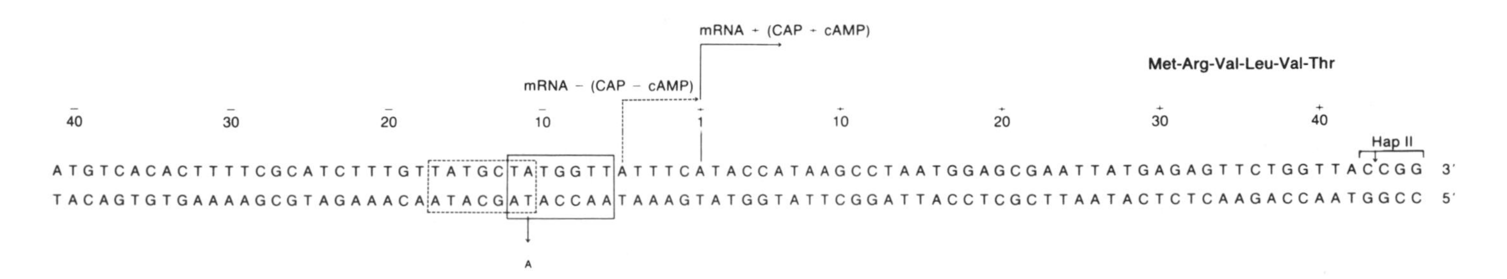

FIGURE 19.13
DNA sequence of the operator–promoter region of the galactose operon. The +1 corresponds to the start of the CAP–cAMP dependent mRNA. Two similar overlapping heptamers preceding each start site are boxed.

by the CAP binding site and the promoters. In the induced state (when galactose is present), transcription starts at promoter P1 if the cAMP–CAP complex is bound to the CAP site. If the CAP binding site is not occupied, however, transcription begins at a lower rate at a second promoter, P2, that is five base pairs in front of P1. These two overlapping promoters are close enough to the CAP binding site that CAP must interact directly with RNA polymerase when transcription begins at P1. Furthermore, because the operator and the promoters are separated by the CAP site, it is possible that the repressor does not directly block binding of the RNA polymerase at the promoters. Perhaps repressor binding distorts the double helix in the vicinity of the promoters so that they are not recognized by RNA polymerase.

Synthesis of Ribosomal Proteins Is Regulated in a Coordinated Manner

Another interesting group of operons are those that contain the structural genes for the 70 or more proteins that comprise the ribosome (Figure 19.14). Each ribosome contains one copy of each ribosomal protein (except for protein L7-L12, which is probably present in four copies). Therefore, all 70 proteins are required in equimolar amounts, and it makes sense that their synthesis is regulated in a coordinated fashion. The characterization of this set of operons is not yet complete, but six operons, containing about one-half of the ribosomal protein genes, have been identified in two major gene clusters. One cluster contains four adjacent operons (*str, Spc, S10,* and α), and the other two operons are near each other elsewhere in the *E. coli* chromosome. There seems to be no real pattern to the distribution of these genes among the different operons. Some operons contain genes for proteins of just one ribosomal subunit; others code for proteins of both subunits. In addition to the structural genes for the ribosomal proteins, these operons also contain genes for other (related) proteins. For example, the *str* operon contains genes for the two soluble elongation factors, EF-Tu and EF-G, as well as genes for some proteins in the 30S subunit. The α operon has genes for proteins of both the 30S and 50S subunits plus a gene for one of the subunits of RNA polymerase. The *rif* operon has genes for two other protein subunits of RNA polymerase and ribosomal protein genes.

A common theme among the six ribosomal operons is that their expression is regulated by one of their own structural gene products, that is, they are self-regulated. The precise mechanism of this self-regulation varies

Operon	Regulator	Proteins specified by the operon
Spc	S8	L14-L24-L5-S14-S8-L6-L18-S5-L15-L30
S10	L4	S10-L3-L2-L4-L23-S19-L22-S3-S17-L16-L29
str	S7	S12-S7-Ef · G-EF · Tu
α	S4	S13-S11-S4-α-L17
L11	L1	L11-L1
rif	L10	L10-L7-β-β'

FIGURE 19.14
Location of genes for ribosomal proteins of *E. coli.*
Genes for the protein components of the small (S) and large (L) ribosomal subunits of *E. coli* are clustered on several operons. Some of these operons also contain genes for RNA polymerase subunits ($\alpha\beta\beta'$) and protein synthesis factors (EF·G and EF·Tu). At least one of the protein products of each operon usually regulates expression of that operon (see text).

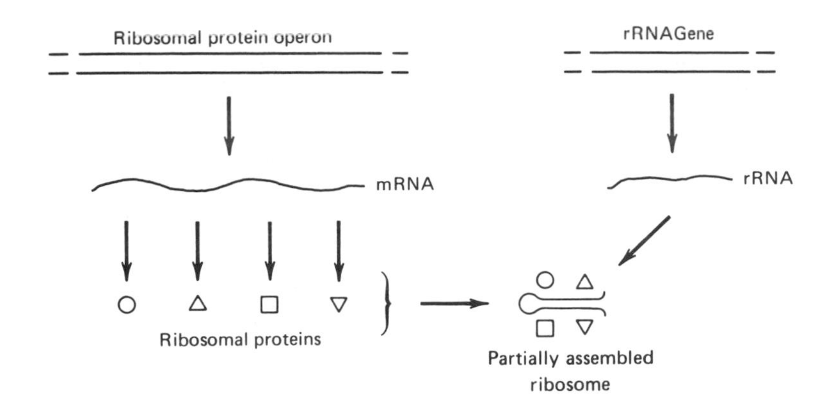

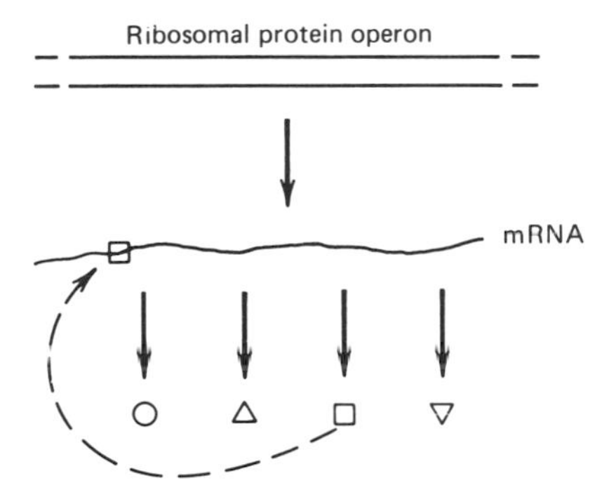

FIGURE 19.15
Self-regulation of ribosomal protein synthesis.
Individual ribosomal proteins bind to the polycistronic mRNA from their own operon, blocking further translation, if free rRNA is not available for the assembly of new ribosomal subunits.

considerably with each operon and is not yet understood in detail. However, in at least some cases the regulation occurs at the level of translation not transcription as discussed for the other operons. After the polycistronic mRNA is made, the "regulatory" ribosomal protein determines which regions, if any, are translated. In general, the ribosomal protein that regulates expression of its own operon, or part of its own operon, is a protein that is associated with one of the ribosomal RNAs (rRNAs) in the intact ribosome. This ribosomal protein has a high affinity for the rRNA and a lower affinity for one or more regions of its own mRNA. Therefore, a competition between the rRNA and the operon's mRNA for binding with the ribosomal protein occur. As the ribosomal protein accumulates to a higher level than the free rRNA, it binds to its own mRNA and prevents the initiation of protein synthesis at one or more of the coding sequences on this mRNA (Figure 19.15). As more ribosomes are formed, the excess of this particular ribosomal protein is used up and translation of its coding sequence on the mRNA can begin again.

The Stringent Response Controls the Synthesis of rRNAs and tRNAS

Bacteria have several ways in which to respond molecularly to emergency situations, that is, times of extreme general stress. One of these situations is when the cell does not have a sufficient pool of amino acids to maintain protein synthesis. Under these conditions the cell invokes what is called the **stringent response**, a mechanism that reduces the synthesis of the rRNAs and tRNAs about 20-fold. This places many of the activities within the cell on hold until conditions improve. The mRNAs are less affected, but there is also about a threefold decrease in their synthesis.

The stringent response is triggered by the presence of an uncharged tRNA in the A site of the ribosome. This occurs when the concentration

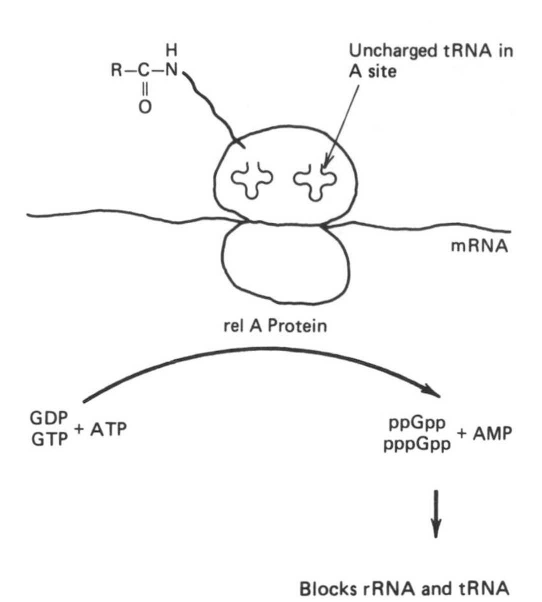

FIGURE 19.16
Stringent control of protein synthesis in *E. coli.*
During extreme amino acid starvation, an uncharged tRNA in the A site of the ribosome activates the rel A protein to synthesize ppGpp and pppGpp which, in turn, are involved in decreasing transcription of the genes coding for rRNAs and tRNAs.

of the corresponding charged tRNA is very low. The first result, of course, is that further peptide enlongation by the ribosome stops. This event causes a protein called the *stringent factor,* the product of the *relA* gene, to synthesize two small molecules, **guanosine tetraphosphate** (ppGpp) and **guanosine pentaphosphate** (pppGpp), from ATP and GTP or GDP as shown in Figure 19.16. The stringent factor is loosely associated with a few, but not all, ribosomes of the cell. Perhaps a conformational change in the ribosome is induced by the occupation of the A site by an uncharged tRNA, which, in turn, activates the associated stringent factor. The exact functions of ppGpp and pppGpp are not known. However, they seem to inhibit transcription initiation of the rRNA and tRNA genes. In addition they affect transcription of some operons more than others.

19.6 BACTERIAL TRANSPOSONS

Transposons Are Mobile Segments of DNA

So far we have only discussed the regulation of bacterial genes that are fixed in the chromosome. Their positions relative to the neighboring genes do not change. The vast majority of bacterial genes are of this type. In fact the genetic maps of *E. coli* and *Salmonella typhimurium* are quite similar, indicating the lack of much evolutionary movement of most genes within the bacterial chromosome. There is a class of bacterial genes, however, in which newly duplicated gene copies "jump" to another genomic site with a frequency of about 10^{-7} per generation, the same rate as spontaneous point mutations occur. The mobile segments of DNA containing these genes are called **transposable elements or transposons** (Figure 19.17).

Transposons were first detected as rare insertions of foreign DNA into structural genes of bacterial operons. Usually, these insertions interfere with the expression of the structural gene (into which they have inserted) and all downstream genes of the operon. This is not surprising since they can potentially destroy the translation reading frame, introduce transcription termination signals, affect the mRNA stability, and so on. A number of transposons and the sites into which they insert have now been isolated using recombinant DNA techniques and have been extensively characterized. These studies have revealed many interesting features about the mechanisms of transposition and the nature of genes located within transposons.

Different transposons vary tremendously in length. Some are a few thousand base pairs and contain only two or three genes; others are many thousands of base pairs long, possessing several genes. Several small transposons can occur within a large transposon. All transposons contain at least one gene that codes for a **transposase,** an enzyme required for the transposition event. Often they also contain genes that code for resistance to antibiotics or heavy metals. All transpositions involve the generation of an addition copy of the transposon and the insertion of this copy into another location. The original transposon copy is the same after the duplication as before; that is, the donor copy is unaffected by insertion of its duplicate into the recipient site. All transposons contain short **inverted terminal repeat sequences** that are essential for the insertion mechanism, and in fact these inverted repeats are often used to define the two boundaries of a transposon. The multiple target sites into which most transposons can insert seem to be fairly random in sequence; other transposons have a propensity for insertion at specific "hot spots." The duplicated transposon can be located in a different DNA molecule than its donor. Frequently, transposons are found on plasmids that pass from one bacte-

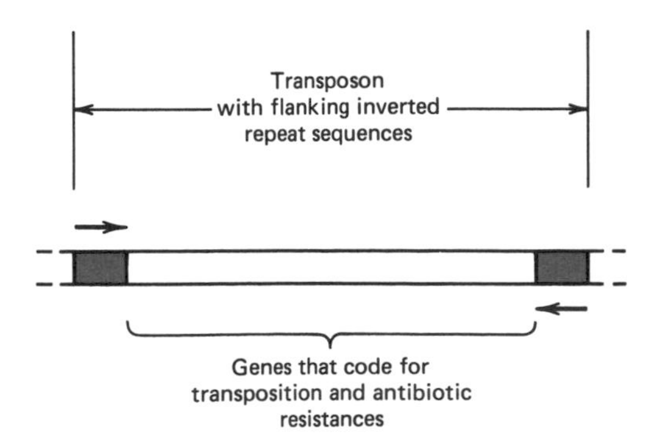

FIGURE 19.17
The general structure of transposons.
Transposons are relatively rare mobile segments of DNA that contain genes coding for their own rearrangement and (usually) genes that specify resistance to various antibiotics.

rial strain to another and are the source of a suddenly acquired resistance to one or more antibiotics by a bacterium (Clin. Corr. 19.1).

As with bacterial operons, each transposon or set of transposons has its own distinctive characteristics. It is beyond the scope of this chapter to compare different transposons, so one well-characterized transposon called *Tn3* will be discussed as an example of their general properties.

The *Tn3* Transposon Contains Three Structural Genes

The transposon *Tn3* has been cloned using recombinant DNA techniques and its complete sequence has been determined. Its general structure is shown in Figure 19.18. It contains 4957 base pairs including 38 base pairs at one end that occur as an inverted repeat at the other end. Three genes are present on *Tn3*. One gene codes for the enzyme β-lactamase, which hydrolyzes ampicillin and renders the cell resistant to this antibiotic. The other two genes, *tnpA* and *tnpR,* code for a transposase and a repressor protein, respectively. The transposase is composed of 1021 amino acids and it binds to single-stranded DNA. Little else is known about its action, but it is thought to recognize the repetitive ends of the transposon and to participate in the cleavage of the recipient site into which the new transposon copy inserts. The *tnpR* gene product is a protein of 185 amino acids. In its role as a repressor it controls transcription of both the transposase gene and its own gene. The *tnpA* and *tnpR* genes are transcribed divergently from a 163 base pair control region located between the two genes that is recognized by the repressor. In addition to serving as a repressor of these two genes, the *tnpR* product also participates in the recombination process that results in the insertion of the new transposon. Transcription of the ampicillin-resistance gene is not affected by the *tnpR* gene product.

Mutations in the transposase gene generally decrease the frequency of *Tn3* transposition, demonstrating its direct role in the transposition process. Mutations that destroy the repressor function of the *tnpR* product cause an increased frequency of transposition. These mutations derepress the *tnpA* gene resulting in more molecules of the transposase, which increases the formation of more transposons. They also derepress the *tnpR* gene but since the repressor is inactive, this has no effect on the system.

CLIN. CORR. 19.1 TRANSMISSIBLE MULTIPLE DRUG RESISTANCES

Pathogenic bacteria are becoming increasingly resistant to a large number of antibiotics, which is viewed with alarm by many physicians. Many cases have been documented in which a bacterial strain in a patient being treated with one antibiotic suddenly becomes resistant to that antibiotic and, simultaneously, to several other antibiotics even though the bacterial strain had never been previously exposed to these other antibiotics. This occurs when the bacteria suddenly acquire from another bacterial strain a plasmid that contains several different transposons, each containing one or more antibiotic-resistance genes. Examples include the genes encoding β-lactamase, which inactivates penicillins and cephalosporins, chloramphenicol acetyltransferase, which inactivates chloramphenicol, and phosphotransferases that modify aminoglycosides such as neomycin and gentamicin.

Olarte, J., Filloy, L., and Galindo, E. Resistance of Shigella dysenteriae type 1 to ampicillin and other antimicrobial agents: Strains isolated during a dysentery outbreak in a hospital in Mexico City. *J. Infect. Dis.* 133:572, 1976.

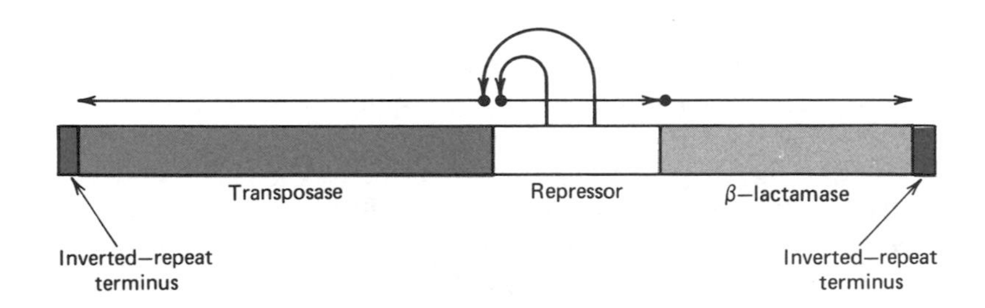

FIGURE 19.18
Functional components of the transposon *Tn3*.
Genetic analysis shows there are at least four kinds of regions: the inverted-repeat termini; a gene for the enzyme β-lactamase, which confers resistance to ampicillin and related antibiotics; a gene encoding an enzyme required for transposition (transposase); and a gene for a repressor protein that controls the transcription of the genes for transposase and for the repressor itself. The arrows indicate the direction in which DNA of various regions is transcribed.
Redrawn from S. N. Cohen and J. A. Shapiro, *Sci. Am.* 242:40, 1980. W. H. Freeman and Company, Copyright © 1980.

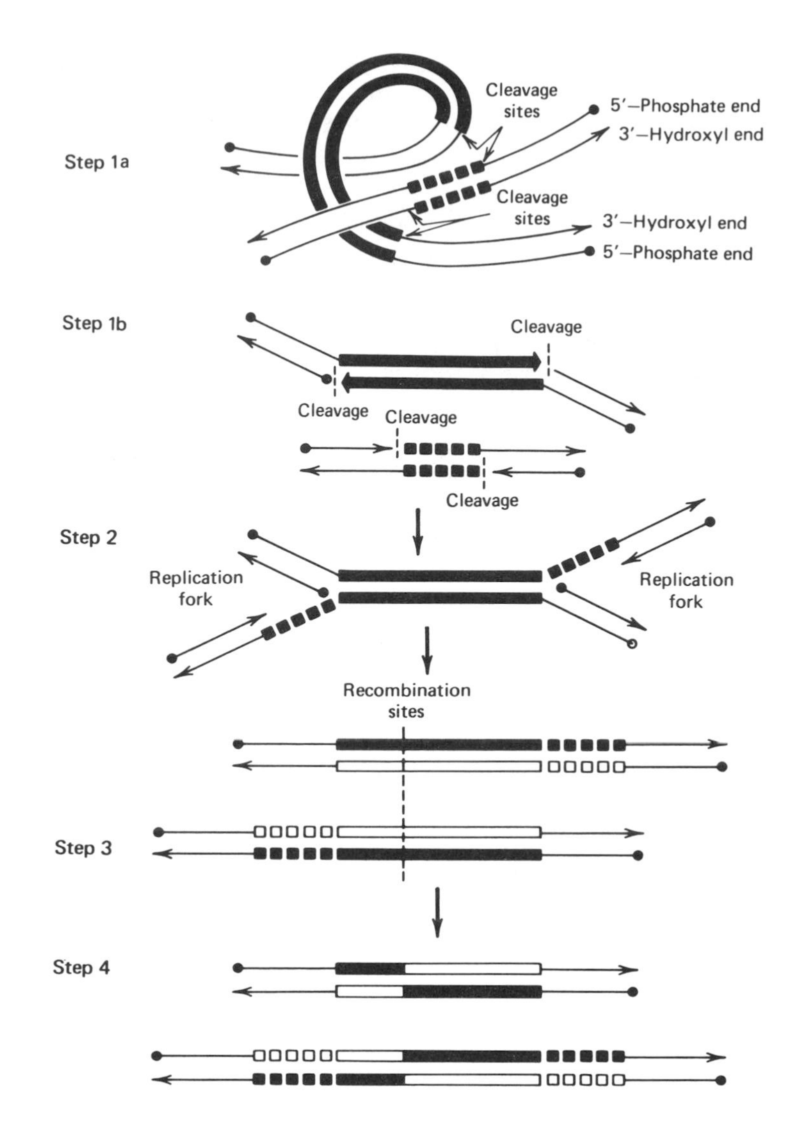

FIGURE 19.19
Proposed molecular pathway for transposition and chromosome rearrangements.
The donor DNA, including the transposon, is in solid bars; the recipient DNA is in squares. The pathway has four steps, beginning with single-strand cleavage (1a) at each end of the transposable element and at each end of the "target" nucleotide sequence to be duplicated. The cleavages expose (1b) the chemical groups involved in the next step: the joining of DNA strands from donor and recipient molecules in such a way that the double-stranded transposable element has a DNA-replication fork at each end (2). DNA synthesis (3) replicates the transposon (open bars) and the target sequence (□) accounting for the observed duplication. This step forms two new complete double-stranded molecules; each copy of the transposable element joins a segment of the donor molecule and a segment of the recipient molecules. (The copies of the element serve as linkers for the recombination of two unrelated DNA molecules.) In the final Step (4), reciprocal recombination between copies of the transposable element inserts the element at a new genetic site and regenerates the donor molecule. The mechanism of this recombination is not known.
Redrawn from S. N. Cohen and J. A. Shapiro, *Sci. Am.* 242:40, 1980. W. H. Freeman and Company, Copyright © 1980.

When a transposon, containing its terminal inverted repeats, inserts into a new site, it generates short (5–10 bp) direct repeats of the sequence at the recipient site that flank the new transposon. This is due to the mechanism of recombination that occurs during the insertion process, as illustrated in Figure 19.19. The first step in the process is the generation of staggered nicks at the recipient sequence. These staggered single-strand, protruding 5′ ends then join covalently to the inverted repeat ends of the transposon. The resulting intermediate then resembles two replicating forks pointing toward each other and separated by the length of the transposon. The replication machinery of the cell fills in the gaps and continues the divergent elongation of the two primers through the transposon region. This ultimately results in two copies of the transposon sequence. Reciprocal recombination within the two copies regenerates the original transposon copy at its old (unchanged) position and completes the process of forming a new copy at the recipient site that is flanked by direct repeats of the recipient sequence.

In recent years the practical importance of transposons located on plasmids has taken on increased significance for the use of antibiotics in the treatment of bacterial infections. Plasmids that have not been altered in the laboratory usually contain genes that facilitate their transfer from one bacterium to another. As the plasmids transfer (e.g., between different infecting bacterial strains), their transposons containing antibiotic-resistance genes are moved into new bacterial strains. Once inside a new bacterium, the transposon can be duplicated onto the chromosome and become permanently established in that cell's lineage. The result is that more and more pathogenic bacterial strains become resistant to an increasing number of antibiotics.

19.7 INVERSION OF GENES IN *SALMONELLA*

A different mechanism of differential gene regulation has been discovered for one set of genes in *Salmonella*. Since similar control mechanisms exist for the expression of other genes in other procaryotes (e.g., the bacteriophage μ), this gene system will be described briefly as an example of still another mechanism of gene regulation in procaryotes.

Bacteria move by waving their flagella. The flagella are composed predominantly of subunits of a protein called flagellin. Many *Salmonella* species possess two different flagellin genes and express only one of these genes at a time. The bacteria are said to be in *phase 1* if they are expressing the H1 flagellin gene and in *phase 2* if they are expressing the H2 flagellin gene. A bacterial clone in one phase switches to the other phase about once every 1000 divisions. This switch is called *phase variation,* and its occurrence is controlled at the level of transcription of the *H1* and *H2* genes.

The organization of the flagellin genes and their regulatory elements is shown in Figure 19.20. A 995 bp segment of DNA flanked by 14 bp repeats is adjacent to the *H2* gene and a *rhl* gene that codes for a repressor of *H1*. The *H2* and *rhl* genes are coordinately transcribed. Therefore, when *H2* is expressed, the repressor is also made and turns off *H1* expression. When *H2* protein and the repressor are not made, the *H1* gene is derepressed and H1 synthesis occurs.

The promoter for the operon containing *H2* and *rhl* lies near one end of the 995 base pair segment, just inside one copy of the 14 bp repeats. Furthermore, this segment can undergo inversions about the 14 base pair

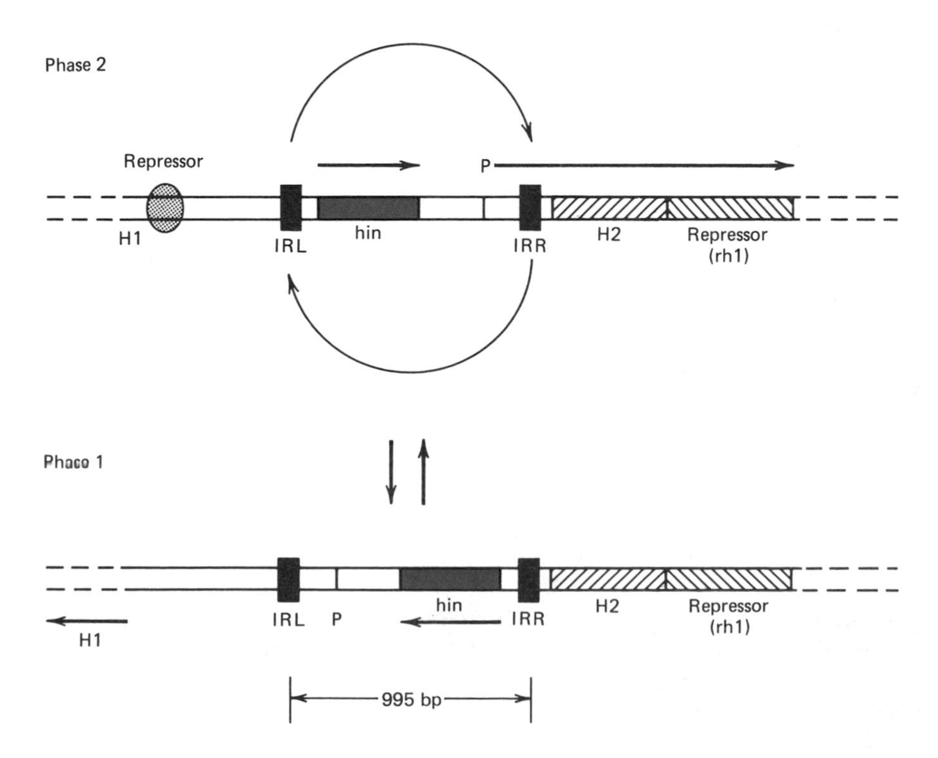

FIGURE 19.20
Organization of the flagellin genes.
The orientation of a 995 bp DNA segment flanked by 14 base pairs inverted repeats (IRL and IRR) controls the expression of the *H1* and *H2* flagellin genes. In phase 2, transcription initiates at promoter P within the invertable segment and continues through the *H2* and *rh1* genes. In phase 1, the orientation is reversed so that transcription of the *H2* and *rh1* genes does not occur.

repeats. In one orientation of the segment, the promoter is upstream of the *H2-rhl* transcription unit; in the other orientation it points toward the opposite direction so that *H2* and *rhl* are not transcribed. In addition to containing this promoter, the invertible segment of DNA possesses the *hin* gene whose product is an enzyme that catalyzes the inversion event itself. The *hin* gene seems to be transcribed constitutively at a low rate. Mutations in *hin* reduce the rate of inversion by 10,000-fold.

Therefore, phase variation is controlled by the physical inversion of the segment of DNA that removes a promoter from its position in front of the *H2-rhl* operon. When the promoter is in the opposite direction, it presumably still serves to initiate transcription, but the fate of that RNA is not known. It does not code for the *H1* that maps in this direction. That gene apparently has its own promoter that is controlled directly by the presence of the *rhl* repressor.

The inversion of the *hin* segment probably occurs via recombination between the 14 bp inverted repeats that is similar to recombination events involved in the transposition of a transposon. In fact, transposons have been shown to invert relative to their flanking sequences in a fashion exactly analogous to the *hin* inversion. Furthermore, the amino acid sequence of the *hin* product shows considerable homology with the *tnpR* product of the *Tn3* transposon, which, as described above, participates in the integration of the transposon into a new site. Thus, it is possible, and even likely, that the two processes are evolutionarily related.

19.8 ORGANIZATION OF GENES IN MAMMALIAN DNAs

The past 15 years have seen a virtual explosion of new information about the organization, structure, and regulation of genes in eucaryotic organisms. The reason for this enormous increase in our knowledge about eucaryotic genes has been the concurrent development of recombinant DNA techniques and the development of techniques for determining nucleotide sequences of DNA. These techniques are described in Chapter 18. Experiments undreamed of a few years ago are now routine accomplishments because of the use of these laboratory procedures.

The mammalian genome contains 4×10^9 bp of DNA, about 1000 times more DNA than the *E. coli* chromosome. Each mammalian cell contains virtually a complete copy of this genome, and all except the haploid germ line cells contain two copies. Different types of mammalian cells express widely different proteins even though each contains the same complement of genes. In addition, widely different patterns of protein synthesis occur at different developmental stages of the same type of cells. Therefore, extremely intricate and complicated mechanisms of regulation for these genes must exist, and, in fact, these mechanisms are not understood for even one mammalian gene to the extent that they are understood for many bacterial operons. Despite the great advances of the past 15 years, our understanding of gene regulation in mammals, and indeed all eucaryotes, remains fragmentary at best and most likely is still very naive.

Eucaryotic DNA Does Not Correlate With the Complexity or Function of an Organism

It was appreciated even before the advent of recombinant DNA methodology that eucaryotic cells, including mammalian cells, contain far more DNA than seems necessary to code for all of the required proteins. Furthermore, organisms that appear rather similar in complexity can have a several-fold difference in cellular DNA content. A housefly, for example, has about 6 times the cellular DNA content of a fruit fly. Some plant cells have almost 10 times more DNA than human cells. Therefore, DNA content does not always correlate with the complexity and diversity of functions of the organism.

It is difficult to obtain an accurate estimate of the number of different proteins, and therefore genes, that are found in a mammalian cell or in the entire mammalian organism. However, nucleic acid hybridization procedures have been used to determine that a maximum of 5000–10,000 different mRNAs may be present in a mammalian cell at a given time. Most of these mRNAs code for proteins that are common to many cell types. Therefore, a generous estimate might be that there are approximately 40,000 genes for the entire mammalian genome. If the average coding sequence is 1500 nucleotides (specifying a 500 amino acid protein), this accounts for 6% of the genome. Controlling elements, repetitive genes for rRNAs, and so on, may account for another 5–10%. However, as much as 80–90% of the mammalian genome may not have a direct genetic function. This is in contrast to the bacterial genome in which virtually all of the DNA is consumed by genes and their regulatory elements.

Eucaryotic Genes Usually Contain Intervening Sequences (Introns)

As discussed in Chapter 16, the coding sequences **(exons)** of eucaryotic genes are frequently interrupted by *intervening sequences* or **introns** that do not code for a product. These intervening sequences are transcribed into a precursor RNA species found in the nucleus and are removed by **RNA splicing** events in the processing of the precursor RNA to the mature mRNA found in the cytoplasm. The number and length of the intervening sequences in a gene can vary tremendously. Histone genes and interferon genes do not have intervening sequences; they contain a continuous coding sequence for the protein as do bacterial genes. The mammalian collagen gene, on the other hand, has over 50 different intervening sequences that collectively consume 10 times more DNA than the coding sequence for collagen. On the basis of the mammalian genes analyzed to date, it appears that most have one or more intervening sequences but that 50 introns in one gene is an extreme case. Therefore, intervening sequences of genes can account for some of the "excess" DNA present in eucaryotic genomes. The evolutionary significance of intervening sequences and their potential biological functions, if any, are the subject of much current speculation and experimentation. In many ways, they remain as big an enigma as they were when first discovered.

19.9 REPETITIVE DNA SEQUENCES IN EUCARYOTES

Another curiosity about mammalian DNA, and the DNA of most higher organisms, is that, in contrast to bacterial DNA, it contains repetitive sequences in addition to single copy sequences. This repetitive DNA falls into two general classes—highly repetitive simple sequences and moderately repetitive longer sequences of several hundred to several thousand base pairs.

The Importance of the Highly Repetitive Sequences Is Unknown

The **highly repetitive sequences** range from 5 to about 100 bp in length and occur in tandem. Their contribution to the total genomic size is extremely

Genome (%)	Number of copies in genome	Predominant sequence
25	1×10^7	5′ -ACAAACT- 3′ 3′ -TGTTTGA- 5′
8	3.6×10^6	5′ -A[T]AAACT- 3′ 3′ -T[A]TTTGA- 5′
8	3.6×10^6	5′ -ACAAA[T]T- 3′ 3′ -TGTTT[A]A- 5′

FIGURE 19.21
Main repeat units of repetitive sequences of *Drosophilia virilis*.
Approximately 41% of the genomic DNA of the fruit fly, *Drosophilia virilis*, is comprised of three related repeat sequences of seven base pairs. The bottom two sequences differ from the top sequence at one base pair shown in the box.

variable, but in most organisms these sequences are tandemly repeated millions of times and in some organisms they consume 50% or more of the total DNA. They are concentrated at the ends of chromosomes, that is, the telomeres, and at the chromosomal centromeres. Figure 19.21 shows the three main repeat units of the highly repetitive sequences of the fruit fly, *Drosophila virilis*. Repeats of these three sequences of seven base pairs comprise 41% of the organism's DNA. They are obviously related evolutionarily since two of the repeats can be derived from the third by a single base pair change. Relatively little transcription occurs from the highly repetitive sequences, and their biological importance remains, for the most part, a mystery. The repetitive sequences that occur near the telomeres are probably required for the replication of the ends of the linear DNA molecules. The ones that occur at the centromeres might play a structural role since this is the region of attachment during chromosome pairing and segregation in mitosis and meiosis.

A Variety of Repeating Units Are Defined as Moderately Repetitive Sequences

The **moderately repetitive sequences** comprise a large number of different sequences that are repeated to such different extents that it is somewhat misleading to group them under one heading. Some moderately repetitive sequences are clustered in one region of the genome; others are scattered throughout the DNA. Some moderate repeats are several thousand base pairs in length; other repeats come in a unit size of only a hundred base pairs. Sometimes the sequence is highly conserved from one repeat to another; in other cases, different repeat units of the same basic sequence will have undergone considerable divergence. To illustrate the diversity in the structures and functions of moderately repetitive sequences, two examples from the human genome will be briefly described.

In mammalian cells the 18S, 5.8S, and 28S rRNAs are transcribed as a single precursor transcript that is subsequently processed to yield the mature rRNAs. In humans the length of this precursor is 13,400 nucleotides, about one-half of which is comprised of the mature rRNA sequences. Several sequential posttranscriptional cleavage steps remove the extra sequences from the ends and the middle of the precursor RNA, releasing the mature rRNA species. The DNA that contains the corresponding rRNA genes is a moderately repetitive sequence of about 43,000 base pairs of which 30,000 bp are nontranscribed spacer DNA. Clusters of this entire DNA unit occur on five human chromosomes. In total there are about 280 repeats of this unit, which comprise about 0.3% of the total

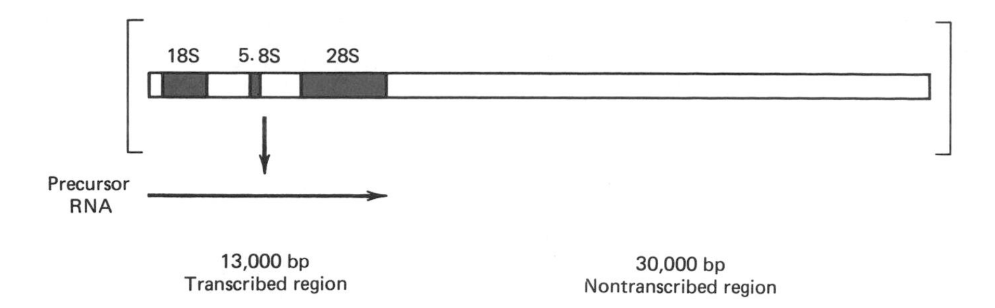

FIGURE 19.22
Repetitive sequence in human DNA for rRNA.
In human cells a single transcription unit of 13,000 nucleotides is processed to yield the 18S, 5.8S, and 28S rRNAs. About 280 copies of the corresponding rRNA genes are clustered on five chromosomes. Each repeat contains a nontranscribed spacer region of about 30,000 bp.

genome (Figure 19.22). The 5S *rRNA* genes are also repeated but in different clusters. The human genome contains about 2000 repeats of the 5S *rRNA* genes. The need for so many rRNA genes is because the rRNAs are structural RNAs. Each transcript from the gene yields only one rRNA molecule. On the other hand, each transcript from a ribosomal protein gene can be repeatedly translated to give many protein molecules.

In contrast to the repetitive rRNA genes that are clustered at only a few sites, most moderately repetitive sequences in the mammalian genome do not code for a stable gene product and are interspersed with nonrepetitive sequences that occur only once or a few times in the genome. The average size of these interspersed repetitive sequences is about 300 bp. Almost one-half of these sequences are members of a general family of moderately repetitive sequences called the **Alu family** because they can be cleaved by the restriction enzyme Alu I. There are about 300,000 Alu sequences scattered throughout the human haploid genome (on the high side of being moderately repetitive). Individual members are related in sequence but frequently are not identical. Their average homology with a consensus sequence is about 87%.

Interestingly, there is some repeat symmetry within an individual Alu sequence itself. The sequence appears to have arisen by a tandem duplication of a 130 base sequence with a 31-bp insertion in one of the two adjacent repeats. Furthermore, some members of the *Alu* family resemble bacterial transposons in that they are flanked by short direct repeats. This does not prove that an Alu repeat can be duplicated and transposed to another site like true transposons, but it suggests that such events may occur.

The biological function, if any, of Alu sequences is not known. One interesting suggestion is that they serve as the multiple origins for the DNA replication during the S phase of the cell, but more of these sequences occur than seem necessary for this function. Some Alu sequences appear in the intervening sequences of some genes and are transcribed as part of large precursor RNAs in which the Alu sequences are removed during RNA splicing to form the mature mRNA. Other Alu sequences are transcribed into small RNA molecules whose function is unknown. A final point is that all mammalian genomes appear to have a counterpart to the human interspersed Alu sequence family although the size of the repeat and its distribution can vary considerably from one species to another.

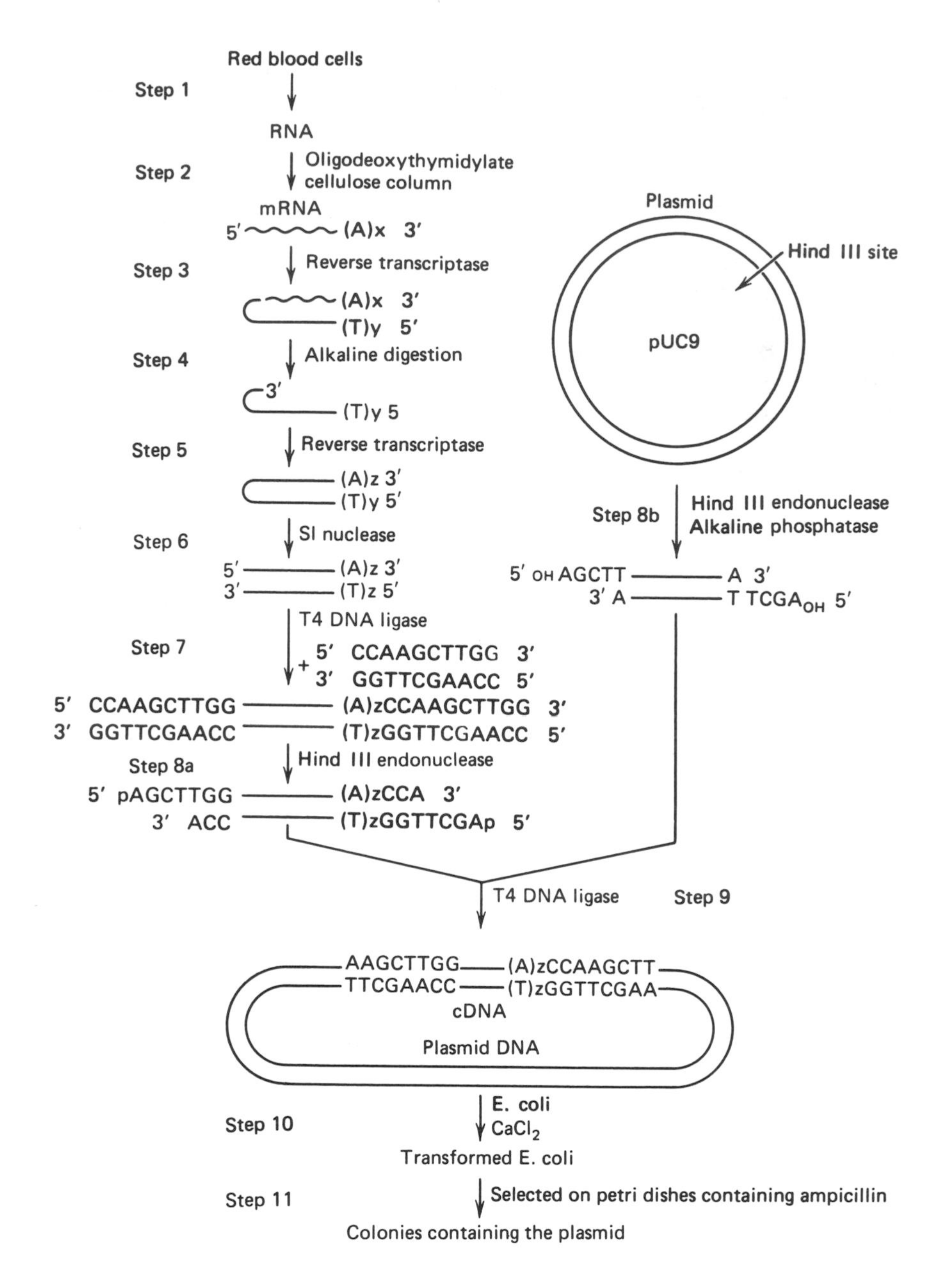

FIGURE 19.23
Cloning of globin cDNA.
Step 1: Total RNA is extracted from red cells. Step 2: The total RNA is passed through an oligodeoxythymidylate cellulose column. This column separates the polyadenylated mRNA (see Chapter 16) from rRNA and tRNA. The polyadenylated RNA is thought to contain significant amounts of mRNA coding for hemoglobin. Step 3: The mRNA is transcribed into cDNA using reverse transcriptase, the viral enzyme that transcribes DNA on RNA templates (see Chapter 16). Step 4: The mRNA is hydrolyzed with alkali. Step 5: The single-stranded cDNA is converted into a double-stranded DNA helix by using reverse transcriptase. Step 6: The resulting double helix contains a single-stranded hairpin loop that is removed by SI nuclease, an enzyme that hydrolyzes single-stranded DNA. Step 7: The cDNA is now a double helix with an unknown number of A–T base pairs at one end. In order to produce cohesive ends for the introduction of this cDNA into a plasmid, a chemically synthesized decanucleotide is attached to both ends using DNA ligase. This decanucleotide contains the palindromic symmetry recognized by the Hind III restriction nuclease. Step 8a: Treatment with Hind III restriction nuclease produces a cDNA molecule with Hind III cohesive ends. Step 8b: The plasmid pUC9, which contains an ampicillin-resistance gene, is cleaved with Hind III restriction endonuclease and exposed to bacterial alkaline phosphatase, an enzyme that

19.10 GENES FOR GLOBIN PROTEINS

Recombinant DNA Technology Has Been Used to Clone Genes for the Globin Molecules

Many of the mammalian structural genes that have been cloned by recombinant DNA techniques specify proteins that either occur in large quantity in a specific cell type, such as the globins of the red blood cell, or after induction of a specific cell type, e.g., growth hormone or prolactin in the pituitary. As a result, more is understood about the regulation of these genes than of other genes whose protein products occur at lower levels in many different cell types. We will discuss the organization, structure, and regulation of the related members of two gene families—the globin and the growth hormone-like proteins.

The first step in characterizing a eucaryotic gene is usually to use recombinant DNA techniques to clone a complementary DNA (cDNA) copy of that gene's corresponding mRNA. In fact, this is the reason that the most extensively studied mammalian genes are ones that code for the major proteins of specific cells; a large fraction of the total mRNA isolated from these cells codes for the protein of interest. The main constituent of red cells in human adults is hemoglobin, which is comprised of two α subunits (141 amino acids) and two β subunits (146 amino acids), each of which is complexed with a heme group. Almost all of the mRNA isolated from immature red cells (reticulocytes) codes for these two subunits of hemoglobin. Thus, the study of the structure and regulation of the globin genes began with the identification of cloned cDNA molecules that contain the coding sequences for α and β globin.

There are several experimental variations of the procedure for synthesizing double-stranded cDNA copies of isolated mRNA in vitro. As discussed in Chapter 18, several different plasmid and viral DNA vectors are available for cloning the (passenger) cDNA molecules. Figure 19.23 shows one protocol for constructing and cloning cDNAs prepared from mRNA of reticulocytes.

A synthetic oligonucleotide composed of 12–18 residues of deoxythymidine is hybridized to the 3′-polyadenylate tail of the mRNA and serves as a primer for reverse transcriptase, an enzyme that copies the RNA sequence into a DNA strand in the presence of the four deoxynucleoside triphosphates. The resulting RNA–DNA heteroduplex is treated with NaOH, which degrades the RNA strand and leaves the DNA strand intact. The 3′ end of the remaining DNA strand can then fold back and serve as a primer for initiating synthesis of a second DNA strand at random locations by *E. coli* DNA polymerase I. The hairpin loop is then nicked by S1 nuclease, an enzyme that cleaves single-stranded DNA but has little activity against double-stranded DNA. The 3′ ends of the resulting double-stranded cDNAs are ligated to small synthetic "linker" oli-

---

removes the phosphates from the cleaved 5′-terminal ends of the plasmids at the Hind III site. This prevents the cleaved plasmid from recircularizing without the insertion of the cDNA. Step 9: The linear plasmid and the cDNA molecules are mixed, and formation of circular, dimeric, "recombinant" DNA molecules is allowed to take place. DNA ligase is used to repair the breaks. Step 10: This mixture is used to transform an *E. coli* strain. Step 11: Individual *E. coli* cells that took up the plasmid were selected by their ability to grow on ampicillin. The globin cDNA is confirmed by determining the nucleotide sequence of the small DNA fragment released from the plasmid DNA by Hind III restriction endonuclease; if the observed nucleotide sequences corresponded to those expected based on the known amino acid sequence of α and β globin, then the cDNA is identified.

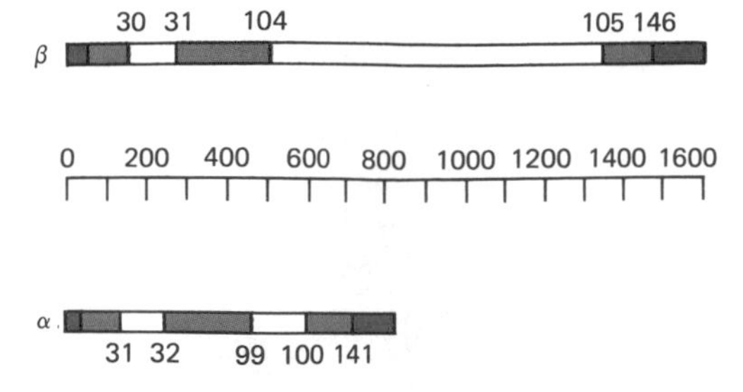

FIGURE 19.24
Structures of human globin genes.
Structures for the human α-like and β-like globin genes are drawn to approximate scale. Dark colored and open boxes represent coding (exon) and noncoding (intron) sequences, respectively. Light colored boxes indicate the (5′) upstream and (3′) downstream nontranslated regions in the RNA. The α-like globin genes contain introns of approximately 95 and 125 bp, located between codons 31 and 32 and 99 and 100, respectively. The β-like globin genes contain introns of approximately 125–150 and 800–900 bp, located between codons 30 and 31 and 104 and 105, respectively.

gonucleotides that contain the recognition site for the restriction enzyme Hind III. Digestion of the resulting DNA with Hind III generates DNA fragments that contain Hind III-specific ends. These fragments can be ligated into the Hind III site of a plasmid, and when the resulting circular "recombinant" DNA species are incubated with *E. coli* in the presence of cations such as calcium or rubidium, a few molecules will be taken up by the bacteria. The incorporated recombinant DNAs will be replicated and maintained in the progeny of the original transformed bacterial cell.

The collection of cloned cDNAs synthesized from the total mRNA isolated from a given tissue or cell type is called a **cDNA library,** for example, a liver cDNA library or a reticulocyte cDNA library. Since most of the mRNAs of a reticulocyte code for either α or β globin, it is relatively easy to identify these globin cDNAs in a reticulocyte cDNA library using procedures discussed in Chapter 16. Once identified, the nucleotide sequences of the cDNAs can be determined to confirm that they do code for the known amino acid sequences of the α and β globins. In cases in which the amino acid sequence of the protein is not known, other procedures (usually immunological) are used to confirm the identification of the desired cDNA clone.

Among the first things to be noticed in comparing the α and β globin cDNA sequences with the corresponding chromosomal globin genes, which have also been cloned using recombinant DNA techniques, is that all members of both sets of genes contain two introns at approximately the same positions relative to the coding sequences (Figure 19.24). The α- (and α-like) genes have an intron of 95 base pairs between codons 31 and 32 and a second intron of 125 bp between codons 99 and 100. The β- (and β-like) genes have introns of 125–150 and 800–900 base pairs located between codons 30 and 31 and codons 104 and 105, respectively. Although the function of introns is presently unknown, they do separate the coding sequences of different functional domains of *some* proteins, including the globins. The coding region between the two globin introns specifies the region of the protein that interacts with the heme group. The final coding region (after the second intron) encodes the region of the protein that provides the interface with the opposite subunit, that is the α-β protein–protein interaction. This separation of the coding sequences for functional domains of a protein by introns is not a general phenomenon, however. The positioning of introns in other genes seems to bear little relationship to the final three-dimensional structure of the encoded protein.

Different α-like and β-like globin subunits are synthesized at different developmental stages. For example, embryonic red cells contain a different hemoglobin tetramer than do adult red cells. These developmentally distinct subunits have slightly different amino acid sequences and oxygen affinities but are closely related to each other. In humans there are two α-like chains—ζ, which is expressed in the embryo during the first 8 weeks and α itself, which replaces ζ in the fetus and continues through adulthood. There are four β-like chains. Epsilon and γ are expressed in the embryo, γ in the fetus, and δ plus β in the adult. Adults possess 97% $\alpha_2\beta_2$, 2% $\alpha_2\delta_2$, and 1% $\alpha_2\gamma_2$, which persists from the fetus.

Each of the different globin chains is coded by at least one gene in the haploid genome. The α-like genes are clustered on human chromosome 16, and the β-like genes are clustered on the short arm of chromosome 11. The gene organization within these two clusters is shown in Figure 19.25. Interestingly, the genes within a cluster are positioned relative to one another in the order of both their transcriptional direction and their developmental expression; that is, 5′ — embryonic—fetal—adult—3′.

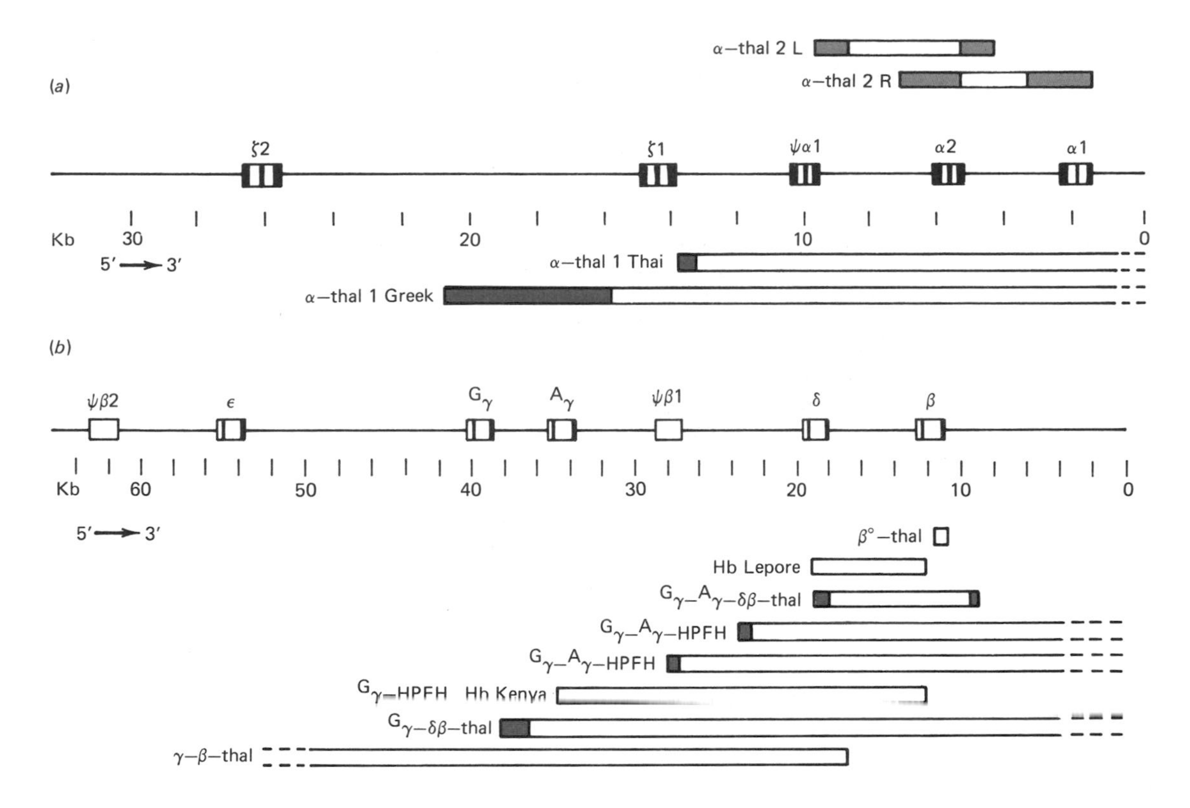

FIGURE 19.25
Gene organization for α-like and β-like genes of human hemoglobin.
(*a*) Linkage arrangement of the human α-like globin genes and locations of deletions within the α-like gene cluster. The positions of the adult (α1, α2) and embryonic (ζ1, ζ2) α-like globin genes and the α-like pseudogene (ψα1) are shown. For each gene the black and white boxes represent coding (exon) and noncoding (intron) sequences respectively. The introns in ζ2 are assumed to exist by analogy with the other α-like genes. The locations of deletions associated with the leftward and rightward types of α-thalassemia 2 are indicated by the rectangles labeled α-thal 2 L and α-thal 2 R. The crosshatched boxes at the ends of these rectangles indicate regions of sequence homology. The breakpoints of each type of α-thalassemia 2 deletion can occur anywhere within the regions of homology. The locations of deletions associated with two cases of α-thalassemia 1 (*α-thal 1 Thai* and *α-thal 1 Greek*) are shown below the linkage map. The stippled boxes indicate uncertainty in the extent of each deletion. (*b*) Linkage arrangement of the human β-like globin genes and locations of deletions within the β-like gene cluster. The positions of the embryonic (ε), fetal ($^{G}\gamma^{A}\gamma$), and adult (δ, β) β-like globin genes and the two β-like pseudogenes (ψβ1, ψβ2) are shown. For each gene the black and white boxes represent the coding (exon) and noncoding (intron) sequences, respectively. The distribution of coding and noncoding sequences within ψβ1 and ψβ2 is not known. The locations of various deletions within the gene cluster are presented below the map. Open boxes represent areas known to be deleted; dashed lines indicate that the endpoint of the deletion has not been determined; and stippled boxes represent uncertainty in the extent of the deletions. For δβ-thalassemia and HPFH, the type of fetal globin chain that is produced ($^{G}\gamma$ and/or $^{A}\gamma$) is indicated in the name of each syndrome (e.g., in $^{G}\gamma$-$^{A}\gamma$-δβ-thalassemia, the $^{G}\gamma$- and $^{A}\gamma$-globin chains are produced). The percentage of HbF observed in heterozygotes is given to the right of each deletion.
Redrawn from T. Maniatis, E. F. Fritsch, J. Lauer, and R. M. Lawn. *Annu. Rev. Genet.* 14:145, 1980. Copyright © 1980 *Annu. Rev. Inc.*

The α gene cluster encompasses about 28 kb and includes one ζ gene, one ζ pseudogene, two α genes, and one α pseudogene. The pseudogenes contain sequences that are very similar to the active genes, but a determination of their nucleotide sequences has revealed that they do not code for a functional globin subunit found in the red cell. Pseudogenes are, in fact, rather common in eucaryotic genomes. In most cases they do not seem to be deleterious to the organism and probably arose via a duplication of a segment of DNA followed by evolutionary mutation. Both α genes are active and code for identical proteins.

The β gene cluster contains five active genes and one pseudogene (for the γ subunit). Of the five functional genes, two are for the γ subunit and specify proteins that differ only at position 136, which is a glycine in the G variant and an alanine in the A variant. Only a single haploid gene exists for the ε-, δ-, and β-globin subunits.

Other mammalian species often have a different number of globin-like genes within the two clusters. For example, rabbits have only four β-like genes, goats have seven, and mice have as many as nine. At least some of these additional genes are pseudogenes similar to the gamma pseudogene in the human genome.

Many patients have been identified who have abnormalities in hemoglobin structure or expression. Furthermore, in many cases the precise molecular defect that is responsible for these abnormalities is known. The two that have been the most extensively studied are sickle cell anemia and a family of diseases collectively called *thalassemias*.

CLIN. CORR. **19.2** PRENATAL DIAGNOSIS OF SICKLE-CELL ANEMIA

Sickle-cell anemia can be diagnosed using fetal DNA obtained by amniocentesis. This genetic disease is caused by a single base pair change that converts a glutamate codon to a valine codon in the sixth position of β globin. In the normal β-globin gene, the sequence that specifies amino acids 5, 6, and 7 (Pro-Glu-Glu) is CCT-GAG-GAG. In a carrier of sickle-cell anemia, this sequence is CCT-GTG-GAG. An A in the middle of the sixth codon has been changed to a T. The restriction enzyme MstII recognizes and cleaves the sequence CCT-GAG-G, which is present at this position in normal DNA but not the mutated DNA. Therefore, digestion of the fetal DNA with *MstII* followed by the Southern hybridization technique (Chapter 18) using β-globin cDNA as the radioactive probe will reveal if this restriction site is present in one or both allelic copies of the gene. If it is absent in both copies, the fetus will likely be homozygous for the sickle trait; if it is missing in only one copy, the fetus will be heterozygous for the trait. There are many additional examples of other genetic diseases (e.g., Huntington's disease, muscular dystrophy, and hemophilias) that can be diagnosed by analysis of restriction fragment length polymorphisms (RFLP).

Other methods are required if the disease mutation does not cause a change in a restriction site or is not linked to another RFLP. For example, the DNA carrying the mutation can be amplified by polymerase chain reaction, and then the alleles can be detected by hybridization with allele-specific oligonucleotides (ASOs). Two ASOs differing at usually one nucleotide are made so that one ASO matches the normal allele perfectly while the other ASO matches the abnormal allele. Hybridization conditions can be set up so that only the ASO matching perfectly remains bound to the DNA.

Sickle-Cell Anemia Is Due to a Single Base Pair Change

A single base pair change within the coding region for the β chain appears to be responsible for sickle-cell anemia. This change occurs in the second position of the codon for position 6 of the β chain. In the corresponding mRNA the codon, GAG, which specifies glutamate in normal β chains, is converted to the codon, GUG, which specifies valine. The resultant hemoglobin, called hemoglobin S, has altered surface charge properties (because the polar side group of glutamate has been replaced by valine's distinctly nonpolar group), which is responsible for clinical symptoms of the sickle-cell trait. Carriers of the sickle-cell mutation can be detected by restriction enzyme digestion of a sample of the potential carrier's DNA followed by Southern hybridization technique with the β-globin cDNA as described in Clin. Corr. 19.2.

Thalassemias Are Caused by Mutations in the Genes for the α or β Chains of Globin

Thalassemias are a family of related genetic diseases that occur in people who are frequently from or originate from populations living in the Mediterranean areas. If there is a reduced synthesis or a total lack of synthesis of α-globin mRNA, the disease is classified as α-thalassemia; if the β-globin mRNA level is affected, it is called β-thalassemia. Thalassemias can be due to the deletion of one or more globin-like genes in either of the globin gene clusters or be caused by a defect in the transcription or processing of a globin gene's mRNA.

Since each human chromosome 16 contains two adjacent α-globin genes, a normal diploid individual has four copies of this gene. α-Thalassemic patients have been identified who are missing one to four α-globin genes. The condition in which one α gene is missing is referred to as *α-thal-1;* when two α genes are gone, the condition is *α-thal-2*. In both cases the individuals can experience mild to moderate anemia but may have no additional symptoms. When three α genes are missing, many more β-chain molecules are synthesized than α-chain molecules resulting in the formation a globin tetramer of four β chains, which causes HbH disease and accompanying anemia. When all four α genes are absent, the disease hydrops fetalis occurs, which is fatal at or before birth. Some chromosomal deletions that have been mapped in the α-gene cluster are shown in Figure 19.25.

The different β-thalassemias also exhibit differing degrees of severity and can be caused by a variety of defects or deletions. In one case the β gene is present but has undergone a mutation in the codon 17, which generates a termination codon. In another case the β gene is transcribed in the nucleus but no β-globin mRNA occurs in the cytoplasm. Thus, a defect has occurred in the processing and/or transport of the primary transcript of the gene.

Other β-thalassemias are clearly caused by deletions within the β gene cluster on chromosome 11, as illustrated in Figure 19.25. In some cases these deletions remove the DNA between two adjacent genes resulting in a new fusion gene. For example, in the normal person the linked δ and β genes differ in only about 7% of their positions. In Hb *Lepore* a deletion has placed the front portion of the δ gene in register with the back portion of the β gene. From this fusion gene a new β-like chain is produced in which the N-terminal sequence of δ is joined to the C-terminal sequence of β. Several variants of Hb Lepore are now known, and in each case the globin product is a composite of the δ and β sequence, but the actual fusion junction is different.

Another fusion β-like globin is produced in Hb Kenya. This deletion results in a gene product that contains the N-terminal sequence of the A_γ gene and the C-terminal sequence of the β gene. Still another series of deletions has been found in which both the δ and β genes are removed, causing HPFH (hereditary persistence of fetal hemoglobin). Frequently there are no clinical symptoms of this condition because fetal hemoglobin (α_2, γ_2) continues to be synthesized after the time at which γ gene expression is normally turned off (see Clin. Corrs. 19.2 and 19.3.)

CLIN. CORR. 19.3
PRENATAL DIAGNOSIS OF THALASSEMIA

If a fetus is suspected of being thalassemic because of its genetic background, recombinant DNA techniques can now be used to determine if one or more globin genes are missing from its genome. Fetal DNA can be easily obtained (in relatively small quantities) from amniotic fluid cells aspirated early during the second trimester of pregnancy. This DNA is digested with one of several specific restriction enzymes that divides the globin genes on DNA restriction fragments that are several thousand base pairs in length. These fragments are separated by electrophoresis through an agarose gel and hybridized with radioactive cDNA for α and/or β globin using the Southern hybridization technique described in Chapter 18. If one or more globin genes are missing, the corresponding restriction fragment will not be detected or its hybridization to the radioactive cDNA probe will be reduced (in the case when only one of two diploid genes is absent). The disadvantage of this technique is that it requires time for the growth of amniotic cells in culture in order to obtain sufficient DNA to perform the test. Polymerase-chain reaction methods require less DNA.

Benz, E. J. The Hemoglobinopathies. *Textbook of Internal Medicine*. W. N. Kelly (Eds.). Philadelphia: Lippincott, 1989, pp. 1423–1432.

19.11 GENES FOR HUMAN GROWTH HORMONE-LIKE PROTEINS

Human growth hormone (somatotropin) is a single polypeptide of 191 amino acids. A larger precursor of the protein is synthesized in the somatotrophs of the anterior pituitary, and the mature form is secreted into the circulatory system. Growth hormone induces liver (and perhaps other) cells to produce other hormones called somatomedins, which are insulin-like growth factors that stimulate proliferation of mesodermal tissues such as bone, cartilage, and muscle. Infants with a deficiency in growth hormone become dwarfs, whereas those who produce too much become giants.

A closely related protein, displaying 85% homology with growth hormone, is human **chorionic somatomammotropin** (also called placental lactogen), which is synthesized in the placenta. The complete role of this hormone in normal fetal–maternal physiology is still unclear, but it participates in placental growth and contributes to mammary gland preparation for lactation during pregnancy.

Human growth hormone (hGH) and chorionic somatomammotropin (hCS) are an example of two very similar proteins that serve different biological roles and are synthesized in different tissues. Therefore, it was expected that their genes would also be closely related but expressed in a different tissue-specific fashion. Analysis of the cloned cDNAs and corre-

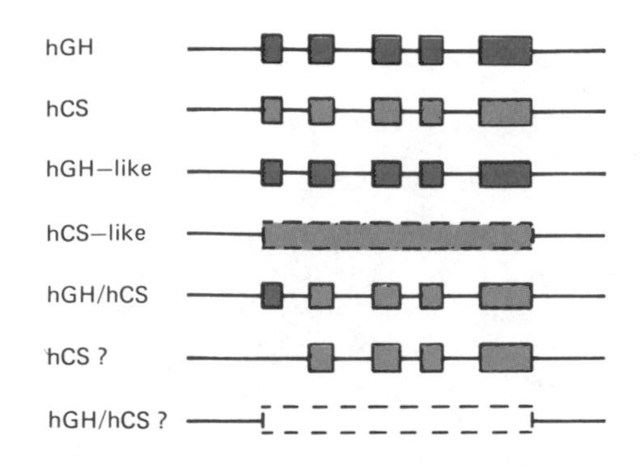

FIGURE 19.26
Members of the *hGH* gene family.
Various cloned members of the *hGH* family are diagrammed. Dark colored boxes: sequences most similar to *hGH*. Light colored boxes: sequences most similar to *hCS*. Open boxes: undefined sequences. Solid outlines: sequenced exons. Dashed outlines: sequences defined by restriction and hybridization analysis. The structure of the *hCS* gene, which has not been conclusively identified, is defined by analogy with *hGH*.
Redrawn from D. D. Moore, M. A. Conkling, and H. M. Goodman. *Cell* 29:285, 1982.

sponding chromosomal genes for hGH and hCS has revealed that, indeed, the two DNA coding sequences are very similar. Furthermore, in addition to the *hGH* and *hCS* genes, there are at least five other closely related DNA regions in the human genome, as shown in Figure 19.26. It is not known if all of these regions are linked in the genomes.

These seven (and perhaps other) related DNA regions comprise the **human growth hormone gene family.** Some of these regions have been extensively characterized; others require more analysis. The *hGH* and *hCS* genes are each comprised of five coding regions (exons) interrupted by four intervening sequences (introns). The introns occur at exactly the same positions relative to the exons in the two genes. On the basis of partial analysis, the seven genes can be divided into three classes: *hGH* type, *hCS* type, and hybrids between *hGH* and *hCS*. The *hGH* class contains two members: the *hGH* gene itself and a *hGH*-like variant, which has a very similar but nonidentical sequence. The *hCS* class has three members: the *hCS* gene, a closely related *hCS* variant, and another variant that appears to have lost the first exon. The hybrid class contains at least one member in which the first exon is nearly identical to the first exon of *hGH,* whereas at least the next three exons are identical to the *hCS* cDNA. Thus, this gene may be the result of a crossover event between two other genes that occurs in or near the first intron.

Only the *hGH* and *hCS* genes are known to be transcribed. Transcription of the other related genes has not been shown. They may be pseudogenes; alternatively, they may be transcribed at a very low level or in other unsuspected tissues. A surprising observation is that about 10% of the *hGH* produced in all examined pituitaries is missing 15 internal amino acids. These 15 amino acids are specified by the beginning of the second exon. Determination of the DNA sequence of the second exon revealed a sequence very similar to the upstream boundary between the first intron and second exon. This sequence occurs 15 codons in from this boundary. Therefore, the RNA splicing event that removes the first intron (between the first and second exon) may not occur with complete faithfulness. If the splicing reaction occurs at the alternative interior site 10% of the time, the shortened version of *hGH* would occur in the detected proportion. However, direct evidence for this alternative splicing event is not yet available. Therefore, the shortened *hGH* could be the result of the expression of another as yet unidentified member of the *hGH* gene family.

The expression of both the *hGH* and *hCS* genes is under the regulation of other hormones. Both thyroxine and cortisol stimulate increased transcription of these genes. Studies using cultured rat pituitary tumor cells reveal that these hormones act in a synergistic fashion in inducing growth hormone mRNA synthesis. Pituitary cells that have only about two molecules of growth hormone mRNA per cell can be stimulated to a level of 1000 growth hormone mRNA molecules per cell—a 500-fold range that is comparable to the magnitude of induction of many bacterial operons.

The precise mechanism by which thryoxine and cortisol stimulate this increased transcription is not known. The regulation is clearly more complicated molecularly than the control of bacterial operon transcription. The regulatory hormones may be transported into the nucleus and either directly or in association with a binding protein affect transcription initiation of the gene. Alternatively, they may interact with other factors in the cell that in turn regulate the level of transcription. There is some evidence that DNA regulatory site for the glucocorticoid influence is upstream of the site at which transcription of the gene begins. Other evidence hints that it may also be located within the first intron. Clearly, our understanding of the mechanisms of eucaryotic gene regulation is still in its infancy, and activity in this research area will continue for years to come. An example of the many transcription initiation protein factors that can interact with the DNA in the vicinity of eucaryotic genes is shown in Figure 16.21. In addition, Chapter 16 discusses the three classes of RNA polymerases that transcribe eucaryotic genes.

19.12 GENES IN THE MITOCHONDRION

Not all of the genetic information of eucaryotic cells is encoded by DNA in the nucleus. About 0.3% of the DNA of human cells occurs in the mitochondria. Human mitochondrial DNA (mtDNA) is a double-stranded circular molecule of 16,569 bp whose sequence has been completely determined. As many as 100 molecules of mtDNA can occur in a metabolically active cell. Each mtDNA codes for 2 rRNAs, 22 tRNAs, and 13 proteins, most of which are subunits of multisubunit complexes in the mitochondrial inner membrane that catalyze oxidative phosphorylation (Fig. 19.27). For example, Complex I (NADH dehydrogenase), the first of three proton-pumping complexes involved in oxidative phosphorylation, is comprised of 26 proteins. Seven of these proteins are encoded by the mtDNA and the other 19 are coded for by genes in the nucleus. Mitochondrial DNA also contains genes for three cytochrome oxidase subunits, 2 ATP synthase subunits, and cytochrome *b*. In contrast to the nucleus where much of the chromosomal DNA seems to have no genetic function, virtually every base pair in human mtDNA is essential. The regions between the protein-coding genes usually encode tRNAs and sometimes the last nucleotide of one gene will be the first nucleotide of the adjacent gene. Polyadenylation at the 3′ ends of some of the mitochondrial mRNAs adds the last two A residues of the termination codon, UAA, to create the end of the open translation reading frame.

Even more remarkable, the genetic code of mtDNA is not identical to the genetic code of nuclear or procaryotic DNA. The codon UGA codes

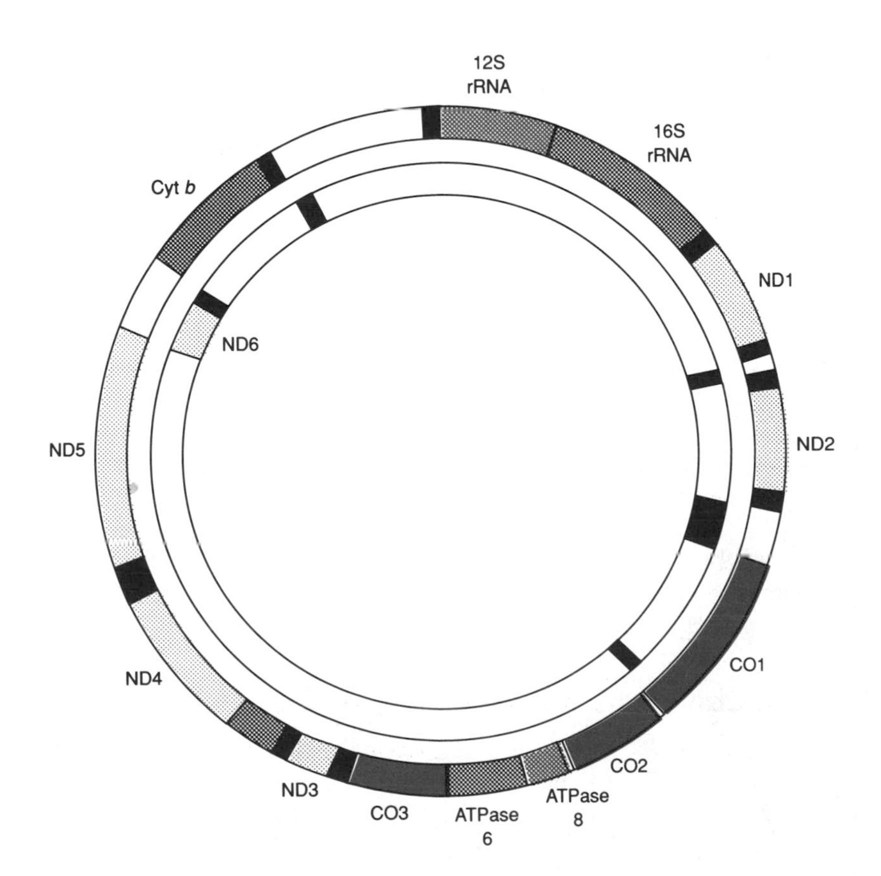

FIGURE 19.27
Human mitochondrial DNA.
The 16,569 bp human mtDNA molecule codes for two ribosomal RNAs (rRNA), some of the subunits for NADH dehydrogenase (ND), cytochrome oxidase (CO), ATP synthase (ATPase) and cytochrome *b* (cyt *b*), and 22 tRNAs (not shown). Most of the genes occur on the outer DNA strand but the genes for ND6 and a few tRNAs are on the inner strand.

for tryptophan instead of for termination, AUA codes for methionine rather than isoleucine, and AGA and AGG serve as stop codons instead of specifying arginine. It is not clear why a mitochondrion has to have its altered small genetic system. Perhaps mtDNA is an evolutionary vestige of an early symbiotic relationship between a bacterium and the progenitor of eucaryotic cells. What is clear is that the cell makes a large investment to express the 13 mitochondria-encoded proteins. To produce those proteins a whole set of nuclear-encoded ribosomal proteins and associated translation factors must be imported into the mitochondrion and assembled, as well as all of the enzymes and binding proteins required for mtDNA replication and transcription. More than 100 different nuclear-encoded proteins are probably necessary to maintain the mtDNA and express its gene products.

Since mitochondria are in the cytoplasm, mtDNA molecules are maternally inherited. This fact has led to the use of mtDNA sequences as markers for maternal lineages. Recently, a single base pair change in mtDNA has been found to be responsible for Leber's hereditary optic neuropathy (see Clin. Corr. 19.4). Similar mtDNA mutations also may be the cause of two other maternally inherited genetic diseases, myoclonic epilepsy and infantile bilateral striatal necrosis.

CLIN. CORR. 19.4
LEBER'S HEREDITARY OPTIC NEUROPATHY (LHON)

Leber's hereditary optic neuropathy, first described in 1871, is a maternally inherited genetic disease that usually strikes young adults and results in complete or partial blindness from optic nerve degeneration. Other neurological disorders such as cardiac dysrhythmia can also be associated with the disease. The cause of this defect in many patients has been traced to a single base pair mutation in the mtDNA that changes an arginine to a histidine at amino acid 340 in the NADH dehydrogenase subunit 4 of Complex I in the inner mitochondrial membrane. Although it is not clear why this mutation leads to blindness, the eyes require a high level of mitochondrial activity and perhaps become sensitive over time to a small decrease in ATP synthesis by oxidative phosphorylation.

19.13 BACTERIAL EXPRESSION OF FOREIGN GENES

Recombinant DNA techniques are now frequently used to construct bacteria that are ''factories'' for making large quantities of specific human proteins that are useful in the diagnosis or treatment of disease. The two examples to be illustrated here are the construction of bacteria that synthesize human insulin and human growth hormone.

Many factors must be considered in designing recombinant plasmids that contain a eucaryotic gene to be expressed in bacteria. First, the cloned eucaryotic gene cannot have any introns since the bacteria do not have the RNA-splicing enzymes that correctly remove introns from the initial transcript. Thus, the actual eucaryotic chromosomal gene is usually not used for these experiments; instead, the cDNA or a synthetic equivalent of the coding sequence, or a combination of both, is placed in the bacterial plasmid.

Another consideration is that different nucleotide sequences comprise the binding sites for RNA polymerase and ribosomes in bacteria and eucaryotes. Therefore, to achieve expression of the desired protein it is necessary to insert the eucaryotic coding sequence directly behind a set of bacterial controlling elements. This has the advantage that the foreign gene is now under the regulation of the bacterial control elements, but its disadvantage is that considerable recombinant DNA manipulation is required to make the appropriate plasmid. Still other factors to be considered are that the foreign gene product must not be degraded by bacterial proteases or require modification before it is active (e.g., specific glycosylation events that the bacteria cannot perform) and must not be toxic to the bacteria. Furthermore, even when the bacteria do synthesize the desired product, it must be isolated from the 1000 or more endogenous bacterial proteins.

Bacteria Can Synthesize Human Insulin

Insulin is produced by the β cells of the pancreatic islets of Langerhans. It is initially synthesized as preproinsulin, a precursor polypeptide that possesses an N-terminal signal peptide and an internal C-peptide of 33 amino acids that are removed during the subsequent maturation and secretion of insulin. The A-peptide (21 amino acids) and B-peptide (30 amino acids) of mature insulin are both derived from this initial precursor and are held together by two disulfide bridges. Bacteria do not have the processing enzymes that convert the precursor form to mature insulin. Therefore, the initial strategy for the bacterial synthesis of human insulin involved the production of the A chain and B chain by separate bacteria followed by purification of the individual chains and the subsequent formation of the proper disulfide linkages.

The first step was to use synthetic organic chemistry methods to prepare small single-stranded oligonucleotides (between 11 and 18 nucleotides in length) that were both complementary and overlapping with each other. When these oligonucleotides were mixed together in the presence of DNA ligase under the proper conditions, they formed a double-stranded fragment of DNA with termini equivalent to those that are formed by specific restriction enzymes (Figure 19.28). Furthermore, the sequences of the oligonucleotides were carefully chosen so that one of the two strands contained a methionine codon followed by the coding sequence of the A chain of insulin and a termination codon. A second set of overlapping complementary oligonucleotides were prepared and ligated

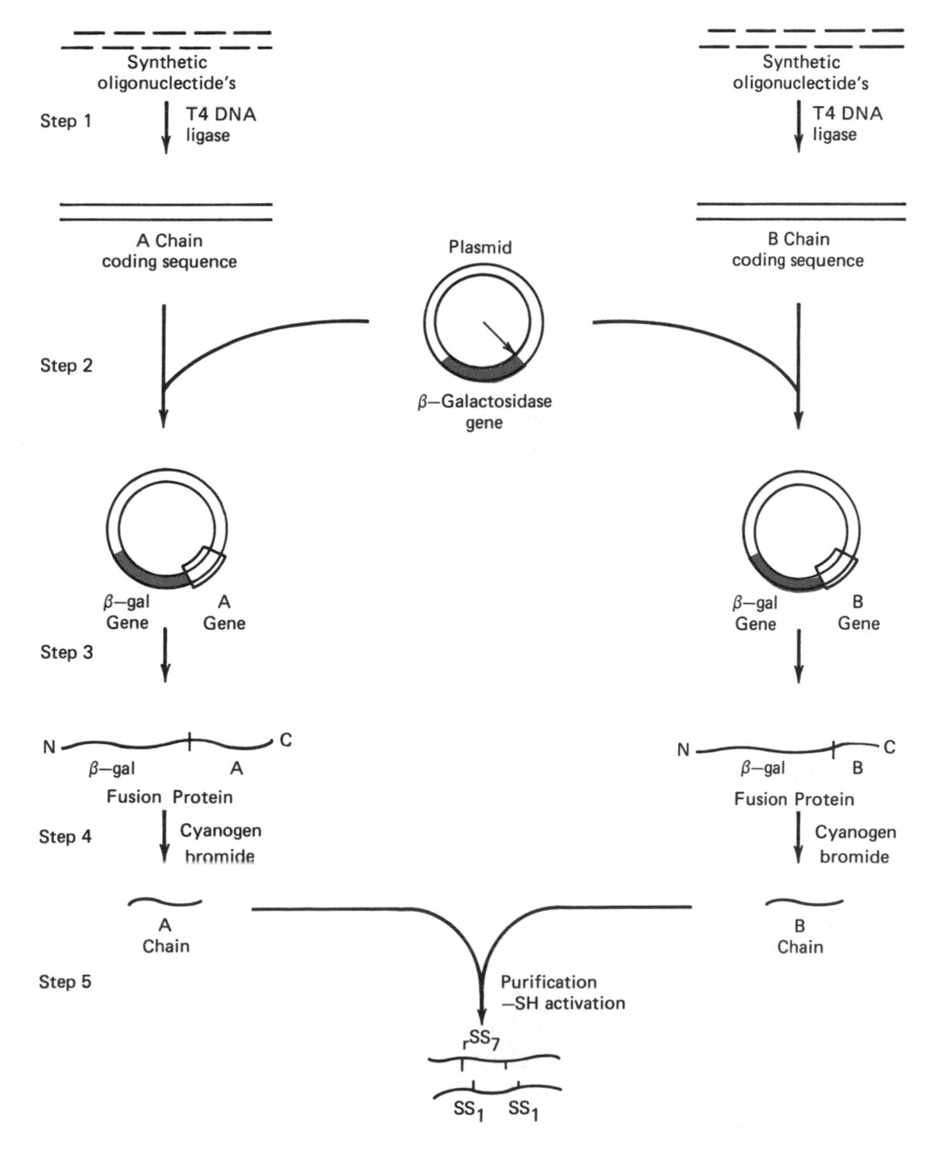

FIGURE 19.28
Bacterial expression of the A and B chains of human insulin.
Step 1: A series of overlapping, complementary oligonucleotides (11 for the A chain and 18 for the B chain) were synthetically prepared and ligated together. One strand of the resulting small DNA fragments contained a methionine codon followed by the coding sequence for the A chain and B chain, respectively. Step 2: The small DNA fragments were ligated into a restriction site near the end of the β-galactosidase gene of the lactose operon that was on a plasmid. Step 3: The recombinant plasmids were introduced into *E. coli* and the β-galactosidase gene induced with IPTG, an inducer of the lactose operon. A fusion protein was produced that contained most of the β-galactosidase sequence at the N terminus and the A chain (or B chain) at the C terminus. Step 4: Bacterial cell lysates containing the fusion protein were treated with cyanogen bromide, which cleaves peptide bonds following methionine residues. Step 5: The A and B chains were purified away from all of the other cyanogen bromide peptides using biochemical and immunological separation techniques. The SH groups were activated and reacted to form the intra- and interchain disulfide bridges found in mature human insulin.
Redrawn from R. Crea, A. Krazewski, T. Hirose, and K. Itakura. *Proc. Natl. Acad. Sci. U.S.A.* 75:5765, 1980.

together to form another double-stranded DNA fragment that contained a methionine codon followed by 30 codons specifying the B chain of insulin and a termination codon.

These two double-stranded fragments were then individually cloned at a restriction site in the β-galactosidase gene of the lactose operon that was on a plasmid. These two recombinant plasmids were introduced into bacteria. The bacteria could now produce a fusion protein of β-galactosidase and the A chain (or B chain) that is under the control of the lactose operon. In the absence of lactose in the bacterial medium, the lactose operon is repressed and only very small amounts of the fusion protein are synthesized. Using induction with IPTG and some additional genetic tricks, the bacteria can be forced to synthesize as much as 20% of their protein as the fusion protein. The A peptide (or B peptide) can be released from this fusion protein by treatment with cyanogen bromide, which cleaves on the carboxyl side of methionine residues. Since neither the A nor B peptide contains a methionine, they will be liberated intact and can be subsequently purified to homogeneity. The final steps involve chemically activating the free SH groups on the cysteines and mixing the activated A and B chains together in a way that the proper disulfide linkages form to generate molecules of mature human insulin.

Bacteria Can Synthesize Human Growth Hormone

The strategy for generating a recombinant DNA plasmid from which bacteria can synthesize human growth hormone is somewhat different than for insulin synthesis. First, human growth hormone is 191 amino acids long so the total synthetic construction of the corresponding DNA coding sequence is more difficult (although certainly not impossible) than in the above insulin case. On the other hand, growth hormone is a single polypeptide so it is not necessary to deal with the production of two chains and their subsequent dimerization to form a protein with biological activity.

Because of these considerations, the growth hormone coding sequence was initially cloned into a bacterial expression plasmid using part of a cloned growth hormone cDNA and several synthetic oligonucleotides (Figure 19.29). The overlapping oligonucleotides were prepared so that, when ligated together, they would form a small double-stranded DNA containing the codon for the first 24 amino acids of mature human growth hormone. One end of this DNA fragment was designed so that the fragment could be ligated in front of a restriction fragment of growth hormone cDNA that provided the rest of the coding sequence, including the termination codon. The other end of the synthetic fragment was chosen so that the composite coding sequence could be easily inserted into a site immediately downstream of the promoter–operator–ribosome binding site of the lactose operon cloned on a plasmid. After the introduction into bacteria, the bacteria were induced with IPTG to transcribe this foreign coding region and the greatly overproduced human growth hormone subsequently purified away from the bacterial proteins.

19.14 INTRODUCTION OF THE RAT GROWTH HORMONE GENE INTO MICE

The previous section described examples of the use of bacteria to produce large quantities of a human protein that is used to treat a disease. It is also possible to microinject molecules of purified RNA or DNA directly into eucaryotic cells. This provides a very powerful approach for identifying

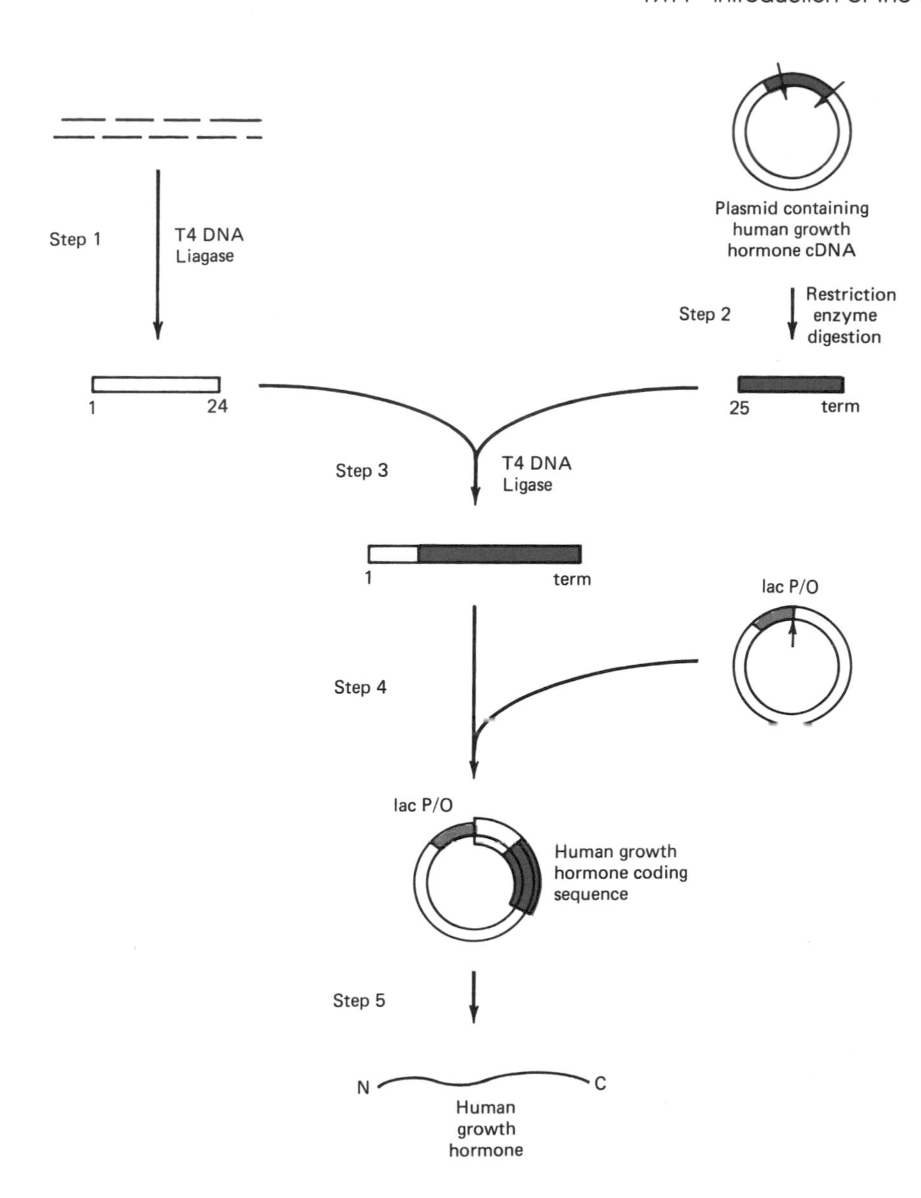

FIGURE 19.29
Expression of human growth hormone in *E. coli.*
Step 1: Several overlapping, complementary, oligonucleotides were synthesized and ligated together. One strand of the resulting small DNA fragment contains the coding sequence for the first 24 amino acids of *mature* human growth hormone (after removal of the N-terminal signal peptide). Step 2: A recombinant plasmid with a full length human growth hormone cDNA, which is not expressed, is cleaved with restriction enzymes that release a fragment containing the complete growth hormone coding sequence after codon 24. Step 3: The synthetic fragment and the partial cDNA—containing fragment are ligated together to yield a new fragment containing the complete coding sequence of mature hGH. Step 4: The new fragment is ligated into a restriction site just downstream from the lactose promoter–operator region cloned on a plasmid. Step 5: The resulting recombinant DNA plasmid is introduced into bacteria in which synthesis of hGH can be induced with IPTG, an inducer of the lactose operon.

conditions under which specific genes are expressed in eucaryotic cells. One of the most dramatic illustrations of this approach is the microinjection of a chromosomal DNA fragment containing the structural gene for rat growth hormone into the pronuclei of fertilized mouse eggs. The eggs were then reimplanted into the reproductive tracts of foster mouse mothers. Some of the mice that developed from this procedure were **transgenic;** one or more copies of the microinjected growth hormone gene integrated into a host mouse chromosome at an early stage of embryo development. These foreign genes were transmitted through the germ line and became a permanent feature in the host chromosomes of the progeny (Figure 19.30).

Analysis of these transgenic mice revealed that in some cases several tandem copies of the rat growth hormone gene had integrated into a mouse chromosome; in other cases only one gene copy was present. In all cases at least some transcription occurred from the integrated gene(s), and in a few cases a dramatic overproduction of rat growth hormone resulted. In these latter cases, as much as 800 times more growth hormone was present in the transgenic mice than in normal mice, resulting in animals more than three times the size and weight of their unaffected littermates.

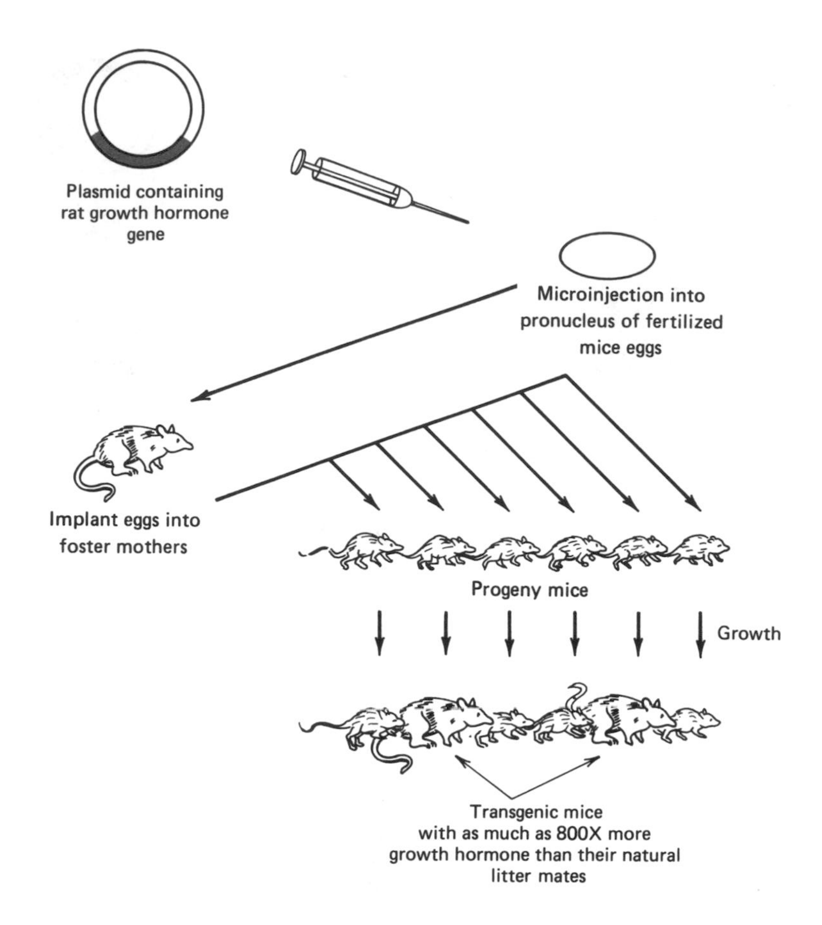

FIGURE 19.30
Schematic illustration of the introduction of rat growth hormone gene into mice.
Copies of a recombinant plasmid DNA containing the rat growth hormone gene were microinjected into fertilized mouse eggs that were reimplanted into foster mothers. Some of the progeny that resulted contained the foreign gene integrated into their own genome and overexpressed growth hormone, growing much larger than their normal-sized litter mates.
Redrawn from R. D. Palmiter, et al. *Nature (London)* 300:611, 1982.

These results present many potential experimental possibilities for the future and raise a number of issues. One implied possibility is the use of similar growth hormone gene insertions to stimulate rapid growth of commercially valuable animals. This could result in a shorter production time and increased efficiency of food utilization. Another long-term possibility is the use of this approach to correct certain human genetic diseases or mimic the diseases in experimental animals so that they can be studied more carefully. One obvious human disease that is a candidate for this **"gene therapy"** approach is thalassemia. For example, an individual with two to three missing α-globin genes might benefit tremendously from receiving bone marrow transplants of his/her own cells that have been established in culture and microinjected with additional copies of the normal α-globin gene. At the moment this approach to gene therapy is not technically feasible, but in the future it may become possible. A final point is that insertion of normal genes into human somatic cells of a defective tissue or organ (instead of germ line cells) does not result in transmission of these genes to the progeny. This lessens the ethical considerations that enter into the design of experiments intended to alter germ line characteristics.

BIBLIOGRAPHY

Procaryotic Gene Expression

Cohen, S. N. and Shapiro, J. A. Transposable genetic elements. *Sci. Am.* 242:40, 1980.

Miller, J. H. The *lac* gene: Its role in *lac* operon control and its use as a genetic system, J. H. Miller and W. S. Resnikoff (Eds.). *The Operon.* Cold Spring Harbor Laboratory, 1978, p. 31.

Platt, T. Regulation of gene expression in the tryptophan operon of *Escherichia coli,* J. H. Miller and W. S. Resnikoff (Eds.), *The Operon.* Cold Spring Harbor Laboratory, 1978, p. 263.

Simon, M., Zieg, J., Silverman, M., Mandel, G., and Doolittle, R. Phase variation: Evolution of a controlling element. *Science* 209:1370, 1980.

Eucaryotic Gene Expression

Brown, D. D. Gene expression in eukaryotes. *Science* 211:667, 1981.

Maniatis, T., Fritsch, E. F., Laurer, J., and Lawn, R. M. The molecular genetics of human hemoglobins. *Annu. Rev. Genet.* 14:145, 1980.

Mitchell, P. J. and Tjian, R. Transcription regulation in mammalian cells by sequence-specific DNA binding proteins. *Science* 245:371, 1989.

Moore, D. D., Conkling, M. A., and Goodman, H. M. Human growth hormone: A multigene family. *Cell* 29:285, 1982.

Orkin, S. H. and Kazazian, H. H., Jr. The mutation and polymorphism of the human beta-globin gene and its surrounding DNA. *Annu. Rev. Genet.* 18:131, 1984.

Palmiter, R. D., Brinster, R. L., Hammer, R. E., et al. Dramatic growth of mice that develop from eggs microinjected with metallothionein-growth hormone fusion genes. *Nature* (*London*) 300:611, 1982.

Singh, G., Lott, M. T., and Wallace, D. C. A mitochondrial DNA mutation as a cause of Leber's Hereditary Optic Neuropathy. *N. Eng. J. Med.* 320:1300, 1989.

QUESTIONS

J. BAGGOTT AND C. N. ANGSTADT

*Q*uestion *T*ypes are described on page xxiii.

A. Induction | C. Both
B. Repression | D. Neither

1. (QT4) Occur(s) in response to the presence of a specific low molecular weight organic compound.

2. (QT3) A. Affinity of the *lac* repressor for the *lacO* sequence when lactose is present in the growth medium of *E. coli*.
   B. Affinity of the *lac* repressor for the *lacO* sequence when lactose is absent.

3. (QT2) The *E. coli lacZYA* region will be transcribed at a rate greater than the low basal rate if:
   1. there is a defect in binding of the inducer to the product of the *lacI* gene.
   2. glucose and lactose are both present in the growth medium, but there is a defect in the cell's ability to synthesize cAMP.
   3. glucose and lactose are both readily available in the growth medium.
   4. the operator has mutated so it can no longer bind repressor.

4. (QT2) An operon:
   1. includes structural genes.
   2. is expected to code for polycistronic mRNA.
   3. contains control sequences such as an operator.
   4. can have only a single promoter.

A. Tryptophan synthesis in *E. coli*. | C. Both
B. Lactose utilization by *E. coli*. | D. Neither

5. (QT4) Controlled by a repressor.
6. (QT4) Controlled by feedback inhibition.
7. (QT4) Affected by a secondary promotor.

8. (QT3) A. Basal level of mRNA for the first gene in the *E. coli* tryptophan operon.
   B. Basal level of mRNA for the last gene in the *E. coli* tryptophan operon.

A. *trp* operator site | C. Both
B. *trp* attenuator site | D. Neither

9. (QT4) Site at which tryptophan availability plays a role in preventing transcription.

10. (QT1) Ribosomal operons:
    A. all contain genes for proteins of just one ribosomal subunit.
    B. all contain genes for proteins of both ribosomal subunits.
    C. all contain genes for only ribosomal proteins.
    D. expression can be regulated at the level of translation.
    E. are widely separated in the *E. coli* chromosome.

11. (QT2) Transposons:
    1. are a means for the permanent incorporation of antibiotic resistance into the bacterial chromosome.
    2. contain short inverted terminal repeat sequences.
    3. include at least one gene that codes for a transposase.
    4. code for an enzyme that synthesizes guanosine tetraphosphate and guanosine pentaphosphate, which inhibit further transposition.

12. (QT1) Repetitive DNA:
    A. is common in bacterial and mammalian systems.
    B. is all uniformly distributed throughout the genome.
    C. includes DNA that codes for rRNA.
    D. consists mostly of DNA that codes for enzymes catalyzing major metabolic processes.
    E. is resistant to the action of restriction endonucleases.

A. α-Globin gene | C. Both
B. β-Globin gene | D. Neither

13. (QT4) Free of introns.
14. (QT4) Defective in sickle-cell anemia.
15. (QT4) Four copies are found in the normal human.

16. (QT1) In designing a recombinant DNA for the purpose of synthesizing an active eucaryotic polypeptide in bacteria all of the following should be true *except:*
    A. the eucaryotic gene may contain its usual complement of introns.
    B. the foreign polypeptide should be resistant to degradation by bacterial proteases.
    C. glycosylation of the polypeptide should be unnecessary.
    D. the foreign polypeptide should be nontoxic to the bacteria.
    E. bacterial controlling elements are necessary.

ANSWERS

1. C Induction is an increase in transcription of a structural gene in response to a low molecular weight substrate (p. 807). Repression is a decrease in transcription in response to a low molecular weight product (p. 808).
2. B The repressor protein has a high affinity for the *lacO* (operator) sequence of DNA unless the repressor has previously bound the inducer, allolactose (p. 810). The complex of inducer with repressor, however, does not bind to the operator site.
3. D Only 4 true. 1: The product of the *lacI* gene is the repressor protein. When this protein binds an inducer, it changes its conformation, no longer binds to the operator site of DNA, and transcription occurs at an increased rate. Failure to bind an inducer prevents this sequence. 2 and 3: In the presence of glucose catabolite repression occurs. Glucose lowers the intracellular level of cAMP. The catabolite activator protein (CAP) then cannot complex with cAMP, so there is no CAP–cAMP complex to activate transcription. The same would occur if the cell had lost its capacity to synthesize cAMP (p. 814). 4: If the operator is unable to bind repressor, the rate of transcription is greater than the basal level (p. 811).
4. A 1–3 true. 1, 2, 3: An operon is the complete regulatory unit of a set of clustered genes, including the structural genes (which are transcribed together to form a polycistronic mRNA), regulatory genes, and control elements, such as the operator (p. 807). 4: An operon may have more than one promoter, as does the tryptophan operon of *E. coli* (p. 817).
5. C Both are controlled by a repressor. A is "turned on" unless tryptophan is present, whereas B is "turned off" unless lactose is present (p. 815).
6. A Tryptophan synthesis is controlled both by control of enzyme synthesis and by regulation of enzyme activity through feedback (p. 815).
7. A The tryptophan operon has a secondary promoter (p. 817).
8. B The tryptophan operon has two promoters. The first one is controlled by the repressor, whereas the second one (located in the *trpD* gene) is not. Although the *trpP2* promoter is not very efficient, it does promote transcription, so more mRNA for the last three genes of the *trp* operon are made than for the entire *trp* operon (p. 817).
9. C At the *trp* operator site the repressor–tryptophan complex binds, preventing transcription. Tryptophan serves as a corepressor (p. 815). At the *trp* attenuator site, the presence of trp-tRNA, which can be made only if tryptophan is available, permits synthesis of leader peptide, which contains two *trp* residues. If this occurs, the mRNA on which it occurs folds into a secondary structure that causes termination of its own synthesis (p. 818). Figure 19.10 shows the secondary structure, a hairpin loop with many CG pairs in its stem, followed by an oligo-U sequence (bases 114–140). This is a standard termination sequence.
10. D A, B, C, E: The genes for one-half of the ribosomal proteins are in two major clusters. There is no pattern to the distribution of genes for the proteins of the two ribosomal subunits, and they are intermixed with genes for other proteins involved in protein synthesis. D: Excess ribosomal protein binds to its own mRNA, preventing initiation of further synthesis of that protein (p. 823).
11. A 1, 2, 3 true. 1: See p. 827. 2 and 3: See p. 824. 4: These guanosine phosphates are synthesized by the product of the *relA* gene; they inhibit initiation of transcription of the rRNA and tRNA genes, shutting off protein synthesis in general. This is the stringent response (p. 824).
12. C A and B: Highly repetitive and moderately repetitive DNA are found only in eucaryotes. Highly repetitive sequences tend to be clustered, as are some moderately repetitive sequences (p. 830). C: This makes sense, since many copies of these structural elements are needed (p. 830). D: Most repetitive DNA does not code for a stable gene product (p. 831). E: The *Alu* family of moderately repetitive DNA is named for the restriction endonuclease that cleaves them (p. 831).
13. D Both genes contain two introns (p. 834). Introns are very common in eucaryotic genes.
14. B Sickle-cell anemia is due to a genetic error in the β-chain gene (p. 836).
15. A The normal human diploid cell contains four copies of the α-globin gene and two copies of the β-globin gene (p. 836).
16. A A and C: The bacterial system has no mechanism for posttranscriptional modification of mRNA or for posttranslational (or cotranslational) modification of protein. E: Bacterial systems need bacterial promoters, and so on (p. 840).

20

Biochemistry of Hormones I: Peptide Hormones

GERALD LITWACK

20.1 OVERVIEW 848

20.2 HORMONES AND THE HORMONAL CASCADE SYSTEM 849

The Cascade System Is a Means to Amplify a Specific Signal 849

Polypeptide Hormones of the Anterior Pituitary 852

20.3 MAJOR POLYPEPTIDE HORMONES AND THEIR ACTIONS 854

20.4 GENES AND FORMATION OF POLYPEPTIDE HORMONES 858

Proopiomelanocortin Is a Precursor Peptide for Eight Hormones 859

Many Polypeptide Hormones Are Encoded Together in a Single Gene 859

Multiple Copies of a Hormone Can Be Encoded on a Single Gene 861

20.5 SYNTHESIS OF AMINO ACID DERIVED HORMONES 862

Epinephrine Is Synthesized From Phenylalanine/Tyrosine 862

Synthesis of Thyroid Hormone Requires Incorporation of Iodine Into a Tyrosine of Thyroglobulin 865

20.6 INACTIVATION AND DEGRADATION OF HORMONES 868

20.7 CELL REGULATION AND HORMONE SECRETION 868

G Proteins Serve as Cellular Transducers of the Hormone Signal 870

Cyclic AMP Activates a Protein Kinase A Pathway 873

Inositol Trisphosphate Formation Leads to Release of Calcium From Intracellular Stores 873

Diacylglycerol Activates a Protein Kinase C Pathway 875

20.8 CYCLIC HORMONAL CASCADE SYSTEMS 876

Melatonin and Serotonin Synthesis Are Controlled by Light and Dark Cycles 876

The Ovarian Cycle Is Controlled by Gonadotropic Releasing Hormone 876

20.9 HORMONE–RECEPTOR INTERACTIONS 881

Scatchard Analysis Permits Determination of the Number of Receptor Binding Sites and the Association Constant for Ligand 882

Some Hormone–Receptor Interactions Involve Multiple Hormone Subunits 883

20.10 STRUCTURE OF RECEPTORS: β-ADRENERGIC RECEPTOR 884

20.11 INTERNALIZATION OF RECEPTORS 886

Clathrin Forms a Lattice Structure to Direct Internalization of Hormone–Receptor Complexes From the Cell Membrane 886

20.12 INTRACELLULAR ACTION: PROTEIN KINASES 889

The Insulin Receptor: Transduction Through Tyrosine Kinase 890

Activity of Vasopressin: Protein Kinase A 891

Gonadotropin Releasing Hormone (GnRH): Protein Kinase C 892

Activity of Atrionatriuretic Factor (ANF): Protein Kinase G 894

20.13 ONCOGENES AND RECEPTOR FUNCTIONS 897
BIBLIOGRAPHY 899
QUESTIONS AND ANNOTATED ANSWERS 899
CLINICAL CORRELATIONS
20.1 Testing the Activity of the Anterior Pituitary 852
20.2 Hypopituitarism 854
20.3 Lithium Treatment of Manic Depressive Illness: The Phosphatidylinositol Cycle 876

20.1 OVERVIEW

It is now well known that cells are regulated by many hormones, growth factors, neurotransmitters, and certain toxins through interactions with cognate receptors at the cell surface. This system of receptors in the cell membrane is the major mechanism through which peptide hormones and amino acid derived hormones exert their effects at the cellular level. Another important mechanism involves the permeation of the cell membrane by steroid hormones that interact with intracellular cognate receptors (Chapter 21). These two sites, the cell membrane and the intracellular milieu, represent the principal locations of initial interaction of ligand and cellular receptor and are diagrammed in Figure 20.1. Polypeptide hormones and amino acid derived hormones, in general, bind cognate receptors at the cell membrane with some exceptions, such as the thyroid hormone receptor, which appears to reside in the nucleus and resembles certain steroid hormone receptors.

The hormonal cascade system is applicable to many, but not all, hormones. It begins with signals in the central nervous system (CNS), followed by the hypothalamus, pituitary, and end target organ. Major polypeptide hormones are summarized and formation of specific hormones is described. The synthesis of the amino acid derived hormones epinephrine and triiodo-L-thyronine is outlined. Examples of hormone inactivation and degradation are given. The remainder of this chapter focuses on receptors, signal transduction, and second messenger pathways. Receptor internalization is described. Examples of cyclic hormonal cascade systems are introduced. Finally, there is a discussion of oncogenes and receptor function.

In terms of receptor mechanism, aspects of hormone–receptor interac-

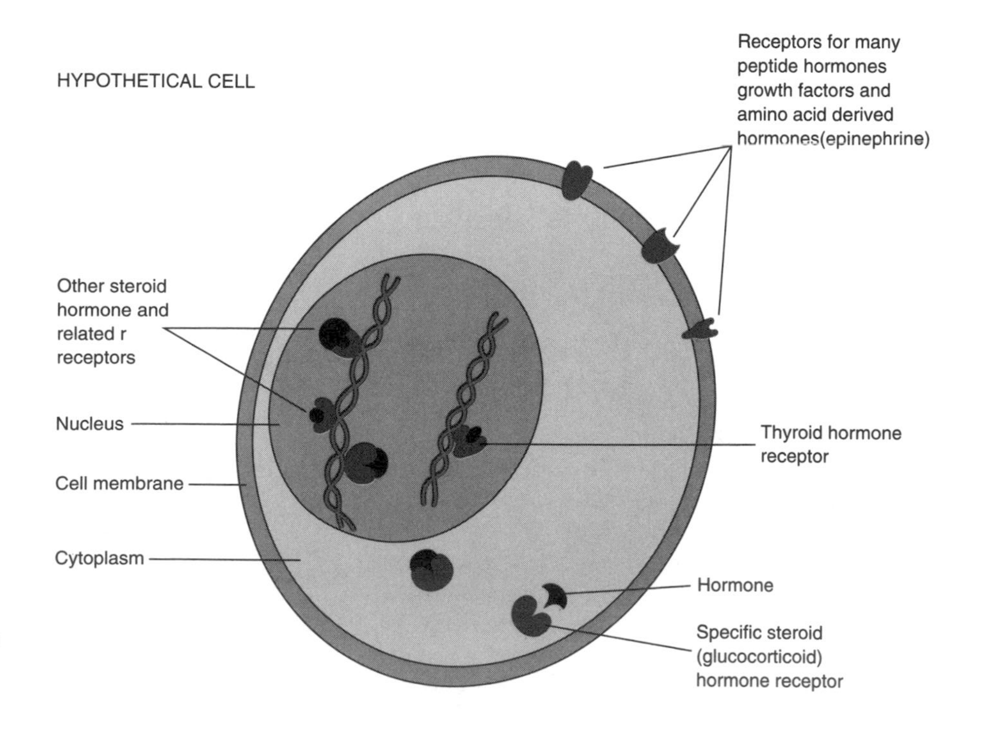

FIGURE 20.1
Diagram showing the different locations of classes of receptors in the cell.

tion are presented with a brief mathematical analysis. Signal transduction is considered, especially in reference to GTP binding proteins. Second messenger systems include cAMP and the protein kinase A pathway, inositol trisphosphate–diacylglycerol and the Ca^{2+}–protein kinase C pathway, and cGMP and the protein kinase G pathway. These pathways are discussed in the context of representative hormone action. In addition, insulin action and its tyrosine kinase and second messenger pathways are considered.

20.2 HORMONES AND THE HORMONAL CASCADE SYSTEM

The definition of a hormone has been expanded over the last several decades. **Endocrine hormones** originally were considered to define all of the hormones. Now, the term *hormone* essentially means any substance in an organism that carries a signal to generate some alteration at the cellular level. Thus, *endocrine hormones* represent a class of hormones that arise in one tissue, or "gland," and travel a considerable distance through the circulation to reach a target cell bearing its cognate receptor. **Paracrine hormones** arise from a cell and travel a relatively small distance to interact with their cognate receptors on another cell. **Autocrine hormones** are produced by the same cell that is also a target (nearby cells may be targets). Thus, we can classify hormones based on their radii of action. Often, endocrine hormones that travel further to the target cell may be more stable than autocrine hormones that operate over very small distances.

The Cascade System Is a Means to Amplify a Specific Signal

For many hormonal systems in higher animals, the signal pathway originates with the brain and culminates with the ultimate target cell. Figure 20.2 outlines the sequence of events. A stimulus may originate in the environment or inside the organism mediated by a specific neuron. This signal may be transmitted as an electrical pulse or as a chemical signal or both. In many cases, but not all, such signals are forwarded to the limbic system, the hypothalamus, the pituitary, and the target gland that secretes

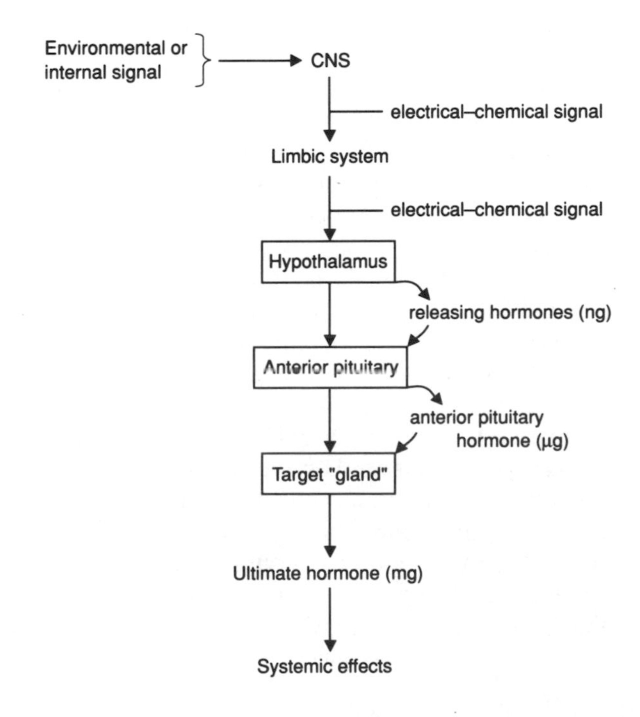

FIGURE 20.2
Hormonal cascade of signals from CNS to ultimate hormone.
The target "gland" refers to the last hormone-producing tissue in the cascade, which is stimulated by an appropriate anterior pituitary hormone. Examples would be thyroid gland, adrenal cortex, and liver. Amounts [nanogram (ng), microgram (μg), and milligram (mg)] represent approximate quantities of hormone released.
Redrawn from A. W. Norman and G. Litwack. *Hormones*. New York: Academic, 1987, p. 38.

the final hormone, which then affects various target cells to a degree proportional to the number of cognate cellular receptors. This may be a true **cascade** in the sense that increasing amounts of hormones at successive levels (hypothalamus, pituitary, and target gland) are generated and also because the half-lives of these hormones in blood tend to become greater in progression from the hypothalamus to the ultimate gland. In the case of environmental stress, for example, there is a single stressor (change in temperature, noise, trauma, etc.). This results in a signal to the hippocampal structure in the limbic system, which signals the hypothalamus to release a hypothalamic releasing hormone, corticotrophic releasing hormone (CRH), usually secreted in nanogram amounts and may have a $t_{1/2}$ in blood of several minutes. This hormone travels down a closed portal system to gain access to the **anterior pituitary,** where it binds to its cognate receptor in the cell membrane of the corticotrophic cell and initiates a set of metabolic changes resulting in the release of adrenocorticotropic hormone (ACTH) (and β-lipotropin). This hormone, released in microgram amounts with an increased $t_{1/2}$ over CRH, circulates in the blood until it binds its cognate receptor in the cell membranes of the inner layer of cells of the adrenal gland (target gland), where it affects metabolic changes leading to the synthesis and release of the ultimate hormone, cortisol, in multimilligram amounts in 24 h with a substantial $t_{1/2}$ in blood. Cortisol is taken up by a wide variety of cells that contain varying amounts of the glucocorticoid receptor. Cellular changes follow and the individual hormonal effects at the cellular levels summate to produce the systemic effect of the hormone. The cascade is represented in this example by a single environmental stimulus generating a series of hormones, in relay, in progressively larger amounts with increasing stability that finally affect most of the cells in the body. Many other systems operate similarly, there being different specific **releasing hormones,** anterior pituitary hormones, and ultimate hormones involved in the process and the final number of cells affected may be large or small depending on the distribution of receptors for the ultimate hormone.

A related system involves the **posterior pituitary hormones,** oxytocin and vasopressin, which are stored in the posterior pituitary but are synthesized in neuronal cell bodies located in the hypothalamus. This system is represented in Figure 20.3; elements of Figure 20.2 appear in the central vertical pathway. The posterior pituitary system branches to the right from the hypothalamus. Oxytocin and vasopressin are synthesized in separate cell bodies located in the supraoptic nucleus and the paraventricular nucleus of the hypothalamus. More cell bodies dedicated to synthesis of vasopressin are located in the supraoptic nucleus and more cell bodies dedicated to synthesis of oxytocin are located in the paraventricular nucleus. Their release from the posterior pituitary together with *neurophysin,* a stabilizing protein, occurs by individual controls on each type of neuronal cell.

There are highly specific signals dictating the release of polypeptide hormones along the cascade of this system. Thus, there are a variety of **aminergic neurons** (secreting amine-containing substances like dopamine, serotonin, etc.), which connect to neurons involved in the synthesis and release of the releasing hormones of the hypothalamus. Releasing hormones are summarized in Table 20.1. These aminergic neurons fire depending on various types of internal or external signals and their activities account for pulsatile release patterns of certain hormones, such as the gonadotropic releasing hormone (GnRH), and **biorhythms,** such as in the **rhythmic cycling releases** of many hormones, like cortisol.

Another prominent feature of the hormonal cascade (Figure 20.3) is the negative feedback system operating when sufficient terminal hormone has

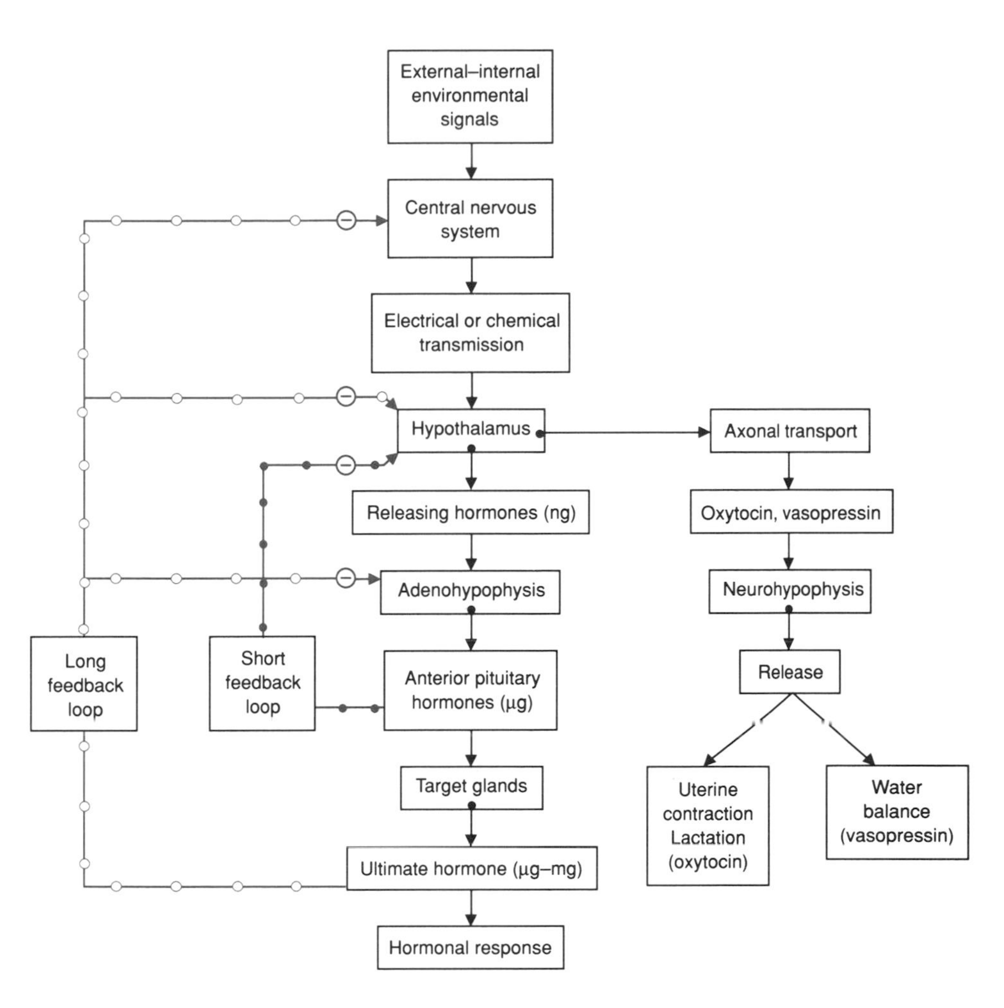

FIGURE 20.3
Many hormonal systems involve the hypothalamus.
This figure shows a cascade of hormonal signals starting with an external or internal environmental signal. This is transmitted first to the CNS and may involve components of the limbic system, such as the hippocampus and amygdala. These structures innervate the hypothalamus in a specific region, which responds with secretion of a specific releasing hormone, usually in nanogram amounts. Releasing hormones are transported down a closed portal system connecting the hypothalamus and the anterior pituitary cell membrane receptor and cause the secretion of specific anterior pituitary hormones, usually in microgram amounts. These access the general circulation through fenestrated local capillaries and bind to specific target gland receptors. The interactions trigger release of an ultimate hormone in microgram to milligram daily amounts, which generate the hormonal response by binding to receptors in several target tissues. In effect, this overall system is an amplifying cascade. Releasing hormones are secreted in nanogram amounts and they have short half-lives of the order of a few minutes. Anterior pituitary hormones are produced often in microgram amounts and have longer half-lives than releasing hormones. Ultimate hormones can be produced in daily milligram amounts with much longer half-lives. Thus, the products of mass × half-life constitute an amplifying cascade mechanism. With respect to differences in mass of hormones produced from hypothalamus to target gland, the range is nanograms to milligrams, or as much as 1 millionfold. When the ultimate hormone has receptors in nearly every cell type, it is possible to affect the body chemistry of virtually every cell by a single environmental signal. Consequently, the organism is in intimate association with the external environment, a fact that we tend to underemphasize. Arrows with a black dot at the origin indicate a secretory process. Long arrows studded with black dots indicate feedback pathways.
Redrawn from, A. W. Norman and G. Litwack, *Hormones*. New York: Academic, 1987. p. 102.

TABLE 20.1 Hypothalamic Releasing Hormones

Releasing hormone	Number of amino acids in structure	Anterior pituitary hormone released or inhibited
Thyrotropin releasing hormone (TRH)	3	Thyrotropin (TSH); can also release prolactin (PRL) experimentally
Gonadotropin releasing hormone (GnRH)	10	Lactogenic (LH) and Follicle stimulating hormones (FSH) from the same cell type; leukotriene C_4 (LTC_4) can also release LH & FSH by a different mechanism
Gonadotropin release inhibiting factor (GnRIF)	12.2-kDa molecular weight	LH and FSH release inhibited
Corticotropin releasing hormone (CRH)	41	ACTH, β-lipotropin (β-LPH) and some β-endorphin
Arginine vasopressin (AVP)	9	Stimulates CRH action in ACTH release; releases ACTH with 5-HT
Angiotensin II (AII)	8	Stimulates CRH action in ACTH release; releases ACTH weakly
Somatocrinin (GRH)	44	Growth hormone (GH) release
Somatostatin (GIH)	14	GH release inhibited
Hypothalamic gastrin releasing peptide		Inhibits release of GH and PRL
Prolactin releasing factor (PRF)		Releases prolactin (PRL)
Prolactin release inhibiting factor (PIF)		Evidence that a new peptide may inhibit PRL release; dopamine also inhibits PRL release and was thought to be PIF for some time; dopamine may be a secondary PIF; oxytocin may inhibit PRL release

Melanocyte stimulating hormone (MSH) is a major product of the pars intermedia (Figure 20.5) in the rat and is under the control of aminergic neurons. The human may also secrete α-MSH from pars intermedia-like cells although this structure is anatomically indistinct in the human.

been formed and secreted into the circulation. Generally, there are two feedback loops, the long feedback and the short feedback loops. In the long feedback loop, the final hormone binds a cognate receptor in/on cells of the anterior pituitary, hypothalamus, and CNS to prevent further elaboration of hormones from those cells that are involved in the cascade. The short feedback loop is accounted for by the pituitary hormone that feeds back negatively on the hypothalamus operating through a cognate receptor. These mechanisms provide tight controls on the operation of the cascade responding to stimulating signals and negative feedback to render this system highly responsive to the hormonal milieu. Clin. Corr. 20.1 describes approaches for testing the responsiveness of the anterior pituitary gland.

Polypeptide Hormones of the Anterior Pituitary

The polypeptide hormones of the anterior pituitary are shown in Figure 20.4 together with their controlling hormones from the hypothalamus. The major hormones of the anterior pituitary are growth hormone (GH),

CLIN. CORR. 20.1 TESTING THE ACTIVITY OF THE ANTERIOR PITUITARY

Considerable progress has been made in synthesizing releasing hormones and chemical analogs, particularly of the smaller peptides. The gonadotropin releasing hormone, a decapeptide, is available for use in assessing the function of the anterior pituitary. This is of importance when a disease situation may involve either the hypothalamus, the anterior pituitary, or the end organ. Infertility is an example of such a situation. What needs to be assessed is which organ is at fault in the hormonal cascade. Consequently, the end organ, in this case the gonads must be considered. This can be done by injecting the anterior pituitary hormone LH or FSH. If sex hormone is elicited, then the ultimate gland would appear to be functioning properly. Next, the anterior pituitary would need to be analyzed. This can be done by i.v. administration of synthetic GnRH; by this route GnRH can gain access to the gonadotropic cells of the anterior pituitary and elicit secretion of LH and FSH. Routinely, LH levels are measured in the blood as a function of time after the injection. These levels are measured by radioimmunoassay (RIA) in which radioactive LH or hCG is displaced from binding to an LH-binding protein by LH in the serum sample. The extent of the competition is proportional to the amount of LH in the serum. In this way a progress of response is measured that will be within normal limits or clearly deficient. If the response is deficient, the anterior pituitary cells are not functioning normally and are the cause of the syndrome. On the other hand, normal pituitary response to GnRH would indicate that the hypothalamus was nonfunctional. Such a finding would prompt examination of the hypothalamus for conditions leading to insufficient availability/production of releasing hormones. Obviously, the knowledge of hormone structure and the ability to synthesize specific hormones permits the diagnosis of these disease states.

Marshall, J. C. and Barkan, A. L., Disorders of the hypothalamus and anterior pituitary. Internal Medicine, W. N. Kelley, ed., J. B. Lippincott, New York, 1989, p. 2159; and Conn, P. M. The molecular basis of gonadotropin-releasing hormone action. Endocrine Rev. 7:3, 1986.

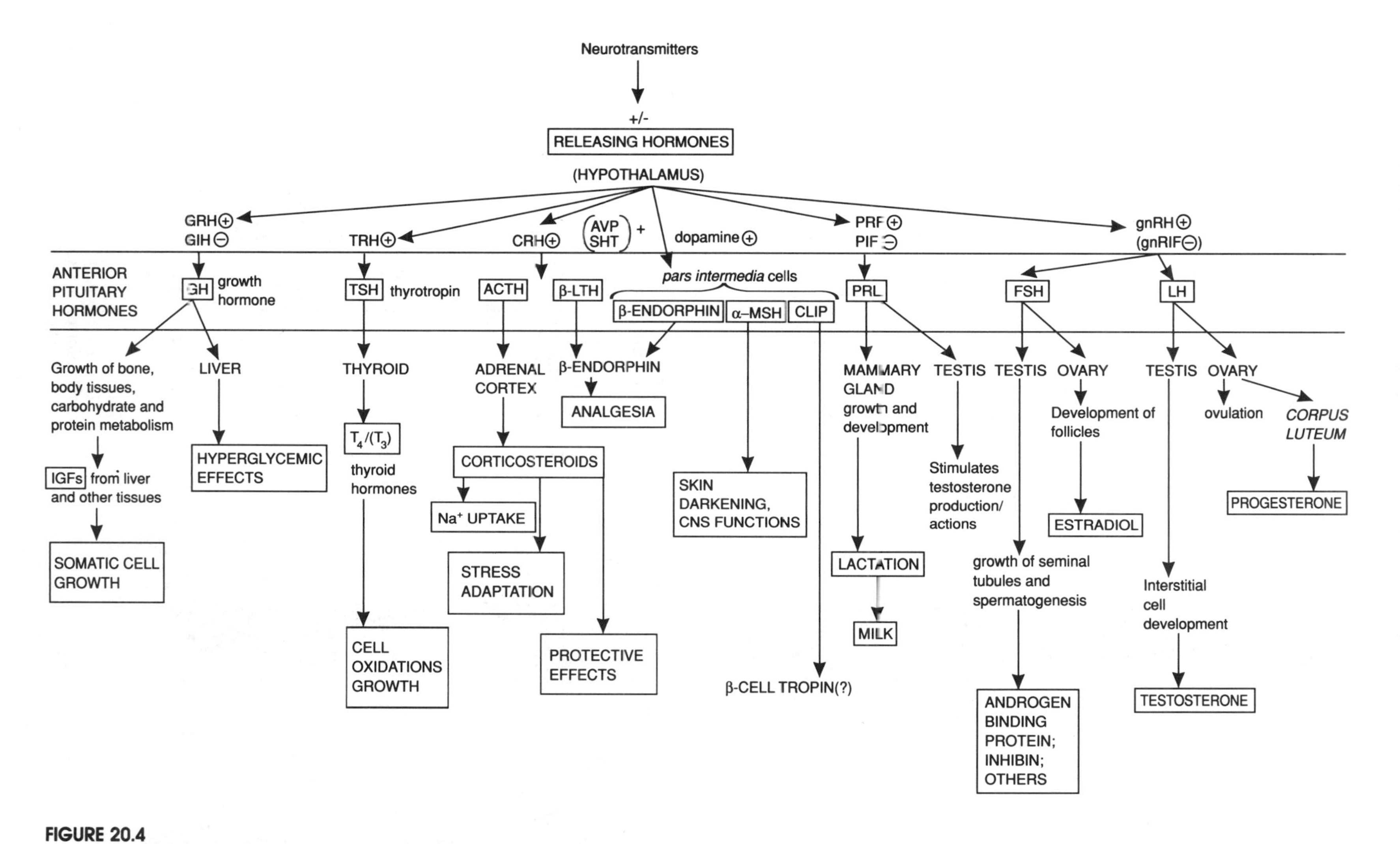

FIGURE 20.4
Overview of anterior pituitary hormones and their actions together with hypothalamic releasing hormones.

CLIN. CORR. 20.2
HYPOPITUITARISM

The hypothalamus is connected to the anterior pituitary by a delicate stalk that contains the portal system through which releasing hormones, secreted from the hypothalamus, gain access to the anterior pituitary cells. In the cell membranes of these cells are specific receptors for releasing hormones. In most cases, different cells contain different releasing hormone receptors. The connection between the hypothalamus and anterior pituitary can be broken by trauma or tumors. Trauma can occur in automobile accidents or other local damaging events that may result in severing the stalk and preventing the releasing hormones from reaching their target anterior pituitary cells. When this happens, the anterior pituitary cells no longer have their signaling mechanism for the release of the anterior pituitary hormones. In the case of tumors of the pituitary gland, all of the anterior pituitary hormones may not be shut off to the same degree or the secretion of some may disappear sooner than others. In any case, hypopituitarism occurs that may result in a life-threatening situation in which the clinician must determine the extent of loss of pituitary hormones, especially ACTH. Posterior pituitary hormones oxytocin and vasopressin may also be lost, precipitating problems of excessive urination (vasopressin deficiency) that must be addressed. The usual therapy involves administration of the end organ hormones, such as thyroid hormone, cortisol, sex hormones and progestin; with female patients it is also necessary to maintain the ovarian cycle. These hormones can be easily administered in oral form. Growth hormone is not a problem in the adult but would be an important problem in a growing child. The patient must learn to anticipate needed increases of cortisol in the face of stressful situations. Fortunately, these patients are usually maintained in reasonably good condition.

Marshall, J. C. and Barkan, A. L. Disorders of the hypothalamus and anterior pituitary. Internal Medicine, W. N. Kelley, ed., J. B. Lippincott, New York: 1989, p 2159; and Robinson, A. G. Disorders of the posterior pituitary. Internal Medicine, W. N. Kelley, ed. J. B. Lippincott, New York: 1989, p 2172.

thyrotropin or thyroid stimulating hormone (TSH), adrenocorticotropic hormone (ACTH), β-lipotropin (β-LTH), β-endorphin (*pars intermedia*-like cells), α-MSH (*pars intermedia*-like cells), corticotropin-like intermediary peptide (CLIP), from *pars intermedia*-like cells), prolactin (PRL), follicle-stimulating hormone (FSH), and luteinizing hormone (LH). Of these, all are single polypeptide chains, except TSH, FSH, and LH, which are dimers and they share a similar or identical subunit, the α subunit. The consequences of hypopituitarism is presented in Clin. Corr. 20.2.

20.3 MAJOR POLYPEPTIDE HORMONES AND THEIR ACTIONS

Since cellular communication is so specific, it is not surprising that there is a large number of hormones in the body with new ones still being discovered. Limitations of space permit a summary of only a few of the hormones. Table 20.2 presents some major polypeptide hormones and their actions. By inspection of Table 20.2 it becomes evident that many hormones cause the release of other substances, some of which may themselves be hormones. This is particularly the case for hormonal sys-

TABLE 20.2 Some Important Polypeptide Hormones in the Body and Their Actions[a]

Location	Hormone	Action
Hypothalamus	Thyrotropin releasing hormone (TRH)	Acts on thyrotrope to release TSH
	Gonadotropic releasing hormone (GnRH)	Acts on gonadotrope to release LH and FSH from the same cell
	Growth hormone releasing hormone or Somatocrinin (GRH)	Acts on sommatotrope to release GH
	Growth hormone release inhibiting hormone or somatostatin (GIH)	Acts on sommatotrope to prevent release of GH
	Corticotropin releasing hormone (CRH) Vasopressin is a helper hormone to CRH in releasing ACTH; Angiotensin II also stimulates CRH action in releasing ACTH)	Acts on corticotrope to release ACTH and β-lipotropin
	Prolactin releasing factor (PRF) (not well established)	Acts on lactotrope to release PRL
	Prolactin release inhibiting factor (PIF) (not well established; may be a peptide hormone under control of dopamine or may be dopamine itself)	Acts on lactotrope to inhibit release of PRL
Anterior pituitary	Thyrotropin (TSH)	Acts on thyroid follicle cells to bring about release of $T_4(T_3)$
	Luteinizing hormone (LH) (human chorionic gonadotropin, hCG, is a similar hormone from the placenta)	Acts on Leydig cells of testis to increase testosterone synthesis and release; acts on corpus luteum of ovary to increase progesterone production and release

TABLE 20.2 Continued

Location	Hormone	Action
	Follicle stimulating hormone (FSH)	Acts on Sertoli cells of semiferous tubule to increase proteins in sperm and other proteins; acts on ovarian follicles to stimulate maturation of ovum and production of estradiol
	Growth hormone (GH)	Acts on a variety of cells to produce IGFs, cell growth, and bone sulfation
	Adrenocorticotropic hormone (ACTH)	Acts on cells in the adrenal gland to increase cortisol production and secretion
	β-Endorphin	Acts on cells and neurons to produce analgesic and other effects
	Prolactin (PRL)	Acts on mammary gland to cause differentiation of secretory cells (with other hormones) and to stimulate synthesis of components of milk
	Melanocyte stimulating hormone (MSH)	Acts on skin cells to cause the dispersion of melanin (skin darkening)
Ultimate gland hormones	Insulin-like growth factors (IGF)	Respond to GH and produce growth effects by stimulating cell mitosis
	Thyroid hormone (T_4/T_3) (amino acid-derived hormone)	Responds to TSH and stimulates oxidations in many cells
	Opioid peptides	May derive as breakdown products of γ-lipotropin or β-endorphin or from specific gene products; can respond to CRH or dopamine and may produce analgesia and other effects
	Inhibin	Responds to FSH in ovary and in Sertoli cell; regulates secretion of FSH from anterior pituitary. Second form of inhibin (activin) may stimulate FSH secretion
	Corticotropin-like intermediary peptide (CLIP)	Derives from intermediate pituitary by degradation of ACTH; contains β-cell tropin activity which stimulates insulin release from β-cells in presence of glucose
Peptide hormones responding to other signals than anterior pituitary hormones	Arginine Vasopressin (AVP antidiuretic hormone, ADH)	Responds to increase in osmoreceptor which senses extracellular $[Na^+]$; increases water reabsorption from distal kidney tubule
	Oxytocin	Responds to suckling reflex and estradiol; causes milk ejection in lactating female, involved in

TABLE 20.2 Continued

Location	Hormone	Action
Peptide hormones responding to other signals than anterior pituitary hormones (continued)	Oxytocin (continued)	uterine contractions of labor; luteolytic factor produced by *corpus luteum;* decreases steroid synthesis in testis
	Insulin	β-cells of pancreas respond to glucose and other blood constituents to release insulin; increases tissue utilization of glucose
	Glucagon	α cells of pancreas respond to low levels of glucose falling serum calcium; decreases tissue utilization of glucose to elevate blood glucose
	Angiotensin II and III (AII and AIII)	Derived from circulating blood protein by actions of renin and converting enzyme; renin initially responds to decreased blood volume or decreased [Na^+] in the *macula densa* of the kidney. AII/AIII stimulate outer layer of adrenal cells to synthesize & release aldosterone
	Atrionatriuretic factor (ANF) or Atriopeptin	Released from heart atria in response to hypovolemia; regulated by other hormones; acts on outer adrenal cells to decrease aldosterone release; has other effects also
	Bradykinin	Generates from plasma, gut or other tissues; modulates extensive vasodilation resulting in hypotension
	Neurotensin	From hypothalamus and intestinal mucosa; effects on gut; may have neurotransmitter actions
	Substance P	From hypothalamus, CNS and intestine; pain transmitter, increases smooth muscle contractions of GI tract
	Bombesin	From nerves and endocrine cells of gut; hypothermic hormone; increases gastric acid secretion
	Cholecystokinin (CCK)	Stimulates gallbladder contraction and bile flow; increases secretion of pancreatic enzymes
	Gastrin	From stomach antrum; increases secretion of gastric acid and pepsin

TABLE 20.2 Continued

Location	Hormone	Action
Peptide hormones responding to other signals than anterior pituitary hormones (continued)	Secretin	From duodenum at pH values below 4.5; stimulates pancreatic acinar cells to release bicarbonate and water to elevate duodenal pH
	Vasointestinal peptide (VIP)	From hypothalamus and GI tract; acts as a neurotransmitter in peripheral autonomic nervous system; relaxes smooth muscles of circulation; increases secretion of water and electrolytes from pancreas and gut.
	Erythropoietin	From kidney; acts on bone marrow for terminal differentiation and initiates hemoglobin synthesis
	Relaxin	Ovarian corpus luteum; inhibits myometrial contractions; its secretion increases during gestation
	Human placental lactogen (hPL)	Acts like PRL and GH because of large amount of hPL produced
	Epidermal growth factor	Salivary gland; stimulates proliferations of cells derived from ectoderm and mesoderm together with serum; inhibits gastric secretion
	Thymopoietin (α-thymosin)	From thymus; stimulates phagocytes; stimulates differentiation of precursors into immune competent T cells
	Calcitonin (CT)	From parafollicular C cells of thyroid gland; lowers serum calcium
	Parathyroid hormone (PTH)	From parathyroid glands; stimulates bone resorption; stimulates phosphate excretion by kidney; raises serum calcium levels

SOURCE: Part of this table is reproduced from Norman, A. W. and Litwack, G. *Hormones*, Orlando, FL: Academic, 1987.

[a] This table gives only a partial list of polypeptide hormones in humans. TSH = thyroid-stimulating hormone or thyrotropin; LH = luteinizing hormone; FSH = follicle-stimulating hormone; GH = growth hormone; ACTH = adrenocorticotropic hormone; PRL = prolactin; T_4 = thyroid hormone (also T_3); IGF = insulin-like growth factor. For the releasing hormones and for some hormones in other categories, the abbreviation may contain "H" at the end when the hormone has been well characterized, and "F" in place of H to refer to "factor" when the hormone has not been well characterized. Names of hormones may contain "tropic" or "trophic" endings; tropic is mainly used here. Tropic refers to a hormone generating a change, whereas trophic refers to growth promotion. Both terms can refer to the same hormone at different stages of development. Many of these hormones have effects in addition to those listed here.

tems that are included in cascades like that presented in Figures 20.2 and 20.3. Other activities of receptor–hormone complexes located in cell membranes are to increase the flux of ions into cells, particularly calcium ions, and to activate or suppress activities of enzymes in contact with the receptor or a transducing protein with which the receptor interacts. Examples of these kinds of activities are discussed later. In the functioning of most membrane receptor complexes, stimulation of enzymes or flux of ions is followed by a chain of events, which may be described as intracellular cascades, where a high degree of amplification is obtained.

20.4 GENES AND FORMATION OF POLYPEPTIDE HORMONES

Genes for polypeptide hormones contain the information for the hormone and the control elements upstream of the transcriptionally active sequence. In some cases, more than one hormone is encoded in a gene. For example, systems like proopiomelanocortin, a gene product that encodes

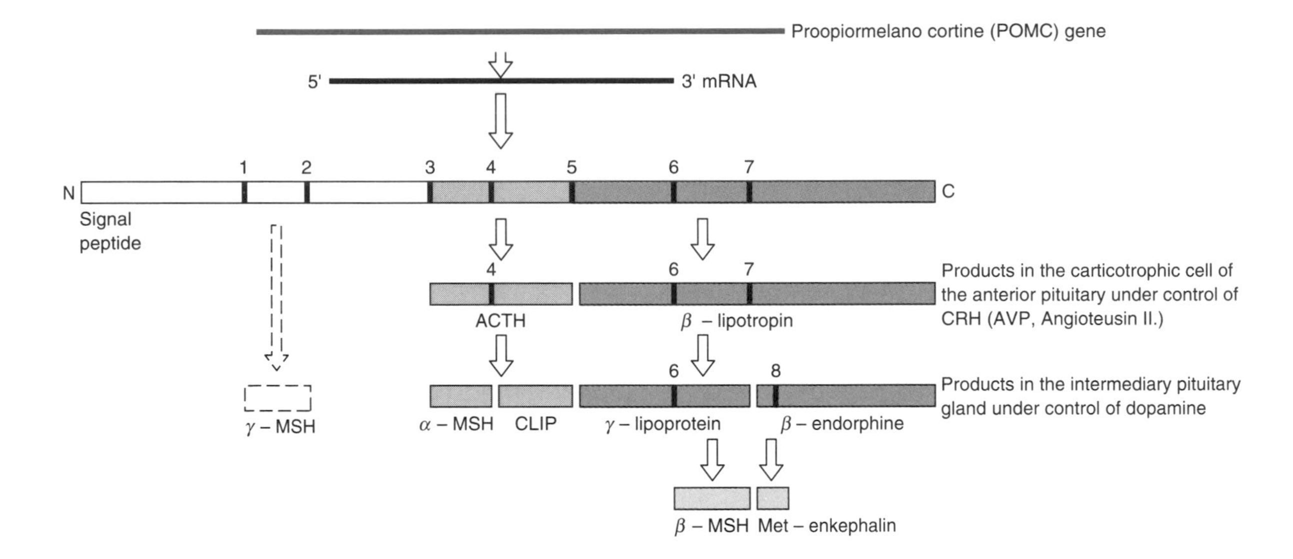

FIGURE 20.5

Proopiomelanocortin is a polypeptide product encoded by a single gene. The dark vertical bars represent proteolytic cleavage sites for specific enzymes. The cleavage sites are Arg-Lys, Lys-Arg, or Lys-Lys. Some specificity also may be conferred by neighboring amino acid residues. In the corticotrophic cell of the anterior pituitary, enzymes are present that cleave at sites 3 and 5, releasing the major products, ACTH and β-lipotropin into the general circulation. In the pars intermedia, especially in vertebrates below humans, these products are further cleaved at major sites 4, 6, and 7 to release α-MSH, CLIP, γ-lipotropin, and β-endorphin into the general circulation. Some β-lipotropin arising in the corticotroph may be further degraded to form β-endorphin. These two cell types appear to be under separate controls. The corticotrophic cell of the anterior pituitary is under the positive control of the CRH and its auxiliary helpers, arginine vasopressin (AVP), and Angiotensin II. AVP by itself does not release ACTH but enhances the action of CRH in this process. The products of the intermediary pituitary, α-MSH, CLIP (corticotropin-like intermediary peptide), γ-lipotropin, and β-endorphin are under the positive control of dopamine, rather than CRH, for release. Obviously there must exist different proteases in these different cell types in order to generate a specific array of hormonal products. β-Endorphin also contains a pentapeptide, enkephalin, which potentially could be released at some point (hydrolysis at 8).

TABLE 20.3 Summary of Stimuli and Products of Proopiomelanocortin

Cell type	Corticotroph[a]	pars intermedia[a]
Stimulus	CRH	Dopamine
Auxiliary stimulus	AVP,AII	
Major products	ACTH β-lipotropin (β-endorphin)	α-MSH, CLIP, γ-lipotropin, β-endorphin

[a] CRH = corticotropin releasing hormone; AVP = arginine vasopressin; AII = angiotensin II; ACTH = adrenocorticotropin; α-MSH = alpha melanocyte stimulating hormone; CLIP = corticotropin-like intermediary peptide.

the following hormones: ACTH, β-lipotropin, and other hormones like γ-lipotropin, γ-MSH, α-MSH, CLIP, β-endorphin and potentially, enkephalins. A gene active in the adrenal medulla encodes multiple copies of enkephalins. In the case of the posterior pituitary hormones, oxytocin and vasopressin, information for these hormones are each encoded on a separate gene together with information for each respective **neurophysin,** a protein that binds to the completed hormone and stabilizes it.

Proopiomelanocortin Is a Precursor Peptide for Eight Hormones

Proopiomelanocortin, as schematized in Figure 20.5, can generate at least eight hormones from a single gene product. All products do not appear at once in a single-cell type, but occur in separate cells based on their content of specific proteases, needed to cleave the propeptide, specific metabolic controls and based on different positive regulators. Thus, while proopiomelanocortin occurs in both the corticotropic cell of the anterior pituitary and the *pars intermedia* cell, the stimuli and products are different as summarized in Table 20.3. The *pars intermedia* is a discrete anatomical structure located between the anterior and posterior pituitary in the rat (Figure 20.6). In the human, however, the *pars intermedia* is not a discrete anatomical structure, although the cell type may be present in the equivalent location. More needs to be discovered about the controls of MSH release in the human.

Many Polypeptide Hormones Are Encoded Together in a Single Gene

An example of another gene and gene products encoding more than one protein are the genes for vasopressin and oxytocin and their accompanying neurophysin proteins, products that are released from the posterior pituitary upon specific stimulation. In much the same manner that ACTH and β-lipotropin (β-LPH) are split out of the proopiomelanocortin precursor peptide, so are the products vasopressin, neurophysin II, and a glycoprotein of as yet unknown function split out of the vasopressin precursor. A similar situation exists for oxytocin and neurophysin I (Figure 20.7).

Vasopressin and **neurophysin II** are released by baroreceptors and osmoreceptors, which sense a fall in blood pressure or a rise in extracellular sodium ion concentration, respectively. Generally, **oxytocin** and **neurophysin I** are released by the suckling response in lactating females or by other stimuli mediated by a specific cholinergic mechanism. Oxytocin–neurophysin I release can be triggered by injection of estradiol. Release of

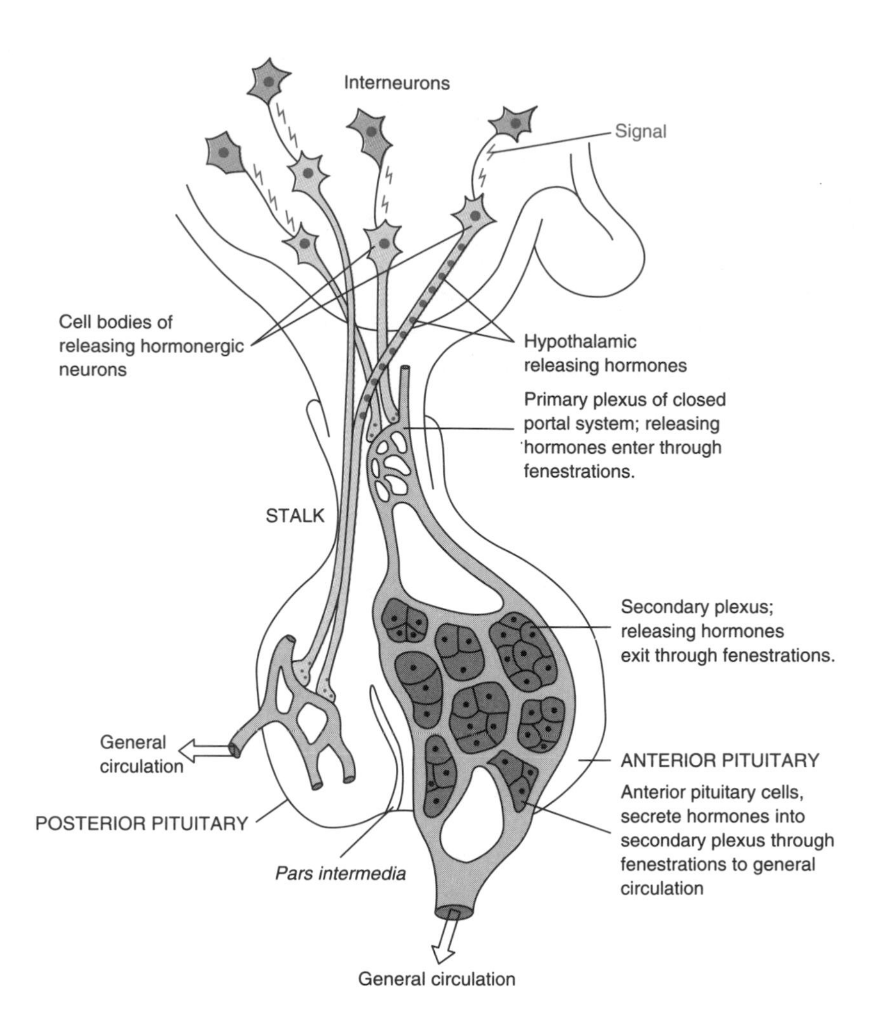

FIGURE 20.6
The hypothalamus with nuclei in various locations in which the hypothalamic releasing hormones are synthesized.
Shown is the major vascular network consisting of a primary plexus where releasing hormones enter its circulation through fenestrations and the secondary plexus in the anterior pituitary where the releasing hormones are transported out of the circulation again through fenestrations in the vessels to the region of the anterior pituitary target cells. This figure also shows the resultant effects of the actions of the hypothalamic releasing hormones causing the secretion into the general circulation of the anterior pituitary hormones.
Adapted from A. W. Norman and G. Litwack, *Hormones*, New York: Academic, 1987, p. 104.

vasopressin–neurophysin II can be stimulated by administration of nicotine. The two separate and specific releasing agents, estradiol and nicotine, prove that oxytocin and vasopressin together with their respective neurophysins, are synthesized and released from different cell types. Although oxytocin is well known for its milk-releasing action in the lactating female, in the male it seems to have a separate role connected with a reduction in testosterone synthesis in the testes.

Other polypeptide hormones are being discovered that are encoded together by a single gene. An example is the discovery of the gene encod-

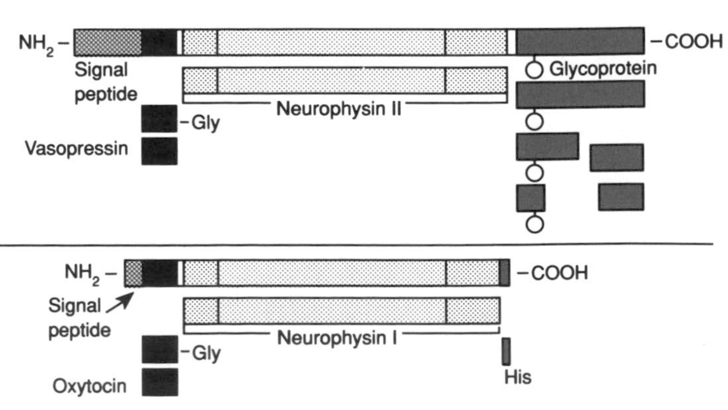

FIGURE 20.7
Prepro-vasopressin and prepro-oxytocin.
Proteolytic maturation proceeds from top to bottom for each precursor. The organization of the gene translation products is similar in either case except that a glycoprotein is included on the proprotein of vasopressin in the C-terminal region. Heavily stippled bars of the neurophysin represent conserved amino acid regions; lightly stippled bars represent variable C and N termini.
Redrawn with permission from D. Richter, VP and OT are expressed as polyproteins. *Trends Biochem.* 8:278, 1983.

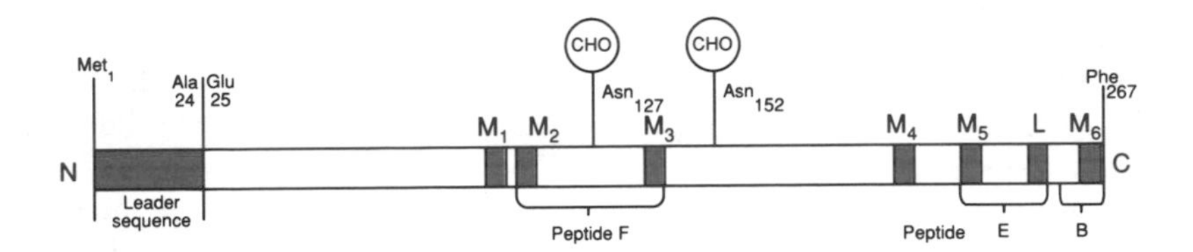

FIGURE 20.8
Model of enkephalin precursor.
The distribution of Met-enkephalin sequences (M_1–M_6) and Leu-enkephalin (L) are presented.
Redrawn from M. Comb, P. H. Seeburg, J. Adelman, L. Eiden, and E. Herbert, *Nature* (*London*) 295:663, 1982.

ing the GnRH, a decapeptide that appears to reside to the left of a gene for the GnRH-associated peptide (GAP), which may be a hormone whose action is to inhibit the release of prolactin. Thus it would be the prolactin release inhibiting factor (PRIF), that is PRIF = GAP. Since the putative PRIF is encoded by the same gene encoding the information of GnRH, it is apparent that both hormones must be secreted by the same cell.

Multiple Copies of a Hormone Can Be Encoded on a Single Gene

An example of multiple copies of a single hormone encoded on a single gene is the gene product for enkephalins located in the chromaffin cell of the adrenal medulla. **Enkephalins** are pentapeptides with opioid activity; methionine-enkephalin (Met-ENK) and leucine-enkephalin (Leu-ENK) have the structures:

Tyr-Gly-Gly-Phe-Met (Met-ENK)

Tyr-Gly-Gly-Phe-Leu (Leu-ENK)

A model of enkephalin precursor in adrenal medulla is presented in Figure 20.8, which encodes several met-ENK(M) molecules and a molecule of leu-ENK(L). Again, the processing sites to release enkephalin molecules from the protein precursor involve Lys-Arg, Arg-Arg, and Lys-Lys bonds.

Many genes for hormones are constructed to encode only one hormone and this may be the general situation. An example of a single hormone gene is shown in Figure 20.9. In this case the information for the hormone is contained in the second exon and the information in the first exon is not expressed. Having cDNAs for use as probes that contain the information

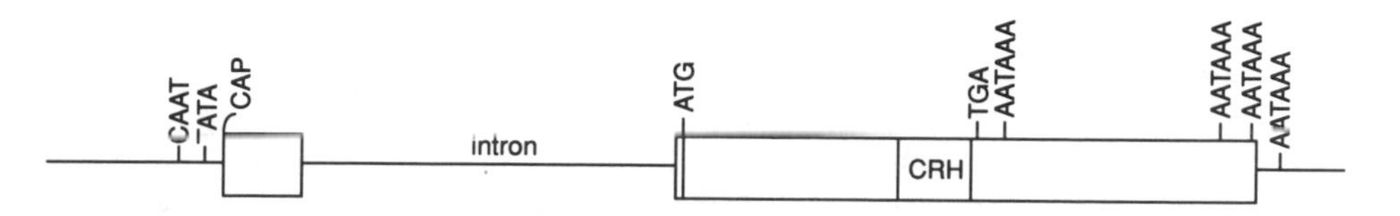

FIGURE 20.9
Nucleic acid sequence for rat *proCRH* genes.
A schematic representation of the rat *proCRH* gene. The exons are shown as *blocks* and the intron by a *line*. The TATA and CAAT sequence, putative cap site, translation initiation ATG, translation terminator UGA, and poly(A) addition signals (AATAAA) are indicated. The location of the CRH peptide is indicated by CRH.
Redrawn from R. D. Thompson, A. F. Seasholz, and E. Herbert, *Molec. Endocrinol.*, 1:363, 1987.

for expression of CRH allows for the localization of the hormone in tissues. Previously it was thought that the hormone should be restricted to the hypothalamus, the anterior pituitary, and the stalk, which contains the closed vascular transporting system (Figure 20.6). However, RNA extracts from different tissues probed with this DNA revealed the location of CRH mRNA in testis, brainstem, and adrenal in addition to pituitary and hypothalamus. The presence of the hormone in extrahypothalamic pituitary axis tissues and its functions there are subjects of active investigation.

20.5 SYNTHESIS OF AMINO ACID DERIVED HORMONES

A large number of hormones and neurotransmitters are derived from amino acids, principally from tyrosine and phenylalanine. Glutamate, aspartate, and other compounds are important neurotransmitter substances as well. Although there may be some confusion about which compounds are neurotransmitters and which are hormones, it is clear that epinephrine from the adrenal medulla is a hormone, whereas norepinephrine is a neurotransmitter. This section considers epinephrine and thyroxine or triiodothyronine. The other biogenic amines, such as dopamine, which are considered to be neurotransmitters, are discussed in Chapter 22.

Epinephrine Is Synthesized From Phenylalanine/Tyrosine

The formation of epinephrine occurs in the adrenal medulla. A number of steroid hormones, aldosterone, cortisol, and dehydroepiandrosterone (sulfate), are produced in the adrenal cortex, which are discussed in Chapter 21.

The biochemical reactions leading to the formation of **epinephrine** from tyrosine or phenylalanine are presented in Figure 20.10. Epinephrine is a principal hormone secreted from the adrenal medulla chromaffin cell along with some norepinephrine, enkephalins, and some of the enzyme *dopamine-β-hydroxylase*. Secretion of epinephrine is signaled by the neural response to stress, which is transmitted to the adrenal medulla by way of a preganglionic acetylcholinergic neuron (Figure 20.11). Release of acetylcholine by the neuron increases the availability of intracellular calcium ion, which stimulates exocytosis and release of the material stored in the **chromaffin granules** (Figure 20.11*b*). This overall system of epinephrine synthesis, storage, and release from the adrenal medulla is regulated by neuronal controls and also by glucocorticoid synthesized in and secreted from the adrenal cortex in response to stress. Since the products of the adrenal cortex are transported to the adrenal medulla on their way out to the general circulation, cortisol becomes elevated in the medulla and induces **phenylethanolamine *N*-methyltransferase (PNMT),** a key enzyme catalyzing the conversion of norepinephrine to epinephrine. Thus, in biochemical terms, the stress response at the level of the adrenal cortex insures the production of epinephrine from the adrenal medulla (Figure 20.12). Presumably, epinephrine once secreted into the bloodstream not only affects α receptors of hepatocytes to ultimately increase blood glucose levels as indicated, but also interacts with α receptors on vascular smooth muscle cells and on pericytes to cause cellular contraction and increase blood pressure.

FIGURE 20.10
Biochemical steps in the synthesis of epinephrine by the chromaffin cell of the adrenal medulla.

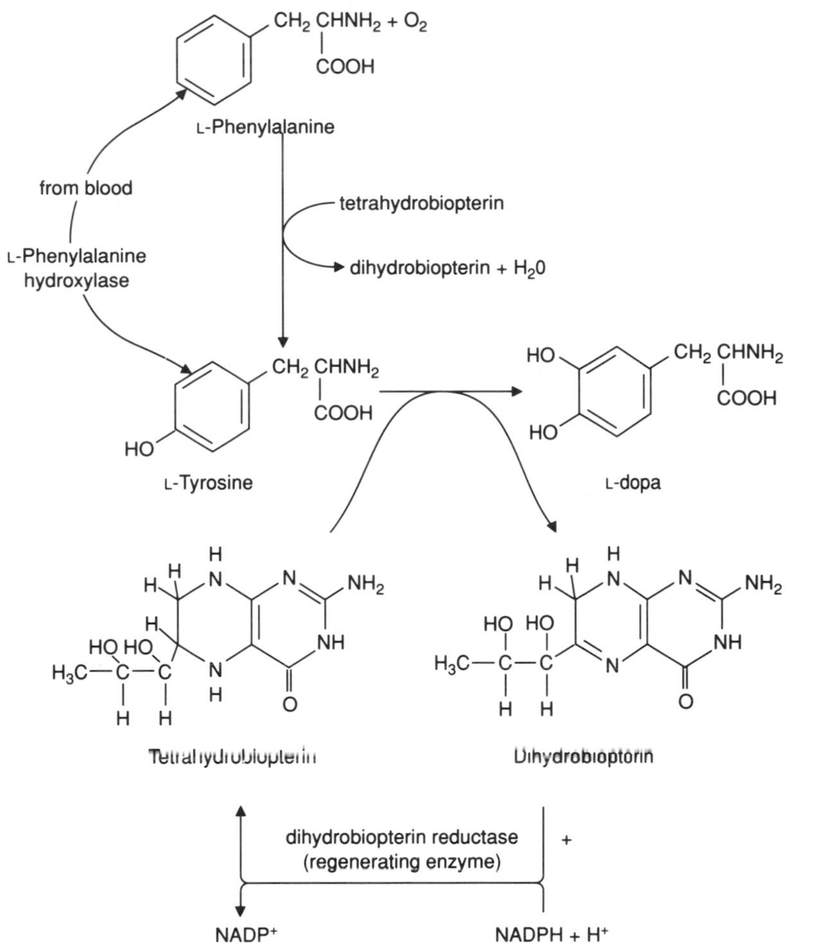

$CH_2\,CHNH_2 + O_2$
COOH
L-Phenylalanine
from blood
L-Phenylalanine hydroxylase
tetrahydrobiopterin
dihydrobiopterin + H_20
$CH_2\,CHNH_2$
COOH
HO
L-Tyrosine
HO
$CH_2\,CHNH_2$
COOH
HO
L-dopa
Tetrahydrobiopterin
Dihydrobiopterin
dihydrobiopterin reductase (regenerating enzyme)
+
$NADP^+$
$NADPH + H^+$

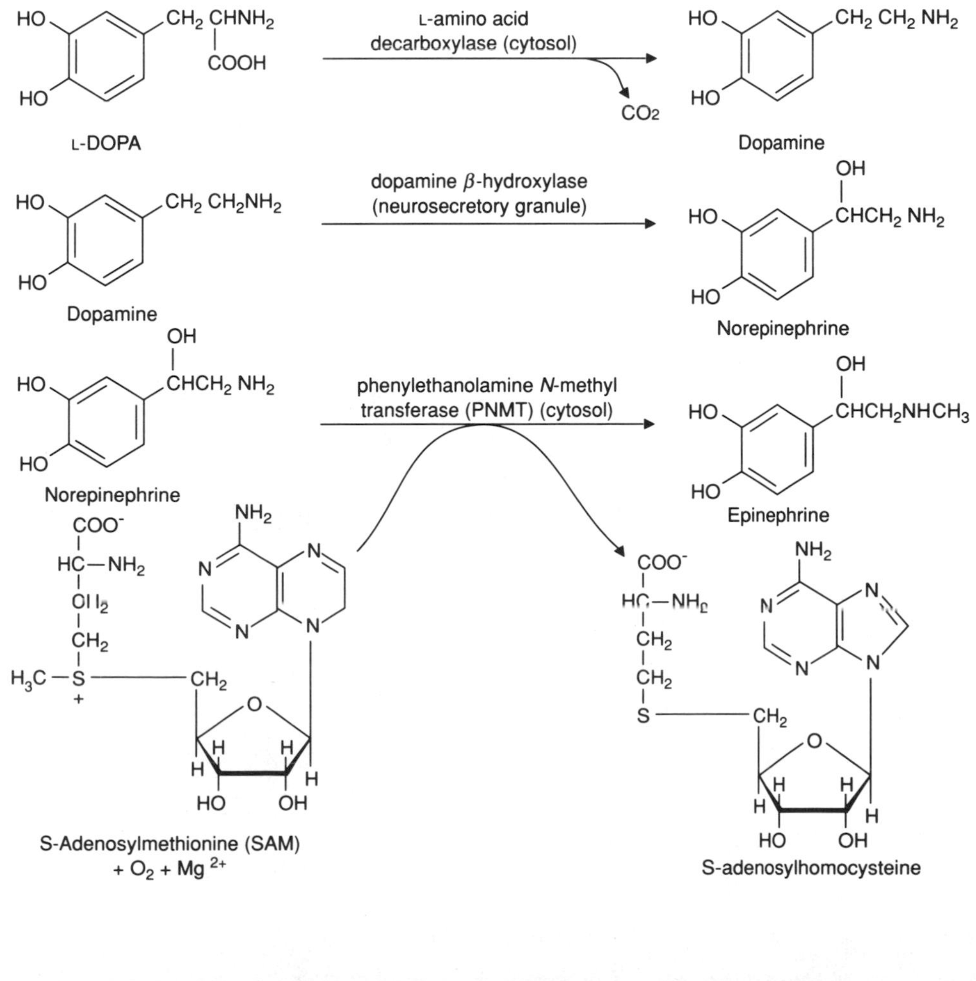

L-DOPA → L-amino acid decarboxylase (cytosol) → Dopamine + CO_2

Dopamine → dopamine β-hydroxylase (neurosecretory granule) → Norepinephrine

Norepinephrine → phenylethanolamine *N*-methyl transferase (PNMT) (cytosol) → Epinephrine

S-Adenosylmethionine (SAM) + O_2 + Mg^{2+} → S-adenosylhomocysteine

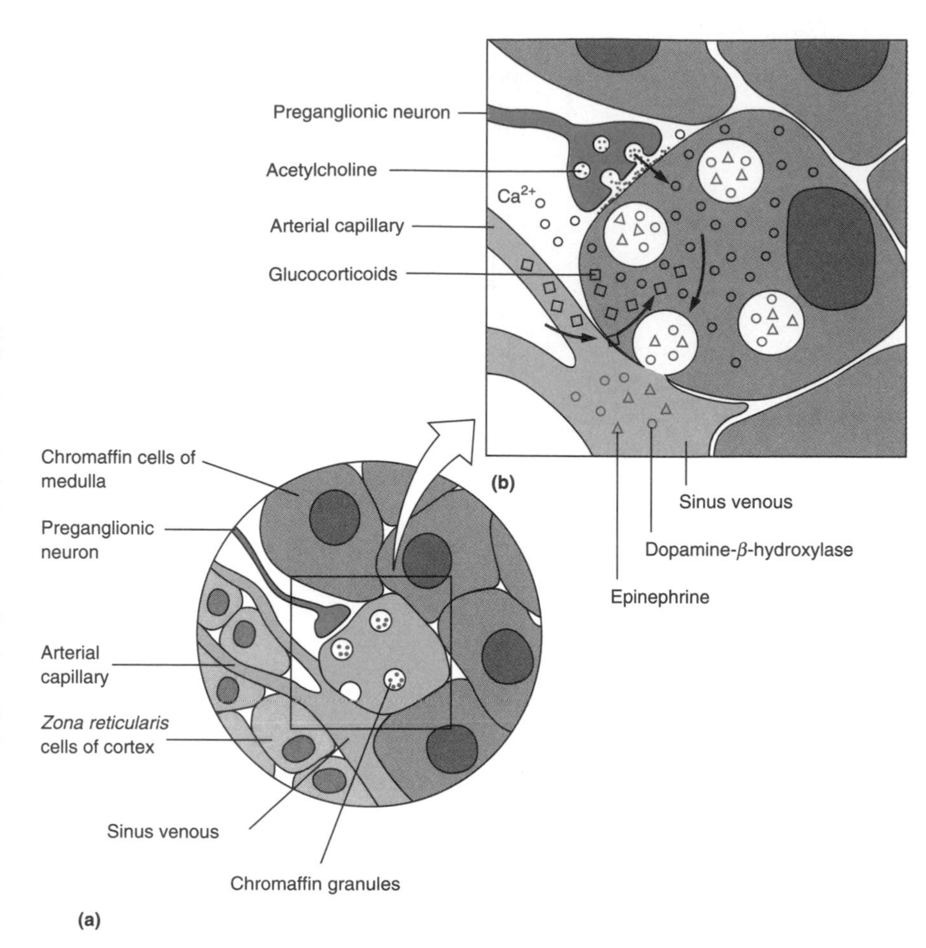

FIGURE 20.11
Diagrams showing the relationship of adrenal medulla chromaffin cells to preganglionic neuron innervation and the structural elements involved in the synthesis of epinephrine and the discharge of catecholamines in response to acetylcholine.
(*a*) Functional relationship between cortex and medulla for the control of synthesis of adrenal catecholamines. Glucocorticoids that stimulate enzymes catalyzing the conversion of norepinephrine to epinephrine reach the chromaffin cells from capillaries shown in (b). (*b*) Discharge of catecholamines from storage granules in chromaffin cells after nerve fiber stimulation resulting in the release of acetylcholine. Calcium enters the cells as a result, causing the fusion of granular membranes with the plasma membrane and exocytosis of the contents.
Reprinted with permission from D. T. Krieger and J. C. Hughes (Eds.). *Neuroendocrinology*. Sunderland, MA: Sinauer Associates, 1980.

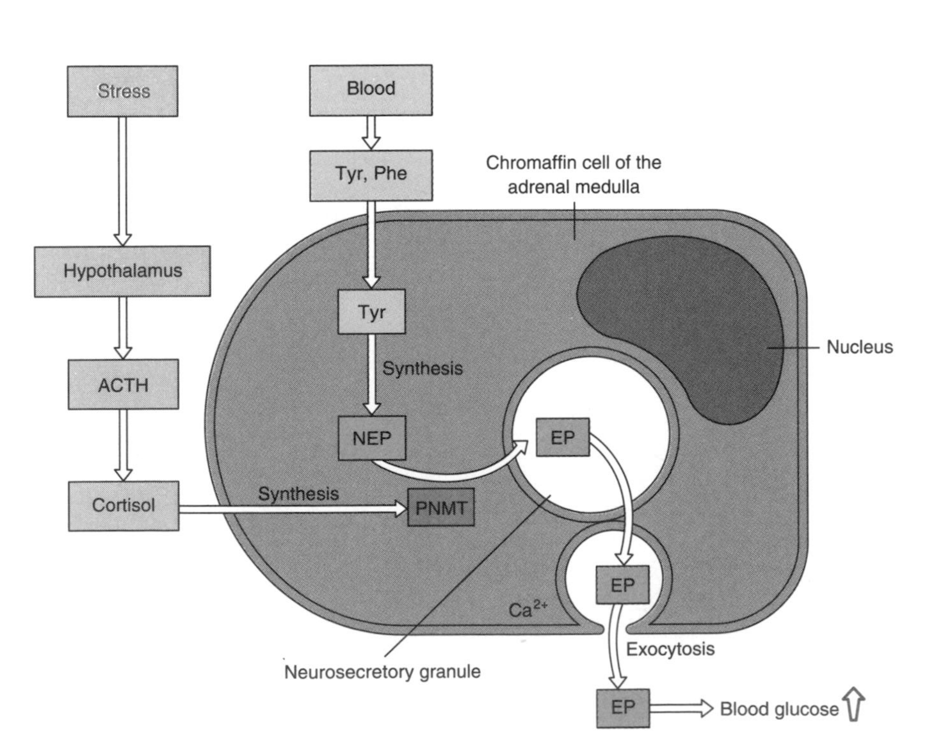

FIGURE 20.12
Biosynthesis, packaging, and release of epinephrine in the adrenal medulla chromaffin cell.
PNMT = phenylethanolamine *N*-methyltransferase; EP = epinephrine; NEP = norepinephrine. Arrows pointing upward following the name of a compound refer to an increase in the concentration of that substance. Neurosecretory granules contain epinephrine, dopamine β-hydroxylase, ATP, Met- or Leu-enkephalin, as well as larger enkephalin-containing peptides or norepinephrine in place of epinephrine. Epinephrine and norepinephrine are contained in different cells. Enkephalins could also be contained in separate cells, although that is not completely clear.
Adapted from Norman, A. W. and Litwack, G. *Hormones*, New York: Academic, 1987, p. 464.

Synthesis of Thyroid Hormone Requires Incorporation of Iodine Into a Tyrosine of Thyroglobulin

An outline of the biosynthesis of thyroid hormone, **tetraiodo-L-thyronine (T_4),** also referred to as **thyroxine,** and its active cellular counterpart, **triodo-L-thyronine (T_3)** (Figure 20.13) is presented in Figure 20.14. The thyroid gland is differentiated to concentrate iodide from the blood and through the series of reactions shown horizontally in the center of Figure 20.14, monoiodotyrosine (MIT), diodotyrosine (DIT), T_4 and T_3 are produced and associated with **thyroglobulin** (TG). Thus, the iodinated amino acids and thyronines are stored in the thyroid follicle with thyroglobulin. Monoiodotyrosine (MIT) can be formed by two theoretical mechanisms shown in Figure 20.15a and b. Presumably an additional iodide can be added to the ring by similar mechanisms to form diiodotyrosine. It is believed that the thyroid hormones are synthesized from DIT. In Figure 20.16a a hypothetical coupling sequence is shown for the formation of thyroxine. Figure 20.16b presents a scheme by which T_4 can be incorporated into thyroglobulin. Recent work indicates that there are hot spots (regions for very active iodination) in the thyroglobulin sequence for the incorporation of iodine. Apparently, the sequences around iodotyrosyls occur in three concensus groups: Glu/Asp-Tyr associated with the synthesis of thyroxine or iodotyrosines; Ser/Thr-Tyr-Ser, associated with the synthesis of iodothyronine and iodotyrosine and Glu-X-Tyr, the remaining iodotyrosyls in the sequence.

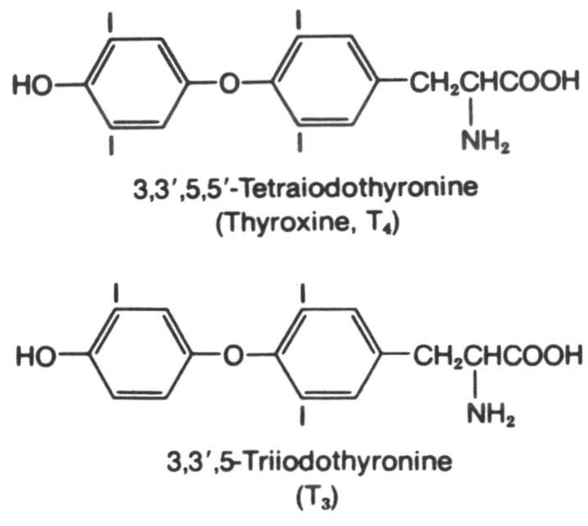

3,3′,5,5′-Tetraiodothyronine (Thyroxine, T_4)

3,3′,5-Triiodothyronine (T_3)

FIGURE 20.13
Structures of thyroid hormones.

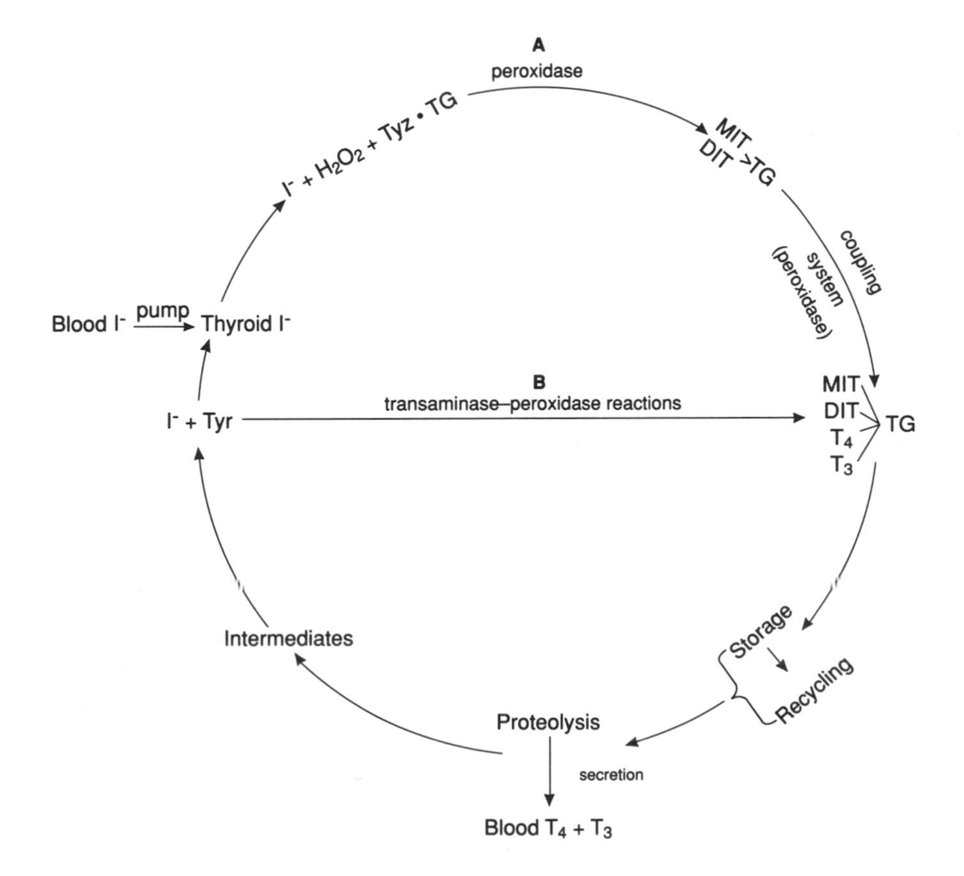

FIGURE 20.14
Pathways of iodide metabolism by the thyroid gland.
Thyroidal iodide is oxidized by H_2O_2 in the presence of peroxidase to a higher valence state and bound enzymatically to tyrosyl residues of thyroglobulin (TG) in pathway A. The coupling of monoiodotyrosine (MIT) and diiodotyrosine (DIT) might be intramolecular or intermolecular. In pathway A, peroxidase probably catalyzes the coupling reaction in the molecule of thyroglobulin. In pathway B, a series of steps producing free intermediates such as *p*-hydroxyphenylpyruvic acid (DIHPPA) could lead to the formation of DIHPHPO, a hydroperoxide derivative of DIHPPA, that reacts with a DIT molecule of thyroglobulin to synthesize T_4 in an intermolecular coupling process.
Redrawn from L. J. DeGroot and H. Niepomniszcze, *Metab. Clin. Exp.* 26:666, 1977.

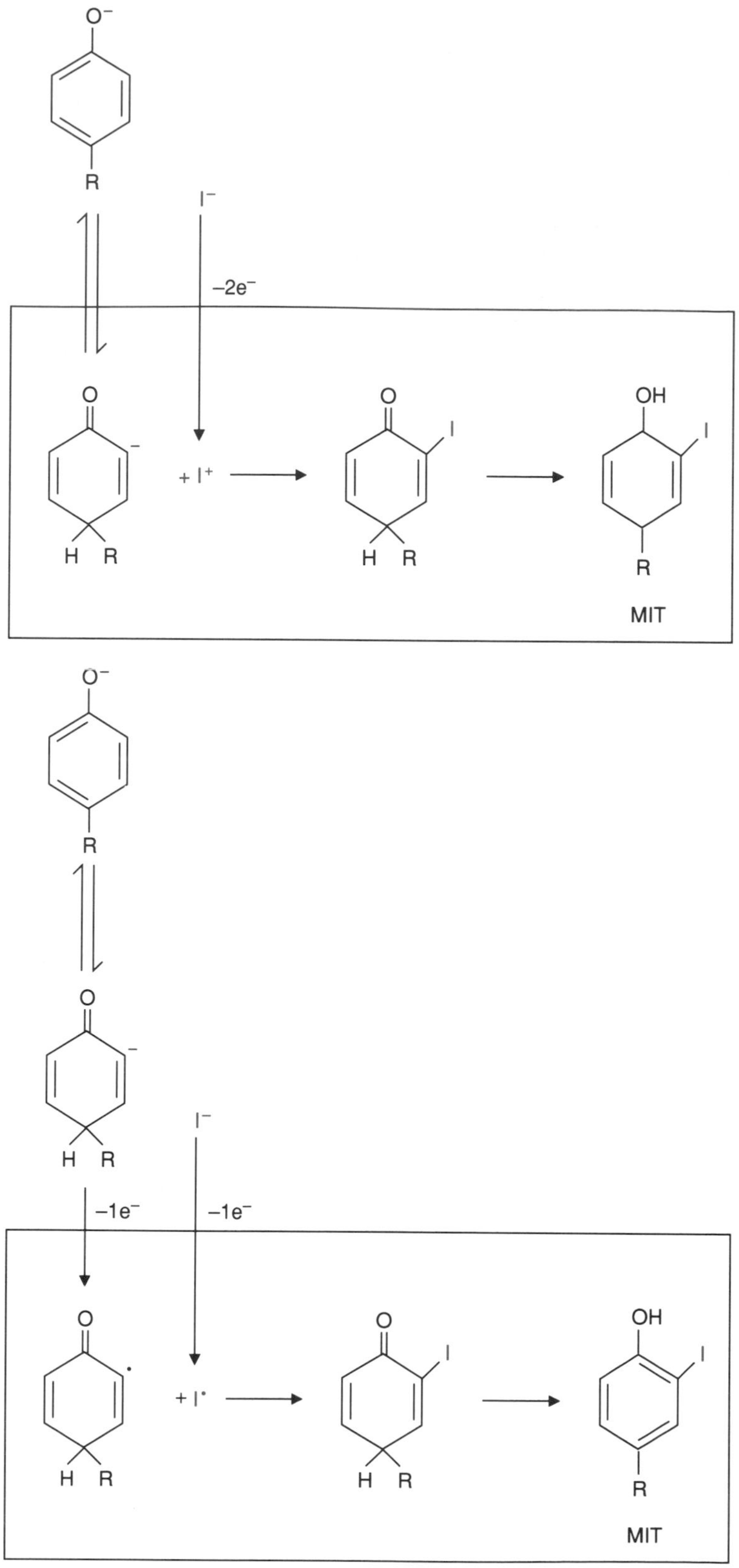

FIGURE 20.15
Iodination of tyrosine.
(*a*) Iodide is oxidized by thyroid peroxidase in a $2e^-$ oxidation step, forming I^+. The phenolate anion of tyrosine is in equilibrium with its quinoid form. Iodinium and tyrosine quinoid react to form an iodinated quinoid intermediate that forms MIT by electronic rearrangement. (*b*) Iodine is oxidized by thyroid peroxidase in a $1e^-$ step forming a free radical (I^0). The quinoid anion of tyrosine is oxidized to a free radical by peroxidase or by another I^0. The quinoid free radical reacts with I^0 to form the iodinated quinoid intermediate that forms MIT by electronic rearrangement.
Redrawn from L. J. DeGroot and H. Niepomniszcze, *Metab. Clin. Exp.* 26:666–681, 1977 as published in A. W. Norman and G. Litwack, *Hormones,* New York: Academic, 1987, p. 232, 234.

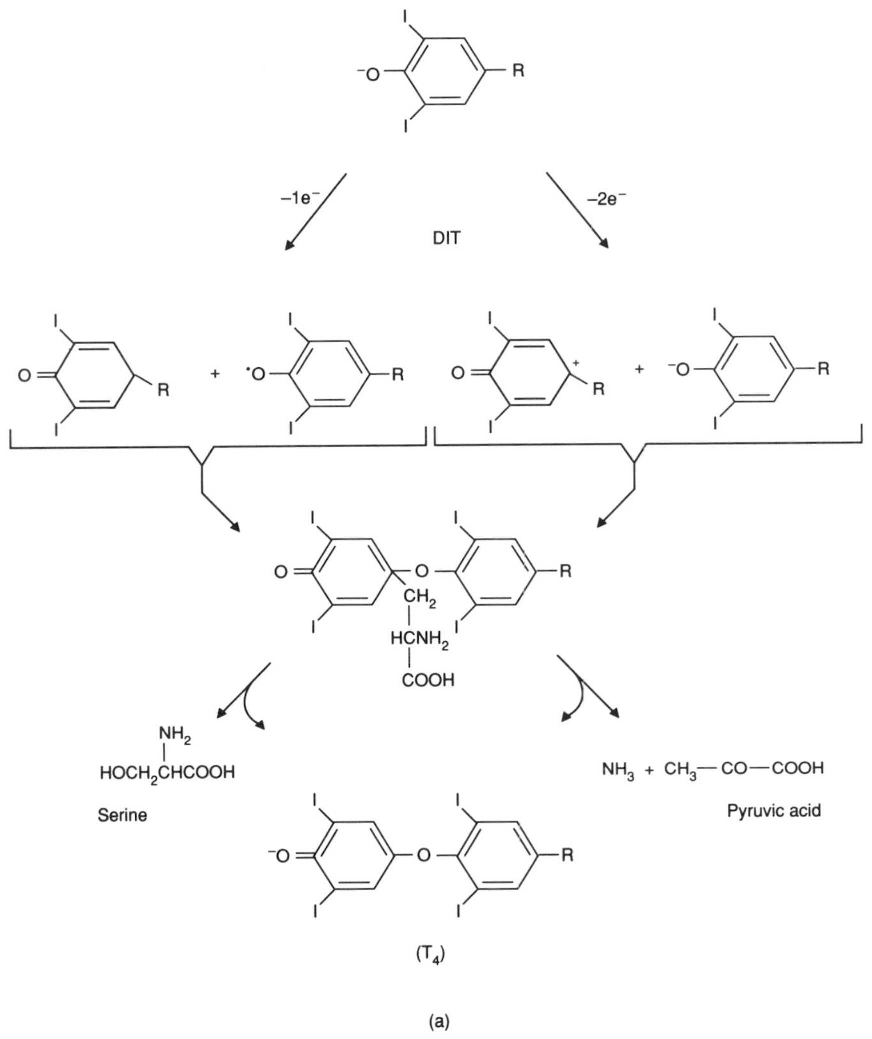

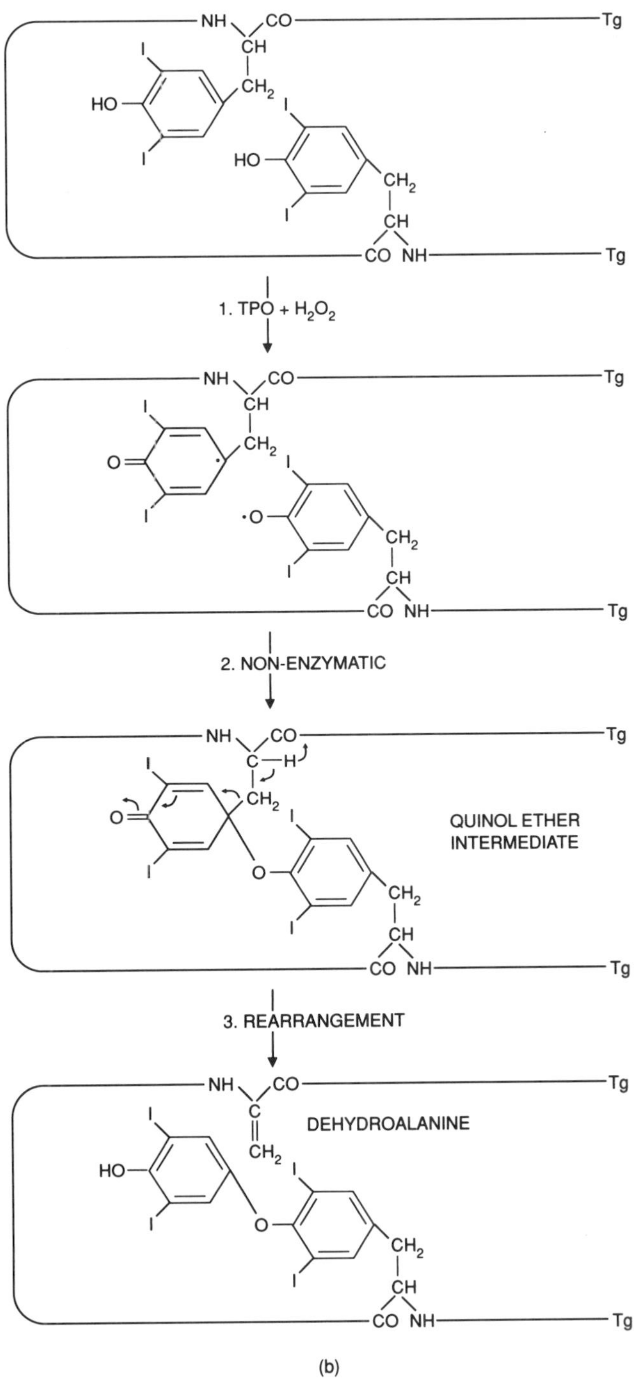

FIGURE 20.16
Proposed intramolecular coupling mechanisms for formation of T_4 from DIT (a) and hypothetical coupling scheme for intramolecular formation of T_4 in thyroglobin (b).
(a) Is modified from L. J. DeGroot and H. Niepomniszcze, *Metab. Clin. Exp.* 26, 694, 1977. (b) Is modified from A. Taurog, *Endocrinology*, Vol. 1. L. J. DeGroot et al. (Eds.). New York: Grune & Stratton, 1979, pp. 331–346.
Redrawn from A. W. Norman and G. Litwack, *Hormones*, New York: Academic, 1987, pp. 237–238.

TABLE 20.4 Hypothalamic Releasing Hormones Containing a C-Terminal Pyroglutamate[a] or an N-Terminal Amino Acid Amide or Both

Hormone	Sequence[b]
Thyrotropin releasing hormone (TRH)	pGlu[a]-H-[b]-Pro-NH_2
Gonadotropin releasing hormone (GnRH)	pGlu-HWSYGLRP-Gly-NH_2
Corticotropin releasing hormone (CRH)	SQEPPISLDLTFHLLREVLEMTKADQLAQQAHSNRKLLDI-Ala-NH_2
Somatocrinin (GRH)	YADAIFTNSYRKVLGQLSARKLLQDIMSRQQGESNQERGARAR-Leu-NH_2

[a] The pyroglutamate structure is

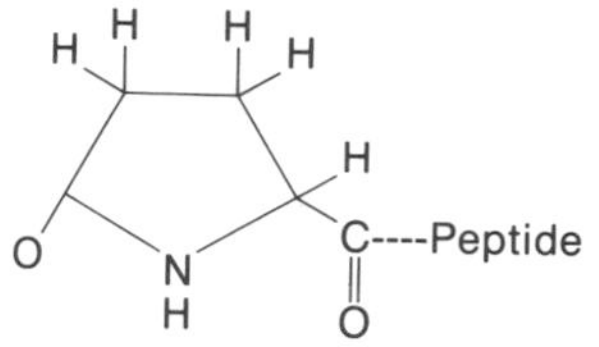

[b] Denotes single-letter abbreviation for an amino acid. Abbreviations are as follows: Ala = A; Arg = R; Asn = N; Asp = D; Cys = C; Glu = E, Gln = Q; Gly = G; His = H; Ile = I; Leu = L; Lys = K; Met = M; Phe = F; Pro = P; Ser = S; Thr = T; Trp = W; Tyr = Y; Val = V.

20.6 INACTIVATION AND DEGRADATION OF HORMONES

Most polypeptide hormones are degraded to amino acids by hydrolysis presumably in the lysosome. Partial hydrolysis by proteinases is a principal pathway for degradation. Certain hormones, however, contain modified amino acids; for example, among the hypothalamic releasing hormones, the N-terminal amino acid can be cycloglutamic acid (or pyroglutamic acid) (Table 20.4) and a C-terminal amino acid amide. Some of the releasing hormones that have either or both of these amino acid derivatives are listed in Table 20.4. Apparently, breakage of the cyclic glutamate ring or cleavage of the C-terminal amide can lead to inactivation of many of these hormones and such enzymic activities have been reported in blood. This probably accounts, in part, for the short half-life of many of these hormones.

Some hormones contain a ring structure joined by a cystine disulfide bridge. A few examples are given in Table 20.5. Peptidc hormones, such as those shown in Table 20.5 may be degraded initially by the random action of **cystine aminopeptidase** and **glutathione transhydrogenase** as shown in Figure 20.17. Alternatively, as has been suggested in the case of oxytocin, the peptide may be broken down through partial proteolysis to shorter peptides, some of which may have hormonal actions on their own. Maturation of prohormones in many cases involves proteolysis, which may be considered as a degradation process in the sense that the prohormone is degraded to active forms (e.g., Figure 20.5), whereas degradation is usually thought of as the reduction of active peptides to inactive ones.

20.7 CELL REGULATION AND HORMONE SECRETION

Hormonal secretion is under specific control. In the cascade system displayed in Figures 20.2 and 20.3, hormones must emanate from one

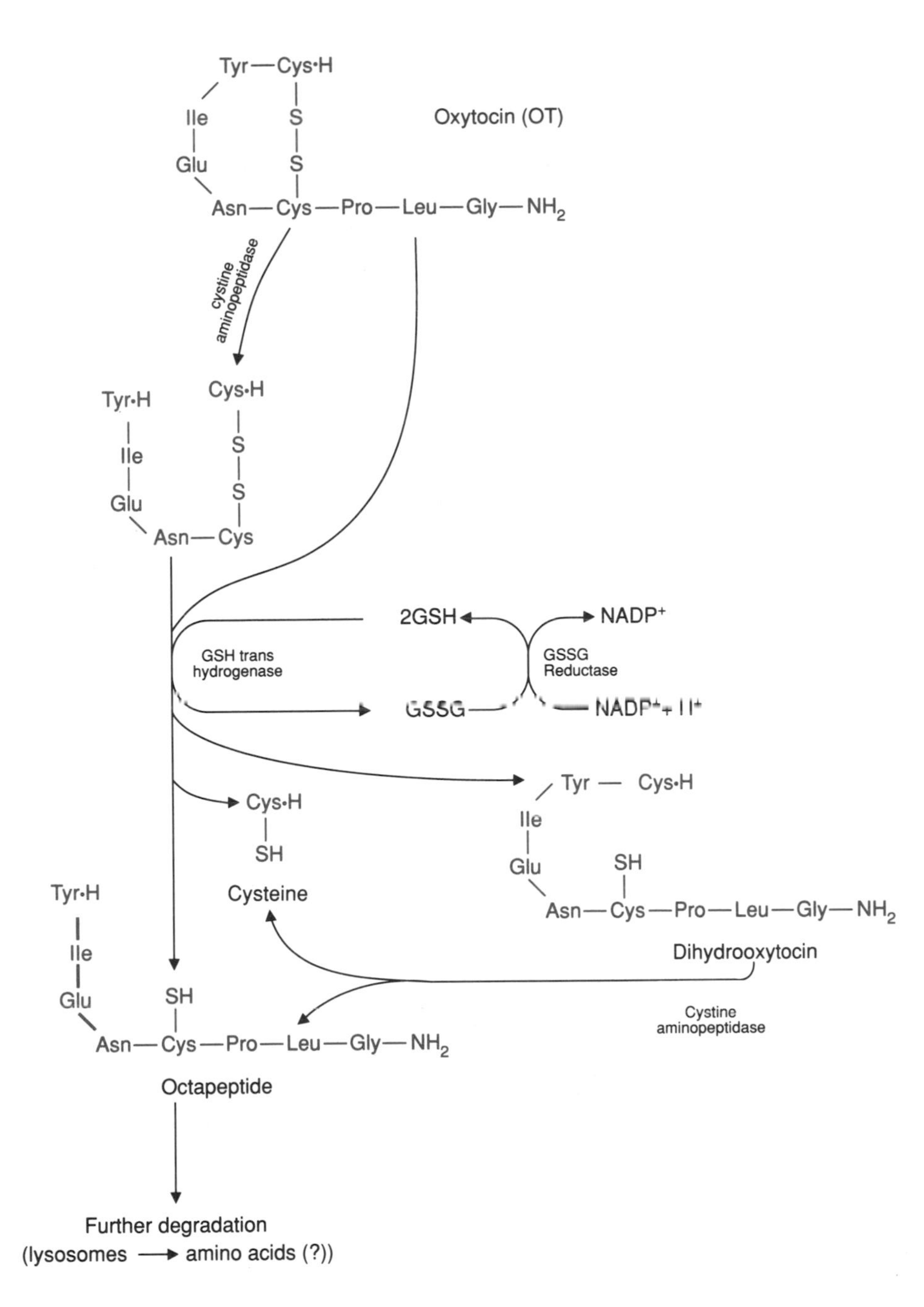

TABLE 20.5 Examples of Hormones Containing a Cystine Disulfide Bridge Structure

Hormone	Sequence[a]
Somatostatin (GH)	FFNKCGA[1] / W / K / TFTSC[14] (ring closed by S—S bridge between C and C)
Oxytocin	YC[1] / I / E / NCPLG—NH$_2$ (ring closed by S—S bridge between C and C)
Arg vasopressin	YC / F / Q / NCPRG—NH$_2$ (ring closed by S—S bridge between C and C)

[a] Letters refer to single-letter amino acid abbreviations (Table 20.4)

FIGURE 20.17
Degradation of posterior pituitary hormones.
Oxytocin transhydrogenase is similar to degrading enzymes for insulin; presumably, these enzymes also degrade vasopressin.
Redrawn from A. W. Norman and G. Litwack, *Hormones*, New York: Academic, 1987, p. 167.

source, cause hormonal release from the next cell type in line, and so on, down the cascade system. The right responses must follow from a specific stimulus. The precision of these signals is defined by the hormone and the receptor as well as by the activities of the CNS, preceding the first hormonal response, in many cases. Certain generalizations can be made. Polypeptide hormones generally bind to their cognate receptors in cell membranes. The receptor recognizes structural features of the hormone to generate a high degree of specificity and affinity. The affinity constants for these interactions are in the range of 10^9–10^{11} M^{-1}, representing tight binding. This interaction usually activates or complexes with a transducing protein in the membrane, such as a **G protein** (GTP binding protein), or other transducer and causes an activation of some enzymatic function

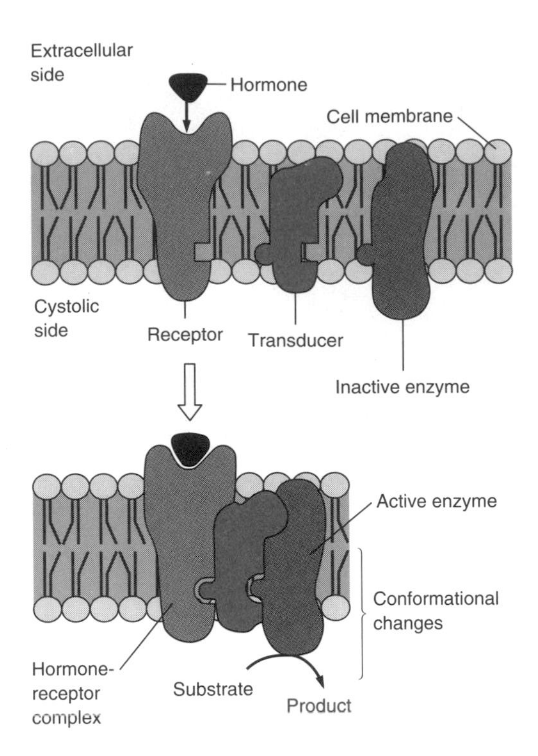

FIGURE 20.18
Hypothetical events following from the interaction of a polypeptide hormone with its cognate membrane receptor.

on the cytoplasmic side of the membrane. In some cases receptors undergo **internalization** to the cell interior; these receptors may or may not (e.g., the insulin receptor) be coupled to transducing proteins in the cell membrane. A discussion on internalization of receptors is presented in Section 20.11. The ''activated'' receptor complex could, physically, open a membrane ion channel or have other profound impacts on membrane structure and function. This signal transfer is schematized in Figure 20.18. Binding of the hormone to the receptor causes conformational changes in the receptor molecule enabling the association with transducer in which further conformational changes may occur to permit interaction with an enzyme on the cytoplasmic side of the cell membrane. This interaction causes conformational changes in an enzyme so that its catalytic site becomes active.

G Proteins Serve as Cellular Transducers of the Hormone Signal

Most transducers of receptors in the cell membrane are GTP binding proteins and are referred to as G proteins. G Proteins consist of three types of subunits, α, β, and γ (Figure 20.19). The **α subunit** is the guanine nucleotide binding component and is thought to interact with the receptor indirectly through the **β,γ subunits** and then directly with an enzyme, such as adenylate cyclase, resulting in enzyme activation. Actually there are two forms of the α subunit, designated α_s for α stimulatory subunit and α_i for the α inhibitory subunit. Two types of receptors, and thus hormones, control the adenylate cyclase reaction; hormone–receptors that lead to a stimulation of the adenylate cyclase and those that lead to an inhibition of the cyclase. This is depicted in Figure 20.20 with an indication of the role of α_s and α_i and some of the hormones that interact with the stimulatory and inhibitory receptor.

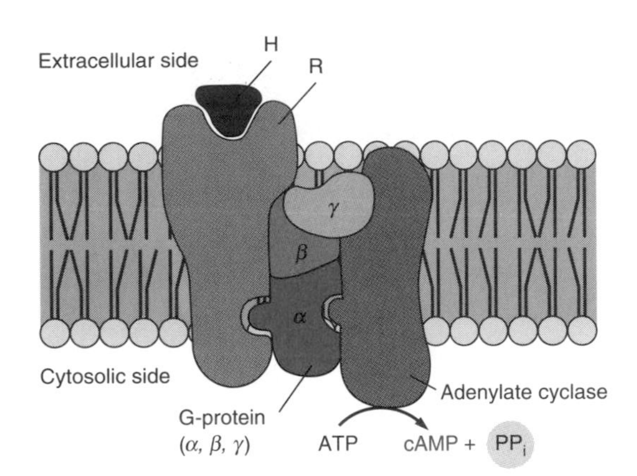

FIGURE 20.19
Artist's conception of how the hormone receptor–G protein–adenylate cyclase system might appear in the cell membrane.
H = hormone; R = hormone receptor.

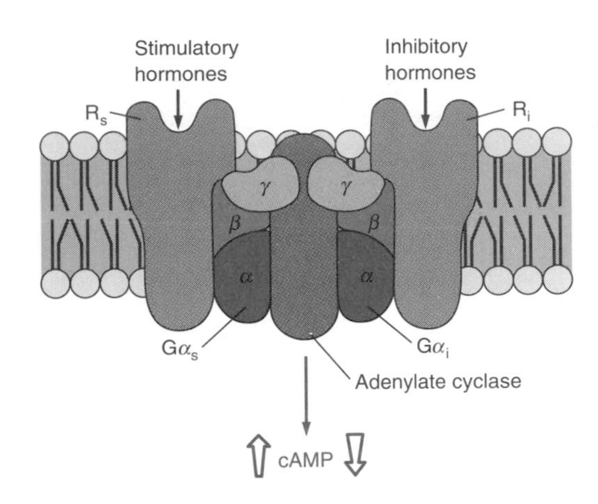

FIGURE 20.20
Components that constitute a hormone-sensitive adenylate cyclase system and the subunit composition.
Adenylate cyclase is responsible for conversion of ATP to cAMP. The occupancy of R_s by stimulatory hormones stimulates adenylate cyclase via formation of an active dissociated $G_{\alpha s}$ subunit. The occupancy of R_i by inhibitory hormones results in the formation of an "active" $G_{\alpha i}$ complex and concomitant reduction in cyclizing activity. The fate of β and γ subunits in these dissociation reactions is not yet known. R_s = stimulatory hormone receptor; R_i = inhibitory hormone receptor.

The sequence of events when hormone and receptor interact is presented in Figure 20.21 and is as follows: receptor binds hormone in the membrane (Step 1) which produces a conformational change in receptor to expose a site for G protein (β,γ subunit) attachment (Step 2); G protein can be either stimulatory, G_s, or inhibitory, G_i, referring to the ultimate effects on the activity of adenylate cyclase; the receptor interacts with β,γ subunit of G protein enabling α subunit to exchange bound GDP to GTP (Step 3); dissociation of GDP by GTP causes separation of G protein α subunit from β,γ subunit and a binding site for interaction with adenylate cyclase appears on the surface of the G protein α subunit (Step 4); α subunit binds to adenylate cyclase and activates the catalytic center, so that ATP is converted to cAMP; GTP is hydrolyzed by the GTPase activity of the α subunit returning it to its original conformation and allowing its interaction with β,γ subunit once again (Step 5); GDP associates with the α subunit and the system is returned to the unstimulated state awaiting another cycle of activity.

In the case where an inhibitory G protein is coupled to the receptor, the events are similar but inhibition of adenylate cyclase activity may arise by direct interaction of the inhibitory α subunit with adenylate cyclase or, alternatively, the inhibitory α subunit may interact directly with the stimulatory α subunit on the other side, and prevent the stimulation of adenylate cyclase activity indirectly.

Table 20.6 lists some activities transduced by G proteins.

TABLE 20.6 Some Activities Transduced by G Proteins

		Transducer	
Enzyme activity	Products	Stimulatory	Inhibitory
Adenylate cyclase	cAMP	$G_s(\alpha_s\beta\gamma)$	$G_i(\alpha_i\beta\gamma)$
Phospholipase	IP_3, DAG	$G_p(\alpha_p\beta\gamma)$	$G_i(\alpha_i\gamma)$
cGMP Phosphodiesterase	5′-GMP	Transducin (heterotrimeric G protein)	

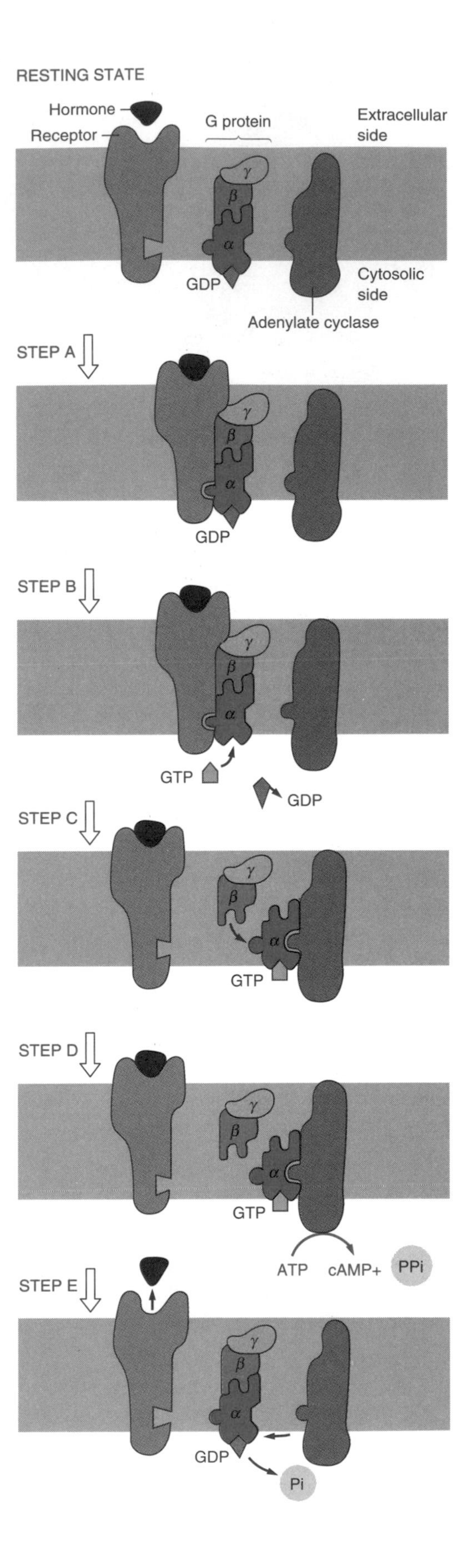

FIGURE 20.21
The activation of adenylate cyclase by binding of a hormone to its receptor.
The cell membrane is depicted, which contains on its outer surface a receptor protein for a hormone. On the inside surface of the membrane is adenylate cyclase protein and the transducer protein G. In the resting state GDP is bound to the α subunit of the G protein. When a hormone binds to the receptor, a conformational change occurs (Step A). The activated receptor binds to the G protein (Step B), which activates the latter so that it releases GDP and binds GTP, causing the α and the complex of β and γ subunits to dissociate (Step C). Free G_α subunit binds to the adenylate cyclase and activates it so that it catalyzes the synthesis of cAMP from ATP (Step D); this step may involve a conformational change in G_α. When GTP is hydrolyzed to GDP, a reaction most likely catalyzed by G_α itself, G_α is no longer able to activate C (Step E), and G_α and $G_{\beta,\gamma}$ reassociate. The hormone dissociates from the receptor and the system returns to its resting state.
Redrawn from J. Darnell, H. Lodish, and D. Baltimore, *Molecular Cell Biology,* New York: Scientific American Books, Inc., 1986, p. 682.

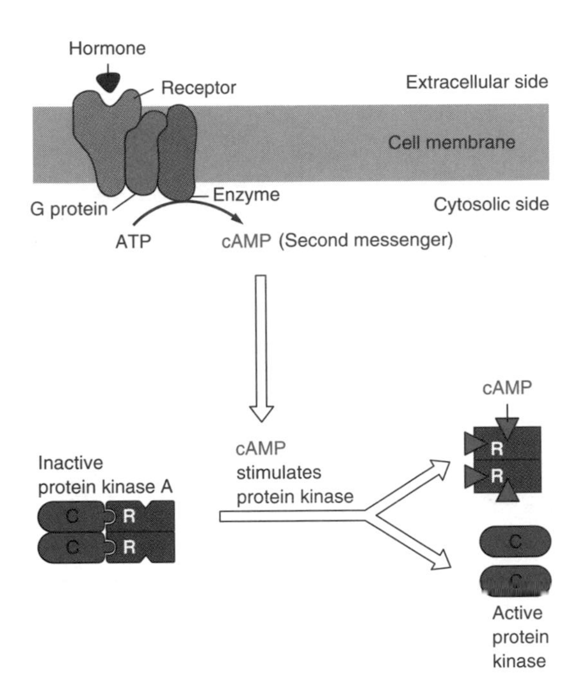

FIGURE 20.22
Activation of protein kinase A.
Hormone–receptor modulated stimulation of adenylate cyclase and subsequent activation of protein kinase A.

Cyclic AMP Activates a Protein Kinase A Pathway

The generation of cAMP in the cell usually activates protein kinase A, referred to as the **protein kinase A pathway.** The overall pathway is presented in Figure 20.22.

Four cAMP molecules are used in the reaction to complex two regulatory subunits (R) liberating two protein kinase catalytic subunits (C). The liberated catalytic subunits are able to phosphorylate proteins to produce a cellular effect.

In many cases the cellular effect leads to the release of preformed hormones. For example, ACTH releases cortisol from the *zona fasciculata* cell of the adrenal gland by this general mechanism. Part of the mechanism of release of thyroid hormones from the thyroid gland involves the cAMP pathway as outlined in Figure 20.23, as is the release of testosterone, presented in Figure 20.24. There are many other examples of hormonal actions mediated by cAMP and the protein kinase A pathway.

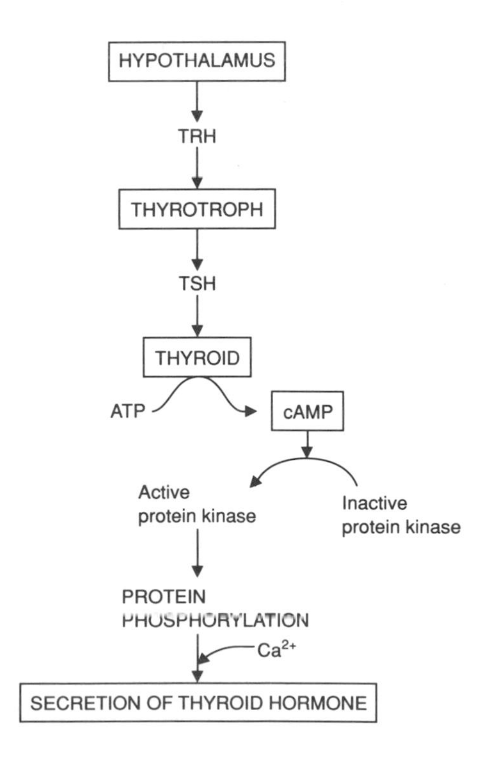

FIGURE 20.23
Overview of secretion controls of thyroid hormone.

Inositol Trisphosphate Formation Leads to Release of Calcium From Intracellular Stores

Uptake of calcium from the cell exterior through calcium channels may be affected directly by hormone–receptor interaction at the cell membrane. In some cases, ligand–receptor interaction is thought to open calcium channels directly in the cell membrane (Chapter 5, Section 5). Another system to increase intracellular Ca^{2+} concentration derives from hormone–receptor activation of phospholipase C activity transduced by a G protein (Figure 20.25).

A hormone, operating through this system binds to a specific cell membrane receptor, which interacts with a G protein in a mechanism similar to that of the protein kinase A pathway and transduces the signal resulting in

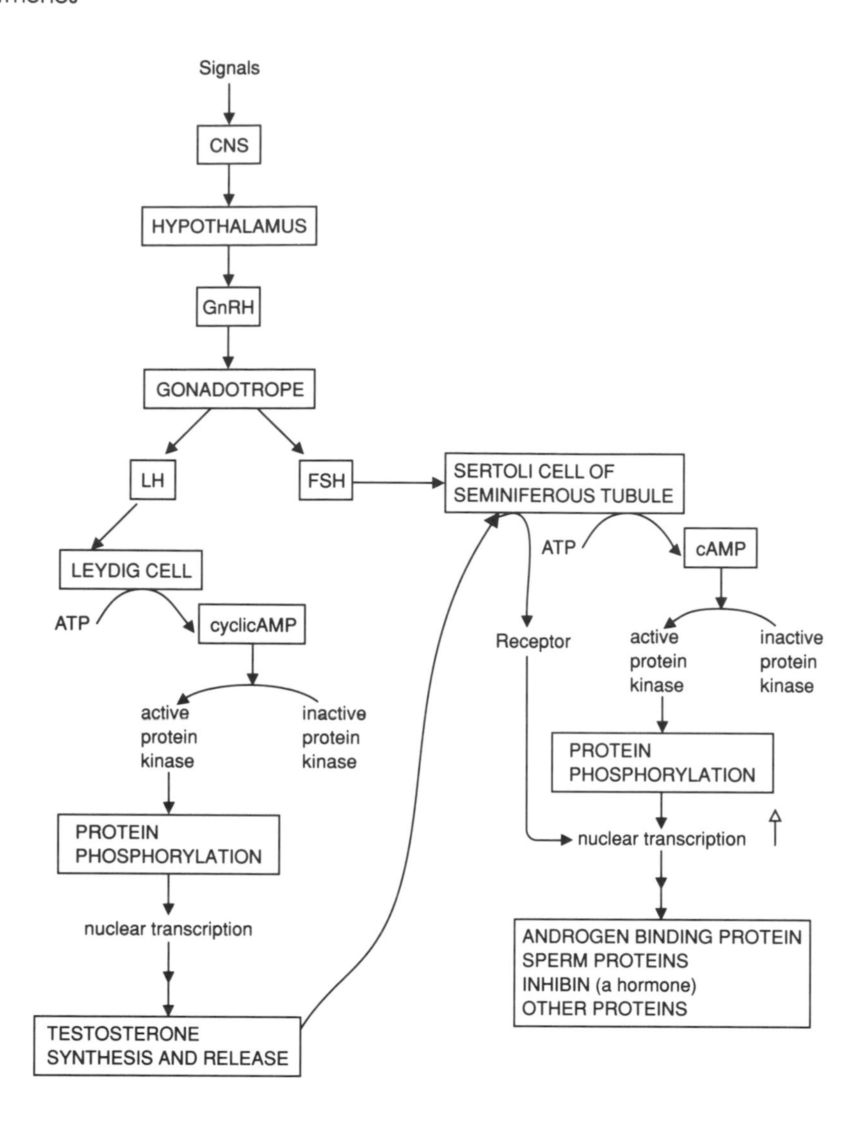

FIGURE 20.24
Overview of the secretion controls and some general actions of the gonadotropes and testosterone release in the male.

stimulation of phospholipase C. This enzyme catalyzes the hydrolysis of phosphatidylinositol-4,5-bisphosphate (PIP_2) to form two **second messengers, diacylglycerol (DAG),** and **inositol-1,4,5-trisphosphate (IP_3).**

Inositol-1,4,5-triphosphate diffuses to the cytoplasm and binds to an IP_3 receptor on the membrane of a particulate **calcium store,** either separate from or part of the endoplasmic reticulum. Its binding results in the release of calcium ions contributing to the large increase in cytoplasmic Ca^{2+}. Calcium ions may be important to the process of exocytosis by taking part in the fusion of secretory granules to the internal cell membrane, in microtubular aggregation or in the function of contractile proteins, which may be part of the structure of the exocytotic mechanism, or all of these.

The IP_3 is metabolized by stepwise removal of phosphate groups (Figure 20.25) to form inositol. This combines with phosphatidic acid (PA) to form phosphatidylinositol (PI) in the cell membrane. It is phosphorylated twice by a kinase to form PIP_2, which is ready to undergo another round of hydrolysis and formation of second messengers (DAG and IP_3) upon hormonal stimulation, unless the receptor is still occupied by hormone, in which case several rounds of the cycle could occur until the hormone–receptor complex dissociated or some other feature of the cycle becomes limiting. It is interesting that the conversion of inositol phosphate to inositol is inhibited by **lithium** ion (Figure 20.25). This could be the metabolic

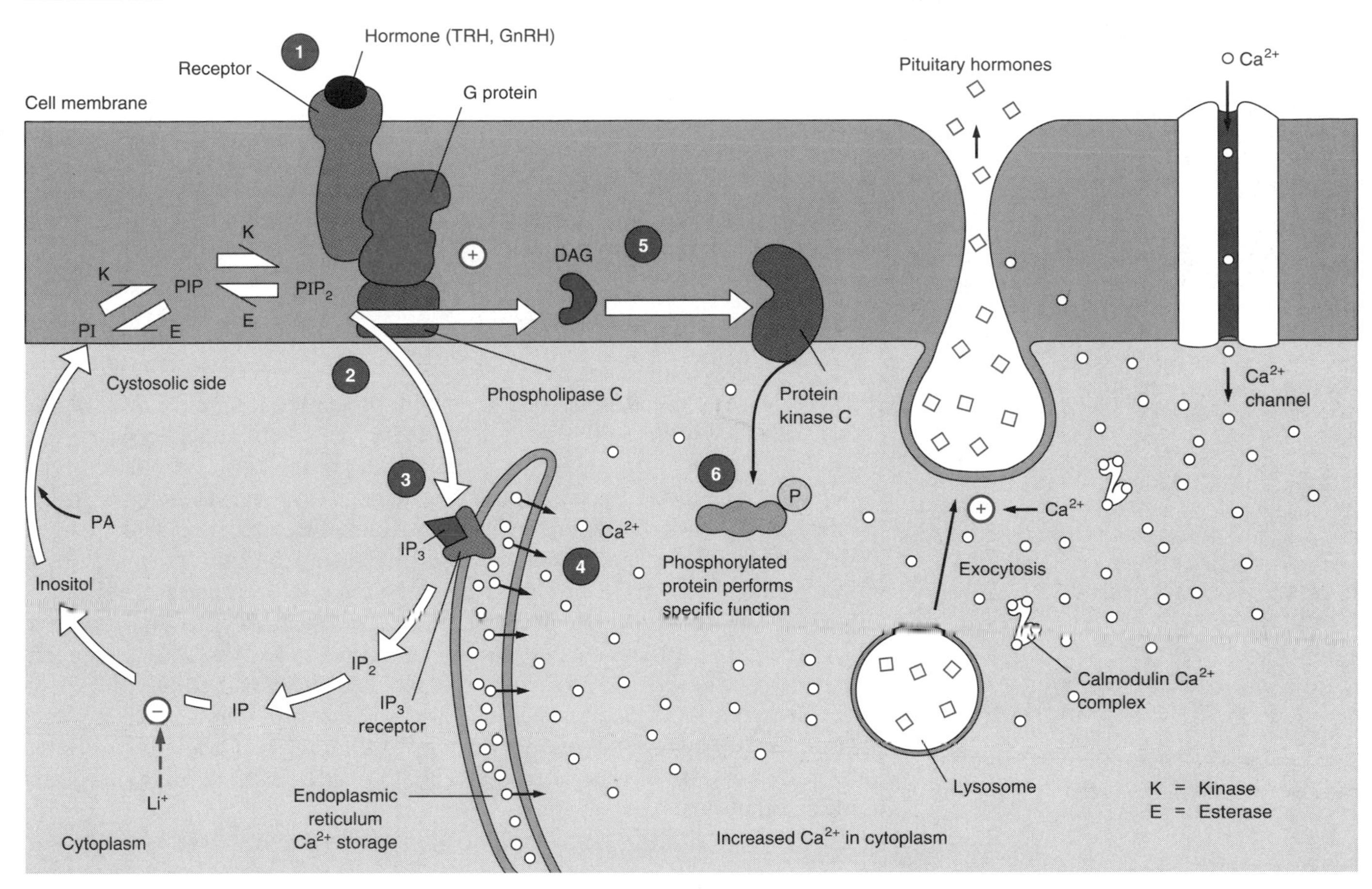

FIGURE 20.25
Overview of hormonal signaling through the phosphatidylinositol system generating the second messengers, inositol-1,4,5-trisphosphate (IP_3) and diacylglycerol (DAG).
The action of IP_3 is to increase cytoplasmic Ca^{2+} levels by a receptor-mediated event in the cellular calcium store. DAG, diacylglycerol; PA, phosphatidic acid; IP, inositol phosphate; IP_2, inositol bisphosphate; IP_3, inositol 1,4,5-trisphosphate; IP_4, inositol 1,3,4,5-tetraphosphate; PIP, phosphatidyl inositol phosphate; PIP_2, phosphatidyl inositol 4,5-bisphosphate; K, Kinase; E, Esterase.

basis for the beneficial effects of Li^+ in manic depression (Clin. Corr. 20.3).

Diacylglycerol Activates a Protein Kinase C Pathway

Diacylglycerol binds to a regulator site on protein kinase C in the cell membrane; some of the enzyme is in the cytoplasm either with its regulatory site intact or with it cleaved off; in the latter form, it is called *protein kinase M*. (Details of protein kinase C are discussed on p. 892.) Binding of DAG stimulates the activity of protein kinase C in the presence of phosphatidylserine and Ca^{2+}, which then phosphorylates specific proteins in the cytoplasm, or in some cases, in the membrane. These phosphorylated proteins perform specific functions that they could not do in the nonphosphorylated state. For example, a phosphorylated protein could migrate to the nucleus and increase mitosis and growth. It is also possible that a phosphorylated protein could play a role in the secretion of cellular substances, such as preformed hormones.

20.8 CYCLIC HORMONAL CASCADE SYSTEMS

Hormonal cascade systems can be generated by external signals as well as by internal signals. Examples of this are the **diurnal variations** in levels of cortisol secreted from the adrenal gland probably initiated by serotonin and vasopressin, the day and night variations in secretion of melatonin from the pineal gland and the internal regulation of the ovarian cycle. Some of these systems operate on a cyclic basis, often dictated by daylight and darkness, and consequently become a biorhythm.

Melatonin and Serotonin Synthesis Are Controlled by Light and Dark Cycles

The release of melatonin from the pineal presented in overview in Figure 20.26, is an example of a biorhythm. Here, as in other such systems, the internal signal is provided by a neurotransmitter, in this case, norepinephrine produced by an adrenergic neuron. In this system, control is exerted by light entering the eyes and is transmitted to the pineal gland by way of the CNS. The adrenergic neuron, innervating the pinealocyte, is inhibited by light transmitted through the eyes. Norepinephrine released from the neurotransmitter in the dark stimulates cAMP formation through a β receptor in the pinealocyte cell membrane, which leads to the enhanced synthesis of *N*-acetyltransferase activity causing the conversion of serotonin to *N*-acetylserotonin and then to melatonin, which is secreted in the dark hours but not during light hours. Melatonin is circulated to cells containing receptors that generate effects on reproductive and other functions.

The Ovarian Cycle Is Controlled by Gonadotropic Releasing Hormone

An example of a pulsatile release mechanism is the regulation of the periodic release of the GnRH. A periodic control regulates the release of this substance at definitive periods (of about 1 h in higher animals) and is

CLIN. CORR. 20.3
LITHIUM TREATMENT OF MANIC DEPRESSIVE ILLNESS: THE PHOSPHATIDYLINOSITOL CYCLE

Lithium has been used for years in the treatment of manic depression. Our newer knowledge suggests that lithium therapy involves the phosphatidylinositol (PI) pathway. This pathway generates the second messengers inositol-1,4,5-trisphosphate (IP_3) and diacylglycerol following the hormone/neurotransmitter membrane receptor interaction and involves the G protein complex and activation of phospholipase C. IP_3 and its many phosphorylated derivatives are ultimately dephosphorylated in a stepwise fashion to free inositol. Inositol is then used for the synthesis of phosphatidylinositol monophosphate. The phosphatases that dephosphorylate IP_3 to inositol are inhibited by Li^+. In addition Li^+ may also interfere directly with G-protein function. The result of Li^+ inhibition is that the PI cycle is greatly slowed even in the face of continued hormonal/neurotransmitter stimulation and the cell becomes less sensitive to these stimuli. Manic depressive illness may occur through the overactivity of certain CNS cells, perhaps as a result of abnormally high levels of hormones or neurotransmitters whose actions are to stimulate the PI cycle. The chemotherapeutic effect of the Li^+ could be to decrease the cellular responsiveness to elevated levels of agents which would promote high levels of PI cycle and precipitate manic depressive illness.

Avissar, S. and Schreiber, G. Muscarinic receptor subclassification and G-proteins: Significance for lithium action in affective disorders and for the treatment of the extrapyramidal side effects of neuroleptics. Biol. Psychiatry 26:113, 1989; and Hallcher, L. M. and Sherman, W. R. The effects of lithium ion and other agents on the activity of myoinositol 1-phosphatase from bovine brain. J. Biol. Chem. 255:896, 1980.

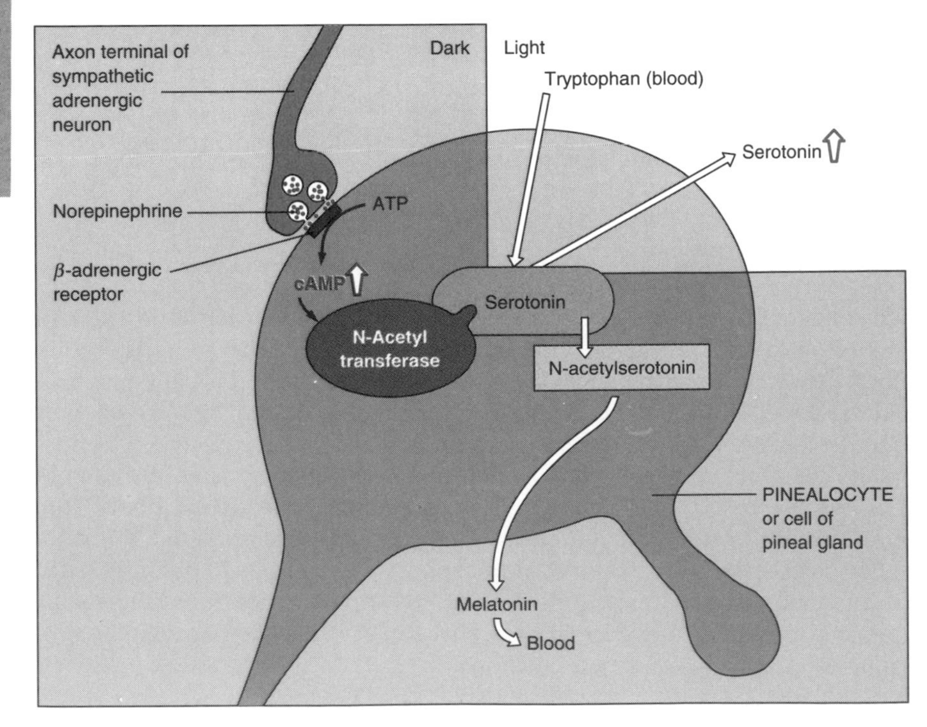

FIGURE 20.26
Biosynthesis of melatonin in the pinealocyte.
Numbers refer to the sequence of events in the light. Numbers with (A) refer to dark-induced sequences of events. HIOMT = hydroxyindole *O*-methyl transferase.
Redrawn from A. W. Norman and G. Litwack, *Hormones*, New York: Academic, 1987, p. 710.

controlled by aminergic neurons, which may be adrenergic (norepinephrine secreting) in nature. The initiation of this function occurs at puberty and is important in both the male and female. While the male system functions continually, the female system is periodic, known as the **ovarian cycle.** This system is presented in Figure 20.27. In the male, the cycling center in the CNS does not develop because its development is blocked by androgens before birth.

In the female, a complicated set of signals needs to be organized in the CNS before the initial secretion of GnRH occurs at puberty. The higher

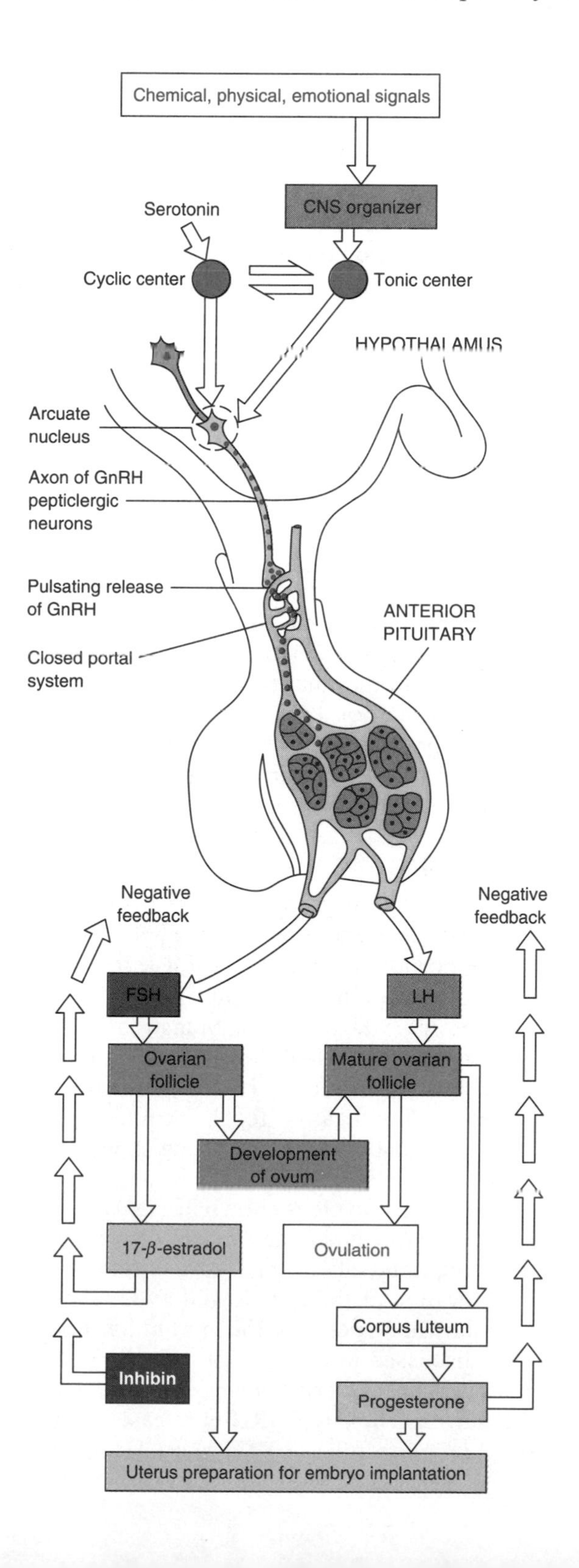

FIGURE 20.27
Overview of the ovarian cycle in terms of generation of hypothalamic hormone, pituitary gonadotropes, and sex hormones.
In order to begin the cycle at puberty, several centers in the CNS coordinate with the hypothalamus so that hypothalamic GnRH can be released in a pulsatile fashion. This causes the release of the gonadotropes, LH and FSH, which in turn affects the ovarian follicle, ovulation, and the *corpus luteum.* Products of the follicle and *corpus luteum,* respectively, are β-estradiol and progesterone. GnRH = gonadotropin releasing hormone; FSH = follicle stimulating hormone; LH = luteinizing hormone.

centers (CNS organizer) must harmonize with the tonic and cycling centers and these interact with each other in order to prime the hypothalamus. The pulsatile system, which innervates the arcuate nucleus of the hypothalamus, must also function for GnRH to be released and this system apparently must be functional throughout life for these cycles to be maintained. Release of GnRH from the axon terminals of the cells that manufacture this hormone is followed by entry of the hormone into the primary plexus of the closed portal system connecting the hypothalamus and the anterior pituitary (Figure 20.27). The blood–brain barrier preventing peptide transport is overcome in this process by allowing GnRH to enter the vascular system through fenestrations, or openings in the blood vessels, that permit such transport. The GnRH is then carried down the portal system and leaves the secondary plexus through fenestrations, again, in the region of the target cells (gonadotropes) of the anterior pituitary. The hormone binds to its cognate membrane receptor and the signal, mediated by the phosphatidylinositol metabolic system, causes the release of both FSH and LH from the same cell. The **FSH** binds to its cognate membrane receptor on the ovarian follicle and, operating through the protein kinase A pathway via cAMP elevation, stimulates synthesis and secretion of 17β-estradiol, the female sex hormone, together with a maturing effect on the follicle and ovum. Other proteins, such as inhibin are also synthesized. Inhibin is a negative feedback regulator of FSH production in the gonadotrope. When the follicle reaches full maturation and the ovum also is matured, **LH** binds to its cognate receptor and plays a role in ovulation together with other factors, such as prostaglandin $F_{2\alpha}$. The residual follicle, after ovulation, becomes the functional corpus luteum under primary control of LH (Figure 20.27). The LH binds to its cognate receptor in the corpus luteum cell membrane and, through stimulation of the protein kinase A pathway, stimulates the synthesis of progesterone, the progestational hormone. **Estradiol** and **progesterone,** bind to intracellular receptors (Chapter 21) in the uterine endometrium and cause major changes resulting in the thickening of the wall and vascularization in preparation for implantation of the fertilized egg. Estradiol, which is made in large amount prior to the production of progesterone, induces the progesterone receptor as one of its inducible phenotypes at a time before the large rise in progesterone generated from the corpus luteum.

Absence of Fertilization

If fertilization of the ovum does not occur, the corpus luteum dies from diminished LH supply. Progesterone levels fall sharply in the blood with the collapse of the corpus luteum. Estradiol levels also fall due to its production by the corpus luteum; the stimuli for a thickened and vascularized uterine endometrial wall are lost. Menstruation occurs through a process of programmed cell death of the uterine endometrial cells until the endometrium reaches its unstimulated state. Ultimately the fall in blood steroid levels releases the negative feedback inhibition on the gonadotrope and hypothalamus and the cycle starts again with release of gonadotropes in response to GnRH.

The course of the ovarian cycle is shown in Figure 20.28 with respect to the relative blood levels of hormones released from the hypothalamus, anterior pituitary, ovarian follicle, and corpus luteum. In addition, the changes in the maturation of the follicle and ovum as well as the uterine endometrium are shown. Aspects of the steroid hormones, estradiol and progesterone, are discussed in Chapter 21.

The cycle first begins at puberty when GnRH is released, corresponding to day 1 in Figure 20.28. The GnRH is released in a pulsatile fashion causing the gonadotrope to release FSH and LH; there is a rise in their

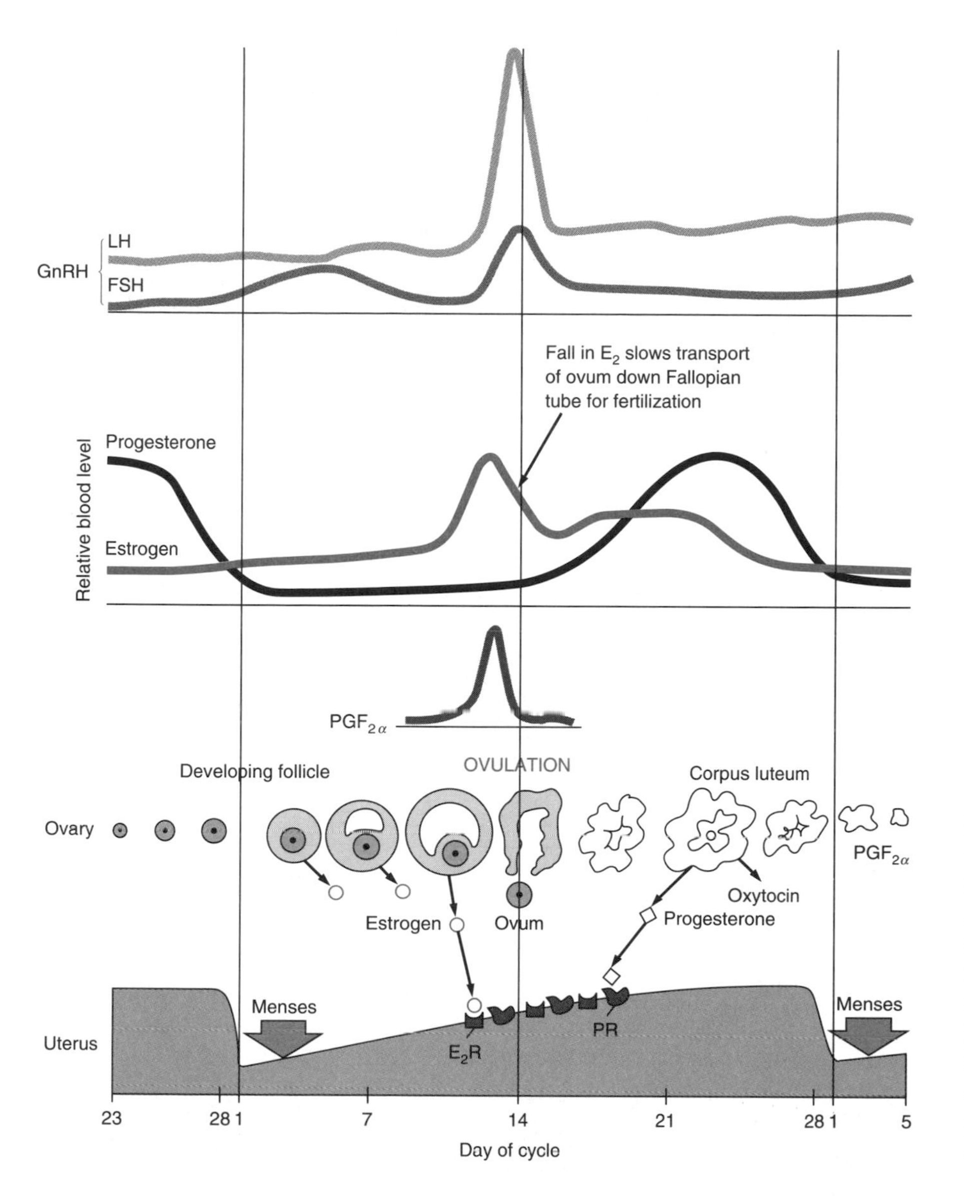

FIGURE 20.28
The ovarian cycle.
In the upper diagram, relative blood levels of GnRH, LH, FSH, progesterone, estrogen, and $PGF_{2\alpha}$ are shown. In the lower diagram, events in the ovarian follicle, *corpus luteum,* and uterine endometrium are diagrammed. GnRH = gonadotropic releasing hormone; LH = luteinizing hormone; FSH = follicle stimulating hormone; $PGF_{2\alpha}$ = prostaglandin $F_{2\alpha}$; E_2 = estradiol; E_2R = estrogen receptor; and PR = progesterone receptor.

blood levels in subsequent days. Under the stimulus of FSH the follicle begins to mature (lower section of Figure 20.28) and estradiol (E_2) is produced. In response to estradiol the uterine endometrium begins to thicken (there would have been no prior menstruation in the very first cycle). Eventually, under the continued action of FSH, the follicle matures with the maturing ovum, and extraordinarily high levels of estradiol are produced (around day 13 of the cycle). These levels of estradiol, instead of causing feedback inhibition, now generate, through feedback *stimulation,* a huge release of LH and to a lesser extent FSH from the gonadotrope as if there were two types of estradiol receptor in the gonadotrope; a high-affinity receptor operating at low levels of E_2 causing a feedback inhibition and a low-affinity receptor operating only at high levels of estradiol causing gonadotropic hormone synthesis (probably mainly of the common α subunit of FSH and LH) and release. The FSH responds to a smaller extent due to the ovarian production of inhibin under the influence of FSH. **Inhibin** is a specific negative feedback inhibitor of FSH but not of LH probably by suppressing the synthesis of the β subunit of FSH. The high mid-cycle peak of LH is referred to as the "LH spike." Ovulation then occurs at about day 14 (mid-cycle) through the effects of high LH concentration together with other factors, such as $PGF_{2\alpha}$. Both LH and $PGF_{2\alpha}$ act on cell membrane receptors. After ovulation, the function of the follicle declines as reflected by the fall in blood estrogen levels. The spent follicle now differentiates into the functional corpus luteum driven by the still high levels of blood LH (Figure 20.27, top). Under the influence of prior high levels of estradiol (estrogen) and the high levels of progesterone produced by the now functional corpus luteum, the uterine endometrial wall reaches its greatest development in preparation for implantation of the fertilized egg should fertilization occur. Note that the previous availability of estradiol in combination with the estrogen receptor (E_2R) produce elevated levels of progesterone receptor (PR) in the uterine wall. The blood levels of estrogen fall with the loss of function of the follicle but some estrogen is produced by the corpus luteum in addition to the much greater levels of progesterone. In the absence of fertilization the corpus luteum continues to function for about 2 weeks then dies because of the loss of high levels of LH. The production of oxytocin by the corpus luteum itself and the production or availability of $PGF_{2\alpha}$ cause inhibition of progesterone synthesis and luteolysis by a process of programmed cell death (Chapter 21). With the death of the corpus luteum, there is a profound decline in blood levels of estradiol and progesterone so that the thickened endometrial wall can no longer be maintained and menstruation occurs, followed by the start of another cycle with a new developing follicle.

Fertilization

The situation changes if fertilization occurs as shown in Figure 20.29. The corpus luteum, which would have ceased function by 28 days (dashed line in progesterone level) remains viable for about 2 weeks longer, driven by the production of **chorionic gonadotropin,** which resembles and acts like LH, from the trophoblast. Eventually, the production of **human chorionic**

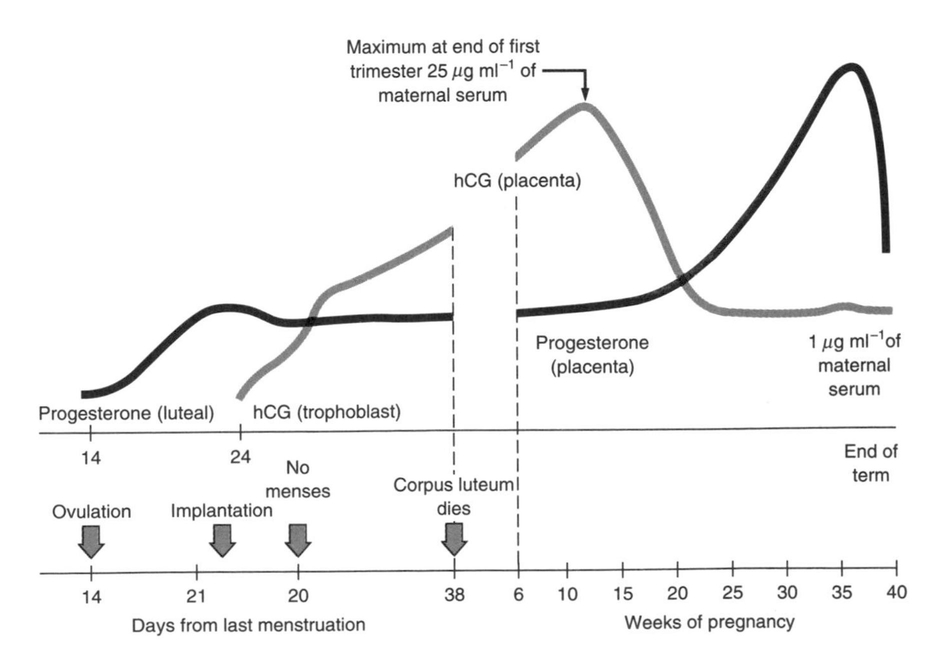

FIGURE 20.29
Effect of fertilization on the ovarian cycle in terms of progesterone and secretion of human chorionic gonadotropin (HCG).
CL = *corpus luteum;* PROG = progesterone.

gonadotropin (hCG) is taken over by the placenta, which continues to produce the hormone at very high levels throughout most of the gestational period (hatched line). Nevertheless, the corpus luteum, referred to as the "corpus luteum of pregnancy," eventually dies and by about 6 weeks of pregnancy, the placenta has taken over the production of progesterone, which is secreted at high levels throughout pregnancy under the influence of high levels of hCG. Near the termination of pregnancy, hCG production falls dramatically and progesterone production also falls. This dramatic fall in progesterone production may result from the release of CRH from the fetal hypothalamus, through a timed signal, which causes ACTH and cortisol to be released. Cortisol, thus released, induces surfactant in the fetal lung in preparation for respiratory function and induces estrogen and $PGF_{2\alpha}$ levels in the placenta and inhibits progesterone release. The marked fall in progesterone levels cause oxytocin to be released from fetal and maternal pituitaries and oxytocin, in concert with $PGF_{2\alpha}$ and the action of estradiol on the uterine myometrium, bring about the final contractions of labor.

As mentioned before, the system in the male is similar, but less complex in that cycling is not involved and it progresses much as outlined in Figure 20.29. This is only one example of biorhythmic and pulsatile systems. Biology is full of them.

20.9 HORMONE–RECEPTOR INTERACTIONS

Receptors are proteins and differ by their specificity for ligands and by their location in the cell (Figure 20.1). The interaction of ligand with receptor essentially resembles a semienzymatic reaction:

$$\text{Hormone} + \text{receptor} \rightleftharpoons \text{hormone–receptor complex}$$

The **hormone–receptor complex** usually undergoes conformational changes resulting from interaction with the hormonal ligand. These changes allow for a subsequent interaction with a transducing protein (G protein) in the membrane for activation to a new state in which active domains become available on the surface, and so on. The mathematical treatment of the interaction of hormone and receptor is a function of the concentrations of the reactants, hormone (H) and receptor (R) in the formation of the hormone–receptor complex (RH) and the rates of formation and reversal of the reaction.

$$[\mathrm{H}] + [\mathrm{R}] \underset{k_{-1}}{\overset{k_{+1}}{\rightleftharpoons}} [\mathrm{RH}]$$

The reaction can be studied under conditions, such as low temperature, that will reduce further reactions involving the hormone–receptor complex. The equilibrium can thus be expressed in terms of the association constant, K_a, which is equal to 1/dissociation constant, K_d,

$$K_a = [\mathrm{RH}]/[\mathrm{H}][\mathrm{R}] = k_{+1}/k_{-1} = 1/K_d$$

The concentrations are equilibrium concentrations that can be restated in terms of the forward and reverse velocity constants, k_{+1} being the on-rate and k_{-1} being the off-rate (on refers to hormone association with the receptor and off refers to hormone dissociation). Experimentally, the equilibrium, under given conditions, is determined by a progress curve of

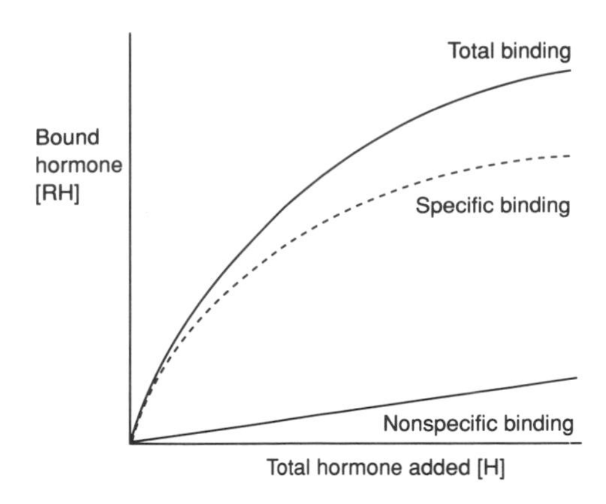

FIGURE 20.30
Typical plot showing specific hormone binding.

binding that reaches saturation. A saturating amount of hormone is determined using variable amounts of free hormone and measuring amount bound with some convenient assay. The half-maximal value of a plot of receptor bound hormone (ordinate) versus total free hormone concentration (abscissa) approximates the dissociation constant, which will have a specific hormone concentration in molarity as its value. Hormone bound to receptor is corrected for nonspecific binding of the hormone to the membrane. This can be measured conveniently if the hormone is radiolabeled, by measuring receptor plus labeled hormone (hot) and receptor plus labeled hormone after the addition of an excess (100–1000 times) of unlabeled hormone (hot + cold). The excess of unlabeled hormone will displace the high-affinity hormone binding sites but not the low-affinity nonspecific binding sites. Thus, when the "hot plus cold" curve is subtracted from the "hot" curve, Figure 20.30, an intermediate curve will represent specific binding to receptor. This is of critical importance when receptor is measured in a system containing other proteins. As an approximation, 20 times the K_d value of hormone is usually enough to saturate the receptor.

Scatchard Analysis Permits Determination of the Number of Receptor Binding Sites and the Association Constant for Ligand

Most measurements of K_d are made using **Scatchard analysis,** which is a manipulation of the equilibrium equation. The equation can be developed by a number of routes but can be envisioned from mass action analysis of the equation presented above. At equilibrium the total possible number of binding sites (B_{max}) = the unbound plus the bound sites, so that B_{max} = R + RH and the unbound sites (R) will be equal to R = B_{max} − RH. To consider the sites left unbound in the reaction the equilibrium equation becomes

$$K_a = \frac{[RH]}{[H](B_{max} - [RH])}$$

Thus,

$$\frac{\text{bound}}{\text{free}} = \frac{[RH]}{[H]} = K_a(B_{max} - [RH]) = \frac{1}{K_d}(B_{max} - [RH])$$

The Scatchard plot of bound/free = [RH]/[H] on the ordinate versus bound [RH] on the abscissa yields a straight line as shown in Figure 20.31. When the line is extrapolated to the abscissa, the intercept gives the value of B_{max} (the total number of specific receptor binding sites). The slope of the negative straight line is $-K_a$ or $-1/K_d$.

These analyses are sufficient for most systems, but become more complex when there are two components in the Scatchard plot. In this case the straight line usually turns as it approaches the abscissa and a second phase is observed somewhat assymptotic to the abscissa while still retaining a negative slope (Figure 20.32*a*). In order to obtain the true value of K_d for the steeper, higher-affinity sites, the low-affinity curve must be subtracted from the first set, which also corrects the extrapolated value of B_{max}. From these analyses information is concluded on the K_d, the number of classes of binding sites (usually one or two), and the maximal number of high-affinity receptor sites (receptor number) in the system. From application to a wide variety of systems it appears the K_d values for many hormone receptors range from 10^{-9} to 10^{-11} M indicating very tight binding. These interactions are generally marked by a high degree of

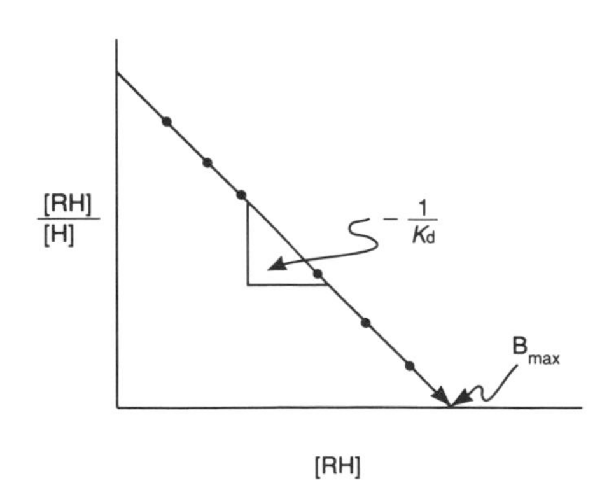

FIGURE 20.31
Typical plot of Scatchard analysis of specific binding of ligand to receptor.

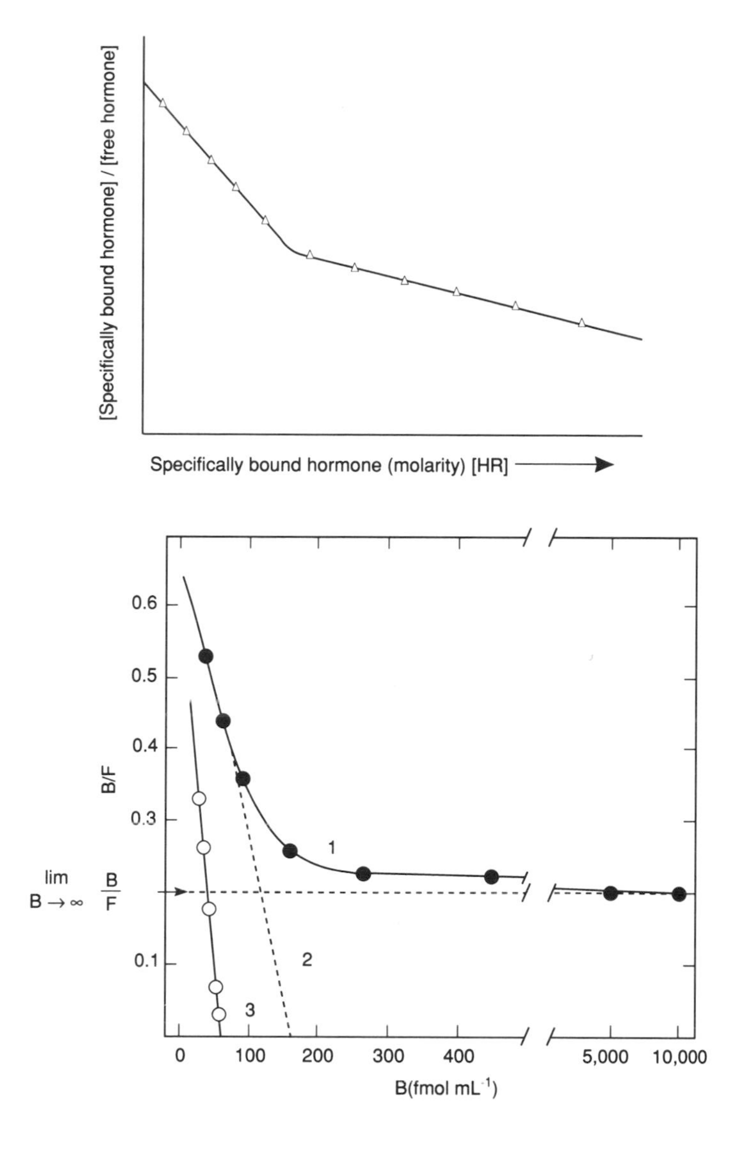

FIGURE 20.32
Scatchard analysis of curves representing two components.
(*a*) Scatchard curve showing two components. (*b*) Scatchard plot with correction of high-affinity component by subtraction of nonspecific binding attributable to the low-affinity component. Curve 1: total binding. Curve 2: Linear extrapolation of high-affinity component that includes contribution from low-affinity component. Curve 3: Specific binding of high-affinity component after removal of nonspecific contribution representing the true high-affinity curve.
Redrawn from G. C. Chamness and W. L. McQuire, *Steroids* 26:538, 1975.

specificity so that both parameters describe interactions of a high order indicating the uniqueness of receptors and the selectivity of signal reception.

Some Hormone–Receptor Interactions Involve Multiple Hormone Subunits

Interaction of hormone and receptor can be exemplified by the anterior pituitary hormones, thyrotropin, luteinizing hormone and follicle9stimulating hormone. These hormones each contain two subunits, an α and a β subunit. The α subunit for all three hormones is nearly identical and the α subunit of any of the three can substitute for the other two. Consequently, the α subunit performs some function in common to all three hormones in their interaction with receptor, but is obviously not responsible for the specificity required for each cognate receptor. The hormones cannot replace each other in binding to their specific receptor, thus, the specificity of receptor recognition is imparted by the β subunit whose structure is unique for the three hormones.

On the basis of topological studies with monoclonal antibodies, a picture of the interaction of LH with its receptor has been suggested as shown in Figure 20.33. In this model, the receptor recognizes both sub-

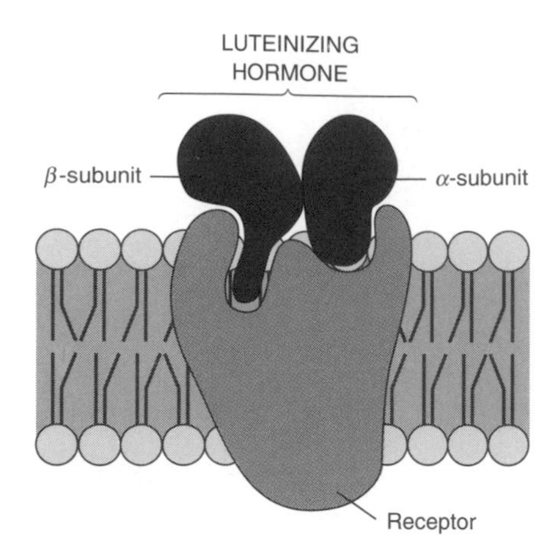

FIGURE 20.33
The interaction of the α and β subunits of LH with the LH receptor of rat Leydig cells.
The interaction was determined by topological analysis with monoclonal antibodies directed against epitopes on the α and β subunits of the hormone.
Diagram adapted from C. Alonoso-Whipple, M. L. Couet, R. Doss, J. Koziarz, E. A. Ogunro, and W. E. Crowley, Jr., *Endocrinology,* 123:1854, 1988.

units of the hormonal ligand, but the β subunit is specifically recognized by the receptor to lead to a response. With the TSH receptor complex there may be more than one second message generated. In addition to the stimulation of adenylate cyclase and the increased intracellular level of cAMP, the phosphatidylinositol pathway (Figure 20.25) is also turned on. The preferred model is one in which there is a single receptor whose interaction with hormone activates both the adenylate cyclase and the phospholipid second messenger systems as shown in Figure 20.34. Thus, a variety of reactions could follow the hormone–receptor interaction through the subsequent stimulation of cAMP levels (protein kinase A pathway) and stimulation of phosphatidylinositol turnover (protein kinase C pathway).

20.10 STRUCTURE OF RECEPTORS: β-ADRENERGIC RECEPTOR

The structures of receptors are conveniently discussed in terms of functional domains. Consequently, for membrane receptors, there will be **transmembrane domains** and the **ligand binding domains,** functional domains which for many membrane receptors involve protein kinase activities. In addition, specific **immunological domains** contain primary epitopes of antigenic regions. Several membrane receptors have been cloned and studied with regard to structure and function, including the **β receptors** (β_1 and β_2), which recognize catecholamines, principally norepi-

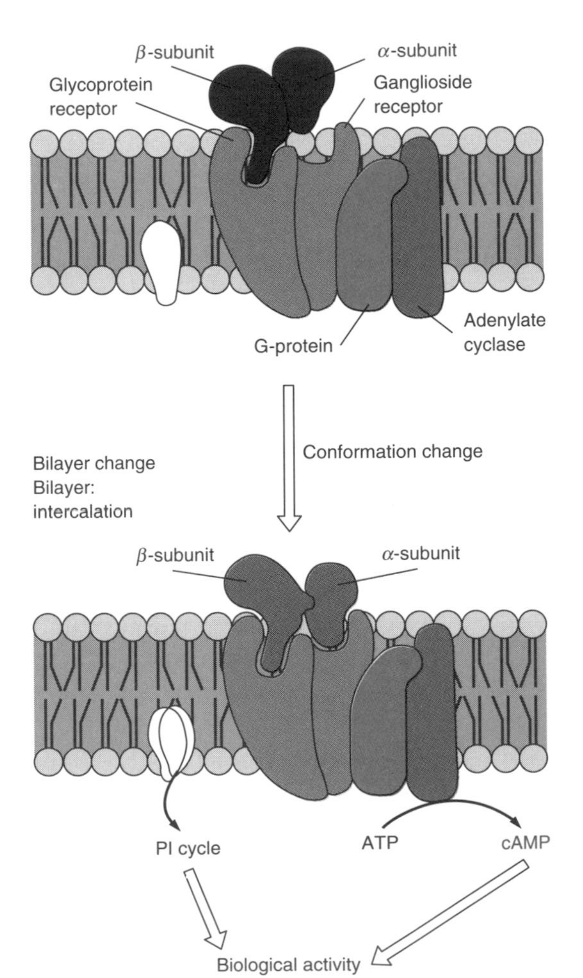

FIGURE 20.34
Model of TSH receptor, the receptor is composed of glycoprotein and ganglioside component.
After the TSH β subunit interacts with the receptor, the hormone changes its conformation and α subunit is brought into the bilayer where it interacts with other membrane components. The β subunit of TSH may carry the primary determinants recognized by the glycoprotein receptor component. It is suggested that the TSH signal to the adenylate cyclase is via the ganglioside; the glycoprotein component appears more directly linked to phospholipid signal system. PI = phosphatidylinositol.
Adapted with modifications from L. D. Kohn, et al. *Biochemical Actions of Hormones,* 12. G. Litwack (Ed.). Academic, 1985, p. 466.

nephrine, and stimulate adenylate cyclase. The β_1- and β_2-receptors are subtypes that differ in affinities for norepinephrine and for synthetic antagonists. Thus, β_1-adrenergic receptor binds norepinephrine with a higher affinity than epinephrine, whereas the order of affinities is reversed for the β_2-adrenergic receptor. The drug isoproterenol has a greater affinity for both receptors than the two hormones. In Figure 20.35 the amino acid sequence is shown (with single letter abbreviations for amino acids; see Table 20.4 for list) for the β_2-adrenergic receptor. A polypeptide stretch extending from α helix I extends to the extracellular space. There

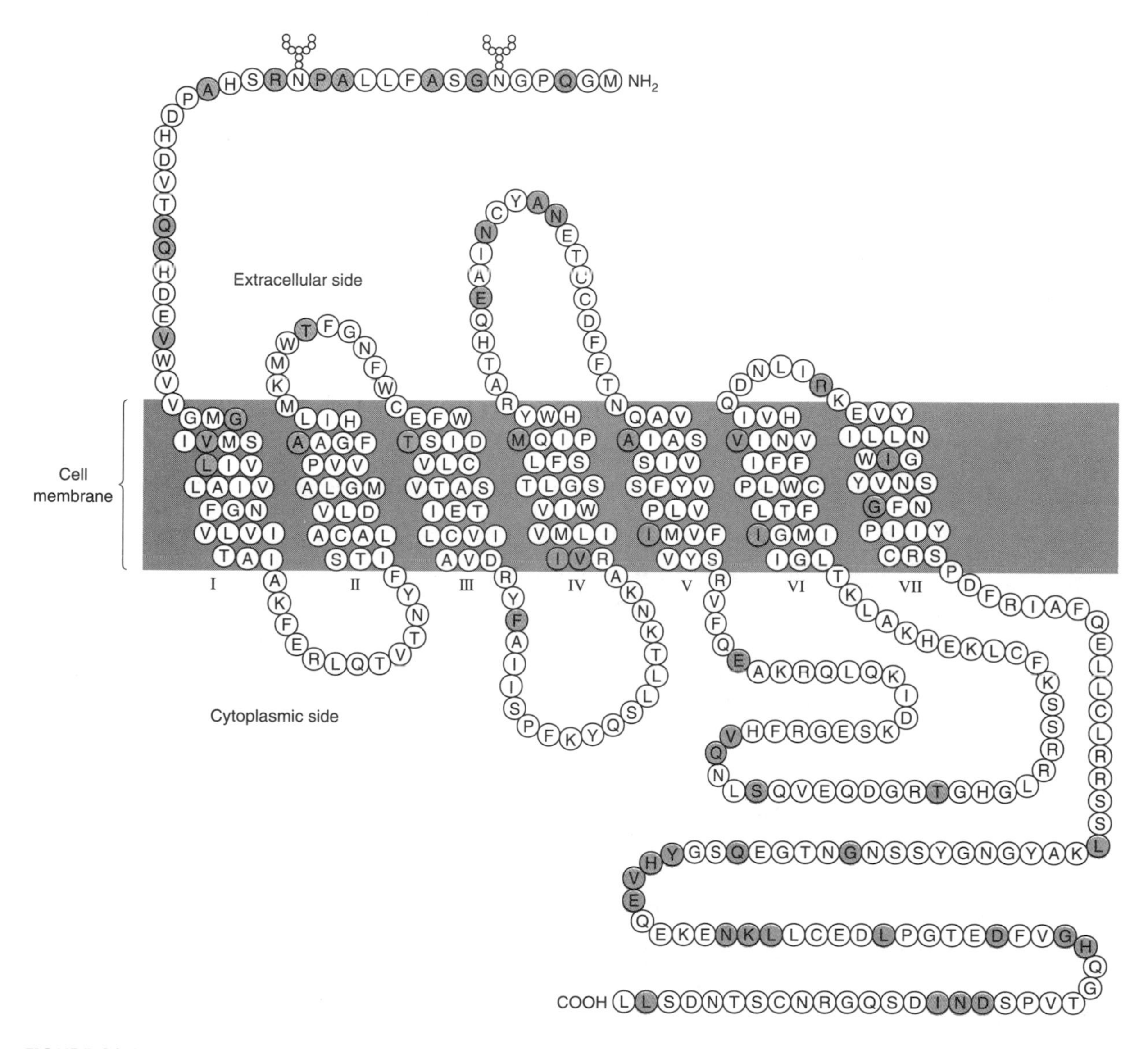

FIGURE 20.35
Proposed model for insertion of the β_2 adrenergic receptor (AR) in the cell membrane.
The model is based on hydropathicity analysis of the human β_2AR. The standard one-letter code for amino acid residues is used. Hydrophobic domains are represented as transmembrane helices. Black circles with white letters indicate residues in the human sequence that differ from those in hamster. Also noted are the potential sites of *N*-linked glycosylation.
Redrawn from B. K. Kobilka, et al. *Proc. Natl. Acad. Sci. U.S.A.* 84:46, 1987.

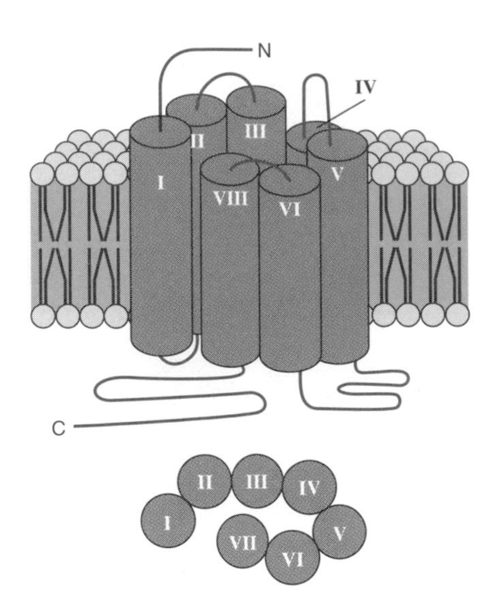

FIGURE 20.36
Proposed arrangement of the β-adrenergic receptor helices in the membrane.
The lower portion of figure is a view from above the plane of the plasma membrane. It is proposed that helices IV, VI, and VII reside in the membrane so as to delineate a ligand binding pocket, with helix VII centrally located.
Adapted from T. Frielle, K. W. Daniel, M. G. Caron, and R. J. Lefkowitz, *Proc. Natl. Acad. Sci. U.S.A.* 95:9494, 1988.

are seven membrane-spanning domains and these appear also in the β_1 receptor where there is extensive homology with the β_2 receptor. Cytoplasmic peptide regions extend to form loops from I to II, III to IV, V to VI and an extended chain from VII. The long extended chain from VII may contain sites of phosphorylation (serine residues) of the receptor, which is part of the receptor regulation process involving receptor desensitization. Cell exterior peptide loops extend from II to III, IV to V, and VI to VII, but mutational analysis suggests that the external loops do not take part in ligand binding. It appears that ligand binding may occur in a pocket arranged by the location of the membrane-spanning cylinders I–VII, which for the β_1 receptor appear to form a ligand pocket as shown from a top view in Figure 20.36. Recently, reported work suggests that the sixth transmembrane domain may play a role in the stimulation of adenylate cyclase activity. By substitution of a specific cysteine residue in the sixth transmembrane domain, a mutant was produced that displays normal ligand binding properties but a decreased ability to stimulate the cyclase.

20.11 INTERNALIZATION OF RECEPTORS

Up to now we have described receptor systems that transduce signals through other membrane proteins, such as G proteins, which move about in the fluid cell membrane, but many types of cell membrane hormone–receptor complexes are internalized, that is, moved from the cell membrane to the cell interior by a process called **endocytosis.** This would represent the opposite of exocytosis in which components within the cell are moved to the cell exterior. The process of endocytosis as presented in Figure 20.37 involves the polypeptide–receptor complex bound in **coated pits,** which are indentations in the plasma membrane that invaginate into the cytoplasm and pinch off the membrane to form **coated vesicles**. The vesicles shed their coats, fuse with each other and form vesicles called **receptosomes.** The receptors and ligands on the inside can have different fates. Receptors can be returned to the cell surface following fusion with the Golgi apparatus. Alternatively, the vesicles can fuse with lysosomes for degradation of both the receptor and hormone. In addition, some hormone–receptor complexes are separated in the lysosome and only the hormone degraded, with the receptor returned intact to the membrane. In some systems, the receptor may also be concentrated in coated pits in the absence of exogenous ligand and cycle in and out of the cell in a constitutive nonligand-dependent manner.

Clathrin Forms a Lattice Structure to Direct Internalization of Hormone–Receptor Complexes From the Cell Membrane

The major protein component of the coated vesicle is **clathrin** a nonglycosylated protein of 180,000 molecular weight whose amino acid sequence is highly conserved. The coated vesicle contains 70% clathrin, 5% polypeptides of about 35 kDa and other polypeptides of 50–100 kDa are also present. Aspects of the structure of a coated vesicle are shown in Figure 20.38. Coated vesicles have a latticelike surface structure comprised of hexagons and pentagons. Three clathrin molecules generate each polyhedral vertex and two clathrin molecules contribute to each edge. The smallest such structure would contain 12 pentagons with 4–8 hexagons and 84 or 108 clathrin molecules. A 200-nm diameter coated vesicle contains about 1000 clathrin molecules. Clathrin can form flexible lattice structures

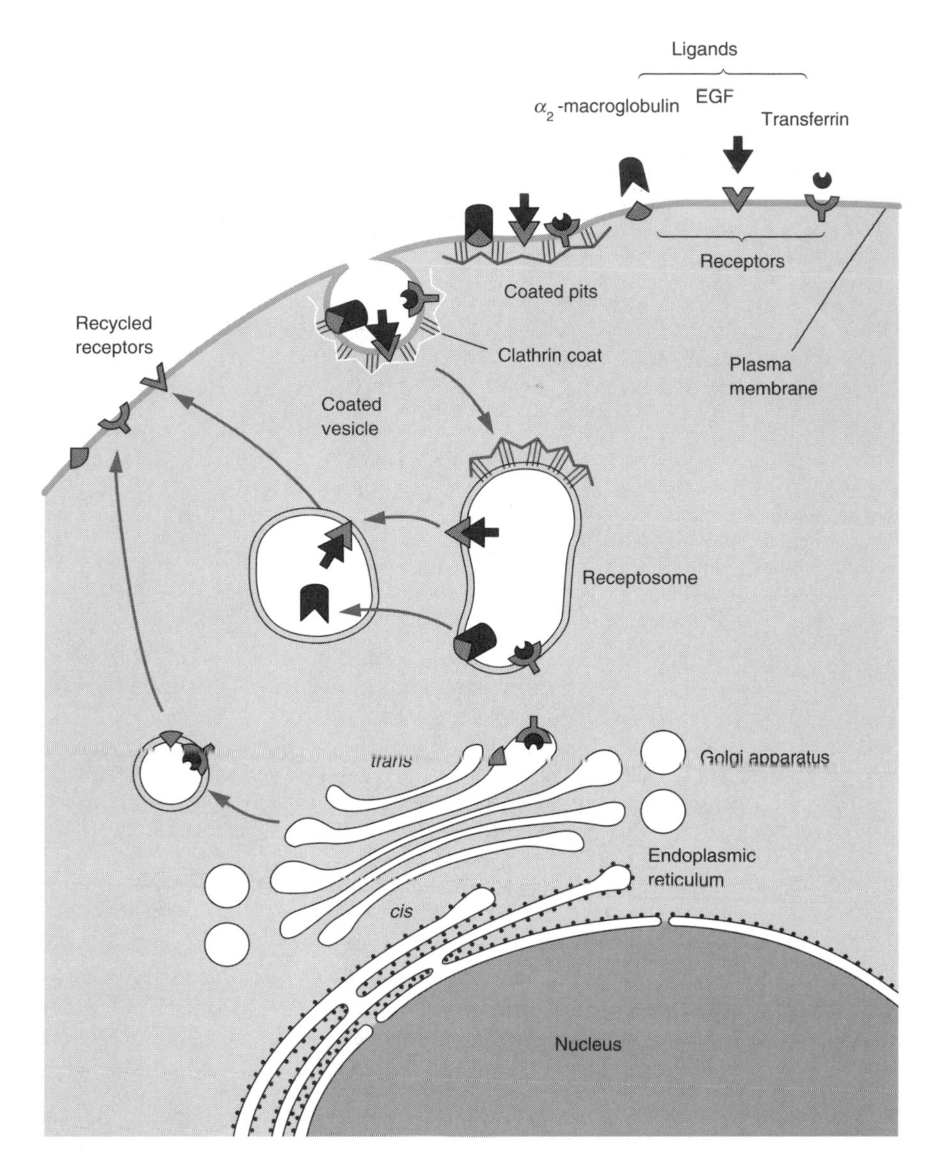

FIGURE 20.37
A diagrammatic summary of the morphological pathway of endocytosis in cells. The morphological elements of the pathway of endocytosis are not drawn to scale. The ligands shown as examples are EGF, transferrin, and α_2-macroglobulin. EGF is shown as an example of a receptor system in which both the ligand and the receptor are delivered to lysosomes; transferrin is shown as an example of a system in which both the ligand and receptor recycle to the surface; α_2-macroglobulin is shown as an example of a system in which the ligand is delivered to lysosomes but the receptor recycles efficiently back to the cell surface.
Adapted from I. Pastan and M. C. Willingham (Eds.). *Endocytosis*, New York: Plenum, 1985, p. 3.

that can act as scaffolds for vesicular budding. Completion of the budding process results in the completed vesicle able to enter the cycle.

The events following endocytosis are not always clear with respect to a specific membrane receptor system. This process can be a means to introduce the intact receptor or ligand to the cell interior where the nucleus, in some cases, is thought to contain a receptor or ligand binding site. Consider, for example, growth factor hormones that are known to bind to a cell membrane receptor but trigger events leading to mitosis. It is possible that signal transmission occurs by the alteration of a specific cytoplasmic protein, perhaps by membrane growth factor receptor associated protein kinase activity, resulting in the nuclear translocation of the covalently modified cytoplasmic protein. In case of internalization, the delivery of an intact ligand (or portion of the ligand) could interact with a nuclear receptor. Such mechanisms are speculative. Nevertheless, these ideas could constitute a rationale for the participation of endocytosis in signal transmission to intracellular components. In Table 20.7 are listed some hormones known to bind to cell membrane receptors and have in common abilities to undergo internalization and stimulate mitosis.

Endocytosis is a means to make a cell less responsive to hormone, by removal of the receptor to the interior, or by cycling membrane components, thereby altering responsiveness or metabolism (e.g., glucose recep-

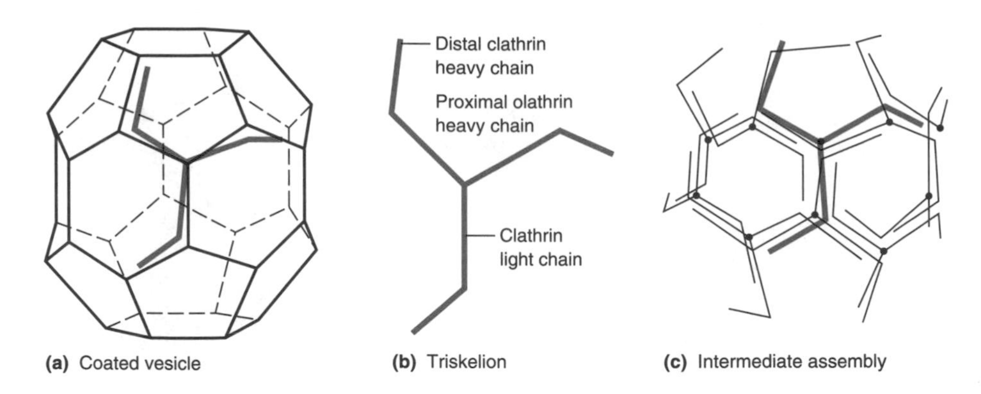

FIGURE 20.38
Structure and assembly of a coated vesicle.
(*a*) A typical coated vesicle contains a membrane vesicle about 40 nm in diameter surrounded by a fibrous network of 12 pentagons and 8 hexagons. The fibrous coat is constructed of 36 clathrin triskelions. One clathrin triskelion is centered on each of the 36 vertices of the coat. Coated vesicles having other sizes and shapes are believed to be constructed similarly: each vesicle contains 12 pentagons but a variable number of hexagons.
See R. A. Crowther and B. M. F. Pearse, *J. Cell. Biol.* 91:790, 1981.
(*b*) Detail of a clathrin triskelion. Each of three clathrin heavy chains is bent into a proximal arm and a distal arm. A clathrin light chain is attached to each heavy chain, most likely near the center. (*c*) An intermediate in the assembly of a coated vesicle, containing 10 of the final 36 triskelions, illustrates the packing of the clathrin triskelions. Each of the 54 edges of a coated vesicle is constructed of two proximal and two distal arms intertwined. The 36 triskelions contain $36 \times 3 =$ 108 proximal and 108 distal arms, and the coated vesicle has precisely 54 edges.
Redrawn from J. Darnell, H. Lodish, and D. Baltimore, *Molecular Cell Biology*. New York: Scientific American Books, 1986, p. 647.

tors can be shuttled between the cell interior and the cell membrane under the control of hormones in certain cells. In another type of **down-regulation,** a message to the nucleus can repress mRNA for receptor by interacting with a specific DNA sequence. More about this form of receptor down regulation is mentioned in Chapter 21.

TABLE 20.7 Characteristics of Some Internalized Polypeptide Hormones

Hormone	Action at cell membrane	Known intracellular receptors[a]
Insulin	Glucose transporter of adipocyte; insulin second messenger in liver and other cells; may enhance activity of IGF receptor; receptor autophosphorylates	Perinuclear membrane
EGF	Phosphorylation of tyrosyl peptides	(Mitogenic effect)
TSH	Increases adenylate cyclase activity and iodide transport in thyroid follicular cell membrane	(Mitogenic effect)
IGF_1	Increases glucose transport in some cells	(Mitogenic effect)
IGF_2		(Mitogenic effect)
NGF		(Mitogenic effect)

SOURCE: A. W. Norman and G. Litwack, *Hormones*, Orlando, FL: Academic, 1987, p. 13.
[a] Although there are known intracellular binding sites resembling receptors, physiological actions cannot yet be ascribed to them.

20.12 INTRACELLULAR ACTION: PROTEIN KINASES

Many amino acid derived hormones or polypeptides bind to cell membrane receptors (except for thyroid hormone) and transmit their signal by: (a) elevation of cAMP and transmission through the **protein kinase A pathway;** (b) triggering of the hydrolysis of phosphatidylinositol 4,5-bisphosphate and stimulation of the **protein kinase C pathway** and IP_3–Ca^{2+} pathways; or (c) stimulation of intracellular levels of cGMP and activation of the **protein kinase G pathway.** There are also other less prevalent systems for signal transfer, which, for example, affect molecules in the membrane like phosphatidylcholine. In the case of TSH-receptor signaling it may be possible that two of these pathways are activated.

The cAMP system operating through protein kinase A activation has been described. Specific proteins are expected to be phosphorylated by this kinase compared to other protein kinases, such as protein kinase C. Both protein kinase A and C phosphorylate proteins on serine or threonine residues. An additional protein kinase system involves phosphorylation of tyrosine, which occurs in cytoplasmic domains of some membrane receptors especially growth factor receptors. This system is important for insulin receptor, IGF receptor, and certain oncogenes discussed below. The cellular location of these protein kinases is presented in Figure 20.39.

The catalytic domain in the protein kinases is similar in amino acid sequence, suggesting that they have all evolved from a common primordial kinase. The three **tyrosine-specific kinases** shown in Figure 20.39 are transmembrane receptor proteins that, when activated by the binding of specific extracellular ligands, phosphorylate proteins (including themselves) on tyrosine residues inside the cell. Both chains of the insulin receptor are encoded by a single gene, which produces a precursor protein that is cleaved into the two disulfide-linked chains. The extracellular domain of the PDGF receptor is thought to be folded into five immunoglobulin (Ig)-like domains, suggesting that this protein belongs to the Ig superfamily.

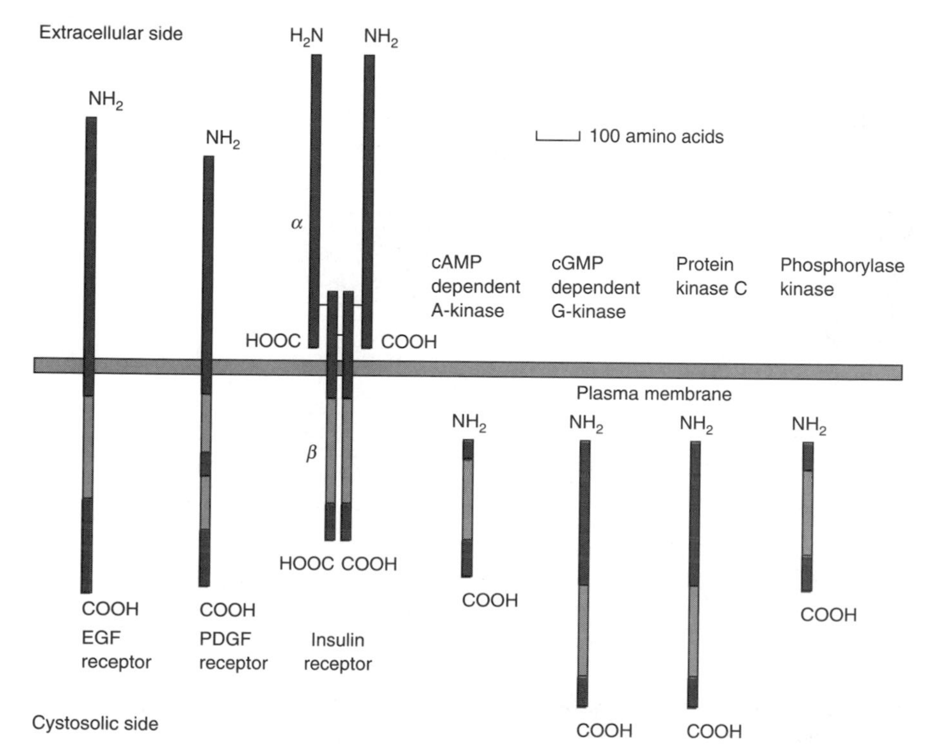

FIGURE 20.39
Protein kinases showing the size and location of their catalytic domain.
In each case the catalytic domain (*filled-in*) is about 250 amino acid residues long. The regulatory subunits normally associated with A-kinase and with phosphorylase kinase are not shown.
Redrawn from B. Alberts, D. Bray, J. Lewis, M. Raff, K. Roberts, and J. D. Watson, *Molecular Biology of the Cell,* 2nd ed. New York: Garland Publishing, p. 707.

Proteins that are regulated by phosphorylation–dephosphorylation can have multiple phosphorylation sites and may be phosphorylated by more than one class of protein kinase.

The Insulin Receptor: Transduction Through Tyrosine Kinase

From Figure 20.39 it is seen that the α subunits of the **insulin receptor** are located outside the cell membrane and apparently serve as the insulin binding site. The insulin–receptor complex undergoes an activation sequence probably involving conformational changes and phosphorylations (autophosphorylations) on the cytoplasmic portion of the receptor (β subunits). This results in activation of the tyrosine kinase activity located in the β subunit, which is now able to phosphorylate cytoplasmic proteins that may carry the insulin signal to the interior of the cell. The substrates of the insulin–receptor tyrosine kinase are an important current research effort since phosphorylated proteins could produce the long-term effects of insulin. On the other hand, there is evidence that an insulin second messenger may be developed at the cell membrane to account for the short-term metabolic effects of insulin. The substance released as a result of insulin–insulin receptor interaction may be a glycoinositol derivative that, when released from the membrane into the cytoplasm, could be a stimulator of phosphoprotein phosphatase. This activity would dephosphorylate a variety of enzymes either activating or inhibiting them and producing effects already known to be associated with the action of insulin. In addition, this second messenger, or the direct phosphorylating activity of the receptor tyrosine kinase might explain the movement of glucose receptors from the cell interior to the surface to account for enhanced cellular glucose utilization in cells that utilize this mechanism to control glucose uptake. These possibilities are reviewed in Figure 20.40. Activation of the enzymes indicated in this figure leads to increased metabolism of glucose while inhibition of enzymes indicated leads to decreased breakdown of glucose or fatty acid stores.

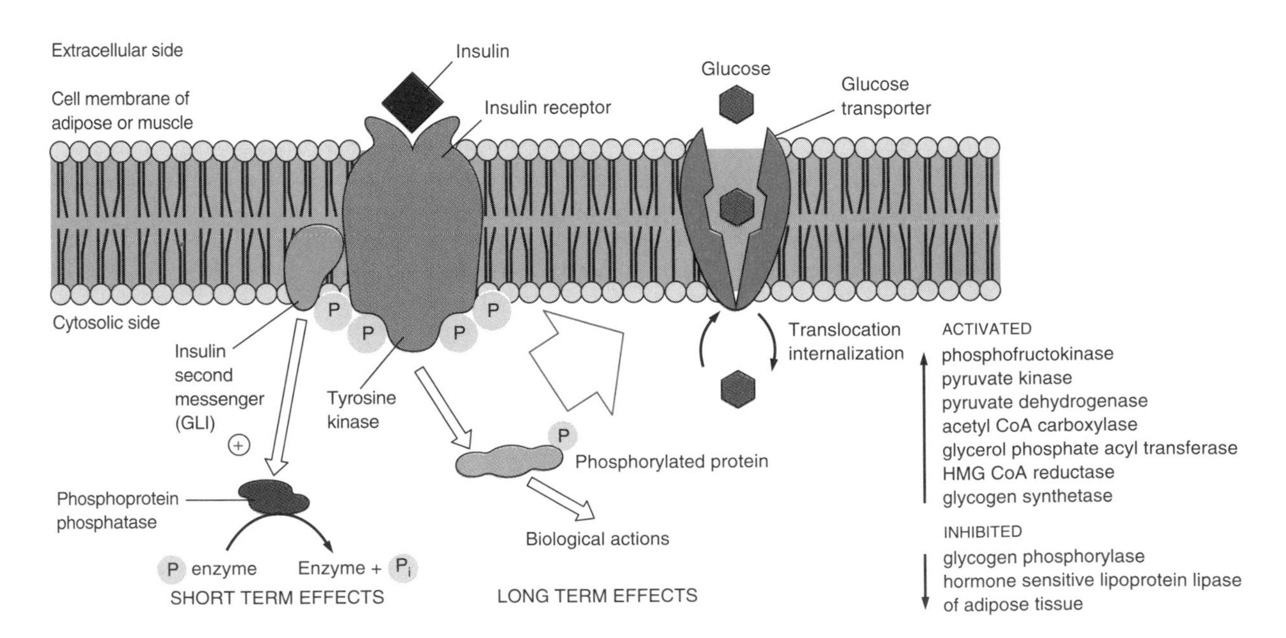

FIGURE 20.40
Possible mechanisms for short-term metabolic effects and long-term actions of insulin.

Activity of Vasopressin: Protein Kinase A

An example of the activation of the protein kinase A pathway by a hormone is the activity of arg-vasopressin (AVP) on the distal kidney cell. Here the action of **vasopressin** also called the antidiuretic hormone (Table 20.5) is to cause increased water reabsorption from the urine in the distal kidney. A mechanism for this system is shown in Figure 20.41. Neurons synthesizing AVP (vasopressinergic neurons) are signaled to release AVP

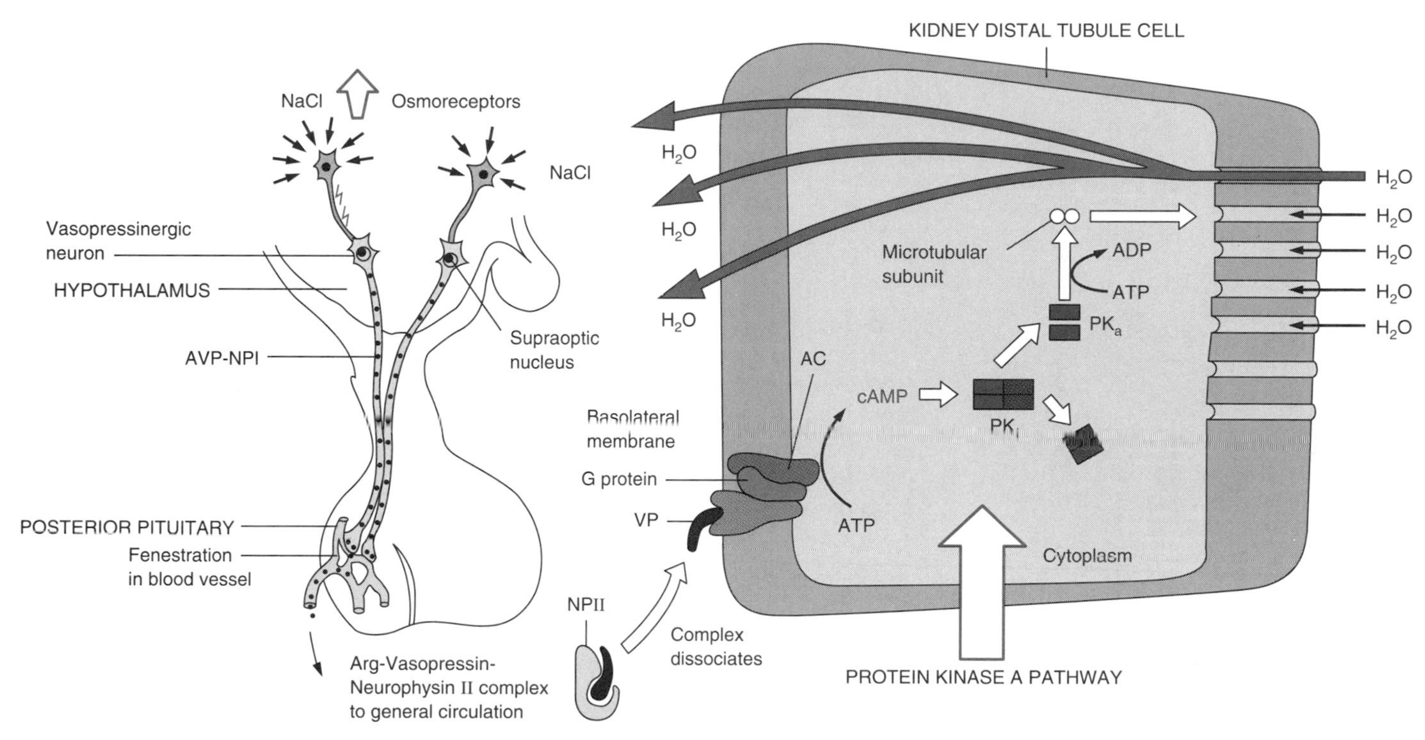

FIGURE 20.41
Secretion and action of arginine vasopressin in the distal kidney.
The release of arginine vasopressin (AVP or VP) from the posterior pituitary begins with a signal from the osmoreceptor, or baroreceptor (not shown), in the upper right-hand corner of figure. The signal can be an increase in the extracellular concentration of sodium chloride, which causes the osmoreceptor neuron to shrink and send an electric message down its axon, which interfaces with the vasopressinergic cell body. This signal is transmitted down the long axon of the vasopressinergic neuron and depolarizes the nerve endings causing the release, by exocytosis, of the VP–neurophysin complex stored there. They enter the local circulation through fenestrations in the vessels and perfuse the general circulation. Soon after release, neurophysin dissociates from VP and VP binds to its cognate receptor in the cell membrane of the kidney distal tubule cell (other VP receptors are located on the corticotrope of the anterior pituitary and on the hepatocytes and their mechanisms in these other cells are different than the one for the kidney tubule cell).

NPII = Neurophysin II; VP = vasopressin; R = receptor; AC = adenylate cyclase; MF = myofibril; GP = glycogen phosphorylase; PK_i = inactive protein kinase; PK_a = active protein kinase; R-Ca = regulatory subunit-cyclic AMP complex; TJ = tight junction; PD = phosphodiesterase. Vasopressin–neurophysin complex dissociates at some point and free VP binds to its cell membrane receptor in the plasma membrane surface. Through a G protein adenylate cyclase is stimulated on the cytoplasmic side of the cell membrane, generating increased levels of cAMP from ATP. Cyclic AMP-dependent protein kinases are stimulated and phosphorylate various proteins (perhaps including microtubular subunits) which, through aggregation, insert as water channels in the luminal plasma membrane, thus increasing the reabsorption of water by free diffusion.

Redrawn in part with permission from T. P. Dousa and H. Valtin, Cellular actions of vasopressin in the mammalian kidney. *Kidney Int.* 10:45, 1975.

TABLE 20.8 Examples of Hormones that Operate Through the Protein Kinase A Pathway

Hormone	Location of action
CRH	Corticotrope of anterior pituitary
TSH (also phospholipid metabolism?)	Thyroid follicle
LH	Leydig cell of testis Mature follicle at ovulation and *corpus luteum*
FSH	Sertoli cell of seminiferous tubule Ovarian follicle
ACTH	Inner layers of cells of adrenal cortex
Opioid peptides	Some in CNS function on inhibitory pathway through G_i.
AVP	Kidney distal tubular cell (the AVP hepatocyte receptor causes phospholipid turnover and calcium ion uptake; the AVP receptor in anterior pituitary causes phospholipid turnover)
PGI_2 (prostacyclin)	Blood platelet membrane
Norepinephrine/epinephrine	β-Receptor

from their nerve endings by interneuronal firing from a *baroreceptor* responding to a fall in blood pressure or from an *osmoreceptor* (probably an interneuron), which responds to an increase in extracellular salt concentration; the trigger could be elevated [Na^+]. The high extracellular salt concentration apparently causes shrinkage of the osmoreceptor cell volume and generates an electrical signal transmitted down the axon of the osmoreceptor to the cell body of the VP neuron generating an action potential. This signal is then transmitted down the long axon from the VP cell body to its nerve ending where, by depolarization, the VP–neurophysin II complex is released to the extracellular space. The complex enters local capillaries through fenestrations and progresses to the general circulation. The complex dissociates and free VP is able to bind to its cognate membrane receptors in the distal kidney, anterior pituitary, hepatocyte, and perhaps other cell types. After binding to the kidney receptor, VP causes stimulation of adenylate cyclase through the stimulatory G protein and activates protein kinase A. The protein kinase phosphorylates microtubular subunits that aggregate to form new channels in the cell membrane for admission of larger volumes of water by free diffusion. Water is transported across the cell to the basolateral side and to the general circulation causing a dilution of the original high salt (signal) and an increase in blood pressure.

A number of hormones operate through the protein kinase A pathway and some are listed in Table 20.8.

Gonadotropin Releasing Hormone (GnRH): Protein Kinase C

Table 20.9 presents examples of polypeptide hormones, which stimulate the phospatidylinositol pathway.

An example of a system operating through stimulation of the phosphatidylinositol pathway and subsequent activation of the protein kinase C system is **GnRH** action, shown in Figure 20.42. Probably, under aminergic interneuronal controls, a signal is generated to stimulate the cell body of the GnRH-ergic neuron where GnRH is synthesized. The signal is transmitted down the long axon to the nerve ending where the hormone is stored. The hormone is released from the nerve ending by

TABLE 20.9 Examples of Polypeptide Hormones that Stimulate the Phosphatidylinositol Pathway

Hormone	Location of action
TRH	Thyrotrope of the anterior pituitary-releasing TSH
GnRH	Gonadotrope of the anterior pituitary-releasing LH & FSH
AVP	Corticotrope of the anterior pituitary; assists CRH in releasing ACTH; hepatocyte—causes increase in cellular Ca^{2+}
TSH	Thyroid follicle—releasing thyroid hormones causes increase in phosphatidylinositol cycle as well as increase in protein kinase A pathway
Angiotensin II/III	*Zona glomerulosa* cell of adrenal cortex—releases aldosterone
Epinephrine (thrombin)	Platelet—releasing ADP/serotonin; hepatocyte via α receptor-releasing intracellular Ca^{2+}

exocytosis resulting from depolarization caused by signal transmission. The GnRH enters the primary plexus of the closed portal system connecting the hypothalamus and anterior pituitary through fenestrations. Then GnRH exits the closed portal system through fenestrations in the secondary plexus and binds to cognate receptors in the cell membrane of the gonadotrope (see exploded view in Figure 20.42). The signal from the

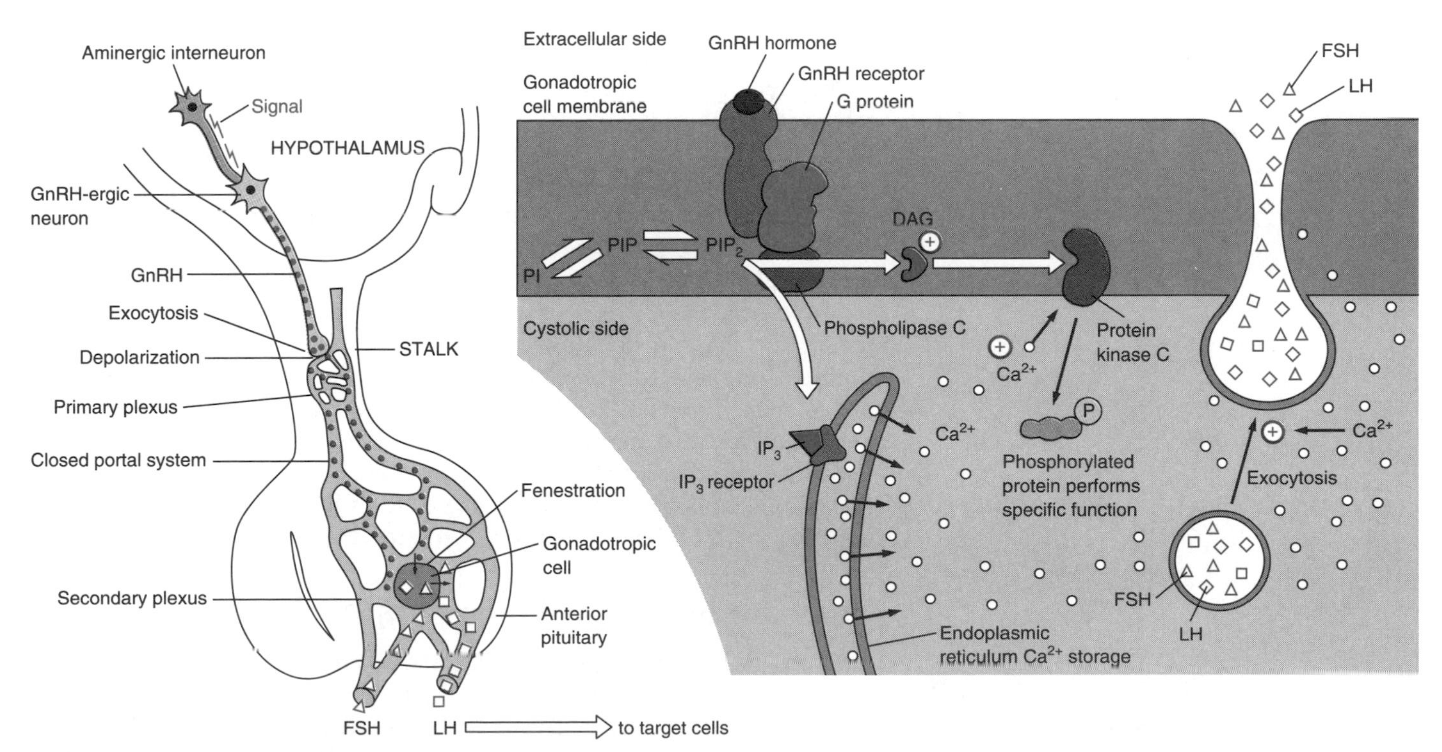

FIGURE 20.42
Overview of regulation of the secretion of the gonadotropes, LH and FSH.
A general mode of action of GnRH to release the gonadotropes from the gonadotropic cell of the anterior pituitary is presented.
GnRH = gonadotropic releasing hormone; FSH = follicle stimulating hormone; LH = luteinizing hormone; PKC = protein kinase C.

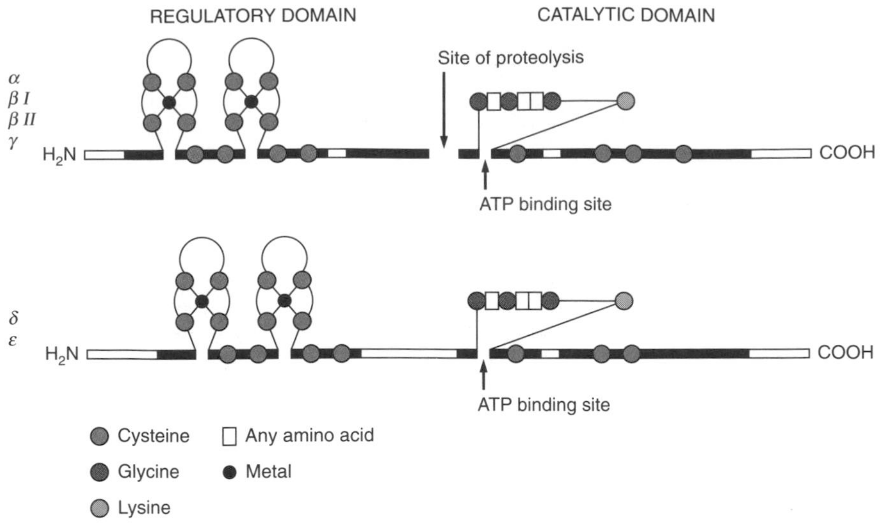

FIGURE 20.43
Common structure of protein kinase C subspecies.
C, G, K, X, and M represent cysteine, glycine, lysine, any amino acid, and metal, respectively.
Modified from U. Kikkawa, A. Kishimoto, and Y. Nishizuka, *Annu. Rev. Biochem.* 58:31, 1989.

hormone–receptor complex is transduced (through a G protein) and phospholipase C is activated. This enzyme catalyzes the hydrolysis of PIP_2 to form DAG and IP_3. Diacylglycerol activates protein kinase C, which phosphorylates specific proteins, some of which may participate in the resulting secretory process to transport LH and FSH to the cell exterior. The product IP_3, which binds to a receptor on the membrane of the calcium storage particle, probably located near the cell membrane, stimulates the release of calcium ion. Elevated cytoplasmic Ca^{2+} causes increased stimulation of protein kinase C and participates in the exocytosis of LH and FSH from the cell.

Much recent work has focused on protein kinase C. It has been shown to have a number of subspecies; such heterogeneity may indicate that there are multiple functions for this critical enzyme (Figure 20.43). The enzyme consists of two domains, a regulatory and a catalytic domain, which can be separated by proteolysis at a specific site. The free catalytic domain, formerly called **protein kinase M,** can phosphorylate proteins free of the regulatory components. The free catalytic subunit, however, may be degraded. More needs to be learned about the dynamics of this system and the translocation of the enzyme from one compartment to another. The regulatory domain contains two Zn^{2+} fingers usually considered to be hallmarks of DNA binding proteins (see Chapter 3). This activity has not yet been demonstrated for protein kinase C and metal fingers may participate in other types of interactions. The ATP binding site in the catalytic domain contains the G box GXGXXG, which is a consensus sequence for ATP binding with a downstream lysine (K) residue.

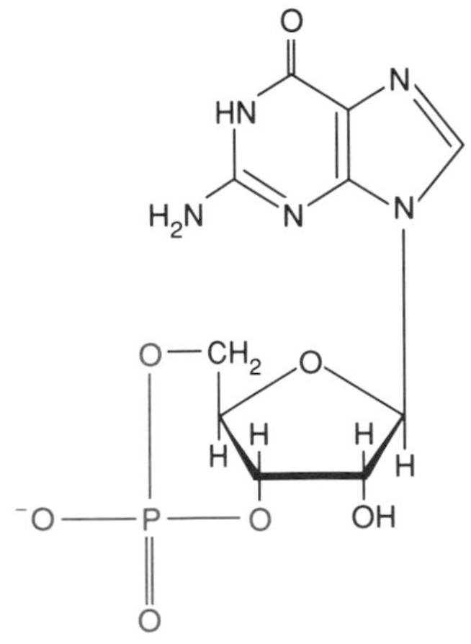

FIGURE 20.44
Structure of cGMP.

Activity of Atrionatriuretic Factor (ANF): Protein Kinase G

The third system is the protein kinase G system, which is stimulated by the elevation of cytoplasmic cGMP (Figure 20.44). **Cyclic GMP** is synthesized by guanylate cyclase from GTP. Like adenylate cyclase, guanylate cyclase is linked to a specific biological signal through a membrane receptor. The guanylate cyclase extracellular domain may serve the role of the hormone receptor. This is directly coupled to the cytoplasmic domain through one membrane spanning domain (Figure 20.45), which may be applicable to the atrionatriuretic factor (ANF; also referred to as atriopeptin) receptor–**guanylate cyclase system.** Thus, the hormone binding site,

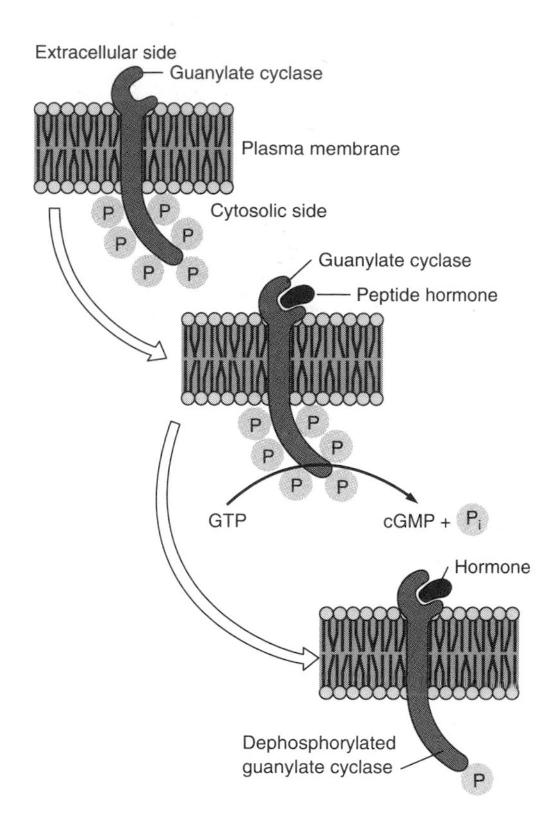

FIGURE 20.45
Model for the regulation of guanylate cyclase activity after peptide hormone binding.
The enzyme exists in a highly phosphorylated state under normal conditions. The binding of hormone markedly enhances enzyme activity, followed by a rapid dephosphorylation of guanylate cyclase and a return of activity to the basal state despite the continued presence of the hormonal peptide.
Redrawn from S. Schultz, M. Chinkers, and D. L. Garbers, *FASEB J.* 3:2026, 1989.

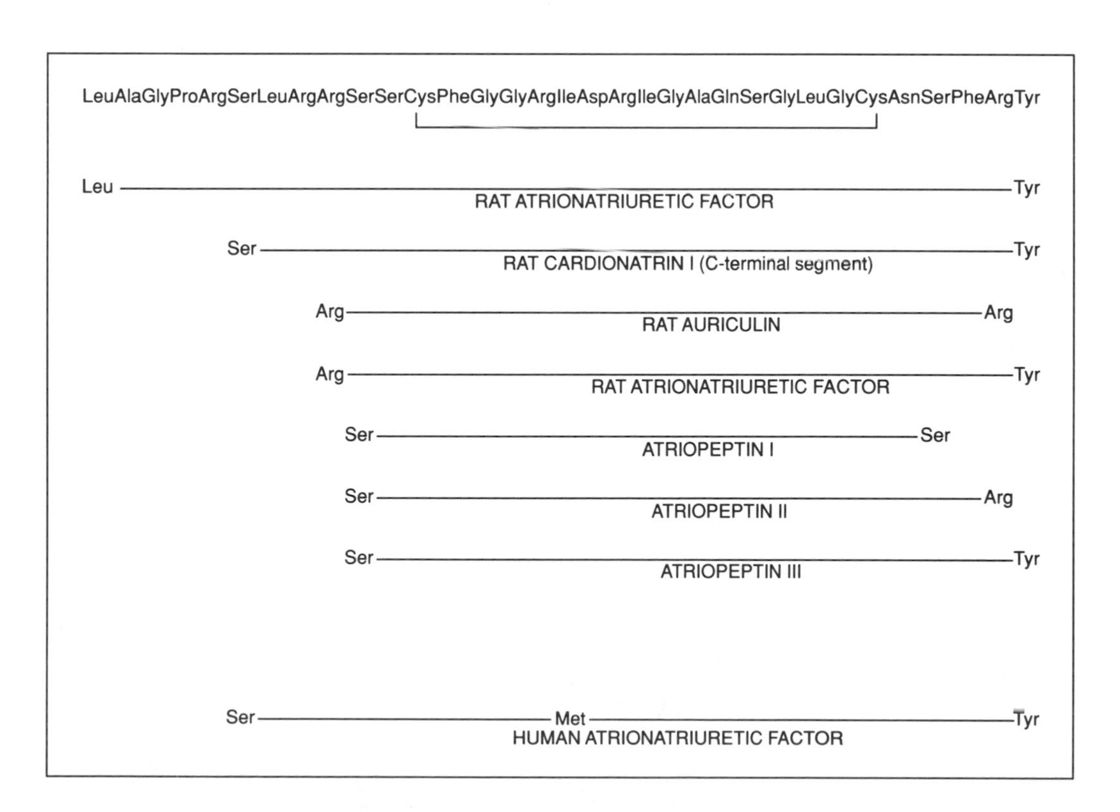

FIGURE 20.46
Atrial natriuretic peptides.
These peptides relax vascular smooth muscle and produce vasodilation and have other effects discussed in the text.
Modified from Chemolog hi-lites, Vol. 11, May 1987, South Plainfield, N.J.: Chemical Dynamics Corp.

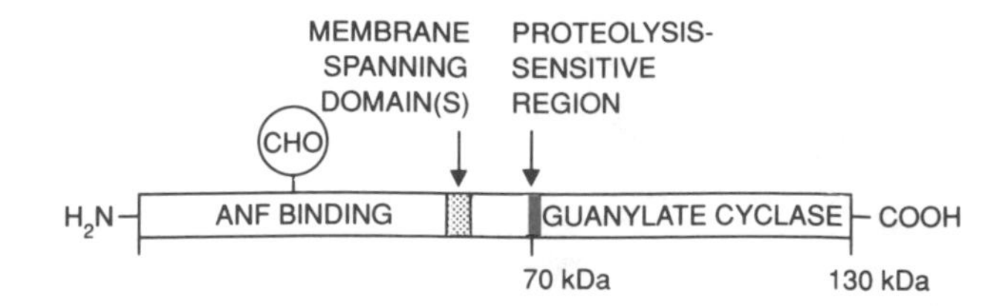

FIGURE 20.47
Functional domains of ANF-R_1 receptor.
The hypothetical model shows the sequence of an ANF binding domain, a membrane-spanning domain(s), a proteolysis-sensitive region, a guanylate cyclase catalytic domain, glucosylation sites (CHO), and amino (H_2H) and carboxyl terminals (COOH) of the receptor.
Redrawn from B. Liu, S. Meloche, N. McNicoll, C. Lord, and A. DeLéan, *Biochemistry* 28:5599, 1989.

transmembrane domain, and guanylate cyclase activities are all served by a single polypeptide chain.

This hormone is a family of peptides as shown in Figure 20.46; sequence of human ANF is shown at the bottom. The functional domains of the ANF receptor are illustrated in Figure 20.47. There are two functional domains of the receptor, a ligand binding domain and a catalytic center. There is also a proteolysis-sensitive region, which separates the ANF binding site from the guanylate cyclase.

Atrionatriuretic factor is released from atrial cells of the heart under control of several hormones. Data from atrial cell culture suggest that ANF secretion is stimulated by activators of protein kinase C and decreased by activators of protein kinase A. These opposing actions may be mediated by the actions of α- and β-adrenergic receptors, respectively. An overview of the secretion of ANF and its general effects is shown in Figure 20.48. The ANF is released by a number of signals, such as blood volume expansion, elevated blood pressure directly induced by vasoconstrictors, high salt intake, and increased heart pumping rate. Atrionatriuretic factor is secreted as a dimer that is inactive for receptor interac-

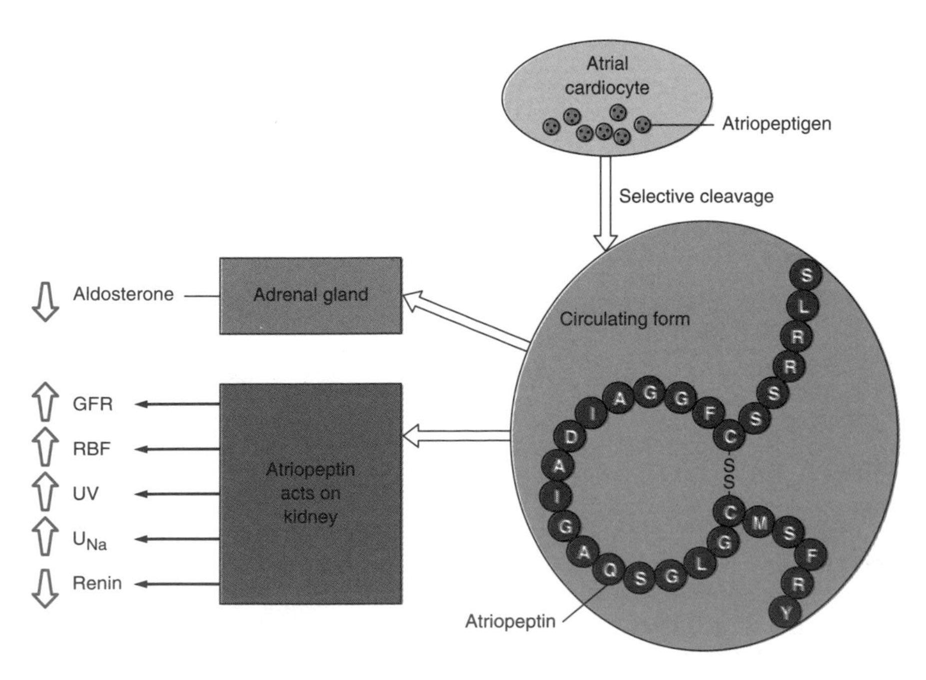

FIGURE 20.48
Schematic diagram of the atrial natriuretic factor–atriopeptin hormonal system.
The prohormone is stored in granules located in perinuclear atrial cardiocytes. An elevated vascular volume results in cleavage and release of atriopeptin, which acts on the kidney (glomeruli and papilla) to increase the glomerular filtration rate (GFR), to increase renal blood flow (RBF), to increase urine volume (UV), and sodium excretion (U_{Na}) and to decrease plasma renin activity. Natriuresis and diuresis are also enhanced by the suppression of aldosterone and its actions and the release from the posterior pituitary of arginine vasopressin (AVP). Diminution of vascular volume provides a negative feedback signal that suppresses circulating levels of atriopeptin.
Redrawn from P. Needleman and J. E. Greenwald, Atriopeptin: A cardiac hormone intimately in fluid, electrolyte, and blood pressure homeostasis. *N. Engl. J. Med.* 314:828, 1986.

tion and is converted in plasma to a monomer capable of interacting with receptor. The actions of ANF (Figure 20.48) are to increase the glomerular filtration rate without increasing renal blood flow leading to increased urine volume and excretion of sodium ion. Renin secretion is also reduced and aldosterone secretion is lowered. This action reduces aldosterone mediated sodium reabsorption. The ANF inhibits the vasoconstriction produced by angiotensin II and relaxes the constriction of the renal vessels, other vascular beds, and large arteries. Atrionatriuretic factor operates through its membrane receptor, which appears to be the extracellular domain of guanylate cyclase (Figure 20.47). The cGMP produced activates a protein kinase G, which further phosphorylates cellular proteins to express many of the actions of this pathway. More needs to be learned about protein kinase G.

20.13 ONCOGENES AND RECEPTOR FUNCTIONS

Oncogenes are genes that are expressed by cancerous transformed cells. A cancer cell may express few or many oncogenes that dictate the aberrant uncontrolled behavior of the cell. There are three mechanisms by which oncogenes allow a cell to escape dependence on exogenous growth factors. These are presented in Figure 20.49. Some are genes for parts of receptors, most often related to growth factor hormone receptors, which can function in the absence of the hormonal ligand since the oncogene may represent a truncated gene where the ligand binding domain is missing. This would result in the production of this protein, insertion into the cell membrane and continuous constitutive function in the absence or presence of ligand (Figure 20.49*b*,*c*). The second messengers would be produced constitutively at a high rate, instead of being regulated by ligand, and the result would be uncontrolled growth of the cell. Some may have tyrosine protein kinase activity and therefore function like tyrosine kinase normally related to certain cell membrane receptors. Other oncogenes relate to thyroid and steroid hormone receptors (see Chapter 21) while still others are DNA binding proteins, some of which may be

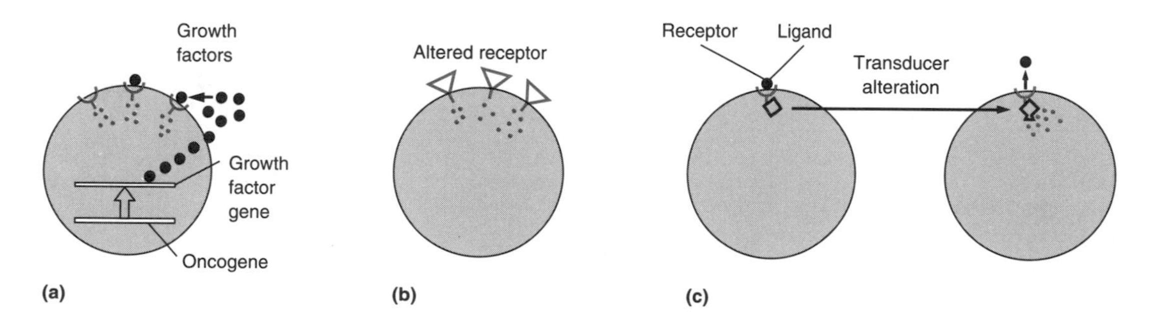

FIGURE 20.49
Mechanisms by which oncogenes can allow a cell to escape dependence on exogenous growth factors.
(*a*) By autocrine mechanism where the oncogene product is a growth factor active ligand and overstimulates its receptor; (*b*) by receptor alteration so that receptor is "permanently turned on" without a requirement for growth factor binding; and (*c*) by transducer alteration where the intermediate between the receptor and its resultant activity, that is, the GTP stimulatory protein, is permanently turned on uncoupling the normal requirement of ligand–receptor binding.
Redrawn from R. A. Weinberg, The action of oncogenes in the cytoplasm and nucleus, *Science* 230:770, 1985.

TABLE 20.10 Known Oncogenes, Their Products and Functions[a]

Name of oncogene	Retrovirus	Tumor	Oncogenic protein: Cellular location	Oncogenic protein: Function
src	Chicken sarcoma		Plasma membrane	
yes	Chicken sarcoma		Plasma membrane (?)	
fgr	Cat sarcoma		(?)	
abl	Mouse leukemia	Human leukemia	Plasma membrane	Tyrosine-specific protein kinase
fps	Chicken sarcoma		Cytoplasm (plasma membrane?)	
fes	Cat sarcoma		Cytoplasm (cytoskeleton?)	
ros	Chicken sarcoma		(?)	
erb-B	Chicken leukemia		Plasma and cytoplasmic membranes	EGF receptor's cytoplasmic tyrosine-specific protein-kinase domain
fms	Cat sarcoma		Plasma and cytoplasmic membranes	Cytoplasmic domain of a growth-factor receptor (?)
mil	Chicken carcinoma		Cytoplasm	(?)
raf	Mouse sarcoma		Cytoplasm	(?)
mos	Mouse sarcoma	Mouse leukemia	Cytoplasm	(?)
sis	Monkey sarcoma		Secreted	PDGF-like growth factor
Ha-*ras*	Rat sarcoma	Human carcinoma, rat carcinoma	Plasma membrane	
Ki-*ras*	Rat sarcoma	Human carcinoma, leukemia and sarcoma	Plasma membrane	GTP binding
N-*ras*	—	Human leukemia and carcinoma	Plasma membrane	
fos	Mouse sarcoma		Nucleus	Transcriptional control
myc	Chicken leukemia	Human lymphoma	Nucleus	DNA binding; related to cell proliferation; transcriptional control
myb	Chicken leukemia	Human leukemia	Nucleus	(?)
B-*lym*	—	Chicken lymphoma, human lymphoma	Nucleus (?)	(?)
ski	Chicken sarcoma		Nucleus (?)	(?)
rel	Turkey leukemia		(?)	(?)
erb-A	Chicken leukemia		(?)	Thyroid hormone receptor (c-erb-Aα1); related to steroid hormone receptors, retinoic acid receptor, and vitamin D_3 receptor
ets	Chicken leukemia		(?)	DNA binding
elk (ets-like)				DNA binding protein
jun				Transcription factor, AP-1-like
int-1				Wingless = Drosophila homolog

[a] The second column gives the source from which each viral oncogene was first isolated and the cancer induced by the oncogene. Some names, such as *fps* and *fes*, may be equivalent genes in birds and mammals. The third column lists human and animal tumors caused by agents other than viruses in which the *ras* oncogene or an inappropriately expressed protooncogene has been identified. Modified from T. Hunter, The proteins of oncogenes. *Sci. Am.* 251:70, 1984.

transactivating factors or related to them, or are known currently as DNA binding proteins. Oncogene phenotypes that bind to DNA may be identical with or related to transactivation factors. The oncogene, *Jun*, for example, is related to activator protein 1 (AP1), a transactivating factor regulating transcription. In Table 20.10 are reviews of some of the oncogenes together with their functions.

BIBLIOGRAPHY

Alberts, B., Bray, D., Lewis, J., Raff, R., Roberts, K., and Watson, J. D. *Molecular Biology of the Cell,* 2nd ed., New York: Garland, 1989.

Cuatrecasas, P. Hormone receptors, membrane phospholipids, and protein kinases. *The Harvey Lectures,* Series 80, 89, 1986.

DeGroat, L. J., et al. (Eds.). *Endocrinology,* New York: Grune and Stratton, 1979.

Hunter, T. The proteins of oncogenes. *Sci. Am.* 251:70, 1984.

Krieger, D. T. and Hughes, J. C. (Eds.). *Neuroendocrinology,* Sunderland, MA: Sinauer Associates, 1980.

Litwack, G. (Ed.). *Biochemical Actions of Hormones,* Vol. 1–14, New York: Academic, 1973–1987.

Norman, A. W. and Litwack, G. *Hormones,* Orlando: Academic, 1987.

Richter, D. Molecular events in expression of vasopressin and oxytocin and their cognate receptors. *Am. J. Physiol.* 255:F207, 1988.

Ryan, R. J., Charlesworth, M. C., McCormick, D. J., Milius, R. P., and Keutmann, H. T. *FASEB J.* 2:2661, 1988.

Weinberg, R. A. The action of oncogenes in the cytoplasm and nucleus. *Science* 230:770, 1985.

QUESTIONS

J. BAGGOTT AND C. N. ANGSTADT

*Q*uestion *T*ypes are described on page xxiii.

A. Quantity of hormone C. Both
B. Half-life of hormone D. Neither

1. (QT4) Can be expected to increase in successive steps of a cascade.

2. (QT1) All of the following have an identical (or very similar) α subunit EXCEPT:
   A. growth hormone.
   B. thyroid stimulating hormone.
   C. luteinizing hormone.
   D. follicle stimulating hormone.
3. (QT1) If a single gene contains information for the synthesis of more than one hormone molecule
   A. all the hormones are produced by any tissue that expresses the gene.
   B. all of the hormone molecules are identical.
   C. cleavage sites in the gene product are pairs of basic amino acids.
   D. all of the peptides of the gene product have well-defined biological activity.
   E. the hormones all have similar function.
4. (QT1) In the following sequence of events associated with signal transduction, which one is out of place?
   Receptor binds hormone.
   A. Conformational change occurs in receptor.
   B. Receptor interacts with G protein.
   C. α subunit of G protein hydrolyses GTP.
   D. α subunit of G protein dissociates from β and γ subunits.
   E. α subunit of G protein binds to adenylate cyclase.
5. (QT1) The direct effect of cAMP in the protein kinase A pathway is to:
   A. activate adenylate cyclase.
   B. dissociate regulatory subunits from protein kinase.
   C. phosphorylate certain cellular proteins.
   D. phosphorylate protein kinase A.
   E. release hormones from a target tissue.
6. (QT1) Activation of phospholipase C initiates a sequence of events including all of the following EXCEPT:
   A. release of inositol 4,5-bisphosphate from a phospholipid.
   B. increase in intracellular Ca^{2+} concentration.
   C. release of diacylglycerol (DAG) from a phospholipid.
   D. activation of protein kinase C.
   E. phosphorylation of certain cytoplasmic proteins.

7. (QT2) In the ovarian cycle:
   1. GnRH enters the vascular system via transport by a specific membrane carrier.
   2. the *corpus luteum* dies only if fertilization does not occur.
   3. inhibin works by inhibiting the synthesis of the α subunit of FSH.
   4. FSH activates a protein kinase A pathway.

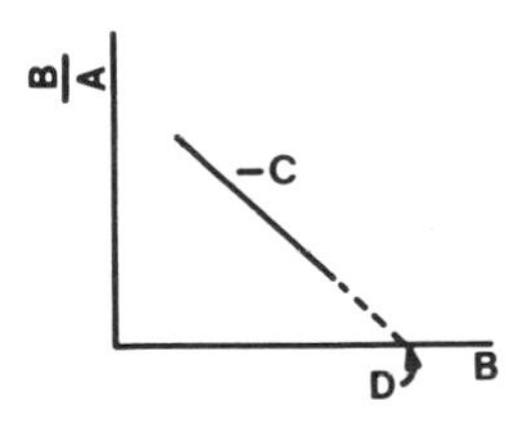

8. (QT1) The Scatchard plot, shown in the accompanying figure, could be used to determine kinetic parameters of an enzyme. Which letter in the graph corresponds to total binding sites in a Scatchard plot or V_{max} in an enzyme kinetic plot?
9. (QT1) With the anterior pituitary hormones, TSH, LH, and FSH:
   A. the α subunits are all different.
   B. the β subunits are specifically recognized by the receptor.
   C. the β subunit alone can bind to the receptor.
   D. hormonal activity is expressed through activation of protein kinase B.
   E. intracellular receptors bind these hormones.

10. (QT2) During the interaction of a hormone with its receptor:
    1. more than one polypeptide chain may be necessary.
    2. more than one second messenger may be generated.
    3. an array of transmembrane helices may form the binding site for the hormone.

4. receptors have a greater affinity for hormones than for synthetic agonists or antagonists.

A. Protein kinase A
B. Tyrosine kinase
C. Protein kinase C
D. Protein kinase G

In the following questions, match the numbered hormone with the lettered kinase it stimulates.

11. (QT5) Atrionatriuretic factor
12. (QT5) Gonadotrophin releasing hormone
13. (QT5) Insulin
14. (QT5) Vasopressin

ANSWERS

1. C Each successive step typically releases a larger amount of a longer lived hormone (p. 850).
2. A All of these are anterior pituitary hormones, but only the last three, the glycoprotein hormones, have an α subunit that is similar or identical from hormone to hormone (p. 854).
3. C One or more trypsin-like proteases catalyzes the reaction (Figure 20.5). A: The POMC gene product is cleaved differently in different parts of the anterior pituitary (p 859). B: Multiple copies of a single hormone may occur (p. 861), but not necessarily (Figure 20.5, p. 858). D: Some fragments have no known function. E: ACTH and β-endorphin, for example, hardly have similar functions (p. 859).
4. C Hydrolysis of GTP returns the α subunit to its original conformation, and allows it to associate with the β and γ subunits (p. 871).
5. B cAMP binding causes a conformational change in the regulatory subunits, resulting in the release of active protein kinase A (p. 873).
6. A Inositol 1,4,5-trisphosphate (IP_3) is released from the phospholipid, phosphatidylinositol 4,5-bisphosphate (PIP_2) (p. 874).
7. D Only 4 is correct. 1: GnRH enters the vascular system through fenestrations (p. 878). 2. The *corpus luteum* is replaced by the placenta if fertilization occurs (p. 881). 3: The glycoprotein hormones share a common α subunit. Specific control of them would not involve a subunit they share.
8. D A is free ligand concentration (analogous to substrate concentration), B is bound ligand concentration (analogous to v), C is the equilibrium constant (analogous to K_m), and D is the extrapolated maximum number of binding sites (analogous to V_{max}) (p. 882).
9. B A: The α subunits are identical or nearly so (p. 883). B and C: Although specificity is conferred by the β subunits, which differ among the three hormones, binding to the receptor requires both subunits (p. 883). D: It is protein kinase A, and perhaps also protein kinase C in the case of TSH (p. 884). E: These large glycoprotein hormones do not penetrate the cell membrane; they bind to receptors on the cell surface (p. 884).
10. A 1–3 correct. 1 and 2: These are true of the anterior pituitary hormones (p. 886). 3: This appears to be true for the β_1 receptor (Figure 20.36). 4: β receptors bind isoproterenol more tightly than their hormones (p. 885).
11. D See p. 894.
12. C See p. 892.
13. B See p. 890.
14. A See p. 891.

21

Biochemistry of Hormones II: Steroid Hormones

GERALD LITWACK

21.1 OVERVIEW 901
21.2 STRUCTURES OF STEROID HORMONES 902
21.3 BIOSYNTHESIS OF STEROID HORMONES 905
Steroid Hormones Are Synthesized From Cholesterol 905
21.4 METABOLIC INACTIVATION OF STEROID HORMONES 907
21.5 CELL–CELL COMMUNICATION AND THE CONTROL OF SYNTHESIS AND RELEASE OF STEROID HORMONES 910
Steroid Hormone Synthesis Is Controlled by Specific Hormones 911
21.6 TRANSPORT OF STEROID HORMONES IN BLOOD 915
Steroid Hormones Are Bound to Specific Proteins or Albumin in Blood 915
21.7 STEROID HORMONE RECEPTORS 916
Steroid Hormones Are Bound to Specific Intracellular Protein Receptors 916
Some Steroid Receptors Are Part of the c*Erb*A Family; c*Erb*A Is a Protooncogene 918
21.8 RECEPTOR ACTIVATION AND UP- AND DOWN-REGULATION 920
Steroid Receptors Can Be Up- and Down-Regulated Depending on Exposure to the Hormone 921
21.9 SPECIFIC EXAMPLES OF STEROID HORMONE ACTION AT THE CELLULAR LEVEL 921
Programmed Death of Specific Cells Is Controlled by Some Steroid Hormones 921
The Adrenal Gland Has an Important Role in Stress 922
BIBLIOGRAPHY 924
QUESTIONS AND ANNOTATED ANSWERS 924
CLINICAL CORRELATIONS
21.1 Oral Contraception 912
20.2 Programmed Cell Death in the Ovarian Cycle 922

21.1 OVERVIEW

Steroid hormones in the human involve cortisol as the major glucocorti coid or antistressing hormone, aldosterone as an important regulator of Na^+ uptake, and the sex and progestational hormones. The sex hormones are 17β-estradiol in the female and testosterone in the male. Progesterone is the major progestational hormone. Testosterone is converted to dihydrotestosterone, a higher affinity ligand for the androgen receptor. Vitamin D is converted to the steroid hormone, dihydroxy vitamin D_3. The genes in the steroid receptor supergene family also include the retinoic acid binding protein and the thyroid hormone receptor although their

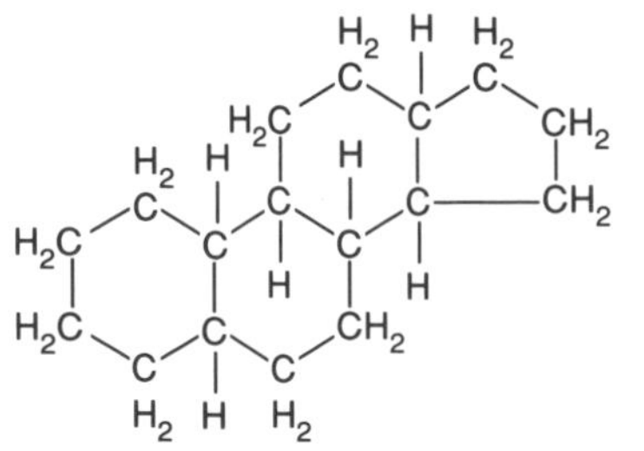

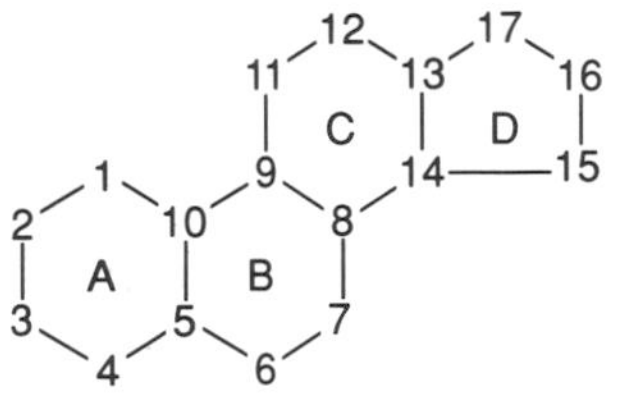

Cyclopentanoperhydrophenanthrene nucleus

Numbering system of carbons

FIGURE 21.1
The steroid nucleus.

ligands are not steroid hormones. Retinoic acid and thyroid hormone, however, have a six-membered ring in their structures that could be thought to resemble the A ring of a steroid.

Steroidal structure will be reviewed along with the synthesis and inactivation of steroid hormones. Regulation of the synthesis of steroid hormones is reviewed with respect to the renin–angiotensin system for aldosterone, the gonadotropes, especially follicle-stimulating hormone for 17β-estradiol, and the vitamin D_3 mechanism. Steroid hormone transport is reviewed with respect to the transporting proteins in blood. A general model for steroid hormone action at the cellular level is presented with information on receptor activation and regulation of receptor levels. Finally, specific examples of steroid hormone action for programmed cell death and for stress are presented.

21.2 STRUCTURES OF STEROID HORMONES

Steroid hormones are derived from specific tissues in the body and are divided into two classes: the **sex hormones,** including the progestational hormones, and the **adrenal hormones.** They are synthesized from cholesterol and all of these hormones pass through the required intermediate, Δ^5-pregnenolone. The structure of steroid hormones is related to thc **cyclopentanoperhydrophenanthrene** nucleus. The numbering of the cyclopentanoperhydrophenanthrene ring system (sterane) and the lettering of the rings is presented in Figure 21.1. The ring system of the steroid hormones is stable and not broken down by mammalian cells. Conversion of active hormones to less active or inactive forms involves alteration of ring substituents rather than the ring structure itself. The parental precursor of the steroid is **cholesterol,** shown in Figure 21.2. The biosynthesis of cholesterol is given in Chapter 10, p. 439.

The major steroid hormones of humans and their actions are shown in Table 21.1. Many of these hormones are similar in gross structure, however, the specific receptor for each hormone is able to distinguish the cognate ligand. In the case of cortisol and aldosterone, however, there is overlap in ability of each specific receptor to bind both ligands. Thus, the availability and levels of each receptor and the relative amounts of each hormone in a given cell become paramount considerations. The steroid hormones listed in Table 21.1 can be described as classes based on the carbon number in their structures. Thus, a C-27 steroid is the vitamin D_3 hormone; C-21 steroids are progesterone, cortisol and aldosterone; C-19 steroids are testosterone and dehydroepiandrosterone; and the C-18 steroid is 17β-estradiol. Classes, such as sex hormones, can be distinguished easily by the carbon number, C-19 being androgens, C-18 being estrogens,

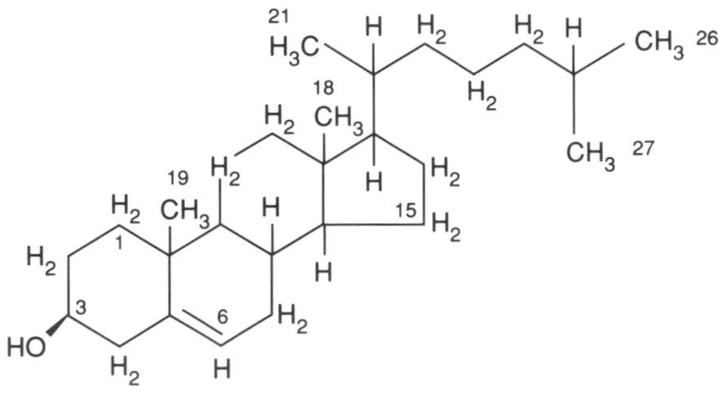

Cholesterol

FIGURE 21.2
Structure of cholesterol.

TABLE 21.1 Major Steroid Hormones of Humans

Hormone	Structure	Secretion from	Secretion signal[a]	Functions
Progesterone	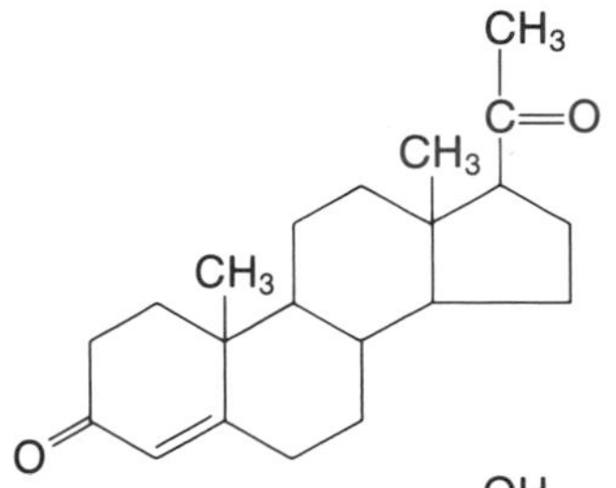	*Corpus luteum*	LH	Maintains (with estradiol) the uterine endometrium for implantation; differentiation factor for mammary glands
17β-Estradiol	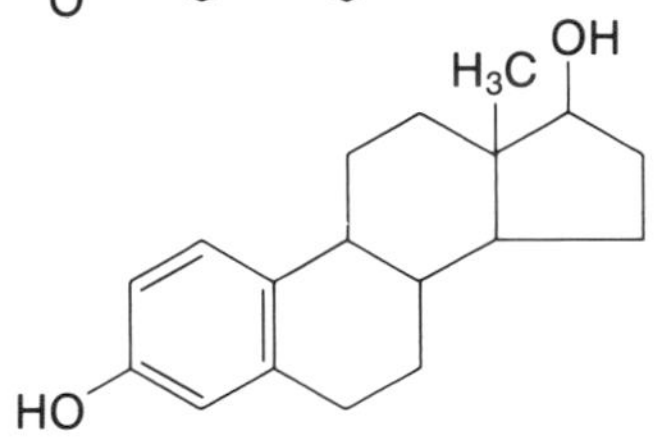	Ovarian follicle; *corpus luteum;* (Sertoli cell)	FSH	Female: regulates gondotrope secretion in ovarian cycle (see Chapter 20); maintains (with progesterone) uterine endometrium; differentiation of mammary gland; Male: negative feedback inhibitor of Leydig cell synthesis of testosterone.
Testosterone	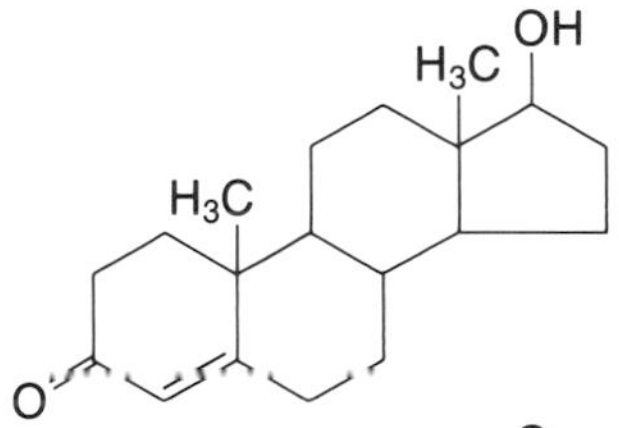	Leydig cells of testis; (adrenal gland); ovary	LH	Male: after conversion to dihydrotestosterone, production of sperm proteins in Sertoli cells; secondary sex characteristics
Dehydroepiandro-sterone	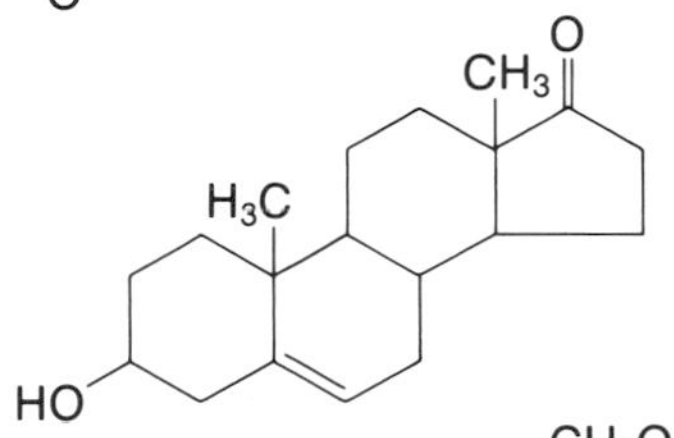	Reticularis cells	ACTH	Various protective effects; weak androgen; can be converted to estrogen; no receptor yet found; inhibitor of G 6-PDH; regulates NAD^+ coenzymes.
Cortisol	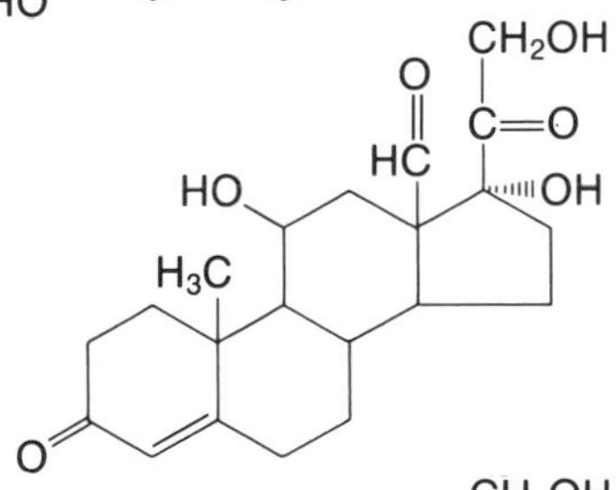	Fasciculata cells	ACTH	Stress adaptation through various cellular phenotypic expression; slight elevation of liver glycogen; killing effect on certain T cells in high doses; elevates blood pressure; sodium uptake in lumenal epithelia
Aldosterone	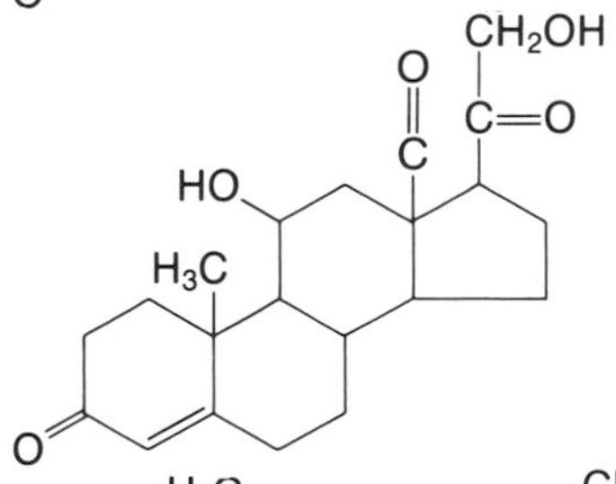	Glomerulosa cells of adrenal cortex	Angiotensin II/III	Causes sodium ion uptake via conductance channel; occurs in high levels during stress; raises blood pressure; fluid volume increased
1,25-Dihydroxy-vitamin D_3	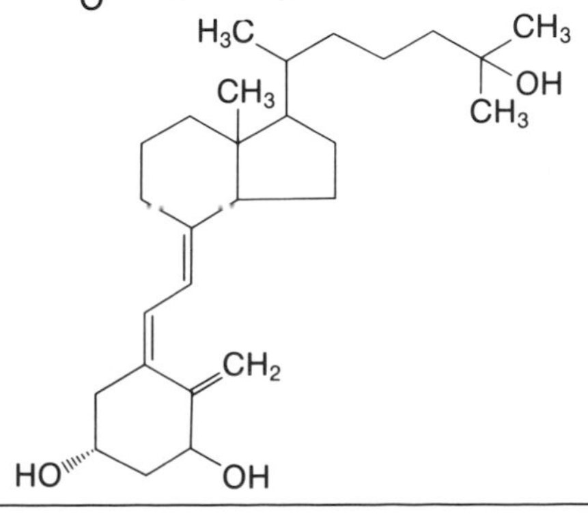	Vitamin D arises in skin cells after irradiation and successive hydroxylations occur in liver and kidney	PTH operates on kidney proximal tubule hydroxylation system	Causes synthesis of Ca^{2+} transport protein

[a] LH = luteinizing hormone; FSH = follicle stimulating hormone; ACTH = adrenocorticotropic hormone; PTH = parathyroid hormone

Phenolic OH on C-3
C-18 methyl group
Hydroxyl OH on C-17 (β-oriented)
Aromatic A-ring
ESTRADIOL

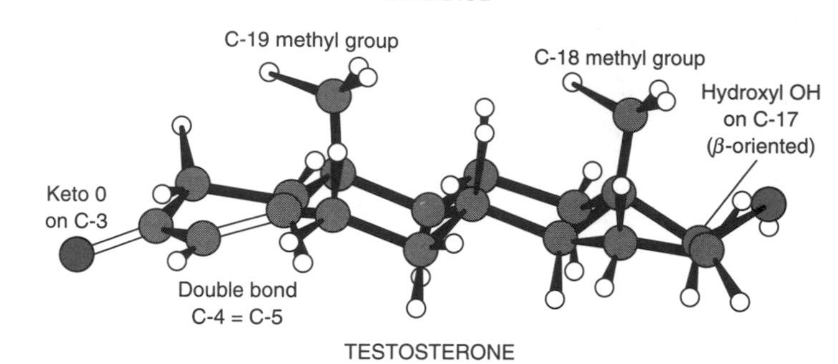

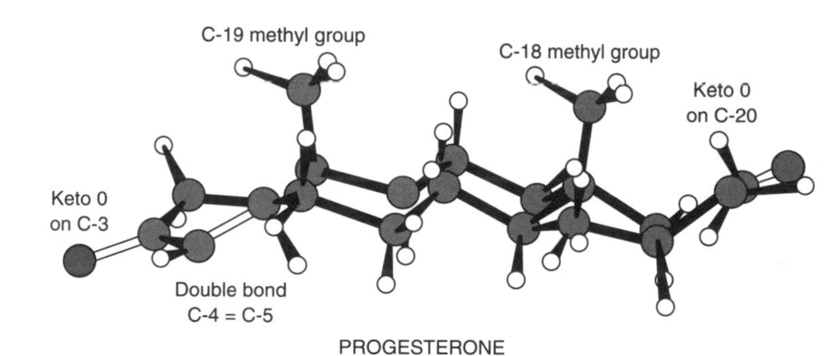

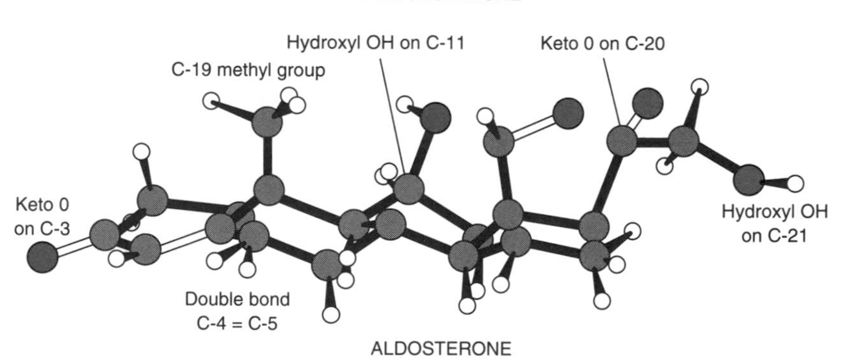

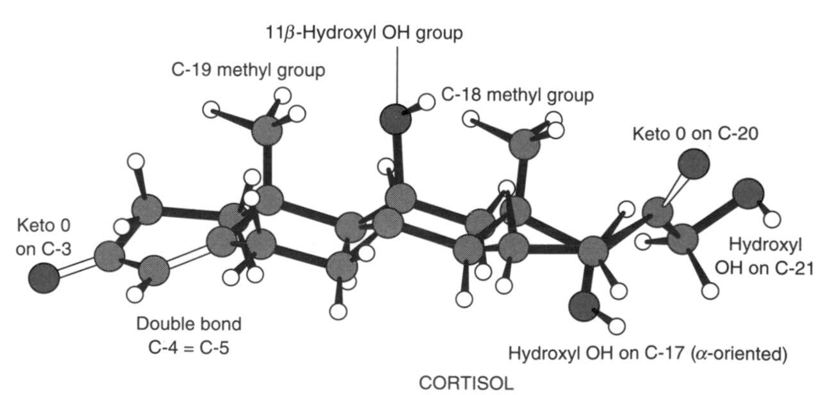

FIGURE 21.3

"Ball-and-stick" representations of the structures of some steroid hormones determined by X-ray crystallographic methods. Details of each structure are labeled. In aldosterone the acetal grouping is

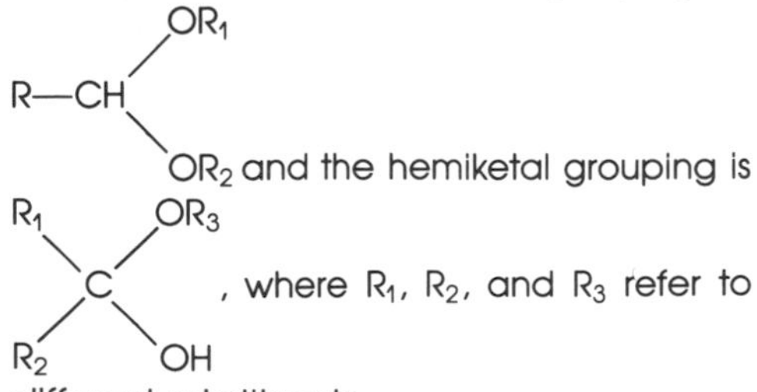

$R—CH(OR_1)(OR_2)$ and the hemiketal grouping is $R_1R_2C(OR_3)(OH)$, where R_1, R_2, and R_3 refer to different substituents.

Reprinted with permission from J. P. Glusker, *Biochemical Actions of Hormones* Vol. 6. G. Litwack (Ed.). Academic, 1979, pp. 121–204.

and C-21 being progestational or adrenal steroids. Aside from the number of carbon atoms in a class structure, certain substituents in the ring system are characteristic. For example, glucocorticoids and mineralocorticoids (typically aldosterone) possess a C-11 OH or oxygen function. In rare exceptions, certain synthetic compounds can elicit a response without a C-11 OH group but they require a new functional group in proximity within the A–B ring system. Estrogens do not have a C-19 methyl group and the A ring is contracted by the content of three double bonds. Since many receptors recognize the ligand A ring primarily, the estrogen receptor can distinguish the A ring of estradiol stretched out of the plane of the B–C–D rings compared to other steroids in which the A ring is in the same plane as the B–C–D rings. These relationships are emphasized in Figure 21.3.

21.3 BIOSYNTHESIS OF STEROID HORMONES

Steroid Hormones Are Synthesized From Cholesterol

All of the steroid hormones are formed from cholesterol. A critical step in the ability to synthesize steroid hormones is the cell's activity in mobilizing cholesterol, which may be stored in a droplet, and transport of cholesterol to the mitochondrion. The rate-limiting step in steroid hormone biosynthesis is the rate of **cholesterol side chain cleavage** in the mitochondrion, accomplished by enzymes (possibly four) collectively known as the **cytochrome P450 side chain cleavage enzyme complex** (See Chapter 23 for

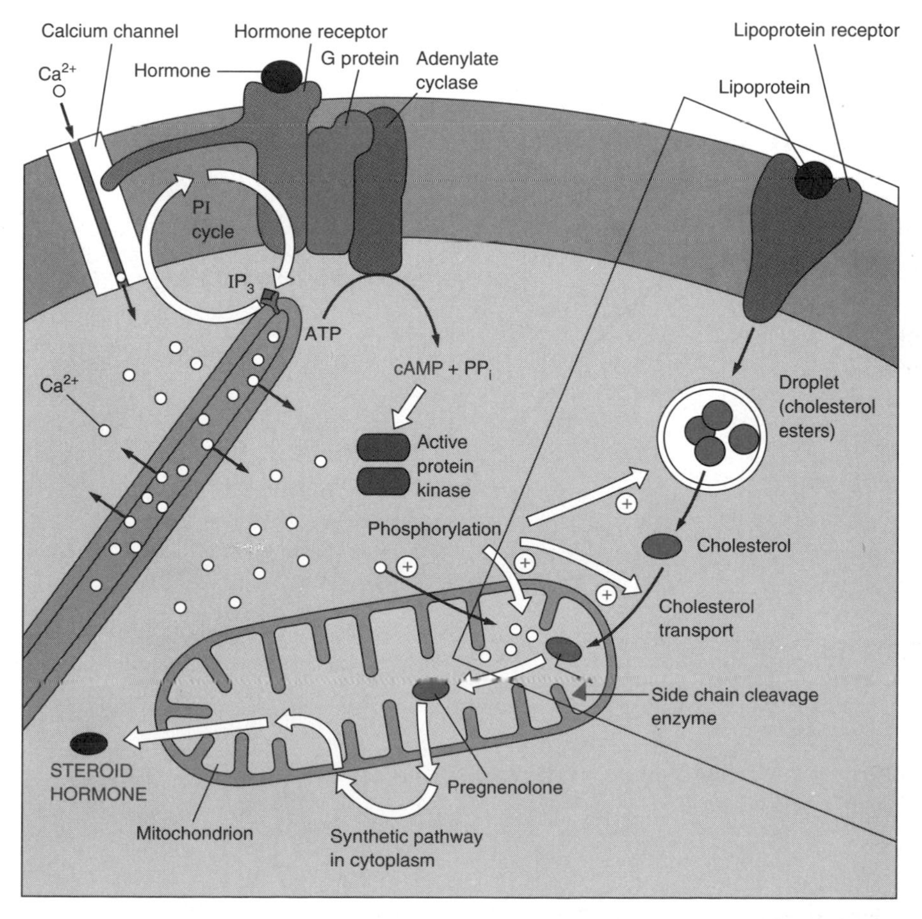

FIGURE 21.4
General overview of hormonal stimulation of steroid hormone biosynthesis.
The nature of the hormone (top of figure) depends on the cell type and receptor (ACTH for cortisol synthesis; FSH for estradiol synthesis; LH for testosterone synthesis, etc., as given in Table 21.1). It binds to cell membrane receptor and activates adenylate cyclase mediated by a stimulatory G protein. Receptor, activated by hormone, may directly stimulate a calcium channel or indirectly stimulate it by activating the phosphatidylinositol cycle (PI cycle) as shown in Figure 20.25. If the PI cycle is concurrently stimulated, IP_3 could augment cytoplasmic Ca^{2+} levels from the intracellular calcium store. The increase in cAMP activates protein kinase (Figure 20.21) whose phosphorylations cause increased hydrolysis of cholesteryl esters from the droplet to free cholesterol and increase cholesterol transport into the mitochondrion. The combination of elevated Ca^{2+} levels and protein phosphorylation bring about induced levels of the side chain cleavage reaction. These combined reactions overcome the rate-limiting steps in steroid biosynthesis and more steroid is produced, which is secreted into the extracellular space and circulated to the target tissues in the bloodstream.

a discussion of P450 enzymes.) These enzymes can be induced as a consequence of hormonal action on the cell eliciting the steroid hormone. The primary signals are usually elevations of cAMP and Ca^{2+}, but inositol trisphosphate as a second messenger may be involved. An overview of steroid hormone biosynthesis and its control is presented in Figure 21.4.

Pathways for the conversion of cholesterol to the adrenal cortical steroid hormones are presented in Figure 21.5. Cholesterol is the major

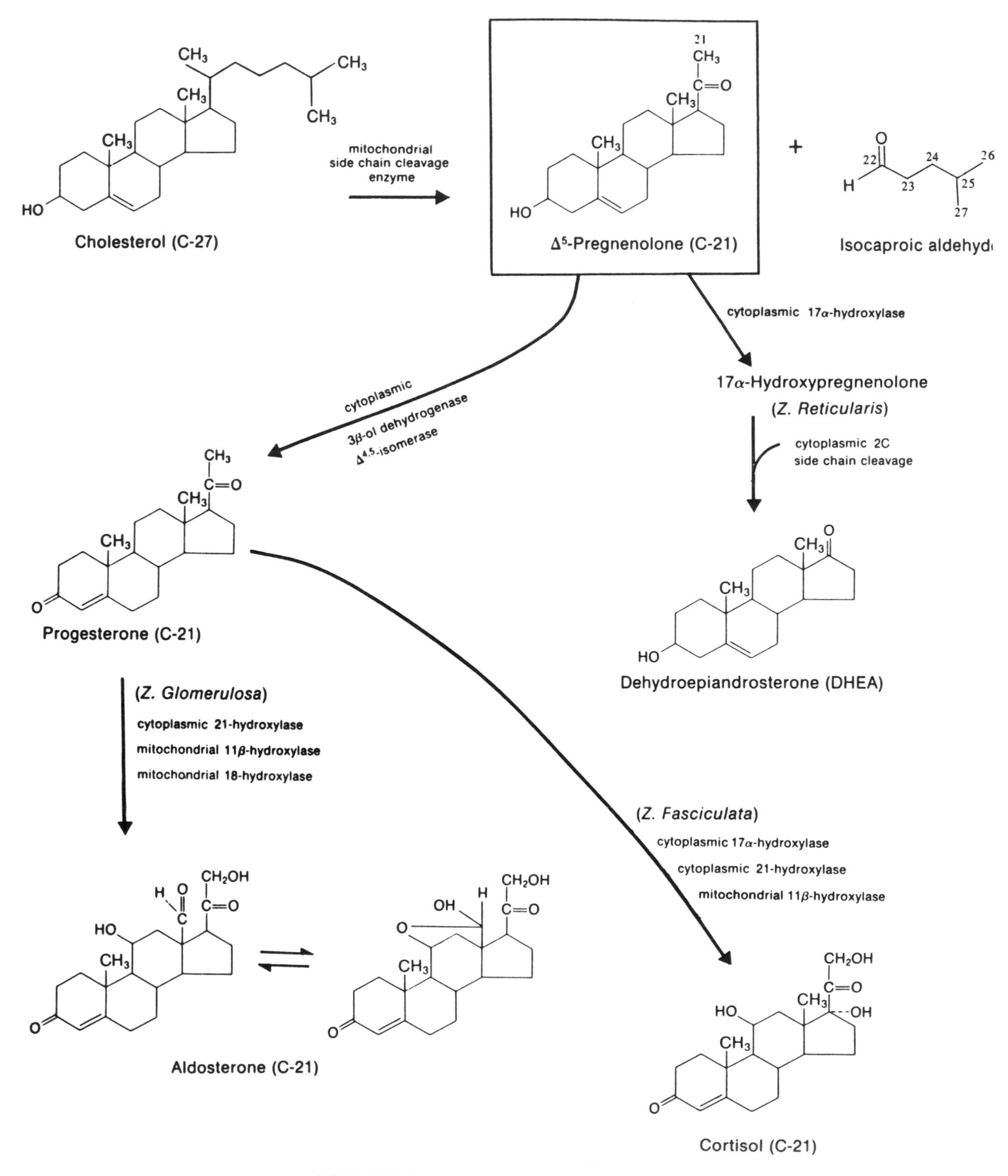

FIGURE 21.5
Conversion of cholesterol to adrenal cortical hormones.

precursor and undergoes side chain cleavage to form Δ^5-pregnenolone releasing a C_6 aldehyde, isocaproaldehyde. **Δ^5-Pregnenolone** is mandatory in the synthesis of all steroid hormones. As shown in Figure 21.5, pregnenolone can be converted directly to progesterone, which requires two cytoplasmic enzymes, **3β-ol dehydrogenase** and **$\Delta^{4,5}$-isomerase.** The dehydrogenase converts the 3-OH group of pregnenolone to a 3-keto group and the isomerase moves the double bond from the B ring to the A ring to produce progesterone. In the *corpus luteum* the bulk of steroid synthesis stops at this point. Progesterone is further converted to aldosterone or cortisol. Conversion to **aldosterone,** which occurs in the adrenal *zona glomerulosa* cells, requires cytoplasmic 21-hydroxylase, and mitochondrial 11β-hydroxylase and 18-hydroxylase. To form cortisol, primarily in adrenal *z. fasciculata* cells, cytoplasmic **17-hydroxylase** and **21-hydroxylase** are required together with mitochondrial **11β-hydroxylase.** The cytoplasmic hydroxylases are all cytochrome P450 linked enzymes of the endoplasmic reticulum (ER) (see Chapter 23). Δ^5-Pregnenolone is also converted to **dehydroepiandrosterone** in the adrenal zona reticularis cells by the action of 17α-hydroxylase of the endoplasmic reticulum to form 17α-hydroxypregnenolone and then by the action of a carbon side chain cleavage system to form dehydroepiandrosterone.

Cholesterol is also converted to the sex hormones by way of Δ^5-pregnenolone (Figure 21.6). Progesterone can be formed as described above and further converted to testosterone by the action of the cytoplasmic enzymes and 17-dehydrogenase. **Testosterone,** so formed, is a major secretory product in the Leydig cell of the testis and undergoes conversion to dihydrotestosterone in androgen target cells before binding to the androgen receptor. This conversion requires the activity of **5α-reductase** located in the ER and nuclear fractions. Pregnenolone can enter an alternative pathway to form dehydroepiandrosterone as described above. This compound can be converted to 17β-estradiol via the aromatase enzyme system and the action of 17-reductase. Also estradiol can be formed from testosterone by the action of the aromatase system. In Figures 21.5 and 21.6 the cellular locations of the requisite enzymes for the synthesis of a given steroidal hormone are denoted.

The hydroxylases of endoplasmic reticulum involved in steroid hormone synthesis are cytochrome P450 enzymes (Chapter 23). Molecular oxygen (O_2) is a substrate with one oxygen atom incorporated into the steroidal substrate (as an OH) and the second atom incorporated into a water molecule. Electrons are generated from NADH or NADPH through a flavoprotein to ferredoxin or similar nonheme protein. Various agents can induce the levels of cytochrome P450.

Another feature, which may be obvious from the discussion of Figures 21.5 and 21.6, is that there is movement of intermediates in and out of the mitochondrial compartment during the synthetic process.

21.4 METABOLIC INACTIVATION OF STEROID HORMONES

A feature of the steroid ring system is its great stability. For the most part, inactivation of steroid hormones involves reduction. Testosterone is initially reduced to a more active form by the enzyme 5α-reductase to form **dihydrotestosterone,** the preferred ligand for the androgen receptor. However, further reduction similar to the other steroid hormones results in inactivation. The inactivation reactions predominate in liver and generally render the steroids more water soluble, as marked by subsequent conjugation with glucuronides or sulfates (see Chapter 23) that are excreted in the urine. Table 21.2 summarizes reactions leading to inactivation and excretory forms of the steroid hormones.

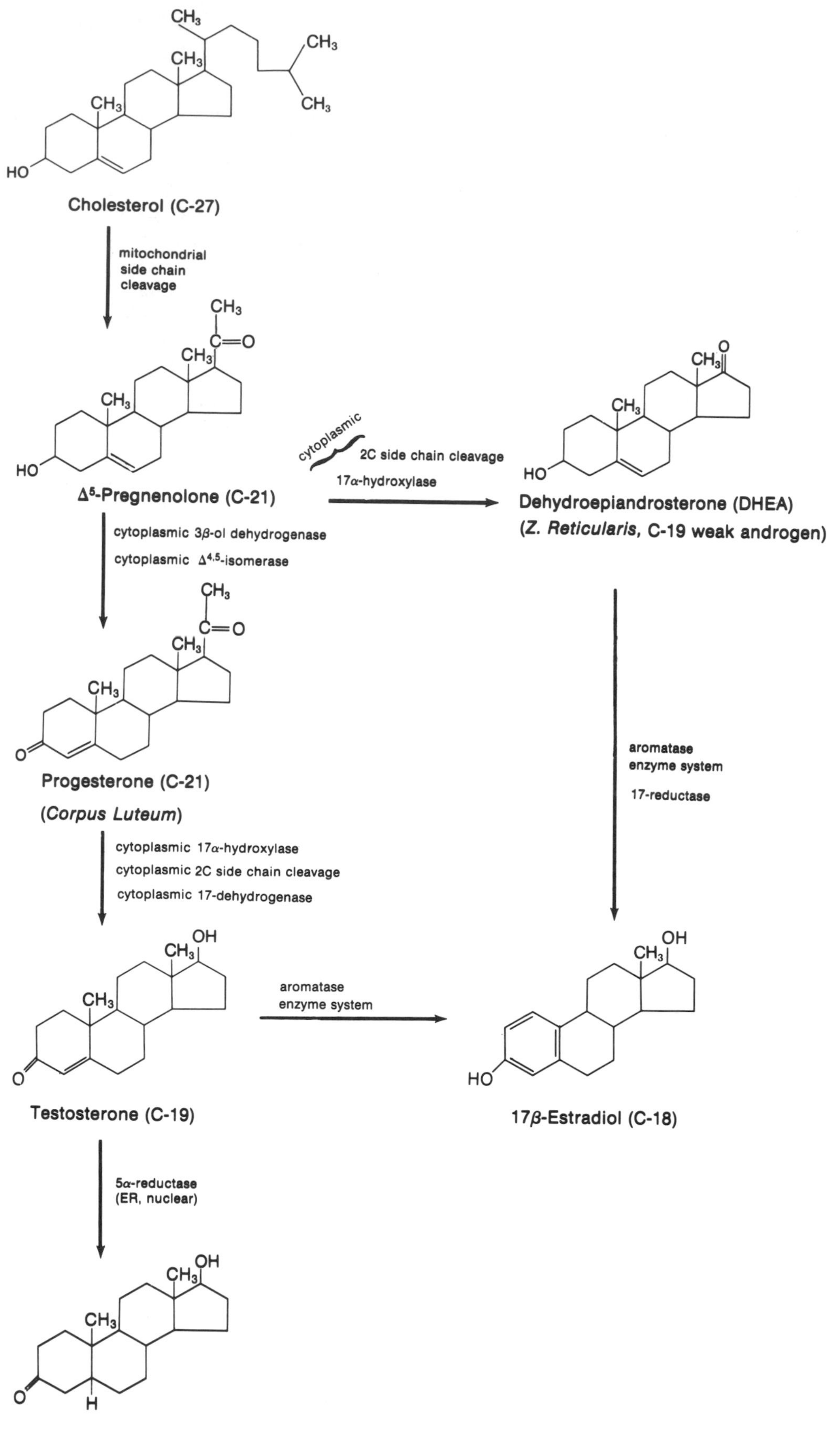

FIGURE 21.6
Conversion of cholesterol to sex hormones.
Mt = mitochondrial, cyto = cytoplasmic.

TABLE 21.2 Excretion Pathways for Steroid Hormones

Steroid class	Starting steroid	Inactivation steps	A:B ring junction	Steroid structure representations of excreted product	Principal conjugate present[a]
Progestins	Progesterone	1. Reduction of C-20 2a. Reduction of 4-ene-3-one or 2b. 3β-steroid dehydrogenase	(cis)	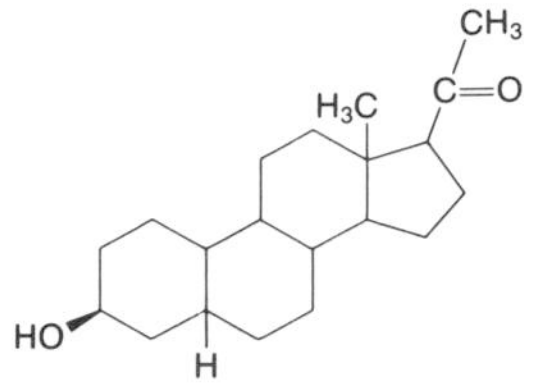 Pregnanodiol (5β-pregnane-3α,20-diol)	G[a]
Estrogens	Estradiol	1. Oxidation of 17β-OH 2. Hydroxylation at C-2 with subsequent methylation 3. Further hydroxylation or ketone formation at a variety of positions, e.g., C-6, C-7, C-14, C-15, C-16, C-18		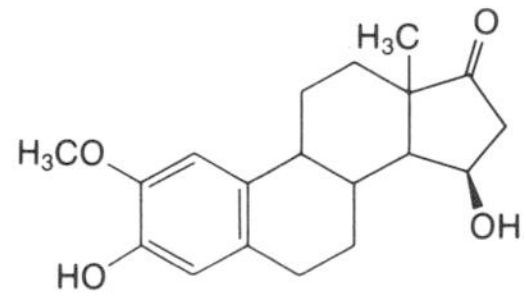 One of many possible compounds	G
Androgens	Testosterone	1. Reduction of 4-ene-3-one 2. Oxidation of C-17 oxo	(cis and trans)	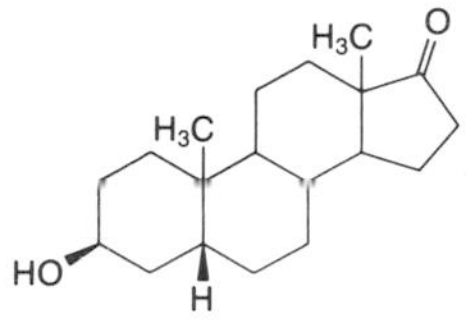 Androsterone 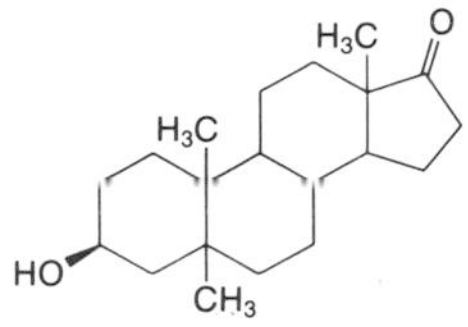 + Etiocholanolone	G, S[a]
Glucocorticoids	Cortisol	1. Reduction of 4-ene-3-one 2. Reduction of 20-oxo group 3. Side chain cleavage	(trans)	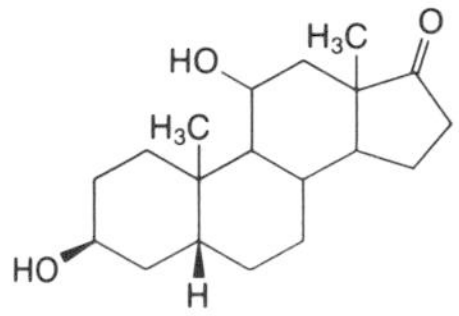 11β-OH-androsterone 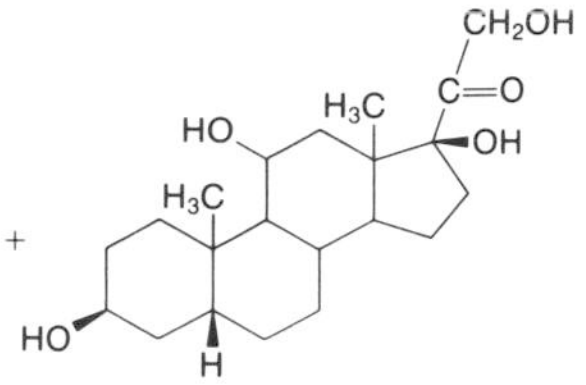 + Allo tetrahydrocortisone	G
Mineralocorticoids	Aldosterone	1. Reduction of 4-ene-3-one	(trans)	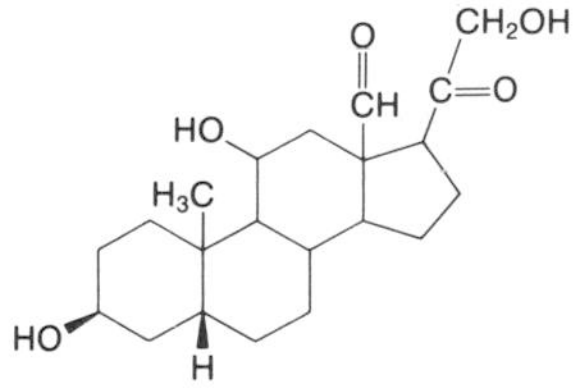 3α,11β,21-$(OH)_3$-20-oxo-5β-pregnane-18-al	G
Vitamin D metabolites	$1,25(OH)_2D_3$	1. Side chain cleavage between C-23 and C-24	—	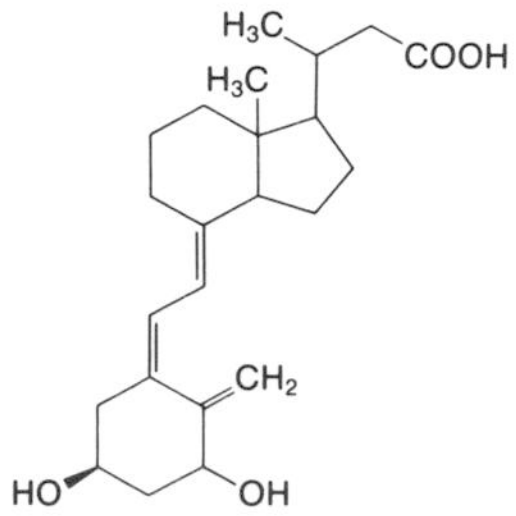 Calcitroic acid	?

[a] G = Glucoronide; S = sulfate.
Reproduced from A. W. Norman and G. Litwack, *Hormones*. Orlando, FL: Academic, 1987.

TABLE 21.3 Hormones that Directly Stimulate Synthesis and Release of Steroid Hormones

Steroid hormone	Steroid producing cell	Signal	Second messenger	Signal system
Cortisol	Adrenal zona fasciculata	ACTH	cAMP, PI cycle, Ca^{2+}	Hypothalamic–pituitary cascade
Aldosterone	Adrenal zona glomerulosa	Angiotensin II	PI cycle, Ca^{2+}	Renin–angiotensin system
Testosterone	Leydig cell	LH	cAMP (PI cycle?)	Hypothalamic–pituitary cascade
17β-Estradiol	Ovarian follicle	FSH	cAMP (PI cycle?)	Hypothalamic-pituitary–ovarian cycle
Progesterone	Corpus luteum	LH	cAMP	Hypothalamic-pituitary–ovarian cycle

ACTH = adrenocorticotropic hormone; LH = luteinizing hormone; FSH = follicle stimulating hormone; PI = phosphatidylinositol

21.5 CELL–CELL COMMUNICATION AND THE CONTROL OF SYNTHESIS AND RELEASE OF STEROID HORMONES

Secretion of steroid hormones from cells where they are synthesized is elicited by other hormones. Many, but not all, such systems are described in Figures 20.2 and 20.3. The hormones that directly stimulate the biosynthesis and secretion of the steroid hormones are summarized in Table 21.3. The signals for stimulation of biosynthesis and secretion of steroid hormones are polypeptide hormones operating through cognate cell membrane receptors. It is not always clear the extent to which the phosphati-

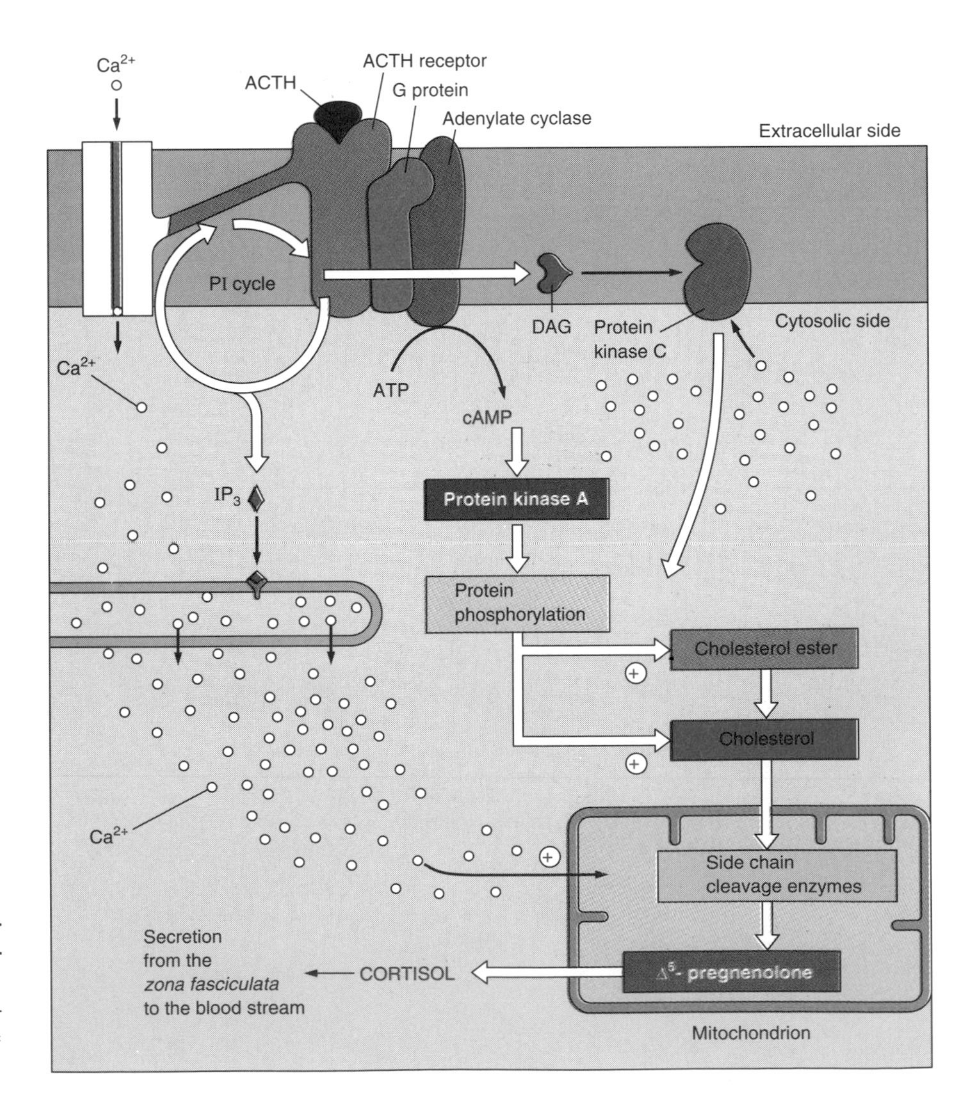

FIGURE 21.7
System for stimulation of cortisol biosynthesis and release initiated by ACTH in the adrenal zona fasciculata.
PI = phosphatidyl inositol; IP_3 = inositol trisphosphate; DAG = diacylglycerol; cAMP = cyclic AMP.

dylinositol (PI) turnover system is the major second messenger compared to cAMP. In many such systems, for example, aldosterone synthesis and secretion, probably several components are involved.

Steroid Hormone Synthesis Is Controlled by Specific Hormones

The general mechanism for hormonal stimulation of steroid hormone synthesis is presented in Figure 21.4. Figure 21.7 presents the system for stimulation of cortisol biosynthesis and release. There are some uncertainties in the pathway in that the role of elevated Ca^{2+} may be larger than presented here. For example, Ca^{2+} could play some role in steroid secretion about which little is known. It is also not clear what is the extent of participation of protein kinase A and protein kinase C. The rate-limiting steps in the biosynthetic process involve the availability of cholesterol from cholesteryl esters in the droplet, the transport of cholesterol to the mitochondrion and the up regulation of the otherwise rate-limiting side chain cleavage reaction.

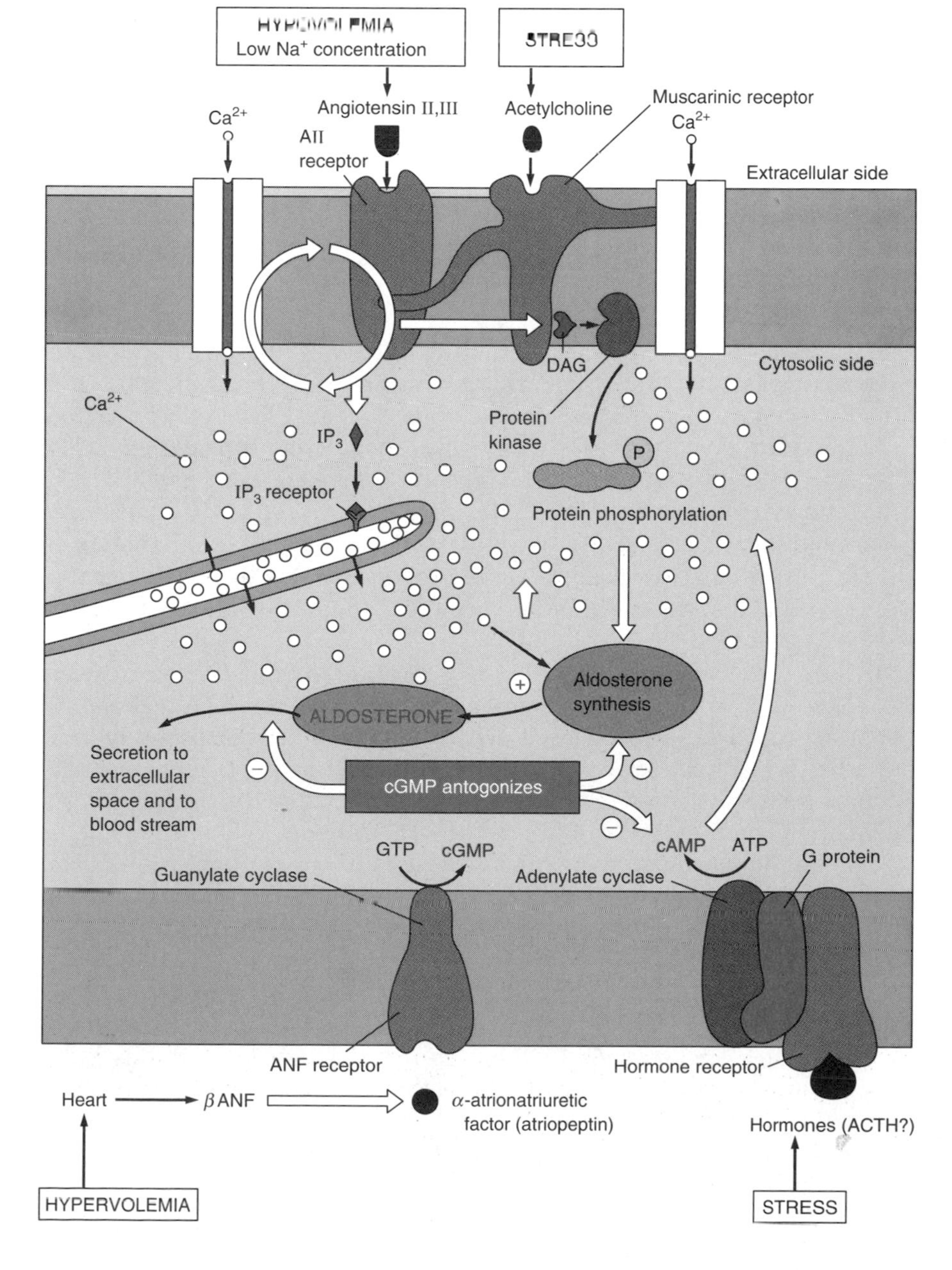

FIGURE 21.8
Reactions leading to the secretion of aldosterone in the adrenal zona glomerulosa cell.
cGMP = cyclic GMP; ANF = factor atrionatriuretic; see Figure 21.7 for additional abbreviations.

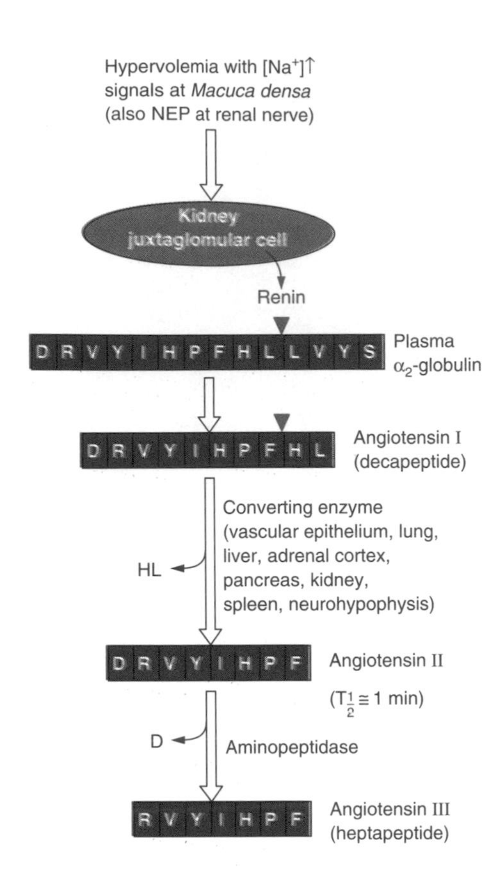

FIGURE 21.9
Renin–angiotensin system.
Amino acid abbreviations are found on p. 30. NEP = norepinephrine.

CLIN. CORR. 21.1
ORAL CONTRACEPTION

The oral contraceptives usually contain an estrogen and a progestin. Taken orally, the levels of these steroids increase in blood to a level where the secretion of the gonadotropins, FSH and LH are repressed. Consequently, the gonadotropic hormone levels in blood fall and there is insufficient FSH to drive the development of the ovarian follicle. As a result the follicle does not mature and ovulation cannot occur. In addition any corpora lutea cannot survive because of the low LH levels. In sum, the ovarian cycle ceases. The uterine endometrium thickens and remains in this state, however, because of the elevated levels of estrogen and progestin. Pills without the steroids (placebos) are usually inserted in the regimen at about the 28th day and, as a result, the blood levels of the steroids fall dramatically and menstruation occurs. When the oral contraceptive steroids are resumed, the blood levels of estrogen and

Aldosterone

Figure 21.8 shows the overall reactions leading to the secretion of aldosterone in the adrenal zona glomerulosa cell. This set of regulatory controls on aldosterone synthesis and secretion is complicated. The main driving force is **angiotensin II** generated from the signaling to the **renin-angiotensin system** shown in Figure 21.9. Essentially, the signal is generated under conditions when blood [Na^+] and blood pressure (blood volume) are required to be increased. The N-terminal decapeptide of circulating α_2 globulin is cleaved by **renin,** a protease. This decapeptide is the hormonally inactive precursor, angiotensin I. It is converted to the octapeptide hormone, angiotensin II, by the action of converting enzyme. Angiotensin II is converted to the heptapeptide, angiotensin III by an aminopeptidase. Both angiotensins II and III can bind to the angiotensin receptor (Figure 21.8), which activates the phosphatidylinositol cycle to generate IP_3 and DAG. The IP_3 stimulates release of calcium ions from the intercellular calcium storage vesicles. In addition, the activity of the Ca^{2+} channel is stimulated by the angiotensin–receptor complex. The K^+ ion is also required to stimulate the Ca^{2+} channel and these events lead to a greatly increased level of cytoplasmic Ca^{2+}. The enhanced cytoplasmic [Ca^{2+}] has a role in aldosterone secretion and together with diacylglycerol stimulates protein kinase C. **Acetylcholine** released through the neuronal stress signals has similar effects mediated by the muscarinic acetylcholine receptor to further reinforce Ca^{2+} uptake by the cell and stimulation of protein kinase C. Enhanced protein kinase C activity leads to protein phosphorylations that stimulate the rate-limiting steps of aldosterone synthesis leading to elevated levels of aldosterone, which are then secreted into the extracellular space and finally into the blood. Once in the blood aldosterone enters the distal kidney cell, binds to its receptor, which may be cytoplasmic, and stimulates expression of mediators that increase the transport of Na^+ from the glomerular filtrate to the blood (see p. 1046).

A set of signals opposite to those that activate the formation of angiotensin generates the **atrionatriuretic factor** (ANF) or atriopeptin from the heart atria (Figure 21.8; see also Figure 20.48). The ANF binds to a specific zona glomerulosa cell membrane receptor and activates guanylate cyclase, which is part of the same receptor polypeptide (Figure 20.47), so that the cytoplasmic level of cGMP increases. Cyclic GMP antagonizes the synthesis and secretion of aldosterone as well as the formation of cAMP by adenylate cyclase; it has been proposed that the involvement of ACTH in aldosterone synthesis and release may involve an adenylate cyclase system.

Aldosterone should be regarded as a stress hormone since its presence in elevated levels in blood occurs as a result of stressful situations. In contrast, cortisol, also released in stress has an additional biorhythmic release (possibly under control of serotonin and vasopressin), which accounts for a substantial reabsorption of Na^+ probably through glucocorticoid stimulation of the Na^+ of the Na^+–H^+ antiport in luminal epithelial cells in addition to the many other activities of cortisol (e.g., antiinflammatory action, control of T-cell growth factors, and synthesis of glycogen and effects on carbohydrate metabolism).

Estradiol

The controls on the formation and secretion of 17β-estradiol, the female sex hormone, are shown in Figure 21.10. During development, a steady-state center and a cycling center arise in the central nervous system. Their functions are required to initiate the ovarian cycle at puberty. These centers must harmonize with the firing of other neurons, such as those producing a clocklike mechanism via release of catecholamines or other amines to generate the pulsatile release of gonadotropic releasing hormone (GnRH), probably at hourly intervals. Details of these reactions are presented on page 876, Chapter 20. The FSH circulates to, and binds its cognate receptor on the cell membrane of the ovarian follicle cell and through its second messengers, primarily cAMP and the activation of cAMP dependent protein kinase, there is a stimulation of the synthesis and secretion of the female sex hormone, 17β-estradiol. At normal stimulated levels of 17β-estradiol there is a negative feedback on the **gonadotrope** (anterior pituitary) suppressing further secretion of FSH. Near ovarian mid-cycle, however, there is a superstimulated level of 17β-estradiol produced that has a positive rather than a negative feedback effect on the gonadotrope. This causes very high levels of LH to be released, referred to as the *LH spike,* and elevated levels of FSH. The level of FSH released under this is substantially lower than LH because the follicle produces **inhibin,** a polypeptide hormone that specifically inhibits FSH release without affecting LH release. The elevation of LH in the LH spike participates in the process of ovulation. After ovulation, the remnant of the follicle is differentiated into the functional corpus luteum, which now synthesizes *progesterone* (and also some estradiol), under the influence of elevated LH levels. Progesterone, however, is a feedback inhibitor of LH synthesis and release (operating through a progesterone receptor in the gonadotropic cell) and eventually the corpus luteum dies owing to a fall in the level of available LH and the production of oxytocin, a luteolytic agent, by the corpus luteum. Prostaglandin $F_{2\alpha}$ may also be involved. With the death of thc corpus luteum, the blood levels of progesterone and estradiol fall, causing menstruation as well as a decline in the negative feedback effects of these steroids on the anterior pituitary and hypothalamus and the cycle begins all over again. Clin. Corr. 21.1 describes how oral contraceptives interrupt this sequence.

The situation is similar in the male with respect to the regulation of gonadotrope secretion but LH acts principally on the Leydig cell for the stimulated production of testosterone and FSH acts on the Sertoli cells to stimulate thc production of inhibin and sperm proteins. The production of testosterone is subject to the negative feedback effect of 17β-estradiol synthesized in the Sertoli cell. The 17β-cstradiol so produced operates through a nuclear estrogen receptor in the Leydig cell to produce inhibition of testosterone synthesis at the transcriptional level. In all cases of steroid hormone production, the synthetic system resembles that shown in Figure 21.4.

progestin increase again and the uterine endometrium thickens. This sequence creates a false "cycling" because of the occurrence of menstruation at the expected time in the cycle. The mechanism by which the ovarian cycle and ovulation is suppressed by the oral contraceptive is based on the negative feedback effects of estrogen and progestin on the secretion of the anterior pituitary gonadotropes. It is also possible to provide contraception by implanting in the skin silicone tubes containing progestins. The steroid is slowly released, providing contraception for up to 3 to 5 years.

Zatuchni, G. I. Female contraception, Principles and Practice of Endocrinology and Metabolism, K. L. Becker, ed., J. B. Lippincott, NY 1990, p. 861; and Shoupe, D. and Mishell, D. R. Norplant: Subdermal implant system for long term contraception. Am. J. Obstet. Gyn. 160:1286, 1988.

Vitamin D_3

Upon activation of vitamin D to vitamin D_3, it becomes a hormone and has the general features of a steroid hormone. The active form of vitamin D stimulates the intestinal absorption of dietary calcium and phosphorous, the mineralization of bone matrix, bone resorption, and reabsorption of calcium and phosphate in the renal tubule. The **vitamin D endocrine system** is diagrammed in Figure 21.11. **7-Dehydrocholesterol** is activated in the skin by sunlight to form vitamin D_3. This form is subsequently hydroxylated in the liver and kidney to form the **1α,25-dihydroxy vitamin D_3,** a steroid hormone. The hormone can bind to nuclear vitamin D_3 receptors in intestine, bone, and kidney to transcriptionally activate genes encoding calcium binding proteins whose actions lead to absorption and reabsorption of Ca^{2+} and phosphorus. The subcellular mode of action is presented in Figure 21.12. In this scheme the active form of vitamin D_3 enters the intestinal cell from the blood side and migrates to the nucleus. Once inside it binds to the high-affinity vitamin D_3 receptor, probably undergoes an activation event, and associates with a vitamin D_3 respon-

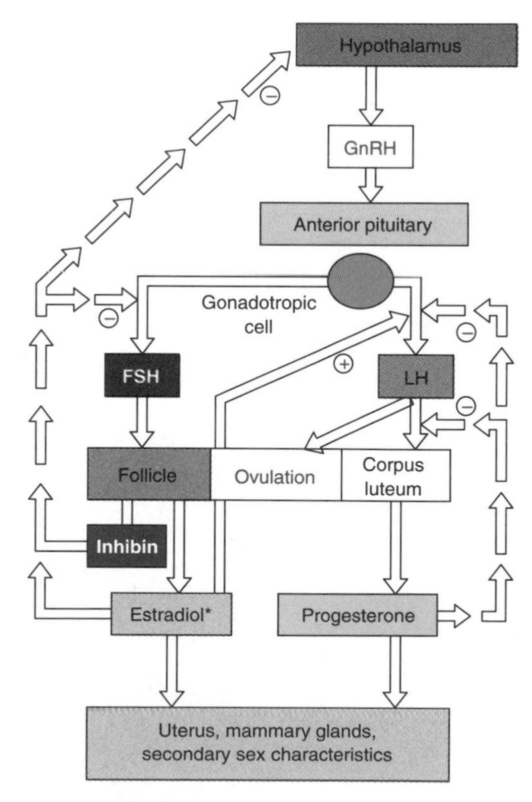

FIGURE 21.10
Formation and secretion of 17β-estradiol and progesterone.

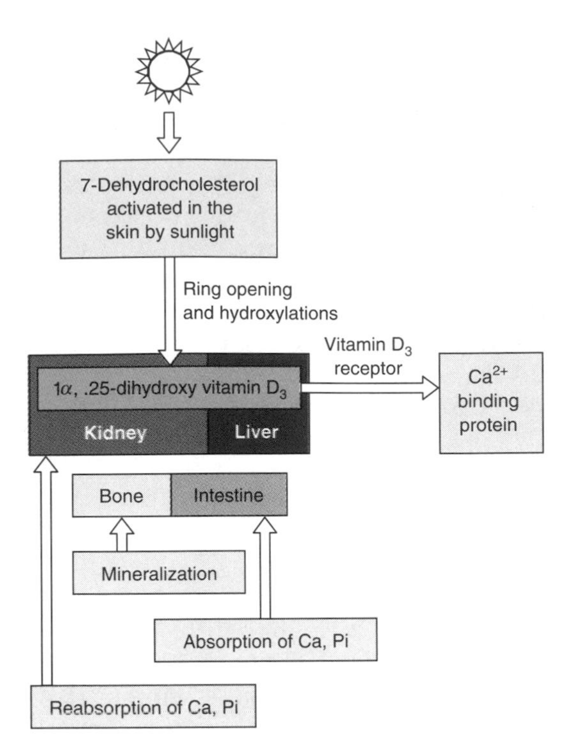

FIGURE 21.11
The vitamin D endocrine system.
P_i = inorganic phosphate.
Adapted from A. W. Norman and G. Litwack, *Hormones*. Orlando, FL: Academic, 1987, p. 379.

sive element to activate genes responsive to the hormone. Messenger RNA is produced and translated in the cytoplasm; these RNAs encode calcium binding proteins, Ca^{2+}-ATPase, other ATPases, membrane components, and facilitators of vesicle formation. Increased levels of calcium binding proteins cause increased uptake of Ca^{2+} from the intestine.

With each of the steroid-producing systems discussed, feedback controls are operative whereby sufficient amounts of the circulating steroid

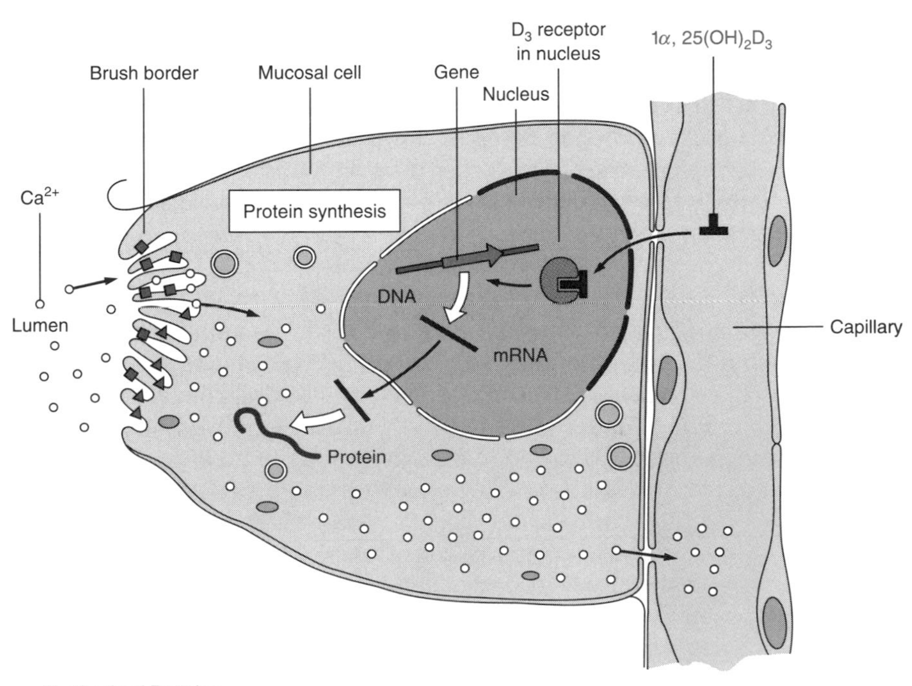

FIGURE 21.12
Schematic model to describe the action of 1,25-$(OH)_2D_3$ in the intestine in stimulating intestinal calcium transport.
Redrawn from I. Nemere and A. W. Norman, *Biochim. Biophys. Acta*. 694:307, 1982.

hormone inhibit the further production and release of intermediate hormones in the pathway at the levels of the pituitary and hypothalamus, as viewed in Figure 20.3. In the case of the vitamin D systems the controls are different since the steroid production is not stimulated by the cascade process applicable to estradiol. The same is true for the aldosterone system where the originating signal (low [Na^+] or hypovolemia) is neutralized by the action of aldosterone (elevated [Na^+] and elevated blood volume). Thus, as with all biological homeostatic systems, there are close controls that maintain the equilibrium environment.

21.6 TRANSPORT OF STEROID HORMONES IN BLOOD

Steroid Hormones Are Bound to Specific Proteins or Albumin in Blood

There are four major proteins in the circulation that account for much of the steroid hormones bound in the blood. These bound forms apparently serve several purposes. They assist in maintaining a level of these hormones in the circulation. In addition, by being bound the hormone is protected from metabolism and inactivation. The binding proteins of importance are corticosteroid binding globulin protein, sex hormone binding protein, androgen binding protein, and albumin.

The **corticosteroid binding globulin (CBG)** or **transcortin** is about 52 kDa, is 3–4 mg% in human plasma, and binds about 80% of the total 17-hydroxysteroids in the blood. In the case of cortisol, which is the principal antistress corticosteroid in humans, about 75% is bound by CBG, 22% is bound in a loose manner to albumin, and 8% is in free form. The unbound cortisol is the form that can permeate cells and bind to intracellular receptors to produce biological effects. The CBG has a high affinity for cortisol with a binding constant (K_a of $2.4 \times 10^7 M^{-1}$). Critical structural determinants for steroid binding to CBG are the Δ^4-3-ketone and 20-ketone structures. Aldosterone binds weakly to CBG but is also bound by albumin and other plasma proteins. Normally 60% of aldosterone is bound to albumin and 10% is bound to CBG. In human serum, albumin is 1000-fold the concentration of CBG, and binds cortisol with an affinity of $10^3 M^{-1}$ compared to the affinity of CBG for cortisol of $2.4 \times 10^7 M^{-1}$. Thus cortisol will always fill CBG binding sites first. During stress when secretion of cortisol is very high, CBG sites will be filled but there will always be sufficient albumin to accommodate excess cortisol.

Sex hormone binding globulin (SHBG) is about 40 kDa and binds androgens with an affinity constant of about $10^9 M^{-1}$ which is much tighter than albumin binding of androgens. One to three percent of testosterone is unbound in the circulation and 10% is bound to SHBG with the remainder bound to albumin. The level of SHBG is probably important in controlling the balance between circulating androgens and estrogens along with the actual amounts of these hormones produced in given situations. About 97–99% of testosterone in blood is bound reversibly to SHBG but much less estrogen is bound to this protein in the female. As mentioned above, only the unbound steroid hormone can permeate cells and bind to intracellular receptors, thus expressing its activity. The level of SHBG before puberty is about the same in males and females, but at puberty, when the functioning of the sex hormones becomes important, there is a small decrease in the level of circulating SHBG in females and a larger decrease in males, insuring a relatively greater amount of the unbound, biologically active sex hormones, testosterone and 17β-estradiol. In adults, males have about one-half as much circulating SHBG as females, so that the unbound testosterone in males is about 20 times greater than in females. In addition, the total concentration of testosterone is about 40 times greater in males. Testosterone itself lowers SHBG levels in blood, whereas 17β-estradiol raises SHBG levels in blood. These effects have important ramifications in pregnancy and in other conditions.

Androgen binding protein (ABP) is produced by the Sertoli cells in response to testosterone and FSH both of which stimulate protein synthesis in these cells. Androgen binding protein is doubtless not of great importance in the entire blood circulation but is vital for the region of the testes and Sertoli cells where it maintains a ready supply of testosterone for the production of protein constituents of spermatozoa. Its role may be to maintain a high local concentration of testosterone in the vicinity of the developing germ cells within the tubules.

From a variety of studies it is clear that these transport proteins and perhaps others, protect the circulating pool of steroid hormones and avail free steroids that can enter cellular targets by dissociation of increasing amounts of the bound forms as the free forms are utilized, thus serving the needs of target cells by a mass action effect.

21.7 STEROID HORMONE RECEPTORS

Steroid Hormones Are Bound to Specific Intracellular Protein Receptors

A general model for steroid hormone action is shown in Figure 21.13. This model takes into account the differences among steroid receptors in terms

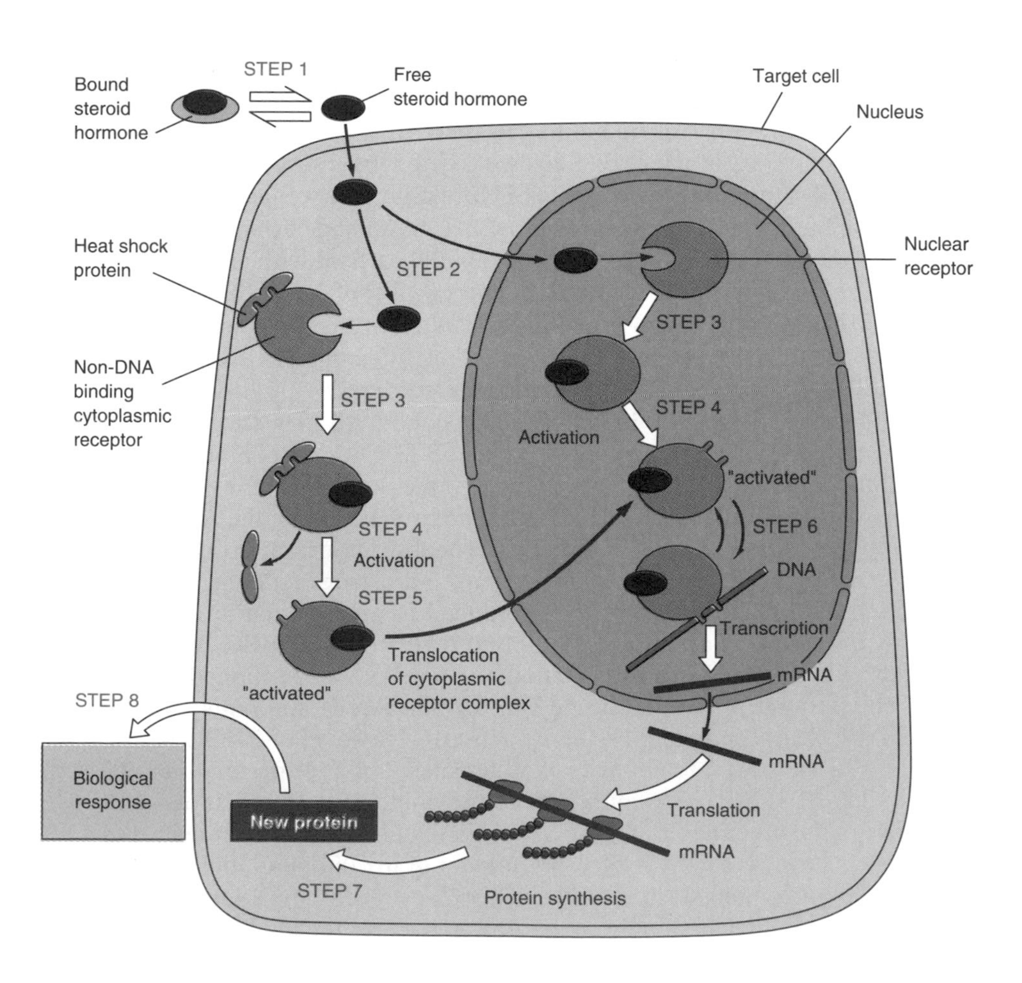

FIGURE 21.13
Model of steroid hormone action.

of their location within the cell. As mentioned in Chapter 20, in contrast to polypeptide hormone receptors that are generally located on/in the cell surface, steroid hormone related receptors are located in the cell interior. Among the steroid receptors there appear to be some differences as to the subcellular location of the **non-DNA binding forms** of these **receptors.** The glucocorticoid receptor and possibly the aldosterone receptor appear to reside in the cytoplasm, whereas the other receptors, for which suitable data have been collected, may be located within the nucleus, presumably in association with DNA, although not necessarily, at productive acceptor sites on the DNA. Starting at the top of Figure 21.13 with Step 1 is shown a bound and a free form of a steroid hormone(s). The free form may enter the cell by a process of free diffusion. In the case of glucocorticoids, like cortisol, the steroid would bind to an unactivated–nontransformed receptor with an open ligand binding site (Step 3). The binding constant for this reaction is of the order of 10^9 M^{-1} compared to about 10^7 M^{-1} for the binding to CBG (see above). The non-DNA binding form also referred to as the unactivated or nontransformed receptor is about 300 kDa because other proteins have been claimed to be associated in the complex. Many investigators believe that a dimer of the 90-kDa protein formed when cells are stressed **(heat shock proteins)** is associated with the receptor in this form and occludes its DNA binding domain, accounting for its non-DNA binding activity. The dimer of heat shock 90-kDa protein is depicted by the pair of darkened ovals attached to the cytoplasmic receptor that block the DNA binding domain pictured as a pair of "fingers" in the subsequently activated form. Activation or transformation to the **DNA binding form** is accomplished by release of the heat shock 90-kDa proteins (Step 4). It is not clear what actually drives the activation step(s). Clearly, the binding of the steroidal ligand is important but there may be an involvement of other factors. A low molecular weight component has been proposed to be part of the cross-linking between the nonhomologous proteins and the receptor in the DNA binding complex. The unactivated complex, however, has yet to be reassembled from purified components. In the case of the glucocorticoid receptor, only the non-DNA binding form has a high affinity for binding the steroidal ligand. Following activation and exposure of the DNA binding domain the receptor translocates to the nucleus, possibly through the nucleopore (Step 5), binds to DNA and "searches" the DNA for a high-affinity acceptor site. At this site the bound receptor complex acts as a transactivation factor, which together with other transactivators, allows for the starting of RNA polymerase and the stimulation of transcription. In some cases the binding of the receptor may lead to repression of transcription and this effect is less well understood. New mRNAs are translocated to the cytoplasm and assembled into translation complexes for the synthesis of proteins (Step 7) that alter metabolism and functioning of the cell (Step 8).

When the unoccupied (nonliganded) steroid hormone receptor is located in the nucleus, as may be the case with the estradiol receptor, the progesterone receptor, the androgen receptor, and the vitamin D_3 receptor (see Figure 21.12), the steroid must travel through a cytoplasm and cross the perinuclear membrane. It is not clear whether this transport through the cytoplasm (aqueous environment) requires a transport protein for the hydrophobic steroid molecules. Once inside the nucleus the steroid can bind to the high-affinity, unoccupied receptor, presumably already on DNA, and cause it to be "activated" to a form bound to the acceptor site. The ligand might promote a conformation that decreases the off-rate of the receptor from its acceptor, if it is located on or near its acceptor site, or might cause the receptor to initiate searching if the unoccupied receptor is on DNA at a locus remote from the acceptor site.

TABLE 21.4 Steroid Hormone Receptor Responsive DNA Elements—Consensus Acceptor Sites

Element	DNA sequence[a]
POSITIVE	
Glucocorticoid responsive element (GRE)	5′-GGTACAnnnTGTTCT-3′ (applies to GRE, MRE, PRE, ARE)
Mineralocorticoid responsive element (MRE)	
Progesterone responsive element (PRE)	
Androgen responsive element (ARE)	
Estrogen responsive element (ERE)	5′-AGGTCAnnnTCACT-3′
NEGATIVE	
Glucocorticoid responsive element	5′-ATYACNnnnTGATCW-3′

[a] n = any nucleotide; Y = a purine; W = a pyrimidine
These data are summarized from work of M. Beato. *Cell* 56:355, 1989.

Consequently, activation (Step 4 in nucleus) is less clear than the mechanism in the cytoplasm. After binding of the activated receptor complex to the DNA acceptor site, the same steps as described above lead to enhancement or repression of transcription, as the case may be. The evidence regarding the cellular location of the aldosterone receptor is not yet convincing and it may be cytoplasmic like the glucocorticoid receptor.

Consensus DNA sequences defining specific acceptor sites for the binding of various activated steroid hormone-receptor complexes are now known and are summarized in Table 21.4. Receptors for glucocorticoids, mineralocorticoids, progesterone, and androgen can all bind to the same **responsive element** on the DNA. Thus, in a given cell type the extent and type of receptor expressed will determine the hormone sensitivity. For example, it is known that the sex hormone receptors are expressed in only a few cell types and the progesterone receptor is likewise restricted to certain cells, whereas the glucocorticoid receptor is expressed in a large number of cell types. In cases where aldosterone and cortisol receptors are expressed, only one form may predominate depending on the cell type and aldosterone is produced mainly in response to stress, whereas cortisol is produced in nonstress conditions in addition to stress. The estrogen receptor has a response element distinct from the others, in particular the androgen receptor. Recently, it has been appreciated that there is a somewhat distinct element for the glucocorticoid receptor that results in down regulation of the effected gene product.

Thus, the changes produced in different cells by the activation of steroid hormone receptors may be different in different cells that contain the relevant receptor in suitable concentration. The whole process is triggered by the entry of the steroidal ligand in amounts that supercede the dissociation constant of the receptor. The different phenotypic changes in different cell types in response to a specific hormone then summate to give the systemic or organismic response to the hormone.

Some Steroid Receptors Are Part of the c*Erb*A Family; c*Erb*A Is a Protooncogene

As a model of steroid receptor structure, the **functional domains** of the glucocorticoid receptor are presented in Figure 21.14. The protein is conveniently divided into three major domains. Starting at the C terminus the steroid binding domain is indicated and has 30–60% homologies with the

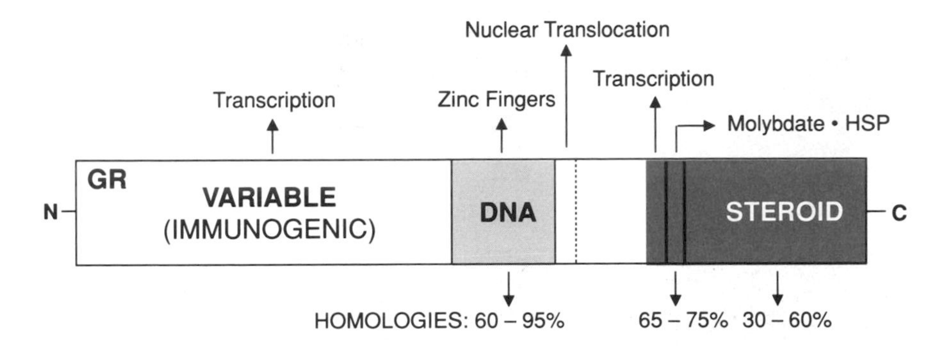

FIGURE 21.14
Model of a typical steroid hormone receptor.
The results are derived from studies on cDNA in various laboratories, especially those of R. Evans and K. Yamamoto.

ligand binding domains of other receptors in the steroid receptor family. The more alike two steroids that bind different receptors are, the greater the extent of homology to be anticipated in this domain. The steroid binding domain contains a sequence that may be involved in the binding of molybdate and a dimer of the 90-kDa heat shock protein whose function would theoretically result in the assembly of the high molecular weight unactivated–nontransformed steroid–receptor complex. To the left of that domain is a region that modifies transcription. In the center of the molecule is the **DNA binding domain.** Among the steroid receptors there is 60–95% homology in this domain. There are two zinc fingers (Chapter 3) that interact in the complex with DNA. The structure of the zinc finger DNA binding motif is shown in Figure 21.15. Further to the left in Figure 21.14 is the N-terminal domain, which contains the principal **antigenic domains** and a site that modulates transcriptional activation. The amino acid sequences in this site are highly variable among the steroid receptors. These are features that are common to the other steroid receptors. The family of steroid receptors is diagrammed in Figure 21.16. The ancestor to which these receptor genes are related is **v-*erb*A** or **c-*erb*A** (see Chapter 20, p. 797). v-*Erb*A is an oncogene that binds to DNA but has no ligand binding domain. In some cases the DNA binding domains are homologous enough that more than one receptor will bind to a common responsive element (consensus sequence on DNA) as shown in Table 21.4. In addition to those genes pictured in Figure 21.16 the **aryl hydrocarbon receptor** (Ah) may also be a member of this family. The Ah receptor binds carcinogens with increasing affinity paralleling increasing carcinogenic potency

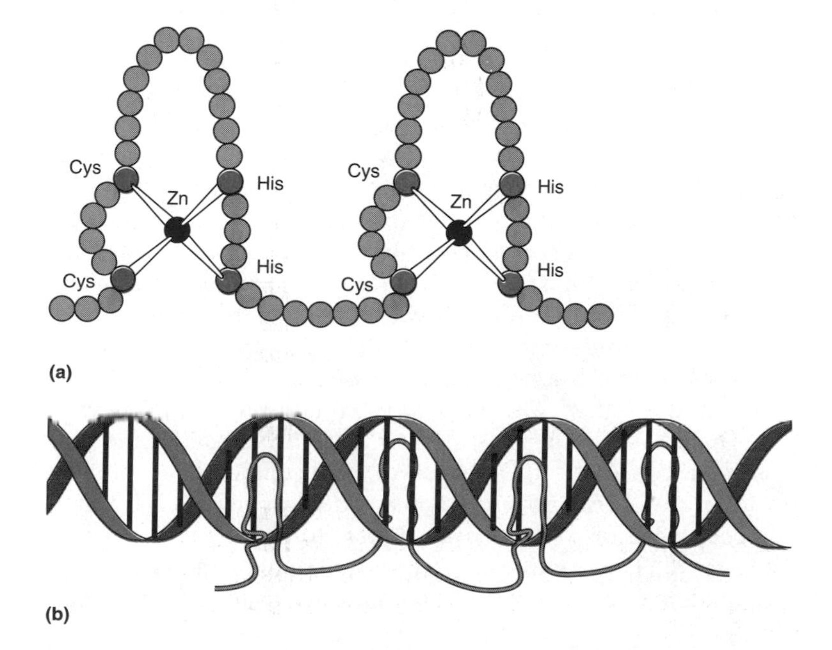

FIGURE 21.15
The "zinc finger" family of sequence-specific DNA-binding proteins.
A highly schematic model for the general conformation of the DNA binding domain is shown in (*a*), with each amino acid represented by a sphere; the polypeptide chain in each finger is actually thought to be folded into a complex globular conformation. (*b*) Schematic view of how four such zinc fingers might bind to a specific DNA sequence. Each zinc finger is postulated to recognize a specific sequence of about five nucleotide pairs.
Adapted from A. Dlug and D. Rhodes, *Trends Biochem. Sci.* 12:464, 1987.

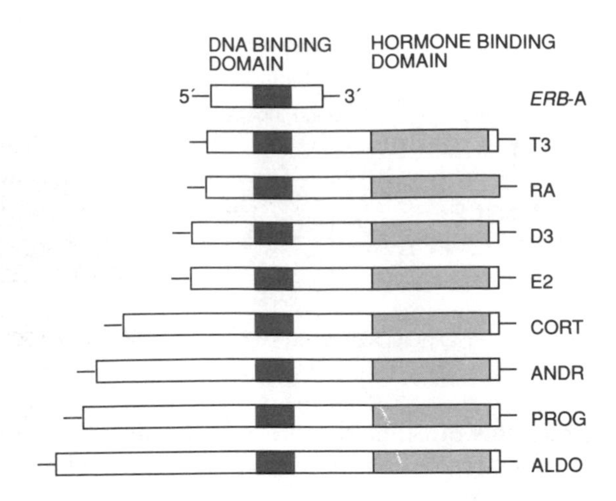

FIGURE 21.16
Steroid receptor gene superfamily.
T3 = triiodothyronine; RA = retinoic acid; D3 = dihydroxyvitamin D3; E2 = estradiol; CORT = cortisol; ANDR = androgen; PROG = progesterone; ALDO = aldosterone. This figure shows roughly the relative sizes of the genes for these receptors. The information derives from the laboratories of R. Evans, K. Yamamoto, P. Chambon, and others. In some cases there is high homology in DNA binding domains and lower homologies in ligand binding domains.

and either translocates the carcinogen to the cellular nucleus or the receptor is already located in the nucleus. The N-terminal portion of the receptors usually contains major antigenic sites and may also contain a site that is active in modulating binding of the receptor to DNA.

It is interesting that the thyroid hormone receptors and the retinoic acid receptors are also members of this class of hormones, since the structures of their ligands are not steroidal. They do, however, contain six-membered rings as shown in Figure 21.17. Clearly, for some of the steroid receptors the A ring of the steroidal ligand is the prominent site of recognition by the receptor, presenting the likelihood that the A ring inserts into the binding pocket of the receptor. In some cases derivatives with a six-membered ring have been shown to bind to the estradiol and glucocorticoid receptors. Thus, it may be that the ring structures of thyroid hormone and retinoic acid have structural similarities not unlike many of the steroidal ligands involved in binding.

The receptors in this large gene family may act as transcriptional activators that together with other transcriptional regulators bring about gene activation at the promotor and cause the binding of RNA polymerase to its start site.

21.8 RECEPTOR ACTIVATION AND UP- AND DOWN-REGULATION

Little is known about activation or transformation of steroid receptors. Activation or transformation involves the reactions that convert a non-DNA binding form (unactivated–nontransformed) of the receptor to a form (activated–transformed) that is able to bind nonspecific DNA, or specific DNA (hormone responsive element) to nuclei. The likelihood that certain receptors are cytoplasmic (glucocorticoid receptor and possibly the mineralocorticoid receptor) while others seem to be nuclear (progesterone, estradiol, vitamin D_3, and androgen receptors) may have a bearing on the significance of the activation phenomena. Most information is available for cytoplasmic receptors. The current view is that the non-DNA binding form is a heteromeric trimer consisting of one molecule of receptor and a dimer of 90-kDa heat shock protein as shown in Figure 21.18. The DNA binding site of the receptor is blocked by the heteromeric proteins or by some other factor or by a combination of both. Upon activation–transformation a stepwise *disaggregation* of this complex could occur leading to the activated receptor having its DNA binding site fully exposed. The reaction may be started by the binding of steroid to the ligand binding site that produces a conformational change in the receptor

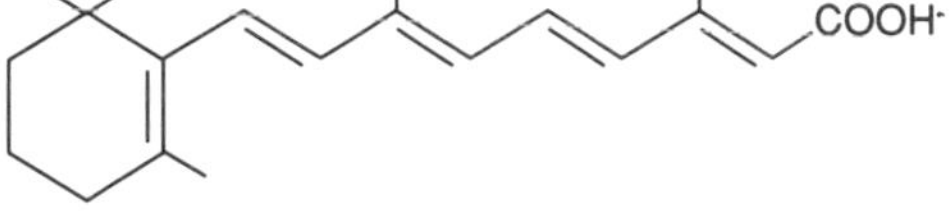

COOH

Retinoic acid
(vitamin A acid)

HO — O — $CH_2CHCOOH$ — NH_2 — I

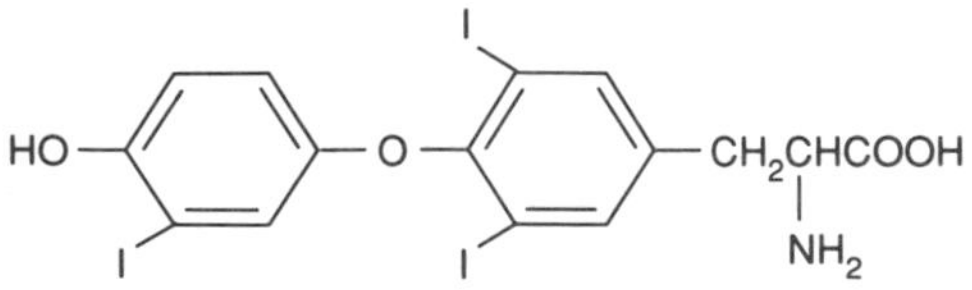

3,5,3′ - Triiodothyronine

FIGURE 21.17
Structures of retinoic acid (Vitamin A acid) and 3, 5, 3′-Triiodothyronine.

protein. Something happens in the cell to cause the other factors and hsp 90 proteins to dissociate from the receptor.

Although the conditions to bring about activation in vitro are well known, the primary signal within the cell is not. Many believe that the binding of ligand is not sufficient to cause the activation process. Once the liberated receptor is free in the cytoplasm, it crosses the perinuclear membrane, perhaps through the nucleopore, to gain access to the nucleus. It binds nonspecifically and specifically to chromatin probably as a dimer, presumably in search of the specific response element (Table 21.4). Thus, these receptors are transactivating factors and are presumed to act in concert with other transactivating factor to provide the appropriate structure to allow RNA polymerase II to bind to its start site and initiate the process of transcription. Most steroid receptors have in their DNA binding domains an SV-40 like sequence known to code for nuclear translocation. The classical signal is Pro-Lys-Lys-Lys-Arg-Lys-Val. Steroid receptors have variants of this signal sequence as some degeneracy is permitted but there is probably a specific lysine residue that cannot be altered. This signal may provide recognition for the nucleopore, although sparse information is available on this point.

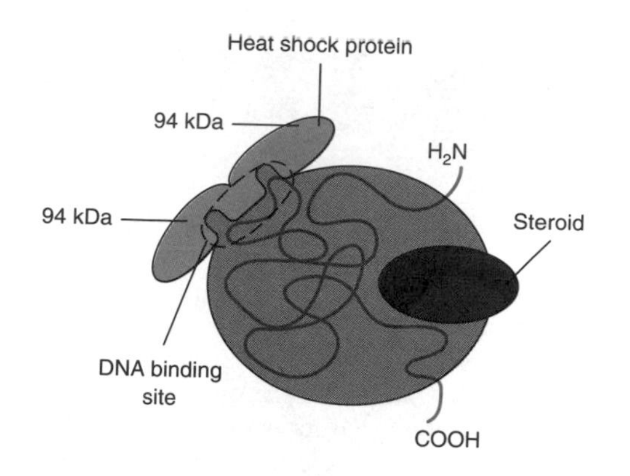

FIGURE 21.18
Hypothetical minimal model of a non-DNA binding form of a steroid receptor.
This form of the receptor cannot bind to DNA because the DNA binding site is blocked by the 94 kDa hsp proteins or by some other constituent. Molecular weight of this complex is approximately 300 kDa.

Steroid Receptors Can Be Up- and Down-Regulated Depending on Exposure to the Hormone

In general, many receptors, whether on the cell membrane or located in the cell interior are down regulated when the cell has been exposed to a certain amount of the hormonal ligand. In some cases, the ensuing down regulation is called *"desensitization."* **Down-regulation** can take many forms. For cell membrane receptors the mechanism may be internalization by endocytosis of the receptors on the surface of the cell after exposure to hormone (see Chapter 20). Internalization reduces the number of receptors available on the cell surface and renders the cell less responsive to hormone; that is, desensitizes the cell. In the case of intracellular steroid receptors, down-regulation generally takes the form of reducing the level of receptor mRNA, which causes a fall in the number of receptor molecules synthesized. This means that the receptor gene may have on its promotor a specific responsive element whose action results in an inhibition of transcription of receptor mRNA or that the receptor may stimulate the mRNA for a protein that degrades the mRNA of the receptor. In reference to the first possibility, sequences are now being recognized on receptor gene promotors that may result in inhibition of transcription (Table 21.4).

In at least one type of reaction, **up-regulation** of a steroid receptor is followed by exposure of a cell to hormone. This occurs in T cells undergoing "programmed cell death" in response to glucocorticoids. It is possible that the up-regulation of receptor mRNA in response to hormone plays an important role in the suicide reaction of the cell, as discussed in more detail below.

21.9 SPECIFIC EXAMPLES OF STEROID HORMONE ACTION AT THE CELLULAR LEVEL

Programmed Death of Specific Cells Is Controlled by Some Steroid Hormones

Programmed cell death is a suicide process by which cells die according to a program that may be beneficial for the organism. Programmed cell death

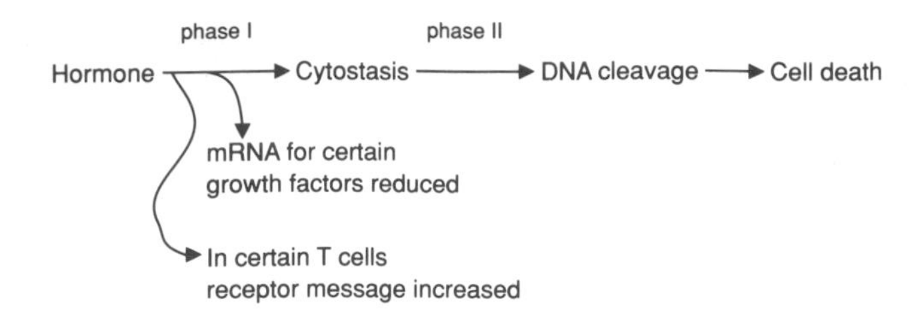

FIGURE 21.19
Two phases of glucocorticoid-initiated programmed cell death.

can result from the fall or the rise in the level of a hormone(s). The death of uterine endometrial cells at the beginning of menstruation is an example and in this case programmed cell death is initiated by the *fall* in the levels of progesterone and estradiol in the blood (see Clin. Corr. 21.2). Another case is the killing of thymus cells during development when the adrenal cortex becomes functional and begins to synthesize and secrete relatively large amounts of cortisol. A newborn has a large thymus, covering an extensive area of the chest. When cortisol is synthesized and released the thymus cortical cells begin to be killed by the hormone until a resistant core of cells is reached and the gland achieves the size of the adult. Thus, programmed cell death is a mechanism used in development for the maturation of certain organs as well as in cyclic systems where cells are proliferated and then the proliferation is regressed until another cycle is initiated to begin the proliferation all over again as is the case with the ovarian cycle.

In the case of glucocorticoid-initiated programmed cell death that occurs in the thymus and also in a variety of T cells, the process is mediated by the glucocorticoid receptor. There appear to be two phases of this complex process that culminate in the cleavage of cellular DNA and finally death of the cell as shown in Figure 21.19.

Cytostasis, or inhibition of cell proliferation, may be distinguished as a separate phase from DNA cleavage and cell death because certain cell mutants respond to hormone through only phase I, cessation of cell growth, and do not proceed further in the mechanism to DNA cleavage and cell death. DNA cleavage may occur through conformational changes in chromatin resulting from the action of the hormone and exposing the linker DNA to attack by an endonuclease. In thymocytes, which are also killed by glucocorticoids, the endoclease responsible for **internucleosomal DNA cleavage** is Ca^{2+} dependent, whereas in T cells the endonuclease is Ca^{2+} independent. Furthermore, glucocorticoid receptor message is induced in certain T cells but down regulated in thymocytes and one important growth factor, c-*myc,* is repressed in its messenger level in certain T cells but is increased in thymocytes. Since both cell types demonstrate internucleosomal DNA cleavage and cell death, but differ in enzymes and message responses, there may be two distinct pathways leading to the same end results. The future will bring more work on this fascinating phenomenon.

CLIN. CORR. 21.2 PROGRAMMED CELL DEATH IN THE OVARIAN CYCLE

During the ovarian cycle, the ovarian follicle expels the mature ovum at day 14 and the residue of the follicle is differentiated into a functional corpus luteum. The corpus luteum produces some estradiol to partially replace that provided earlier by the maturing follicle. But its principal product is progesterone. Estradiol and progesterone are the main stimulators of uterine endometrial wall thickening in preparation for implantation. One of the products of estradiol action in the endometrium is the progesterone receptor, so that the uterine endometrial cells become exquisitely sensitive to estradiol and progesterone. The corpus luteum supplies the latter, but in the absence of fertilization and development of an embryo, the corpus luteum lives only for a short while and then dies because of lack of LH or chorionic gonadotropin, a hormone produced by the early embryo. The production of oxytocin and PGF_{2a} in the ovary may bring about the destruction of the corpus luteum (luteolysis). The blood levels of estradiol and progesterone fall dramatically after luteolysis and the stimulators of uterine endometrial cells disappear, causing the breakdown of this thickened, vascularized layer of tissue and precipitating menstruation. These cells die by programmed cell death due to the withdrawal of steroids. The hallmark of programmed cell death is internucleosomal cleavage of DNA. Thus, programmed cell death appears to play specific roles in development and in tissue cycling either due to a specific hormonal stimulus or to withdrawal of hormone(s) as in the case described here.

Erickson, G. F. and Schreiber, J. R. Morphology and physiology of the ovary. Principles and Practice of Endocrinology and Metabolism, K. L. Becker, ed., J. B. Lippincott, NY 1990, p. 776; Rebar, R. W., Kenigsberg, D. and Hogden, G. D. The normal menstrual cycle and the control of ovulation. Principles and Practice of Endocrinology and Metabolism, K. Becker, ed., J. B. Lippincott, NY 1990, p. 788; and Hamburger, L., Hahlin, M., Hillensjo, T., Johanson, C., and Sjogren, A. Luteotropic and luteolytic factors regulating human corpus luteum function. Ann. NY Acad. Sci. 541:485, 1988.

The Adrenal Gland Has an Important Role in Stress

Stress can be caused by a number of environmental factors, such as trauma, extremes of temperature, emotional states, loud noises, and so on. These environmental signals are filtered by the reticular formation in the brain and "alarm" and other stress signals are transmitted from the central nervous system to the limbic system (e.g., hippocampus), which further transmits a signal to the hypothalamus. In response, the hypothalamus generates signals through the peripheral nervous system (neuronal system) and chemical signals through the anterior pituitary (ACTH, β-lipotropin, and β-endorphin). The neuronal system ends in the adrenal medulla mediated by a cholinergic neuron and causes the secretion of epinephrine, enkephalins, and some norepinephrine into the general circulation. The hormonal system through ACTH ends in the adrenal cortex and causes the output of the stress adaptational hormone, cortisol, and also plays a role in the formation and release of aldosterone (and dehydroepiandrosterone). **β-Endorphin** acts on receptors in the CNS to produce analgesia. These responses limit the deleterious effects of stress, provide immediate energy sources, and if stress continues for a prolonged period of time, produce pathological changes.

Figure 21.20 shows an overview of the **humoral stress pathway.** An environmental stress event is detected in the CNS and signals the limbic system (hippocampal structure) which, in turn, signals the hypothalamus to release the CRH. The CRH courses down the closed portal system, having entered it through fenestrated capillaries in the primary plexus, and exits by way of fenestrations of the secondary plexus in the vicinity of corticotrophic cells. There, CRH binds to a membrane receptor and promotes release of ACTH, β-lipotropin (β-LTH), and β-endorphin (derived from β-LTH), the proopiomelanocortin products. β-Endorpin acts on cells of the CNS to promote analgesia, possibly by lowering the level of cellular cAMP in certain cells through a β-endorphin receptor coupled to an inhibitory G protein transducer and adenylate cyclase. β-Endorphin also binds to receptors on sommatotrophs and lactotrophs of the anterior pituitary causing the secondary release of growth hormone and prolactin, which may play some role in the stress response by virtue of the hyperglycemic actions of these hormones in liver. The ACTH binds to the zona fasciculata (middle layer) cells of the adrenal cortex and enhances the rate of synthesis of cortisol and its secretion (see Figure 21.7). Cortisol circulates in the blood bound to transcortin or corticosteroid binding globulin (CBG) with a finite dissociation to the free form. Free cortisol is taken up by most cells of the body that contain variable amounts of the glucocorticoid receptor. Cortisol enters the cell by free diffusion and binds to the cytoplasmic unoccupied, unactivated–nontransformed receptor. The unactivated receptor complex may consist of one molecule of receptor, a dimer of heat shock 90-kDa protein, and other factors. Activation results in the disaggregation of this complex and the production of the activated receptor, which now has its DNA binding domain available for interaction with DNA (see Table 21.4) and is also able to translocate to the nucleus, possibly through the nucleopore. The receptor performs its transactivation function along with other transactivating proteins and the phenotypic products expressed in the cell alter cellular functions and represent the cellular actions of the hormone. The many different cellular actions combine into the systemic effects that constitute stress adaptation. In the short run, this process is useful to the body because the actions of the hormone may limit the deleterious effects of stress. If the process continues without adaptation, however, the effects of this system can be harmful. Nevertheless, this system is vital for survival in the face of stress.

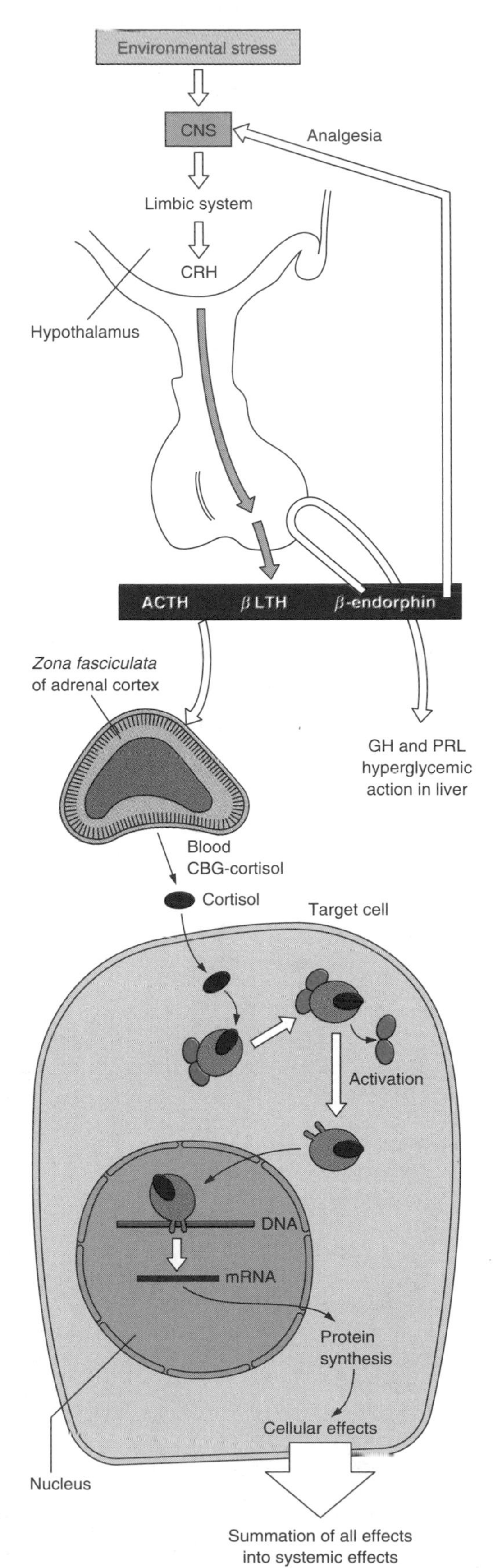

FIGURE 21.20
Overview of the humoral stress pathway.

BIBLIOGRAPHY

Baulieu, E.-E. Steroid hormone antagonists at the receptor level: a role for heat-shock protein MW 90,000 (hsp 90). *J. Cell Biochem.* 35:161, 1987.

Beato, M. Gene regulation by steroid hormones. *Cell* 56:335, 1989.

Carson-Jurica, M. A., Schrader, W. T., and O'Malley, B. W. Steroid receptor family: structure and functions. *Endocrine Rev.* 11:201, 1990.

Chrousos, G. P., Loriaux, D. L., and Lipsett, M. B. (Eds.). *Steroid Hormone Resistance*. New York: Plenum, 1986.

Evans, R. M. The steroid and thyroid hormone receptor superfamily. *Science* 240:889, 1988.

Giguere, V., Hollenberg, S. M., Rosenfeld, M. G., and Evans, R. M. Functional domains of the human glucocorticoid receptor. *Cell* 46:645, 1986.

Green, S., Kumar, V., Theulaz, I., Wahli, W., and Chambon, P. The N-terminal DNA-binding "zinc-finger" of the estrogen and glucocorticoid receptors determines target gene specificity. *EMBO J.* 7:3037, 1988.

Gustafsson, J. A., et al. Biochemistry, molecular biology, and physiology of the glucocorticoid receptor. *Endocrine Rev.* 8:185, 1987.

Huft, R. W. and Pauerstein, C. J. *Human Reproduction: Physiology and Pathophysiology*. New York: Wiley, 1979.

King, R. J. B. and Mainwaring, W. I. P. *Steroid-Cell Interactions,* Baltimore: University Park Press, 1974.

Litwack, G. (Ed.). *Biochemical Actions of Hormones,* Vols. 1–14. New York: Academic, 1973–1987.

Norman, A. W. and Litwack, G. *Hormones,* Orlando, FL: Academic, 1987.

Rusconi, S. and Yamamoto, K. R. Functional dissection of the hormone and DNA binding activities of the glucocorticoid receptor. *EMBO J.* 6:1309, 1987.

Schwabe, J. W. R. and Rhodes, D. Beyond zinc fingers: steroid hormone receptors have a novel structural motif for DNA recognition. *Trend. Biochem. Sci.* 16:291, 1991.

QUESTIONS

J. BAGGOTT AND C. N. ANGSTADT

*Q*uestion *T*ypes are described on page xxiii.

1. (QT1) The C-21 steroid hormones include:
   A. aldosterone
   B. dehydroepiandrosterone
   C. estradiol
   D. testosterone
   E. vitamin D_3

2. (QT2) side chain cleavage enzyme complex activity may be stimulated by:
   1. cAMP
   2. Ca^{2+} released via stimulation of the IP_3 pathway
   3. Ca^{2+} entering the cell through a channel
   4. 5′-AMP

3. (QT1) Δ^5-pregnenolone is a precursor of all the following EXCEPT:
   A. aldosterone
   B. cortisol
   C. 17β-estradiol
   D. progesterone
   E. vitamin D_3

4. (QT1) Major steps in the inactivation and excretion of ALL classes of steroid hormones (except vitamin D_3) include:
   A. conjugation to glucuronic acid
   B. conjugation to sulfuric acid
   C. hydroxylation
   D. oxidation
   E. side chain cleavage

A. Aldosterone C. Both
B. Cortisol D. Neither

5. (QT4) A stress hormone.
6. (QT4) Synthesis and secretion stimulated by angiotensin II and III, but not by angiotensin I.

7. (QT2) Reactions in the pathway of synthesis of active vitamin D involve which of the following organs?
   1. Skin
   2. Kidney
   3. Liver
   4. Parathyroid gland

A. Cortisol binding globulin
B. Serum albumin
C. Sex hormone binding protein
D. Androgen binding protein

8. (QT5) Major aldosterone carrier in blood.
9. (QT5) Supplies testosterone to the Sertoli cells.
10. (QT5) Binds about 20% of the cortisol in the plasma.

11. (QT2) Receptors for steroid hormones are found in the:
    1. cell membranes.
    2. cytoplasm.
    3. ribosomes.
    4. nucleus.
12. (QT2) Which of the following involve(s) the same response element of DNA?
    1. Glucocorticoid
    2. Mineralocorticoid
    3. Progesterone
    4. Estrogen
13. (QT2) Which of the following receptors belong(s) to the same class as the steroid receptors?
    1. Thyroid hormone receptor
    2. Epinephrine receptor
    3. Retinoic acid receptor
    4. α-Tocopherol receptor

A. Programmed cell death C. Both
B. Stress response D. Neither

14. (QT4) Mechanism for the maturation of certain organs.
15. (QT4) Neurohormonal process.
16. (QT4) Involves cortisol.

ANSWERS

1. A B and D: These are C-19 androgens. C: Estradiol is a C-18 estrogen. E: Vitamin D_3 is a C-27 compound (p. 902).
2. A 1,2,3 correct. See Figure 21.4, p. 905.
3. E See Figure 21.5 (p. 906) and Figure 21.6 (p. 908) for the synthesis of A–D. The synthesis of vitamin D_3 is summarized in Figure 21.11 (p. 914).
4. A Oxidation (including hydroxylation) and reduction are common in steroid hormone degradation. Glucocorticoids undergo side chain cleavage. Conjugation to sulfate is important in the excretion of androgens. But conjugation to glucuronide is significant for all steroid hormones except vitamin D_3 (Table 21.2, p. 909).
5. C See p. 912.
6. A Only the octapeptide (angiotensin II) and the heptapeptide (angiotensin III) are active (p. 912).
7. A 1,2,3 correct. 1: Light-induced cleavage of 7-dehydrocholesterol occurs in the skin. 2: Hydroxylation of 25-$(OH)D_3$ occurs in the kidney. 3: Hydroxylation of D_3 occurs in the liver. 4: The parathyroid gland produces a hormone involved in control, but no reaction of the 1,25-$(OH)_2D_3$ synthetic pathway occurs there. See Figure 21.11, p. 914.
8. B See p. 915.
9. D See p. 916.
10. B Cortisol binding globulin carries most of the cortisol. Serum albumin, however, nonspecifically binds a large number of hydrophobic substances, including cortisol (p. 915).
11. C 2,4 correct. 1: Membrane receptors are generally associated with nonhydrophobic hormones, such as epinephrine and peptide hormones (p. 917). 2 and 4: See Figure 21.13, p. 916.
12. A 1,2,3 correct. The *positive* glucocorticoid response element is the same as the mineralocorticoid response element and the progesterone response element. 4: The estrogen response element differs (Table 21.4, p. 918).
13. B 1,3 correct. See p. 920.
14. A Thymus cortical cells are killed by cortisol (p. 922).
15. B Nerve impulses and circulating cortisol are both involved. See Figure 21.20, p. 923.
16. C See answers 14 and 15 above.

22

Molecular Cell Biology

THOMAS E. SMITH

22.1 OVERVIEW 927
22.2 STIMULI RECOGNITION AND TRANSMEMBRANE SIGNALING 928
22.3 NERVOUS TISSUE: METABOLISM AND FUNCTION 929
ATP Maintains a Transmembrane Electrical Potential in Neurons 930
Neuron–Neuron Interaction Occurs Through Synapses 931
Synthesis, Storage, and Release of Neurotransmitters Are Part of Nerve Transmission 933
Neuropeptides Are Cleaved From Precursor Proteins 937
22.4 THE EYE: METABOLISM AND VISION 938
The Cornea Derives Its ATP From Aerobic Metabolism 940
The Lens Is Mostly Water and Protein 940
The Retina Derives ATP From Anaerobic Glycolysis 941
The Process of Visual Transduction Involves Photochemical, Biochemical, and Electrical Events 942
Photoreceptor Cells Are Rods and Cones 945
Transducin Is a G Protein Involved in Signal Transduction 948
Color Vision Resides in the Cones 951
22.5 MUSCLE CONTRACTION 954
Skeletal Muscle Contraction Follows an Electrical to Chemical to Mechanical Path 954
The Thick Filament of Muscle Is Myosin 957
Actin Is a Protein of the Thin Filament 959
Mechanism of Muscle Contraction Begins With Ca^{2+} Release 962
Energy for Muscle Contraction Is Supplied by ATP Hydrolysis 963
Calcium Flux in the Sarcomere Is Controlled by Several Events 963
Calcium Regulates Smooth Muscle Contraction 965
22.6 MECHANISM OF BLOOD COAGULATION 966
Rupture of the Endothelium Initiates Platelet Clumping 966
Clot Formation Occurs at the Site of Injury 967
Thrombin Converts Fibrinogen to Fibrin 967
Factors III and VII Are Unique to the Extrinsic Pathway 969
Several Factors Are Restricted to the Intrinsic Pathway 970
Properties of Many of the Proteins of Coagulation Are Known 970
Fibrinolysis Requires Plasminogen and Tissue Plasminogen Activator (t-PA) to Produce Plasmin for Fibrinolysis 974
Blood Coagulation Occurs By a Cascade of Reactions 974
BIBLIOGRAPHY 976
QUESTIONS AND ANNOTATED ANSWERS 977
CLINICAL CORRELATIONS
22.1 Myasthenia Gravis: A Neuromuscular Disorder 935
22.2 Chromosomal Location of Genes for Vision 953
22.3 Classic Hemophilia 970

22.1 OVERVIEW

Animals know the existence of their environment through the responses of certain organs to stimuli: touch, pain, heat, cold, intensity (light or noise), color, shape, position, pitch, quality, acid, sweet, bitter, salt, alkaline, fragrance, and undefined others. Externally, these generally reflect the responses of stimulus to the skin, eye, ear, tongue, and nose. Some of these signals are localized to the point at which they occur, others—sound and sight—are projected in space, that is, the environment outside of and distant to the animal.

Discrimination of the signals occurs at the point of reception, but acknowledgment of what they are occurs as a result of secondary stimulation of the nervous system and the transmission of the signals to the brain. In many instances, a physical response is indicated and results in a change in muscular activity, either voluntary or involuntary. Common to these events are the electrical activity associated with the signal transmission along the nerve and the chemical activity associated with the signal transmission across the synaptic junctions. In all cases, the stimuli received from the environment in the form of pressure (skin, feeling), light (sight, eye), noise (ear, hearing), taste (tongue), or smell (nose) must be converted (transduced) into an electrical impulse and to some other form of energy in order to effect the desired terminal response as dictated by the brain. There is a biochemical component associated with each of these events.

In this chapter, the general biochemical mechanisms of signal transduction and amplification will be discussed as they relate to the biochemical events involved in nerve transmission, vision, and muscular contraction. Finally, a specialized case of biochemical signal amplification will be discussed, namely, blood coagulation. This is a process that is initiated on the surface of membranes as a result of the exposure of specific proteins that act as receptors and form nucleation sites for formation of the multienzyme complexes that lead to the amplification of the process through a cascade mechanism.

22.2 STIMULI RECOGNITION AND TRANSMEMBRANE SIGNALING

Recognition by cells of external stimuli requires **transmembrane signaling,** which generically involves the intervention of at least three proteins: a surface or initial receptor, a signal transduction protein, and an effector protein. The signal pathway can be illustrated as follows:

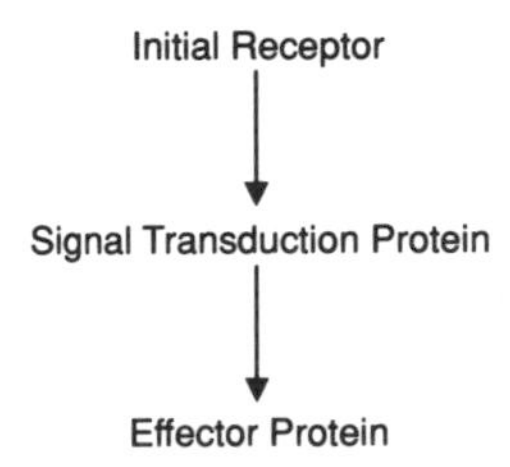

The initial receptor is the *discriminator* and could be sensitive to a physical stimulus or to a chemical stimulus. The physical stimulus is generally pressure or light, while the chemical stimulus may be one of two types of biological molecules: lipophilic molecules that may penetrate the membrane and interact with internal receptors (see p. 916) or water-soluble molecules that interact with receptors at the cell surface. In this chapter only the latter type of chemical stimulus, that is, water-soluble, is applicable to the systems discussed.

The receptor for water-soluble molecules is embedded in or spans the cell membrane. When the stimulus interacts with the receptor, the receptor undergoes a conformation change that starts a series of events that amplifies the signal and transmits it across the membrane where specific proteins are activated. The action or reaction of these proteins is the final response of the cell or tissue to the initiating stimulus.

The **signal transduction protein** could be a protein kinase, an ion channel as is the case with nerve cells, or a protein that undergoes some other energy-dependent change in which the energy is supplied by the hydrolysis of a high-energy compound such as guanosine triphosphate (GTP). If the high-energy compound is in fact GTP, the class of proteins with which it interacts is referred to as G proteins.

G proteins, whether involved in hormone action or transmission of external stimuli, have certain properties and characteristics in common. They require both a specific ligand for activation of the receptor and GTP to elicit the response. In the absence of specific ligand–receptor interaction, nonhydrolyzable analogs of GTP such as GTP-S or Gpp(NH)p provoke the response because the termination of the response requires hydrolysis of the GTP. Binding of guanine nucleotides to a G protein reduces binding of the G protein to the specific receptor that controls its action. Cholera or pertussis toxins catalyze ADP ribosylation of G proteins, which alters the ability of the G proteins to participate in the signal transduction. A detailed discussion of the G proteins is presented on page 869.

These properties have been demonstrated for G proteins in relation to hormone action and will be shown in Section 22.4 to occur in the process of signal transduction in the visual cycle. This chapter will begin with a discussion of the nervous system since it is central to the other processes emphasized here.

22.3 NERVOUS TISSUE: METABOLISM AND FUNCTION

The nervous system provides the communications network between the senses, the environment, and all parts of the body. The brain is the command center. This system is always at work and requires a large amount of energy to keep it operational. Under normal conditions, the only *metabolic fuel* for the brain is *glucose*. It has been estimated that the human brain uses approximately 103 g of glucose per day. This would correspond to a rate of utilization of 0.3 μmol min^{-1} g^{-1} of tissue. The equivalent amount and rate of oxygen consumption are 3.4 L day^{-1} or 1.7 μmol min^{-1} g^{-1} of tissue, respectively. Similarly, the amount and rate of ATP consumption are 20.4 mol day^{-1} and 10.2 μmol min^{-1} g^{-1} of tissue, respectively. Most of the ATP used by the brain and other nervous tissue is generated through the operation of the tricarboxylic acid (TCA cycle), which operates at near maximum capacity. Glycolysis functions at approximately 20% capacity. Much of the energy used by the brain is to maintain the ionic gradients across the plasma membranes and for the synthesis of neurotransmitters and other cellular components.

Two features of brain macromolecules are worth noting. There are complex and specialized lipids in the brain, but they appear to function to maintain membrane integrity (see Chapter 5) rather than metabolic roles. Brain proteins have a rapid turnover rate relative to other body proteins in spite of the fact that the cells do not divide after they have differentiated.

The cells of the nervous system, including the brain, that are responsible for collecting and transmitting messages are the **neurons** and are very highly specialized (Figure 22.1). Each neuron consists of a cell body, dendrites that are short antenna-like protrusions that function to receive signals from other cells, and an axon that extends from the cell body and functions to transmit signals to other cells.

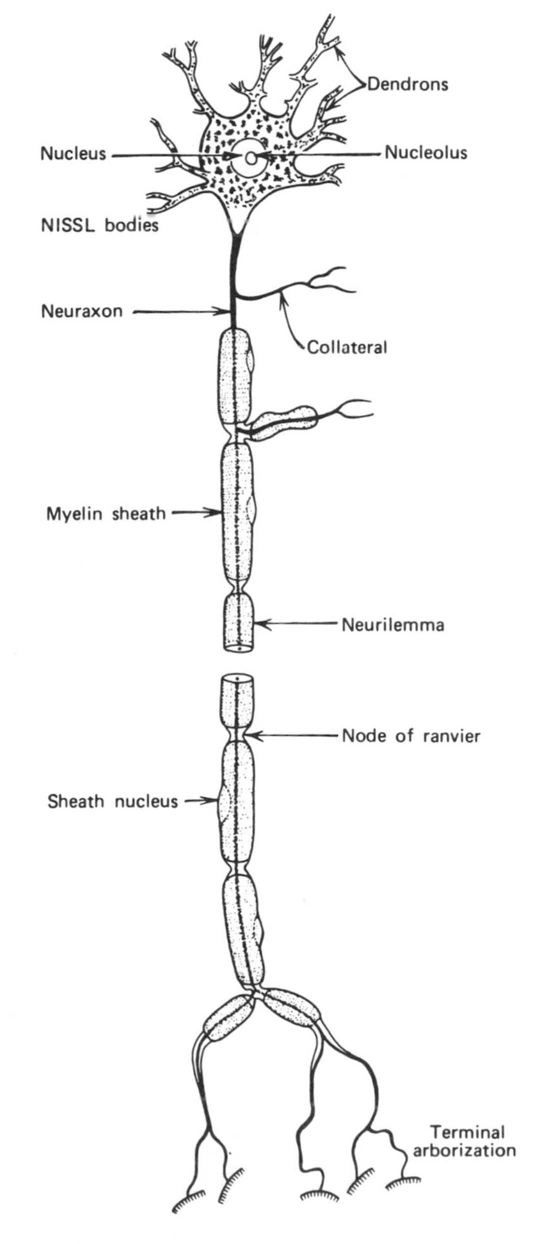

FIGURE 22.1
A motor nerve cell and investing membranes.

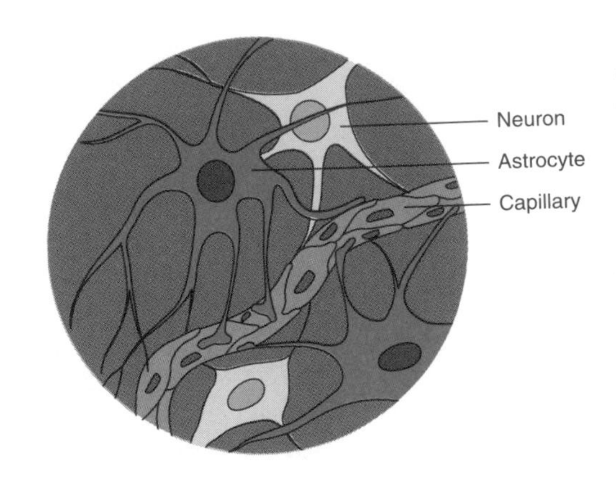

FIGURE 22.2
Diagrammatic representation of an astrocyte.
Redrawn from B. Alberts, D. Bray, J. Lewis, M. Raff, K. Roberts, and J. Watson, *Molecular Biology of the Cell,* 2nd ed., New York: Garland Publishing, Inc., 1989, Figure 19.8, p. 1065.

There are other cells in the CNS besides neurons. In the brain, for example, there are about 10 times more **glial cells** than there are neurons. Glial cells occupy the spaces between neurons and provide some electrical insulation between them. Glial cells are generally not electrically active, but they are capable of division. There are basically five types of glial cells: Schwann cells, oligodendrocytes, microglia, ependymal cells, and astrocytes. Each of the types of glial cells has a specialized function, but because the interest here is focused on transport of metabolites in and out of cells, the function of only one type will be discussed here, namely, the **astrocyte** (Figure 22.2). The astrocytes send out processes at the external surfaces of the central nervous system (CNS). These processes are linked to form anatomical complexes that provide sealed barriers and isolate the CNS from the external environment. The astrocytes also send out similar processes to the circulatory system to seal it off in a selective manner. The astrocytes induce the endothelial cells of the capillaries to become sealed by forming tight junctions that prevent the passive entry into the brain of water-soluble molecules. These tight junctions form what is commonly known as the **blood–brain barrier.** Water-soluble compounds enter the brain only if there are specific membrane transport systems for them.

The normal individual has between 10^{11} and 10^{13} neurons, and communication between the neurons of the nervous system is by electrical and chemical signals. Electrical signals transmit the nerve impulse down the axon and chemicals transmit the signal across the gap between cells. Some of the biochemical events that give the cell its electrical properties and are involved in the propagation of an impulse will be discussed.

ATP Maintains a Transmembrane Electrical Potential in Neurons

Adenosine triphosphate generated from the metabolism of glucose is used to help maintain an *equilibrium electrical potential* across the membrane of the neuron of approximately −70 mV, with the inside of the cell being more negative than the outside. This potential is maintained by the action of the Na^+,K^+ ion pump (p. 226), the energy for which is derived from the hydrolysis of ATP to give ADP and phosphate. This system pumps Na^+ out of the cell in an antiport mechanism where K^+ is moved into the cell. The channels through which Na^+ enters the cell are voltage gated, that is the proteins of the channel undergo a *charge-dependent conformation change* and open when the electrical potential across the membrane decreases (specifically, becomes less negative) by a value greater than some threshold value. When the membrane becomes depolarized, Na^+, whose concentration is higher outside the cell than inside, flows into the cell and K^+, whose concentration is greater inside the cell, flows out of the cell, both going down their respective concentration gradients. The channels are open in a particular geographical region of the cell for only milliseconds. The *localized depolarization* (voltage change) causes a conformation change in the neighboring proteins that make up the **"voltage-gated" ion channels.** These channels open momentarily to allow more ions in and, thus, by affecting adjacent channel proteins, allow the process to continue down the axon. There is a finite recovery time during which the proteins that form the channel cannot repeat the process of opening. Thus, the charge propagation proceeds in one direction. This process is diagrammed in Figure 22.3. This is an example of ion-dependent (charge) protein conformation change that propagates the nerve impulse (electrical). It is the progressive depolarization and repolarization along the length of the axon that allows the electrical impulse to be propagated undiminished in amplitude. Electrical impulse transmission is a continuous process in nervous tissue, and it is the ATP generated from the metabolism of glucose that keeps the system operational.

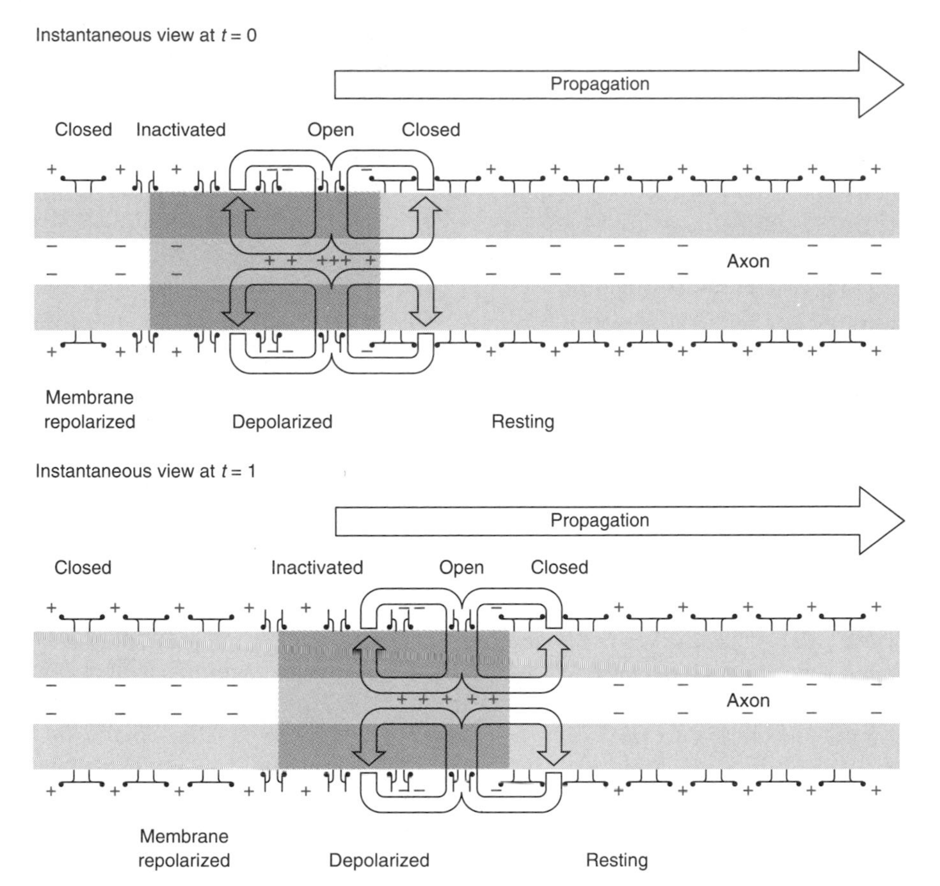

FIGURE 22.3
Schematic of the mechanism of nerve impulse transmission.
Redrawn from B. Alberts, D. Bray, J. Lewis, M. Raff, K. Roberts, and J. Watson, *Molecular Biology of the Cell,* 2nd ed., New York: Garland Publishing, 1989, p. 1071.

A current area of active research in biochemistry involves the use of gene cloning techniques to isolate the proteins involved in ion channels, in order to determine their structures and elucidate their mechanisms of action.

Neuron–Neuron Interaction Occurs Through Synapses

There are generally two mechanisms for neuron–neuron interaction: electrical synapses and chemical synapses. **Electrical synapses** permit the more rapid transfer of signals from cell to cell. **Chemical synapses** allow for various levels of versatility in cell–cell communication and are themselves of two types: those that bind directly to an ion channel and cause it to open or to close, and those that bind to a receptor that releases a second messenger to react with the ion channel to cause it to open or to close. The primary emphasis here is on chemical synapses, and the remainder of this section will focus on the synthesis and degradation of chemical neurotransmitters.

Chemical neurotransmitters may be **excitatory** or **inhibitory.** Some excitatory neurotransmitters are compounds such as acetylcholine and the catecholamines. Some inhibitory neurotransmitters are compounds such as γ-aminobutyric acid (also known as GABA or 4-aminobutyric acid), glycine, and taurine. The names of some of these compounds are listed in Table 22.1.

These compounds are classified as **neurotransmitters** because they fit the following criteria: they are found in the presynaptic axon terminal; the enzymes necessary for their syntheses are present in the presynaptic neuron; stimulation under physiological conditions results in their release; mechanisms exist (within the synaptic junction) for the rapid termination of their action; and their direct application to the postsynaptic terminal mimics the action of nervous stimulation. A sixth criterion, as a

TABLE 22.1 Some of the Neurotransmitters Found in Nervous Tissue

EXCITATORY
Acetylcholine
Aspartate
Dopamine
Histamine
Norepinephrine
Epinephrine
ATP
Glutamate
5-Hydroxytryptamine
INHIBITORY
4-Aminobutyrate
Glycine

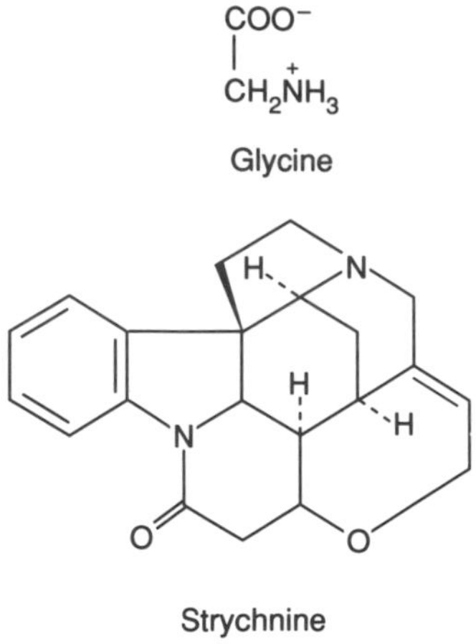

Glycine

Strychnine

FIGURE 22.4
Structures of glycine and strychnine.

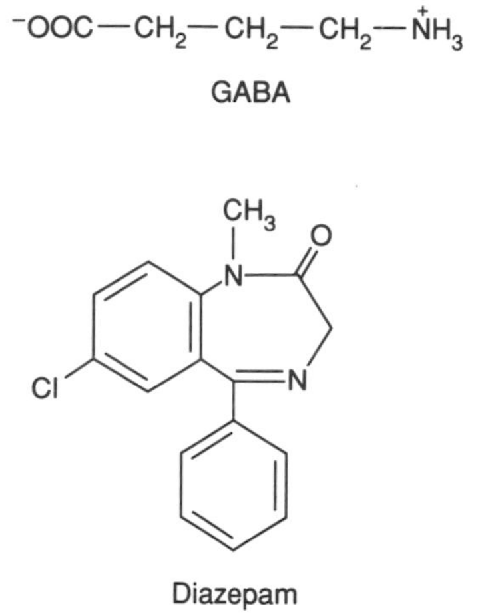

$^-OOC-CH_2-CH_2-CH_2-\overset{+}{N}H_3$

GABA

Diazepam

FIGURE 22.5
Structures of GABA and diazepam.

corollary of the five criteria listed above, is that drugs that modify the metabolism of the neurotransmitter should have predictable physiological effects in vivo, assuming that the drug was transported to the appropriate site where the neurotransmitter acts.

All of the neurotransmitters are made and stored in the *presynaptic neuron*. They are released from it after stimulation of the neuron. They traverse the synapse and bind to a specific receptor on the postsynaptic junction to excite the next cell, causing one of the responses mentioned above. If the neurotransmitter is an excitatory one, it causes depolarization of the membrane as described above. If it is an inhibitory neurotransmitter, it will bind to a channel-linked receptor, causing a conformation change in the proteins of which the channel is composed, opening the pores to admit into the cell small negatively charged ions, specifically Cl^-. The net effect of this is to increase the chloride conductance of the postsynaptic membrane, making it more difficult for the cell to become depolarized, that is, the inhibitory transmitter effectively causes hyperpolarization.

The two major inhibitory neurotransmitters in the central nervous system are **glycine,** which acts predominantly in the spinal cord and the brain stem, and **gamma-aminobutyric acid** (GABA), which acts predominantly in all other parts of the brain. Strychnine (Figure 22.4), a highly poisonous alkaloid obtained from *nux vomica* and related plants of the genus *Strychnos,* binds to the glycine receptors of the CNS. This poison has been used in very small amounts as a CNS stimulant. Can you propose how it works? The receptor to which GABA binds also reacts with a variety of pharmacologically significant agents such as benzodiazepines (Figure 22.5) and barbiturates. As is the case with strychnine and glycine, there is little structural similarity between GABA and benzodiazepines.

The genes for the acetylcholine receptor, which also binds nicotinic acid, the glycine receptor, and the GABA receptor have been cloned and amino acid sequences inferred. There is a relatively high degree of homology between the amino acid sequences of these proteins. Three-dimensional structures of these receptor proteins, however, have not been ob-

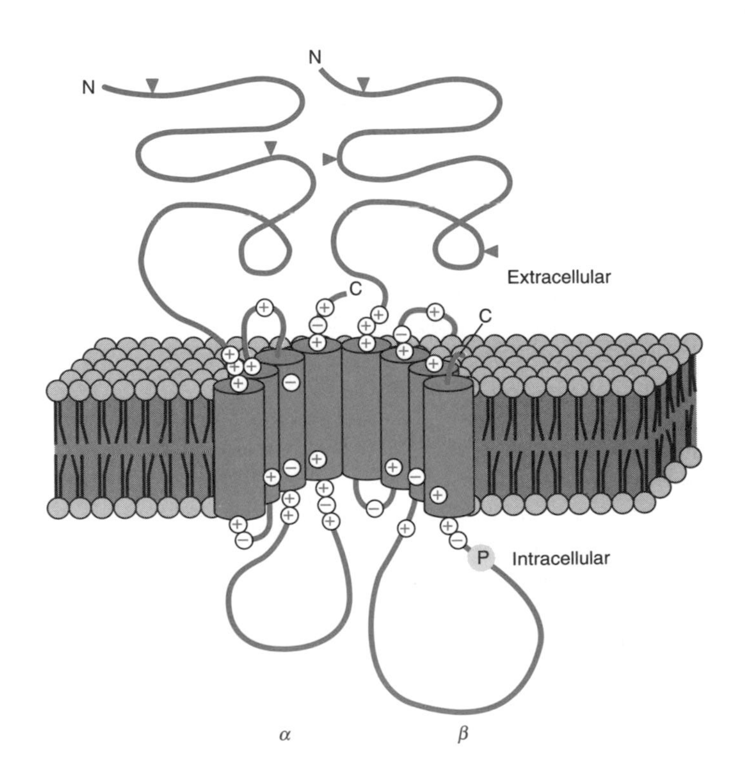

FIGURE 22.6
A schematic model of one-half of the GABA receptor in the cell membrane.
Two of the subunit structures shown form the complete receptor and ion channel molecule. The site labeled P is a serine that is the site for a cAMP-dependent phosphorylation. Redrawn from P. R. Schofield, et al. *Nature (London)* 328:221, 1987.

tained. Schematic models have been constructed using knowledge of how proteins should fold based on their amino acid sequences.

A model for one-half of the GABA receptor is shown in Figure 22.6. This receptor has an $\alpha_2\beta_2$ structure. The peptides are synthesized with "signal peptides" that direct their transport to the membrane. The α chain has 456 amino acids and the β chain has 474. The signal peptides are cleaved, leaving α and β subunits of 429 and 449 amino acid residues, respectively. Interestingly, the pharmaceutical agents bind to the α subunit, whereas GABA, the natural inhibitory neurotransmitter, binds to the β subunit. The protrusion of an extended length of the amino terminal end of each peptide to the extra-cellular side of the membrane suggests that the residues to which the channel regulators bind are at the N terminal. A smaller segment consisting of the C terminal residues is also on the extracellular side of the membrane. The two halves of the receptor form a channel through which small negative ions (Cl^-) can flow, depending on what is bound to the receptor end of the molecule.

The amino acid sequences of several of these ion–channel proteins are similar. This, perhaps, is not too surprising since a common function they all have is to form channels to permit the flow of ions.

Synthesis, Storage, and Release of Neurotransmitters Are Part of Nerve Transmission

The nonpeptide neurotransmitters may be synthesized in almost any part of the cell, in the cytoplasm in the vicinity of the nucleus as well as in the axon. Most of the nonpeptide neurotransmitters are amino acids, derivatives of amino acids, or other intermediary metabolites. Synthesis and degradation of many of them have been discussed elsewhere in earlier chapters, but some aspects of their metabolism relative to nerve transmission will be discussed later.

Storage of neurotransmitters occurs in vesicles in the presynaptic terminal. As the electrical impulse reaches the presynaptic terminal, voltage-gated Ca^{2+} channels in addition to the Na^+ channels open, thereby increasing the concentration of Ca^{2+} in the cell (Figure 22.7). The Ca^{2+} binds to the synaptic vesicles and initiates a series of events that convert the electrical impulse back into a chemical event. It is believed that Ca^{2+} also activates a calcium–calmodulin-dependent protein kinase that phosphorylates a protein called **synapsin I,** which is attached to the surface of the presynaptic membrane. Upon phosphorylation, this protein dissociates from the membrane, and the synaptic vesicles attach to the presynaptic membrane, releasing their neurotransmitters by a process of exocytosis. Presumably, the synapsin I is dephosphorylated and displaces the synaptic vesicles so that the vesicles can take up more neurotransmitters and start the cycle over again.

The neurotransmitters that are released travel rapidly across the synaptic junction (~20 nm), bind to receptors on the postsynaptic side, induce a conformation change on that membrane, and start the process of electrical impulse propagation in the postsynaptic cell.

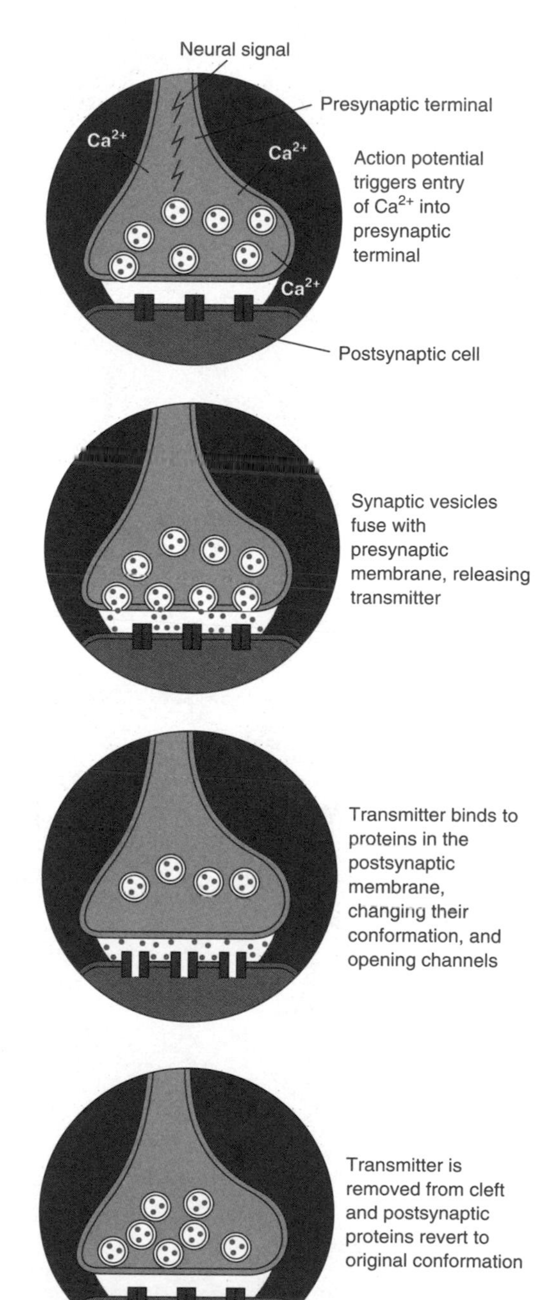

FIGURE 22.7
A schematic diagram of the synapse and some of its associated reactions.
Redrawn from B. Alberts, D. Bray, J. Lewis, M. Raff, K. Roberts, and J. Watson, *Molecular Biology of the Cell,* 2nd ed., New York: Garland Publishing, 1989, Figure 19.19, p. 1077.

Termination of Signal at the Synaptic Junction

There are generally three ways by which neurotransmitter action may be terminated: *metabolism, reuptake,* and *diffusion.* The chemical substances responsible for the fast responses are generally disposed of by one or both of the first two mechanisms. The following sections will outline some of the biochemical pathways involved in the synthesis and degradation of representative neurotransmitters from the fast-acting group, spe-

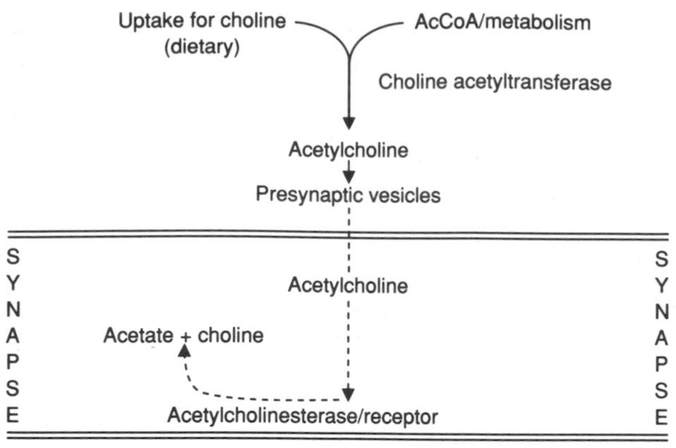

FIGURE 22.8
Summary of the reactions of acetylcholine at the synapse.
AcCoA = acetylcoenzyme A; AcChEase = acetylcholinesterase.

cifically, acetylcholine, the catecholamines, 5-hydroxytryptamine, and 4-aminobutyrate.

Acetylcholine

A summary of the reactions involving acetylcholine at the synapse is presented in Figure 22.8. Acetylcholine is synthesized by the condensation of choline and acetyl CoA catalyzed by the enzyme **choline acetyltransferase.** The reaction is

$$\underset{\text{Choline}}{(CH_3)_3\overset{+}{N}CH_2CH_2OH} + CH_3CO-SCoA \longrightarrow \underset{\text{Acetylcholine}}{(CH_3)_3\overset{+}{N}CH_2CH_2OCOCH_3} + CoASH$$

Choline is derived mainly from the diet; however, some of it may come from its reabsorption from the synaptic junction or from other metabolic sources (see p. 519). The major source of acetyl CoA is the decarboxylation of pyruvate by the pyruvate dehydrogenase complex. Remember that the major metabolic fuel is glucose and that the major pathway for energy production is the tricarboxylic acid cycle. Acetyl CoA for acetyl choline synthesis is obtained by shunting some of it from the TCA cycle. Enzymes for making acetyl CoA from acetate and CoA do not appear to be present. Acetyl CoA is synthesized inside the mitochondria and choline acetyltransferase is present in the cytosol. The same mechanism previously discussed (see p. 397) for getting acetyl CoA across the mitochondrial membrane in other tissues is operational in the presynaptic neuron.

Acetylcholine is released and reacts with the nicotinic-acetylcholine receptor located in the postsynaptic membrane (see Clin. Corr. 22.1). The action of acetylcholine at the postsynaptic membrane is terminated by the action of the enzyme **acetylcholinesterase,** which hydrolyzes the acetylcholine to acetate and choline.

$$\text{Acetylcholine} + H_2O \longrightarrow \text{acetate} + \text{choline}$$

Choline is taken up by the presynaptic membrane and reutilized for synthesis of more acetylcholine. Acetate probably gets reabsorbed into the blood and is metabolized by tissues other than nervous tissue.

CLIN. CORR. 22.1 MYASTHENIA GRAVIS: A NEUROMUSCULAR DISORDER

Myasthenia gravis is an acquired autoimmune disease characterized by muscle weakness due to decreased neuromuscular transmission. The neurotransmitter involved is acetylcholine. More than 90% of patients with myasthenia gravis have in their circulation antibodies to the nicotinic–acetylcholine receptor (AChR), which is located on the postsynaptic membrane of the neuromuscular junction. Antibodies against the AChR are presumed to interact with the receptor and to inhibit its function, that is, either its reaction with acetylcholine or its ability to undergo the requisite conformation change leading to ion transport. Evidence in support of myasthenia gravis being an autoimmune disease affecting the AChR is the finding that the amount of the AChR is reduced in patients with the disease, and the fact that experimental models of myasthenia gravis have been made by either immunizing animals with the AChR or by injecting them with antibodies against it.

It is not known what event(s) trigger the onset of the disease. It has been demonstrated that there are a number of environmental antigens that have epitopes like those on the AChR. A rat monoclonal antibody of the IgM type prepared against the AChR has been shown to react with two proteins from the intestinal bacterium, *E. coli*. Both of the proteins are membrane proteins of 38 and 55 kDa, the smaller of which is located in the outer membrane. This does *not* prove, however, that exposure to *E. coli* proteins is likely to trigger the disease. The sera of both normal individuals and myasthenia gravis patients have antibodies against a large number of *E. coli* proteins. Some environmental antigens from other sources also react with antibodies against the AChR.

The thymus gland, which is involved in antibody production, is also implicated in this disease. Antibodies found in the thymus gland of myasthenia gravis patients have been shown to react with the AChR and in some cases with environmental antigens. The relationship, if there is any, between environmental antigens, thymus antibodies against AChR, and the onset of myasthenia gravis is not clear.

Myasthenia gravis patients may be subjected to one or a combination of several therapies. Drugs, preferably those that do not cross the blood–brain barrier and are reversible inhibitors of acetylcholine esterase (AChE), have been used. The inhibition of AChE within the synapse increases the half-time for acetylcholine hydrolysis. This leads to an increase in its concentration, stimulation of remaining AChR, and some signal transmission. Other treatments include the use of immunosuppressant drugs, steroids, and surgical removal of the thymus gland to decrease further production of the antibodies. Future treatment may include the use of antiidiotype antibodies to the AChR antibodies, and/or the use of small nonantigenic peptides that compete with the AChR epitopes for binding to the AChR antibodies.

Stefansson, K., Dieperink, M. E., Richman, D. P., Gomez, C. M., and Marton, L. S. *N. Engl. J. Med.* 312:221, 1985; Drachman, D. B. (Ed.). "Myasthenia Gravis: Biology and Treatment." *Ann. N.Y. Acad. Sci.* 505:1, 1987; and Steinman, L. and Mantegazza, R. *FASEB J.* 4:2726, 1990.

Catecholamines

The catecholamine neurotransmitters are **epinephrine, norepinephrine,** and **dopamine** (3,4-dihydroxyphenylethylamine) (Figure 22.9). Their biosynthesis has been discussed on page 507.

There are primarily three methods by which the action of the catecholamine neurotransmitters may be terminated: methylation of one of the OH groups of the catechol, oxidative deamination of the side chain, and reuptake into the presynaptic neuron. These reactions are shown in Figure 22.10. The enzyme **catechol-*O*-methyltransferase** is found in the cytoplasm of the cell and catalyzes the transfer of a methyl group from *S*-adenosylmethionine to one of the phenolic OH groups. **Monoamine oxidase** catalyzes the oxidative deamination of the amines to aldehydes

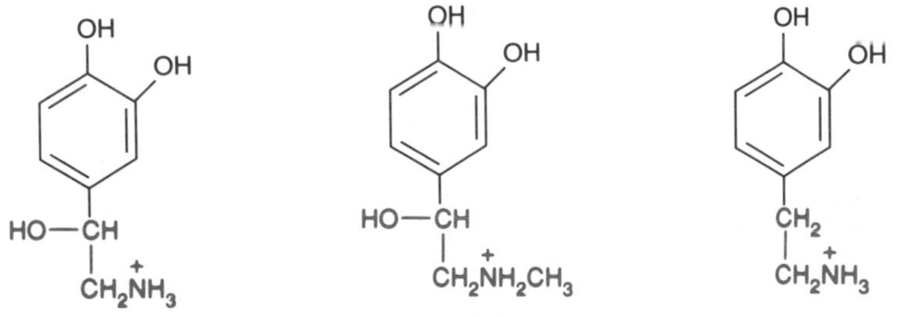

Norepinephrine — Epinephrine — Dopamine

FIGURE 22.9
Catecholamine neurotransmitters.

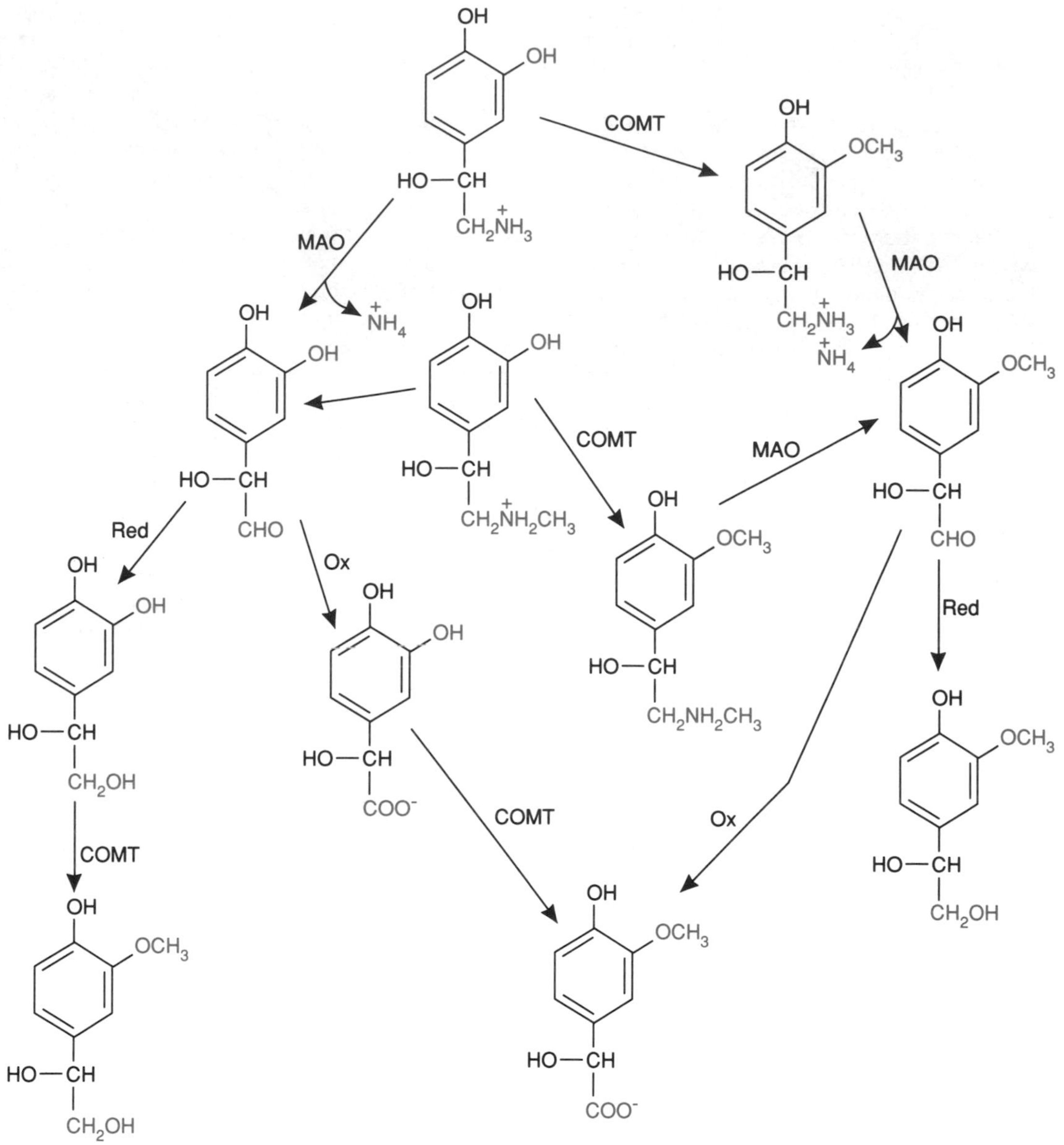

FIGURE 22.10
Pathways of catecholamine degradation.
COMT = catechol-*O*-methyl transferase (requires *S*-adenosylmethionine); MAO = monoamine oxidase; Ox = oxidation; Red = reduction.

and ammonium ions. This enzyme can act on amines whether or not they have been subjected to alteration by the methyltransferase reaction.

The fact that the degradative enzymes are located in the cytosol indicates that the amines must be taken up by the cells prior to degradation. For dopamine, as for others, there is a transporter protein that is responsible for reuptake into the presynaptic neuron. One of the actions of *cocaine* is to bind to the dopamine transporter and block the reuptake of dopamine. This permits dopamine to remain within the synapse for a prolonged period of time and to continue to stimulate the receptors of the postsynaptic neuron.

5-Hydroxytryptamine (Serotonin)

The biosynthesis of 5-hydroxytryptamine (Figure 22.11) involves some steps similar to those for the synthesis of the catecholamines (see p. 510).

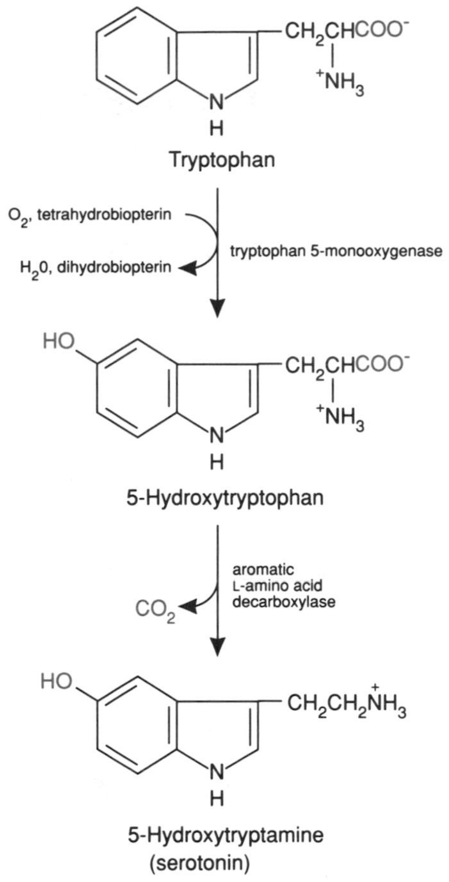

Tryptophan

O_2, tetrahydrobiopterin
H_2O, dihydrobiopterin
tryptophan 5-monooxygenase

5-Hydroxytryptophan

CO_2
aromatic L-amino acid decarboxylase

5-Hydroxytryptamine (serotonin)

FIGURE 22.11
Synthesis of 5-hydroxytryptamine (serotonin) from tryptophan.

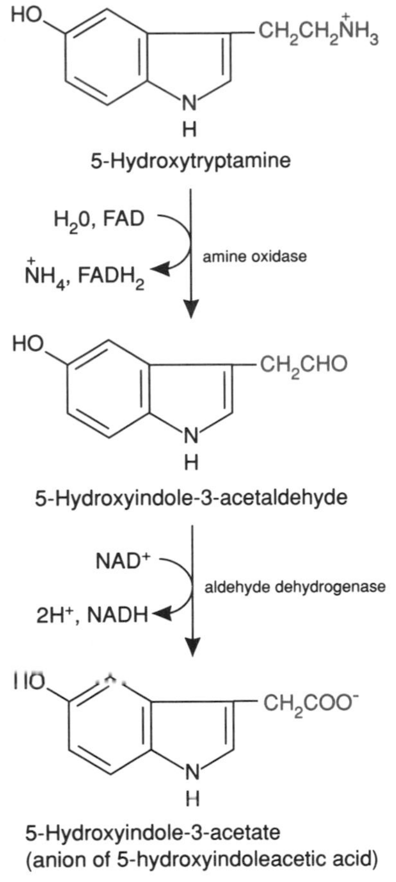

5-Hydroxytryptamine

H_2O, FAD
$\overset{+}{N}H_4$, $FADH_2$
amine oxidase

5-Hydroxyindole-3-acetaldehyde

NAD^+
$2H^+$, NADH
aldehyde dehydrogenase

5-Hydroxyindole-3-acetate
(anion of 5-hydroxyindoleacetic acid)

FIGURE 22.12
Degradation of 5-hydroxytryptamine (serotonin).

Unlike the catecholamines, the primary route for its degradation is by oxidative deamination to the corresponding acetaldehyde catalyzed by the enzyme monoamine oxidase (Figure 22.12). The aldehyde is further oxidized to 5-hydroxyindole-3-acetate by an aldehyde dehydrogenase.

4-Aminobutyrate (γ-Aminobutyrate)

Gamma-aminobutyrate (GABA) can be made and degraded through a series of reactions commonly known as the **GABA shunt.** In these reactions, α-ketoglutarate of the tricarboxylic acid cycle is transaminated to produce glutamate. Glutamate decarboxylase removes the 1-carboxyl group to produce GABA. The GABA is catabolized by deamination and oxidation reactions (see p. 501) to produce succinate that can enter the TCA cycle.

Gamma-aminobutyrate, the inhibitory neurotransmitter, and glutamate, the excitatory neurotransmitter, may share some common routes of metabolism involving the astrocytes (Figure 22.13). Both glutamate and GABA are taken up by astrocytes and converted to glutamine, which is then transported back into the presynaptic vesicles. In the excitatory nerve, glutamine is converted to glutamate. In the inhibitory nerve, the glutamine is converted to GABA with glutamate as an intermediate.

Neuropeptides Are Cleaved From Precursor Proteins

The peptides that are involved in neurotransmitter activity are generally synthesized as larger proteins that are cleaved specifically to produce the

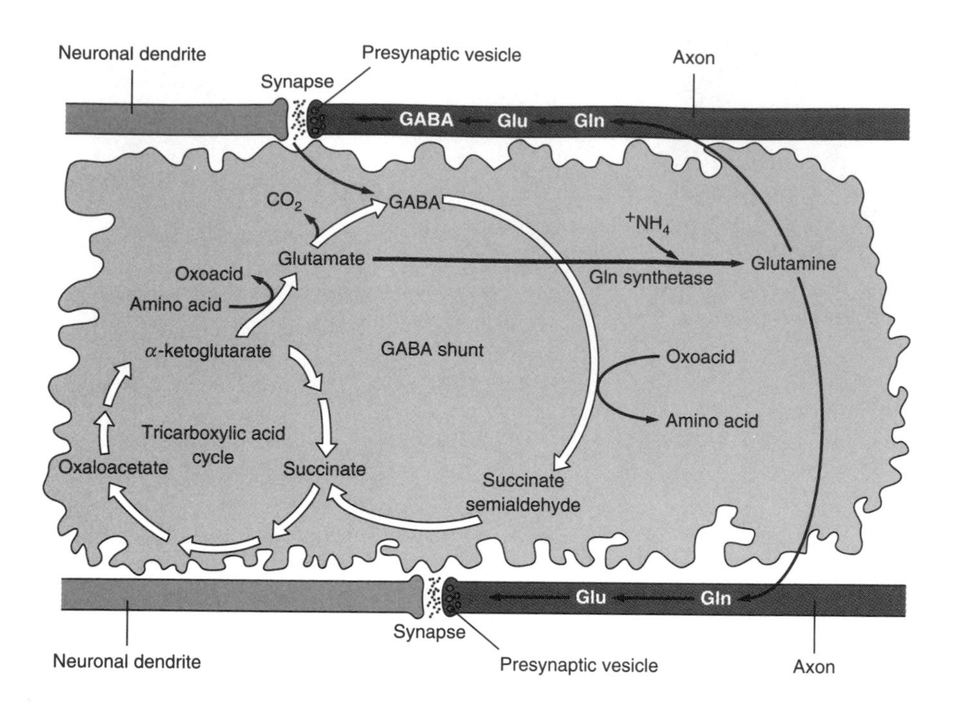

FIGURE 22.13
Involvement of the astrocytes in the metabolism of GABA and glutamate.

smaller peptide molecules. Their synthesis, therefore, require the same biochemical machinery as does any protein synthesis and must take place in the cell body, not the axon. Once they are made, they travel down the axon to its presynaptic region. There are two generic mechanisms for neuropeptide transport down the axon: fast axonal transport at a rate of about 400 millimeters per day and slow axonal transport at a rate of one to five millimeters per day. Since axons may vary in length from one millimeter to one meter, theoretically the total transit time could vary from 150 milliseconds to 200 days. It is highly unlikely that the latter transit time would ever occur under normal physiological conditions, and the upper limit is probably more like hours than days.

Neuropeptides are involved in the mediation of sensory and emotional responses such as those associated with hunger, thirst, sex, pleasure, pain, etc. Included in this category are peptides such as the enkephalins, endorphins, and substance P. **Substance P** is an excitatory neurotransmitter that has a role in pain transmission, whereas the **endorphins** have roles in eliminating the sensation of pain. Some of the peptides found in brain tissue are shown in Table 22.2. Note that *Met*-**enkephalin** is derived from the N-terminal region of β-endorphin. In many of the peptide neurotransmitters, either the N-terminal or both the N-terminal and C-terminal amino acids are modified. For a further discussion of these peptides, see Chapter 20.

22.4 THE EYE: METABOLISM AND VISION

The eye, our window to the outside world, allows us to view the beauties of nature, the beauties of life and, vide this textbook, the beauties of biochemistry. What are the features of this organ that permits this view? A view through any window, through any camera lens, is clearest when unobstructed by objects that result in distortion of the objects being

TABLE 22.2 Peptides Found in Brain Tissue[a]

Peptide	Structure
Beta-endorphin	Y G G F M T S E K S Q T P L V T L F K N A I I K N A Y K K G E
Met-enkephalin	Y G G F M
Leu-enkephalin	Y G G F L
Somatostatin	A G C K N F F W C S T F T K
Leutinizing hormone-releasing hormone	p-E H W S Y G L R P G-NH_2
Thyrotropin-releasing hormone	p-E H P-NH_2
Substance P	R P K P E E F F G L M-NH_2
Neurotensin	p-E L Y E N K P R R P Y I L
Angiotensin I	D R V Y I H P F H L
Angiotensin II	D R V Y I H P F
Vasoactive intestinal peptide	H S D A V F T D N Y T R L R K E M A V K K Y L N S I L N-NH_2

[a] Peptides with p preceding the structure indicate that the N terminal is pyroglutamate. Those with NH_2 at the end indicate that the C terminal is an amide.

viewed. The eye has evolved in such a way that a similar objective has been achieved. It is composed of live tissues that require continuous nourishment for survival. Yet the mechanisms by which this metabolic energy is derived do not place particles in the light path to interfere with the process of vision. The tissues through which light must pass do not have high numbers of mitochondria or other subcellular organelles that would scatter light. For this reason, glycolysis, for which the enzymes are soluble, plays a major role in energy production. An enhanced appreciation for the organ in which these processes take place may be obtained by viewing the schematic of a cross section of the eye shown in Figure 22.14.

Light entering the eye passes progressively through the *cornea;* the *anterior chamber,* which consists of the aqueous humor; the lens; the vitreous body, which consists of the vitreous humor; and finally focuses on the *retina,* which contains the visual sensing apparatus. The exterior of

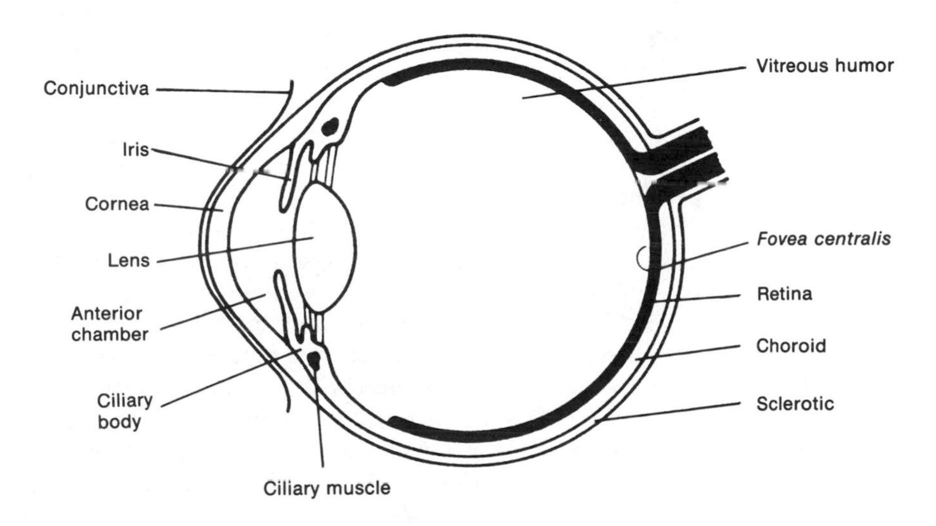

FIGURE 22.14
Schematic of a horizontal section of the left eye.

the cornea is bathed by tears, while the interior is bathed by the aqueous humor. The aqueous humor is an isoosmotic fluid containing salts, albumin, globulin, glucose, and other constituents. The aqueous humor brings nutrients to the cornea and to the lens and removes end products of metabolism from these tissues. The vitreous humor is a collagenous or gelatinouslike mass that helps maintain the shape of the eye, but also allows it to retain some pliability.

The Cornea Derives Its ATP From Aerobic Metabolism

Like tissues of the central nervous system, the major metabolic fuel for the tissues of the eye is glucose. The cornea, which is not a homogeneous tissue, obtains a relatively large percentage of its ATP from aerobic metabolism. About 30% of the glucose used by the cornea is metabolized by glycolysis and about 65% is metabolized by the hexose monophosphate pathway. On a relative weight basis, the cornea has the distinction of having the highest activity of the hexose monophosphate pathway of any other mammalian tissue. It also has a high activity of glutathione reductase, an activity that requires NADPH, a product of the hexose monophosphate pathway. The corneal epithelium is permeable to atmospheric oxygen, which participates in various oxidative reactions. These reactions can result in the formation of various active oxygen species that are harmful to the tissues, perhaps in some cases by oxidizing protein sulfhydryl groups to disulfides. Reduced glutathione (GSH) is used to reduce those disulfides back to their original native states while GSH itself is converted to oxidized glutathione (GSSG). Furthermore, oxidized glutathione (GSSG) itself may be formed by autooxidation. Glutathione reductase, in the presence of NADPH, catalyses the reduction of GSSG to 2GSH.

$$\text{GSSG} + \text{NADPH} + \text{H}^+ \xleftrightarrow{\text{GSH reductase}} 2\text{GSH} + \text{NADP}^+$$

The high activity of the hexose monophosphate pathway and of the glutathione reductase provides a system for maintaining this tissue in an appropriately reduced state by effectively neutralizing the active oxygen species.

The Lens Is Mostly Water and Protein

The lens of the eye is bathed on one side by the aqueous humor and supported on the other side by the vitreous humor. The lens has no blood supply but it is an active metabolizing tissue. It gets its nutrients from the aqueous humor and eliminates its waste into the aqueous humor. The lens is mostly water and proteins. The majority of the proteins are the α, β, and γ crystallins. There are also albuminoids, enzymes, and membrane proteins. All of these are synthesized within the lens, occurring mostly in an epithelial layer around the edge of the lens. The center area of the lens, the core, consists of the lens cells that were present at birth. The lens grows from the periphery (Figure 22.15). The human lens increases in weight and thickness with age and becomes less elastic. This is accompanied by a loss of near vision (Table 22.3); a condition referred to as *presbyopia*. On average the lens may increase threefold in size and approximately $1\frac{1}{2}$-fold in thickness from birth to about age 80.

TABLE 22.3 Changes in Focal Distance With Age

Age	Focal distance (in.)
10	2.8
20	4.4
35	9.8
45	26.2
70	240.0

SOURCE: Adapted from J. F. Koretz and G. H. Handelman, *Sci. Am.* 92, July 1988.

The proteins of the lens must be maintained in a native unaggregated state. These proteins are sensitive to various insults such as changes in the oxidation–reduction state of the cells, the osmolarity of the cells, excessively increased concentrations of metabolites, and various physical insults such as UV irradiation. Reactions of the lens that help maintain its

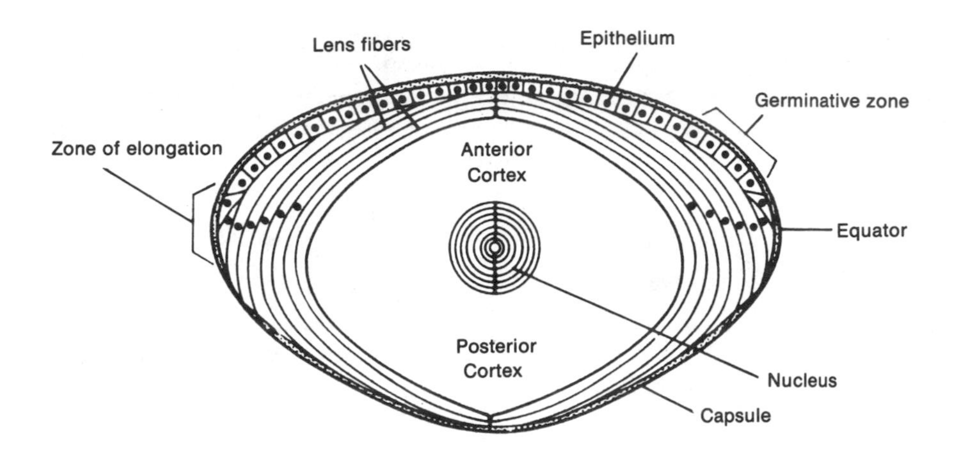

FIGURE 22.15
Schematic representation of a meridonal section of a mammalian lens.

structural integrity are the Na^+,K^+-ATPase for osmotic balance, glutathione reductase for redox state balance, and protein synthesis for growth and maintenance. The energy for these processes comes from the metabolism of glucose. About 85% of the glucose metabolized by the lens is by glycolysis, 10% by the hexose monophosphate pathway, and 3% by the tricarboxylic acid cycle, presumably mainly by the cells located at the periphery.

Cataract is the only known disease of the lens. Cataracts are opacities of the lenses brought about by a loss of osmolarity and a change in solubility of some of the proteins of the lens. This results in the occurrence of regions of high light scatter. Cataracts affect about 1 million people per year and there are no known cures or preventative measures. The remedy is lens replacement, which is a very common operation in the United States. There are basically two types of cataracts: senile cataracts and diabetic cataracts. The problem with the lens is the same in both cases: changes in the solubility and aggregation states of the lens crystallins. In senile cataracts, changes in the architectural arrangement of the lens crystallins are age-related and due to such changes as breakdown of the protein molecules starting at the C-terminal ends, deamidation, and racemization of aspartyl residues. Diabetic cataracts are a result of a loss in osmolarity of the lens due to the activity of the polyol metabolic pathway. The lens contains the enzymes **aldose reductase** and **polyol (aldose) dehydrogenase.** When the glucose concentration of the lens is high, aldose reductase reduces some of it to **sorbitol** (Figure 22.16). Sorbitol may be converted to fructose by polyol dehydrogenase. However, in the human lens the ratio of activities of these two enzymes favors sorbitol accumulation, especially since sorbitol is not used otherwise and it diffuses out of the lens rather slowly. The accumulation of sorbitol in the lens leads eventually to an increase in osmolarity of the lens. This change in osmolarity affects the structural organization of the crystalline proteins within the lens cells and enhances the rate of their aggregation and denaturation. The areas where this occurs will have increased light scattering properties—which is the definition of cataracts. Normally, sorbitol formation is not a problem because the K_m of glucose for aldose reductase is about 200 mM. Thus, very little sorbitol would be formed under normal conditions. In diabetics, where the circulating concentration of glucose is high, significant activity of this enzyme can be realized.

The Retina Derives ATP From Anaerobic Glycolysis

The *retina,* like the lens, depends heavily on anaerobic glycolysis for ATP production. Unlike the lens, the retina is a vascular tissue, but there are essentially no blood vessels in the area where visual acuity is greatest, the

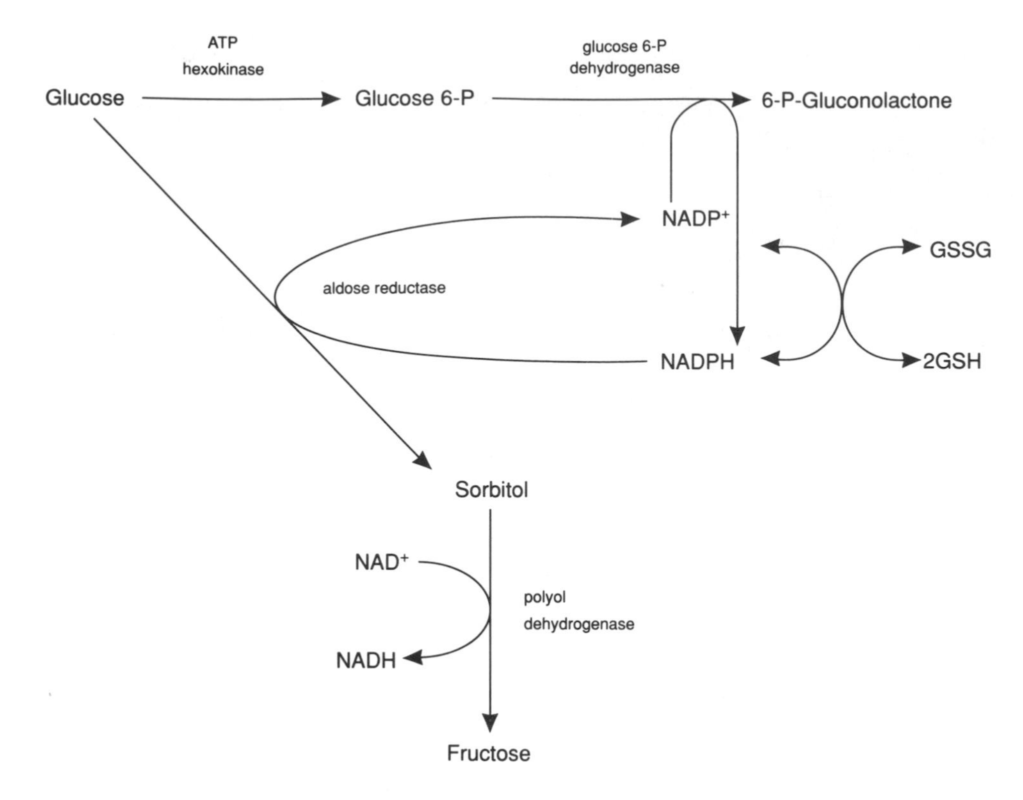

FIGURE 22.16
Some metabolic interrelationships of lens metabolism.

fovea centralis. There are also mitochondria in the retina, including the rods and the cones. There are no mitochondria in the outer segments of the rods and cones, however, where the visual pigments are located.

Excess NADH produced during glycolysis can be used to reduce pyruvate to lactate. The lactate dehydrogenase of the retina can use either NADH or NADPH, the latter being formed from the hexose monophosphate pathway. It is not clear whether lactate dehydrogenase of the retina plays any substantial role in mediating the regulation of glucose metabolism through either of these pathways by its selective use of NADH or NADPH.

The Process of Visual Transduction Involves Photochemical, Biochemical, and Electrical Events

It appears important that the visual system should have evolved to the point that there would be no structures or subcellular particles between the light source and the photoreceptors. The electron micrograph and schematic of the retinal membrane shown in Figure 22.17 appear to represent a design flaw and to be contradictory to what was said previously. If seeing is believing, it certainly can be concluded that there is not a design flaw. Light entering the eye from the lens passes the optic nerve fibers, the ganglion neurons, the bipolar neurons, and the nuclei of the rods and cones before reaching the outer segments of the rods and cones where the visual pigments are located, and hence where the initial reactions to light images occur. In spite of the presence of these structures, the more visually sensitive region of the retina is practically transparent.

It may be more appropriate to compare the eye with a video camera. The camera collects images, converts them to an electrical pulse, records them on a magnetic tape, and allows their visualization by decoding the

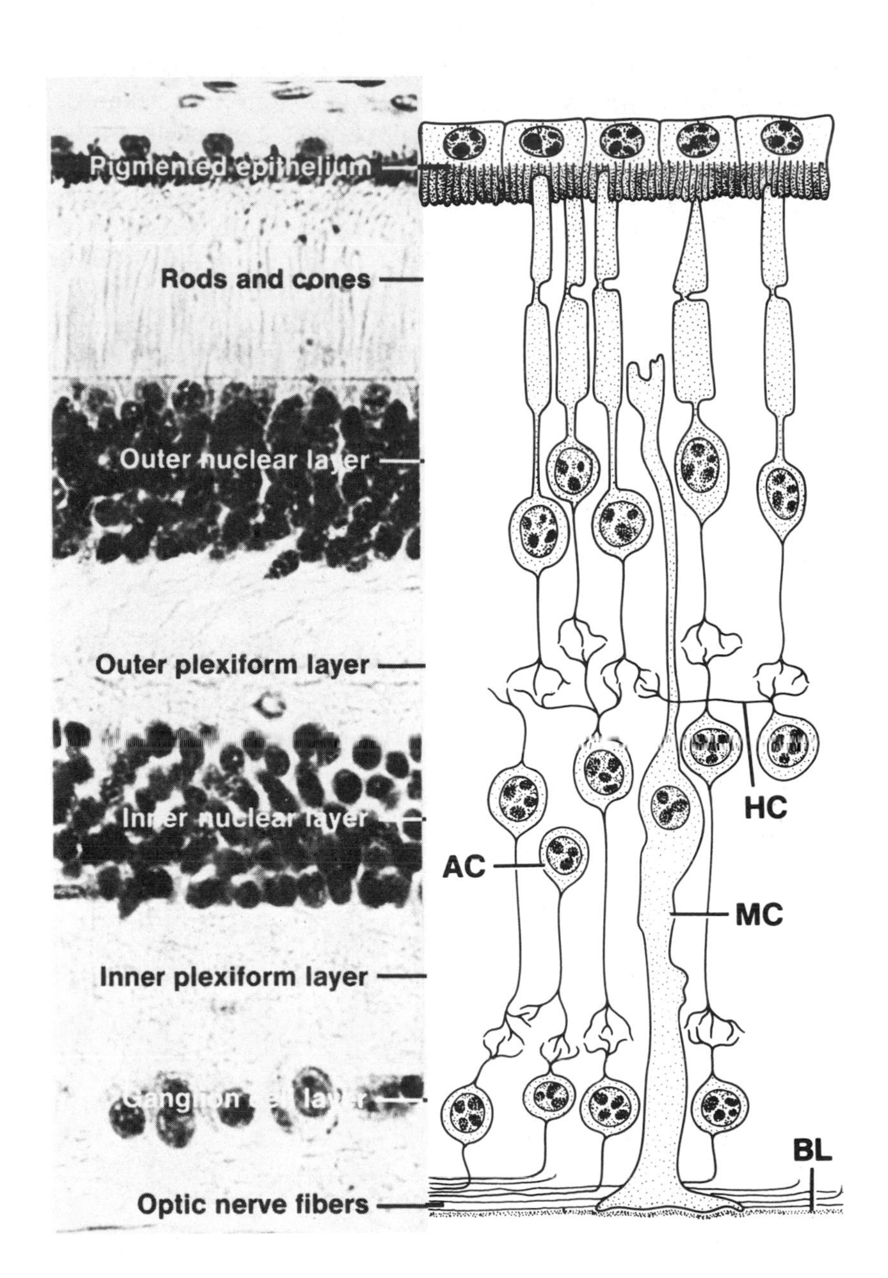

FIGURE 22.17
Electron micrograph and schematic representation of the cells of the human retina.
The tips of the rods and cones are buried in the pigmented epithelium of the outermost layer. The rods and cones form synaptic junctions with many bipolar neurons, which in turn form synapses with cells in the ganglion layer that send axons through the optic nerve to the brain. The synapse of a rod or cone with many cells is important for the integration of information. HC = Horizontal cells; AC = Amacrine cell; MC = Müller cell; BL = Basal lamina.
Reprinted with permission from R. G. Kessel and R. H. Kardon, *Tissues and Organs: A Text Atlas of Scanning Electron Microscopy*, New York: W. H. Freeman and Company, p. 87, 1979.

information on the tape. The eye focuses on an image by projecting that image onto the retina within the *fovea centralis* region. At that point, a series of events begins, the first of which is photochemical, followed by a series of biochemical events that amplify the signal, and finally a series of electrical impulses is sent to the brain where the image is reconstructed in "the mind's eye." During this process, the initial event has been transformed from a physical event to a chemical event, through a series of biochemical events, to an electrical event, to a conscious acknowledgment of the presence of an object in the environment outside of the body.

The events described above occur in the following manner. When a photon of light energy enters the eye and is absorbed by the photoreceptor, the rod outer segment, for example, it causes an isomerization of the visual pigment, **retinal** of rhodopsin, from the 11-cis form to the all-trans

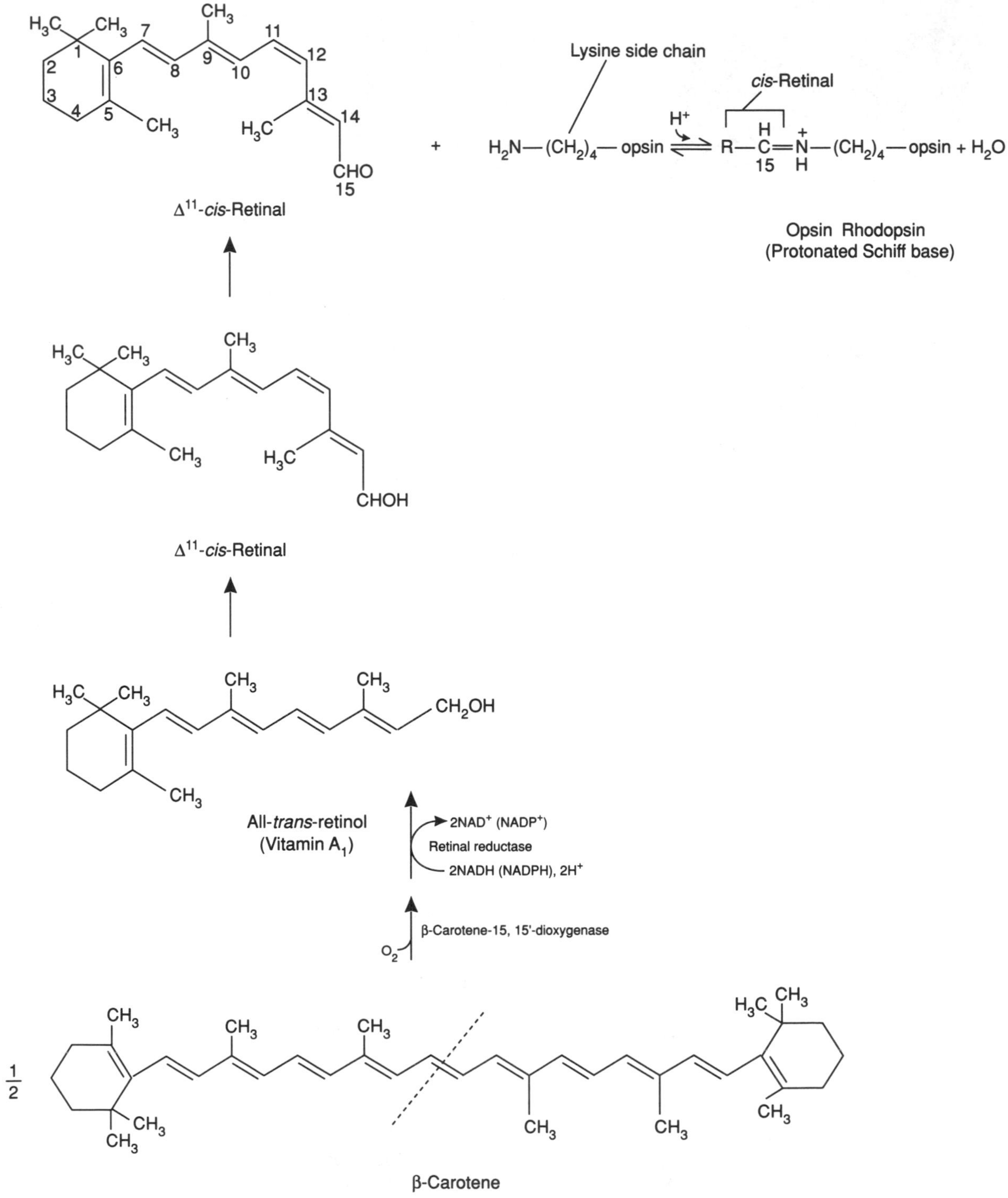

Lysine side chain

cis-Retinal

Δ^{11}-*cis*-Retinal + $H_2N—(CH_2)_4—$opsin $\overset{H^+}{\rightleftharpoons}$ $R—\overset{H}{C}=\overset{+}{N}(H)—(CH_2)_4—$opsin + H_2O

Opsin Rhodopsin
(Protonated Schiff base)

Δ^{11}-*cis*-Retinal

All-*trans*-retinol
(Vitamin A_1)

$2NAD^+$ ($NADP^+$)
Retinal reductase
2NADH (NADPH), $2H^+$

O_2
β-Carotene-15, 15'-dioxygenase

$\frac{1}{2}$ β-Carotene

FIGURE 22.18
Formation of 11-*cis*-retinal and rhodopsin from *β*-carotene.

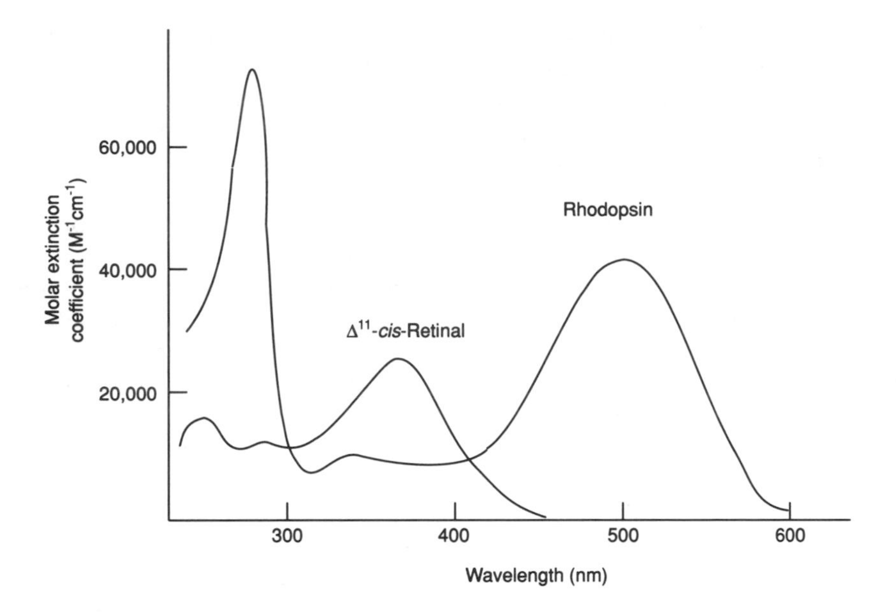

FIGURE 22.19
Absorption spectra of 11-*cis*-retinal in hexane and of rhodopsin in aqueous solution.
The absorption of rhodopsin in the 280-nm region is due mainly to the protein opsin.

form. This isomerization in turn causes a conformation change in the protein moiety of the rhodopsin complex that affects the resting membrane potential of the rod cell, and results in an electrical signal being sent to the optic nerve and transmitted to the brain. These processes will be viewed in some detail, but first let us look at the structures of the molecules and the organelles involved.

Photoreceptor Cells Are Rods and Cones

The photoreceptor cells of the eye are the rods and the cones. They are arranged as shown in Figure 22.17 above. There are flattened disks in both the rods and the cones that contain a photoreceptor pigment. In the rod cells, this pigment is rhodopsin. **Rhodopsin** is a transmembrane protein to which is bound a prosthetic group, **11-*cis*-retinal.** The protein minus its prosthetic group is opsin. There are three other opsin-like proteins in cone cells, which are discussed on p. 952. Focus for now will be on the rod cells and rhodopsin.

Reactions involved in the formation of 11-*cis*-retinal from **β-carotene** and in the formation of rhodopsin from opsin and 11-*cis*-retinal are shown in Figure 22.18. Rhodopsin is a protein of approximately 40 kDa to which is attached 11-*cis*-retinal. Retinal is attached to the ε-amino group of lysine at position 296 through a protonated Schiff base. The 11-*cis*-retinal is derived from vitamin A and/or β-carotene of the diet. Cleavage of β-carotene yields two molecules of all-*trans*- retinol. There is an enzyme in the pigmented epithelial cell layer of the retina that catalyzes the isomerization of the **all-*trans*-retinol** to the **11-*cis*-retinol.** Oxidation of the 11-*cis*-retinol to the 11-*cis*-retinal occurs in the rod outer segment.

The absorption spectra of 11-*cis*-retinal and of rhodopsin are shown in Figure 22.19. Note the shift in the wavelength of maximum absorption of 11-*cis*-retinal upon binding to opsin. The broad absorption band that peaks at about 500 nm is coincident with the light-sensitive response of the rod cells. The second peak at low wavelength for the rhodopsin is due to the absorption of the protein moiety of the complex.

Based on knowledge of the amino acid sequence and the mechanistic requirements for protein folding, a three-dimensional model for rhodopsin has been constructed (Figure 22.20). Rhodopsin contains seven α helices that are embedded within the membrane. The lysine residue to which

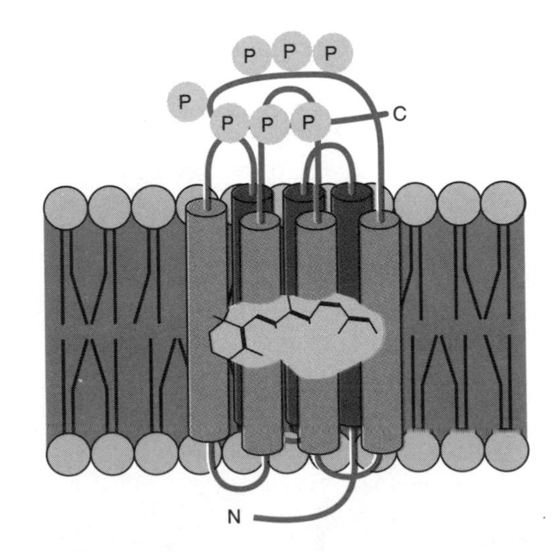

FIGURE 22.20
A model of the structure of the vertebrate rhodopsin.
Redrawn from L. Stryer, *Ann. Rev. Neurosci.* 9:87, 1986.

retinal is attached is on the seventh of these helices approximately midway between the two faces of the membrane.

The initial events, absorption of a photon and the subsequent isomerization reaction, are rapid, requiring only picoseconds. Following this, however, a series of changes take place in rhodopsin, leading to various conformational states, each of which has specific absorption characteristics. These are referred to as "activated" rhodopsin molecules, one of which (R*), plays an important role in the activation of other proteins in the visual cascade. Finally, rhodopsin undergoes a dissociation reaction giving opsin and all-*trans*-retinal. The processes outlined in Figure 22.18 can start over again.

Following exposure of the rod cell to a light pulse, the electrical potential of the cell changes, which is different in magnitude from that of neuron depolarization. The rod cell membrane's resting potential is approximately −30 mV instead of the −70 mV seen with neurons. Excitation of the rod cell causes a *hyperpolarization* of the membrane; it goes from an approximately −30 mV to approximately −35 mV or more before returning to the resting potential (Figure 22.21). It takes a significant fraction of a second or more for the potential to reach its maximum state of hyperpolarization. A number of biochemical events take place during this time interval, beginning with a reaction involving the activated rhodopsin molecule, R*.

The intermediate conformational stages that rhodopsin goes through after light activation are shown in Figure 22.22. Note that at 37°C, the activated rhodopsin has decayed in slightly more than 1 ms through several intermediates to **metarhodopsin II.** Metarhodopsin II has a half-life of approximately 1 min. The active species, R*, that is involved in the biochemical reactions of interest is the *metarhodopsin II* species. This does not mean that it takes up to 1 min for the biochemical events to be initiated. That process can commence after metarhodopsin I is formed and begins to decay to metarhodopsin II, in the hundredths of microsecond time range.

A list of the major proteins involved in the **phototransduction cascade** is given in Table 22.4. The relationship of each protein to the membrane, the

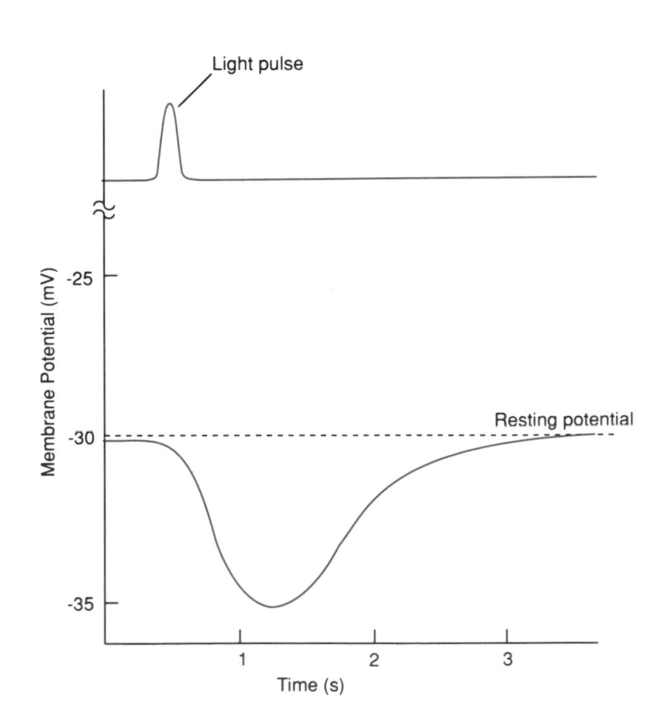

FIGURE 22.21
Changes in the potential of a rod cell membrane after a light pulse.
Redrawn from J. Darnell, H. Lodish, and D. Baltimore, *Molecular Cell Biology*, New York: Scientific American Books, Figure 17-58, 1986, p. 763.

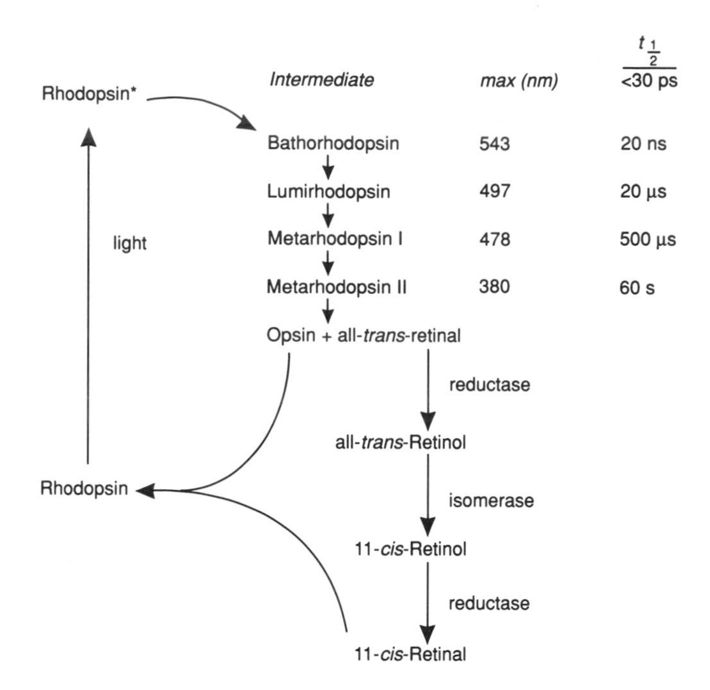

FIGURE 22.22
Light activation of rhodopsin.

molecular mass, and the equivalent cytoplasmic concentration (where applicable) is given, and the approximate number of molecules involved in a single photon event is also given. A scheme of how these proteins may interact with each other is shown in Figure 22.23.

The dark-adapted rods are very sensitive to light, being capable of detecting a single photon. The rod is a specialized type of neuron. As indicated above, the membrane potential of the rod (and other photoreceptor cells) is -30 mV instead of -70 mV. The reason for this difference is that the Na^+ channels of the photoreceptor cells are maintained in a partially opened state by the direct and kinetically dynamic interaction of cyclic GMP (cGMP) with an ion–channel protein. Thus, the state of polarization of the membrane has a direct relationship to the concentration of cGMP. The biochemical events that are involved in the transduction of the light signal to electrical impulse revolve around the metabolic perturbations that result in changes in the concentration of this nucleotide.

The *disks* or *rod cells* are completely separated from the outer plasma membrane. Thus, the signal generated as a result of light activation of the rhodopsin within the disk membrane must travel through the cytosol to the plasma membrane to effect the hyperpolarization event. The signal

TABLE 22.4 Major Proteins Involved in the Phototransduction Cascade

Protein	Relation to membrane	Molecular mass (kDa)	Concentration in cytoplasm (μM)
Rhodopsin	Intrinsic	39	—
Transducin ($\alpha + \beta + \gamma$)	Peripheral or soluble	80	500
Phosphodiesterase	Peripheral	200	150
Rhodopsin kinase	Soluble	65	5
Arrestin	Soluble	48	500
Guanylate cyclase	Attached to cytoskeleton	?	?
cGMP-activated channel	Intrinsic	66	?

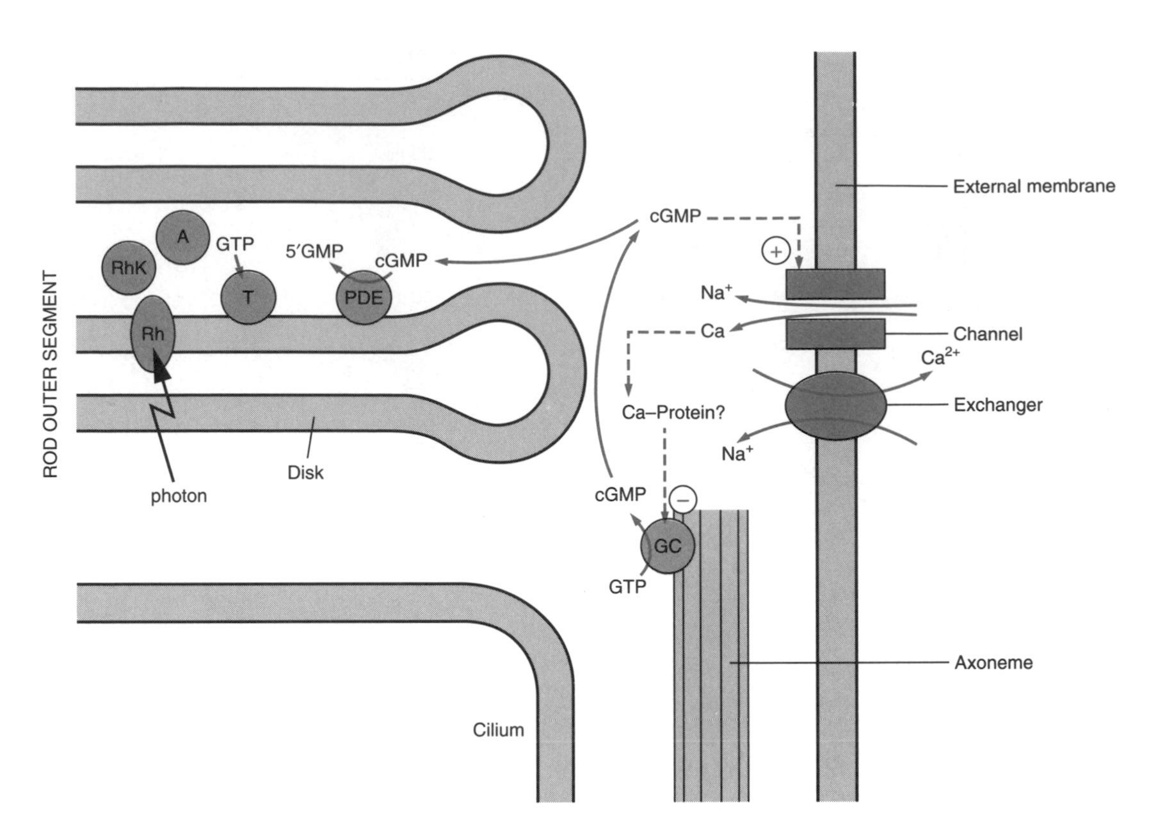

FIGURE 22.23
Schematic representation of the location of the proteins shown in Table 21.4 relative to intracellular compartments.
Rh = rhodopsin; RhK = rhodopsin kinase; A = arrestin; GC = guanylate cyclase; T = transducin; P = phosphodiesterase.
Redrawn from M. Chabre and P. Deterre, *Eur. J. Biochem.* 179:255, 1989.

that is generated by the rod cells, unlike neurons, is not an all-or-none type response. The signal may be graded in intensity. The net effect of light is to reduce the rate of opening of the sodium channels with no effect on the rate of closing of the channels. Thus, if the rate of sodium channel opening is momentarily decreased, the concentration of sodium within the rods will decrease, leading to hyperpolarization. The extent to which the channels are open is dependent on the concentration of cGMP. Thus, a decrease in the concentration of cGMP directly affects the number of channels that are open. The opening rate for the sodium channels is $1.8 \times 10^3\ s^{-1}$, which is within an order of magnitude of the diffusion control limit of approximately $10^4\ s^{-1}$. The closing rate is approximately $0.2 \times 10^3\ s^{-1}$, and, as stated above, is unaffected by cGMP.

Transducin Is a G Protein Involved in Signal Transduction

Chemical Properties of Transducin

Transducin, a protein found in the retinal rods and cones, is a G protein that plays a major role in signal transduction. Its function in the visual process parallels that of other G proteins associated with hormone action (Figure 22.24). Transducin, like other G proteins, is composed of three dissimilar subunits called alpha (α), beta (β), and gamma (γ). The α subunit of transducin contains 350 amino acids and is approximately 44 kDa in size. The amino acid sequence is significantly different from many other G proteins, yet it is sufficiently homologous that the functional

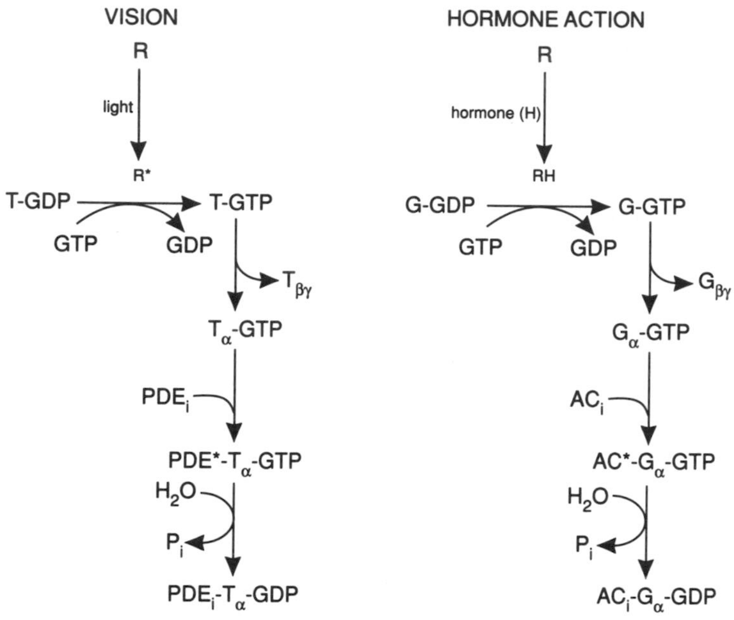

VISION

R → (light) → R*

T-GDP → T-GTP (GTP → GDP)

T-GTP → $T_{\beta\gamma}$ + T_α-GTP

PDE_i + T_α-GTP → PDE*-T_α-GTP

PDE*-T_α-GTP (H_2O → P_i) → PDE_i-T_α-GDP

HORMONE ACTION

R → (hormone (H)) → RH

G-GDP → G-GTP (GTP → GDP)

G-GTP → $G_{\beta\gamma}$ + G_α-GTP

AC_i + G_α-GTP → AC*-G_α-GTP

AC*-G_α-GTP (H_2O → P_i) → AC_i-G_α-GDP

FIGURE 22.24
Similarities between transducin (T) function in the visual cycle and G protein (G) function in hormone action.
In the visual cycle, R = the visual pigment, and for hormone action, it is the membrane receptor; PDE = phosphodiesterase; AC = adenylate cyclase. Inactive forms are denoted by the subscript i, and active forms by the superscript *. The final complexes shown dissociate and the T_α-GDP and G_α-GDP reassociate with their other subunits to start the cycle over again.

criteria for classifying it as a G protein are met. The amino acid sequence of bovine transducin inferred from its cDNA is shown in Figure 22.25. The residues shown in color are in the area where ADP ribosylation occurs under the influence of cholera or pertussis toxins.

Two forms of the ***α* subunit** of transducin have been identified, $G_{t\alpha1}$ and $G_{t\alpha2}$. Form $G_{t\alpha1}$ obtained from rods is a 350 amino acid protein, whereas form $G_{t\alpha2}$ obtained from cones is a 354 amino acid protein. The first 350 amino acids of the $G_{t\alpha2}$ form from cones are identical to those of the $G_{t\alpha1}$ form from rods. Antibodies raised against the rod α subunit ($G_{t\alpha1}$) or against the first 350 amino acids of the cone $G_{t\alpha2}$ react only with rod $G_{t\alpha1}$, whereas antibodies raised against the cone α subunit ($G_{t\alpha2}$) show specificity for the cone photoreceptor. This suggests that the orientation of the α subunit in the complex is such that the exposed epitope(s) are in the region of the C termini. The function of these two isoforms of the α subunit is the same, that is, activation of the phosphodiesterase.

The ***β* subunit** of transducin has 340 amino acid residues and is approximately 37.4 kDa in size (Figure 22.26). The amino acid sequence is nearly identical to the β subunit of all other G proteins. Experiments performed in vitro show that they can be interchanged with no apparent effect on the function of the protein.

The ***γ* subunit** of transducin has 69 amino acids (Figure 22.27) and an approximate molecular weight of 8 kDa. It is a very hydrophilic and acidic protein. There is a total of 5 more acidic than basic amino acids, and 29 of the amino acids are hydrophobic. There are no alanine, histidine, or tryptophan residues. The γ subunit of transducin appears to be identical in amino acid sequence to the γ subunit of other G proteins. The conformations of the different proteins in their cellular environment may be different, since antibodies prepared against the γ subunit of transducin fail to cross-react with γ subunits of other G proteins.

Transducin, sometimes referred to as G_t, exists in either a peripheral or soluble state relative to the disk membrane. In the dark-adapted state, transducin has GDP bound to its α subunit. The transducin–GDP complex binds to the activated rhodopsin, R*, to give an activated-rhodopsin–G_{ta} complex, R*–$G_{t\alpha}$–GDP. This complex undergoes an exchange reac-

```
M G A G A S A E E K
H S R E L E K K L K
E D A E K D A R T V
K L L L L G A G E S
G K S T I V K Q M K
I I H Q D G Y S L E
E C L E F I A I I Y
G N T L Q S I L A I
V R A M T T L N I Q
Y G D S A R Q D D A
R K L M H M A D T I
E E G T M P K E M S
D I I Q R L W K D S
G I Q A C F D R A S
E Y Q L N D S A G Y
Y L S D L E R L V T
P G Y V P T E Q D V
L R S R V K T T G I
I E T Q F S F K D L
N F R M F D V G G Q
R S E R K K W I H C
F E G V T C I I F I
A A L S A Y D M V L
V E D D E V N R M H
E S L H L F N S I C
N H R Y F A T T S I
V L F L N K K D V F
S E K I K K A H L S
I C F P D Y N G P N
T Y E D A G N Y I K
V Q F L E L N M R R
D V K E I Y S H M T
C A T D T Q N V K F
V F D A V T D I I I
K E N L K D C G L F
```

FIGURE 22.25
Amino acid sequence of the α subunit of transducin as determined from its cDNA.

```
M S E L E Q L R Q E
A E Q L R N Q I R D
A R K A C G D S T L
T Q I T A G L D P V
G R I Q M R T R R T
L R G H L A K I Y A
M H W G T D S R L L
V S A S Q D G K L I
I W D S Y T T N K V
H A I P L R S S W V
M T C A Y A ? S G N
F V A C G G L D N I
C S I Y S L K T R E
G N V R V S R E L P
G H T G Y L S C C R
F L D D N Q I I T S
S G D T T C A L W D
I E T G Q Q T V G F
A G H S G D V M S L
S L A P N G R T F V
S G A C D A S I K L
W D V R D S M C R Q
T F I G H E S D I N
A V A F F P N G Y A
F T T G S D D A T C
R L F D L R A D Q E
L L M Y S H D N I I
C G I T S V A F S R
S G R L L L A G Y D
D F N C N I W D A M
K G D R A G V L A G
H D N R V S C L G V
T D D G M A V A T G
S W D S F L K I W N
```

FIGURE 22.26
Amino acid sequence of the β subunit of transducin as determined from its cDNA.

```
P V I N I E D L T E
K D K L K M E V D Q
L K K E V T L E R M
L V S K C C E E F R
D Y V E E R S G E D
P L V K G I P E D K
N P F K E L K G G
```

FIGURE 22.27
Amino acid sequence of the γ subunit of transducin as determined by chemical sequencing.
Note that it contains no Ala, His, or Trp. There are two Cys that are next to each other.

tion with GTP to give the R*–$G_{t\alpha}$–GTP complex, which binds to and activates a **phosphodiesterase.** The phosphodiesterase hydrolyzes some of the cGMP, reducing the cellular concentration of free cGMP and thereby effecting sodium channel closing as described above. These events are summarized in the diagram of Figure 22.28.

The actual concentration of cGMP in normal rod cells changes very little during the light event. However, the flux of nucleotides through the cGMP pool changes over a fivefold range in response to a light pulse. In vitro experiments with specially prepared rod cell preparations have shown that light flashes capable of exciting 10^6 rhodopsin molecules led to a substantial decrease in cGMP concentration. The decrease in cGMP concentration preceded the change in current of the membrane, and 10% of the cGMP was hydrolyzed within 50 ms. This allowed the conclusion that the decrease (or change) in cGMP concentration occurs fast enough for it to be the mediator of the electrical activity of this cell.

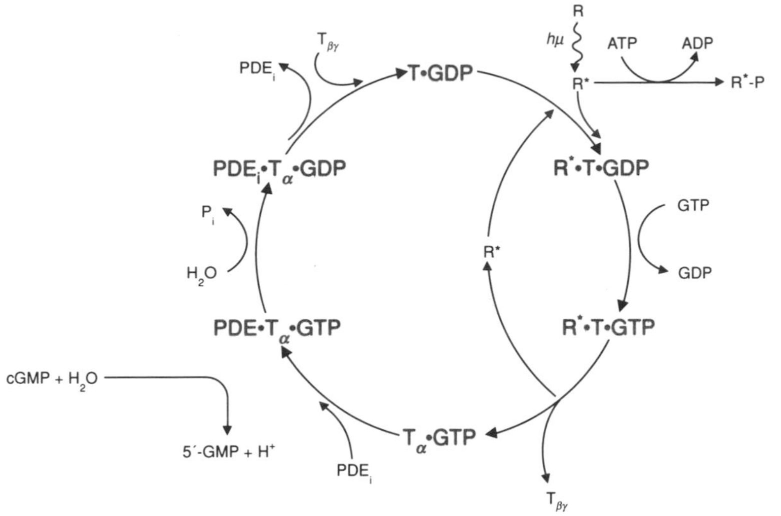

FIGURE 22.28
Diagrammatic representation of the phototransduction cascade.
Redrawn from M. Chabre and P. Deterre, *Eur. J. Biochem.* 179:255, 1989.

The cGMP concentration is regulated by the intracellular Ca^{2+} concentration. The activities of the enzymes responsible for the synthesis and degradation of cGMP are sensitive to Ca^{2+} concentration: at low Ca^{2+} concentration, the guanylate cyclase activity increases whereas the phosphodiesterase activity decreases. Calcium enters the rod cells in the dark through the sodium channels, but is pumped out of the cell, thus maintaining the Ca^{2+} concentration at a steady-state level. When the sodium channels are closed, Ca^{2+} entry is inhibited but the efflux is unchanged. This results in a decrease in the intracellular Ca^{2+} concentration, which in turn activates the guanylate cyclase and inhibits the phosphodiesterase.

There are other important events that lead to a cessation of the cascade. The $G_{t\alpha}$ subunit has GTPase activity. The bound GTP is hydrolyzed to GDP and leads to a dissociation of the phosphodiesterase–$G_{t\alpha}$ GDP complex, yielding the inactive phosphodiesterase and $G_{t\alpha}$ GDP complex that can reassociate with the $G_{t\beta} : G_{t\gamma}$ complex to give back the trimeric transducin. Active rhodopsin, R*, is a substrate for **rhodopsin kinase. Arrestin,** a 48-kDa soluble protein binds to the phosphorylated R* and inhibits its ability to activate further the phosphodiesterase. The concentration of transducin is about 100 times greater than that of the kinase. Thus, R* first activates many molecules of transducin before it is inactivated by the combined action of the kinase and arrestin.

The relative contribution of each of these activators to termination of the visual cascade has not been quantified. Other mechanisms have been proposed for explaining the decrease in the action of the phosphodiesterase at the end of the activation cycle. However, they do not fit the appropriate time frame in in vitro experiments. It appears that inactivation of the phosphodiesterase–$G_{t\alpha}$–GTP complex only as a result of GTP hydrolysis by transducin ($G_{t\alpha}$) does not occur fast enough, requiring tens of seconds, to stop the cascade under physiological conditions. The real situation probably involves the intervention of all of these steps to stop the cascade, namely, Ca^{2+}-dependent changes in enzyme activity, hydrolysis of GTP in the complex, and the combined action of rhodopsin kinase and arrestin.

Color Vision Resides in the Cones

Even though there are photographic artists, such as the late Ansel Adams, who make the world look beautiful in black and white, the intervention of

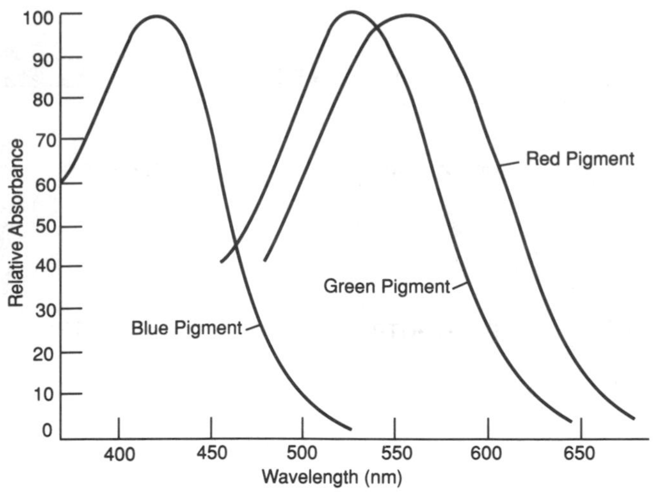

FIGURE 22.29
Relative spectral sensitivity curves of the three color pigments of the human cones.
Adapted from J. Nathans, *Sci. Am.*, February, 1989, p. 42.

colors in the spectrum of life's pictures brings another degree of beauty to the wonders of nature and the beauty of life . . . even the ability to make a distinction between tissues from histological staining. This ability of humans to distinguish colors resides within a relatively small portion of the visual system: the *cones*. The number of cones are few compared with the number of rods within the human eye. Some animals like dogs have even fewer cones, and other animals, like birds, have many more.

The general mechanism for the stimulation of the cone cells is exactly the same as it is for the rod cells. There are three types of cone cells, each with a different visual pigment: one sensitive to blue light, one to green light, and one to red light. The blue pigment has optimum absorbance at 420 nm, the green pigment at 535 nm, and the red pigment at 565 nm (Figure 22.29). Each of these pigments has 11-*cis*-retinal as the prosthetic group, and, when activated by light, the 11-*cis*-retinal isomerizes to all-*trans*-retinal in exactly the same manner as it does in the rod cells. Colors other than those of the visual pigments are distinguished by graded stimulation of the different cones and the comparative analysis by the brain. Color vision in mammals is trichromatic.

The characteristic of **color discrimination** by the cone cells is an inherent property of the proteins to which the 11-*cis*-retinal is attached. As shown in Figure 22.18, 11-*cis*-retinal is attached to the protein through a protonated Schiff base. The 11-*cis*-retinal also has a conjugated double-bond system that contributes to its absorption spectrum (Figure 22.19). When 11-*cis*-retinal is bound to different visual proteins, amino acid residues in the vicinity of the protonated base and in the vicinity of the conjugated π bond system influence the energy levels of the conjugated bond system of the 11-*cis*-retinal and impart different characteristics to the absorption spectra. Thus, the absorption maxima are different for the different color pigments, and more or less energy, in the form of different wavelengths of light, is required to effect the cis–trans isomerization. On the basis of structural comparison of the amino acid sequences and from models, it appears that the 11-*cis*-retinal in the green and the red pigments is in very similar environments.

The genes for the color pigments have been cloned and their amino acid sequences inferred from the gene sequence. A structural comparison of the sequences of the visual pigments is shown in Figure 22.30. Open circles represent amino acids that are the same, and closed circles represent amino acids that are different. A string of closed circles at either end may represent an extension of the chain of one protein relative to the other. The red and the green pigments show the greatest degree of homol-

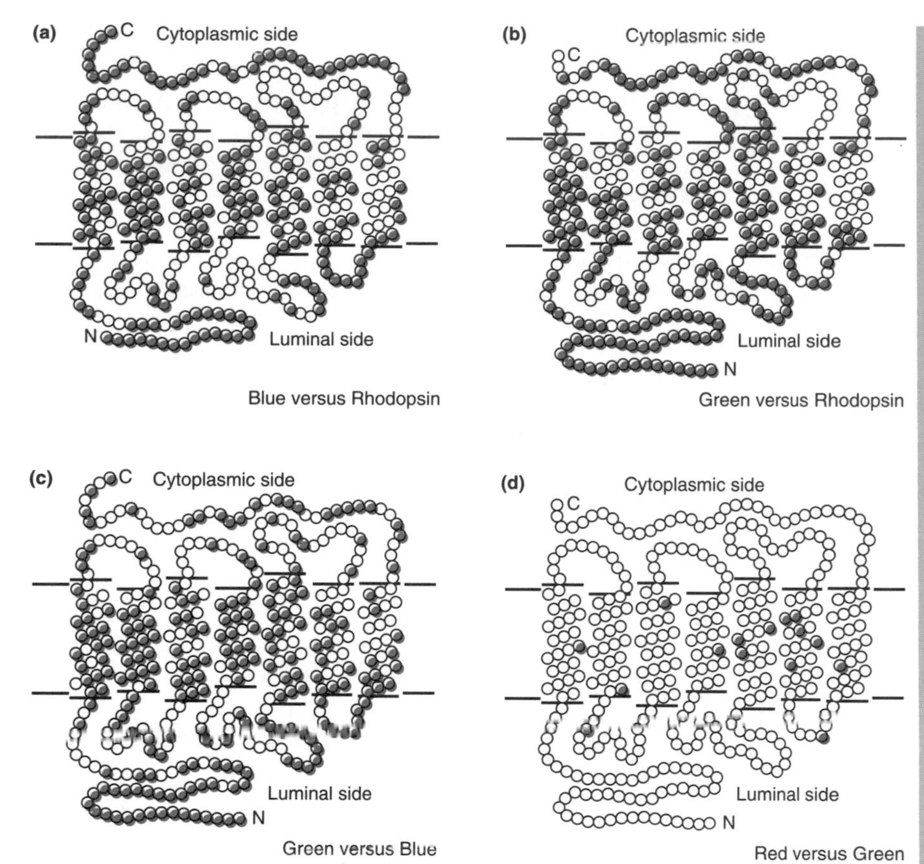

FIGURE 22.30
Comparisons of the amino acid sequences of the human visual pigments. Each black dot indicates an amino acid difference.
Adapted from J. Nathans, *Annu. Rev. Neurosci.* 10:163, 1987.

ogy, about 96% identity, whereas the degree of homology between different pairs of the others is between 40 and 45%.

The genes encoding the visual pigments have been mapped to specific chromosomes (Clin. Corr. 22.2). The rhodopsin gene resides on the third chromosome, the gene encoding the blue pigment resides on the seventh chromosome, and the two genes for the red and green pigments reside on the X chromosome. In spite of their great similarity, the red and green pigments are distinctly different proteins. Individuals have been identified with inherited variations that affect one but not both pigments simultaneously.

There are some differences between the rods and the cones structurally as well as in the proteins involved in signal transduction. There is no evidence, however, that the mechanism for the chemical or the biochemical reactions are different. The disks of the cone cells do show some attachment to the outer plasma membrane, which may contribute to the somewhat quicker response of the plasma membrane to hyperpolarization than in the rod cells. As shown in Table 22.4, there is an isoform of the G-α-subunit of the transducin molecule that occurs in the cone cells. The phosphodiesterase of the cones also appears to react differently immunologically from that of the rods. There is no evidence to date, however, to suggest that their functions are different or that the mechanism of their actions differ from their counterproteins in the rods.

The sensitivity and the response time of the rods is different from that of the cones. The absorption of a single photon by the photoreceptors in

CLIN. CORR. 22.2 CHROMOSOMAL LOCATION OF GENES FOR VISION

The chromosomal arrangement of the genes for vision, very interestingly, precludes the inheritance of a single defective gene from a single parent that would render the recipient sightless. All of the genes for expression of the visual pigments occur on chromosomes that exist in pairs except in males where there is only a single X chromosome containing the genes for the red and the green pigments. In females, there is a pair of X chromosomes and, therefore, color abnormalities in females are rare, affecting only about ½ of 1% of the population. By contrast, about 8% of males have abnormal color vision affecting red or green perception, and, on rare occasions, both. For the sake of simplicity, the proteins coded for by the different genes will be referred to as pigments in spite of the fact that they become pigments only when they form complexes with 11-*cis*-retinal.

The gene that codes for the protein moiety of rhodopsin, the rod pigment, is located on the third chromosome. The genes that code for the three opsin proteins of the cone cells and are responsible for color vision are located on two different chromosomes. The gene for the blue pigment is on the seventh chromosome. The genes for the red and green pigments are tightly linked and are on the X chromosome, which normally contains one gene for the red pigment and from one to three genes for the green pigment. In a given set of cones, however, only one of these gene types is expressed, that is, either the gene for the red pigment or the gene for the green pigment.

Genetic mutations may cause structural abnormalities in the proteins that influence the binding of retinal or the environment in which retinal resides. In addition, the gene for a specific opsin protein may not be expressed. If 11-*cis*-retinal does not bind or one of the proteins is not expressed, the individual will have dichromatic color vision and be color blind for the color of the missing pigment. If the mutation causes a change in the environment around the 11-*cis*-retinal that causes a shift in the absorption spectrum of the pigment, the individual will have abnormal trichromatic color vision, that is, have different degrees of stimulation of the three cone pigments, resulting in a different integration of the signal and hence different interpretation of the color.

Vollrath, D., Nathans, J., and Davis, R. W. *Science* 240:1669, 1988. Also, see other references by J. Nathans at the end of this chapter.

the rods generates a current of approximately 1–3 pA (1–3 $\times$ 10^{-12}), whereas the same event in the cones generates a current of approximately 10 fA (10 $\times$ 10^{-15}), about one one-hundredth of the rod response. The cone response, however, is about four times faster than the rod response. Thus, the cones are better suited for discerning rapidly changing visual events and the rods are better suited for low-light visual acuity. A direct correlation between these functional differences and the structural differences between the organelles and/or some of the proteins involved in the transduction and signal amplification has not yet been made.

22.5 MUSCLE CONTRACTION

On the basis of an extensive evaluation of electron micrographs of skeletal muscle tissue, the **sliding filament model** for muscle contraction was proposed. This simple but eloquent model has weathered the test of time. During most recent years, the genes for many of the proteins found in muscle tissue have been cloned and their amino acid sequences inferred from the sequences of their cDNA molecules. Yet the story on muscle contraction is far from over. The three-dimensional structures of the proteins are just beginning to appear in the literature; thus, details of the mechanisms of interactions between them are still speculative, even though based on sound physicochemical principles. Many aspccts of the regulation and control of the genes expressing these proteins during development remain mysteries. The mechanism by which calcium is released to start the process of muscle contraction is not clearly understood either. There are, however, several plausible theories of how the process works. In this section, a fundamental approach will be taken to explain in biochemical terms the mechanism of muscle contraction.

There are three types of muscles: skeletal, cardiac, and smooth. This discussion will focus primarily on skeletal muscle. Differences in contractile mechanism or regulation of the contraction process for the others will be highlighted as appropriated.

Skeletal Muscle Contraction Follows An Electrical to Chemical to Mechanical Path

For skeletal muscle contraction, the signal begins with an electrical impulse from a nerve. This is followed by a chemical change occurring within the unit cell of the muscle, and is followed by contraction, a mechanical process. Thus, **the signal transduction process** goes from *electrical* to *chemical* to *mechanical.*

A schematic diagram of the structural organization of skeletal muscle is shown in Figure 22.31. The muscle consists of bundles of fibers (diagram c). Each of these bundles is called a fasciculus (diagram b). The fibers themselves are made up of myofibrils (diagram d), and each myofibril is a continuous series of units called sarcomeres. The muscle cell is multinucleated and is no longer capable of division. Most of the muscle cells survive for the life of the animal, but they can be replaced or lengthened by the fusion of myoblast cells.

A muscle cell is shown diagrammatically in Figure 22.32. Note that the myofibrils are surrounded by a membranous structure called a sarcoplasmic reticulum. At discrete intervals along the bundles of myofibrils and connected to the terminal cisterna of the sarcoplasmic reticulum are structures called transverse tubules. The transverse tubules are connected to the external plasma membrane that surrounds the entire structure. The nuclei and the mitochondria lie just inside the plasma membrane.

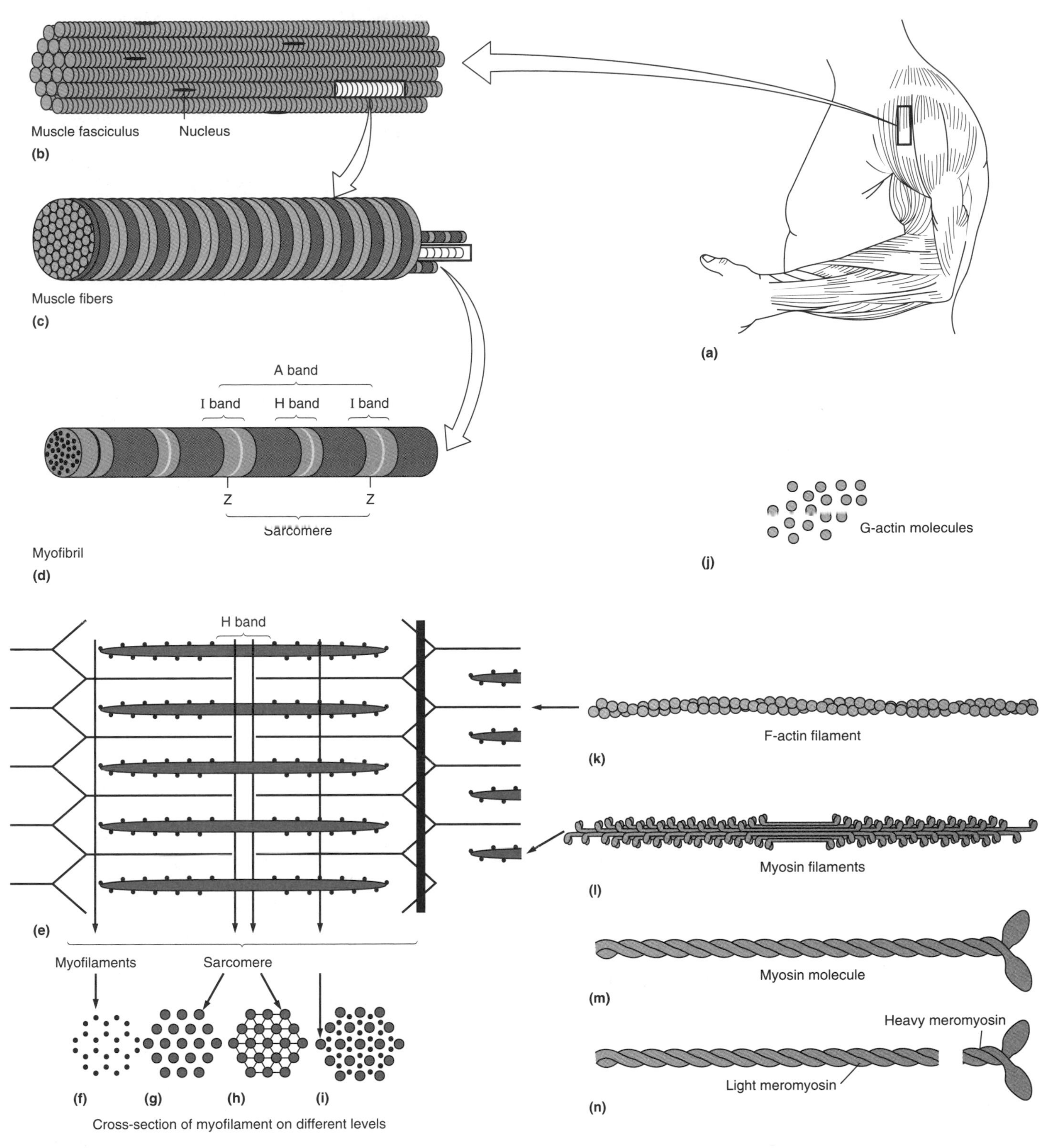

FIGURE 22.31
Structural organization of skeletal muscle.
Redrawn from W. D. Bloom and D. W. Fawcett, *Textbook of Histology,* 10th ed. Philadelphia: Saunders, 1975.

The contractile mechanism of the sarcomere consists of the fibrous material and bands as seen in the electron microscope. The single **contractile unit** extends from Z line to Z line as shown in both Figures 22.31d and 22.32. The bands seen in the sarcomere are due to the arrangements

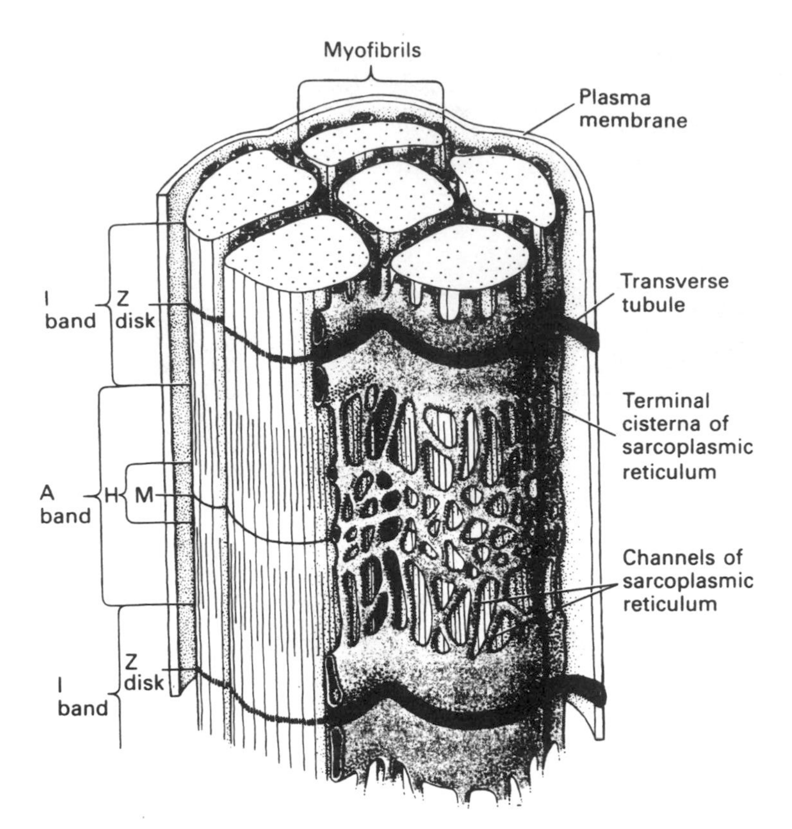

FIGURE 22.32
A three-dimensional representation of six myofibrils.
The lumen of the transverse tubules may connect with the extra-cellular medium and enter the fibers at the Z disk.
Reprinted with permission from J. Darnell, H. Lodish, and D. Baltimore, *Molecular Cell Biology,* New York: Scientific American Books, Figure 19-16, 1986, p. 827.

of specific proteins within the contractile element. These are shown in Figure 22.31E. Two types of fibers are apparent: long thick ones that lie near the center of the sarcomere and have protrusions on both ends, and long thin ones that are attached to the area called the Z line. There is an area called the **I band (isotropic)** that extends for a short distance on both sides of the Z line. Within this region, there are only thin filaments attached to a protein band within the Z line. In the center of the sarcomere there is another band called the **H band.** Within this region there are no thin filaments, just the center regions of the heavy fibers. In the middle of the H band, there is a somewhat diffuse band resulting from the presence of other proteins that assist in cross-linking the fibers of the heavy filaments (Figure 22.31, pattern H). The region that includes the thick fibers and is located between the inner edges of the I bands is called the **A band (anisotropic).** When the muscle contracts, the H and I bands shorten, but the distance between the Z line and the near edge of the H band remains constant. Similarly, the distance between the innermost edges of the I

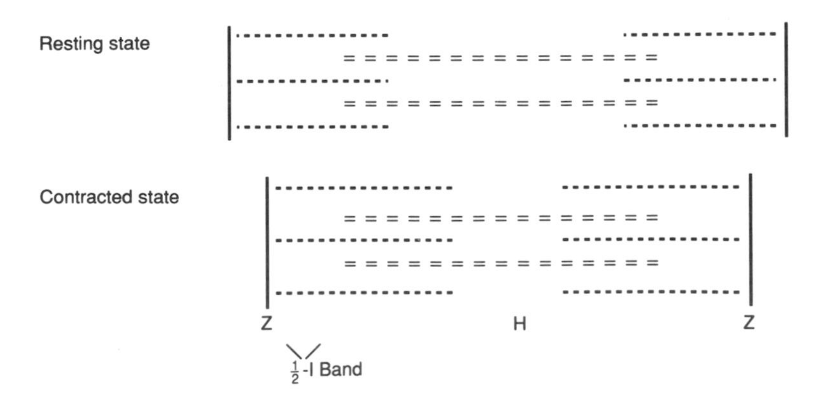

FIGURE 22.33
Sliding filament model of muscle contraction.
Note that the H and the I band have become shorter in the contracted state compared with their width in the resting state. The lengths of the large and small filaments have not changed.

bands on both ends of the sarcomere also remains constant. This occurs because neither the thin filaments nor the thick filaments change in length during contraction (Figure 22.33). Contraction, therefore, results when these filaments "slide" past each other.

The contractile elements of muscle consist of many different proteins, eight of which are listed in Table 22.5. The two most abundant proteins in the contractile element are myosin and actin. About 60–70% of the muscle protein is myosin and about 20–25% is actin. The thick filament is mostly myosin and the thin filament is mostly actin. However, three of the other proteins listed in Table 22.5 are also associated with the thin filaments and two are associated with the thick filaments.

TABLE 22.5 Molecular Weights of Skeletal Muscle Contractile Proteins

Myosin	500,000
Heavy chain	200,000
Light chain	20,000
Actin monomer (G-actin)	42,000
Tropomyosin	70,000
Troponin	76,000
Tn-C subunit	18,000
Tn-I subunit	23,000
Tn-T subunit	37,000
α-Actinin	200,000
C-protein	150,000
β-Actinin	60,000
M-protein	100,000

The Thick Filament of Muscle Is Myosin

A schematic drawing of the myosin molecule is shown in Figure 22.34, which is a representation of the electron micrographs shown in Figure 22.35. Myosin is a long molecule with two globular heads on one end. It is composed of two heavy chains of about 230,000 mol wt and two dissimilar pairs of light chains of about 20,000 mol wt each; each heavy chain is associated with a dissimilar pair of light chains. The light chains are "calmodulin-like" proteins that bind calcium. Two of them can be removed easily, the other two with difficulty. Their removal in vitro does not appear to affect the essential activity of the heavy chains. The light chains bind the heavy chains in the vicinity of the head groups of the myosin.

The tail section of myosin is the carboxyl end of the chain. It is coiled around itself in an α-helical arrangement (Figure 22.34). Treatment with

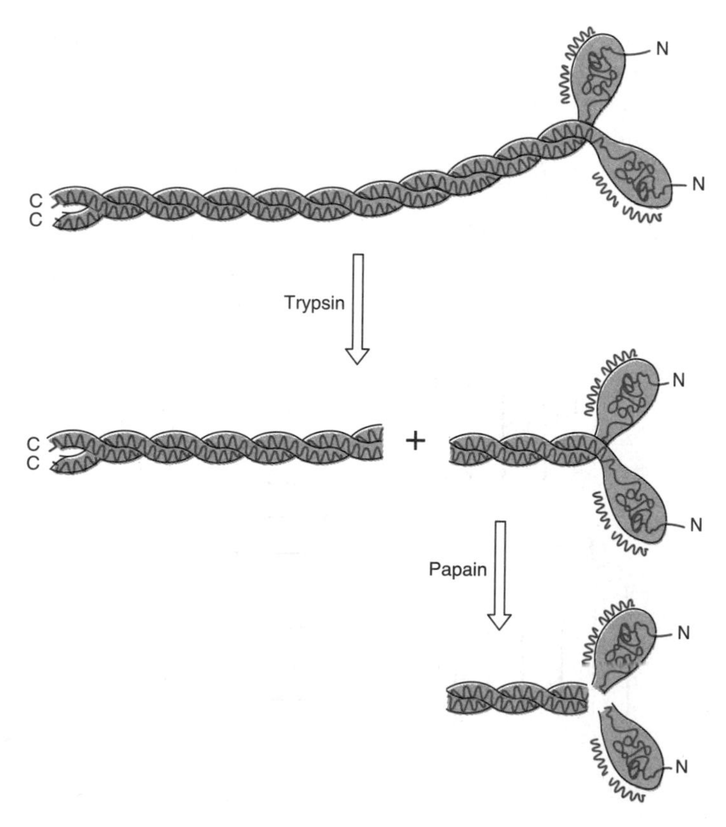

FIGURE 22.34
Schematic drawing of a myosin molecule.
Diagram shows the two heavy chains and the two light chains of myosin. Also shown are the approximate positions of cleavage by trypsin and papain.

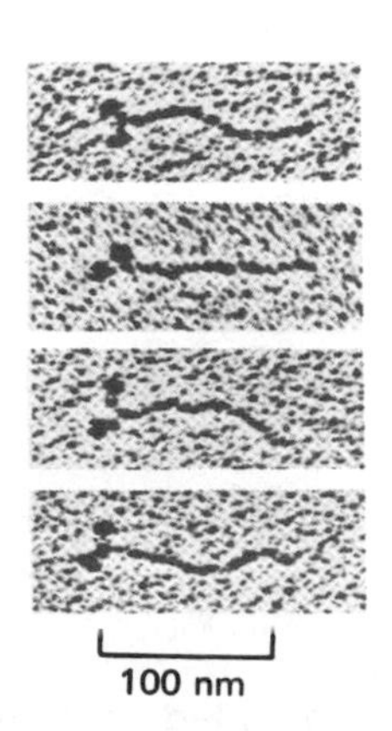

FIGURE 22.35
Electron micrographs of the myosin molecule.
Reprinted with permission from B. Alberts, D. Bray, J. Lewis, M. Raff, K. Roberts, and J. Watson, *Molecular Biology of the Cell,* 2nd ed., New York: Garland Publishing, 1983, Figure 10-9, pp. 554.

trypsin results in cleavage of the tail section about one-third of the way down from the head junction to produce **heavy meromyosin** (the head group and a short tail) and **light meromyosin** (the remainder of the tail section). The latter still has the ability to aggregate under physiological conditions, whereas the former does not have that ability. This suggests that the aggregation of myosin tail sections is a function of the extreme carboxyl end of the molecule. The head section can also be separated from the remainder of the tail section by treatment with papain. The results of studies with these proteases demonstrate that the molecule has at least two hinge points, at the head–tail junction and about one-third of the way down from the head group (Figure 22.34).

The cDNAs for myosin from many different species and from different types of muscle have been cloned and the amino acid sequences for the different myosin molecules inferred. There is sufficient similarity between these molecules of myosin that examples of any one can be used to illustrate essential points about myosin structure and function. Myosin has evolved very slowly, with a very high degree of homology, particularly within the head, or globular, region. Even though there is somewhat less absolute amino acid homology within the tail region, functional homology still exists to an extraordinarily high degree in spite of its various lengths, which range in different species from about 86 to about 150 nm. The head, as pointed out, is less variable. It varies in amino acid number from about 839 in the rat skeletal muscle to about 850 in the nematode and 849 in chicken gizzard smooth muscle heavy chain. The myosin heavy chain of mammalian species has about 1950 amino acids, with nearly one-half of them in the globular head piece.

Myosin, the major protein of the large filament of the contractile element, as shown from electron microscopy, forms a symmetrical aggregate around the M line of the H zone in the sarcomere. For this to occur, myosin tail sections must be aligned in both a parallel and an antiparallel manner. Also, the head groups on each end must be polarized in the same direction, that is, away from the center. About 400 molecules of myosin make up a single filament. The **C protein** (Table 22.5) is involved in their assembly. The **M protein** is also involved, presumably in holding the tail

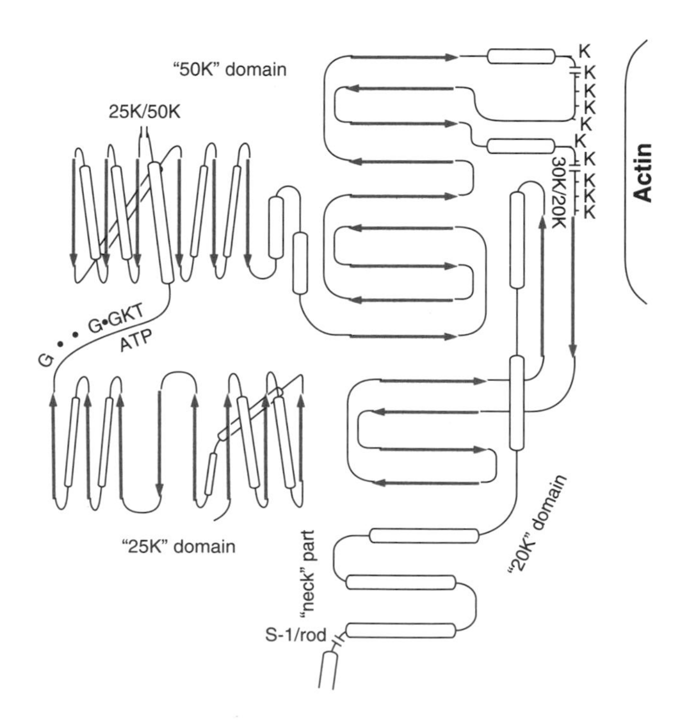

FIGURE 22.36
Diagram of the folded structure of myosin heavy chain.
Alpha helices and β strands are shown as open boxes and closed arrows, respectively. This is not geometrically accurate.
Redrawn from M. Yanagisawa, Y. Hamada, Y. Katsuragawa, M. Imamura, T. Mikawa, and T. Masaki, *J. Mol. Biol.* 198:143, 1987.

sections together as well as anchoring them to the center M line of the H zone.

The globular head section of myosin is the working unit. It contains the ATPase activity that provides the energy for the contractile process, and it contains the actin binding site. A schematic diagram of the relative positions of these various activities of the head group is shown in Figure 22.36. The head group of myosin, also referred to as subfragment 1 or S–1, consists of several domains that have been defined by their production from the controlled digestion of native S–1. It is believed that these domains are capable of folding independently into their proper stable tertiary structures. What is shown in Figure 22.36 is that the binding site for actin is between a "20K" domain and a "50K" domain. The "20K" domain is the neck part of the beginning of the rod section. The ATP binding region is between the "50K" domain and the "25K" domain. In a strict sense, ATP does not provide the energy directly for contraction, but its binding to myosin and its hydrolysis by the **myosin ATPase** induce the reversible conformation changes of the head group that allows its binding to and dissociation from actin during the process of pulling the Z-lines toward the center of the sarcomere. The conformation of myosin that has ATP bound to it has an affinity for actin that is 1/10,000 that of the conformation of myosin that does not have ATP bound to it! Thus, the process of chemical energy transduction to mechanical work depends on the primary event of protein conformation changes.

Actin Is a Protein of the Thin Filament

Actin is a major protein of the thin filament and makes up about 20–25% of muscle protein. It is synthesized as a globular protein with a molecular weight of about 42,000 Da. Even though there may be six or more genes that code for actin, it is a protein with a very highly conserved amino acid sequence. The amino acid sequence of rabbit skeletal muscle actin, shown in Figure 22.37, has 374 amino acid residues.

Table 22.6 shows the amino acid differences between skeletal muscle, smooth muscle, and cardiac muscle actin in three different species of animals. Differences are observed at most in about seven different positions, meaning that the sequence homology shows a better than 90% conservation. In fact, the primary amino acid sequences of more than 30 different actin isotypes, with the longest being 375 amino acids, reveal that a maximum of only 32 residues in any of them had been substituted. A significant number of them occurred at the N terminal, which may not be too surprising, if one considers that all actin molecules are post-transcriptionally modified by N-terminal acetylation. The N-terminal methionine is acetylated, removed, and the next amino acid is acetylated. The process may end here for some actins, but it may proceed through

Ac-
D E T E D T A L V C
D D G S G L V K A G
F A G D D A P R A V
F P S I V G R P R H
Q G V M V G M G Q K
D S Y V G D E A Q S
K R G I L T L K Y P
I E **H** W G I I T N D
D M E K I W H H T F
Y N E L R V A P E E
H P T L L T E A P L
N P K A N R E K M T
Q I M F E T F N V P
A M Y V A I Q A V L
S L Y A S G R T T G
I V L D S G D G V T
H N V P I T E G Y A
L P H A I M R L D L
A G R D L T D Y L M
K I L T E R G Y S F
V T T A E R E I V R
D I K Q K L C Y V A
L D F E N E M A T A
A S S S L E K S Y E
L P D G E V I T I G
N E R F R C P E T L
F Q P S F I G M E S
A G I H E T T Y N S
I M K C D I D I R K
D L Y A N N V M S G
G T T M Y P G T A D
R M Q K E I T A L A
P S T M K I K I I A
P P E R K Y S V W I
G G S I L A S L S T
F Q Q M W I T K Q E
Y D E A G P S I V H
R K C F

FIGURE 22.37
Amino acid sequence of actin.
The N-terminal residue is *N*-acetyl-aspartate. Histidine at position 73 is methylated.

TABLE 22.6 Summary of the Amino Acid Differences Between Chicken Gizzard Smooth Muscle Actin, Skeletal Muscle Actin, and Bovine Cardiac Actin

	Residue number						
Actin type	1	2	3	17	89	298	357
Skeletal muscle[a]	Asp	Glu	Asp	Val	Thr	Met	Thr
Cardiac muscle[b]		Asp	Glu			Leu	Ser
Smooth muscle[c]	absent		Glu	Cys	Ser	Leu	Ser

SOURCE: Adapted from J. Vandekerckhove and K. Weber, FEBS Lett. 102:219, 1979.

[a] From rabbit, bovine, and chicken skeletal muscle.
[b] From bovine heart.
[c] From chicken gizzard.

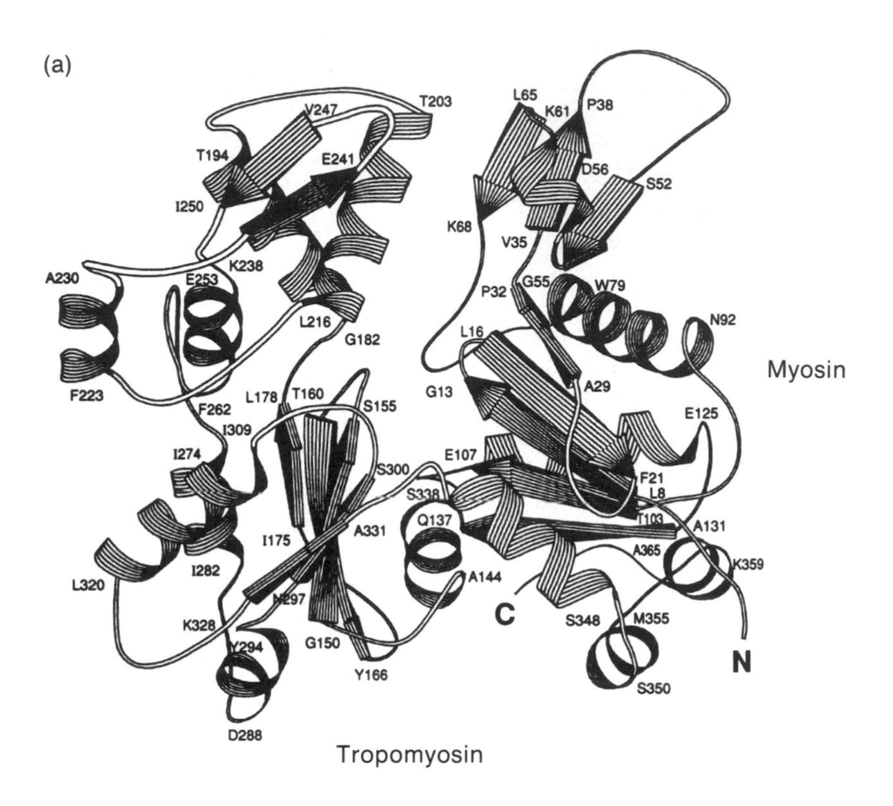

FIGURE 22.38
Actin molecules.
(*a*) A schematic representation of the three-dimensional structure of actin. The domain on the right is historically known as the small domain. The diagram has been altered to show general areas where ATP, Ca^{2+}, tropomyosin, and myosin interact. Aggregation to form F-actin involves interaction on the back side of the large domain in the vicinity of the alpha-helix containing residues 266-269. (*b*) A space-filling model of eight monomers of the F-actin helix. Each monomer is shown in contrasting colors. The small domain of the G-actin monomer is at large radius. The amino acid residues that interact with myosin in the actinomyosin complex are highlighted.
Reprinted with permission from W. Kabsch, H. G. Mannherz, D. Suck, E. F. Pai, and K. C. Holmes, *Nature* 347:37, 1990; and K. C. Holmes, D. Popp, W. Gebhard, and W. Kabsch, *Nature* 347:44, 1990.

one or two additional steps for other actins. In all cases, however, the N-terminal amino acid will be acetylated.

Actin as first synthesized is called **G-actin** for globular actin. In fact, as shown in Figure 22.38, it is not strictly globular. Actin has two distinct domains of approximately equal size that, historically, have been designated as large (left) and small (right) domains. Each of these domains consists of two sub-domains. Both the N terminal and the C terminal are located within the small domain. The molecule has polarity, and when it aggregates to form **F-actin,** or fibrous actin, it does so with a specific directionality. This is important for the "stick and pull" processes involved in sarcomere shortening during muscular contraction.

The process of formation of F-actin is interesting, but it is not fully understood. Some of the important points about the aggregation of G-actin to form F-actin are outlined here. G-Actin contains a specific binding site for ATP and a high-affinity binding site for divalent metal ions. The Mg^{2+} ion is most likely the physiologically important cation, but Ca^{2+} also binds tightly and competes with Mg^{2+} for the same tight binding site. It is the G-actin–ATP–Mg^{2+} complex that aggregates to form the F-actin polymer. Aggregation can occur from either direction, but kinetic data indicate that the preferred direction of aggregation is by addition to the barbed end of the molecule (where the rate is diffusion controlled) rather than to the more pointed end. ATP hydrolysis occurs by orders of magnitude faster in the aggregate than it does in the monomeric complex, but its hydrolysis has not yet been shown to have a cause-and-effect relationship to the process. G-Actin–ADP–Mg^{2+} also aggregates to form F-actin but at a somewhat slower rate.

There are a number of proteins in the cytosol that can bind to actin. Some of them form "capped" complexes that may limit the length of the aggregate formed. It is not clear what the roles of many of these proteins are, but it does appear that in the sarcomere ***β*-actinin** binds to F-actin and plays a major role in limiting the length of the thin filament. ***α*-Actinin,** a protein with two subunits of 90–110 kDa each, also binds actin, but its main role appears to be to anchor the actin filament to the Z line of the sarcomere. There are two other major proteins associated with the thin filament, tropomyosin and troponin.

Tropomyosin is a rod-shaped protein associated with the F-actin filament. It consists of two dissimilar α-helical peptide chains coiled around each other in a head–tail configuration. Its subunits are about 35,000 Da. Tropomyosin is associated with the thin filament for its entire length. However, each tropomyosin molecule binds strongly to about seven monomers in the F-actin molecule. Tropomyosin lies within the groves of the F-actin filament (Figure 22.39). Tropomyosin may help to stabilize and strengthen the filament as well as compete for the myosin binding sites during the relaxation phase of contraction.

Troponin consists of three dissimilar subunits with molecular weights of about 18,000, 21,000, and 37,000 Da, and are designated Tn-C, Tn-I, and Tn-T. The Tn-T subunit binds to tropomyosin. The Tn-I subunit is involved in the inhibition of the binding of actin to myosin under resting conditions. A calmodulin like protein, Tn-C, binds calcium. The interaction of this subunit with calcium induces a conformation change that presumably alters the conformation of tropomyosin, resulting in the exposure of the actin–myosin binding sites and initiating the contraction process by permitting the free myosin (or the ADP–myosin complex) to bind to the actin.

Most of the pictures of muscle components shown so far have been two dimensional. Before discussing further the mechanism of muscular con-

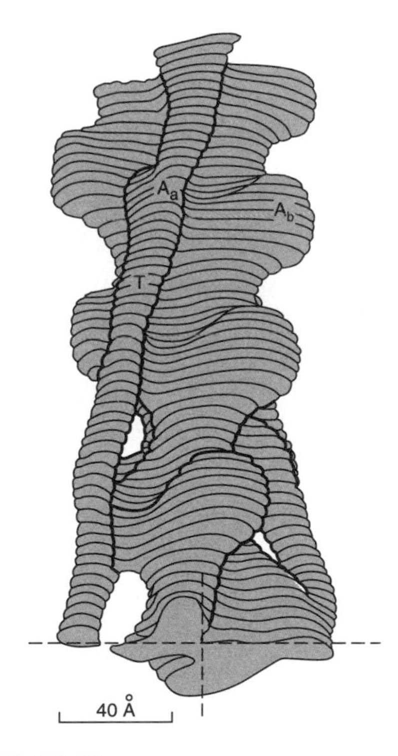

FIGURE 22.39
A three-dimensional image of actin–tropomyosin complex.
The image has been modified to highlight the tropomyosin (T) molecule. The locations of the two domains of actin are labeled Aa and Ab.
Modified from L. A. Amos, *Annu. Rev. Biophys. Biophys. Chem.* 14:291, 1985.

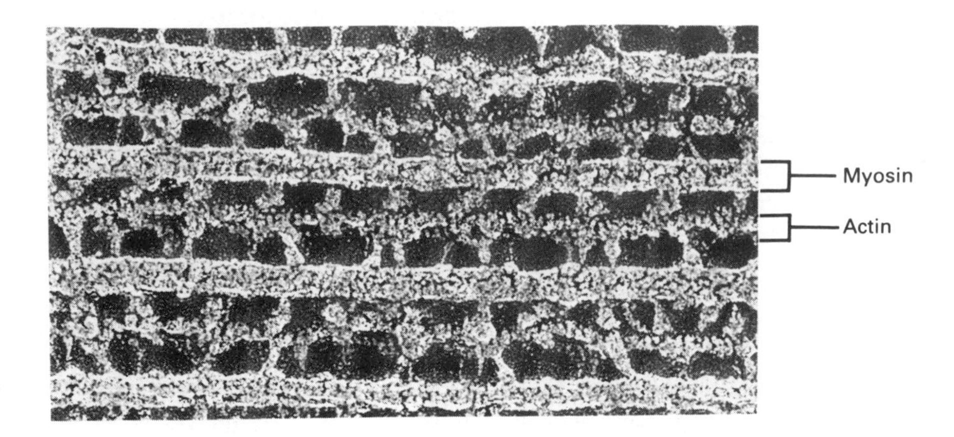

FIGURE 22.40
An electron micrograph of actin–myosin cross-bridges in a striated insect flight muscle.
Reproduced with permission from J. Darnell, H. Lodish, and D. Baltimore, *Molecular Cell Biology,* New York: Scientific American Books, Figure 19-10, 1986, p. 823.

traction, refer back to diagrams f–i of Figure 22.31. Specific attention to diagram H is requested where extensive cross-linking is seen in the vicinity of the M line of the H zone. This cross-linking is due to the interactions of the tails of the myosin heavy chain together with the M protein. These interactions are involved in helping to keep the large fibers both together and centered within the sarcomere. The cross section designated I shows the relative arrangement of the light and heavy chains. Note that throughout the center region there are six light chains surrounding each heavy chain. The arrangement of head groups around the heavy chain together with the flexible regions of the heavy chain make it possible for each heavy chain to interact with multiple light chains. When cross-bridges are formed between the heavy and light chains, they may be formed in patterns consistent with the protrusions of head groups from the thick myosin filaments as shown in the electron micrograph of Figure 22.40. This figure shows a two-dimensional view of myosin interacting with actin filaments lying on either side of it. Similar interactions occur with the other four actin filaments that surround the myosin filament (see Figure 22.31, Diagram i).

Mechanism of Muscle Contraction Begins With Ca^{2+} Release

When a nerve impulse reaches the neuromuscular junction of skeletal muscle, Ca^{2+} is released into the sarcomere. The Ca^{2+} ion binds Tn-C of troponin, causing a conformation change that results in the exposure of the myosin binding sites on the actin molecule. Myosin binds to actin and undergoes a conformation change that allows the actin–myosin complex to bind ATP. The ATP–myosin–actin ternary complex, however, is less stable than the actin–myosin binary complex by a factor of approximately 10^4. Thus, the binding of ATP facilitates the dissociation of myosin from actin. The hydrolysis of ATP by the myosin ATPase and subsequent dissociation of products produces a myosin molecule that is again capable of binding actin. This process of myosin–ATP interaction and hydrolysis triggers a series of conformation changes that allows the myosin molecules to "walk" along the actin molecule and pull it towards the middle of the sarcomere, which is equivalent to the "sliding" of the filaments past each other as seen in the electron micrographs. Tension can develop progressively as the myosin–actin bonds are asynchronously made and broken with the surrounding actin molecules. The flexibility of the myosin head structures and the number of actin molecules surrounding each myosin heavy chain allow this to be an efficient process.

When Ca^{2+} is extruded from the sarcomere and again approaches its resting concentration, it dissociates from the Tn-C molecules. The resulting conformation change in troponin leads to changes in the conformation of Tn-I and tropomyosin with either mechanical or functional blockage of the actin–myosin binding sites. The ATP displaces ADP on myosin, but the ATPase activity is inhibited and the actin–myosin complex dissociates. The muscle is at rest.

Energy for Muscle Contraction Is Supplied by ATP Hydrolysis

The main aspects of metabolism in muscle cells have been dealt with in previous chapters. It is sufficient to state here that ATP is an absolute requirement for the process of muscular contraction to occur. ATP hydrolysis by the myosin–ATPase to give the myosin–ADP complex leads to a myosin conformation that has a very high binding affinity for actin. Additional ATP is required for the dissociation of the myosin–actin complex.

In a live muscle cell, the concentration of ATP remains fairly constant even during strenuous muscle activity. This is brought about, first, because of increased metabolic activity and, second, because of the action of two enzymes. The most important of the enzymes is the **creatine phosphokinase** that catalyzes the transfer of phosphate from phosphocreatine to ADP in an energetically favored manner to make ATP.

$$\text{Phosphocreatine} + \text{ADP} \rightleftharpoons \text{ATP} + \text{creatine}$$

Thus, if the metabolic process is insufficient to keep up with the energy demand, the creatine phosphokinase system serves as a "buffer" for maintenance of the ATP level. The second enzyme is **adenylate kinase** that catalyzes the reaction

$$2\text{ADP} \rightleftharpoons \text{ATP} + \text{AMP}$$

Under strained conditions this enzyme helps maintain the concentration of ATP relatively constant.

ATP depletion brings about rather rigid consequences to muscle cells. When the ATP supply of the muscle is exhausted and the intracellular Ca^{2+} concentration is no longer controlled, the myosin will exist exclusively bound to actin, a condition called *rigor mortis*. The function of ATP in the muscular contraction process, therefore, is to promote the dissociation of the actin–myosin complex, not to promote its formation.

Calcium Flux in the Sarcomere Is Controlled by Several Events

Contraction of skeletal muscle is initiated by a nerve impulse being transmitted across the neuromuscular junction through the intervention of the neurotransmitter acetylcholine. The interaction of acetylcholine with its receptors on the postsynaptic membrane sets in motion a series of events that lead to an increased Ca^{2+} concentration within the sarcomere, the interaction of Ca^{2+} with Tn-C, and the resulting processes of contraction. A question of current interest is the mechanism by which Ca^{2+} is released or transported into the sarcomeres. This process must occur rapidly enough to cause contraction to occur approximately simultaneously in all of the cells of a stimulated muscle.

The electron micrograph and accompanying diagrams of Figure 22.41 will help to demonstrate the complexity of the problem. The acetylcholine receptors are associated with the plasma membrane of the cell. There are transverse tubules (T tubules) along the membrane in the vicinity of the Z lines. These tubules are connected also to the terminal cisternae of the

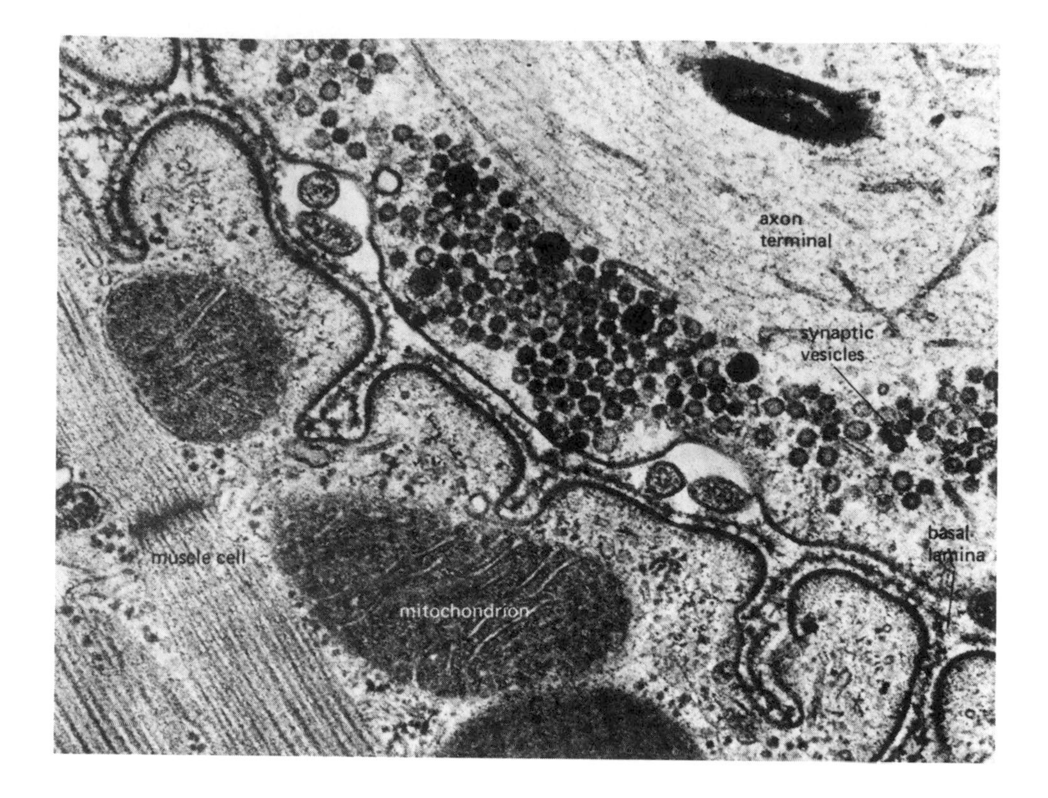

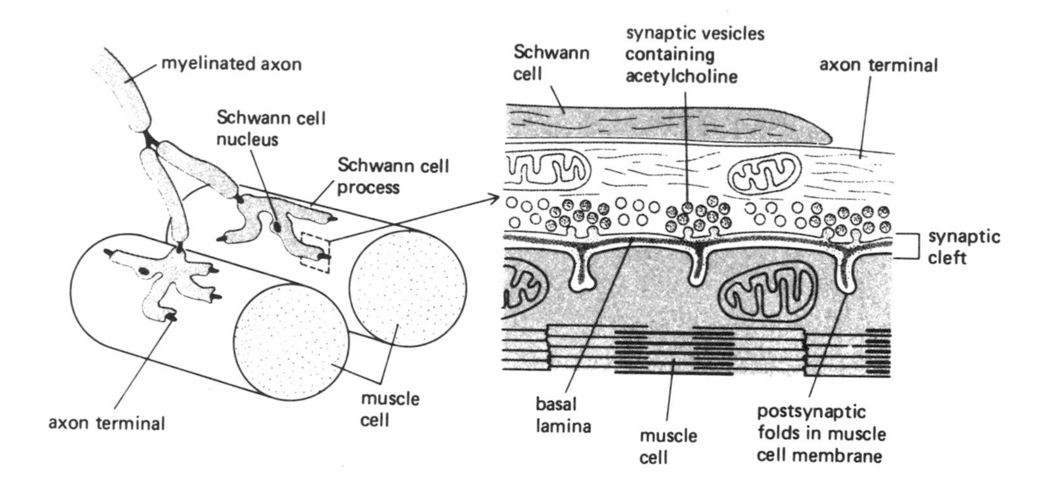

FIGURE 22.41
(*a*) Electron micrograph of a neuromuscular junction. (*b*) Schematic diagram of the neuromuscular junction shown in *a*.
Reproduced with permission from B. Alberts, D. Bray, J. Lewis, M. Raff, K. Roberts, and J. Watson, *Molecular Biology of the Cell,* New York: Garland Publishing, Figures 18-24 and 18-25, 1983, pp. 1036–1037.

sarcoplasmic reticulum. The nerve impulse results in a depolarization of the plasma membrane and the transverse tubules. The question is whether depolarization of the transverse tubules causes direct conformational alteration of the terminal cisternae that results in opening of the Ca^{2+} channels or whether depolarization causes the release of a transmitter that binds to a receptor and causes opening of the Ca^{2+} channels. An attractive mechanism for the latter possibility involves stimulation of the enzyme phospholipase C in the plasma membrane with accompanying production of inositol 1,4,5-trisphosphate (IP_3) from phosphatidylinositol 4,5-bisphosphate (see p. 875). The IP_3 could diffuse across the approximately 20-nm space to the sarcoplasmic reticulum to interact with receptors that open Ca^{2+} channels. The components appear to be present for the IP_3 system to operate, but it has not been proven that it could operate fast enough to accommodate the known rapidity of skeletal muscle response to stimuli. On the other hand, the IP_3 pathway could operate fast enough to control smooth muscle contraction since the rate of force development in smooth muscles is much slower than in skeletal muscle.

Calcium Regulates Smooth Muscle Contraction

The sliding filament model also explains the mechanism of smooth muscle contraction. Calcium ions play an important role in smooth muscle contraction also, but there are some important differences in the mechanism by which its acts. A mechanism for calcium regulation of smooth muscle contraction is shown in Figure 22.42. Key elements of this mechanism are (a) A phosphorylated form of myosin light chain binds to actin to stimulate the MgATPase, which supplies the energy for the contractile process. (b) Myosin light chain is phosphorylated by a **myosin light chain kinase** (MLCK). (c) The MLCK is activated by a Ca^{2+}–calmodulin (CM) complex. (d) The formation of the Ca^{2+}–CM complex is dependent on the concentration of intracellular Ca^{2+}. Thus, the release of Ca^{2+} from its intracellular stores or its flux across the plasma membrane is important for control of the process. (e) Contraction is stopped by the action of a myosin phosphatase or the transport of Ca^{2+} out of the cells. It is apparent that in smooth muscle, many more biochemical steps are involved in the regulation of contraction, steps that can be regulated in a progressive

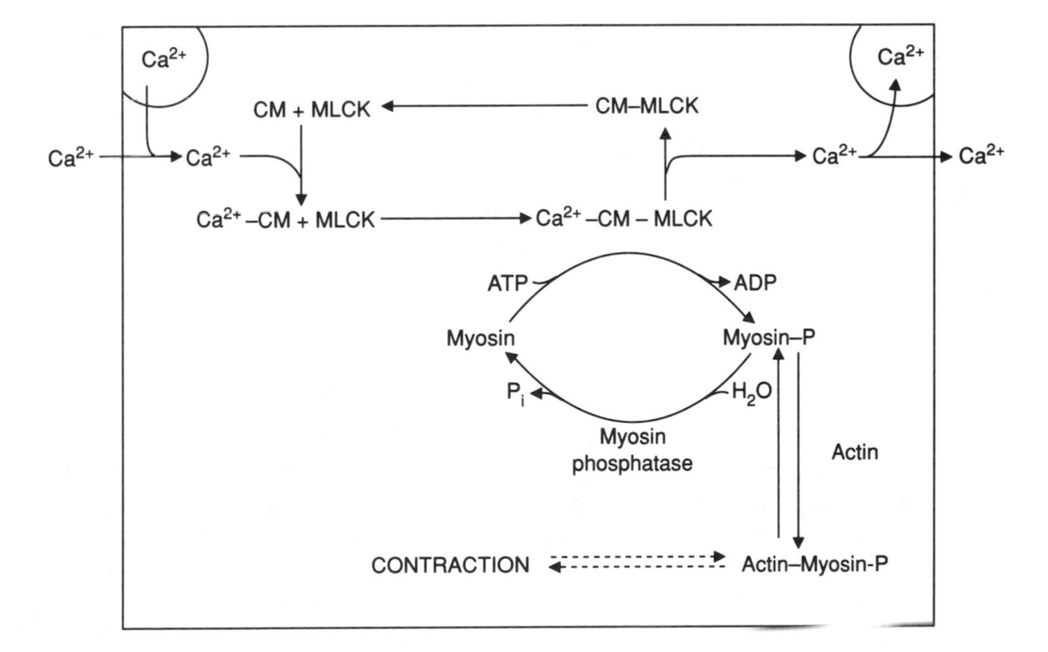

FIGURE 22.42
A schematic representation of the mechanism of regulation of smooth muscle contraction.
Heavy arrows show the pathway of tension development and light arrows show the pathway for release of tension. The MgATPase activity is highest in the Actin–Myosin-P complex. CM = calmodulin and MLCK = myosin light chain kinase.
Adapted from K. E. Kramm and J. T. Stull, *Ann. Rev. Pharmacol. Toxicol.* 25:593, 1985.

manner, which serves the function of smooth muscles well, namely, to be able to develop various degrees of tension and to retain it for some prolonged time period.

22.6 MECHANISM OF BLOOD COAGULATION

The importance of blood in maintaining pH, in the transport of oxygen and nutrients to cells, and in the transport of carbon dioxide and waste products from cells has been discussed in other chapters. The circulation of blood, therefore, is essential for life, and the integrity of the process must be maintained. Circulation occurs in a very specialized type of closed system in which the volume of circulating fluid is maintained fairly constant, but it is also one in which the transfer of solutes across the boundaries that define inside and outside is a necessary function. Like any system of pipes and tubes, leaks can occur and must be repaired. The process of blood clotting addresses the question of stopping the leaks.

The purpose of this section is to give a general picture of the mechanism of blood clotting from a biochemical viewpoint. To this end, this section will focus on the relationship between blood clotting and enzyme activation, control, and inhibition. It is not a process of signal transduction in the same sense as are other topics of this chapter. Instead, it is a process of signal amplification.

Some of the primary questions to be addressed are (a) What initiates the clotting process? (b) What substances and reactions are responsible for forming the clot? (c) What factors prevent complete intravascular clotting once the process is initiated? (d) How is the clot dissolved?

It is important to maintain a state of hemostasis, that is, no bleeding. Thus the process of blood clotting is designed to stop as rapidly as possible the loss of blood following a vascular injury of some sort. When such an injury occurs, three major events take place in an effort to maintain hemostasis: (1) clumping of blood platelets at the site of injury in an effort to form a physical plug to stop the leak; (2) vasoconstriction in an effort to reduce the blood flow through the area; and (3) aggregation of a protein, fibrin, into an insoluble network, or clot, to cover the ruptured area and to prevent the further loss of blood.

All of the processes mentioned above are emergency mechanisms for stopping the loss of blood. More permanent repair of the system occurs when the ruptured vessel itself is repaired. The clot then dissolves and the system is restored to normal.

Rupture of the Endothelium Initiates Platelet Clumping

Contact with collagen fibers that become exposed when the endothelial lining of the blood vessels is ruptured initiates **platelet clumping.** The platelets adhere to the collagen, undergo a morphological change, release ADP, serotonin, some phospholipids, lipoproteins, and other proteins that aid in coagulation and the repair process (Figure 22.43). One of the circulating proteins, a glycoprotein called **von Willebrand factor** (vWF), concentrates in the area of the injury and helps to form a link between the exposed surface and the platelets, facilitating formation of the platelet plug. The von Willebrand factor also serves as a carrier for one of the protein cofactors of blood clotting, **factor VIII.**

The process of platelet aggregation becomes autocatalytic with the release of ADP and thromboxane A_2. Platelet factor IV, heparin binding protein, prevents heparin–antithrombin III complexes from inhibiting the serine proteinase coagulation factors and it attracts cells with antiinflam-

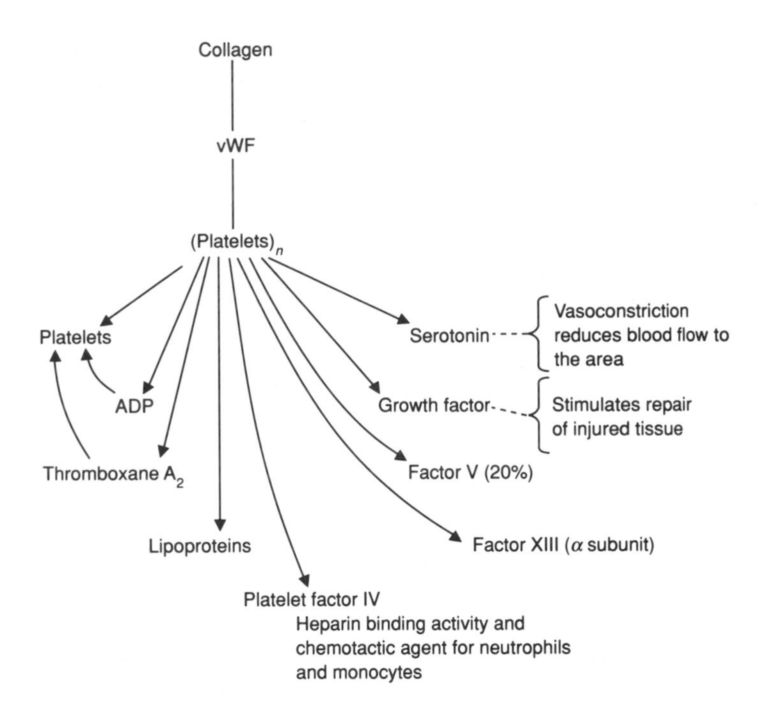

FIGURE 22.43
Action of platelets in blood coagulation.

matory activity to the site of injury. Factor V, about 20% of which exists in platelets, is a protein cofactor necessary for the expression of activity of factor Xa. Factor XIII is a transglutamidase and is involved in cross-linking fibrin fibers. These factors will be described in more detail later.

Intact vascular endothelium normally does not initiate platelet aggregation since connective tissue elements are not exposed and activators such as ADP are rapidly degraded or are not in blood in sufficient concentration to be effective in this activity. The endothelium also appears to secrete prostacyclin (PGI_2), a potent inhibitor of platelet aggregation and secretion. The combination of these activities normally prevents the aggregation of platelets.

Clot Formation Occurs at the Site of Injury

Clot formation is initiated at the site of injury. Specifically exposed extravascular membrane proteins serve as receptors for the various zymogens and protein cofactors involved in the process. These multienzyme complexes catalyze the formation of fibrin, which aggregates into an insoluble matrix. This insoluble matrix is stabilized by cross-linking of the fibrin monomers through the action of the enzyme **transglutamidase,** or factor XIII.

Thrombin Converts Fibrinogen to Fibrin

The enzyme responsible for the conversion of fibrinogen to fibrin is thrombin. **Thrombin** circulates in plasma as the **zymogen prothrombin,** a protein with about 582 amino acid residues (Figure 22.44). Prothrombin contains five to six **γ-carboxyglutamate** (Gla) residues in its N terminal region. The binding of calcium ions to these residues facilitates the binding of prothrombin to its activation complex on the membrane surface at the site of injury. Prothrombin is activated by an enzyme complex formed on the surface of the injured membrane. The active proteinase of this complex is **factor Xa.** This prothrombinase complex activates prothrombin by making two proteolytic cleavages in the prothrombin molecule, one on the carboxyl side of arginine 274, an arginine–threonine bond, and

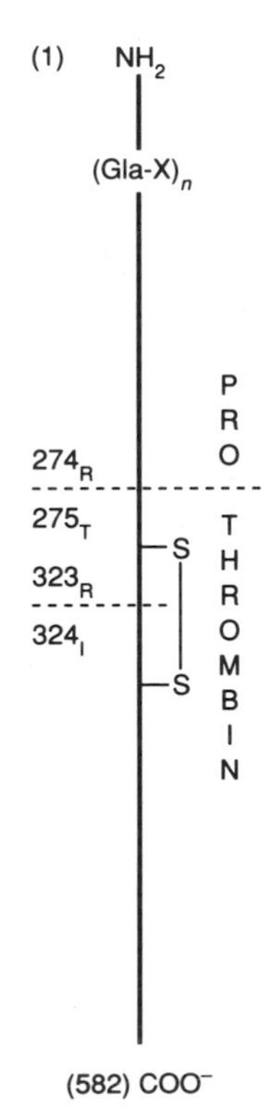

FIGURE 22.44
Diagrammatic representation of the prothrombin molecule showing the points of cleavage by factor Xa.
The peptide that is lost upon activation is the one in the PRO region of the chain.

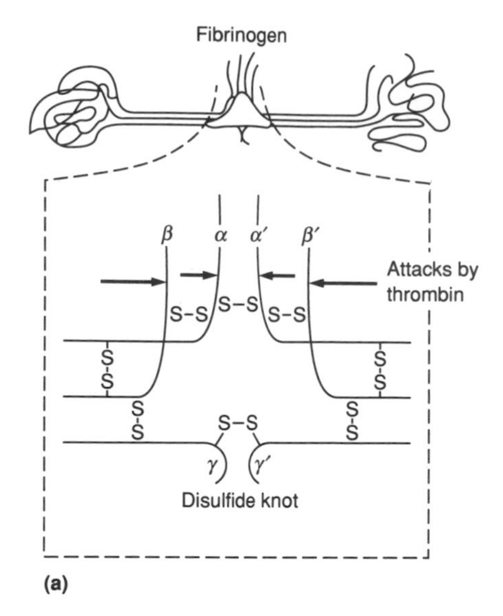

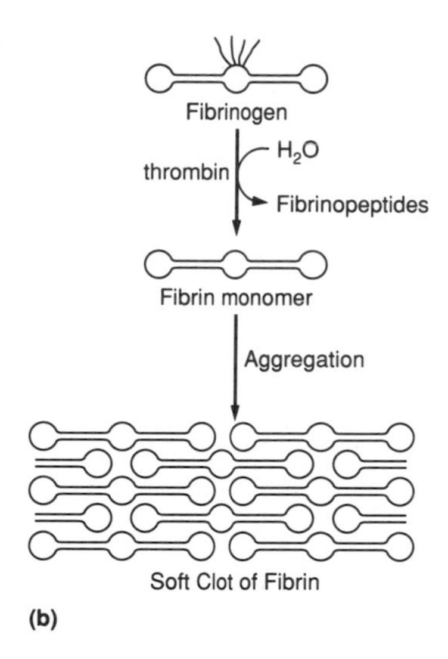

FIGURE 22.45
Diagrammatic representation of the fibrinogen molecule and its conversion to the soft clot of fibrin.

another on the carboxyl side of arginine 323, an arginine–isoleucine bond. The active thrombin molecule (*alpha*-thrombin) consists of amino acid residues 275–582 and has two chains of 49 and 259 amino acid residues. These are covalently connected by a disulfide bridge. The substrate for this active thrombin molecule is fibrinogen.

Fibrinogen is a large molecule of approximately 340 kDa. It consists essentially of two tripeptide units with $\alpha\beta\gamma$ structure. The complete molecule, therefore, has an $\alpha\alpha'$, $\beta\beta'$, $\gamma\gamma'$ subunit configuration (Figure 22.45). These two tripeptide structures are "tied" together at their N-terminal regions by a group of disulfide bridges. Fibrinogen has three globular domains, one on each end and one in the middle where the chains are joined. The domains are separated from each other by rod-like configurations. A short segment of the free N-terminal regions project out from the central globular domain. The $\alpha\alpha'$ and the $\beta\beta'$ N-terminal regions, through *charge-charge repulsion,* play a role in preventing the aggregation of fibrinogen from occurring normally. Thrombin cleaves these groups and allows the resulting fibrin molecules to aggregate and form the **"soft" clot.** This soft clot is stabilized and strengthened by the action of factor XIII, transglutamidase. This enzyme catalyzes a reaction in which an isopeptide linkage is formed by replacing the γ-amide group of glutamine residues of one chain with the ε-amino group of lysine residues of another chain (Figure 22.46).

Activation of prothrombin occurs as a result of the action of another hydrolase, blood clotting factor Xa. Factor Xa is formed by the activation of Factor X through proteolysis. There are two pathways that lead to the activation of factor X to give Xa. They have generally been referred to as the **extrinsic** and the **intrinsic pathways** for blood clotting. Both of these pathways originate with the interaction of enzymes involved in the blood clotting process with surface proteins, reactions that can be considered to be analogous to ligand–receptor interactions. The net result of these interactions is the formation of multienzyme complexes with greatly enhanced catalytic activity. The rate enhancement within these complexes is the result of protein–protein interaction and a catalytic cascade.

Major proteins that are involved in the blood clotting process are listed in Table 22.7. Four of these proteins have Gla residues in the N-terminal regions. The structure of the Gla residues resembles that of other organic

FIGURE 22.46
Reactions catalyzed by transglutamidase.

TABLE 22.7 Factors Involved in Blood Clotting

Factor	Trivial name	Pathway	Characteristic
I	Fibrinogen	Both	
II	Prothrombin	Both	Contains N-terminal Gla segment
III	Tissue factor	Extrinsic	
IV	Calcium	Both	
V	Proaccelerin	Both	Protein cofactor
VI	Accelerin		This is Va, redundant to Factor V.
VII	Proconvertin	Extrinsic	Endopeptidase with Gla residues.
VIII	Antihemophiliac	Intrinsic	Protein cofactor
IX	Christmas factor	Intrinsic	Endopeptidase with Gla residues.
X	Stuart factor	Both	Endopeptidase with Gla residues.
XI	Plasma thromboplastin antecedent	Intrinsic	Endopeptidase
XII	Hageman factor	Intrinsic	Endopeptidase
XIII	Protransglutamidase	Both	Transpeptidase

compounds like ethylenediamine tetraacetic acid (EDTA) that are used to chelate metal ions. The Gla residues are also good chelators of divalent metal ions, specifically calcium. The binding of calcium to these residues within the protein facilitate the formation of complexes with the membranes. The four blood clotting factors that contain Gla residues are II (prothrombin), VII, IX, and X; each is an endopeptidase.

Factors III and VII Are Unique to the Extrinsic Pathway

The reactions of the extrinsic pathway are shown in Figure 22.47. The two factors that are unique to the extrinsic pathway are tissue factor (factor III or TF) and factor VII. Tissue factor is a membrane-bound protein whose receptor-like activity initiates this process. Tissue factor is exposed only upon rupture of the blood vessels. Factor VII is a Gla-containing protein that binds to tissue factor only in the presence of Ca^{2+}. The resulting **TF–VII–Ca^{2+}** complex is the catalytically active species, and it has high affinity for factor X. This complex catalyzes the formation of factor Xa from X. Factor Xa remains attached to the membrane. The initial activation of the zymogen form of factor VII takes place on the membrane as a result of protein–protein interaction at the site of injury and possibly without initial cleavage of any peptide bonds in factor VII. This is still somewhat of a controversial point since factor VIIa has a long half-life in the circulation. Thus, formation of the initial complex with TF could involve some of the

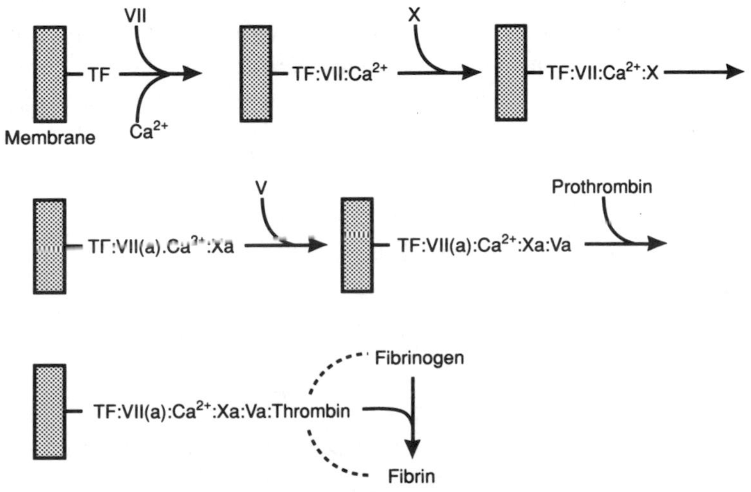

FIGURE 22.47
Extrinsic pathway.
The notation VIIa is to indicate that it has catalytic activity either as the zymogen or after activation by proteolysis.

already preformed factor VIIa. In any case, activation of either involves protein–protein interaction, and more factor VII is activated to factor VIIa by the action of factor Xa in the multi-enzyme complex.

Factor Xa is the active endopeptidase required for the activation of prothrombin to thrombin. The protein cofactor, **factor Va,** is required for this activity. Factor V, which exists in platelets and in plasma, is released at the site of injury, binds to the **TF–VII–Ca^{2+}–Xa complex,** and is activated by factor Xa to give factor Va that is now capable of serving as the cofactor for the activation of prothrombin to thrombin. Thrombin, in addition to catalyzing the conversion of fibrinogen to fibrin, also activates factors V, VII, VIII, and XIII.

Several Factors Are Restricted to the Intrinsic Pathway

The intrinsic pathway was initially thought to require only factors *internal* to the circulating blood. The factors that are restricted to the intrinsic pathway for blood coagulation are **VIII, IX, XI, and XII.** All except factor VIII are endopeptidases. As with the extrinsic pathway, the process starts with interactions and complex formation on the surface of the exposed tissues at the site of injury (Figure 22.48). Factor XII is activated in a surface complex involving **high molecular weight kininogen** (HMK) and **prekallikrein.** The surface proteins that serve as receptors for this complex are different from the tissue factor protein and probably also include collagen molecules as well as the HMK that have been exposed as a result of tissue damage. In this complex, prekallikrein is activated to **kallikrein,** which activates factor XII. Factor XIIa, anchored to HMK, activates more prekallikrein in an autocatalytic cycle. The kallikrein–HMK complex activates factor XI, and the XIa–receptor complex activates factor IX. (Factor IX can also be activated by the TF VIIa complex.) Those are the initial reactions of the intrinsic cascade (Figure 22.48). For factor IXa to be active, it must form a complex with the activated form of the protein cofactor VIIIa. Factor VIII is activated to VIIIa by the action of thrombin. Since the extrinsic pathway involving tissue factor operates faster initially than the intrinsic pathway, some thrombin will be present for factor VIII activation. Factor VIII circulates in plasma tightly associated with von Willebrand factor (vWF). The vWF is a glycoprotein that, in addition to serving as a carrier for factor VIII, also functions to facilitate platelet binding to the subendothelial components of the injured vascular tissue, thus promoting platelet aggregation. Clinical Correlation 22.3 describes a major disease of blood clotting, Classic Hemophilia.

CLIN. CORR. 22.3
CLASSIC HEMOPHILIA

Hemophilia is an inherited disorder characterized by a permanent tendency to hemorrhages, spontaneous or traumatic, due to a defective blood clotting system. Classic hemophilia is hemophilia A, an X-linked recessive disorder, in which there is a deficiency of blood clotting factor VIII. About 1 in 10,000 males is born with a deficiency of factor VIII. Of the approximate 25,000 hemophiliacs in the USA, more than 80% are of the A type. Hemophilia B is due to a dysfunction in factor IX.

Some hemophilia A patients may have a normal prothrombin time if the concentration of tissue factor is high. One possible explanation for this focuses on the fact that factor V in human plasma is much lower in concentration than factor X. Activation of an amount of factor X to Xa in excess of that required to bind to all of the factor Va available would initiate blood clotting by the extrinsic pathway and give normal prothrombin time. The intrinsic pathway would not function normally due to the deficiency in factor VIII. Without the two pathways operating in concert, the overall process of blood clotting would be impaired. Both factor Xa and thrombin activate factor V. They are involved in a number of other reactions also, and the kinetics of their interaction with the normally low concentration of factor V must be considered in this regard. Thus, if the overall process cannot show the acceleration expected at its onset by the intervention of the intrinsic pathway, the clotting disorder is still expressed.

Severely affected hemophilia A patients have blood levels of factor VIII that are less than 5% of normal. These patients are generally treated by blood transfusion. There are several dangers associated with transfusion, including the possibility of contraction of hepatitis or AIDS, and also a significant number of patients (~6%) receiving frequent blood transfusions develop autoantibodies. Treatment of these patients has been made much safer as a result of the cloning and expression of the gene for factor VIII. The pure protein can be administered to patients with none of the associated dangers mentioned above.

Nemerson, Y. *Blood* 71:1, 1988.

Properties of Many of the Proteins of Coagulation Are Known

Calcium ions serve at least two important functions in the blood coagulation process. They form complexes with the four factors that contain γ-carboxyglutamate (Gla) residues and induce a conformational and possi-

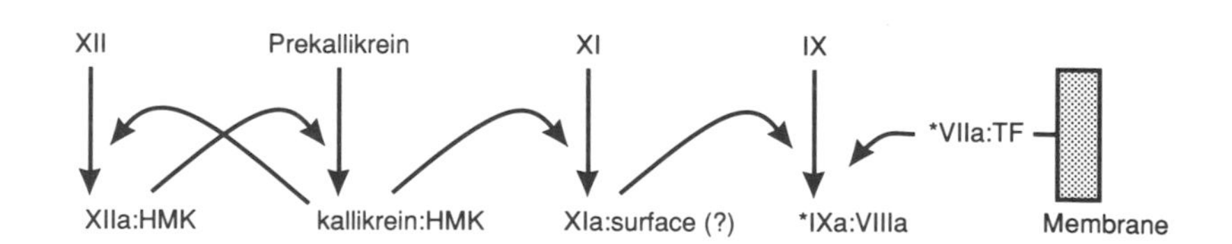

FIGURE 22.48
Intrinsic pathway.
The (*) mark proteins that contain γ-carboxyglutamate (Gla) residues and require calcium ions for activity.

bly an electronic state that permits their interaction with membrane "receptors" that form the site for the initiation and localization of the reactions. They are also involved in some other processes, probably conformational to enhance activity, since at least one of the enzymes involved in the process, thrombin, has other calcium binding sites distinct from those associated with the Gla residues. The activation of thrombin leads to the cleavage and elimination of the N-terminal region containing the Gla residues, yet calcium ions are still required for the reaction.

The modification of prothrombin, and factors VII, IX, and X, to form the Gla residues occurs during synthesis by a **carboxylase enzyme** located on the luminal side of the rough endoplasmic reticulum. **Vitamin K** (phytonadione, the "koagulation" vitamin) is an essential cofactor for this carboxylase enzyme. During the course of the reaction, the dihydroquinone or reduced form of vitamin K (Figure 22.49), vit K(H_2), is oxi-

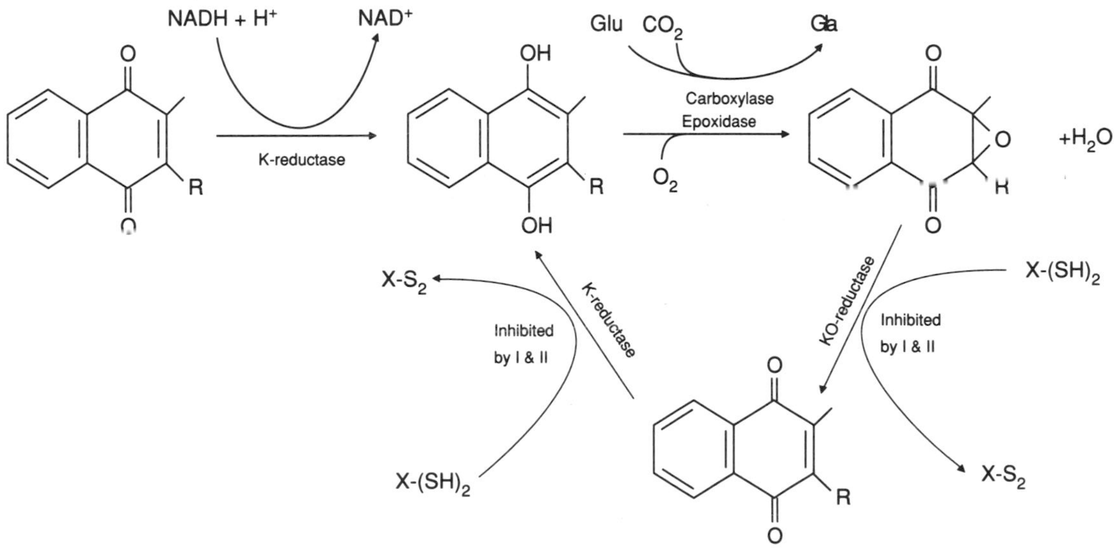

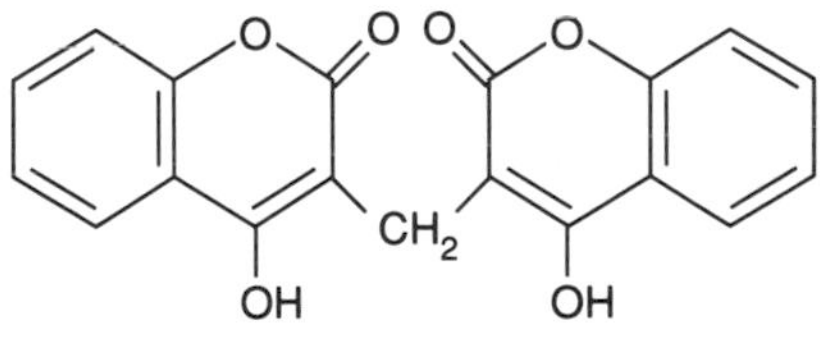

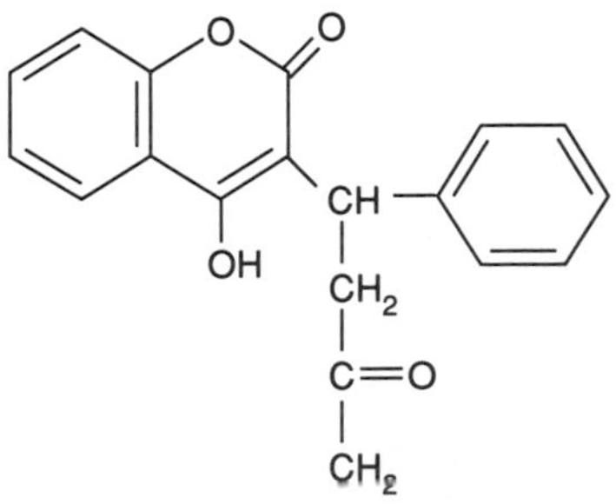

FIGURE 22.49
The vitamin K cycle as it functions in protein glutamyl carboxylation reaction. X-$(SH)_2$ and X-S_2 represent the reduced and oxidized forms, respectively, of a thioredoxin. The NADH-dependent and the dithiol-dependent vitamin K reductases are different enzymes. The dithiol-dependent K and KO reductases are inhibited by dicoumarol (I) and warfarin (II).
Redrawn and modified from C. Vermeer, *Biochem. J.* 266:625–636, 1990.

dized to the epoxide form, vit K(O). The vitamin K epoxide is converted back to the dihydroquinone enzymatically by enzymes requiring dithiols like thioredoxin as cofactors. Analogs of vitamin K inhibit the dithiol-requiring vitamin K reductases and result in the conversion of all available vitamin K to the epoxide form, a form that is not functional with the protein carboxylase. The overall carboxylation reaction is

$$\mathrm{HOOC{-}CH_2{-}CH_2{-}CH(NHR_1){-}C(=O){-}R_2} \xrightarrow[\text{Protein carboxylase}]{\text{vitamin K}(H_2) + O_2 + CO_2} \mathrm{(HOOC)_2CH{-}CH_2{-}CH(NHR_1){-}C(=O){-}R_2} + \text{vitamin K(O)} + H_3O^+$$

Some analogs that interfere with the action of vitamin K are dicoumarol and warfarin (Figure 22.49). In animals treated with these compounds, prothrombin, factors VII, IX, and X are not modified during their synthesis to contain the Gla residues, and are therefore deficient in Ca^{2+} binding and cannot participate in the blood coagulation process. It follows then that dicoumarol and warfarin do not have any effect on blood coagulation *in the test tube*.

Tissue factor **(factor III)** is a membrane protein that serves as the primary receptor for the initiation of the clotting process following injury. The cDNA for mature human tissue factor has been obtained and sequenced. Its structure is shown in Figure 22.50. It is a transmembrane protein of 263 amino acids. Residues 243–263 are located on the inside of the cell. Residues 220–242 are hydrophobic residues and represent the transmembrane sequence. Residues 1–219 are on the outside of the membrane and are the residues that become exposed after injury and form the receptor for formation of the initial complex of the extrinsic pathway. The extracellular domain is glycosylated, apparently at three sites. There are four cysteine residues within this domain that most likely form two disulfide bridges.

The **protransglutamidase** exists in both plasma and platelets. The structural form of the platelet enzyme is α_2, whereas that of the plasma form is

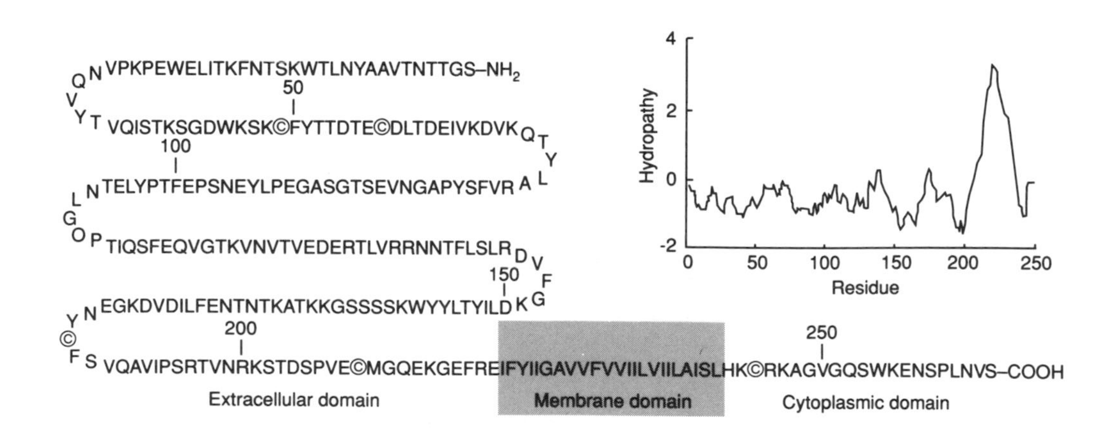

FIGURE 22.50
Amino acid sequence of human tissue factor derived from its cDNA sequence.
Redrawn from E. K. Spicer, et al. *Proc. Natl. Acad. Sci. U.S.A.* 84:5148, 1987.

$\alpha_2\beta_2$. Conversion to the active transglutamidase form is catalyzed by thrombin (Figure 22.51) through a proteolytic cleavage of the α subunit of either the platelet or the plasma enzyme. In the case of the plasma enzyme, cleavage of the α subunit leads to dissociation of the β subunit. These reactions occur on the surface and within the multienzyme complex. The platelet enzyme is released at the site and is activated more rapidly.

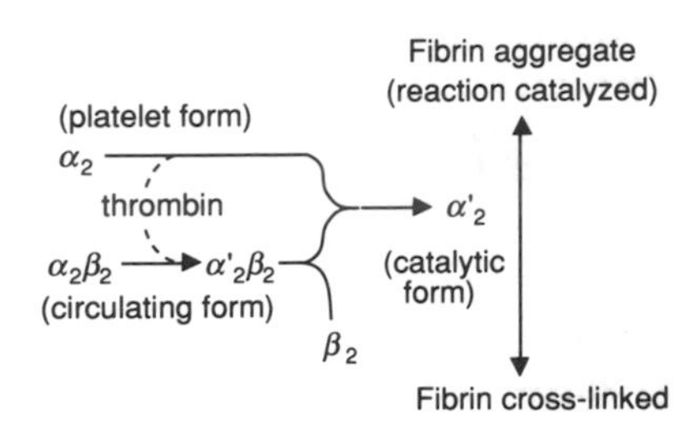

FIGURE 22.51
Activation of transglutamidase by thrombin.

The process of blood coagulation is self-controlling. One of the proteins that is involved in this process is **protein C.** Protein C, a Gla-containing protein, is activated in a membrane-bound complex of thrombin, thrombomodulin, and calcium. **Thrombomodulin** is an integral protein of the endothelial cell membrane. It is a glycoprotein containing 560 amino acid residues. Thrombomodulin shows homology with the low-density lipoprotein receptor but very little with tissue factor. There is, however, a great deal of similarity in functional domains between tissue factor and thrombomodulin. Thrombomodulin serves as a cofactor for thrombin for the activation of protein C. Protein C inhibits the coagulation process by inactivating, through proteolysis, the protein cofactors Va and VIIIa that are required for the activity of factors IXa and Xa. Important reactions of thrombin and protein C are summarized in Figure 22.52.

There are also a number of **proteinase inhibitors** in the blood that interact with the enzymes of the blood coagulation cascade. The half-times for the interactions of some of these inhibitors are listed in Table 22.8. The process of inhibition of the proteinases involved in coagulation is a kinetic one and can begin almost as soon as the process of coagulation itself begins. Initially, inhibitor complex formation is a slow process because the concentrations of the enzymes with which the inhibitors interact are low. As the activation process proceeds, the reactions leading to inhibition will increase and become more prominent. These reactions along with those that lead to destruction of protein cofactors, such as the inactivation of factors Va and VIIIa by protein C, will eventually stop the coagulation process completely. In general, the complexes formed with the proteinases and the inhibitors do not dissociate readily and are removed intact from the circulation by the liver and become substrates for catabolic reactions.

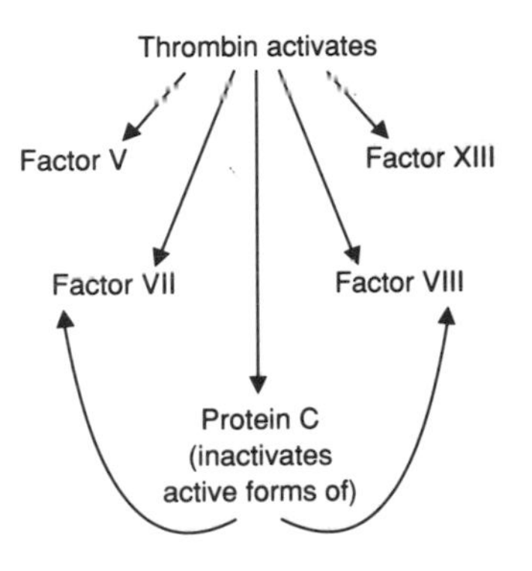

FIGURE 22.52
Summary of the reactions of thrombin.

The mechanism of inhibition of the extrinsic pathway, i.e., the TF–VIIa–Ca^{2+}–Xa complex, appears to be unique and to involve specific interaction with a **lipoprotein-associated coagulation inhibitor** (LACI), formerly known as anticonvertin. LACI is a 32 kd protein that contains three tandem domains. Each domain is a functionally homologous protease inhibitor that resembles other individual protease inhibitors such as the bovine pancreas trypsin inhibitor. LACI inhibits the extrinsic pathway by interacting specifically with TF–VIIa–Ca^{2+}–Xa complex. Domain 1

TABLE 22.8 Half-Time (in s) of the Association of Selected Proteinases With Human Plasma Proteinase Inhibitors

	Enzyme[a]			
Inhibitor	Kallikrein	Thrombin	Plasmin	Factor Xa
Alpha1-PI	+	830	210	170
Alpha2-M	6	500	+	350
Cl-Inh	8	?	+	?
Alpha2-AP	+	–	0.029	+
ATIII	+	0.847	+	230

SOURCE: Adapted from J. Travis and G. S. Salvesen, *Annu. Rev. Biochem.* 52:655, 1983.

[a] A + indicates that interaction occurred but kinetic data not available; a – indicates no interaction; a ? indicates no data available from this source.

binds to Xa and domain 2 binds to VIIa. The latter interaction occurs only in the presence of Xa. The uniqueness of this reaction is that LACI itself is a multi-enzyme inhibitor in which each separate domain inhibits the action of one of the enzymes of the multi-enzyme complex of the extrinsic pathway.

Fibrinolysis Requires Plasminogen and Tissue Plasminogen Activator (t-PA) to Produce Plasmin for Fibrinolysis

The reactions involved in fibrinolysis are shown in Figure 22.53. Lysis of the fibrin clot occurs through the action of the enzyme plasmin. **Plasmin** exists in the blood in its zymogen form, plasminogen. **Plasminogen** has a high affinity for the fibrin clot and forms complexes with fibrin throughout various regions of the porous fibrin network. Plasminogen is activated to plasmin by the action of another protease, **tissue plasminogen activator,** frequently referred to as **t-PA or TPA.** The t-PA also binds to the fibrin clot and, in the complex, activates plasminogen by cleavage of specific bonds within the zymogen. The clot is then solubilized by the action of plasmin.

The cDNA for human t-PA has been obtained and expressed. The t-PA is a 72-kDa protein with several functional domains. It has a growth factor activity near its N terminal, two adjacent Kringle regions that are involved in the interaction of t-PA with fibrin, and closer to its C terminal end is its protease activity. **Kringle domains** are conserved sequences of amino acids that fold into large loops stabilized by disulfide bridges. These domains are thought to be important structural features for the protein–protein interactions that occur with the blood coagulation factors. A diagram of the structure of t-PA is shown in Figure 22.54. This diagram shows the complete molecule with its leader sequence attached. The number 1 between the letters B and C is the N terminal of the mature protein. The molecule consists of a heavy and a light chain resulting from cleavage at the triangle between the letters I and J. The serine protease activity is located within the light chain of the molecule.

The activity of t-PA is also regulated through the action of other protein inhibitors. Approximately four immunologically distinct types have been identified, two of which are of greater physiological significance because they react rapidly with t-PA and are specific for t-PA. These are plasminogen activator-inhibitor type 1 (PAI-1) and plasminogen activator-inhibitor type 2 (PAI-2). The cDNAs for each of these inhibitors have been cloned and found to code for proteins of 402 amino acid residues for PAI-1 (rat tissue) and for a protein of 415 amino acid residues for human PAI-2.

Blood Coagulation Occurs By a Cascade of Reactions

The blood coagulation process may be divided into three parts: the terminal phase in which the clot is formed, the extrinsic pathway, and the

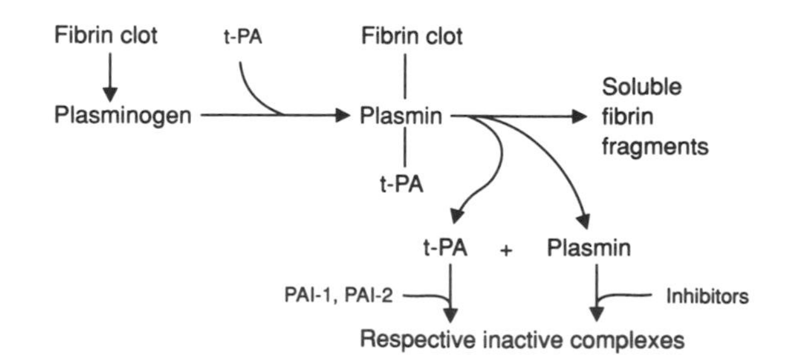

FIGURE 22.53
Reactions involved in the dissolution of the clot.

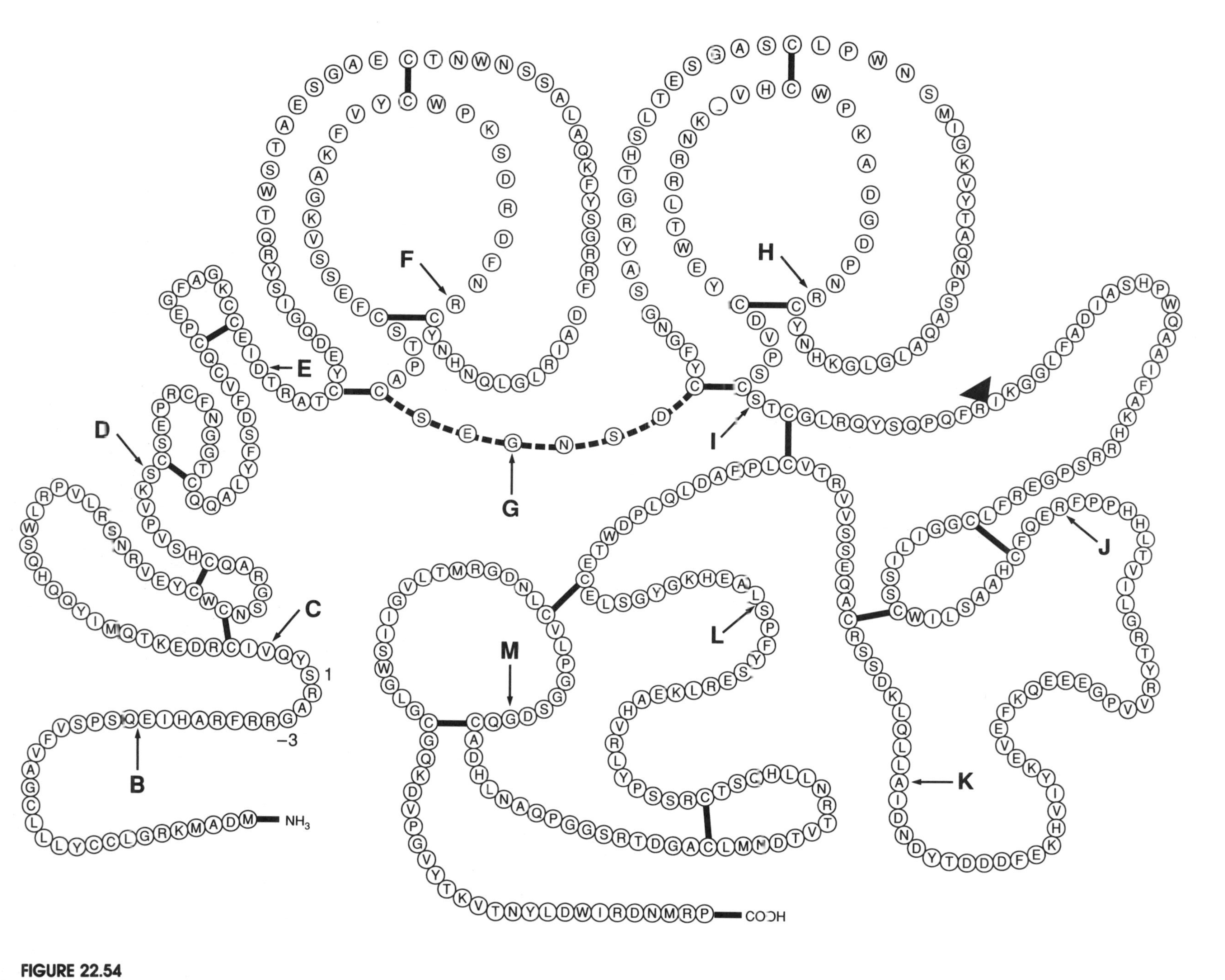

FIGURE 22.54
Schematic model of the human tissue plasminogen activator including the signal peptide and the prosequence.
Redrawn from T. Ny, F. Elgh, and B. Lund, *Proc. Natl. Acad. Sci. U.S.A.* 81:5355, 1984.

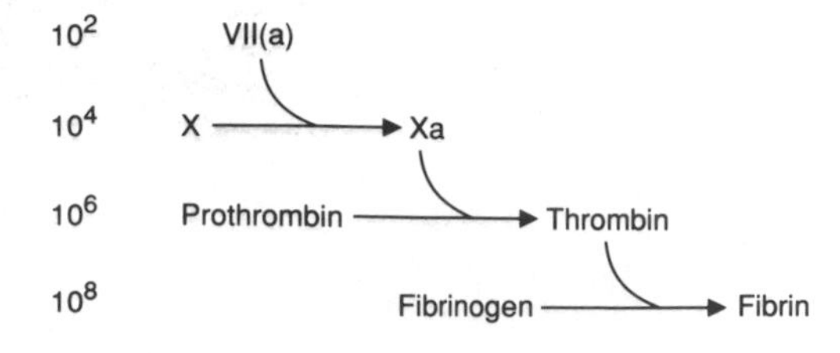

FIGURE 22.55
Example of amplification by the cascade mechanism.

intrinsic pathway, the latter two of which lead to formation of the proteases that catalyze the terminal reactions that lead to clot formation. The initiation of clot formation, that is, the initial reactions of the extrinsic and the intrinsic pathways occur on the surface of membranes at the sites of injury. The injury leads to the exposure of certain proteins, tissue factor for the extrinsic pathway, and collagen–high molecular weight kininogen for the intrinsic pathway, which function as *receptors* for the respective zymogens and their cofactors.

Multienzyme complexes are formed at these receptor sites. The activity of these proteinases is enhanced through the protein–protein interactions that occur either as a result of the initial interactions of the additional binding of specific protein cofactors. The process of clot formation is enhanced by the amplification of the *cascade* process resulting from successive zymogen activation. For example, in the following scheme (Figure 22.55), if *each molecule* of enzyme catalyzes the formation of just 100 molecules of the next enzyme, the process is amplified 1 millionfold.

Blood clotting is self-regulated by the interaction of specific inhibitors with the proteinases and the activation of other proteinases that inactivate the essential protein cofactors involved in the clotting process. Dissolution of the clot is also a similarly controlled process. It too occurs within a protein–protein complex that begins with the binding of plasminogen to the clot matrix.

BIBLIOGRAPHY

General

Alberts, B., Bray, D., Lewis, J., Raff, M., Roberts, K., and Watson, J. *Molecular Biology of the Cell,* 2nd ed., New York: Garland Publishing, 1989.

Berridge, M. J. Regulation of ion channels by inositol trisphosphate and diacylglycerol. *J. Exp. Biol.* 124:323, 1986.

Darnell, J., Lodish, H., and Baltimore, D. *Molecular Cell Biology,* New York: Scientific American Books, 1986.

Rawls, R. L. G-Proteins. *Chem. Eng. News,* December 21, 26, 1987.

Nerve

Grenningloh, G. et al. The strychnine-binding subunit of the glycine receptor shows homology with nicotinic acetylcholine receptors. *Nature* (*London*) 328:215, 1987.

Schofield, P. B. et al. Sequence and functional expression of the GABA receptor shows a ligand-gated receptor super-family. *Nature* (*London*) 328:221, 1987.

Stühmer, W. Structure-function studies of voltage-gated ion channels. *Annu. Rev. Biophys. Biophys. Chem.* 20:65, 1991.

Sussman, J. L., Harel, M., Frolow, F., Oefner, C., Goldman, A., Toker, L., and Silman, I. Atomic structure of acetylcholinesterase from *Torpedo californica*: A prototypic acetylcholine-binding protein. *Science* 253:872, 1991.

Taylor, P. The cholinesterases. *J. Biol. Chem.* 266:4025, 1991.

Vision

Abrahamson, E. W. and Ostroy, S. E. (Eds.). *Molecular Processes in Vision, Benchmark Papers in Biochemistry/3,* Stroudsburg, PA: Hutchinson Ross Publishing Company, 1981.

Chabre, M. and Deterre, P. Molecular mechanisms of visual transduction. *Eur. J. Biochem.* 179:255, 1989.

deJong, W. W., Hendriks, W., Mulders, J. W. M., and Bloomendal, H. Evolution of eye lens crystallins: The stress connection. *Trends Biochem. Sci.* 14:365, 1989.

Gao, B., Gilman, A. G., and Robishaw, J. D. A second form of the beta subunit of signal-transducing G-Proteins. *Proc. Natl. Acad. Sci. U.S.A.* 84:6122, 1987.

Koretz, J. F. and Handelman, G. H. How the human eye focuses. *Sci. Am.,* 92, July 1988.

Medynski, D. C. et al. Amino acid sequence of the alpha-subunit of transducin deduced from the cDNA sequence. *Proc. Natl. Acad. Sci. U.S.A.* 82:4311, 1985.

Nathans, J. Molecular biology of visual pigments. *Ann. Rev. Neurosci.* 10:163, 1987.

Nathans, J. The genes for color vision. *Sci. Am.* 42, February 1989.

Nathans, J. et al. Molecular genetics of human blue cone monochromacy. *Science* 245:831, 1989.

Ovchinnikov, Y. A., Lipkin, V. M., Shuvaeva, T. M., Bogachuk, A. P., and Shemyakin, V. V. Complete amino acid sequence of gamma subunit of the GTP-binding protein from cattle retina. *FEBS Lett.* 179:107, 1985.

Schrapf, J. L. and Baylor, D. A. How photo–receptor cells respond to light. *Sci. Am.* 40, April 1987.

Stryer, L. Cyclic GMP cascade of vision. *Ann. Rev. Neurosci.* 9:87, 1986.

Stryer, L. Visual excitation and recovery. *J. Biol. Chem.* 266:10711, 1991.

Zigler, J. S., Jr., and Goosey, J. Aging of protein molecules: Lens crystallins as a model system. *Trends Biochem. Sci.* 7:133, 1981.

Muscle

Ames, L. A. Structure of muscle filaments studied by electron microscopy. *Ann. Rev. Biophys. Biophys. Chem.* 14:291, 1985.

Carlier, M.-F. Actin: Protein structure and filament dynamics. *J. Biol. Chem.* 266:1, 1991.

da Silva, A. C. R. and Reinach, F. C. Calcium binding induces conformational changes in muscle regulatory proteins. *Trends Biochem. Sci.* 16:53, 1991.

Ebashi, S. Excitation-contraction coupling and the mechanism of muscle contraction. *Annu. Rev. Physiol.* 53:1, 1991.

Gerisch, G., Noegel, A. A., and Schleicher, M. Genetic alteration of proteins in actin-based motility systems. *Annu. Rev. Psychol.* 53:607, 1991.

Hambly, B. D., Barden, J. A., Miki, M., and dos Remedies, C. G. Structural and functional domains on actin. *BioEssays* 4:124, 1986.

Huxley, H. E. The mechanism of muscular contraction. *Science* 164:1356, 1969.

Kramm, K. E. and Stull, J. T. The function of myosin and myosin light chain kinase phosphorylation in smooth muscle. *Ann. Rev. Pharmacol. Toxicol.* 25:593, 1985.

Matsudaira, P. Modular organization of actin crosslinking proteins. *Trends Biochem. Sci.* 16:87, 1991.

McLachlan, A. D. Structural implications of the myosin amino acid sequence. *Ann. Rev. Biophys. Bioeng.* 13:167, 1984.

Pollard, T. D. and Cooper, J. A. Actin and Actin Binding Proteins. A Critical Evaluation of Mechanisms and Functions. *Annu. Rev. Biochem.* 55:987, 1986.

Rimm, D. L., Sinard, J. H., and Pollard, T. D. Location of the head–tail junction of myosin. *J. Cell. Biol.* 108:1783, 1989.

Vanderkerckhove, J. and Weber, K. The amino acid sequence of actin from chicken skeletal muscle actin and chicken gizzard smooth muscle actin. *FEBS Lett.* 102:219, 1979.

Yanagisawa, M., Hamada, Y., Katsuragawa, Y., Imamura, M., Mikawa, T., and Masaki, T. Complete primary structure of vertebrate smooth muscle myosin heavy chain deduced from its complementary DNA sequence: Implications on topography and function of myosin. *J. Mol. Biol.* 198:143, 1987.

Blood

Andrews, B. S. Is the WKS motif the tissue-factor binding site for coagulation factor VII? *Trends Biochem. Sci.* 16:31, 1991.

Antalis, T. M. et al. Cloning and expressing of a cDNA coding for a human monocyte-derived plasminogen activator inhibitor. *Proc. Natl. Acad. Sci. U.S.A.* 85:985, 1988.

Colombatti, A. and Bonaldo, P. The superfamily of proteins with von Willebrand factor type A-like domains: One theme common to components of extracellular matrix, hemostasis, cellular adhesion, and defense mechanisms. *Blood* 77:2305, 1991.

Cooper, D. N. The molecular genetics of familial venous thrombosis. *Blood Reviews* 5:55, 1991.

Hessing, M. The interaction between complement component C4b-binding protein and the vitamin K-dependent protein S forms a link between blood coagulation and the complement system. *Biochem. J.* 277:581, 1991.

Kalyan, N. K., Lee, S. G., Cheng, S. M., Hartzell, R., Urbano, C., and Hung, P. P. Construction and expression of a hybrid plasminogen activator gene with sequences from non-protease region of tissue-type plasminogen activator (t-PA) and protease region of urokinase (u-PA). *Gene* 68:205, 1988.

Mann, K. G., Jenny, R. J., and Krishnaswamy, S. Co-factor proteins in the assembly and expression of blood clotting enzyme complexes. *Annu. Rev. Biochem.* 57:915, 1988.

Nemerson, Y. Tissue factor and hemostasis. *Blood* 71:1, 1988.

Ny, T., Elgh, F., and Lund, B. The structure of the human tissue-type plasminogen activator gene: correlation of introns and exon structures to functional and structural domains. *Proc. Natl. Acad. Sci. U.S.A.* 81:5355, 1984.

Roth, G. J. Developing relationships: Arterial platelet adhesion, glycoprotein Ib, and leucine-rich glycoproteins. *Blood* 77:5, 1991.

Steiner, H., Pohl, G., Gunne, H., Hellers, M., Elhammer, A., and Hansson, L. Human tissue-type plasminogen activator synthesized by using a baculovirus vector in insect cells compared with human plasminogen activator produced in mouse cells. *Gene* 73:449, 1988.

Vermeer, C. γ-Carboxyglutamate-containing proteins and the vitamin K-dependent carboxylase. *Biochem. J.* 266:625, 1990.

Zeheb, R. and Gelehrter, T. D. Cloning and sequencing of cDNA for the rat plasminogen activator inhibitor-1. *Gene* 73:459, 1988.

QUESTIONS

C. N. ANGSTADT AND J. BAGGOTT

*Q*uestion *T*ypes are described on page xxiii.

1. (QT1) G Proteins involved in transmembrane signaling:
   A. must recognize and interact with the stimulus.
   B. form an ion channel during the signaling process.
   C. are not susceptible to cholera or pertussis toxins.
   D. undergo a conformational change, dependent on binding of GTP.
   E. are involved in transmission of signals only from lipophilic molecules.

2. (QT2) In the propagation of a nerve impulse by an electrical signal:
   1. the electrical potential across the membrane maintained by the ATP-driven Na^+,K^+ ion pump becomes less negative.
   2. local depolarization of the membrane causes protein conformational changes that allow Na^+ and K^+ to move down their concentration gradients.
   3. charge propagation is unidirectional along the axon because the "voltage-gated" ion channels have a finite recovery time.
   4. astrocytes are the antennalike protrusions that receive signals from other cells.

3. (QT1) All of the following are characteristics of nonpeptide neurotransmitters EXCEPT:
   A. they transmit the signal across the synapse between cells.
   B. they must be made in the cell body and then travel down the axon to the presynaptic terminal.
   C. electrical stimulation increasing Ca^{2+} in the presynaptic terminal fosters their release from storage vesicles.
   D. binding to receptors on the postsynaptic terminal induces a conformational change in proteins of that membrane.

E. their actions are terminated by specific mechanisms within the synaptic junction.

A. Acetylcholine
B. 4-Aminobutyrate (GABA)
C. Catecholamines
D. 5-Hydroxytryptamine (Serotonin)

4. (QT5) Binding to its receptor opens a channel for Cl^-, causing hyperpolarization of the cell.
5. (QT5) Termination of the signal typically involves the actions of both methyltransferase and monoamine oxidase, as well as reuptake into the presynaptic neuron.

6. (QT2) Which of the following statements about biochemical events occurring in the eye is (are) true?
   1. The high rate of the hexose monophosphate pathway in the cornea is necessary to provide NADPH as a substrate for glutathione reductase.
   2. Glucose in the lens is metabolized by the TCA cycle in order to provide ATP for the Na^+,K^+-ATPase.
   3. Controlling the blood glucose level might reduce the incidence of diabetic cataracts by minimizing the production of sorbitol.
   4. The retina contains mitochondria so it depends on the TCA cycle for its production of ATP.

7. (QT1) Which of the following statements about rhodopsin is true?
   A. Rhodopsin is the primary photoreceptor of both rods and cones.
   B. The prosthetic group of rhodopsin is all-*trans*-retinol derived from cleavage of β-carotene.
   C. Conversion of rhodopsin to activated rhodopsin, R*, by a light pulse requires depolarization of the cell.
   D. Rhodopsin is located in the cytosol of the cell.
   E. Absorption of a photon of light by rhodopsin causes an isomerization of 11-*cis*-retinal to all-*trans*-retinal.
8. (QT1) All of the following statements about the transduction of the light signal on rhodopsin are true EXCEPT:
   A. cGMP is involved in the transmission of the signal between the disk membrane and the plasma membrane.
   B. it involves the G protein, transducin.
   C. cGMP concentration is increased in the presence of an activated rhodopsin–transducin–GTP complex.
   D. the signal is turned off, in part, by the GTPase activity of the α subunit of transducin.
   E. both guanylate cyclase and phosphodiesterase are regulated by calcium concentration.
9. (QT1) The cones of the retina:
   A. are responsible for color vision.
   B. are much more numerous than the rods.
   C. have red, blue, and green light-sensitive pigments that differ because of small differences in the retinal prosthetic group.
   D. do not use transducin in signal transduction.
   E. are better suited for discerning rapidly changing visual events because a single photon of light generates a stronger current than it does in the rods.

10. (QT1) When a muscle contracts, the:
    A. transverse tubules shorten, drawing the myofibrils and sarcoplasmic reticulum closer together.
    B. thin filaments and the thick filaments of the sarcomeres shorten.
    C. light chains dissociate from the heavy chains of myosin.
    D. H bands and I bands of the sarcomeres shorten because the thin filaments and thick filaments slide past each other.
    E. cross linking of proteins in the heavy filaments increases.

11. (QT2) Which of the following statements is(are) true about actin and myosin?
    1. The globular head section of myosin has domains for binding ATP and actin.
    2. The binding of ATP to the actin–myosin complex promotes dissociation of actin and myosin.
    3. F-actin, formed by aggregation of G-actin–ATP–Mg^{2+} complex, is stabilized when tropomyosin is bound to it.
    4. Binding of calcium to the calmodulin-like subunit of troponin induces conformational changes that permit myosin to bind to actin.

A. Adenylate kinase
B. Creatine phosphokinase
C. Both
D. Neither

12. (QT4) Involved in maintaining the [ATP] relatively constant during muscle contraction.

13. (QT2) The nerve impulse which initiates muscular contraction:
    1. begins with the binding of acetylcholine to plasma membrane receptors.
    2. causes both the plasma membrane and the transverse tubules to undergo depolarization.
    3. causes opening of calcium channels, which leads to an increase in calcium concentration within the sarcomere.
    4. is not likely to use cAMP as a second messenger for opening calcium channels.
14. (QT2) Platelet aggregation:
    1. is initiated at the site of an injury by conversion of fibrinogen to fibrin.
    2. is inhibited in uninjured blood vessels by the secretion of prostacyclin by intact vascular endothelium.
    3. causes morphological changes and a release of the vasodilator, serotonin.
    4. is facilitated by the release of ADP and thromboxane A_2.

15. (QT1) In the formation of a blood clot:
    A. proteolysis of γ-carboxyglutamate residues from fibrinogen to form fibrin is required.
    B. the clot is stabilized by the cross-linking of fibrin molecules by the action of factor XIII, transglutamidase.
    C. antagonists of vitamin K inhibit the formation of γ-carboxyglutamate residues in various proteins, thus facilitating the clotting process.

D. Tissue factor, factor III, must be inactivated for the clotting process to begin.
E. the role of calcium is primarily to bind fibrin molecules together to form the clot.

A. Extrinsic pathway for blood clotting C. Both
B. Intrinsic pathway for blood clotting D. Neither

16. (QT4) Factor Xa, necessary for conversion of prothrombin to thrombine, is formed by the action of the TF–VII–Ca^{+2} complex on factor X.

17. (QT1) Lysis of the fibrin clot once the blood vessel has been repaired occurs by the action of:
A. Kallikrein
B. PAI-1
C. Plasmin
D. Plasminogen
E. t-PA (TPA)

ANSWERS

1. D Definition of this type of signal transduction protein. A: This is the receptor. B: Another type of signal transduction protein. C: G proteins are ADP-ribosylated by these agents. E: Lipophilic molecules do not require this mechanism because their receptors are internal (p. 928).
2. A 1,2,3, correct. 1: This is a major role for ATP in the neuron. 2 and 3: This is the mechanism for impulse propagation. 4: This describes dendrites. Astrocytes are glial cells that are involved in processes isolating this CNS from the external environment (p. 930).
3. B This is true for neuropeptides, but many non peptide neurotransmitters are synthesized in the presynaptic terminal (pp. 931,933). A: This is a difference between electrical and chemical signals (p. 932). C: What is the role of synapsin I in this process (p. 933)? E: Make sure you know the three types of processes involved (p. 933).
4. B GABA is an inhibitory neurotransmitter. All the others are excitatory ones that cause depolarization of the cells (p. 932).
5. C Methylation by catecholamine-*O*-methyltransferase is an important part of the metabolism of the catecholamines. A: Acetylcholinesterase terminates the action of this (p. 934). B: GABA is converted into an intermediate of the TCA cycle (p. 937). D: Monoamine oxidase is the primary enzyme responsible for terminating serotonin's action (p. 936).
6. B 1,3 correct. 1: Make sure you understand the role of glutathione in protecting against harmful byproducts from atmospheric oxygen (p. 940). 3: Increased sorbitol results in aggregation of crystalline proteins in the lens, increasing their light-scattering properties (p. 941). 2: Most of the ATP (85%) in the lens is generated by glycolysis (p. 941). 4: Its metabolism is similar to that of other eye tissues directly involved in the visual process. Thus, its major source of energy is from glycolysis (p. 941).
7. E This causes the conformational change of the protein that affects the resting membrane potential and initiates the rest of the events (p. 944). A: Cones have the same prosthetic group but different proteins, so rhodopsin is in rods only (p. 945). B: This is the precursor of the prosthetic group 11-*cis*-retinal (p. 945). C: Isomerization of the prosthetic group leads to hyperpolarization (p. 946, Figure 21.21). D: Rhodopsin is a transmembrane protein (p. 945).
8. C The transducin complex *activates* the phosphodiesterase, thus lowering [cGMP] (p. 950). A: This is an example of a second messenger type chemical synapse (p. 947). B and D: Transducin meets the criteria for a typical G protein. E: The enzymes are regulated in opposite directions by Ca^{2+}, thus controlling [cGMP] (p. 951).
9. A Rods are responsible for low light vision. C: All three pigments have 11-*cis*-retinal; the proteins differ and are responsible for the slightly different spectra (p. 952). D: The biochemical events are believed to be the same in rods and cones, although there are some immunological differences in the transducins (p. 953). E: Cones are better suited for rapid events because their response rate is about four times faster than rods, even though their sensitivity to light is much less (p. 954).
10. D This occurs because of association–dissociation of actin and myosin (p. 962). A: Depolarization in the transverse tubules may be involved in transmission of the signal but not directly in the contractile process (p. 964). B: The filaments do not change in length, but slide past each other (p. 962). C: This is not physiological. E: Cross-linking occurs in the H band of the sarcomeres but does not change during the contractile process (p. 962).
11. E All four correct. 1: (see Figure 22.42). 2: Note that the role of ATP in contraction is to favor dissociation, not formation, of the actin–myosin complex (p. 963). 3 and 4: Tropomyosin, tropinin and actin are the three major proteins of the filament. Their actions are closely interconnected (pp. 959,961).
12. C Both. Make sure you know the reactions both of these enzymes catalyze (p. 963).
13. A 1–3 correct. 4: IP_3, released by the action of phospholipase C, is the most likely second messenger to react with calcium channels. The system is present but it has not yet been proven to be sufficiently rapid (p. 965).
14. C 2,4 correct. 1: Initiation is by contact with collagen fibers exposed by the injury. Clot formation requires activation of various enzymes (p. 966). 2 and 4: the "ying–yang" nature of PGI_2 and TXA_2 help to control platelet aggregation to a need for it. 3: Serotonin is a vasoconstrictor. Vasodilation would be contraindicated in this situation (p. 966).

15. B The cross-linking occurs between a glutamine and a lysine (Figure 22.46). A and E: Gamma-carboxyglutamate residues are on various enzymes; they bind calcium and facilitate the interaction of these proteins with membranes that form the sites for initiation of reaction (pp. 967,970). C: Vitamin K is an activator for the γ-carboxylation reaction, which is a *necessary* posttranslational modification of some of the enzymes involved in clot formation (p. 971). D: TF, factor III, is the primary receptor for initiation of the clotting process (p. 969).

16. A Tissue factor and factor VII are unique to the extrinsic pathway. The membrane interaction with the intrinsic pathway is with high-molecular-weight kininogen and pre-kallikrein (p. 970).

17. C A: kallikrein activates factor XII in clot formation (p. 970). B,D,E: Plasminogen is activated to plasmin by tissue plasminogen activator, whose activity is regulated by protein inhibitors (e.g., PAI—1) (p. 974).

23

Biotransformations: The Cytochromes P450

RICHARD T. OKITA
BETTIE SUE SILER MASTERS

23.1 OVERVIEW 981
23.2 CYTOCHROME P450: NOMENCLATURE AND OVERALL REACTION 982
23.3 CYTOCHROME P450: MULTIPLE FORMS 984
Multiplicity of Genes Produces Many Forms of Cytochrome P450 984
23.4 INHIBITORS OF CYTOCHROME P450 986
23.5 CYTOCHROME P450 ELECTRON TRANSPORT SYSTEMS 987
NADPH-Cytochrome P450 Reductase Is the Intermediate Electron Donor in the Endoplasmic Reticulum 987
Mitochondria Use NADPH-Adrenodoxin Reductase as the Intermediate Electron Donor 989
23.6 PHYSIOLOGICAL FUNCTIONS OF CYTOCHROMES P450 990
Cytochromes P450 Participate in the Synthesis of Steroid Hormones and Oxygenation of Eicosanoids 990
Cytochromes P450 Oxidize Exogenous Lipophilic Substrates 995
BIBLIOGRAPHY 997
QUESTIONS AND ANNOTATED ANSWERS 998
CLINICAL CORRELATIONS
23.1 Consequences of Induction of Drug-Metabolizing Enzymes 986
23.2 Deficiency of Cytochrome P450–21-Hydroxylase 994
23.3 Steroid Hormone Production During Pregnancy 996

23.1 OVERVIEW

The term **cytochrome P450** refers to a family of heme proteins present in all mammalian cell types, except mature red blood cells and skeletal muscle cells, which catalyze the oxidation of a wide variety of structurally diverse compounds. Cytochrome P450 also occurs in procaryotes. Substrates for these enzyme systems include endogenously synthesized compounds such as steroids, fatty acids (including prostaglandins and leukotrienes), and compounds such as drugs, food additives, or industrial byproducts that enter the body through food sources, injection, inhalation from the air, or absorption through the skin. The cytochrome P450 system has far reaching effects in medicine. It is involved in: (a) inactivation or activation of therapeutic agents, (b) conversion of chemicals to highly

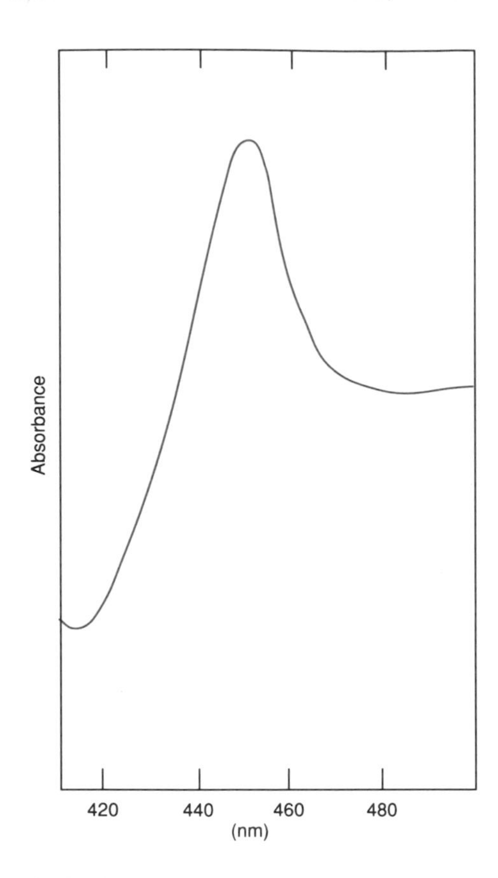

FIGURE 23.1
The absorbance spectrum of the carbon monoxide-bound cytochrome P450.

reactive molecules, which may produce unwanted cellular damage, cell death, or mutations, (c) participation in several steps in steroid hormone biosynthesis, and (d) metabolism of fatty acids and their derivatives.

23.2 CYTOCHROME P450: NOMENCLATURE AND OVERALL REACTION

The designation of a protein as cytochrome P450 originated from its spectral properties before its catalytic function was known. This group of proteins has a unique absorbance spectrum that is obtained by adding a reducing agent, such as sodium dithionite, to a suspension of endoplasmic reticulum vesicles, frequently referred to as microsomes, followed by the bubbling of carbon monoxide gas into the solution. The carbon monoxide is bound to the reduced heme protein and produces an absorbance spectrum with a peak at 450 nm (Figure 23.1); thus the name P450 for a *pigment* with an absorbance at 450 nm. Specific forms of cytochromes P450 differ in their maximum absorbance wavelength, with a range between 446 and 452 nm. The many forms of cytochrome P450s are classified, according to their sequence similarities, into various gene subfamilies; this system of nomenclature is being adopted almost universally.

The general reaction catalyzed by cytochrome P450 is written as follows:

$$\text{NADPH} + \text{H}^+ + \text{O}_2 + \text{SH} \longrightarrow \text{NADP}^+ + \text{H}_2\text{O} + \text{S—OH}$$

where the substrate (S) may represent a steroid, fatty acid, drug, or other chemical that has an alkane, alkene, aromatic ring, or heterocyclic ring substituent that can serve as a site for oxygenation. The reaction is referred to as a **monooxygenation** and the enzyme as a **monooxygenase** because only one of the two oxygen atoms is incorporated into the substrate.

In mammalian cells, cytochromes P450 serve as terminal electron acceptors and monooxygenases in electron transport systems, which are present either in the endoplasmic reticulum or inner mitochondrial membrane. The cytochrome P450 protein contains a single iron protoporphyrin IX prosthetic group (see p. 1011), and the resulting heme protein contains binding sites for both an oxygen molecule and the substrate. The heme iron of all known cytochromes P450 is bound to the four pyrrole nitrogen atoms of the porphyrin ring and two axial ligands, one of which is a sulfhydryl group from a cysteine residue located toward the carboxyl end of the molecule (Figure 23.2). The heme iron may exist in two different **spin states:** (1) a hexa-coordinated low-spin iron, or (2) penta-coordinated high-spin state. The low- and high-spin states are descriptions of the electronic shells around the iron atom. When a cytochrome P450 molecule binds a substrate, there is a perturbation of these electronic shells and the heme iron atom changes from its hexa-coordinated to the penta-coordinated state. The substrate bound, penta-coordinated state has a more positive reduction potential (−170 mV) than the hexa-coordinated unbound state (−270 mV), accelerating the rate at which cytochrome P450 may be reduced by electrons donated from NADPH (Figure 23.3). In order for the **hydroxylation** (monooxygenation) reaction to occur, the heme iron must be reduced from the ferric (Fe^{3+}) to its ferrous (Fe^{2+}) state so that oxygen may bind to the heme iron. A total of $2e^-$ are required for the monooxygenation reaction. The electrons are transferred to the cytochrome P450 molecule individually, the first to allow oxygen binding and

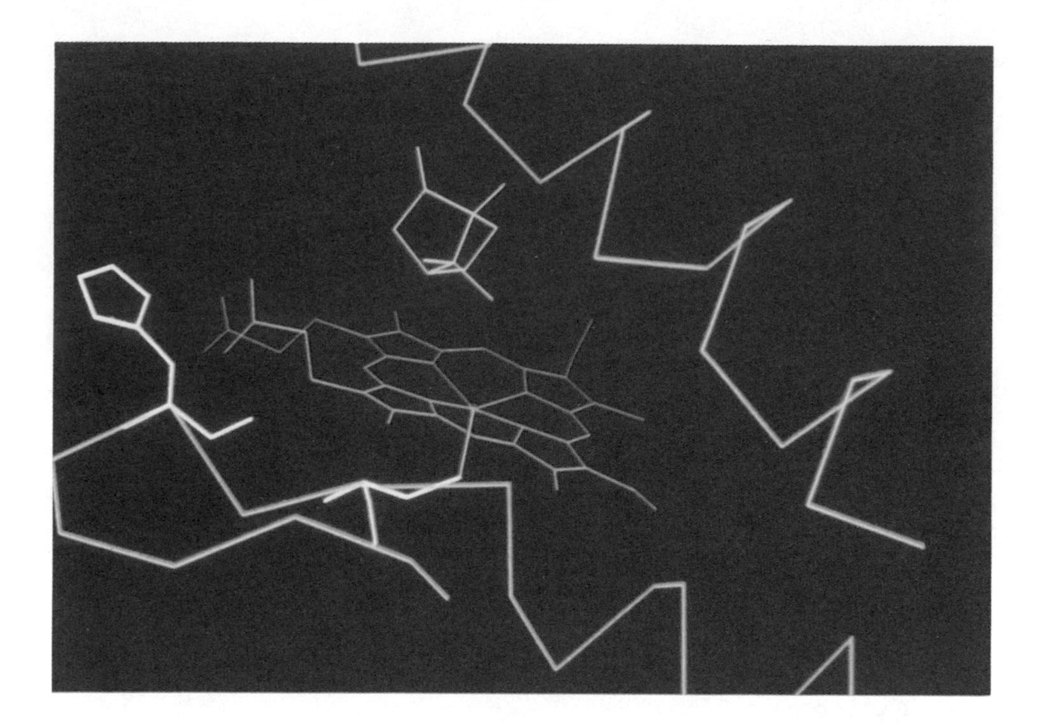

FIGURE 23.2
The penta-coordinated heme prosthetic group of bacterial cytochrome P450. The iron is held by four pyrrole nitrogen atoms and the axial thiolate ligand is provided by cysteine 357, which is located below the iron atom. Also, histidine residue 355 is shown to the left of the iron atom. The substrate camphor is shown above the iron atom. This figure was drawn using the program TOM on a Silicon Graphics computer based on studies of Dr. Thomas Poulos (P. R. Ortiz de Montellano (Ed.). *Cytochrome P450. Structure, Mechanism, and Biochemistry*, Chapter 13, New York: Plenum, pp. 505–523). TOM is a version of FRODO, a molecular graphics program written by A. Jones and modified for Silicon Graphics by C. Cambillau. The coordinates are from the Brookhaven Protein Data Bank.

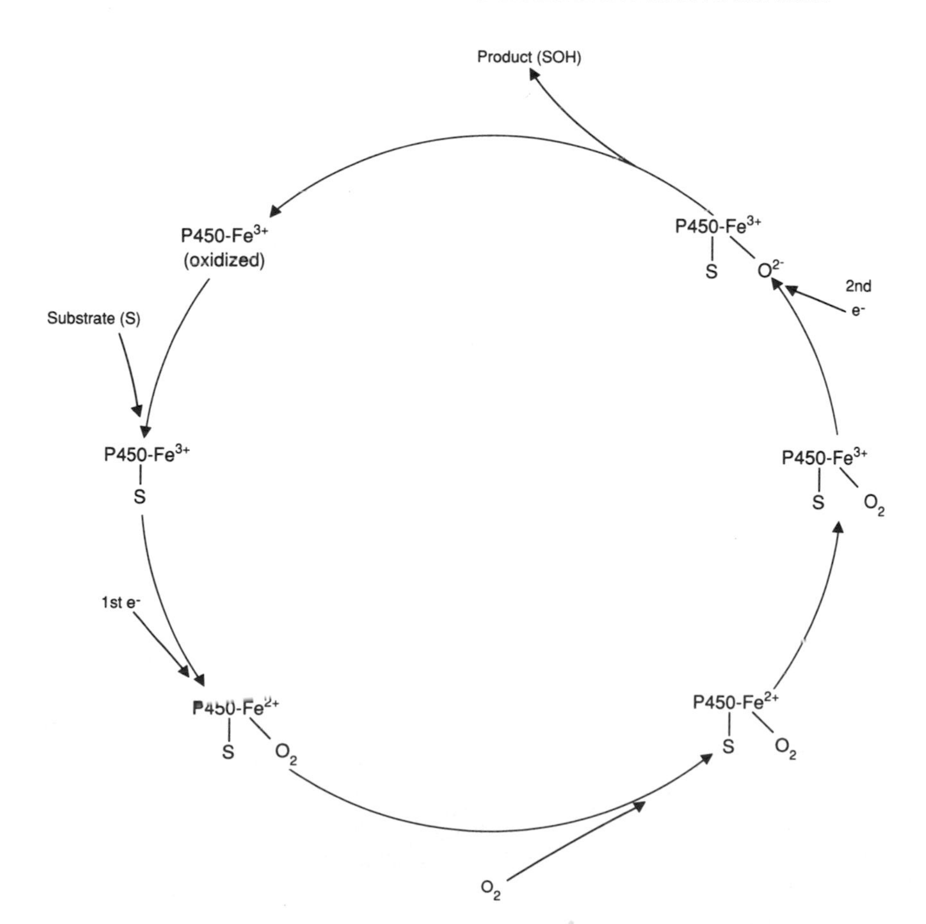

FIGURE 23.3
The sequence of reactions at cytochrome P450.
The diagram demonstrates the binding of substrate, transfer of the first and second electrons, and binding of molecular oxygen.

the second to cleave the oxygen molecule to generate the active oxygen species for insertion into the reaction site of the substrate.

23.3 CYTOCHROME P450: MULTIPLE FORMS

Since the mid-1950s it has been known that one of the atoms of molecular O_2 can be inserted into a substrate being metabolized. It should be emphasized that this process of *monooxygenation* is also performed by other specialized proteins such as flavoprotein monooxygenases (hydroxylases). None of the other proteins classified as oxygenases, however, displays the versatility of the members of the cytochrome P450 family. In the past decade, a plethora of information has been assimilated on the sequence and structure of cytochromes P450 that has led to a further understanding of their evolution and regulation.

Multiplicity of Genes Produces Many Forms of Cytochrome P450

It has become abundantly clear that many forms of cytochromes P450 have emerged due to **gene duplication** events occurring in the last 5–50 million years, resulting in the divergence of various species. The different forms of cytochrome P450, which are now known among the various animal species, have likely arisen from the selective pressure of environmental influences, such as dietary habits or exposure to environmental agents. It is logical that the primordial genes gave rise to those cytochromes P450 that metabolized endogenous substrates. An examination of the phylogenetic tree, generated by comparing amino acid sequences and assuming a constant evolutionary change rate, leads to the conclusion that the earliest cytochromes P450 evolved to metabolize cholesterol and fatty acids. Therefore, they may have played a role in the maintenance of *membrane integrity* in early eucaryotes.

Substrate Specificity

By the end of 1990, over 150 cytochrome P450 genes, coding for different proteins catalyzing the oxygenation of a variety of endogenous and exogenous substrates, had been characterized. There are many other members of this gene superfamily for which sequences have not yet been determined and, therefore, classification is not possible. Until such sequence information is available on each protein, it is inappropriate to designate a particular cytochrome P450 as new and unique. One of the ways of characterizing these enzymes is the determination of *substrate specificity*. While this has been possible with many of the members of this family, the similarity of molecular weights and other molecular properties has made purification of individual cytochromes P450 from the same organ or even the same subcellular organelle very difficult, if not impossible. One way of determining the substrate specificity of a cytochrome P450 has been to express the cDNA for the particular protein via an expression vector in an appropriate cellular expression system. To date, this has been achieved in bacterial, insect, yeast, and mammalian cell systems, and permits the unequivocal determination of substrate specificity uncomplicated by impurities of protein purification. The assumption is that, if nucleotide sequences can be determined from the cloned genes appropriately characterized prior to transfection into the cultured cells, there can be little doubt as to the source of enzyme activity expressed in those cells. This approach has been used, but a complication can be that a single cytochrome P450 protein can metabolize different substrates that do not bear structural resemblance to one another.

Induction of Cytochrome P450

The induction of various cytochromes P450 by both endogenous and exogenous compounds has been known since the mid-1960s. The mechanisms of induction of cytochromes P450 have been demonstrated to be at either the transcriptional or posttranscriptional level and it is not possible to predict the mode of induction based on the inducing compound. For example, ethanol and acetone both induce members of two different gene subfamilies of cytochrome P450 by different mechanisms. In one case, induction occurs at the transcriptional level and, in the other, it involves posttranscriptional events, that is, stabilization of mRNA. An example of the complexity of the induction process occurs with cytochrome P450IIE1 in rats as a result of ethanol, acetone or pyrazole pretreatment, fasting, or diabetes. Administration of the small organic compounds produces larger amounts of the cytochrome P450IIE1 protein without affecting the levels of mRNA. While the mechanism is not completely understood, pyrazole may stabilize this specific cytochrome P450 from proteolytic degradation. On the other hand, in diabetic rats the sixfold induction of cytochrome P450IIE1 is accompanied by a 10-fold increase in mRNA in the absence of an increase in transcription of the gene, suggesting stabilization of the mRNA. Again, it is difficult to predict the mechanism of induction under various conditions, even with a single cytochrome P450.

The role of specific receptor proteins has been indicated in the case of some of the known inducing agents. One of the most extensively studied is the interaction of **2,3,7,8-tetrachlorodibenzo-*p*-dioxin (TCDD)** with its cytosolic receptor, which is also referred to as the aryl hydrocarbon (or

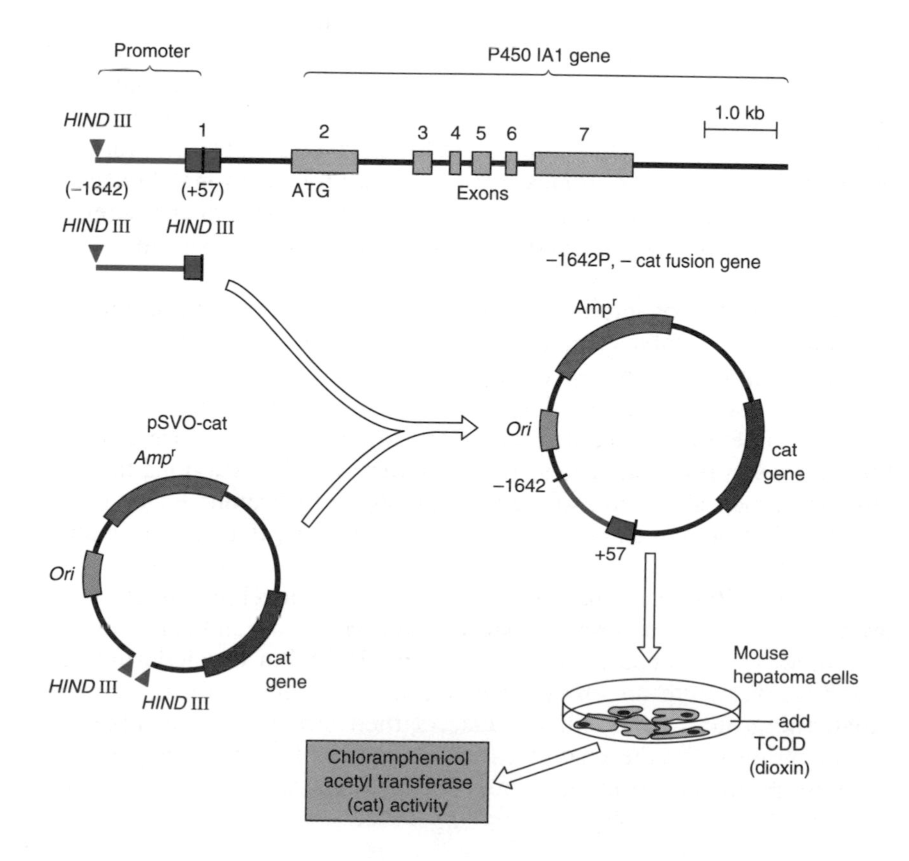

FIGURE 23.4
Diagram of the process for identifying a regulatory sequence of a P450 gene.
Nucleotides −1642 to +57 of *P450IA1* gene are excised at HIND III restriction enzyme sites and inserted into vector pSVO-cat containing the chloramphenicol acetyl transferase (CAT) gene, which is then transfected into mouse hepatoma cells. The TCDD is added to culture media, and if TCDD–receptor complex can initiate transcription of the CAT gene and synthesis of CAT protein, this region of the gene is identified as containing a regulatory site for the inducer. Amp^r = ampicillin resistant gene.
Adapted from F. J. Gonzalez. The molecular biology of cytochrome P450s, *Pharmacological Rev.* 40: 243–248, 1989. See Figure 6.

Ah) receptor. Polycyclic aromatic hydrocarbons bind to the receptor, producing a complex that may be involved in binding to the upstream regulatory regions of specific cytochrome P450 genes. Utilizing cytochrome P450 gene transfection and expression vector technology (Figure 23.4), it has been possible to express those portions of the cytochrome P450 genomic DNA representing the RNA polymerase II promoter region and the upstream DNA sequences in conjunction with another gene coding for an enzyme that is not expressed in eucaryotes. In an all-or-none assay of the procaryotic enzyme activity (chloramphenicol acetyl transferase, CAT) in the expression system, it is possible to determine which sequences on the DNA are involved in regulating these genes (**xenobiotic regulatory elements** or XREs). It is still not known, however, whether TCDD is actually required for DNA binding or for modulation or receptor partitioning between the cytoplasm and the nucleus. It is important to emphasize that there is no evidence for a receptor-mediated process in the case of other inducers such as phenobarbital, nor have cis-acting (regulatory) DNA sequence elements been identified. Some consequences of induction of drug-metabolizing enzymes are presented in Clin. Corr. 23.1.

CLIN. CORR. 23.1 CONSEQUENCES OF INDUCTION OF DRUG-METABOLIZING ENZYMES

Induction of the cytochrome P450 system may result in altered efficacy of therapeutic drugs, as the accelerated rate of hydroxylation will increase the inactivation and/or enhance the excretion rate of drugs. Induction of the cytochrome P450 system may also produce unexpected and unwanted side effects of therapeutic agents due to increased formation of toxic metabolites that may cause cell injury if produced in large enough concentrations.

The induction of cytochromes P450 by a drug may stimulate the metabolism of itself or other drugs that are substrates for the induced cytochromes P450. The consequence of accelerated metabolism is, in general, higher clearance of the drug from the body and reduction of the therapeutically active form of the drug. Clinical problems, however, may also develop as a consequence of cytochrome P450 induction. The increase in clearance of oral contraceptives by rifampicin, an antituberculosis agent and cytochrome P450 inducer, has been shown to decrease the effectiveness of the contraceptive agent and increase the incidence of pregnancy in women who are prescribed both drugs.

Fatalities have been reported in patients who are simultaneously treated with phenobarbital, a long-acting sedative and potent cytochrome P450 inducer, and warfarin, an anticoagulant, which is prescribed to patients with clotting disorders. Higher doses of warfarin are required in these patients to maintain the same effective concentration of the drug to delay coagulation because warfarin is a substrate for the cytochrome P450 induced by phenobarbital. Consequently, the drug is metabolized and cleared at a faster rate, which reduces its therapeutic efficacy. Clinical problems are created when phenobarbital is removed from the treatment regimen with no corresponding decrease in the concentration of warfarin prescribed. With time, cytochrome P450 levels decrease to the noninduced state but the high concentrations of warfarin, proper under conditions of accelerated metabolism and clearance, are in excess and produce unwanted hemorrhaging and result in death in some cases. This represents a classic example of drug–drug interactions that can lead to unwanted and unexpected sequelae if antagonistic drugs are prescribed unknowingly.

23.4 INHIBITORS OF CYTOCHROME P450

Due to the many forms of cytochrome P450, it is of interest to examine the metabolic roles of these various enzymes in the organs in which they function. Several inhibitors have been utilized to demonstrate that cytochrome(s) P450 may be involved in a metabolic pathway, for example, the metabolism of steroids in the adrenal or specific reproductive organs. As has been discussed, the detection of cytochrome P450 in most tissues can be ascertained by the reduced-carbon monoxide difference spectrum. Carbon monoxide binds to the heme iron, in lieu of oxygen, with a much higher binding affinity and thereby is a potent inhibitor of its function. The identity of cytochrome P450 as the terminal oxygenase in the catalysis of a putative substrate in a metabolic pathway rested on the reversal of CO inhibition by light at 450 nm, corresponding to the reduced-CO absorption maximum. This was first demonstrated for steroids as substrates for adrenal mitochondrial cytochromes P450 and later for drugs metabolized by liver microsomes. However, this is a nonspecific inhibition characteristic of most cytochromes P450 and does not differentiate among the various forms.

Therefore, it is of interest to design more specific inhibitors that can determine the role of a specific cytochrome P450 in a particular metabolic pathway. Although monospecific polyclonal and monoclonal antibodies have been developed to a number of cytochromes P450, it is not always possible to determine that a single form is responsible because of inhibition of a given reaction. The strong structural homology among the various forms may allow cross-reactivity among cytochromes P450. This is particularly true of members of the same gene family that exhibit immune cross-reactivity.

Recently, efforts have been directed toward the development of **mechanism-based inhibitors,** so-called **suicide substrates,** which bear strong resemblance to the substrate(s) of the specific P450, but which during catalytic turnover form an irreversible inhibition product with the enzyme prosthetic group or protein. Because of their structural resemblance to the substrate(s) for the enzyme, these inhibitors become highly specific for that particular form of P450. These inhibitors contain functional groups that result in their covalent binding to the enzymes, thereby accounting for their irreversibility.

23.5 CYTOCHROME P450 ELECTRON TRANSPORT SYSTEMS

Although the cytochrome P450-catalyzed reaction requires $2e^-$ to accomplish its task of heme iron reduction, oxygen binding, and oxygen cleavage, a basic mechanistic problem is the direct and simultaneous transfer of electrons from NADPH to the cytochrome P450. Pyridine nucleotides are $2e^-$ donors (see p. 151), but cytochrome P450, with its single heme prosthetic group, may only accept $1e^-$ at a time. Thus, a protein that serves to transfer electrons from NADPH to the cytochrome P450 molecule must have the capacity to accept $2e^-$ but serve as a $1e^-$ donor. This problem is solved by the presence of a NADPH-dependent flavoprotein reductase, which accepts $2e^-$ from NADPH simultaneously but transfers the electrons individually to an intermediate iron–sulfur protein or directly to cytochrome P450. The active redox group of the flavin molecule is the isoalloxazine ring (see p. 152). The isoalloxazine nucleus is uniquely suited to perform this chemical task since it can exist in oxidized and 1- and $2e^-$-reduced states (Figure 23.5). The transfer of electrons from NADPH to cytochrome P450 is accomplished by two distinct electron transport systems that reside almost exclusively in either mitochondria or endoplasmic reticulum.

NADPH-Cytochrome P450 Reductase Is the Intermediate Electron Donor in the Endoplasmic Reticulum

In the endoplasmic reticulum, NADPH donates electrons to a flavoprotein called **NADPH–cytochrome P450 reductase.** This enzyme has a mass of approximately 78,000 and contains both flavin adenine dinucleotide (FAD) and flavin mononucleotide (FMN) as prosthetic groups. It is the only mammalian flavoprotein known to contain both FAD and FMN. A significant number of residues at the amino end of the molecule are hydrophobic, and this portion of the molecule is embedded in the endoplasmic reticulum (Figure 23.6). The FAD serves as the entry point for electrons from NADPH, and FMN serves as the exit point, transferring electrons individually to cytochrome P450. Because the flavin molecule may exist as 1- or $2e^-$-reduced forms and two flavin molecules are bound per reductase molecule, the enzyme may receive electrons from NADPH and store them between the two flavin molecules before transferring them individu-

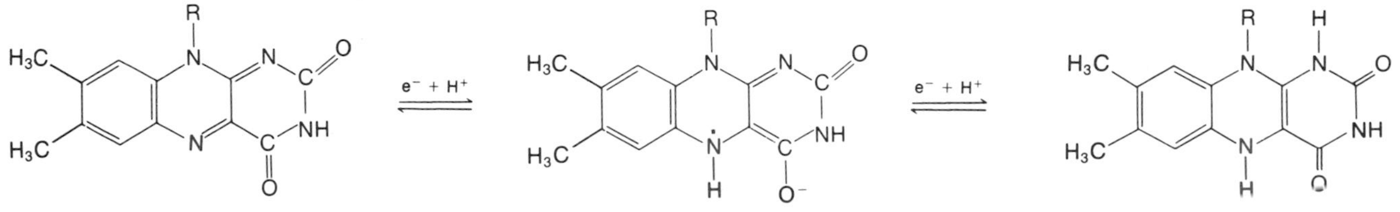

FAD or FMN (oxidized) $\underset{}{\overset{e^- + H^+}{\rightleftharpoons}}$ FAD′ or FMN′ (semiquinone) $\underset{}{\overset{e^- + H^+}{\rightleftharpoons}}$ $FADH_2$ or $FMNH_2$ (reduced)

FIGURE 23.5
The isoalloxazine ring of FMN or FAD in its oxidized, semiquinone ($1e^-$ reduced form), or fully reduced ($2e^-$ reduced) states.

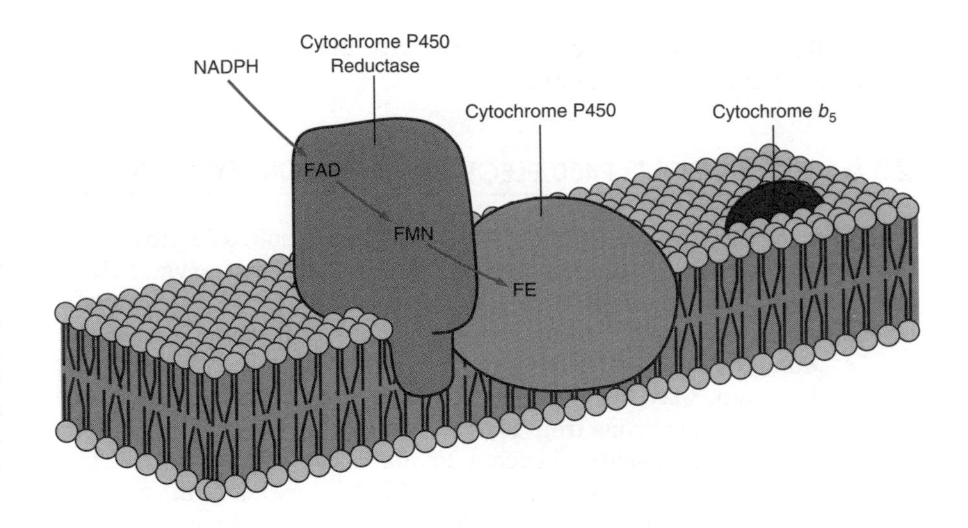

FIGURE 23.6
The components of the endoplasmic reticulum (microsomal) cytochrome P450 system.
NADPH–cytochrome P450 reductase is bound by its hydrophobic tail to the membrane whereas cytochrome P450 is deeply embedded in the membrane. Also shown is cytochrome b_5, which may participate in selected cytochrome P450-mediated reactions.

ally to the heme iron for oxygen binding (first electron) and cleavage of the oxygen molecule (second electron) (Figure 23.7).

In certain reactions catalyzed by the microsomal P450, the transfer of the second electron may not be directly from NADPH–cytochrome P450 reductase, but may occur from **cytochrome b_5,** a small heme protein of about molecular weight 16,000, which is also present in the endoplasmic reticulum (Figure 23.7). Cytochrome b_5 is reduced either by NADPH–cytochrome P450 reductase or another microsome-bound flavoprotein, **NADH-cytochrome b_5 reductase,** which is specific for NADH. It is not known why certain reactions catalyzed by specific cytochromes P450 apparently require cytochrome b_5 to transfer the second electron to cytochrome P450 for expression of maximal enzymatic activity. In addition, NADH–cytochrome b_5 reductase and cytochrome b_5 constitute the elec-

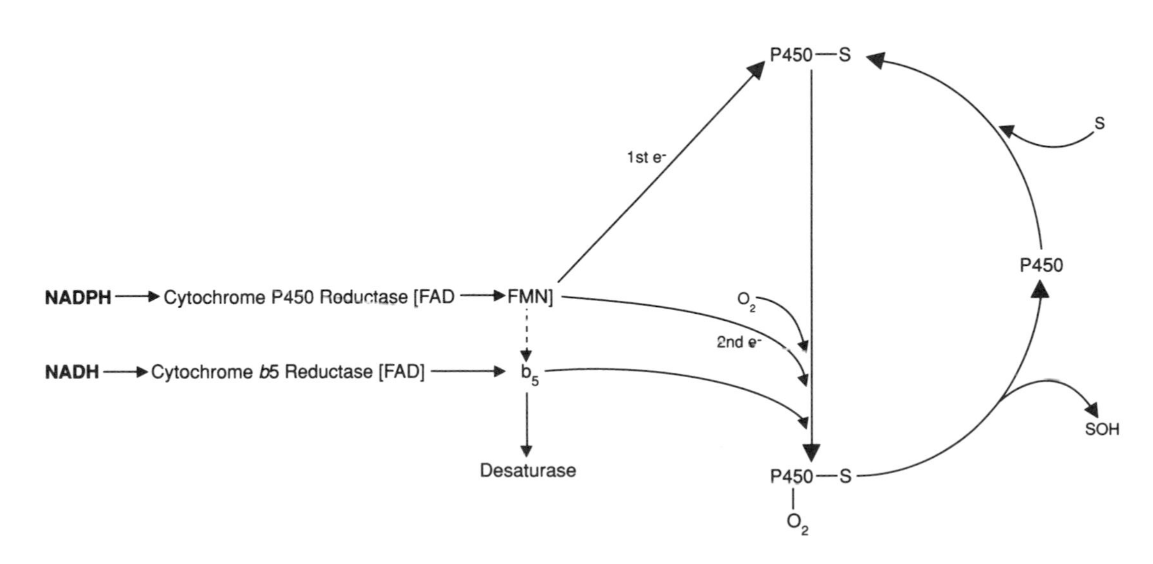

FIGURE 23.7
The endoplasmic reticulum electron transport pathway.
The first electron must be donated by the NADPH–cytochrome P450 reductase, but the second electron may be donated by the cytochrome P450 reductase or, with certain cytochromes P450, by cytochrome b_5. S = substrate.

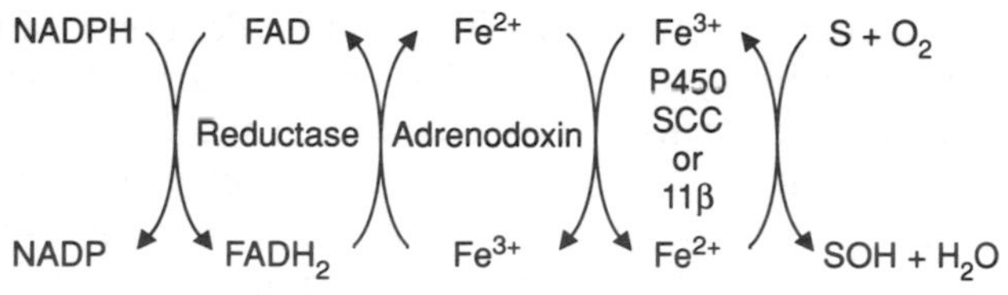

NADPH → NADP; FAD ⇄ $FADH_2$ (Reductase); Fe^{2+} ⇄ Fe^{3+} (Adrenodoxin); Fe^{3+} ⇄ Fe^{2+} (P450 SCC or 11β); $S + O_2$ → $SOH + H_2O$

FIGURE 23.8
The process of electron transfer from NADPH to cytochrome P450.
Electrons are donated from NADPH to adrenodoxin reductase, which reduces adrenodoxin. Adrenodoxin serves as the reductant of the mitochondrial cytochrome P450.

tron transfer system for NADH to the iron–sulfur protein, desaturase, which catalyzes the formation of double bonds in fatty acids (see p. 399).

Mitochondria Use NADPH-Adrenodoxin Reductase as the Intermediate Electron Donor

In mitochondria, a flavoprotein reductase also acts as the electron acceptor from NADPH. This protein is referred to as NADPH–**adrenodoxin reductase** because its characteristics were described for the flavoprotein first isolated from the adrenal gland. This protein contains only FAD and has a molecular weight of approximatley 51,000. The adrenodoxin reductase is only weakly associated with its membrane milieu, unlike its flavoprotein counterpart in endoplasmic reticulum membranes or cytochrome P450 molecules. This reductase cannot directly transfer either the first or second electron to the heme iron of cytochrome P450 (Figure 23.8). A small molecular weight protein, called **adrenodoxin** (12,500 Da) serves as an intermediate between the adrenodoxin reductase and cytochrome P450. As in the case of the adrenodoxin reductase, the adrenodoxin molecule is also weakly associated with the inner mitochondrial membrane. Adrenodoxin contains two iron–sulfur clusters, which serve as redox centers for this molecule and function as an electron shuttle between the adrenodoxin reductase and the mitochondrial cytochromes P450. Adreno-

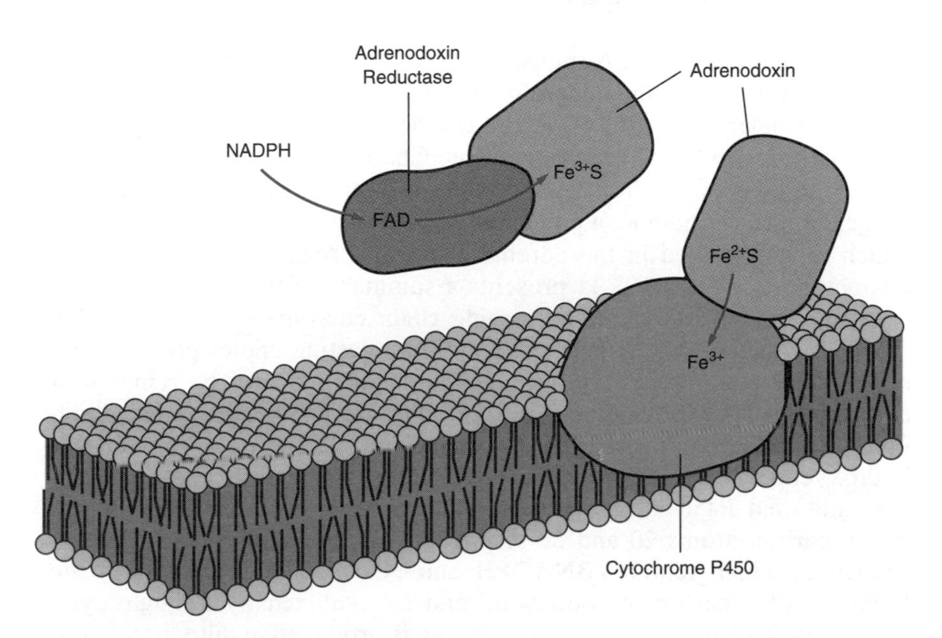

FIGURE 23.9
The components of the mitochondrial cytochrome P450 system.
Cytochrome P450 is an integral protein of the inner mitochondrial membrane. NADPH–adrenodoxin reductase and adrenodoxin (ADR) are peripheral proteins and are not embedded in the membrane.

doxin receives an electron from its mitochondrial flavoprotein reductase and then interacts with cytochrome P450 embedded in the inner mitochondrial membrane to transfer this electron to the heme iron (Figure 23.9). The components of the mitochondrial cytochrome P450 system, including the flavoprotein and iron–sulfur protein, are synthesized in the cytosol as larger molecular weight precursors, transported into mitochondria, and processed by proteases into smaller molecular weight, mature proteins.

23.6 PHYSIOLOGICAL FUNCTIONS OF CYTOCHROMES P450

Cytochromes P450 constitute a family of enzymes that metabolize a variety of lipophilic compounds of endogenous or exogenous origin. Examples are presented in Figure 23.10. These enzymes may catalyze simple hydroxylations of the carbon atom of a methyl group, insertion of an OH group into a methylene carbon of an alkane, hydroxylation of an aromatic ring to form a phenol, or addition of an oxygen atom across a carbon–carbon double bond to form an *epoxide*. The epoxide may nonenzymatically decompose to an alcohol group, or a second enzyme called the **epoxide hydrolase** may add water to the epoxide group to form vicinal alcohol derivatives. In dealkylation reactions, the oxygen is inserted into the carbon–hydrogen bond, but the resulting product is unstable and rearranges to the primary alcohol, amine, or sulfhydryl compound. Oxidation of nitrogen, sulfur, and phosphorus atoms and dehalogenation reactions are also catalyzed by cytochromes P450, as shown in Figure 23.10, Cytochrome P450-catalyzed reactions may also be complex. These enzymes cleave carbon–carbon bonds or oxidize alcohols to aldehydes.

Cytochromes P450 Participate in the Synthesis of Steroid Hormones and Oxygenation of Eicosanoids

An illustration of the importance of cytochrome P450-catalyzed reactions is the synthesis of steroid hormones from cholesterol in the adrenal gland and sex organs. Both mitochondrial and microsomal cytochrome P450 systems are required to metabolize cholesterol stepwise into aldosterone and cortisol in adrenal glands, testosterone in testes, and estradiol in ovaries.

Cytochromes P450 are responsible for several steps in the adrenal synthesis of aldosterone, the mineralocorticoid responsible for regulating salt and water balance, and cortisol, the glucocorticoid that governs protein, carbohydrate, and lipid metabolism (see Chapter 21, p. 905). In addition, cytochromes P450 mediate the production of small quantities of the androgen, androstenedione, a precursor of both estrogens and testosterone, which is synthesized in the adrenal gland and regulates secondary sex characteristics. Figure 23.11 presents a summary of these pathways.

In the adrenal mitochondria, a **side chain cleavage cytochrome P450** ($P450_{scc}$) catalyzes the complex reaction converting cholesterol to pregnenolone, a committed step in cholesterol metabolism. The removal of isocaproic aldehyde results from a cytochrome P450-catalyzed reaction involving sequential hydroxylation at C-22 and C-20 atoms to produce 22-hydroxycholesterol and then 20,22-dihydroxycholesterol (Figure 23.12). An additional P450-catalyzed step is necessary to cleave the bond between carbon atoms 20 and 22 to produce pregnenolone. This reaction sequence, which requires 3NADPH and $3O_2$ molecules, results in the breakage of a carbon–carbon bond and is catalyzed by a single cytochrome P450 enzyme. After pregnenolone is produced in mitochondria, it is transported into the cytoplasm where it is oxidized by **3β-hydroxysteroid dehydrogenase/$\Delta^{4,5}$-isomerase** to progesterone. Progesterone is hydroxylated at carbon atom 21 to form 11-deoxycorticosterone (DOC) by an endoplasmic reticulum cytochrome P450, the **21-hydroxylase.** The DOC is transported back into the mitochondria to be hydroxylated by an additional mitochondrial cytochrome P450 which catalyzes the **11β-hydroxylase** and **18-hydroxylase** activities in the *zona glomerulosa* of the adrenal gland to form aldosterone, a mineralocorticoid.

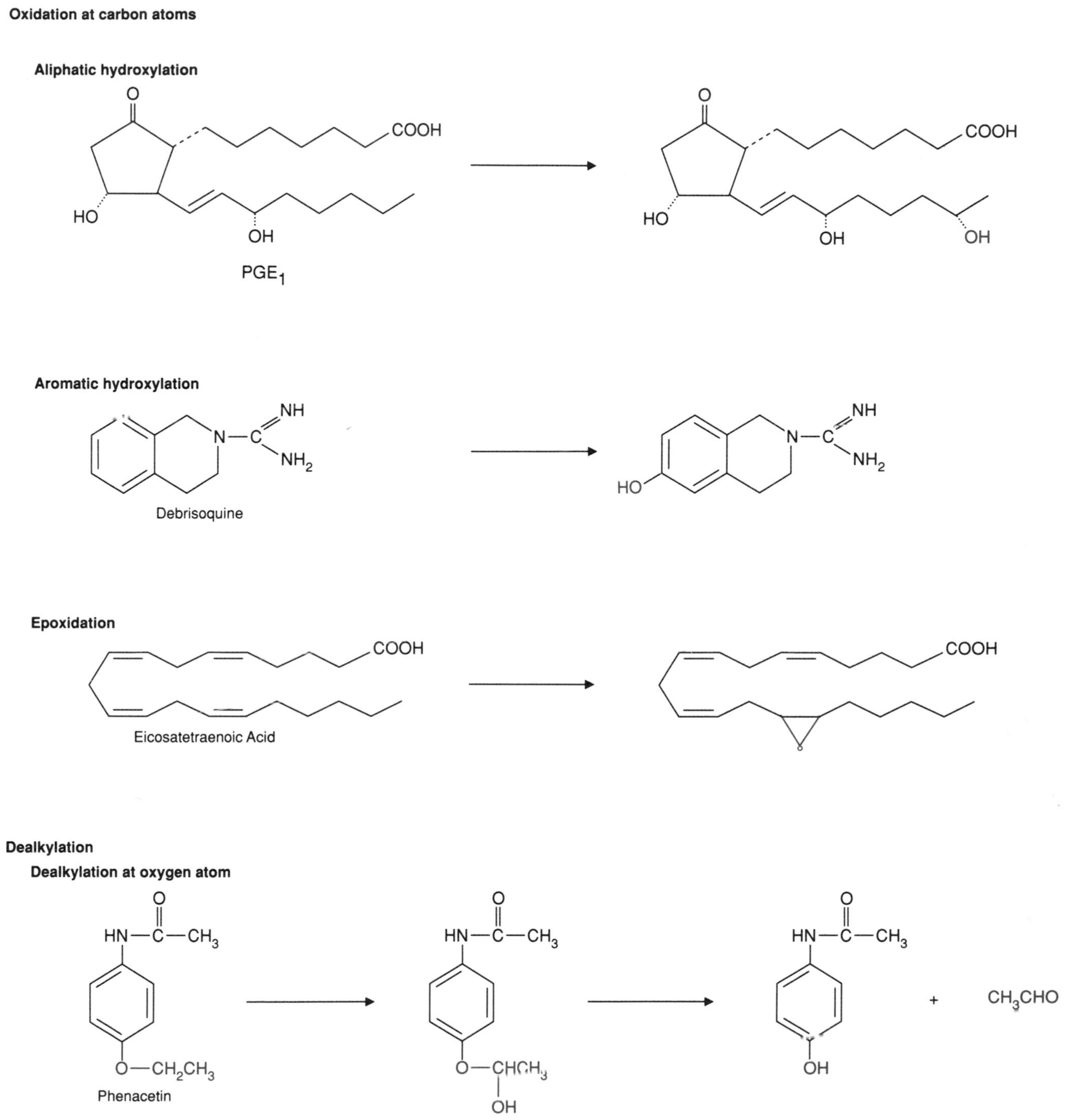
Oxidation at carbon atoms
Aliphatic hydroxylation
COOH
HO
OH
PGE1
Aromatic hydroxylation
NH
N—C
NH2
Debrisoquine
HO
Epoxidation
COOH
Eicosatetraenoic Acid
Dealkylation
Dealkylation at oxygen atom
HN—C—CH3
O—CH2CH3
Phenacetin
O—CHCH3
OH
+
CH3CHO

FIGURE 23.10
Reaction types catalyzed by cytochromes P450.
Continued on page 992.

Dealkylation at nitrogen atom

Aminopyrine

+ HCHO

Dealkylation at sulfur atom

S-Methylthiopurine

+ HCHO

Oxidation at nitrogen atom

2-Acetylaminofluorene

Oxidation at sulfur atom

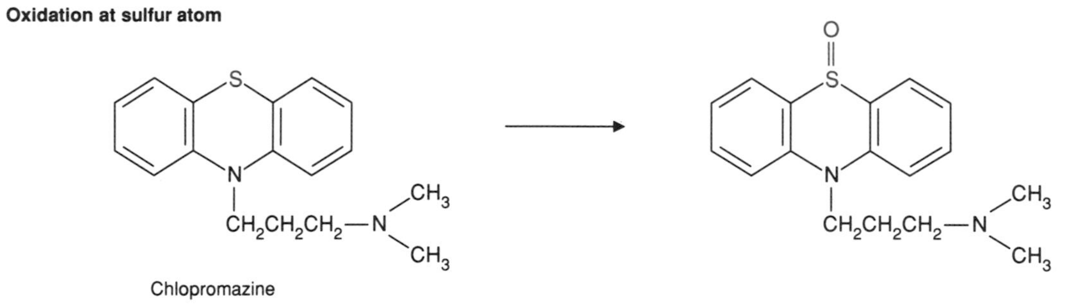

Chlopromazine

Oxidation at phosphorus atom

Parathion

Dehalogenation

Halothane

FIGURE 23.10
Continued from page 991.

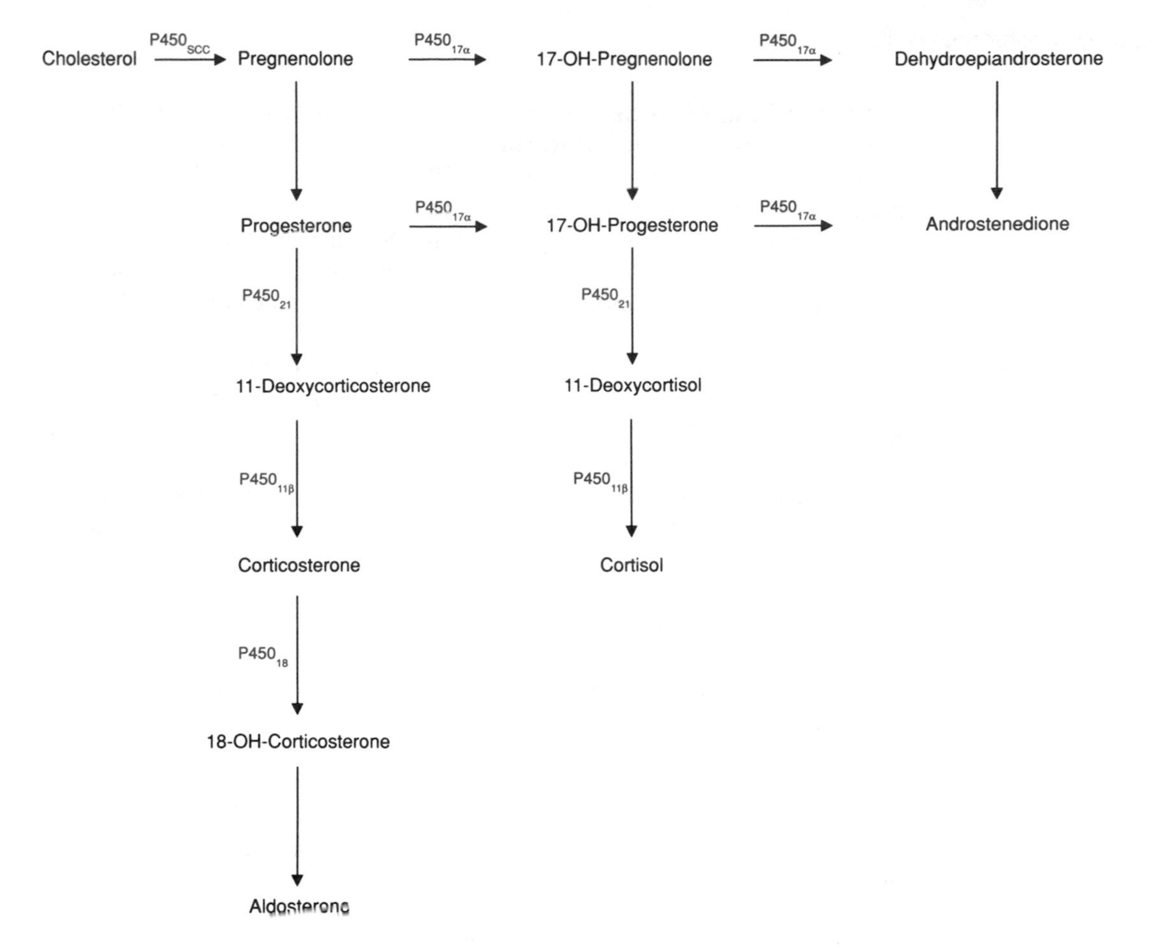

FIGURE 23.11
Steroid hormone synthesis in the adrenal gland. The reactions catalyzed by cytochromes P450 are indicated.

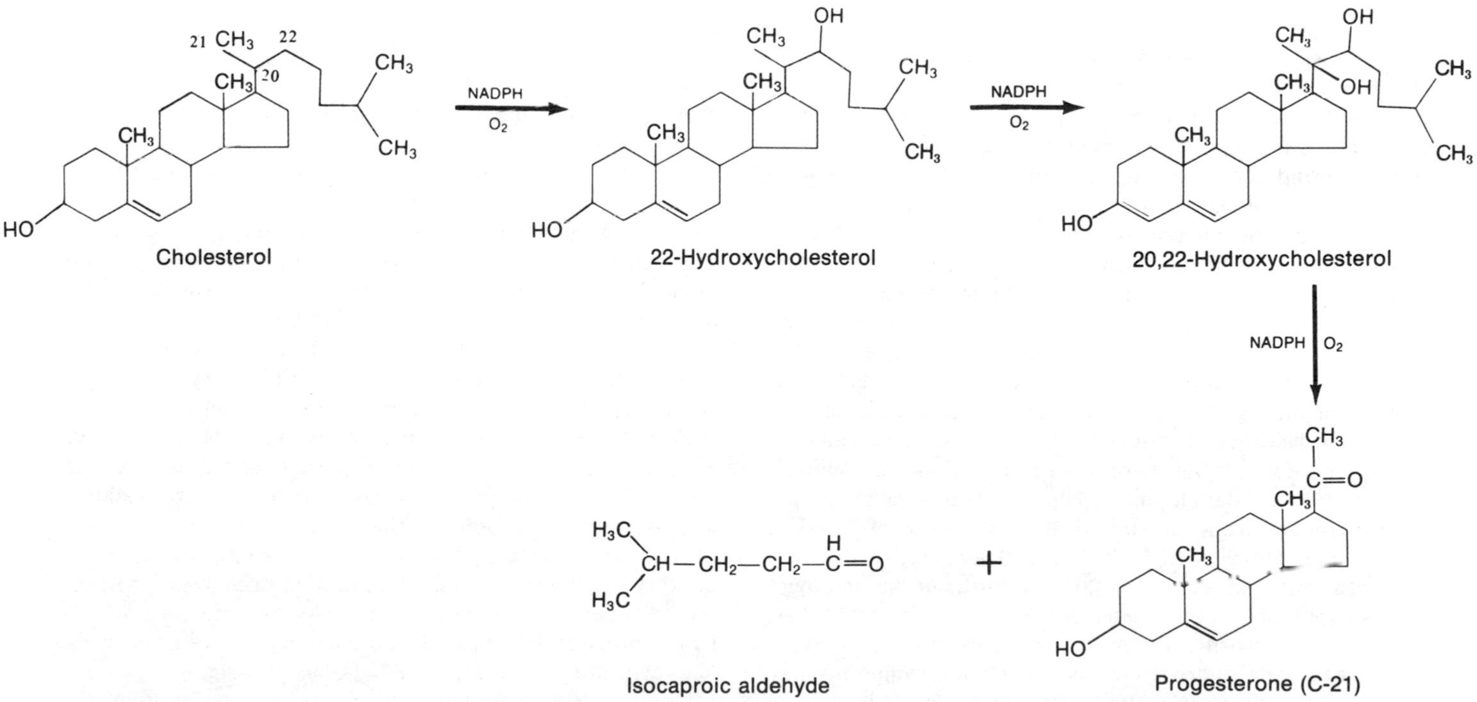

Cholesterol → (NADPH, O_2) → 22-Hydroxycholesterol → (NADPH, O_2) → 20,22-Hydroxycholesterol → (NADPH, O_2) → Isocaproic aldehyde + Progesterone (C-21)

FIGURE 23.12
The side chain cleavage reaction of cholesterol.
Three sequential reactions are catalyzed by cytochrome P450 to produce pregnenolone and isocaproic aldehyde.

The synthesis of cortisol proceeds from either pregnenolone or progesterone and involves the **cytochrome P450 17α-hydroxylase** ($P450_{17\alpha}$), an endoplasmic reticulum cytochrome P450. Hydroxylation of 17α-hydroxyprogesterone by cytochrome P450 21-hydroxylase ($P450_{21}$) produces 11-deoxycortisol, which is transported into the mitochondrion where it is hydroxylated by the 11β-hydroxylase to form cortisol. This reaction occurs primarily in the *zona fasciculata*. The consequences of a deficiency of cytochrome $P450_{21}$ is presented in Clin. Corr. 23.2.

The synthesis of C_{19} steroids from 17α-hydroxypregnenolone or 17α-hydroxyprogesterone is the result of removal of the acetyl group at C-17 and is catalyzed by the **P450 17,20-lyase,** identified as the same cytochrome P450 that hydroxylates the C-17 atom. Thus, cleavage of the carbon–carbon bond between C-17 and C-20 with loss of the acetyl group is also catalyzed by a cytochrome P450 molecule. The factors that determine whether this cytochrome P450 performs only a single hydroxylation step to produce the 17-OH product or proceeds further to cleave the 17,20 carbon–carbon bond has not been determined. The reaction products are dehydroepiandrosterone (DHEA) from 17α-hydroxypregnenolone or androstenedione from 17α-hydroxyprogesterone. The DHEA may be metabolized by dehydrogenation of the 3-OH group to androstenedione, a potent androgenic steroid that serves as the immediate precursor of testosterone.

Another physiologically important reaction catalyzed by cytochrome P450 is the synthesis of estrogens from androgens, collectively called **aromatization** because an aromatic ring is introduced into the product.

CLIN. CORR. 23.2
DEFICIENCY OF CYTOCHROME P450–21-HYDROXYLASE

The adrenal cortex is a major site of steroid hormone production during fetal and adult life. The adrenal gland is a far more active biosynthetic system in fetal life than in the adult and may produce 100–200 mg of steroids per day in comparison to the 20–30 mg produced per day in the nonstressed adult adrenal gland. A number of enzymes are required for the production of steroid hormones, and enzyme deficiencies have been reported at all steps of cortisol production. Diseases associated with insufficient cortisol production are referred to as congenital adrenal hyperplasias (CAH). There are two types of CAH, classical and nonclassical. In the former, clinical consequences of 21-hydroxylase deficiency are recognizable at birth, particularly in females, because the excessive buildup of C_{19}-androgenic steroids may cause obvious irregular development of their genitalia. In male newborns, a deficiency in 21-hydroxylase may be overlooked, because male genitalia will appear normal, but there will be precocious masculinization and physical development. High levels of circulating 17α-hydroxyprogesterone are found. The incidence of classical CAH varies from as high as 1 in 490 live births in Yupik Eskimos in Alaska to 1 in 67,000 live births of the heterogeneous population in the United States. In nonclassical CAH, individuals are born without obvious signs of prenatal exposure to excessive androgen levels, and clinical symptoms may vary considerably from early development of pubic hair, early fusion of epiphyseal growth plates causing premature cessation of growth, or male baldness patterns in females. Circulating levels of 17α-hydroxyprogesterone are elevated but are lower than those found in classical CAH. Nonclassical CAH occurs more frequently and the disease frequency varies for ethnic populations. Significant information has been obtained about the physiological functions of the hormones that are produced by the adrenal gland and that regulate the adrenal cortex because of these enzyme deficiencies.

The formation of cortisol from cholesterol requires five enzymatic steps as presented in Figure 23.11. Enzyme deficiencies have been associated at each step and are inherited as autosomal recessive traits. The reaction that is most commonly involved is hydroxylation at C-21. Approximately 90% of all cases of congenital adrenal hyperplasia can be accounted for by a defective 21-hydroxylase gene. Failure to metabolize 17α-hydroxyprogesterone to 11-deoxycortisol and subsequently to cortisol causes an increase in secretion of ACTH, the pituitary hormone that regulates adrenal cortex production of cortisol. Prolonged periods of elevated ACTH levels will cause adrenal hyperplasia and an increase in androgenic steroid production, for this steroid does not require the 21-hydroxylase activity for its synthesis. Patients with 21-hydroxylase deficiency will excrete in their urine pregnanetriol, the principal metabolite of 17α-hydroxyprogesterone, and 17-keto steroids, DHEA, and androstenedione, Clinical problems will arise because the additional production of androgenic C_{19} steroids will cause virilization in females, precocious sex organ development in prepubertal males, or diseases related to salt imbalance because of decreased levels of aldosterone.

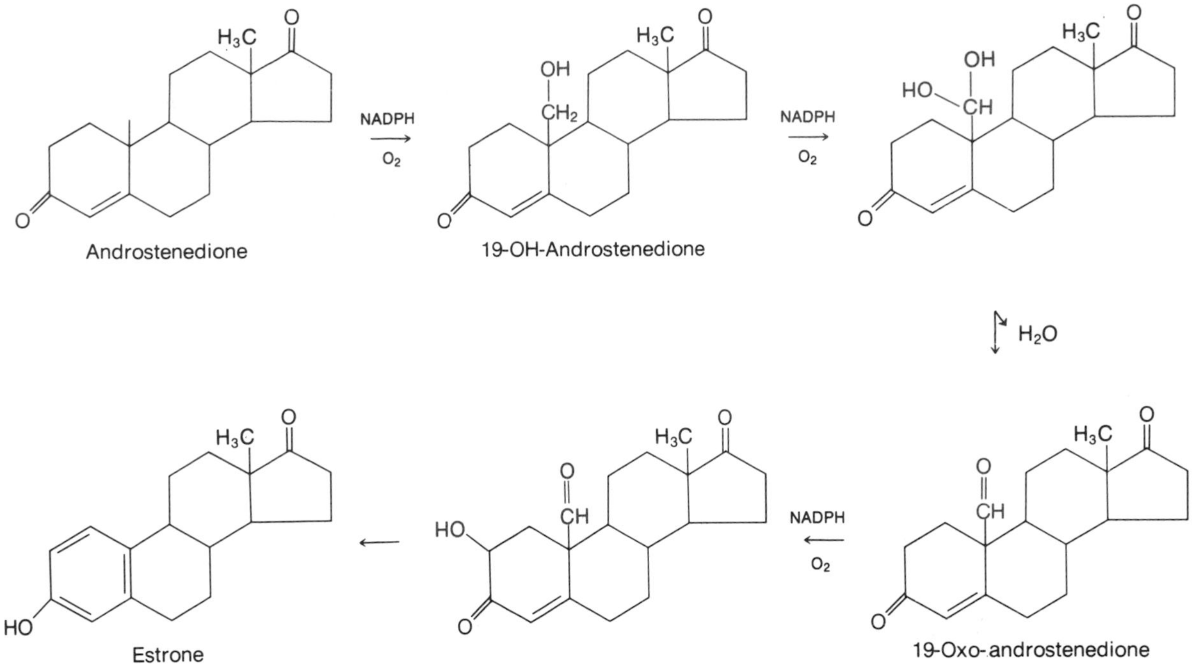

Androstenedione
NADPH O_2
19-OH-Androstenedione
NADPH O_2
H_2O
Estrone
NADPH O_2
19-Oxo-androstenedione

FIGURE 23.13
The sequence of reactions leading to the aromatization of androgens to estrogens.

This is a complex reaction in which three molecules of NADPH and three molecules of oxygen are required for formation of the aromatic ring and to remove the methyl group at C-19. Figure 23.13 outlines the aromatization reaction. Two cytochrome P450-mediated hydroxylation reactions at the C-19 methyl carbon atom introduce an aldehyde group. A hydroxylation at the C_2 atom of the cyclohexenone ring is the final enzymatic reaction catalyzed by this cytochrome P450. The resulting structure is unstable and nonenzymatically converts to the aromatic ring of the estrogen molecule with loss of the C-19 carbon atom. All three reaction steps of this sequence are catalyzed by the same cytochrome P450, and the enzyme is referred to as the **aromatase.** The complexity of steroid hormone production and the role of cytochromes P450 is illustrated in Clin. Corr. 23.3.

Other cytochromes P450 metabolize vitamin D_3 to produce the 1,25-dihydroxy vitamin D_3, which is the active form of this important hormone (see p. 448), leukotriene B_4 to 20-OH-LTB_4, which is the less active form of this chemotactic agent (see p. 468), and arachidonic acid to produce epoxides, hydroxy and dihydroxy derivatives of arachidonic acid, which may have important regulatory functions (see p. 462).

Cytochromes P450 Oxidize Exogenous Lipophilic Substrates

Exogenous substrates are often referred to as **xenobiotics,** meaning "foreign to life," and this term includes drugs that are used therapeutically and nontherapeutically, chemicals used in the work place, industrial by-products that become environmental contaminants, and food additives. The cytochromes P450 are able to oxidize a variety of xenobiotics, particularly lipophilic cyclic compounds. The addition of an OH group makes the compound more polar and thus more soluble in the aqueous environment of the cell. Many exogenous compounds are highly lipophilic and accumulate within cells, interfering with cellular function over a period of

CLIN. CORR. 23.3 STEROID HORMONE PRODUCTION DURING PREGNANCY

Steroid hormone production increases dramatically during pregnancy and, at term, the pregnant woman produces 15–20 mg of estradiol, 50–100 mg of estriol, and approximately 250 mg of progesterone per 24-h period. The amount of estrogens synthesized during pregnancy far exceeds the amount synthesized by the nonpregnant woman. For example, the pregnant woman at the end of gestation produces in 24 h 1000 times more estrogen than premenopausal women.

The production of progesterone and estrogens in the pregnant woman is decidedly different from that in the nonpregnant woman. The corpus luteum of the ovary is the major site for estrogen production in the first few weeks of pregnancy, but at approximately 4 weeks of gestation, the placenta begins synthesizing and secreting progesterone and estrogens. After 8 weeks of gestation, the placenta becomes the dominant source for the synthesis of progesterone, a C_{21}-steroid, and certain C_{18}-estrogenic steroids. An interesting difference between the steroid hydroxylating systems in the placenta and the ovary is that the human placenta lacks the cytochrome P450 that catalyzes the 17α-hydroxylation reaction and the cleavage of the 17,20 carbon–carbon bond (see Chapter 21, p. 906, for details of synthesis of steroid hormones). Thus, the placenta cannot, by itself, synthesize estrogens from cholesterol. The placenta catalyzes the side chain cleavage reaction to form pregnenolone from cholesterol and oxidizes pregnenolone to progesterone but releases this hormone into the maternal circulation. How then does the placenta produce estrogens if it cannot synthesize DHEA or androstenedione from progesterone? This is accomplished in the fetal adrenal gland, which represents a significant proportion of the total fetal weight compared to its adult state. The fetal adrenal gland catalyzes the synthesis of DHEA from cholesterol and releases it into the fetal circulation. A large proportion of the fetal DHEA is metabolized by the fetal liver to 16α-OH-DHEA, and this product is aromatized in the placenta to the estrogen, estriol. The DHEA produced by the fetal adrenal that is not hydroxylated by the fetal liver to 16α-OH-DHEA is also aromatized in the placenta to form estradiol-17β. This is an elegant demonstration of the cooperativity of the cytochrome P450-mediated hydroxylating systems in the fetal and maternal organ systems leading to the progressive formation of estrogens during the gestational development of the human fetus.

TABLE 23.1 Xenobiotics Metabolized by Cytochromes P450

Reaction	Examples
Aliphatic hydroxylation	Valproic acid, pentobarbital
Aromatic hydroxylation	Debrisoquine, acetanilide
Epoxidation	Benzene, benzo[*a*]pyrene
Dealkylation	Aminopyrine, phenacetin, 6-methylthiopurine
Oxidative deamination	Amphetamine
Nitrogen or sulfur oxidation	2-Acetylaminofluorene, chlorpromazine
Dehalogenation	Halothane
Alcohol oxidation	Ethanol

time. Examples of xenobiotics that are oxidized by cytochromes P450 are presented in Table 23.1 and Figure 23.10. In many cases the action of the cytochromes P450 leads to a compound with reduced pharmacological activity or toxicity, which can be readily excreted in the urine or bile. Modified and unmodified xenobiotics can be chemically altered by a variety of conjugating enzyme systems forming products that are even less toxic and that can be readily eliminated from the body. A list of enzymes that metabolize xenobiotics is presented in Table 23.2; many of these reactions occur primarily in the liver.

One xenobiotic that has received considerable attention is benzo[*a*]pyrene, a common environmental contaminant produced from the burning of coal, from the combustion of plant materials in tobacco, from food barbecued on charcoal, and as an industrial byproduct. Benzo[*a*]pyrene binds to the **aryl hydrocarbon** receptor and induces cytochromes P450, thus increasing its own metabolism. Several sites of the molecule may be hydroxylated by different forms of cytochrome P450. It is known that benzo[*a*]pyrene is metabolized to a product that is a carcinogen in animals and a mutagen in bacteria, prompting considerable work in identifying the enzymes involved in this process. The product found to represent the ultimate carcinogen is benzo[*a*]pyrene-7,8-dihydrodiol-9,10-epoxide, the formation of which is illustrated in Figure 23.14. The initial step involves a cytochrome P450-catalyzed epoxidation at the 7,8 position, hydrolysis by epoxide hydrolase to the vicinal hydroxylated compound, benzo[*a*]pyrene-7,8-dihydrodiol, and then another epoxidation reaction to form benzo[*a*]pyrene-7,8-dihydrodiol-9,10-epoxide. Benzo[*a*]pyrene is a weak carcinogen and, like most carcinogens that have been characterized, requires metabolic activation to the carcinogenic form.

TABLE 23.2 Xenobiotic Metabolizing Enzymes

Type of reaction	Enzyme	Representative substrate
Oxidation	Cytochrome P450	Toluene
	Alcohol dehydrogenase	Ethyl alcohol
	Flavin containing monooxygenase	Dimethylaniline
Reduction	Ketone reductase	Metyrapone
Hydration	Epoxide hydrolase	Benzo[*a*]pyrene-7,8-epoxide
Hydrolysis	Esterase	Procaine
Conjugation	UDP glucuronyl transferase	Acetaminophen
	Sulfotransferase	β-Naphthol
	N-acetyltransferase	Sulfanilamide
	Methyltransferase	Thiouracil
	Glutathione transferase	Acetaminophen

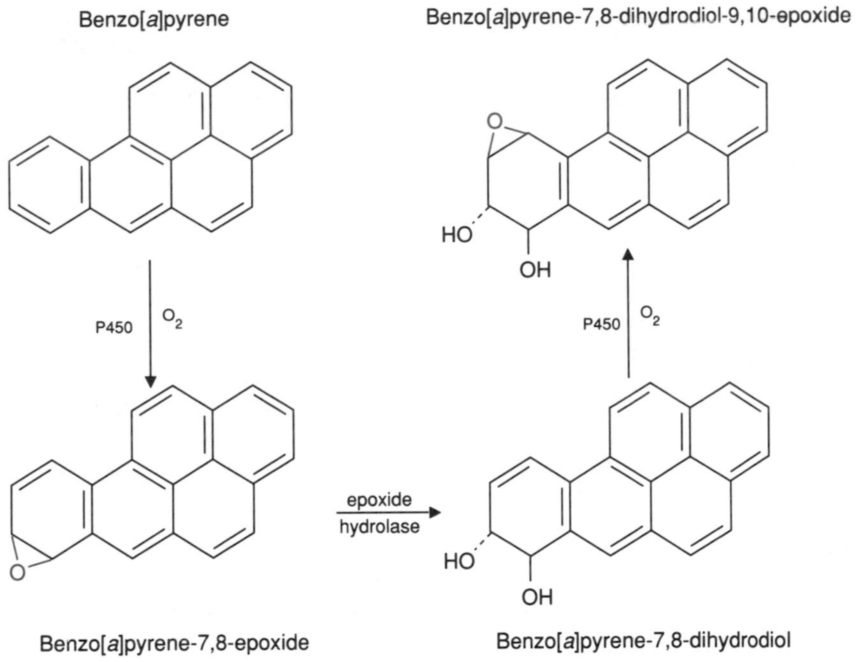

Benzo[*a*]pyrene

Benzo[*a*]pyrene-7,8-dihydrodiol-9,10-epoxide

P450 O_2

P450 O_2

epoxide hydrolase

HO OH

Benzo[*a*]pyrene-7,8-epoxide

Benzo[*a*]pyrene-7,8-dihydrodiol

FIGURE 23.14
The metabolism of benzo[*a*]pyrene by cytochrome P450 and epoxide hydrolase to form benzo[*a*]pyrene-7,8-dihydrodiol-9,10-epoxide.

In a number of cases, the cytochrome P450 system is responsible for the generation of the ultimate carcinogen. The formation of toxic compounds by the cytochrome P450 system, however, does not mean that cell damage or cancer will occur, because many other factors will determine whether or not the toxic metabolite will cause cell injury. These include the involvement of detoxification enzyme systems, the status of the immune system, nutritional state, genetic predisposition, and environmental factors. One may ask why the body should possess an enzyme system that would create highly toxic compounds? As indicated, the purpose of the cytochrome P450 system is to add or expose functional groups making the molecule more polar and/or more susceptible to attack by additional detoxification enzyme systems. In addition, many of these compounds resemble hormones that are our natural communication signals and would interfere with cell–cell or organ–organ communication.

Thus, the cytochrome P450 system plays a significant role in the health and disease of humans. Different cytochromes P450 are responsible for the generation of essential steroid hormones, the regulation of blood levels of therapeutic agents, the removal of unwanted chemicals that would accumulate because of their lipophilicity, and the generation of potentially toxic metabolites that may cause acute cell injury or damage to genetic material and lead to the production of tumors.

BIBLIOGRAPHY

Drug Metabolism and Cytochrome P450

Gibson, G. G. and Skett, P. *Introduction to Drug Metabolism*. New York: Chapman and Hall, 1986.

Gonzalez, F. J. "The molecular biology of cytochrome P450s." *Pharmacol. Rev.* 40:243, 1988.

Guengerich, F. P. "Cytochromes P-450." *Comp. Biochem. Physiol.* 89C:1, 1988.

Guengerich, F. J. Reactions and significance of cytochrome P-450 enzymes. *J. Biol. Chem.* 266:10019, 1991.

MacDonald, M. G. and Robinson, D. S. "Clinical observations of possible barbiturate interference with anticoagulation." *J. Am. Med. Assoc.* 204:97, 1968.

Park, B. K. and Breckenridge, A. M. "Clinical implications of enzyme induction and enzyme inhibition." *Clin. Pharm.* 6:1, 1981.

Porter, T. D. and Coon, M. J. Cytochrome P-450. Multiplicity of isoforms, substrates, and catalytic and regulatory mechanisms. *J. Biol. Chem.* 266:13469, 1991.

Vessey, D. A. "Hepatic metabolism of drugs and toxins." D. Zakim and T. D. Boyer (Eds.). *Hepatology. A Textbook of Liver Disease*. Philadelphia: Saunders, 1982, Chapter 8, p. 197.

Watkins, P. B. Role of cytochromes P450 in drug metabolism and hepatotoxicity. *Seminars Liver Dis.* 10:235, 1990.

Electron Transport and Cytochrome P450

Peterson, J. A. and Prough, R. A. "Cytochrome P-450 reductase and cytochrome b_5 in cytochrome P-450 catalysis." P. R. Ortiz de Montellano (Ed.). *Cytochrome P-450 Structure, Mechanism, and Biochemistry.* New York: Plenum, 1986, Chapter 4, p. 89.

Induction of Cytochrome P450

Eisen, H. J. "Induction of hepatic P-450 isozymes." P. R. Ortiz de Montellano (Ed.). *Cytochrome P-450. Structure, Mechanism, and Biochemistry.* New York: Plenum, 1986, Chapter 9, p. 315.

Gonzalez, F. J. "The molecular biology of cytochrome P450s." *Pharmacol. Rev.* 40:243, 1988.

Endogenous Substrates and Cytochrome P450

Cunningham, F. G., MacDonald, P. C., and Gant, N. F. "The placental hormones." *Williams Obstetrics.* Connecticut: Appleton and Lange, 18th ed., 1989, Chapter 5, p. 67.

Estabrook, R. W., Cooper, D. Y., and Rosenthal, O. "The light reversible carbon monoxide inhibition of the steroid C21-hydroxylase system of the adrenal cortex." *Biochem. Zeit.* 338:741, 1963.

Masters, B. S. S., Myerhoff, A. S., and Okita, R. T. Enzymology of extrahepatic cytochromes P450. F. P. Guengerich (Ed.). *Mammalian Cytochromes P450.* Boca Raton, FL: CRC Press, Inc., 1987, Chapter 3, p. 107.

New, M. I., White, P. C., Pang, S., DuPont, B., and Speiser, P. W. "The adrenal hyperplasias." C. R. Scriver, A. L. Beaudet, W. S. Sly, and D. Valle (Eds.). *The Metabolic Basis of Inherited Disease.* New York: McGraw-Hill, 1989, Vol. II, Chapter 74, p. 1881.

Simpson, E. R. and MacDonald, P. C. "Endocrine physiology of the placenta." *Ann. Rev. Physiol.* 43:163, 1981.

Takemori, S. and Kominami, S. "The role of cytochromes P-450 in adrenal steroidogenesis." *Trends Biochem. Sci.* 9:393, 1984.

Waterman, M. R. and Simpson, E. R. "Regulation of the biosynthesis of cytochromes P-450 involved in steroid hormone synthesis." *Mol. Cell. Endocrinol.* 39:81, 1985.

Waterman, M. R., John, M. E., and Simpson, E. R. "Regulation of synthesis and activity of cytochrome P-450 enzymes in physiological pathways." P. R. Ortiz de Montellano (Ed.). *Cytochrome P-450. Structure, Mechanism, and Biochemistry.* New York: Plenum, 1986, Chapter 10, p. 315.

QUESTIONS

C. N. ANGSTADT AND J. BAGGOTT

*Q*uestion *T*ypes are described on page xxiii.

1. (QT2) A molecule designated as a cytochrome P450:
   1. contains a heme as a prosthetic group.
   2. catalyzes the hydroxylation of a hydrophobic substrate.
   3. may accept electrons from a substance such as NADPH.
   4. undergoes a change in the heme iron upon binding a substrate.

2. (QT1) Known roles for cytochromes P450 include all of the following EXCEPT:
   A. synthesis of steroid hormones.
   B. conversion of some chemicals to mutagens.
   C. hydroxylation of an amino acid.
   D. inactivation of some hydrophobic drugs.
   E. metabolism of fatty acid derivatives.
3. (QT1) The induction of cytochromes P450:
   A. occurs only by endogenous compounds.
   B. occurs only at the transcriptional level.
   C. necessarily results from increased transcription of the appropriate mRNA.
   D. necessitates the formation of an inducer–receptor protein complex.
   E. may occur by posttranscriptional processes.
4. (QT1) Flavoproteins are usually intermediates in the transfer of electrons from NADPH to cytochrome P450 because:
   A. NADPH cannot enter the membrane.
   B. flavoproteins can accept two electrons from NADPH and donate them one at a time to cytochrome P450.
   C. they have a more negative reduction potential than NADPH so accept electrons more readily.
   D. as proteins, they can bind to cytochrome P450 while the nonprotein NADPH cannot.

   A. Cyto P450 electron transport in the endoplasmic reticulum
   B. Cyto P450 electron transport in the mitochondria
   C. Both
   D. Neither
5. (QT4) The flavoprotein reductase uses both FAD and FMN as prosthetic groups.
6. (QT4) System(s) necessary for the formation of double bonds in fatty acids.
7. (QT4) Transport chain contains a protein with iron–sulfur clusters that transfer electrons to the heme iron.
8. (QT4) Cytochrome b_5 may be an intermediate between NADPH–cytochrome P450 reductase and cytochrome P450.

9. (QT1) Cytochrome P450 systems are able to oxidize:
   A. —CH_2— groups.
   B. benzene rings.
   C. nitrogen atoms in an organic compound.
   D. sulfur atoms in an organic compound.
   E. all of the above.
10. (QT1) In the conversion of cholesterol to steroid hormones in the adrenal gland:
    A. all of the cytochrome P450 oxidations occur in the endoplasmic reticulum.
    B. all of the cytochrome P450 oxidations occur in the mitochondria.
    C. side chain cleavage of cholesterol to pregnenolone is one of the cytochrome P450 systems that uses adrenodoxin reductase.
    D. cytochrome P450 is necessary for the formation of aldosterone and cortisol but not for the formation of the androgens and estrogens.
    E. aromatization of the first ring of the steroid does not use cytochrome P450 because it involves removal of a methyl group, not a hydroxylation.

11. (QT1) Many xenobiotics (exogenous substrates) are oxidized by cytochromes P450 in order to:
    A. make them carcinogenic.
    B. increase their solubility in an aqueous environment.
    C. enhance their deposition in adipose tissue.
    D. increase their pharmacological activity.
    E. all of the above.

12. (QT2) Benzo[*a*]pyrene, a xenobiotic produced by combustion of a variety of substances:
    1. induces the synthesis of cytochrome P450.
    2. undergoes epoxidation by a cytochrome P450.
    3. is converted to a carcinogen in animals by cytochrome P450.
    4. would be rendered more water-soluble after the action of cytochrome P450.

13. (QT1) Phenobarbital is a potent inducer of cytochrome P450. Warfarin, an anticoagulant, is a substrate for cytochrome P450 with the result that the drug is metabolized and cleared from the body more rapidly than normal. If phenobarbital is added to the therapeutic regimen of a patient, with no change in the dosage of warfarin, the expected consequence would be
    A. no change in the clinical results.
    B. an increased possibility of clot formation.
    C. an increased possibility of hemorrhaging.

ANSWERS

1. E All four correct. 2: The types of substrates are hydrophobic. It is classified as a monooxygenase. 3: See Figure 23.3. 4: The change from hexa to penta coordinated gives the compound a more positive reduction potential (p. 982).
2. C Cytochromes P450 are not the only hydroxylases and other types are active with amino acids (p. 990).
3. E There may be a stabilization of mRNA (as seen in diabetic rats) or decrease in the degradation of the protein, which may be a mechanism for pyrazole (p. 985). A: One of the roles of cytochromes P450 is in the metabolism of exogenous substances. B,C: Transcriptional modification is only one of the mechanisms of induction (see E). D: This has been shown with induction by some compounds, but with others, like phenobarbital, this is not so (p. 986).
4. B Heme can accept only one electron at a time while NADPH always donates two at a time. A: NADPH passes only electrons; it does not have to enter the membrane. C: If this were true, the flow of electrons would not occur in the way it does. D: Protein–protein binding is not known to play a role here (p. 987).
5. A NADPH–cytochrome P450 reductase is the only mammalian protein known to do so (p. 987).
6. D Cytochrome b_5 may reduce *either* desaturase *or* cytochrome P450, but desaturase does not react with cytochrome P450 (Figure 23.7).
7. B The mitochondrial system uses adrenodoxin as an intermediate between adrenodoxin reductase and cytochrome P450 (Figure 23.8).
8. A Notice that while the usual path is for NADPH–cytochrome P450 reductase to react with cytochrome P450, some systems of the endoplasmic reticulum have cytochrome b_5 interposed (Figure 23.7).
9. E See Figure 23.10.
10. C This is a mitochondrial process (Figure 23.8, p. 990). A,B: Hormone synthesis involves a series of reactions that move back and forth between mitochondria and endoplasmic reticulum (p. 991). D,E: Removal of side chains frequently begins with oxidation reactions (p. 994).
11. B The types of xenobiotics oxidized by cytochrome P450 are usually highly lipophilic but must be excreted in the aqueous urine or bile. A: This may happen but is certainly not the purpose. C: They do that prior to oxidation. D: Oxidation tends to reduce pharmacological activity (p. 995).
12. E All four correct. 1: It is not uncommon for xenobiotics to induce synthesis of something that will enhance their own metabolism. 2,3: Epoxidation is the first step in the conversion of this compound to one that is carcinogenic—again, a common occurrence (p. 996). 4: Benzo[*a*]pyrene, with its four fused benzene rings, is highly hydrophobic; introducing oxygens increases water solubility (p. 997).
13. B If warfarin is metabolized and cleared more rapidly by cytochrome P450, its therapeutic efficiency is decreased. Therefore, at the same dosage, it will be less effective as an anticoagulant. Think what would happen if the warfarin dosage were adjusted for a proper response, and then phenobarbital were withdrawn without adjusting the warfarin dose (Clin. Corr. 23.1).

24

Iron and Heme Metabolism

MARILYN S. WELLS
WILLIAM M. AWAD, JR.

24.1 IRON METABOLISM: OVERVIEW 1001
24.2 IRON-CONTAINING PROTEINS 1002
Transferrin Transports Iron in Serum 1003
Lactoferrin Binds Iron in Milk 1003
Ferritin Is a Protein Involved in Storage of Iron 1004
Other Nonheme Iron-Containing Proteins Are Involved in Enzymatic Processes 1004
24.3 INTESTINAL ABSORPTION OF IRON 1005
24.4 MOLECULAR REGULATION OF IRON UTILIZATION 1006
24.5 IRON DISTRIBUTION AND KINETICS 1007
24.6 HEME BIOSYNTHESIS 1009
Enzymes in Heme Biosynthesis Occur in Both Mitochondria and Cytosol 1011
ALA Synthase Catalyzes Rate-Limiting Step of Heme Biosynthesis 1016
24.7 HEME CATABOLISM 1017
Bilirubin Is Conjugated to Form Bilirubin Diglucuroxide in Liver 1017
Intravascular Hemolysis Requires Scavenging of Iron 1020
BIBLIOGRAPHY 1021
QUESTIONS AND ANNOTATED ANSWERS 1022
CLINICAL CORRELATIONS
24.1 Iron-Deficiency Anemia 1008
24.2 Hemochromatosis and Iron-Fortified Diet 1009
24.3 Acute Intermittent Porphyria 1012
24.4 Neonatal Isoimmune Hemolysis 1019

24.1 IRON METABOLISM: OVERVIEW

Iron is closely involved in the metabolism of oxygen; the properties of this metal permit the transportation and participation of oxygen in a variety of biochemical processes. The common oxidation states are either ferrous (Fe^{2+}) or ferric (Fe^{3+}); higher oxidation levels occur as short-lived intermediates in certain redox processes. Iron has an affinity for electronegative atoms such as oxygen, nitrogen, and sulfur, which provide the electrons that form the bonds with iron. These can be of very high affinity when favorably oriented on macromolecules. In forming complexes, no bonding electrons are derived from iron. There is an added complexity to the structure of iron: The nonbonding electrons in the outer shell of the metal (the incompletely filled 3*d* orbitals) can exist in two states. Where bonding interactions with iron are weak, the outer nonbonding electrons will avoid pairing and distribute throughout the 3*d* orbitals. Where bonding electrons interact strongly with iron, however, there will be pairing of the outer nonbonding electrons, favoring lower energy $3d$ orbitals. These two different distributions for each oxidation state of iron can be determined by electron spin resonance (ESR) measurements. Dispersion of $3d$ electrons to all orbitals leads to the high-spin state, whereas restriction of $3d$ electrons to lower energy orbitals, because of electron pairing, leads to a low-spin state. Some iron–protein complexes reveal changes in spin state without changes in oxidation during chemical events (e.g., binding and release of oxygen by hemoglobin).

At neutral and alkaline pH ranges, the redox potential for iron in aqueous solutions favors the ferric state; at acid pH values, the equilibrium favors the ferrous state. In the ferric state iron slowly forms large polynuclear complexes with hydroxide ion, water, and other anions that may be present. These complexes can become so large as to exceed their solubility products, leading to their aggregation and precipitation with pathological consequences.

Iron can bind to and influence the structure and function of a variety of macromolecules, with deleterious results to the organism. To protect against such reactions, several iron-binding proteins function specifically to store and transport iron. These proteins have both a very high affinity for the metal and, in the normal physiological state, also have incompletely filled iron-binding sites. The interaction of iron with its ligands has been well characterized in some proteins (e.g., hemoglobin and myoglobin), whereas for others (e.g., transferrin) it is presently in the process of being defined. The major area of ignorance in the biochemistry of iron lies in the in vivo transfer processes of iron from one macromolecule to another. Several mechanisms have been proposed to explain the process of iron transfer. Two of these have been supported by excellent model studies but with varying degrees of relevance to the physiological state. The proposed processes are the following: First, the redox change of iron has been an attractive mechanism because it is supported by selective in vitro studies and because in many cases macromolecules show a very selective affinity for ferric ions, binding ferrous ions poorly. Thus reduction of iron would permit ferrous ions to dissociate, and reoxidation would allow the iron to redistribute to appropriate macromolecules. Redox mechanisms have only been defined in a very few settings, some of which will be described below. An alternative hypothesis involves chelation of ferric ions by specific small molecules with high affinities for iron; this mechanism has been supported also by selective in vitro studies. The chelation mechanism suffers from the lack of a demonstrably specific in vivo chelator. Because the redox potential strongly favors ferric ion at almost all tissue sites and because Fe^{3+} binds so strongly to liganding groups, the probability is that there are cooperating mechanisms regulating the intermolecular transfer of iron.

24.2 IRON-CONTAINING PROTEINS

Iron binds to proteins either by incorporation into a **protoporphyrin IX** ring (see below) or by interaction with other protein ligands. Ferrous- and ferric-protoporphyrin IX complexes are designated **heme** and **hematin,** respectively. Heme-containing proteins include those that transport (e.g., hemoglobin) and store (e.g., myoglobin) oxygen; and certain enzymes that contain heme as part of their prosthetic groups (e.g., tryptophan pyrrolase). Discussions on structure–function relationships of heme proteins are presented in Chapters 6 and 25.

Nonheme proteins include transferrin, ferritin, a variety of redox enzymes that contain iron at the active site, and iron–sulfur proteins. A significant body of information has been acquired that relates to the structure–function relationships of some of these molecules.

Transferrin Transports Iron in Serum

The protein in serum involved in the transport of iron is **transferrin,** a β_1-glycoprotein synthesized in the liver, consisting of a single polypeptide chain of 78,000 Da with two iron-binding sites. Sequence studies indicate that the protein is a product of gene duplication derived from a putative ancestral gene coding for a protein binding only one atom of iron. Although several other metals bind to transferrin, the highest affinity is for ferric ion. Transferrin does not bind ferrous ion. The binding of each ferric ion is absolutely dependent on the coordinate binding of an anion, which in the physiological state is carbonate as indicated below:

$$\text{Transferrin} + Fe^{3+} + CO_3^{2-} \longrightarrow \text{transferrin} \cdot Fe^{3+} \cdot CO_3^{2-}$$

$$\text{Transferrin} \cdot Fe^{3+} \cdot CO_3^{2-} + Fe^{3+} + CO_3^{2-} \longrightarrow \text{transferrin} \cdot 2(Fe^{3+} \cdot CO_3^{2-})$$

In experimental settings, other organic polyanions can substitute for carbonate. Estimates of the association constants for the binding of ferric ion to transferrins from different species range from 10^{19} to 10^{31} M^{-1}, indicating for practical purposes that wherever there is excess transferrin free ferric ions will not be found. There is no evidence for cooperativity in the binding of iron at the two sites. In the normal physiological state, approximately one-ninth of all transferrin molecules are saturated with iron at both sites; four-ninths of transferrin molecules have iron at either site; and four-ninths of circulating transferrin are free of iron. The two iron-binding sites, although homologous, are not completely identical; they show some differences in sequences and also some differences in affinities for other metals (especially the lanthanides). Transferrin binds to specific cell surface receptors that mediate the internalization of the protein.

The **transferrin receptor** is a transmembrane protein consisting of two subunits of 90,000 Da each, joined by a disulfide bond; each subunit can bind a transferrin molecule, favoring the diferric form. Internalization of the receptor–transferrin complex is dependent on receptor phosphorylation by a Ca^{2+}–calmodulin–protein kinase C complex. Release of the iron occurs within the acidic milieu of the lysosome after which the receptor–apotransferrin complex returns to the cell surface where the apotransferrin is released to be reutilized in the plasma.

Lactoferrin Binds Iron in Milk

Milk contains iron that is bound almost exclusively to a protein that is closely homologous to transferrin, with two sites binding the metal. The iron content of **lactoferrin** varies, but the protein is never saturated. The role of lactoferrin in facilitating the transfer of iron to intestinal receptor sites in the infant has now begun to be studied. Other major studies on the function of lactoferrin have been directed toward its antimicrobial effect, protecting the newborn from gastrointestinal infections. Microorganisms require iron for replication and function. The presence of incompletely saturated lactoferrin results in the rapid binding of any free iron leading to the inhibition of microbial growth by preventing a sufficient amount of iron from entering these microorganisms. Other microbes, such as *E. coli*,

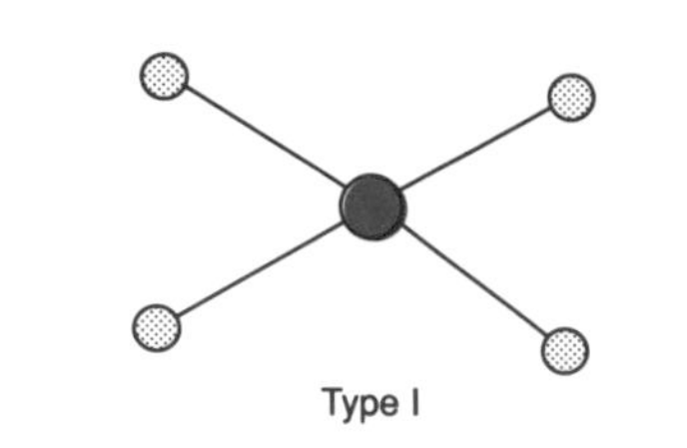

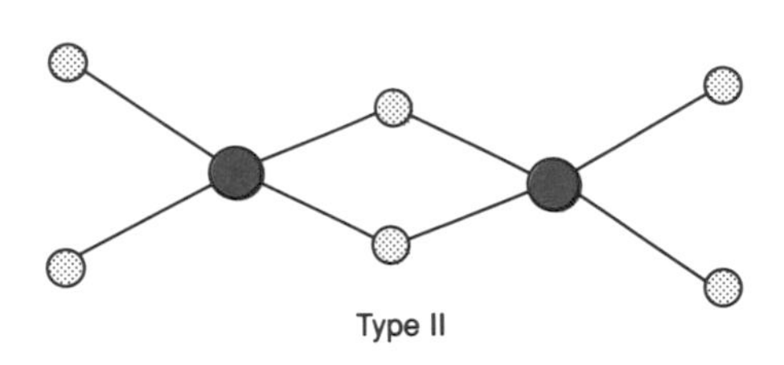

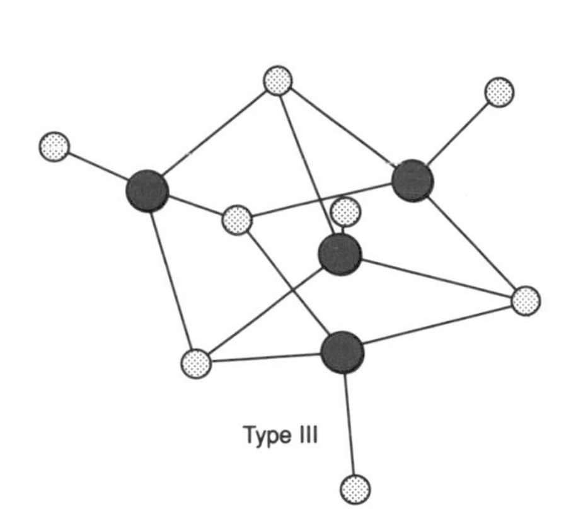

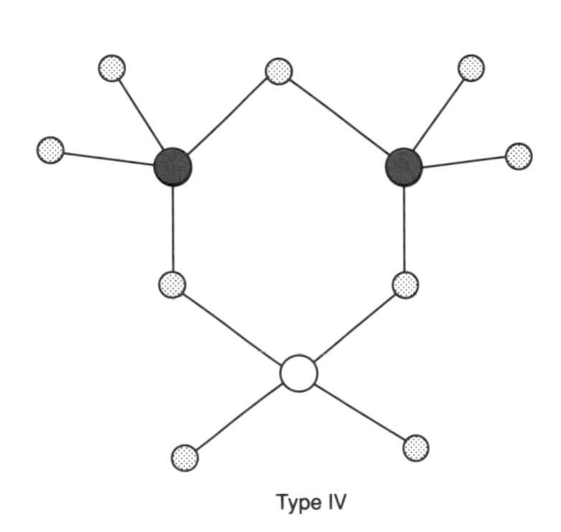

FIGURE 24.1
Structure of ferredoxins.
The dark colored circles represent the iron atoms; the light colored circles represent the inorganic sulfur atoms; and the small stippled circles represent the cysteinyl sulfur atoms derived from the polypeptide chain. Variation in type IV ferredoxins can occur where one of the cysteinyl residues can be substituted by a solvent oxygen atom of an OH group.

which release competitive iron chelators, are able to proliferate despite the presence of lactoferrin, since the chelators transfer the iron specifically to the microorganism.

Ferritin Is a Protein Involved in Storage of Iron

Ferritin is the major protein involved in the storage of iron. The protein consists of an outer polypeptide shell 130 Å in diameter with a central ferric–hydroxide–phosphate core 60 Å across. The apoprotein, **apoferritin,** consists of 24 subunits of a varying mixture of H chains (21,000 Da) and L chains (19,000 Da) providing a variety of isoprotein forms. The ratio of iron to polypeptide is not constant, since the protein has the ability to gain and release iron according to physiological needs. With a capacity of 4500 iron atoms, the molecule contains usually less than 3000. Channels from the surface permit the accumulation and release of iron. When iron is in excess, the storage capacity of newly synthesized apoferritin may be exceeded. This leads to iron deposition adjacent to ferritin spheres. Histologically such amorphous iron deposition is called **hemosiderin.**

The H chains of ferritin oxidize ferrous ions to the ferric state. Ferritins derived from different tissues of the same species demonstrate differences in electrophoretic mobility in a fashion analogous to the differences noted with isoenzymes. In some tissues ferritin spheres form latticelike arrays, which are identifiable by electron microscopy.

Other Nonheme Iron-Containing Proteins Are Involved in Enzymatic Processes

Many iron-containing proteins are involved in enzymatic processes, most of which are related to oxidation mechanisms. The structural features of the ligands binding the iron are not well known, except for a few components involved in mitochondrial electron transport. These latter proteins are characterized by iron being bonded, with one exception, only to sulfur atoms. Four major types of such proteins, termed **ferredoxins,** are known (see Figure 24.1). The smallest, type I, (e.g., nebredoxin) found only in microorganisms, consists of a small polypeptide chain with a molecular weight of about 6000, containing one iron atom bound to four cysteine residues. Type II consists of ferredoxins found in both plants and animal tissues where two iron atoms are found, each liganding to two separate cysteine residues and sharing two sulfide anions. The most complicated of the iron–sulfur proteins are the bacterial ferredoxins, type III, which contain four atoms of iron, each of which is linked to single separate cysteine residues but also shares three sulfide anions with neighboring iron molecules to form a cubelike structure. In some anerobic bacteria a family of ferredoxins may contain two type III iron–sulfur groups per macromolecule. Type IV ferredoxins contain structures with three atoms of iron each linked to two separate cysteine residues and each sharing two sulfide anions, forming a planar ring. In one example of this ferredoxin type, an exception of iron atoms being liganded only to sulfur atoms was found where the sulfur of a cysteinyl residue was substituted by a solvent oxygen atom. The redox potential afforded by these different ferredoxins varies widely and is in part dependent on the environment of the surrounding polypeptide chain that envelops these iron–sulfur groups. In nebredoxin the iron undergoes ferric–ferrous conversion during electron transport. With the plant and animal ferredoxins (type II iron–sulfur proteins) both irons are in the ferric form in the oxidized state; upon reduction only one iron goes to the ferrous state. In the bacterial ferredoxin (type III iron–sulfur protein) the oxidized state can be either $2Fe^{3+} \cdot 2Fe^{2+}$ or $3Fe^{3+} \cdot Fe^{2+}$, with corresponding reduced forms of $Fe^{3+} \cdot 3Fe^{2+}$ or $2Fe^{3+} \cdot 2Fe^{2+}$.

24.3 INTESTINAL ABSORPTION OF IRON

The high affinity of iron for both specific and nonspecific macromolecules leads to the absence of significant formation of free iron salts, and thus this metal is not lost via usual excretory routes. Rather, excretion of iron occurs only through the normal sloughing of tissues that are not reutilized (e.g., epidermis and gastrointestinal mucosal cells). In the healthy adult male the loss is about 1 mg day^{-1}. In premenopausal women, the normal physiological events of menses and parturition substantially augment iron loss. A wide variation of such loss exists, depending on the amounts of menstrual flow and the multiplicity of births. In the extremes of the latter settings, a premenopausal woman may require an amount of iron that is four to five times that needed in an adult male for prolonged periods of time. The postmenopausal woman who is not iron-deficient has an iron requirement similar to that of the adult male. Children and patients with blood loss naturally have increased iron requirement.

Cooking of food facilitates the breakdown of ligands attached to iron, increasing the availability of the metal in the gut. The low pH of stomach contents permits the reduction of ferric ion to the ferrous state, facilitating dissociation from ligands. The latter requires the presence of an accompanying reductant, which is usually achieved by adding ascorbate to the diet. The absence of a normally functioning stomach reduces substantially the amount of iron that is absorbed. Some iron-containing compounds bind the metal so tightly that it is not available for assimilation. Contrary to popular belief, spinach is a poor source of iron because of an earlier erroneous record of the iron content and because some of the iron is bound to phytate (inositol hexaphosphate), which is resistant to the chemical actions of the gastrointestinal tract. Specific protein cofactors derived from the stomach or pancreas have been suggested as being facilitators of iron absorption in the small intestine.

The major site of absorption of iron is in the small intestine, with the largest amount being absorbed in the duodenum and a gradient of lesser absorption occurring in the more distal portions of the small intestine. The metal enters the mucosal cell either as the free ion or as heme; in the latter case the metal is split off from the porphyrin ring in the mucosal cytoplasm. The large amount of bicarbonate secreted by the pancreas neutralizes the acidic material delivered by the stomach and thus favors the oxidation of ferrous ion to the ferric state. The major barrier to the absorption of iron is not at the lumenal surface of the duodenal mucosal cell. Whatever the requirements of the host are, in the face of an adequate delivery of iron to the lumen, a substantial amount of iron will enter the mucosal cell. Regulation of iron transfer occurs between the mucosal cell and the capillary bed (see Figure 24.2). In the normal state, certain processes define the amount of iron that will be transferred. Where there is iron deficiency, the amount of transfer increases; where there is iron overload in the host, the amount transferred is curtailed substantially. One mechanism that has been demonstrated to regulate this transfer of iron across the mucosal–capillary interface is the synthesis of apoferritin by the mucosal cell. In situations in which little iron is required by the host, a large amount of apoferritin is synthesized to trap the iron within the mucosal cell and prevent transfer to the capillary bed. As the cells

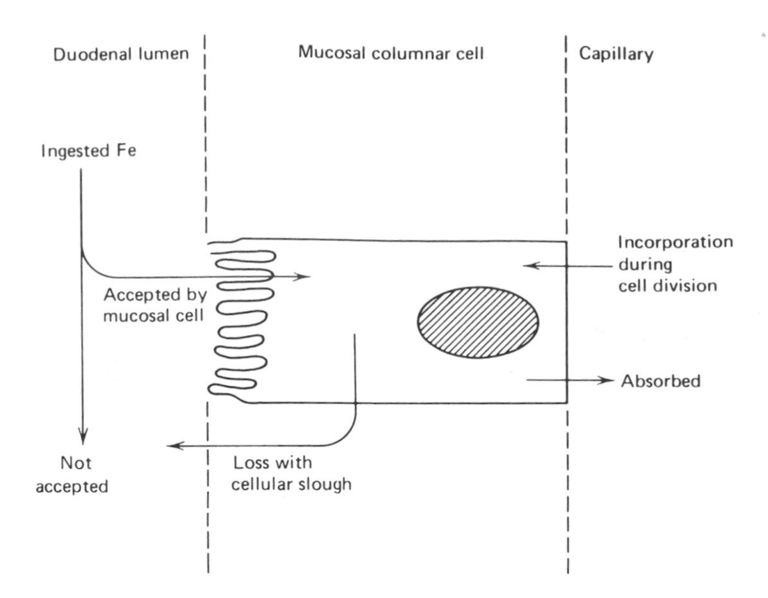

FIGURE 24.2
Intestinal mucosal regulation of iron absorption.
The flux of iron in the duodenal mucosal cell is indicated. A fraction of the iron that is potentially acceptable is transferred from the intestinal lumen into the epithelial cell. A large portion of ingested iron is not absorbed, in part because it is not presented in a readily acceptable form. Some iron is retained within the cell, bound by apoferritin to form ferritin. This iron is sloughed into the intestinal lumen with the normal turnover of the cell. A portion of the iron within the mucosal cell is absorbed and transferred to the capillary bed to be incorporated into transferrin. During cell division, which occurs at the bases of the intestinal crypts, iron is incorporated for cellular requirements. These fluxes change dramatically in iron-depleted or iron-excess states.

turn over (within a week), their contents are extruded into the intestinal lumen without absorption occurring. In situations in which there is iron deficiency, virtually no apoferritin is synthesized so as not to compete against the transfer of iron to the deficient host. There are other as yet undefined positive mechanisms that increase the rate of iron absorption in the iron-deficient state. Iron transferred to the capillaries is trapped exclusively by transferrin.

24.4 MOLECULAR REGULATION OF IRON UTILIZATION

Resting nonproliferating cells have a modest number of transferrin receptors on the cell surface, whereas proliferating cells have a variable number depending on the iron content in the cell. A high iron content reduces the number of receptors to a baseline level; a low content can increase the number up to sevenfold. Two iron-dependent mechanisms regulate the synthesis of receptor protein in proliferating cells. One acts at the DNA level where a 5′ flanking sequence of the receptor gene responds to iron deprivation by increasing gene transcription two- to threefold. The other acts at the mRNA level by a posttranscriptional mechanism wherein low iron concentrations lead to stabilization of the receptor mRNA. A 3′ untranslated region of the mRNA contains five stem–loop structures, which constitute an **iron-responsive element** (see Figure 24.3). In the iron-deprived state this region binds a 90,000-Da regulatory protein, which stabilizes the mRNA leading to its reduced turnover and therefore an increased number of receptor-specific mRNA molecules, thereby leading

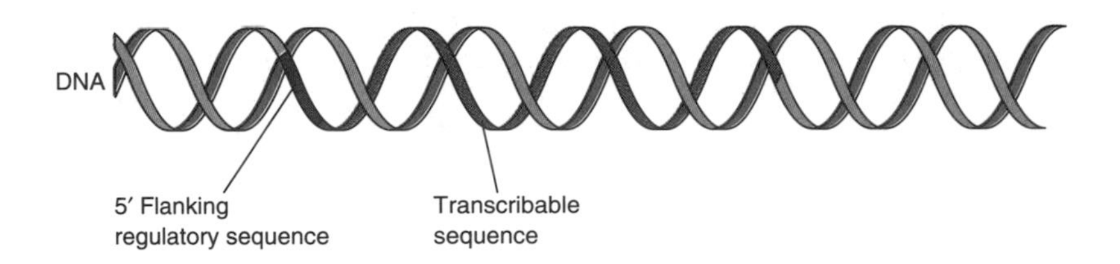

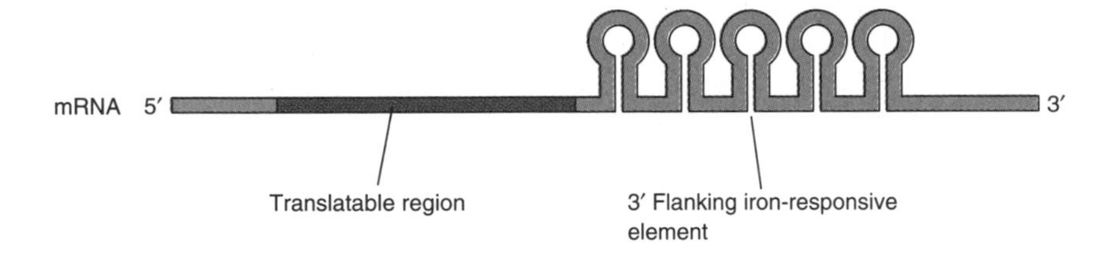

FIGURE 24.3
Control elements of transferrin receptor gene and mRNA.

to the increased synthesis of receptor protein. Deletion analysis suggests that all five stem–loops are required for this part of regulation.

Hepatocytes are a major site for ferritin subunit synthesis, which is regulated by the iron content of the cell. A high iron content increases whereas a low iron level decreases the synthesis. A postranscriptional mechanism regulates this process. The 5′ untranslated region of ferritin mRNA contains a single stem–loop structure, homologous to the 3′ stem–loop structures of the transferrin receptor mRNA (see Figure 24.4). In addition, this stem–loop structure of ferritin mRNA is an iron-responsive element that binds the 90,000-Da regulatory protein. However, in this case binding of the protein leads to a decreased rate of translation of the mRNA and thereby to a decreased concentration of apoferritin molecules. Note that the molecular events that are controlled are different for the syntheses of transferrin receptor and apoferritin.

Sulfhydryl groups on the regulatory protein appear to be crucial for its activity. In the presence of iron an inactive protein is generated through the formation of an intramolecular disulfide bond. With a low iron concentration, cytosolic reduction of the disulfide bond occurs, generating the essential sulfhydryl groups of the active protein.

In summary, low iron concentrations lead to the activation of a 90,000-Da regulatory protein that binds to the mRNAs of transferrin receptor and ferritin. In the former case, more receptor is synthesized, while in the latter case less apoferritin is synthesized. The net effect is the utilization of iron by proliferating cells. In contrast, high iron concentrations lead to the inactivation of the regulatory protein with a shift of iron from uptake by proliferating cells to storage in the liver.

24.5 IRON DISTRIBUTION AND KINETICS

A normal 70-kg male has 3–4 g of iron, of which only 0.1% (3.5 mg) is in the plasma. Approximately 2.5 g are in hemoglobin. Table 24.1 lists the distribution of iron in humans.

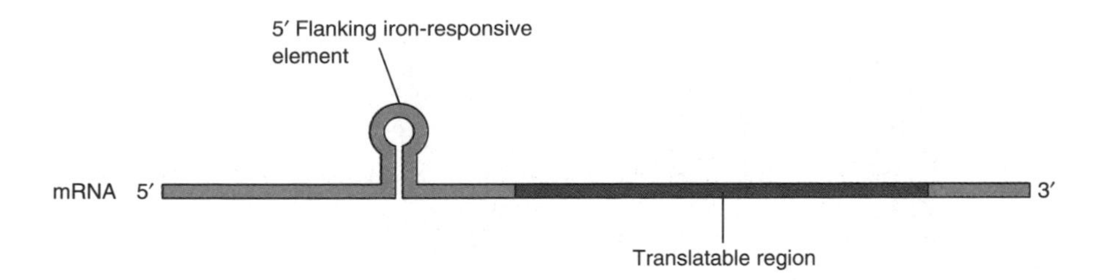

FIGURE 24.4
Structure of apoferritin H-chain mRNA.

TABLE 24.1 Approximate Iron Distribution: 70-kg Man

	g	%
Hemoglobin	2.5	68
Myoglobin	0.15	4
Transferrin	0.003	0.1
Ferritin, tissue	1.0	27
Ferritin, serum	0.0001	0.004
Enzymes	0.02	0.6
Total	3.7	100

Normally about 33% of the sites on transferrin contain iron. Iron picked up from the intestine is delivered primarily to the marrow for incorporation into the hemoglobin of red blood cells. The mobilization of iron from the mucosa and from storage sites involves in part the reduction of iron to the ferrous state and its reoxidation to the ferric form. The reduction mechanisms have not been well described. On the other hand, conversion of the ferrous ion back to the ferric state is regulated by serum enzymes called ferroxidases as indicated below:

$$Fe^{2+} + \text{ferroxidase} \longrightarrow Fe^{3+} + \text{reduced ferroxidase}$$

Ferroxidase I is also known as **ceruloplasmin.** A deficiency of this protein is associated with **Wilson's disease,** in which there is progressive hepatic failure and degeneration of the basal ganglia associated with a characteristic copper deposition in the cornea (Kayser–Fleischer rings). The ferroxidase activity of ceruloplasmin is not clinically important, since there is no evidence for significant impairment of mobilization of iron in Wilson's disease. Another serum protein, **ferroxidase II,** appears to be the major serum component that oxidizes ferrous ions. If an inappropriately large amount of iron is administered by injection to a subject who is not iron-deficient, this iron will be transported to the liver for storage in the form of ferritin.

In any disease process in which iron loss exceeds iron repletion, a sequence of physiological responses occurs. The initial events are without symptoms to the subject and involve depletion of iron stores without compromise of any physiological function. This depletion will be manifested by a reduction or absence of iron stores in the liver and in the bone marrow and also by a decrease in the content of the very small amount of ferritin that is normally present in plasma. Serum ferritin levels reflect slow release from storage sites during the normal cellular turnover that occurs in the liver; measurements are made by radioimmune assays. Serum ferritin is mostly apoferritin in form, containing very little iron. During this early phase, the level and percentage saturation of serum transferrin is not distinctly abnormal. As the iron deficiency progresses, the level of hemoglobin begins to fall and morphological changes appear in the red blood cells. Concurrently, the serum iron falls with a rise in the level of total serum transferrin, the latter reflecting a physiological adaptation in an attempt to absorb more iron from the gastrointestinal tract. At this state of iron depletion a very sensitive index is the percentage saturation of serum transferrin with iron (normal range 21–50%). At this point the patient usually comes to medical attention, and the diagnosis of iron deficiency is made. In countries in which iron deficiency is severe without available corrective medical measures, a third and severe stage of iron deficiency can occur, where there begins a depletion of iron-containing enzymes leading to very pronounced metabolic effects (see Clin. Corr. 24.1).

Iron overload can occur in patients so that the iron content of the body can be elevated to values as high as 100 g. This may happen for a variety of reasons. Some patients have a recessive heritable disorder associated with a marked inappropriate increase in iron absorption. In such cases the serum transferrin can be almost completely saturated with iron. This state, which is known as **idiopathic hemochromatosis,** is more commonly seen in men because women with the abnormal gene are protected somewhat by menstrual and childbearing events. The accumulation of iron in the liver, the pancreas, and the heart can lead to cirrhosis and liver tumors, diabetes mellitus, and cardiac failure, respectively. The treatment for these patients is periodic withdrawals of large amounts of blood where

CLIN. CORR. 24.1 IRON-DEFICIENCY ANEMIA

Microscopic examination of a blood smear in patients with iron-deficiency anemia usually reveals the characteristic findings of microcytic (small in size) and hypochromic (underpigmented) red blood cells. These changes in the red cell result from decreased rates of globin synthesis when heme is not available. A bone marrow aspiration will reveal no storage iron to be present, and serum ferritin values are virtually zero. The serum transferrin value (expressed as the total iron-binding capacity) will be elevated (upper limits of normal: 410 $\mu g\ dL^{-1}$) with a serum iron saturation of less than 16%. Common causes for iron deficiency include excessive menstrual flow, multiple births, and gastrointestinal bleeding that may be occult. The common causes of gastrointestinal bleeding include medications which can cause ulcers or erosion of the gastric mucosa (especially aspirin or cortisone-like drugs), hiatal hernia, peptic ulcer disease, gastritis associated with chronic alcoholism, and gastrointestinal tumors. The management of such patients must include both a careful examination for the cause and source of bleeding and supplementation with iron. The latter is usually provided in the form of oral ferrous sulfate tablets; occasionally intravenous iron therapy may be required. Where the iron deficiency is severe, transfusion with packed red blood cells may also be indicated.

Finch, C. A. and Huebers, H. Perspectives in iron metabolism. *N. Engl. J. Med.* 306:1520, 1982.

the iron is contained in the hemoglobin. Another group of patients has severe anemias, among the most common of which are the thalassemias, a group of hereditary **hemolytic anemias.** In these cases the subjects require transfusions throughout their lives, leading to the accumulation of large amounts of iron derived from the transfused blood. Clearly bleeding would be an inappropriate measure in these cases; rather, the patients are treated by the administration of iron chelators, such as desferrioxamine, which leads to the excretion of large amounts of complexed iron in the urine. Rarely, a third group of patients will acquire excess iron because they ingest large amounts of both iron and ethanol, the latter promoting iron absorption. In these cases excess stored iron can be removed by bleeding (see Clin. Corr. 24.2).

24.6 HEME BIOSYNTHESIS

Heme is produced in virtually all mammalian tissues. Its synthesis is most pronounced in the bone marrow and liver because of the requirements for incorporation into hemoglobin and the cytochromes, respectively. As depicted in Figure 24.5, heme is largely a planar molecule. It consists of one ferrous ion and a tetrapyrrole ring, **protoporphyrin IX.** The diameter of the iron atom is a little too large to be accommodated within the plane of the porphyrin ring, and thus the metal puckers out to one side as it coordinates with the apical nitrogen atoms of the four pyrrole groups. Heme is one of the most stable of compounds, reflecting its strong resonance features.

Figure 24.6 depicts the pathway for heme biosynthesis. The following are the important aspects to be noted. First, the initial and last three enzymatic steps are catalyzed by enzymes that are in the mitochondrion, whereas the intermediate steps take place in the cytoplasm. This is important in considering the regulation by heme of the first biosynthetic step; this aspect is discussed below. Second, the organic portion of heme is derived totally from eight residues each of glycine and succinyl CoA. Third, the reactions occurring on the side groups attached to the tetrapyrrole ring involve the colorless intermediates known as **porphyrinogens.** The latter compounds, though exhibiting resonance features within each pyrrole ring, do not demonstrate resonance between the pyrrole groups. As a consequence, the porphyrinogens are unstable and can be readily oxidized, especially in the presence of light, by nonenzymatic means to their stable **porphyrin** products. In the latter cases resonance between pyrrole groups is established by oxidation of the four methylene

CLIN. CORR. 24.2 HEMOCHROMATOSIS AND IRON-FORTIFIED DIET

For many years there was a controversy as to whether food should be fortified with iron because of the prevalence of iron-deficiency anemia, especially among premenopausal women. Proponents suggested that if at least 50 mg of iron was incorporated per pound of enriched flour, dietary iron deficiency would be reduced markedly. Opponents stated that the possibility of toxicity from excess iron absorption through iron fortification was too great and thus such a measure should not be sponsored. Recent studies have indicated that the heterozygote prevalence for hemochromatosis is extraordinarily high, about 3–5% in the general population. Since the disease is expressed primarily in the homozygous state, about 0.25% of all individuals are at risk for hemochromatosis. A study in Sweden measured the serum iron and iron-binding capacity in a group of 347 subjects. No women showed evidence of iron overload. However, 5% of the men had persistent elevation of serum iron values, with 2% of the men having increased iron stores consonant with the distribution found in early stages of hemochromatosis. Of relevance is the fact that Sweden has had mandated iron fortification for the past 30 years, and approximately 42% of the average daily intake of iron is derived from fortified sources. This study points out the danger of iron-fortified diets. In other settings, however, such as in countries in which iron deficiency is widespread, fortification may still be the most appropriate measure.

Bothwell, T. H., Derman, D., Bezwoda, W. R., Torrance, J. D., and Charlton, R. W. Can iron fortification of flour cause damage to genetic susceptibles (idiopathic hemochromatosis and β-thalassemia major)? *Human Genet.* 1:131, 1978.

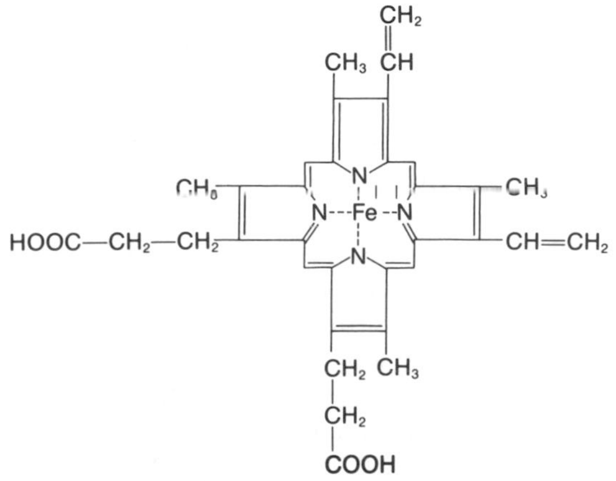

FIGURE 24.5
Structure of heme.

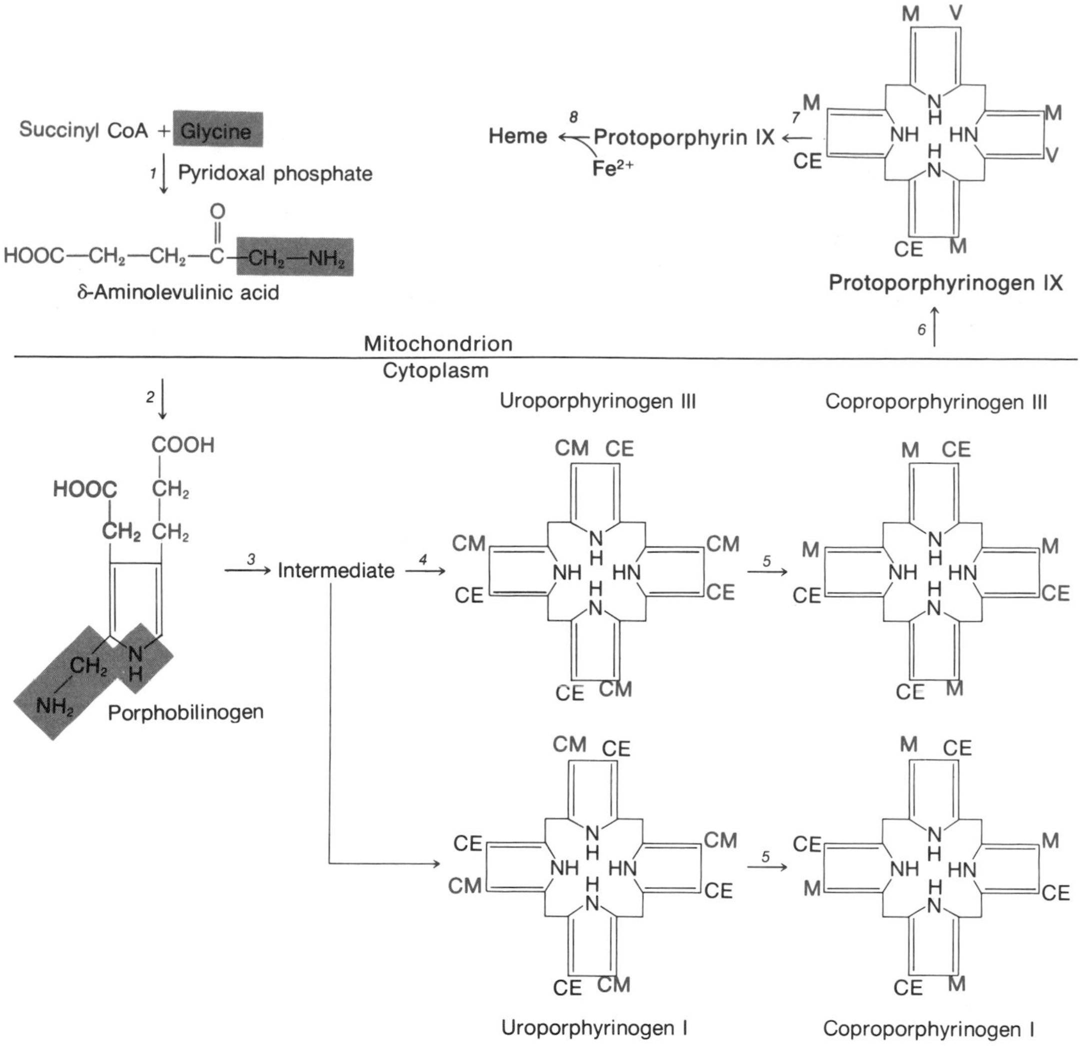

FIGURE 24.6
Pathway for heme biosynthesis.
The numbers indicate the enzymes involved in each of the biochemical steps according to the following code: 1, Ala synthase; 2, Ala dehydratase; 3, uroporphyrinogen I synthase; 4, uroporphyrinogen III cosynthase; 5, uroporphyrinogen decarboxylase; 6, coproporphyrinogen III oxidase; 7, protoporphyrinogen IX oxidase; 8, ferrochelatase. The pyrrole ligands are indicated by the following abbreviations: CE = β-carboxyethyl (propionic); CM = carboxymethyl (acetic); M = methyl; V = vinyl.

bridges. Figure 24.7 depicts the enzymatic conversion of protoporphyrinogen to protoporphyrin by this oxidation mechanism. This is the only known porphyrinogen oxidation that is enzyme regulated in humans; all other porphyrinogen–porphyrin conversions are nonenzymatic and catalyzed by light rather than catalyzed by specific enzymes. Fourth, once the tetrapyrrole ring is formed, the order of the R groups as one goes clockwise around the tetrapyrrole ring defines which of the four possible types of **uro-** or **coproporphyrinogens** are being synthesized. These latter compounds have two different substituents, one each for every pyrrole group. Going clockwise around the ring, the substituents can be arranged as ABABABAB (where A is one substituent and B the other), forming a type I porphyrinogen, or the arrangement can be ABABABBA, forming a type III porphyrinogen. In principle, two other arrangements can occur to

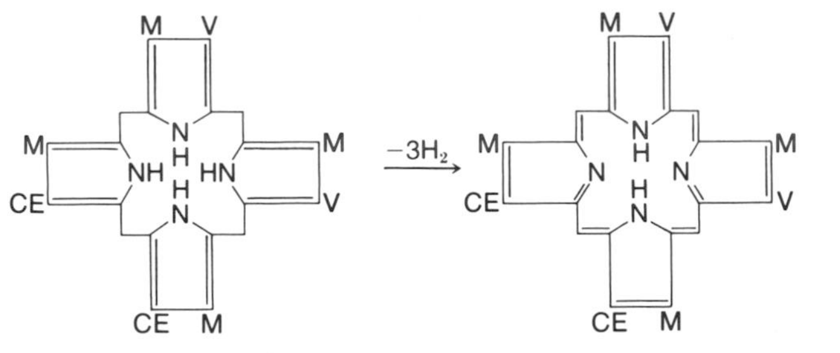

Protoporphyrinogen IX $\xrightarrow{-3H_2}$ Protoporphyrin IX

FIGURE 24.7
Action of protoporphyrinogen IX oxidase, an example of the conversion of a porphyrinogen to a porphyrin.

form porphyrinogens II and IV, and these can be synthesized chemically; however, they do not occur naturally. In protoporphyrinogen and protoporphyrin there are three types of substituents, and the classification becomes more complicated; type IX is the only form that is synthesized naturally.

Derangements of porphyrin metabolism are known clinically as the **porphyrias.** This family of diseases is of great interest because it has revealed that the regulation of heme biosynthesis is complicated. The clinical presentations of the different porphyrias provide a fascinating exposition of biochemical regulatory abnormalities and their relationship to pathophysiological processes. Table 24.2 lists the details of the different porphyrias (see Clin. Corr. 24.3).

Enzymes in Heme Biosynthesis Occur in Both Mitochondria and Cytosol

Aminolevulinic Acid Synthase

Aminolevulinic acid (ALA) synthase controls the rate-limiting step of heme synthesis in all tissues studied. The synthesis of the enzyme is not directed by mitochondrial DNA but occurs rather in the cytosol, being directed by mRNA derived from the nucleus. The enzyme is incorporated into the matrix of the mitochondrion. Succinyl CoA is one of the substrates and is found only in the mitochondrion. This protein has been purified to homogeneity from rat liver mitochondria. The cytosolic pro-

TABLE 24.2 Derangements in Porphyrin Metabolism

Disease state	Genetics	Tissue	Enzyme	Activity	Organ pathology
Acute intermittent porphyria	Dominant	Liver	1. ALA synthase 2. Uroporphyrinogen I synthase 3. Δ^4-5α-Reductase	Increase Decrease Decrease	Nervous system
Hereditary coproporphyria	Dominant	Liver	1. ALA synthase 2. Coproporphyrinogen oxidase	Increase Decrease	Nervous system; skin
Variegate porphyria	Dominant	Liver	1. ALA synthase 2. Protoporphyrinogen oxidase	Increase Decrease	Nervous system; skin
Porphyria cutanea tarda	Dominant	Liver	1. Uroporphyrinogen decarboxylase	Decrease	Skin, induced by liver disease
Hereditary protoporphyria	Dominant	Marrow	1. Ferrochelatase	Decrease	Gallstones, liver disease, skin
Erythropoietic porphyria	Recessive	Marrow	1. Uroporphyrinogen III cosynthase	Decrease	Skin and appendages; reticuloendothelial system
Lead poisoning	None	All tissues	1. ALA dehydrase 2. Ferrochelatase	Decrease Decrease	Nervous system; blood; others

CLIN. CORR. 24.3
ACUTE INTERMITTENT PORPHYRIA

A 40-year-old single, white woman appears in the emergency room in an agitated state, weeping, and complaining of severe abdominal pain. She states that she has been constipated for several days and has noted a feeling of marked weakness in her arms and legs and that "things do not appear to be quite right." Physical examination reveals a slightly rapid heart rate (110/min) and moderate hypertension (blood pressure of 160/110 mmHg). The only other significant findings are two well-healed abdominal operative scars. When queried, she relates that there have been earlier episodes of severe abdominal pain, in fact more severe than what she is presently experiencing. Exploratory abdominal operations undertaken on two of those past occasions revealed no abnormalities.

The usual laboratory tests are obtained and appear to be largely normal. None of her neurological complaints are accompanied by physical findings or have any localized anatomical focus. The decision is made that the present symptoms are largely psychiatric in origin and have a functional rather than an organic basis. Because of her agitated state, the patient is sedated with 60 mg of phenobarbital; a consultant psychiatrist agrees by telephone to see the patient in about 4 h. During the ensuing interval, the emergency room staff notices marked deterioration in the patient's status: Generalized weakness rapidly appears, progressing to a compromise of respiratory function. This ominous development leads to immediate incorporation of a ventilatory assistance regimen, with transfer of the patient to the intensive care unit for close physiological monitoring. Despite these measures the patient's condition deteriorates further and she dies 48 h later. A short time before death a urine sample of the patient is found to have a markedly elevated level of porphobilinogen.

This patient had acute intermittent porphyria, a disease of incompletely understood derangement of heme biosynthesis. This entity must be considered in any postpubertal patient who develops unexplained neurological, abdominal, or psychiatric symptoms. There is a dominant pattern of inheritance associated with an overproduction of the porphyrin precursors, ALA and porphobilinogen. Several associated well-defined enzyme abnormalities are noted in the cases that have been studied carefully. These include (1) a marked increase in ALA synthase, (2) a reduction by one-half of the activity of uroporphyrinogen I synthase, and (3) a reduction of one-half of the activity of steroid Δ^4-5α-reductase. The change in content of the second enzyme is consonant with a dominant expression. The change in content of the third enzyme is acquired and not apparently a heritable expression of the disease. It is believed that a decrease in uroporphyrinogen I synthase leads to a minor decrement in the content of heme in the liver. The low concentration of heme leads to a failure both to repress the synthesis and to inhibit the activity of ALA synthase. Since this disease is almost never manifested before puberty, it is thought that only with the induction of Δ^4-5β-reductase at adolescence does the disease become manifest. Since these patients do not have a sufficient amount of Δ^4-5α-reductase, it is assumed that the observed increase in the 5β steroids is due to a shunting of Δ^4 steroids into the 5β-reductase pathway. The importance of abnormalities of this metabolic pathway in the pathogenesis of porphyria is not known.

Pathophysiologically, the disease poses a great riddle: The derangement of porphyrin metabolism is confined to the liver, which anatomically appears normal, whereas the pathological findings are restricted to the nervous system. In the present case, involvement of (1) the brain led to the agitated and confused state, (2) the autonomic system led to the hypertension, increased heart rate, constipation, and abdominal pain, and (3) the peripheral nervous system and spinal cord led to the weakness and sensory disturbances. These observations suggest that there must be a metabolic intermediate that circulates to reach and affect neural tissue. However, experimentally, no known intermediate of heme biosynthesis can cause the pathology noted in acute intermittent porphyria.

In the present case there should have been a greater suspicion of the possibility of porphyria early in the patient's presentation. The analysis for porphobilinogen in the urine is a relatively simple test. The treatment would have been glucose infusion, the exclusion of any drugs that could cause elevation of ALA synthase (e.g., barbituates), and, if her disease failed to respond satisfactorily despite these measures, the administration of intravenous hematin to inhibit the synthesis and activity of ALA synthase.

Acute intermittent porphyria is of historic political interest. The disease has been diagnosed in two descendants of King George III, suggesting that the latter's deranged personality preceding and during the American Revolution could possibly be ascribed to porphyrias.

Meyer, U. A., Strand, L. J., Doss, M., Rees, A. C., and Marver, H. S. Intermittent acute porphyria: demonstration of a genetic defect in porphobilinogen metabolism. *N. Engl. J. Med.* 286:1277, 1972; Stein, J. A. and Tschudy, D. P. Acute intermittent porphyria: a clinical and biochemical study of 46 patients. *Medicine* 49:1, 1970.

tein is a dimer of a 71,000-Da subunit, containing a basic N-terminal signaling sequence that directs the enzyme into the mitochondrion. An ATP-dependent 70,000-Da cytosolic component, known as a chaperone protein, maintains ALA synthase in the unfolded extended state, the only form that can pass through the mitochondrial membrane. Thereafter, the N-terminal signaling sequence is cleaved by a metal-dependent protease in the mitochondrial matrix, yielding an ALA synthase with subunits of 65,000 Da each. Within the matrix an oligomeric protein, of 14 subunits of 60,000 Da each, catalyzes the correct folding of the protein in a second

ATP-dependent process (see Figure 24.8). The ALA synthase has a short biological half-life (~60 min). Both the synthesis and the activity of the enzyme are subject to regulation by a variety of substances; 50% inhibition of activity occurs in the presence of 5 μM of hemin, and virtually complete inhibition is noted at a 20-μM concentration. The enzymatic reaction involves the condensation of a glycine residue with a residue of succinyl CoA (Figure 24.9). The reaction has an absolute requirement for **pyridoxal phosphate;** the latter interacts with the nitrogen of glycine to form a Schiff's base. This generates a carbanion intermediate on the α carbon of the glycine, allowing a nucleophilic attack and condensation with the succinyl group from succinyl CoA. The bound intermediate, α-amino-β-ketoadipic acid, decarboxylates to form the released product, ALA. Pyridoxal deficiency and drugs competing with pyridoxal phosphate lead to a decrease in enzyme activity.

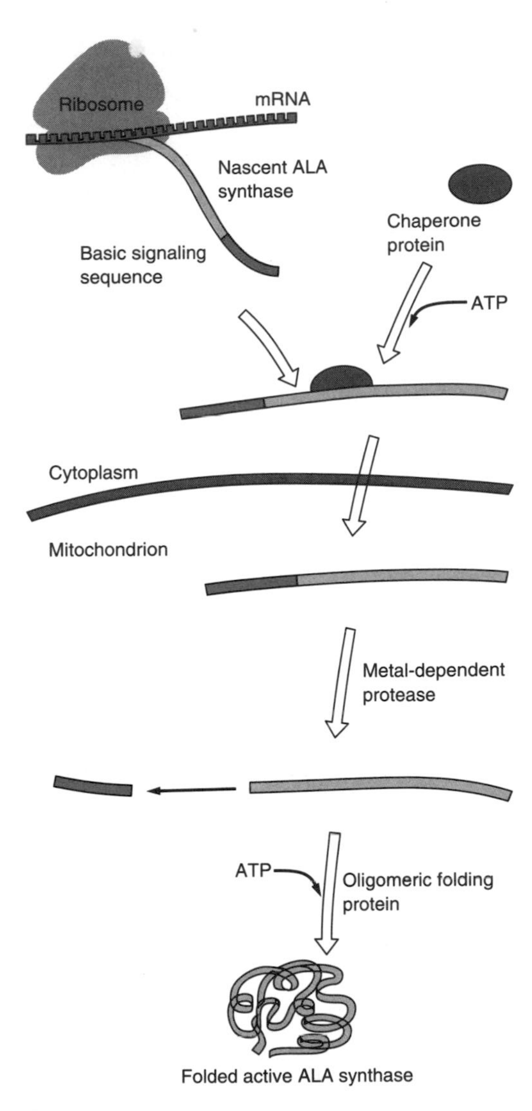

FIGURE 24.8
Synthesis of ALA synthase and its transfer into the mitochondrion.

ALA Dehydrase

Aminolevulinic acid dehydrase is a soluble cytosol component with a molecular weight of 280,000 and consisting of eight subunits, of which only four interact with the substrate. This protein also interacts with the substrate to form a Schiff's base, but in this case the ε-amino group of a lysine residue binds to the ketonic carbon of the substrate molecule (Figure 24.10). Two molecules of ALA condense asymmetrically to form **porphobilinogen.** The ALA dehydrase is a sulfhydryl enzyme and is very sensitive to inhibition by heavy metals. A characteristic finding of **lead poisoning** is the elevation of ALA in the absence of an elevation of porphobilinogen.

Uroporphyrinogen I Synthase

The synthesis of the porphyrin ring is a complicated process that has recently been defined. A sulfhydryl group on uroporphyrinogen I synthase forms a thioether bond with a porphobilinogen residue through a deamination reaction. Thereafter, five additional porphobilinogen residues are deaminated successively to form a linear hexapyrrole adduct with the enzyme. The adduct is cleaved hydrolytically to form both an enzyme pyrromethane complex and the linear tetrapyrrole, hydroxymethylbilane. The enzyme pyrromethane complex is then ready for an-

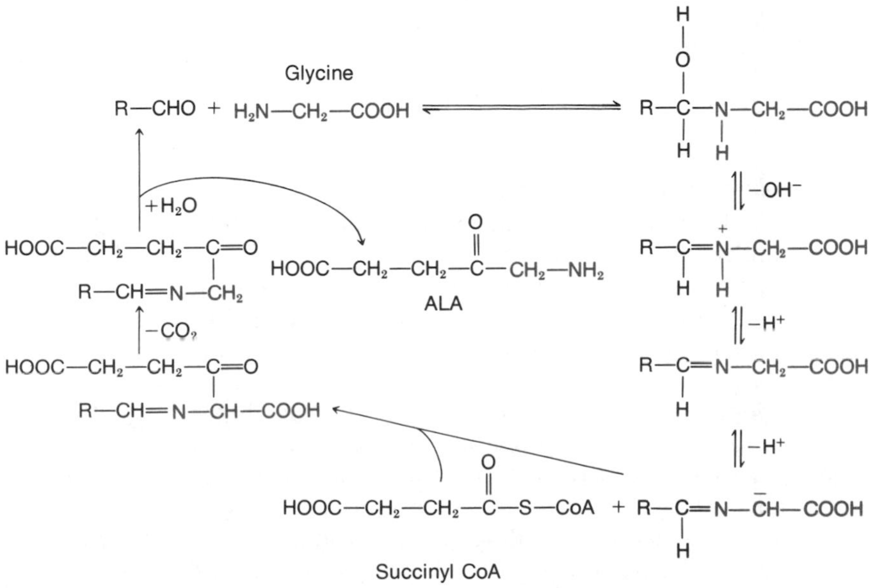

$$\text{R—CHO} + \underset{\text{Glycine}}{\text{H}_2\text{N—CH}_2\text{—COOH}} \rightleftharpoons \text{R—C(H)(OH)—N(H)—CH}_2\text{—COOH}$$

$$\xrightarrow{-\text{OH}^-} \text{R—CH=N}^+\text{H—CH}_2\text{—COOH} \xrightarrow{-\text{H}^+} \text{R—CH=N—CH}_2\text{—COOH} \xrightarrow{-\text{H}^+} \text{R—CH=N—}\bar{\text{C}}\text{H—COOH}$$

$$\underset{\text{Succinyl CoA}}{\text{HOOC—CH}_2\text{—CH}_2\text{—C(=O)—S—CoA}} + \text{R—CH=N—}\bar{\text{C}}\text{H—COOH} \rightarrow \text{HOOC—CH}_2\text{—CH}_2\text{—C(=O)—CH(COOH)—N=CH—R}$$

$$\xrightarrow{-\text{CO}_2} \text{HOOC—CH}_2\text{—CH}_2\text{—C(=O)—CH}_2\text{—N=CH—R} \xrightarrow{+\text{H}_2\text{O}} \text{R—CHO} + \underset{\text{ALA}}{\text{HOOC—CH}_2\text{—CH}_2\text{—C(=O)—CH}_2\text{—NH}_2}$$

FIGURE 24.9
Synthesis of δ-aminolevulinic acid (ALA).

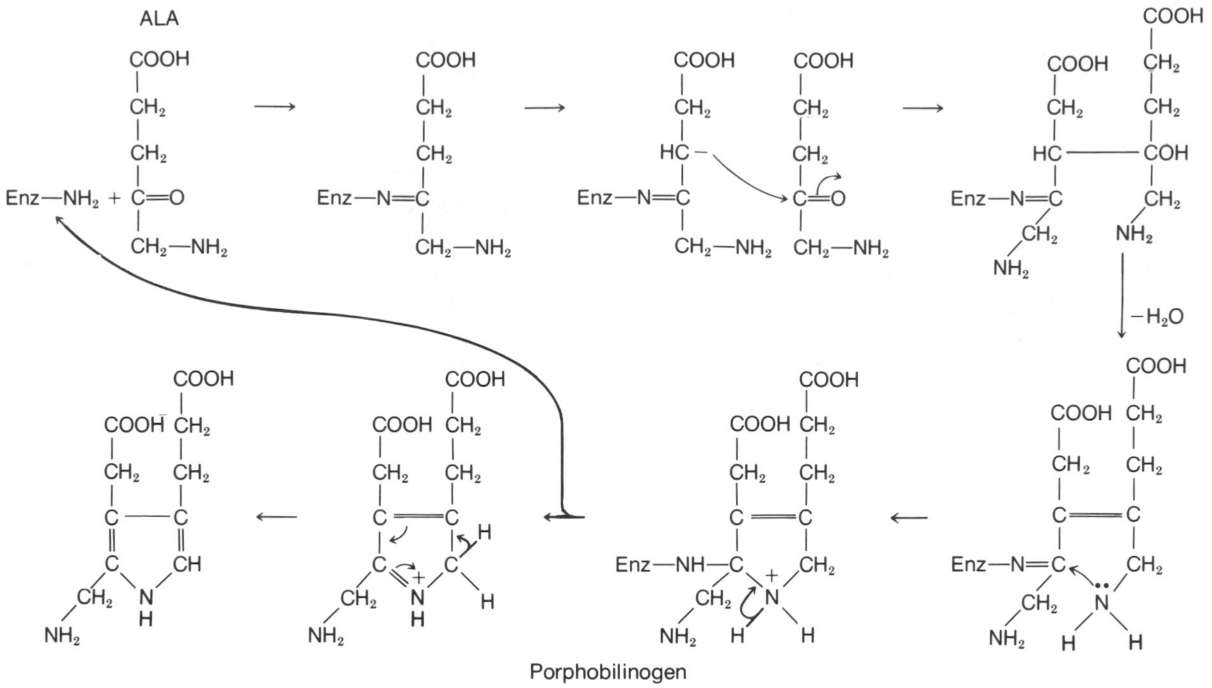

FIGURE 24.10
Synthesis of porphobilinogen.

other cycle of addition of four porphobilinogen residues to generate another tetrapyrrole. Thus, pyrromethane is the covalently attached novel cofactor for the enzyme. Uroporphyrinogen I synthase has no ring-closing function; hydroxymethylbilane closes in an *enzyme-independent* step to form uroporphyrinogen I if no additional factors are present. However, the synthase is closely associated with a second protein, **uroporphyrinogen III cosynthase,** which directs the synthesis of the III isomer. The formation of the latter involves a spiro intermediate generated from hydroxymethylbilane; this allows the inversion of one of the pyrrole groups (see Figure 24.11). In the absence of the cosynthase, uroporphyrinogen I is synthesized slowly; in its presence, the III isomer is synthesized rapidly. A rare recessively inherited disease, erythropoietic porphyria, associated with marked cutaneous light sensitization, is due to an abnormality of red blood cell cosynthase. Here large amounts of the type I isomers of uroporphyrinogen and coproporphyrinogen are synthesized in the bone marrow.

Uroporphyrinogen Decarboxylase

This enzyme acts on the side chains of the uroporphyrinogens to form the coproporphyrinogens. The protein catalyzes the conversion of both I and III isomers of uroporphyrinogen to the respective coproporphyrinogen isomers. Uroporphyrinogen decarboxylase is inhibited by iron salts. Clinically the most common cause of porphyrin derangement is associated with patients who have a single gene abnormality for this enzyme, leading to 50% depression of the enzyme's activity. This disease, which shows cutaneous manifestations primarily with sensitivity to light, is known as **porphyria cutanea tarda.** The condition is not expressed unless patients either take drugs that cause an increase in porphyrin synthesis or drink large amounts of alcohol, leading to the accumulation of iron, which then acts to inhibit further the activity of uroporphyrinogen decarboxylase.

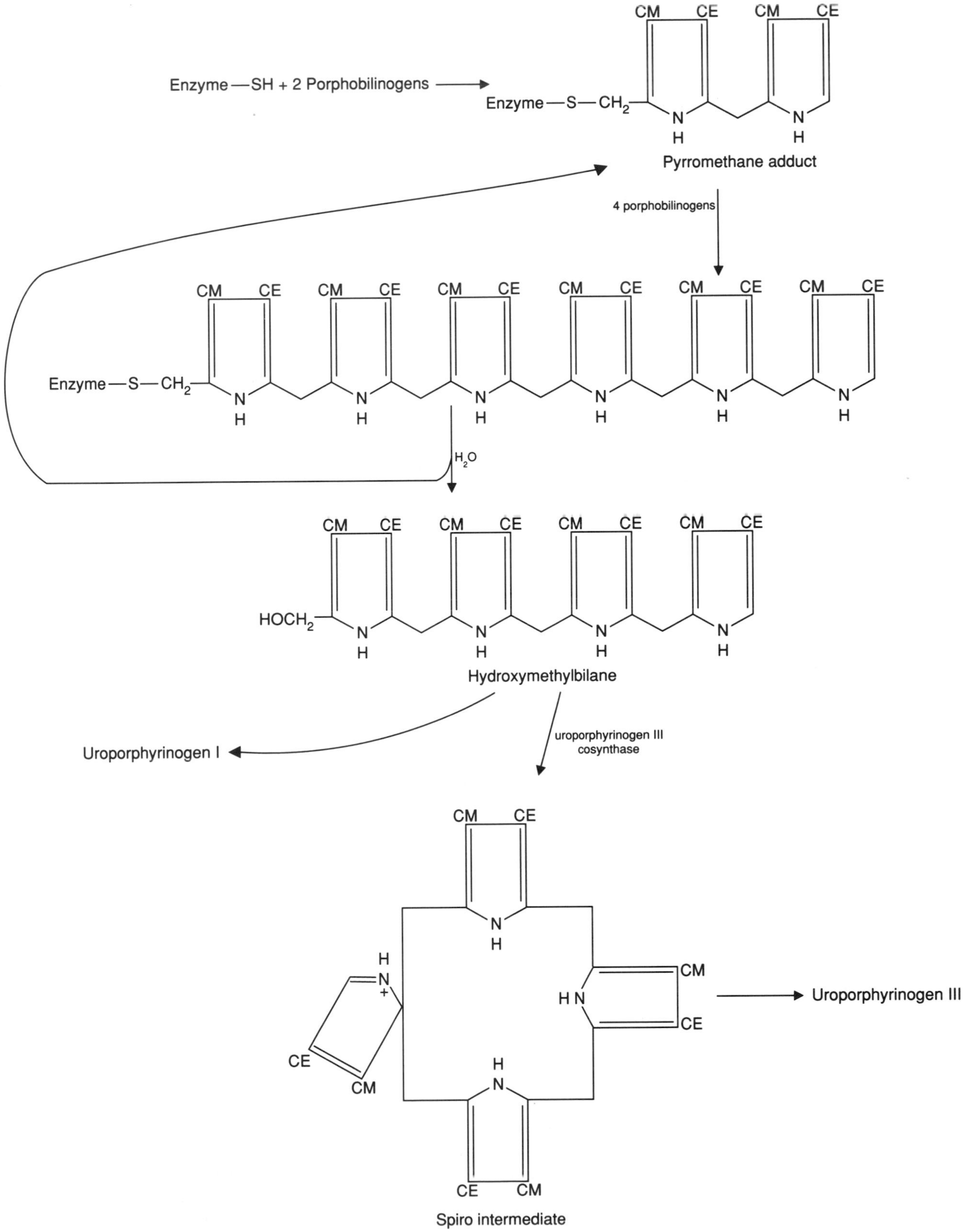

Enzyme—SH + 2 Porphobilinogens → Enzyme—S—CH_2—

Pyrromethane adduct

4 porphobilinogens

Enzyme—S—CH_2—

H_2O

$HOCH_2$—

Hydroxymethylbilane

Uroporphyrinogen I

uroporphyrinogen III cosynthase

Spiro intermediate

→ Uroporphyrinogen III

FIGURE 24.11
Synthesis of uroporphyrinogens I and III.

Coproporphyrinogen Oxidase

This mitochondrial enzyme is specific for the type III isomer of coproporphyrinogen, not acting on the type I isomer. Coproporphyrinogen III enters the mitochondrion and is converted to protoporphyrinogen IX. The mechanism of action is not understood. A dominant disease associated with a deficiency of this enzyme leads to a form of hereditary hepatic porphyria, known as **hereditary coproporphyria.**

Protoporphyrinogen Oxidase

This mitochondrial enzyme generates a product, protoporphyrin IX, which, in contrast to the other heme precursors, is very water-insoluble. Excess amounts of protoporphyrin IX that are not converted to heme are excreted by the biliary system into the intestinal tract. A dominant disease, **variegate porphyria,** is due to a deficiency of protoporphyrinogen oxidase.

Ferrochelatase

Ferrochelatase inserts ferrous iron into protoporphyrin IX in the final step of the synthesis of heme. Reducing substances are required for its activity. The protein is sensitive to the effects of heavy metals (especially lead) and, of course, to iron deprivation. In these latter instances, zinc instead of iron is incorporated to form a zinc–protoporphyrin IX complex. In contrast to heme, the zinc–protoporphyrin IX complex is brilliantly fluorescent and easily detectable in small amounts.

ALA Synthase Catalyzes Rate-Limiting Step of Heme Biosynthesis

Aminolevulinic acid synthase controls the rate-limiting step of heme synthesis in all tissues. Succinyl CoA and glycine are substrates for a variety of reactions. The modulation of the activity of ALA synthase determines the quantity of the substrates that will be shunted into heme biosynthesis. Heme (and also hematin) acts both as a repressor of the synthesis of ALA synthase and as an inhibitor of its activity. Since heme resembles neither the substrates nor the product of the enzyme's action, it is probable that the latter inhibition occurs at an allosteric site. Almost 100 different drugs and metabolites can cause induction of ALA synthase; for example, a 40-fold increase is noted in the rat after treatment with 3,5-dicarbethoxy-1,4-dihydrocollidine. The effect of pharmacological agents has led to the important clinical feature where some patients with certain kinds of porphyria have had exacerbations of their condition following the inappropriate administration of certain drugs (e.g., barbiturates). The ALA dehydrase is also inhibited by heme; but this is of little physiological consequence, since the maximal activity of the total amount of ALA dehydrase present is about 80-fold greater than that of ALA synthase, and thus heme-inhibitory effects are reflected first in the activity of ALA synthase.

Glucose or one of its proximal metabolites serves to inhibit heme biosynthesis in a mechanism that is not yet defined. This is of clinical relevance, since some patients manifest their porphyric state for the first time when placed on a very low caloric (and therefore glucose) intake. Other regulators of porphyrin metabolism include certain steroids. Steroid hormones (e.g., oral contraceptive pills) with a double bond in ring A between C–4 and C–5 atoms can be reduced by two different reductases. The product of 5α-reduction has little effect on heme biosynthesis; however, the product of 5β-reduction serves as a stimulus for the synthesis of

ALA synthase. The observation that **5β-reductase** appears at puberty suggests this to be the reason why, in a few of the porphyrias, manifestations are not present at an earlier age.

24.7 HEME CATABOLISM

The catabolism of heme-containing proteins presents two requirements to the mammalian host: (a) the development of a means of processing the hydrophobic products of porphyrin ring cleavage and (b) the retention and mobilization of the contained iron so that it may be reutilized.

Red blood cells have a life span of approximately 120 days. Senescent cells are recognized by their membrane changes and removed and engulfed by the reticuloendothelial system at extravascular sites. The globin chains denature, releasing heme into the cytoplasm. The globin is degraded to its constituent amino acids, which are reutilized for general metabolic needs.

Figure 24.12 depicts the sequence of events of heme catabolism. Heme is degraded primarily by a microsomal enzyme system in reticuloendothelial cells that requires molecular oxygen and NADPH. Cytochrome *c* serves as the major vehicle for regenerating the NADPH utilized in the reaction. **Heme oxygenase** is substrate inducible. The enzyme specifically catalyzes the cleavage of the α-methene bridge, which joins the two pyrrole residues containing the vinyl substituents. The α-methene carbon is converted quantitatively to carbon monoxide. The only endogenous source of **carbon monoxide** in humans is the α-methene carbon. A fraction of the carbon monoxide is released via the respiratory tract. Thus the measurement of carbon monoxide in an exhaled breath provides an index to the quantity of heme that is degraded in an individual. The oxygen present in the carbon monoxide and in the newly derivatized lactam rings are generated entirely from molecular oxygen. The stoichiometry of the reaction requires 3 mol of oxygen for each ring cleavage. Heme oxygenase will only use heme as a substrate, with the iron possibly participating in the cleavage mechanism. Thus, free protoporphyrin IX is not a substrate. The linear tetrapyrrole **biliverdin IX** is the product formed by the action of heme oxygenase. Biliverdin IX is reduced by **biliverdin reductase** to bilirubin IX.

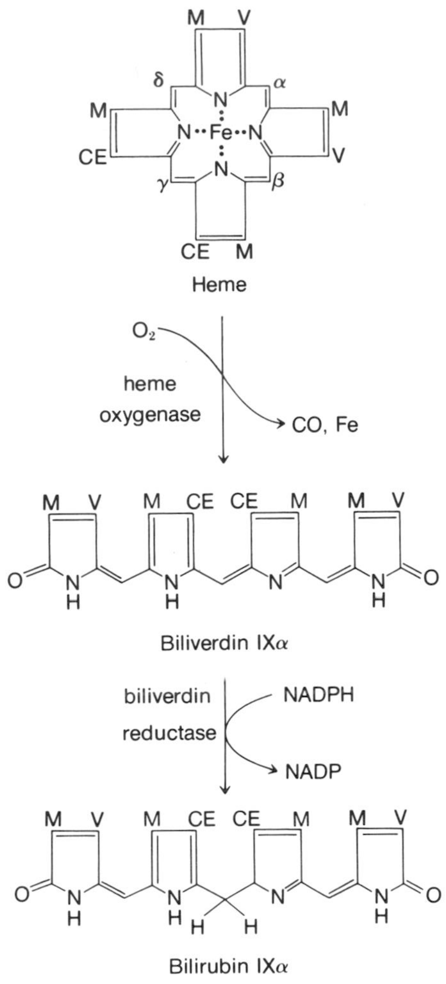

FIGURE 24.12
Formation of bilirubin from heme.
The Greek letters indicate the labeling of the methene carbon atoms in heme.

Bilirubin Is Conjugated to Form Bilirubin Diglucuroxide in Liver

Bilirubin is derived not only from senescent red cells but also from the turnover of other heme-containing proteins, such as the cytochromes. Studies with labeled glycine as a precursor have revealed that an early-labeled bilirubin, with a peak amount present within 1–3 h, appears a very short time after a pulsed administration of the labeled precursor. A larger amount of bilirubin appears much later at about 120 days, reflecting the turnover of heme in red blood cells. Early-labeled bilirubin can be divided into two parts: an early–early part, which reflects the turnover of heme proteins in the liver, and a late-early part, which consists of both the turnover of heme-containing hepatic proteins and the turnover of bone marrow heme, which is either poorly incorporated or easily released from red blood cells. The latter is a measurement of ineffective erythropoiesis and can be very pronounced in disease states such as pernicious anemia (see Chapter 28) and the thalassemias.

Bilirubin is poorly soluble in aqueous solutions at physiological pH values. When transported in plasma, it is bound to serum albumin with an association constant greater than 10^6 M^{-1}. Albumin contains one such high-affinity site and another with a lesser affinity. At the normal albumin concentration of 4 g dL^{-1}, about 70 mg of bilirubin per deciliter of plasma can be bound on the two sites. However, bilirubin toxicity **(kernicterus),** which is manifested by the transfer of bilirubin to membrane lipids, commonly occurs at concentrations greater than 25 mg dL^{-1}. This suggests that the weak affinity of the second site does not allow it to serve effectively in the transport of bilirubin. Bilirubin on serum albumin is rapidly cleared by the liver, where there is a free bidirectional flux of the tetrapyrrole across the sinusoidal–hepatocyte interface. Once in the hepatocyte, bilirubin is bound to several cytosolic proteins, of which only one has been well characterized. The latter component, **ligandin,** is a small basic component making up to 6% of the total cytoplasmic protein of rat liver. Ligandin has been purified to homogeneity from rat liver and characterized as having two subunits of molecular weights 22,000 and 27,000. Each subunit contains glutathione *S*-epoxide transferase activity, a function important in detoxification mechanisms of aryl groups. The stoichiometry of binding is one bilirubin residue per complete ligandin molecule. The functional role of ligandin and other hepatic bilirubin-binding proteins remains to be defined.

Once in the hepatocyte the β-carboxyethyl side chains of bilirubin are conjugated to form a diglucuronide (see Figure 24.13). The reaction mechanism includes the utilization of uridine diphosphoglucose, which is oxidized by a dehydrogenase to uridine diphosphoglucuronate. The latter serves as a glucuronate donor to bilirubin; different specific transferases form sequentially the mono- and diglucuronide adducts of bilirubin. The mammalian liver contains several **uridine diphosphoglucuronyl transferases,** each of which is substrate-specific for the acceptor molecule. In normal bile the diglucuronide is the major form of excreted bilirubin, with only small amounts present of the monoglucuronide or other glycosidic adducts. **Bilirubin diglucuronide** is much more water-soluble than free bilirubin, and thus the transferase facilitates the excretion of the bilirubin into bile. Bilirubin diglucuronide is poorly absorbed by the intestinal mu-

$$\text{UDP-Glucose} + 2\text{NAD}^+ \longrightarrow \text{UDP-glucuronate} + 2\text{NADH} + 2\text{H}^+$$

$$2\ \text{UDP-glucuronate} + \text{bilirubin IX}\alpha$$

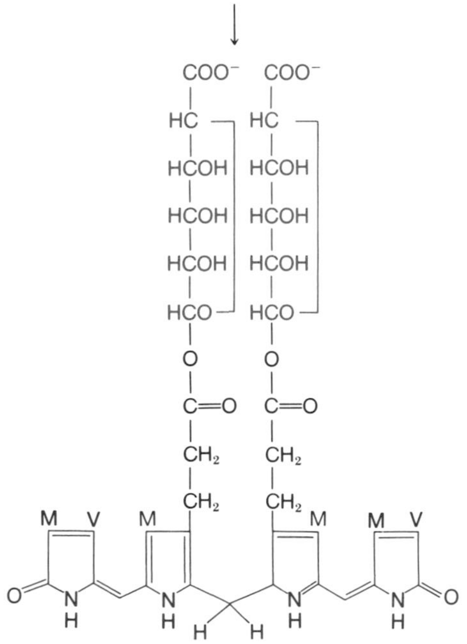

FIGURE 24.13
Biosynthesis of bilirubin diglucuronide.

cosa. The glucuronide residues are released in the terminal ileum and large intestine by bacterial hydrolases; the released free bilirubin is reduced to the colorless linear tetrapyrroles known as **urobilinogens.** Urobilinogens can be oxidized to colored products known as urobilins, which are excreted in the feces. A small fraction of urobilinogen can be reabsorbed by the terminal ileum and large intestine to be removed by hepatic cells and resecreted in bile. When urobilinogen is reabsorbed in large amounts in certain disease states, the kidney serves as a major excretory site.

In the normal state plasma bilirubin concentrations are 0.3–1 mg dL^{-1}, and this is almost all in the unconjugated state. In the clinical setting conjugated bilirubin is expressed as **direct bilirubin** because it can be coupled readily with diazonium salts to yield azo dyes; this is the direct **van den Bergh reaction.** Unconjugated bilirubin is bound noncovalently to albumin and will not react until it is released by the addition of an organic solvent such as ethanol. The reaction with diazonium salts yielding the azo dye after the addition of ethanol is the indirect van den Bergh reaction, and this measures the **indirect bilirubin** or the unconjugated bilirubin. Unconjugated bilirubin binds so tightly to serum albumin and lipid that it does not diffuse freely in plasma and therefore does not lead to an elevation of bilirubin in the urine. Unconjugated bilirubin has a high affinity for membrane lipids, which leads to the impairment of cell membrane function, especially in the nervous system. In contrast, conjugated bilirubin is relatively water-soluble, and elevations of this bilirubin form lead to high urinary concentrations with the characteristic deep yellow–brown color. The deposition of conjugated and unconjugated bilirubin in skin and the sclera gives the yellow to yellow–green color seen in patients with jaundice.

Recently a third form of plasma bilirubin has been described. This occurs only with hepatocellular disease in which a fraction of the bilirubin binds so tightly that it is not released from serum albumin by the usual techniques and is thought to be linked covalently to the protein. In some cases up to 90% of total bilirubin can be in this newly discovered covalently bound form.

The normal liver has a very large capacity to conjugate and mobilize the bilirubin that is delivered. As a consequence, hyperbilirubinemia due to excess heme destruction, as in hemolytic diseases, rarely leads to bilirubin levels that exceed 5 mg dL^{-1}, except in situations in which functional derangement of the liver is present (see Clin. Corr. 24.4). Thus, marked elevation of unconjugated bilirubin reflects primarily a variety of hepatic diseases, including those that are heritable and those that are acquired. A severe jaundice in infants occurs with the **Crigler–Najjar syndrome,** in which there exists a homozygous complete functional deficiency of the specific uridine diphosphoglucuronyl transferase for bilirubin. A low-grade mild hyperbilirubinemia known as **Gilbert's syndrome** occurs in adults. Although this disease has not been as well characterized as the Crigler–Najjar syndrome, one of the findings is a moderate reduction in bilirubin–uridine diphosphoglucuronyl-transferase activity.

Elevations of conjugated bilirubin level in plasma are attributable to liver and/or biliary tract disease. In simple uncomplicated biliary tract obstruction, the major component of the elevated serum bilirubin is the diglucuronide form, which is released by the liver into the vascular compartment. Biliary tract disease may be extrahepatic or intrahepatic, the latter involving the canaliculi and biliary ductules. The **Dubin-Johnson syndrome** is an autosomal recessive disease involving a defect in the biliary secretory mechanism of the liver. The excretion through the biliary tract of a variety of (but not all) organic anions is affected. The retention

CLIN. CORR. 24.4
NEONATAL ISOIMMUNE HEMOLYSIS

Rh negative women pregnant with Rh positive fetuses will develop antibodies to Rh factors. These antibodies will cross the placenta to hemolyze fetal red blood cells. Usually this is not of clinical relevance until about the third Rh positive pregnancy, in which the mother has had antigenic challenges with earlier babies. Antenatal studies will reveal rising maternal levels of IgG antibodies against Rh positive red blood cells, indicating that the fetus is Rh positive. Before birth, placental transfer of fetal bilirubin occurs with excretion through the maternal liver. Because hepatic enzymes of bilirubin metabolism are poorly expressed in the newborn, infants may not be able to excrete the large amounts of bilirubin that can be generated from red cell breakdown. At birth these infants usually appear normal; however, the unconjugated bilirubin in the umbilical cord blood is elevated up to 4 mg dL^{-1} due to the hemolysis initiated by maternal antibodies. During the next 2 days the serum bilirubin rises, reflecting continuing isoimmune hemolysis, leading to jaundice, hepatosplenomegaly, ascites, and edema. If untreated, signs of central nervous system damage can occur, with the appearance of lethargy, hypotonia, spasticity, and respiratory difficulty, constituting the syndrome of kernicterus. Treatment involves exchange transfusion with whole blood, which is serologically compatible with both the infant's blood and maternal serum. The latter requirement is necessary to prevent hemolysis of the transfused cells. Additional treatment includes external phototherapy, which facilitates the metabolism of bilirubin.

The entire problem can be prevented by treating Rh negative mothers with anti-Rh globulin. These antibodies recognize the fetal red cells, block the RH antigens, and cause them to be destroyed without stimulating an immune response in the mothers.

Mauer, H. M., Shumway, C. N., Draper, D. A., and Hossaini, A. A. Controlled trial comparing agar, intermittent phototherapy, and continuous phototherapy for reducing neonatal hyperbilirubinemia. *J. Pediatr.* 82:73, 1973; Bowman, J. J. Management of Rh-isoimmunization. *Obstet. Gynecol.* 52:1, 1978.

of undefined pigments in the liver in this disorder leads to a characteristic gray–black color of this organ. This pigment is apparently not derived from heme. A second heritable disorder associated with elevated levels of plasma-conjugated bilirubin is **Rotor syndrome.** In this poorly defined disease no hepatic pigmentation occurs, and the associated finding of increased secretion of urinary coproporphyrins I and III is found. The relationship of the derangement of coproporphyrin metabolism to bilirubin metabolism is not well understood.

Intravascular Hemolysis Requires Scavenging of Iron

In certain diseases destruction of red blood cells occurs in the intravascular compartment rather than in the extravascular reticuloendothelial cells. In the former case the appearance of free hemoglobin and heme in the plasma potentially could lead to the excretion of these substances through the kidney with a substantial loss of iron. To prevent this occurrence, specific plasma proteins are involved in scavenging mechanisms. Transferrin binds free iron and thus permits the reutilization of the metal. Free hemoglobin in the plasma leads to the following sequence of events. After oxygenation in the pulmonary capillaries, plasma oxyhemoglobin dissociates into α,β dimers, which are bound to a family of circulating plasma proteins, the **haptoglobins,** having a high affinity for the oxyhemoglobin dimer. Since deoxyhemoglobin does not dissociate into dimers in physiological settings, it is not bound by haptoglobin. The stoichiometry of binding is two α,β-oxyhemoglobin dimers per haptoglobin molecule. Interesting studies have been made with rabbit antihuman–hemoglobin antibodies on the haptoglobin–hemoglobin interaction. Human haptoglobin interacts with a variety of hemoglobins from different species. The binding of human haptoglobin with human hemoglobin is not affected by the binding of rabbit antihuman-hemoglobin antibody. These studies suggest that haptoglobin binds to sites on hemoglobin that are highly conserved in evolution and therefore are not sufficiently antigenic to generate antibodies. The most likely site for the molecular interaction of hemoglobin and haptoglobin is the interface of the α and β chains of the tetramer that dissociates to yield α,β dimers. Sequence determinations have indicated that these contact regions are highly conserved in evolution.

The haptoglobins are α_2-globulins. Made in the liver, they consist of two pairs of polypeptide chains (α being the lighter and β the heavier). The genes for the α and β chains are linked so that a single mRNA is synthesized, generating a single polypeptide chain that is cleaved to form the two different chains. The β chains are glycopeptides of 39,000 Da and are invariant in structure; α chains are of several kinds. The shorter α^{1S} chains each consist of 84 residues, with a molecular weight of about 9000 varying only in the residues at positions 54 where Glu is found in α^{1S} and Lys in α^{1F}. The α^2 chain consists of an incomplete fusion duplication of the genes for α^{1F} and α^{1S}, leading to a polypeptide chain with 143 residues. The haptoglobin peptide chains are joined by disulfide bonds between the α and β chains and between the two α chains. Thus, in contrast to the immunoglobulins, the disulfide bond between similar chains occurs with the smaller polypeptides. Because the α^2 chain has one more half-cystine residue than the α^1 chain, haptoglobins with the formula ${\alpha^2}_2\beta_2$ can polymerize into larger aggregates through disulfide linkages. There may not be symmetry in any single haptoglobin molecule; thus haptoglobins are known that have the following molecular structures: $\alpha^{1F}\alpha^{1S}\beta_2$ or $\alpha^{1F}\alpha^2\beta_2$. These variations in the structure of haptoglobins have been useful in analyses of population genetics.

The interaction of haptoglobin with hemoglobin leads to a complex that is too large to be filtered through the renal glomerulus. Free hemoglobin (appearing in the renal tubules and in the urine) will occur during intravascular hemolysis only when the binding capacity of circulating haptoglobin has been exceeded. Haptoglobin delivers hemoglobin to the reticuloendothelial cells. The heme in free hemoglobin is relatively resistant to the action of heme oxygenase, whereas the heme residues in an α,β dimer of hemoglobin when attached to haptoglobin are very susceptible. This enhancement of oxygenase activity is especially pronounced with the $\alpha^2{}_2\beta_2$ haptoglobin type. It has been suggested that the high proportion of haptoglobin $\alpha^2{}_2\beta_2$ in certain populations (such as southeast Asia), where there is a high incidence of hemolytic disease, is a reflection of a selective genetic advantage. The measurement of serum haptoglobin is used clinically as an indication of the degree of intravascular hemolysis. Patients who have significant intravascular hemolysis will have low or absent levels of haptoglobin because of the removal of haptoglobin–hemoglobin complexes by the reticuloendothelial system. Haptoglobin levels can also be low in severe extravascular hemolysis, in which the large load of hemoglobin in the reticuloendothelial system leads to the transfer of free hemoglobin into plasma.

Free heme and hematin appearing in plasma are bound by a β-globulin, **hemopexin,** which has a molecular weight of 57,000. One heme residue binds per hemopexin molecule. Hemopexin transfers heme to liver, where further metabolism by heme oxygenase occurs. In the normal state, hemopexin contains very little bound heme, whereas in intravascular hemolysis, the hemopexin is almost completely saturated by heme and is cleared with a half-life of about 7 h. In the latter instance excess heme will bind to albumin, with newly synthesized hemopexin serving as a mediator for the transfer of the heme from albumin to the liver. Hemopexin also binds free protoporphyrin.

BIBLIOGRAPHY

Battersby, A. R. The Bakerian Lecture, 1984. Biosynthesis of the pigments of life. *Proc. R. Soc. London B* 225:1, 1985.

Bothwell, T. H., Charlton, R. W., and Motulsky, A. G. Hemochromatosis, in C. R. Scriver, A. L. Beaudet, W. S. Sly, and D. Valle (Eds.). *The metabolic basis of inherited disease,* Vol. 1. New York: McGraw-Hill, 1989, p. 1433.

Casey, J. L., Hentze, M. W., Koeller, D. H., Caughman, S. Rouault, T. A., Klausner, R. D., and Harford, J. B. Iron-responsive elements: regulatory RNA sequences that control mRNA levels and translation. *Science* 240:924, 1988.

Chowdhury, J. R., Wolkoff, A. W., and Arias, I. M. Hereditary jaundice and disorders of bilirubin metabolism. C. R. Scriver, A. L. Beaudet, W. S. Sly, and D. Valle (Eds.). *The metabolic basis of inherited disease,* Vol. 1. New York: McGraw-Hill, 1989, p. 1367.

Gordon, E. R., Sommerer, U., and Goresky, C. A. The hepatic microsomal formation of bilirubin diglucuronide. *J. Biol. Chem.* 258:15028, 1983.

Hentze, M. W., et al. Identification of the iron-responsive element for translational regulation of human ferritin mRNA. *Science* 238:1570, 1987.

Hentze, M. W., Rouault, T. A., Harford, J. B., and Klausner, R. D. Oxidation-reduction and the molecular mechanism of a regulatory RNA–protein interaction. *Science* 244:357, 1989.

Huebers, H. A. and Finch, C. A. Transferrin: Physiologic behavior and clinical implications. *Blood* 64:763, 1984.

Kappas, A., Sassa, S., Galbraith, R. A., and Nordmann, V. The porphyrias. C. R. Scriver, A. L. Beaudet, W. S. Sly, and D. Valle (Eds.). *The metabolic basis of inherited disease,* Vol. 1. New York: McGraw-Hill, 1989, p. 1305.

Koeller, D. M., et al. A cytosolic protein binds to structural elements within the iron regulatory region of the transferrin receptor mRNA. *Proc. Natl. Acad. Sci. USA* 86:3574, 1989.

Lustbader, J. W., Arcoleo, J. P., Birken, S., and Greer, J. Hemoglobin-binding site on haptoglobin probed by selective proteolysis *J. Biol. Chem.* 258:1227, 1983.

Maeda, N., Yang, F., Barnett, D. R., Bowman, B. H., and Smithies, O. Duplication within the haptoglobin Hp^2 gene. *Nature* (*London*). 309:131, 1984.

May, W. S., Sahyoun, N., Jacobs, S., Wolf, M., and Cuatracasas, P. Mechanism of phorbol-diester-induced regulation of surface transferrin receptor involves the action of activated protein kinase C and an intact cytoskeleton. *J. Biol. Chem.* 260:9419, 1985.

Mullner, E. W., and Kuhn, L. C. A stem–loop in the 3′ untranslated region mediates iron–dependent regulation of transferrin receptor mRNA stability in the cytoplasm. *Cell* 53:815, 1988.

Osterman, J., Horwich, A. L., Neupert, W., and Hartl, F.-U. Protein folding in mitochondria requires complex formation with hsp60 and ATP hydrolysis. *Nature* (*London*). 341:125, 1989.

Rouault, T. A., Hentze, M. W., Haile, D., Harford, J. B., and Klausner, R. D. The iron-response element binding protein: A method for the affinity purification of a regulatory RNA-binding protein. *Proc. Natl. Acad. Sci. USA* 86:5768, 1989.

Weiss, J. S., et al. The clinical importance of a protein-bound fraction of serum bilirubin in patients with hyperbilirubinemia. *N. Engl. J. Med.* 309:147, 1983.

Yamashiro, D. J., Tycko, B., Fluss, S. R., and Maxfield, F. R. Segregation of transferrin to a mildly acidic (pH 6.5) para-golgi compartment in the recycling pathway. *Cell* 37:789, 1984.

QUESTIONS

C. N. ANGSTADT AND J. BAGGOTT

*Q*uestion *T*ypes are described on page xxiii.

A. Ferritin
B. Ferredoxin
C. Hemosiderin
D. Lactoferrin
E. Transferrin

1. (QT5) A type of protein in which iron is specifically bound to sulfur
2. (QT5) Exhibits an antimicrobial effect in the intestinal tract of newborns because of its ability to bind iron
3. (QT5) Delivers iron to tissues by binding to specific cell surface receptors

4. (QT2) In the intestinal absorption of iron:
   1. the presence of ascorbate enhances the availability of the iron.
   2. the regulation of uptake occurs between the lumen and the mucosal cells.
   3. the amount of apoferritin synthesized in the mucosal cell is related inversely to the need for iron by the host.
   4. iron bound tightly to a ligand, such as phytate, is more readily absorbed than free iron.

5. (QT1) Iron overload:
   A. cannot occur because very efficient excretory mechanisms are available.
   B. occurs because cells cannot regulate their uptake of iron with changing iron content.
   C. would be accompanied by an increase in total serum transferrin.
   D. might be caused by the ingestion of large amounts of iron along with alcohol (ethanol).
   E. might be caused by the ingestion of iron chelators.

6. (QT2) The biosynthesis of heme requires:
   1. succinyl CoA.
   2. glycine.
   3. ferrous ion.
   4. propionic acid.

7. (QT1) Uroporphyrin III:
   A. is an intermediate in the biosynthesis of heme.
   B. does not contain a tetrapyrrole ring.
   C. differs from coproporphyrin III in the substituents around the ring.
   D. is formed from uroporphyrinogen III by an oxidase.
   E. formation is the primary control step in heme synthesis.

8. (QT2) Aminolevulinic acid synthase:
   1. requires pyridoxal phosphate for activity.
   2. is allosterically inhibited by heme.
   3. synthesis can be induced by a variety of drugs.
   4. is synthesized in the cytoplasm but catalyzes a mitochondrial reaction.

9. (QT1) Lead poisoning would be expected to result in an elevated level of:
   A. aminolevulinic acid.
   B. porphobilinogen.
   C. protoporphyrin I.
   D. heme.
   E. bilirubin.

10. (QT1) Ferrochelatase:
    A. is an iron-chelating compound.
    B. releases iron from heme in the degradation of hemoglobin.
    C. binds iron to sulfide ions and cysteine residues.
    D. is inhibited by heavy metals.
    E. is involved in the cytoplasmic portion of heme synthesis.

11. (QT2) Heme oxygenase:
    1. can oxidize the methene bridge between any two pyrrole rings of heme.
    2. requires molecular oxygen.
    3. produces bilirubin.
    4. produces carbon monoxide.

A. Direct bilirubin   C. Both
B. Indirect bilirubin   D. Neither

12. (QT4) Deposited in skin and sclera in jaundice
13. (QT4) Hepatic disease leads to major elevation of blood level of ___
14. (QT4) Biliary obstruction leads to major elevation of blood level of ___
15. (QT4) Acute intermittent porphyria is accompanied by increased blood level of ___

16. (QT1) Haptoglobin binds:
    A. a globin monomer.
    B. an oxyhemoglobin molecule.
    C. α,β-oxyhemoglobin dimers.
    D. a deoxyhemoglobin molecule.
    E. α,β-deoxyhemoglobin dimers.

17. (QT2) Haptoglobin:
    1. helps prevent loss of iron following intravascular red blood cell destruction.
    2. levels in serum are elevated in severe intravascular hemolysis.
    3. facilitates the action of heme oxygenase.
    4. binds heme and hematin as well as hemoglobin.

ANSWERS

1. B Animal ferredoxins, also known as nonheme iron-containing proteins, have two irons bound to two cysteine residues and sharing two sulfide ions (p. 1004).
2. D As long as lactoferrin is not saturated, its avid binding of iron diminishes the amount available for growth of microorganisms (p. 1003).
3. E Internalization of the receptor–transferrin complex is mediated by a Ca^{+2}–calmodulin–protein kinase C complex. Internalization is followed by release of the iron and recycling of the apotransferrin to the plasma (p. 1003). Ferritin and hemosiderin (p. 1004) are storage forms of iron.
4. B 1, 3 correct. 1, 4: Ascorbate facilitates reduction to the ferrous state and, therefore, dissociation from ligands and absorption. 2: Substantial iron enters the mucosal cell regardless of need, but the amount transferred to the capillary beds is controlled. 3: Iron bound to apoferritin is trapped in mucosal cells and not transferred to the host (p. 1005).
5. D Ethanol enhances the absorption of iron and could lead to an overload. A: The high affinity of many macromolecules for iron prevents efficient excretion B: A regulatory protein, whose activity responds to changes in iron content, affects the mRNAs for transferrin receptor and ferritin (p. 1007). C: Transferrin increases in iron deficiency to improve absorption. E: This is the treatment for iron overload when bleeding is inappropriate (p. 1007, 1009).
6. A 1–3 correct. 1, 2, 4: The organic portion of heme comes totally from glycine and succinyl CoA; the propionic acid side chain comes from the succinate. 3: The final step of heme synthesis is the insertion of the ferrous ion (p. 1009, Figure 24.6).
7. C A, B, D: The tetrapyrrole porphyrins (except for protoporphyrin IX) are not intermediates but end products formed from the porphyrinogens nonenzymatically. E: The synthesis of aminolevulinic acid is the rate-limiting step (p. 1009), Figure 24.6).
8. E All four correct. 1: The mechanism involves a Schiff's base with glycine. 2, 3: Heme both allosterically inhibits and suppresses synthesis of the enzyme, which can also be induced in response to need (many drug detoxications are cytochrome P450 dependent). 4: The gene for this enzyme is on nuclear DNA (p. 1011).
9. A A–D: Lead inhibits ALA dehydrase so it inhibits synthesis of porphobilinogen and subsequent compounds. Heme certainly would not be elevated, because lead also inhibits ferrochelatase. E: Bilirubin is a breakdown product of heme, not an intermediate in synthesis (p. 1013).
10. D This enzyme, in the mitochondria, catalyzes the last step of heme synthesis, the insertion of Fe^{2+}, and is sensitive to the effects of heavy metals (p. 1016).
11. C 2, 4 correct. 1: The enzyme is specific for the methene between the two rings containing the vinyl groups (α-methene bridge). 2–4: It uses O_2 and the products are biliverdin and CO; the measurement of CO in the breath is an index of heme degradation (p. 1017).
12. C Both conjugated (direct) and unconjugated (indirect) bilirubin are deposited (p. 1019).
13. B Since the liver is responsible for conjugating bilirubin, hepatic disease leads to the elevation of unconjugated (indirect) bilirubin in blood (p. 1019).
14. A Conjugated (direct) bilirubin is excreted in the bile (p. 1019).
15. D Bilirubin reflects heme catabolism; porphyrias reflect a derangement of synthesis (Clin. Corr. 24.3).
16. C Haptoglobin binds dimers, two per haptoglobin molecule, specifically the oxyhemoglobin dimers since deoxyhemoglobin does not dissociate to dimers physiologically (p. 1020).
17. B 1, 3 correct. 1, 2: Haptoglobin is part of the scavenging mechanism to prevent urinary loss of heme and hemoglobin from intravascular degradation of red blood cells. Since the complex is taken up by the reticuloendothelial system, the haptoglobin levels in serum are low. It also prevents clogging of the glomerular filter by hemoglobin. 3: Heme residues in the dimers bound to haptoglobin are more susceptible than free heme to oxidation by heme oxygenase. 4: Heme and hematin are bound by a β-globulin, while haptoglobin is an α-globulin (p. 1020).

25

Gas Transport and pH Regulation

JAMES BAGGOTT

25.1 INTRODUCTION TO GAS TRANSPORT 1025
25.2 NEED FOR A CARRIER OF OXYGEN IN THE BLOOD 1026
25.3 HEMOGLOBIN AND ALLOSTERISM: EFFECT OF 2,3-DIPHOSPHOGLYCERATE 1029
25.4 OTHER HEMOGLOBINS 1030
25.5 PHYSICAL FACTORS THAT AFFECT OXYGEN BINDING 1032
High Temperature Weakens Hemoglobin's Oxygen Affinity 1032
Low pH Weakens Hemoglobin's Oxygen Affinity 1032
25.6 CARBON DIOXIDE TRANSPORT 1033
25.7 INTERRELATIONSHIPS AMONG HEMOGLOBIN, OXYGEN, CARBON DIOXIDE, HYDROGEN ION, AND 2,3-DIPHOSPHOGLYCERATE 1038
25.8 INTRODUCTION TO pH REGULATION 1039
25.9 BUFFER SYSTEMS OF PLASMA, INTERSTITIAL FLUID, AND CELLS 1039
25.10 THE CARBON DIOXIDE–BICARBONATE BUFFER SYSTEM 1041
25.11 ACID–BASE BALANCE AND ITS MAINTENANCE 1045
The Kidney Plays a Critical Role in Acid–Base Balance 1046
25.12 COMPENSATORY MECHANISMS 1049
25.13 ALTERNATIVE MEASURES OF ACID–BASE IMBALANCE 1052
25.14 THE SIGNIFICANCE OF Na^+ AND Cl^- IN ACID–BASE IMBALANCE 1054
BIBLIOGRAPHY 1056
QUESTIONS AND ANNOTATED ANSWERS 1057
CLINICAL CORRELATIONS
25.1 Cyanosis 1029
25.2 Chemically Modified Hemoglobins: Methemoglobin and Sulfhemoglobin 1031
25.3 Hemoglobins With Abnormal Oxygen Affinity 1032
25.4 The Case of the Variable Constant 1042
25.5 The Role of Bone in Acid–Base Homeostasis 1046
25.6 Acute Respiratory Alkalosis 1051
25.7 Chronic Respiratory Acidosis 1052
25.8 Salicylate Poisoning 1053
25.9 Evaluation of Clinical Acid–Base Data 1055
25.10 Metabolic Alkalosis 1056

25.1 INTRODUCTION TO GAS TRANSPORT

Large organisms, especially terrestrial ones, require a relatively tough, impermeable outer covering to help ward off dust, twigs, nonisotonic fluids like rain and seawater, and other elements of the environment that would be harmful to living cells. One of the consequences of being large and having an impermeable covering is that individual cells of the organism cannot exchange gases directly with the atmosphere. Instead there must exist a specialized exchange surface, such as a lung or a gill, and a system to circulate the gases (and other materials, such as nutrients and waste products) in a manner that will meet the needs of every living cell in the body.

The existence of a system for the transport of gases from the atmosphere to cells deep within the body is not merely necessary, it has definite advantages. Oxygen is a good oxidizing agent, and at its partial pressure in the atmosphere, about 160 mmHg or 21.3 kPa [1 mmHg = 1 torr = 0.133 kPa (kilopascal)], it would oxidize and inactivate many of the components of the cells, such as essential sulfhydryl groups of enzymes. By the time O_2 gets through the transport system of the body its partial pressure is reduced to a much less damaging 20 mmHg (2.67 kPa) or less. In contrast, CO_2 is relatively concentrated in the body and becomes diluted in transit to the atmosphere. In the tissues, where it is produced, its partial pressure is 46 mmHg (6.13 kPa) or more. In the lungs it is 40 mmHg (5.33 kPa), and in the atmosphere only 0.2 mmHg (0.03 kPa), less abundant than the rare gas, argon. Its relatively high concentration in the body permits it to be used as one component of a physiologically important buffering system, a system that is particularly useful because, upon demand, the concentration of CO_2 in the extracellular fluid can be varied over a rather wide range. This is discussed in more detail later in this chapter.

Oxygen and CO_2 are carried between the lungs and the other tissues by the blood. In the blood some of each gas is present in simple physical solution, but mostly each is involved in some sort of interaction with hemoglobin, the major protein of the red blood cell. There is a reciprocal relation between hemoglobin's affinity for O_2 and CO_2, so that the relatively high level of O_2 in the lungs aids the release of CO_2, which is to be expired, and the high CO_2 level in other tissues aids the release of O_2 for use by those tissues. Thus, a description of the physiological transport of O_2 and CO_2 is the story of the interaction of these two compounds with hemoglobin.

25.2 NEED FOR A CARRIER OF OXYGEN IN THE BLOOD

An O_2 carrier is needed in the blood simply because O_2 is not soluble enough in blood plasma to meet the body's needs. At 38°C 1 L of plasma will dissolve only 2.3 mL of O_2. Whole blood, with its **hemoglobin,** has a much greater oxygen capacity. One liter of blood normally contains about 150 g of O_2 hemoglobin, and each gram of hemoglobin can combine with 1.34 mL of O_2. Thus the hemoglobin in 1 L of blood can carry 200 mL of O_2, 87 times as much as plasma alone would carry. Without an O_2 carrier, the blood would have to circulate 87 times as fast to provide the same capacity to deliver O_2. As it is, the blood makes a complete circuit of the body in 60 s under resting conditions, and in the aorta it flows at the rate of about 18.6 m s^{-1}. An 87-fold faster flow would require a fabulous high-pressure pump, would produce tremendously turbulent flow and high shear forces in the plasma, would result in uncontrollable bleeding from wounds, and would not even allow the blood enough time in the lungs to take up O_2. The availability of a carrier not only permits us to avoid these impracticalities, but also gives us a way of controlling oxygen delivery, since the O_2 affinity of the carrier is responsive to changing physiological conditions.

The respiratory system includes the trachea, in the neck, which bifurcates in the thorax into right and left bronchi, as shown schematically in

Figure 25.1. The bronchi continue to bifurcate into smaller and smaller passages, ending with tiny bronchioles, which open into microscopic gas-filled sacs called alveoli. It is in the alveoli that gas exchange takes place with the alveolar capillary blood.

As we inhale and exhale, the alveoli do not appreciably change in size. Rather, it is the airways that change in length and diameter as the air is pumped into and out of the lungs. Gas exchange between the airways and the alveoli then proceeds simply by diffusion. These anatomical and physiological facts have two important consequences. In the first place, since the alveoli are at the ends of long tubes that constitute a large dead space, and the gases in the alveoli are not completely replaced by fresh air with each breath, the gas composition of the alveolar air differs from that of the atmosphere, as shown in Table 25.1. Oxygen is lower in the alveoli because it is removed by the blood. Carbon dioxide is higher because it is added. Since we do not usually breathe air that is saturated with water vapor at 38°C, water vapor is generally added in the airways. The level of nitrogen is lower in the alveoli, not because it is taken up by the body, but simply because it is diluted by the CO_2 and water vapor.

A second consequence of the existence of alveoli of essentially constant size is that the blood that flows through the pulmonary capillaries during expiration, as well as the blood that flows through during inspiration, can exchange gases. This would not be possible if the alveoli collapsed during expiration and contained no gases, in which case the composition of the blood gases would fluctuate widely, depending on whether the blood passed through the lungs during an inspiratory or expiratory phase of the breathing cycle.

We have seen that an O_2 carrier is necessary. Clearly this carrier would have to be able to bind oxygen at an O_2 tension of about 100 mmHg (13.3 kPa), the partial pressure of oxygen in the alveoli. The carrier must also be able to release O_2 to the extrapulmonary tissues. The O_2 tension in the capillary bed of an active muscle is about 20 mmHg (2.67 kPa). In resting muscle it is higher, but during extreme activity it is lower. These O_2 tensions represent the usual limits within which an oxygen carrier must work. An efficient carrier would be nearly fully saturated in the lungs, but should be able to give up most of this to a working muscle.

Let us first see whether a carrier that binds O_2 in a simple equilibrium represented by

$$\text{Oxygen} + \text{carrier} \rightleftharpoons \text{oxygen} \cdot \text{carrier}$$

would be satisfactory. For this type of carrier the dissociation constant would be given by the simple expression

$$K_d = \frac{[\text{oxygen}][\text{carrier}]}{[\text{oxygen} \cdot \text{carrier}]}$$

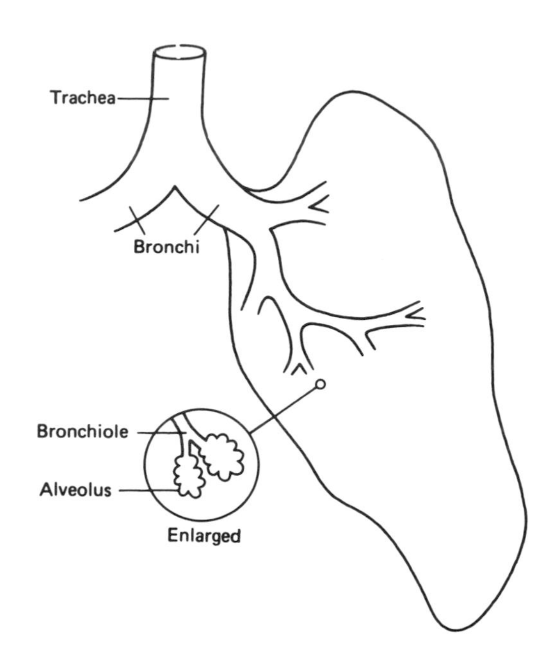

FIGURE 25.1
Diagram showing the respiratory tract.

TABLE 25.1 Partial Pressures of Important Gases Given in Millimeters of Hg, (kPa)

	In the atmosphere		In the alveoli of the lungs	
Gas	mmHg	kPa	mmHg	kPa
O_2	159	21.2	100	13.3
N_2	601	80.1	573	76.4
CO_2	0.2	0.027	40	5.33
H_2O	0	0	47	6.27
Total	760	101	760	101

and the saturation curve would be a **rectangular hyperbola.** This model would be valid even for a carrier with several oxygen-binding sites per molecule, which we know is the case for hemoglobin, as long as each site were independent and not influenced by the presence or absence of O_2 at adjacent sites.

If such a carrier had a dissociation constant that permitted 90% saturation in the lungs, then, as shown in Figure 25.2, Curve *A*, at a partial pressure of 20 mmHg (2.67 kPa) it would still be 66% saturated, and would have delivered only 27% of its O_2 load. This would not be very efficient.

What about some other simple carrier, one that bound O_2 less tightly and therefore released most of it at low partial pressure, so that the carrier was, say, only 20% saturated at 20 mmHg (2.67 kPa)? Again, as shown in Figure 25.2, Curve *B*, it would be relatively inefficient; in the lungs this carrier could fill only 56% of its maximum O_2 capacity, and would deliver only 36% of what it could carry. It appears then that the mere fivefold change in O_2 tension between the lungs and the unloading site is not compatible with efficient operation of a simple carrier. Simple carriers are not sensitive enough to respond massively to a signal as small as a fivefold change.

Figure 25.2 also shows the oxygen-binding curve of hemoglobin in normal blood. The curve is **sigmoid,** not hyperbolic, and it cannot be described by a simple equilibrium expression. Hemoglobin, however, is a very good physiological O_2 carrier. It is 98% saturated in the lungs and only about 33% saturated in the working muscle. Under these conditions it delivers about 65% of the O_2 it can carry.

It can be seen in Figure 25.2 that hemoglobin is 50% saturated with O_2 at a partial pressure of 27 mmHg (3.60 kPa). The partial pressure corresponding to 50% saturation is called the $\mathbf{P_{50}}$. The term P_{50} is the most common way of expressing hemoglobin's O_2 affinity. In analogy with K_m for enzymes, a relatively high P_{50} corresponds to a relatively low O_2 affinity.

It is important to notice that the steep part of hemoglobin's saturation curve lies in the range of O_2 tensions that prevail in the extrapulmonary tissues of the body. This means that relatively small decreases in oxygen tension in these tissues will result in large increases in O_2 delivery. Furthermore, small shifts of the curve to the left or right will also strongly influence O_2 delivery. In Sections 25.3, 25.5, and 25.6 we see how physiological signals effect such shifts and result in enhanced delivery under conditions of increased O_2 demand. Small decreases of O_2 tension in the

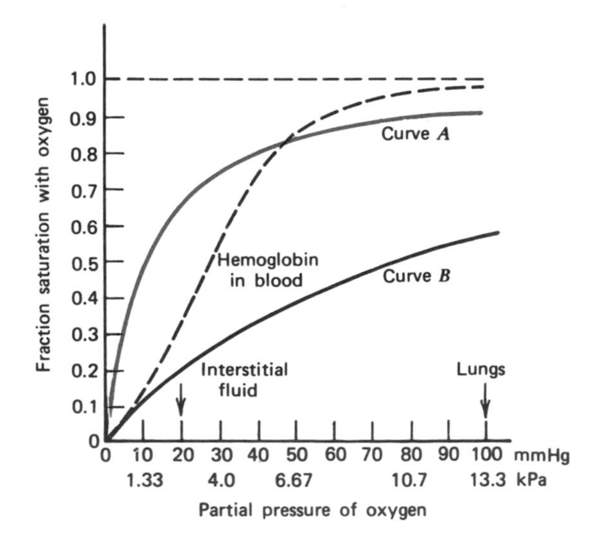

FIGURE 25.2
Oxygen saturation curves for two hypothetical oxygen carriers and for hemoglobin.
Curve *A*: Hypothetical carrier with hyperbolic saturation curve (a simple carrier) 90% saturated in the lungs and 66% saturated at the partial pressure found in interstitial fluid. Curve *B*: Hypothetical carrier with hyperbolic saturation curve (another simple carrier), 56% saturated in the lungs and 20% saturated at the partial pressure found in interstitial fluid. Dotted curve: Hemoglobin in whole blood.

lungs, however, such as occur at moderately high altitudes, do not seriously compromise hemoglobin's ability to bind oxygen. This will be true as long as the alveolar partial pressure of O_2 remains in a range that corresponds to the relatively flat region of hemoglobin's O_2 dissociation curve (Clin. Corr. 25.1).

Finally, we can see from Figure 25.2 that the binding of oxygen by hemoglobin is cooperative. At very low O_2 tension the hemoglobin curve tends to follow the hyperbolic curve which represents relatively weak O_2 binding, but at higher tensions it actually rises above the hyperbolic curve that represents tight binding. Thus it can be said that hemoglobin binds O_2 weakly at low oxygen tension and tightly at high tension. *The binding of the first O_2 to each hemoglobin molecule enhances the binding of subsequent O_2 atoms.*

Hemoglobin's ability to bind O_2 cooperatively is reflected in its **Hill coefficient,** which has a value of about 2.7. (The Hill equation is derived and interpreted on p. 127.) Since the maximum value of the Hill coefficient for a system at equilibrium is equal to the number of cooperating binding sites, a value of 2.7 means that hemoglobin, with its four oxygen-binding sites, is more cooperative than would be possible for a system with only two cooperating binding sites, but it is not as cooperative as it could be.

> **CLIN. CORR. 25.1**
> **CYANOSIS**
>
> Cyanosis is a condition in which a patient's skin or mucous membrane appears gray or (in severe cases) purple-magenta. It is due to an abnormally high concentration of deoxyhemoglobin below the surface, which is responsible for the observed color. The familiar blue of superficial veins is due to their deoxyhemoglobin content and is a normal manifestation of this color effect.
>
> Cyanosis is most commonly caused by diseases of the cardiac or pulmonary systems, resulting in inadequate oxygenation of the blood. It can also be caused by certain hemoglobin abnormalities. Severely anemic individuals cannot become cyanotic; they do not have enough hemoglobin in their blood for the characteristic color of its deoxy form to be apparent.
>
> Albert, R. K. Approach to the patient with cyanosis and/or hypoxemia. *Textbook of internal medicine,* W. N. Kelley (Ed.). Philadelphia: J. B. Lippincott, 1989, pp. 2041–2044.

25.3 HEMOGLOBIN AND ALLOSTERISM: EFFECT OF 2,3-DIPHOSPHOGLYCERATE

Hemoglobin's binding of O_2 was the original example of a **homotropic effect** (cooperativity and allosterism are discussed in Chapter 4), but hemoglobin also exhibits a **heterotropic effect** of great physiological significance. This involves its interaction with **2,3-diphosphoglycerate** (DPG). Figure 25.3 shows the structure of DPG; it is closely related to the glycolytic intermediate, 1,3-diphosphoglycerate, from which it is in fact biosynthesized.

It had been known for many years that hemoglobin in the red cell bound oxygen less tightly than purified hemoglobin could (Figure 25.4). It had also been known that the red cell contained high levels of DPG, nearly equimolar with hemoglobin. Finally, the appropriate experiment was done to demonstrate the relationships between these two facts. It was shown that the addition of DPG to purified hemoglobin produced a shift to the right of its oxygen-binding curve, bringing it into congruence with the curve observed in whole blood. Other organic polyphosphates, such as ATP and inositol pentaphosphate, also have this effect. Inositol pentaphosphate is the physiological effector in birds, where it replaces DPG, and ATP plays a similar role in some fish.

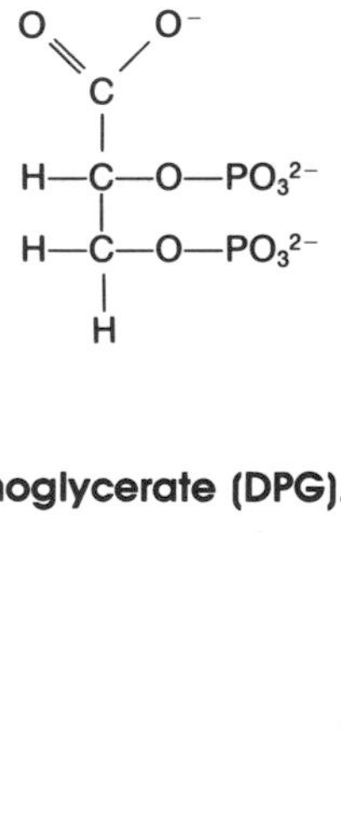

O   O⁻
C
H—C—O—PO_3^{2-}
H—C—O—PO_3^{2-}
H

FIGURE 25.3
2,3-Diphosphoglycerate (DPG).

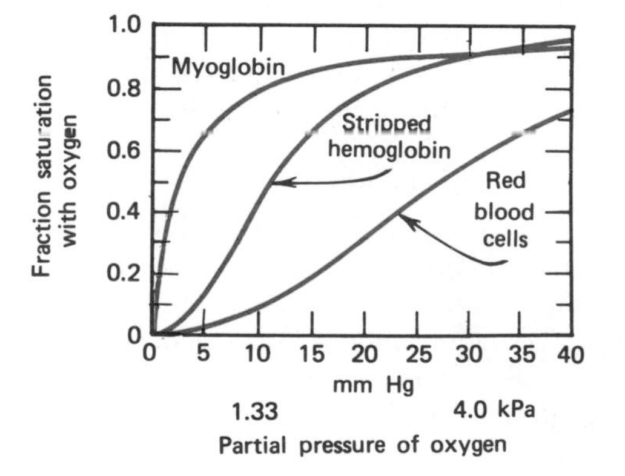

FIGURE 25.4
Oxygen dissociation curves for myoglobin, for hemoglobin that has been stripped of CO_2 and organic phosphates, and for whole red blood cells.
Data from O. Brenna, et al. *Adv. Exp. Biol. Med.* 28:19, 1972. Adapted from R. W. McGilvery. *Biochemistry: A functional approach,* 2nd ed. Philadelphia: W. B. Saunders, 1979, p. 236.

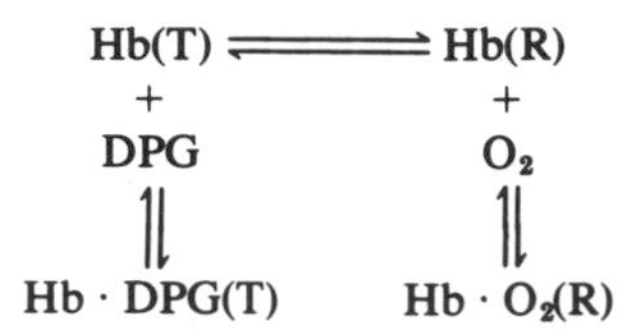

FIGURE 25.5
Schematic representation of equilibria among DPG, O_2, and the T and R states of hemoglobin.

Monod's model of allosterism explains heterotropic interaction. Applying this model to hemoglobin, in the deoxy conformation (the **T state**) a cavity large enough to admit DPG exists between the β chains of hemoglobin. This cavity is lined with positively charged groups and firmly binds one molecule of the negatively charged DPG. In the oxy conformation (the **R state**) this cavity is smaller, and it no longer accommodates DPG as easily. The result is that the binding of DPG to oxyhemoglobin is much weaker. Since DPG binds preferentially to the T state, the presence of DPG shifts the R–T equilibrium in favor of the T state; the deoxyhemoglobin conformation is thus stabilized over the oxyhemoglobin conformation (Figure 25.5). For oxygen to overcome this and bind to hemoglobin, a higher concentration of oxygen is required. Oxygen tension in the lungs is sufficiently high under most conditions to saturate hemoglobin almost completely, even when DPG levels are high. The physiological effect of DPG, therefore, can be expected to be upon release of oxygen in the extrapulmonary tissues, where O_2 tensions are low.

The significance of a high DPG concentration is that the efficiency of O_2 delivery is increased. Levels of DPG in the red cell rise in conditions associated with **tissue hypoxia,** such as various anemias, cardiopulmonary insufficiency, and high altitude. These high levels of DPG enhance the formation of deoxyhemoglobin at low partial pressures of oxygen; hemoglobin then delivers more of its O_2 to the tissues. This effect can result in a substantial increase in the amount of O_2 delivered because the venous blood returning to the heart of a normal individual is (at rest) at least 60% saturated with O_2. Much of this O_2 can dissociate in the peripheral tissues if the DPG concentration rises.

The DPG mechanism works very well as a compensation for tissue hypoxia as long as the partial pressure of oxygen in the lungs remains high enough that oxygen binding in the lungs is not compromised. Since, however, the effect of DPG is to shift the oxygen-binding curve to the right, the mechanism will not compensate for tissue hypoxia when the partial pressure of O_2 in the lungs falls too low. Then the increased efficiency of O_2 unloading in the tissues is counterbalanced by a decrease in the efficiency of loading in the lungs. This may be a factor in determining the maximum altitude at which people choose to establish permanent dwellings, which is about 18,000 ft (~5500 m). There is evidence that a better adaptation to extremely low ambient partial pressures of O_2 would be a shift of the curve to the left.

25.4 OTHER HEMOGLOBINS

Although hemoglobin A is the major form of hemoglobin in adults and in children over 7 months of age, accounting for about 90% of their total hemoglobin, it is not the only normal hemoglobin species. Normal adults also have 2–3% of **hemoglobin A_2,** which is composed of two α chains like those in hemoglobin A and two δ chains. It is represented as $\alpha_2\delta_2$. The δ chains are distinct from the β chains, and are under independent genetic control. Hemoglobin A_2 does not appear to be particularly important in normal individuals.

Several species of modified hemoglobin A also occur normally. These are designated A_{Ia1}, A_{Ia2}, A_{Ib}, and A_{Ic}. They are adducts of hemoglobin with various sugars, such as glucose, glucose 6-phosphate, and fructose 1,6-bisphosphate. The quantitatively most significant of these is **hemoglobin A_{Ic}.** It arises from the covalent binding of a glucose residue to the N terminal of the β chain. The reaction is not enzyme catalyzed, and its rate depends on the concentration of glucose. As a result, hemoglobin A_{Ic}

forms more rapidly in uncontrolled diabetics and can comprise up to 12% of their total hemoglobin under some circumstances. Hemoglobin A_{1c} levels or total glycosylated hemoglobin levels have become a useful measure of how well diabetes has been controlled during the days and weeks before the measurement is taken; measurement of blood glucose only indicates how well it is under control at the time the blood sample is taken. Chemical modification of hemoglobin A can also occur due to interaction with drugs or environmental pollutants (Clin. Corr. 25.2).

Fetal hemoglobin, **hemoglobin F,** is the major hemoglobin component of the newborn. It contains two γ chains in place of the β chains, and is represented as $\alpha_2\gamma_2$. Shortly before birth γ chain synthesis diminishes and β chain synthesis is initiated, and by the age of 7 months well over 90% of the infant's hemoglobin is hemoglobin A.

Hemoglobin F is adapted to the environment of the fetus, who must get oxygen from the maternal blood, a source that is far poorer than the atmosphere. In order to compete with the maternal hemoglobin for O_2, fetal hemoglobin must bind O_2 more tightly; its oxygen-binding curve is thus shifted to the left relative to hemoglobin A. This is accomplished not through an intrinsic difference in the oxygen affinities of these hemoglobins but through a difference in the influence of DPG upon them. In hemoglobin F two of the groups that line the DPG-binding cavity have neutral side chains instead of the positively charged ones that occur in hemoglobin A. Consequently, hemoglobin F binds DPG less tightly and thus binds oxygen more tightly than hemoglobin A does. Furthermore, about 15–20% of the hemoglobin F is acetylated at the N terminals; this is referred to as hemoglobin F_1. Hemoglobin F_1 does not bind DPG, and its affinity for oxygen is not affected at all by DPG. The postnatal change from hemoglobin F to hemoglobin A, combined with a rise in red cell DPG that peaks 3 months after birth, results in a gradual shift to the right of the infant's oxygen-binding curve (Figure 25.6). The result is greater delivery of oxygen to the tissues at this age than at birth, in spite of a 30% decrease in the infant's total hemoglobin concentration.

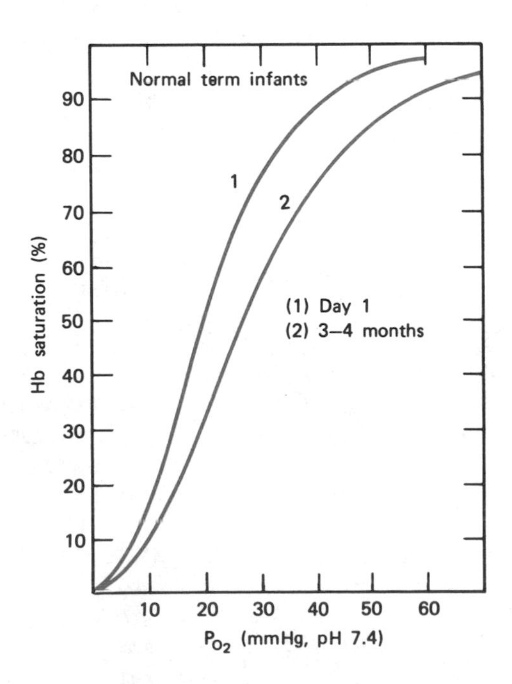

FIGURE 25.6
Oxygen dissociation curves after birth.
Adapted from F. A. Oski and M. Delivoria-Papadopoulos, *J. Pediatr.* 77:941, 1970.

CLIN. CORR. 25.2
CHEMICALLY MODIFIED HEMOGLOBINS: METHEMOGLOBIN AND SULFHEMOGLOBIN

Methemoglobin is a form of hemoglobin in which the iron is oxidized from the iron (II) state to the iron (III) state. A tendency for methemoglobin to be present in excess of its normal level of about 1% may be due to a hereditary defect of the globin chain or to exposure to oxidizing drugs or chemicals.

Sulfhemoglobin is a species that forms when a sulfur atom is incorporated into the porphyrin ring of hemoglobin. Exposure to certain drugs or to soluble sulfides produces it. Sulfhemoglobin is green.

Hemoglobin subunits containing these modified hemes do not bind oxygen, but they change the oxygen-binding characteristics of the normal subunits in hybrid hemoglobin molecules containing some normal subunits and one or more modified subunits. The accompanying figure shows the oxygen-binding curve of normal HbA, 15% methemoglobin and 12% sulfhemoglobin. The presence of methemoglobin shifts the curve to the left, impairing the delivery of the decreased amount of bound oxygen. In contrast, the sulfhemoglobin curve is shifted to the right, a DPG-like effect. As a result, oxygen delivery is enhanced, partially compensating for the inability of the sulfur-modified hemes to bind oxygen.

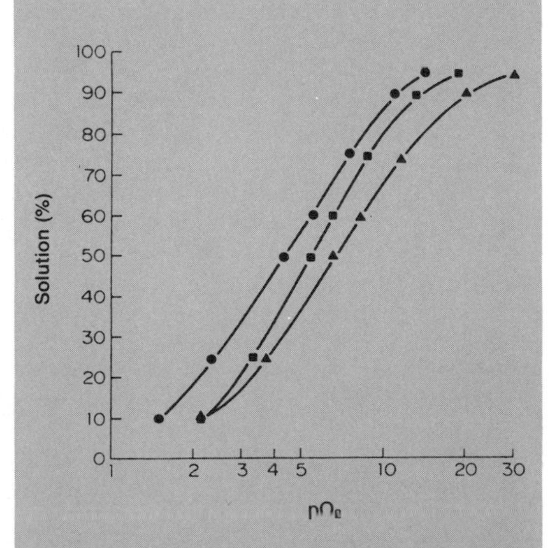

Oxygenation curves of unmodified hemoglobin A (squares) of a 15% oxidized hemolysate (circles) and of a hemolysate containing 12% sulfhemoglobin (triangles) in 0.1 M phosphate, pH 7.35, at 20°C.
Data from C. M. Park and R. L. Nagel, *N. Engl. J. Med.* 310:1579, 1984.

CLIN. CORR. 25.3
HEMOGLOBINS WITH ABNORMAL OXYGEN AFFINITY

Some abnormal hemoglobins have an altered affinity for oxygen. If oxygen affinity is increased (P_{50} decreased), oxygen delivery to the tissues will be diminished unless some sort of compensation occurs. Typically the body responds by producing more erythrocytes (polycythemia) and more hemoglobin. Hb Rainier is an abnormal hemoglobin in which the P_{50} is 12.9 mmHg, far below the normal value of 27 mmHg.

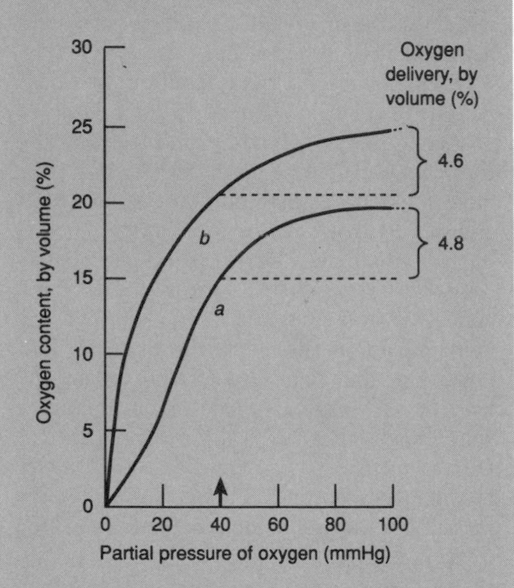

Oxygen content plotted against partial pressure of oxygen.
Curve *a* shows the oxygen dissociation curve of normal blood with a hemoglobin of 15 g dL^{-1}, P_{50} 27 mmHg, *n* 2.8, at pH 7.4, 37°C. Curve *b* shows that of blood from a patient with Hb Rainier, having a hemoglobin of 19.5 g dL^{-1}, P_{50} 12.9 mmHg, *n* 1.2, at the same pH and temperature. (1 mmHg ≈ 133.3 Pa.) On the right is shown the oxygen delivery. The compensatory polycythemia and hyperbolic curve of Hb Rainier results in practically normal arterial and venous oxygen tensions. Arrow indicates normal mixed venous oxygen tension.
From A. J. Bellingham, Hemoglobins with altered oxygen affinity. *Br. Med. Bull.*, 32:234,1976.

In the accompanying figure the oxygen content in volume percent (mL of O_2 per 100 mL of blood) is plotted versus partial pressure of oxygen, both for normal blood (curve *a*) and for the blood of a patient with Hb Rainier (curve *b*). Obviously the patient's blood carries more oxygen; this is because it

There are many inherited anomalies of hemoglobin synthesis in which there is formation of a structurally abnormal hemoglobin; these are called **hemoglobinopathies.** They may involve the substitution of one amino acid in one type of polypeptide chain for some other amino acid or they may involve absence of one or more amino acid residues of a polypeptide chain. In some cases the change is clinically insignificant, but in others it causes serious disease (Clin. Corr. 25.3).

25.5 PHYSICAL FACTORS THAT AFFECT OXYGEN BINDING

High Temperature Weakens Hemoglobin's Oxygen Affinity

Temperature has a significant effect on O_2 binding by hemoglobin, as shown in Figure 25.7. At below-normal temperatures the binding is tighter, resulting in a leftward shift of the curve; at higher temperatures the binding becomes weaker, and the curve is shifted to the right.

The effect of elevated temperature is like that of high levels of DPG, in that both enhance unloading of oxygen. The temperature effect is physiologically useful, as it makes additional O_2 available to support the high metabolic rate found in fever or in exercising muscle with its elevated temperature. The relative insensitivity to temperature of O_2 binding at high partial pressure of oxygen minimizes compromise of O_2 uptake in the lungs under these conditions.

The tighter binding of O_2 that occurs in hypothermic conditions is not consequential in hypothermia induced for surgical purposes. The decreased O_2 utilization by the body and increased solubility of O_2 in plasma at lower temperatures, as well as the increased solubility of CO_2, which acidifies the blood, compensate for hemoglobin's diminished ability to release oxygen.

Low pH Weakens Hemoglobin's Oxygen Affinity

Hydrogen ion concentration influences hemoglobin's O_2 binding. As shown in Figure 25.8, low pH shifts the curve to the right, enhancing O_2 delivery, whereas high pH shifts the curve to the left. It is customary to express oxygen binding by hemoglobin as a function of the pH of the plasma because it is this value, not the pH within the erythrocyte, that is usually measured. The pH of the erythrocyte cell sap is lower than the

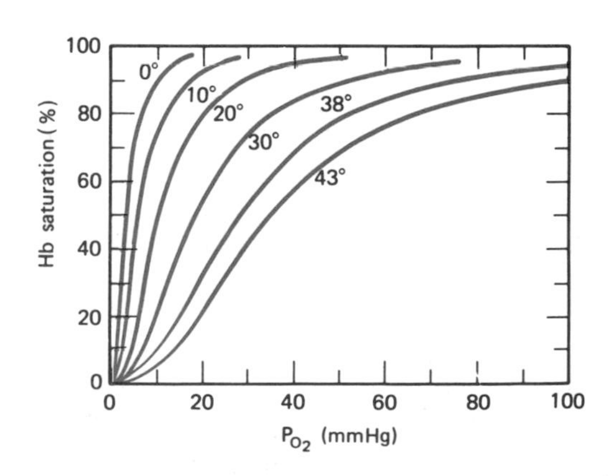

FIGURE 25.7
Oxygen dissociation curve for whole blood at various temperatures.
From C. J. Lambertson, *Medical physiology*, 11th ed., P. Bard (Ed.). St. Louis, MO.: Mosby, 1961, p. 596.

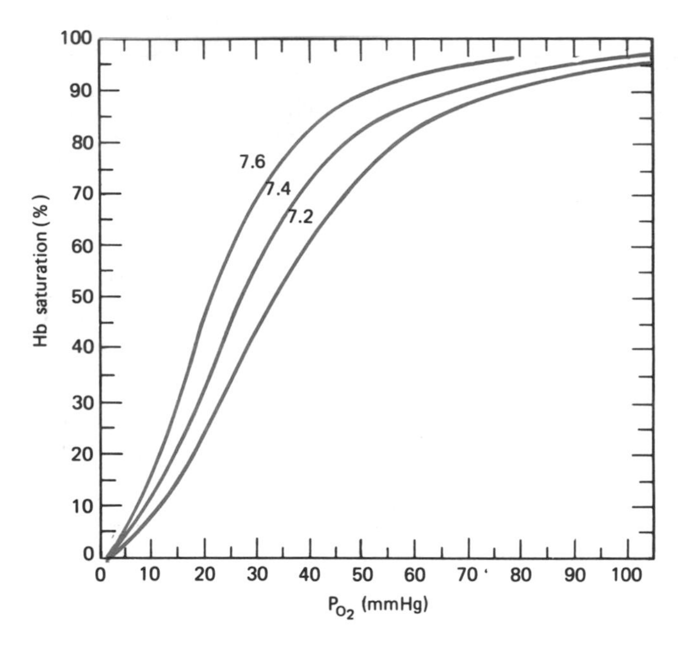

FIGURE 25.8
Oxygen dissociation curve for whole blood at various values of plasma pH.
Adapted from C. J. Lambertson, *Medical physiology,* 11th ed., P. Bard (Ed.). St. Louis, MO.: Mosby, 1961, p. 596.

contains 19.5 g of Hb per 100 mL instead of the usual 15 g per 100 mL.

Since the partial pressure of oxygen in mixed venous blood is about 40 mmHg, the volume of oxygen the blood of each individual can deliver may be obtained from the graph by subtracting the oxygen content of the blood at 40 mmHg from its oxygen content at 100 mmHg. As shown in the figure, the blood of the patient with Hb Rainier delivers nearly as much oxygen as normal blood does, although Hb Rainier delivers a significantly smaller fraction of the total amount it carries. Evidently polycythemia is an effective compensation for this condition, at least in the resting state.

plasma pH, but since these two fluids are in equilibrium, changes in the one reflect changes in the other.

The influence of pH upon O_2 binding is physiologically significant, since a decrease in pH is often associated with increased oxygen demand. An increased metabolic rate results in increased production of carbon dioxide and, as in muscular exercise, lactic acid. Lactic acid is also produced by hypoxic tissue. These acids produced by metabolism help release oxygen to support that metabolism.

The increase in acidity of hemoglobin as it binds O_2 is known as the **Bohr effect;** an equivalent statement is that the Bohr effect is the increase in basicity of hemoglobin as it releases oxygen. The effect may be expressed by the equation

$$HHb + O_2 \rightleftharpoons HbO_2 + H^+$$

Clearly this equation gives the same information as Figure 25.8, namely, that increases in hydrogen ion concentration will favor the formation of free oxygen from oxyhemoglobin, and conversely, that oxygenation of hemoglobin will lower the pH of the solution.

25.6 CARBON DIOXIDE TRANSPORT

The carbon dioxide we produce is excreted by the lungs, to which it must be transported by the blood. Carbon dioxide transport is closely tied to hemoglobin and to the problem of maintaining a constant pH in the blood, a problem that will be discussed subsequently.

Carbon dioxide is present in the blood in three major forms, as dissolved CO_2, as HCO_3^- (formed by ionization of H_2CO_3 produced when CO_2 reacts with H_2O), and as carbamino groups (formed when CO_2 reacts with amino groups of protein). Each of these is present both in arterial blood and in venous blood, as shown in the top three lines of Table 25.2.

TABLE 25.2 Properties of Blood of Humans at Rest[a]

	Arterial			Venous			Δ		
	Serum	Cells	Blood	Serum	Cells	Blood	Serum	Cells	Blood
Hb carbamino groups (meq L^{-1} of blood)		1.13	1.13		1.42	1.42		*+0.29*	*+0.29*
HCO_3^- (meq L^{-1} of blood)	13.83	*5.73*	*19.56*	14.84	*6.41*	*21.25*	+1.01	*+0.68*	*+1.69*
Dissolved CO_2 (meq L^{-1} of blood)	0.71	0.48	1.19	0.82	0.56	1.38	+0.11	+0.08	+0.19
Total CO_2 (meq L^{-1} of blood)	14.54	7.34	21.88	15.66	8.39	24.05	+1.12	+1.05	+2.17
Free O_2 (mmol L^{-1} of blood)			0.10			0.04			−0.06
Bound O_2 (mmol L^{-1} of blood)			8.60			6.01			−2.59
Total O_2 (mmol L^{-1} of blood)			8.70			6.05			−2.65
P_{O_2} (mmHg)			88.0			37.2			−50.8
P_{CO_2} (mmHg)			41.0			47.5			+6.5
pH	7.40	7.19		7.37	7.17		−0.03	−0.02	
Volume (cc L^{-1} of blood)	551.7	448.3	1000	548.9	451.1	1000	−2.8	+2.8	0.0
H_2O (cc L^{-1} of blood)	517.5	322.8	840.0	514.7	325.6	840.0	−2.8	+2.8	0.0
Cl^- (meq L^{-1} of blood)	57.71	24.30	82.01	56.84	25.17	82.01	−0.88	+0.88	0.0

SOURCE: From J. Baggott, The contribution of carbamate to physiological carbon dioxide transport. *Trends Biochem. Sci.* 3:N207, 1978, with permission of the publisher.

[a] Hemoglobin, 9 mM; serum protein, 39.8 g L^{-1} of blood; respiratory quotient, 0.82.

Net transport to the lungs for excretion is represented by the concentration difference between arterial and venous blood, shown in the last column. Notice that for each form of carbon dioxide the arterial–venous difference is only a small fraction of the total amount present; venous blood contains only about 10% more **total carbon dioxide** (total CO_2 is the sum of HCO_3^-, dissolved CO_2, and carbamino hemoglobin) than arterial blood does.

Carbon dioxide, after it enters the bloodstream for transport, generates hydrogen ions in the blood. Most come from bicarbonate ion formation, which occurs in the following manner.

Carbon dioxide entering the blood diffuses into the erythrocytes. The erythrocyte membrane, like most other biological membranes, is freely permeable to dissolved CO_2. Within the erythrocytes most of the carbon dioxide is acted upon by the intracellular enzyme, **carbonic anhydrase,** which catalyzes the reaction

$$CO_2 + H_2O \underset{\text{anhydrase}}{\overset{\text{carbonic}}{\rightleftharpoons}} H_2CO_3$$

This reaction will proceed in the absence of a catalyst, as is well known to all who drink carbonated beverages. Without the catalyst, however, it is too slow to meet the body's needs, taking over 100 s to reach equilibrium. Recall that at rest the blood makes a complete circuit of the body in only 60 s. Carbonic anhydrase is a very active enzyme, having a turnover number of the order of 10^6, and inside the erythrocytes the reaction reaches equilibrium within 1 s, less than the time spent by the blood in the capillary bed. The enzyme is zinc requiring, and accounts for a portion of our dietary requirement for this metal.

The ionization of carbonic acid, $H_2CO_3 \rightleftharpoons H^+ + HCO_3^-$, is a rapid, spontaneous reaction. It results in the production of equivalent amounts of H^+ and HCO_3^-. Since, as shown in the last column of line 2 in Table 25.2, 1.69 meq of bicarbonate was added to each liter of blood by this process, 1.69 meq of H^+ must also have been generated per liter of blood. The addition of this much acid, over 10^{-3} equiv of H^+, to 1 L of water would give a final pH below 3. Since the pH of venous plasma has an average value of 7.37, clearly most of the H^+ generated during HCO_3^-

production must be consumed by buffer action and/or other processes. This is discussed below.

Because of the compartmentalization of carbonic anhydrase, essentially all of the conversion of CO_2 to carbonic acid, and ultimately to HCO_3^-, occurs inside the erythrocyte. Negligible amounts of CO_2 react nonenzymatically in the plasma. This means that virtually all of the increase in HCO_3^- in venous as compared to arterial blood comes from intraerythrocyte HCO_3^- generation. To be sure, most of this diffuses into the plasma, so that venous plasma HCO_3^- is higher than arterial, but the erythrocyte was the site of its formation.

It has been observed that in the presence of carbonic anhydrase inhibitors, such as acetazolamide or cyanide, blood will still take up a certain amount of carbon dioxide rapidly. This is due to the reaction of carbon dioxide with amino groups of proteins within the erythrocyte to form **carbamino groups.**

$$R{-}NH_2 + CO_2 \rightleftharpoons R{-}NH{-}C(=O){-}O^- + H^+$$

Hemoglobin is quantitatively the most important protein involved in this reaction. Deoxyhemoglobin forms **carbamino hemoglobin** more readily than oxyhemoglobin does, and oxygenation causes the release of CO_2 that had been bound in carbamino hemoglobin.

Carbamino hemoglobin formation occurs only with uncharged aliphatic amino groups, not with the charged form, $R{-}NH_3^+$. The pH within the erythrocyte is normally about 7.2, somewhat more acidic than the plasma. Since amino groups of proteins have pK values well to the alkaline side of 7.2, they will be mostly in the charged (undissociated acid) form. Removal of some of the uncharged form via carbamino group formation will shift the equilibrium, generating more uncharged amino groups and an equivalent amount of H^+.

$$R{-}NH_3^+ \rightleftharpoons R{-}NH_2 + H^+$$

Clearly the formation of a carbamino group is, like HCO_3^- formation, a process that generates H^+.

The fact that only uncharged groups can form carbamino groups severely limits the number that can potentially participate in this reaction. Typical amino groups, such as the ε-amino groups in the side chains of lysyl residues, have pK values about 9.5–10.5. If the pK were 10.2, then at an intracellular pH of 7.2 only one ε-amino group in 1000 would be uncharged and able to react with CO_2. The α-amino groups at the N terminals of proteins, however, have much lower pK values, in the range of 7.6–8.4. This occurs because of the electron-withdrawing effect of the nearby oxygen of the peptide linkage. For an amino group with pK = 8.2, 1 out of every 10 molecules would be uncharged inside the cell and able to react with CO_2. A lower pK (or a higher intracellular pH) would result in an even greater availability of the group. Because of their lower pK values the α-amino groups at the N terminals of hemoglobin's polypeptide chains are the principal sites of carbamino group formation. If all four N-terminal amino groups of hemoglobin are blocked chemically by reaction with cyanate, carbamino formation does not occur.

The N-terminal amino groups of the β chains form part of the binding site of DPG. Since the N terminals cannot bind DPG and simultaneously

TABLE 25.3 Major Forms of Carbon Dioxide Transport

Species	Transport (%)
HCO_3^-	78
CO_2 (dissolved)	9
Carbamino hemoglobin	13

form carbamino groups, a competition arises. Carbon dioxide diminishes the effect for DPG and, conversely, DPG diminishes the ability of hemoglobin to form carbamino hemoglobin. Ignorance of the latter interaction led to a major overestimation of the role of carbamino hemoglobin in carbon dioxide transport. Prior to the discovery of the DPG effect, careful measurements were made of the capacity of purified hemoglobin (no DPG present) to form carbamino hemoglobin. The results were assumed to be applicable to hemoglobin in the erythrocyte, leading to the erroneous conclusion that carbamino hemoglobin accounted for 25–30% or more of CO_2 transport. It now appears that 13–15% of CO_2 transport is via carbamino hemoglobin. Table 25.3 summarizes the contribution of each major form of blood carbon dioxide to overall CO_2 transport.

Hemoglobin, in addition to being the primary O_2 carrier and a transporter of CO_2 in the covalently bound form of a carbamino group, also plays the major role in handling the hydrogen ions produced in CO_2 transport. It does this by buffering and by a second mechanism, which is discussed below. **Hemoglobin's buffering** power is due to its ionizable groups with pK values in the neighborhood of the intracellular pH of the erythrocyte. These include the four α-amino groups of the N-terminal amino acids and the imidazole side chains of the histidine residues. Hemoglobin has 38 histidines per tetramer; these therefore provide the bulk of hemoglobin's buffering ability.

In whole blood, buffering absorbs about 60% of the acid generated in normal carbon dioxide transport. Although hemoglobin is by far the most important nonbicarbonate buffer in blood, the organic phosphates in the erythrocytes, the plasma proteins, and so on, also make a significant contribution. Buffering by these compounds accounts for about 10% of the acid, leaving about 50% of acid control specifically attributable to buffering by hemoglobin. These buffer systems minimize the change in pH that occurs when acid or base is added, but do not altogether prevent that change. A small difference in pH between arterial and venous blood is therefore observed.

The remainder of the acid arising from carbon dioxide is absorbed by hemoglobin via a mechanism that has nothing to do with buffering. Recall that when hemoglobin became oxygenated it became a stronger acid and released H^+ (the Bohr effect). In the capillaries, where O_2 is released, the opposite occurs:

$$HbO_2 + H^+ \rightleftharpoons HHb + O_2$$

Simultaneously, CO_2 enters the capillaries and is hydrated:

$$CO_2 + H_2O \rightleftharpoons H^+ + HCO_3^-$$

Addition of these two equations gives

$$HbO_2 + CO_2 + H_2O \rightleftharpoons HHb + HCO_3^- + O_2$$

revealing that to some extent this system can take up H^+ arising from CO_2, and can do so with no change in H^+ concentration (that is, with no change in pH). Hemoglobin's ability to do this, through the operation of the Bohr effect, is referred to as the **isohydric carriage of CO_2.** As already pointed out, there is a small A-V difference in plasma pH. This is because the isohydric mechanism cannot handle all the acid generated during normal CO_2 transport; if it could, no such difference would occur. Figure 25.9 is a schematic representation of O_2 transport and the isohydric mechanism, showing what happens in the lungs and in the other tissues.

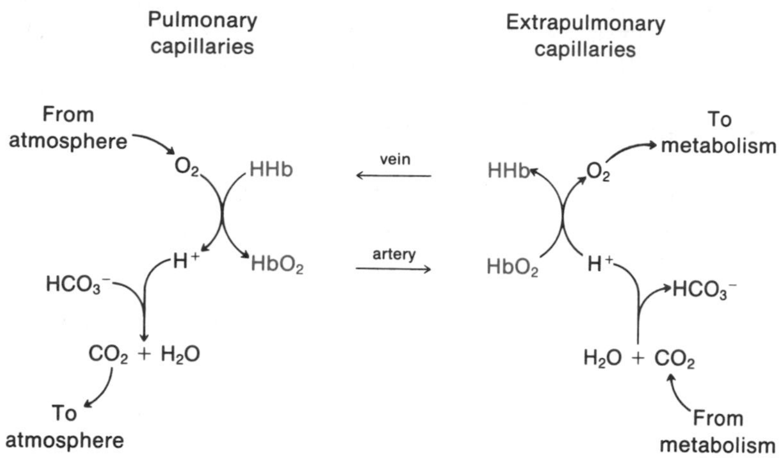

Pulmonary capillaries
From atmosphere
O_2
HHb
H^+
HbO_2
HCO_3^-
$CO_2 + H_2O$
To atmosphere
vein
artery
Extrapulmonary capillaries
HHb
O_2
To metabolism
HbO_2
H^+
HCO_3^-
$H_2O + CO_2$
From metabolism

FIGURE 25.9
Schematic representation of oxygen transport and the isohydric carriage of CO_2 by hemoglobin.
In the lungs (left) O_2 from the atmosphere reacts with deoxyhemoglobin, forming oxyhemoglobin and H^+. The H^+ combines with the HCO_3^- to form H_2O and CO_2. The CO_2 is exhaled. Oxyhemoglobin is carried to extrapulmonary tissues (right), where is dissociates in response to low P_{O_2}. The O_2 is used by metabolic processes, and CO_2 is produced. CO_2 combines with H_2O to give HCO_3^- and H^+. H^+ can then react with deoxyhemoglobin to give HHb, which returns to the lungs, and the cycle repeats.

Estimates of the importance of the isohydric mechanism in handling normal respiratory acid production have changed upward and downward over the years. The older, erroneous estimates arose out of a lack of knowledge of the multiple interactions in which hemoglobin participates. The earliest experiments, titrations of purified oxyhemoglobin and purified deoxyhemoglobin, revealed that oxygenation of hemoglobin resulted in release of an average of 0.7 H^+ for every O_2 bound. This figure still appears in textbooks, and much is made of it. Authors point out that with a Bohr effect of this magnitude the isohydric mechanism alone could handle all of the acid produced by the metabolic oxidation of fat (RQ of fat is 0.7), and buffering would be unnecessary. Unfortunately the experimental basis for this interpretation is physiologically unrealistic; the titrations were done in the total absence of carbon dioxide, which we now know binds to some of the Bohr groups, forming carbamino groups and diminishing the effect. When later experiments were carried out in the presence of physiological amounts of carbon dioxide, there was a drastic diminution of the Bohr effect, so much so that at pH 7.45 the isohydric mechanism was able to handle only the amount of acid arising from carbamino group formation. This work, however, was done prior to our appreciation of the competition between DPG and CO_2 for the same region of the hemoglobin molecule (see Table 25.4). Finally, in 1971, careful titrations of whole blood under presumably physiological conditions were carried out, yielding a value of 0.31 H^+ released per O_2 bound. This value is the basis of the present assertion that the isohydric mechanism accounts for about 40% of the acid generated during normal carbon dioxide transport. The quantitative contributions of various mechanisms to the handling of acid arising during carbon dioxide transport are summarized in Table 25.5. The major role of hemoglobin in handling this acid is obvious.

We have seen that essentially all of HCO_3^- formation is intracellular, catalyzed by carbonic anhydrase, and that the vast bulk of the H^+ generated by CO_2 is handled within the erythrocyte. These two observations bear upon the final distribution of HCO_3^- between plasma and the erythrocyte.

Intracellular formation of HCO_3^- increases its intracellular concentration. Since HCO_3^- and Cl^- freely exchange across the erythrocyte membrane, HCO_3^- will diffuse out of the erythrocyte, increasing the plasma HCO_3^- concentration. Electrical neutrality must be maintained across the membrane as this happens. Maintenance of neutrality can be accomplished in principle either by having a positively charged ion accompany

TABLE 25.4 Processes Occurring at the N Terminals of the α Chains and β Chains of Hemoglobin

	N terminals	
Process	α Chains	β Chains
Carbamino formation	Yes	Yes
DPG binding	No	Yes
H^+ Binding in the Bohr effect	Yes	No

TABLE 25.5 Control of the Excess H^+ Generated During Normal Carbon Dioxide Transport

Buffering	
By hemoglobin	50%
By other buffers	10%
Isohydric mechanism (hemoglobin)	40%

HCO_3^- out of the cell or by having some other negatively charged ion enter the cell in exchange for the HCO_3^-. Since the distribution of the major cations, Na^+ and K^+, is under strict control, it is the latter mechanism that is seen, and the ion that is exchanged for HCO_3^- is Cl^-. Thus as HCO_3^- is formed in red cells during their passage through the capillary bed, it moves out into the plasma and Cl^- comes in to replace it. The increase in intracellular $[Cl^-]$ is shown in the last line of Table 25.2. In the lungs, where all events of the peripheral tissue capillary beds are reversed, HCO_3^- migrates into the cells to be converted to CO_2 for exhalation, and Cl^- returns to the plasma. The exchange of Cl^- and HCO_3^- between the plasma and the erythrocyte is called the **chloride shift.**

The intracellular buffering of H^+ from carbon dioxide causes the cells to swell, giving venous blood a slightly (0.6%) higher hematocrit than arterial blood. (The hematocrit is the volume percent of red cells in the blood.) This occurs because the charge on the hemoglobin molecule becomes more positive with every H^+ that binds to it. Each bound positive charge requires an accompanying negative charge to maintain neutrality. Thus as a result of buffering there is a net accumulation of HCO_3^- or Cl^- inside the erythrocyte. An increase in the osmotic pressure of the intracellular fluid results from this increase in concentration of particles. As a consequence, water migrates into the cells, causing them to swell slightly. Typically, an arterial hematocrit might be 44.8 and a venous hematocrit 45.1, as shown in Table 25.2 by the line labeled "volume (cc L^{-1} of blood)."

25.7 INTERRELATIONSHIPS AMONG HEMOGLOBIN, OXYGEN, CARBON DIOXIDE, HYDROGEN ION, AND 2,3-DIPHOSPHOGLYCERATE

By now it should be clear that multiple interrelationships of physiological significance exist among the ligands of hemoglobin. These interrelationships may be summarized schematically as follows:

$$HHb\begin{matrix} \diagup DPG \\ \diagdown CO_2 \end{matrix} + O_2 \rightleftharpoons HbO_2 + CO_2 + DPG + H^+$$

This equation shows that changes in the concentration of H^+, DPG, or CO_2 have similar effects on O_2 binding. The equation will help you remember the effect of changes in any one of these variables upon hemoglobin's O_2 affinity.

DPG levels in the red cell are controlled by product inhibition of its synthesis and by pH. Hypoxia results in increased levels of deoxyhemoglobin on a time-averaged basis. Since deoxyhemoglobin binds DPG more tightly, in hypoxia there is less free DPG to inhibit its own synthesis, and so DPG levels will rise due to increased synthesis. The effect of pH is that high pH increases DPG synthesis and low pH decreases DPG synthesis; this reflects the influence of pH on the activity of **DPG mutase,** the enzyme that catalyzes DPG formation. Since changes in DPG levels take many hours to become complete, this means that the immediate effect of a decrease in blood pH is to enhance oxygen delivery by the Bohr effect. If the acidosis is sustained (most causes of chronic metabolic acidosis are not associated with a need for enhanced oxygen delivery), diminished DPG synthesis leads to a decrease in intracellular DPG concentration, and hemoglobin's oxygen affinity returns toward normal. Thus we have a system that can respond appropriately to acute conditions, such as vigorous exercise, but which when faced with a prolonged abnormality of pH readjusts to restore normal (and presumably optimal) oxygen delivery.

25.8 INTRODUCTION TO pH REGULATION

When we considered carbon dioxide transport we noted the large amount of H^+ generated by this process, and we considered the ways in which the blood pH was kept under control. Control of blood pH is important because changes in blood pH will cause changes in intracellular pH, which in turn may profoundly alter metabolism. Protein conformation is affected by pH, as is enzyme activity. In addition, the equilibria of important reactions that consume or generate hydrogen ions, such as any of the oxidation–reduction reactions involving pyridine nucleotides, will be shifted by changes in pH.

The normal arterial plasma pH is 7.40 ± 0.05; the pH range compatible with life is about 7.8–6.8. Intracellular pH varies with the type of cell. The pH of the erythrocyte is nearly 7.2, whereas most other cells are lower, about 7.0. Values as low as 6.0 have been reported for skeletal muscle.

It is fortunate for both diagnosis and treatment of diseases that the acid–base status of the intracellular fluid influences and is influenced by the acid–base status of the blood. Blood is readily available for analysis, and when alteration of body pH becomes necessary, intravenous administration of acidifying or alkalinizing agents is efficacious.

25.9 BUFFER SYSTEMS OF PLASMA, INTERSTITIAL FLUID, AND CELLS

Each body water compartment is defined spatially by one or more differentially permeable membranes. Each type of compartment contains characteristic kinds and concentrations of solutes, some of which are buffers at physiological pH values. Although the solutes in the cytoplasm of each type of cell are different, most cells are similar enough that they are considered together for purposes of acid–base balance. Thus there are, from this point of view, three major body water components: plasma, which is contained in the circulatory system; interstitial fluid, the fluid that bathes the cells; and intracellular fluid.

The compositions of these fluids are given in Figure 25.10. In plasma the major cation is Na^+; small amounts of K^+, Ca^{2+}, and Mg^{2+} are also present. The two dominant anions are HCO_3^- and Cl^-. Smaller amounts of protein, phosphate, and SO_4^{2-} are also found, along with a mixture of organic anions (amino acids, etc.), each of which would be insignificant if taken separately. The sum of the anions equals, of course, the sum of the cations. It is apparent at a glance that the composition of interstitial fluid is very similar. The major difference is that interstitial fluid contains much less protein than plasma contains (the capillaries are not normally permeable to the plasma proteins) and, correspondingly, a lower cation concentration. Plasma and interstitial fluid taken together are called the extracellular fluid, and low molecular weight components equilibrate fairly rapidly between the two. For example, H^+ equilibrates between the plasma and interstitial fluid within about $\frac{1}{2}$ h. The composition of intracellular fluid is strikingly different. The major cation is K^+, while organic phosphates (ATP, DPG, glycolytic intermediates, etc.) and protein are the major anions.

As a result of the differences among these fluid compartments, each fluid makes a different contribution to buffering. The major buffer of the extracellular fluid, for example, is the HCO_3^-/CO_2 system. Since the pK

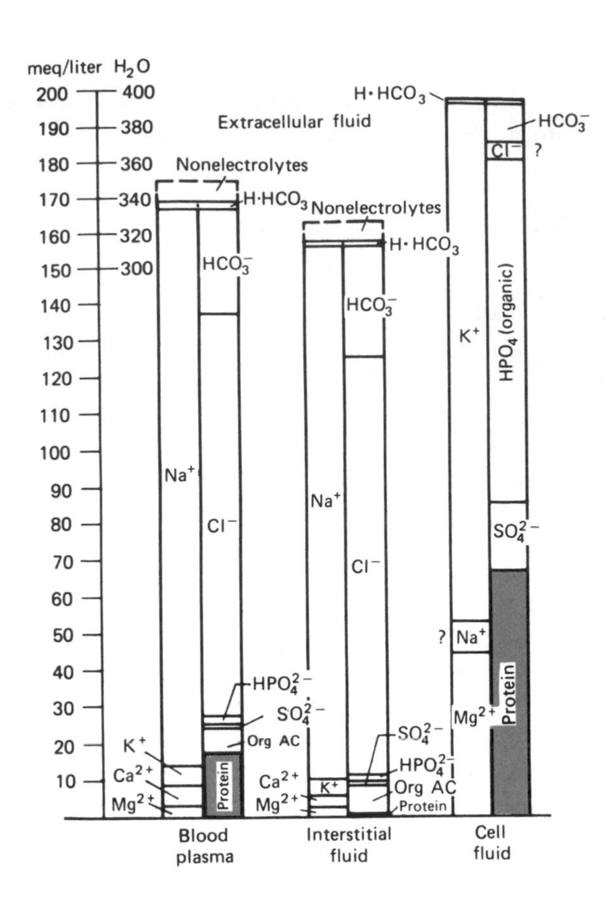

FIGURE 25.10
Diagram showing chief chemical constituents of the three fluid compartments.
Height of left half of each column indicates total concentration of cations; that of right half, concentration of anions. Both are expressed in milliequivalents per liter (meq L^{-1}) of water. Note that chloride and sodium values in cell fluid are questioned. It is probable that at least in muscle, the intracellular phase contains some sodium but no chloride.
Modified from Gamble. From M. I. Gregersen, *Medical physiology*, 11th ed., P. Bard (Ed.). St. Louis, MO.: Mosby, 1961, p. 307.

of the HCO_3^-/CO_2 system is 6.1 (Table 25.6 lists the major physiological buffers and their pK values), extracellular fluid at a pH of 7.4 is not very effective in resisting changes in pH arising from changes in P_{CO_2}. Intracellular fluid, with its high levels of protein and organic phosphates, is responsible for most of the buffering that occurs when P_{CO_2} changes. We have already seen the importance of buffering by hemoglobin and organic phosphates within the red cell. On the other hand, for reasons that will be explained in Section 25.10, the bicarbonate buffer system is quite effective in controlling pH changes due to causes other than changes in P_{CO_2}. Extracellular fluid and intracellular fluid share almost equally in the buffering of strong organic or inorganic acids (see Table 25.7). The plasma $[HCO_3^-]$ is, therefore, an excellent indicator of the whole body's capacity to handle additional loads of these acids.

Since acid–base imbalance arising from metabolic production of organic acids is a common and potentially life-threatening condition, and since the plasma $[HCO_3^-]$ is such a good indicator of the whole body's capacity to handle further metabolic acid loads, plasma $[HCO_3^-]$ is of major clinical concern. It is hydrogen ion concentration that must be kept within acceptable limits, but measuring pH alone is like walking on thin ice while observing merely whether or not you are still on the surface. Knowledge of $[HCO_3^-]$ tells you how close the ice is to the breaking point and how deep the water is underneath.

Because of the importance of the bicarbonate buffer system and its interaction with the other buffers of blood and other tissues, we will consider blood as a buffer in some detail. We will begin with a brief consideration of a model buffer.

Every buffer consists of a weak acid, HA, and its **conjugate base,** A^-. Examples of conjugate base/weak acid pairs include acetate$^-$/acetic acid, NH_3/NH_4^+, and $HPO_4^{2-}/H_2PO_4^-$. Notice that the weak acid may be neutral, positively charged or negatively charged, and that its conjugate base

TABLE 25.6 Acid Dissociation Constants of Major Physiological Buffers

Buffer system	pK
HCO_3^-/CO_2	6.1
Phosphate	
$HPO_4^{2-}/H_2PO_4^-$	6.7–7.2
Organic phosphate esters	6.5–7.6
Protein	
Histidine side chains	5.6–7.0
N-terminal amino groups	7.6–8.4

TABLE 25.7 Buffering of Metabolic Acids

Tissue	Buffering (%)
Extracellular fluids	42
Red cells	6
Tissue cells	52

must (since a H^+ has been lost) have one less positive charge (or one more negative charge) than the weak acid.

The degree of ionization of a weak acid depends on the concentration of free hydrogen ions. This may be expressed in the form of the **Henderson–Hasselbalch equation** (derived on p. 11) as follows:

$$pH = pK + \log \frac{[\text{conjugate base}]}{[\text{acid}]}$$

This is a mathematical rearrangement of the fundamental equilibrium equation. It states that there is a direct relationship between the pH and the ratio of [conjugate base]/[acid]. It is important to realize that this *ratio,* not the absolute concentration of any particular species, is the factor that is related to pH. Use of this equation will help you to understand the operation of and to predict the effects of various alterations upon acid–base balance in the body.

Blood plasma is a mixed buffer system; in the plasma the major buffers are HCO_3^-/CO_2, $HPO_4^{2-}/H_2PO_4^-$, and protein/Hprotein. The pH is the same throughout the plasma, so each of these buffer pairs distributes independently according to its own Henderson–Hasselbalch equation:

$$\begin{aligned} pH &= pK_1 + \log \frac{[HCO_3^-]}{[CO_2]} \\ &= pK_2 + \log \frac{[HPO_4^{2-}]}{[H_2PO_4^-]} \\ &= pK_3 + \log \frac{[\text{protein}^-]}{[\text{Hprotein}]} \end{aligned}$$

Because each pK is different, the [conjugate base]/[acid] ratio is also different for each buffer pair. Notice, though, if the ratio is known for any given buffer pair, one automatically has information about the others (assuming the pK values are known).

25.10 THE CARBON DIOXIDE–BICARBONATE BUFFER SYSTEM

As we have seen, the major buffer of the plasma (and of the interstitial fluid as well) is the **bicarbonate buffer system.** The bicarbonate system has two peculiar properties that make its operation unlike that of typical buffers. We will examine this important buffer in some detail, since a firm understanding of it is the key to a grasp of acid–base balance.

In the first place, the component that we consider to be the acid in this buffer system is CO_2, which is not truly an acid, but an acid anhydride. It reacts with water to form carbonic acid, which is indeed a typical weak acid.

$$CO_2 + H_2O \rightleftharpoons H_2CO_3$$

Carbonic acid then rapidly ionizes to give H^+ and HCO_3^-.

$$H_2CO_3 \rightleftharpoons H^+ + HCO_3^-$$

If these two equations are added, H_2CO_3 cancels out, and the sum is

$$CO_2 + H_2O \rightleftharpoons H^+ + HCO_3^-$$

Elimination of H_2CO_3 from formal consideration is realistic since, not only does it simplify matters, but H_2CO_3 is, in fact, quantitatively insignificant. Because the equilibrium of the reaction,

$$CO_2 + H_2O \rightleftharpoons H_2CO_3$$

lies far to the left, H_2CO_3 is present only to the extent of 1/200 of the concentration of dissolved CO_2. Since the concentration of water is virtually constant, it need not be included in the equilibrium expression for the reaction, and one may write:

$$K = \frac{[H^+][HCO_3^-]}{[CO_2]}$$

The value of K is 7.95×10^{-7}.

The concentration of a gas in a solution is proportional to the partial pressure of the gas. Thus we measure the partial pressure of CO_2 (P_{CO_2}). Then P_{CO_2} is multiplied by a **conversion factor, α,** to get the millimolar concentration of dissolved CO_2.

$$\alpha P_{CO_2} = \text{meq L}^{-1}$$

α has a value of 0.03 meq liter^{-1} mmHg^{-1} [or 0.225 meq liter^{-1} kPa^{-1}] at 37°C. The equilibrium expression thus becomes

$$K = \frac{[H^+][HCO_3^-]}{0.03 \cdot P_{CO_2}}$$

and the Henderson–Hasselbalch equation for this buffer system becomes

$$pH = 6.1 + \log \frac{[HCO_3^-]}{0.03 \cdot P_{CO_2}}$$

with $[HCO_3^-]$ expressed in units of milliequivalents per liter (Clin. Corr. 25.4).

We said earlier that the bicarbonate buffer system, with a pK of 6.1, was not effective against carbonic acid in the pH range of 7.8–6.8, but that it was effective against noncarbonic acids. The usual rules of chemical equilibrium dictate that a buffer is not very useful in a pH range more than about one unit beyond its pK. Thus what needs to be explained is how the bicarbonate system can be effective against noncarbonic acids; its failure to buffer carbonic acid is expected. The manner in which it buffers noncarbonic acids in a pH range far from its pK is the second unusual property of this buffer system. Notice that the explanation of this property in the following paragraph involves the flow of materials in a living system, and so departs from mere equilibrium considerations.

Consider first a typical buffer, consisting of a mixture of a weak acid and its conjugate base. When a strong acid is added, most of the added H^+ combines with the conjugate base. As a result, [weak acid] increases and simultaneously [conjugate base] diminishes. The ratio of [conjugate base]/[weak acid] therefore changes, and so does the pH. Of course, the pH changes much less than if there were no buffer present. Now imagine a system in which the weak acid, as it is generated by the reaction of the added strong acid with the conjugate base, is somehow removed so that while [conjugate base] diminishes, [weak acid] remains nearly constant. In this case the ratio of [conjugate base]/[weak acid] would change much

CLIN. CORR. 25.4
THE CASE OF THE VARIABLE CONSTANT

In clinical laboratories plasma pH and P_{CO_2} are commonly measured with suitable electrodes, and plasma $[HCO_3^-]$ is then calculated from the Henderson–Hasselbalch equation using pK = 6.1. Although this procedure is generally satisfactory, there have been several reports of severely erroneous results in patients whose acid–base status was changing rapidly.[1] Clinicians who are attuned to this phenomenon urge that *direct* measurements of all three variables be made in acutely ill patients.

The clinical literature discusses this problem in terms of departure of the value of pK from 6.1. Studies of model systems suggest that this interpretation is incorrect; pK does change with ionic strength, temperature, and so on, and so does α, but not enough to account for the magnitude of the clinical observations.

Astute commentators have speculated that the real basis of the phenomenon is disequilibrium. The detailed nature of the putative disequilibrium has not yet been established, but it is probably related to the difference in pH across the erythrocyte membrane. Normally the pH of the erythrocyte is about 7.2, and the plasma pH is 7.4. If the plasma pH changes rapidly in an acute illness, the pH of the erythrocyte will also change, but the rate of change within the erythrocyte is not known. If the change within the erythrocyte lags sufficiently behind the change in the plasma, the system would indeed be in gross disequilibrium, and equilibrium calculations would not apply.

[1] See Hood, I. and Campbell, E. J. M. *N. Engl. J. Med.* 306:864, 1982.

less for a given addition of strong acid, and the pH would also change much less. This is exactly what happens with the bicarbonate buffer system in the body. As strong acid is added, $[HCO_3^-]$ diminishes and CO_2 is formed. But the excess CO_2 is exhaled, so that the ratio of $[HCO_3^-]/\alpha P_{CO_2}$ does not change so dramatically. In like manner, if strong base is added to the body, it will be neutralized by carbonic acid, but CO_2 will be replaced by metabolism, and again, the ratio of $[HCO_3^-]/P_{CO_2}$ will not change as much as would be expected. The bicarbonate buffer system in the body is thus an **open system** in which the P_{CO_2} term is adjusted to meet the body's needs. If respiration should be unable to accomplish this adjustment, then P_{CO_2} would change strikingly, and the bicarbonate system would be relatively ineffective, in keeping with the prediction of chemical equilibrium.

A graphical representation of the Henderson–Hasselbalch equation for the bicarbonate buffer system is a valuable aid to learning and understanding how this system reflects the acid–base status of the body. One of the most common of these representations is the **pH–bicarbonate diagram,** shown in Figure 25.11. $[HCO_3^-]$ of up to 40 meq L^{-1} is shown on the ordinate; this is adequate to deal with most situations. Similarly, since plasma pH does not exceed 7.8 or (except transiently) fall below 7.0 in living patients, the abscissa of the graph is limited to the range of 7.0–7.8. The normal plasma $[HCO_3^-]$, 24 meq L^{-1}, and the normal plasma pH, 7.4, are indicated. The third variable, CO_2, can be shown on a two-dimensional graph by assigning a fixed value to P_{CO_2} and then showing, *for that value,* the relationship between pH and $[HCO_3^-]$. Figure 25.11 shows that relationship when P_{CO_2} has its normal physiological value of 40 mmHg (5.33 kPa). The line is called the 40 mmHg (5.33 kPa) isobar. Whenever P_{CO_2} is 40 mmHg (5.33 kPa), pH and $[HCO_3^-]$ must be somewhere on that line.

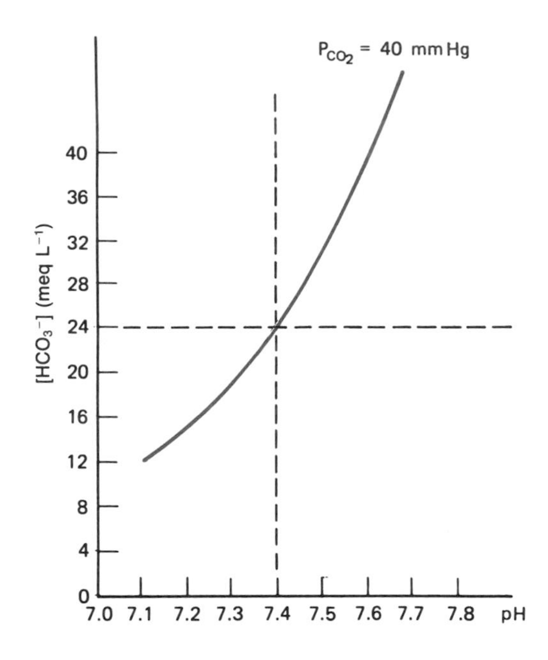

FIGURE 25.11
pH–Bicarbonate diagram including the 40 mmHg (5.33 kPa) CO_2 isobar, and showing the normal values of plasma pH and bicarbonate ion concentration.

In a similar manner we can plot isobars for various abnormal values of P_{CO_2}. These curves are shown in Figure 25.12. The range of values given covers those found in patients. Any point on the graph gives the values of the three variables of the Henderson–Hasselbalch equation for the bicarbonate system at that point. Since all you need to locate a point are any two of these variables, the third can be read directly from the graph.

Let us now see how the bicarbonate buffer system behaves when it is in the presence of other buffers, as it is in whole blood. First, let us acidify the system by increasing the concentration of the acid-producing component, CO_2. For every CO_2 that reacts with water to produce a H^+, one HCO_3^- will also form. Most of the H^+, however, will be buffered by protein and phosphate. As a result, $[HCO_3^-]$ will rise much more than $[H^+]$. Similarly, if acid is removed from this system by decreasing P_{CO_2}, $[HCO_3^-]$ will decrease. The $[H^+]$ will not decrease by an equivalent amount, though, because the other buffers will dissociate to resist the pH change. The results of these processes as they occur in whole blood, with its various intracellular and extracellular buffers, are shown in Figure 25.13. Let us start at the point that represents the normal values: pH of 7.4, $[HCO_3^-]$ of 24 meq L^{-1}, and P_{CO_2} of 40 mmHg (5.33 kPa). As P_{CO_2} rises to 80 mmHg (10.7 kPa), bicarbonate goes up to 28 meq L^{-1}, an increase of 4 meq L^{-1}. This means H_2CO_3 must have increased by 4 meq L^{-1}, and that it immediately ionized, giving H^+ and HCO_3^-. The pH, however, drops to 7.18; this represents an increase in $[H^+]$ of only 26×10^{-6} meq L^{-1}. The other 3.999974 meq L^{-1} of H^+ produced by the ionization of carbonic acid were taken up by the phosphate, hemoglobin, plasma protein, and other buffer systems. If P_{CO_2} were to decrease, the opposite would occur. Thus by altering P_{CO_2} in the presence of HCO_3^- and other buffers, a line is generated with a definite nonzero slope. For the blood

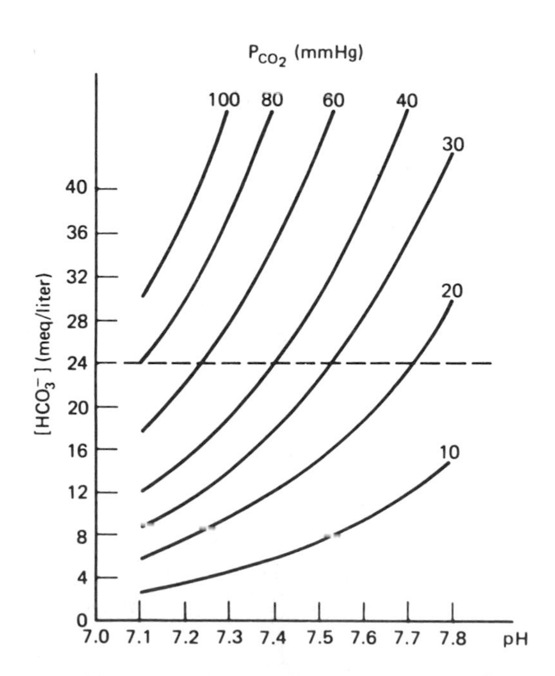

FIGURE 25.12
pH–Bicarbonate diagram, showing CO_2 isobars from 10 to 100 mmHg.

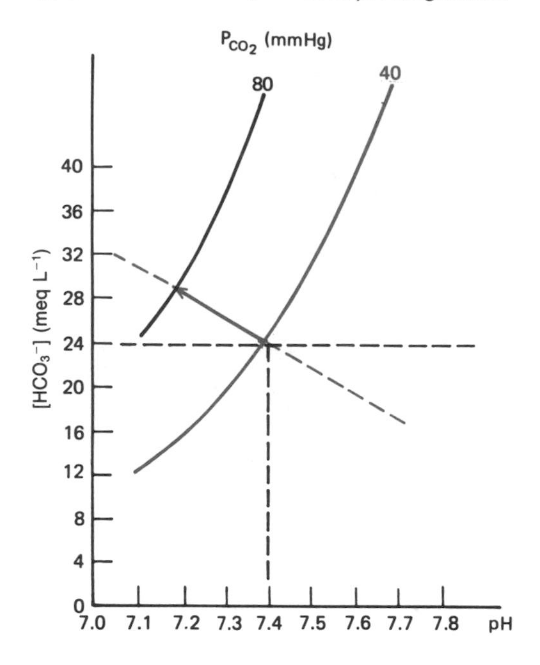

FIGURE 25.13
The buffering line of blood.
This pH–bicarbonate diagram shows the changes in pH and [HCO_3^-] that occur in whole blood in vitro when P_{CO_2} is changed. Notice that the relationship between pH and [HCO_3^-] is described by a straight line with a nonzero slope.

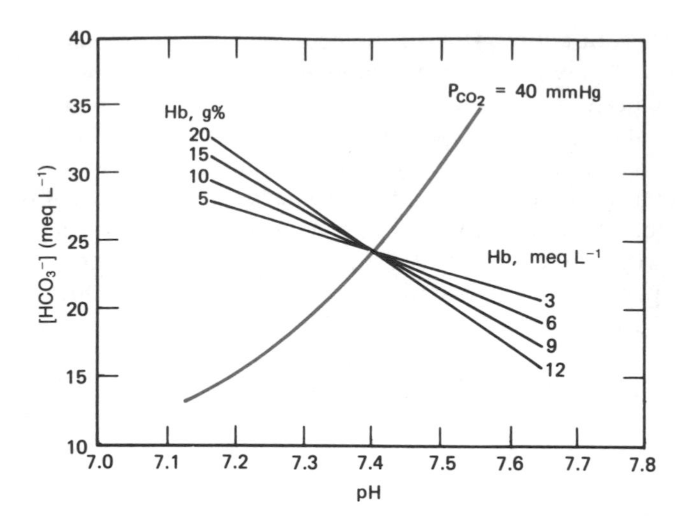

FIGURE 25.14
Slope of the buffering line of blood as it varies with hemoglobin concentration.
From H. W. Davenport, *The ABC of acid–base chemistry*, 6th ed. revised. Chicago: University of Chicago Press, 1974, p. 55.

system, this is called the **buffering line of blood.** Notice that if P_{CO_2} is the only variable that is changed, the response of the system is confined to movements along this line.

The slope of the buffering line depends on the concentration of the nonbicarbonate buffers. If they were more concentrated, they would better resist changes in pH. An increase in P_{CO_2} to 80 mmHg (10.7 kPa) would then cause a smaller drop in pH, and since the more concentrated buffers would react with more hydrogen ions (produced by the ionization of carbonic acid), [HCO_3^-] would rise higher. Thus the slope of the buffering line would be steeper.

Hemoglobin is quantitatively the second most important blood buffer, exceeded only by the bicarbonate buffer system. Since hemoglobin concentration in the blood can fluctuate widely in various disease states, it is the most important physiological determinant of the slope of the blood buffer line. Figure 25.14 shows how the slope of the blood buffer line varies with hemoglobin concentration.

Having now seen how the bicarbonate buffer system in blood responds to changes in P_{CO_2} and how this response is modified by changing the hemoglobin concentration, let us examine the response of blood to the addition of noncarbonic acids such as HCl, acetoacetic acid, and so on. We will continue to analyze the situation in terms of the pH–bicarbonate diagram. The starting point will again be the normal state: pH = 7.4, [HCO_3^-] = 24 meq L^{-1}, and P_{CO_2} = 40 mmHg (5.33 kPa). As acid is added, it will react with all the blood buffers, and the concentrations of all of the conjugate bases will decrease. Since the bicarbonate system is the major blood buffer, the decrease in [HCO_3^-] will be substantial. If P_{CO_2} is held constant at 40 mmHg (5.33 kPa) as a noncarbonic acid is added, the changes in the system can be represented by a point sliding down the 40 mmHg (5.33 kPa) isobar, as shown in Figure 25.15. If alkali is added to the blood, all the undissociated acids of the various buffer systems will participate in neutralizing it. Again, if this occurs at a fixed P_{CO_2} of 40 mmHg (5.33 kPa), the changes in the system will be represented by a point sliding up the 40 mmHg (5.33 kPa) isobar. Notice that, just as changes in P_{CO_2} were represented by points confined to the blood buffer line, changes due

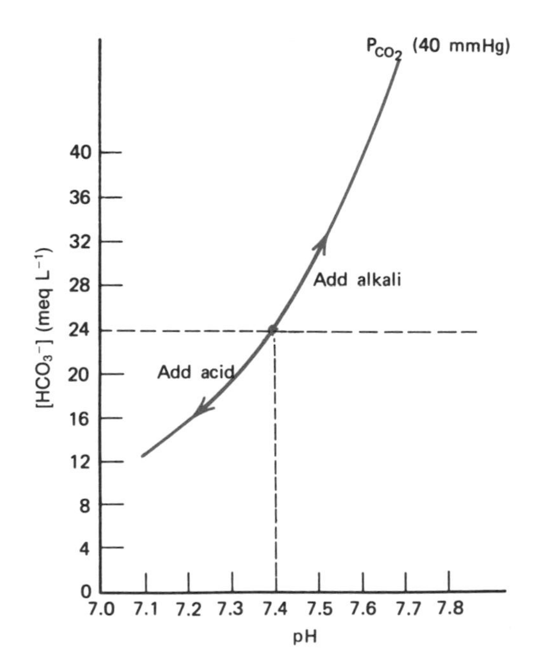

FIGURE 25.15
Effect of adding noncarbonic acid or alkali to whole blood with P_{CO_2} fixed at 40 mmHg.

to the addition of acid or base at a fixed P_{CO_2} are represented by points confined to the CO_2 isobar.

The effects on blood of changing P_{CO_2} or of adding acid or alkali, as we have just described, are realistic qualitative models of what can happen in certain disease states. In the following sections we will see how these changes occur in the body and how the body compensates for them.

25.11 ACID–BASE BALANCE AND ITS MAINTENANCE

It should come as no surprise that mechanisms exist whereby the body normally rids itself of excess acid or alkali. The physiological implication is that if a patient is in a state of continuing **acidosis** (excess acid or deficiency of alkali in the body) or **alkalosis** (excess alkali or deficiency of acid in the body), there must be a continuing cause of the imbalance. In such a situation the body's first task is to somehow compensate so plasma pH does not exceed the limits compatible with life. Assistance from the physician is sometimes necessary. The body's second task is to eliminate the primary cause of the imbalance, that is, to cure the disease, so that a normal acid–base status can be reestablished. Again, intervention by the physician may be needed.

All individuals, in sickness or in health, produce large amounts of acids every day. The major acid is CO_2. The amount of CO_2 produced depends on the individual's caloric expenditure; CO_2 production ranges from 12,500 to nearly 50,000 meq day^{-1}, and in an average young adult male, about 22,000 meq of CO_2 are produced daily. This acid is volatile, and is normally excreted by the lungs. Inability of the lungs to perform this task adequately leads to respiratory acidosis or alkalosis. **Respiratory acidosis** is the result of hypoventilation of the alveoli, so that CO_2 accumulates in the body. Alveolar hypoventilation occurs when the depth or rate of respiration is diminished. Obstruction of the airway, neuromuscular disorders, and diseases of the central nervous system are common causes of acute respiratory acidosis. Chronic respiratory acidosis is seen in patients with chronic obstructive lung disease, such as emphysema. Obviously, since the common element in all these conditions is increased alveolar P_{CO_2}, inhalation of a gas mixture with a high P_{CO_2} could also cause respiratory acidosis.

Respiratory alkalosis, on the other hand, arises from decreased alveolar P_{CO_2}. Hyperventilation due to anxiety is probably the most common cause. Central nervous system injury involving the respiratory center, salicylate poisoning, fever, and artificial ventilation are other causes. At high altitude, due to the decrease in total atmospheric pressure, alveolar P_{CO_2} also falls, producing chronic respiratory alkalosis.

Various amounts of nonvolatile acids are also produced by the body. The diet and the physiological state of the individual determine the kinds and amounts of these acids. Oxidation of sulfur-containing amino acids produces H^+ and SO_4^{2-}, the equivalent of sulfuric acid. Hydrolysis of phosphate esters is equivalent to the formation of phosphoric acid. The contribution of these processes depends on the amount of acid precursors ingested; for an individual consuming an average American diet there is a net daily acid production of about 60 meq.

Metabolism normally produces certain amounts of lactic acid, acetoacetic acid, and β-hydroxybutyric acid. In some physiological or pathological states these are produced in excess, and accumulation of the excess causes acidosis. When an ammonium salt of a strong acid, such as ammonium chloride, or when arginine hydrochloride or lysine hydrochloride is administered, it is converted to urea, and the corresponding strong acid (hydrochloric acid in these examples) is synthesized. Ingestion of salicylates, methyl alcohol, or ethylene glycol results in production of strong organic acids. Accumulation of any of these nonvolatile acids leads to **metabolic acidosis.**

While it is obvious that excess acid production can cause acidosis, the same net effect can arise from abnormal loss of base, as could be predicted from the Henderson–Hasselbalch equation for the bicarbonate buffer system. Renal tubular acidosis is a condition in which this occurs. Abnormal amounts of HCO_3^- escape from the blood into the urine, leaving the body acidotic (Clin. Corr. 25.5). A more common cause of bicarbonate depletion is severe diarrhea. In this chapter it will be assumed that kidney function is normal.

Mammals do not synthesize alkaline compounds from neutral starting materials. **Metabolic alkalosis** therefore arises from intake of excess alkali or abnormal loss of acid. An alkali commonly taken by many people is sodium bicarbonate. A less obvious source of alkali is the salt of any metabolizable organic acid. Sodium lactate is often administered to combat acidosis; normal metabolism converts it to sodium bicarbonate. The net reaction is as follows:

$$Na^+ + CH_3CHOHCOO^- + 3O_2 \rightleftharpoons Na^+ + HCO_3^- + 2CO_2 + 2H_2O$$

Most dietary fruits and vegetables have a net alkalinizing effect on the body for this reason. They contain a mixture of organic acids, which are metabolized to CO_2 and H_2O, and therefore have no long-term effect on acid–base balance, and salts of organic acids, which give rise to bicarbonate. Abnormal loss of acid, as can occur with prolonged vomiting or gastric lavage, causes alkalosis. Alkalosis may also be produced by rapid loss of body water, as in diuresis, which may temporarily increase $[HCO_3^-]$ in the plasma and extracellular fluid.

CLIN. CORR. 25.5 THE ROLE OF BONE IN ACID–BASE HOMEOSTASIS

The average adult skeleton contains 50,000 meq of Ca^{2+} in the form of salts that are alkaline relative to the pH of plasma. In chronic acidosis this large reservoir of base is drawn upon to help control the plasma pH. Thus people with chronic kidney disease and severely impaired renal acid excretion do not experience a continuous decline in plasma pH and $[HCO_3^-]$. Rather, the pH and $[HCO_3^-]$ stabilize at some below-normal level. The resulting change in bone composition is not inconsequential, and clinical and roentgenologic evidence of rickets or osteomalacia often appear. Bone healing has been shown in these patients after prolonged administration of alkali in the form of sodium bicarbonate or citrate sufficient to restore plasma $[HCO_3^-]$.

Lemann, J., Jr. and Lennon, E. J. *Kidney Intern.* 1:275, 1972.

The Kidney Plays a Critical Role in Acid–Base Balance

Excess nonvolatile acid and excess bicarbonate are excreted by the kidney. As a result, the pH of the urine varies as a function of the body's need to excrete these materials. For an individual on a typical American diet, urine pH is about 6, indicating a net acidification as compared to plasma. This is consistent with our knowledge that the typical diet results in a net production of acid. The pH of the urine can range from a lower limit of 4.4 up to 8.0.

A typical daily urine volume is about 1.2 L. At the minimum urine pH of 4.4, $[H^+]$ is only 0.04 meq L^{-1}, and it would take 1250 L of urine to excrete 50 meq of acid as free hydrogen ions. Clearly most of the acid we excrete must be in some form other than H^+. A form that can be excreted in a reasonable concentration, such as $H_2PO_4^-$ or NH_4^+, is needed.

Let us now see how the kidney accomplishes the excretion of acid or base. Figure 25.16 shows the fundamental functioning unit of the kidney, the nephron. Each human kidney contains at least a million of these. They serve first to filter the blood and then to modify the filtrate into urine.

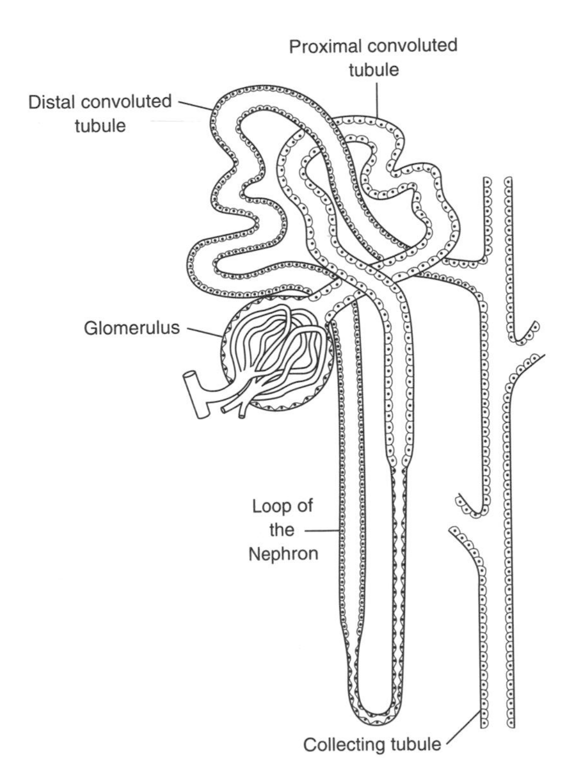

FIGURE 25.16
Diagram showing the essential features of a typical nephron in the human kidney.
Reprinted with permission from H. W. Smith, *The Physiology of the Kidney,* London: Oxford University Press, 1937, p. 6.

Filtration occurs in the glomerulus, which consists of a tuft of capillaries enclosed by an epithelial envelope called the glomerular capsule (formerly Bowman's capsule). Water and low-molecular-weight solutes, such as inorganic ions, urea, sugars, and amino acids (but not normally substances with molecular weights above 70,000, such as plasma proteins), escape from these capillaries and collect in the capsular space. This ultrafiltrate of plasma then passes through the proximal convoluted tubule, where most of the water and solutes are reabsorbed. The tubule fluid continues through the loop of the nephron (loop of Henle) and through the distal convoluted tubule, where further reabsorption of some solutes or secretion of others occurs. The tubule fluid then passes into the collecting tubule, where concentration can occur if necessary. The fluid may now be called urine; it contains 1% or less of the water and solutes of the original glomerular filtrate.

The kidney regulates acid–base balance by controlling bicarbonate reabsorption and by secreting acid. Both of these processes depend on formation of H^+ and HCO_3^- from CO_2 and H_2O within the tubule cells, shown in Figure 25.17*a*. The H^+ formed in this reaction is then actively secreted into the tubule fluid in exchange for Na^+. Sodium uptake by the tubule cell is partly passive, with Na^+ flowing down the electrochemical gradient, and partly active, via a Na^+, H^+-antiport system. At this point sodium has been reabsorbed in exchange for H^+, and sodium bicarbonate has been generated within the tubule cell. The sodium bicarbonate is then pumped out of the cell into the interstitial fluid, which equilibrates with the plasma.

The H^+ that has been secreted into the tubule fluid can now experience one of three fates. First, it can react with a HCO_3^-, as shown in Figure 25.17*b*, to form CO_2 and H_2O. The overall net effect of this process is to move sodium bicarbonate from the tubule fluid back into the interstitial fluid. The name given to this is **reabsorption of sodium bicarbonate.**

As reabsorption of sodium bicarbonate proceeds, the tubule fluid becomes depleted of HCO_3^-, and the pH drops from its initial value, which was identical to the pH of the plasma from which it was derived. As HCO_3^- becomes less available and the pH comes closer to the pK of the $HPO_4^{2-}/H_2PO_4^-$ buffer system, more and more of the H^+ will be taken up by this buffer. Buffering is the second fate the H^+ can experience, and it is represented in Figure 25.17*c*. The $H_2PO_4^-$ is not readily reabsorbed by the kidney. It passes out in the urine, and its loss represents net excretion of H^+.

Although phosphate is normally the most important buffer in the urine, other ions can become significant. For example, in diabetic ketoacidosis, plasma levels of acetoacetate and β-hydroxybutyrate are elevated. They pass through the glomerular filter, and appear in the tubule fluid. Since acetoacetic acid has a pK = 3.6 and β-hydroxybutyric acid has a pK = 4.7, as the urine pH approaches its minimum of 4.4, these will begin to serve as buffers.

The effect of buffering is not only to excrete acid, but to regenerate the bicarbonate that was lost when the acid was first neutralized. Let us consider a situation in which the metabolic defect of a diabetic patient has produced the elements of β-hydroxybutyric acid. The protons are neutralized by sodium bicarbonate, leaving sodium β-hydroxybutyrate. In the kidney, then, β-hydroxybutyrate appears in the filtrate, is converted to β-hydroxybutyric acid, which is excreted, and sodium bicarbonate is returned to the extracellular fluid. Net acid excretion and bicarbonate regeneration occur no matter what anion in the tubule fluid acts as the H^+ acceptor.

The amount of acid excreted as the acid component of a urinary buffer can be measured easily. One merely titrates the urine back to the normal pH of the plasma, 7.4. The amount of base required is identical to the amount of acid excreted in this form, and is referred to as the **titratable acidity** of the urine.

The formation of titratable acidity accounts for about one-third to one-half of our normal daily acid excretion. It is thus an important mechanism for acid excretion, and is capable of putting out as much as 250 meq of

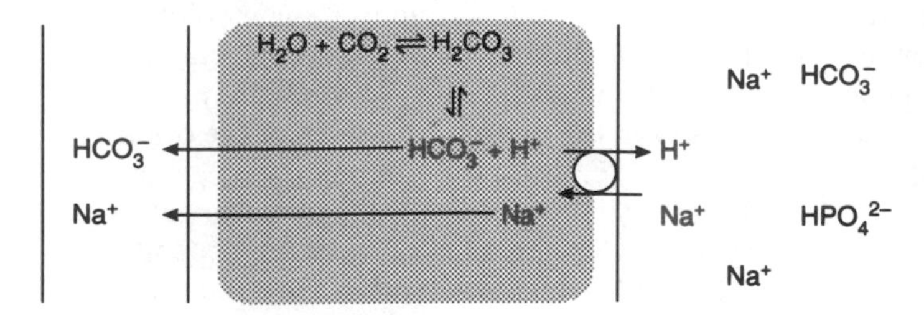

(*a*) Basic ion exchange mechanism

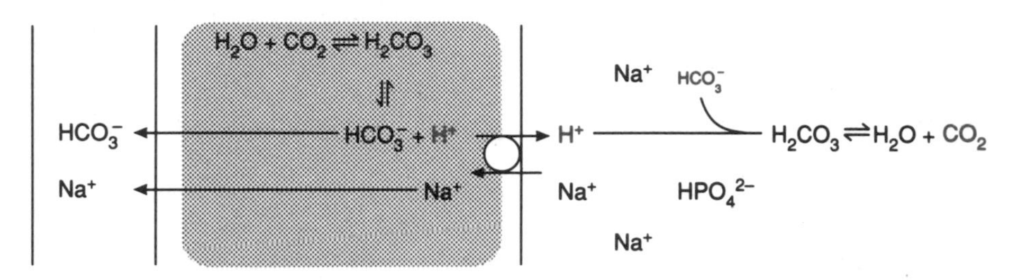

(*b*) Reabsorption of bicarbonate

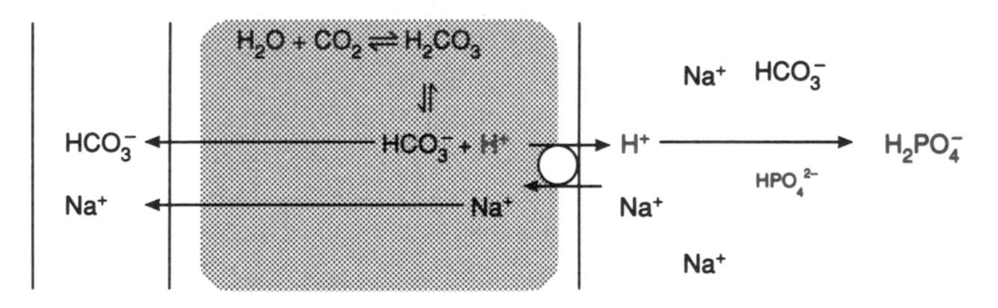

(*c*) Excretion of titratable acidity

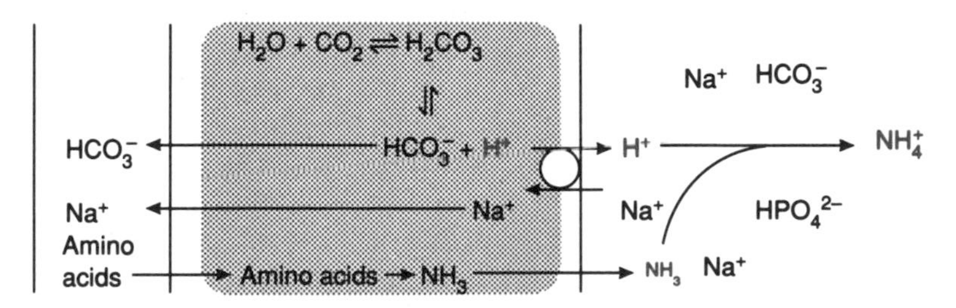

(*d*) Excretion of ammonia

FIGURE 25.17
Role of the exchange of tubular cell H^+ ions in tubular fluid in renal regulation of acid–base balance.
(*a*) Basic ion exchange mechanism. (*b*) Reabsorption of bicarbonate. (*c*) Excretion of titratable acid. (*d*) Excretion of ammonia.
Adapted from R. E. Pitts, Role of ammonia production and excretion in regulation of acid–base balance. *N. Engl. J. Med.* 284:32, 1971, with permission of the publisher.

acid daily. There is, however, a limit to the amount of acid that can be excreted in this manner. Titratable acidity can be increased only by lowering the pH of the urine or by increasing the concentration of buffer in the urine, and neither of these processes can proceed indefinitely. The urine pH cannot go below about 4.4; evidently the sodium-for-hydrogen exchange mechanism is incapable of pumping H^+ out of the tubule cells against more than a 1000-fold concentration gradient. Buffer excretion is limited not only by the solubility of the buffer, but by limitations to the supply of the buffer ion and of the cations that are necessarily part of the important buffer systems. If, for example, a 600 meq day^{-1} of acid load were excreted as NaH_2PO_4, the body would be totally depleted of sodium in less than 1 week.

The third fate the H^+ can experience in the tubule fluid is neutralization by NH_3. The tubule cells produce NH_4^+ from amino acids, particularly glutamine, as shown in Figure 25.17*d*. **Elimination of NH_4^+** in the urine contributes to net acid excretion.

The ammonium ion is normally a major urinary acid. Typically, one-half to two-thirds of our daily acid load is excreted as NH_4^+. For three reasons it becomes even more important in acidosis. In the first place, since the pK of NH_4^+ is 9.3, acid can be excreted in this form without lowering the pH of the urine, whereas formation of titratable acidity requires a decrease in urine pH. Second, enormous amounts of acid can be excreted in this form. Ammonia is readily available from amino acids, and in prolonged acidosis the NH_4^+ excretion system becomes activated. This activation, however, takes several days; it does not begin to adapt until after 2–3 days, and the process is not complete until 5–6 days after the onset of acidosis. Once complete, though, amounts of acid in excess of 500 meq can be excreted daily as NH_4^+. The third role of NH_4^+ in acidosis is that it spares the body's stores of Na^+ and K^+. Excretion of titratable acid, such as $H_2PO_4^-$, and of the anions of strong acids, such as acetoacetate, requires simultaneous excretion of a cation to maintain electrical neutrality. At the onset of acidosis this role is filled by Na^+, and as the body's Na^+ stores become depleted, K^+ excretion rises. If NH_4^+ did not then become available, even a moderate acidosis could quickly become fatal.

Total acid excretion, the **total acidity of the urine,** is the sum of the titratable acidity and NH_4^+. Strictly speaking, one should subtract from this sum the urinary HCO_3^-, but this correction is seldom made in practice. Obviously, in severe metabolic acidosis, where the total acid excretion would be of greatest interest, the urine would be so acidic that HCO_3^- would be nil.

In alkalosis the role of the kidney is simply to allow HCO_3^- to escape. Metabolic alkalosis is therefore seldom long-lasting unless alkali is continuously administered or HCO_3^- elimination is somehow prevented. HCO_3^- elimination may be restricted if the kidney receives a strong signal to conserve Na^+ at a time when there is a deficiency of an easily reabsorbable anion, such as Cl^-, to be reabsorbed with it. Some diuretics cause this. The first renal response is to put out K^+ in exchange for Na^+ from the tubule fluid, and when K^+ stores are depleted, H^+ is exchanged for Na^+. This results in the production of an acidic urine by an alkalotic patient. If NaCl is administered, alkalosis associated with volume and Cl^- depletion may correct itself.

25.12 COMPENSATORY MECHANISMS

We have defined four primary types of acid–base imbalances and we have seen their chemical causes. Respiratory acidosis arises from an increased plasma P_{CO_2}. Respiratory alkalosis is caused by a decreased plasma P_{CO_2}. In metabolic acidosis addition of strong organic or inorganic acid (or loss of HCO_3^-) results in a decreased plasma [HCO_3^-]. Conversely, in metabolic alkalosis loss of acid from the body or ingestion of alkali raises the plasma [HCO_3^-]. Recall that in an acute respiratory acid–base imbalance, as long as the body has not attempted to compensate, the pH will be abnormal, and the [HCO_3^-] will be somewhere on the buffer line. In an acute metabolic acid–base imbalance, if the patient has made no attempt to compensate, the pH will be abnormal and the [HCO_3^-] will be somewhere on the 40-mm isobar.

When the plasma pH deviates from the normal range, various compensatory mechanisms begin to operate. The general principle of compensation is that, since an abnormal condition has directly altered one of the terms of the [HCO_3^-]/[CO_2] ratio, the plasma pH can be readjusted back toward normal by a compensatory alteration of the other term. For example, if a diabetic patient becomes acidotic due to excess production of ketone bodies, plasma [HCO_3^-] will decrease. Compensation would involve decreasing the plasma [CO_2] so that the [HCO_3^-]/[CO_2] ratio, and therefore the pH, is readjusted back toward normal. Notice that compensation does not involve a return of [HCO_3^-] and [CO_2] toward normal. Rather, compensation is a secondary alteration in one of these, an alteration that has the effect of counteracting the primary alteration in the other. The result is that the plasma pH is readjusted toward normal. That this is necessarily so is evident from the Henderson–Hasselbalch equation.

$$\mathrm{pH} = 6.1 + \log \frac{[\mathrm{HCO_3^-}]}{0.03\ P\mathrm{CO_2}}$$

If [HCO_3^-] changes, the only way to restore the original [HCO_3^-]/[CO_2] ratio is to change $P\mathrm{CO_2}$ in the *same direction*. If the primary change is in $P\mathrm{CO_2}$, the original ratio can be restored only by altering [HCO_3^-] in the *same direction*.

Although some compensatory mechanisms begin to operate rapidly and produce their effects rapidly, others are slower. Several stages of compensation may therefore be seen. First is the acute stage, before any significant degree of compensation could possibly occur. After the acid–base imbalance has been in effect for a period of time the patient may become **compensated.** This means the compensatory mechanisms have come into play in a normal manner, as expected on the basis of experience with other individuals with an acid–base imbalance of similar type and degree. The "compensated state" does not necessarily imply that the plasma pH is within the normal range. Alternatively, the patient may show no sign of compensation, even though compensation is expected. This state is referred to as **uncompensated;** it arises because compensation cannot occur due to some other abnormality. Finally, there is an intermediate state where compensation is occurring but is not yet as complete as it should be. This is the **partially compensated** state. Factors that limit the compensatory processes will be discussed at the end of this section.

Let us now follow the course of acute onset of each type of acid–base imbalance and of the compensatory process. Each of these will be schematically illustrated in a pH–bicarbonate diagram.

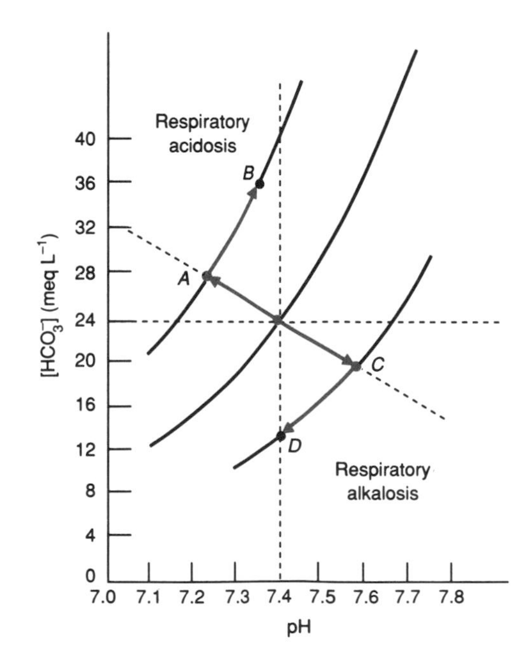

FIGURE 25.18
pH–Bicarbonate diagram showing compensation for respiratory acidosis (normal state to point *B*) and for respiratory alkalosis (normal state to point *D*).

Imagine an individual in normal acid–base balance who goes into an acute respiratory acidosis as a result of breathing a gas mixture containing a high level of CO_2. As $P\mathrm{CO_2}$ rises, plasma pH will drop and [HCO_3^-] will rise. (If a decrease in pH and a simultaneous rise in [HCO_3^-] suddenly seems anomalous, turn back to Figure 25.13 and the text on p. 1043, and review the blood buffer line.) The point describing his or her condition will follow the buffer line to point A, as shown in Figure 25.18. Eventually a new steady-state $P\mathrm{CO_2}$ will be established in the alveoli and in the blood, and no further change in $P\mathrm{CO_2}$ will occur. The abnormal condition has fixed this patient on an abnormally high CO_2 isobar. If the condition is returned to normal, he can drop back to the 40 mmHg (5.33 kPa) isobar and all will be well, but until that time all compensatory processes are confined to the higher CO_2 isobar. Compensation will, of course, consist of renal excretion of H^+. Since this is a bicarbonate-producing process, $[HCO_3^-]$ should rise, even though it is already above normal. This could have been predicted from the pH–HCO_3^- diagram with no knowledge of the renal mechanism of compensation. Since it is assumed that the individual is fixed on the high CO_2 isobar by the abnormal condition, the only way the pH can possibly be adjusted toward normal is by sliding up the isobar to point *B* in Figure 25.18. This movement is necessarily linked to an increase in $[HCO_3^-]$. Thus the correct analysis of this compensation could be made either from an understanding of the nature of the compensatory mechanism or from an appreciation of the physical chemistry of the bicarbonate buffer system as expressed in the pH–HCO_3^- diagram.

Although the path we have described, up the buffer line to point *A* and then up the isobar to point *B*, is a real possibility, it is also possible that a respiratory acidosis would develop gradually, with compensation occurring simultaneously. The points describing this progress would fall on the curved line from the normal state to point *B*.

In sudden onset respiratory alkalosis P_{CO_2} drops rapidly. The pH rises and $[HCO_3^-]$ falls, following the buffer line to point *C* in Figure 25.18. Clin. Corr. 25.6 describes a case of acute respiratory alkalosis. As with respiratory acidosis, unless the cause of the decreased alveolar P_{CO_2} is removed, the patient is fixed on an abnormal CO_2 isobar. Compensation consists of renal excretion of HCO_3^-; plasma $[HCO_3^-]$ diminishes (at a fixed, subnormal P_{CO_2}), and the plasma pH decreases toward normal. This is described in Figure 25.18 by movement along the isobar from point *C* to point *D*. With a gradual onset of respiratory alkalosis, the bicarbonate buffer system would follow points along the curved line from the normal state to point *D*.

In metabolic acidosis two mechanisms are usually available for dealing with the excess acid. The kidneys increase their H^+ excretion, but this takes time, and is not adequate to return $[HCO_3^-]$ and the pH to normal. The other mechanism, which begins to operate almost instantly, is respiratory compensation. The acidosis stimulates the respiratory system to hyperventilate, decreasing the P_{CO_2}. Thus, if onset of a primary metabolic acidosis is represented in Figure 25.19 by a fall in plasma $[HCO_3^-]$ along the 40 mmHg (5.33 kPa) isobar from the normal state to point *E*, the compensatory decrease in P_{CO_2} and the concomitant rise in pH will be along the line from *E* to *F*. Notice that this line is parallel to the buffer line, and so compensation for a metabolic acidosis involves not only the expected decrease in P_{CO_2} but also a *further* small decrease in $[HCO_3^-]$. This is due to the same factor that causes the buffer line itself to have a slope: titration of the nonbicarbonate buffers. The inevitability and the magnitude of the further decrease in $[HCO_3^-]$ can be seen clearly in the pH–bicarbonate diagram.

The principles governing compensation for metabolic alkalosis are like those for metabolic acidosis, but everything is operating in the opposite direction. In metabolic alkalosis the primary defect is an increase in plasma $[HCO_3^-]$; it rises from the normal state to point *G* in Figure 25.19. The immediate physiological response is hypoventilation, followed by increased renal excretion of HCO_3^-. As a result of the hypoventilation the P_{CO_2} increases along the line from *G* to *H*, and a further small rise in $[HCO_3^-]$ occurs.

The respiratory response to a metabolic acid–base imbalance is rapid, and the bicarbonate buffer system would in most cases be expected to follow points along the curved line from the normal state to the compen-

CLIN. CORR. 25.6
ACUTE RESPIRATORY ALKALOSIS

An anesthetized surgical patient with a uretheral catheter in place was hyperventilated as an adjunct to the general anesthesia. Prior to hyperventilation normal values of plasma P_{CO_2} and pH were obtained. Alveolar ventilation was then increased mechanically, and a new steady state was reached, in which the plasma P_{CO_2} was 25 mmHg and the pH was 7.55. Plasma HCO_3^- was not directly measured, but interpolation from a pH–bicarbonate diagram (e.g., Figure 25.12) or calculation from the Henderson–Hasselbalch equation reveals that the plasma $[HCO_3^-]$ decreased to 21.2 meq L^{-1}. Analysis of the urine showed negligible loss of HCO_3^- through the kidneys. It can be concluded that the decrease in $[HCO_3^-]$ was due to titration of bicarbonate by the acid components of the body's buffer systems. The point representing the patient's new steady-state condition clearly must be on the buffering line that represents whole-body buffering. (Since the buffers of the whole body are not identical in type or concentration to the blood buffers, the buffer line for the whole body will be analogous, but not identical, to the blood buffer line.)

Magarian, G. J. *Medicine* 61:219, 1982.

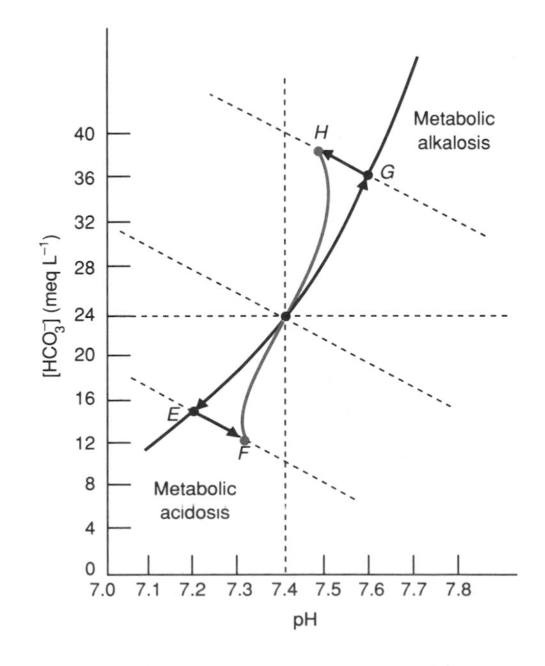

FIGURE 25.19
pH–Bicarbonate diagram showing compensation for metabolic acidosis (normal state to point *F*) and for metabolic alkalosis (normal state to point *H*).

CLIN. CORR. 25.7
CHRONIC RESPIRATORY ACIDOSIS

H.W. was admitted to the hospital with marked dyspnea, cyanosis, and signs of mental confusion. As his acute problems were relieved by appropriate treatment, his symptoms disappeared except for a continuing dyspnea. Blood gas analysis performed 8 days later yielded the following data: pH, 7.32; $P\text{CO}_2$, 70 mmHg; $[HCO_3^-]$, 34.9 meq L^{-1}. This is a typical compensation for this degree of chronic respiratory acidosis.

Another patient, C.Q., with chronic obstructive lung disease was found to have arterial plasma pH, 7.40; $[HCO_3^-]$, 35.9 meq L^{-1}; and $P\text{CO}_2$, 60 mmHg. A return of the plasma pH to 7.4 is not expected in the average patient with respiratory acidosis. Some patients are capable of attaining it, although it becomes less and less likely as $P\text{CO}_2$ increases. In this case, with a $P\text{CO}_2$ of 60 mmHg, a plasma pH of 7.4 lies outside the 95% probability range. Close questioning of the patient revealed that he had surreptitiously been taking a relative's thiazide diuretic, which superimposed a metabolic alkalosis upon respiratory acidosis.

Rastegar, A. and Thier, S. O. *Chest* 63:355, 1972.

sated state. An acute metabolic imbalance will not generally be seen outside the experimental laboratory. Indeed, if a physician sees a patient whose plasma pH, $[HCO_3^-]$, and $P\text{CO}_2$ are consistent with an acute metabolic imbalance, he concludes that the patient's compensatory mechanisms are impaired and that the patient cannot compensate. The patient would be suffering a mixed respiratory and metabolic acidosis or a mixed respiratory and metabolic alkalosis. Obviously, if a patient had a primary acidosis of one type (respiratory or metabolic) and a primary alkalosis of the other, both caused by independent diseases, the effects of the two on plasma pH would tend to cancel. But even if the pH were within the normal range due to such a circumstance, $[HCO_3^-]$ and $P\text{CO}_2$ would be abnormal.

How complete can the process of compensation be? Can the body totally compensate (bring the pH back to the normal range) for any imbalance? Generally, the answer is no. The organs used in compensation, the lungs and the kidneys, do not exist exclusively to deal with acid–base imbalance. There is a limit to how much one can hyperventilate; it is simply impossible to move air into and out of the lungs at an indefinitely high rate for an indefinitely long time. Also, one cannot suspend respiration merely to raise $P\text{CO}_2$ to some desired level. The kidney, too, has limits. As the $P\text{CO}_2$ rises above 70 mmHg (9.33 kPa) in respiratory acidosis, renal mechanisms for reabsorbing HCO_3^- fail to keep pace, and further increases in plasma $[HCO_3^-]$ are only about what could be expected from titration of the nonbicarbonate buffers (Clin. Corr. 25.7). In respiratory alkalosis renal excretion of excess HCO_3^- can, with time, be sufficient to return the plasma pH to within the normal range. Individuals who dwell at high altitude are typically in compensated respiratory alkalosis, with their plasma pH within the normal range. For the other types of acid–base imbalance the exact degree of compensation expected of a patient with a given clinical picture is well worked out, but a detailed discussion is beyond the scope of this chapter. Suffice it to say that if a patient is compensating, but not as well as expected, this is taken to mean that the patient cannot compensate appropriately and must therefore have a mixed acid–base disturbance.

25.13 ALTERNATIVE MEASURES OF ACID–BASE IMBALANCE

Modern clinical laboratories generally report plasma bicarbonate concentration, and the value is used by the physician just as we have used it here. Some laboratories however, report **total plasma CO_2.** Total plasma CO_2 as reported by the clinical laboratory is the sum of bicarbonate and dissolved CO_2, and so is always slightly higher than $[HCO_3^-]$. At pH 7.4, for example, the ratio of $[HCO_3^-]$ to $[CO_2]$ is 20 : 1 (dissolved CO_2 is only 1 : 21 of the total CO_2); if $[HCO_3^-]$ is 24 meq L^{-1}, $[CO_2]$ is 1.2 meq L^{-1} and total CO_2 is 25.2 meq L^{-1}. At pH 7.1, HCO_3^- is still 10 times as concentrated as dissolved CO_2. Because the major contributor to total CO_2 is HCO_3^-, total CO_2 is often used in the same manner as bicarbonate to make clinical judgments. Strictly speaking, total CO_2 also includes carbamino groups, but current clinical laboratory practice is to ignore carbamino groups when making a blood gas and pH report. If, however, they were included in a total CO_2 measurement, it would not change the interpretation of the measurement, since carbamino groups, like dissolved CO_2, represent only a small fraction of the total CO_2.

The clinical importance of bicarbonate as a gauge of the whole body's ability to buffer further loads of metabolic acid (see Clin. Corr. 25.8) has given rise to several ways of expressing what the [HCO_3^-] would be if there were no respiratory component or respiratory compensation involved in a patient's condition. The **base excess** is one of the more common of these expressions. It is defined as the amount of acid that would have to be added to the blood to titrate it to pH 7.4 at a $P\text{CO}_2$ of 40 mmHg (5.33 kPa) at 37°C. Since the titration is carried out at the normal $P\text{CO}_2$, only the metabolic contribution to acid–base imbalance (primary metabolic imbalance *and* nonrespiratory compensatory processes) would be measured. If a blood sample were acidic under the conditions of the titration, alkali would have to be added instead of acid, and the base excess would be negative.

The concept and the quantitation of base excess are most easily understood if we refer to the pH–bicarbonate diagram. In our discussion of the blood buffer line we saw how increasing the $P\text{CO}_2$ in blood, where other buffers are present, would result in a rise in [HCO_3^-] and a virtually identical decrease in the concentration of other buffer bases. This was because equivalent amounts of the other buffer bases were consumed as they buffered carbonic acid. Since virtually all the carbonic acid formed was buffered, for every HCO_3^- formed one conjugate base of some other system was consumed. In this situation the *total* base in the blood is not measurably changed; only the distribution of HCO_3^- and nonbicarbonate buffer conjugate base is changed. Thus, as long as one remains on the blood buffer line, [HCO_3^-] can change but total base will not. There will be no positive or negative base excess.

If, however, renal activity, diet, or some metabolic process adds or removes HCO_3^-, then a positive or negative base excess will be seen. The patient's status will no longer be described by a point on the buffer line, and the base excess will be the difference between the observed plasma [HCO_3^-] and the [HCO_3^-] on the buffer line at the same pH. This is shown in Figure 25.20. In order to calculate this difference, the position of the

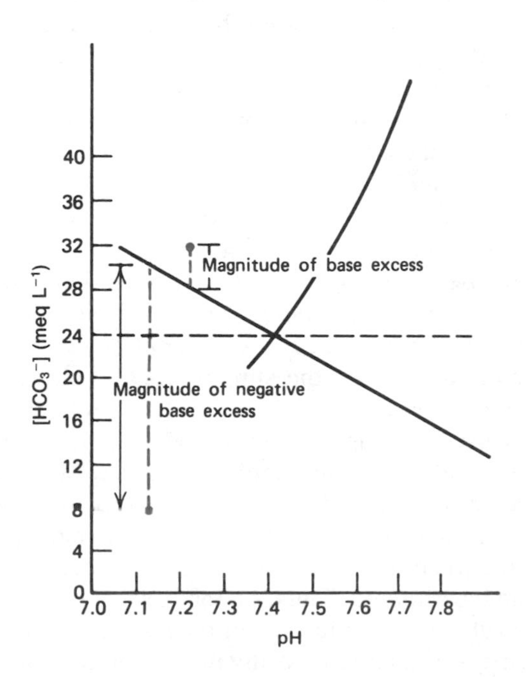

FIGURE 25.20
Calculation of base excess for a point above the blood buffer line, and calculation of negative base excess for a point below the blood buffer line.
The base excess is 32 − 28 = 4 meq L^{-1}. The negative base excess is 30 − 8 = 22 meq L^{-1}.

CLIN. CORR. 25.8
SALICYLATE POISONING

Salicylates are the most common cause of poisoning in children. A typical pathway of salicylate intoxication is plotted in the accompanying figure. The first effect of salicylate overdose is stimulation of the respiratory center, resulting in respiratory alkalosis. Renal compensation occurs, lowering the plasma [HCO_3^-]. A second, delayed effect of salicylate may then appear, metabolic acidosis. Since [HCO_3^-] had been lowered by the previous compensatory process, the victim is at a particular disadvantage in dealing with the metabolic acidosis. In addition, but not shown in the graph, respiratory stimulation sometimes persists after the acidosis has run its course. Rational management of salicylate intoxication requires knowledge of the plasma pH and the plasma [HCO_3^-] or its equivalent throughout the course of the condition.

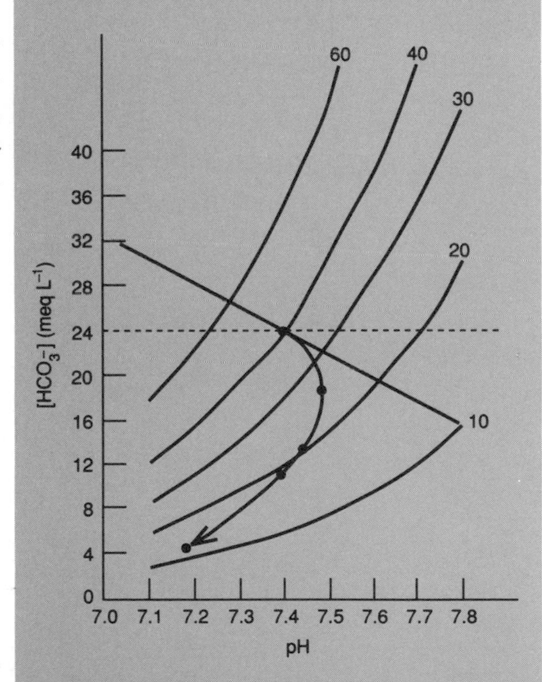

A typical pathway of salicylate intoxication.
Data replotted from R. B. Singer, The acid–base disturbance in salicylate intoxication. *Medicine* 33:1–13, 1954.

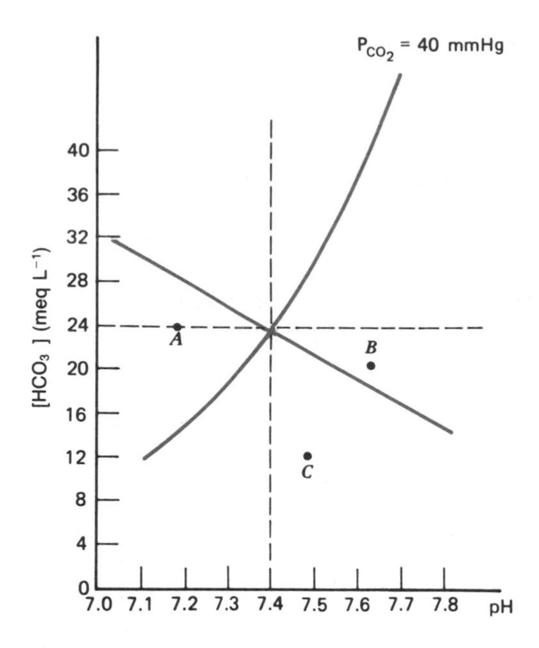

FIGURE 25.21
Examples showing the sign of the base excess at various points.
At points *A* and *C* there is a negative base excess. At point *B* the base excess is positive.

buffer line, which can be determined from knowledge of the slope and the point representing the normal state, must be known. In the clinical laboratory it can be estimated by measuring the hemoglobin concentration and assuming that it is the major nonbicarbonate buffer.

The buffer line, then, is the dividing line between positive and negative base excess. Any point above it is in the region of positive base excess, and any point below it is in the region of negative base excess. This gives rise to situations that may seem peculiar at first. In Figure 25.21 the [HCO_3^-] at point *A* is normal, but the patient has a negative base excess. A positive or negative base excess occurs as a result of *compensation* for a respiratory acid–base imbalance or *directly* from a metabolic one. Respiratory compensation for a metabolic acid–base imbalance, since it involves movement along a line parallel to the buffer line (Figure 25.19), would cause no further change in the value of the base excess. Clin. Corr. 25.9 involves consideration of base excess.

25.14 THE SIGNIFICANCE OF Na^+ AND Cl^- IN ACID–BASE IMBALANCE

An important concept in diagnosing certain acid–base disorders is the so-called anion gap. Most clinical laboratories routinely measure plasma Na^+, K^+, Cl^-, and HCO_3^-. A glance back at the graph in Figure 25.10 will confirm that in the plasma of a normal individual the sum of Na^+ and K^+ is greater than the sum of Cl^- and HCO_3^-. This difference is called *A*, the **anion gap;** it represents the other plasma anions (Figure 25.10), which are not routinely measured. It is calculated as follows:

$$A = (Na^+ + K^+) - (Cl^- + HCO_3^-)$$

The normal value of *A* is in the range of 12–16 meq L^{-1}. In some clinical laboratories K^+ is not measured; then the normal value is 8–12 meq L^{-1}. The gap is changed only by conditions that change the sum of the cations or the sum of the anions, or by conditions that change both sums by different amounts. Thus administration or depletion of sodium bicarbonate would not change the anion gap because [Na^+] and [HCO_3^-] would be affected equally. Metabolic acidosis due to HCl or NH_4Cl administration would also leave the anion gap unaffected; here [HCO_3^-] would decrease, but [Cl^-] would increase by an equivalent amount, and the sum of [HCO_3^-] plus [Cl^-] would be unchanged. In contrast, diabetic ketoacidosis or methanol poisoning involves production of organic acids, which react with HCO_3^-, decreasing its concentration. But since the [HCO_3^-] is replaced by some organic anion, the sum of [HCO_3^-] plus [Cl^-] decreases, and the anion gap increases.

The anion gap is most commonly used to establish a differential diagnosis for metabolic acidosis. In a metabolic acidosis with an increased anion gap, H^+ must have been added to the body with some anion other than chloride. Metabolic acidosis without an increased anion gap must be due either to accumulation of H^+ with chloride or to a decrease in the concentration of sodium bicarbonate. Thus, on the basis of the anion gap, certain diseases can be ruled out, while certain others would have to be considered. This information can be especially important in dealing with patients who cannot give good histories.

The electrolytes of the body fluids interact with each other in a multitude of ways. One of the most important of these involves the capacity for K^+ and H^+ to substitute for one another under certain circumstances. This can occur in the cell, where as we have seen, K^+ is the major cation.

CLIN. CORR. 25.9
EVALUATION OF CLINICAL ACID–BASE DATA

In a 1972 study of total parenteral nutrition of infants, it was found that infants who received amino acids in the form of a hydrolyzate of the protein fibrin maintained normal acid–base balance. In contrast, infants receiving two different mixtures of synthetic amino acids, FreAmine and Neoaminosol, became acidotic. Both synthetic mixtures contained adequate amounts of all the essential amino acids, but neither contained aspartate or glutamate. The fibrin hydrolyzate contained all of the common amino acids.

The accompanying figure shows the blood acid–base data from these infants. Notice that the normal values for infants, given by the dotted lines, are not quite the same as normal values for adults. (A child is not a small adult.)

The blood pH data show that the infants receiving synthetic mixtures were clearly acidotic. The low $[HCO_3^-]$ of the Neoaminosol group immediately suggests a metabolic acidosis, and the P_{CO_2} and base excess data are compatible with this interpretation. The FreAmine group, however, shows nearly normal $[HCO_3^-]$, and all of these infants have elevated P_{CO_2} values. The P_{CO_2} values indicate respiratory acidosis, but a simple respiratory acidosis should be associated with a slightly elevated $[HCO_3^-]$. The absence of this finding in most of the infants indicates that the acidosis must also have a metabolic component. This is confirmed by the observation that all the infants receiving FreAmine have a significant negative base excess.

The infants with mixed acid–base disturbances did, in fact, have pneumonia or respiratory distress syndrome. The metabolic acidosis, which all the infants receiving synthetic mixtures experienced, was due to synthesis of aspartic acid and glutamic acid from a neutral starting material (presumably glucose). Subsequent incorporation of these acids into body protein imposed a net acid load upon the body. Addition of aspartate and/or glutamate to the synthetic mixtures was proposed as a solution of the problem.

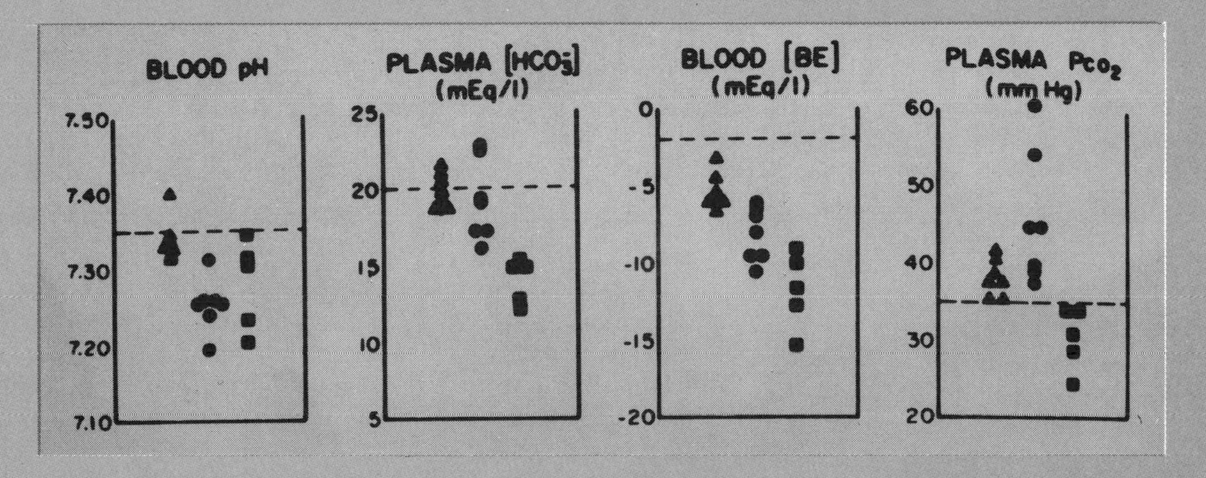

Blood acid–base data of patients receiving fibrin hydrolysate (▲) and of those receiving synthetic L-amino acid mixtures, FreAmine (●) and Neoaminosal (■).
Values are those observed at the time of the lowest blood base excess. Dashed lines represent accepted normal values for infants.
Adapted from W. C. Heird, *N. Engl. J. Med.* 287:943, 1972.

In acidosis intracellular $[H^+]$ rises, and it replaces some of the intracellular K^+. The displaced K^+ appears in the plasma, and in time is excreted by the kidneys. This leaves the patient with normal plasma $[K^+]$ (normokalemia), but with seriously depleted body K^+ stores (hypokalia). Subsequent excessively rapid correction of the acidosis may then reverse events. As the plasma pH rises, K^+ flows back into the cells, and plasma $[K^+]$ may decline to the point where muscular weakness sets in and respiratory insufficiency may become life threatening.

In the kidney the reciprocal relationship between K^+ and H^+ results in an association between metabolic alkalosis and hypokalemia. If hypokalemia arises from long-term insufficiency of dietary potassium or long-term diuretic therapy, K^+ levels in the cells will diminish, and intracellular $[H^+]$ will rise. This leads to increased acid excretion, acidic urine, and an

CLIN. CORR. 25.10
METABOLIC ALKALOSIS

Prolonged gastric lavage produces a metabolic alkalosis that is a good experimental model of the metabolic alkalosis that results from repeated vomiting. The following table gives plasma and urine acid–base and electrolyte data from a healthy volunteer on a low sodium diet who, after a control period, was subjected to gastric lavage for two days. After a 5-day recovery period he was placed on a low-potassium diet and given a sodium (130 meq day^{-1}) and chloride (121 meq day^{-1}) supplement.

During the control period the data are within normal limits. After gastric lavage that selectively removed HCl (Na^+, K^+, and H_2O lost with the gastric juice were restored), an uncomplicated metabolic alkalosis developed. Notice that the subject excreted an alkaline urine, containing a substantial amount of HCO_3^-. The Na^+ excretion increased, depleting the body's Na^+ stores. Plasma P_{CO_2} was not measured, but plotting the values of pH and HCO_3^- on a pH–bicarbonate diagram (e.g., Figure 25.12) allows one to interpolate a value of about 47 mmHg. Clearly, respiratory compensation was occurring. Plasma [K^+] was decreased. Plasma [Cl^-] decreased, but no more than would be expected on the basis of the changes in [Na^+], [K^+], and [HCO_3^-].

When the subject was placed on a low-potassium diet the alkalosis grew worse, and plasma [HCO_3^-] rose. Additional compensatory hypoventilation evidently prevented a further rise in plasma pH. Notice, though, that the urine became acid, in spite of the increased severity of the alkalosis. The Na^+ was conserved, not in exchange for K^+, but in exchange for H^+. After several days of Na^+ and Cl^- administration, however, the subject was able to restore the depleted Cl^-, excrete the excess HCO_3^- and repair the acid–base imbalance with no other treatment.

	Control	After lavage	Low KCl	After NaCl
PLASMA				
pH	7.4	7.50	7.48	7.41
HCO_3^- (meq L^{-1})	29.3	35.3	38.1	26.1
Na^+ (meq L^{-1})	138	134	141	144
K^+ (meq L^{-1})	4.2	3.2	2.9	3.2
Cl^- (meq L^{-1})	101	88	85	108
URINE				
pH	6.12	7.48	5.70	7.19
HCO_3^- (meq L^{-1})	3	51	1	17
NH_4^+ (meq L^{-1})	22	4	36	14
Titratable acidity (meq L^{-1})	10	0	14	1
Total acidity (meq L^{-1})	29	−49	49	−2
Na^+ (meq L^{-1})	2	28	1	95

SOURCE: Data from J. P. Kassirer and W. B. Schwartz, *Am. J. Med.*, 40:10, 1966.

alkaline arterial plasma pH. We have already seen how in an alkalotic individual a hormonal signal to absorb Na^+ can lead to K^+ loss and then to an exacerbation of the metabolic alkalosis (p. 1049). The association also operates in the opposite direction, with alkalosis leading to hypokalemia. In this case increased amounts of $Na^+ + HCO_3^-$ are presented to the distal convoluted tubule, where all K^+ secretion normally takes place (all *filtered* K^+ is reabsorbed; K^+ loss is due to distal tubular secretion). The distal tubule takes up some of the Na^+, but since HCO_3^- does not readily follow across that membrane, the increased Na^+ uptake is linked to increased K^+ secretion. The K^+ excretion is complicated, with its control under the influence of a variety of hormones and other factors. The end result, however, is that metabolic alkalosis and hypokalemia go hand in hand, so much so that in some circles the term, "hypokalemic alkalosis," is used synonymously with metabolic alkalosis. Clin. Corr. 25.10 discusses a case of experimental metabolic alkalosis in which this occurred.

BIBLIOGRAPHY

Gas Transport

Bunn, H. F. and Forget, B. G. *Hemoglobin: molecular, genetic, and clinical aspects*. Philadelphia: Saunders, 1986.

Bunn, H. F., Gabbay, K. H., and Gallop, P. M. The glycosylation of hemoglobin: relevance to diabetes mellitus. *Science* 200:21, 1978.

Kilmartin, J. V. Interaction of haemoglobin with protons, CO_2 and 2,3-diphosphoglycerate. *Br. Med. Bull.* 32:209, 1976.

Perutz, M. F. and Lehmann, H. Molecular pathology of human haemoglobin. *Nature (London)*. 219:902, 1968.

Steffes, M. W. and Mauer, S. M. Toward a basic understanding of diabetic complications. *N. Engl. J. Med.* 325:883, 1991.

pH Regulation

Davenport, H. W. *The ABC of acid–base chemistry,* 6th ed. Chicago: University of Chicago Press, 1974.

Gabow, P. A., et al. Diagnostic importance of an increased serum anion gap. *N. Engl. J. Med.* 303:854, 1980.

Gamble, J. L., Jr., and Bettice, J. A. Acid-base relationships in the different body compartments: the basis for a simplified diagnostic approach. *Johns Hopkins Med. J.* 140:213, 1977.

Masoro, E. J. and Siegel, P. D. *Acid–base regulation: its physiology, pathophysiology and the interpretation of blood–gas analysis,* 2nd ed. Philadelphia: Saunders, 1977.

Siggaard-Andersen, O. *The acid–base status of the blood,* 4th ed. Baltimore: Williams & Wilkins, 1974.

QUESTIONS

J. BAGGOTT AND C. N. ANGSTADT

Question Types are described on page xxiii.

1. (QT1) During a breathing cycle:
   A. the alveolar gases are completely exchanged for atmospheric gases.
   B. gas exchange between the alveoli and the capillary blood can occur at all times.
   C. gas exchange with the capillary blood occurs at the surface of all the airways.
   D. there is net uptake of nitrogen by the blood.
   E. atmospheric water vapor is taken up by the lungs.
2. (QT1) From an oxygen saturation curve for normal blood we can determine that:
   A. P_{50} is in the Po_2 range found in extrapulmonary tissues.
   B. oxygen binding is hyperbolic.
   C. an oxygen carrier is necessary.
   D. tighter oxygen binding occurs at lower Po_2.
   E. shifts of the curve to the left or right would have little effect on oxygen delivery.

A. α-Chains of hemoglobin C. Both
B. β-Chains of hemoglobin D. Neither

3. (QT4) Found in HbA, HbA_2, and HbF.
4. (QT4) Modified in the formation of HbA_{1c}.

5. (QT2) At a Po_2 of 30 mmHg hemoglobin's percent saturation will:
   1. increase with increasing temperature.
   2. decrease with decreasing pH.
   3. increase with increasing Pco_2.
   4. decrease with increasing 2,3-diphosphoglycerate concentration.
6. (QT2) Significant contributor(s) to the total carbon dioxide of whole blood include(s):
   1. bicarbonate ion.
   2. dissolved carbon dioxide (CO_2).
   3. carbamino hemoglobin.
   4. carbonic acid (H_2CO_3).

7. (QT1) 2,3-Diphosphoglycerate (DPG):
   A. is absent from the normal erythrocyte.
   B. is a homotropic effector for hemoglobin.
   C. binds more tightly to HbF than to HbA.
   D. synthesis increases when hemoglobin's T ⇌ R equilibrium is shifted in favor of the T state.
   E. synthesis decreases when the erythrocyte pH rises.

8. (QT3) A. Contribution of intracellular buffers to minimizing changes in pH due to increased Pco_2.
   B. Contribution of extracellular buffers to minimizing changes in pH due to increased Pco_2.
9. (QT3) A. Increase in blood $[H^+]$ when Pco_2 rises from 40 to 75 mmHg.
   B. Increase in blood $[HCO_3^-]$ when Pco_2 rises from 40 to 75 mmHg.
10. (QT3) A. Ability of normal blood to buffer excess CO_2.
    B. Ability of anemic blood to buffer excess CO_2.

11. (QT2) Which of the following produce(s) H^+?
    1. Formation of bicarbonate ion from CO_2 and water.
    2. Formation of carbamino hemoglobin from CO_2 and hemoglobin.
    3. Binding of oxygen by hemoglobin.
    4. Oxidation of sulfur-containing amino acids.
12. (QT2) A substantial fraction of the urinary acid of a normal individual consists of:
    1. H_2CO_3.
    2. NH_4^+.
    3. acetoacetate.
    4. $H_2PO_4^-$.

13. (QT1) In a patient with diabetic ketoacidosis of long duration:
    A. the major urinary acid is $H_2PO_4^-$.
    B. hemoglobin's oxygen dissociation curve would be shifted to the right.
    C. the distribution of hemoglobin species would be the same as in a normal individual.
    D. 1 mol of bicarbonate is regenerated for every mole of $H_2PO_4^-$ formed in the renal tubule.
    E. hypoventilation would be expected.

14. (QT2) The following laboratory data are obtained from a patient:
    Pco_2 = 60 mmHg, HCO_3^- = 27 meq L^{-1}, pH = 7.28; these values define a point on the patient's blood buffer line.
    1. The patient has an acute condition.
    2. The condition would lead to production of an alkaline urine.
    3. Of the blood buffers, hemoglobin is the most important in resisting this pH change.
    4. Increasing the alveolar Pco_2 could restore the plasma $[HCO_3^-]$ to normal.
15. (QT2) During the process of compensation for a metabolic acid–base imbalance, which of the following would become increasingly abnormal?
    1. Plasma pH 3. Base excess
    2. Blood Pco_2 4. Plasma HCO_3^-

16. (QT2) In respiratory alkalosis:
    1. the acute state is associated with an abnormally low plasma [HCO_3^-].
    2. the mechanism of compensation causes an increase in the plasma [HCO_3^-].
    3. the plasma pH may be within the normal range in the fully compensated state.
    4. in the partially compensated state, there will be a negative base excess equal to the difference between 24 meq L^{-1} and the actual plasma [HCO_3^-].
17. (QT2) Hypokalemia can be expected to:
    1. occur if the plasma pH is rapidly raised.
    2. lead to increased urine acidity.
    3. be associated with a high plasma [HCO_3^-].
    4. raise the value of the anion gap.

ANSWERS

1. B A and C: The alveoli, where gas exchange with the blood occurs, are of constant size, and exchange gases with the airways by diffusion. D,E: Water vapor and CO_2 are added to the alveolar gases by the lung tissue, diluting the nitrogen (p. 1027).
2. A Po_2 of tissues is typically in the neighborhood of 20 mm in active muscle and is higher in less active situations. The normal P_{50}, 27 mm, is in this range (p. 1028). B and D: The curve is sigmoid, with tighter binding at higher Po_2 (p. 1028, Figure 25.2). C: If O_2 were soluble enough in plasma, no carrier would be necessary (p. 1026). E: Shifts profoundly affect delivery (p. 1028).
3. A It is the non-α chain that differs (p. 1030).
4. B The β chains are nonenzymatically glycosylated (p. 1030).
5. C 2 and 4 true. 1–3: High temperature, low pH (and therefore high Pco_2) favor dissociation; that is, decreased saturation (p. 1032). 4: High DPG has the same effect (p. 1029).
6. A 1–3 true. Carbonic acid is present in very small amounts; the equilibrium strongly favors CO_2 and H_2O (p. 1034, Table 25.2; see also p. 1042).
7. D A and B: DPG is a normal component of the red cell, where it serves as a heterotropic effector of HbA (p. 1029, Figure 25.4). C: It binds weakly or not at all to the HbF (p. 1031). D and E: DPG binds to the T state, relieving product inhibition of DPG synthesis; DPG synthesis is also inhibited by low pH (p. 1038).
8. A The bicarbonate system is a major extracellular buffer; with a pK of 6.1 it is ineffective toward CO_2. The intracellular buffers (phosphates and protein) are, however, effective (p. 1040, Table 25.6).
9. B $CO_2 + H_2O \rightleftharpoons H^+ + HCO_3^-$, but most of the H^+ is taken up by buffers (p. 1043).
10. A Hemoglobin is the major nonbicarbonate blood buffer (see Question 8) (p. 1044, Figure 25.14).
11. E 1–3, and 4 true. 1 and 2 are reactions whose products include H^+ (pp. 1034–1035). 3 is the Bohr effect (p. 1033). 4 is a major source of acid in the typical American diet (p. 1045).
12. C 2 and 4 true. 1: The level of H_2CO_3 is very low. 3: Acetoacetate is not an acid; acetoacetic acid would appear only in some kinds of severe acidosis. 2 and 4 are true (pp. 1047–1049).
13. D A: After adaption to acidosis NH_4^+ excretion rises enormously (p. 1049). B: True only in acidosis of short duration; decreasing DPG in prolonged acidosis tends to restore the normal position (p. 1038). C: Large amounts of HbA_{1c} would be expected (p. 1030). D: See p. 1047. E: Hyperventilation, to expel CO_2, would be expected (p. 1051).
14. B 1 and 3 true. 1: High Pco_2, low pH, point on the blood buffer line define an acute respiratory acidosis. 2: An acid urine would be produced in compensation. 3: The nonbicarbonate buffers would be most important, and hemoglobin is the major one in blood. 4: Increasing Pco_2 would exacerbate the condition (p. 1050).
15. C 2 and 4 true. 1: Plasma pH would be restored. 2: Pco_2 would fall below normal in acidosis or rise above normal in alkalosis. 3: Base excess would be unchanged. 4: Bicarbonate would decrease in acidosis or increase in alkalosis (pp. 1051–1053, Figures 25.19 and 25.20).
16. B 1 and 3 true. 1 and 2: See p. 1050, Figure 25.18. 3: This is the only acid–base abnormality in which compensation is expected to restore the plasma pH to 7.4 (p. 1052). 4: There is a negative base excess equivalent to the difference between the patient's [HCO_3^-] and the [HCO_3^-] of the point on the blood buffer line at the same pH, a point that will be less than 24 meq L^{-1} (p. 1053, Figure 25.20).
17. A 1–3 true. 1, 2, and 3: See p. 1055. 4: Decreasing K^+ would lower the anion gap, but only by a small amount (p. 1054).

26

Digestion and Absorption of Basic Nutritional Constituents

ULRICH HOPFER

26.1 OVERVIEW 1060
Digestive Processes Were Among the Earliest Biochemical Events Studied 1060
Gastrointestinal Organs Have Multiple Functions in Digestion 1061
26.2 DIGESTION: GENERAL CONSIDERATIONS 1062
Pancreas Supplies Enzymes for Intestinal Digestion 1062
Digestive Enzymes Are Secreted as Zymogens 1063
Regulation of Secretion Occurs Through Secretagogues 1064
26.3 EPITHELIAL TRANSPORT 1067
Solute Transport May Be Transcellular or Paracellular 1067
NaCl Absorption Has Both Active and Passive Components 1068
NaCl Secretion Depends on Contraluminal Na^+,K^+-ATPase 1069
Concentration Gradients or Electrical Potentials Drive Transport of Nutrients 1071
Gastric Parietal Cells Secrete HCl 1072
26.4 DIGESTION AND ABSORPTION OF PROTEINS 1073
Mixture of Peptidases Assures Efficient Protein Digestion 1073
Pepsins Catalyze Gastric Digestion of Protein 1074
Pancreatic Zymogens Are Activated in Small Intestine 1075
Intestinal Peptidases Digest Small Peptides 1076
Free Amino Acids Are Absorbed by Carrier-Mediated Transport 1076
The Fetus and Neonate Can Absorb Intact Proteins 1077
26.5 DIGESTION AND ABSORPTION OF CARBOHYDRATES 1077
Di- and Polysaccharides Require Hydrolysis 1077
Monosaccharides Are Absorbed by Carrier-Mediated Transport 1080
26.6 DIGESTION AND ABSORPTION OF LIPIDS 1081
Lipid Digestion Requires Overcoming the Limited Water Solubility of Lipids 1081
Lipids Are Digested By Gastric and Pancreatic Lipases 1082
Bile Acid Micelles Solubilize Lipids During Digestion 1083
Most Absorbed Lipids Are Incorporated Into Chylomicrons 1087
26.7 BILE ACID METABOLISM 1088
BIBLIOGRAPHY 1090
QUESTIONS AND ANNOTATED ANSWERS 1090
CLINICAL CORRELATIONS
26.1 Electrolyte Replacement Therapy in Cholera 1073
26.2 Neutral Amino Aciduria (Hartnup Disease) 1077
26.3 Disaccharidase Deficiency 1079
26.4 Cholesterol Stones 1086
26.5 A-β-lipoproteinemia 1087

26.1 OVERVIEW

Digestive Processes Were Among the Earliest Biochemical Events Studied

Secretion of digestive fluids and digestion of food were some of the earliest biochemical events to be investigated at the beginning of the era of modern science. Major milestones were the discovery of hydrochloric acid secretion by the stomach and enzymatic hydrolysis of protein and starch by gastric juice and saliva, respectively. The discovery of gastric HCl production goes back to the American physician William Beaumont (1785–1853). In 1822 he treated a patient with a stomach wound. The patient recovered from the wound, but retained a gastric fistula (abnormal opening through the skin). Beaumont seized the opportunity to obtain and study gastric juice at different times during and after meals. Chemical analysis revealed, to the surprise of chemists and biologists, the presence of the inorganic acid HCl. This discovery established the principle of unique secretions into the gastrointestinal tract, which are elaborated by specialized glands.

Soon thereafter, the principle of enzymatic breakdown of food was recognized. Theodor Schwann, a German anatomist and physiologist (1810–1882), noticed in 1836 the ability of gastric juice to degrade albumin in the presence of dilute acid. He recognized that a new principle was involved and coined for it the word *pepsin* from the Greek *pepsis,* meaning digestion. Today the process of secretion of digestive fluids, digestion of food, and absorption of nutrients and of electrolytes can be described in considerable detail.

The basic nutrients fall into the classes of proteins, carbohydrates, and fats. Many different types of food can satisfy the nutritional needs of humans, even though they differ in the ratios of proteins to carbohydrates and to fats and in the ratio of digestible to nondigestible materials. Unprocessed plant products are especially rich in fibrous material that can be neither digested by human enzymes nor easily degraded by intestinal bacteria. The fibers are mostly carbohydrates, such as cellulose (β-1,4-glucan) or pectins (mixtures of methyl esters of polygalacturic acid, polygalactose, and polyarabinose). High-fiber diets enjoy a certain popularity nowadays because of a postulated preventive effect on development of colonic cancer.

Table 26.1 describes average contributions of different food classes to the diet of North Americans. The intake of individuals may substantially deviate from the average, as food consumption depends mainly on availability and individual tastes. The ability to utilize a wide variety of food is possible because of the great adaptability and digestive reserve capacity of the gastrointestinal tract.

TABLE 26.1 Contribution of Major Food Groups to Daily Nutrient Supplies in the United States

	Total consumption (g)	Dairy products, except butter (%)	Meat, poultry, fish (%)	Eggs (%)	Fruits, nuts, vegetables (%)	Flour, cereal (%)	Sugar, sweeteners (%)	Fats, oils (%)
Protein	100	22	42	6	12	18	0	0
Carbohydrate	381	7	0.1	0.1	19	36	37	0
Fat	155	13	35	3	4	1	0	42

Knowledge of the nature of proteins and carbohydrates in the diet is important from a clinical point of view. Certain proteins and carbohydrates, although good nutrients for most humans, cannot be properly digested by some individuals and produce gastrointestinal ailments. Omission of the offending material and change to another diet can eliminate the gastrointestinal problems for these individuals. Examples of food constituents that can be the cause of gastrointestinal disorders are gluten, one of the protein fractions of wheat, and lactose, the disaccharide in milk.

Gastrointestinal Organs Have Multiple Functions in Digestion

The bulk of ingested nutrients consists of large polymers that have to be broken down to monomers before they can be absorbed and made available to all cells of the body. The complete process from food intake to absorption of nutrients into the blood consists of a complicated sequence of events, of which, at the minimum, the following steps are discernible (see Figure 26.1):

1. Mechanical homogenization of food and mixing of ingested solids with fluids secreted by the glands of the gastrointestinal tract

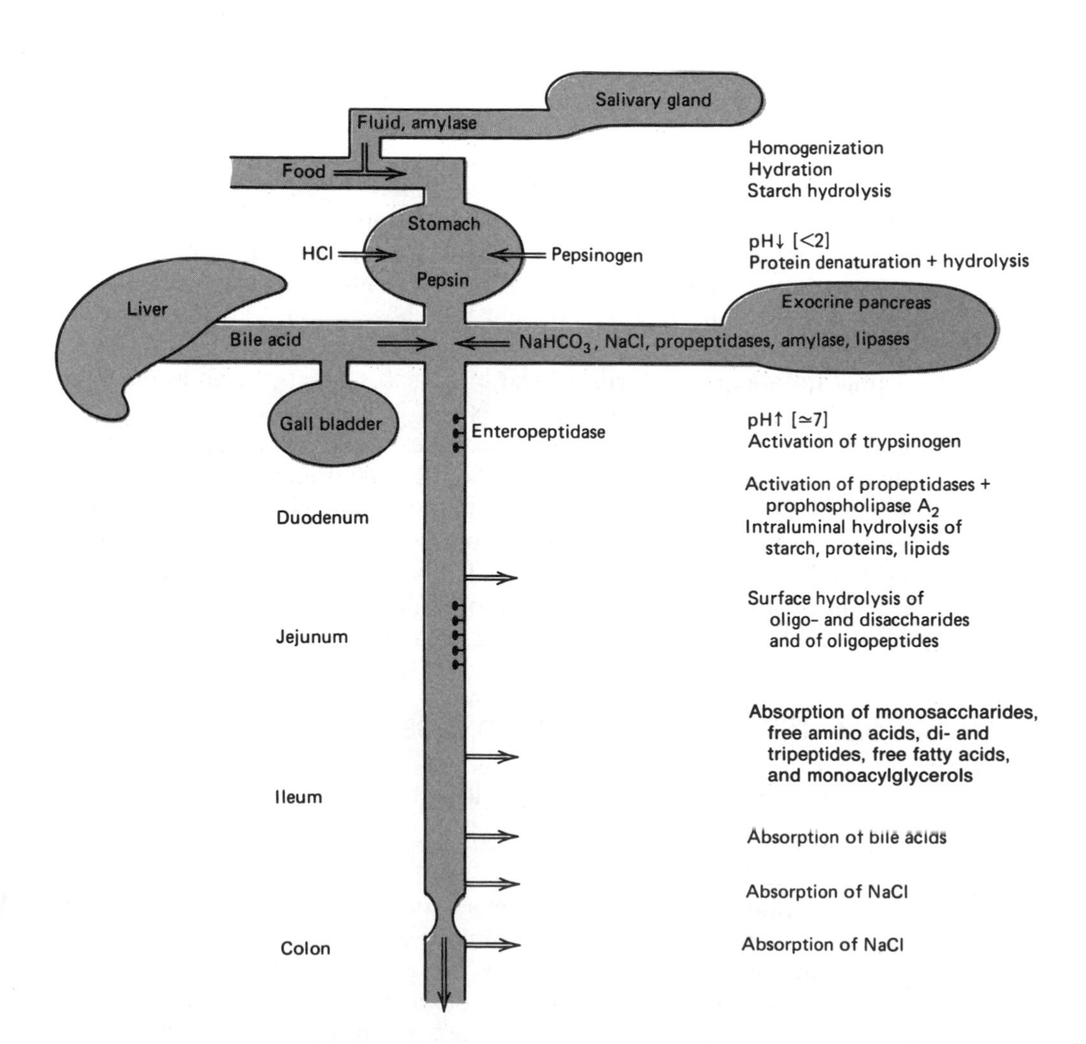

FIGURE 26.1
Gastrointestinal organs and their functions.

2. Secretion of digestive enzymes that hydrolyze macromolecules to oligomers, dimers, or monomers
3. Secretion of electrolytes, acid, or base to provide an appropriate environment for optimal enzymatic digestion
4. Secretion of bile acids as detergents to solubilize lipids and facilitate their absorption
5. Hydrolysis of nutrient oligomers and dimers by intestinal surface enzymes
6. Transport of nutrient molecules and of electrolytes from the intestinal lumen across the epithelial cells into blood or lymph

To accomplish these diverse functions, the gastrointestinal tract contains specialized glands and surface epithelia:

Organ	Major function in digestion and absorption
Salivary glands	Elaboration of fluid and digestive enzymes
Stomach	Elaboration of HCl and proteases
Pancreas	Elaboration of $NaHCO_3$ and enzymes for intraluminal digestion
Liver	Elaboration of bile acids
Gallbladder	Storage and concentration of bile
Small intestine	Terminal digestion of food, absorption of nutrients and electrolytes
Large intestine	Absorption of electrolytes

The pancreas and the small intestine are essential for digestion and absorption of all basic nutrients. Fortunately, both organs have large reserve capacities. For example, maldigestion due to pancreatic failure becomes a problem only when the pancreatic secretion rate of digestive enzymes drops below one-tenth of the normal rate. The secretion of the liver (bile) is important for efficient lipid absorption, which depends on the presence of bile acids. In contrast, gastric digestion of food is nonessential for adequate nutrition, and loss of this function can be compensated for by the pancreas and the small intestine. Yet, normal gastric digestion greatly increases the smoothness and efficiency of the total digestive process. The stomach aids in the digestion through its reservoir function, its churning ability, and initiation of protein hydrolysis, which, although small, is important for stimulation of pancreatic and gallbladder output. The peptides and amino acids liberated in the stomach serve as stimuli for the coordinated release of pancreatic juice and bile into the lumen of the small intestine, thereby ensuring efficient digestion of food.

26.2 DIGESTION: GENERAL CONSIDERATIONS

Pancreas Supplies Enzymes for Intestinal Digestion

Since Schwann's discovery of gastric pepsin, it has been recognized that most of the breakdown of food is catalyzed by soluble enzymes and occurs within the lumen of the stomach or small intestine. However, the pancreas, not the stomach, is the major organ that synthesizes and secretes the large amounts of enzymes needed to digest the food. Secreted enzymes amount to at least 30 g of protein per day in a healthy adult. The pancreatic enzymes together with bile are poured into the lumen of the second (descending) part of the duodenum, so that the bulk of the intraluminal digestion occurs distal to this site in the small intestine. However,

TABLE 26.2 Digestive Enzymes of the Small Intestinal Surface

Enzyme (common name)	Substrate
Maltase	Maltose
Sucrase/isomaltase	Sucrose/α-limit dextrin
Glucoamylase	Amylose
Trehalase	Trehalose
β-Glucosidase	Glucosylceramide
Lactase	Lactose
Endopeptidase 24.11	Protein (cleavage at internal hydrophobic amino acids)
Aminopeptidase A	Oligopeptide with acidic NH_2-terminal amino acid
Aminopeptidase N	Oligopeptide with neutral NH_2-terminal amino acid
Dipeptidyl aminopeptidase IV	Oligopeptide with X-Pro or X-Ala at NH_2 terminus
Leucine aminopeptidase	Peptides with NH_2-terminal neutral amino acids
γ-Glutamyltransferase	Glutathione + amino acid
Enteropeptidase (enterokinase)	Trypsinogen
Alkaline phosphatase	Organic phosphates

pancreatic enzymes cannot completely digest all the nutrients to forms that can be absorbed. Even after exhaustive contact with pancreatic enzymes, a substantial portion of the carbohydrates and amino acids are present as dimers and oligomers that depend for final digestion on enzymes present on the luminal surface or within the lining of chief epithelial cells of the small intestine **(enterocytes).**

The luminal plasma membranes of enterocytes is enlarged by a regular array of projections, termed **microvilli,** which gives it the appearance of a brush and have lead to the name **brush border** for the luminal pole of enterocytes. This membrane contains many di- and oligosaccharidases, amino- and dipeptidases, as well as esterases (Table 26.2). Many of these enzymes protrude up to 100 Å into the intestinal lumen, attached to the plasma membrane by an anchoring polypeptide that itself has no role in the hydrolytic activity. The substrates for these enzymes are the nutrient oligomers and dimers that result from the pancreatic digestion of food. The surface enzymes are glycoproteins that are relatively stable toward digestion by pancreatic proteases or toward detergents.

A third site of digestion is the cytoplasm of enterocytes. Intracellular digestion is of some importance for the hydrolysis of di- and tripeptides, which can be absorbed across the luminal plasma membrane.

Digestive Enzymes Are Secreted as Zymogens

The salivary glands, the gastric mucosa, and the pancreas contain specialized cells for the synthesis, packaging, and release of enzymes into the lumen of the gastrointestinal tract (see Figure 26.2). This secretion is termed **exocrine** because of its direction toward the lumen.

Proteins destined for secretion are synthesized on the polysomes of the rough endoplasmic reticulum, which is particularly abundant in exocrine cells. The NH_2-terminal amino acid sequence of nascent secretory proteins contains a signal sequence that results in anchorage of ribosomes to the membrane and release of the peptide chain into the cisternal space of the endoplasmic reticulum. The amino acids forming the signal sequence may be clipped off during further processing. Secretory proteins are then transported from the endoplasmic reticulum to the Golgi complex in small membrane-bounded vesicles. In the endoplasmic reticulum and also the

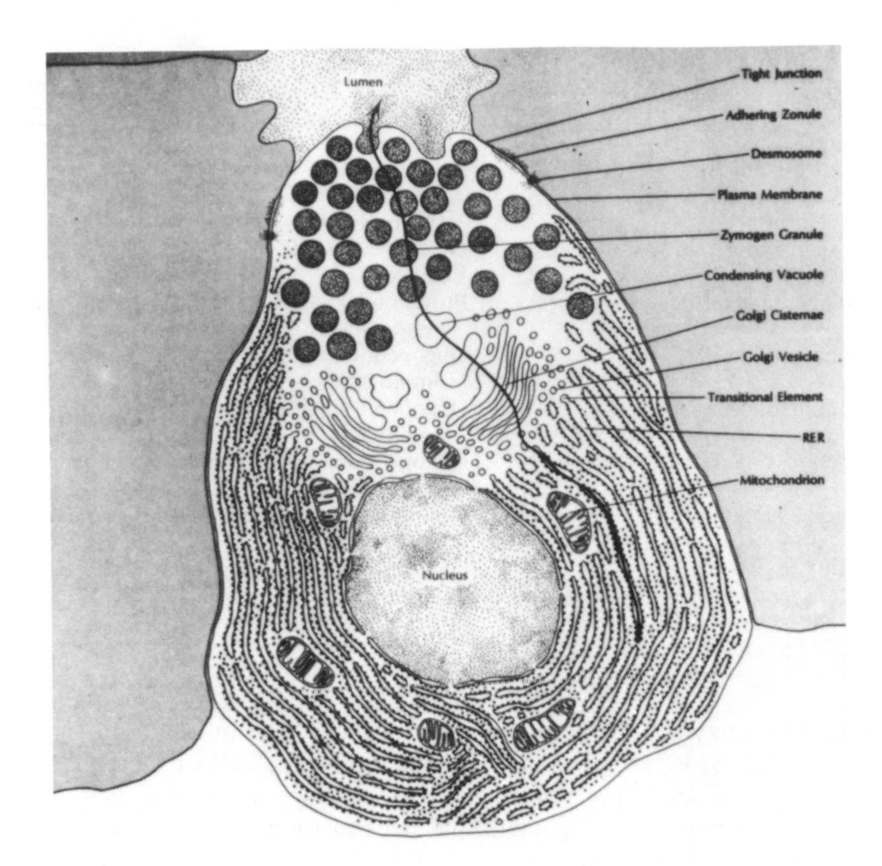

FIGURE 26.2
Exocrine secretion of digestive enzymes.
Reproduced with permission from J. D. Jamieson, *Membrane and Secretion, in Cell Membranes: Biochemistry, Cell Biology and Pathology*. G. Weissmann and R. Claiborne (Eds.). New York: HP Publishing Co., Inc., 1975. Figure by B. Tagawa.

Golgi apparatus, glycosylation may occur. Subsequent to processing, the secretory proteins are packaged into larger vesicles about 1 μm in diameter, which serve as storage forms until the stimulus for secretion is received. Proteases and phospholipase A are produced and stored as inactive precursors, also termed proenzymes or **zymogens.** Therefore the storage vesicles are also called **zymogen granules.** These zymogen granules are bounded by a typical cellular membrane with trilaminar appearance in conventional electron microscopy. When an appropriate stimulus for secretion is received by the cell, the granules move to the luminal plasma membrane, where their membranes fuse with the plasma membrane releasing the contents into the lumen. The process of fusion of granule membrane with plasma membrane and of release of secretory proteins is termed **exocytosis.** Activation of proenzymes occurs only after they are released from the cells.

Regulation of Secretion Occurs Through Secretagogues

The processes involved in the secretion of enzymes and electrolytes are regulated and coordinated. Elaboration of electrolytes and fluids simultaneously with that of enzymes is required to flush any discharged digestive enzymes out of the gland into the gastrointestinal lumen. The physiological regulation of secretion occurs through **secretagogues** that interact with receptors on the contraluminal surface of the exocrine cells (Table 26.3). Neurotransmitters, hormones, pharmacological agents, and certain bacterial toxins can be secretagogues. Different exocrine cells, for example, in different glands, usually possess different sets of receptors. Interaction of the secretagogues with the receptors sets off a chain of events that ends with fusion of intracellular membrane-bounded granules with the plasma membrane and release of the granular material into the extracellular

TABLE 26.3 Physiological Secretagogues

Organ	Secretion	Secretagogue
Salivary gland	NaCl, amylase	Acetylcholine (catecholamines?)
Stomach	HCl, pepsinogen	Acetylcholine, histamine, gastrin
Pancreas	NaCl, enzymes	Acetylcholine, cholecystokinin (secretin)
	$NaHCO_3$,NaCl	Secretin

space. Two major signaling pathways have been identified (Figure 26.3): (1) Activation of phosphatidylinositol-specific phospholipase C with liberation of inositol-1,4,5-trisphosphate and diacylglycerol; which in turn triggers Ca^{2+} release into the cytosol and activate protein kinase C, respectively; and (2) activation of adenylate cyclase, resulting in elevated cAMP levels. Secretagogues appear to activate either pathway.

Cholera toxin, which is secreted by the pathogenic bacterium *Vibrio cholerae,* bypasses the receptors and becomes a secretagogue by catalyzing ADP-ribosylation of the stimulatory GTP binding protein, which in turn provides tonic stimulation of adenylate cyclase.

Acetylcholine (Figure 26.4) elicits salivary, gastric, and pancreatic enzyme and electrolyte secretion. It appears to be the major neurotransmitter for stimulating secretion, with input from the central nervous system in salivary and gastric glands, or via local reflexes in gastric glands and the pancreas. The acetylcholine receptor of exocrine cells is of the *mus-*

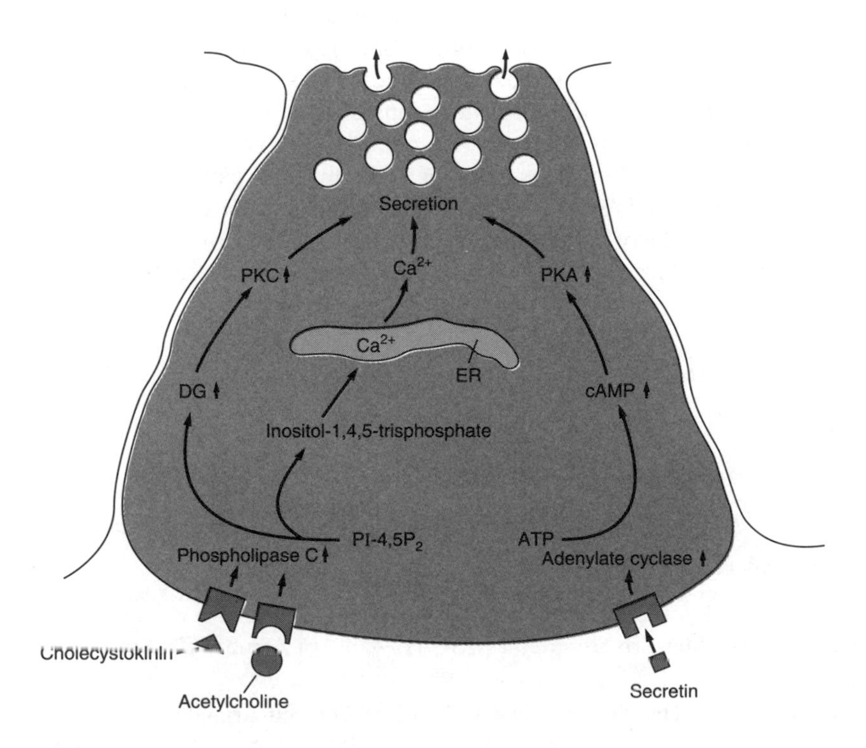

FIGURE 26.3
Cellular regulation of exocrine secretion in the pancreas.
Abbreviations: PI-4,5P_2 = phosphatidylinositol-4,5-bisphosphate; DG = diacylglycerol; ER = endoplasmic reticulum; PKC = protein kinase C; PKA = protein kinase A.
Adapted from J. D. Gardner, *Annu. Rev. Physiol.* 41:63, 1979. Copyright © 1979 by Annual Reviews Inc.

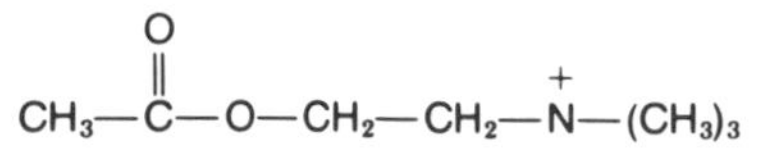

FIGURE 26.4
Acetylcholine.

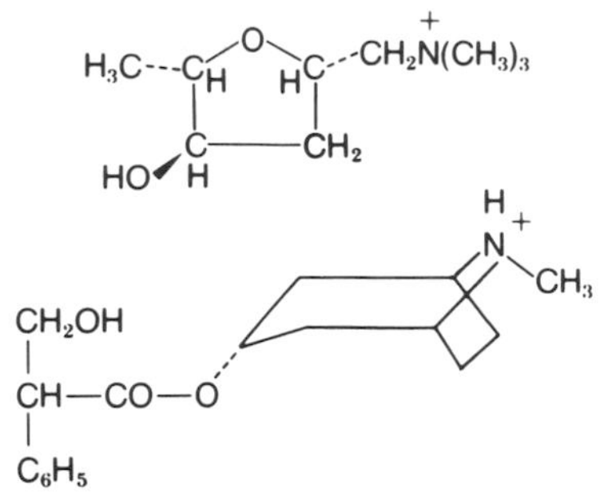

FIGURE 26.5
(*a*) L(+)-Muscarine and (*b*) atropine.

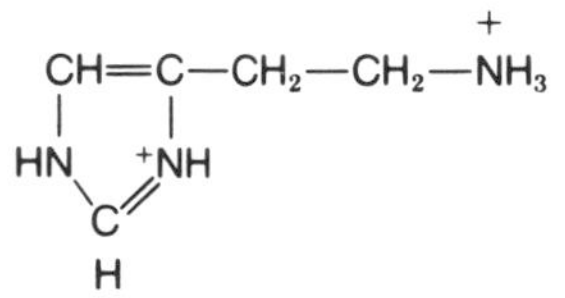

FIGURE 26.6
Histamine.

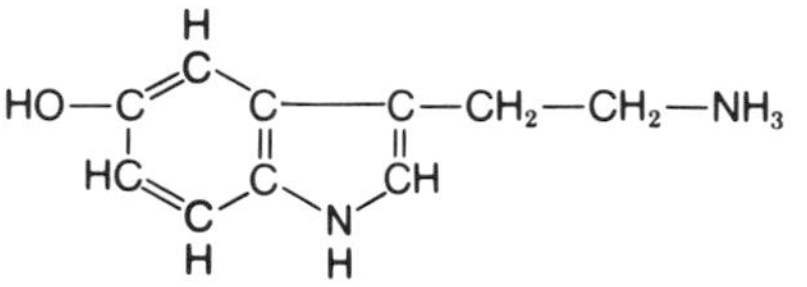

FIGURE 26.7
5-OH-Tryptamine (serotonin).

carinic type, that is, it can be blocked by atropine (Figure 26.5). Most people have experienced the effect of atropine because it is used by dentists to "dry up" the mouth for dental work.

A second class of secretagogues consists of certain **biogenic amines.** For example, histamine (Figure 26.6) is a potent stimulator of HCl secretion. It interacts with a gastric-specific histamine receptor, also referred to as the H_2 receptor, on the contraluminal plasma membrane of parietal cells. The cellular origin of **histamine** involved in the regulation of HCl secretion is not exactly known. Histamine as well as analogs, which act as antagonists at the H_2 receptor, are used medically to increase or decrease HCl output. **5-Hydroxytryptamine** (serotonin) is another biogenic amine that is present in relatively high amounts in the gastrointestinal tract (Figure 26.7). It probably is involved in stimulation of NaCl secretion by the small intestinal mucosa.

A third class of secretagogues consists of peptide hormones. The gastrointestinal tract is rich in specialized epithelial cells, containing a large number of different biologically active amines or peptides. The peptides are localized in granules, usually close to the contraluminal pole of these cells and are released into the blood. Hence these epithelial cells are classified as endocrine cells. Of particular importance are the peptides gastrin, cholecystokinin (pancreozymin), and secretin (Table 26.4).

Gastrin occurs predominantly as either a large peptide of 34 amino acids (G-34) or a smaller one of 17 residues (G-17) from the COOH-terminus of G-34. The functional portion of gastrin resides mainly in the last five amino acids of the COOH terminus. Thus pentagastrin, an artificial pentapeptide containing only the last five amino acids, can be used specifically to stimulate gastric HCl and pepsin secretion. Gastrin as well as cholecystokinin have an interesting chemical feature, a sulfated tyrosine, which considerably enhances the potency of both hormones.

Cholecystokinin and **pancreozymin** denote the same peptide. The different names allude to the different functions elicited by the peptide and had been coined before purification. The peptide stimulates gallbladder contraction (cholecystokinin) as well as secretion of pancreatic enzymes (pancreozymin). It is secreted by epithelial endocrine cells of the small intestine, particularly in the duodenum, and this secretion is stimulated by luminal amino acids and peptides, usually derived from gastric proteoly-

TABLE 26.4 Structure of Human Gastrin, Cholecystokinin, and Secretin

Peptide	Sequence
Gastrin G-34-II[a] (↓ start) … **G-17-II** (↓ at Gln)	[b]Glp-Leu-Gly-Pro-Gln-Gly-Pro-Pro-His-Leu-Val-Ala-Asp-Pro-Ser-Lys-Lys-Gln
	[c]NH_2-Phe-Asp-Met-Trp-Gly-Tyr(SO_3H)-Ala-(Glu)$_5$-Leu-Trp-Pro-Gly
Cholecystokinin	Lys-Ala-Pro-Ser-Gly-Arg-Met-Ser-Ile-Val-Lys-Asn-Leu-Gln-Asn-Leu-Asp-Pro
	[c]NH_2-Phe-Asp-Met-Trp-Gly-Met-Tyr(SO_3H)-Asp-Arg-Asp-Ser-Ile-Arg-His-Ser
Secretin	His-Ser-Asp-Gly-Thr-Phe-Thr-Ser-Glu-Leu-Ser-Arg-Leu-Arg-Glu
	[c]NH_2-Val-Leu-Gly-Gln-Leu-Leu-Arg-Gln-Leu-Arg-Ala-Gly

SOURCE: C. Yanaihara, Handbook of Physiology. Section 6: Alimentary Canal. Vol. II: Neural and Endocrine Biology. B. B. Rauner, G. M. Makhlouf, and S. G. Schultz (Eds.) Bethesda, MD: American Physiology Society, 1989, pp. 45–62.

[a] Gastrin I is not sulfated.
[b] Glp = pyrrolidino carboxylic acid, derived from Glu through internal amide formation.
[c] NH_2 = amide of carboxy terminal amino acid.

sis, by fatty acids, and by an acid pH. Cholecystokinin and gastrin are thought to be related in an evolutionary sense, as both share an identical amino acid sequence at the COOH terminus.

Secretin is a polypeptide of 27 amino acids. This peptide is secreted by yet other endocrine cells of the small intestine. Its secretion is stimulated particularly by luminal pH less than 5. The major biological activity of secretin is stimulation of secretion of pancreatic juice rich in $NaHCO_3$. Pancreatic $NaHCO_3$ is essential for the neutralization of gastric HCl in the duodenum. Secretin also enhances pancreatic enzyme release, acting synergistically with cholecystokinin.

26.3 EPITHELIAL TRANSPORT

Solute Transport May Be Transcellular or Paracellular

Solute movement across an epithelial cell layer is determined by the properties of epithelial cells, particularly their plasma membranes, as well as by the intercellular tight junctional complexes (see Figure 26.8). The **tight junctions** extend in a beltlike manner around the perimeter of each epithelial cell and connect neighboring cells. Therefore, the tight junctions constitute part of the barrier between the two extracellular spaces on either side of the epithelium, that is, the lumen of the gastrointestinal tract and the intercellular (interstitial) space on the other side. The tight junction also marks the boundary between the luminal and the contraluminal region of the plasma membrane of epithelial cells.

Two potentially parallel pathways for solute transport across epithelial cell layers can be distinguished: through the cells **(transcellular)** and through the tight junctions between cells **(paracellular)** (Figure 26.8). The transcellular route in turn consists mainly of two barriers in series, which are formed by the luminal and by the contraluminal plasma membrane. Because of this combination of different barriers in parallel (cellular and paracellular pathways) and in series (luminal and contraluminal plasma membranes), biochemical and biophysical information on all three barriers as well as their mutual influence is required for understanding the overall transport properties of the epithelium.

One of the main functions of epithelial cells in the gastrointestinal tract is active transport of nutrients, electrolytes, and vitamins from one side of the epithelium to the other. The cellular basis for this vectorial solute movement must lie in the different properties of the luminal and contraluminal regions of the plasma membrane. The small intestinal cells provide a prominent example of the differentiation and specialization of the two types of membrane. The luminal and the contraluminal plasma membranes differ in morphological appearance, enzymatic composition,

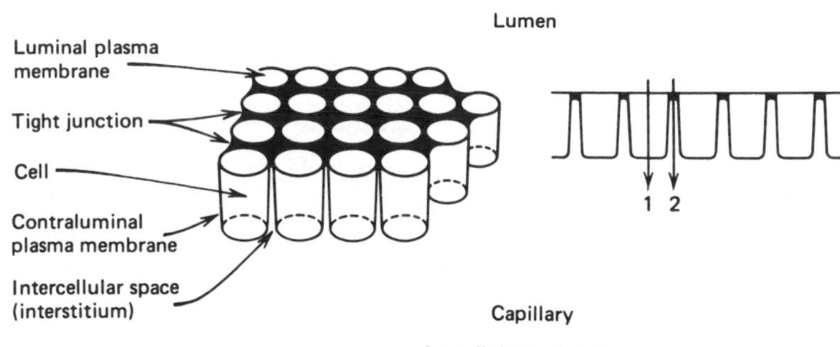

FIGURE 26.8
Pathways for transport across epithelia.

TABLE 26.5 Characteristic Differences Between Luminal and Contraluminal Plasma Membrane of Small Intestinal Epithelial Cells

	Luminal	Contraluminal
Morphological appearance	Microvilli in ordered arrangement (= brush border)	Few microvilli
Enzymes	Di- and oligosaccharidases Aminopeptidase Dipeptidases γ-Glutamyltransferase Alkaline phosphatase	Na^+,K^+-ATPase Adenylate cyclase
Transport systems	Na^+-monosaccharide cotransport Na^+-neutral amino acid cotransport Na^+-bile acid cotransport	Facilitated monosaccharide transport Facilitated neutral amino acid transport

chemical composition, and transport functions (Table 26.5). The luminal membrane is in contact with the nutrients in the chyme (the semifluid mass of partially digested food) and is specialized for terminal digestion of nutrients through its digestive enzymes and for nutrient absorption through transport systems that accomplish concentrative uptake. Such transport systems are well known for monosaccharides, amino acids, peptides, and electrolytes. In contrast, the contraluminal plasma membrane, which is in contact with the intercellular fluid, capillaries, and lymph, has properties similar to the plasma membrane of most cells. It possesses receptors for hormonal or neuronal regulation of cellular functions, a Na^+,K^+-ATPase for removal of Na^+ from the cell, and transport systems for the entry of nutrients for its own consumption. In addition, the contraluminal plasma membrane contains the transport systems necessary for exit of the nutrients derived from the lumen so that the digested food can become available to all cells of the body. Some of the transport systems in the contraluminal plasma membrane may fulfill both the function of catalyzing exit when the intracellular nutrient concentration is high after a meal and that of mediating their entry when the blood levels are higher than those within the cell.

NaCl Absorption Has Both Active and Passive Components

The transport of Na^+ plays a crucial role not only for epithelial NaCl absorption or secretion, but also in the energization of nutrient uptake. The Na^+,K^+-ATPase provides the dominant mechanism for transduction of chemical energy in the form of ATP into osmotic energy of a concentration (chemical) or a combined concentration and electrical (electrochemical) ion gradient across the plasma membrane. In epithelial cells this enzyme is located exclusively in the contraluminal plasma membrane (Figure 26.9). The stoichiometry of the Na^+,K^+-ATPase reaction is 1 mol of ATP coupled to the outward pumping of 3 mol of Na^+ and the simultaneous inward pumping of 2 mol of K^+. The Na^+,K^+-ATPase maintains the high K^+ and low Na^+ concentrations in the cytosol and is directly or indirectly responsible for an electrical potential of about −60 mV of the cytoplasm relative to the extracellular solution. The direct contribution comes from the charge movement when $3Na^+$ ions are replaced by $2K^+$; the indirect contribution is by way of the K^+ gradient, which becomes the dominant force for establishing the potential when the K^+ conductance of the membrane is the major ion conductance.

Transepithelial NaCl movements are produced by the combined actions of the Na^+,K^+-ATPase and additional "passive" transport systems in the

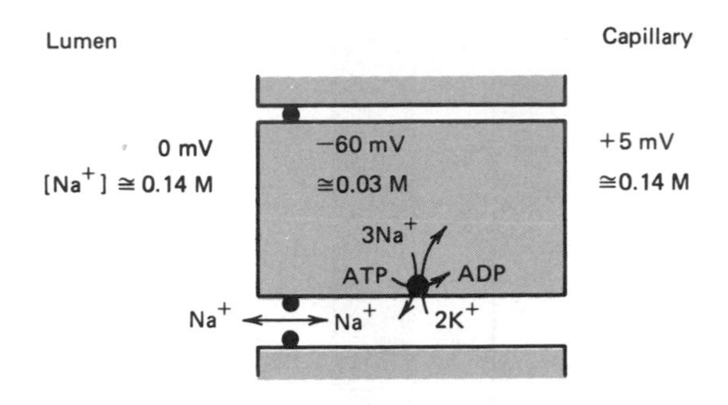

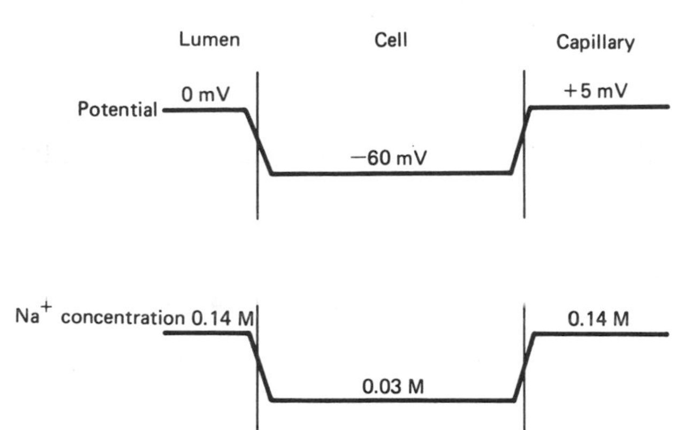

FIGURE 26.9
Na^+ concentrations and electrical potentials in enterocytes.

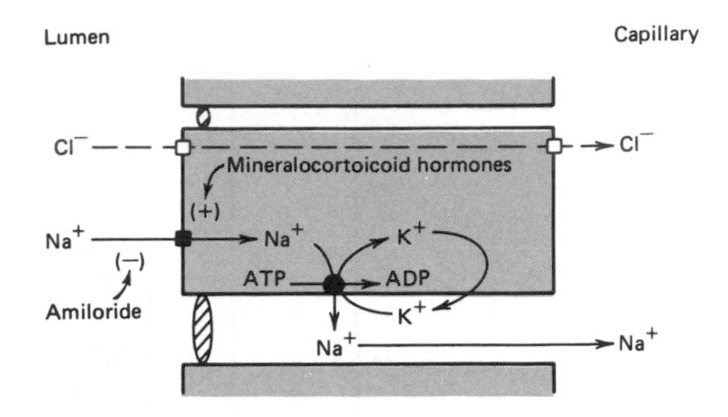

FIGURE 26.10
Model for electrogenic NaCl absorption in the lower intestine.

FIGURE 26.11
Amiloride.

plasma membrane, which allow the entry of Na^+ or Cl^- into the cell. Sodium chloride absorption results from Na^+ entry into the cell across the luminal plasma membrane and its extrusion by the Na^+,K^+-ATPase across the contraluminal membrane. The epithelial cells of the lower portion of the large intestine possess a luminal Na^+ channel that allows the uncoupled entry of Na^+ down its electrochemical gradient (Figure 26.10). This Na^+ flux is electrogenic, that is, it is associated with an electrical current, and it can be inhibited by the drug *amiloride* at micromolar concentrations (Figure 26.11). The presence of this transport system, and hence NaCl absorption, is regulated by mineralocorticoid hormones.

Epithelial cells of the upper portion of the intestine possess a transport system in the brush border membrane, which catalyzes an electrically neutral **Na^+/H^+ exchange** (see Figure 26.12). The exchange is not affected by low concentrations of amiloride and not regulated by mineralocorticoids. The Na^+ absorption secondarily drives Cl^- absorption through a specific **Cl^-/HCO_3^--exchanger** in the luminal plasma membrane, as illustrated in Figure 26.12. The necessity for two types of NaCl absorption may arise from the different functions of upper and lower intestine, which require different regulation. The upper intestine reabsorbs the bulk of NaCl from diet and from secretions of the exocrine glands after each meal, while the lower intestine participates in the fine regulation of NaCl retention, depending on the overall electrolyte balance of the body.

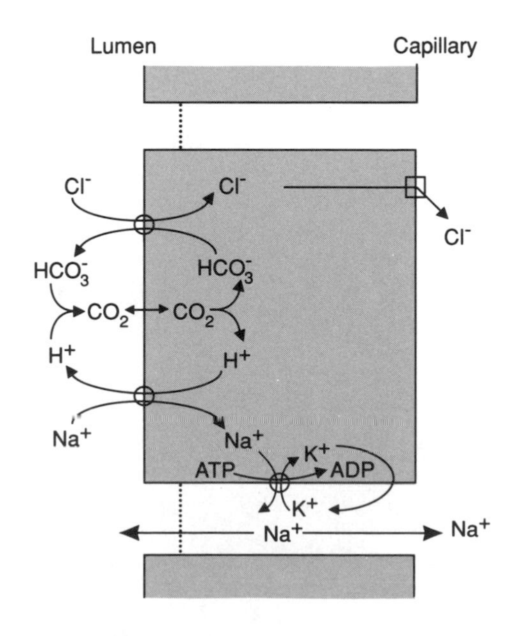

FIGURE 26.12
Model for electrically neutral NaCl absorption in the small intestine.

NaCl Secretion Depends on Contraluminal Na^+,K^+-ATPase

The epithelial cells of most regions of the gastrointestinal tract have the potential for electrolyte and fluid secretions. The major secreted ions are Na^+ and Cl^-. Water follows passively because of the osmotic forces

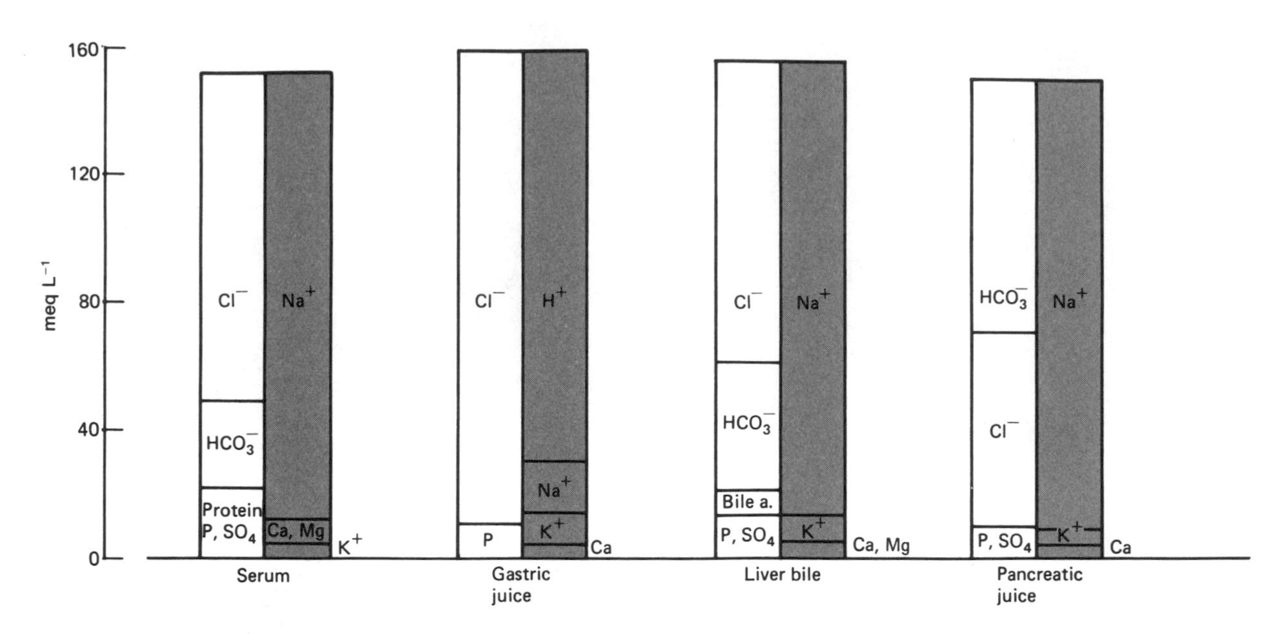

FIGURE 26.13
Ionic composition of secretions of the gastrointestinal tract.
Serum is included to facilitate comparison between fluids. Note the high H^+ concentration in gastric juice (pH < 1) and the high HCO_3^- concentration in pancreatic juice. P = organic and inorganic phosphate; SO_4 = inorganic and organic sulfate; Ca = calcium; Mg = magnesium; bile a. = bile acids.
Adapted from *Biological Handbooks. Blood and Other Body Fluids*. Federation of American Societies for Experimental Biology, 1961.

exerted by any secreted solute. Thus, NaCl secretion secondarily results in fluid secretion. The fluid may be either hypertonic or isotonic, depending on the contact time of the secreted fluid with the epithelium and the tissue permeability to water. The longer the contact and the greater the water permeability, the closer the secreted fluid gets to osmotic equilibrium, that is, isotonicity (see Figure 26.13).

The cellular mechanisms for NaCl secretion involve the Na^+,K^+-ATPase located in the contraluminal plasma membrane of epithelial cells (Figure 26.14). The enzyme is implicated because cardiac glycosides, inhibitors of this enzyme, abolish salt secretion. However, the involvement of Na^+,K^+-ATPase does not provide a straightforward explanation for a NaCl movement from the capillary side to the lumen because the enzyme extrudes Na^+ from the cell toward the capillary side. Thus the active step of Na^+ transport across one of the plasma membranes has a direction opposite to that of overall transepithelial NaCl movements. The apparent paradox has been resolved by an electrical coupling of Cl^- secretion across the luminal plasma membrane and the finding of Na^+ movements

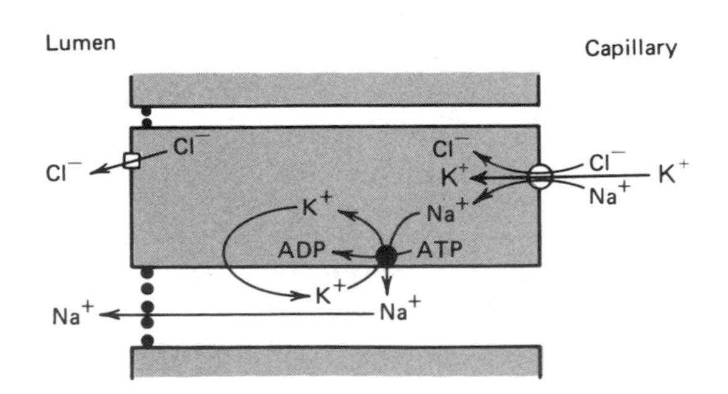

FIGURE 26.14
Model for epithelial NaCl secretion.

via the paracellular route, illustrated in Figure 26.14. The Cl^- secretion depends on coupled uptake of $2Cl^-$ ions with Na^+ and K^+ via a specific cotransporter in the contraluminal plasma membrane and specific luminal Cl^- channels. The **$Na^+,K^+,2Cl^-$-cotransporter** can be identified by specific inhibitors, such as the common diuretic **furosemide** (Figure 26.15) it utilizes the energy of the Na^+ gradient to accumulate Cl^- within the cytoplasmic compartment above its electrochemical equilibrium concentration. Subsequent opening of luminal Cl^- channels allow efflux of Cl^- together with a negative charge.

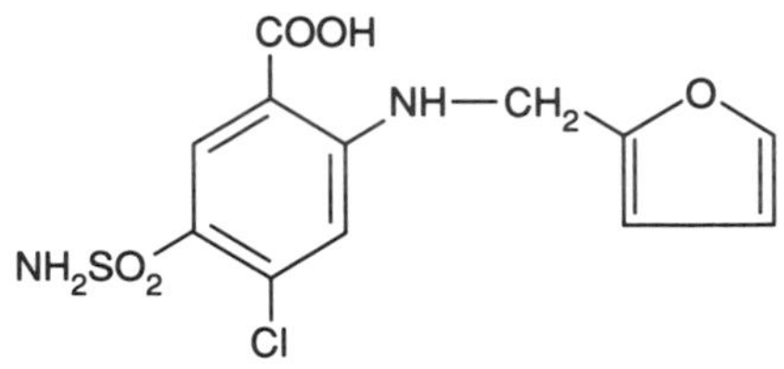

FIGURE 26.15
Furosemide.

In the pancreas a fluid rich in Na^+ and Cl^- is secreted by acinar cells. This fluid provides the vehicle for the movement of digestive enzymes from the acini, where they are released, to the lumen of the duodenum. The fluid is modified in the ducts by the additional secretion of $NaHCO_3$ (Figure 26.16). The HCO_3^- concentration in the final pancreatic juice can reach concentrations of up to 120 mM.

The permeability of the tight junction to H_2O, Na^+, or other ions modifies active transepithelial solute movements. For example, a high permeability is necessary to allow Na^+ to equilibriate between extracellular solutions of the intercellular and the luminal compartments during NaCl or $NaHCO_3$ secretion. Different regions of the gastrointestinal tract differ not only with respect to the transport systems that determine the passive entry (see above for amiloride-sensitive Na^+ channel and Na^+–H^+ exchange), but also with respect to the permeability characteristics of the tight junction. The distal portion (colon) is much tighter to prevent leakage of Na^+ from blood to lumen, in accordance with its function of scavenging of NaCl from the lumen.

Concentration Gradients or Electrical Potentials Drive Transport of Nutrients

Many solutes are absorbed across the intestinal epithelium against a concentration gradient. The energy for this "active" transport is directly derived from the Na^+ concentration gradient or the electrical potential across the luminal plasma membrane, rather than from the chemical energy of a covalent bond change, such as ATP hydrolysis. Glucose transport provides an example of uphill solute transport that is driven directly by the electrochemical Na^+ gradient and only indirectly by ATP (Figure 26.17).

In vivo, glucose is absorbed from the lumen into the blood against a concentration gradient. This vectorial transport is the combined result of

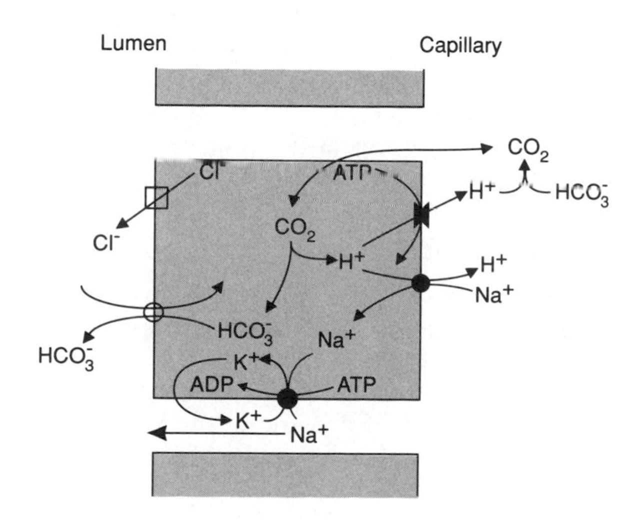

FIGURE 26.16
Model for epithelial $NaHCO_3$ secretion.
Note that two different mechanisms for H^+ secretion exist in the contraluminal plasma membrane: (1) Na^+/H^+ exchange; (2) H^+-ATPase.

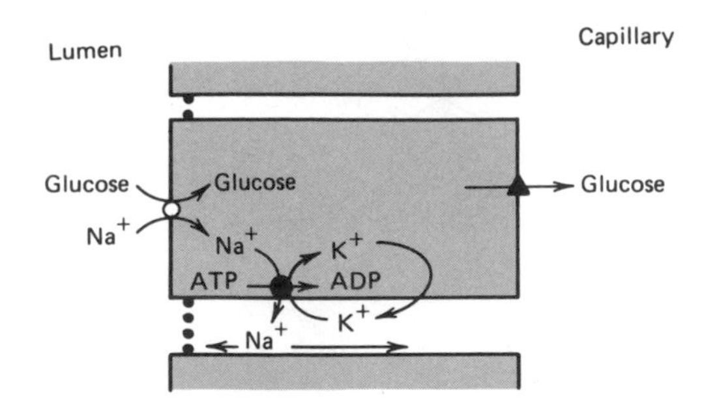

Glucose concentration profile normal to epithelial plane

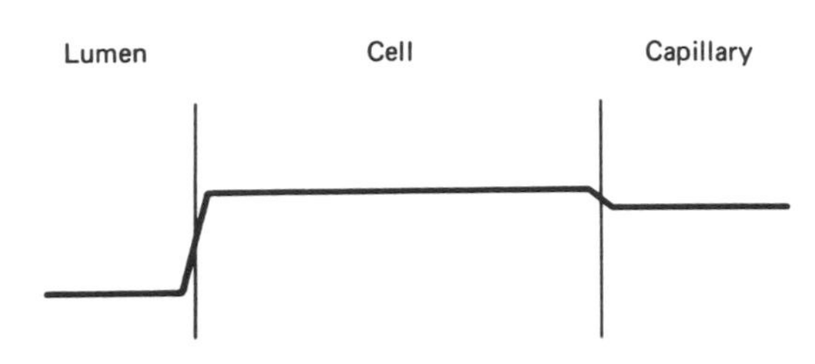

FIGURE 26.17
Model for epithelial glucose absorption.

$$3Na^+_{lumen} + 3Glc_{lumen} \rightleftharpoons 3Na^+_{cell} + 3Glc_{cell}$$
$$3Na^+_{cell} + 2K^+_{interstitium} + ATP_{cell} \longrightarrow 3Na^+_{interstitium} + 2K^+_{cell} + ADP_{cell} + P_{cell}$$
$$2K^+_{cell} \rightleftharpoons 2K^+_{interstitium}$$
$$3Na^+_{interstitium} \rightleftharpoons 3Na^+_{lumen}$$
$$3Glc_{cell} \rightleftharpoons 3Glc_{interstitium}$$
$$\text{Sum:} \quad 3Glc_{lumen} + ATP_{cell} \longrightarrow 3Glc_{interstitium} + ADP_{cell} + P_{cell}$$

FIGURE 26.18
Transepithelial glucose transport as translocation reactions across the plasma membranes and the tight junction.

several separate membrane events: (a) ATP-dependent Na^+ transport out of the cell at the contraluminal pole that establishes an electrochemical Na^+ gradient across the plasma membrane (first three reactions in Figure 26.18); (b) an asymmetric insertion of two different transport systems for glucose into the luminal and the contraluminal plasma membrane; and (c) the coupling of Na^+ and glucose transport across the luminal membrane.

The luminal plasma membrane contains a transport system that facilitates a tightly coupled movement of Na^+ and D-glucose (or structurally similar sugars). The transport system mediates glucose and Na^+ transport equally well in both directions. However, because of the higher Na^+ concentration in the lumen and the negative potential within the cell, the observed direction is from lumen to cell, even if the cellular glucose concentration is higher than the luminal one. In other words, downhill Na^+ movement normally supports concentrative glucose transport. Concentration ratios of up to 20-fold between intracellular and extracellular glucose have been observed in vitro under conditions of blocked efflux of cellular glucose. In some situations Na^+ uptake via this route is actually more important than glucose uptake (Clin. Corr. 26.1).

The contraluminal plasma membrane contains another type of transport system for glucose, which allows glucose to exit. This transport system facilitates equilibration of glucose across the membrane, whereby the direction of the net flux is determined by the glucose concentration gradient. The two glucose transport systems in the luminal and contraluminal plasma membrane share glucose as substrate, but otherwise differ considerably in terms of Na^+ as cosubstrate, specificity for other sugars, sensitivity to inhibitors, or biological regulation. Since both Na^+-glucose cotransport and the simple, facilitated diffusion are not inherently directional, "active" transepithelial glucose transport can be maintained under steady-state conditions only if the Na^+,K^+-ATPase continues to move Na^+ out of the cell. Thus the active glucose transport is indirectly dependent on a supply of ATP and an active Na^+,K^+-ATPase.

The advantage of an electrochemical Na^+ gradient serving as intermediate in the energization is that the Na^+,K^+-ATPase can drive many different nutrients. The only requirement is presence of a transport system catalyzing cotransport of the nutrient with Na^+.

Gastric Parietal Cells Secrete HCl

The parietal (oxyntic) cells of gastric glands are capable of secreting HCl into the gastric lumen. Luminal H^+ concentrations of up to 0.14 M (pH 0.8) have been observed (see Figure 26.19). As the plasma pH = 7.4, the parietal cell transports protons against a concentration gradient of $10^{6.6}$. The free energy required for HCl secretion under these conditions is minimally 9.1 kcal mol^{-1} of HCl (=38 J mol^{-1} of HCl), as calculated from

$$\Delta G' = RT\, 2.3 \log 10^{6.6} \qquad RT = 0.6 \text{ kcal mol}^{-1} \text{ at } 37°\text{C}$$

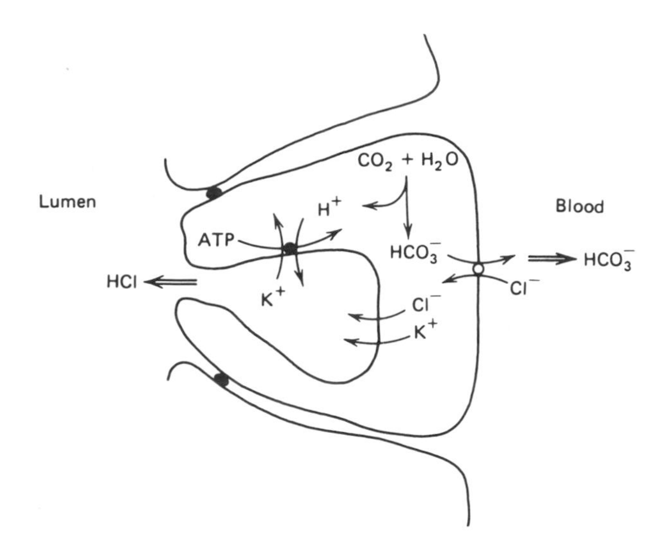

FIGURE 26.19
Model for secretion of hydrochloric acid.

A **K^+-activated ATPase** (K^+,H^+-ATPase) is intimately involved in the mechanism of active HCl secretion. This enzyme is unique to the parietal cell and is found only in the luminal region of the plasma membrane. It couples the hydrolysis of ATP to an electrically neutral obligatory exchange of K^+ for H^+, secreting H^+ and taking K^+ into the cell. The stoichiometry appears to be 1 mol of transported H^+ and K^+ for each mole of ATP:

$$ATP_{cell} + H^+_{cell} + K^+_{lumen} \rightleftharpoons ADP_{cell} + P_{i_{cell}} + H^+_{lumen} + K^+_{cell}$$

In the steady state, HCl can be elaborated by this mechanism only if the luminal membrane is permeable to K^+ and Cl^- and the contraluminal plasma membrane catalyzes an exchange of Cl^- for HCO_3^-. The exchange of Cl^- for HCO_3^- is essential to refurnish the cell with Cl^- and to prevent accumulation of base within the cell. Thus, under steady-state conditions, secretion of HCl into the gastric lumen is coupled to movement of HCO_3^- into the plasma.

26.4 DIGESTION AND ABSORPTION OF PROTEINS

Mixture of Peptidases Assures Efficient Protein Digestion

The total daily protein load to be digested consists of about 70–100 g of dietary proteins and 35–200 g of endogenous proteins from digestive enzymes and sloughed-off cells. Digestion and absorption of proteins are very efficient processes in healthy humans, since only about 1–2 g of nitrogen are lost through feces each day, which is equivalent to 6–12 g of protein.

With the exception of a short period after birth, oligo- and polypeptides (proteins) are not absorbed intact in appreciable quantities by the intestine. Proteins are broken down by hydrolases with specificity for the peptide bond, that is, by peptidases. This class of enzymes is divided into **endopeptidases** (proteases) which attack internal bonds and liberate large peptide fragments, and **exopeptidases,** which cleave off one amino acid at a time from either the COOH or the NH_2 terminus. Thus exopeptidases are further subdivided into **carboxy- and aminopeptidases.** Endopeptidases are important for an initial breakdown of long polypeptides into smaller products, which can then be attacked more efficiently by the exopeptidases. The final products are free amino acids and di- and tripeptides, which are absorbed by the epithelial cells (see Figure 26.20).

CLIN. CORR. 26.1
ELECTROLYTE REPLACEMENT THERAPY IN CHOLERA

Voluminous, life-threatening intestinal electrolyte and fluid secretion (diarrhea) occurs in patients with cholera, an intestinal infection by *Vibrio cholerae*. The secretory state is a result of cholera enterotoxin produced by the bacterium. The toxin activates adenylate cyclase by causing ADP-ribosylation of the $G_{\alpha s}$ protein involved in regulating the cyclase (see p. 871) and thereby turns on electrolyte secretion. Modern, oral treatment of cholera takes advantage of the presence of Na^+-glucose cotransport in the intestine, which is not regulated by cAMP and remains intact in this disease. In this case, the presence of glucose allows uptake of Na^+ to replenish body NaCl. Composition of solution for oral treatment of cholera patients is glucose 110 mM, Na^+ 99 mM, Cl^- 74 mM, HCO_3^- 39 mM, and K^+ 4 mM. The major advantages of this form of therapy are its low cost and ease of administration when compared with intravenous fluid therapy.

Carpenter, C. C. J., *Secretory diarrhea*, M. Field, J. S. Fordtran, and S. G. Schultz (Eds.). *American Physiol. Society*. MD: Bethesda, pp. 67–83, 1980.

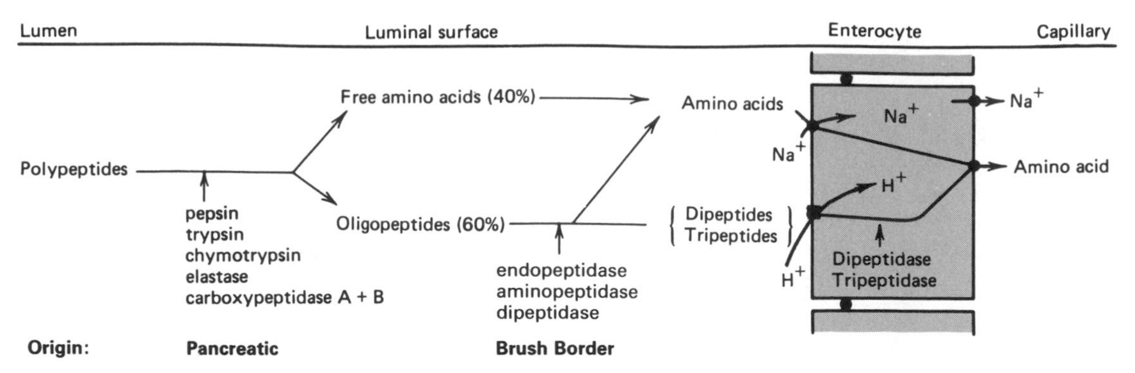

FIGURE 26.20
Digestion and absorption of proteins.

The process of protein digestion can be divided into a gastric, a pancreatic, and an intestinal phase, depending on the source of peptidases.

Pepsins Catalyze Gastric Digestion of Protein

Gastric juice is characterized by the presence of HCl and therefore a low pH less than 2 as well as the presence of proteases of the pepsin family. The acid serves to kill off microorganisms and also to denature proteins. Denaturation makes proteins more susceptible to hydrolysis by proteases. **Pepsins** are unique in that they are acid stable; in fact, they are active at acid but not at neutral pH. The catalytic mechanism that is effective for peptide hydrolysis at the acid pH depends on two carboxylic groups at the active site of the enzymes. Pepsin A, the major gastric protease, prefers peptide bonds formed by the amino group of aromatic acids (Phe, Tyr) (see Table 26.6).

TABLE 26.6 Gastric and Pancreatic Peptidases

Enzyme	Proenzyme	Activator	Reaction catalyzed	R
CARBOXYL PROTEASES				
Pepsin A	Pepsinogen A	Autoactivation, pepsin	CO↓—NH(R)CHCO↓—NH(R′)CHCO	Tyr, Phe, Leu
SERINE PROTEASES				
Trypsin	Trypsinogen	Enteropeptidase, trypsin	CO—NH(R)CHCO↓—NH(R′)CHCO	Arg, Lys
Chymotrypsin	Chymotrypsinogen	Trypsin	CO—NH(R)CHCO↓—NH(R′)CHCO	Tyr, Trp, Phe, Met, Leu
Elastase	Proelastase	Trypsin	CO—NH(R)CHCO↓—NH(R′)CHCO	Ala, Gly, Ser
ZN-PEPTIDASES				
Carboxypeptidase A	Procarboxypeptidase A	Trypsin	CO↓—NH(R)CHCO$_2^-$	Val, Leu, Ile, Ala
Carboxypeptidase B	Procarboxypeptidase B	Trypsin	CO↓—NH(R)CHCO$_2^-$	Arg, Lys

Active pepsin is generated from the proenzyme **pepsinogen** by the removal of 44 amino acids from the NH_2 terminus (pig enzyme). Cleavage of the peptide bond between residues 44 and 45 of pepsinogen can occur as either an intramolecular reaction (autoactivation) below pH 5 or by active pepsin (autocatalysis). The liberated peptide from the NH_2 terminus remains bound to pepsin and acts as "pepsin inhibitor" above pH 2. This inhibition is released either by a drop of the pH below 2 or further degradation of the peptide by pepsin. Thus, once favorable conditions are reached, pepsinogen is converted to pepsin by autoactivation and subsequent autocatalysis at an exponential rate.

The major products of pepsin action are large peptide fragments and some free amino acids. The importance of gastric protein digestion does not lie so much in its contribution to the breakdown of ingested macromolecules, but rather in the generation of peptides and amino acids that act as stimulants for cholecystokinin release in the duodenum. The gastric peptides therefore are instrumental in the initiation of the pancreatic phase of protein digestion.

Pancreatic Zymogens Are Activated in Small Intestine

The pancreatic juice is rich in proenzymes of endopeptidases and carboxypeptidases (see Figure 26.21). These proenzymes are activated only after they reach the lumen of the small intestine. The key to activation is **enteropeptidase** (old name: enterokinase), a protease produced by duodenal epithelial cells. Enteropeptidase activates pancreatic **trypsinogen** to **trypsin** by scission of a hexapeptide from the NH_2 terminus. Trypsin in turn autocatalytically activates more trypsinogen to trypsin and also acts on the other proenzymes, thus liberating the endopeptidases chymotrypsin and elastase and the carboxypeptidases A and B. Since trypsin plays a pivotal role among pancreatic enzymes in the activation process, pancreatic juice normally contains a small-molecular-weight peptide that acts as a trypsin inhibitor and neutralizes any trypsin formed prematurely within the pancreatic cells or pancreatic ducts.

Trypsin, chymotrypsin, and elastase are endopeptidases with different substrate specificity, as shown in Table 26.6. They are all active only at neutral pH and depend on pancreatic $NaHCO_3$ for neutralization of gastric HCl. The mechanism of catalysis of all three enzymes involves an essential *serine* residue. Thus reagents that interact with serine and modify it, inactivate the enzymes. A prominent example of such a reagent is diisopropylphosphofluoridate (Figure 26.22).

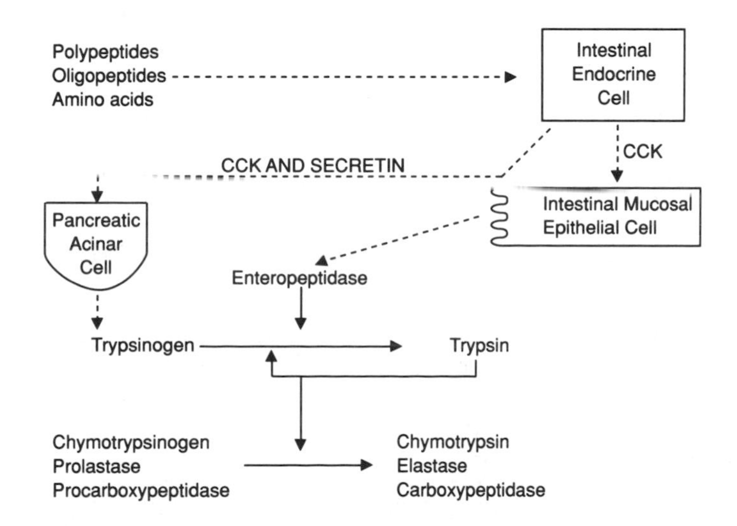

FIGURE 26.21
Secretion and activation of pancreatic enzymes.
Abbreviation: CCK = cholecystokinin.
Reproduced with permission from H. J. Freeman and Y. S. Kim, *Annu. Rev. Med.* 29:102, 1978, Copyright © 1978 by Annual Reviews Inc.

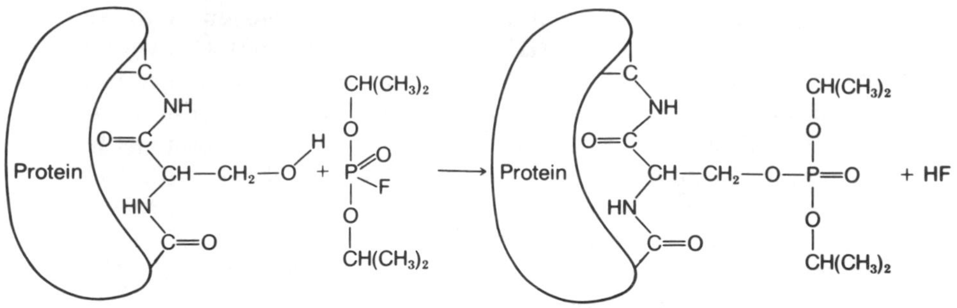

FIGURE 26.22
Inactivation of "serine" proteases by diisopropylphosphofluoridate.

The polypeptides generated from ingested proteins by the action of gastric and pancreatic endopeptidases are degraded further within the small intestinal lumen by **carboxypeptidase A and B.** The pancreatic carboxypeptidases are metalloenzymes that require Zn^{2+} for activity and thus possess a different type of catalytic mechanism than the "carboxyl" or "serine" peptidases.

The combined action of pancreatic peptidases results in the formation of free amino acids and small peptides of 2–8 residues. Peptides account for about 60% of the amino nitrogen at this point.

Intestinal Peptidases Digest Small Peptides

Since pancreatic juice does not contain appreciable aminopeptidase activity, final digestion of di- and oligopeptides depends on small intestinal enzymes. The luminal surface of intestinal epithelial cells is particularly rich in **endopeptidase** and **aminopeptidase** activity, but also contains dipeptidases (Table 26.2). The end products of the cell surface digestion are free amino acids and di- and tripeptides. Amino acids and small peptides are then absorbed by the epithelial cells via specific **amino acid** or **peptide transport systems.** The di- and tripeptides are generally hydrolyzed within the cytoplasmic compartment before they leave the cell. The cytoplasmic dipeptidases explain why practically only free amino acids are found in the portal blood after a meal. The virtual absence of peptides had previously been taken as evidence that luminal protein digestion had to proceed all the way to free amino acids before absorption could occur. However, it is now established that a large portion of dietary amino nitrogen is absorbed in the form of small peptides with subsequent intracellular hydrolysis. Exception to this general rule are di- and tripeptides containing proline and hydroxyproline or unusual amino acids, such as β-alanine in carnosine (β-alanylhistidine) or anserine [β-alanyl(1-methyl)histidine] after ingestion of chicken meat. These peptides are not good substrates for the intestinal cytoplasmic dipeptidases and therefore are available for transport out of the cell into the portal blood.

Free Amino Acids Are Absorbed by Carrier-Mediated Transport

The small intestine has a high capacity to absorb free amino acids. Most L-amino acids can be transported across the epithelium against a concentration gradient, although the need for concentrative transport in vivo is not obvious, since luminal concentrations are usually higher than the plasma levels of 0.1–0.2 mM. Amino acid transport in the small intestine has all the characteristics of carrier-mediated transport, such as discrimination between D- and L-amino acids and energy and temperature dependence. In addition, genetic defects are known to occur in humans (Clin. Corr. 26.2).

On the basis of genetics and of transport experiments, at least six brush border specific transport systems for the uptake of L-amino acids from the luminal solution can be distinguished:

1. For neutral amino acids with short or polar side chains (Ser, Thr, Ala).
2. For neutral amino acids with aromatic or hydrophobic side chains (Phe, Tyr, Met, Val, Leu, Ile).
3. For imino acids (Pro, Hyp).
4. For β-amino acids (β-Ala, taurine).
5. For basic amino acids and cystine (Lys, Arg, Cys-Cys).
6. For acidic amino acids (Asp, Glu).

The mechanisms for concentrative transepithelial transport of L-amino acids appear to be similar to those discussed for D-glucose (see Figure 26.17). The Na^+-dependent transport systems have been identified in the luminal (brush border) membrane and Na^+-independent ones in the contraluminal plasma membrane of small intestinal epithelial cells. Similarly, as for active glucose transport, the energy for concentrative amino acid transport is derived directly from the electrochemical Na^+ gradient and only indirectly from ATP. The amino acids are not chemically modified during membrane transport, although they may be metabolized within the cytoplasmic compartment.

The Fetus and Neonate Can Absorb Intact Proteins

The fetal and neonatal small intestines can absorb intact proteins. The uptake occurs by endocytosis, that is, the internalization of small vesicles of plasma membrane, which contain ingested macromolecules. The process is also termed "pinocytosis" because of the small size of vesicles. The small intestinal pinocytosis of protein is thought to be important for the transfer of maternal antibodies (γ-globulins) to the offspring, particularly in rodents. The pinocytotic uptake of proteins is not important for nutrition, and its magnitude usually declines after birth. Persistance of low levels of this process beyond the neonatal period may, however, be responsible for absorption of sufficient quantities of macromolecules to induce antibody formation.

CLIN. CORR. 26.2 NEUTRAL AMINO ACIDURIA (HARTNUP DISEASE)

Transport functions, like enzymatic functions, are subject to modification by mutations. An example of a genetic lesion in epithelial amino acid transport is Hartnup disease, named after the family in which the disease entity resulting from the defect was first recognized. The disease is characterized by the inability of renal and intestinal epithelial cells to absorb neutral amino acids from the lumen. In the kidney, in which plasma amino acids reach the lumen of the proximal tubule through the ultrafiltrate, the inability to reabsorb amino acids manifests itself as excretion of amino acids in the urine (amino aciduria). The intestinal defect results in malabsorption of free amino acids from the diet. Therefore the clinical symptoms of patients with this disease are mainly those due to essential amino acid and nicotinamide deficiencies. The latter is explained by a deficiency of tryptophan, which serves as precursor for nicotinamide. Investigations of patients with Hartnup disease revealed the existence of intestinal transport systems for di- or tripeptides, which are different from the ones for free amino acids. The genetic lesion does not affect transport of peptides, which remains as a pathway for absorption of protein digestion products.

Silk, D. B. A. Disorders of nitrogen absorption. *Clinics in Gastroenterology: Familial Inherited Abnormalities*, Harries, J. T. (Ed.). Vol. 11: London: Saunders, pp. 47–73, 1982.

26.5 DIGESTION AND ABSORPTION OF CARBOHYDRATES

Di- and Polysaccharides Require Hydrolysis

Dietary carbohydrates provide a major portion of the daily caloric requirement. From a digestive point of view, it is important to distinguish between mono-, di-, and polysaccharides (see Table 26.7). Monosaccharides need not be hydrolyzed prior to absorption. Disaccharides require the small intestinal surface enzymes for breakdown into monosaccharides, while polysaccharides depend additionally on **pancreatic amylase** for degradation (see Figure 26.23).

Starch is a major nutrient. It is a plant polysaccharide with a molecular weight of more than 100,000. It consists of a mixture of linear chains of glucose molecules linked by α-1,4-glucosidic bonds (amylose) and of branched chains with branch points made up by α-1,6 linkages (amylopec-

TABLE 26.7 Dietary Carbohydrates

Carbohydrate	Typical source	Structure
Amylopectin	Potatoes, rice, corn, bread	α-Glc(1 ⟶ 4)$_n$Glc with α-Glc(1 ⟶ 6) branches 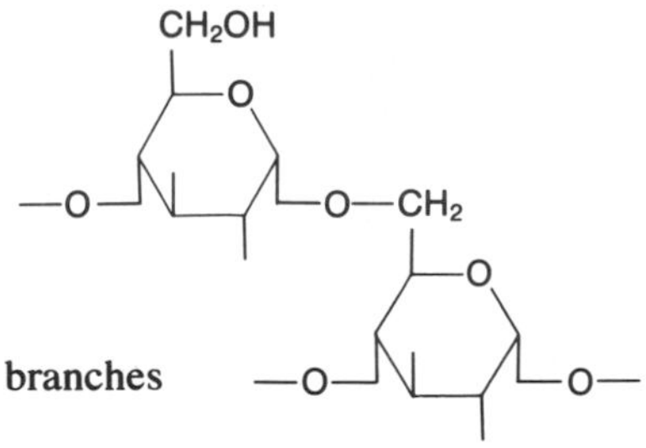
Amylose	Potatoes, rice, corn, bread	α-Glc(1 ⟶ 4)$_n$Glc 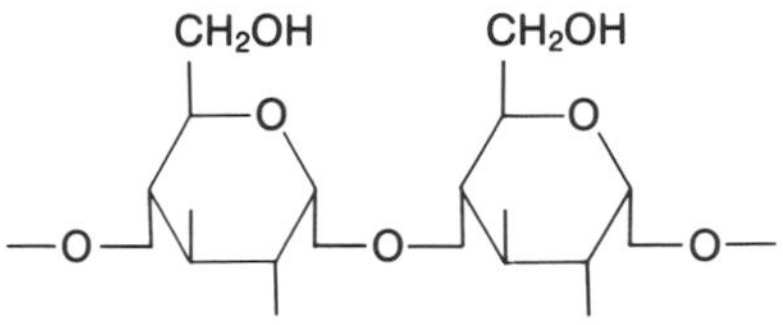
Sucrose	Table sugar, desserts	α-Glc(1 ⟶ 2)β-Fru 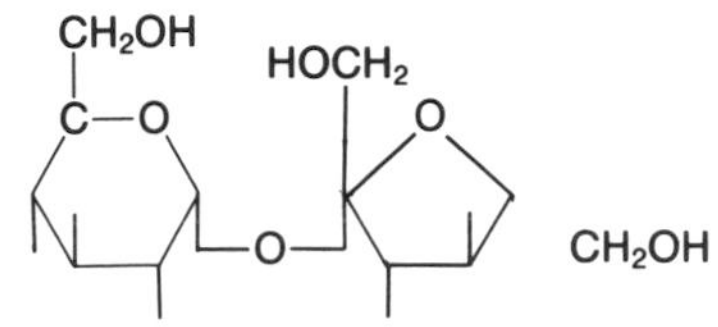
Trehalose	Young mushrooms	α-Glc(1 ⟶ 1)α-Glc 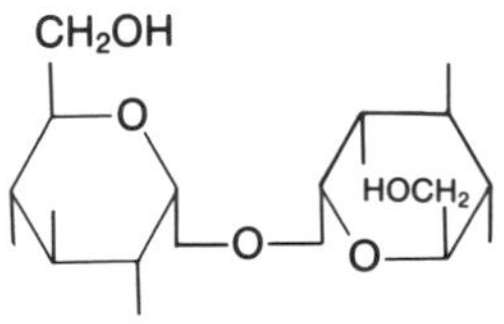
Lactose	Milk, milk products	β-Gal(1 ⟶ 4)Glc 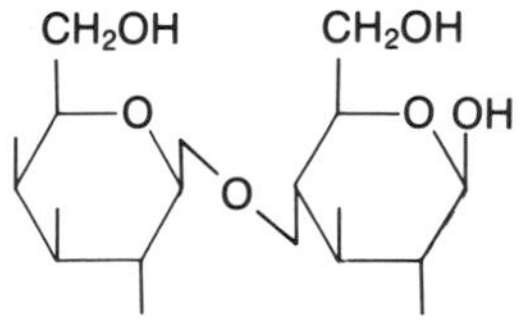
Fructose	Fruit, honey	Fru 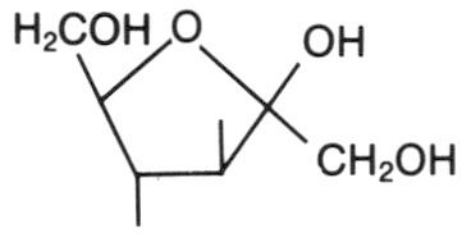
Glucose	Fruit, honey, grape	Glc 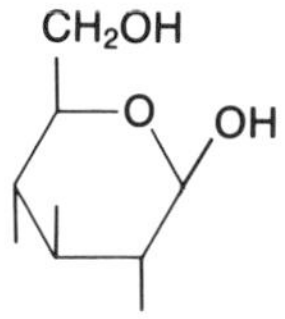
Raffinose	Leguminous seeds	α-Gal(1 ⟶ 6)α-Glc (1 ⟶ 2)β-Fru 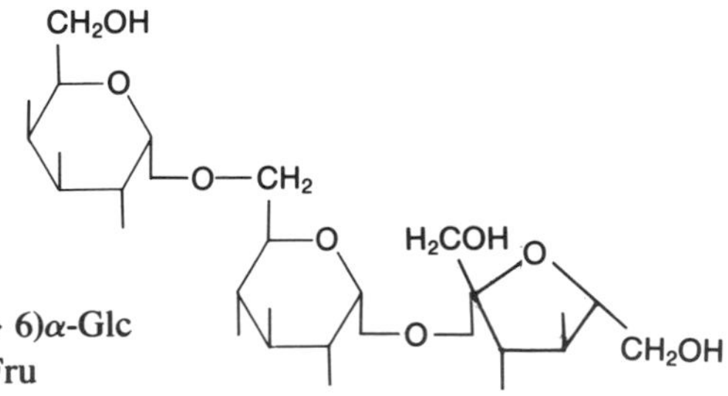

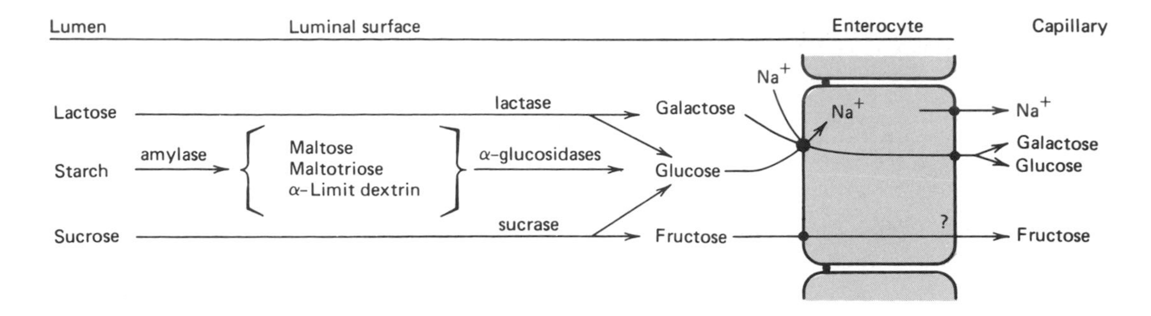

FIGURE 26.23
Digestion and absorption of carbohydrates.

tin). The ratio of 1,4- to 1,6-glucosidic bonds is about 20 : 1. Glycogen is an animal polysaccharide similar in structure to amylopectin. The two compounds differ in terms of the number of branch points, which occur more frequently in glycogen.

Hydrated starch and glycogen are attacked by the endosaccharidase **α-amylase,** which is present in saliva and pancreatic juice (see Figure 26.24). Hydration of the polysaccharides occurs during heating, which is essential for efficient digestion of starch. Amylase has specificity for internal α-1,4-glucosidic bonds; α-1,6 bonds are not attacked, nor are α-1,4 bonds of glucose units that serve as branch points. The pancreatic isoenzyme is secreted in large excess relative to starch intake and is more important than the salivary enzyme from a digestive point of view. The products of the digestion by α-amylase are mainly the disaccharide maltose [α-Glc(1 → 4)Glc], the trisaccharide **maltotriose** [α-Glc(1 → 4)α-Glc(1 → 4)Glc], and so-called **α-limit dextrins** containing on the average eight glucose units with one or more α-1,6-glucosidic bonds.

Final hydrolysis of di- and oligosaccharides to monosaccharides is carried out by surface enzymes of the small intestinal epithelial cells (Table 26.8). Most of the surface oligosaccharidases are exoenzymes that clip off one monosaccharide at a time from the nonreducing end. The capacity of the α-glucosidases normally is much greater than that needed for completion of the digestion of starch. Similarly, there usually is excess capacity for sucrose (table sugar) hydrolysis relative to dietary intake. In contrast, **β-galactosidase (lactase)** can be rate-limiting in humans for hydrolysis and utilization of lactose, the major milk carbohydrate (Clin. Corr. 26.3).

Di-, oligo-, and polysaccharides that are not hydrolyzed by α-amylase and/or intestinal surface enzymes cannot be absorbed; therefore they

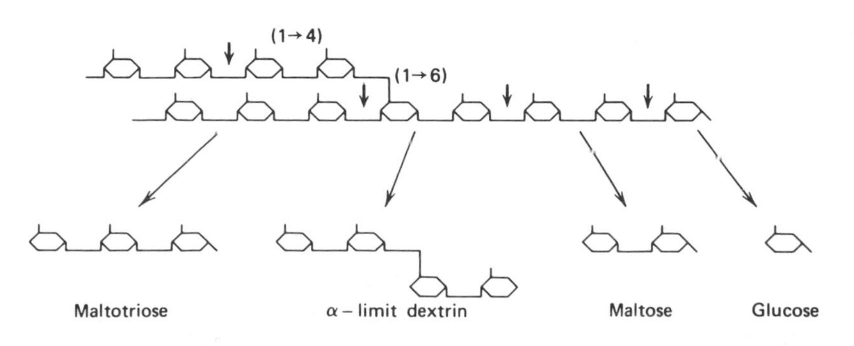

FIGURE 26.24
Digestion of amylopectin by salivary and pancreatic α-amylase.

CLIN. CORR. 26.3
DISACCHARIDASE DEFICIENCY

Intestinal disaccharidase deficiencies are encountered relatively frequently in humans. Deficiency can be present in either a single enzyme or several enzymes for a variety of reasons (genetic defect, physiological decline with age, or as result of "injuries" to the mucosa). Of the disaccharidases, lactase is the most common enzyme with an absolute or relative deficiency, which is experienced as milk intolerance. The consequences of an inability to hydrolyze lactose in the upper small intestine are (a) inability to absorb lactose, and (b) bacterial fermentation of ingested lactose in the lower small intestine. Bacterial fermentation results in the production of gas (distension of gut and flatulence) and osmotically active solutes that draw water into the intestinal lumen (diarrhea). The lactose in yogurt has already been partially hydrolyzed during the fermentation process of making yogurt. Thus, individuals with lactase deficiency can often tolerate yogurt than unfermented dairy products. The enzyme lactase has also been purified and is commercially available to pretreat milk so that the lactose is hydrolyzed.

Büller, H. A. and Grant, R. G. Lactose intolerance. *Annu. Rev. Med.* 41:141, 1990.

TABLE 26.8 Di- and Oligosaccharidases of the Luminal Plasma Membrane in the Small Intestine

Enzyme	Specificity	Natural substrate	Product
exo-1,4-α-Glucosidase (glucoamylase)	α-(1 ⟶ 4)Glucose	Amylose	Glucose
Oligo-1,6-glucosidase (isomaltase)	α-(1 ⟶ 6)Glucose	Isomaltose, α-dextrin	Glucose
α-Glucosidase (maltase)	α-(1 ⟶ 4)Glucose	Maltose, maltotriose	Glucose
Sucrose-α-Glucosidase (sucrase)	α-Glucose	Sucrose	Glucose, fructose
α,α-Trehalase	α-(1 ⟶ 1)Glucose	Trehalose	Glucose
β-Glucosidase	β-Glucose	Glucosyl-ceramide	Glucose, ceramide
β-Galactosidase (lactase)	β-Galactose	Lactose	Glucose, galactose

reach the lower tract of the intestine, which from the lower ileum on contains bacteria. Bacteria can utilize many of the remaining carbohydrates because they possess many more types of saccharidases than humans. The monosaccharides that are released as a result of bacterial enzymes are predominantly anaerobically metabolized by the bacteria themselves, resulting in such degradation products as short-chain fatty acids, lactate, hydrogen gas (H_2), methane (CH_4), and carbon dioxide (CO_2). These compounds can cause fluid secretion, increased intestinal motility, and cramps, either because of increased intraluminal osmotic pressure, distension of the gut, or because of a direct irritant effect of the bacterial degradation products on the intestinal mucosa.

The well-known problem of flatulence after ingestion of leguminous seeds (beans, peas, and soya) can be traced to oligosaccharides, which cannot be hydrolyzed by human intestinal enzymes. The leguminous seeds contain modified sucrose to which one or more galactose moieties are linked. The glycosidic bonds of galactose are in the α configuration, which can only be split by bacterial enzymes. The simplest sugar of this family is **raffinose** [α-Gal(1 → 6)α-Glc (1 → 2) β-Fru] (see Table 26.7).

Trehalose [α-Glc(1 → 1)α-Glc] is a disaccharide that occurs in young mushrooms. The digestion of this sugar requires a special disaccharidase, **trehalase.**

Monosaccharides Are Absorbed by Carrier-Mediated Transport

The major monosaccharides that result from the digestion of di- and polysaccharide hydrolysis are D-glucose, D-galactose, and D-fructose. Absorption of these and other minor monosaccharides are carrier-mediated processes that exhibit such features as substrate specificity, stereospecificity, saturation kinetics, and inhibition by specific inhibitors.

At least two types of monosaccharide transporters are known to catalyze monosaccharide uptake from the lumen into the cell: (1) a **Na^+-monosaccharide cotransporter,** existing probably as a tetramer of 75-kDa peptides, has high specificity for D-glucose and D-galactose and catalyzes "active" sugar absorption, and (2) a **Na^+-independent, facilitated-diffusion** type of monosaccharide transport system with specificity for D-fructose. In addition, a **Na^+-independent** monosaccharide transporter, con-

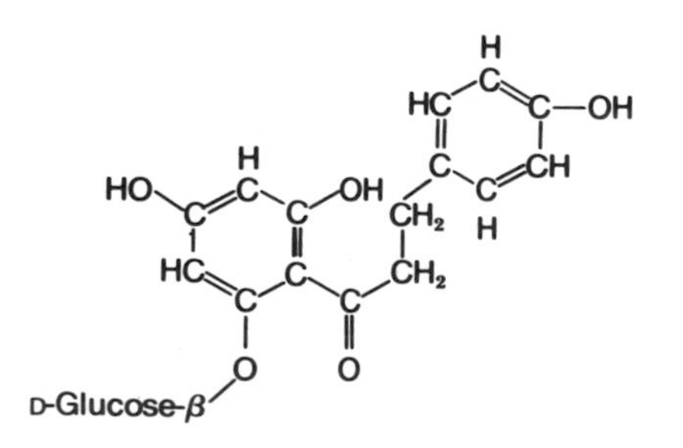

FIGURE 26.25
Phlorizin (phloretin-2′-β-glucoside).

TABLE 26.9 Characteristics of Glucose Transport Systems in the Plasma Membranes of Enterocytes

Characteristic	Luminal	Contraluminal
Effect of Na^+	Cotransport with Na^+	None
Good substrates	D-Glc, D-Gal, α-methyl-D-Glc	D-Glc, D-Gal, D-Man, 2-deoxy-D-Glc
Molecular wt of subunit (kDa)	75	57
Inhibition by	Phlorizin (Figure 26.25)	Cytochalasin B (Figure 26.26)

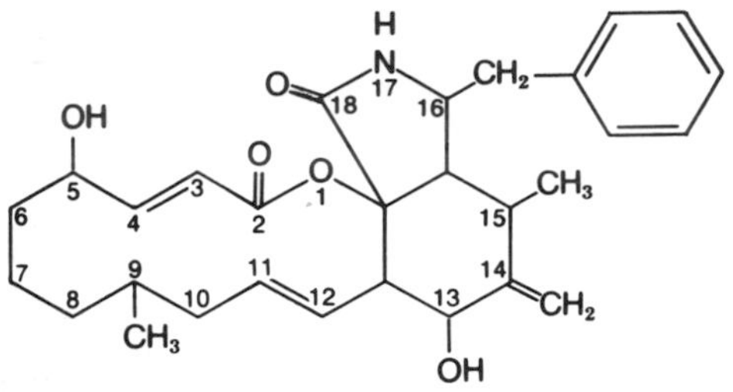

FIGURE 26.26
Cytochalasin B.

sisting of 57-kDa peptide(s), with specificity for D-glucose and D-galactose, is present in the contraluminal plasma membrane. The same type of transporter is also located in the liver and the kidney, and a similar one in brain, erythrocytes, adipocytes, and cardiac muscle cells. These Na^+-independent transporters mediate uncoupled D-glucose flux down its concentration gradient. The Na^+-independent transporter of the gut, liver, and kidney moves D-glucose out of the cell into the blood under physiological conditions, while in other tissues (e.g., erythrocytes and brain) it is mainly involved in D-glucose uptake. The properties of the Na^+ dependent and the Na^+-independent intestinal glucose transporters are compared in Table 26.9, and their role in transepithelial glucose absorption is illustrated in Figure 26.18.

26.6 DIGESTION AND ABSORPTION OF LIPIDS

Lipid Digestion Requires Overcoming the Limited Water Solubility of Lipids

An adult man ingests about 60–150 g of fat per day. Triacylglycerols constitute more than 90% of the dietary fat. The rest is made up of phospholipids, cholesterol, cholesterol esters, and free fatty acids. In addition, 1–2 g of cholesterol and 7–22 g of phosphatidylcholine (lecithin) are secreted into the small intestinal lumen as constituents of bile.

Lipids are defined by their good solubility in organic solvents. Conversely, they are sparingly or not at all soluble in aqueous solutions. The poor water solubility presents problems for digestion because the substrates are not easily accessible to the digestive enzymes in the aqueous phase. In addition, even if ingested lipids are hydrolyzed into simple constituents, the products tend to aggregate to larger complexes that make poor contact with the cell surface and therefore are not easily absorbed. These problems are overcome by (a) increases in the interfacial area between the aqueous and the lipid phase, and (b) "solubilization" of the hydrolysis products with detergents. Thus changes in the physical state of lipids are intimately connected to chemical changes during digestion and absorption (see Figure 26.27).

At least five different phases can be distinguished:

1. Hydrolysis of triacylglycerols to free fatty acids and monoacylglycerols.
2. Solubilization of free fatty acids and monoacylglycerols by detergents (bile acids) and transportation from the intestinal lumen toward the cell surface.

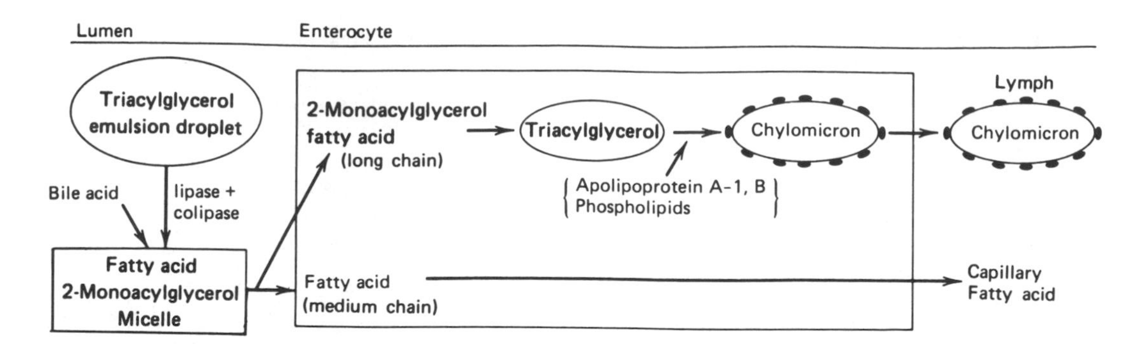

FIGURE 26.27
Digestion and absorption of lipids.

3. Uptake of free fatty acids and monoacylglycerols into the cell and resynthesis to triacylglycerols.
4. Packaging of newly synthesized triacylglycerols into special lipid-rich globules, called chylomicrons.
5. Exocytosis of chylomicrons from cells and release into lymph.

Lipids Are Digested By Gastric and Pancreatic Lipases

The digestion of lipids is initiated in the stomach by an **acid-stable lipase,** most of which is thought to originate from glands at the back of the tongue. However, the rate of hydrolysis is slow because the ingested triacylglycerols form a separate lipid phase with a limited water–lipid interface. The lipase adsorbs to that interface and converts triacylglycerols into fatty acids and diacylglycerols (see Figure 26.28). The importance of the initial hydrolysis is that some of the water-immiscible triacylglycerols are converted to products that possess both polar and nonpolar groups. Such products spontaneously adsorb to water–lipid interfaces, and therefore are said to be *surfactive*. In effect, surfactants confer a hydrophilic surface to lipid droplets and thereby provide a stable interface with the aqueous environment. Among the dietary lipids, free fatty acids, monoacylglycerols, and phospholipids are the major surfactants. One of the effects of the action of lipase is then a release of surfactive molecules and through these an increase in interfacial area between lipid

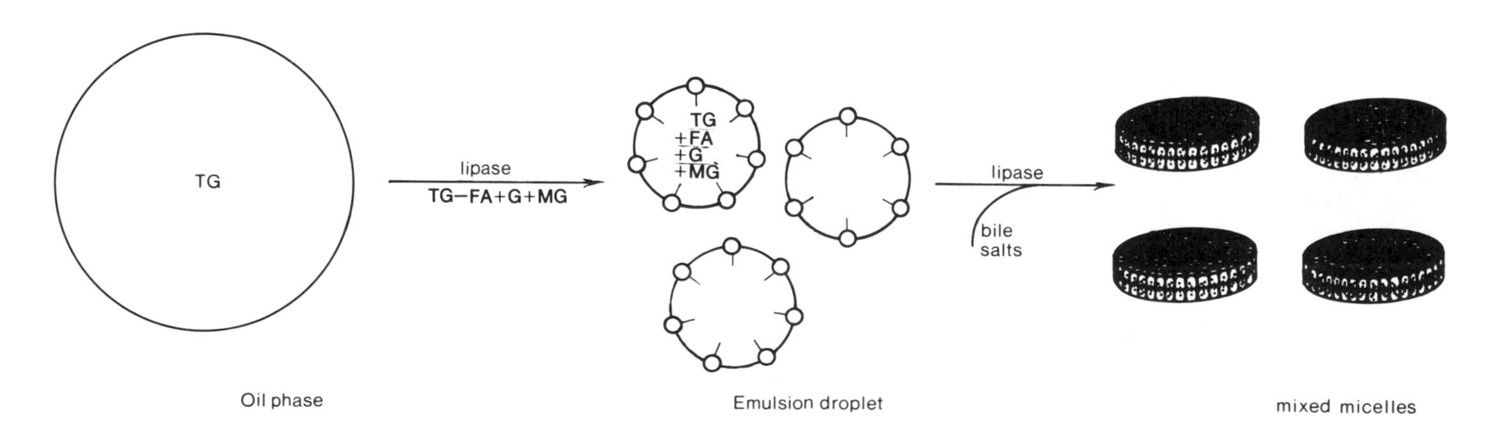

FIGURE 26.28
Changes in physical state during triacylglycerol digestion.
Abbreviations: TG = triacylglycerol, DG = diacylglycerol, MG = monoacylglycerol, FA = fatty acid.

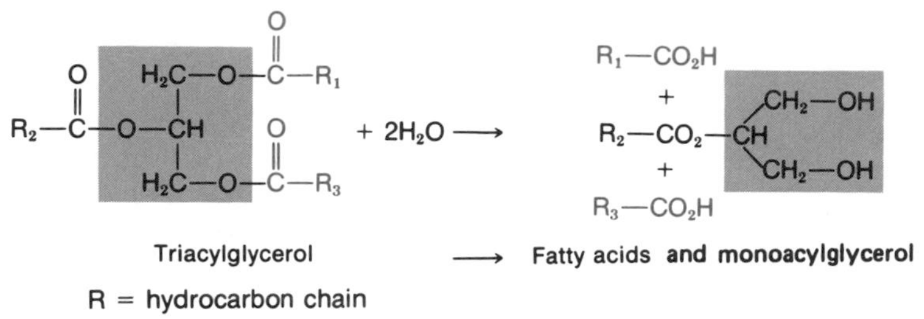

Triacylglycerol $\longrightarrow$ Fatty acids and monoacylglycerol

R = hydrocarbon chain

FIGURE 26.29
Mechanism of action of lipase.

and water. At constant volume of the lipid phase, any increase in interfacial area produces dispersion of the lipid phase into smaller droplets (emulsification). Thus lipase autocatalytically enhances the availability of more triacylglycerol substrate through an increase in interfacial area.

The major enzyme for triacylglycerol hydrolysis is the **pancreatic lipase** (Figure 26.29). This enzyme is specific for esters in the α-position of glycerol and prefers long-chain fatty acids of more than 10 carbon atoms. Triacylglycerol hydrolysis by the pancreatic enzyme also occurs at the water–lipid interface of emulsion droplets. The products are free fatty acids and β-monoacylglycerols. The purified form of the enzyme is strongly inhibited by the bile acids that normally are present in the small intestine during lipid digestion. The problem of inhibition is overcome by the addition of colipase, a small protein with a molecular weight of 12,000 Da. Colipase binds to both the water–lipid interface and to lipase, thereby anchoring and activating the enzyme. It is secreted by the pancreas as procolipase and depends on tryptic removal of a NH_2-terminal decapeptide for full activity.

In addition to lipase, pancreatic juice contains another less specific **lipid esterase.** This enzyme acts on cholesterol esters, monoglycerides, or other lipid esters, such as esters of vitamin A with carboxylic acids. In contrast to triacylglycerol lipase, the less specific lipid esterase requires bile acids for activity.

Phospholipids are broken down by specific phospholipases. Pancreatic secretions are especially rich in the proenzyme for **phospholipase A_2** (Figure 26.30). As other pancreatic proenzymes, this one, too, is activated by trypsin. Phospholipase A_2 requires bile acids for activity.

Bile Acid Micelles Solubilize Lipids During Digestion

Bile acids are biological detergents that are synthesized by the liver and secreted with the bile into the duodenum. At physiological pH values, the acids are present as anions, which exhibit detergent properties. Therefore, the terms **bile acids** and **bile salts** are often used interchangeably (Figure 26.31).

Bile acids at pH values above the pK (see Table 26.10) reversibly form aggregates at concentrations above 2–5 mM. These aggregates are called

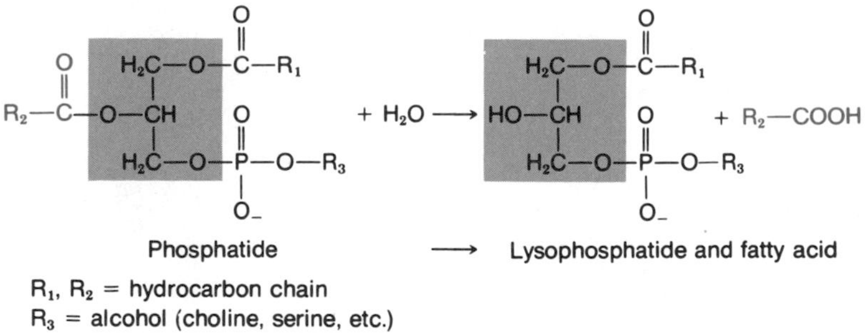

Phosphatide $\longrightarrow$ Lysophosphatide and fatty acid

R_1, R_2 = hydrocarbon chain
R_3 = alcohol (choline, serine, etc.)

FIGURE 26.30
Mechanism of action of phospholipase A_2.

Cholic acid

Stereochemistry of cholic acid

FIGURE 26.31
Cholic acid, a bile acid.

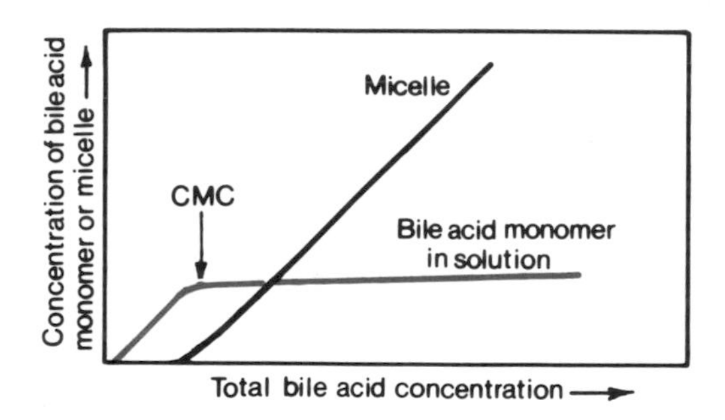

FIGURE 26.32
Solubility properties of bile acids in aqueous solutions.
Abbreviation: CMC = critical micellar concentration.

micelles, and the minimal concentration necessary for micelle formation is the **critical micellar concentration** (Figure 26.32). The bile acids in micelles are in equilibrium with those free in solution. Thus, micelles, in contrast to emulsified lipids, are equilibrium structures with well-defined sizes that are much smaller than emulsion droplets. Micelle sizes typically range between 40 and 600 Å depending on bile acid concentration and the ratio of bile acids to lipids.

The arrangements of bile acids in micelles is such that the hydrophobic portions of the molecule are removed from contact with water, while hydrophilic groups remain exposed to the water molecules. The hydrophobic region of bile acids is formed by one surface of the fused ring system, while the carboxylate or sulfonate ion and the hydroxyl groups on the other side of the ring system are hydrophilic. Since the major driving forces for micelle formation are the removal of apolar, hydrophobic groups from and the interaction of polar groups with water molecules, the distribution of polar and apolar regions places some constraints on the

TABLE 26.10 The Effect of Conjugation on the Acidity of Cholic, Deoxycholic, and Chenodeoxycholic Acids

Bile acid	Ionized group	pK_a
Unconjugated bile acids	$—CO_2^-$ of cholestanoic acid	≃5
Glycoconjugates	$—CO_2^-$ of glycine	≃3.7
Tauroconjugates	$—SO_3^-$ of taurine	≃1.5

Primary: Cholic; Chenic

Secondary: Deoxycholic

+ glycine ($NH_3^+CH_2COO^-$) → $R—C(=O)—NH—CH_2COO^-$: Cholylglycine, Chenylglycine, Deoxycholylglycine

+ taurine ($NH_3^+(CH_2)_2SO_2O^-$) → $R—C(=O)—NH—(CH_2)_2SO_2O^-$: Cholyltaurine, Chenyltaurine, Deoxycholyltaurine

SOURCE: Reproduced with permission from Hofmann, A. F. *Handbook of Physiology* 5:2508, 1968.

stereochemical arrangements of bile acid molecules within a micelle. Four bile acid molecules are sufficient to form a very simple micelle as shown in Figure 26.33.

Bile salt micelles can solubilize other lipids, such as phospholipids and fatty acids. These **mixed micelles** have disklike shapes, whereby the phospholipids and fatty acids form a bilayer and the bile acids occupy the edge positions, rendering the edge of the disk hydrophilic (Figure 26.34). Within the mixed phospholipid–bile acid micelles, other water-insoluble lipids, such as cholesterol, can be accommodated and thereby "solubilized" (for potential problems see Clin. Corr. 26.4).

During triacylglycerol digestion, free fatty acids and monoacylglycerols are released at the surface of fat emulsion droplets. In contrast to triacylglycerols, which are water-insoluble, free fatty acids and monoacylglycerols are slightly water-soluble, and molecules at the surface equilibrate with those in solution. The latter in turn become incorporated into bile acid micelles. Thus the products of triacylglycerol hydrolysis are continuously transferred from emulsion droplets to the micelles (Figure 26.28).

Micelles provide the major vehicle for moving lipids from the intestinal lumen to the cell surface where absorption occurs. Because the fluid layer next to the cell surface is poorly mixed, the major transport mechanism for solute flux across this "unstirred" fluid layer is diffusion down the concentration gradient. With this type of transport mechanism, the delivery rate of nutrients at the cell surface is proportional to their concentration difference between luminal bulk phase and cell surface. Obviously, the unstirred fluid layer presents problems for sparingly soluble or insoluble nutrients, in that reasonable delivery rates cannot be achieved. Bile acid micelles overcome this problem for lipids by increasing their effec-

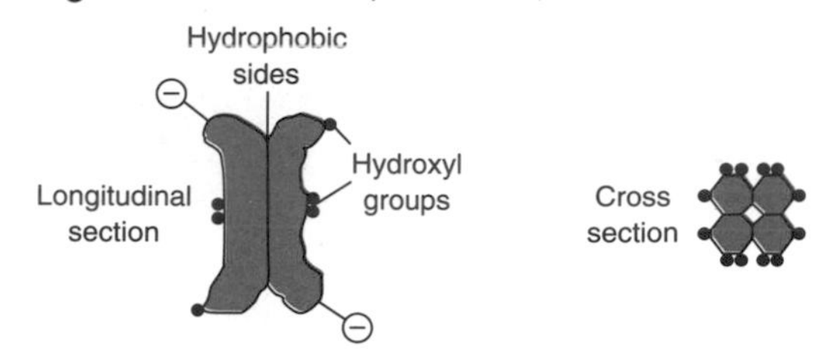

FIGURE 26.33
Diagrammatic representation of a Na^+ cholate micelle.
Adapted from D. M. Small, *Biochim. Biophys. Acta* 176:178, 1969.

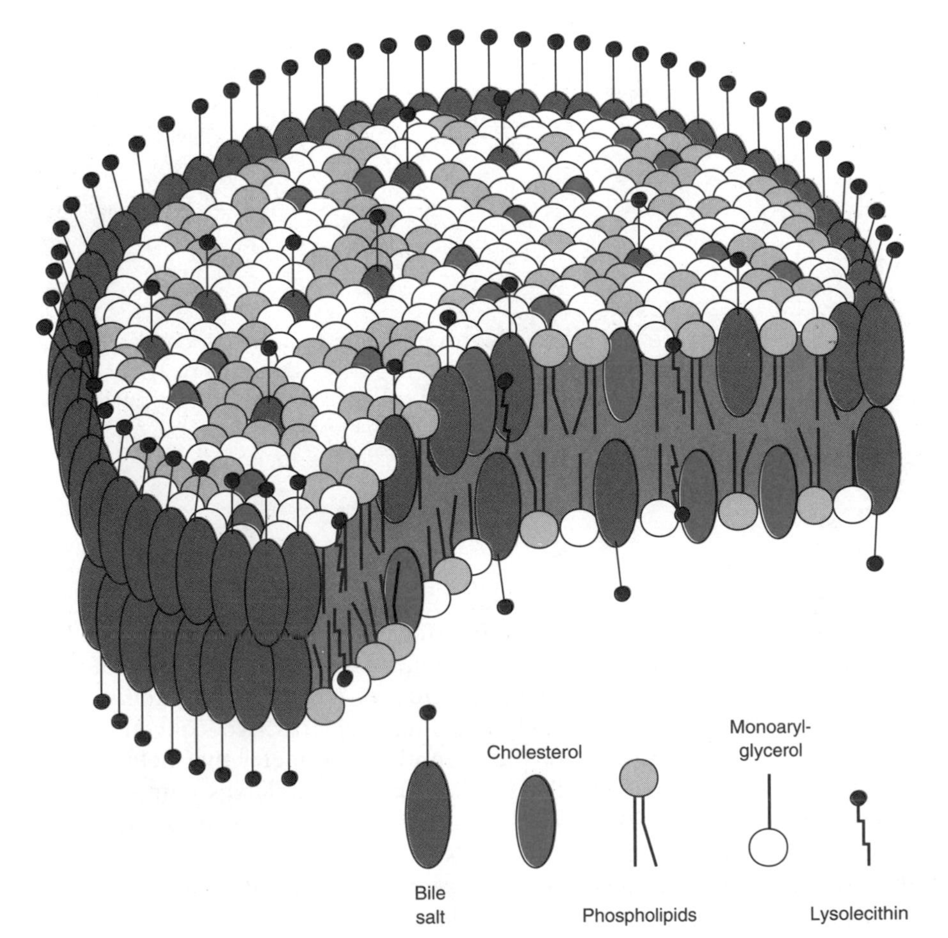

FIGURE 26.34
Proposed structure of the intestinal mixed micelle.
The bilayer disk has a band of bile salt at its periphery and other, more hydrophobic components (fatty acids, monoacylglycerol, phospholipids, and cholesterol) protected within its interior.
Redrawn based on figure from M. C. Carey, *The liver: Biology and pathology*, A. M. Arias, H. Popper, D. Schachter, et al. (Eds.). New York: Raven, 1982.

CLIN. CORR. 26.4
CHOLESTEROL STONES

Liver secretes phospholipids and cholesterol together with bile acids into the bile. Because of the limited solubility of cholesterol, its secretion in bile can result in cholesterol stone formation in the gallbladder. Stone formation is a relatively frequent complication; up to 20% of North Americans will develop stones during their lifetime.

Cholesterol molecules are practically insoluble in aqueous solutions. However, they can be incorporated into mixed phospholipid–bile acid micelles up to a mole ratio of 1 : 1 for cholesterol/phospholipids and thereby "solubilized" (see accompanying figure). However, the liver can produce supersaturated bile with a higher ratio than 1 : 1 of cholesterol/phospholipid. This excess cholesterol has a tendency to crystallize out of solution. Such bile with excess cholesterol is considered lithogenic, that is, stoneforming. The crystal formation usually occurs in the gallbladder, rather than the hepatic bile ducts, because contact times between bile and any crystallization nuclei are greater in the gallbladder. In addition, the gallbladder concentrates bile by absorption of electrolytes and water and the gallbladder bile is not moving except after meals. The bile salts chenodeoxycholate and ursodeoxycholate are now available for oral use to dissolve gallstones. Ingestion of these bile salts reduces cholesterol excretion into the bile and allows cholesterol in stones to be solubilized.

The tendency to secrete bile supersaturated with respect to cholesterol is inherited and found more frequently in females than in males, often associated with obesity. Supersaturation also appears to be a function of the size and nature of the bile acid pool as well as the secretion rate.

Schoenfield, L. J. and Lachin, J. M. Chenodiol (chenodeoxycholic acid) for dissolution of gallstones: The National Cooperative Gallstone Study. A controlled trial of safety and efficacy. *Ann. Int. Med.* 95:257, 1981; and Carey, M. C. and Small, D. M. The Physical Chemistry of Cholesterol Solubility in Bile. *J. Clin. Invest.* 61:998, 1978.

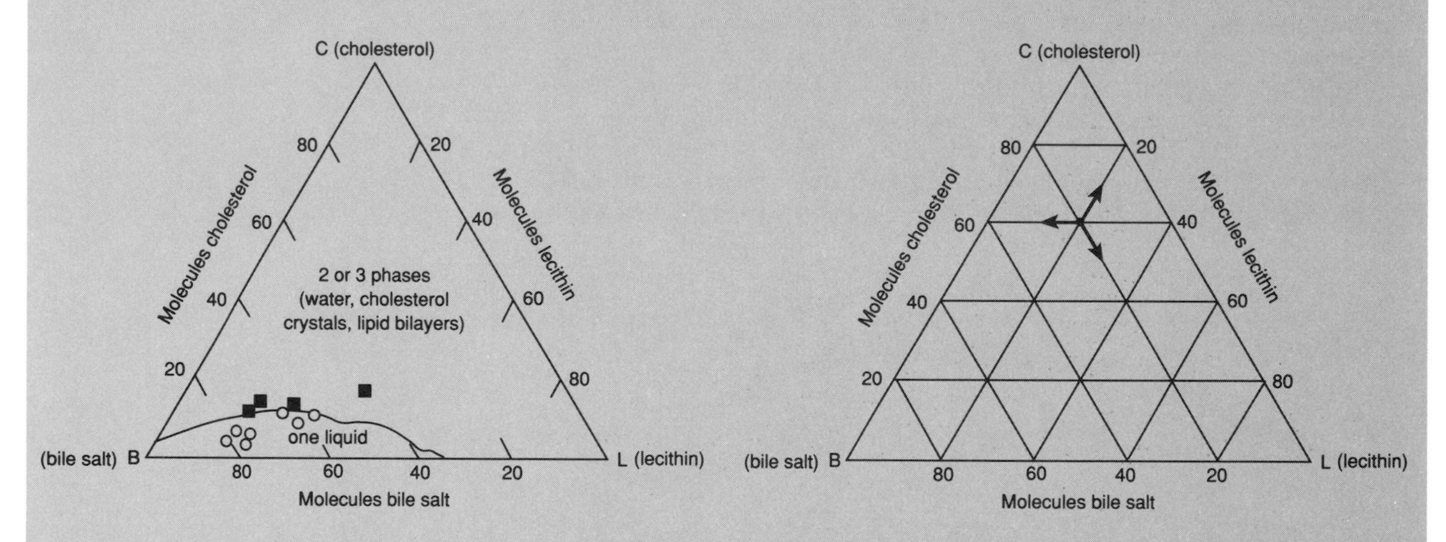

Diagram of the physical states of mixtures of 90% water and 10% lipid.
The 10% lipid is made up of bile acids, lecithin, and cholesterol, and the triangle represents all possible ratios of the three lipid constituents. Each point within the triangle corresponds to a particular composition of the three components, which can be read off the graph as indicated; each point on one of the sides corresponds to a particular composition of just two components. The left triangle contains the composition of gallbladder bile samples from patients without stones (O) and with cholesterol stones (■). Lithogenic bile has a composition that falls outside the "one liquid" area in the lower left corner.
Redrawn from A. F. Hofmann and D. M. Small, *Annu. Rev. Med.* 18:362, 1967.

tive concentration in the unstirred layer. The increase in transport rate is nearly proportional to the increase in effective concentration and can be 1000-fold over that of individually solubilized fatty acids, in accordance with the different solubility of fatty acids as micelles or as individual molecules. This relationship between flux and effective concentration holds because the diffusion constant, another parameter that determines the flux, is only slightly smaller for the mixed micelles as compared to lipid molecules free in solution. Thus, efficient lipid absorption depends on the presence of sufficient bile acids to "solubilize" the ingested and hydrolyzed lipids in micelles. In the absence of bile acids, the absorption of triacylglycerols does not completely stop, although the efficiency is drastically reduced. The residual absorption depends on the slight water solubility of the free fatty acids and monoacylglycerols. Unabsorbed lipids reach the lower intestine where a small part can be metabolized by bacteria. The bulk of unabsorbed lipids, however, is excreted with the stool (this is called **steatorrhea**).

Micelles also serve as transport vehicles through the unstirred fluid layers for those lipids that are even less water-soluble than fatty acids, such as cholesterol or the vitamins A and K. For these compounds, bile acid secretion is absolutely essential for absorption by the intestine.

Most Absorbed Lipids Are Incorporated Into Chylomicrons

The uptake of lipids by the epithelial cells of the small intestine can be explained on the basis of diffusion through the plasma membrane. Absorption is virtually complete for fatty acids and monoacylglycerols, which are slightly water-soluble. It is less efficient for water insoluble lipids. For example, only 30–40% of the dietary cholesterol is absorbed.

Within the intestinal cell, the fate of absorbed fatty acids depends on chain length. Fatty acids of medium chain length (6–10 carbon atoms) pass through the cell into the portal blood without modification. In contrast, the long-chain fatty acids (>12 carbon atoms) become bound to a soluble, fatty acid binding protein in the cytoplasm and are transported to the endoplasmic reticulum, where they are resynthesized into triacylglycerols. Glycerol for this process is derived from the absorbed 2-monoacylglycerols and, to a minor degree, from glucose. The resynthesized triacylglycerols form lipid globules to which surface-active phospholipids and special proteins, termed **apolipoproteins,** adsorb. The lipid globules migrate within membrane-bounded vesicles through the Golgi to the basolateral plasma membrane. They are finally released into the intercellular space by fusion of the vesicles with the basolateral plasma membrane. The final size can be several micrometers in diameter. Because the lipid globules can be so large and because they leave the intestine via lymph vessels, they are called **chylomicrons** (chyle = milky lymph that is present in the intestinal lymph vessels, lacteals, and the thoracic duct after a lipid meal; the word chyle is derived from the Greek *chylos*, which means juice). The intestinal apolipoproteins can be distinguished from those of the liver by antisera. The major intestinal ones are called A-1 and B apolipoproteins. **Apolipoprotein B** is essential for chylomicron release from enterocytes (see Clin. Corr. 26.5).

While medium-chain fatty acids from the diet reach the liver directly with the portal blood, the long-chain fatty acids bypass the liver by being released in the form of chylomicrons into the lymphatics. The intestinal lymph vessels drain into the large body veins via the thoracic duct. Blood from the large veins first reaches the lungs and then the capillaries of the peripheral tissues, including adipose tissue and muscle, before it comes into contact with the liver. Fat and muscle cells in particular take up large amounts of dietary lipids for storage or metabolism. The bypass of the liver may have evolved to protect this organ from a lipid overload after a meal.

The differential handling of medium- and long-chain fatty acids by intestinal cells can be specifically exploited to provide the liver with fatty acids, which constitute high-caloric nutrients. Short- and medium-chain fatty acids are not very palatable; however, triacylglycerols synthesized from these fatty acids are quite palatable and can be used as part of the diet. In the small intestine the triacyglycerols would be hydrolyzed by pancreatic lipase and thus provide fatty acids that reach and can be utilized by the liver.

CLIN. CORR. 26.5
A-β-LIPOPROTEINEMIA

A-β-lipoproteinemia is a human, recessively inherited disorder. It is characterized by the absence of all lipoproteins containing apo-β-lipoprotein, that is, chylomicrons, very low density lipoproteins (VLDL), and low density lipoprotein (LDL). The clinical features are triacylglycerol malabsorption and deficiency of fat-soluble vitamins, notably vitamin E. The enterocytes fail to synthesize apo-β-lipoprotein and accumulate lipid droplets at their apex.

Kane, J. P. Apolipoprotein B: Structural and metabolic heterogeneity. *Annu. Rev. Physiol.* 45:673, 1983.

26.7 BILE ACID METABOLISM

All bile acids are synthesized initially within the liver from cholesterol, but they can be modified by bacterial enzymes during passage through the intestinal lumen. The **primary bile acids** synthesized by the liver are cholic and chenodeoxycholic (or chenic) acid. The **secondary bile acids** are derived from the primary bile acids by bacterial dehydroxylation in position 7 of the ring structure, resulting in deoxycholate and lithocholate, respectively (Figure 26.35).

Primary and secondary bile acids are reabsorbed by the intestine into the portal blood, taken up by the liver, and then resecreted into bile. Within the liver, primary as well as secondary bile acids are linked to either glycine or taurine via an isopeptide bond. These derivatives are called **glyco- and tauroconjugates,** respectively, and constitute the forms

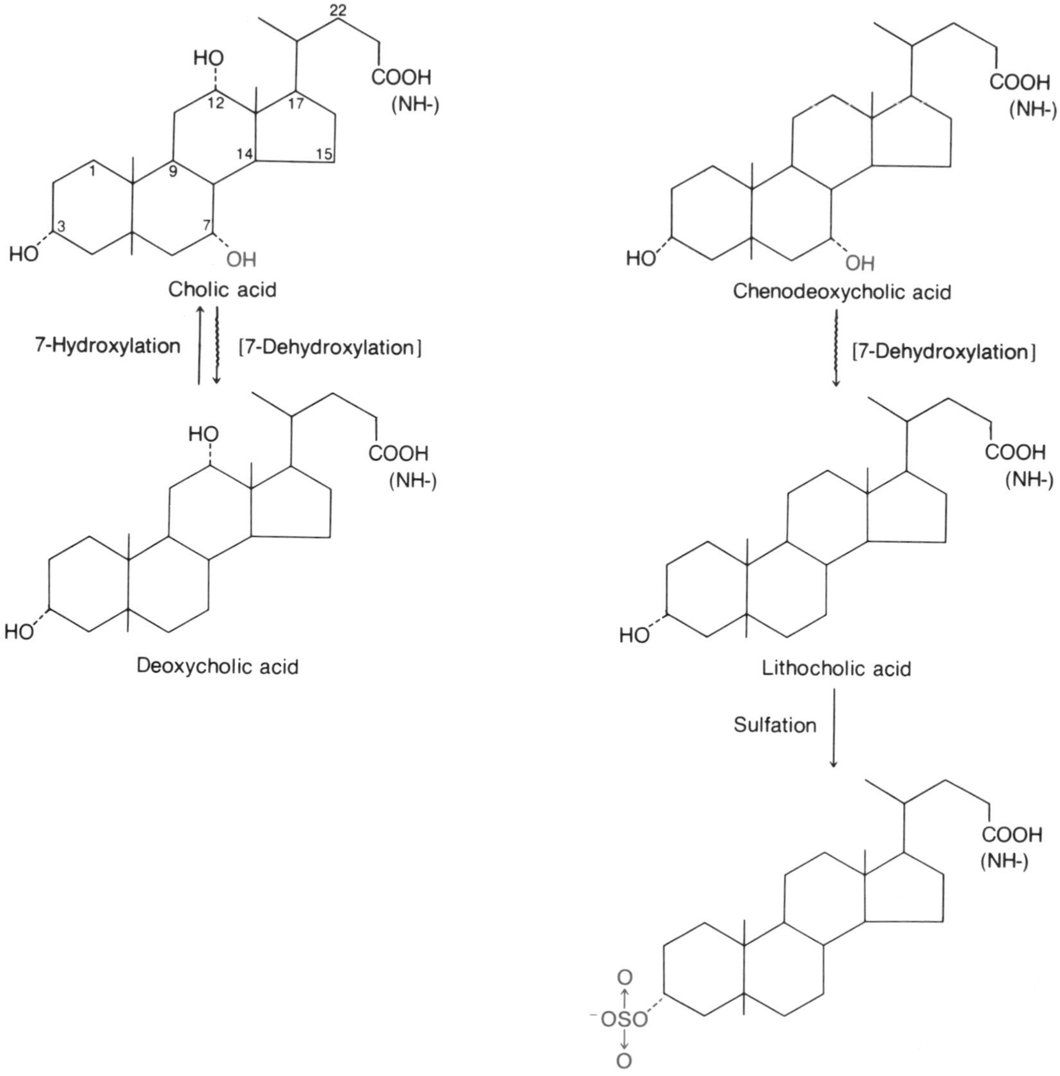

FIGURE 26.35
Bile acid metabolism in the rat.
Straight arrows indicate reactions catalyzed by liver enzymes; bent arrows indicate those of bacterial enzymes within the intestinal lumen. (NH -) = glycine or taurine conjugate of the bile acids.

that are secreted into bile. With the conjugation, the carboxyl group of the unconjugated acid is replaced by an even more polar group. The p*K* values of the carboxyl group of glycine and of the sulfonyl group of taurine are lower than that of unconjugated bile acids, so that conjugated bile acids remain ionized over a wider pH range (Table 26.10). The conjugation is partially reversed within the intestinal lumen by hydrolysis of the isopeptide bond.

The total amount of conjugated and unconjugated bile acids secreted per day by the liver is 16–70 g for an adult. As the total body pool is only 3–4 g, bile acids have to recirculate 5–14 times each day between the intestinal lumen and the liver. Reabsorption of bile acids is important to conserve the pool. Most of the uptake is probably by passive diffusion along the entire small intestine. In addition, the lower ileum contains a specialized **Na^+-bile acid cotransport system** for concentrative reuptake. Thus during a meal bile acids from the gallbladder and the liver are released into the lumen of the upper small intestine, pass with the chyme down the small intestinal lumen, are reabsorbed by the epithelium of the lower small intestine into the portal blood, and are then extracted from the portal blood by the liver parenchymal cells. The process of secretion and reuptake is referred to as the enterohepatic circulation (Figure 26.36). The reabsorption of bile acids by the intestine is quite efficient as only about 0.5 g of bile acids escape reuptake each day and are secreted with the feces. Serum levels of bile acids normally vary with the rate of reabsorption and therefore are highest during a meal when the enterohepatic circulation is most active.

Cholate, deoxycholate, chenodeoxycholate, and their conjugates continuously participate in the enterohepatic circulation. In contrast, most of the **lithocholic acid** that is produced by the action of bacterial enzymes within the intestine is sulfated during the next passage through the liver. The sulfate ester of lithocholic acid is not a substrate for the bile acid transport system in the ileum, and therefore is excreted with the feces.

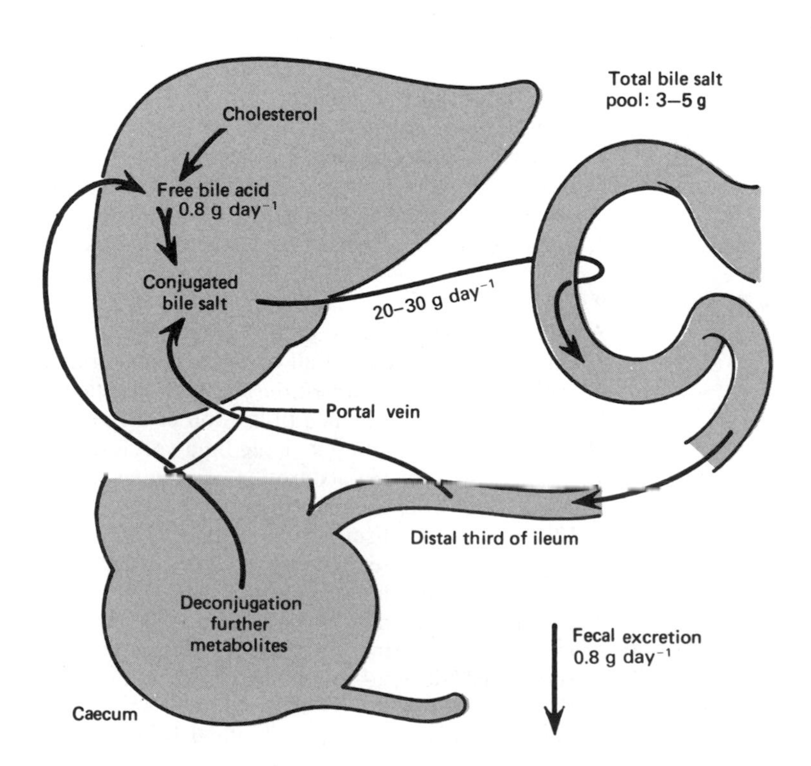

FIGURE 26.36
Enterohepatic circulation of bile acids.
Redrawn from M. L. Clark and J. T. Harries, *Intestinal absorption in man,* McColl, I. and Sladen, G. E. (Eds.). New York: Academic, 1975, p. 195.

BIBLIOGRAPHY

Alvarado, F. and Van Os, C. H. (Eds.). *Ion gradient-coupled transport. Inserm Symposium No. 26*. New York: Elsevier, 1986.

Cristofaro, E., Mottu, F., and Wuhrmann, J. J. Involvement of the raffinose family of oligosaccharides in flatulence, H. L. Sipple and K. W. McNutt (Eds.). *Sugars in nutrition*. New York: Academic, 1974, p. 314.

Davison, J. S. and Shaffer, E. A. (Eds.). Gastrointestinal & hepatic secretions: mechanism and control. Canada: University Calgary Press, 1988.

Erickson, R. H. and Kim, Y. S. Digestion and absorption of dietary protein. *Annu. Rev. Med.* 41:133, 1990.

Field, M., Rao, M. C., and Chang, E. B. Intestinal electrolyte transport and diarrheal disease. *N. Engl. J. Med.* 321:800 and 879, 1989.

Harris, T. J. (Eds.). Familial inherited abnormalities. *Clin. Gastroenterol.* 11:17 and 87, 1982.

Johnson, L. R. *Physiology of the gastrointestinal tract,* Vols. 1 and 2, 2nd ed., New York: Raven, 1987.

Porter, R. and Collins, G. M. *Brush border membranes*. Vol. 95. Ciba Foundation Symposium, London: Pitman, 1983.

Sleisenger, M. H. (Ed.). Malabsorption and nutritional support. *Clin. Gastroenterol.* 12:323, 1983.

QUESTIONS

J. BAGGOTT AND C. N. ANGSTADT

*Q*uestion *T*ypes are described on page xxiii.

1. (QT1) All of the following are involved in the digestive process *except:*
   A. mouth.
   B. pancreas.
   C. liver.
   D. spleen.
   E. gallbladder.

2. (QT2) Active forms of enzymes that digest food may normally be found:
   1. in soluble form in the lumen of the stomach.
   2. attached to the luminal surface of the plasma membrane of intestinal epithelial cells.
   3. dissolved in the cytoplasm of intestinal epithelial cells.
   4. in zymogen granules of pancreatic exocrine cells.

3. (QT1) Histamine is a potent secretagogue of:
   A. amylase by the salivary glands.
   B. HCl by the stomach.
   C. gastrin by the stomach.
   D. hydrolytic enzymes by the pancreas.
   E. $NaHCO_3$ by the pancreas.

4. (QT2) The luminal and contraluminal membranes of small intestinal epithelial cells:
   1. both contain systems for mediated transport of Na^+.
   2. differ in that only the contraluminal membrane contains Na^+,K^+-ATPase.
   3. both contain active transport systems that involve Na^+.
   4. both contain an energy-linked (active) transport system for glucose.

A. Stomach C. Both
B. Small intestine D. Neither

5. (QT4) Site of protein digestion.
6. (QT4) Site of ATP-linked K^+,H^+-antiport.
7. (QT4) Site of chymotrypsinogen synthesis.
8. (QT4) Site of amino acid absorption.

9. (QT1) Oral administration of large amounts of tyrosine could be expected to interfere with the intestinal absorption of:
   A. leucine.
   B. lysine.
   C. glycine.
   D. aspartate.
   E. none of the above.

A. Starch C. Both
B. Protein D. Neither

10. (QT4) Digestion is more efficient after heating.
11. (QT4) Final hydrolysis to monomers may occur in cytoplasm of enterocytes.

A. Autoactivation C. Both
B. Autocatalysis D. Neither

12. (QT4) Pepsin.
13. (QT4) Gastric lipase.

14. (QT2) During the digestion and absorption of triacylglycerols:
    1. a gastric lipase is involved.
    2. hydrolysis occurs at the interface between lipid droplets and the aqueous phase.
    3. most of the triacylglycerol hydrolysis is carried out by a pancreatic enzyme that is inhibited by bile acids.
    4. efficiency is greatly decreased if bile acids are absent.

15. (QT1) Micelles:
    A. are the same as emulsion droplets.
    B. form from bile acids at all bile acid concentrations.
    C. although they are formed during lipid digestion, do not significantly enhance utilization of dietary lipid.
    D. always consist of only a single lipid species.
    E. are essential for the absorption of vitamins A and K.

16. (QT2) In the metabolism of bile acids:
    1. the liver synthesizes the primary bile acids, cholic, and chenodeoxycholic acids.
    2. primary and secondary bile acids may be conjugated to glycine or taurine.
    3. secondary bile acids are formed from primary bile acids by intestinal bacteria.
    4. daily bile acid secretion by the liver is approximately equal to daily bile acid synthesis.

ANSWERS

1. D A: Mechanical homogenization, fluid, amylase. B: HCO_3^-, amylase, proteases, lipase. C: Bile acid synthesis. E: Bile acid storage (p. 1062).
2. A 1–3 true. Zymogen granules contain inactive proenzymes or zymogens, which are not activated until after release from the cell (pp. 1062–1064).
3. B Stimulation of H_2 receptors of the stomach causes HCl secretion (p. 1066).
4. A 1–3 true. Only the luminal surface contains a system for the Na^+-linked active uptake of glucose. Only the contraluminal surface contains the Na^+, K^+-ATPase. The contraluminal surface contains a transport system for glucose that equilibrates glucose across the membranes (p. 1068, Table 26.5, p. 1069).
5. C Pepsin acts in the stomach, producing peptides that stimulate cholecystokinin release, which activates the pancreatic stage of protein digestion. Pancreatic proteases and intestinal peptidases are active in the small intestine (p. 1074).
6. A This system is responsible for H^+ secretion into the gastric lumen by the oxyntic cells (p. 1073). See also item 4 above.
7. D Chymotrypsinogen is synthesized in the pancreas and is activated by trypsin in the small intestine (p. 1074, Table 26.6).
8. B The small intestine has at least six amino acid transport systems (p. 1077).
9. A Tyrosine shares a transport system with Val, Leu, Met, Phe, and Ile (p. 1077).
10. C α-Amylase attacks hydrated starch more readily than unhydrated; heating hydrates the starch granules (p. 1079). Proteolytic enzymes act more readily on denatured proteins (p. 1074).
11. B Cytoplasmic peptidases hydrolyze di- and tripeptides that are absorbed by specific peptide transporters (p. 1076).
12. C Autoactivation is an intramolecular reaction, which produces active pepsin. The pepsin thus formed can, by autocatalysis, activate other pepsinogen molecules (p. 1075).
13. B Gastric lipase digestion products are surface active agents, which stabilize small emulsion droplets. This process enlarges the surface area upon which the lipase acts. Thus the process is autocatalytic (p. 1083).
14. E 1–4 true. 1 and 2: See item 13. 3: Pancreatic lipase carries out the bulk of lipid digestion but is inhibited by bile acids (p. 1083). 4: A protein, colipase, prevents bile acid inhibition in vivo (p. 1083). Bile acid micelles are the major vehicles for moving lipids to the cell surface, where absorption occurs (p. 1087).
15. E A: Micelles are of molecular dimensions and are highly ordered structures; emulsion droplets are much larger and are random (p. 1082, Figure 26.28; p. 1085, Figure 26.34). B: Micelle formation occurs only above the critical micellar concentration (CMC); below that concentration the components are in simple solution (p. 1084, Figure 26.32). C: See item 14. D: Micelles may consist of only one component, or they may be mixed (p. 1085). E: The lipid-soluble vitamins must be dissolved in mixed micelles as a prerequisite for absorption (p. 1087).
16. A 1–3 true. The primary bile acids are synthesized in the liver. In the intestine they may be modified to form the secondary bile acids. Both are reabsorbed and recirculated (enterohepatic circulation). Both are conjugated to glycine or taurine. Only a small fraction of the bile acid escapes reuptake; this must be replaced by synthesis (p. 1089).

27

Principles of Nutrition I: Macronutrients

STEPHEN G. CHANEY

27.1 OVERVIEW 1093
27.2 ENERGY METABOLISM 1094
Energy Content of Food Is Measured in Kilocalories 1094
Energy Expenditure Is Influenced by Four Factors 1095
Recommended Dietary Allowances (RDA) Are General Guidelines 1095
27.3 PROTEIN METABOLISM 1096
Dietary Protein Serves Many Roles Including Energy Production 1096
Essential Amino Acids Must Be Present in the Diet 1098
Protein Sparing Is Related to Dietary Content of Carbohydrate and Fat 1098
Normal Adult Protein Requirements Depend on Diet 1098
Protein Requirements Are Increased During Growth and Recovery Following an Illness 1099
27.4 PROTEIN–ENERGY MALNUTRITION 1100
27.5 EXCESS PROTEIN–ENERGY INTAKE 1101
Obesity May Be Related in Part to a Defect in Thermogenesis 1101
Metabolic Consequences of Obesity May Be Related to Decreased Insulin Receptors 1102
27.6 CARBOHYDRATES 1103
27.7 FATS 1104
27.8 FIBER 1105
27.9 COMPOSITION OF MACRONUTRIENTS IN THE DIET 1106
Lipid Composition of the Diet Affects Serum Cholesterol 1106
Effects of Refined Carbohydrate in the Diet Are Not Straightforward 1107
Mixed Vegetable and Animal Protein Meet Nutritional Protein Requirements 1108
An Increase in Fiber From Varied Sources Is Desirable 1109
Current Recommendations Are for a "Prudent Diet" 1109
BIBLIOGRAPHY 1111
QUESTIONS AND ANNOTATED ANSWERS 1111
CLINICAL CORRELATIONS
27.1 Vegetarian Diets and Protein–Energy Requirements 1098
27.2 Low-Protein Diets and Renal Disease 1099
27.3 Providing Adequate Protein and Calories for the Hospitalized Patient 1100
27.4 Fad Diets: High Protein–High Fat 1102
27.5 Carbohydrate Loading and Athletic Endurance 1103
27.6 High-Carbohydrate–High-Fiber Diets and Diabetes 1104
27.7 Polyunsaturated Fatty Acids and Risk Factors for Heart Disease 1106
27.8 Metabolic Adaptation: The Relationship Between Carbohydrate Intake and Serum Triacylglycerols 1108

27.1 OVERVIEW

Nutrition is best defined as the utilization of foods by living organisms. Since the process of food utilization is clearly biochemical in nature, the major thrust of Chapters 27 and 28 is a discussion of basic nutritional concepts in biochemical terms. However, simply understanding basic nutritional concepts is no longer sufficient. Nutrition appears to attract more than its share of controversy in our society, and a thorough understanding of nutrition almost demands an understanding of the issues behind these controversies. Thus these chapters also explore the biochemical basis for some of the most important nutritional controversies.

Why so much controversy in the first place? Part of the problem is simply that nonscientists feel competent to be "experts." After all, nutrition is concerned with food, and everyone knows about food. However, part of the problem is scientific. Both food itself and our utilization of it are very complex. Quite often the ideal scientific approach of examining only one variable at a time—isolated from other, related variables—is simply inadequate to handle this complex situation. Human variability raises further problems. In studying biochemistry the tendency is to study metabolic pathways and control systems as if they were universal, yet probably no two people utilize nutrients in exactly the same manner. Finally, it is important to realize that much of our knowledge of nutrient requirements and functions comes from animal studies. The days of using prison "volunteers" for nutritional experimentation are behind us. Yet, animals are seldom completely adequate models for human beings, since their biochemical makeup almost always differs (the ability of most other animals to synthesize their own ascorbic acid is just one illustration). Thus, despite the best efforts of many reputable scientists, some of the most important nutritional questions have yet to be answered.

The study of human nutrition can logically be divided into three areas: undernutrition, overnutrition, and ideal nutrition. With respect to **undernutrition,** the primary concern in this country is not with the nutritional deficiency diseases, which are now quite rare. Today more attention is directed toward potential, or subclinical, nutritional deficiencies. **Overnutrition,** on the other hand, is a particularly serious problem in developed countries. Current estimates suggest that between 15 and 30% of the U.S. population are obese, and obesity is known to have a number of serious health consequences. Finally, there is increasing interest today in the concept of ideal, or **optimal nutrition.** This is a concept that has meaning only in an affluent society. Only when the food supply becomes abundant enough so that deficiency diseases are a rarity does it become possible to consider the long-range effects of nutrients on health. This is probably the more exciting and least understood area of nutrition today.

27.2 ENERGY METABOLISM

Energy Content of Food Is Measured in Kilocalories

By now you should be well acquainted with the energy requirements of the body. The energy of much of the foods we eat is converted to ATP and other high-energy compounds, which are in turn utilized by the body to drive biosynthetic pathways, generate nerve impulses, and power muscle contraction. We generally describe the energy content of foods in terms of calories. Technically speaking, we are actually referring to the kilocalories of heat energy released by combustion of that food in the body. Some nutritionists today prefer to use the term kilojoule (a measure of mechanical energy), but since the American public is likely to be counting calories rather than joules in the foreseeable future, we will restrict ourselves to that term here. Experimentally, **calories** can be measured as the heat given off when a food substance is completely burned in a bomb calorimeter, although the caloric value of foods in our body is slightly less, due to incomplete digestion and metabolism. The actual caloric values of protein, fat, carbohydrate, and alcohol are roughly 4, 9, 4, and 7 kcal g^{-1}, respectively. Given these data and the composition of the food, it is simple to calculate the caloric content (input) of the foods we eat. Calculating caloric content of foods does not appear to be a major problem in this country. Millions of Americans appear to be able to do that with ease. The problem lies in balancing caloric input with caloric output. Where do these calories go?

Energy Expenditure Is Influenced by Four Factors

In practical terms, there are four principal factors that affect individual energy expenditure: surface area (which is related to height and weight), age, sex, and activity levels. (1) The effects of surface area are thought to be simply related to the rate of heat loss by the body—the greater the surface area, the greater the rate of heat loss. While it may seem surprising, a lean individual actually has a greater surface area, and thus a greater energy requirement, than an obese individual of the same weight. (2) Age, on the other hand, may reflect two factors: growth and lean muscle mass. In infants and children more energy expenditure is required for rapid growth, and this is reflected in a higher **basal metabolic rate** (rate of energy utilization in the resting state). In adults (even lean adults), muscle tissue is gradually replaced with fat and water during the aging process, resulting in a 2% decrease in basal metabolic rate (BMR) per decade of adult life. (3) As for sex, women tend to have a lower BMR than men due to a smaller percentage of lean muscle mass and the effects of female hormones on metabolism. (4) The effect of activity levels on energy requirements is obvious. However, most of us overemphasize the immediate, as opposed to the long-term, effects of exercise. For example, one would need to jog for over 1 h to burn up the calories found in one piece of apple pie.

Yet, the effect of a regular exercise program on energy expenditure can be quite beneficial. Regular exercise will increase basal metabolic rate, allowing one to burn up calories more rapidly 24 h a day. A regular exercise program should be designed to increase lean muscle mass and should be repeated 4 or 5 days a week but need not be aerobic exercise to have an effect on basal metabolic rate. For an elderly or infirm individual, even daily walking may, with time, help to increase basal metabolic rate slightly.

Recommended Dietary Allowances (RDA) Are General Guidelines

The effect of all of these variables on energy requirements can be readily calculated and, assuming light activity levels (a safe assumption for most Americans) and ideal body weight (not such a safe assumption), are presented in tabular form as **Recommended Dietary Allowances (RDAs)** for energy in Table 27.1.

The RDAs are average values and are widely quoted, but they tell us little about the energy needs of individuals. For example, body composition is known to affect energy requirements, since lean muscle tissue has a higher basal metabolic rate than adipose tissue. This, in part, explains the higher energy requirements of athletes and the lower energy requirements of obese individuals compared to lean individuals of the same weight. Hormone levels are important also, since thyroxine, sex hormones, growth hormones, and, to a lesser extent, epinephrine and corti-

TABLE 27.1 Recommended Dietary Allowances for Energy Intake[a]

Age (year) and sex group	Weight (lb)	Height (in)	Energy needs (kcal)
Infants			
0.0–0.5	13	24	650
0.5–1.0	20	28	850
Children			
1–3	29	35	1300
4–6	44	44	1800
7–10	62	52	2000
Males			
11–14	99	62	2500
15–18	145	69	3000
19–22	160	70	2900
23–50	174	70	2900
51+	170	68	2300
Females			
11–14	101	62	2200
15–18	120	64	2200
19–22	128	65	2200
23–50	138	64	2200
51+	143	63	1900
Pregnancy 1st trimester			+0
2nd trimester			+300
3rd trimester			+300
Lactation			+500

SOURCE: From Recommended Dietary Allowances, Revised 1989, Food and Nutrition Board, National Academy of Sciences–National Research Council, Washington, DC.

[a] The data in this table have been assembled from the observed median heights and weights as surveyed in the U.S. population (NHANES II data).

sol are known to increase BMR. The effects of epinephrine and cortisol probably explain in part why severe stress and major trauma significantly increase energy requirements. Finally, energy intake itself has an inverse relationship to expenditure in that during periods of starvation or semi-starvation BMR can decrease up to 50%. This is of great survival value in cases of genuine starvation, but not much help to the person who wishes to lose weight on a calorie-restricted diet. Clearly then, the above tables are only general guidelines and may bear little resemblance to the energy needs of a given individual—a fact that is important to remember in the treatment of obesity.

27.3 PROTEIN METABOLISM

Dietary Protein Serves Many Roles Including Energy Production

Protein carries a certain mystique as a "body-building" food. While it is true that protein is an essential structural component of all cells, protein is equally important for maintaining the output of essential secretions such as digestive enzymes and peptide hormones. Protein is also needed to synthesize the plasma proteins, which are essential for maintaining osmotic balance, transporting substances through the blood, and maintaining immunity. However, the average adult in this country consumes far more protein than needed to carry out these essential functions. The excess protein is simply treated as a source of energy, with the glucogenic amino acids being converted to glucose and the ketogenic amino acids being converted to fatty acids and keto acids. Both kinds of amino acids will of course eventually be converted to triacylglycerol in the adipose tissue if fat and carbohydrate supplies are already adequate to meet energy requirements. Thus for most of us the only body building obtained from high-protein diets is adipose tissue.

It has always been popular to say that the body has no storage depot for protein, and thus adequate dietary protein must be supplied with every meal. However, in actuality, this is not quite accurate. While there is no separate class of "storage" protein, there is a certain percentage of protein that undergoes a constant process of breakdown and resynthesis. In the fasting state the breakdown of this store of body protein is enhanced, and the resulting amino acids are utilized for glucose production, the synthesis of nonprotein nitrogenous compounds, and for the synthesis of the essential secretory and plasma proteins described above (see also Chapter 14). Even in the fed state, some of these amino acids will be utilized for energy production and as biosynthetic precursors. Thus the turnover of body protein is a normal process—and an essential feature of what is called nitrogen balance.

Nitrogen Balance Relates Intake of Nitrogen to Its Excretion

Nitrogen balance (Figure 27.1) is simply a comparison between the intake of nitrogen (chiefly in the form of protein) and the excretion of nitrogen (chiefly in the form of undigested protein in the feces and urea and ammonia in the urine). The normal adult will be in nitrogen equilibrium, with

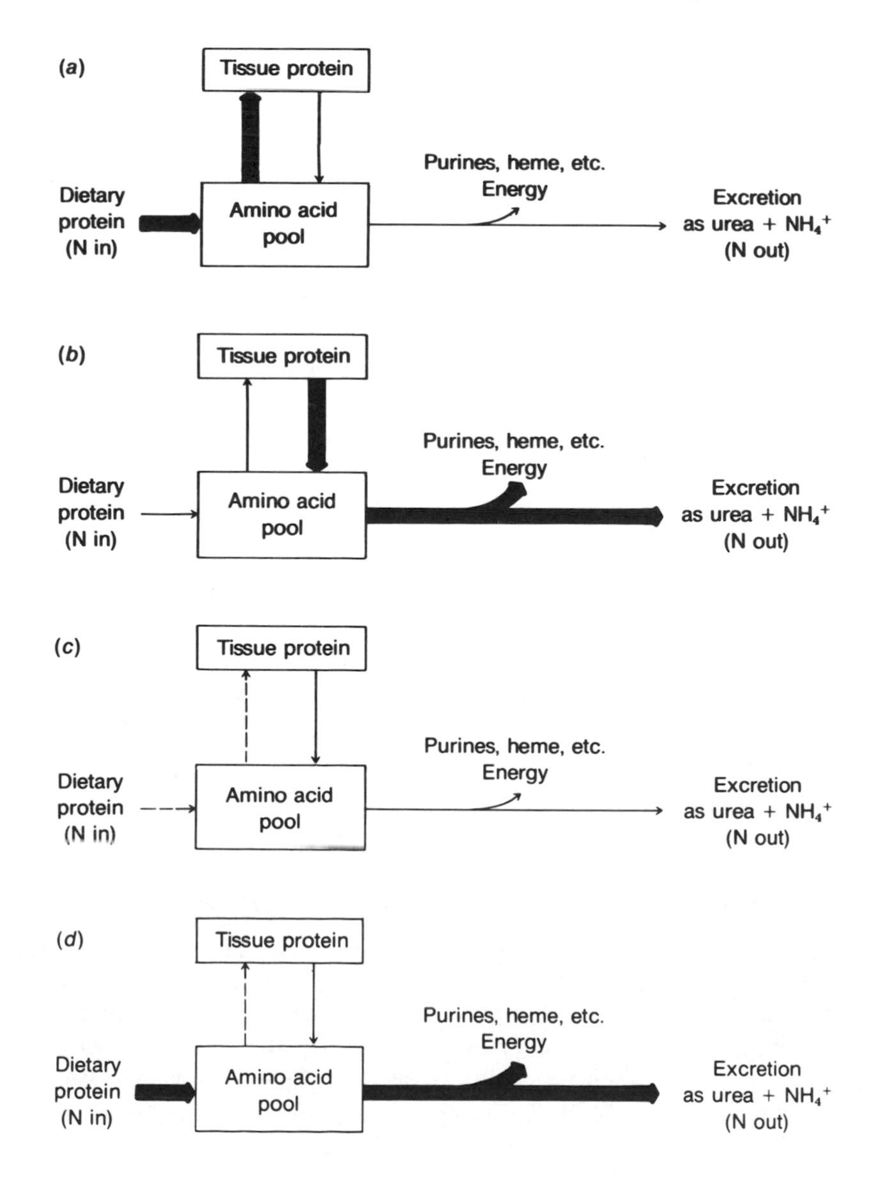

FIGURE 27.1
Factors affecting nitrogen balance.
Schematic representations of the metabolic interrelationship involved in determining nitrogen balance. (*a*) Positive nitrogen balance (growth, pregnancy, lactation, and ro covery from metabolic stress). (*b*) Negative nitrogen balance (metabolic stress). (*c*) Negative nitrogen balance (inadequate dietary protein). (*d*) Negative nitrogen balance (lack of an essential amino acid). Each figure represents the nitrogen balance resulting from a particular set of metabolic conditions. The dominant pathways in each situation are indicated by heavy arrows.

CLIN. CORR. **27.1**
VEGETARIAN DIETS AND PROTEIN–ENERGY REQUIREMENTS

One of the most important problems of a purely vegetarian diet (as opposed to a lacto-ovo vegetarian diet) is the difficulty in obtaining sufficient calories and protein. The potential caloric deficit results from the fact that the caloric densities of fruits and vegetables are much less than the meats they replace (30–50 cal per 100 g vs 150–300 cal per 100 g). The protein problem is generally threefold: (1) most plant products are much lower in protein (1–2 g of protein per 100 g vs 15–20 g per 100 g); (2) most plant protein is of low biological value; and (3) some plant proteins are incompletely digested. Actually, any reasonably well-designed vegetarian diet will usually provide enough calories and protein for the average adult. In fact, the reduced caloric intake may well be of benefit because strict vegetarians do tend to be lighter than their nonvegetarian counterparts.

However, whereas an adult male may require about 0.8 g of protein and 40 cal kg^{-1} of body weight, a young child may require 2–3 times that amount. Similarly, a pregnant woman needs an additional 10 g of protein and 300 cal day^{-1} and a lactating woman an extra 15 g of protein and 500 cal. Thus both young children and pregnant and lactating women run a risk of protein–energy malnutrition.

It is possible to provide sufficient calories and protein even for these high-risk groups provided the diet is adequately planned. There are three principles that can be followed to design a calorie–protein–sufficient vegetarian diet for young children: (1) whenever possible, include eggs and milk in the diet; they are both excellent sources of calories and high-quality protein; (2) include liberal amounts of those vegetable foods with high-caloric density in the diet, including nuts, grains, dried beans, and dried fruits; and (3) include liberal amounts of high-protein vegetable foods that have complementary amino acid patterns. It used to be felt that these complementary proteins must be present in the same meal. Recent animal studies, however, suggest that a meal low in (but not devoid of) an essential amino acid may be supplemented by adding the limiting amino acid at a subsequent meal.

Shultz, T. D., Craig, W. J., Johnson, P. K., Sanchez, A., and Register, U. D. Vegetarianism and Health, J. Weininger and G. M. Briggs (Eds.). *Nutrition Update*, Vol. 2, New York: Wiley, 1985, p. 129; and First International Congress on Vegetarian Nutrition. *Proc. Am. J. Clin. Nutr.* 48 (Suppl. 1):707, 1988.

losses just balanced by intake. Negative nitrogen balance can result from an inadequate dietary intake of protein, since the amino acids utilized for energy and biosynthetic reactions are not replaced. However, it also is observed in injury when there is net destruction of tissue and in major trauma or illness in which the body's adaptive response causes increased catabolism of body protein stores. A positive nitrogen balance will be observed whenever there is a net increase in the body protein stores, such as in a growing child, a pregnant woman, or a convalescing adult.

Essential Amino Acids Must Be Present in the Diet

In addition to the amount of protein in the diet, several other factors must be considered. One of these is the complement of essential amino acids present in the protein. *The* **essential amino acids** *are those amino acids that cannot be synthesized by the body* (Chapter 11). If just one of these essential amino acids is missing from the diet, the body cannot synthesize new protein to replace the protein lost due to normal turnover, and a negative nitrogen balance results (Figure 27.1). Obviously then, the complement of essential amino acids in any given protein will determine how well it can be used by our body.

Generally, most animal proteins contain all of the essential amino acids in about the quantities needed by the human body. Vegetable proteins, on the other hand, often lack one or more essential amino acids and may, in some cases, be more difficult to digest. Even so, vegetarian diets can provide adequate protein provided (a) enough extra protein is consumed to provide sufficient quantities of the essential amino acids and/or (b) two or more proteins are consumed together, which complement one another in amino acid content. For example, if corn, which is deficient in lysine, is combined with legumes, which are deficient in methionine but rich in lysine, the efficiency of utilization for the combination of the two vegetable proteins approaches that of animal protein. The adequacy of vegetarian diets with respect to protein and calories is discussed more fully in Clin. Corr. 27.1, and the need for high-quality protein in low-protein diets in renal disease is discussed in Clin. Corr. 27.2.

Protein Sparing Is Related to Dietary Content of Carbohydrate and Fat

Another factor that must be considered in determining protein requirements is the dietary intake of fat and carbohydrate. If these components are present in insufficient quantities, some of the dietary protein must be used for energy generation and is unavailable for building and replacing tissue. Thus, as the energy (calorie) content of the diet from carbohydrate and fat increases, the need for protein decreases. This is referred to as **protein sparing.** Carbohydrate is somewhat more efficient at protein sparing than fat—presumably because carbohydrate can be used as an energy source by almost all tissues, whereas fat cannot.

Normal Adult Protein Requirements Depend on Diet

Assuming adequate calorie intake and 75% efficiency of utilization, which is typical of the mixed protein in the average American diet, the **recommended protein intake** is 0.8 g per kg body wt per day. This amounts to about 58 g of protein per day for a 72-kg (160-lb) man and about 44 g per day^{-1} for a 55-kg (120-lb) woman. These recommendations would obviously need to be increased on a vegetarian diet if the overall efficiency of utilization were less than 75%.

CLIN. CORR. 27.2
LOW-PROTEIN DIETS AND RENAL DISEASE

Chronic renal failure is characterized by the buildup of the end products of protein catabolism, mainly urea. Some degree of dietary protein restriction is usually necessary because these toxic end products are responsible for many of the symptoms associated with renal failure. The amount of protein restriction is dependent on the severity of the disease. It is easy to maintain patients in nitrogen equilibrium for prolonged periods on diets containing as little as 40 g of protein per day if the diet is calorically sufficient. Diets containing less than 40 g day^{-1} pose problems. Protein turnover continues and one is forced to walk a tightrope between providing enough protein to avoid negative nitrogen balance, but little enough to avoid buildup of waste products.

The strategy employed in such diets is twofold: (1) to provide a minimum of protein, primarily protein of high BV, and (2) to provide the rest of the daily calories in the form of carbohydrates and fats. The goal is to provide just enough essential amino acids to maintain positive nitrogen balance. In turn, the body should be able to synthesize the nonessential amino acids from other nitrogen-containing metabolites. Enough carbohydrate and fat are provided so that essentially all of the dietary protein can be spared from energy metabolism. With this type of diet, it is possible to maintain a patient on 20 g of protein per day for considerable periods. Because of the difficulty in maintaining nitrogen equilibrium at such low-protein intakes, the patient's protein status should be monitored. This can be done by measuring parameters such as serum albumin and transferrin.

Moreover, such diets are extremely monotonous and difficult to follow. A typical 20 g of protein diet is shown below:

1. One egg plus ¾ cup milk or 1 additional egg or 1 oz of meat.
2. One-half pound of deglutenized (low protein) wheat bread; all other breads and cereals must be avoided—this includes almost all baked goods.
3. A limited amount of certain low-protein, low-potassium fruits and vegetables.
4. Sugars and fats to fill the rest of the needed calories; however, cakes, pies, and cookies would need to be avoided.

The palatability of this diet would be improved considerably if it were possible to include more foods containing protein of low BV (vegetables, cereals, and bread) for variety, yet still maintain a positive nitrogen balance. One approach to this problem is to use dietary supplements (usually in liquid or tablet form), which provide only the essential amino acids. Another approach is to use supplements containing the carbon skeletons (in the form of α-OH and α-keto acids) of the essential amino acids. Preliminary studies indicate that both of these techniques will help preserve renal function when used in conjunction with low protein diets containing a somewhat greater variety of foods.

Goodship, T. H. J. and Mitch, W. E. Nutritional Approaches to Preserving Renal Function. *Adv. Intern. Med.* 33:377, 1988.

Protein Requirements Are Increased During Growth and Recovery Following an Illness

Because dietary protein is essential for the synthesis of new body tissue, as well as for maintenance and repair, the need for protein increases markedly during periods of rapid growth. Such growth occurs during pregnancy, infancy, childhood, and adolescence. These recommendations are summarized in Table 27.2.

Once growth requirements have been considered, age does not seem to have much effect on protein requirements. If anything, the protein requirement may slightly decrease with age. However, older people need and generally consume fewer calories, so high-quality protein should provide a larger percentage of their total calories. Furthermore, some older people may have special protein requirements due to malabsorption problems.

One of the myths of popular nutrition is that athletes need more protein. While there may be a slight increase in protein requirements during periods of intensive body building, vigorous exercise alone increases only energy needs—not protein requirements. Thus the protein needs of the athlete are only slightly greater than those of the nonathlete and can be met from a well-balanced diet that provides for the additional caloric needs.

Illness, major trauma, and surgery all cause a major **catabolic response** in the body. Both energy and protein needs are very large, and the body responds by increasing production of glucagon, glucocorticoids, and epinephrine (see Chapter 14). In these situations the breakdown of body

TABLE 27.2 Recommended Dietary Allowances for Protein

Age (year) and sex group	Weight (lb)	Height (in)	Protein (g)
Infants			
0.0–0.5	13	24	13
0.5–1.0	20	28	14
Children			
1–3	29	35	16
4–6	44	44	24
7–10	62	52	28
Males			
11–14	99	62	42
15–18	145	69	59
19–22	160	70	58
23–50	174	70	63
51+	170	68	63
Females			
11–14	101	62	46
15–18	120	64	44
19–22	128	65	46
23–50	138	64	50
51+	143	63	50
Pregnancy			+10
Lactation 1st 6 months			+15
2nd 6 months			+12

SOURCE: From Recommended Dietary Allowances, Revised 1989, Food and Nutrition Board, National Academy of Sciences, National Research Council, Washington, DC.

protein is greatly accelerated and a negative nitrogen balance results unless protein intake is increased (Figure 27.1). Although this increased protein requirement is of little significance in short-term illness, it can be vitally important in the recovery of hospitalized patients as discussed in Section 27.4 (see also Clin. Corr. 27.3).

27.4 PROTEIN–ENERGY MALNUTRITION

The most common form of malnutrition in the world today is **protein–energy malnutrition (PEM).** In the developing countries inadequate intake of protein and energy is all too common, and it is usually the infants and young children who suffer most. While the actual symptoms of protein–energy insufficiency vary widely from case to case, it is common to classify most cases as either marasmus or kwashiorkor. **Marasmus** is usually defined as inadequate intake of both protein and energy. **Kwashiorkor** is defined as inadequate intake of protein in the presence of adequate energy intake. More often than not, the diets associated with marasmus and kwashiorkor may be similar, with the kwashiorkor being precipitated by conditions of increased protein demand such as infection. The marasmic infant will have a thin, wasted appearance and will be small for his/her age. If the PEM continues long enough, the child will be permanently stunted in both physical and mental development. In kwashiorkor, on the other hand, the child will often have a deceptively plump appearance due to edema. Other telltale symptoms associated with kwashiorkor are dry, brittle hair, diarrhea, dermatitis of various forms, and retarded growth. Perhaps the most devastating result of both marasmus and kwashiorkor is reduced ability of the afflicted individuals to fight off infection. They have a reduced number of T lymphocytes (and thus diminished cell-mediated immune response) as well as defects in the generation of phagocytic cells and production of immunoglobulins, interferon, and other components of the immune system. Many of these indi-

CLIN. CORR. 27.3 PROVIDING ADEQUATE PROTEIN AND CALORIES FOR THE HOSPITALIZED PATIENT

The normal metabolic response to infection, trauma, and surgery is a complex and carefully balanced catabolic state. As discussed in the text, epinephrine, glucagon, and cortisol are released, greatly accelerating the rates of lipolysis, proteolysis, and gluconeogenesis. The net result is an increased supply of fatty acids and glucose to meet the increased energy demands of such major stress. The high serum glucose results in elevation of circulating insulin levels, which is more than counterbalanced by the increased levels of epinephrine and other hormones. Skeletal muscle, for example, uses very little of the serum glucose, but continues instead to rely on free fatty acids and its own catabolized protein as a primary source of energy. It also continues to export amino acids, primarily alanine, for use elsewhere in the body, resulting in a very rapid depletion of body protein stores.

A highly catabolic hospitalized patient may require 35–45 kcal kg^{-1} day^{-1} and 2–3 g of protein kg^{-1} day^{-1}. A patient with severe burns may require even more. The physician has a number of options available to provide this postoperative patient with sufficient calories and protein to insure optimal recovery. When the patient is simply unable to take in enough food, it may be adequate to supplement the diet with high-calorie–high-protein preparations, which are usually mixtures of homogenized cornstarch, egg, milk protein, and flavorings. When the patient is unable to take in solid food or unable to digest complex mixtures of foods adequately, elemental diets are usually administered via a nasogastric tube. Elemental diets consist of small peptides or purified amino acids, glucose and dextrins, some fat, vitamins, and electrolytes. These diets are generally low residue and can be used in patients with lower gastrointestinal tract disturbances. They are also very efficiently digested and absorbed in the absence of pancreatic enzymes or bile salts. These diets are sometimes sufficient to meet most of the short-term caloric and protein needs of a moderately catabolic patient.

When the patient is severely catabolic or unable to digest and absorb foods normally, parenteral (intravenous) nutrition is necessary. The least invasive method is to use a peripheral, slow-flow vein in a manner similar to any other iv infusion. The main limitation of this method is hypertonicity. If the infusion fluid is too hypertonic, there is endo-

viduals die from secondary infections, rather than from the starvation itself. In the United States, classical marasmus and kwashiorkor are exceedingly rare, but milder forms of protein–energy malnutrition are seen.

The most common form of PEM seen in the United States today occurs in the hospital setting. A typical course of events is as follows: The patient is not eating well for several weeks or months prior to entering the hospital due to chronic or debilitating illness. He/she enters the hospital with major trauma, severe infection, or for major surgery, all of which cause a large negative nitrogen balance. This is often compounded by difficulties in feeding the patient or by the necessity of fasting the patient in preparation for surgery or diagnostic tests. The net result is PEM as measured by low levels of serum albumin and other serum proteins or by decreased cellular immunity tests. Recent studies have shown that hospitalized patients with demonstrable PEM have delayed wound healing, decreased resistance to infection, increased mortality, and increased length of hospitalization. Currently, most major hospitals have instituted programs to monitor the nutritional status of their hospitalized patients and to intervene where necessary to maintain a positive nitrogen and energy balance (Clin. Corr. 27.3).

thelial cell damage and thrombosis. However, a solution of 5% glucose and 4.25% purified amino acids can be used safely. This solution will usually provide enough protein to maintain positive nitrogen balance, but will rarely provide enough calories for long-term maintenance of a catabolic patient.

The most aggressive nutritional therapy is total parenteral nutrition. Usually an indwelling catheter is inserted into a large fast-flow vessel such as the superior vena cava, so that the very hypertonic infusion fluid can be rapidly diluted. This allows solutions of up to 60% glucose and 4.25% amino acids to be used, providing sufficient protein and most of the calories for long-term maintenance. Intravenous lipid infusion is often added to boost calories and provide essential fatty acids. All of these methods can be used to prevent or minimize the negative nitrogen balance associated with surgery and trauma. The actual choice of method depends on the patient's condition. As a general rule it is preferable to use the least invasive technique.

Hopkins, B. S., Bistrain, B. R., and Blackburn, G. L. Protein–Calorie Management in the Hospitalized Patient, H. A. Schneider, C. E. Anderson, and D. B. Coursin (Eds.). *Nutritional Support of Medical Practice,* 2nd ed., New York: Harper & Row, 1983, pp. 140–159; and Streat, S. J. and Hill, G. L. Nutritional Support in the Management of Critically Ill Patients in Surgical Intensive Care. *World J. Surg.* 11:194, 1987.

27.5 EXCESS PROTEIN–ENERGY INTAKE

Much has been said in recent years about the large quantities of protein that the average American consumes. Certainly most of us do consume far more than needed to maintain positive nitrogen balance. The average American currently consumes 99 g of protein, 68% of it from animal sources. However, most studies seem to show that a healthy adult can consume that quantity of protein with no apparent harm. Concern has been raised about the possible effect of high-protein intake on calcium requirements. Some studies suggest that high-protein intakes increase urinary loss of calcium and thus may accelerate the bone demineralization associated with the aging process. However, this issue is far from settled.

Obesity May Be Related in Part to a Defect in Thermogenesis

Perhaps the more serious nutritional problem in this country is excessive energy consumption. In fact, **obesity** is the most frequent nutritional disorder in the United States. It would, however, be unfair to label obesity as simply a problem of excess consumption. Overeating plays an important role in many individuals, as does inadequate exercise, but there is a strong genetic component as well. While the biochemical mechanism(s) for this genetic predisposition are still unclear, one popular theory holds that predisposition to obesity may result from a breakdown in thermogenesis. Whenever we consume calories in excess of those needed for immediate energy production, a portion of those excess calories is converted to heat rather than being stored as fat, a process known as **diet-induced thermogenesis.** In part, this may be influenced by exercise and composition of the diet, with both exercise and high-carbohydrate diets favoring greater thermogenesis. One theory is that the hyperinsulinemia that follows ingestion of a carbohydrate-rich meal activates the sympathetic nervous system, which, in turn, stimulates metabolic processes that dissipate energy as heat. In rats and mice, most of this thermogenesis appears to occur in specialized brown adipose tissue, but in humans skeletal muscle appears to be the major site of thermogenesis, with adipose tissue making only a small contribution. Some individuals with a predisposition to obesity appear to have a partial genetic defect in this diet-induced thermogenesis. Simply put, they are more efficient at converting excess calories to fat

CLIN. CORR. **27.4**
FAD DIETS: HIGH PROTEIN–HIGH FAT

Among the most popular fad diets over the years have been the high-protein–high-fat diets such as the Stillman and Atkins diets. Actually, at first glance, it seems as if these diets have a sound metabolic basis. The basic premise is that if one severely restricts carbohydrate intake, it is possible to eat large amounts of high-protein–high-fat foods because the body will not be able to utilize the fat efficiently. This hypothesis is primarily based on the fact, made abundantly clear in any biochemistry textbook, that glucose is needed to replenish the intermediates of the citric acid cycle. Thus, in the absence of glucose, fat should simply be converted to ketone bodies and be disposed of.

There are several problems with this oversimplified hypothesis. First, it ignores the fact that many amino acids can be readily converted to citric acid cycle intermediates. Second, many tissues in the body are perfectly able to use ketone bodies for energy generation. Third, the loss of ketones in the urine cannot possibly lead to any significant caloric deficit. Maximum ketone excretion is about 20 g (100 kcal) day^{-1}.

It is important, however, to realize that this diet does appear to "work" for many patients. The apparent success of the diet is related primarily to two factors. In the fist place, any low-carbohydrate diet results in a significant initial water loss. This is primarily due to depletion of glycogen reserves, since 3 g of water is bound for every 1 g of glycogen. It is this rapid initial weight loss that makes the diet so appealing. Second, while a high-protein–high-fat diet sounds appealing initially, it is relatively unpalatable and expensive, leading ultimately to decreased caloric intake.

This diet is also not without its health risks. The high-fat content may contribute to atherosclerosis and heart disease. The lack of fruits and vegetables may lead to vitamin deficiencies. Finally, ketosis should be avoided by pregnant women (ketone bodies can be harmful to the developing fetal brain), and high-protein intakes should be avoided by anyone with a history of kidney disease.

Today the most popular diets tend to be low fat–high carbohydrate. There may be some metabolic basis for these diets since dietary carbohydrates appear to be less efficiently converted to fat stores than dietary fat. Carbohydrate intake also results in greater diet-induced thermogenesis than fat intake although it is not clear how significant this effect is for people consuming a mixed diet. Finally, a low-fat diet is consistent with

than others in the population. It is not yet clear whether the defect is a lack of response of the sympathetic nervous system to the consumption of excess calories, or an inability of the target tissue(s) to respond to stimulation by norepinephrine.

The biochemical mechanisms for thermogenesis in the target tissues are also not known. However, several of the proposed mechanisms are instructive to the student of biochemistry because they review pathways you have learned previously. For example, the α-glycerol phosphate shuttle has been proposed as one possible source of thermogenesis. As you recall (p. 404), the α-glycerol phosphate shuttle is one mechanism for getting reducing equivalents into the mitochondria, but it is less efficient than the malate–aspartate shuttle because cytoplasmic NADH is converted to mitochondrial $FADH_2$. For each turn of the α-glycerol phosphate shuttle, the equivalent of 1 ATP is lost as heat. Thus, stimulation of the α-glycerol phosphate shuttle and/or inhibition of the malate–aspartate shuttle would lead to decreased efficiency of energy utilization and increased thermogenesis. Alternatively, oxidative phosphorylation may simply be uncoupled during diet-induced thermogenesis. Mitochondria in brown adipose tissue of animals contain an **"uncoupling protein"** that acts to dissipate the proton gradient when the brown adipose tissue is stimulated by the sympathetic nervous system.

Metabolic Consequences of Obesity May Be Related to Decreased Insulin Receptors

A discussion of the treatment of obesity is clearly beyond the scope of this chapter, but it is worthwhile to consider some of the metabolic consequences of obesity. One striking clinical feature of overweight individuals is a marked elevation of serum free fatty acids, cholesterol, and triacylglycerols irrespective of the dietary intake of fat. Why is this? Obesity is obviously associated with an increased number and/or size of adipose cells. Furthermore, these cells contain fewer insulin receptors and thus respond more poorly to insulin, resulting in increased activity of the hormone-sensitive lipase. The increased lipase activity along with the increased mass of adipose tissue is probably sufficient to explain the increase in circulating free fatty acids. These excess fatty acids are, of course, carried to the liver, where they are broken down to acetyl CoA, which is a precursor for both triacylglycerol and cholesterol synthesis. The excess triacylglycerol and cholesterol are released as very low density lipoprotein (VLDL) particles, leading to higher circulating levels of both triacylglycerol and cholesterol (for more detail on these metabolic interconversions, see Chapters 9 and 10).

A second striking finding in obese individuals is higher fasting blood sugar levels and decreased glucose tolerance. Fully 80% of adult onset diabetics are overweight. Again the culprit appears to be the decrease in insulin receptors, since many adult onset diabetics have higher than normal insulin levels. This hyperinsulinemia also appears to stimulate the sympathetic nervous system, leading to sodium and water retention and vasoconstriction, all of which tend to increase blood pressure. Because of these metabolic changes, obesity is one of the primary risk factors in coronary heart disease, hypertension, and diabetes. This is nutritionally significant because all of these metabolic changes are reversible. Quite often reduction to ideal weight is the single most important mode of nutritional therapy. Furthermore, when the individual is at ideal body weight, the composition of the diet becomes a less important consideration in maintaining normal serum lipid and glucose levels.

Any discussion of weight reduction regimens should include a mention of one other metabolic consequence of obesity. As discussed above, obesity can lead to increased retention of both sodium and water. Thus, in effect as the fat stores are metabolized, they are converted to water, which is denser than the fat, and the water may be largely retained. In fact, some individuals may actually observe short-term weight gain on certain diets, even though the diet is working perfectly well in terms of breaking down their adipose tissue. This metabolic fact of life can be psychologically devastating to the dieters, who expect to see quick results for all their sacrifice. This is one major reason for the frequent popularity of the low-carbohydrate diets, which decrease water retention (Clin. Corr. 27.4).

27.6 CARBOHYDRATES

The chief metabolic role of carbohydrates in the diet is for energy production. Any carbohydrate in excess of that needed for energy is converted to glycogen and triacylglycerol for long-term storage. The human body can adapt to a wide range of carbohydrate levels in the diet. Diets high in carbohydrate result in higher steady-state levels of glucokinase and some of the enzymes involved in the hexose monophosphate shunt and triacylglycerol synthesis. Diets low in carbohydrate result in higher steady-state levels of some of the enzymes involved in gluconeogenesis, fatty acid oxidation, and amino acid catabolism. Glycogen stores can also be affected by the carbohydrate content of the diet (Clin. Corr. 27.5). Very low

most current dietary recommendations. There is nothing magical about carbohydrates, however. Claims that "you can eat all you want as long as you cut down on fat" are almost certainly false. In addition, diets containing less than 30% fat are radically different from the typical American diet. Most patients find it difficult to follow such diets for an extended period of time. Thus, the best dietary advice is still to reduce total caloric intake by limiting portion size and frequency of snacking, and choosing low calorie foods. Since carbohydrate-rich foods have a lower caloric density than foods high in fat, this will be easiest to accomplish if low-fat–high-carbohydrate foods are selected whenever possible.

Council on Foods and Nutrition. A Critique of Low-Carbohydrate Ketogenic Weight Reduction Regimens. *J. Am. Med. Assoc.* 224:1415, 1973; Friedman, R. B. Fad Diets: Evaluation of Five Common Types. *Postgrad. Med.* 79:249, 1986; and Truswell, A. S. Pop Diets for Weight Reduction. *Br. Med. J.* 285:1519, 1982.

CLIN. CORR. 27.5
CARBOHYDRATE LOADING AND ATHLETIC ENDURANCE

The practice of carbohydrate loading dates back to observations made in the early 1960s that endurance during vigorous exercise was limited primarily by muscle glycogen stores. Of course, the glycogen stores are not the sole energy source for muscle. Free fatty acids are present in the blood during vigorous exercise and are utilized by muscle along with the glycogen stores. Once the glycogen stores have been exhausted, however, muscle cannot rely entirely on free fatty acids without tiring rapidly. This is probably related to the fact that muscle becomes partially anaerobic during vigorous exercise. Oxygen simply cannot be supplied to the muscle tissue rapidly enough to keep up with the demand for ATP production. While the glycogen stores can be utilized equally well aerobically or anaerobically, fatty acids can only be utilized aerobically. Under those conditions, fatty acids can not provide ATP rapidly enough to serve as the sole energy source.

Thus, the practice of carbohydrate loading to increase glycogen stores was devised for track and other endurance athletes. Originally, it was thought that it would be necessary to trick the body into increasing glycogen stores. The original carbohydrate loading regimen consisted of a 3–4-day period of heavy exercise while on a low-carbohydrate diet, followed by 1–3 days of light exercise while on a high-carbohydrate diet. The initial low-carbohydrate–high-energy demand period caused a depletion of muscle glycogen stores. Apparently, the subsequent change to a high carbohydrate diet resulted in a slight rebound effect, with the production of higher than normal levels of insulin and growth hormone. Under these conditions glycogen storage was favored and glycogen stores reached almost twice the normal amounts. This practice did increase significantly endurance. For example, in one study, test subjects on a high-fat and high-protein diet had less than 1.6 g of glycogen per 100 g of muscle and could perform a standardized work load for only 60 min. When the same subjects then consumed a high-carbohydrate diet for 3 days, their glycogen stores increased to 4 g per 100 g of muscle and the same workload could be performed for up to 4 h.

Thus, the technique clearly worked, but the athletes often felt lethargic and irritable during the low-carbohydrate phase of the regimen, and the high-fat diet ran counter to current health recommendations. Fortunately, recent studies show that regular consumption of a high complex carbohydrate–low-fat diet during training increases glycogen stores without the need for tricking the body with sudden dietary changes. Thus, the current recommendations are for endurance athletes to consume a high-carbohydrate diet (with emphasis on complex carbohydrates) during training. Then carbohydrate intake is increased further (to 70% of calories) and exercise tapered off during the 2–3 days just prior to an athletic event. This procedure has been shown to increase muscle glycogen stores to levels comparable to the original carbohydrate loading regimen.

Conlee, R. K. Muscle Glycogen and Exercise Endurance: A Twenty-Year Perspective. *Exer. Sports Sci. Rev.* 15:1, 1987; Costill, D. L., Sherman, W. M., Fink, W. J., Moresh, C., Witten, M., and Miller, J. M. The Role of Dietary Carbohydrate in Muscle Glycogen Resynthesis After Strenuous Running. *Am. J. Clin. Nutr.* 34:1831, 1981; and Ivey, J. L., Katz, A. L., Cutler, C. L., Sherman, W. M., and Cayle, E. F. Muscle Glycogen Synthesis After Exercise: Effect of Time of Carbohydrate Ingestion. *J. Appl. Physiol.* 64:1480, 1988.

levels of carbohydrate result in a permanent state of ketosis similar to that seen during starvation. The mechanism has been discussed earlier (Chapter 14). If continued over a period of time this ketosis may, in some instances, be detrimental to the patient's health. Most Americans, of course, do consume more than adequate carbohydrate levels.

The most common nutritional problems involving carbohydrates are seen in those individuals with various **carbohydrate intolerances.** The most common form of carbohydrate intolerance is diabetes mellitus, which is caused either by lack of insulin production or lack of insulin receptors. This causes an intolerance to glucose and those simple sugars that can be readily converted to glucose. The dietary treatment of diabetes usually involves limiting intake of most simple sugars, and increasing intake of those carbohydrates that are better tolerated, mostly complex carbohydrates, but including some simple sugars such as fructose and sorbitol (Clin. Corr. 27.6). Lactase insufficiency is also a common disorder of carbohydrate metabolism affecting over 30 million people in the United States alone. It is most prevalent among blacks, Asians, and South Americans. Without the enzyme lactase, the lactose is not significantly hydrolyzed or absorbed. It remains in the intestine where it acts osmotically to draw water into the gut and serves as a substrate for conversion to lactic acid, CO_2, and H_2O by intestinal bacteria. The end result is bloating, flatulence, and diarrhea—all of which can be avoided simply by eliminating milk and milk products from the diet.

CLIN. CORR. 27.6
HIGH-CARBOHYDRATE–HIGH-FIBER DIETS AND DIABETES

The American Diabetes Association recently recommended diets that are low in fat and high in complex carbohydrates and fiber for diabetes. The logic of such a recommendation seemed to be inescapable. Diabetics are prone to hyperlipidemia with the attendent risk of heart disease, and low-fat diets appear likely to reduce the risk of both hyperlipidemia and heart disease. In addition, numerous clinical studies had suggested that the high-fiber content of these diets resulted in improved control of blood sugar. Yet this recommendation has proven to be controversial. An understanding of the controversies involved illustrates the difficulties in making dietary recommendations for population groups rather than individuals. In the first place, it is very difficult to make any major changes in dietary composition without changing other components of the diet. Changing from the typical American diet to a high-carbohydrate–high-fiber diet results in changes in protein sources, micronutrient intake, and caloric density. In fact, most of the clinical trials of the high-carbohydrate–high-fiber diets have resulted in significant weight reduction, either by design or because of the lower caloric density of the diet. Since weight reduction is known to improve diabetic control, it is not entirely clear whether the improvements seen in the treated group were due to the change in diet composition per se or because of the weight loss. Second, there is significant individual variation in how diabetics respond to these diets. A few diabetic patients appear to show poorer control (as

27.7 FATS

Triacylglycerols, or fats, can be directly utilized by many tissues of the body as an energy source and, as phospholipids, are an important part of membrane structure. Any excess fat in the diet can be stored as triacylglycerol only. As with carbohydrate, the human body can adapt to a wide range of fat intakes. However, some problems can develop at the extremes (either high or low) of fat consumption. At the low end, **essential fatty acid (EFA)** deficiencies may become a problem. The fatty acids linoleic, linolenic, and arachidonic acid cannot be made by the body and thus are essential components of the diet. These EFA are needed for maintaining the function and integrity of membrane structure, for fat metabolism and transport, and for synthesis of prostaglandins. The most characteristic symptom of EFA deficiency is a scaly dermatitis. Essential fatty acid deficiency is very rare in the United States, being seen primarily in low-birth-weight infants fed on artificial formulas lacking EFA and in hospitalized patients maintained on total parenteral nutrition for long periods of time.

At the other end of the scale, there is legitimate concern that excess fat in the diet does cause elevation of serum lipids and thus an increased risk of heart disease. Most experts agree with that general conclusion. Unfortunately, there is no firm consensus as to how much is too much. However, our understanding of metabolism tells us that any fat consumed in excess of energy needs (except for the small amount needed for membrane formation) has nowhere to go but our adipose tissue, and obesity is correlated with an increased risk of heart disease, diabetes, and stroke. Recent studies also suggest that high-fat intakes are associated with increased risk of colon, breast, and prostate cancer, but it is not yet certain whether the cancer risk is associated with fat intake per se or with the excess calories associated with a high-fat diet. To the extent that fat intake is associated with cancer risk, animal studies suggest that polyunsaturated fatty acids of the ω-6 series may be more tumorigenic than other unsaturated fatty acids. The reason for this is not known, but it has been suggested that prostaglandins derived from the ω-6 fatty acids may stimulate tumor progression.

evidenced by higher blood glucose levels, elevated VLDL and/or LDL levels, and reduced HDL levels) on the high-carbohydrate–high-fiber diets. Thus, this diet may not be an equally appropriate recommendation for all diabetics. The "glycemic index" concept (Table 27.4) may also turn out to be difficult to apply to the diabetic population as a whole, because of individual variation. Thus, at present, many experts have concluded that the high-carbohydrate–high-fiber diet may be useful for diabetics when it is part of a diet that enables the patient to reach and maintain normal weight. The safety and efficacy of such a diet may vary considerably from patient to patient.

Anderson, J. W., Gustafson, N. J., Bryant, C. A., and Tietyen-Clark, J. Dietary Fiber and Diabetes: A Comprehensive Review and Practical Application. *J. Am. Diet Assoc.* 87:1189, 1987; Jenkins, D. J. A., Wolener, T. M. S., Jenkins, A. L., and Taylor, R. H. Dietary Fiber, Carbohydrate Metabolism and Diabetes. *Molec. Aspects Med.* 9:97, 1987; and Nutall, F. Q. The High-Carbohydrate Diet in Diabetes Management. *Ann. Intern. Med.* 33:165, 1988.

27.8 FIBER

Dietary fiber is defined as those components of food that cannot be broken down by human digestive enzymes. It is incorrect, however, to assume that fiber is indigestible. Some fibers are, in fact, at least partially broken down by intestinal bacteria. Our knowledge of the role of fiber in human metabolism has expanded significantly in the past decade. Our current understanding of the metabolic roles of dietary fiber is based on three important observations: (1) there are several different types of dietary fiber, (2) they each have different chemical and physical properties, and (3) they each have different effects on human metabolism, which can be understood, in part, from their unique properties.

The major types of fiber and their properties are summarized in Table 27.3. **Cellulose** and most **hemicelluloses** increase stool bulk and decrease transit time. These are the types of fiber that should most properly be associated with the effects of fiber on regularity. They also decrease intracolonic pressure and appear to play a beneficial role with respect to diverticular diseases. By diluting out potential carcinogens and speeding their transit through the colon, they may also play a role in reducing the risk of colon cancer. **Lignins,** on the other hand, play a slightly different role. In addition to their bulk-enhancing properties, they adsorb organic substances such as cholesterol and appear to play a role in the cholesterol-lowering effects of fiber. The **mucilaginous fibers,** such as pectin and gums, have a very different mode of action. They tend to form viscous

TABLE 27.3 Major Types of Fiber and Their Properties

Type of fiber	Major source in diet	Chemical properties	Physiological effects
Cellulose	Unrefined cereals Bran Whole wheat	Nondigestible Water-insoluble Absorbs water	Stool bulk Intestinal transit time Intracolonic pressure
Hemicellulose	Unrefined cereals Some fruits and vegetables Whole wheat	Partially digestible Usually water-insoluble Absorbs water	Stool bulk Intestinal transit time Intracolonic pressure
Lignin	Woody parts of vegetables	Nondigestible Water-insoluble Adsorbs organic substances	Stool bulk Binds cholesterol Binds carcinogens
Pectin	Fruits	Digestible Water-soluble Mucilaginous	Rate of gastric emptying Rate of sugar uptake Serum cholesterol
Gums	Dried beans Oats	Digestible Water-soluble Mucilaginous	Rate of gastric emptying Rate of sugar uptake Serum cholesterol

gels in the stomach and intestine and slow the rate of gastric emptying, thus slowing the rate of absorption of many nutrients. The most important clinical role of these fibers is to slow the rate at which carbohydrates are digested and absorbed. Thus, both the rise in blood sugar and the subsequent rise in insulin levels are significantly decreased if these fibers are ingested along with carbohydrate-containing foods. The water-soluble fibers (pectins, gums, some hemicelluloses, and storage polysaccharides) also help to lower serum cholesterol levels in most people. Whether this is a consequence of their effect on insulin levels (insulin stimulates cholesterol synthesis and export) or to other metabolic effects (perhaps caused by end products of partial bacterial digestion) is not known at present. Vegetables, wheat, and most grain fibers are the best sources of the water-insoluble cellulose, hemicellulose, and lignin. Fruits, oats, and legumes are the best source of the water-soluble fibers. Obviously, a balanced diet should include food sources of both soluble and insoluble fiber.

27.9 COMPOSITION OF MACRONUTRIENTS IN THE DIET

From the foregoing discussion it is apparent that there are relatively few instances of macronutrient deficiencies in the American diet. Thus, much of the interest in recent years has focused more on the question of whether there is an ideal diet composition consistent with good health. It would be easy to pass off such discussions as purely academic, yet our understanding of these issues could well be vital. Heart disease, stroke, and cancer kill many Americans each year, and if some experts are even partially correct, many of these deaths could be preventable with prudent diet. So it is only fitting that we now turn to the question of diet composition.

Lipid Composition of the Diet Affects Serum Cholesterol

Much of the current discussion centers around two key issues: (1) Can serum cholesterol and triacylglycerol levels be controlled by diet? (2) Does lowering serum cholesterol and triacylglycerol levels protect against heart disease? The controversies centered around dietary control of cholesterol levels illustrate perfectly the trap one falls into by trying to look too closely at each individual component of the diet instead of the diet as a whole. For example, there are at least four components that can be identified as having an effect on serum cholesterol: cholesterol itself, **polyunsaturated fatty acids (PUFA), saturated fatty acids (SFA),** and fiber. It would appear obvious that the more cholesterol one eats, the higher the serum cholesterol would be. However, cholesterol synthesis is tightly regulated via a feedback control at the hydroxymethylglutaryl CoA reductase step, so decreases in dietary cholesterol have relatively little effect on serum cholesterol levels (Chapter 10). One can obtain a more significant reduction in cholesterol and triacylglycerol levels by increasing the ratio of PUFA/SFA in the diet. Finally, some plant fibers, especially the water-soluble fibers, appear to decrease cholesterol levels significantly.

While the effects of various fats in the diet can be dramatic, the biochemistry of their action is still uncertain. Saturated fats have been shown to inhibit receptor-mediated uptake of low density lipoprotein (LDL), but the mechanism of this effect is obviously complex. Palmitic acid (saturated, C_{16}) raises cholesterol levels while stearic acid (saturated, C_{18}) is neutral. Polyunsaturated fatty acids lower both LDL and HDL cholesterol levels, while oleic acid (monounsaturated, C_{18}) appears to lower

CLIN. CORR. 27.7 POLYUNSATURATED FATTY ACIDS AND RISK FACTORS FOR HEART DISEASE

The recent NIH study confirming that reduction of elevated serum cholesterol levels can reduce the risk of heart disease has rekindled interest in the effects of diet on serum cholesterol levels and other risk factors for heart disease. We have known for years that one of the most important dietary factors regulating serum cholesterol levels is the ratio of polyunsaturated fats (PUFAs) to saturated fats (SFAs) in the diet.

One of the most interesting recent developments is the discovery that different types of polyunsaturated fatty acids have different effects on lipid metabolism and on other risk factors for heart disease. As discussed in Chapter 9, there are two families of polyun-

LDL without affecting high density lipoprotein (HDL) levels. Furthermore, the ω-3 and ω-6 polyunsaturated fatty acids have slightly different effects on lipid profiles (see Clin. Corr. 27.7). However, these mechanistic complexities do not significantly affect dietary recommendations. Most foods high in saturated fats contain both palmitic and stearic acid, and are still atherogenic. The data showing that oleic acid lowers LDL levels mean olive oil, and possibly peanut oil, may be considered as beneficial as the polyunsaturated oils.

Actually, there is very little disagreement with respect to these data. The question is, what can be done with the information? Much of the disagreement arises from the tendency to look at each dietary factor in isolation. For example, it is indeed debatable whether it is worthwhile placing a patient on a highly restrictive 300-mg cholesterol diet (1 egg = 213 mg of cholesterol) if his serum cholesterol is lowered only 5–10%. Likewise, changing the **PUFA/SFA ratio** from 0.3 (the current value) to 1.0 would either require a radical change in the diet by elimination of foods containing saturated fat (largely meats and fats) or an addition of large amounts of rather unpalatable polyunsaturated fats to the diet. For many Americans this would be unrealistic. Fiber is another good example. One could expect, at the most, a 5% decrease in serum cholesterol by adding any reasonable amount of fiber to the diet. (Very few people would eat the 10 apples per day needed to lower serum cholesterol by 15%.) Are we to conclude then that any dietary means of controlling cholesterol levels are useless? Only if each element of the diet is examined in isolation. For example, recent studies have shown that vegetarians, who have lower cholesterol intakes plus higher PUFA/SFA ratios and higher fiber intakes, may average 25–30% lower cholesterol levels than their nonvegetarian counterparts. Perhaps, more to the point, diet modifications of the type acceptable to the average American have been shown to cause a 10–15% decrease in cholesterol levels in long-term studies. The second question has been much more difficult to answer. However, a recent 7-year clinical trial sponsored by the National Institutes of Health (NIH) has proved conclusively that lowering serum cholesterol levels will reduce the risk of heart disease in men. It is, of course, important to keep in mind that serum cholesterol is just one of many risk factors. Unfortunately, some nutritionists, and certainly the popular press, may have overemphasized the importance of this one risk factor in preventing heart disease.

Effects of Refined Carbohydrate in the Diet Are Not Straightforward

Much of the current dispute in the area of carbohydrates centers around the amount of *refined carbohydrate* in the diet. In the past simple sugars (primarily sucrose) have been blamed for almost every ill from tooth decay to heart disease and diabetes. In the case of tooth decay, these assertions were clearly correct. In the case of heart disease and diabetes, however, the linkage is more obscure.

It is evident that much of the excess dietary carbohydrate in the American diet is converted to triacylglycerol in the liver, exported as VLDL, and stored in the adipose tissue. It is even somewhat logical to assume that simple sugars, which are absorbed and metabolized very rapidly, might cause a slightly greater elevation of triacylglycerols than complex carbohydrates. In actual studies with human volunteers, an isocaloric switch from a diet high in starch to one high in simple sugars, does cause a transient rise in triacylglycerol levels. Over a period of 2–3 months, however, adaptation occurs and the triacylglycerol levels return to normal.

saturated essential fatty acids—the ω-6, or linoleic family [18:2(9,12)], and the ω-3, or linolenic family [18:3(9,12,15)].

Recent clinical studies have shown that the ω-6 PUFAs (chief dietary source is linoleic acid from plants and vegetable oils) primarily decrease serum cholesterol levels, with only modest effects on triacylglycerol levels. The ω-3 PUFAs (chief dietary source is eicosapentaenoic acid from certain ocean fish and fish oils), on the other hand, cause modest decreases in serum cholesterol levels and significantly lower triacylglycerol levels as well. The biochemical mechanism behind these different effects on serum lipid levels is unknown at present.

The ω-3 PUFAs have yet another unique physiological effect that may decrease the risk of heart disease—they decrease platelet aggregation. The mechanism of this effect is a little clearer. Arachidonic acid (ω-6 family) is known to be a precursor of thromboxane A_2 (TXA_2), which is a potent proaggregating agent, and prostaglandin I_2 (PGI_2), which is a weak antiaggregating agent (see Chapter 10). The ω-3 PUFAs are thought to act by one of two mechanisms: (1) Eicosapentaenoic acid (ω-3 family) may be converted to thromboxane A_3 (TXA_3), which is only weakly proaggregating, and prostaglandin I_3 (PGI_3), which is strongly antiaggregating. Thus, the balance between proaggregation and antiaggregation would be shifted toward a more antiaggregating condition as the ω-3 PUFAs displace ω-6 PUFAs as a source of precursors to the thromboxanes and prostaglandins. (2) The ω-3 PUFAs may also act by simply inhibiting the conversion of arachidonic acid to TXA_2. Since the conversion of eicosapentaenoic acid to TXA_3 appears to be relatively inefficient, the second mechanism is considered the most likely at present.

The unique potential of eicosapentaenoic acid and other ω-3 PUFAs in reducing the risk of heart disease is currently being tested in numerous clinical trials. Although the results of these clinical studies may affect dietary recommendations in the future, it is well to keep in mind that no long-term clinical studies of the ω-3 PUFAs have yet been carried out. No major health organization is yet recommending that we attempt to replace ω-6 with ω-3 PUFAs in the American diet.

Carroll, K. K. Biological Effects of Fish Oils in Relation to Chronic Diseases. *Lipids* 21:731, 1986; Holub, B. J. Dietary Fish Oils Containing Eicosapentaenoic Acid and the Prevention of Atherosclerosis and Thrombosis. *Can. Med. Assoc. J.* 139:377, 1988; and Sanders, T. A. B. Nutritional and Physiological Implications of Fish Oils. *J. Nutr.* 116:1857, 1986.

CLIN. CORR. 27.8 METABOLIC ADAPTATION: THE RELATIONSHIP BETWEEN CARBOHYDRATE INTAKE AND SERUM TRIACYLGLYCEROLS

In evaluating the nutrition literature, it is important to be aware that most clinical trials are of rather short duration (2–6 weeks), while some metabolic adaptations may take considerably longer. Thus, even apparently well-designed clinical studies may lead to erroneous conclusions that will be repeated in the popular literature for years to come. For example, there were several studies carried out in the 1960s and 1970s that tried to assess the effects of carbohydrate intake on serum triacylglycerol levels. Typically, young college age males were given a diet in which up to 50% of their fat calories were replaced with sucrose or other simple sugars for a period of 2–3 weeks. In most cases serum triacylglycerol levels increased markedly (up to 50%). This led to the tentative conclusion that high intake of simple sugars, particularly sucrose, might increase the risk of heart disease, a notion that was popularized by nutritional best sellers such as "sugar blues" and "sweet and dangerous." Unfortunately, while the original conclusions were being promoted in the lay press, the experiments themselves were being called into question. It turned out in subsequent studies that if these trials were continued for longer periods of time (3–6 months), the triacylglycerol levels usually normalized. The nature of this slow metabolic adaptation is unknown at present.

It should be noted that while the interpretation of the original clinical trials may have been faulty, the ensuing dietary recommendations may not have been entirely off base. Many of the snack and convenience foods in the American diet that are high in sugar are also high in fat and in caloric density. Thus, removing some of these foods from the diet can aid in weight control, and overweight is known to contribute to hypertriacylglycerolemia. Also, there are some individuals who do exhibit carbohydrate-induced hypertriacylglycerolemia. Triacylglycerol levels in these individuals will respond dramatically to diets that substitute foods containing complex carbohydrates and fiber for these foods containing primarily simple sugars as a carbohydrate source.

MacDonald, I. Effects of Dietary Carbohydrates on Serum Lipids. *Prog. Biochem. Pharmacol.* 8:216, 1973; and Vrana, A. and Fabry, P. Metabolic Effects of High Sucrose or Fructose Intake. *World Rev. Nutr. Diet* 42:56, 1983.

Thus, for most individuals, there is no evidence that simple sugars can cause a permanent elevation of serum triacylglycerols (Clin. Corr. 27.8).

The situation with respect to diabetes is probably even less direct. Whereas restriction of simple sugars is often desirable in a patient who already has diabetes, there is little direct evidence that an excess of simple sugars in the diet is a direct cause of diabetes.

In fact, recent studies show less than expected correlation between the type of carbohydrate ingested and the subsequent rise in serum glucose levels (Table 27.4). Ice cream, or example, causes a much smaller increase in serum glucose levels than either potatoes or whole wheat bread. It turns out that other components of food—such as protein, fat, and the soluble fibers—are much more important than the type of carbohydrate present in determining how rapidly glucose will enter the bloodstream. Foods rich in simple sugars, however, are often also high in fat and thus have a very high **caloric density,** contributing to overeating and obesity. Obesity, as discussed earlier, does have a direct relationship to heart disease and diabetes. This may go a long way toward explaining some of the epidemiologic studies linking consumption of simple sugars with heart disease and diabetes.

Mixed Vegetable and Animal Protein Meet Nutritional Protein Requirements

Much concern has also been voiced recently about the type of protein in the American diet. Epidemiologic data and animal studies suggest that consumption of animal protein is associated with increased incidence of heart disease and various forms of cancer. One would assume that it is probably not the animal protein itself that is involved, but the associated fat and cholesterol. What sort of protein should we consume? Although the present diet may not be optimal, a strictly vegetarian diet is not without some health risks of its own, unless the individual is nutritionally well informed (Clin. Corr. 27.1). Perhaps a middle road is best. Clearly,

TABLE 27.4 Glycemic Index of Some Selected Foods[a]

Food	Glycemic index	Food	Glycemic index
Grain and cereal products		Root vegetables	
Bread (white)	69 ± 5	Beets	64 ± 16
Bread (whole wheat)	72 ± 6	Carrots	92 ± 20
Rice (white)	72 ± 9	**Potato (white)**	70 ± 8
Sponge cake	46 ± 6	Potato (sweet)	48 ± 6
Breakfast cereals		Dried legumes	
All bran	51 ± 5	Beans (kidney)	29 ± 8
Cornflakes	80 ± 6	Beans (soy)	15 ± 5
Oatmeal	49 ± 8	Peas (blackeye)	33 ± 4
Shredded wheat	67 ± 10	Fruits	
Vegetables		Apple (golden delicious)	39 ± 3
Sweet corn	59 ± 11	Banana	62 ± 9
Frozen peas	51 ± 6	Oranges	40 ± 3
Dairy products		Sugars	
Ice cream	36 ± 8	Fructose	20 ± 5
Milk (whole)	34 ± 6	Glucose	100
Yogurt	36 ± 4	Honey	87 ± 8
		Sucrose	59 ± 10

SOURCE: Data from D. A. Jenkins et al. Glycemic index of foods: A physiological basis for carbohydrate exchange. *Am. J. Clin. Nutr.* 34:362, 1981.

[a] Glycemic index is defined as the area under the blood glucose response curve for each food expressed as a percentage of the area after taking the same amount of carbohydrate as glucose (means 5–10 individuals).

there are no known health dangers associated with a mixed diet that is lower in animal protein than the current American standard.

An Increase in Fiber From Varied Sources Is Desirable

Because of our current knowledge about the effects of fiber on human metabolism, most suggestions for a prudent diet recommend an increase in dietary fiber. The main question is "How much is enough?" Because of possible effects of high-fiber diets on nutrient absorption, it would seem wise not to increase fiber levels too much. The current fiber content of the American diet is about 14–15 g day^{-1}.

Most experts feel that an increase to at least 30 g would be safe and beneficial. Since we know that different fibers have different metabolic roles, this increase in fiber intake should come from a wide variety of fiber sources—including fresh fruits, vegetables, and legumes as well as the more popular cereal sources of fiber, which are primarily cellulose and hemicellulose.

Current Recommendations Are for a "Prudent Diet"

In the midst of all of this controversy, it would seem to be premature to make specific recommendations with respect to the ideal dietary composition for the American public. Yet, this is just what several private and government groups have done in recent years. This movement was spearheaded by the Senate Select Committee on Human Nutrition, which first published its *Dietary Goals for the United States* in 1977. The Senate

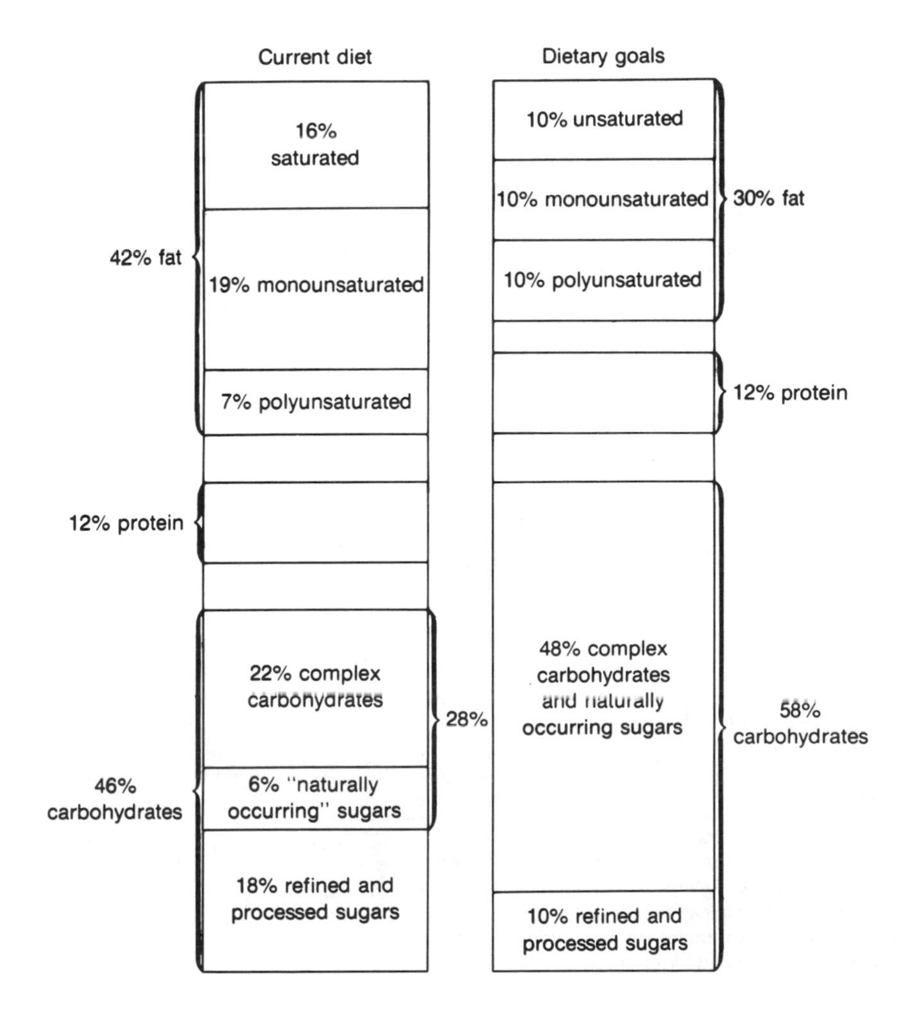

FIGURE 27.2
United States dietary goals.
Graphical comparison of the composition of the current U.S. diet and the dietary goals for the U.S. population suggested by the Senate Select Committee on Human Nutrition.
From *Dietary goals for the United States, 2nd ed.*, U.S. Government Printing Office, Washington, DC, 1977.

Select Committee recommended that the American public reduce consumption of total calories, total fat, saturated fat, cholesterol, simple sugars, and salt to "ideal" goals more compatible with good health (Figure 27.2). In recent years the USDA, the American Heart Association, the American Diabetes Association, the National Research Council, and the Surgeon General all have published similar recommendations (Table 27.5). These recommendations have become popularly known as the **"prudent diet."** How valid is the scientific basis of the recommendations for a prudent diet? Is there evidence that a prudent diet will improve the health of the general public? These remain controversial questions at present.

The most important argument against such recommendations is that we presently do not have enough information to set concrete goals. We might be creating some problems while solving others. For example, the goals of reducing total fat and saturated fat in the diet are best met by replacing animal protein with vegetable protein. This, in turn, might reduce the amount of available iron and vitamin B_{12} in the diet. It is also quite clear that the same set of guidelines does not apply for every individual. For example, exercise is known to raise serum HDL cholesterol and obesity is known to elevate cholesterol, triacylglycerols, and reduce glucose tolerance. Thus the very active individual who maintains ideal body weight can likely tolerate higher fat and sugar intakes than an obese individual.

On the "pro" side, however, it clearly can be argued that all of the dietary recommendations are in the right direction for reducing nutritional risk factors in the general population. Furthermore, similar diets have been consumed by our ancestors and by people in other countries with no apparent harm. Whatever the outcome of this debate in the years ahead, it will undoubtedly shape much of our ideas concerning the role of nutrition in medicine.

TABLE 27.5 Dietary Recommendations of Major Health Organizations

	American Heart Association[a]	National Research Council[b]	Senate Select Committee on Human Nutrition[c]	USDA[d] and Surgeon General[e]
Obesity	Reduce to ideal weight		Reduce to ideal weight	Maintain ideal weight
Fat	30% of total calories	30% of total calories	30% of total calories	Avoid too much fat
Type of fat	10% SFA 10% MUFA 10% PUFA	10% SSFA	10% MUFA 10% PUFA	Avoid too much saturated fat
Cholesterol	< 300 mg day^{-1}		300 mg day^{-1}	Avoid too much cholesterol
Carbohydrate	55% of total calories		58% of total calories	
Type of carbohydrate	Increase complex carbohydrates and fiber-rich foods	Fiber may be beneficial	48% complex CHO 10% simple sugars	Eat foods with starch and fiber Avoid too much sugar
Vitamins/minerals		Include fruits, vegetables, and whole grain foods in diet		Eat a variety of foods
Alcohol	If you drink alcohol, do so in moderation	Avoid excess alcohol consumption		If you drink do so in moderation
Sodium		Avoid salt-cured foods	Limit sodium intake to 5 g day^{-1}	Avoid too much sodium

[a] Recommendation to decrease risk of heart disease. Source: American Heart Association. *Recommendation for treatment of hyperlipidemia in adults*. American Heart Association, 1984.
[b] Recommendation to decrease risk of cancer. Source: National Research Council. *Diet, nutrition, and cancer*. National Academy Press, 1982.
[c] Senate Select Committee on Human Nutrition. *Dietary goals for the United States*, 2nd ed. U.S. Government Printing Office, 1977.
[d] U.S. Department of Agriculture. *Nutrition and your health*. U.S. Government Printing Office, 1980.
[e] The Surgeon General's Report on Nutrition and Health, U.S. Government Printing Office, 1988.

BIBLIOGRAPHY

Protein Energy Malnutrition in Hospitalized Patients

Bistrian, B. R. Interactions of Nutrition and Infection in the Hospital Setting. *Am. J. Clin. Nutr.* 30:1228, 1977.

The Veterans Administration Cooperative Trial of Perioperative Total Parenteral Nutrition in Malnourished Surgical Patients. *Am. J. Clin. Nutr.* 47:351, 1988.

Metabolic Consequences of Obesity

Hershcopf, R. J. and Bradlow, H. L. Obesity, Diet, Endogenous Estrogens, and the Risk of Hormone-Sensitive Cancer. *Am. J. Clin. Nutr.* 45:283, 1987.

Maxwell, M. H. and Waks, A. U. Obesity and Hypertension. *Bibl. Cardiol.* 41:29, 1987.

Mancini, M., Lewis, B., and Cantaldo, F. (Eds.). *Medical Complications of Obesity*. New York: Academic, 1979.

Simopoulos, A. P. Obesity and Carcinogenesis: Historical Perspective. *Am. J. Clin. Nutr.* 45:271, 1987.

Metabolic Predisposition to Obesity

Bjorntorp, P. Fat Cell Distribution and Metabolism. *Ann. N.Y. Acad. Sci.* 499:66, 1987.

Bray, G. A. Obesity—A Disease of Nutrient or Energy Balance? *Nutr. Rev.* 45:33, 1987.

Dulloo, A. G. and Miller, D. S. Obesity: A Disorder of the Sympathetic Nervous System. *World Rev. Nutr. Diet.* 50:1, 1987.

Landsberg, L. Diet, Obesity and Hypertension: An Hypothesis Involving Insulin, the Sympathetic Nervous System, and Adaptive Thermogenesis. *Quart. J. Med.* 61:1081, 1986.

Complex Carbohydrates and Fiber

Anderson, J. W. Fiber and Health: An Overview. *Am. J. Gastroenterol.* 81:892, 1986.

Eastwood, M. Dietary Fiber and the Risk of Cancer. *Nutr. Rev.* 45:193, 1987.

Jenkins, D. J. A. and Jenkins, A. L. The Glycemic Index, Fiber, and the Dietary Treatment of Hypertriglyceridemia and Diabetes. *J. Am. Cell. Nutr.* 6:11, 1987.

Spiller, G. A. and Kay, R. M. *Medical Aspects of Dietary Fiber*. New York: Plenum, 1980.

Macronutrient Composition and Health

Glueck, C. J. and Conner, W. E. Diet-Coronary Heart Disease Relationships. *Am. J. Clin. Nutr.* 31:727, 1978.

Gorlin, R. The Biological Actions and Potential Clinical Significance of Dietary ω-3 Fatty Acids. *Arch. Intern. Med.* 148:2043, 1988.

Grundy, S. M. Monounsaturated Fatty Acids, Plasma Cholesterol, and Coronary Heart Disease. *Am. J. Clin. Nutr.* 45:1168, 1987.

Grundy, S. M. et al. Rationale of the Diet–Heart Statement of the American Heart Association, *News From the American Heart Association. Circulation 65,* 839A, 1982.

Kisselbah, A. and Schetman, G. Polyunsaturated and Saturated Fat, Cholesterol, and Fatty Acid Supplementation. *Diabetes Care.* 11:129, 1988.

Kritchevsky, D. and Klurfeld, D. M. Caloric Effects in Experimental Mammary Tumorigenesis. *Am. J. Clin. Nutr.* 45:236, 1987.

Sanders, T. A. B. Nutritional and Physiological Implications of Fish Oils. *J. Nutr.* 116:1857, 1986.

Welsh, C. W. Enhancement of Mammary Tumorigenesis by Dietary Fat: Review of Potential Mechanisms. *Am. J. Clin. Nutr.* 45:191, 1987.

Dietary Recommendations

American Heart Association. *Recommendations for treatment of hyperlipidemia in adults*. Dallas: American Heart Association. 1984.

Food and Nutrition Board of the National Academy of Sciences. *Towards healthful diets*. Washington, DC: U.S. Government Printing Office, 1980.

National Research Council. *Diet, nutrition and cancer*. Washington: National Academy Press, 1982.

Senate Select Committee on Human Nutrition. *Dietary goals for the United States,* 2nd ed., Stock No. 052-070-04376-8. Washington, DC: U.S. Government Printing Office, 1977.

Truswell, A. S. Evolution of Dietary Recommendations, Goals, and Guidelines. *Am. J. Clin. Nutr.* 45:1060, 1987.

U.S. Department of Agriculture. *Nutrition and your health, dietary guidelines for Americans,* Stock No. 017-001-00416-2. Washington, DC: U.S. Government Printing Office, 1980.

U.S. Department of Health and Human Services. *The Surgeon General's Report on Nutrition and Health,* Stock No. 017-001-00465-1. Washington, DC: U.S. Government Printing Office, 1988.

QUESTIONS

C. N. ANGSTADT AND J. BAGGOTT

Question Types are described on page xxiii.

1. (QT2) Of two people with approximately the same weight, the one with the higher basal energy requirement would most likely be:
   1. taller.
   2. female if the other were male.
   3. younger.
   4. under less stress.

2. (QT3) A. Basal metabolic rate of a person consuming about 2000 kcal day^{-1}
   B. Basal metabolic rate of a person consuming about 600 kcal day^{-1}

3. (QT2) Diet-induced thermogenesis:
   1. accounts for the conversion of a portion of excess calories to heat rather than to fat.
   2. may be defective in some obese people.
   3. in humans, occurs primarily in skeletal muscle.
   4. can be increased by both exercise and a high-fat diet.

4. (QT1) The primary effect of the consumption of excess protein beyond the body's immediate needs will be:
   A. excretion of the excess as protein in the urine.
   B. an increase in the "storage pool" of protein.
   C. an increased synthesis of muscle protein.
   D. an enhancement in the amount of circulating plasma proteins.
   E. an increase in the amount of adipose tissue.
5. (QT1) Which of the following individuals would most likely be in nitrogen equilibrium?
   A. A normal, adult male
   B. A normal, pregnant female
   C. A growing child
   D. An adult male recovering from surgery
   E. A normal female on a very low protein diet
6. (QT1) Vegetarian diets:
   A. cannot meet the body's requirements for all of the essential amino acids.
   B. contain only protein that is very readily digestible.
   C. are adequate as long as two different vegetables are consumed in the same meal.
   D. would require less total protein to meet the requirement for all of the essential amino acids.
   E. require that proteins consumed have essential amino acid contents that complement each other.

7. (QT2) In which of the following circumstances would a protein intake of 0.8 g of protein kg^{-1} (body weight) day^{-1} probably not be adequate?
   1. the diet is a vegetarian one.
   2. infancy.
   3. severe burn.
   4. about 85–90% of total calories from carbohydrate and fat.

8. (QT1) Kwashiorkor is:
   A. the most common form of protein–calorie malnutrition in the United States.
   B. characterized by a thin, wasted appearance.
   C. an inadequate intake of food of any kind.
   D. an adequate intake of total calories but a specific deficiency of protein.
   E. an adequate intake of total protein but a deficiency of the essential amino acids.

9. (QT2) An excessive intake of calories:
   1. is likely to have adverse metabolic consequences if continued for a long period of time.
   2. leads to metabolic changes that are usually irreversible.
   3. is frequently associated with an increased number or size of adipose cells.
   4. is frequently associated with an increased number of insulin receptors.

10. (QT1) A diet very low in carbohydrate:
    A. would cause weight loss because there would be no way to replenish citric acid cycle intermediates.
    B. would result in no significant metabolic changes.
    C. could lead to a chronic ketosis.
    D. would lead to water retention.
    E. would be the diet of choice for a diabetic.
11. (QT1) Lactase insufficiency:
    A. is a more serious disease than diabetes mellitus.
    B. has no clinical symptoms.
    C. causes an intolerance to glucose.
    D. causes an intolerance to milk and milk products.
    E. affects utilization of milk by the liver.
12. (QT1) Dietary fat:
    A. is usually present, although there is no specific need for it.
    B. if present in excess, can be stored as either glycogen or adipose tissue triacylglycerol.
    C. should include linoleic and linolenic acids.
    D. should increase on an endurance training program in order to increase the body's energy stores.
    E. if present in excess, does not usually lead to health problems.
13. (QT1) Which one of the following dietary regimens would be most effective in lowering serum cholesterol?
    A. Restrict dietary cholesterol
    B. Increase the ratio of polyunsaturated to saturated fatty acids
    C. Increase fiber content
    D. Restrict cholesterol and increase fiber
    E. Restrict cholesterol, increase PUFA/SFA, increase fiber
14. (QT1) Most nutrition experts currently agree that an excessive consumption of sugar causes:
    A. tooth decay.
    B. diabetes.
    C. heart disease.
    D. permanently elevated triacylglycerol levels.
    E. all of the above.

A. 10% of total calories
B. 12% of total calories
C. 30% of total calories
D. 48% of total calories
E. 58% of total calories

The dietary goal recommended by the Senate Select Committee on Human Nutrition for:

15. (QT5) Polyunsaturated fatty acids is
16. (QT5) Complex carbohydrates and naturally occurring sugars is ____

17. (QT1) A complete replacement of animal protein in the diet by vegetable protein:
    A. would be expected to have no effect at all on the overall diet.
    B. would reduce the total amount of food consumed for the same number of calories.
    C. might reduce the total amount of iron and vitamin B_{12} available.
    D. would be satisfactory regardless of the nature of the vegetable protein used.
    E. could not satisfy protein requirements.

ANSWERS

1. B 1, 3 correct. 1: A taller person with the same weight would have a greater surface area. 2: Males have higher energy requirements than females. 3: Energy requirements decrease with age. 4: Stress, probably because of the effects of epinephrine and cortisol, increase energy requirements (Section 27.2).
2. A The BMR can decrease up to 50% on an inadequate caloric intake (p. 1096).
3. A 1–3 correct. 2: Some individuals predisposed to obesity have a partial genetic defect in this process. 3: In rats and mice this occurs in brown adipose tissue but the location is different in humans. 4: Exercise and a high-carbohydrate diet favor thermogenesis (p. 1101).
4. E A: Protein is not found in normal urine except in very small amounts. The excess nitrogen is excreted as NH_4^+ and urea, whereas the excess carbon skeletons of the amino acids are used as energy sources. B–D: There is no discrete storage form of protein, and although some muscle and structural protein is expendable, there is no evidence that increased intake leads to generalized increased protein synthesis. E: Excess protein is treated like any other excess energy source and stored (minus the nitrogen) eventually as adipose tissue fat (Section 27.3).
5. A B–D: Although normal, pregnancy is also a period of growth, requiring positive balance as does a period of convalescence. E: Inadequate protein intake leads to negative balance (Section 27.3).
6. E A–E: It is possible to have adequate protein intake on a vegetarian diet provided enough is consumed (protein content is generally low and may be difficult to digest) and there is a *mixture* of proteins that supplies all of the essential amino acids since individual proteins are frequently deficient in one or more (Clin. Corr. 27.1, p. 1098).
7. A 1–3 correct. 1: Essential amino acids are low in vegetable protein. 2, 3: Periods of rapid growth require extra protein, as does major trauma. 4: This level of calories from carbohydrate and fat is more than adequate for protein sparing (Section 27.3).
8. D A: The most common protein–calorie malnutrition occurs in severely ill, hospitalized patients who would be more likely to have generalized malnutrition. B, C: These are the characteristics of marasmus. E: This would lead to negative nitrogen balance but does not have a specific name (Section 27.4).
9. B 1, 3 correct. 1: Excess caloric intake will lead to obesity if continued long enough. 2: Fortunately most of the changes accompanying obesity can be reversed if weight is lost. 3, 4: Many of the adverse effects of obesity are associated with an increased number of adipocytes that are deficient in insulin receptors (Section 27.5).
10. C A: This is a popular myth but untrue because many amino acids are glucogenic. B, C, E: The liver adapts by increasing gluconeogenesis, fatty acid oxidation, and ketone body production, contraindicated for a diabetic. D: Low carbohydrate leads to a depletion of glycogen with its stored water, accounting for rapid initial weight loss on this kind of diet (Section 27.6, Clin. Corr. 27.8).
11. D B, D, E: Lactase insufficiency is an inability to digest the sugar in milk products, causing intestinal symptoms, but is easily treated by eliminating milk products from the diet. A, C: Diabetes, caused by inadequate insulin or insulin receptors, inhibits appropriate utilization of glucose (Section 27.6).
12. C A, C: Linoleic and linolenic acids are essential fatty acids and so must be present in the diet. B, D: Excess carbohydrate can be stored as fat but the reverse is not true. D: Carbohydrate loading has been shown to increase endurance. E: High-fat diets are associated with many health risks (Section 27.7, Clin. Corr. 27.5).
13. E Any of the measures alone would decrease serum cholesterol slightly, but to achieve a reduction of more than 15% requires all three (Section 27.9).
14. A This is the only direct linkage shown. B, C: There may be an association with these conditions but not a direct cause–effect relationship. D: Transient elevations may occur on an isocaloric switch from a high starch to a high simple sugar diet but not a permanent elevation (Section 27.9).
15. A See Figure 27.2, p. 1109.
16. D See Figure 27.2, p. 1109.
17. C A, C: This would reduce the amount of fat, especially saturated fat, but could also reduce the amount of necessary nutrients that come primarily from animal sources. B: The protein content of vegetables is quite low, so much larger amounts of vegetables would have to be consumed. D, E: It is possible to satisfy requirements for all of the essential amino acids completely if vegetables with complementary amino acid patterns, in proper amounts, are consumed (Section 27.9, Clin. Corr. 27.1).

28

Principles of Nutrition II: Micronutrients

STEPHEN G. CHANEY

28.1 ASSESSMENT OF MALNUTRITION 1116
28.2 RECOMMENDED DIETARY ALLOWANCES 1117
28.3 FAT-SOLUBLE VITAMINS 1118
Vitamin A Is Derived From Plant Carotenoids 1118
Vitamin D Synthesis in the Body Requires Sunlight 1121
Vitamin E Is a Mixture of Tocopherols 1124
Vitamin K Is a Quinone Derivative 1125
28.4 WATER-SOLUBLE VITAMINS 1126
28.5 ENERGY-RELEASING WATER-SOLUBLE VITAMINS 1127
Thiamine (Vitamin B_1) Forms the Coenzyme Thiamine Pyrophosphate (TPP) 1127
Riboflavin Is Part of FAD and FMN 1128
Niacin Forms the Coenzymes NAD and NADP 1129
Pyridoxine (Vitamin B_6) Forms the Coenzyme Pyridoxal Phosphate 1129
Pantothenic Acid and Biotin Are Also Energy Releasing Vitamins 1132
28.6 HEMATOPOIETIC WATER-SOLUBLE VITAMINS 1132
Folic Acid (Folacin) Functions as Tetrahydrofolate in One-Carbon Metabolism 1132
Vitamin B_{12} (Cobalamine) Contains Cobalt in a Tetrapyrrole Ring 1134
28.7 OTHER WATER-SOLUBLE VITAMINS 1136
Ascorbic Acid Functions in Reduction and Hydroxylation Reactions 1136
28.8 MACROMINERALS 1137
Calcium Has Many Physiological Roles 1137
Other Macrominerals Are Phosphorus and Magnesium 1138
28.9 TRACE MINERALS 1139
Iron Is Efficiently Reutilized 1139
Iodine Is Incorporated Into Thyroid Hormones 1140
Zinc Binds to Many Enzymes 1141
Copper Functions as a Cofactor for Many Enzymes 1142
Chromium Is a Component of Glucose Tolerance Factor 1142
Selenium Is a Scavenger of Peroxides 1143
Manganese, Molybdenum, and Fluoride Are Other Trace Minerals 1143
28.10 THE AMERICAN DIET: FACT AND FALLACY 1143
28.11 ASSESSMENT OF NUTRITIONAL STATUS IN CLINICAL PRACTICE 1144
BIBLIOGRAPHY 1146
QUESTIONS AND ANNOTATED ANSWERS 1147
CLINICAL CORRELATIONS
28.1 Nutritional Considerations for Cystic Fibrosis 1120
28.2 Renal Osteodystrophy 1122
28.3 Nutritional Considerations in the Newborn 1126
28.4 Anticonvulsant Drugs and Vitamin Requirements 1126
28.5 Nutritional Considerations in the Alcoholic 1128
28.6 Vitamin B_6 Requirements for Users of Oral Contraceptives 1131
28.7 Diet and Osteoporosis 1138
28.8 Calculation of Available Iron 1140
28.9 Iron-Deficiency Anemia 1141
28.10 Nutritional Considerations for Vegetarians 1145
28.11 Nutritional Needs of the Elderly 1145

The micronutrients play a vital role in human metabolism since they are involved in almost every biochemical reaction and pathway known. The biochemistry of these nutrients, however, is of little interest unless we also know if dietary deficiencies are likely. Alarming reports of nutritional deficiencies in the American diet continually appear in the popular press. Is there any truth to these reports? On the one hand, the American diet is undoubtedly the best that it ever has been. Our current food supply provides us with an abundant variety of foods all year long—a luxury not available in the ''good old days''—and deficiency diseases have become medical curiosities. On the other hand, our diet is far from optimal. The old adage that we get everything we need from a balanced diet is true only if in fact we eat a balanced diet. Unfortunately, most Americans do not know how to select a balanced diet. Foods of high caloric density and low nutrient density (often referred to elsewhere as empty calories or junk food) are an abundant and popular part of the American diet, and our nutritional status suffers because of these food choices. Obviously then, neither alarm nor complacency are fully justified. We need to know how to best evaluate the adequacy of our diet.

28.1 ASSESSMENT OF MALNUTRITION

Why does one see so many reports of vitamin and mineral deficiencies in the popular press when deficiency diseases are so rare? In many cases these reports result from the misinterpretation of valid scientific data. One needs to be aware that there are three increasingly stringent criteria for measuring **malnutrition.**

1. Dietary surveys, which are usually based on a 24-h recall, are the least stringent. In the first place, 24-h recalls almost always tend to overestimate the number of people with deficient diets. Also, poor dietary intake alone is usually not a problem in this country unless the situation is compounded by increased need. Thus dietary surveys are not indicative of malnutrition by themselves, although they are often quoted as if they were.

2. Biochemical assays, either direct or indirect, are a more useful indicator of the nutritional status of an individual. At their best, they indicate **subclinical nutritional deficiencies** that can be treated before actual deficiency diseases develop. However, all biochemical assays are not equally valid—an unfortunate fact that is not sufficiently recognized. Furthermore, changes in biochemical parameters due to stress need to be interpreted with caution. The distribution of many nutrients in the body changes dramatically in a stress situation such as illness, injury, and pregnancy. A drop in level of a nutrient in one tissue compartment (usually blood) need not signal a deficiency or an increased requirement. It could simply reflect a normal metabolic adjustment to stress.

3. The most stringent criterion is, of course, the appearance of clinical symptoms. It is desirable, however, to intervene long before clinical symptoms became apparent.

The question remains: When should dietary surveys or biochemical assays be interpreted to indicate the necessity of nutritional intervention? At what level should we become concerned? Obviously, the situation is complex and controversial, but the following general guidelines are probably useful. Dietary surveys are seldom a valid indication of general malnutrition unless the average intake for a population group falls significantly below the standard (usually two-thirds of the Recommended Dietary Allowance) for one or more nutrients. However, by looking at the percentage of people within a population group who have suboptimal intake, it is possible to identify high-risk population groups that should be monitored more closely. This is the real value of dietary surveys. Biochemical assays, on the other hand, can definitely identify subclinical cases of malnutrition where nutritional intervention is desirable provided (a) the assay has been shown to be reliable, (b) the deficiency can be verified by a second assay, and (c) there is no unusual stress situation that may alter micronutrient distribution. In evaluating nutritional claims and counterclaims it is well to keep an open, but skeptical, mind and evaluate each issue on the basis of its scientific merit. While serious vitamin and mineral deficiencies are rare in this country, mild to moderate deficiencies can be found in many population groups. It is important for the clinician to be aware of these population groups at risk and their most probable symptoms.

28.2 RECOMMENDED DIETARY ALLOWANCES

One hears a lot about the Recommended Dietary Allowances (RDAs). What are they and how are they determined? Briefly, the RDAs are the levels of intake of essential nutrients considered in the judgment of the Food and Nutrition Board of the National Research Council on the basis of available scientific knowledge, to be adequate to meet the known nutritional need of practically all healthy persons. Optimally, the RDAs are based on daily intake sufficient to prevent the appearance of nutritional deficiency in at least 95% of the population. This determination is relatively easy to make for those nutrients associated with dramatic deficiency diseases, such as vitamin C and scurvy. In other instances more indirect measures must be used, such as tissue saturation or extrapolation from animal studies. In some cases, such as vitamin E, in which no deficiency symptoms are known to occur in the general population, the RDA is simply defined as the normal level of intake in the American diet. Obviously, there is no one set of criteria that can be used for all micronutrients, and there is always some uncertainty and debate as to the correct criteria. Furthermore, the criteria are constantly changed by new research. The Food and Nutrition Board normally meets every 6 years to consider currently available information and update their recommendations. Because of recent information suggesting that optimal intake of certain micronutrients (e.g., vitamins A, C, and E) may reduce cancer risk, some experts feel that the focus of the RDAs should shift from preventing nutritional deficiencies to defining optimal levels that might reduce the risk of other diseases. This controversy prevented the publication of revised RDAs in 1986, and required the formation of a new scientific committee which formulated the most recent version of the RDAs in 1989.

The RDAs serve as a useful general guide in evaluating the adequacy of diets and (as the USRDA) the nutritional value of foods. The RDAs, however, have several limitations that should be kept in mind:

1. The RDAs represent an ideal average intake for groups of people and are best used for evaluating nutritional status of population groups. The RDAs are not meant to be standards or requirements for individuals. Some individuals would have no problem with intakes below the RDA, whereas a few might develop deficiencies on intakes above the RDA.

2. The RDAs were designed to meet the needs of healthy people and do not take into account any special needs arising from infections, metabolic disorders, or chronic diseases.

3. Since present knowledge of nutritional needs is incomplete, there may be unrecognized nutritional needs. To provide for these needs, the RDAs should be met from as varied a selection of foods as possible. No single food can be considered complete, even if it meets the RDA for all

known nutrients. This is an important consideration, especially in light of the current practice of fortifying foods of otherwise low nutritional value.

4. As currently formulated, the RDAs make no effort to define the "optimal" level of any nutrient, since optimal levels are difficult to define on the basis of current scientific information.

28.3 FAT-SOLUBLE VITAMINS

Vitamin A Is Derived From Plant Carotenoids

The active forms of vitamin A are **retinol, retinal** (retinaldehyde), and **retinoic acid.** These substances are synthesized by plants as the more complex **carotenoids** (Figure 28.1), which are cleaved to retinol by most animals and stored in the liver as retinol palmitate. Liver, egg yolk, butter, and whole milk are good sources of the preformed retinol. Dark green and yellow vegetables are generally good sources of the carotenoids. The conversion of carotenoids to retinol is rarely 100%, so that the vitamin A potency of various foods is expressed in terms of retinol equivalents (1 RE = 1 μg of retinol, 6 μg of β-carotene, and 12 μg of other carotenoids). β-Carotene and other carotenoids are the major sources of vita-

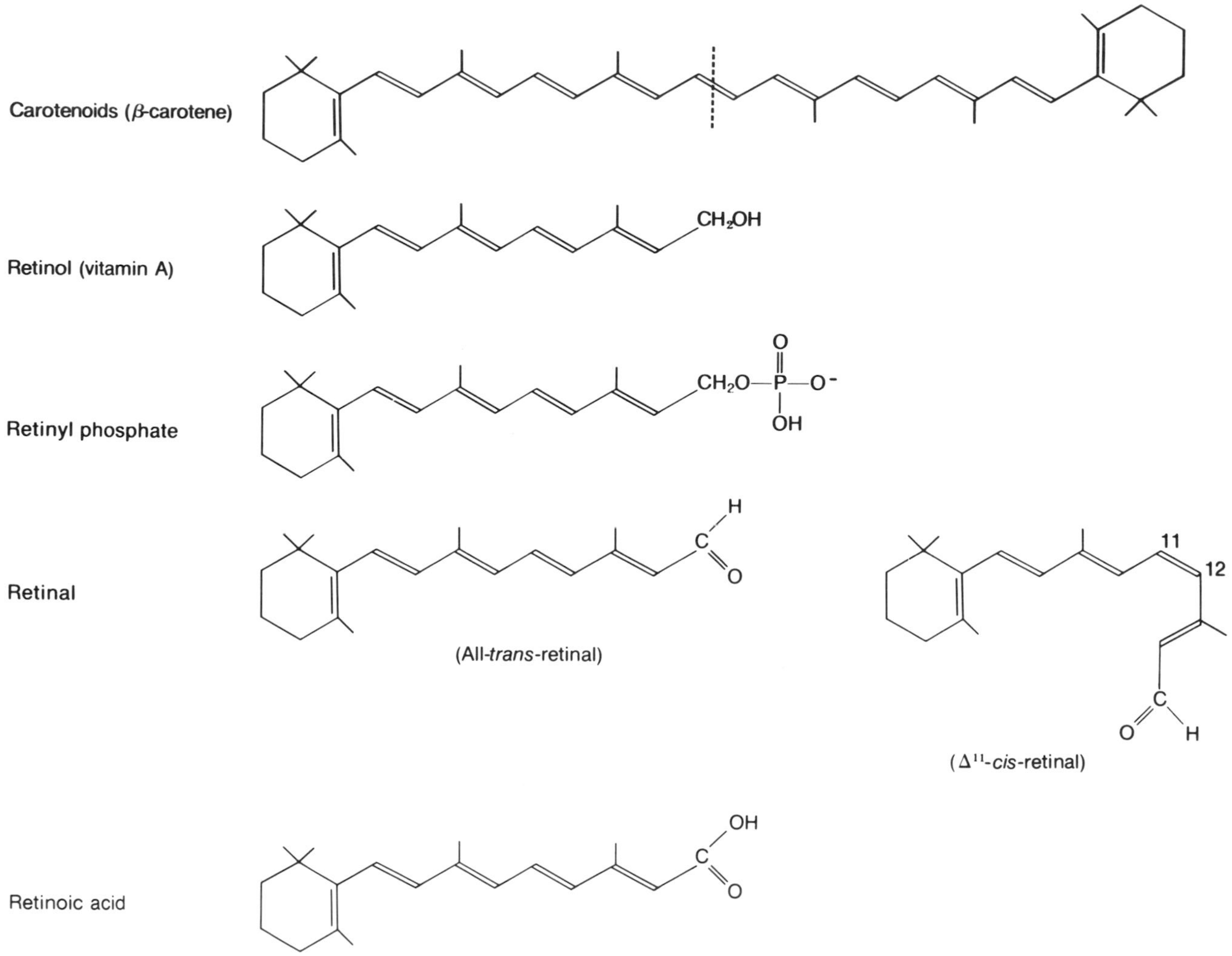

Carotenoids (β-carotene)

Retinol (vitamin A) — CH_2OH

Retinyl phosphate — $CH_2O{-}P(=O)(OH){-}O^-$

Retinal — (All-*trans*-retinal); (Δ^{11}-*cis*-retinal)

Retinoic acid

FIGURE 28.1
Structures of vitamin A and related compounds.

min A in the American diet. These carotenoids are first cleaved to retinol and then converted to other vitamin A metabolites in the body (Figure 28.1).

Vitamin A serves a number of functions in the body. However, only in recent years has its biochemistry become well understood (Figure 28.2). **β-Carotene** and some of the other carotenoids have recently been shown to play an important role as antioxidants. At the low oxygen tensions prevalent in the body, β-carotene is a very effective antioxidant and might be expected to reduce the risk of those cancers initiated by free radicals and other strong oxidants. Several restrospective clinical studies have suggested that adequate dietary β-carotene may play an important role in reducing the risk of lung cancer, especially in people who smoke. The National Institutes of Health (NIH) is currently testing this hypothesis in a multicenter prospective study.

Retinol can be converted to **retinyl phosphate** in the body. The retinyl phosphate appears to serve as a glycosyl donor in the synthesis of some glycoproteins and mucopolysaccharides in much the same manner as dolichol phosphate, which has been discussed previously (Chapter 20, Figure 28.3). It appears that retinyl phosphate is essential for the synthesis of certain glycoproteins needed for normal growth regulation and for mucous secretion.

Both retinol and retinoic acid bind to specific receptor proteins, which then bind to chromatin and affect the synthesis of proteins involved in the regulation of cell growth and differentiation. Thus, both retinol and retinoic acid can be considered to act like steroid hormones in regulating growth and differentiation.

Finally, in the Δ^{11}-cis-retinal form, vitamin A becomes reversibly associated with the visual pigments. When light strikes the retina, a number of complex biochemical changes take place, resulting in the generation of a nerve impulse, conversion of the retinal to the all-trans form, and its dissociation from the visual pigment (Chapter 22). Regeneration of more visual pigment requires isomerization back to the Δ^{11}-cis form (Figure 28.4).

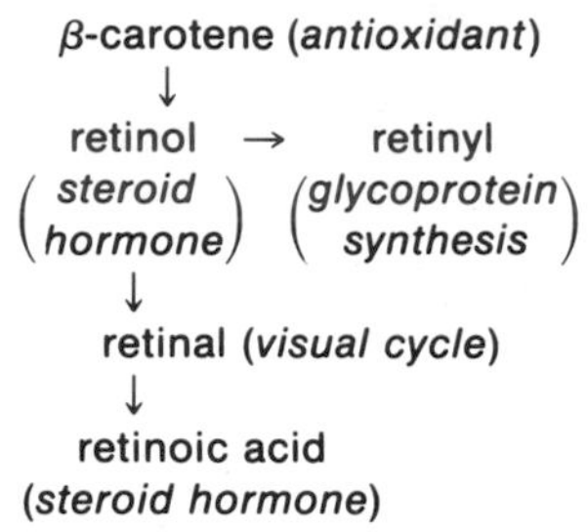

β-carotene (*antioxidant*)
↓
retinol → retinyl
(*steroid hormone*) (*glycoprotein synthesis*)
↓
retinal (*visual cycle*)
↓
retinoic acid
(*steroid hormone*)

FIGURE 28.2
Vitamin A metabolism and function.

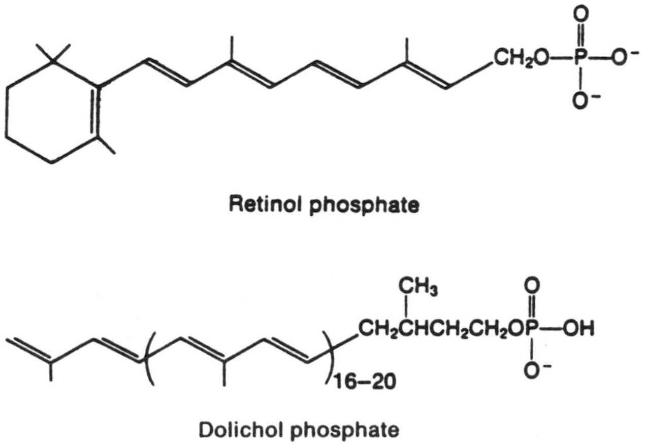

$CH_2O—P(=O)(O^-)—O^-$

Retinol phosphate

CH_3, $CH_2CHCH_2CH_2OP(=O)(O^-)—OH$, 16–20

Dolichol phosphate

FIGURE 28.3
Carriers involved in glycoprotein synthesis.

Rhodopsin
$h\nu$
opsin
nerve impulse
opsin
Δ^{11}-*cis*-Retinal ← *trans*-Retinal

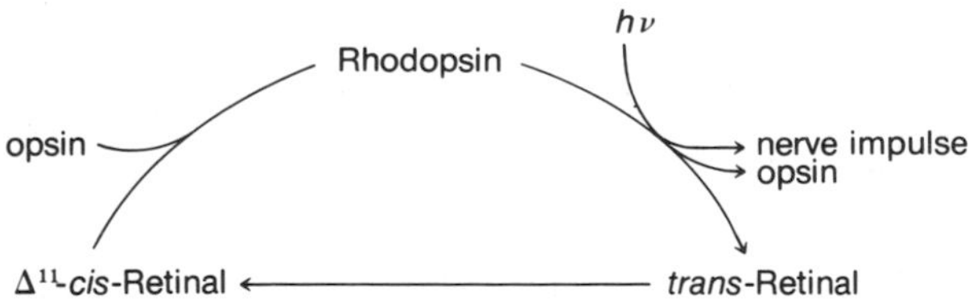

FIGURE 28.4
The role of vitamin A in vision.

Based on what is now known about the biochemical mechanisms of vitamin A action, its biological effects are easier to understand. For example, vitamin A is known to be required for the maintenance of healthy epithelial tissue. We now know that retinol and/or retinoic acid are required to prevent the synthesis of high molecular weight forms of keratin and that retinyl phosphate is required for the synthesis of mucopolysaccharides (an important component of the mucus secreted by many epithelial tissues). The lack of mucus secretion leads to a drying of these cells, and the excess keratin synthesis leaves a horny keratinized surface in place of the normal moist and pliable epithelium. It has also been observed that vitamin A deficiency can lead to anemia caused by impaired mobilization of iron from the liver. Recent studies have shown that retinol and/or retinoic acid are required for the synthesis of the iron transport protein transferrin.

Finally, vitamin A-deficient animals have been shown to be more susceptible to both infections and cancer. The decreased resistance to infections is thought to be due to the keratinization of the mucosal cells lining the respiratory, gastrointestinal, and genitourinary tract. Under these conditions fissures readily develop in the mucosal membranes, allowing microorganisms to enter. Some evidence suggests that vitamin A deficiency may impair the immune system as well. The protective effect of vitamin A against many forms of cancer probably results from the antioxidant potential of β-carotene and the effects of retinol and retinoic acid in regulating cell growth.

Since vitamin A is stored by the liver, deficiencies of this vitamin can develop only over prolonged periods of inadequate uptake. Mild **vitamin A deficiencies** are characterized by follicular hyperkeratosis (rough kertinized skin resembling *goosebumps*), anemia (biochemically equivalent to iron deficiency anemia, but in the presence of adequate iron intake), and increased susceptibility to infection and cancer. Night blindness is also an early symptom of vitamin A deficiency. Severe vitamin A deficiency leads to a progressive keratinization of the cornea of the eye known as xerophthalmia in its most advanced stages. In the final stages, infection usually sets in, with resulting hemorrhaging of the eye and permanent loss of vision.

The severe symptoms of vitamin A deficiency are generally seen only in developing countries. In this country even mild vitamin A deficiencies are rare, but the potential for deficiencies does exist. For most people (unless they happen to eat liver) the dark green and yellow vegetables are the most important dietary source of vitamin A. Unfortunately, these are the foods most often missing from the American diet. Nationwide, dietary surveys indicate that between 40–60% of the population consumes less than two-thirds of the RDA for vitamin A. How significant are these dietary surveys? While plasma vitamin A levels are low in a significant number of individuals, the clinical symptoms of vitamin A deficiency are rare. Follicular hyperkeratosis is occasionally seen and is the most characteristic symptom of vitamin A deficiency. The decreased resistance to infection and cancer are obviously of great concern in preventive medicine, but they are too nonspecific to be useful indicators of vitamin A status. Night blindness is seldom seen in the general population. While clinically detectable vitamin A deficiency is rare in the general population, it is a fairly common consequence of severe liver damage or diseases that cause fat malabsorption (Clin. Corr. 28.1).

Since vitamin A does accumulate in the liver, large amounts of this vitamin over prolonged periods of time can be toxic. Doses of 15,000–50,000 RE day^{-1} over a period of months or years will prove to be toxic for many children and adults. The usual symptoms include bone pain,

CLIN. CORR. 28.1
NUTRITIONAL CONSIDERATIONS FOR CYSTIC FIBROSIS

Patients with malabsorption diseases often develop malnutrition. As an example, let us examine the nutritional consequences of one disease with malabsorption components. Cystic fibrosis (CF) involves a generalized dysfunction of the exocrine glands that leads to formation of a viscid mucus, which progressively plugs their ducts. Obstruction of the bronchi and bronchioles leads to pulmonary infections, which are usually the direct cause of death. In many cases, however, the exocrine glands of the pancreas are also affected, leading to a deficiency of pancreatic enzymes and sometimes a partial obstruction of the common bile duct.

The deficiency (or partial deficiency) of pancreatic lipase and bile salts leads to severe malabsorption of fat and fat-soluble vitamins. Calcium tends to form insoluble salts with the long-chain fatty acids, which accumulate in the intestine. While these are the most severe problems, some starches and proteins are also trapped in the fatty bolus of partially digested foods. This physical entrapment, along with the deficiencies of pancreatic amylase and pancreatic proteases, can lead to severe protein–calorie malnutrition as well. Excessive mucus secretion on the luminal surfaces of the intestine may also interfere with the absorption of several nutrients, including iron.

Fortunately, microsphere preparations of pancreatic enzymes are now available that can greatly alleviate many of these malabsorption problems. With these preparations, protein and carbohydrate absorption are returned to near normal. Fat absorption is improved greatly but not normalized, since deficiencies of bile salts and excess mucus secretion persist. Because dietary fat is a major source of calories, these patients have difficulty obtaining sufficient calories from a normal diet. This is complicated by increased protein and energy needs resulting from the chronic infections often seen in these patients. Thus, many experts recommend energy intakes ranging from 120–150% of the RDA.

scaly dermatitis, enlargement of liver and spleen, nausea, and diarrhea. It is, of course, virtually impossible to ingest toxic amounts of vitamin A from normal foods unless one eats polar bear liver (6000 RE g^{-1}) regularly. Most instances of **vitamin A toxicity** are due to the use of massive doses of this vitamin to treat acne or prevent colds. Fortunately, this practice, while once common, is now relatively rare because of increased public awareness of vitamin A toxicity.

Whereas vitamin A itself is too toxic to be used therapeutically, certain **synthetic retinoids** (chemical derivatives of retinoic acid) are being used or tested because they often have lower toxicity. For example, 13-*cis*-retinoic acid is widely used in the treatment of acne and etretinate (an aromatic analog of all *trans*-retinoic acid) is being used in the treatment of psoriasis and related disorders (Figure 28.5). Other synthetic retinoids have been shown to be useful in the prevention and treatment of cancers in laboratory animals.

Vitamin D Synthesis in the Body Requires Sunlight

Technically, vitamin D could be considered a hormone rather than a vitamin (see Chapter 21). **Cholecalciferol (D_3)** is produced in the skin by UV irradiation of 7-dehydrocholesterol (see Figure 28.6). Thus, as long as the body is exposed to adequate sunlight, there is little or no dietary requirement for vitamin D. The best dietary sources of vitamin D_3 are saltwater fish (especially salmon, sardines, and herring), liver, and egg yolk. Milk, butter, and other foods are routinely fortified with **ergocalciferol (D_2)** prepared by irradiating ergosterol from yeast. Vitamin D potency is measured in terms of μg cholecalciferol (1 μg of cholecalciferol or ergocalciferol = 40 IU).

Both cholecalciferol and ergocalciferol are metabolized identically. They are carried to the liver where the 25-OH derivative is formed. **25-hydroxycholecalciferol** [25-(OH)D] is the major circulating derivative of vitamin D, and it is in turn converted into the biologically active **1α,25-dihydroxycholecalciferol** in the proximal convoluted tubules of the kidney (Clin. Corr. 28.2).

In the past, low-fat diets were often used to minimize fat malabsorption problems, but it was difficult for patients to obtain sufficient calories on such diets because of their low caloric density. Since inadequate energy intake results in poor growth and increased susceptibility to infection, inadequate caloric intake is of great concern for cystic fibrosis patients. Thus, the current recommendations are for high-energy–high-protein diets without any restriction of dietary fat (50% carbohydrate, 15% protein, and 35% fat). Where symptoms of fat malabsorption (steatorrhea and gastrointestinal distress) are a problem, they are controlled by increasing the dose of pancreatic extract. If caloric intake from the normal diet is inadequate, dietary supplements or enteral feedings may be used. The dietary supplements most often contain easily digested carbohydrates and milk protein mixtures. Medium-chain triglycerides are sometimes used as a partial fat replacement since they can be absorbed directly through the intestinal mucosa in the absence of bile salts and pancreatic lipose.

Since some fat malabsorption is present, deficiencies of the fat soluble vitamins often occur. These deficiencies can be prevented by supplementing with slightly greater than RDA levels of these nutrients. The usual recommendations are for 1,600–2,000 μg of vitamin A, 20 μg of vitamin D, and 100–200 mg of vitamin E. Iron deficiency is fairly common in cystic fibrosis patients but iron supplementation is not usually recommended because of concern that higher iron levels in the blood might encourage systemic bacterial infections. Calcium levels in the blood are usually normal. However, since calcium absorption is probably suboptimal, it is important to make certain that the diet provides at least RDA levels of calcium.

Littlewood, J. M. and MacDonald, A. Rationale of Modern Dietary Recommendations in Cystic Fibrosis. *J. R. Soc. Med.* 80 (Suppl. 15):16, 1987.

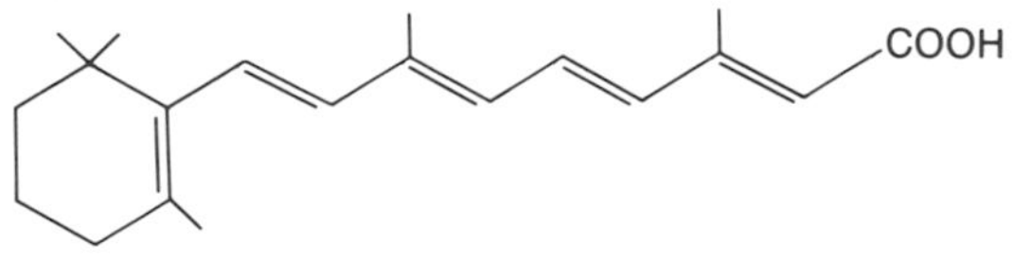

COOH

All-*trans*-Retinoic acid

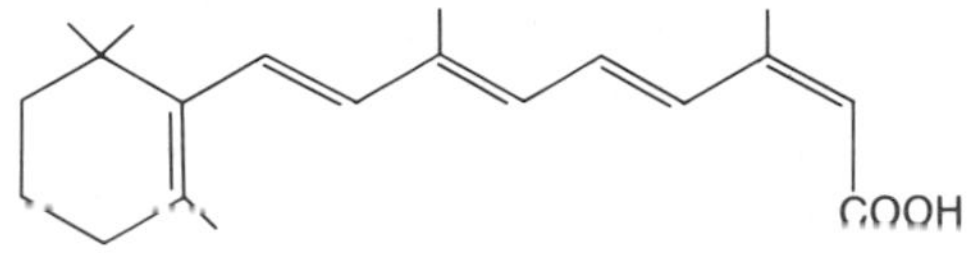

COOH

13-*cis*-Retinoic acid

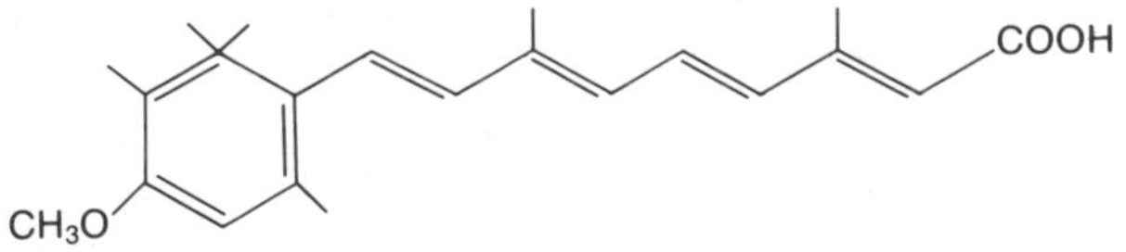

COOH

CH_3O

Etretinate

FIGURE 28.5
Therapeutically useful analogs of vitamin A.

CLIN. CORR. 28.2
RENAL OSTEODYSTROPHY

In patients with chronic renal failure, a complicated chain of events is set in motion, leading to a condition known as renal osteodystrophy. The renal failure results in an inability to produce 1,25-$(OH)_2D$, and thus bone calcium becomes the only important source of serum calcium. The situation is complicated further by increased renal retention of phosphate and resulting hyperphosphatemia. The serum phosphate levels are often high enough to cause metastatic calcifications (i.e., calcification of soft tissue), which tends to lower serum calcium levels further (the solubility product of calcium phosphate in the serum is very low and a high serum level of one component necessarily causes a decreased concentration of the other). The hyperphosphatemia and hypocalcemia stimulate parathyroid hormone secretion, and the resulting hyperparathyroidism further accelerates the rate of bone loss. One ends up with both bone loss and metastic calcifications. In this case, simple administration of high doses of vitamin D or its active metabolites would not be sufficient since the combination of hyperphosphatemia and hypercalcemia would only lead to more extensive metastatic calcification. The readjustment of serum calcium levels by high calcium diets and/or vitamin D supplementation must be accompanied by phosphate reduction therapies. The most common technique is to use phosphate binding antacids that make phosphate unavailable for absorption. Orally administered 1,25-$(OH)_2D$ is effective at stimulating calcium absorption in the mucosa but does not enter the peripheral circulation in significant amounts. Thus, patients with severe hyperparathyroidism may need to be treated with intravenous 1,25-$(OH)_2D$.

Johnson, W. J. Use of Vitamin D Analogs in Renal Osteodystrophy. *Sem. Nephrol.* 6:31, 1986; and McCarthy, J. T. and Kumar, R. Behavior of the Vitamin D Endocrine System in the Development of Renal Osteodystrophy. *Semin. Nephrol.* 6:21, 1986.

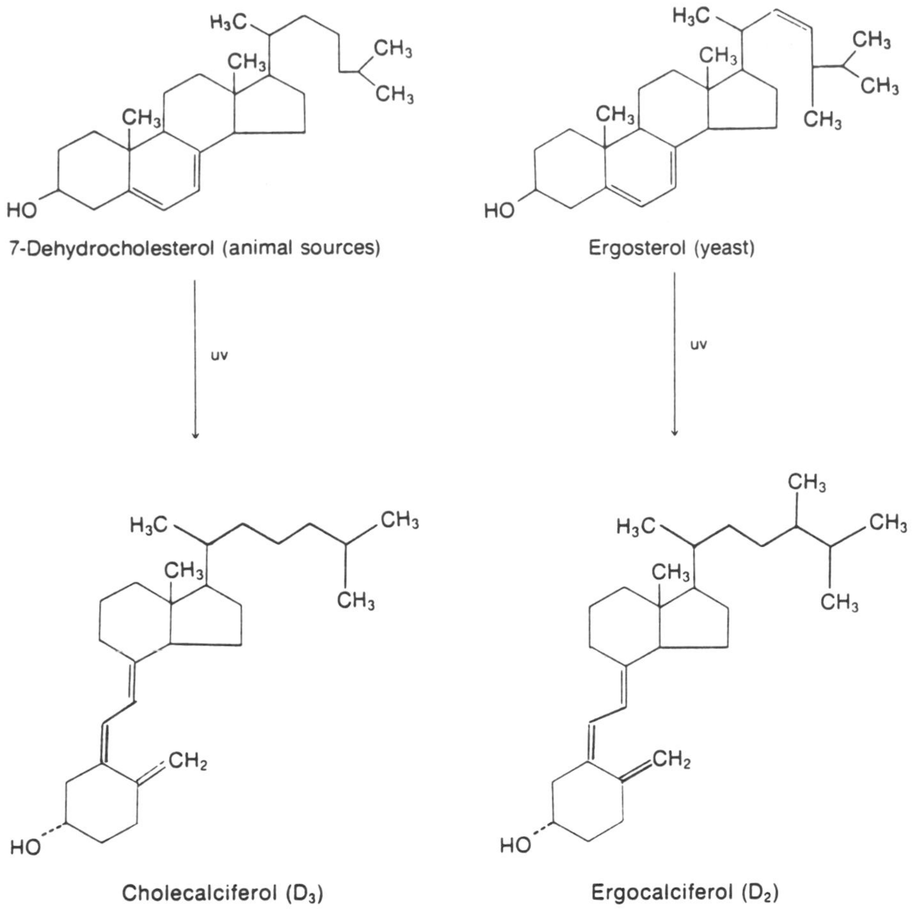

7-Dehydrocholesterol (animal sources) — uv → Cholecalciferol (D_3)

Ergosterol (yeast) — uv → Ergocalciferol (D_2)

FIGURE 28.6
Structures of vitamin D and related compounds.

The compound 1,25-$(OH)_2D$ acts in concert with **parathyroid hormone (PTH),** which is also produced in response to low serum calcium. Parathyroid hormone plays a major role in regulating the activation of vitamin D. High PTH levels stimulate the production of 1,25-$(OH)_2D$, while low PTH levels induce formation of an inactive 24,25-$(OH)_2D$. Once formed, the 1,25-$(OH)_2D$ acts alone as a typical steroid hormone in intestinal mucosal cells, where it induces synthesis of a protein required for calcium absorption. In the bone 1,25-$(OH)_2D$ and PTH act synergistically to promote bone resorption (demineralization) by stimulating osteoblast formation and activity. Finally, PTH and 1,25-$(OH)_2D$ inhibit calcium excretion in the kidney by stimulating calcium reabsorption in the distal renal tubules. The overall response of calcium metabolism to several different physiological situations is summarized in Figure 28.7. The response to low serum calcium levels is characterized by elevation of PTH and 1,25-$(OH)_2D$, which act to enhance calcium absorption and bone resorption and to inhibit calcium excretion (Figure 28.7*a*). High serum calcium levels block production of PTH. The low PTH levels allow 25-(OH)D to be metabolized to 24,25-$(OH)_2D$ instead of 1,25-$(OH)_2D$. In the absence of PTH and 1,25-$(OH)_2D$ bone resorption is inhibited and calcium excretion is enhanced. Furthermore, the high levels of serum calcium and phosphate increase the rate of bone mineralization (Figure 28.7*b*). Thus bone serves as a very important reservoir of the calcium and phosphate needed to maintain homeostasis of serum levels. When vitamin D and dietary calcium are adequate, no net loss of bone calcium occurs. However, when dietary calcium is low, PTH and 1,25-$(OH)_2D$ will cause net

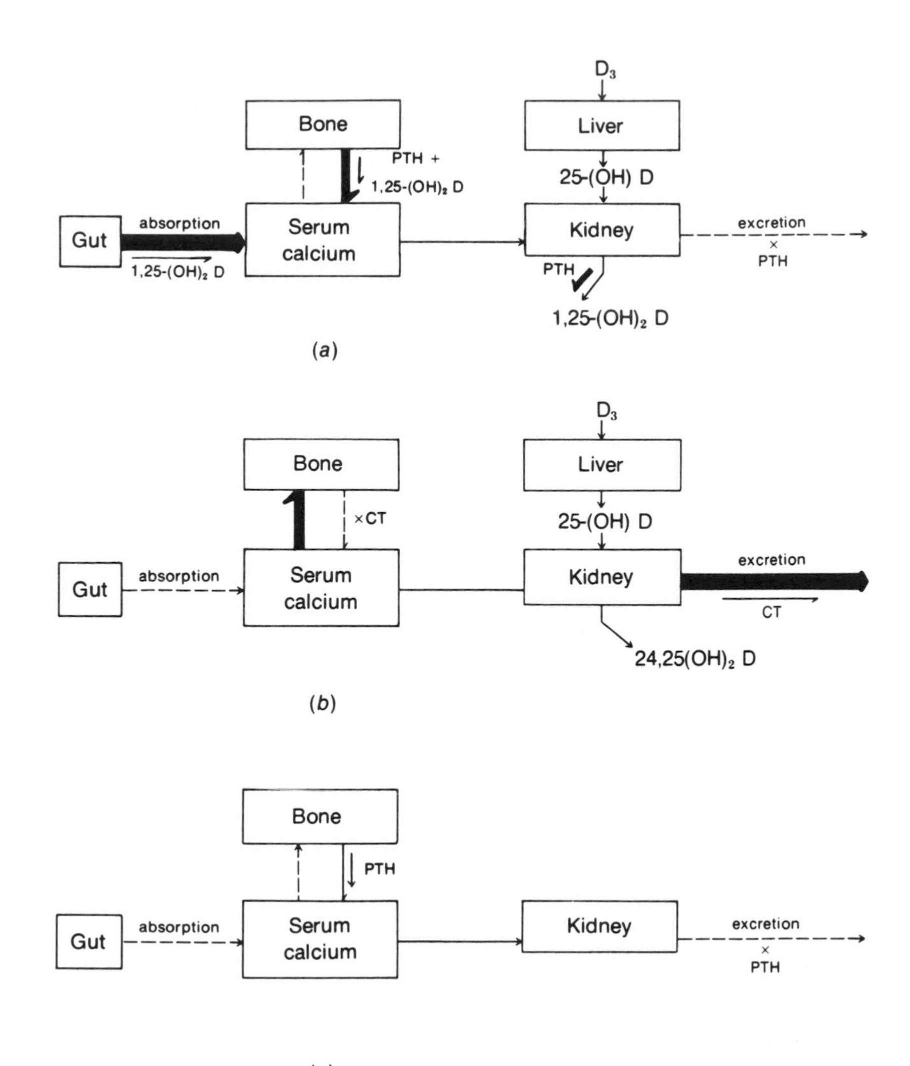

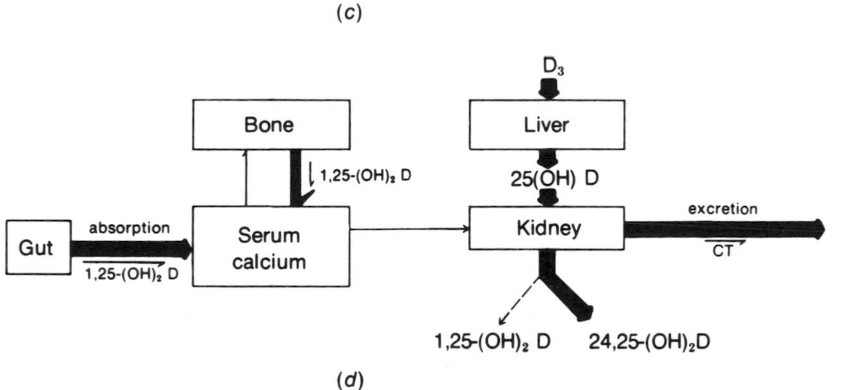

FIGURE 28.7
Vitamin D and calcium homeostasis.
(*a*) Low serum calcium. (*b*) High serum calcium. (*c*) Low vitamin D. (*d*) Excess vitamin D. The dominant pathways of calcium metabolism under each set of metabolic conditions are shown with heavy arrows. The effect of various hormones on these pathways is shown by → (stimulation) or X (repression). PTH = parathyroid hormone; D = cholecalciferol; 25-(OH)D = 25-hydroxycholecalciferol; 1,25-$(OH)_2$D = 1α,25-dihydroxycholecalciferol.

demineralization of bone to maintain normal serum calcium levels. Vitamin D deficiency also causes net demineralization of bone due to elevation of PTH (Figure 28.7*c*).

The most common symptoms of **vitamin D deficiency** are **rickets** in young children and *osteomalacia* in adults. Rickets is characterized by continued formation of osteoid matrix and cartilage, which are improperly mineralized resulting in soft, pliable bones. In the adult demineralization of preexisting bone takes place, causing the bone to become softer and more susceptible to fracture. This osteomalacia is easily distinguishable from the more common osteoporosis, by the fact that the osteoid matrix remains intact in the former, but not in the latter. Vitamin D may be involved in more than regulation of calcium homeostasis. Receptors for 1,25-$(OH)_2$D have been found in many tissues in the body including the parathyroid gland, the islet cells of the pancreas, keratinocytes of skin, and myeloid stem cells in the bone marrow. The role of vitamin D in these tissues is the subject of active investigation.

Because of fortification of dairy products with vitamin D, dietary deficiencies are very rare. The cases of dietary vitamin D deficiency that do occur are most often seen in low-income groups, the elderly (who often also have minimal exposure to sunlight), strict vegetarians (especially if their diet is also low in calcium and high in fiber), and chronic alcoholics. Most cases of vitamin D deficiency, however, are a result of diseases causing fat malabsorption or severe liver and kidney disease (Clin. Corrs. 28.1 and 28.2). Certain drugs also interfere with vitamin D metabolism. For example, corticosteroids stimulate the conversion of vitamin D to inactive metabolites and have been shown to cause bone demineralization when used for long periods of time.

Vitamin D can also be toxic in doses 10–100 times the RDA. The mechanism of **vitamin D toxicity** is summarized in Figure 28.7*d*. Enhanced calcium absorption and bone resorption cause hypercalcemia, which can lead to metastatic calcifications. The enhanced bone resorption also causes bone demineralization similar to that seen in vitamin D deficiency. Finally, the high serum calcium leads directly to hypercalciuria, which predisposes the patient to formation of renal stones.

Vitamin E Is a Mixture of Tocopherols

For many years **vitamin E** was described as the "vitamin in search of a disease." While vitamin E deficiency diseases are still virtually unknown, its metabolic role in the body has become better understood in recent years. Vitamin E occurs in the diet as a mixture of several closely related compounds, called tocopherols. **α-Tocopherol** is considered the most potent of these and is used as the measure of vitamin E potency (1α-tocopherol equiv = 1 mg of α-tocopherol).

First and foremost, vitamin E appears to play an important role as a naturally occurring *antioxidant*. Due to its lipophilic structure it tends to accumulate in circulating lipoproteins, cellular membranes, and fat deposits, where it reacts very rapidly with molecular oxygen and free radicals. It acts as a scavenger for these compounds, protecting unsaturated fatty acids (especially those in the membranes) from peroxidation reactions. Vitamin E appears to play a role in cellular respiration, perhaps by stabilizing coenzyme Q. It also appears to enhance heme synthesis by increasing the levels of δ-aminolevulinic acid (ALA) synthetase and ALA dehydratase. Most of these vitamin E effects are thought to be an indirect effect of its antioxidant potential, rather than its actual participation as a coenzyme in any biochemical reactions.

Symptoms of **vitamin E deficiency** vary widely from one animal species to another. In various animals vitamin E deficiencies can be associated with sterility, muscular dystrophy, central nervous system changes, and megaloblastic anemia. In humans, however, the symptoms are usually limited to increased fragility of the red blood cell membrane (presumably due to peroxidation of membrane components), although neurological symptoms have been reported following prolonged vitamin E deficiency associated with malabsorption diseases.

Premature infants fed on formulas low in vitamin E sometimes develop a form of hemolytic anemia that can be corrected by vitamin E supplementation. Adults suffering from fat malabsorption show a decreased red blood cell survival time. Hence, vitamin E supplementation may be necessary with premature infants and in cases of fat malabsorption.

Studies on the recommended levels of vitamin E in the diet have been hampered by the difficulty in producing severe vitamin E deficiency in humans. In general it is assumed that the vitamin E levels in the American diet are sufficient, since no major vitamin E deficiency diseases have been found. However, vitamin E requirements do increase as the intake of polyunsaturated fatty acids (PUFA) increases. The propensity of PUFA to form free radicals on exposure to oxygen may lead to an increased cancer risk. Thus it appears only prudent to increase vitamin E intake along with high PUFA diets.

As a fat-soluble vitamin, E has the potential for toxicity. However, it does appear to be the least toxic of the fat-soluble vitamins. No instances of toxicity have been reported at doses of 600 mg day^{-1} or less. A few scattered reports of malaise and easy fatigability have been reported at doses of 800 mg day^{-1}.

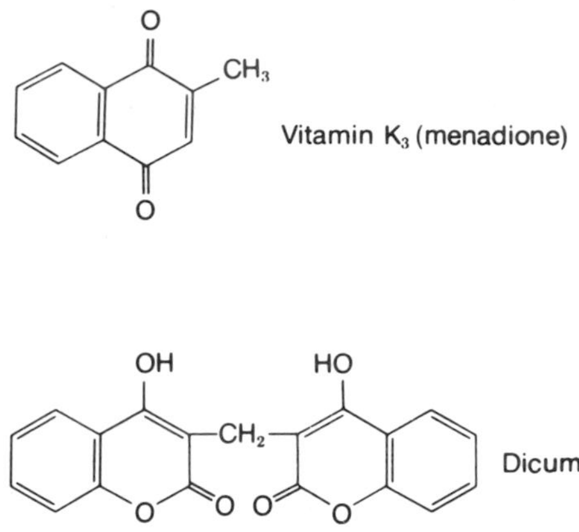

FIGURE 28.8
Structures of vitamin K and related compounds.

Vitamin K Is a Quinone Derivative

Vitamin K is found naturally as $\mathbf{K_1}$ (phytylmenaquinone) in green vegetables and $\mathbf{K_2}$ (multiprenylmenaquinone), which is synthesized by intestinal bacteria. The body is also able to convert synthetically prepared menaquinone (menadione) and a number of water-soluble analogs to a biologically active form of vitamin K (see Figure 28.8). Dietary requirements are measured in terms of micrograms of vitamin K_1 with the RDA for adults being in the range of 60–80 μg day^{-1}.

Vitamin K_1 has been shown to be required for the conversion of several clotting factors and **prothrombin** to the active state. The mechanism of this action has been most clearly delineated for prothrombin. Prothrombin is synthesized as an inactive precursor called preprothrombin. Conversion to the active form requires a vitamin K-dependent carboxylation of specific glutamic acid residues to **γ-carboxyglutamic acid** (Figure 28.9). The γ-carboxyglutamic acid residues are good chelators and allow prothrombin to bind calcium. The prothrombin–Ca^{2+} complex in turn binds to the phospholipid membrane, where proteolytic conversion to thrombin can occur in vivo. The mechanism of the carboxylation reaction has not been fully clarified, but appears to involve the intermediate formation of a 2,3-epoxide derivative of vitamin K. **Dicumarol,** a naturally occurring anticoagulant, inhibits the reductase that converts the epoxide back to the active vitamin.

Recently, proteins containing γ-carboxyglutamic acid have been reported in a variety of tissues including bone, kidney, and placenta, suggesting that vitamin K-dependent carboxylation reactions may plat a role in metabolic processes other than blood clotting. However, the only known symptom of **vitamin K deficiency** in humans is increased coagulation time. Since vitamin K is relatively abundant in the diet and synthesized in the intestine, deficiencies are rare. The most common deficiency is seen in newborn infants (Clin. Corr. 28.3), especially those whose mothers have been on anticonvulsant therapy (Clin. Corr. 28.4). Vitamin K deficiency is also seen in patients with obstructive jaundice and other

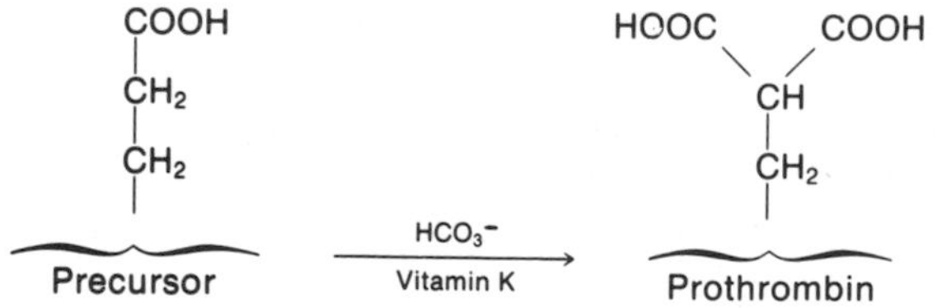

FIGURE 28.9
Function of vitamin K.

CLIN. CORR. **28.3**
NUTRITIONAL CONSIDERATIONS IN THE NEWBORN

Newborn infants are at special nutritional risk. In the first place, this is a period of very rapid growth, and needs for many nutrients are high. Some micronutrients (such as vitamins E and K) do not cross the placental membrane well and tissue stores are low in the newborn infant. The gastrointestinal tract may not be fully developed, leading to malabsorption problems (particularly with respect to the fat-soluble vitamins). The gastrointestinal tract is also sterile at birth and the intestinal flora that normally provide significant amounts of certain vitamins (especially vitamin K) take several days to become established. If the infant is born prematurely, the nutritional risk is slightly greater, since the gastrointestinal tract will be less well developed and the tissue stores will be less.

The most serious nutritional complications of newborns appears to be hemorrhagic disease. Newborn infants, especially premature infants, have low tissue stores of vitamin K and lack the intestinal flora necessary to synthesize the vitamin. Breast milk is also a relatively poor source of vitamin K. Approximately 1 out of 400 live births shows some signs of hemorrhagic disease. One milligram of the vitamin at birth is usually sufficient to prevent hemorrhagic disease.

Iron is another potential problem. Most newborn infants are born with sufficient reserves of iron to last 3–4 months (although premature infants are born with smaller reserves). Since iron is present in low amounts in both cow's milk and breast milk, iron supplementation is usually begun at a relatively early age by the introduction of iron-fortified cereal. Vitamin D levels are also somewhat low in breast milk and supplementation with vitamin D is usually recommended. However, some recent studies have suggested that iron in breast milk is present in a form that is particularly well utilized by the infant and that earlier studies probably underestimated the amount of vitamin D available in breast milk. Other vitamins and minerals appear to be present in adequate amounts in breast milk as long as the mother is getting a good diet. Recent studies have suggested that in situations in which infants must be maintained on assisted ventilation with high oxygen concentrations, supplemental vitamin E may reduce the risk of bronchopulmonary dysplasia and retrolental fibroplasia, two possible side effects of oxygen therapy.

In summary, most infants are provided with supplemental vitamin K at birth to prevent hemorrhagic disease. Breast-fed infants are usually provided with supplemental vitamin D, with iron being introduced along with solid foods. Bottle-fed infants are provided with supplemental iron. If infants must be maintained on oxygen, supplemental vitamin E may be beneficial.

Barness, L. A. Pediatrics, in H. Schneider, C. E. Anderson, and D. B. Coursin (Eds.). *Nutritional Support of Medical Practice*, 2nd ed. New York: Harper & Row, 1983, pp. 541–561.

CLIN. CORR. **28.4**
ANTICONVULSANT DRUGS AND VITAMIN REQUIREMENTS

Anticonvulsant drugs such as phenobarbital or diphenylhydantoin (DPH) present an excellent example of the type of drug–nutrient interactions that are of concern to the physician. Metabolic bone disease appears to be the most significant side effect of prolonged anticonvulsant therapy. Whereas children and adults on these drugs seldom develop rickets or severe osteomalacia, as many as 65% of those on long-term therapy will have abnormally low serum calcium and phosphorus and abnormally high serum alkaline phosphatase. Some bone loss is usually observed in these cases. While the cause of the hypocalcemia and bone loss was originally thought to be an effect of the anticonvulsant drugs on vitamin D metabolism, more recent studies have shown normal levels of 25-(OH)D and 1,25-$(OH)_2D$ in most patients on these drugs. In spite of this evidence suggesting that vitamin D metabolism may not be directly involved in this disorder, supplemental vitamin D in the range of 2000 to 10,000 units per day appears to correct both the hypocalcemia and osteopenia. Anticonvulsants

diseases leading to severe fat malabsorption (Clin. Corr. 28.4) and patients on long-term antibiotic therapy, which may destroy vitamin K-synthesizing organisms in the intestine. Finally, vitamin K deficiency is occasionally seen in the elderly, who are prone to poor liver function (reduced preprothrombin synthesis) and fat malabsorption. Clearly vitamin K deficiency should be suspected in any patient demonstrating easy bruising and prolonged clotting time.

28.4 WATER-SOLUBLE VITAMINS

The water-soluble vitamins differ from the fat-soluble vitamins in several important aspects. In the first place, most of these compounds are readily excreted once their concentration surpasses the renal threshold. Thus toxicities are rare. It is popular to speak of these vitamins as "not being stored by the body." While that is not quite accurate, the metabolic stores are labile and depletion can often occur in a matter of weeks or months. Since the water-soluble vitamins are coenzymes for many common biochemical reactions, it is often possible to assay vitamin status by measuring one or more enzyme activities in isolated red blood cells. These assays are especially useful if one measures both the endogenous enzyme activity and the stimulated activity following addition of the active coenzyme derived from that vitamin.

Most of the water-soluble vitamins are converted to coenzymes, which are utilized either in the pathways for energy generation or hematopoiesis. Deficiencies of the energy-releasing vitamins produce a number of overlapping symptoms. In many cases the vitamins participate in so many biochemical reactions that it is impossible to pinpoint the exact biochemical cause of any given symptom. It is possible, however, to generalize that because of the central role these vitamins play in energy metabolism, deficiencies show up first in rapidly growing tissues. Typical symptoms include dermatitis, glossitis (swelling and reddening of the tongue), cheilitis at the corners of the lips, and diarrhea. In many cases nervous tissue is also involved due to its high-energy demand or specific effects of the vitamin. Some of the common neurological symptoms include peripheral neuropathy (tingling of nerves at the extremities), depression, mental confusion, lack of motor coordination, and malaise. In some cases demyelination and degeneration of nervous tissues also takes place. These deficiency symptoms are so common and overlapping that they can be considered as properties of the energy-releasing vitamins as a class, rather than being specific for any one.

also tend to increase needs for vitamin K, leading to an increased incidence of hemorrhagic disease in infants born to mothers on anticonvulsants. In addition, anticonvulsants appear to increase the need for folic acid and B_6. Low serum folate levels are seen in 75% of patients on anticonvulsants and megaloblastic anemia may occur in as many as 50% without supplementation. By biochemical parameters, 30–60% of the children on anticonvulsants exhibit some form of B_6 deficiency. Clinical symptoms of B_6 deficiency are rarely seen, however. From 1 to 5 mg of folic acid and 10 mg of vitamin B_6 appear to be sufficient for most patients on anticonvulsants. Since folates may speed up the metabolism of some anticonvulsants, it is important that excess folic acid not be given.

Keith, D. A., Gundberg, C. M., Japaur, A., Aronoff, J., Alvarez, N., and Gallop, P. M. Vitamin K-Dependent Proteins and Anticonvulsant Medication. *Clin. Pharmacol. Ther.* 34:529, 1983; Theuer, R. C. and Vitale, J. L. Drug and Nutrient Interactions, H. Schneider, C. E. Anderson, and D. B. Coursin (Eds.). *Nutritional Support of Medical Practice*. New York: Harper & Row, 1977, pp. 297–305; and Weinstein, R. S., Bryce, G. F., Sappington, L. J., King, D. V. and Gallagher, B. B. Decreased Serum Ionized Calcium and Normal Vitamin D Metabolite Levels with Anticonvulsant Drug Treatment. *J. Clin. Endocrinol. Metabol.* 58:1003, 1984.

28.5 ENERGY-RELEASING WATER-SOLUBLE VITAMINS

Thiamine (Vitamin B_1) Forms the Coenzyme Thiamine Pyrophosphate (TPP)

Thiamine (Figure 28.10) is rapidly converted to the coenzyme thiamine pyrophosphate (TPP), which is required for the key reactions catalyzed by pyruvate and α-ketoglutarate dehydrogenases (see Figure 28.11). Thus, the cellular capacity for energy generation is severely compromised in thiamine deficiency. Thiamine pyrophosphate is also required for the transketolase of the pentose phosphate pathway. While the pentose phosphate pathway is not quantitatively important in terms of energy generation, it is the sole source of ribose for the synthesis of nucleic acid precursors and the major source of NADPH for fatty acid biosynthesis and other biosynthetic pathways. The red blood cell transketolase is also the enzyme most commonly used for measuring thiamine status in the body. Finally, TPP appears to play an important role in the transmission of nerve impulses. The TPP (or a related metabolite, thiamine triphosphate)

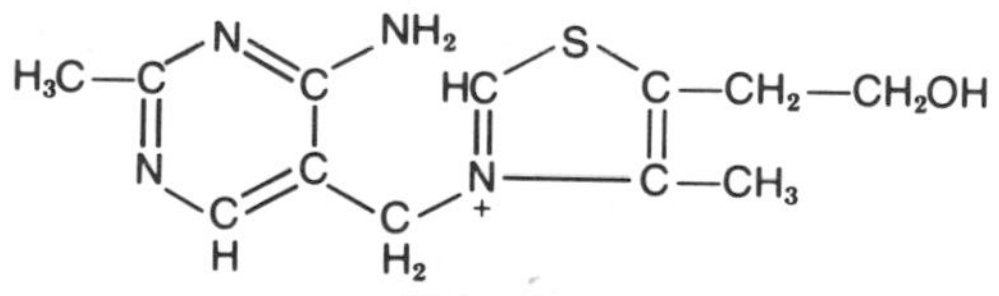

FIGURE 28.10
Structure of thiamine.

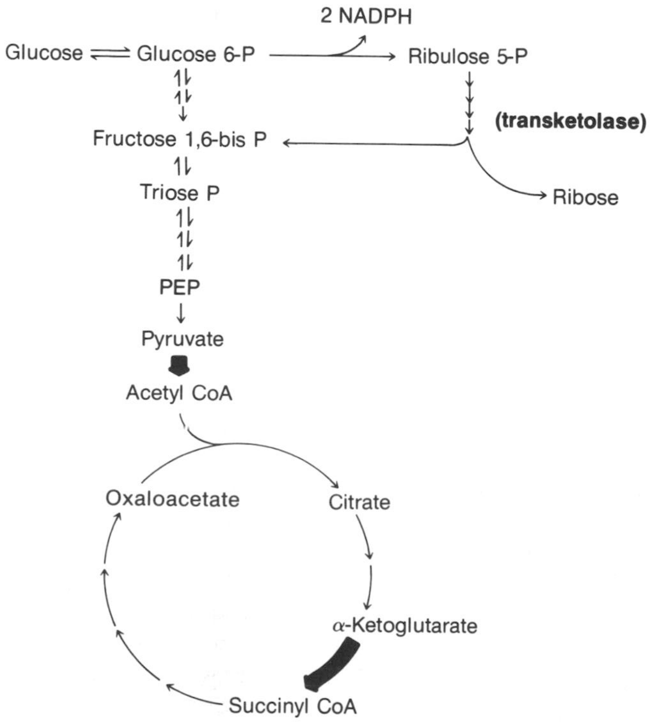

FIGURE 28.11
Summary of important reactions involving thiamine pyrophosphate.
The reactions involving thiamine pyrophosphate are indicated in boldface type.

CLIN. CORR. **28.5**
NUTRITIONAL CONSIDERATIONS IN THE ALCOHOLIC

Chronic alcoholics run considerable risk of nutritional deficiencies. The most common problems are neurologic symptoms associated with thiamine or pyridoxine deficiencies and hematological problems associated with folate or pyridoxine deficiencies. The deficiencies seen with alcoholics are not necessarily due to poor diet alone, although it is often a strong contributing factor. Alcohol causes pathological alterations of the gastrointestinal tract that often directly interfere with absorption of certain nutrients. The liver is the most important site of activation and storage of many vitamins. The severe liver damage associated with chronic alcoholism appears to interfere directly with storage and activation of certain nutrients.

Up to 40% of hospitalized alcoholics are estimated to have megaloblastic erythropoiesis due to folate deficiency. Alcohol appears to directly interfere with folate absorption and alcoholic cirrhosis impairs storage of this nutrient. Another 30% of hospitalized alcoholics have sideroblastic anemia or identifiable sideroblasts in erythroid marrow cells characteristic of pyridoxine deficiency. Some alcoholics also develop a peripheral neuropathy that responds to pyridoxine supplementation. This problem appears to result from impaired activation and increased degradation of pyridoxine. In particular, acetaldehyde (an end product of alcohol metabolism) displaces pyridoxal phosphate from its carrier protein in the plasma. The free pyridoxal phosphate is then rapidly degraded to inactive compounds and excreted.

The most dramatic nutritionally related neurological disorder is the Wernicke–Korsakoff syndrome. The symptoms include mental disturbances, ataxia (unsteady gait and lack of fine motor coordination), and uncoordinated eye movements. Congestive heart failure similar to that seen with beriberi is also seen in a small number of these patients. While this syndrome may only account for 1–3% of alcohol-related neurologic disorders, the response to supplemental thiamine is so dramatic that it is usually worth consideration. The thiamine deficiency appears to arise primarily from impaired absorption, although alcoholic cirrhosis may also affect the storage of thiamine in the liver.

While those are the most common nutritional deficiencies associated with alcoholism, deficiencies of almost any of the water-soluble vitamins can occur and cases of alcoholic scurvy and pellagra are occasion-

is localized in peripheral nerve membranes. It appears to be required for acetylcholine synthesis and may also be required for ion translocation reactions in stimulated neural tissue.

Although the biochemical reactions involving TPP are fairly well characterized, it is not clear how these biochemical lesions result in the symptoms of **thiamine deficiency.** The pyruvate dehydrogenase and transketolase reactions are the most sensitive to thiamine levels. Thus thiamine deficiency appears to selectivity inhibit carbohydrate metabolism, causing a buildup of pyruvate. The cells may be directly affected by the lack of available energy and NAPDH or may be poisoned by the accumulated pyruvate. Other symptoms of thiamine deficiency involve the neural tissue and probably result from the direct role of TTP in nerve transmission.

Loss of appetite, constipation, and nausea are among the earliest symptoms of thiamine deficiency. Mental depression, peripheral neuropathy, irritability, and fatigue are other early symptoms and probably directly relate to the role of thiamine in maintaining healthy nervous tissue. These symptoms of thiamine deficiency are most often seen in the elderly and low-income groups on restricted diets. Symptoms of moderately severe thiamine deficiency include mental confusion, ataxia (unsteady gait while walking and general inability to achieve fine control of motor functions), and ophthalmoplegia (loss of eye coordination). This set of symptoms is usually referred to as **Wernicke–Korsakoff syndrome** and is most commonly seen in chronic alcoholics (Clin. Corr. 28.5). Severe thiamine deficiency is known as **beriberi.** Dry beriberi is characterized primarily by advanced neuromuscular symptoms, including atrophy and weakness of the muscles. When these symptoms are coupled with edema, the disease is referred to as wet beriberi. Both forms of beriberi can be associated with an unusual type of heart failure characterized by high cardiac output. Beriberi is found primarily in populations relying exclusively on polished rice for food, although cardiac failure is sometimes seen in alcoholics as well.

The thiamine requirement is proportional to the caloric content of the diet and will be in the range of 1.0–1.5 mg day^{-1} for the normal adult. This requirement should be raised somewhat if carbohydrate intake is excessive or if the metabolic rate is elevated (due to fever, trauma, pregnancy, or lactation). Coffee and tea both contain substances that destroy thiamine, but this is not a problem for individuals consuming normal amounts of these beverages. The routine enrichment of cereals has assured that most Americans have an adequate intake of thiamine on a normal mixed diet.

Riboflavin Is Part of FAD and FMN

Riboflavin (Figure 28.12) is converted to the coenzymes flavin adenine dinucleotide (FAD) and flavin mononucleotide (FMN), both of which are involved in a wide variety of redox reactions. The flavin coenzymes are essential for energy production and cellular respiration. The most characteristic symptoms of riboflavin deficiency are angular cheilitis, glossitis, and scaly dermatitis (especially around the nasolabial folds and scrotal areas). The best flavin-requiring enzyme for assaying riboflavin status appears to be erythrocyte glutathione reductase. The recommended riboflavin intake is 1.2–1.7 mg day^{-1} for the normal adult. Foods rich in riboflavin include milk, meat, eggs, and cereal products. Riboflavin deficiencies are quite rare in this country. When riboflavin deficiency does occur, it is usually seen in chronic alcoholics. Hypothyroidism has recently been shown to slow the conversion of riboflavin to FMN and FAD. It is not known whether this affects riboflavin requirements, however.

Niacin Forms the Coenzymes NAD and NADP

Niacin (Figure 28.13) is not a vitamin in the strictest sense of the word, since the body is capable of making some niacin from tryptophan. However, the conversion of tryptophan to niacin is relatively inefficient (60 mg of tryptophan are required for the production of 1 mg of niacin) and occurs only after all of the body requirements for tryptophan (protein synthesis and energy production) have been met. Since the synthesis of niacin requires thiamine, pyridoxine, and riboflavin, it is also very inefficient on a marginal diet. Thus, in practical terms, most people require dietary sources of both tryptophan and niacin. Niacin (nicotinic acid) and niacinamide (nicotinamide) are both converted to the ubiquitous oxidation–reduction coenzymes NAD and NADP in the body.

Borderline **niacin deficiencies** are first seen as a glossitis of the tongue, somewhat similar to riboflavin deficiency. Pronounced deficiencies lead to **pellagra,** which is characterized by the three D's: dermatitis, diarrhea, and dementia. The dermatitis is characteristic in that it is usually seen only in skin areas exposed to sunlight and is symmetric. The neurologic symptoms are associated with actual degeneration of nervous tissue. Because of food fortification, pellagra is a medical curiosity in the developed world. Today it is primarily seen in alcoholics, patients with severe malabsorption problems, and elderly on very restricted diets. Pregnancy, lactation, and chronic illness lead to increased needs for niacin, but a varied diet will usually provide sufficient amounts.

Since tryptophan can be converted to niacin, and niacin itself can exist in a free or bound form, the calculation of available niacin for any given food is not a simple matter. For this reason, niacin requirements are expressed in terms of niacin equivalents (1 niacin equiv = 1 mg of free niacin). The current recommendation of the Food and Nutrition Board for a normal adult is 13–19 niacin equivalents (NE) per day. The richest food sources of niacin are meats, peanuts and other legumes, and enriched cereals.

When **nicotinic acid** (but not nicotinamide) is used in pharmacologic doses (2–4 g day^{-1}) it appears to cause a number of metabolic effects in the body not related to its normal function as a vitamin. For example, vasodilation (flushing) is a very immediate reaction. Over the longer term there is a decreased mobilization of fatty acids from adipose tissue, a marked decrease in circulating cholesterol and lipoproteins (especially LDL), and an elevation of serum glucose and uric acid. These effects can be explained in part by an effect of nicotinic acid on cAMP levels. While the cholesterol lowering effects of nicotinic acid may be desirable in certain controlled clinical situations, there are potential side effects of pharmacologic doses of this vitamin. The reduced mobilization of fatty acids from adipose tissue causes depletion of the glycogen and fat reserves in skeletal and cardiac muscle. The tendency toward elevated glucose and uric acid could cause problems if someone is borderline for diabetes or gout. Finally, continued use of nicotinic acid in those doses is sometimes associated with elevated levels of serum enzymes suggestive of liver damage.

ally reported. Chronic ethanol consumption causes an interesting redistribution of vitamin A stores in the body. Vitamin A stores in the liver are rapidly depleted while levels of vitamin A in the serum and other tissues may be normal or slightly elevated. Apparently, ethanol causes both increased mobilization of vitamin A from the liver and increased catabolism of liver vitamin A to inactive metabolites by the hepatic P450 system. Alcoholic patients have decreased bone density and an increased incidence of osteoporosis. This probably relates to a defect in the 25-hydroxylation step in the liver as well as an increased rate of metabolism of vitamin D to inactive products by an activated cytochrome P450 system. Dietary calcium intake is also often poor. In fact, alcoholics generally have decreased serum levels of zinc, calcium, and magnesium due to poor dietary intake and increased urinary losses. Iron-deficiency anemia is very rare unless there is gastrointestinal bleeding or chronic infection. In fact, excess iron is a more common problem with alcoholics. Many alcoholic beverages contain relatively high iron levels, and alcohol appears to enhance iron absorption.

Hayumpa, A. M. Mechanisms of Vitamin Deficiencies in Alcoholism. *Alcoholism Clin. Exper. Res.* 10:573, 1986; and Lieber, C. S. The Influence of Alcohol on Nutritional Status. *Nutr. Rev.* 46:241, 1988.

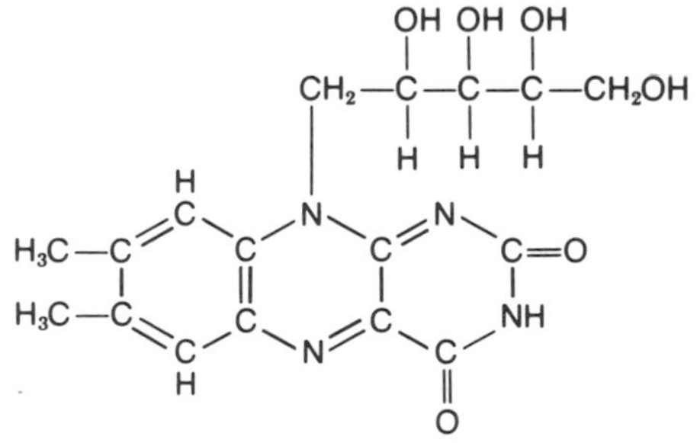

Riboflavin [7,8-dimethyl-10-(1′-D-ribityl)isoalloxazine]

FIGURE 28.12
Structure of riboflavin.

Niacin   Niacinamide

FIGURE 28.13
Structure of niacin and niacinamide.

Pyridoxine (Vitamin B_6) Forms the Coenzyme Pyridoxal Phosphate

Pyridoxine, pyridoxamine, and **pyridoxal,** are all naturally occurring forms of vitamin B_6 (see Figure 28.14). All three forms are efficiently converted by the body to **pyridoxal phosphate,** which is required for the synthesis, catabolism, and interconversion of amino acids. The role of

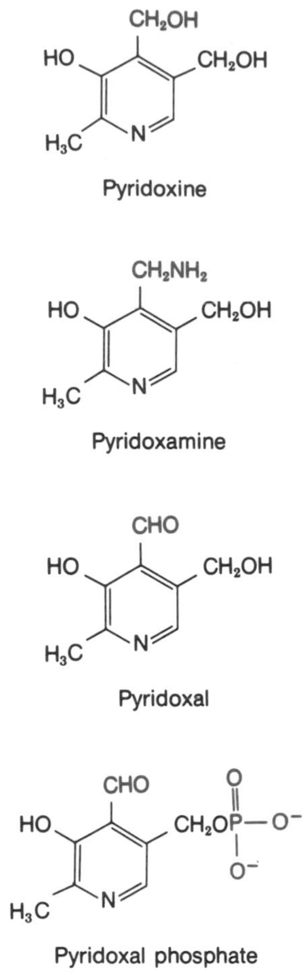

Pyridoxine

Pyridoxamine

Pyridoxal

Pyridoxal phosphate

FIGURE 28.14
Structures of vitamin B_6.

pyridoxal phosphate in amino acid metabolism has been discussed in Chapter 11 and will not be considered here. While pyridoxal phosphate-dependent reactions are legion, there are a few instances in which the biochemical lesion seems to be directly associated with the symptoms of **B_6 deficiency.** Some of these more important reactions are summarized in Figure 28.15. Obviously, pyridoxal phosphate is essential for energy production from amino acids and can be considered an energy-releasing vitamin. Thus some of the symptoms of severe B_6 deficiency are similar to those of thc other energy-releasing vitamins. Pyridoxal phosphate is also required for the synthesis of the neurotransmiters serotonin and norepinephrine and appears to be required for the synthesis of the sphingolipids necessary for myelin formation. These effects are thought to explain the irritability, nervousness, and depression seen with mild deficiencies and the peripheral neuropathy and convulsions observed with severe deficiencies. Pyridoxal phosphate is required for the synthesis of δ-aminolevulinic acid, a precursor of heme. Vitamin B_6 deficiencies occasionally cause **sideroblastic anemia,** which is characteristically a microcytic anemia seen in the presence of high serum iron. Pyridoxal phosphate is also an essential component of the enzyme glycogen phosphorylase. It is covalently linked to a lysine residue and stabilizes the enzyme. This role of B_6 may explain the decreased glucose tolerance associated with deficiency, although B_6 appears to have some direct effects on the glucocorticord receptor as well. Finally, pyridoxal phosphate is one of the cofactors required for the conversion of tryptophan to NAD^+. While this may not be directly related to the symptomatology of B_6 deficiency, a tryptophan load test is one of the most sensitive indicators of vitamin B_6 status (Clin. Corr. 28.6).

The amount of B_6 required in the diet is roughly proportional to the protein content of the diet. Assuming that the average American consumes close to 100 g of protein per day, the RDA for vitamin B_6 has been set at 1.4–2.0 mg day^{-1} for a normal adult. This requirement is increased during pregnancy and lactation and may increase somewhat with age as well. Vitamin B_6 is fairly widespread in foods, but meat, vegetables, whole grain cereals, and egg yolks are among the richest sources.

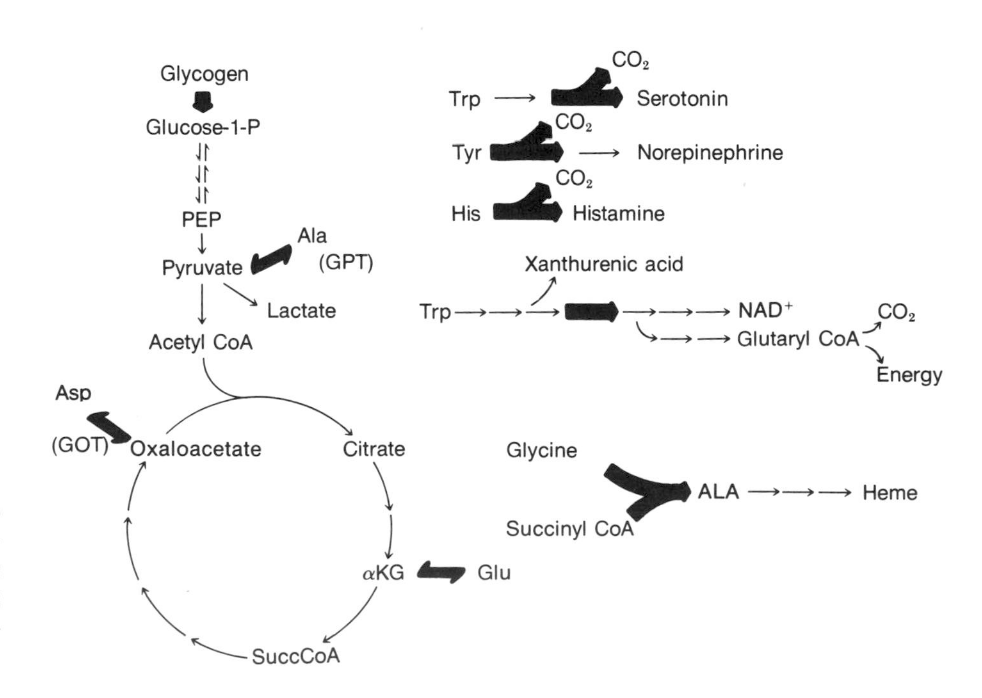

FIGURE 28.15
Some important metabolic roles of pyridoxal phosphate.
The reactions requiring pyridoxal phosphate are indicated with heavy arrows. ALA = δ-aminolevulinic acid; αKG = α-ketoglutarate; GPT = glutamate pyruvate transaminase; and GOT = glutamate oxaloacetate transaminase.

The evaluation of B_6 nutritional status has become a controversial topic in recent years. Both questions of dietary intake and nutritional needs have been clouded by the issue of how to adequately measure B_6 status. Some of this controversy is discussed in Clin. Corr. 28.6. In terms of dietary intake, it has usually been assumed that the average American diet is adequate in B_6 and it is not routinely added to flour and other fortified foods. Recent nutritional surveys, however, have cast doubt on that assumption. In several instances, a significant fraction of the survey population was found to consume less than two-thirds of the RDA for B_6. Although dietary intake of B_6 may be marginal for many individuals, this does not necessarily cause adverse effects unless coupled with increased demand. Pregnancy, for example, increases the need for B_6. The usual recommendation is for an additional intake of 0.6 mg day^{-1}.

CLIN. CORR. **28.6**
VITAMIN B_6 REQUIREMENTS FOR USERS OF ORAL CONTRACEPTIVES

The controversy over B_6 requirements for users of oral contraceptives best illustrates the potential problems associated with biochemical assays. For years, one of the most common assays for vitamin B_6 status had been the tryptophan load assay. This assay is based on the observation that when tissue pyridoxal phosphate levels are low, the normal catabolism of tryptophan is impaired and most of the tryptophan is catabolized by a minor pathway leading to synthesis of xanthurenic acid. Under many conditions, the amount of xanthurenic acid recovered in a 24-h urine sample following ingestion of a fixed amount of tryptophan is a valid indicator of vitamin B_6 status. When the tryptophan load test was used to assess the vitamin B_6 status of oral contraceptive users, however, alarming reports started appearing in the literature. Not only did oral contraceptive use increase the excretion of xanthurenic acid considerably but the amount of pyridoxine needed to return xanthurenic acid excretion to normal was 20 mg day^{-1}. This amounts to 10 times the RDA and almost 20 times the level required to maintain normal B_6 status in control groups. As might be expected, this observation received much popular attention in spite of the fact that most classical symptoms of vitamin B_6 deficiency were not observed in oral contraceptive users.

More recent studies using other measures of vitamin B_6 have painted a slightly different picture. For example, erythrocyte glutamate pyruvate transaminase and erythrocyte glutamate oxaloacetate transaminase are both pyridoxal phosphate-containing enzymes. One can also assess vitamin B_6 status by measuring the endogenous activity of these enzymes and the degree of stimulation by added pyridoxal phosphate. These types of assays show a much smaller difference between nonusers and users of oral contraceptives. The minimum level of pyridoxine needed to maintain normal vitamin B_6 status as measured by these assays was only 2.0 mg day^{-1}, which is slightly greater than the RDA and about twice that needed by nonusers.

Why the large discrepancy? For one thing, it must be kept in mind that enzyme activity can be affected by hormones as well as vitamin cofactors. Kynureninase is the key pyridoxal phosphate containing enzyme of the tryptophan catabolic pathway. The activity of kynureninase is regulated both by pyridoxal phosphate availability and by estrogen metabolites. Even with normal vitamin B_6 status most of the enzyme exists in the inactive apoenzyme form. However, this does not affect tryptophan metabolism because tryptophan oxygenase, the first enzyme of the pathway, is rate limiting. Thus, the small amount of active holoenzyme is more than sufficient to handle the metabolites produced by the first part of the pathway. However, kynureninase is inhibited by estrogen metabolites. Thus, with oral contraceptive use its activity is reduced to a level where it becomes rate limiting and excess tryptophan metabolites are shunted to xanthurenic acid. Higher than normal levels of vitamin B_6 overcome this problem by converting more apoenzyme to holoenzyme, thus increasing the total amount of enzyme. Since the estrogen was having a specific effect on the enzyme used to measure vitamin B_6 states in this assay, it did not necessarily mean that pyridoxine requirements were altered for other metabolic processes in the body.

Does this mean that vitamin B_6 status is of no concern to users of oral contraceptives? Oral contraceptives do appear to increase vitamin B_6 requirements slightly. Several dietary surveys have shown that a significant percentage of women in the 18–24-year age group consume diets containing less than 1.3 mg of pyridoxine per day. If these women are also using oral contraceptives, they are at some increased risk for developing a borderline deficiency. This is of some concern since recent data suggest that borderline vitamin B_6 deficiency may increase the risk of hormone-sensitive breast cancers. It should also be noted that there are documented cases of depression and decreased glucose tolerance in oral contraceptive users that respond to pyridoxine supplementation. This appears to be due to metabolic effects of pharmacologic doses of vitamin B_6 rather than suggestive of vitamin B_6 deficiency. While the tryptophan load test was clearly misleading in a quantitative sense, it did alert the medical community to a previously unsuspected nutritional risk.

Bender, D. A. Oestrogens and Vitamin B_6—Actions and Interactions. *World Rev. Nutr. Diet.* 51:140, 1987; and Kirksey, A., Keaton, K., Abernathy, R. P., and Grager, J. L. Vitamin B_6 Nutritional Status of a Group of Female Adolescents. *Am. J. Clin. Nutr.* 31:946, 1978.

There are, however, a few instances where B_6 deficiencies are clear-cut and noncontroversial. For example, the drug **isoniazid** (isonicotinic acid hydrazide), which is commonly used in the treatment of tuberculosis, reacts with pyridoxal or pyridoxal phosphate to form a hydrazone derivative, which inhibits pyridoxal phosphate-containing enzymes. Patients on long-term isoniazid treatment develop a peripheral neuropathy, which responds well to B_6 therapy. Also, **penicillamine** (β-dimethylcysteine), which is used in the treatment of patients with Wilson's disease, cystinuria, and rheumatoid arthritis, reacts with pyridoxal phosphate to form an inactive thiazolidine derivative. Patients treated with penicillamine occasionally develop convulsions that can be prevented by B_6 supplementation.

Pantothenic Acid and Biotin Are Also Energy Releasing Vitamins

Pantothenic acid is an essential component of coenzyme A (CoA) and of phosphopantetheine of fatty acid synthase and, thus, is required for the metabolism of all fat, protein, and carbohydrate via the citric acid cycle. In short, more than 70 enzymes have been described to date that utilize CoA or ACP derivatives. In view of the importance of these reactions, one would expect pantothenic acid deficiencies to be a serious concern in humans. This, however, does not appear to be the case, and the reasons are essentially twofold: (1) pantothenic acid is very widespread in natural foods, probably reflecting its widespread metabolic role, and (2) most symptoms of panthothenic acid deficiency are vague and mimic those of other B vitamin deficiencies.

Biotin is the prosthetic group for a number of carboxylation reactions, the most notable being pyruvate carboxylase (needed for synthesis of oxaloacetate for gluconeogenesis and replenishment of the citric acid cycle) and acetyl CoA carboxylase (fatty acid biosynthesis). Biotin is found in peanuts, chocolate and eggs, and is usually synthesized in adequate amounts by intestinal bacteria. Biotin deficiency is generally seen only following long-term antibiotic therapy or excessive consumption of raw egg white. The raw egg white contains a protein, **avidin**, which binds biotin in a nondigestible form. However, in humans raw egg white must comprise 30% of the caloric intake (~20 egg whites per day) to precipitate a biotin deficiency.

28.6 HEMATOPOIETIC WATER-SOLUBLE VITAMINS

Folic Acid (Folacin) Functions as Tetrahydrofolate in One-Carbon Metabolism

The simplest form of **folic acid** is pteroylmonoglutamic acid. However, folic acid usually occurs as polyglutamate derivatives with from two to seven additional glutamic acid residues (Figure 28.16). These compounds are taken up by intestinal mucosal cells and the extra glutamate residues removed by **conjugase,** a lysosomal enzyme. The free folic acid is then reduced to H_4folate by the enzyme H_2folate reductase and circulated in the plasma primarily as N^5-methyl H_4folate (Figure 28.16). Inside cells H_4folates are found primarily as polyglutamate derivatives, and these appear to be the biologically most potent forms. Folic acid is also stored as a polyglutamate derivative of H_4folate in the liver.

Various 1-carbon H_4folate derivatives are used in biosynthesis reactions (see Figure 28.17). They are required, for example, in the synthesis

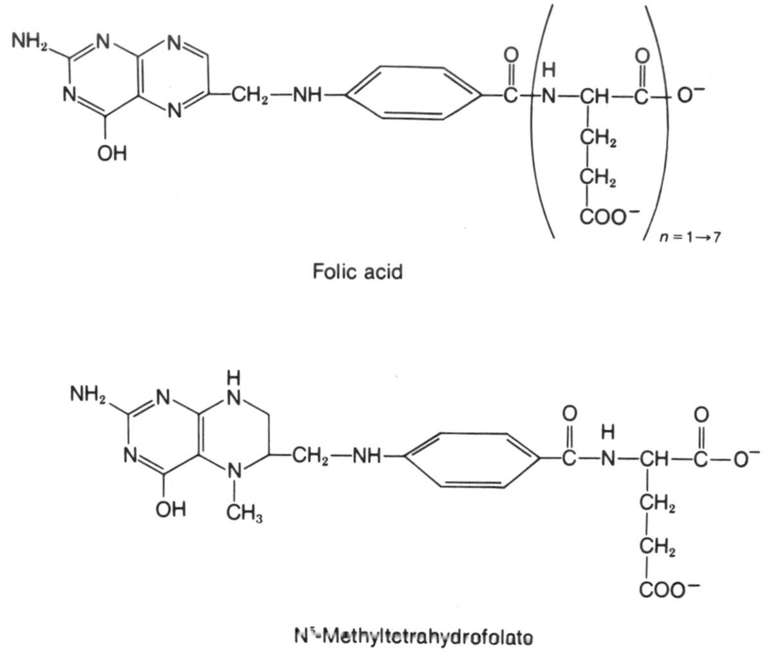

Folic acid

N^5-Methyltetrahydrofolate

FIGURE 28.16
Structure of folic acid and N^5-methyltetrahydrofolate.

of choline, serine, glycine, methionine, purines, and dTMP. Since adequate amounts of choline and the amino acids can usually be obtained from the diet, the participation of folates in purine and dTMP synthesis appears to be metabolically most significant. However, under some conditions the folate-dependent conversion of homocysteine to methionine can make a significant contribution to the available pool. Methionine, of course, is also converted to *S*-adenosylmethionine, which is used in a number of biologically important methylation reactions.

The most pronounced effect of folate deficiency is inhibition of DNA synthesis due to decreased availability of purines and dTMP. This leads to

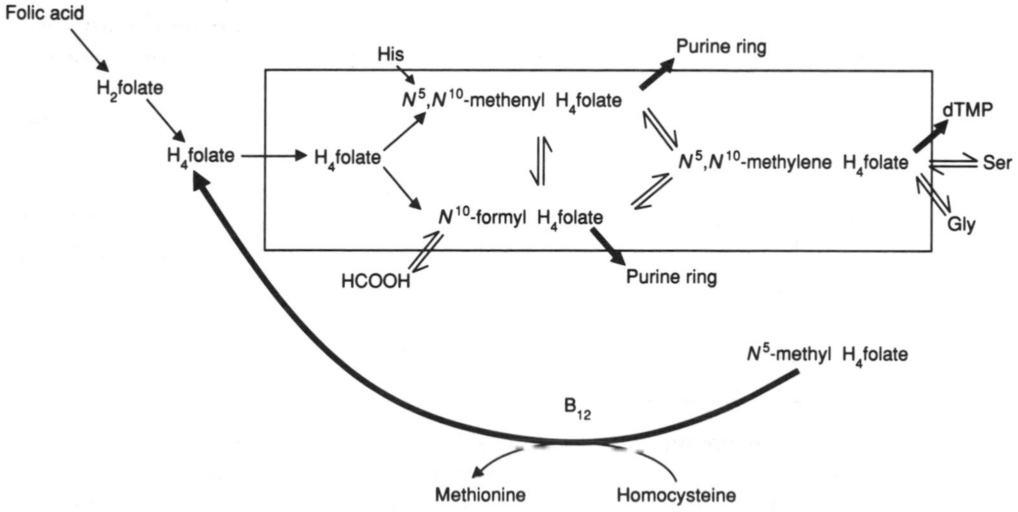

FIGURE 28.17
Metabolic roles of folic acid and vitamin B_{12} in one-carbon metabolism.
The metabolic interconversions of folic acid and its derivatives are indicated with light arrows. Pathways relying exclusively on folate are shown with heavy arrows. The important B_{12}-dependent reaction converting N^5-methyl H_4folate back to H_4folate is also shown with a heavy arrow. The box encloses the "pool" of C_1 derivatives of H_4folate.

an arrest of cells in S phase and a characteristic "megaloblastic" change in the size and shape of the nuclei of rapidly dividing cells. The block in DNA synthesis also slows down the maturation of red blood cells, causing production of abnormally large "macrocytic" red blood cells with fragile membranes. Thus a **macrocytic anemia** associated with megaloblastic changes in the bone marrow is fairly characteristic of folate deficiency.

There is some uncertainty over the incidence of folate deficiencies. Folates occur very widely in foods, especially meats in a variety of different forms that are utilized with varying efficiencies. Assessment of nutritional status is further complicated by the use of different biochemical methods to measure folate levels. For example, the serum folate level decreases rapidly on restricted folate intake and is a very sensitive indicator of folate status. Levels of polyglutamate folate derivatives in red blood cells decrease much more slowly and are more indicative of tissue levels. Symptoms of folate deficiency appear only as tissue folates are depleted. There can be many causes of folate deficiency, including inadequate intake, impaired absorption, increased demand, and impaired metabolism. Some dietary surveys have suggested that inadequate intake may be more common than previously supposed. However, as with most other vitamins, inadequate intake is probably not sufficient to trigger symptoms of folate deficiency in thc absence of increased requirements or decreased utilization.

Perhaps the most common example of increased need occurs during pregnancy and lactation. As the blood volume and the number of rapidly dividing cells in the body increase, the need for folic acid increases. By the third trimester the folic acid requirement has almost doubled. In the United States almost 20–25% of otherwise normal pregnancies are associated with low serum folate levels but actual megaloblastic anemia is rare and is usually seen only after multiple pregnancies. Normal diets seldom supply the 400 μg of folate needed during pregnancy, so most physicians routinely recommend supplementation. Folate deficiency is common in alcoholics (Clin. Corr. 28.5). Folate deficiencies are also seen in a number of malabsorption diseases and are occasionally seen in the elderly due to a combination of poor dietary habits and poor absorption.

There are a number of drugs that also directly interfere with folate metabolism. Anticonvulsants and oral contraceptives may interfere with folate absorption and anticonvulsants appear to increase catabolism of folates (Clin. Corr. 28.4). Oral contraceptives and estrogens also appear to interfere with folate metabolism in their target tissue. Long-term use of any of these drugs can lead to folate deficiencies unless adequate supplementation is provided. For example, 20% of patients using oral contraceptives develop megaloblastic changes in the cervicovaginal epithelium, and 20–30% show low serum folate levels.

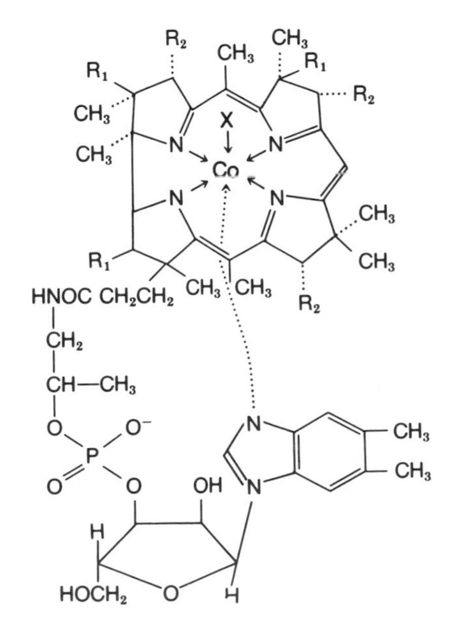

FIGURE 28.18
Structure of vitamin B_{12} (cobalamine).

Vitamin B_{12} (Cobalamine) Contains Cobalt in a Tetrapyrrole Ring

Pernicious anemia, a megaloblastic anemia associated with neurological deterioration, was invariably fatal until 1926 when liver extracts were shown to be curative. Subsequent work showed the need for both an extrinsic factor present in liver and an intrinsic factor produced by the body. Vitamin B_{12} was the **extrinsic factor.** Chemically, vitamin B_{12} consists of **cobalt** in a coordination state of six, coordinated in four positions by a tetrapyrrol (or corrin) ring, in one position by a benzimidazole nitrogen, and in the sixth position by one of several different ligands (Figure 28.18). The crystalline forms of B_{12} used in supplementation are usually hydroxycobalamine or cyanocobalamine. In foods B_{12} usually occurs bound to protein in the methyl or 5′-deoxyadenosyl forms. To be utilized the B_{12} must first be removed from the protein by acid hydrolysis in the stomach or trypsin digestion in the intestine. It then must combine with **intrinsic factor,** a protein secreted by the stomach, which carries it to the ileum for absorption.

In humans there are two major symptoms of B_{12} deficiency, hematopoietic and neurological, and only two biochemical reactions in which B_{12} is known to participate (Figure 28.19). Thus it is very tempting to speculate on exact cause and effect mechanisms. The methyl derivative of B_{12} is required for the conversion of homocysteine to methionine and the 5-deoxyadenosyl derivative is required for the methylmalonyl CoA mutase reaction (methylmalonyl CoA → succinyl CoA), which is a key step in the catabolism of some branched-chain amino acids. The neurologic disorders seen in B_{12} deficiency are due to progressive demyelination of nervous tissue. It has been proposed that the methylmalonyl CoA that accumulates interferes with myelin sheath formation in two ways:

1. Methylmalonyl CoA is a competitive inhibitor of malonyl CoA in fatty acid biosynthesis. Since the myelin sheath is continually turning over, any severe inhibition of fatty acid biosynthesis will lead to its eventual degeneration.

2. In the residual fatty acid synthesis that does occur, methylmalonyl CoA can substitute for malonyl CoA in the reaction sequence, leading to branched-chain fatty acids, which might disrupt normal membrane structure. There is some evidence supporting both mechanisms.

The megaloblastic anemia associated with B_{12} deficiency is thought to be due to the effect of B_{12} on folate metabolism. The B_{12}-dependent homocysteine to methionine conversion (homocysteine + N^5-methyl H_4folate → methionine + H_4folate) appears to be the only major pathway by which N^5-methyl H_4folate can return to the H_4folate pool (Figure 28.17). Thus in B_{12} deficiency there is a buildup of N^5-methyl H_4folate and a deficiency of the H_4folate derivatives needed for purine and dTMP biosynthesis. Essentially all of the folate becomes "trapped" as the N^5-methyl derivative. Vitamin B_{12} also may be required for uptake of folate by cells and for its conversion to the biologically more active polyglutamate forms. High levels of supplemental folate can overcome the megaloblastic anemia associated with B_{12} deficiencies but not the neurological problems. Hence caution must be utilized in using folate to treat megaloblastic anemia.

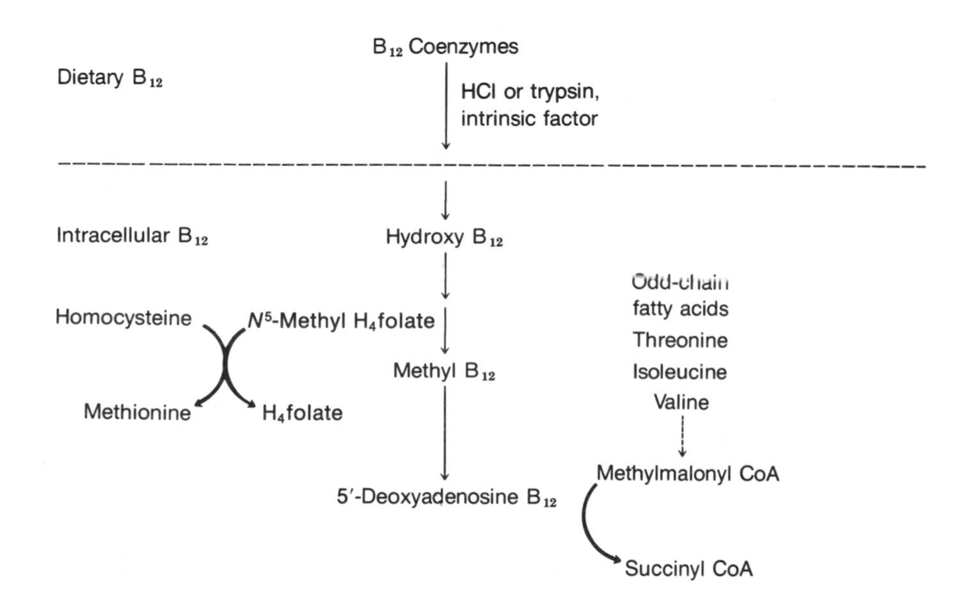

FIGURE 28.19
Metabolism of vitamin B_{12}.
The metabolic interconversions of B_{12} are indicated with light arrows, and B_{12} requiring reactions are indicated with heavy arrows. Other related pathways are indicated with dashed arrows.

Vitamin B_{12} is widespread in foods of animal origin, especially meats. Furthermore, the liver stores up to a 6-year supply a vitamin B_{12}. Thus, deficiencies of B_{12} are extremely rare. They are occasionally seen in older people due to insufficient production of intrinsic factor and/or HCl in the stomach. Vitamin B_{12} deficiency can also be seen in patients with severe malabsorption diseases and in long-term vegetarians.

28.7 OTHER WATER-SOLUBLE VITAMINS

Ascorbic Acid Functions in Reduction and Hydroxylation Reactions

Vitamin C or **ascorbic acid** is a C_6 compound closely related to glucose. Its main biological role appears to be as a reducing agent in a number of important hydroxylation reactions in the body. For example, there is clear evidence that ascorbic acid is required for the hydroxylation of lysine and proline in procollagen. Without the hydroxylation of these amino acids, the procollagen is unable to properly cross-link into normal collagen fibrils. Thus vitamin C is obviously important for maintenance of normal connective tissue and for wound healing, since the connective tissue is laid down first. Vitamin C is also necessary for bone formation because bone tissue has an organic matrix containing collagen as well as the inorganic, calcified portion. Finally, collagen appears to be a component of the ground substance surrounding capillary walls, so vitamin C deficiency is associated with capillary fragility.

Since vitamin C is concentrated in the adrenal gland, especially in periods of stress, it has also been postulated to be required for the hydroxylation reactions involved in the synthesis of some corticosteroids. Ascorbic acid has other important properties as a reducing agent that appear to be nonenzymatic. For example, it aids in the absorption of iron by reducing it to the ferrous state in the stomach. It spares vitamin A, vitamin E, and some B vitamins by protecting them from oxidation. Also, it enhances the utilization of folic acid, either by aiding the conversion of folate to H_4folate or the formation of polyglutamate derivatives of H_4folate. Finally, vitamin C appears to be a biologically important antioxidant. The National Research Council has recently concluded that adequate amounts (RDA levels) of antioxidants such as β-carotene and vitamin C in the diet reduce the risk of cancer. The data for other naturally occurring antioxidants such as vitamin E and selenium are not yet conclusive.

Most of the symptoms of vitamin C deficiency can be directly related to its metabolic roles. Symptoms of mild vitamin C deficiency include easy bruising and the formation of petechiae (small, pinpoint hemorrhages in the skin) due to increased capillary fragility. Mild vitamin C deficiencies are also associated with decreased immunocompetence. **Scurvy** itself is associated with decreased wound healing, osteoporosis, hemorrhaging, and anemia. The osteoporosis results from the inability to maintain the organic matrix of the bone, followed by demineralization. The anemia results from extensive hemorrhaging coupled with defects in iron absorption and folate metabolism.

Since vitamin C is readily absorbed, vitamin C deficiencies almost invariably result from poor diet and/or increased need. There is some uncertainty over the need for vitamin C in periods of stress; in severe stress or trauma there is a rapid drop in serum vitamin C levels. In these situations most of the body's supply of vitamin C is mobilized to the adrenals and/or the area of the wound. These facts are clear but their interpretation is variable. Does this represent an increased demand for vitamin C or merely a normal redistribution of vitamin C to those areas where it is needed most? Do the lowered serum levels of vitamin C impair its functions in other tissues in the body? The current consensus appears to be that the lowered serum vitamin C levels do indicate an increased demand but there is little agreement as to how much.

A similar situation exists with respect to the effect of various drugs on vitamin C status. Smoking has been shown to cause lower serum levels of vitamin C. In fact, the 1989 RDAs recommend that smokers consume 100 mg of vitamin C per day instead of the 60 mg day^{-1} needed by nonsmoking adults. Aspirin appears to block uptake by white blood cells. Oral contraceptives and corticosteroids also lower serum levels of vitamin C. While there is no universal agreement as to the seriousness of these effects of vitamin C requirements, the possibility of marginal vitamin C deficiencies should be considered with any patient using these drugs over a long period of time, especially if dietary intake is less than optimal.

Of course, the most controversial question surrounding vitamin C is its use in megadoses to prevent and cure the common cold. Ever since this use of vitamin C was first popularized by Linus Pauling in 1970, the issue has generated considerable controversy. Some double-blind studies appear to have substantiated the claim in part; the number of colds experienced by vitamin C supplemented groups appeared to be about the same as for control groups but the severity and duration of the colds were decreased. Thus, while vitamin C does not appear to be useful in preventing the common cold, it may moderate its symptoms. There is no clear indication at present as to how much vitamin C is required to achieve this effect. In the original experiment the control group had a balanced diet providing at least 45 mg of ascorbic acid per day while the experimental group received 1–4 g day^{-1}. Subsequent experiments suggested that considerably less vitamin C may achieve the same result. The mechanism by which vitamin C ameliorates the symptoms of the common cold is not known; it has been suggested that vitamin C is required for normal leukocyte function or for synthesis and release of histamine during stress situations.

While megadoses of vitamin C are probably no more harmful than the widely used over-the-counter cold medications, there are some potential side effects of high vitamin C intake that should be considered. For example, oxalate is a major metabolite of ascorbic acid, thus, high ascorbate intakes could lead to the formation of oxalate kidney stones in predisposed individuals. Pregnant women taking megadoses of vitamin C may give birth to infants with abnormally high vitamin C requirements. Earlier suggestions that megadoses of vitamin C interfered with B_{12} metabolism have proven to be controversial.

28.8 MACROMINERALS

Calcium Has Many Physiological Roles

Calcium is the most abundant mineral in the body. Most of the calcium is in the bone but a small amount of calcium outside of the bone functions in a number of essential processes. It is required for many enzymes, mediates some hormonal responses, and is essential for blood coagulation. It is also essential for muscle contractility and normal neuromuscular irritability. In fact only a relatively narrow range of serum calcium levels is compatible with life. Since maintenance of constant serum calcium levels is so vital, an elaborate homeostatic control system has evolved. Part of

CLIN. CORR. 28.7 DIET AND OSTEOPOROSIS

The controversies raging over the relationships between calcium intake and osteoporosis illustrate the difficulties we face in making simple dietary recommendations for complex biological problems. Based on the TV ads and wide variety of calcium-fortified foods on the market, it would be easy to assume that all an older woman needs to prevent osteoporosis is a diet rich in calcium. However, that may be like closing the barn door after the horse has left. There is strong consensus that the years from age 10 to 35, when the bone density is reaching its maximum, are the most important for reducing the risk of osteoporosis. The maximum bone density obtained during these years is clearly dependent on both calcium intake and exercise and dense bones are less likely to become seriously depleted of calcium following menopause. Unfortunately, most American women are consuming far too little calcium during these years. The RDA for calcium is 1200 mg day^{-1} (4 glasses of milk per day) for women from age 11 to 24 and 800 mg day^{-1} ($2\frac{2}{3}$ glasses of milk per day) for women over 24. The median calcium intake for women in this age range is only about 500 mg day^{-1}. Thus, it is clear that increased calcium intake should be encouraged in this group.

But what about postmenopausal women? After all, many of the advertisements seem to be targeted at this group. Do they really need more calcium? The 1984 NIH consensus panel on osteoporosis recommended that postmenopausal women consume 1500 mg of calcium per day, but this recommendation has been vigorously disputed by other experts in the field. Lets examine the evidence. Calcium balance studies have shown that many postmenopausal women need 1200–1500 mg of calcium per day to maintain a positive calcium balance (more calcium coming in than is lost in the urine), but that does not necessarily mean that the additional calcium will be stored in their bones. In fact, some recent studies have failed to find a correlation between calcium intake and loss of bone density in postmenopausal women while others have reported a protective effect. All of those studies have been complicated by the discovery that calcium intake may have different effects on different types of bones. Calcium intakes in the range of 1000–1500 mg day^{-1} appear to slow the decrease in density of cortical bone, such as that found in the hip, hand, and some parts of the forearm. Similar doses, however, appear to have little or no effect on loss of density

this was discussed earlier in the section on vitamin D metabolism. Low serum calcium stimulates formation of 1,25-dihydroxycholecalciferol, which enhances calcium absorption. If dietary calcium intake is insufficient to maintain serum calcium, 1,25-dihydroxycholecalciferol and parathyroid hormone stimulate bone resorption. Long-term dietary calcium insufficiency, therefore, almost always results in net loss of calcium from the bones.

Dietary calcium requirements, however, are very difficult to determine because of the existence of other factors that affect availability of calcium. One important factor is vitamin D. As discussed in Chapter 27, excess protein in the diet may upset calcium balance by causing more rapid excretion of calcium. Finally, exercise increases the efficiency of calcium utilization for bone formation. Thus calcium balance studies carried out on Peruvian Indians, who have extensive exposure to sunlight, get extensive exercise, and subsist on low-protein vegetarian diets, indicate a need for only 300–400 mg of calcium per day. However, calcium balance studies carried out in this country consistently show higher requirements and the RDA has been set at 800–1200 mg day^{-1}.

The chief symptoms of **calcium deficiency** are similar to those of vitamin D deficiency, but other symptoms such as muscle cramps are possible with marginal deficiencies. Dietary surveys indicate that a significant portion of certain population groups in this country do not have adequate calcium intake, especially low-income children and adult women. This is of particular concern because these are the population groups with particularly high needs for calcium. For this reason, the U.S. Congress has established the WIC (Women and Infant Children) program to assure adequate protein, calcium, and iron for indigent families with pregnant–lactating women or young infants.

Dietary surveys also show that 34–47% of the over-60 population consume less than one-half of the RDA for calcium. This is also the age group most at risk of developing **osteoporosis,** which is characterized by loss of the organic matrix as well as progressive demineralization of the bone. The causes of osteoporosis are multifactorial and largely unknown but it appears likely that part of the problem has to do with calcium metabolism (Clin. Corr. 28.7). Recent studies have also suggested that inadequate intake of calcium may result in elevated blood pressure. Although this hypothesis has not been conclusively demonstrated, it is of great concern because most of the low sodium diets recommended for patients with high blood pressure severely limit dairy products, which are the main dietary source of calcium for Americans.

Other Macrominerals Are Phosphorus and Magnesium

Phosphorous is a universal constituent of living cells and, for that reason, is almost always present in adequate amounts in the diet. Hypophosphatemia, a symptom of vitamin D deficiency, can occasionally be seen following excessive use of antacids containing aluminum hydroxide or calcium carbonate that form insoluble precipitates with phosphate. Uncontrolled metabolic acidosis can lead to excessive phosphate loss in the urine. The initial symptom of hypophosphatemia is muscle weakness but eventually a form of rickets can develop.

Magnesium is also ubiquitous in living tissue. It is required for many enzyme activities and for neuromuscular transmission. Magnesium deficiency is most often observed in conditions of alcoholism, use of certain diuretics, and metabolic acidosis. The main symptoms of magnesium deficiency are weakness, tremors, and cardiac arrhythmia. The discovery that heart muscle from patients with myocardial infarctions was low in magnesium led to the hypothesis that magnesium deficiency might be a predisposing condition for various forms of heart disease. There is no evidence, however, that the patients with heart disease had, in fact, consumed diets inadequate in magnesium. The possibility must be considered that a redistribution of tissue magnesium takes place as a result of the heart attack. There is some evidence that supplemental magnesium may help prevent the formation of calcium oxalate stones in the kidney.

28.9 TRACE MINERALS

Iron Is Efficiently Reutilized

Iron metabolism is unique in that it operates largely as a closed system, with iron stores being efficiently reutilized by the body. Not only are iron losses normally minimal (<1 mg day^{-1}) but iron absorption is also minimal under the best of conditions. Iron usually occurs in foods in the ferric form bound to protein or organic acids. Before absorption can occur, the iron must be split from these carriers, a process that is facilitated by the acid secretions of the stomach, and reduced to the ferrous form, a process that is enhanced by ascorbic acid. Only 10% of the iron in an average mixed diet is usually absorbed but the efficiency of absorption can be increased to 30% by severe iron deficiency. Iron absorption and metabolism have been discussed in Chapter 24 and are summarized in Figure 28.20.

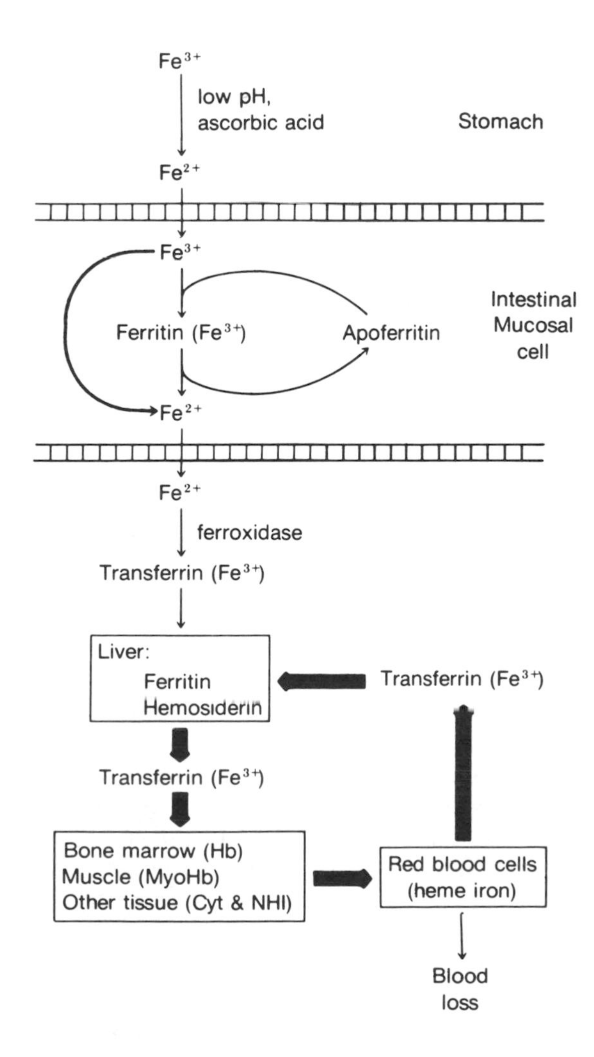

FIGURE 28.20
Overview of iron metabolism.
This figure reviews some of the features of iron metabolism discussed previously in Chapter 24. The heavy arrows indicate that most of the body's iron is efficiently reutilized by the pathway shown. Hb = hemoglobin; Cyt = cytochromes; and NHI = nonheme iron.

from the trabecular bone found in the spine, wrist, and other parts of the forearm. At least some of the confusion in the earlier studies appears to have resulted from differences in the site used for measurement of bone density. Thus, the effect of high calcium intakes alone on slowing bone loss in postmenopausal women remains controversial at present. It is clear that elderly women should be getting the RDA for calcium in their diet. With the recent concern about the fat content of dairy products, calcium intakes in this group appear to be decreasing rather than increasing. Furthermore, even with estrogen replacement therapy, calcium intake should not be ignored. Recent studies have shown that with calcium intakes in the range of 1000–1500 mg day^{-1}, the effective dose of estrogen can be significantly reduced.

While the advertisements and much of the popular literature focus on calcium intake, we also need to remember that bones are not made of calcium alone. If the diet is deficient in other nutrients, the utilization of calcium for bone formation will be impaired. Vitamin C is needed to form the bone matrix and the macrominerals magnesium and phosphorus are an important part of bone structure. Recent research has also shown that a variety of trace minerals, including copper, zinc, manganese, and possibly boron, are important for bone formation. Thus, calcium supplements may be of little use if the overall diet is inadequate. Vitamin D deserves special mention since it may be a particular problem for the elderly (see Clin. Corr. 28.11). Finally, an adequate exercise program is just as important as estrogen replacement therapy and an adequate diet for preventing the loss of bone density.

Heaney, R. P. The Role of Nutrition in Prevention and Management of Osteoporosis. *Clin. Obstet. Gynecol.* 50:833, 1987; and Schaafsma, G., Van Berensteyn, E. C. H., Raymakers, J. A., and Dursma, S. A. Nutritional Aspects of Osteoporosis. *World Rev. Nutr. Diet.* 49:121, 1987.

Iron, of course, plays a number of important roles in the body. As a component of hemoglobin and myoglobin, it is required for O_2 and CO_2 transport. As a component of cytochromes and nonheme iron proteins, it is required for oxidative phosphorylation. As a component of the essential lysosomal enzyme myeloperoxidase, it is required for proper phagocytosis and killing of bacteria by neutrophils. The best-known symptom of iron deficiency is a microcytic hypochromic anemia and a deficiency is also associated with decreased immunocompetence.

Assuming a 10% efficiency of absorption, the Food and Nutrition Board has set a recommended dietary allowance of 10 mg day^{-1} for a normal adult male and 15 mg day^{-1} for a menstruating female (see Clin. Corr. 28.8). For pregnant women this allowance is raised to 30 mg day^{-1}. While 10 mg of iron can easily be obtained from a normal diet, 15 mg is marginal at best and 30 mg can almost never be obtained. The best dietary sources of iron are meats, dried legumes, dried fruits, and enriched cereal products.

Iron-deficiency anemia has long been considered the most prevalent nutritional disorder in the United States (Clin. Corr. 28.9). Young children need enough iron to allow for a continuing increase in blood volume as do pregnant women; menstruating women lose iron through blood loss. Thus, iron deficiency anemia is a problem for these groups. This is reflected in dietary surveys, which indicate that 95% or more of children and menstruating women are not obtaining adequate iron in their diet. It is also reflected in biochemical measurements of a 10–25% incidence of iron-deficiency anemia in this same group. Iron-deficiency anemia is also occasionally a problem with the elderly due to poor dietary intake and increased frequency of achlorhydria.

Because of the widespread nature of iron-deficiency anemia, government programs of nutritional intervention such as the WIC program have emphasized iron rich foods. A more extensive iron fortification of foods has also been discussed but there is a concern among some nutritionists that iron deficiency has been overemphasized. Since iron excretion is very limited, it is possible to build up toxic levels of iron. The excess iron leads to a condition called **hemochromatosis** in which iron deposits are found in abnormally high levels in many tissues. This can lead to liver, pancreatic, and cardiac dysfunction as well as pigmentation of the skin. This condition is usually only seen in hemolytic anemias and liver disease but concern has been voiced that if iron fortification of foods were to become more widespread, iron overload could become more prevalent.

Iodine Is Incorporated Into Thyroid Hormones

Dietary **iodine** is very efficiently absorbed and transported to the thyroid gland, where it is stored and used for the synthesis of the thyroid hormones triiodothyronine and thyroxine. These hormones play a major role in regulating the basal metabolic rate of the adult and the growth and development of the child. Saltwater fish are the best natural food sources of iodine and in earlier years population groups living in inland areas suffered from the endemic deficiency disease **goiter**. The most characteristic symptom of goiter is the enlargement of the thyroid gland to the point where a large nodule is visible on the neck. Since iodine has been routinely added to table salt, goiter has become relatively rare in this country. In some inland areas, however, mild forms of goiter may still be seen in up to 5% of the population. It is a possible deficiency for individuals not using iodinized salt provided that they have no other dietary sources of iodine.

CLIN. CORR. 28.8
CALCULATION OF AVAILABLE IRON

For many years the RDA for iron for women of childbearing age has been estimated at 18 mg day^{-1}, based on a need for 1.5–2.0 mg of absorbed iron and an average absorption of 10%. While this calculation made meal planning simple enough, it has long been regarded as unsatisfactory. While it is almost impossible to design a diet containing more than 6 mg of iron per 1000 kcal, most American women of childbearing age do not suffer from iron-deficiency anemia. In recognition of this fact, the RDA has recently been lowered to 15 mg day^{-1}. However, since the absorption of iron can vary from 2% to over 30% depending on the food source and need, even the revised RDA may not be completely adequate as a dietary standard. Another method that has been proposed is to calculate the actual amount of dietary iron available for absorption in the diet and compare that to the 1.5–2.0 mg needed. This calculation is based on the following data: (a) 40% of the iron in meat is heme iron and the efficiency of absorption of heme iron is 23%. This does not depend on the presence of other food factors. (b) All other dietary iron is nonheme iron, and absorption is dependent on other food, primarily the presence of ascorbic acid and/or some factor(s) present in meat.

Availability	Consumed in presence of	Absorption (%)
Low	<1 oz of meat or <25 mg of vitamin C	3
Medium	1–3 oz of meat or 25–75 mg of vitamin C	5
High	>3 oz of meat or >75 mg of vitamin C or >1 oz of meat + >25 mg of vitamin C	8

Calculations based on this information provide a much more rational estimate of dietary iron. For example, 3 oz of soybeans contain more than twice as much iron as a 3-oz steak. However, at 3% availability, they would provide only 60% as much available iron. By improving availability to 5%, one could obtain an equivalent amount of iron. This adds a new dimension to planning vegetarian diets. Not only should the protein be balanced and include ample quantities of iron rich sources (dried beans and dried fruit), but these meals should be consumed along with fresh fruit

Zinc Binds to Many Enzymes

Zinc absorption appears to be proportional to metaleothionein levels in the intestinal mucosa cells. The exact role of metallothoinein in zinc transport is uncertain but it may serve as a buffer for zinc ions as the metal transverses the intestinal cells. Over 100 zinc metalloenzymes have been described to date, including a number of regulatory proteins and both RNA and DNA polymerases. **Zinc deficiencies** in children are usually marked by poor growth and impairment of sexual development. In both children and adults zinc deficiencies result in poor wound healing. Zinc is also present in gustin, a salivary polypeptide, that appears to be necessary for normal development of taste buds. Thus zinc deficiencies also lead to decreased taste acuity.

The few dietary surveys that have been carried out in this country have indicated that zinc intake may be marginal for some individuals. However, few symptoms of zinc deficiency other than decreased taste acuity can be demonstrated in these individuals. Severe zinc deficiency is seen primarily in alcoholics, especially if they have developed cirrhosis, patients with chronic renal disease or severe malabsorption diseases and occasionally in patients on long-term parenteral nutrition (TPN). The most characteristic early symptom of zinc deficient patients on TPN is dermatitis. Zinc is occasionally used therapeutically to promote wound healing and may be of some use in treating gastric ulcers.

and/or fruit juices to provide enough vitamin C for efficient absorption. While 75 mg of vitamin C might be difficult to obtain in a single meal without supplementation, many fresh fruits and fruit juices will supply 50 mg.

These calculations are also useful in planning diets containing iron-fortified food products. While $FeSO_4$ taken on an empty stomach is well absorbed, it is very poorly utilized in some fortified foods. A serving of iron-fortified cereal with milk alone may actually provide very little available iron. Practical application of this type of information has helped solve problems such as the mysterious occurrence of iron-deficiency anemia in some children on "well-balanced" school meal programs. As long as an iron-fortified cereal was served at 8:00 *AM* and orange juice at 10:00 *AM*, there was a continued incidence of iron deficiency. By moving the orange juice to 8:00 *AM* with the cereal, the problem virtually disappeared.

Monsen, E. R., et al. Estimation of Available Dietary Iron. *Am. J. Clin. Nutr.* 31:134, 1978.

CLIN. CORR. **28.9**
IRON-DEFICIENCY ANEMIA

While iron-deficiency anemia is one of the most common nutritional deficiencies in this country, there is some confusion as to just how common it is. The problem is that iron depletion occurs very gradually over a period of time. Iron deficiency occurs in at least three distinct stages.

1. In the first stage, iron stores (usually as ferritin and hemosiderin) are depleted. Depleted tissue stores can be determined by measuring plasma ferritin levels.
2. Once the iron stores have been depleted, serum iron levels fall and the total iron binding capacity (TIBC) of transferrin increases. As the supply of iron for hemoglobin synthesis becomes limiting, red cell protoporphyrin levels increase as well.
3. In the final stage, hemoglobin levels fall and anemia becomes evident.

Iron deficiency anemia is associated with a decrease in hemoglobin concentration and a decrease in red cell size (mean corpuscular size, MCV).

Since there is a large range of normal hemoglobin concentrations in the population, a diagnosis of iron-deficiency anemia is rarely made on the basis of hemoglobin and MCV alone. The TIBC and serum ferritin assays are widely used in conjunction with hemoglobin and MCV to determine iron status and the red cell protoporphyrin assay is coming into increasing usage. The estimates for the incidence of iron deficiency anemia in various studies will depend on the biochemical assays used. Obviously, serum ferritin or TIBC will be more sensitive indicators of iron deficiency than hemoglobin or MCV.

Just how serious are the symptoms associated with iron-deficiency anemia? In most cases the symptoms are mild enough that the anemia is seldom the reason that the patient comes to see the physician. The most common symptoms are mild fatigue, weakness, and anorexia. These symptoms seem to be associated with depleted tissue stores rather than the anemia itself. As the iron-deficiency anemia becomes more severe, the fingernails become thin and flat with a characteristic spoon-shaped appearance (koilonychia). If these were the only possible symptoms of iron-deficiency anemia, a major public health effort to prevent iron deficiency would not appear to be warranted. However, there is the possibility that iron deficiency may lead to an increased susceptibility to infection. When tissue stores of iron are depleted, there is an impairment of cell mediated immunity and phagocytic activity. There also appear to be effects of iron deficiency on attention span and behavior in small children. Unfortunately, it is very difficult to correlate accurately learning deficits or the frequency of infection with the iron status in a human population. Since both possibilities exist and a significant portion of the population evidences iron deficiency by one measure or another, several major public health measures have been undertaken to improve iron availability in the diet. These include iron fortification of flour and, more recently, the Women and Infant Children (WIC) program.

Dallman, P. R. Biochemical Basis for the Manifestations of Iron Deficiency. *Ann. Rev. Nutr.* 6:13, 1986; and Lanzkowsky, P. Problems in Diagnosis of Iron Deficiency Anemia. *Pediatr. Ann.* 14:618, 1985.

Copper Functions as a Cofactor for Many Enzymes

Copper absorption may also be dependent on the protein metallothionein since excess intake of either copper or zinc interferes with the absorption of the other. Copper is contained in a number of important metalloenzymes, including cytochrome *c* oxidase, dopamine β-hydroxylase, superoxide dismutase, lysyl oxidase, and C_{18},Δ-9 desaturase. The lysyl oxidase is necessary for the conversion of certain lysine residues in collagen and elastin to allysine, which is needed for cross-linking. Some of the symptoms of **copper deficiency** in humans are **hypercholesterolemia,** leukopenia, demineralization of bones, anemia, fragility of large arteries, and demyelination of neural tissue. The anemia appears to be due to a defect in iron metabolism. The copper-containing enzyme ferroxidase is necessary for conversion of iron from the Fe^{2+} state, in which form it is absorbed, to the Fe^{3+} state, in which form it can bind to the plasma protein transferrin. The bone demineralization and blood vessel fragility can be directly traced to defects in collagen and elastin formation. The hypercholesterolemia may be related to increases in the ratio of saturated to monounsaturated fatty acids of the C_{16} and C_{18} series due to reduced activity of the C_{18},Δ-9 desaturase. The causes of the other symptoms are not known.

Copper balance studies carried out with human volunteers seem to indicate a minimum requirement of 1.0–2.6 mg day^{-1}. Thus the RDA has been set at 1.5–3 mg day^{-1}. Most dietary surveys find that the average American diet provides only 1 mg at less than or equal to 2000 cal day^{-1}. At present, this remains a puzzling problem. No symptoms of copper deficiency have been identified in the general public. It is not known whether there exist widespread marginal copper deficiencies, or whether the copper balance studies are inaccurate. Recognizable symptoms of copper deficiency are usually seen only in **Menke's syndrome,** a relatively rare hereditary disease. Menke's syndrome is associated with a defect in copper transport. **Wilson's disease,** on the other hand, is associated with abnormal accumulation of copper in various tissues and can be treated with the naturally occurring copper chelating agent penicillamine.

Chromium Is a Component of Glucose Tolerance Factor

Chromium probably functions in the body primarily as a component of **glucose tolerance factor (GTF),** a naturally occurring substance that appears to be a coordination complex between chromium, nicotinic acid, and the amino acids glycine, glutamate, cysteine, or glutathione. The GTF potentiates the effects of insulin presumably by facilitating its binding to cell receptor sites. The chief symptom of chromium deficiency is impaired glucose tolerance, a result of the decreased insulin effectiveness.

The frequency of occurrence of chromium deficiency is virtually unknown at present. The RDA for chromium has been set at 50–200 μg for a normal adult. The best current estimate is that the average consumption of chromium is around 30 μg day^{-1} in the United States. Unfortunately, the range of intakes is very wide (5–100 μg) even for individuals otherwise consuming balanced diets. Those most likely to have marginal or low intakes of chromium are individuals on low caloric intakes or consuming large amounts of processed foods. Some concern has been voiced that many Americans may be marginally deficient in chromium. However, it is difficult to assess the extent of this problem, if it exists, until better chromium analyses of food become available.

The situation is further confused by individual differences in chromium absorption and utilization. For some individuals, GTF appears to be a hormone-like substance in that they can utilize dietary chromium salts, niacin, and amino acids to synthesize GTF. Other individuals, however, utilize chromium salts very poorly and appear to need preformed GTF in the diet. Unfortunately, very little is known about the requirements for preformed GTF in the general population or about its distribution in natural foods. While it is clear that most diabetics do not respond significantly to either chromium or GTF, there are well-documented cases in which GTF has been useful in treating cases of diabetes.

Selenium Is a Scavenger of Peroxides

Selenium appears to function primarily in the metalloenzyme **glutathione peroxidase,** which destroys peroxides in the cytosol. Since the effect of vitamin E on peroxide formation is limited primarily to the membrane, both selenium and vitamin E appear to be necessary for efficient scavenging of peroxides. Selenium is one of the few nutrients not removed by the milling of flour and is usually thought to be present in adequate amounts in the diet. The selenium levels are very low in the soil in certain parts of the country and foods raised in these regions will be low in selenium. Fortunately, this effect is minimized by the current food distribution system, which assures that the foods marketed in any one area are derived from a number of different geographical regions.

Manganese, Molybdenum, and Fluoride Are Other Trace Minerals

Manganese is a component of pyruvate carboxylase and probably other metalloenzymes. **Molybdenum** is a component of xanthine oxidase and sulfite oxidase. Deficiencies of both of these trace minerals are virtually unknown in humans. **Fluoride** is known to strengthen bones and teeth and is usually added to drinking water.

28.10 THE AMERICAN DIET: FACT AND FALLACY

Much has been said lately about the supposed deterioration of the American diet. How serious a problem is this? Clearly Americans are eating much more processed food than their ancestors. These foods differ from simpler foods in that they have a higher caloric density and a lower nutrient density than the foods they replace. However, these foods are almost uniformly enriched with iron, thiamine, riboflavin, and niacin. In many cases they are even fortified (usually as much for sales promotion as for nutritional reasons) with as many as 11–15 vitamins and minerals. Unfortunately, it is simply not practical to replace all of the nutrients lost, especially the trace minerals. Imitation foods present a special problem in that they are usually incomplete in more subtle ways. For example, the imitation cheese and imitation milkshakes that are widely sold in this country usually do contain the protein and calcium one would expect of the food they replace, but often do not contain the riboflavin, which one should also obtain from these items. Fast food restaurants have also been much maligned in recent years. Some of the criticism has been undeserved, but fast food meals do tend to be high in calories and fat and low in certain vitamins and trace minerals. For example, the standard fast food meal provides over 50% of the calories the average adult needs for the entire day, while providing less than 5% of the vitamin A and less than 30% of biotin, folic acid, and pantothenic acid. Unfortunately, much of the controversy in recent years has centered around whether these trends are "good" or "bad." This simply obscures the issue at hand. Clearly it is possible to obtain a balanced diet that includes processed, imitation, and fast foods if one compensates by selecting foods for the other meals that are low in caloric density and rich in nutrients. However, without such compensation the "balanced diet" becomes a myth. Unfortunately, few nutritionists have taken the initiative in pointing out that such a compensation is both necessary and possible if one wishes to consume a balanced diet.

What do dietary surveys tell us about the adequacy of the American diet? The most comprehensive dietary survey presently available is the 1977–1978 Nationwide Food Consumption Survey carried out by the USDA, which analyzed 3-day dietary reports from 38,000 Americans. It showed that many Americans may be consuming suboptimal amounts of iron, calcium, vitamin A, vitamin B_6, and vitamin C. Less extensive dietary surveys have suggested that a significant fraction of the population might have inadequate intakes of folic acid and certain trace minerals. Data are not yet available from the third nationwide health and nutrition survey (NHANES-III) begun in 1988, but there is no reason to expect that it will be significantly different from previous nutrition surveys. How are these data to be interpreted? In every instance, biochemical measurements show significantly fewer individuals with marginal nutritional status and clinical symptoms of these deficiencies are rare indeed. Thus a physician need not be alarmed by reports of potential dietary deficiencies but should be aware of them when dealing with patients with increased nutrient requirements. Clearly, these dietary surveys tell us that the diet provides no surplus of most nutrients. Thus, if a patient has increased nutritional needs, it is unlikely that he or she will meet these needs from diet alone without proper nutritional counseling and/or dietary supplementation.

28.11 ASSESSMENT OF NUTRITIONAL STATUS IN CLINICAL PRACTICE

Having surveyed the major micronutrients and their biochemical roles, it might seem that the process of evaluating the nutritional status of an individual patient would be an overwhelming task. It is perhaps best to recognize that there are three factors that can add to nutritional deficiencies: poor diet, malabsorption, and increased nutrient need. Only when two or three components overlap in the same person (Figure 28.21) do the risks of symptomatic deficiencies become significant. For example, infants and young children have increased needs for iron, calcium, and protein. Dietary surveys show that many of them consume diets inadequate in iron and some consume diets that are low in calcium. Protein is seldom a problem unless the children are being raised as strict vegetarians (see Clin. Corr. 28.10). Thus, the chief nutritional concerns for most children are iron and calcium. Teenagers tend to consume diets low in calcium, magnesium, vitamin A, vitamin B_6, and vitamin C. Of all of these nutrients, their needs are particularly high for calcium and magnesium during the teenage years, so these are the nutrients of greatest concern. Young women are likely to consume diets low in iron, calcium, magnesium, vitamin B_6, folic acid, and zinc and all of these nutrients are needed in greater amounts during pregnancy and lactation. Adult women often consume diets low in calcium, yet they may have a particularly high need for calcium to prevent rapid bone loss. Finally, the elderly have unique nutritional needs (Clin. Corr. 28.11) and tend to have poor nutrient intake due to restricted income, loss of appetite, and loss of the ability or interest to prepare a wide variety of foods. They are also more prone to

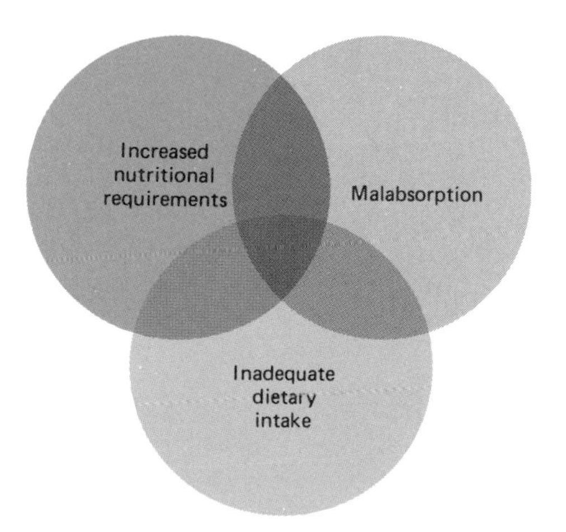

FIGURE 28.21
Factors affecting individual nutritional status.
Schematic representation of three important risk factors in determining nutritional status. A person on the periphery would have very low risk of any nutritional deficiency, whereas people within the interior areas would be much more likely to experience some symptoms of nutritional deficiencies.

CLIN. CORR. 28.10
NUTRITIONAL CONSIDERATIONS FOR VEGETARIANS

A vegetarian diet poses certain problems in terms of micronutrient intake that need to be recognized in designing a well-balanced diet. Vitamin B_{12} is of special concern, since it is found only in foods of animal origin. Vitamin B_{12} should be obtained from fortified foods (such as some brands of soybean milk) or in tablet form. However, surprisingly few vegetarians ever develop pernicious anemia, perhaps because an adult who has previously eaten meat will have a 6–10-year store of B_{12} in their liver.

Iron is another problem. The best vegetable sources of iron are dried beans, dried fruits, whole grain or enriched cereals, and green leafy vegetables. Vegetarian diets can provide adequate amounts of iron provided that these foods are regularly selected and consumed with vitamin C rich foods to promote iron absorption (Clin. Corr. 28.9). However, iron supplementation is usually recommended for children and menstruating women.

When milk and dairy products are absent from the diet, certain other problems must be considered as well. Normally, dietary vitamin D is obtained primarily from fortified milk. While some butters and margarines are fortified with vitamin D, they are seldom consumed in sufficient quantities to supply significant amounts of vitamin D. Although adults can usually obtain sufficient vitamin D from exposure to sunlight, dietary sources are often necessary during periods of growth and for adults with little exposure to sunlight. Vegetarians may need to obtain their vitamin D from fortified foods such as cereals, certain soybean milks, or in tablet form. Riboflavin is found in a number of vegetable sources such as green leafy vegetables, enriched breads, and wheat germ. However, since none of these sources supply more than 10% of the RDA in normal serving sizes, fortified cereals or vitamin supplements may become an important source of this nutrient. The important sources of calcium for vegetarians include soybeans, soybean milk, almonds, and green leafy vegetables. Those green leafy vegetables without oxalic acid (mustard, turnip, and dandelion greens, collards, kale, romaine, and loose leaf lettuce) are particularly good sources of calcium. None of these sources, however, is equivalent to cow's milk in calcium content, so calcium supplements are usually recommended during periods of rapid growth.

CLIN. CORR. 28.11
NUTRITIONAL NEEDS OF THE ELDERLY

If current trends continue one out of five Americans will be over the age of 65 by the year 2030. With this projected aging of the American population, there has been increased interest in defining the nutritional needs of the elderly.

Recent research shows altered needs for several essential nutrients. For example, the absorption and utilization of vitamin B_6 has been shown to decrease with age. Dietary surveys have consistently shown that B_6 is a problem nutrient for many Americans and the elderly appear to be no exception. Many older Americans get less than 50% of the RDA for B_6 from their diet. While clinical symptoms of B_6 deficiency are still rare in the elderly population, biochemical tests of B_6 status indicate that marginal deficiencies may be fairly widespread. Vitamin B_{12} deficiency is also more prevalent in the elderly. Many older adults develop a condition called atrophic gastritis, which results in decreased acid production in the stomach. That along with a tendency towards decreased production of intrinsic factor leads to poor absorption of B_{12}. Vitamin D can be a problem as well. Many elderly do not spend much time in the sunlight and to make matters worse both the conversion of 7-dehydrocholesterol to vitamin D in the skin and 25-(OH)D to 1,25-$(OH)_2$D in the kidney decrease with age. These factors often combine to produce significant deficiencies of 1,25-$(OH)_2$D in the elderly, which can in turn lead to negative calcium balance. These changes do not appear to be the primary cause of osteoporosis but they certainly may contribute to it.

There is some evidence for increased need for chromium and zinc as well. Chromium is not particularly abundant in the American diet and many elderly appear to have difficulty converting dietary chromium to the biologically active glucose tolerance factor. The clinical relevance of these observations is not clear but chromium deficiency could contribute to adult-onset diabetes. Similarly, dietary surveys show that most elderly consume between one-half and two-thirds the RDA for zinc. Conditions such as atrophic gastritis can also interfere with zinc absorption. Symptoms of zinc deficiency include loss of taste acuity, dermatitis, and a weakened immune system. All of these symptoms are common in the elderly population and it has been suggested that zinc deficiency might contribute.

To make matters worse caloric needs decrease with age. Thus, older Americans must eat better quality food. This is further compounded by socioeconomic factors such as decreased income, loss of mobility, and isolation that often result in poorer food choices. Finally, drug–nutrient interactions can become particularly significant in the elderly because they are often taking multiple medications.

Not all of the news is bad, however. Atrophic gastritis causes an overgrowth of folate-producing bacteria in the intestine. Thus, many elderly appear to have reduced folate requirements. Vitamin A absorption actually increases as we age and the ability of the liver to clear vitamin A from the blood decreases, so it remains in the circulation for a longer time. In fact, not only does the need for vitamin A decrease as we age, but the elderly also need to be particularly careful to avoid vitamin A toxicity. While this does not restrict their choice of foods or multivitamin supplements, they should generally avoid separate vitamin A supplements.

Munro, H. N., Suter, P. M., and Rusell, R. M. Nutritional Requirements of the Elderly. *Ann. Rev. Nutr.* 7:23, 1987; and Suter, P. M. and Rusell, R. M. Vitamin Requirements of the Elderly. *Am. J. Clin. Nutr.* 45:501, 1987.

TABLE 28.1 Drug–Nutrient Interactions

Drug	Potential nutrient deficiencies
Alcohol	Thiamine
	Folic acid
	Vitamin B_6
Anticonvulsants	Vitamin D
	Folic acid
	Vitamin K
Cholestyramine	Fat soluble vitamins
	Iron
Corticosteroids	Vitamin D and calcium
	Zinc
	Potassium
Diuretics	Potassium
	Zinc
Isoniazid	Vitamin B_6
Oral contraceptives and estrogens	Vitamin B_6
	Folic acid and B_{12}

suffer from malabsorption problems and to use multiple prescription drugs that increase nutrient needs (Table 28.1).

Illness and metabolic stress often cause increased demand or decreased utilization of certain nutrients. For example, diseases leading to fat malabsorption cause a particular problem with absorption of calcium and the fat-soluble vitamins. Other malabsorption diseases can result in deficiencies of many nutrients depending on the particular malabsorption disease. Liver and kidney disease can prevent activation of vitamin D and storage or utilization of many other nutrients including vitamin A, vitamin B_{12}, and folic acid. Severe illness or trauma increase the need for calories, protein, and possibly some micronutrients such as vitamin C and certain B vitamins. Long-term use of many drugs in the treatment of chronic disease states can affect the need for certain micronutrients. Some of these are summarized in Table 28.1.

Who then is at a nutritional risk? Obviously, this depends on many factors. Nutritional counseling will be an important part of the treatment for infants, young children, and pregnant–lactating women. A brief analysis of a dietary history and further nutritional counseling will also be important when dealing with many other high-risk patients.

BIBLIOGRAPHY

Recommended Dietary Allowances

Food and Nutrition Board of the National Academy of Sciences. *Recommended Daily Allowances*. 10th ed. Washington, DC: National Academy of Sciences, 1989.

Micronutrients–General Information

The Nutrition Foundation. *Present Knowledge in Nutrition,* 5th ed. New York, 1985.

Symposium on Metal Metabolism and Disease. *Clin. Physiol. Biochem.* 4:1, 1986.

Vitamin A

Goodman, D. S. Vitamin A and Retinoids in Health and Disease. *N. Eng. J. Med.* 310:1023, 1984.

Wolf, G. Multiple Functions of Vitamin A. *Physiol. Rev.* 64:873, 1984.

Vitamin D

DeLuca, H. F. The Vitamin D Story. A Collaborative Effort of Basic Science and Clinical Medicine. *FASEB J.* 2:224, 1988.

Vitamin E

Bieri, J. G., Coresh, L., and Hubbard, V. S. Medical Uses of Vitamin E. *N. Engl. J. Med.* 308:1063, 1983.

Howitt, M. K. Vitamin E: A Reexamination. *Am. J. Clin. Nutr.* 29:569, 1976.

Vitamin B_6

Merril, A. H. Jr. and Henderson, J. M. Diseases Associated with Defects in Vitamin B_6 Metabolism or Utilization. *Ann. Rev. Nutr.* 7:137, 1987.

Folate

Blakely, R. L. and Whitehead, V. M. (Eds.). *Folate and Pterins,* Vol. 3. New York: Wiley, 1986.

Vitamin C

Third International Conference on Vitamin C. *Ann. N.Y. Acad. Sci.* 498:1, 1987.

Calcium

Parrot-Garcia, M. and McCarron, D. A. Calcium and Hypertension. *Nutr. Rev.* 42:205, 1984.

Chromium

Saner, G. *Chromium in Nutrition and Disease*. New York: Alan R. Liss, 1980.

Dietary Surveys

Pao, E. M. and Mickle, S. J. Problem Nutrients in the United States. *Food Technol.* 35:58, 1981.

Rizek, R. L. and Posati, L. Continuing Survey of Food Intakes by Individuals. *Fam. Econ. Rev.* 1:16, 1985.

USDA Human Nutrition Information Service. Nationwide Food Consumption Survey, Continuing Survey of Food Intakes by Individuals, Women 19–50 Years and Their Children 1–5 Years, 4 Days, 1985. NFCS, CSFII Report No. 85-4, 1987, pp. 182.

Yetty, E. and Johnson, C. Nutritional Applications of the Health and Nutritional Examination Surveys (HANES). *Ann. Rev. Nutr.* 7:441, 1987.

QUESTIONS

J. BAGGOTT AND C. N. ANGSTADT

*Q*uestion *T*ypes are described on page xxiii.

1. (QT1) The effects of vitamin A may include all of the following EXCEPT:
   A. prevention of anemia.
   B. serving as an antioxidant.
   C. cell differentiation.
   D. the visual cycle.
   E. induction of certain cancers.

A. Visual cycle C. Both
B. Cholecalciferol metabolism D. Neither

2. (QT4) Kidney plays a critical role.

A. Fat-soluble vitamins C. Both
B. Water-soluble vitamins D. Neither

3. (QT4) Can generally be synthesized in adequate amounts by healthy people who have an adequate protein–calorie intake.

A. Vitamin A
B. Biotin
C. Niacin
D. Vitamin D
E. Vitamin B_{12} (cobalamine)

4. (QT5) Requirement may be totally supplied by intestinal bacteria.
5. (QT5) Precursor is synthesized by green plants.
6. (QT5) Tryptophan is a precursor.
7. (QT5) Deficiency may be seen in long-term adherence to a strict vegetarian diet.
8. (QT5) Is required for normal regulation of calcium metabolism.

9. (QT2) Ascorbic acid:
   1. is a biological reducing agent.
   2. is absorbed with difficulty.
   3. aids in iron absorption.
   4. is harmless in high doses.
10. (QT2) In assessing the adequacy of a person's diet:
    1. intake of the RDA (recommended dietary allowance) of every nutrient assures adequacy.
    2. a drop in the plasma concentration of a nutrient is a clear sign of an increased requirement.
    3. a 24-h dietary intake history provides an adequate basis for making a judgment.
    4. currently administered medications must be considered.

A. Calcium
B. Iron
C. Iodine
D. Copper
E. Selenium

11. (QT5) Absorption is inhibited by excess dietary zinc.
12. (QT5) Excess dietary protein causes rapid excretion.
13. (QT5) Risk of nutritional deficiency is high in young children.
14. (QT5) Unsupplemented diets of populations living in inland areas may be deficient.
15. (QT5) Essential component of glutathione peroxidase.

ANSWERS

1. E Vitamin A *deficiency* is linked to increased susceptibility to certain cancers. A. Retinyl phosphate serves as a glycosyl donor in the synthesis of certain glycoproteins (p. 1119), including transferrin (p. 1120). B. See p. 1119. C. Retinol and retinoic acid may function like steroid hormones (p. 1119). D. Retinol cycles between the Δ^{11}-cis and all-trans forms (p. 1119).
2. B Here 25-hydroxycholecalciferol is converted to the active 1,25-dihydroxy compound (p. 1121).
3. D Although some niacin is synthesized in the body (p. 1129), it is not synthesized in adequate amounts. When there is sufficient exposure to sunlight for the conversion of 7-dehydrocholesterol to cholecalciferol, vitamin D is not a vitamin, but a hormone (p. 1121).
4. B See p. 1132.
5. A β-Carotene, from green plants, is converted to vitamin A (p. 1118).
6. C See p. 1129.
7. E This vitamin is from animal sources (p. 1136).
8. D 1,25-Dihydroxyvitamin D is required for calcium absorption and, along with parathyroid hormone, regulates bone reabsorption and calcium excretion.
9. B 1 and 3 true. 1, 2, and 3: See p. 1136. 4: See p. 1137.
10. D Only 4 true. 1: RDAs meet the known needs of most healthy people. Individual variations may occur, and certainly stress or disease can be expected to change some requirements. 2: This could reflect a mere redistribution, not deficiency. 3: Can you be sure that any 24-h diet history is either accurate or representative of the individual's typical diet? (p. 1116) 4: Isoniazid and penicillamine affect B_6 (p. 1132), and antacids interfere with phosphate absorption (p. 1138).
11. D See p. 1142.
12. See p. 1138.
13. B Rapid growth in children causes high demands for iron (p. 1140).
14. C The problem is rare in the United States due to the common use of iodized salt (p. 1140).
15. E See p. 1143.

APPENDIX

Review of Organic Chemistry

CAROL N. ANGSTADT

FUNCTIONAL GROUPS

Alcohols

The general formula of **alcohols** is R—OH, where R equals an alkyl or aryl group. They are classified as *primary, secondary,* or *tertiary,* according to whether the hydroxyl (OH) bearing carbon is bonded to no carbon or one, two, or three other carbon atoms:

$$\underset{\text{Primary}}{(\text{H or C})\text{—}\overset{\text{H}}{\underset{\text{H}}{\text{C}}}\text{—OH}} \qquad \underset{\text{Secondary}}{\text{—}\overset{|}{\underset{|}{\text{C}}}\text{—}\overset{\text{—C—}}{\underset{\text{H}}{\text{C}}}\text{—OH}} \qquad \underset{\text{Tertiary}}{\text{—}\overset{|}{\underset{|}{\text{C}}}\text{—}\overset{\text{—C—}}{\underset{\text{—C—}}{\text{C}}}\text{—OH}}$$

Aldehydes and Ketones

Aldehydes and ketones contain a carbonyl group $\text{—}\overset{\text{O}}{\overset{\|}{\text{C}}}\text{—}$; *aldehydes* are $\text{R—}\overset{\text{O}}{\overset{\|}{\text{C}}}\text{—H}$, and a *ketone* has two alkyl groups at the carbonyl group $\text{R—}\overset{\text{O}}{\overset{\|}{\text{C}}}\text{—R}'$.

Acids and Acid Anhydrides

Carboxylic acids contain the functional group $\text{—}\overset{\text{O}}{\overset{\|}{\text{C}}}\text{—OH}$ (—COOH). Dicarboxylic and tricarboxylic acids contain two or three carboxyl groups. A carboxylic acid ionizes in water to a negatively charged carboxylate ion:

$$\underset{\text{Carboxylic acid}}{\text{R—}\overset{\text{O}}{\overset{\|}{\text{C}}}\text{—OH}} \longrightarrow \underset{\text{Carboxylate ion}}{\text{R—}\overset{\text{O}}{\overset{\|}{\text{C}}}\text{—O}^- + \text{H}^+}$$

Names of carboxylic acids usually end in -ic and the carboxylate ion in -ate. **Acid anhydrides** are formed when two molecules of acid react with loss of a molecule of water. An acid anhydride may form between two organic acids, two inorganic acids, or an organic and an inorganic acid:

$$\underset{\text{Organic anhydride}}{\text{R—}\overset{\text{O}}{\overset{\|}{\text{C}}}\text{—O—}\overset{\text{O}}{\overset{\|}{\text{C}}}\text{—R}} \qquad \underset{\text{Inorganic anhydride}}{\text{HO—}\overset{\text{O}}{\overset{\|}{\underset{\text{OH}}{\text{P}}}}\text{—O—}\overset{\text{O}}{\overset{\|}{\underset{\text{OH}}{\text{P}}}}\text{—OH}}$$

$$\underset{\text{Organic–inorganic anhydride}}{\text{R—}\overset{\text{O}}{\overset{\|}{\text{C}}}\text{—O—}\overset{\text{O}}{\overset{\|}{\underset{\text{OH}}{\text{P}}}}\text{—OH}}$$

Esters

Esters form in the reaction between a carboxylic acid and an alcohol:

$$\text{R—COOH} + \text{R}'\text{—OH} \longrightarrow \text{R—}\overset{\text{O}}{\overset{\|}{\text{C}}}\text{—OR}' + H_2O$$

Esters may form between an inorganic acid and an organic alcohol, for example, glucose 6-phosphate.

Hemiacetals, Acetals, and Lactones

A reaction between an aldehyde and an alcohol gives a **hemiacetal,** which may react with another molecule of alcohol to form an *acetal:*

$$\text{R—CHO} \xrightarrow{\text{R}'\text{—OH}} \underset{\text{Hemiacetal}}{\text{R—}\overset{\text{OH}}{\underset{\text{H}}{\text{C}}}\text{—OR}'} \xrightarrow{\text{R}''\text{—OH}} \underset{\text{Acetal}}{\text{R—}\overset{\text{OR}''}{\underset{\text{H}}{\text{C}}}\text{—OR}'}$$

Lactones are cyclic esters formed when an acid and an alcohol group on the same molecule react and usually requires that a five- or six-membered ring be formed.

Unsaturated Compounds

Unsaturated compounds are those containing one or more carbon–carbon multiple bonds, for example, a double bond: —C═C—

Amines and Amides

Amines, $R—NH_2$, are organic derivatives of NH_3 and are classified as *primary, secondary,* or *tertiary,* depending on the number of alkyl groups (R) bonded to the nitrogen. When a fourth substituent is bonded to the nitrogen, the species is positively charged and called a *quaternary ammonium ion:*

$R—NH_2$	R—NH—R′	R—NR″—R′	R—N⁺(H (or R‴))(R″)—R′
Primary amine	Secondary amine	Tertiary amine	Quaternary ammonium ion

Amides contain the functional group —C(═O)—N(H)—X; where X can be H (simple) or R (*N* substituted). The carbonyl group is from an acid, and the *N* is from an amine. If both functional groups are from amino acids, the amide bond is referred to as a **peptide bond.**

TYPES OF REACTIONS

Nucleophilic Substitutions at an Acyl Carbon

R(L)C═O: + :X—H ⇌ R—C(XH⁺)(L)—O:⁻ ⇌ R—C(X)(LH⁺)—O:⁻ ⟶ X(R)C═O + LH

New compound   Leaving group

If the acyl carbon is on a carboxylic group, the leaving group is water. Nucleophilic substitution on carboxylic acids usually requires a catalyst or conversion to a more reactive intermediate; biologically this occurs via enzyme catalysis. X—H may be an alcohol (R—OH), ammonia, amine ($R—NH_2$), or another acyl compound. Types of nucleophilic substitutions include *esterification, peptide bond formation,* and *acid anhydride* formation.

Hydrolysis and Phosphorolysis Reactions

Hydrolysis is the cleavage of a bond by water:

R—C(═O)—OR′ + H_2O ⇌ R—C(═O)—OH + R′—OH

Hydrolysis is often catalyzed by either acid or base. *Phosphorolysis* is the cleavage of a bond by inorganic phosphate:

Glucose-glucose + HO—P(═O)(O⁻)—O⁻ ⟶ Glucose 1-phosphate + glucose

Oxidation–Reduction Reactions

Oxidation is the loss of electrons; **reduction** is the gain of electrons. Examples of oxidation are as follows:

1. Fe^{2+} + acceptor ⟶ Fe^{3+} + acceptor · e^-
2. S(ubstrate) + O_2 + DH_2 ⟶ S—OH + H_2O + D
3. S—H_2 + acceptor ⟶ S + acceptor · H_2

Some of the group changes that occur on oxidation–reduction are

1. —CH_2OH ⇌ —C(H)═O
2. >C—OH ⇌ >C═O
3. —C(H)═O ⇌ —C(═O)—OH
4. —CH_2NH_2 ⇌ —C(H)═O + NH_3
5. —CH_2—CH_2— ⇌ —CH═CH—

STEREOCHEMISTRY

Stereoisomers are compounds with the same molecular formulas and order of attachment of constituent atoms but with different arrangements of these atoms in space.

Enantiomers are stereoisomers in which one isomer is the mirror image of the other and requires the presence of a chiral atom. A *chiral carbon* (also called an **asymmetric carbon**) is one that is attached to four different groups:

D◄C►A (B above, E below)   A◄C►D (B above, E below)

Enantiomers

Enantiomers will be distinguished from each other by the designations *R* and *S* or *D* and *L*. The maximum number of stereoisomers possible is 2^n, where n is the number of chiral carbon atoms. A molecule with more than one chiral center will be an achiral molecule if it has a point or plane of symmetry.

Diastereomers are stereoisomers that are not mirror images of each other and need not contain chiral atoms. **Epimers** are diastereomers that contain more than one chiral carbon and differ in configuration about *only one* asymmetric carbon.

Anomers are a special form of carbohydrate epimers in which the difference is specifically about the anomeric carbon (see p. 1152). Diastereomers can also occur with molecules in which there is restricted rotation about carbon–carbon bonds. Double bonds exhibit **cis–trans isomerism.** The double bond is in the cis-configuration if two similar groups are on the same side and is trans if two similar groups are on opposite sides. Fused ring systems, such as those found in steroids (see p. 1156), also exhibit cis–trans isomerism.

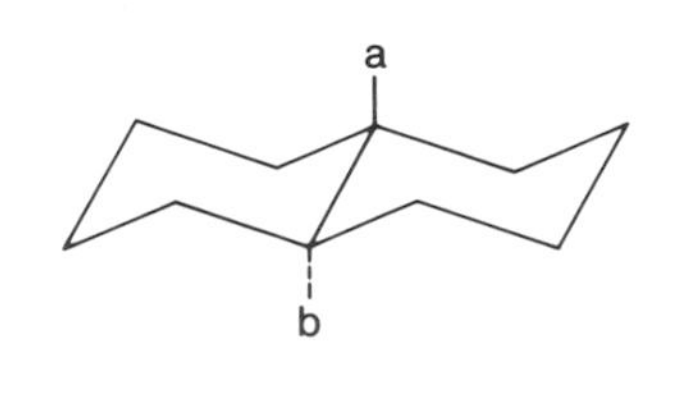

trans-rings

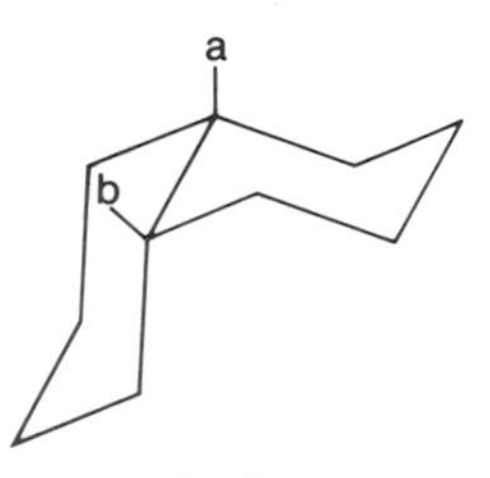

cis-rings

TYPES OF FORCES INVOLVED IN MACROMOLECULAR STRUCTURES

A **hydrogen bond** is a dipole–dipole attraction between a hydrogen atom attached to an electronegative atom and a nonbonding electron pair on another electronegative atom:

$$\underset{\delta^-}{:\ddot{X}}\text{—}\underset{\delta^+}{H}\ \cdots\cdots\ \underset{\delta^-}{:\ddot{X}}\text{—}\underset{\delta^+}{H}$$

Hydrogen bonds of importance in macromolecular structures occur between two nitrogen atoms, two oxygen atoms, or an oxygen and a nitrogen atom.

A **hydrophobic interaction** is the association of nonpolar groups in a polar medium. *van der Waals* forces consist of dipole and induced dipole interactions between two nonpolar groups. A nonpolar residue dissolved in water induces a highly ordered, thermodynamically unfavorable, solvation shell. Interaction of nonpolar residues with each other, with the exclusion of water, increases the entropy of the system and is thermodynamically favorable.

Ionic (electrostatic) interactions between charged groups can be attractive if the charges are of opposite signs or repulsive if they are of the same sign. The strength of an electrostatic interaction in the interior of a protein molecule may be high. Most charged groups on the surface of a protein molecule interact with water rather than with each other.

A **disulfide bond** (S—S) is a covalent bond formed by the oxidation of two sulfhydryl (SH) groups.

CARBOHYDRATES

Carbohydrates are polyhydroxy aldehydes or ketones or their derivatives. **Monosaccharides** (simple sugars) are those carbohydrates that cannot be hydrolyzed into simpler compounds. The name of a type of monosaccharide includes the type of function, a Greek prefix indicating the number of carbon atoms, and the ending -ose; for example, *aldohexose* is a six-carbon aldehyde and *ketopentose* a five-carbon ketone. Monosaccharides may react with each other to form larger molecules. With fewer than eight monosaccharides, either a Greek prefix indicating the number or the general term *oligosaccharide* may be used. **Polysaccharide** refers to a polymer with more than eight monosaccharides. Oligo- and polysaccharides may be either homologous or mixed.

Most *monosaccharides* are asymmetric, an important consideration since enzymes usually work on only one isomeric form. The simplest carbohydrates are glyceraldehyde and dihydroxyacetone whose structures, shown as Fischer projections, are as follows:

```
H—C=O            H—C=O           CH2OH
  |                |               |
H—C—OH          HO—C—H            C=O
  |                |               |
  CH2OH            CH2OH           CH2OH
```

D-Glyceraldehyde   L-Glyceraldehyde   Dihydroxyacetone

D-Glyceraldehyde may also be written as follows:

```
     H
    /
   C=O
   |
H◄─C─◄OH
   |
   CH2OH
```

In the Cahn–Ingold–Prelog system, the designations are (*R*) (rectus; right) and (*S*) (sinister, left).

The configuration of monosaccharides is determined by the stereochemistry at the asymmetric carbon furthest from the carbonyl carbon (number 1 for an aldehyde; lowest possible number for a ketone). Based on the *position* of the OH on the highest number asymmetric carbon, a monosaccharide is D if the OH projects to the *right* and L if it projects to the *left*. The D and L monosaccharides with the same name are *enantiomers*, and the substituents on all asymmetric carbon atoms are reversed as in

```
  H—C=O          H—C=O
    |              |
  H—C—OH        HO—C—H
    |              |
 HO—C—H          H—C—OH
    |              |
  H—C—OH        HO—C—H
    |              |
  H—C—OH        HO—C—H
    |              |
    CH2OH          CH2OH
```

D-Glucose   L-Glucose

Epimers (e.g., glucose and mannose) are stereoisomers that differ in the configuration about *only one* asymmetric carbon. The relationship of OH groups to *each other* determines the specific monosaccharide. Three aldohexoses and three pentoses of importance are

```
  H—C=O          H—C=O          H—C=O
    |              |              |
  H—C—OH        HO—C—H          H—C—OH
    |              |              |
 HO—C—H         HO—C—H         HO—C—H
    |              |              |
  H—C—OH         H—C—OH        HO—C—H
    |              |              |
  H—C—OH         H—C—OH         H—C—OH
    |              |              |
    CH2OH          CH2OH          CH2OH
```

D-Glucose   D-Mannose   D-Galactose

```
  H—C=O            CH2OH           CH2OH
    |              |               |
  H—C—OH           C=O             C=O
    |              |               |
  H—C—OH         H—C—OH         HO—C—H
    |              |               |
  H—C—OH         H—C—OH          H—C—OH
    |              |               |
    CH2OH          CH2OH           CH2OH
```

D-Ribose   D-Ribulose   D-Xylulose

Fructose, a ketohexose, differs from glucose only on carbon atoms 1 and 2:

```
   1 CH2OH
     |
   2 C=O
     |
HO—3 C—H
     |
 H—4 C—OH
     |
 H—5 C—OH
     |
   6 CH2OH
```

Five- and six-carbon monosaccharides form **cyclic hemiacetals** or *hemiketals* in solution. A new asymmetric carbon is generated so two isomeric forms are possible:

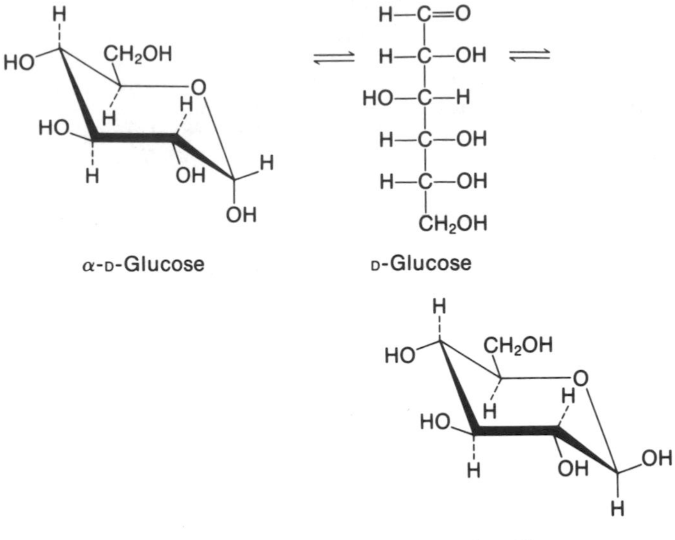

α-D-Glucose   D-Glucose

β-D-Glucose

Both five (furanose)- and six-membered (pyranose) ring structures are possible, although pyranose rings are more common. A furanose ring is written as follows:

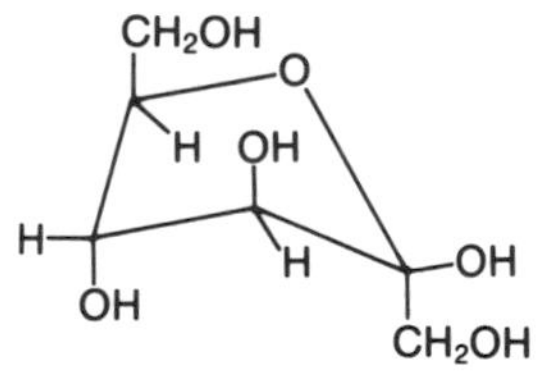

β-D-Fructose

The isomer is designated α if the OH group and the CH_2OH group on the two carbon atoms linked by the oxygen are trans to each other and β if they are cis. The hemiacetal or hemiketal forms may also be written as modified *Fischer projection formulas:* α if OH on the acetal or ketal carbon projects to the same side as the ring and β if on the opposite side:

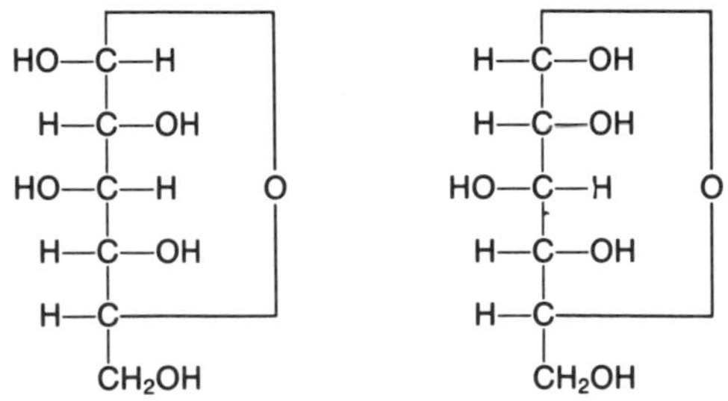

β-D-Glucose   α-D-Glucose

Haworth formulas are used most commonly:

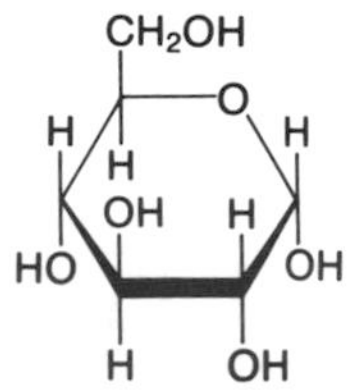

α-D-Glucose 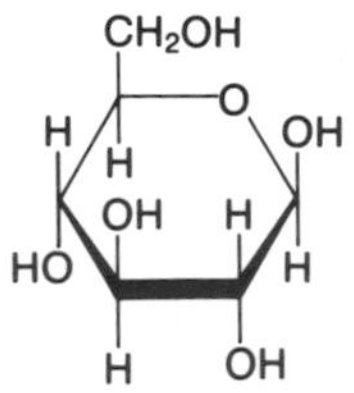   β-D-Glucose

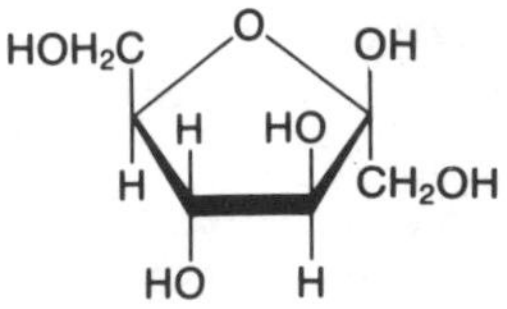

β-D-Fructose

The ring is perpendicular to the plane of the paper with the oxygen written to the back (upper) right, C-1 to the right, and substituents above or below the plane of the ring. The OH at the acetal or ketal carbon is below in the α isomer and above in the β. Anything written to the right in the Fischer projection is written down in the Haworth formula.

In this text, a *modified Haworth formula,* not indicating ring orientation, will be used:

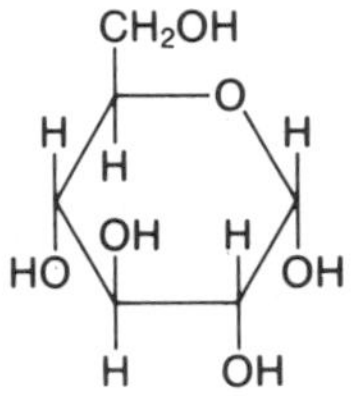

α-D-Glucose

The α and β forms of the same monosaccharide are special forms of epimers called *anomers,* differing only in the configuration about the anomeric (acetal or ketal) carbon. Monosaccharides exist in solution primarily as a mixture of the hemiacetals (or hemiketals) but react chemically as aldehydes or ketones. *Mutarotation* is the equilibration of α and β forms through the free aldehyde or ketone. Substitution of the H of the anomeric OH prevents mutarotation and fixes the configuration in either the α or β form.

Monosaccharide Derivatives

A **deoxymonosaccharide** is one in which an OH has been replaced by H. In biological systems, this occurs at C-2 unless otherwise indicated. An **amino monosaccharide** is one in which an OH has been replaced by NH_2, again at C-2 unless otherwise specified. The amino group of an amino sugar may be *acetylated:*

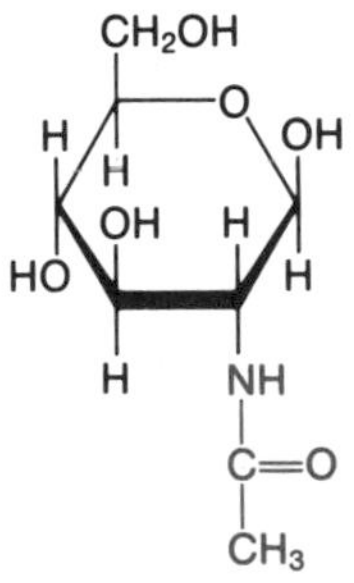

β-*N*-Acetylglucosamine

An aldehyde is reduced to a primary and a ketone to a secondary **monosaccharide alcohol** (**alditol**). Alcohols are named with the base name of the sugar plus the ending *-itol* or with a trivial name (glucitol = sorbitol). Monosaccharides that differ around only two of the first three carbon atoms yield the same alditol. D-Glyceraldehyde and dihydroxyacetone give glycerol:

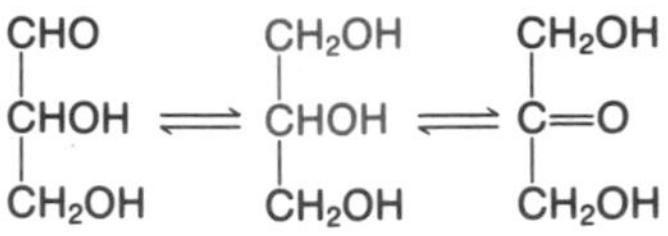

$CHO-CHOH-CH_2OH \rightleftharpoons CH_2OH-CHOH-CH_2OH \rightleftharpoons CH_2OH-C{=}O-CH_2OH$

D-Glucose and D-fructose give D-sorbitol; D-fructose and D-mannose give D-mannitol. Oxidation of the terminal CH_2OH, but not of the CHO, yields a **-uronic acid,** a *monosaccharide acid:*

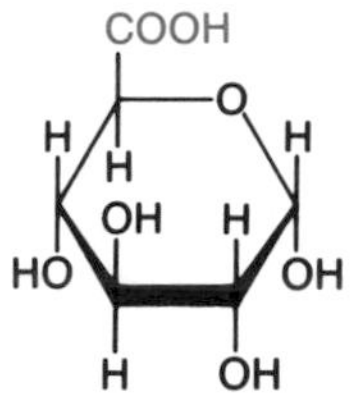

D-Glucuronic acid

Oxidation of the CHO, but not the CH_2OH, gives an **-onic acid:**

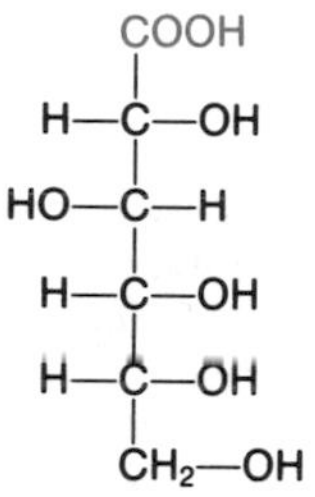

D-Gluconic acid

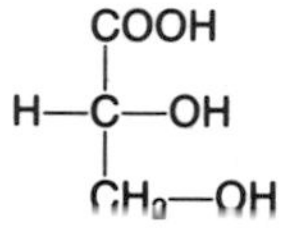

D-Glyceric acid

Oxidation of both the CHO and CH_2OH gives an **-aric acid:**

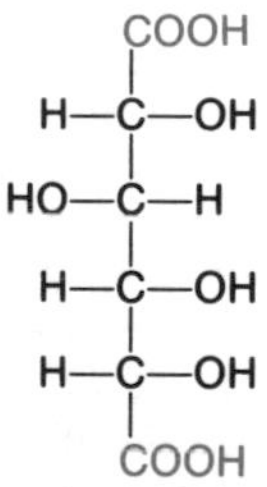

D-Glucaric acid

Ketones do not form acids. Both -onic and -uronic acids can react with an OH in the same molecule to form a **lactone** (see p. 1150):

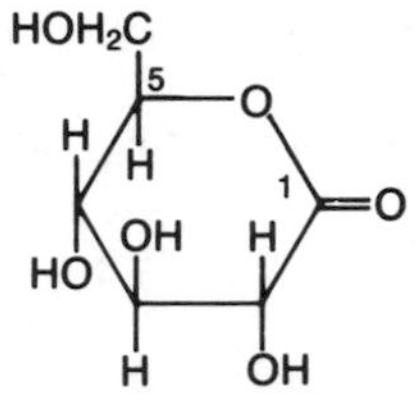

D-Glucono-S-lactone

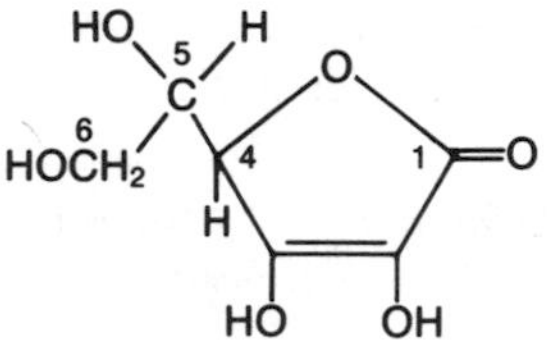

L-Ascorbic acid (Derivative of L gulose)

Reactions of Monosaccharides

The most common *esters* of monosaccharides are phosphate esters at carbon atoms 1 and/or 6:

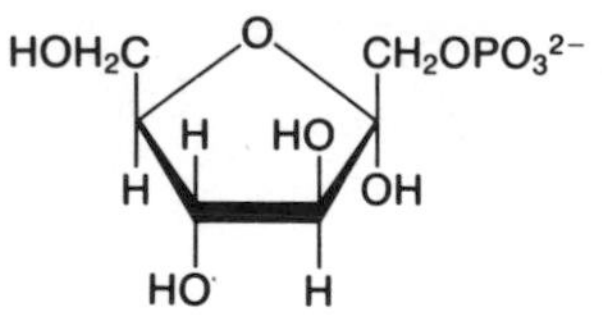

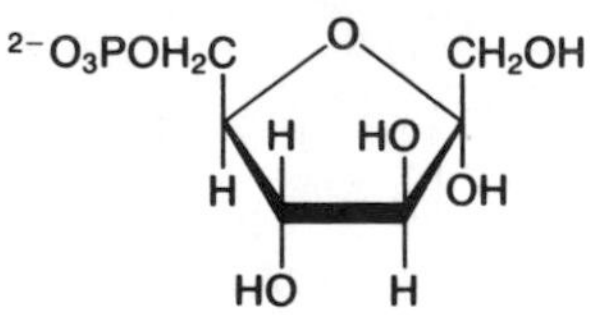

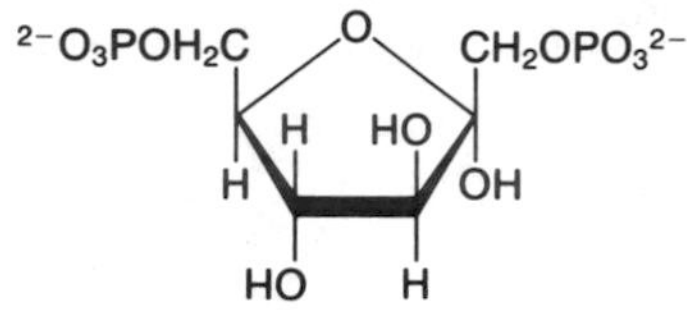

To be a **reducing sugar,** mutarotation must be possible. In alkali, enediols form that may migrate to 2,3 and 3,4 positions:

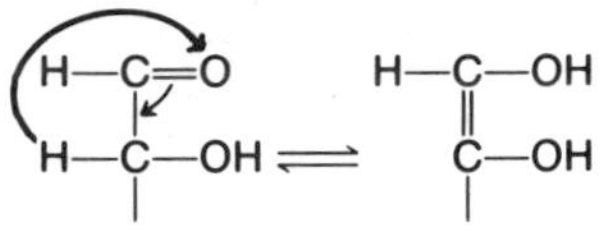

Enediols may be oxidized by O_2, Cu^{2+}, Ag^+, Hg^{2+}. Reducing ability is more important in the laboratory than physiologically. A hemiacetal or hemiketal may react with the OH of another monosaccharide to form a disaccharide (*acetal*: *glycoside*) (see p. 1154):

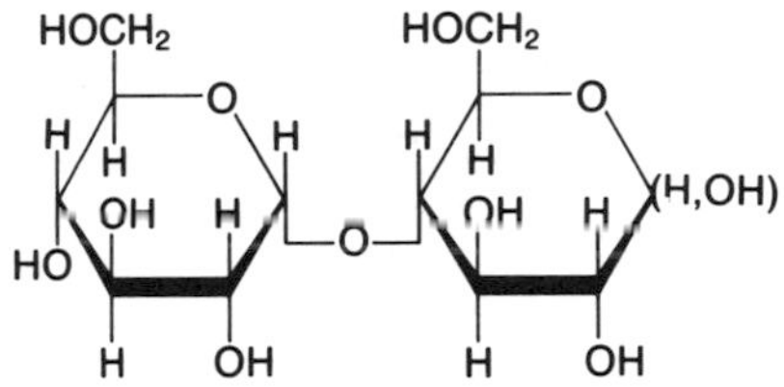

(May be either α or β)

α-1,4-Glycosidic linkage

One monosaccharide still has a free anomeric carbon and can react further. Reaction of the anomeric OH may be with any OH on the other monosaccharide, including the anomeric one. The anomeric OH that has reacted is fixed as either α or β and cannot mutarotate or reduce. If the glycosidic bond is not between two anomeric carbon atoms, one of the units will still be free to mutarotate and reduce.

Oligo- and Polysaccharides

Disaccharides have two monosaccharides, either the same or different, in glycosidic linkage. If the glycosidic linkage is between the two anomeric carbon atoms, the disaccharide is nonreducing:

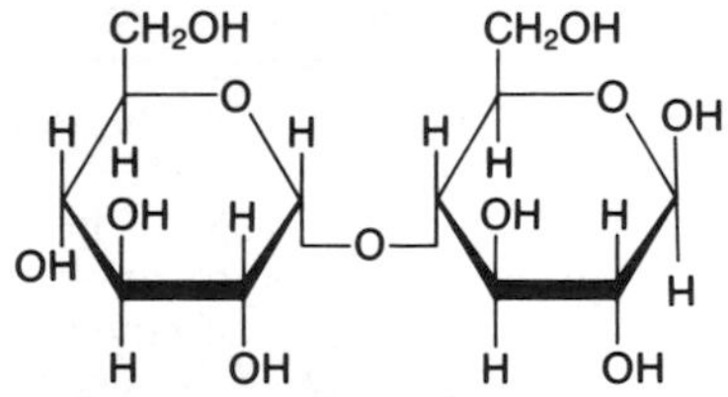

Maltose

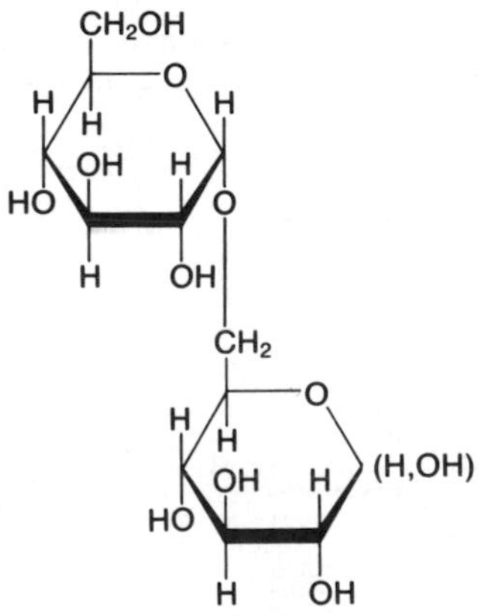

Isomaltose

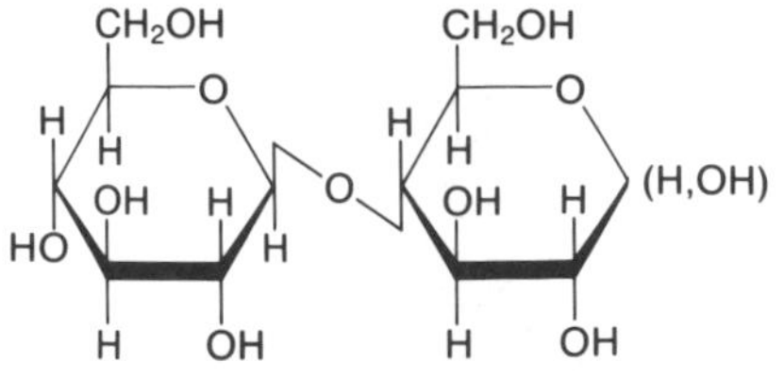

Cellobiose

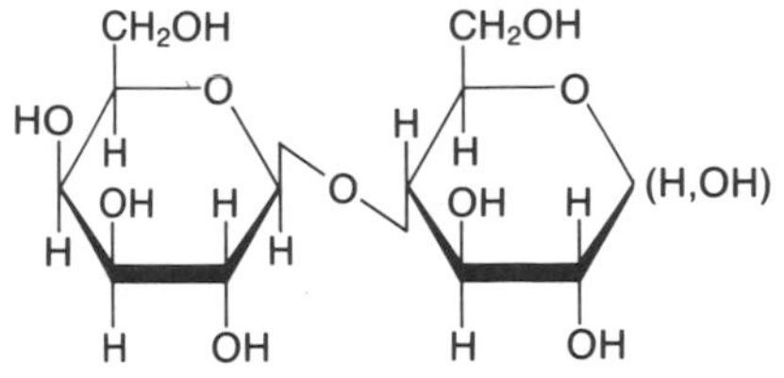

Lactose

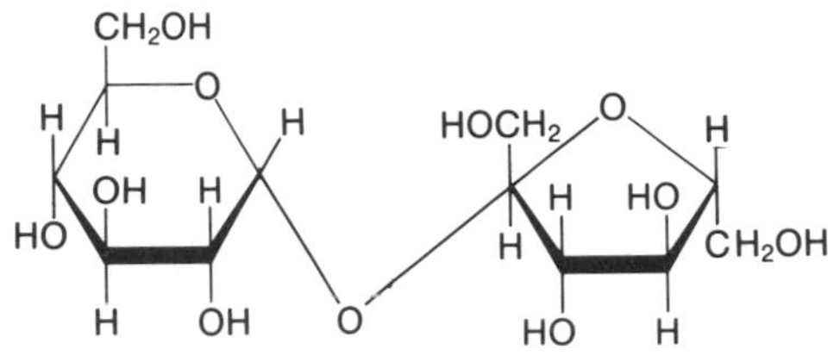

Sucrose

Maltose = 4-*O*-(α-D-glucopyranosyl)D-glucopyranose; reducing
Isomaltose = 6-*O*-(α-D-glucopyranosyl)D-glucopyranose; reducing
Cellobiose = 4-*O*-(β-D-glucopyranosyl)D-glucopyranose; reducing
Lactose = 4-*O*-(β-D-galactopyranosyl)D-glucopyranose; reducing
Sucrose = α-D-glucopyranosyl-β-D-fructofuranoside; nonreducing

As many as thousands of monosaccharides, either the same or different, may be joined by glycosidic bonds to form *polysaccharides*. The anomeric carbon of one unit is usually joined to C-4 or C-6 of the next unit. The ends of a polysaccharide are not identical (reducing end = free anomeric carbon; nonreducing = anomeric carbon linked to next unit; branched polysaccharide = more than one nonreducing end). The most common carbohydrates are homopolymers of glucose; for example, starch, glycogen, and cellulose. Plant starch is a mixture of **amylose,** a linear polymer of maltose units, and **amylopectin,** branches of repeating maltose units (glucose–glucose in α-1,4 linkages) joined via isomaltose linkages. **Glycogen,** the storage form of carbohydrate in animals, is similar to amylopectin, but the branches are shorter and occur more frequently. **Cellulose,** in plant cell walls, is a linear polymer of repeating cellobioses (glucose–glucose in β-1,4 linkages). *Mucopolysaccharides* contain amino sugars, free and acetylated, uronic acids, sulfate esters, and sialic acids in addition to the simple monosaccharides. ***N*-Acetylneuraminic acid,** a sialic acid, is

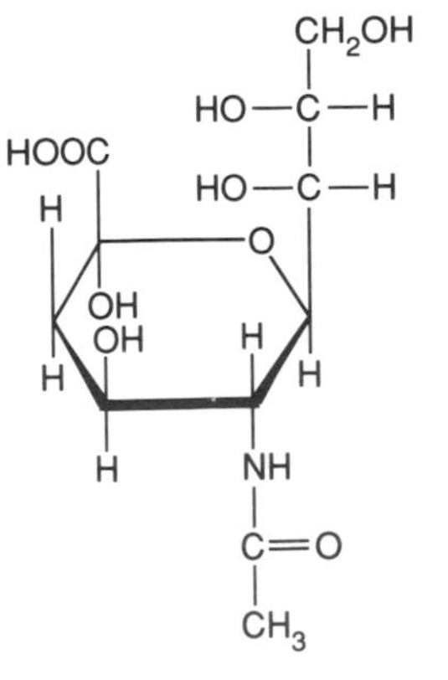

LIPIDS

Lipids are a diverse group of chemicals related primarily because they are insoluble in water, soluble in nonpolar solvents, and found in animal and plant tissues.

Saponifiable lipids yield salts of fatty acids upon alkaline hydrolysis. *Acylglycerols* = glycerol + fatty acid(s); *phosphoacylglycerols* = glycerol + fatty acids + HPO_4^{2-} + alcohol; *sphingolipids* = sphingosine + fatty acid + polar group (phosphorylalcohol or carbohydrate); *waxes* = long-chain alcohol + fatty acid. *Nonsaponifiable lipids* (*terpenes, steroids, prostaglandins,* and related compounds) are not usually subject to hydrolysis. *Amphipathic* lipids have both a polar "head" group and a nonpolar "tail." Amphipathic molecules can stabilize emulsions and are responsible for the lipid bilayer structure of membranes.

Fatty acids are monocarboxylic acids with a short (<6 carbon atoms), medium (8–14 carbon atoms), or long (>14 carbon atoms) aliphatic chain. Biologically important ones are usually linear molecules with an even number of carbon atoms (16–20). Fatty acids are numbered using either arabic numbers (COOH is 1) or the Greek alphabet (COOH is not given a symbol; adjacent carbon atoms are α, β, γ, etc.). **Saturated fatty acids** have the general formula $CH_3(CH_2)nCOOH$. (*Palmitic acid* = C_{16}; *stearic acid* = C_{18}.) They tend to be extended chains and solid at room temperature unless the chain is short. Both trivial and systematic (prefix indicating number of carbon atoms + *anoic acid*) names are used. $CH_3(CH_2)_{14}COOH$ = palmitic acid or hexadecanoic acid.

Unsaturated fatty acids have one or more double bonds. Most naturally occurring fatty acids have cis double bonds and are usually liquid at room temperature. Fatty acids with trans double bonds tend to have higher melting points. A double bond is indicated by Δ^n, where n is the number of the first carbon of the bond. *Palmitoleic* = Δ^9-hexadecenoic acid; *oleic* = Δ^9-octadecenoic acid; *linoleic* = $\Delta^{9,12}$-octadecadienoic acid; *linolenic* = $\Delta^{9,12,15}$-octadecatrienoic acid; *arachidonic* = $\Delta^{5,8,11,14}$-eicosatetraenoic acid. Since fatty acids are elongated in vivo from the carboxyl end, biochemists use alternate terminology to assign these fatty acids to families. Omega (ω) minus x (or $n - x$), where x is the number of carbon atoms from the methyl end where a double bond is first encountered. *Palmitoleic* and *oleic* are ω-9 acids, *linoleic* and *arachidonic* are ω-6 acids, and *linolenic* is an ω-3 acid. Addition of carbon atoms does not change the family to which an unsaturated fatty acid belongs.

Since the p*K* values of fatty acids are about 4–5, in physiological solutions, they exist primarily in the ionized form, called salts or "soaps." Long-chain fatty acids are insoluble in water, but soaps form micelles. Fatty acids form esters with alcohols and thioesters with CoA.

Biochemically significant reactions of unsaturated fatty acids are

1. *Reduction* —CH=CH— + XH_2 ⟶ —CH_2CH_2— + X
2. *Addition of water* —CH=CH— + H_2O ⟶ —CH(OH)—CH_2—
3. *Oxidation* R—CH=CH—R′ ⟶ R—CHO + R′CHO

Prostaglandins, thromboxanes, and *leukotrienes* are derivatives of C_{20}, polyunsaturated fatty acids, especially arachidonic acid. **Prostaglandins** have the general structure:

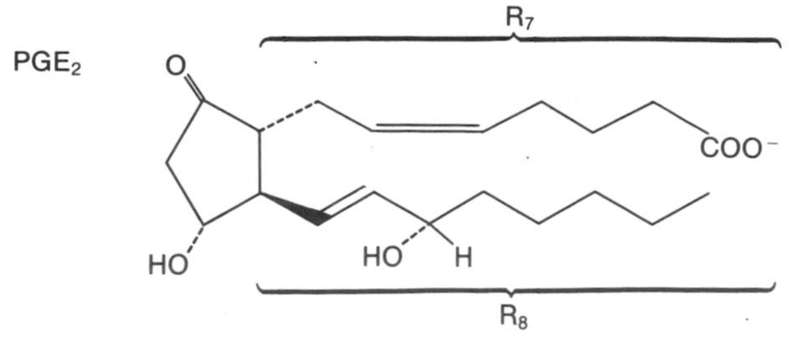

The series differ from each other in the substituents on the ring and whether C-15 contains an OH or O · OH group. The subscript indicates the number of double bonds in the side chains. Substituents indicated by ——(β) are above the plane of the ring; ···(α) below:

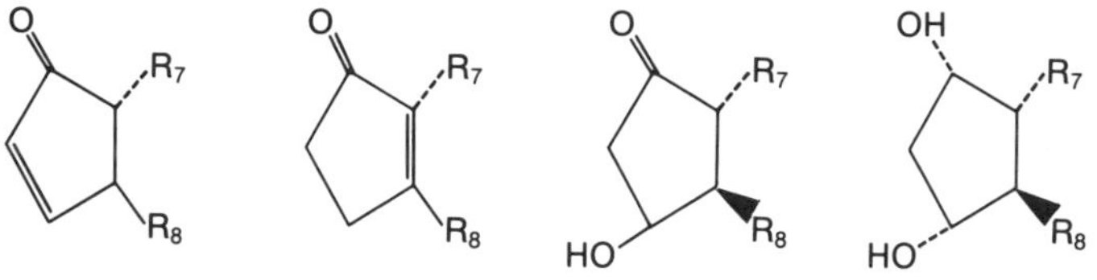

PGA PGB PGE PGF

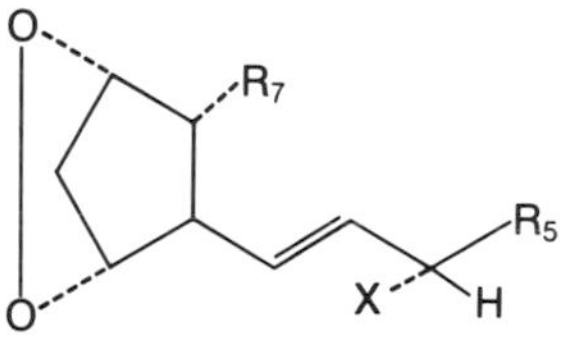

PGG(X = OH); PGH(X = OOH) 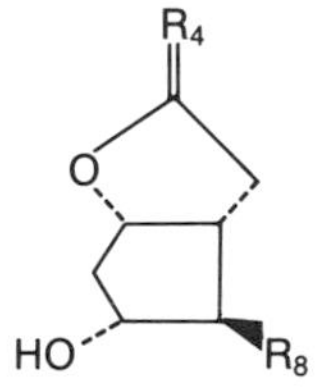 PGI

Thromboxanes have an oxygen incorporated to form a six-membered ring:

TXA_2

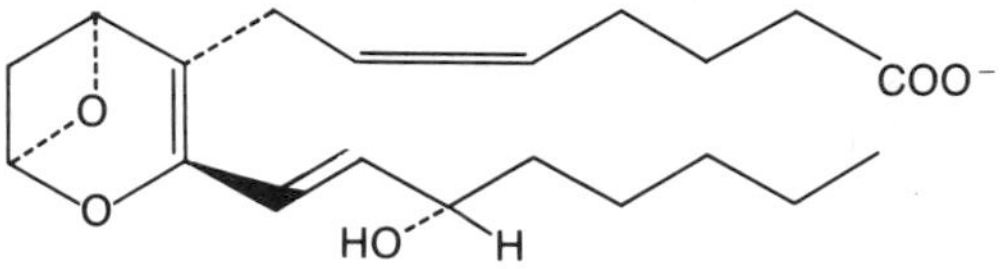

Leukotrienes are substituted derivatives of arachidonic acid in which no internal ring has formed: R is variable:

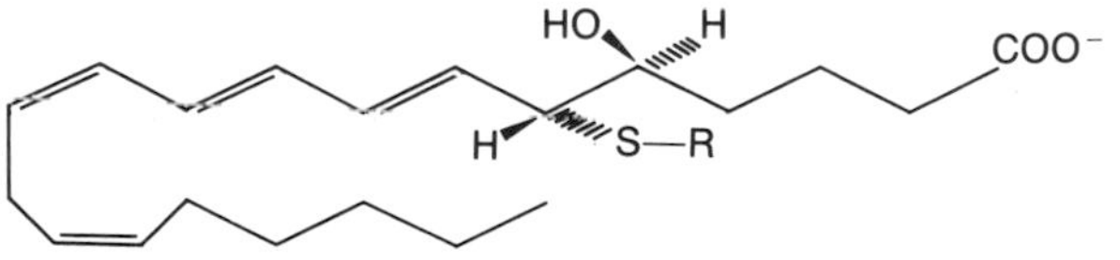

Leukotriene C, D, or E

Acylglycerols are compounds in which one or more of the three OH groups of glycerol is esterified. In **triacylglycerols** (triglycerides) all three OH groups are esterified to fatty acids. At least two of the three R groups are usually different. If R_1 is not equal to R_3, the molecule is asymmetric and of the L configuration:

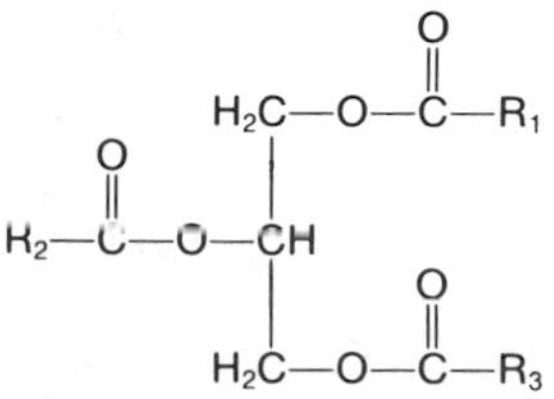

The properties of the triacylglycerols are determined by those of the fatty acids they contain; with *oils,* liquids at room temperature (preponderance of short-chain and/or cis-unsaturated fatty acids), and *fats,* solid (preponderance of long-chain and/or saturated).

Triacylglycerols are hydrophobic and do not form stable micelles. They may be hydrolyzed to glycerol and three fatty acids by strong alkali or enzymes (lipases). *Mono-* [usually with the fatty acid in the β (2) position] and *diacylglycerols* also exist in small amounts as metabolic intermediates. Mono- and diacylglycerols are slightly more polar than triacylglycerols. *Phosphoacylglycerols* are derivatives of L-α-glycerolphosphate (L-glycerol 3-phosphate):

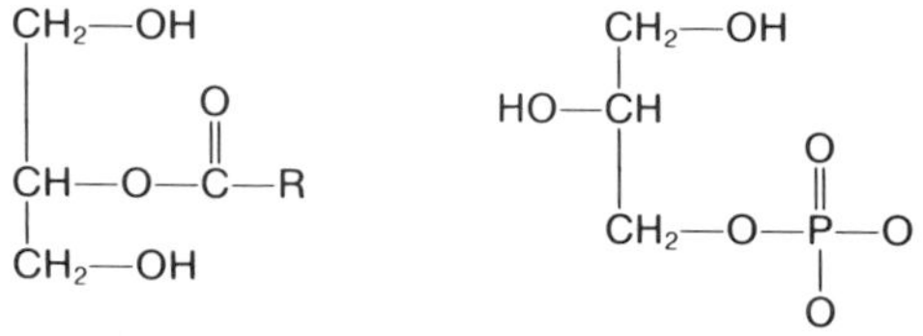

The parent compound, **phosphatidic acid** (two OH groups of L-α-glycerolphosphate esterified to fatty acids), has its phosphate esterified to an alcohol (X—OH) to form several series of phosphoacylglycerols. These are amphipathic molecules, but the net charge at pH 7.4 depends on the nature of X—OH.

X—OH	Phosphoacylglycerol
$HO—CH_2—CH_2—\overset{+}{N}—(CH_3)_3$	Phosphatidylcholines (lecithins)
$HO—CH_2—CH_2—\overset{+}{N}H_3$	Phosphatidylethanolamines (cephalins)
$HO—CH_2—CH(\overset{+}{N}H_3)—COO^-$	Phosphatidylserines
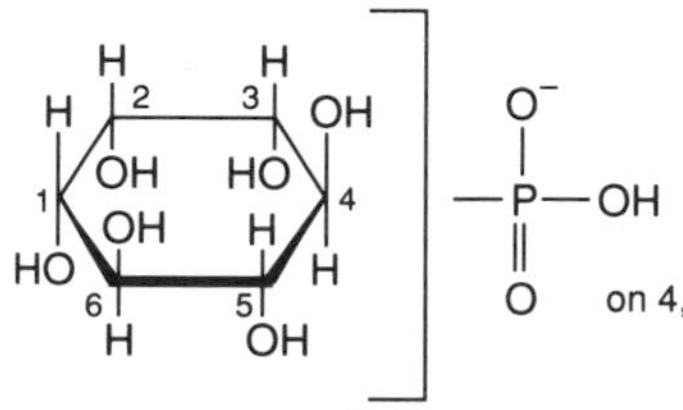 inositol; —P(O)(O⁻)—OH on 4, or 4 and 5	Phosphatidylinositols

In **plasmalogens,** the OH on C-1 is in *ether,* rather than ester, linkage to an alkyl group. If *one* fatty acid (usually β) has been hydrolyzed from a phosphoacylglycerol, the compound is a *lyso*-compound; for example, lysophosphatidylcholine (lysolecithin):

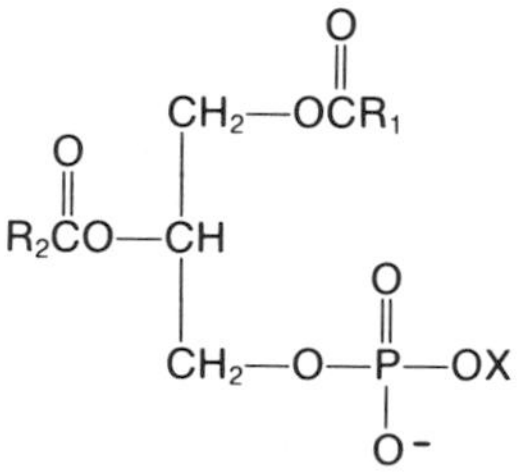

A phosphoacylglycerol 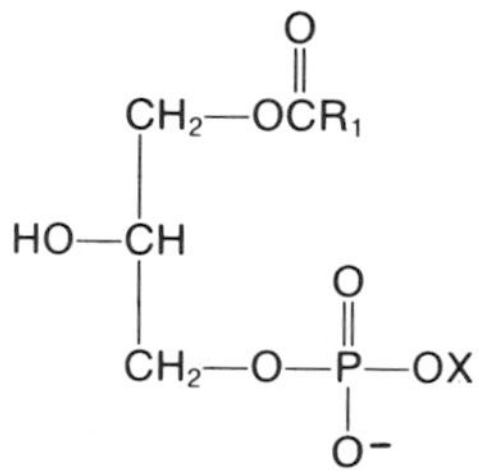 A lysocompound

Sphingolipids are complex lipids based on the C-18, unsaturated alcohol, sphingosine. In *ceramides,* a long-chain fatty acid is in amide linkage to sphingosine:

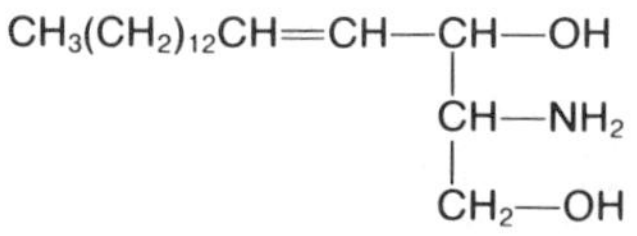

$CH_3(CH_2)_{12}CH{=}CH—CH(OH)—CH(NH_2)—CH_2—OH$

Sphingosine

$CH_3(CH_2)_{12}CH{=}CH—CH(OH)—CH(NH—CO—R)—CH_2—OH$

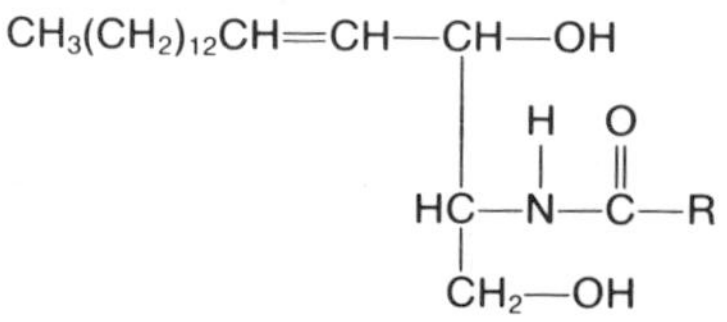

A ceramide

Sphingomyelins, the most common sphingolipids, are a family of compounds in which the primary OH group of a ceramide is esterified to phosphorylcholine (phosphorylethanolamine):

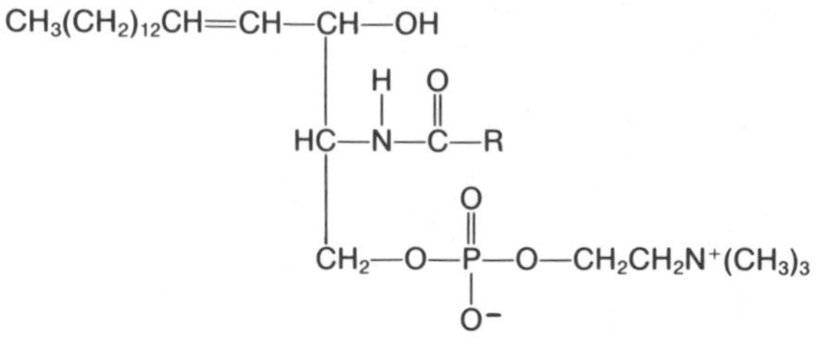

They are amphipathic molecules, existing as zwitterions at pH 7.4 and the only sphingolipids that contain phosphorus. *Glycosphingolipids* do not contain phosphorus but contain carbohydrate in glycosidic linkage to the primary alcohol of a ceramide. They are amphipathic and either neutral or acidic if the carbohydrate moiety contains an acidic group. **Cerebrosides** have a single glucose or galactose linked to a ceramide. *Sulfatides* are galactosylceramides esterified with sulfate at C-3 of the galactose:

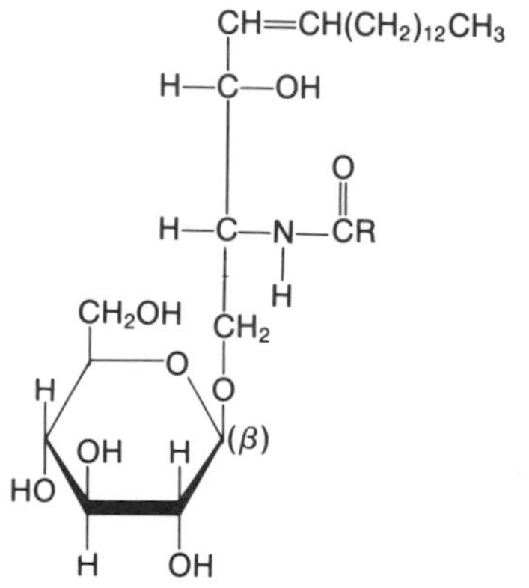

Glucosylceramide (glucocerebroside)

Globosides (*ceramide oligosaccharides*) are ceramides with two or more neutral monosaccharides, whereas **gangliosides** have an oligosaccharide containing one or more sialic acids.

Steroids are derivatives of cyclopentanoperhydrophenanthrene. The steroid nucleus is a rather rigid, essentially planar structure with substituents above or in the plane of the rings designated β (solid line) and those below called α (dotted line):

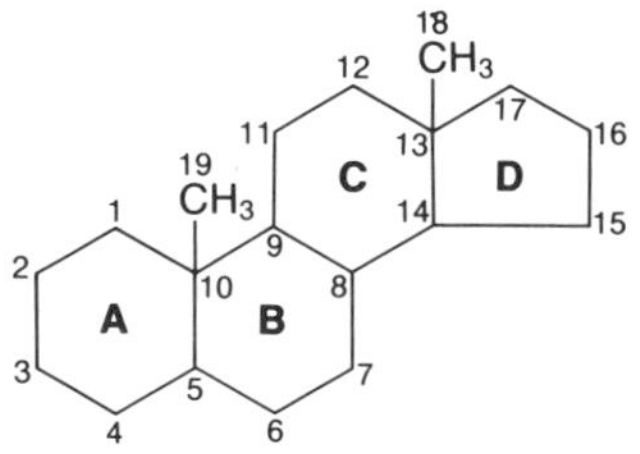

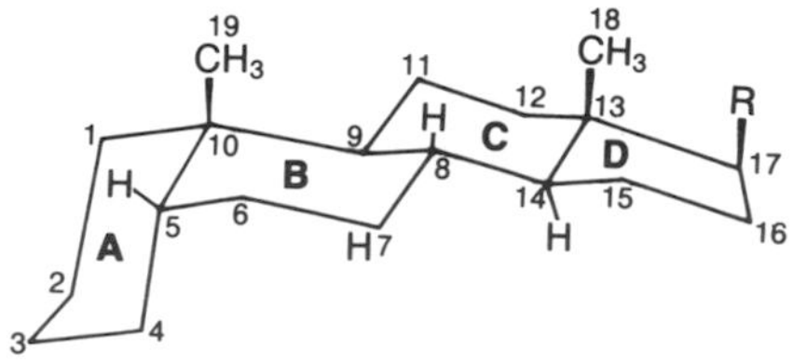

A and B rings—cis
the others—trans

Most steroids in humans have methyl groups at positions 10 and 13 and frequently a side chain at position 17. Sterols contain one or more OH groups, free or esterified to a fatty acid. Most steroids are nonpolar. In a liposome or cell membrane, **cholesterol** orients with the OH toward any polar groups; cholesterol esters do not. **Bile acids** (e.g., cholic acid) have a polar side chain and so are amphipathic:

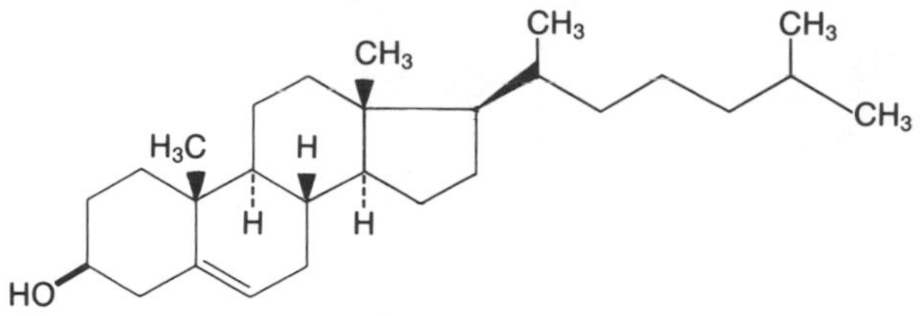

Cholesterol

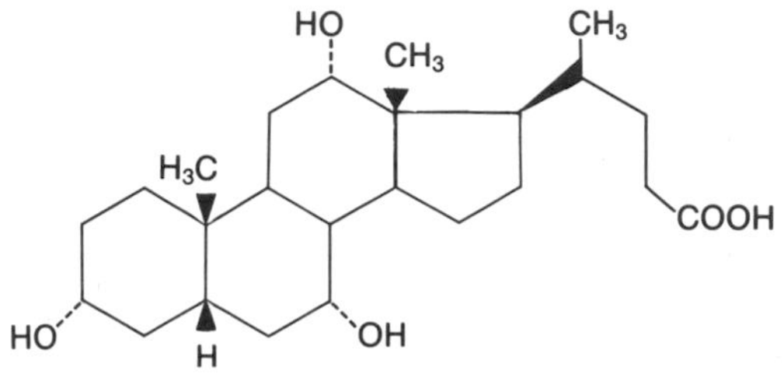

Cholic acid

Steroid hormones are oxygenated steroids of 18–21 carbon atoms. *Estrogens* have 18 carbon atoms, an aromatic ring A, and no methyl at C-10. *Androgens* have 19 carbon atoms and no side chain at C-17. *Glucocorticoids* and *mineralocorticoids* have 21 carbon atoms including a C_2, oxygenated side chain at C-17. *Vitamin D_3* (*cholecalciferol*) is not a sterol but is derived from 7-dehydrocholesterol in humans:

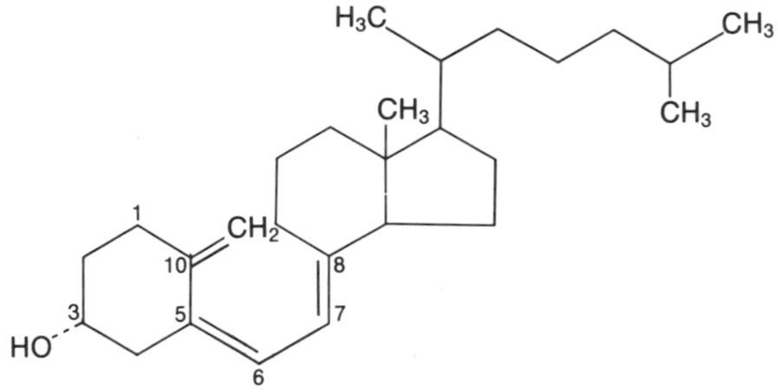

Cholecalciferol

Terpenes are polymers of two or more isoprene units. **Isoprene** is

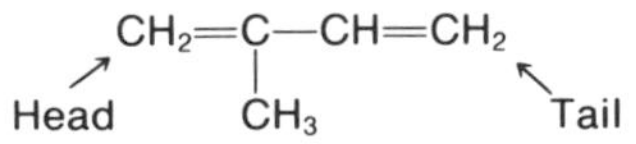

$$CH_2{=}C(CH_3){-}CH{=}CH_2$$

Head ... Tail

Terpenes may be linear or cyclic, with the isoprenes usually linked head to tail and most double bonds trans (but may be cis as in vitamin A). *Squalene,* the precursor of cholesterol, is a linear terpene of six isoprene units. Fat-soluble *vitamins* (A, D, E, and K) contain polymers of isoprene units:

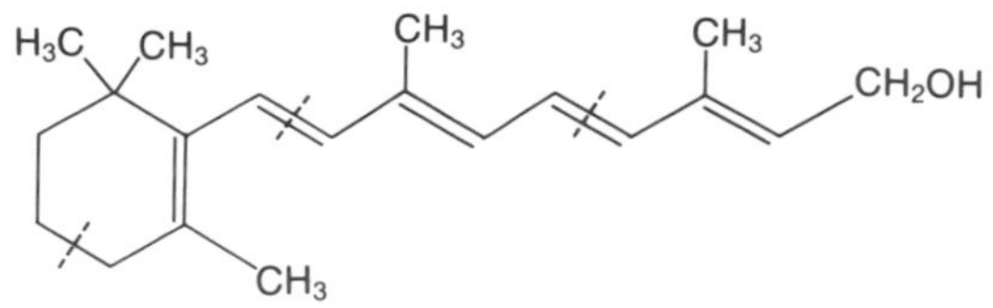

Vitamin A

Vitamin E (α-tocopherol)

Vitamin K_2

AMINO ACIDS

Amino acids contain both an *amino* (NH_2) and a *carboxylic acid* (COOH) group. Biologically important amino acids are usually α-amino acids with the formula

L-α-Amino acid

The amino group, with an unshared pair of electrons, is basic, with a pK_a of about 9.5, and exists primarily as $—NH_3^+$ at pH values near neutrality. The carboxylic acid group ($pK \sim 2.3$) exists primarily as a carboxylate ion. If R is anything but H, the molecule is asymmetric with most naturally occurring ones of the L-configuration (same relative configuration as L-glyceraldehyde; see p. 1151).

The *polarity* of amino acids is influenced by their side chains (R groups) (see p. 28 for complete structures). *Nonpolar* amino acids include those with large, aliphatic, aromatic, or undissociated sulfur groups (aliphatic = Ala, Ile, Leu, Val; aromatic = Phe, Trp; sulfur = Cys, Met). *Intermediate* polarity amino acids include Gly, Pro, Ser, Thr, and Tyr (undissociated).

Amino acids with ionizable side chains are *polar*. The pK values of the side groups of arginine, lysine, glutamate, and aspartate are such that these are nearly always charged at physiological pH, whereas the side groups of histidine ($pK = 6.0$) and cysteine ($pK = 8.3$) exist as both charged and uncharged species at pH 7.4 (acidic = Glu, Asp, Cys; basic = Lys, Arg, His). Although undissociated cysteine is nonpolar, cysteine in its dissociated form is polar.

All amino acids are at least *dibasic acids* because of the presence of both the α-amino and α-carboxyl groups, the ionic state being a function of pH. The presence of another ionizable group will give a tribasic acid, as shown for cysteine.

$$H_3\overset{+}{N}CH(CH_2SH)COOH \underset{pK_1}{\rightleftharpoons} H_3\overset{+}{N}CH(CH_2SH)COO^- \underset{pK_2}{\rightleftharpoons}$$

pK_1 (α-COOH) = 1.7–2.6; pK_2 (—SH) = 8.3

$$H_3\overset{+}{N}CH(CH_2S^-)COO^- \underset{pK_3}{\rightleftharpoons} H_2NCH(CH_2S^-)COO^-$$

pK_3 (α-NH_3^+) = 8.8–10.8

The **zwitterionic form** is the form in which the *net* charge is zero. The *isoelectric point* is the average of the two pK values involved in the formation of the zwitterionic form. In the above example this would be the average of $pK_1 + pK_2$.

PURINES AND PYRIMIDINES

Purines and pyrimidines, often called *bases,* are nitrogen-containing heterocyclic compounds with the structures

Purine

Pyrimidine

Major bases found in nucleic acids and as cellular nucleotides are

Purines		Pyrimidines	
Adenine:	6-amino	Cytosine:	2-oxy, 4-amino
Guanine:	2-amino, 6-oxy	Uracil	2,4-dioxy
		Thymine:	2,4-dioxy, 5-methyl

Other important bases found primarily as intermediates of synthesis and/or degradation are

Hypoxanthine:	6-oxy	Orotic acid: 2,4-dioxy, 6-carboxy
Xanthine:	2,6-dioxy	

Oxygenated purines and pyrimidines exist as *tautomeric* structures with the keto form predominating and involved in hydrogen bonding between bases in nucleic acids:

Keto ⇌ Enol

Nucleosides have either β-D-ribose or β-D-2-deoxyribose in an *N*-glycosidic linkage between C-1 of the sugar and N-9 (purine) or N-1 (pyrimidine).

Nucleotides have one or more phosphate groups esterified to the sugar. Phosphates, if more than one is present, are usually attached to each other via phosphoanhydride bonds. Monophosphates may be designated as either the base monophosphate or as an *-ylic acid* (AMP: adenylic acid):

A pyrimidine nucleotide

By conventional rules of *nomenclature,* the atoms of the base are numbered 1–9 in purines or 1–6 in pyrimidines and the carbon atoms of the sugar 1′–5′. A nucleotide with an unmodified name indicates that the sugar is ribose and the phosphate(s) is/are attached at C-5′ of the sugar. Deoxy forms are indicated by the prefix d (dAMP = deoxyadenylic acid). If the phosphate is esterified at any position other than 5′, it must be so designated [3′-AMP; 3′-5′-AMP (cyclic AMP = cAMP)]. The nucleosides and nucleotides (ribose form) are named as follows:

Base	Nucleoside	Nucleotide
Adenine	Adenosine	AMP, ADP, ATP
Guanine	Guanosine	GMP, GDP, GTP
Hypoxanthine	Inosine	IMP
Xanthine	Xanthosine	XMP
Cytosine	Cytidine	CMP, CDP, CTP
Uracil	Uridine	UMP, UDP, UTP
Thymine	dThymidine	dTMP, dTTP
Orotic acid	Orotidine	OMP

Minor (modified) bases and nucleosides also exist in nucleic acids (see p. 688 for a list). *Methylated* bases have a methyl group on an amino group (*N*-methyl guanine), a ring atom (1-methyl adenine) or on an OH group of the sugar (2′-*O*-methyl adenosine). *Dihydrouracil* has the 5–6 double bond saturated. In *pseudouridine,* the ribose is attached to C-5 rather than to N-1.

In **polynucleotides** (*nucleic acids*), the mononucleotides are joined by phosphodiester bonds between the 3′-OH of one sugar (ribose or deoxyribose) and the 5′-OH of the next (see p. 611 for the structure).

Index

HOW TO USE THIS INDEX: The letter "f" after a page number designates a figure; "t" designates tabular material; "cc" designates a Clinical Correlation. **Boldface** page numbers show where terms are explained. "*See*" cross-references direct the reader to the synonymous term. "*See also*" cross-references direct the reader to (1) related topics or (2) the synonym under which a more detailed topic breakdown may be found.

A (anisotropic) bands, **956**
A codon base pairing, 730t
AAUAAA sequence (polyadenylation signal sequence), **714**
Abetalipoproteinemia, 405cc
ABO blood group, 373cc
Absolute configuration, **31**
Absorption of nutrients, 1059–1089, *see also* Digestion and absorption
Abzymes, **177**, 177–178
ACAT (fatty acyl CoA:cholesterol acyltransferase), **445**
Accelerin (Factor VI), 969t
Acetals, 1149
Acetoacetate, **406,** *see also* Ketone bodies
Acetoacetyl CoA, 416
Acetokinase, **440**
Acetone formation, 414–415, *see also* Ketone bodies
Acetylcholine, **520**, **934**
  in blood pressure regulation, 912
  as secretagogue, 1065
Acetylcholinesterase, **934**
Acetyl CoA
  in cholesterol synthesis, 439–442
  fatty acid oxidation and, 409–410
  in fatty acid synthesis, 392–395
  in gluconeogenesis, 588
  mitochondrial transport, 397–398
  in nitrogen elimination, 488
  in oxidative metabolism, 260
  intramitochondrial, 267
Acetyl CoA carboxylase, 393, 394t
N-Acetylgalactosaminyltransferase, **381**
N-Acetylglucosamine, in sialic acid formation, 371–372
N-Acetylglucosaminylpyrophosphoridolichol, **377**
N-Acetylglutamate, **486**
N-Acetylglutamate synthetase, **488**
  deficiency, 487cc
N-Acetylneuraminic acid, **371**, 455, 1154
Acid-base balance
  alternative laboratory measures, 1052–1054
  of amino acids and proteins, 34–36, 35t
  basic principles, 8–13
  catalysis, 172–174
  clinical correlations, *see under* pH
  compensatory mechanisms, 1049–1052
  disorders of, 1045–1046
  glutamine in, 602–603
  Henderson-Hasselbalch equation, 10–11, 13cc, 35–36, 1041, 1042cc
  kidneys in, 1046–1049
  Na^+ and Cl^- in, 1054–1056
  see also pH
Acidic amino acids, **35**
Acidity
  and oxygen-hemoglobin binding, 1032–1033
  titratable of urine, **1047**
  total of urine, **1049**
Acidosis, **1045**
  metabolic, **1046**, 1051–1052
  respiratory, **1045**, 1050–1051
    chronic, 1052cc
Acids, **8**, 8–18
  acid-base pairs, 9t
  branched-chain, **390**
  buffering of, 11–13, 1039–1045
  chemistry of, 1149
  see also Acid-base balance; Buffering; *and individual acids*
Acid-stable lipase, **1082**
Aciduria, orotic, 548cc
Aconitase reaction, **256**
Acquired immune deficiency syndrome. *See* AIDS
ACTH, 855t
Actin, 958–962
  F-, **n961**
  G (monomer), 957t
Actin filaments, **21**
α-Actinin, 957t, **961**
β-Actinin, 957t, **n961**
Activation energy, **171**
Active site direction, **107**
Acute intermittent porphyria, 1011t, 1012cc
Acute respiratory alkalosis, 1051cc
Acyclovir (acycloguanosine), **569**
Acyl carrier protein (ACP), **393**
Acyl CoA dehydrogenases, **409**
  deficiency, 410cc
1-Acylglycerol phosphate:acyltransferase, **432**
Adenine nucleotide translocator, **267**, 267–268
Adenine phosphoribosyltransferase (APRTase), **542**
Adenosine deaminase, **544**, 544cc, 544–545
Adenosine 5′-diphosphate. *See* ADP
Adenosine 5′-triphosphate. *See* ATP
S-Adenosylmethionine (AdoMet), **433**, 501–504
Adenylate cyclase, **189**, 190f
Adenylate kinase, **315**, **963**
Adenylosuccinate synthetase, **541**
Adenylsuccinase, **538**
ADH. *See* Vasopressin
Adipose tissue
  brown, 1101–1102
  in starve-feed cycle, 582, 591
  in triacylglycerol synthesis, storage, and utilization, 401–404
  white, 401–404
AdoMet, **433**, 501–504
ADP
  oxidative phosphorylation and, 282–286
  structure and function, 239f, 240
  transport mechanisms, 268
ADP-ribosylation, **745**
Adrenal gland hormones
  in stress, 922–923
  synthesis, 991–993f
Adrenal hyperplasia, 994cc
Adrenaline. *See* Epinephrine
α-Adrenergic receptors, **353**
β-Adrenergic receptors, **353**, **884**, 884–886
Adrenocorticotropic hormone. *See* ACTH
Adrenodoxin, 156, **989**
Adrenodoxin reductase, **989**
Aerobic vs. anaerobic exercise, 593–595
Affinity chromatography, **87**

Age pigment (lipofuscin), **20**, 20–21, 486
Agonists, **221**
Agricultural applications of recombinant techniques, 801–802
AIDS, 79
  enzyme-linked immunoassay (ELISA) test for, 183–184
  purine and pyrimidine analog drugs, 569–570
  recombinant techniques in screening and treatment, 799cc, 799–800
A 23187 ionophore, 232t, 233f
ALA (aminoevulinic acid synthase), **1011**, 1011–1013, 1016–1017
ALA dehydrase, **1013**
Alamethicin, 232t, 233f
Alanine, **27**, 493
Alanine cycle, **324**, 324–327
Albinism, 508cc
Albumin titration curve of human serum, 38f
Albuminemia, 405cc
Alcaptonuria, 508, 508cc
Alcohol
  carbohydrate metabolism and, 307cc
  chemistry, 1149
  fatty, 401
  fermentation of, **294**, 295
  hypoglycemia and intoxication, 336cc, 336–337
  metabolism, 306, 307cc, 600–601
  sensitivity in Asians, 147cc
Alcohol dehydrogenase, 137, 180cc, **306**
Alcoholism
  and nutritional intake, 1128cc
  phosphorus deficiency in, 1138
Aldehyde chemistry, 1149
Aldolases in fructose intolerance, 311cc
Aldose reductase, **941**
Aldosterone, **907**, 909t, 910t, 911–912
  structure and function, 903t
  synthesis, 991
Alkaline phosphatase, 1063t
Alkalosis, **1045**
  metabolic, **1046**, 1048–1049
  respiratory, **1045**
    acute, 1051cc
Allergy and monohydroxyeicosatetraenoic acids in, 470
Allosteric mechanisms, **162**, 162–167
  enzymes, **131**, 131–132
  nucleotides in, 531
  in hemoglobin–oxygen binding
  in starve–feed cycle, 586–587
Allysine, **65**
Alpha, beta structure, **58**, 58–60
Alpha-carbon atom, **27**
Alpha-complementation, 780–781
Alpha-folded domain, 58f
Alpha helix conformation, **51**, 51–54
Alpha keratin, **66**, 66–67
Alpha-limit dextrins, **1079**
Alpha oxidation of fatty acids, 412–414
Alpha subunit, **870**
Alu family of repetitive sequences, **831**
Amanita phalloides poisoning, 699cc
American diet, 1143–1144
Ames test, 658cc
Amides, chemistry, 1151
Aminergic hormones, **850**
Amines, chemistry, 1151
Amino acid derived hormones. *See under* Hormones
Amino acid metabolism
  clinical correlations
    branched chain amino acid-related diseases, 492
    cytine-related diseases, 503
    folate-related diseases, 518
    folic acid deficiency, 517
    glutathione deficiency, 523
    histinemia, 512
    hyperammonemia and hepatic coma, 483
    lysine- and ornithine-related diseases, 515
    nonketotic hyperglycinemia, 498
    Parkinson's disease, 509
    phenylketonuria, 506
    propionate- and methylmalonate-related diseases, 496
    transulfuration defects, 503
    tyrosine-related diseases, 508
    urea cycle enzyme deficiency diseases, 487
  general reactions
    alternative nitrogen-eliminating reactions, 488
    ammonia reactions, 480–481
    glutamate dehydrogenase catalysis, 478–480
    transamination, 477–478
    urea cycle and its regulation, 481–488
  glucogenicity and ketogenicity, 492–493
  of individual amino acids
    alanine, 493
    γ-aminobutyrate (GABA), 501
    arginine, 501
    aspartate, 509
    branched-chain types, 493–496
    carnitine, 514–516
    cysteine, 502f, 504–506
    glutathione, 522–524
    glycine, 496–498
    histidine catabolism, 512–513, 513cc
    hydroxyamine acids, 496–498
    lysine, 513–514
    nitrogenous derivatives, 518–522
    ornithine, 499–501
    phenylalanine, 506
    proline, 498–499
    serine and threonine, 496–498
    tryptophan, 510–512
    tyrosine, 506
  overview, 475–477, 492
  see also Amino acids; Proteins
L-Amino acid oxidase, **480**
Amino acids
  abbreviations for, 30t
  acid–base properties, 35–37
  activation (aminoacylation), 733–735
  assay techniques, 38–42, 44–49
  basic, 35
  branched chain, 492–493cc, **493**, 493–496, **579**
  charge properties, 37–38
  chemical reactions, 41–42, 42t, 1157
  clinical correlations
    diagnostic applications of amino acid analysis, 42
    insulin synthesis, 49
    see also under Amino acid metabolism
  common, 27–33
  configurations, 31
  derived, 33–35
  dietary requirements, 476t, 1098, 1098cc, 1099cc
  essential, 476, **1098**
    vs. nonessential, 476
  fasting state and, 580–582
  free, 1076–1077
  glucagenic, 330t, 329–332, 492–493
  ionizable side chain groups, 34–36, 35t
  ketogenic, 330t, 330–331, 492–493
  membrane transport, 231
  metabolism
    general pathways, 475–489
    individual amino acids, 491–525
    see also Amino acid metabolism
  modified, 756t, 756–757
  oxidation rates, 480t
  physical and observed properties, 30t
  polymerization, 31–33
  residues, **32**
    invariant, **49**
  sequence determination, 44–49
  serine proteases, 111t, 111–112
  side chain groups, 27–31
  structures of common, 28f
  transport systems, **1076**
  transcription control, 820–821
  tRNA activation, 687–688
  see also Amino acid metabolism; Proteins
Aminoaciduria, 42cc
Aminoacylation, 733–735
Aminoacyl-tRNA synthetases, **733**
γ-Aminobutyrate, *See* GABA
Aminolevulinic acid synthase (ALA), **1011**, 1011–1013, 1016–1017
Aminopeptidases, 1063t, **1073**, **1076**
Amino sugars, **371**
Aminotransferases, **138**, 185–186
  glutamate-oxaloacetate transaminase, 257
Ammonia, 480, **485**
AMP
  IMP as precursor, 539–540
  Pasteur effect and, 313–315
  as signal of energy status, 315–316
  see also cAMP
AMP deaminase (5′AMP aminohydrolase), **544**
Amphipathicity, **200**
Amphipathic molecules, **6**
Ampicillin resistance, 778–780, *see also* Drugs, resistance
Amplification, **190**, **348**
Amylo-α-[1,6]-glucosidase, **342**
Amylopectin, 1078t, 1154

Amylose, 1078t, 1154
Amytal, 281
Anabolic reactions, **238**
Anaplerotic reactions, **330**
Androgen binding protein, **916**
Androgens, 902, 905, 905t, 909t
  cytochrome O450 in synthesis, 994–995
  see also Hormones, steroid
Anemia
  hemolytic, 179cc, 324cc, 362cc
    vitamin E supplementation in, 1124
  iron deficiency, 1008cc, 1141cc, **1140**
  macrocytic, **1134**
  megaloblastic, 1135
    and orotic aciduria, 548cc
  non-iron deficiency, 1120
  pernicious, **1134**, 1134–1136
  sickle-cell. *See* Sickle-cell anemia
  sideroblastic, **1130**
Anesthetics
  and malignant hyperthermia, 316cc
  and membrane fluidity, 214
Angina pectoris, 318cc
Angiotensin I, 938t
Angiotensin II, 852t, 854t, 856t, **912**, 938t
  horse, 34t
Angiotensin III, 856t
Anserine, **512**
Anterior pituitary gland, functional assessment, 852cc
Anterior pituitary hormones, **850**, 852–854, 854cc, 854t–855t
  functional assessment, 851cc
Antibiotics
  amino sugars in, 371
  ionophore, 231–234, 232t
  protein synthesis inhibitors, 744t, 744–746
  resistance, 749, 778–781, 787, 825cc
Antibody class functions, 96cc
Antibody molecules, 93–102; *see also* Immunoglobulins
Anticoagulant drugs, 375cc
Anticodon–codon interactions, 730–731
Anticodons, degenerate, **730**
Anticodon triplet, **687**
Antidiuretic hormone (ADH). *See* Vasopressin
Antifolate drugs, 567–568
Antigenic determinants, **93**
Antigens, **93**
Antihemophiliac factor (Factor VIII), 969t
Anti-inflammatory agents and prostaglandin inhibition, 465
Antimetabolite chemotherapy, **159**, 159–162, 566–569
Antimycin A, 281
Antioncogenes, 707cc
Antiparallelism, **618**
α_2-Antiplasmin, 109t
Antisense nucleic acids, **799**
Antithrombin III, 109t
Antitumor drugs, 566–569
  growth-inhibiting, 569
  see also Cancer; Chemotherapy
AP endonuclease, **660**
ApoCII deficiency, 67cc
Apoenzymes, **136**
Apoferritin, **1004**
Apolar edge of polypeptide chain, **64**
Apolipoproteins, **67**, 68t, **1087**
  A-I, 67
  B, **443**
  B-100, 34, 67, **445**
Apoproteins, **120**, 405
  B-100, **445**
  E, **445**
APRTase, **542**
Apurinic-apyrimidinic (AP) endonuclease, **660**
AraC (cytosine arabinoside), 567
Arachidonic acid, **462**
Arginase, **484**, 484–486
  deficiency, 487cc
Arginine, **29**, 35t, 501
  synthesis, 483, 483–484
Arginine vasopressin (AVP), 852t, 855t, 869t, 891–892
Argininosuccinate lyase, **483**
  deficiency, 487cc
Arrestin, **951**
Arsenate, inhibition of glycolysis, 308
Arsenic poisoning, 309cc
Aryl hydrocarbon receptor, **919**
Ascorbic acid (vitamin C). *See* Vitamins, C
Asn-X-Thr sequence, **71**
Asparaginase in leukemia, 186
Asparagine, **29**, 376, **481**, **489**
Aspartate, **29**, 509
  in purine synthesis, 538–539
  in pyrimidine synthesis, 547–550
  in urea cycle, 483
Aspartate carbamoyl transferase, **548**
Aspartate transaminase, **478**
Aspartylglycosylaminuria, 374t, 374cc
Aspirin
  and prostaglandin inhibition, 465
  salicylate poisoning, 1053cc
Astrocytes, **930**
Asymmetry in binding sites, 170
Atherosclerosis, 445–446
  polyunsaturated fatty acids and heart disease, 1106cc
ATP
  arsenate inhibition, 308
  in corneal metabolism, 940
  function, 150–151, 238–240
  in gluconeogenesis, 334
  glycolytic synthesis, 302
  oxidative phosphorylation and, 282–286
  structure, 150–151, 238–240
  in transmembrane signaling, 930–931
  transport mechanisms, 268
ATP-ADP translocase, **225**
ATPases
  F_1F_0, **285**
  K^+,H^+-, **1073**
  myosin, **958**
  see also Na^+,K^+-ATPase pump
Atrionatriuretic factor (ANF), 856t, 894–897, **912**
Atriopeptin. *See* Atrionatriuretic factor
Attenuation, **817**, 817–821
AUG codon, **724**, **730**
Autoimmune diseases and RNA, 717cc
Autoimmunity in connective tissue disease, 717cc
Autolysis, **111**
Autoxidation and free radical formation, 419–420
AVP (arginine vasopressin), 852t, 855t, 869t, 891-892
Azaserine, **569**
Azide, 281
3′-Azido-3′-deoythymidine (AZT), **569**

B-100 apolipoprotein, 34, 67, **445**
Bacterial transformation, 777–780
Bacteriophage λ vectors, **785**, 785–786
Bacteriophage M13, **796**
Barbiturates, 307cc
Basal metabolic rate (BMR), **1095**, 1096
Base excess, **1053**
Base pairing, 9t
  intramolecular, **684**
  of rRNAs, 689
  of tRNAs, 687–688
  wobble, 731t
Bases, **8**
  acid-base pairs, 9t
  conjugate, **1040**
  modified, **1158**
  symbols and abbreviations, 534t
  see also Purines; Pyrimidines
Base substitution mutations, **653**
Basic amino acids, **35**
Beriberi, **1128**
Beta barrels, **58**, 99–100, 115
Beta-carotene, **1118**
Beta-endorphin, 938t
Beta-folded domains, **58**, 58–60, 59f
Betaine, **520**
Beta oxidation of fatty acids, 409–412
Beta structure conformation, **51**, 54–55
Beta subunit, **870**
Bicarbonate
  membrane transport, 224
  renal resorption, **1047**
Bicarbonate buffering, 9, 13cc, 1041–1045
Bifunctional enzymes, **302**
Bile, 438–440, *see also* Cholesterol
Bile acids, **439**, 447–448, **1083**, **1156**
  absorption, 1083–1087, 1086cc
  metabolism, 1088–1089
  primary, **1088**
  secondary, **1088**
Bile salts, **1083**
Bilirubin, **1017**, 1017–1020
  plasma types, **1019**
Bilirubin diglucuronide, **1018**
Biliverdin IX, **1017**
Binding element of DNA, **116**
Bioenergetics. *See* Energy transfer; Oxidative metabolism
Biogenic amines, **1066**
Biorhythms, **850**
Biotin, 1132
2,3-Bisphosphoglycerate (DPG), 302–303, **303**
Blood
  buffering line of, **1045**, 1053–1054

buffering of glucose, **338**
carbon dioxide transport, 1033–1038
chemical constituents, 16f
lipid transport, 404–407
oxygen transport, 1026–1029, *see also* Hemoglobin
properties of human at rest, 1034t
steroid hormone transport, 915–916
see also Blood coagulation; Plasma
Blood bicarbonate in metabolic acidosis, 13cc
Blood-brain barrier, **930**
Blood buffer line, **1045**, 1053–1054
Blood cells. *See* Erythrocytes; Leukocytes
Blood coagulation
clinical correlations
classic hemophilia, 970
cytochrome P450 and warfarin therapy, 986cc
heparin therapy, 375
clotting cascade, 974–976
clotting factors, 969t, 1125–1126
extrinsic pathway, 969–970
fibrinolysis, 974
intrinsic pathway, 970
proteins involved in, 970–974
stages of, 966–969
vitamin K deficiency and, 1125–1126
Blood groups, 376
and carbohydrate metabolism, 373cc
sugars determining, 373, 373cc
Blood pressure regulation, **849**, 859-861
aldosterone in, 911–912
prostaglandins and, 466
see also Vasopressin
Bohr effect, **131**, 131–132, **1033**
Bombesin, 856t
Bonding
chemistry review, 1152
5′-phosphate-5′-phosphate, **692**
high-energy, **243**, 244t
strength of typical protein bonds, 72t
Bone in acid-base homeostasis, 1046cc
Bovine pancreas deoxyribonuclease (DNase I), 613t
Bovine spleen phosphodiesterase, 613t
Bradykinins, 34t, 856t
Brain
glucose uptake mechanisms, 295–296
peptides found in, 938t
Branched-chain amino acids, 492–496, 493cc, **579**
Branched-chain fatty acids, **390**
Branching enzyme, **344**
Branch migration, **676**
Brush border, **1063**
Bubble form of DNA, 621
Buffering (buffers), **11**, 11–13
bicarbonate, 1041–1045
in metabolic acidosis, 13cc
blood buffer line, 1053–1054
dissociation constants, 1040t
hemoglobin, **1036**
problems of, 12f
Buffering line of blood, **1045**, 1053–1054
Buffering of blood glucose, **338**
Butyryl CoA in fatty acid synthesis, 392–395
B vitamin coenzymes, 151–153

C codon base pairing, 730t
C_1 inhibitor, 109t
C peptides, **48**, **755**
C proteins, 957t, **958**
C regions, **94**, 98–101
Ca^{2+}
in blood cocagulation (as Factor IV), 969
in cytostasis, 922
dietary and osteoporosis, 1138cc
in glycogenolysis, 353–355
in glycogen phosphorylation, 350
mitochondrial transport, 269
in muscle contraction, 962–966
neural control, 355
parathyroid hormone and, 1122–1123
in renin–angiotensin system, 912
as second messenger, 228–229
uptake and release mechanisms, 873–875
vitamin D and homeostasis, 913–914, 1121–1122
Ca^{2+}-ATPase, **228**
Cachexia in cancer, 602cc
Calcitonin, 857t
Calcium. *See* Ca^{2+}
Calf thymus deoxyribonuclease (DNase II), 613t
Calmodulin, **228**, **347**
Caloric density, **1108**
Caloric homeostasis, **583**
Caloric value, **243**
Calpains, **763**
cAMP, 589–591
energy-rich bonding properties, 245
in gluconeogenesis, 588
in glycogenolysis, 353–354
in glycogen phosphorylation, 189–190, 349–350
as protein kinase activator, 319–322, 349–350, 590–591, 873
repression of, 814–815
as second messenger, 318–319
cAMP-CAP complex, 814–815, 821–822
cAMP-dependent protein kinase, **319**, 319–322, 349–350, 590–591
Cancer
Ames test of carcinogenicity, 658cc
antioxidants and risk reduction, 1136
chemotherapy and purine–pyrimidine metabolism, 566–569
clinical correlations
cachexia in, 602cc
carcinogenesis and RNA transcription, 707cc
DNA mutagens and, 658cc
DNA repair diseases and susceptibility, 662cc
oncogenes and antioncogenes, 707cc
recombinant tumor identification techniques, 783cc
vitamin A deficiency and, 1119, 1120
CAP binding protein, **813**, 815, 821–822
Cap structures of RNA, **692**
Carbamino groups, **1035**
Carbamino hemoglobin, **1035**
Carbamoyl phosphate synthetases, **482**, II, **547**, 550
deficiency, 487cc
Carbohydrate intolerance, **1104**, *see also* Diabetes mellitus
Carbohydrate loading, 1103cc
Carbohydrate metabolism
clinical correlations
alcohol and barbiturate metabolism, 307
angina pectoris and myocardial infarction, 318
arsenic poisoning, 309
aspartylglycosylaminuria, 374
blood group substances, 373
diabetes mellitus, 312
fructose intolerance, 309, 366
galactosemia, 368
glycogen storage disease, 340–341
hemolytic anemia and pyruvate kinase deficiency, 324
hemolytic anemia (drug-induced), 362
heparin as anticoagulant, 375
hypoglycemia and alcohol intoxication, 336
hypoglycemia in premature infants, 325
lactic acidosis, 316
malignant hyperthermia, 316
mucopolysaccaridoses, 379
non-hemolytic jaundice, 369
pentosuria, 368
complex carbohydrate synthesis, 372–373
gluconeogenesis, 324–337
glycogenolysis and glycogenesis, **337**
debranching enzyme in, 341–343
glycogen as energy source, 337–340
glycogenin as primer, 344–346
glycogen phosphorylase in, 340–341
glycogen synthase in, 343–344
regulation, 346–356, *see also* Insulin
glycolysis, **293**, 293–297
inhibition, 308
NADH reduction and reoxidation, 304–306
shuttle mechanisms in, 306–308
stages of, **293**, 297–304
glycoproteins in, 373–378
metabolic pathways
overview, 292–293
special, 360
pentose phosphate pathway, 361–365
proteoglycans in, 378–383
sugar interconversions and nucleotide sugar formation, 365–372
see also Carbohydrates
Carbohydrates
chemistry review, 1152–1154
complex, 372–373, 1107–1108
dietary, 1078t, 1103–1104, 1104cc

metabolic adaptation and, 1108cc
refined vs. unrefined, 1107–1108
digestion and absorption, 1077–1081
in glycoproteins, 70–71, 374–375
linkages with proteins, 71
membrane, 205–206
metabolism. *See* Carbohydrate metabolism
Carbon dioxide
in pyrimidine synthesis, 547–550
total plasma, **1034, 1052**
transport, 1033–1038
isohydric carriage, **1034**
Carbon dioxide-bicarbonate buffer system, 9, 1041–1045
Carbonic acid formation, 10, 1034–1036
Carbonic anhydrase, **1034**
Carbon monoxide, 1017, 282cc
Carbonylcyanide-*p*-trifluromethoxyphenylhydrazone (FCCP), **283**
γ-Carboxyglutamic acid, **109**, **967, 1125**
Carboxylases in blood coagulation, 971
Carboxylic acid chemistry, 1149
Carboxyl proteases, 1074t
Carboxypeptidases, **1073**, 1074t
Cardiolipin (phosphatidylglycerol phosphoglyceride), 197, 199, **426**
Carnitine, **408**, 408-409, 514–515
deficiency, 409cc
Carnitine palmitoyltransferase I (CPT I), **408**
Carnitine palmitoyltransferase II (CPT II), **409**
Carnosine, **512**
Carotenoids, **1118**
Catabolic reactions, **238**
Catabolite repression, **814**
Catalase, **21,** 21t, **138**, 428
Catalysis, 172–178
Catalytic constant (turnover number), **144**
Catecholamines
metabolism, 521–522
as neurotransmitters, **935**, 935–936
synthesis, 509f
see also individual catacholamines
Catechol-*O*-methyl transferase (COMT), **521, 935**
Cation-exchange chromatography, 43–44
—CCA_{OH} terminus, **687**
cDNA library, **834**
Cell–cell communication and steroid hormones, 909–911
Cell cycle, nucleotide metabolizing enzymes in, 557–559
Cell membrane. *See* Membranes; Plasma membrane
Cell nucleus. *See* Nucleus
Cells
cardiac, 262–263
chemical composition, 4t, 14–16, 627t
enzyme activity, 187–188
eucaryotic, structure and function, 1–22, *see also* Eucaryotic cell structure and function
procaryotic. *See* Procaryotic cells
programmed death in ovarian cycle, 878–879
transmembrane signaling, **928**
tumor. *See* Cancer; Tumorigenesis structure
Cellulose, **1105**, 1105t, 1154
Centrifugation, 15f, 628–629
Cephalin (phosphatidylethanoloamine), 197, 199
Ceramide, **450**, 450–451
Cerebrosides, **202**, **1156**
Ceremide trihexoside, **454**
Ceruloplasmin (ferrodoxidase I), **1008**
cGMP, **894**, 894–897
Chain termination and release, **696**
Channels, membrane, **220**, *see also* Membrane transport
Chelates, 153
Chemical coupling hypothesis, **284**
Chemiostatic coupling hypothesis, **285**, 285–286
Chemotherapeutic drugs
antifolates, 567–568
antimetabolites, 566–567
glutamine antagonists, 569
Chenodeoxycholic acid, **447**
Chloride. *See* Cl^-
Chloride shift, **1038**
Cholecalciferol. *See* Vitamins, D_3
Cholecystokinin (CCK), 856t, **1066**
Cholera, 1072cc
Cholera toxin, 456, 1065
Cholesterol, **204**, **902**, **1156**
in cell membrane, 204
clinical correlations
atherosclerosis, 446
gallstones, 1086
hypercholesterolemia, 445
polyunsaturated fatty acids and heart disease, 1106
excretion, 447–448
and membrane fluidity, 214
in plasma, 446–447
lipid consumption and, 1100–1107
side chain cleavage, **905**, 993f
in steroid hormone synthesis, 905–907, 908f
structure and function, 438–439
synthesis, 439–446
in vitamin D synthesis, 448–449
see also Lipoproteins
Cholic acid, **447**
Choline, **519**, 519–520
Choline acetyltransferase, **934**
Chondroitin sulfate, **380**, 380–384
Chorionic gonadotropin, **880**
Christmas factor (Factor X), 969t
Chromaffin granules, **862**
Chromatin, **638**
Chromatography techniques, 41, 43–44, 86
Chromatosomes, **638**
Chromium, dietary, 1142–1143
Chromosomes
interspersion pattern, **647**
structure, 638–641
SV (simian virus), 696
yeast artificial (YAC), **787**, 787–789
Chromosome walking, **788**
Chylomicrons, **67**, **405**, **1087**, 1087–1088
Chymotrypsin, 46, 1074t, 176–177
α, 106t, 109t, 112–115
UV absorption spectrum, 81f
Circular dichroism spectroscopy, **83**
Cis-acting mutations, 820
Cis conformation, **32**, 33–34
Cis-dominant (operator-constitutive) mutations, **811**
Citrate cleavage enzyme, **397**
Citrate synthase, **253,** 255–256
Citric acid cycle. *See* Tricarboxylic acid cycle
Citrulline, 483, **579**
Cl^-
in acid-base balance, 1054–1056
membrane transport, 224
Clathrin, **445**, **763**, **886**, 886–888
Cl^-/HCO_3^- exchanger, 224, **1069**
Clinical correlations
acute intermittent porphyria, 1012
acyl CoA debydrogenase deficiency, 410
albinism, 508
alcoholism, 1128 and nutrition
alcohol metabolism, 307
allergy, 96
Amanita phalloides poisoning, 699
amino acid analysis in diagnosis, 42
amino acid metabolism
branched chain amino acid-related diseases, 492
cystine-related diseases, 503
folate-related diseases, 518
folic acid deficiency, 517
glutathione deficiency, 523
histidinemia, 512
hyperammonemia and hepatic coma, 483
lysine- and ornithing-related diseases, 515
nonketotic hyperglycinemia, 498
Parkinson's disease, 509
phenylketonuria, 506
propionate- and methylmalonate-related diseases, 496
transulfuration defects, 503
urea cycle enzyme deficiency diseases, 487
antigen–antibody responses, 70
aspartylglycosylaminuria, 374cc
atherosclerosis, 446cc
autoimmune connective tissue diseases, 717
barbiturate metabolism, 307
blood groups, 373cc
branched chain amino acid related diseases, 492
cancer
carcinogenesis and RNA transcription, 707
DNA mutations and, 658
recombinant identification techniques, 783
serine proteases in tumor cell metastasis, 105

carbohydrate metabolism
  alcohol and barbiturate metabolism, 307
  angina pectoris and myocardial infarction, 318
  arsenic poisoning, 309
  aspartylglycosylaminuria, 374
  blood group substances, 373
  diabetes mellitus. *See* diabetes mellitus
  fructose intolerance, 309, 366
  galactosemia, 368
  glycogen storage disease, 340–341
  glucose tolerance tests, 312
  hemolytic anemia and pyruvate kinase deficiency, 324
  hemolytic anemia (drug-induced), 362
  heparin as anticoagulant, 375
  hypoglycemia and alcohol intoxication, 336
  hypoglycemia in premature infants, 325
  malignant hyperthermia, 316
  mucopolysaccharidoses, 379
  non-hemolytic jaundice, 369
  pentosuria, 368
carnitine deficiency, 409
cholera, 1072
collagen diseases, 61, 717, 760–762
congenital adrenal hyperplasias, 994
cyanide poisoning and electron transport, 282
cyanosis, 1029
cystic fibrosis, 1120
cystine-related diseases, 503
cytochromes P450
  21-hydroxylase deficiency, 994
  induction of drug-metabolizing enzymes, 986
  pregnancy and steroid hormone production, 996
diabetes mellitus, 312
  complications and polyol pathway, 598
  glycosylated hemoglobin monitoring, 71
  high carbohydrate-high fiber diets, 1104
  hyperglycemic, hyperosmolar coma, 584
  insulin-dependent disease, 597
  ketoacidosis (diabetic coma), 417
  noninsulin-dependent disease, 600
  primary structure of insulins, 49
diagnostic use of amino acid analysis, 42
digestion and absorption
  cholera and electrolyte replacement, 1072
  disaccharidase deficiency (milk intolerance), 1079
  gallstones, 1086
  Hartnup disease (neutral amino aciduria), 1077
  β-lipoproteinemia, 1087
DNA
  cancer and mutations, 658
  diagnostic use of probes, 626
  genetic mapping, 677
  xeroderma pigmentosum, 662
drug resistance, 825
drugs
  anticonvulsant drugs and vitamin requirements, 1126
  heparin therapy, 375
  oral contraceptives and vitamin B_6 requirements, 1131
Ehlers-Danlos syndromes, 760–761cc
folate metabolic diseases, 518
folic acid deficiency, 517
fructose intolerance, 366
gallstones, 1086
gas transport
  abnormal hemoglobins, 1032
  chemically modified hemoglobins, 1031
  cyanosis, 1029
Gaucher's disease, 461
gene expression
  Leber's hereditary optic neuropathy, 840
  transmissible multiple drug resistance, 825
glutathione deficiency, 523
glycogen storage diseases, 340–341
glycoprotein function, 70
gout, 19, 546
Hartnup disease (neutral amino aciduria), 1077
heart disease and polyunsaturated fatty acids, 1106
heme and neonatal isoimmune hemolysis, 1019
hemochromatosis, 1008
hemolytic anemia
  drug-induced, 362
  and pyruvate kinase deficiency, 324
hemophilia, 970
heparin as anticoagulant, 375
histidinemia, 512
hormones
  anterior pituitary activity tests, 852
  hypopituitarism, 854
  lithium treatment of manic depressive illness, 876
  oral contraception, 912
  ovarian cycle and programmed cell death, 922
21-hydroxylase deficiency, 994
hyperammonemia and hepatic coma, 483
hypercholesterolemia, 445
hyperglycemia and protein glycosylation, 585
hyperlipidemias, 67, 445
hyperlysinemia, 515
hyperproinsulinemia, 755
hypoglycemia and alcohol intoxication, 336
hypolipoproteinemias, 68
hypopituitarism, 854
I-cell disease, 753
immune response
  antibody class functions, 96
  complement proteins, 96
  immunization, 97
iron metabolism
  acute intermittent porphyria, 1012
  hemochromatosis, 1008
  iron-deficiency anemia, 1008
ketoacidosis
  diabetic, 417
  methylmalonate-related, 496
  proprionate related, 496
lactic acidosis, 252, 315
Leber's hereditary optic neuropathy, 840
Lesch–Nyhan syndrome, 544
lipid metabolism
  acyl CoA dehydrogenase deficiency, 410
  atherosclerosis, 446
  carnitine deficiency, 409
  diabetic ketoacidosis, 417
  Gaucher's disease, 461
  genetic lipid-transport abnormalities, 405
  hypercholesterolemia, 445
  obesity, 404
  Refsum's disease, 414
  respiratory distress syndrome, 428
A-β-lipoproteinemia, 1087
lysinuric protein intolerance, 515
malignant hyperthermia, 316
manic depressive illness and lithium treatment, 876
Marfan syndrome, 761
McKees Rocks hemoglobin disorder, 732
membrane fluidity in disease states, 214
membrane transport, liposomes as drug and enzyme carriers, 208
metabolic acidosis, 13
metabolic alkalosis, 1056
metabolic interrelationships
  diabetes mellitus, 584, 597, 598
  hyperglycemia and protein glycosylation, 585
  hyperglycemic, hyperosmolar coma, 584
  obesity, 576
  protein malnutrition, 577
  Reye's syndrome, 582
  starvation, 577
milk intolerance, 1079
mitochondrial myopathies, 268–269
mucopolysaccharidoses, 379
myasthenia gravis
Mycobacterium tuberculosis, 699
myocardial infarction, 104
neonatal isoimmune hemolysis, 1019
nonhemolytic jaundice, 369cc
nonketotic hyperglycinemia, 498
nucleotide metabolism
  immunodeficiency disease and purine nucleotides, 544
  Lesch-Nyhan syndrome, 544
  orotic aciduria, 548

nutrition
anticonvulsant drugs and vitamin requirements, 1126
cancer cachexia, 602
carbohydrate loading, 1103
cystic fibrosis, 1120
diabetes and high carbohydrate-high fiber diets, 1104
diabetes mellitus, 597, 598, 600
fad diets, 1102
heart disease and polyunsaturated fatty acids, 1106
hypertriglycerolemia and metabolic adaptation, 1108
iron availability calculations, 1140
iron deficiency anemia, 1141
low-protein diets and renal disease, 1099
neonatal nutrition, 1126
oral contraceptives and vitamin B_6 requirements, 1131
renal osteodystrophy, 1122
total parenteral nutrition and acidosis in infants, 1055
vegetarian diets, 1098, 1145
obesity, 404, *see also under* nutrition
oral contraception, 912
ornithine-related diseases, 515
orotic aciduria, 548
ovarian cycle and programmed cell death, 922
Parkinson's disease, 509
pentosuria, 368
pH
bone in acid-base homeostasis, 1046
laboratory errors, 1042
metabolic alkalosis, 1056
respiratory acidosis, 1052
respiratory alkalosis, 1051
salicylate poisoning, 1053
total parenteral nutrition and acidosis in infants, 1055
phenylketonuria, 506
protein synthesis
collagen diseases, 760–762
familial hyperproinsulinemia, 755
I-cell disease, 753
thalessemia, 733
recombinant techniques
HIV screening, 799
HSV I site-directed mutagenesis, 795
restriction mapping and evolution, 773
transgenic animal models, 800
tumor identification, 783
Refsum's disease, 414
renal failure, 1099, 1122
respiratory distress syndrome, 428
Reye's syndrome, 582cc
RNAs
Amanita phalloides poisoning, 699
autoimmunity in connective tissue diseases, 717
carcinogenesis and transcriptional factors, 707
Mycobacterium tuberculosis, 699
staphylococcal erythromycin resistance, 699
thalassemias, 714
salicylate poisoning, 1053
sickle cell anemia
hemoglobin and gene mutation, 50
prenatal diagnosis, 836
steroid hormones in pregnancy and cytochromes P450, 996
thalassemias, 714, 733
prenatal diagnosis, 837
transsulfuration defects, 503
tyrosinemias, 508
urea cycle enzyme deficiency diseases, 487
visual defects and gene arrangement, 953
xeroderma pigmentosum, 662cc
xylitol dehydrogenase deficiency, 368
Zellweger syndrome and peroxisome function, 21cc
CLIP (corticotropin-like intermediary peptide), 855t
Cloning techniques, 767–802, *see also* Recombinant techniques
Cloning vectors, **775**, 776–777
Closed DNA complexes, **701**
Clotting factors, 966–976, 969t, *see also* Blood coagulation
CO_2. *See* Carbon dioxide
CoA. *See* Acetyl CoA; Coenzyme A
Coagulation of blood, 966–976, *see also* Blood coagulation
Coated pits, **886**
Cobalamine (vitamin B_{12}), 1134–1136
Coding sequences, 712–714
Codons, **729**, 729t, 729–730
stop, 742–743
terminator disorders, 732cc
tRNA in mRNA recognition, 687–688
wobble base pairing rules, 731t
Coenzyme A
acetoacetyl, 416
acetyl. *See* Acetyl CoA
butyryl, 392–395
fatty acyl, 408–409
HMG, **414**, 414–415, **440**
linoleoyl, 412
malonyl, **393**
methylglutaryl (HMG), 414–415, **440**
propionyl, 411, **494**
succinyl, **258**, 258–259
synthesis, 562–563
Coenzymes
adenosine triphosphate (ATP), 150–151
cystathioninuria, 149
FAD (flavin adenine dinucleotide), 152–153, 182–183, 274–275, 987–990, 1128
FNM (flavin mononucleotide), 152–153, 182–183, 274–275, 1128
NAD (nicotinamine adenine nucleotide), 151–152, 182–183
NADP (nicotinamide adenine dinucleotide phosphate), 151–152, 182–183
nucleotides as components, 531
of pyruvate dehydrogenase complex, 251t
Q (ubiquinone), **277**, 277–279
water-soluble vitamins as, 1125–1132
Cofactors, **136**
Collagen, **61**
amino acid composition and sequence, 61–62, 62t
compared with elastin, 62t
covalent cross-link formation, 65–66
distribution in humans, 61
pro-, 758–759
structure, 62–65
synthesis
abnormal, 61cc, 760t, 761–762cc
posttranslational modifications in, 757–762
types, 63t, 757
Color vision, 951–954
Committed steps, **187**, **312**
in fatty acid synthesis, 393
in purine nucleotide synthesis, 536, 540–542
Compensated and uncompensated acid base states, **1050**
Competitive inhibitors, **157**
Complementarity, **615**
Complementary determining (C) regions, **94**
Complementary DNA (cDNA), 784
α-Complementation, 780–781
Complement proteins, 96cc
COMT (catechol-*O*-methyl transferase), **521**
Concatemers, **672**
Cone cells of the eye, 951–954
Conformational coupling hypothesis, **285**
Conformational states, **130**
Congenital adrenal hyperplasias, 994cc
Conjugase, **1132**
Conjugate acids and bases, 8–13
Conjugate bases, **1040**
Connective tissue diseases and RNA, 717cc, *see also* Collagen diseases
Consensus (conserved) sequences, **696**, 697t
Conservative replication, **663**
Conservative substitution, **49**
Conserved (consensus) sequences, **696**, 697t
Contraceptives. *See* Oral contraceptives
Contractile unit, **955**
Control elements, **807**
Convergent evolution, **112**
Cooperativity index, **126**
Cooperativity of ligands, **165**, 165–167
concerted model, 166
sequential-induction model, 166
Coordinate expression, **809**
Copper, 156, 157, **1142**
Coproporphyrinogen oxidase, **1016**
Coproporphyrinogens, **1010**
Core enzymes, **696**
Cori cycle, **324**, 324–327, 579–580
Cori's disease, 341cc
Coronary artery bypass, 319cc

Correlated spectroscopy (COSY), **85**
Corticosteroid binding globulin (transcortin), **915**
Corticotropin-like intermediary peptide (CLIP), 855t
Corticotropin releasing hormone, 852t, 854t
Cortisol, 903t, 909t, 910t
COS cells, 793
Cotransfection, **792**
Covalent bonds, **50**
Covalent catalysis, 175–176
Creatine, 518–519, **519**
Creatine kinase, 155, 184–185, **963**
Creatinine, 518–519
CRH (corticotropin releasing hormone), 852t, 854t
Crigler–Najjar syndrome, **1019**
Critical micellar concentration, **207**, **1084**
Cross-regulation, **189**
Crossover theorem in glycolysis, **313**, 313–316
C_0t-a-half value, **623**
CTP (cytidine 5′-triphosphate), 549–550
CTP synthetase, **549**
Cushing's syndrome, 576cc
Cutis laxa (Ehlers-Danlos syndrome type V), 761cc
Cyanide poisoning, 281, 282cc
Cyanogen bromide, **46**
3′,5′-Cyclic adenosine monophosphate. *See* cAMP
Cyclic GMP. *See* cGMP
Cycloheximide, 744t
Cyclooxygenases, **462**, 461–465
Cyclopentanoperhydrophenanthrene, **902**
Cystathionase, **505**
Cystathionine, **501**
Cystathioninuria, 149cc
Cysteine, **29**, 35t, 501, 504–506
Cysteine sulfinate, **505**
Cystic fibrosis and diet, 1120cc
Cystine, **33**, 33–34
Cystine aminopeptidase, **868**
Cystinosis, 503cc
Cystinuria, 42cc, 232cc, 503cc
Cytidine 5′triphosphate (CTP), 549–550
Cytochrome *c*, 1017, 278f
Cytochrome P450 17α-hydroxylase, **994**
Cytochrome P450 17,20-lyase, **994**
Cytochromes, **157**, **276**, 276–279
Cytochromes P450, 658, **981**
  clinical correlations
    21-hydroxylase deficiency, 994
    induction of drug-metabolizing enzymes, 986
    pregnancy and steroid hormone production, 996
  electron transport systems, 270, 987–990
  functions
    steroid synthesis and eicosanoid oxygenation, 990–994
    xenobiotic oxidation, 995–997, 996t
  inhibitors, 986–987
  nomenclature and overall reaction, 982–984
  side chain cleavage enzyme complex, 905, **990**, 990–994
  types, 984–986
Cytoplasmic enzymes and metabolic pathways, 188t
Cytosine arabinoside (AraC), 567
Cytoskeleton functions, 21–22
Cytosol, **22**
Cytostasis, **922**

Dansyl chloride assays, 44–45
Davison–Danielli membrane model, 208
Debranching enzyme, **341**
Decarboxylases, **140**
Decarboxylation of nucleotide sugars, 370–371
Dehydratases, **140**
7-Dehydrocholesterol, **913**
Dehydrogenation, **407**
Dehydropiandrosterone, 903t, **907**
Deletion mutations, **794**
Denaturation, 75–76, 620–622, **621**
2-Deoxyglucose, **308**
Deoxyribodipyrimidine photolyase, **662**
DephosphoCoA kinase, **563**
DephosphoCoA pyrophosphorylase, **563**
Derepression, **815**
Dermatan sulfate, 379cc, **381**
Desaturation of fatty acids, **399**
Desmosine, **66**
Dextrin, phosphorylase-limit, **341**
DHAP (dihydroxyacetone phosphate), **300**
dhfr (dihydrofolate reductase) gene, 792
Diabetes mellitus
  clinical correlations
    complications and polyol pathway, 598
    diabetic ketoacidosis, 417
    glucose tolerance test, 312
    glycosylated hemoglobin HbA_{1c} test, 71
    high carbohydrate, high fiber diets, 1104
    hyperglycemic, hyperosmolar coma, 584
    insulin-dependent (juvenile onset; type I) disease, 597
    noninsulin-dependent (maturity onset; type II) disease, 600
  compensatory mechanisms for acid-base regulation, 1050
  diabetic ketoacidosis, 1047
  glucose tolerance factor (GTF) and chromium, 1142–1143
  insulin-dependent (juvenile onset; type I), 597
  noninsulin-dependent (maturity onset; type II), 597–598
    and obesity, 1102–1103
  in pregnancy, 596
  refined carbohydrates and, 1108
Diacylglycerol (DAG), **874**, 874-875
6-Diazo-5-oxo-L-norleucine, **569**
Dicarboxylic acid, 414
Dicumarol, **1125**
Dielectric constant, **74**
Dietary Goals for the United States, 1106–1110, 1110t
Diet-induced thermogenesis, **1101**
Diet
  clinical assessment, 1116–1117, 1144
  clinical correlations
    carbohydrate loading, 1103
    cystic fibrosis, 1120
    fad diets, 1102
    fiber, 1104
    high carbohydrate-high fiber diets in diabetes, 1104
    iron-fortified and hemochromatosis, 1009
    low-protein and renal disease, 1099cc
    osteoporosis, 1138
    total parenteral nutrition, 1100
    vegetarian diets, 1098, 1145
  fatty acid source, 392
  fiber in, **1105**, 1105t, 1109
  membrane fluidity and, 314
  "prudent", 1106–1110,1110t
  in renal failure, 600
  vegetarian, 1108
  see also Digestion and absorption; Nutrition
Diffusion, **215**, *see also* Membrane transport
  Fick's first law of, **216**
Digestion and absorption
  bile acid metabolism, 1088–1089
  carbohydrates, 1077–1081
  clinical correlations
    cholera and electrolyte replacement, 1072
    cholesterol stones, 1086
    disaccharidase deficiency (milk intolerance), 1079
    Hartnup disease (neutral amino aciduria), 1077
    A-β-lipoproteinemia, 1087
  epithelial transport, 1067–1073
  gastrointestinal organs and functions, 1061–1062
  historical aspects of study, 1060–1061
  iron absorption, 1005–1006
  lipids, 1081–1088
    overview
    gastrointestinal organs and functions, 1061–1062
  pancreatic enzymes, 1062–1063, 1063t
    secretagogues, 1064–1067, 1065t
    zymogens, 1063–1064
  protein, 1073–1077
  see also Diet; Nutrition
Dihydrofolate reductase, **517**
  deficiency, 518cc
Dihydroorotase, **548**
Dihydroorotate dehydrogenase, **548**
Dihydrotestosterone, **907**
Dihydroxyacetone phosphate (DHAP), **300**
1α,25-Dihydroxycholecalciferol, **1121**, 903t
3,4-Dihydroxyphenylalanine, *See* DOPA
Dimethyllysine, **514**

2,4-Dinitrophenol, **283**
Dioxygenases, **138**
Dipalmitoyllecithin, **427**, 427–428
Dipeptidyl aminopeptidase IV, 1063t
Diphosphatidylglycerol (phosphatidylglycerol phosphoglyceride), 197, 199
2,3-Diphosphoglycerate (DPG), **1029**
Diphthamide, 757
Diphtheria toxin, 744t, **745**
Dipole, 5
Directional cloning, **776**
Disaccharide digestion and absorption, 1077–1080
Dispersive replication, **664**
Dissociation, 6–8
  Bohr effect, 131–132
Dissociation constants, 10t
Distance matrix, **51**, 52f, 125
Disulfide isomerase, **756**
Disulfite bonds, 1152
Diuretic drugs, furosemide, **1071**
DNA
  A, B, and C forms, **615**, 615–617
  antisense, 799–800
  chromosome structure and, 638–641
  clinical correlations
    cancer and mutations, 658
    genetic mapping, 677
    probe techniques, 662
    xeroderma pigmentosum, 662
  closed and open complexes, **701**
  complementary (cDNA), **784**, 784–785
  control elements, **807**
  denaturation, **621**, 621–622
  functions, 644–649
  general properties, 608–609
  heteroduplex techniques, **626**
  hybridization experiments, 624–625
  internucleosomal cleavage, **922**
  in vivo phosphodiester bond formation, 649–653
  leading and lagging strands, **667**
  mammalian gene expression, 828–829
  measurement techniques, 627–629
  mutations, 653–657, *see also* Mutations
  nascent (Okazaki fragments), **666**, 671, 673
  nucleotide sequences, 677
    classification, 646
    in encoding, 642–644
    flanking regions, 793–794
    highly reiterated, **646**, 648–649, 828–829
    interspersion pattern, **647**
    inverted repeats, **646,** 649
    inverted terminal, **824**
    linker, 638–639
    moderately reiterated, **646**, 647–648, 830–832
    operator regions, **810**, 821
    promoter, **811**
    repetitive, **646**, **648**, 829–832
    satellite, 647
    segregated tandem arrays, **648**
    single copy, **646**, 647
  primer and template strands, **650**
  probe techniques, 625–626, 781–782, *see also* Recombinant techniques
  reassociation kinetics
    eucaryotic, 624f
    procaryotic, 623f
  recombinant, 675–677, 773–778, *see also* Recombinant techniques
  relaxed, **631**
  renaturation, **622**, 622–624
  repair, 657–659
    error-prone (SOS), 662
    eucaryotic, 662
    excision, 659–661
    mismatched, **657**
    postreplication, **652**
    reversal mechanisms, 662
    responsive element of, **918**
  sequencing
    Maxam–Gilbert procedure, 771–772
    Sanger procedure, 772–773
  structure, 609–610
    circular, **630**, 630–635
    double helix, 614–619
    double stranded circles, 630
    eucaryotic vs. procaryotic, 637–641
    histones and nonhistone proteins, 637–641
    linearity and circularity, 630–637
    nuclease hydrolysis, 612–613
    nucleotide-phosphodiester bonds, 610–612
    periodic (secondary) structures, 613–619
    single-stranded, 630
    size, 627–629
    stabilization of, 619–627
    superhelical, **631**, 631–635
    types, 627–631
  synthesis
    bidirectional, **667**
    complementary strands and, 662–669
    conservative, **663**
    dispersive, **664**
    E. coli as model, 669–673
    eucaryotic, 673–675, 674–675
    nucleotides in, 557–559
    nucleus in, 17–19
    rolling circle model, 672–673
    semiconservative, **664**
    template function, 682
  transcription to RNA, **695**, 695–717, *see also* Protein synthesis
  Z form, 616–618, **617**
DNA binding proteins, **92**, 116–118
DNA glycosylases, **660**, 660–661
DNA polymerases
  DNA-dependent, 649–683
  mutagenic errors, 656–657
  as primers, 664–665
DNA probe techniques, 662cc, 625–626, 781–782
DNase I, 613t
DNAse II, 613t
Docking proteins, **747**
Dolichol phosphate, **376**, 376–378, **749**
Domains
  antigenic, **919**
  DNA binding, **919**
  of hormone receptors, 884–886
  ligand binding, **919**
  of membrane proteins, 210
  of polypeptide strands, **57**, 57–60
  of protein kinases, 894–897
  of steroid receptors, 918–920
  of t-PA, 974
DOPA (3,4-dihydroxyphenylalanine), **507**, 507–509, 509cc
Dopamine, **507**, 507–509, 509cc, 854t
Dopamine β-hydroxylase, **157**
Double helix, 614–619
  A, B, and C forms, **615**, 615–617
  antiparallelism of, **618**
  elongation, 702
  stability factors
    denaturation, 620–622
    hybridization experiments, 624–625
    renaturation, **622**, 622–624
  stretching (intercalation), **657**, 659
  Z form, 616–618, **617**
  see also DNA, structure; Helix formations
Double sieve mechanism, **735**
Down-regulation, **445**, **921**
  of steroid receptors, 920–921
DPG (2,3-diphosphoglycerate), **1029**, 1038–1039
Drugs
  ALA synthase induction and porphyria, 1016
  anesthetics and membrane fluidity, 214
  antibiotics, 371
    erythromycin resistance, 692cc
    inhibition of protein synthesis, 749
    ionophore and membrane transport, 231–234, 232t
    penicillamine, 1132
    protein synthesis inhibiting, 744t, 744–746
    resistance and bacterial transformation, 778–781
    transmissible multiple resistance, 825cc
  anticoagulant, 375cc
  anticonvulsant drugs and vitamin requirements, 1126
  anti-inflammatory agents
    prostaglandin inhibition, 465
    salicylate poisoning, 1053cc
  antitumor, 569
    resistance to, 570cc
    purine–pyrimidine metabolism and, 566–569
  barbiturate metabolism, 307cc
  chemotherapeutic and purine–pyrimidine metabolism, 566–569
  diuretics, furosemide, **1071**
  as enzyme inhibitors, 159–162
  enzymes as, 186
  heparin, 375cc
  liposomes as carriers, 208cc
  lithium in manic depressive illness, 876cc

nitroglycerin in angina pectoris, 318cc
nutrient interactions, 1134, 1146t
oral contraceptive, 912cc, 1131cc
prostaglandins as, 465–466
salicylate poisoning, 1053cc
Dubin–Johnson syndrome, **1019**
Ductus arteriosus, 466
Dynamic aspects of protein structure, 76–78
Dynamic disorder, **79**
Dynamic equilibrium, **478**
Dynesin, **21**

E apoprotein, **445**
Edman reaction, **45**
Effector control of glycogen metabolism, 351
EF hand, **229**
Ehlers-Danlos syndrome
type IV, 760cc
type VI, 761
type VII, 761cc
Ehrlich tumor cells, 532
Eicosanoids, **462**
Eicosapentaeinoic acids, 470, 1106cc
eIF-4D (eucaryotic initiation factor 4D), 522
eIF2 (eucaryotic initiation factor 2), **737**
eIF3 (eucaryotic initiation factor 3), **737**
18 SRNA, **688**
Elastase, 106t, 1074t
Elastin, **65**, 66f
compared with collagen, 62t
Elderly persons, nutritional needs, 1145cc
Electrogenicity, **219**
Electrolyte balance in cholera, 1072cc
Electrolytes, **6**, 6–7
non-, **6**
weak, **7**
Electron acceptor (oxidant), **270**
Electron-density distance mapping, 80f, **81**
Electron donor (reductant), **270**
Electron transport, **270**, 272–283, 306
clinical correlation: cyanide poisoning, 282
components involved in, 272–279
cytochrome P450 in, 987–989
inhibition, 279–281
reversibility, 281–283
site, 277–279
see also Mitochondrial membranes
Electrophoresis techniques, **38**, 38–41, 86
Electrostatic interactions in proteins, 74
ELISA (enzyme-linked immunoassay), 183–184, 460–461
Elongation factors (Elfs), **741**, 741–742, 757
Elongation of fatty acids, **398**
Elongation steps in protein synthesis, 737–739
Emden-Meyerhof pathway, 293–324, *see also* Glycolysis
Endergonic and exergonic reactions, **241**
Endocrine hormones, **849**
Endocytosis, **886**, 886–888
receptor-mediated, **445**
Endonucleases, **613**, **718**, 719t
apurinic-apyrimidinic (AP), **660**
restriction, **642**, 642–644, **769**, *see also* Recombinant techniques
specificities, 613t
Endopeptidase, 1063t, 1073, 1076, *see also* Proteases
Endoplasmic reticulum, **18**
in collagen synthesis, 758–759
cytochrome P450 system, 987–988
enzymes and metabolic pathways, 188t
function, 18
in gluconeogenesis, 329
in protein synthesis, 746–753
β-Endorphin, 855t, **923**, 938t
Endorphins, **939**
Energy production
fatty acids in, 407–417
measures
basal metabolic rate, **1095**
kilocalorie, **1094**, 1094–1095
nutrition and, 584–585
Recommended Dietary Allowances (RDAs), **1095**, 1096
starve-feed cycle and, 584–585
see also Nutrition; Oxidative metabolism
Energy reserves, **584**, 584t
Energy transfer, 246–247
Enkephalins, 34t, **861**, 938t
Enolase, **303**
Enterokinase, 1075
Enteropeptidase (enterokinase), 1063t, **1075**
Entropy, **241**
Entropy effect, **176**, 176–177
Enzyme-linked immunoassay (ELISA), 183–184
in Tay-Sachs disease, 460–461
Enzymes
allosteric control mechanisms, 162–167
amplification of regulatory signals, 190–191
basic terminology, 136–137
bifunctional, **302**
binding specificity, 167–170
catalytic mechanisms, 171–179
cell cycle and nucleotide metabolizing, 557–559
classification, 137–141, 141t
clinical applications, 179–186
clinical correlations
alcohol dehydrogenase activity, 180
alcohol sensitivity in Asians, 147
cystathioninuria, 149
drug-metabolizing enzymes, 986
enzyme deficiencies, 181
gout, 19, 146, 163, 183
hemolytic anemia, 179
immune disease, 544–545
liposomes as carriers, 208
myocardial infarction enzyme assay, 180
coenzyme structure and function, 149–157, *see also* Coenzymes *and individual coenzymes*
core, **696**
coupled reactions and free energy, 245–246
dephosphorylation, 320
as drugs, 186
function, 26
general concepts, 136–137
in hepatic adaptive metabolism, 589–593
holo-, **696**
immunoglobulins, artificial (abzymes), **177**, 177–178, *see also* Immunoglobulins
inducible, **312**
induction, **807**
inhibition, 157–162
drugs in, 159–162
kinetics, 141–144
reversible reactions, 148–149
saturation, 144–148
in lipid bilayer, 210–211
in long-term adaptation, 591–592
lysosomal, 19–22
metal cofactors, 153–157
metallo-, **155**
multidomain, **57**
NAD^+-linked dehydrogenases, 272–274, 273t
pancreatic, 1062–1063, 1063t
phospholipid activation, 427
phosphorylation, 320
pro-, **109**
processive, **650**
proteolytic, 102–105, 105t
rate-limiting, **187**
regulatory, **309**
in glycolysis, 309–324
regulatory mechanisms, 186–190
repression, **808**
salvage pathways, 542, 551
serine proteases, 102–116
terminology of measurement, 144
see also Proteins *and individual enzymes and pathways*
Epidermal growth factor, 857t
Epimerases (racemases), **140**
Epimerization, 367–369
Epinephrine, **862**
cAMP and, 190f
in glycogenolysis, 353–355
in glycogen phosphorolysis, 190–191
as neurotransmitter, **935**
synthesis, 508–509, 862–864
Epithelial transport in digestion and absorption, 1067–1073
Epoxide hydrolase, **990**
Equilibrium
dynamic, **478**
near-equilibrium reaction, **309**
nonequilibrium reaction, **309**
Equilibrium constants, **7**, **242**, 242–243
for hepatic glycolysis and gluconeogenesis, 310t
for oxygen binding of myoglobin, 124–126
ErbA protooncogene, 918–920
Ergocalciferol. *See* Vitamins, D_2
Error-prone DNA repair, **662**
Erythrocyte ghosts, **227**

Erythrocytes
  buffer systems, 1039–1041
  2,3-diphosphoglycerate (DPG) in, 1038–1039
  genetic variation and G6P deficiency, 362cc
  glucose metabolism in, 303
  glutathione in, 365, 522–523
  intravascular hemolysis, 1020–1021
  membrane structure, 212, 212f
  nucleotide synthesis in, 542
  PRPP utilization, 565–566
Erythromycin, 744t, **745**
  resistance and RNA, 692cc
Erythropoietic porphyria, 1011t
Erythropoietin, 857t
Escherichia coli
  cellular structure, 3f
  DNA polymerases, 649–652, 650t
  nucleoprotein structure, 641
  operons in, 807–808
    lactose, 808–815
    tryptophan, 815–817
  plasmid pS101, **775**
  pyruvate dehydrogenase complex in, 249–250
  RNA polymerases in, 696–698
  see also Gene expression
Esters, chemistry of, 1149
Estradiol, **878**, 878–880, 909t, 910t, 912–913
  17β-, 903t
Estrogens, 902, 905t, 909t, 994–995, *see also* Ovarian cycle; Hormones, steroid
Ethanol
  hepatic oxidation, 600–601, 600–603
  inhibition of gluconeogenesis, 336cc, 336–337
  metabolism, 306
Ethanolamine (phosphatidylethanoloamine), 197, 199
Ethanolamine phosphotransferase, **434**
Eucaryotic cells
  expression vectors in, 790–792
  glycosyltransferases in, 748t
  RNA synthesis, 703–708
  structure and function, 1–13, 4t, 17t
    lysosomal enzymes and gout, 19cc
    metabolic acidosis, 13cc
    organization and composition, 13–16
    subcellular organelles and membranes, 16–22, 17t
  transfected, 792–793
Eucaryotic gene expression, 828–840, *see also* Gene expression
Eucaryotic initiation factors, **522**, **737**, 737–739
Eukaryotic transcription, 703
Excision DNA repair, 660–661
Excitation energy transfer, **83**
Exercise, aerobic vs. anaerbic, 593–595
Exergonic and endergonic reactions, **241**
exo-1,4-α-glucosidase (glucoamylase), 1063t, 1080t
Exocrine secretion, **1063**, 1063–1067
Exoglycosidases, 377–378
Exons, **101**, **643**, **714**, 829
Exonucleases, **612**, **718**, 719t
  specificities, 613t
Exopeptidases, **1073**
Expressed sequences (exons), **712**
Expression vectors, 788–791, 794f
Extrinsic and intrinsic clotting pathways, **968**, 969–970
Extrinsic factor, **1134**
Eye and visual system
  clinical correlation: genetic visual disorders, 953
  color vision, 951–954
  night blindness and vitamin A deficiency, 1120
  photoreception: rods and cones, 945–948
  structure and general function, 939–942
  transducin and signal transduction, 948–951
  visual transduction process, 942–945

Fab fragments, **97**, 97–98
Fabry's disease, 459t
Facilitated diffusion. *See* Membrane transport, passive mediated
F-actin, **n961**
Factors, clotting, 966–976, 969t
Fad diets, 1102cc
FAD (flavin adenine dinucleotide), **152**, 152–153, 182–183, 274–275, 306–307, 561, 987–990, 1128
  synthesis, 562
Fanconi's syndrome, 42cc
Farnesyl pyrophosphate in cholesterol synthesis, 441–443
Fast atom bombardment mass spectrometry (FABMS), **46**
Fasting state, 579–586, *see also* Starve-feed cycle
Fat, body. *See* Adipose tissue
Fats. *See* Lipid metabolism; Lipids
Fatty acids, **1154**
  asymmetric distribution in phospholipids, 434–436
  chemical nature, 389–392
  desaturation, **399**
  diet as source, 392
  elongation of, **398**
  in gluconeogenesis, 332–333
  interorgan transport, 404–407
  linoleic, **392**, 418
  linolenic, **392**, 461
  metabolism, 389–417, 597–598, *see also* Lipid metabolism
  nomenclature, 390t
  oxidation
    α, 412–414
    β, 419–420
    ω, 414
  polyunsaturated, **392**, 417–420, **1106**
    and heart disease, 1106cc
  saturated, **1106**, **1154**
  sources, 392–401
  storage, 401–403
  synthesis, 392–401
  unsaturated, **1154**
Fatty acid synthase, **392**, 394t
Fatty acyl CoA, 408–409
Fatty acyl CoA:cholesterol acyltransferase (ACAT), **445**
Fatty alcohols, 401
FCCP (carbonylcyanide-*p*-trifluromethoxyphenylhydrazone), **283**
Fc fragments, **97**, 97–98
Fe^+. *See* Iron
Feedback mechanisms, **189**, 351–352
Feedback inhibition, **189**
Ferredoxins, **1004**
Ferritin, **1004**
Ferrochelase, **1016**
Ferroxidase I and II, **1008**
Fetal protein absorption, 1077
Fever and prostaglandins, 466
F_1F_0-ATPase, **285**
Fiber, dietary, 1104cc, **1105**, 1105t, 1105–1106, 1109
Fibrinogen (Factor I), **968**, 969t
Fibrous proteins, **61**, 61–67
Fick's first law of diffusion, **216**
Fingerprinting of proteins, 46
First law of thermodynamics, **241**
First-order reactions, **142**, 142–143
Fish oil diets, 470, 1106cc
5.8S ribosomal particle, **689**
5′-phosphate-5′-phosphate bonding, **692**
5′-phosphate–yielding endonucleases, 719t
Flagellin genes, 827–828
Flanking regions, 793–794
Flavin-linked dehydrogenases, 274–275, 275t
Flavokinase (riboflavin kinase), **562**
Flavoproteins, 548, **987**
Fluid mosaic model of membranes, **208**
Fluorescence spectroscopy, 82–83
Fluoride, **1143**
  inhibition of glycolysis, 308
5-Flurouracil (FUra), 161, 567
FMN (flavin mononucleotide), **152**, 152–153, 274–275
Folacin, 1132–1134
Folate
  antagonist drugs in cancer therapy, 567–568
  deficiency, 517cc, 1134–1136
  H_4, **516**, 518cc
  synthesis, 159–161
  N^5,N^{10}-methylene H_4, 552–556
Folic acid, 516–518, 517cc
Follicle stimulating hormone (FSH), 854t, **878**, 878–880, 893f, 912cc, 911–912
Folylpolyglutamate synthetase, **567**
Formiminotransferase deficiencies, 518cc
Formylkynurinine, **510**
40S ribosomal particle, **688**
Fourier synthesis, 80–81
Four-letter genetic code alphabet, **729**, 729t
Frame shift mutations, **653**, **732**
Free catalytic domain, **894**
Free energy, **241**, 272
Free radicals, **419**, 419–420
  selenium as scavenger, 1143
Fructokinase, **366**

Fructolysis, **333**
Fructose, **312**, 333–334, 1078t
Fructose 2,6-bisphosphatase, **319**
Fructose 1,6-bisphosphate aldolase, **300**
Fructose 1,6-bisphosphatase, **299**, **315**, **329**
Fructose intolerance, 309cc, 311cc, 366cc
Fructose-6-phosphate, **299**
Fructose 1-phosphate aldolase, **366**
Fructose 6-phosphate-glutamine transamidase reaction, 366f
FSH. *See* Follicle stimulating hormone
Fucosidosis, 374t, 374cc, 459t
Fumarase, 140, **259**
Furosemide, **1071**
Fusidic acid, 744t
Fusion proteins, 790

G codon base pairing, 731t
G factor, **741**
G proteins, 190f, 351, **869, 929**
  as signal transducers, 870–873, 929
  transducin and, **948,** 948–951
GABA (gamma-aminobutyric acid), 501, **932**, 932–933, **936**, 937
GABA shunt, **937**
Galactocerebrosides, **453**, 453–454
Galactokinase, **366**
Galactose, **334**, 367–369, 452
  β-, 1063t
Galactosemia, 368cc
Galactose 1-phosphate uridylyl transferase, **334**, **367**
β-Galactosidase. *See* Lactose
Galactose 4-epimerase, **368**
Gallbladder, 1062t
Gallstones, 1086cc
Gamma subunit, **870**
Gangliosides, **203**, 455–456, 456t, **1156**
Gangliosidosis, 374t, 374cc, 459t
GAP (glyceraldehyde-3-phosphate), **300**
Gap junctions, **222**
GAP proteins, 860
Gas transport
  carbon dioxide, 1033–1038
  clinical correlations
    abnormal hemoglobins, 1032
    chemically modified hemoglobins, 1031
    cyanosis, 1029
  hemoglobin ligand interrelationships, 1038–1039
  oxygen
    in blood, 1026–1029
    hemoglobin and allosterism, 1029–1030
    hemoglobin types, 1030–1032
    physical factors affecting binding, 1032–1033
    see also Hemoglobin; Iron metabolism
Gastric lavage and metabolic alkalosis, 1056cc
Gastric peptidases, 1074t
Gastrin, 34t, 856t, **1066**
Gastrointestinal organs, 1061–1062, 1062t
Gaucher's disease, 459t, 461t
Gel electrophoresis, 39–41, 86
Gene duplications, **101**
Gene expression
  clinical correlations
    Leber's hereditary optic neuropathy, 840
    sickle cell anemia prenatal diagnosis, 836
    thalassemia prenatal diagnosis, 837
    transmissible multiple drug resistance, 825
  eucaryotic
    gene organization of mammalian DNA, 828–829
    for globin proteins, 833–837
    for human growth hormone-like proteins, 837–839
    mitochondrial genes, 839–848
    repetitive sequences, 829–832
  procaryotic
    galactose operon, 821–822
    gene inversion in *Salmonella*, 827–828
    lactose operon of *E. coli*, 808–815
    operon as unit of transcription, 807–808
    ribosomal protein synthesis, 822–823
    stringent response, **823**, 823–824
    transposons in, 824–827
    tryptophan operon of *E. coli*, 815–821
  recombinant techniques
    bacterial human growth hormone synthesis, 841–842
    bacterial insulin synthesis, 840–842
    transgenic species, 842–844
  see also Genes
Gene libraries, 777–778, **778**, 834
Genes
  duplication and cytochromes P450, 984–986
  flagellin, 827–828
  human immunoglobulin, 100–101
  inversion in *Salmonella*, 827–828
  jumping (transposons), **675**
  in peptide hormone formation, 858–862
  pseudo-, **119**, **647**
  regulatory, **807**
  structural, **644**, 644–649, **807**, 822–823
  of *Tn3* transposon, 825–827
  transposition complexes (transposons), **675**
  vision-related, 953, 953cc
  see also Gene expression; Mutations; RNA
Gene therapy, **844**
Genetic code
  breaking of, 731, 732t
  four-letter alphabet, **729**, 729t
  see also Gene expression; Mutations
Genetic mapping, 677cc
Genetics, reverse, **799**
Gentamicins, **744**, 744t
GIH (somatostatin), 852t, 854t, 869t, 938t
Gilbert's syndrome, **1019**
Glial cells, **930**
Globins, hapto-, **1020**
Globoid leukodystrophy (Krabbe's disease), 453, 459t
Globosides, 454
Globulins
  corticosteroid binding (transcortin), **915**
  sex hormone binding, **915**
  see also Immunoglobulins
Glucagon, **321**, 321–322, 856t
  bovine, 34t
  function, **336**
  hepatic function, 352–353
Glucagon:insulin ratio in caloric homeostasis, 583–584
4-α-D-Glucanotransferase, **341**
Glucocerebroside, **453**
Glucocorticoids, 909t
Glucogenic amino acids, 330t, 330–331
Glucokinase, **311**, 311–312
Glucoamylase, 1063t
Gluconeogenesis
  amino acids in, 329–332
  ATP in, 334–335
  Cori and alanine cycles, 324–327, 579–580
  in fasting state, 580–582
  fatty acids in, 332–333
  glycolytic enzymes in, 328–329
  homeostasis and hormonal control, 335–336
  importance to survival, 324, 325cc
  lactate in, 326–327
  nonglucose sugars in, 333–334
  pyruvate carboxylase and PEP in, 326–327, 327–328
  regulatory sites, 334–335
  tricarboxylic acid cycle, 330–332
Gluconolactonases, **362**
Glucose, 1078t
  binding characteristics, 170
  cellular transport, 1071–1072
  heme biosynthesis inhibition, 1016–1017
  interconversion to galactose, 367–369
  mutarotation of, 172–173
  Na^+-dependent transport, 229–230
  oxidative metabolism, 243
  passive mediated transport, 224
  UDP-, **367,** 383
  see also Carbohydrate metabolism; Diabetes mellitus
Glucose homeostasis, 585–586
Glucose-6-phosphatase, **299**, **329,** 341
Glucose-6-phosphate, **299**
Glucose 6-phosphate dehydrogenase, **362**
  deficiency, 362cc
Glucose-1-phosphate uridylyltransferase, **343**
Glucose reserve, **345**
Glucose tolerance factor (GTF), **1142**, 1142–1143
Glucose tolerance test, 312cc
α-Glucosidase (maltase), 1063t, 1080t
β-Glucosidase, 1080t
Glucotransferases, **139**
Glucuronic acid, **306**, 369–370, 370cc
Glucuronides, **306**
Glucuronosyltransferases, **381**

Glutamate, **29**
Glutamate dehydrogenase, 257, **478**, 478–480
Glutamate-oxaloacetate transaminase, 257
Glutamic acid, charge and pH relationships, 37
Glutaminase, **480**
Glutamine, **29**, **480**, 480–481
  in acid-base regulation, 602–603
  antagonist drugs in cancer therapy, 569
  in pyrimidine synthesis, 547–550
Glutamine synthetase, **480**
γ-Glutamyl cycle, **230**, 523cc, **524**
γ-Glutamylcystineglycine, **522**
γ-Glutamyltransferase, 1063t
Glutathione, **365,** 522–525, 523cc
  deficiency, 523cc
Glutathione peroxidase, **523**
Glutathione *S* transferases, **523**
Glutathione transhydrogenase, **868**
Glycemic index, 1108t
Glyceraldehyde 3-phosphate dehydrogenase, **300**
Glyceraldehyde-3-phosphate (GAP), **300**
Glycerol in gluconeogenesis, 580–582
Glycerol ether phospholipids, **200**
Glycerol kinase, **402**
Glycerol phosphate:acyltransferase, **432**
Glycerol 3-phosphate dehydrogenase, **333**
Glycerol phosphate shuttle, **266**, **304**
Glycerophospholipids, **424**, *see also* Phospholipids
Glycine, **27**, 496–498, 536, **932**
Glycocholic acid, **447**
Glycoconjugates, **1088**
Glycogen-binding protein, **351**
Glycogenesis, 139, **337**, 337–356, 1154; *see also* Carbohydrate metabolism
Glycogen granules, 338–339
Glycogenin, **345**
Glycogenolysis, **337**, 337–356; *see also* Carbohydrate metabolism
Glycogen phosphorylase regulation, 346–349
Glycogen storage diseases, 340–341cc
Glycogen synthase, **343**, 343–344
  D form, **349**
  I form, **349**
  regulation, 349–351
Glycolipids, 205–206
Glycolysis, **293**, 293–297
  equilibrium constants and mass-action ratios, 310t
  inhibition, 308
  NADH reduction and reoxidation, 304–306
  regulation, 309–324
  shuttle mechanisms in, 306–308
  stages of, **293**, 297–304
Glycophorin, **209**
Glycoproteins, **70**, 70–71, 205–206, **373**, 373–378
  dolichol in synthesis, 376–378
  peptide bond structure, 375f
  see also Proteoglycans
Glycosaminoglycans
  chondroitin sulfate, **380**, 380–384
  dermatan sulfate, 379t, 379cc, 381, **381**
  heparin and heparan sulfate, 375cc, 379cc, **381**
  hyaluronate, **378**
  keratan sulfate, **381**, 381–382
  mucopolysaccharidoses, 379cc
Glycosidases, **373**, 374cc, 375cc
N- and *O*-Glycosidic linkages, **71**
Glycosphingolipids, **202**
Glycosylation
  hyperglycemia and protein, 585cc
  N-linked, **749**, 749–751
  O-linked, **748**
  of proteins, 748–751
Glycosyl-phosphatidylinositol anchor, **211**, 429–430
Glycosyltransferases, **372**, **748**, 748t
G_{M1} and G_{M2} gangliosides, **456**, 456t
G_{M1} and G_{M2} gangliosidosis (Sandhoff-Jatzkewitz disease), 374t, 374cc, 479t
Gm_{M2} gangliosidosis, **460**
GMP, synthesis, 539–540
GnRH. *See* Gonadotropin releasing hormone (GnRH)
GnRIF. *See* Gonadotropin release inhibiting factor (GnRIF)
Goiter, **1140**
Golgi apparatus, 18
  in collagen synthesis, 758–759
  enzymes and metabolic pathways, 188t
  in glycoprotein synthesis, 377
  in protein synthesis, 31–33, 748–751, 752–753
Gonadotropic releasing hormone (GnRH)
Gonadotropin release inhibiting factor (GnRIF), 852t
Gonadotropin releasing hormone (GnRH), 852t, 854t
  in ovarian cycle, 876–881
  and protein kinase C, 892–894
Gout
  enzyme activity and, 146cc, 163cc, 183cc
  PRPP formation and, 565
  uric acid formation and, 546
Gramicidin A, 232t, **234**
Gratuitous inducers, **810**
GRH (somatocrinin), 852t
Group translocation, **217**, **230**
Growth hormone, 855t
  recombinant techniques, 842–844
Growth hormone-like proteins, gene expression, 837–839
Growth hormone release inhibiting factor (somatostatin), 852t, 854t, 869t, 938t
Growth hormone releasing hormone (somatocrinin), 854t
Growth-inhibiting antitumor drugs, 569
GSH. *See* Glutathione
GTP in purine synthesis, 539–540
GTP binding protein. *See* G proteins
Guanidinium groups, **29**
Guanosine 5′-monophosphate. *See* GMP
Guanosine pentophosphate, **824**
Guanosine tetraphosphate, **824**
Guanylate cyclase system, **894**, 894–897
Gum-type fiber, 1105, 1105t
Gyrase, 636, **637**

H bands, **956**
Hageman factor (Factor XII), 969t
Hairpin forms, **684**
Hammerhead RNA structure, **709**
Haptens, **93**
Haptoglobins, **1020**
Hartnup's disease (neutral amino aciduria), 42cc, 232cc, 1077cc
hCG. *See* Human chorionic gonadotropin
HCl secretion, 1072–1073
HCO^{-3}. *See* Bicarbonate
HDLs (high-density lipoproteins). *See* Lipoproteins
Heart disease
  polyunsaturated fats and risk, 1106cc
  prostaglandins and congenital, 466
  see also Atherosclerosis; Myocardial infarction
Heat shock protein, **917**
Heavy chains, **93**, 93–94, 100
Helix formations
  alpha, 51–54
  cloverleaf, **687**
  collagen, 759
  DNA types, 634f
  double helix, 614–622, *see also* Double helix
  hairpin, **684**
  helix-loop-helix motif, **229**
  helix-turn-helix motif, **116**
  negative superhelix, **684**
  polyproline type II, 63
  superhelix, 64
  topoisomeres in formation, 635–637
Helix-to-coil transition. *See* Denaturation
Hematin, **1002**
Heme, **1002**
  catabolism, 1017–1021
  clinical correlation: neonatal isoimmune hemolysis, 1019
  in cytochrome *c*, 276–277
  cytochrome P450 in binding, 982–984
  prosthetic group, 120–121
  spin states, **982**
  structure, 120f, 276f
  synthesis, 1009–1017
  see also Gas transport, oxygen; Hemoglobin; Iron metabolism
Heme iron proteins, 157
Heme oxygenase, **1017**
Hemiacetal chemistry, 1149
Hemicellulose, **1105**, 1105t
Hemochromatosis, **1008**, 1008cc, 1009cc, **1140**
Hemoglobin, **1026**
  A (types), **1030**, 1030–1031
  A_1 (HbA_1), 50, **119**
  A_2 (HbA_2), **119**
  allosteric effect, 1029–1030
  binding characteristics, 120–121, 126–132
  carbamino, **1035**
  clinical correlations
    abnormal oxygen affinity, 1032

chemically modified hemoglobins, 1031
hemoglobinopathies, 1032
McKees-Rocks disorder, 732
missense mutation disorders, 732
mutation in sickle-cell anemia, 50
F (HbF), **119**, **1031**
gene organization of human, 835f, 835–836
mutations
frameshift, 733t
read-through, 732t
oxygen binding, 126–132, 1027–1032
S, 50
structure, 122–123, 123t
types, 119t, 119–120
see also Gas transport, oxygen; Heme; Iron metabolism; Myoglobin
Hemolysis
intravascular, 1020–1021
neonatal isoimmune, 1019cc
Hemolytic anemia, 179cc
drug-induced, 362cc
pyruvate kinase deficiency and, 324cc
vitamin E supplementation and, 1124
Hemopexin, **1021**
Hemophilia, 970cc
Hemosiderin, **1004**
Henderson–Hasselbalch equation, **10**, 10–11, 13cc, 35–36, **1041**, 1042cc
Heparan sulfate, 379cc, **380**
Heparin, 375cc, **380**
Hepatic coma, 483cc
Hepatic disease and membrane fluidity, 214cc
Hepatic metabolism, adaptive mechanisms, 586–593
Hepatocytes, 3f, 15f, 262–263
Hereditary coproporphyria, 1011t, **1016**
Hereditary protoporphyria, 1011t
5-HETE (5-hydroxyeicosatetraenoic acid), **468**, 468–470
Heteroduplexes, **626**
Heterogeneous nuclear RNA (hnRNA), 683t
Heterotropic effect of hemoglobin, **1029**
Heterotropic ligand interactions, **163**
Hexokinase, 57, 58f, **299**, 309–312, 311cc
Hexosaminidases, **458**
Hexose monophosphate shunt (pentose phosphate pathway), 361–365, *see also* Carbohydrate metabolism
H_4 folate, **515**, 515–517, 537
antagonist drugs in cancer therapy, 567–568
hGh genes, **838**
HGPRTase, **542**
deficiency and gout, 546cc
deficiency and Lesch-Nyhan syndrome, 544cc
Hierarchical phosphorylation, **350**
High density lipoproteins (HDLs). *See* Lipoproteins
High-energy bonds, **243**, 244t
High molecular weight kininogen, **970**
High performance liquid chromatography (HPLC), **44**, **87**, 535
Hill coefficient, **126**, **1029**
Hill equation, **126**
Histamine, **512**, 520–522, **1066**
Histidase, **512**
Histidinemia, 512
Histidines, **29**, 35t, 512–513
distal, **121**
invariant sequences, 113t
proximal, **121**
Histone acetylase, **641**
Histone deacylase, **641**
Histones, **637**, 637t, 639–641
HMG CoA, **414**, 414–415, **440**
HMG CoA reductase, **441**
HMG CoA synthase, **415**, **440**
HMK (high molecular weight kininogen), **970**
HnRNA, 683t
Holoenzymes, **136**, **696**
Holoproteins, **120**
Homocysteine, **501**
Homogentisate, **507**
Homology, **44**
Homotropic ligand interactions, **163**
Hormone–receptor complex, **881**
Hormones
adrenal, **902**, 905t, *see also* Hormones, steroid
amino acid derived, 862–865
cell regulation and secretion, 869–875
clinical correlations
anterior pituitary activity tests, 852
hypopituitarism, 854
lithium treatment of manic depressive illness, 876
oral contraception, 912
ovarian cycle and programmed cell death, 922
endocrine, **849**
hormonal cascade systems
anterior pituitary polypeptide hormones, 852–854
cyclic, 876–881
as signal amplification mechanism, 849–852
human growth hormone-like proteins, 837–839
human growth (somatotropin). *See* Human growth hormone
hypothalamic releasing, 852t
inactivation and degradation, 868–869
ovarian cycle, 876–881
peptide, *see also* Peptide hormones
genes and formation, 858–862
major hormones and their functions, 854–857t
receptors
hormone–receptor interactions, 881–884
internalization of, 886–888
intracellular action: protein kinases, 889–897
oncogenes and, 897–898
steroid hormone, 916–921
structure of β-adrenergic, 884–886
in gluconeogenesis, 335–336
sex, **902**, 905t, 994–995, *see also* Hormones, steroid
synthesis, 994–995
starve-feed cycle levels, 584t
steroid, **1156**
blood transport, 915–916
cellular communication and control, 909–915
cErbA protooncogene and, 918–920
cholesterol as precursor, 439
cytochromes P450 in synthesis, 990–995
examples of cellular activity, 921–923
heme biosynthesis inhibition, 1017
hormones controlling synthesis, 911–915
metabolic inactivation, 907–909
production in pregnancy, 996cc
and programmed cell death, 921–922, 922cc
and stress, 922–923
structure, 902–905, 903t
synthesis, 905–907, 991–993f
in triacylglycerol regulation, 407t
types, **849**, 849–852, 854–857t
see also Peptides
Hormone-sensitive triacylglycerol lipase, **406**, 407t
Hospital diets, 1100cc
HPETES (ionohydroperoxeicosatetraenoic acids), **467**, 467–470
hPL. *See* Human placental lactogen
HPLC. *See* High performance liquid chromatography
HPRT gene, 783cc
Human chorionic gonadotropin, **880**, 880–881
Human growth hormone, recombinant techniques, 842–844
Human growth hormone-like proteins, 837–839
Human immunodeficiency virus (HIV), 799cc, 799–800
Human placental lactogen (hPL), 857t
Humoral stress pathway, **923**
Hunter's syndrome, 379t, 379cc
Hurler's syndrome, 379t, 379cc
Hyaluronate, **378**
Hybridization, **624**
Hydration, **407**
Hydrochloric acid. *See* HCl
Hydrogen bonding, **5**
chemistry review, 1152
in DNA, 614–615
in proteins, 51–55, 73–74
representative, 5f
of water molecules, 4–6, 5f
Hydrogen peroxide, 21, 480
Hydrolases, **140**, 141t
Hydrolysis, 403–404, 1151
Hydronium ion, **7**
Hydrophobic interaction forces, **73**, 1152
Hydrophobicity, **30**
of amino acids and side chains, 30–31
of fatty acids, 391–392
Hydrophobic signal peptide, **747**
4-Hydroxyproline, **498**
Hydroxyamino acids, 496–498
β-Hydroxy-β-methylglutaryl coenzyme A (HMG CoA), **414**,
β-Hydroxybutyrate, **406**, *see also* Ketone bodies

β-Hydroxybutyrate dehydrogenase, **415**
25-Hydroxycholecalciferol, **1121**, *see also* Vitamins, D
5-Hydroxyeicosatetraenoic acid (5-HETE), **468**, 468–470
α-Hydroxy fatty acids, **400**
Hydroxylases
  11β, **907**, **991**
  17-, **907**
  18-, **991**
  21-, **907**, **991**
    deficiency, 994cc
  in steroid hormone synthesis, 907
Hydroxylation, **982**
3β-Hydroxysteroid dehydrogenase/$\delta^{4,5}$-Isomerase, **991**
5-Hydroxytryptamine (serotonin), **536**, 876, **936**, **1066**
Hydroxyurea, **569**
Hyperammonemia, 483cc
Hyperbilirubinemia, 1019–1020
Hypercholesterolemia, 67cc, 445–446
  copper deficiency and, 1142
Hyperglycemia
  hyperglycemic hyperosmolar coma, 584cc
  nonketotic, 498
  and protein glycosylation, 585cc
  see also Diabetes mellitus
Hyperlipidemia, 1104–1105
Hyperlysinemia, 515cc
Hyperproinsulinemia, 755cc
Hypersensitivity, *See* Allergy
Hypertriclyceridemia, 597
Hypertriglycerolemia, 1108cc
Hypervariable regions, **94**
Hypochromic effect, **621**
Hypoglycemia, alcohol intoxication and, 336cc, 336–337
  in premature infants, 325cc
Hypolipoproteinemias, 69cc
Hypopituitarism, 851cc, 854cc
Hypothalamic gastrin releasing peptide, 852t
Hypothalamic hormones, 851f, 852t, 854t
Hypoxanthine-guanine phosphoribosyltransferase. *See* HGPRTase
Hypoxia, 284cc, 1038–1039, *see also* Gas transport, oxygen
Hypusine, 522

I (isotropic) bands, **956**
I-cell disease, 753, 753cc
Idiopathic hemochromatosis, **1008**, 1008–1009, 1009cc
Imidazolium groups, 29
Immunization, 97cc
Immunodeficiency diseases and enzyme deficiencies, 544cc
Immunoglobulin fold motif, **102**
Immunoglobulin light and heavy chains, 93–94
Immunoglobulins, **92**
  artificial (abzymes), **177**, 177–178
  classes, 94t
  functions of immunoglobulin fold, 102
  genetics, 100–102
  IgA, deficiency, 96cc
  IgE, 426
  IgF (insulin-like growth factor), 855t
  IgG, 94f, 95f, 98f
  IgG isotypes, **101**
  IgM, 96cc
  structure, 93–100
Immunological domains, **884**
IMP dehydrogenase, **541**
IMP (inosine 5′monophosphate), **536**, 536–540
Induced fit model of enzyme binding, **169**
Inducers, **809**
  gratuitous, **810**
Inducible enzymes, **312**
Induction, **807**
Inflammation and prostaglandins, 465–466
Inhibin, 855t, **879**, **913**
Inhibition
  competitive, **157**
  cross regulation, **189**
  drugs in, 159
  feedback, **189**
  irreversible, 158
  noncompetitive, **158**
  uncompetitive, **158**
Inhibitor-2 protein, **351**
Initial velocity, **145**
Initiation sites, **72**
Injury, metabolic changes in, 598
Inosine 5′-monophosphate. *See* IMP
Inosinicase, **539**
Inositides in signal transduction, 428–429
Inositol 1,4,5-trisphosphate (IP_3), **353**, **428**, 428–429, **874**
  in Ca^{2+} release, 873–875
Insertional inactivation, **778**
Insulin, **336**, 856t
  in caloric homeostasis, 583–584
  and glucokinase synthesis, 312
  in glycogen synthesis, 355–356
  mode of action, 321–323
  prepro-, **755**
  primary structure, 46–49, 49t, 49cc
  pro-, 46, **755**
  receptor deficiency and obesity, 1102–1103
  receptor structure and function, **356**, **890**
  recombinant, 840–842
  resistance to, 597–598
    in cancer, 602cc
  synthesis, 755, 755cc
  see also Diabetes mellitus; Gluconeogenesis
Insulin : glucogen ratio in caloric homeostasis, 583–584
Insulin-like growth factor (IgF), 855t
Insulin receptors, **356**, **890**
Insulin resistance, 597–598
  in cancer, 602cc
Integral (intrinsic) proteins, **205**, **209**
Integrase, **675**
Inter-α-trypsin inhibitor, 109t
Intercalation, **657**, 659
Interleukins, **598**
Intermediate density lipoproteins (IDLs), **67**
Internalization of receptors, **870**
Internal promoters, **696**
Interspersion pattern, **647**
Interstitial fluid buffer systems, 1039–1041
Intervening sequences (introns), **714**
Interwound turns, **634**
Intracellular fluid, chemical composition vs. blood, 16f
Intramolecular base pairing, **684**
Intravascular hemolysis, 1020–1021
Intrinsic and extrinsic clotting pathways, **968**, 969–970
Intrinsic factor, **1135**
Intron–exon junction, 715–717
Introns, **101**, **643**, **714**, 828, 829
  cytoplasmic, 753
Invariant amino acid residues, **49**
Inverted repeat sequences (palindromes), **642**, 642–644, **649**
Inverted terminal repeat sequences, **824**
Iodine
  metabolism, 865f
  in thyroid hormone synthesis, 864–865
Ion-exchange chromatography, **41**
Ionic interaction chemistry, 1152
Ionizable side chain groups, **34**, 34–36
Ionophore antibiotics, 231–234, 232t
IP_3 (inositol 1,4,5-trisphosphate) *See* Inositol 1,4,5-trisphosphate
Iron
  availability calculations, 1140cc
  dietary, 1139–1140
  nonheme, **275**
  see also Heme; Hemoglobin; Gas transport, oxygen
Iron metabolism
  clinical correlations
    acute intermittent porphyria, 1012
    hemochromatosis, 1008
    iron-deficiency anemia, 1008
  distribution and kinetics, 1007–1009
  intestinal absorption, 1005–1006
  iron-containing proteins, 1002–1005
  regulatory mechanisms, 1006–1007
  see also Iron
Iron-sulfur (nonheme iron) enzymes, **156**, 156–157, 275
Ischemia, 284cc
Isocitrate dehydrogenase, **256**
Isodesmosine, **66**
Isoelectric focusing, **40**
Isoelectric pH, **36**, 36–41
Isohydric CO_2 carriage, **1036**, 1036–1037
Isoleucine, **27**, 493–496
Isomaltase, 1063t, 1080t
Isomerases, **140**, 141t
  disulfide, **756**
  $\delta^{4,5}$-, **907**
Isomerization in carbohydrate interconversion, 365–366
Isoprene, **1156**
Isozymes (isoenzymes), **184**
  assays, 184–186
IUB enzyme classification system, 137–141, 141t

Jaundice, nonhemolytic, 369cc
Jumping genes (transposons), **675**
Junctions, tight, **1067**

K^+, in enzyme activity, 154–155, *see also* Na^+,K^+-ATPase pump

K$^+$-ATPase, **1073**
Kalikrein, **970**
  pre-, **970**
Kasugamycin, **744**
K class allosteric enzymes, **162**, 162–163
Keratan sulfate, **381**, 381–382
Keratins, α, **66**, 66–67
Kernicterus, **1018**
Ketoacidosis
  diabetic. *See* Diabetes mellitus; Diabetic ketoacidosis
  methylmalonate related, 496cc
  proprionate related, 496cc
β-Ketoacyl-ACP synthase, **394**
Ketogenesis in stress and trauma, 598
Ketogenic amino acids, 330t, 330–331
α-Ketoglutarate dehydrogenase, **256**, 256–257
Ketone bodies, **406**, 414, 597
  nonhepatic utilization, 416
  in starvation, 416–417
Ketone chemistry, 1149
β-Ketothiolase, **414**
Kidneys
  in acid-base balance, 1046–1049
  in bilirubin excretion, 1119
  disease and metabolic derangements, 600
  gluconeogenesis in, 324–326
  NH_4^+ elimination, **1049**
  see also Renal
Kilocalorie, **1094**
Kinases, **139**
  in carbohydrate interconversion, 365–366
  in FAD synthesis, 562–563
  in glucose metabolism, 299–300
  in glycogenolysis, 353–354
  in glycogen synthase phosphorylation, 349–351
  in glycolysis, 302, 303
  in muscle contraction, 963
  nucleotide di- and monophosphate reactions, 240f
  see also individual kinases
Kinesin, **21**
K_m constant, **146**, 146–147
Krabbe's disease (globoid leukodystrophy), 453, 459t
Krebs cycle. *See* Tricarboxylic acid cycle
Kringle domain, **974**
    Kwashiokor, 577cc
Kwashiorkor, 577cc, **1100**
Kynurenate, **510**
Kynureninase, **510**

lac operon, 644f, 808–815, *see also* Operons
lac repressor, **809**
Lactase, 1063t, **1079**, 1080t
Lactate in gluconeogenesis, 326–327
Lactate dehydrogenase, 185, **303**, 303–304
Lactate synthesis, 315–317
Lactation, metabolic changes in, 596–597
Lactic acid
  acid–base titration curves, 11f
  dissociation, 8–9
Lactic acidosis, 252–253cc, 315cc
Lactoferrin, **1002**
Lactone chemistry, 1149
Lactose, **334**, 596–597, 1078t
Lactose intolerance, 186, 1079cc
Lanosterol, **441**
LAPC (lipoprotein-associated coagulation inhibitor), **973**
Lattice structure of clathrin, 886–888
LCAT (lecithin:cholesterol aminotransferase), **446**
LDL receptors, **445**
LDLs (low density lipoproteins). *See* Lipoproteins
Leader sequences, **817**
Lead poisoning, 1011t
Leber's hereditary optic neuropathy, 840cc
Lecithin in pulmonary surfactant, 427–428, 428cc
Lecithin:cholesterol aminotransferase (LCAT), **446**
Lecithin deficiency, 69cc
Lens of eye, 940–941, *see also* Eye and visual system
Lesch–Nyhan syndrome, 542-544, 543cc
Leucine, 493–496
Leucine aminopeptidase, 1063t
Leucine zipper, **117**, 117–118
Leu-enkaphalin, 938t
Leukemia, 186
Leukocytes, polymorphonuclear, 426–427
Leukodystrophies, 459t
Leukotrienes, **468**, 468–470, 523**1155**
Lewis acids, 153
Lewis blood group, 373cc
LH. *See* Luteinizing hormone
Ligand binding domains, **884**, **919**
Ligandin, **1018**
Ligands, **162**
  cooperativity, 165–167
  homotropic and heterotropic interactions, **163**
Ligases, **140**, 141t
Light chains, **93**, 93–94
Light-dark cycles, 876
Lignins, **1105**, 1105t
Lineweaver-Burk double-reciprocal plot, **147**
Linoleic fatty acids, **392**, 461
Linolenic fatty acids, **392**, 418
Linoleoyl CoA, 412
Lipases
  acid-stable, **1002**
  hormone-sensitive triacylglycerol, **406**, 407t
  lipoprotein, **406**
    deficiency, 405cc
  pancreatic, **1003**
  in triacylglycerol mobilization, 404
Lipid esterase, **1083**
Lipid hydroperoxide, **419**
Lipid metabolism
  cholesterol, 438–449
  clinical correlations
    acyl CoA dehydrogenase deficiency, 410
    carnitine deficiency, 409
    diabetic ketoacidosis, 417
    Gaucher's disease, 461
    genetic lipid-transport abnormalities, 405
    obesity, 404
    polyunsaturated fatty acids and heart disease, 1106cc
    Refsum's disease, 414
  in fasting state, 582–583
  fatty acids
    chemical nature of, 389–392
    interorgan transport, 404–407
    polyunsaturated, 417–420
    sources, 392–401
    storage, 401–404
    utilization in energy production, 407–417
    see also Fatty acids
  hypertriglycerolemia and metabolic adaptation, 1108cc
  lipoxygenase and oxy-eicosatetraenoic acids, 466–470
  phospholipids, 424–438, 428cc
  polyunsaturated fatty acids and heart disease, 1106cc
  prostaglandins and thromboxanes, 461–466
  sphingolipids, 449–461
  see also Fatty acids; Lipids *and individual lipids and enzymes*
Lipids, **1154**
  amphipathic, **200**
  chemistry review, 1154–1157
  dietary, 1104–1107
  digestion and absorption, 1081–1088
  lipid bilayer, 206f, 207–212
  lipid membrane, 427
  in membrane structure and transport, 204, 208–209, 213–214
  metabolism. *See* Lipid metabolism
  sphingo-, **201**, 201–206
  vesicular structure of, 206–207
  see also Fatty acids; Lipid bilayer; Phospholipids
Lipofuscin (age pigment), **20**, 20–21
Lipoprotein-associated coagulation inhibitor, **973**
A-β-Lipoproteinemia, 1087cc
Lipoprotein lipases, **406**
  deficiency, 405cc
Lipoproteins, **67**, 67–69
  chemical composition, 69t
  classes, 67t
  clinical correlation: hyperlipidemias, 67
  high density (HDLs), **67**, **405**
  intermediate density (IDLs), **67**
  low density (LDLs), **67**, **405**, 443, 595, 1087cc
    apolipoprotein B-100, 34
  structure, 69f
  triacylglycerol transport, 405cc, 405–407
  very low density (VLDLs), **67**, **405**, 443, 579, 597–598, 1087cc
Liposomes, 206–208, 208cc
Lipoxygenases, **466**, 466–470
Lithium, **874**
  in manic depressive illness, 876cc

- Lithocholic acid, **1089**
- Liver
  - adaptive metabolism, 586–593
  - bile acid formation, 448
  - in bilirubin metabolism, 1017–1020
  - in detoxification, 550, 600–601
  - digestive and absorptive functions, 1062t
  - disease and metabolic derangements, 214cc, 598–599
  - epinephrine in glycogenolysis, 353–354
  - ethanol oxidation, 600–601
  - in gluconeogenesis, 324–326
  - glycogen granules in, 338–339
  - inducible enzymes, 591–593
  - insulin in glycogen synthesis, 355–356
  - in nitrogen metabolism, 482, 483cc
  - phosphorylase as "glucose receptor," 352
  - in triacylglycerol synthesis, storage, and utilization, 401–404
  - tryptophan catabolism, 510–512
  - *see also* Hepatic
- LLAT (lysolecithin:lecithin acyltransferase), **437**
- Lock-and-key model of enzyme binding, **169**
- Low birth weight/prematurity and hypoglycemia, 325cc, *see also* Neonates
- Low density lipoproteins (LDLs). *See* Lipoproteins
- LTA_4 synthase, **468**, 468–470
- LTB_4 synthase, **469**
- Luteinizing hormone (LH), 854t, **878**, 878–880, 893f, 911–912, 912cc
- Luteinizing hormone releasing hormone, 938t
- Lyases, **140**, 141t
- Lysine, **29**, 513–514, 515cc
- Lysinuric protein intolerance, 515cc
- Lysis, auto-, **111**
- Lysolecithin:lecithin acyltransferase, **437**
- Lysosomal enzymes, 20t, 188t
  - clinical correlation: gout, 19cc
  - functional role, 19–22, 20t
- Lysosomes, **19**
  - functional role, 19–22, 20f
  - in protein degradation and turnover, 762
  - protein targetting, 752–753
- Lysozymes, 174–175
- Lysyl hydroxylase, **758**
- Lysylhydroxylase deficiency (Ehlers-Danlos syndrome type VI), 761cc

- M proteins, 957t, **958**
- Macrocytic anemia, **1134**
- α_2-Macroglobulin, 109t
- Macromolecular binding specificity, **115**
- Magnesium, *See* Mg^{2+}
- Malate–aspartate shuttle, **266**, **304**
- Malate dehydrogenase, **260**
- Malignant hyperthermia, 316cc
- Malnutrition. *See* Digestion and Absorption; Metabolic interrelationships; Nutrition
- Malonyl CoA, **393**
- Maltase, 1063t
- Maltotriose, **1079**
- Mammary glands, fatty acids in milk production, 400
- Manganese, **1143**
- Manic depressive illness, 876cc
- Mannokinase, **366**
- Mannose, **334**
- Mannose phosphate isomerase, **334**
- α-Mannosidosis, 374t, 374cc
- β-Mannosidosis, 374t, 374cc
- Mapping
  - electron-density distance, **80**, 80f
  - genetic, 677cc
  - genome, 733f
  - restriction, **677**, **769**, 769–770
- Marasmus, 577cc, **1100**
- Marfan syndrome, 761cc
- Maroteaux-Lamy syndrome, 379t, 379cc
- Mass-action ratios, **309**, 319t
- Maxam-Gilbert procedure, **771**, 771–772
- Maximum velocity (V_{max}), **144**
- M13 bacteriophage, **796**
- McArdle's disease, 341cc
- McKees Rocks hemoglobin disorder, 732cc
- Megaloblastic anemia, 1135
- Melanin, **509**
- Melanocyte stimulating hormone, 855t
- Melanocyte stimulating hormone (MSH), 852t
- Melatonin synthesis, 876
- Membranes, **13**, 196–197
  - artificial systems, 207–208
  - channels and pores, 221–223
  - chemical composition, *see also* Lipid bilayer
    - carbohydrates, 205–206
    - cholesterol, 204
    - lipids, 197–205
    - phosphoglycerides, 197–201
    - proteins, 205
    - spingolipids, 201–203
  - cholesterol and fluidity changes, 214
  - diffusion of molecules, 215–222, *see also* Membrane transport
  - micelles and liposomes, 206–208, 208cc
  - mitochondrial, 261–269, 264t
  - phosphatidylinositol protein anchor, 429–430
  - plasma. *See* Plasma membrane
  - semipermeability of, 13–14
  - structure
    - erythrocytes as model, 212
    - fluid mosaic model, 208
    - integral proteins in, 208–211
    - lipid distribution, 212–213
    - peripheral proteins in, 211–212
    - protein and lipid diffusion, 213–215
  - *see also* Membrane transport
- Membrane transport
  - active
    - mediated, 225–226
    - primary, 226–229
    - secondary, 229–230
    - substrate modification in, 230–231
  - clinical correlations
    - diseases due to loss, 232
    - liposomes as drug and enzyme carriers, 208
  - energetics of, 219–220
  - four steps of, 218–219, 219f
  - ionophore antibiotics, 231–234, 232t
  - major systems, 232t
  - mechanism of diffusion, 215–216
  - passive mediated, 223–225
  - translocation systems, 216t, 216–217
  - transporter characteristics, 217–218, 218t
- Menke's syndrome, **1142**
- Menstrual cycle. *See* Ovarian cycle
- Mercaptopuric acid formation, 523
- 6-Mercaptopurine, 161–162, 566–567
- Meromyosins, **958**
- Messenger RNAs. *See* RNAs, messenger
- Metabolic acidosis, 1051–1052
- Metabolic alkalosis, 1048–1049, 1056cc
- Metabolic interrelationships, 575–603
  - clinical correlations
    - cancer cachexia, 602
    - diabetes mellitus, 584, 597, 598, 600
    - hyperglycemia and protein glycosylation, 585
    - hyperglycemic, hyperosmolar coma, 584
    - obesity, 576
    - protein malnutrition, 577
    - Reye's syndrome, 582
    - starvation, 577
  - hepatic regulatory mechanisms, 586–593
  - starve–feed cycle, 576–586
  - tissue relationships, pregnancy, and disease states, 593–603
  - *see also* Digestion and absorption; Nutrition; Starve-feed cycle
- Metabolic pathways
  - committed steps, **187**, **312**
    - in fatty acid synthesis, 393
    - in purine nucleotide synthesis, 536, 540–542
  - compartmentation, **187**
  - Emden–Meyerhof (glycolytic), 293–324
  - enzymes important to, 188t
  - glycolytic, 293–324
  - pentose phosphate pathway, **295**
  - tricarboxylic acid (Krebs) cycle, 253–255, 481–486
  - urea cycle, 480–488, **481**
- Metabolic wear and tear, 486
- Metachromic leukodystrophy, 459t
- Metal enzyme cofactors, 153–157
- Metalloenzymes, **155**
- Metals
  - as enzyme cofactors, 153–157
  - as structural elements, 155–156
  - *see also* Iron metabolism *and specific metals*
- Metarhodopsin II, **946**
- Met-enkephalin, 938t, **939**
- Methemoglobin, 1031cc
- Methionine, **29**
- Methionine enkephalin, 34t
- Methotrexate, 159–160, **567**
- Methylases, **643**

Methylation of amines and polyamines, 520–522
N^5,N^{10}-Methylene H_4 folate, 552–556
Methylene tetrahydrofolate reductase deficiency, 518cc
Methylglutaryl CoA, **440**
Methylmalonyl CoA, **400**
Methylmalonyl CoA mutase, **494**
Met-tRNAs, **737**
Mevalonic acid n cholesterol synthesis, 440–442
Mg^{2+}
in enolase inhibition, 308
in enzyme activity, 155
as substrate for creatine kinase, 155
Micelles, **1084**, 1084–1087
critical micelle concentration, **207**
mixed, **1085**
structure and function, 206–208
Michaelis–Menten equation, **145**, 145–147
linear form (Lineweaver-Burk plot), **148**
Microbodies. *See* Peroxisomes
Microfilaments. *See* Actin filaments
Microsomes, **18**
Microtubules, **21**
Microvilli, **1063**
Mineralocorticoids, 909t
Mineral nutrition
macrominerals
calcium, 1137–1138, 1138cc
phosphorus and magnesium, 1138–1139
nutritional needs of the elderly, 1145cc
trace minerals
chromium, 1142–1143
copper, 1142
dietary iron, 1139–1140
iodine, 1140
iron, 1140cc, *see also* Hemoglobin; Iron metabolism
manganese, molybdenum, and fluoride, 1143
selenium, 1143
zinc, 153–154, 1016, 1140
see also individual minerals
Missense mutations, **732**, 732cc
Mitochondria, **18**
aspartate transport, 305–306
cytochrome P450 system in, 989–990
electron transport, 224–225, 270–283, 306, 987–989
clinical correlation: cyanide poisoning, 282
enzymes and metabolic pathways, 188t
functional role, 18–19
gene expression, 839–840
heme biosynthesis, 1011–1016
protein synthesis in, 743–744
protein targetting, 753–754
structure and compartmentation, 261–269, 264t
see also Mitochondrial membranes
Mitochondrial membranes
acetyl CoA transport, 397–398
acyl group transport, 408–409
inner, 265–266
outer, 263–265
substrate shuttle systems, 266
Mitochondrial myopathies, 268–269cc
Mitochondrial RNAs (mtRNAs), 694
Mitochondrial targetting signal, **753**
Mitoplasts, **265**
Mn^{2+}, dietary requirements, 1138
Molar extinction coefficients, **81**
Molecular cell biology
blood coagulation, 966–976
clinical correlations
genetic visual defects, 953
hemophilia, 970
myasthenia gravis, 935
muscular contraction, 954–966
nervous tissue structure and function, 929–939
stimuli recognition and transmembrane signaling, 928–929
visual system: ocular metabolism and vision, 939–954
see also Blood coagulation; Eye and visual system; Muscle contraction; Nervous system
Molecular exclusion chromatography, **86**
Molecular weight
of contractile proteins, 957t
of proteins, 85–86
pyruvate dehydrogenase complex, 249t
Molecules
as acids or bases, 8–13
amphipathic, **6**
Molybdenum, **1143**
Monensin, 232t
Monoacylglycerol hydrolase, **406**
Monoamine oxidase, **520**, 520–522, **935**
Monocistronic mRNAs, **692**
Monoenoic acids, 399–400
Monohydroperoxy-eicosatetraenoic acids (HPETES), **467**, 467–470
Monomethyllysine, **514**
Monooxygenases, **138**, **982**
Monooxygenation reaction and cytochrome P450, **982**, 982–984
Monosaccharides
chemistry review, 1152–1153
digestion and absorption, 1080–1081
Morquito syndrome, 379cc
mRNAs. *See* RNAs, messenger
Mucilaginous fibers, **1105**
Mucolipidosis VII, 379t, 379cc
Mucolipidosis (sialidosis), 374t, 374cc
Mucopolysaccharides, *See* Proteoglycans
Mucopolysaccharidoses, 379cc, 379t, **383**
Multidomain enzymes, **57**
Muscle
glycogenolysis and gluconeogenesis, 337–340, 354–355
insulin in glycogen synthesis, 355–356
red and white fibers, 338–339
Muscle contraction
actin as thin muscle filament, 958–962
ATP hydrolysis in, 963
Ca^{2+} in, 962–966
electrochemical-mechanical pathway, 954–957
myosin as thick muscle filament, 957–958
sarcomeric Ca^{2+} flux, 963–965
sliding filament model, **954**
smooth muscle contraction and Ca^{2+}, 965–966
Mutagenesis, site-directed, **793**, 793–797
Mutagens
chemical, 654–656
DNA polymerase errors, 656–657
radiation, 656
Mutarotation of glucose, 172–173
Mutases, **140**, 302
Mutations
cis-acting, 820
deletion, **794**
β-globin genes and thalassemia, 836–837, 837cc
missense, 732cc
operator-constitutive (cis-dominant), **811**
repressor-constitutive, **811**
trans-dominant, **812**
types, 732–733, 735
Myasthenia gravis, 935cc
Mycobacterium tuberculosis, 699cc
Myocardial infarction
alcohol dehydrogenase enzyme assay in, 180cc
angina pectoris and, 318cc
lactate dehydrogenase assays in, 185f
magnesium therapy in, 1138–1139
see also Heart disease
Myofibrils, 954–957
Myoglobin binding characteristics, 124–126
Myokinase. *See* Adenylate kinase
Myopathies, mitochondrial, 269cc
Myosin, 22, **958**, 957–959
ATPase, **958**
mero-, **958**
molecular weights, 957t
tropo-, 66, 957t, **961**
Myosin light chain kinase, **965**

Na^+ in acid–base balance, 1054–1056
Na^+-bile acid cotransport system, **1089**
Na^+ channel, 220, **221**, *see also* Na^+,K^+-ATPase pump
NAD^+ (nicotinamide adenine dinucleotide), **150**, 182–183
degradation, 561–562
lactate dehydrogenase and in mitochondrial electron, 304
synthesis, 559–560
transport, 273–274
Na^+-dependent transport, 229–230
NAD^+-linked dehydrogenases, 272–274, 273t
NAD^+ malate dehydrogenase, **397**
NAD-glycohydrolase, **561**
NAD-kinase, 561
NAD-phosphorylase, **561**
NAD-pyrophosphorylase, **561**
NAD-synthetase, **561**
NADH, 300–306, 397–398
NADH-cytochrome b_5 reductase, **988**

NADP, (nicotinamie adenine dinucleotide phosphate), **150**, 182–182, 256, 273–274, 561
NADP-linked malic enzyme, **397**
NADPH
  in corneal metabolism, 940
  in glucuronic acid reduction, 370
  in heme catbolism, 1017
  in monoenoic acid synthesis, 399
  of ocular retina, 942
  synthesis, 362–365
NADPH-cytochrome P450 reductase, **987**
Na$^+$/H$^+$ exchange, **1069**
Na$^+$-independent, facilitated diffusion, **1080**
Na$^+$,K$^+$ ATPase pump, **226**, 226–230
  in digestion and absorption, 1068–1072
  erythrocyte ghost studies, 227–228
  in glucose and amino acid transfer, 229–230
Na$^+$,K$^+$,2Cl-cotransporter, **1071**
Na$^+$-monosaccharide transporter, **1080**
NANA (*N*-acetylneuraminic acid), 455
Nascent DNA (Okazaki fragments), **666**, 671, 673
Near-equilibrium reaction, **309**
N-end rule, **763**
Neomycin, **744**, 744t
Neonates
  hemolytic anemia in, 1124
  hypoglycemia in, 325cc
  nutritional considerations, 1126cc
  protein absorption, 1077
  respiratory distress syndrome, 428cc
  Rh incompatibility and neonatal isoimmune hemolysis, 1019cc
Nernst equation, **271**
Nervous system
  ATP and transmembrane potential, 930–931
  clinical correlation: myasthenia gravis, 935
  components and function, 929–930
  α-hydroxy fatty acids in, **400**
  neurotransmitters, **931**, 931t, 933–939
  synapses, **931**
  see also Neurotransmitters
Neuraminic acid (Neu), **455**
Neurons, **929**
Neurophysin I, **859**
Neurophysin II, **859**
Neurotensin, 856t, 938t
Neurotransmitters, **931**, 931t, 933–937
  in brain tissue, 938t
  excitatory and inhibitory, **931**
Newborns. *See* Neonates
Niacin (vitamin B$_3$), 151–152, 1129, *see also* Nicotinamide; Nicotinic acid
Nick translation, **781**
Nicotinamide, 273–274, 559–562, *see also* Niacin
Nicotinamine phosphoribosyltransferase, **561**
Nicotinate in NAD synthesis, 559–560
Nicotinic-acetylcholine channel, **221**
Nicotinic acid, **510**, 510–512, **1129**
Niemann-Pick disease, 459t, 459–460
Nigericin, 232t, 233–234, **234**
Nitrogen balance, **477**, 598, **1097**, 1097–1098, *see also* Amino acid metabolism; Urea cycle
Nitroglycerin in angina pectoris, 318cc
N-linked glycosylation, **749**
NOESY spectroscopy, **85**
Nonactin, 232t
Noncompetitive inhibition, **158**
Nonconservative substitution, **49**
Noncovalent association, **60**
Noncovalent interactions, **60**, **72**, 72–76, 128
Nonelectrolytes, **6**
Nonequilibrium reaction, **309**
Nonheme iron proteins, **156**, 156–157, **275**, 1004–1005
Nonhemolytic jaundice, 369cc
Nonketotic hyperglycinemia, 498
Nonsense mutations, **732**
Non-steroidal anti-inflammatory agents (NSAIDs) and prostaglandin inhibition, 465
Noradrenaline. *See* Norepinephrine
Norepinephrine, 508–509, **935**
Nuclear magnetic resonance spectroscopy (NMR), **83**, 83–85
Nuclear Overhauser effect, **83**
Nuclear Overhauser effect spectroscopy (NOESY), **84**
Nuclear pores, **223**
Nucleases, 614t, 717–719, 719t
Nucleic acids, **1158**
  antisense, **799**
  helix parameters, 617t
  nucleotides as monomeric units, 531
  see also DNA; RNA
Nucleofilaments, **639**, 639–641
Nucleophilic substitution chemistry, 1151
Nucleoplasmin, **754**
Nucleosides, **1157**, 1158t
  modified tRNA, 710–712
  in RNA, 688t
  spectrophotometric constants, 535t
  symbols and abbreviations, 534t
Nucleosomes, **638**
Nucleotide diphosphate kinase reactions, 240f
Nucleotide-linked sugars, **366**
Nucleotides, **1157**, 1158t
  adenine transport, 267–268
  cellular distribution, 532
  chemistry, 532–535
  chemotherapy and, 566–571
  clinical correlations
    gout, 546
    immunodeficiency disease and purines, 544
    Lesch–Nyhan syndrome, 544
    orotic aciduria, 548
  coenzyme synthesis, 559–563
  deoxyribonucleotide formation, 551–555
  drug resistance and, 570–571
  enzyme activity and the cell cycle, 557–559
  metabolism, 530–532
  modified, **684**
  nicotinamide, 273–274
  nucleoside and nucleotide kinases, 555–557
  5-phosphoribosyl 1-pyrophosphate, 563–566
  polynucleotide synthesis, 610–612
  purine
    analogs as antiviral agents, 569–570
    metabolism, 536–547, 566–567
  pyrimidine
    analogs as antiviral agents, 569–570
    metabolism, 547–551, 566–567
  site-directed metagenesis of single, 794–797
  symbols and abbreviations, 534t
  see also DNA
Nucleotide sequences. *See* DNA, nucleotide sequences
Nucleotide sugar formation, 365–372
Nucleus, **17**, 188t, 754
Nutrition
  clinical correlations
    alcoholism, 1128
    anticonvulsant drugs and vitamin requirements, 1126
    carbohydrate loading, 1103
    cystic fibrosis, 1120
    diabetes mellitus and high carbohydrate–high fiber diets, 1104
    fad diets, 1102
    heart disease and polyunsaturated fatty acids, 1106
    hospital diets, 1055, 1100
    hypertriglycerolemia and metabolic adaptation, 597
    iron availability calculations, 1140
    iron deficiency anemia, 1141
    low-protein diets and renal disease, 1099
    nutritional needs of the elderly, 1145
    nutrition in neonates, 1126
    oral contraceptives and vitamin B$_6$ requirements, 1131
    renal osteodystrophy, 1122
    total parenteral nutrition, 1055, 1100
    vegetarian diets, 1098, 1145
  controversial aspects, 1094
  drug-nutrient interactions, 1146t
  macronutrients
    carbohydrates, 1103–1104
    composition of dietary, 1106–1110
    energy metabolism, 1094–1096
    fats, 1104–1105
    fiber, 1105–1106
    protein-energy excess, 1101–1103
    protein-energy malnutrition, 1100–1101
    protein metabolism, 1096–1100
  major food groups and U.S. diet, 1060t
  micronutrients
    American diet, 1144
    clinical assessment, 1116–1117, 1144–1146

minerals, 1137–1143, *see also* Minerals
Recommended Dietary Allowances (RDAs), 1117–1118
vitamins, 1118–1137, *see also* Vitamins
under-, over-, and optimal, **1094**
see also Diets; Digestion and Absorption; Metabolic interrelationships, clinical correlations; Obesity
Obesity, 593, 1108
exercise and, 593–595
insulin receptor deficiency and, 1102–1103
insulin resistance in, 598
and lipid metabolism, 404cc
and metabolic interrelationships, 576cc
thermogenetic defect and, 1101–1102
see also Diabetes mellitus
Ocular myopathies, 269cc
Ocular system. *See* Eye and visual system
Okazaki (precursor) fragments, **666**, 671, 673
3β-Ol-dehydrogenase, **907**
Oligo-1,6-Glucosidase (isomaltase), 1080t
Oligomycin, **283**
Oligonucleotides, **610**
Oligosaccharides, 749–751, 1153–1154
O-linked glycosylation, **748**
Omega oxidation of fatty acids, 414
Omega protein, **635**
OMP-decarboxylase, **549**
OMP synthesis, 548–549
Oncogenes, 707cc, 897–898, 898t
Open DNA complexes, **701**
Operator-constitutive (cis-dominant) mutations, **811**
Operator regions, 821
Operons
galactose, 821–822
lac, 644f, 808–809
catabolite activator protein, 813–815
lac repressor protein, 809–811
operator sequence, 811–813
promoter sequence, 813–815
in ribosomal protein synthesis, 822–823
tryptophan
attenuation of transcription, 817–821
trp repressor protein, 815–817
Opioid peptides, 855t
Optical rotation spectroscopy, **83**
Oral contraceptives, 912cc
heme biosynthesis inhibition, 1017
nutrient interactions, 1134
vitamin B_6 requirements and, 1131cc
Order characterization of reactions, 142–143
Ordered mechanism, **149**
Organelles, 16–22, 746–748, 753–755, *see also individual structures*
Ornithine, 483–486, 499–501, 515cc
Ornithine cycle. *See* Urea cycle
Ornithine transcarbamoylase, **483**
deficiency, 487cc
Orotate phosphoribosyltransferase, **548**
Orotic aciduria, 548cc
Orotidine 5′-monophosphate (OMP) synthesis, 548–549
Osteogenesis imperfecta, 761cc
Osteomalacia, **1123**, 1123–1124
Osteoporosis, **1138**, 1138cc
Ouabain, 227f
Ovarian cycle, **877**, 877–881
and programmed cell death, 922cc
Overhauser effects, **83**
Oxidants (electron acceptors), **270**
Oxidases, **138**
auto- and free radical formation, 419–420
metals in, 156–157
Oxidation-reduction potentials, **270**, 270–272, 271t
Oxidation-reduction reactions, 270–272, 1151
Oxidative metabolism
acetyl CoA-producing pathways, 247–253
anabolic and catabolic and catabolic pathways, **238**
ATP in, 238–240
clinical correlations
cyanide poisoning, 282
hypoxic injury, 284
myopathies, 268–269
pyruvate dehydrogenase deficiency/lactic acidosis, 252–253
of fatty acids, 156–157, 407–420, *see also* Fatty acids; Lipid metabolism
mitochondrial membranes in
electron transport, 270–283
structure and function, 261–269, 264t
nucleotides in, 531
oxidative phosphorylation, 284–286
thermodynamic relationships
caloric value, 243
coupled enzyme reactions, 245–246
energy-rich compounds, 243–245, 244t
energy transfer, 246–247
free energy, 241–243
tricarboxylic acid cycle, 253–261
Oxidative phosphorylation, 282–286, 299
chemical coupling hypothesis, 284, **284**
chemiostatic coupling hypothesis, **285**, 285–286
conformational coupling hypothesis, **285**
Oxidoreductases, **137**, 138, 141t
Oxy-eicosatetraenoic acids, **467**, 467–470
Oxygenases, **138**
Oxygen binding
of hemoglobins, 126–132
of myoglobin, 120
Oxygen saturation curve, **125**
Oxygen transport, 1026–1032, *see also* Hemoglobin; Iron metabolism
Oxytocin, 855t, **859**, 869t
P (peptidyl) site, **741**
PAF (platelet activating factor), **426**
Pain and prostaglandins, 466
Palindromes (inverted repeats), **642**, 642–644, 649
Palmitic acid, **392**, 392–400, 410
Pancreas, 1062t
Pancreatic amylase, **1077**, 1077–1080
Pancreatic beta cells, 579, 597, *see also* Diabetes mellitus
Pancreatic enzymes, 1062–1063, 1063t, 1065, *see also* Enzymes, pancreatic
Pancreatic lipase, **1083**
Pancreatic peptidases, 1074t
Pancreatic trypsin inhibitor, 108
Pancreozymin, **1066**
Pantothenic acid, 1132
Papovirus, 792
PAPS (3′-phosphoadenosine 5′-phosphosulfate), **383**
Paracellular transport, **1067**
Paracrine hormones, **849**
Parathyroid hormone, 857t, 1122–1123, **1123**
Parkinson's disease, 509cc
Pasteur effect, **313**, 313–316
pBR322 plasmid, **775**, 778
Pectin, 1105, 1105t
Pellagra, **1129**
Penicillamine side effects and vitamin B_6 therapy, 1132
Pentose phosphate pathway, **295**, 361–365
Pentosuria, 368cc
PEP (phosphenolpyruvate), **303**, 327–328
Pepsinogen, **1075**
Pepsins, **1074**
pepsin A, 1074t
Peptic ulcer and prostaglandins, 466
Peptidases, 1071t
endo-, 1073
exo-, **1073**
signal, **748**
Peptide bond conformations, **32**, 33–34
Peptides, **26**, 34t
C, **48**, **755**
domains, **57**
hydrophobic signal, **747**
initiation sites, **72**
neuro-, *See* Neurotransmitters
opioid, 855t
reagents for cleaving, 46f
synthesis, 31–33
telo-, **65**
transport systems, **1076**
see also Hormones, peptide
Peptidyl site (P site), **741**
Peptidyl transferase, **739**
Periodic structures of DNA, 613–619
Peripheral proteins, **205**, **211**
Pernicious anemia, **1134**, 1134–1136
Peroxidases, **138**, 420, 464, **523**
Peroxide dismutases, 420
Peroxisomes, **21**, 188t, **417**
clinical correlation: Zellweger syndrome, 21cc
PEST sequences, **763**
PG. *See* Prostaglandins

$PGF_{2\alpha}$, 879–880
PG-peroxidases, 464
pH, **8**
  of biological fluids, 8t
  calculation of, 8
  catalytic activity and, 179
  and charge properties of amino acids, 37–38
  clinical correlations
    acute respiratory alkalosis, 1051
    bone in acid-base homeostasis, 1046
    chronic respiratory acidosis, 1052
    laboratory errors, 1042
    metabolic alkalosis, 1056
    salicylate poisoning, 1053
    total parenteral nutrition and acidosis in infants, 1055
  conjugate base/acid ratio of function of, 10f
  glutamine regulation, 602–603
  glycolysis and intracellular, 316–317
  $[H^+]$, $[OH^-]$, pOH relationships, 8t
  isoelectric, **36**, 36–41
  and oxygen–hemoglobin binding, 1032–1033
  regulation of
    acid–base balance, 8–13, 1045–1056
    bicarbonate buffer system, 1041–1045
    buffer systems of plasma, interstitial fluid, and cells, 1038–1039
  see also Acid-base balance
Phase transition temperature, **213**
pH-bicarbonate diagram, **1044**
Phenylalanine, 27–28, 42cc, **506**, 506cc, 506–509
  in epinephrine synthesis, 862–864
Phenylethanolamine *N*-methyltransferase, **508**, **862**
Phenylketonuria, 42cc, 506cc
Phi bond, **50**
Phorbol esters, **350**
Phosphatase inhibitor-1, **348**
Phosphate transport, 268
3′ Phosphate yielding endonucleases, 719t
5′ Phosphate yielding endonucleases, 719t
Phosphatidate phosphatase, **402**
Phosphatidic acid, **402**, 432–435, **1155**
Phosphatidylcholine, 197, 199, **425**
Phosphatidylethanolamine, 197, 199, **425**
Phosphatidylethanolamine *N*-methyltransferase, **433**
Phosphatidylglycerol, **426**
Phosphatidylinositol, 198–199, **426**, 429–430
Phosphatidylinositol 4,5-bisphosphate (PIP_2), **428**, 428–429
Phosphatidylinositol synthase, **434**
Phosphatidylserine, **425**
3′-Phosphoadenosine 5′-phosphosulfate (PAPS), **383**, **505**
Phosphocholine cytidylyltransferase, **432**
Phosphocreatine, 518–519
Phosphodiesterases, **613**, 613t, **950**
Phosphodiester bond formation, 649–653, *see also* DNA, repair
Phosphoenolpyruvate (PEP), **231**, 245, **303**
Phosphoethanolamine cytidylyltransferase reaction, **434**
6-Phosphofructo-1-kinase, **299**, **312**, 313–317
6-Phosphofructo-2-kinase, **319**
Phosphoglucomutase, **340**
6-Phosphogluconate dehydrogenase, **362**
6-Phosphogluconate pathway (pentose phosphate pathway), 361–365; *see also* Carbohydrate metabolism
Phosphoglucose isomerase, **299**
Phosphoglycerate kinase, **302**
Phosphoglycerate mutase, **302**
Phosphoglycerides, **197**, 197–200
Phospholipase A_2, **1083**
Phospholipase C, **353**
Phospholipids, **197**, 197–201
  clinical correlations: respiratory distress syndrome, 428
  function, 427–431
  glycerol ether, **200**
  major fatty acids in, 199t
  structure, 425–427
  synthesis, 431–438
  types, 424–425
Phosphomannose isomerase, **365**
Phosphopantetheine, **393**
Phosphopentose epimerase, **362**
Phosphoprotein phosphatase, 320, **348**, 351
5-Phosphoribose pyrophosphokinase (PRPP synthetase), **564**
Phosphoribosylaminomidazole carboxylase, **538**
Phosphoribosylaminomidazole succinocarboxamide synthetase, **538**
Phosphoribosylaminomidazole synthetase, **537**
Phosphoribosylglycinamide formyltransferase, **537**
Phosphoribosylglycinamide synthetase, **536**
Phosphorolysis chemistry, 1151
Phosphorus, dietary requirements, 1138
Phosphorylase
  cascade, 190–191
  b forms, **346**
  a forms, **346**
  as hepatic "glucose receptor," 352
Phosphorylase kinase, **347**, 347–351
Phosphorylase-limit dextrin, **341**
Phosphorylation
  in carbohydrate interconversion, 365–366
  hierarchical, **350**
  oxidative, 282–286
  synergistic, **350**
Phototransduction cascade, 946–947, 947t
Phytonadione (vitamin K), **971**, 971–972
pI. *See* Isoelectric pH
Pi electron interaction, 75
PIF. *See* Prolactin release inhibiting factor
Ping-pong mechanism, **149**
PIP_2 (phosphatidylinositol 4,5-bisphosphate), **428**, 428–429
Pituitary hormones. *See* Hormones
pK′ and dissociation constant of biochemically important compounds, 10t
Placental enzyme (PRPP amidotransferase), 536, 541
Plasma
  bradykinin, 34t, 856t
  buffer systems, 1039–1041
  chemical constituents, 16f
  cholesterol in, 446–447
  enzyme assays in clinical diagnosis, 179–186
  lipoproteins. *See* Lipoproteins
  protein classical (Tiselius) electrophoresis patterns, 39f
  thromboplastin antecedent (Factor XI), 969t
Plasmalogens, **200**, 438, **1155**
Plasma membrane, **2**
  in digestion and absorption, 1068t, 1068–1072
  Na^+,K^+ ATPase transport system, **226**, 226–227
  structure and function, 2–4, 16
Plasmids, **775**, 775–780
Plasmin, **974**
Plasminogen, **974**
Plasminogen activator inhibitors I and II, 109t
Plasmalogens, **426**
Platelet activating factor (PAF), **426**
Platelet clumping, 466, **966**
Point mutations, **732**
Poisoning
  Amanita phalloides, 699cc
  arsenic, 309cc
  carbon monoxide, 1017
  cyanide, 282cc
  lead, 1011t
  salicylate, 1053cc
  vitamin A, 1120–1121
  vitamin D, **1124**
  vitamin E, potential for, 125
Polyacrylamide gel electrophoresis, 86
Polyadenylation signal sequence (AAUAAA), **714**
Poly(ADP-ribose) synthetase, **561**
Polyamines, **503**
Poly(A) synthesis, 714–715
Poly(A) tail, **693**
Polycistronic mRNAs, **692, 807**
Polycloning sites, **776**
Polylinker sequences, **776**
Polymerases, 673–674
Polymerization of amino acids, 31–33
Polymorphonuclear leukocytes, 426–427
Polynucleosomes, **638**
Polynucleotide phosphorylase, **731**
Polynucleotides, **609**, **1158**, *see also* DNA; Nucleic acids; RNAs
Polyol (aldose) dehydrogenase, **941**
Polypeptides. *See* Peptides
Polyproline type II helix, **63**
Polysaccharides, 1077–1080, 1153–1154

Polyunsaturated fatty acids, **392**, 417–420, **1106**
  and heart disease, 1106cc
Pompe's disease, 341cc
Pores, membrane, **221**, *see also* Membrane transport
Porphyria cutanea tarda, 1011t, **1014**
Porphyrias, **1011**, 1011t, 1012cc
Porphyrinogens, **1009**
Positive cooperation, **126**
Posterior pituitary hormones, **850**
Posttranscriptional RNA processing, 708–717
  mRNA, 712–717
  ribozymes in, 708–710
  tRNA, 710–712
Potassium. *See* K^+; Na^+,K^+-ATPase pump
Precursor (Okazaki) fragments, **666**, 671, 673
Pregnancy
  cytochromes P450 and steroid hormone production, 996cc
  folic acid deficiency, 1134
  iron requirements, 1140
  metabolic changes in, 595–596
  Rh incompatibility, 1019cc
  vitamin C intake, 1137
δ^5-Pregnenolone, **907**
Prematurity/low birth weight and hypoglycemia, 325cc
Prepriming complexes, **670**
Prepriming proteins, **666**
PRF. *See* Prolactin releasing factor (PRF)
Pribnow boxes, **699**, 701
Primary transporters, **226**
Primases, **665**
Primers, **345**
Proaccelerin (Factor V), 969t
Probe techniques, 625–626, 626cc, **781**, *see also* Recombinant techniques
Procaryotes
  definition, **2**, 3f
  gene expression, 807–828, *see also* Gene expression, procaryotic
  nucleotide interactions vs. eucaryotic, 530
  RNA synthesis, 249–250, 698–703
  serine proteases, 112
Procollagen, 758–759
Proconvertin (Factor VII), 969t
Proenzymes, **109**
Progesterone, **878**, 878–880, **907**, 903t, 909t, 910t
Progestins, 909t
Programmed cell death, 921–922, 922cc
Proinsulin, **46**
Prolactin, 855t
  release inhibiting factor (PIF), 852t, 854t, 860
  releasing factor (PRF), 852t, 854t
Proline, **29**, 498–499
Prolyl hydroxylases, **758**
Promoter regions, **696**
Promoters, **811**
Proofreading sites, **651**, **735**
Proopiomelanocortin, **858**
Propionate, 332
Propionyl CoA, 411, **494**
Prostaglandins
  function, 465–466
  nomenclature, 461
  structures, 461–462
  synthesis, 462–464
Prostaglandin synthase complex, **462**
Prosthetic groups, **120**, **136**
Protease nexins I and II, 109t
Proteases, 1063t, 1073, 1074t
  serine, 102–116
Proteinase inhibitors, **973**
  α_1–, 109t
Proteinase inhibitors, **973**
Protein C, **973**
Protein–energy excess, 1101–1103, *see also* Obesity
Protein–energy malnutrition, 577, 1100–1101
Protein kinases
  A, 891–892
  C, 892–894
  cAMP activation, 190f
  Ca^{2+}- and phospholipid-dependent (protein kinase C), **350**
  calmodulin-dependent, 350
  cAMP-dependent, **319**, 319–322, 349–350, 590–591
  G, 894–897
  M (free catalytic domain), 875, **894**
  types, **889**
Proteins, **26**
  acid–base relationships, 34–36, 35t
  acyl carrier, **393**
  amino acid composition, 27–34
  assay techniques, 38–42, 44–49
  binding, 72t, 668, 813, 815
  C, 957t, **958**, 973
  CAP, **813**, 815, 821–822
  charge and chemical properties, 34–44
  clinical correlations
    amino acid sequences in diagnosis, 42
    collagen diseases, 61
    complement proteins, 96
    glycoprotein functions, 70
    glycosylated hemoglobin HbA_{1c} in diabetes mellitus, 71
    hyperglycemia and glycosylation, 585
    hyperlipidemias, 67
    immunization, 97
    immunoglobulin (antibody) functions, 96
    insulin structure and diabetes mellitus treatment, 49
    myocardial infarction and fibrin formation, 104
    protein malnutrition, 577
    sickle hemoglobin, 50
    tumor metastasis and serine proteases, 105
  cytochromes, **275**, 275–277
    P450, 982–997
  degradation and turnover, 761–763
  dietary, 1096–1100, 1108–1109
  digestion and absorption, 1073–1077
  docking, **747**
  extrinsic, **205**, **211**
  families
    DNA binding proteins, 116–118
    hemoglobin and myoglobin, 119–132
    immunoglobulins, 93–102
    serine proteases, 102–116
  functional roles, 26–27
  fusion, 790
  G (GTP-binding). *See* G Proteins
  gene expression
    globin proteins, 833–836
    human growth hormone-like proteins, 837–839
  globular domains, 57f
  glycosyl-phosphatidylinositol anchor, **211**
  glycogen-binding, 190f, **351**
  heat shock, **917**
  heme, 1003–1005
  histone, **637**, 639–641
  inhibitor-2, **351**
  integral (intrinsic), **205**, **209**
  ion-channel in neurotransmission, 932–933
  ionizable side chain groups, 34–36, 35t
  iron-containing, 1003–1005
  lac repressor, **809**
  lactation requirements, 596–597
  M, 957t, **958**
  membrane, 205, 208–214
  molecular weight of contractile, 957t
  nonheme iron (iron-sulfur enzymes), 156–157, 275
  nonhistone, **638**
  of ocular lens, 940–941
  omega, **635**
  operon
    lactose, 809–813
    tryptophan, 815–817
  peripheral, **205**, **211**
  of phototransduction cascade, 947t
  pI values, 37t
  plasma, Tiselius patterns, 39f
  prepriming, **666**
  protein–energy excess, 1101–1103
  protein–energy malnutrition, 1100–1101
  serine-protease-inhibiting (serpins), **109**, 109t
  signal transduction, **929**
  single-strand binding, **666**
  steroid hormone receptors, 915–916
  structural and heterogenous
    fibrous proteins, 61–67, *see also* Collagen
    glycoproteins, 70–72
    lipoproteins, 67–70
  structure
    analogy, **44**
    dynamic aspects, 76–78
    folding (noncovalent interactions), 72–76
    homology, **44**
    primary, *32*, 44–49

quaternary, 60
secondary, 32f, 50f, 50–55
study methods, 78–87
tertiary, 55–60, 57f, 58f, 59f
synthesis, 31–33, 723–763, *see also* Protein synthesis
trifunctional, **539**
zinc finger, 116
see also Amino acids; Enzymes; Peptides
Protein sparing, **1098**
Protein suicide, 761
Protein synthesis, 31–33
antibiotic and toxin inhibition, 744–746
apparatus
aminoacylation, 723–736
codon-anticodon interactions, 730–733
mRNA, 724–725
nucleotide 4-letter alphabet, 729–730
ribosomes, 725–727
tRNA, 727–729
clinical correlations
collagen diseases, 760–762
familial hyperproinsulinemia, 755
I-cell disease, 753
McKees Rocks hemoglobin disorder, 732
thalessemia, 733
degradation and turnover, 761–763
Golgi apparatus in, 18
initiation complexes, 737–739
maturation, 748–752
mitochondrial vs. cytoplasmic, 743–744
mRNA in, 692–694, 724–725
organelle biogenesis and targetting, 753–757
ribosomal proteins in, 822–823
rRNAs in, 688–692
steps
1: initiation, 737–739
2: elongation, 739–742
3: termination, 742–743
temporal sequence of proteins, 736
see also Amino acid synthesis; Proteins
Proteoglycans, **378**, 378–383, *see also*, Glycoproteins; Glycosaminoglycans
Proteolipids, **205**
Proteolysis, 755–756
Proteolytic enzymes, 102–105, 105t, 140
Prothrombin (Factor II), 969t, **1125**
Protooncogene *cErbA*, 918–920
Protophyrinogen oxidase, **1016**
Protoporphyrins, **1009**
IX, **1002**
Protransglutamidase (Factor XIII), 969t, 972
Proximal histidine, **121**
PRPP
in NAD synthesis, 560–561
in pyrimidine synthesis, 548
synthesis and utilization, 563–566
PRPP amidotransferase (placental enzyme), **536**, 541
PRPP synthase, **564**, 546cc
Prudent diet, 1106–1110, **1110**, 1110t
pSC101 plasmid, **775**
Pseudogenes, **119**, **647**
Psi bond, **50**
Pulmonary surfactant, 427–428, 428cc
Purine nucleoside phosphorylase, **544**
Purine nucleotides, 536–571, **1157**, *see also* Nucleotides, purine
Purine ring, 536
Puromycin, 744t, **745**
Putrescine, **503**
Pyridoxal phosphate, **478**, **1129**
Pyridoxine (vitamin B_6), 149, 1129–1132, 1131cc
Pyridoxol phosphate, **478**, **1013**
Pyrimidine nucleoside kinases, **551**
Pyrimidine nucleotides, 536–571, **1157**, *see also* Nucleotides, pyrimidine
Pyrimidine phosphoribosyltransferase, **550**
Pyrophosphatase, **366**
Pyruvate, 248–252, 295–296
Pyruvate carboxylase, 140, 141
Pyruvate dehydrogenase, **249**, 249–252
deficiency, 252cc
Pyruvate kinase, 155, 323–324, 324cc

Quaternary structure
of hemoglobin, 129f
of proteins, **49**, 60
Quencher function, **83**
Quinolinic acid, **560**

R conformational state, **130**
R groups in amino acids, 27–29, 72
Racemases (epimerases), **140**
Radiation and free radical formation, 419
Raffinose, 1078t, **1080**
Random mechanism, **149**
Rate equations, 141–142
Michaelis-Menten equation, **145**, 145–148
Rate-limiting enzymes, **187**
RDA. *See* Recommended Dietary Allowances
Reading frames, **731**
Reagents
double helix effects, 620t
nuclease specificities, 613t
polypeptide cleaving, 46f
sulfhydryl, **308**
Reannealing, *see* Renaturation
Receptor-mediated endocytosis, **445**
Receptors
functional domains, 884–886
insulin, 598, 1102–1103
oncogenes and, 897–898
peptide hormone, 881–898, *see also* Hormones, peptide
steroid hormones
activation and up-.and down-regulation, 920–921
DNA-binding and non-DNA binding, **917**
functional domains, **918,** 918–920
Receptosomes, **886**
Recombinant techniques
applications, 799–802
bacteriophage vectors, 785–786
chromosome walking, 788
clinical correlations
HIV-screening, 799
HSV I site-directed mutagenesis, 795
restriction mapping and evolution, 773
transgenic animal models, 800
tumor identification, 783
complementary DNA and DNA libraries,784–785
cosmid and yeast artificial chromosome vectors, 786–788
DNA sequencing, 770–771
chemical cleavage (Maxam–Gilbert procedure), 771–772
interrupted enzyme method (Sanger procedure), 772–773
expression vectors and fusion proteins, 790–793
nucleic acid detection and identification
probe techniques, 781–782
Southern blot technique, **782**, 782–784
polymerase chain reaction, 797–798
recombinant DNA, 775–781
restriction mapping, 769–770
site-directed mutagenesis, 793–797
subcloning, 788
Recombinase, **676**
Recommended Dietary Allowances (RDAs), **1095**, 1096, 1100t, 1117–1118
Red blood cells. *See* Erythrocytes
Redox couples, **274**
Redox reactions, 272
Reductant (electron donor), **270**
5α-Reductase, **907**
Reduction, metals in, 156–157
Refsum's disease, 413cc, 414cc
Regulatory genes, **807**
Regulatory subunits of allosteric enzymes, 167
Reiteration of DNA sequences, **648**, *see also* DNA, nucleotide sequences
Relaxed plasmids, **775**
Relaxin, 857t
Release factors, **742**, 742–743
Releasing hormones, **850**
Renal failure, 600
low-protein diet in, 1099cc
renal osteodystrophy, 1046cc, 1122cc
Renin, **912**
Renin–angiotensin system, **912**
Replication. *See* DNA synthesis
Replicons, **608**, **775**
Replisomes, **669**, 669t
Repression, **887**
catabolite, **814**
de-, **815**
Repressor-constitutive mutations, **811**
Respiratory acidosis, **1045**, 1050–1051, 1052cc

Respiratory alkalosis, **1045**, 1051cc
Respiratory control, **261**
Respiratory control ratio, **283**
Respiratory distress syndrome, 428cc
Respiratory tract, 1027f
Restriction digest, **643**
Restriction endonucleases, **642**, 642–644, **769**, *see also* Recombinant techniques
Restriction fragments, **677**
  length polymorphisms, **782**
    clinical correlation: tumor origin detection, 783
Restriction mapping, **677**, **769**, 769–770
Restriction site banks, **776**
Retina. *See* Eye and visual system
Retinal (retinaldehyde), **1118**
Retinal (visual pigment), **944**
Retinoic acid, **1118**
  receptors, 920
Retinoids, synthetic, **1121**
Retinol, **1118**
  all-trans-retinol, **945**
  11-cis-retinol, **945**
Retinyl phosphate, **1118**
Reverse genetics, **799**
Reverse transcriptase, 785
Reversibility of chemical reactions, 143–144, 148–149
Reye's syndrome, 582cc
Rh incompatibility, 1019cc
Rhodanese, **505**
Rhodopsins, **945**, 945–948
Rhodopsin kinase, **951**
Rhythmic cycling hormone release, **850**
Riboflavin kinase (flavokinase), **562**
Riboflavin (vitamin B_2), 152–153, 1127–1128
Ribonucleic acids. *See* RNAs
Ribonucleoprotein particles, 694, **708**
Ribonucleoproteins, small nuclear (snRNPs), **715**
Ribose isomerase, **362**
Ribosomal RNAs. *See* RNAs, ribosomal
Ribosomes, 688–689, 725t, 725–727
Ribozymes, **708**, 708–710
Ricin, 744t, **745**
Rickets, 232cc, **1123**, 1123–1124
RNAs
  antisense, **799**
  attenuator region, **817**, 817–822
  cap structures, **692**
  clinical correlations
    Amanita phalloides poisoning, 699
    autoimmunity in connective tissue diseases, 717
    carcinogenesis and transcriptional factors, 707
    Mycobacterium tuberculosis, 699
    staphylococcal erythromicin resistance, 692
    thalassemia, 714
  hammerhead structure, **709**
  heterogeneous nuclear (hnRNA), 683t
  leader sequences, **693**, **817**
  messenger (mRNAs), **682**, 683t, **692**, 692–694
    in cDNA synthesis, 784
    codon-anticodon interactions, 730–733
    cytoplasmic, 683t
    eucaryotic, 692–693
    intron removal, 715–716
    iron-responsive elements, **1006**, 1006–1007
    mitochondrial (mt mRNA), 683t
    monocistronic, **692**
    posttranscriptional processing, 712–717
    in protein synthesis, 692–694, 724–725, 736–737
    RNA polymerase II in synthesis, 698
    splicing, 715–717, 716cc
  mitochondrial (mtRNAs), **694**, 694–695
  nucleases and turnover, 717–719
  nucleosides in, 688t
  probe techniques, 781, *see also* Recombinant techniques
  postranscriptional processing, 708–717
  reverse transcribed, **682**
  ribosomal (rRNAs), **682**, 683t
    cytoplasmic, 683t
    5S, **689**
    18S, **688**
    mitochondrial (mt rRNA), 683t
  in ribonucleoprotein particles, 694
  sizes, 683t
  small cytoplasmic [scRNA; 7S(l)RNA], 683t
  small nuclear (snRNA), 683t, 698
  splicing, **829**
  stringent response, **823**, 823–824
  structure, 683–684
    secondary, 684–685
    tertiary, 685–686
  synthesis. *See* RNA synthesis
  transfer (tRNA), **682**, 683t, **688**
    in amino acid activation and codon recognition, 687–688
    aminoacylation, 733–735
    cloverleaf structure, 690f
    cytoplasmic, 683t
    L-shaped conformation, 686
    formylmethionyl, 738
    methionyl, **737**
    mitochondrial (mt tRNA), 683t
    posttranscriptional processing, 710–712
    regional characteristics, 691t
    synthesis, 727–728, 823–824
  types, 681–683, 683t
RNA synthesis, **695**, 695–696
  DNA as template, 696
  eucaryotic, 703–708
  nucleotides in, 558–559
  nucleus in, 17–18
  posttranscriptional processing, 708–717
  procaryotic, 698–703
  regulation of, 822–823
    stringent response, **823**, 823–824
  RNA polymerase catalysis, 696–698
  as strand-selected process, **698**
  see also RNAs
Rods and cones in vision, 945–948
Rolling circle replication, **672**
Rotenone, 281
Rotor syndrome, **1020**
Rough endoplasmic reticulum. *See* Endoplasmic reticulum
Rous sarcoma virus, 792

Salicylate poisoning, 1053cc
Salivary gland secretagogues, 1065t
Salivary glands, 1062t
Salmonella gene expression, 827–828
Salvage pathways, 542, 551
Sandhoff-Jatzkewitz disease (G_{M2} gangliosidosis variant O), 374t, 374cc, 459t
Sanfilippo syndromes, 379t, 379cc
Sanger procedure, **772**, 772–773
α-Sarcin, **745**
Saturated fatty acids, **1106**, **1154**
Saturation kinetics of enzymes, 144–148
Scatchard analysis, **882**
Scurvy, 761cc, **1136**
SDS-polyacrylamide gel electrophoresis, 86
Secondary structure
  of proteins, **49**, 50–55
  of RNA, 684–685
Secondary transporters, **226**
Second messengers, **318, 874**
Second-order reactions, **143**
Secretagogues, **1064**, 1064–1067, 1065t
Secretin, 857t, **1067**
Selenium, **1143**
Semen fructose content, 333–334
Semiconservative replication, **664**
Semipermeable membranes, 13–14
Sequential mechanisms, **149**
Serine, **29**, 113t, 498
Serine hydroxymethyl transferase, **496**
Serine proteases, **92**, 1074t
  binding characteristics, 105–108
  catalysis in, 176
  classification, 102–105, 105t
  clinical correlations
    myocardial infarction, 104
    tumor metastasis, 105
  genetic organization, 110f
  inhibitor proteins, 109t, 109–110
  invariant sequences, 113t
  structure-function relationships, 110–116
  X-ray crystallographic structures, 113t
  zymogen form synthesis, 108–109
Serotonin (5-hydroxytryptamine), **512**, 876, 936, 1066
Serpins (serine-protease-inhibiting proteins), **109**, 109t
Sertoli cells, 916
Serum albumin titration curve, 38f
7S(L) RNA, 683t
Sex hormone binding globulin, **915**
Sex hormones. *See* Hormones, sex
SGOT (serum glutamate–oxaloacetate transaminase) assay, 185–186
SGPT (serum glutamate–pyruvate transaminase) assay, 185–186
Shine-Delgarno sequence, **739**, 789–790

Shuttle systems in glycolysis, **304**, 304–308
Shuttle vectors, **791**
Sialic acids, **371**, 371–372, 454–455
Sialisosis (mucoliposis), 374t, 374cc
Sickle cell anemia, 50cc, 836cc, *see also* Hemoglobin; Thalassemias
Side chain cleavage cytochrome P450 ($P450_{scc}$), **990**, 990–994
Side chain groups of amino acids, 27–31, 34–36
Sideroblastic anemia, **1130**
Sigmoidal kinetics of allosteric enzymes, 164–165
Signal peptidase, **748**
Signal recognition particles, **747**
Signal transduction proteins, 428–429, **929**
Singer-Nicolson membrane model, 208–211
Site-directed mutagenesis, **793**, 793–797, 795cc
Site-specific recombination, **675**
60S ribosomal particle, **688**
Skeletal muscle. *See* Muscle; Muscle contraction
Sliding filament model of muscle contraction, **954**, 954–957
Small cytoplasmic mRNA (scRNA; 7s(L)RNA), 683t
Small nuclear ribonucleoproteins (snRNPs), **715**
Small nuclear RNA (snRNA), 683t, 698
Smoking and vitamin C levels, 1137
Smooth muscle, Ca^{2+} in contraction, 965
Snake venom phosphodiesterase, 613t
Sodium. *See* Na^+; Na^+-K^+ ATPase pump
Solute transport in digestion and absorption, 1067–1068
Somatocrinin (growth hormone releasing hormone), 852t, 854t
Somatostatin (GIH), 852t, 869t, 938t
Somatotropin. *See* Human growth hormone
Sorbitol, **941**
Southern blot technique, **782**
Specific activity of enzymes, **144**
Spectrometry techniques, 46, 81–85
Spermatozoa, fructose as energy source, 333–334
Spermidine, 522
Spermine, **503**, 522
Sphinganine, **450**
Sphingolipidoses, 456–461, 459t
Sphingolipids, **201**, 201–206, **1155**
  clinical correlation: Gaucher's disease, 461
  storage diseases, 456t, 456–461
  structure and function, 450–456
  synthesis, 449–450
Sphingomyelin, 200t, **451**, 451–452, **1155**
Sphingomyelin synthase, **452**
Spin states of heme, **982**
Spleen phosphodiesterase, 719t
Spliceosomes, **715**
Splicing of mRNA, 715–717, 716cc
Squalene, **441**
Squalene oxidocyclase, **442**
Squalene synthase, **441**
Starvation, 416–417, 577cc
Starvation studies of glucose homeostasis, 585–586
Starve–feed cycle, **576**
  clinical correlations
    hyperglycemia, hyperosmolar coma, 584
    hyperglycemia and protein glycosylation, 585
    obesity, 404, 576
    protein malnutrition, 577
    Reye's syndrome, 582
    starvation, 577
  early fasting state, 579
  early refed state, 582
  energy requirements and reserves, 584–585
  fasting state, 580–582
  glucose homeostasis, 585–586
  hepatic metabolism and, 586–593
  insulin:glucagon ratio, 583–584
  well-fed state, 577–579
  see also Metabolic interrelationships
Static disorder, **79**
Stearoyl CoA desaturase, 394t, **399**, **418**, 418–419
Steatorrhea, **1087**
Sterochemistry review, 1151
Steroid hormones, 901–923; *see also* Hormones, steroid
Stomach, 1062t
  secretagogues, 1065t
Stop codons, 742–743
Streptomycin, 744t
Stress, 598, 912, **923**, 1136
Stringent response, **823**, 823–824
String of beads nucleofilament structure, 640f, 641
Structural genes, **644**, 644–649, **807**, 822–823
Structural motifs, **55**
Subcellular organelles. *See* Organelles
Subcloning, **788**
Substance P, 34t, 856t, 938t, **939**
Substrate binding site, **137**, **167**
Substrates, **136**
  suicide, 161, **986**
Substrate shuttles, **266**
Substrate specificity, **984**
Substrate-strain catalysis, 174–175
Succinate dehydrogenase, **258**, 258–259
Succinyl CoA, **258**, 258–259
Sucrase, 1063t, 1080t
Sucrose, 1078t
Sucrose-α-glucosidase (sucrase), 1063t, 1080t
Sugars
  amino, **371**
  glucose synthesis from, 333–334
  interconversions and nucleotide sugar formation, 365–372
  membrane transport, 231
  see also Carbohydrates; Carbohydrate metabolism
Suicide substrates, 161, **986**
Sulfa drugs as enzyme inhibitors, 159
Sulfatide (sulfaglactocerebroside), **202**, **453**
Sulfhemoglobin, 1031cc
Sulfhydryl reagents, **308**
Sulfur amino acids, 501–506
Superfamily, **102**
Superhelical DNA, **64**, **631**, 631–635
Super-secondary structure, **55**
Suppressor mutations, **735**
SV40, 733f, 792
SV chromosome, 696
Svedberg coefficients, **85**, 86t
Svedberg units, **85**
Synapses, neural, **931**, *see also* Neurotransmitters
Synapsin I, **933**
Synapsis of DNA molecules, **676**
Synergistic phosphorylation, **350**
Synthases, **140**

Tangier disease, 69cc
Targetting, **752**
Taurine, **505**
Taurocholic acid, **447**
Tauroconjugates, **1088**
Tay-Sachs disease, 455, 459t, 460–461
TCDD (1,2,7,8-tetrachlorodibenzo-*p*-dioxin), **985**, 985–986
T conformational state, **130**
Telopeptides, **65**
Temperature
  catalytic activity and, 178–179
  and denaturation, 621–622
  and oxygen-hemoglobin binding, 1032
  phase transition, **213**
Temporal sequence of protein synthesis, 736
—10 sequences (Pribnow boxes), **699**
Terpenes, 1156
Tertiary structure
  of proteins, **49**, 55–60
  of RNA, 685–686
  in serine proteases, 112–116
Testosterone, 903t, **907**, 909t, 910t
1,2,7,8-Tetrachlorodibenzo-*p*-dioxin, **985**, 985–986
Tetracyclines, 744t, **745**
Tetrahedral configuration, **31**
Tetrahydrobiopterin, **506**
Tetrahydrofolate methyltransferase deficiency, 519cc
Tetraiodo-L-thyronine. *See* Thyroid hormone, T_4
Thalassemias, 717, 714cc, 733cc, 836–837, 837cc
  gene mutations and, 836–837
Thermodynamics, laws of, 241
Thermogenesis, diet-induced, **1101**
Thiamine (vitamin B_1), 1127–1128
Thioesterases, **400**
Thiohemiacetal, **301**
Thiolysis, **407**
—35 sequence, **699**
3′ phosphate yielding endonucleases, 719t
Threonine, **29**, 496–498
Thrombin, **967**, 967–969
Thrombomodulin, **973**
Thrombosis, 466, *see also* Blood coagulation
Thromboxane A_2, 1106cc
Thromboxane A synthase, **465**

Thromboxanes, **465**, **1155**, *see also* Prostaglandins
Thrombus formation, *See* Blood coagulation
Thymopoietin (alpha-thymosin), 857t
Thyroglobulin, **865**
Thyroid hormone
  dietary iodine and, 1140
  synthesis, 865
  T_3, 855t
  T_4, (thyroxine), 855t, **865**
Thyroid hormone receptors, 920
Thyroid-stimulating hormone (TSH), 884
Thyrotropin, 854t
Thyrotropin releasing hormone (TRH), 34t, 852t, 854t, 938t
Tiazofurin, **569**
Tight junctions, **1067**
Tiselius patterns of plasma proteins, 39f
Tissue factor (Factor IV), 969t, **972**
Tissue factor (TF)-VII-Ca^{2+} complex, **969**, 969–970
Tissue factor (TF)-VII-Ca^{2+} -Xa complex, **970**
Tissue hypoxia, 1029cc, **1030**
Tissue plasminogen activator (t-PA), 186, **974**
tk (thymidine kinase) gene, 792
Tn3 transposon, 825–826, 825–827
α-Tocopherol, **1124**
Tomato bushy stunt virus quaternary structure, 60f
TOPA (trihydroxyphenyalanine), 521
Topoisomerases, **635**, 635–637, 668
Toroidal turns, **634**
Total parenteral nutrition, 1100cc, 1141
  and acidosis in infants, 1055cc
Total plasma carbon dioxide, **1052**
Toxicity. *See* Poisoning
Toxins as protein synthesis inhibitors, 744t, 744–746
t-PA (tissue plasminogen activator), 186, **974**
Transaldolase, **362**
Transamidation, **371**
Transamidinase reaction, **500**
Transaminases. *See* Aminotransferases
Transamination, **477**, 477–478
Transcellular transport, **1067**
Trans conformation, **32**, 33–34
Transcortin (corticosteroid binding globulin), **915**
Transcription. *See* RNA synthesis
Trans-dominant mutations, **812**
Transducin, **948**, 948–951
  α-subunit, **949**
  β-subunit, **949**
  γ-subunit, **949**
Transfection, **791**
  co-, **792**
Transferases, **138**, 138–140, 141t
  uridine diphosphoglucuronyl, **1018**
Transferrin, **1002**
Transferrin receptor, **1002**
Transfer RNAs. *See* RNAs, transfer
Transformation, 608–609, **609**
  bacterial, 777–780
Transgenic species, 800cc, **801**, 842–844
Transglutaminase (clotting Factor XIII), **967**
Transition mutations, **653**
Transition state, **171**, 171–172
Transition state analogs, **172**
Transketolase, **362**
Translational initiation signal (AUG codon), **724**
Translocation group, **217**
Transmembrane domains, **884**
Transmembrane potential, **221**
Transmembrane signaling, **928**
  ASTP and, 930–931
Transporters
  membrane, **217**, 218t
  tricarboxylate, **267**
Transposase, **824**
Transpositions, **675**
Transposons, **675**, **824**, 824–825
  Tn3, 825–827
Transsulfuration defects, 503cc
Transversion mutations, **653**
Trauma, metabolic changes in, 598
Trehalase, 1063t
  αα-, 1063t, 1080t
Trehalose, 1078t
TRH, *see* Thyrotropin releasing hormone (TRH)
Triacylglycerols, 332–333, **390**, 401–404, **424**, **1155**
  hormones regulating, 407t
  in lactation, 596–597
  see also, Phospholipids
Tricarboxylate transporter, **267**
Tricarboxylic acid cycle, **253**
  in gluconeogenesis, 330–332
  reactions, 253–261, 254f
  regulation, 260–261
  urea cycle and, 485–486
Trimethyllysine, **514**
Triodo-L-thyronine. *See* Thyroid hormone, T_3
Triose phosphate isomerase, **300**
tRNA. *See* RNAs, transfer
Tropocollagen, **61**
Tropomyosin, **66**, 66–67, 957t, **961**
Troponin, 957t, **961**
trp genes, 815–817
Trypsin, 1074t, **1075**
  as example of tertiary structure, 55–57
  structure–function relationships, 112–115
Trypsinogen, 77–78, 80f, **1075**
Trypsin reagent techniques, 46, 47f
Tryptophan, **29**, 510–512
  deficiency, 42cc
  in NAD synthesis, 559–560
Tryptophan oxygenase, **510**
Tubulin, **21**
Tumor cells
  drug resistance, 570–571
  growth inhibiting drugs, 569
  life cycle and nucleotide metabolism, 559
  recombinant identification techniques, 783cc
  see also Cancer; Chemotherapeutic drugs
Tumorigenicity and fat intake, 1104–1105
Tumor necrosis factor (TNF), **598**, 602cc
Tunicamycin, **749**
Turnover number (catalytic constant), **144**
28S ribosomal particle, **689**
Two-dimensional nuclear magnetic resonance spectroscopy, **83**,
TXA_2, 465, 466
Tyrosine, **29**, 35t, **506**
  in epinephrine synthesis, 862–864
  iodination, 866f
Tyrosinemias, 508cc
Tyrosine-specific kinases, **889**

U (standard unit of enzyme activity), **144**
U codon base pairing, 731t
UAA codon, **730**, 742–743
UAG codon, 742–743
Ubiquinone (coenzyme Q), **277**, 277–279
Ubiquitin, **762**, 762–763
UDP-glucoronic acid, 306–307
UDP-glucose, **367**, 383
UDP-glucose epimerase, **367**
UDP-glucose pyrophosphorylase, **367**
UDP-xylose, **370**, 383
UGA codon, **730**, 742–743
Ultimate gland hormones, 855t
Ultracentrifugation, 85–86
Ultraviolet light spectroscopy, 81–82
UMP, 549–550
Uncompetitive inhibition, **158**
Unsaturated fatty acids, **1154**
Upregulation, **921**
Uracil nucleotides, 534
Urea cycle, 480–486, **481**, 481–486, **482**
  enzyme deficiency diseases, 487cc
  regulation, 486–488
Urease, **484**, 484–486
Urea synthesis, 330–332
  starve-feed cycle and, 581–582, 586
Uric acid formation, 544–547, 546cc, 565
Uridine diphosphoglucuronyl transferases, **1018**
Uridine 5′-triphosphate (UMP), 549–550
Urine
  titratable acidity, **1047**
  total acidity, **1049**
Urobilinogens, **1019**
Urocoporphyrinogens, **1010**
Uroporphyrinogen decarboxylase, **1014**
Uroporphyrinogens I and II, **1013**, **1014**, 1015f

V class allosteric enzymes, **162**
V regions, **94**, 94–95, 98–101
Valine, **27**, 493–496
Valinomycin, **231**, 232–234
van den Bergh reaction, **1019**
van der Waals contact distance, **74**
van der Waals-London dispersion interaction, **74**, 75t
Variable (V) regions, **94**, 94–95, 98–101
Variegate porphyria, 1011t, **1016**
Vasoactive intestinal peptide (VIP), 857t, 938t

Vasopressin, 34t, 854t, **859**, 859–861, 859t, **891**
arginine, 852t
and protein kinase A, 890–892
see also Blood pressure regulation
Vectors
bacteriophage lambda, **785**, 785–786
cloning. *See* Cloning vectors
cosmid, **787**, 787–789
expression, 788–791, 794f
shuttle, **791**
yeast artificial chromosomes (YAC), **787**, 787–789
Vegetarian diets, 1098cc, 1108, 1145cc
Velocity, 141
initial, **145**
maximum (V_{max}), **144**
see also Rate equations
Very low density lipoproteins (VLDLs), **67**, **405**, 443, 579, 597–598, 1087cc
Vesicular structure of lipids, 206–207
Viral coat proteins, 60
Viruses, 569–570, 792
Virus transformed cells, 791–793
Viscoelastic retardation, 629
Visual disorders and gene arrangement, 953cc
Visual system, 939–954, *see also* Eye and visual system
Vitamins
A, 1118–1121, 1128cc
B complex
B_1 (thiamine), 1127–1128
B_2 (riboflavin), 1127–1128
B_3 (niacin), 151–152, 1129
B_6 (pyridoxine), 149, 150, 1129–1132, 1131cc
B_{12} (cobalamine), 1134–1136
coenzymes, 151–153
biotin, 1132
C (ascorbic acid), 378, 761cc, 1136–1137
D, 1121–1124
alcoholism and, 1128cc
and Ca^{2+} binding, 913–914
cholesterol in synthesis, 448–449
cytochromes P450 in synthesis, 995
D_2 (ergocalciferol), **1121**
D_3 (cholecalciferol), **448**, 913–915, **1121**
deficiency in newborns, 1126cc
1,25-dihydroxy, 903t
endocrine system, 913–914
metabolites, 909t
precursors, 913–914
E, 1124–1125
fat-soluble, 1118–1126
folacin, 1132–1134
deficiency in alcoholics, 1128cc
K (phytonadione), 98f, **971**, 971–972, 1125–1126
deficiency in newborns, 1126cc
nutritional needs of the elderly, 1145cc
pantothenic acid, 1132
water-soluble, 1126–1127
energy-releasing, 1127–1132
hematopoietic, 1132–1136
Voltage-gated channels, **221**, **931**, 931–932, *see also* Electron transport
von Gierke's disease, 340cc
von Willebrand factor, **966**, 969t, 970, *see also* Blood coagulation

Warfarin therapy and cytochrome P450, 986cc
Water, **4**, 4–8, *see also* Electrolytes
Wayne hemoglobin frameshift mutations, 733t
Weak electrolytes, **7**
Wear and tear pigment (lipofuscin), **20**, 20–21, 486
Wernicke-Korsakoff syndrome, **1128**, 1128cc
Wilson's disease, **1142**
Wobble hypothesis, **730**, 730–731

Xanthine oxidase, **546**, 546–547cc
Xanthomas, 405cc
Xanthosine 5′monophosphate (XMP), 540–544
Xenobiotics, **986**, 996t
Xenopus laevis mRNA, 692
Xeroderma pigmentosum, 662cc
X-linked agammaglobinemia, 96cc
X-ray diffraction crystallography. **78**, 78–81
Xylitol dehydrogenase deficiency, 368cc
Xylose, UDP-, **370**, 383

YAC. *See* Yeast artificial chromosomes
Yeast artificial chromosomes (YAC), **787**, 787–789

Z-DNA, 616–618, **617**
Zellweger syndrome and peroxisome function, 21cc
Zero-order reactions, **143**
Zinc
as cofactor, 153–154
deficiency, 1141
Zinc finger motif, **116**
Zinc finger proteins, 919
Zinc-peptidases, 1074t
Zinc–protoporphytin IX complex, 1016
Zipper structure, leucine, **117**, 117–118
Zwitterion form, **36**, **1157**
Zymogen prothrombin, **967**
Zymogens, 108–109, 755–756, **1064**

NORMAL CLINICAL VALUES: BLOOD*

INORGANIC SUBSTANCES

Ammonia	12–15 μmol/L
Calcium	8.5–10.5 mg/dl
Carbon dioxide	24–30 meq/L
Chloride	100–106 meq/L
Copper	100–200 μg/dl
Iron	50–150 μg/dl
Lead	50 μg/dl or less
Magnesium	1.5–2.0 meq/L
P_{CO_2}	35–40 mmHg 4.7–6.0 kPa
pH	7.35–7.45
Phosphorus	3.0–4.5 mg/dl
P_{O_2}	75–100 mmHg 10.0–13.3 kPa
Potassium	3.5–5.0 meq/L
Sodium	135–145 meq/L

ORGANIC MOLECULES

Acetoacetate	negative
Ascorbic acid	0.4–15 mg/dl
Bilirubin	
Direct	0.4 mg/dl
Indirect	0.6 mg/dl
Carotenoids	0.8–4.0 μg/ml
Creatinine	0.6–1.5 mg/dl
Glucose	70–110 mg/dl
Lactic acid	0.6–1.8 meq/L
Lipids	
Total	450–1000 mg/dl
Cholesterol	120–220 mg/dl
Phospholipids	9–16 mg/dl as lipid P
Total fatty acids	190–420 mg/dl
Triglycerides	40–150 mg/dl
Phenylalanine	0–2 mg/dl
Pyruvic acid	0–0.11 meq/L
Urea nitrogen (BUN)	8–25 mg/dl
Uric acid	3.0–7.0 mg/dl
Vitamin A	0.15–0.6 μg/ml

PROTEINS

Total	6.0–8.4 g/dl
Albumin	3.5–5.0 g/dl
Ceruloplasmin	27–37 mg/dl
Globulin	2.3–3.5 g/dl
Insulin	6–20 μU/ml

ENZYMES

Aldolase	1.3–8.2 mU/ml
Amylase	4–25 U/ml
Cholinesterase	0.5 pH U or more/h
Creatine phosphokinase (CK)	10–148 U/L
Lactic dehydrogenase	45–90 U/ml
Lipase	2 U/ml or less
Nucleotidase	1–11 U/L
Phosphatase (acid)	0.1–0.63 Sigma U/ml
Phosphatase (alkaline)	13–39 U/L
Transaminase (SGOT)	7–27 U/ml

PHYSICAL PROPERTIES

Blood pressure	120/80 mmHg
Blood volume	8.5–9.0% of body weight in kg
Iron binding capacity	250–410 μg/dl
Osmolality	280–296 mOsm/kg H_2O
Hematocrit	37–52%

NORMAL CLINICAL VALUES: URINE*

Acetoacetate (acetone)	0
Amylase	24–76 U/ml
Calcium	300 mg/d or less
Copper	0–100 μg/d
Coproporphyrin	50–250 μg/d
Creatine	under 0.75 mmol/d
Creatinine	15–25 mg/kg body weight/d
5-Hydroxyindoleacetic acid	2–9 mg/d
Lead	120 μg/d or less
Phosphorus (inorganic)	varies; average 1 g/d
Porphobilinogen	0
Protein (quantitative)	less than 150 mg/d
Sugar	0
Titratable acidity	20–40 meq/d
Urobilinogen	up to 1.0 Ehrlich U
Uroporphyrin	0–30 μg/d

* Selected values are taken from normal reference laboratory values in use at the Massachusetts General Hospital and published in the *New England Journal of Medicine* **314:**39, 1986. The reader is referred to the complete list of reference laboratory values in the literature citation for references to methods and units. dl, deciliters (100 ml); d, day.